The Vascular Flora of Pennsylvania:
Annotated Checklist and Atlas

THE VASCULAR FLORA
Annotated Checklist

Virginia bluebells (*Mertensia virginica*) and Bloodroot (*Sanguinaria canadensis*) are common spring woodland flowers.

OF PENNSYLVANIA
and Atlas

ANN FOWLER RHOADS
Chair of Botany,
Morris Arboretum
of the University of Pennsylvania

WILLIAM McKINLEY KLEIN, JR.
Director, Fairchild Tropical Garden

Illustrations by **Janet E. Klein**

Compiled and produced
at the Morris Arboretum
of the University of Pennsylvania
with the assistance of **Alfred E. Schuyler**
and **Carl S. Keener**

Published by the
American Philosophical Society
Philadelphia, PA
1993

The Vascular Flora of Pennsylvania: Annotated Checklist and Atlas
Ann Fowler Rhoads & William McKinley Klein, Jr.

104 South Fifth Street
Philadelphia, PA 19106

For its Memoirs series, *Vol. 207*

ISBN # 0-87169-207-4

ISSN # 0065-9738

Library of Congress # 92-85316

Cover illustration: **Janet E. Klein**
Mountain laurel (*Kalmia latifolia*), the state flower of Pennsylvania, common in woods throughout the state.

Text and cover design: **Rodelinde Albrecht**

Maps: **J/B Woolsey Associates**

Data transformation and page formats: **McMunn Associates**

CONTENTS

INTRODUCTION

This volume is the first published product of the Pennsylvania Flora Database, created and maintained at the Morris Arboretum of the University of Pennsylvania. The database has its roots in the work of Edgar T. Wherry, John M. Fogg, Jr. and Herbert A. Wahl, the *Atlas of the Flora of Pennsylvania* (Wherry et al. 1979), published by the Morris Arboretum. Over a period of 40 years, Wherry and his colleagues gathered data from the major Pennsylvania herbaria and manually placed a quarter of a million dots on over 3500 maps (Fogg 1944.)

Throughout the development of the Pennsylvania Flora Database we have retained the stress they placed on specimen-based, site-specific data. Since publication of the *Atlas*, the checklist of included taxa has undergone extensive review to reflect recent taxonomic and nomenclatural revisions. Questionable specimens have been re-evaluated with the result that several taxa included in earlier works were dropped. Recent discoveries have been added.

Distribution data has also been updated. A published list of new county records from the herbarium of the Carnegie Museum of Natural History in Pittsburgh (Thompson et al. 1989) indicated that although the CM herbarium was surveyed for the original atlas project, many recent collections were not included. Therefore we launched a massive effort to update species distribution records through the addition of herbarium specimen label data from the collections of the Carnegie Museum of Natural History (CM), Pennsylvania State University (PAC), Academy of Natural Sciences of Philadelphia (PH), Morris Arboretum (MOAR), Albright College (ALBR), Shippensburg University (SHIP), Millersville University (MVSC), Wilkes College (WILK), Pennsylvania Department of Agriculture (PAM), Cleveland Museum of Natural History (CLM), US Forest Service Laboratory at Warren, PA and from several private collections as well.

This volume also includes collections made during the past 10 years in conjunction with the Pennsylvania Natural Diversity Inventory (PNDI), the state heritage program. Field work stimulated by PNDI has greatly increased knowledge of the current status of endangered, threatened and rare taxa. Many new records have resulted, and several erroneous identifications and questionable taxonomic decisions have been reevaluated.

The maps present the accumulated collection information for each taxon as represented in the herbaria. Dates of collection range from the early 1800s to 1992. Clearly the floristic makeup of many parts of the state has been greatly altered, therefore not all dots represent currently extant populations.

Numerous individuals assisted in reviewing early drafts of the checklist, however, the authors accept full responsibility for taxonomic and nomenclatural decisions. Key references utilized in arriving at the final treatments are listed under family headings at the end of the book. Subspecies or varieties were recognized where it seemed appropriate to do so; however, we decided not to recognize any taxa at the level of forma. For the most part, hybrid taxa are listed by the progenitor species; with nothospecies epithets, if they exist, shown as synonyms. Hybrids represented by only a few collections are mentioned, but not mapped.

The organization of the checklist follows Cronquist (1981) for the flowering plants and *Flora of North America Vol.2* (Flora of North America Editorial Committee, in press) for the Pteridophyta and Pinophyta. Within each family the arrangement of genera and species is alphabetical. No attempt was made to include full synonymy. However, when alternate nomenclature was used in one of the standard regional manuals (Fernald 1950, Gleason 1952, Gleason and Cronquist 1963, Gleason and Cronquist 1991), the earlier atlas (Wherry et al. 1979) or Kartesz and Kartesz (1980), it is indicated for ease of cross reference. Nomenclature of Porter (1903) is also indicated for the purpose of continuity and comparison, however, the synonymy derived from Porter's treatment includes some (now) very obscure names.

A variety of published and unpublished sources provided common names. The format for common names follows Kartesz and Thieret (1991).

Habitat information was drawn from herbarium label notations and is specific to Pennsylvania. Ballast, unless otherwise indicated, refers to filled areas at the port of Philadelphia in which ships' ballast was dumped. Such areas were rich hunting grounds for 19th century botanists seeking unusual species (Smith 1867, Burk 1877).

Federal conservation status indicates that a taxon is listed or proposed for listing (LE, LT or C2 status) under the Endangered Species Act (US Department of the Interior 1990 and 1991). State conservation status indicates taxa listed as extirpated, endangered, threatened, rare or undetermined under the state program for the protection of native wild plants (Department of Environmental Resources 1987) and includes changes in species status scheduled for final approval early in 1993. Species believed to be extirpated in Pennsylvania are noted.

Noxious weed status at the federal and state levels was defined by the Federal Noxious Weed Act and the Pennsylvania Noxious Weed Control Act respectively. Wetland indicator status was based on the National Wetland Inventory list (Reed 1988).

Native plants are defined as those present before settlement by Europeans or those migrating into the state independent of human activity. Non-native, or introduced, plants include those which have become widely naturalized as well as others which have a very limited occurrence. Plants known only from cultivated situations were not included.

Several cases in which there are reliable early reports of species occurrence but no corroborating herbarium specimens include: *Chamaecyparis thyoides* (L.) BSP in southeastern Bucks County (Thomas and Moyer 1876) and Philadelphia County (Smith 1886); *Picea mariana* in Philadelphia Co. (Barton 1818); *Triadenum walteri* (Gmel.) Gleason in Philadelphia County (Barton 1818). These occurrences are not mapped.

The computer generated maps in this volume use latitude and longitude coordinates for specimen collection sites to plot the data. Because most herbarium specimen labels do not include latitude and longitude, coordinates were obtained from the Geographic Names Information System of the US Geological Survey (USGS 1988), or manually from USGS 71/2 minute maps for others not included in the USGS gazetteer. Specimens are thus mapped at the nearest named place.

The color illustrations are part of a larger set of original water color paintings of native Pennsylvania plants by Janet E. Klein, botanical illustrator.

ACKNOWLEDGEMENTS

Alfred E. Schuyler of the Academy of Natural Sciences of Philadelphia provided frequent and invaluable counsel during the preparation of this work. Carl S. Keener of Pennsylvania State University advised on nomenclatural and taxonomic issues and proofread all entries.

Robert Noland and Robert Wirtshafter of the Center for Energy and the Environment of the University of Pennsylvania assisted in the development of the database structure and created the programs making it possible.

Kathryn Fogarasi served as a research assistant on the project for 6 years. Flora of Pennsylvania interns who assisted in data gathering and entry included: Lynn Faust, Jeffrey Walck, Pamela White and Stephanie Neid.

Sue A. Thompson, Frederick Utech and Bonnie Isaac of the Carnegie Museum of Natural History, Section on Botany cooperated in the entry of label data from the 120,000 Pennsylvania specimens in the Carnegie Museum collection. James Bissell made data available from the herbarium of the Cleveland Museum of Natural History. Other individuals who provided specimen label data included: John R. Kunsman, Theodore Grisez, Thomas L. Smith, James Parks, Larry Klotz, Susan Munch, James Montgomery.

The following gave generously of their time to review portions of the checklist and/or examine questionable specimens: George W. Argus, Daniel F. Austin, T.M. Barkley, Mary E. Barkworth, I.J. Bassett, Joseph Beitel, Will H. Blackwell, Jr., Orland J. Blanchard, Jr., James G. Bruce, R.K. Brummitt, Robert Bye, Paul M. Catling, Janice Coffey-Swab, Lincoln Constance, Tom S. Cooperrider, William Crins, Arthur Cronquist, William D'Arcy, Lauramay T. Demster, Theodore Dudley, James E. Eckenwalder, Richard H. Eyde, Robert B. Faden, Jan Farmer, Shirley A. Graham, James W. Hardin, Robert R. Haynes, R. James Hickey, Peter A. Hypio, Duane Isely, Miles F. Johnson, Almut G. Jones, John T. Kartesz, Carl S. Keener, Robert Krall, Walter Lewis, Carlyle A. Luer, Elbert L. Little, Jr., Landon E. McKinney, John T. McNeil, John T. Mickel, Richard S. Mitchell, Robert H. Mohlenbrook, Michael

O. Moore, Nancy Morin, Larry Morse, J.K. Morton, Gerald B. Ownbey, James C. Parks, Howard Wm. Pfeifer, James Phipps, Richard W. Pohl, James S. Pringle, Richard K. Rabeler, James L. Reveal, Kenneth R. Robertson, W. Ann Robinson, C.M. Rogers, Reed C. Rollins, Paul Rothrock, A.E. Schuyler, Gerald Seiler, Charles Sheviak, Richard W. Spellenberg, G. Ledyard Stebbins, Constance Taylor, John W. Thieret, Sue A. Thompson, Arthur O. Tucker, Gordon C. Tucker, Frederick H. Utech, Edward G. Voss, Warren L. Wagner, Warren H. Wagner, Marcia Waterway, Grady L. Webster, Dieter H. Wilbur, Jean Wooten.

The Pew Charitable Trusts, Mrs. Lammot du Pont Copeland, the Estate of Mary Tyson Janney and the Gladys K. Delmas Fund of the American Philosophical Society generously provided a part of the support required to complete this book.

Ann Fowler Rhoads
December 1992

THE SETTING

Pennsylvania lies between 39°43′13″ and 42° north latitude and 74°37′30″ and 80°31′13″ west longitude. The state is divided into 67 counties (Figure 1). Elevation ranges from sea level to 979 M. Average annual precipitation ranges from 97 to 112 cm, with the central part of the state receiving the least moisture. Slightly more precipitation falls during the spring and summer than during the fall and winter seasons. Average July maximum temperature ranges from 24.4°C in the northeast to 28.8°C in the southwest. Average January minimum temperature ranges from -6.6°C in the southeast to -10°C in the northeast and northcentral regions. The length of the growing season ranges from 180 days in the southeast to 120 days in the northcentral region (Yarnal 1989).

The state includes portions of seven physiographic provinces extending from the Atlantic Coastal Plain along the Delaware River estuary in the southeast through the Piedmont, Valley and Ridge and Allegheny Plateaus to the Central Lowland

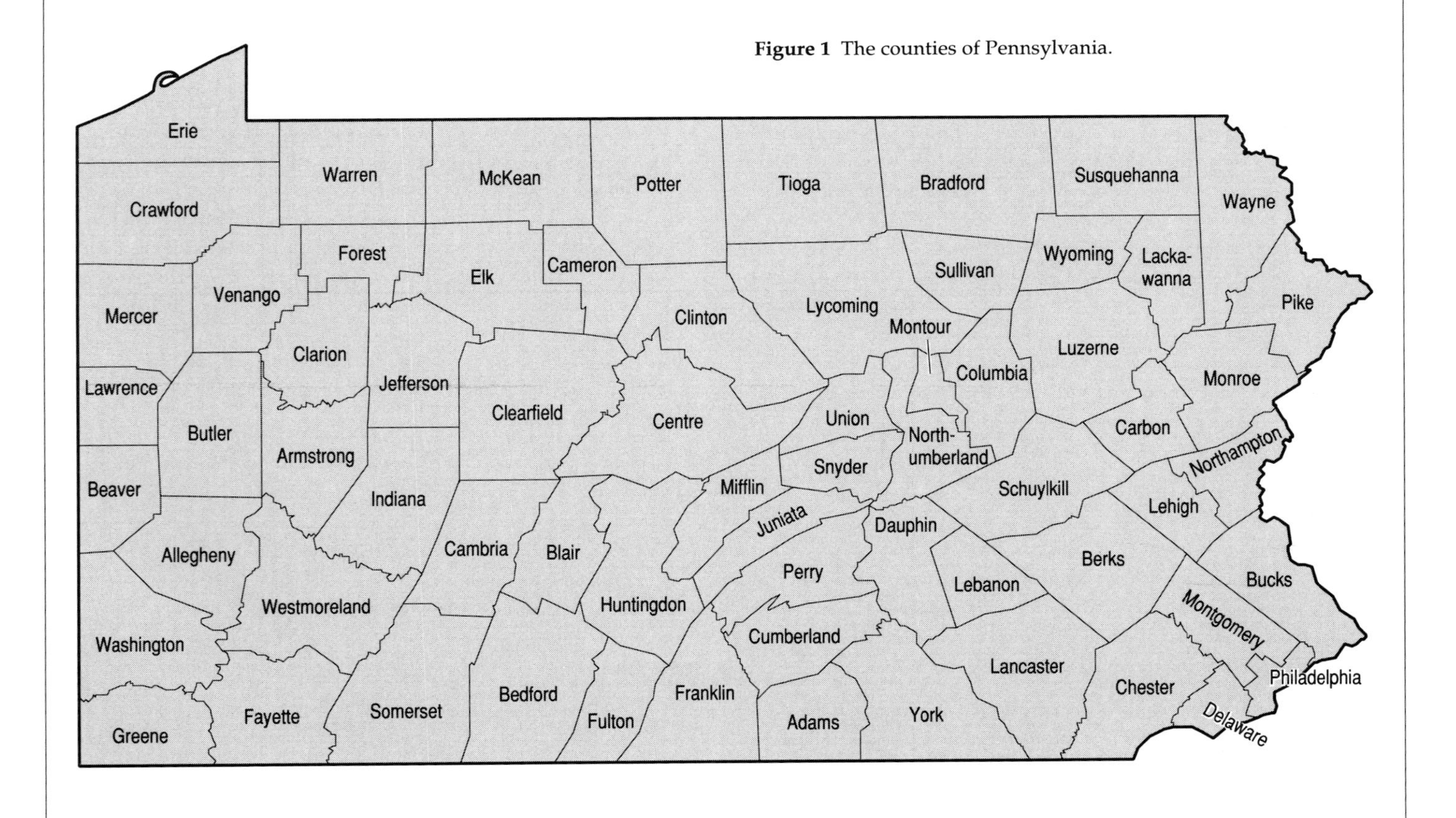

Figure 1 The counties of Pennsylvania.

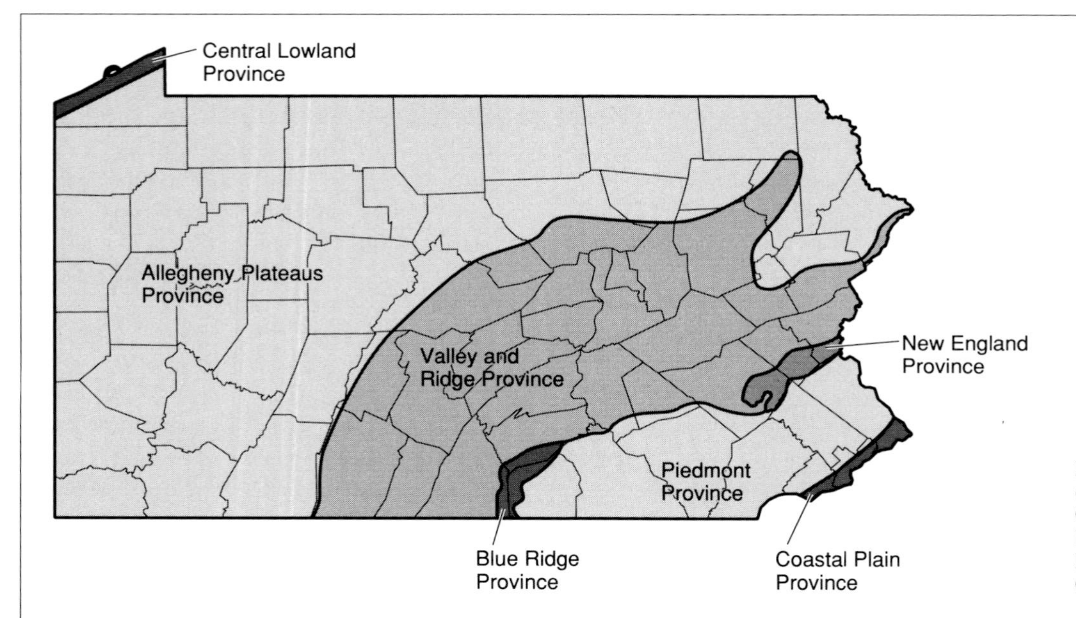

Figure 2 Physiographic Provinces of Pennsylvania, after Map 13, Commonwealth of Pennsylvania, Department of Environmental Resources, Topographic and Geologic Survey, undated.

Province along the shore of Lake Erie (Figure 2). The southern tip of the New England Province extends into eastern Pennsylvania where it is known as the Reading Prong. Similarly, South Mountain, in the southcentral portion of the state, represents the northern extent of the Blue Ridge Province (Marsh and Marsh 1989).

GEOLOGY

North and west of the Allegheny Front the surficial geology is primarily sedimentary, consisting of horizontal strata of sandstone and shale of Devonian, Mississippian, Pennsylvanian and Permian age. The Valley and Ridge Province is characterized by folded and highly deformed rock layers forming a series of parallel ridges that arc across the state from northeast to southwest. Sandstone ridges alternate with limestone or shale valleys.

Within the Piedmont, the Triassic Lowlands are characterized by shales and sandstones into which diabase dikes and sills have intruded. The Piedmont Uplands consist of low, rolling hills formed by metamorphic schists and quartzites, and broad limestone valleys. Outcrops of serpentine rock contribute to the geological and floristic diversity of the Piedmont Uplands (Figure 3). The

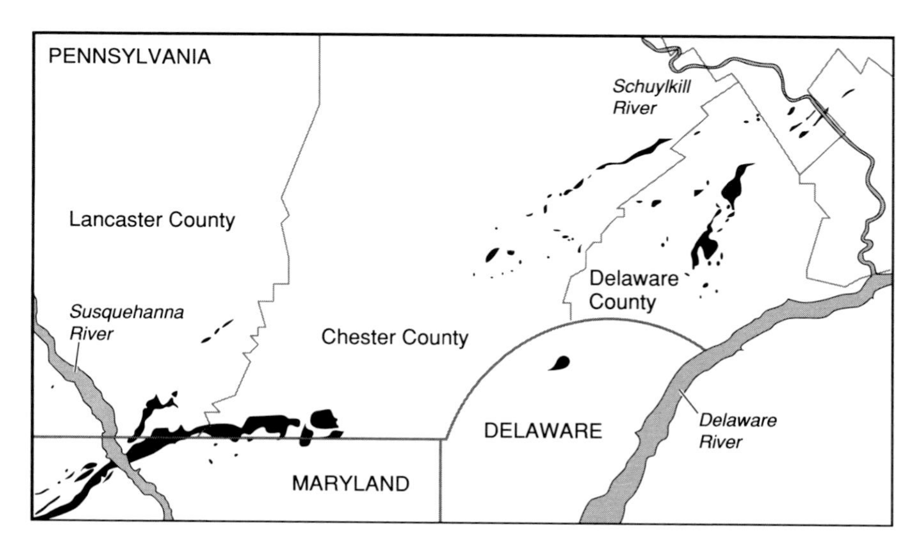

Figure 3 Serpentine Rock Outcrops of Pennsylvania, adapted from Nancy C. Pearre and Allen V. Heyl, Jr. 1960. Chromite and Other Mineral Deposits in Serpentine Rocks of the Piedmont Upland, Maryland, Pennsylvania and Delaware. Geological Survey Bulletin 1082-K.

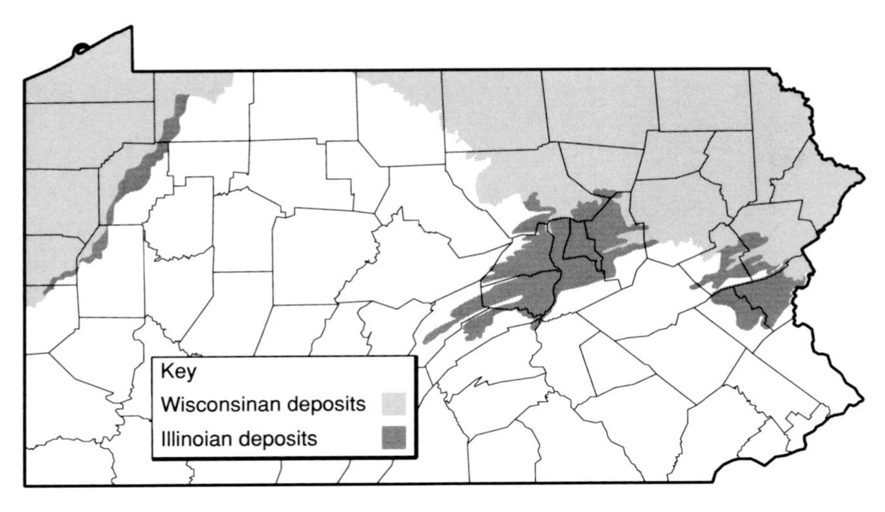

Figure 4 Glacial Deposits of Pennsylvania, after Map 59, Commonwealth of Pennsylvania, Department of Topographic and Geologic Survey, undated.

Coastal Plain is a narrow belt of unconsolidated sands and gravels deposited during the Tertiary and Quaternary periods (Myer 1989).

Glaciation has also contributed to shaping the landscape of Pennsylvania. Most recently, the Wisconsinan Glaciation covered the northeast and northwest corners of the state leaving surficial deposits of till which disrupted drainage patterns and created kettlehole lakes, bogs and extensive wetlands. In the northwest, the till has a calcareous component which is much less frequent in the east. Deposits of earlier Illinoian till occur south of the Wisconsinan border in several areas (Figure 4).

VEGETATION

Pennsylvania lies within the eastern deciduous forest as described by Braun (1950). According to Kuchler (1964) the major forest types include the northern hardwood forest, which occupies the higher elevations on the Allegheny Plateau, and the Appalachian oak forest, which covers the southern two-thirds of the state. In addition, the mid-western beech-maple forest extends into northwest Pennsylvania, and the mixed mesophytic forest of the southern Appalachians reaches the southwest corner of the state. The oak-hickory-pine forest reaches its northern limit in southcentral Pennsylvania (Figure 5).

Little remains of the original forest cover due to early clearing for agriculture followed by extensive clear cutting of the remote forested areas which took place between 1880 and 1930 (Whitney 1990). Remnants of old growth total only a few thousand acres at scattered sites. Forest composition has been further altered by introduced pests and diseases such as the gypsy moth, chestnut blight, beech bark disease, Dutch elm disease and, most recently, excessive

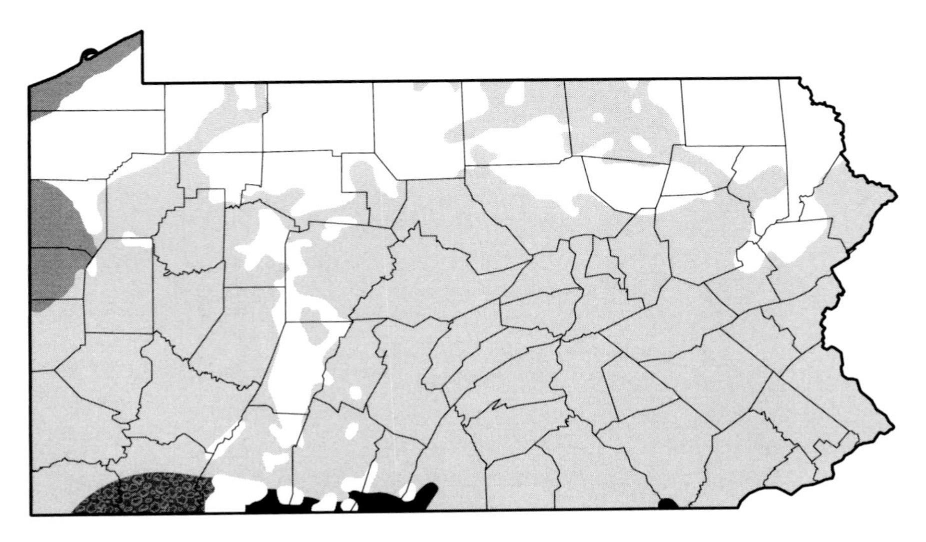

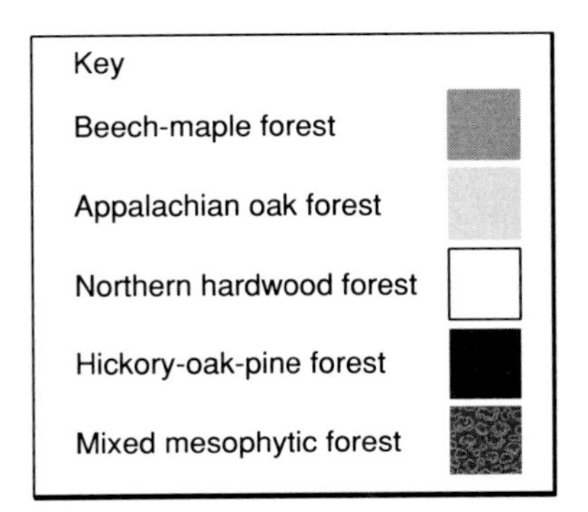

Figure 5 Major Forest Types of Pennsylvania, adapted from A.W. Kuchler. 1964. Potential Natural Vegetation of the Conterminous United States. American Geographical Society, Special Publication No.36.

browsing by white-tail deer (Marquis 1981).

Pennsylvania's floristic diversity is derived from the meeting and mixing of northern and southern forest types within the state. Further variety is provided by specialized habitats including freshwater tidal marshes; serpentine, shale, limestone and till barrens; wet and dry cliffs and rock outcrops; seeps; fens; bogs; lake bluffs and sand plains.

CHANGES IN THE FLORA

EXCLUDED SPECIES

The taxa listed below were included in earlier treatments of the Pennsylvania flora but have been omitted from this work.

Agalinis obtusifolia Raf. - specimens reidentified
Carex complanata Torr. & Hook. - specimens reidentified
Carex nigra (L.) Reichard - specimens reidentified
Cerastium pumillum Curtis - specimen reidentified
Chaenomeles japonica (Thunb.) Lindl. ex Spach - cultivated only
Crataegus arnoldiana Sarg. - cultivated only
Crataegus laevigata DC. - no specimen
Cyperus brevifolius (Rottb.) Hassk. - specimens reidentified
Dichromena colorata (L.) Pfeiffer (syn. *Rhyncospora c.*) - specimen reidentified
Fraxinus excelsior L. - cultivated only
Gymnocarpium robertianum (Hoffman) Newman - specimens reidentified
Helonias bullata L. - cultivated only
Hydrangea quercifolia Bartram - cultivated only
Juncus longii Fern. - no confirmed PA material
Lycopodium complanatum L. - specimens reidentified
Matelea carolinense (Jacq.) Woodson - specimens reidentified
Panicum calliphyllum Ashe - specimens reidentified
Sesuvium maritimum (Walt.) BSP - no specimen
Triglochin maritimum L. - no specimen
Tradescantia subaspera Ker-Gwal. - cultivated only

ADDITIONS

The plants listed below were not included in earlier published treatments of the Pennsylvania flora. They represent taxa new to the state, or newly discovered in the field or the herbaria; rather than the result of taxonomic splitting. Hybrids are not included.

Native species

Aster borealis (Torr. & A.Gray) Prov.
Aster drummondii Lindl.
Aster nemoralis Ait.
Carex adusta Boott.
Carex alopecoidea Tuckerman
Carex crawfordii Fern.
Carex formosa Dewey
Carex gravida Bailey
Carex wiegandii Mackenzie
Echinacea laevigata (C.L.Boynt. & Beadle) Blake
Eleocharis caribaea (Rottb.) Blake
Eleocharis parvula (Roemer & Schultes) Link ex Buff. & Fingerh.
Elodea schweinitzii (Planch) Caspary
Epilobium parviflorum Schreber
Equisetum laevigatum A.Braun
Fraxinus profunda (Bush) Bush
Gentianopsis procera (Holm) Ma
Hieracium traillii Greene
Hydrophyllum macrophyllum Nutt.
Hypericum sphaerocarpum Michx.
Juncus brachycarpus Engelm.
Leiophyllum buxifolium (Berg.) Ell.
Lemna obscura (Austin) Daubs
Lemna turionifera Landolt
Lespedeza angustifolia (Pursh) Ell.
Linum medium (Planchon) Britton var. *medium*
Ludwigia decurrens Walt.
Lysimachia quadriflora Sims.
Myriophyllum verticillatum L.
Ophioglossum engelmannii Prantl.
Panicum laxiflorum Lam.
Panicum tuckermanii Fern.
Polygala curtissii A. Gray
Prunus angustifolia Marshall
Schoenoplectus robustus (Pursh) Strong (syn. *Scirpus r.*)
Spiranthes casei Catling & Cruise
Spiranthes magnicamporum Sheviak

Thalictrum dasycarpum Fisch. & Ave.-Lall.
Trichomanes intricatum Farrar
Vittaria appalachiana Farrar & Mickel
Wolffiella gladiata (Hegelm.) Hegelm.

Non-native species

Acanthospermum australe (Loefl.) Kuntze
Agastache foeniculum (Pursh) Kuntze
Albizia julibrissin Durazz
Ammophila arenaria (L.) Link
Apium leptophyllum (Pers.) F. Muell.
Betula pendula Roth.
Calamovilfa longifolia (Hook.) Scribn.
Carex nebraskensis Dewey
Carex praegracilis W. Boott.
Chorispora tenella (Pallas) DC.
Cirsium undulatum (Nutt.) Spreng.
Cleome spinosa L.
Colutea arborescens L.
Cornus drummondii C.A. Mey
Crepis vesicaria L.
Critesion brachyantherum (Nevski) Barkw. & D.R.Dewey
Cyperus aggregatus (Willd.) Endl.
Cyperus difformis L.
Cyperus serotinus Rottb.
Epilobium parviflorum Schreber
Eriochloa villosa (Thunb.) Kunth
Erodium moschatum (L.) L'Her.
Erysimum hieraciifolium L.
Euonymus fortunei (Turcz.) Hand.-Mazz.
Euonymus yedoensis Koehne
Eupatorium cannabinum L.
Galanthus elwesii Hook. f.
Geranium pratense L.
Glyceria fluitans (L.) R.Br.
Hyacinthus orientalis L.
Iva annua L.
Lactuca pulchella (Pursh) DC.
Larix kaempferi (Lamb) Carr.
Leptochloa fascicularis (Lam.) A.Gray v. *acuminata* (Nash) Gleason
Linum catharticum L.
Lobelia chinensis Lour.
Lonicera fragrantissima Lindl. & Paxt.
Pachysandra procumbens Michx.- naturalized from cultivated sources
Pachysandra terminalis Siebold & Zucc.
Picea pungens Engelm.
Polemonium caeruleum L.
Polygonum nepalense Meissner
Quercus acutissima Carruth.
Spergularia marina (L.) Griseb.
Spergularia media (L.) Presl ex Griseb.
Sporobolis pyramidatus (Lam.) A.S.Hitchcock
Triodia stricta (Nutt.) Benth.
Trapa natans L.
Vigna luteola (Jacquin) Bentham
Wisteria floribunda (Willd.) DC.
Xanthorhiza simplicissima Marsh.
Zantedeschia albomaculata (Hook.) Baill.
Zinnia elegans Jacq.

SPECIES WHICH ARE SPREADING

This list includes plants which have expanded their ranges in Pennsylvania significantly in recent years and are probably more frequent than the record of herbarium specimens indicates. Several are plants of brackish or alkaline habitats which are spreading along highways in response to the use of de-icing salts. Others are invading old fields, disturbed woods and edges. Two of the species are relatively recent additions to the flora of the freshwater tidal marshes along the Delaware estuary.
Acer platanoides L. - disturbed woods
Alliaria petiolata (Bieb.) Cavara & Grande - disturbed woods
Anthriscus sylvestris (L.) Hoffm. - alluvial areas
Baccharis halimifolia L. - highway ditches
Bidens polylepis Blake - moist fields
Carduus nutans L. - roadsides and old fields
Celastrus orbiculatus Thunb. - disturbed woods
Elaeagnus umbellata Thunb. - old fields
Eleocharis parvula (Roemer & Schultes) Link. ex Buff. & Fingerh. - tidal marshes
Eupatorium serotinum Michx. - sandy fields
Hesperis matronalis L. - moist woods, floodplains
Hutera cheiranthos (Vill.) Gomez-Campo - highway berms
Leptochloa fascicularis (Lam.) A.Gray v. *acuminata* (Nash) Gleason - highway berms
Lobelia chinensis Luer. - tidal marshes
Lonicera maackii (Rupr.) Maxim. - old fields and disturbed woods
Lonicera morrowii A.Gray - old fields and disturbed woods
Lythrum salicaria L. - wetlands
Magnolia tripetala (L.) L. - disturbed woods

Microstegium vimineum (Trin.) A.Camus. - disturbed woods
Myriophyllum spicatum L. - lakes and rivers
Phragmites australis (Cav.) Trin. - highway ditches and wetlands
Polygonum cuspidatum Sieb. & Zucc. - floodplains
Polygonum perfoliatum L. - floodplains and moist fields
Puccinella distans (Jacq.) Parl. - highway berms
Ranunculus ficaria L. - floodplains
Rosa multiflora Thunb. - old fields
Senecio viscosus L. - roadsides
Solidago sempervirens L. - highway berms
Spergularia marina (L.) Griseb. - highway berms
Spergularia media (L.) L'Her. - highway berms
Triplasis purpurea (Walter) Chapman - sandy fields
Typha angustifolia L. - highway ditches

STATISTICAL SUMMARY

The first statewide treatment of the Pennsylvania flora (Porter 1903) listed 2201 species. Wherry et al. (1979) included distribution maps for 3574 taxa. This volume treats 3319 taxa segregated into 179 families and 982 genera. An additional 62 hybrids, for which there are very limited collections, are mentioned, but not mapped.

	native	**non-native**	**total**
Pteridophyta	107	1	108
Pinophyta	15	11	26
Magnoliopsida	1266	1016	2282
Liliopsida	687	215	902
Totals	2076	1242	3318
Largest families:			
Asteraceae	215	155	370
Poaceae	177	142	319
Cyperaceae	265	20	285
Rosaceae	95	62	157
Fabaceae	55	68	123
Brassicaceae	27	72	99

	native	**non-native**	**total**
Largest genus:			
Carex	169	9	178
Trees	118	68	186
Shrubs	188	100	288
Woody vines	21	14	35
Herbs:			
Annual or biennial	295	602	897
Perennial	1454	458	1912

Autotrophic	3263
Heterotrophic:	
Parasitic or hemiparasitic	37
Sparophytic	6
Carnivorous	12
Federal endangered, threatened or candidate	21
State extirpated, endangered, threatened, rare or undetermined	607
Federal noxious weeds	4
State noxious weeds	13

KEY TO CHECKLIST FORMAT

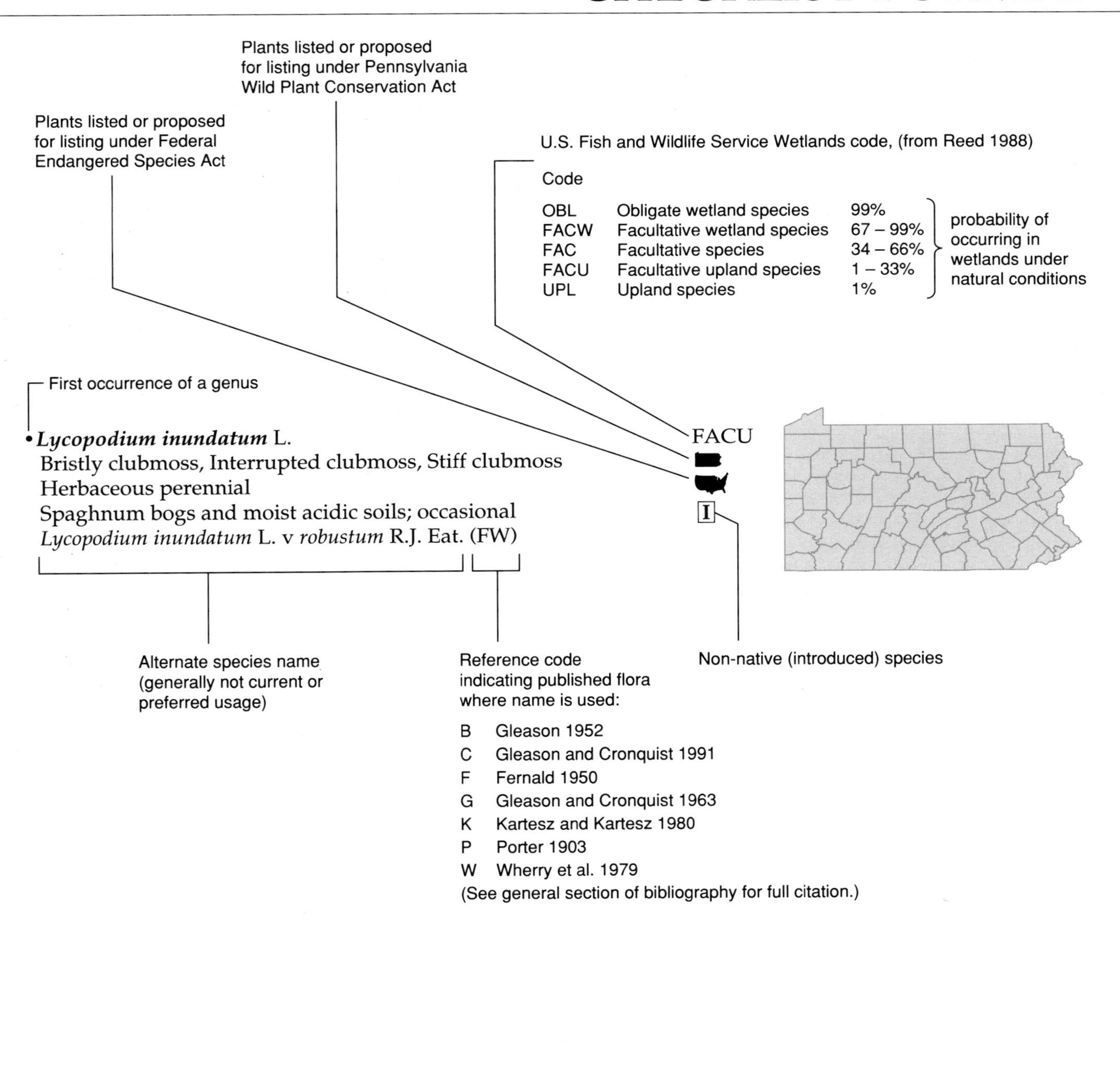

PTERIDOPHYTA
(Ferns and Fern Allies)

Christmas fern (*Polystichum acrostichoides*) grows on moist, wooded slopes throughout the state.

LYCOPODIACEAE

•***Diphasiastrum digitatum*** (A.Braun) Holub
Southern ground-cedar
Herbaceous perennial
Open woods and thickets, in acidic to subacidic soil.
Lycopodium complanatum L. var. *flabelliforme* Fern. FGB; *Lycopodium flabelliforme* (Fern.) Blanch. W; *Lycopodium digitatum* Dillen. C

FACU-

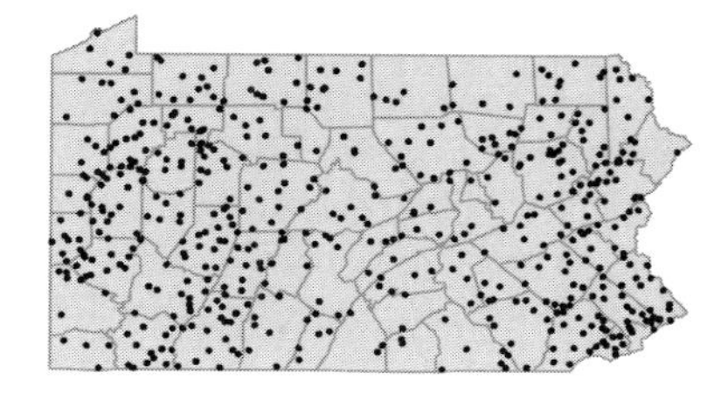

Diphasiastrum digitatum x tristachyum
Ground-pine
Herbaceous perennial
Open woodlands or barrens, in acidic soils.
Lycopodium x habereri House W; *Diphasiastrum x harbereri* (House) Holub

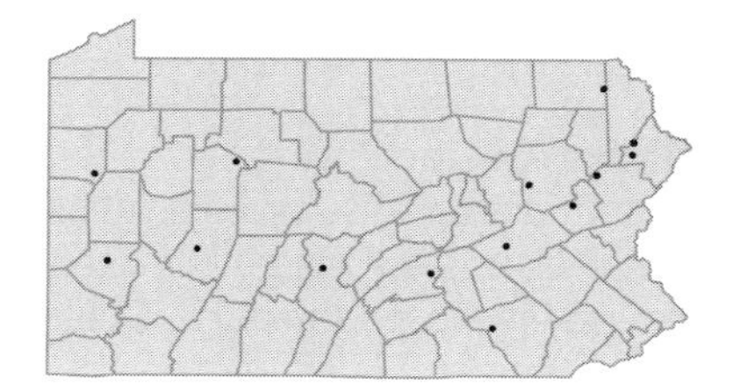

Diphasiastrum sabinifolium (Willd.) Holub
Fir clubmoss; Ground-fir
Herbaceous perennial
Open woods and meadows, in acidic soil. Believed to be extirpated, last collected in 1942.
Lycopodium sabinifolium Willd. CWGFBK

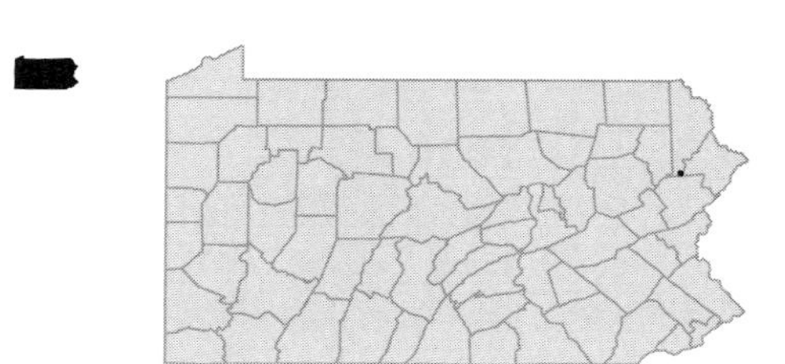

Diphasiastrum tristachyum (Pursh) Holub
Ground-cedar
Herbaceous perennial
Open coniferous woods and sandy or rocky barrens, in acidic soil.
Lycopodium tristachyum Pursh CFWBGK

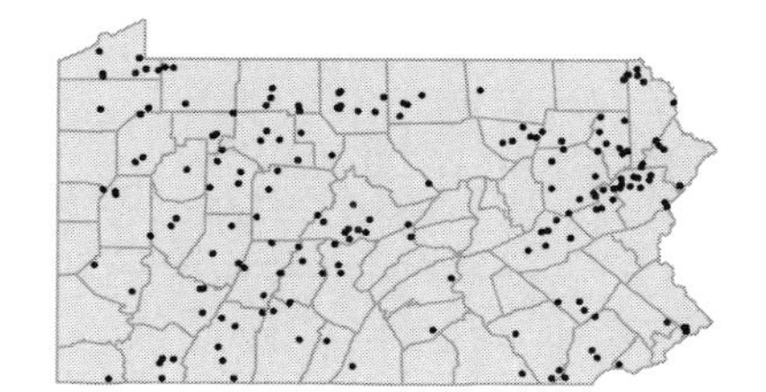

•***Huperzia lucidula*** (Michx.) Trevisan
Shining clubmoss
Herbaceous perennial
Cool, moist woods in humus-rich soils.
Lycopodium lucidulum Michx. BFGKWC

FACW-

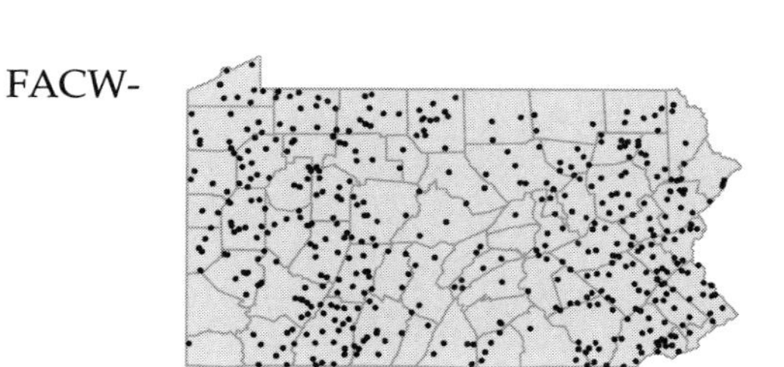

Huperzia porophila (Lloyd & Underw.) Holub
Cliff clubmoss; Rock clubmoss
Herbaceous perennial
Moist sandstone cliffs adjacent to waterfalls.
Lycopodium selago L. var. *patens* (Beauv.) Desv. GB; *Lycopodium porophilum* Lloyd & Underw. C

FACU-

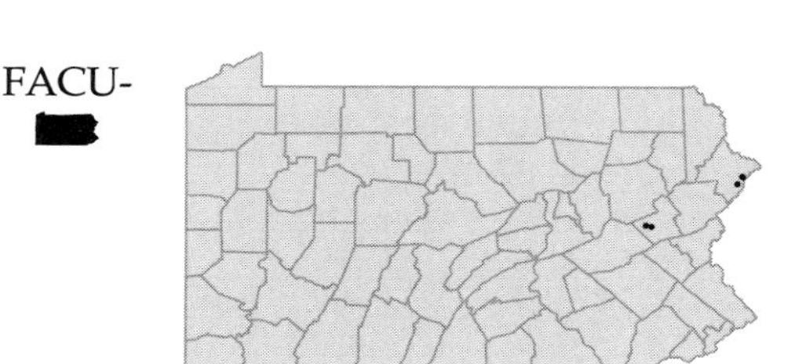

Huperzia selago (L.) C. Martius & Schrank
Fir clubmoss; Mountain clubmoss
Herbaceous perennial
Rocky slopes and ledges, in acidic soil. Believed to be extirpated, last collected in 1870.
Lycopodium selago L. BFGKWC

FAC

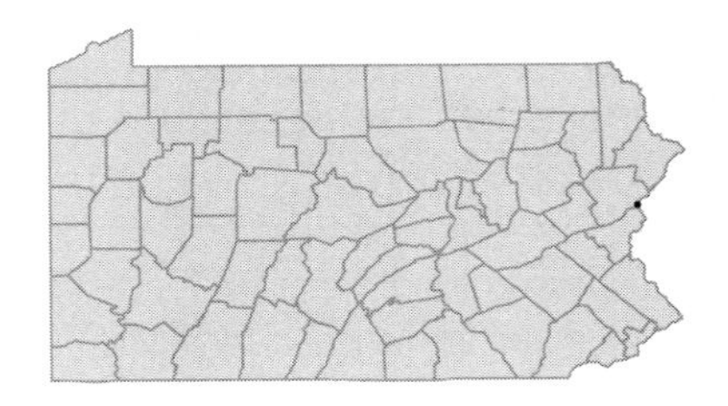

• ***Lycopodiella alopecuroides*** (L.) Cranfill
Fox-tail clubmoss
Herbaceous perennial
Openings in moist, coastal plain woods. A hybrid with *L. appressa* (*L. copelandii* Eig.) has been collected occasionally.
Lycopodium alopecuroides L. BFGKWC

FACW+

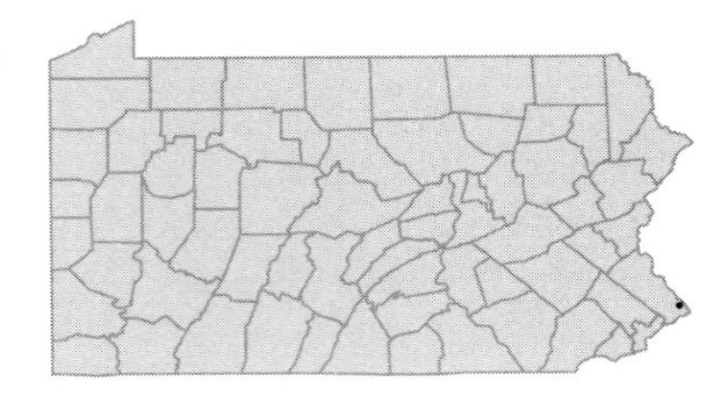

Lycopodiella appressa (Lloyd & Underw.) Cranfill
Bog clubmoss
Herbaceous perennial
Moist, acidic soil on or near the Coastal Plain. A hybrid with *L. inundata* has been collected occasionally.
Lycopodium inundatum L. var. *bigelovii* Tuckerman FGB; *Lycopodium appressum* (Chapm.) Lloyd & Underw. C

FACW+

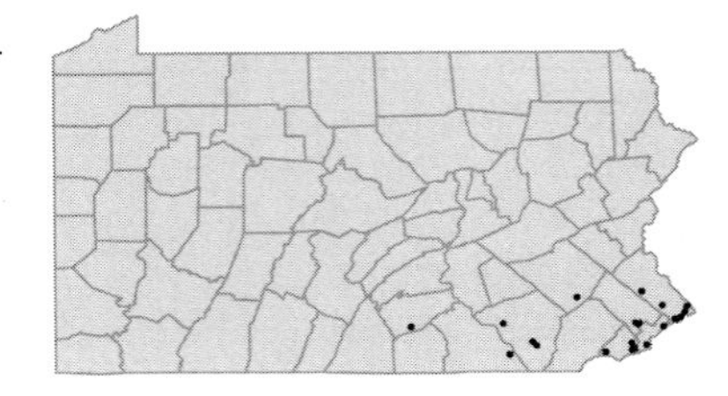

Lycopodiella inundata (L.) Holub
Bog clubmoss
Herbaceous perennial
Sphagnum bogs and moist, acidic soils.
Lycopodium inundatum L. BFGKWC

OBL

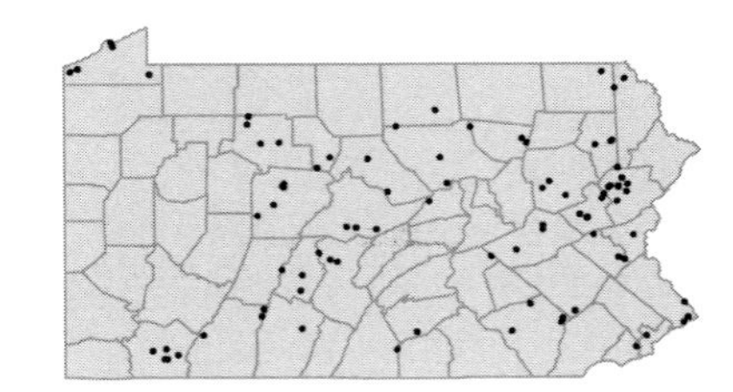

• ***Lycopodium annotinum*** L.
Bristly clubmoss; Interrupted clubmoss; Stiff clubmoss
Herbaceous perennial
Cool, moist, coniferous forests in acidic soils.

FAC

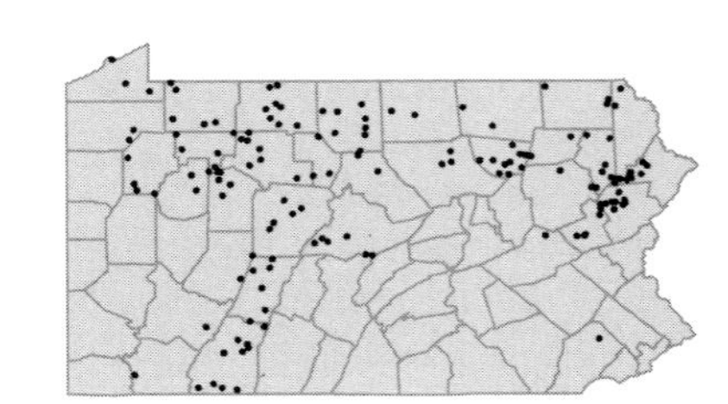

Lycopodium clavatum L. var. ***clavatum***
Common clubmoss; Running-pine
Herbaceous perennial
Open woods, bog margins, or rocky barrens, in acidic or subacidic soils.

FAC

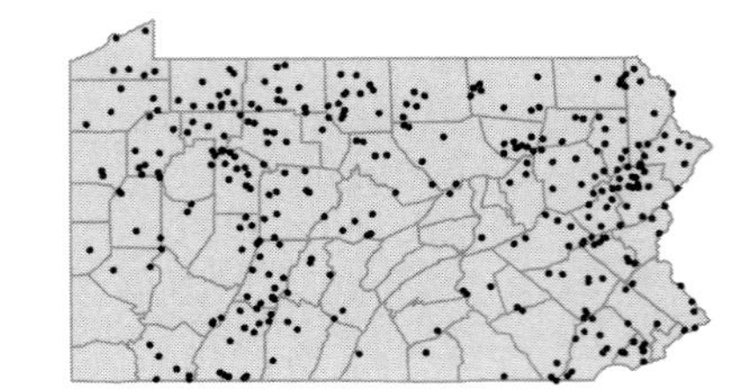

Lycopodium clavatum L. var. ***monostachyon*** Grev. & Hook.
Common clubmoss; Running-pine
Herbaceous perennial
Open woods and edges.

FAC

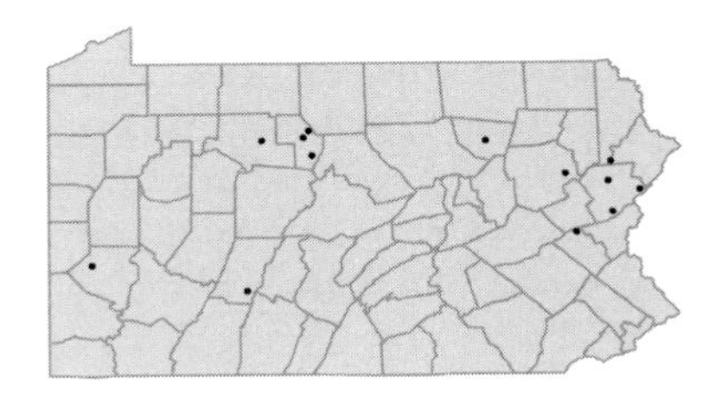

Lycopodium dendroideum Michx.
Round-branch ground-pine
Herbaceous perennial
Bogs and barrens, in moist, acidic soils.
Lycopodium obscurum L. var. *dendroideum* (Michx.) D.C.Eat. FBW

FACU

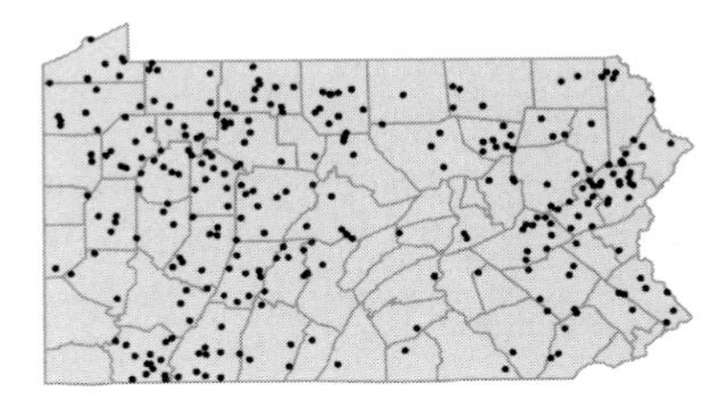

Lycopodium hickeyi W.Wagner, Beitel & R.C.Moran
Ground-pine
Herbaceous perennial
Moist woods or marshes.
Lycopodium obscurum L. var. *isophyllum* Hickey K

FACU

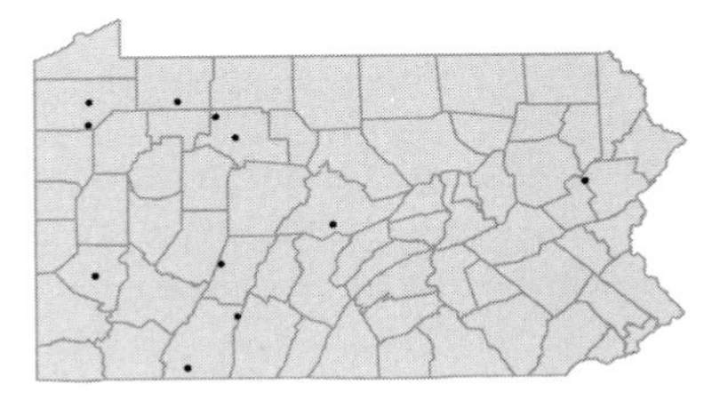

Lycopodium obscurum L.
Princess-pine; Tree clubmoss
Herbaceous perennial
Moist to dry woods, in acidic to subacidic soils.

FACU

SELAGINELLACEAE

• ***Selaginella apoda*** (L.) Fern.
Creeping spikemoss; Meadow spikemoss
Herbaceous perennial
Moist meadows and stream banks, in circumneutral soils.

FACW

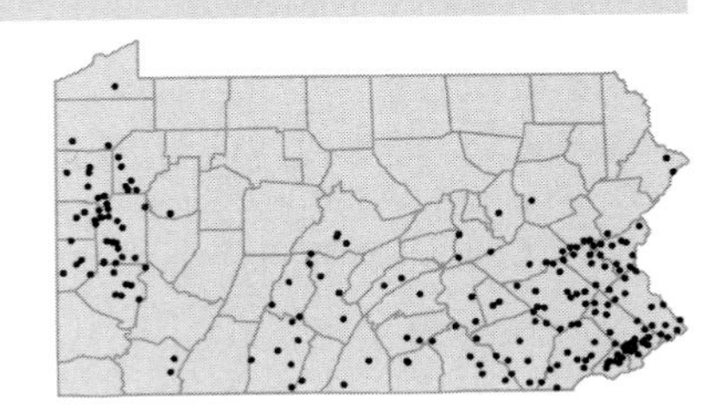

Selaginella rupestris (L.) Spring
Rock spikemoss
Herbaceous perennial
Dry, exposed ledges and slopes.

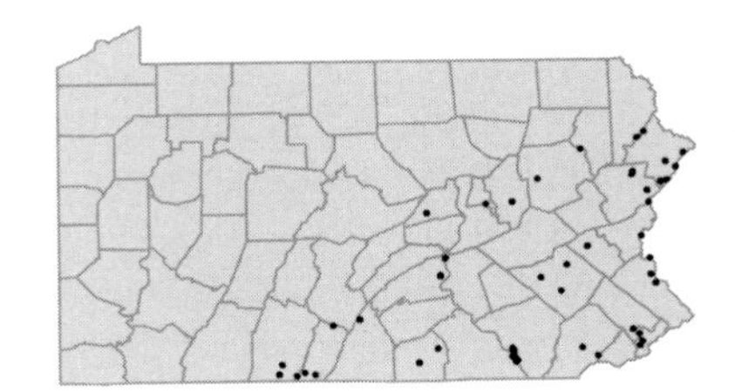

ISOETACEAE

• ***Isoetes echinospora*** Durieu
Spiny-spored quillwort
Herbaceous perennial, rooted submergent aquatic
Shallow water of cold lakes, ponds and slow-moving streams.
Isoetes muricata Durieu FW

OBL

Isoetes engelmannii A.Braun
Engelmann's quillwort
Herbaceous perennial, rooted submergent aquatic
Shallow water of lakes, ponds, slow-moving streams, and river shores.

OBL

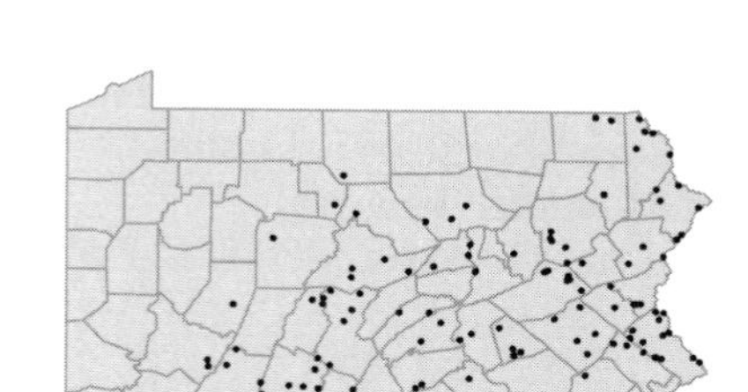

Isoetes engelmannii x riparia
Quillwort
Herbaceous perennial, rooted submergent aquatic
Shallow water, shorelines of reservoirs.
Isoetes x brittonii D.F.Brunton & W.C.Taylor

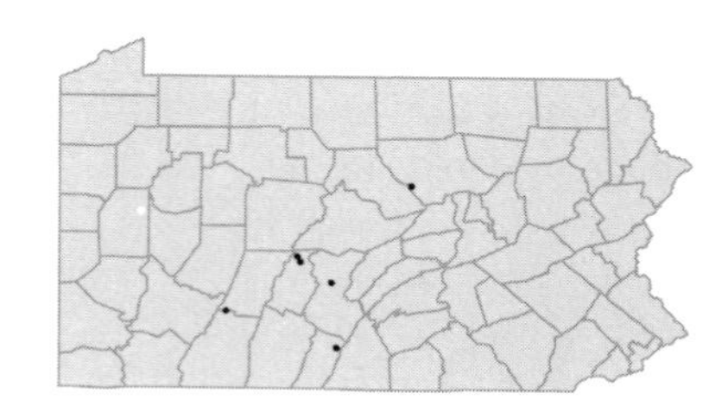

Isoetes riparia Engelm. ex A.Braun
Riverbank quillwort
Herbaceous perennial, rooted submergent aquatic
Shallow water of slow-moving rivers and streams, also intertidal mudflats.

OBL

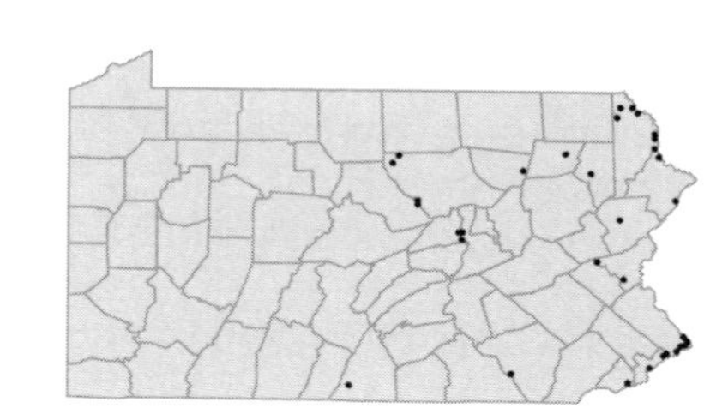

EQUISETACEAE

• ***Equisetum arvense*** L.
Common horsetail; Devil's-guts
Herbaceous perennial
Stream banks, meadows, roadsides and railroad embankments.

FAC

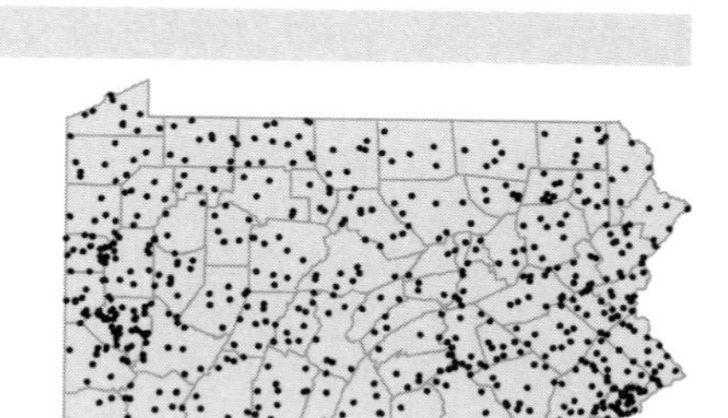

Equisetum arvense x fluviatile
Shore horsetail; Water horsetail
Herbaceous perennial
Moist stream banks.
Equisetum x litorale Kuehlew. ex Rupr. FGBW

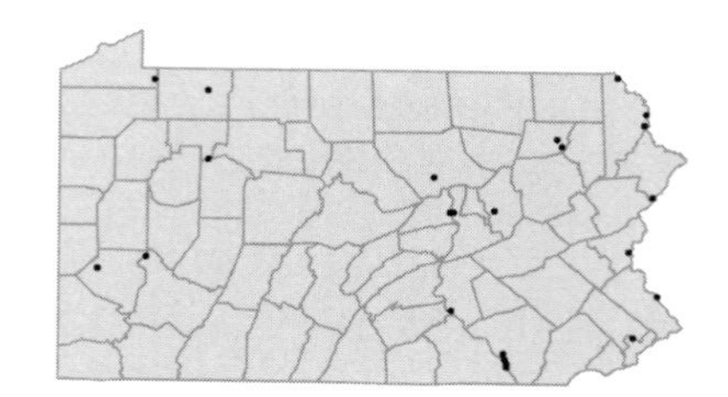

Equisetum fluviatile L.
Water horsetail
Herbaceous perennial, emergent aquatic
Shallow water of ponds, swamps and sluggish streams.

OBL

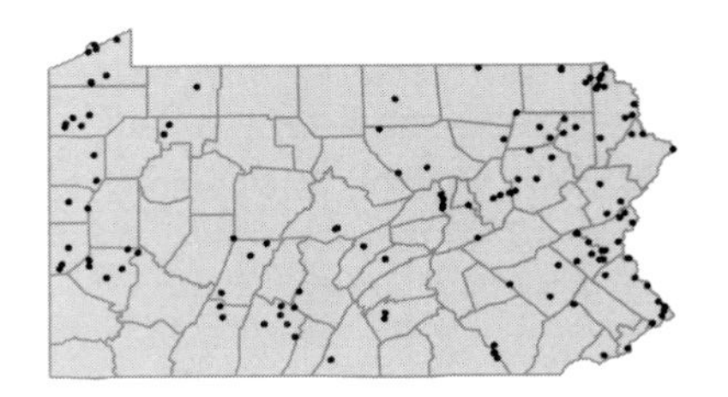

Equisetum hyemale L. var. ***affine*** (Engelm.) A.A.Eat.
Scouring-rush
Herbaceous perennial
Sandy, circumneutral shores, banks and roadsides.
Equisetum hyemale L. var. *pseudohiemale* (Farw.) Morton GB

FACW

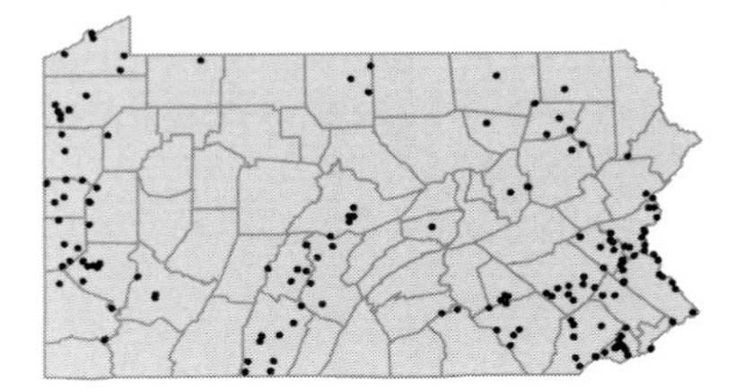

Equisetum hyemale var. ***affine x laevigatum***
Scouring-rush
Herbaceous perennial
Moist, sandy or clayey circumneutral soils.
Equisetum x ferrissii Clute W

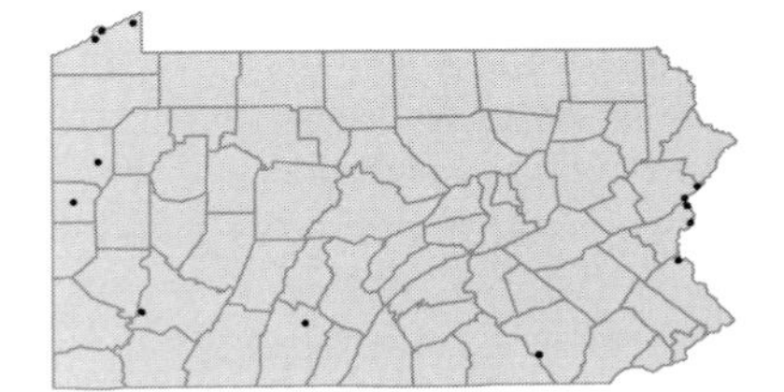

Equisetum laevigatum A.Braun
Smooth horsetail
Herbaceous perennial
Sandy shores.
Equisetum hyemale L. var. *intermedium* A.A.Eat. F

FACW

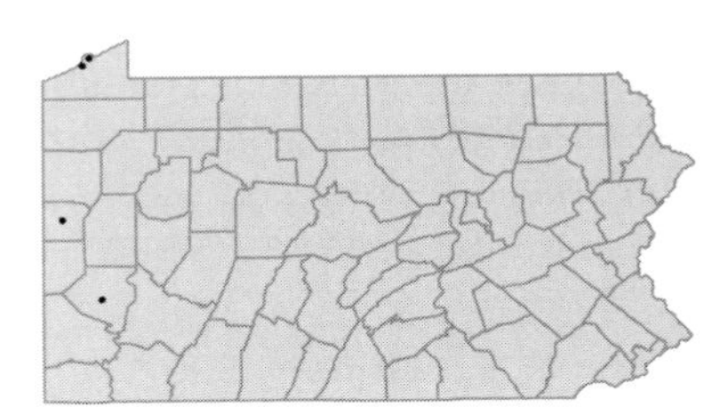

Equisetum sylvaticum L.
Woodland horsetail
Herbaceous perennial
Moist, open woods and wet meadows, in subacidic soils.

FACW

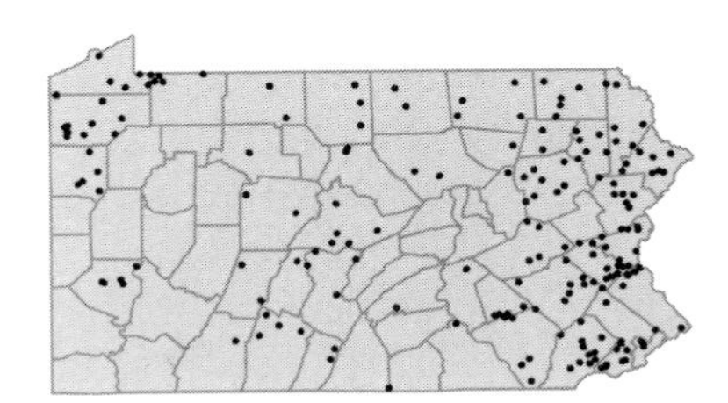

Equisetum variegatum Schleich. ex Weber & C.Mohr
Variegated horsetail
Herbaceous perennial
Damp, sandy flats and stream banks, in circumneutral to alkaline soils.

FACW

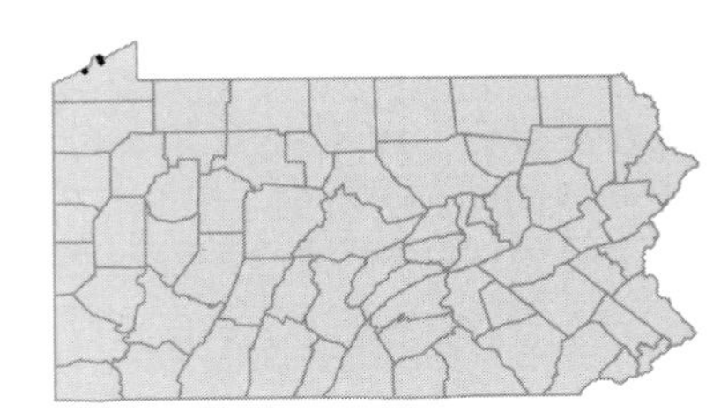

OPHIOGLOSSACEAE

• ***Botrychium dissectum*** Spreng. FAC
Cut-leaved grape-fern
Herbaceous perennial
Moist, open woods, meadows and barrens, in moderately acidic soils.
Botrychium obliquum Muhl. W

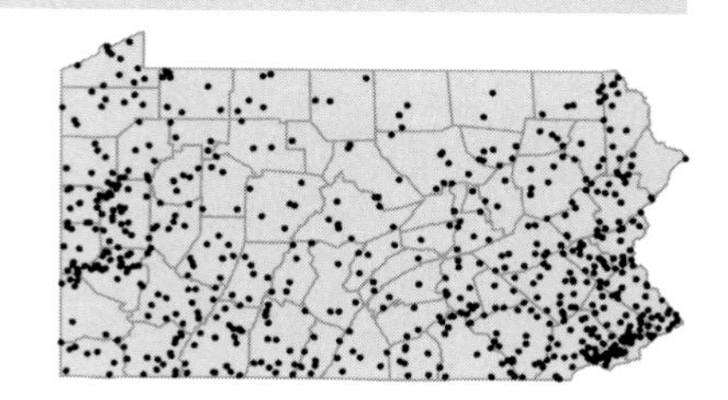

Botrychium lanceolatum (Gmel.) Angstr. FACW
Lance-leaved grape-fern; Lance-leaved moonwort
Herbaceous perennial
Cool, humus-rich woods and hummocks.

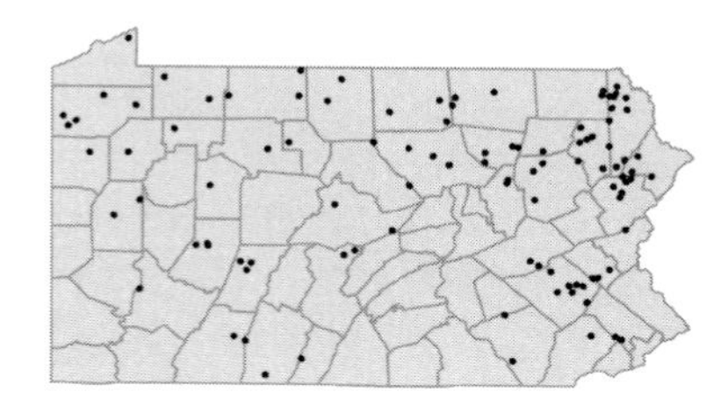

Botrychium matricariifolium (A.Braun ex Doll) A.Braun ex Koch
Matricary grape-fern
Herbaceous perennial
Moist, humus-rich woods and edges, in subacidic to circumneutral soils.

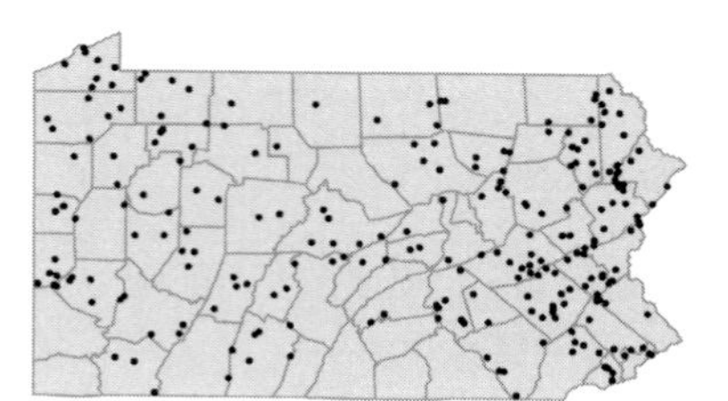

Botrychium multifidum (Gmel.) Rupr.
Broad grape-fern; Leathery grape-fern; Northern grape-fern
Herbaceous perennial
Moist thickets, meadows and barrens, in subacidic soils.

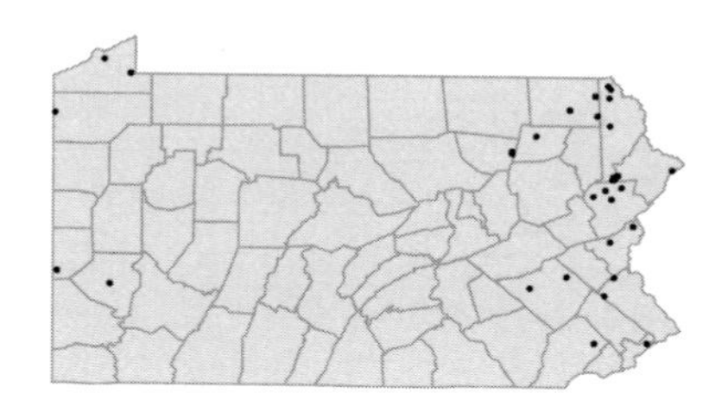

Botrychium oneidense (Gilbert) House
Blunt-lobed grape-fern
Herbaceous perennial
Moist, rich woods.

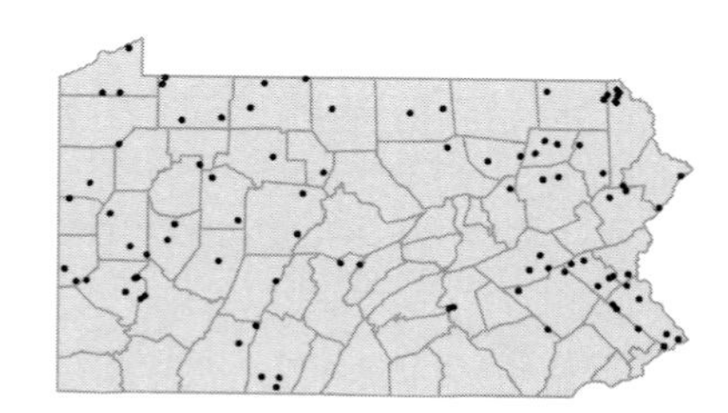

Botrychium simplex E.Hitchc.
Least grape-fern; Least moonwort
Herbaceous perennial
Moist woods, meadows and barrens, in subacidic to circumneutral soils.

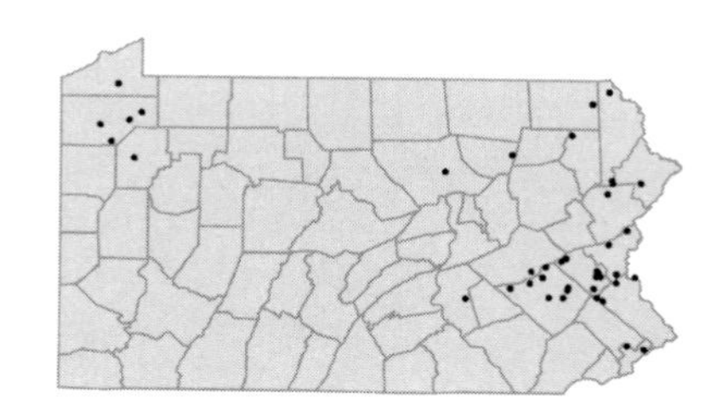

Botrychium virginianum (L.) Swartz
Rattlesnake fern
Herbaceous perennial
Dry or moist, humus-rich woods, in subacidic to circumneutral soils.

FACU

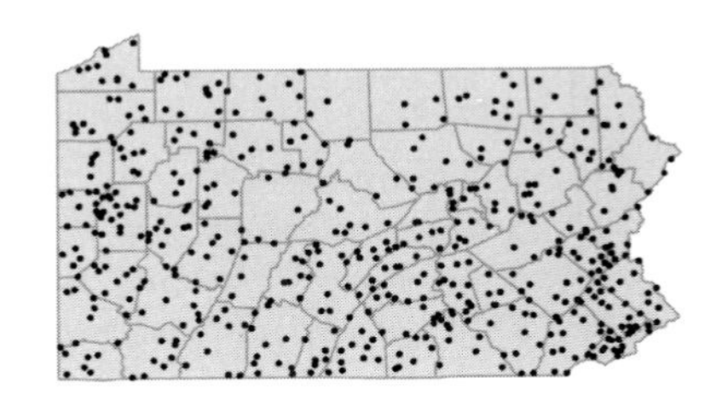

• ***Ophioglossum engelmannii*** Prantl
Adder's-tongue
Herbaceous perennial
Limestone glade.

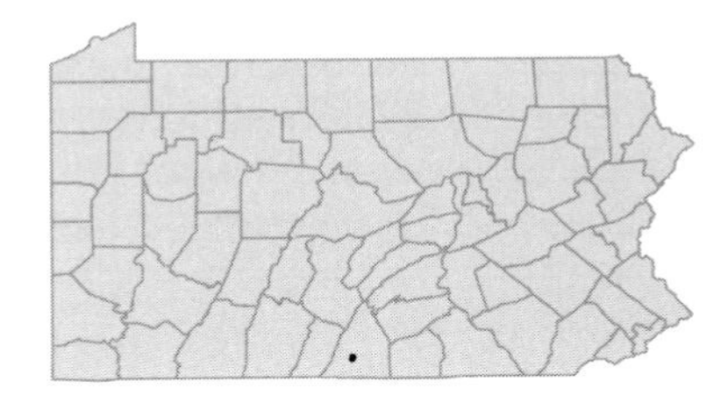

Ophioglossum pusillum Raf.
Northern adder's-tongue
Herbaceous perennial
Moist to dry, open woods and meadows.
Ophioglossum vulgatum L. var. *pseudopodum* (Blake) Fern. FWC

FACW

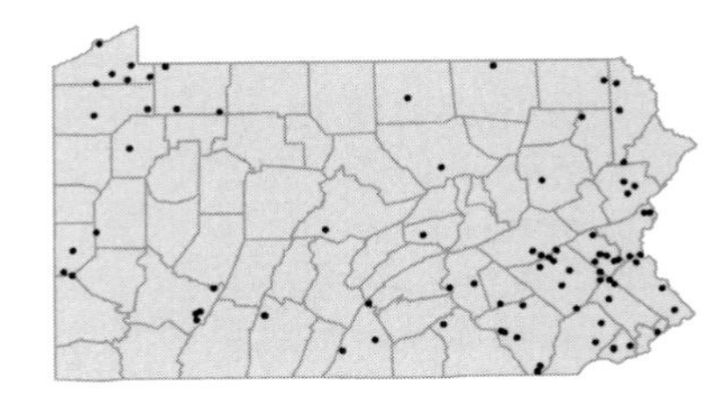

Ophioglossum pycnostichum (Fern.) Loeve & Loeve
Southeastern adder's-tongue
Herbaceous perennial
Moist to dry, open woods, meadows and floodplains, in acidic to circumneutral soils. Believed to be extirpated, last collected in 1952.
Ophioglossum vulgatum L. var. *pycnostichum* Fern. FWC

FACW

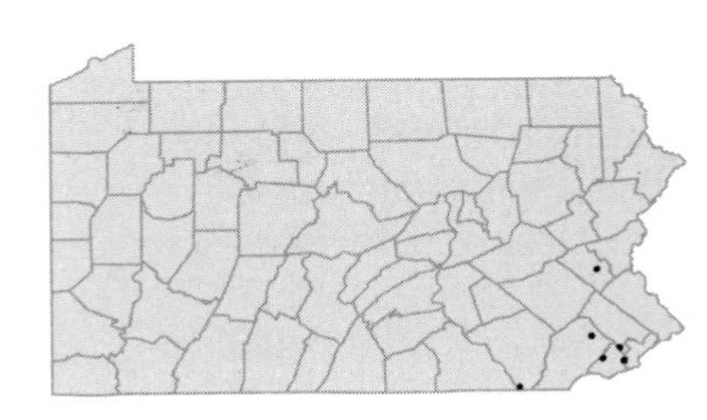

OSMUNDACEAE

• ***Osmunda cinnamomea*** L.
Cinnamon fern
Herbaceous perennial
Swamps, bog margins and stream banks, in wet acidic soils.

FACW

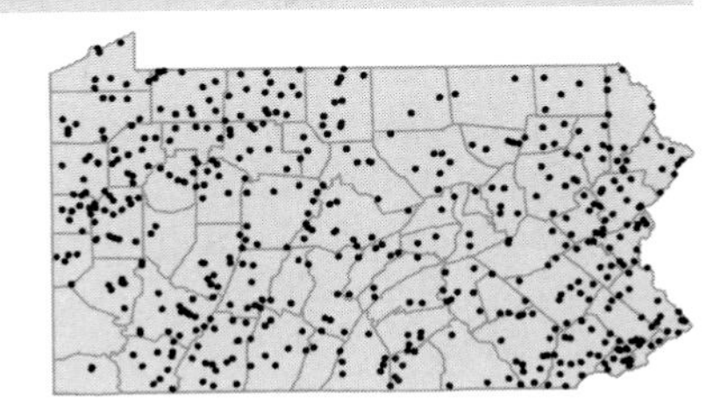

Osmunda claytoniana L.
Interrupted fern
Herbaceous perennial
Moist woodlands, bog edges and hummocks, in subacidic to neutral soils.

FAC

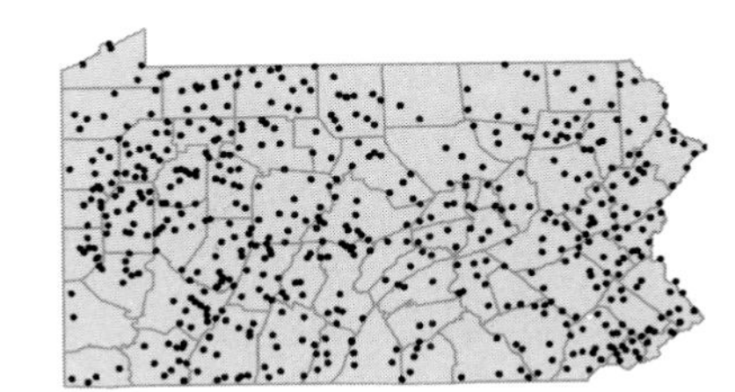

OSMUNDACEAE

Osmunda regalis L. var. ***spectabilis*** (Willd.) A.Gray
Royal fern
Herbaceous perennial
Bogs and swamps in moist acidic soils.

OBL

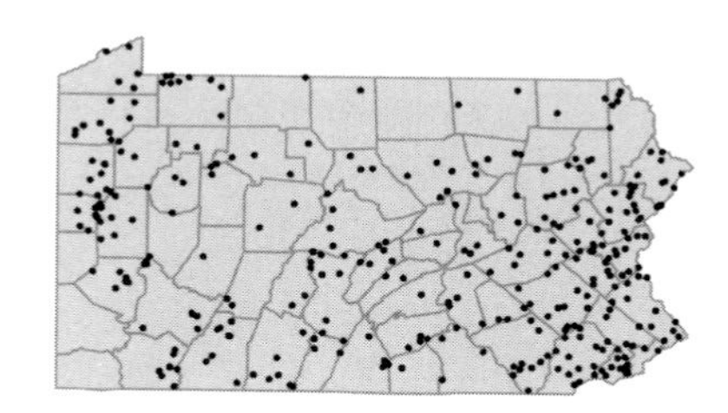

LYGODIACEAE

• ***Lygodium palmatum*** (Bernh.) Swartz
Climbing fern; Hartford fern
Herbaceous perennial vine
Moist, sandy-peaty thickets and edges, in acidic soils.

FACW

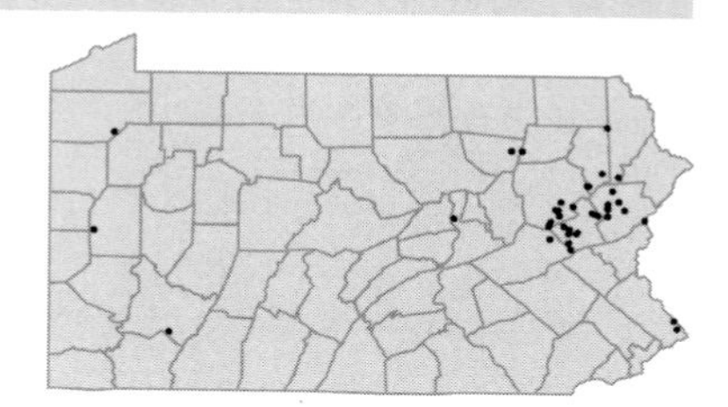

ADIANTACEAE

• ***Adiantum aleuticum*** (Ruprecht) Paris
Aleutian maidenhair fern
Herbaceous perennial
Serpentine barrens.
Adiantum pedatum L. ssp. *calderi* Cody KC; *Adiantum pedatum* L. var. *aleuticum* Ruprecht FBK

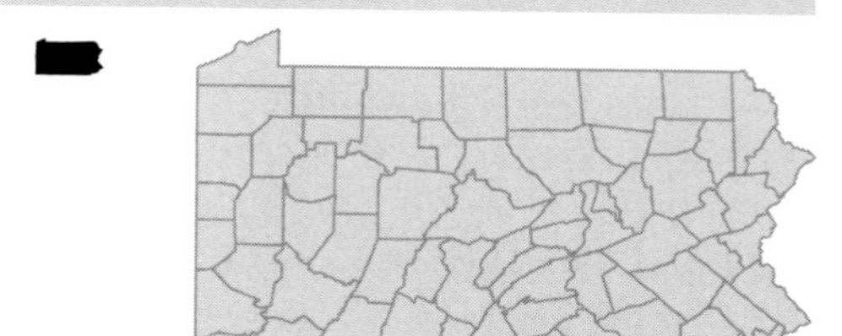

Adiantum pedatum L.
Maidenhair fern
Herbaceous perennial
Moist, shaded, humus-rich woods, in subacidic to neutral soils.

FAC-

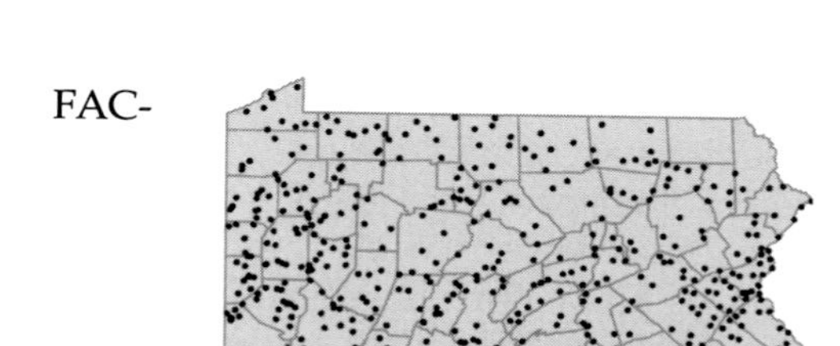

• ***Cheilanthes lanosa*** (Michx.) D.C.Eat.
Hairy lip fern
Herbaceous perennial
Dry cliffs and rock outcrops, in subacidic to circumneutral soils.

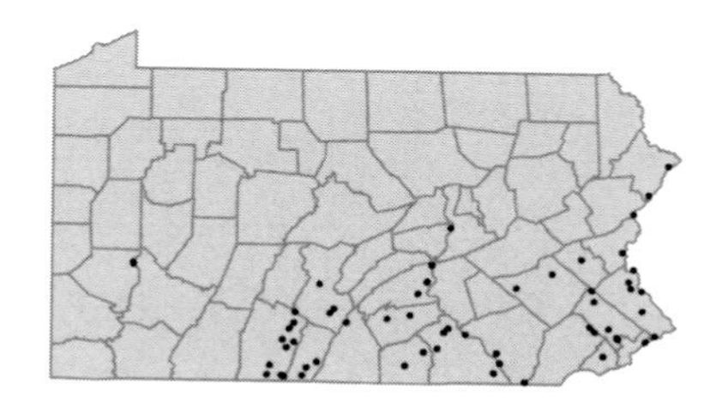

• ***Cryptogramma stelleri*** (Gmel.) Prantl
Slender cliff-brake
Herbaceous perennial
Cool, moist calcareous cliffs and ravines.

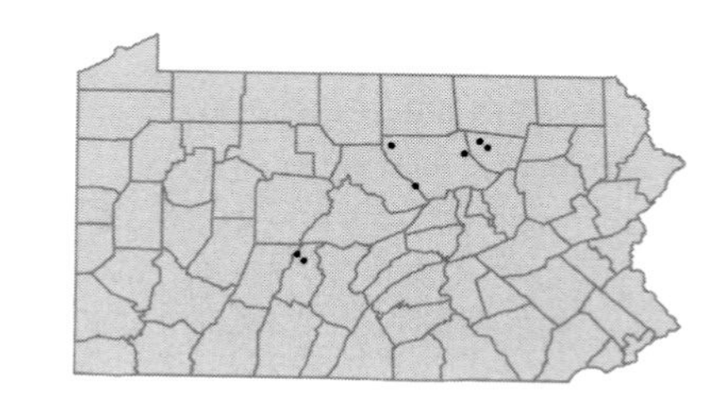

• ***Pellaea atropurpurea*** (L.) Link
Purple cliff-brake
Herbaceous perennial
Calcareous cliffs, ledges and talus slopes, also masonry cracks.

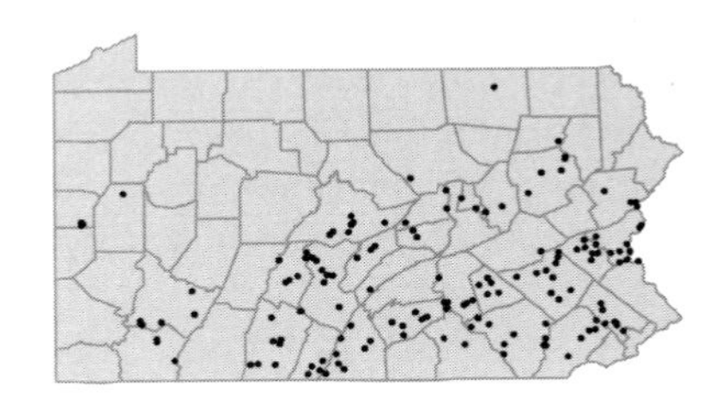

Pellaea glabella Mett. ex Kuhn var. ***glabella***
Smooth cliff-brake
Herbaceous perennial
Moist to dry, exposed calcareous cliffs and ledges, also masonry cracks.
Pellaea atropurpurea (L.) Link var. *bushii* Mackenzie B

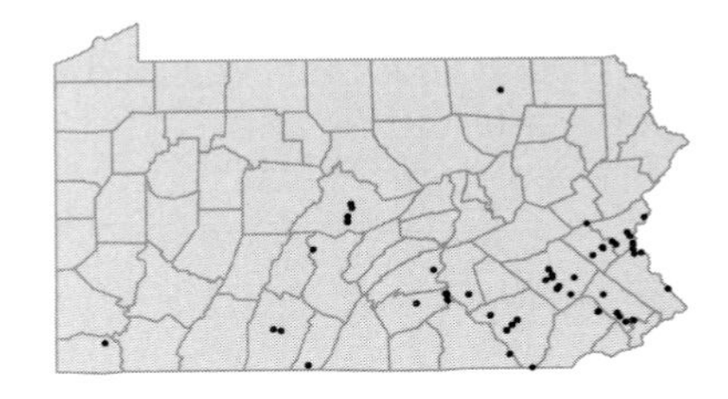

VITTARIACEAE

• ***Vittaria appalachiana*** Farrar & Mickel
Appalachian grass fern
Herbaceous perennial
Heavily shaded, moist crevices and overhangs in noncalcareous rock. Occurs in PA as gametophytes only.

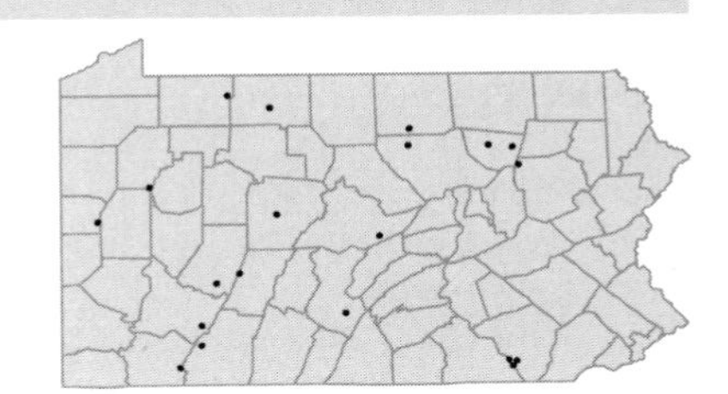

POLYPODIACEAE

• ***Polypodium virginianum*** L.
Rock polypody; Rock-cap fern
Herbaceous perennial
Dry rocks and ledges, in subacidic to neutral soils. The tetraploid, *P. virginianum*, and the recently described diploid, *P. appalachianum* Haufler & Windham, are both included here. Critical examination of existing specimens is needed to map them separately.
Polypodium vulgare L. var. *virginianum* (L.) Eat. GB

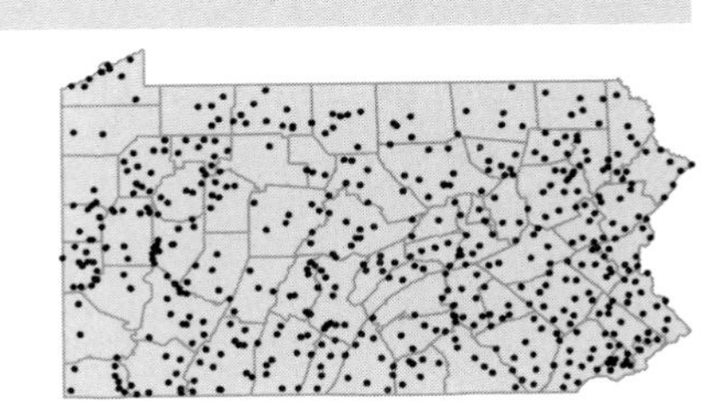

HYMENOPHYLLACEAE

• ***Trichomanes intricatum*** Farrar
Filmy fern
Herbaceous perennial
Heavily shaded, moist crevices and overhangs in noncalcareous rock. Occurs in PA as gametophytes only.

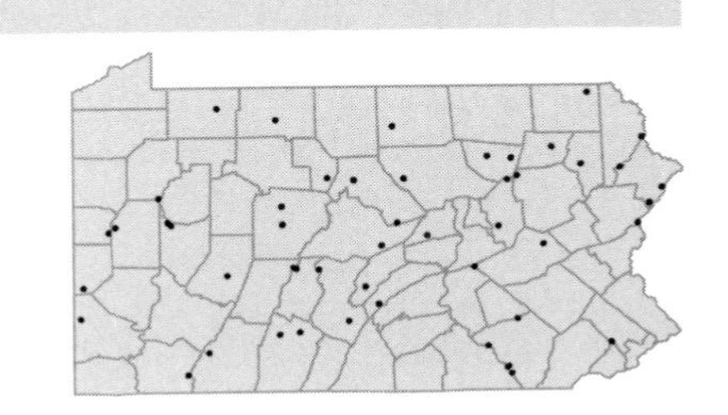

DENNSTAEDTIACEAE

• ***Dennstaedtia punctilobula*** (Michx.) T.Moore
Hayscented fern
Herbaceous perennial
Open woods, meadows and slopes, in acidic soils.

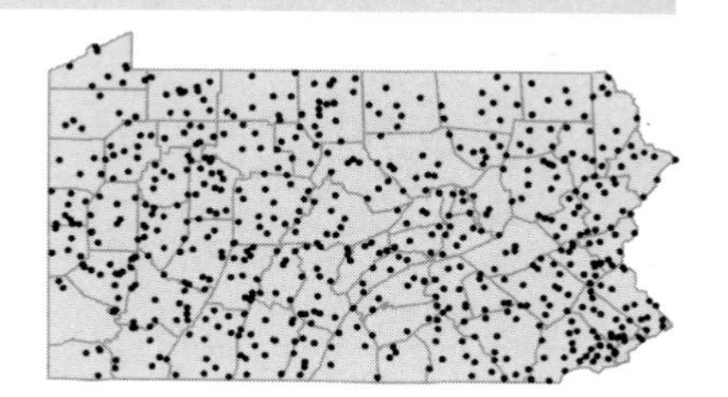

• ***Pteridium aquilinum*** (L.) Kuhn var. ***latiusculum*** (Desv.) Underw. ex Heller — FACU
Bracken fern
Herbaceous perennial
Open woods and barrens, in sandy acidic soils.

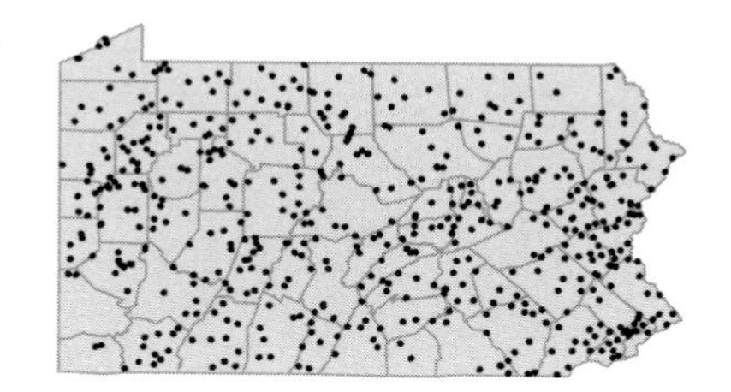

THELYPTERIDACEAE

• ***Thelypteris hexagonoptera*** (Michx.) Weatherby — FAC
Broad beech fern; Southern beech fern
Herbaceous perennial
Rich, moist wooded slopes and swamp margins, in humus-rich, moderately acidic soils.
Dryopteris hexagonoptera (Michx.) C.Christens. F; *Phegopteris hexagonoptera* (Michx.) Fee W

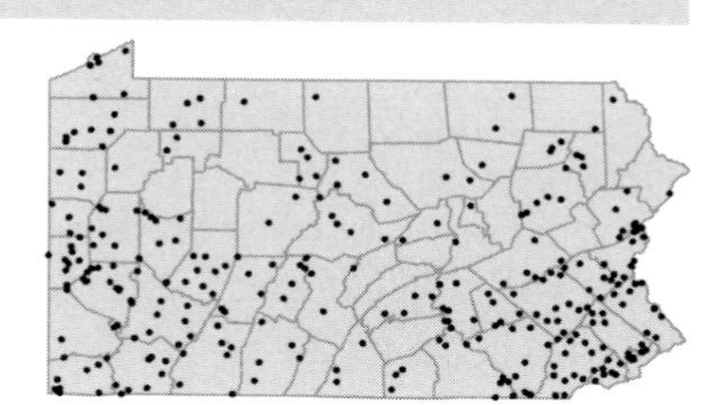

Thelypteris noveboracensis (L.) Nieuwl. — FAC
New York fern
Herbaceous perennial
Moist woods, thickets and swamps, in humus-rich, moderately acidic soils.
Dryopteris noveboracensis (L.) A.Gray F

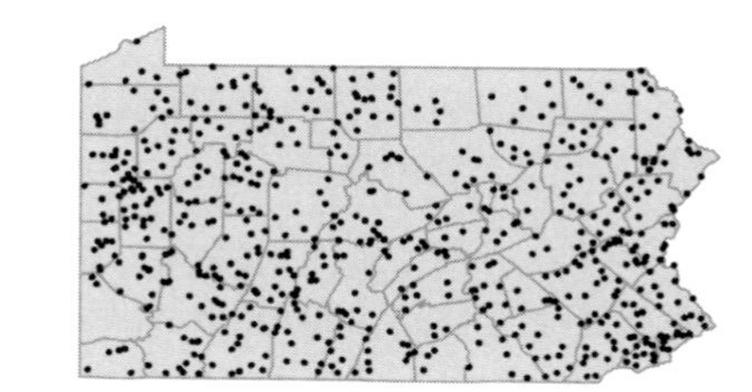

Thelypteris palustris Schott var. ***pubescens*** (Lawson) Fern. — FACW
Marsh fern
Herbaceous perennial
Marshes, wet meadows and bog margins, in acidic soils.
Dryopteris thelypteris (L.) A.Gray var. *pubescens* (Lawson) Nakai F; *Thelypteris palustris* Schott var. *haleana* Fern. W

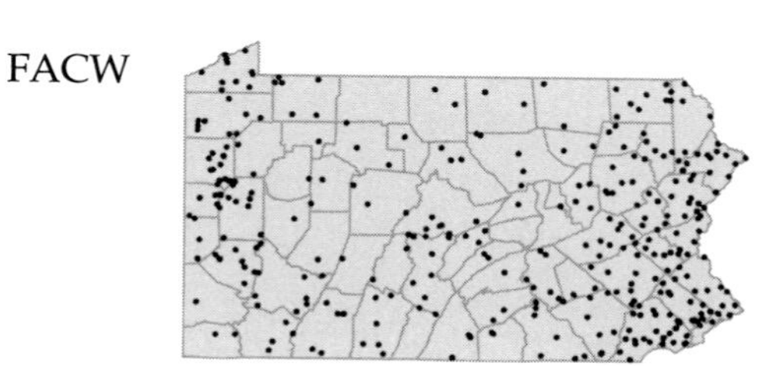

Thelypteris phegopteris (L.) Slosson
Long beech fern; Northern beech fern
Herbaceous perennial
Cool woods, damp thickets and shaded rock crevices, in acidic soils.
Dryopteris phegopteris (L.) C.Christens. F; *Phegopteris connectilis* (Michx.) Watt W

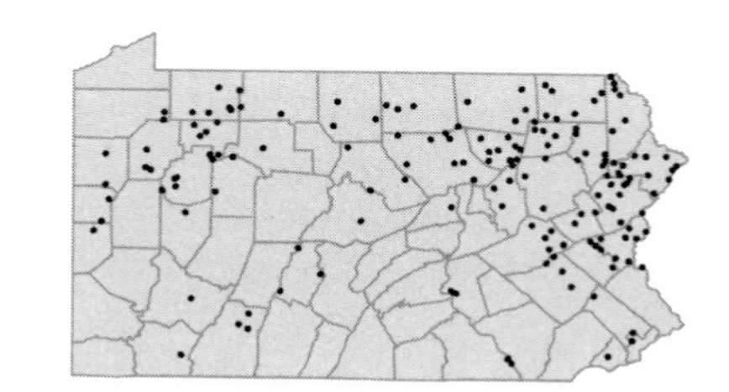

Thelypteris simulata (Davenp.) Nieuwl.
Bog fern; Massachusetts fern
Herbaceous perennial
Bogs and swamp hummocks, in acidic soils.
Dryopteris simulata Davenp. F

FACW

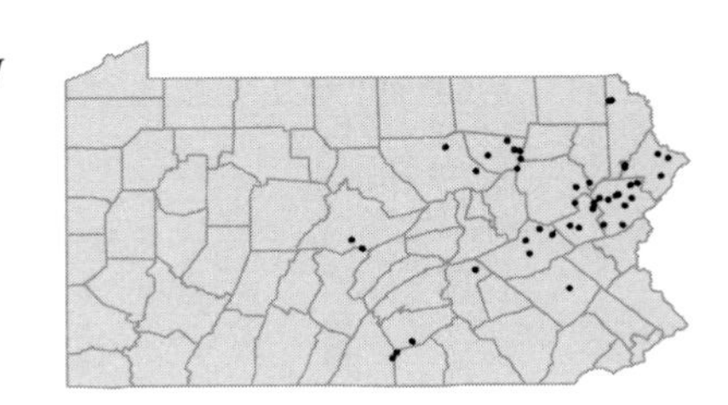

ASPLENIACEAE

• ***Asplenium bradleyi*** D.Eaton
Bradley's spleenwort
Herbaceous perennial
Dry, shaded crevices of noncalcareous rock.
Asplenium montanum x platyneuron

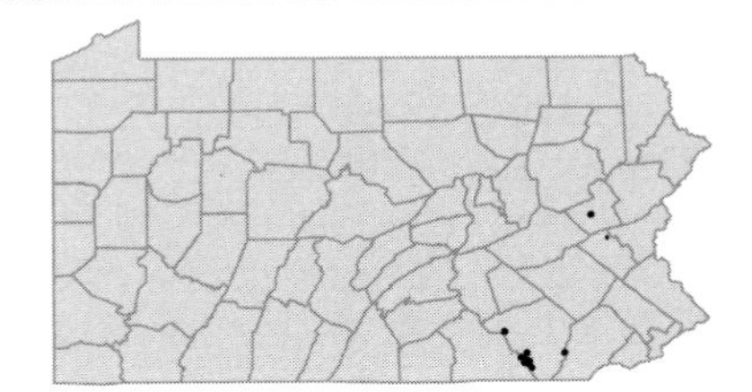

Asplenium bradleyi x pinnatifidum
Graves' spleenwort; Sand-mountain spleenwort
Herbaceous perennial
Crevices of noncalcareous rock.
Asplenium x gravesii Maxon FGBWC

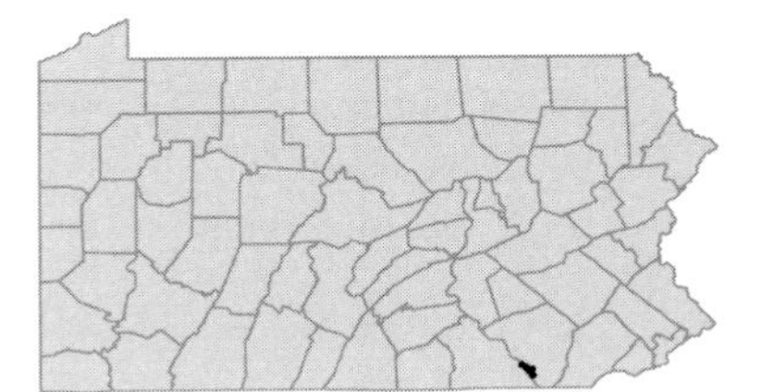

Asplenium montanum Willd.
Mountain spleenwort
Herbaceous perennial
Damp, shaded crevices of noncalcareous rock.

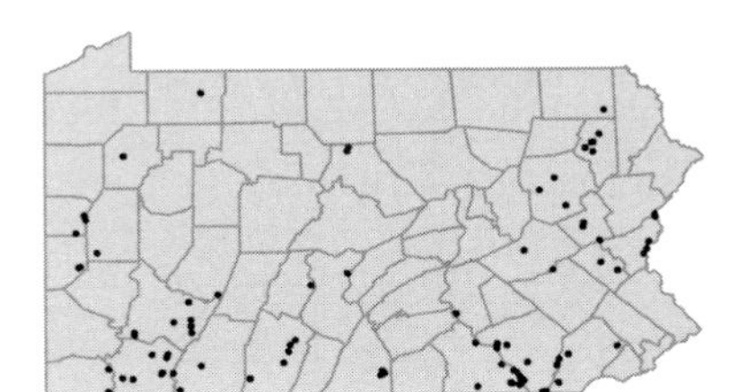

Asplenium montanum x pinnatifidum
Trudell's spleenwort
Herbaceous perennial
Crevices of acidic rock.
Asplenium x trudellii Wherry FGWC

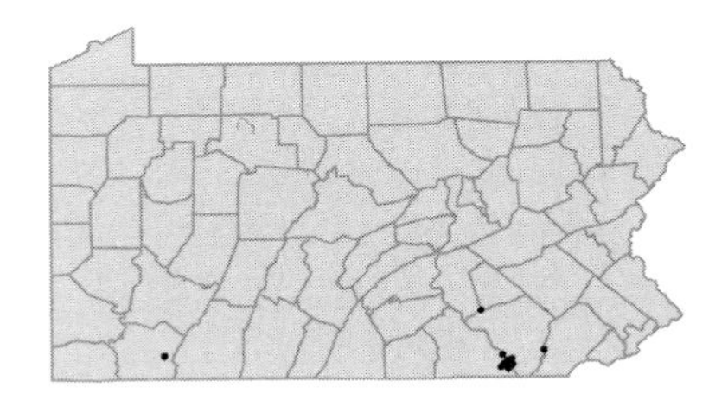

Asplenium pinnatifidum Muhl.
Lobed spleenwort
Herbaceous perennial
Dry, shaded rock crevices, in subacidic to circumneutral soils.
Asplenium montanum x rhizophyllum

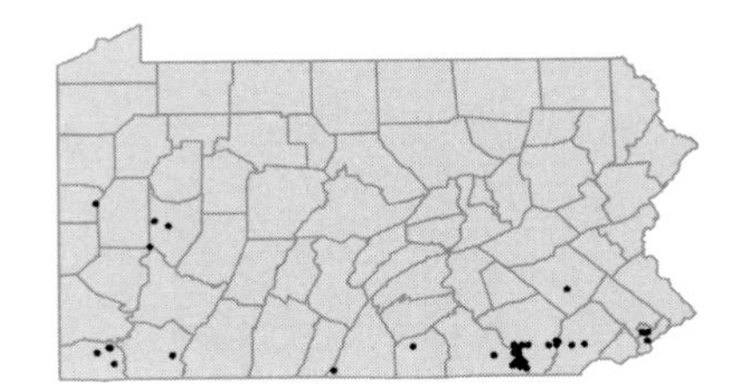

Asplenium platyneuron (L.) BSP
Ebony spleenwort
Herbaceous perennial
Dry to moist, wooded slopes and rock ledges, in subacidic soils.

FACU

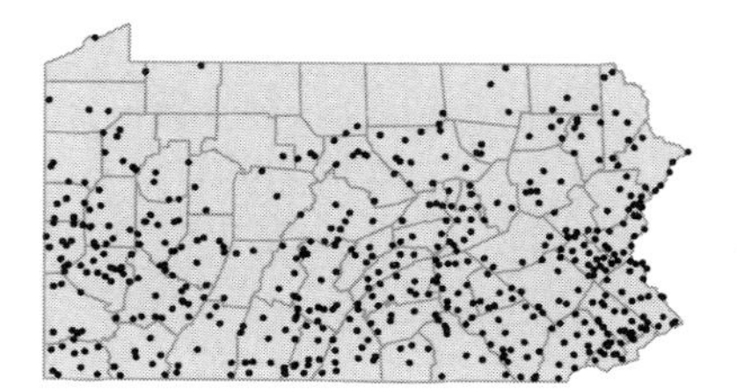

Asplenium platyneuron x rhizophyllum
Scott's spleenwort
Herbaceous perennial
Crevices of shaded rocks.
Asplenium x ebenoides R.R.Scott WC; *Asplenosorus x ebenoides* (R.R.Scott) Wherry FGB

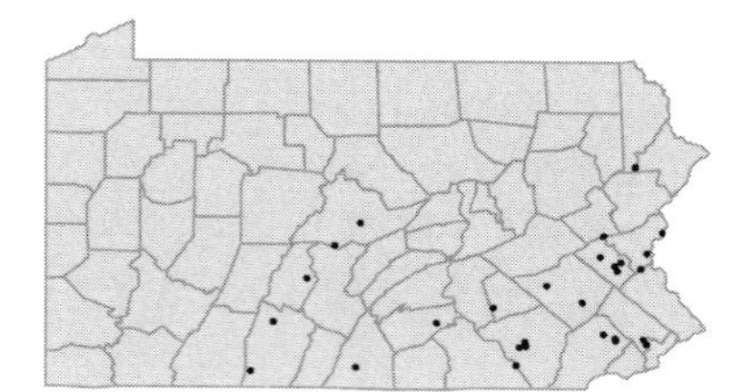

Asplenium resiliens Kunze
Black-stemmed spleenwort; Little ebony spleenwort
Herbaceous perennial
Shaded limestone ledges.

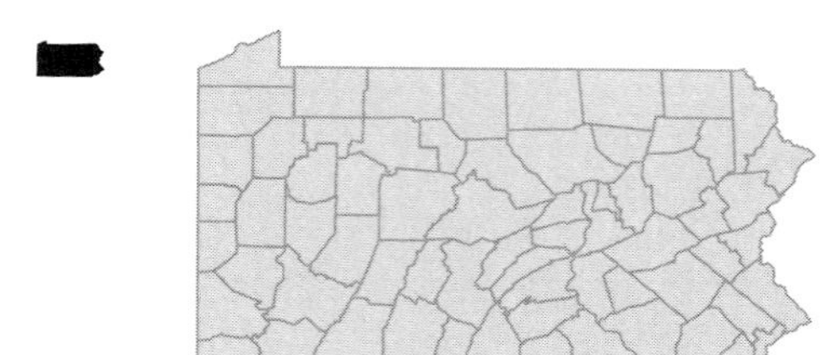

Asplenium rhizophyllum L.
Walking fern
Herbaceous perennial
Ledges and crevices in calcareous rock.
Camptosorus rhizophyllus (L.) Link FGBW

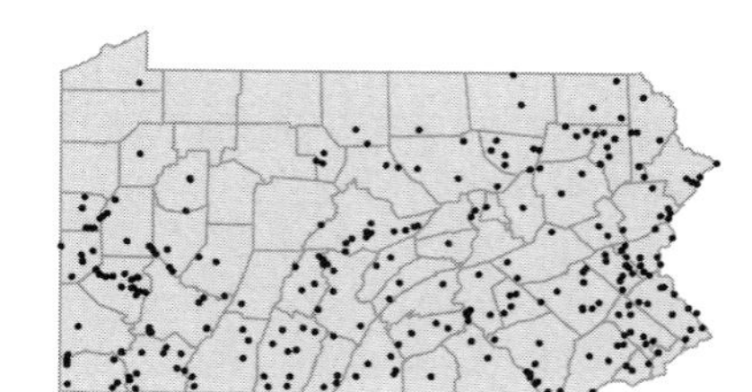

Asplenium ruta-muraria L.
Wall-rue spleenwort
Herbaceous perennial
Shaded, calcareous ledges and talus.

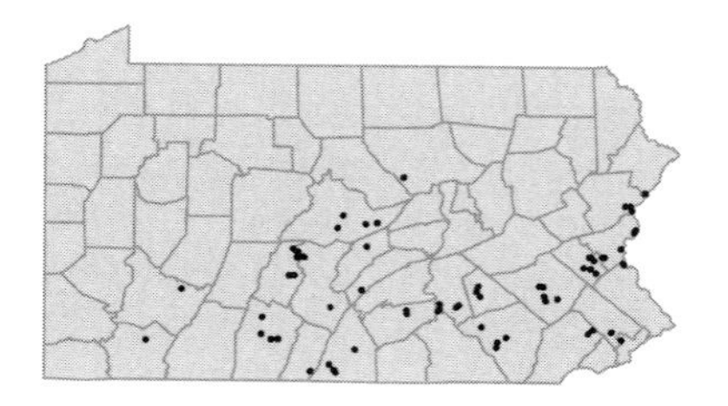

Asplenium trichomanes L.
Maidenhair spleenwort
Herbaceous perennial
Moist rock crevices, in neutral to moderately acidic soils.

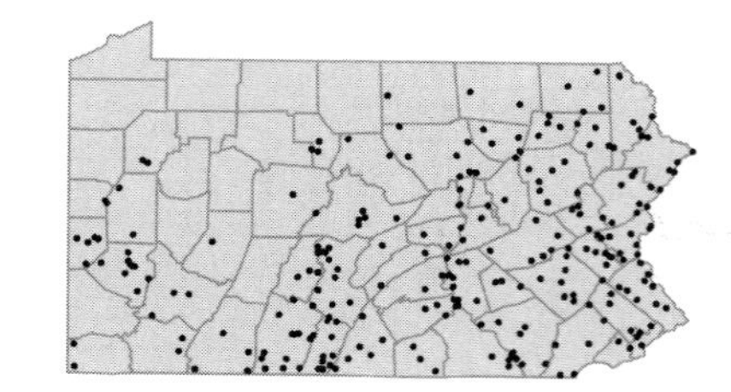

DRYOPTERIDACEAE

• ***Athyrium filix-femina*** (L.) Roth var. ***angustum*** (Willd.) Lawson
Northern lady fern
Herbaceous perennial
Damp woods, swamps and thickets, in subacidic soils.
Athyrium angustum (Willd.) Presl WFGBC

FAC

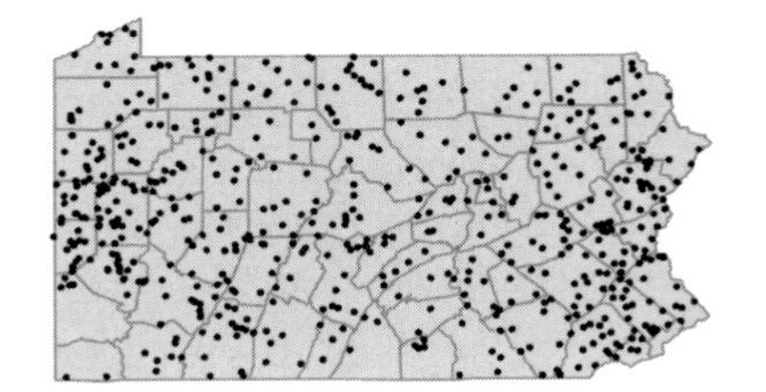

Athyrium filix-femina (L.) Roth var. ***asplenioides*** (Michx.) Farw.
Southern lady fern
Herbaceous perennial
Swamps, damp thickets and wooded ravines, in subacidic soils.
Athyrium asplenioides (Michx.) Eat. WC

FAC

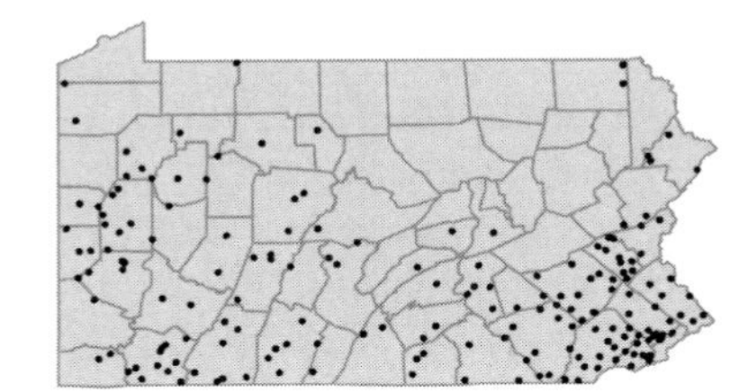

• ***Cystopteris bulbifera*** (L.) Bernh.
Bladder fern; Bulblet fern
Herbaceous perennial
Calcareous cliffs and ledges, also limy swamps.

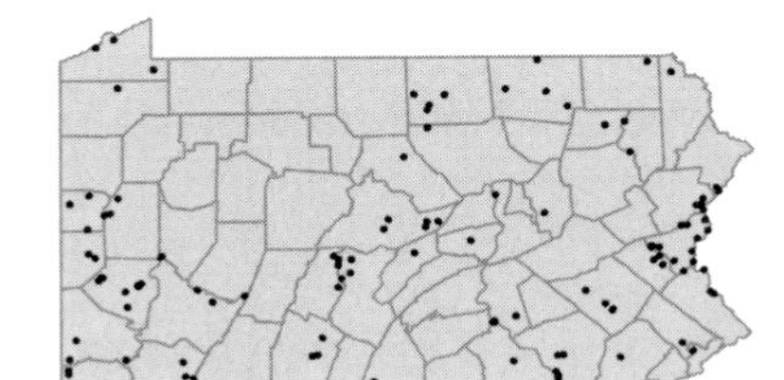

Cystopteris bulbifera x fragilis
Laurentian bladder fern
Herbaceous perennial
Calcareous rocks.
Cystopteris fragilis (L.) Bernh. var. *laurentiana* Weatherby FGB; *Cystopteris x laurentiana* (Weatherby) Blasdell

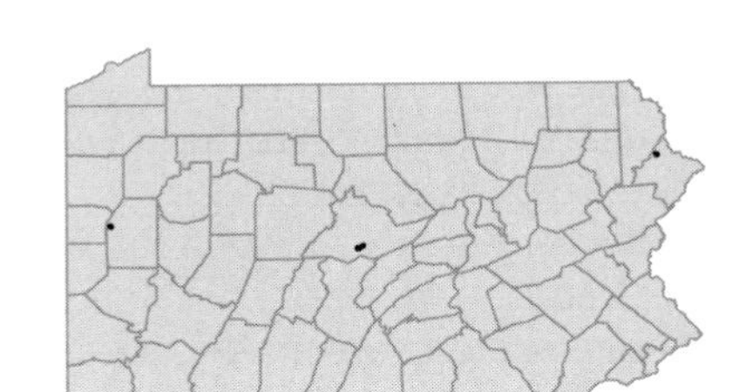

Cystopteris bulbifera x protrusa
Bladder fern; Fragile fern
Herbaceous perennial
Limestone or sandstone cliffs.
Cystopteris x tennesseensis Shaver GW

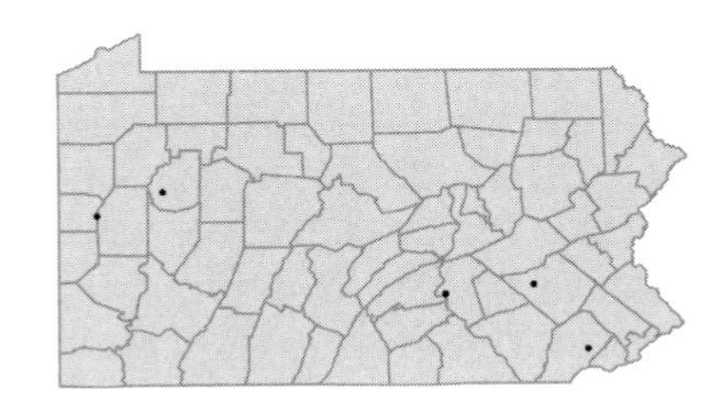

Cystopteris fragilis (L.) Bernh.
Common bladder fern; Fragile fern
Herbaceous perennial
Cool rock crevices and talus slopes, in neutral to subacidic soils.

FACU

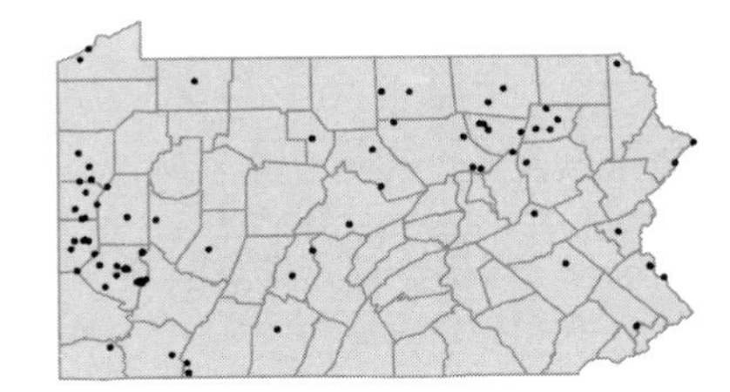

Cystopteris protrusa (Weatherby) Blasdell
Bladder fern; Fragile fern
Herbaceous perennial
Wooded, alluvial flats, in humus-rich, circumneutral soils.
Cystopteris fragilis (L.) Bernh. var. *protrusa* Weatherby FGB

FACU

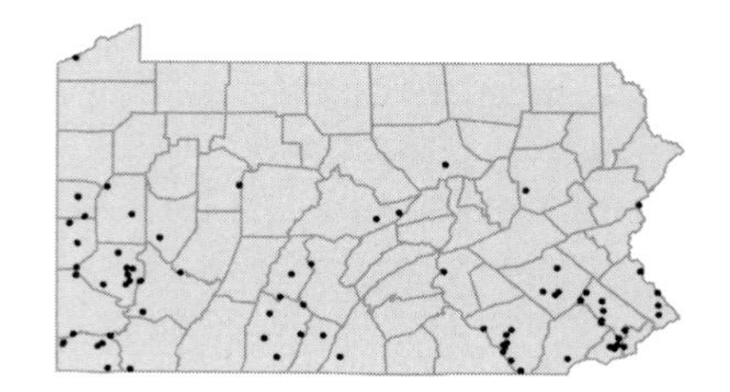

Cystopteris tenuis (Michx.) Desv.
Fragile fern; MacKay's brittle fern
Herbaceous perennial
Cool rock crevices and talus slopes, in neutral to subacidic soils.
Cystopteris fragilis (L.) Bernh. var. *mackayi* Lawson FGBW

FACU

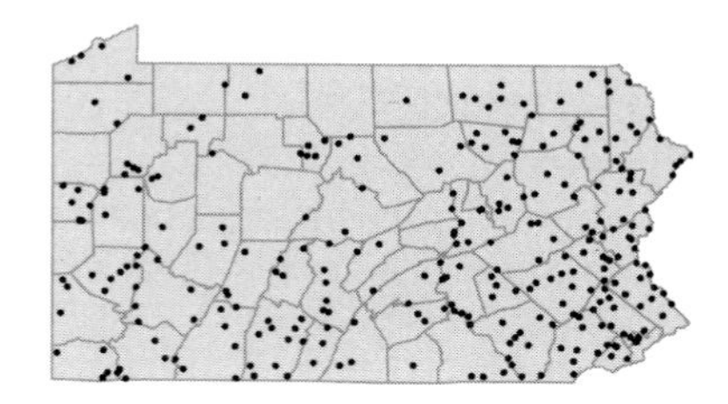

• ***Deparia acrostichoides*** (Swartz) M.Kato
Silvery glade fern; Silvery spleenwort
Herbaceous perennial
Damp woods and shaded slopes, in humus-rich, subacidic to circumneutral soils.
Athyrium thelypteroides (Michx.) Desv. FWC

FAC

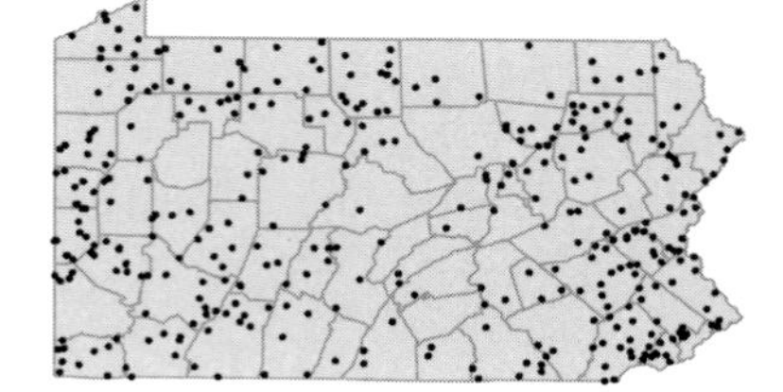

• ***Diplazium pycnocarpon*** (Spreng.) Broun
Glade fern; Narrow-leaved spleenwort
Herbaceous perennial
Woodland glades, alluvial thickets and rocky slopes, in rich, circumneutral soils.
Athyrium pycnocarpon (Spreng.) Tidestr. FGBWC

FAC

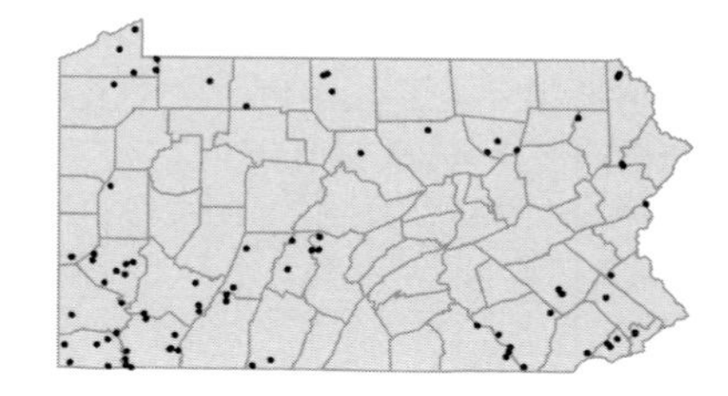

Dryopteris—In addition to the taxa treated below, limited collections exist representing the following *Dryopteris* hybrids:*carthusiana x clintoniana* (*D. x benedictii* Wherry); *carthusiana x marginalis* (*D. x pittsfordensis* Slossen); *celsa x marginalis* (*D. x leedsii* Wherry); *clintoniana x intermedia* (*D. x dowellii* [Farw.] Wherry); *clintoniana x marginalis* (*D. x burgessii* Boivin); *goldiana x marginalis* (*D. x neowherryi* W.H.Wagner); *intermedia x marginalis.*

• ***Dryopteris campyloptera*** (Kunze) Clarkson
Mountain wood fern
Herbaceous perennial
Cool, rocky, wooded slopes, in humus-rich, subacidic soils.
Dryopteris austriaca (Jacq.) Woynar var. *austriaca* GB; *Dryopteris spinulosa* (O.F.Muell.) Watt var. *americana* (Fisch.) Fern. F

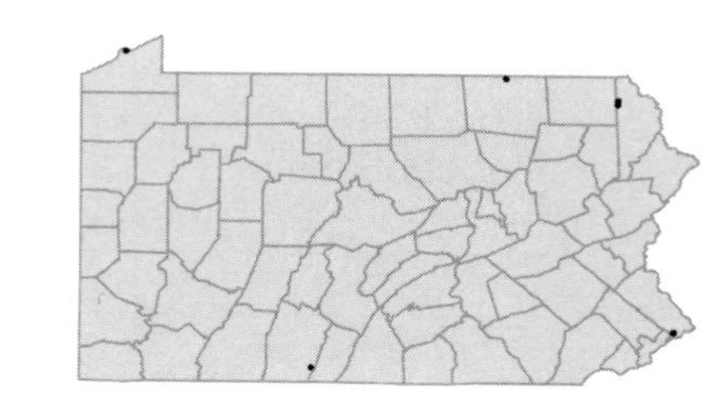

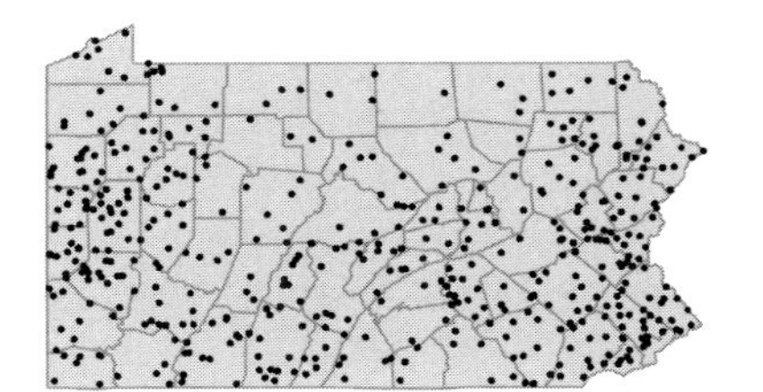

Dryopteris carthusiana (Villars) H.P.Fuchs FAC+
Fancy fern; Spinulose wood fern
Herbaceous perennial
Moist to wet woods and swamps.
Dryopteris austriaca (Jacq.) Woynar var. *spinulosa* (O.F.Muell.) Fiori GB;
Dryopteris spinulosa (O.F.Muell.) Watt F

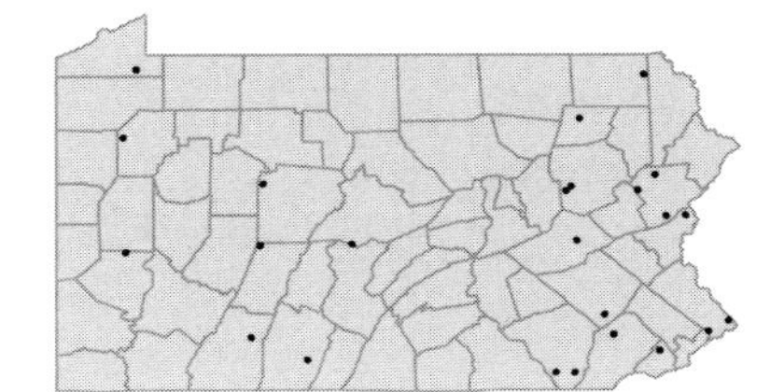

Dryopteris carthusiana x cristata
Braun's wood fern
Herbaceous perennial
Moist woods, swamps and thickets.
Dryopteris x uliginosa (A.Braun. ex Dowell) Druce W

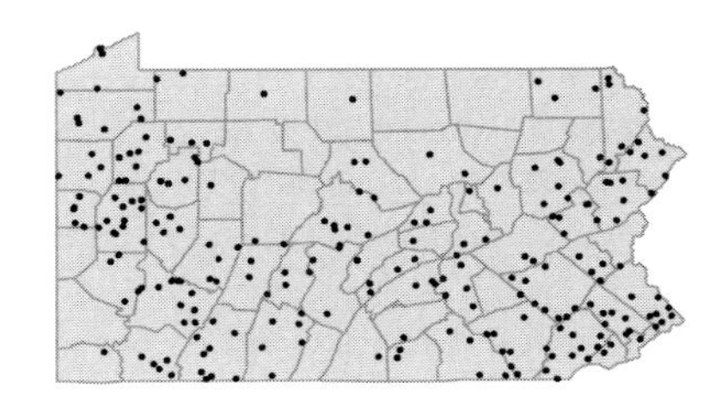

Dryopteris carthusiana x intermedia
Triploid wood fern
Herbaceous perennial
Moist woods, swamps and thickets.
Dryopteris x triploidea Wherry W

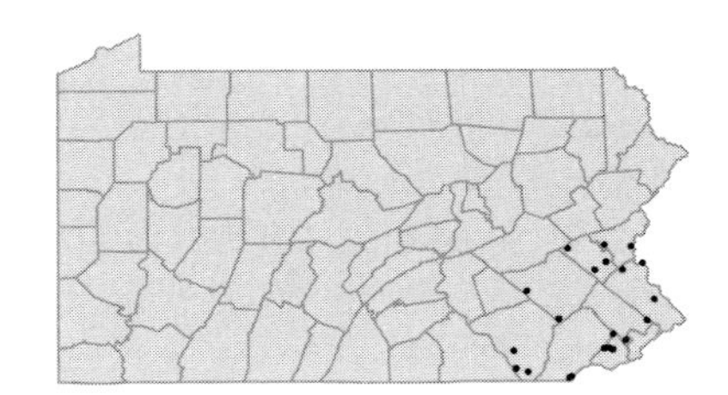

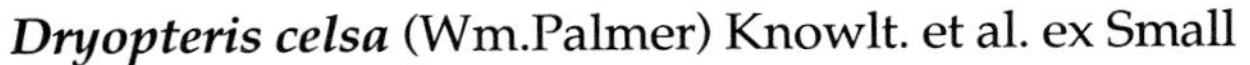

Dryopteris celsa (Wm.Palmer) Knowlt. et al. ex Small FAC+
Log fern
Herbaceous perennial
Rotting logs and swampy hummocks, in acidic soils.
Dryopteris goldiana (Hook.) A.Gray ssp. *celsa* Wm. Palmer GB

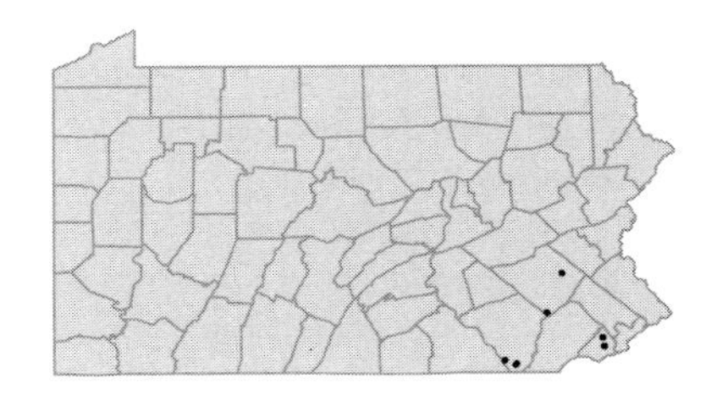

Dryopteris celsa x cristata
Wood fern
Herbaceous perennial
Ravines and damp woods.

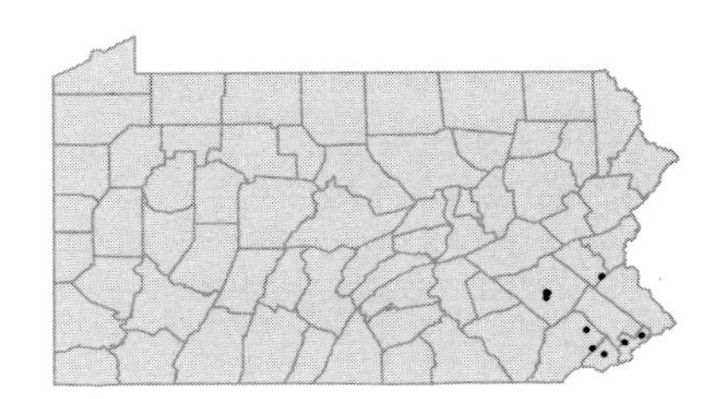

Dryopteris celsa x goldiana
Wood fern
Herbaceous perennial
Wet, rocky woods and stream edges.

Dryopteris clintoniana (D.Eaton) Dowell
Clinton's shield fern; Clinton's wood fern
Herbaceous perennial
Swamps and damp woods, in subacidic soils.
Dryopteris cristata (L.) A.Gray var. *clintoniana* (D.C.Eat.) Underw. F

FACW-

Dryopteris clintoniana x cristata
Deceptive wood fern
Herbaceous perennial
Swamps and moist hummocks.

FACW+

Dryopteris cristata (L.) A.Gray
Crested shield fern; Crested wood fern
Herbaceous perennial
Swamps and wet thickets, in subacidic soils.

FACW+

Dryopteris cristata x intermedia
Boott's wood fern; Glandular swamp fern
Herbaceous perennial
Swamps and moist woodland edges, in subacidic soils.
Dryopteris x boottii (Tuckerman) Underw. FGBW

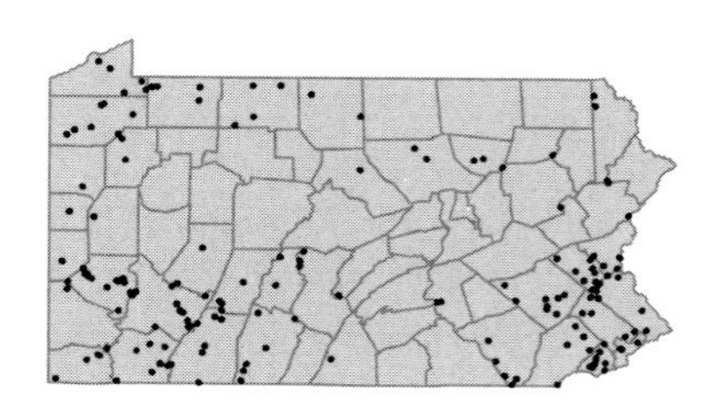

Dryopteris cristata x marginalis
Miss Slosson's wood fern
Herbaceous perennial
Swampy woods, stream banks and springheads.
Dryopteris x slossoniae Wherry ex Lellinger W

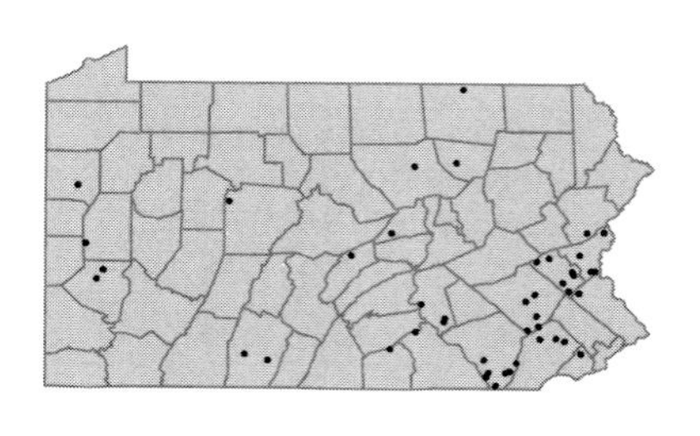

Dryopteris goldiana (Hook.) A.Gray
Goldie's wood fern
Herbaceous perennial
Moist woods or shaded talus or seepage slopes, in humus-rich, circumneutral soils.

FAC+

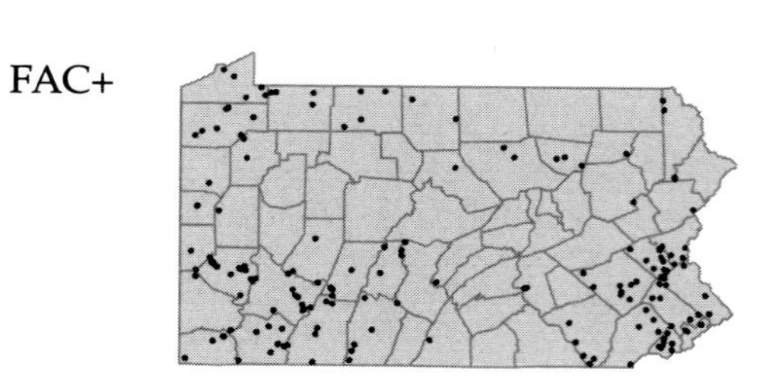

Dryopteris carthusiana (Villars) H.P.Fuchs FAC+
Fancy fern; Spinulose wood fern
Herbaceous perennial
Moist to wet woods and swamps.
Dryopteris austriaca (Jacq.) Woynar var. *spinulosa* (O.F.Muell.) Fiori GB;
Dryopteris spinulosa (O.F.Muell.) Watt F

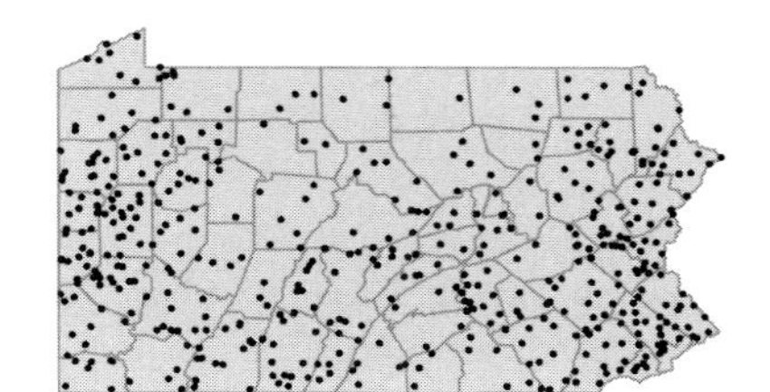

Dryopteris carthusiana x cristata
Braun's wood fern
Herbaceous perennial
Moist woods, swamps and thickets.
Dryopteris x uliginosa (A.Braun. ex Dowell) Druce W

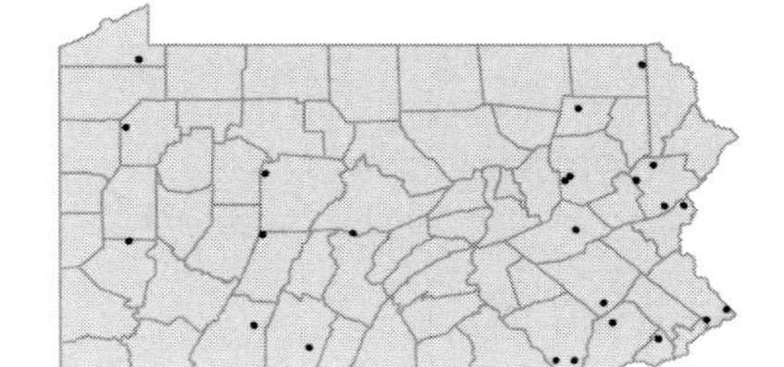

Dryopteris carthusiana x intermedia
Triploid wood fern
Herbaceous perennial
Moist woods, swamps and thickets.
Dryopteris x triploidea Wherry W

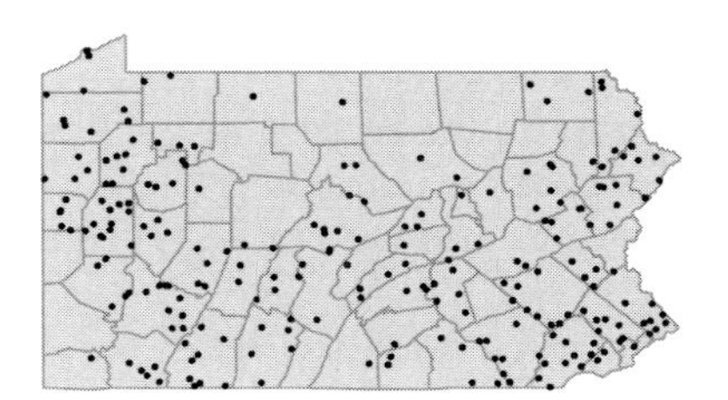

Dryopteris celsa (Wm.Palmer) Knowlt. et al. ex Small FAC+
Log fern
Herbaceous perennial
Rotting logs and swampy hummocks, in acidic soils.
Dryopteris goldiana (Hook.) A.Gray ssp. *celsa* Wm. Palmer GB

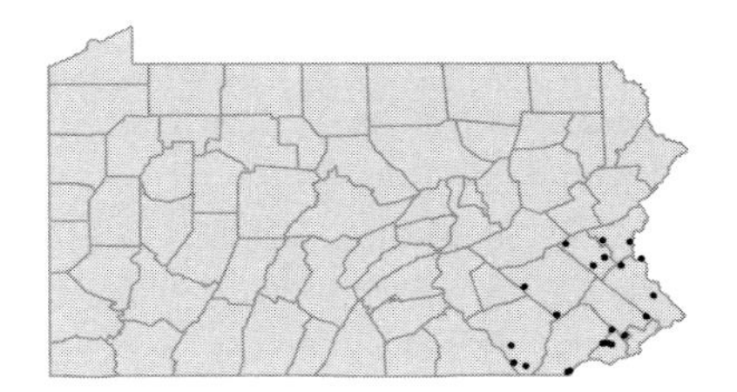

Dryopteris celsa x cristata
Wood fern
Herbaceous perennial
Ravines and damp woods.

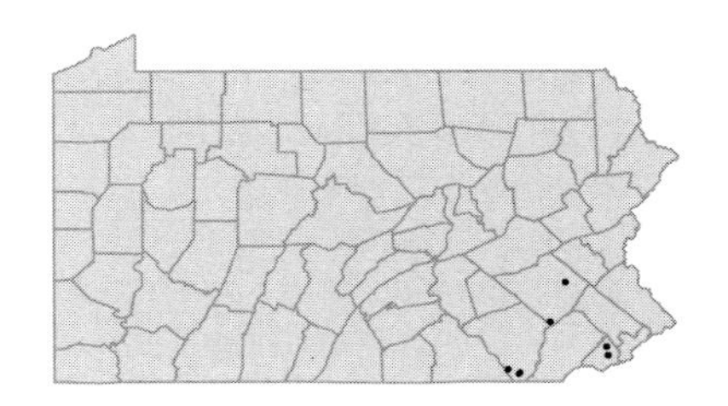

Dryopteris celsa x goldiana
Wood fern
Herbaceous perennial
Wet, rocky woods and stream edges.

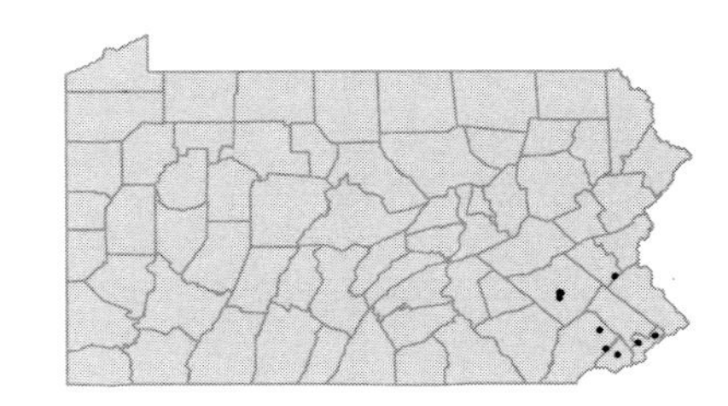

Dryopteris clintoniana (D.Eaton) Dowell
Clinton's shield fern; Clinton's wood fern
Herbaceous perennial
Swamps and damp woods, in subacidic soils.
Dryopteris cristata (L.) A.Gray var. *clintoniana* (D.C.Eat.) Underw. F

FACW-

Dryopteris clintoniana x cristata
Deceptive wood fern
Herbaceous perennial
Swamps and moist hummocks.

FACW+

Dryopteris cristata (L.) A.Gray
Crested shield fern; Crested wood fern
Herbaceous perennial
Swamps and wet thickets, in subacidic soils.

FACW+

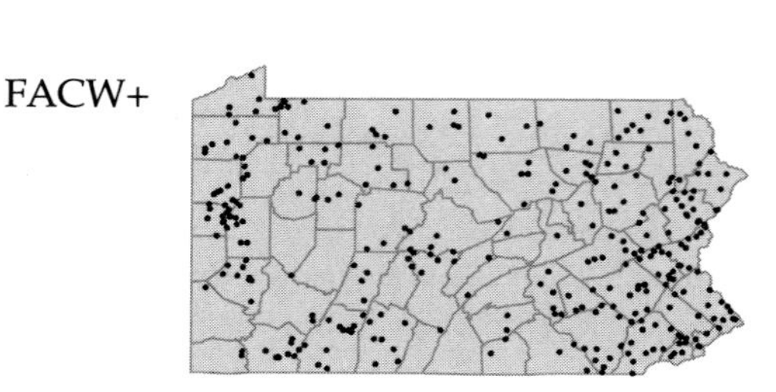

Dryopteris cristata x intermedia
Boott's wood fern; Glandular swamp fern
Herbaceous perennial
Swamps and moist woodland edges, in subacidic soils.
Dryopteris x boottii (Tuckerman) Underw. FGBW

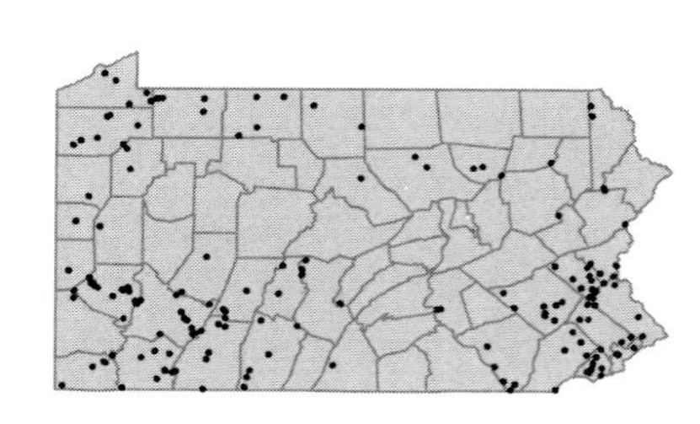

Dryopteris cristata x marginalis
Miss Slosson's wood fern
Herbaceous perennial
Swampy woods, stream banks and springheads.
Dryopteris x slossoniae Wherry ex Lellinger W

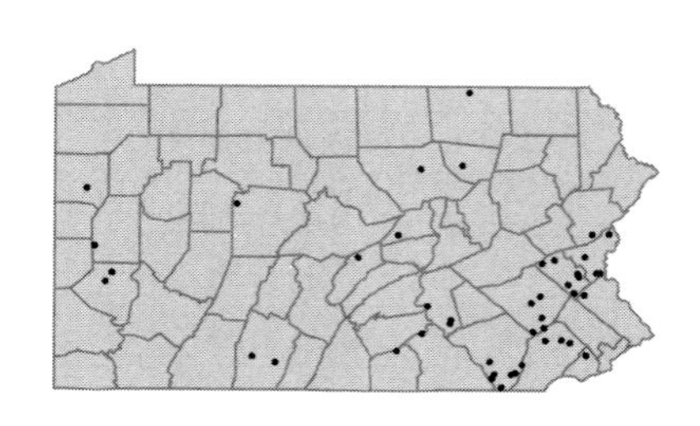

Dryopteris goldiana (Hook.) A.Gray
Goldie's wood fern
Herbaceous perennial
Moist woods or shaded talus or seepage slopes, in humus-rich, circumneutral soils.

FAC+

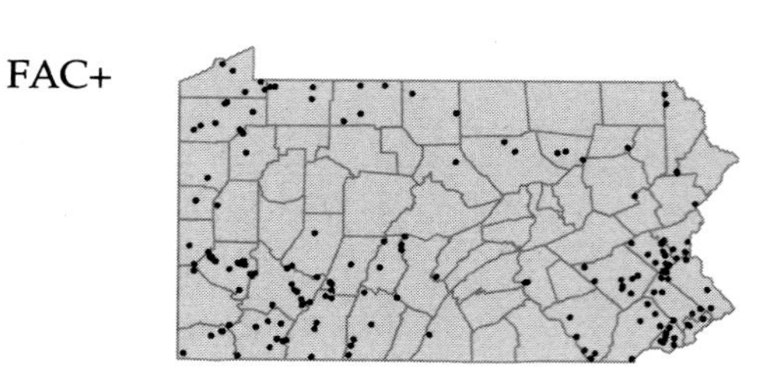

Dryopteris intermedia (Muhl. ex Willd.) A.Gray
Common wood fern; Fancy fern
Herbaceous perennial
Moist woods, shaded slopes and swamp hummocks, in humus-rich, acidic to neutral soils.
Dryopteris austriaca (Jacq.) Woynar var. *intermedia* (Muhl. ex Willd.) Morton GB; *Dryopteris spinulosa* (O.F.Muell.) Watt var. *intermedia* (Muhl. ex Willd.) Underw. F

FACU

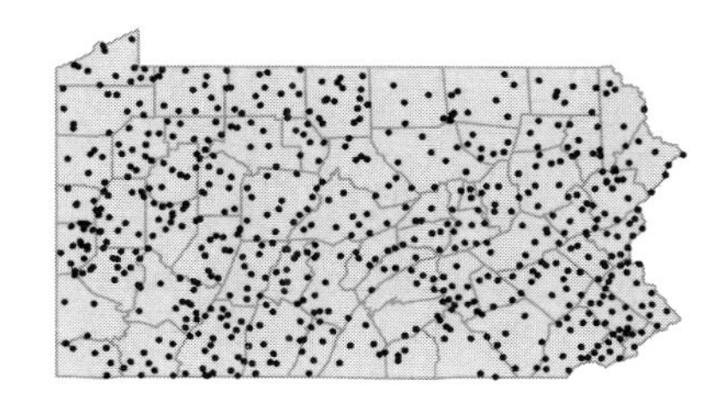

Dryopteris marginalis (L.) A.Gray
Evergreen wood fern; Marginal shield fern
Herbaceous perennial
Rocky ledges, talus slopes and shaded edges, in subacidic to circumneutral soils.

FACU-

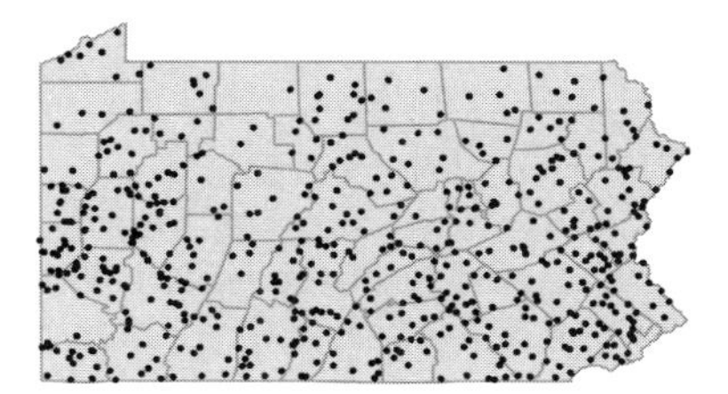

• ***Gymnocarpium appalachianum*** Pryer & Haufler
Oak fern
Herbaceous perennial
Rocky woods.

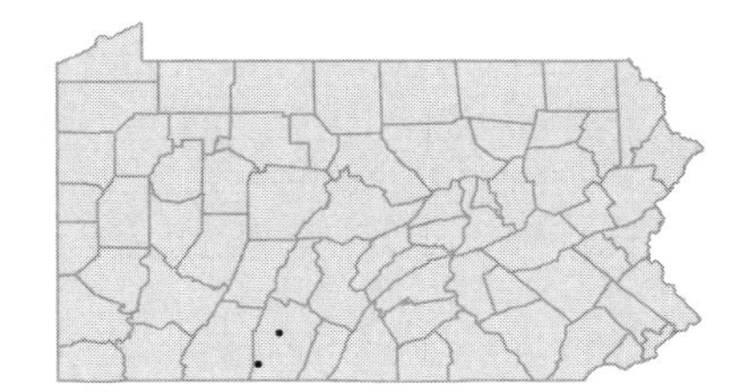

Gymnocarpium dryopteris (L.) Newman
Oak fern
Herbaceous perennial
Cool rocky woods, shaded talus slopes and swamp margins, in humusy, subacidic soils.
Dryopteris disjuncta (Ledeb.) C.V.Mort. F

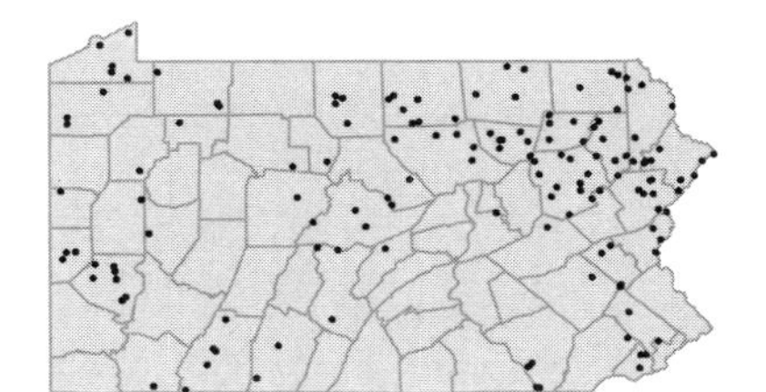

Gymnocarpium dryopteris x robertianum
Oak fern
Herbaceous perennial
Limestone outcrop on wooded hillside. Believed to be extirpated in Pennsylvania, last collected in 1957.
Gymnocarpium x heterosporum W.H.Wagner

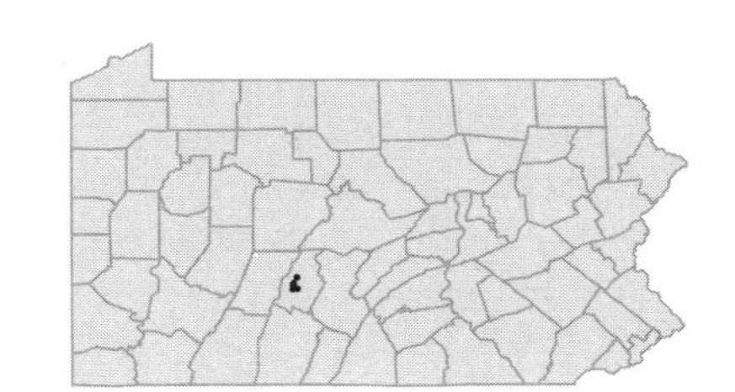

• ***Matteuccia struthiopteris*** (L.) Todaro
FACW
Ostrich fern
Herbaceous perennial
Moist, alluvial flats and swamps, in circumneutral soils.
Matteuccia pensylvanica (Willd.) Raymond W; *Pteretis pensylvanica* (Willd.) Fern. F

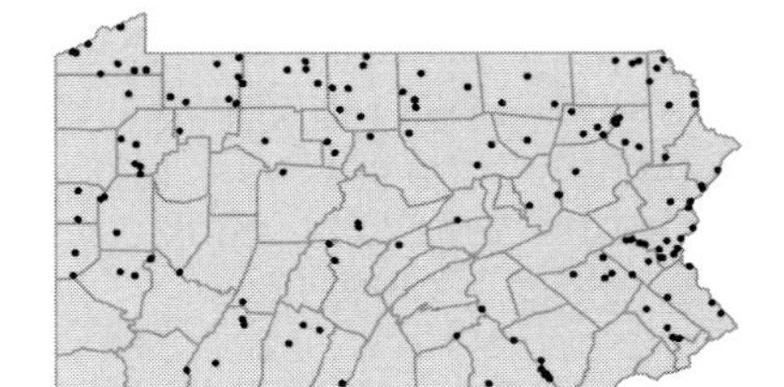

• ***Onoclea sensibilis*** L.
FACW
Sensitive fern
Herbaceous perennial
Marshes, swamps, moist open woods and wet meadows, in subacidic soils.

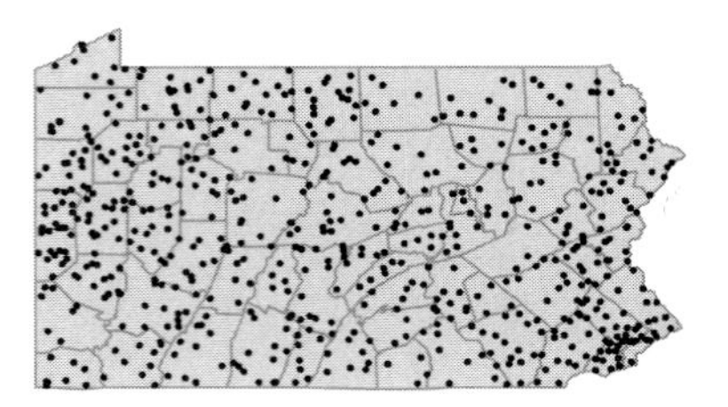

• ***Polystichum acrostichoides*** (Michx.) Schott
Christmas fern
Herbaceous perennial
Shaded slopes and well-drained flats, in subacidic to circumneutral soils. A hybrid with *P. braunii* (*P. x potteri* Barrington) has been collected at a site in Wayne Co. where both parent species occur.

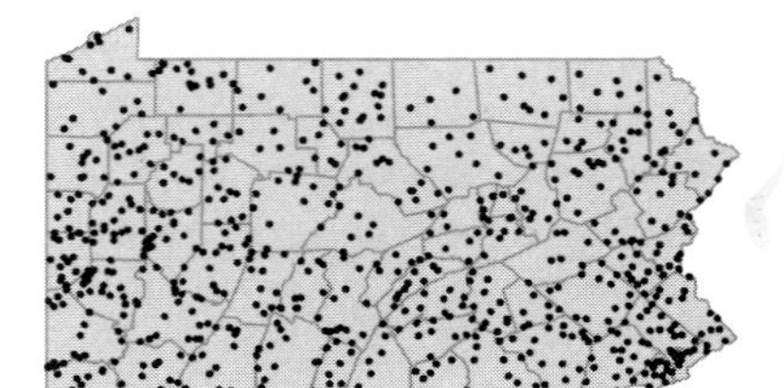

Polystichum braunii (Spenner) Fee
Braun's holly fern
Herbaceous perennial
Cool, rocky slopes and shaded ravines, in humus-rich, circumneutral soils.

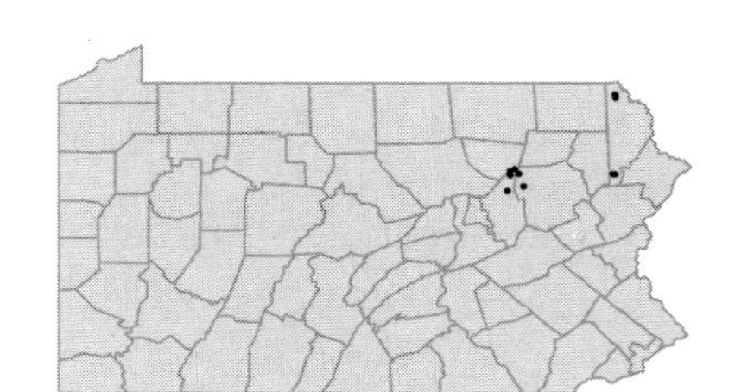

• ***Woodsia ilvensis*** (L.) R.Br.
Rusty woodsia
Herbaceous perennial
Dry, sunny crevices of cliffs, ledges and talus slopes, in acidic to subacidic soil.

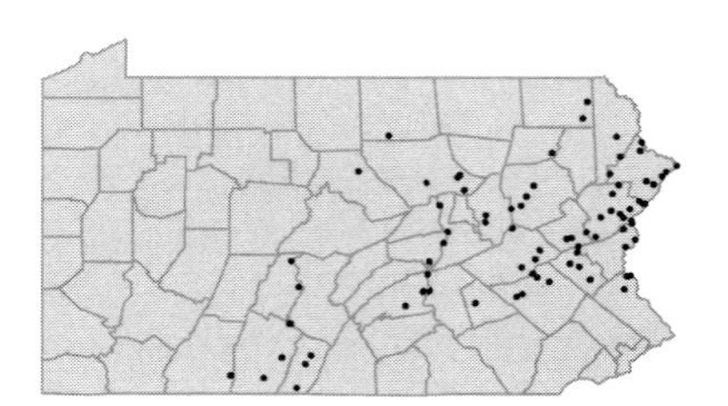

Woodsia obtusa (Spreng.) Torr.
Blunt-lobed woodsia
Herbaceous perennial
Shaded rock crevices, talus slopes, sandy banks or masonry, in acidic to neutral soil.

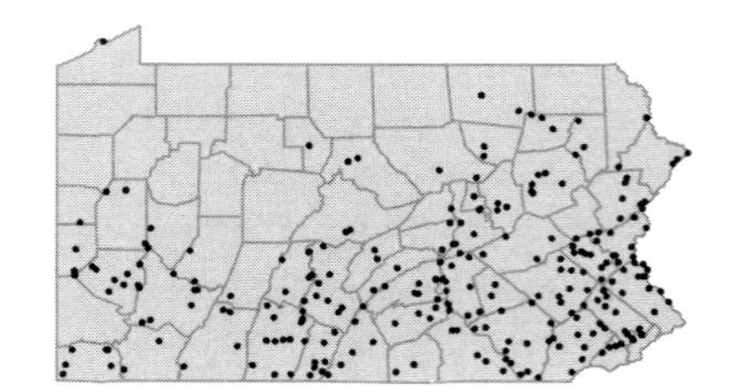

BLECHNACEAE

• ***Woodwardia areolata*** (L.) T.Moore — FACW+
Netted chain fern
Herbaceous perennial
Moist woods; in humus-rich, strongly acidic soils.

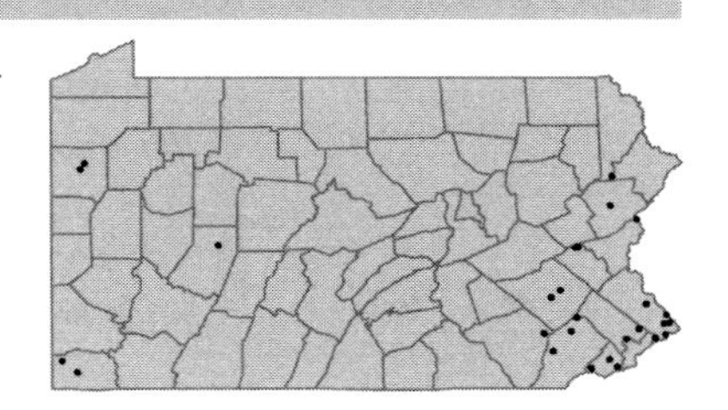

Woodwardia virginica (L.) Smith — OBL
Virginia chain fern
Herbaceous perennial
Bogs, swamps, marshes and shallow ponds, in strongly acidic to neutral soils.

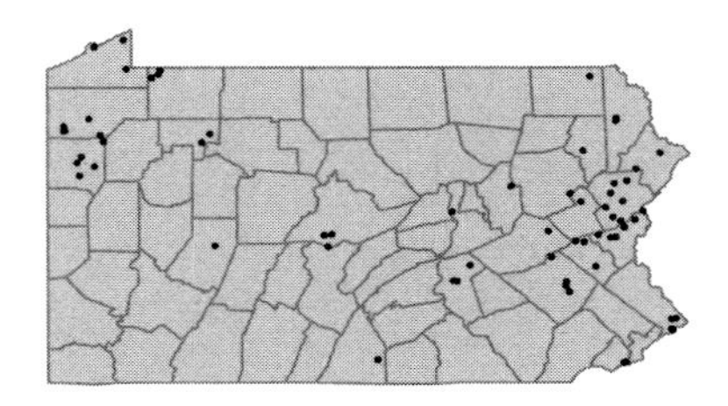

MARSILEACEAE

• ***Marsilea quadrifolia*** L. — OBL [I]
European water-clover; Pepperwort
Herbaceous perennial, floating-leaf aquatic
Shallow water of ponds and sluggish streams.

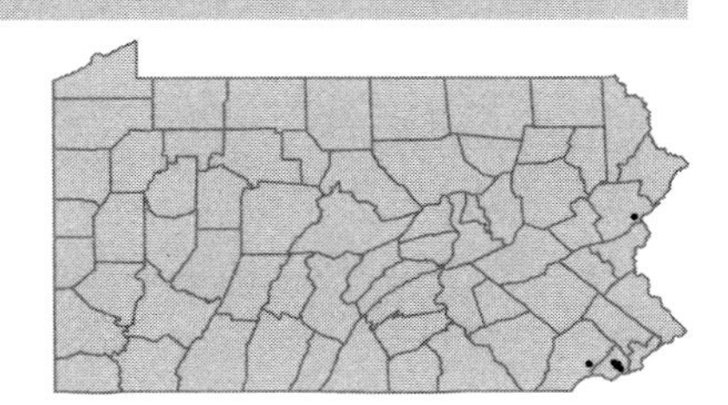

PINOPHYTA
(Gymnosperms)

Canada hemlock (*Tsuga canadensis*), the state tree of Pennsylvania, grows in cool, moist woods throughout the state.

GINKGOACEAE

• ***Ginkgo biloba*** L.
Ginkgo
Deciduous tree
An occasional garden escape, naturalizing in disturbed woods.

[I]

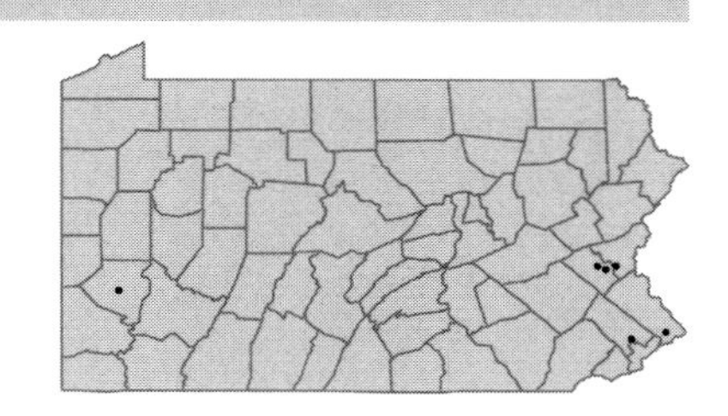

PINACEAE

• ***Abies balsamea*** (L.) P.Mill.
Balsam fir
Evergreen tree
Cool swamps or bogs, in peaty soils.

FAC

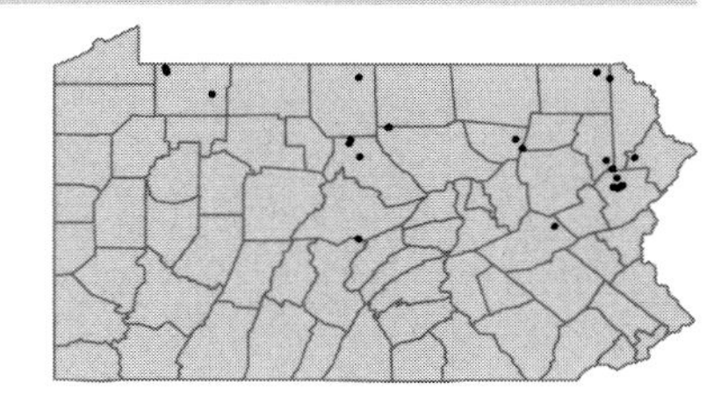

• ***Larix decidua*** P.Mill.
European larch
Deciduous tree
Forest plantations. Planted by the Bureau of Forestry and the Game Commission.

[I]

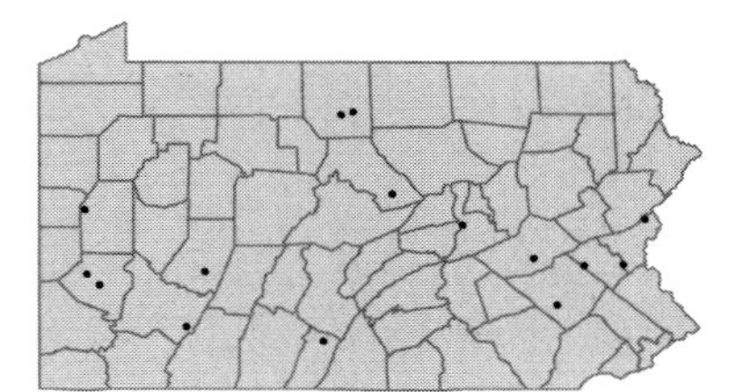

Larix kaempferi (Lamb.) Carr.
Japanese larch
Deciduous tree
Forest plantations. Planted by the Bureau of Forestry and the Game Commission.

[I]

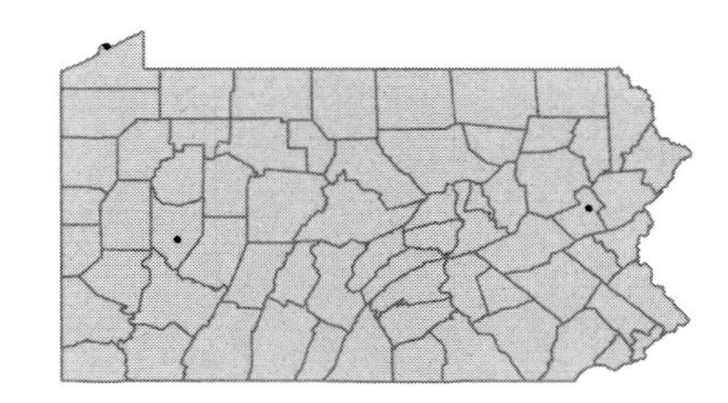

Larix laricina (DuRoi) K.Koch
American larch; Tamarack
Deciduous tree
Margins of sphagnum bogs and peatlands.

FACW

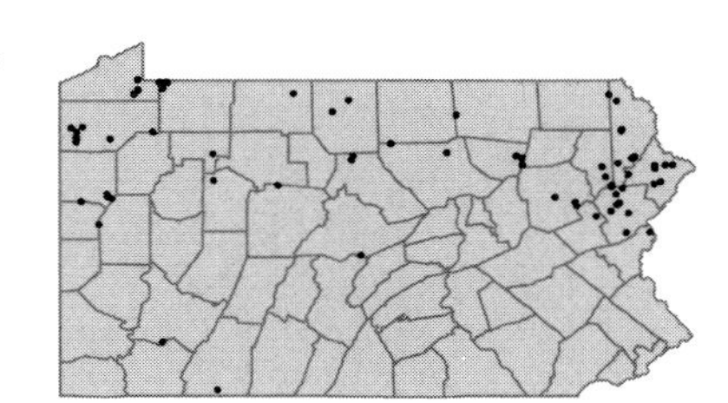

• ***Picea abies*** (L.) Karst.
Norway spruce
Evergreen tree
Forest plantations. Planted by the Bureau of Forestry and the Game Commission.

[I]

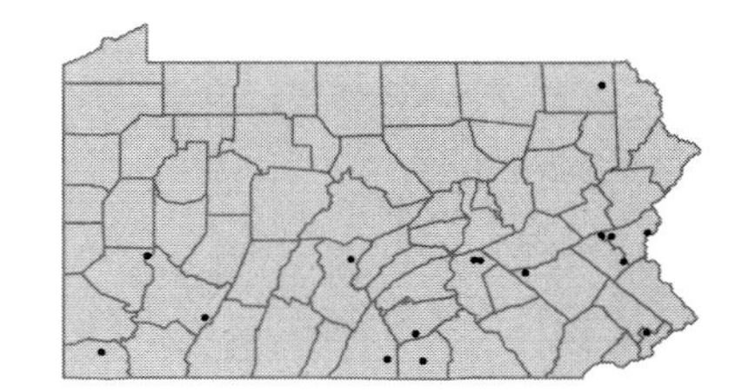

Picea glauca (Moench) Voss
White spruce
Evergreen tree
Forest plantations. Planted by the Bureau of Forestry and the Game Commission.

FACU
[I]

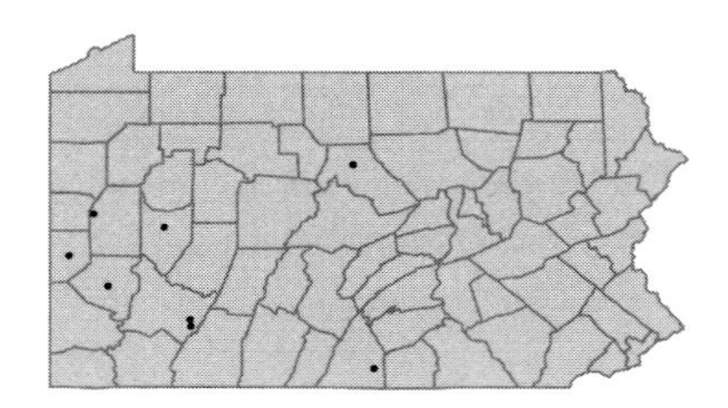

Picea mariana (P.Mill.) BSP
Black spruce; Bog spruce
Evergreen tree
Sphagnum bogs.
Picea brevifolia Peck P

FACW-

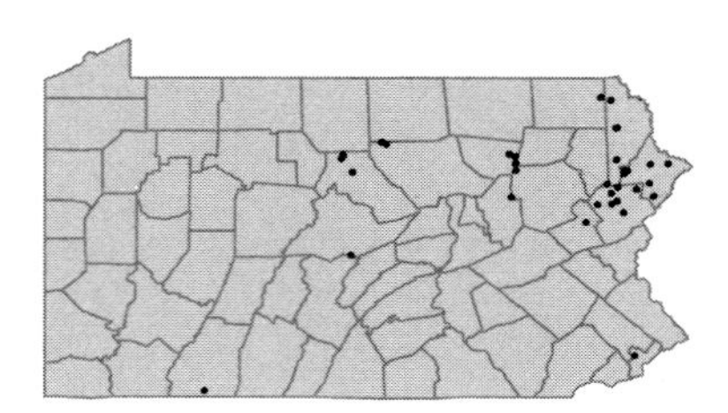

Picea pungens Engelm.
Colorado blue spruce
Evergreen tree
Forest plantations, old homesteads or other cultivated sites. Planted by the Bureau of Forestry and the Game Commission.

[I]

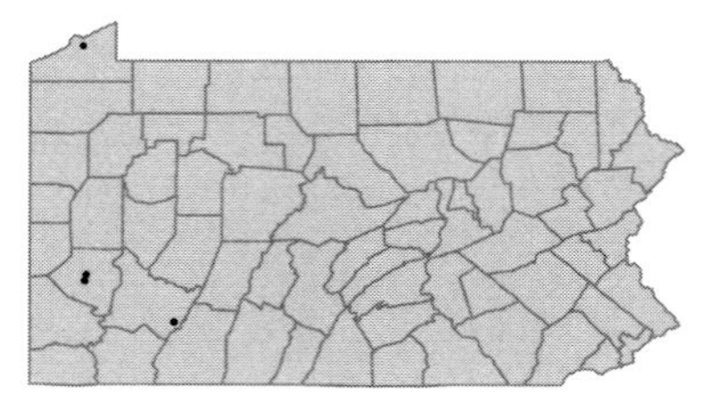

Picea rubens Sarg.
Red spruce
Evergreen tree
Cool, moist woodlands or margins of bogs and swamps.

FACU

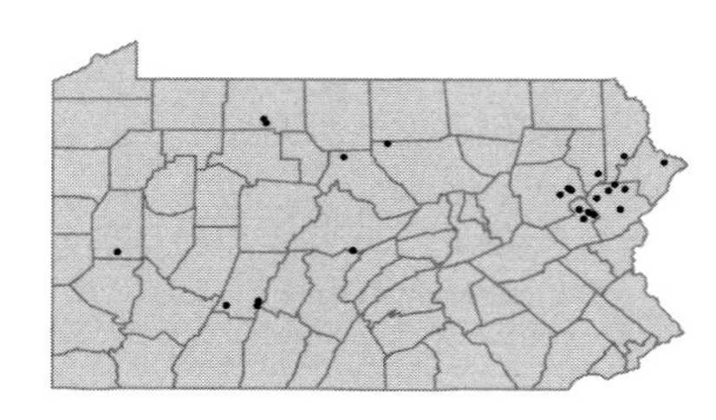

• ***Pinus banksiana*** Lamb.
Jack pine
Evergreen tree
Forest plantations. Planted by the Bureau of Forestry and the Game Commission.

FACU
[I]

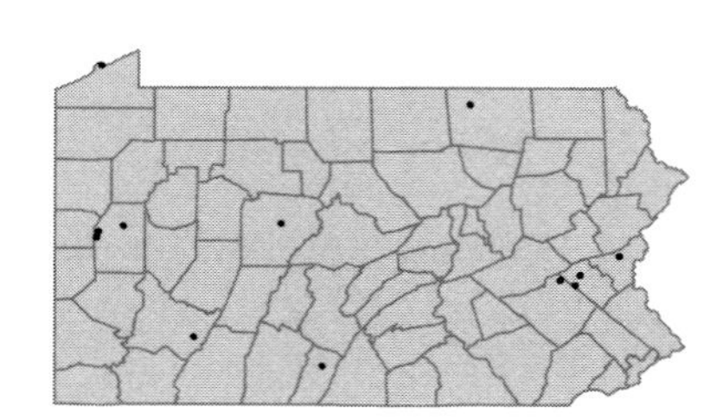

Pinus echinata P.Mill.
Short-leaf pine; Yellow pine
Evergreen tree
Slopes and ridges, in dry, sterile soils.

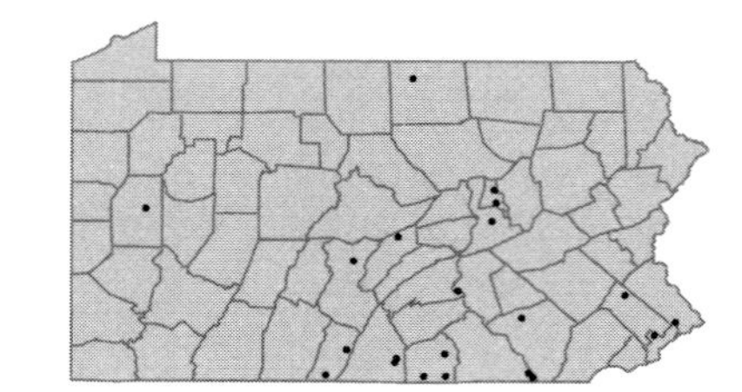

Pinus nigra Arnold
Austrian pine
Evergreen tree
Forest plantations and other cultivated sites. Planted by the Bureau of Forestry and the Game Commission.

I

Pinus pungens Lamb.
Table-mountain pine
Evergreen tree
Dry, rocky and gravelly slopes and ridge tops.

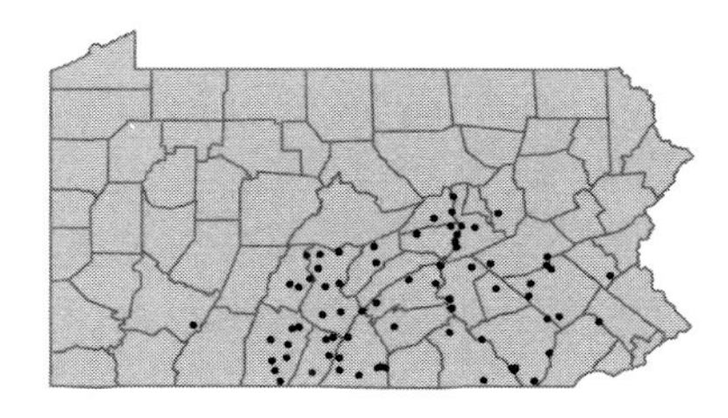

Pinus resinosa Ait.
Norway pine; Red pine
Evergreen tree
Native on dry slopes in Luzerne, Wyoming, Tioga and Centre Cos. Also planted extensively by the Bureau of Forestry and the Game Commission.

FACU

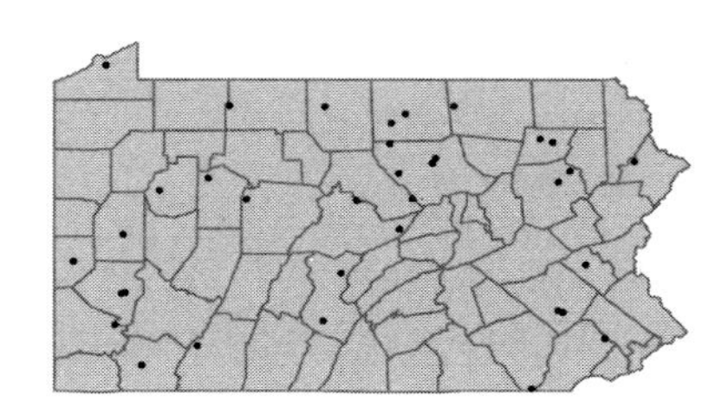

Pinus rigida P.Mill.
Pitch pine
Evergreen tree
Moist to dry, sterile soils including serpentine barrens.

FACU

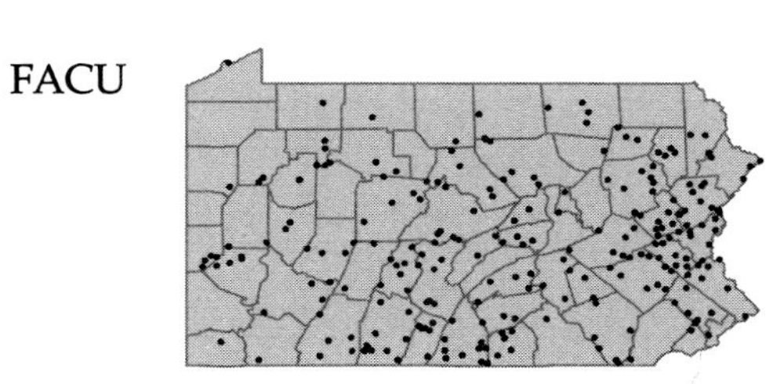

Pinus strobus L.
Eastern white pine
Evergreen tree
Moist to dry woodlands and forested slopes.

FACU

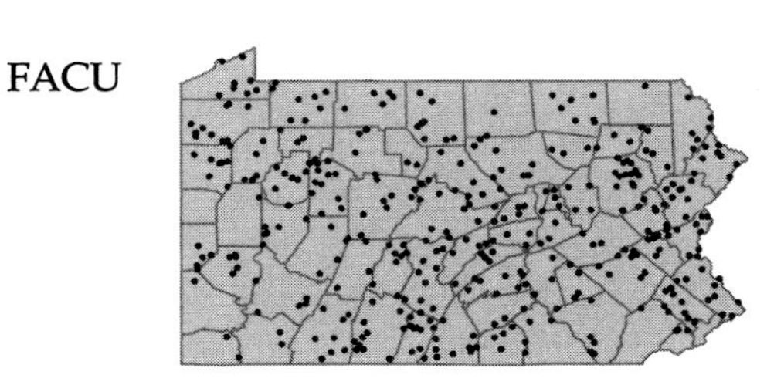

Pinus sylvestris L.
Scots pine
Evergreen tree
Forest plantations and other cultivated sites. Planted by the Bureau of Forestry and the Game Commission.

I

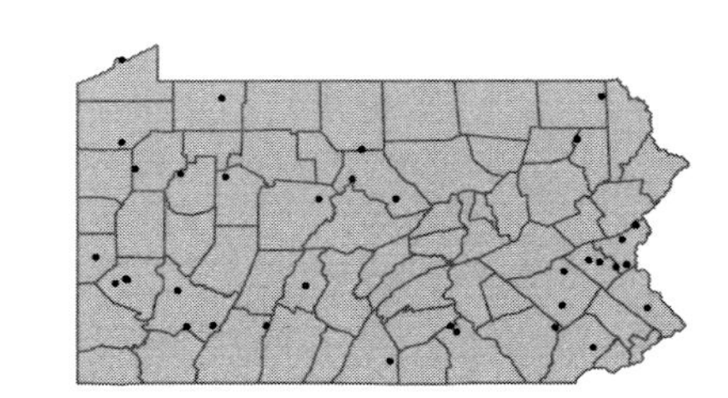

Pinus virginiana P.Mill.
Virginia pine
Evergreen tree
Barrens and ridgetops, in dry sandy or rocky soils.

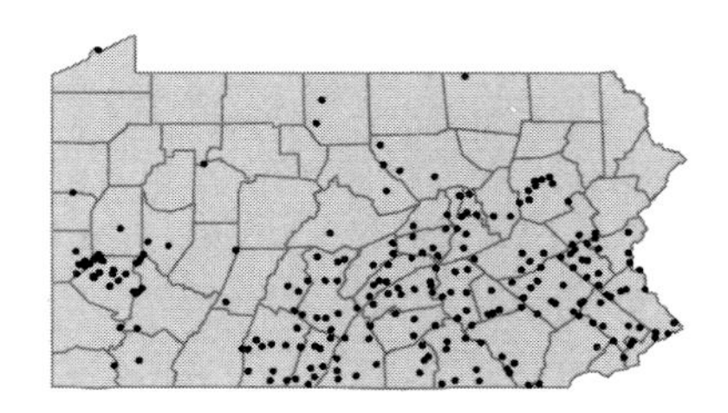

• ***Tsuga canadensis*** (L.) Carr. FACU
Canada hemlock
Evergreen tree
Cool, moist woods and shaded slopes. The state tree of Pennsylvania.

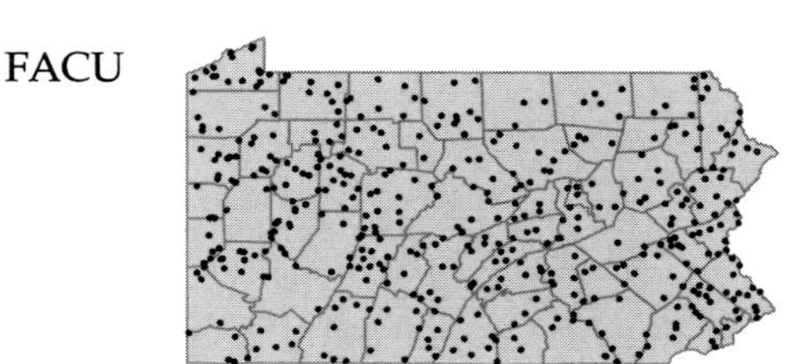

CUPRESSACEAE

• ***Chamaecyparis thyoides*** (L.) BSP OBL

Atlantic white-cedar
Evergreen tree
Sphagnum bogs. Early Bucks and Phila. Co. occurrences reported but not substantiated by herbarium specimens. Site on Laurel Mt, Westmoreland Co. probably represents a naturalized population.

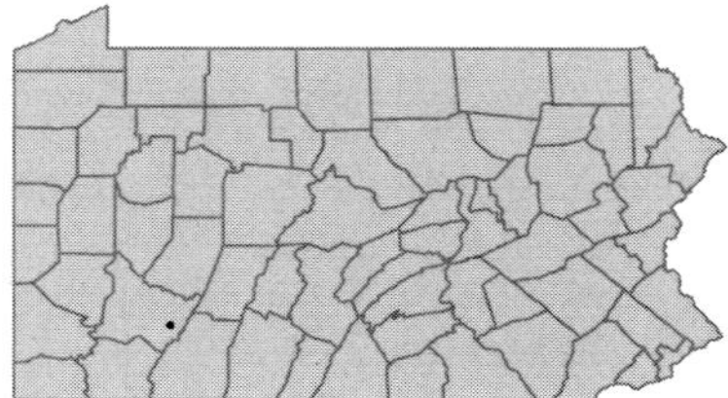

• ***Juniperus communis*** L.
Common juniper
Evergreen shrub
Dry slopes or pastures.

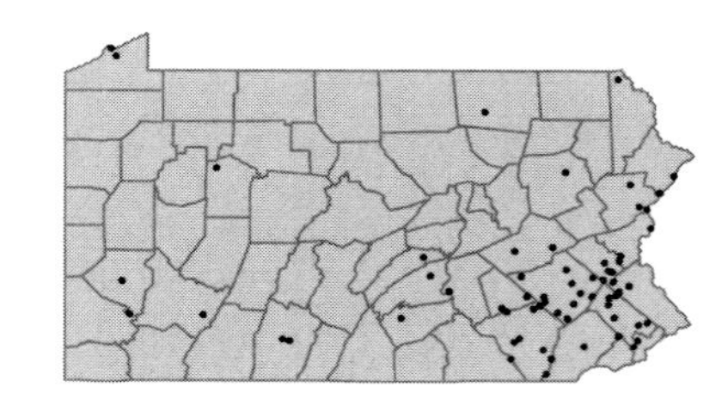

Juniperus virginiana L. FACU

Red-cedar
Evergreen tree
Old fields, serpentine barrens and other moist to dry, sterile soils.

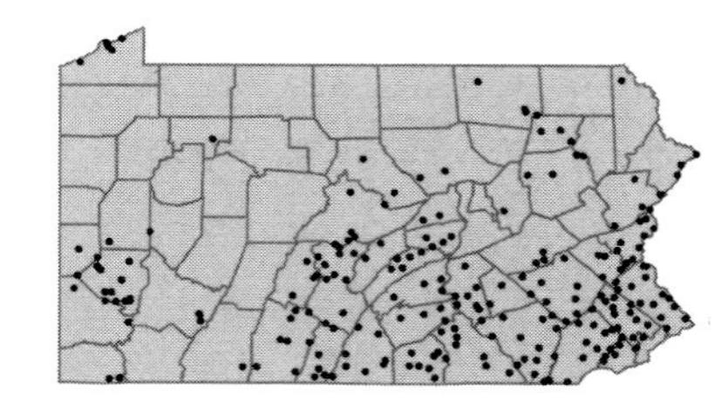

• ***Taxodium distichum*** (L.) L.C.Rich. OBL

Bald cypress
Deciduous tree
Persisting at old nurseries or other sites of cultivation.

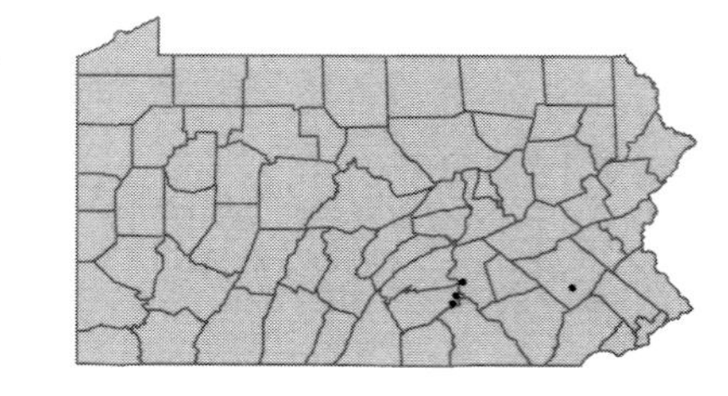

• ***Thuja occidentalis*** L.
Arbor-vitae; Northern white-cedar
Evergreen tree
Old homesites, abandoned nurseries or plantations.

FACW
I

TAXACEAE

• ***Taxus canadensis*** Marshall
American yew
Evergreen shrub
Cool, moist, rocky slopes and ravines. Declining in some areas due to heavy browsing by deer.

FAC

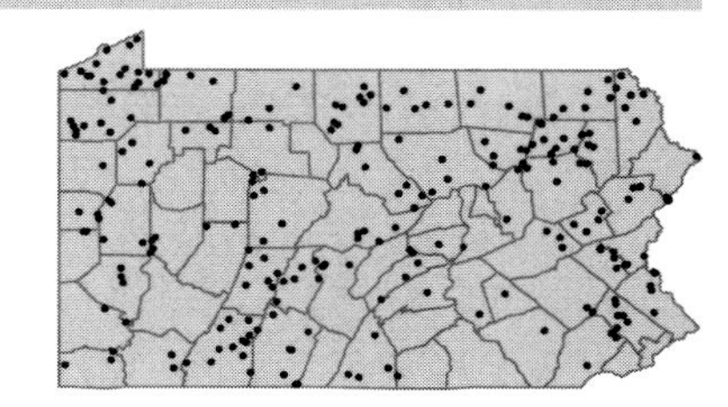

MAGNOLIOPSIDA

(Dicotyledons)

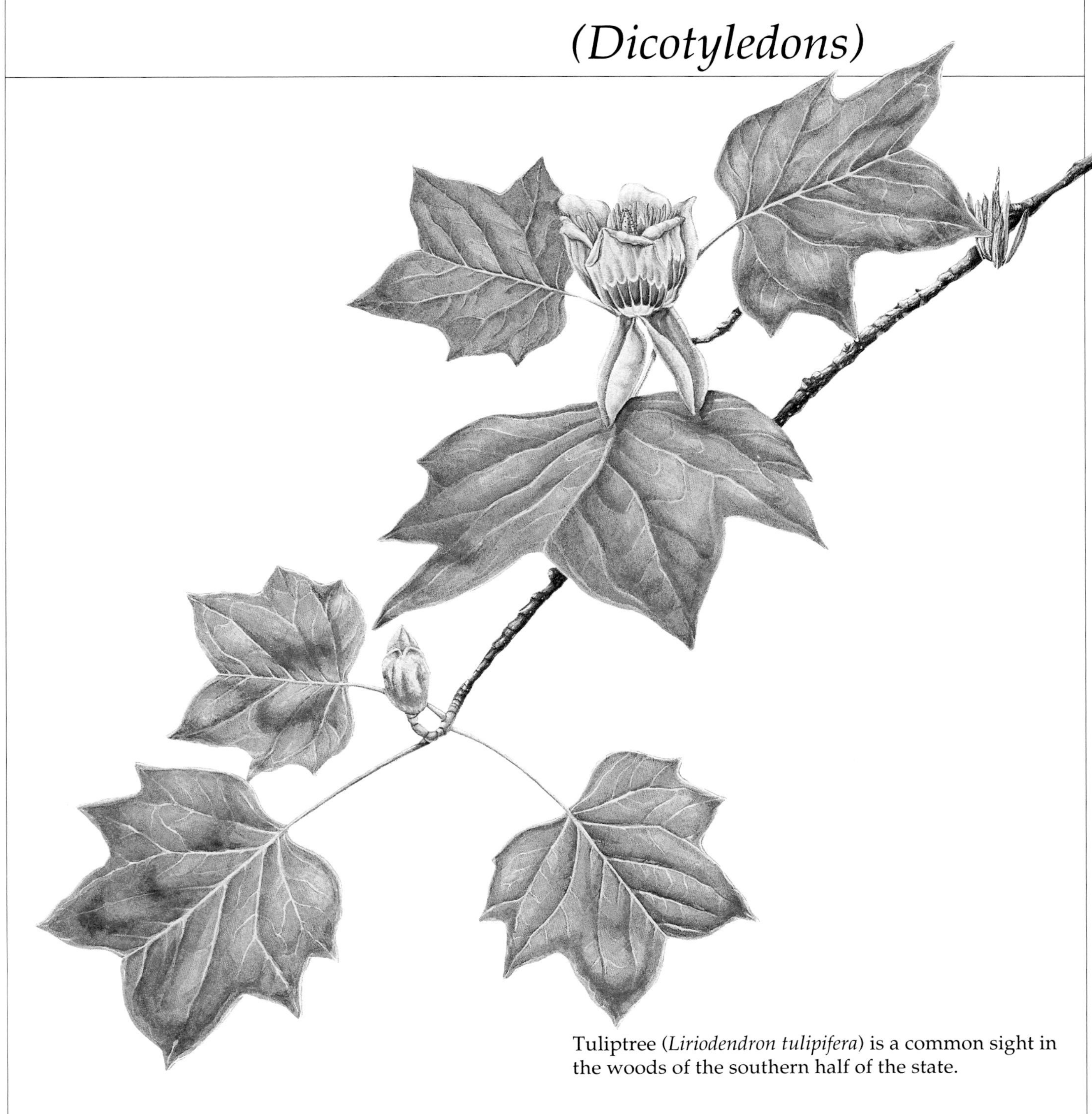

Tuliptree (*Liriodendron tulipifera*) is a common sight in the woods of the southern half of the state.

MAGNOLIACEAE

• ***Liriodendron tulipifera*** L.
Tuliptree; Yellow poplar
Deciduous tree
A common forest tree of rich woods.

FACU

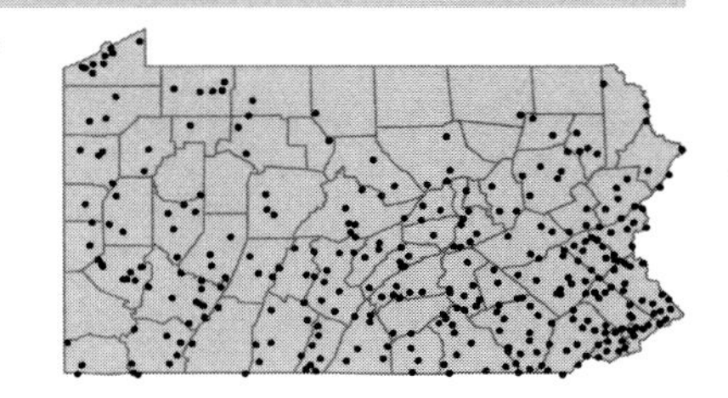

• ***Magnolia acuminata*** (L.) L.
Cucumber-tree
Deciduous tree
Rich upland woods and slopes. Native in central and western counties, planted in the east.

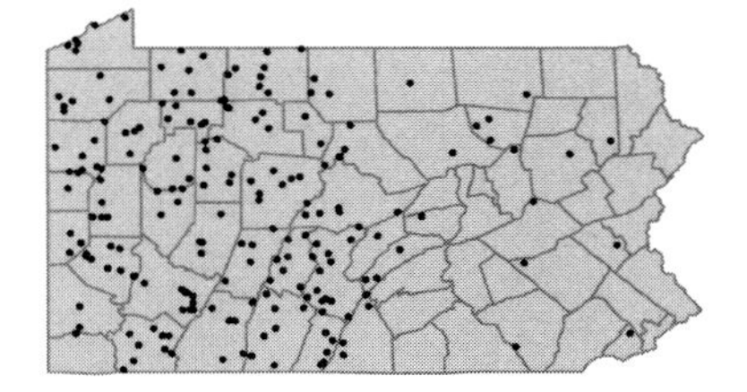

Magnolia tripetala L.
Umbrella magnolia
Deciduous tree
Rich wooded slopes and floodplains. Native in southcentral and southwest counties, apparently introduced in the east. Spreading rapidly at several sites in southeast PA.

FACU

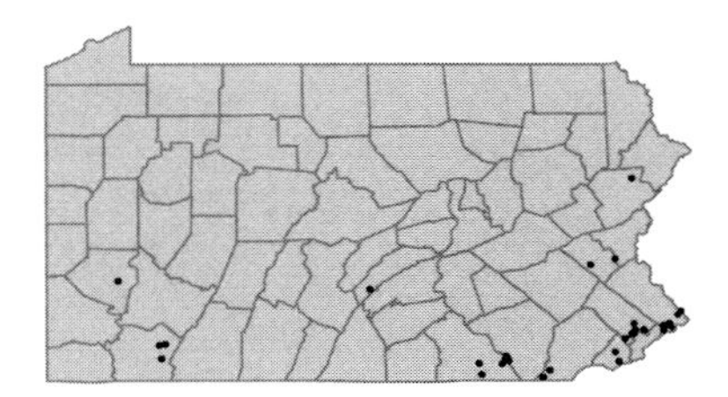

Magnolia virginiana L.
Sweet-bay magnolia
Deciduous tree
Moist woods and swamps, in sandy, peaty soils.

FACW+

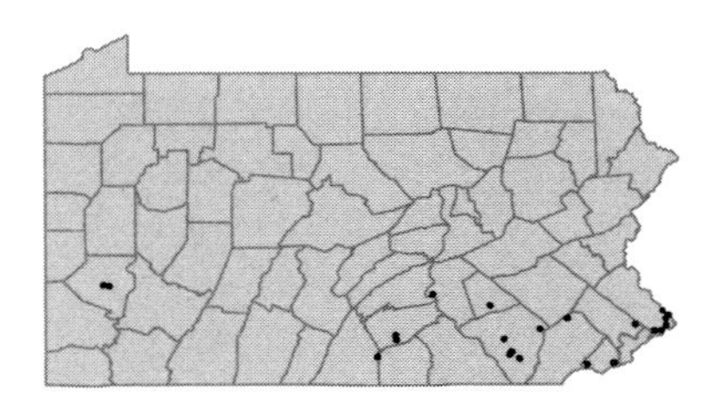

ANNONACEAE

• ***Asimina triloba*** (L.) Dunal
Pawpaw
Deciduous tree
Moist, rich woodlands.

FACU+

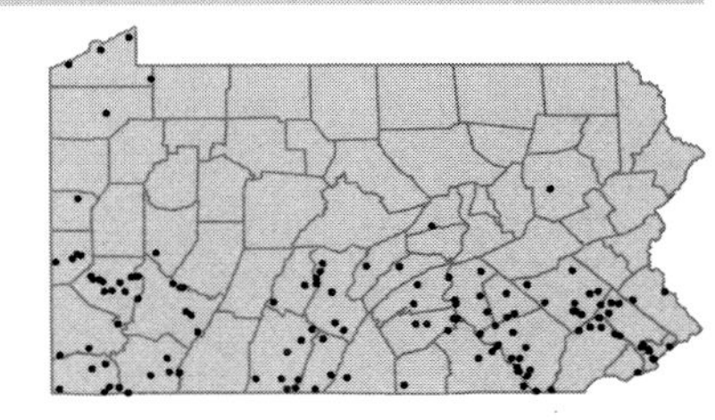

CALYCANTHACEAE

• ***Calycanthus floridus*** L. var. ***floridus***
Carolina allspice; Strawberry-shrub; Sweetshrub
Deciduous shrub
Cultivated and occasionally escaped.

I

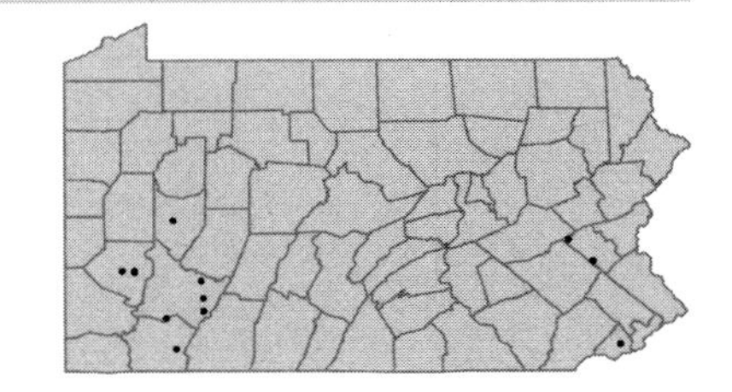

CALYCANTHACEAE

Calycanthus floridus L. var. ***laevigatus*** (Willd.) Torr. & A.Gray
Carolina allspice; Strawberry-shrub; Sweetshrub
Deciduous shrub
Although this plant is native as far north as PA, most collections represent cultivated sources.
Butneria fertilis (Walt.) Kearney P; *Calycanthus fertilis* Walt. FGB

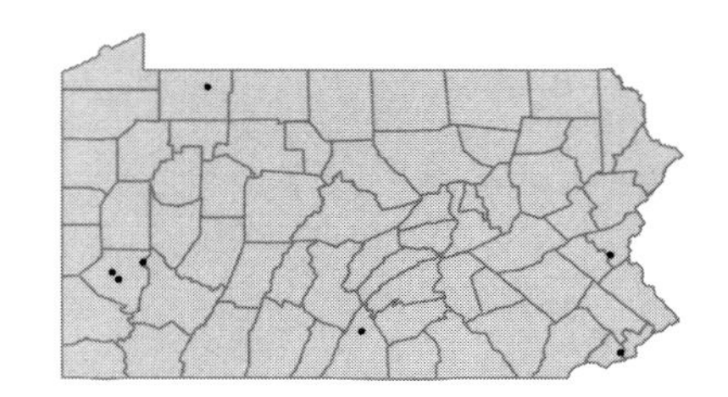

LAURACEAE

• ***Lindera benzoin*** (L.) Blume — FACW-
Spicebush
Deciduous shrub
A common component of moist, rich woods.

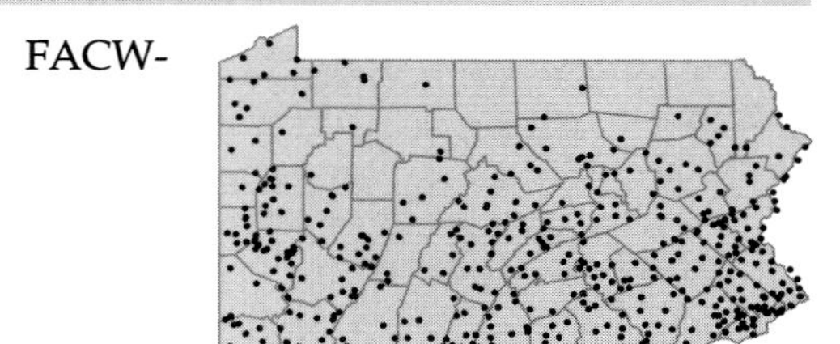

• ***Sassafras albidum*** (Nutt.) Nees — FACU-
Sassafras
Deciduous tree
Old fields, hedgerows and woods edges.
Sassafras sassafras (L.) Karst. P

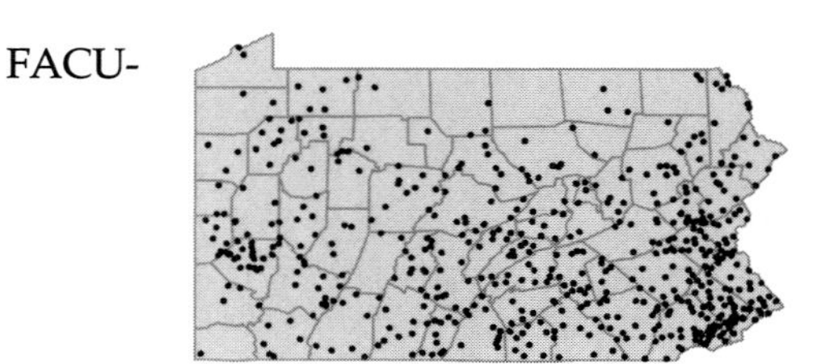

SAURURACEAE

• ***Saururus cernuus*** L. — OBL
Lizard's-tail
Herbaceous perennial, emergent aquatic
Swamps and shallow water along the edges of streams.

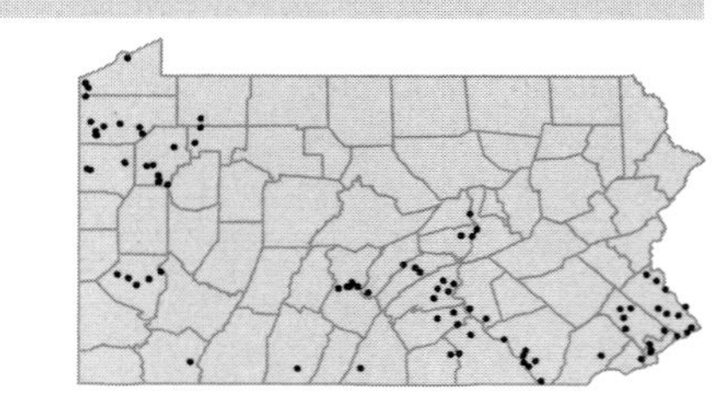

ARISTOLOCHIACEAE

• ***Aristolochia clematitis*** L. — I
Birthwort
Herbaceous perennial
Escaped from cultivation.

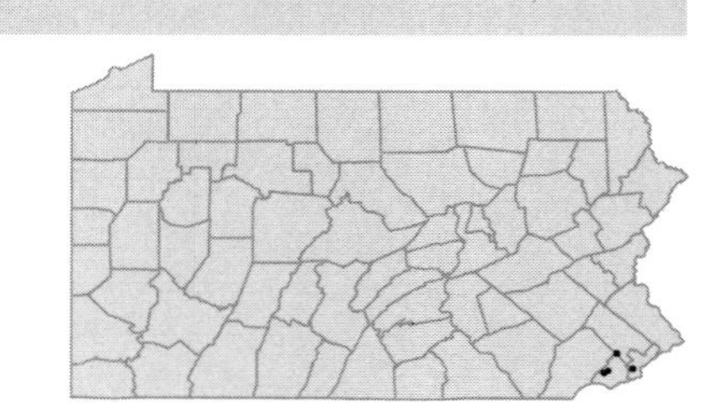

Aristolochia macrophylla Lam.
Dutchman's-pipe; Pipe-vine
Woody vine
Rich woods and rocky slopes. Native in western counties, introduced in the east.
Aristolochia durior Hill FGBW

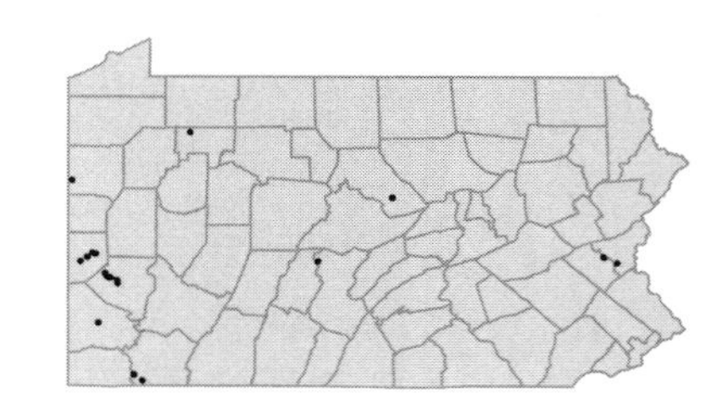

Aristolochia serpentaria L.
Virginia snakeroot
Herbaceous perennial
Moist to dry, wooded slopes.

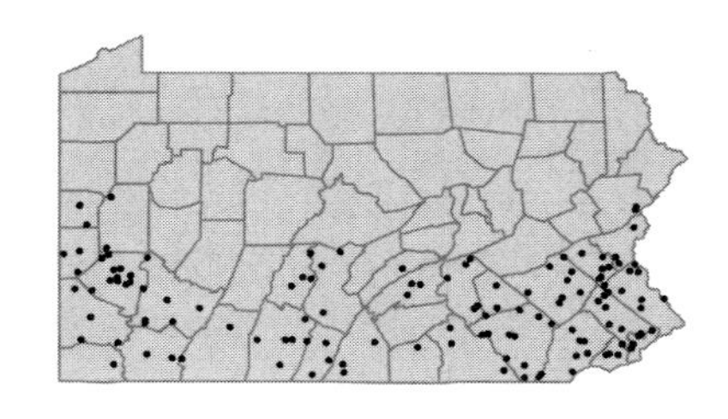

• ***Asarum canadense*** L. var. ***canadense***
Wild ginger
Herbaceous perennial
Moist, rich woods.

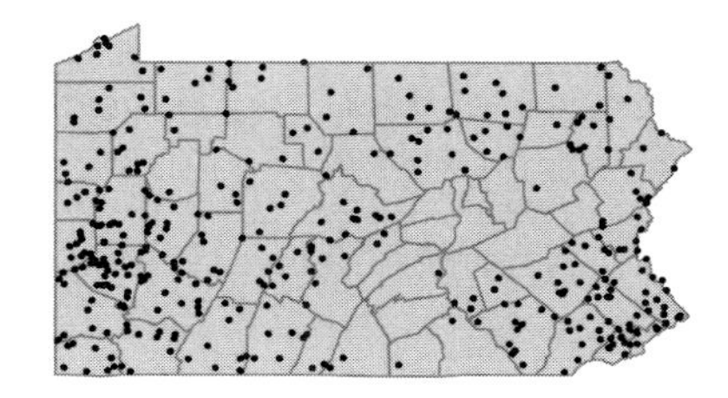

Asarum canadense L. var. ***acuminatum*** Ashe
Wild ginger
Herbaceous perennial
Moist, rich woods.

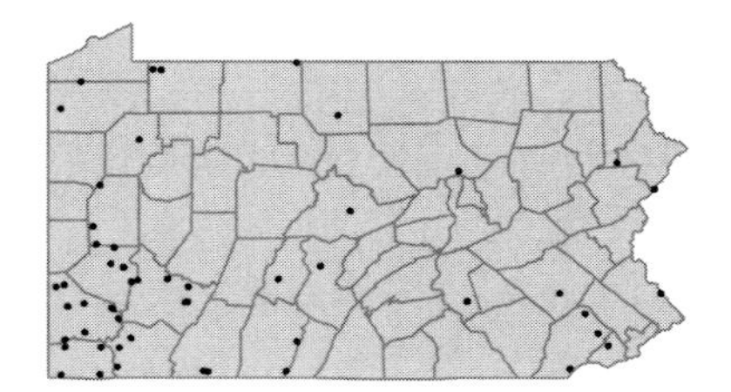

Asarum canadense L. var. ***reflexum*** (Bickn.) B.L.Robins.
Short-lobed wild ginger
Herbaceous perennial
Moist, rich woods.
Asarum reflexum Bickn. P

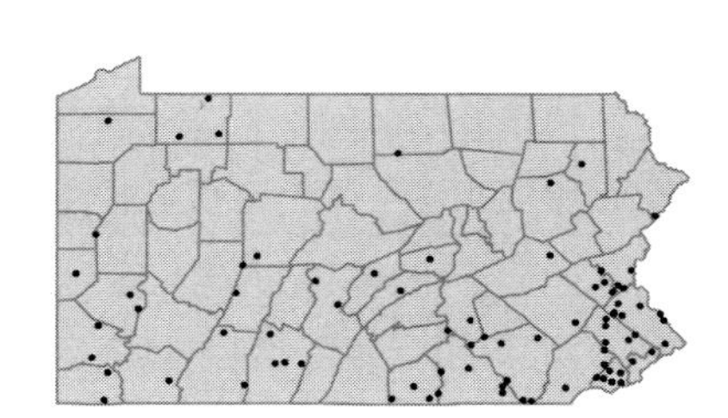

NELUMBONACEAE

• ***Nelumbo lutea*** (Willd.) Pers. OBL
American lotus
Herbaceous perennial, emergent aquatic
Ponds and other quiet water.

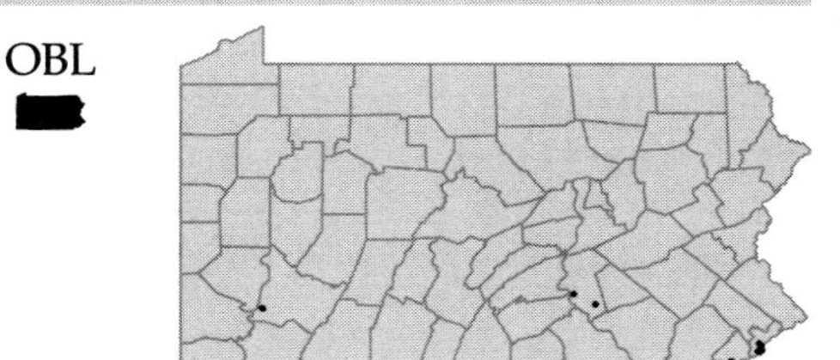

NYMPHAEACEAE

• ***Nuphar lutea*** (L.) Sibth. & Smith *sensu lato* OBL
Spatterdock; Yellow pond-lily; Cow-lily
Herbaceous perennial, floating-leaf aquatic
Ponds, lake margins, slow-moving streams, swamps and tidal marshes. A polymorphic species complex needing further clarification.
Nuphar advena (Ait.) Ait.f. *pro parte* WFGC; *Nuphar variegata* Durand *pro parte* C; *Nuphar x rubrodisca* Morong *pro parte* WFGBC; *Nuphar microphyllum* (Pers.) Fern. *pro parte* FGBC; *Nuphar variegatum* Engelm. *pro parte* FGBW

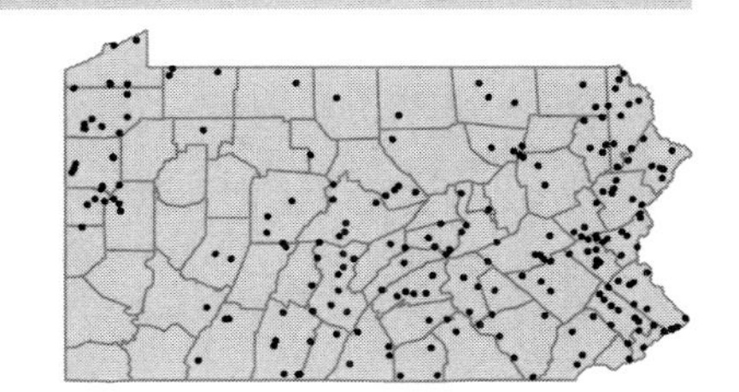

• ***Nymphaea odorata*** Ait.
Fragrant water-lily
Herbaceous perennial, floating-leaf aquatic
Quiet water of lakes and ponds.
Castalia odorata (Dryand) Woodr. & Wood P; *Castalia tuberosa* (Paine) Greene *pro parte* P; *Nymphaea tuberosa* Paine *pro parte* FGBW

OBL

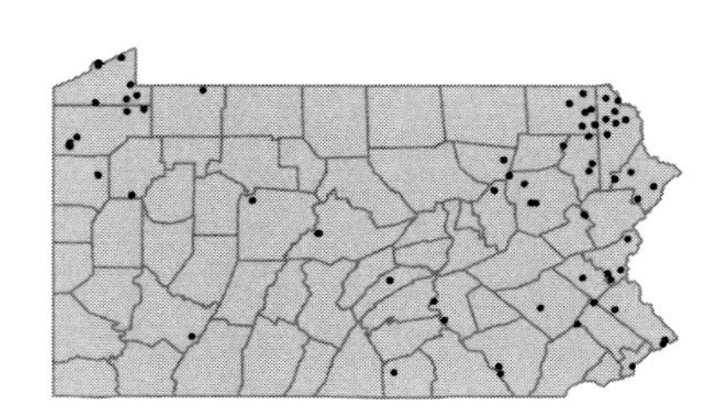

CABOMBACEAE

• ***Brasenia schreberi*** J.F.Gmel.
Purple wen-dock; Water-shield
Herbaceous perennial, floating-leaf aquatic
Quiet water of lakes and streams.
Brasenia peltata (Michx.) Caspary P

OBL

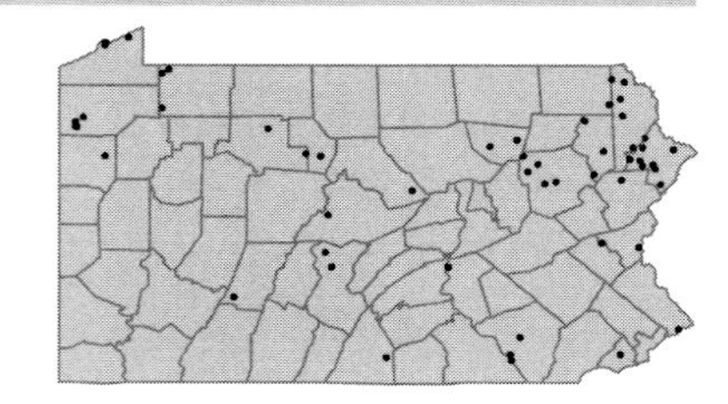

• ***Cabomba caroliniana*** A.Gray
Fanwort
Herbaceous perennial, rooted submergent aquatic
Lakes.

OBL
[I]

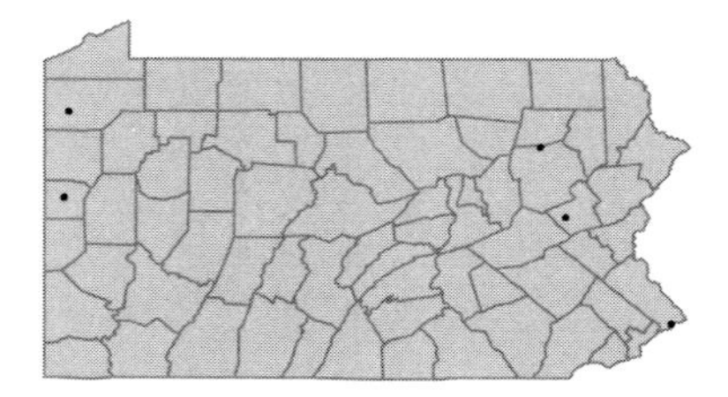

CERATOPHYLLACEAE

• ***Ceratophyllum demersum*** L.
Coontail; Hornwort
Herbaceous perennial, rooted submergent aquatic
Quiet water of lakes, ponds and rivers.

OBL

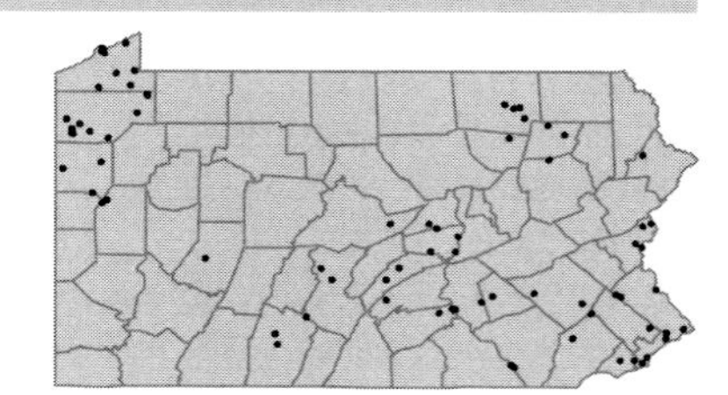

Ceratophyllum muricatum Cham.
Hornwort
Herbaceous perennial, rooted submergent aquatic
Lakes and ponds.
Ceratophyllum echinatum A.Gray FGBWC

OBL

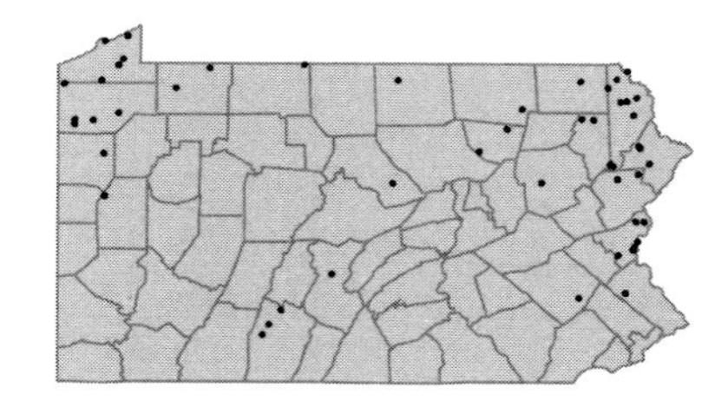

RANUNCULACEAE

• ***Aconitum reclinatum*** A.Gray
Trailing wolfsbane; White monkshood
Herbaceous perennial
Mountain woods.

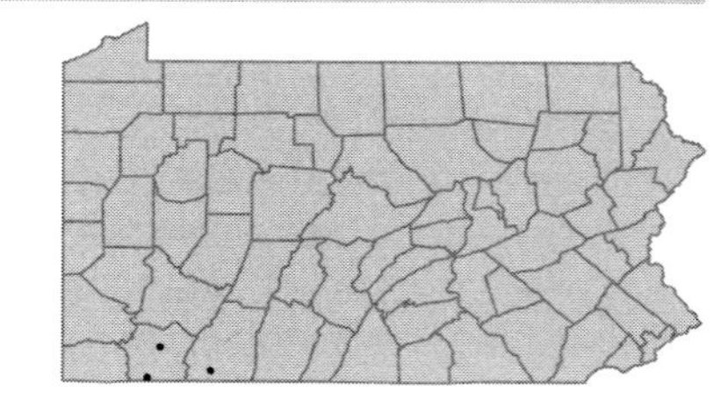

Aconitum uncinatum L.
Blue wild monkshood
Herbaceous perennial
Moist woods and slopes.

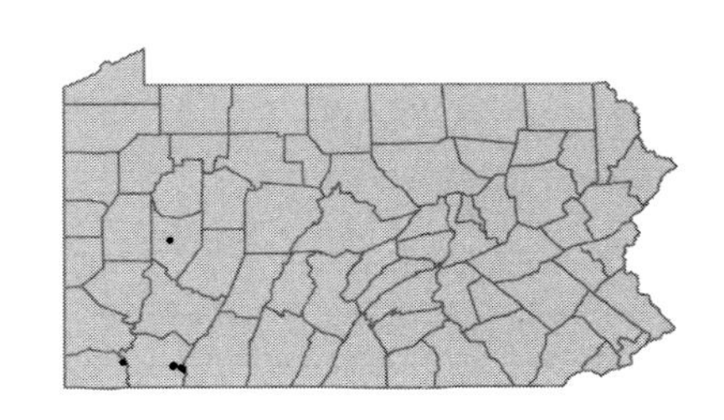

• ***Actaea pachypoda*** Ell.
Doll's-eyes; White baneberry
Herbaceous perennial
Rich woods.
Actaea alba (L.) P.Mill. PGBC

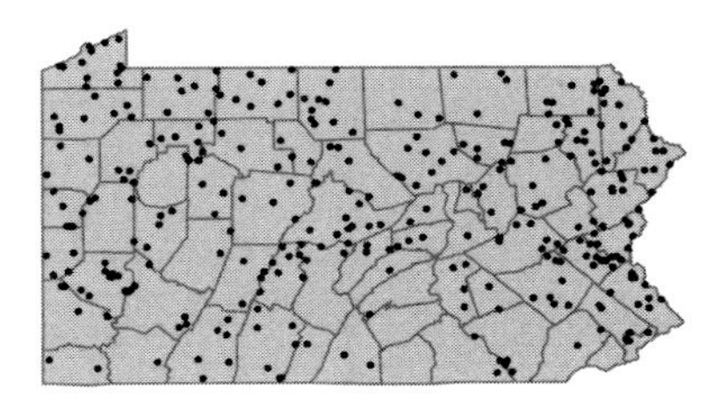

Actaea rubra (Ait.) Willd.
Red baneberry
Herbaceous perennial
Rich woods.

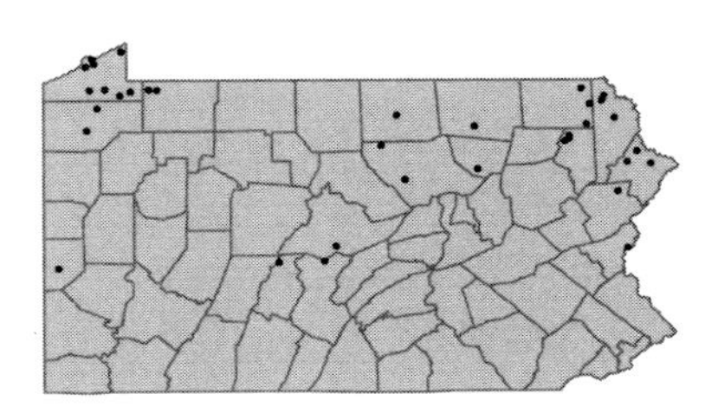

• ***Anemone canadensis*** L.
Canada anemone
Herbaceous perennial
Damp, calcareous meadows and shores.

FACW

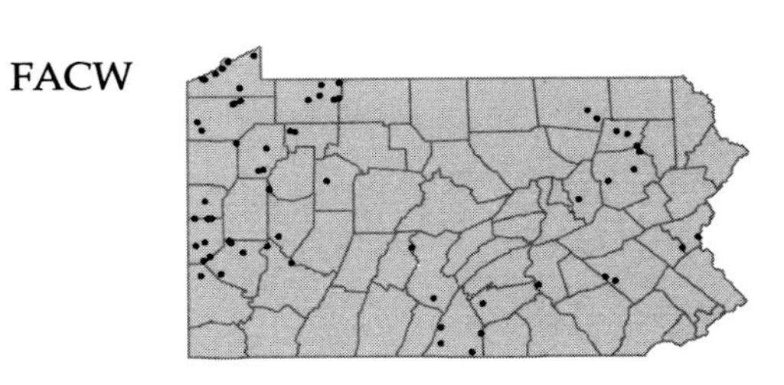

Anemone cylindrica A.Gray
Thimbleweed
Herbaceous perennial
Dry slopes and open fields.

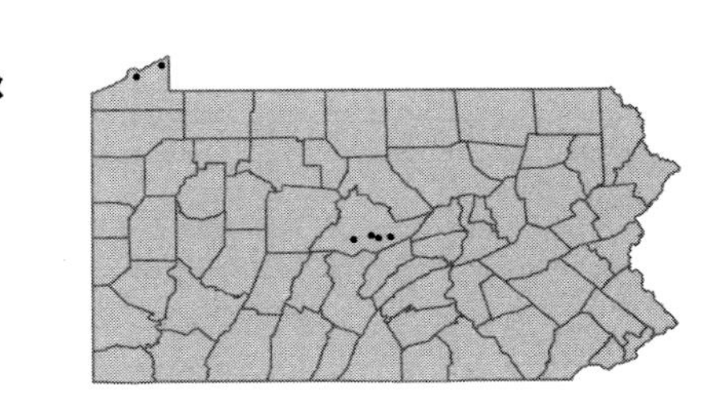

Anemone quinquefolia L.
Wood anemone
Herbaceous perennial
Moist woods and thickets.

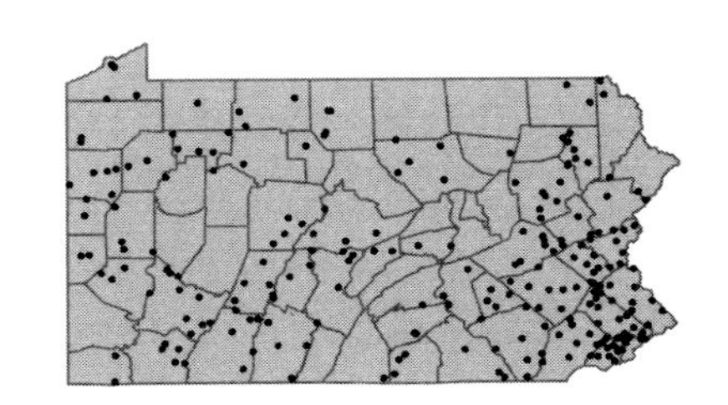

Anemone virginiana L.
Tall anemone; Thimbleweed
Herbaceous perennial
Dry, open woods, slopes and edges.

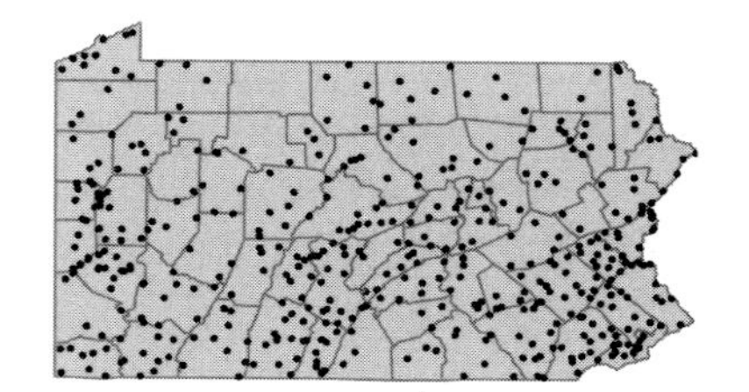

• ***Aquilegia canadensis*** L.
Meetinghouses; Wild columbine
Herbaceous perennial
Cliffs and rocky slopes, usually calcareous.

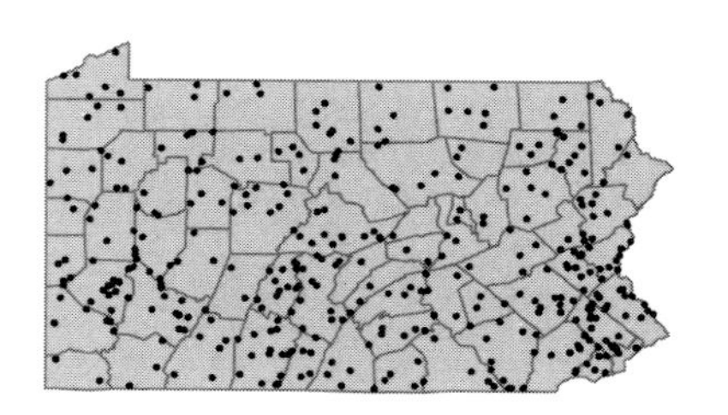

Aquilegia vulgaris L. — [I]
Columbine
Herbaceous perennial
Cultivated and escaped to roadsides and woods edges.

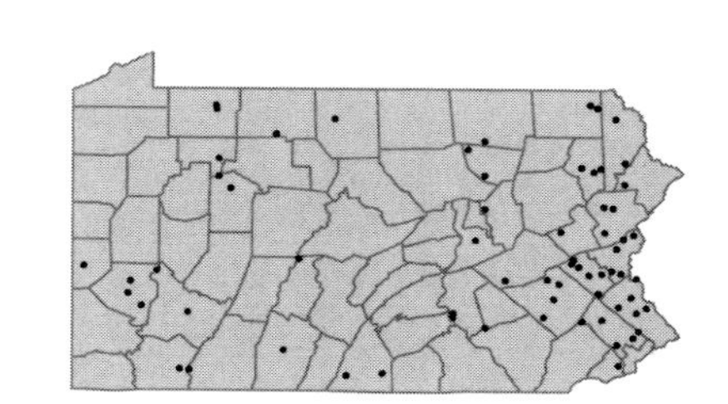

• ***Caltha palustris*** L. var. ***palustris*** — OBL
Cowslip; Marsh marigold
Herbaceous perennial
Swamps, wet meadows and wet woods, often calcareous.

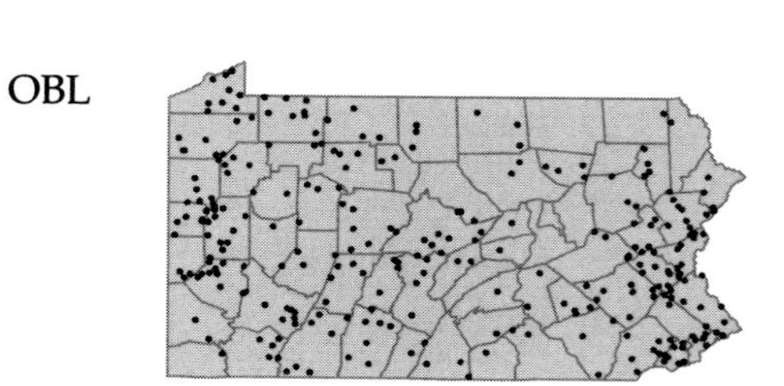

Caltha palustris L. var. ***flabellifolia*** (Pursh) Torr. & A.Gray — OBL
Cowslip; Marsh marigold
Herbaceous perennial
Swamps, wet meadows and wet woods, often calcareous.
Caltha flabellifolia Pursh P

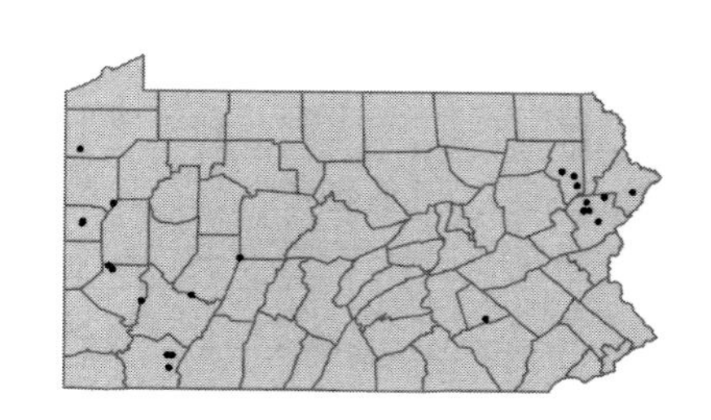

• ***Cimicifuga americana*** Michx.
Mountain bugbane; Summer cohosh
Herbaceous perennial
Moist woods.

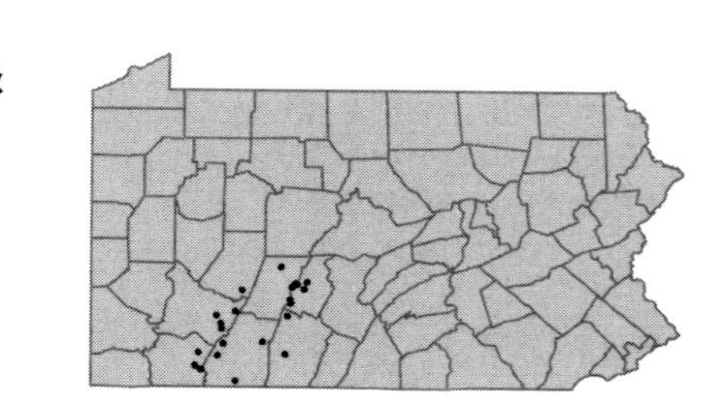

Cimicifuga racemosa (L.) Nutt.
Black snakeroot; Black cohosh
Herbaceous perennial
Rich woods.

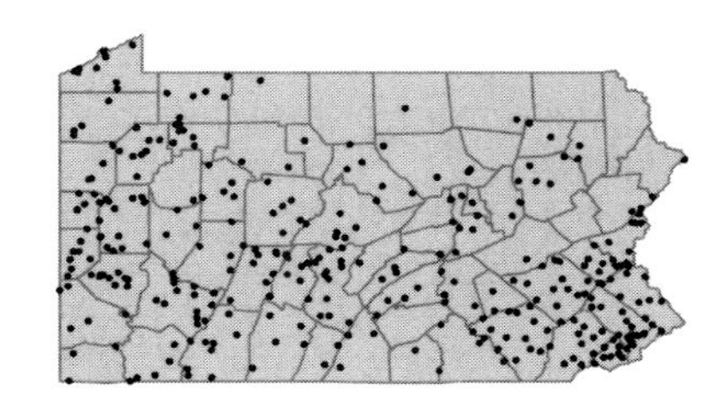

Cimicifuga rubifolia Kearney
Bugbane
Herbaceous perennial
Garden escape. Represented by two collections from Berks Co. in 1935 and 1936.

I

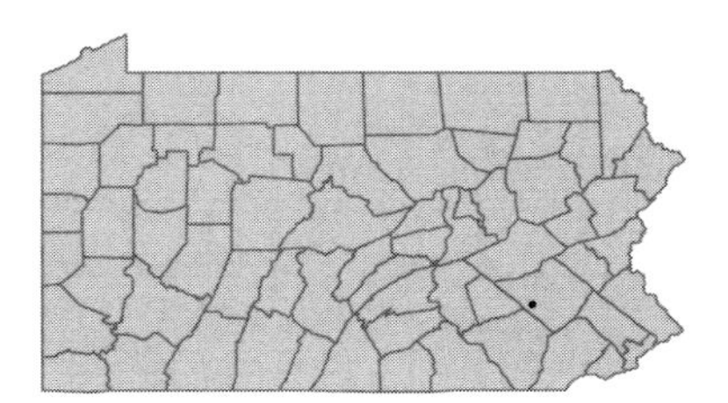

• ***Clematis ligusticifolia*** Nutt.
Western virgin's-bower
Herbaceous perennial vine
Roadside bank. Represented by a single collection from Philadelphia Co. in 1926.

I

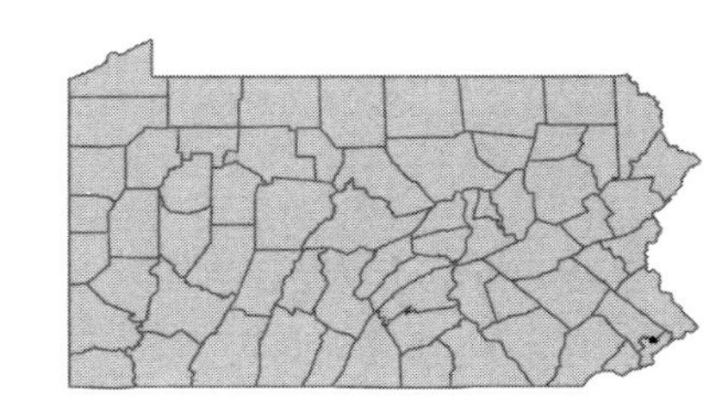

Clematis occidentalis (Hornem.) DC.
Purple clematis
Herbaceous perennial vine
Rocky slopes and open woods, often calcareous.
Atragene americana Sims. P; *Clematis verticillaris* DC. FGB

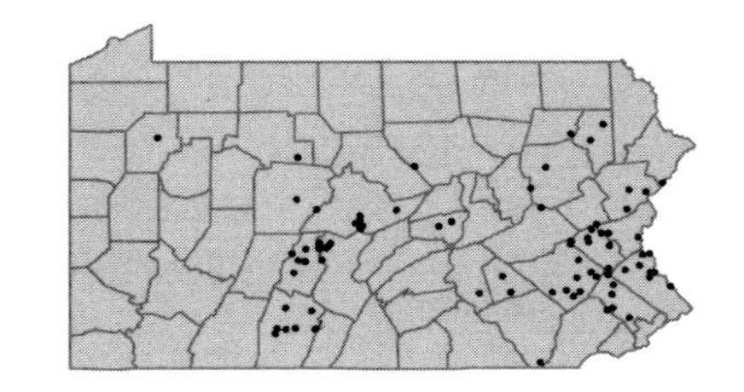

Clematis terniflora DC.
Sweet autumn clematis
Herbaceous perennial vine
Thickets, fencerows and roadsides.
Clematis dioscoreifolia Levl. & Vaniot FGB; *Clematis maximowicziana* Franch. & Savat W

I

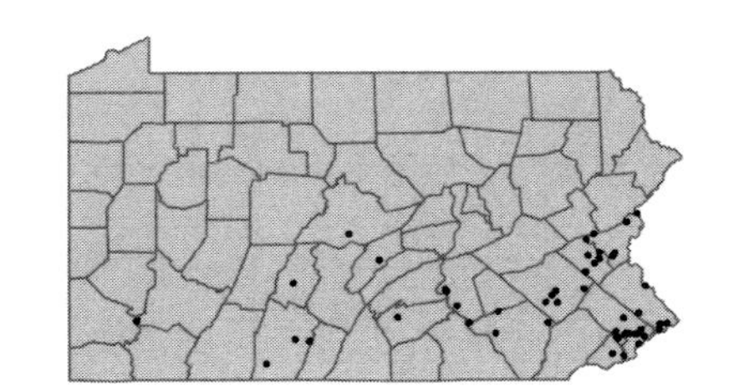

Clematis viorna L.
Leather-flower; Vase-vine
Herbaceous perennial vine
Rich woods and thickets.

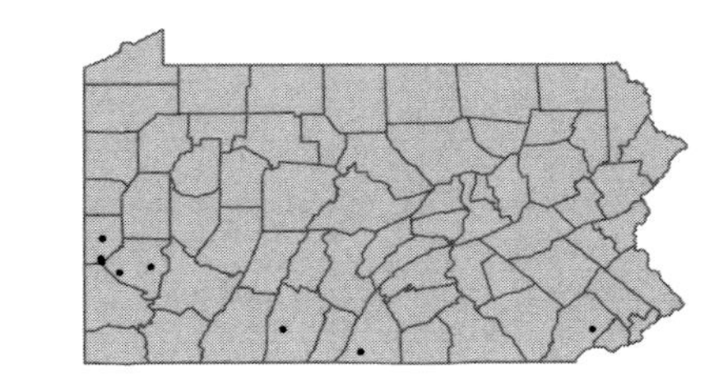

Clematis virginiana L.
Devil's-darning-needle; Virgin's-bower
Herbaceous perennial vine
Thickets and woods edges, in low ground.

FAC

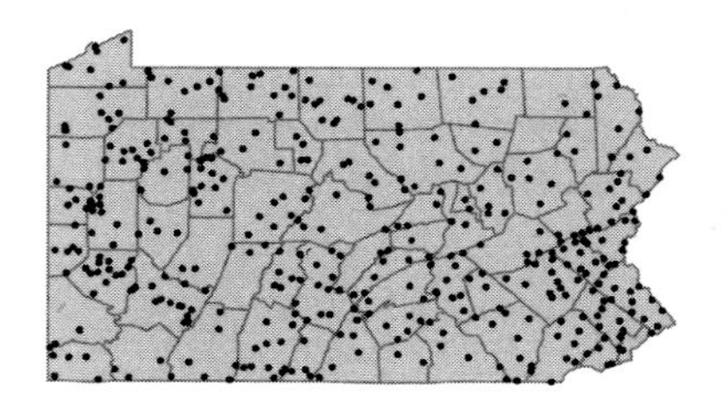

• ***Consolida ambigua*** (L.) Ball & Heywood
Garden larkspur
Herbaceous annual
Roadsides and old fields.
Delphinium ajacis L. PFGBW; *Delphinium ambiguum* L. C

I

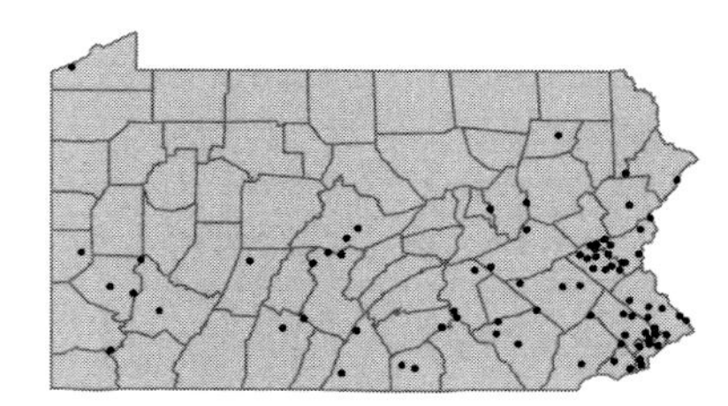

Consolida regalis S.F.Gray
Field larkspur
Herbaceous annual
Waste ground or ballast.
Delphinium consolida L. PGBWC

I

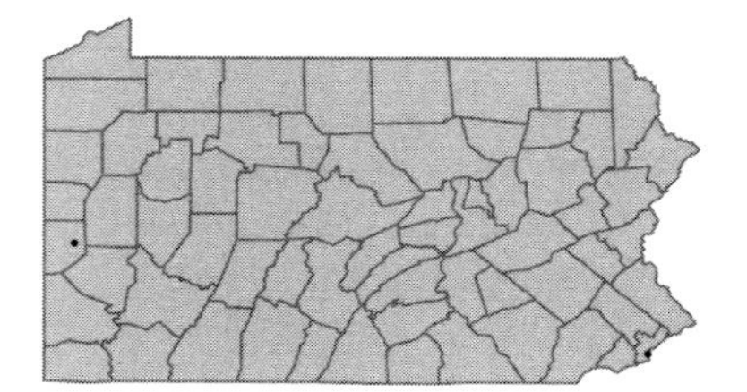

• ***Coptis trifolia*** (L.) Salisb. ssp. ***groenlandica*** (Oeder) Hulten
Goldthread
Herbaceous perennial
Cool, mossy woods and swamps.
Coptis groenlandica (Oeder) Fern. FW

FACW

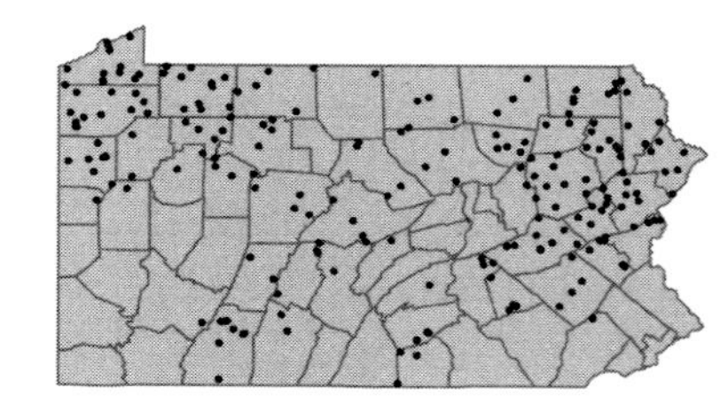

• ***Delphinium exaltatum*** Ait.
Tall larkspur
Herbaceous perennial
Rich woods on rocky slopes.
D. urceolatum Jacq. of Porter (1903) = *D. exaltatum* Ait.

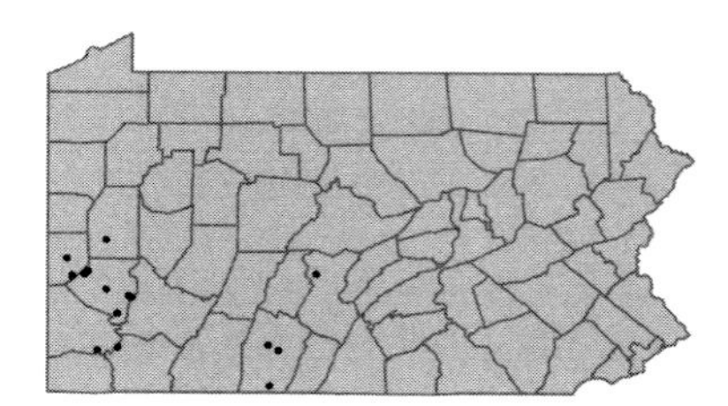

Delphinium tricorne Michx.
Dwarf larkspur
Herbaceous perennial
Rich woods on calcareous slopes.

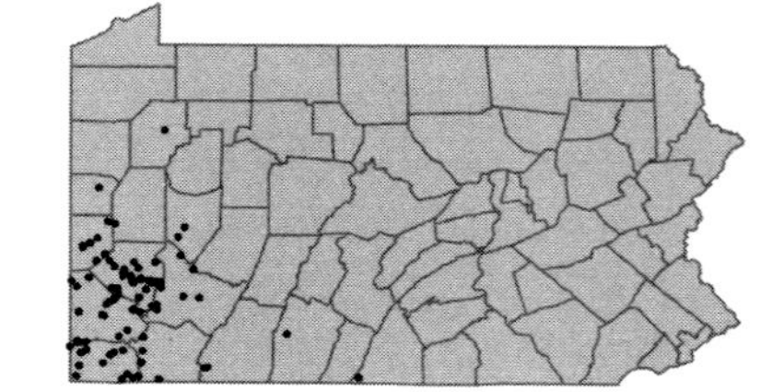

• ***Eranthis hyemalis*** (L.) Salisb.
Winter aconite
Herbaceous perennial
Garden escape, old homesites.
Cammarum hyemale (L.) Greene P

I

• ***Helleborus viridis*** L.
Green hellebore
Herbaceous perennial
Garden escape, old homesites.

I

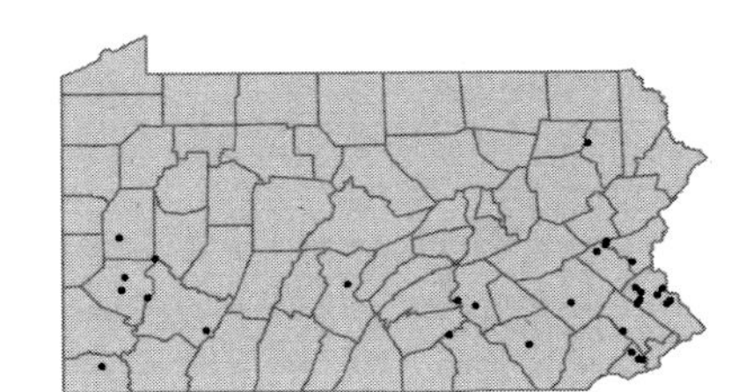

• ***Hepatica nobilis*** P.Mill. var. ***acuta*** (Pursh) Steyerm.
Liverleaf
Herbaceous perennial
Rich, often calcareous, woods.
Hepatica acuta (Pursh) Britt. P; *Hepatica acutiloba* DC. FGBWC

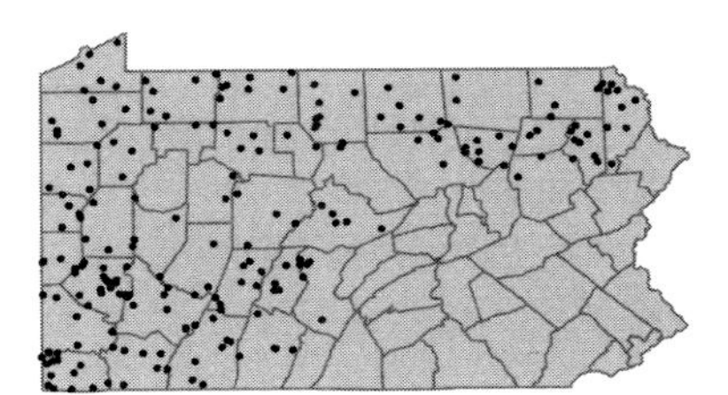

Hepatica nobilis P.Mill. var. ***obtusa*** (Pursh) Steyerm.
Liverleaf
Herbaceous perennial
Rich woods.
Hepatica americana (DC.) Ker-Gawl. FGBWC; *Hepatica hepatica* (L.) Karst. P

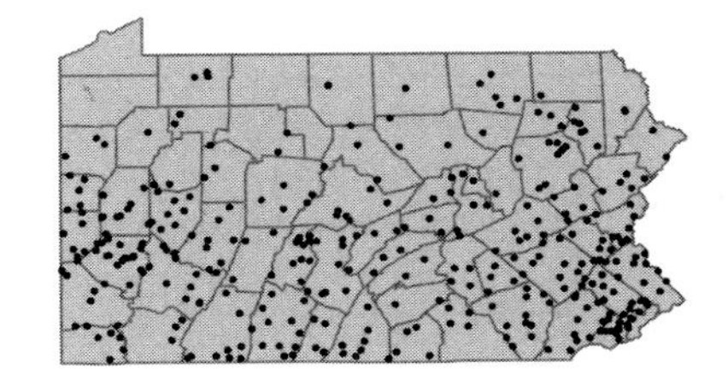

• ***Hydrastis canadensis*** L.
Goldenseal
Herbaceous perennial
Moist, rich woods. Declining due to over-collection.

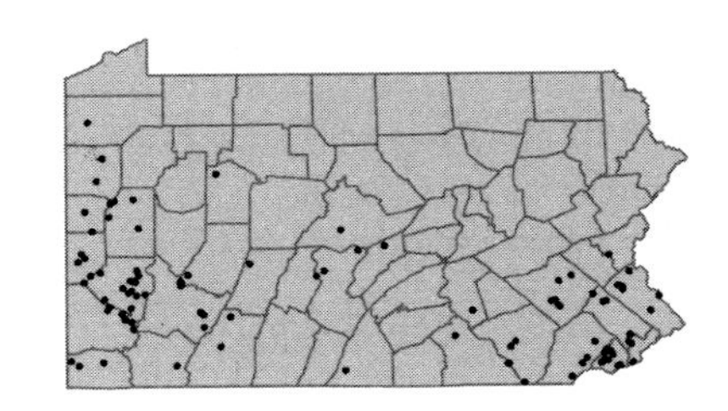

• ***Myosurus minimus*** L.
Mousetail
Herbaceous annual
Moist, sandy soil. Represented by a single collection from Chester Co. in 1822.

FACW+
I

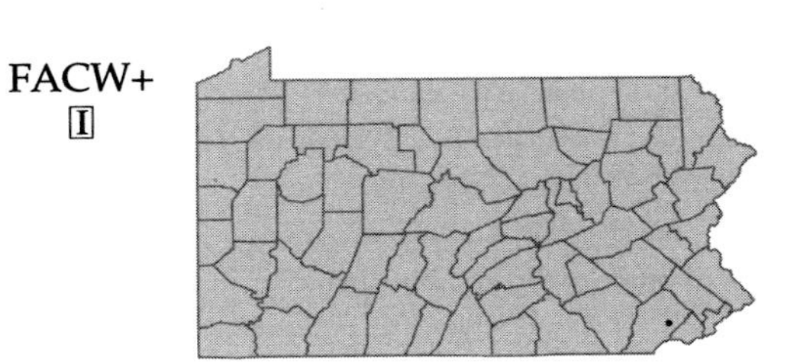

• ***Nigella damascena*** L.
Fennel-flower; Love-in-a-mist
Herbaceous annual
Garden escape, old home sites.

I

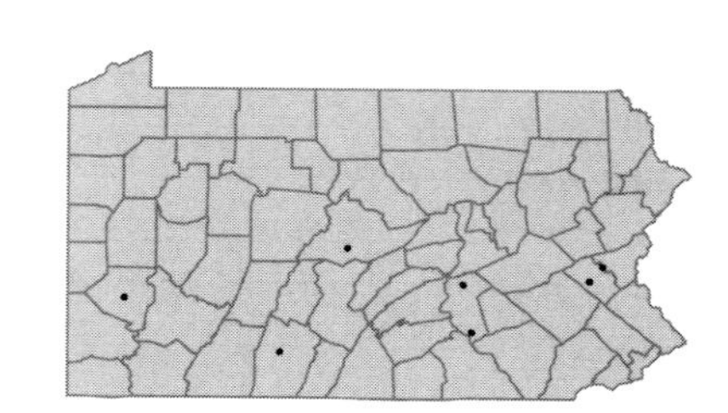

• ***Ranunculus abortivus*** L. var. ***abortivus***
Kidney-leaf buttercup; Small-flowered crowfoot
Herbaceous annual
Low woods, clearings and damp shores.
Ranunculus abortivus L. var. *acrolasius* Fern. FBW

FACW-

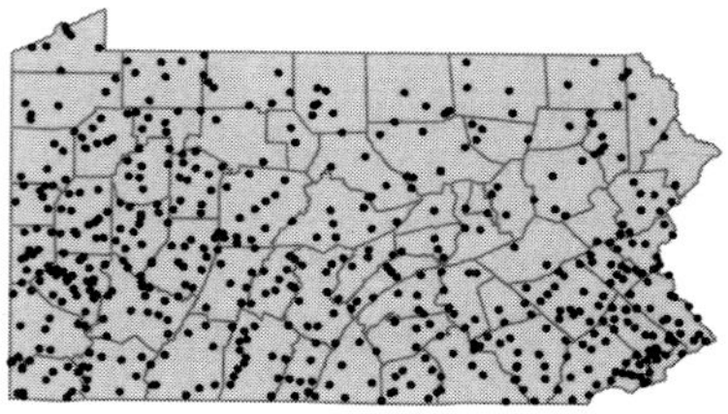

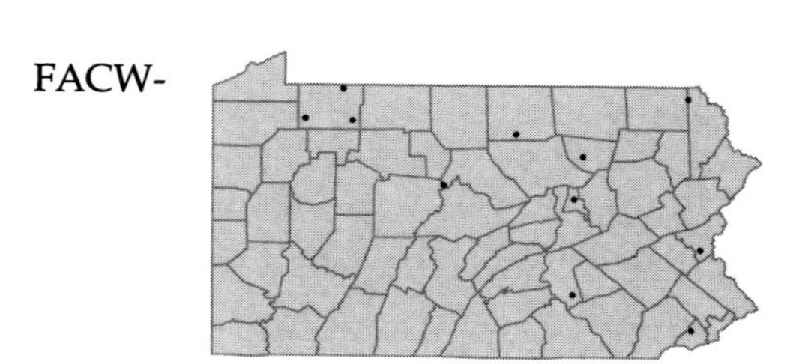

Ranunculus abortivus L. var. ***eucyclus*** Fern.
Kidney-leaf buttercup; Small-flowered buttercup
Herbaceous annual
Low woods, clearings and damp shores.

FACW-

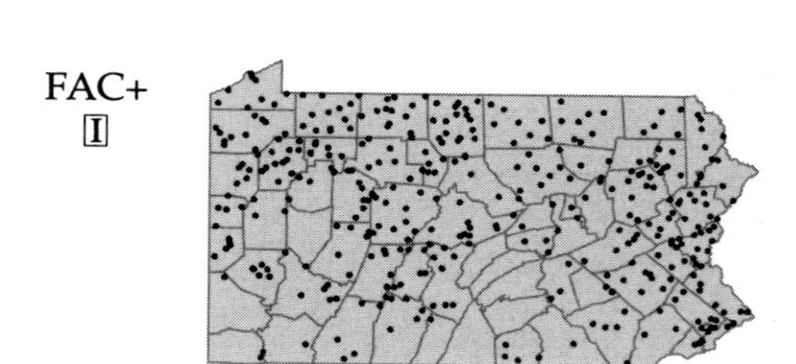

Ranunculus acris L.
Common meadow buttercup
Herbaceous perennial
Pastures, lawns and other open areas.

FAC+
I

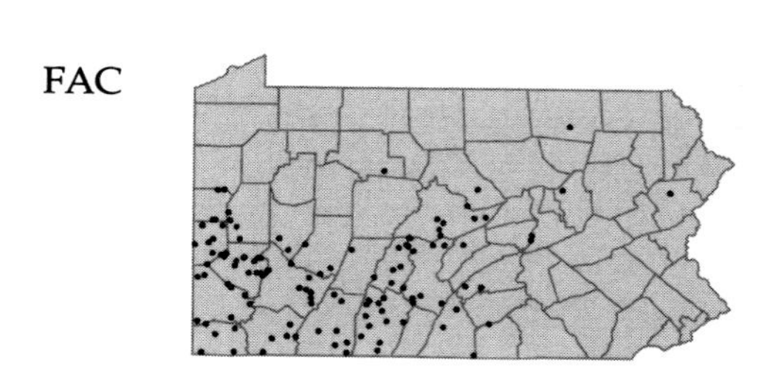

Ranunculus allegheniensis Britt.
Allegheny crowfoot; Mountain crowfoot
Herbaceous annual
Rich woods and rocky, calcareous slopes.

FAC

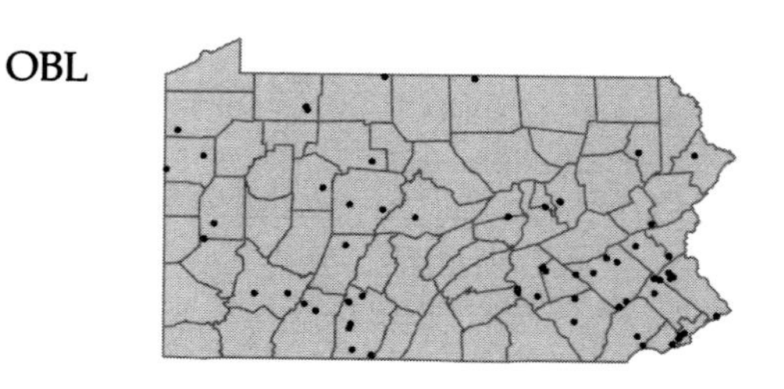

Ranunculus ambigens S.Wats.
Water-plantain spearwort
Herbaceous perennial
Swamps and muddy ditches.
Ranunculus obtusiusculus Raf. P

OBL

Ranunculus arvensis L.
Corn crowfoot
Herbaceous annual
Waste ground and ballast. Represented by two collections from Philadelphia Co. in 1878 and 1880.

I

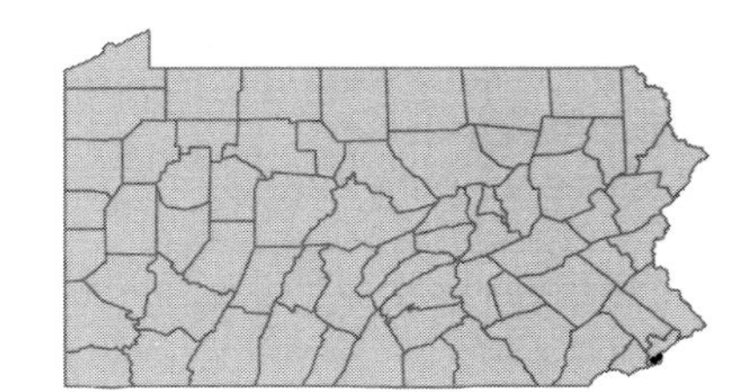

Ranunculus bulbosus L.
Bulbous buttercup
Herbaceous perennial
Open woods, pastures and fields.

I

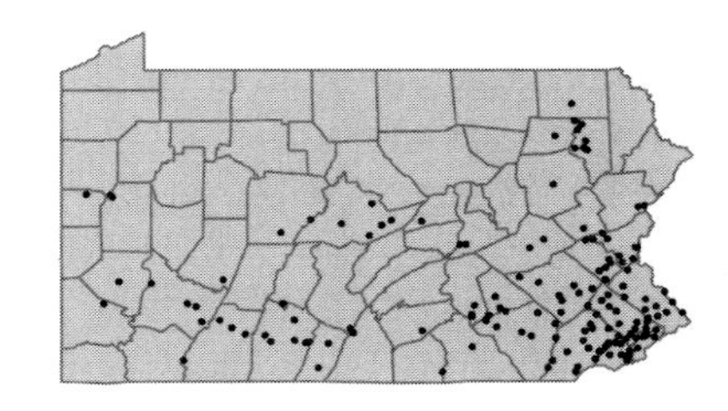

Ranunculus caricetorum Greene
Marsh buttercup; Swamp buttercup
Herbaceous perennial
Woods, meadows and alluvial thickets.
Ranunculus septentrionalis auctt. mult., non Poiret FGS; *Ranunculus septentrionalis* Poir. var. *caricetorum* (Greene) Fern. F; *Ranunculus hispidus* Michx. var. *caricetorum* (Greene) T.Duncan

OBL

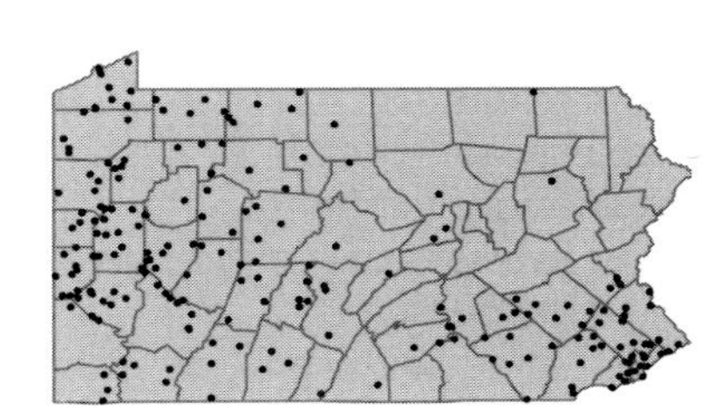

Ranunculus fascicularis Muhl. ex Bigelow
Early buttercup; Tufted buttercup
Herbaceous perennial
Open woods, slopes and edges, often calcareous.

FACU

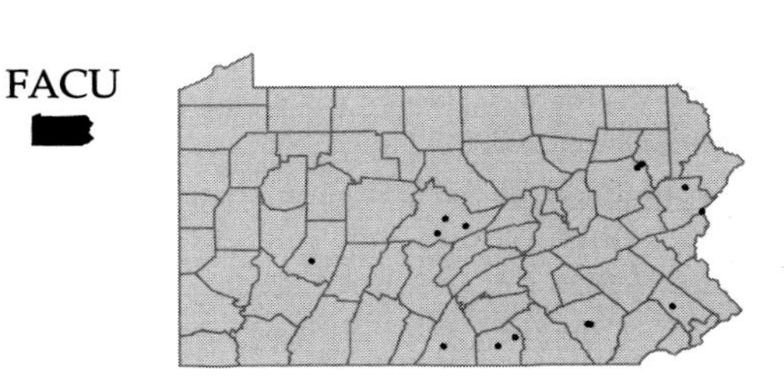

Ranunculus ficaria L.
Lesser celandine; Pilewort
Herbaceous perennial
Naturalized in low woods, floodplains and meadows.
Ficaria ficaria (L.) Karst. P

I

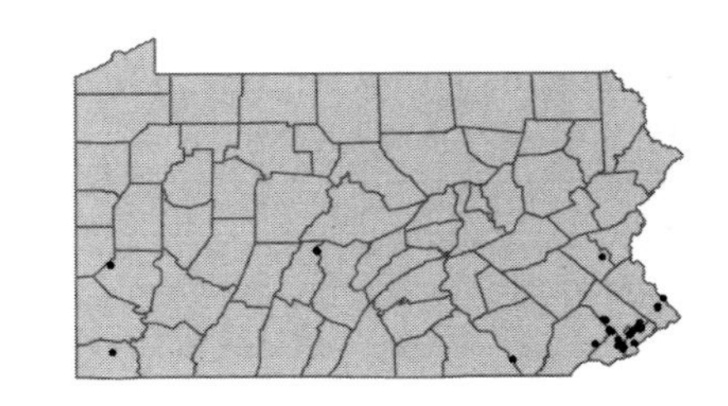

Ranunculus flabellaris Raf.
Yellow water-crowfoot
Herbaceous perennial, rooted submergent aquatic
Quiet water and muddy shores.
Ranunculus delphinifolius Torr. P

OBL

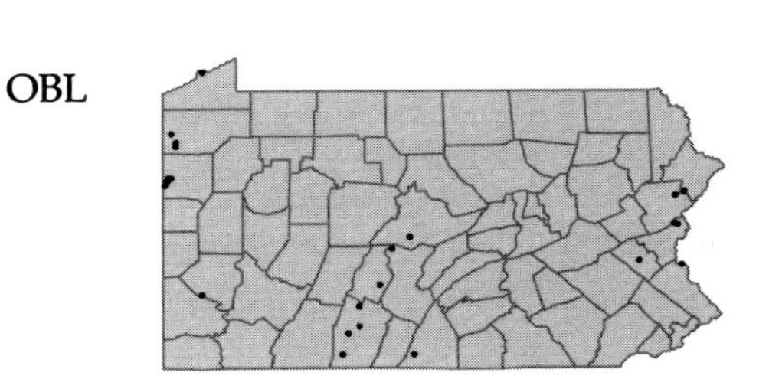

Ranunculus flammula L. var. ***filiformis*** (Michx.) DC.
Creeping spearwort
Herbaceous perennial
Gravelly or sandy shores.
Ranunculus reptans L. var. *reptans* FK

FACW

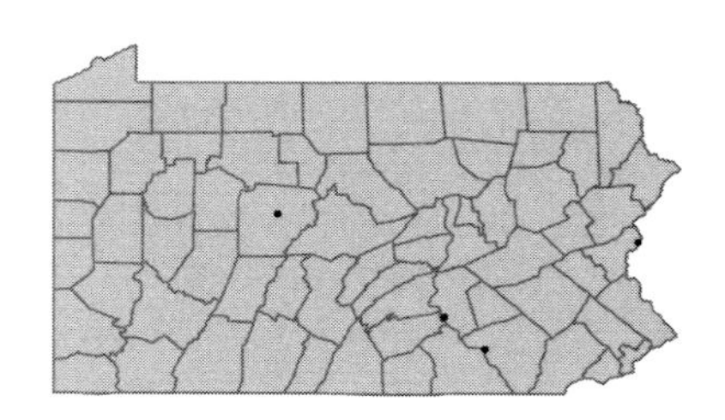

Ranunculus flammula L. var. ***ovalis*** (Bigel.) L.Benson
Creeping spearwort
Herbaceous perennial
Moist pond margins and springy thickets.
Ranunculus reptans L. var. *ovalis* (Bigel.) Torr. & A.Gray FBW

FACW

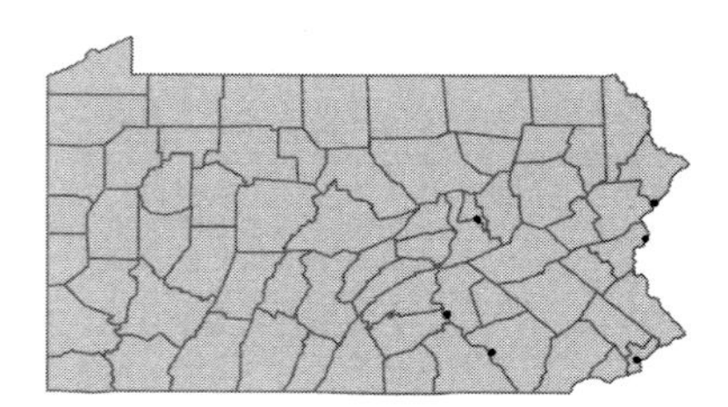

Ranunculus hederaceus L.
Long-stalked crowfoot
Herbaceous perennial
Shallow water or moist depressions. Believed to be extirpated, last collected in 1929.

OBL

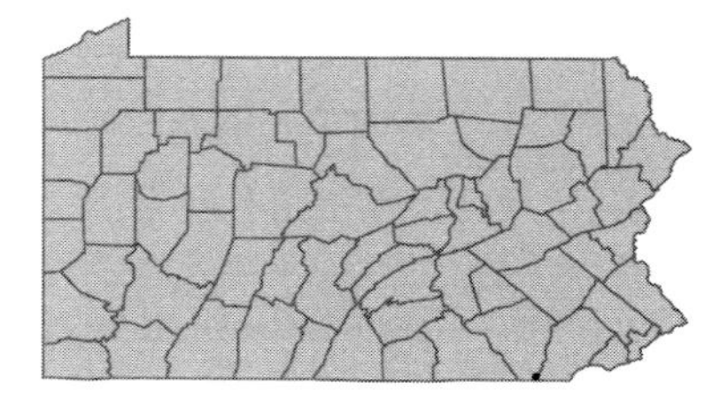

Ranunculus hispidus Michx.
Hairy buttercup
Herbaceous perennial
Rich, moist woods.
Ranunculus carolinianus DC. *pro parte* C

FAC

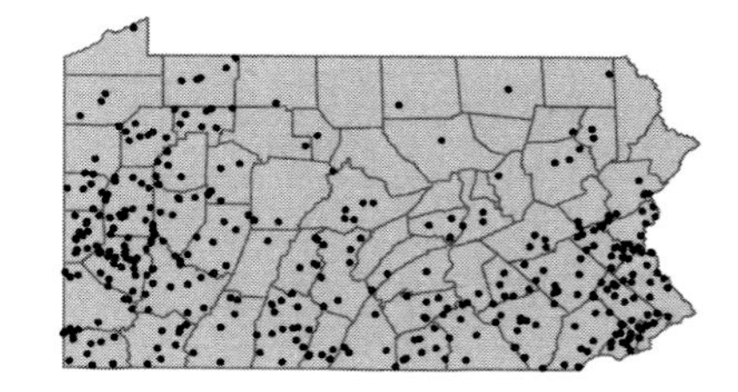

Ranunculus longirostris Godr.
White water-crowfoot
Herbaceous perennial, rooted submergent aquatic
Calcareous water.
Ranunculus circinatus Sibth. B

OBL

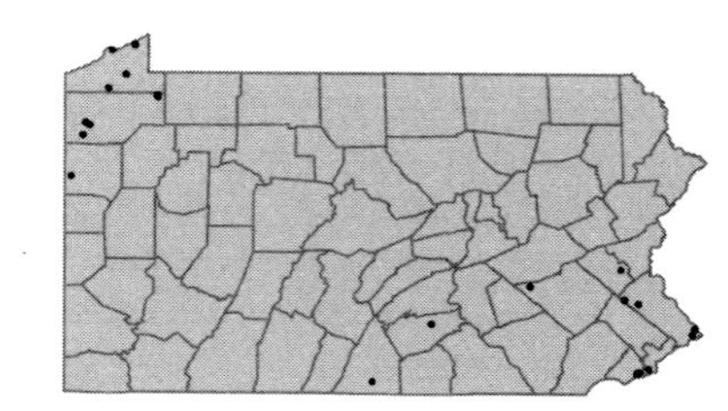

Ranunculus micranthus Nutt. in Torr. & A.Gray
Small-flowered crowfoot
Herbaceous perennial
Moist to dry, shaly slopes and alluvium.

FACU

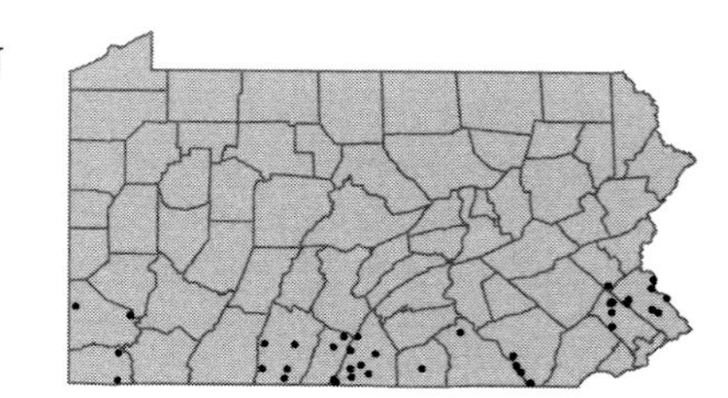

Ranunculus pensylvanicus L.f.
Bristly buttercup
Herbaceous annual
Wet meadows, ditches and marshy bottomland.

OBL

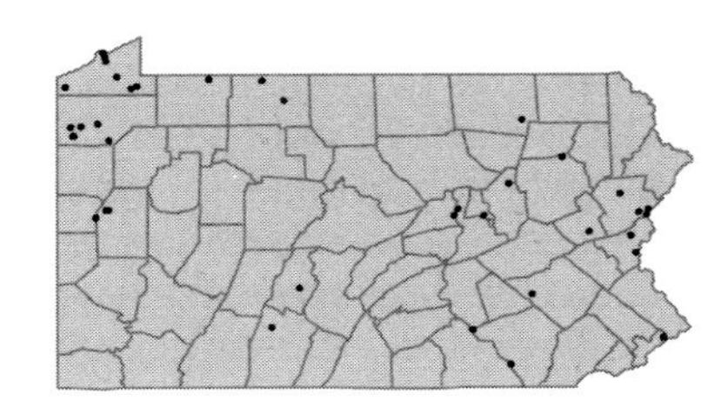

Ranunculus pusillus Poir.
Low spearwort
Herbaceous annual
Swamps, ditches and shallow pools.

OBL

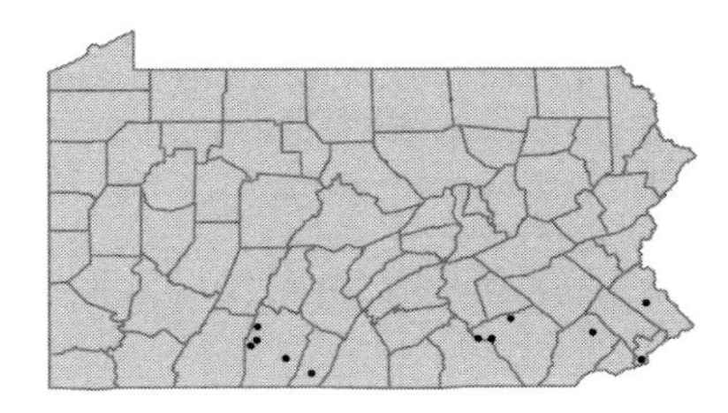

Ranunculus recurvatus Poir.
Hooked crowfoot
Herbaceous perennial
Damp or swampy woods or stream edges.

FAC+

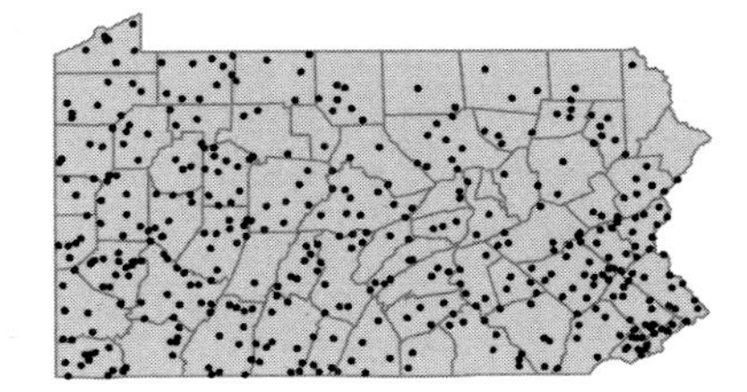

Ranunculus repens L.
Creeping buttercup
Herbaceous perennial
Wet, open ground, ditches and swales.

FAC
[I]

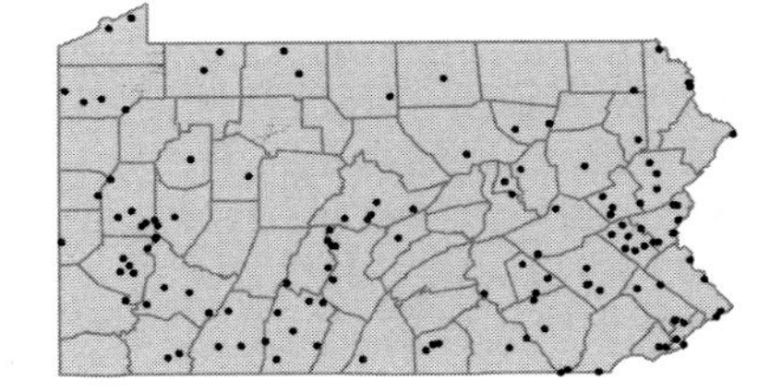

Ranunculus sardous Crantz
Hairy buttercup; Sardinian buttercup
Herbaceous perennial
Roadsides and waste places.

[I]

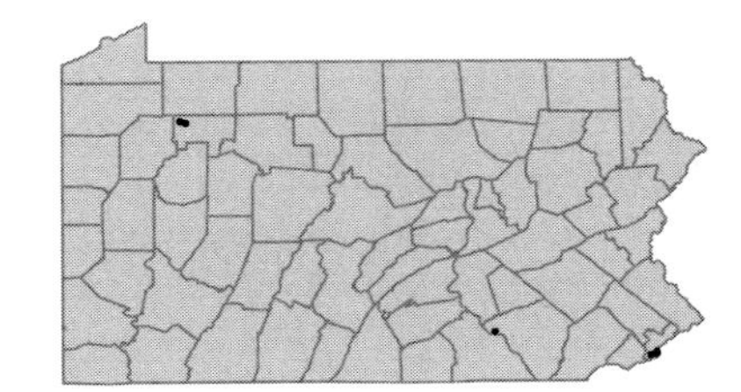

Ranunculus sceleratus L.
Celery-leaved crowfoot; Cursed crowfoot
Herbaceous annual
Ditches, pools and springy areas.

OBL
[I]

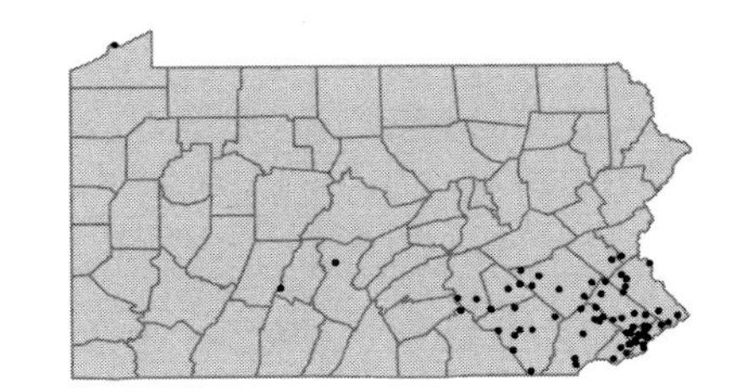

Ranunculus trichophyllus Chaix
White water-crowfoot
Herbaceous perennial, rooted submergent aquatic
Lakes, ponds and other slow-moving water.
Batrachium trichophyllum (Chaix) Bossch. P; *Ranunculus aquatilis* L. GB

OBL

• ***Thalictrum coriaceum*** (Britt.) Small
Thick-leaved meadow-rue
Herbaceous perennial
Upland woods.
Thalictrum steeleanum Boivin *pro parte* FGBW

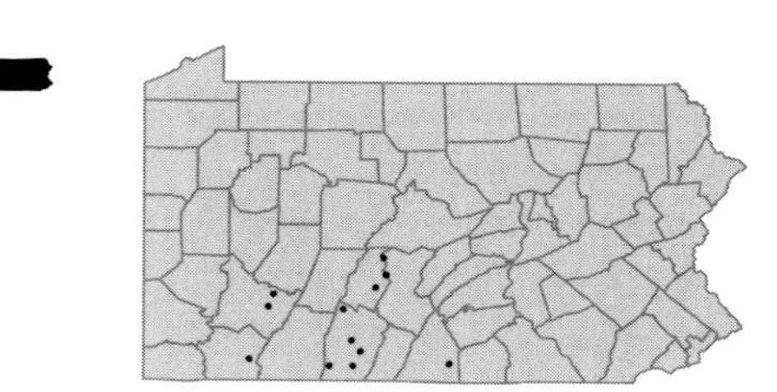

Thalictrum dasycarpum Fisch. & Ave-Lall.
Purple meadow-rue
Herbaceous perennial
Alluvial thickets.

FACW

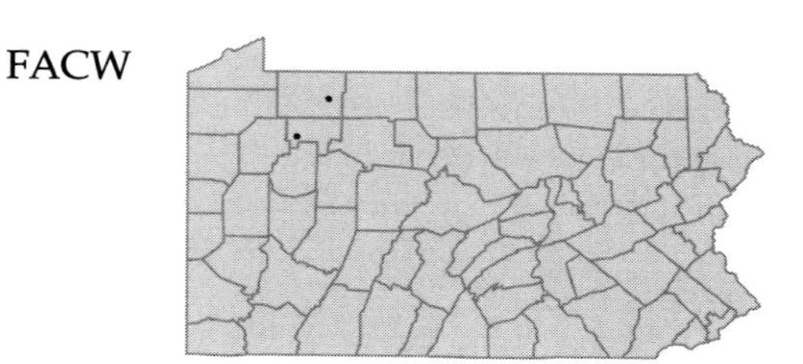

Thalictrum dioicum L.
Early meadow-rue
Herbaceous perennial
Rich, rocky woods and ravines.

FAC

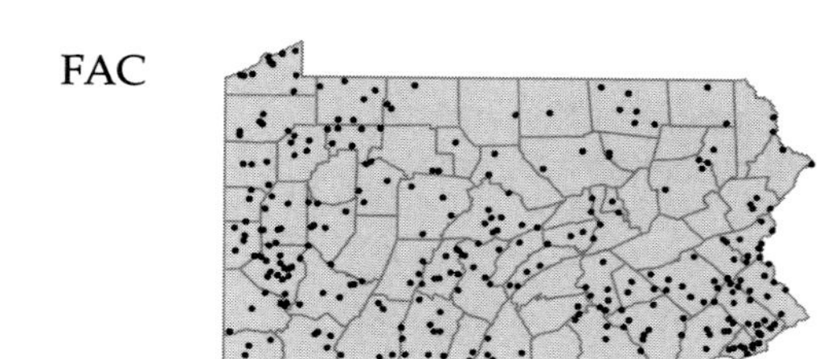

Thalictrum pubescens Pursh
Tall meadow-rue
Herbaceous perennial
Wet meadows, low open woods and swamps.
Thalictrum polygamum Muhl. PFG

FACW+

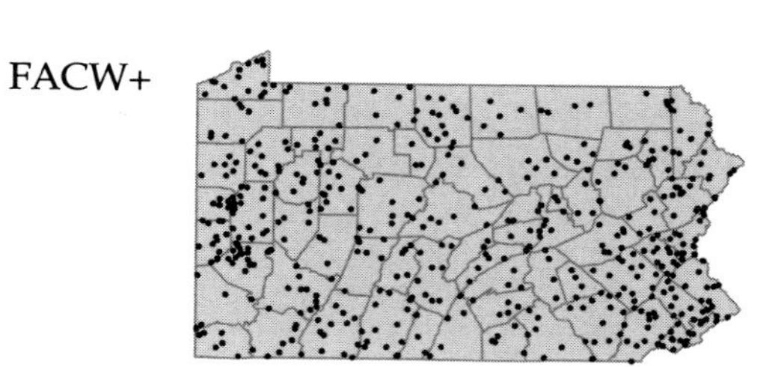

Thalictrum revolutum DC.
Purple meadow-rue; Skunk meadow-rue
Herbaceous perennial
Dry, open woods and barrens.

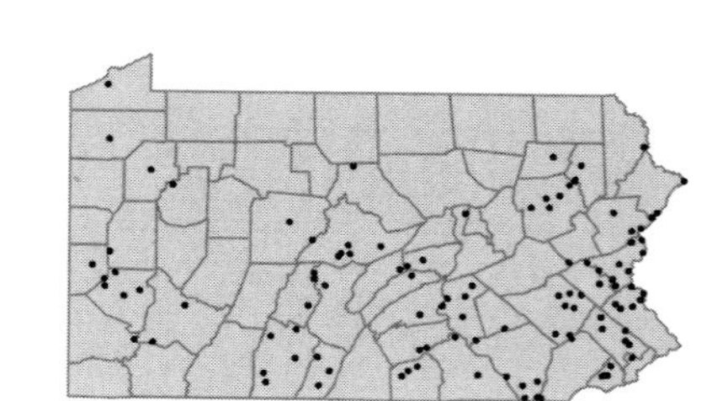

- ***Thalictrum thalictroides*** (L.) Eames & B.Boivin
  Rue-anemone
  Herbaceous perennial
  Rich woods.
  Anemonella thalictroides (L.) Spach FGBWC; *Syndesmon thalictroides* (L.) Hoffmg. P

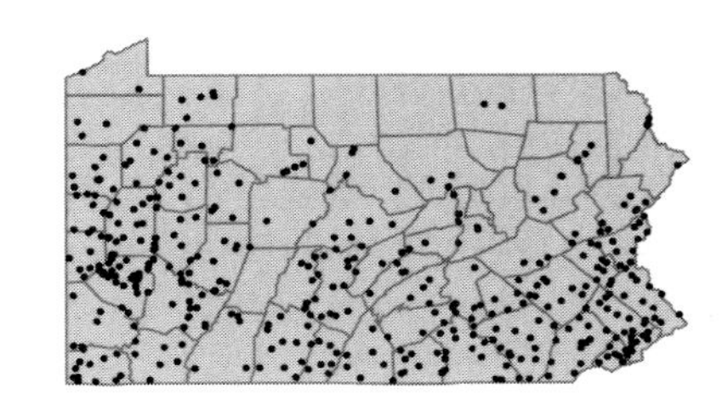

- ***Trautvetteria caroliniensis*** (Walt.) Vail
  Carolina tassel-rue
  Herbaceous perennial
  Wooded slopes and stream banks.

FACW-

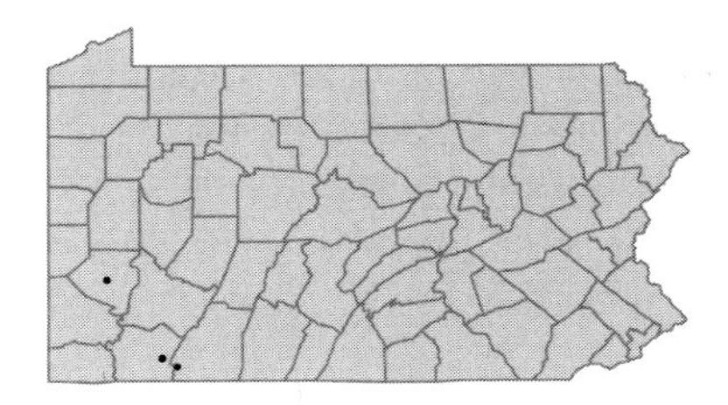

- ***Trollius laxus*** Salisb.
  Spreading globe-flower
  Herbaceous perennial
  Moist calcareous meadows, open woods and swamps.

OBL

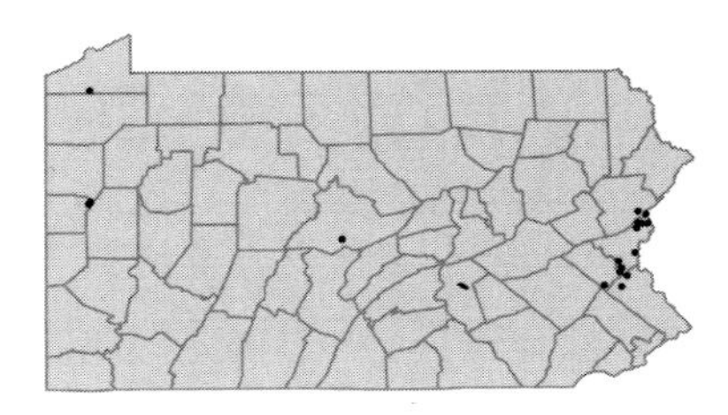

- ***Xanthorhiza simplicissima*** Marshall
  Shrub yellowroot
  Deciduous shrub
  Naturalized in disturbed woods near site of abandoned nursery.

FACW
I

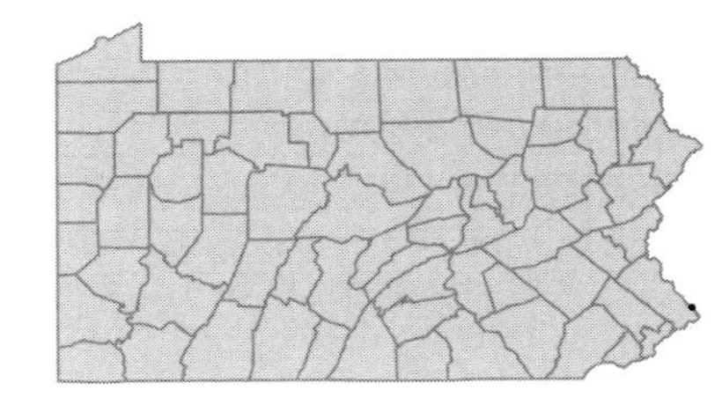

BERBERIDACEAE

- ***Berberis canadensis*** P.Mill.
  Allegheny barberry
  Deciduous shrub
  Woods and waste places. Believed to be extirpated, represented by a single collection from Huntingdon Co. in 1947.

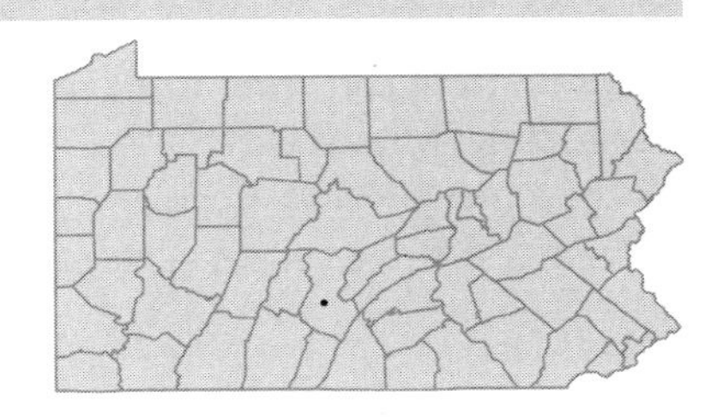

Berberis thunbergii DC.
Japanese barberry
Deciduous shrub
Disturbed woods, roadsides and hedgerows.

I

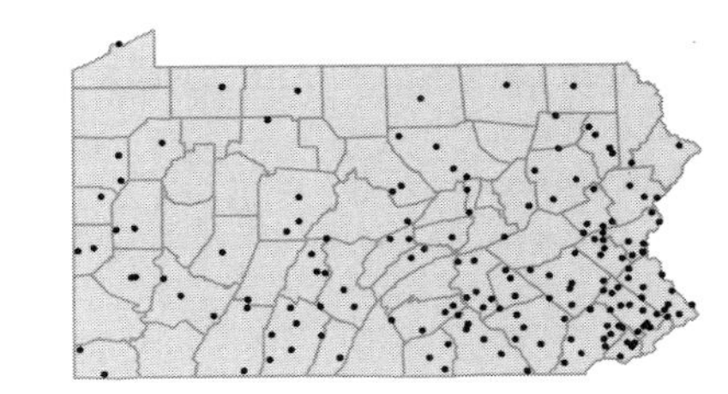

Berberis vulgaris L. [I]
European barberry
Deciduous shrub
Fields, pastures and disturbed woods, escaped from cultivation.

• ***Caulophyllum thalictroides*** (L.) Michx.
Blue cohosh; Squaw-root
Herbaceous perennial
Moist, rich woods.

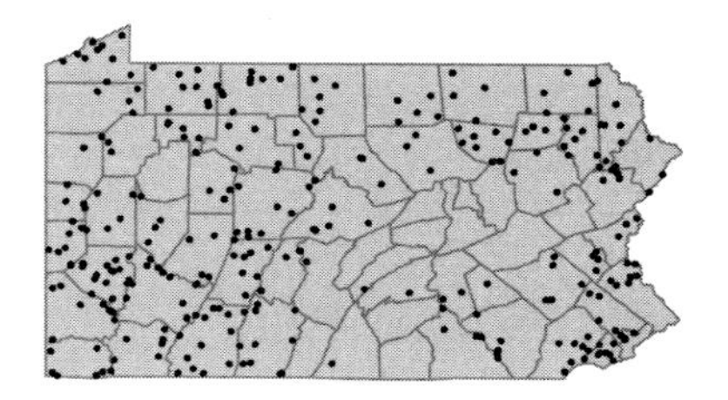

• ***Jeffersonia diphylla*** (L.) Pers.
Twinleaf
Herbaceous perennial
Moist, usually calcareous, woods.

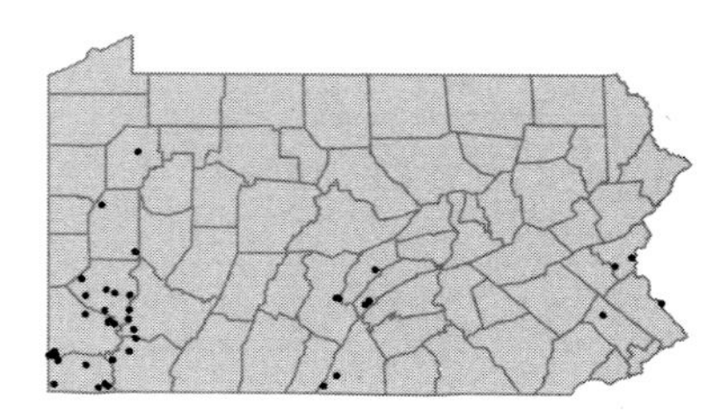

• ***Mahonia repens*** (Lindl.) G.Don [I]
Creeping hollygrape
Evergreen shrub
Naturalized in woods near abandoned home site. Represented by a single collection from Northampton Co. in 1946.

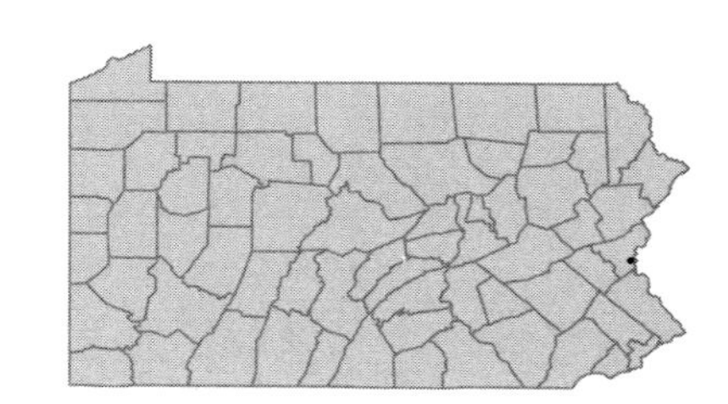

• ***Podophyllum peltatum*** L.
May-apple; Mandrake
Herbaceous perennial
Moist woods.

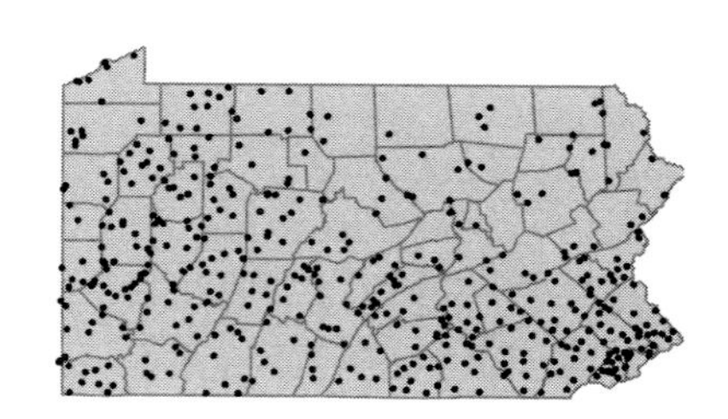

LARDIZABALACEAE

• ***Akebia quinata*** (Houtt.) Decne. [I]
Five-leaf
Woody vine
Disturbed woods, naturalized from cultivated sources.

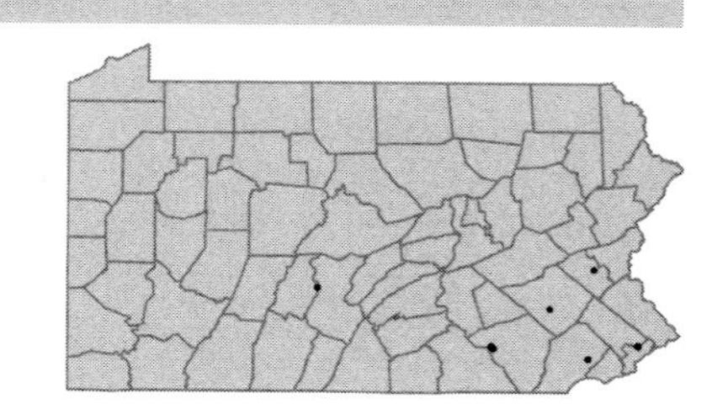

MENISPERMACEAE

• ***Menispermum canadense*** L.
Moonseed; Yellow parilla
Herbaceous perennial vine
Moist ground of stream banks and edges.

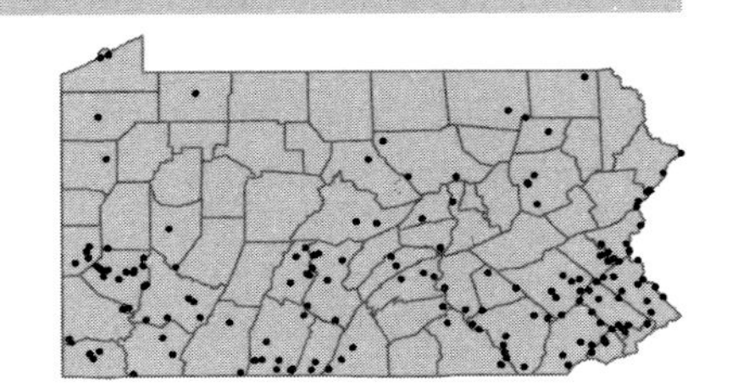

PAPAVERACEAE

• ***Argemone mexicana*** L. [I]
Mexican poppy; Prickly poppy
Herbaceous annual
Ballast and waste ground.

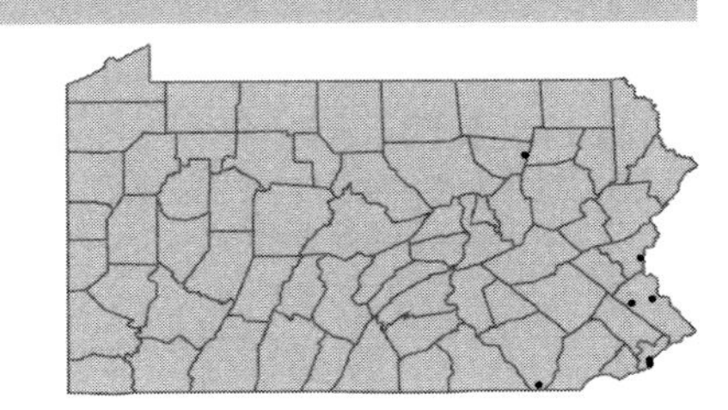

• ***Chelidonium majus*** L. [I]
Greater celandine; Swallowwort
Herbaceous biennial
Moist soil of floodplains, roadsides and gardens.

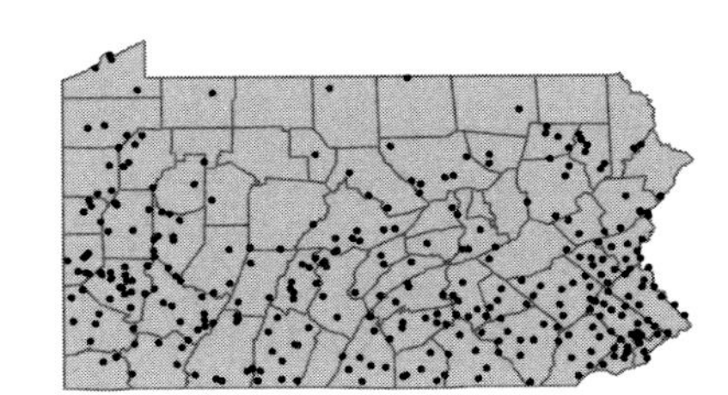

• ***Eschscholzia californica*** Cham. [I]
California poppy
Herbaceous annual
Occasionally escaped from cultivation.

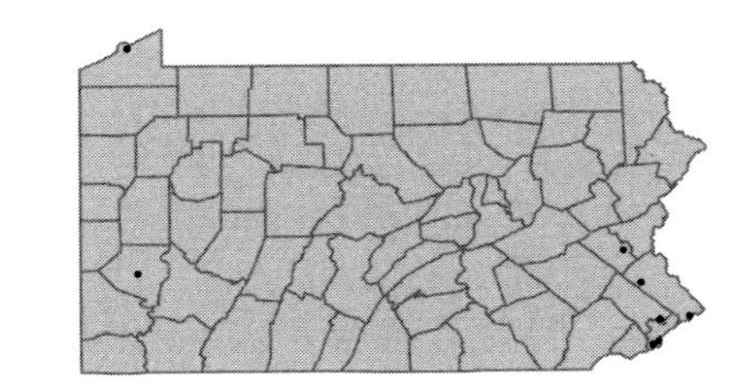

• ***Glaucium corniculatum*** (L.) J.H.Rudolph [I]
Horned poppy
Herbaceous annual
Ore piles. Represented by a single collection from Northampton Co. in 1879.

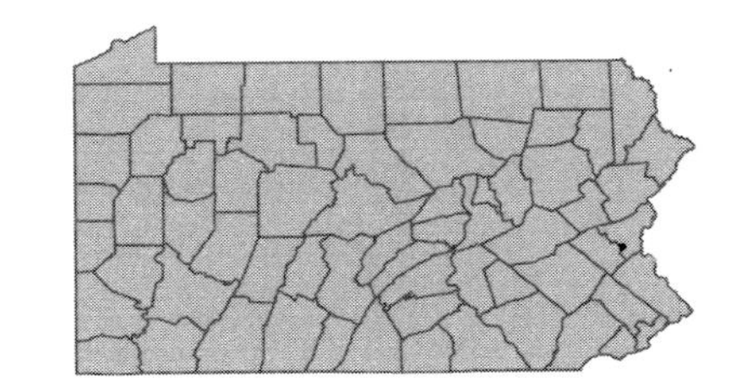

Glaucium flavum Crantz [I]
Yellow horned-poppy
Herbaceous biennial
Ballast. Represented by a single collection from Philadelphia Co. in 1890.
Glaucium glaucium (L.) Karst. P

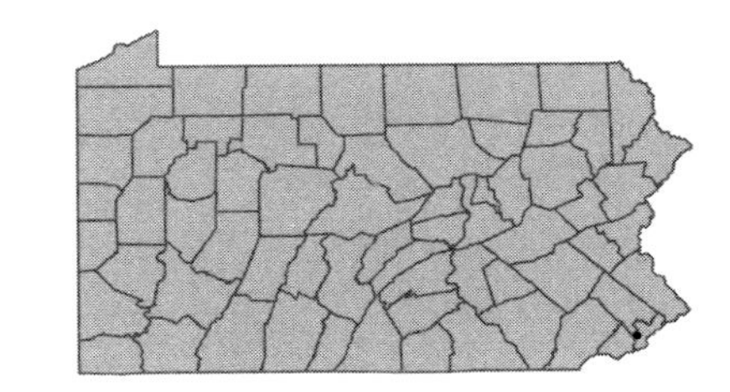

• ***Macleaya cordata*** (Willd.) R.Br.
Plume poppy; Tree celandine
Herbaceous perennial
Railroad banks, waste ground and cultivated areas.

I

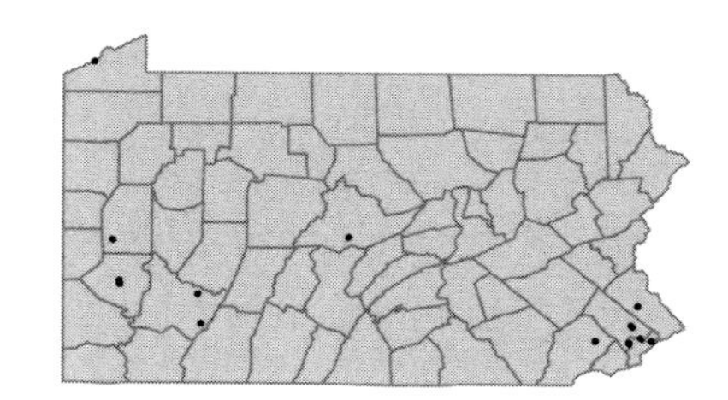

• ***Papaver argemone*** L.
Long rough-fruited poppy
Herbaceous annual
Waste ground and ballast.

I

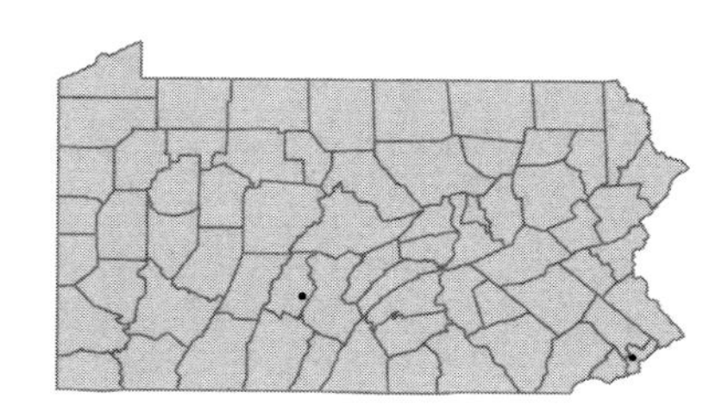

Papaver dubium L.
Long-pod poppy; Smooth-fruited poppy
Herbaceous annual
Occasionally escaped from cultivation.

I

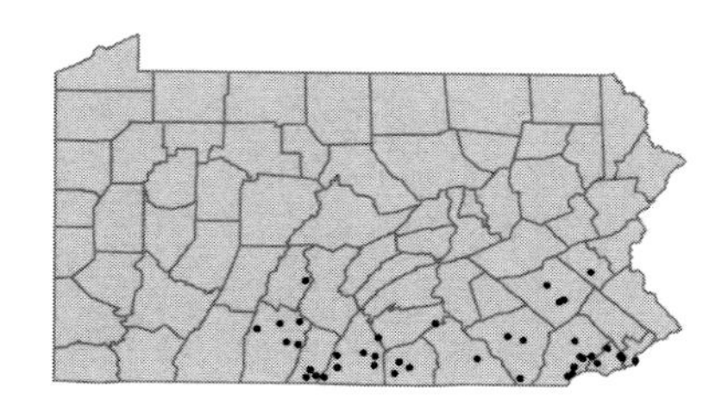

Papaver hybridum L.
Hybrid poppy
Herbaceous annual
Waste ground. Represented by two collections from Philadelphia Co. in 1877.

I

Papaver orientale L.
Oriental poppy
Herbaceous perennial
Frequently cultivated and occasionally persisting in abandoned gardens or rubbish dumps.

I

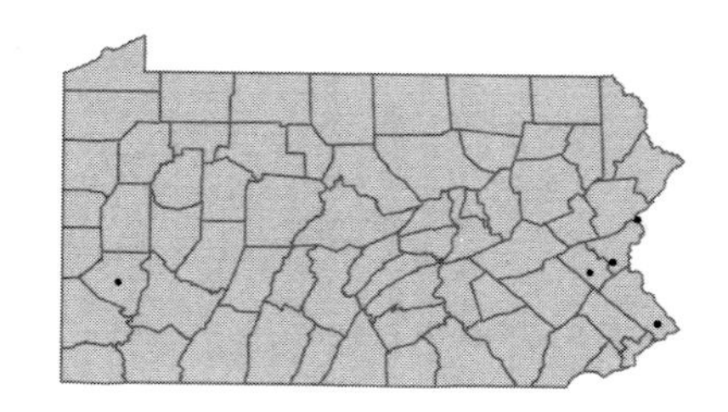

Papaver rhoeas L.
Corn poppy; Field poppy; Red poppy
Herbaceous perennial
Occasionally escaped from cultivation.

I

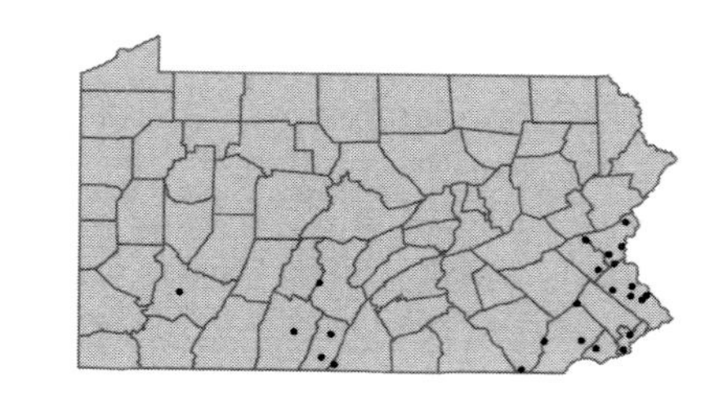

Papaver somniferum L.
Opium poppy
Herbaceous perennial
Cultivated and occasionally escaped.

I

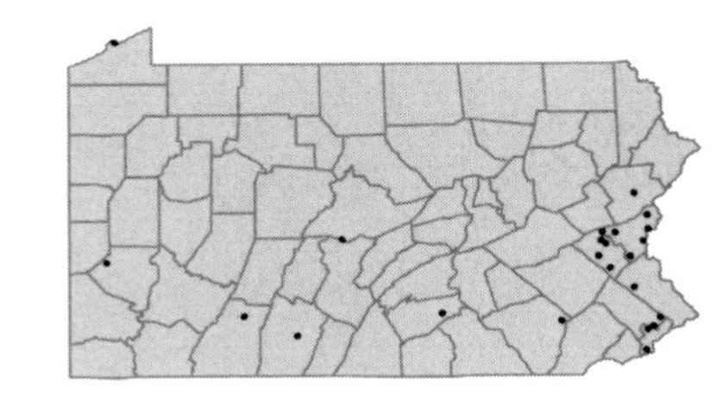

• ***Sanguinaria canadensis*** L.
Bloodroot; Red puccoon
Herbaceous perennial
Rich woods and roadside banks.

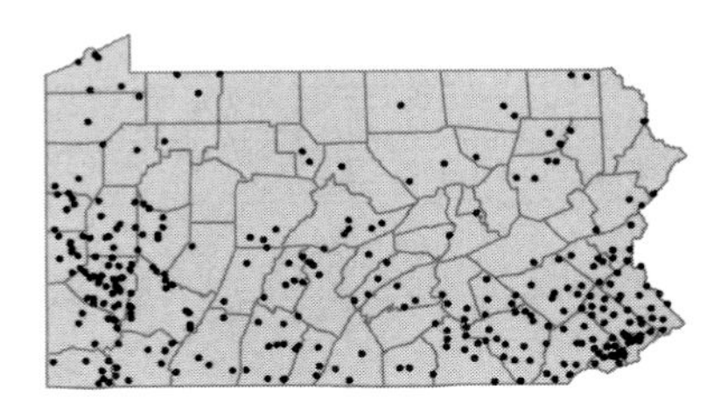

• ***Stylophorum diphyllum*** (Michx.) Nutt.
Celandine poppy
Herbaceous perennial
All confirmed records apparently naturalized from cultivated sources.

I

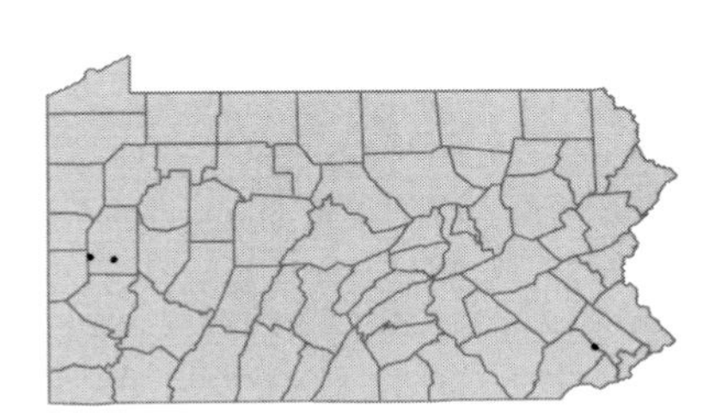

FUMARIACEAE

• ***Adlumia fungosa*** (Ait.) Greene ex BSP
Alleghany-vine; Climbing fumitory
Herbaceous biennial vine
Moist, rocky slopes and woodlands.

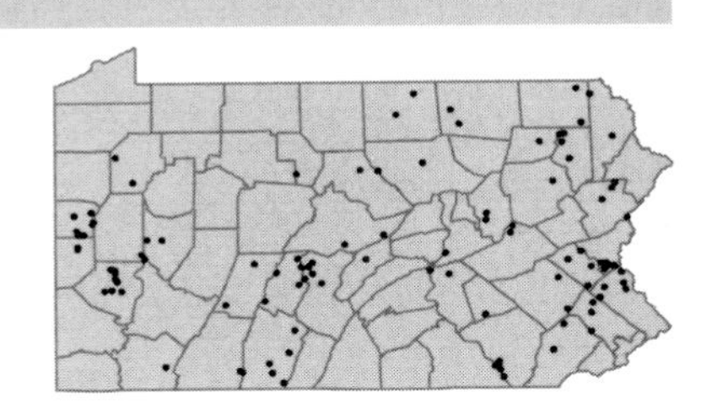

• ***Corydalis aurea*** Willd.
Golden corydalis
Herbaceous biennial
Roadside.

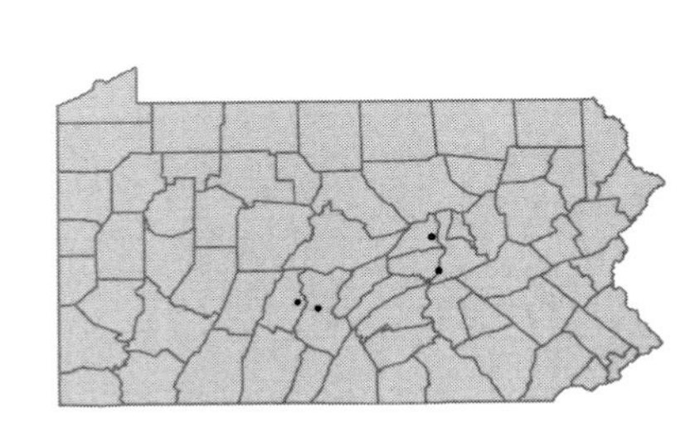

Corydalis flavula (Raf.) DC.
Yellow fumewort; Yellow harlequin
Herbaceous biennial
Moist open woods, slopes and edges.
Capnoides flavulum (Raf.) Kuntze P

FACU

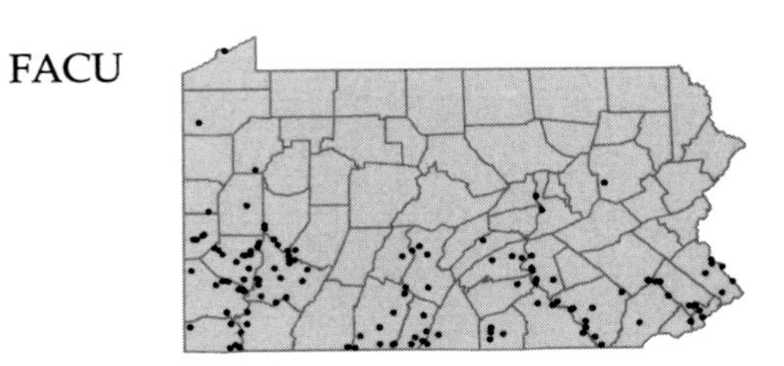

Corydalis sempervirens (L.) Pers.
Rock harlequin
Herbaceous biennial
Dry, rocky woods and rock outcrops.
Capnoides sempervirens (L.) Borck. P

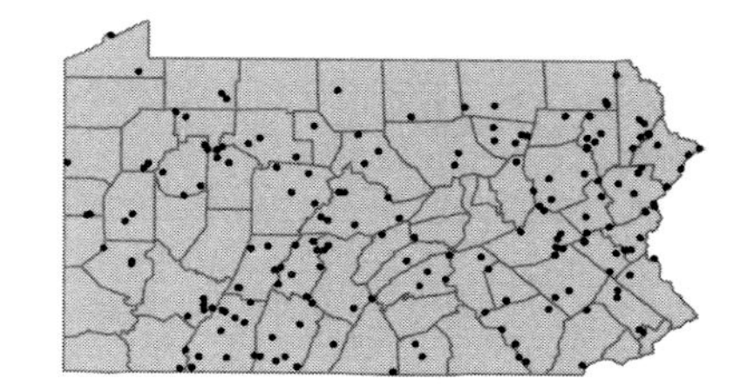

•***Dicentra canadensis*** (Goldie) Walp.
Squirrel corn
Herbaceous perennial
Rich, moist woods.
Bicuculla canadensis (Goldie) Millsp. P

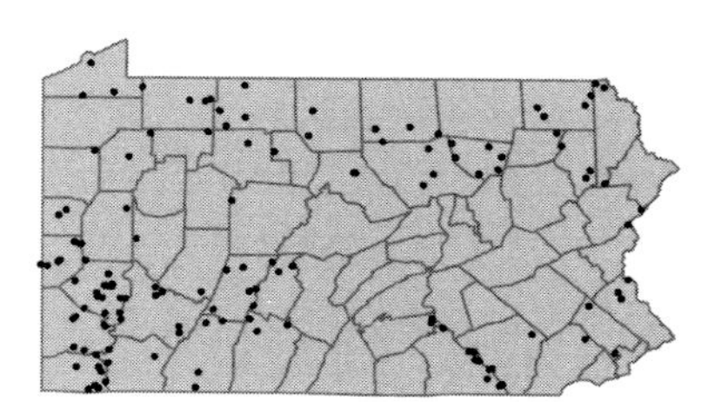

Dicentra cucullaria (L.) Bernh.
Dutchman's-breeches
Herbaceous perennial
Rich woods.
Bicuculla cucullaria (L.) Millsp. P

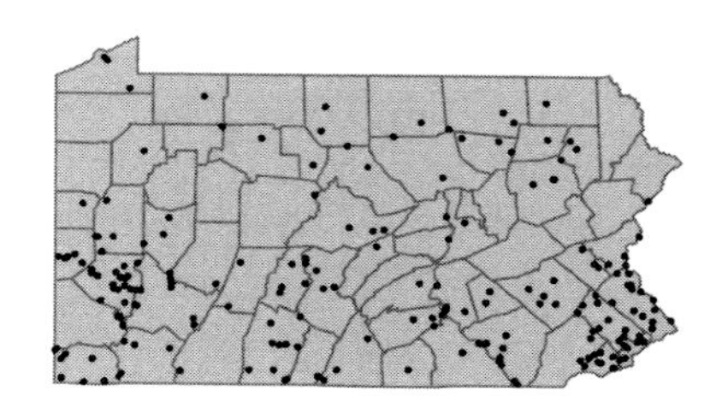

Dicentra eximia (Ker-Gawl.) Torr.
Wild bleeding-heart
Herbaceous perennial
Rocky woods and cliffs. Also cultivated.
Bicuculla eximia (Ker-Gawl.) Millsp. P

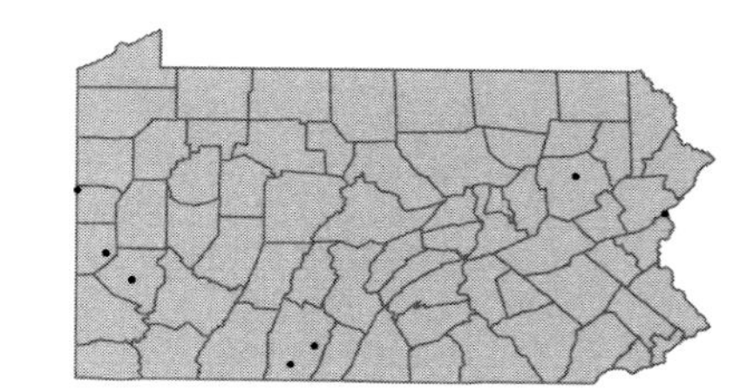

Dicentra spectabilis (L.) Lem. [I]
Tall bleeding-heart
Herbaceous perennial
Cultivated and occasionally persisting in abandoned garden sites or rubbish dumps.

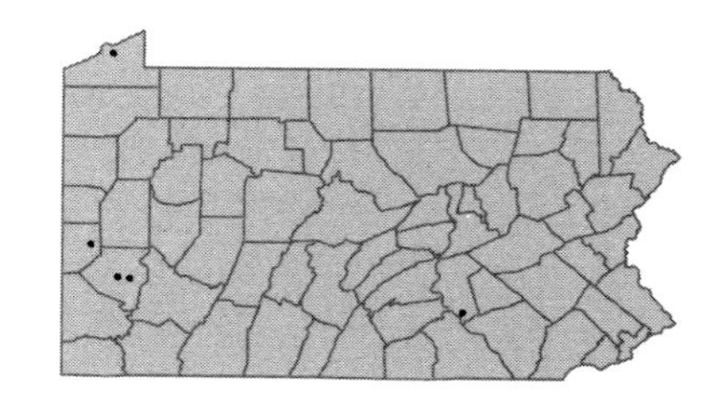

•***Fumaria officinalis*** L. [I]
Common fumitory; Earth-smoke
Herbaceous annual
Cultivated sites and waste ground.

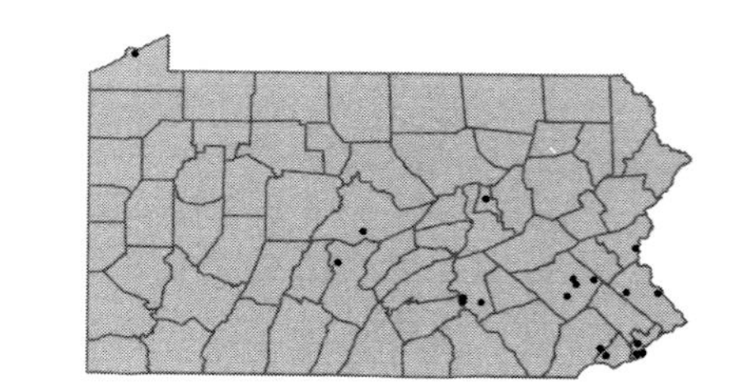

Fumaria parviflora Lam. I
Fumitory
Herbaceous annual
Ballast. Represented by two collections from Philadelphia Co. in 1877.

CERCIDIPHYLLACEAE

• ***Cercidiphyllum japonicum*** Sieb. & Zucc. ex J.Hoffm. & H.Schult. I
Katsura-tree
Deciduous tree
Cultivated and occasionally naturalized in disturbed woods.

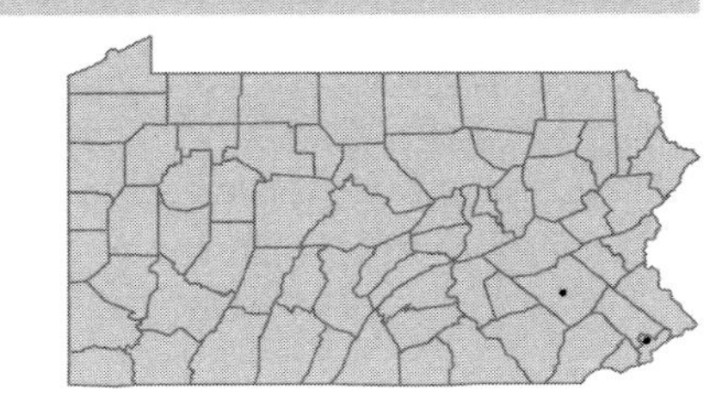

PLATANACEAE

• ***Platanus occidentalis*** L. FACW-
Sycamore; Buttonwood
Deciduous tree
Stream banks, low woods, floodplains and alluvial soils. London planetree, *P. occidentalis x orientalis* (*P. x acerifolia* Willd.), is frequently planted in urban areas.

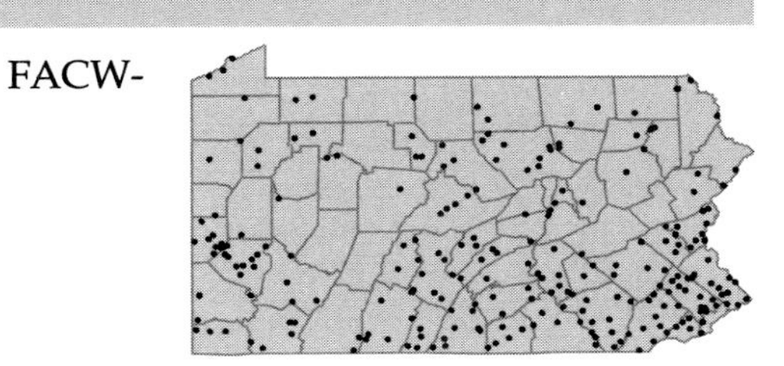

HAMAMELIDACEAE

• ***Hamamelis virginiana*** L. FAC-
Witch-hazel
Deciduous shrub
Rich, rocky woods.

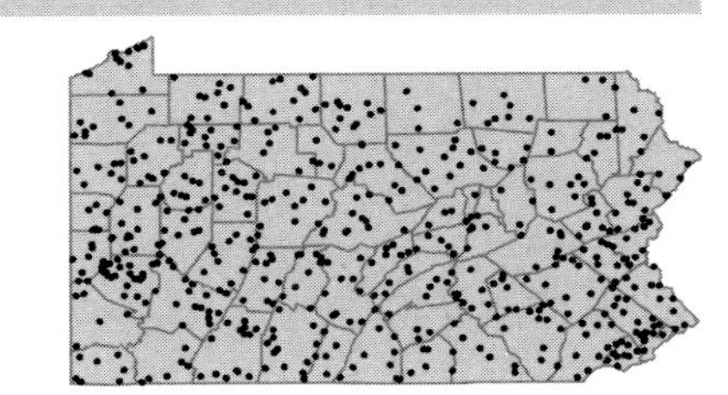

• ***Liquidambar styraciflua*** L. FAC
Sweet-gum
Deciduous tree
Low, wet, coastal plain woods. Also cutivated.

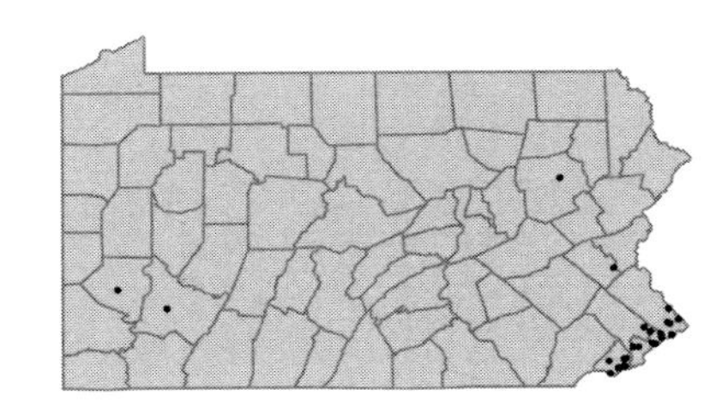

ULMACEAE

• ***Celtis occidentalis*** L. var. ***occidentalis***
Hackberry; Sugarberry
Deciduous tree
Dry to moist woods.
Celtis crassifolia Lam. P

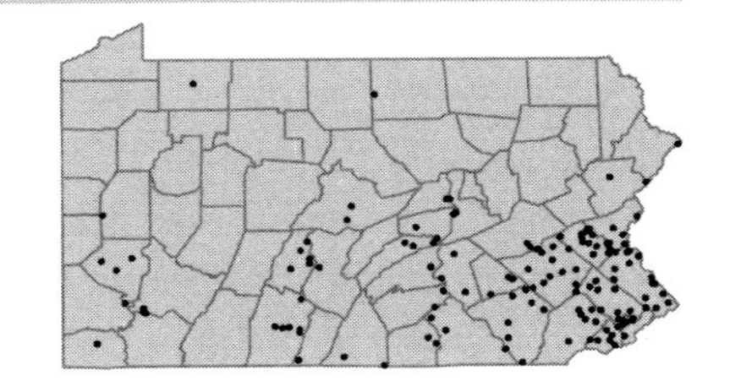

Celtis occidentalis L. var. ***canina*** (Raf.) Sarg.
Dogberry
Deciduous tree
Rocky slopes, rich banks and bottomlands.
Celtis canina Raf. W

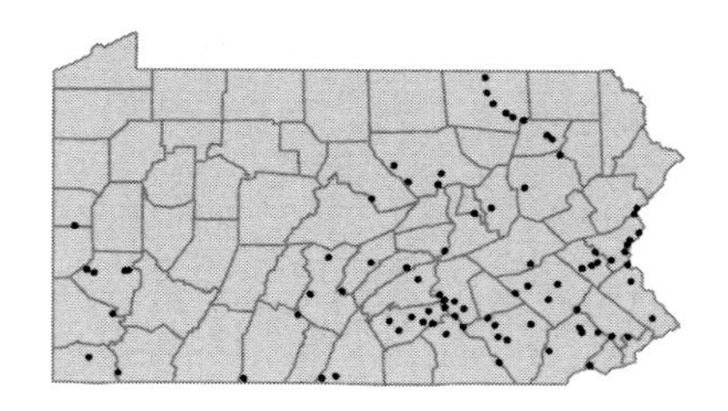

Celtis tenuifolia Nutt. var. ***tenuifolia***
Dwarf hackberry; Georgia hackberry
Deciduous tree
Dry, shaly slopes.

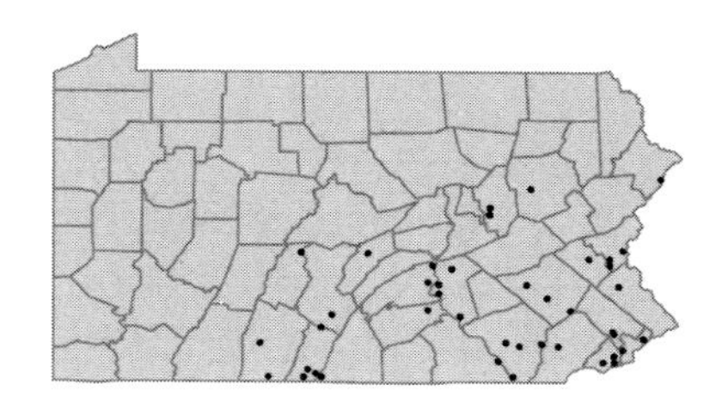

Celtis tenuifolia Nutt. var. ***georgiana*** (Small) Fern. & Schub.
Dwarf hackberry; Georgia hackberry
Deciduous tree
Shale banks, wooded hillsides and limestone cliffs.
Celtis georgiana Small W

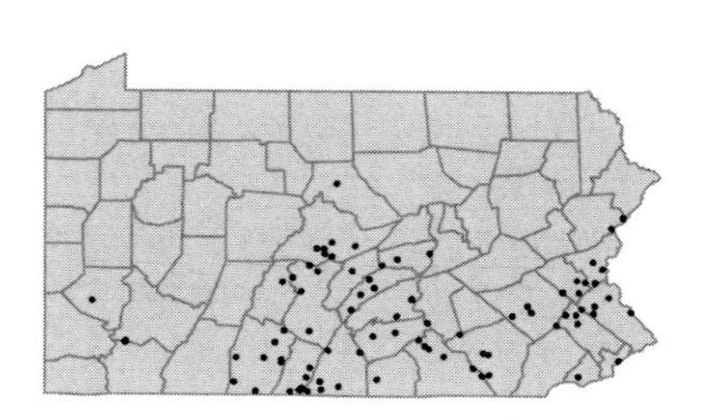

• ***Ulmus americana*** L.
American elm; White elm
Deciduous tree
Stream banks and floodplains, in rich, alluvial soil. Dutch elm disease has reduced numbers of this species severely.

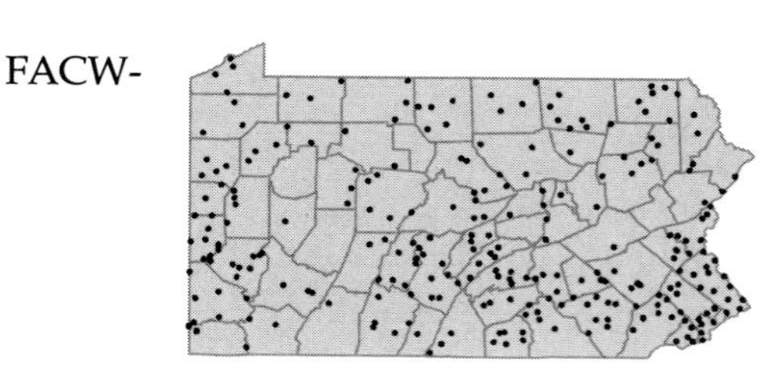

Ulmus pumila L.
Siberian elm
Deciduous tree
Naturalized from planted sources.

I

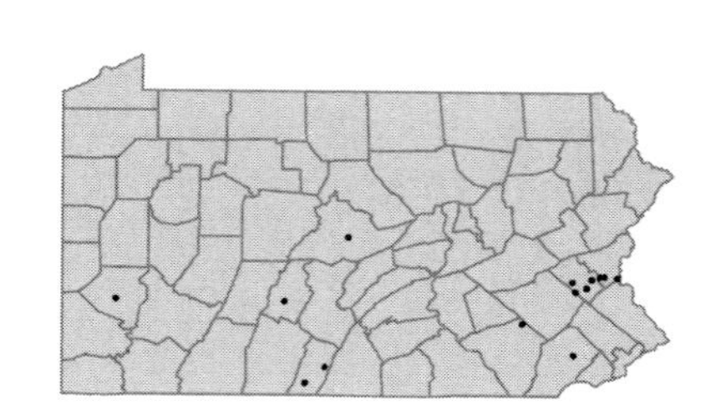

Ulmus rubra Muhl.
Red elm; Slippery elm
Deciduous tree
Moist woods, stream banks and floodplains in circumneutral soils.
Ulmus fulva Michx. P

FAC

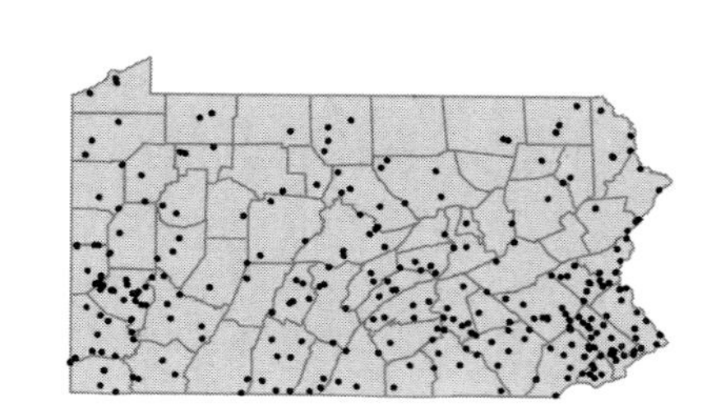

MORACEAE

• ***Broussonetia papyrifera*** (L.) Vent. I
Paper mulberry
Deciduous tree
Escaped from cultivation, waste ground and woods margins.

• ***Ficus carica*** L. I
Fig
Deciduous shrub
Cultivated, may persist briefly at old garden sites.

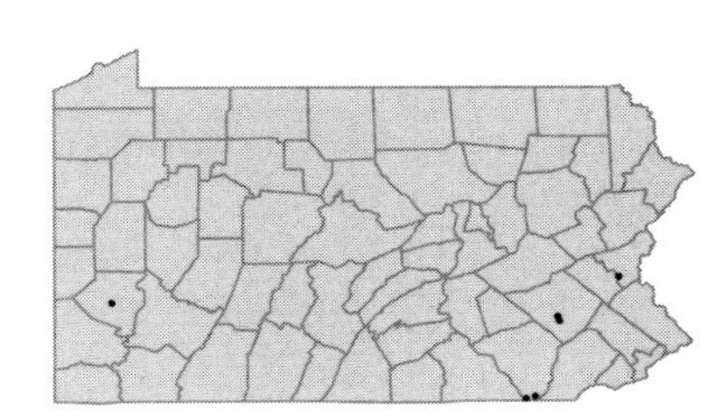

• ***Maclura pomifera*** (Raf. ex Sarg.) Schneid. I
Osage-orange
Deciduous tree
Roadsides, fencerows and abandoned pastures.
Toxylon pomiferum Raf. P

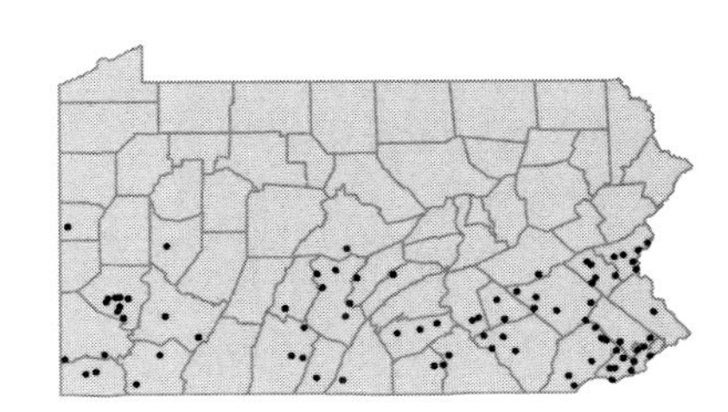

• ***Morus alba*** L. I
White mulberry
Deciduous tree
Fencerows, woods edges and waste ground.

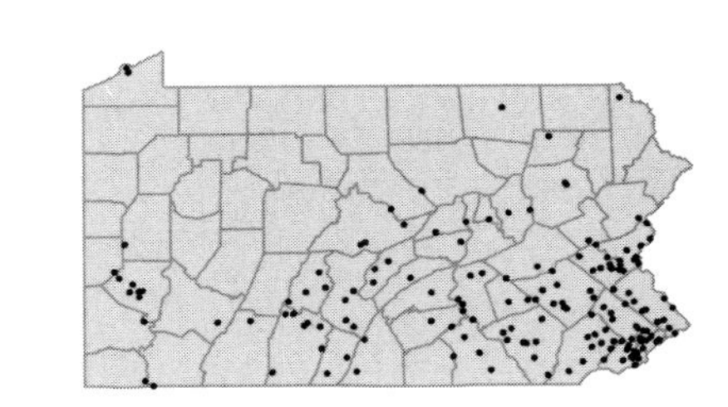

Morus rubra L. FACU
Red mulberry
Deciduous tree
Rich, moist, alluvial soils and wooded slopes.

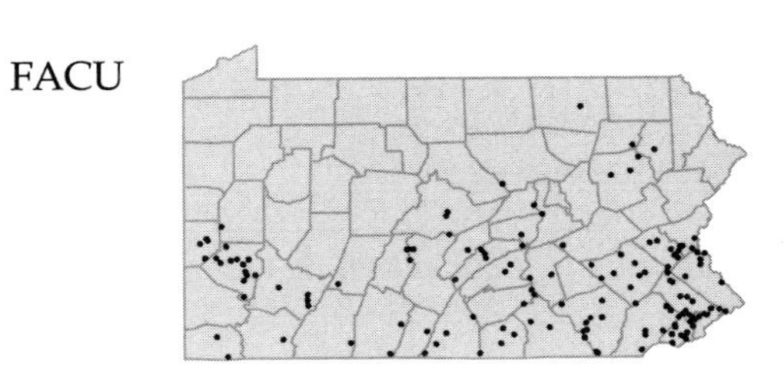

CANNABACEAE

• ***Cannabis sativa*** L. I
Hemp; Marijuana
Herbaceous annual
Waste ground. Designated as a noxious weed in PA.

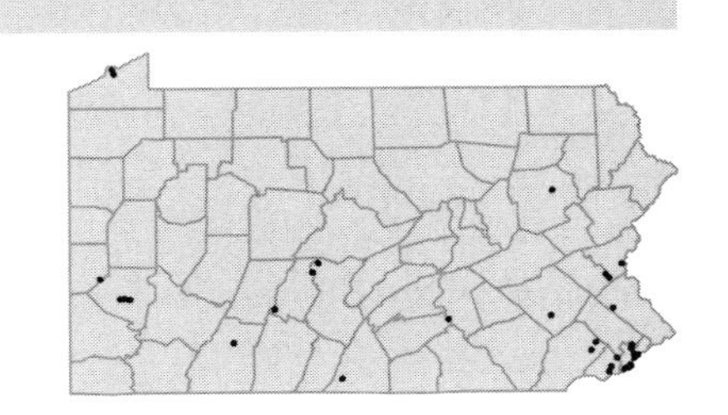

CANNABACEAE

• ***Humulus japonicus*** Sieb. & Zucc. I
Japanese hops
Herbaceous perennial
Meadows, roadsides and waste ground.

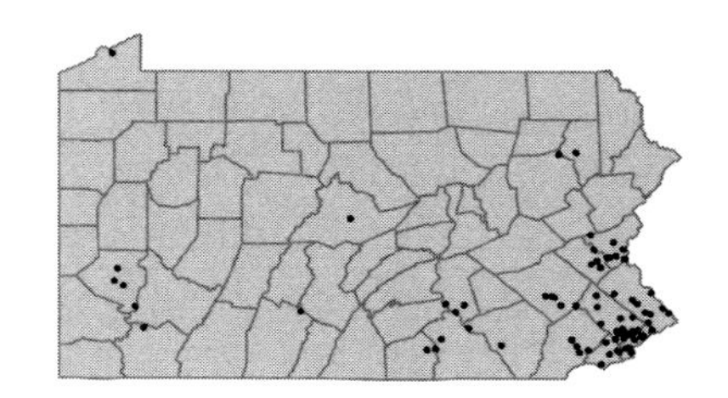

Humulus lupulus L.
Brewer's hops; Common hops
Herbaceous perennial
Moist alluvial soil, woods edges, thickets and waste ground. Native, but also escaped from cultivation.

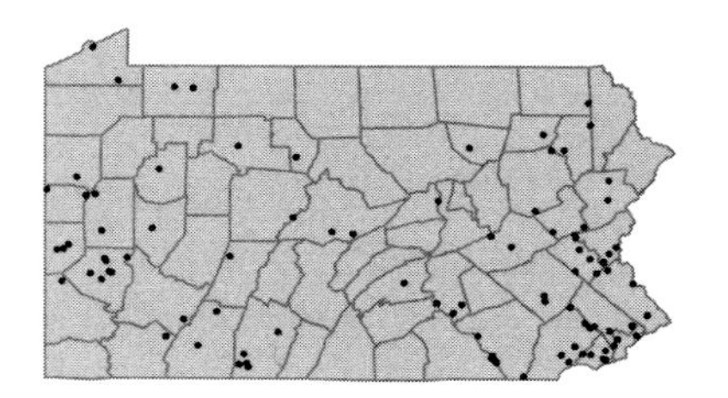

URTICACEAE

• ***Boehmeria cylindrica*** (L.) Swartz var. ***cylindrica*** FACW+
Bog-hemp; False nettle
Herbaceous perennial
Moist, shady ground of wet woods and stream margins.

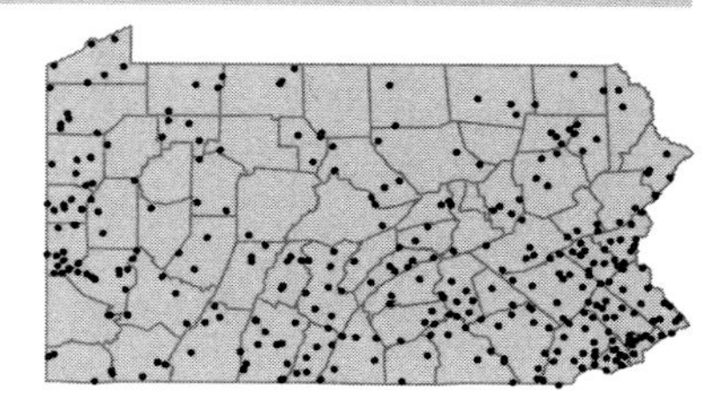

Boehmeria cylindrica (L.) Swartz var. ***drummondiana*** (Weddell) Weddell FACW+
Bog-hemp; False nettle
Herbaceous perennial
Moist, shady ground of wet woods and stream margins.

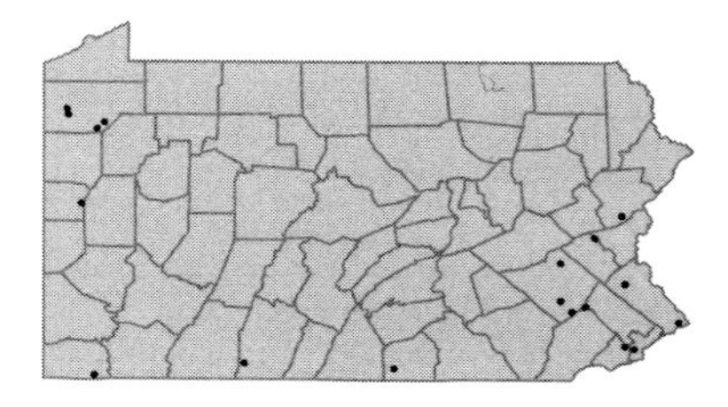

• ***Laportea canadensis*** (L.) Weddell FACW
Wood-nettle
Herbaceous perennial
Low, moist woods and stream banks.
Urticastrum divaricatum (L.) Kuntze P

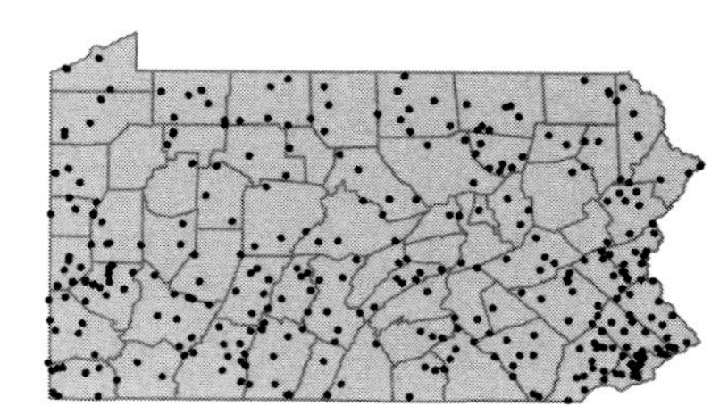

• ***Parietaria judaica*** L. I
Pellitory
Herbaceous annual
Ballast. Represented by several collections from Philadelphia Co. 1879-1921.
Parietaria diffusa Mertens & Koch W

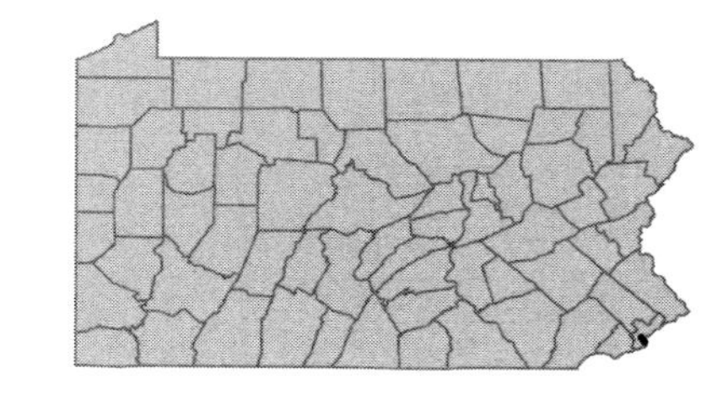

• ***Parietaria pensylvanica*** Muhl. ex Willd.
Pellitory
Herbaceous annual
Dry, rocky or gravelly woods, roadside banks and waste ground.

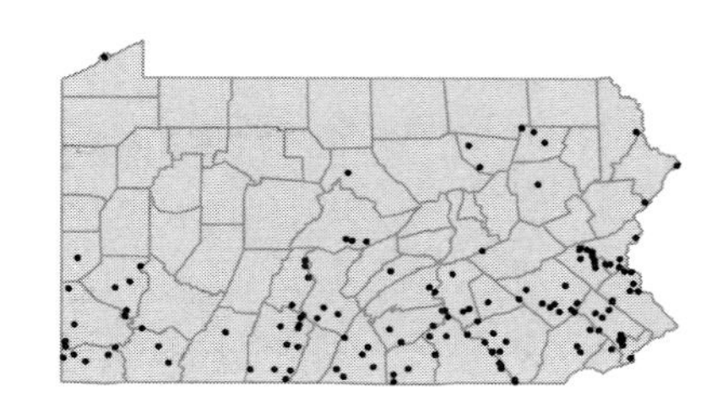

• ***Pilea fontana*** (Lunell) Rydb.
Coolwort; Lesser clearweed
Herbaceous annual
Springheads, wet shores and swamps.

FACW+

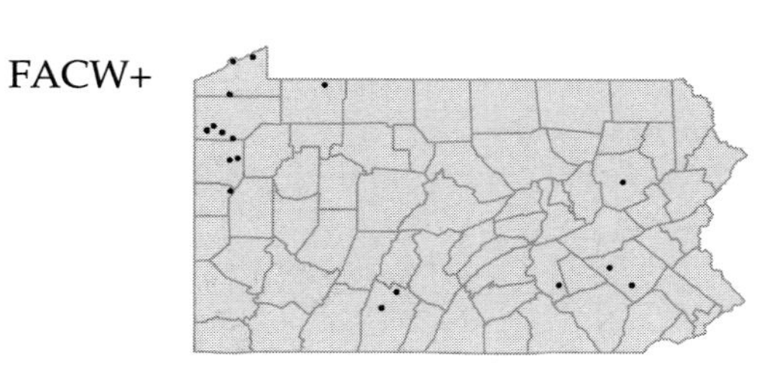

Pilea pumila (L.) A.Gray
Clearweed; Coolwort; Richweed
Herbaceous annual
Cool, moist, shady areas.
Adicia pumila (L.) Raf. P

FACW

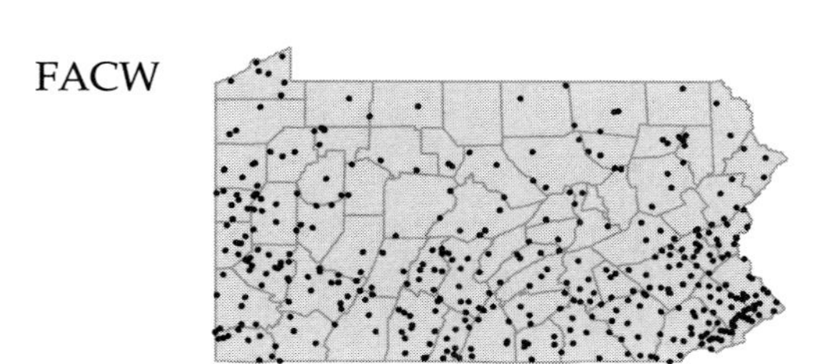

• ***Urtica dioica*** L. ssp. ***dioica***
Great nettle; Stinging nettle
Herbaceous perennial
Disturbed ground of thin woods, floodplains and edges.

FACU
[I]

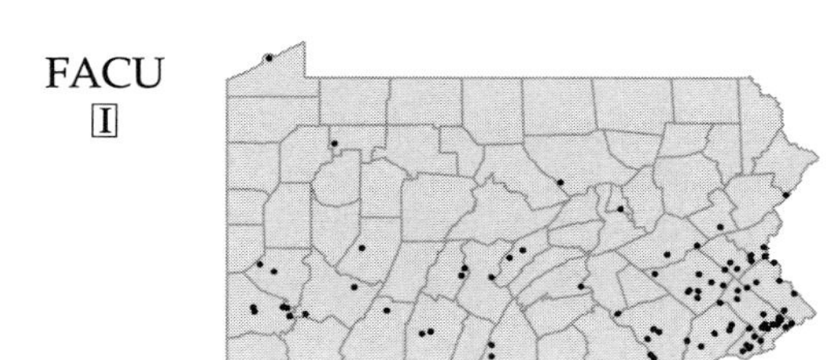

Urtica dioica L. ssp. ***gracilis*** (Ait.) Seland.
Great nettle; Stinging nettle
Herbaceous perennial
Floodplains and thickets, in moist soil.
Urtica dioica L. var. *procera* (Muhl.) Weddell GBC; *Urtica gracilis* Ait. F; *Urtica procera* Muhl. FW

FACU

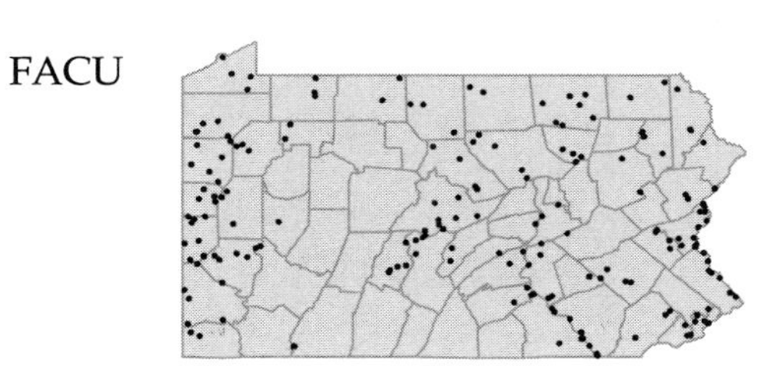

Urtica urens L.
Burning nettle; Dog nettle; Stinging nettle
Herbaceous annual
Disturbed soil.

[I]

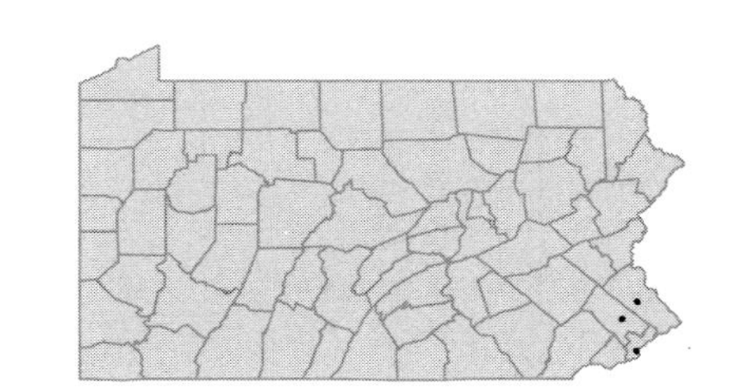

JUGLANDACEAE

• ***Carya cordiformis*** (Wang.) K.Koch — FACU+
Bitternut hickory
Deciduous tree
Moist woods and stream banks. A hybrid with *C. ovata* (*C. x laneyi* Sarg.) was collected at a single location in Crawford Co.
Hicoria minima (Marshall) Britt. P

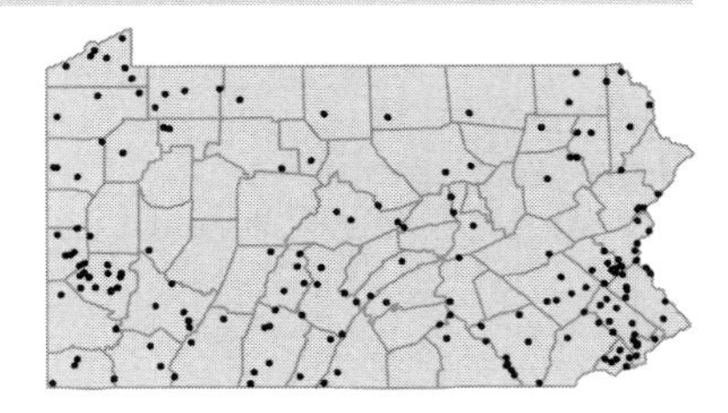

Carya glabra (P.Mill.) Sweet — FACU-
Pignut hickory
Deciduous tree
Upland woods, dry ridge tops and slopes.
Hicoria glabra (P.Mill.) Britt. P

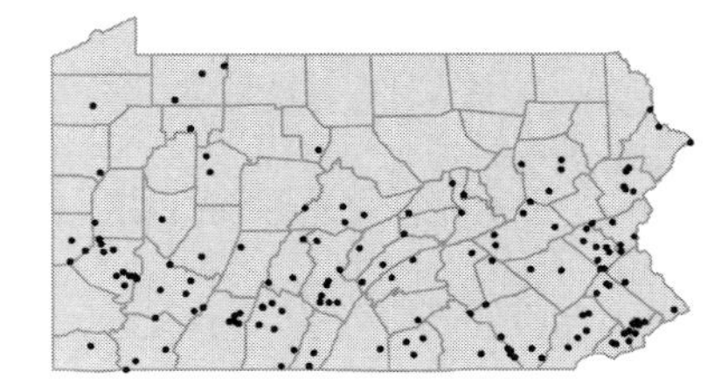

Carya laciniosa (Michx.f.) Loud.
Shellbark hickory
Deciduous tree
Moist, rich bottomlands and slopes.
Hicoria laciniosa (Michx.) Sargent P

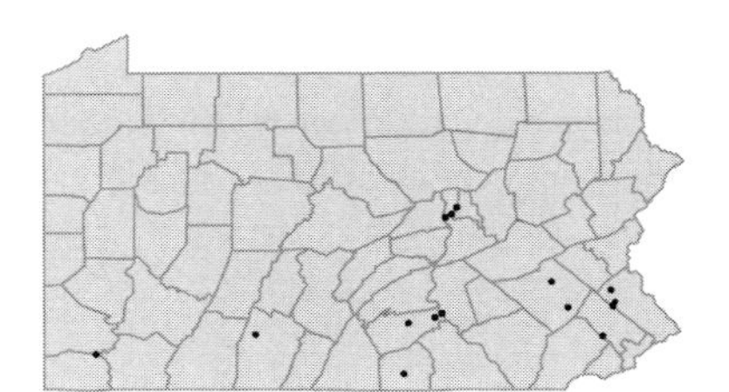

Carya ovalis (Wang.) Sarg.
Sweet pignut hickory
Deciduous tree
Rich, dry woods and bluffs.
Hicoria microcarpa (Nutt.) Britt. P

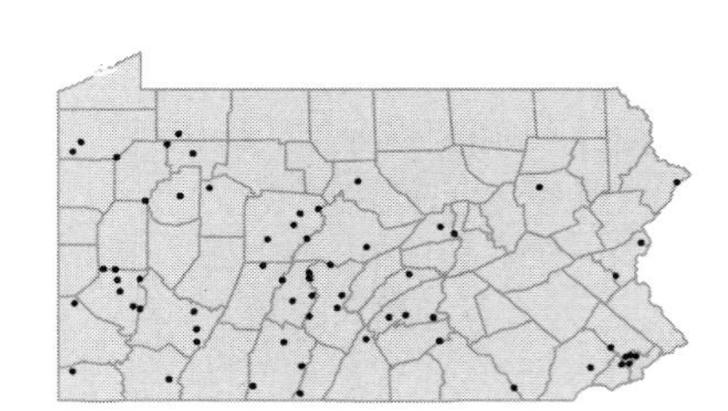

Carya ovata (P.Mill.) K.Koch

Shagbark hickory; Shellbark hickory
Deciduous tree
Low, moist woods and slopes, in rich soil.
Hicoria ovata (P.Mill.) Britt. P

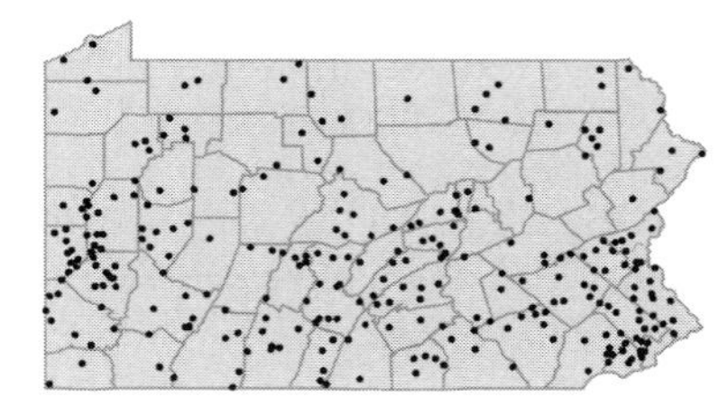

Carya tomentosa (Lam. ex Poir.) Nutt.
Mockernut hickory
Deciduous tree
Moist, open woods and slopes.
Hicoria alba (L.) Britt. P

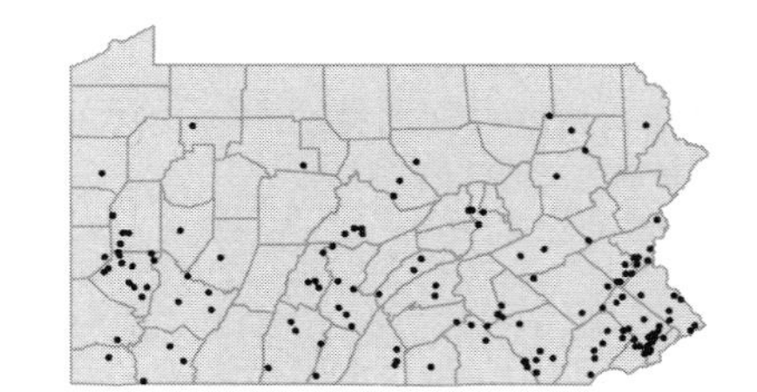

• ***Juglans cinerea*** L. FACU+
Butternut
Deciduous tree
Lowland woods and rich, wooded hillsides.

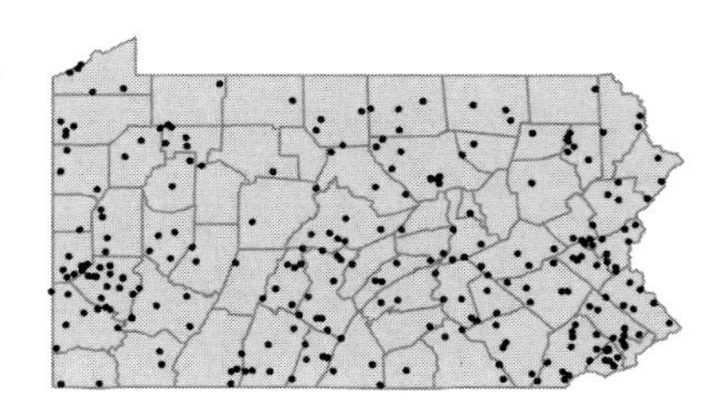

Juglans nigra L. FACU
Black walnut
Deciduous tree
Open woods and meadows in moist, rich, alluvial soils.

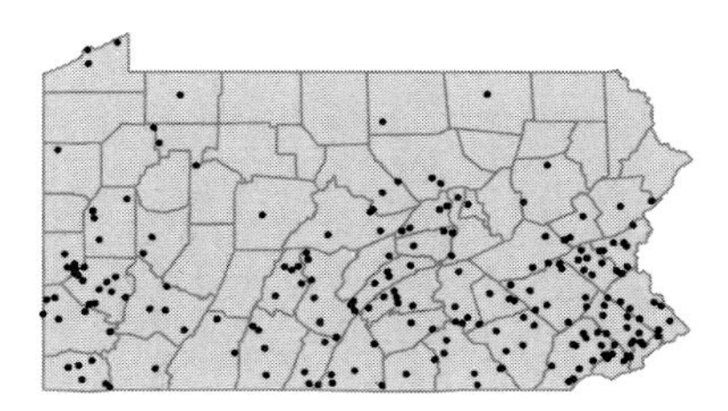

MYRICACEAE

• ***Comptonia peregrina*** (L.) Coult.
Sweet-fern
Deciduous shrub
Dry, sterile soils of open woods and barrens.
Myrica asplenifolia L. GB

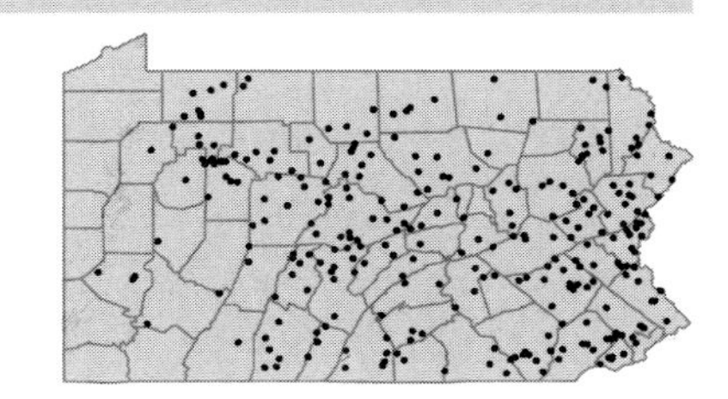

• ***Myrica gale*** L. OBL
Sweet-gale
Deciduous shrub
Bogs and shallow water of lake, pond and stream edges.

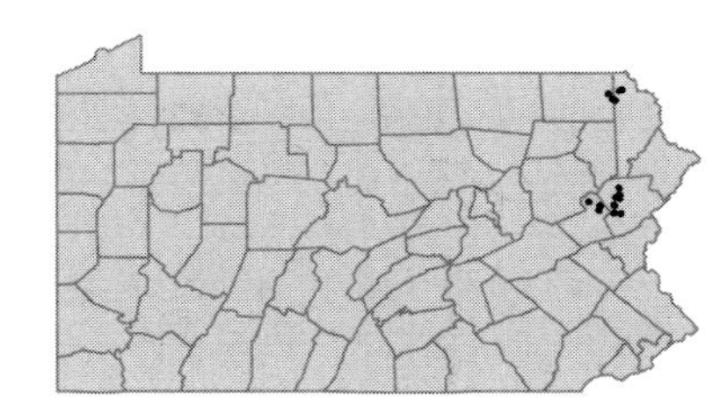

Myrica heterophylla Raf. FAC
Evergreen bayberry
Evergreen shrub
Dry to moist woods or thickets. Believed to be extirpated, last collected in 1946.

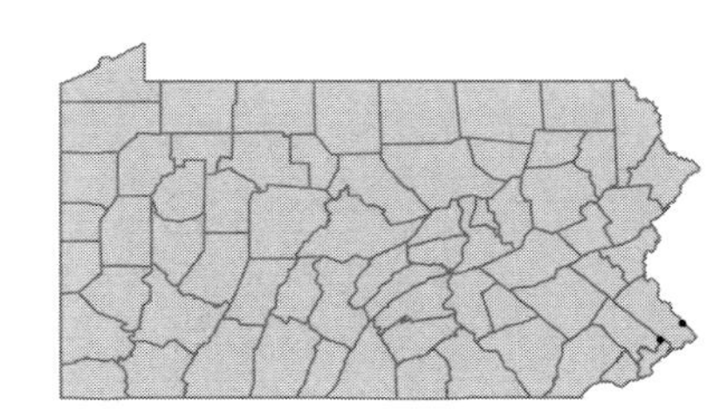

Myrica pensylvanica Loisel. FAC
Bayberry
Deciduous shrub
Old fields or open woods, in dry to moist, sterile, sandy soils.
Myrica carolinensis P.Mill. P

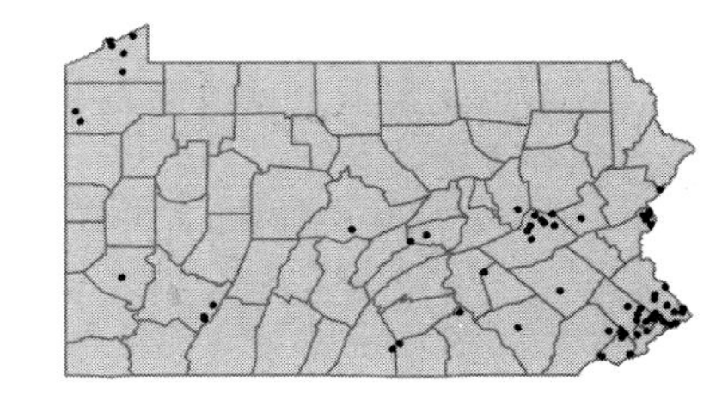

FAGACEAE

• ***Castanea dentata*** (Marshall) Borkh.
American chestnut
Deciduous tree
Wooded slopes and ridges, in dry, acidic soils. Formerly a dominant forest tree but reduced to minor status by Chestnut blight.

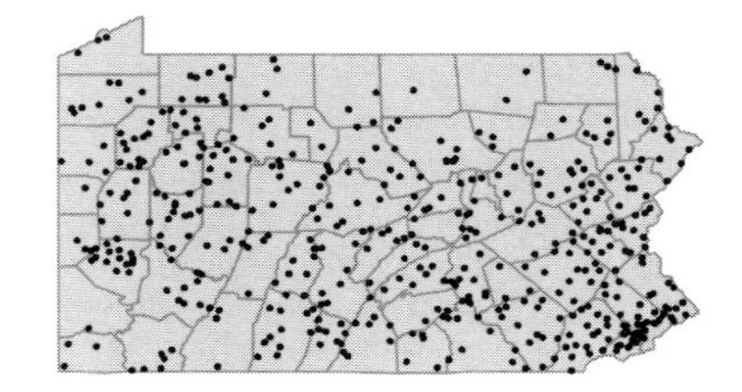

Castanea pumila (L.) P.Mill.
Chinquapin
Deciduous tree
Moist to dry wooded slopes.

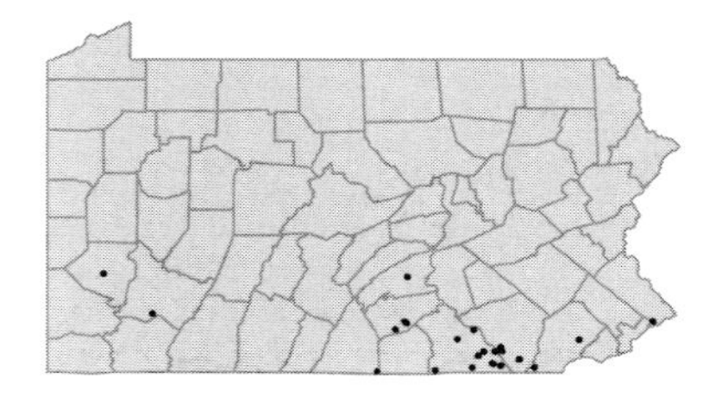

• ***Fagus grandifolia*** Ehrh.
American beech
Deciduous tree
A dominant tree of mature forests on moist, rich soils.

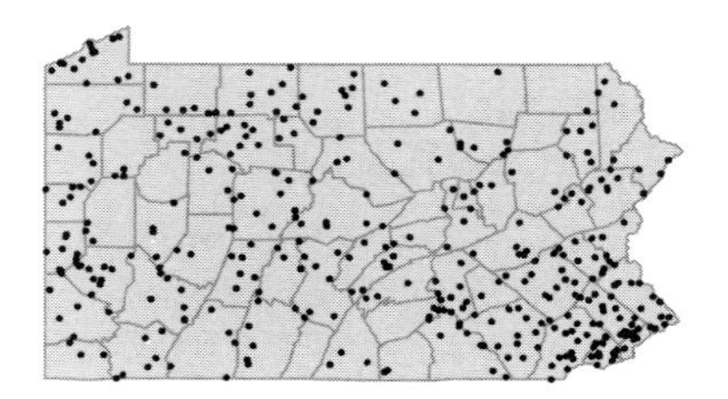

Quercus—In addition to the taxa treated below, limited collections exist representing the following *Quercus* hybrids: *alba x bicolor* (*Q. x jackiana* Schneid.); *alba x macrocarpa* (*Q. x bebbiana* Schneid.); *bicolor x montana*; *coccinea x ilicifolia* (*Q. x robbinsii* Trel.); *coccinea x rubra* (*Q. x benderi* Baenitz); *ilicifolia x marilandica* (*Q. x brittonii* W.T.Davis); *ilicifolia x rubra* (*Q. x fernaldii* Trel.); *imbricaria x marilandica* (*Q. x tridentata* [A.DC.] Engelm.); *imbricaria x palustris* (*Q. x exacta* Trel.); *macrocarpa x muhlenbergii* (*Q. x deamii* Trel.); *marilandica x phellos* (*Q. rudkinii* Britt.); *marilandica x velutina* (*Q. x bushii* Sarg.).

• ***Quercus acutissima*** Carruth. I
Sawtooth oak
Deciduous tree
Escaped from cultivation, fallow fields. Planted by the Game Commission.

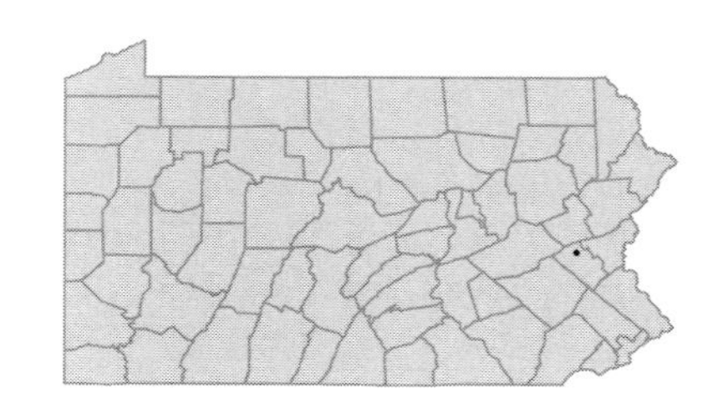

Quercus alba L.
White oak
Deciduous tree
A dominant forest tree on dry to moist sites.

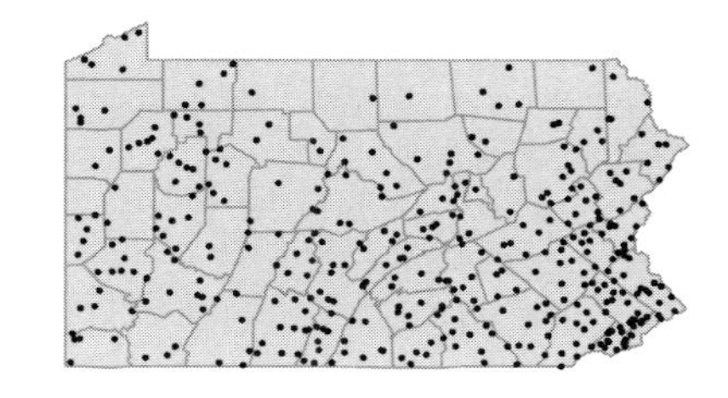

Quercus alba x montana
Saul oak
Deciduous tree
Rocky, upland woods.
Quercus x saulii Schneid.

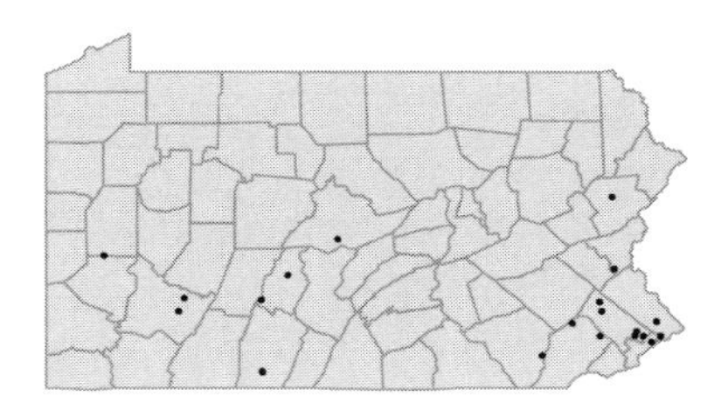

Quercus bicolor Willd.
Swamp white oak
Deciduous tree
Low, moist forests and wooded swamps.
Quercus platanoides (Lam.) Sudworth P

FACW+

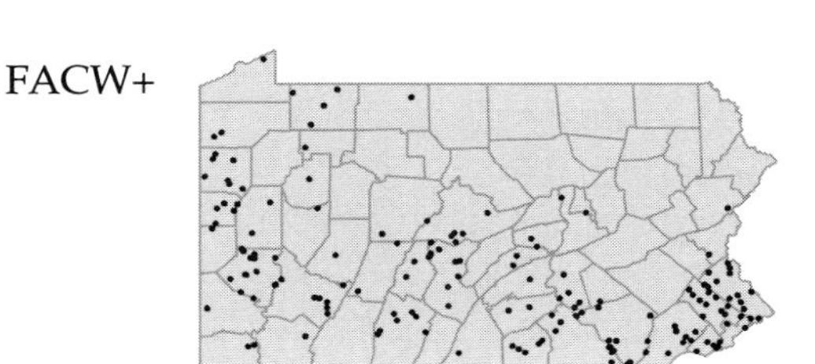

Quercus coccinea Muenchh.
Scarlet oak
Deciduous tree
Dry upper slopes and ridges, in poor soil.

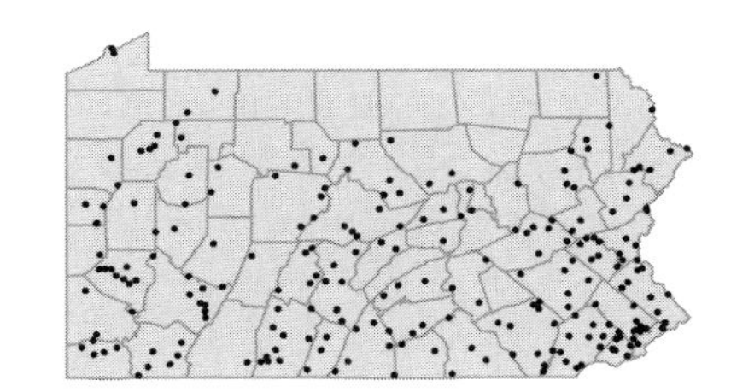

Quercus falcata Michx.
Southern red oak; Spanish oak
Deciduous tree
Dry to moist woodlands on or near the Coastal Plain.
Quercus digitata (Marshall) Sudworth P

FACU-

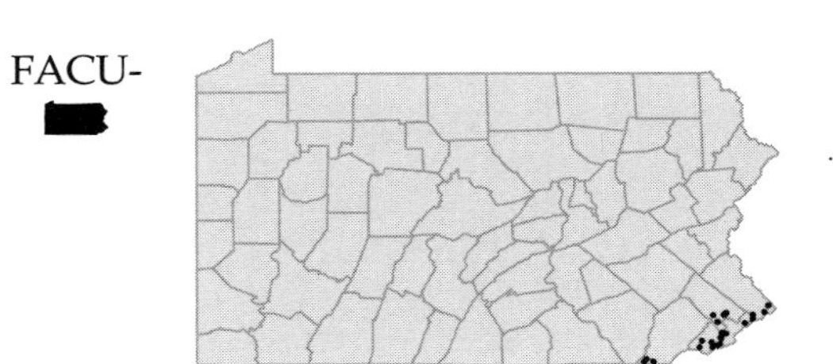

Quercus ilicifolia Wang.
Bear oak; Scrub oak
Deciduous shrub
Dry ridge tops and barrens, in sterile, sandy soil.
Quercus nana (Marshall) Sarg. P

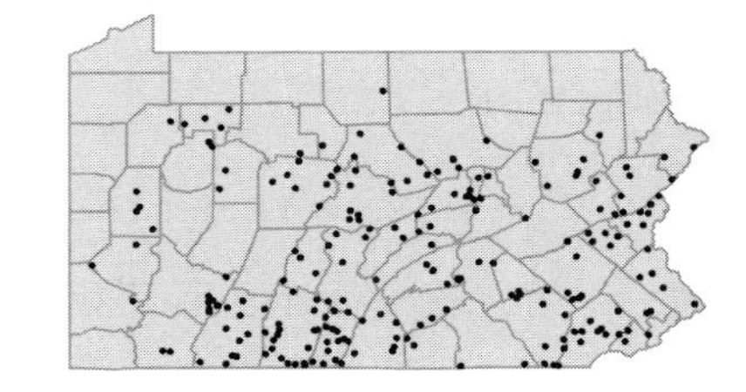

Quercus ilicifolia x velutina
Rehder oak
Deciduous tree
Dry, rocky woods.
Quercus x rehderi Trel.

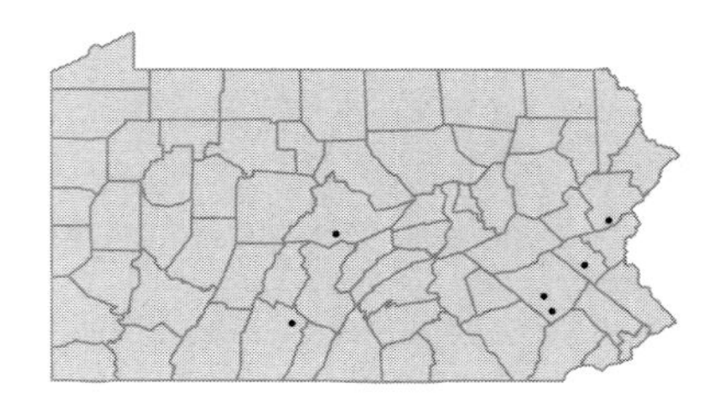

Quercus imbricaria Michx. FAC
Shingle oak
Deciduous tree
Moist, rich bottomlands of southwest counties. Introduced and cultivated elsewhere.

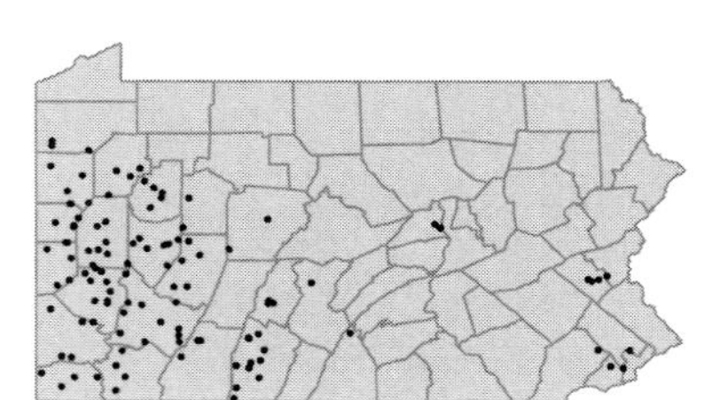

Quercus imbricaria x rubra
Saw-toothed oak
Deciduous tree
Stream valleys and roadsides.
Quercus x runcinata (DC.) Engelm.

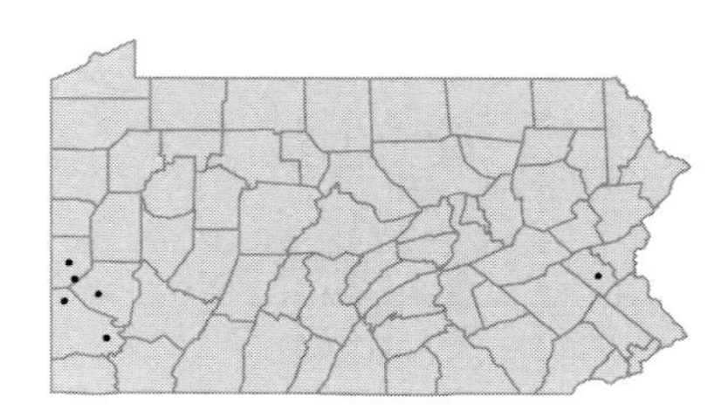

Quercus imbricaria x velutina
Lea oak
Deciduous tree
Woods borders and rich bottomland.
Quercus x leana Nutt.

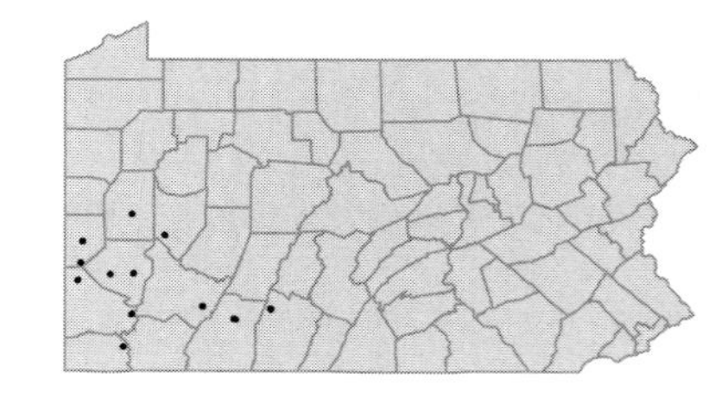

Quercus macrocarpa Michx. FAC-
Bur oak; Mossy-cup oak
Deciduous tree
Dry to moist forests, in neutral or calcareous soils.

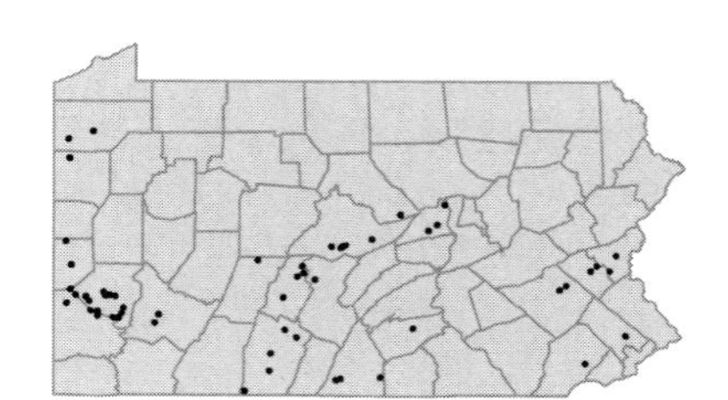

Quercus marilandica Muenchh.
Blackjack oak
Deciduous tree
Dry, sterile soils, serpentine barrens.

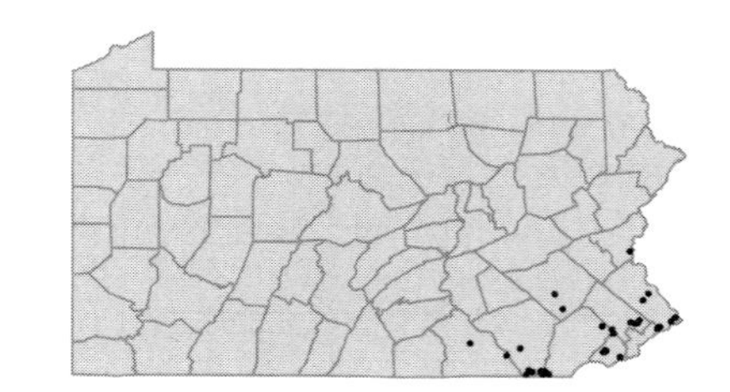

Quercus montana Willd.
Chestnut oak
Deciduous tree
A dominant forest tree on dry slopes and ridgetops, in acidic soils.
Quercus prinus L. PFGBWC

UPL

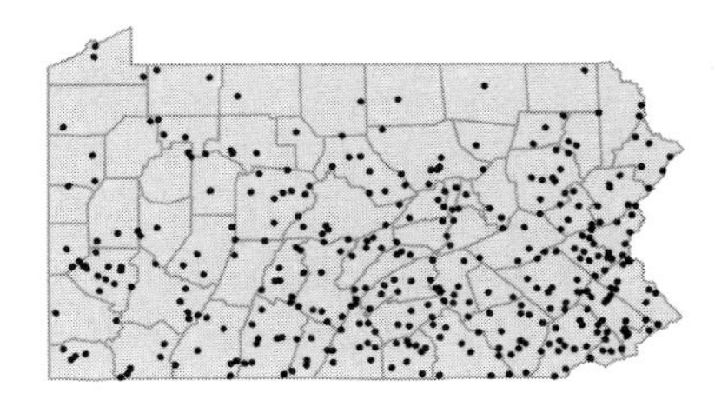

Quercus muhlenbergii Engelm.
Yellow oak
Deciduous tree
Wooded slopes, on limestone.
Quercus acuminata (Michx.) Sarg. P; *Quercus prinoides* Willd. var. *acuminata* (Michx.) Gleason GB

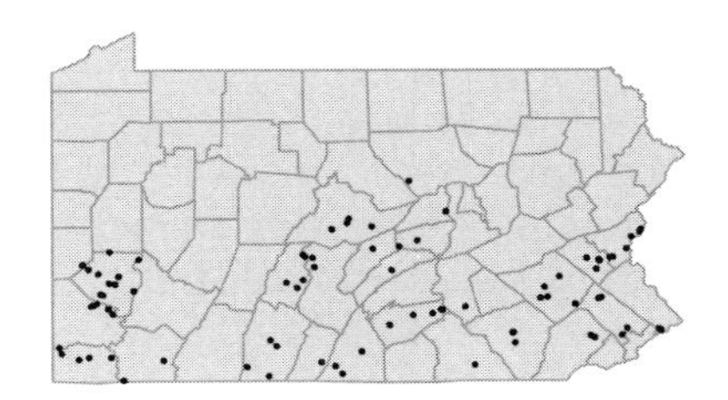

Quercus palustris Muenchh.
Pin oak
Deciduous tree
Low, moist or seasonally wet woods or swamps.

FACW

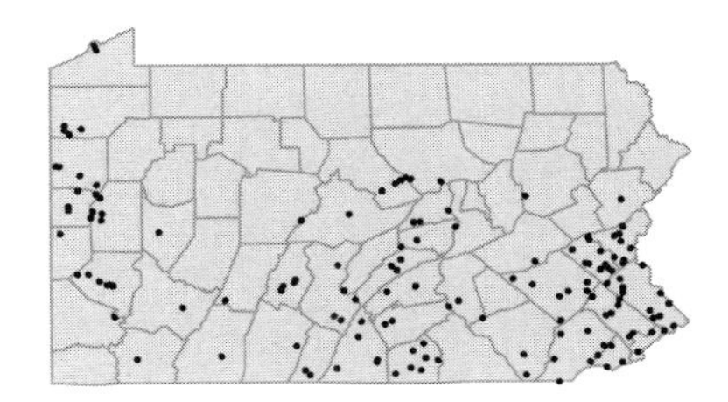

Quercus phellos L.
Willow oak
Deciduous tree
Low, moist or seasonally wet woods.

FAC+

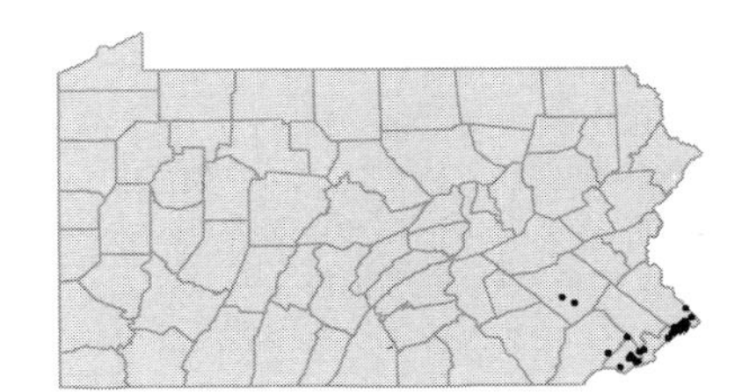

Quercus phellos x rubra
Bartram oak
Deciduous tree
Low, moist woods on or near the Coastal Plain.
Quercus x heterophylla Michx.

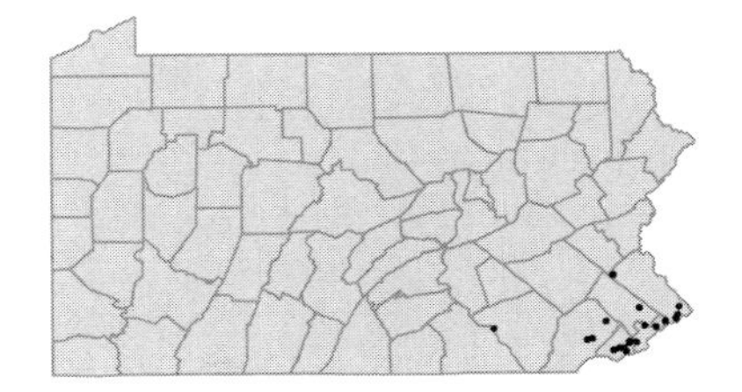

Quercus prinoides Willd.
Chinquapin oak; Dwarf chestnut oak
Deciduous shrub
Dry, rocky ridgetops and slopes and serpentine barrens.

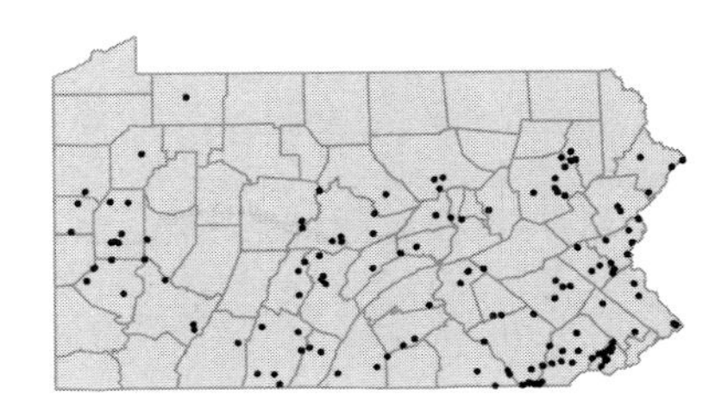

Quercus robur L.
English oak
Deciduous tree
Cultivated and sometimes escaped. Seedlings naturalizing at one Montgomery Co. site.

[I]

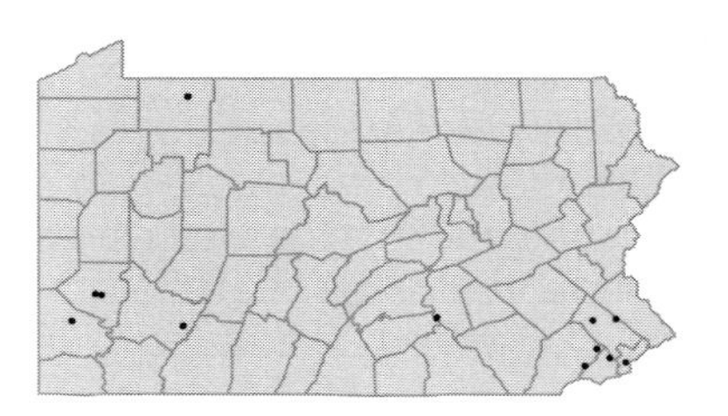

Quercus rubra L.
Northern red oak
Deciduous tree
A dominant forest tree on moist to dry sites.
Quercus borealis Michx.f. GB

FACU-

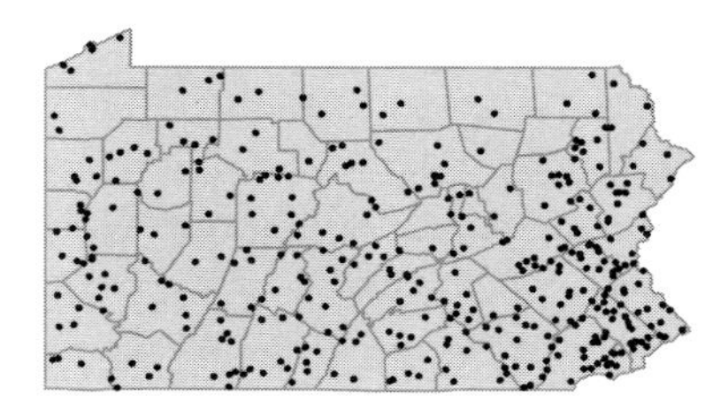

Quercus shumardii Buckl.
Shumard oak
Deciduous tree
Stream banks.

FAC+

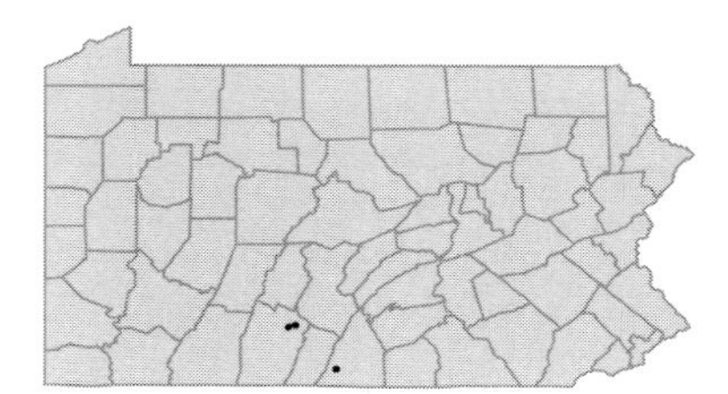

Quercus stellata Wang.
Post oak
Deciduous tree
Dry woods and hillsides, serpentine barrens.
Quercus minor (Marshall) Sarg. P

UPL

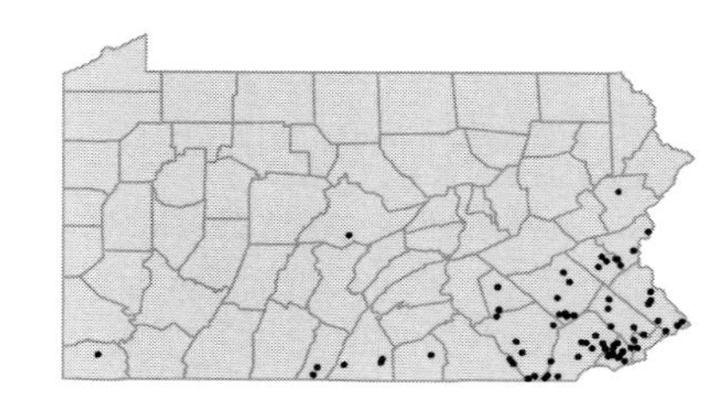

Quercus velutina Lam.
Black oak
Deciduous tree
A dominant forest tree on moist to dry soils.

BETULACEAE

• ***Alnus glutinosa*** (L.) Gaertn.
Black alder; European alder
Deciduous tree
Planted and sometimes naturalized.

FACW-
[I]

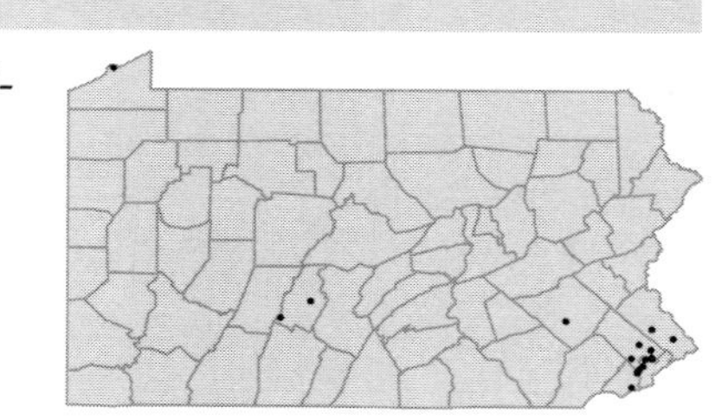

Alnus incana (L.) Moench ssp. ***rugosa*** (DuRoi) Clausen FACW+
Speckled alder
Deciduous shrub
Bogs and swamps.
Alnus rugosa (DuRoi) Spreng. PFGBW

Alnus serrulata (Dryand. ex Ait.) Willd. OBL
Smooth alder
Deciduous shrub
Low, wet woods and swamps.

Alnus viridis (Chaix) DC. ssp. ***crispa*** (Ait.) Turrill
Mountain alder
Deciduous shrub
Cool, rocky, wooded slopes.
Alnus crispa (Ait.) Pursh FGBW

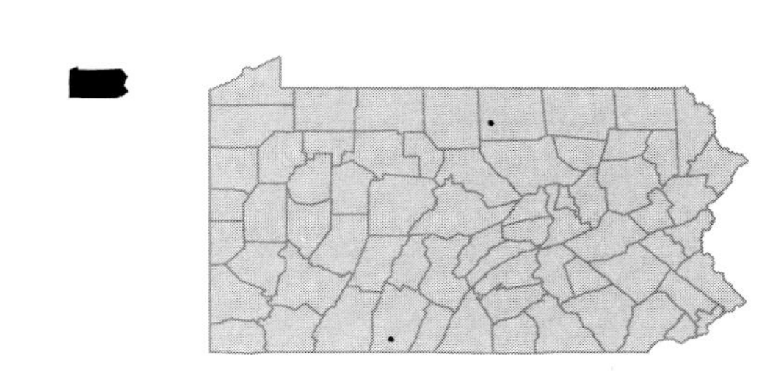

• ***Betula alleghaniensis*** Britt. FAC
Yellow birch
Deciduous tree
Cool, moist, northern woods.
Betula lutea Michx.f. FGBW

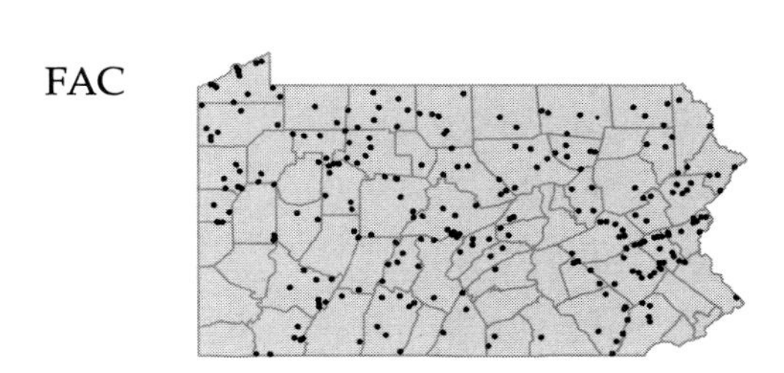

Betula lenta L. FACU
Black birch; Cherry birch; Sweet birch
Deciduous tree
Woods and stream banks.

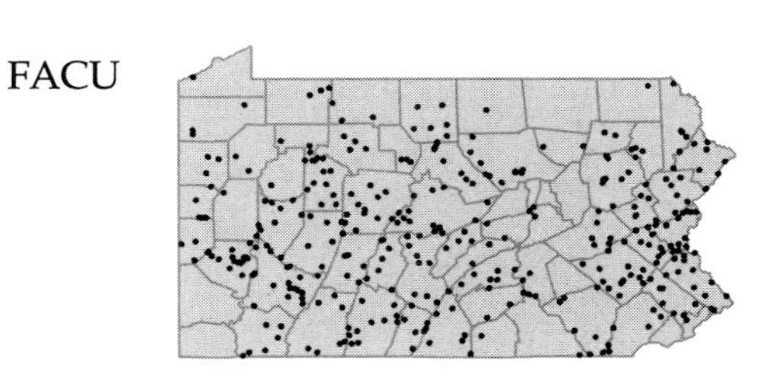

Betula nigra L. FACW
River birch
Deciduous tree
Floodplains, stream banks, wet woods and swamps.

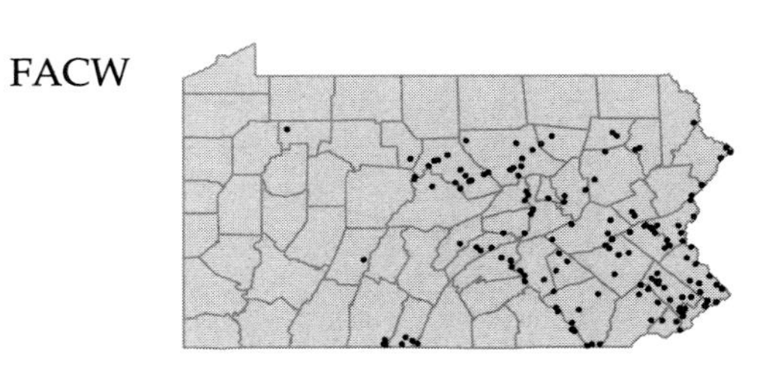

Betula papyrifera Marshall
Canoe birch; Paper birch
Deciduous tree
Upland woods and slopes.

FACU

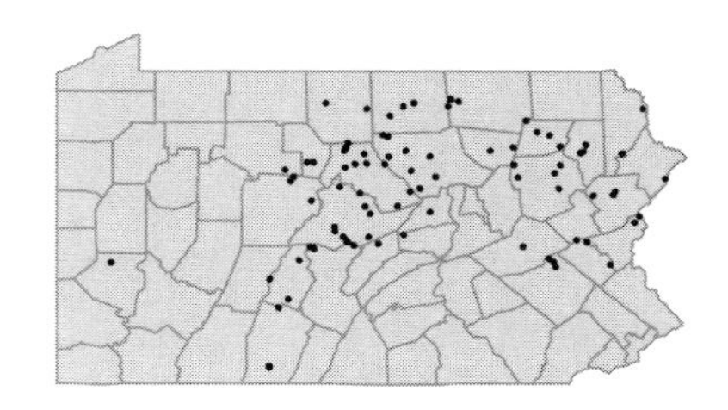

Betula pendula Roth
European white birch
Deciduous tree
Roadside thickets, pastures and waste ground.

I

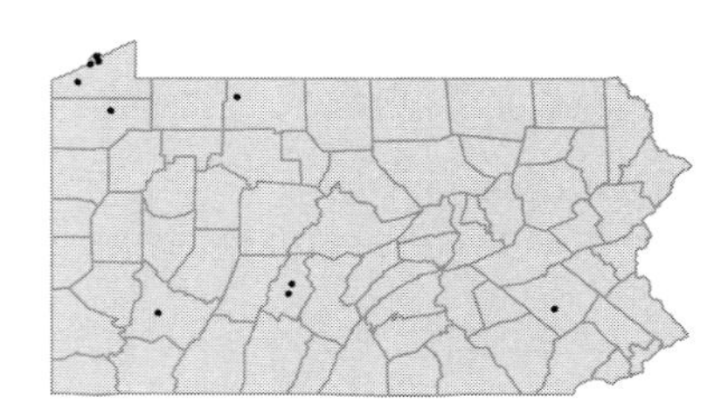

Betula populifolia Marshall
Gray birch
Deciduous tree
Old fields, open woods and disturbed areas, especially on dry, sterile soils.

FAC

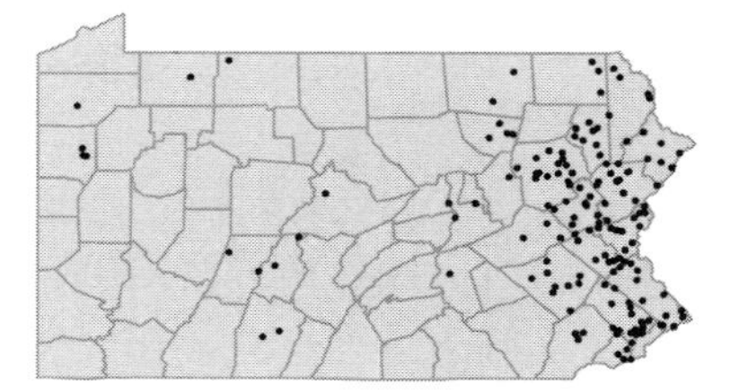

Betula pubescens Ehrh.
European white birch
Deciduous tree
Cultivated and sometimes persisting at abandoned home or nursery sites.
Betula alba L. FW

I

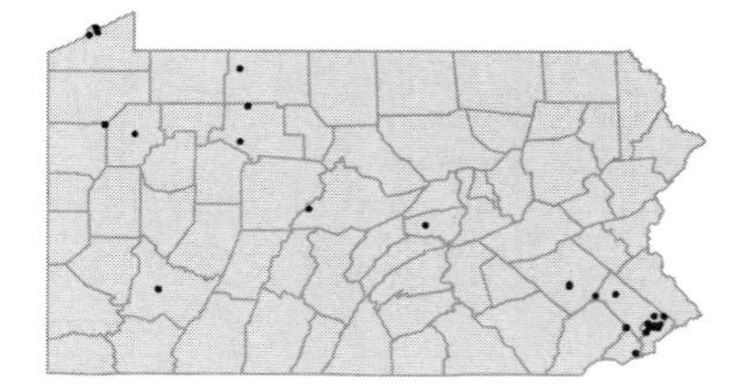

• ***Carpinus caroliniana*** Walt.
Hornbeam; Blue-beech
Deciduous tree
Rich, moist woods and stream edges.

FAC

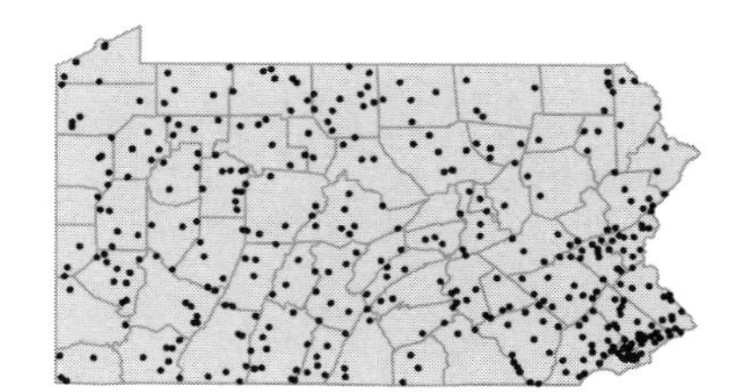

• ***Corylus americana*** Walt.
American filbert; Hazelnut
Deciduous shrub
Rich woods and edges.

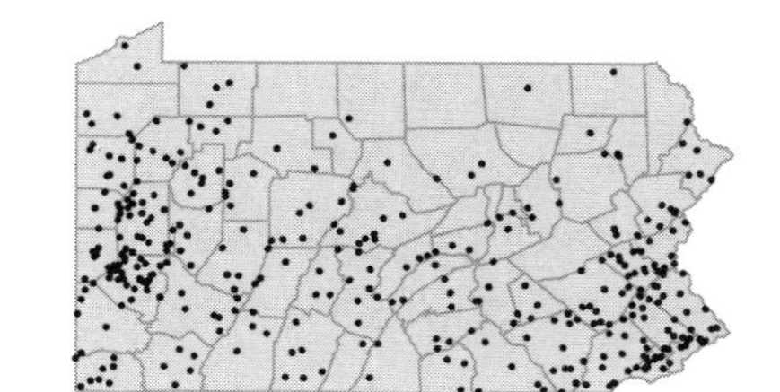

Corylus cornuta Marshall
Beaked hazelnut
Deciduous shrub
Dry, rocky woods and thickets.

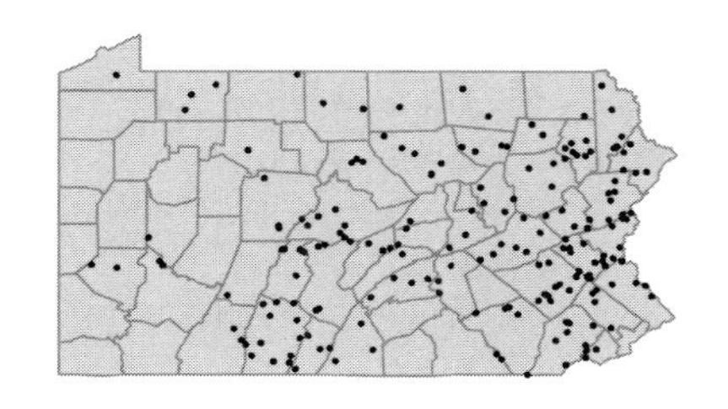

• ***Ostrya virginiana*** (P.Mill.) K.Koch — FACU-
Hop-hornbeam
Deciduous tree
Dry, wooded slopes, often on calcareous soils.

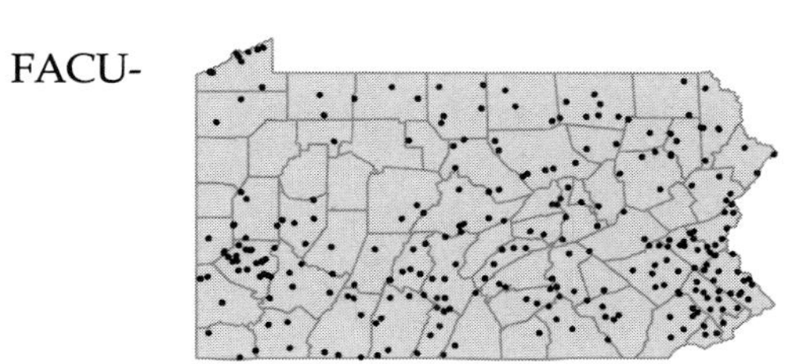

PHYTOLACCACEAE

• ***Phytolacca americana*** L. — FACU+
Pokeweed
Herbaceous perennial
Forest openings, waste ground and gardens.
Phytolacca decandra L. P

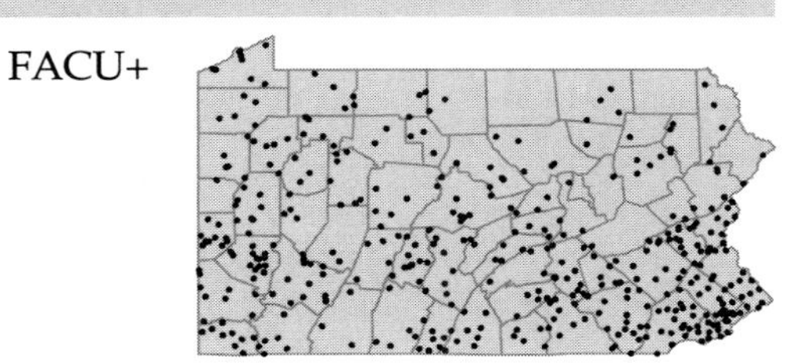

NYCTAGINACEAE

• ***Mirabilis albida*** (Walt.) Heimerl — [I]
Umbrella-wort
Herbaceous perennial
Railroad ballast and streets. Represented by two collections from Bucks Co. in 1898.

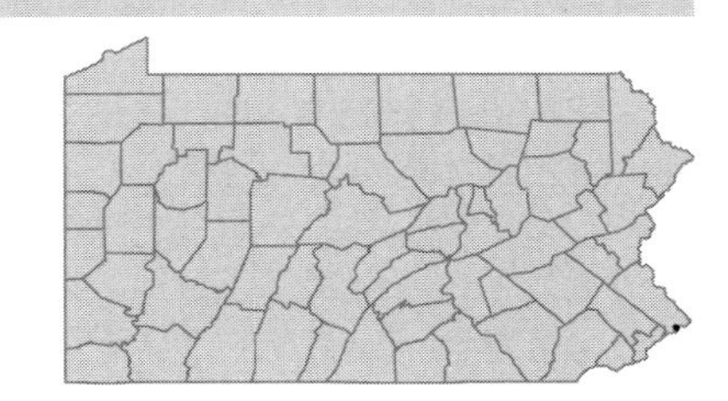

Mirabilis jalapa L. — [I]
Four-o'clock; Marvel-of-Peru
Herbaceous perennial
Cultivated and occasionally spreading to waste ground.

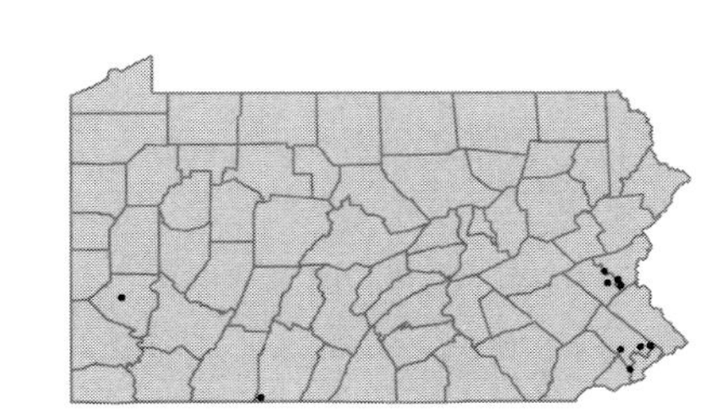

Mirabilis nyctaginea (Michx.) MacM. — [I]
Heart-leaved umbrella-wort; Wild four-o'clock
Herbaceous perennial
Dry soil and waste ground.
Oxybaphus nyctagineus (Michx.) Sweet GBW

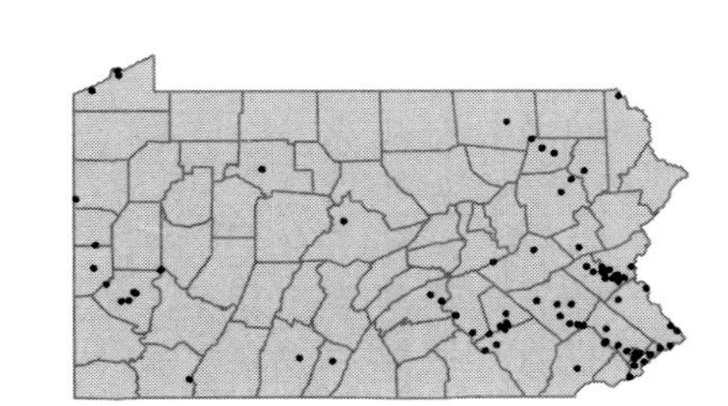

Mirabilis linearis (Pursh) Heimerl
Umbrella-wort
Herbaceous perennial
Railroad embankment. Represented by a single collection from Luzerne Co. in 1937.
Oxybaphus linearis (Pursh) B.L.Robins. GBW

[I]

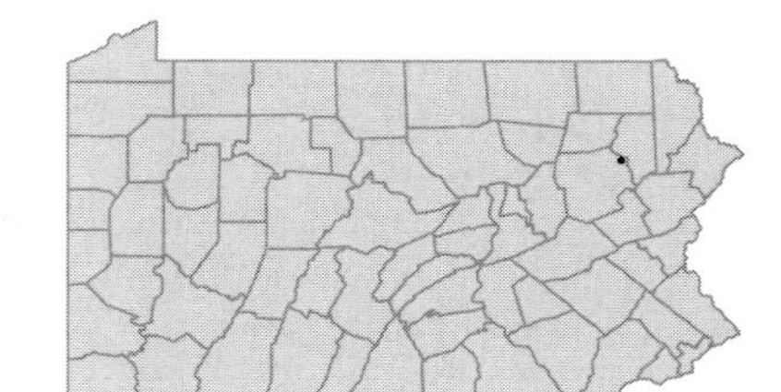

AIZOACEAE

• ***Mesembryanthemum crystallinum*** L.
Ice-plant
Herbaceous annual
Ballast. All collections are from a single location in Philadelphia Co. in 1879.

[I]

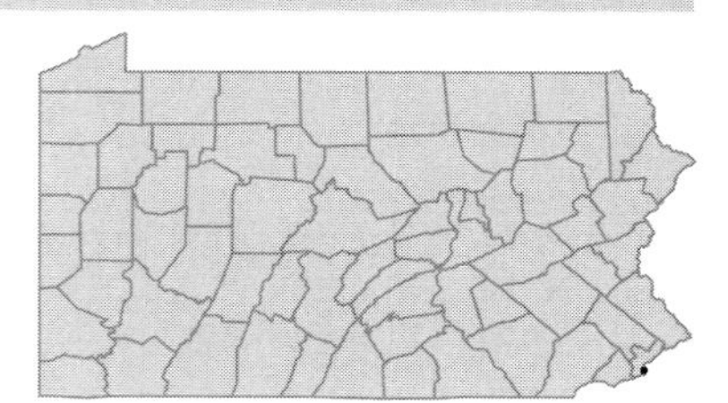

• ***Sesuvium portulacastrum*** (L.) L.
Sea-purslane
Herbaceous perennial
Ballast. Collected at a single location in Philadelphia Co. in 1865.

[I]

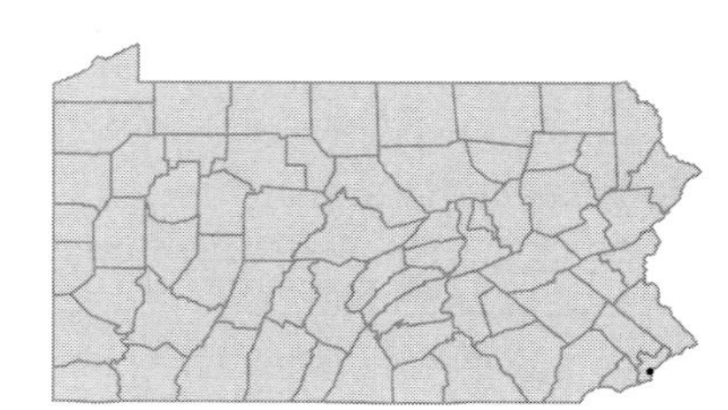

• ***Tetragonia tetragonioides*** (Pallas) Kuntze
New Zealand spinach
Herbaceous annual
Cultivated and very rarely escaped.

[I]

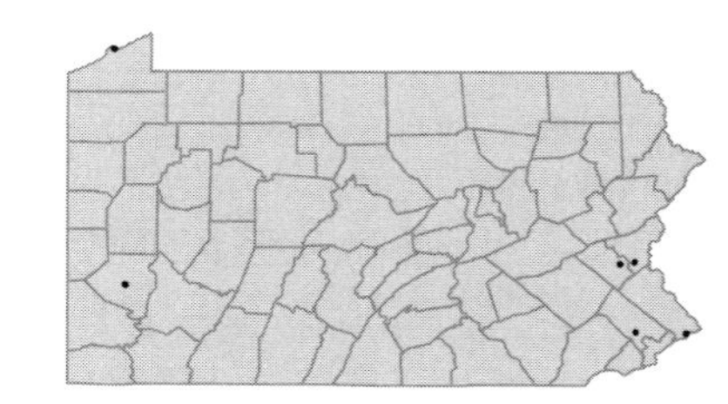

CACTACEAE

• ***Opuntia humifusa*** (Raf.) Raf.
Eastern prickly-pear cactus
Herbaceous perennial
Dry, shaly cliffs and barrens.
Opuntia calcicola Wherry W; *Opuntia compressa* (Salisb.) Macbr. GBW; *Opuntia opuntia* (L.) Coult. P

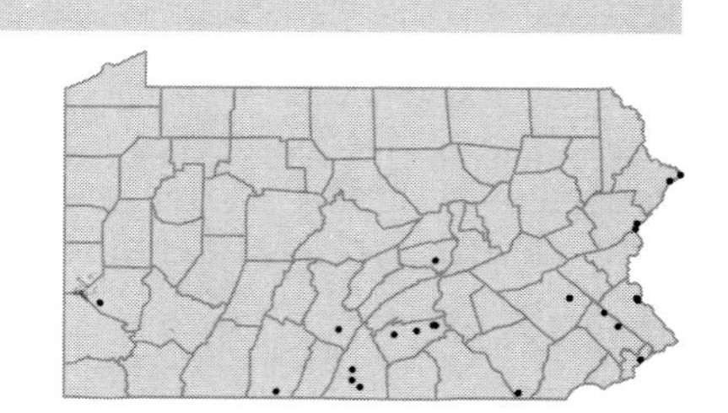

CHENOPODIACEAE

• ***Atriplex arenaria*** Nutt.
Seabeach orach
Herbaceous annual
Ballast and waste ground around salt storage piles.

FAC-
[I]

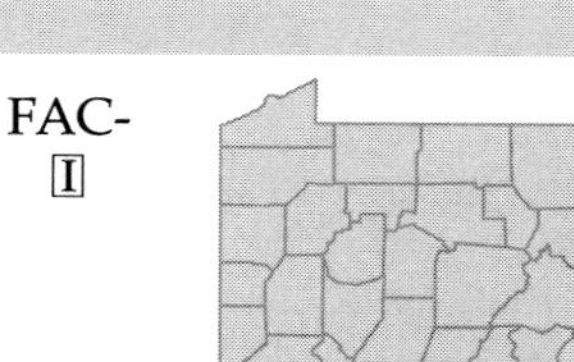

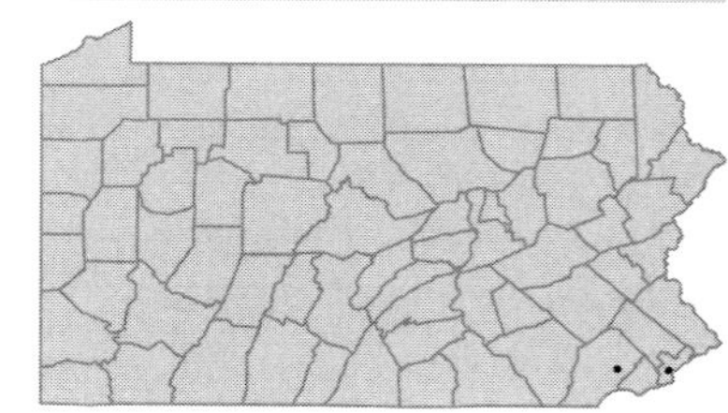

Atriplex littoralis L.
Narrow-leaved atriplex; Seashore orach
Herbaceous annual
Moist areas of waste ground. Believed to be extirpated, last collected in 1967.
Atriplex patula L. var. *littoralis* (L.) A.Gray FGBW

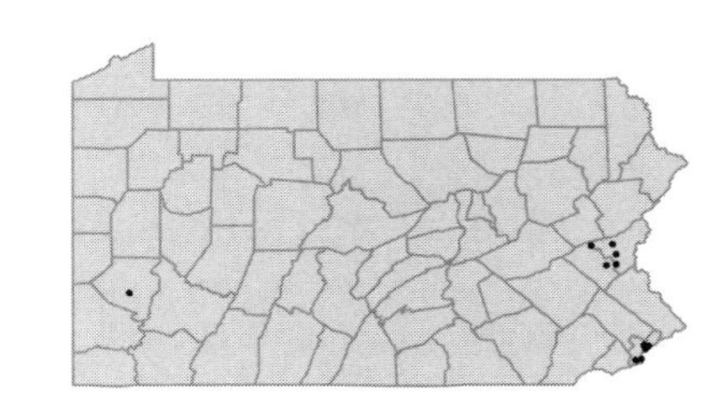

Atriplex patula L.
Spearscale; Spreading orach
Herbaceous annual
Brackish or rich soils, generally in waste ground.

FACW

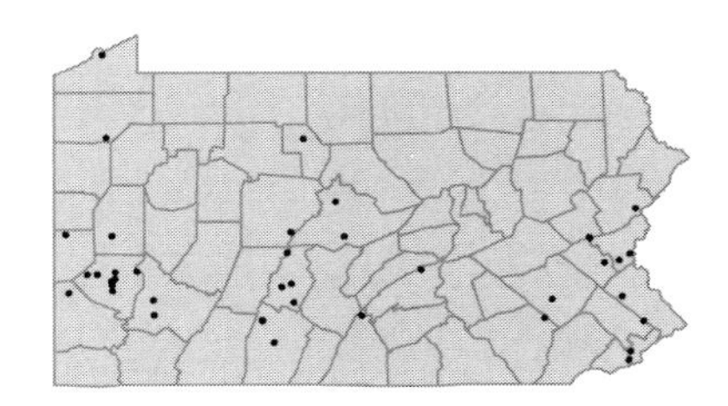

Atriplex prostrata Boucher ex DC.
Halberd-leaved orach
Herbaceous annual
Low, moist areas in brackish or rich soils.
Atriplex hastata sensu Aellen, *non* L. P; *Atriplex patula* L. var. *hastata auct., non* (L.) A.Gray FGBW

FACW

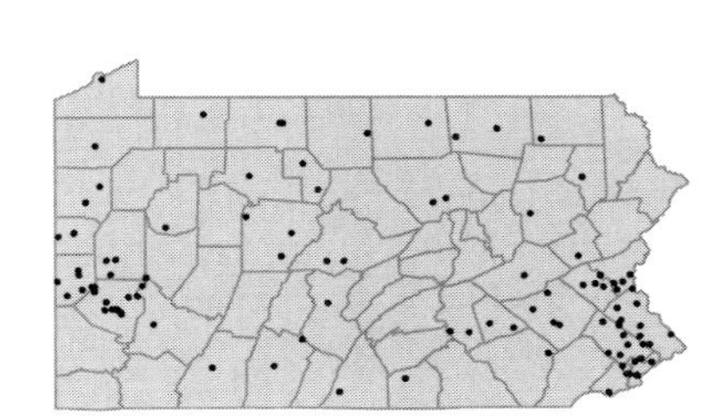

Atriplex rosea L.
Red orach
Herbaceous annual
Waste ground and ballast.

FACU
I

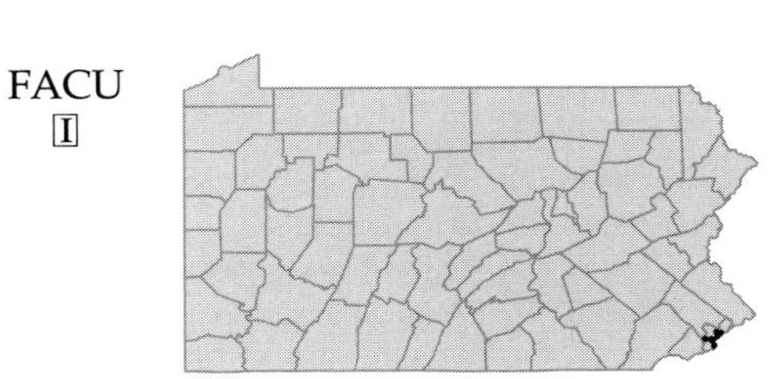

Atriplex tatarica L.
Orach
Herbaceous annual
Ballast. Collected at a single site in Philadelphia Co. in 1874.

I

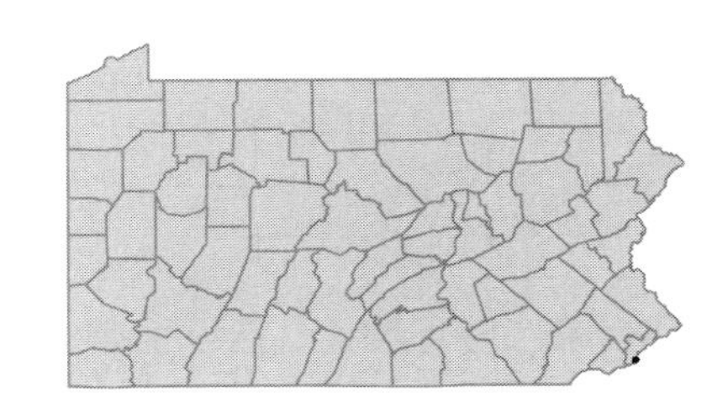

• ***Bassia hirsuta*** (L.) Aschers. ex Schwein.
Bassia
Herbaceous annual
Railroad ballast. Represented by a single collection from Bedford Co. in 1962.

OBL
I

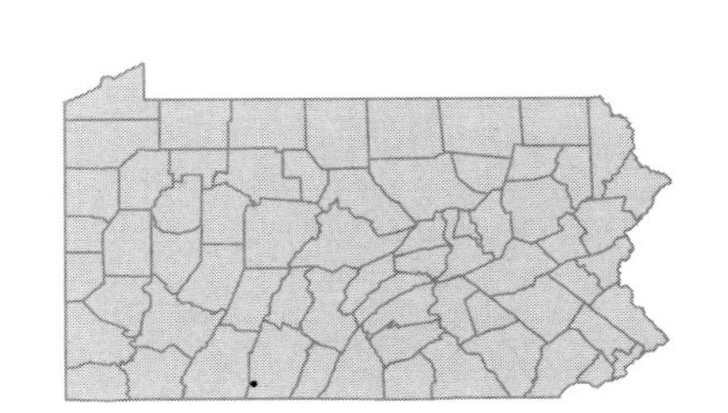

• ***Beta vulgaris*** L.
Beet
Herbaceous biennial
Rare, non-persistent, garden escape.

[I]

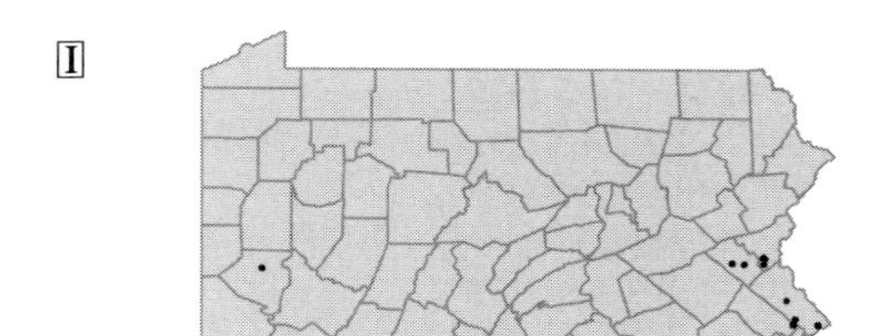

• ***Chenopodium album*** L. var. ***album***
Lamb's-quarters
Herbaceous annual
A weed of fields, gardens and disturbed soil.
Chenopodium lanceolatum Muhl. *pro parte* FGB

FACU+
[I]

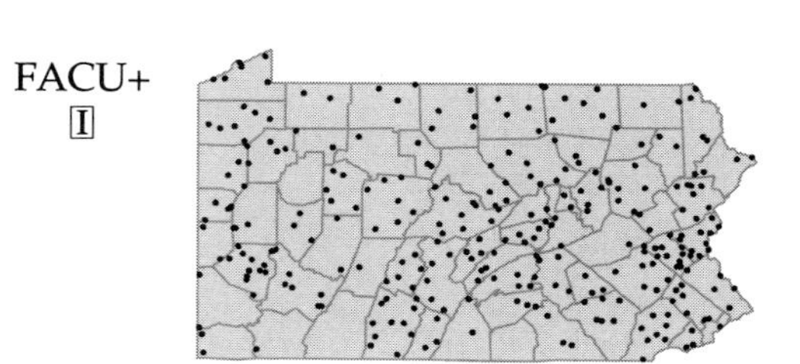

Chenopodium album L. var. ***missouriense*** (Aellen) I.J.Bassett & C.W.Crompton
Goosefoot
Herbaceous annual
Roadsides and waste ground, including heavily compacted soils.
Chenopodium missouriense Aellen WC

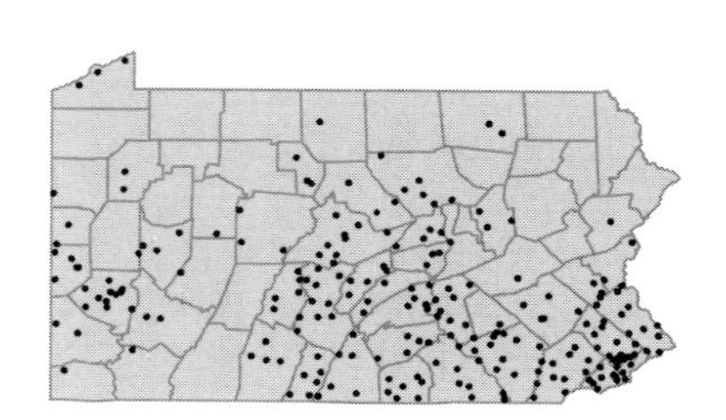

Chenopodium ambrosioides L.
Mexican-tea; Wormseed
Herbaceous annual
Fields and waste ground.
Chenopodium anthelminticum L. P

FACU
[I]

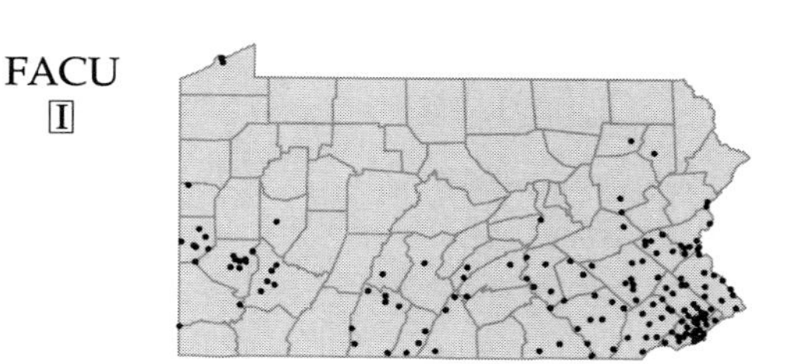

Chenopodium berlandieri Moq.
Goosefoot
Herbaceous annual
Disturbed, open ground.

[I]

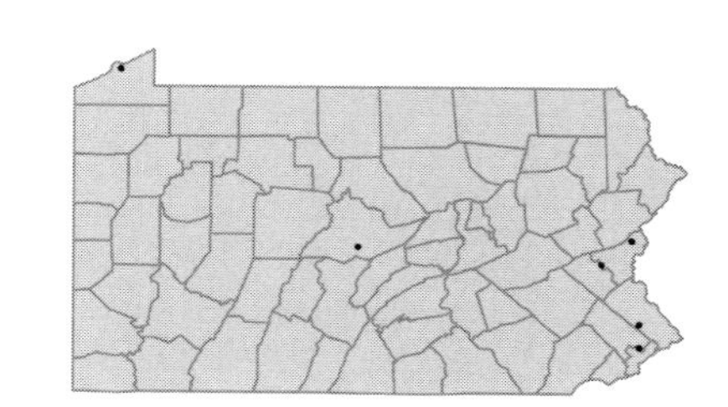

Chenopodium bonus-henricus L.
Good-King-Henry
Herbaceous perennial
Roadsides and disturbed ground.

[I]

Chenopodium botrys L.
Feather-geranium; Jerusalem-oak
Herbaceous annual
Waste ground, especially along rivers.

[I]

Chenopodium bushianum Aellen
Pigweed
Herbaceous annual
Cultivated and waste ground.
Chenopodium paganum Reichenb. F; *Chenopodium berlandieri* Moq. var. *bushianum* (Aellen) Cronq. C

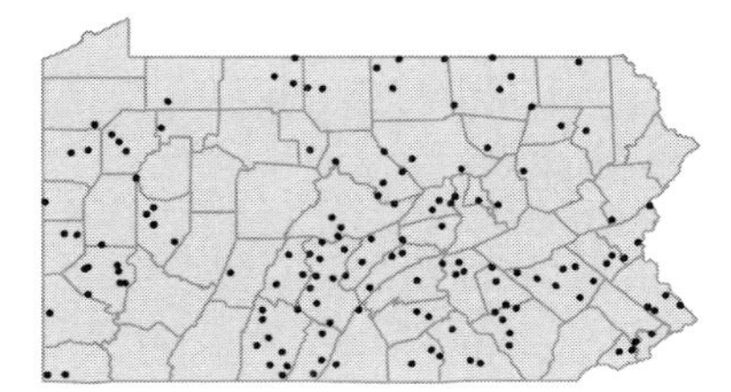

Chenopodium capitatum (L.) Aschers.
Indian-paint; Strawberry-blite
Herbaceous annual
Woodland clearings and burns, in light soil.
Blitum capitatum L. P

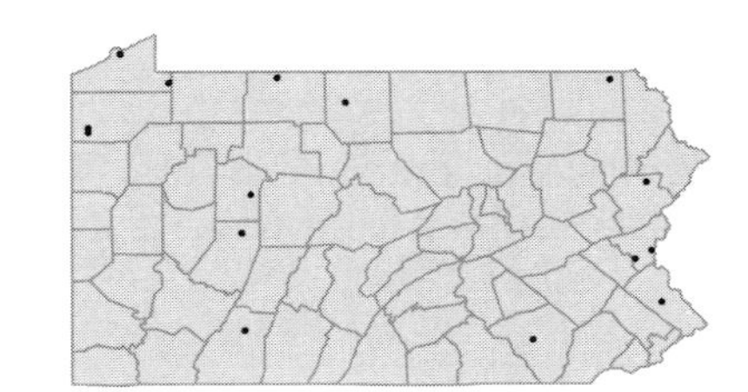

Chenopodium foggii H.A.Wahl
Goosefoot
Herbaceous annual
Dry, shaly slopes.
Chenopodium pratericola Rydb. *pro parte* C

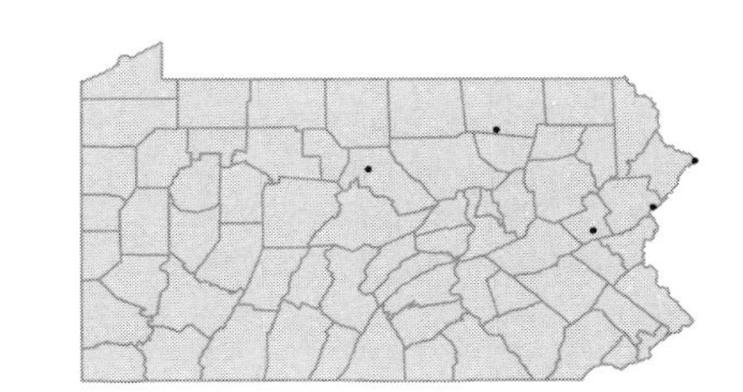

Chenopodium gigantospermum Aellen
Maple-leaved goosefoot
Herbaceous annual
Rocky woods, edges and waste places.
Chenopodium hybridum L. var. *gigantospermum* Aellen F; *Chenopodium hybridum* L. var. *hybridum* GB

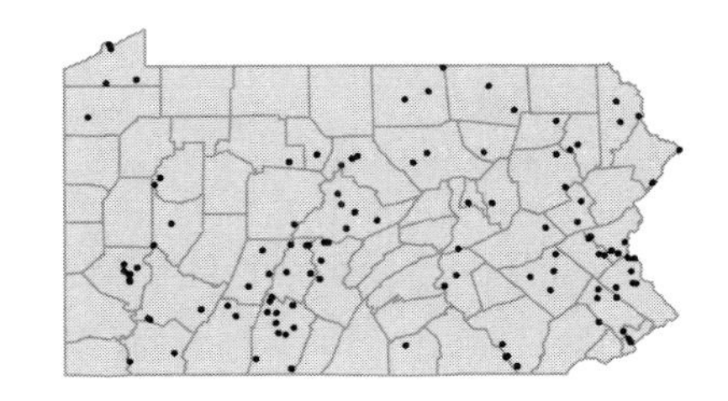

Chenopodium glaucum L.
Oak-leaved goosefoot
Herbaceous annual
Roadsides, gardens and cultivated land.

FACW-
[I]

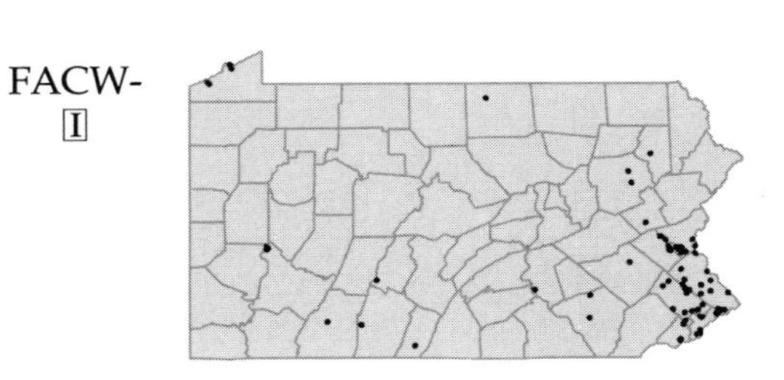

Chenopodium multifidum L. [I]
Cut-leaved goosefoot
Herbaceous annual
Waste ground and ballast.
Roubieva multifida (L.) Moq. PFGBW

Chenopodium murale L. [I]
Nettle-leaved goosefoot; Sowbane
Herbaceous annual
Waste ground.

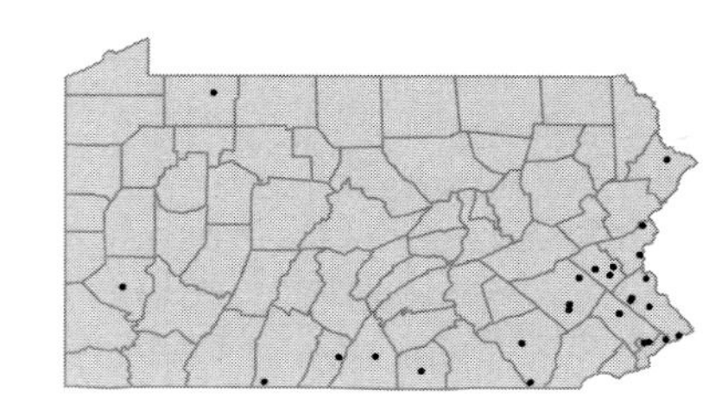

Chenopodium opulifolium Schrad. ex Koch & Ziz [I]
Goosefoot
Herbaceous annual
Ballast and waste ground. Represented by two collections from Philadelphia Co. 1865-1911.

Chenopodium polyspermum L. [I]
Many-seeded goosefoot
Herbaceous annual
Ballast. Represented by four collections from Philadelphia Co. 1877-1921.

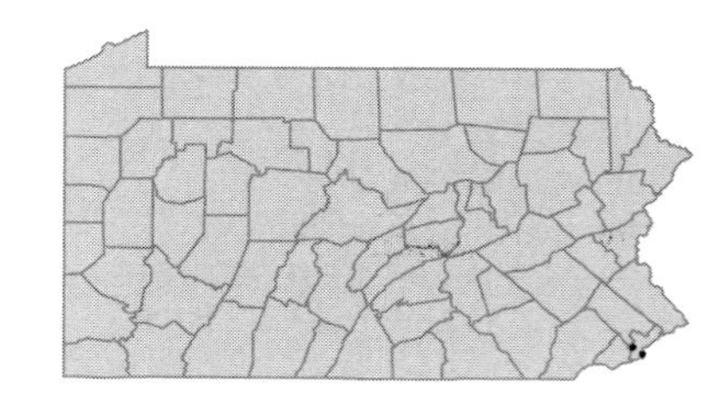

Chenopodium pratericola Rydb. [I]
Goosefoot
Herbaceous annual
Waste ground.
Chenopodium desiccatum A.Nels var. *leptophylloides* (J.Murr.) H.A.Wahl W

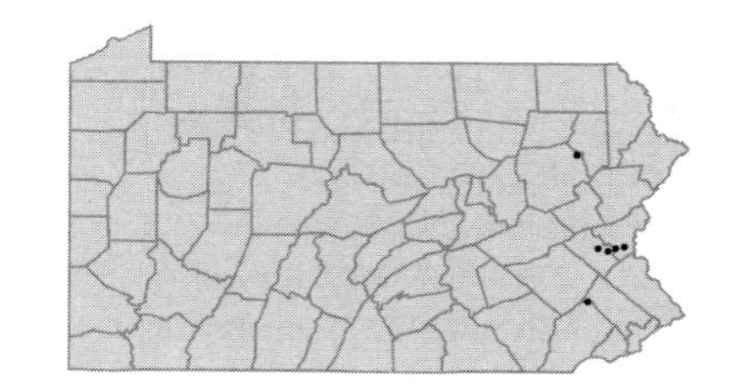

Chenopodium pumilio R.Br. [I]
Goosefoot
Herbaceous annual
Roadsides and waste ground.
Chenopodium carinatum R.Br. PFW

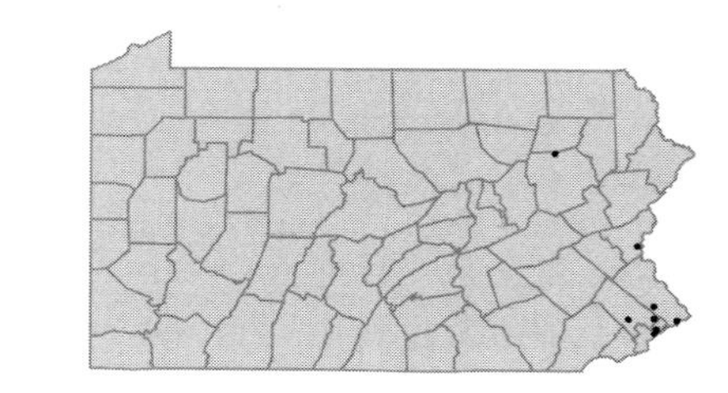

Chenopodium rubrum L.
Coast-blite; Red goosefoot
Herbaceous annual
Ballast or waste ground. Collected at a single site in Philadelphia Co. 1865-1877.

FACW
[I]

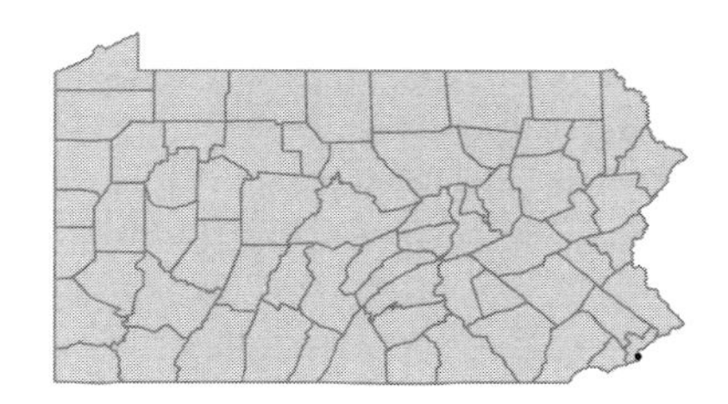

Chenopodium serotinum L.
Goosefoot
Herbaceous annual
Rubbish dumps and waste ground.

[I]

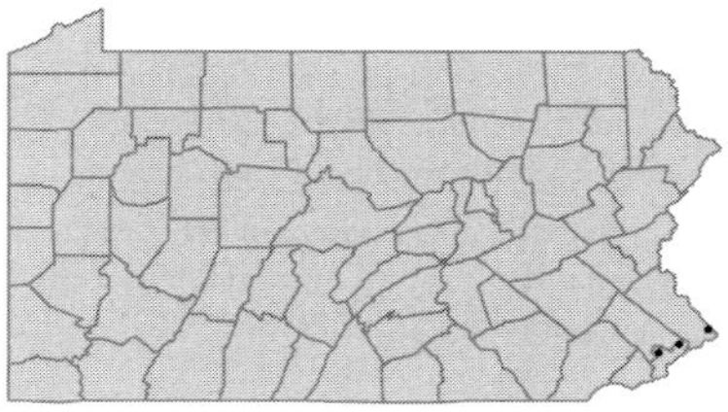

Chenopodium standleyanum Aellen
Goosefoot
Herbaceous annual
Dry, open woods and thickets.
Chenopodium hybridum L. var. *standleyanum* (Aellen) Fern. F

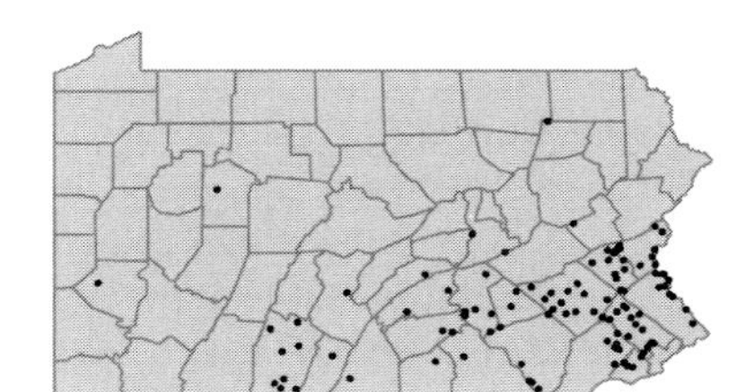

Chenopodium strictum Roth var. ***glaucophyllum*** (Aellen) H.A.Wahl
Goosefoot
Herbaceous annual
Waste ground and cinders.

[I]

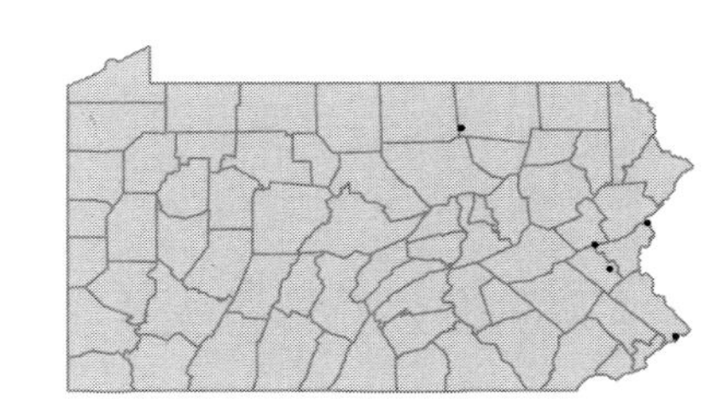

Chenopodium urbicum L.
City goosefoot; Upright goosefoot
Herbaceous annual
Waste ground.

[I]

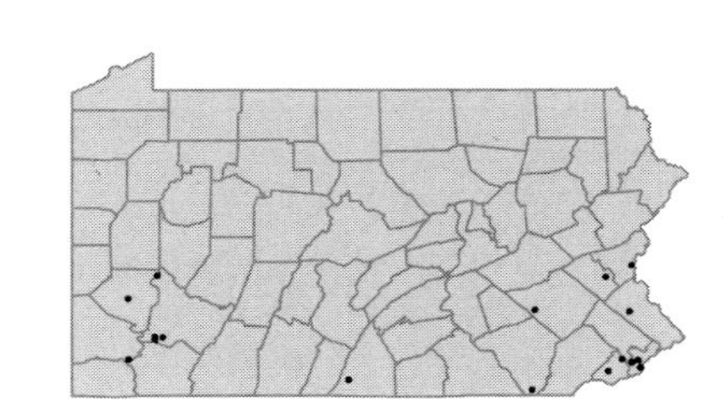

Chenopodium vulvaria L.
Stinking goosefoot
Herbaceous annual
Waste ground and ballast. Represented by four collections from Philadelphia Co. 1870-1920.

[I]

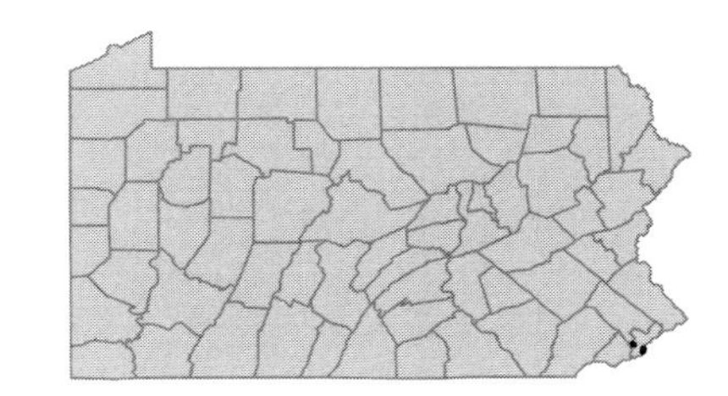

• ***Cycloloma atriplicifolium*** (Spreng.) Coult.
[I]
Winged pigweed
Herbaceous annual
Sandy, disturbed soil.

• ***Kochia scoparia*** (L.) Roth
[I]
Belvedere; Summer-cypress
Herbaceous annual
Garden escape, waste ground.

• ***Salsola kali*** L.
FACU
[I]
Salsola; Barilla
Herbaceous annual
Alluvial meadows, waste ground and railroad ballast. Although native to coastal areas, it also occurs as a waif in PA.
Salsola kali L. var. *caroliniana* (Walt.) Nutt. FGBW

Salsola tragus L.
[I]
Russian thistle
Herbaceous annual
Waste ground, especially in sandy soil.
Salsola kali L. var. *tenuifolia* Tausch FGBW; *Salsola iberica* Sennen & Pau.

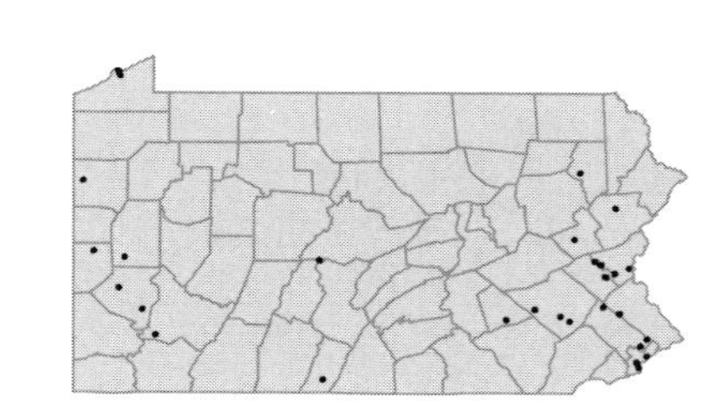

• ***Spinacia oleracea*** L.
[I]
Spinach
Herbaceous annual
Cultivated and rarely escaped. Represented by a single collection from Philadelphia Co. in 1899.

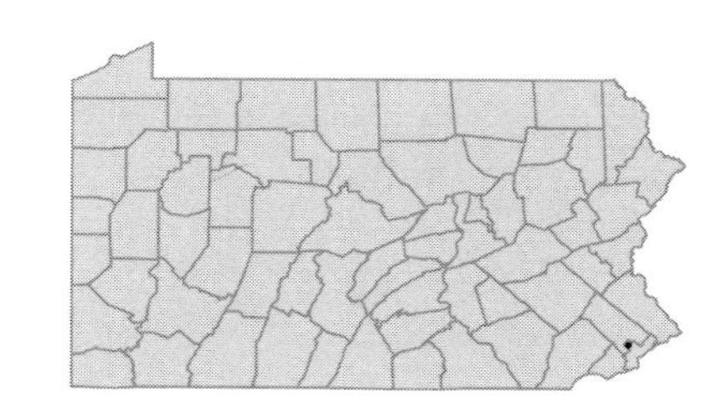

• ***Suaeda maritima*** (L.) Dumort. ssp. ***maritima***
OBL
[I]
Lesser sea-blite
Herbaceous annual
Ballast and waste ground.

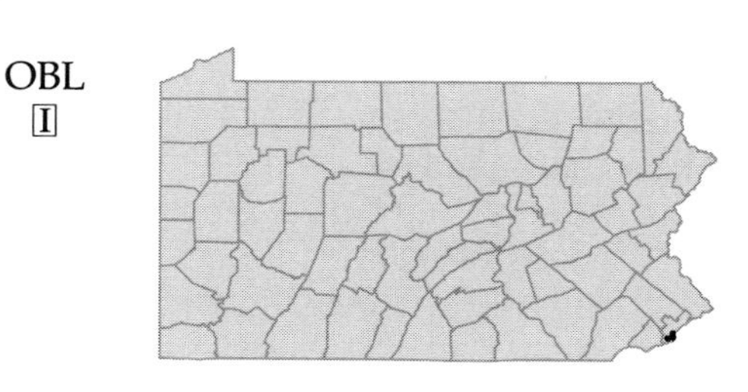

Suaeda maritima (L.) Dumort. ssp. ***richii*** (Fern.) Basset & Crompton
Sea-blite
Herbaceous annual
Ballast. Represented by a single collection from Philadelphia Co. in 1874.
Suaeda richii Fern. FGBW

OBL
[I]

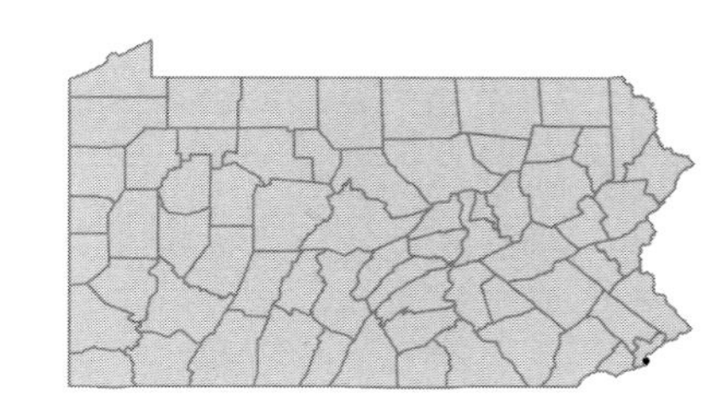

AMARANTHACEAE

• ***Alternanthera paronychioides*** St.Hil.
Alternanthera
Herbaceous annual
Ballast. Represented by a single collection from Philadelphia Co. in 1863.

[I]

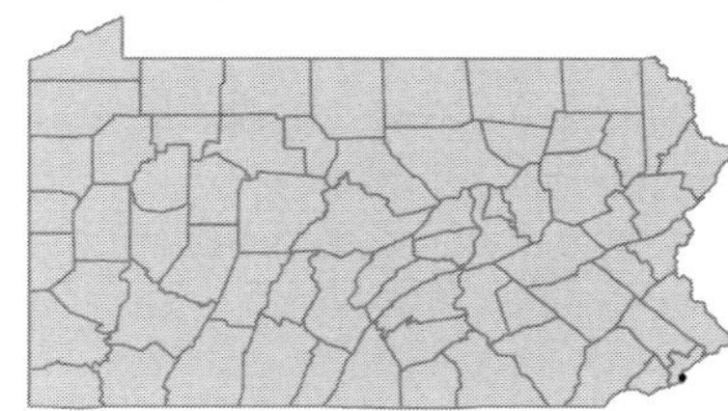

• ***Amaranthus albus*** L.
Tumbleweed
Herbaceous annual
A weed of disturbed ground.
Amaranthus graecizans sensu auctt., non L. FGBW

FACU

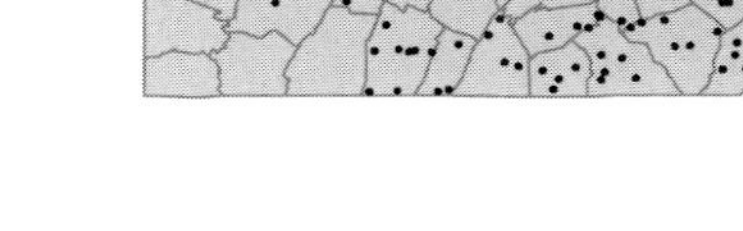

Amaranthus arenicola I.M.Johnston
Amaranth
Herbaceous annual
Waste ground, dump. Represented by a single collection each from Delaware and Allegheny Cos. 1941-1946.
Amaranthus torreyi (A.Gray) Benth. FBW

[I]

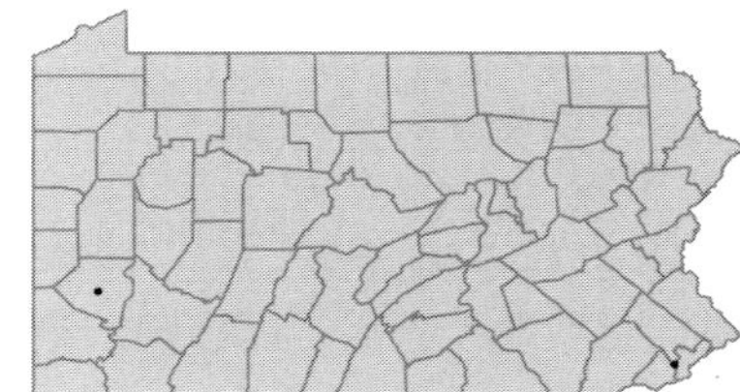

Amaranthus blitum L.
Amaranth
Herbaceous annual
A weed of waste ground.
Amaranthus lividus L. PFGBW

[I]

Amaranthus cannabinus (L.) Sauer
Saltmarsh water-hemp; Waterhemp ragweed
Herbaceous annual
Uppermost zone of freshwater intertidal marsh.
Acnida cannabina L. F

OBL

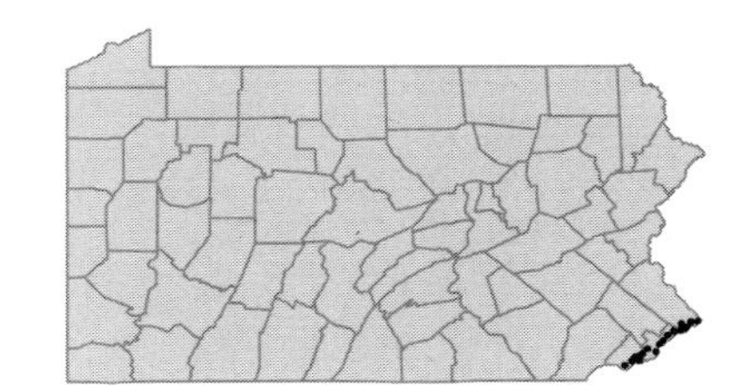

Amaranthus caudatus L.
Love-lies-bleeding
Herbaceous annual
Cultivated and rarely escaped.

[I]

Amaranthus cruentus L.
Blood amaranth; Purple amaranth
Herbaceous annual
An occasional weed of waste places.
Amaranthus hybridus L. ssp. *cruentus* GB

[I]

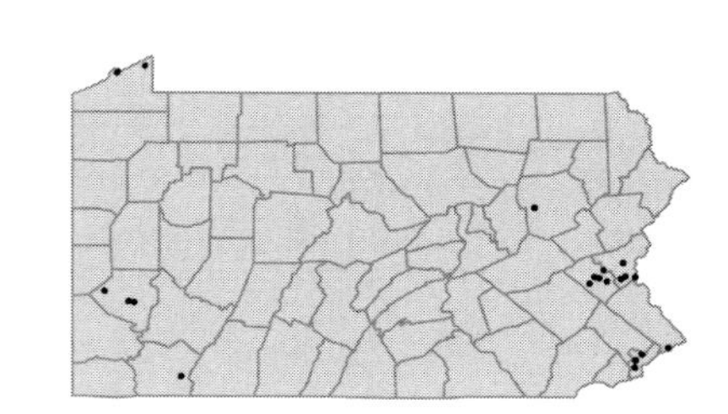

Amaranthus deflexus L.
Low amaranth
Herbaceous annual
Ballast.

[I]

Amaranthus hybridus L.
Pigweed
Herbaceous annual
A weed of cultivated fields, gardens and waste ground.

[I]

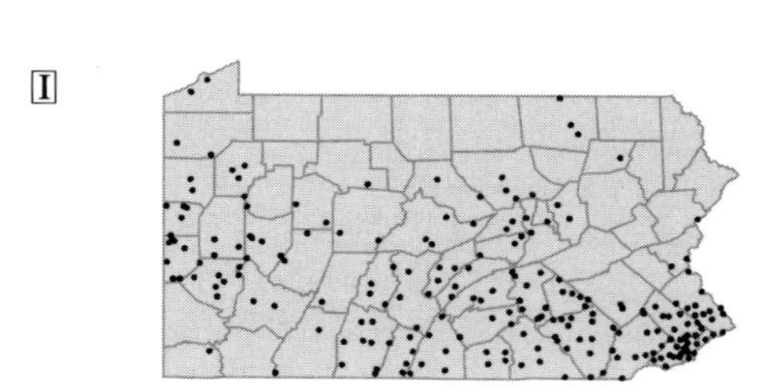

Amaranthus palmeri S.Wats.
Palmer's amaranth
Herbaceous annual
Waste ground and ballast.

FACU
[I]

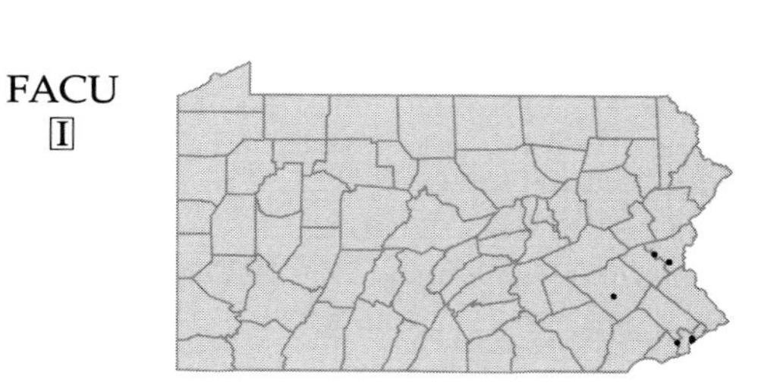

Amaranthus powellii S.Wats.
Amaranth
Herbaceous annual
Waste ground.

[I]

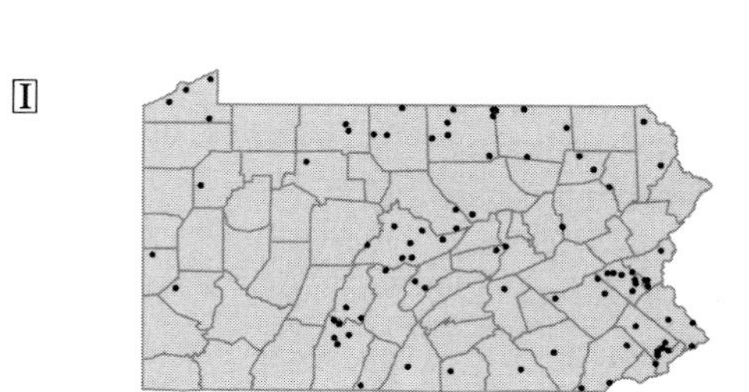

Amaranthus pumilus Raf.
Seabeach amaranth
Herbaceous annual
Ballast. Represented by a single collection from Philadelphia Co. in 1865, although native in coastal areas, adventive in PA.

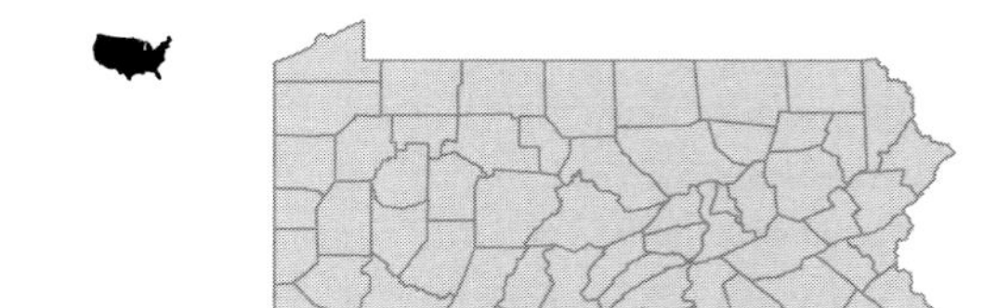

Amaranthus retroflexus L.
Green amaranth; Pigweed; Wild beet
Herbaceous annual
A weed of cultivated fields, gardens and waste ground.

FACU
I

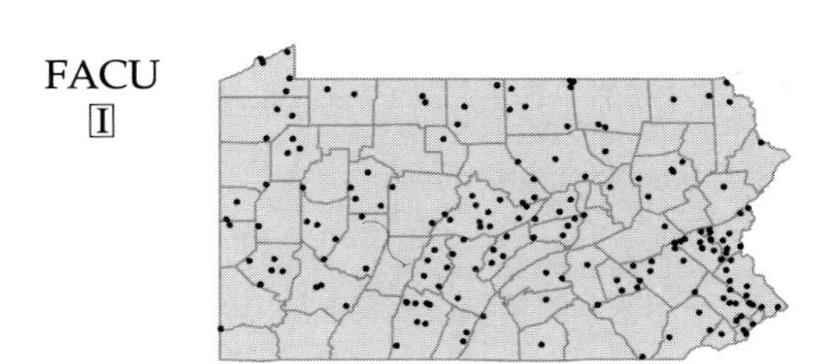

Amaranthus rudis Sauer
Western water-hemp
Herbaceous annual
Waste ground and ballast.
Acnida tamariscina (Nutt.) Wood FB; *Amaranthus tamariscinus* Nutt. GW

I

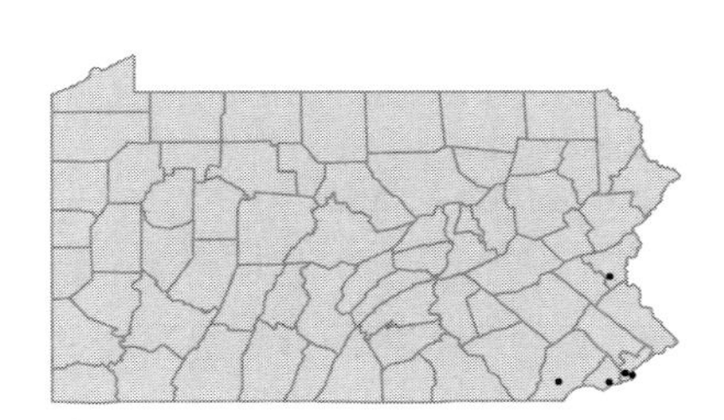

Amaranthus spinosus L.
Spiny amaranth
Herbaceous annual
Cultivated fields, waste ground and ballast.

FACU
I

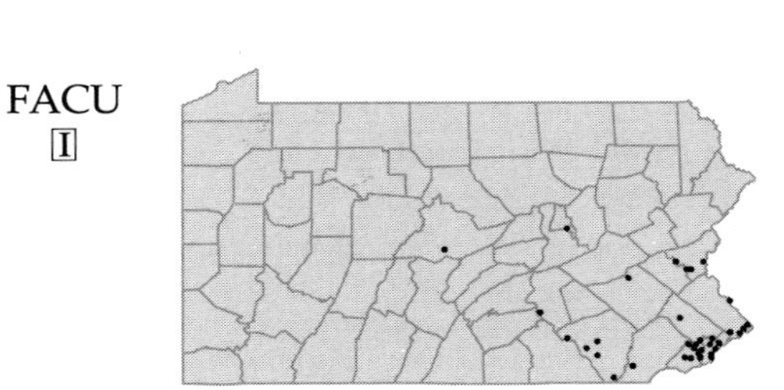

Amaranthus tuberculatus (Moq.) Sauer
Water-hemp
Herbaceous annual
River banks and low, disturbed ground.
Acnida altissima Riddell FB

I

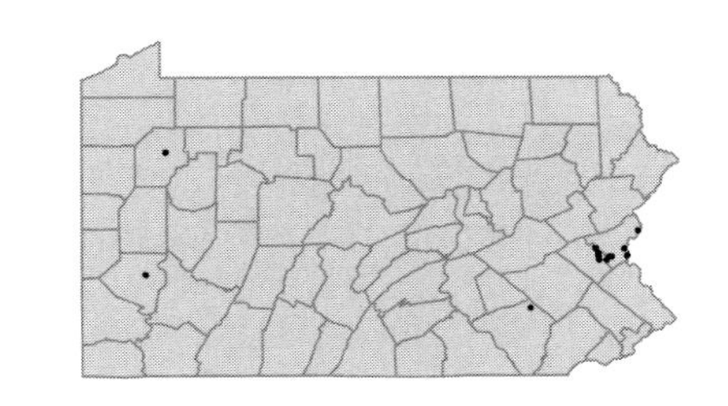

• ***Celosia argentea*** L.
Celosia
Herbaceous annual
Cultivated and occasionally escaped.

I

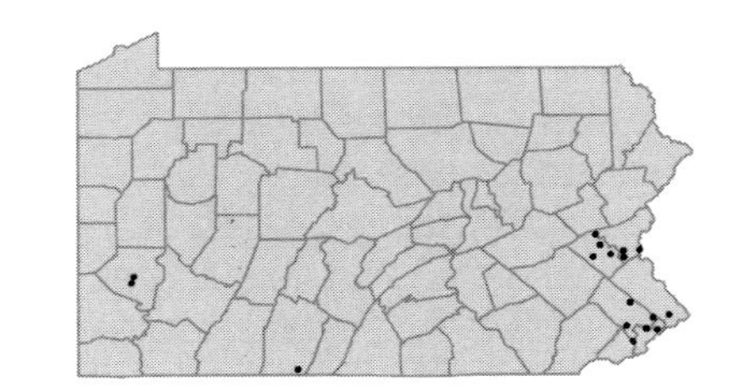

• ***Froelichia gracilis*** (Hook.) Moq. [I]
Cottonweed
Herbaceous annual
Dry, open, sandy soil, often along railroads.

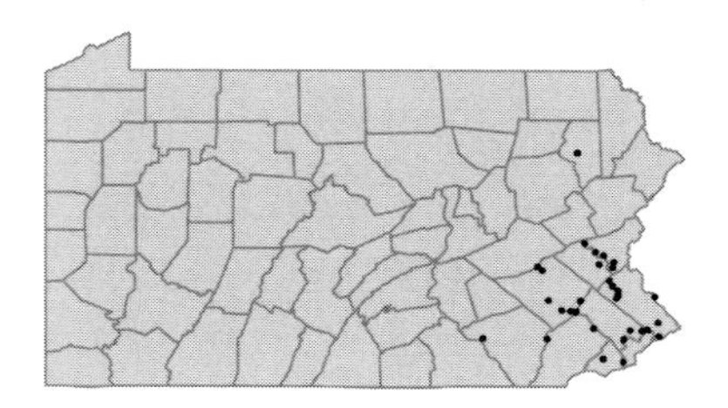

• ***Gomphrena globosa*** L. [I]
Globe-amaranth
Herbaceous annual
Ballast.

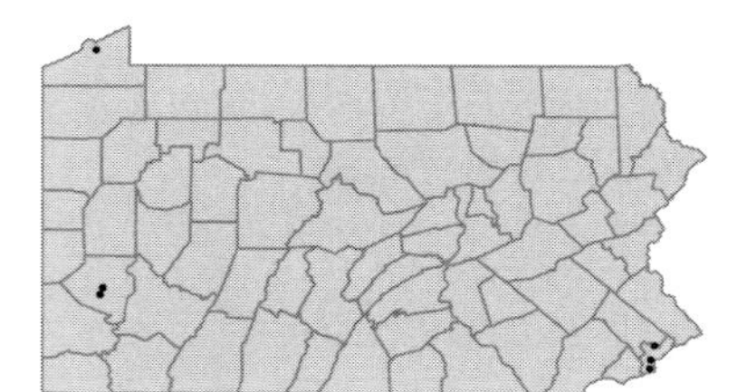

• ***Tidestromia lanuginosa*** (Nutt.) Standl. [I]
Tidestromia
Herbaceous annual
Railroad ballast. Represented by a single collection from Erie Co. in 1914.

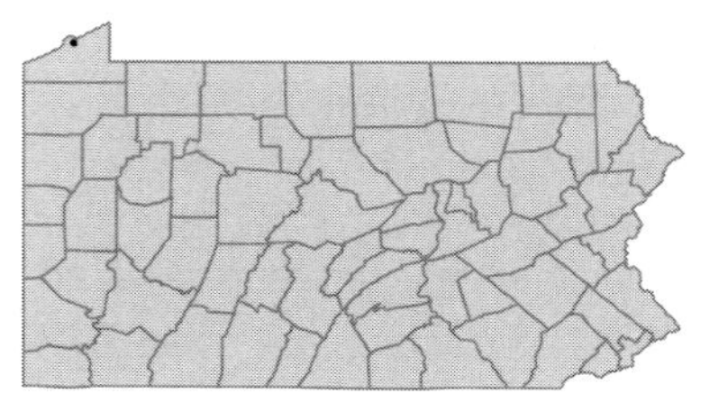

PORTULACACEAE

• ***Claytonia caroliniana*** Michx.
Carolina spring-beauty
Herbaceous perennial
Moist, rocky, wooded slopes.

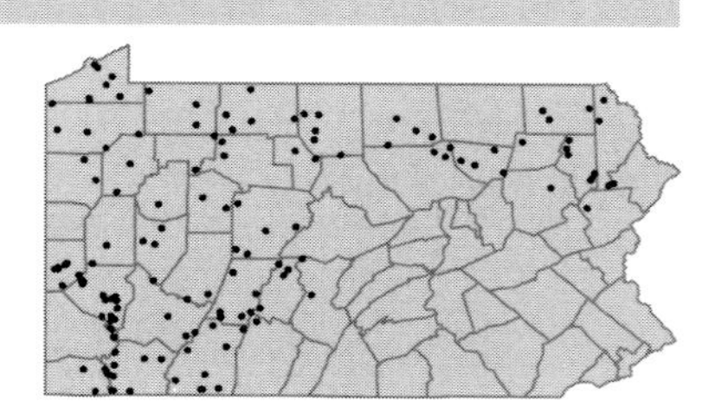

Claytonia virginica L.
Spring-beauty
Herbaceous perennial
Moist woods and meadows, frequently on alluvial soils.

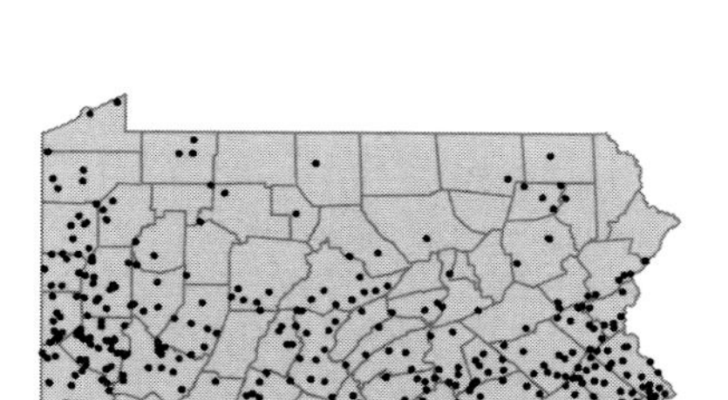

• ***Montia chamissoi*** (Ledeb. ex Spreng.) Greene
Chamisso's miners-lettuce
Herbaceous perennial
Moist, rocky ledges and river banks.

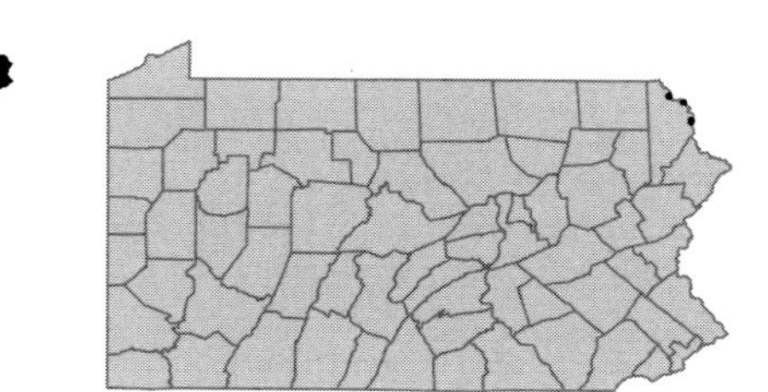

• ***Portulaca grandiflora*** Hook.
Moss-rose
Herbaceous annual
Cultivated and occasionally lingering in abandoned gardens or rubbish dumps.

I

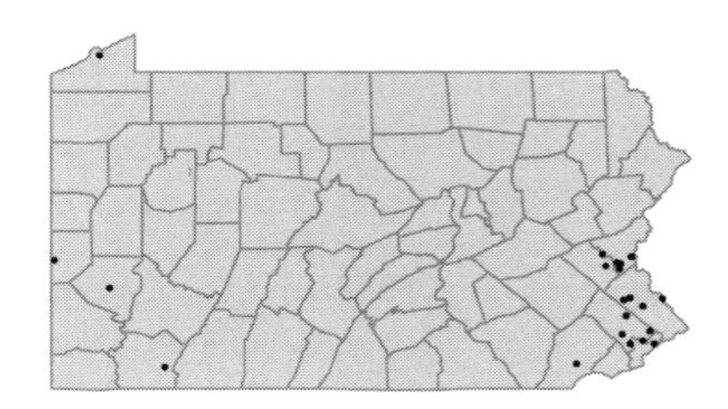

Portulaca oleracea L.
Purslane
Herbaceous annual
A weed of gardens and fields.

FAC

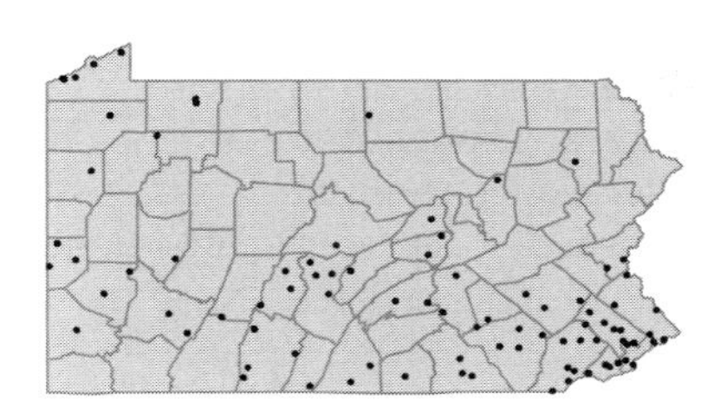

• ***Talinum teretifolium*** Pursh
Round-leaved fame-flower
Herbaceous perennial
Serpentine barrens, on rock outcrops.

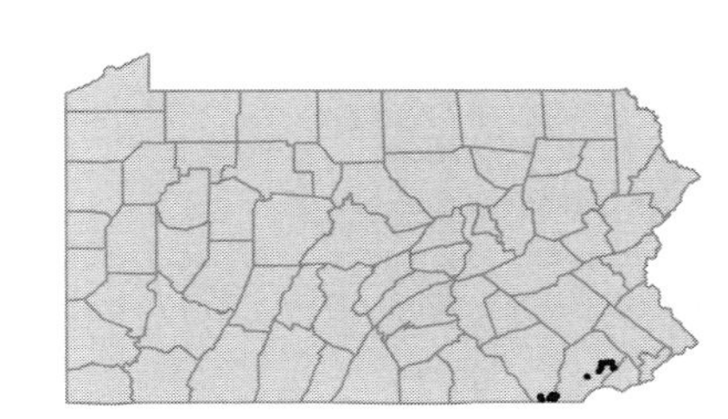

MOLLUGINACEAE

• ***Mollugo cerviana*** (L.) Ser.
Carpet-weed
Herbaceous annual
Ballast. Represented by a single collection from Philadelphia Co. ca 1879.

I

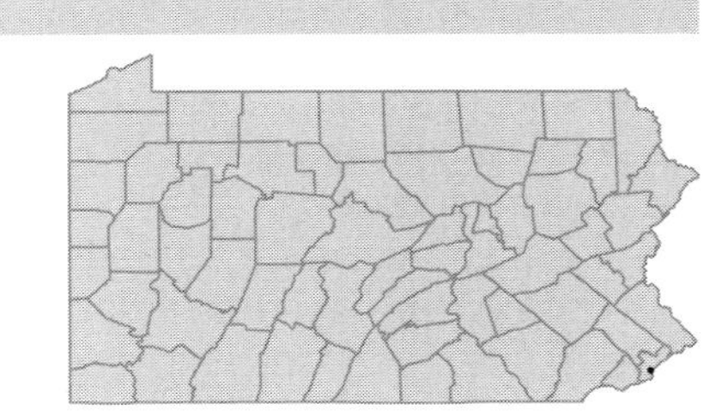

Mollugo verticillata L.
Carpet-weed
Herbaceous annual
A weed of waste ground, roadsides, and pavement cracks.

FAC
I

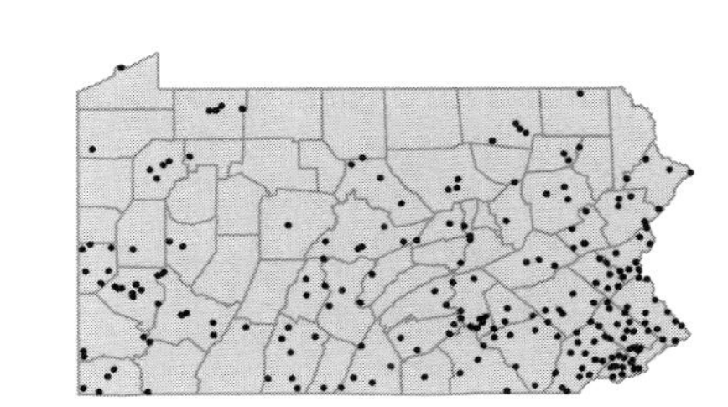

CARYOPHYLLACEAE

• ***Agrostemma githago*** L.
Corn cockle; Purple cockle
Herbaceous annual
Cultivated fields, roadsides and waste ground.

I

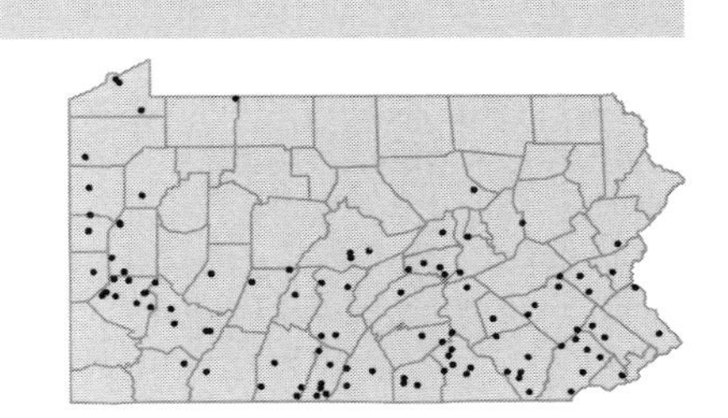

• ***Arenaria leptoclados*** (Reichenb.) Guss.
Slender sandwort
Herbaceous annual
Dry fields and roadsides.
Arenaria serpyllifolia L. var. *tenuior* Mert. & Koch FGBWC

[I]

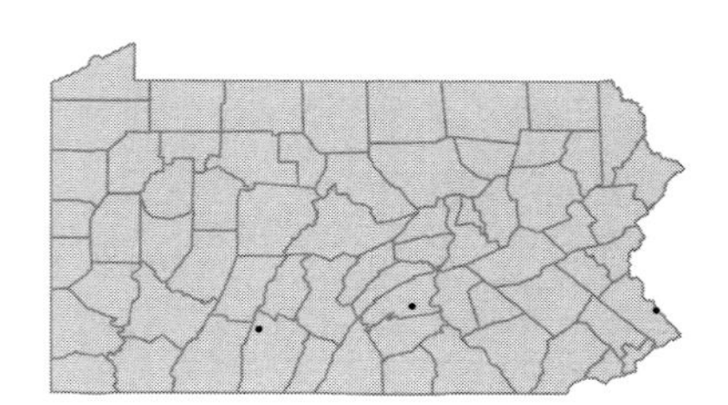

Arenaria serpyllifolia L.
Thyme-leaved sandwort
Herbaceous annual
A weed of dry, sterile fields, cultivated ground and roadsides.

FAC
[I]

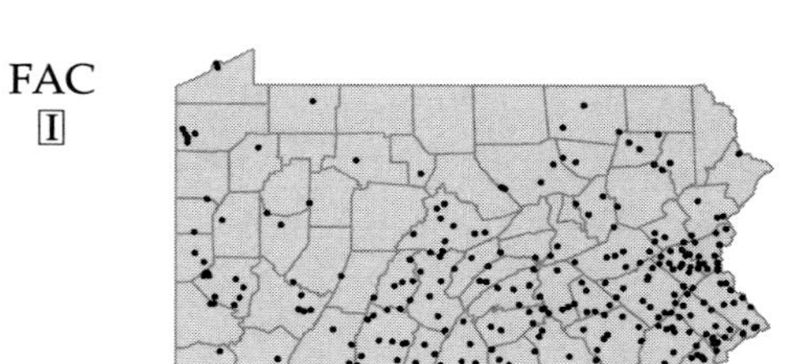

• ***Cerastium arvense*** L. var. ***arvense***
Field chickweed
Herbaceous perennial
Dry, rocky slopes or sandy fields.
Cerastium arvense L. var. *villosum* (Muhl.) Hollick & Britt. *pro parte* FBW

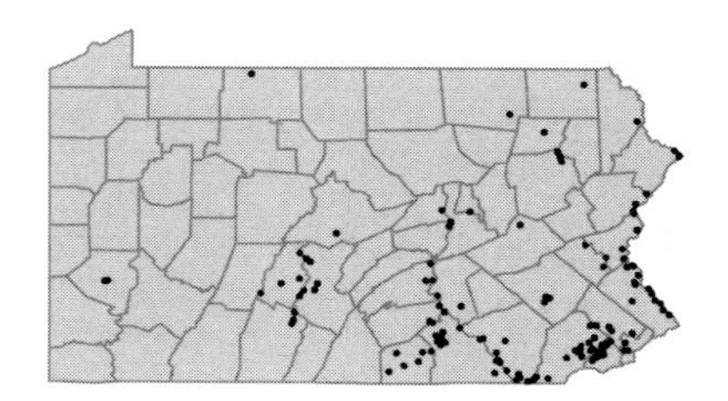

Cerastium arvense L. var. ***villosissimum*** Pennell
Serpentine barrens chickweed
Herbaceous perennial
Grassy openings on serpentine barrens. This taxon should probably be placed in *C. velutinum* Raf. see Morton (1987).

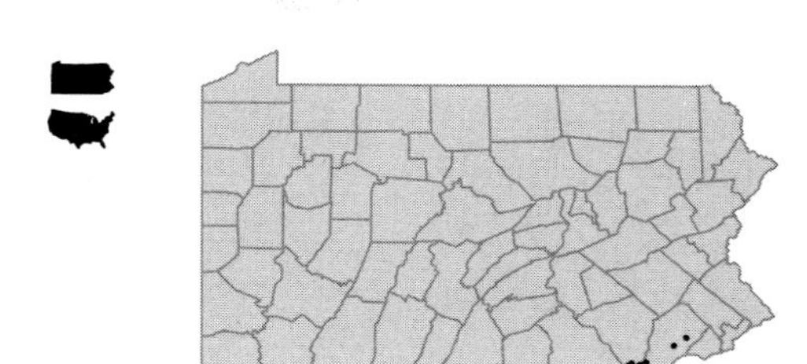

Cerastium arvense L. var. ***viscidulum*** Gremli
Field chickweed
Herbaceous perennial
Waste ground at site of zinc mine. All collections are from a single site in Lehigh Co. 1912-1955.

[I]

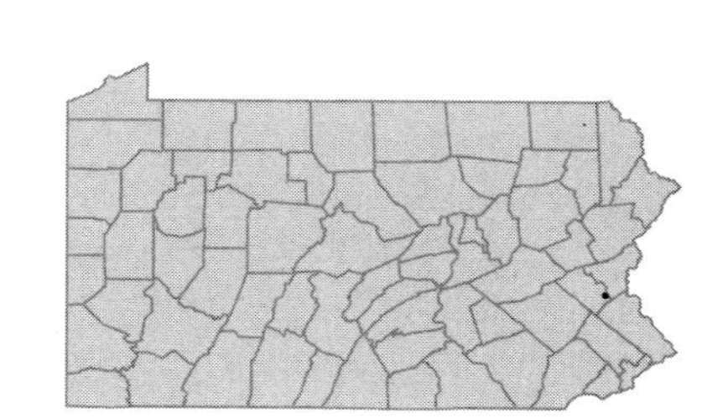

Cerastium fontanum Baumg. ssp. ***triviale*** (Link) Jalas
Common mouse-ear chickweed
Herbaceous perennial
A weed of cultivated ground.
Cerastium holosteoides Fries W; *Cerastium vulgatum* L. FGBC

FACU-
[I]

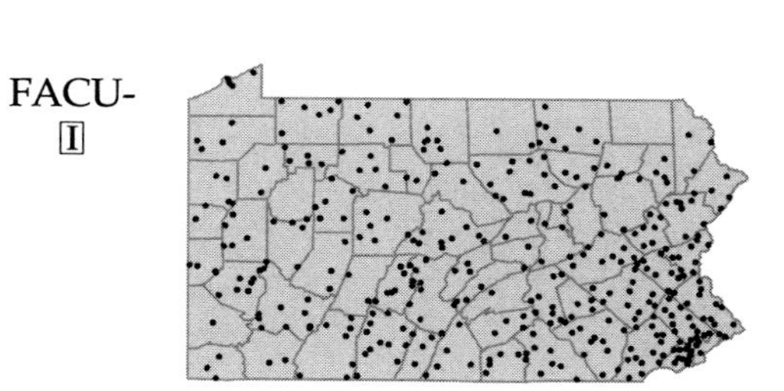

Cerastium glomeratum Thuill.
Mouse-ear chickweed
Herbaceous annual
Fields, roadsides and waste ground.
Cerastium viscosum L. PFGBC

I

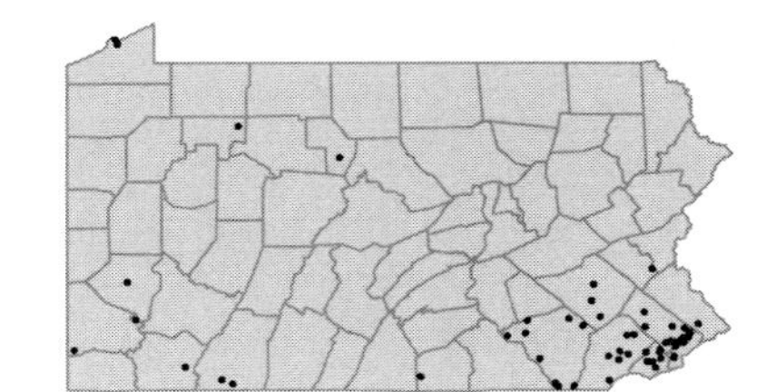

Cerastium nutans Raf.
Nodding chickweed
Herbaceous annual
Rich wooded slopes and alluvium.
Cerastium longipedunculatum Muhl. P

FAC

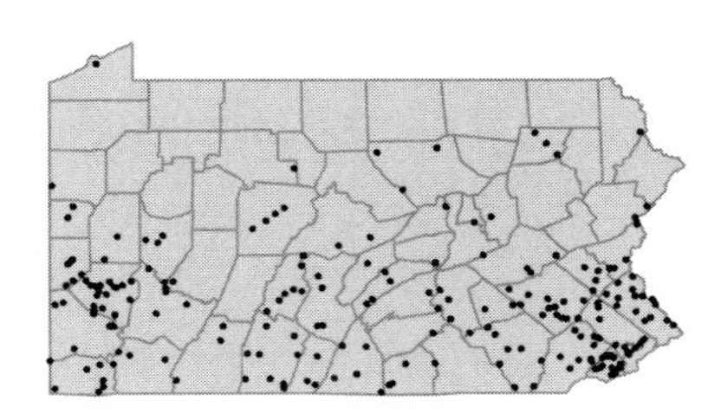

Cerastium semidecandrum L.
Small mouse-ear chickweed
Herbaceous annual
Dry, sandy woodland edges, alluvial shores and beaches.

I

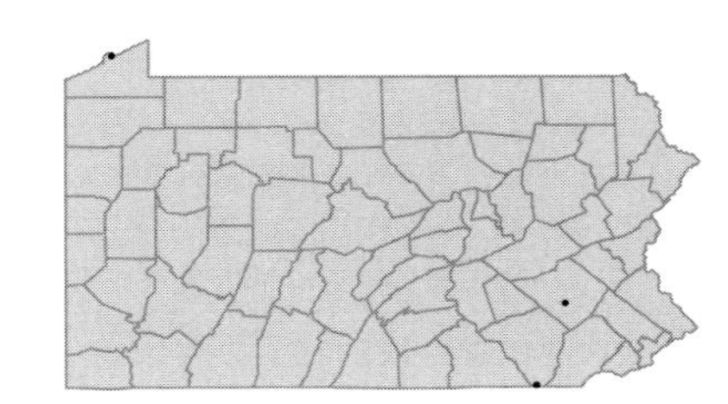

Cerastium tomentosum L.
Snow-in-summer
Herbaceous perennial
Cultivated and sometimes escaped.

I

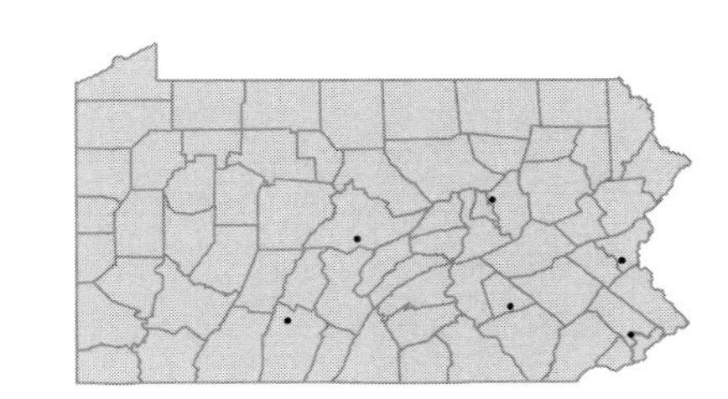

• ***Corrigiola littoralis*** L.
Strapwort
Herbaceous annual
Ballast. Represented by pre-1900 collections from Philadelphia and Delaware Cos.

I

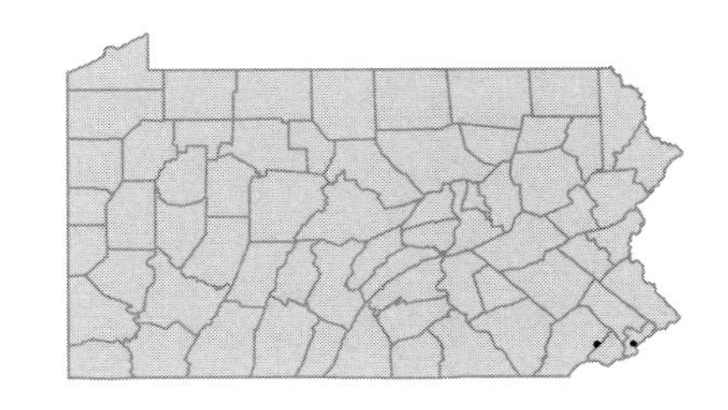

• ***Dianthus armeria*** L.
Deptford pink
Herbaceous biennial
Old fields, waste ground and roadsides.

I

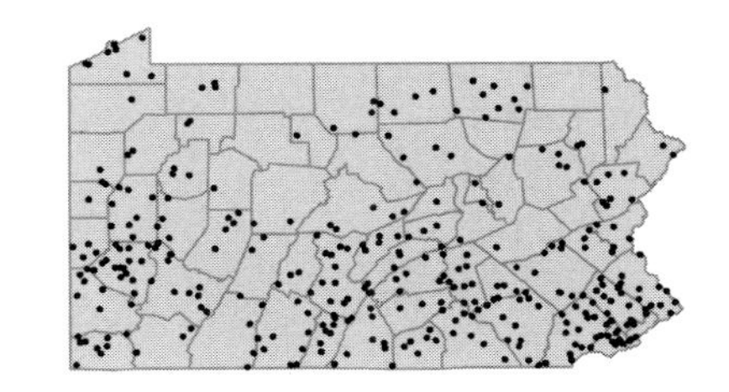

Dianthus barbatus L.
Sweet-william
Herbaceous perennial
Cultivated and sometimes escaped.

I

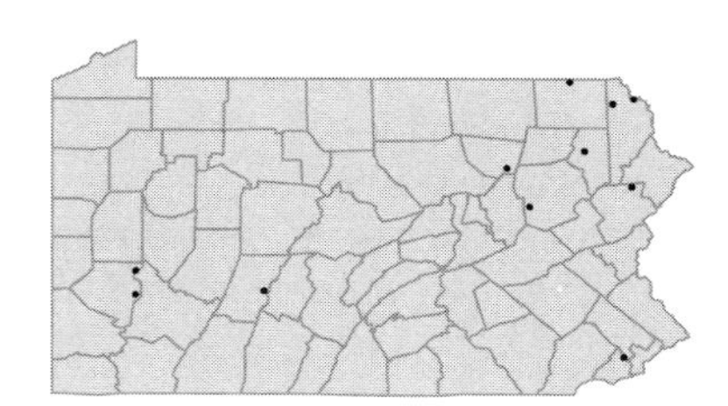

Dianthus deltoides L.
Maiden pink; Meadow pink
Herbaceous perennial
Dry fields and alluvial sand and gravel bars.

I

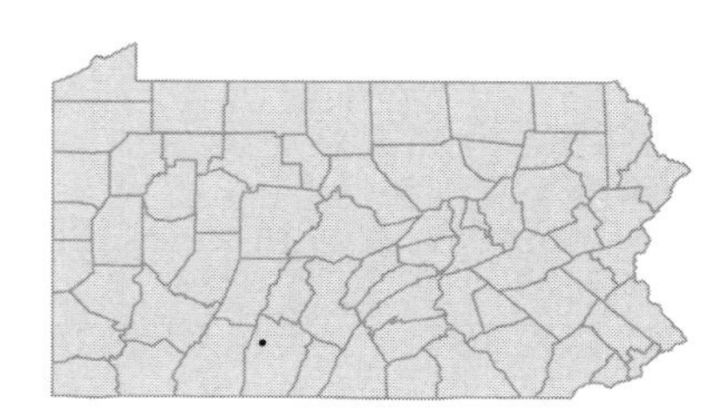

Dianthus plumarius L.
Garden pink
Herbaceous perennial
Roadside bank. Represented by a single collection from Bedford Co. in 1949.

I

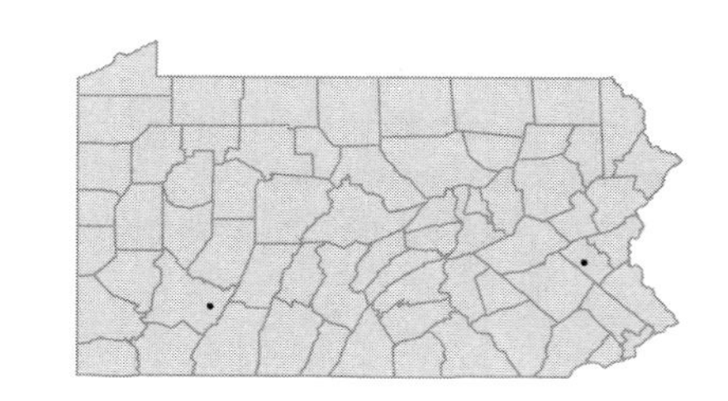

• ***Gypsophila elegans*** Bieb.
Baby's-breath
Herbaceous annual
Waste ground, dump.

I

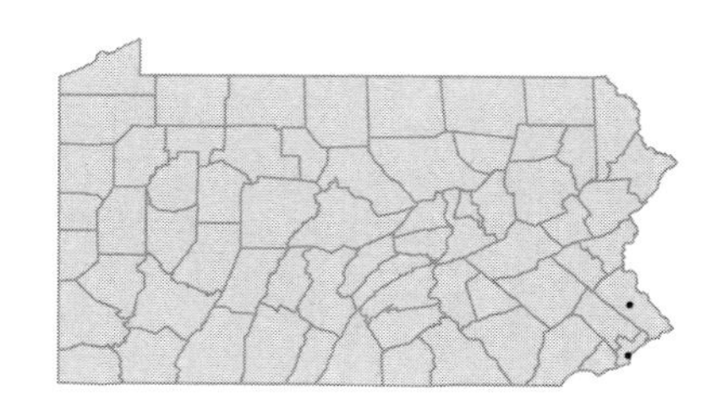

Gypsophila muralis L.
Baby's-breath; Mist
Herbaceous annual
Cultivated, one 1866 collection from waste ground in Philadelphia Co.

I

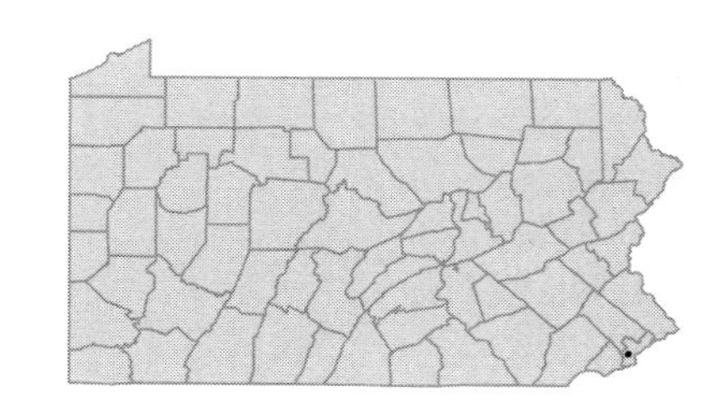

• ***Herniaria glabra*** L.
Herniary
Herbaceous annual
Ballast and waste ground. Represented by two collections from Philadelphia Co. in 1880 and 1960.

I

Herniaria hirsuta L. [I]
Herniary
Herbaceous annual
Ballast and waste ground.

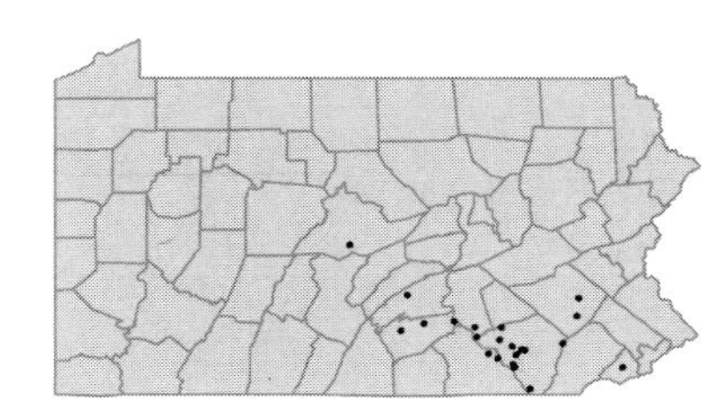

• ***Holosteum umbellatum*** L. [I]
Jagged chickweed
Herbaceous annual
Sandy railroad bank and waste ground.

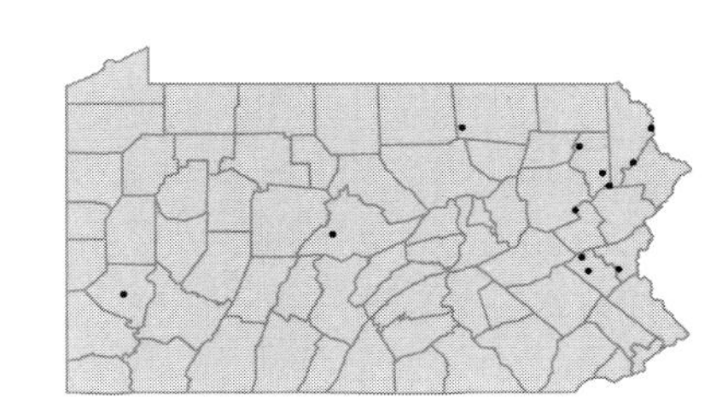

• ***Lychnis chalcedonica*** L. [I]
Maltese-cross
Herbaceous perennial
Cultivated and sometimes persisting in old gardens and fields.

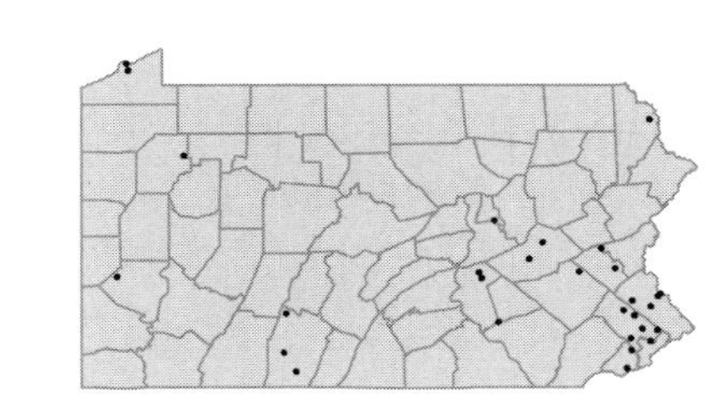

Lychnis coronaria (L.) Desr. [I]
Mullein pink; Rose-campion
Herbaceous perennial
Pastures, fields, cemeteries and dumps.

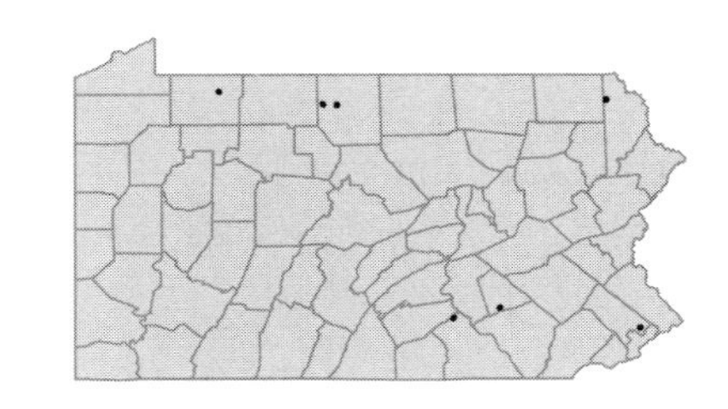

Lychnis flos-cuculi L. [I]
Ragged-robin
Herbaceous perennial
Fields, pastures and roadsides.

Lychnis fulgens Fisch. ex Sims [I]
Brilliant campion
Herbaceous perennial
Alluvial thickets and woods. Represented by two collections from Lackawanna Co. in 1946.

• ***Minuartia glabra*** (Michx.) Mattf.
Appalachian sandwort
Herbaceous annual
Exposed sandstone rocks.
Arenaria glabra Michx. W; *Arenaria groenlandica* (Retz.) Spreng. var. *glabra* (Michx.) Fern. FGB

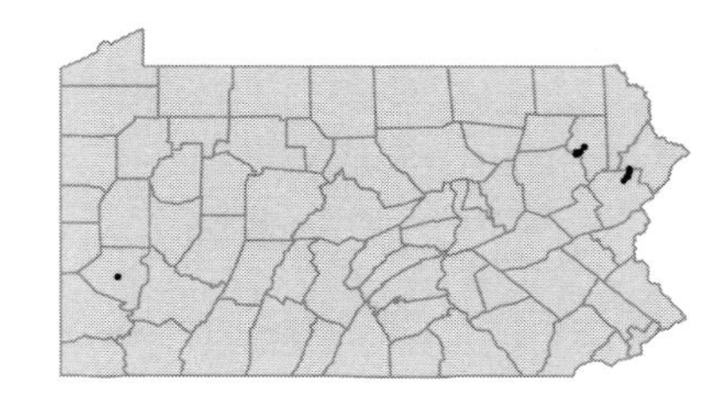

Minuartia michauxii (Fern.) Farw.
Rock sandwort
Herbaceous annual
Dry, open, rocky ledges of limestone or serpentine.
Arenaria stricta Michx. FGBWC; *Arenaria michauxii* (Fenzl) Hook.f. P

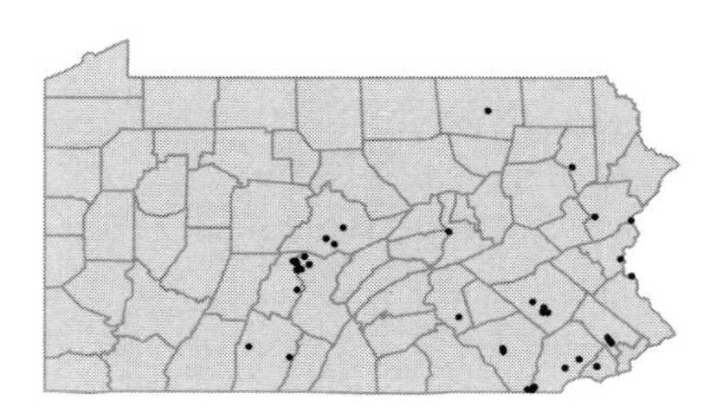

Minuartia patula (Michx.) Mattf.
Sandwort
Herbaceous annual
Rocky slopes, alluvial shores and roadsides.
Arenaria patula Michx. FGBWC

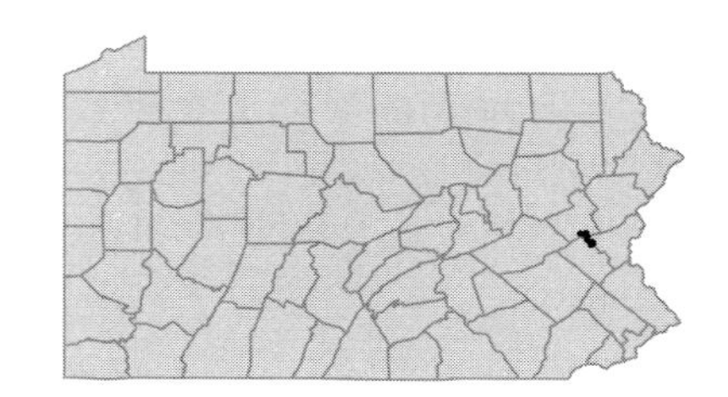

• ***Moehringia lateriflora*** (L.) Fenzl
Blunt-leaved sandwort
Herbaceous perennial
Wet meadows, swamps, swales and low woods.
Arenaria lateriflora L. FGBC

FAC

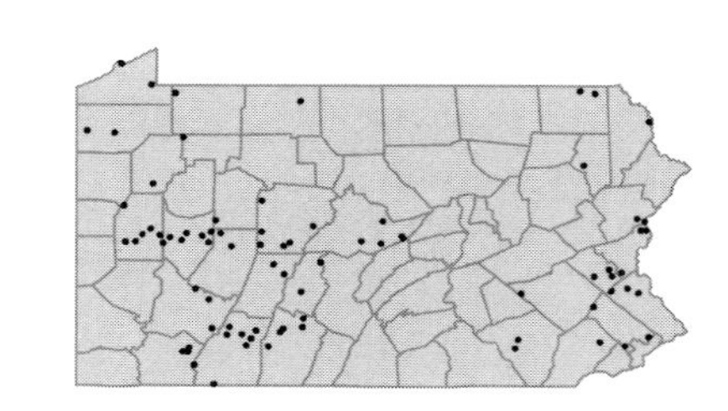

• ***Moenchia erecta*** (L.) Gaertn., Meyer & Scherbius
Moenchia
Herbaceous annual
Waste ground. Represented by a single collection from Philadelphia Co. in 1831.

[I]

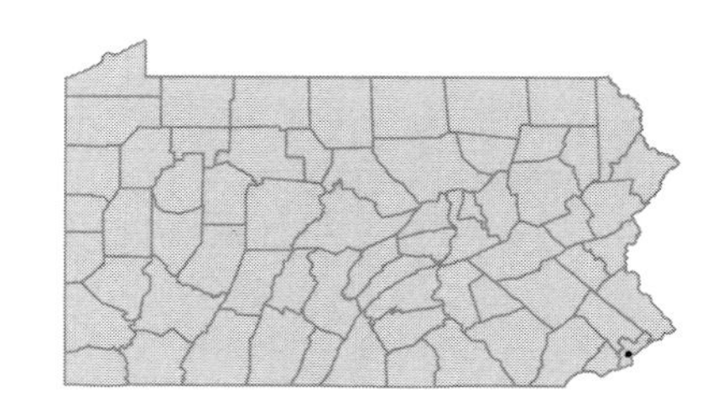

• ***Myosoton aquaticum*** (L.) Moench
Giant chickweed; Water mouse-ear
Herbaceous perennial
Seepy slopes, stream banks, alluvial woods and roadsides.
Alsine aquaticum (L.) Britt. P

FACW
[I]

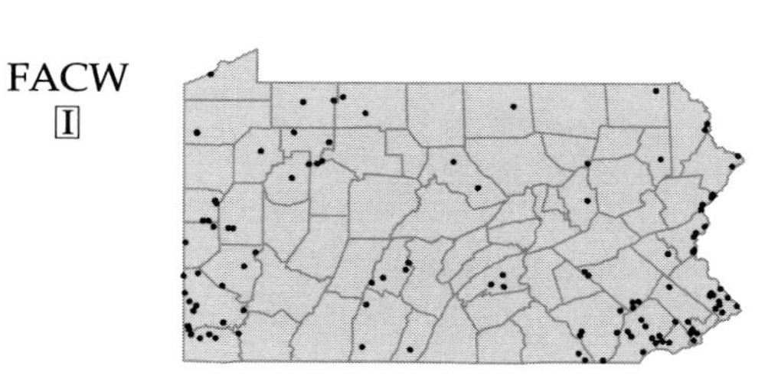

• ***Paronychia canadensis*** (L.) Wood
Forked chickweed
Herbaceous annual
Open woods, in dry rocky or sandy soil.
Anychia canadensis (L.) BSP P

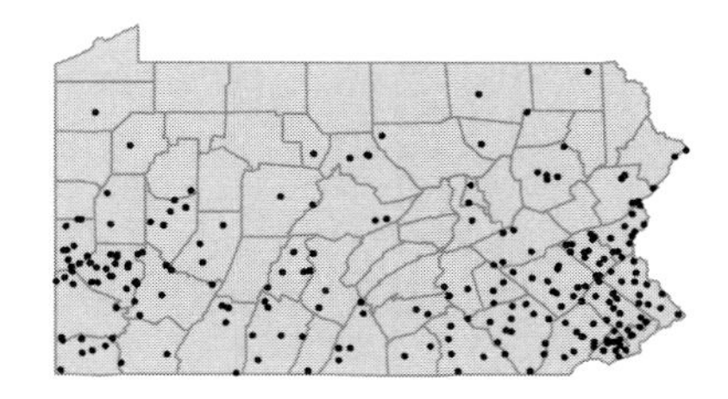

Paronychia fastigiata (Raf.) Fern. var. ***fastigiata***
Forked chickweed; Chaffy whitlow-wort
Herbaceous annual
Dry woods and edges.
Paronychia fastigiata (Raf.) Fern. var. *paleacea* FGBW

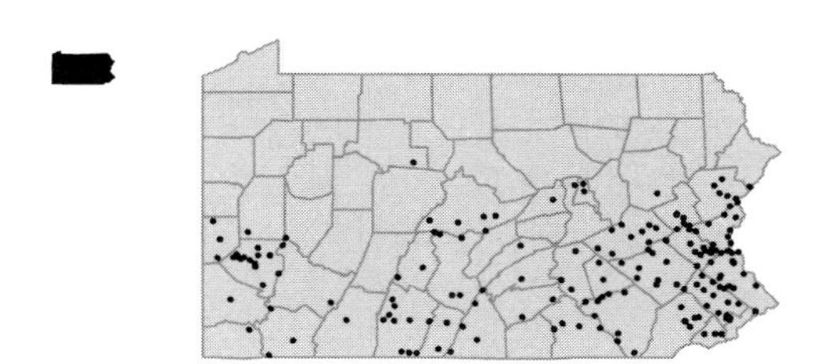

Paronychia fastigiata (Raf.) Fern. var. ***nuttallii*** (Small) Fern.
Forked chickweed; Whitlow-wort
Herbaceous annual
Open woods and edges, in dry rocky or sandy soils.

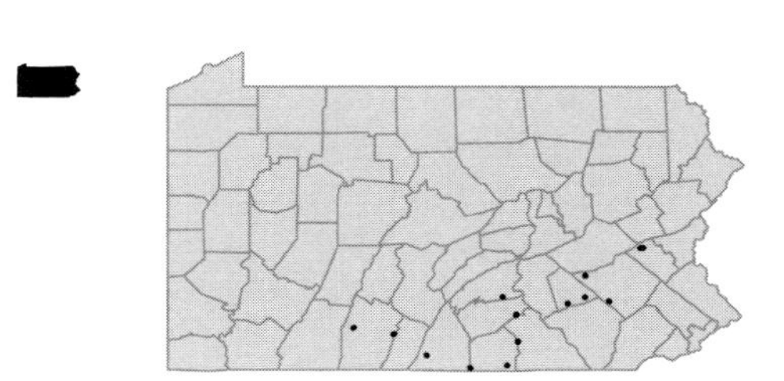

Paronychia montana (Small) Pax & Hoffm.
Forked chickweed
Herbaceous annual
Dry woods and openings, in rocky or sandy soil.
Paronychia fastigiata (Raf.) Fern. var. *pumila* (A.Wood) Fern. FGBC

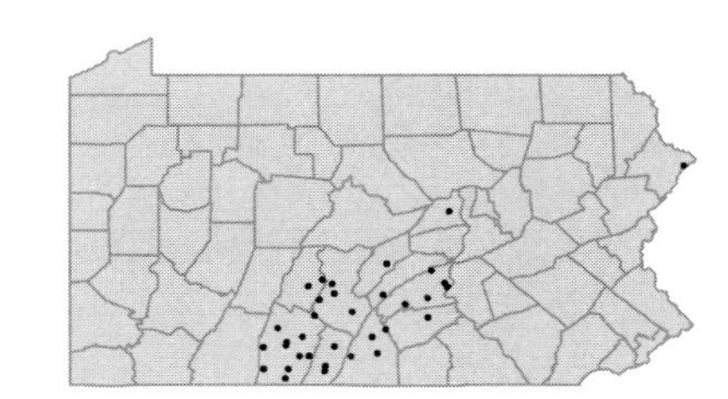

• ***Petrorhagia prolifera*** (L.) Ball & Heywood
I
Childing pink
Herbaceous annual
Dry fields and banks.
Dianthus prolifer L. PF; *Tunica prolifer* (L.) Scop. GB

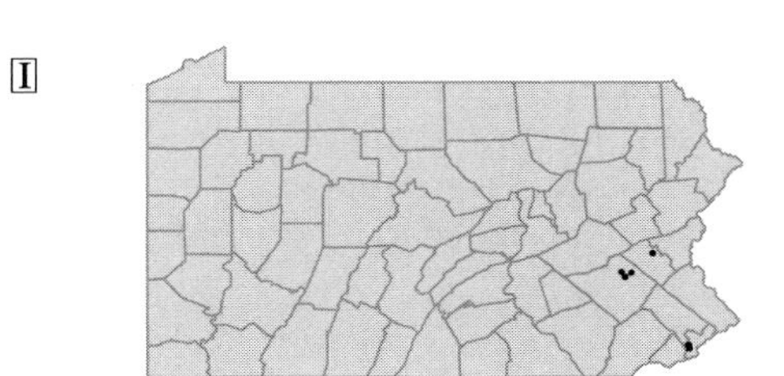

Petrorhagia saxifraga (L.) Link
I
Saxifrage pink
Herbaceous perennial
Cultivated and occasionally escaped.
Tunica saxifraga (L.) Scop. GB

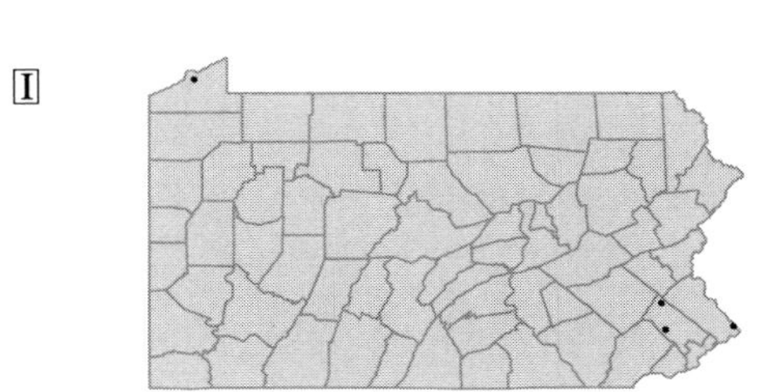

• ***Polycarpon tetraphyllum*** L. [I]
Pink
Herbaceous annual
Ballast. Represented by collections from Philadelphia Co. in 1877 and 1921 and Northampton Co. in 1890.

• ***Sagina decumbens*** (Ell.) Torr. & A.Gray
Pearlwort
Herbaceous annual
Cracks between flagstones.

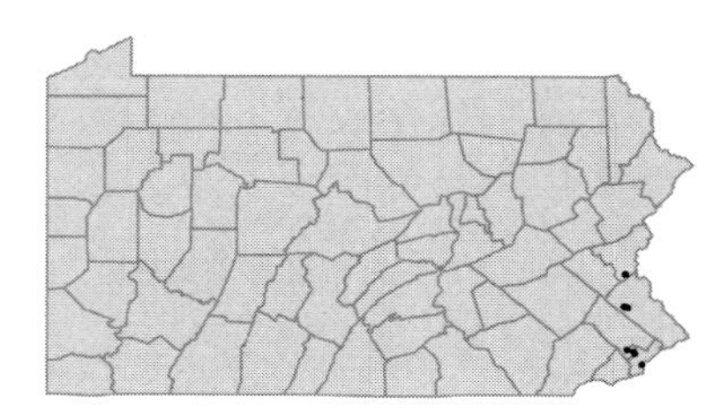

Sagina japonica (Swartz) Ohwi [I]
Japanese pearlwort
Herbaceous annual
Lawns and dry, compacted soil around buildings.

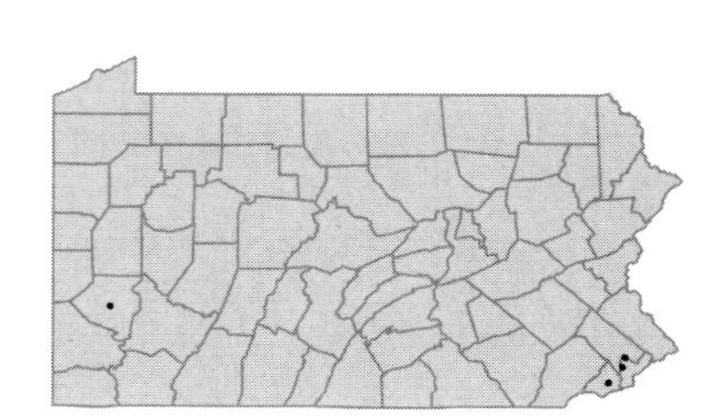

Sagina procumbens L.
Bird's-eye; Pearlwort
Herbaceous perennial
Paving joints, lawns and dooryards.

• ***Saponaria officinalis*** L. FACU- [I]
Bouncing-bet; Soapwort
Herbaceous perennial
Roadsides, railroad banks and other waste ground.

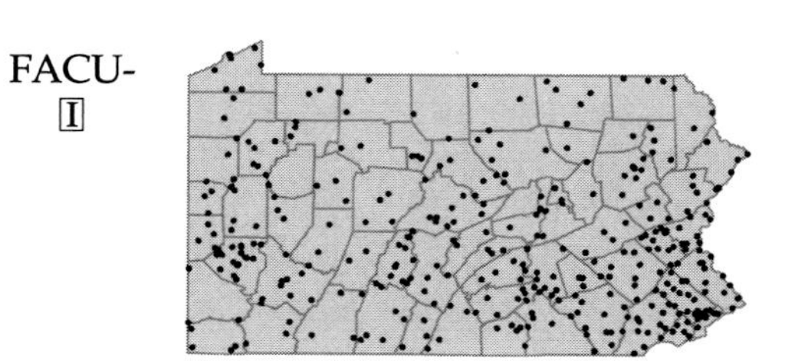

• ***Scleranthus annuus*** L. FACU- [I]
Knawel
Herbaceous annual
Dry, open ground, roadsides and shale barrens.

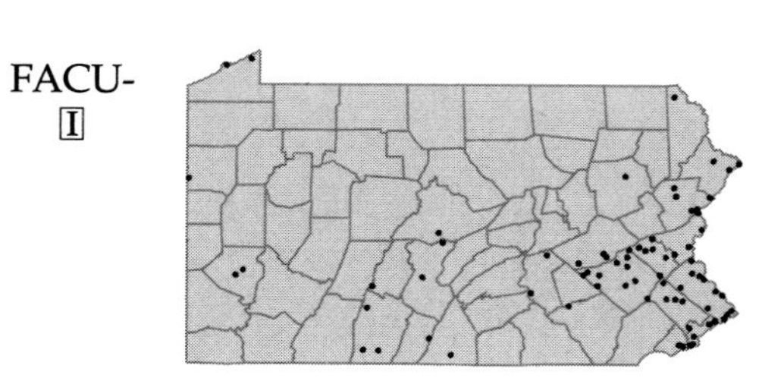

• ***Silene alba*** (Mill.) Krause
White campion; White cockle
Herbaceous annual
Moist ground of woods, fields and roadsides.
Lychnis alba Mill. PFGB; *Silene latifolia* Poir. C

I

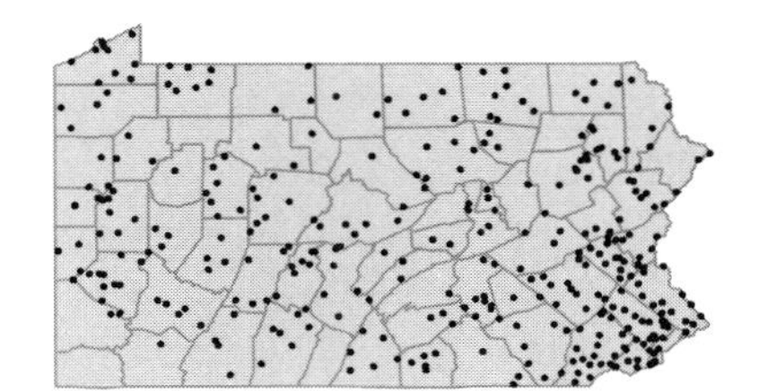

Silene antirrhina L.
Sleepy catchfly
Herbaceous annual
Dry, open woods, fields and waste ground.

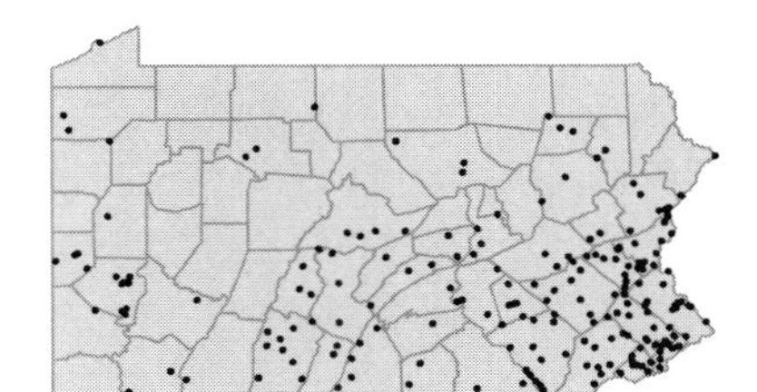

Silene armeria L.
Garden catchfly; None-so-pretty
Herbaceous annual
Old gardens and waste ground.

I

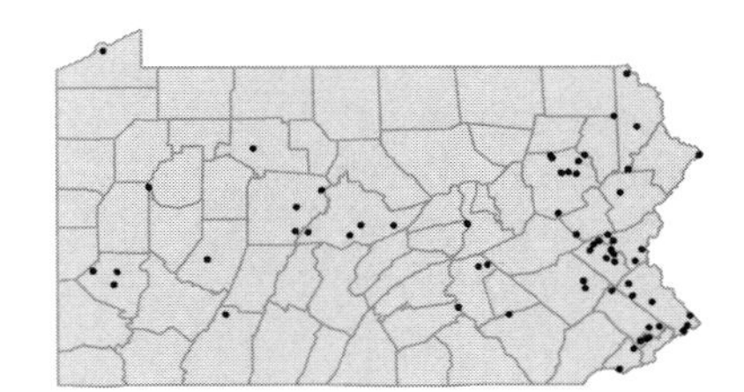

Silene caroliniana Walt. ssp. ***pensylvanica*** (Michx.) Clausen
Sticky catchfly; Wild pink
Herbaceous perennial
Dry open woods, rocky slopes, roadside banks and shale barrens.

Silene cserei Baumg.
Campion
Herbaceous perennial
Alluvium and dry, open soil along railroad tracks.

I

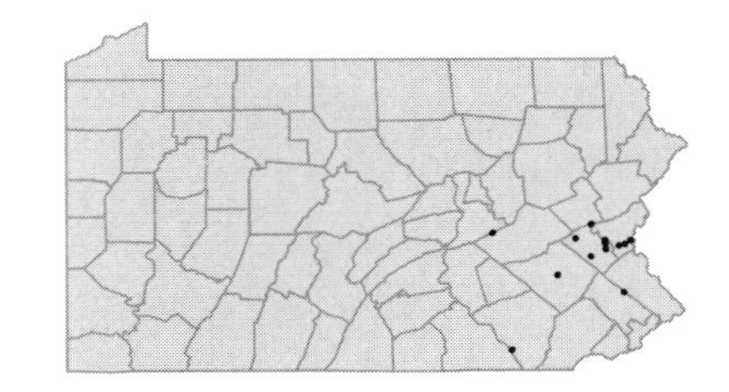

Silene dichotoma Ehrh.
Forked catchfly
Herbaceous annual
Fields and roadsides.

I

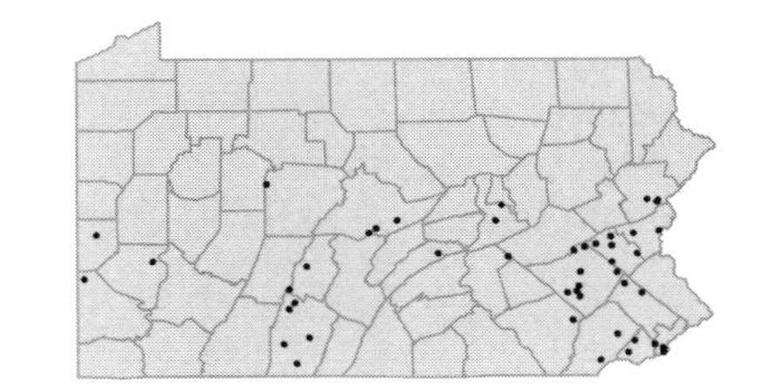

Silene dioica (L.) Clairville
Red campion
Herbaceous annual
Ballast, waste ground, lawns and roadsides.
Lychnis dioica L. PFGB

I

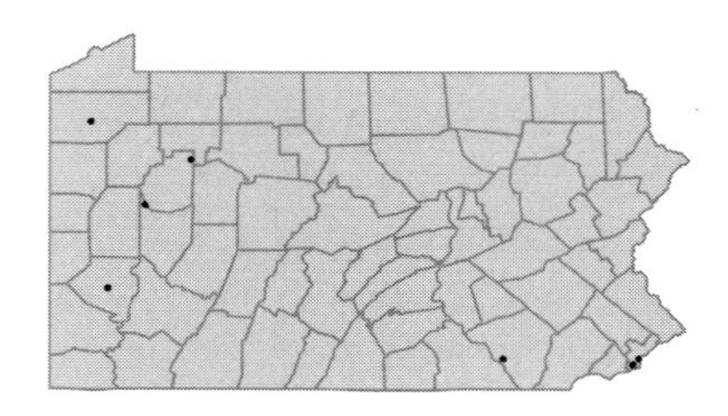

Silene gallica L.
Catchfly
Herbaceous annual
Cultivated and sometimes escaped to waste ground.
Silene anglica L. P

I

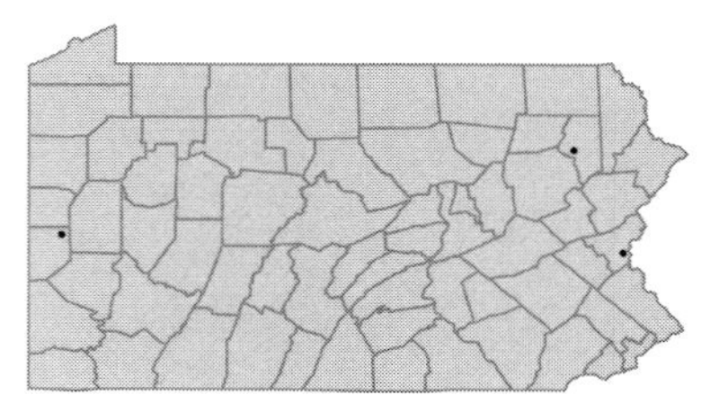

Silene nivea (Nutt.) Otth
Snowy campion
Herbaceous perennial
Alluvial thickets.
Silene alba Muhl. P

FAC

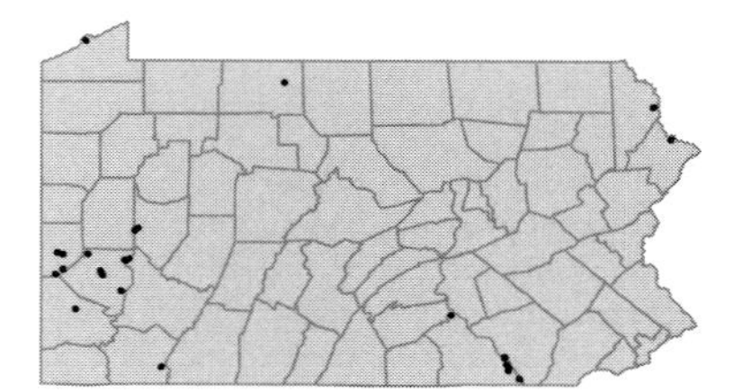

Silene noctiflora L.
Night-flowering catchfly; Sticky campion
Herbaceous annual
Cultivated and escaped to upland woods and fields.

I

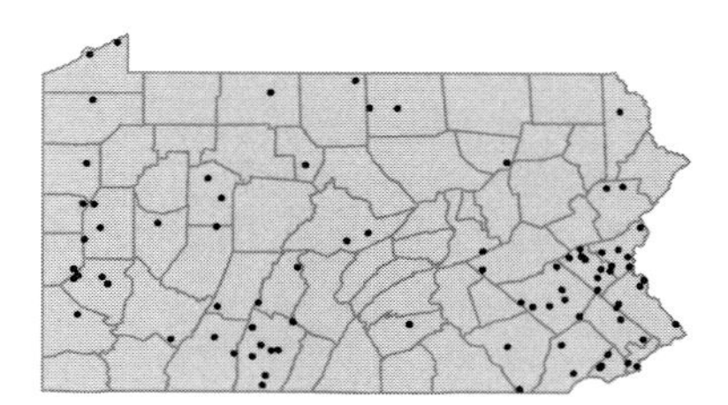

Silene stellata (L.) Ait.f.
Starry campion; Widow's-frill
Herbaceous perennial
Rocky, wooded slopes, roadside banks and barrens.

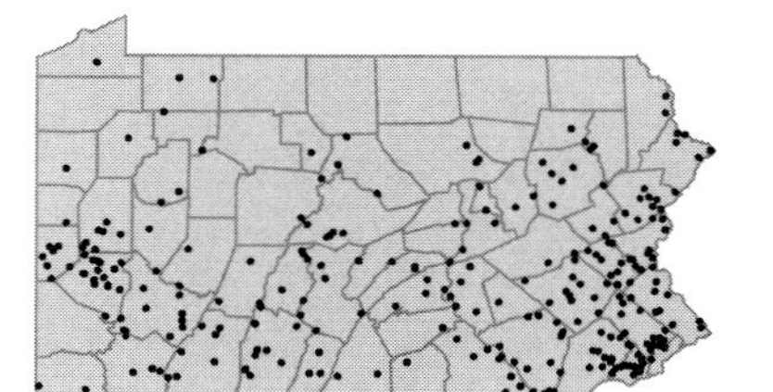

Silene virginica L.
Fire pink
Herbaceous perennial
Upland woods, wooded slopes and stream banks.

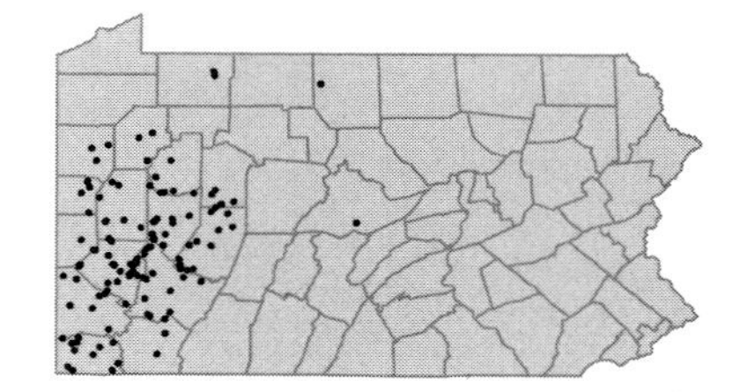

Silene vulgaris (Moench) Garcke
Bladder campion; Maiden's-tears
Herbaceous perennial
Old fields, roadsides, railroad banks and waste ground.
Silene cucubalus Wibel FGB

[I]

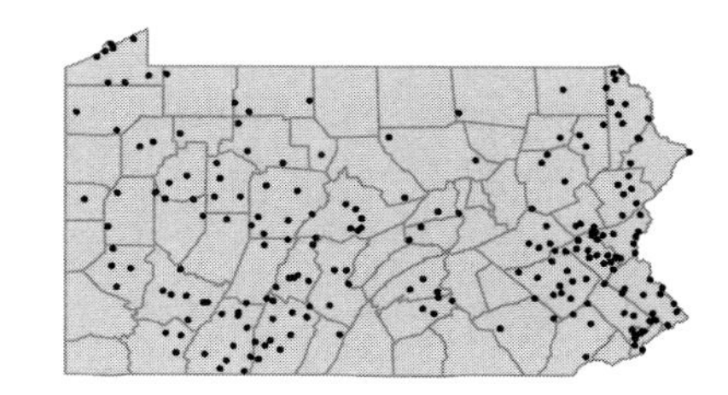

• ***Spergula arvensis*** L.
Corn-spurrey
Herbaceous annual
Fields and dry roadside banks.

[I]

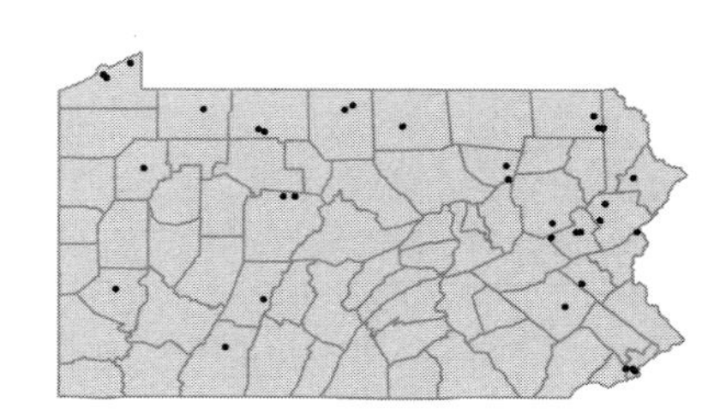

• ***Spergularia marina*** (L.) Griseb.
Saltmarsh sand-spurrey
Herbaceous annual
Roadsides, occurrence probably related to heavy use of de-icing salt.
Represented by a single collection from Somerset Co. in 1987.

OBL
[I]

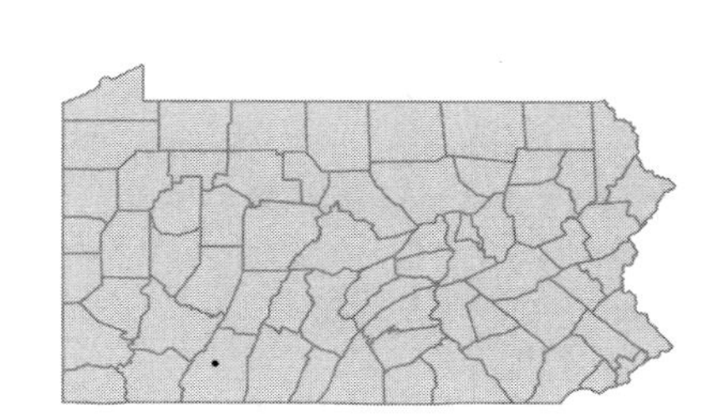

Spergularia media (L.) C. Presl ex Griseb.
Sand-spurrey
Herbaceous perennial
Roadsides, occurrence probably related to heavy use of de-icing salt.

FACW
[I]

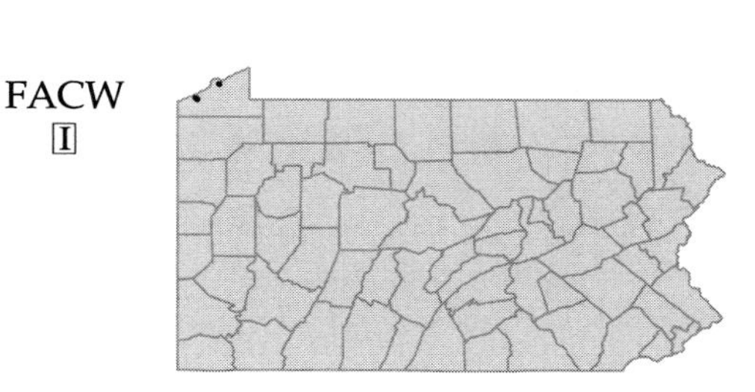

Spergularia rubra (L.) J. & C. Presl
Purple sand-spurrey
Herbaceous annual
Dry, open, sterile soils.

[I]

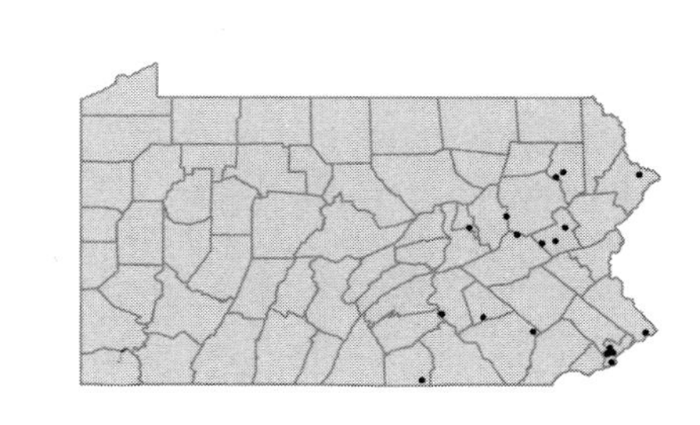

• ***Stellaria alsine*** Grimm
Bog chickweed; Bog starwort
Herbaceous annual
Springs, swamps, seeps and stream edges.
Alsine uliginosa (Murr.) Britt. P

OBL
[I]

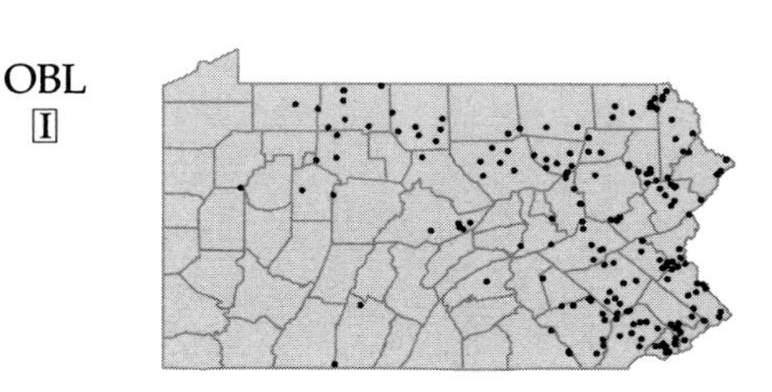

Stellaria borealis Bigelow
Mountain starwort; Northern stitchwort
Herbaceous perennial
Springy wooded slopes, sphagnous swamps and stream banks.
Alsine borealis (Bigel.) Britt. P; *Stellaria calycantha* (Ledeb.) Bong. FGBW

FACW

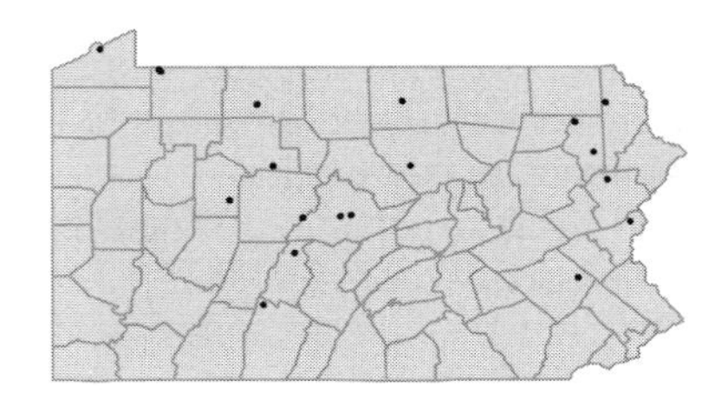

Stellaria corei Shinners
Chickweed
Herbaceous perennial
Rocky hillsides, bluffs and open woods.
Stellaria pubera Michx. var. *sylvatica* (Beguinot) Weatherby FWC; *Stellaria sylvatica* (Beguinot) Maguire GB

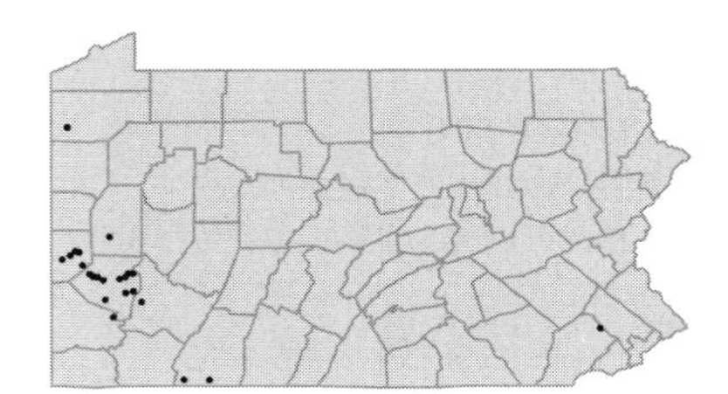

Stellaria graminea L.
Common stitchwort; Lesser stitchwort
Herbaceous perennial
Swampy woods, moist meadows, stream banks or moist shores.
Alsine graminea (L.) Britt. P

FACU-
I

Stellaria holostea L.
Easter-bell; Greater stitchwort
Herbaceous perennial
Roadsides and rocky woods.

I

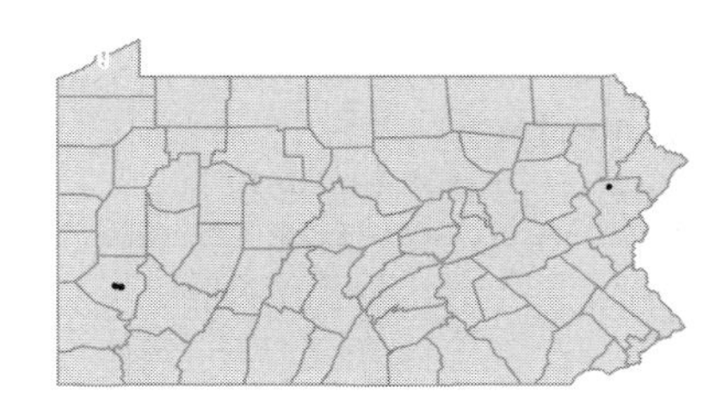

Stellaria longifolia Muhl. ex Willd.
Long-leaved stitchwort
Herbaceous perennial
Marshy open ground, swamps, rich woods and moist roadsides.
Alsine longifolia (Muhl.) Britt. P

FACW

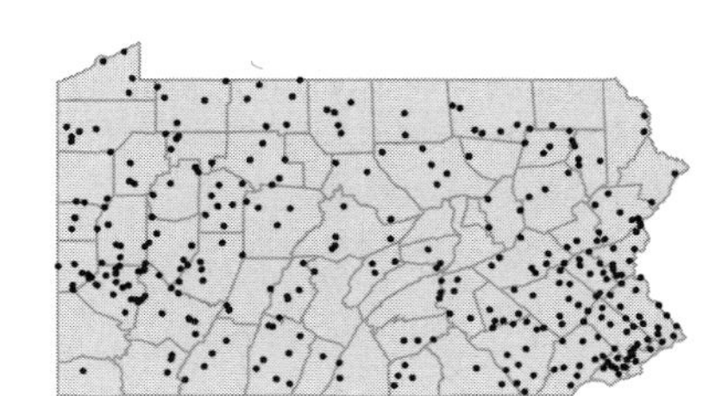

Stellaria media (L.) Vill.
Common chickweed
Herbaceous annual
A common weed of fields and gardens.
Alsine media L. P

I

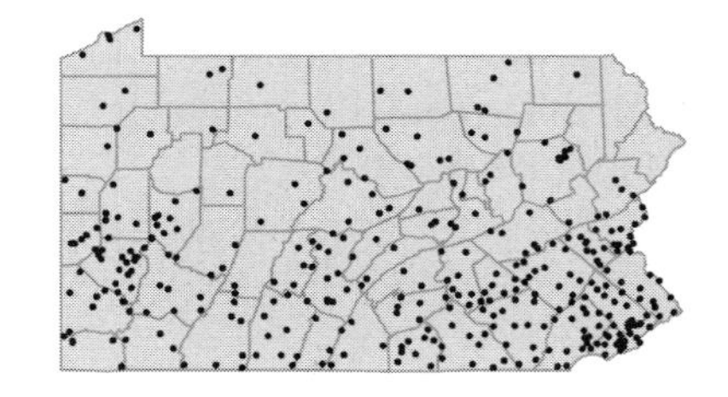

Stellaria pubera Michx.
Great chickweed; Star chickweed
Herbaceous perennial
Moist rocky ground, alluvial woods and mesic forests.

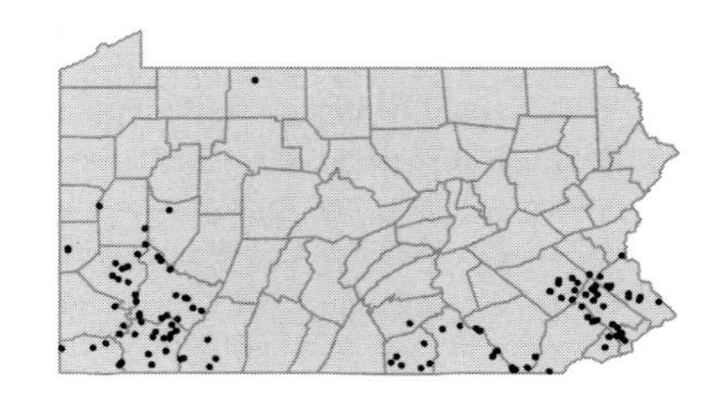

• ***Vaccaria hispanica*** (P.Mill.) Raushcert [I]
Cow-cockle; Cow-herb
Herbaceous annual
Cultivated and escaped to railroad banks and waste ground.
Saponaria vaccaria L. F; *Vaccaria pyramidata* Medic. W; *Vaccaria segetalis* (Neck.) Garcke GB; *Vaccaria vaccaria* (L.) Britt. P

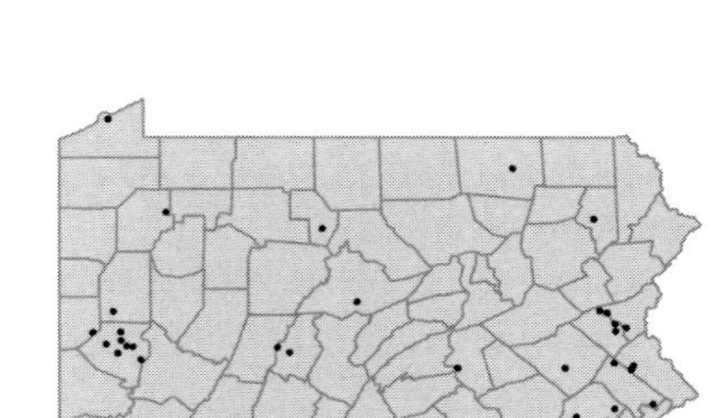

POLYGONACEAE

• ***Fagopyrum sagittatum*** Gilib. [I]
Buckwheat
Herbaceous annual
Cultivated and occasionally escaped to fields, roadsides, railroad banks and dumps.
Fagopyrum fagopyrum (L.) Karst. P; *Fagopyrum esculentum* Moench. KWBGC

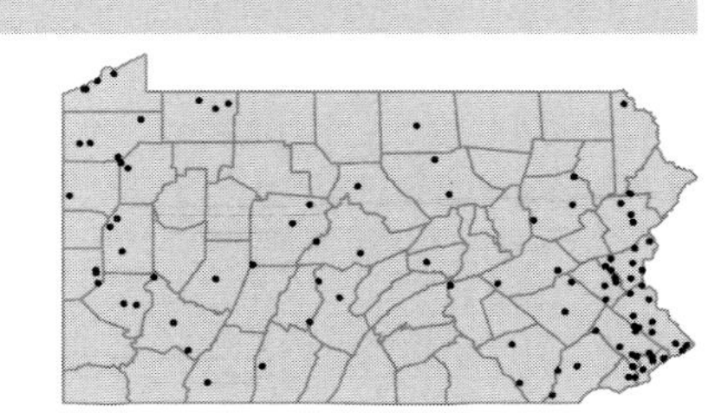

Fagopyrum tataricum (L.) Gaertn. [I]
India-wheat
Herbaceous annual
Cultivated and rarely escaped.

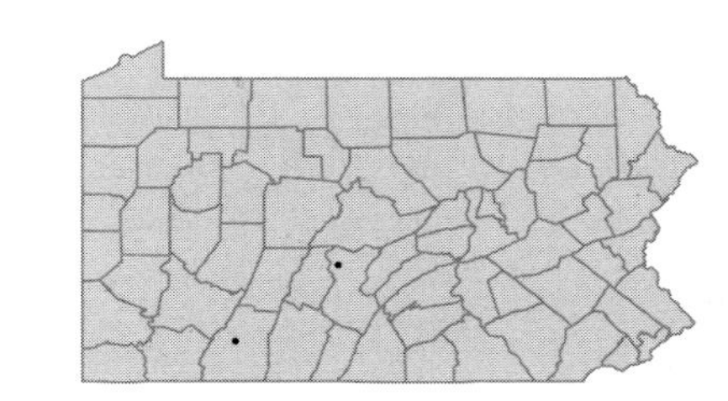

• ***Polygonella articulata*** (L.) Meisn.
Jointweed
Herbaceous annual
Dry, open, sandy acidic soils.

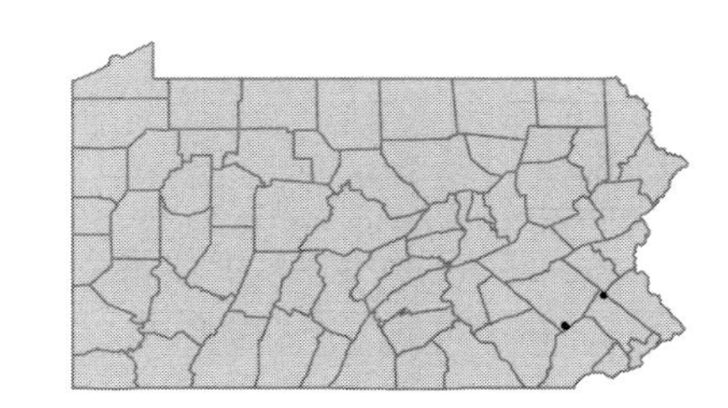

• ***Polygonum amphibium*** L. var. ***emersum*** Michx. OBL
Water smartweed
Herbaceous perennial, floating-leaf aquatic
Muddy shores and margins of ponds, streams or rivers.
Polygonum coccineum Muhl. FGBW; *Polygonum emersum* (Michx.) Britt. P

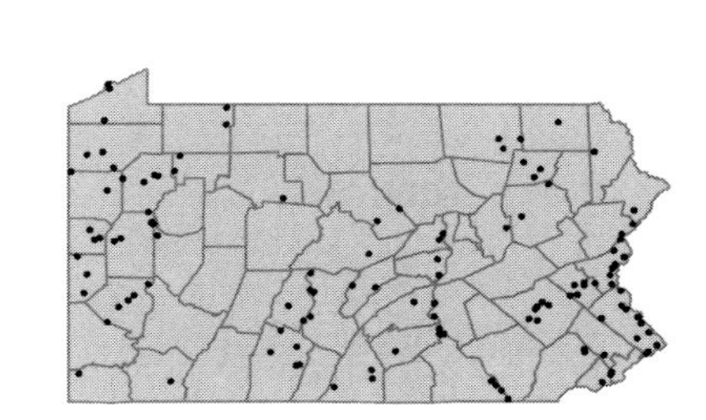

Polygonum amphibium L. var. ***stipulaceum*** Coleman
Floating smartweed; Water smartweed
Herbaceous perennial, floating-leaf aquatic
Edges of lakes, wet meadows, swamps and ditches.
Polygonum hartwrightii A.Gray P; *Polygonum natans* Eat. GB

OBL

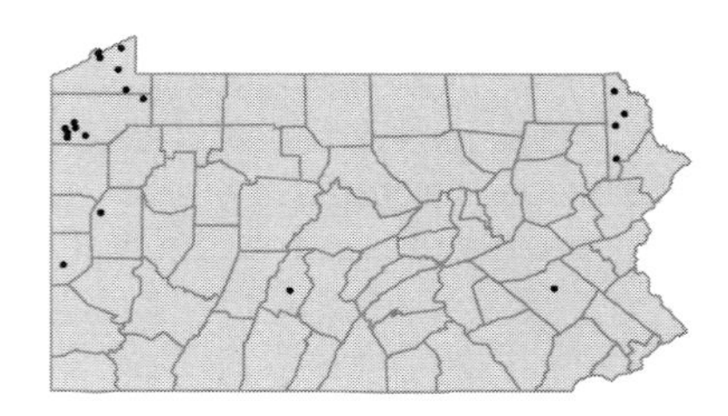

Polygonum arenastrum Jord. ex Boreau
Doorweed; Knotgrass
Herbaceous annual
Disturbed or compacted soil and gravel.
Polygonum aviculare L. var. *arenastrum* (Bor.) Rouy

I

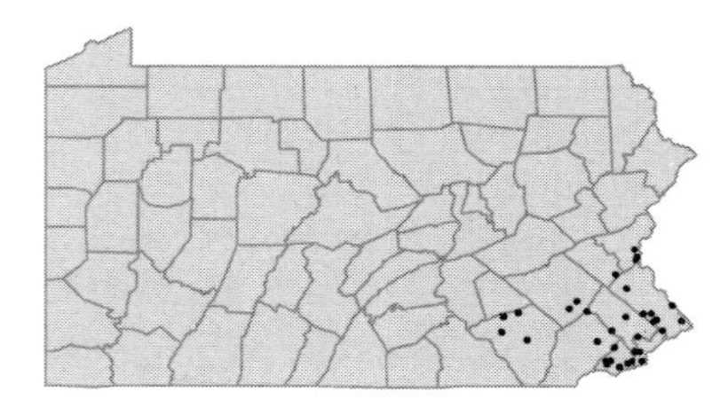

Polygonum arifolium L.
Halberd-leaved tearthumb
Herbaceous annual
Wet woods, boggy thickets, swamps, wet meadows and ditches.

OBL

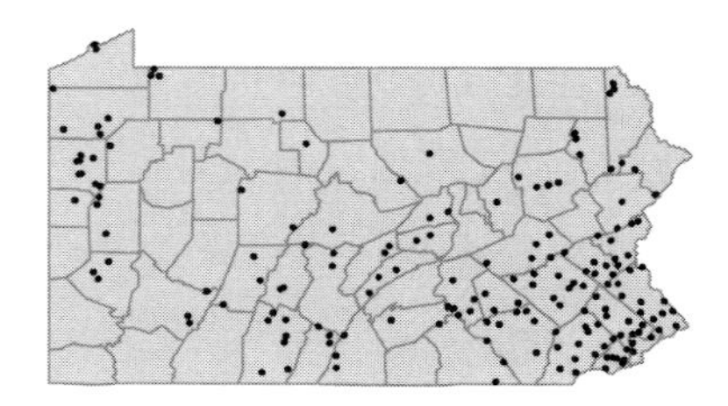

Polygonum aubertii L.Henry
Silver lace-vine
Herbaceous perennial vine
Cultivated and occasionally escaped to roadside thickets.

I

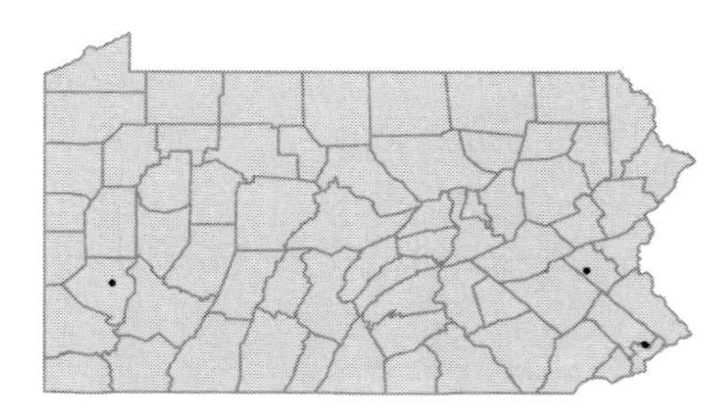

Polygonum aviculare L.
Knotweed
Herbaceous annual
Dry, open ground including yards, pavement, roadsides and trails.
Polygonum monspeliense Pers. K; *Polygonum aviculare* L. var. *vegetum* Ledeb. *pro parte* FW; *Polygonum aviculare* L. var. *erectum* (Roth) W.D.J. Koch *pro parte*

FACU
I

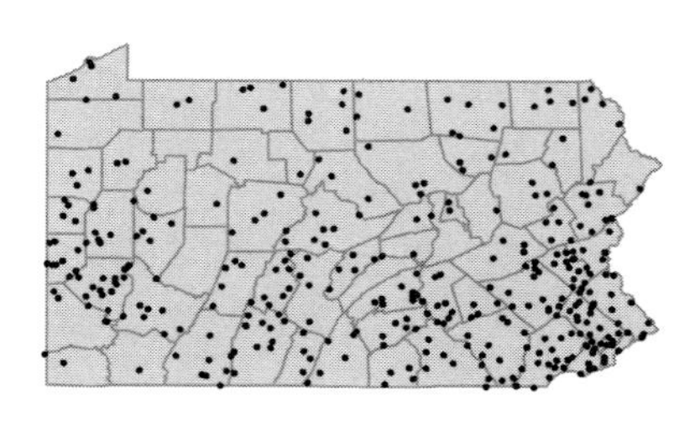

Polygonum buxiforme Small
Knotweed
Herbaceous annual
Ballast and waste ground.
Polygonum aviculare L. var. *aviculare pro parte* GB; *Polygonum aviculare* L. var. *littorale* (Link) W.D.J.Koch FW; *Polygonum littorale* Link P

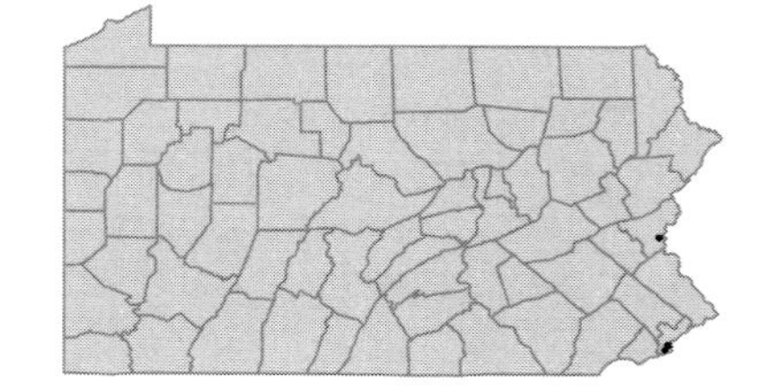

Polygonum caespitosum Blume var. ***caespitosum***
Smartweed
Herbaceous annual
Damp ground of roadsides, shores and waste places.

FACU-
I

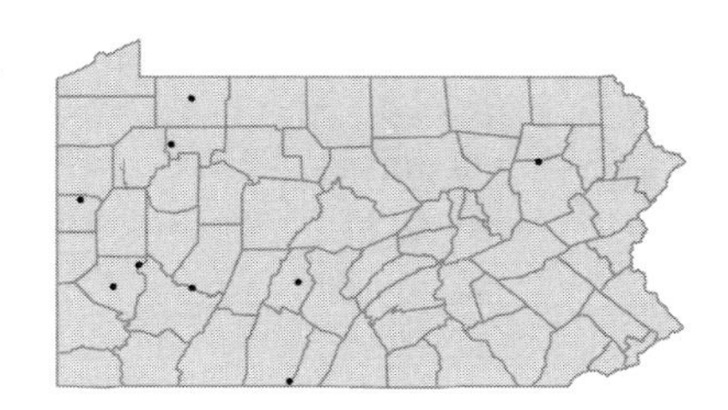

Polygonum caespitosum Blume var. ***longisetum*** (DeBruyn) A.N.Stewart
Smartweed
Herbaceous annual
Damp ground of wooded hillsides, roadsides, fields and waste places.

FACU-
I

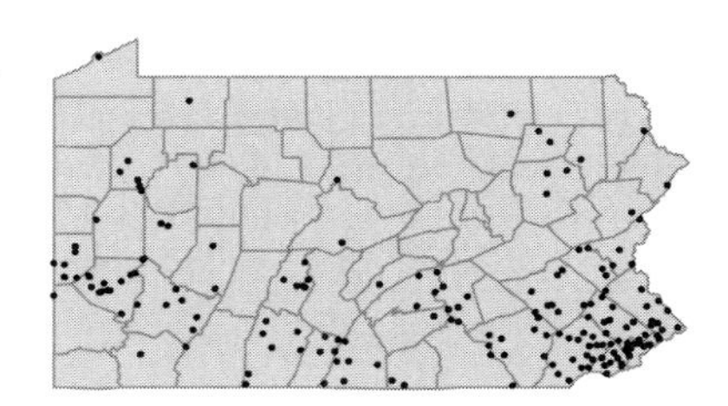

Polygonum careyi Olney
Pinkweed; Smartweed
Herbaceous annual
Wet meadows, moist woodland edges and cultivated ground.

FACW

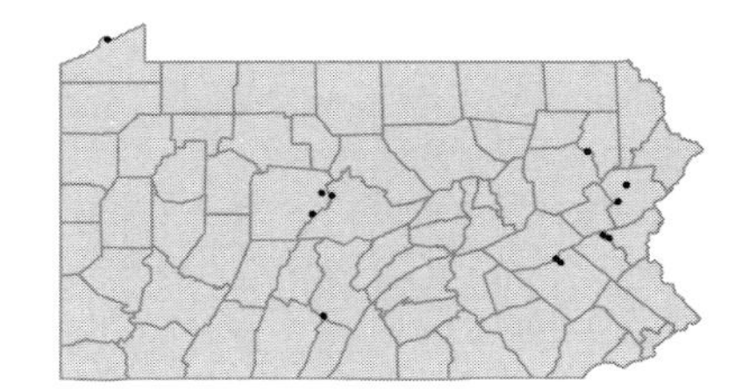

Polygonum cilinode Michx.
Fringed bindweed
Herbaceous perennial vine
Dry woods, rocky slopes, thickets and roadsides.

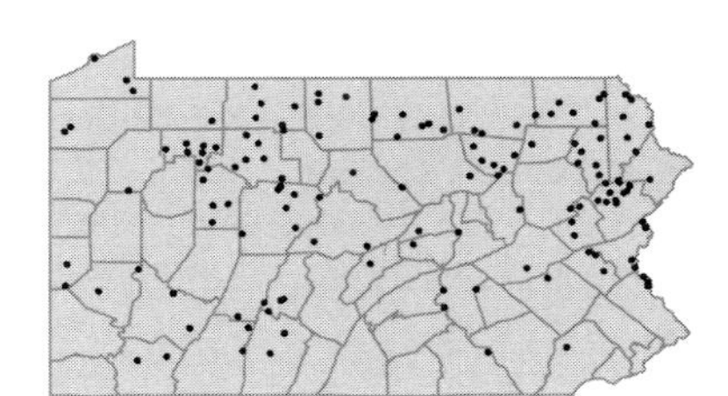

Polygonum convolvulus L.
Black bindweed; Nimble-will
Herbaceous annual vine
Cultivated fields, roadsides and waste ground.

FACU
I

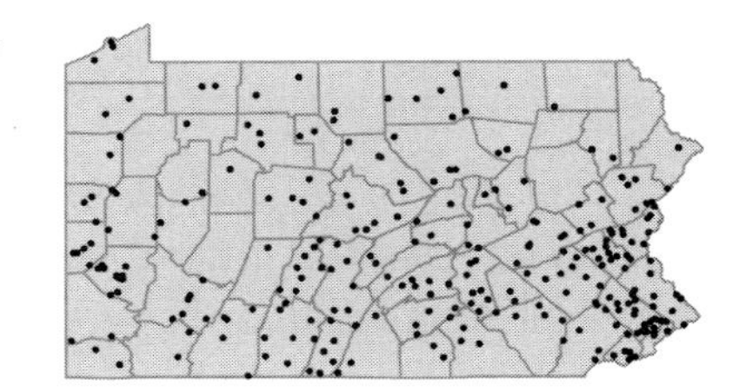

Polygonum cuspidatum Sieb. & Zucc.
Japanese knotweed; Mexican bamboo
Herbaceous perennial
Stream banks, roadsides, railroad banks and waste areas.
Polygonum zuccarinii Small P

FACU-
I

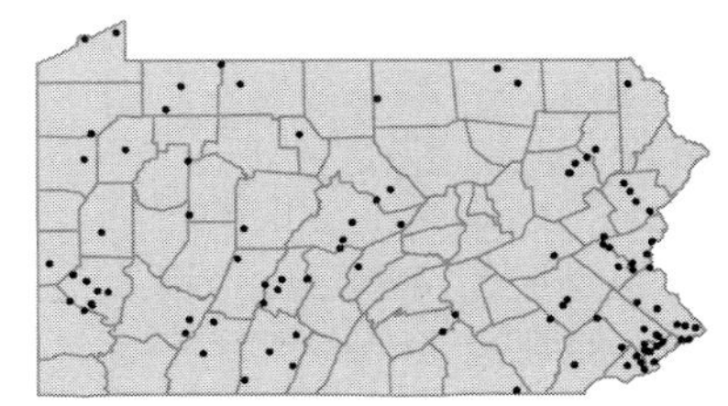

Polygonum erectum L.
Erect knotweed
Herbaceous annual
Moist soil of river banks, abandoned fields, roadsides and waste areas.

FACU

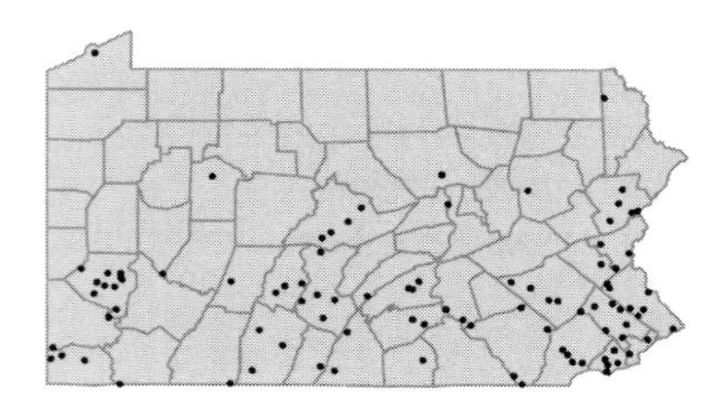

Polygonum hydropiper L.
Common smartweed; Water-pepper
Herbaceous annual
Wet meadows, swamps, stream banks and moist waste ground.

OBL
[I]

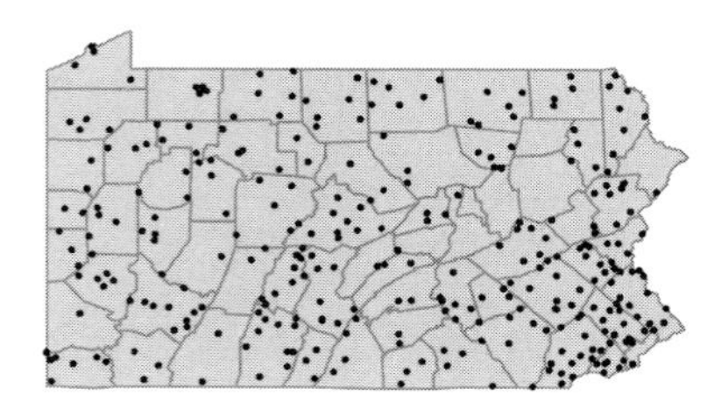

Polygonum hydropiperoides Michx.
Mild water-pepper; Water smartweed
Herbaceous perennial
Openings in wet woods, vernal ponds, stream banks or wet swales.
Polygonum opelousanum Riddell *pro parte* FW

OBL

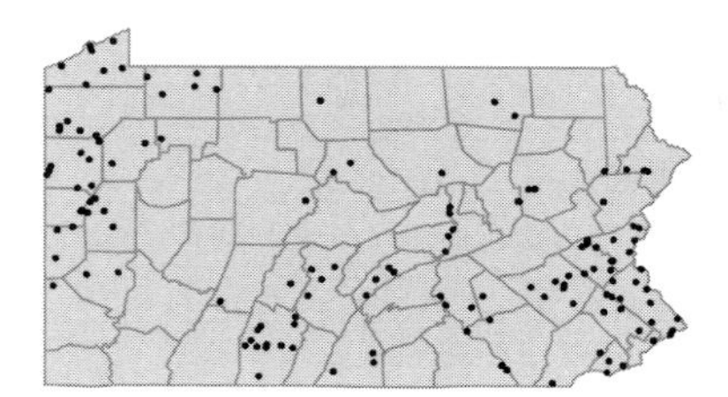

Polygonum lapathifolium L.
Dock-leaved smartweed; Willow-weed
Herbaceous annual
Wet woods, stream banks, ditches, roadsides and cultivated fields.
Polygonum incarnatum Ell. P; *Polygonum scabrum* Moench FW

FACW+
[I]

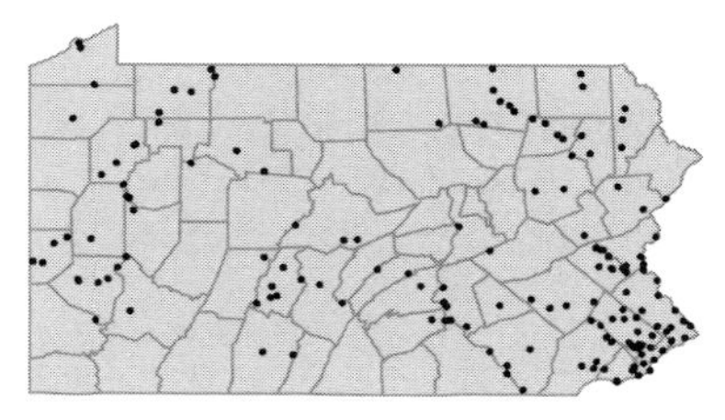

Polygonum neglectum Besser
Needle-leaf knotweed
Herbaceous annual
Dry, sandy roadsides, river banks and railroad ballast.
Polygonum aviculare L. var. *angustissimum* Meisn. F

[I]

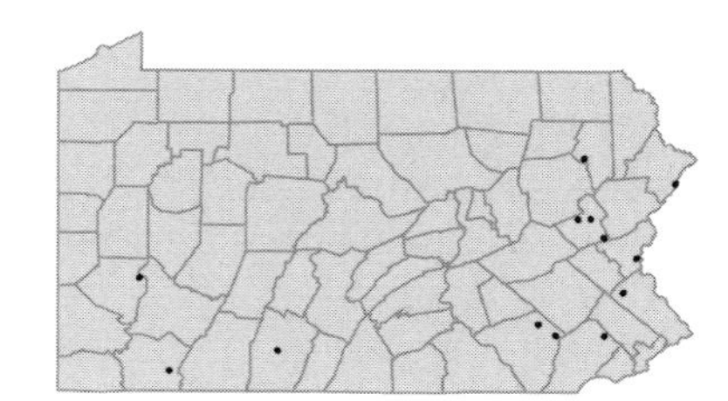

Polygonum nepalense Meissner
Polygonum
Herbaceous annual
River bank.

[I]

Polygonum orientale L. FACU- I
Kiss-me-over-the-garden-gate; Prince's-feather
Herbaceous annual
Cultivated and occasionally escaped to cultivated fields, roadsides and waste ground.
Polygonum spaethii Damm. W

Polygonum pensylvanicum L. FACW
Smartweed
Herbaceous annual
Fields, woodland edges, stream banks, roadsides and waste ground.

Polygonum perfoliatum L. FAC I
Mile-a-minute
Herbaceous annual
Alluvial thickets, moist open woods, abandoned nurseries and roadsides. Designated as a noxious weed in PA.

Polygonum persicaria L. FACW I
Lady's-thumb; Heart's-ease
Herbaceous annual
Wooded slopes, moist thickets, edges of swamps, roadsides and waste ground.
Polygonum dubium Stein FGBW; *Polygonum minus* Huds. FW

Polygonum punctatum Ell. var. ***punctatum*** OBL
Dotted smartweed
Herbaceous perennial, emergent aquatic
Moist pastures, river banks, bogs, swamps and swales.

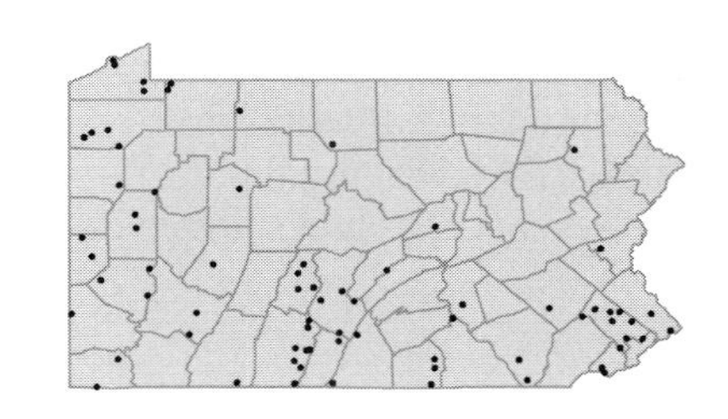

Polygonum punctatum Ell. var. ***confertiflorum*** (Meisn.) Fassett OBL
Dotted smartweed
Herbaceous annual, emergent aquatic
Vernal ponds, stream banks, muddy shores, swamps and swales.
Polygonum punctatum Elliott var. *parvum* Vict. & Rouss. FW

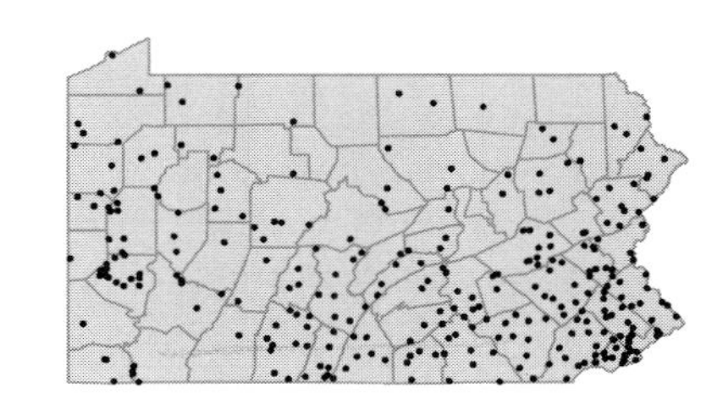

Polygonum ramosissimum Michx.
Bushy knotweed
Herbaceous annual
Sandy shores, railroad ballast, waste ground and rubbish heaps.

FAC

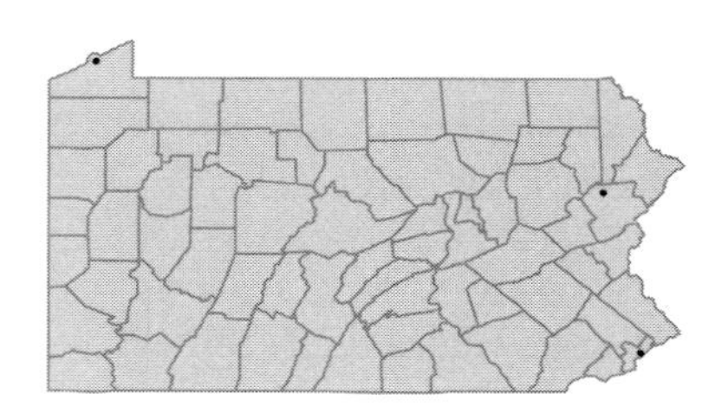

Polygonum robustius (Small) Fern.
Large water-smartweed
Herbaceous perennial
Wet ground or shallow water of stream and pond margins, swamps and bogs.

OBL

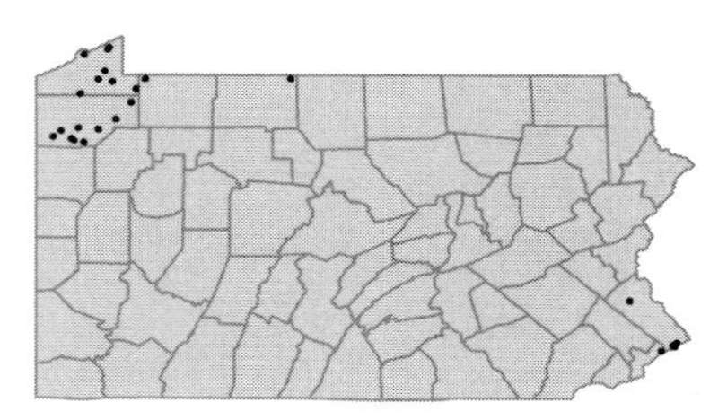

Polygonum sachalinense Schmidt ex Maxim.
Giant knotweed
Herbaceous perennial
Vacant lots, river banks, roadsides and railroad banks.

UPL
I

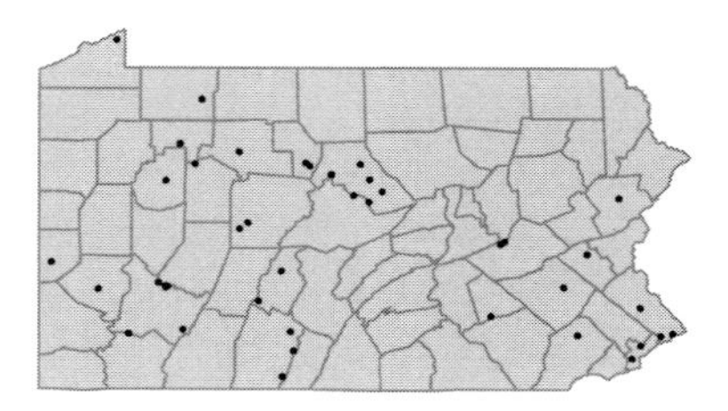

Polygonum sagittatum L.
Arrow-leaved tearthumb; Scratch-grass
Herbaceous annual
Low moist ground, vernal ponds, bogs, swamps or marshes.

OBL

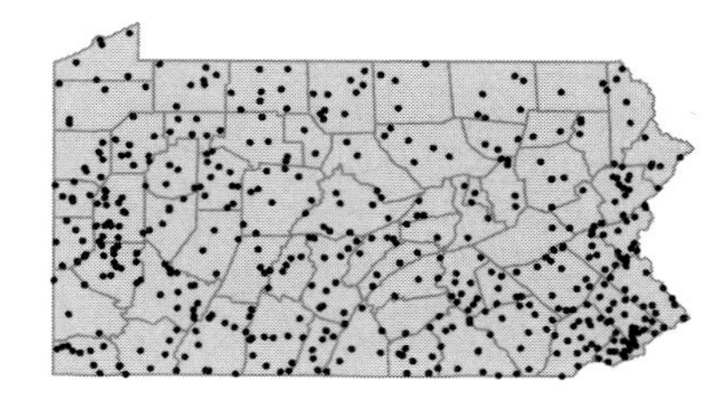

Polygonum scandens L. var. ***scandens***
Climbing false-buckwheat
Herbaceous perennial vine
Moist woods, thickets, roadsides and waste ground.

FAC

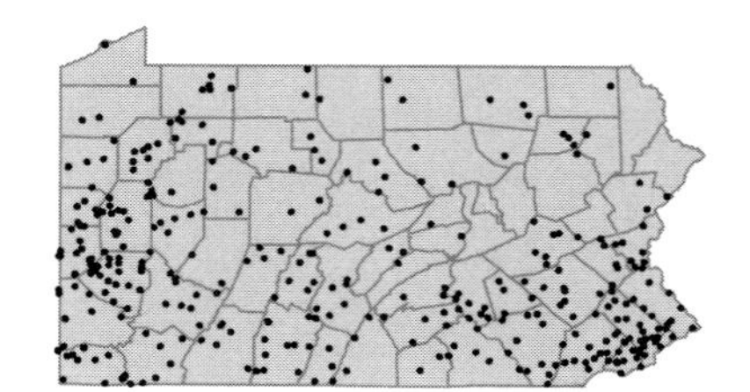

Polygonum scandens L. var. ***cristatum*** (Engelm. & A.Gray) Gleason
Climbing false-buckwheat
Herbaceous perennial vine
Dry rocky woods, wooded banks and thickets.
Polygonum cristatum Engelm. & A.Gray PFW

FAC

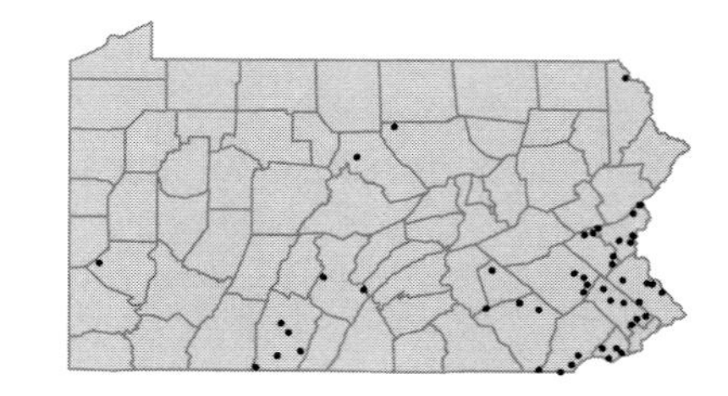

Polygonum scandens L. var. ***dumetorum*** (L.) Gleason
Climbing false-buckwheat
Herbaceous perennial vine
Moist woods, alluvial thickets and roadsides.

FAC
I

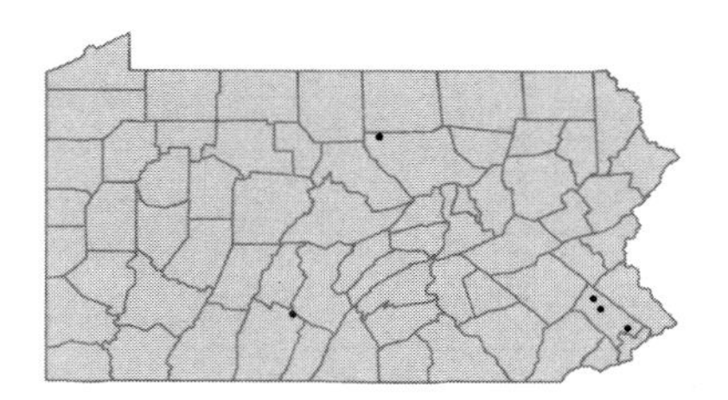

Polygonum setaceum Baldw. ex Ell.
Swamp smartweed
Herbaceous perennial
Shallow water and moist soil.
Polygonum hydropiperoides Michx. var. *setaceum* (Baldw.) Gleason C

OBL

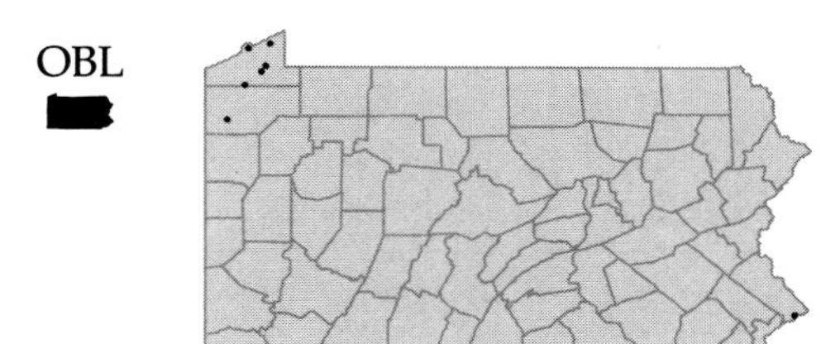

Polygonum tenue Michx.
Slender knotweed
Herbaceous annual
Dry fields in acidic soils, also serpentine barrens and shale barrens.

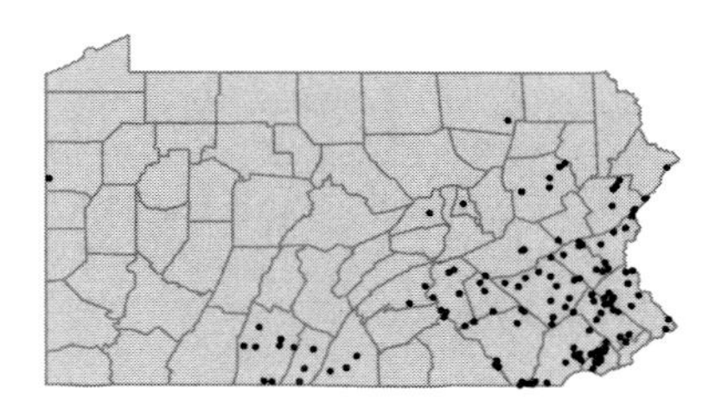

Polygonum virginianum L.
Jumpseed
Herbaceous perennial
Moist open woods, floodplains and roadsides.
Tovara virginiana (L.) Raf. FW

FAC

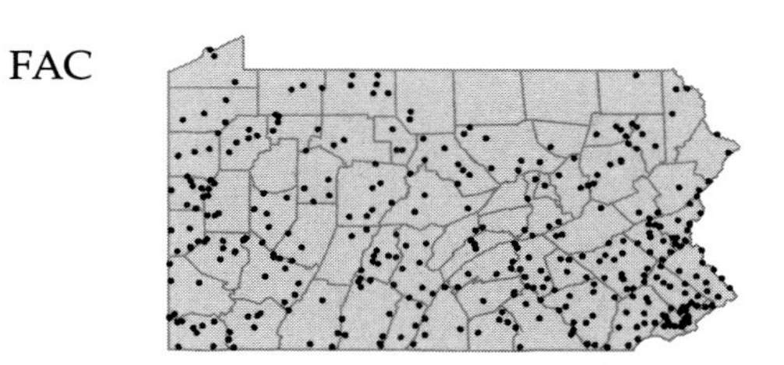

• ***Rheum rhaponticum*** L.
Rhubarb
Herbaceous perennial
Abandoned gardens or old home sites.

I

• ***Rumex acetosa*** L.
Garden sorrel
Herbaceous perennial
Cultivated and occasionally occurring in fallow fields or waste ground.

FACU
I

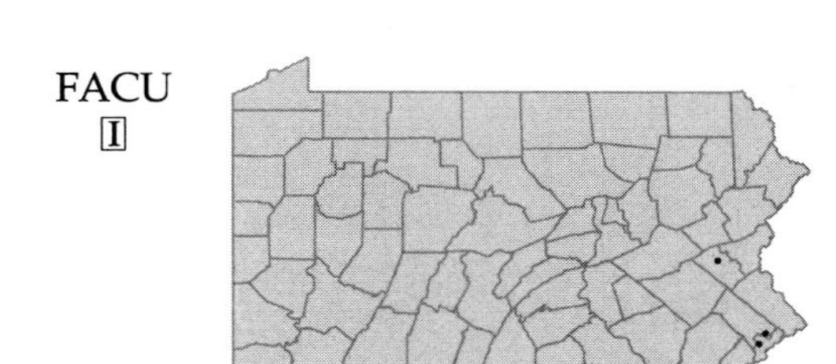

Rumex acetosella L.
Sheep sorrel; Sourgrass
Herbaceous perennial
Fields, lawns, waste places and shale barrens, in acidic soils.

UPL
I

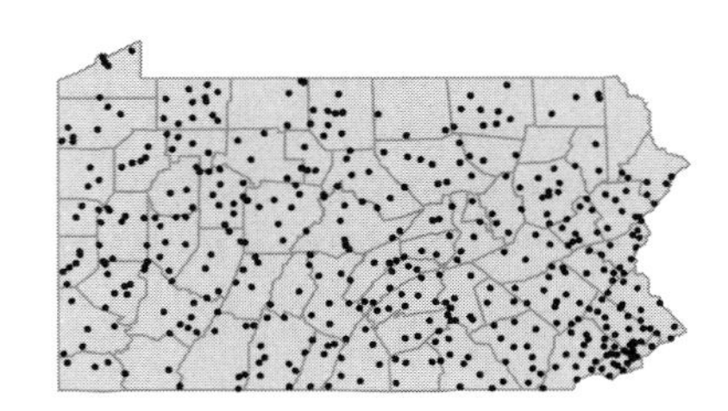

Rumex altissimus A.Wood
Tall dock; Peach-leaved dock
Herbaceous perennial
River banks, alluvial thickets and wet ditches.

FACW-

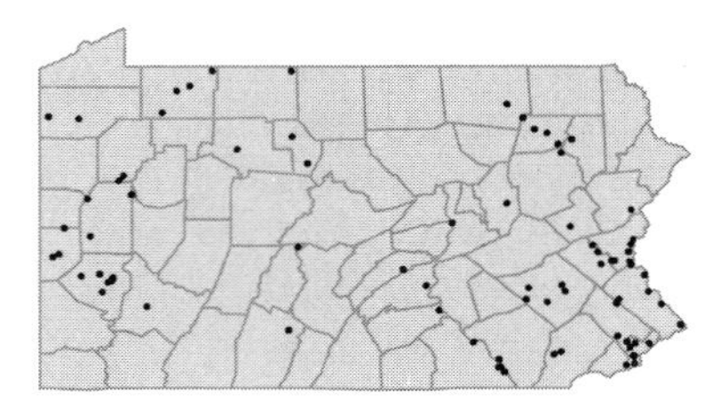

Rumex conglomeratus Murr.
Clustered dock
Herbaceous perennial
Ballast and waste ground. Represented by two collections from Philadelphia Co. ca. 1877.

FAC
I

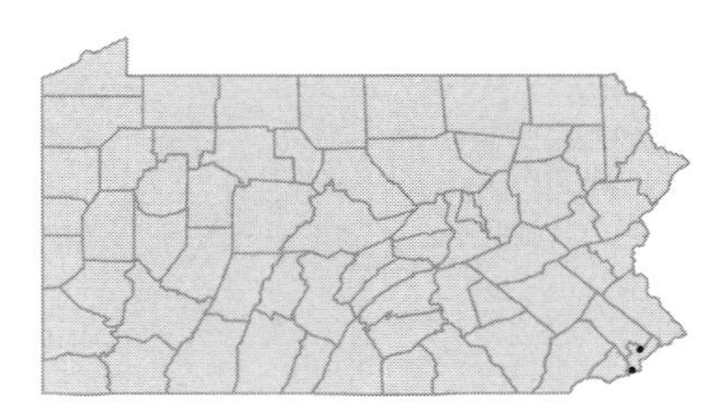

Rumex crispus L.
Curly dock
Herbaceous perennial
Cultivated fields, roadsides and waste ground.

FACU
I

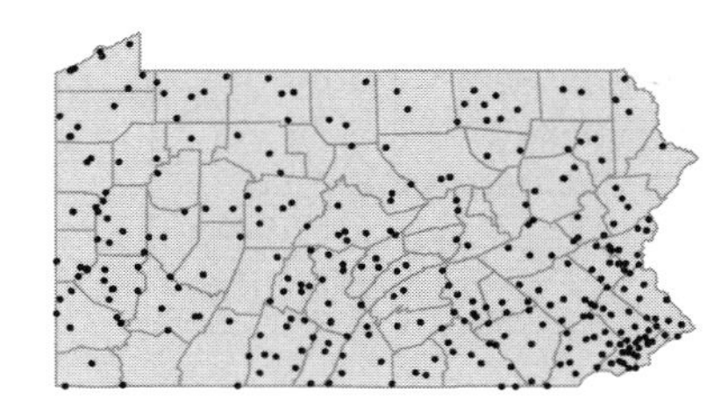

Rumex hastatulus Baldw. ex Ell.
Heart sorrel
Herbaceous perennial
Alluvial meadow. Represented by a single collection from Delaware Co. in 1971.

FACU-

Rumex maritimus L.
Seashore dock
Herbaceous annual
Ballast and waste ground.

FACW
I

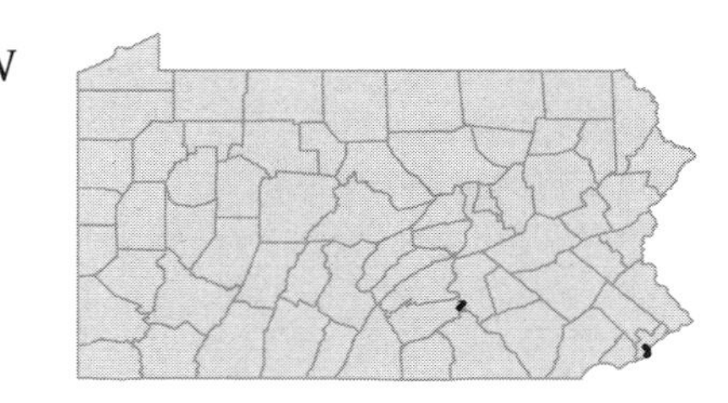

Rumex obtusifolius L.
Bitter dock
Herbaceous perennial
Roadsides, woods and moist, open ground.

FACU-
[I]

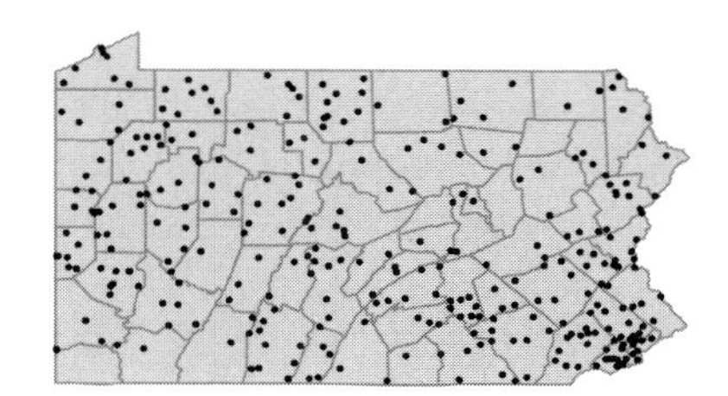

Rumex orbiculatus A.Gray
Great water dock
Herbaceous perennial, emergent aquatic
Wet woods, or shallow water of swamps and bogs.
Rumex britannica L. P

OBL

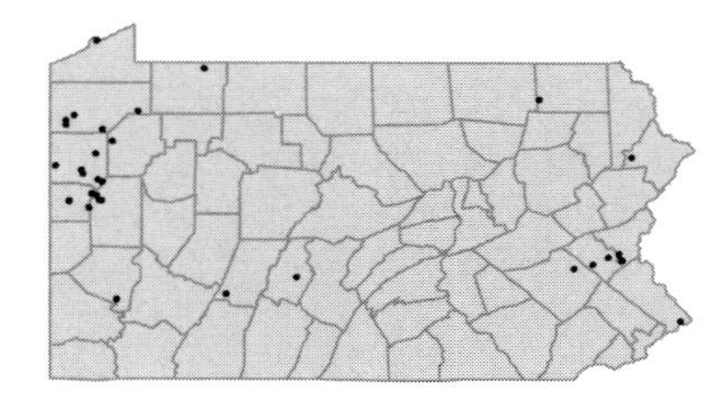

Rumex patientia L.
Monk's-rhubarb; Patience dock
Herbaceous perennial
Fields, railroad banks, roadsides and waste ground.

[I]

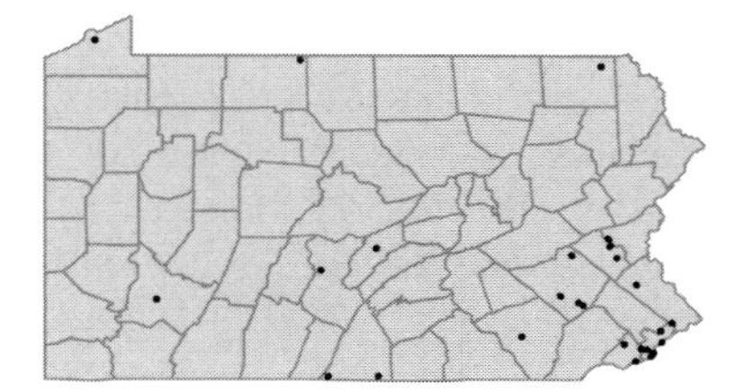

Rumex pulcher L.
Fiddle dock
Herbaceous perennial
Waste ground and rubbish dumps.

FACW-
[I]

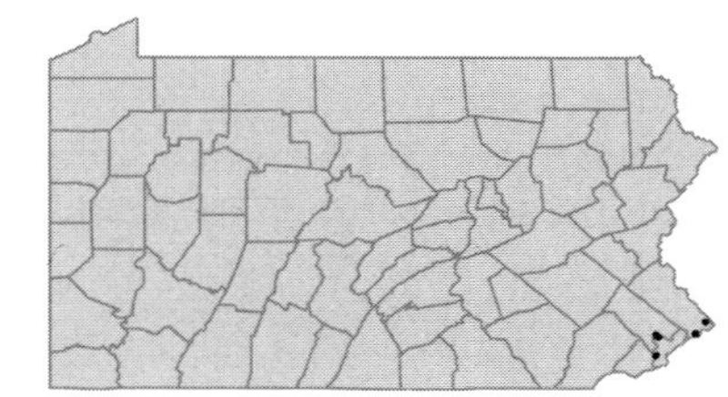

Rumex sanguineus L.
Bloody dock; Red-veined dock
Herbaceous perennial
Waste ground and ballast.

[I]

Rumex triangulivalvis (Danser) Rech.f.
Willow-leaf dock
Herbaceous perennial
Shady banks, railroad ballast and waste ground.
Rumex mexicanus Meisn. *pro parte* FGBW; *Rumex salicifolius* J.A.Weinm. var. *triangulvalvis* (Danser) Hickman. C

FACU
[I]

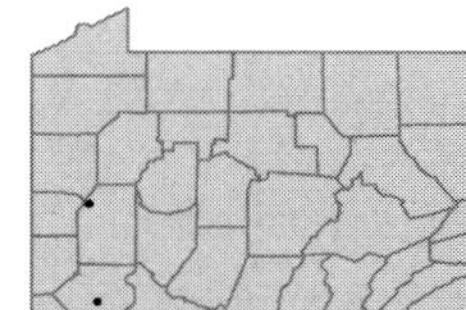

Rumex verticillatus L.
Swamp dock
Herbaceous perennial, emergent aquatic
Swampy fields, floodplains, swamps and swales, in wet soil or shallow water.

OBL

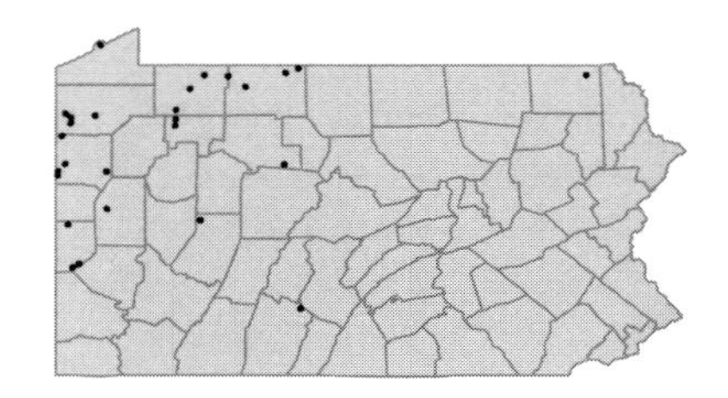

PAEONIACEAE

• ***Paeonia lactiflora*** Pallas
Peony
Herbaceous perennial
Cultivated and occasionally persisting around old home sites.

I

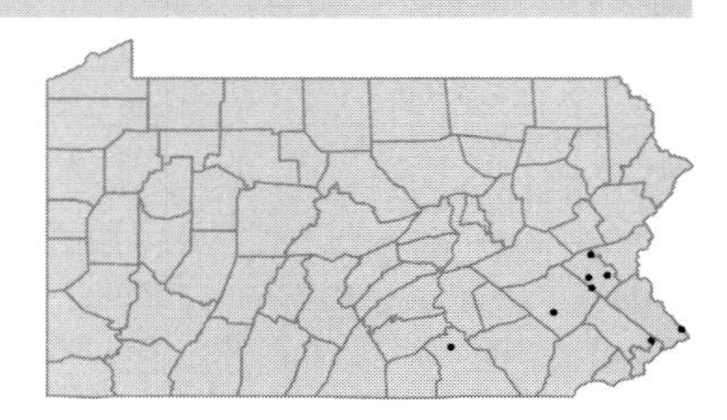

ELATINACEAE

• ***Elatine americana*** (Pursh) Arn.
American waterwort; Long-stemmed waterwort
Herbaceous annual, rooted submergent aquatic
Muddy tidal shores. Believed to be extirpated, last collected in 1868.
Elatine triandra Schkuhr var. *americana* (Pursh) Fassett BGC

OBL

Elatine minima (Nutt.) Fisch. & Mey.
Small waterwort
Herbaceous annual, rooted submergent aquatic
Shallow water of northern lakes and muddy tidal shores.

OBL

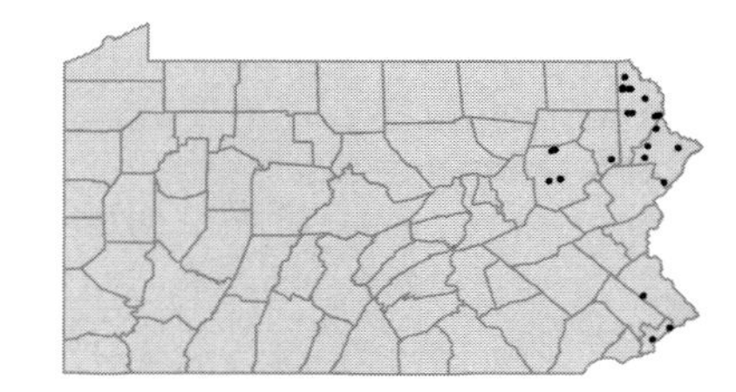

CLUSIACEAE

• ***Hypericum adpressum*** Raf. ex Bart.
Creeping St.-John's-wort
Herbaceous perennial
Swamps and damp ground. Believed to extirpated, last collected in 1908.

OBL

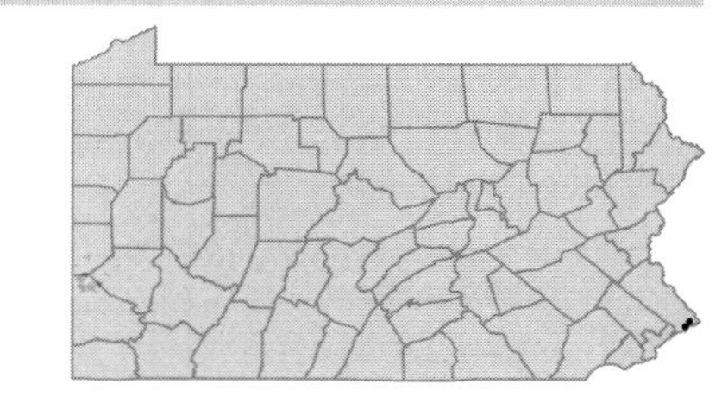

Hypericum boreale (Britt.) Bickn.
Dwarf St.-John's-wort
Herbaceous perennial
Open peat of bog edges, also swampy hummocks and wet meadows.
Hypericum mutilum L. ssp. *boreale* (Britt.) Bickn. K

OBL

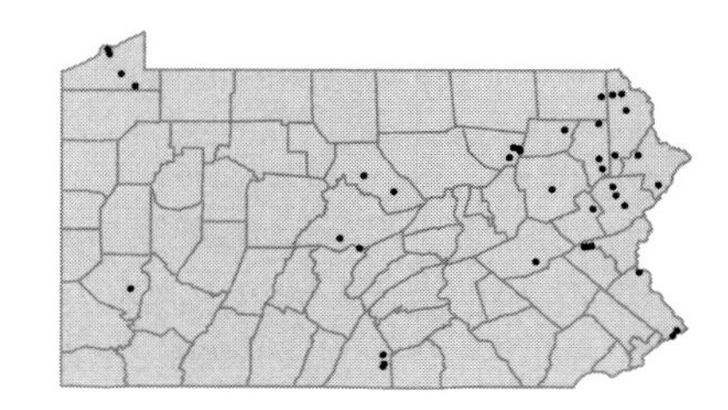

Hypericum canadense L.
Canadian St.-John's-wort
Herbaceous annual
Sandy river banks, moist ground with sphagnum, low thickets and swales.

FACW

Hypericum crux-andreae (L.) Crantz
St.-Peter's-wort
Deciduous shrub
Swamps, in sandy soil. Believed to be extirpated, all collections were from a single site in Bucks Co. in 1864.
Ascyrum stans Michx. PFGBW; *Hypericum stans* (Michx. ex Willd.) Adams & Robson C

FACU

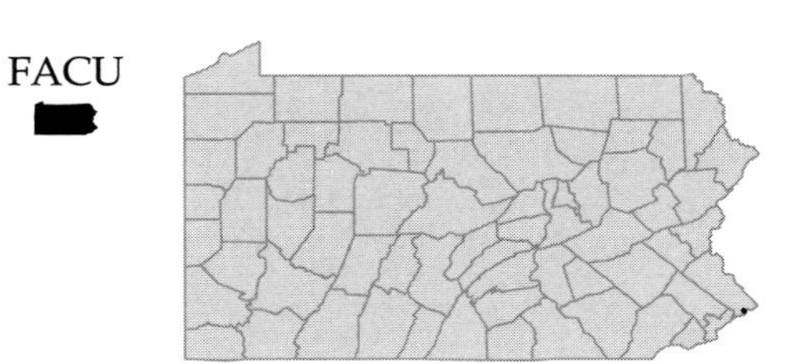

Hypericum densiflorum Pursh
Bushy St.-John's-wort
Deciduous shrub
Rocky river banks, swampy meadows and sphagnum bogs.

FAC+

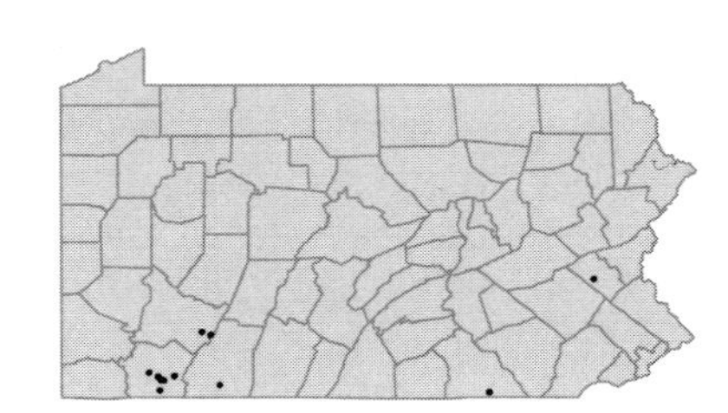

Hypericum denticulatum Walt.
Coppery St.-Johns-wort
Herbaceous perennial
Bogs and wet woods. Believed to be extirpated, all collections were from a single site in Bucks Co. 1865-1866.

FACW-

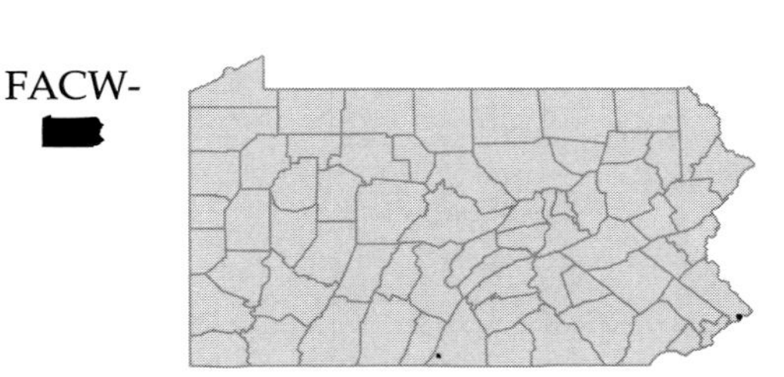

Hypericum dissimulatum Bickn.
St.-John's-wort
Herbaceous perennial
Moist, sandy or peaty soil.

FACW

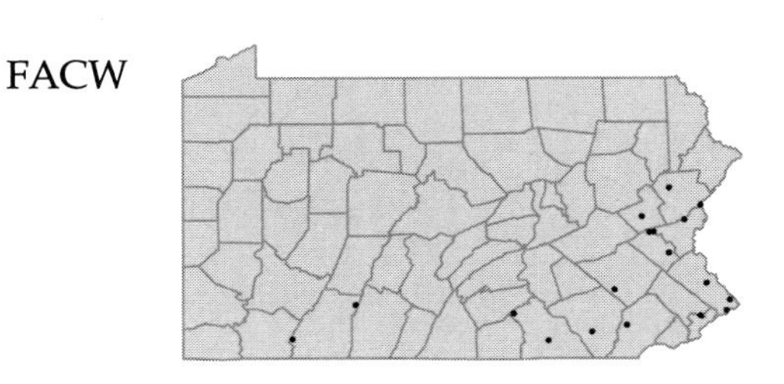

Hypericum drummondii (Grev. & Hook.) Torr. & A.Gray
Nits-and-lice
Herbaceous annual
Dry slopes and stony fields.

UPL

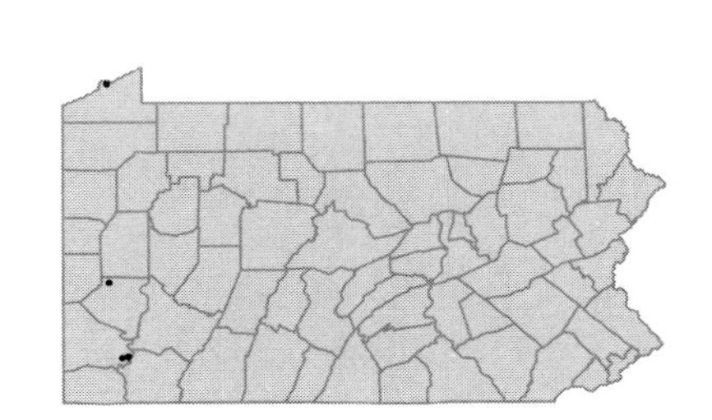

Hypericum ellipticum Hook.
Pale St.-John's-wort
Herbaceous perennial
Sandy river banks, stream borders, marshes, boggy areas and swales.

OBL

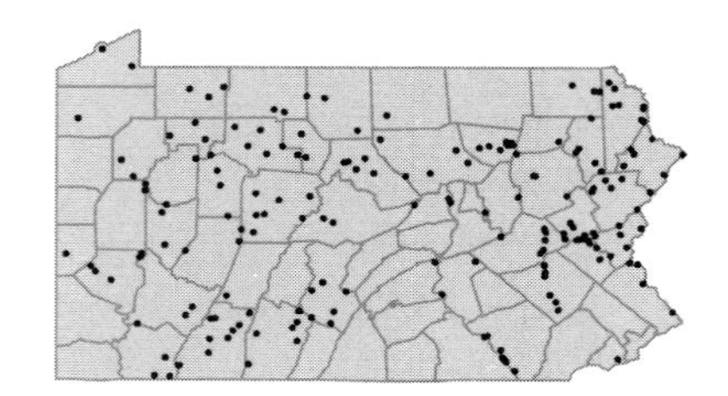

Hypericum gentianoides (L.) BSP
Orange-grass; Pineweed
Herbaceous annual
Abandoned fields, dry hillsides, barrens and open shale slopes.

UPL

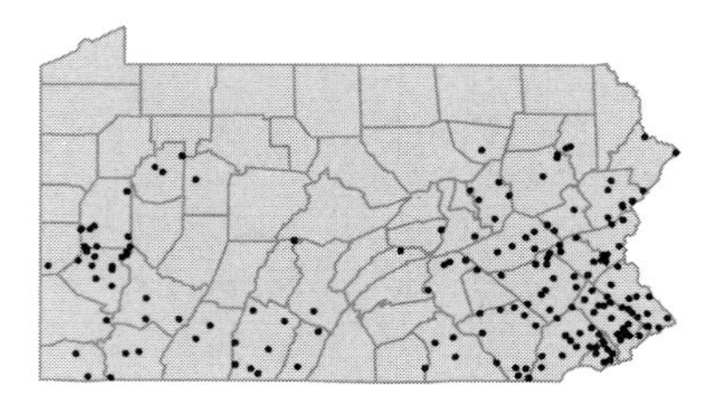

Hypericum gymnanthum Engelm. & A.Gray
Clasping-leaved St.-John's-wort
Herbaceous annual
Muddy shores or edges of intermittent ponds or mud holes. Believed to be extirpated, last collected in 1920.

OBL

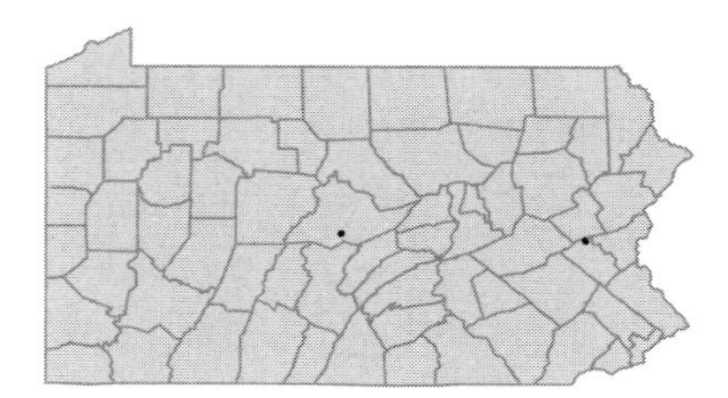

Hypericum hypericoides (L.) Crantz
St.-Andrew's-cross
Deciduous shrub
Open woods, banks, thickets and serpentine barrens, in dry sandy soil.
Ascyrum hypericoides L. PFGBW

FACU

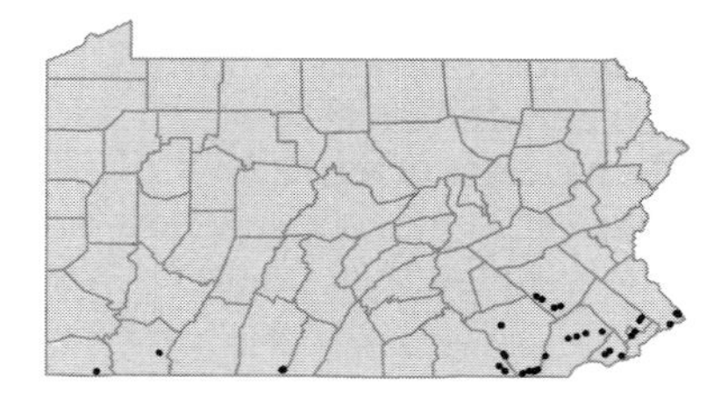

Hypericum majus (A.Gray) Britt.
Canadian St.-John's-wort
Herbaceous annual
Swampy ground and sand plains.

FACW

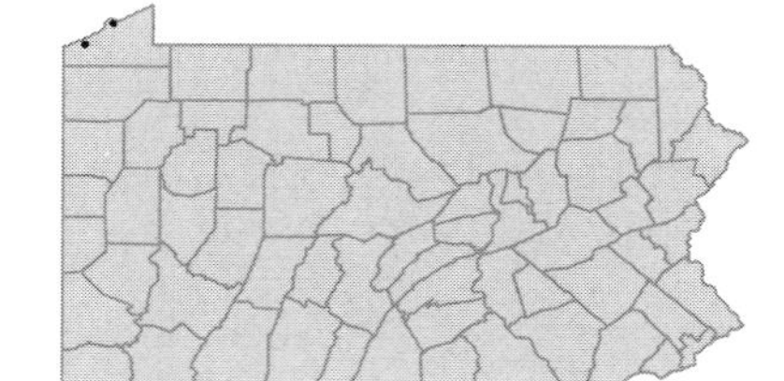

Hypericum mutilum L.
Dwarf St.-John's-wort
Herbaceous perennial
River banks, moist fields, swamps and ditches.

FACW

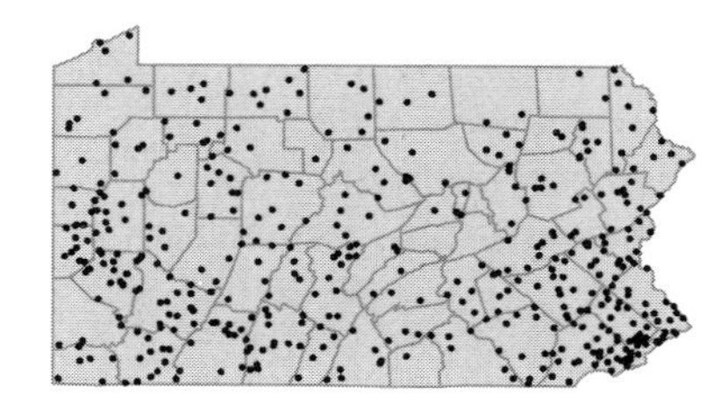

Hypericum perforatum L. I
St.-John's-wort
Herbaceous perennial
Fields, roadsides and waste places.

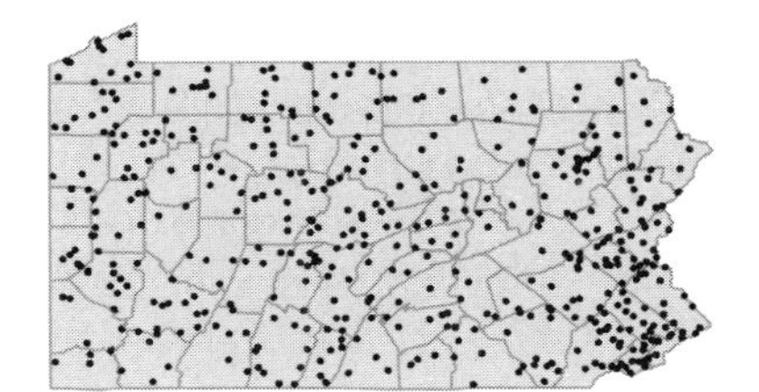

Hypericum prolificum L. FACU
Shrubby St.-John's-wort
Deciduous shrub
Low fields, swamps and thickets.
Hypericum spathulatum (Spach) Steud. FW

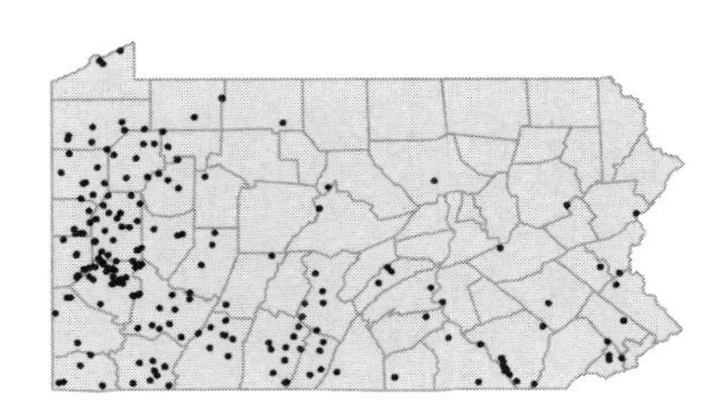

Hypericum punctatum Lam. FAC-
Spotted St.-John's-wort
Herbaceous perennial
Moist fields, floodplains, thickets and roadsides.
Hypericum maculatum Walt. P

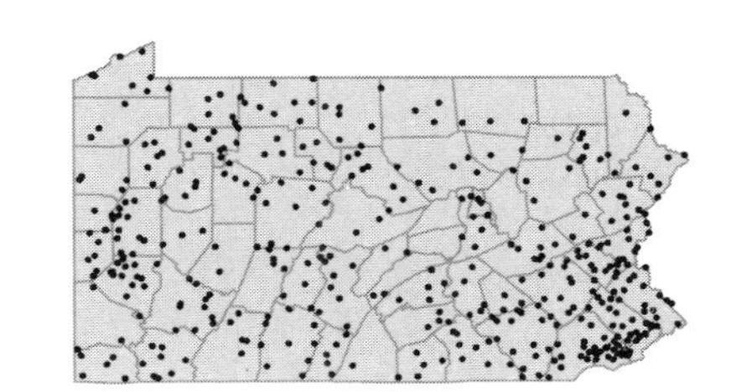

Hypericum pyramidatum Ait. FAC
Great St.-John's-wort
Herbaceous perennial
Alluvial shores, rocky banks, swamps.
Hypericum ascyron L. K

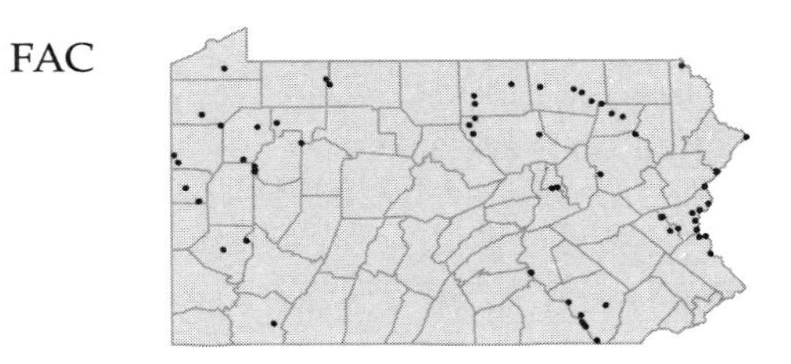

Hypericum sphaerocarpum Michx. FACU
St.-John's-wort
Herbaceous perennial
Rocky shores. Collected several times at a single site in Allegheny Co. 1916-1921.

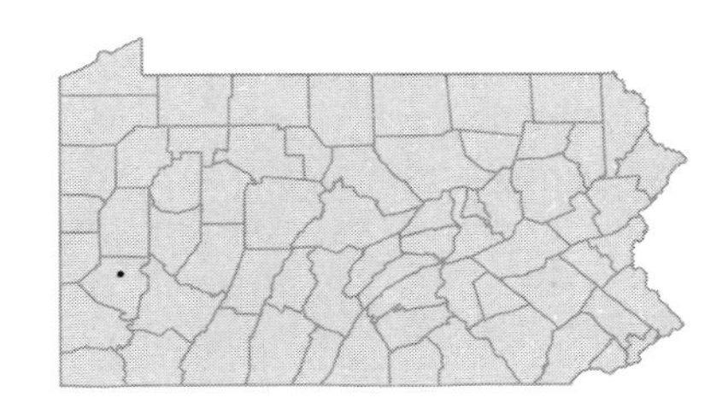

• **_Triadenum fraseri_** (Spach) Gleason OBL
Marsh St.-Johns-wort
Herbaceous perennial
Bogs, swamps, moist meadows and seeps.
Hypericum virginicum L. var. *fraseri* (Spach) Fern. F; *Triadenum virginicum* (L.) Raf. ssp. *fraseri* (Spach) J. Gillett K

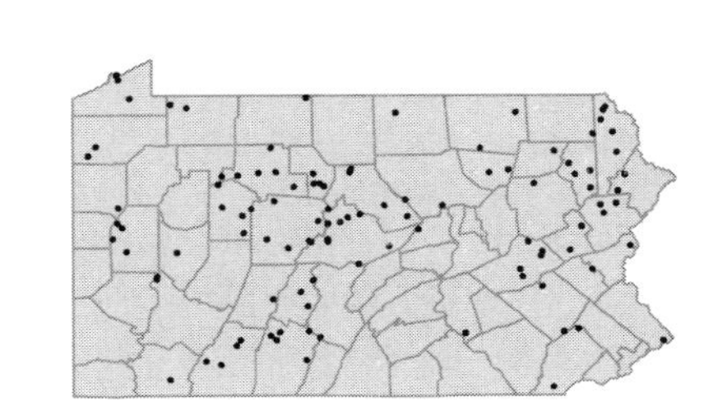

Triadenum virginicum (L.) Raf.
Marsh St.-John's-wort
Herbaceous perennial
Marshes, bogs, swampy woods, river banks and ditches.
Hypericum virginicum L. F

OBL

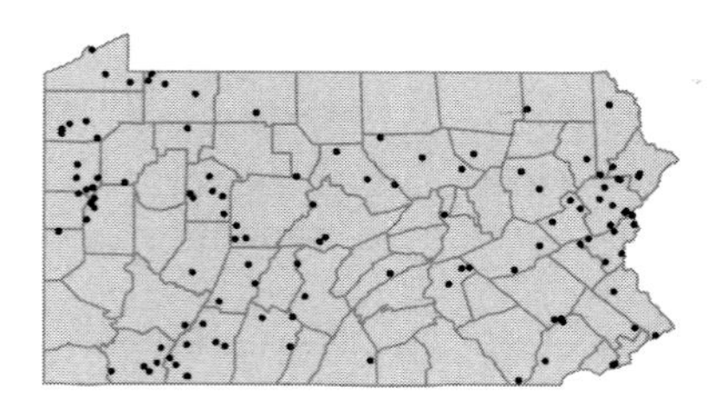

TILIACEAE

• ***Tilia americana*** L. var. ***americana***
Basswood; Linden; Whitewood
Deciduous tree
Rich woods.
Tilia americana L. var. *neglecta* (Spach) Fosberg W

FACU

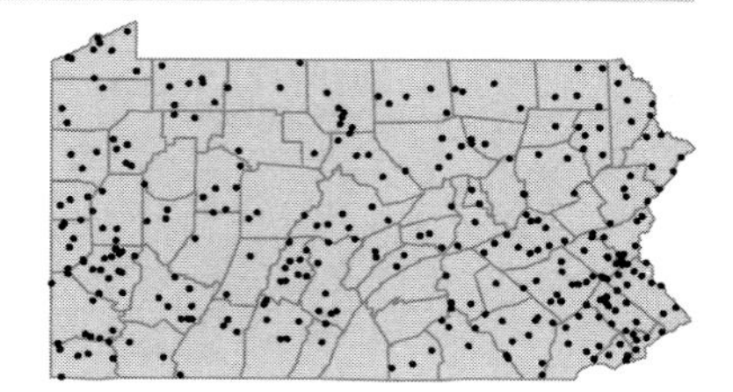

Tilia americana L. var. ***heterophylla*** (Vent.) Loudon
White basswood
Deciduous tree
Rich woods.
Tilia heterophylla Vent. WBGFK

MALVACEAE

• ***Abutilon theophrastii*** Medic.
Butter-print; Pie-marker; Velvet-leaf
Herbaceous annual
Cultivated fields, roadsides and waste ground.

[I]

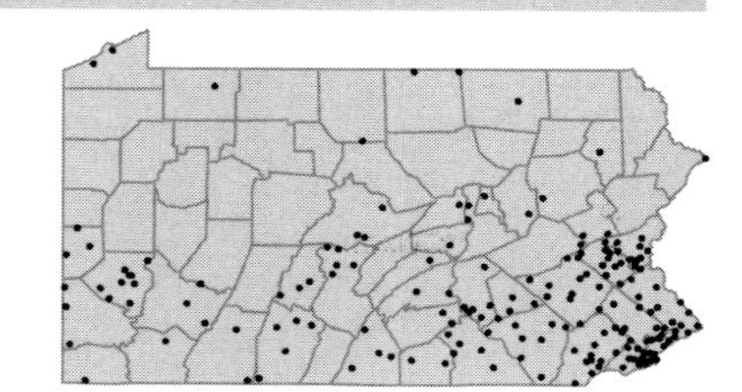

• ***Alcea rosea*** L.
Hollyhock
Herbaceous biennial
Cultivated, and occasionally escaped near gardens, roadsides or rubbish dumps.
Althaea rosea (L.) Cav. PFGBW

[I]

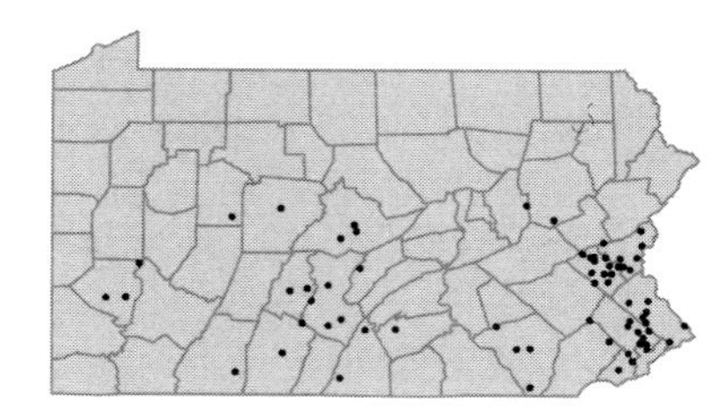

• ***Althaea hirsuta*** L.
Mallow
Herbaceous annual
Ballast and waste ground. Represented by three collections from Philadelphia and Delaware Cos. ca 1878.

[I]

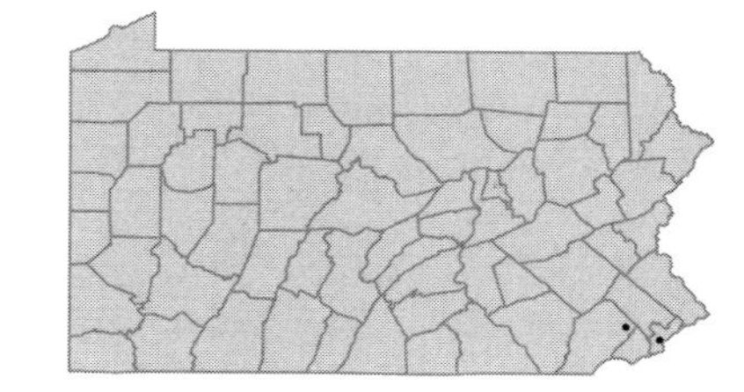

Althaea officinalis L.
Marsh-mallow
Herbaceous perennial
Cultivated and sometimes naturalized in wet ground.

FACW+
[I]

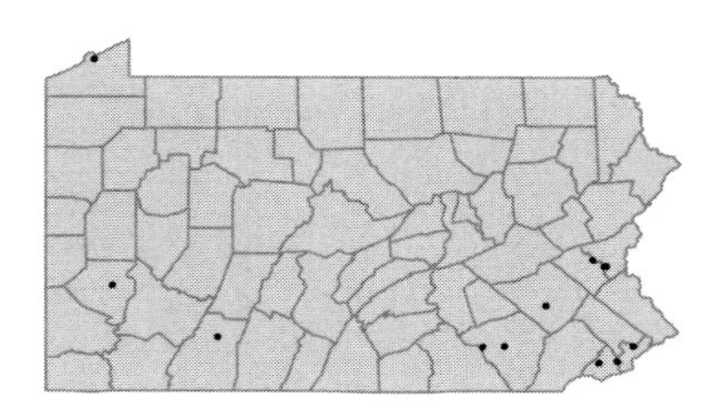

• ***Anoda cristata*** (L.) Schlecht.
Anoda
Herbaceous annual
Waste ground and rubbish dumps.

[I]

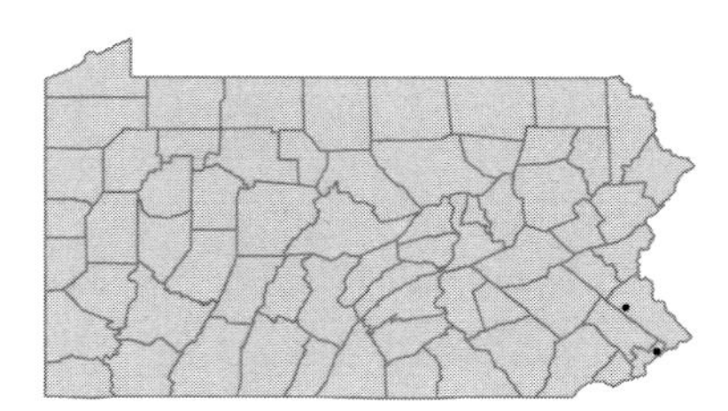

• ***Callirhoe involucrata*** (Torr. & A.Gray) A.Gray
Purple poppy-mallow
Herbaceous perennial
Cultivated and occasionally escaped. Represented by a single collection from Bucks Co. in 1904.

[I]

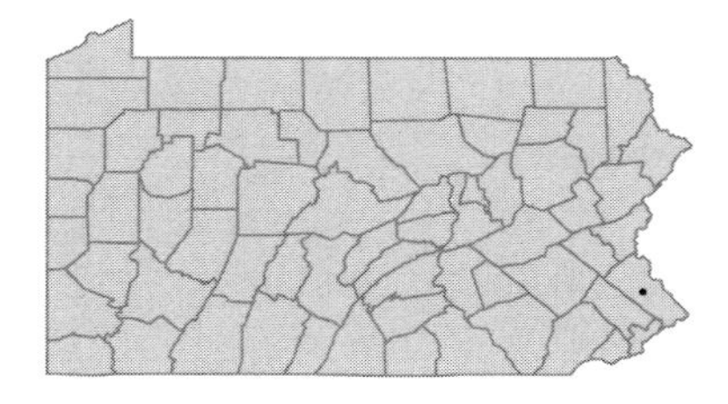

• ***Gossypium herbaceum*** L.
Cotton
Herbaceous annual
Cultivated and also occurring on waste ground and ballast.

[I]

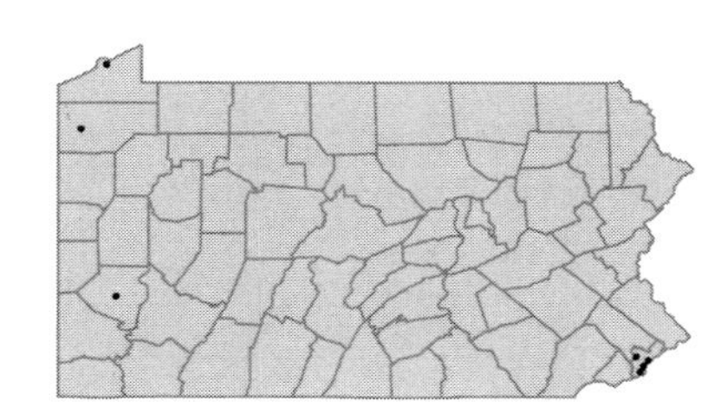

• ***Hibiscus laevis*** All.
Halberd-leaved rose-mallow; Showy hibiscus
Herbaceous perennial
Alluvial shores and marshes, in shallow water.
Hibiscus militaris Cav. PFGBW

OBL

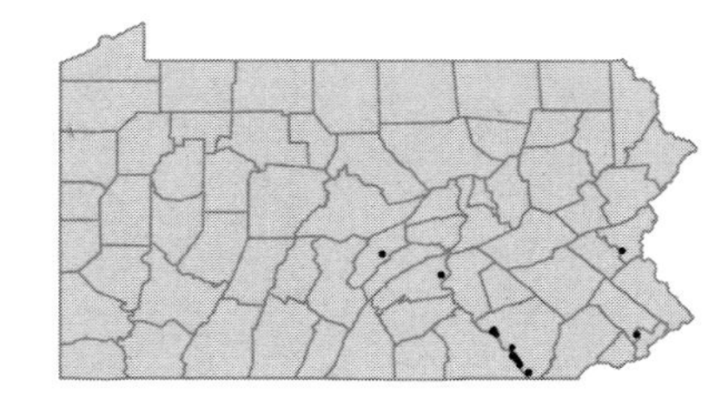

Hibiscus moscheutos L.
Rose-mallow; Swamp-mallow
Herbaceous perennial
Swamps, marshes and ditches, in shallow water.
Hibiscus palustris L. FGBW

OBL

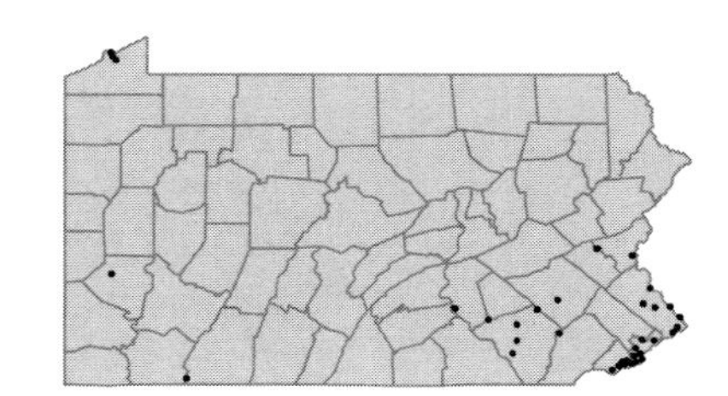

Hibiscus syriacus L.
Rose-of-sharon
Deciduous shrub
Cultivated and occasionally spreading to empty lots or roadsides.

I

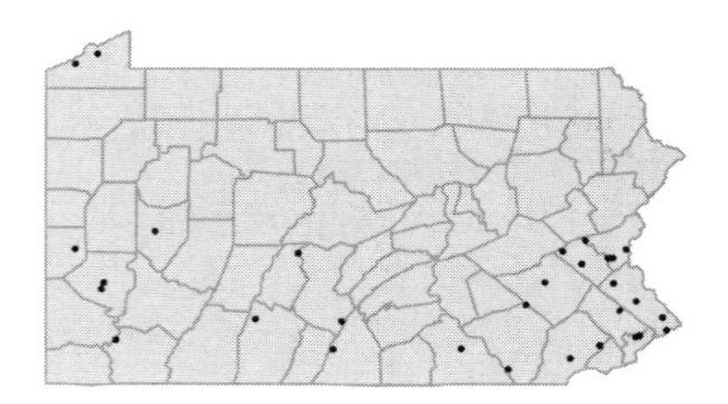

Hibiscus trionum L.
Flower-of-an-hour
Herbaceous annual
Cultivated fields, stream banks and dry, rocky ground.

I

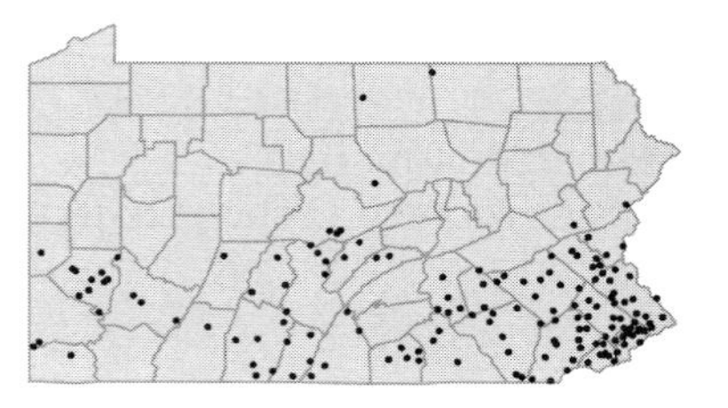

• ***Kosteletzkya virginica*** (L.) Presl
Seaside mallow
Herbaceous perennial
Ballast. Represented by several collections from Philadelphia Co. 1863-1865. Although native in coastal areas, it occurs as a ballast waif in PA.

OBL

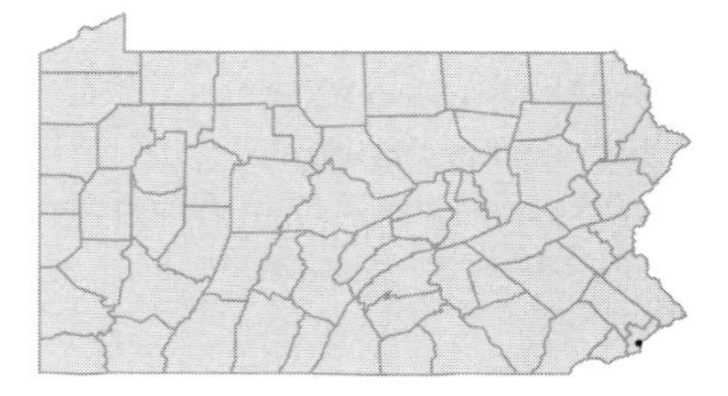

• ***Malva alcea*** L.
Vervain mallow
Herbaceous perennial
Open fields in moist, rocky soil. Collected twice in Tioga Co. 1935-1963.

I

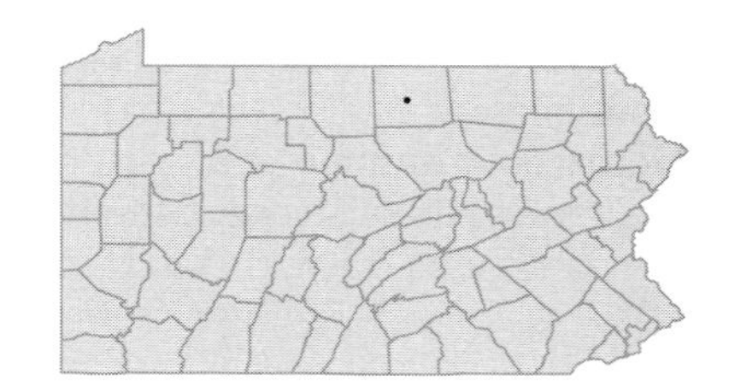

Malva moschata L.
Musk mallow; Rose mallow
Herbaceous perennial
Dry sandy fields, roadsides and fencerows.

I

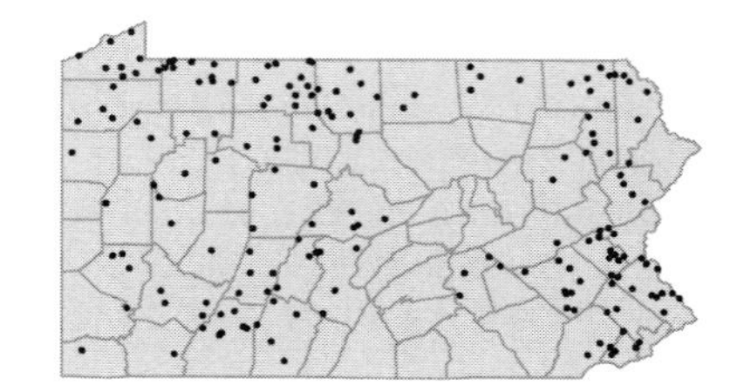

Malva neglecta Wallr.
Cheeses; Common mallow
Herbaceous annual
A weed of gardens, roadsides and waste ground.
M. rotundifolia L. of Porter (1903) = *M. neglecta* Wallr.

I

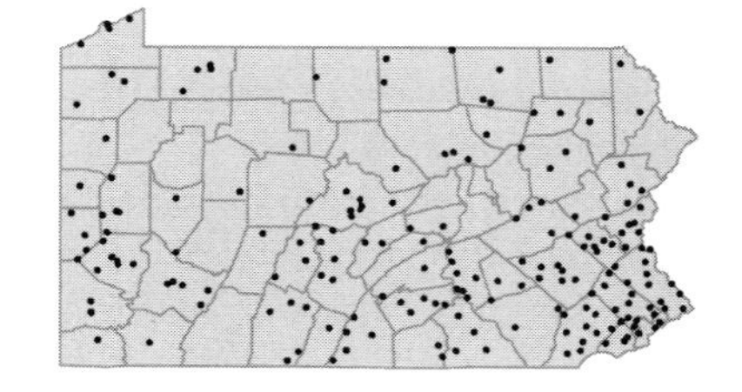

Malva rotundifolia L.
Cheeses
Herbaceous annual
Ballast and waste ground.
Malva pusilla Sm. W

I

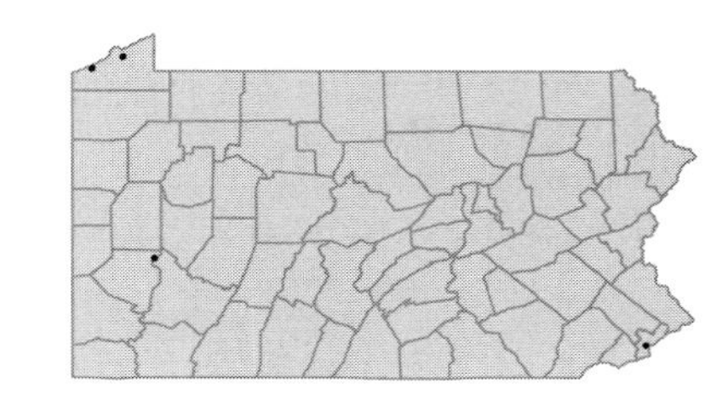

Malva sylvestris L.
Cheeses; High mallow
Herbaceous biennial
Cultivated, also occurring on ballast, roadsides and waste ground.

I

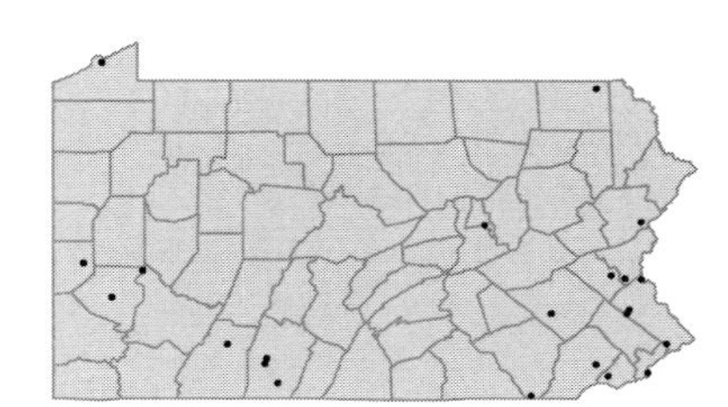

Malva verticillata L.
Whorled mallow
Herbaceous annual
Cultivated and rarely escaped to roadsides and meadows.

I

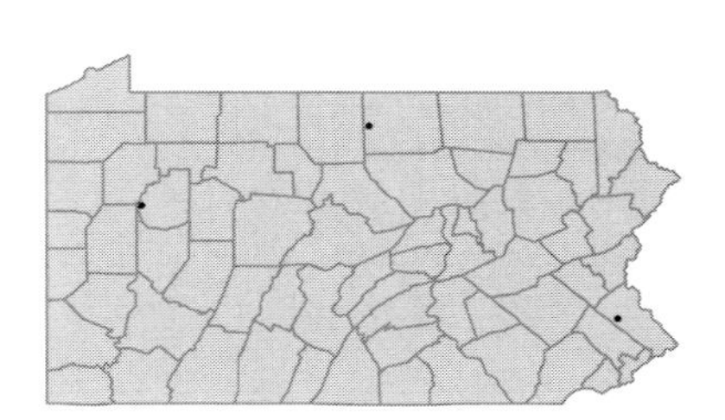

• ***Malvastrum coromandelianum*** (L.) Garcke
False mallow
Herbaceous annual
Ballast and waste ground. Represented by three collections from Philadelphia and Delaware Cos. 1865-1932.

I

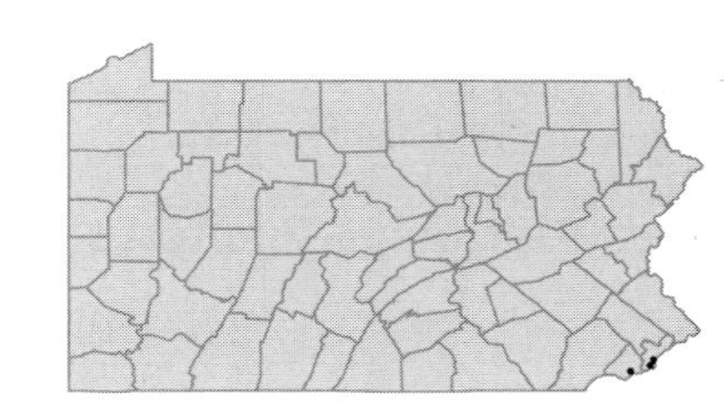

• ***Modiola caroliniana*** (L.) G.Don
Bristly-fruited mallow
Herbaceous annual
Ballast and waste ground. Represented by several collections from Philadelphia Co. ca. 1865.

I

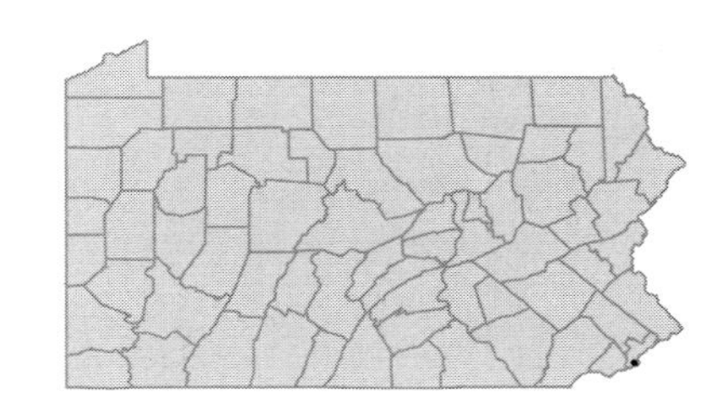

• ***Sida acuta*** Burm. f.
Broomweed
Deciduous shrub
Ballast and waste ground. Represented by several collections from Philadelphia Co. 1864-1869.
Sida stipulacea Cav.

I

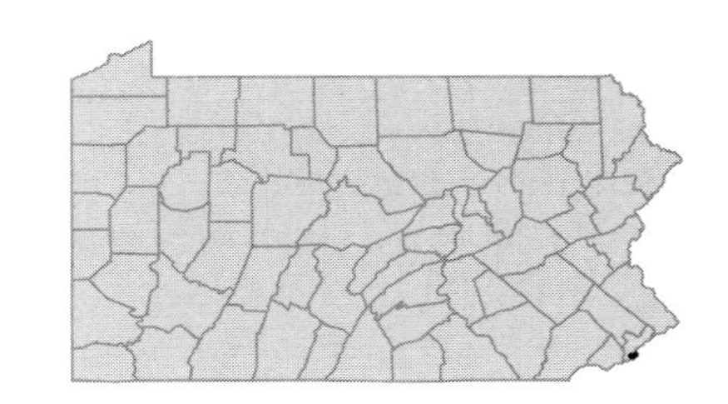

Sida hermaphrodita (L.) Rusby
Virginia mallow
Herbaceous perennial
River banks, in moist alluvial soil.

FAC

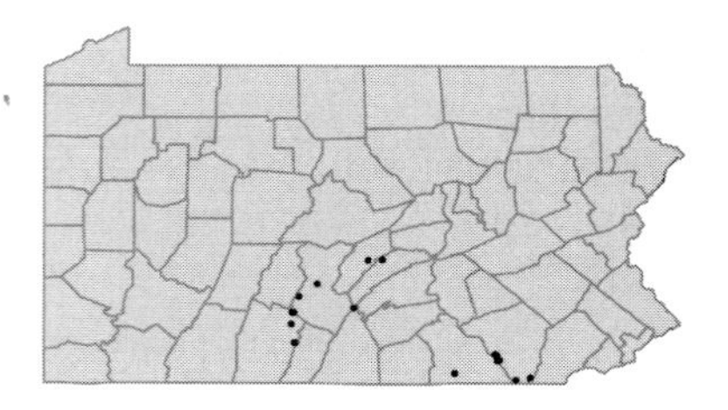

Sida linifolia Cav.
Sida
Herbaceous annual
Ballast. Represented by a single collection from Philadelphia Co. in 1879.

I

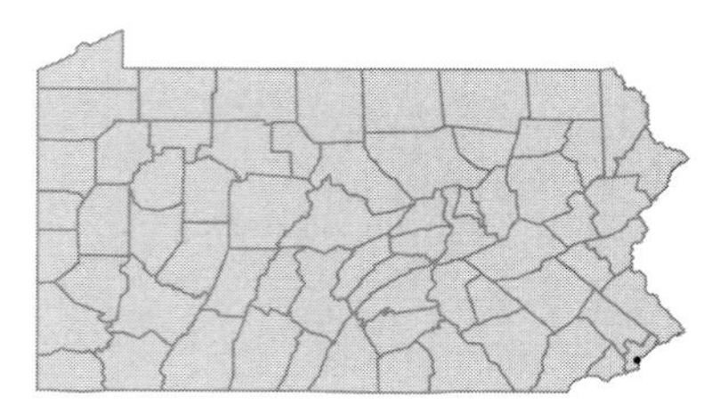

Sida rhombifolia L.
Arrow-leaf sida
Herbaceous annual
Ballast and waste ground. Represented by two collections from Philadelphia and Delaware Cos. 1865-1932.

UPL
I

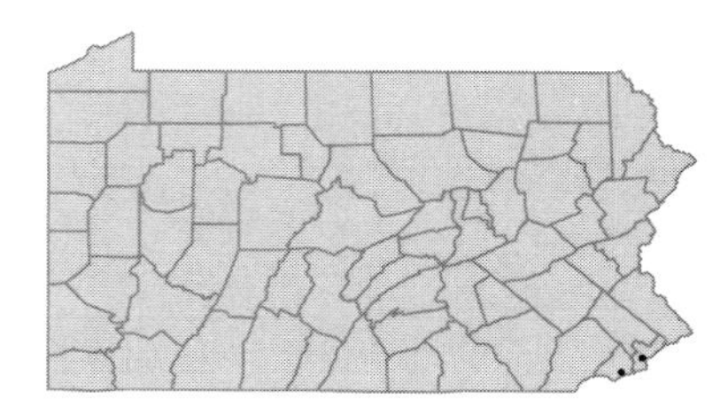

Sida spinosa L.
Prickly sida; False mallow
Herbaceous annual
Cultivated fields, roadsides, stream banks, waste ground and ballast.

UPL
I

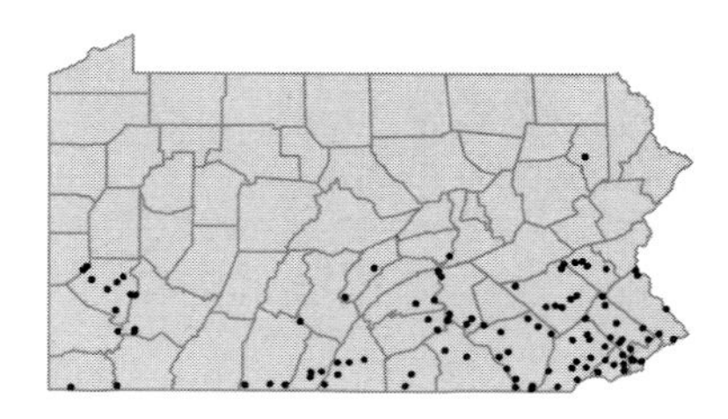

SARRACENIACEAE

• ***Sarracenia purpurea*** L.
Pitcher-plant
Herbaceous perennial
Sphagnum bogs. Carnivorous.

OBL

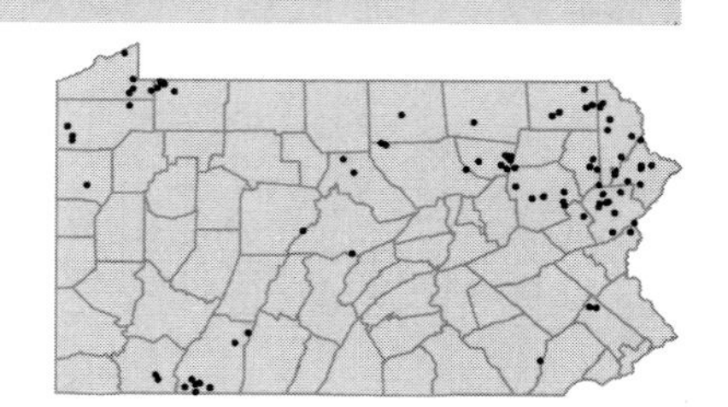

DROSERACEAE

• ***Drosera intermedia*** Hayne
Spatulate-leaved sundew
Herbaceous perennial
Open peat at the edges of bogs and glacial lakes. Carnivorous.

OBL

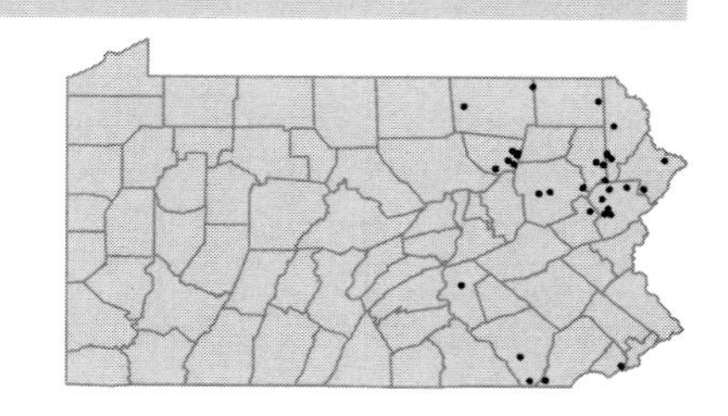

Drosera rotundifolia L.
Round-leaved sundew
Herbaceous perennial
Sphagnum bogs and peaty edges. Carnivorous.

OBL

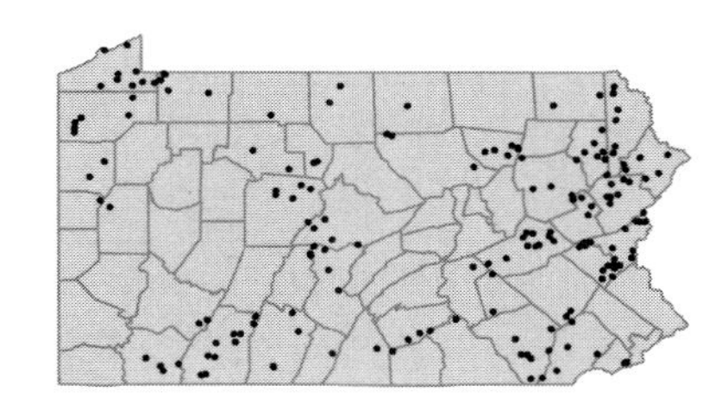

CISTACEAE

• ***Helianthemum bicknellii*** Fern.
Bicknell's hoary rockrose; Frostweed
Herbaceous perennial
Dry, rocky hillsides, open woods and serpentine barrens.
Helianthemum majus (L.) BSP P

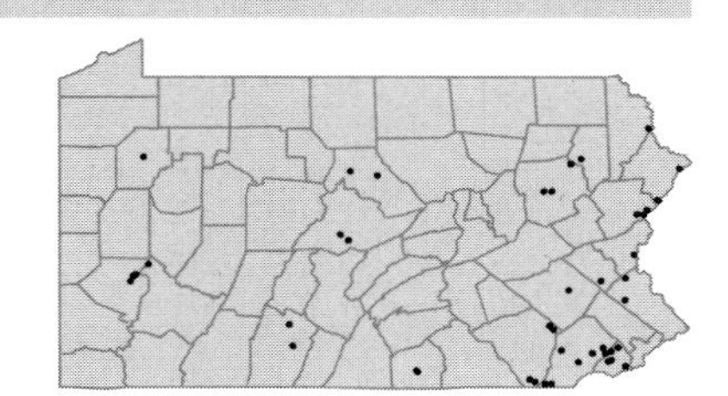

Helianthemum canadense (L.) Michx.
Frostweed
Herbaceous perennial
Dry, sandy or rocky ground, open woods and barrens.

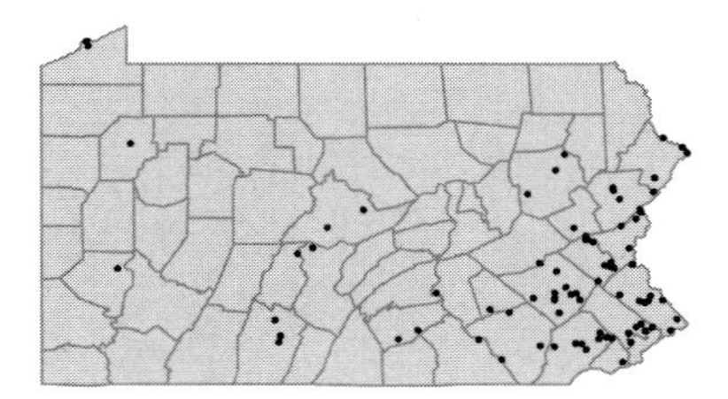

Helianthemum propinquum Bickn.
Frostweed
Herbaceous perennial
Dry, sandy soil and barrens.

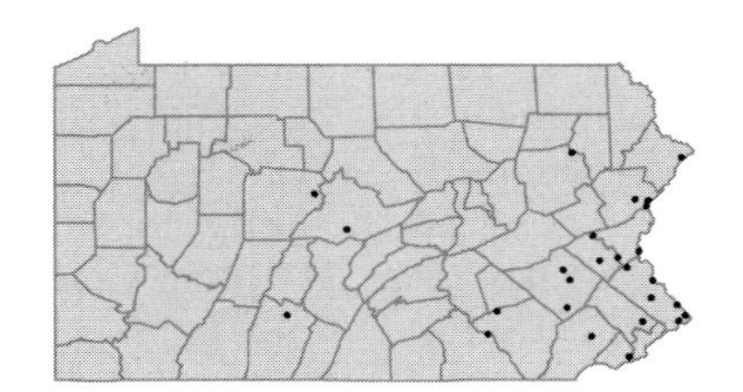

• ***Lechea intermedia*** Leggett ex Britt.
Pinweed
Herbaceous perennial
Dry, sandy fields and dry ridges.

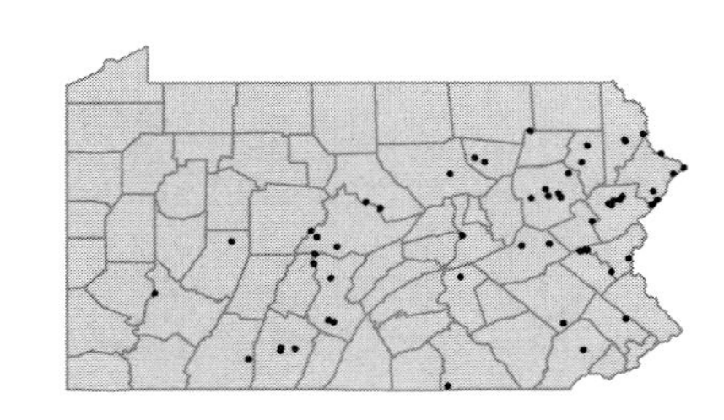

Lechea minor L.
Thyme-leaved pinweed
Herbaceous perennial
Rocky woods, slopes and serpentine barrens, in dry, sandy soil.

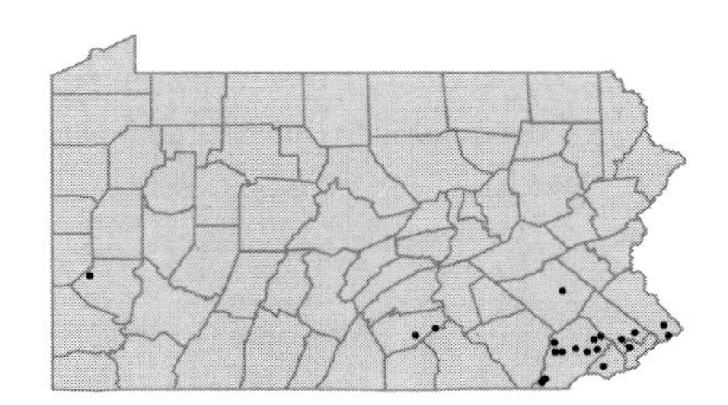

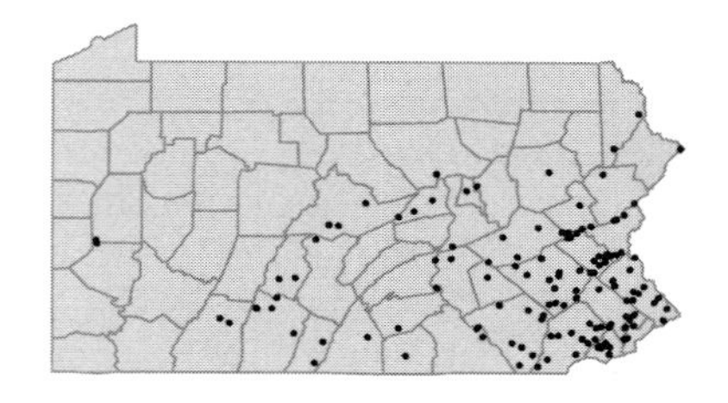

Lechea pulchella Raf.
Pinweed
Herbaceous perennial
Dry open woods, sandy fields and barrens.
Lechea leggettii Britt. & Hollick PFGBW

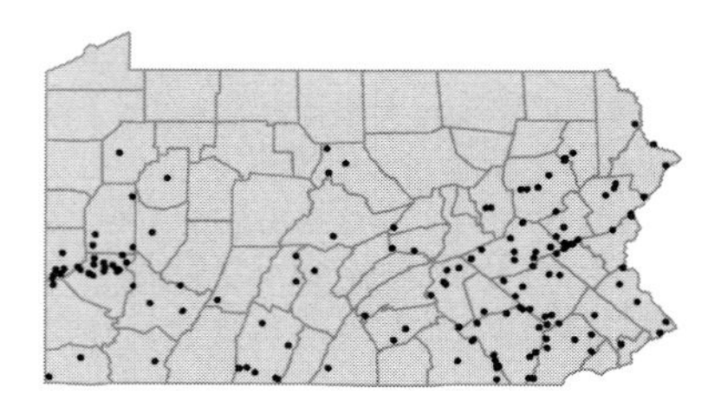

Lechea racemulosa Lam.
Pinweed
Herbaceous perennial
Dry fields and shaly slopes.

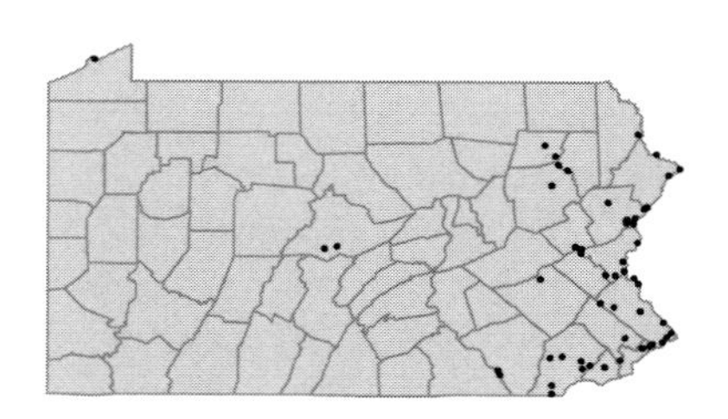

Lechea villosa Ell.
Pinweed
Herbaceous perennial
Dry slopes, open woods and sand plains.
Lechea mucronata Raf. C

VIOLACEAE

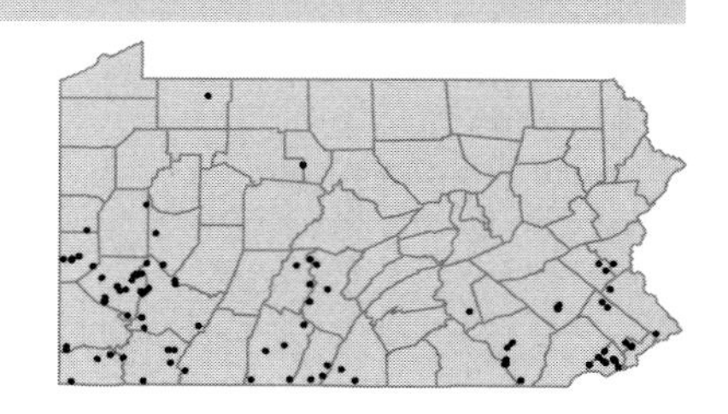

• ***Hybanthus concolor*** (T.F.Forst.) Spreng.
Green violet
Herbaceous perennial
Rich woods and wooded slopes.
Cubelium concolor (T.F.Forst.) Raf. PGB

Viola—In addition to the taxa treated below, limited collections exist representing the following *Viola* hybrids: *affinis x sororia*; *conspersa x rostrata*; *conspersa x striata* (*V. x eclipes* H.Ballard); *cucullata x sororia* (*V. x bissellii* House); *fimbriatula x hirsuta* (*V. x redacta* House); *fimbriatula x sororia* (*V. x aberrans* Greene); *hirsutula x sagittata* (*V. x dimissa* House); *hirsutula x sororia* (*V. x cordifolia* (Nutt.) Schwein.; *palmata x sororia* (*V. x populifolia* Greene); *rostrata x striata* (*V. x brauniae* Grover ex Cooperrider).

• ***Viola affinis*** LeConte
LeConte's violet
Herbaceous perennial
Rich moist woods, especially on alluvial soil.

FACW

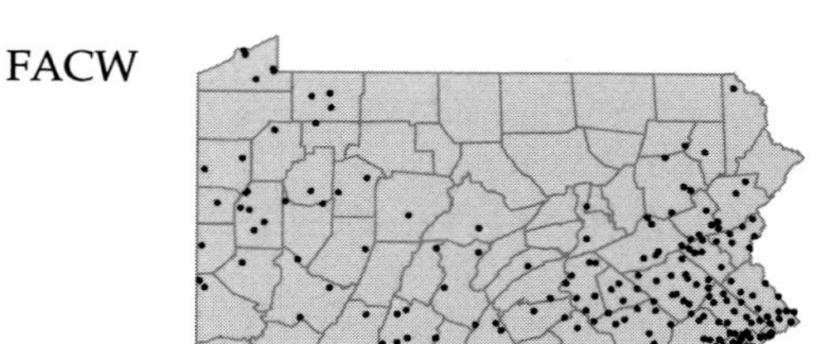

Viola arvensis Murr.
European field pansy
Herbaceous annual
Cultivated fields and waste ground.

I

Viola blanda Willd.
Sweet white violet
Herbaceous perennial
Moist, shady woods and cool ravines.
Viola incognita Brainerd FGBW

FACW

Viola canadensis L. var. ***canadensis***
Canada violet
Herbaceous perennial
Cool, shady woods and ravines.

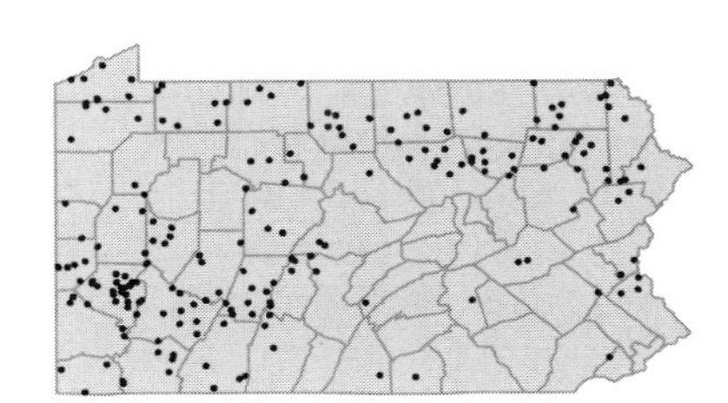

Viola conspersa Reichenb.
American dog violet
Herbaceous perennial
Swamps, meadows and alluvial woods.
Viola labradorica Schrank

FACW

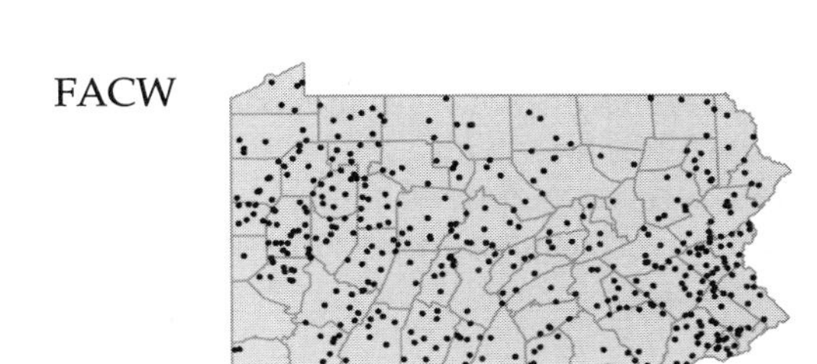

Viola cucullata Ait.
Blue marsh violet
Herbaceous perennial
Swamps, bogs and wet meadows.
Viola obliqua Hill K

FACW+

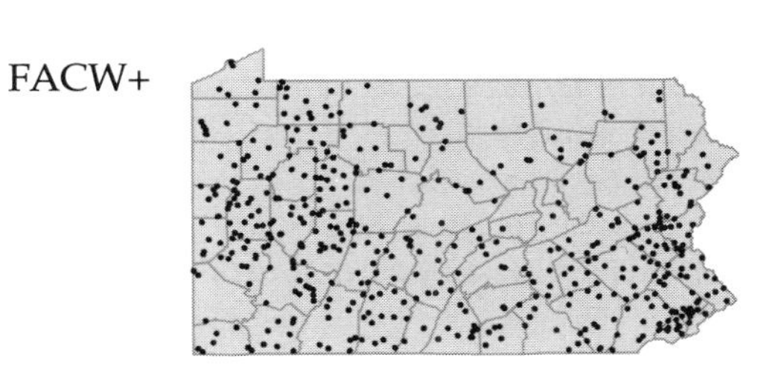

Viola cucullata x fimbriatula
Porter's violet
Herbaceous perennial
Rocky banks and open woods.
Viola x porterana Pollard FW

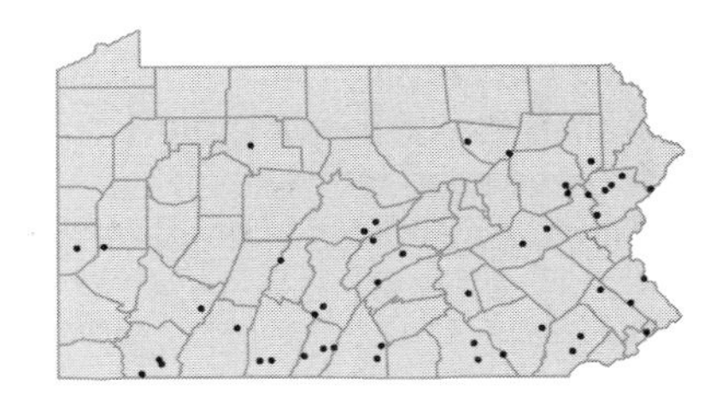

Viola eriocarpa Schwein.
Smooth yellow violet
Herbaceous perennial
Rich woods.
Viola pensylvanica Michx. FW; *Viola pubescens* Ait. var. *eriocarpa* (Schwein.) Russell KC; *Viola scabriuscula* (Torr. & A.Gray) Schwein. P

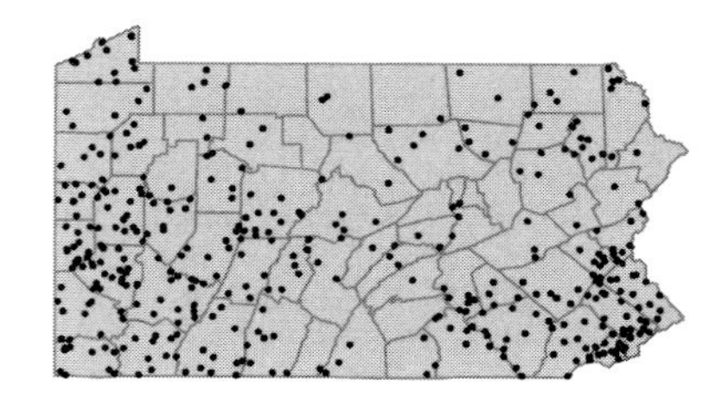

Viola hastata Michx.
Halberd-leaved yellow violet
Herbaceous perennial
Rich deciduous woods.

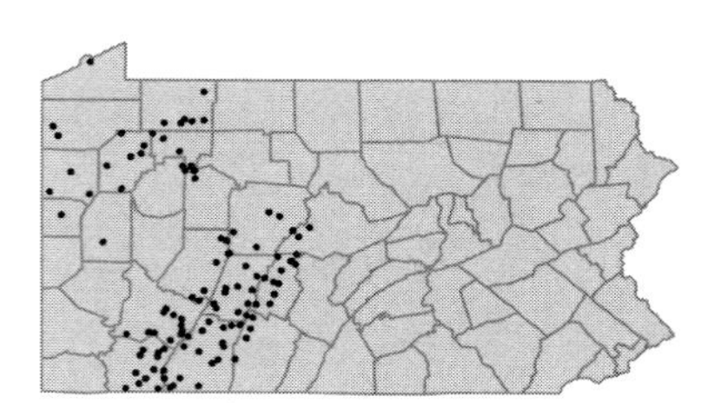

Viola hirsutula Brainerd
Southern wood violet
Herbaceous perennial
Rich, dry, open woods.
Viola villosa Walt. *pro parte* C

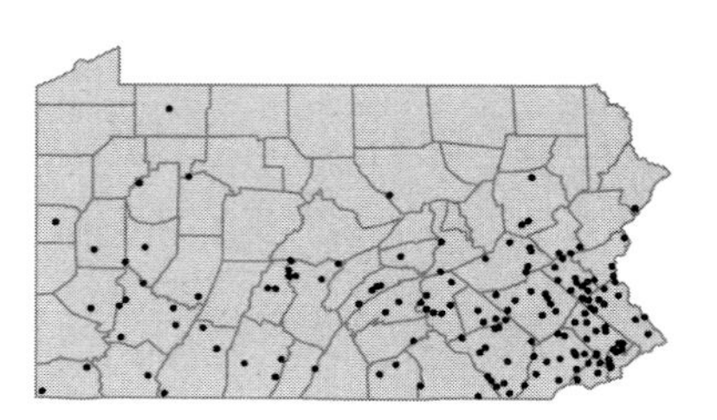

Viola lanceolata L. var. ***lanceolata*** OBL
Lance-leaved violet
Herbaceous perennial
Bogs, wet meadows and moist shores.

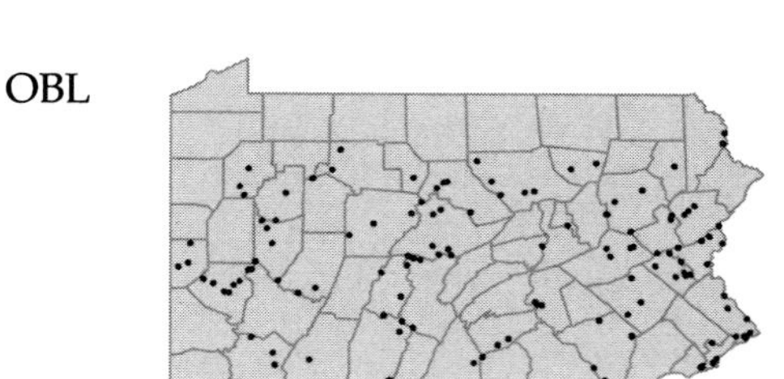

Viola macloskeyi Lloyd ssp. ***pallens*** (Banks ex DC.) M.S.Baker OBL
Sweet white violet
Herbaceous perennial
Bogs, swamps and wet woods.
Viola pallens (Banks) Brainerd FGBW

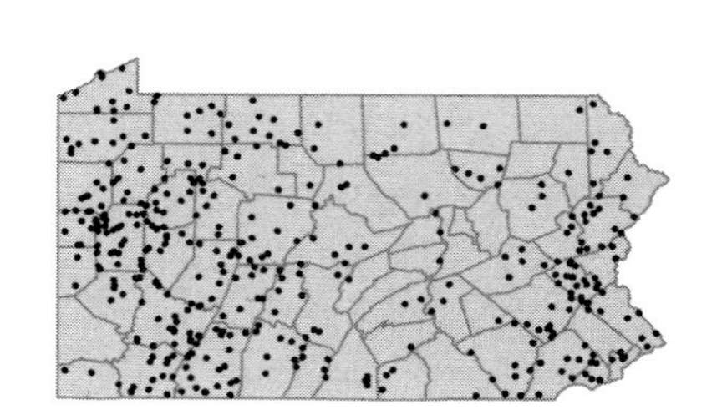

Viola odorata L. I
English violet; Sweet violet
Herbaceous perennial
Cultivated and occasionally naturalized in lawns and woodland edges.

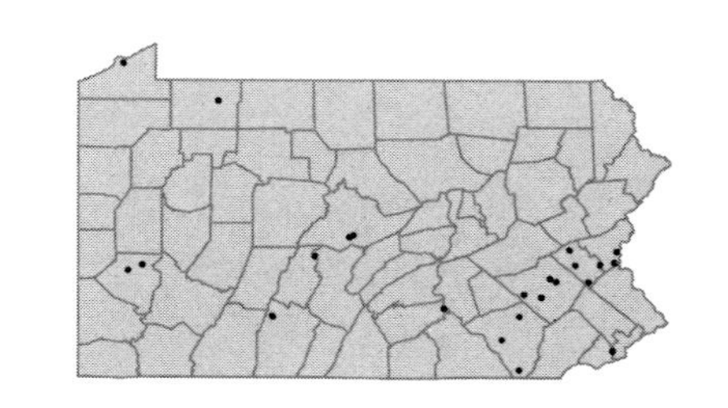

Viola palmata L.
Early blue violet
Herbaceous perennial
Rich, dry, open woods and edges.
Viola triloba Schwein. *pro parte* WFBKG; *Viola stoneana* House *pro parte* WFKBG

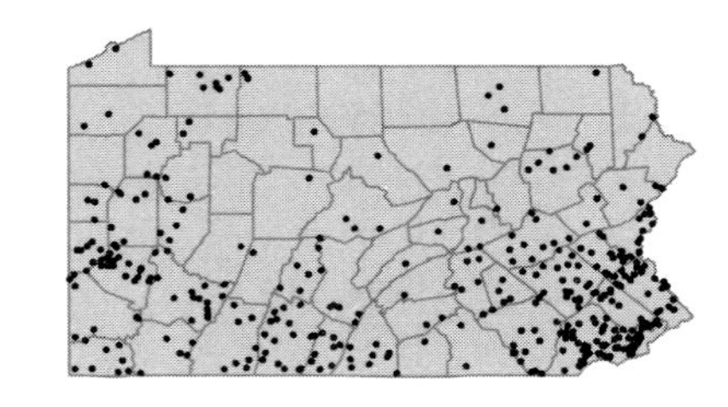

Viola pedata L.
Birdfoot violet
Herbaceous perennial
Dry, rocky or sandy ground, shale barrens.

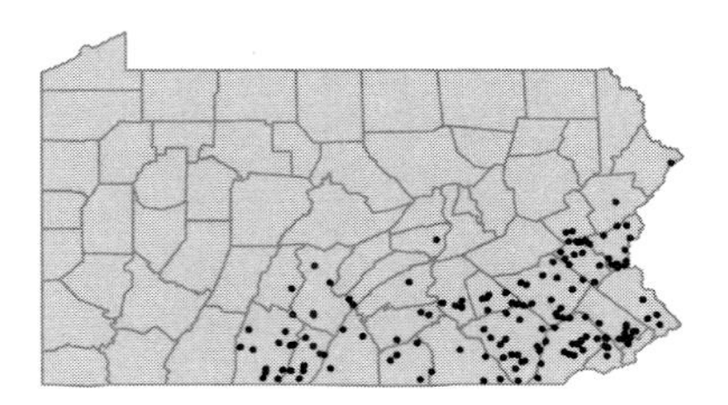

Viola pedatifida G.Don ssp. ***brittoniana*** (Pollard) McKinney
Coast violet
Herbaceous perennial
Moist, sandy soils.
Viola palmata L. var. *palmata pro parte* C; *Viola brittoniana* Pollard WFBGK

FAC

Viola primulifolia L.
Primrose violet
Herbaceous perennial
Moist woods, meadows and swamps.

FAC+

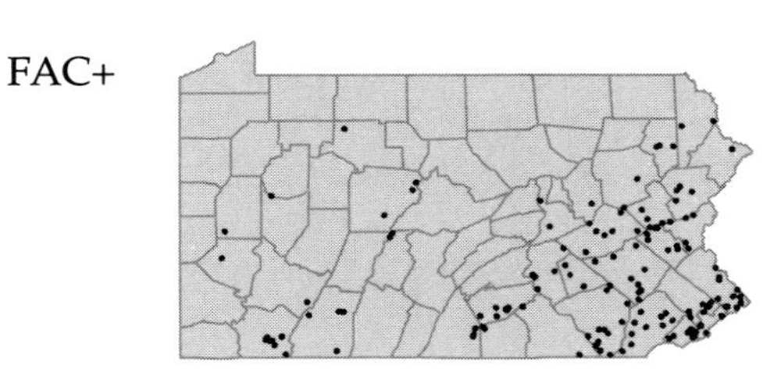

Viola pubescens Ait.
Downy yellow violet
Herbaceous perennial
Dry to moist woods.

FACU-

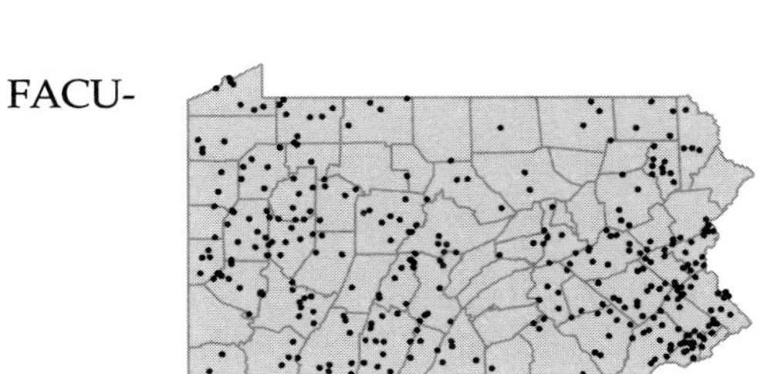

Viola rafinesquii Greene
Field pansy
Herbaceous annual
Fields and open woods.
Viola kitaibeliana Roemer & Schultes var. *rafinesquii* (Greene) Fern. F

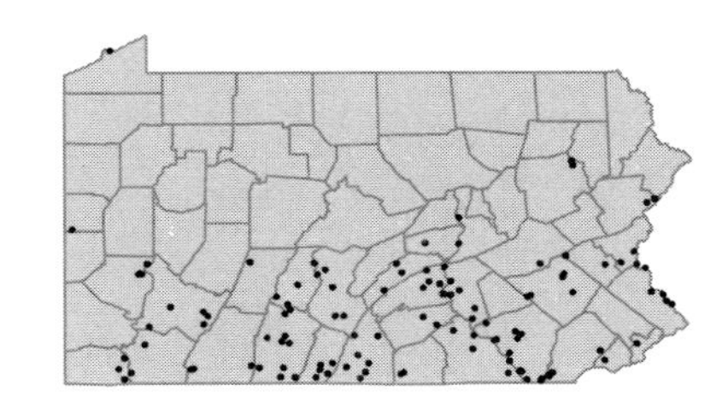

Viola renifolia A.Gray
Kidney-leaved white violet
Herbaceous perennial
Wooded swamps and cool, wooded slopes.

FACW

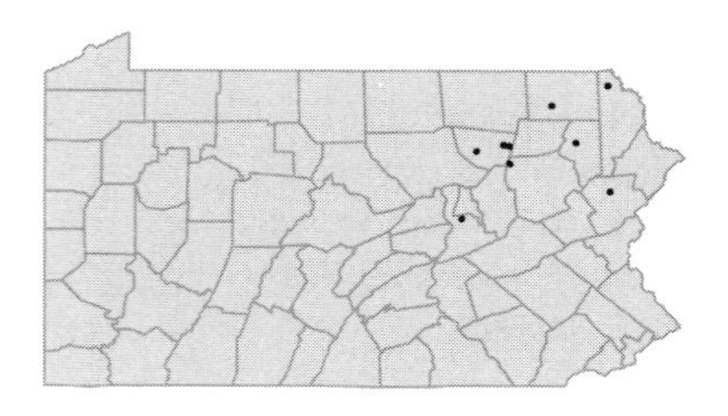

Viola rostrata Pursh
Long-spurred violet
Herbaceous perennial
Rich woods.

FACU

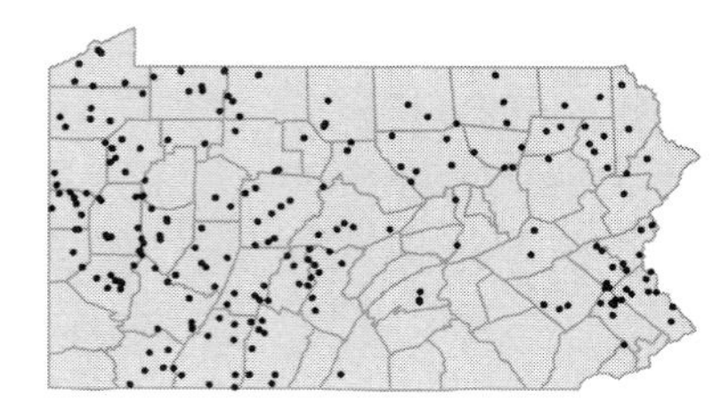

Viola rotundifolia Michx.
Round-leaved violet
Herbaceous perennial
Cool, moist, shady woods and banks.

FAC+

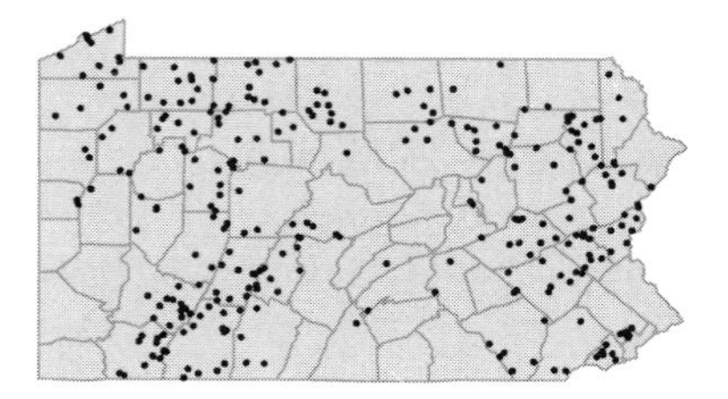

Viola sagittata Ait. var. ***sagittata***
Arrow-leaved violet
Herbaceous perennial
Dry woods, fields and edges.
Viola emarginata (Nutt.) LeConte PFGB

FACW

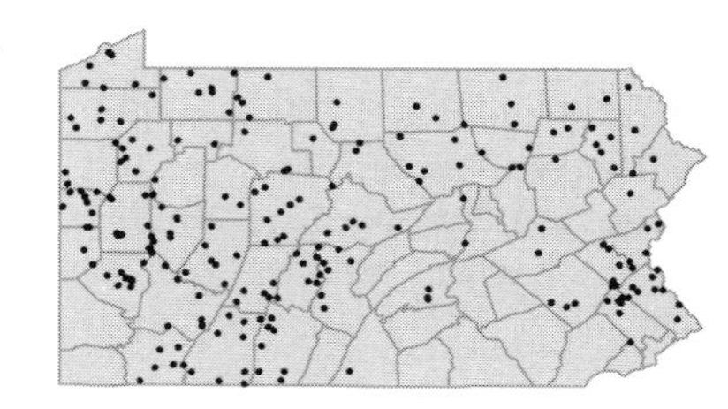

Viola sagittata Ait. var. ***ovata*** (Nutt.) Torr. & A.Gray
Ovate-leaved violet
Herbaceous perennial
Dry fields and woodland edges.
Viola fimbriatula J.E.Smith WBGFK

Viola selkirkii Pursh ex Goldie
Great-spurred violet
Herbaceous perennial
Cool woods and shady ravines.

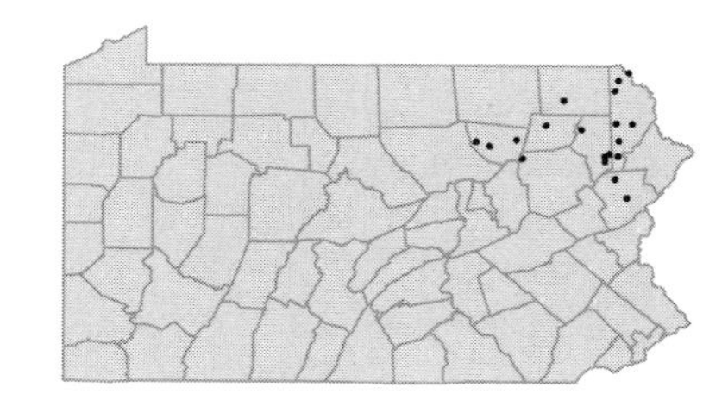

Viola sororia Willd. var. ***sororia*** FAC-
Common blue violet
Herbaceous perennial
Meadows and woods.
Viola papilionacea Pursh *pro parte* PGB; *Viola priceana* Pollard *pro parte* K; *Viola septentrionalis* Greene *pro parte* FGBW; *Viola sororia f. priceana* (Pollard) Cooperrider *pro parte*

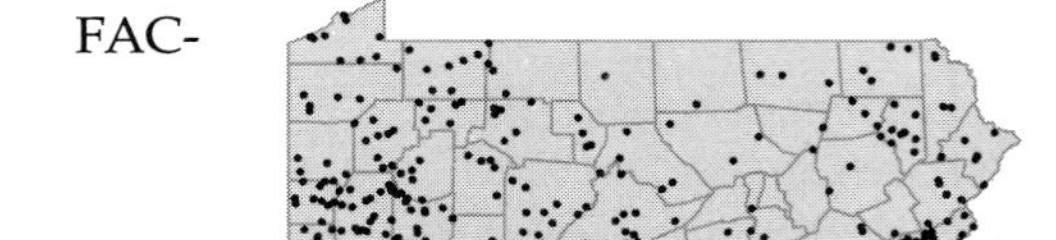

Viola sororia Willd. var. ***missouriensis*** (Greene) McKinney FAC
Missouri violet
Herbaceous perennial
Damp thickets and alluvial woods.
Viola missouriensis Greene

Viola striata Ait. FACW
Striped violet
Herbaceous perennial
Moist, rich woods and floodplains.

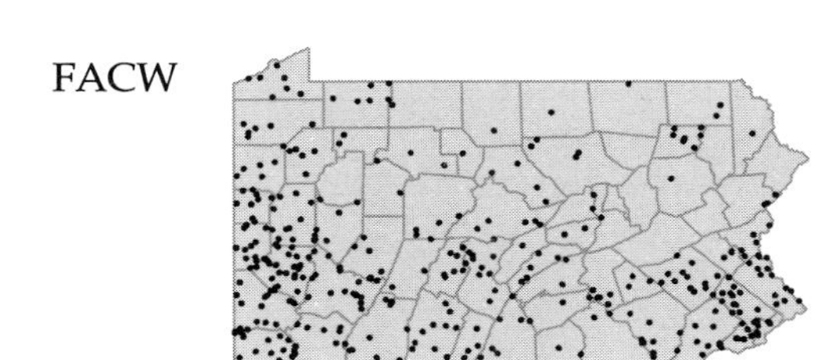

Viola subsinuata Greene
Violet
Herbaceous perennial
Open or rocky woods and cliffs.
Viola palmata L. *pro parte* FGBK

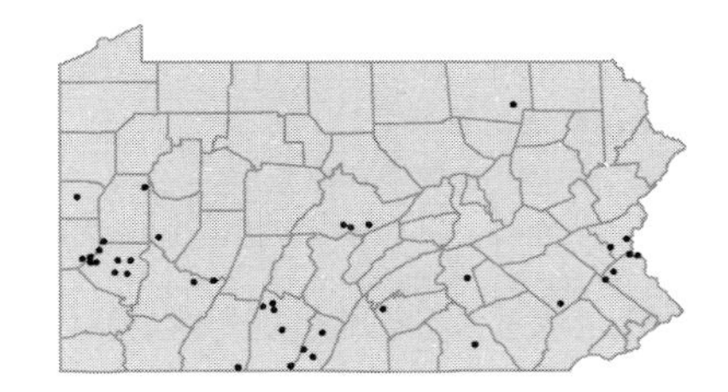

Viola tricolor L. [I]
Johnny-jump-up
Herbaceous annual
Lawns and waste ground, an occasional garden escape.

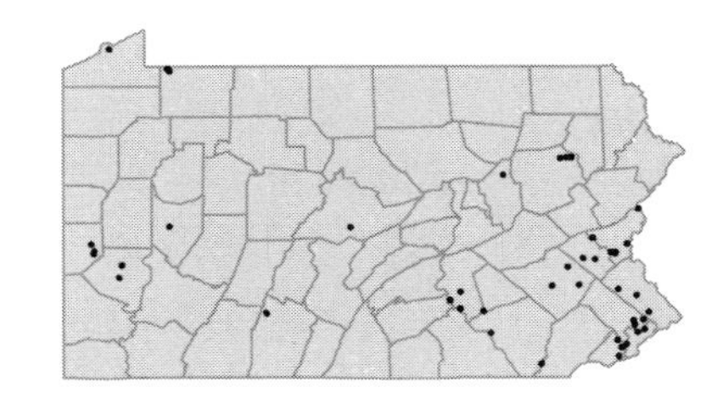

Viola tripartita Ell.
Three-parted violet
Herbaceous perennial
Deciduous woods. Represented by a single collection from Fayette Co. in 1920.

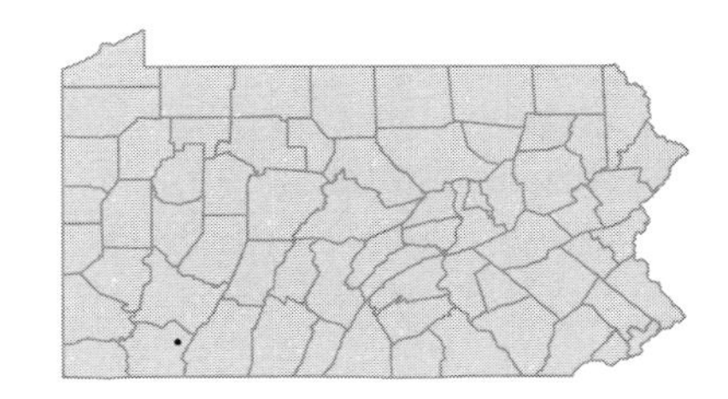

Viola walteri House
Appalachian blue violet
Herbaceous perennial
Dry wooded slopes and rocky ledges.
Viola appalachiensis L.K.Henry W

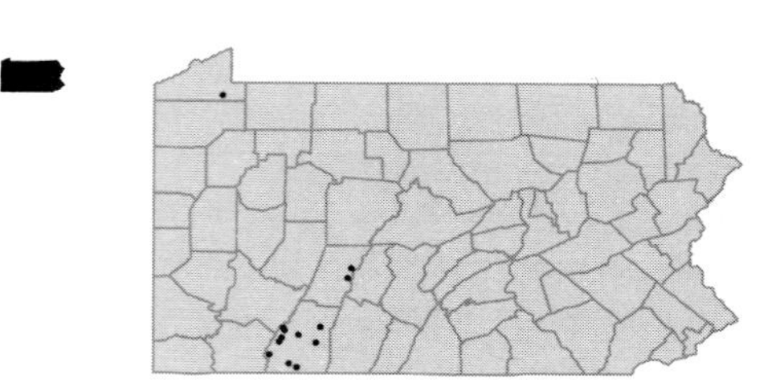

TAMARICACEAE

• ***Tamarix gallica*** L.
French tamarisk
Deciduous shrub
Cultivated and occasionally persisting at old garden sites.

FACW
I

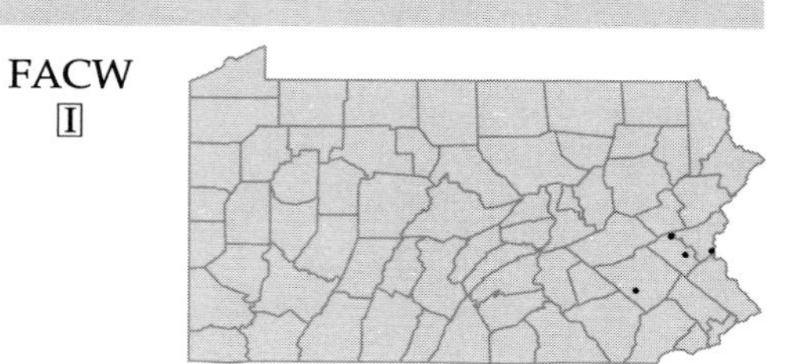

PASSIFLORACEAE

• ***Passiflora lutea*** L.
Passion-flower
Herbaceous perennial vine
Riverbanks and thickets.

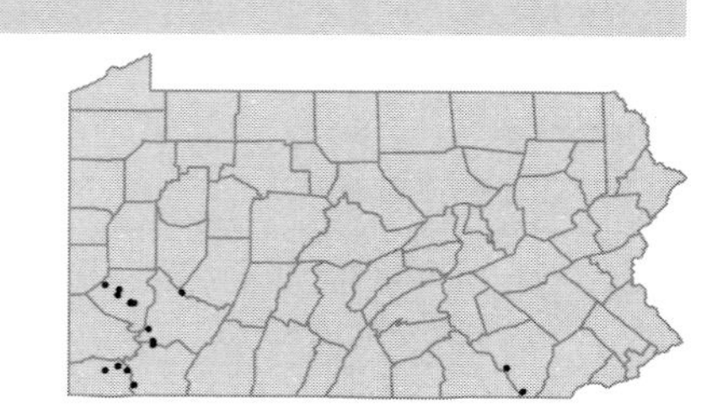

CUCURBITACEAE

• ***Citrullus colocynthis*** (L.) Schrad.
Watermelon
Herbaceous annual vine
Cultivated and occasionally occurring in waste ground or near rubbish dumps.
Citrullus vulgaris Schrad. FGBW

I

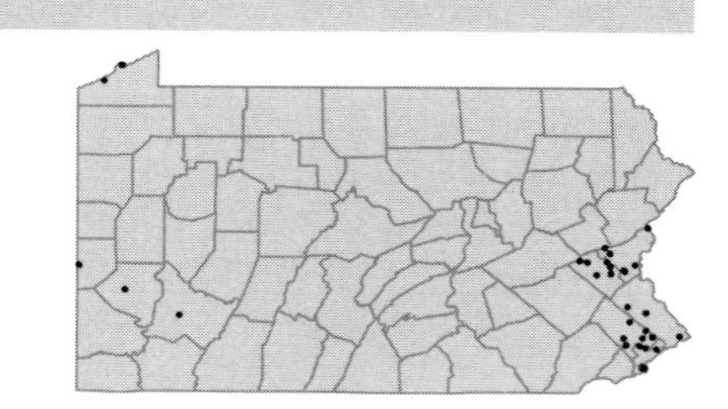

• ***Cucumis melo*** L.
Muskmelon
Herbaceous annual vine
Cultivated and occasionally occurring around rubbish dumps.

I

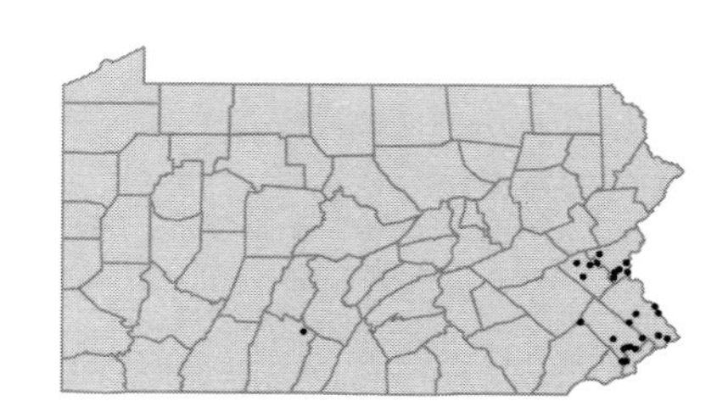

Cucumis sativus L.
Cucumber
Herbaceous annual vine
Cultivated and occasionally occurring in waste ground or rubbish dumps.

I

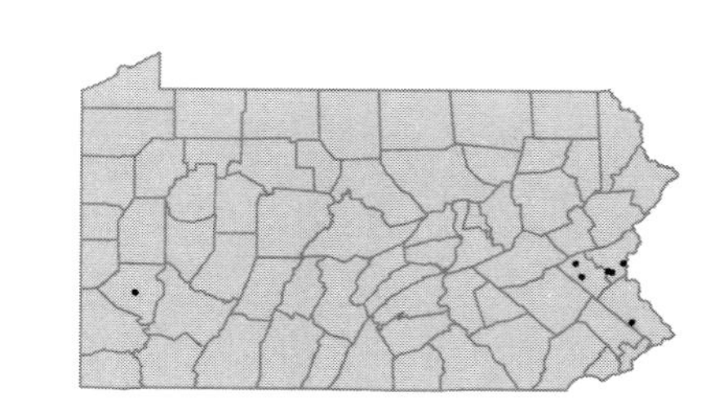

• ***Cucurbita maxima*** Duchesne
Squash
Herbaceous annual vine
Cultivated and rarely escaped to roadsides and woodland edges.
Represented by a single collection from Huntingdon Co. in 1953.

I

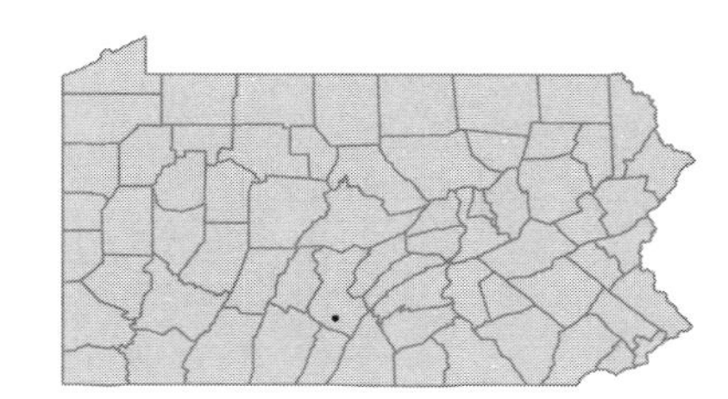

Cucurbita moschata Duchesne ex Poir.
Crookneck squash
Herbaceous annual vine
Cultivated and occasionally occurring in rubbish dumps or waste ground.

I

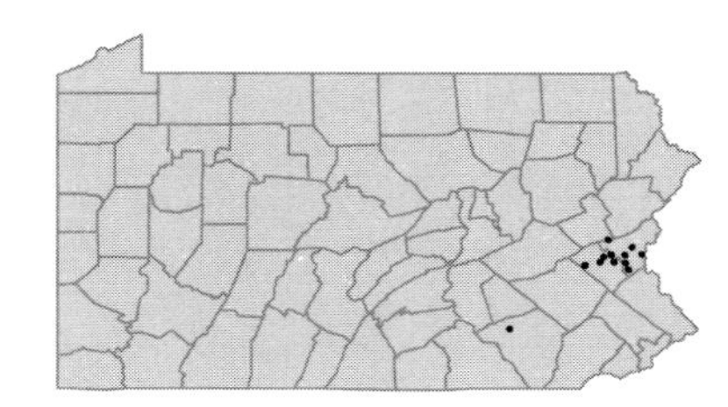

Cucurbita pepo L.
Pumpkin
Herbaceous annual vine
Cultivated and occasionally occurring in rubbish dumps or waste ground.

I

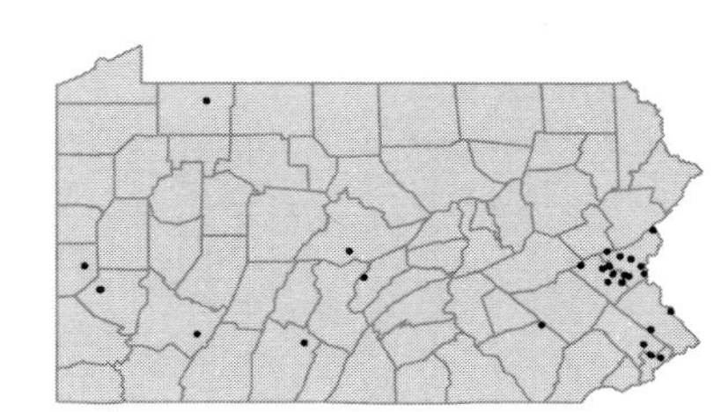

• ***Ecballium elaterium*** (L.) A.Rich.
Squirting cucumber
Herbaceous annual vine
Ballast. Represented by two collections from Philadelphia Co. ca. 1864.

I

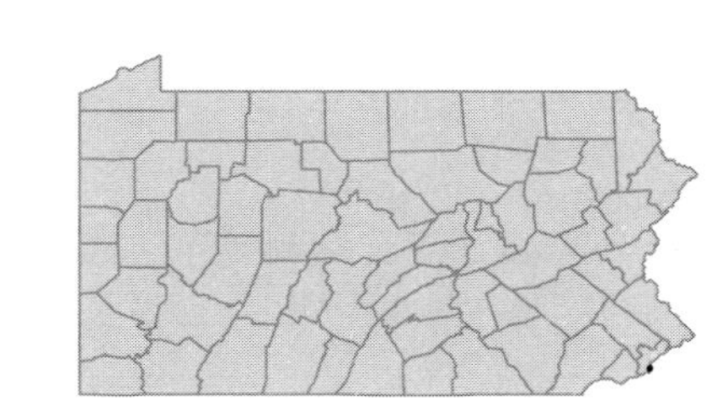

• ***Echinocystis lobata*** (Michx.) Torr. & A.Gray
Prickly cucumber; Wild balsam-apple
Herbaceous annual vine
Moist alluvial soil, stream banks and woods edges.
Micrampelis lobata (Michx.) Greene P

FAC

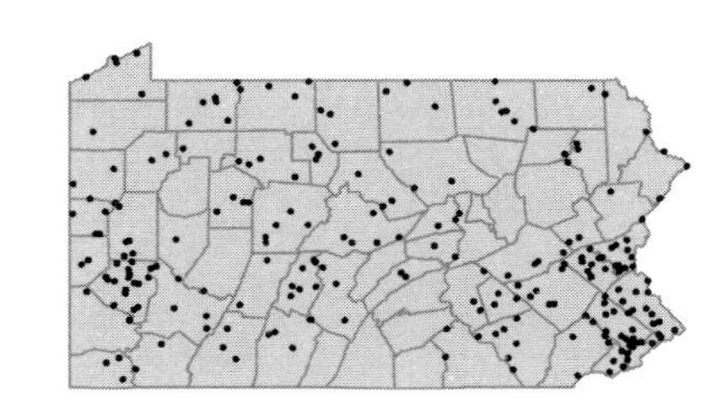

• ***Lagenaria siceraria*** (Molina) Standl.
Gourd
Herbaceous annual vine
Cultivated and rarely occurring in waste ground or rubbish dumps.

I

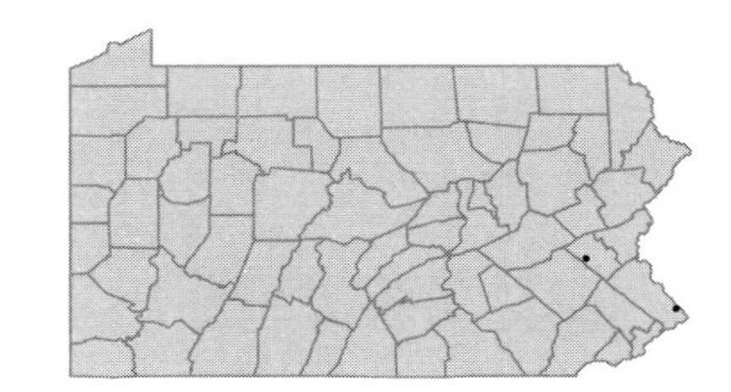

• ***Momordica charantia*** L.
Balsam-pear
Herbaceous annual vine
Cultivated, possibly escaped to waste ground. Represented by a single collection each from Philadelphia and Westmoreland Cos. ca. 1870.

[I]

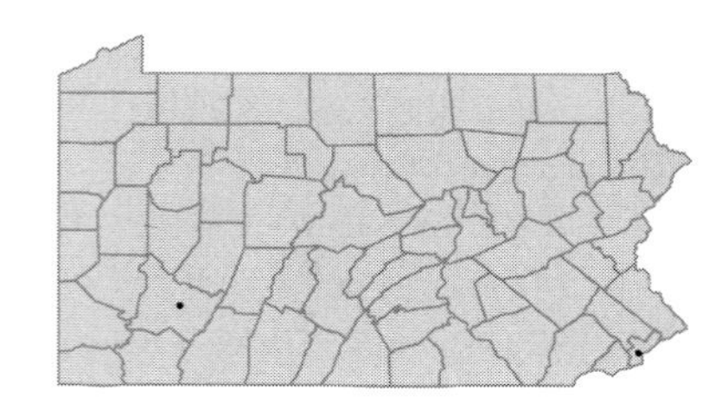

• ***Sicyos angulatus*** L.
Bur cucumber
Herbaceous annual vine
Moist open soil, stream banks, roadsides or waste ground.

FACU

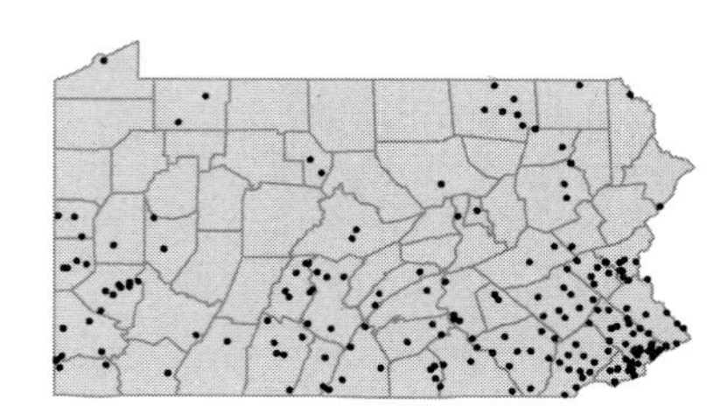

SALICACEAE

• ***Populus alba*** L.
White poplar
Deciduous tree
Cultivated and occasionally escaped to roadsides and old fields.

[I]

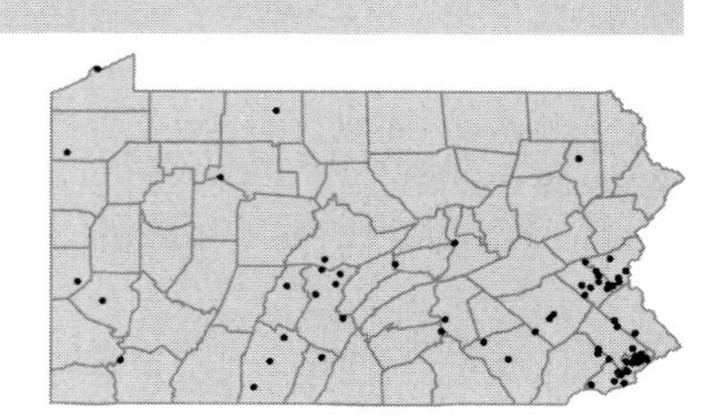

Populus alba x tremula
Gray poplar
Deciduous tree
Cultivated and occasionally escaped to waste ground and edges.
Populus canescens (Ait.) Sm. FGBWC

[I]

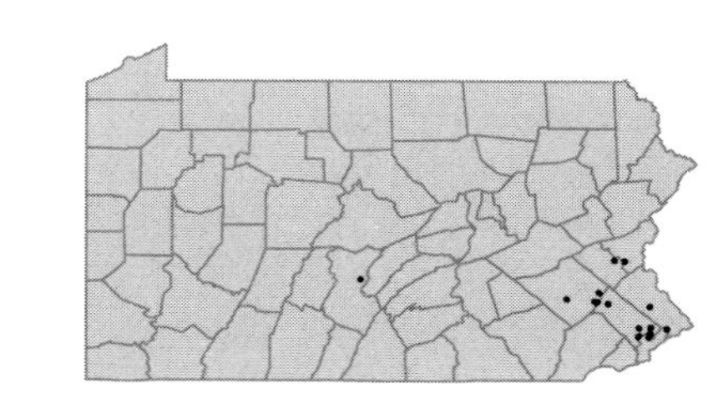

Populus balsamifera L.
Balsam poplar; Hackmatack
Deciduous tree
Swamps, thickets, alluvial gravels and river banks.

FACW

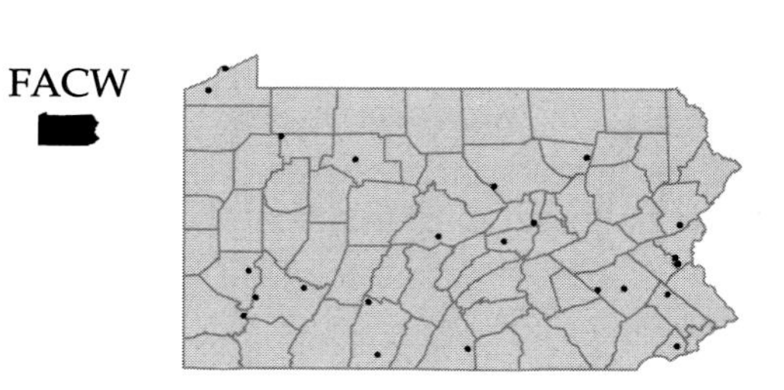

Populus balsamifera x deltoides
Balm-of-Gilead
Deciduous tree
Cultivated and occasionally naturalized.
Populus candicans Ait. P; *Populus x gileadensis* Rouleau FW; *Populus x jackii* Sarg. C

[I]

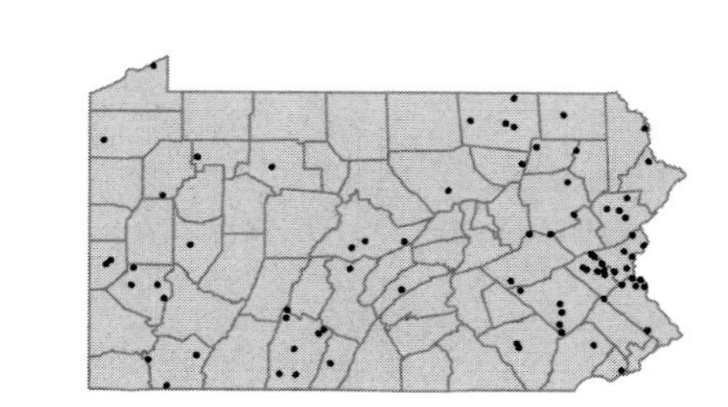

Populus deltoides Bartr. ex Marsh.
Cottonwood
Deciduous tree
River banks and rich alluvial soils.

FACU-

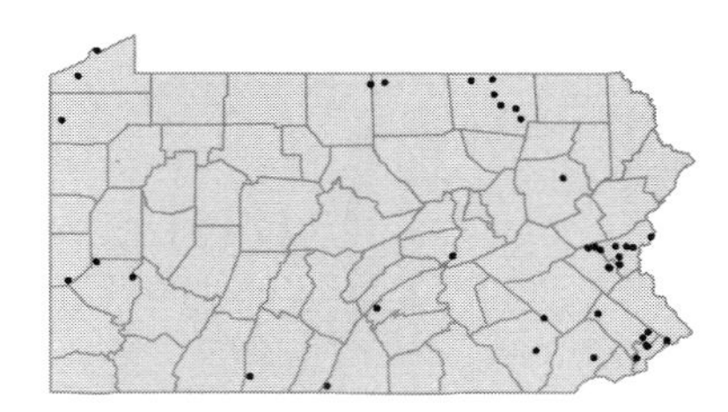

Populus deltoides x nigra
Carolina poplar
Deciduous tree
Cultivated and occasionally escaped.
Populus x canadensis Moench WC

[I]

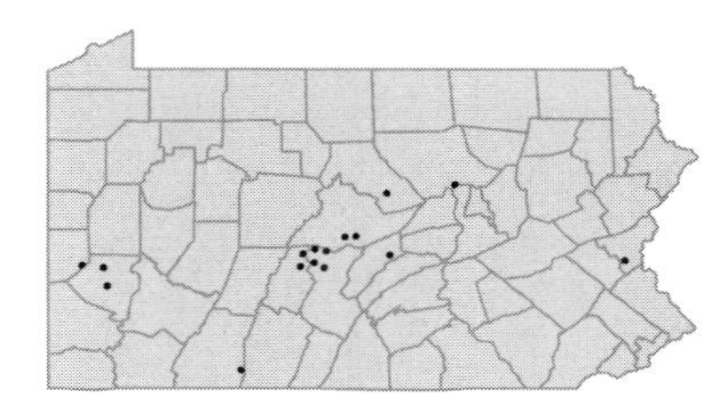

Populus grandidentata Michx.
Large-toothed aspen
Deciduous tree
Early successional woods and floodplains.

FACU-

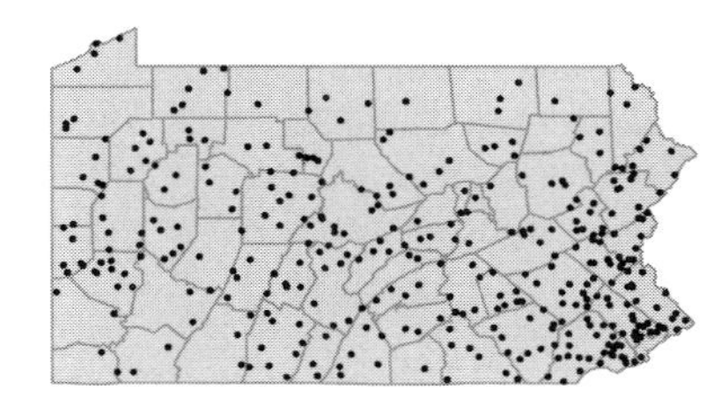

Populus nigra L.
Lombardy poplar
Deciduous tree
Cultivated and rarely escaped to disturbed ground.

[I]

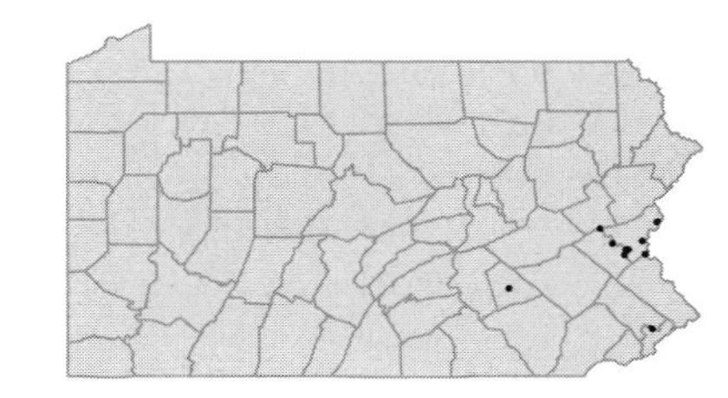

Populus tremuloides Michx.
Quaking aspen; Trembling aspen
Deciduous tree
Old fields, open woods or barrens, usually on sandy or gravelly soils.

• ***Salix alba*** L. var. ***alba***
White willow
Deciduous tree
Naturalized in low ground and along stream banks and roadsides.

FACW
[I]

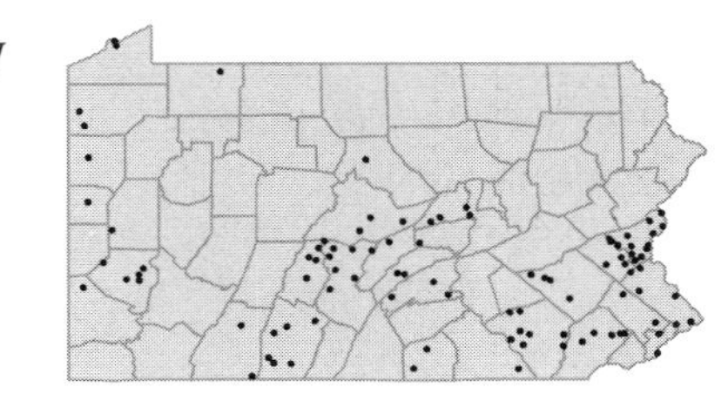

Salix alba L. var. ***vitellina*** (L.) Stokes
White willow
Deciduous tree
Naturalized in low, wet areas.

FACW
I

Salix alba x fragilis
White crack willow
Deciduous tree
Wet shores and roadsides.
Salix x rubens Schrank FW

I

Salix amygdaloides Anderss.
Peach-leaved willow
Deciduous tree
Swamps, bogs and wet shores.

FACW

Salix babylonica L.
Weeping willow
Deciduous tree
Cultivated and occasionally naturalized.

FACW-
I

Salix bebbiana Sarg.
Long-beaked willow; Gray willow
Deciduous shrub
Moist or dry thickets and edges.

FACW

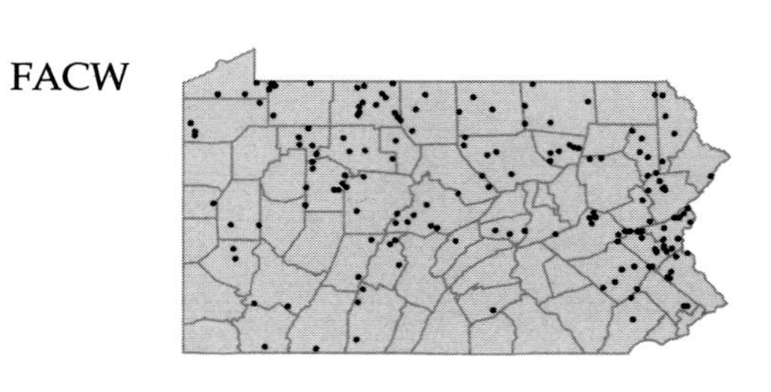

Salix candida Fluegge ex Willd.
Hoary willow; Sage-leaved willow
Deciduous shrub
Fens and wet meadows on calcareous soils.

OBL

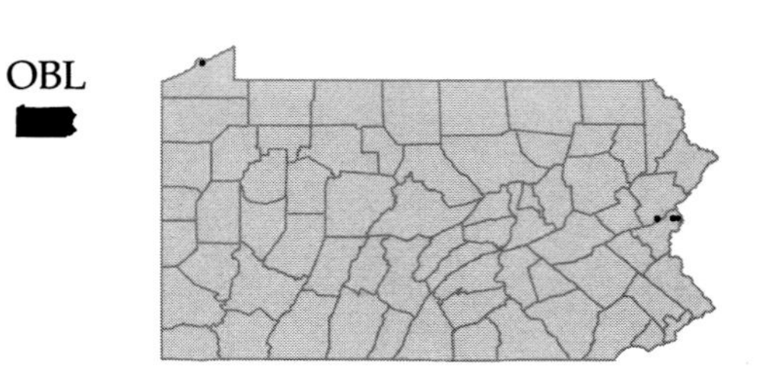

Salix caprea L.
Goat willow
Deciduous shrub
Cultivated and occasionally spreading to thickets and roadsides.

[I]

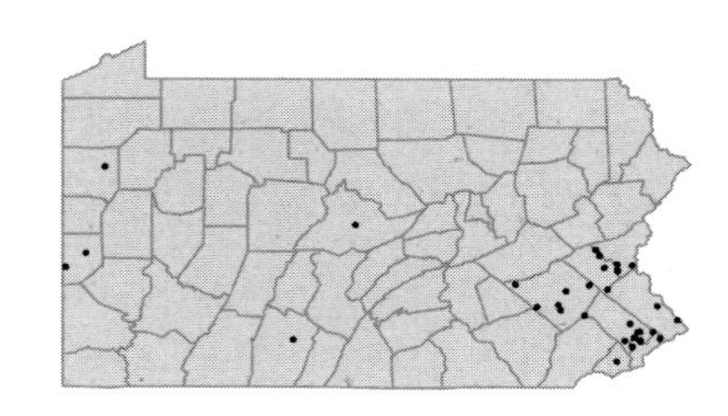

Salix caroliniana Michx.
Carolina willow
Deciduous tree
River banks, shores and low woods.
Salix longipes Shuttlw. var. *wardii* (Bebb) Schneid. BW

OBL

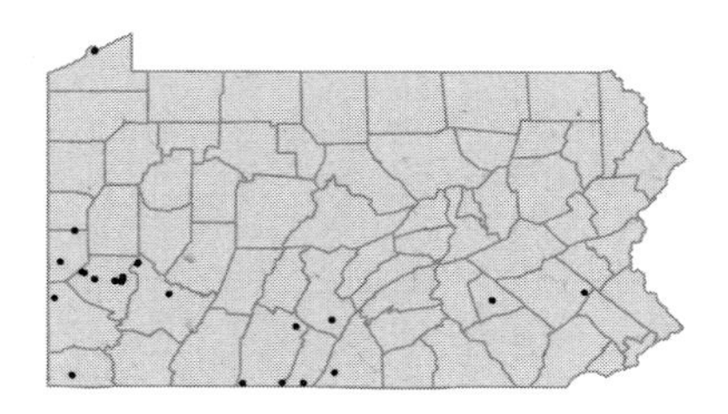

Salix cinerea L.
Gray willow; Pussy willow
Deciduous shrub
Cultivated and rarely escaped.

[I]

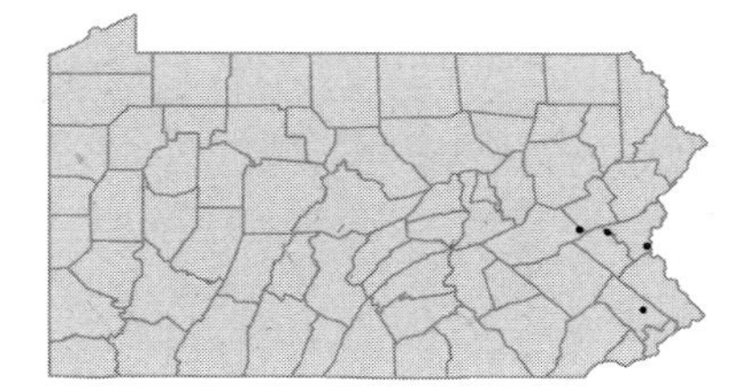

Salix discolor Muhl.
Pussy willow
Deciduous shrub
Swamps and moist or wet woods.

FACW

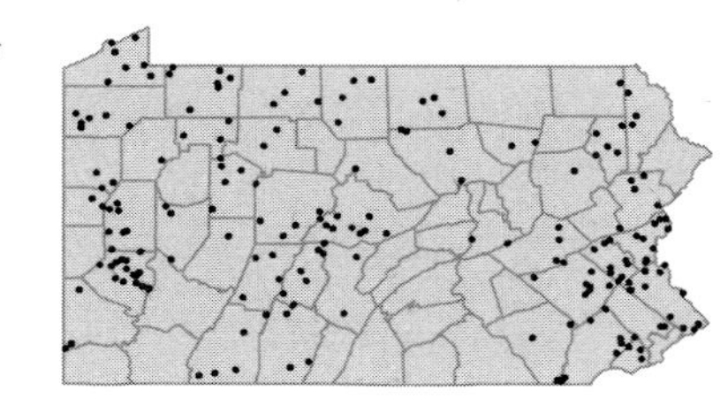

Salix eriocephala Michx.
Heart-leaved willow
Deciduous shrub
Shores and bottomlands.
Salix cordata Muhl. P; *Salix rigida* Muhl. FGBW

FACW

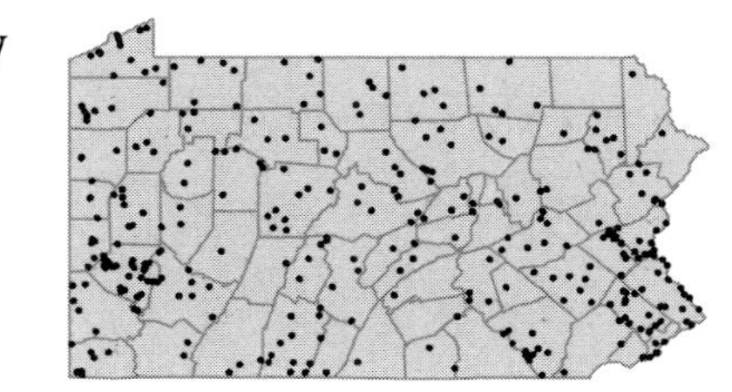

Salix exigua Nutt.
Sandbar willow
Deciduous shrub
Sandy or gravelly alluvial bars and shores.
Salix interior Rowlee PFGBW

OBL

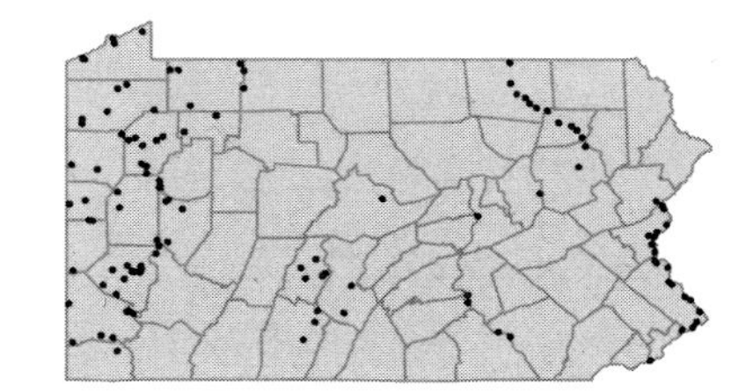

Salix fragilis L.
Crack willow; Brittle willow
Deciduous tree
Cultivated and occasionally escaped to roadsides and woods edges.

FAC+
[I]

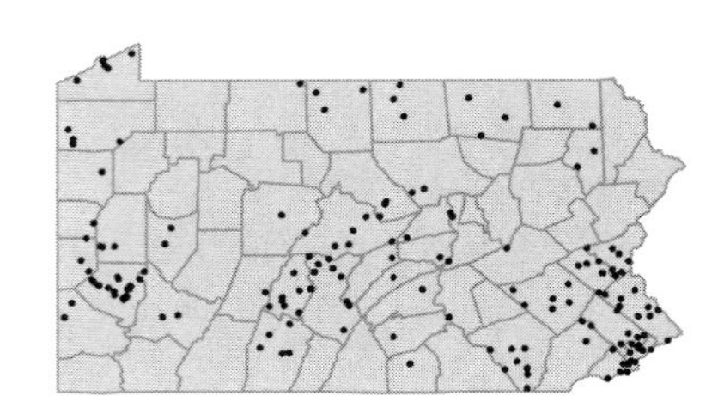

Salix humilis Marshall var. ***humilis***
Upland willow
Deciduous shrub
Dry thickets and barrens, on sandy soils.
Salix humilis Marshall var. *rigidiuscula* (Anderss.) Robins. & Fern. GBW

FACU

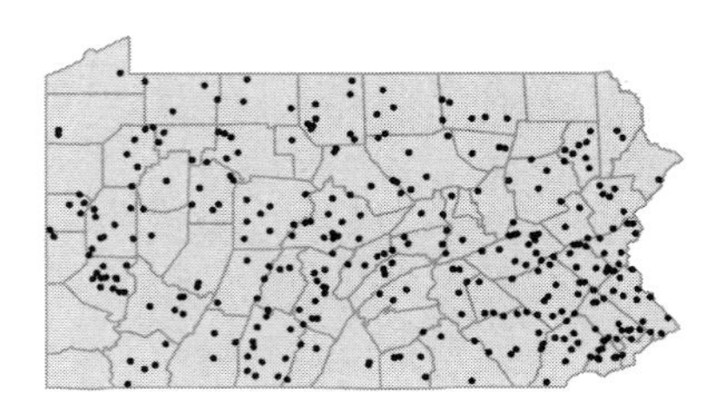

Salix humilis Marshall var. ***microphylla*** (Anderss.) Fern.
Dwarf upland willow; Sage willow
Deciduous shrub
Moist barrens and thickets.
Salix tristis Ait. PGBW

FACU

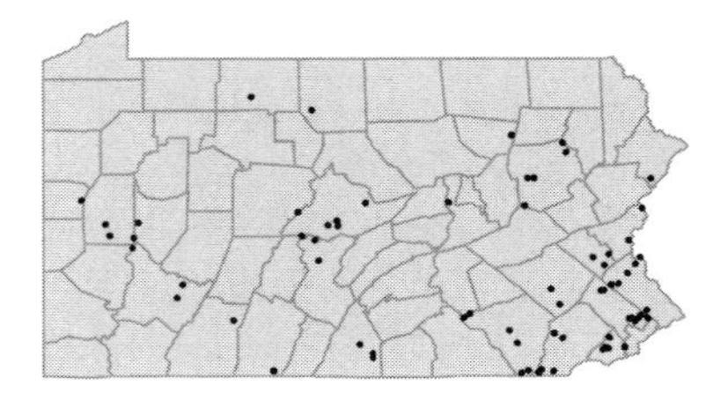

Salix lucida Muhl.
Shining willow
Deciduous shrub
Swamps, low ground and wet shores.

FACW

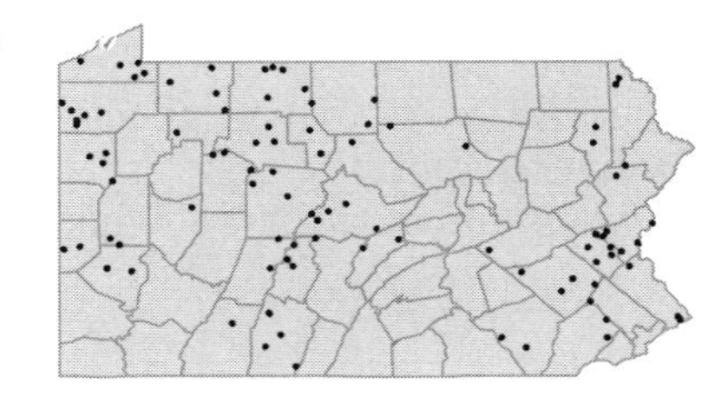

Salix myricoides Muhl. var. ***myricoides***
Broad-leaved willow
Deciduous shrub
Stream banks and swamps.
Salix glaucophylla Bebb P; *Salix glaucophylloides* Fern. var. *glaucophylla* (Bebb) Schneid. FGBW

FAC

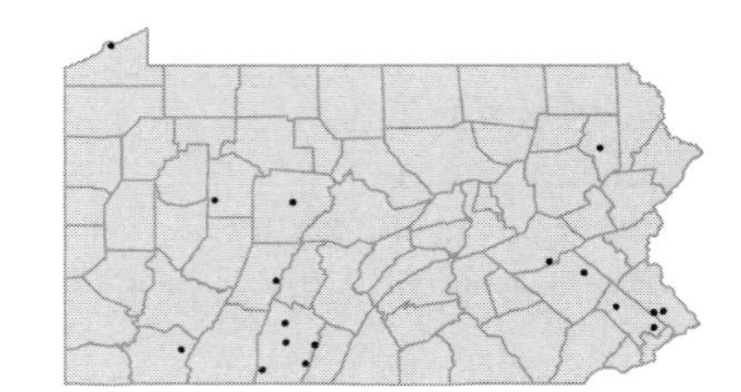

Salix myricoides Muhl. var. ***albovestita*** (Ball) Dorn
Shoreline willow
Deciduous shrub
Sandy shores.
Salix glaucophylloides Fern. var. *albovestita* (Ball) Fern. FGBW

FAC

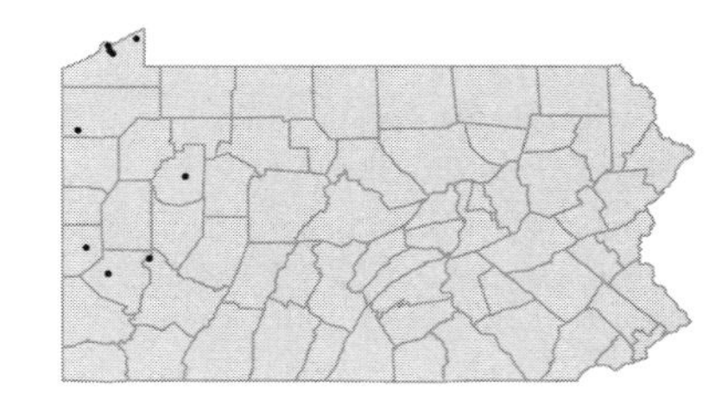

Salix nigra Marshall
Black willow
Deciduous tree
Swamps, wet meadows and rich alluvial soils.

FACW+

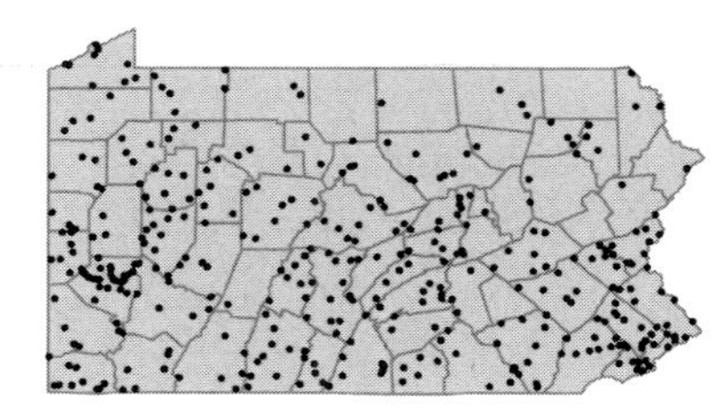

Salix pentandra L.
Bay-leaved willow
Deciduous tree
Cultivated and occasionally escaped.

[I]

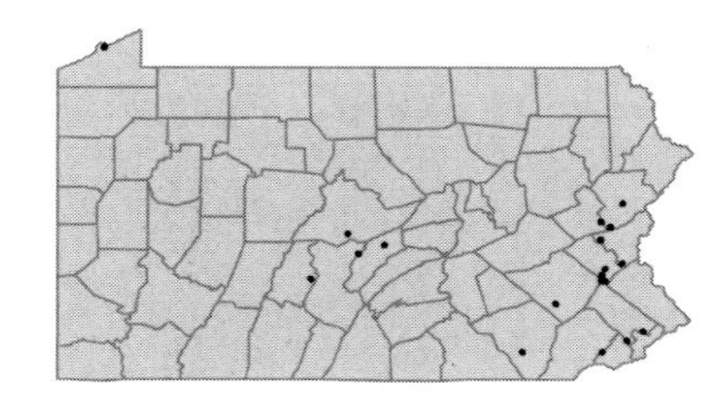

Salix petiolaris Smith
Slender willow
Deciduous shrub
Meadows and swales. A hybrid with *S. sericea* (*S. x subsericea* [Anderss.] Schneid) has been collected at one site in Westmoreland Co.
Salix gracilis Anderss. var. *textoris* Fern. FW

OBL

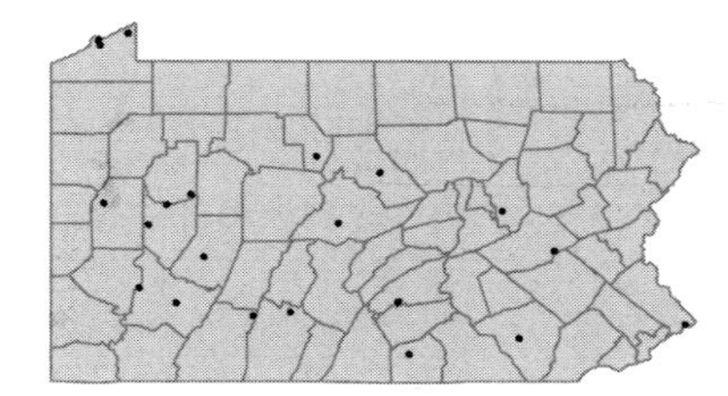

Salix purpurea L.
Basket willow
Deciduous shrub
Cultivated and occasionally naturalized in low ground.

[I]

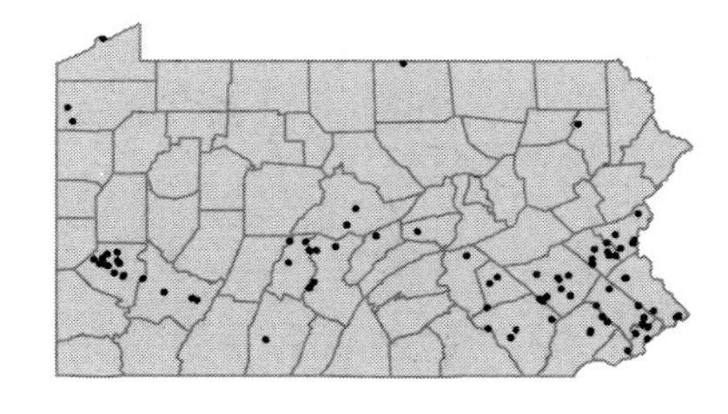

Salix sericea Marshall
Silky willow
Deciduous shrub
Swamps, bogs, stream banks and low woods.

OBL

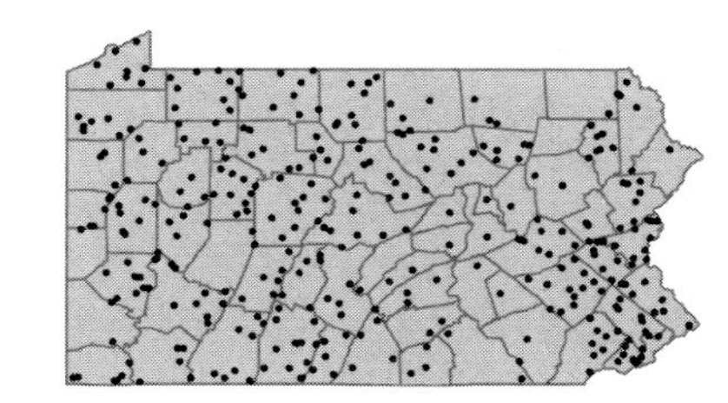

Salix serissima (Bailey) Fern.
Autumn willow
Deciduous shrub
Fens and wet meadows, on calcareous soils.

OBL

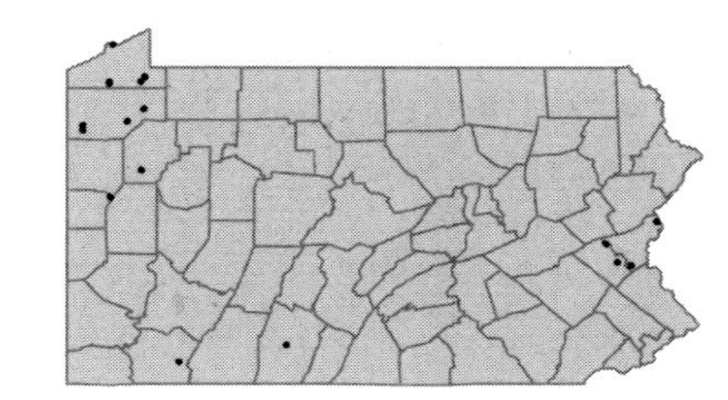

CAPPARACEAE

• ***Cleome hasslerana*** Chodat [I]
Spider-flower
Herbaceous annual
Waste ground.
Cleome houtteana Raf. W

Cleome ornithopodioides L. [I]
Birdfoot cleome
Herbaceous annual
Waste ground. Represented by a single collection from Mercer Co. in 1957.

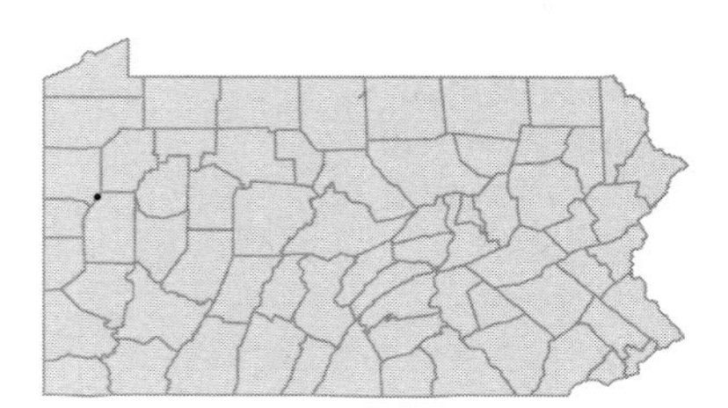

Cleome spinosa L. FACU- [I]
Giant spider-flower
Herbaceous annual
Cultivated and occasionally occurring on waste ground or rubbish dumps.

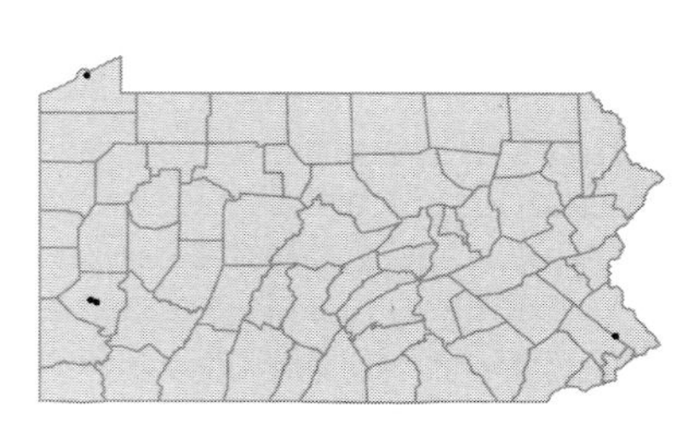

Cleome viscosa L. [I]
Spider-flower
Herbaceous annual
Ballast and waste ground. Represented by three collections from Philadelphia Co. 1865-1879.

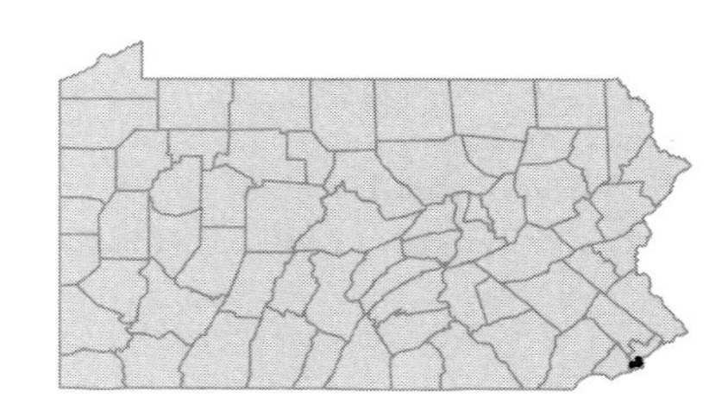

• ***Polanisia dodecandra*** (L.) DC. ssp. ***dodecandra***
Clammyweed
Herbaceous annual
Dry, sandy or gravelly alluvial soils.
Polanisia graveolens Raf. PFB

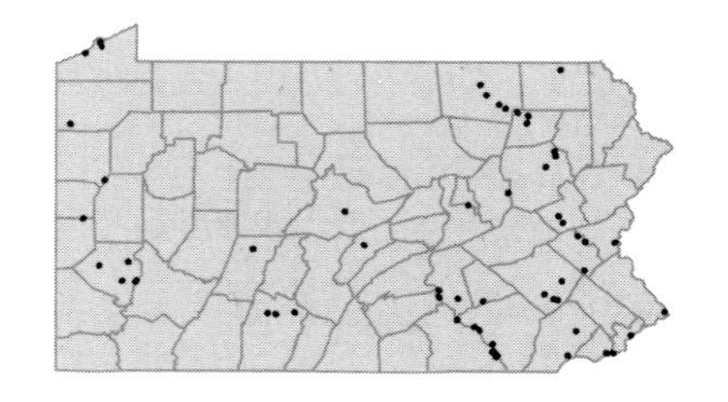

Polanisia dodecandra (L.) DC. ssp. ***trachysperma*** (Torr. & A.Gray) Iltis [I]
Clammyweed
Herbaceous annual
Open ground.
Polanisia trachysperma Torr. & A.Gray FGBW

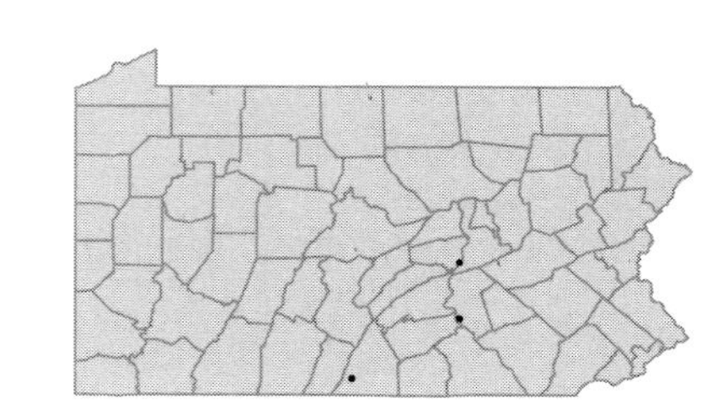

BRASSICACEAE

• ***Alliaria petiolata*** (Bieb.) Cavara & Grande
Garlic-mustard
Herbaceous biennial
Disturbed woods, floodplains and waste ground.
Alliaria officinalis Andrz. ex Bieb. FGB

FACU-
[I]

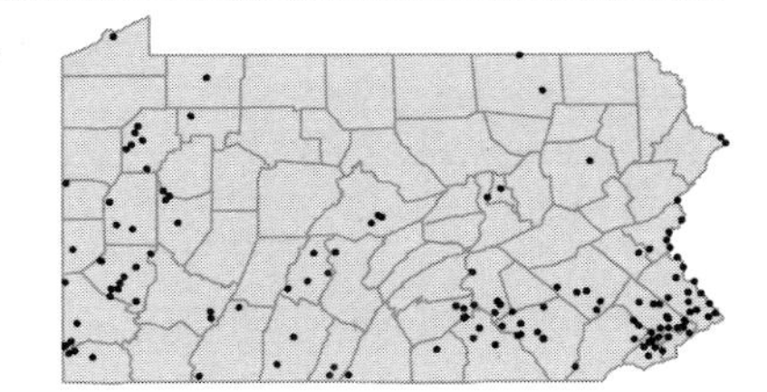

• ***Alyssum alyssoides*** (L.) L.
Alyssum
Herbaceous annual
Waste ground.

[I]

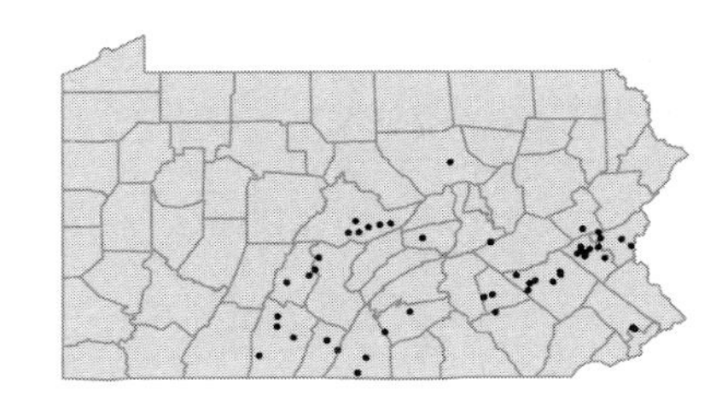

• ***Arabidopsis thaliana*** (L.) Heynh.
Mouse-ear cress
Herbaceous annual
Cultivated fields and waste ground.

[I]

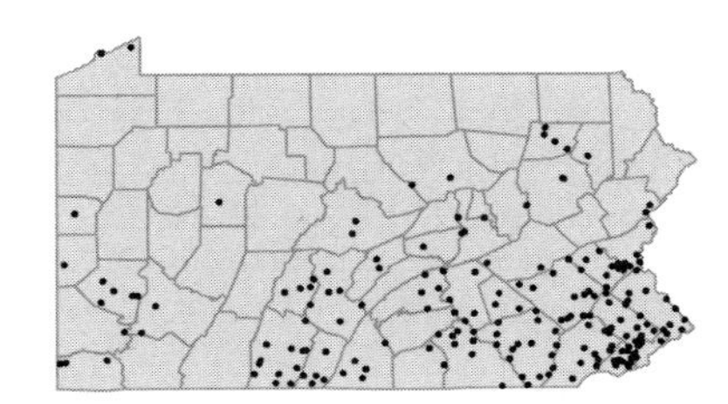

• ***Arabis canadensis*** L.
Sicklepod
Herbaceous biennial
Rocky banks and slopes, in rich soil of woods or edges.

Arabis glabra (L.) Bernh.
Tower cress; Tower mustard
Herbaceous biennial
Dry, rocky soil in open fields and edges.

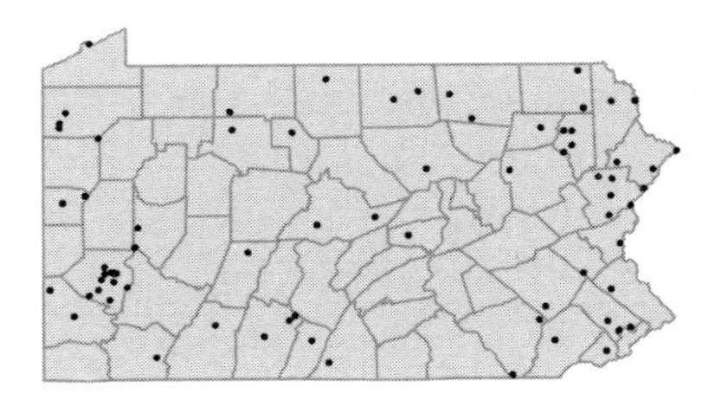

Arabis hirsuta (L.) Scop. var. ***adpressipilis*** (M.Hopkins) Rollins
Hairy rock-cress
Herbaceous biennial
Woods, banks or rock ledges, usually on limestone.

FACU

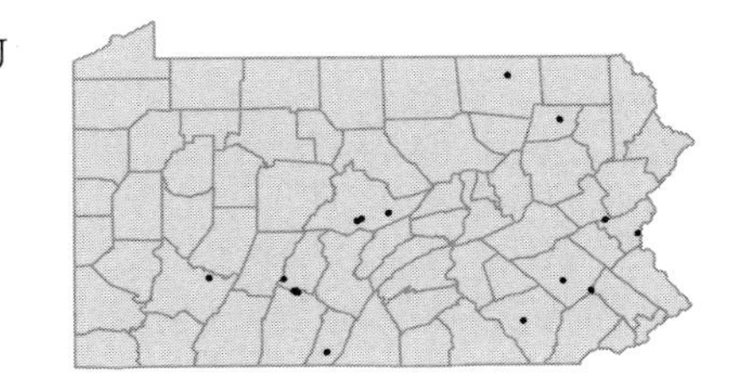

Arabis hirsuta (L.) Scop. var. ***pycnocarpa*** (M.Hopkins) Rollins
Hairy rock-cress
Herbaceous biennial
Dry cliffs and ledges of calcareous shale or limestone.

FACU

Arabis laevigata (Muhl. ex Willd.) Poir. var. ***laevigata***
Smooth rock-cress
Herbaceous biennial
Dry woods and hillsides.

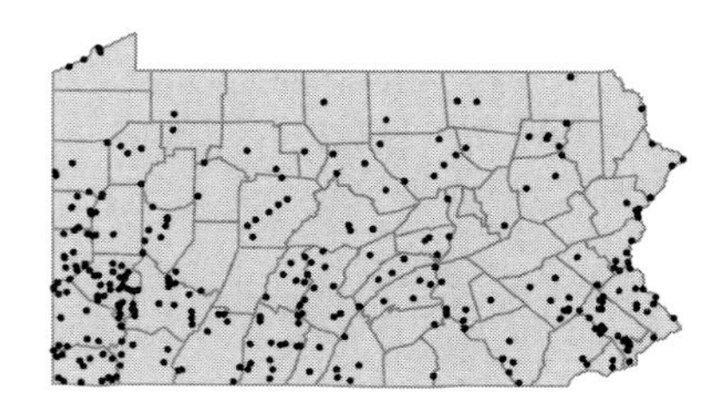

Arabis laevigata (Muhl. ex Willd.) Poir. var. ***burkii*** Porter
Smooth rock-cress
Herbaceous biennial
Dry, wooded slopes.

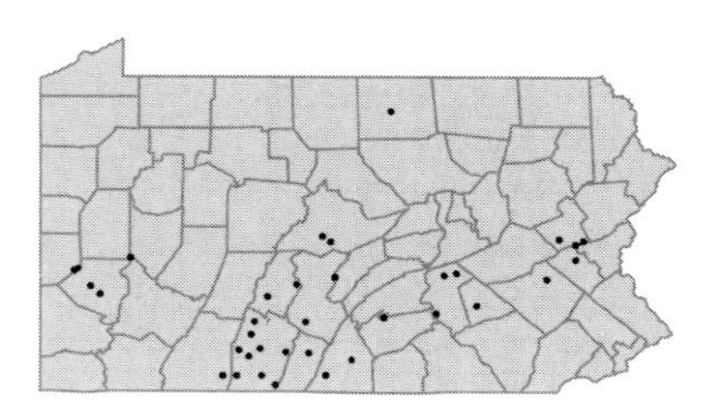

Arabis lyrata L.
Lyre-leaved rock-cress
Herbaceous biennial
Dry, rocky slopes and outcrops, serpentine barrens.

FACU

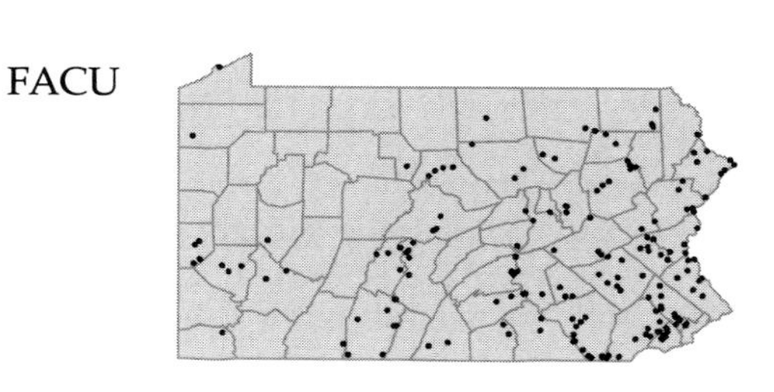

Arabis missouriensis Greene
Missouri rock-cress
Herbaceous biennial
Dry slopes.
Arabis viridis Harger GB

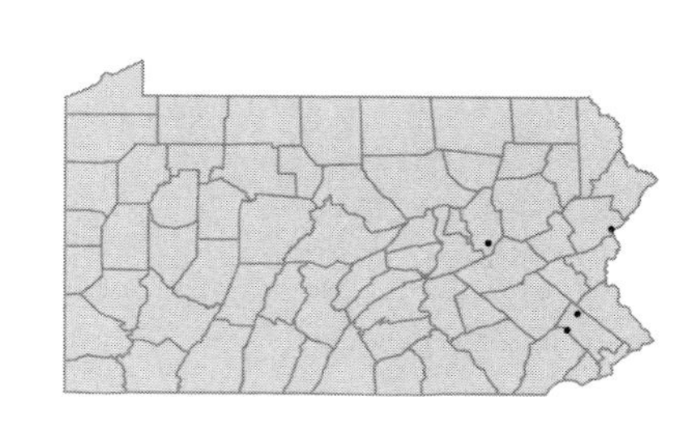

Arabis patens Sullivant
Spreading rock-cress
Herbaceous biennial
Moist, rocky woods.

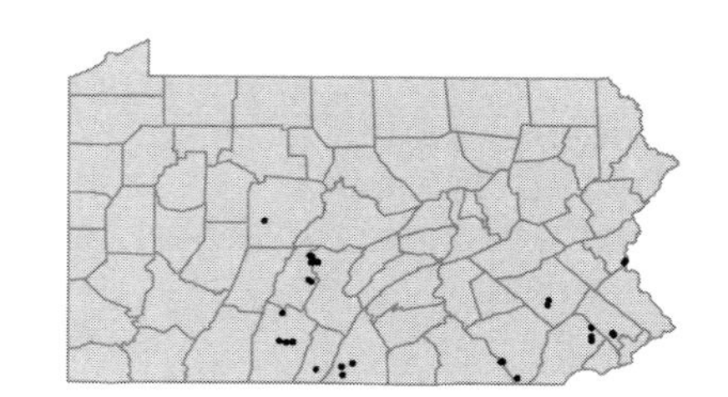

Arabis shortii (Fern.) Gleason var. ***shortii***
Toothed rock-cress
Herbaceous biennial
Rich, moist woods.
Arabis dentata Torr. & A.Gray P; *Arabis perstellata* E.L.Braun var. *shortii* Fern. FW

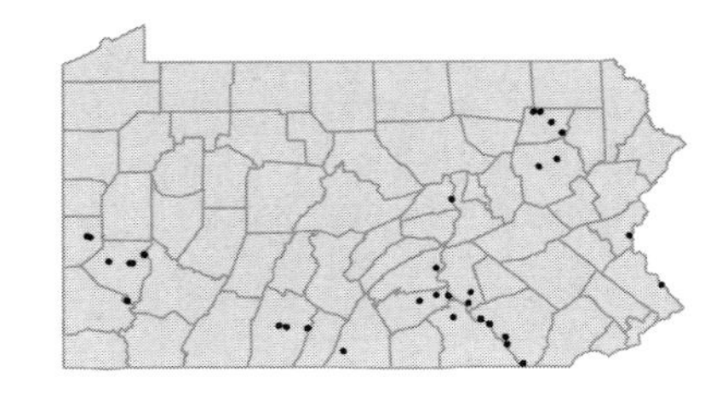

Arabis shortii (Fern.) Gleason var. ***phalacrocarpa*** (M.Hopkins) Steyermark
Toothed rock-cress
Herbaceous biennial
Wooded hillsides.
Arabis perstellata E.L.Braun var. *phalacrocarpa* (M.Hopkins) Fern. FW

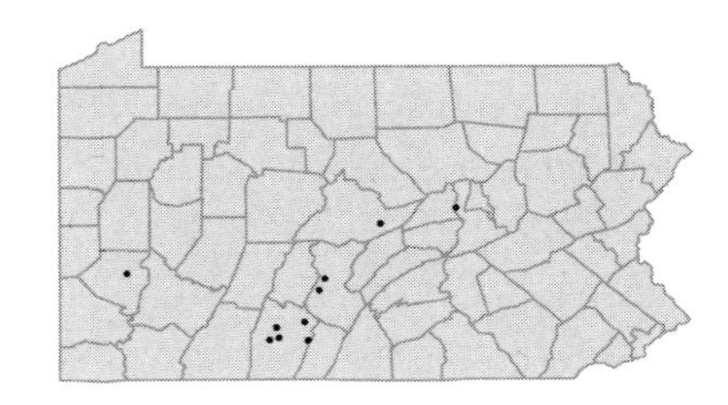

• ***Armoracia rusticana*** (Lam.) Gaertn.,Mey. & Scherb.
I
Horseradish
Herbaceous perennial
Cultivated and occasionally established in old fields and roadsides.
Armoracia lapathifolia Gilib. F; *Roripa armoracia* (L.) A.S.Hitchc. P

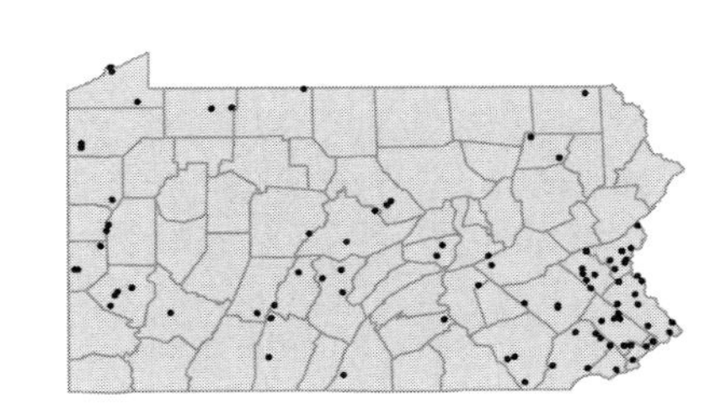

• ***Barbarea verna*** (P.Mill.) Aschers.
I
Early winter-cress
Herbaceous biennial
Fields, roadsides and waste ground.
Barbarea praecox (J.E.Smith) R.Br. P

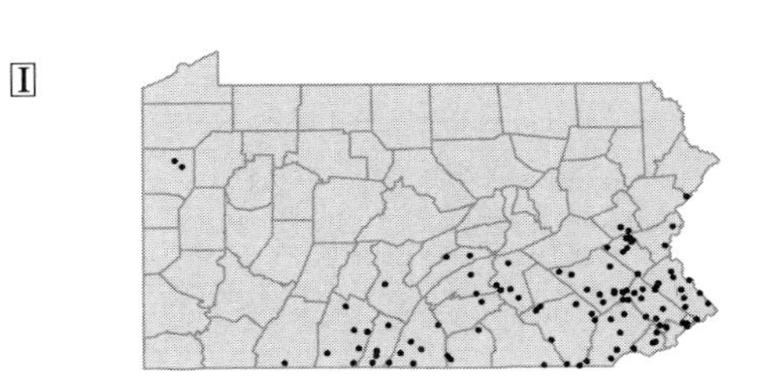

Barbarea vulgaris R.Br. var. ***vulgaris***
FACU
I
Common winter-cress; Yellow rocket
Herbaceous biennial
Moist, open ground.
Barbarea barbarea (L.) MacM. P

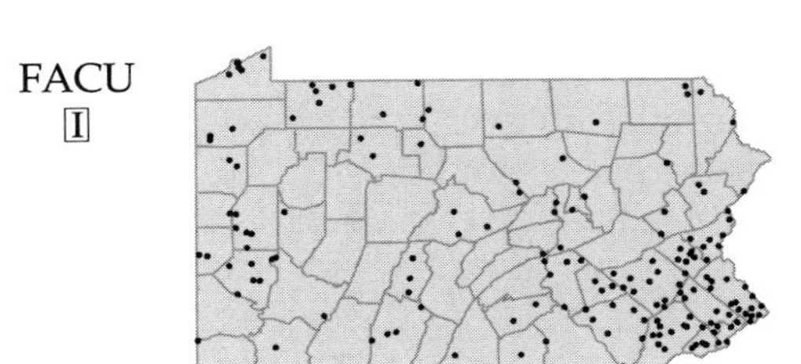

Barbarea vulgaris R.Br. var. ***arcuata*** (Opiz ex J.& C.Presl) Fries
FACU
I
Winter-cress; Yellow rocket
Herbaceous biennial
Moist fields and roadsides.

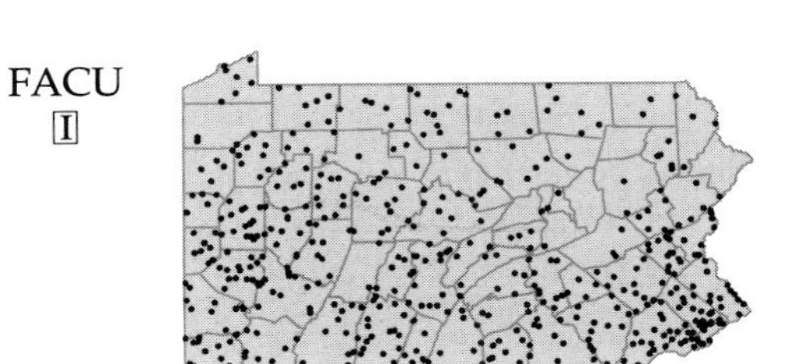

Barbarea vulgaris R.Br. var. ***sylvestris*** Fries
Winter-cress; Yellow rocket
Herbaceous biennial
Fields and waste ground.

FACU
[I]

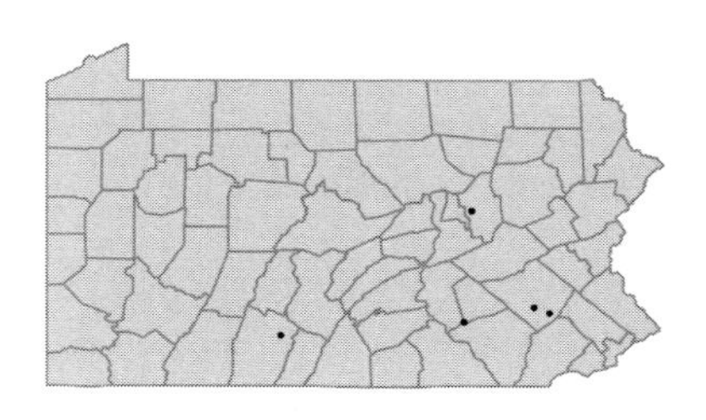

• ***Berteroa incana*** (L.) DC.
Hoary alyssum
Herbaceous annual
Waste ground.

[I]

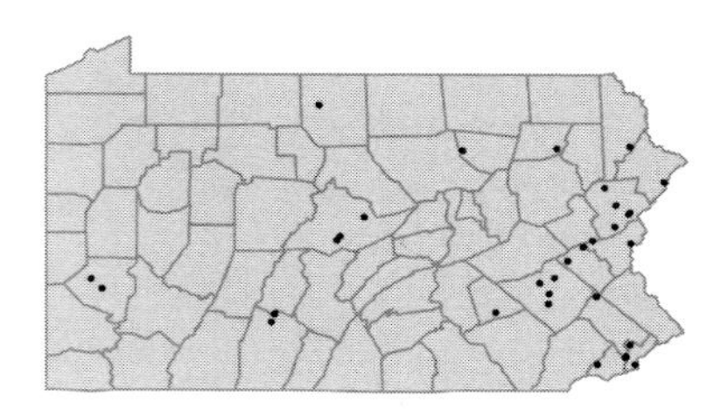

• ***Brassica juncea*** (L.) Czern.
Brown mustard; Chinese mustard; Leaf mustard
Herbaceous annual
Waste ground.

[I]

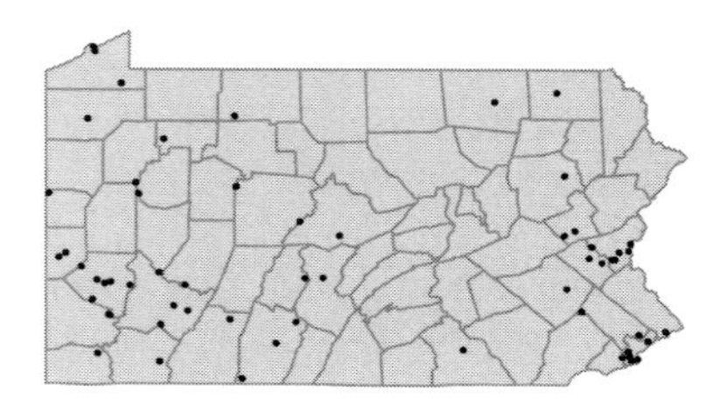

Brassica nigra (L.) W.D.J.Koch
Black mustard
Herbaceous annual
Fields and waste ground.

[I]

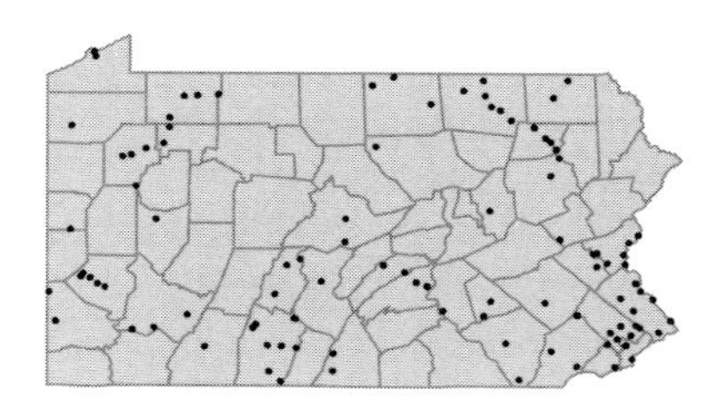

Brassica oleracea L.
Cabbage
Herbaceous biennial
Cultivated and occasionally occurring in rubbish dumps and waste ground.

[I]

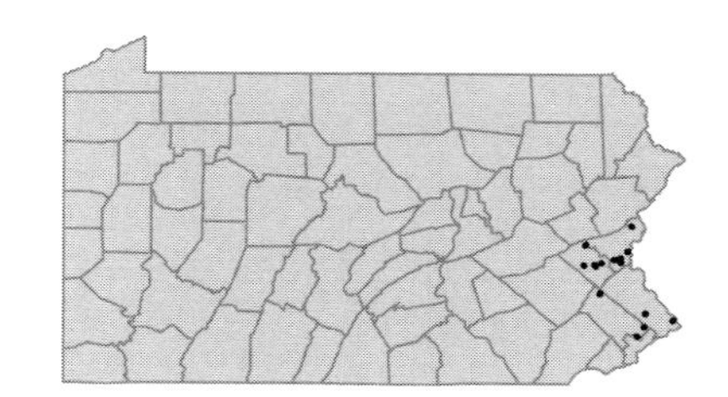

Brassica rapa L. ssp. ***olifera*** DC.
Field mustard
Herbaceous annual
Fields and waste ground.
Brassica campestris L. PGBW

[I]

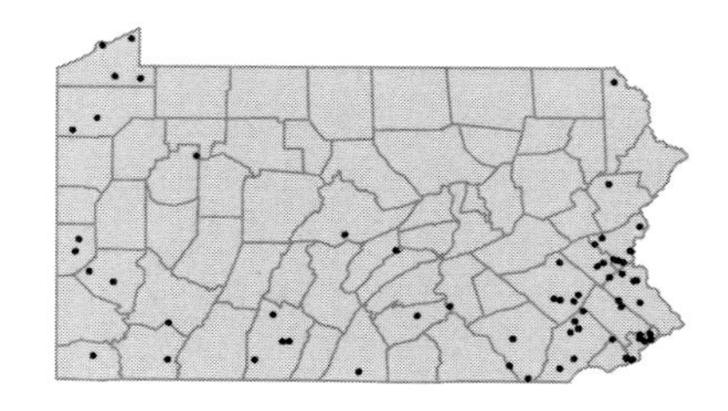

• ***Bunias orientalis*** L.
Hill mustard; Turkish rocket
Herbaceous annual
Ballast.

I

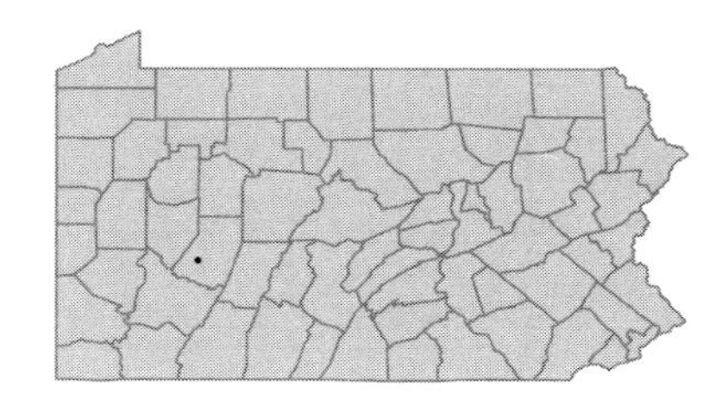

• ***Cakile edentula*** (Bigel.) Hook.
American sea-rocket
Herbaceous annual
Dunes and sand plains along Lake Erie.

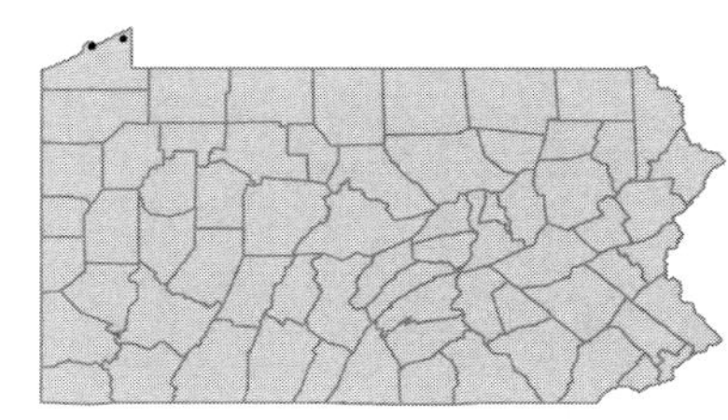

Cakile maritima Scop.
Sea-rocket
Herbaceous annual
Ballast. Represented by several collections from Philadelphia Co. 1878-1880.

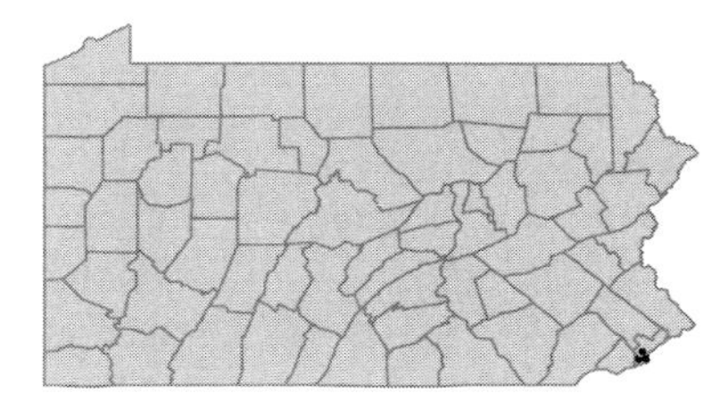

• ***Camelina microcarpa*** Andrz. ex DC.
Small-fruited false-flax
Herbaceous annual
Fields and waste ground, in sandy soil.

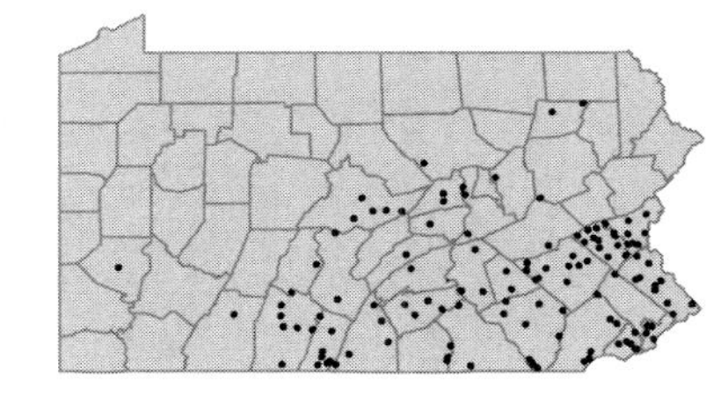

Camelina sativa (L.) Crantz
False flax; Gold-of-pleasure
Herbaceous annual
Open, sandy ground.

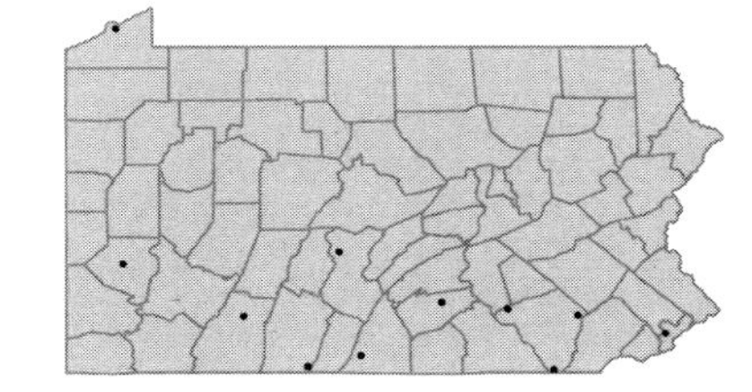

• ***Capsella bursa-pastoris*** (L.) Medic.
Shepherd's-purse
Herbaceous annual
Cultivated fields and waste ground.

FACU

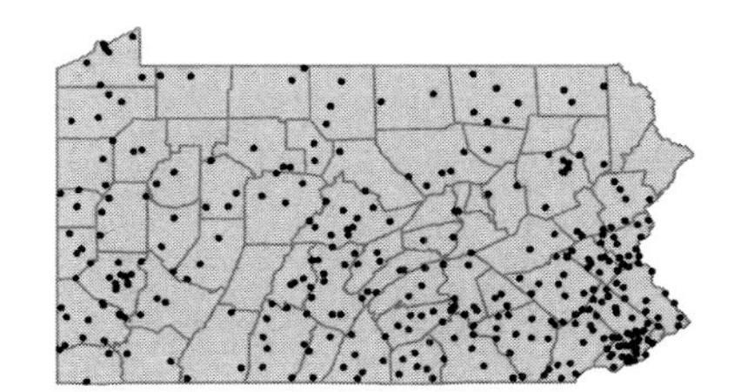

• ***Cardamine angustata*** O.E.Schulz
Toothwort
Herbaceous perennial
Moist woods, thickets and stream banks.
Dentaria heterophylla Nutt. PFGBW

FACU

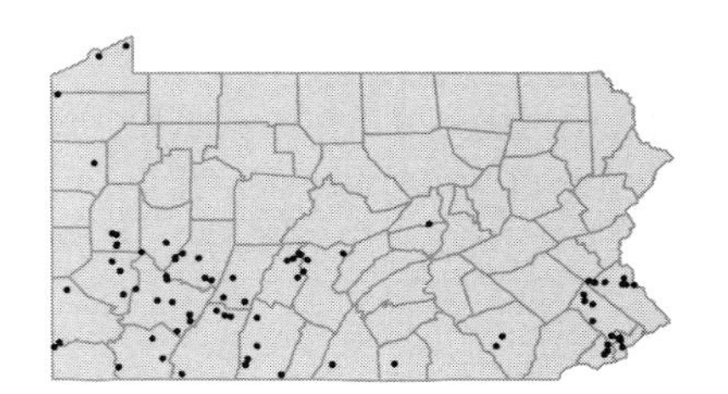

Cardamine bulbosa (Schreb. ex Muhl.) BSP
Bitter-cress; Spring cress
Herbaceous perennial
Low wet ground, shallow water, swamps or springy areas.
Cardamine rhomboidea (Pers.) DC. C

OBL

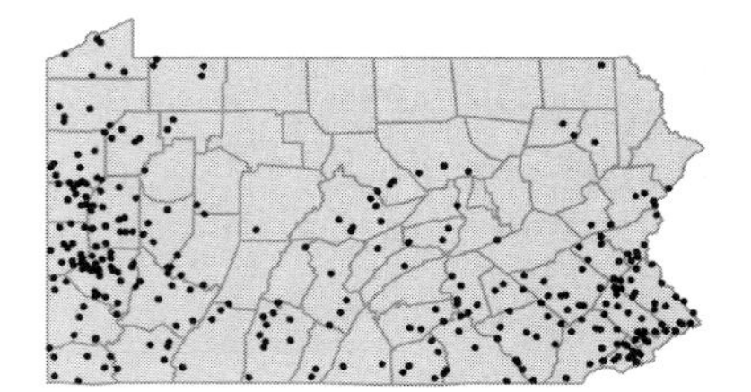

Cardamine concatenata (Michx.) O.E.Schulz
Cut-leaved toothwort; Pepper-root
Herbaceous perennial
Rich, deciduous woods.
Dentaria laciniata Muhl. ex Willd. PFGBW

FACU

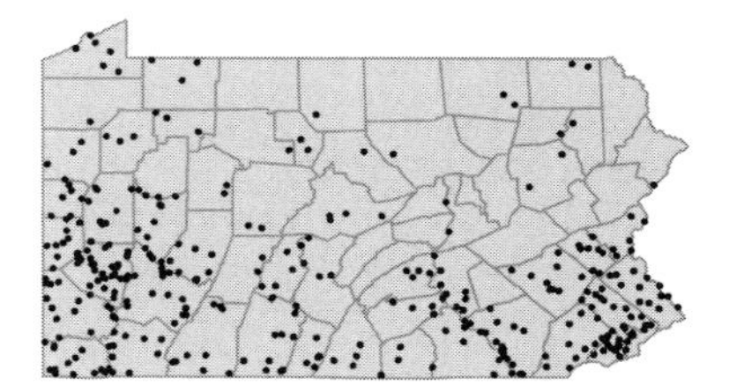

Cardamine diphylla (Michx.) Wood
Pepper-root; Two-leaved toothwort
Herbaceous perennial
Rich woods and floodplains.
Dentaria diphylla Michx. PFGBW

FACU

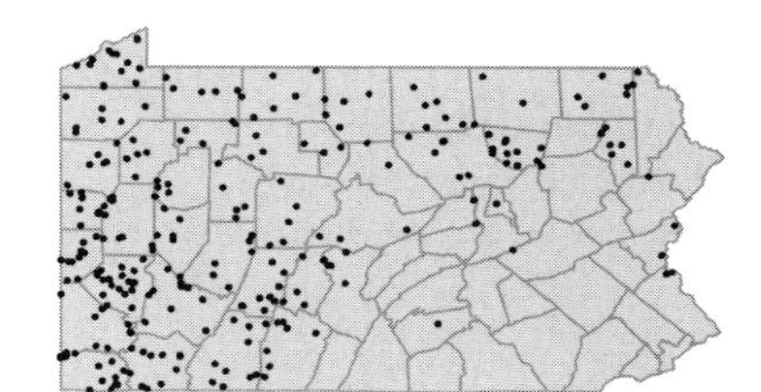

Cardamine douglassii Britt.
Purple cress
Herbaceous perennial
Rich woods and calcareous springs.

FACW+

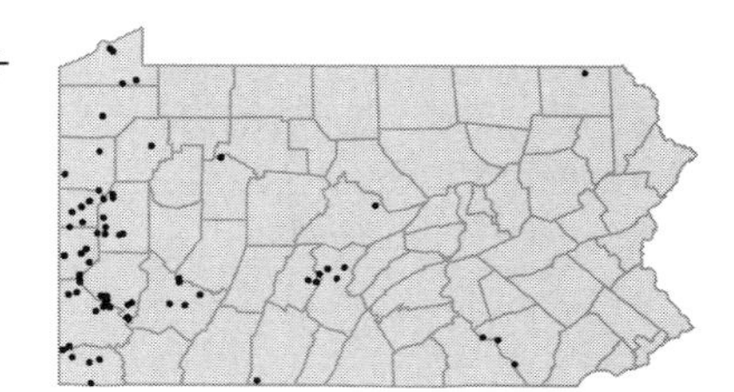

Cardamine flexuosa With.
Bitter-cress
Herbaceous annual
Woods and open areas.

OBL
[I]

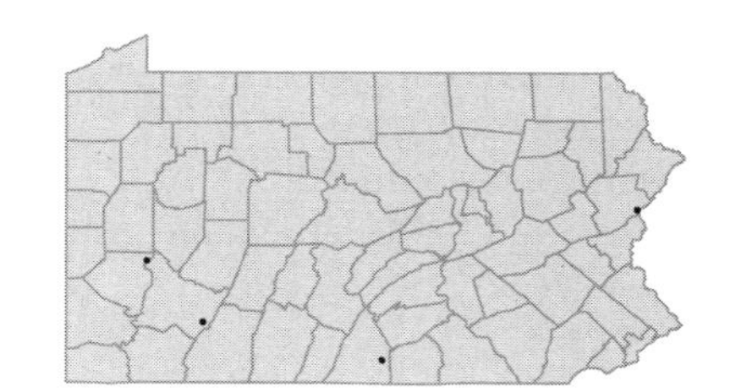

Cardamine hirsuta L.
Hairy bitter-cress; Hairy rock-cress
Herbaceous perennial
Lawns, gardens or waste ground, in moist soil.

FACU
I

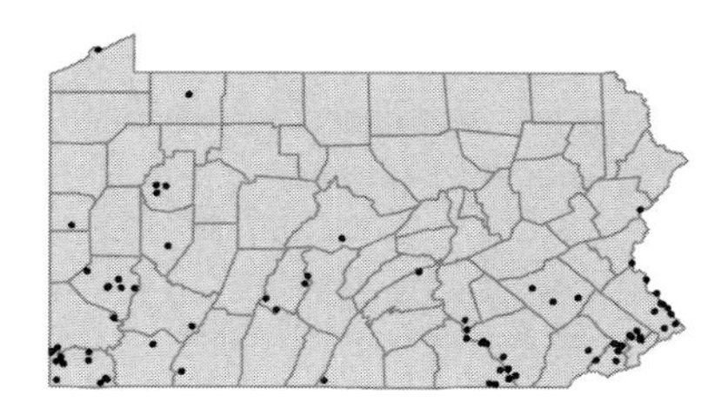

Cardamine impatiens L.
Bitter-cress
Herbaceous perennial
Moist slopes.

I

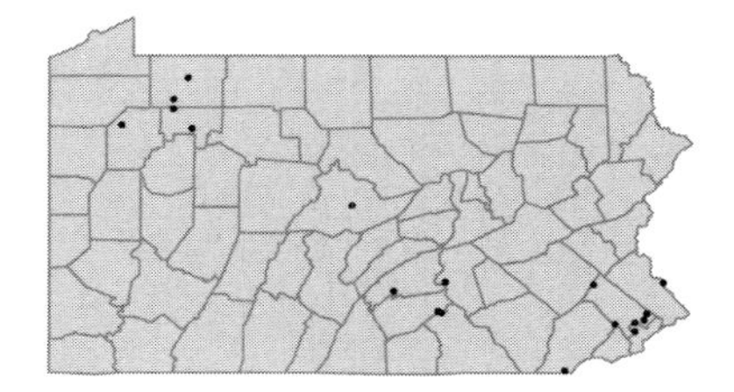

Cardamine maxima (Nutt.) A.Wood
Large toothwort
Herbaceous perennial
Low, wet woods.
Dentaria maxima Nutt. PFGBW

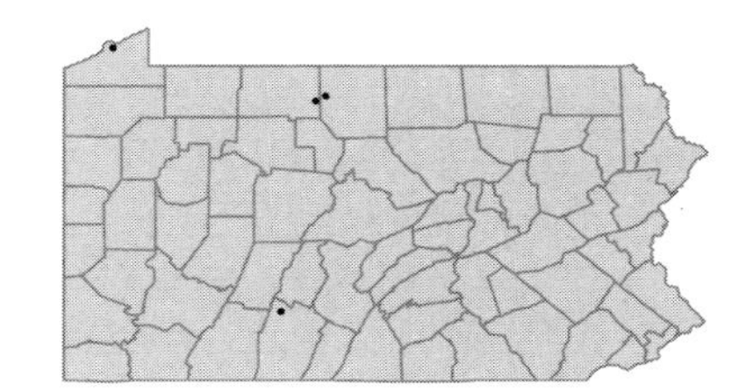

Cardamine parviflora L. var. ***arenicola*** (Britt.) O.E.Schulz
Small-flowered bitter-cress
Herbaceous perennial
Dry, rocky ledges or shaly slopes.
Cardamine arenicola Britt. P

FACU
I

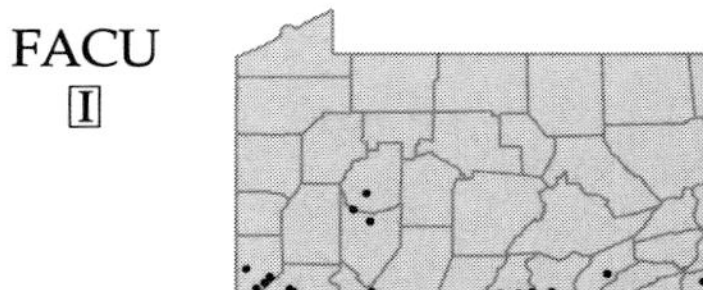

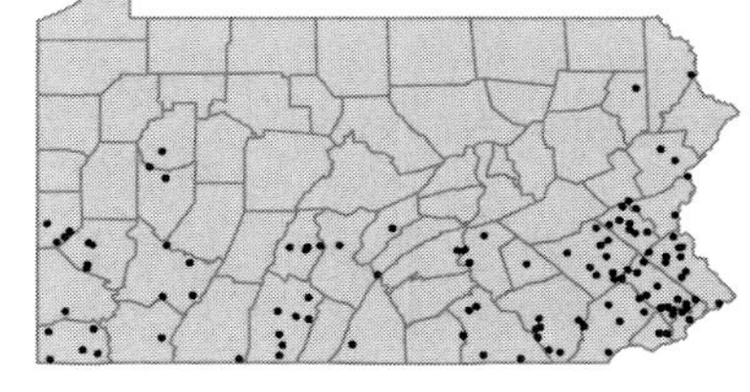

Cardamine pensylvanica Muhl. ex Willd.
Pennsylvania bitter-cress
Herbaceous perennial
Low wet ground, swamps, springs and stream margins.

OBL

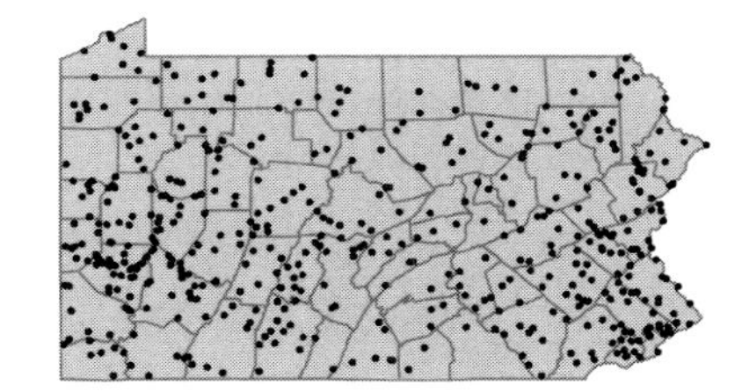

Cardamine pratensis L.
Cuckoo-flower; Lady's-smock
Herbaceous perennial
Swamps, wet meadows and alluvial woods.

OBL

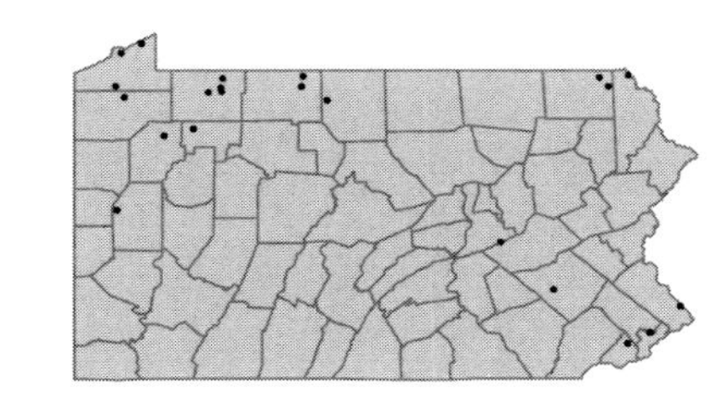

Cardamine rotundifolia Michx. OBL
Mountain watercress; Round-leaved watercress
Herbaceous perennial
Springs and stream edges.

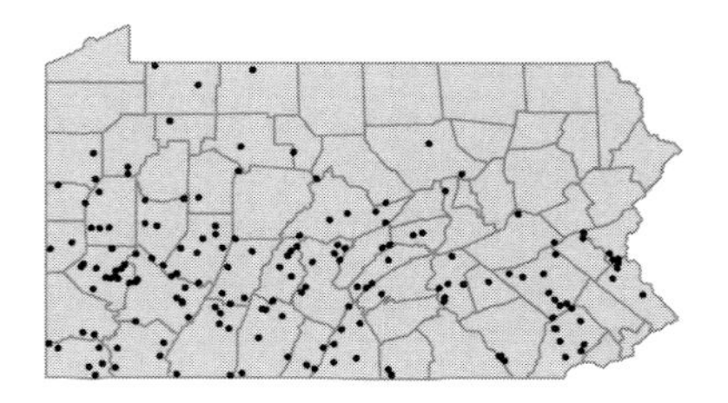

• *Cardaria draba* (L.) Desv. [I]
Hoary cress
Herbaceous perennial
Roadsides, fields and waste ground.

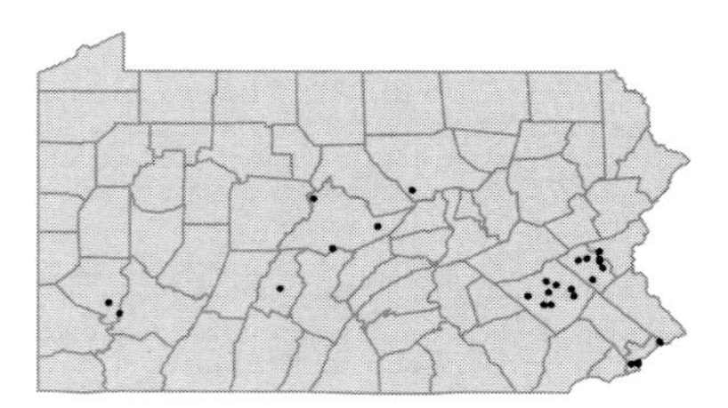

Cardaria pubescens (C.A.Meyer) Jarmolenko [I]
White top
Herbaceous perennial
Dry railroad banks.

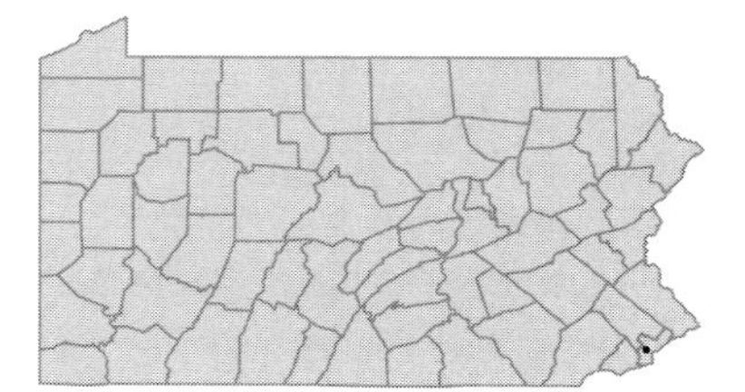

• *Chorispora tenella* (Pallas) DC. [I]
Chorispora
Herbaceous annual
Waste ground, roadsides and edges of cultivated fields.

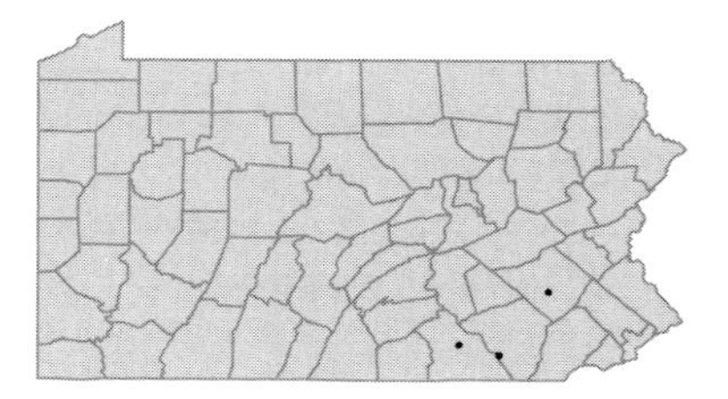

• *Conringia orientalis* (L.) Dumort. [I]
Hare's-ear mustard; Treacle mustard
Herbaceous annual
Waste ground and ballast.

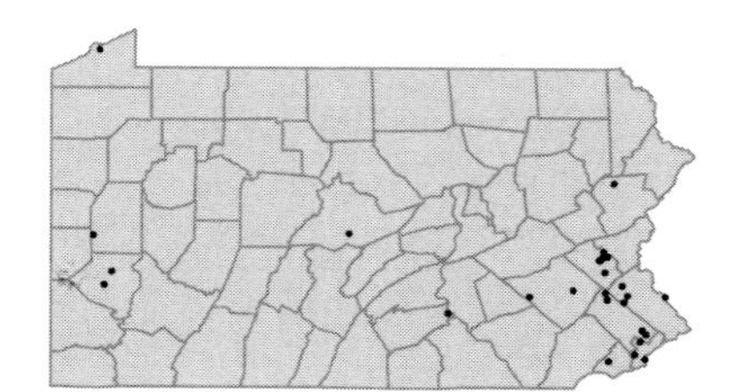

• *Coronopus didymus* (L.) Smith [I]
Swine cress; Wart cress
Herbaceous annual
Waste ground and ballast.

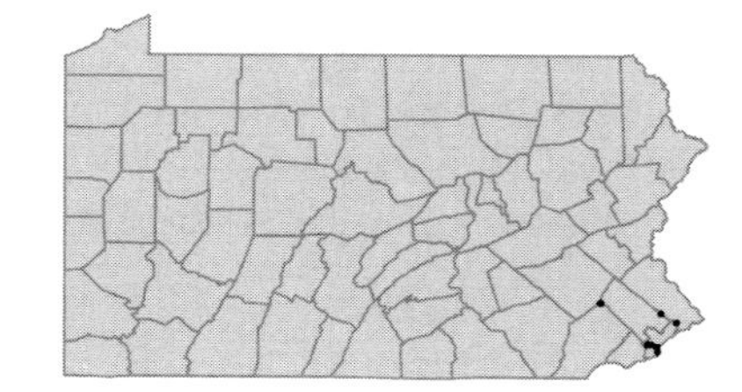

Coronopus squamatus (Forsk.) Aschers.
Swine cress; Wart cress
Herbaceous annual
Waste ground and ballast.
Coronopus coronopus (L.) Karst. P; *Coronopus procumbens* Gilib. FGBW

I

• **_Descurainia pinnata_** (Walt.) Britt.
Tansy mustard
Herbaceous annual
Ballast and waste ground.
Sophia pinnata (Walt.) Britt. P

I

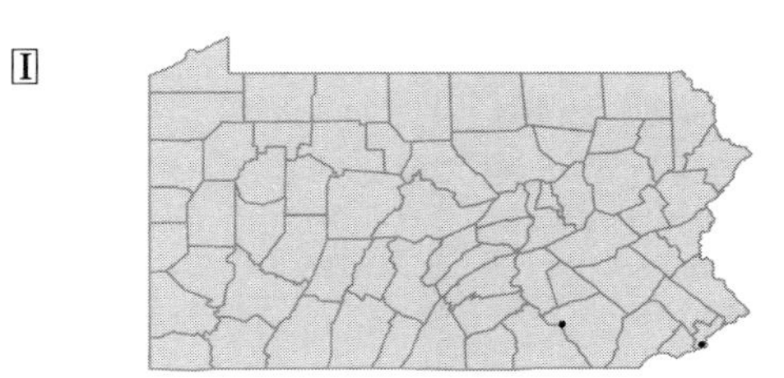

Descurainia sophia (L.) Webb ex Prantl
Herb-sophia
Herbaceous annual
Roadsides, waste ground and ballast.
Sophia sophia (L.) Britt. P

I

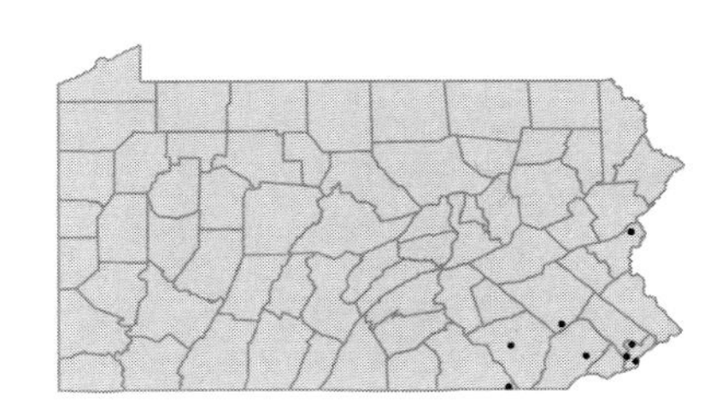

• **_Diplotaxis muralis_** (L.) DC.
Sand-rocket; Wall-rocket
Herbaceous annual
Waste ground and roadsides.

I

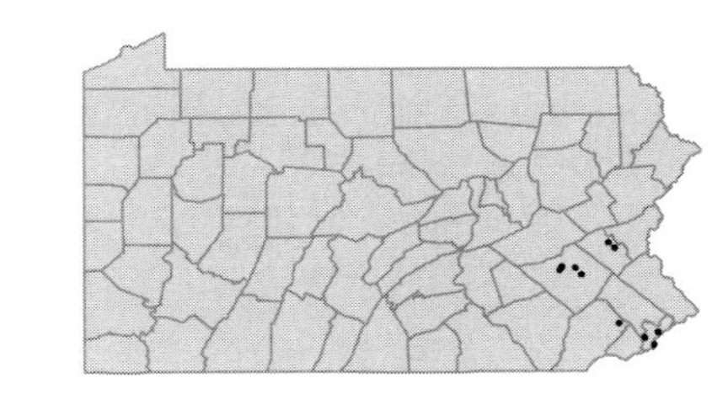

Diplotaxis tenuifolia (L.) DC.
Wall-rocket
Herbaceous perennial
Waste ground and roadsides, in dry, gravelly soil.

I

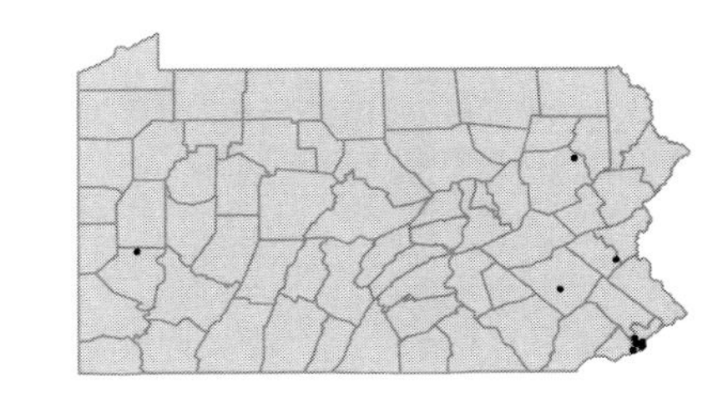

• **_Draba reptans_** (Lam.) Fern.
Whitlow-grass
Herbaceous annual
Dry slopes and ledges. Believed to be extirpated, last collected in 1967.
Draba caroliniana Walt. P

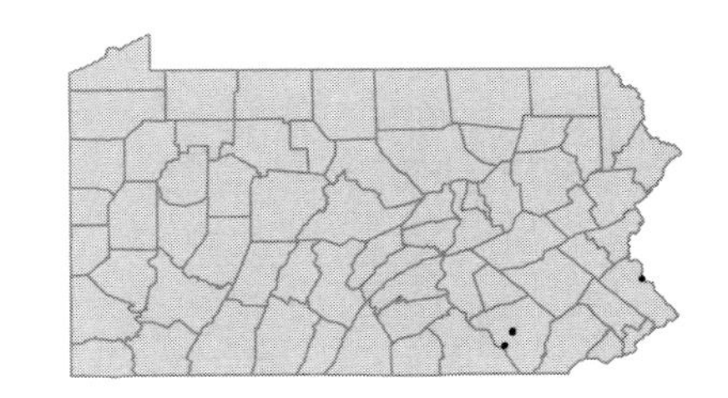

• ***Erophila verna*** (L.) Chev.
Whitlow-grass
Herbaceous annual
Dry fields, roadsides and open woods.
Draba verna L. PFGBWC

[I]

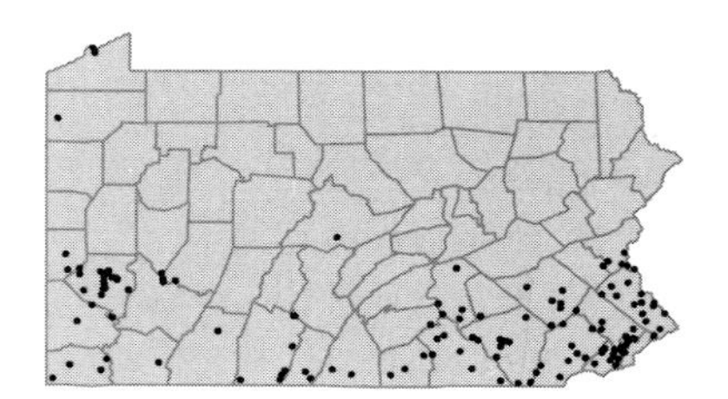

• ***Eruca vesicaria*** (L.) Cav. ssp. ***sativa*** (P.Mill.) Thell.
Garden-rocket
Herbaceous annual
Grassy fields.
Eruca sativa P.Mill FGWC

[I]

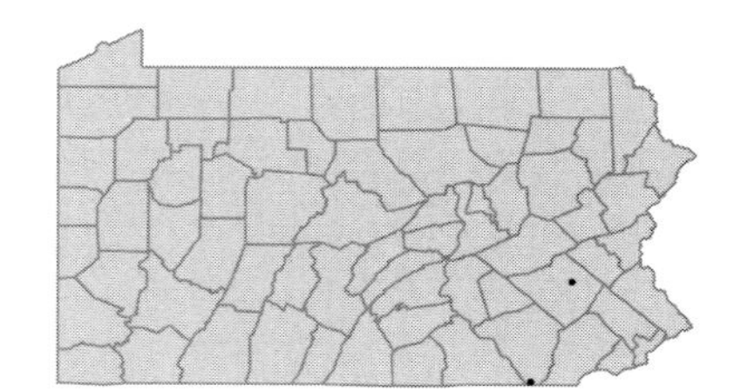

• ***Erucastrum gallicum*** (Willd.) O.E.Schulz
Dog mustard; French rocket
Herbaceous annual
Roadsides and waste ground.

[I]

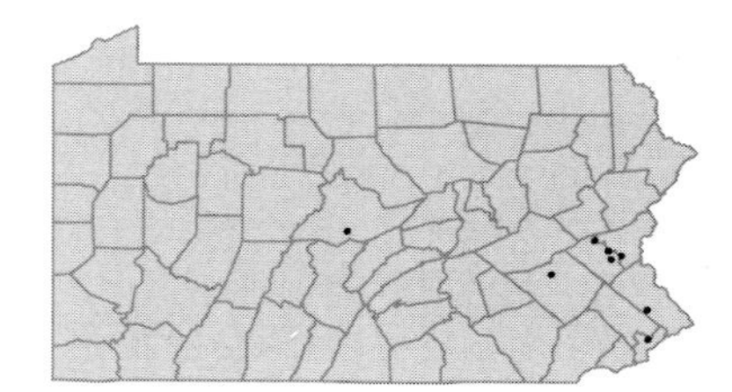

• ***Erysimum cheiranthoides*** L.
Treacle mustard; Wormseed mustard
Herbaceous annual
Roadsides and waste ground.

FAC
[I]

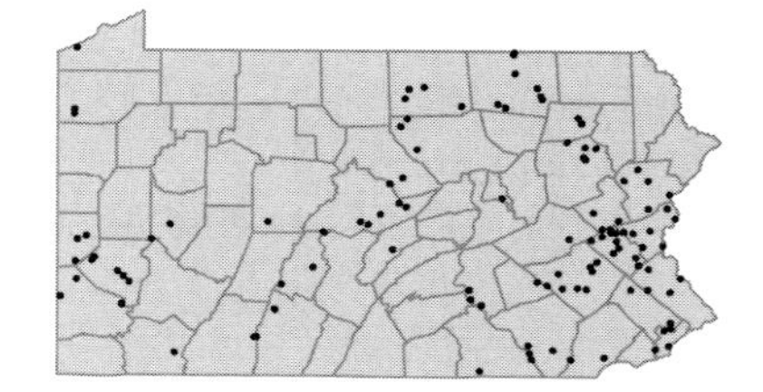

Erysimum hieraciifolium L.
Tall worm-seed mustard
Herbaceous biennial
Dry, rocky bank.

[I]

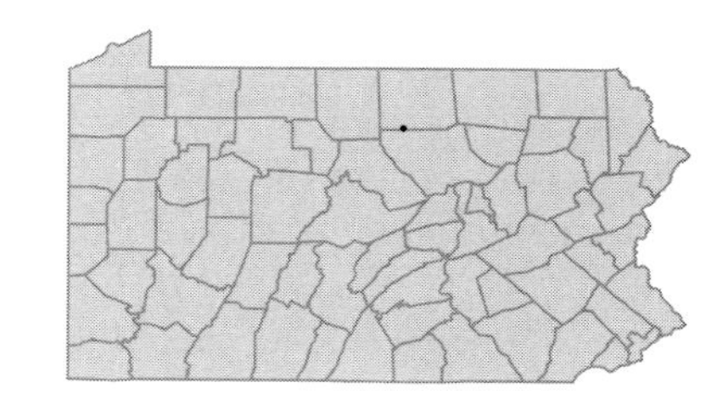

Erysimum inconspicuum (S.Wats.) MacM.
Treacle mustard
Herbaceous perennial
Roadsides, railroad banks and waste ground.

[I]

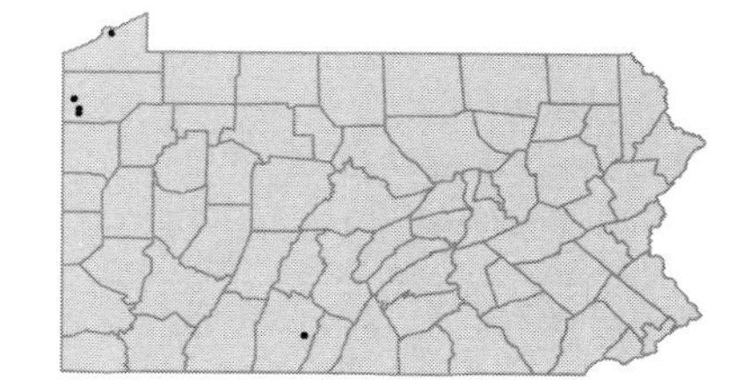

Erysimum repandum L.
Treacle mustard
Herbaceous annual
Roadsides, waste ground and ballast.

I

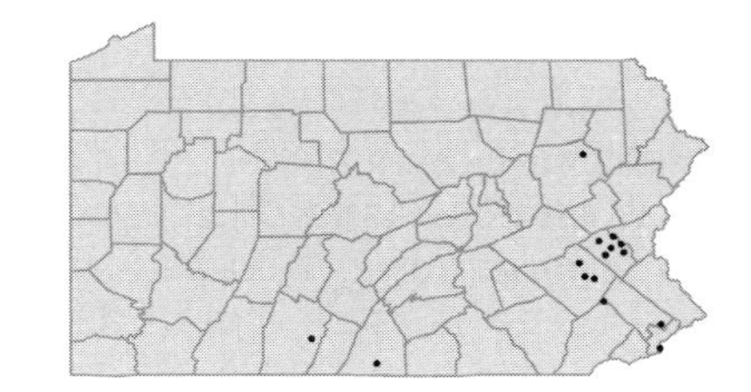

• ***Hesperis matronalis*** L.
Dame's-rocket
Herbaceous perennial
Low woods, wet meadows and roadside ditches.

I

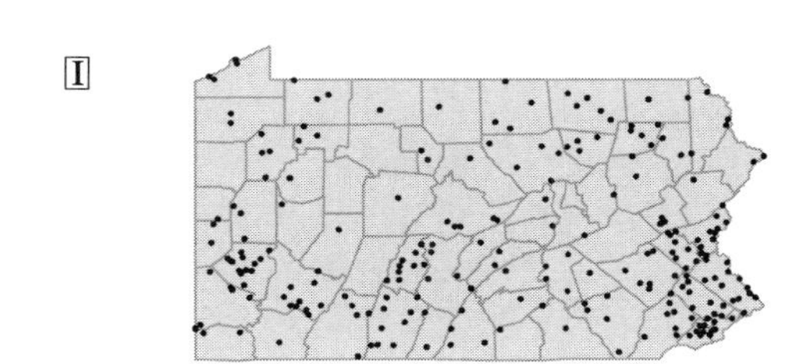

• ***Hutera cheiranthos*** (Vill.) Gomez-Campo
Hutera
Herbaceous annual
Roadsides, apparently spreading along Interstate 80.
Rhyncosinapis cheiranthos (Vill.) Dandy W

I

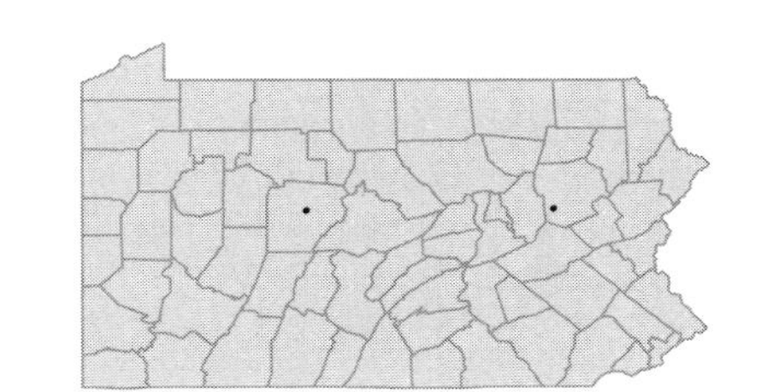

• ***Iberis umbellata*** L.
Candytuft
Herbaceous annual
Cultivated and rarely escaped.

I

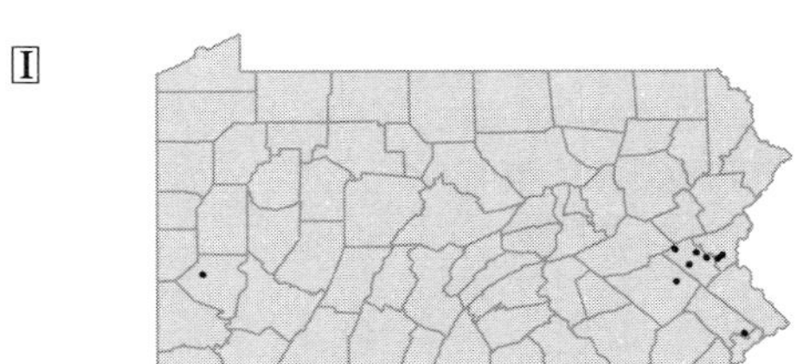

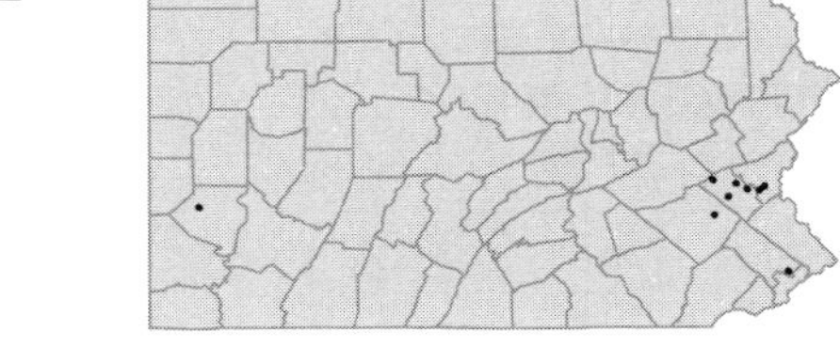

• ***Iodanthus pinnatifidus*** (Michx.) Steud.
Purple rocket
Herbaceous perennial
Moist, alluvial woods and wooded slopes.

FACW

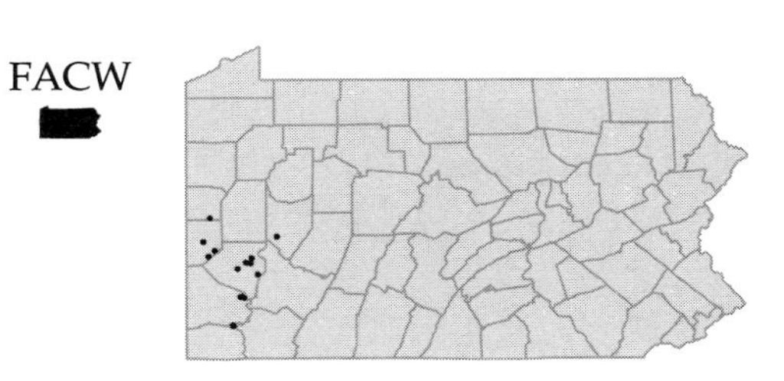

• ***Lepidium campestre*** (L.) R.Br.
Field cress
Herbaceous annual
Fields and waste ground.

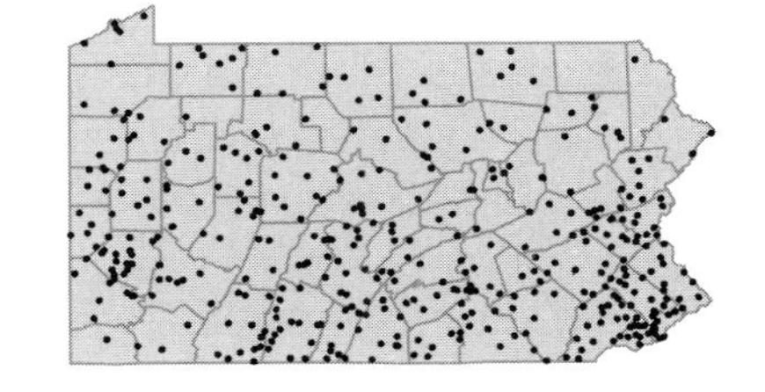

Lepidium densiflorum Schrad.
Wild pepper-grass
Herbaceous annual
Waste ground and roadsides in dry to moist soil.
Lepidium apetalum Willd. P

FAC
I

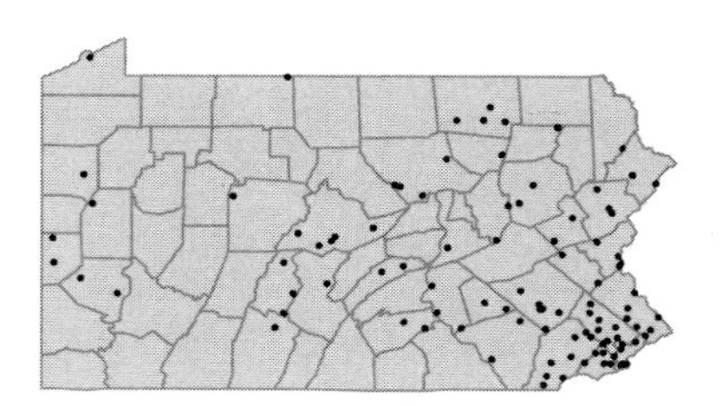

Lepidium graminifolium L.
Pepper-grass
Herbaceous perennial
Ballast. Represented by a single collection from Philadelphia Co. in 1920.

I

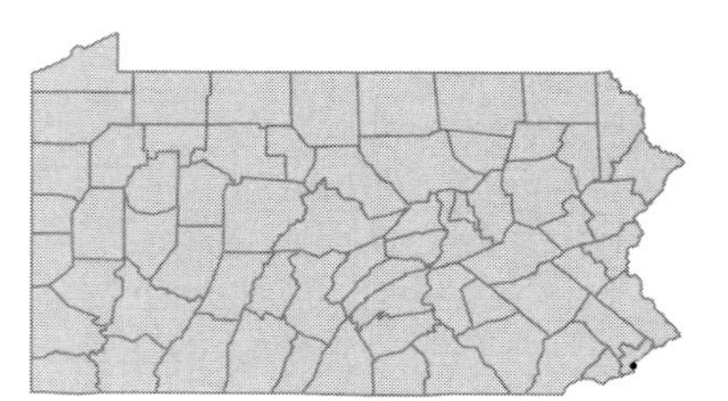

Lepidium heterophyllum Benth.
Pepper-grass; Pepper-wort
Herbaceous perennial
Ballast. Represented by two collections from Philadelphia Co. in the 1870's.
Lepidium smithii Hook. W

I

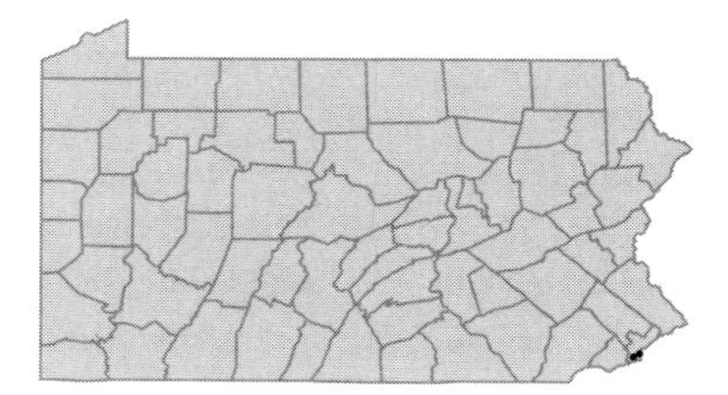

Lepidium perfoliatum L.
Pepper-grass
Herbaceous annual
Waste ground.

I

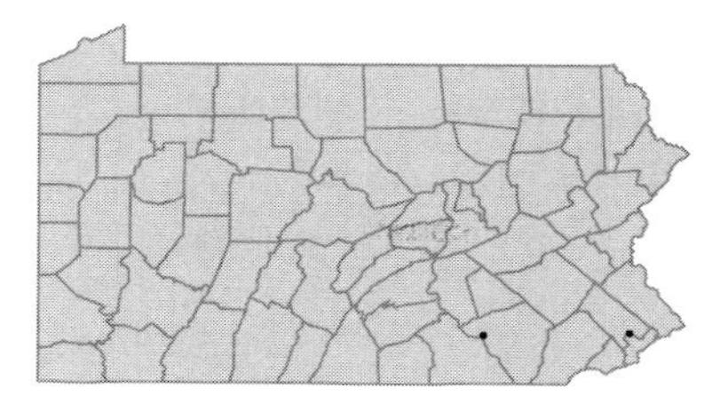

Lepidium ruderale L.
Roadside pepper-grass
Herbaceous annual
Urban waste ground and roadsides. All collections are from Philadelphia Co. 1863-1867.

I

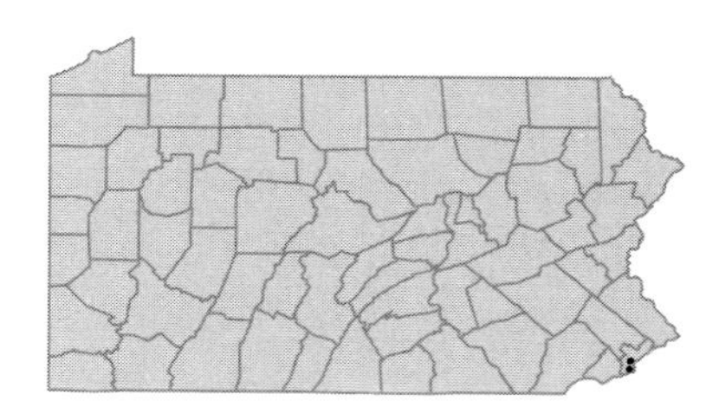

Lepidium sativum L.
Garden cress
Herbaceous annual
Waste ground and roadsides.

I

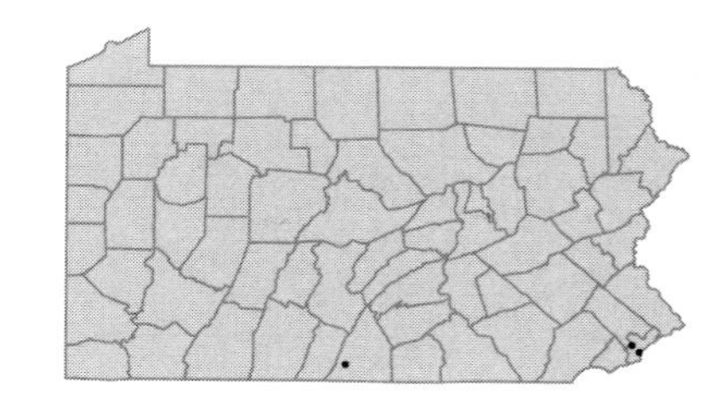

• ***Lepidium virginicum*** L.
Poor-man's pepper; Wild pepper-grass
Herbaceous annual
Roadsides and waste ground, in dry, open soil.

FACU-

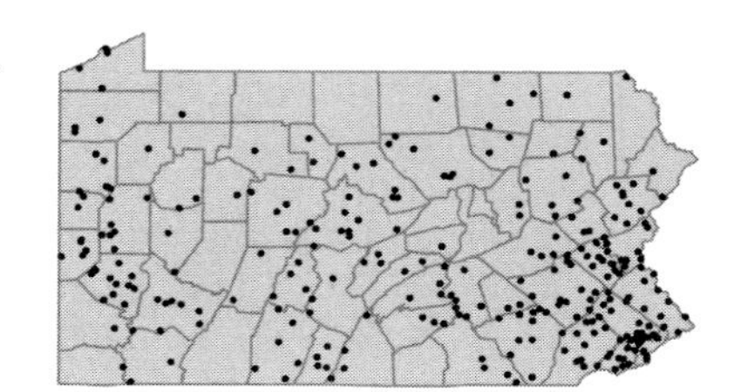

• ***Lobularia maritima*** (L.) Desv.
Sweet alyssum
Herbaceous annual
Cultivated and rarely escaped to waste ground.

I

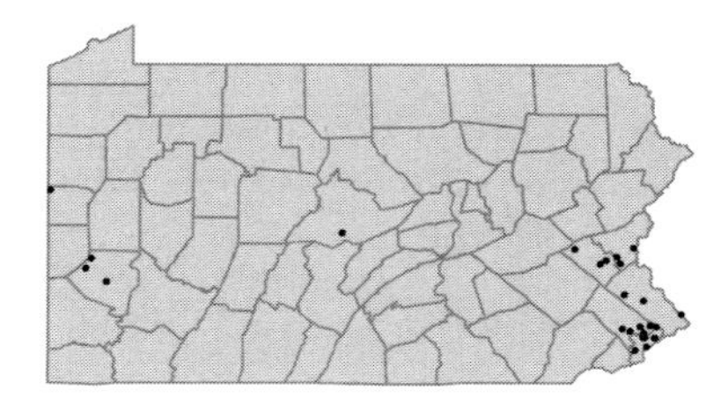

• ***Lunaria annua*** L.
Honesty; Money-plant
Herbaceous biennial
Cultivated and occasionally escaped to roadsides and waste ground.

I

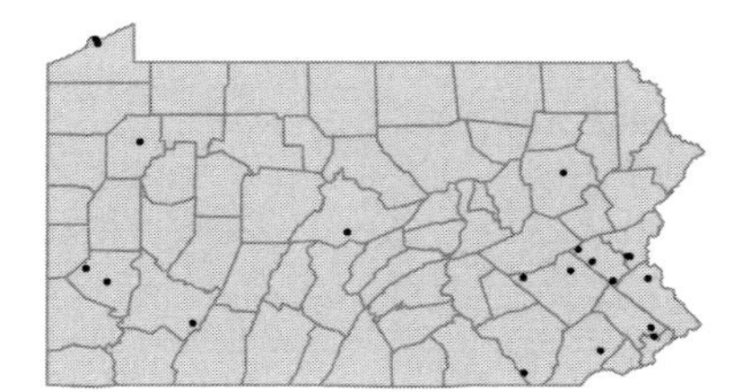

Lunaria rediviva L.
Perennial honesty
Herbaceous perennial
Waste ground.

I

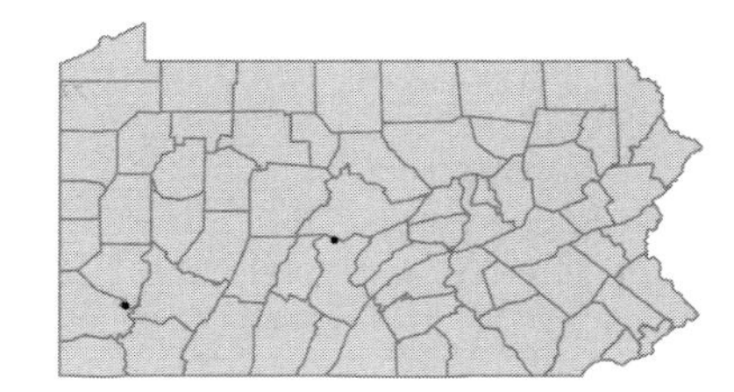

• ***Nasturtium microphyllum*** (Boenn.) Reichenb.
Watercress
Herbaceous perennial, emergent aquatic
Pond edges and river banks. A hybrid with *N. officinale* (*N. x sterilis* [Airy-Shaw] Oefel) has been collected at three sites.
Nasturtium officinale R.Br. var. *microphyllum* (Boenn.) Thell. F

OBL
I

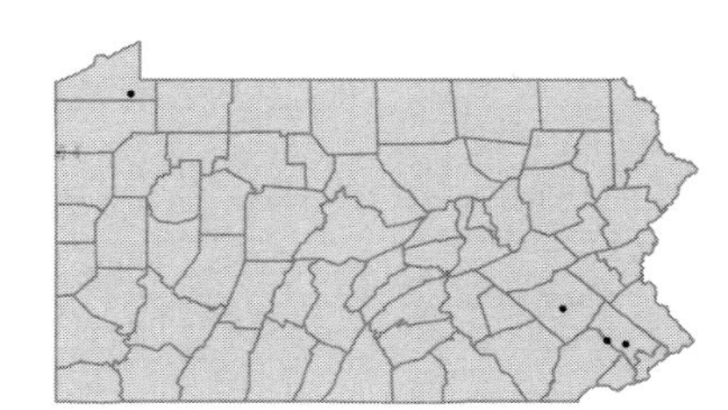

Nasturtium officinale R.Br.
Watercress
Herbaceous perennial, emergent aquatic
Springs and shallow, quiet water.
Roripa nasturtium (L.) Rusby

OBL
I

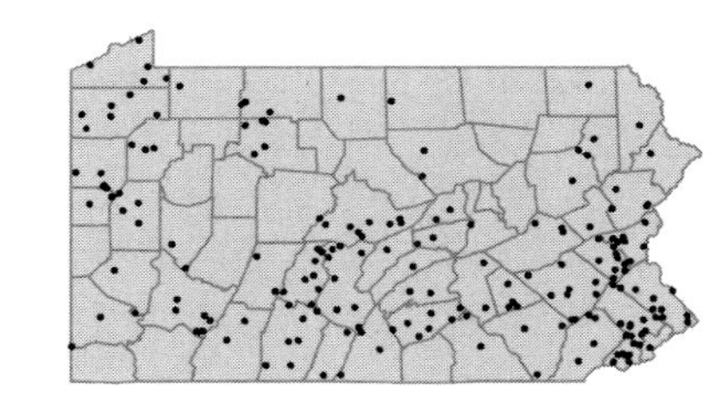

• ***Neslia paniculata*** (L.) Desv. I
Ball mustard
Herbaceous annual
Waste ground along railroad. Represented by a single collection from Lehigh Co. in 1917.

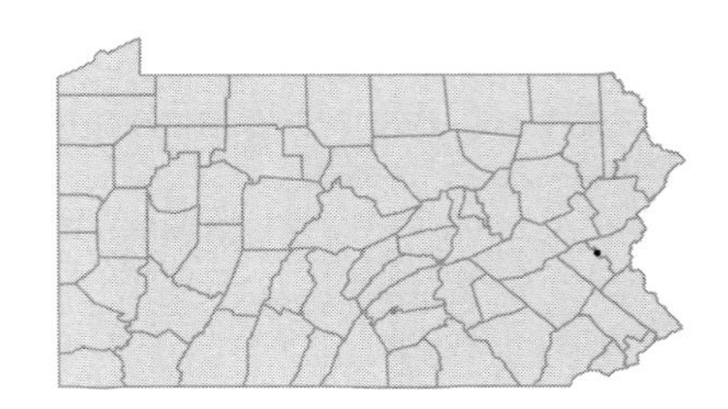

• ***Raphanus raphanistrum*** L. I
White charlock; Wild radish
Herbaceous annual
Waste ground.

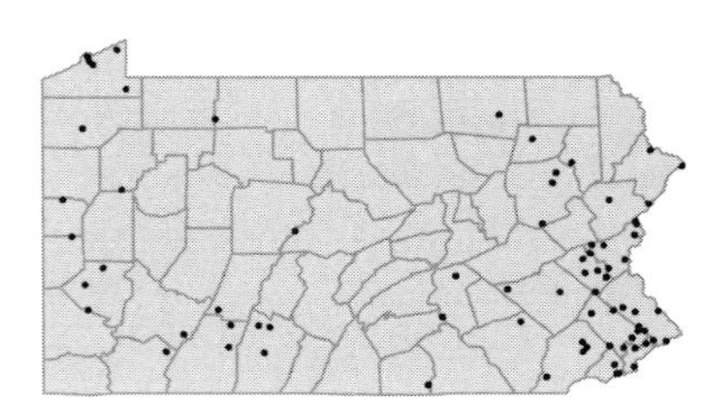

Raphanus sativus L. I
Garden radish
Herbaceous annual
Cultivated and occasionally escaped to roadsides and old fields.

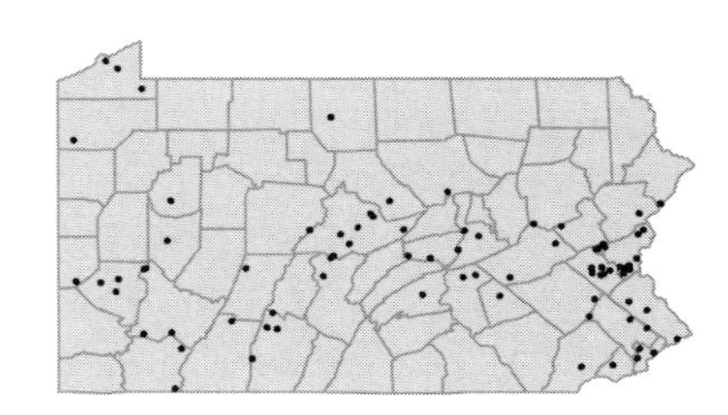

• ***Rapistrum rugosum*** (L.) All. I
Wild rape
Herbaceous annual
Roadsides and ballast.

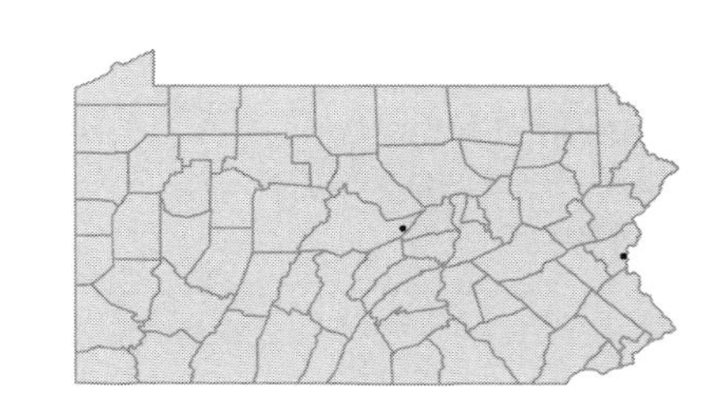

• ***Rorippa amphibia x sylvestris*** I
Yellow cress
Herbaceous perennial
Wet soil.
Rorippa x prostrata (Bergeret) Schniz & Thell. FWKC

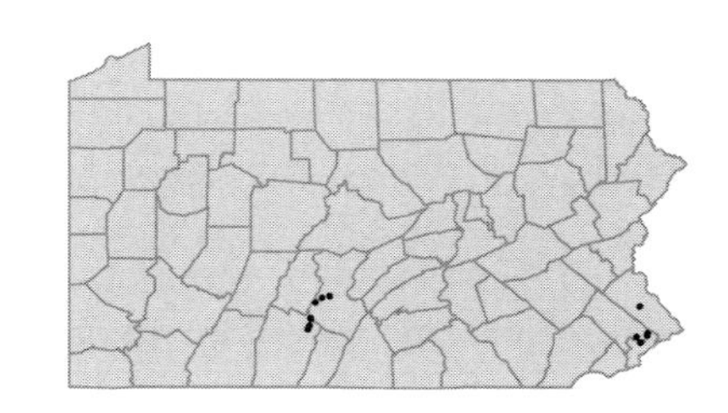

Rorippa austriaca (Crantz) Bess. I
Austrian yellow-cress
Herbaceous perennial
Fields and roadsides.

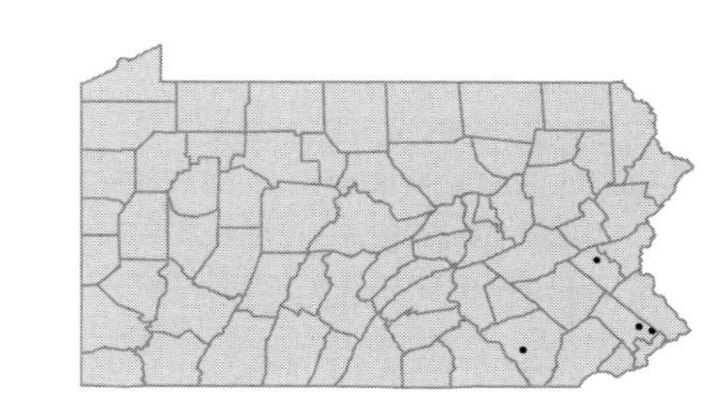

Rorippa palustris (L.) Bess. ssp. ***palustris***
Marsh watercress; Yellow watercress
Herbaceous annual
Wet soil.
Roripa islandica (Oeder) Borbas FGB

OBL

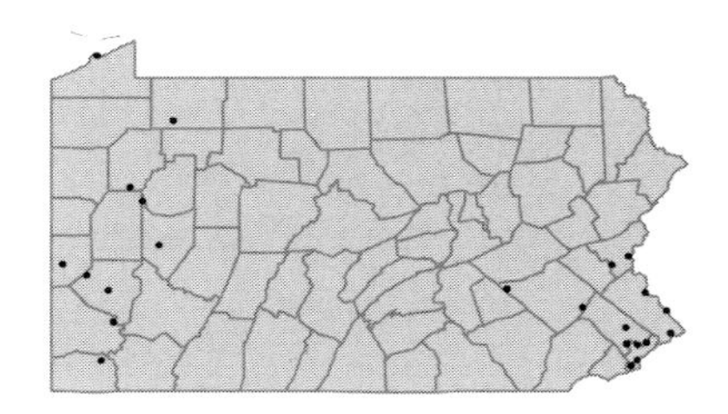

Rorippa palustris (L.) Bess. ssp. ***fernaldiana*** (Butt. & Abbe) Jonsell
Marsh watercress; Yellow watercress
Herbaceous annual
Wet shores and low, open ground.
Rorippa islandica (Oeder) Borbas var. *fernaldiana* Butt. & Abbe FGB

OBL

Rorippa palustris (L.) Bess. ssp. ***hispida*** (Desv.) Jonsell
Marsh watercress; Yellow watercress
Herbaceous annual
Shores and wet ground.
Roripa hispida (Desv.) Britt. P; *Rorippa islandica* (Oeder) Borbas var. *hispida* (Desv.) Butt. & Abbe FGBC

OBL

Rorippa sylvestris (L.) Bess.
Creeping yellow-cress
Herbaceous perennial
Wet soil.

FACW
I

• ***Sinapis alba*** L.
White mustard
Herbaceous annual
Waste ground and ballast.
Brassica hirta Moench FGBW

I

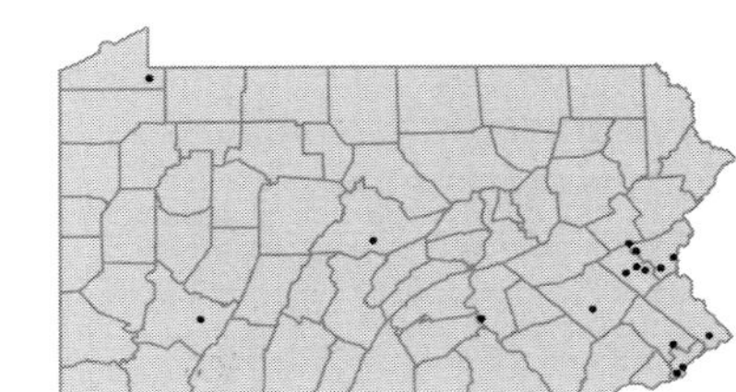

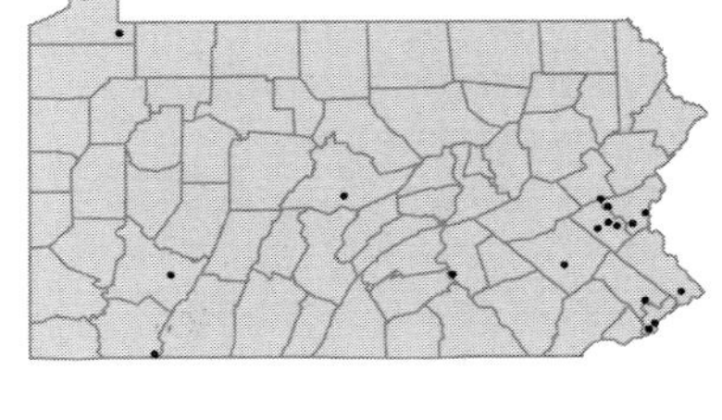

Sinapis arvensis L.
Charlock; Wild mustard
Herbaceous annual
Fields, gardens and waste ground.
Brassica kaber (DC.) L.C.Wheeler FWGB

I

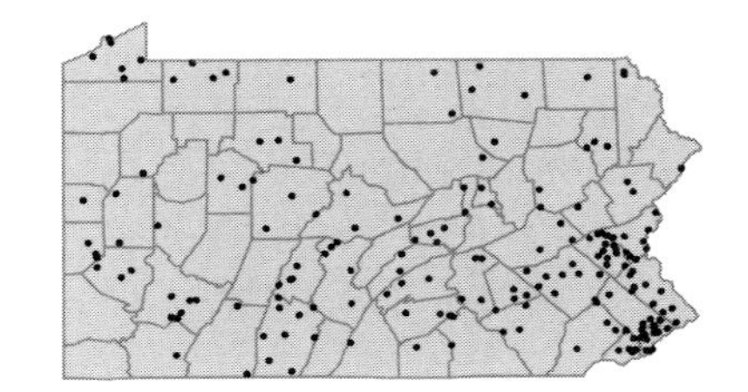

• ***Sisymbrium altissimum*** L.
Tumble mustard
Herbaceous annual
Fields, roadsides and waste ground.

FACU-
I

Sisymbrium irio L.
London rocket
Herbaceous annual
Ballast.

I

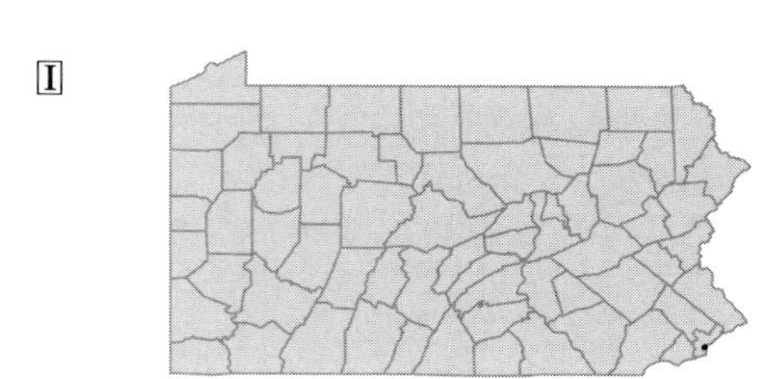

Sisymbrium loeselii L.
Hedge mustard
Herbaceous annual
Roadsides or fields.

I

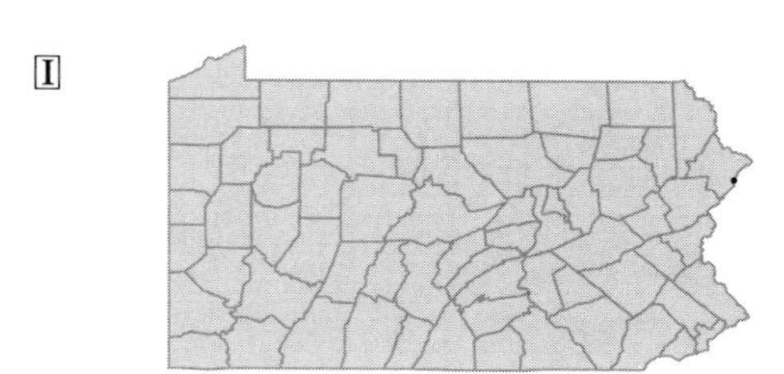

Sisymbrium officinale (L.) Scop. var. ***officinale***
Hedge mustard
Herbaceous annual
Gardens, fields and waste ground.

I

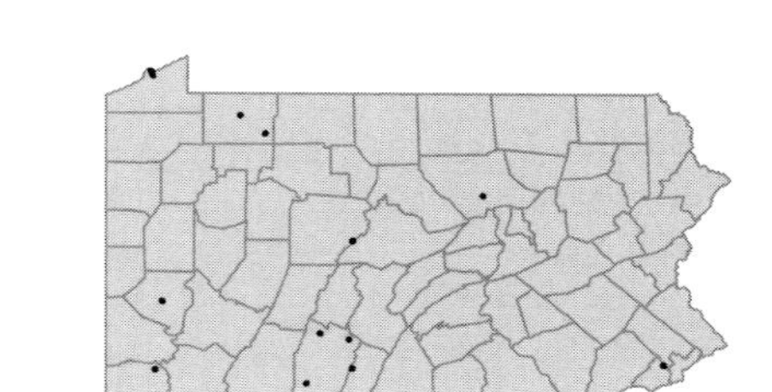

Sisymbrium officinale (L.) Scop. var. ***leiocarpum*** DC.
Bank cress; Hedge mustard
Herbaceous annual
Gardens, fields and waste ground.

I

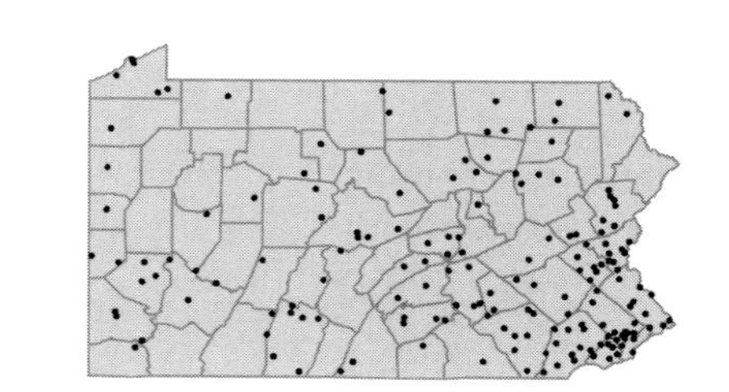

• ***Teesdalia nudicaulis*** (L.) R.Br.
Shepherd cress
Herbaceous annual
Ballast. All collections are from Philadelphia Co. 1877-1880.

I

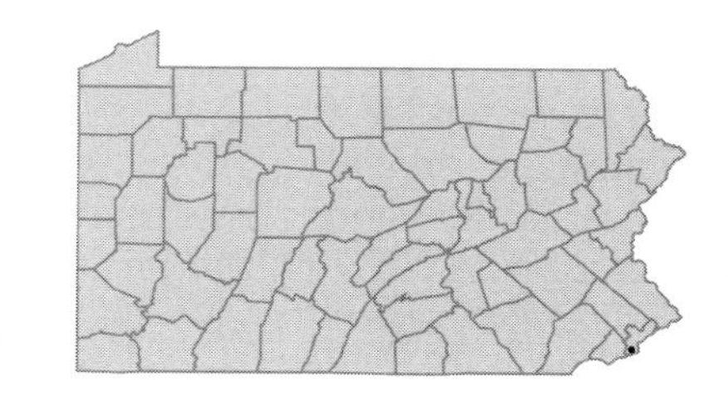

• ***Thlaspi alliaceum*** (L.) Jacq.
Penny-cress
Herbaceous annual
Roadside bank. Represented by a single collection from Bedford Co. in 1947.

I

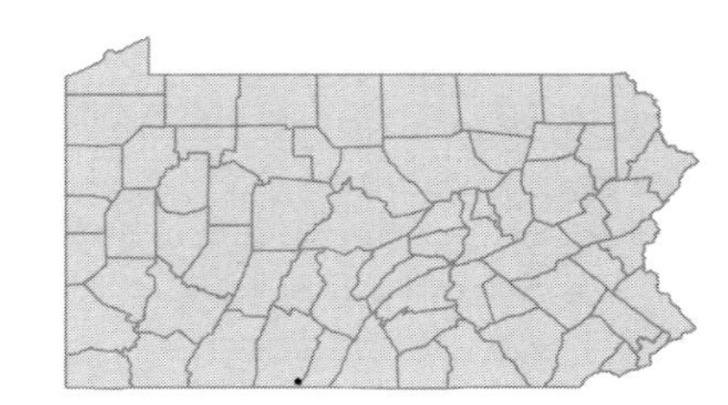

Thlaspi arvense L.
Field penny-cress; Frenchweed
Herbaceous annual
Roadsides and waste ground.

I

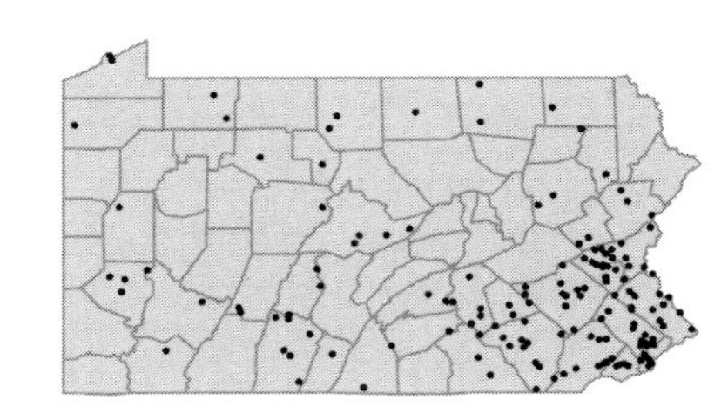

Thlaspi perfoliatum L.
Penny-cress
Herbaceous annual
Fields and roadsides.

I

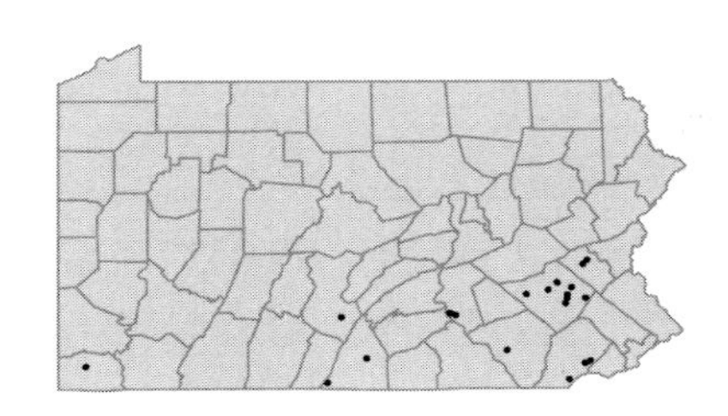

RESEDACEAE

• ***Reseda alba*** L.
White mignonette
Herbaceous annual
Ballast and waste ground. Represented by three collections from Philadelphia and Lehigh Cos. 1875-1912.

I

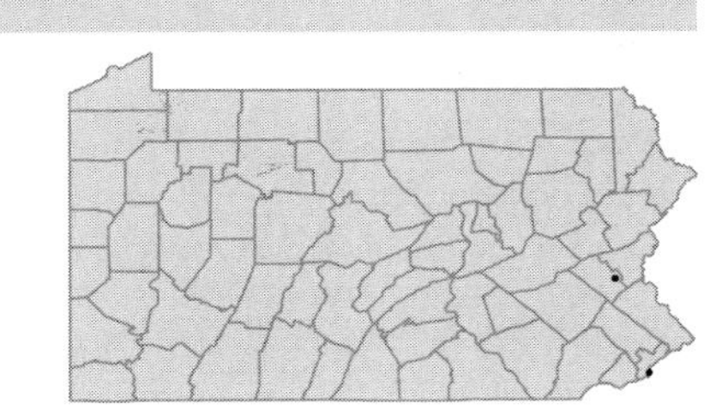

Reseda lutea L.
Yellow mignonette
Herbaceous biennial
Waste ground, often on limestone.

I

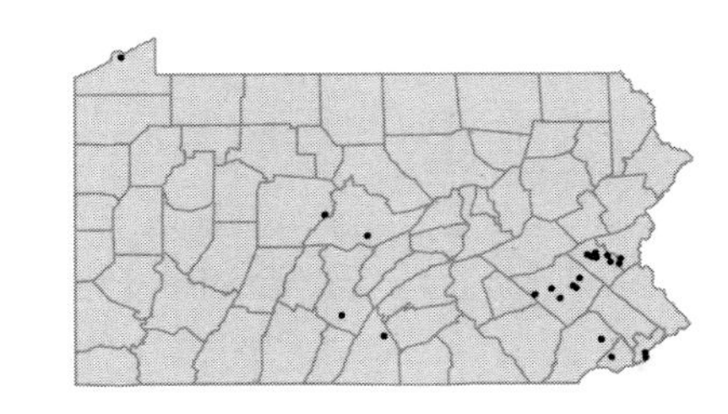

Reseda luteola L.
Dyer's rocket
Herbaceous biennial
Ballast and waste ground, formerly cultivated.

I

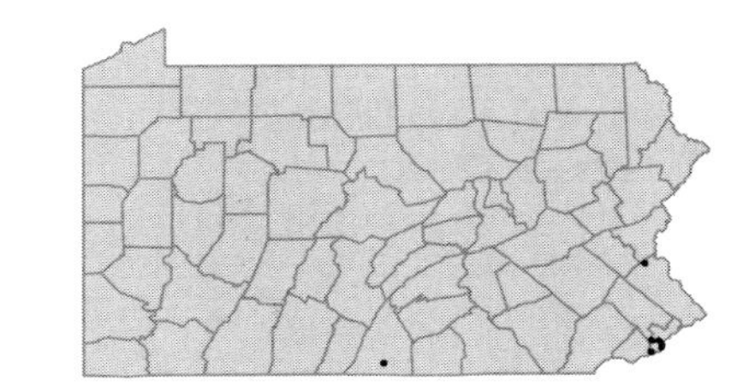

Reseda odorata L. I
Mignonette
Herbaceous annual
Refuse dump. Represented by one collection from Chester Co. in 1884 and one from Lehigh Co. in 1960.

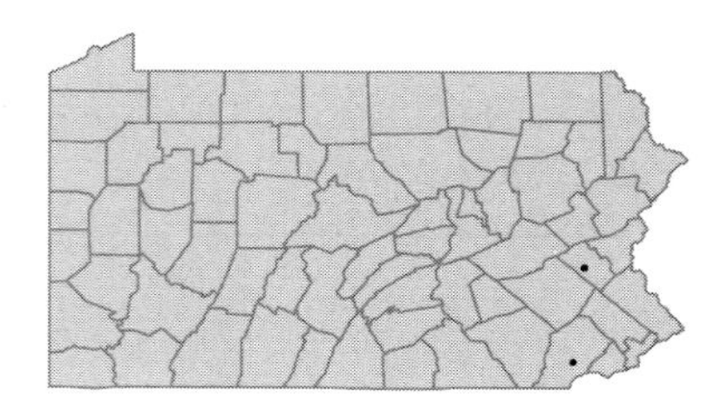

Reseda phyteuma L. I
Mignonette
Herbaceous annual
Ballast and waste ground. Represented by two collections from Philadelphia Co. 1875-1883.

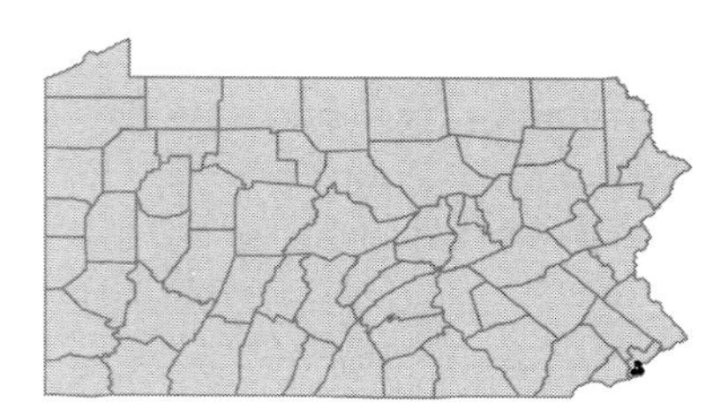

CLETHRACEAE

• ***Clethra acuminata*** Michx.
Mountain pepperbush
Deciduous shrub
Rocky, wooded slopes.

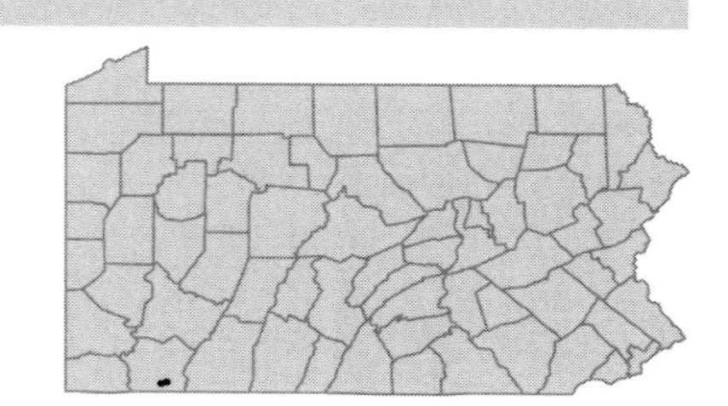

Clethra alnifolia L.
Sweet pepperbush
Deciduous shrub
Low, wet woods and swamps.

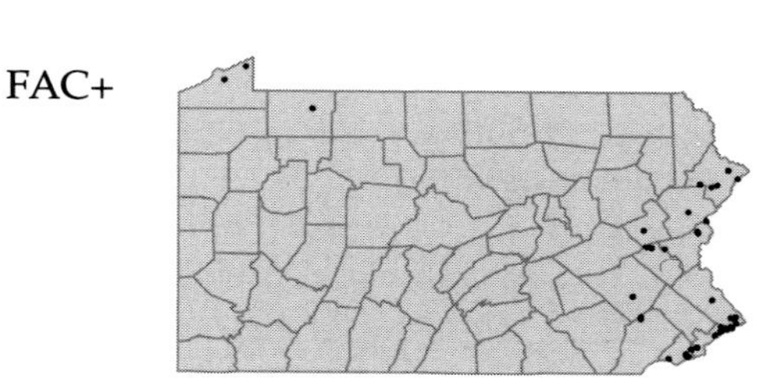

ERICACEAE

• ***Andromeda polifolia*** L. var. ***glaucophylla*** (Link) DC. OBL
Bog-rosemary
Evergreen shrub
Floating sphagnum bog mats.
Andromeda glaucophylla Link FGBWC

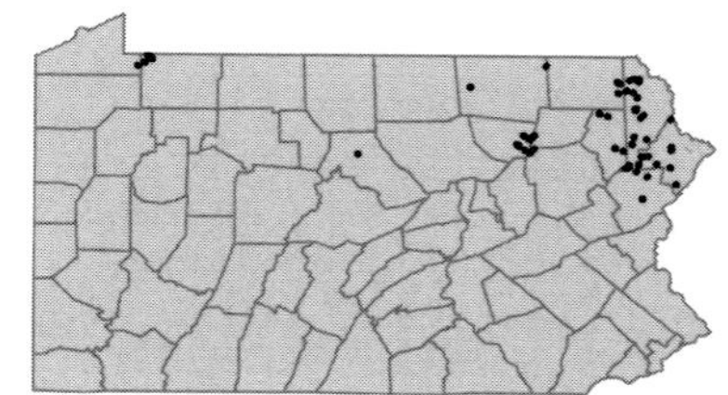

• ***Arctostaphylos uva-ursi*** (L.) Spreng. ssp. ***coactilis*** Fern. & Macbr.
Bearberry; Kinnikinick
Evergreen shrub
Dry, open woods and sand barrens. Believed to be extirpated, last collected in 1959.

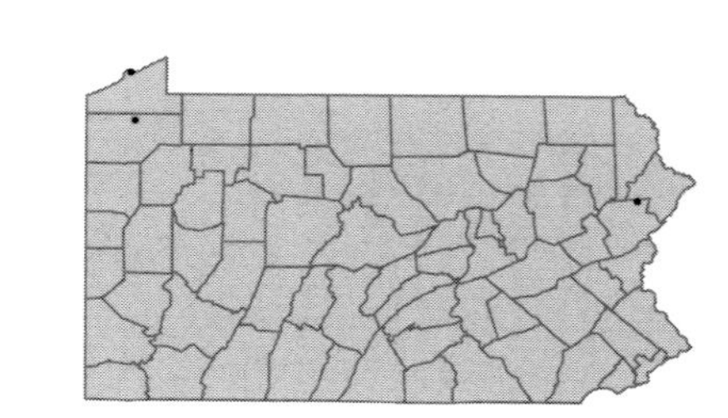

• ***Chamaedaphne calyculata*** (L.) Moench var. ***angustifolia*** (Ait.) Rehd.
Leatherleaf
Evergreen shrub
Sphagnum bogs.

OBL

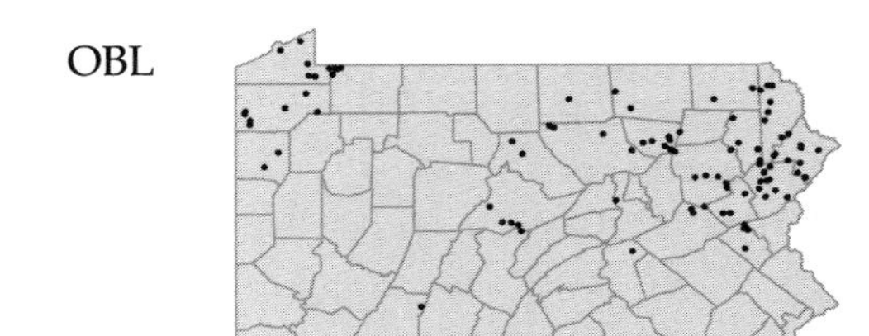

• ***Epigaea repens*** L.
Trailing arbutus; Mayflower
Evergreen sub-shrub
Dry to moist, acidic woods and edges.

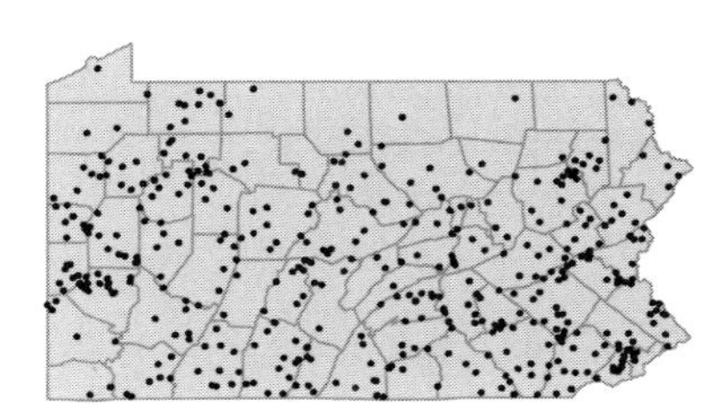

• ***Erica vagans*** L.
Cornish heath
Evergreen shrub
Waste ground. Collected at a single site in Schuylkill Co. in 1944-1945.

I

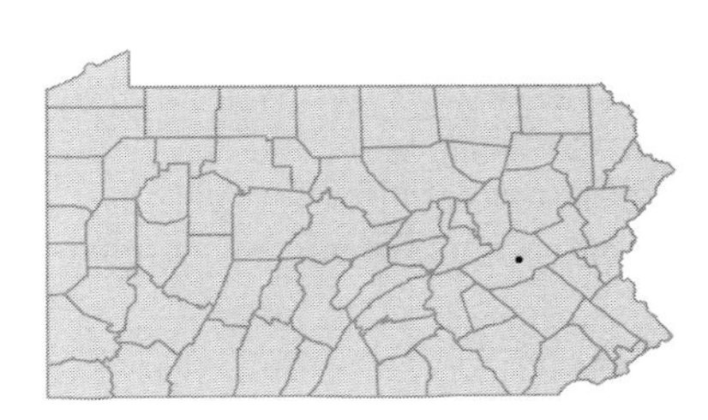

• ***Gaultheria hispidula*** (L.) Muhl. ex Bigel.
Creeping snowberry; Moxie-plum
Evergreen sub-shrub
On hummocks and tree stumps in northern bogs and swamps.
Chiogenes hispidula (L.) Torr. & A.Gray PW

FACW

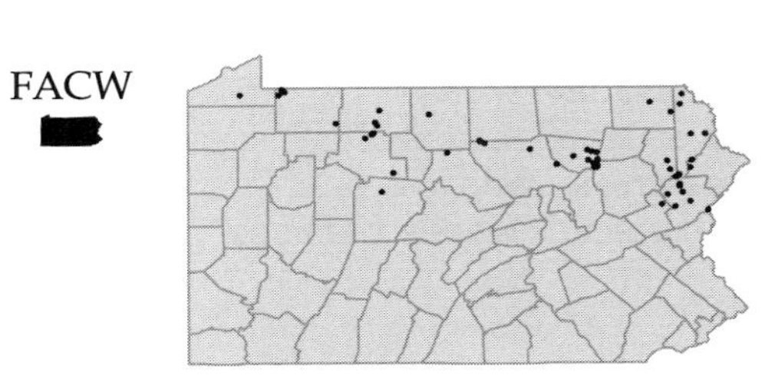

Gaultheria procumbens L.
Teaberry; Wintergreen
Evergreen sub-shrub
Dry to moist, acidic woods.

FACU

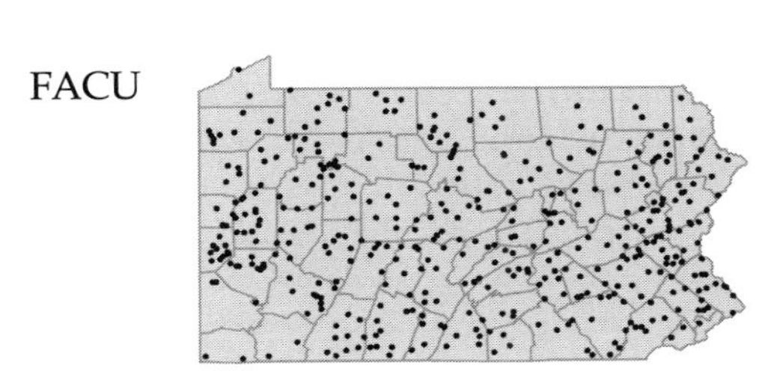

• ***Gaylussacia baccata*** (Wang.) K.Koch
Black huckleberry
Deciduous shrub
Dry to moist, acidic woods and bogs.
Gaylussacia resinosa (Ait.) Torr. & A.Gray P

FACU

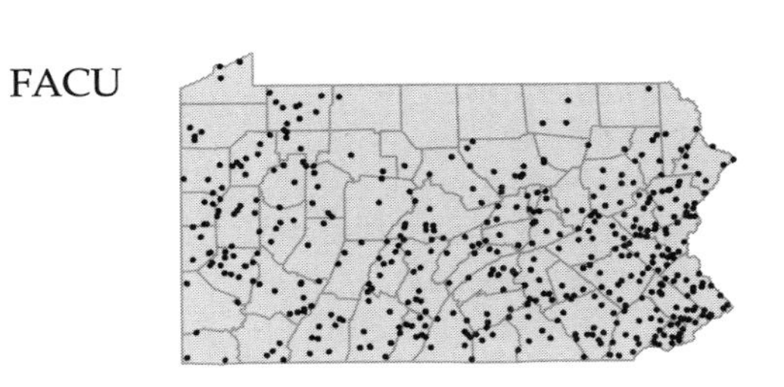

Gaylussacia brachycera (Michx.) A.Gray
Box huckleberry
Evergreen shrub
Dry, wooded slopes.

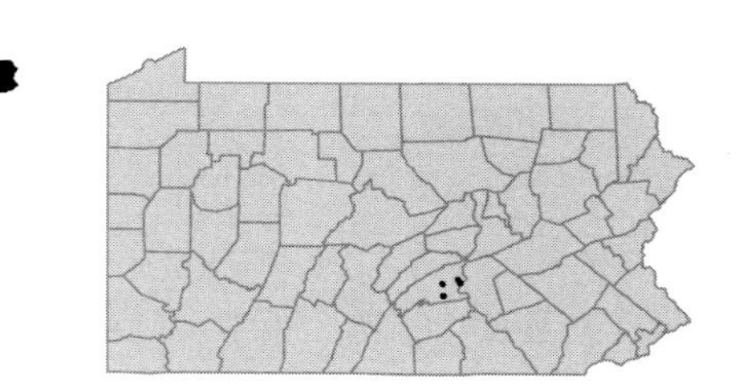

Gaylussacia dumosa (Andr.) Torr. & A.Gray
Dwarf huckleberry
Deciduous shrub
Moist, acidic woods and swamps.

FAC

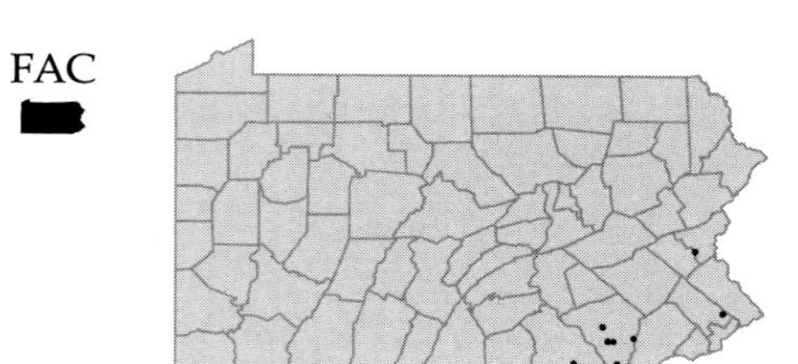

Gaylussacia frondosa (L.) Torr. & A.Gray ex Torr.
Dangleberry
Deciduous shrub
Moist, acidic woods, swamps and bogs.

FAC

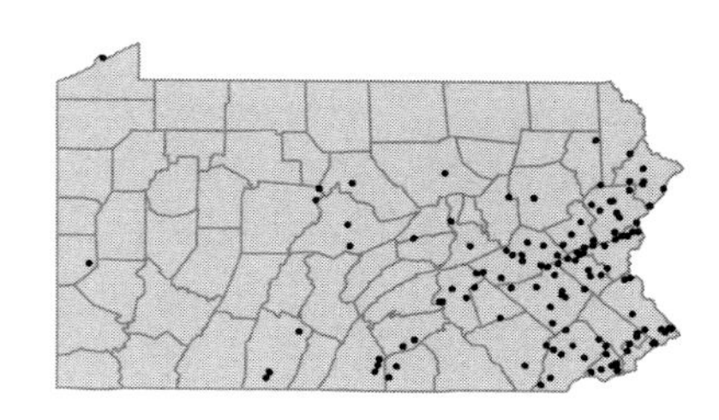

• ***Kalmia angustifolia*** L.
Sheep laurel; Lambkill
Evergreen shrub
Moist, acidic woods and bogs.

FAC

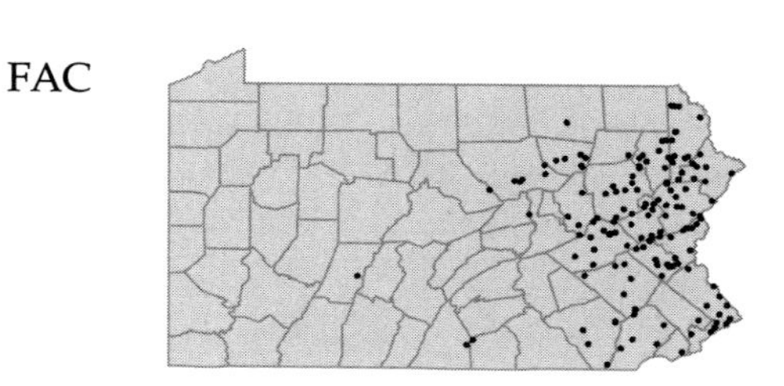

Kalmia latifolia L.
Mountain laurel
Evergreen shrub
Dry to moist, acidic woods and slopes. The state flower of Pennsylvania.

FACU

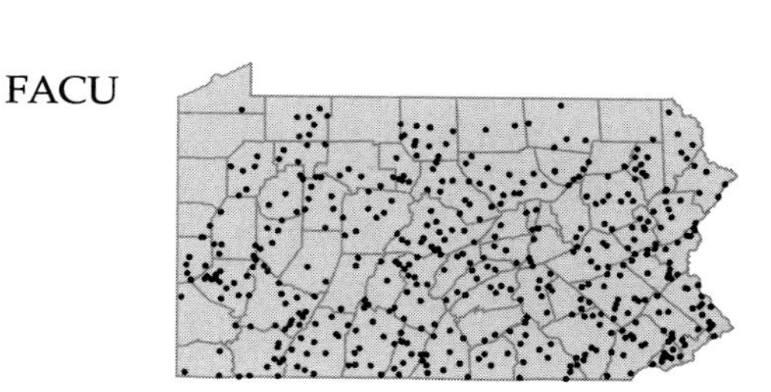

Kalmia polifolia Wang.
Bog laurel
Evergreen shrub
Floating sphagnum bog mats.
Kalmia glauca Ait. P

OBL

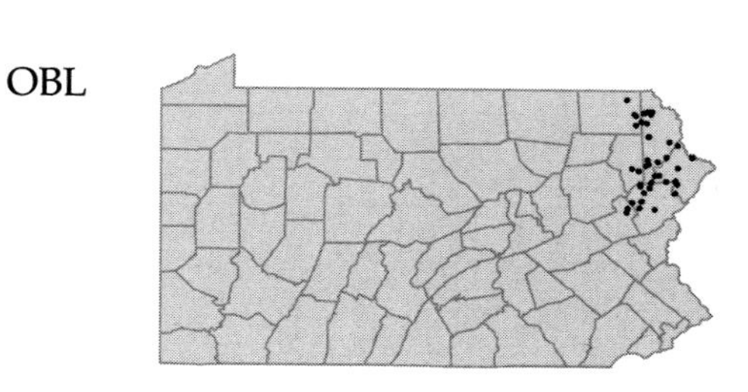

• ***Ledum groenlandicum*** Oeder
Labrador tea
Evergreen shrub
Bogs and swamps.

OBL

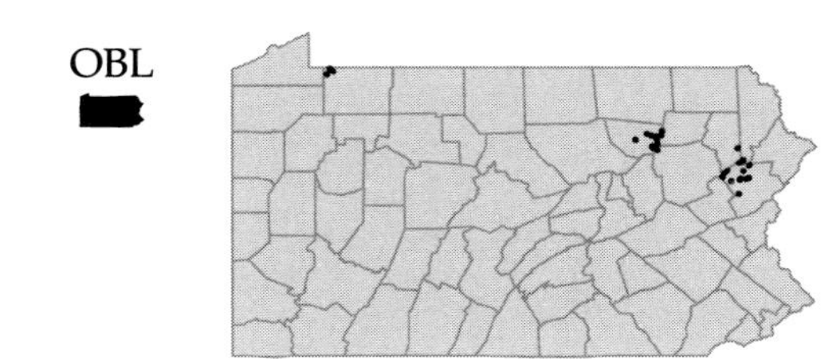

• ***Leiophyllum buxifolium*** (Berg.) Ell.
Sand-myrtle
Evergreen shrub
Dry, acidic woods. Believed to be extirpated, collected once in Monroe Co. in 1880.

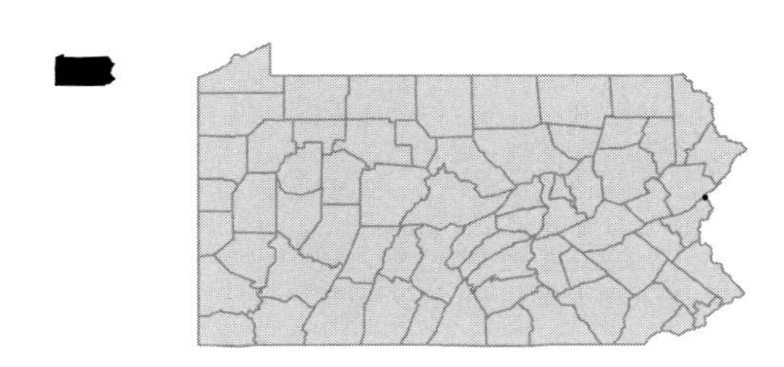

• ***Leucothoe racemosa*** (L.) A.Gray
Fetter-bush; Swamp dog-hobble
Deciduous shrub
Moist, acidic woods.
Eubotrys racemosa (L.) Nutt. GBC

FACW

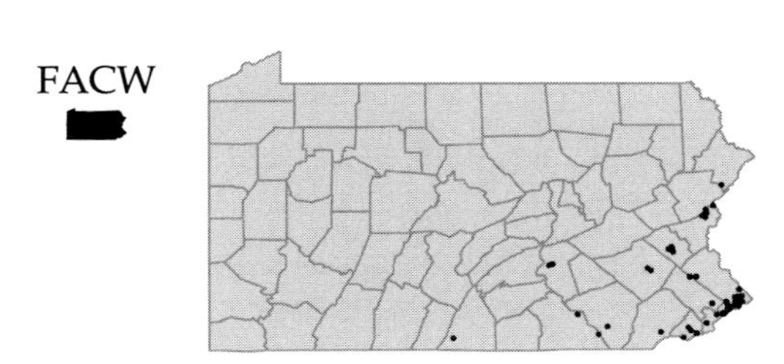

• ***Lyonia ligustrina*** (L.) DC.
Maleberry
Deciduous shrub
Moist woods and swamps.
Xolisma ligustrina (L.) Britt. P

FACW

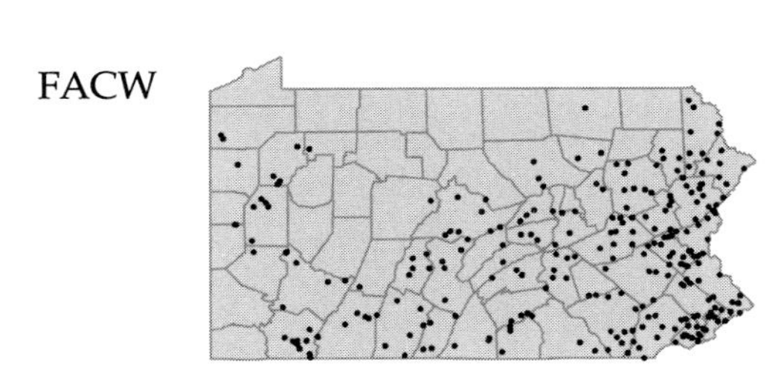

Lyonia mariana (L.) D.Don
Stagger-bush
Deciduous shrub
Dry slopes and barrens.
Pieris mariana (L.) Benth. & Hook. P

FAC-

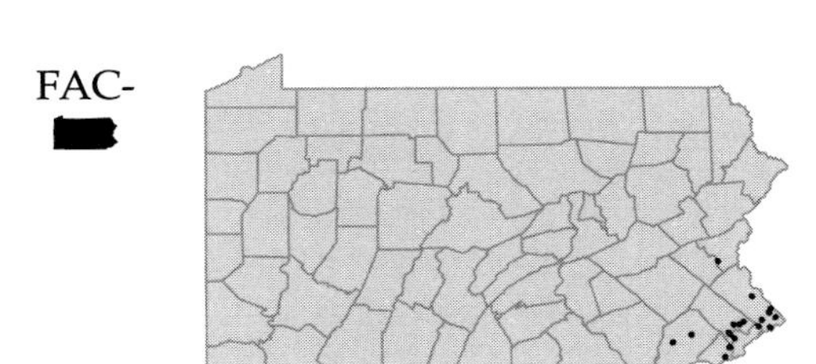

• ***Menziesia pilosa*** (Michx.) Juss.
Minnie-bush
Deciduous shrub
Moist woods, rocky lower slopes and stream banks.

FAC-

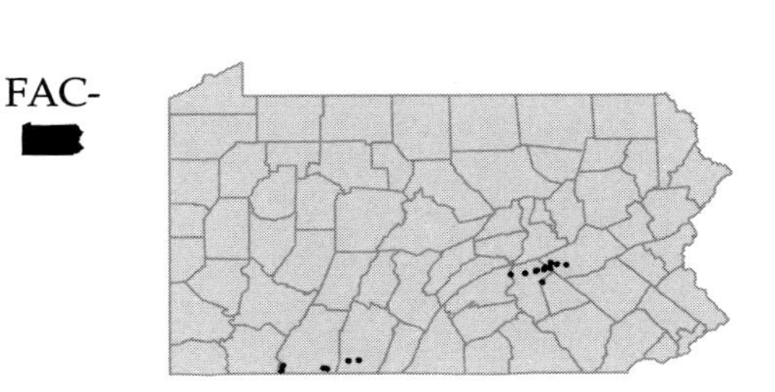

• ***Oxydendrum arboreum*** (L.) DC.
Sourwood
Deciduous tree
Rocky, wooded slopes.

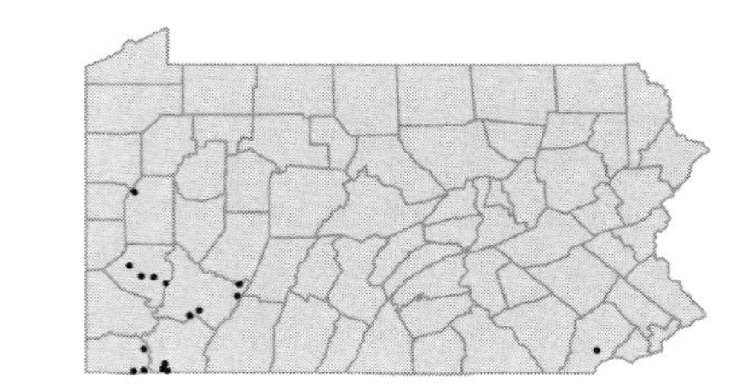

• ***Rhododendron arborescens*** (Pursh) Torr.
Smooth azalea
Deciduous shrub
Moist, rocky woods and swamps. A hybrid with *R. viscosum* has been collected once in Bedford Co.
Azalea arborescens Pursh P

FAC

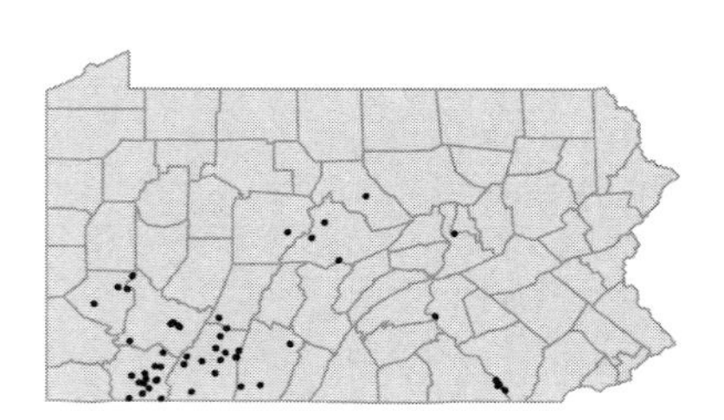

Rhododendron atlanticum (Ashe) Rehd.
Dwarf azalea
Deciduous shrub
Sandy, open woods.

FAC

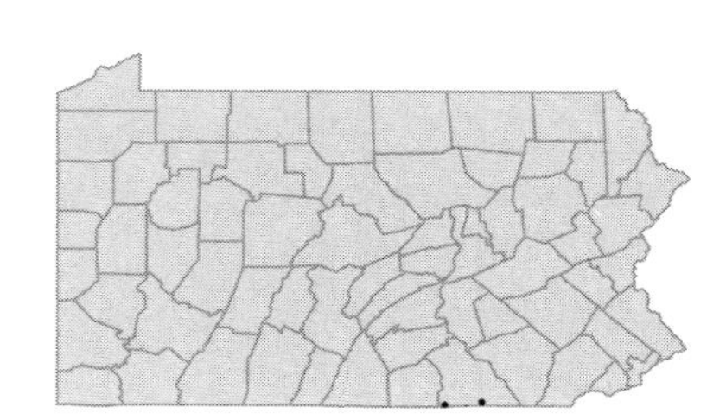

Rhododendron calendulaceum (Michx.) Torr.
Flame azalea
Deciduous shrub
Abandoned pasture land. Believed to be extirpated, last collected in the wild in 1898.
Azalea lutea L. P

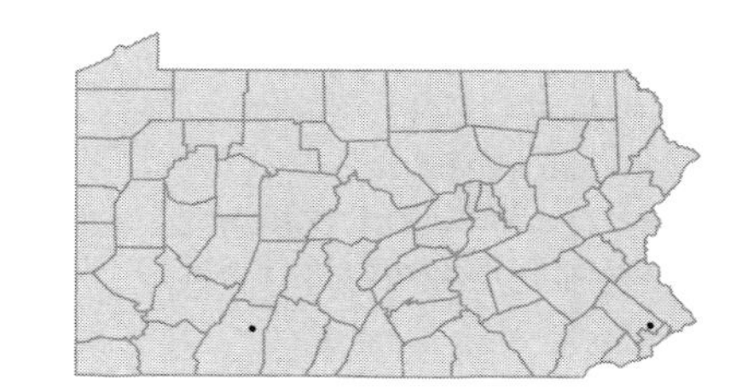

Rhododendron canadense (L.) Torr.
Rhodora
Deciduous shrub
Acidic bogs and barrens.

FACW

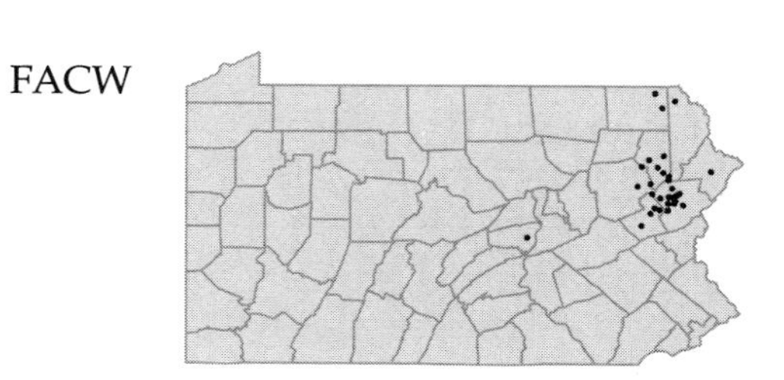

Rhododendron canescens (Michx.) Sweet
Hoary azalea; Mountain azalea
Deciduous shrub
Rich, dry woods and ravines.

FACW

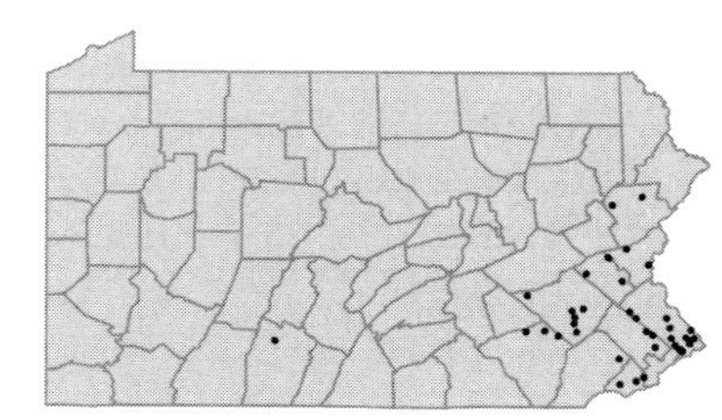

Rhododendron maximum L. FAC
Rosebay
Evergreen shrub
Moist woods, swamps and ravines.

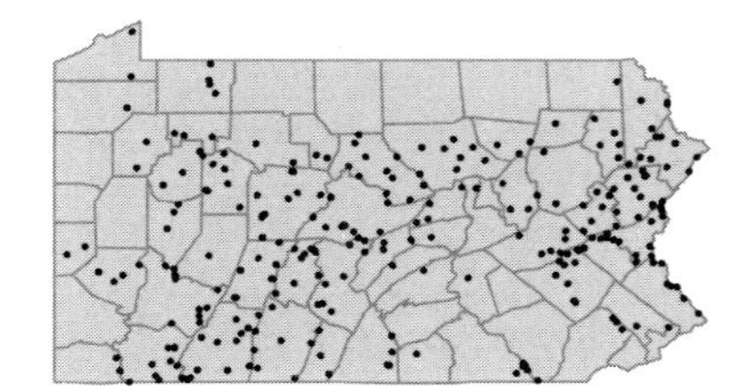

Rhododendron periclymenoides (Michx.) Shinners FAC
Pinxter-flower; Election-pink
Deciduous shrub
Dry to moist, acidic woods. A hybrid with *R. viscosum* has been collected at one location.
Azalea nudiflora L. P; *Rhododendron nudiflorum* (L.) Torr. FGB

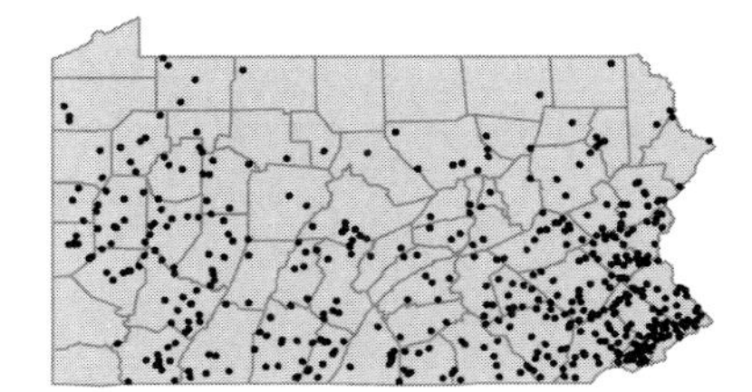

Rhododendron periclymenoides x prinophyllum
Azalea
Deciduous shrub
Open woods and oak barrens.

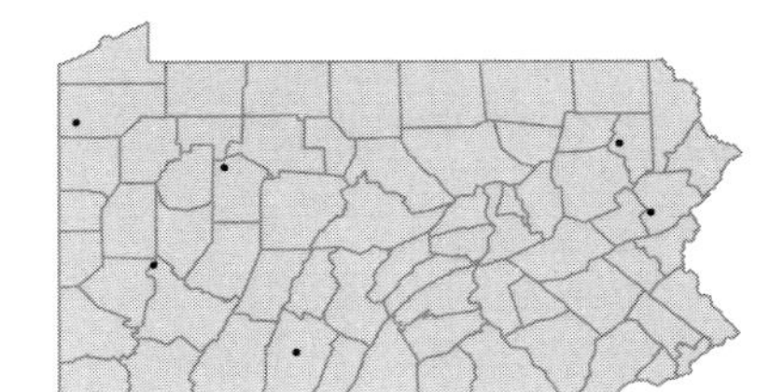

Rhododendron prinophyllum (Small) Millais
Mountain azalea
Deciduous shrub
Open woods, bogs and swamps.
Rhododendron roseum (Loisel.) Rehd. FGB

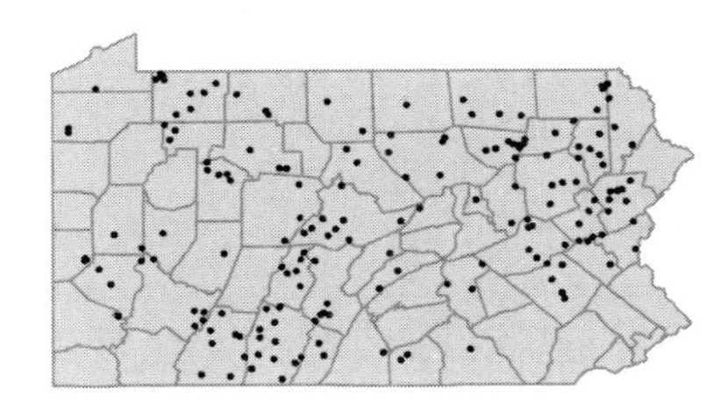

Rhododendron viscosum (L.) Torr. OBL
Swamp azalea; Swamp-honeysuckle
Deciduous shrub
Swamps, bogs and wet woods.
Azalea viscosa L. P

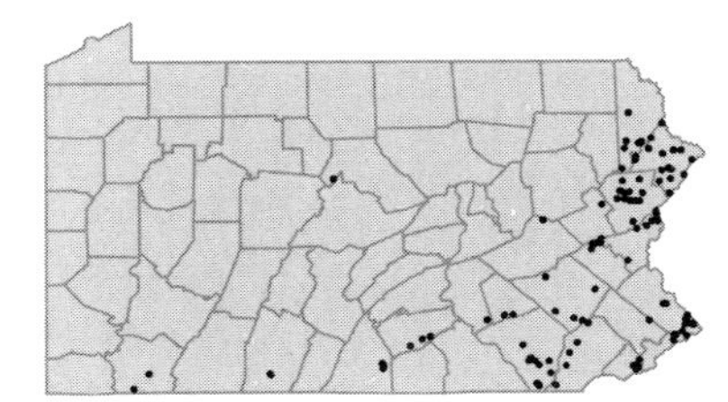

• ***Vaccinium angustifolium*** Ait. FACU-
Low sweet blueberry; Sweet-hurts
Deciduous shrub
Dry, open woods and barrens.
Vaccinium brittonii Porter GBW; *Vaccinium lamarckii* Camp GBW

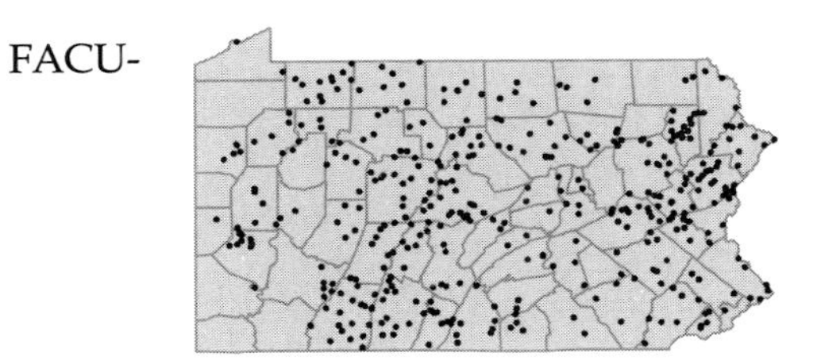

Vaccinium corymbosum L.
Highbush blueberry
Deciduous shrub
Moist woods, bogs and swamps.
Vaccinium atrococcum (A.Gray) Heller *pro parte* PFGBW; *Vaccinium caesariense* Mackenzie *pro parte* FGBW

FACW-

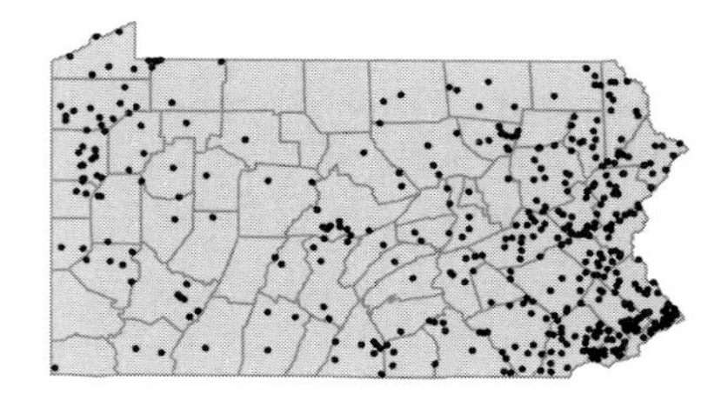

Vaccinium macrocarpon Ait.
Cranberry
Evergreen shrub
Sphagnum bogs.

OBL

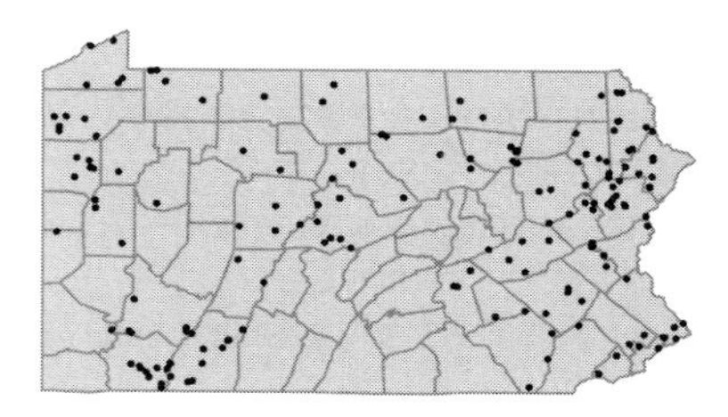

Vaccinium myrtilloides Michx.
Sour-top blueberry; Velvet-leaf blueberry
Deciduous shrub
Bogs and swamps.
Vaccinium canadense Richards P

FAC

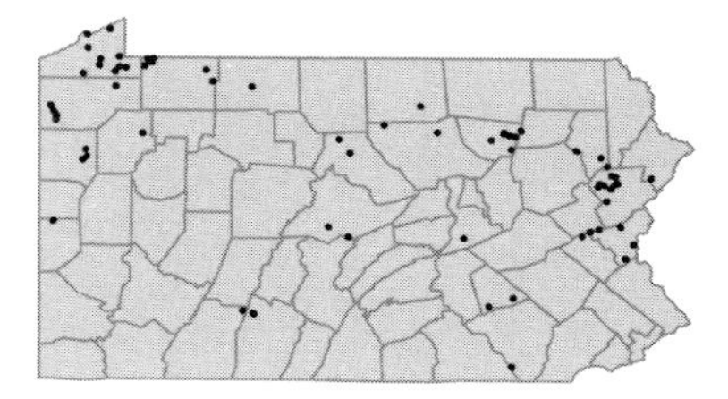

Vaccinium oxycoccos L.
Small cranberry
Evergreen shrub
Floating sphagnum bog mats.
Oxycoccus oxycoccus (L.) MacM. P

OBL

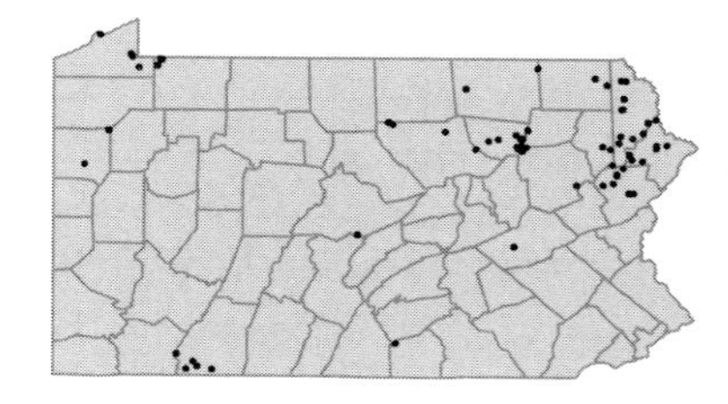

Vaccinium pallidum Ait.
Lowbush blueberry
Deciduous shrub
Dry, acidic woods.
Vaccinium vacillans Torr. FGBW

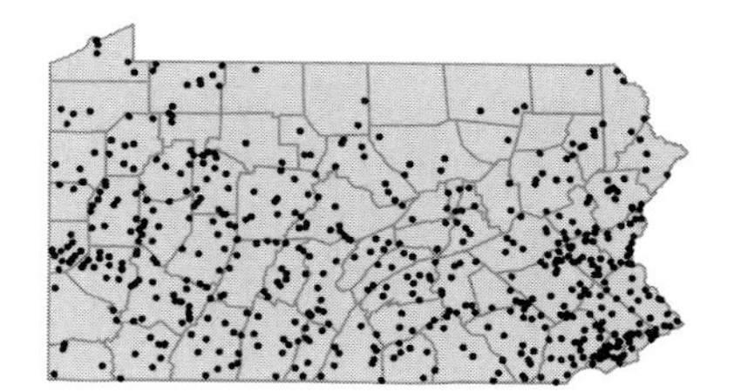

Vaccinium stamineum L.
Deerberry; Squaw huckleberry
Deciduous shrub
Dry, open, acidic woods and slopes.
Vaccinium caesium Greene FW

FACU-

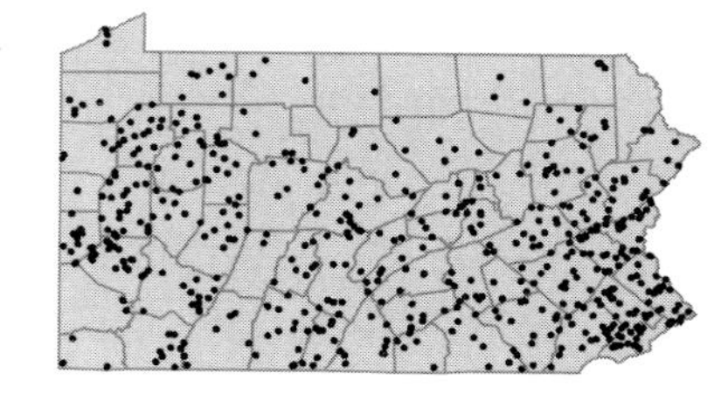

PYROLACEAE

• ***Chimaphila maculata*** (L.) Pursh
Pipsissewa; Spotted wintergreen
Herbaceous perennial
Woods.

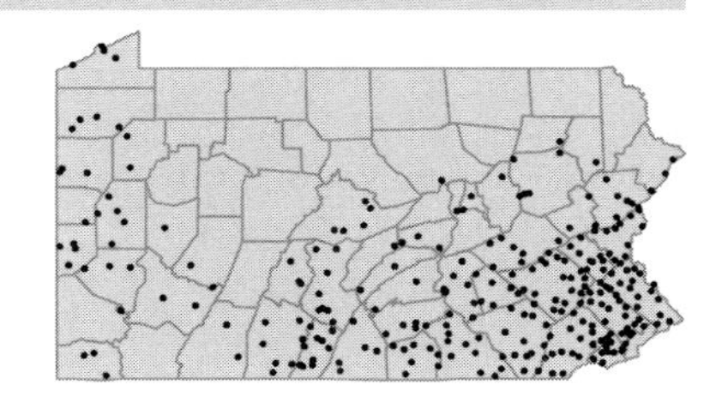

Chimaphila umbellata (L.) Bart. ssp. ***cisatlantica*** (S.F.Blake) Hulten
Pipsissewa; Prince's-pine
Herbaceous perennial
Dry, upland woods and barrens.

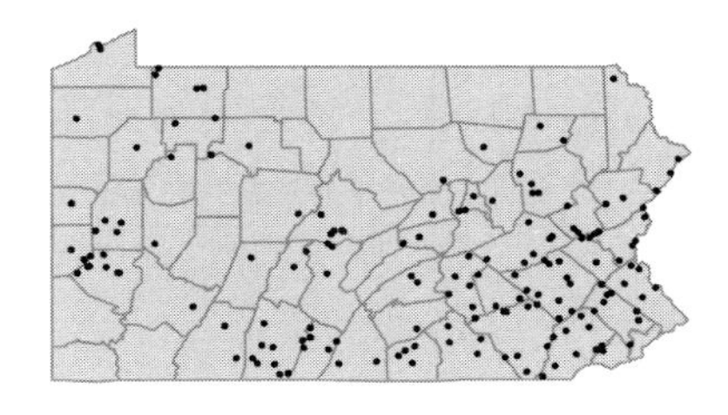

• ***Orthilia secunda*** (L.) House
One-sided shinleaf; Wintergreen
Herbaceous perennial
Dry to moist woods.
Pyrola secunda L. FGBWC

FAC

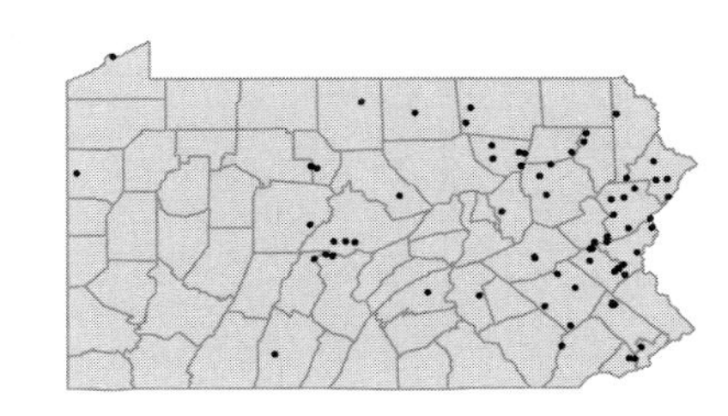

• ***Pyrola americana*** Sweet
Wild lily-of-the-valley
Herbaceous perennial
Woods.
Pyrola rotundifolia L. var. *americana* (Sweet) Fern. FGBWC

FAC

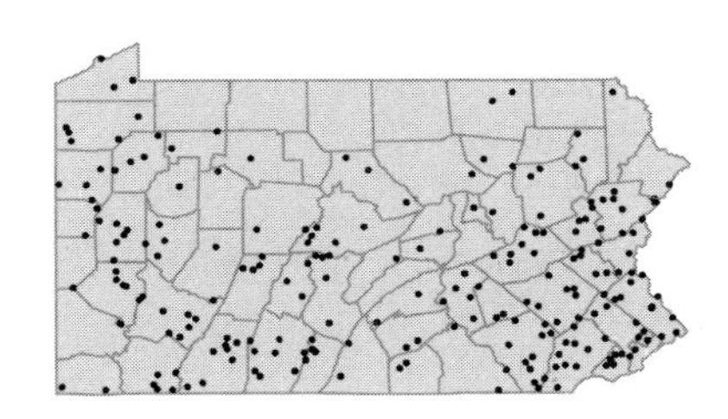

Pyrola chlorantha Swartz
Wintergreen
Herbaceous perennial
Dry woods.
Pyrola virens Schweig. FW

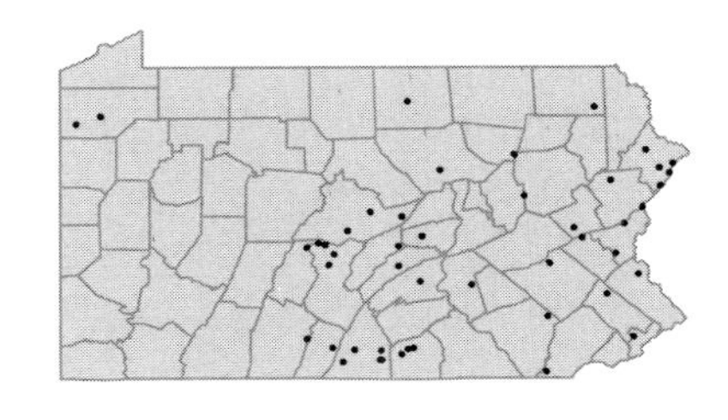

Pyrola elliptica Nutt.
Shinleaf; Wild lily-of-the-valley
Herbaceous perennial
Dry to moist woods, on rich soils.

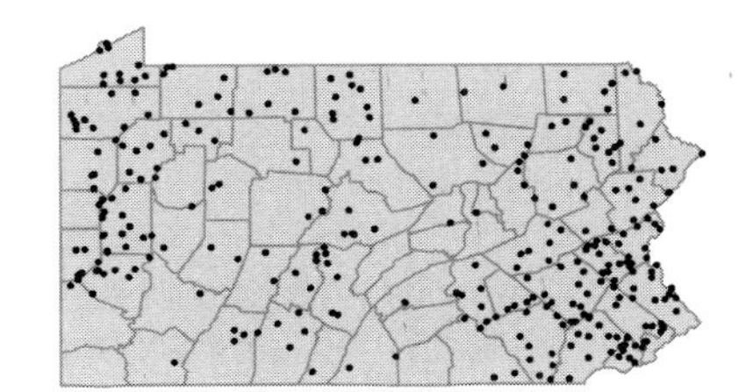

MONOTROPACEAE

• ***Monotropa hypopithys*** L.
Pine-sap
Herbaceous perennial, saprophytic
Dry to moist woods, in humus.
Hypopithys hypopithys (L.) Small P

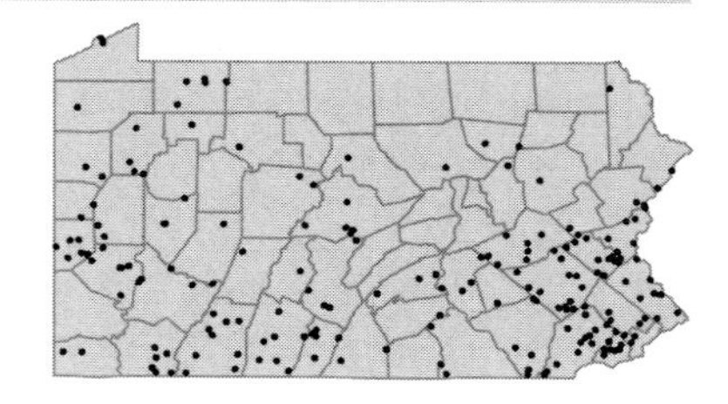

Monotropa uniflora L.
Indian-pipe
Herbaceous perennial, saprophytic
Dry to moist woods, in humus.

FACU-

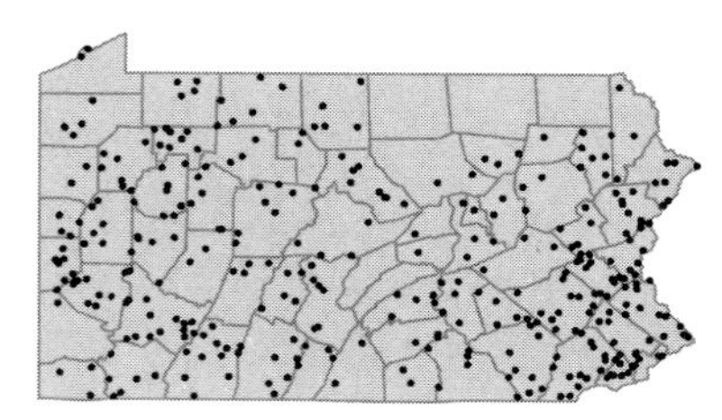

EBENACEAE

• ***Diospyros virginiana*** L.
Persimmon
Deciduous tree
Thin woods, edges, floodplains and old fields.

FAC-

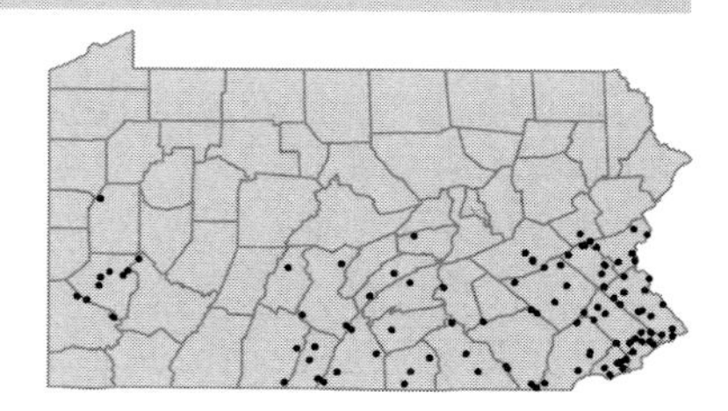

STYRACACEAE

• ***Halesia carolina*** L.
Silverbell-tree
Deciduous tree
Cultivated and rarely spreading to disturbed woods.
Halesia tetraptera Ellis C

I

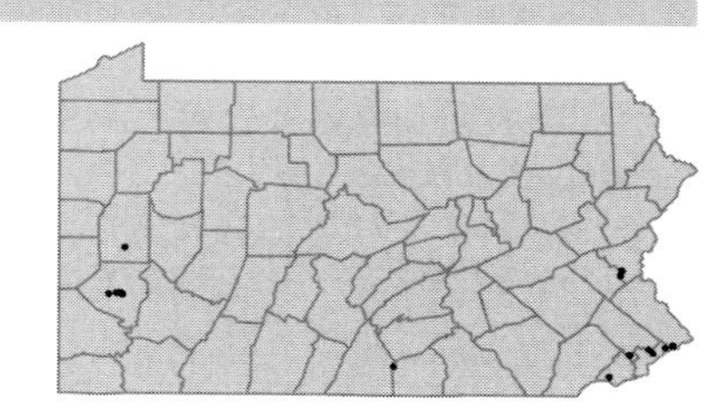

PRIMULACEAE

• ***Anagallis arvensis*** L.
Scarlet pimpernel
Herbaceous annual
Lawns, gardens and waste ground.

I

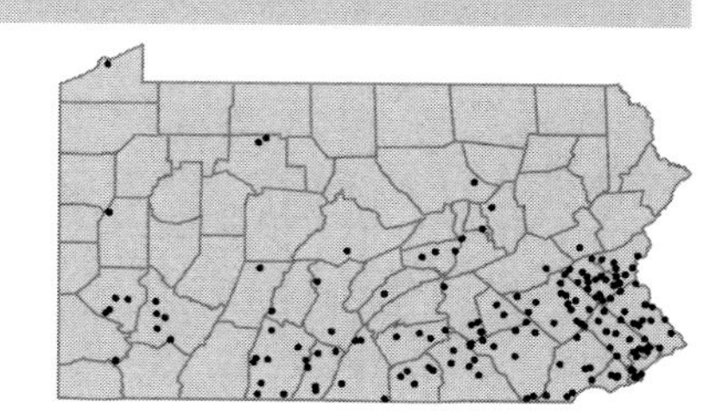

• ***Dodecatheon meadia*** L.
Shooting-star
Herbaceous perennial
Open woods, wooded slopes, bluffs and meadows, on calcareous soils.
Although some PA material has been referred to *D. amethystinum* (Fassett) Fassett, most populations are too variable to separate reliably.
Dodecatheon amethystinum (Fassett) Fassett *pro parte* BFKWC

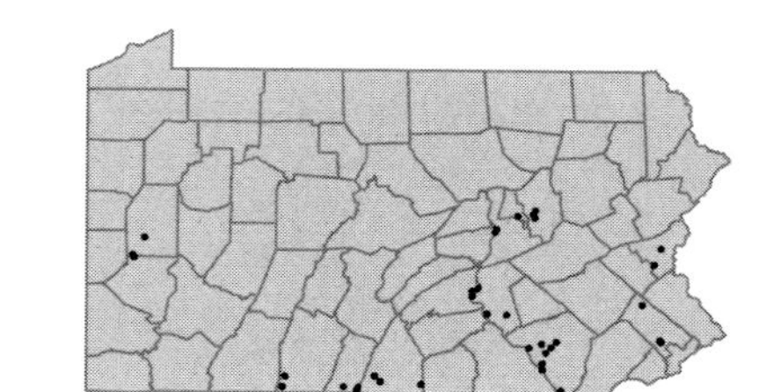

• ***Hottonia inflata*** Ell.
American featherfoil
Herbaceous perennial, floating-leaf aquatic
Ponds and ditches. Believed to be extirpated, last collected in 1931.

OBL

• ***Lysimachia ciliata*** L.
Fringed loosestrife
Herbaceous perennial
Low, moist ground of fields, stream banks and swamp edges.
Steironema ciliata (L.) Raf. PB

FACW

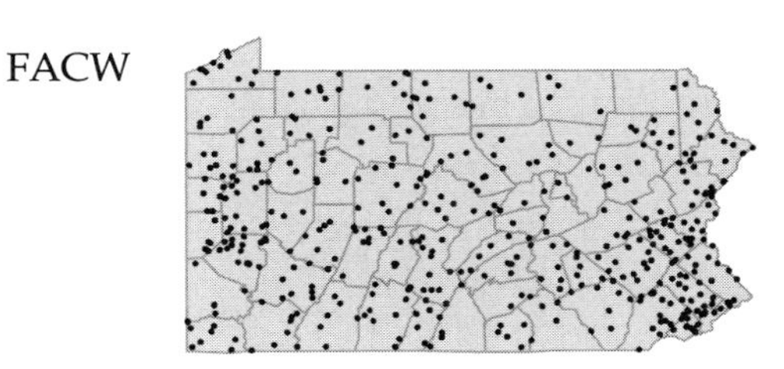

Lysimachia clethroides Duby
Loosestrife
Herbaceous perennial
Cultivated and occasionally spreading from gardens.

I

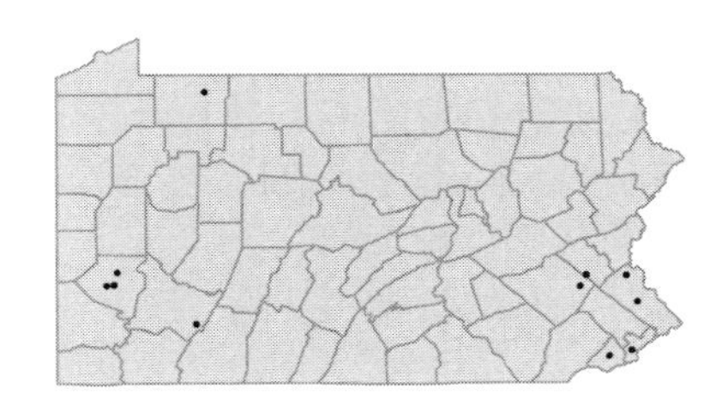

Lysimachia hybrida Michx.
Lance-leaved loosestrife
Herbaceous perennial
Swamps, wet meadows, fens and pond margins.
Steironema hybridum (Michx.) Raf. B

OBL

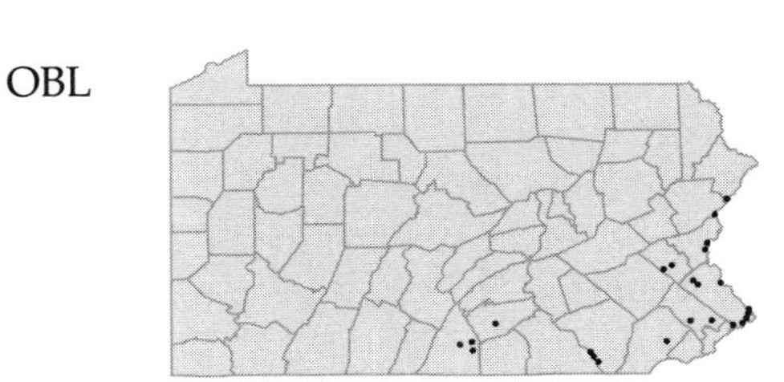

Lysimachia lanceolata Walt.
Loosestrife
Herbaceous perennial
Moist fields, stream banks and sandy or rocky shores.
Steironema lanceolatum (Walt.) A.Gray PB

OBL

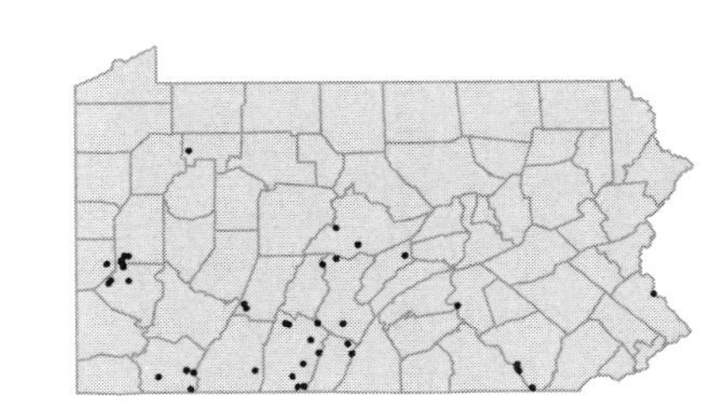

Lysimachia nummularia L.
Creeping-charlie; Moneywort
Herbaceous perennial
Lawns, meadows, wet woods and floodplains.

OBL
I

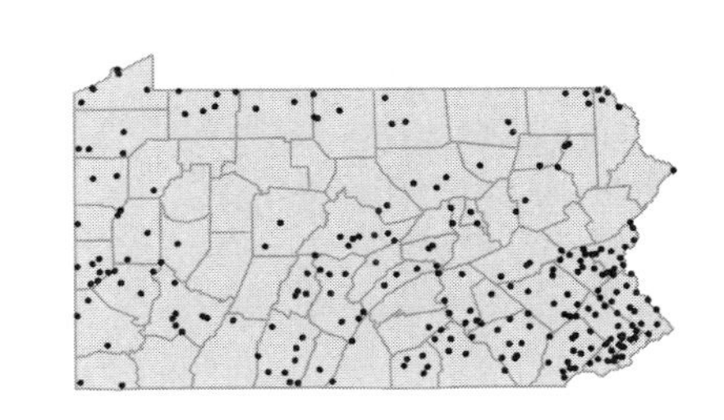

Lysimachia punctata L.
Spotted loosestrife
Herbaceous perennial
Cultivated and occasionally naturalized in wet meadows or roadsides.

OBL
I

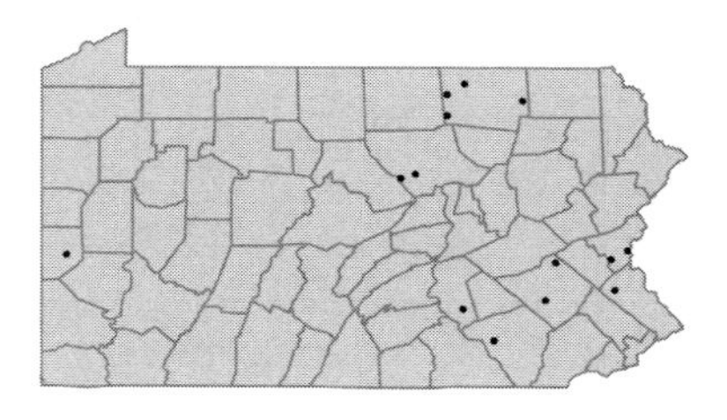

Lysimachia quadriflora Sims
Four-flowered loosestrife
Herbaceous perennial
Calcareous meadows and fens. In addition to the mapped collections, there are two ca. 1890 specimens believed to be from Berks Co.

FACW+

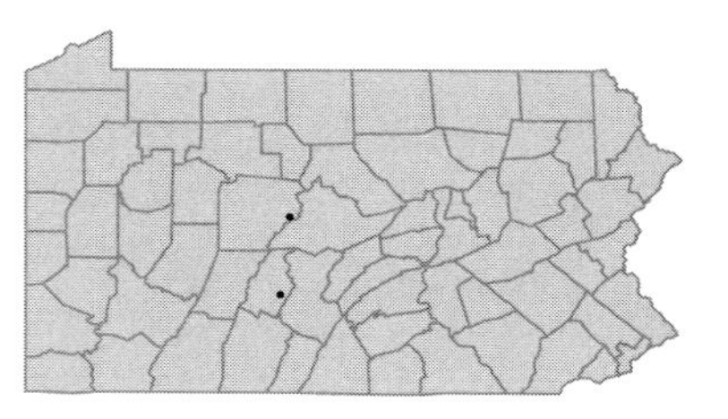

Lysimachia quadrifolia L.
Whorled loosestrife
Herbaceous perennial
Dry, open woods.

Lysimachia quadrifolia x terrestris
Loosestrife
Herbaceous perennial
Open thickets and moist bottomland.
Lysimachia x producta (A.Gray) Fern.

Lysimachia terrestris (L.) BSP
Swamp-candles
Herbaceous perennial
Swamps and bogs.

OBL

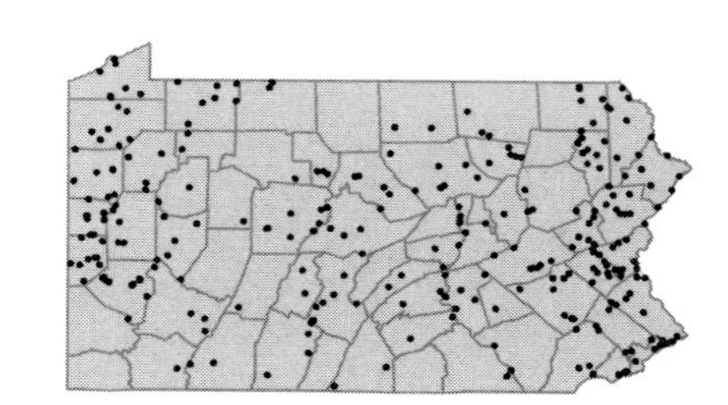

Lysimachia thyrsiflora L.
Tufted loosestrife
Herbaceous perennial
Bogs and swamps.
Naumburgia thyrsiflora (L.) Duby PB

OBL

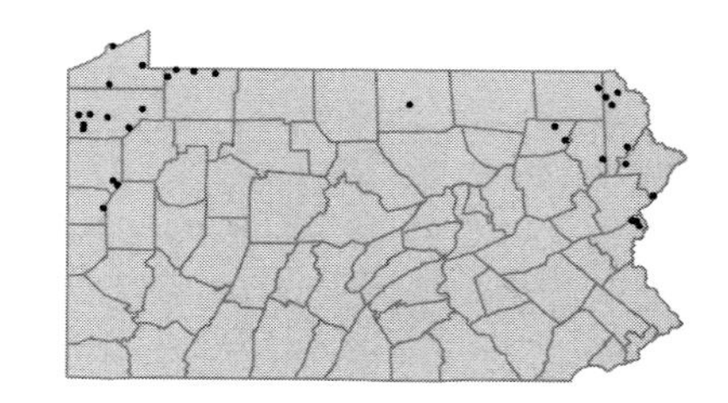

Lysimachia vulgaris L.
Garden loosestrife
Herbaceous perennial
Cultivated and occasionally naturalized in swales or moist shores.

I

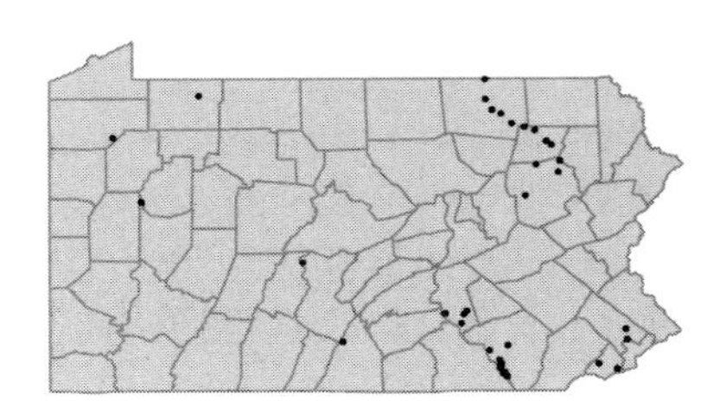

• ***Samolus parviflorus*** Raf.
Brookweed; Water pimpernel
Herbaceous perennial
Ditches and moist shores, also on ballast.
Samolus floribundus HBK PGBC; *Samolus valerandi* L. ssp. *parviflorus* (Raf.) Hulten K

OBL

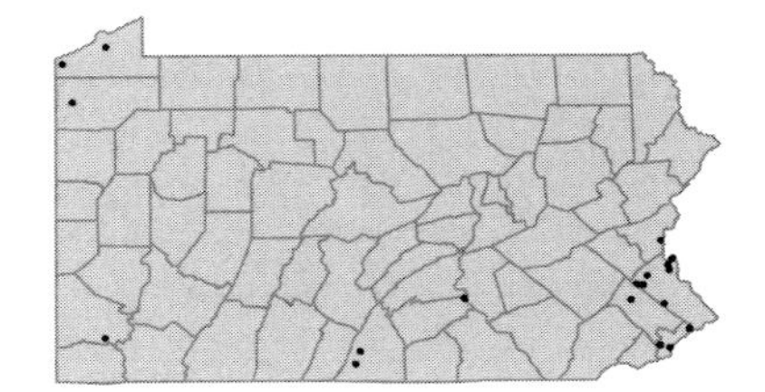

• ***Trientalis borealis*** Raf.
Star-flower
Herbaceous perennial
Moist woods and bogs.
Trientalis americana Pursh P

FAC

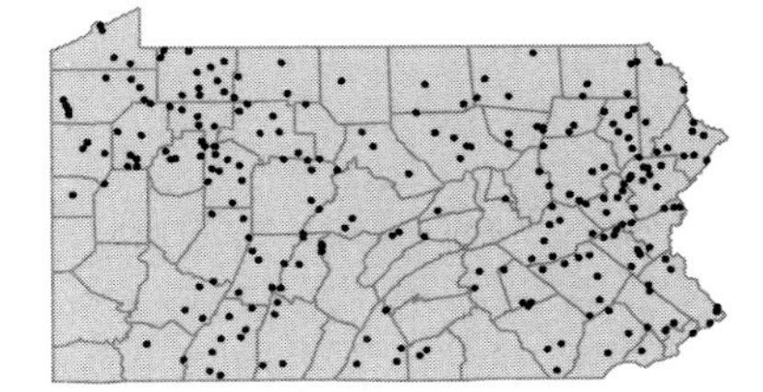

HYDRANGEACEAE

• ***Deutzia scabra*** Thunb.
Deutzia
Deciduous shrub
Cultivated and occasionally persisting on roadside banks or waste ground.

I

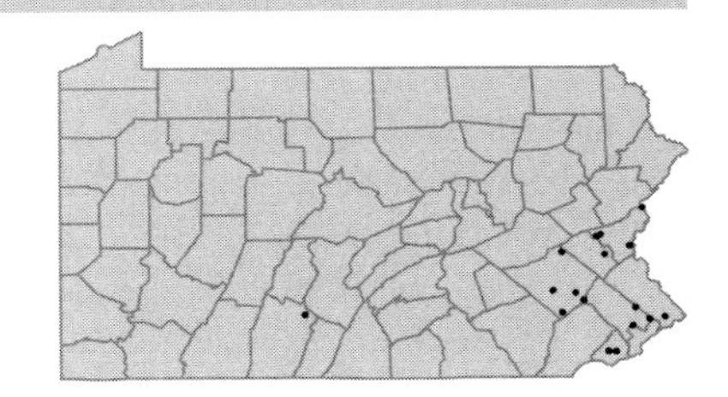

• ***Hydrangea arborescens*** L.
Seven-bark; Wild hydrangea
Deciduous shrub
Rich woods, slopes and stream banks.

FACU

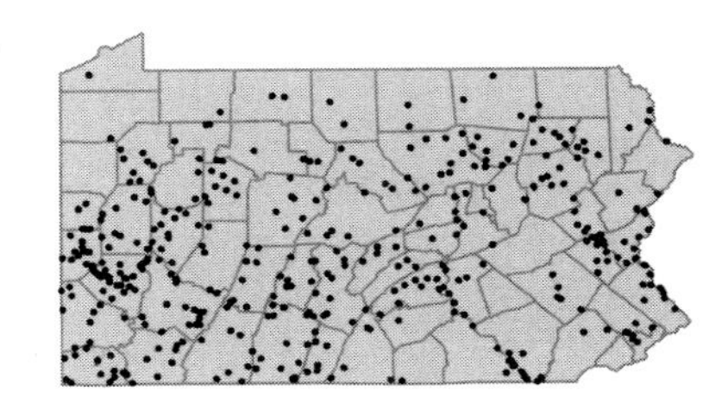

Hydrangea paniculata Sieb.
Peegee hydrangea
Deciduous shrub
Cultivated and occasionally persisting at abandoned home sites or waste ground.

FAC
I

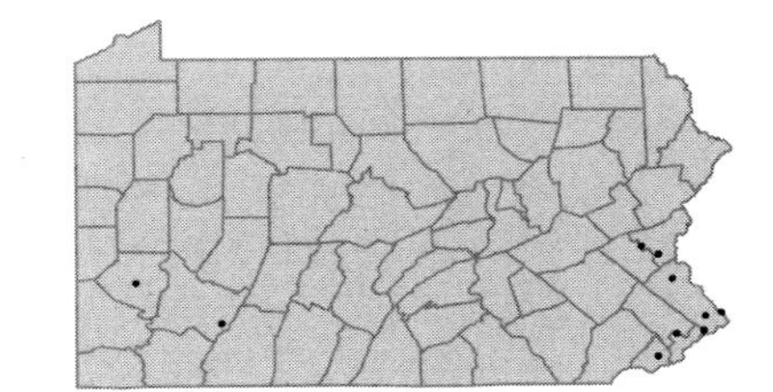

• ***Philadelphus coronarius*** L. [I]
Mock-orange
Deciduous shrub
Cultivated and occasionally spreading to banks, roadsides or alluvial woods.

Philadelphus inodorus L. var. ***grandiflorus*** Willd. [I]
Mock-orange
Deciduous shrub
Cultivated and sometimes persisting near rubbish heaps.
Philadelphus grandiflorus Willd. W

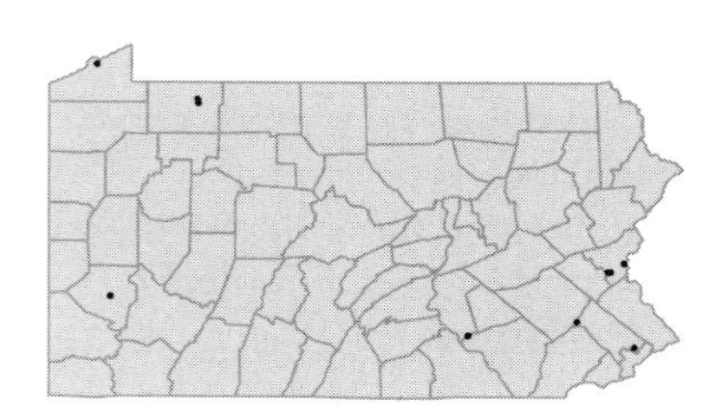

Philadelphus pubescens Loisel. [I]
Mock-orange
Deciduous shrub
Occasionally spreading from cultivation.

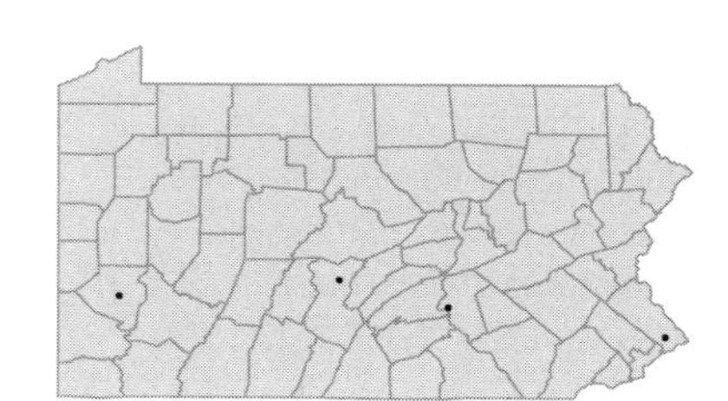

GROSSULARIACEAE

• ***Itea virginica*** L. OBL
Tassel-white; Virginia-willow
Deciduous shrub
River bank. Believed to be extirpated, represented by a single collection from Cumberland Co. in 1841.

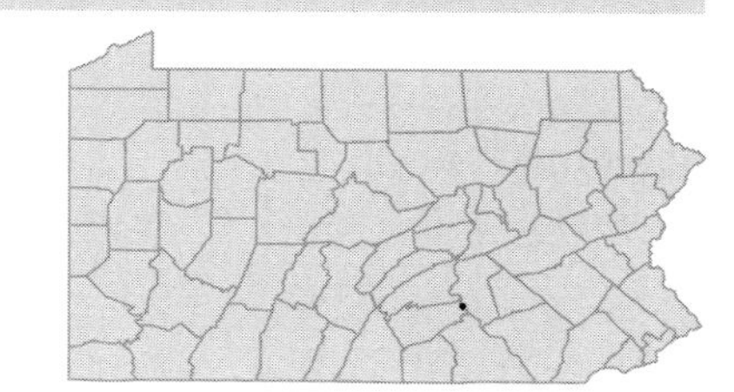

• ***Ribes americanum*** P.Mill. FACW
Wild black currant
Deciduous shrub
Moist woods, swamps and thickets.
Ribes floridum L'Her. P

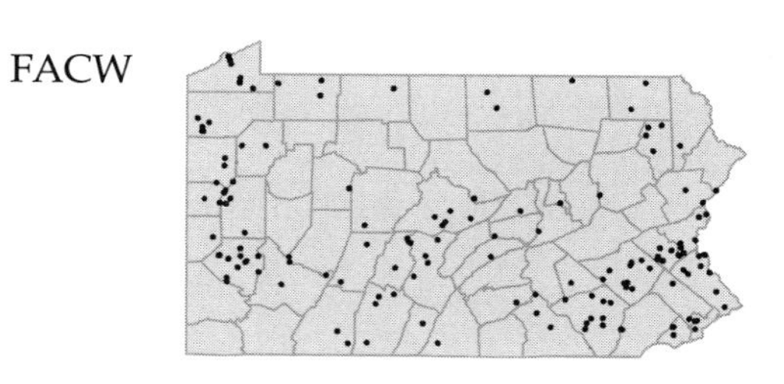

Ribes cynosbati L.
Prickly gooseberry; Dogberry
Deciduous shrub
Thin, moist, often rocky, woods.

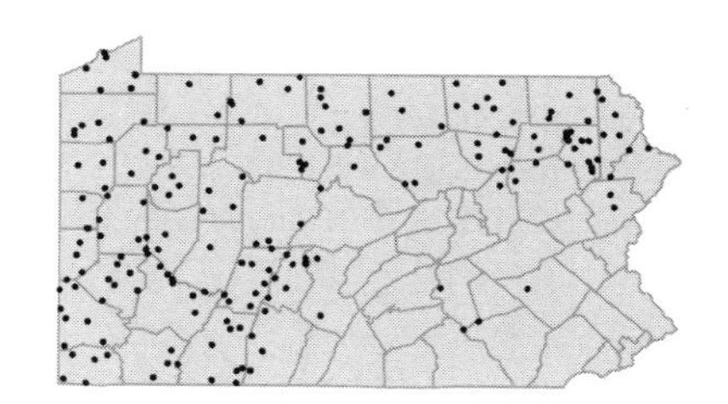

Ribes glandulosum Grauer
Skunk currant
Deciduous shrub
Swamps, bogs, wet woods and moist, rocky slopes.
Ribes prostratum L'Her. P

FACW

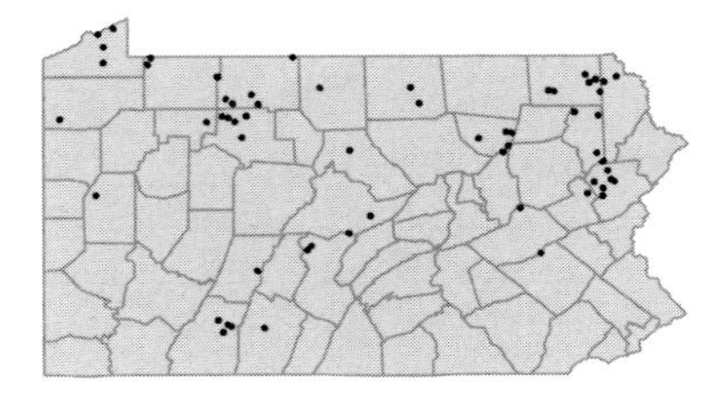

Ribes hirtellum Michx.
Northern wild gooseberry
Deciduous shrub
Calcareous marshes, swamps, rocky woods and cliffs.

FAC

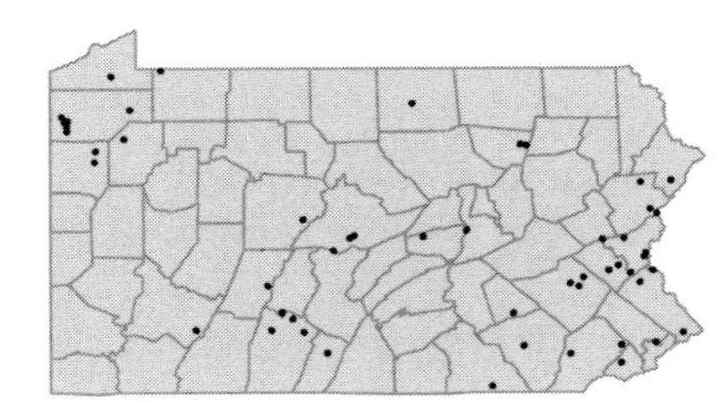

Ribes lacustre (Pers.) Poir.
Bristly black currant; Swamp currant
Deciduous shrub
Swamps and cold, wet woods.

FACW

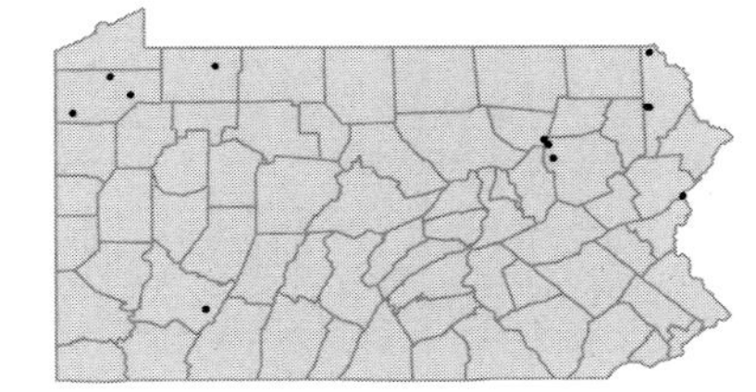

Ribes missouriense Nutt. ex Torr. & A.Gray
Missouri gooseberry
Deciduous shrub
Rich woods.

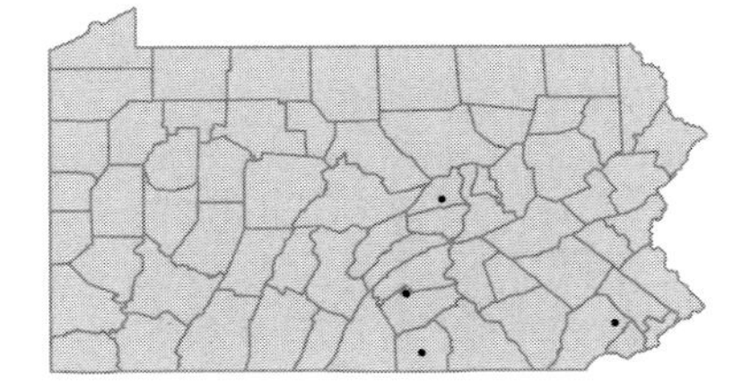

Ribes odoratum H.Wendl.
Buffalo currant
Deciduous shrub
Cultivated and occasionally escaped to meadows and waste ground.

I

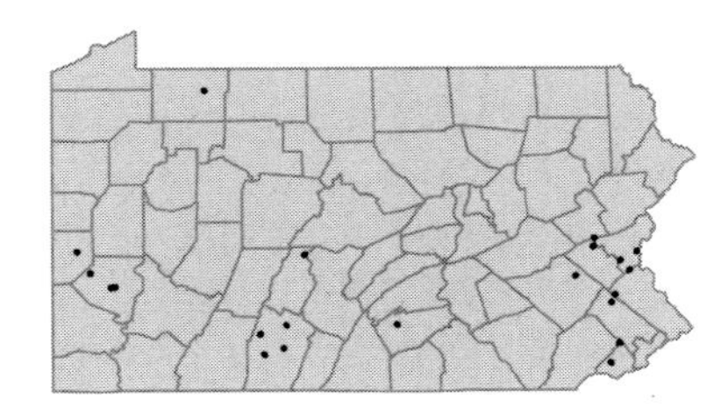

Ribes rotundifolium Michx.
Wild gooseberry
Deciduous shrub
Rocky, upland woods and slopes.

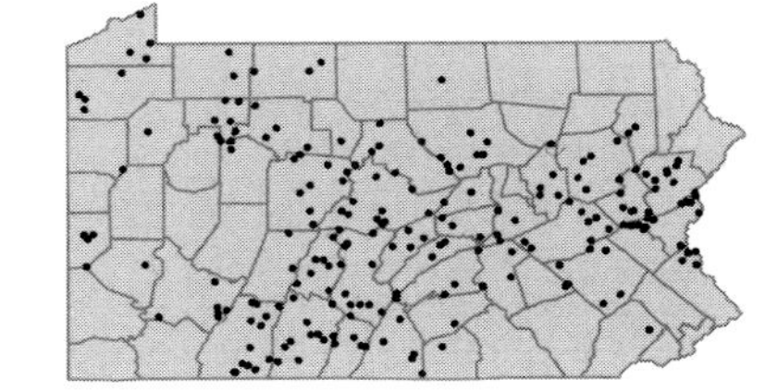

Ribes rubrum L.
Red currant
Deciduous shrub
Cultivated and often naturalized in moist woods, thickets and banks.
Ribes sativum Syme FGBC

I

Ribes triste Pallas
Wild red currant
Deciduous shrub
Wet, rocky woods, swamps and cliffs.

OBL

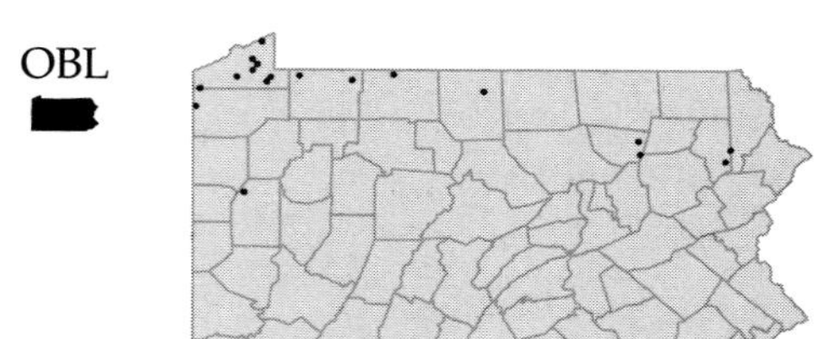

Ribes uva-crispa L. var. ***sativum*** DC.
European garden gooseberry
Deciduous shrub
Cultivated and occasionally escaped.
Ribes grosslularia L. FGBW

I

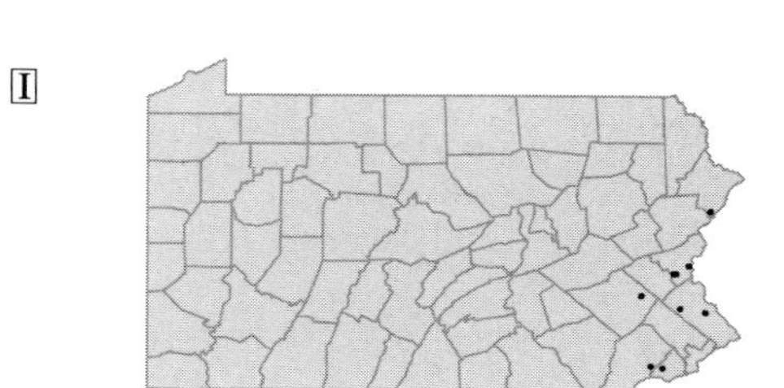

CRASSULACEAE

• ***Crassula aquatica*** (L.) Schoenl.
Water-pigmyweed
Herbaceous annual, rooted submergent aquatic
Fresh water, intertidal marshes. Believed to be extirpated, last collected in 1918.
Tillaea aquatica L. PFGBW

OBL

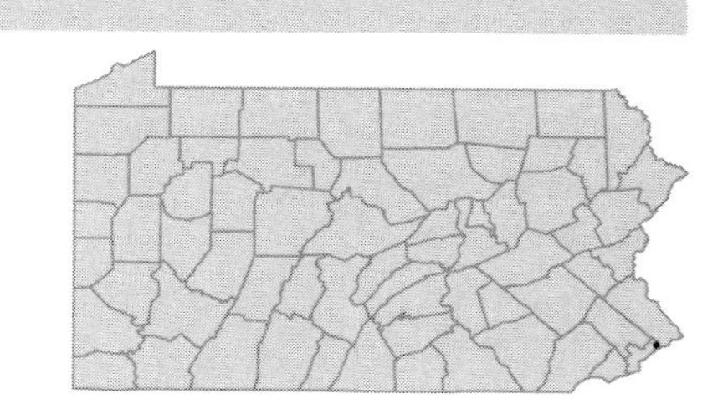

• ***Sedum acre*** L.
Love-entangle; Mossy stonecrop
Herbaceous perennial
Cultivated and frequently naturalized on limestone cliffs, roadsides and waste ground.

I

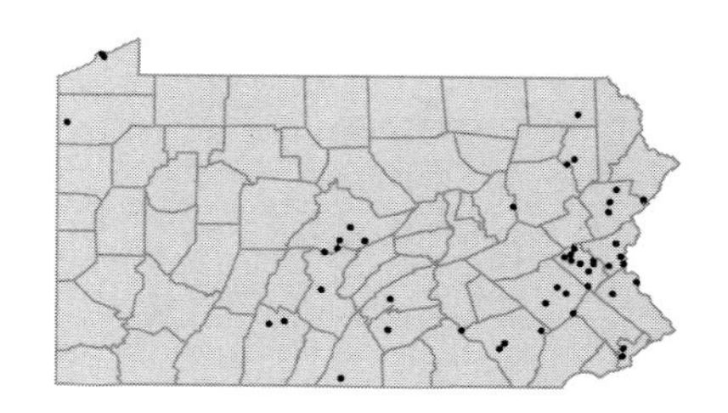

Sedum alboroseum Baker
Garden orpine
Herbaceous perennial
Cultivated and occasionally spreading to waste ground.
Sedum x erythrostictum Miq. C

I

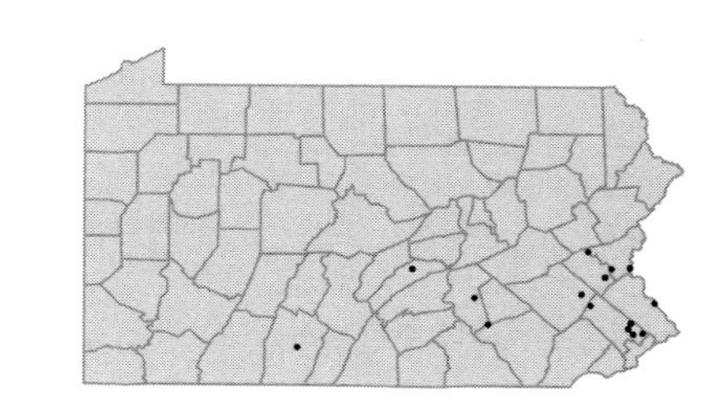

Sedum album L.
White orpine; White stonecrop
Herbaceous perennial
Occasionally spreading from cultivation.

I

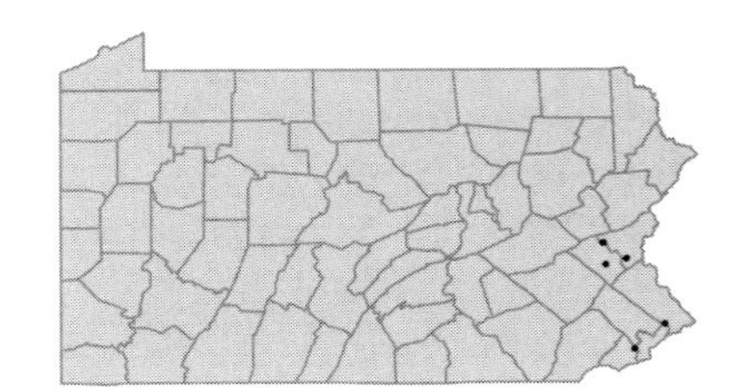

Sedum kamtschaticum Fisch. & Mey. ssp. ***ellacombianum*** (Praeger) R.T.Clausen
Orange stonecrop
Herbaceous perennial
Cultivated and occasionally escaped.
Sedum aizoon L. F; *Sedum ellacombianum* Praeger W

I

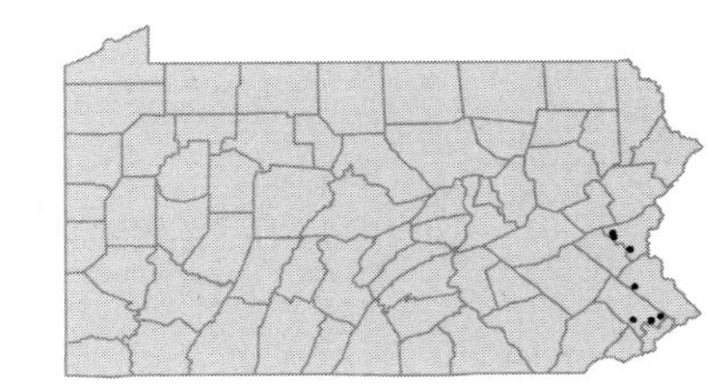

Sedum rosea (L.) Scop.
Roseroot stonecrop
Herbaceous perennial
Moist cliffs and ledges.

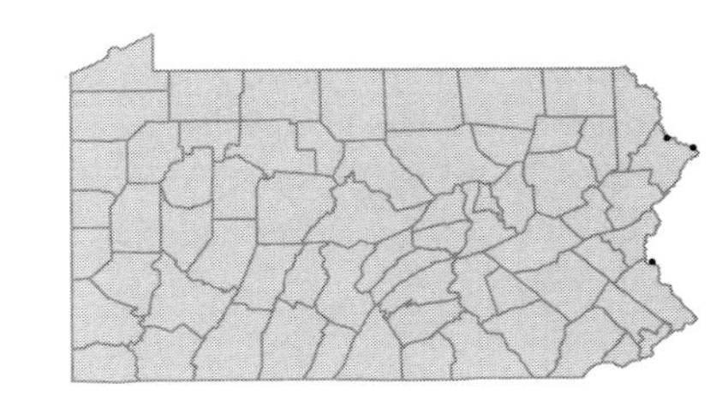

Sedum sarmentosum Bunge
Orpine
Herbaceous perennial
Cultivated and frequently escaped to dry, rocky soils and roadsides.

I

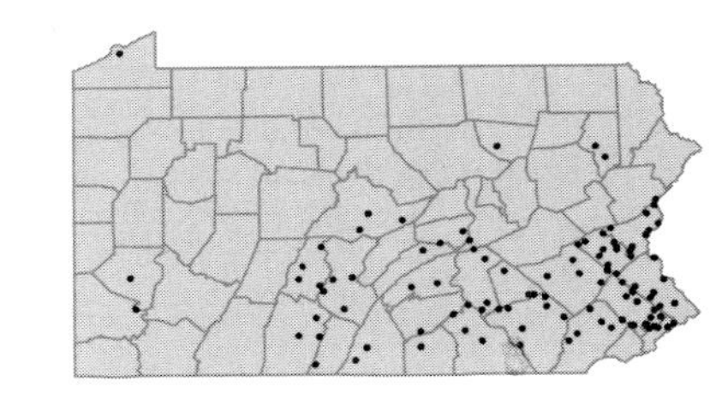

Sedum spectabile Boreau
Orpine
Herbaceous perennial
Cultivated and occasionally persisting at abandoned garden sites or waste ground.

I

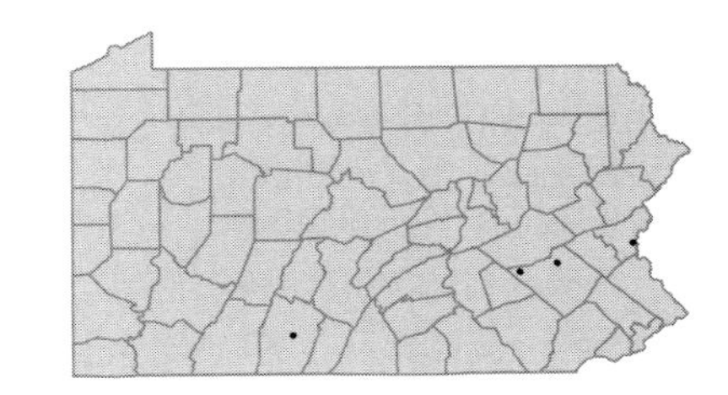

Sedum spurium Bieb.
Orpine
Herbaceous perennial
Cultivated and occasionally established on roadsides or near dump sites.

I

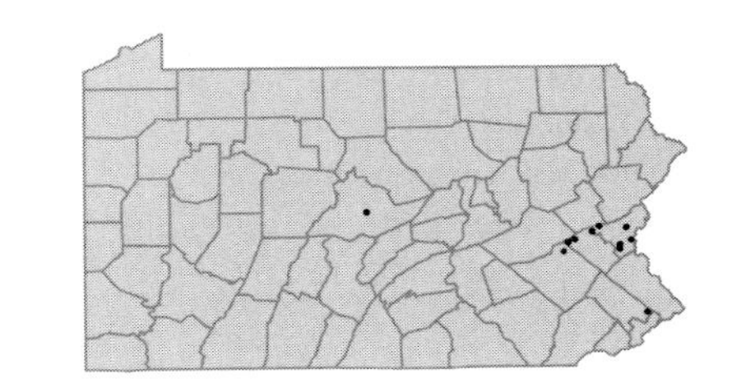

Sedum telephioides Michx.
Allegheny stonecrop
Herbaceous perennial
Rocky woods, dry cliffs, ledges and shale barrens.

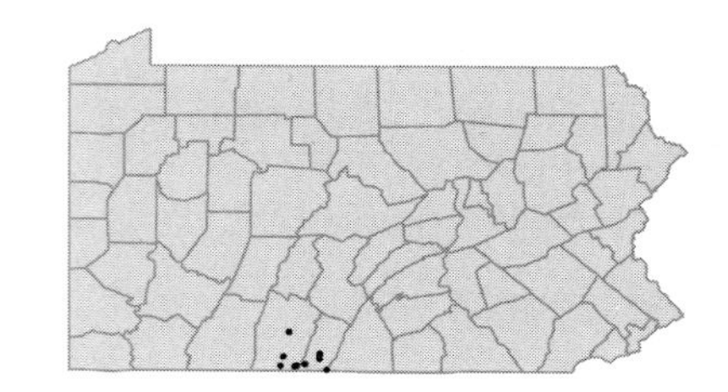

Sedum telephium L. [I]
Garden orpine; Live-forever
Herbaceous perennial
Cultivated and frequently escaped to woods, old fields and moist ditches.

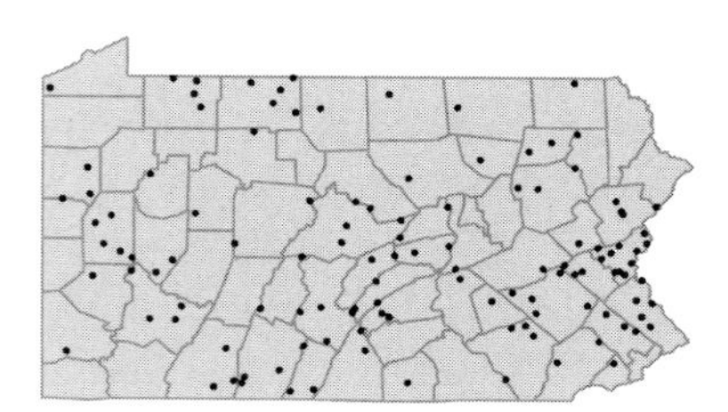

Sedum ternatum Michx.
Wild stonecrop
Herbaceous perennial
Rocky banks, cliffs and woods.

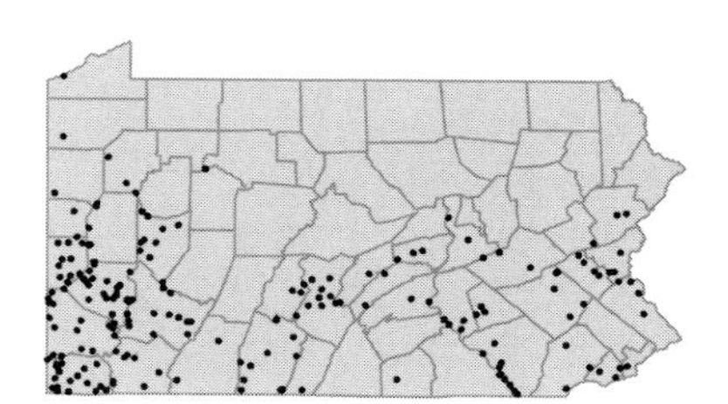

SAXIFRAGACEAE

• ***Chrysosplenium americanum*** Schwein. ex Hook. OBL
Golden saxifrage; Water-carpet
Herbaceous perennial
Wet woods, springs, seeps and cold swamps.

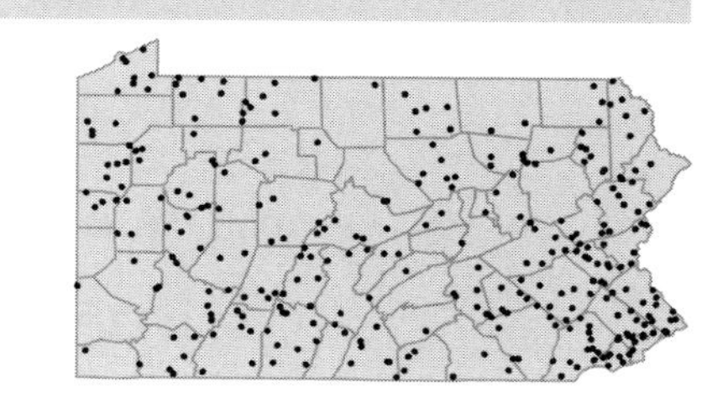

• ***Heuchera americana*** L.
Alum-root; Rock-geranium
Herbaceous perennial
Rich woods, rocky slopes and shaly cliffs. A hybrid with *H. pubescens* has been collected occasionally.

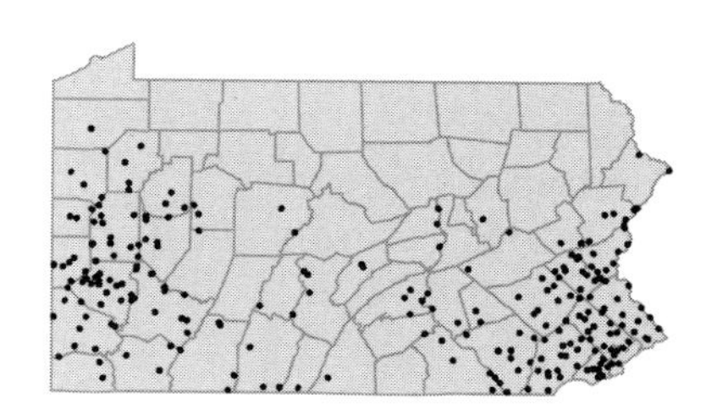

Heuchera pubescens Pursh
Alum-root
Herbaceous perennial
Rocky woods, banks and shale barrens.

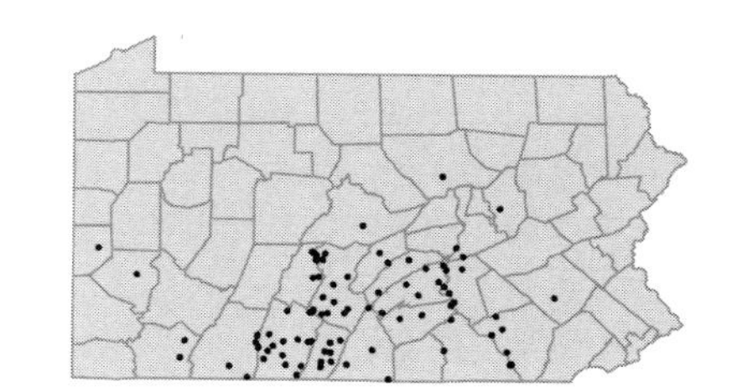

Heuchera sanguinea Engelm.
Coralbells
Herbaceous perennial
Cultivated and occasionally persisting at dump sites. Represented by a single collection from Dauphin Co. in 1952.

[I]

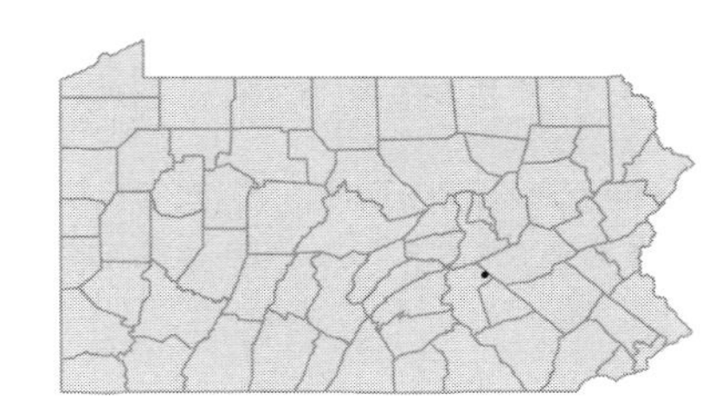

• ***Mitella diphylla*** L.
Bishop's-cap; Miterwort
Herbaceous perennial
Rich, moist woods.

FACU

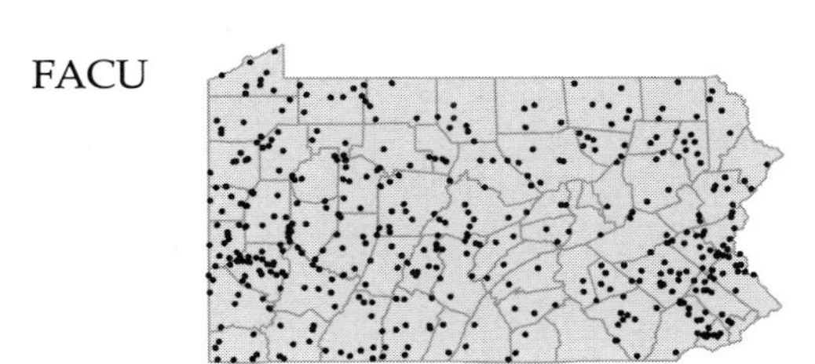

Mitella nuda L.
Bishop's-cap; Miterwort
Herbaceous perennial
Moist woods.

FACW-

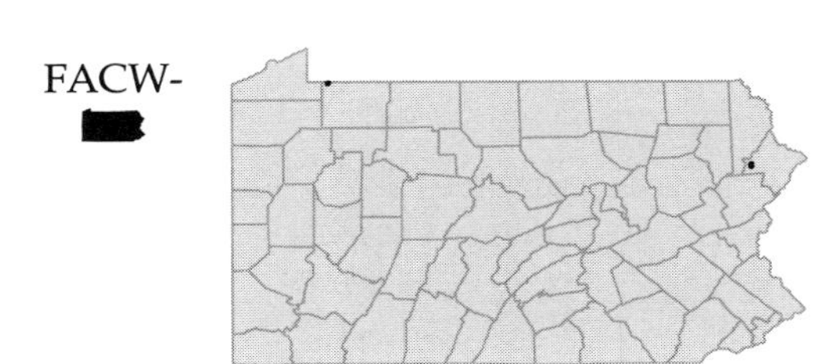

• ***Parnassia glauca*** Raf.
Grass-of-parnassus
Herbaceous perennial
Boggy meadows or seeps, on calcareous soils.

OBL

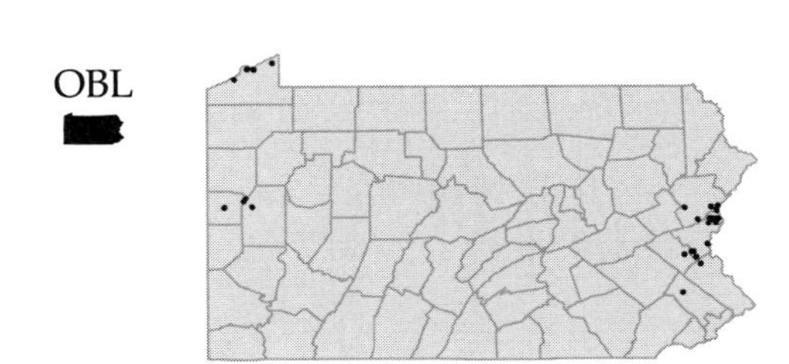

• ***Penthorum sedoides*** L.
Ditch stonecrop
Herbaceous perennial
Low, wet ground and ditches.

OBL

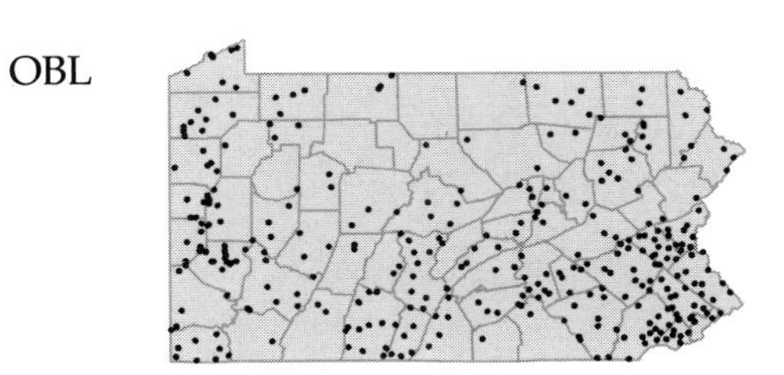

• ***Saxifraga micranthidifolia*** (Haw.) Steud.
Lettuce saxifrage; Mountain-lettuce
Herbaceous perennial
Among mossy rocks of shaded streambeds and seepage areas.

OBL

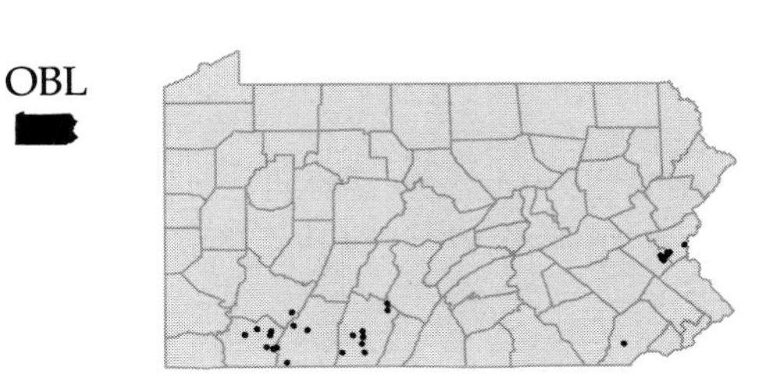

Saxifraga pensylvanica L.
Swamp saxifrage
Herbaceous perennial
Wet woods, seepage areas, bogs and swamps.

OBL

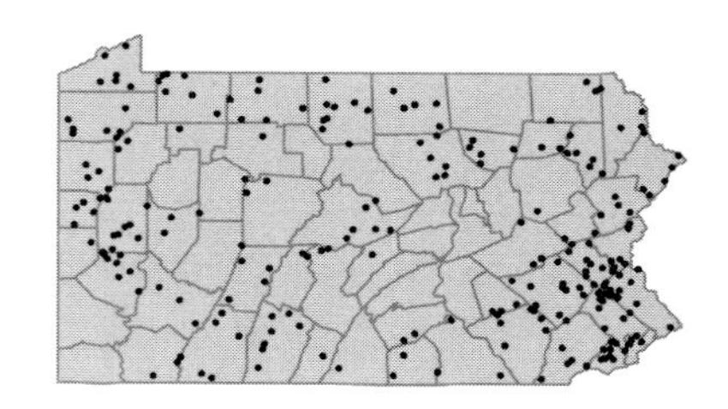

Saxifraga virginiensis Michx.
Early saxifrage
Herbaceous perennial
Moist or dry rock crevices and gravelly slopes.

FAC-

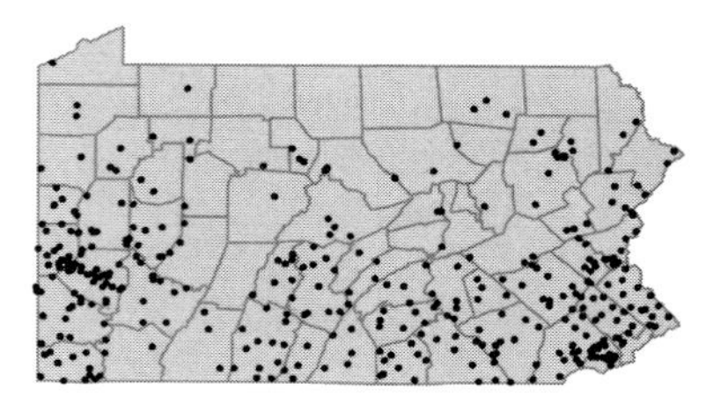

• ***Tiarella cordifolia*** L.
Foamflower
Herbaceous perennial
Moist, rocky, wooded slopes.

FAC-

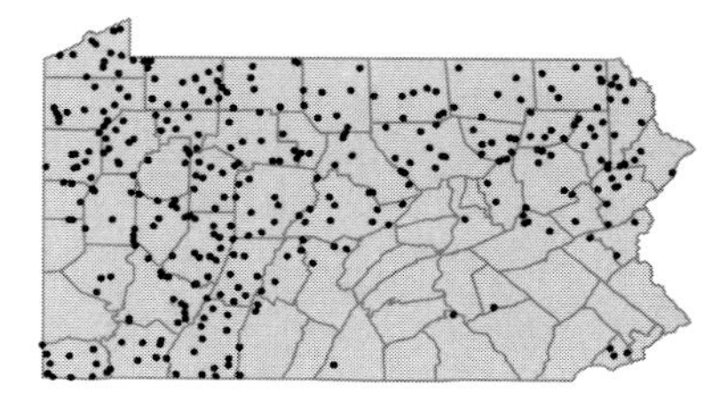

ROSACEAE

• ***Agrimonia gryposepala*** Wallr.
Agrimony; Harvest-lice
Herbaceous perennial
Woods, fields and floodplains.

FACU

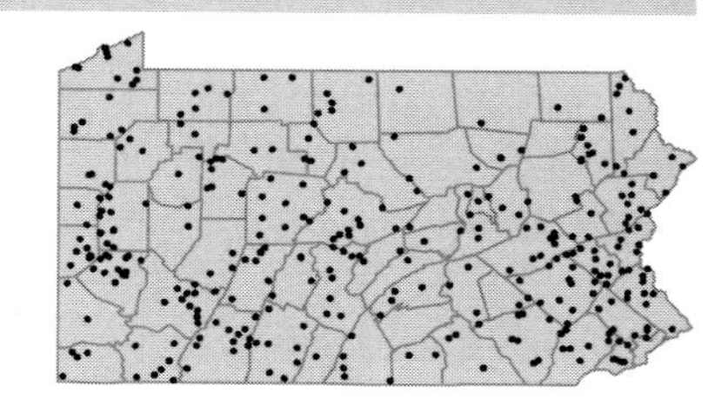

Agrimonia microcarpa Wallr.
Small-fruited agrimony
Herbaceous perennial
Woods.
Agrimonia pumila Muhl. P

Agrimonia parviflora Ait.
Southern agrimony
Herbaceous perennial
Bogs, moist woods and thickets.

FAC

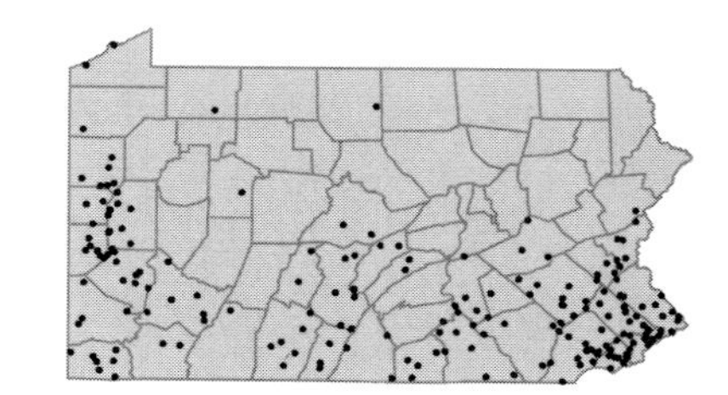

Agrimonia pubescens Wallr.
Downy agrimony
Herbaceous perennial
Rich, rocky woods, woods edges and slopes.
Agrimonia mollis (Torr. & A.Gray) Britt. P

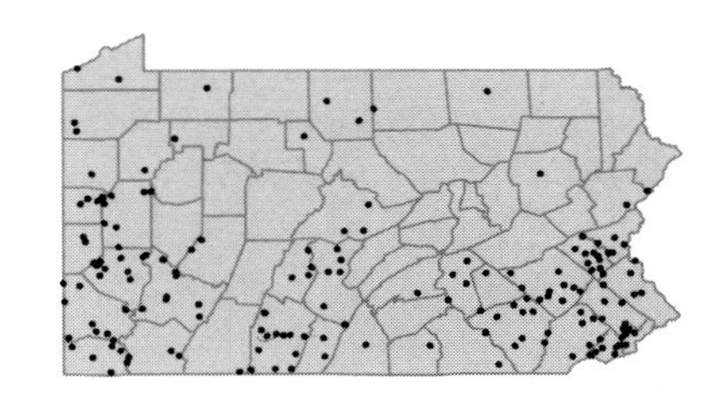

Agrimonia rostellata Wallr.
Woodland agrimony
Herbaceous perennial
Woods, fields and thickets.

FACU

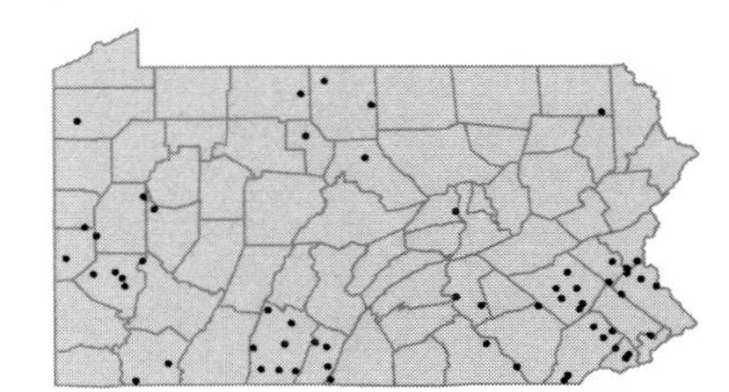

Agrimonia striata Michx.
Roadside agrimony
Herbaceous perennial
Stream banks, open floodplains or woods edges.

FACU-

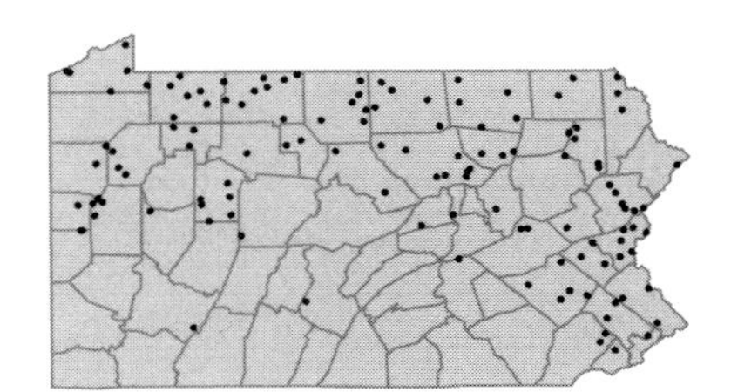

• ***Amelanchier arborea*** (Michx.f.) Fern.
Shadbush; Serviceberry; Juneberry
Deciduous tree
Rocky bluffs and upper slopes. A hybrid with *A. laevis* has been collected at a few locations.

FAC-

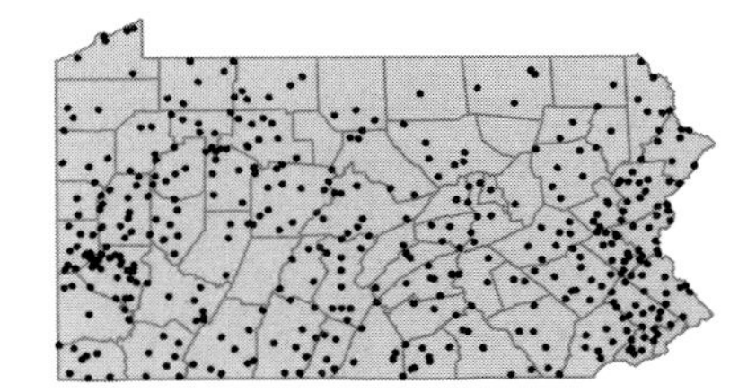

Amelanchier bartramiana (Tausch) M.Roemer
Mountain juneberry; Oblong-fruited serviceberry
Deciduous shrub
Swamps, sphagnum bogs and peaty thickets.
Amelanchier oligocarpa (Michx.) M.Roemer P

FAC

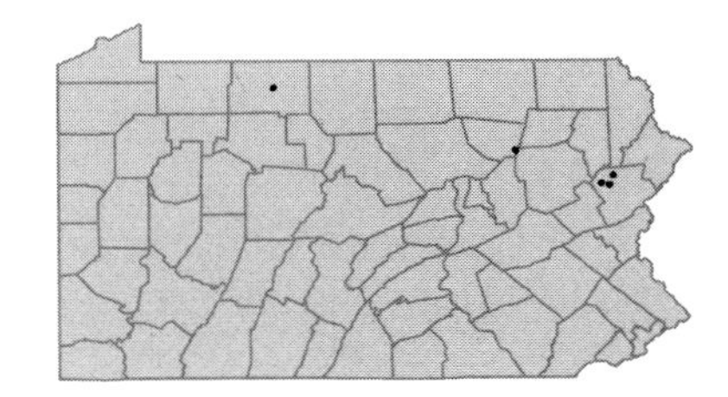

Amelanchier canadensis (L.) Medic.
Shadbush; Serviceberry; Juneberry
Deciduous shrub
Woods and peaty thickets. A hybrid with *A. laevis* has been collected at a few locations.
Amelanchier botryapium (L.f.) DC. P

FAC

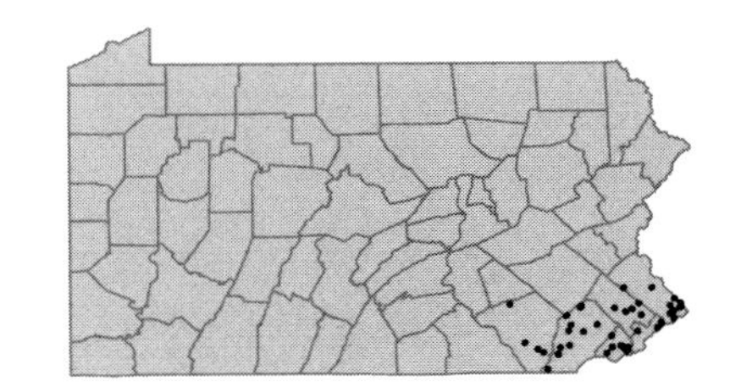

Amelanchier humilis Wieg.
Low juneberry; Low serviceberry
Deciduous shrub
Dry, open, high ground and bluffs.
Amelanchier spicata (Lam.) Decne. PC

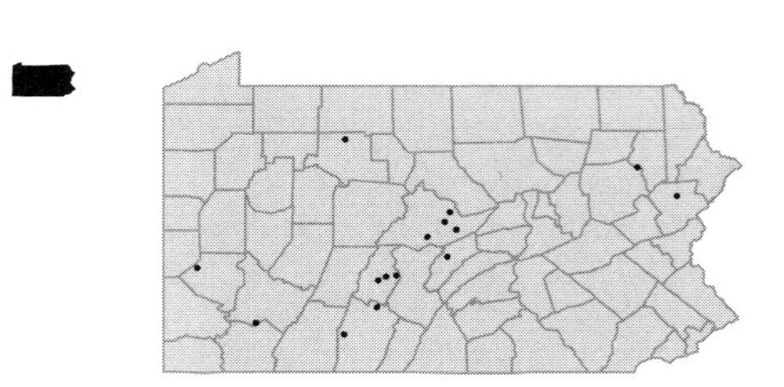

Amelanchier intermedia Spach
Shadbush; Serviceberry
Deciduous shrub
Wet woods, swamps, bogs and river banks.

FACW

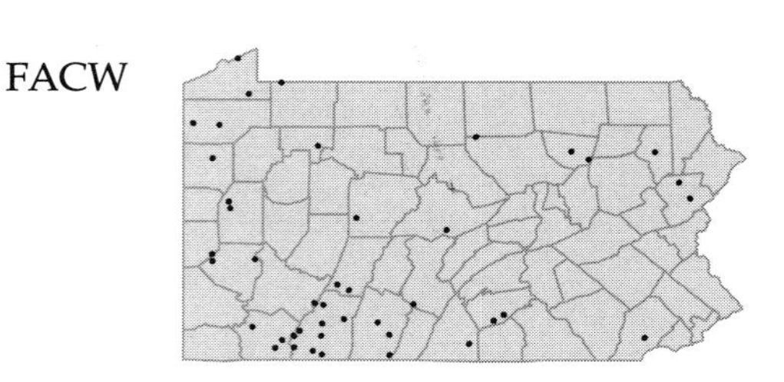

Amelanchier laevis Wieg.
Smooth serviceberry; Smooth shadbush
Deciduous tree
Rocky woods, thickets and roadside banks.

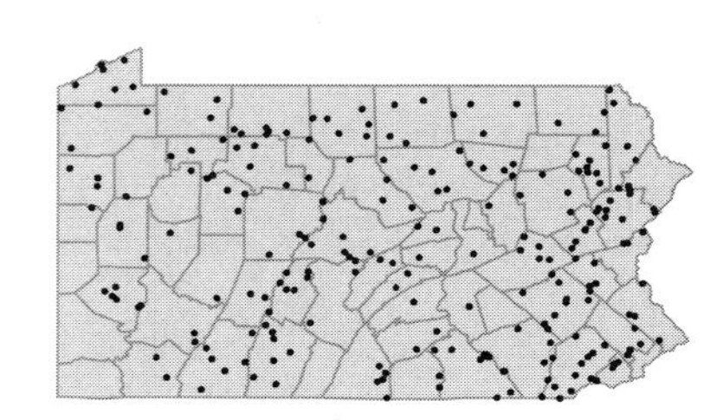

Amelanchier obovalis (Michx.) Ashe
Coastal juneberry; Coastal shadbush
Deciduous shrub
Peaty barrens, thickets and roadsides.

FACU

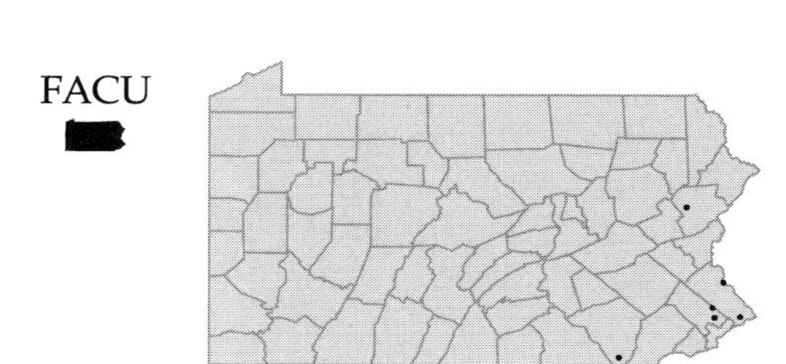

Amelanchier sanguinea (Pursh) DC.
Roundleaf serviceberry; Roundleaf shadbush
Deciduous shrub
Open woods, rocky slopes and barrens.

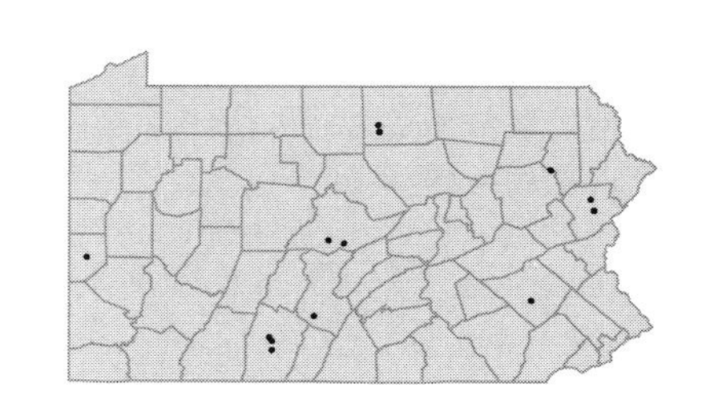

Amelanchier stolonifera Wieg.
Low juneberry; Low shadbush
Deciduous shrub
Woods, old fields, fencerows, roadside banks and serpentine barrens.
Amelanchier spicata (Lam.) K.Koch GC

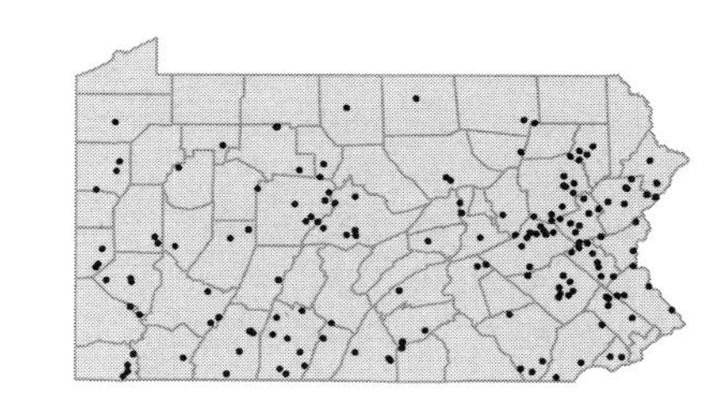

• ***Aronia arbutifolia*** (L.) Ell.
Red chokeberry
Deciduous shrub
Swamps, bogs and moist woods.
Pyrus arbutifolia (L.) L.f. F

FACW

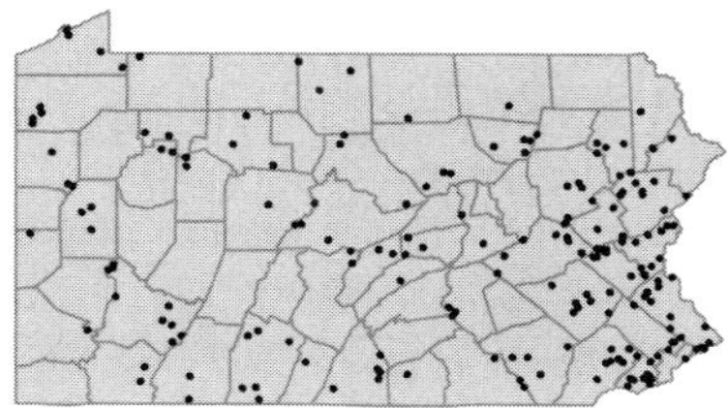

Aronia melanocarpa (Michx.) Ell.
Black chokeberry
Deciduous shrub
Swamps, bogs and wet or dry woods or barrens.
Aronia nigra (Willd.) Britt. P; *Pyrus melanocarpa* (Michx.) Willd. F

FAC

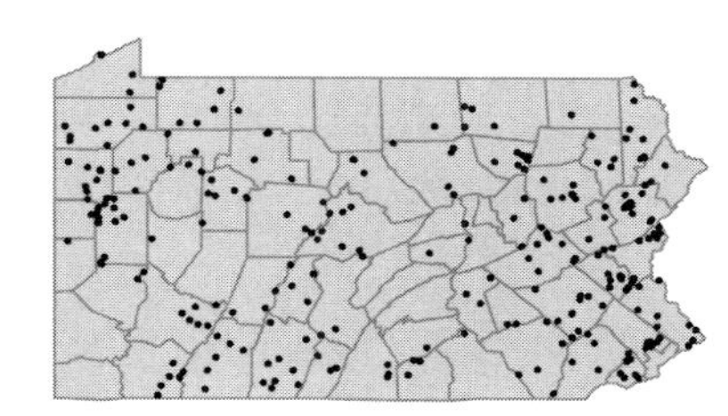

Aronia prunifolia (Marshall) Rehd.
Purple chokeberry
Deciduous shrub
Swampy woods, bogs, rocky ledges and dry, mountaintop thickets.
Pyrus floribunda Lindl. F

FACW

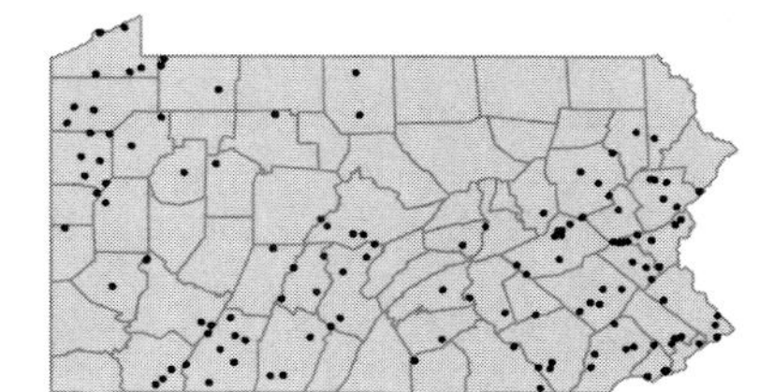

• ***Aruncus dioicus*** (Walt.) Fern.
Goat's-beard
Herbaceous perennial
Rich woods and wooded roadsides.
Aruncus aruncus (L.) Karst. P

FACU

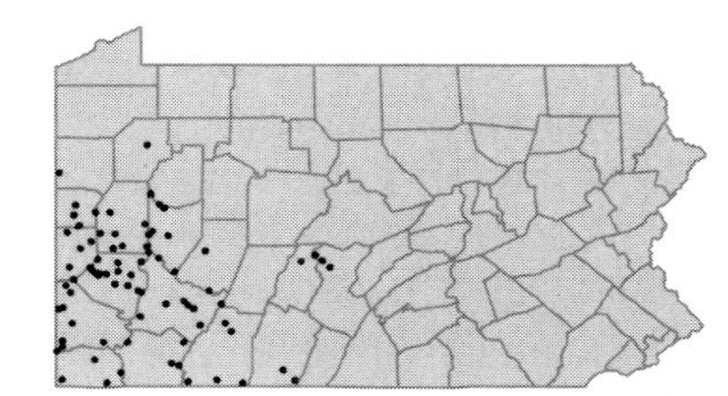

• ***Chaenomeles speciosa*** (Sweet) Nakai
Flowering quince
Deciduous shrub
Cultivated and occasionally persisting at old homesites.

I

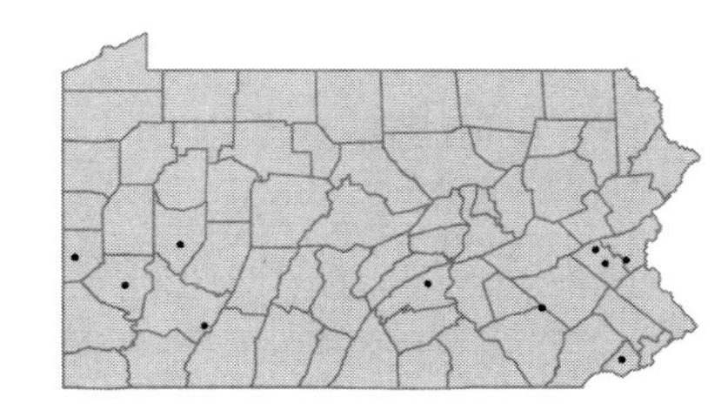

• ***Cotoneaster apiculata*** Rehd. & Wils.
Cranberry cotoneaster
Deciduous shrub
Cultivated and rarely naturalized, wooded river bank.

I

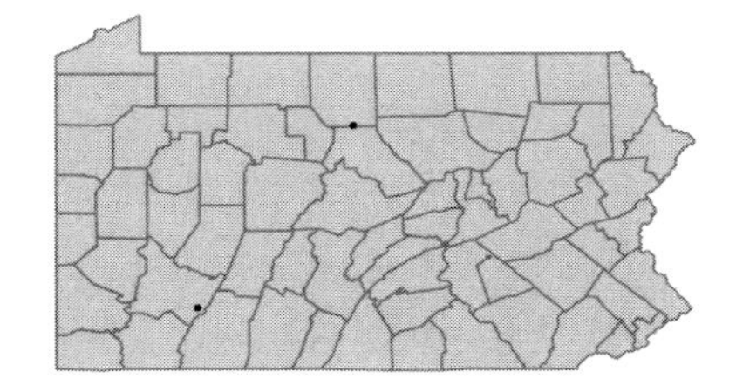

Crataegus—Due to the lack of recent definitive work on the Pennsylvania *Crataegus*, we have chosen to recognize a number of taxa at the level of broad species complexes. The synonyms listed represent an attempt to account for the species mapped in Wherry et al. (1979). Extensive study is needed to determine which of the many segregate species that have been named should be recognized. Type specimens alone from Pennsylvania number 129 and 18 at PH and CM respectively (Thomas and Boufford 1986).

• ***Crataegus brainerdii*** Sarg.
Brainerd's hawthorn
Deciduous shrub
Moist bottomland.

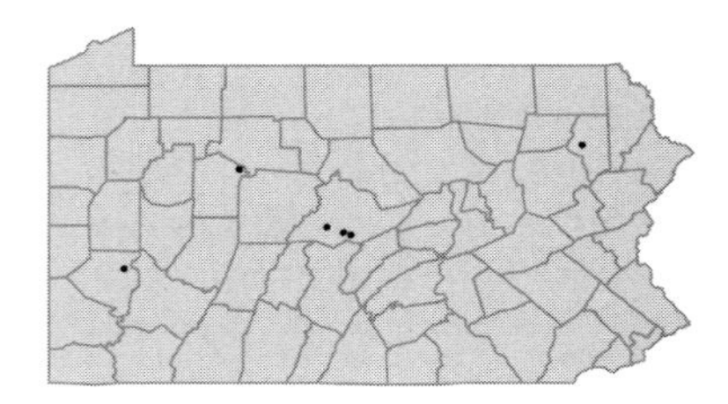

Crataegus calpodendron (Ehrh.) Medic.
Pear hawthorn; Blackthorn hawthorn
Deciduous shrub
Woods, thickets and low meadows.

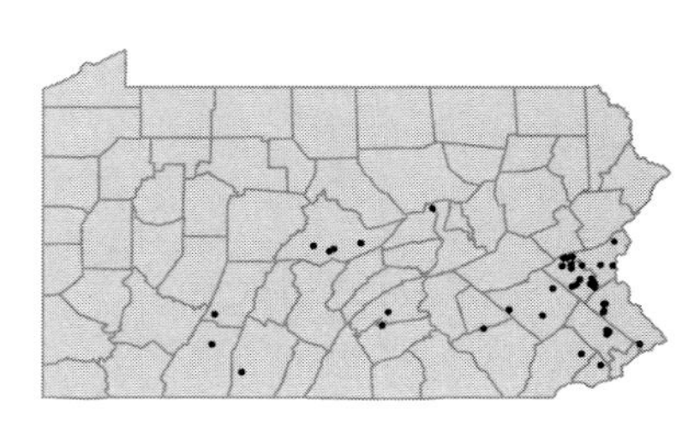

Crataegus coccinea L. *sensu lato*
Red-fruited hawthorn
Deciduous shrub
Open woods, fields, roadsides and stream banks.
Crataegus holmsiana Ashe *pro parte* GW; *Crataegus pensylvanica* Ashe *pro parte* W; *Crataegus pedicellata* Sarg. *pro parte* FBW; *Crataegus tatnalliana* Sarg. *pro parte* W

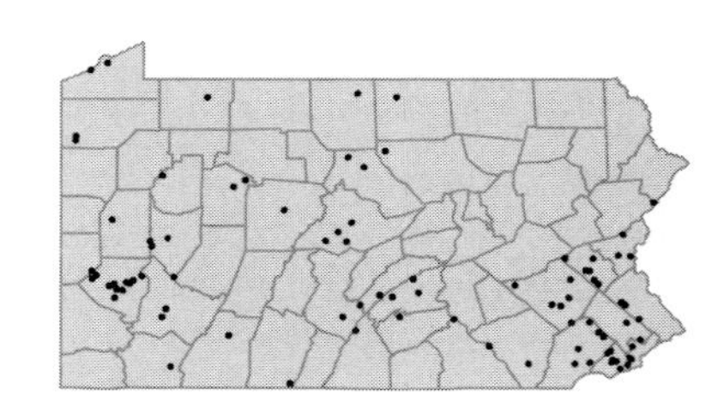

Crataegus crus-galli L. *sensu lato*
Cockspur hawthorn
Deciduous shrub
Woods, meadows, roadsides and thickets.
Crataegus canbyi Sarg. *pro parte* PFBW; *Crataegus disperma* Ashe *pro parte* W; *Crataegus pausiaca* Ashe *pro parte* W; *Crataegus fontanesiana* (Spach) Steud. *pro parte* W; *Crataegus bartramiana* Sarg. *pro parte* W; *Crataegus laetifica* Sarg. *pro parte* W

FACU

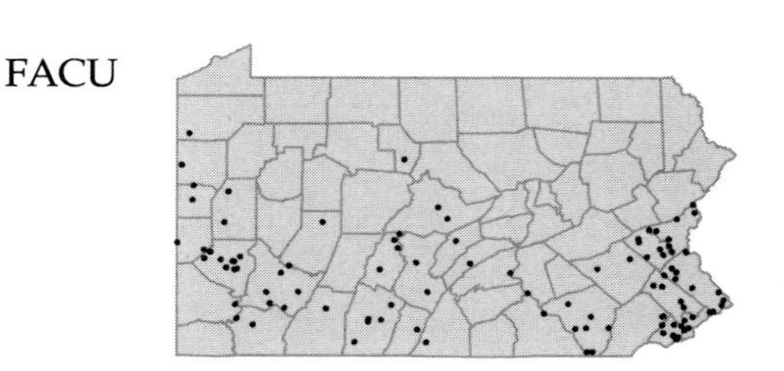

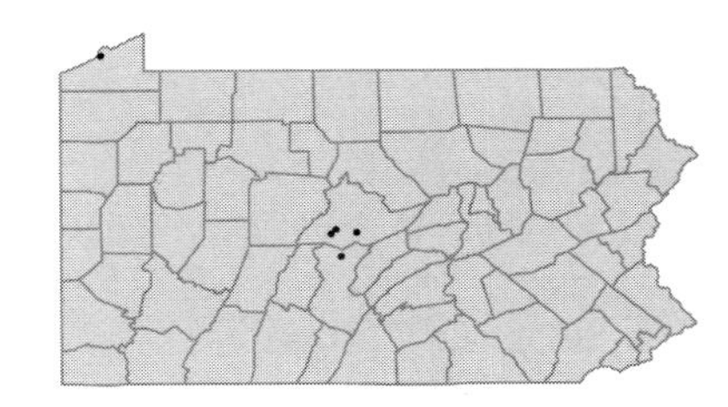

Crataegus dilatata Sarg.
Hawthorn
Deciduous tree
Pastures, thickets and hillsides.

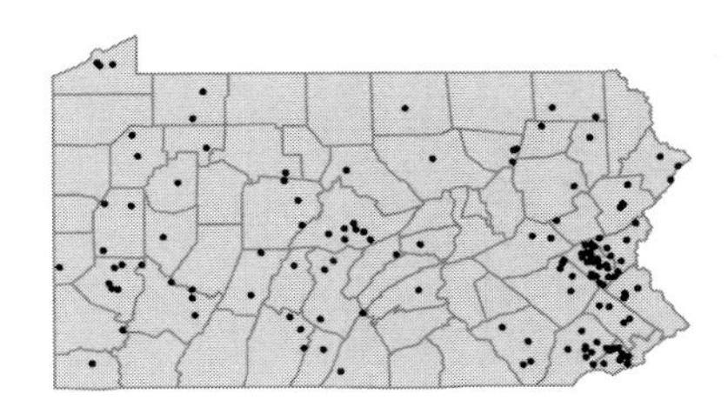

Crataegus flabellata (Spach) Kirchner *sensu lato*
Hawthorn
Deciduous shrub
Open woods, fencerows, abandoned fields and roadsides.
Crataegus brumalis Ashe *pro parte* FBW; *Crataegus populnea* Ashe *pro parte* FBW; *Crataegus stolonifera* Sarg. *pro parte* FBW; *Crataegus macrosperma* Ashe *pro parte* FWG; *Crataegus levis* Sarg. *pro parte* FBW; *Crataegus milleri* Sarg. *pro parte* FBW; *Crataegus vittata* Ashe *pro parte* FBW; *Crataegus compta* Sarg. *pro parte* W; *Crataegus anomala* Sarg. *pro parte* W

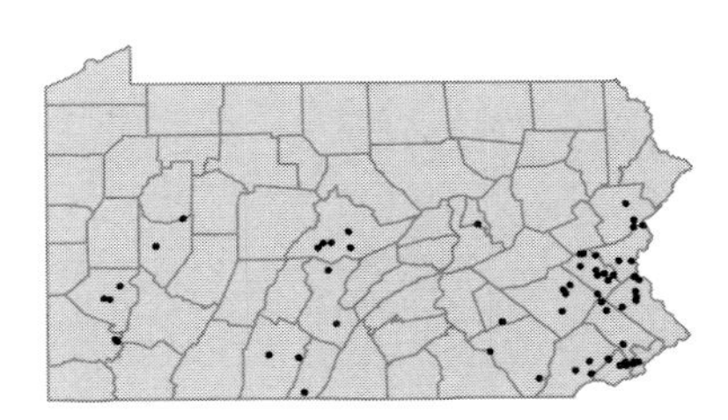

Crataegus intricata Lange *sensu lato*
Hawthorn
Deciduous shrub
Woods, pastures, thickets and barrens.
Crataegus boyntonii Beadle *pro parte* FBW; *Crataegus foetida* Ashe *pro parte* FBW; *Crataegus fortunata* Sarg. *pro parte* FBW; *Crataegus rubella* Beadle *pro parte* FBW; *Crataegus biltmoreana* Beadle *pro parte* W; *Crataegus stonei* Sarg. *pro parte* W; *Crataegus neobushii* Sarg. *pro parte* W

FACU

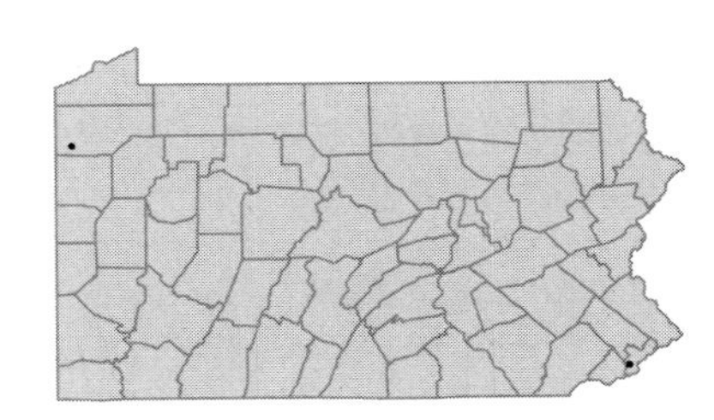

Crataegus mollis (Torr. & A.Gray) Scheele
Downy hawthorn
Deciduous tree
Abandoned field.

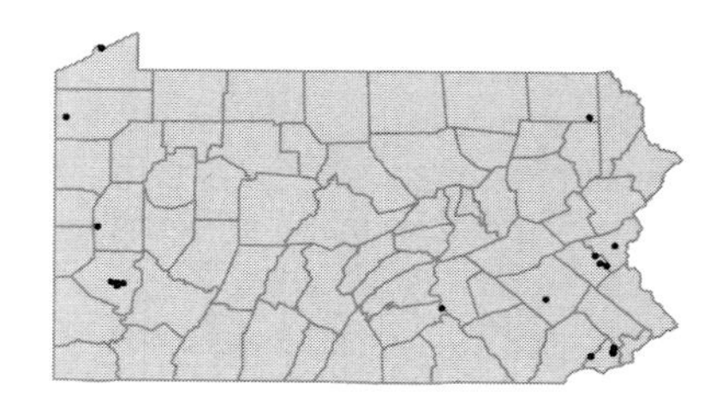

Crataegus monogyna Jacq.
English hawthorn
Deciduous shrub
Escaped from cultivation to roadsides and waste ground.

I

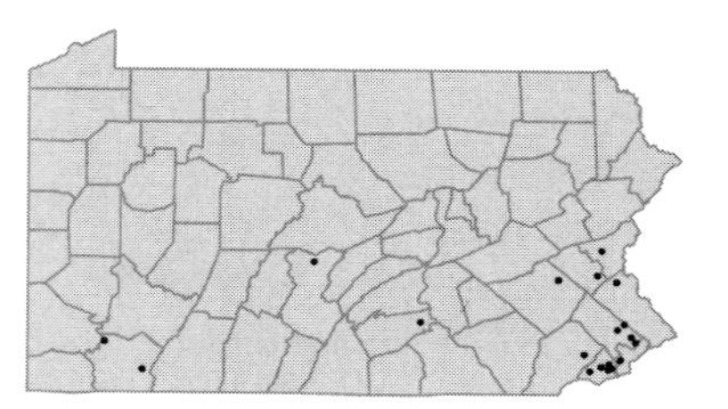

Crataegus phaenopyrum (L.f.) Medic.
Washington hawthorn
Deciduous tree
Cultivated and occasionally escaped to roadsides, hedgerows and open ground.

FAC
I

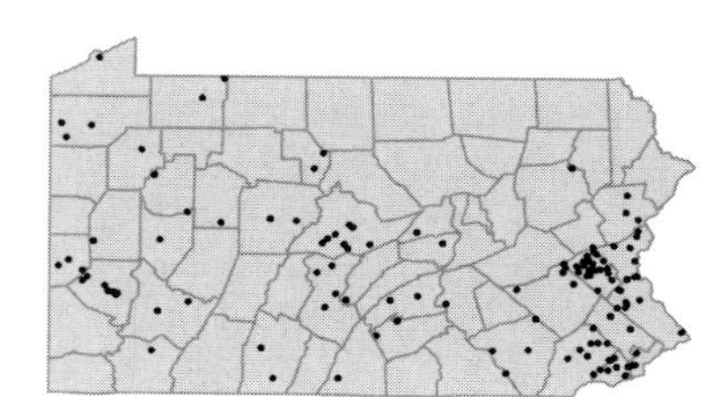

Crataegus pruinosa (Wendl.f.) K.Koch *sensu lato*
Frosted hawthorn
Deciduous shrub
Open woods and thickets.
Crataegus crawfordiana Sarg. *pro parte* FBW; *Crataegus gattingeri* Ashe *pro parte* FBW; *Crataegus gaudens* Sarg. *pro parte* FBW; *Crataegus lecta* Sarg. *pro parte* FBW; *Crataegus leiophylla* Sarg. *pro parte* FBW; *Crataegus porteri* Britt. *pro parte* FBW; *Crataegus compacta* Sarg. *pro parte* W; *Crataegus laetans* Sarg. *pro parte* W; *Crataegus relicta* Sarg. *pro parte* W

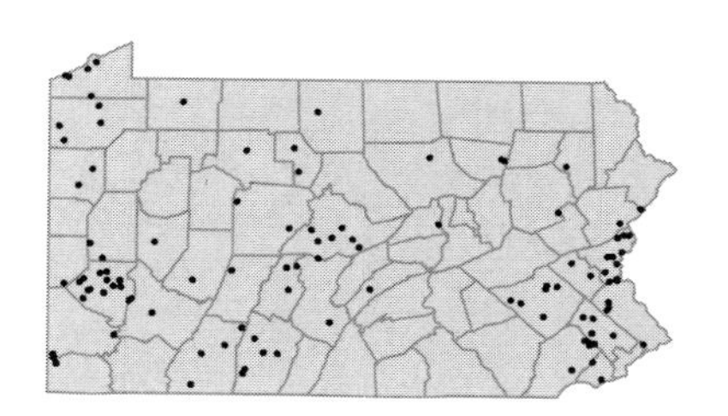

Crataegus punctata Jacq.
Dotted hawthorn; White hawthorn
Deciduous tree
Woods, pastures and alluvial banks.

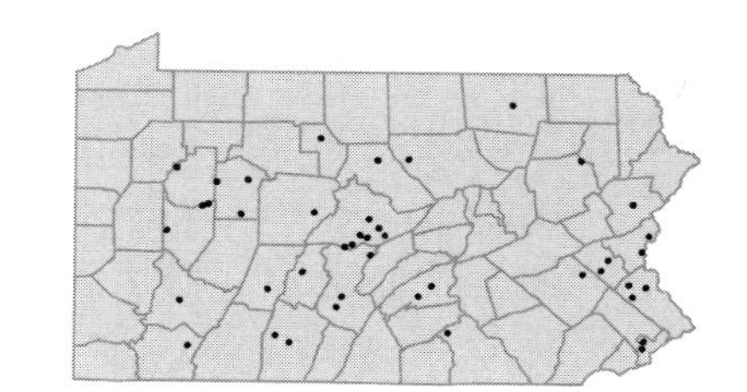

Crataegus rotundifolia Moench.
Hawthorn
Deciduous shrub
Rocky pastures, open woods and roadsides.
Crataegus margaretta Ashe *pro parte* GW; *Crataegus dodgei* Ashe *pro parte* GW; *Crataegus chrysocarpa* Ashe *pro parte* GCW; *Crataegus evansiana* Sarg. *pro parte* W

Crataegus succulenta Schrad. ex Link
Long-spined hawthorn; Fleshy hawthorn
Deciduous tree
Woods, thickets, banks, fencerows and meadows.
Crataegus macracantha Lodd. *pro parte*

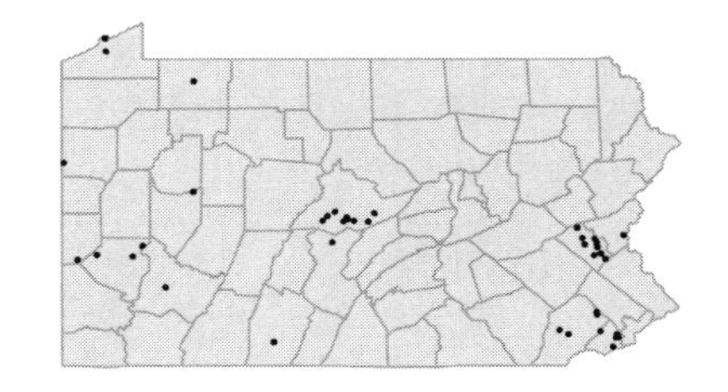

Crataegus uniflora Muenchh.
One-fruited hawthorn
Deciduous shrub
Open woods and dry slopes.
Crataegus tomentosa L. P

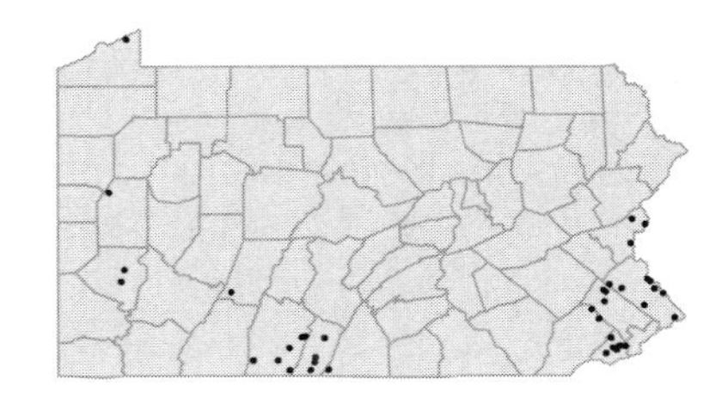

• ***Cydonia oblonga*** P.Mill.
Quince
Deciduous tree
Cultivated and occasionally escaped to woods, hedgerows, thickets and old fields.

[I]

• ***Dalibarda repens*** L.
False-violet; Robin-run-away
Herbaceous perennial
Bogs, peaty barrens and cool mossy woods.

FAC

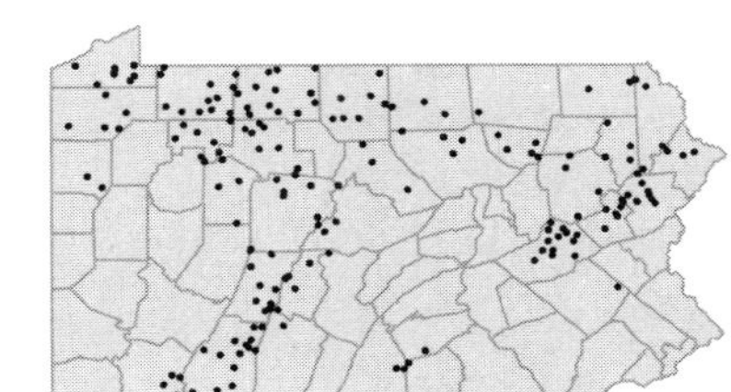

• ***Duchesnea indica*** (Andr.) Focke
Indian strawberry
Herbaceous perennial
Woods, trails and waste ground.

FACU-
[I]

• ***Exochorda racemosa*** (Lindl.) Rehd.
Pearlbush
Deciduous shrub
Cultivated and occasionally persisting in abandoned parks or nurseries.

[I]

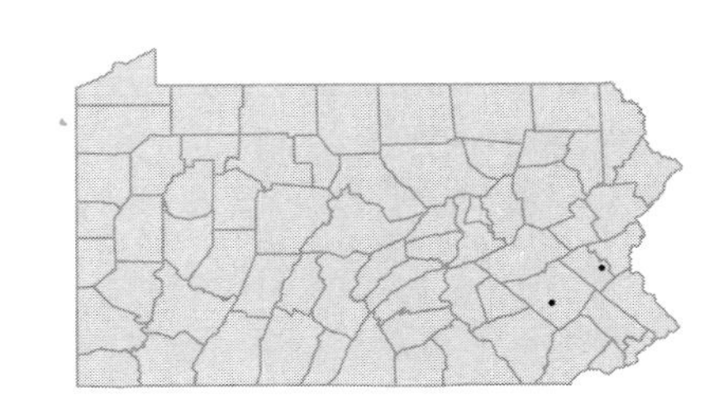

• ***Filipendula rubra*** (Hill) B.L.Robins.
Queen-of-the-prairie
Herbaceous perennial
Moist meadows, thickets and roadsides, sometimes escaped from cultivation.

FACW

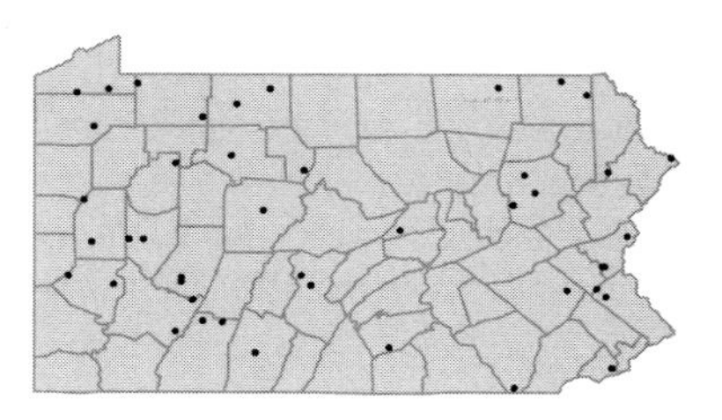

Filipendula ulmaria (L.) Maxim.
Queen-of-the-meadow
Herbaceous perennial
Cultivated and occasionally escaped to swamps or woods edges.

I

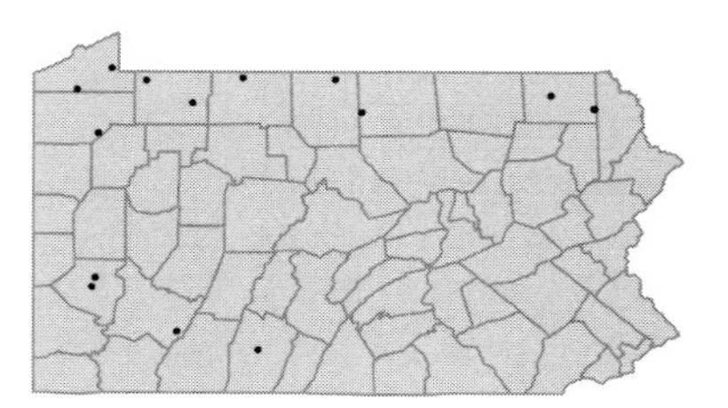

• ***Fragaria chiloensis x virginiana***
Garden strawberry
Herbaceous perennial
Cultivated and occasionally escaped to fields and waste ground.
Fragaria x ananassa Duchesne FWKC; *Fragaria chiloensis* Duchesne var. *ananassa* Bailey B

I

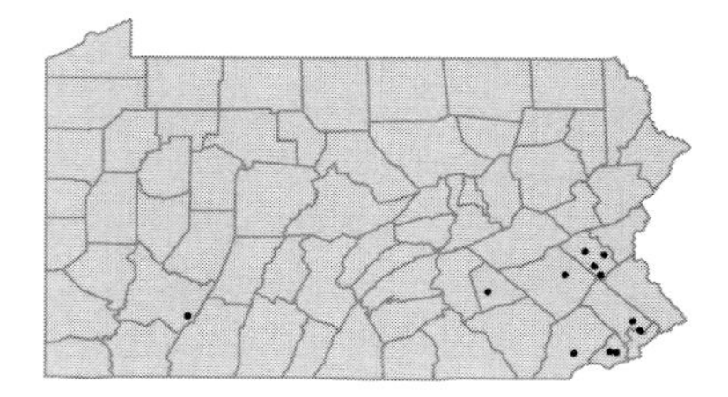

Fragaria vesca L. ssp. ***vesca***
Sow-teat strawberry; Woodland strawberry
Herbaceous perennial
Moist, deciduous woods and banks.

I

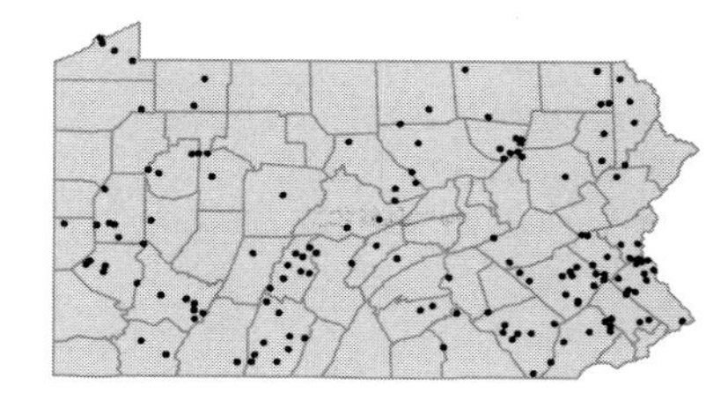

Fragaria vesca L. ssp. ***americana*** (Porter) Staudt
Sow-teat strawberry; Woodland strawberry
Herbaceous perennial
Wooded, upland slopes and rocky ledges.

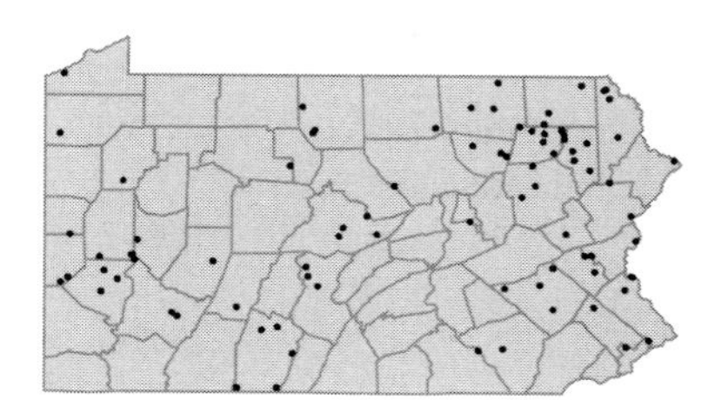

Fragaria virginiana P.Mill. ssp. ***virginiana***
Wild strawberry
Herbaceous perennial
Woods, meadows, old fields and other dry, open ground.

FACU

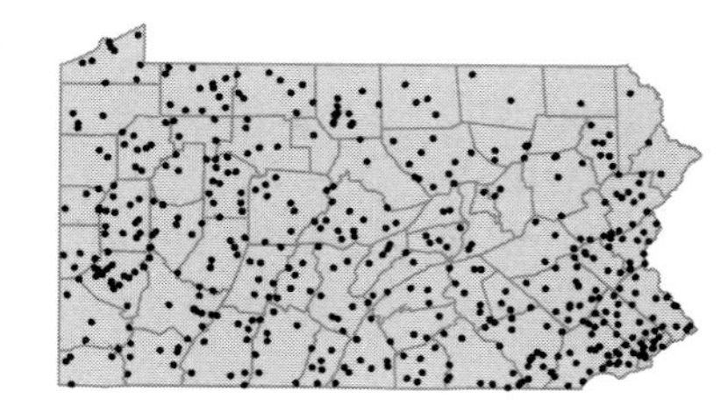

Fragaria virginiana P.Mill. ssp. ***platypetala*** (Rydb.) Staudt
Wild strawberry
Herbaceous perennial
Woods and edges.
Fragaria virginiana P.Mill. var. *illinoensis* (Prince) A.Gray FBW

FACU

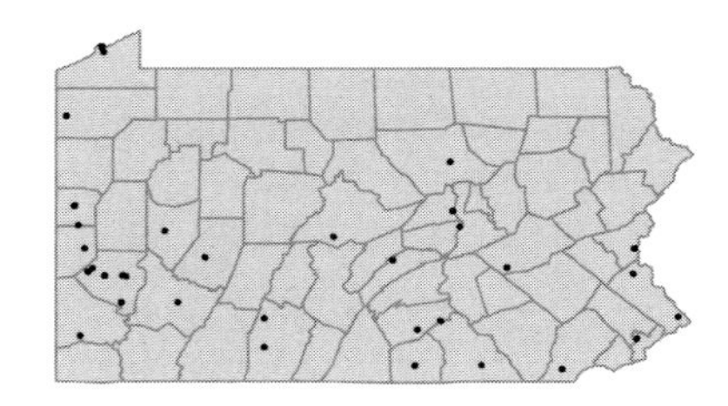

• ***Geum aleppicum*** Jacq.
Yellow avens
Herbaceous perennial
Woods, moist fields, swamps, and roadsides.
Geum strictum Ait. P

FAC
[I]

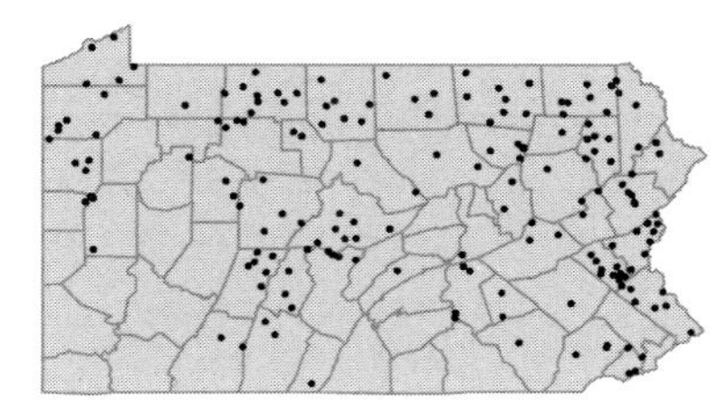

Geum canadense Jacq. var. ***canadense***
White avens
Herbaceous perennial
Woods, stream banks and roadsides.

FACU

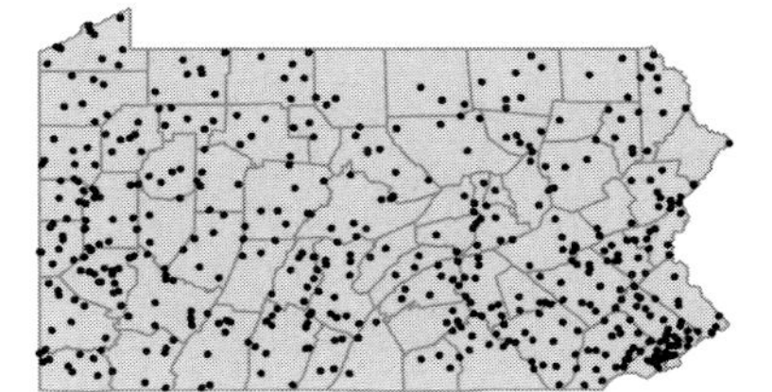

Geum canadense Jacq. var. ***grimesii*** Fern. & Weatherby
White avens
Herbaceous perennial
Stream banks.

FACU

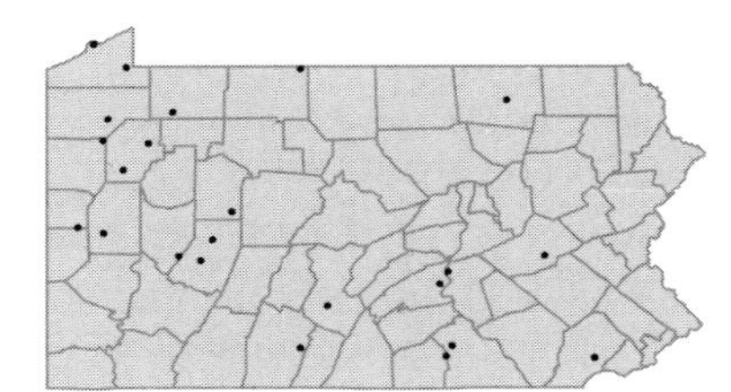

Geum laciniatum Murr. var. ***laciniatum***
Herb-bennet; Rough avens
Herbaceous perennial
Woods, swamps, bogs and wet ditches.

FAC+

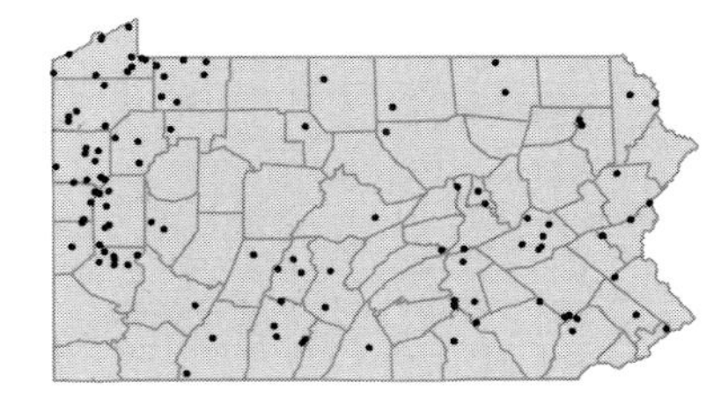

Geum laciniatum Murr. var. ***trichocarpum*** Fern.
Herb-bennet; Rough avens
Herbaceous perennial
Woods, boggy fields and swampy ditches.

FAC+

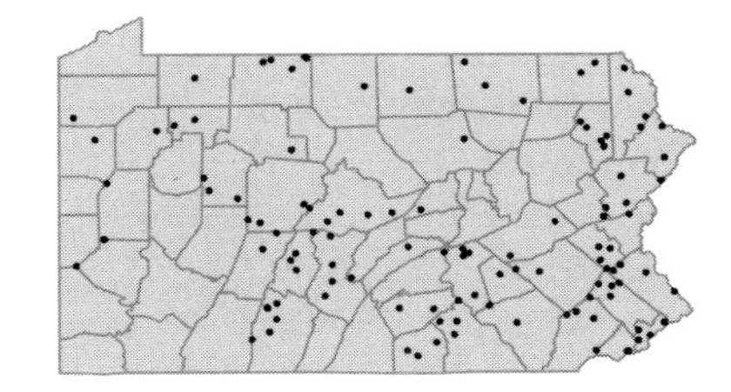

Geum rivale L.
Water avens; Purple avens
Herbaceous perennial
Bogs, calcareous marshes and peaty meadows.

OBL

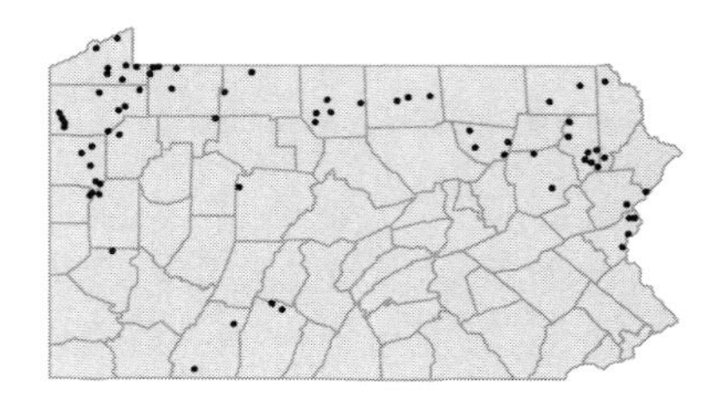

Geum urbanum L.
Herb-bennet; Cloveroot
Herbaceous perennial
Cultivated and occasionally escaped.

[I]

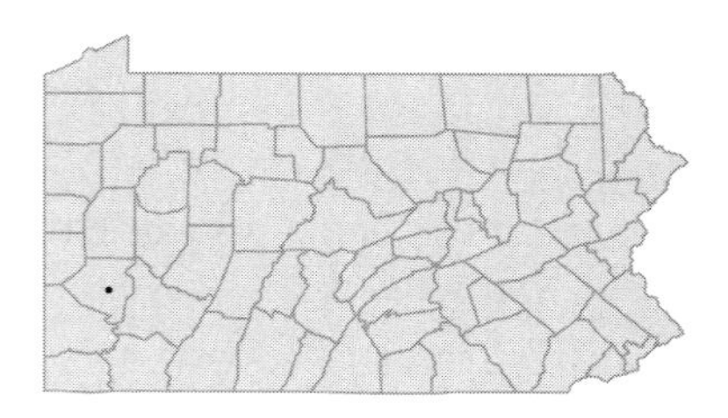

Geum vernum (Raf.) Torr. & A.Gray
Spring avens
Herbaceous perennial
Rich woods and ravines.

FACU

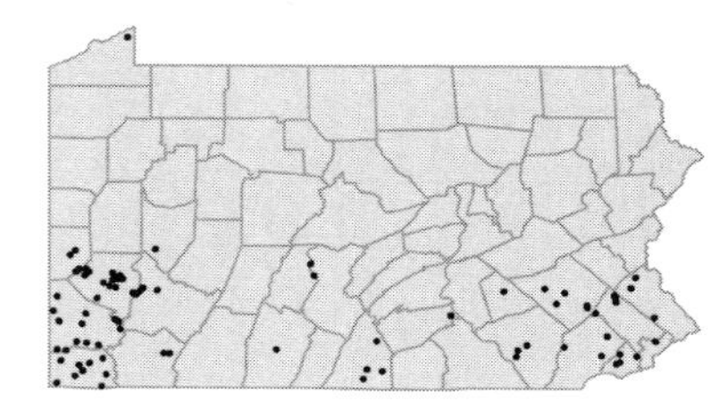

Geum virginianum L.
Cream-colored avens
Herbaceous perennial
Rich woods, open swamps, ravines and bluffs.
Geum flavum (Porter) Bickn. P

FAC-

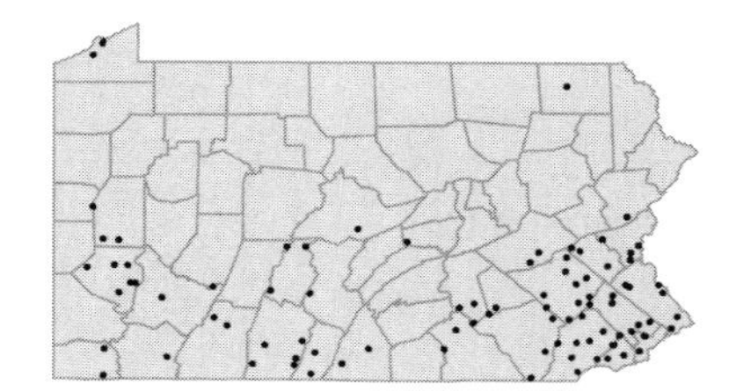

• ***Kerria japonica*** (L.) DC.
Japanese kerria
Deciduous shrub
Cultivated and occasionally escaped.

[I]

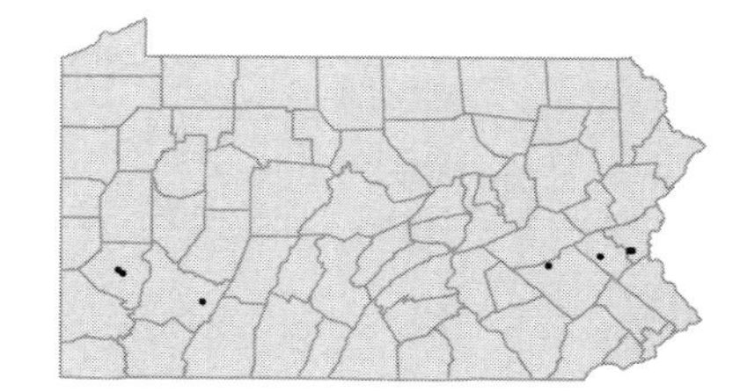

• ***Malus baccata*** (L.) Borkh.
Siberian crabapple
Deciduous tree
Cultivated and occasionally escaped to waste ground or cinders.
Pyrus baccata L. FGBC

[I]

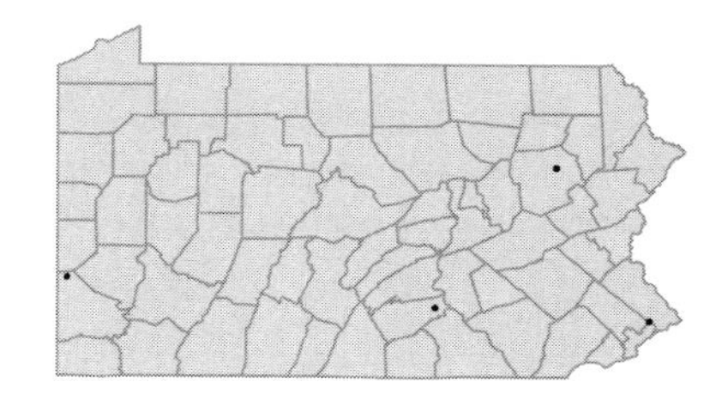

Malus coronaria (L.) P.Mill. var. ***coronaria***
American crabapple; Sweet crab
Deciduous tree
Woods, old fields and thickets.
Pyrus coronaria L. FGBC

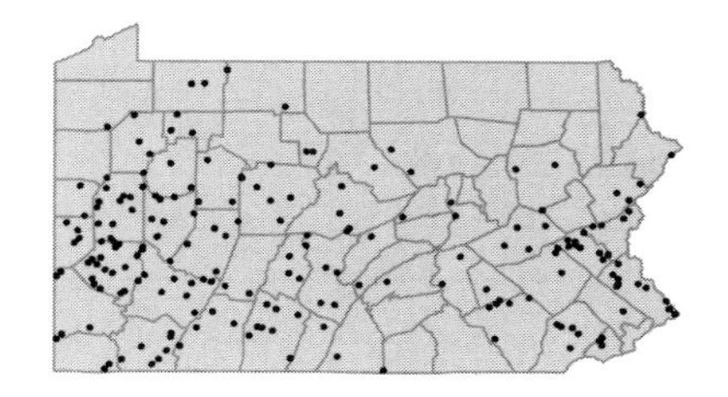

Malus floribunda Sieb. ex Van Houtte
Showy crabapple
Deciduous tree
Cultivated and occasionally escaped.
Malus pulcherrima (Sieb.) Makino W

I

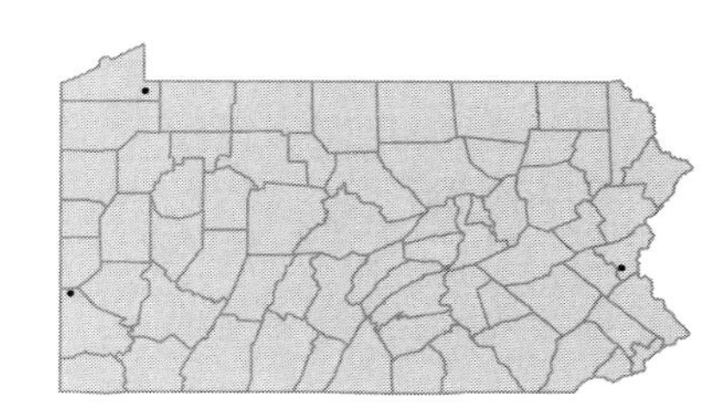

Malus glaucescens Rehd.
American crabapple; Wild crabapple
Deciduous tree
Open woods and wooded slopes.
Pyrus coronaria L. *pro parte* GBC

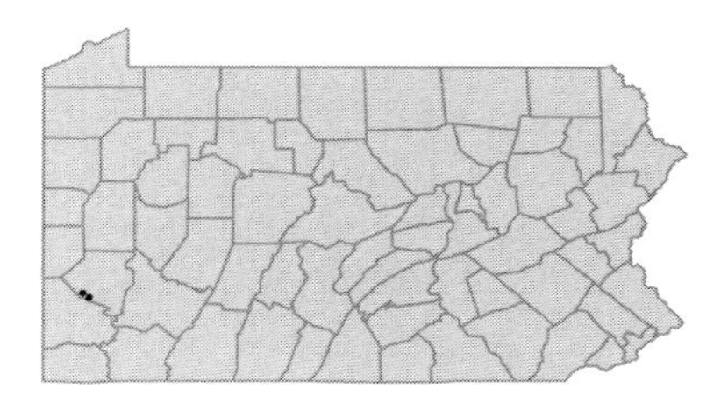

Malus lancifolia Rehd.
Lanceleaf crabapple
Deciduous tree
Dry, open woods, old fields, pastures and edges.
Pyrus coronaria L. *pro parte* CF

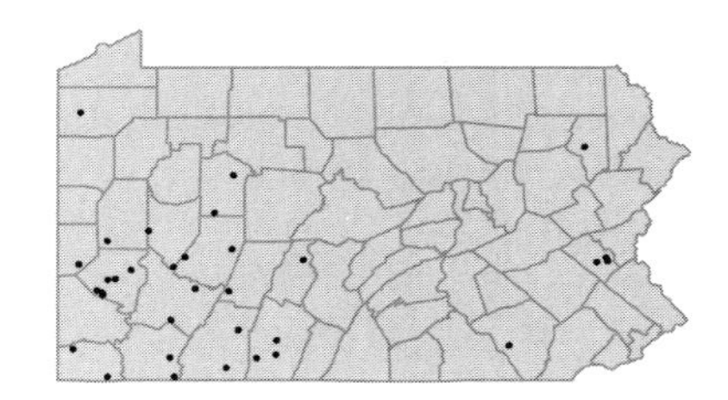

Malus prunifolia (Willd.) Borkh.
Chinese crabapple
Deciduous tree
Parks, waste ground and railroad ballast, escaped from cultivation.
Pyrus prunifolia Willd. FC

I

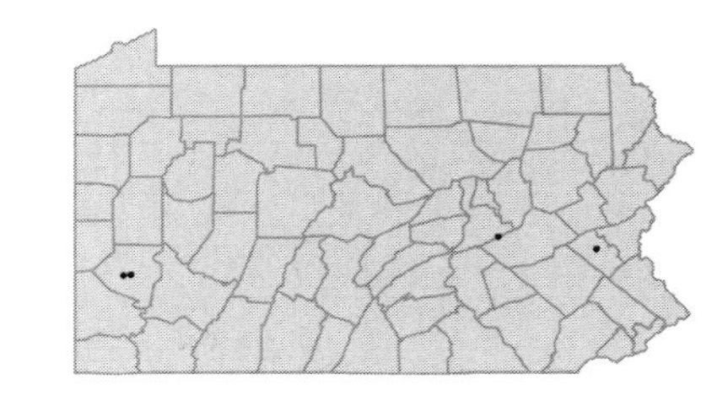

Malus pumila P.Mill.
Apple
Deciduous tree
Cultivated and often persisting at abandoned farm or garden sites, hedgerows and roadsides.
Malus domestica Borkh. W; *Pyrus malus* L. FGBWC

I

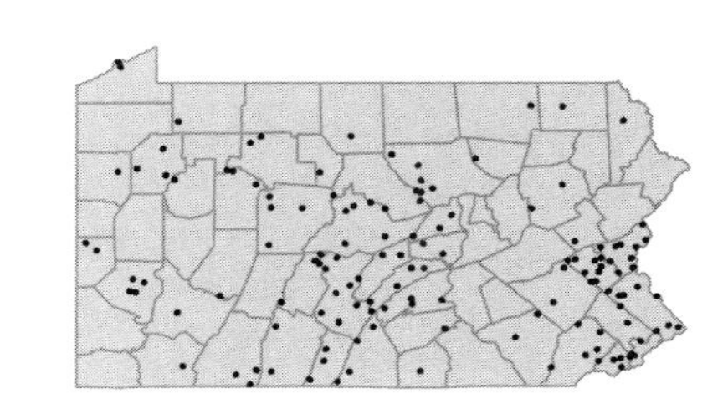

Malus sieboldii Rehd.
Crabapple
Deciduous tree
Cultivated and rarely escaped.
Pyrus sieboldii Regel.

I

• ***Photinia villosa*** (Thunb.) DC.
Photinia
Deciduous tree
Cultivated and rarely escaped.

I

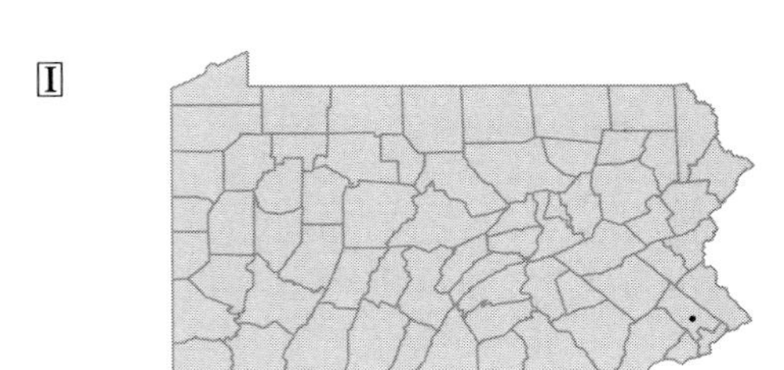

• ***Physocarpus opulifolius*** (L.) Maxim.
Ninebark
Deciduous shrub
Moist cliffs, wet woods, sandy or rocky banks and shores.
Opulaster opulifolius (L.) Kuntze P

FACW-

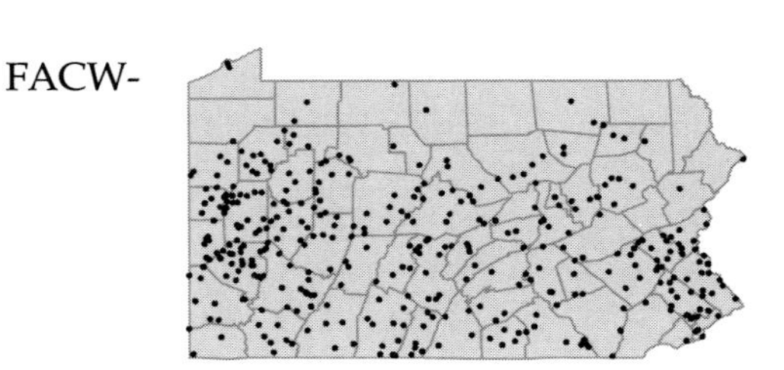

• ***Porteranthus trifoliatus*** (L.) Britt.
Bowman's-root; Mountain Indian-physic
Herbaceous perennial
Upland woods.
Gillenia trifoliata (L.) Moench FGBW

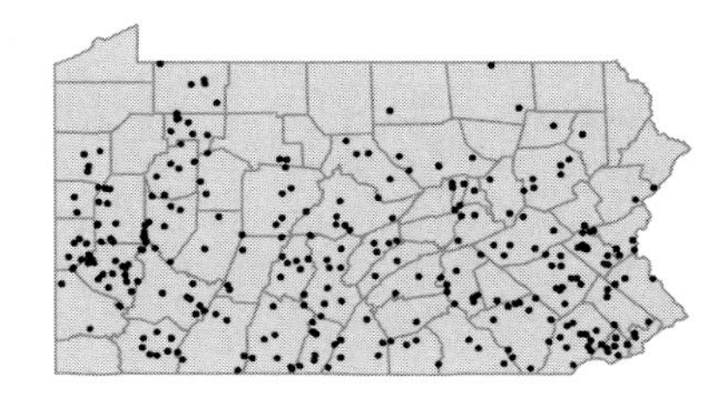

• ***Potentilla anglica*** Laichard.
English cinquefoil
Herbaceous perennial
Rocky pastures.
Potentilla procumbens Sibth. GB

I

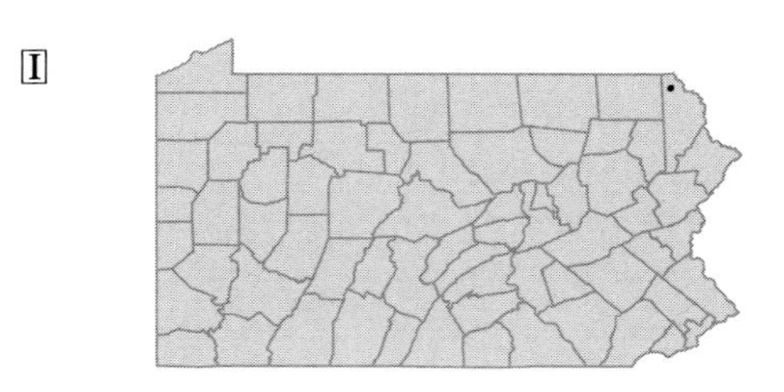

Potentilla anserina L.
Silverweed
Herbaceous perennial
Moist, sandy or gravelly shores, also ballast.

OBL

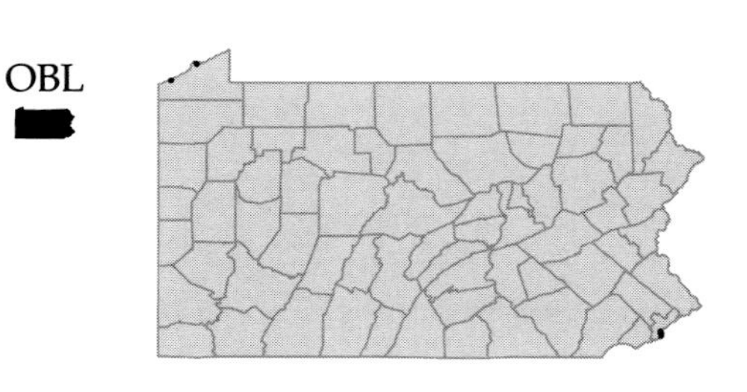

Potentilla argentea L. [I]
Hoary cinquefoil; Silvery cinquefoil
Herbaceous perennial
Dry, open ground.

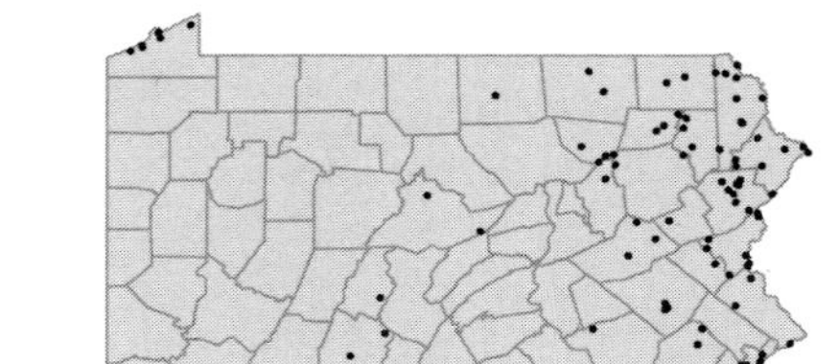

Potentilla arguta Pursh
Tall cinquefoil
Herbaceous perennial
Dry, rocky ledges, fields and woods.

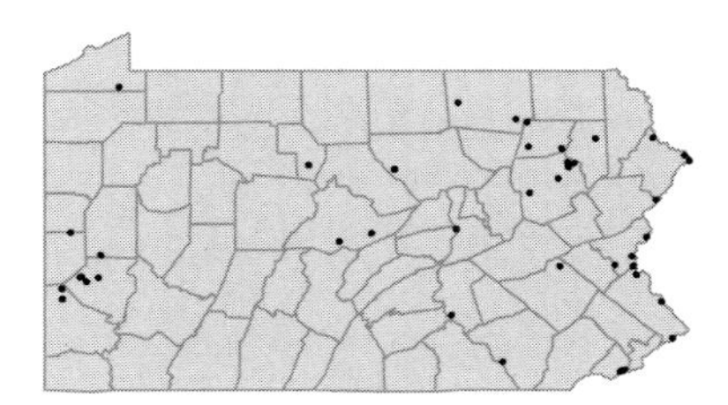

Potentilla canadensis L.
Cinquefoil
Herbaceous perennial
Dry, open woods and fields.
Potentilla pumila Poir. P

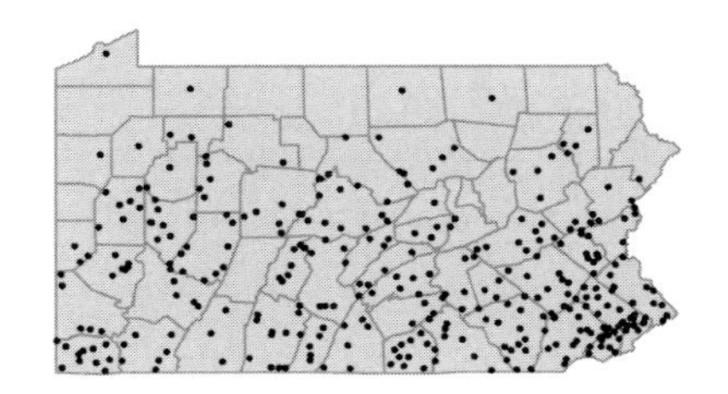

Potentilla fruticosa L. FACW
Shrubby cinquefoil
Deciduous shrub
Calcareous swamps.

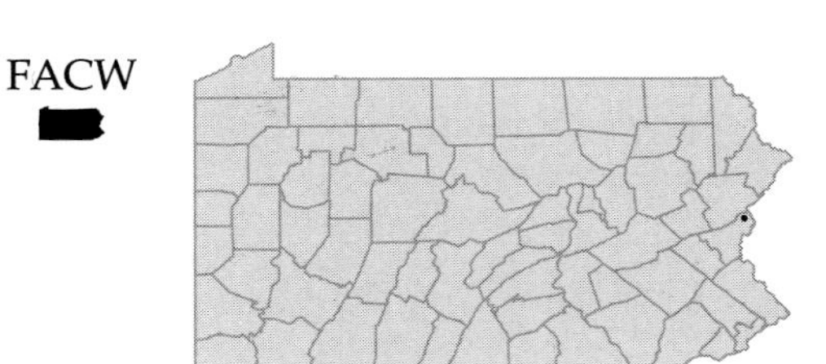

Potentilla intermedia L. [I]
Cinquefoil; Five-fingers
Herbaceous perennial
Roadsides and waste ground.

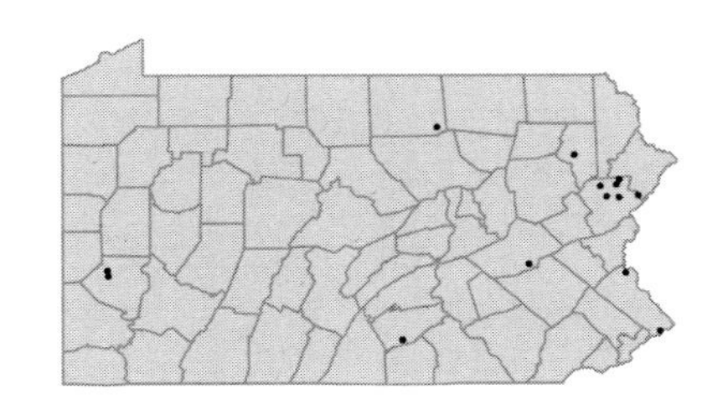

Potentilla norvegica L. ssp. ***monspeliensis*** (L.) Aschers. & Graebn. FACU
Strawberry-weed; Three-leaf cinquefoil
Herbaceous annual
Roadsides and waste ground.
Potentilla monspeliensis L. P; *Potentilla norvegica* L. var. *hirsuta* (Michx.) Torr. & A.Gray C

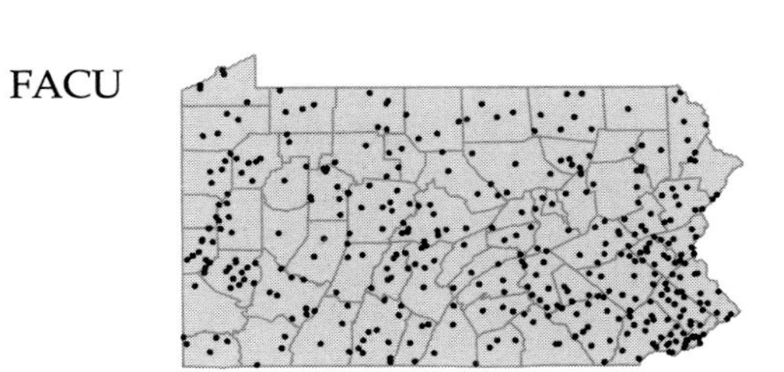

Potentilla palustris (L.) Scop.
Marsh cinquefoil
Herbaceous perennial
Swamps, bogs and peaty lake margins.

OBL

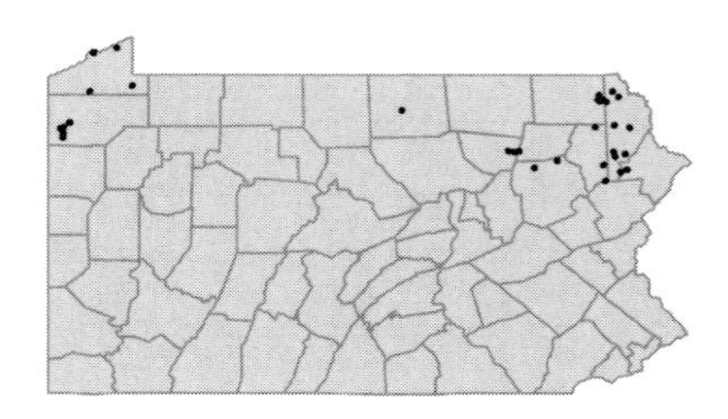

Potentilla paradoxa Nutt.
Bushy cinquefoil
Herbaceous annual
Moist, sandy shores.

OBL

Potentilla recta L.
Sulfur cinquefoil
Herbaceous perennial
Dry fields and waste ground.

I

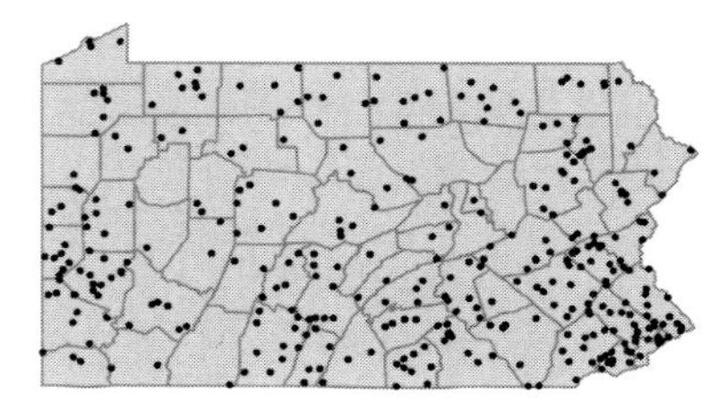

Potentilla reptans L.
Creeping cinquefoil
Herbaceous perennial
Waste ground and ballast.

I

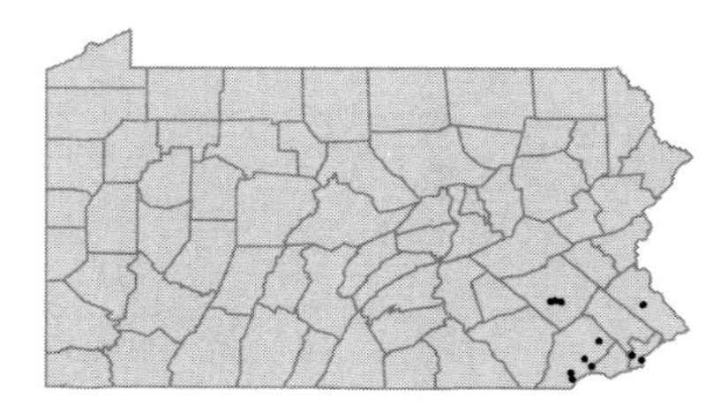

Potentilla simplex Michx.
Old-field cinquefoil
Herbaceous perennial
Dry woods, fields, meadows and roadsides.

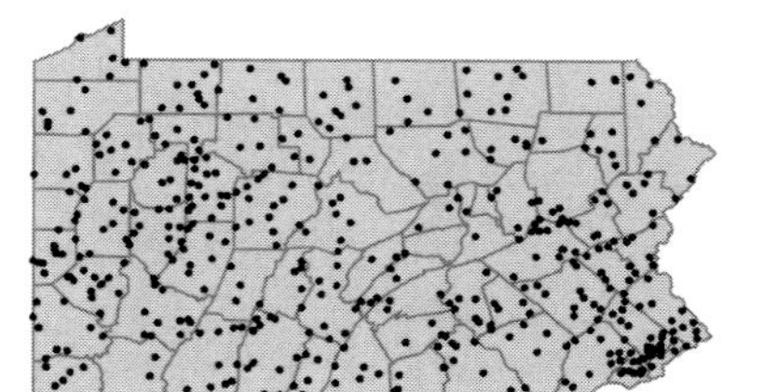

Potentilla tridentata Ait.
Three-toothed cinquefoil
Herbaceous perennial
Dry, exposed, rocky balds and mountain tops.

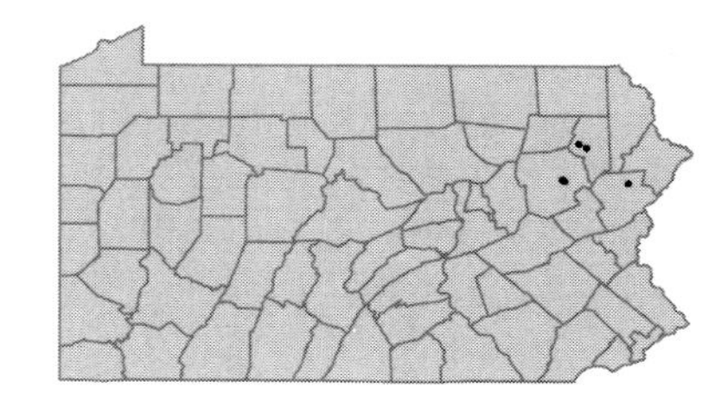

• ***Prunus alleghaniensis*** Porter
Allegheny plum
Deciduous tree
Rocky bluffs, shale barrens, roadsides and floodplains.

UPL

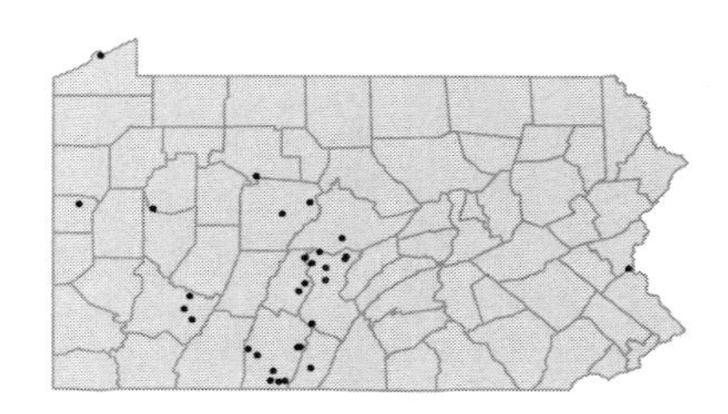

Prunus americana Marshall
Wild plum
Deciduous shrub
Wooded slopes, river banks, hedgerows, and roadside thickets.

FACU-

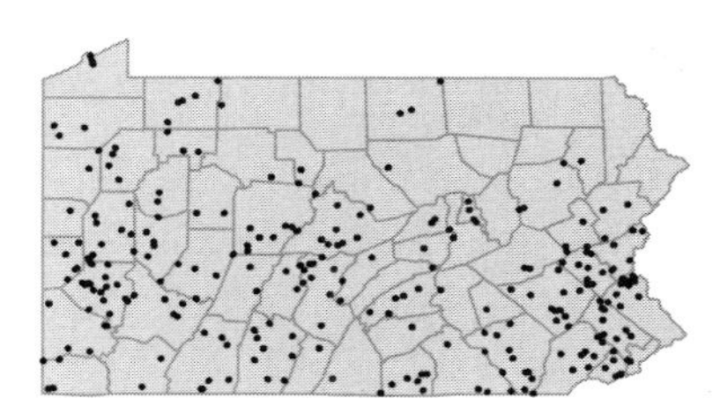

Prunus angustifolia Marshall
Chickasaw plum
Deciduous tree
Roadside thicket.

Prunus armeniaca L.
Apricot
Deciduous tree
Cultivated and occasionally persisting at abandoned homesites.

[I]

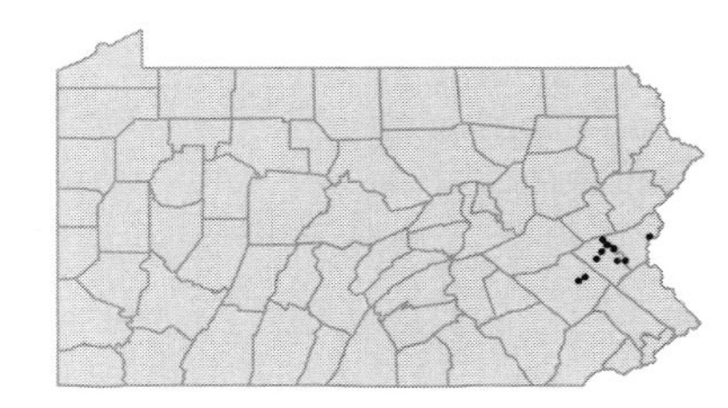

Prunus avium (L.) L.
Sweet cherry
Deciduous tree
Woods margins and fencerows.

[I]

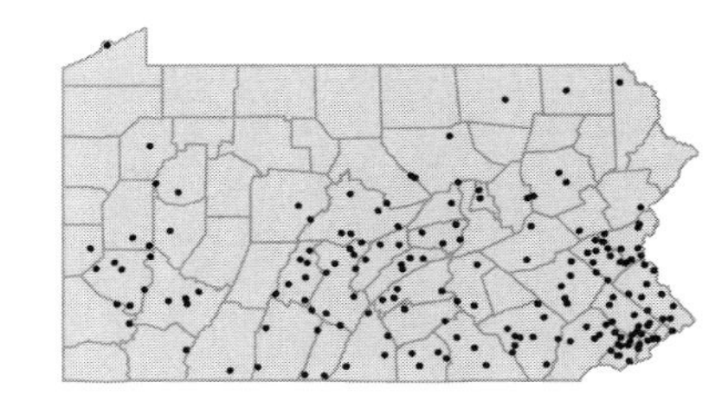

Prunus cerasifera Ehrh.
Plum
Deciduous tree
Occasionally spreading from cultivation.

[I]

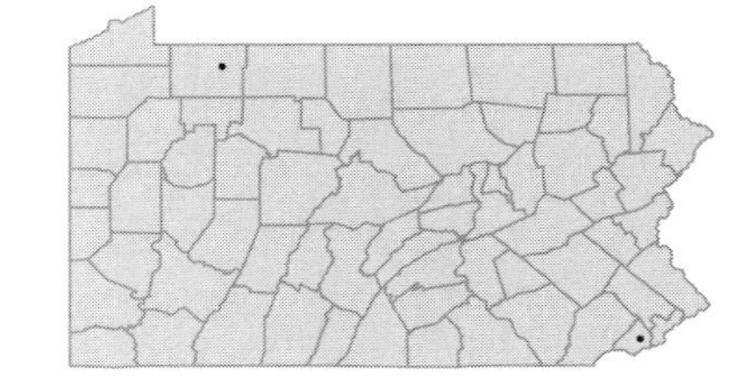

Prunus cerasus L.
Pie cherry
Deciduous tree
Cultivated, and occasionally naturalized in woods edges, thickets or waste ground.

[I]

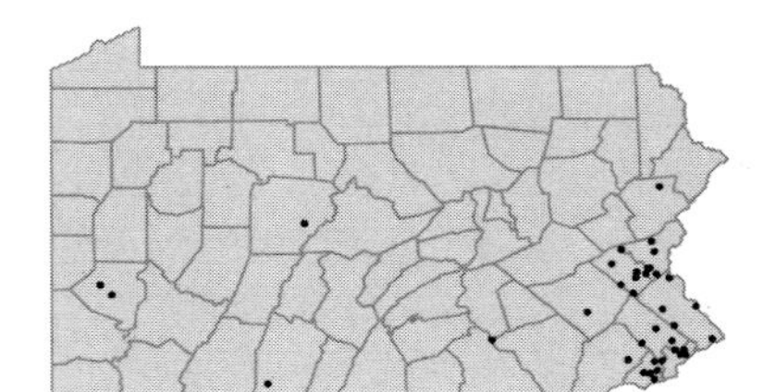

Prunus domestica L. var. ***domestica***
Plum
Deciduous tree
Cultivated and occasionally escaped.

[I]

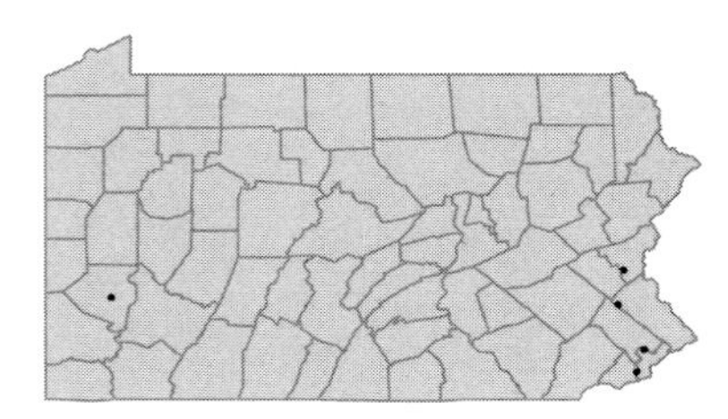

Prunus domestica L. var. ***insititia*** (L.) Fiori & Paol.
Damson plum
Deciduous tree
Waste ground.
Prunus insititia L. FGBW

[I]

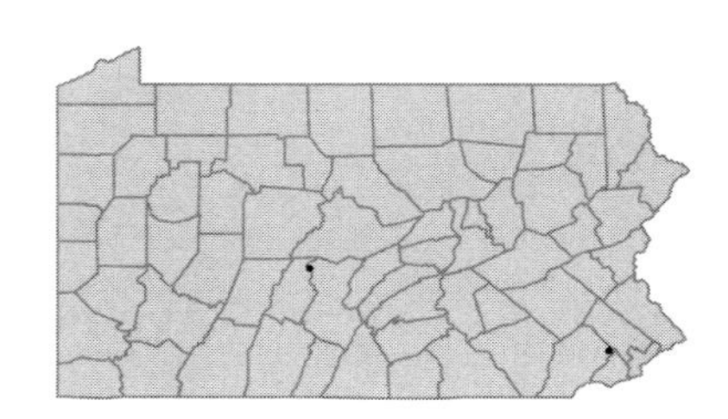

Prunus mahaleb L.
Mahaleb cherry
Deciduous tree
Roadsides, woods and waste ground.

[I]

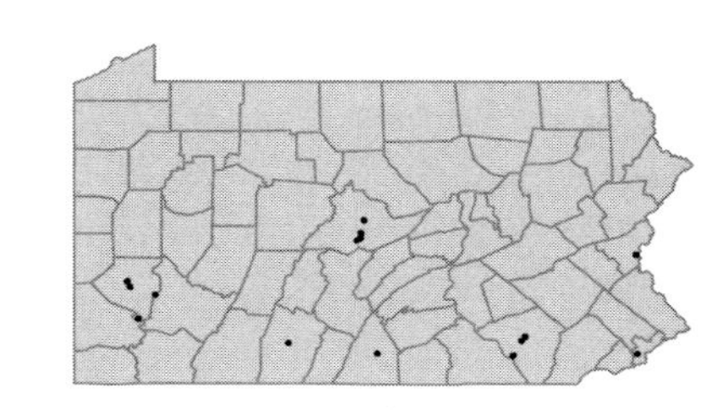

Prunus maritima Marshall
Beach plum
Deciduous shrub
Dry roadside banks and hedgerows.

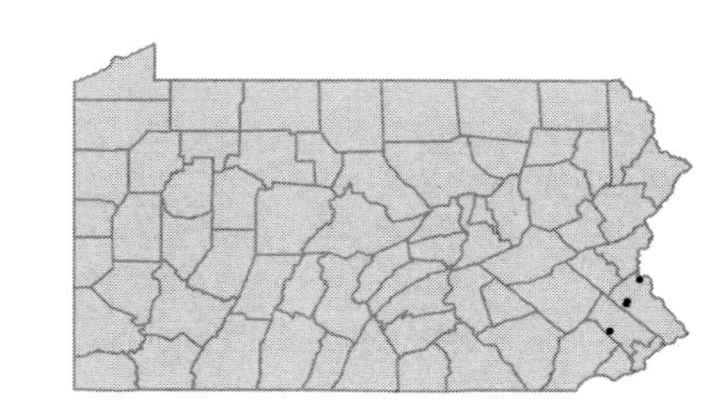

Prunus padus L.
European bird cherry
Deciduous tree
Cultivated and occasionally escaped.

[I]

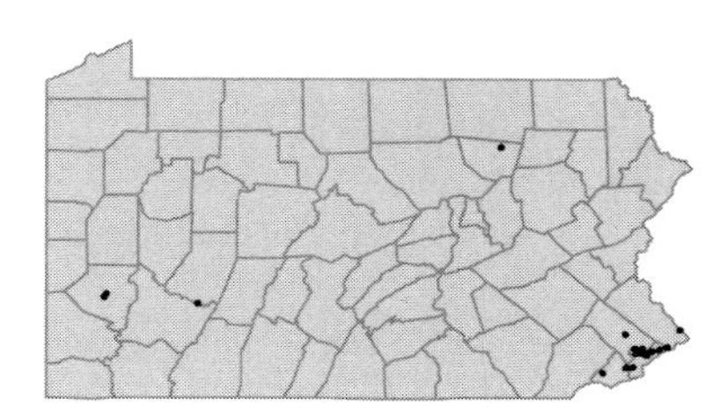

Prunus pensylvanica L.f.
Pin cherry; Fire cherry
Deciduous tree
Dry woods and openings.

FACU-

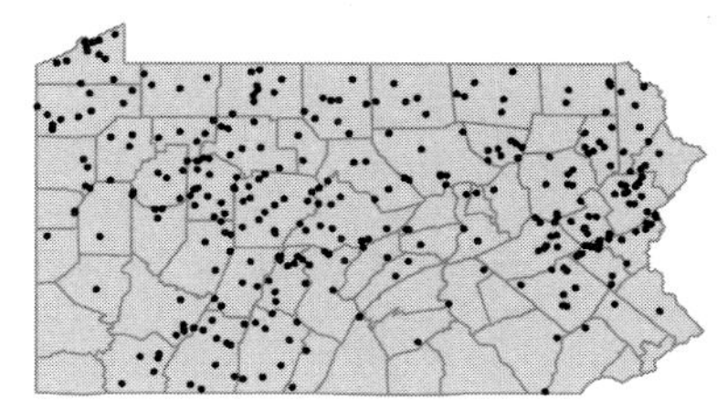

Prunus persica (L.) Batsch
Peach
Deciduous tree
Cultivated and occasionally escaped.

[I]

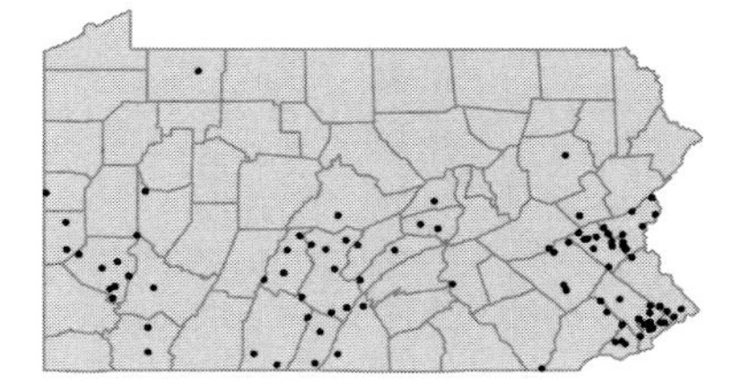

Prunus pumila L. var. ***pumila***
Sand cherry
Deciduous shrub
Dunes and sandy woods.

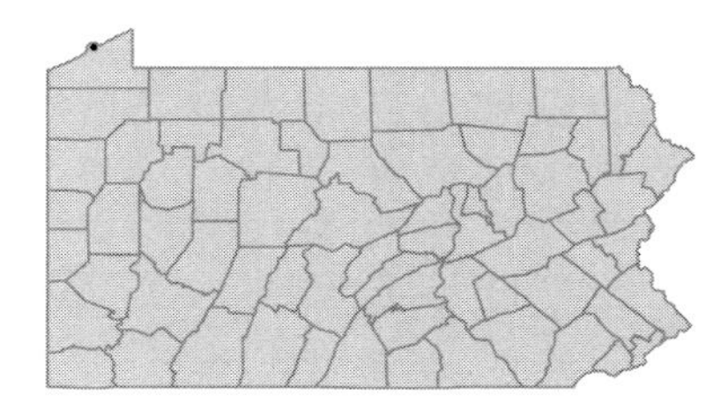

Prunus pumila L. var. ***depressa*** (Pursh) Gleason
Prostrate sand cherry
Deciduous shrub
Alluvial islands and sandy or gravelly shores.
Prunus depressa Pursh FW

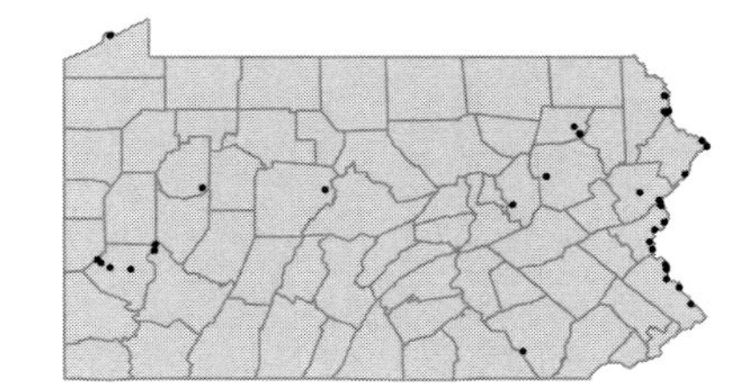

Prunus pumila L. var. ***susquehanae*** (Hortulan ex Willd.) Jaeger
Appalachian sand cherry
Deciduous shrub
Dry, exposed rock outcrops and mountain tops.
Prunus cuneata Raf. P; *Prunus susquehanae* Willd. FW; *Prunus pumila* L. var. *cuneata* (Raf.) Bailey C

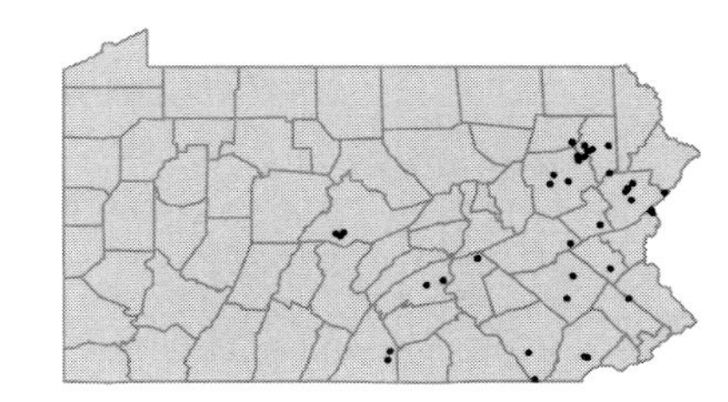

Prunus serotina Ehrh.
Wild black cherry
Deciduous tree
Woods and fencerows.

FACU

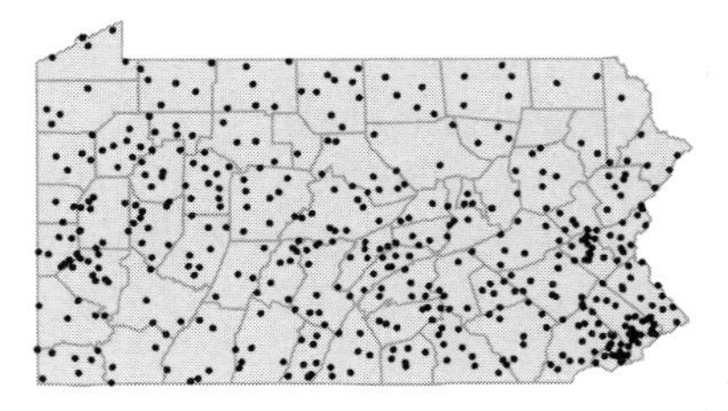

Prunus triloba Lindl.
Flowering almond
Deciduous shrub
Cultivated and occasionally escaped. Represented by a single collection from Lehigh Co. in 1959.

I

Prunus virginiana L.
Choke cherry
Deciduous shrub
Rocky, upland woods.

FACU

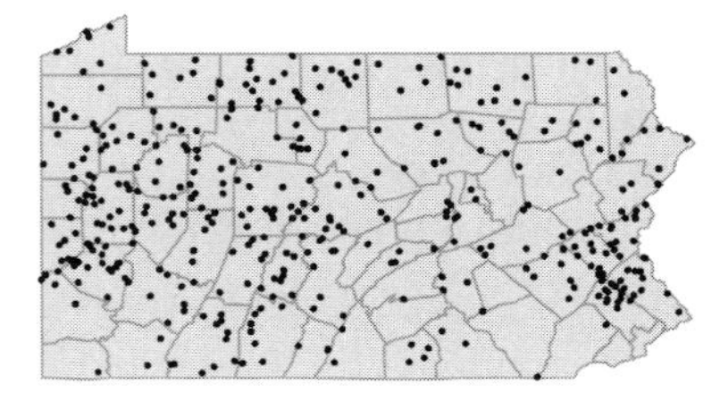

• ***Pyracantha coccinea*** M.Roemer
Firethorn
Evergreen shrub
Cultivated and occasionally escaped.
Cotoneaster pyracantha (L.) Spach PF

I

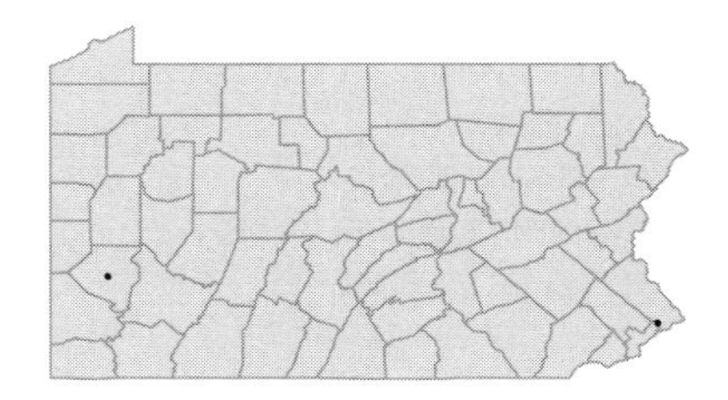

• ***Pyrus communis*** L.
Pear
Deciduous tree
Cultivated and occasionally persisting at abandoned home sites.

I

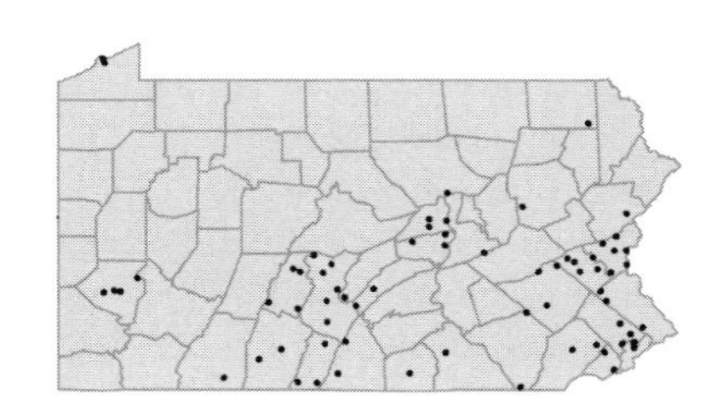

• ***Rhodotypos scandens*** (Thunb.) Makino
Jetbead
Deciduous shrub
Cultivated and occasionally escaped.

I

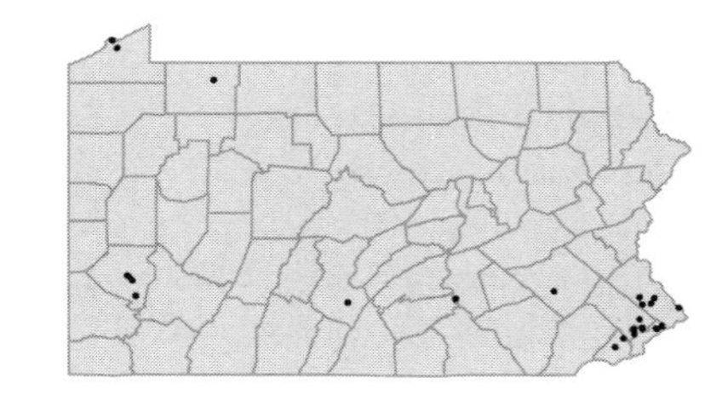

• ***Rosa blanda*** Ait. [I]
Meadow rose
Deciduous shrub
Fencerows and hedges.
Rosa subblanda Rydb. W

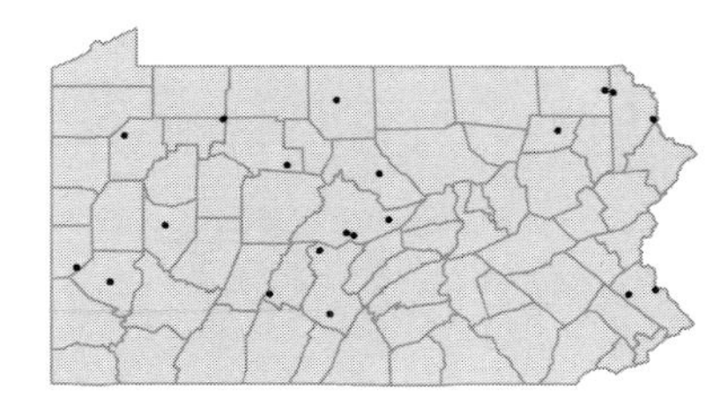

Rosa canina L. [I]
Dog rose
Deciduous shrub
Cultivated and occasionally escaped to roadsides and old fields.

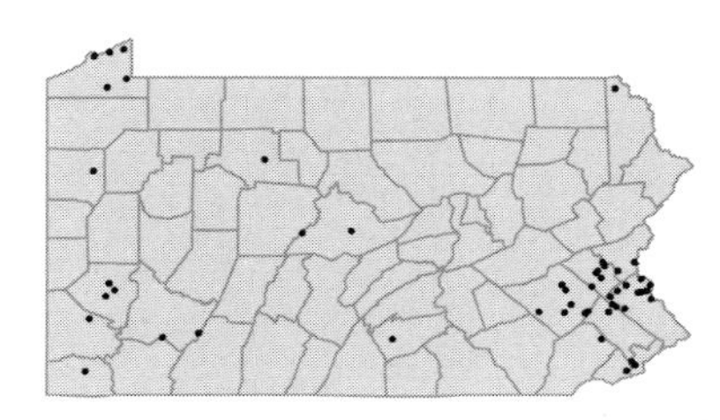

Rosa carolina L. var. ***carolina*** UPL
Pasture rose
Deciduous shrub
Fields, rocky banks, shale barrens and other dry, open ground.

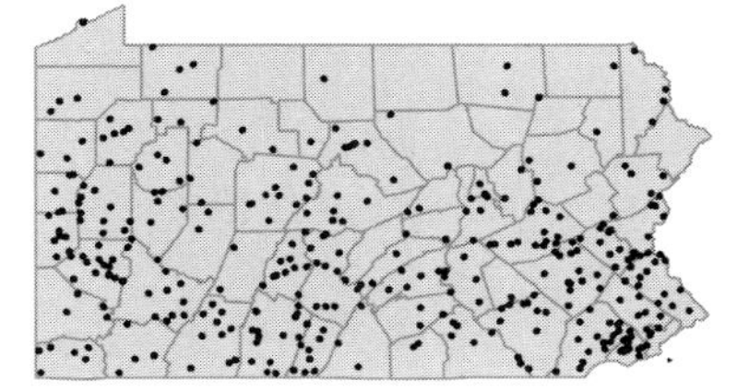

Rosa carolina L. var. ***grandiflora*** (Baker) Rehd.
Rose
Deciduous shrub
Stream banks and rocky slopes.
Rosa obovata Raf.

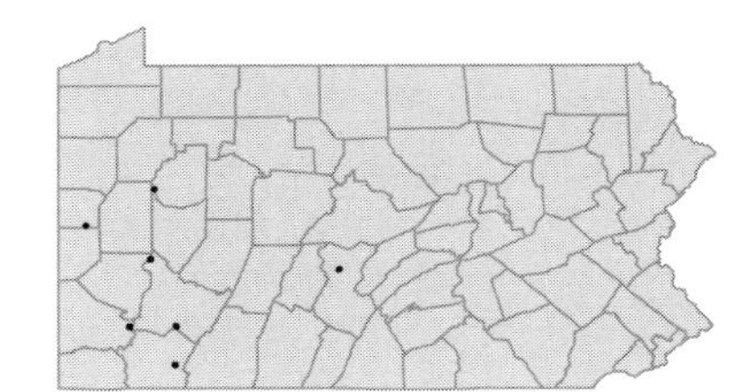

Rosa centifolia L. [I]
Cabbage rose
Deciduous shrub
Cultivated and occasionally escaped.

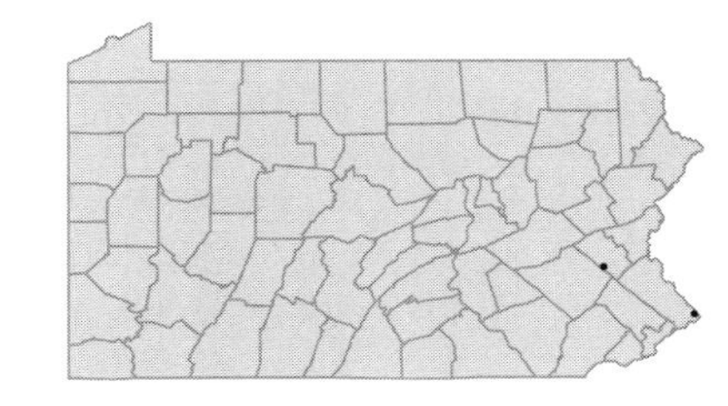

Rosa cinnamomea L. [I]
Cinnamon rose
Deciduous shrub
Cultivated and occasionally escaped to dry, open fields.
Rosa majalis Herrm. C

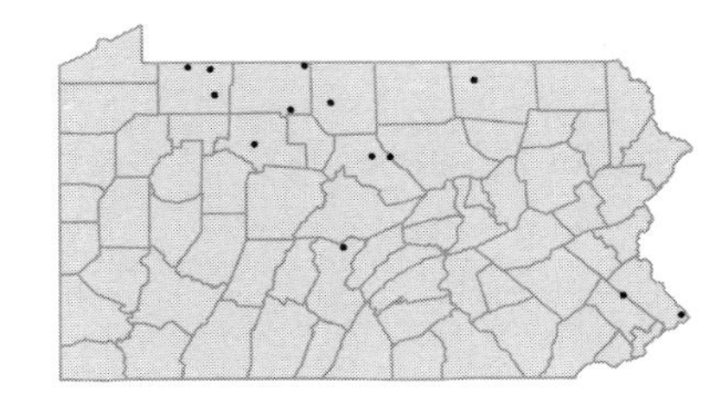

Rosa eglanteria L.
Eglantine; Sweetbrier
Deciduous shrub
Dry fields, shaly hillsides and roadsides.
Rosa rubiginosa L. PFW

[I]

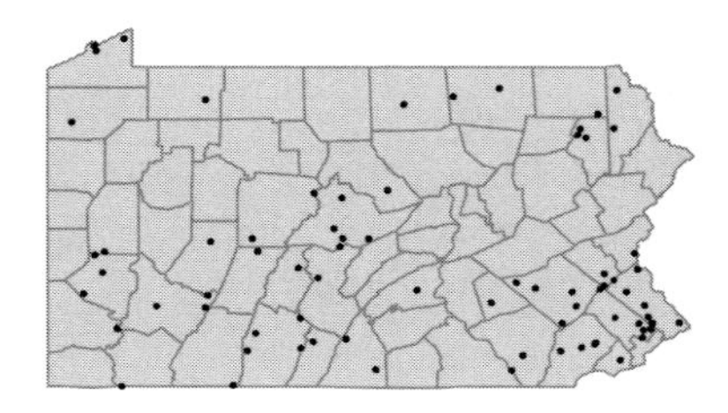

Rosa gallica L.
French rose
Deciduous shrub
Cultivated and occasionally persisting on old homesites or roadsides.

[I]

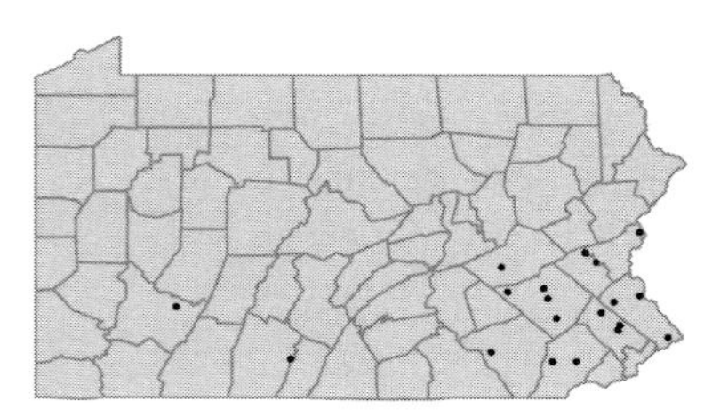

Rosa micrantha Borer ex Sm.
Eglantine; Sweetbrier
Deciduous shrub
Roadsides and old fields.

FACU
[I]

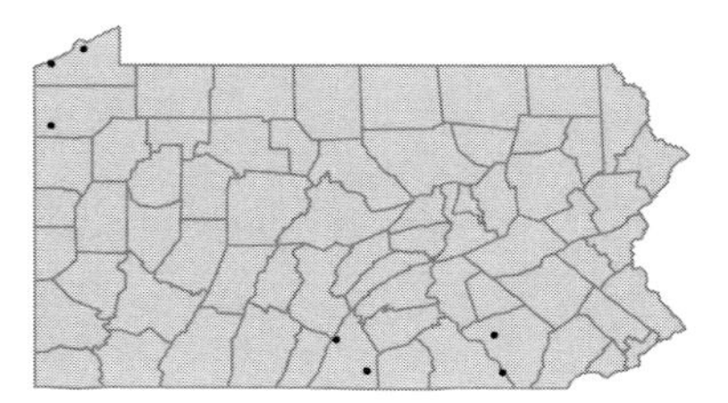

Rosa multiflora Thunb. ex Murr.
Multiflora rose
Deciduous shrub
Disturbed woods, pastures, old fields, roadsides and thickets. Designated as a noxious weed in PA.

FACU
[I]

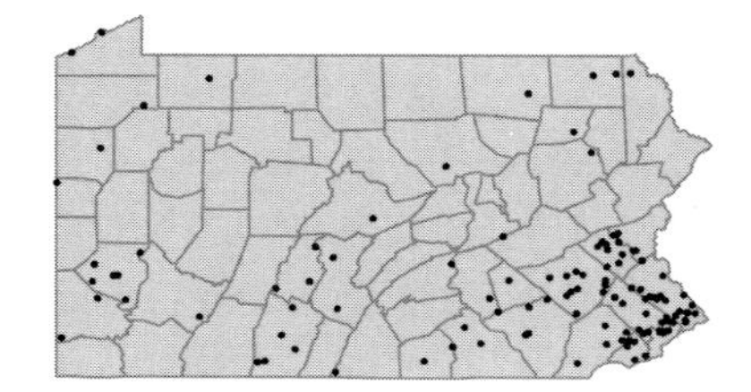

Rosa palustris Marshall
Swamp rose
Deciduous shrub
Swamps and marshes.

OBL

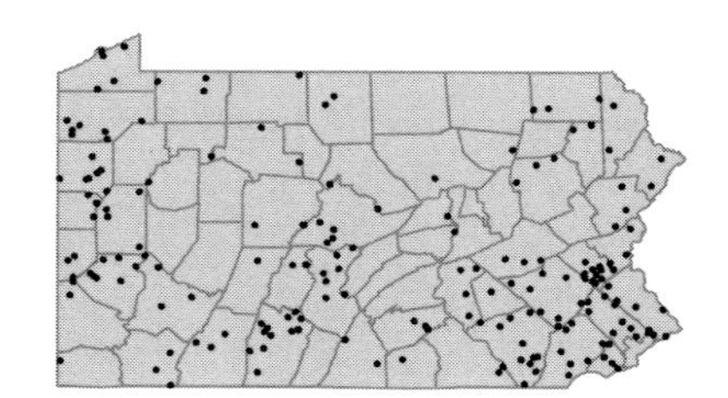

Rosa rugosa Thunb.
Rugosa rose
Deciduous shrub
Cultivated and occasionally escaped.

FACU-
[I]

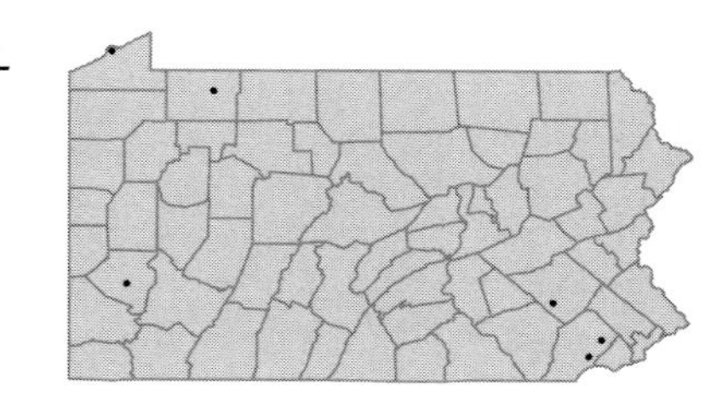

Rosa setigera Michx.
Prairie rose
Deciduous shrub
Sandy old fields, open thickets, roadsides and fencerows.

FACU
[I]

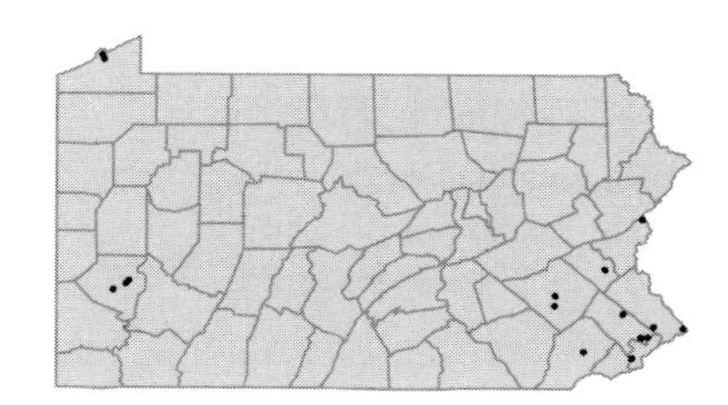

Rosa virginiana P.Mill.
Wild rose; Pasture rose
Deciduous shrub
Pastures, fields, open woods, thickets and roadsides.
Rosa lucida Ehrh. P; *Rosa humilus* Marshall P

FAC

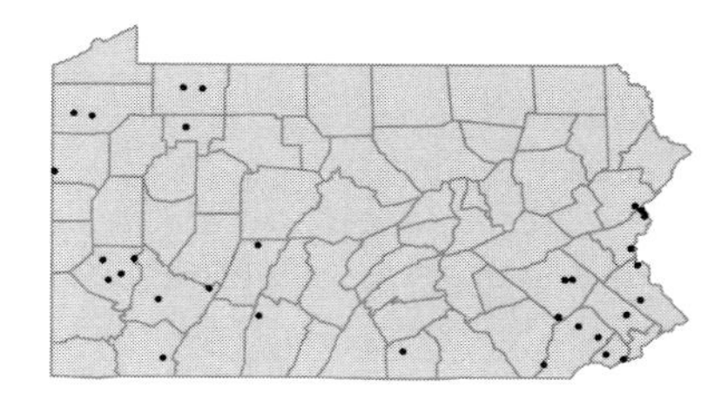

Rosa wichuraiana Crepin
Memorial rose
Deciduous shrub
Fields, floodplains and waste ground.

[I]

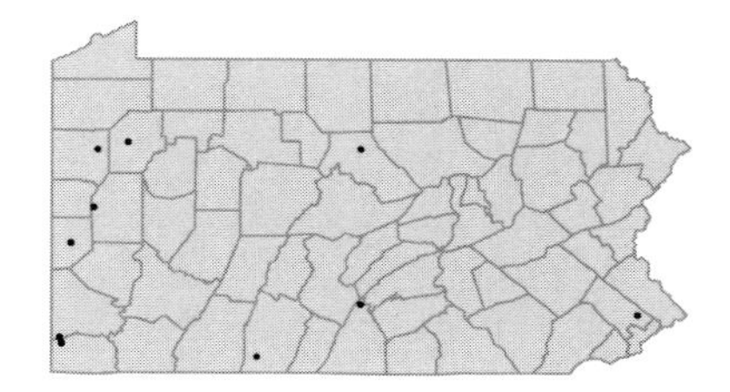

Rubus—The work of Davis et al. on *Rubus* (1968-1970) needs to be updated. Meanwhile, we have chosen to recognize broad species complexes following Gleason and Cronquist (1991). The synonyms listed represent an attempt to account for the species mapped in Wherry et al. (1979).

• ***Rubus allegheniensis*** Porter
Common blackberry; Sow-teat blackberry
Deciduous shrub
Old fields, open woods and clearings.
Rubus nigrobaccus Bailey *pro parte* P; *Rubus pugnax* Bailey *pro parte* FWK

FACU

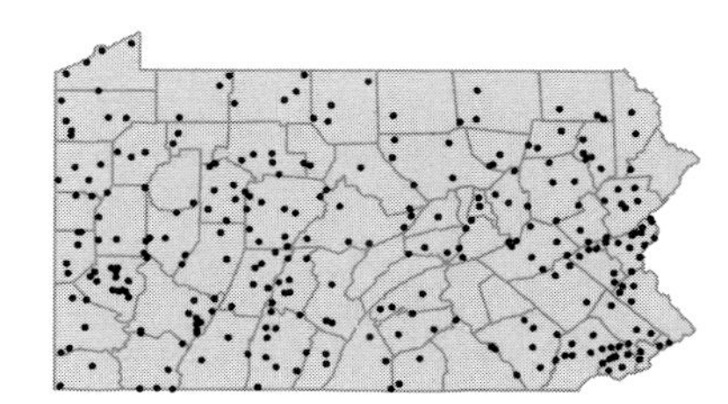

Rubus bifrons Vest ex Tratt.
Blackberry
Deciduous shrub
Dry, open ground.

[I]

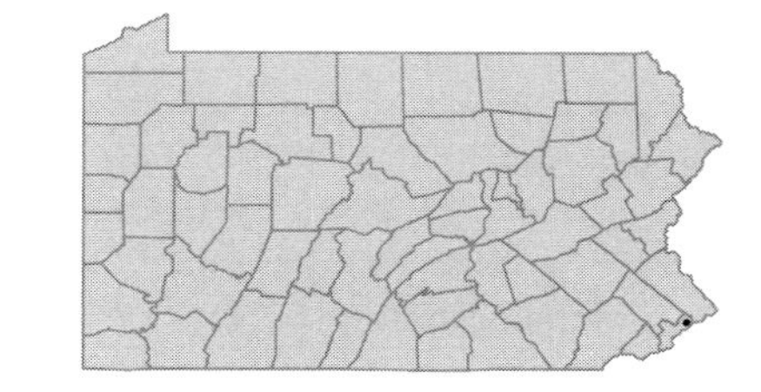

Rubus canadensis L.
Smooth blackberry
Deciduous shrub
Cool, moist woods, rocky slopes and thickets.
Rubus pauperimus Bailey *pro parte* W; *Rubus randii* (Bailey) Rydb. *pro parte* P

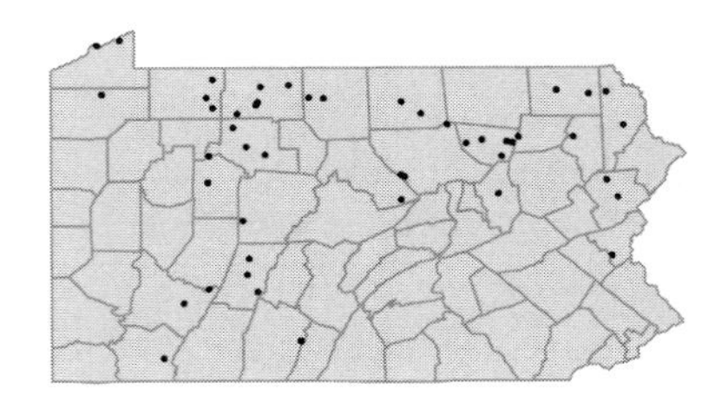

Rubus cuneifolius Pursh
Sand blackberry
Deciduous shrub
Dry, open thickets and roadsides, in sandy soil.
Rubus cacaponensis Davis & Davis *pro parte* WK

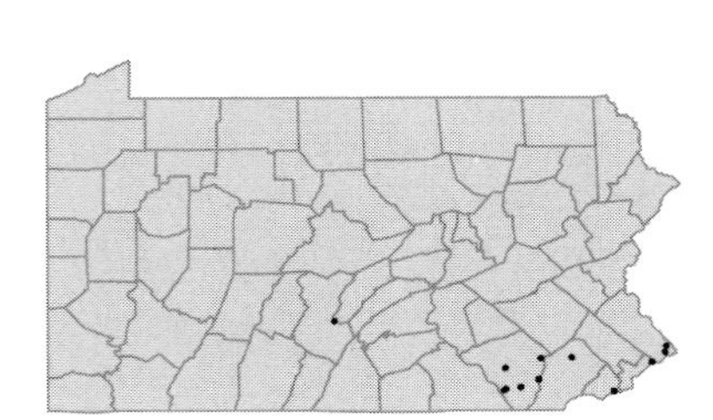

Rubus discolor Weihe & Nees
Himalaya-berry
Deciduous shrub
Cultivated and spreading to roadside thickets and waste ground.
Rubus procerus P.J.Muell. FW

I

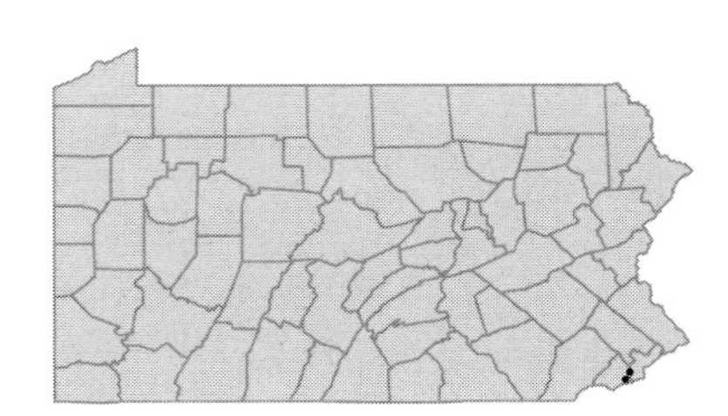

Rubus enslenii Tratt. *sensu lato*
Southern dewberry
Woody vine
Sandy banks.
Rubus celer Bailey *pro parte* FWK; *Rubus centralis* Bailey *pro parte* FK; *Rubus injunctus* Bailey *pro parte* FWK; *Rubus michiganensis* (Card ex Bailey) Bailey *pro parte* FWK; *Rubus multifer* Bailey *pro parte* WK; *Rubus nefrens* Bailey *pro parte* FK; *Rubus vixalacer* Bailey *pro parte* WK

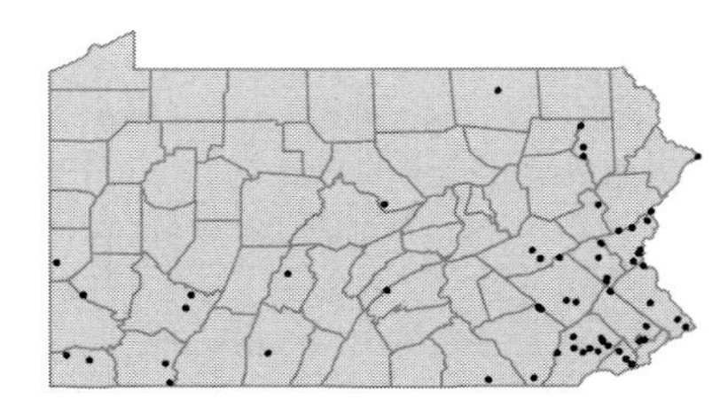

Rubus flagellaris Willd. *sensu lato*
Prickly dewberry; Northern dewberry
Woody vine
Rocky or shaly slopes, cliffs and fields.
Rubus baileyanus Britt. *pro parte* PFWK; *Rubus exsularis* Bailey *pro parte* WK; *Rubus fecundus* Bailey *pro parte* WK; *Rubus invisus* (Bailey) Britt. *pro parte* FWK; *Rubus meracus* Bailey *pro parte* FWK; *Rubus roribaccus* (Bailey) Rydb. *pro parte* FWK; *Rubus sailorii* Bailey *pro parte* WK; *Rubus uvidus* Bailey *pro parte* WK

UPL

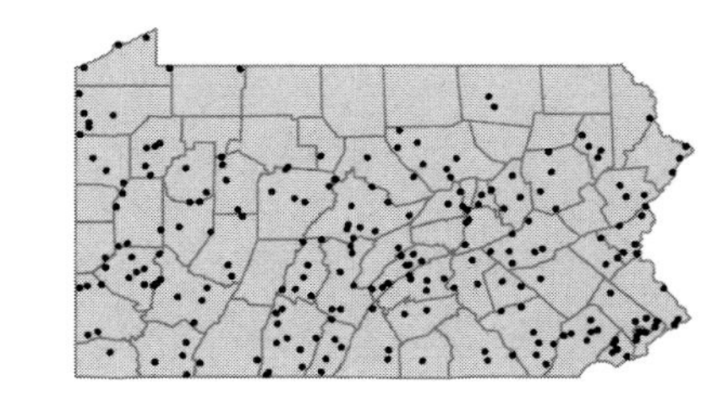

Rubus hispidus L. *sensu lato* FACW
Swamp dewberry
Woody vine
Bogs, swamps, moist woods, thickets and barrens.
Rubus adjacens Fern. *pro parte* FWK; *Rubus ithacanus* Bailey *pro parte* WK; *Rubus montensis* Bailey *pro parte* WK; *Rubus paganus* Bailey *pro parte* FWK; *Rubus permixtus* Blanch. *pro parte* FWK; *Rubus porteri* Bailey *pro parte* FWK; *Rubus vagulus* Bailey *pro parte* FW

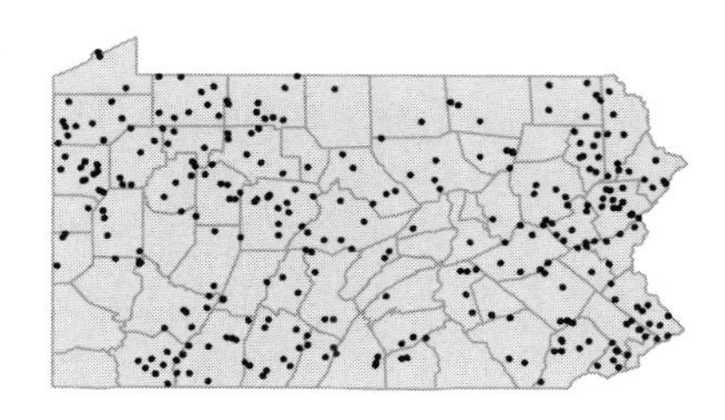

Rubus idaeus L. var. ***strigosus*** (Michx.) Maxim. FAC-
Red raspberry
Deciduous shrub
Rocky woods, clearings and thickets.
Rubus strigosus Michx. PGWK

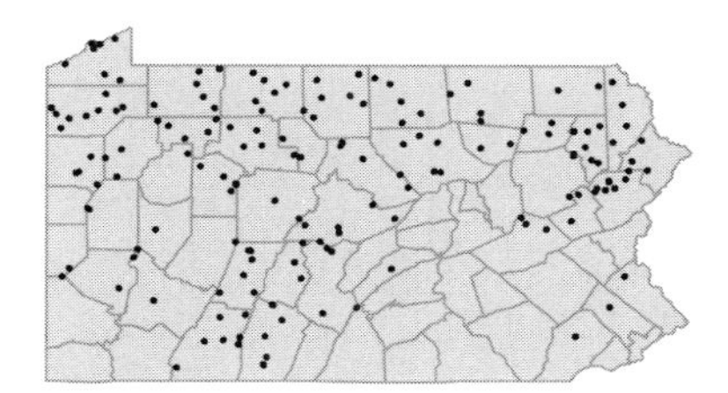

Rubus illecebrosus Focke [I]
Strawberry raspberry
Deciduous shrub
Thin woods, roadsides and fields.

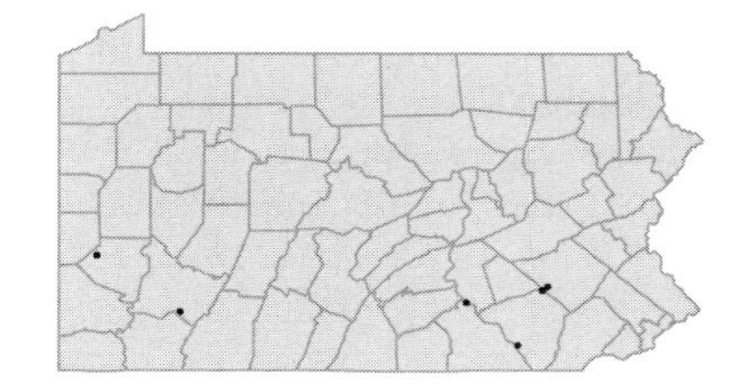

Rubus laciniatus Willd. [I]
Cut-leaved blackberry
Deciduous shrub
Spreading from cultivation to roadsides, fields, sandy woods and waste ground.

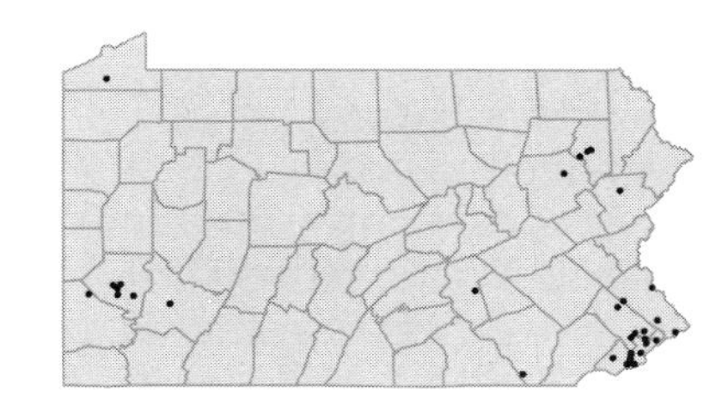

Rubus occidentalis L.
Black-cap; Black raspberry
Deciduous shrub
Sandy or rocky woods, wooded slopes and thickets.

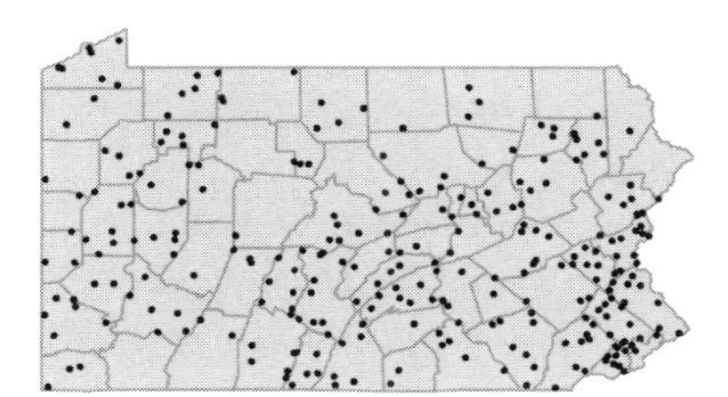

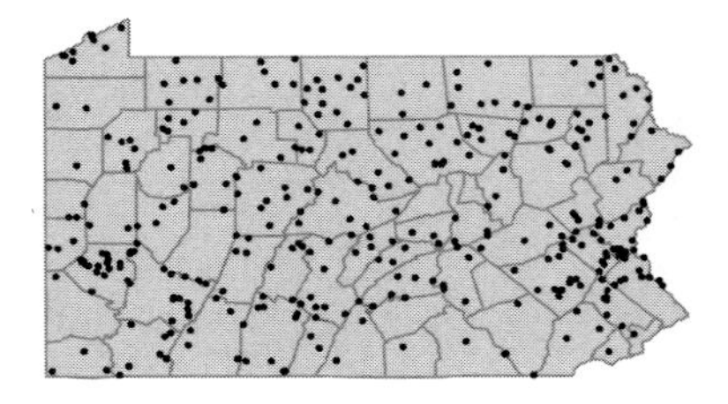

Rubus odoratus L.
Purple-flowering raspberry; Thimbleberry
Deciduous shrub
Cliffs, ledges and rocky, wooded slopes.

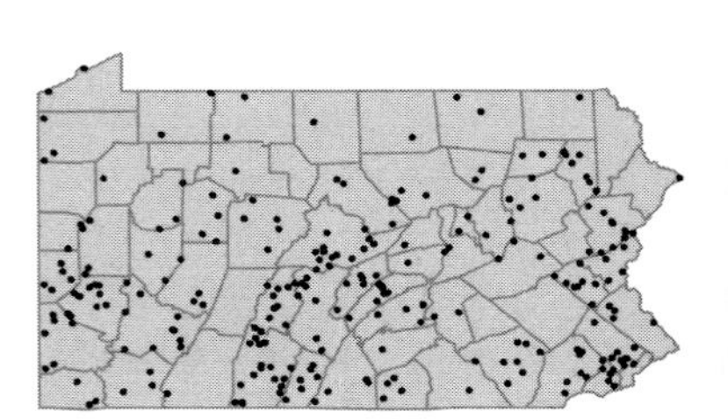

Rubus pensilvanicus Poir. *sensu lato*
Blackberry
Deciduous shrub
Thickets, rocky banks, woods, fields and waste ground.
Rubus alumnus Bailey *pro parte* FWK; *Rubus andrewsianus* Blanch. *pro parte* WK; *Rubus bellobatus* Bailey *pro parte* FWK; *Rubus frondosus* Bigel. *pro parte* PFWK; *Rubus laudatus* Berger *pro parte* WK; *Rubus orarius* Blanch. *pro parte* GK; *Rubus pergratus* Blanch. *pro parte* WK; *Rubus recurvans* Blanch. *pro parte* FWK; *Rubus rosa* Bailey *pro parte* WK

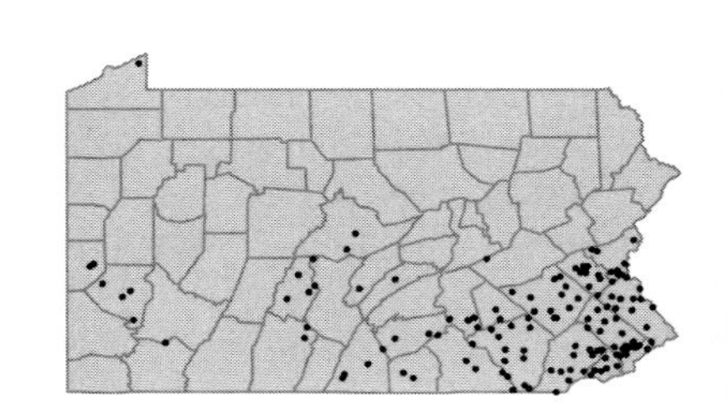

Rubus phoenicolasius Maxim. I
Wineberry
Deciduous shrub
Roadsides, banks and thickets.

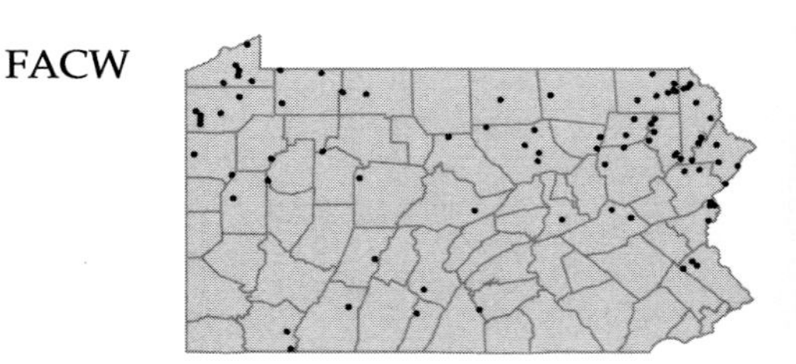

Rubus pubescens Raf. FACW
Dwarf blackberry
Deciduous shrub
Boggy or swampy woods and moist slopes.
Rubus americanus (Pers.) Britt. P

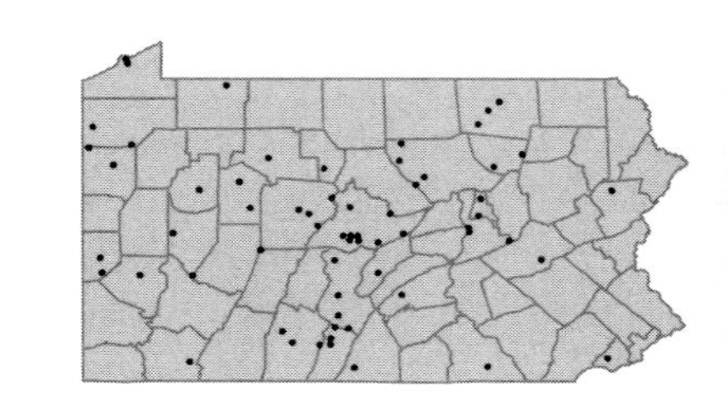

Rubus recurvicaulis Blanch.
Dewberry
Woody vine
Dry, rocky or sandy soil.
Rubus arundelanus Blanch. *pro parte* FGB; *Rubus noveboracus* Bailey *pro parte* FK; *Rubus plicatifolius* Blanch. *pro parte* WK

Rubus setosus Bigel. *sensu lato*
Blackberry
Deciduous shrub
Damp thickets and swamps.
Rubus adirondackensis Bailey *pro parte* W; *Rubus benneri* Bailey *pro parte* FBW; *Rubus elegantulus* Blanch. *pro parte* FK; *Rubus nigricans* Rydb. *pro parte* P; *Rubus notatus* Bailey *pro parte* WK; *Rubus racemiger* Bailey *pro parte* FWK; *Rubus semisetosus* Blanch. *pro parte* FWK; *Rubus wheeleri* (Bailey) Bailey *pro parte* FWK

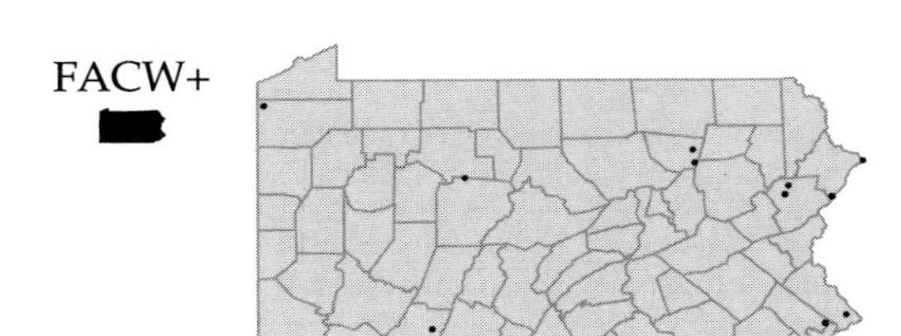

• ***Sanguisorba canadensis*** L.
American burnet
Herbaceous perennial
Moist meadows, swamps, bogs and floodplains.

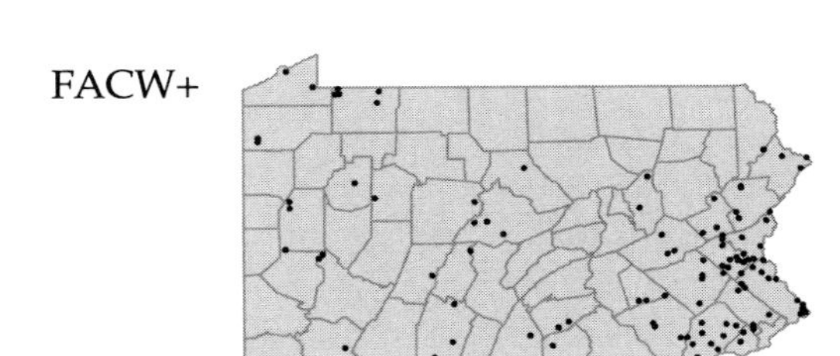

Sanguisorba minor Scop.
Salad burnet
Herbaceous perennial
Cultivated and occasionally escaped to open, rocky ground, roadsides or ballast.
Sanguisorba sanguisorba (L.) Britt. P

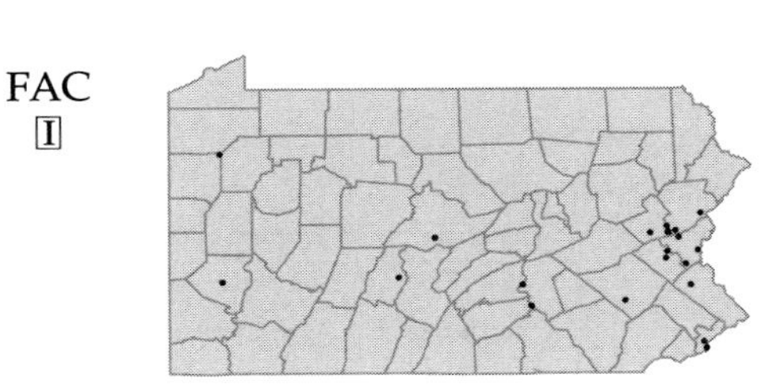

• ***Sorbaria sorbifolia*** (L.) A.Braun
False spiraea
Deciduous shrub
Cultivated and occasionally escaped to roadsides, stream banks or moist thickets.

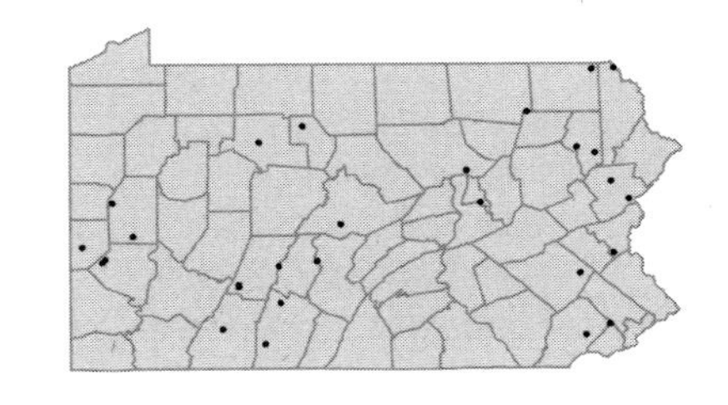

• ***Sorbus americana*** Marshall
American mountain-ash
Deciduous tree
Rocky slopes, bogs and swamps.
Pyrus americana (Marshall) DC. F

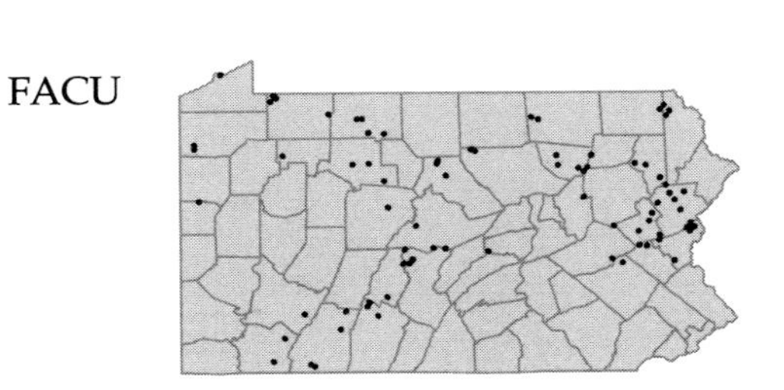

Sorbus aucuparia L.
European mountain-ash; Rowan
Deciduous tree
Cultivated and occasionally escaped.
Pyrus aucuparia (L.) Gaertn. F

I

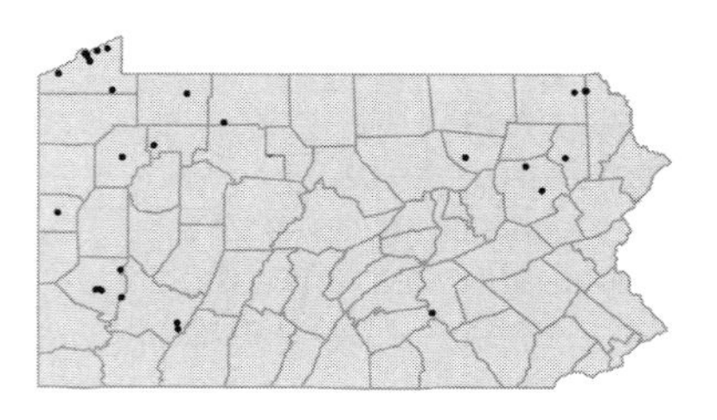

Sorbus decora (Sarg.) Schneid.
Showy mountain-ash
Deciduous tree
Rocky slopes.
Pyrus decora (Sarg.) Hyland F

FAC

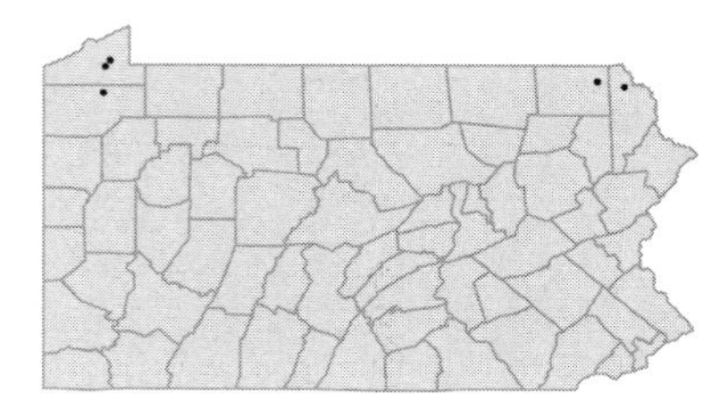

• **Spiraea alba** DuRoi
Meadow-sweet
Deciduous shrub
Bogs and moist, peaty meadows.

FACW+

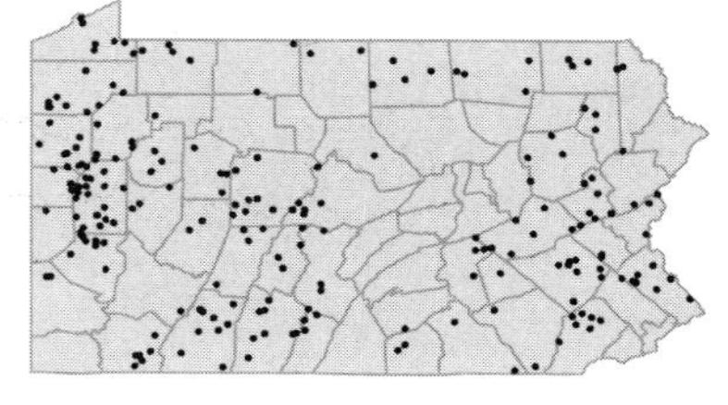

Spiraea alba x latifolia
Meadow-sweet
Deciduous shrub
Swamps, marshes and rocky or boggy shores.

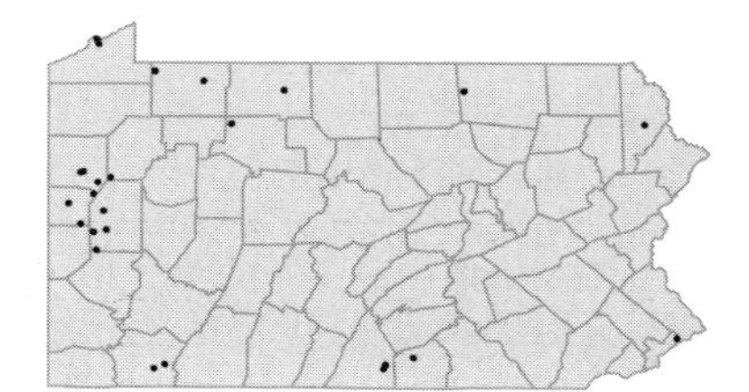

Spiraea albiflora x japonica
Spiraea
Deciduous shrub
Cultivated and rarely escaped to dry roadsides. Represented by a single collection from Berks Co.
Spiraea x bumalda Burven WK

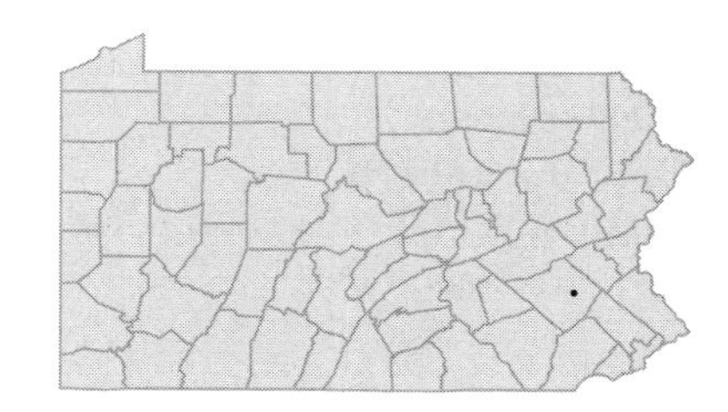

Spiraea betulifolia Pallas var. **corymbosa** (Raf.) Maxim.
Dwarf spiraea
Deciduous shrub
Rocky, wooded slopes.
Spiraea corymbosa Raf. FW

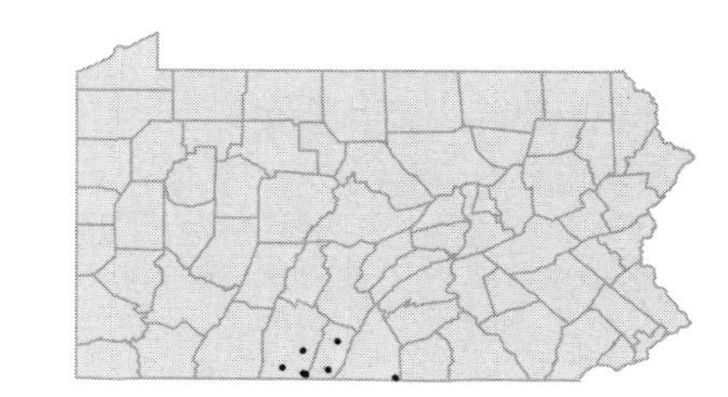

Spiraea douglasii x salicifolia
Spiraea
Deciduous shrub
Cultivated and occasionally escaped to roadsides or stream banks.
Spiraea x billiardii Herincq WK

[I]

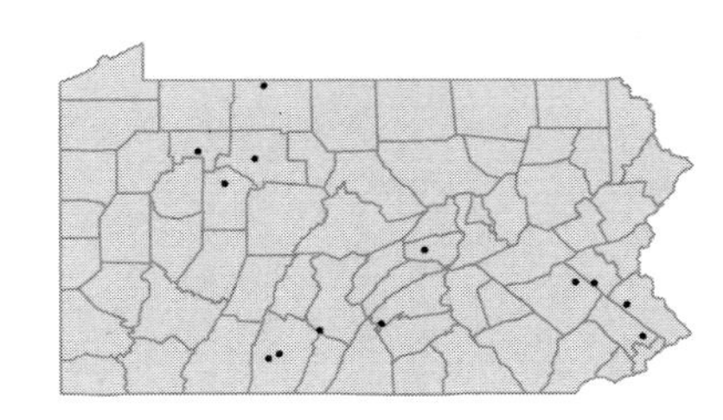

Spiraea japonica L.f.
Japanese spiraea
Deciduous shrub
Cultivated and occasionally escaped.

FACU-
[I]

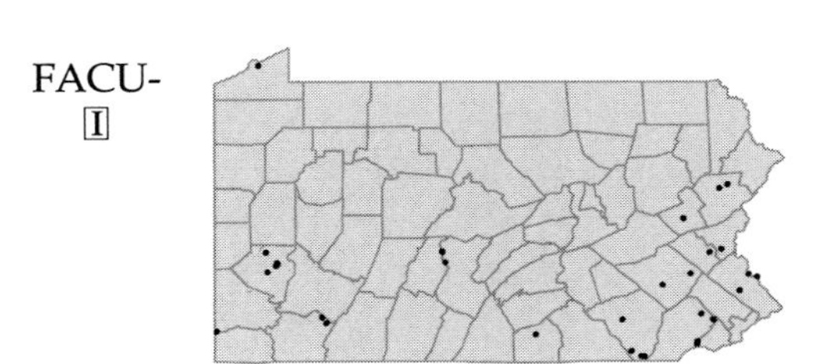

Spiraea latifolia (Ait.) Borkh.
Meadow-sweet
Deciduous shrub
Bogs, moist woods, peaty barrens and swamps.
Spiraea salicifolia L. var. *latifolia* (Ait.) Wieg. P; *Spiraea alba* Duroi var. *latifolia* (Ait.) Dippel. C

FACU+

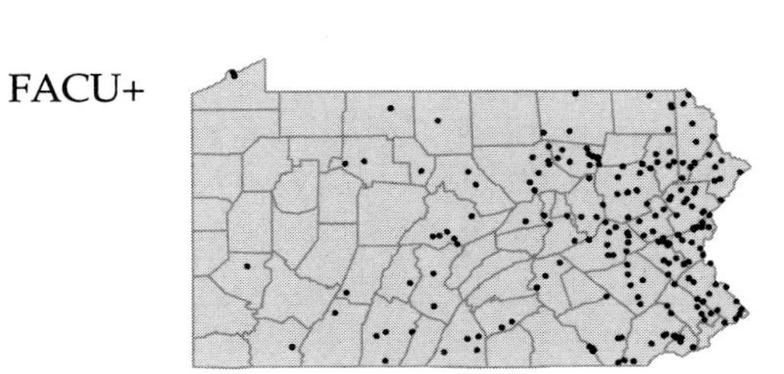

Spiraea prunifolia Sieb. & Zucc.
Bridal-wreath spiraea
Deciduous shrub
Cultivated and occasionally persisting on abandoned homesteads or roadsides.

[I]

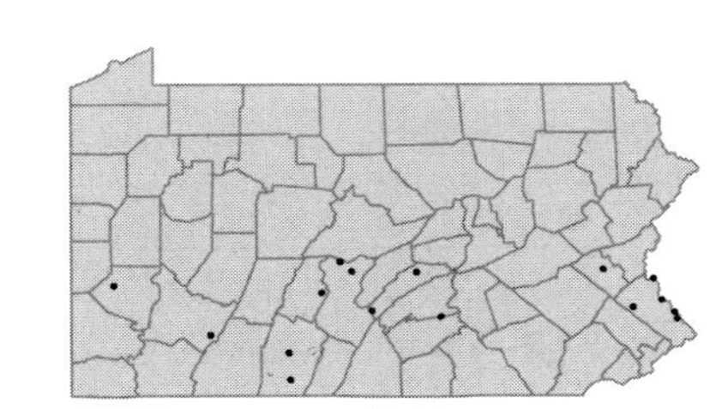

Spiraea thunbergii Sieb. ex Blume
Spiraea
Deciduous shrub
Cultivated and rarely escaped. Represented by a single collection from Bedford Co.

[I]

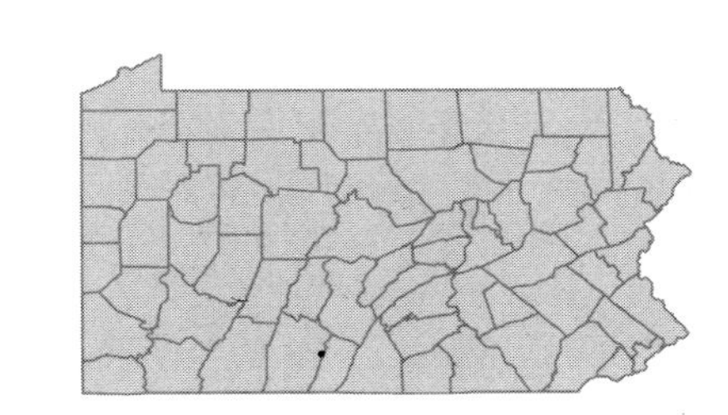

Spiraea tomentosa L.
Hardhack; Steeple-bush
Deciduous shrub
Wet meadows, moist old fields, bogs and swamps.

FACW

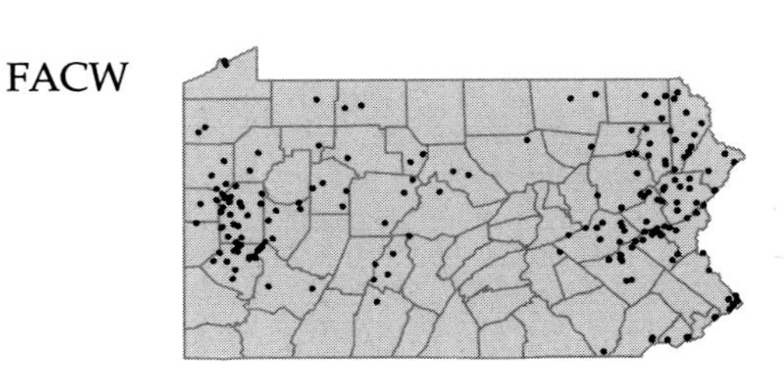

Spiraea virginiana Britt.
Appalachian spiraea; Virginia spiraea
Deciduous shrub
Rocky slopes and stream banks. Believed to be extirpated, last collected in 1901.

FACU

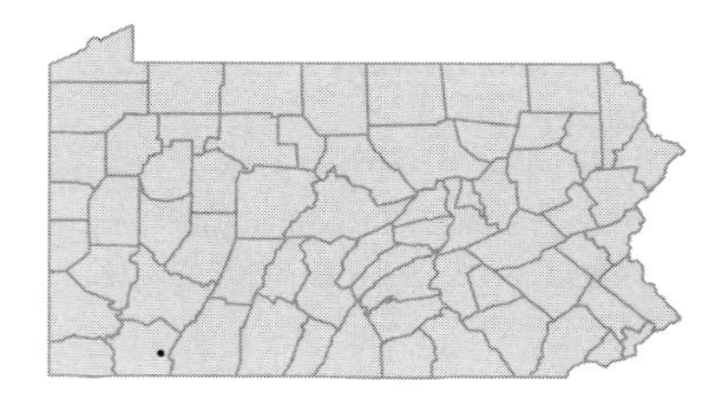

• ***Waldsteinia fragarioides*** (Michx.) Tratt.
Barren-strawberry
Herbaceous perennial
Moist, rich woods and pastures.

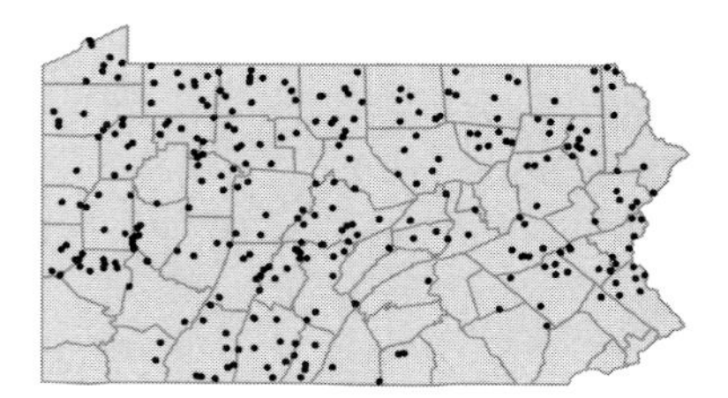

MIMOSACEAE

• ***Albizia julibrissin*** Durazz.
Mimosa; Silk tree
Deciduous tree
Cultivated and occasionally escaped to roadsides and woods edges.

[I]
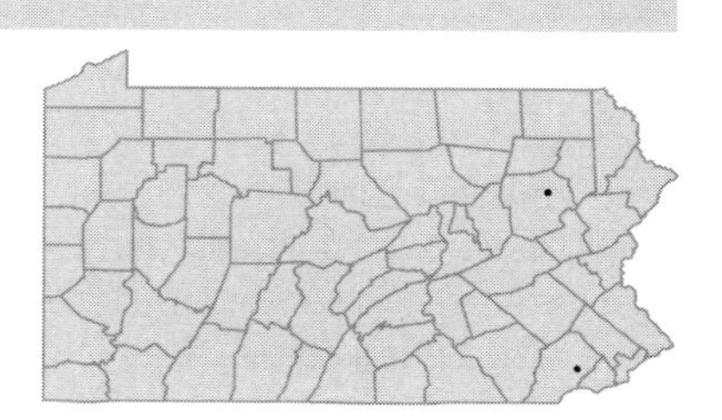

• ***Schrankia nuttallii*** (DC. ex Britt. & Rose) Standl.
Cat-claw; Sensitive-brier
Herbaceous perennial vine
Open field near railroad. Represented by a single collection from Dauphin Co. in 1952.
Mimosa quadrivalis L. C

[I]
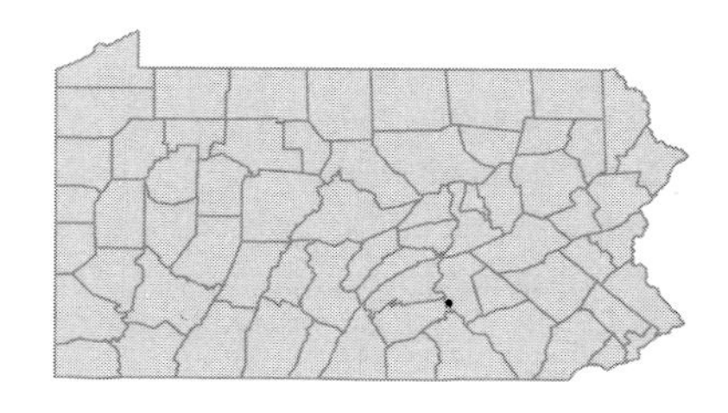

FABACEAE

• ***Aeschynomene virginica*** (L.) BSP
Sensitive joint-vetch
Herbaceous annual, emergent aquatic
Freshwater intertidal marshes. Believed to be extirpated, last collected in 1865.

OBL

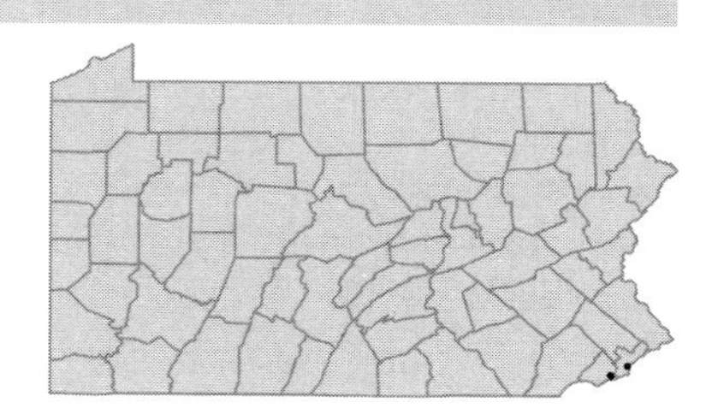

• ***Amorpha fruticosa*** L.
False-indigo
Deciduous shrub
Alluvial soils along streams and rivers and other low, moist areas.

FACW
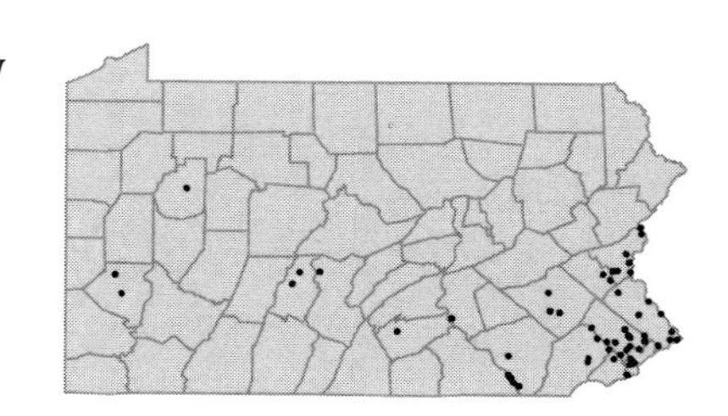

• ***Amphicarpaea bracteata*** (L.) Fern.
Hog-peanut
Herbaceous perennial vine
Moist woods and alluvium.
Amphicarpa comosa (L.) G.Don W; *Falcata comosa* (L.) Kuntze P

FAC

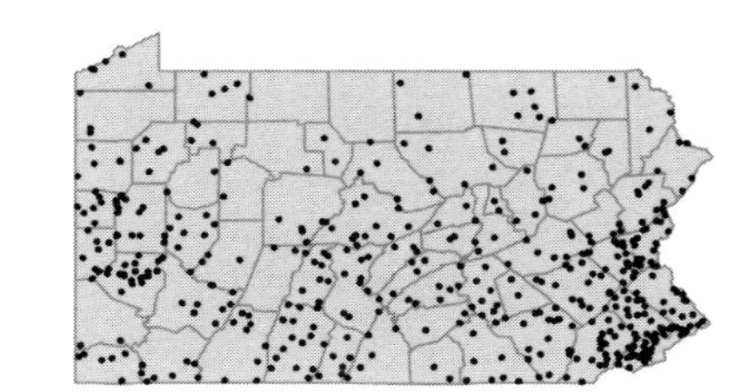

• ***Anthyllis vulneraria*** L.
Lady's-fingers
Herbaceous perennial
Old fields, waste ground and ballast. Represented by two collections from Philadelphia Co. in 1878 and one from Lancaster Co. in 1906.

[I]

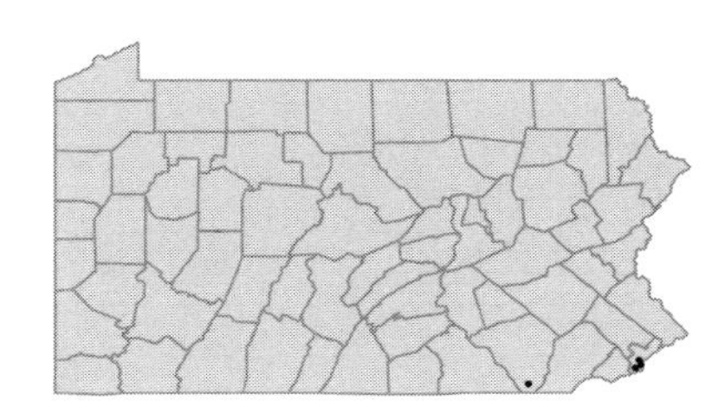

• ***Apios americana*** Medic.
Ground-nut; Wild bean
Herbaceous perennial vine
Low, rich, moist ground and thickets.
Apios apios (L.) MacM. P

FACW

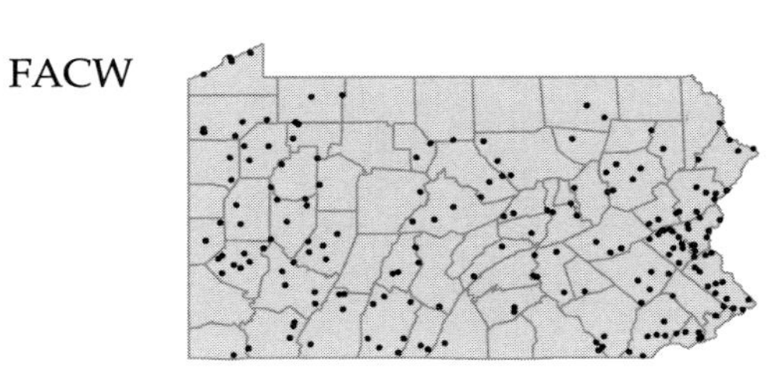

• ***Arachis hypogaea*** L.
Peanut
Herbaceous annual
Cultivated and occasionally escaped.

[I]

• ***Astragalus canadensis*** L.
Milk-vetch
Herbaceous perennial
Rocky roadside banks, limestone ledges and shale barrens.
Astragalus carolinianus L. P

FAC

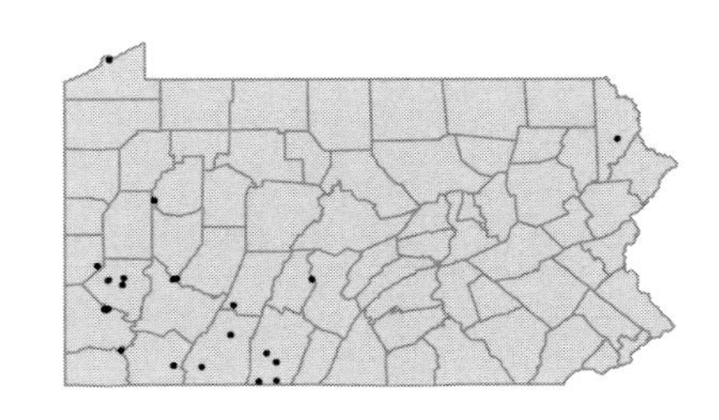

Astragalus neglectus (Torr. & A.Gray) Sheldon
Cooper's milk-vetch
Herbaceous perennial
Gravelly thickets and roadsides. Believed to be extirpated, last collected in 1938.
Astragalus cooperi A.Gray B

FACU

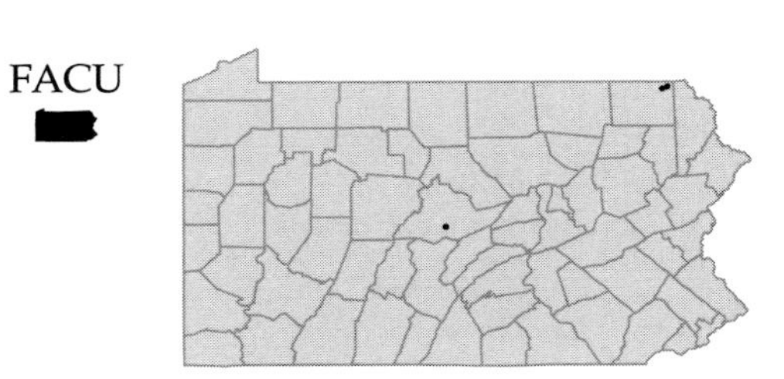

• ***Baptisia australis*** (L.) R.Br.
Blue false-indigo
Herbaceous perennial
Open woods, river banks and sandy floodplains. Native in western counties, introduced elsewhere.

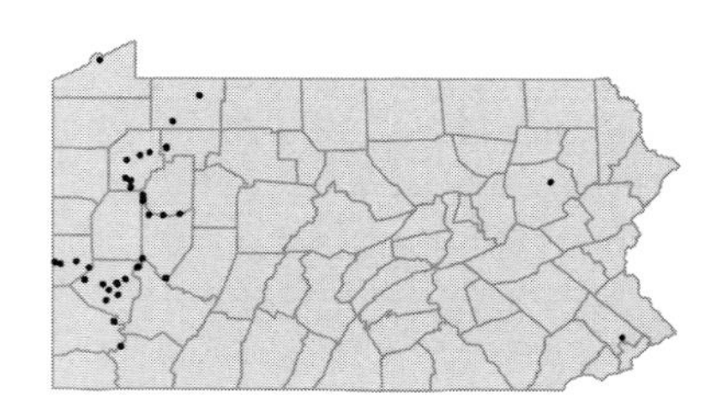

Baptisia tinctoria (L.) Vent.
Wild indigo
Herbaceous perennial
Dry, open woods and clearings, in sandy, acidic soils.

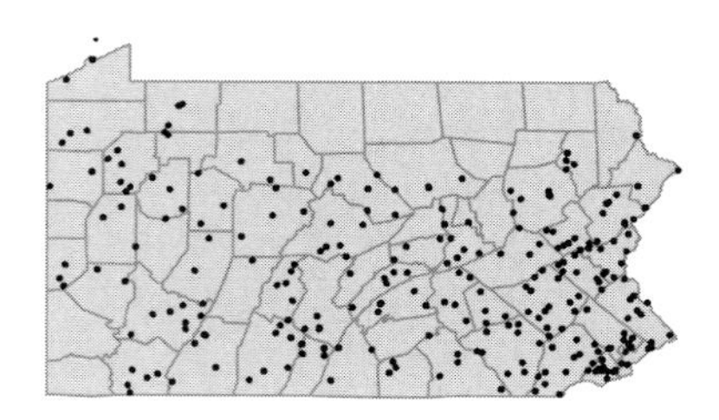

• ***Clitoria mariana*** L.
Butterfly pea
Herbaceous perennial vine
Abandoned railroads and other dry, open areas, on sandy soil.

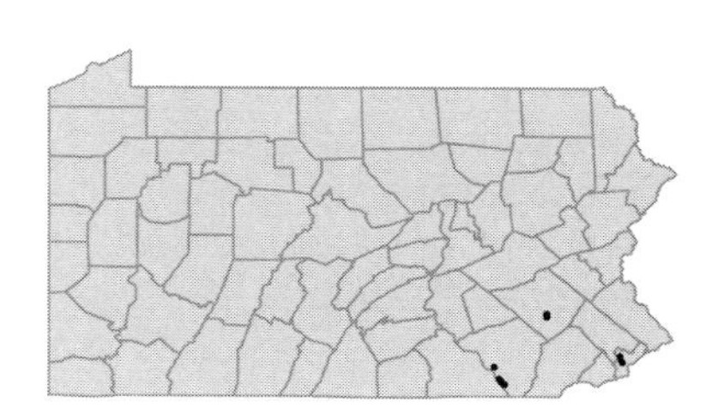

• ***Colutea arborescens*** L. I
Bladder senna
Deciduous shrub
Occasionally escaped from cultivation to fields, roadsides and shores.

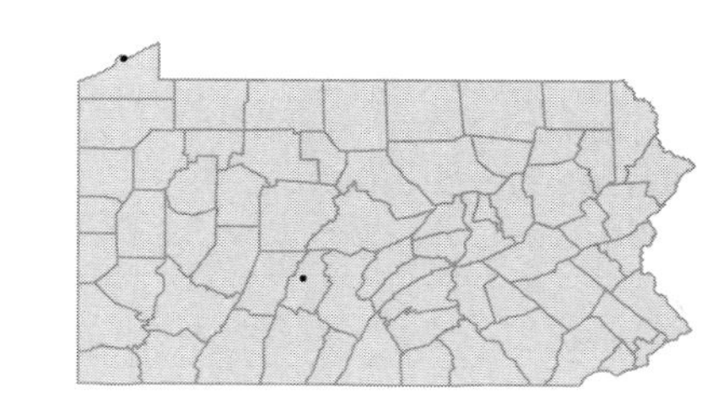

• ***Coronilla varia*** L. I
Crown-vetch
Herbaceous perennial
Planted extensively along highways.

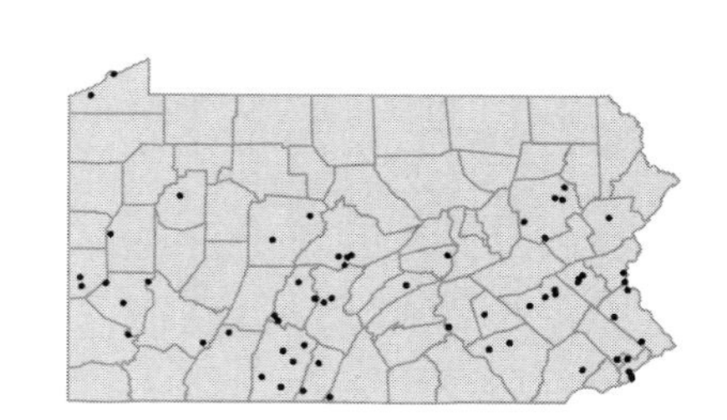

• ***Crotalaria sagittalis*** L.
Rattlebox
Herbaceous annual
Dry, sandy or gravelly soil of woods, old fields or roadsides.

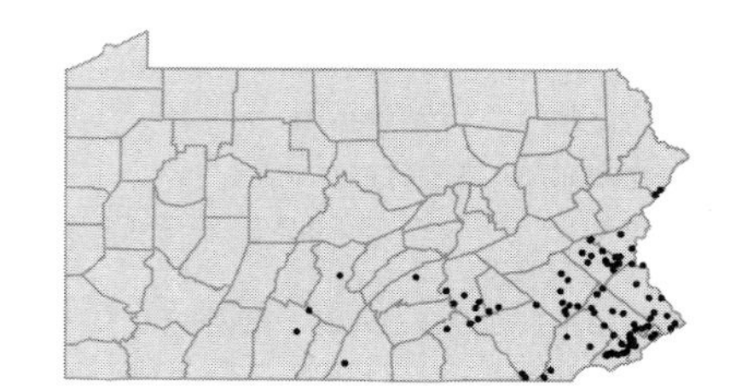

• ***Cytisus scoparius*** (L.) Link — I
Scotch-broom
Deciduous shrub
Cultivated and occasionally escaped to railroad embankments and waste ground.

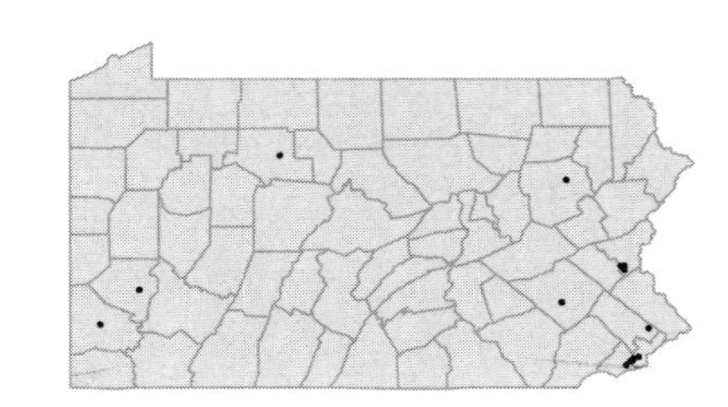

• ***Desmodium canadense*** (L.) DC. — FAC
Beggar-ticks; Showy tick-trefoil
Herbaceous perennial
Open woods.
Meibomia canadensis (L.) Kuntze P

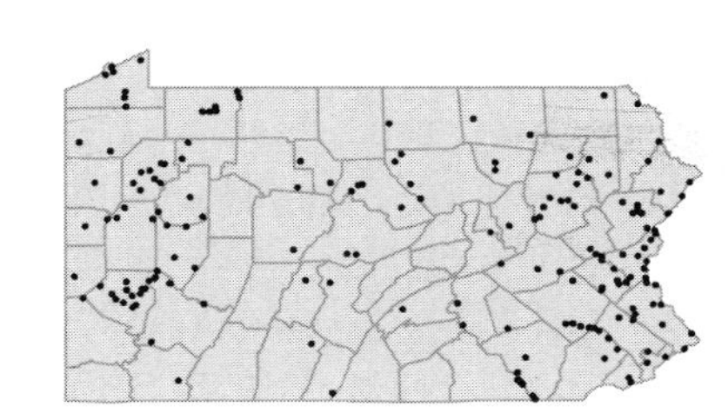

Desmodium canescens (L.) DC.
Hoary tick-trefoil
Herbaceous perennial
Dry, open woods and fields.
Meibomia canescens (L.) Kuntze P

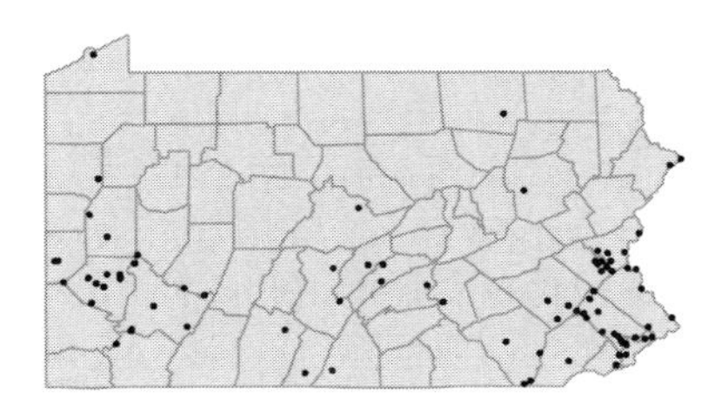

Desmodium ciliare (Muhl. ex Willd.) DC.
Tick-clover; Tick-trefoil
Herbaceous perennial
Dry, sandy woods and edges.

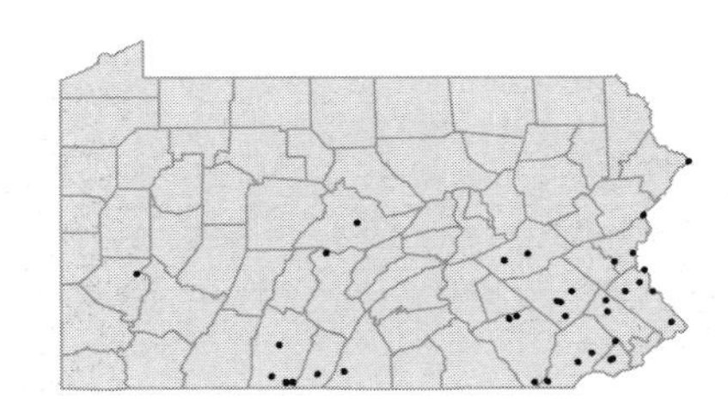

Desmodium cuspidatum (Muhl. ex Willd.) Loud.
Tick-clover; Tick-trefoil
Herbaceous perennial
Rich, rocky woods and banks.
Meibomia bracteosa (Michx.) Kuntze P

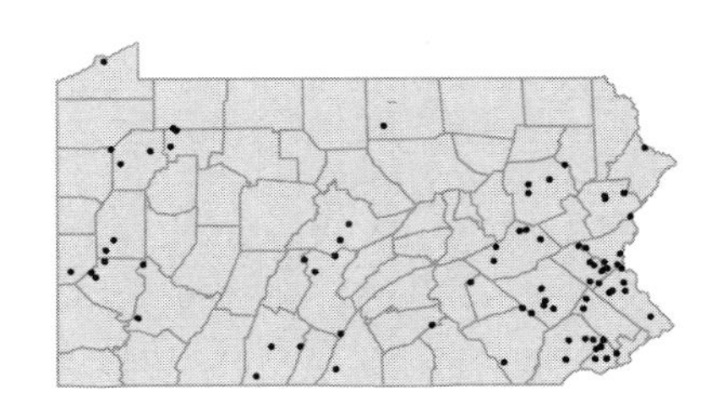

Desmodium glabellum (Michx.) DC.
Tall tick-trefoil
Herbaceous perennial
Wooded roadside banks and open woods.
Meibomia glabella (Michx.) Kuntze P

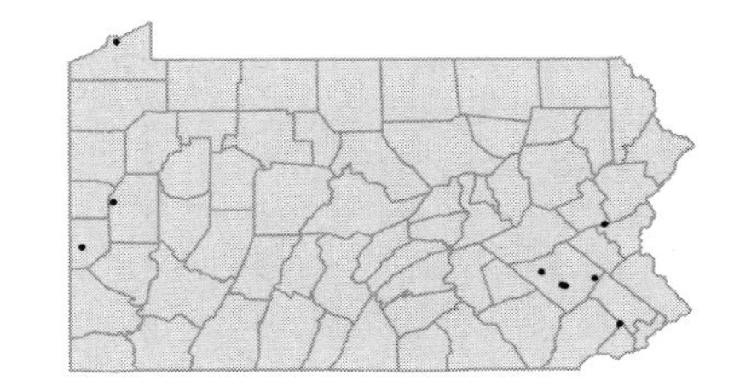

Desmodium glutinosum (Muhl. ex Willd.) A.Wood
Sticky tick-clover
Herbaceous perennial
Rich woods.
Meibomia grandiflora (Walt.) Kuntze P

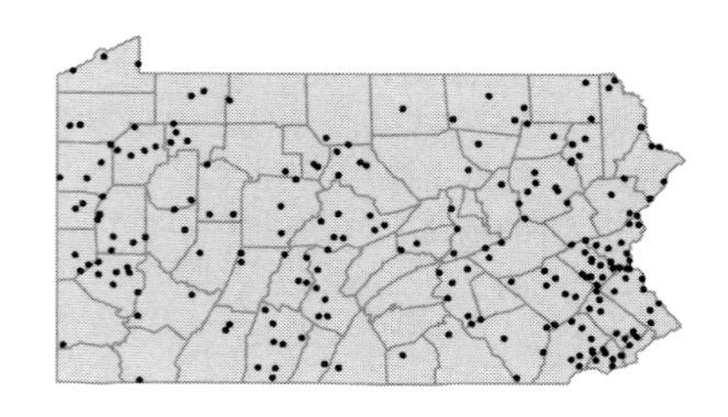

Desmodium humifusum (Muhl. ex Bigel.) Beck
Tick-trefoil
Herbaceous perennial vine
Dry, sandy woods.

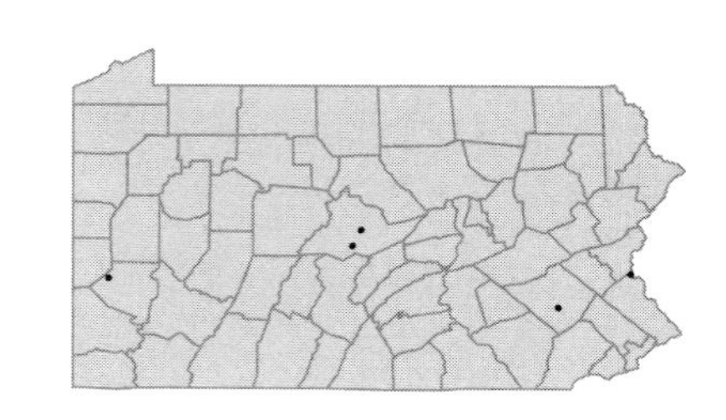

Desmodium laevigatum (Nutt.) DC.
Smooth tick-clover
Herbaceous perennial
Dry, sandy woods and roadsides.
Meibomia laevigata (Nutt.) Kuntze P

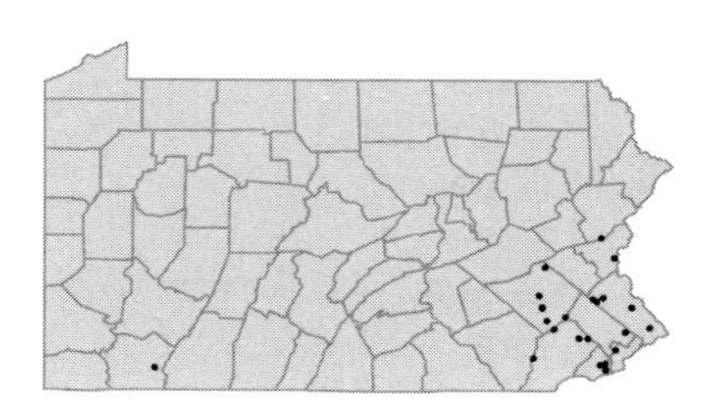

Desmodium marilandicum (L.) DC.
Maryland tick-clover
Herbaceous perennial
Dry, open, upland woods, fields and edges.
Meibomia marylandica (L.) Kuntze P

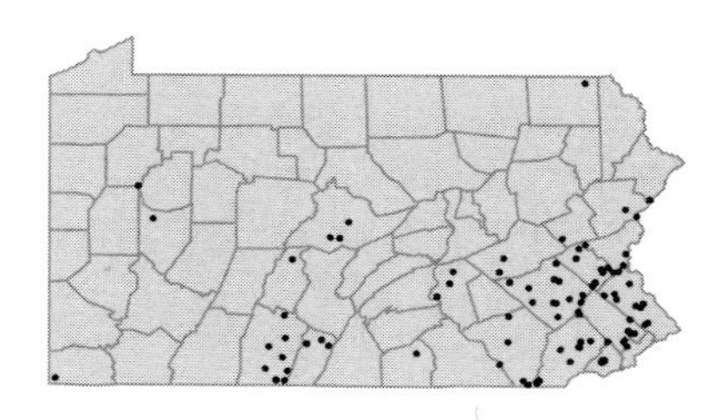

Desmodium nudiflorum (L.) DC.
Naked-flowered tick-trefoil
Herbaceous perennial
Rich, deciduous woods and edges.
Meibomia nudiflora (L.) Kuntze P

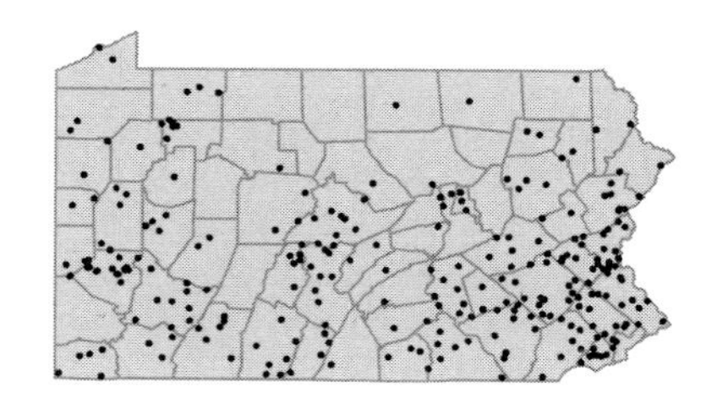

Desmodium nuttallii (Schindl.) Schub.
Nuttall's tick-trefoil
Herbaceous perennial
Open woods and edges.
Desmodium viridiflorum (L.) DC. *pro parte* C

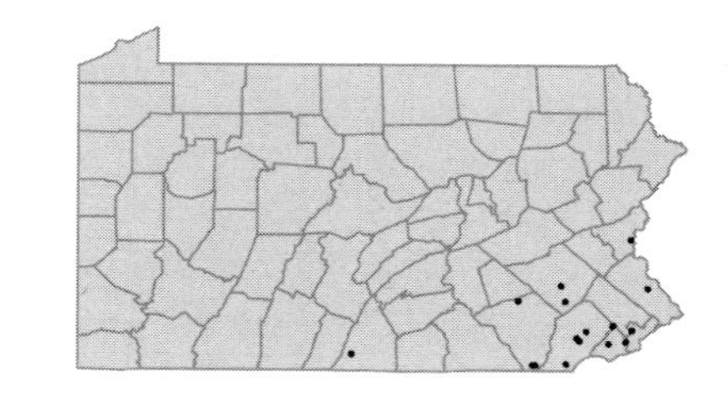

Desmodium obtusum (Muhl. ex Willd.) DC.
Beggar-lice; Tick-trefoil
Herbaceous perennial
Dry, open woods, on sandy soils.
Desmodium rigidum (Ell.) DC. FGBWC; *Meibomia obtusa* (Muhl.) Vail P;
Meibomia rigida (Ell.) Kuntze P

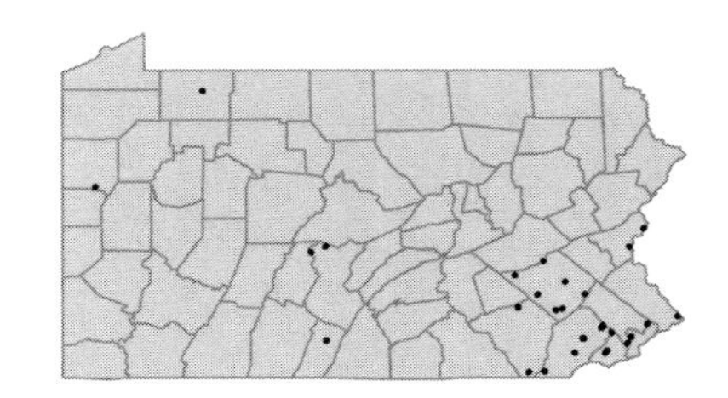

Desmodium paniculatum (L.) DC. UPL
Beggar-ticks; Tick-trefoil
Herbaceous perennial
Clearings and edges of moist or dry woods.
Meibomia dillenii (Darl.) Kuntze P; *Meibomia paniculata* (L.) Kuntze P

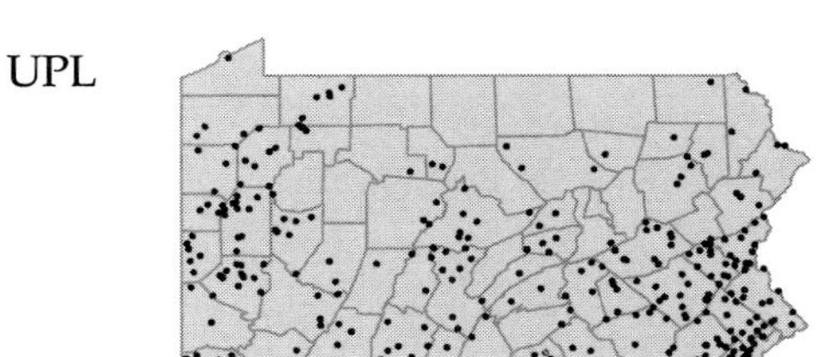

Desmodium perplexum Schub.
Tick-clover; Tick-trefoil
Herbaceous perennial
Dry or moist, open woods.
Desmodium glabellum (Michx.) DC. *pro parte* C

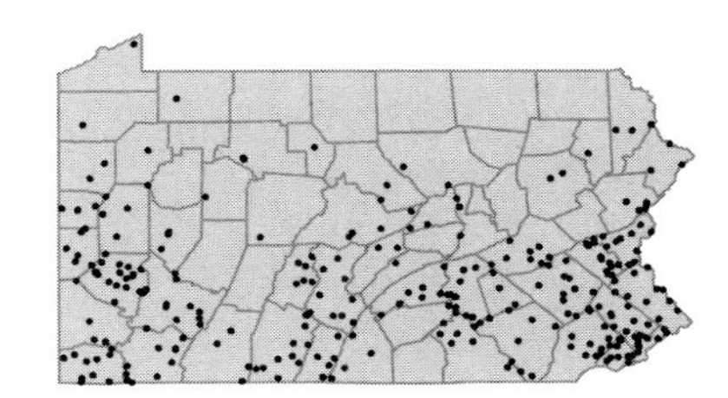

Desmodium rotundifolium DC.
Tick-trefoil
Herbaceous perennial
Dry, open woods.
Meibomia michauxii Vail P

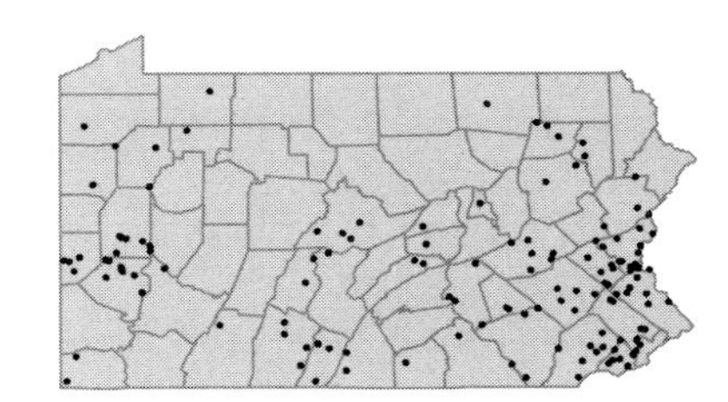

Desmodium sessilifolium (Torr.) Torr. & A.Gray
Sessile-leaved tick-trefoil
Herbaceous perennial
Dry, open woods in sterile, sandy soils. Believed to be extirpated, last collected in 1921.
Meibomia sessilifolia (Torr.) Kuntze P

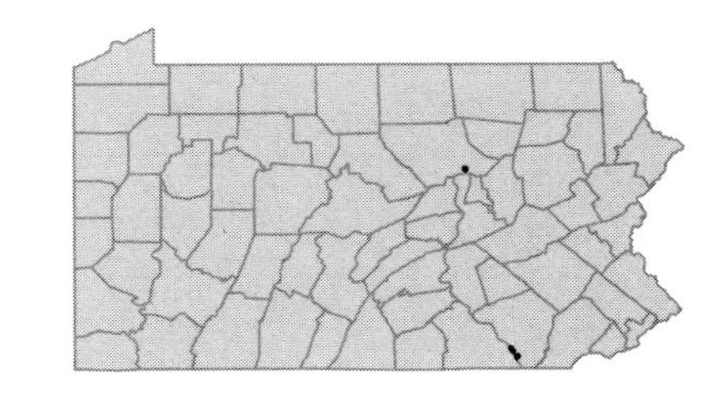

Desmodium viridiflorum (L.) DC.
Velvety tick-trefoil
Herbaceous perennial
Abandoned fields and other dry, open places.

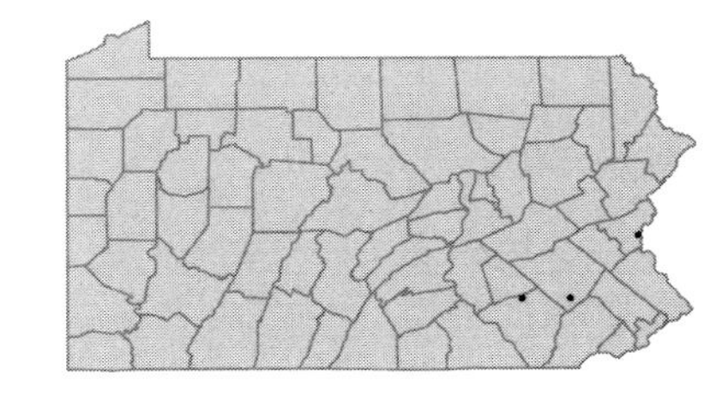

• ***Galactia regularis*** (L.) BSP
Eastern milk-pea
Herbaceous perennial vine
Dry, sandy soil. Believed to be extirpated, last collected in 1901.

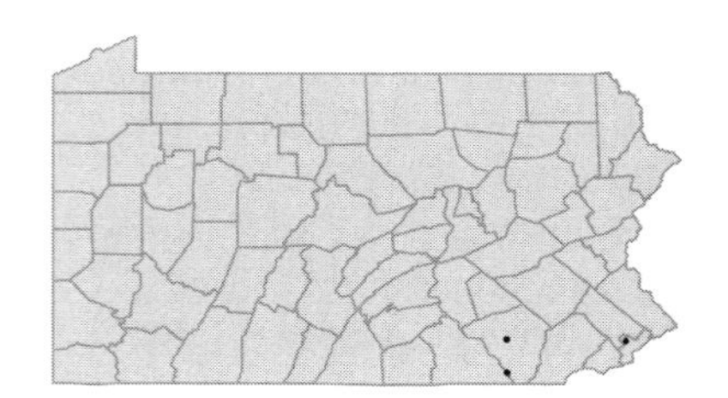

Galactia volubilis (L.) Britt.
Downy milk-pea
Herbaceous perennial vine
Dry thickets and edges. Believed to be extirpated, last collected in 1941.

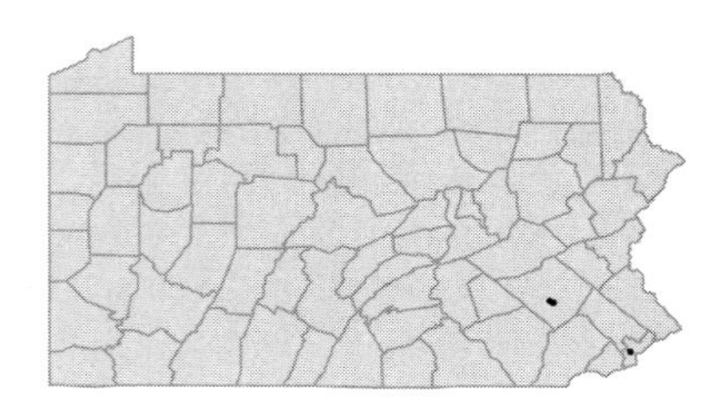

• ***Galega officinalis*** L.
Goat's-rue
Herbaceous perennial
Naturalized in moist, open meadows and along stream banks. A federally designated noxious weed.

I

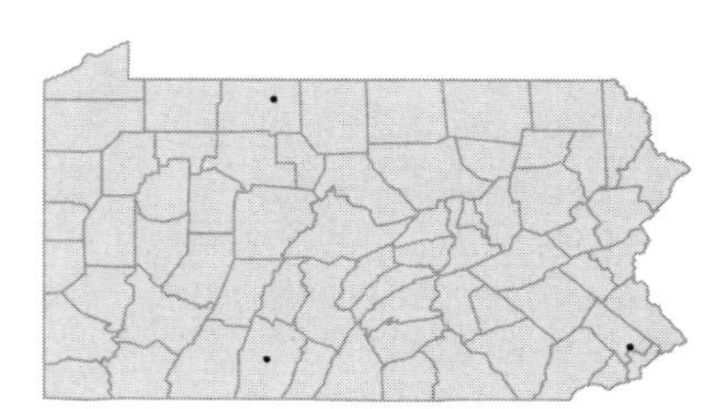

• ***Glycine max*** (L.) Merr.
Soybean
Herbaceous annual
Widely grown crop plant, occasionally escaped to railroad ballast or roadsides.

I

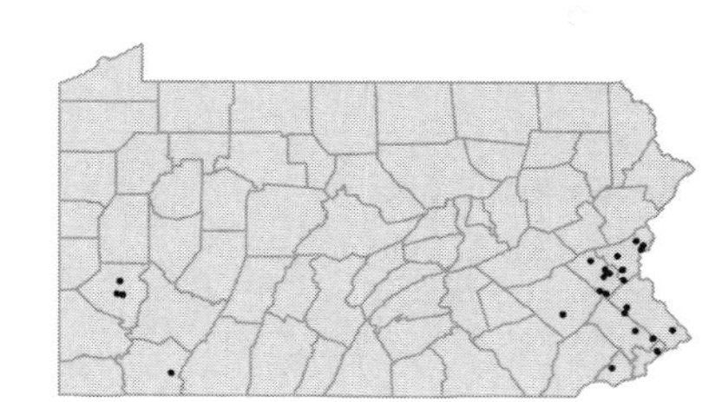

• ***Glycyrrhiza lepidota*** (Nutt.) Pursh
Wild licorice
Herbaceous perennial
Waste ground. Represented by a single collection from Philadelphia Co. in 1950.

I

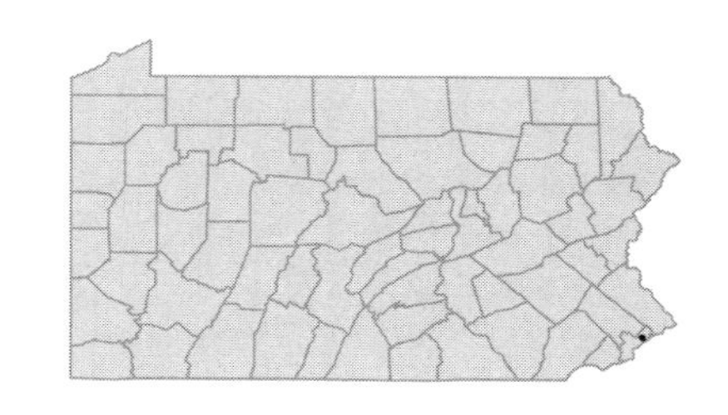

• ***Hippocrepis comosa*** L.
Horseshoe vetch
Herbaceous perennial
Waste ground and ballast. Represented by a single collection from Philadelphia Co. in 1878.

I

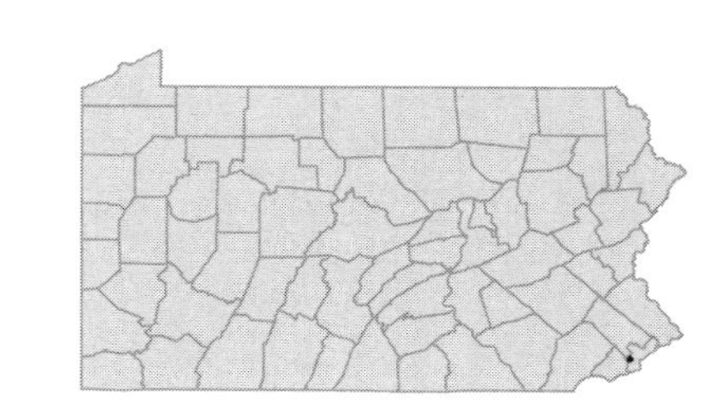

• ***Kummerowia stipulacea*** (Maxim.) Schindler
Korean lespedeza
Herbaceous annual
Dry, open soil of pastures, roadsides and waste ground, planted for erosion control.
Lespedeza stipulacea Maxim. FGBWC

FACU
[I]

Kummerowia striata (Thunb.) Shindl.
Common lespedeza; Japanese clover
Herbaceous annual
Fields, railroad banks, waste ground and shale barrens.
Lespedeza striata (Thunb.) Hook. & Arn. FGBWC

FACU
[I]

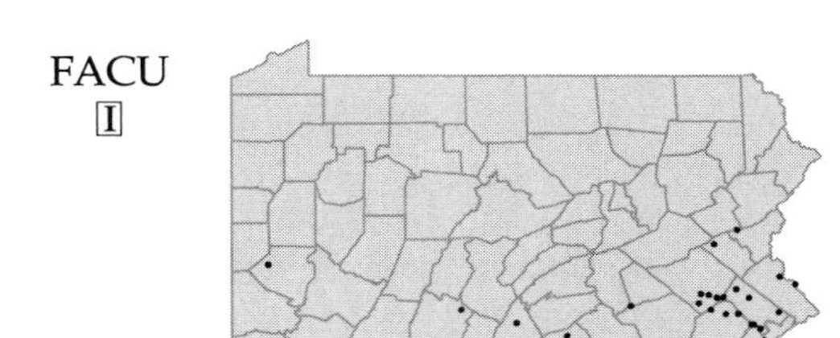

• ***Lablab purpureus*** (L.) Sweet
Hyacinth-bean
Herbaceous annual vine
Cultivated and occasionally occurring in dumps and waste ground.
Dolichos lablab L. FGBWC; *Lablab nigra* Medic. K

[I]

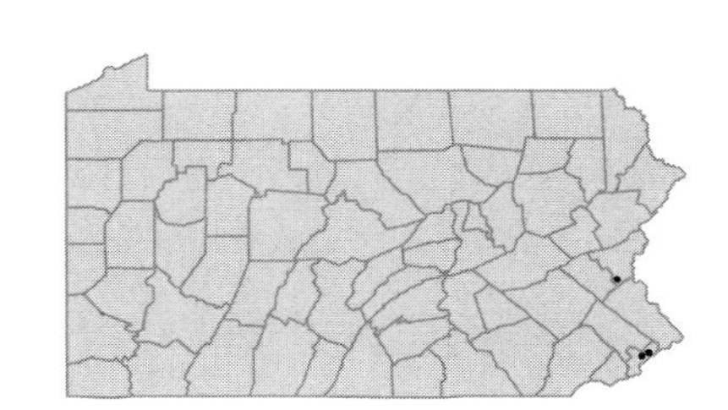

• ***Lathyrus aphaca*** L.
Vetchling
Herbaceous annual
Waste ground and ballast.

[I]

Lathyrus japonicus Willd. var. ***glaber*** (Ser.) Fern.
Beach peavine
Herbaceous perennial
Sandy or gravelly shores, sand plains and dunes.
Lathyrus maritimus (L.) Bigel. GBC

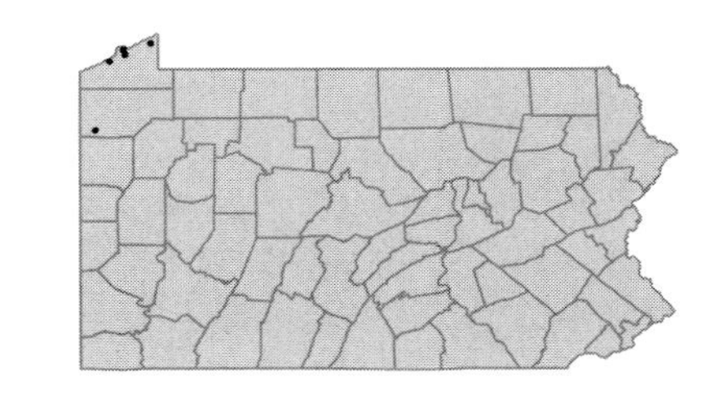

Lathyrus latifolius L.
Perennial sweetpea
Herbaceous perennial vine
Waste ground, old fields, roadsides and disturbed places.

[I]

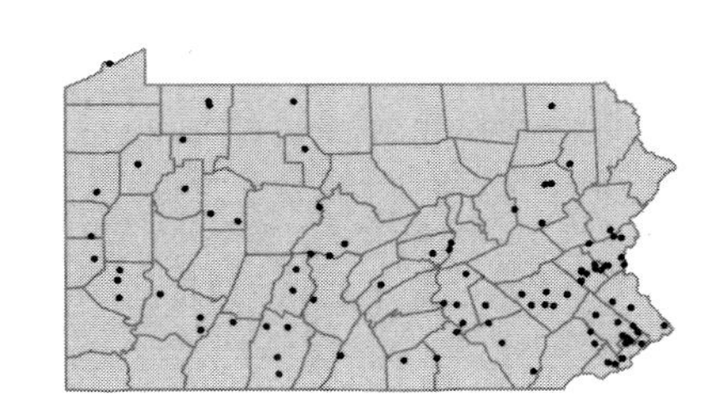

Lathyrus ochroleucus Hook.
Vetchling; Wild pea
Herbaceous perennial
Rocky, limestone woods and slopes.

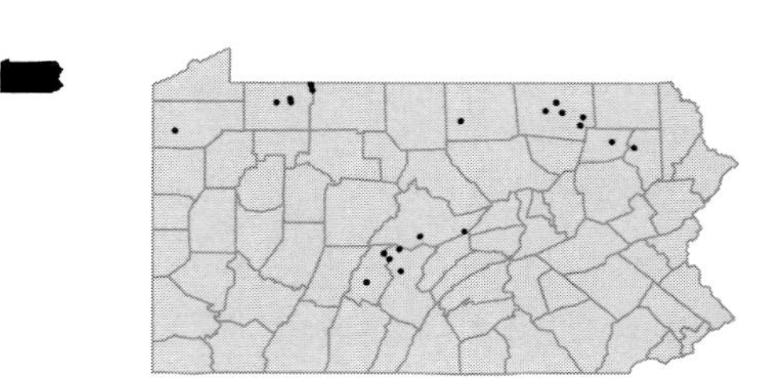

Lathyrus palustris L.
Marsh pea; Marsh-vetch; Vetchling
Herbaceous perennial vine
Shores, moist meadows, sand plains, swamps and thickets.
Lathyrus myrtifolius Muhl. P; *Lathyrus palustris* L. FGB; *Lathyrus palustris* L FGBW

FACW+

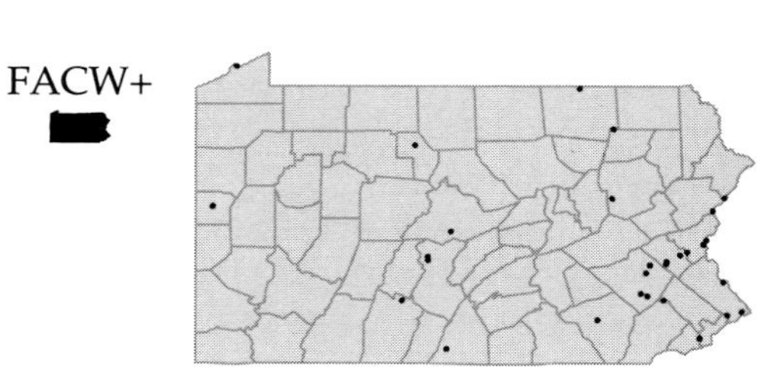

Lathyrus tuberosus L.
Field pea; Yellow vetchling
Herbaceous perennial
Fields, meadows, roadsides and alluvium.

I

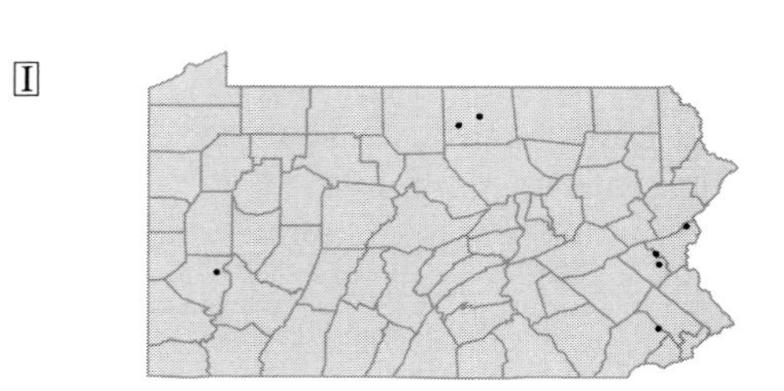

Lathyrus venosus Muhl. ex Willd.
Veiny pea; Veiny vetchling
Herbaceous perennial
Sandy or rocky shores, wooded slopes and railroad banks.

FACW

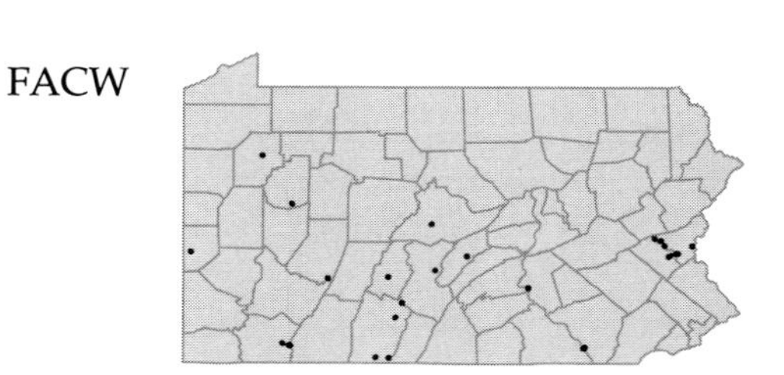

• ***Lens culinaris*** Medic.
Lentil
Herbaceous annual
A rare escape from cultivation.
Lens esculenta Moench GB

I

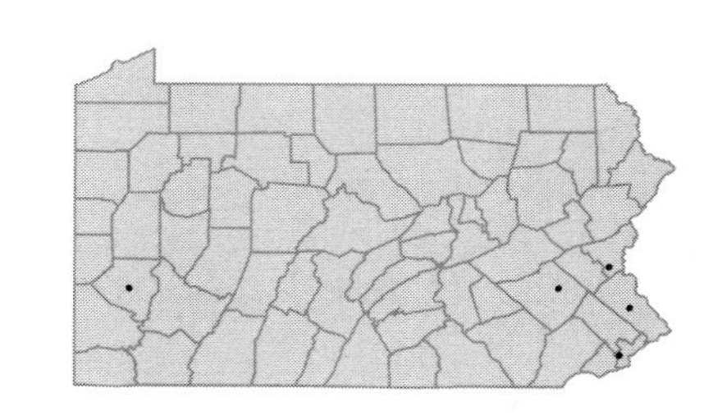

Lespedeza—In addition to the taxa treated below, the following hybrids have been collected at least once in Pennsylvania: *capitata x virginica* (*L. x simulata* Mackenzie & Bush); *hirta x procumbens*; *hirta x violacea*; *hirta x virginica*; *intermedia x virginica*; *procumbens x virginica* (*L. x brittonii* E. Bickn.)

• ***Lespedeza angustifolia*** (Pursh) Ell.
Narrow-leaved bush-clover
Herbaceous perennial
Moist, open, sandy soil of an abandoned gravel pit.

FAC

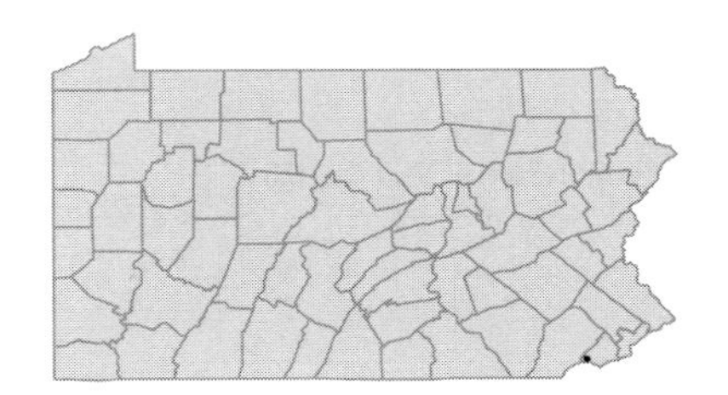

Lespedeza bicolor Turcz.
Bicolor lespedeza
Deciduous shrub
Introduced for erosion control and naturalized in open woods and old fields.

I

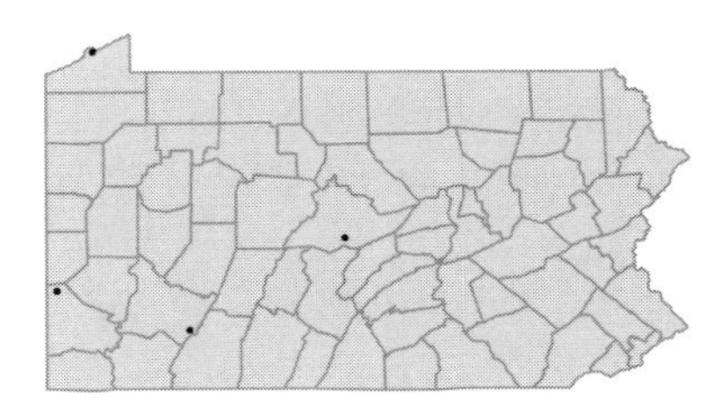

Lespedeza capitata Michx.
Bush-clover
Herbaceous perennial
Fields and thin woods in dry, open ground.

FACU-

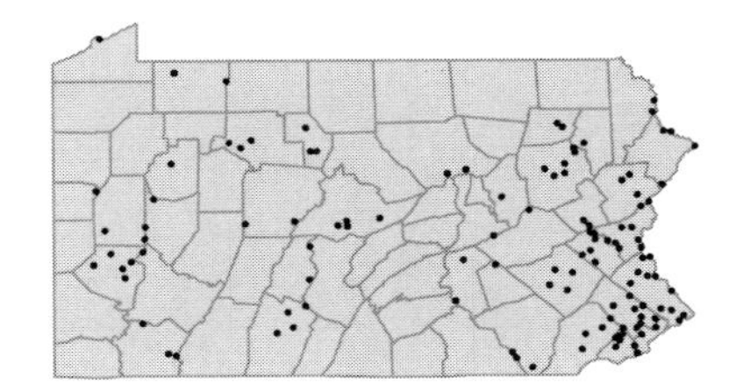

Lespedeza cuneata (Dum.-Cours.) G.Don
Sericea lespedeza
Herbaceous perennial
Fields and grassy roadsides, planted for erosion control.

I

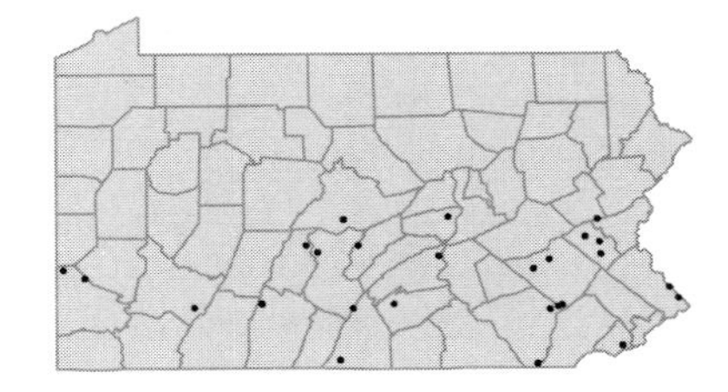

Lespedeza hirta (L.) Hornem.
Bush-clover
Herbaceous perennial
Dry, open soils.

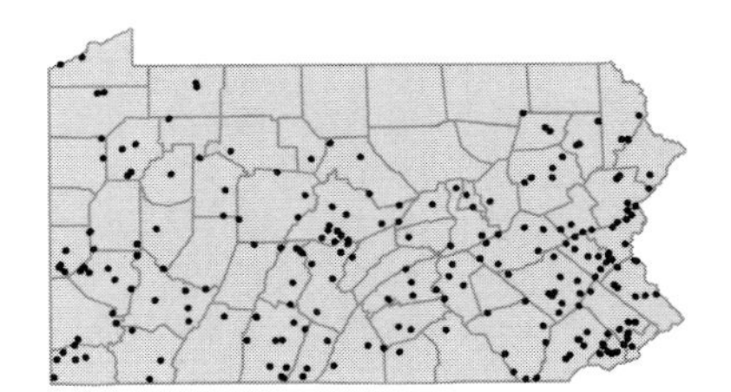

Lespedeza hirta x intermedia
Nuttall's bush-clover
Herbaceous perennial
Dry, open woods and edges, on sandy soils.
Lespedeza x nuttallii Darl. PFGBW

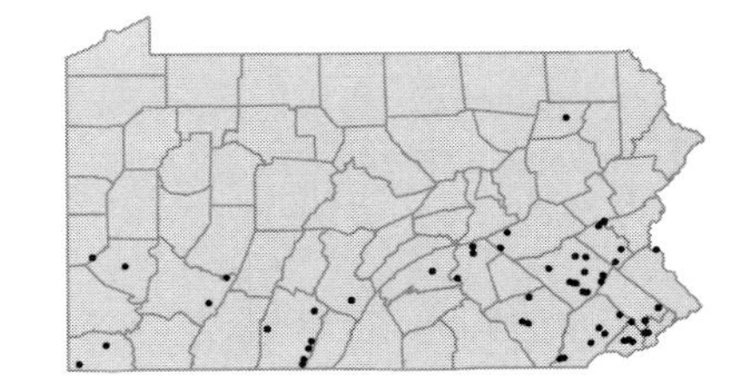

Lespedeza intermedia (S.Wats.) Britt.
Bush-clover
Herbaceous perennial
Dry, open, rocky woods and thickets.
L. frutescens (L.) of Porter (1903) = *L. intermedia* (S.Wats.) Britt.

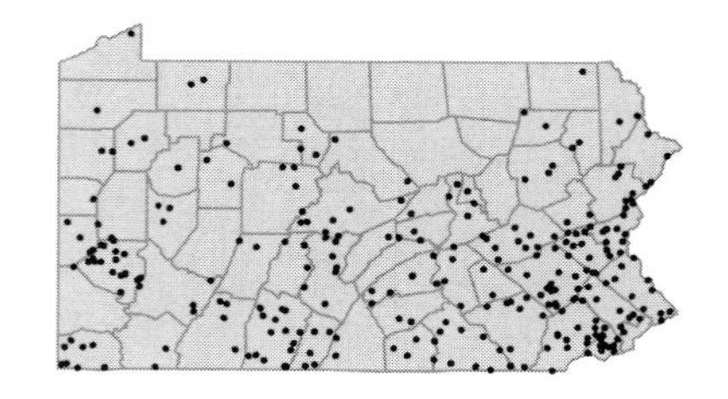

Lespedeza procumbens Michx.
Trailing bush-clover
Herbaceous perennial
Dry woods, fields or roadsides, in sandy or rocky soils.

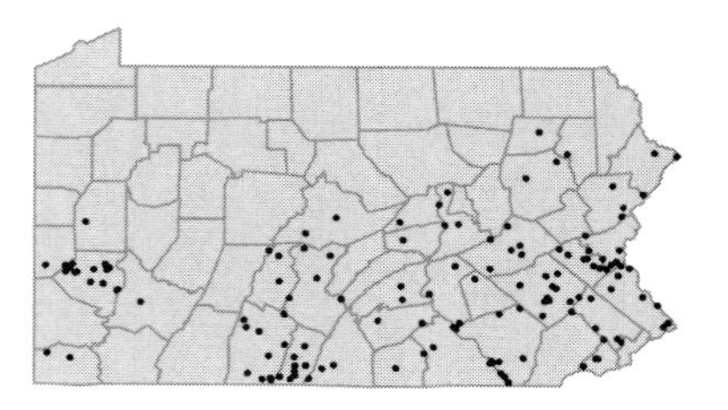

Lespedeza repens (L.) Bart.
Creeping bush-clover
Herbaceous perennial
Woods, banks or edges in dry, sterile, acidic soils.

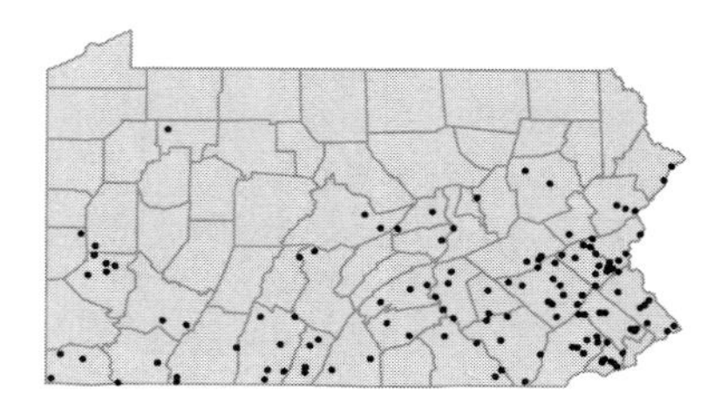

Lespedeza stuevei Nutt.
Tall bush-clover
Herbaceous perennial
Dry, open woods and edges in sterile soil. Believed to be extirpated, last collected in 1933.

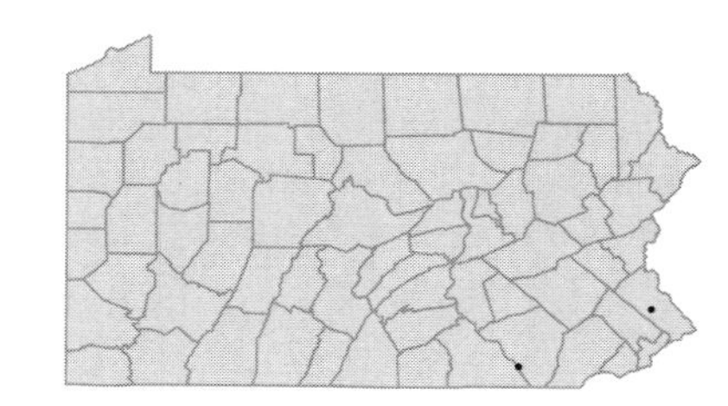

Lespedeza thunbergii (DC.) Nakai
[I]
Thunberg's bush-clover
Deciduous shrub
Dry fields and open woods.

Lespedeza violacea (L.) Pers.
Bush-clover
Herbaceous perennial
Dry upland woods, thickets and openings.

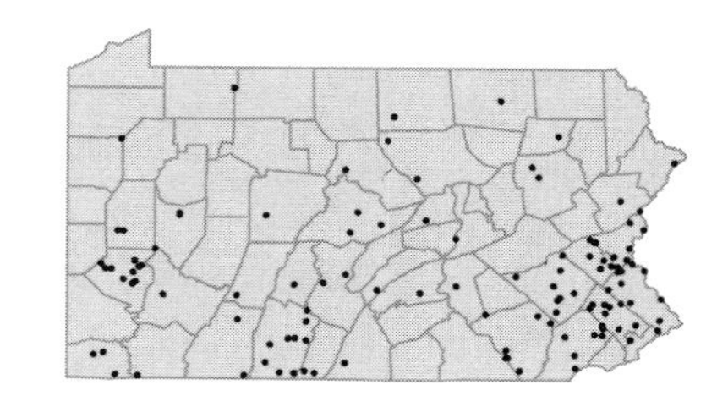

Lespedeza virginica (L.) Britt.
Virginia bush-clover
Herbaceous perennial
Dry fields, stony banks, and serpentine barrens.

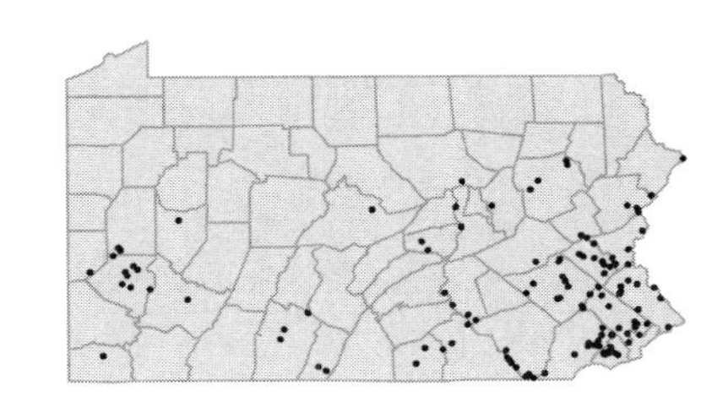

• ***Lotus corniculatus*** L.
Bird's-foot trefoil
Herbaceous perennial
Planted along roadsides and other disturbed sites.

FACU-
I

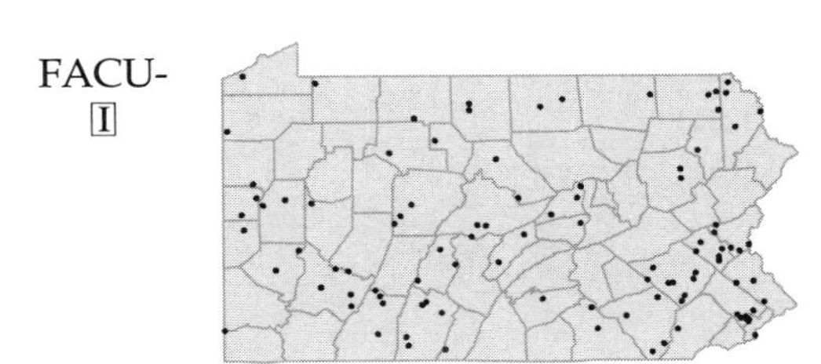

• ***Lupinus perennis*** L.
Wild blue lupine
Herbaceous perennial
Alluvial sand and gravel bars, open fields, woods edges and roadsides in sandy soils.

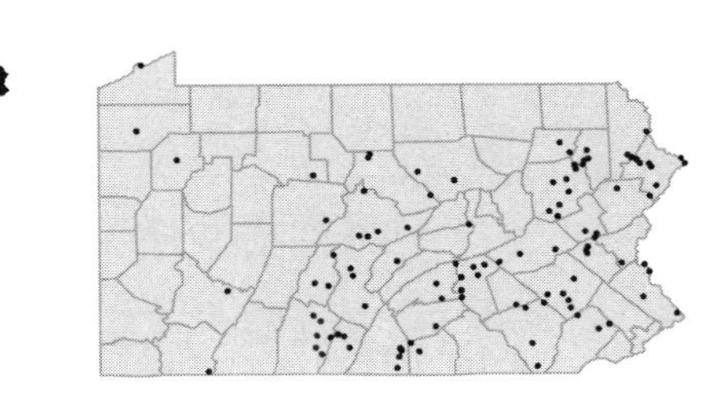

• ***Medicago arabica*** (L.) Huds.
Spotted bur-clover; Spotted medick
Herbaceous annual
Waste ground. Represented by two collections from Philadelphia Co. in 1852 and 1865.

I

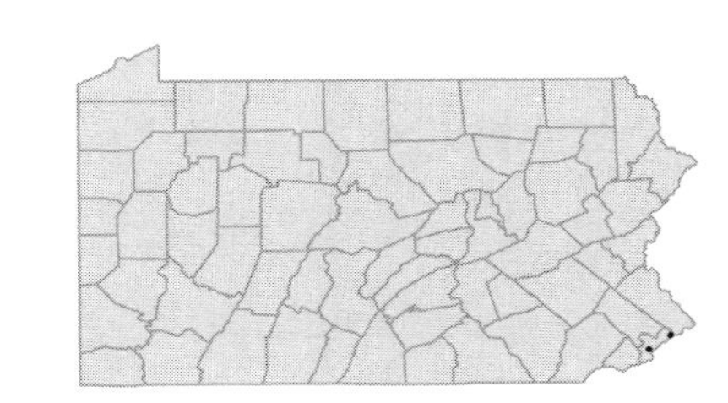

Medicago falcata L.
Yellow alfalfa
Herbaceous perennial
Cultivated and escaped to waste places and roadsides.
Medicago sativa L. ssp. *falcata* (L.) Arcangeli C

I

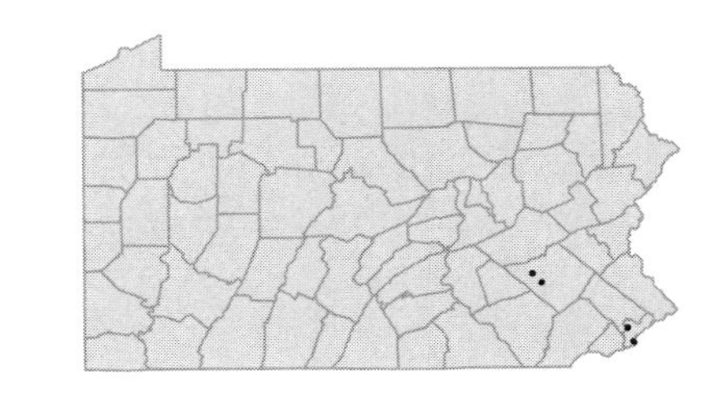

Medicago lupulina L.
Black medick
Herbaceous annual
Roadsides and waste places.

I

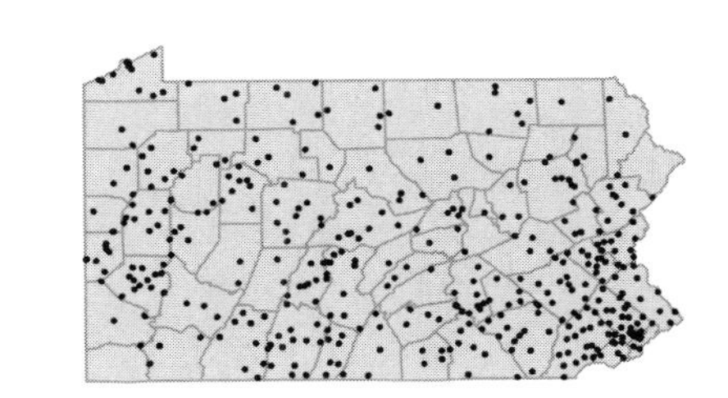

Medicago polymorpha L.
Bur-clover
Herbaceous annual
Ballast and waste ground.
Medicago hispida Gaertn. FGBW

[I]

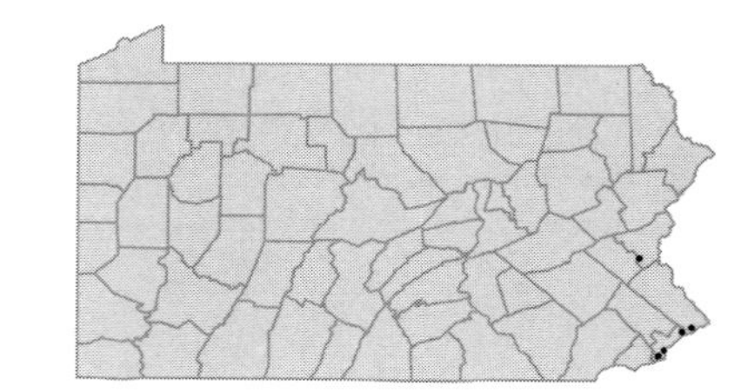

Medicago sativa L.
Alfalfa
Herbaceous perennial
Widely grown crop plant, occasionally escaped to roadsides, vacant lots and railroad banks.

[I]

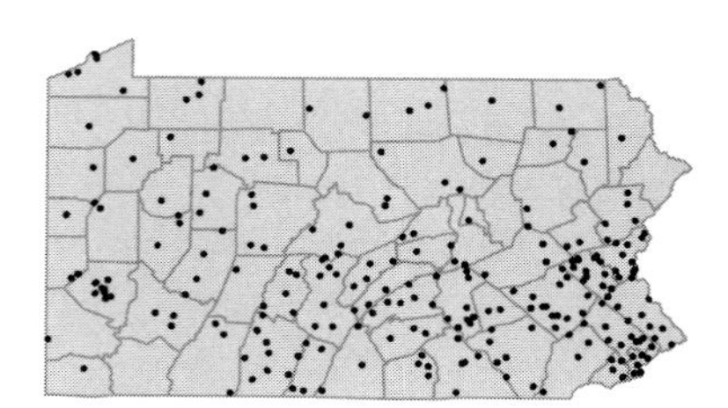

• ***Melilotus alba*** Medic.
White sweet-clover
Herbaceous biennial
Roadsides and old fields in rich soil.

FACU
[I]

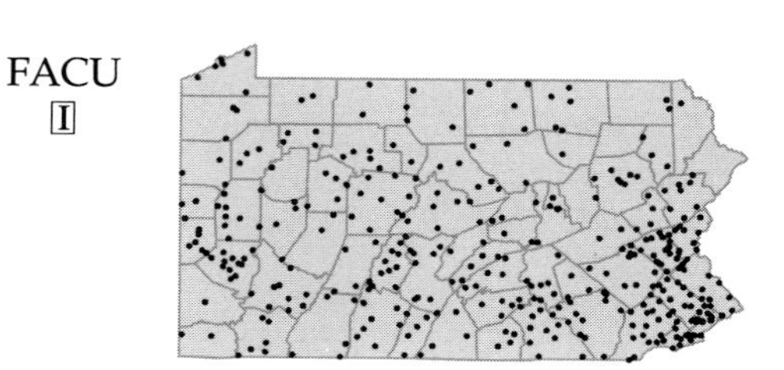

Melilotus altissima Thuill.
Tall sweet-clover
Herbaceous annual
Roadsides and waste places.

[I]

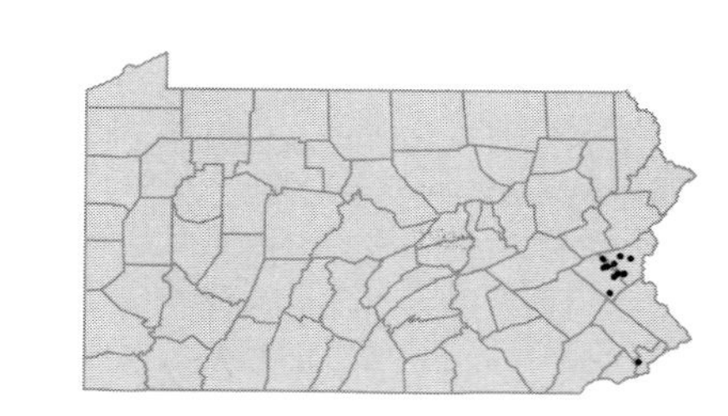

Melilotus officinalis (L.) Pallas
Yellow sweet-clover
Herbaceous biennial
Waste areas, cultivated ground and roadsides.

FACU
[I]

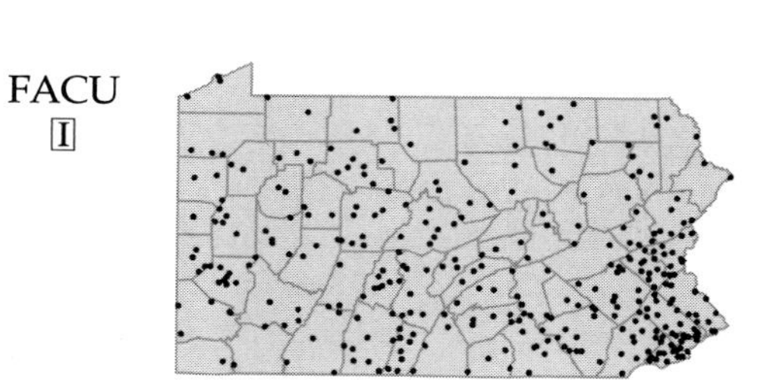

Melilotus sulcata Desf.
Sweet-clover
Herbaceous annual
Waste ground. Represented by two collections from Philadelphia Co. in 1920 and 1921.

[I]

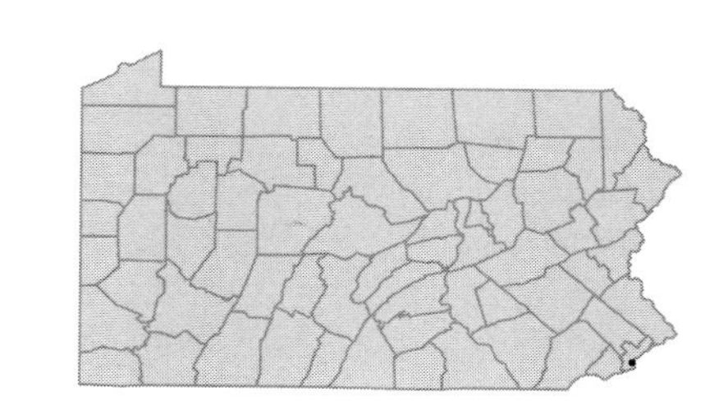

• ***Ononis spinosa*** L. [I]
Rest-harrow; Rest-harvest
Deciduous shrub
Waste ground or ballast. Represented by several collections from Philadelphia Co. 1864-1880.

• ***Ornithopus perpusillus*** L. [I]
Ornithopus
Herbaceous annual
Waste ground or ballast. Represented by three collections from Philadelphia and Chester Cos. in 1878 and 1879.

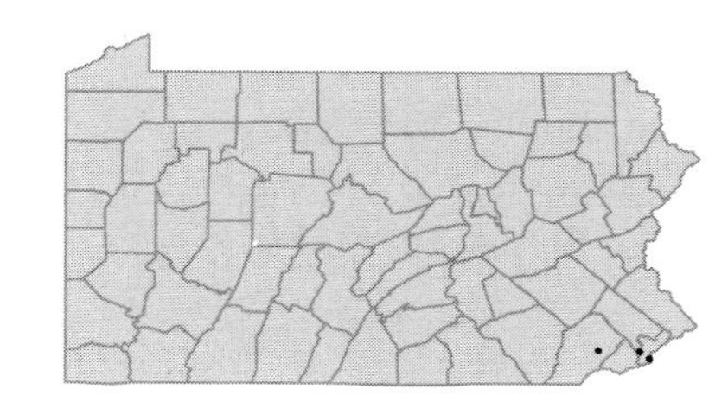

• ***Phaseolus polystachios*** (L.) BSP
Wild bean
Herbaceous perennial vine
Woods, roadside banks and waste ground.

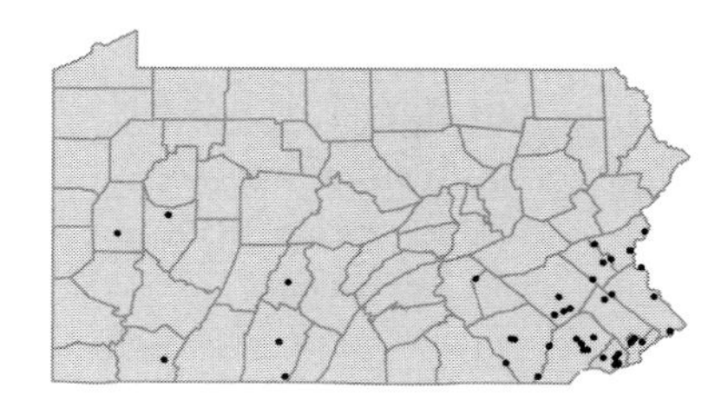

• ***Pisum sativum*** L. [I]
Garden pea
Herbaceous annual vine
Cultivated and occasionally escaped to edges and waste ground.

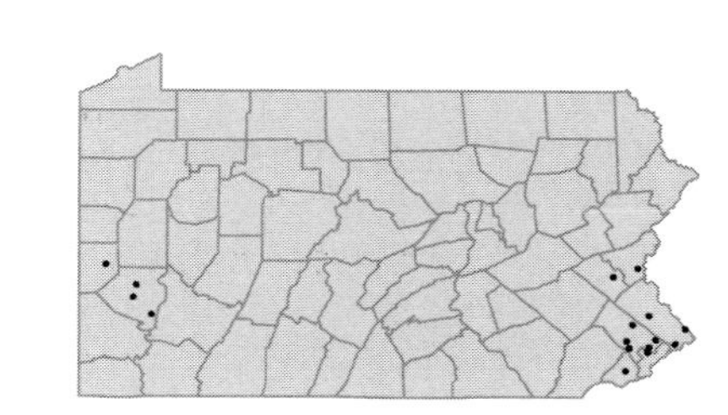

• ***Pueraria lobata*** (Willd.) Ohwi [I]
Kudzu
Herbaceous perennial vine
Waste ground and woods edges, escaped from cultivation. Designated as a noxious weed in PA.
Pueraria thunbergiana (Sieb. & Zucc.) Benth. B

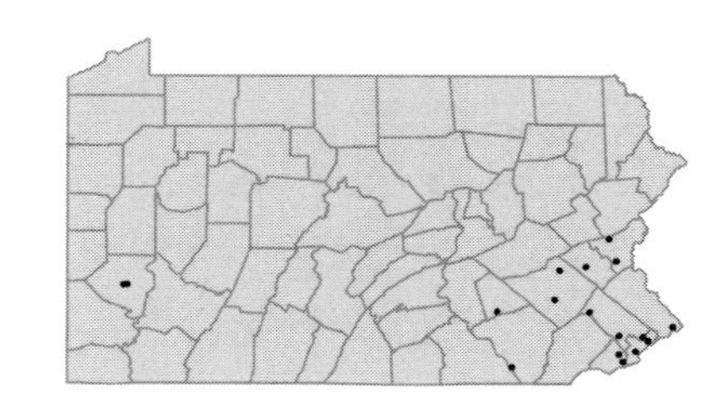

• ***Robinia hispida*** L. [I]
Bristly locust; Mossy locust; Rose-acacia
Deciduous tree
Dry, open woods, slopes and roadsides.

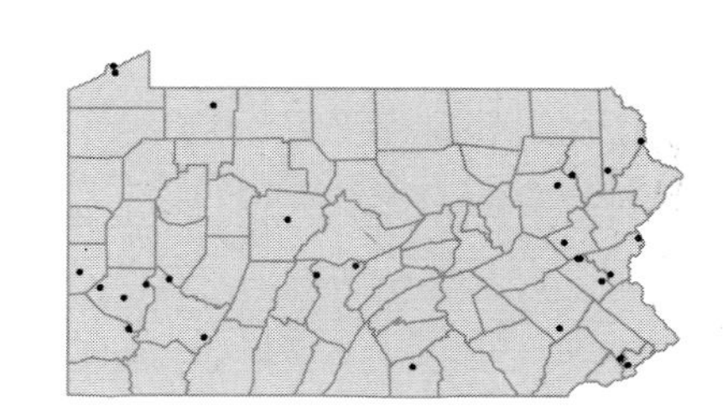

Robinia pseudoacacia L.
Black locust
Deciduous tree
Open woods, floodplains, thickets and fencerows.

FACU-

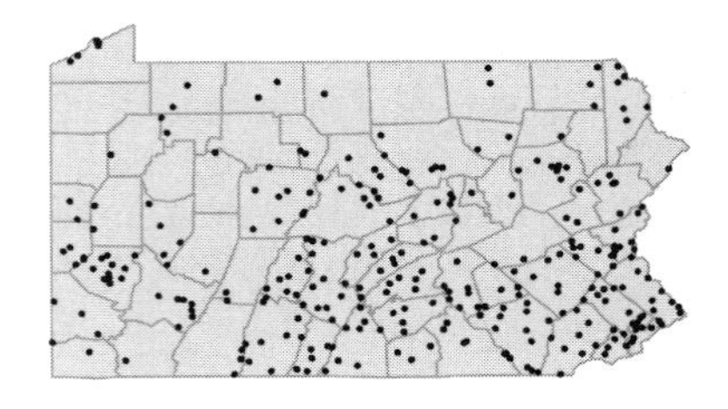

Robinia viscosa Vent.
Clammy locust
Deciduous tree
Dry, open ground, thin woods or slopes.

[I]

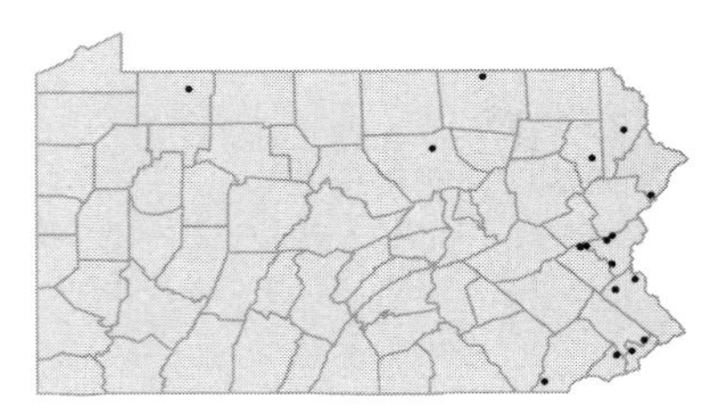

• ***Scorpiurus muricatus*** L.
Caterpiller-plant
Herbaceous annual
Waste ground or ballast. Represented by two collections from Philadelphia Co. in 1879 and 1921.

[I]

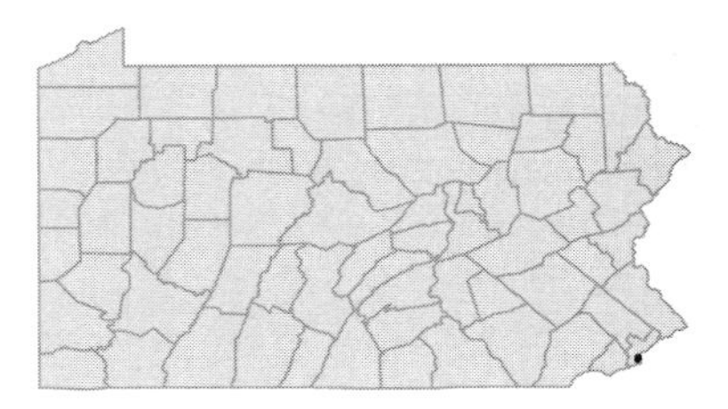

• ***Sesbania exaltata*** (Raf.) Cory
Coffee-weed; Indigo-weed
Herbaceous annual
Waste ground or ballast. Represented by two collections from Philadelphia and Delaware Cos. in 1866 and 1894.
Sesban macrocarpa Muhl. P

[I]

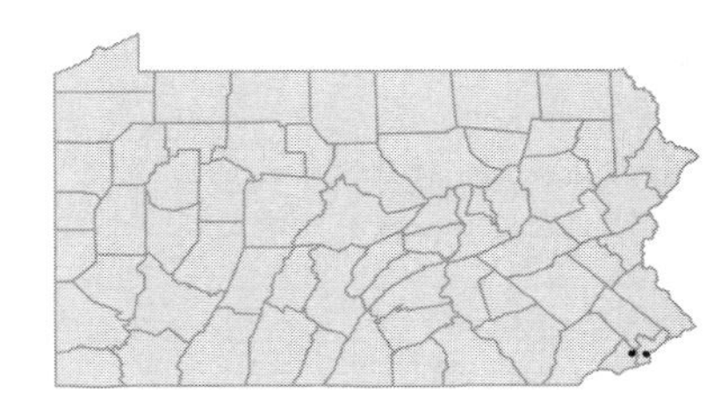

• ***Strophostyles helvola*** (L.) Ell.
Wild bean
Herbaceous annual vine
Sandy fields, railroad embankments and moist roadside ditches.

FACU

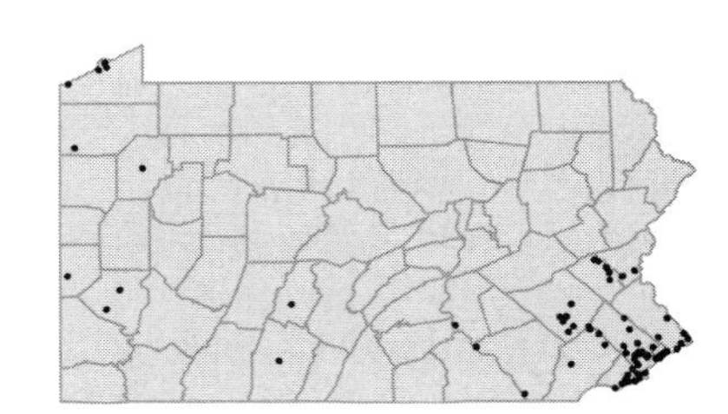

Strophostyles leiosperma (Torr. & A.Gray) Piper
Wild bean
Herbaceous perennial vine
Railroad ballast.
Strophostyles pauciflora (Benth.) S.Wats. W

[I]

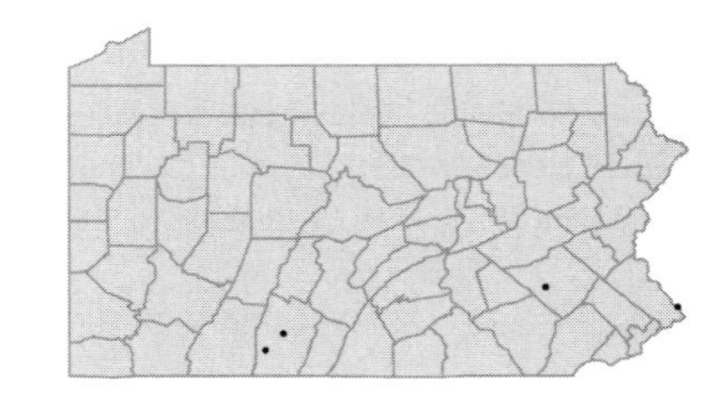

Strophostyles umbellata (Muhl. ex Willd.) Britt. FACU
Wild bean
Herbaceous perennial vine
Woods, clearings and fields on sandy soils, also serpentine barrens.

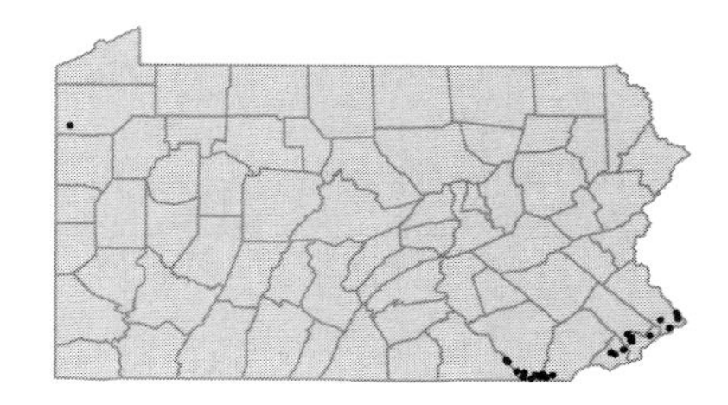

• ***Stylosanthes biflora*** (L.) BSP
Pencil-flower
Herbaceous perennial
River banks, rocky or shaly slopes and sandy fields.
Stylosanthes riparia Kearney *pro parte* FGBW

• ***Tephrosia virginiana*** (L.) Pers.
Goat's-rue
Herbaceous perennial
Dry woods and openings, in sandy, acidic soils.
Cracca virginiana L. P

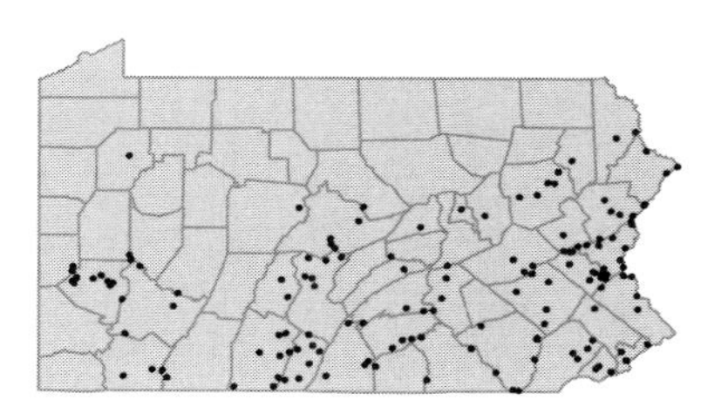

• ***Thermopsis villosa*** (Walt.) Fern. & Schubert I
Buck-bean
Herbaceous perennial
Cultivated and rarely escaped, dry fields or roadsides. Represented by a single collection from McKean Co. in 1950.

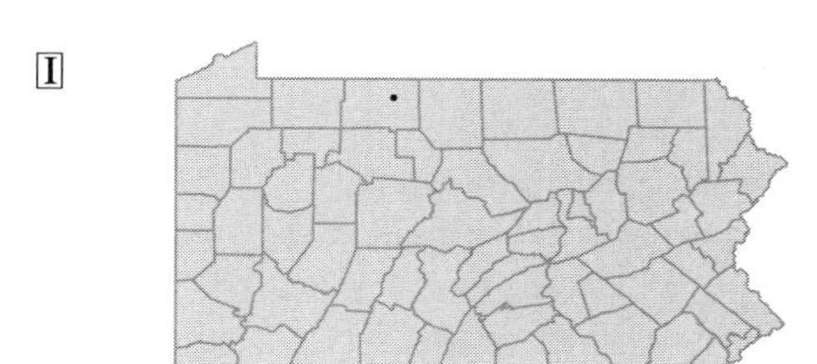

• ***Trifolium arvense*** L. I
Rabbit-foot clover
Herbaceous annual
Dry fields and roadsides.

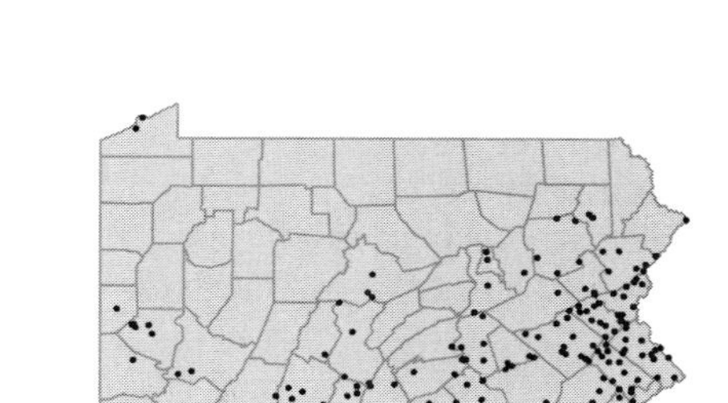

Trifolium aureum Pollich I
Large hop-clover; Yellow clover
Herbaceous annual
Roadsides, dry fields and waste places.
Trifolium agrarium L. FGB

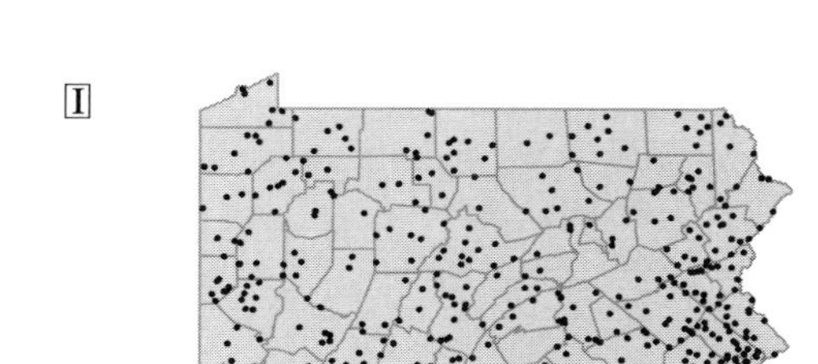

Trifolium campestre Schreb.
Low hop-clover
Herbaceous annual
Roadsides, old fields and waste places.
Trifolium procumbens L. PFGB

I

Trifolium carolinianum Michx.
Wild white clover
Herbaceous annual
Ballast ground. Represented by several collections from Philadelphia Co. in 1865.

I

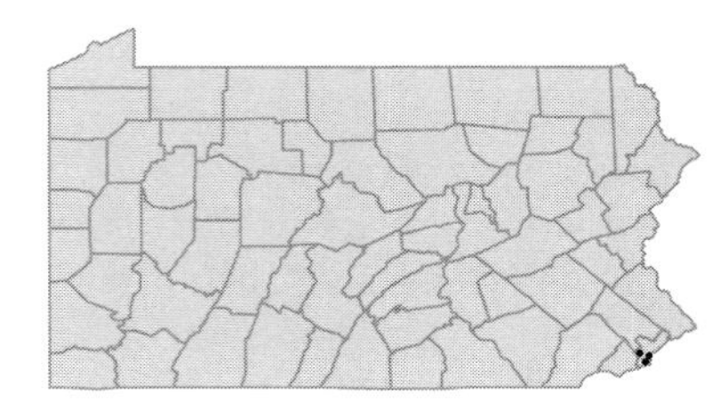

Trifolium dubium Sibth.
Hop-clover
Herbaceous annual
Dry roadsides, old sand pits and lawns.

UPL
I

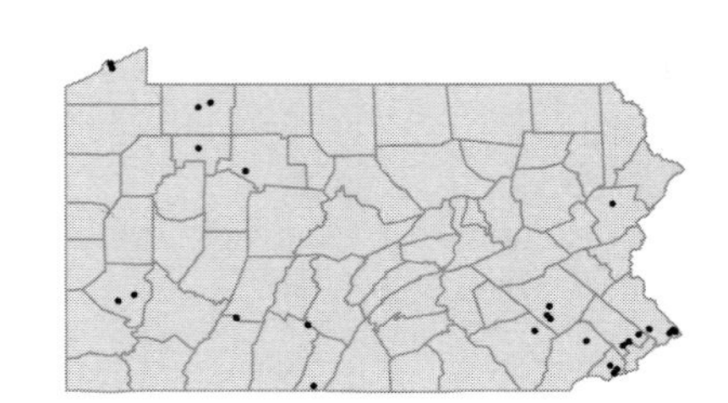

Trifolium fragiferum L.
Strawberry clover
Herbaceous perennial
Waste ground or ballast. Represented by a single collection from Philadelphia Co. in 1878.

FACU
I

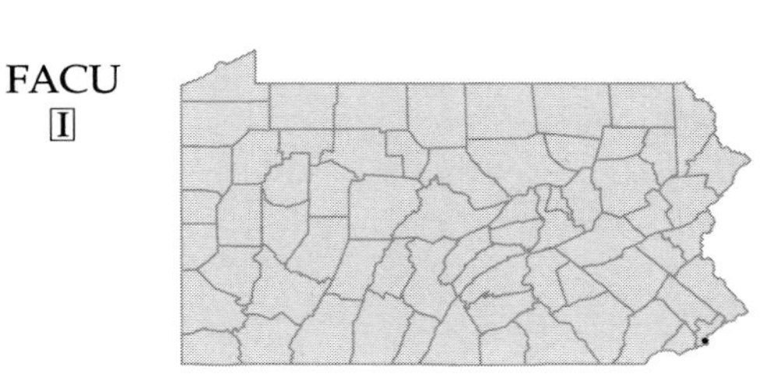

Trifolium hybridum L.
Alsike clover
Herbaceous perennial
Roadsides, clearings and fields.

FACU
I

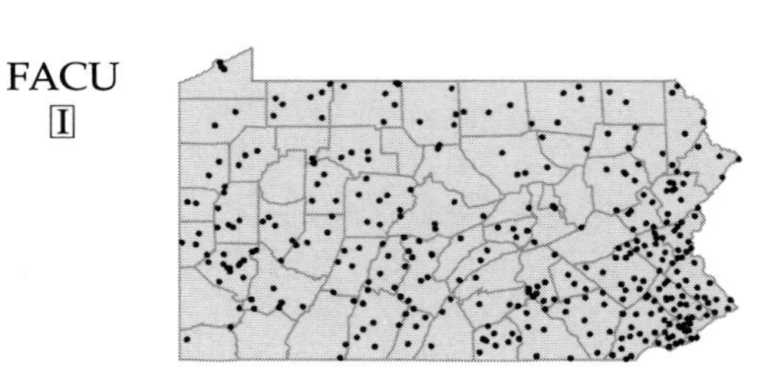

Trifolium incarnatum L.
Crimson clover; Italian clover
Herbaceous annual
Roadsides and waste ground, spreading from cultivation.

I

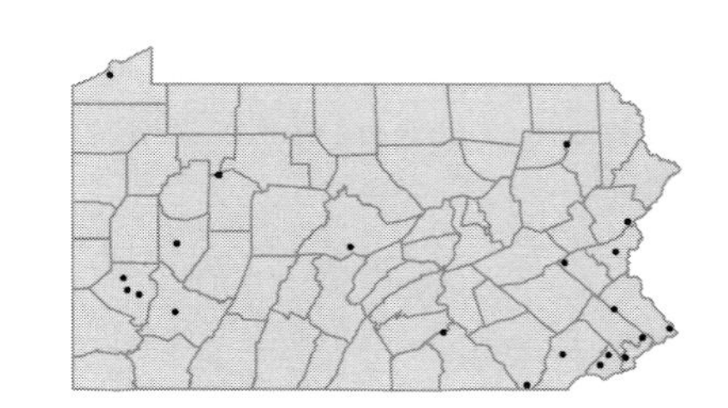

Trifolium lappaceum L.
Lappa clover
Herbaceous annual
Ballast. Represented by a single collection from Philadelphia Co. in 1879.

I

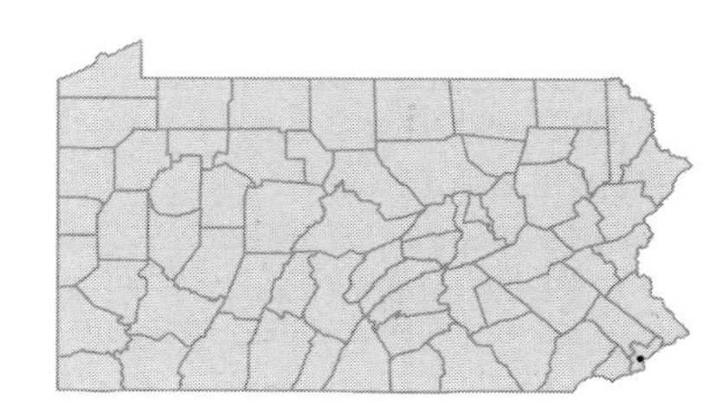

Trifolium pratense L.
Red clover
Herbaceous perennial
Widely grown as a forage plant, also extensively naturalized on roadsides, pastures and fields.

FACU-
I

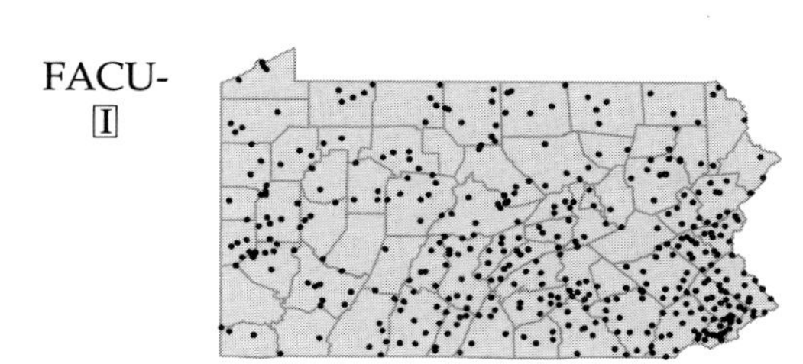

Trifolium reflexum L.
Buffalo clover
Herbaceous annual
Woodland edges, fields and roadsides. Believed to be extirpated, last collected in 1899.

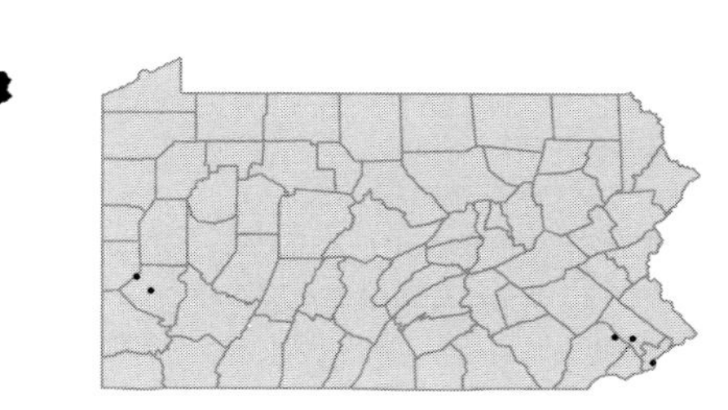

Trifolium repens L.
Dutch clover; White clover
Herbaceous perennial
Roadsides, meadows, old fields and lawns.

FACU-
I

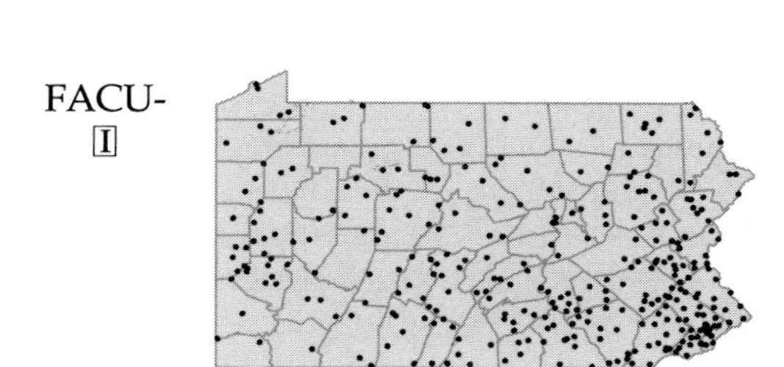

Trifolium resupinatum L.
Persian clover
Herbaceous annual
Lawns, fields, roadsides or stream banks.

UPL
I

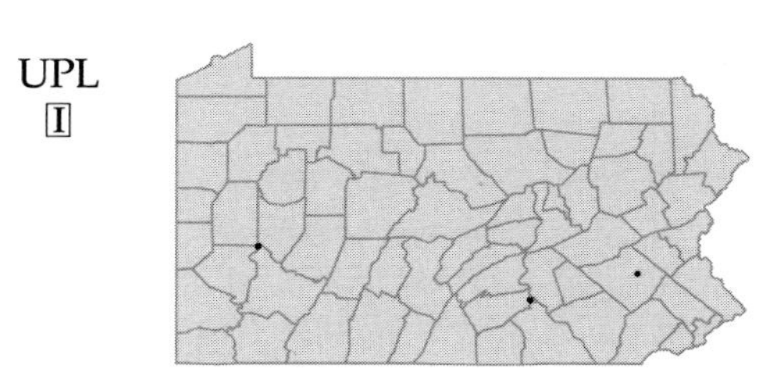

Trifolium squamosum L.
Clover
Herbaceous annual
Waste ground and ballast.

I

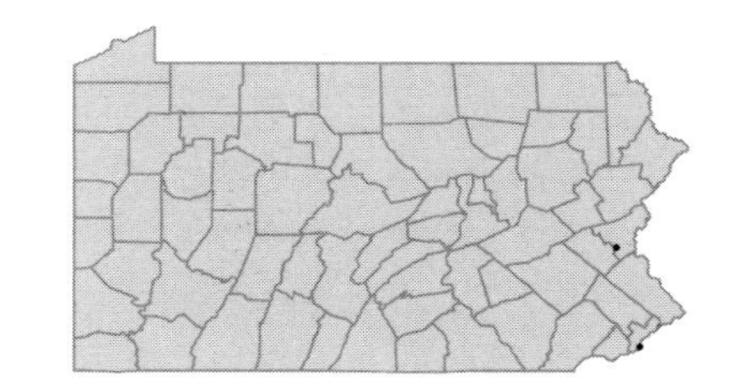

Trifolium striatum L.
Knotted clover
Herbaceous annual
Waste ground and ballast. Represented by a single pre-1900 collection from Philadelphia Co.

I

Trifolium virginicum Small
Kate's-Mountain clover; Shale-barren clover
Herbaceous perennial
Shale barrens.

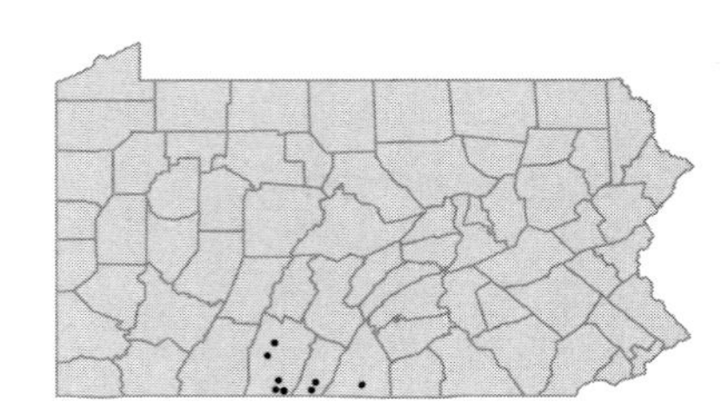

• ***Trigonella procumbens*** (Bess.) Reichenb.
Fenugreek
Herbaceous annual
Waste ground and ballast. Represented by three collections from Philadelphia Co. 1845 to 1876.

I

• ***Ulex europaeus*** L.
Gorse
Deciduous shrub
Waste ground or ballast. Represented by a single collection from Philadelphia Co. in 1878.

I

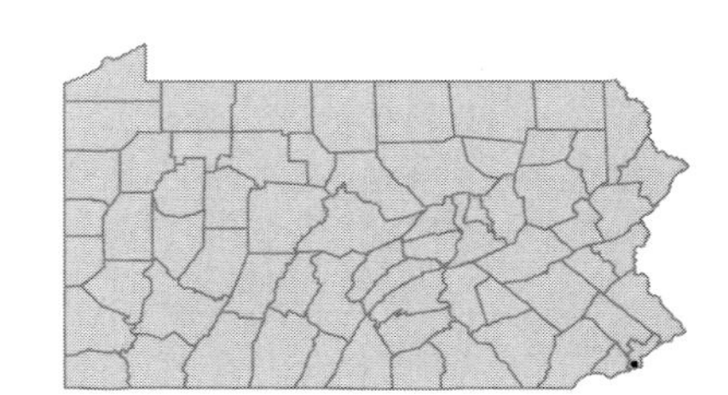

• ***Vicia americana*** Muhl. ex Willd.
Purple vetch; Tare
Herbaceous perennial vine
Moist, gravelly shores, thickets, meadows or roadside banks.

FACU-

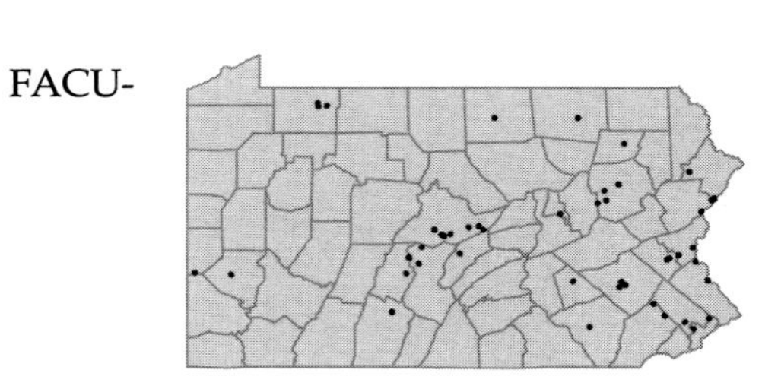

Vicia caroliniana Walt.
Wood vetch
Herbaceous perennial vine
Rich woods and thickets, often on limestone.

FACU-

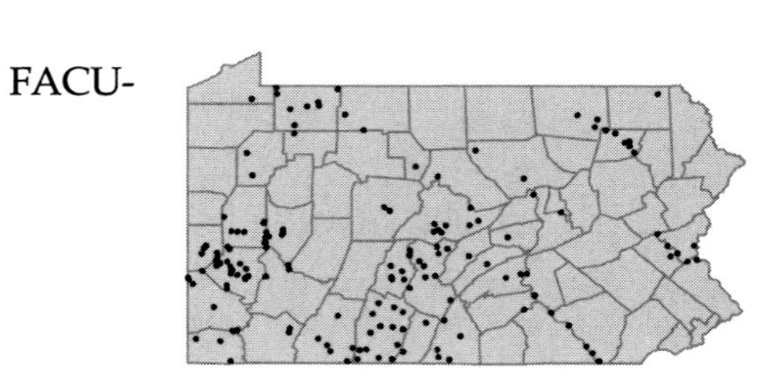

Vicia cracca L. [I]
Canada pea; Cow vetch
Herbaceous perennial vine
Fields, roadsides and floodplains.

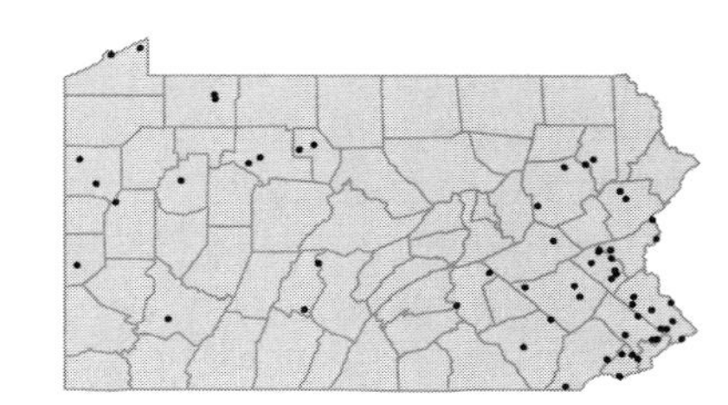

Vicia faba L. [I]
Broad bean; Horse bean
Herbaceous annual
Rubbish dump. Represented by a single collection from Northampton Co. in 1944.

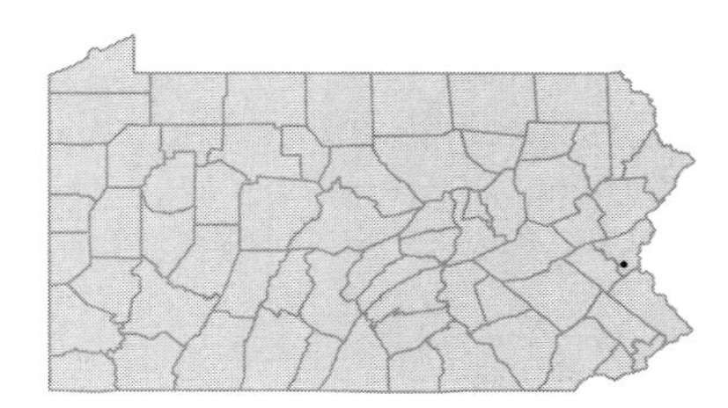

Vicia hirsuta (L.) S.F.Gray [I]
Vetch
Herbaceous annual vine
Roadsides and waste places.

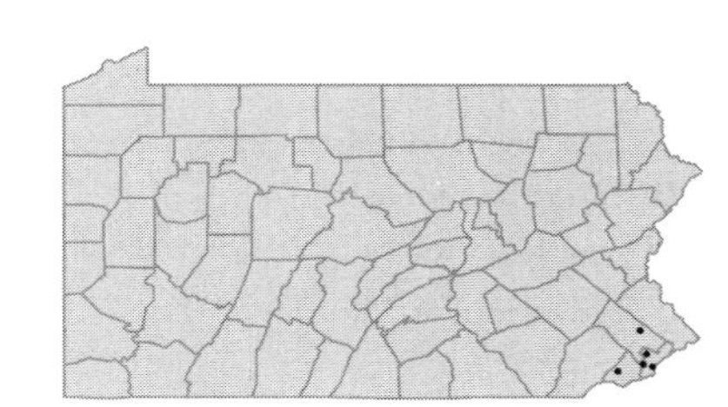

Vicia sativa L. ssp. ***sativa*** FACU- [I]
Common vetch; Tare
Herbaceous annual
Cultivated fields and roadsides.

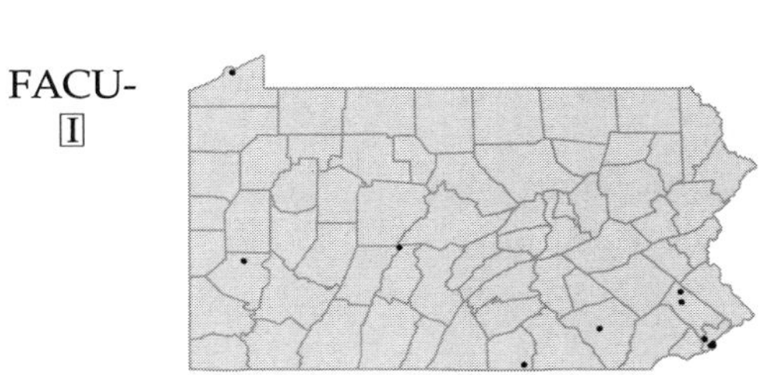

Vicia sativa L. ssp. ***nigra*** (L.) Ehrh. FACU- [I]
Narrow-leaved vetch; Tare
Herbaceous annual
Spreading from cultivation to waste ground and roadsides.
Vicia angustifolia Reichard PFGBWC

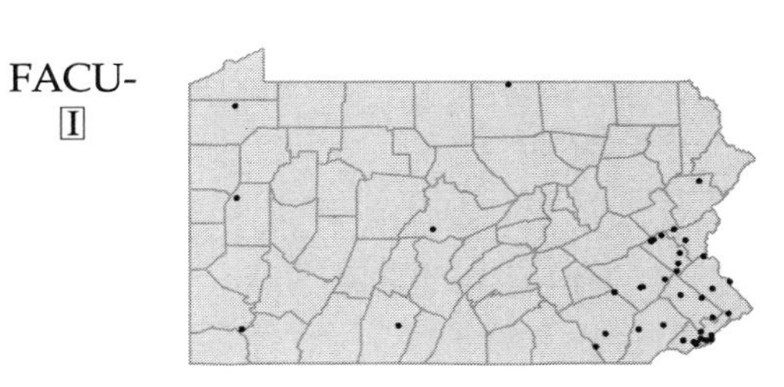

Vicia sepium L. [I]
Wild tare
Herbaceous perennial vine
Roadsides and old fields. Represented by a single collection from Monroe Co. in 1915.

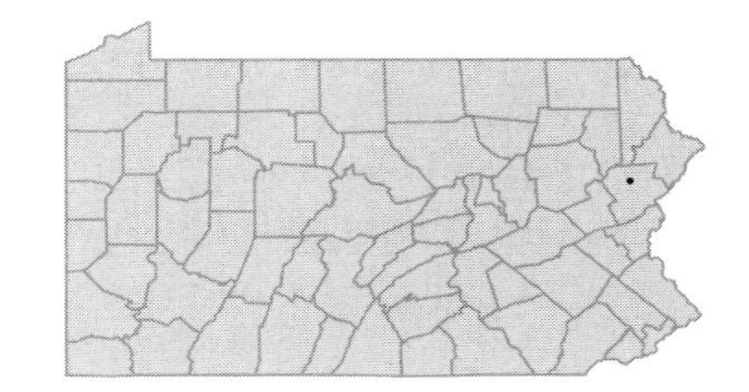

Vicia tetrasperma (L.) Schreb.
Slender vetch
Herbaceous annual
Roadsides and old fields, also moist serpentine barrens.

[I]

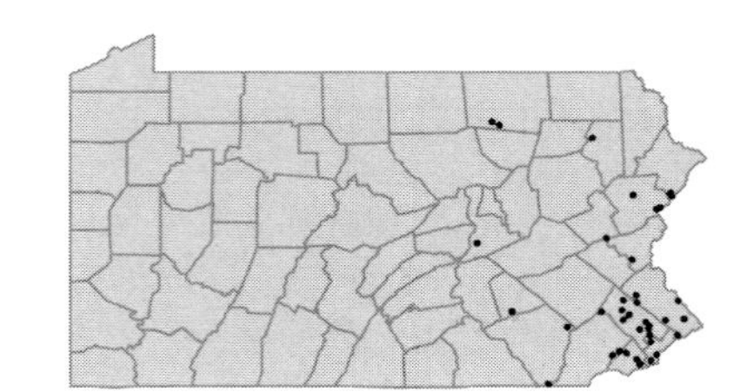

Vicia villosa Roth ssp. ***villosa***
Hairy vetch; Winter vetch
Herbaceous annual
Occasionally spreading from cultivation to fields and roadsides.

[I]

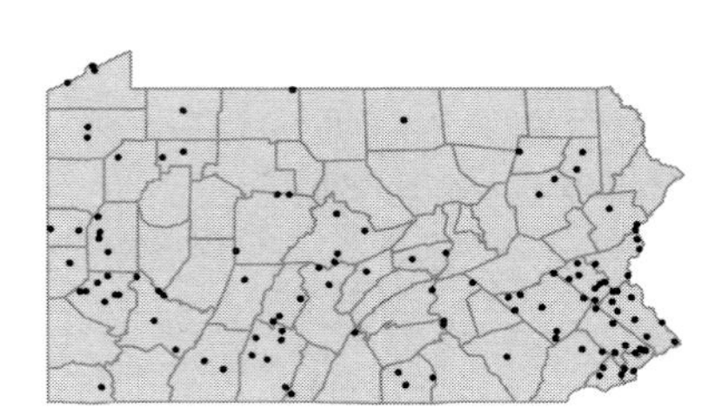

Vicia villosa Roth ssp. ***varia*** (Host) Corb.
Hairy vetch; Winter vetch
Herbaceous annual
Roadsides, fields and disturbed ground.
Vicia dasycarpa Ten. FGBWC

[I]

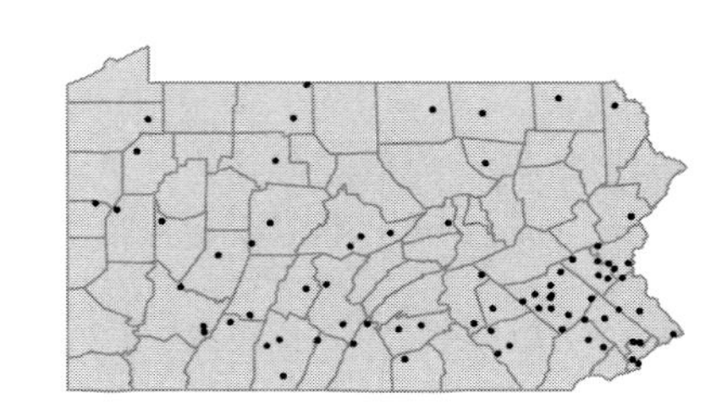

• ***Vigna luteola*** (Jacquin) Bentham
Bean
Herbaceous perennial vine
Swampy ground and ballast.

FACW
[I]

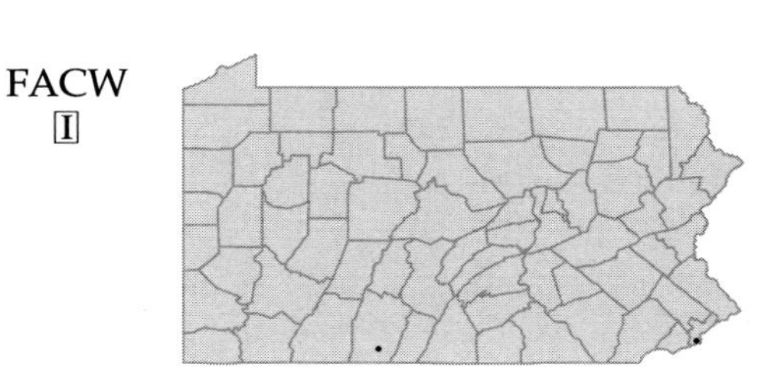

Vigna unguiculata (L.) Walpers
Black-eyed pea
Herbaceous annual
Waste ground or ballast, also cultivated.
Vigna sinensis (L.) Endl. FGB

[I]

• ***Wisteria floribunda*** (Willd.) DC.
Japanese wisteria
Woody vine
Roadside thickets and abandoned gardens.

[I]

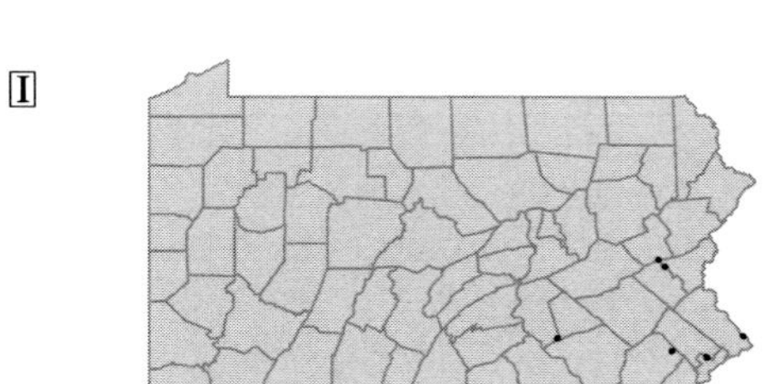

Wisteria frutescens (L.) Poir. FACW-
American wisteria
Woody vine
Alluvial woods and river banks.
Wisteria macrostachya Nutt. FGBW

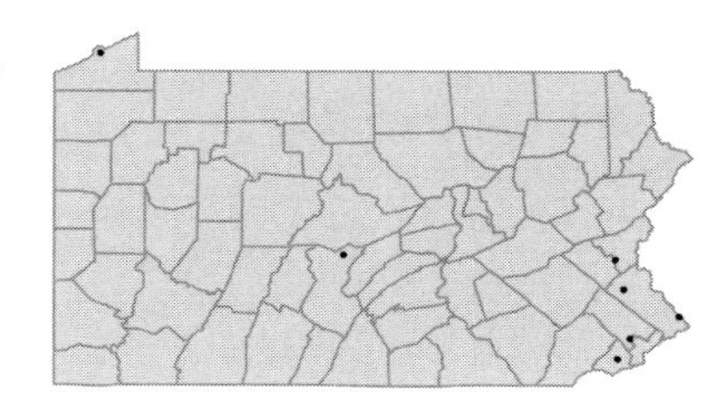

Wisteria sinensis (Sims) Sweet [I]
Chinese wisteria
Woody vine
Disturbed woodlands, abandoned nurseries and gardens.

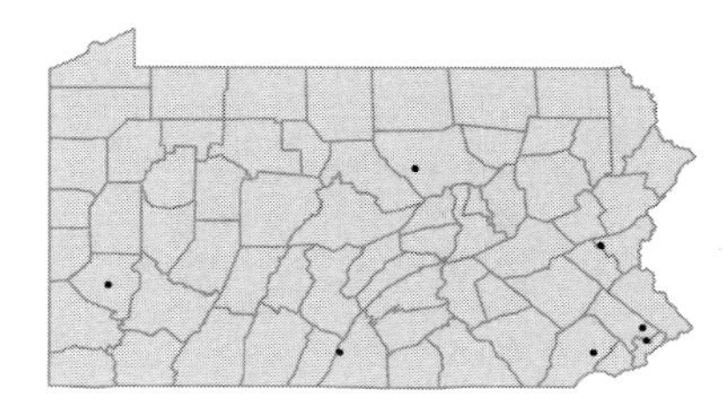

CAESALPINIACEAE

• ***Cercis canadensis*** L.
Redbud; Judas-tree
Deciduous tree
Wooded slopes and ravines in dry to moist, rich soils on limestone or diabase.

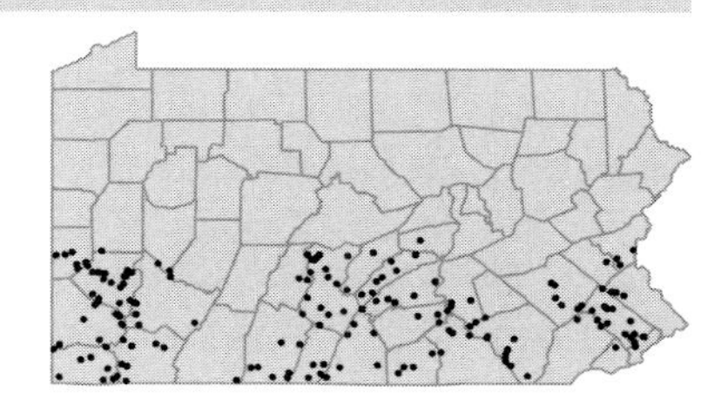

• ***Chamaecrista fasciculata*** (Michx.) Greene FACU
Partridge-pea; Prairie senna
Herbaceous annual
River banks, sandy flats, railroad cinders and serpentine barrens.
Cassia chamaecrista L. P; *Cassia fasciculata* Michx. FGBW

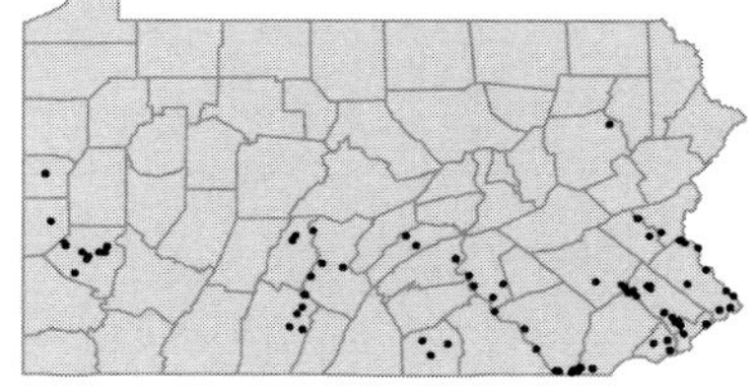

Chamaecrista nictitans (L.) Moench
Wild sensitive-plant
Herbaceous annual
Dry, open, sandy ground of roadsides, railroad embankments and old fields.
Cassia nictitans L. PFGBW

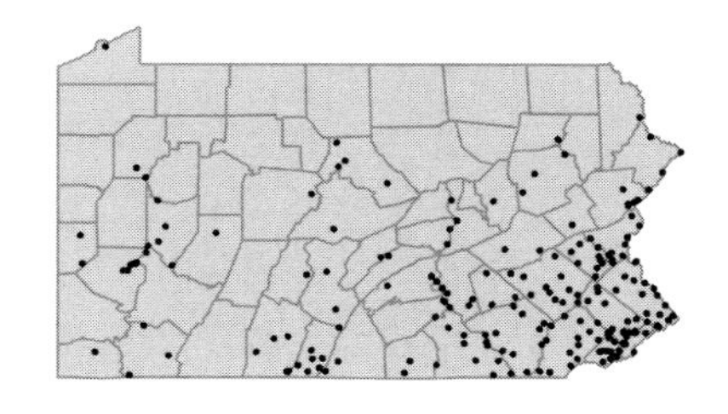

• ***Gleditsia triacanthos*** L. FAC-
Honey-locust
Deciduous tree
Wooded slopes, river banks and floodplains, also frequently planted in urban and suburban areas.

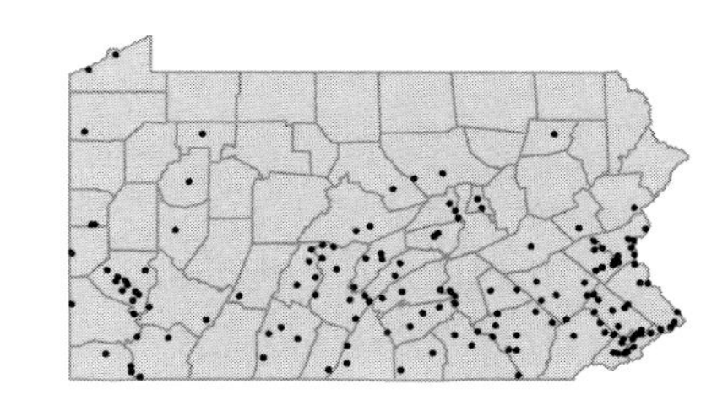

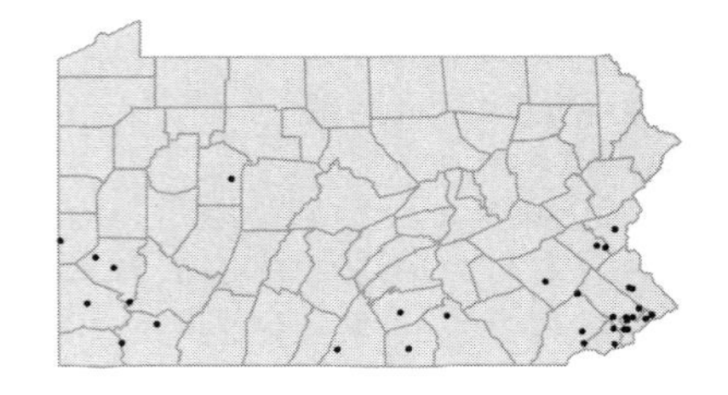

• ***Gymnocladus dioica*** (L.) K.Koch
Kentucky coffee-tree
Deciduous tree
Rich, moist woods and bottomlands, also occasionally planted.

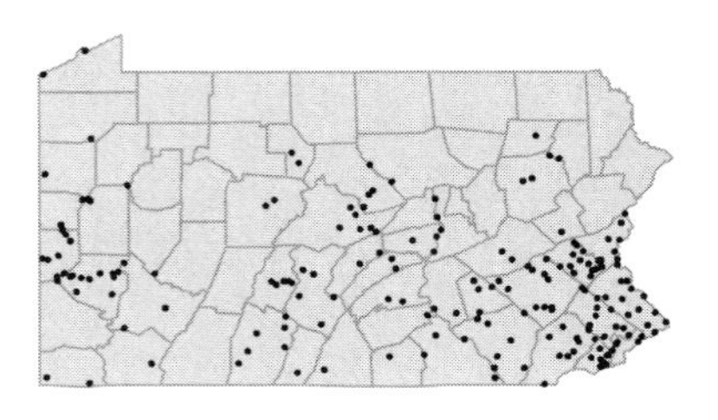

• ***Senna hebecarpa*** (Fern.) Irwin & Barneby
Wild senna
Herbaceous perennial
River banks, sandy shores, and old fields.
Cassia hebecarpa Fern. FGBW

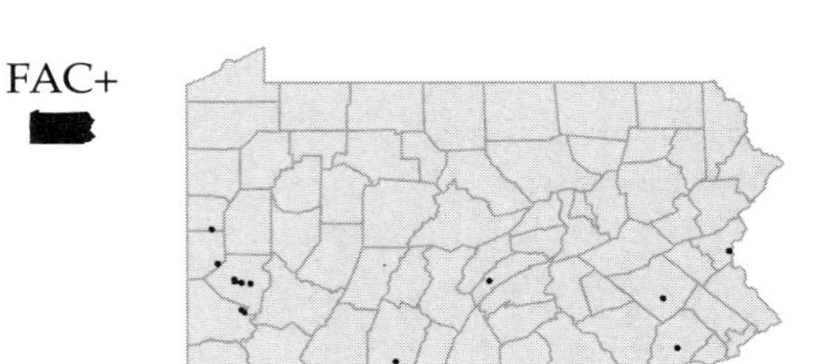

Senna marilandica (L.) Link
FAC+
Wild senna
Herbaceous perennial
Dry roadsides and thickets.
Cassia marilandica L. PFGBW

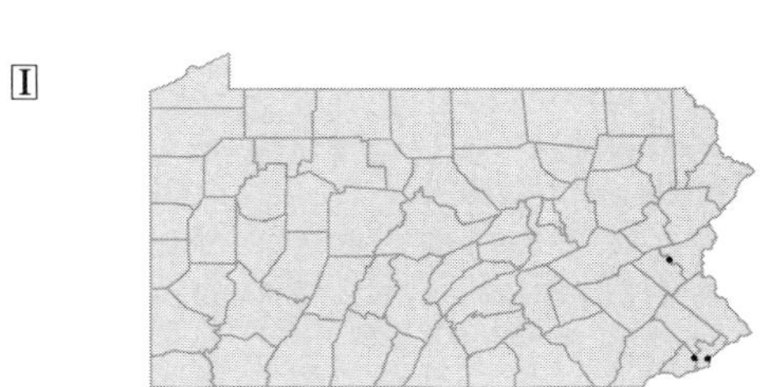

Senna obtusifolia (L.) Irwin & Barneby
I
Coffeeweed; Sicklepod
Herbaceous annual
Rich soil and waste ground.
Cassia tora L. PFGBW; *Cassia tora* L. PFGBW

ELAEAGNACEAE

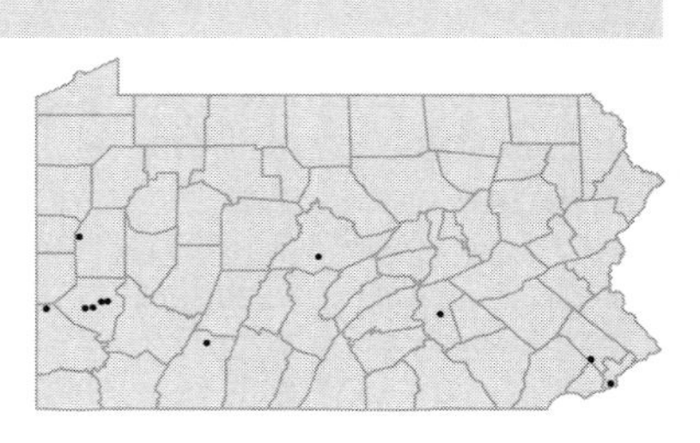

• ***Elaeagnus angustifolia*** L.
FACU
I
Russian olive
Deciduous shrub
Cultivated and escaped to waste ground, also planted for erosion control.

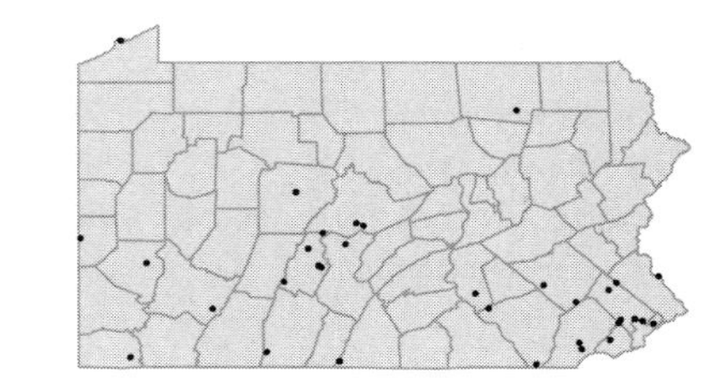

Elaeagnus umbellata Thunb.
Autumn olive
Deciduous shrub
Planted by the Game Commission and extensively naturalized in old fields and abandoned pastures. It has become a serious weed in some parts of the state.

• ***Shepherdia canadensis*** (L.) Nutt.
Buffalo-berry; Soapberry
Deciduous shrub
Wet, shaly banks and open to partly shaded slumps along Lake Erie.

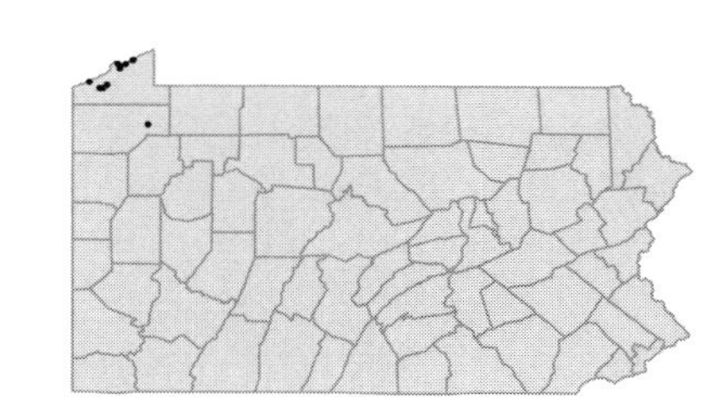

PODOSTEMACEAE

• ***Podostemum ceratophyllum*** Michx.
Riverweed
Herbaceous perennial, rooted submergent aquatic
On rocks in shallow, fast-moving water of rivers and streams, apparently adversely affected by acid mine drainage.

OBL

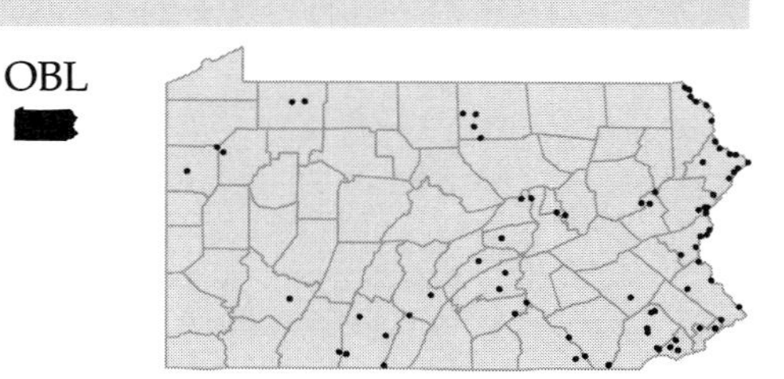

HALORAGACEAE

• ***Myriophyllum aquaticum*** (Vell.) Verdc.
Parrot's-feather; Water-feather
Herbaceous perennial, rooted submergent aquatic
Ponds.
Myriophyllum brasiliense Camb. FGBW

OBL
[I]

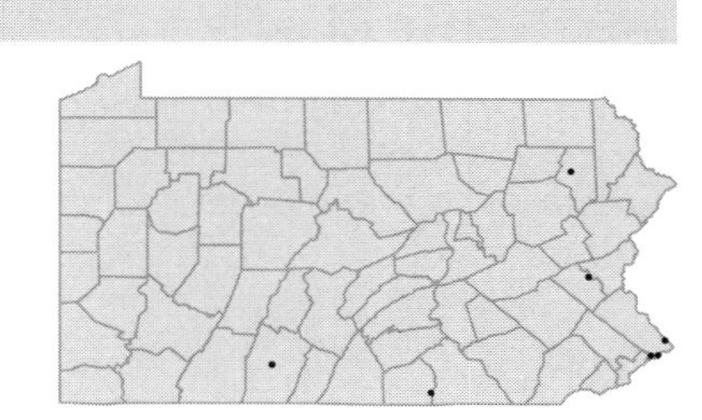

Myriophyllum farwellii Morong
Farwell's water-milfoil
Herbaceous perennial, rooted submergent aquatic
Lakes and ponds.

OBL

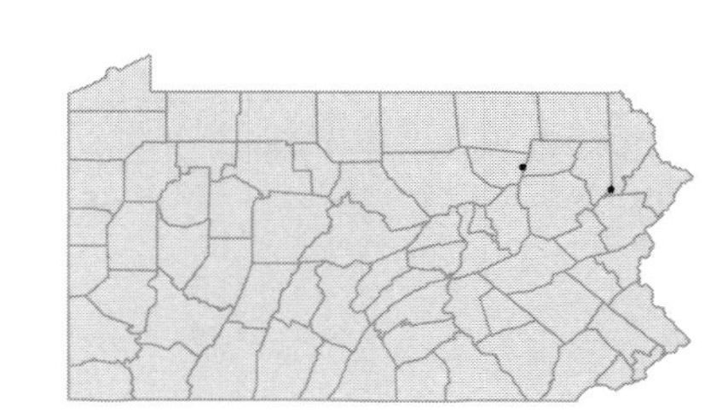

Myriophyllum heterophyllum Michx.
Broad-leaved water-milfoil
Herbaceous perennial, rooted submergent aquatic
Still water of ponds.

OBL

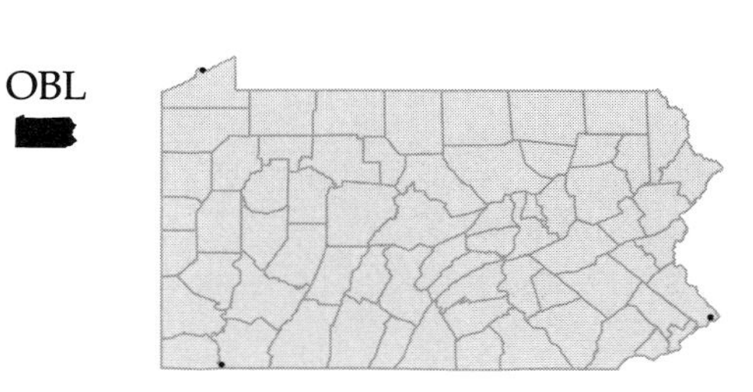

Myriophyllum humile (Raf.) Morong
Water-milfoil
Herbaceous perennial, rooted submergent aquatic
Lakes and ponds. A terrestrial form occurs on soft, exposed peat.

OBL

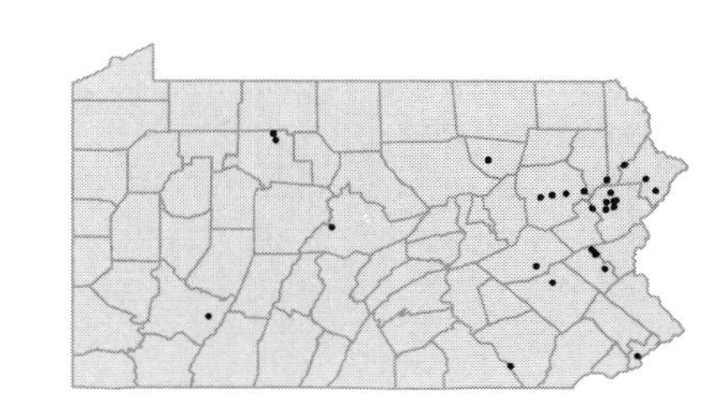

Myriophyllum sibiricum Komarov
Northern water-milfoil
Herbaceous perennial, rooted submergent aquatic
Still water of rivers, lakes, ponds and marshes.
Myriophyllum spicatum L. var. *exalbescens* (Fern.) Jeps. G; *Myriophyllum exalbescens* Fern.

OBL

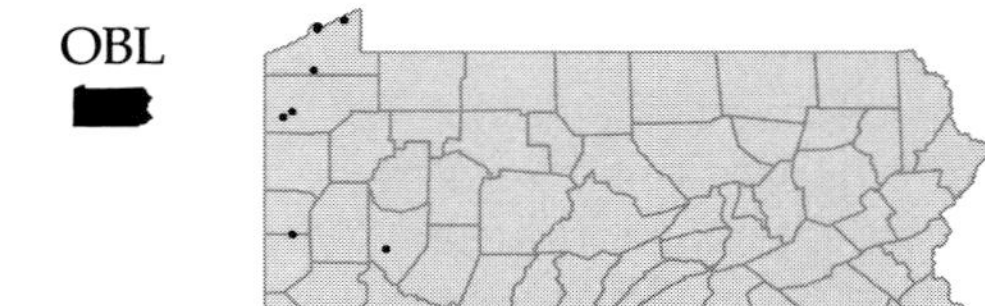

Myriophyllum spicatum L.
Eurasian water-milfoil
Herbaceous perennial, rooted submergent aquatic
Lakes and rivers. A fairly recent arrival, this species is spreading rapidly in the state's waterways. It has sometimes been mistaken for *M. sibiricum* Komarov (syn. *M. exalbescens* Fern.) which is very rare in PA.

OBL
I

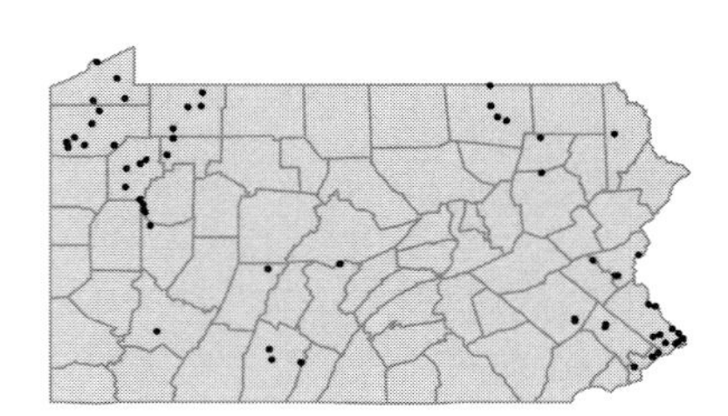

Myriophyllum tenellum Bigel.
Slender water-milfoil
Herbaceous perennial, rooted submergent aquatic
Shallow water of northern lakes, sometimes forming a dense turf on the lake bottom.

OBL

Myriophyllum verticillatum L.
Whorled water-milfoil
Herbaceous perennial, rooted submergent aquatic
Shallow ponds or marshes.

OBL

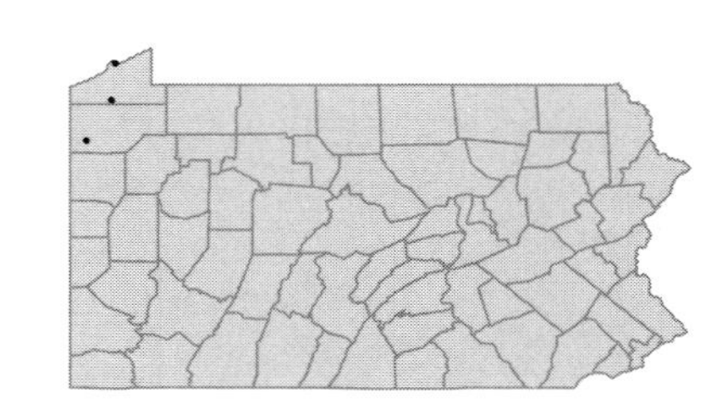

• ***Proserpinaca palustris*** L. var. ***palustris***
Mermaid-weed
Herbaceous perennial, rooted submergent aquatic
Pond and wet swale.

OBL

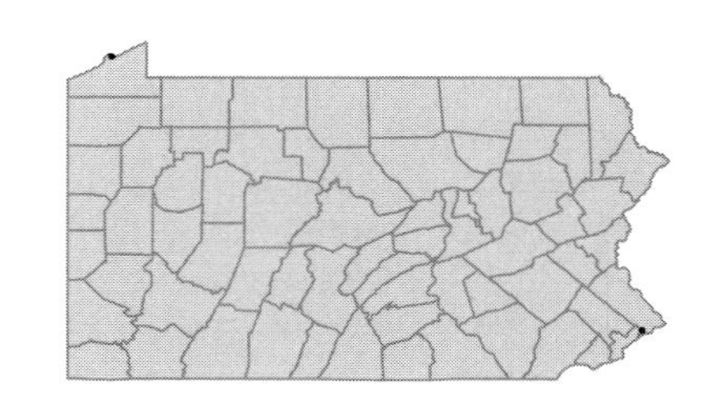

Proserpinaca palustris L. var. ***crebra*** Fern. & Grisc.
Mermaid-weed
Herbaceous perennial, rooted submergent aquatic
Swamps, bogs, marshes, and vernal ponds.
Proserpinaca intermedia Mackenzie C

OBL
I

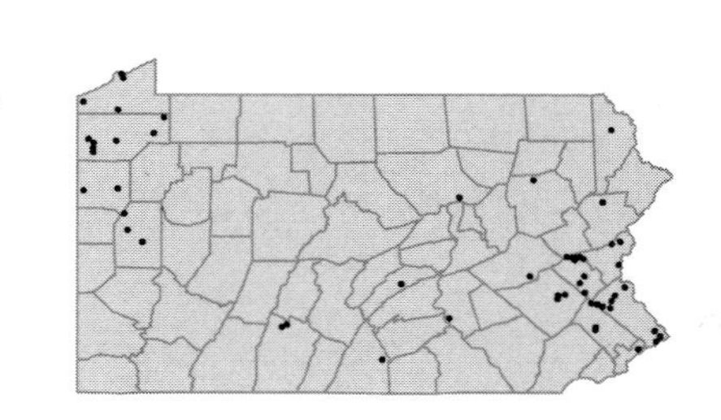

Proserpinaca pectinata Lam.
Comb-leaved mermaid-weed
Herbaceous perennial, rooted submergent aquatic
Swamps or bogs. Believed to be extirpated, represented by a single collection from Bucks Co. in 1865.

OBL

LYTHRACEAE

• ***Ammannia coccinea*** Rottb.
Scarlet ammannia; Tooth-cup
Herbaceous annual
Wet, sandy or silty shores.

OBL

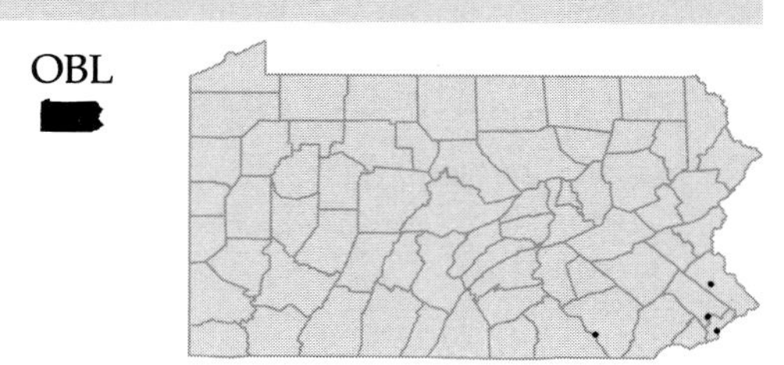

• ***Cuphea viscosissima*** Jacq.
Blue waxweed; Clammy cuphea
Herbaceous annual
Dry open banks, fencerows and fields.
Cuphea petiolata (L.) Koehne FGBW

FAC-

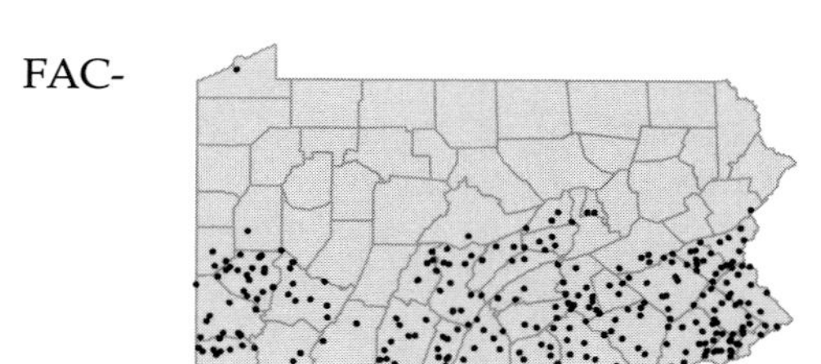

• ***Decodon verticillatus*** (L.) Ell.
Water-willow
Herbaceous perennial
Lakes, swamps and bog margins in shallow water.

OBL

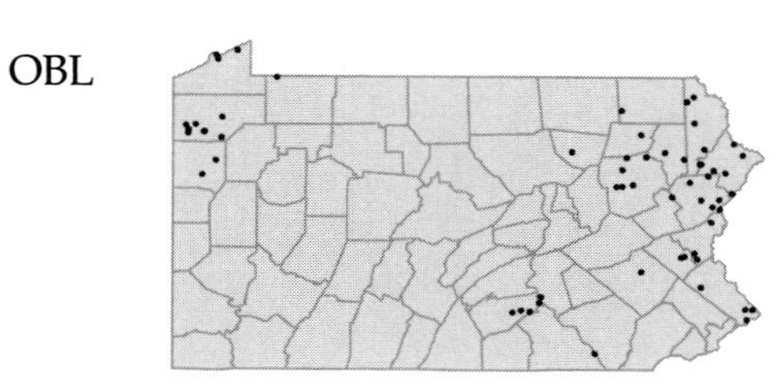

• ***Lythrum alatum*** Pursh
Winged loosestrife
Herbaceous perennial
Swamps, wet meadows, marshy shorelines and ditches.

FACW+

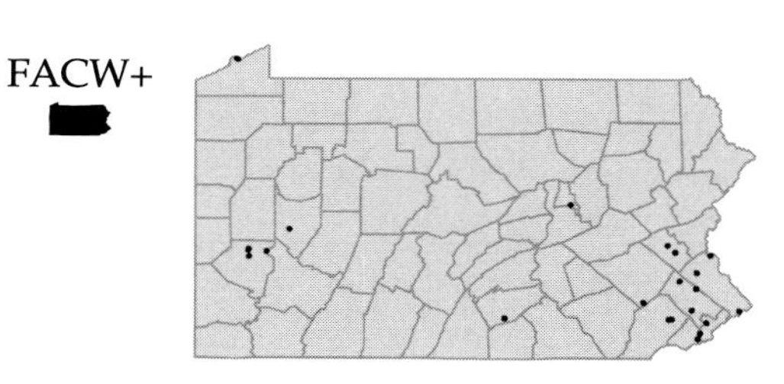

Lythrum hyssopifolia L.
Hyssop loosestrife
Herbaceous annual
Moist roadside ditches.

OBL
I

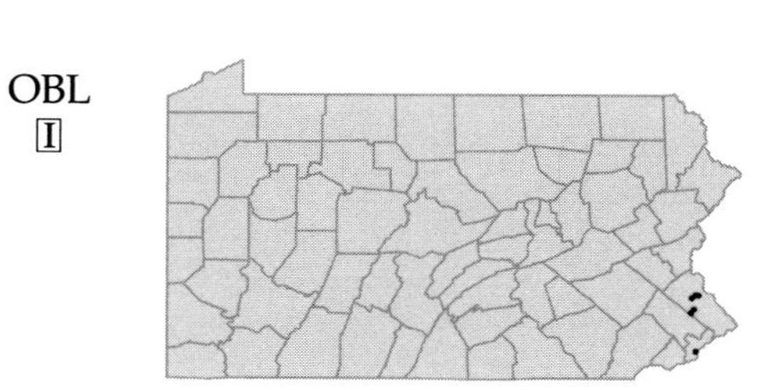

Lythrum salicaria L.
Purple loosestrife
Herbaceous perennial
Swamps, wet meadows and shores. A serious threat to native wetland species in many areas.

FACW+
[I]

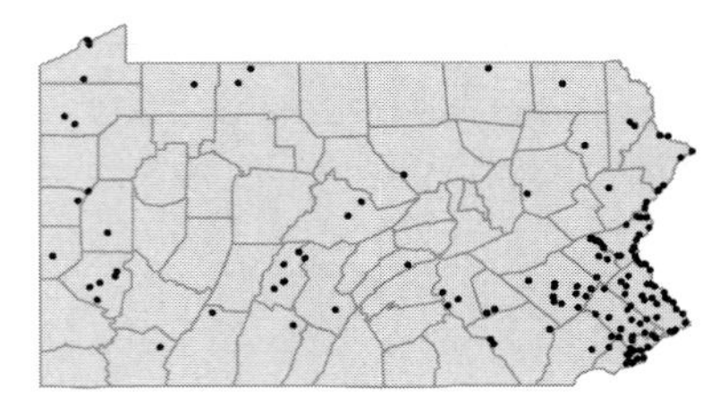

Lythrum virgatum L.
Loosestrife
Herbaceous perennial
Moist, alluvial thickets. Represented by a single collection from Lackawanna Co. in 1946.

[I]

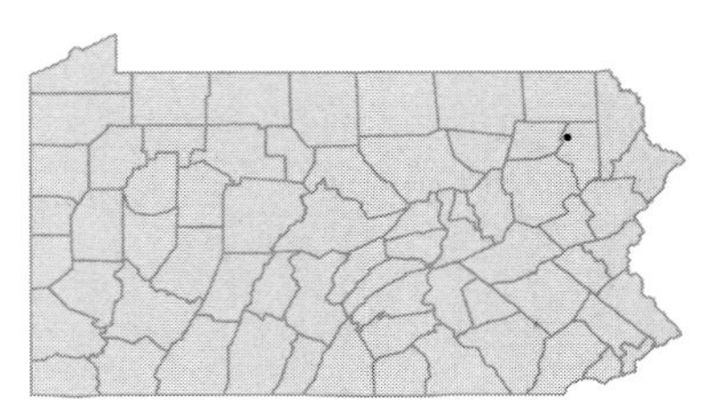

• ***Rotala ramosior*** (L.) Koehne
Tooth-cup
Herbaceous annual
Wet, sandy shores and other swampy, open ground.

OBL

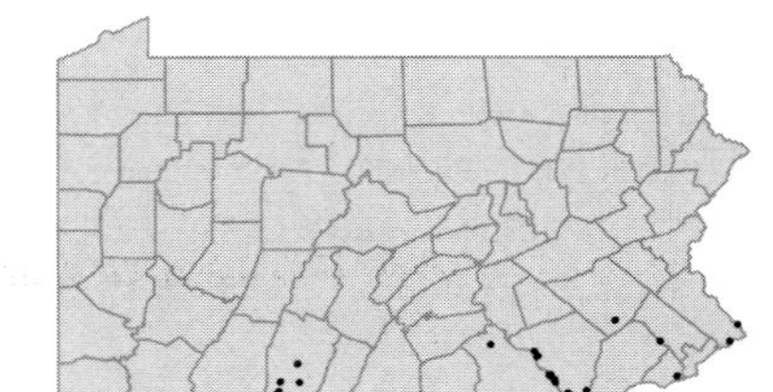

THYMELAEACEAE

• ***Dirca palustris*** L.
Leatherwood
Deciduous shrub
Rich, deciduous woods, rocky banks and thickets.

FAC

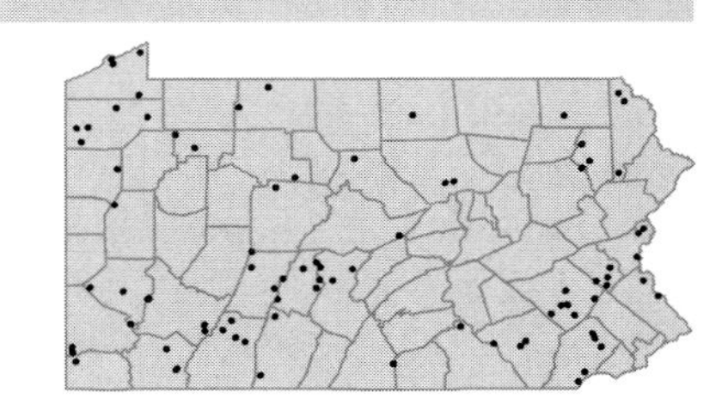

TRAPACEAE

• ***Trapa natans*** L.
Water-chestnut
Herbaceous annual, free-floating aquatic
Shallow water of ponds.

OBL
[I]

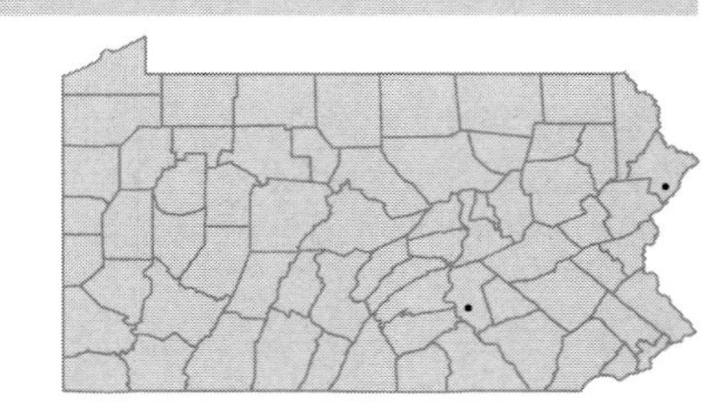

ONAGRACEAE

• ***Circaea alpina*** L.
Enchanter's-nightshade
Herbaceous perennial
Moist hemlock woods, rocky wooded slopes and stream banks.

FACW

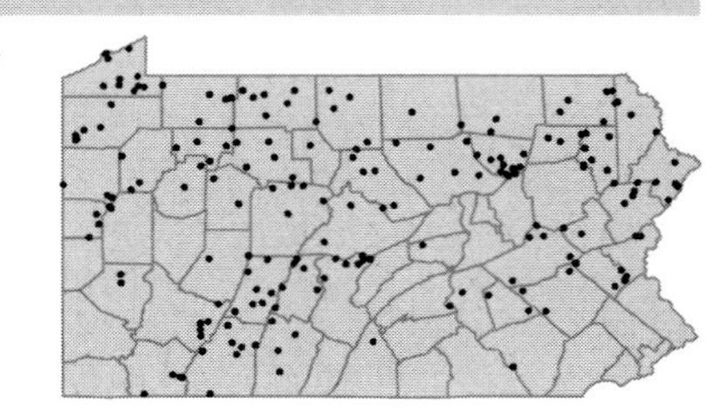

Circaea alpina x lutetiana
Enchanter's-nightshade
Herbaceous perennial
Cool, rocky, mountain woods and stream banks.
Circaea x intermedia Ehrh.

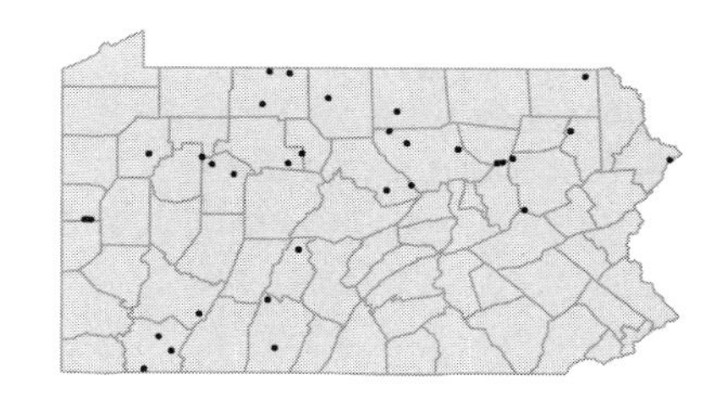

Circaea lutetiana L. ssp. ***canadensis*** (L.) Aschers. & Magnus — FACU
Enchanter's-nightshade
Herbaceous perennial
Rocky, upland woods, damp woods and floodplains.
Circaea quadrisulcata (Maxim.) Franch. & Savat var. *canadensis* (L.) Hara FGB

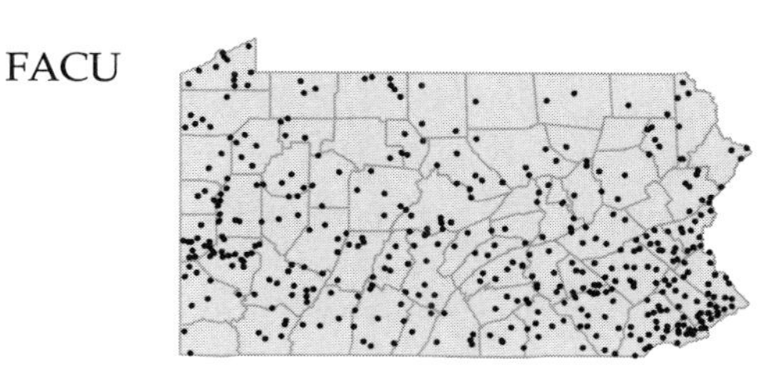

• ***Clarkia purpurea*** (Curtis) A.Nels. & Macb. ssp. ***quadrivulnera*** (Dougl.) Lewis & Lewis — [I]
Farewell-to-spring
Herbaceous annual
Ballast. Represented by a single collection from Philadelphia Co. in 1877.
Godetia quadrivulnera Spach W

• ***Epilobium angustifolium*** L. — FAC
Fireweed
Herbaceous perennial
Woods edges and recent clearings, in open sandy ground.
Chamaenerion angustifolium (L.) Scop. P

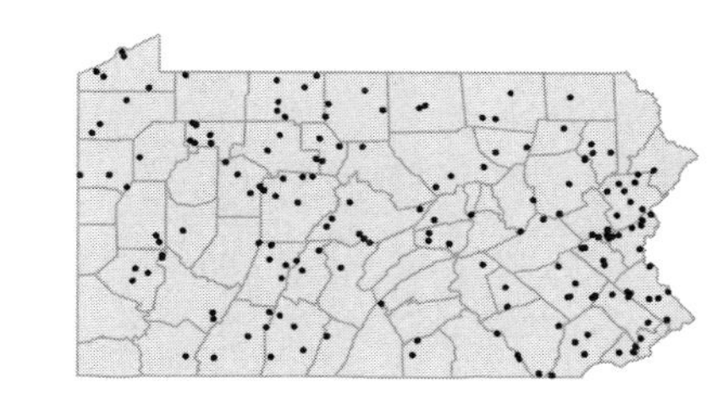

Epilobium ciliatum Raf. — FAC-
Willow-herb
Herbaceous perennial
Moist, springy soil and wet rocks.
Epilobium adenocaulon Haussk. PB; *Epilobium glandulosum* Lehm. var. *adenocaulon* (Haussk.) Fern. FW

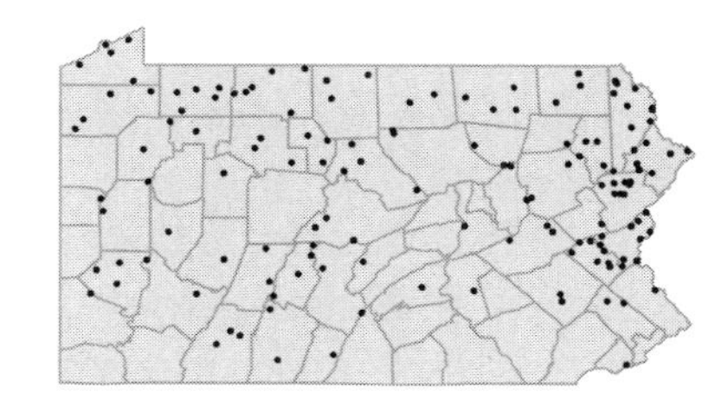

Epilobium coloratum Biehler — OBL
Purple-leaved willow-herb
Herbaceous perennial
Marshes, stream or pond banks and floodplains.

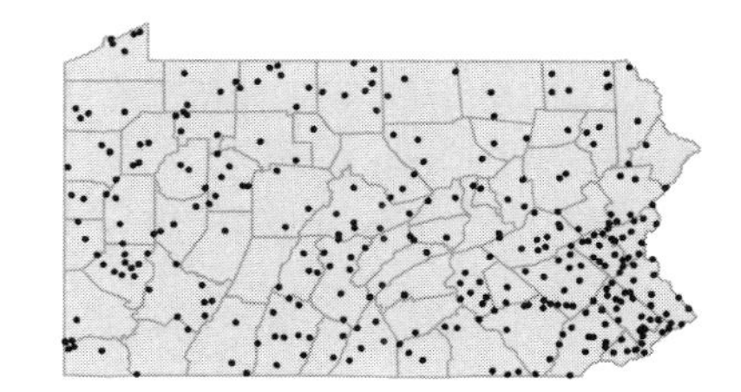

Epilobium hirsutum L.
Hairy willow-herb
Herbaceous perennial
Wet fields, marshes and ditches.

FACW
[I]

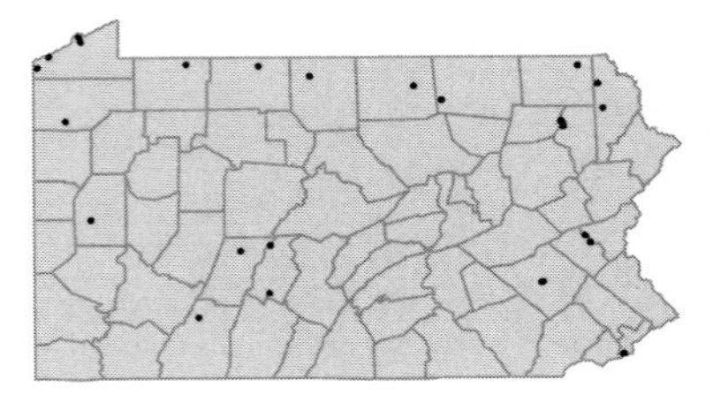

Epilobium leptophyllum Raf.
Willow-herb
Herbaceous perennial
Marshes and boggy pastures.

OBL

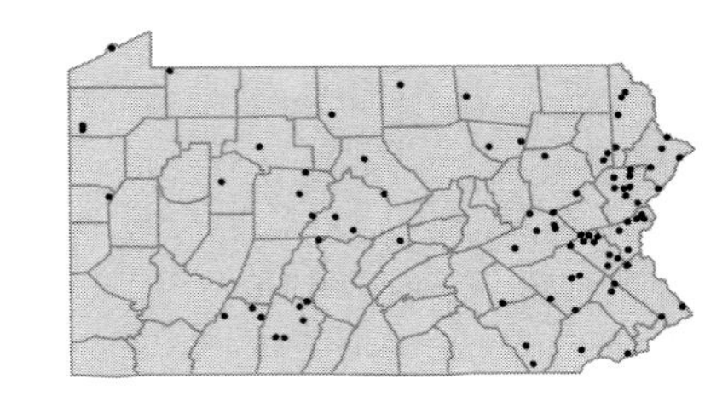

Epilobium palustre L.
Marsh willow-herb
Herbaceous perennial
Bogs and wooded swamps.

OBL

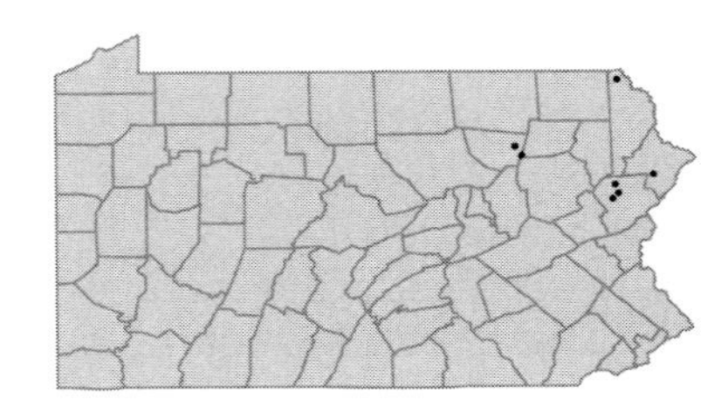

Epilobium parviflorum Schreber
Willow-herb
Herbaceous perennial
Moist shores.

[I]

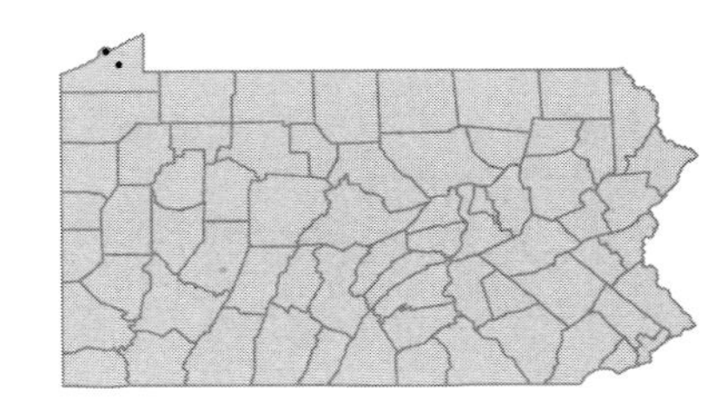

Epilobium strictum Muhl.
Downy willow-herb; Soft willow-herb
Herbaceous perennial
Calcareous marshes, meadows and thickets.

OBL

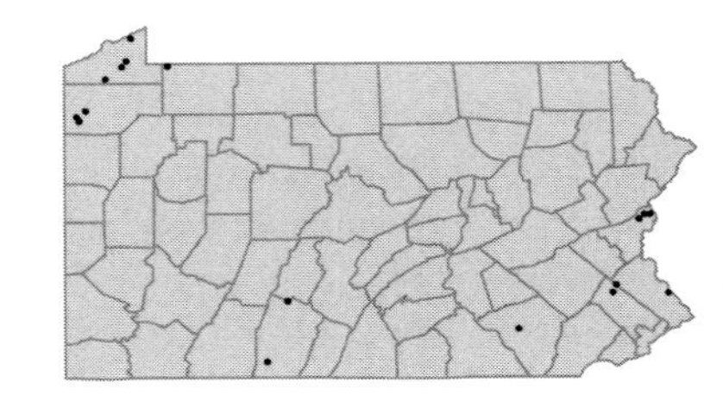

• ***Gaura biennis*** L.
Gaura
Herbaceous annual
Moist meadows, floodplains, stream banks and roadside thickets.

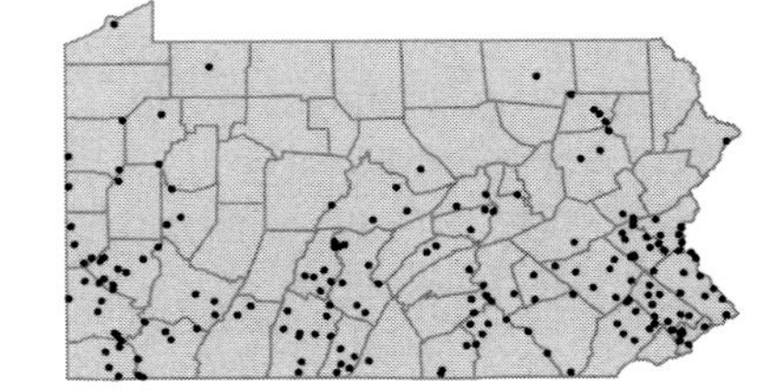

• ***Lopezia coronata*** Andr.
Crown-jewels
Herbaceous annual
Cultivated field. Represented by a single collection from Lehigh Co. in 1953.

[I]

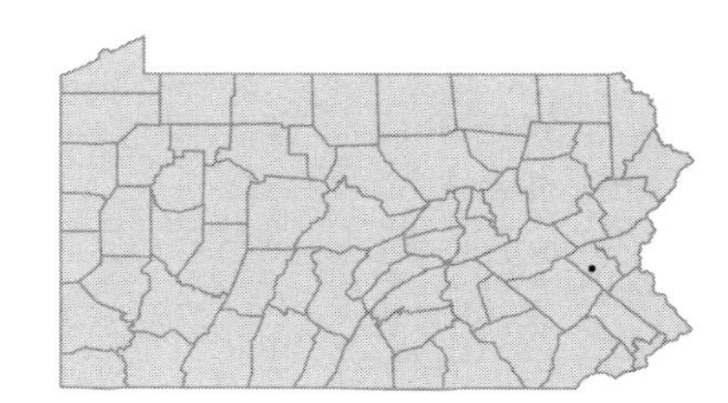

• ***Ludwigia alternifolia*** L.
False loosestrife; Seedbox
Herbaceous perennial
Swampy fields and wet woods.

FACW+

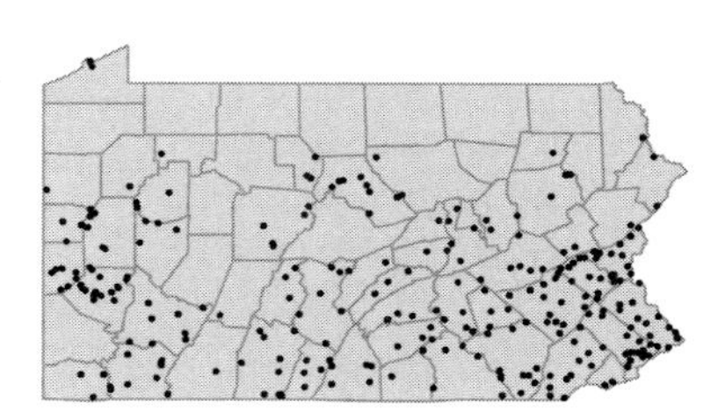

Ludwigia decurrens Walt.
Upright primrose-willow
Herbaceous perennial
Sandy, alluvial beach.

OBL

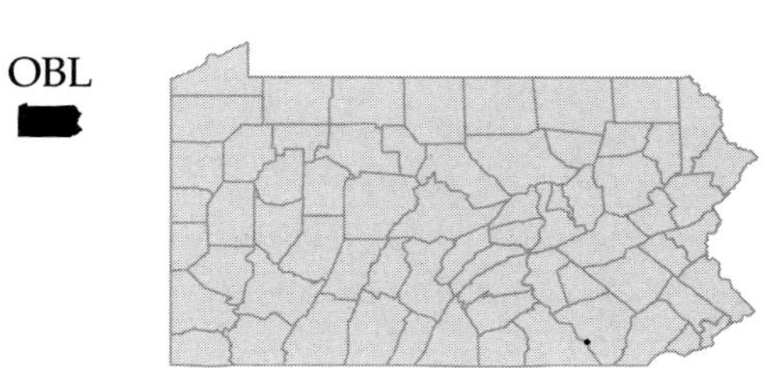

Ludwigia hexapetala (Hook. & Arn.) Zardini, Gu & Raven
Primrose-willow; Water-primrose
Herbaceous perennial
Moist shores or shallow water along streams and canals.
Jussiaea uruguayensis Camb. FGB; *Ludwigia uruguayensis* (Camb.) Hara *pro parte* C; *Jussiaea michauxiana* Fern. W

OBL
[I]

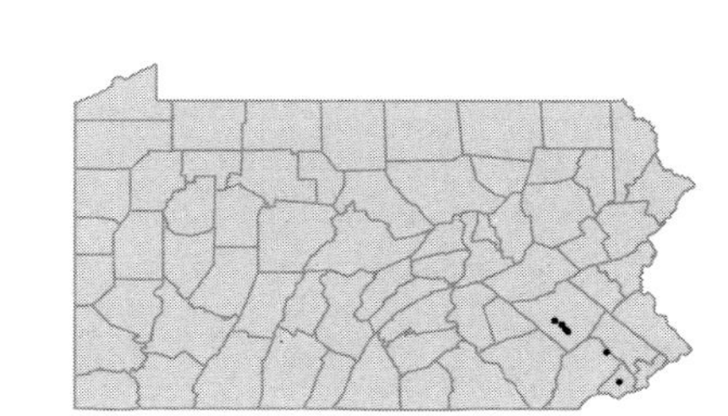

Ludwigia leptocarpa (Nutt.) Hara
Water-willow
Herbaceous annual
Ballast ground. Represented by a single collection from Philadelphia Co. in 1865.
Jussiaea leptocarpa Nutt. F

OBL
[I]

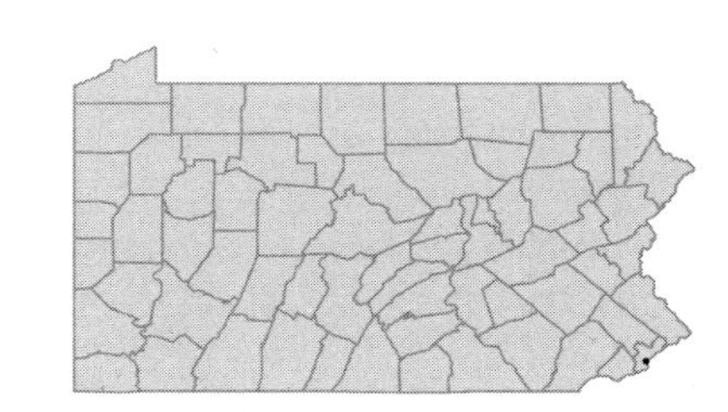

Ludwigia palustris (L.) Ell.
Marsh-purslane
Herbaceous perennial
Swamps, moist meadows, muddy shores, stream banks and ditches.

OBL
[I]

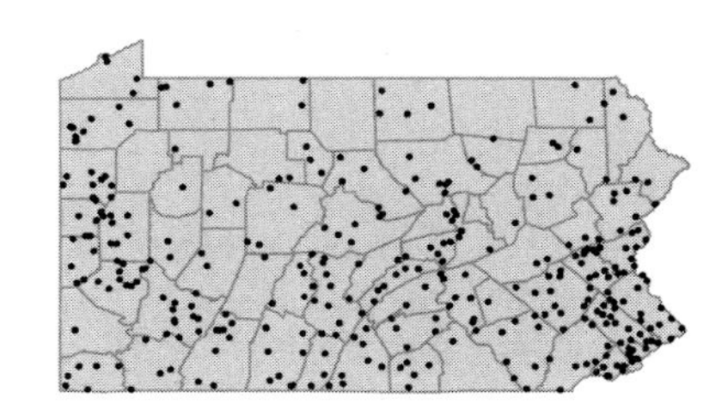

Ludwigia peploides (Kunth) Raven ssp. ***glabrescens*** (Kuntze) Raven OBL
Primrose-willow
Herbaceous perennial
Shallow water or silty, muddy shores.
Jussiaea repens L. var. *glabrescens* Kuntze FGBW

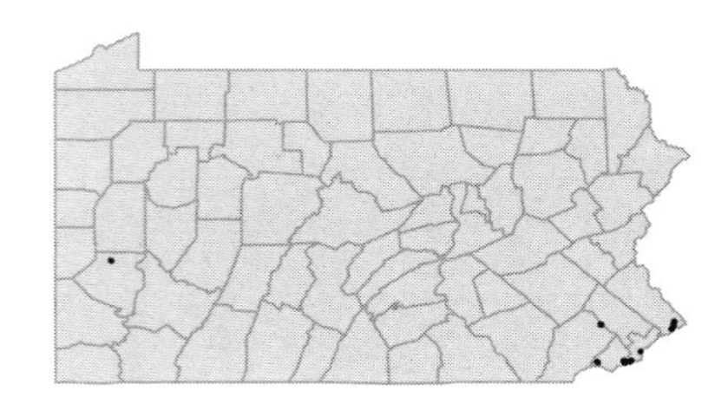

Ludwigia polycarpa Short & Peter OBL
False loosestrife; Seedbox
Herbaceous perennial
Wet meadows and swales.

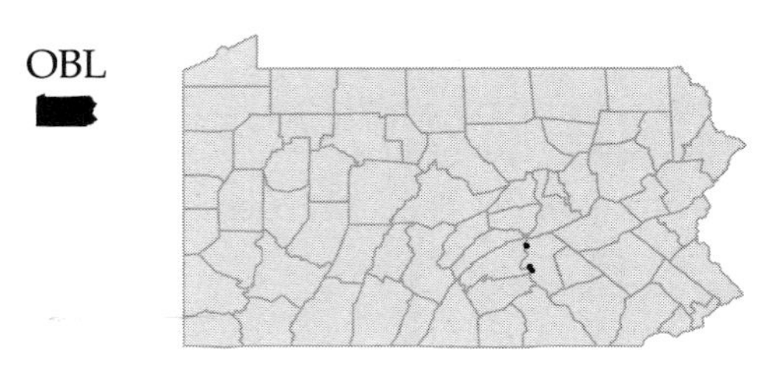

Ludwigia sphaerocarpa Ell. OBL
Spherical-fruited seedbox
Herbaceous perennial
Coastal plain swamps. Believed to be extirpated, last collected in 1866.

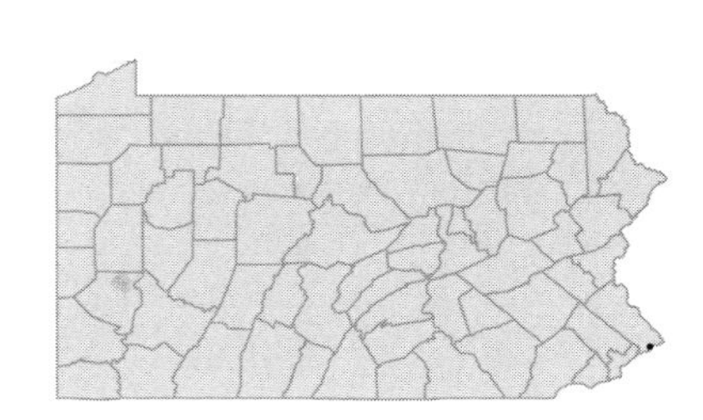

• ***Oenothera argillicola*** Mackenzie
Shale-barren evening-primrose
Herbaceous biennial
Shale barrens.

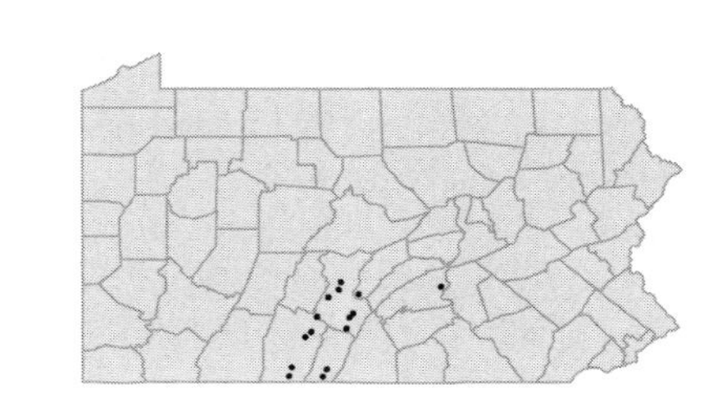

Oenothera biennis L. FACU-
Evening-primrose
Herbaceous biennial
Cultivated fields, waste ground and roadsides.
Onagra biennis (L.) Scop. P

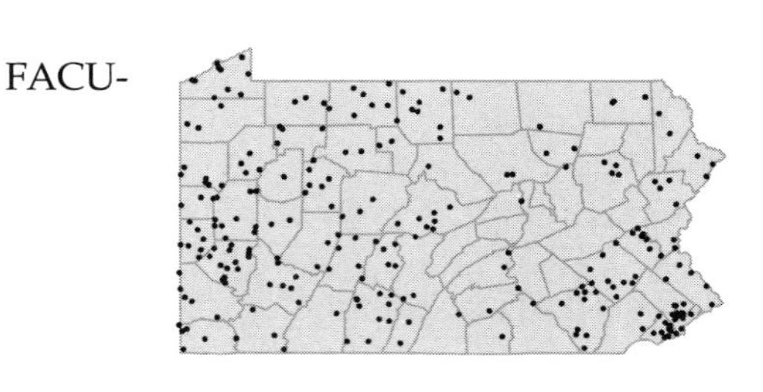

Oenothera fruticosa L. ssp. ***fruticosa*** FAC
Sundrops
Herbaceous perennial
Fields, roadside banks, river flats and pond edges.
Oenothera fruticosa L. var. *linearis* (Michx.) S.Wats. FW; *Oenothera tetragona* Roth var. *longistipata* (Pennell) Munz FGBW

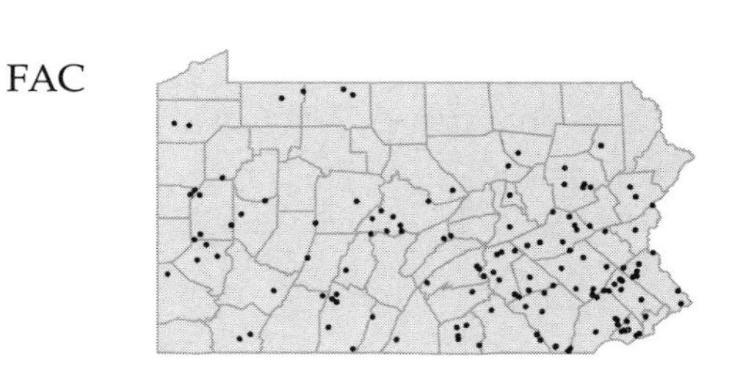

Oenothera fruticosa L. ssp. ***glauca*** (Michx.) Straley FAC
Sundrops
Herbaceous perennial
Meadows, dry fields, roadside banks and barrens.
Oenothera tetragona Roth var. *tetragona* FGBWC; *Oenothera tetragona* Roth var. *latifolia* (Rydb.) Fern. FW

Oenothera glazioviana Micheli I
Evening-primrose
Herbaceous biennial
Waste ground.

Oenothera grandiflora L'Her. I
Evening-primrose
Herbaceous biennial
Dry thickets, cultivated and occasionally escaped.

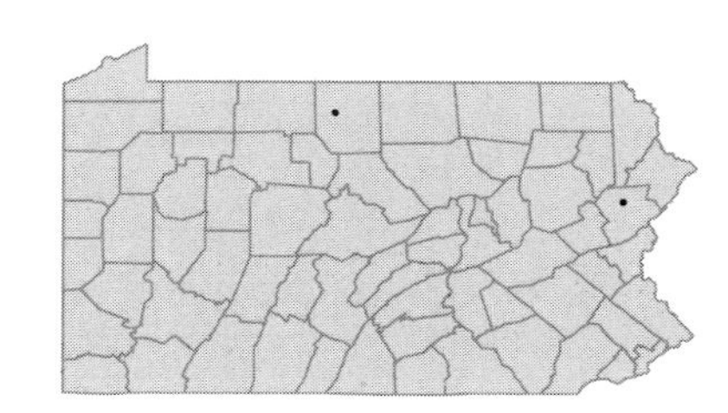

Oenothera laciniata Hill FACU-
Cut-leaved evening-primrose
Herbaceous annual
Dry, sandy or gravelly soil of railroad beds, vacant lots or open, rocky woods.

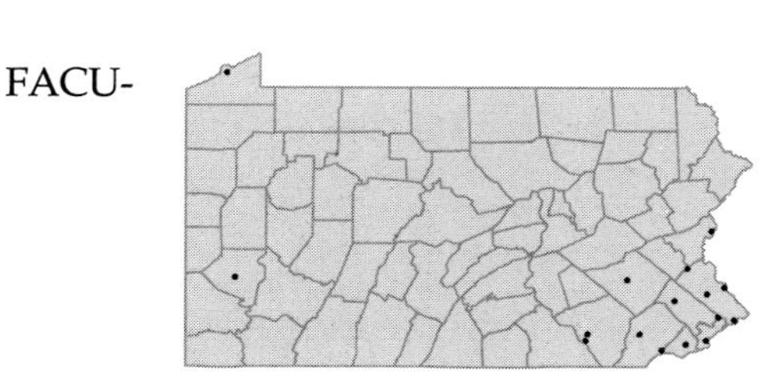

Oenothera nutans Atkinson & Bartlett
Evening-primrose
Herbaceous biennial
Old fields, roadsides and fallow land.
Oenothera biennis L. var. *nutans* (Atkinson & Bartlett) Wieg. F

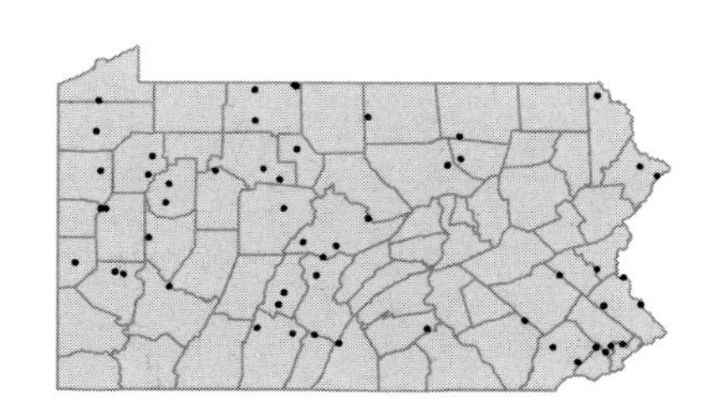

Oenothera oakesiana (A.Gray) Robins ex S.Wats. & Coult. FACU-
Evening-primrose
Herbaceous biennial
Railroad ballast and alluvium.
Oenothera parviflora L. var. *oakesiana* (Robbins) Fern. FC

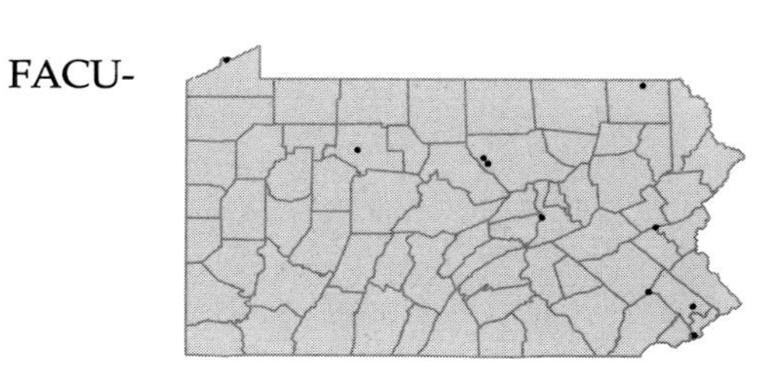

Oenothera parviflora L. var. ***parviflora***
Evening-primrose
Herbaceous biennial
Old fields, roadsides and railroad embankments.

FACU-

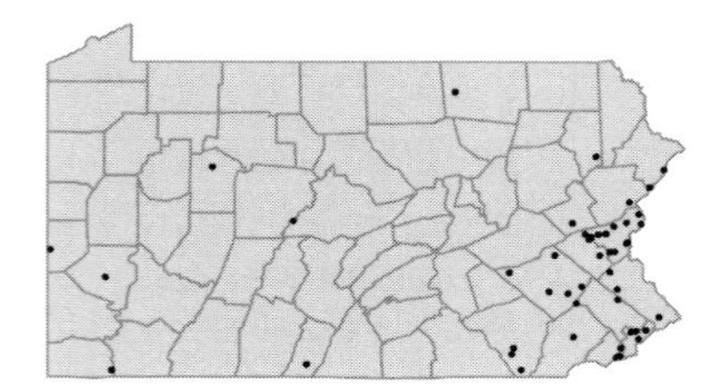

Oenothera perennis L.
Sundrops
Herbaceous perennial
Shaly slopes, dry fields or pastures and roadsides.
Kneiffia pumila (L.) Spach P

Oenothera pilosella Raf.
Evening-primrose; Sundrops
Herbaceous perennial
Open woods, meadows and roadsides.
Kneiffia fruticosa (L.) Raimann var. *pilosella* (Raf.) Britt. P

FAC

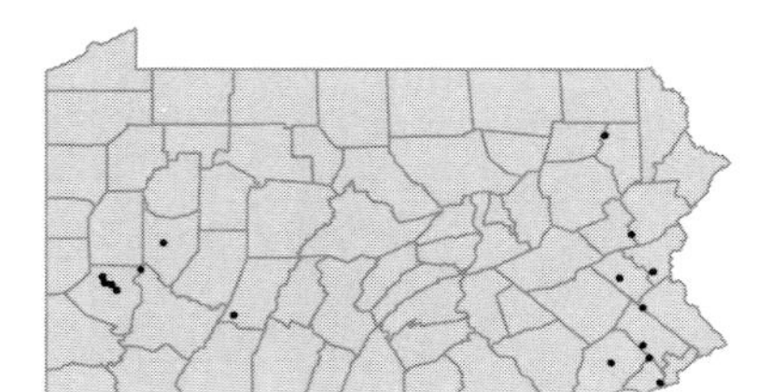

Oenothera speciosa Nutt.
White evening-primrose
Herbaceous perennial
Dry fields and roadsides, escaped from cultivation.

I

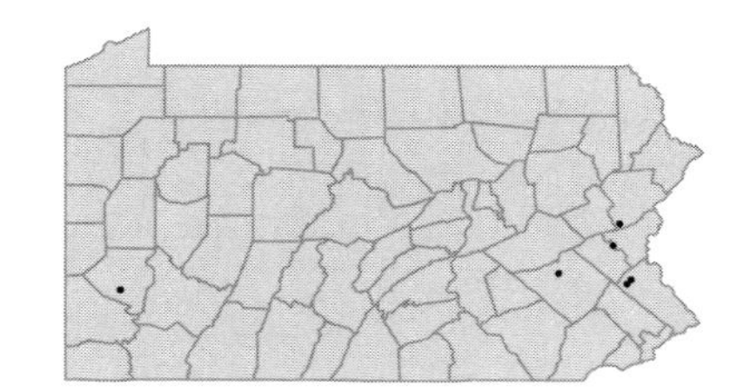

Oenothera triloba Nutt.
Evening-primrose
Herbaceous annual
Dry, often calcareous, soil.

I

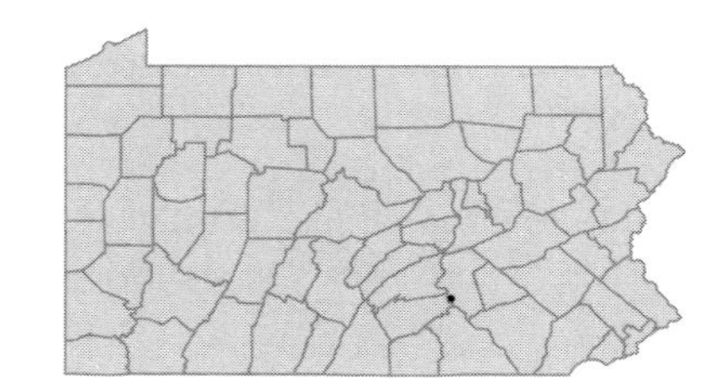

Oenothera villosa Thunb.
Evening-primrose
Herbaceous biennial
Fallow fields and waste ground.
Oenothera biennis L. var. *canescens* Torr. & A.Gray FC

FAC
I

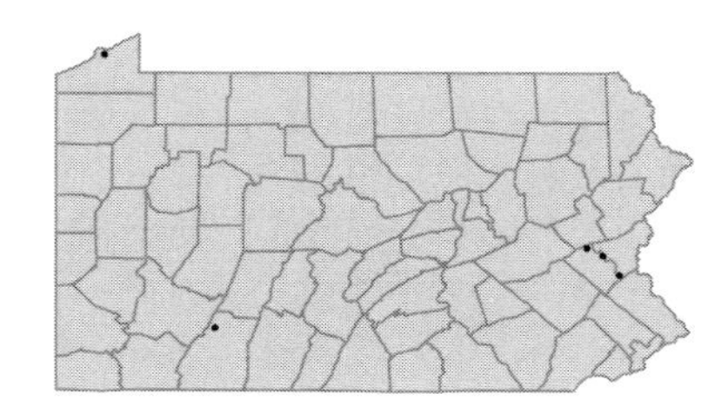

MELASTOMACEAE

• ***Rhexia mariana*** L.
Maryland meadow-beauty
Herbaceous perennial
Moist, open, coastal plain soils.

OBL

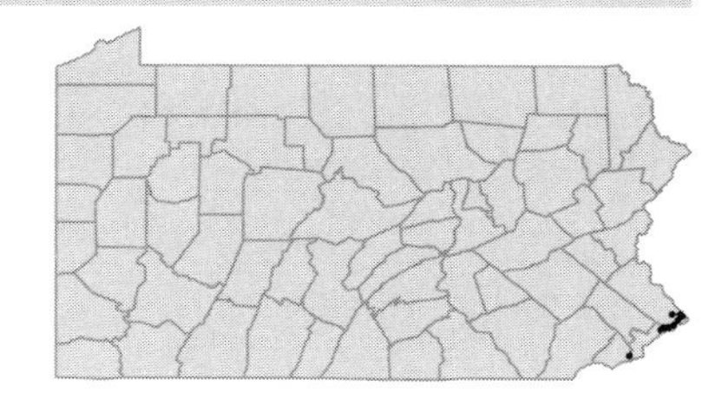

Rhexia virginica L.
Meadow-beauty
Herbaceous perennial
Moist fields, marshes, vernal ponds and other open areas on moist, sandy soils.

OBL

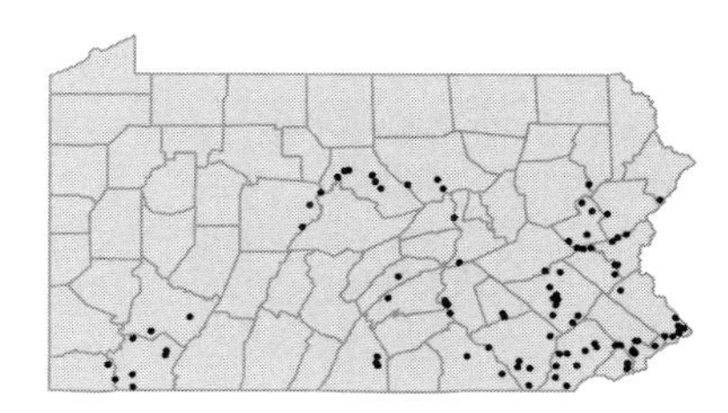

NYSSACEAE

• ***Nyssa sylvatica*** Marshall
Black-gum; Sour-gum; Tupelo
Deciduous tree
Dry to moist woods, rocky slopes and ridge tops.

FAC

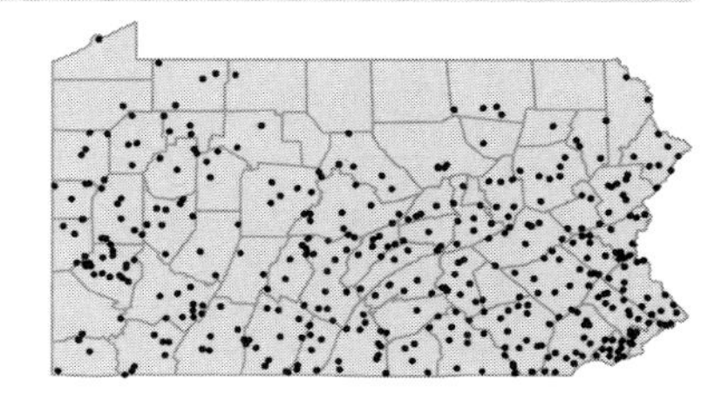

CORNACEAE

• ***Cornus alternifolia*** L.f.
Alternate-leaved dogwood; Pagoda dogwood
Deciduous tree
Low, moist woods and shaded ravines.

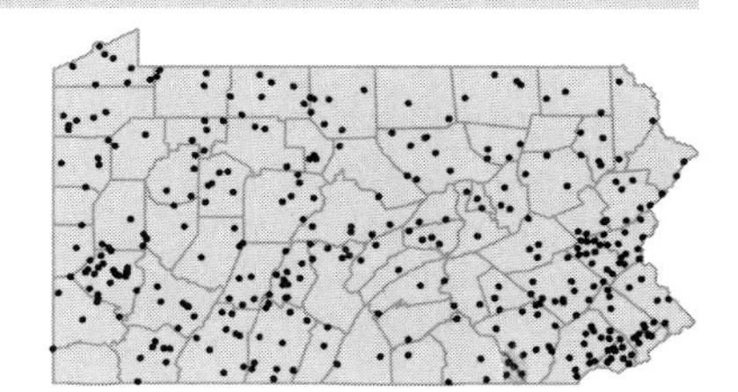

Cornus amomum P.Mill. ssp. ***amomum***
Kinnikinik; Red-willow
Deciduous shrub
Moist woods, meadows, old fields and swamps.

FACW

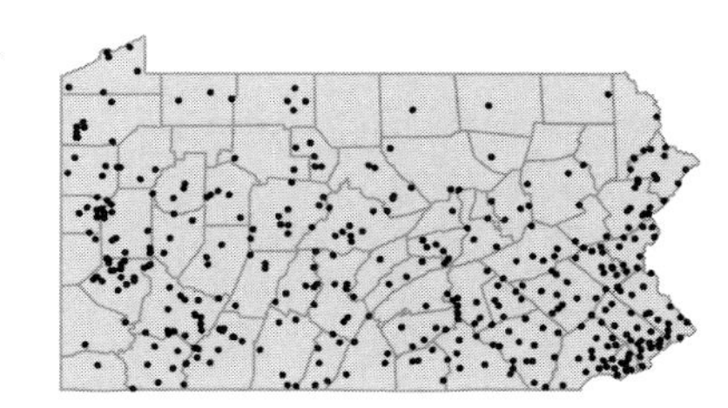

Cornus amomum P.Mill. ssp. ***obliqua*** (Raf.) J.S.Wilson
Kinnikinik; Red-willow
Deciduous shrub
Vernal ponds, swamps and moist thickets.
Cornus amomum P.Mill. var. *schuetziana* (Mey.) Rickett WC; *Cornus obliqua* Raf. F

FACW

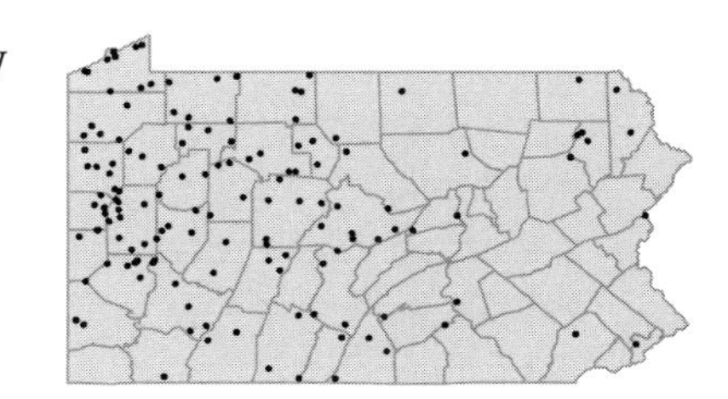

Cornus canadensis L. FAC-
Bunchberry; Dwarf cornel
Herbaceous perennial
Cool, damp woods, bogs and swamp edges.

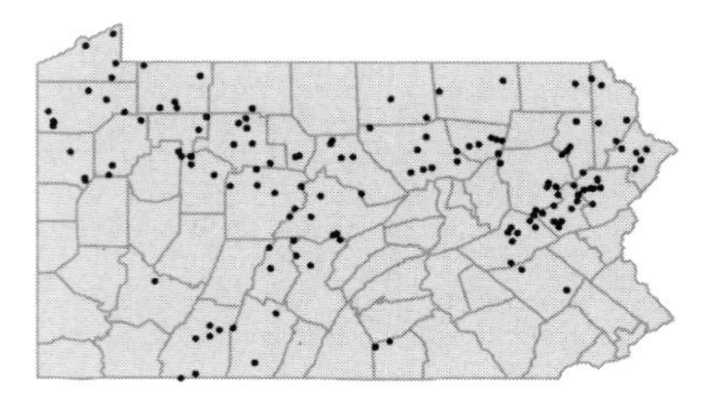

Cornus drummondii C.A.Mey. I
Roughleaf dogwood
Deciduous shrub
Shrub thicket on moist, gravelly ground. Represented by a single collection from Delaware Co. in 1986.

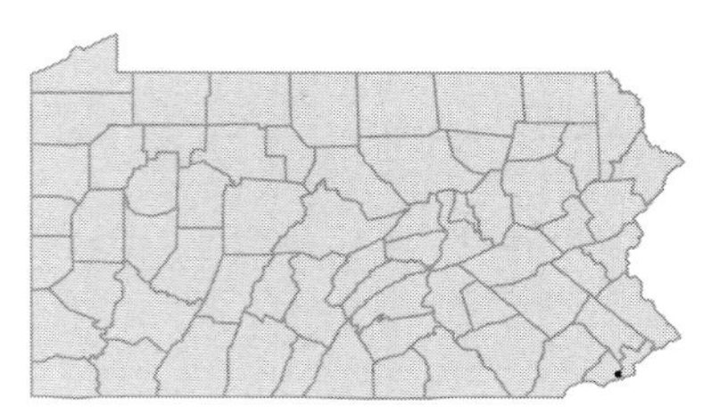

Cornus florida L. FACU-
Flowering dogwood
Deciduous tree
Rich, moist woods and woods edges.

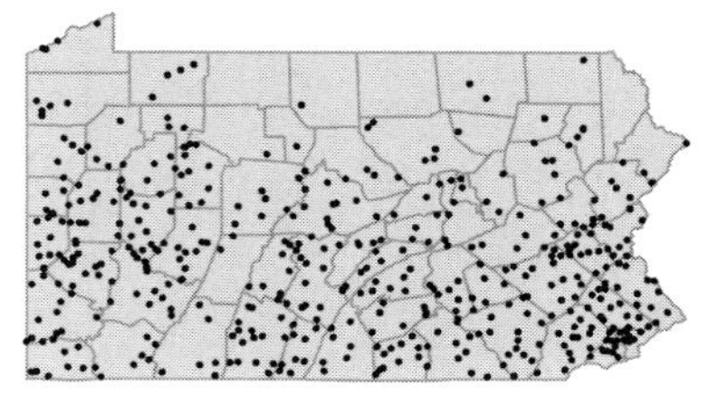

Cornus racemosa Lam. FAC
Swamp dogwood
Deciduous shrub
Swampy meadows, wet woods and thickets.
Cornus candidissima Marshall P; *Cornus foemina* P.Mill. ssp. *racemosa* (Lam.) J.S.Wilson K

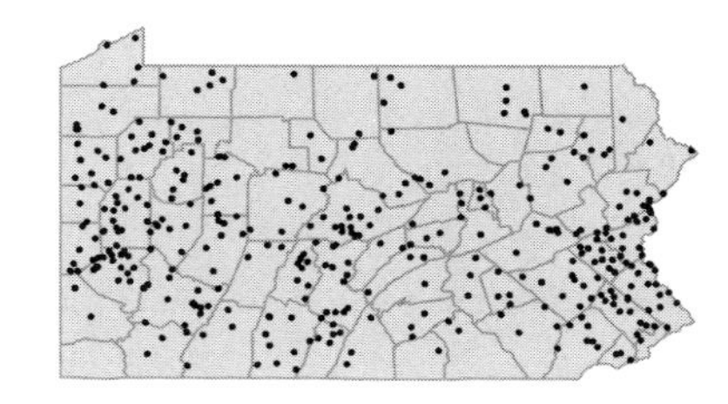

Cornus rugosa Lam.
Round-leaved dogwood
Deciduous shrub
Dry, rocky woods and cliffs.
Cornus circinata L'Her. P

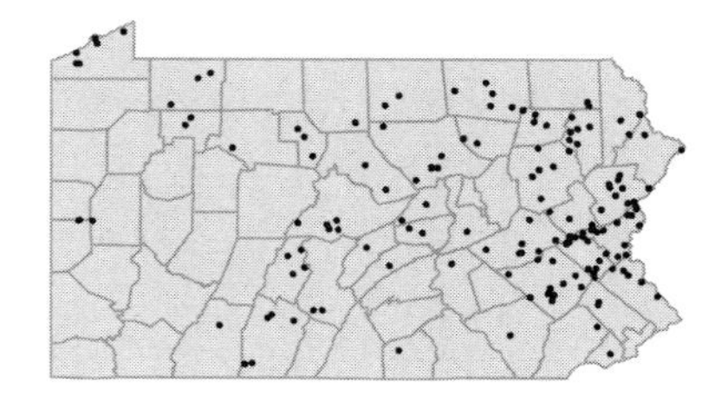

Cornus sanguinea L. I
Bloodtwig dogwood
Deciduous shrub
Old fields, thickets and abandoned nurseries.

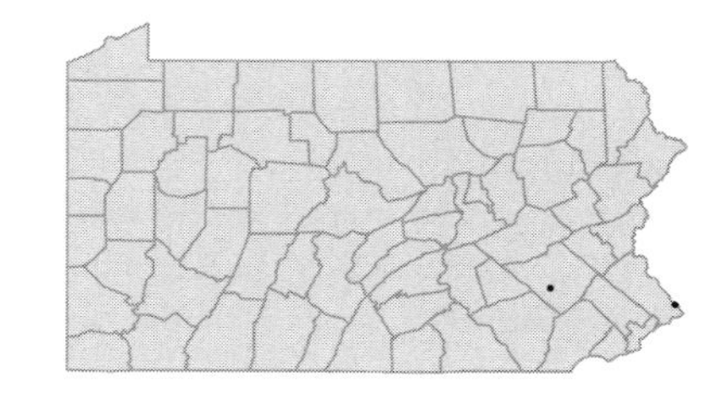

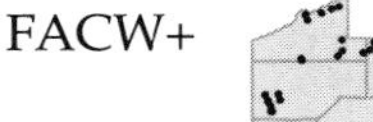

Cornus sericea L.
Red-osier dogwood
Deciduous shrub
Swamps, moist fields and thickets.
Cornus baileyi Coult. & Evans P; *Cornus stolonifera* Michx. PFGBW

FACW+

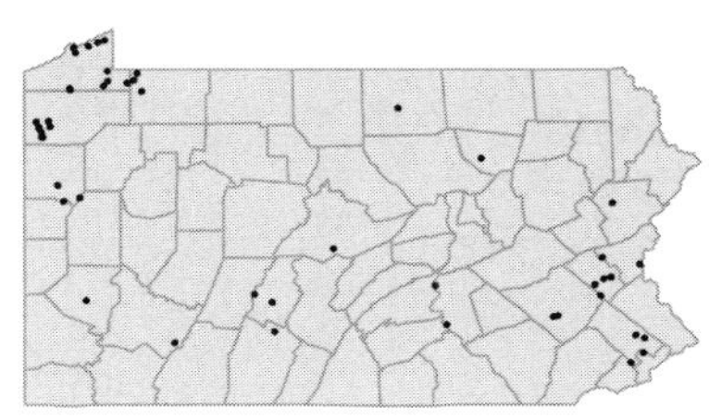

SANTALACEAE

• ***Comandra umbellata*** (L.) Nutt.
Bastard toadflax
Herbaceous perennial
Dry, open oak woods. Parasitic on the roots of oaks.

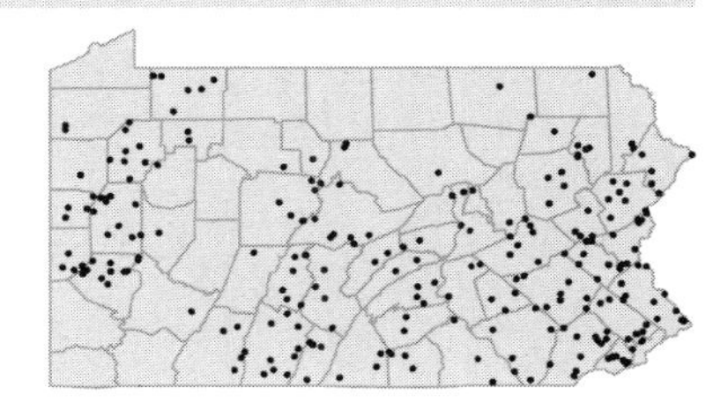

• ***Pyrularia pubera*** Michx.
Buffalo-nut; Oil-nut
Deciduous shrub
Woods and thickets. Parasitic on the roots of deciduous trees and shrubs.

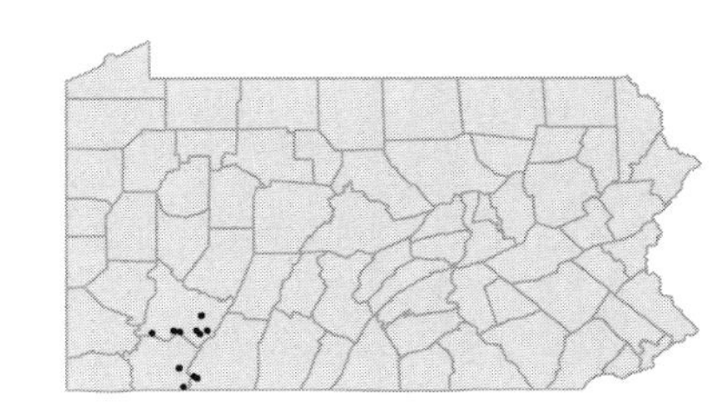

VISCACEAE

• ***Arceuthobium pusillum*** Peck
Dwarf mistletoe
Evergreen shrub
Bogs. Parasitic on the branches of Black spruce.
Razoumofskya pusilla (Peck) Kuntze P

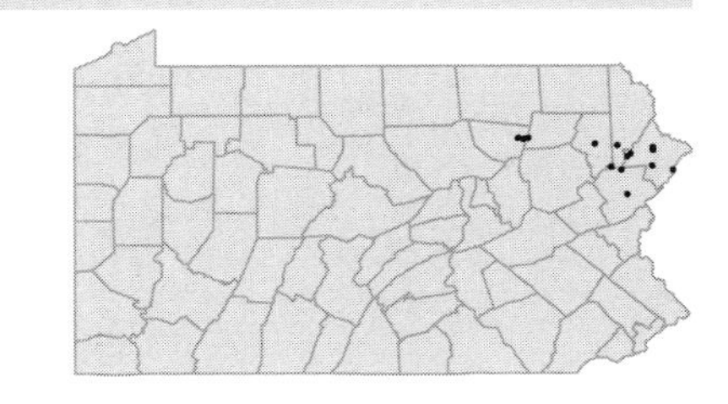

• ***Phoradendron leucarpum*** (Raf.) Rev. & M.C.Johnston
Christmas mistletoe
Evergreen shrub
Woods. Parasitic on the branches of Black-gum and other deciduous trees.
Believed to be extirpated, last collected in 1924.
Phoradendron flavescens (Pursh) Nutt. PFGB; *Phoradendron serotinum* (Raf.) M.C.Johnston KC

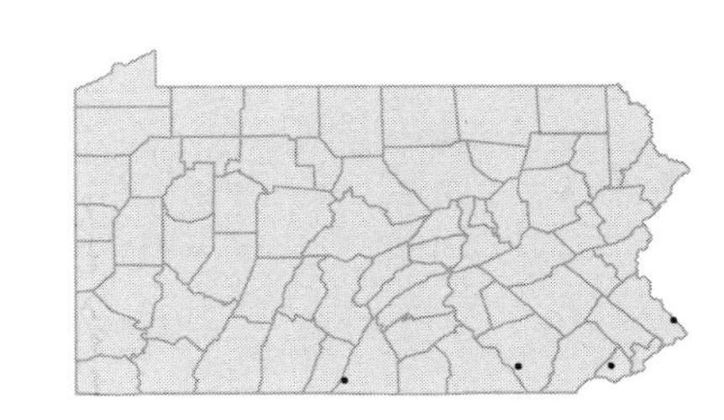

CELASTRACEAE

• ***Celastrus orbiculatus*** Thunb.
Oriental bittersweet
Woody vine
Disturbed woods, fields, fencerows and edges.

UPL
I

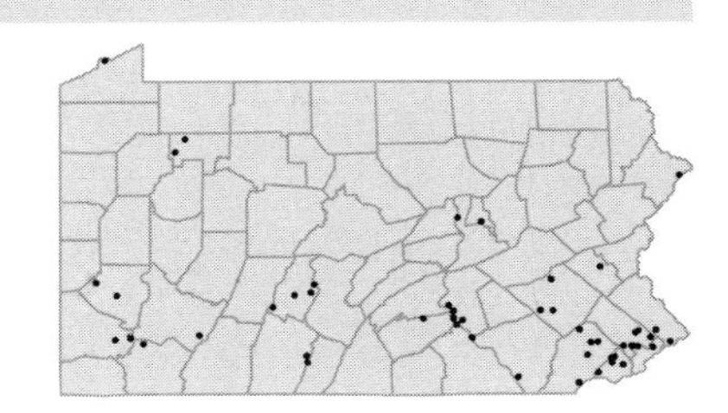

Celastrus scandens L. FACU--
American bittersweet
Woody vine
Dry fields, rocky ledges, woods and thickets.

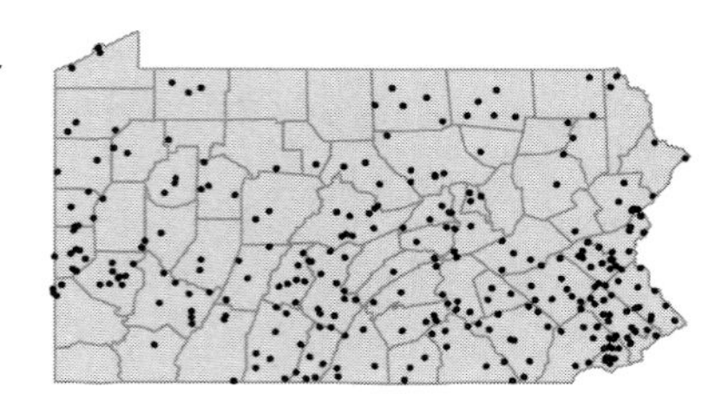

• ***Euonymus alatus*** (Thunb.) Sieb. [I]
Winged euonymous
Deciduous shrub
Cultivated and occasionally naturalized in disturbed woods, stream banks, fencerows and edges.

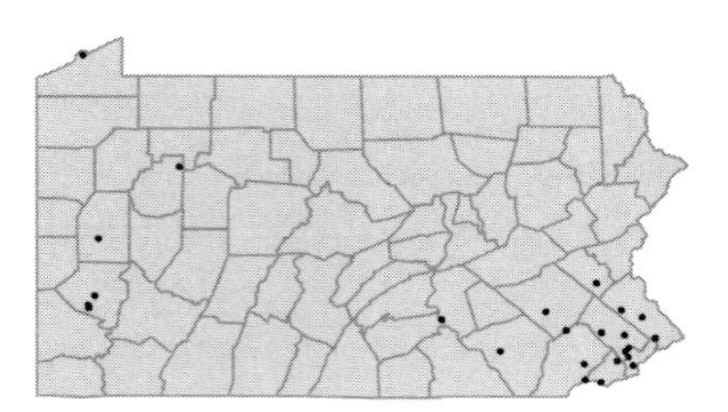

Euonymus americanus L. FAC
Hearts-a-bursting; Strawberry-bush
Deciduous shrub
Moist woods, swamps, floodplains and wet thickets.

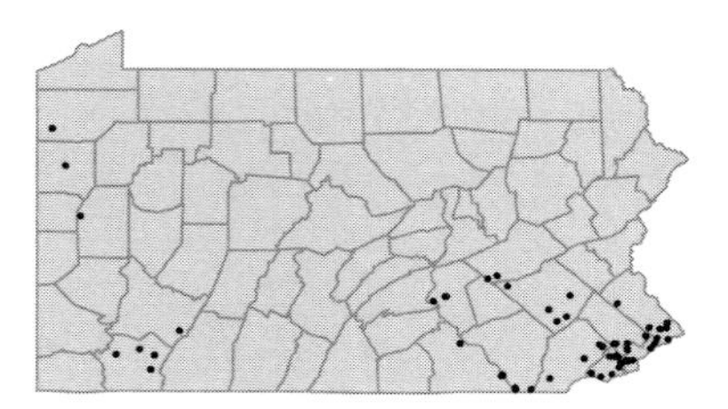

Euonymus atropurpureus Jacq. FACU
Burning-bush; Wahoo
Deciduous shrub
Wooded limestone slopes, rocky bluffs and floodplain thickets.

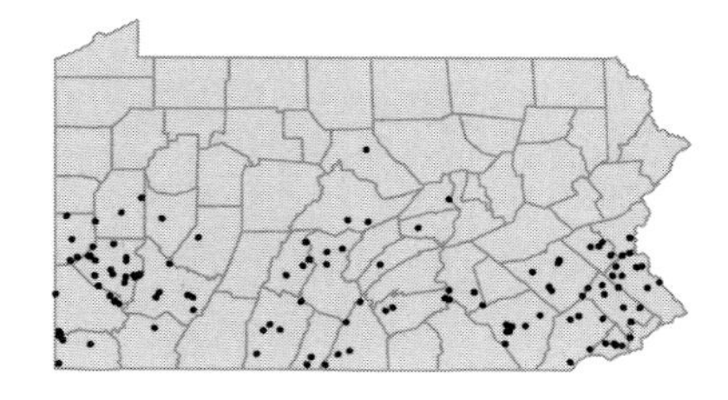

Euonymus europaeus L. [I]
European spindle-tree
Deciduous shrub
Cultivated and occasionally escaped.

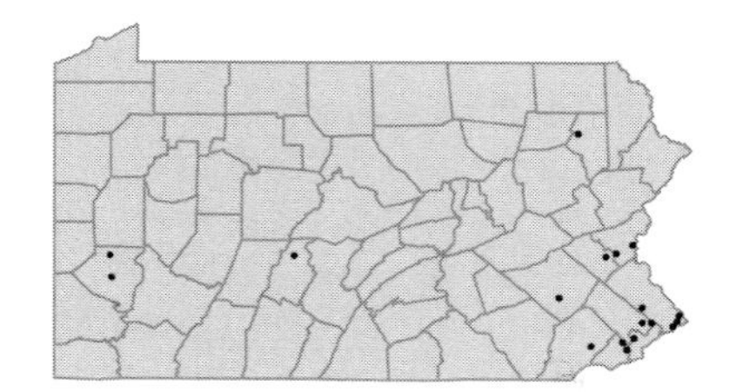

Euonymus fortunei (Turcz.) Hand.-Mazz. [I]
Wintercreeper
Woody vine
Cultivated and occasionally naturalized in woods and wooded floodplains.

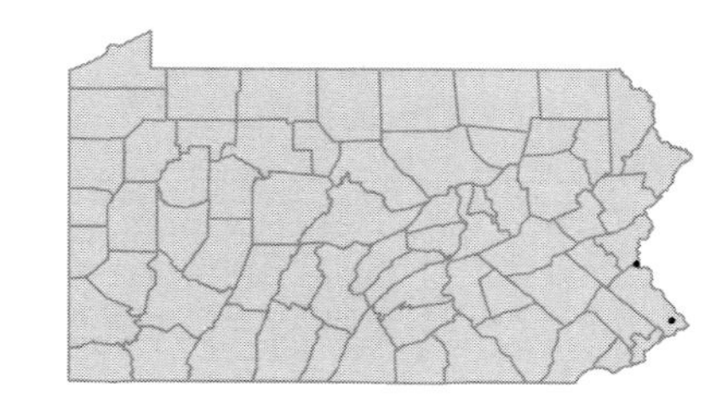

Euonymus obovatus Nutt.
Running strawberry-bush
Deciduous shrub
Wet decidous woods and wooded hillsides.

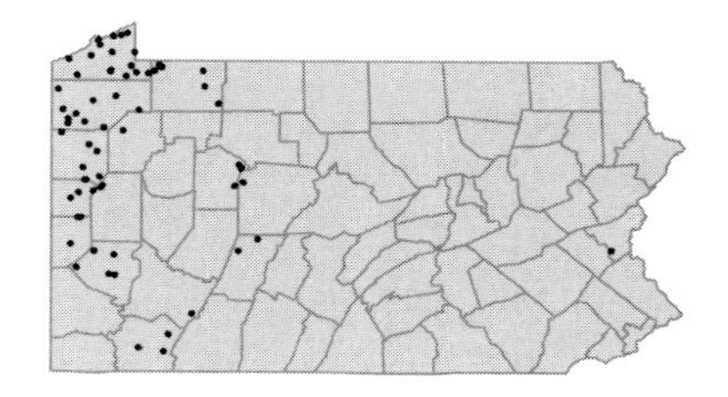

Euonymus yedoensis Koehne
Spindle-tree
Deciduous shrub
Cultivated and rarely escaped to moist thickets.

I

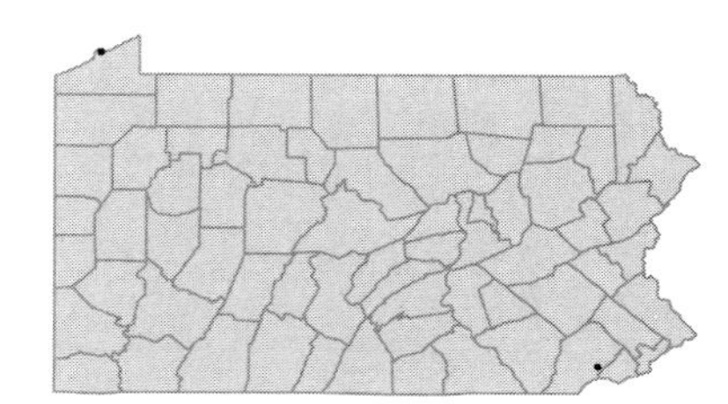

• ***Paxistima canbyi*** A.Gray
Canby's-mountain-lover; Cliff-green
Evergreen shrub
Calcareous cliffs and slopes.
Pachystima canbyi A.Gray FWB

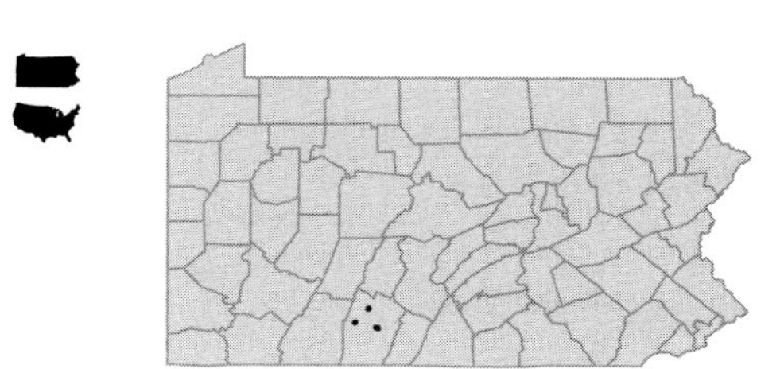

AQUIFOLIACEAE

• ***Ilex beadlei*** Ashe
Mountain holly
Deciduous shrub
Wooded slopes.
Ilex montana Torr. & A.Gray var. *beadlei* (Ashe) Fern. GB; *Ilex montana* Torr. & A.Gray var. *mollis* (A.Gray) Britt. FWC

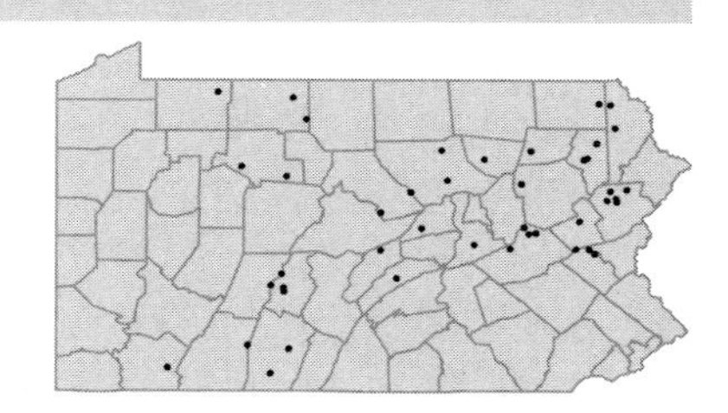

Ilex glabra (L.) A.Gray
Inkberry
Evergreen shrub
Moist, sandy, coastal plain soil. Believed to be extirpated, last collection from a native stand was in 1950, also cultivated.

FACW-

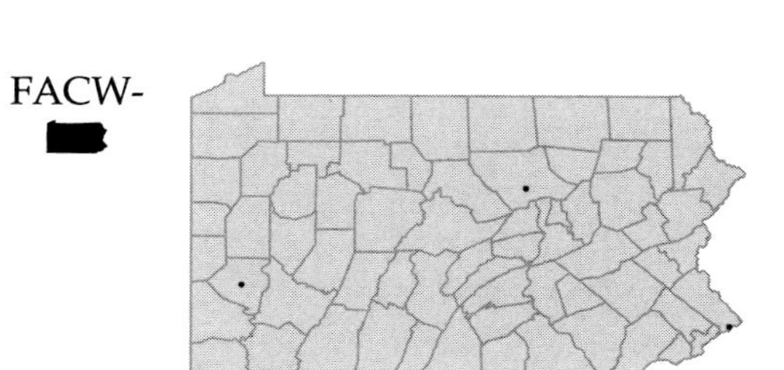

Ilex laevigata (Pursh) A.Gray
Smooth winterberry
Deciduous shrub
Wooded swamps, wet thickets and shores.

OBL

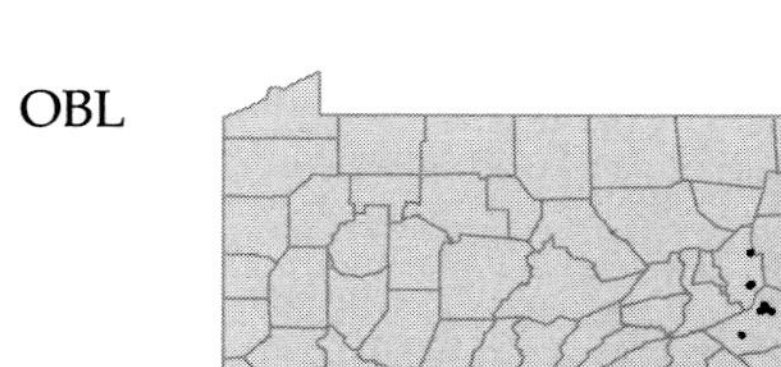

AQUIFOLIACEAE

Ilex montana (Torr. & A.Gray) A.Gray
Mountain holly
Deciduous shrub
Rocky, acidic woods and slopes.

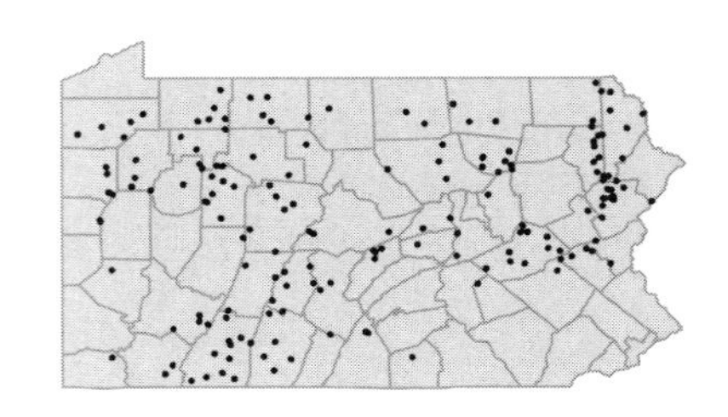

Ilex opaca Ait. — FACU
American holly
Evergreen tree
Moist, alluvial woods and wooded slopes, also cultivated and frequently escaped.

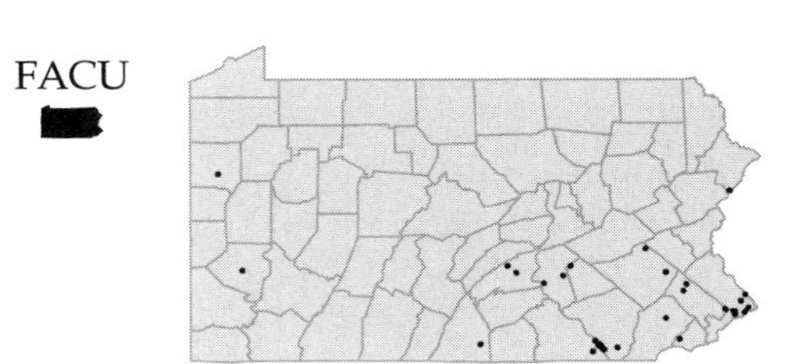

Ilex verticillata (L.) A.Gray — FACW+
Winterberry; Black alder
Deciduous shrub
Swamps, bogs, moist woods and wet shores.

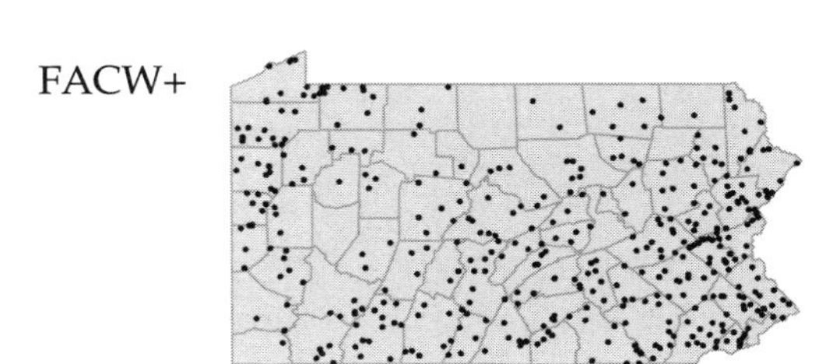

• ***Nemopanthus mucronatus*** (L.) Trel. — OBL
Mountain holly
Deciduous shrub
Swamps, bogs, moist woods and rocky slopes.

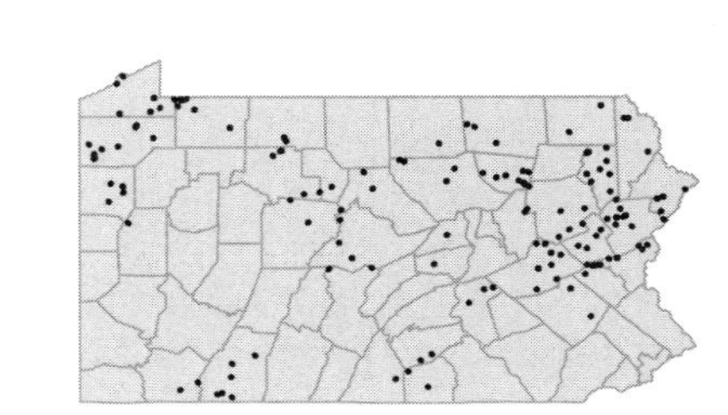

BUXACEAE

• ***Pachysandra procumbens*** Michx. — [I]
Allegheny spurge
Herbaceous perennial
Cultivated in woodland gardens and occasionally naturalized. Although native as far north as WV, there is no evidence that this species is indigenous to PA.

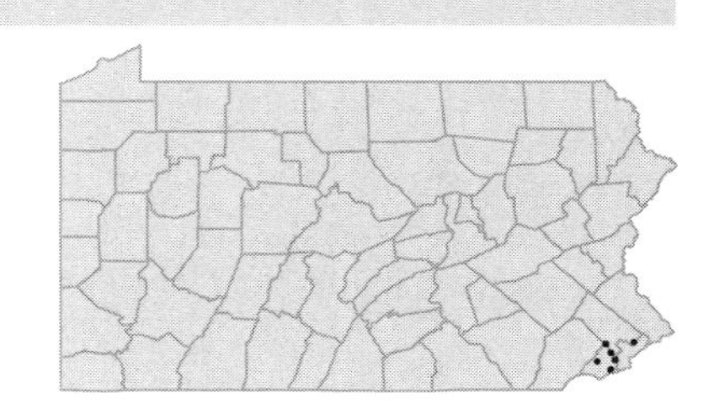

Pachysandra terminalis Siebold & Zucc. — [I]
Japanese spurge
Herbaceous perennial
Frequently cultivated and occasionally naturalized in urban or suburban woods.

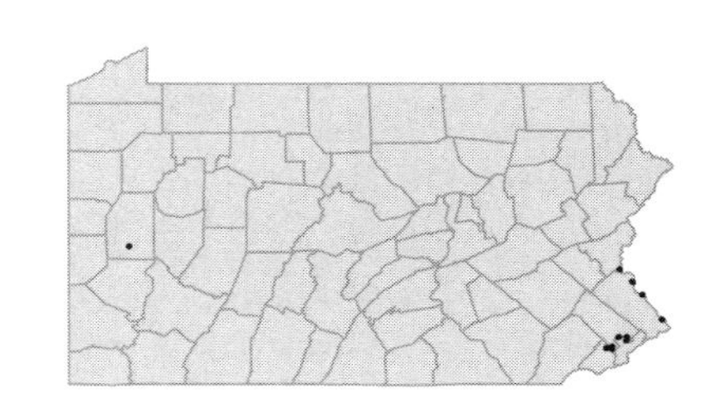

EUPHORBIACEAE

• ***Acalypha deamii*** (Weatherby) Ahles
Three-seeded mercury
Herbaceous annual
Wooded hollow. Represented by a single collection from Allegheny Co. in 1900.
Acalypha rhomboidea Raf. var. *deamii* Weatherby FBW; *Acalypha virginica* L. var. *deamii* Weatherby K

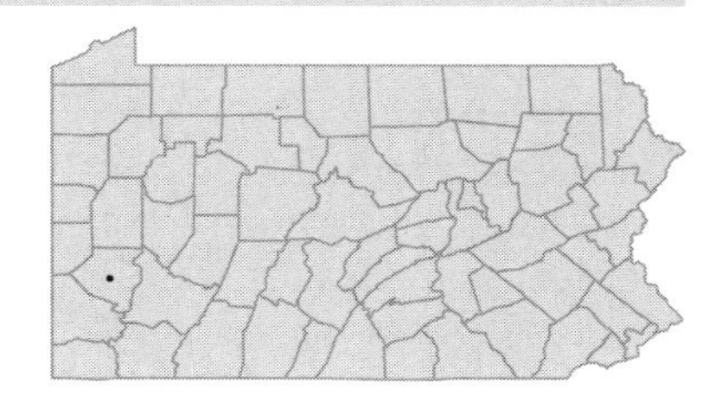

Acalypha gracilens A.Gray
Slender mercury
Herbaceous annual
Dry soils of fields, open woods and shaly slopes.

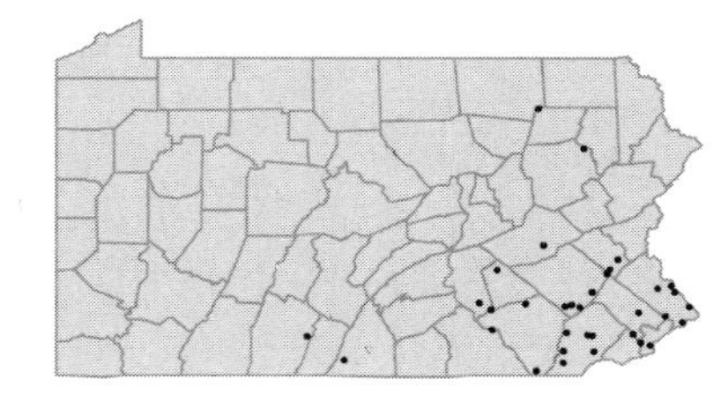

Acalypha rhomboidea Raf.
Three-seeded mercury
Herbaceous annual
Wooded slopes, roadsides, fields and waste ground.

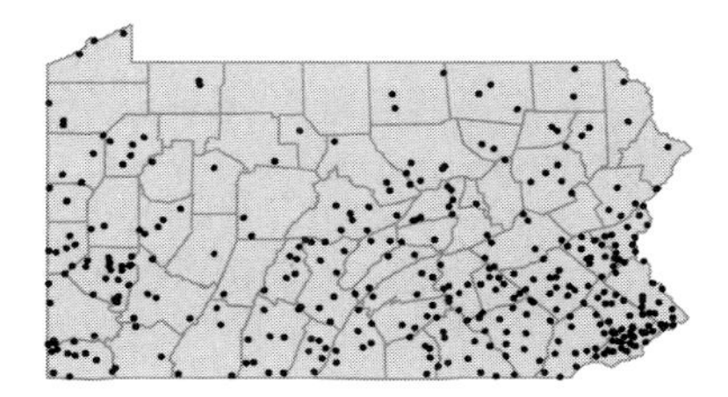

Acalypha virginica L.
Three-seeded mercury
Herbaceous annual
Dry or moist soil of fields, wooded slopes, stream banks and waste ground.

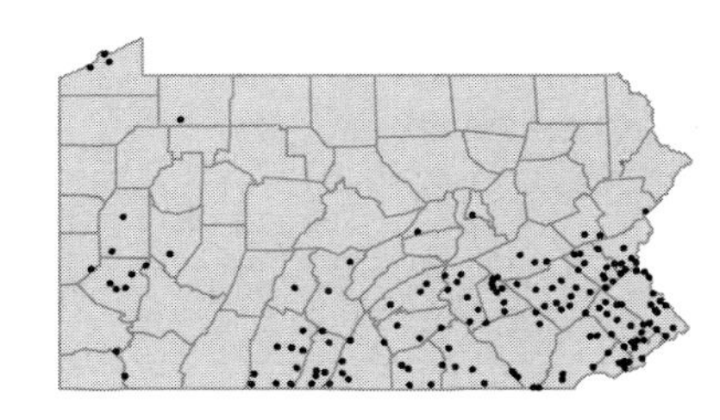

• ***Chamaesyce maculata*** (L.) Small
Wartweed; Spotted spurge
Herbaceous annual
Dry, disturbed ground including sidewalks and gravel ballast.
Euphorbia maculata L. *pro parte* PFG; *Euphorbia supina* Raf. FW

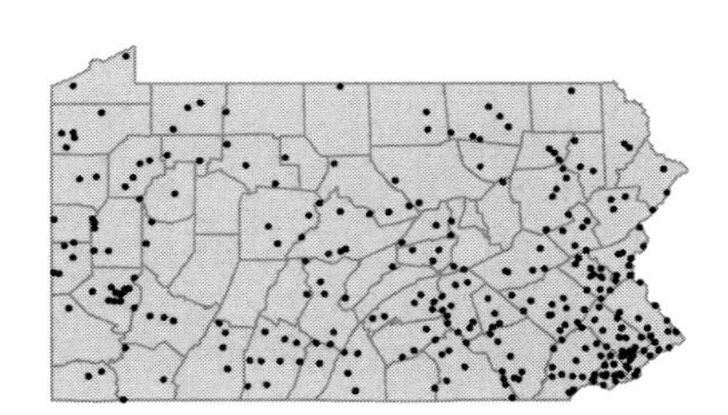

Chamaesyce nutans (Lag.) Small
Eyebane
Herbaceous annual
Dry, open ground, waste areas and cultivated fields.
Euphorbia maculata L. *pro parte* F; *Euphorbia nutans* Lag. PWC; *Euphorbia preslii* Guss. GB

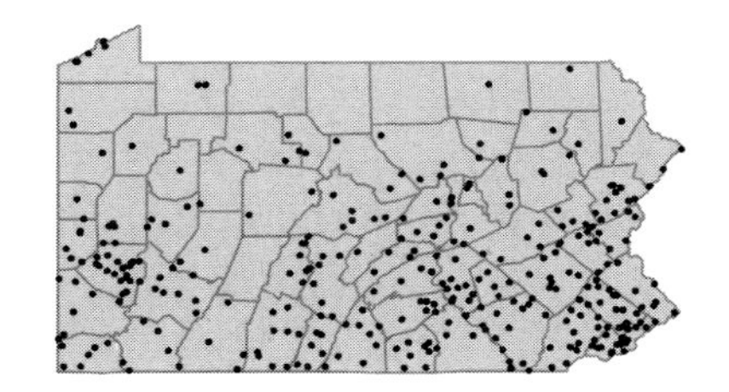

Chamaesyce polygonifolia (L.) Small
Seaside spurge
Herbaceous annual
Dunes and sand plains.
Euphorbia polygonifolia L. PFGBWC

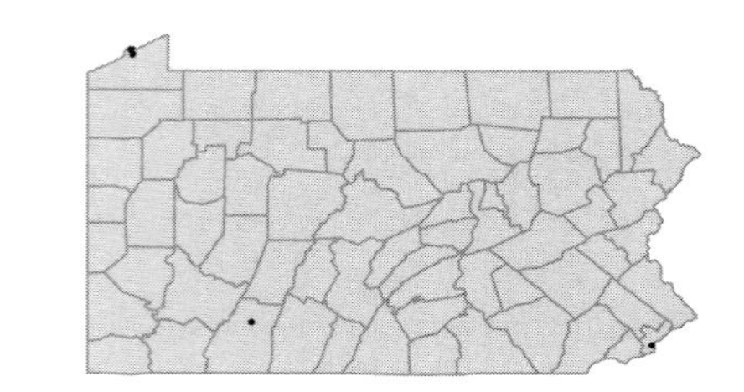

Chamaesyce prostrata (Ait.) Small
Spurge
Herbaceous annual
Weedy margins, wharves and railroad sidings.
Euphorbia chamaesyce L. FGBW; *Euphorbia prostrata* Ait. C

[I]

Chamaesyce serpens (HBK) Small
Crawling spurge
Herbaceous annual
Ballast, waste ground and margins of streamlets.
Euphorbia serpens HBK FGBW

FACW
[I]

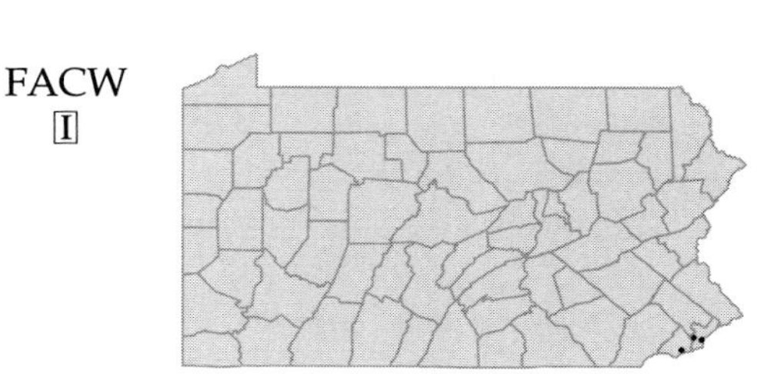

Chamaesyce serpyllifolia (Pers.) Small
Thyme-leaved spurge
Herbaceous annual
Weedy shore. Represented by a single collection from Tioga Co. in 1944.
Euphorbia serpyllifolia Pers. FGBW

[I]

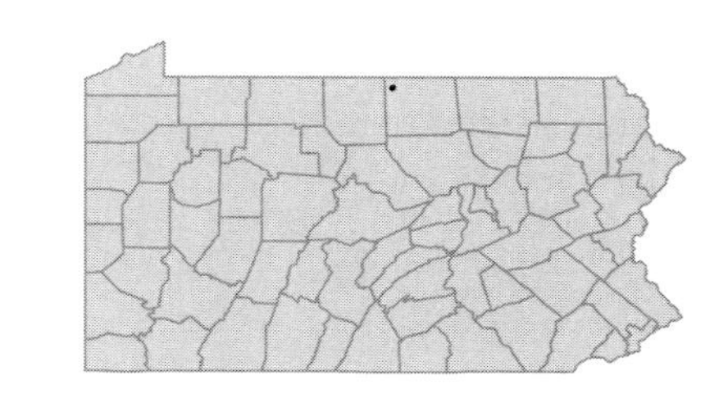

Chamaesyce vermiculata (Raf.) House
Hairy spurge
Herbaceous annual
Dry, open soil and waste ground including walkways, railroad ballast and roadsides.
Euphorbia vermiculata Raf. FGBW

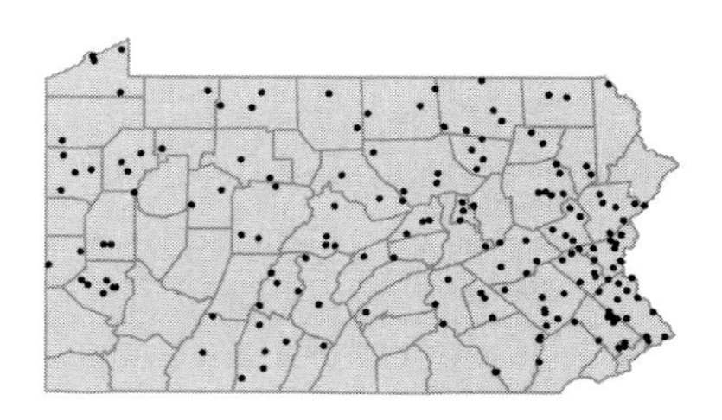

• ***Croton capitatus*** Michx.
Hogwort; Wooly croton
Herbaceous annual
Railroad sidings and waste places.

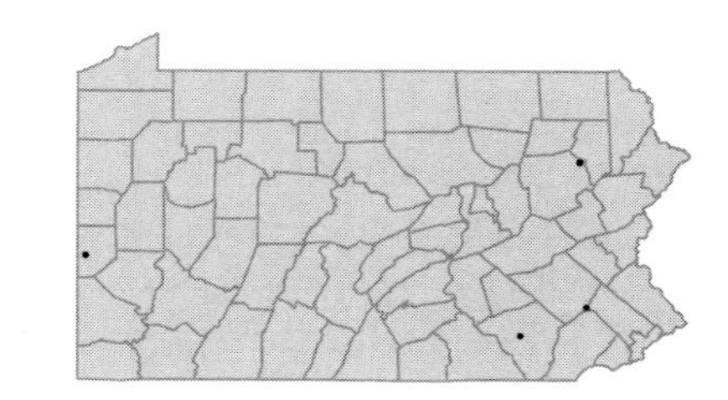

Croton glandulosus L. var. ***septentrionalis*** Muell. Arg.
Croton
Herbaceous annual
Railroad banks, waste ground and ballast.

I

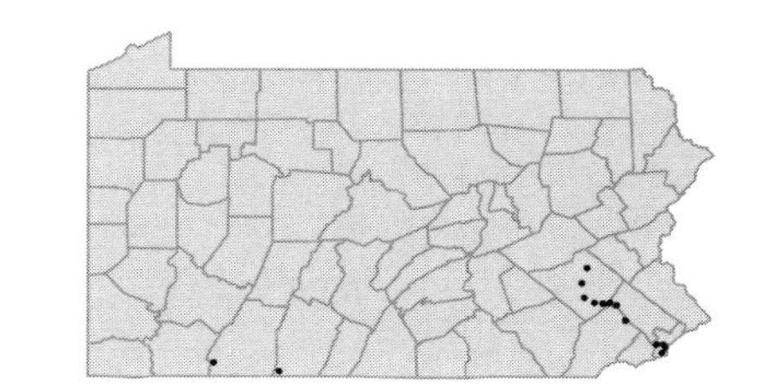

Croton lindheimerianus Scheele
Croton
Herbaceous annual
Ballast, waste ground and pastures.

I

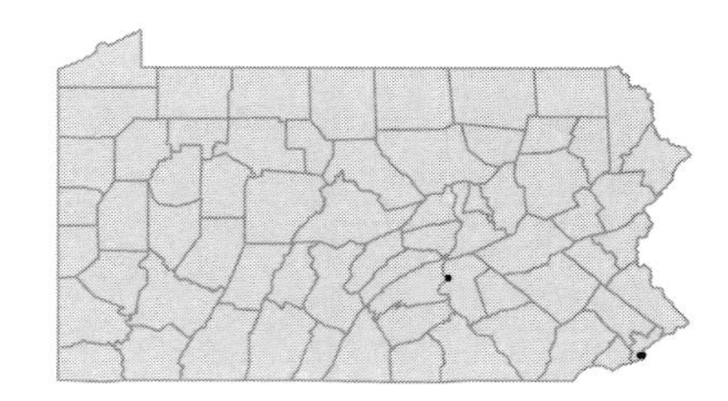

Croton punctatus Jacq.
Croton
Herbaceous annual
Pastures and waste ground.

I

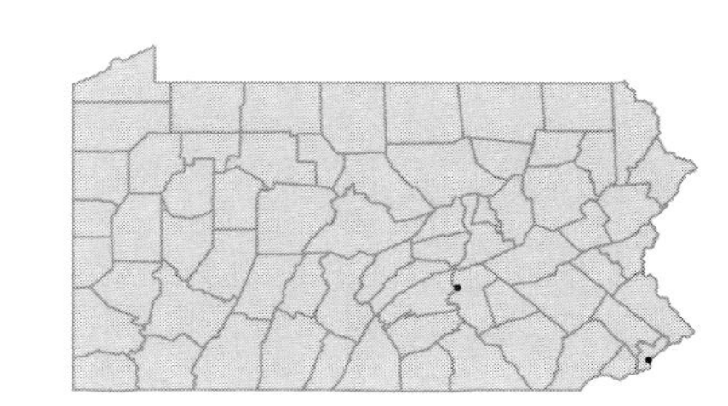

• ***Crotonopsis elliptica*** Willd.
Elliptical rushfoil
Herbaceous annual
Sandy soil in open woods. Believed to be extirpated, last collected in 1892.

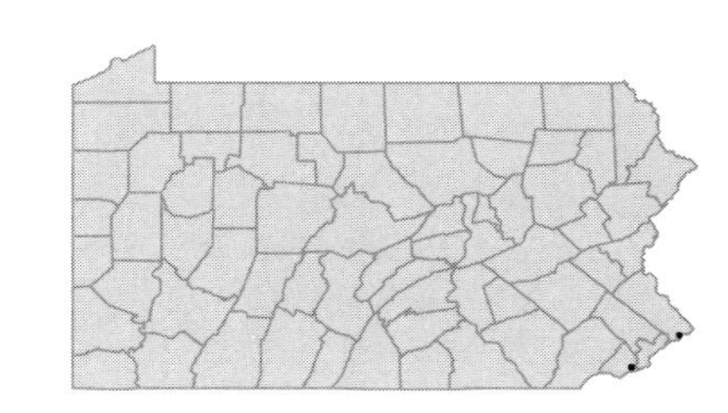

• ***Euphorbia commutata*** Engelm.
Wood spurge
Herbaceous annual
Open woods or wooded slopes on limestone.

FACU

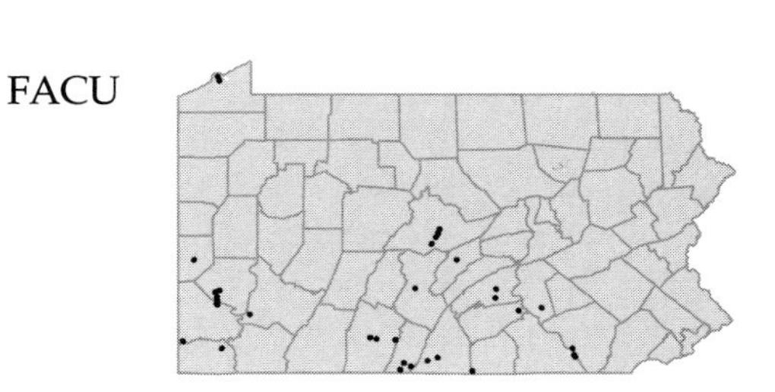

Euphorbia corollata L.
Flowering spurge
Herbaceous perennial
Dry, open woods, shale barrens, fields and sandy waste ground.

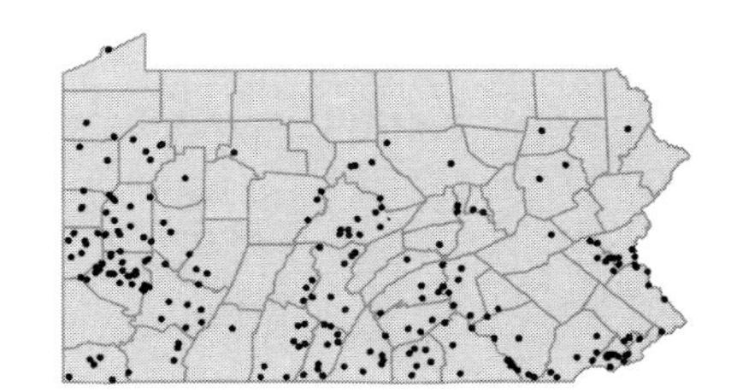

Euphorbia cyparissias L.
Cypress spurge; Cemetery-plant
Herbaceous perennial
Roadsides, old fields, disturbed open woods and abandoned cemeteries.

I

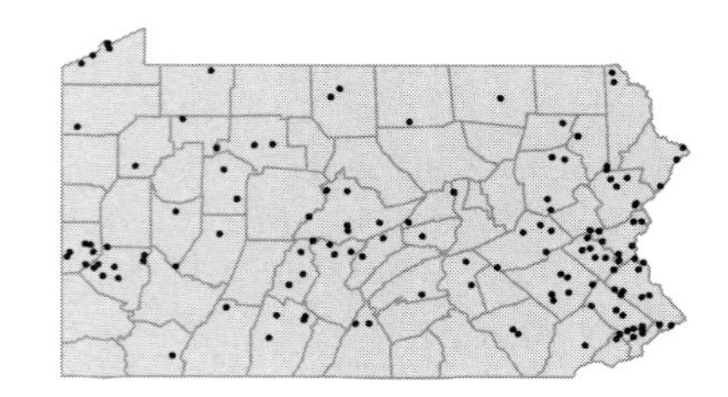

Euphorbia esula L.
Leafy spurge; Wolf's-milk
Herbaceous perennial
Roadsides, openings in woods and waste ground.

I

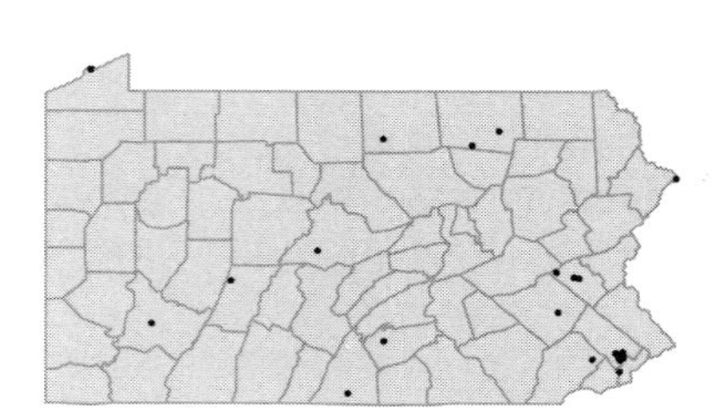

Euphorbia falcata L.
Spurge
Herbaceous annual
Dry, shaly roadside slopes, occasional.

I

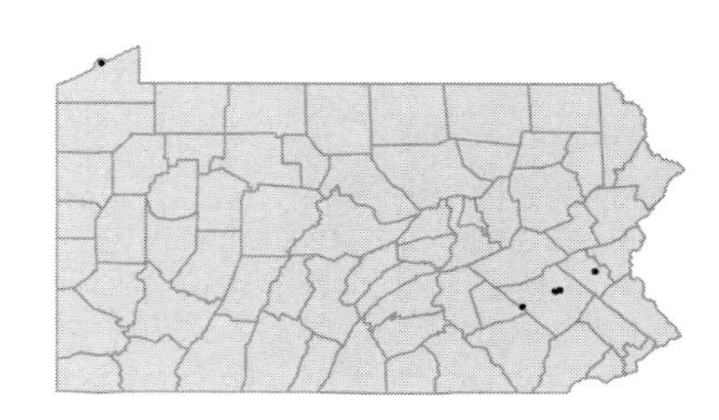

Euphorbia helioscopia L.
Wartweed
Herbaceous annual
Dry, open soil, gardens and waste ground.

I

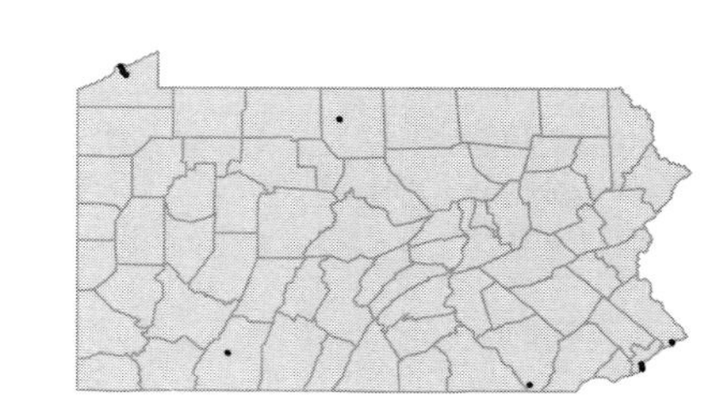

Euphorbia ipecacuanhae L.
Wild ipecac
Herbaceous perennial
Sandy or gravelly soil of the Coastal Plain.

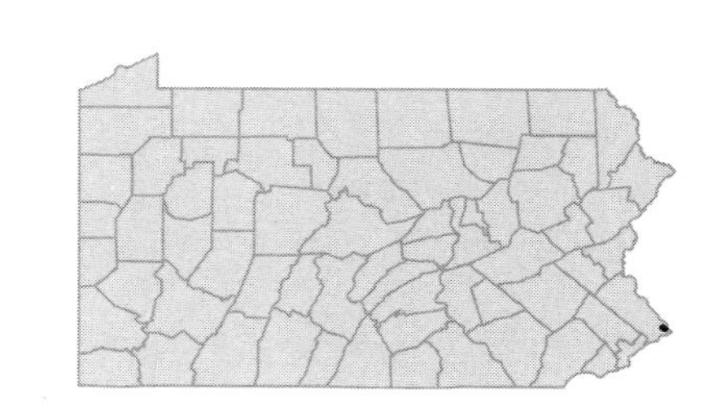

Euphorbia lathyris L.
Caper spurge; Myrtle spurge; Mole-plant
Herbaceous annual
Wooded banks, cultivated fields and waste ground.

I

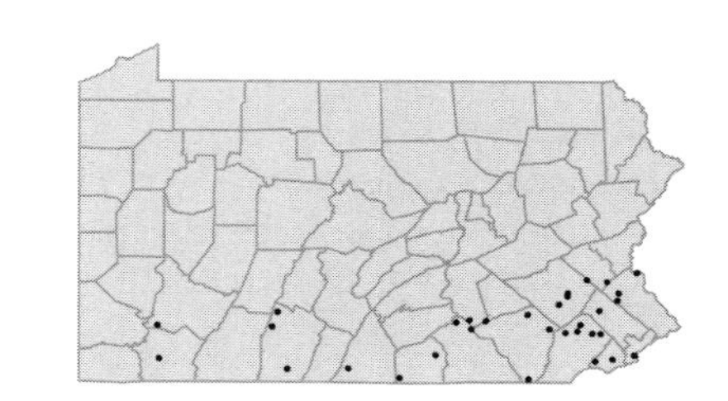

Euphorbia lucida Waldst. & Kit.
Leafy spurge
Herbaceous annual
Dry fields.
Euphorbia agraria Bieb. GB

I

Euphorbia marginata Pursh
Snow-on-the-mountain
Herbaceous annual
Cultivated and occasionally escaped to vacant lots, dumps, railroad cinders and roadsides.

UPL
I

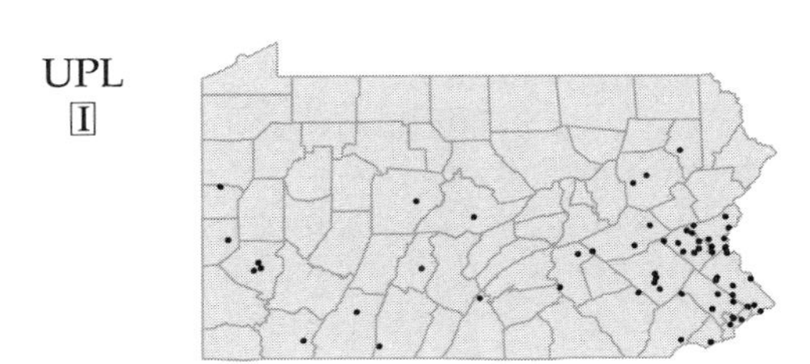

Euphorbia obtusata Pursh
Blunt-leaved spurge
Herbaceous annual
Rich woods or sandy ground along creeks and river banks. Believed to be extirpated, last collected in 1955.

FACU-

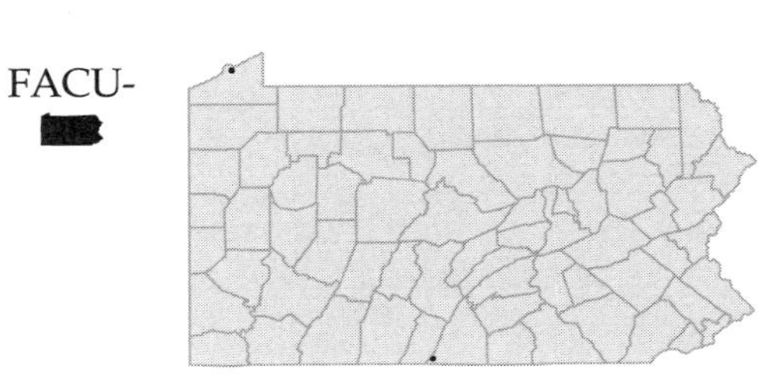

Euphorbia peplus L.
Petty spurge
Herbaceous annual
Yards, gardens and pavements.

I

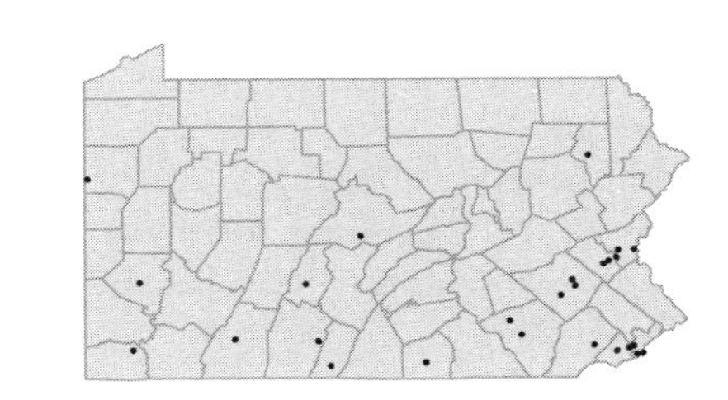

Euphorbia platyphyllos L.
Broad-leaved spurge
Herbaceous annual
Shores and waste ground.

I

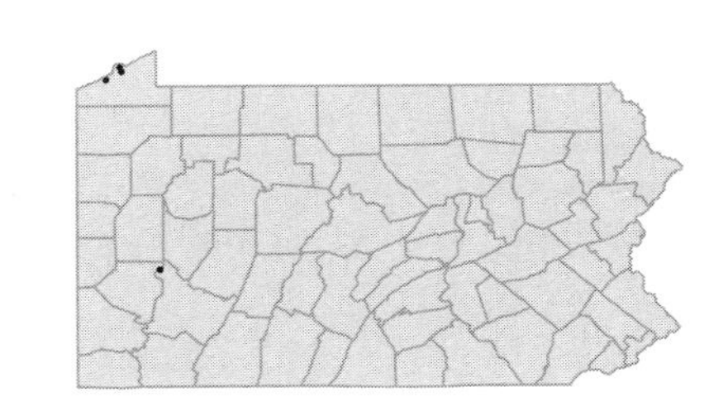

Euphorbia purpurea (Raf.) Fern.
Glade spurge
Herbaceous perennial
Swamps or moist thickets on rich soils.
Euphorbia darlingtonii A.Gray P

FAC

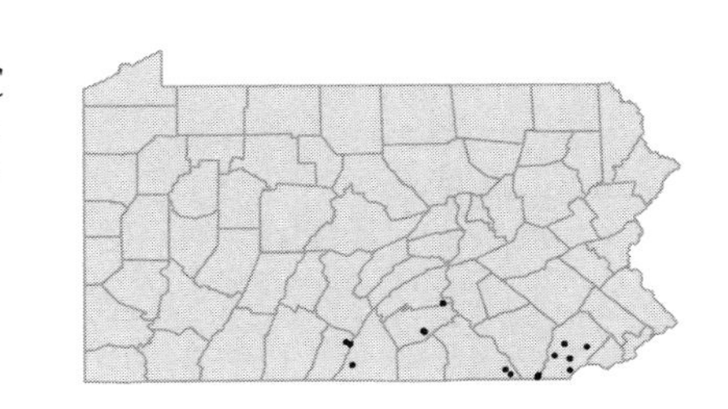

Euphorbia segetalis L.
Spurge
Herbaceous annual
Ballast.

I

• ***Mercurialis annua*** L.
Boys-and-girls; Herb-mercury
Herbaceous annual
Waste ground and ballast.

I

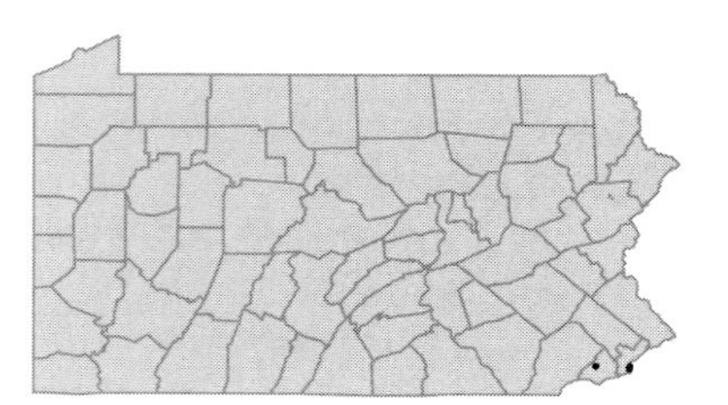

• ***Phyllanthus caroliniensis*** Walt. ssp. ***caroliniensis***
Carolina leaf-flower
Herbaceous annual
Moist soil along creek beds, ravines, roadsides and railroad banks. Believed to be extirpated, last collected in 1940.

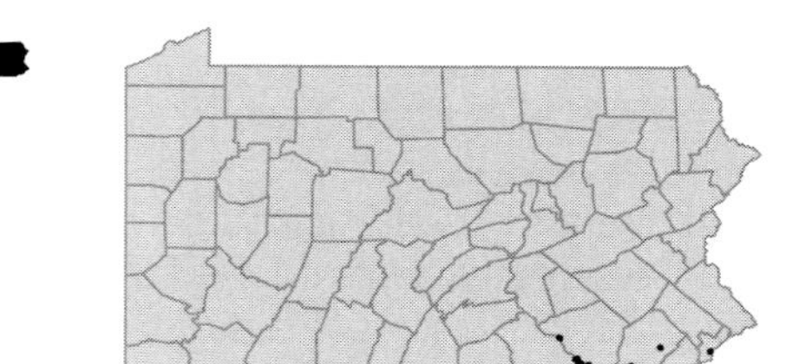

• ***Poinsettia dentata*** (Michx.) Klotzsch & Garcke
Spurge
Herbaceous annual
Wooded slopes, floodplains, railroad banks and waste ground.
Euphorbia dentata Michx. PFGBWC

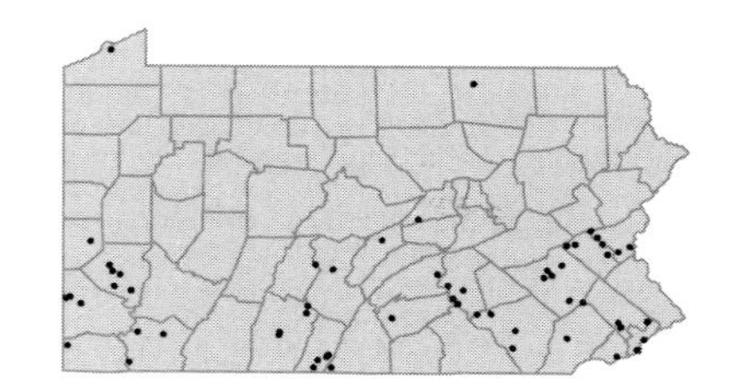

• ***Ricinus communis*** L.
Castor bean
Herbaceous annual
Cultivated and occasionally escaped to ballast or waste ground.

I

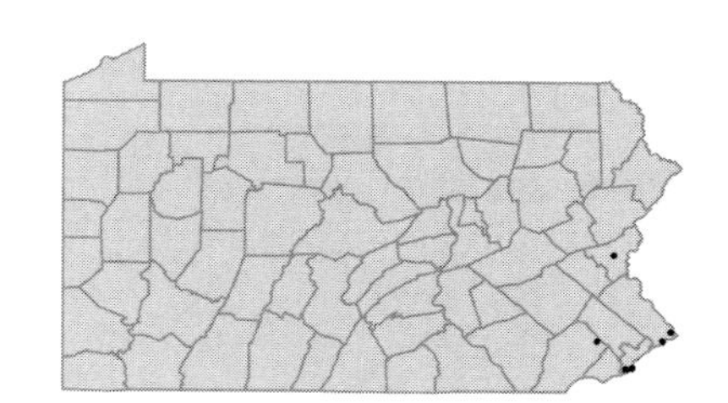

RHAMNACEAE

• ***Ceanothus americanus*** L.
New Jersey tea
Deciduous shrub
Wooded bluffs, roadside banks and shaly slopes.

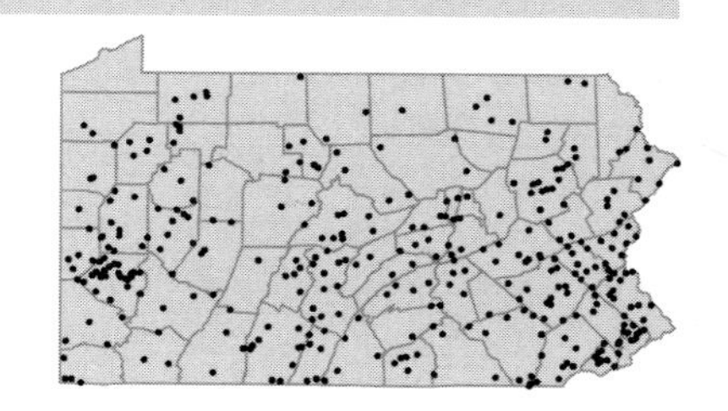

• ***Rhamnus alnifolia*** L'Her.
Alder-leaved buckthorn
Deciduous shrub
Fens, calcareous marshes and wet thickets.

OBL

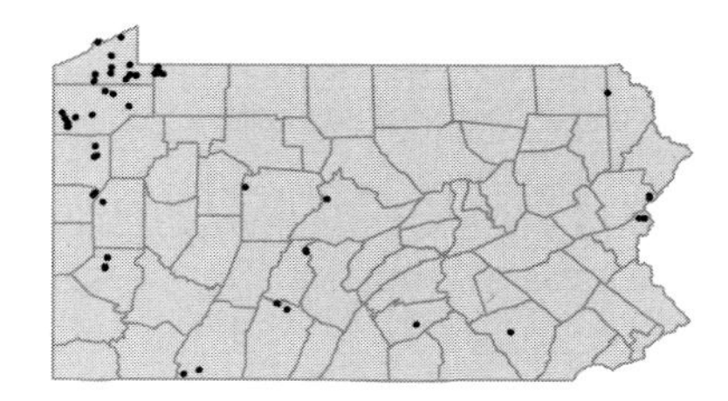

Rhamnus cathartica L.
Buckthorn
Deciduous shrub
Open woods, pastures, fencerows and roadside banks.

I

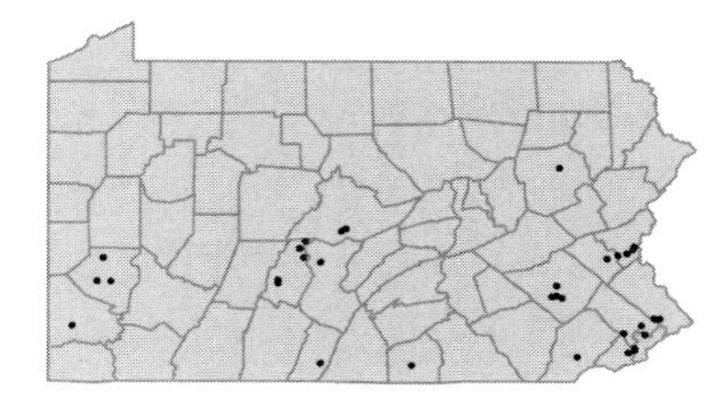

Rhamnus davurica Pallas
Buckthorn
Deciduous tree
Cultivated and occasionally spreading to old fields and roadside thickets.
Rhamnus citrifolia (Weston) W.Hess & Stearn C

I

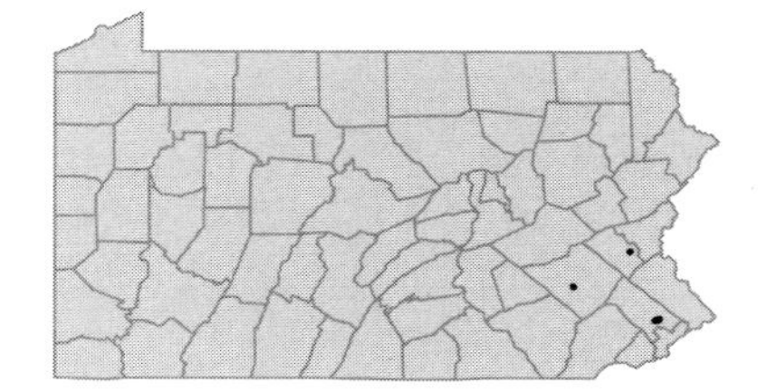

Rhamnus frangula L.
Alder buckthorn
Deciduous shrub
Cultivated and escaped to fields, thickets and roadside banks.

I

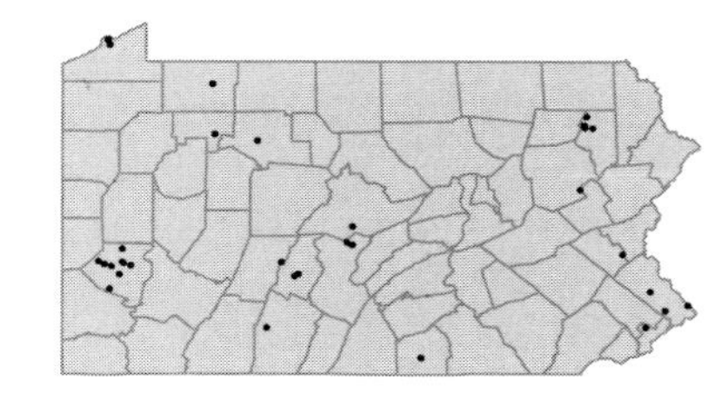

Rhamnus lanceolata Pursh
Lanceolate buckthorn
Deciduous shrub
Boggy fields, stream banks and calcareous woods.

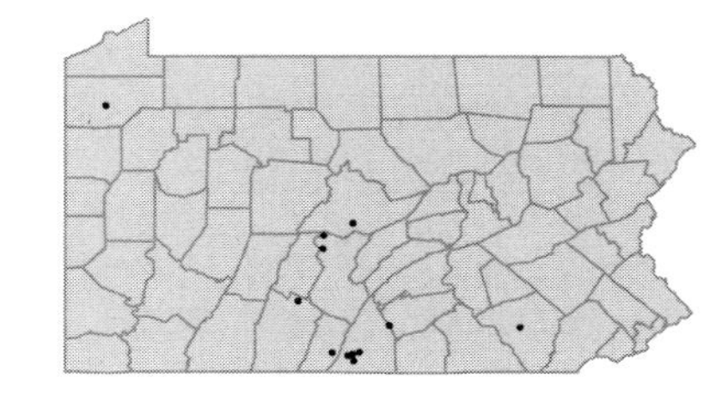

VITACEAE

• ***Ampelopsis brevipedunculata*** (Maxim.) Trautv.
Porcelain-berry
Woody vine
Cultivated and occasionally spreading to rubbish dumps, roadside thickets and railroad banks.

I

• ***Parthenocissus inserta*** (Kern.) Fritsch
Virginia-creeper; Woodbine
Woody vine
Woods, fields and alluvial thickets.
Parthenocissus vitacea (Kern.) A.S.Hitchc. GBWC

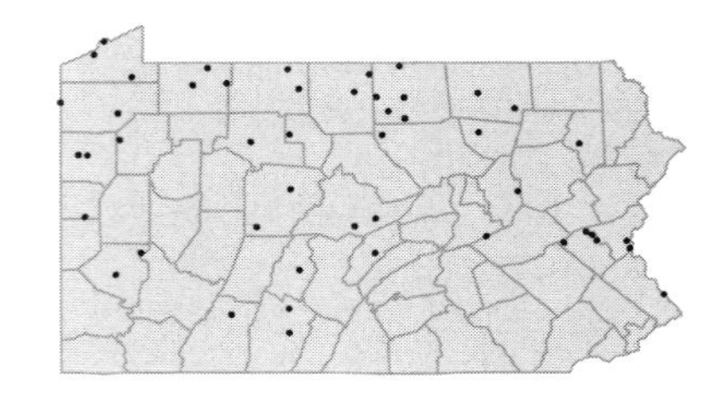

Parthenocissus quinquefolia (L.) Planch.
Virginia-creeper
Woody vine
Woods, fields and edges.

FACU

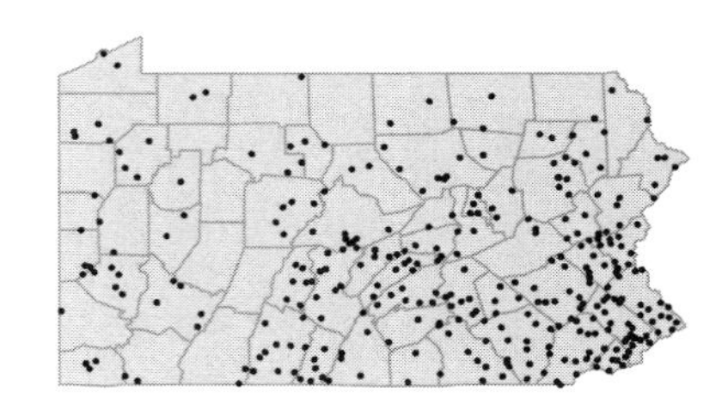

Parthenocissus tricuspidata (Sieb. & Zucc.) Planch.
Boston ivy
Woody vine
Cultivated and occasionally escaped.

I

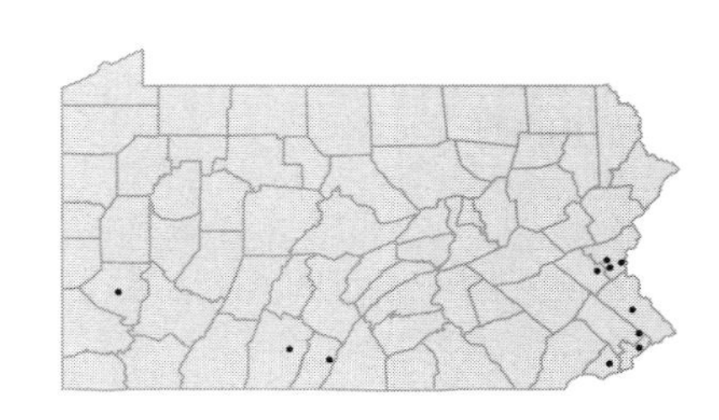

• ***Vitis aestivalis*** Michx.
Summer grape; Pigeon grape
Woody vine
Upland woods and wooded slopes.
Vitis bicolor LeConte P

FACW

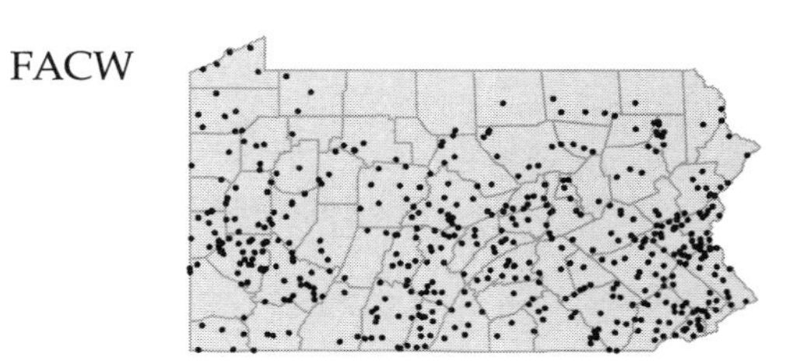

Vitis cinerea (Engelm. in A.Gray) Engelm. ex Millardet var. ***baileyana*** (Munson) Comeaux
Possum grape
Woody vine
Lowland woods.
Vitis baileyana Munson CBGF

FACW

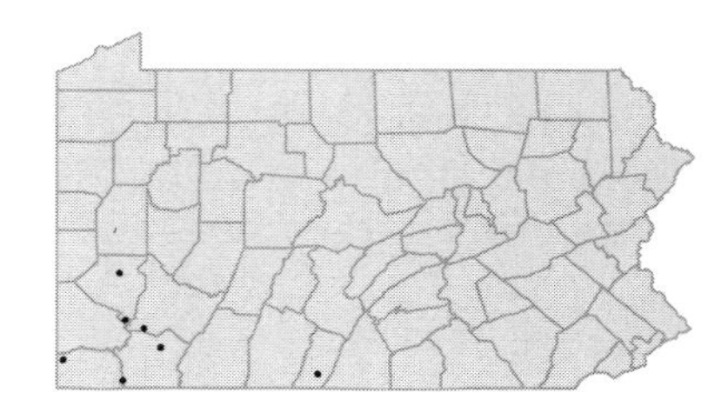

Vitis labrusca L.
Fox grape
Woody vine
Rocky woods, moist thickets and stream banks.

FACU

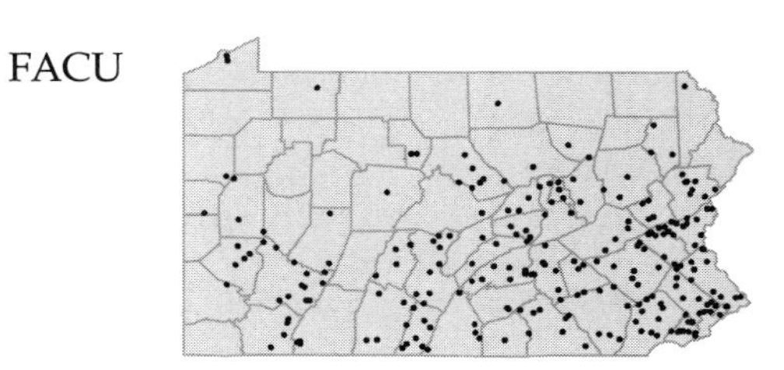

Vitis labrusca x riparia
New England grape
Woody vine
Moist mountain woods, ravines and roadside thickets.
Vitis novae-angliae Fern. FBGKWC

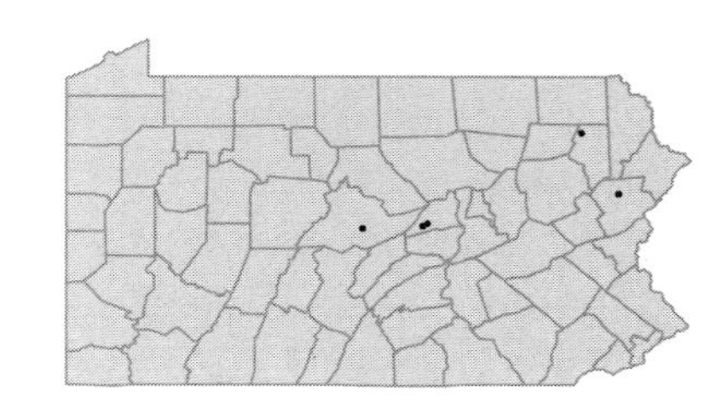

Vitis x labruscana Bailey
Fox grape
Woody vine
Cultivated and occasionally spreading to dumps and roadsides. A collective epithet for cultivars of hybrid origin having *V. labrusca* as one parent. It is distinguished from the native species by the larger and more numerous berries.

I

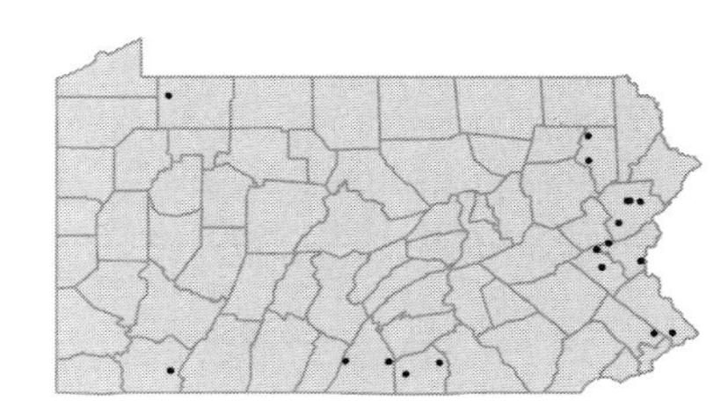

Vitis riparia Michx.
Frost grape
Woody vine
River banks and alluvial thickets.
Vitis syrticola (Fern.& Wieg.) Fern. W

FACW

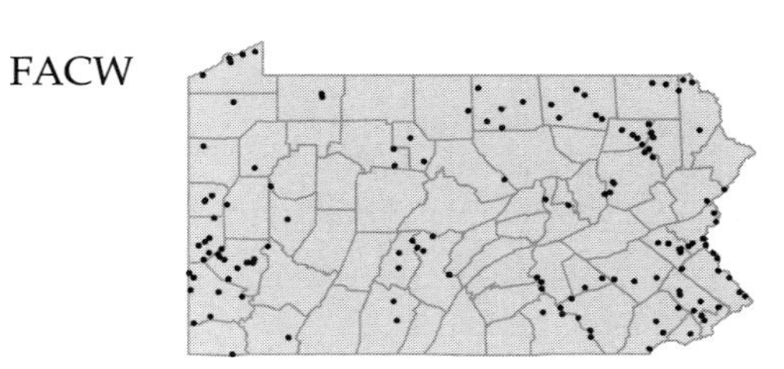

Vitis rupestris Scheele
Sand grape
Deciduous shrub
River banks. Believed to be extirpated.

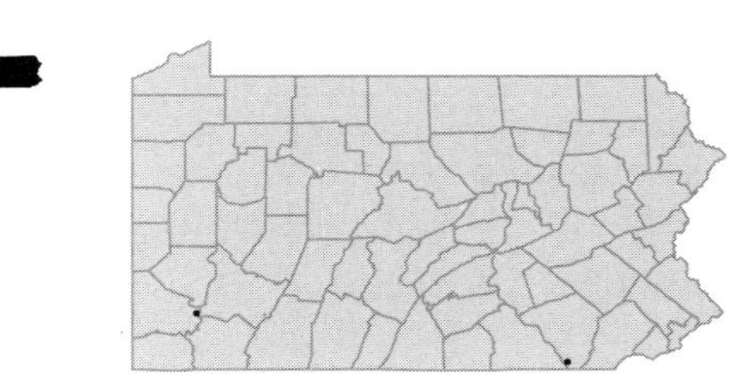

Vitis vinifera L.
European grape
Woody vine
Cultivated and occasionally escaped to rubbish dumps.

I

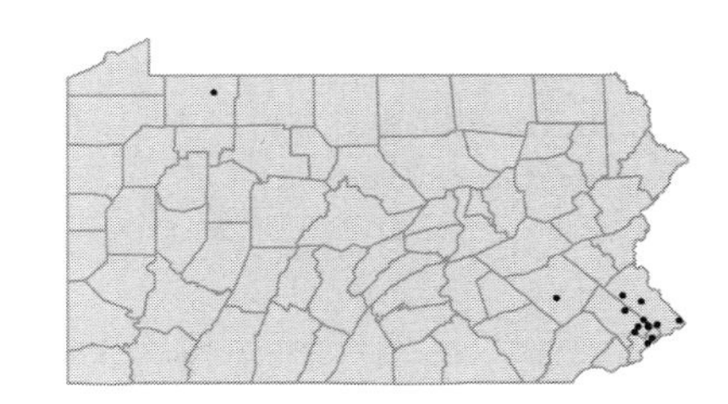

Vitis vulpina L.
Frost grape
Woody vine
Woods, thickets, rocky slopes, roadsides and sand dunes.
Vitis cordifolia Michx. P

FAC

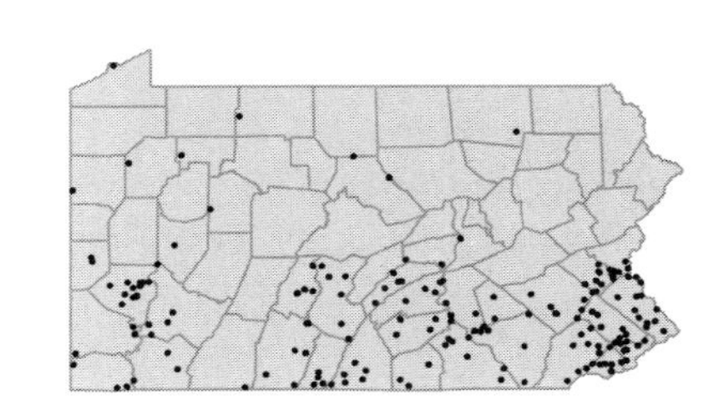

LINACEAE

• ***Linum bienne*** P.Mill.
Flax
Herbaceous annual
Waste ground and ballast.

I

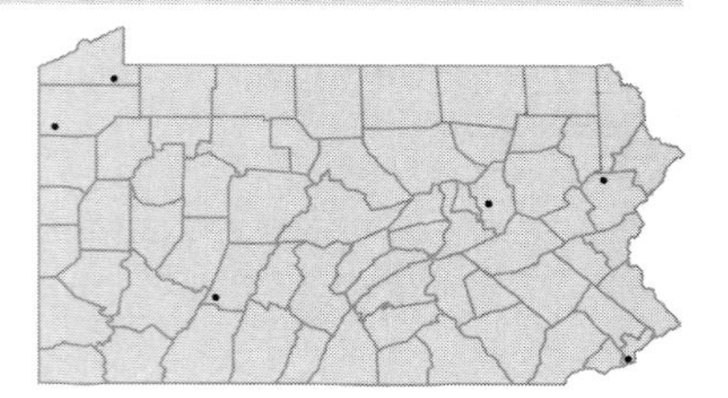

Linum catharticum L.
Fairy flax
Herbaceous annual
Waste ground. Represented by a single collection from Philadelphia Co. in 1865.

I

Linum intercursum Bickn.
Sandplain wild flax
Herbaceous perennial
Moist, clayey, open thickets and serpentine barrens.

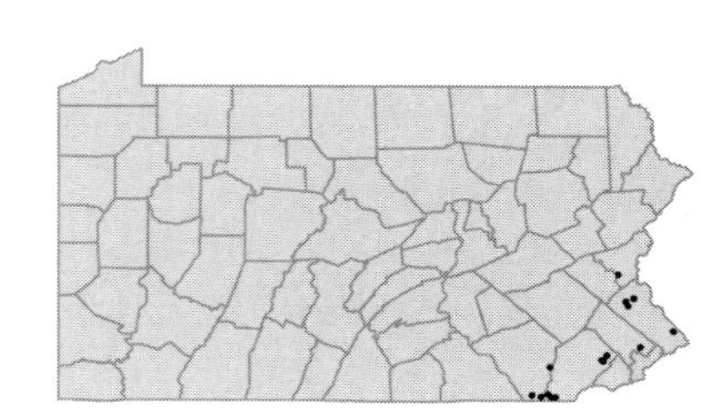

Linum medium (Planch.) Britt. var. ***medium***
Yellow flax
Herbaceous perennial
Moist sand flats. Represented by a single collection from Erie Co. in 1985.

FACU

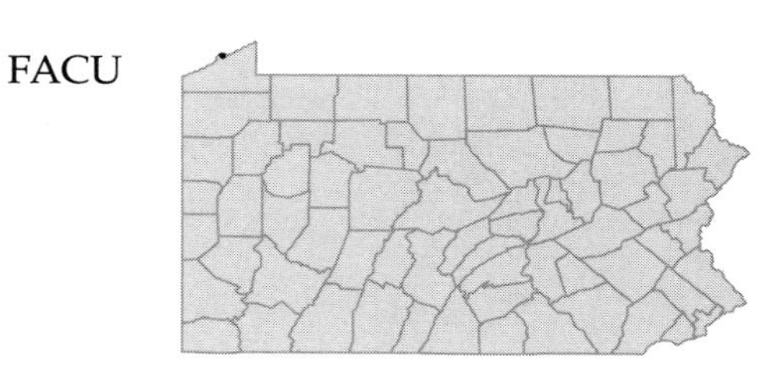

Linum medium (Planch.) Britt. var. ***texanum*** (Planch.) Fern.
Southern yellow flax
Herbaceous perennial
Moist to dry, sandy fields, open thickets or woodland openings.

FACU

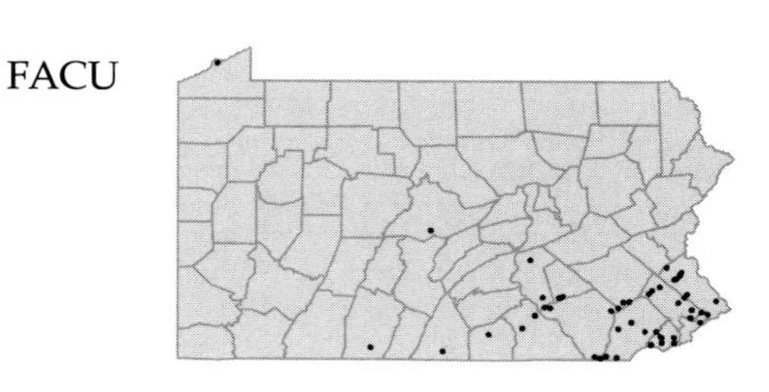

Linum perenne L.
Perennial flax
Herbaceous perennial
Cultivated and rarely naturalized along roadsides.

I

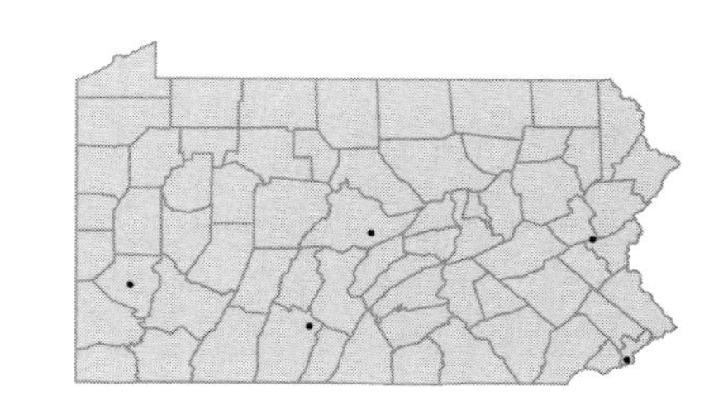

Linum striatum Walt.
Ridged yellow flax
Herbaceous perennial
Moist meadows and wet, open woods.

FACW

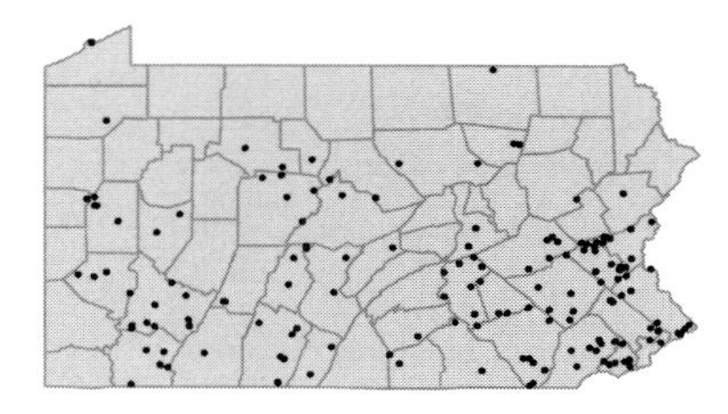

Linum sulcatum Riddell
Grooved yellow flax
Herbaceous annual
Sandy barrens.

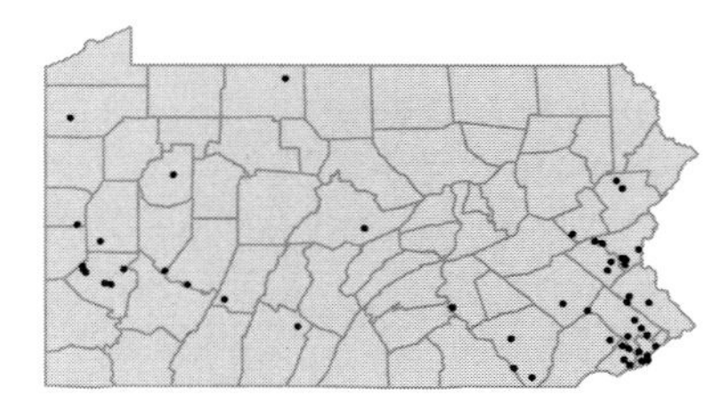

Linum usitatissimum L.
Common flax
Herbaceous annual
Ballast, railroad banks and roadsides.

I

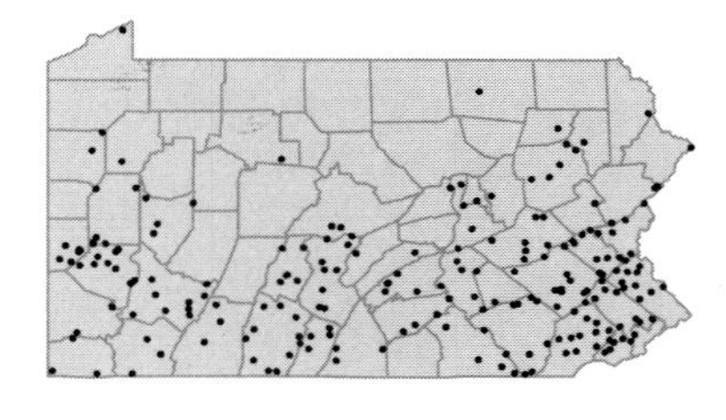

Linum virginianum L.
Slender yellow flax
Herbaceous perennial
Dry, open woods, old fields or shaly slopes.

FACU

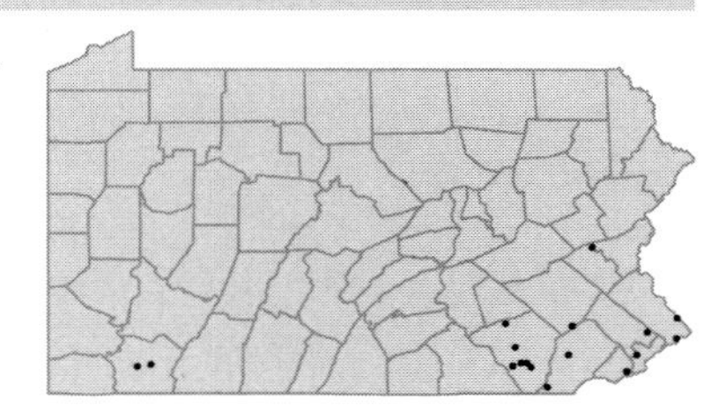

POLYGALACEAE

• ***Polygala cruciata*** L.
Cross-leaved milkwort
Herbaceous annual
Boggy pastures and mountain bogs.

FACW+

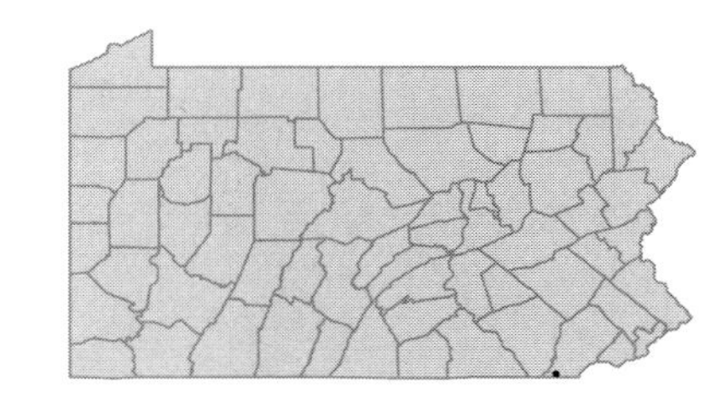

Polygala curtissii A.Gray
Curtis's milkwort
Herbaceous annual
Dry, open serpentine barrens.

Polygala incarnata L.
Pink milkwort
Herbaceous annual
Dry, open soil of serpentine barrens.

UPL

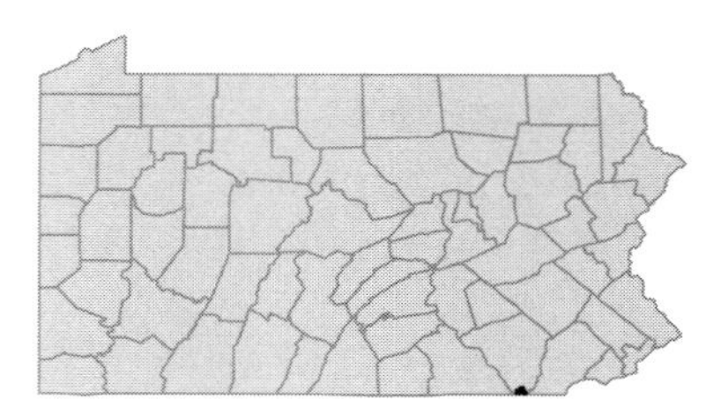

Polygala lutea L.
Yellow milkwort
Herbaceous biennial
Moist, sandy, coastal plain soil. Believed to be extirpated, last collected in 1906.

FACW+

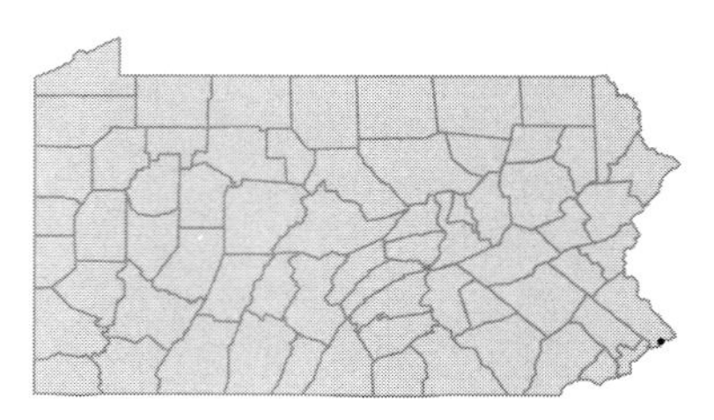

Polygala nuttallii Torr. & A.Gray
Nuttall's milkwort
Herbaceous annual
Open woods, peaty thickets and sphagnum bogs.

FAC

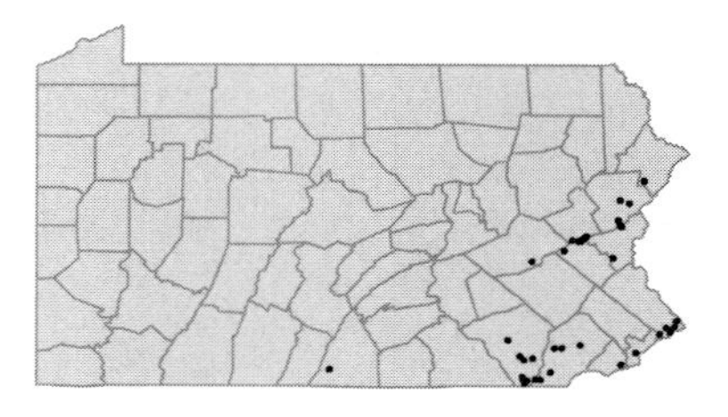

Polygala paucifolia Willd.
Bird-on-the-wing; Fringed milkwort
Herbaceous perennial
Rich, rocky woods and wooded slopes.

FACU

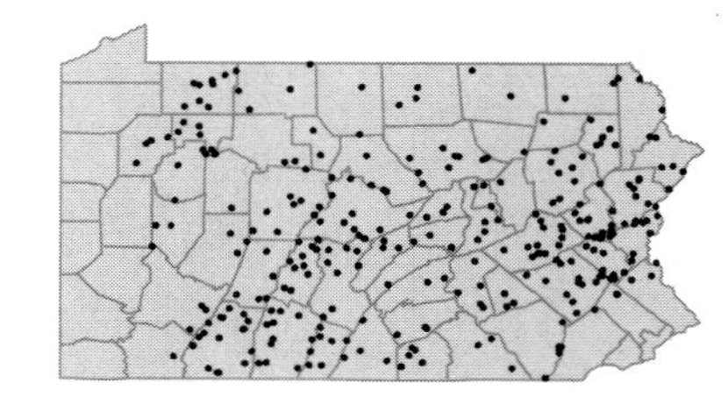

Polygala polygama Walt.
Bitter milkwort; Racemed milkwort
Herbaceous biennial
Abandoned fields and wooded bogs.
Polygala polygama Walt. var. *obtusata* Chodat FGBW

UPL

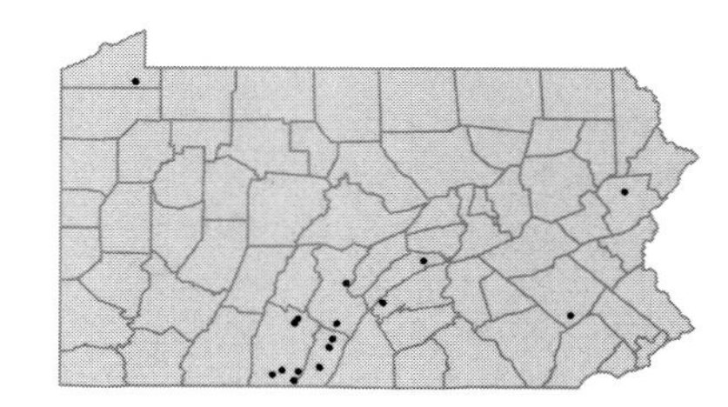

Polygala sanguinea L.
Field milkwort; Rose milkwort
Herbaceous annual
Moist fields on sterile, acidic soils.
Polygala viridescens L. P

FACU

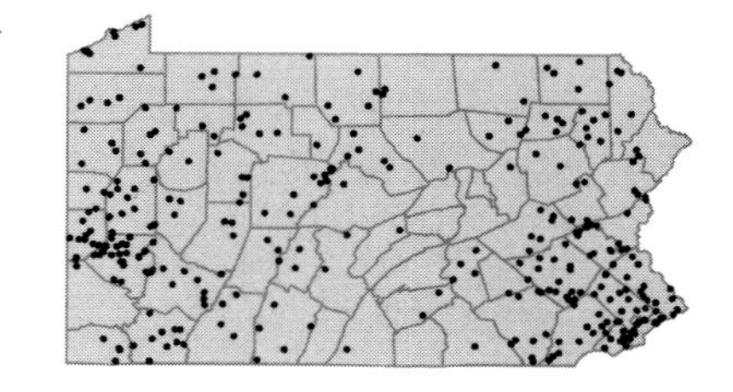

Polygala senega L. var. ***senega***
Seneca snakeroot
Herbaceous perennial
Rocky, wooded, limestone slopes.

FACU

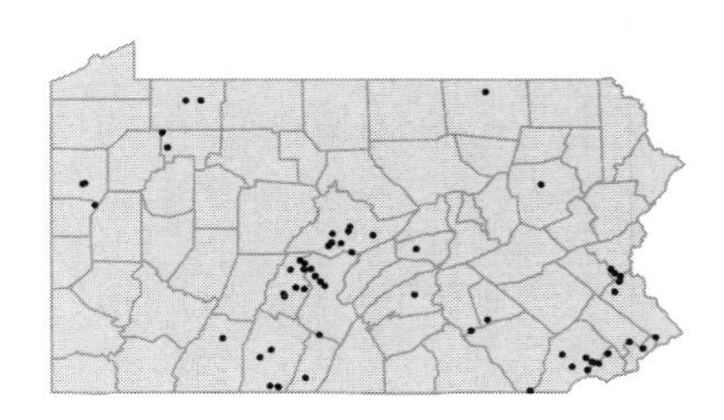

Polygala senega L. var. ***latifolia*** Torr. & A.Gray
Seneca snakeroot
Herbaceous perennial
Scrubby, limestone ridge and moist roadside slopes.

FACU

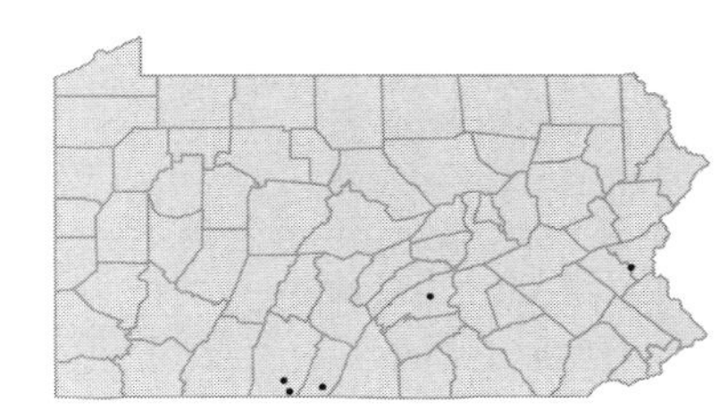

Polygala verticillata L. var. ***verticillata***
Whorled milkwort
Herbaceous annual
Dry, open woods, old fields and roadsides.

UPL

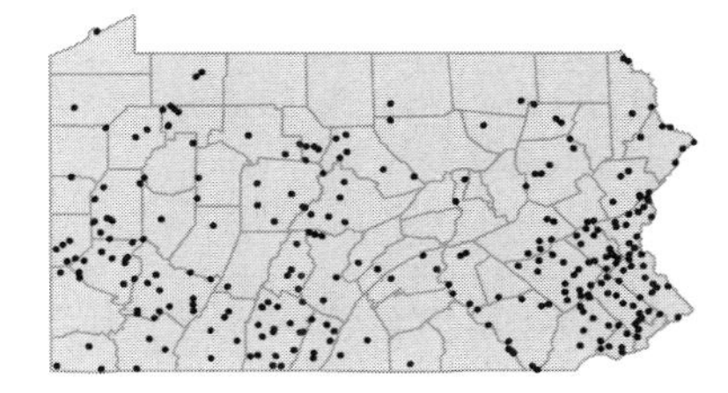

Polygala verticillata L. var. ***ambigua*** (Nutt.) Wood
Whorled milkwort
Herbaceous annual
Moist meadows, marshes and ravines.
Polygala ambigua Nutt. PGBC

UPL

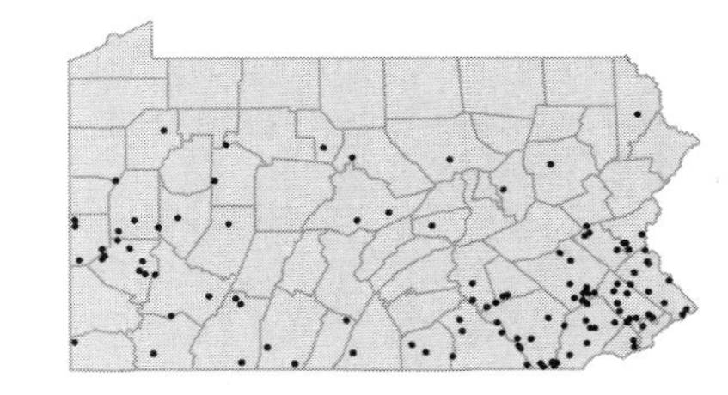

Polygala verticillata L. var. ***isocycla*** Fern.
Whorled milkwort
Herbaceous annual
Waste ground and railroad tracks.

UPL

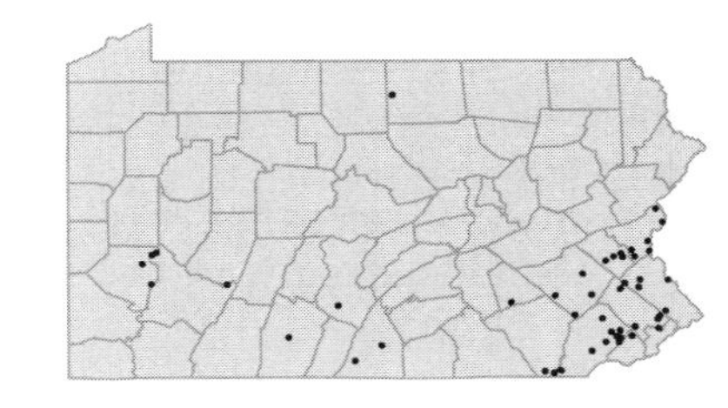

STAPHYLEACEAE

• ***Staphylea trifolia*** L.
Bladdernut
Deciduous shrub
Moist, rocky woods and stream banks.

FAC

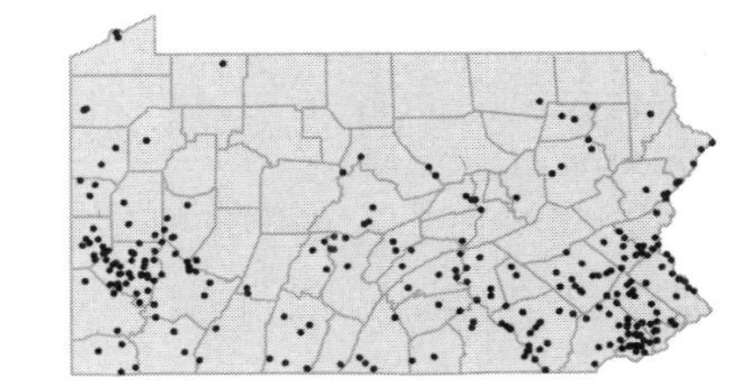

SAPINDACEAE

• ***Cardiospermum halicacabum*** L. [I]
Balloon-vine; Heart-seed
Herbaceous annual vine
Cultivated and occasionally escaped to ballast, railroad sidings, and urban stream valleys.

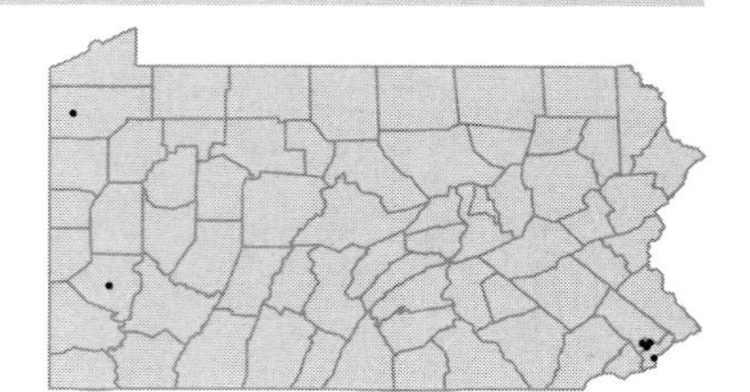

• ***Koelreuteria paniculata*** Laxm. [I]
Goldenrain-tree
Deciduous tree
Cultivated, seedlings occasionally becoming established in waste ground or along roadsides.

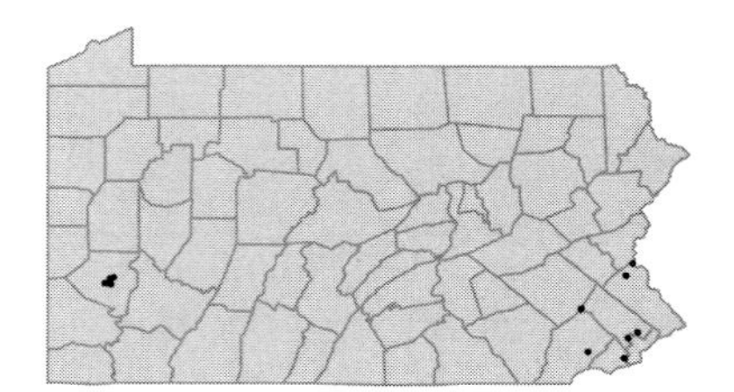

HIPPOCASTANACEAE

• ***Aesculus flava*** Ait.
Yellow buckeye; Sweet buckeye
Deciduous tree
Low woods along streams, native in western counties but introduced in the east.
Aesculus octandra Marshall PFGBW

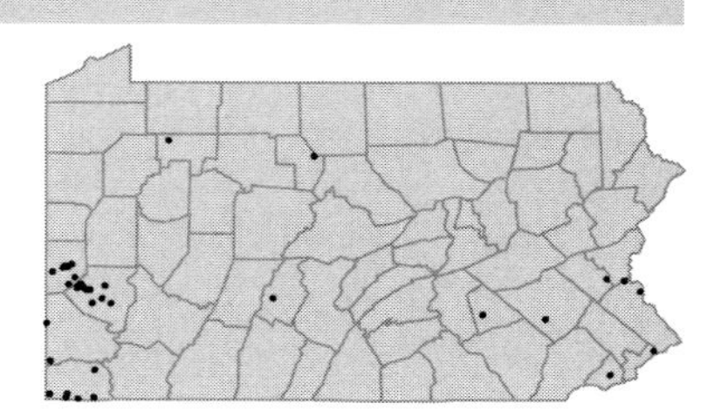

Aesculus glabra Willd. FACU+
Ohio buckeye
Deciduous tree
Moist woods and bottomlands, native in western counties but introduced in the east.

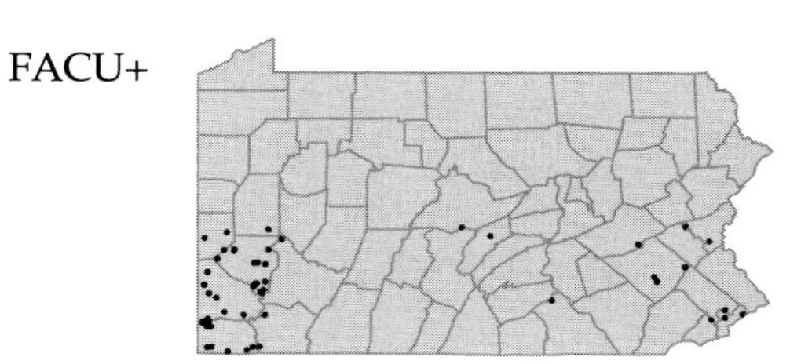

Aesculus hippocastanum L. [I]
Horse-chestnut
Deciduous tree
Cultivated, occasionally escaped to railroad banks or persisting in abandoned gardens.

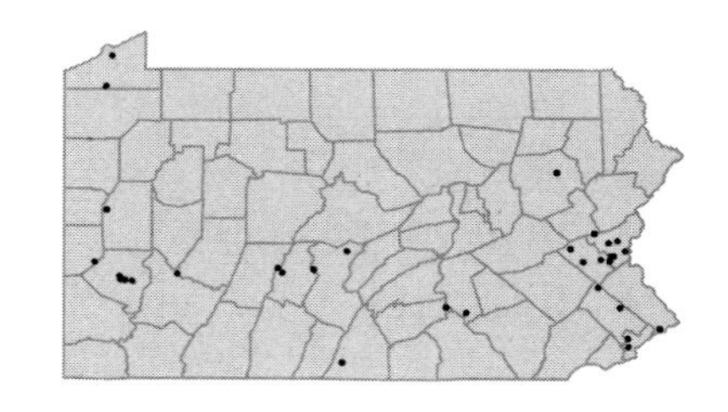

Aesculus parviflora Walt. [I]
Bottlebrush buckeye
Deciduous shrub
Cultivated and occasionally naturalized in disturbed, urban woods.

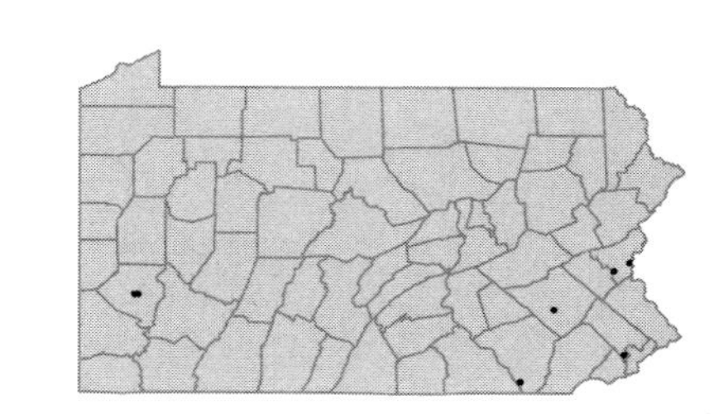

ACERACEAE

• ***Acer campestre*** L. [I]
Hedge maple
Deciduous shrub
Occasionally spreading from cultivation to moist, rocky, disturbed woods.

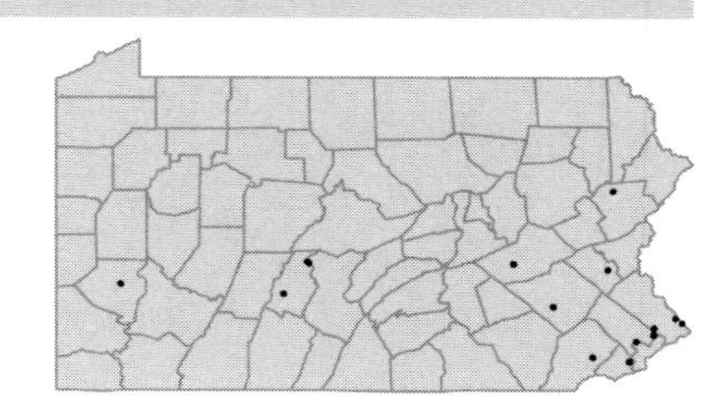

Acer ginnala Maxim. [I]
Amur maple
Deciduous shrub
Cultivated and escaped to a brushy field.

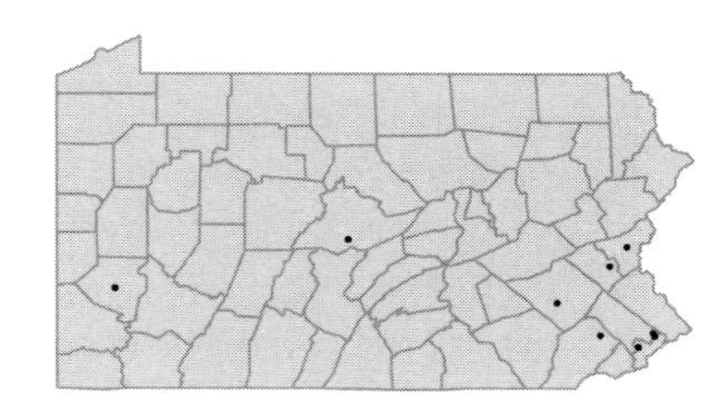

Acer negundo L. FAC+
Box-elder
Deciduous tree
Low, moist areas, stream banks, and floodplains.

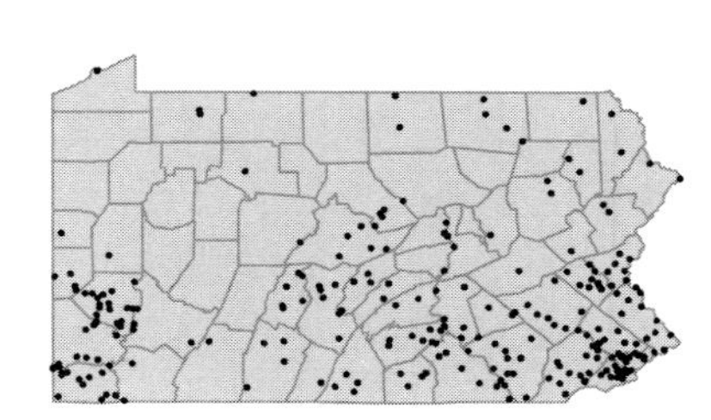

Acer nigrum Michx.f.
Black maple
Deciduous tree
Rich woods, ravines and river banks.

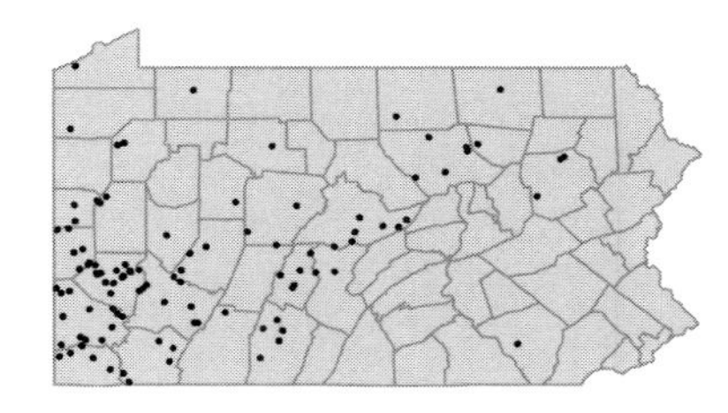

Acer palmatum Thunb. [I]
Japanese maple
Deciduous tree
Cultivated and occasionally escaped to disturbed woods.

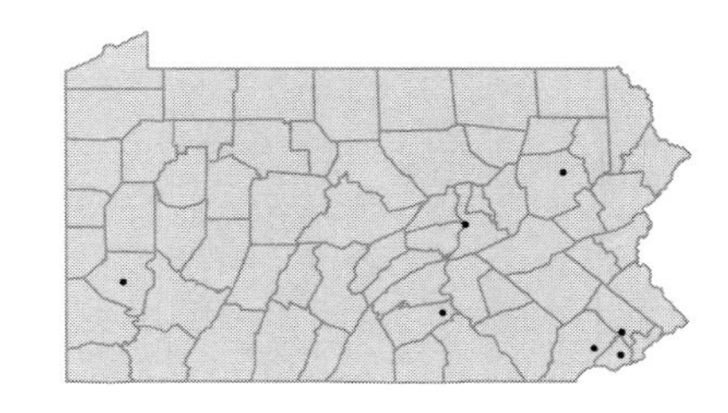

Acer pensylvanicum L. FACU
Moosewood; Striped maple
Deciduous tree
Cool, moist, rocky woods.

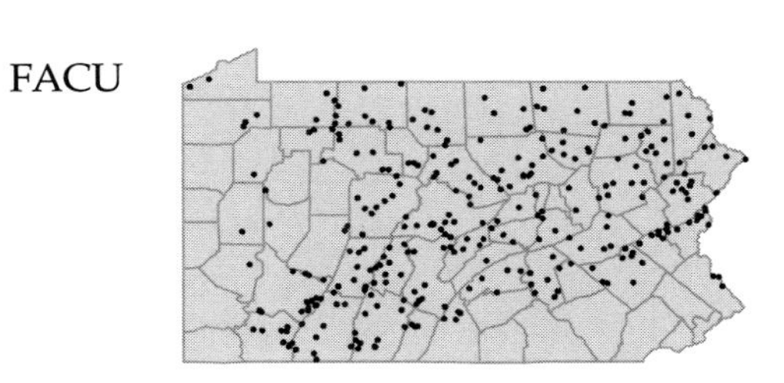

Acer platanoides L. I
Norway maple
Deciduous tree
Cultivated and frequently escaped to disturbed woods, roadsides and edges.

Acer pseudoplatanus L. I
Sycamore maple
Deciduous tree
Occasionally spreading from cultivation to railroad banks, waste ground and urban woods.

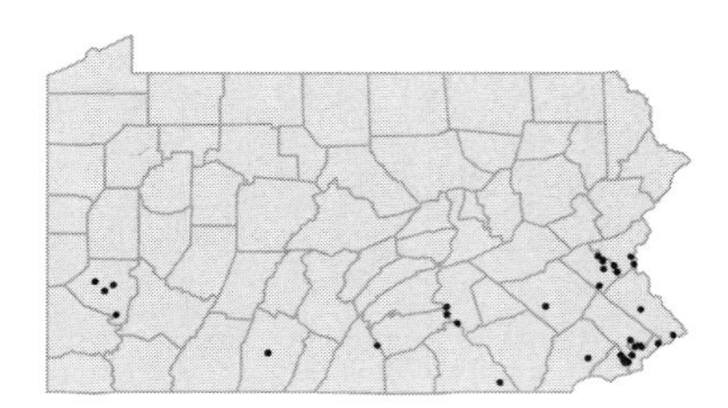

Acer rubrum L. var. ***rubrum*** FAC
Red maple; Swamp maple
Deciduous tree
Dry to moist woods, swamps, and bogs.

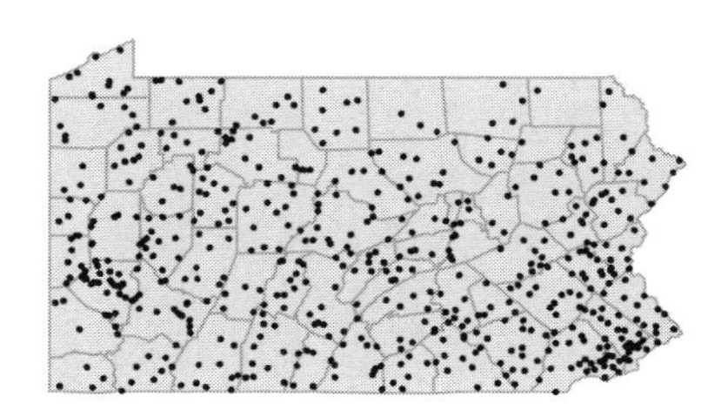

Acer rubrum L. var. ***trilobum*** Torr. & A.Gray ex K.Koch FAC
Trident red maple
Deciduous tree
Wooded slopes, swamps, bogs or moist woods.

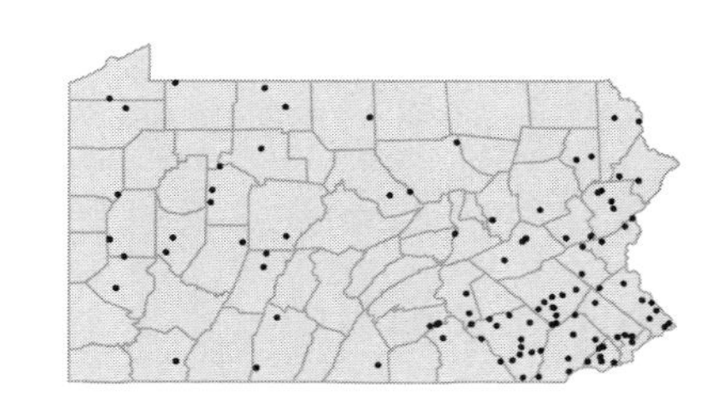

Acer saccharinum L. FACW
Silver maple
Deciduous tree
Moist woods, stream banks and alluvium.

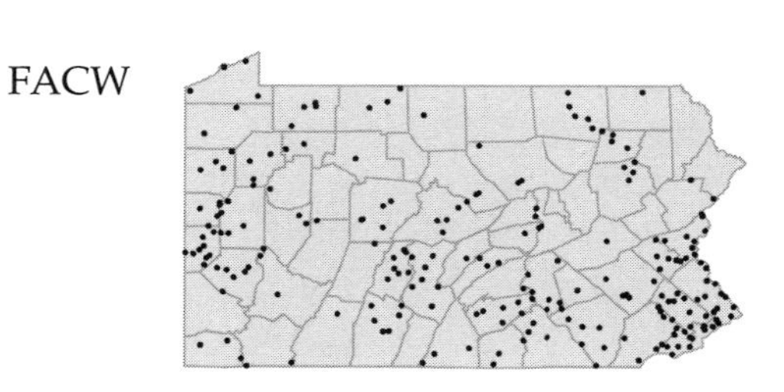

Acer saccharum Marshall var. ***saccharum*** FACU
Sugar maple; Rock maple
Deciduous tree
Moist woods, wooded slopes, ravines and alluvial areas.

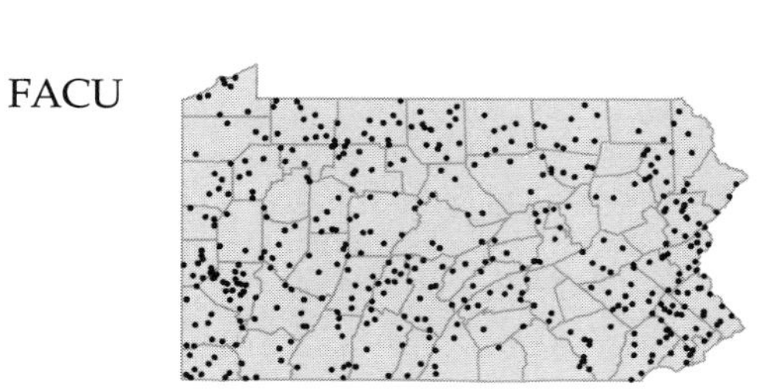

Acer saccharum Marshall var. ***schneckii*** Rehd. FACU
Sugar maple; Rock maple
Deciduous tree
Rich, moist woods.
Acer saccharum Marshall var. *rugelii* (Pax) Palmer & Steyermark FGBW

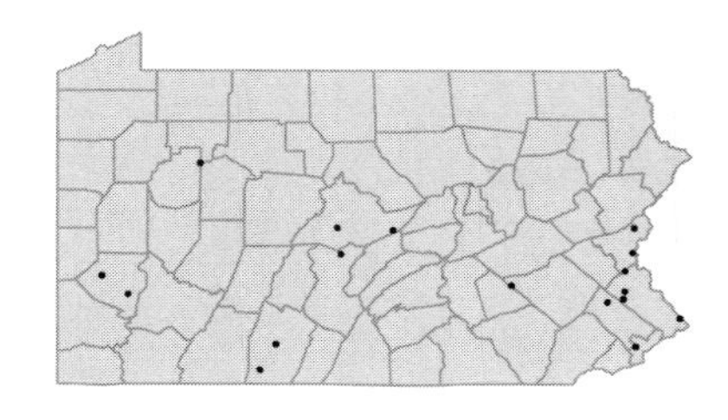

Acer spicatum Lam. FACU-
Mountain maple
Deciduous shrub
Moist, rocky woods.

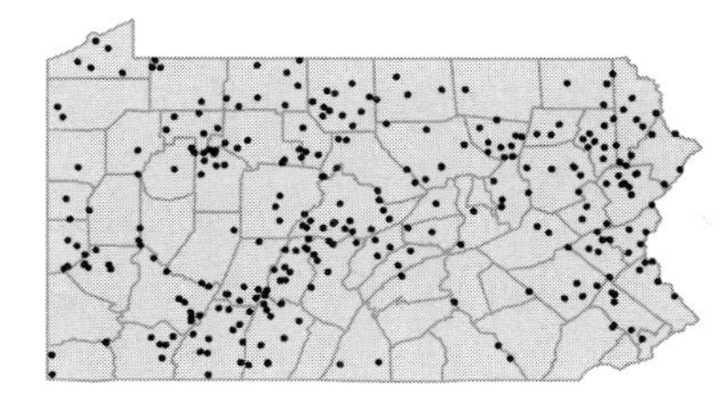

ANACARDIACEAE

• ***Cotinus coggygria*** Scop. [I]
Smoke-tree
Deciduous tree
Cultivated and occasionally persisting at old home sites.

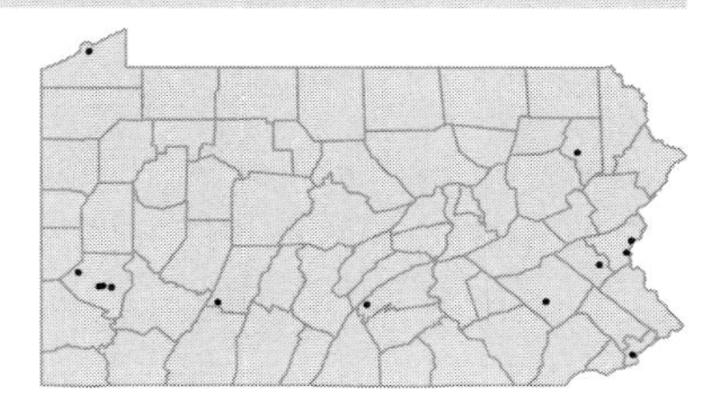

• ***Rhus aromatica*** Ait. var. ***aromatica***
Fragrant sumac
Deciduous shrub
Dry, open woods and shale barrens.

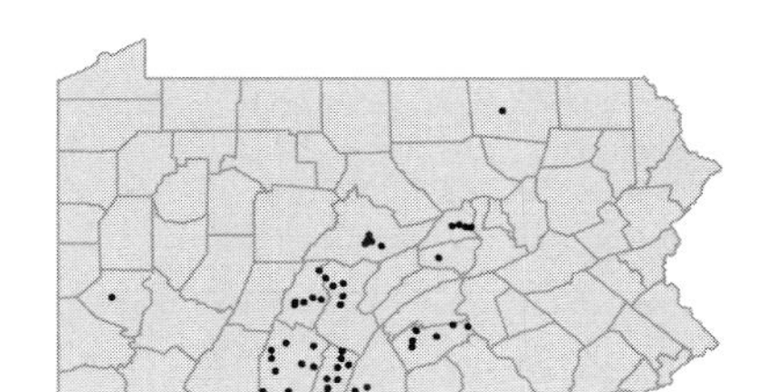

Rhus copallina L. var. ***copallina***
Shining sumac; Dwarf sumac
Deciduous shrub
Dry, open woods, thickets and old fields.

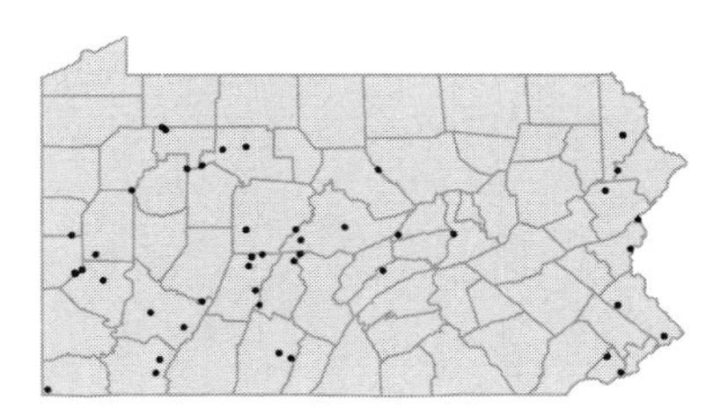

Rhus copallina L. var. ***latifolia*** Engl.
Shining sumac; Dwarf sumac
Deciduous shrub
Serpentine barrens, shale barrens, old fields and rocky slopes.

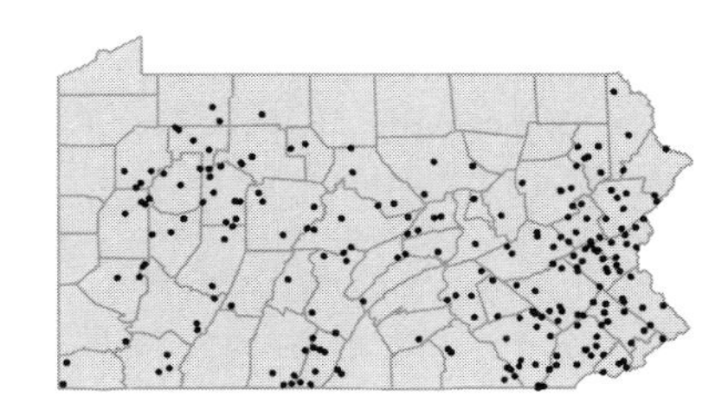

Rhus glabra L.
Smooth sumac
Deciduous shrub
Shale barrens, old fields and dry, open slopes.

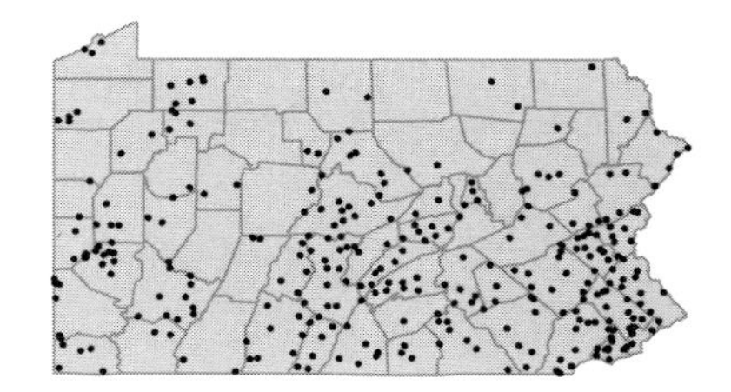

Rhus typhina L.
Staghorn sumac
Deciduous shrub
Dry, open soil of old fields, roadsides and woods edges.
Rhus hirta (L.) Sudworth P

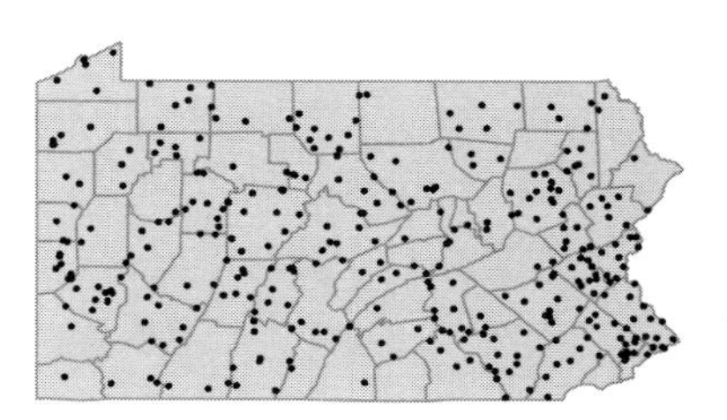

• ***Toxicodendron radicans*** (L.) Kuntze — FAC
Poison-ivy
Woody vine
Open woods, roadside thickets, fencerows and edges.
Rhus radicans L. PFGBW

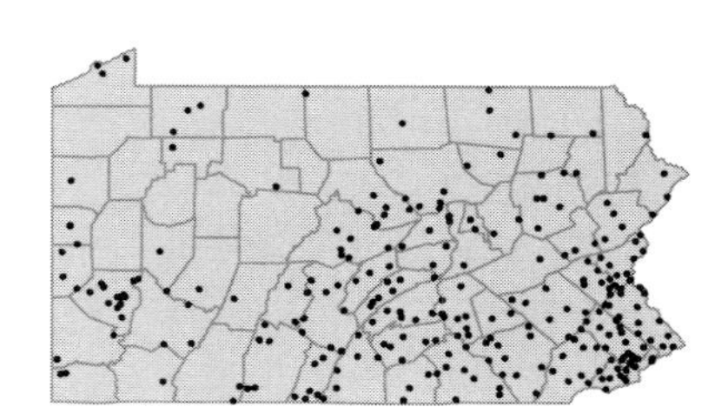

Toxicodendron rydbergii (Small ex Rydb.) Greene — FAC-
Giant poison-ivy
Herbaceous perennial
Dry, rocky woods.
Rhus radicans L. var. *rydbergii* (Small) Rehd. FGBW

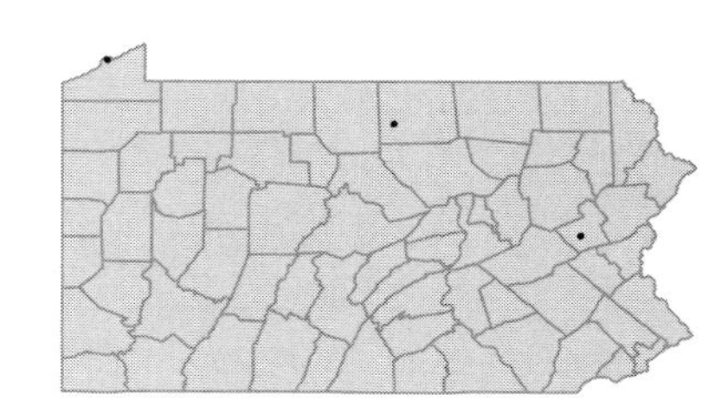

Toxicodendron vernix (L.) Kuntze — OBL
Poison sumac
Deciduous tree
Swamps, fens and marshes.
Rhus vernix L. PFGBW

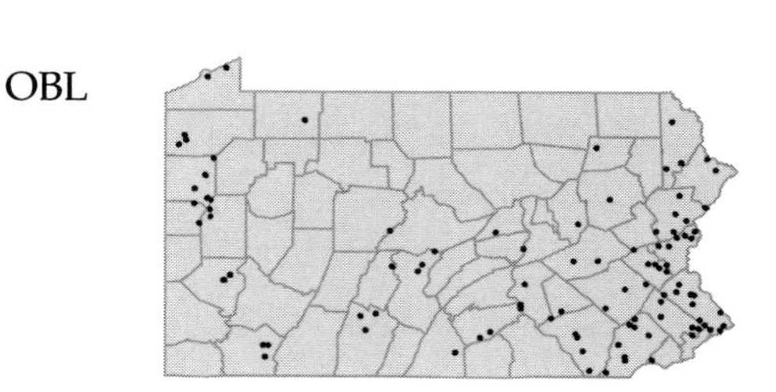

SIMAROUBACEAE

• ***Ailanthus altissima*** (P.Mill.) Swingle — [I]
Tree-of-heaven
Deciduous tree
Disturbed woods, roadsides, fencerows, vacant lots and railroad banks.
Ailanthus glandulosa Desf. P

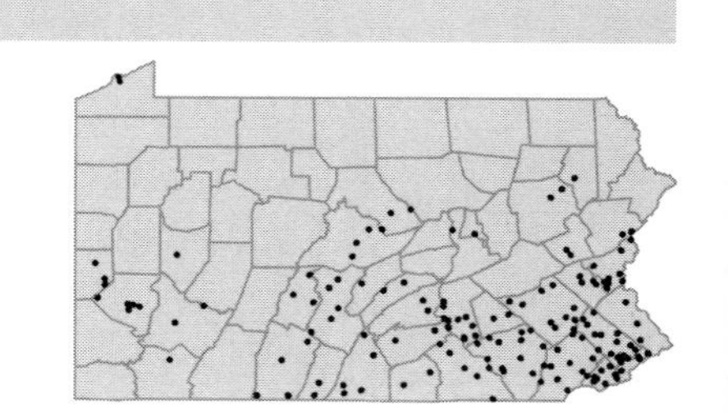

RUTACEAE

• ***Phellodendron japonicum*** Maxim.
Japanese cork-tree
Deciduous tree
Cultivated and occasionally naturalized in disturbed woods.

I

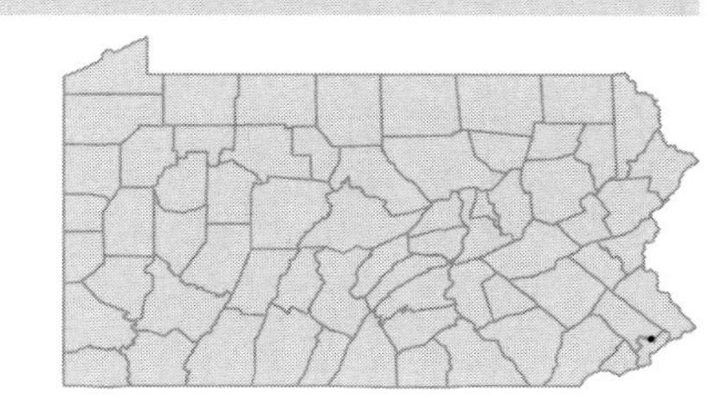

• ***Poncirus trifoliata*** (L.) Raf.
Hardy orange; Trifoliate orange
Deciduous shrub
Cultivated and occasionally escaped to wooded banks or rubbish dumps.

I

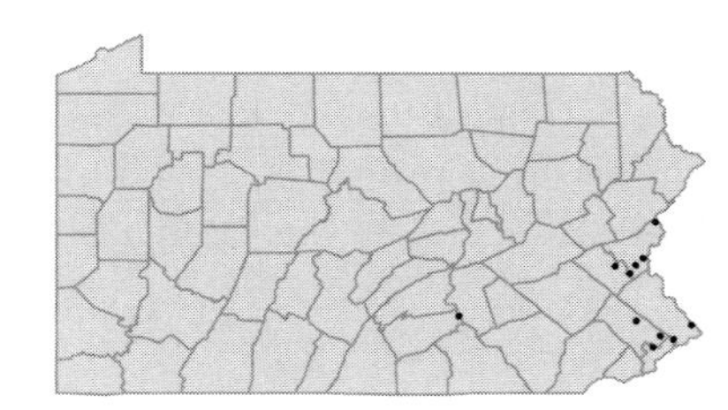

• ***Ptelea trifoliata*** L.
Hop-tree; Wafer-ash
Deciduous shrub
Stream banks and roadside thickets.

FAC

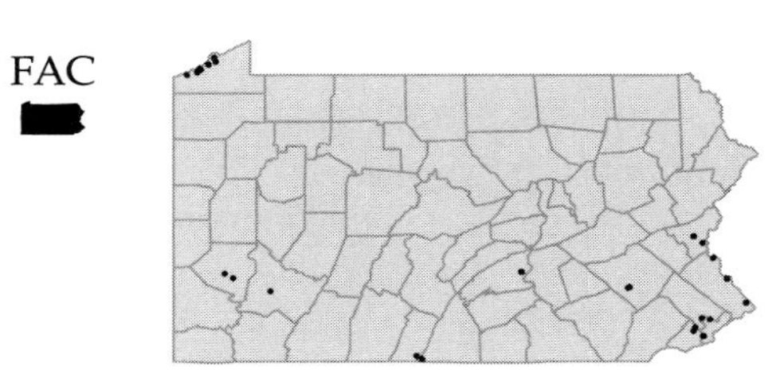

• ***Ruta graveolens*** L.
Common rue; Herb-of-grace
Herbaceous perennial
Cultivated and occasionally escaped to roadsides and old fields.

I

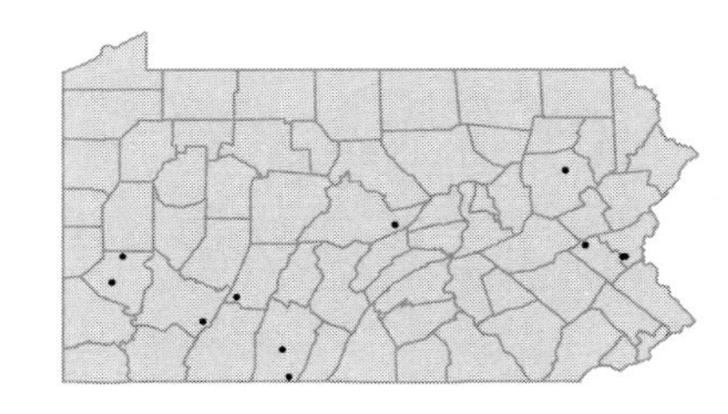

• ***Zanthoxylum americanum*** P.Mill.
Northern prickly-ash; Toothache-tree
Deciduous shrub
Stream banks, river bluffs and roadside thickets, usually on calcareous soils or diabase.

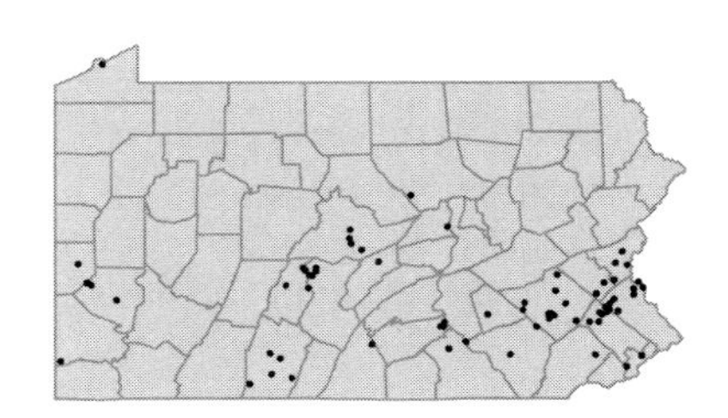

ZYGOPHYLLACEAE

• ***Kallstroemia parviflora*** Norton
Kallstroemia
Herbaceous annual
Wharves and railroad sidings. Represented by a single collection from Delaware Co. in 1932.
Kallstroemia intermedia Rydb. FGB; *Kallstroemia maxima* (L.) Torr. & A.Gray W

I

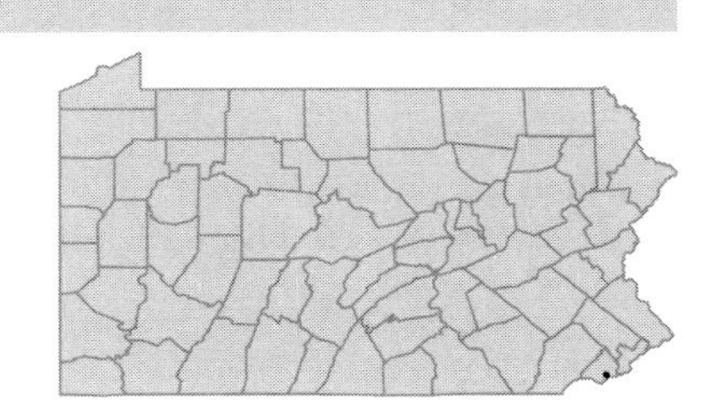

ZYGOPHYLLACEAE

• ***Tribulus terrestris*** L. [I]
Caltrop; Puncture-weed
Herbaceous annual
Ballast, ore heaps and railroad tracks.

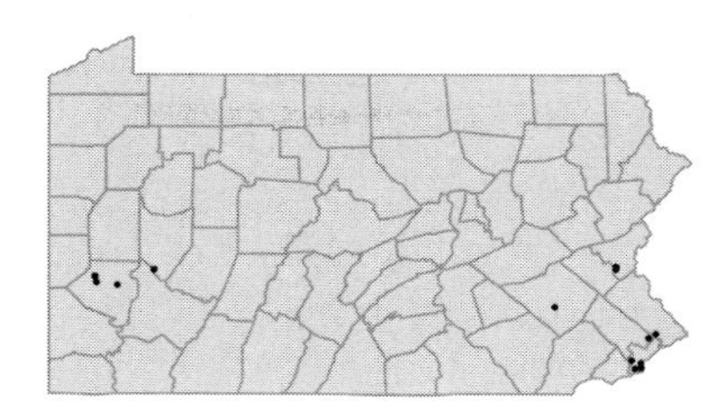

• ***Zygophyllum fabago*** L. [I]
Zygophyllum
Herbaceous annual
Ballast. All collections from a single site in Philadelphia Co. in 1880.

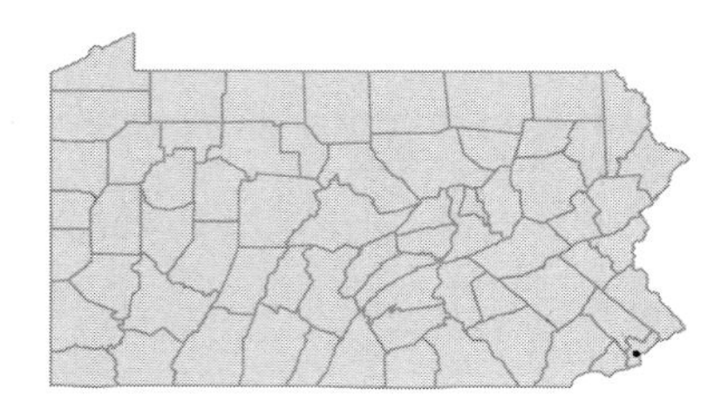

OXALIDACEAE

• ***Oxalis acetosella*** L. FAC-
Common wood-sorrel; Wood-shamrock
Herbaceous perennial
Rich, moist woods, bogs and swamps.
Oxalis montana Raf. FW

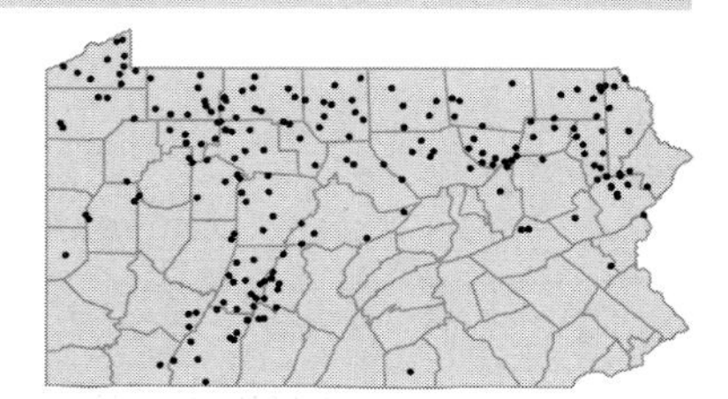

Oxalis corniculata L. FACU [I]
Creeping yellow wood-sorrel
Herbaceous perennial
Roadsides, fields, pavement and gardens.
Oxalis repens Thunb. B

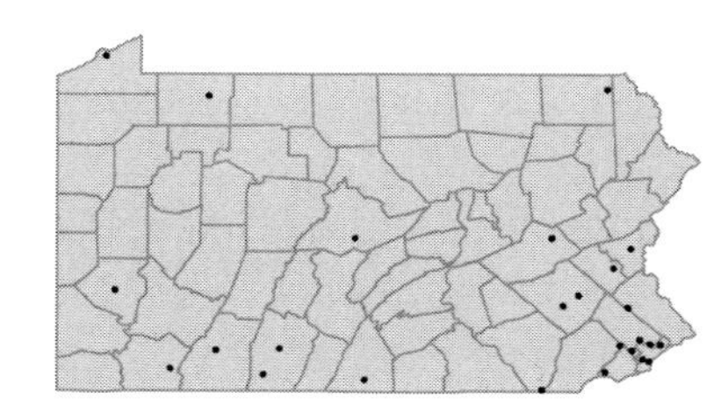

Oxalis dillenii Jacq. ssp. ***filipes*** (Small) Eiten
Wood-shamrock; Southern yellow wood-sorrel
Herbaceous perennial
Rich woods, diabase cliffs, roadsides, waste ground, and woodland edges.
Oxalis brittoniae Small P; *Oxalis filipes* Small FW

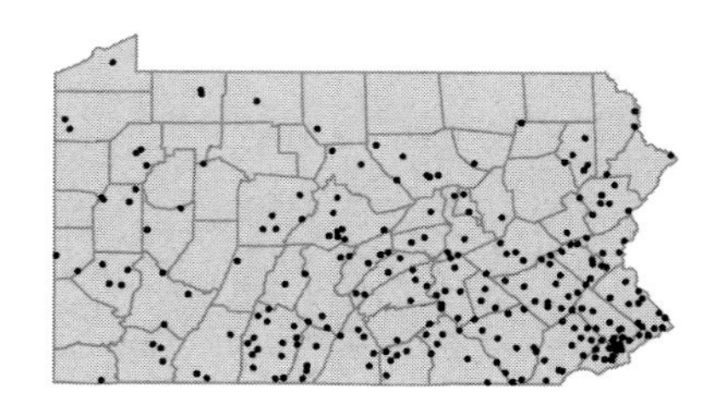

Oxalis grandis Small
Wood-sorrel
Herbaceous perennial
Rich woods, wooded banks and shale barrens.

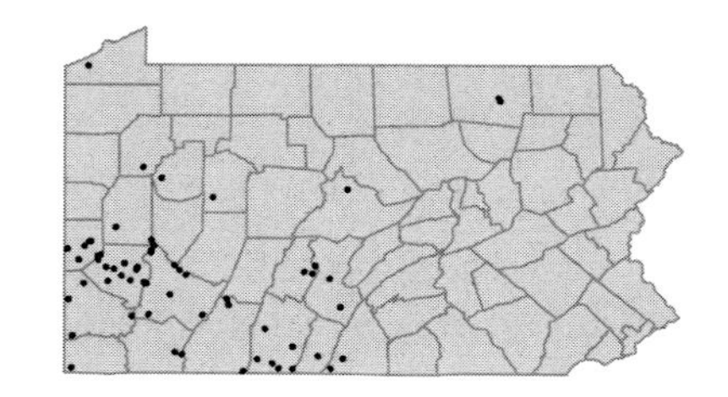

Oxalis stricta L.
Common yellow wood-sorrel
Herbaceous perennial
Lawns, gardens and fields.
Oxalis bushii Small P; *Oxalis cymosa* Small P; *Oxalis europaea* Jord. FBW; *Oxalis rufa* Small P

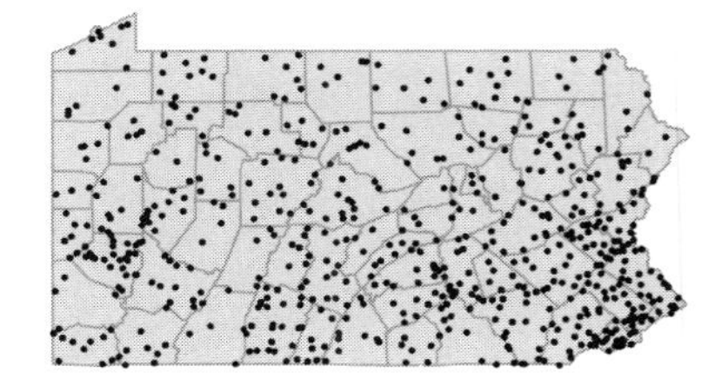

Oxalis violacea L.
Violet wood-sorrel
Herbaceous perennial
Open woods and shaded banks.

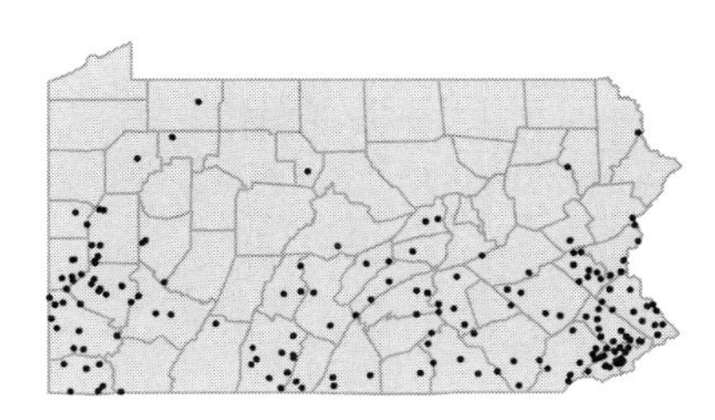

GERANIACEAE

• ***Erodium cicutarium*** (L.) L'Her. ex Soland. [I]
Redstem filaree
Herbaceous annual
Open, roadside banks and weedy fields.

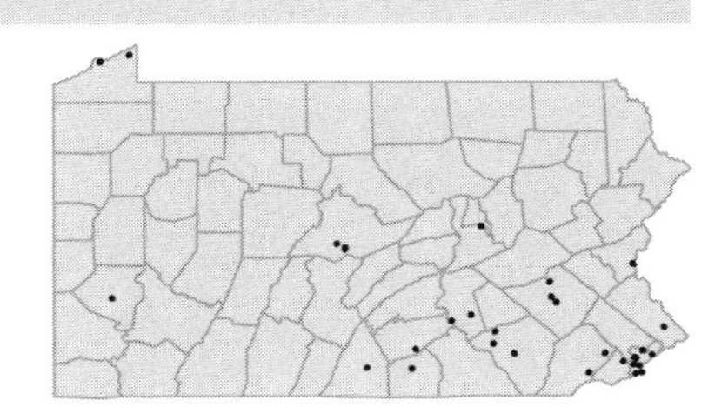

Erodium moschatum (L.) L'Her. [I]
Whitestem filaree
Herbaceous biennial
Waste ground and rubbish dumps.

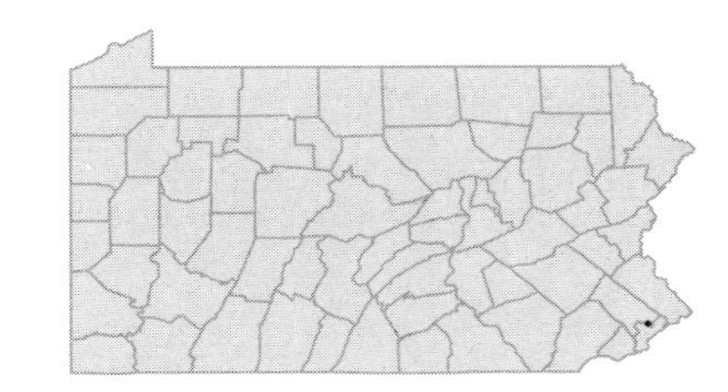

• ***Geranium bicknellii*** Britt.
Cranesbill
Herbaceous annual
Dry, open woods, clearings and rocky ledges.

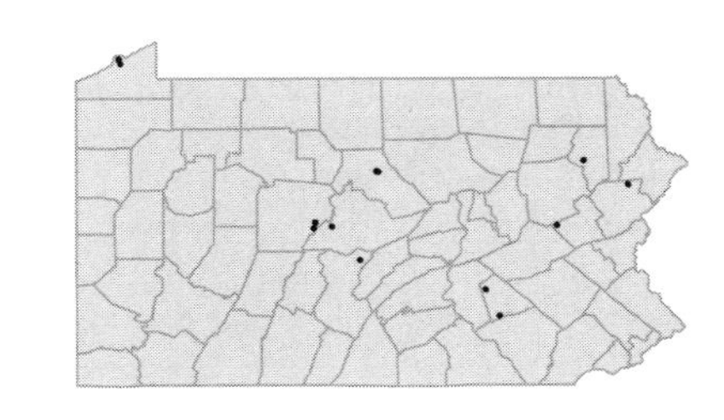

Geranium carolinianum L.
Cranesbill; Wild geranium
Herbaceous annual
Fields, roadsides, and dry woods.

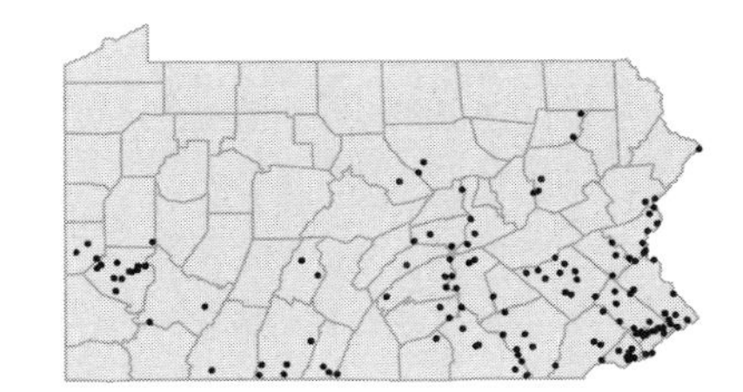

Geranium columbinum L.
Long-stalked cranesbill
Herbaceous annual
Dry, rocky, calcareous or shaly slopes or old fields.

[I]

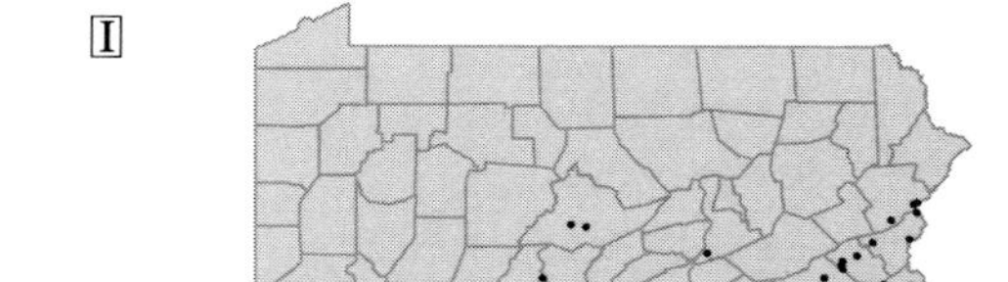

Geranium dissectum L.
Cut-leaved cranesbill
Herbaceous annual
Fields and waste ground.

[I]

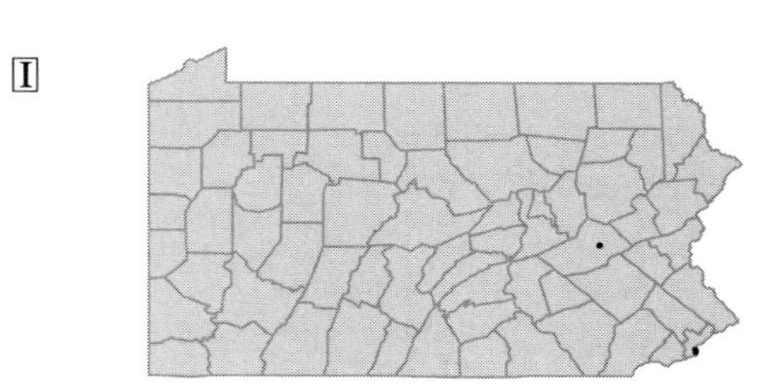

Geranium maculatum L.
Wood geranium
Herbaceous perennial
Rich woods, roadsides and fields.

FACU

Geranium molle L.
Dovesfoot cranesbill
Herbaceous annual
Old fields, roadsides, canal banks, and railroad tracks.

[I]

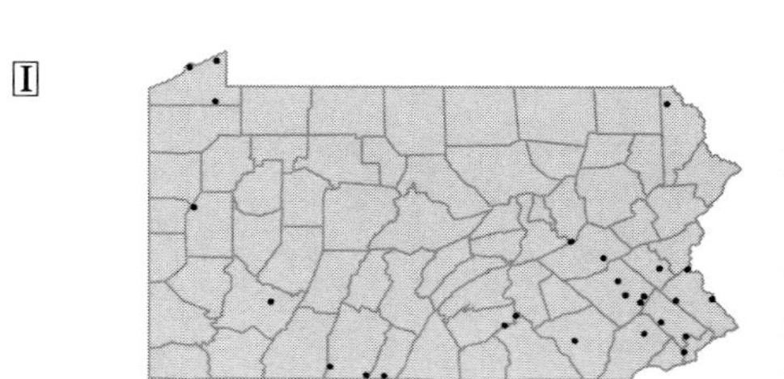

Geranium pratense L.
Meadow geranium
Herbaceous perennial
Cultivated and rarely escaped.

UPL
[I]

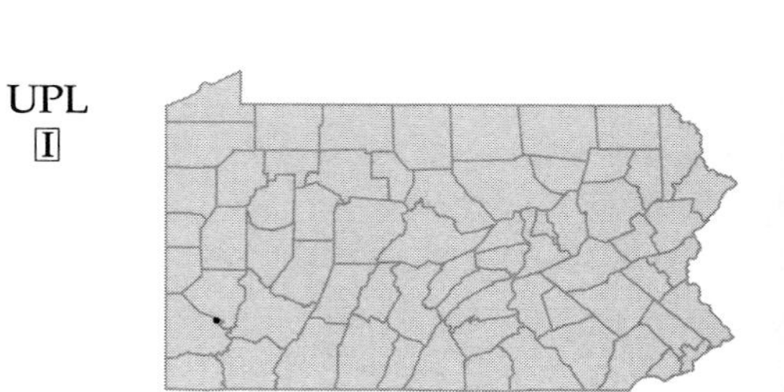

Geranium pusillum L.
Slender cranesbill
Herbaceous annual
Roadsides, fields and waste ground.

[I]

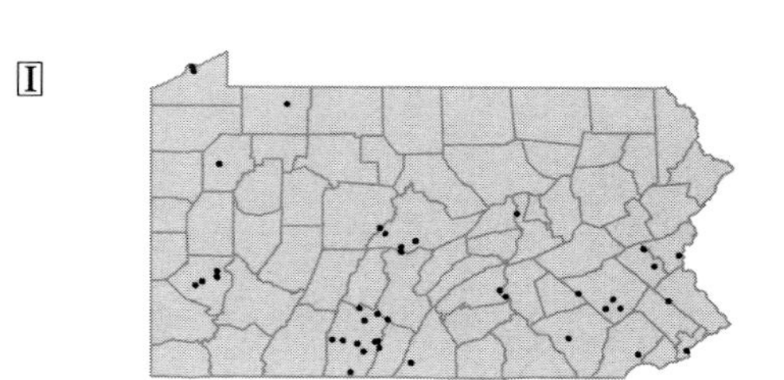

Geranium robertianum L.
Herb-robert
Herbaceous annual
River banks, moist woods, rocky slopes and ravines.

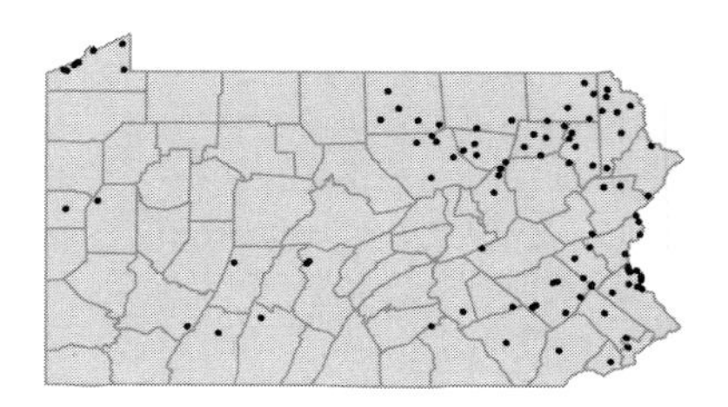

Geranium sanguineum L. [I]
Blood-red cranesbill
Herbaceous perennial
Cultivated and rarely escaped to fields and waste ground.

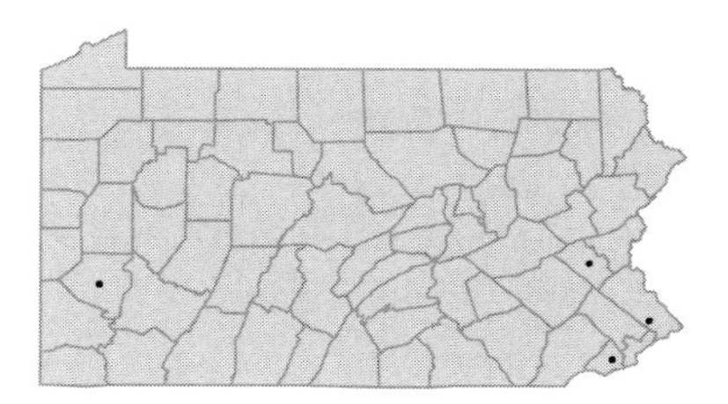

Geranium sibiricum L. [I]
Siberian cranesbill
Herbaceous annual
Roadsides, gardens and waste ground.

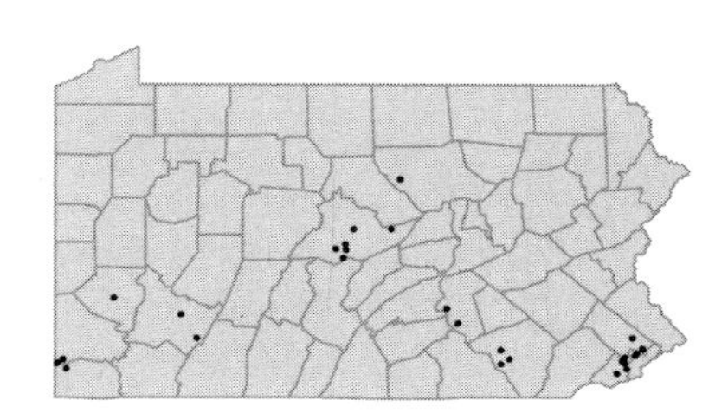

Geranium versicolor L. [I]
Cranesbill
Herbaceous perennial
Fields, gardens, rubbish dumps and low, wet ground.
Geranium striatum L. W

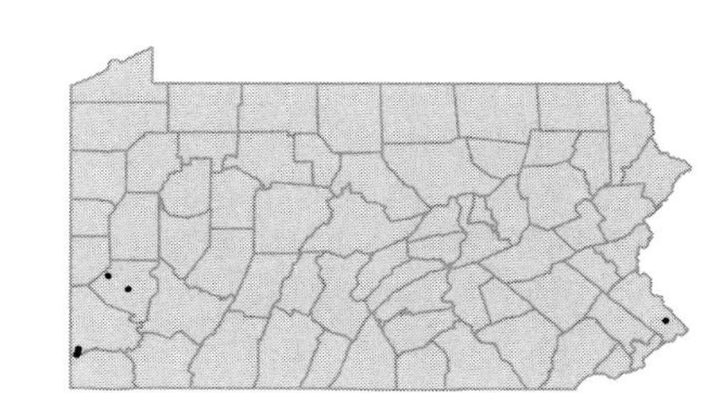

LIMNANTHACEAE

• ***Floerkea proserpinacoides*** Willd. FAC
False-mermaid
Herbaceous annual
Moist or wet woods, often on floodplains.

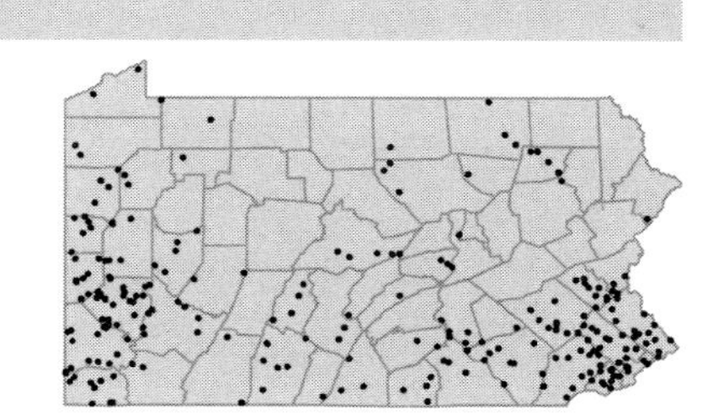

TROPAEOLACEAE

• ***Tropaeolum majus*** L. [I]
Common nasturtium
Herbaceous annual
Cultivated and very rarely escaped to rubbish dumps.

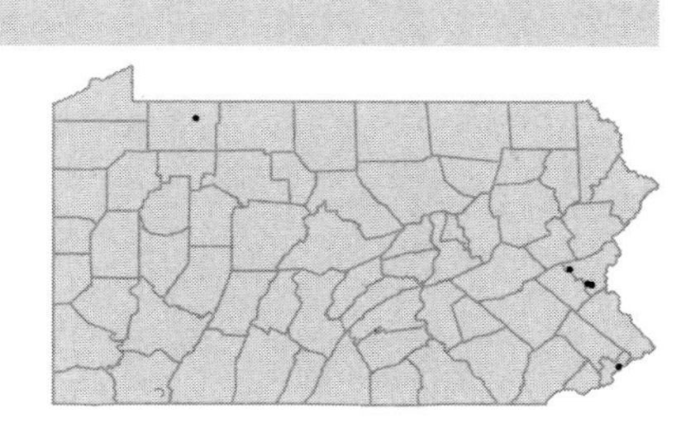

BALSAMINACEAE

• ***Impatiens balsamina*** L.
Garden balsam
Herbaceous annual
Cultivated and occasionally escaped to waste places and old fields.

I

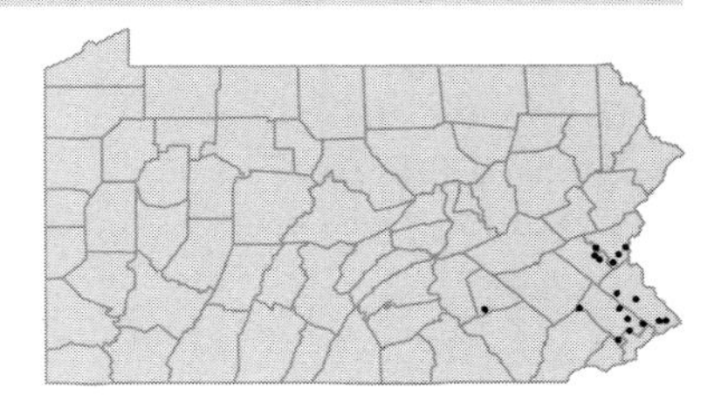

Impatiens capensis Meerb.
Jewelweed; Touch-me-not
Herbaceous annual
Moist meadows, swamps and stream banks.
Impatiens biflora Walt. PGB

FACW

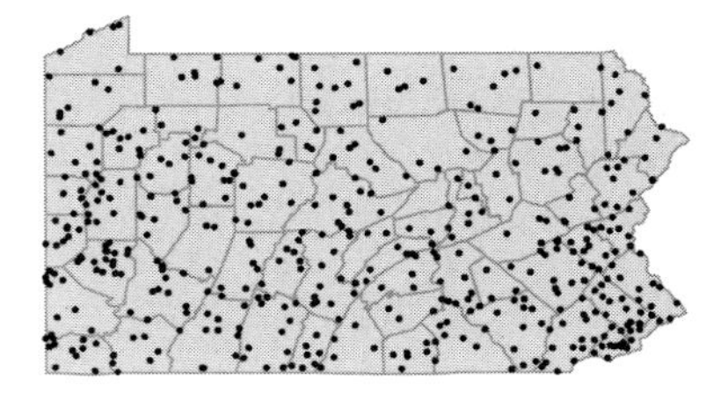

Impatiens pallida Nutt.
Pale jewelweed; Touch-me-not
Herbaceous annual
Swamps, moist woods and stream banks.
Impatiens aurea Muhl. P

FACW

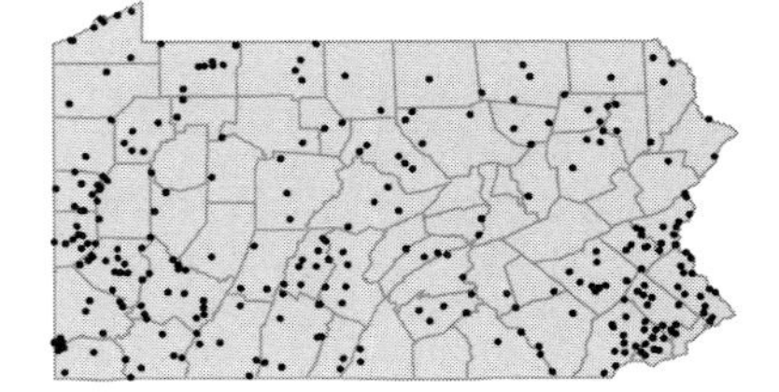

ARALIACEAE

• ***Acanthopanax sieboldianus*** Makino
Fiveleaf aralia
Deciduous shrub
Cultivated and occasionally spreading to waste places or persisting in abandoned garden sites.

I

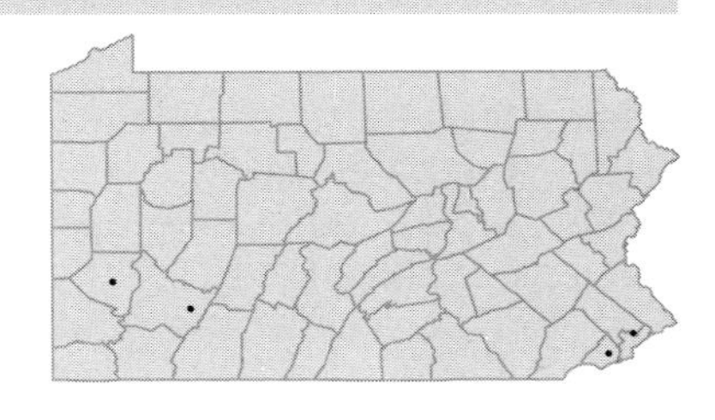

• ***Aralia chinensis*** L.
Chinese aralia
Deciduous shrub
Dry roadside bank. Represented by a single collection from Berks Co. in 1965.

I

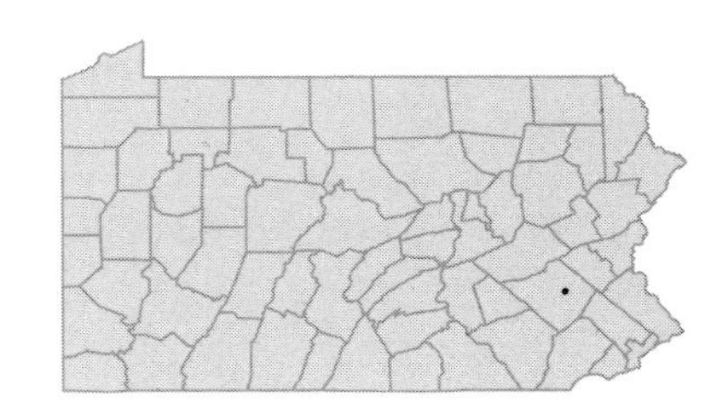

Aralia hispida Vent.
Bristly sarsaparilla
Herbaceous perennial
Dry, rocky woods, damp hollow and roadsides.

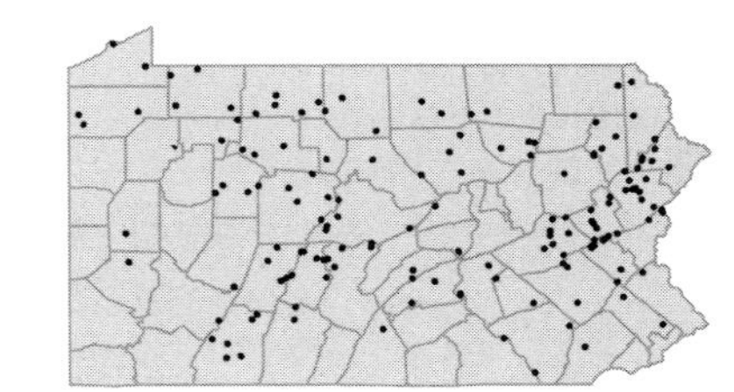

Aralia nudicaulis L.
Wild sarsaparilla
Herbaceous perennial
Dry to moist woods.

FACU

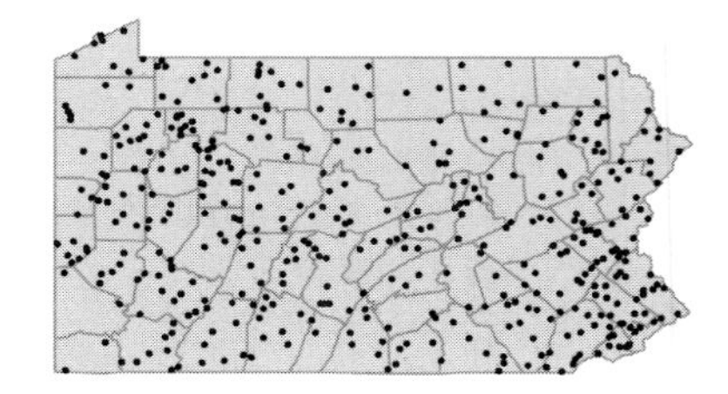

Aralia racemosa L.
Life-of-man; Petty-morrel; Spikenard
Herbaceous perennial
Rich woods, slopes and edges.

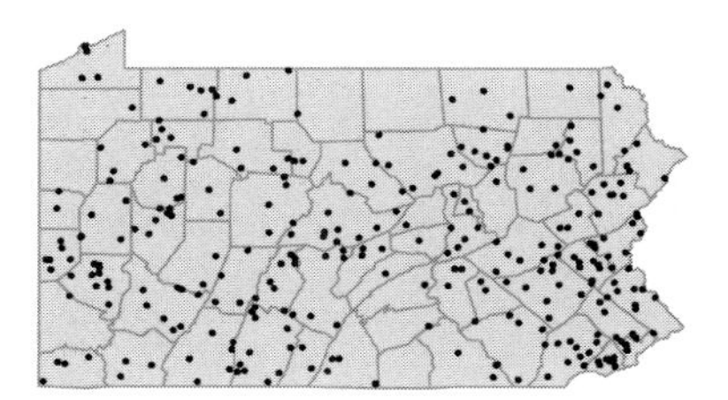

Aralia spinosa L.
Devil's-walking-stick; Hercules'-club
Deciduous tree
Moist woods, river banks and roadsides.

FAC

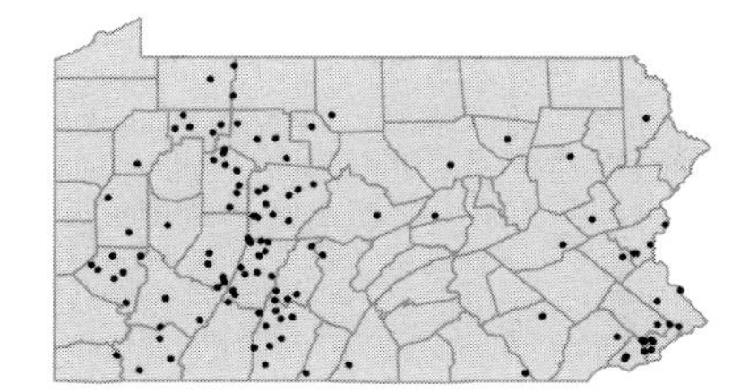

• ***Hedera helix*** L.
English ivy
Woody vine
Cultivated and occasionally naturalized in disturbed woods.

[I]

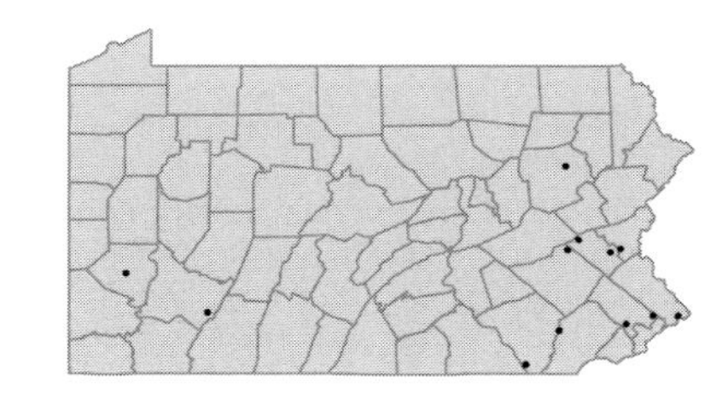

• ***Panax quinquefolius*** L.
Ginseng
Herbaceous perennial
Rich, mesic woods, declining due to over-collection.

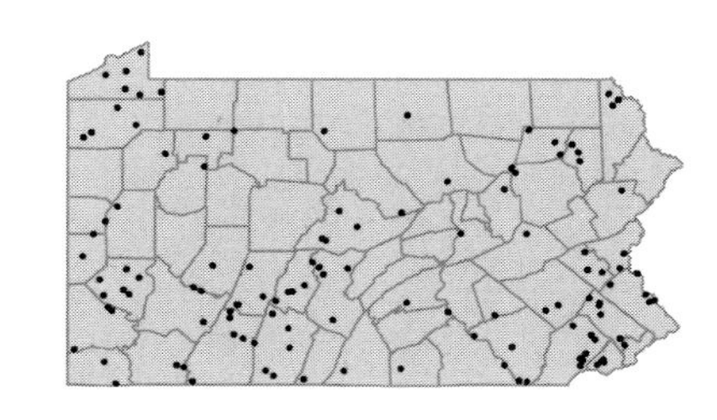

Panax trifolius L.
Dwarf ginseng
Herbaceous perennial
Moist, rich woods.

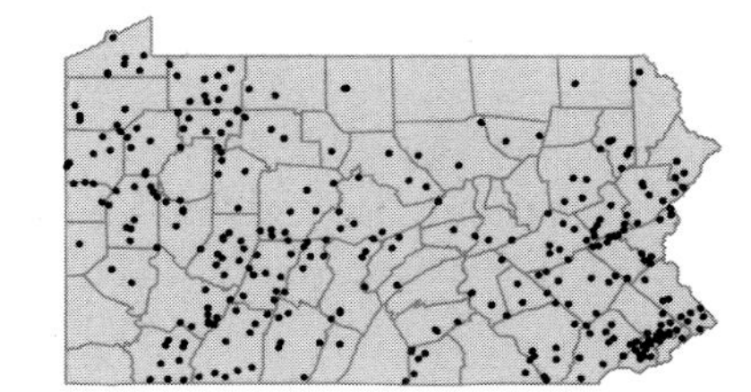

APIACEAE

• ***Aegopodium podagraria*** L. [I]
Goutweed
Herbaceous perennial
Cultivated, also frequently naturalized in fields, thickets, disturbed woods and roadsides.

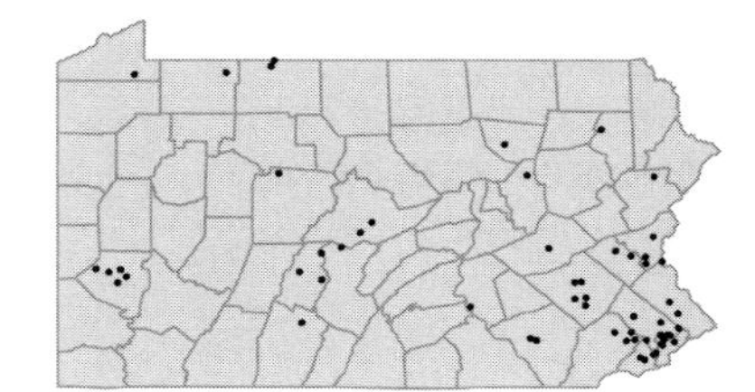

• ***Aethusa cynapium*** L. [I]
Fool's-parsley
Herbaceous annual
Dry roadsides, alluvial terraces, waste ground and ballast.

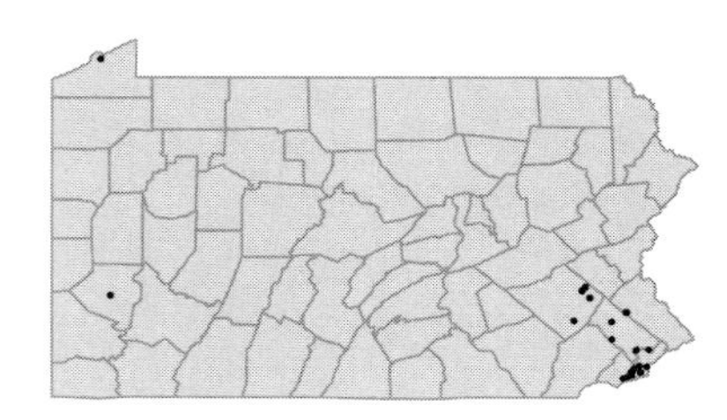

• ***Ammi visnaga*** (L.) Lam. [I]
Ammi
Herbaceous perennial
Ballast. Represented by a single collection from Philadelphia Co. in 1879.

• ***Anethum graveolens*** L. [I]
Dill
Herbaceous annual
Cultivated and occasionally established in fields and rubbish dumps.

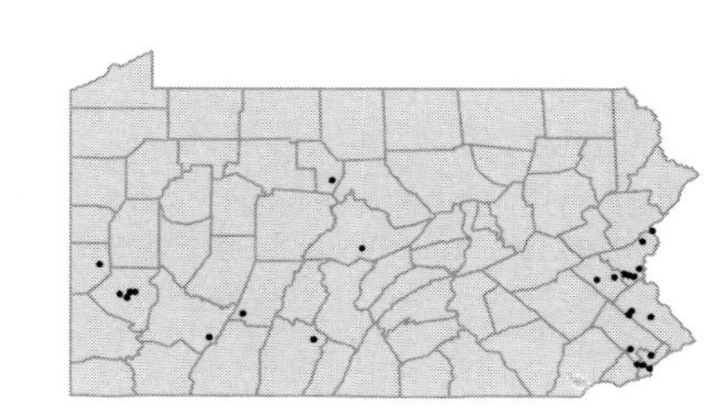

• ***Angelica atropurpurea*** L. OBL
Purple-stemmed angelica
Herbaceous perennial
Swamps, moist meadows, river banks and wet woods.

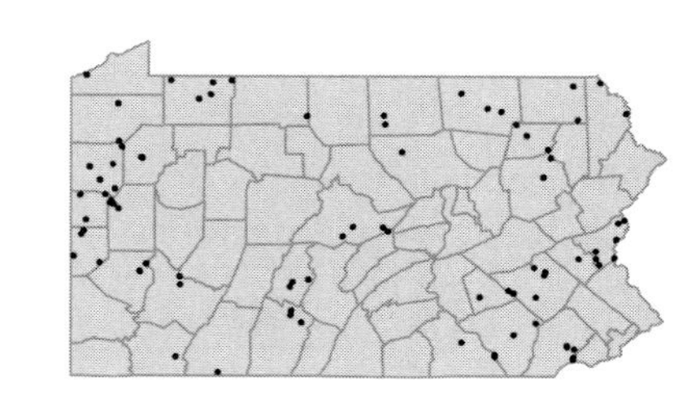

Angelica triquinata Michx. UPL
Angelica
Herbaceous perennial
Banks of mountain streams, wet woods and floodplains.
Angelica curtisii Buckl. P

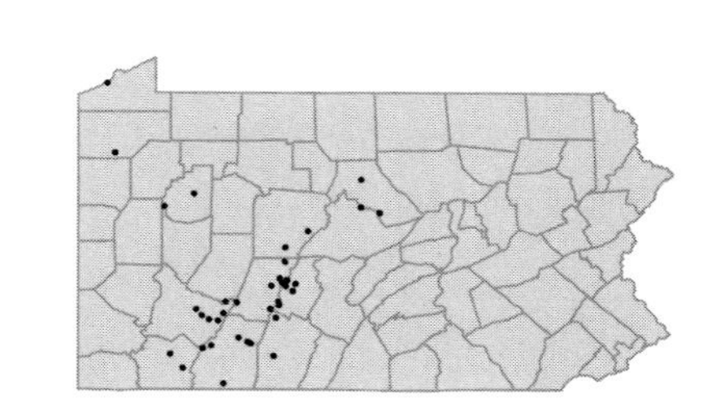

Angelica venenosa (Greenway) Fern.
Deadly angelica; Hairy angelica
Herbaceous perennial
Dry, open woods, roadside banks, serpentine barrens and old fields.
Angelica villosa (Walt.) BSP P

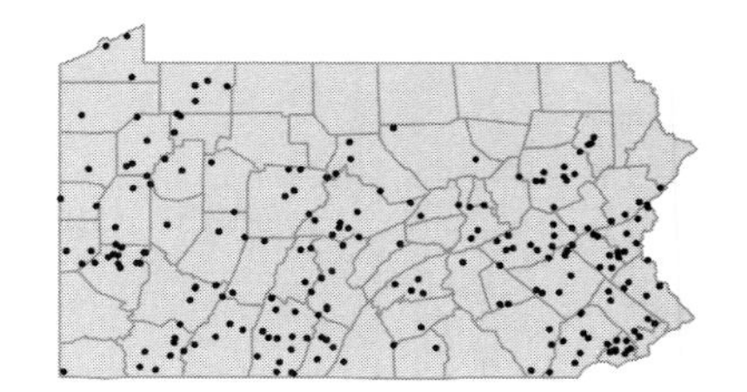

• ***Anthriscus cerefolium*** (L.) Hoffm.
Chervil
Herbaceous annual
Cultivated and occasionally naturalized in woods and roadsides.

I

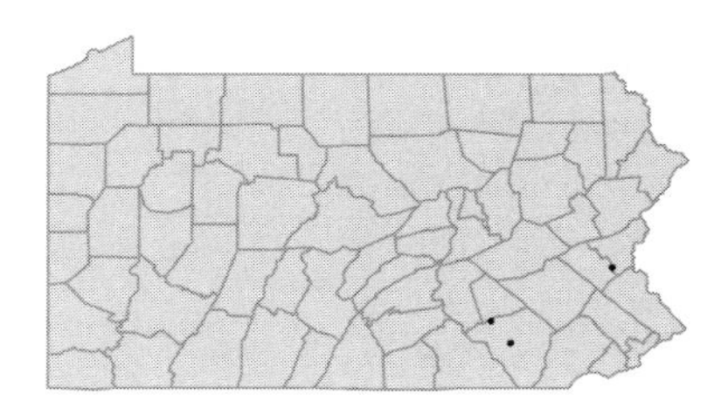

Anthriscus sylvestris (L.) Hoffm.
Chervil
Herbaceous biennial
Gravelly banks, roadsides and alluvial areas.

I

• ***Apium graveolens*** L.
Celery
Herbaceous biennial
Cultivated and occasionally escaped near rubbish dumps or waste ground.

FAC
I

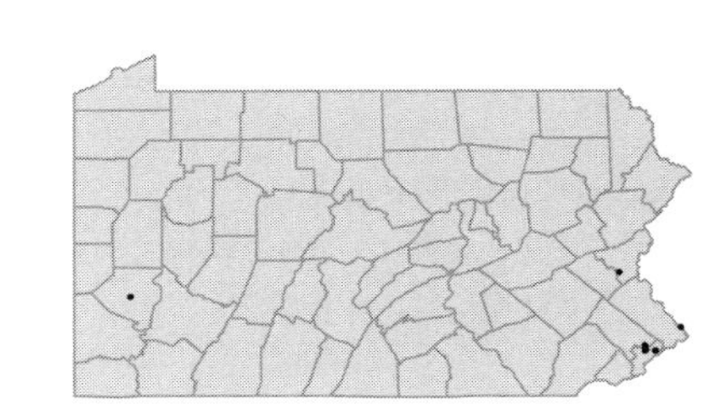

Apium leptophyllum (Pers.) F.Muell.
Marsh parsley
Herbaceous annual
Rubbish dump. Represented by a single collection from Bucks Co. in 1961.

I

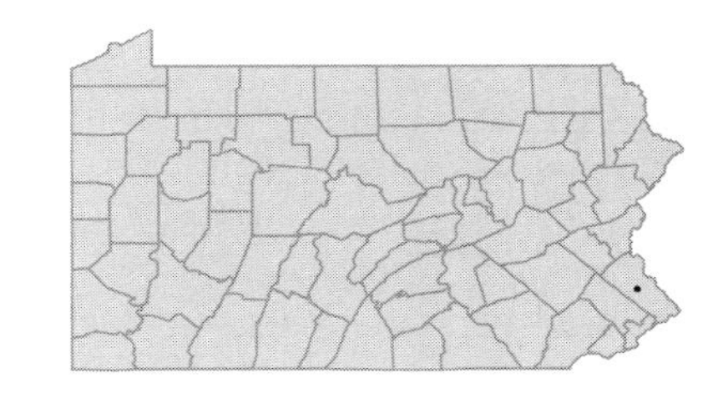

• ***Bupleurum rotundifolium*** L.
Hare's-ear; Thoroughwax
Herbaceous annual
Fallow areas, railroad tracks, wharves and rubbish dumps.

I

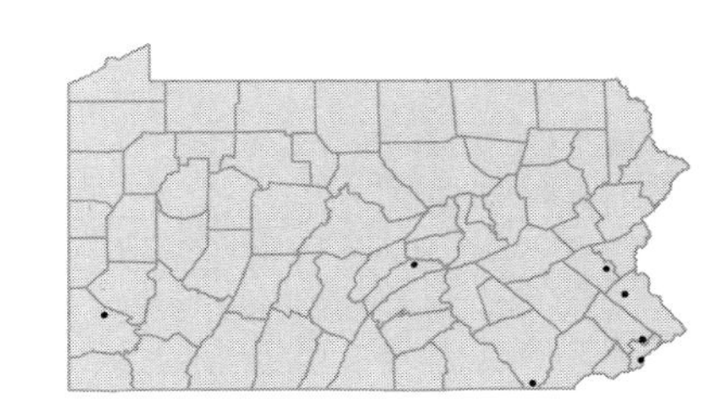

• ***Carum carvi*** L. I
Caraway
Herbaceous biennial
Cultivated and occasionally naturalized in fields, meadows and roadsides.

• ***Caucalis platycarpos*** L. I
Bur-parsley
Herbaceous annual
Cultivated ground.

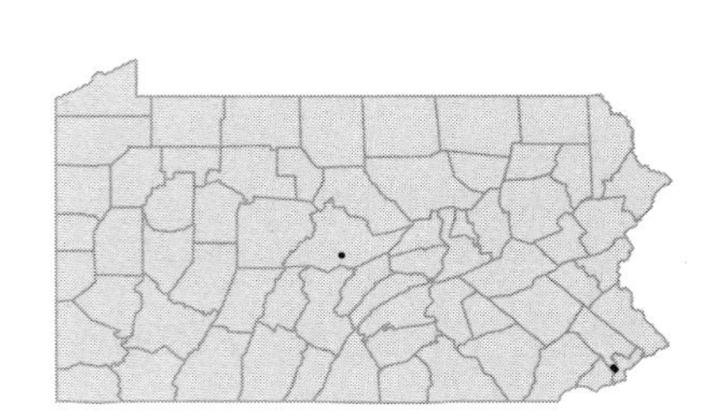

• ***Chaerophyllum procumbens*** (L.) Crantz FACW
Slender chervil; Spreading chervil
Herbaceous annual
Rich woods, wooded slopes and bottomland.

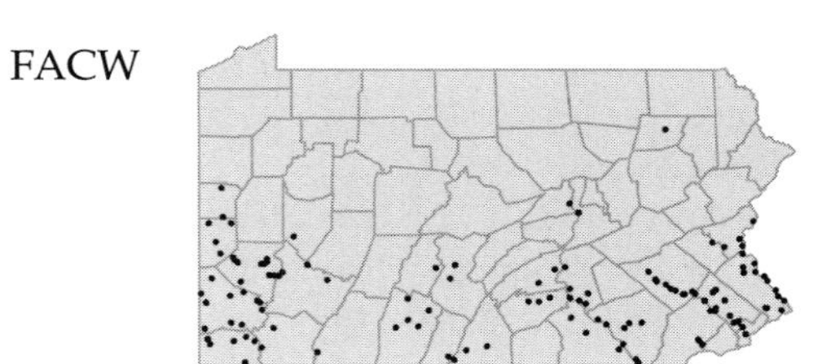

• ***Cicuta bulbifera*** L. OBL
Water-hemlock
Herbaceous perennial
Marshes, swampy meadows, swales and openings in wet, bottomland woods.

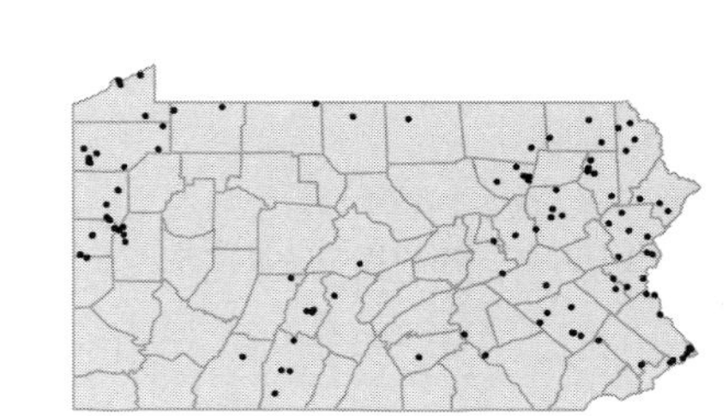

Cicuta maculata L. var. ***maculata*** OBL
Beaver-poison; Musquash-root; Spotted cowbane
Herbaceous perennial
Swamps, marshes, wet meadows, stream banks and ditches.

• ***Conioselinum chinense*** (L.) BSP FACW
Hemlock-parsley
Herbaceous perennial
Moist, rich woods and stream banks.

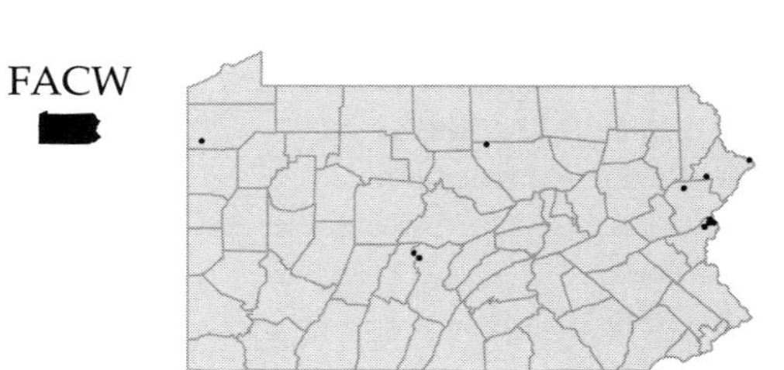

• ***Conium maculatum*** L.
Poison hemlock
Herbaceous biennial
Roadside ditches, floodplains and moist woods.

FACW
[I]

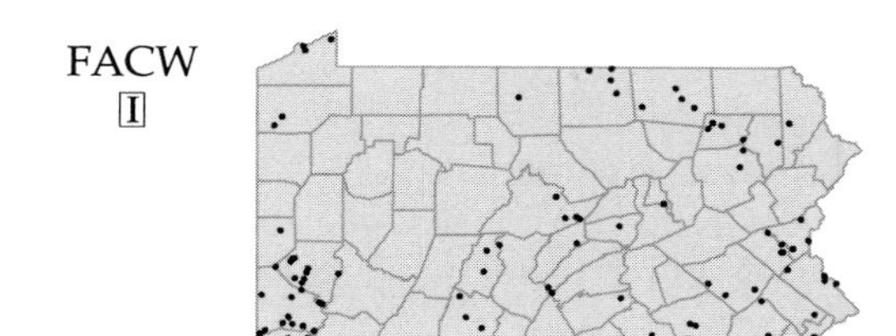

• ***Coriandrum sativum*** L.
Coriander
Herbaceous annual
Ballast, wharves, waste ground and rubbish dumps.

[I]

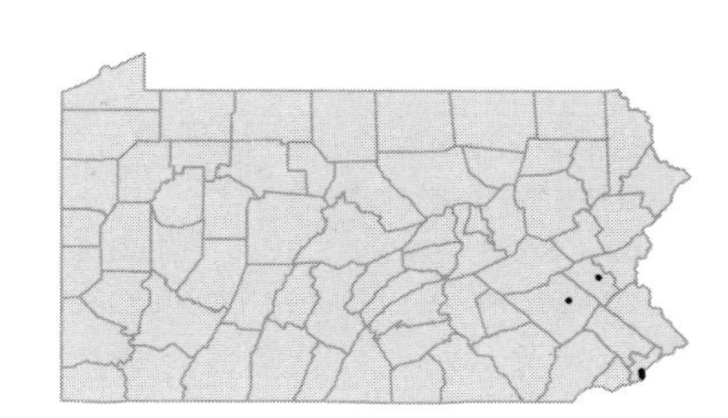

• ***Cryptotaenia canadensis*** (L.) DC.
Honewort; Wild chervil
Herbaceous perennial
Moist woods, wooded stream banks and seeps.
Deringa canadensis (L.) Kuntze P

FAC

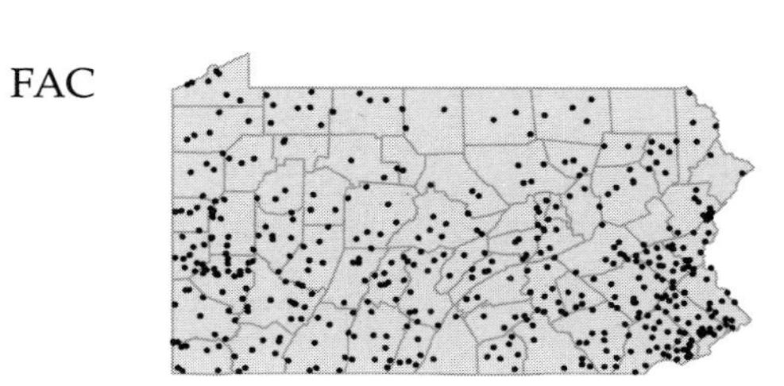

• ***Daucus carota*** L.
Queen-Anne's-lace; Wild carrot
Herbaceous biennial
Roadsides, old fields, gardens and waste ground.

[I]

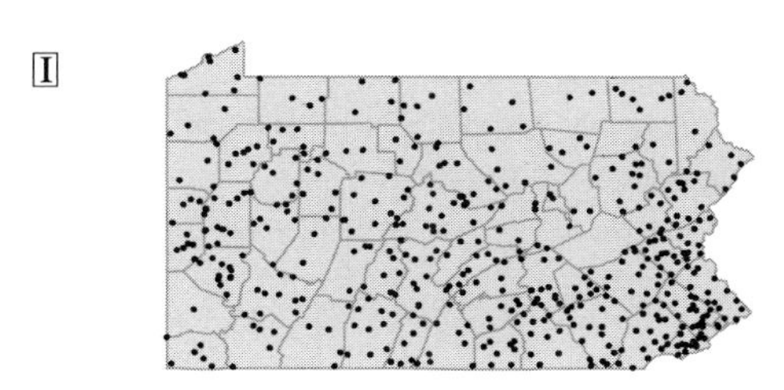

• ***Erigenia bulbosa*** (Michx.) Nutt.
Harbinger-of-spring; Pepper-and-salt
Herbaceous perennial
Seeps and spring heads on wooded slopes.

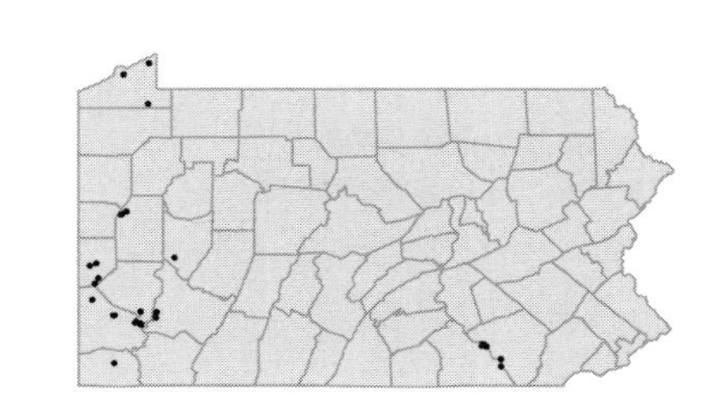

• ***Eryngium aquaticum*** L.
Marsh eryngo; Rattlesnake-master
Herbaceous perennial
River swamps, pond banks and gravelly shores. Believed to be extirpated, last collected in 1866.
Eryngium virginianum Lam. P

OBL

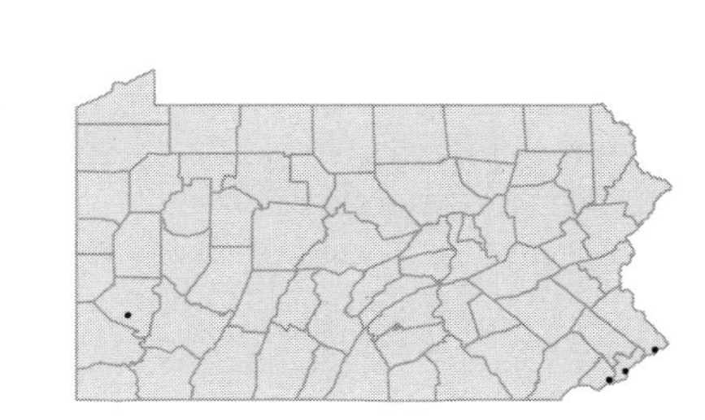

• ***Falcaria vulgaris*** Bernh. [I]
Sickleweed
Herbaceous perennial
Open woods and meadows.
Falcaria sioides (Wibel) Aschers. GB

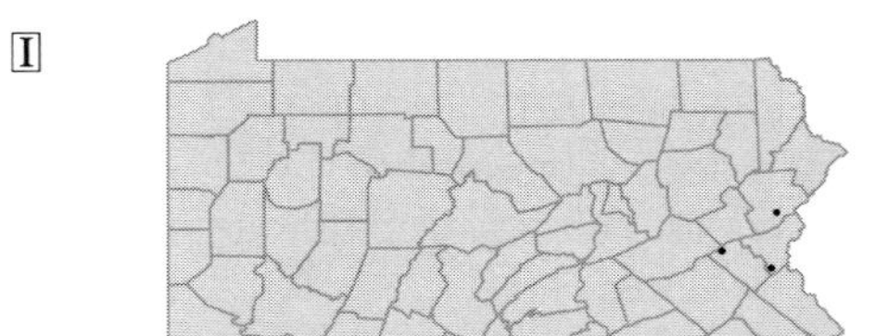

• ***Foeniculum vulgare*** P.Mill. [I]
Fennel
Herbaceous perennial
Cultivated, and occasionally escaped to meadows and waste ground.
Foeniculum foeniculum (L.) Karst. P

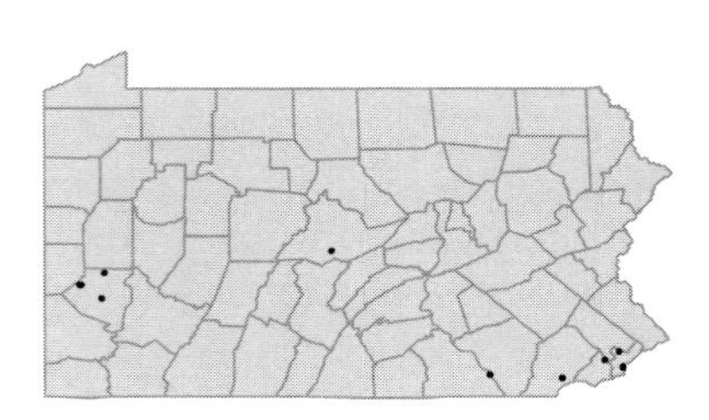

• ***Heracleum lanatum*** Michx. FACU-
Cow-parsnip
Herbaceous perennial
Rich woods, wooded roadside banks, marshy flats, stream banks and ditches.
Heracleum maximum Bartr. PFW

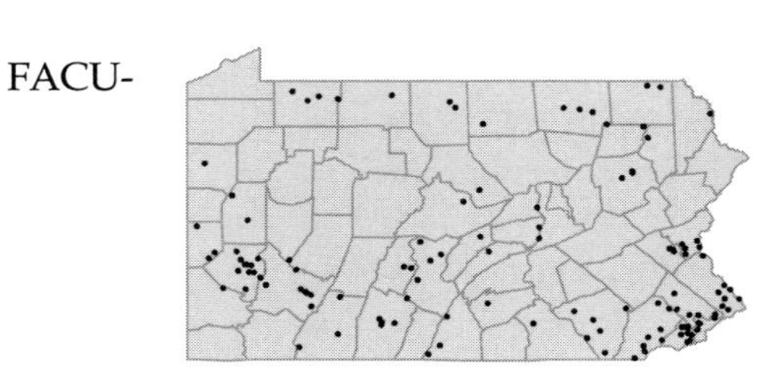

• ***Hydrocotyle americana*** L. OBL
Marsh pennywort; Navelwort
Herbaceous perennial
Swampy thickets, boggy fields, wet woods and lake margins.

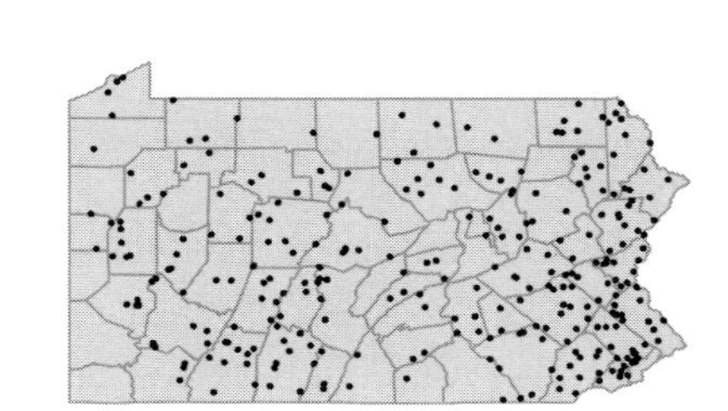

Hydrocotyle ranunculoides L.f. OBL
Floating pennywort
Herbaceous perennial
Shallow water, moist shores and wet meadows.

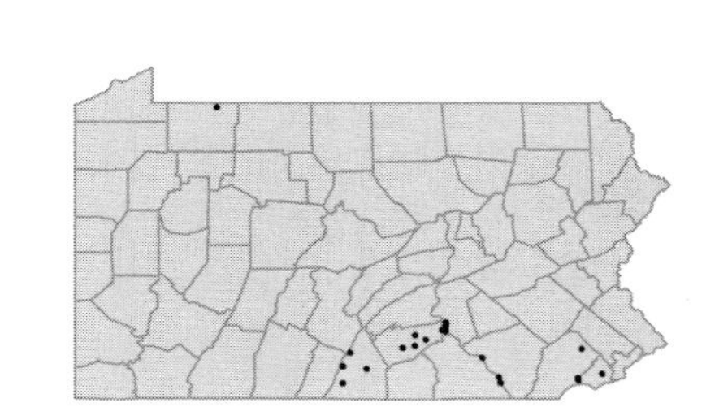

Hydrocotyle sibthorpioides Lam. [I]
Lawn pennywort
Herbaceous perennial
An occasional weed of shady lawns.

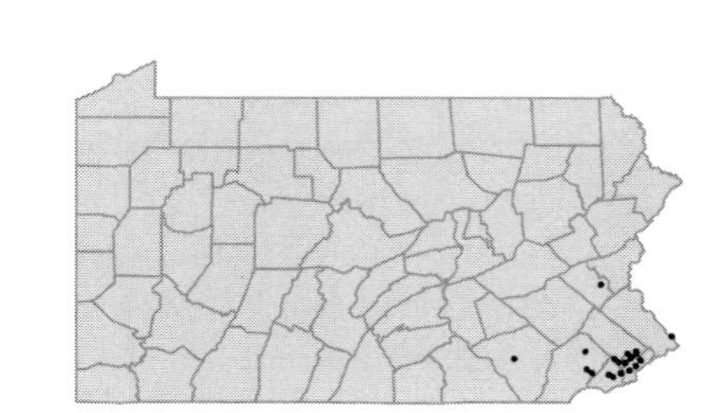

Hydrocotyle umbellata L.
Water pennywort; Navelwort
Herbaceous perennial
Shallow water and muddy shores. Believed to be extirpated, last collected in 1953.

OBL

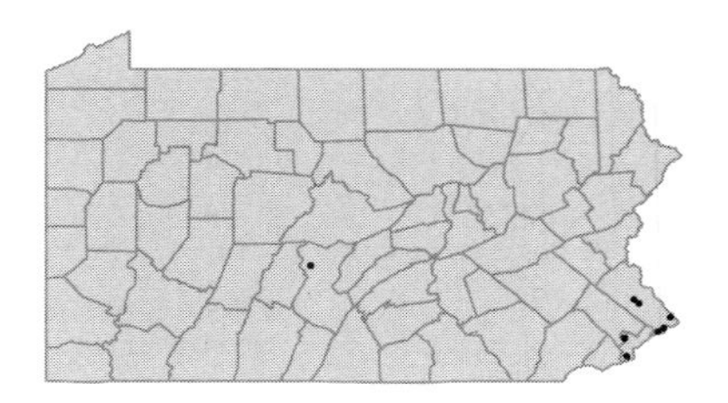

• ***Levisticum officinale*** W.D.J.Koch
Lovage
Herbaceous perennial
Cultivated and occasionally escaped to fields and river banks.
Levisticum levisticum (L.) Karst. P

I

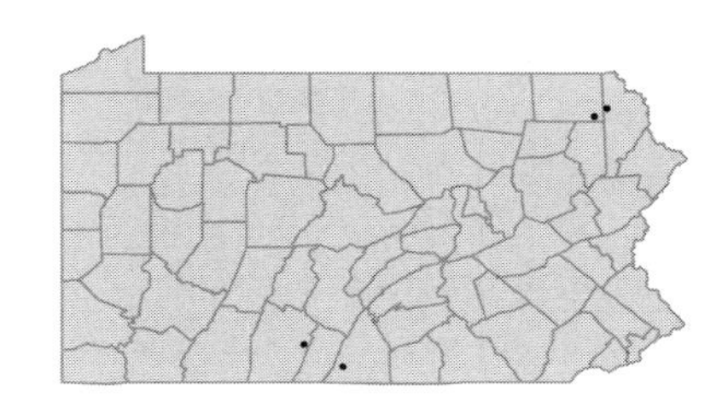

• ***Ligusticum canadense*** (L.) Britt.
Lovage
Herbaceous perennial
Montain woods, stream banks and wooded roadsides.

FAC

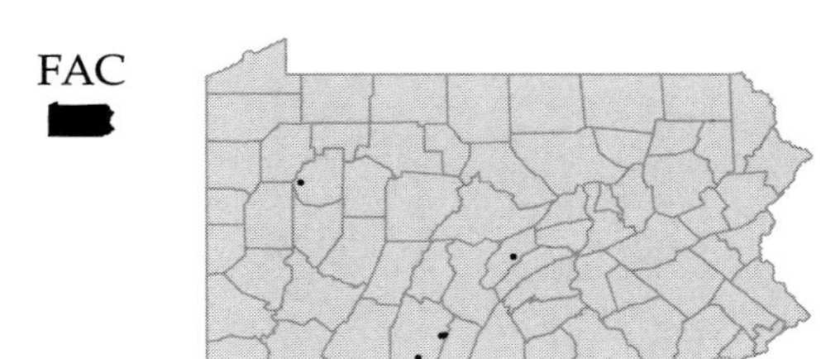

• ***Osmorhiza claytonii*** (Michx.) C.B.Clarke
Sweet-cicely
Herbaceous perennial
Rich woods, wooded stream banks and wet meadows.
Washingtonia claytonii (Michx.) Britt. P

FACU-

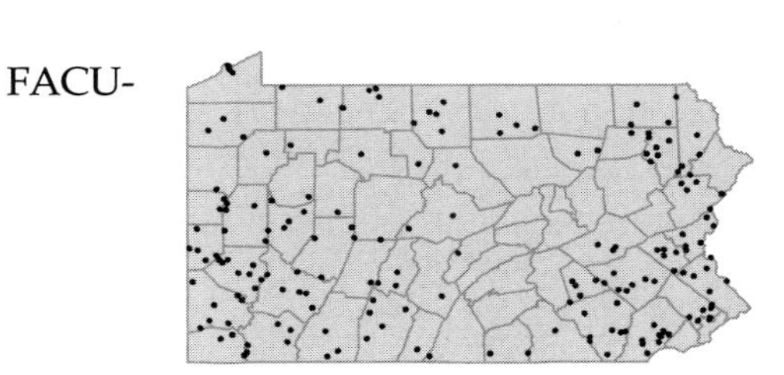

Osmorhiza longistylis (Torr.) DC.
Anise-root
Herbaceous perennial
Rich woods, moist wooded slopes and thickets.
Washingtonia longistylis (Torr.) Britt. P

FACU

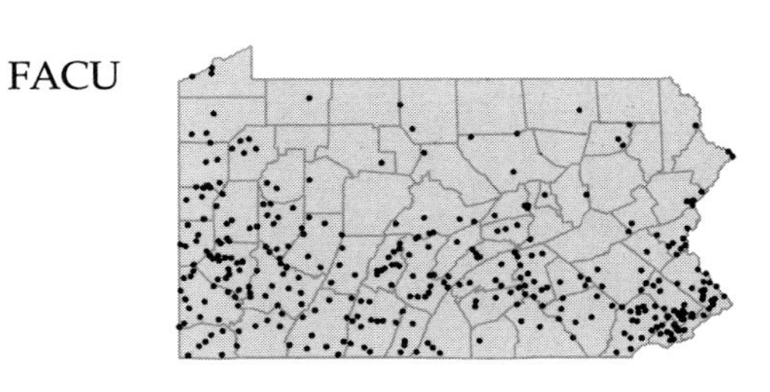

• ***Oxypolis rigidior*** (L.) Raf.
Cowbane; Water-dropwort
Herbaceous perennial
Swamps, bogs, sedge meadows, sandy shores and abandoned railroad bed.

OBL

• ***Pastinaca sativa*** L.
Wild parsnip
Herbaceous biennial
Roadsides, woods edges, fields, meadows and waste ground.

I

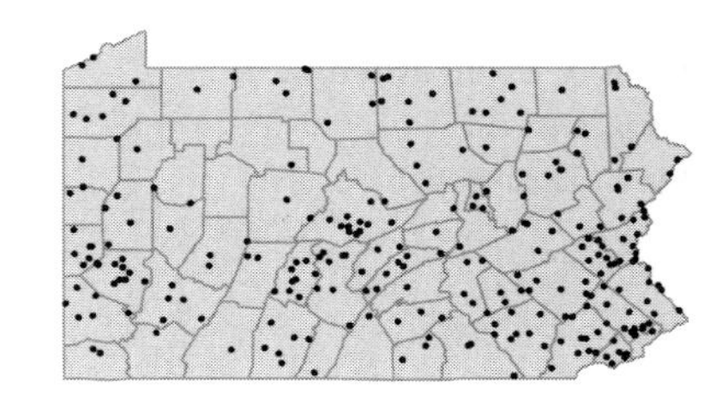

• ***Petroselinum crispum*** (Mill.) Nyman ex A.W.Hill
Parsley
Herbaceous biennial
Cultivated, and occasionally escaped to fallow fields, waste ground and dumps.

I

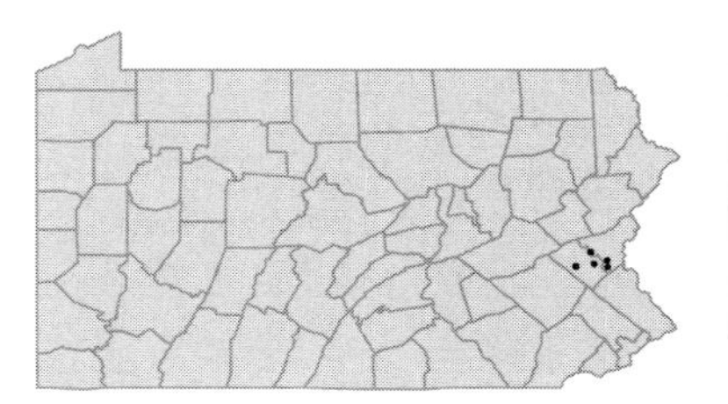

• ***Peucedanum ostruthium*** (L.) W.D.J.Koch
Masterwort
Herbaceous perennial
Cultivated fields and waste ground.
Imperatoria ostruthium L. PGBW

I

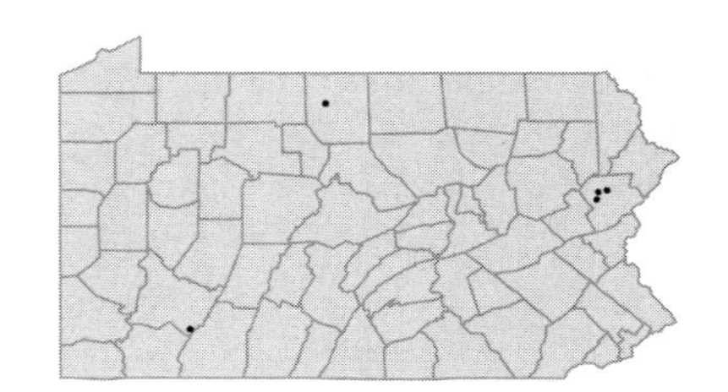

• ***Pimpinella major*** (L.) Huds.
Pimpinella
Herbaceous perennial
Moist thicket. Represented by a single collection from Bucks Co. in 1953.

I

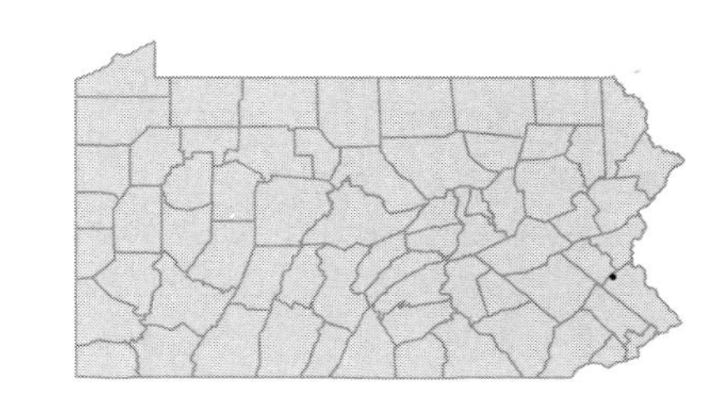

Pimpinella saxifraga L.
Burnet-saxifrage
Herbaceous perennial
Thickets, roadsides and waste ground.

I

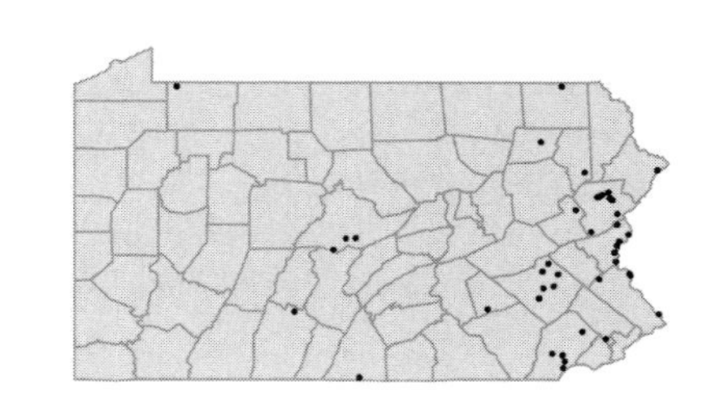

• ***Ptilimnium capillaceum*** (Michx.) Raf.
Mock bishop's-weed
Herbaceous perennial
Wet soil along a creek, also waste ground and ballast.

OBL

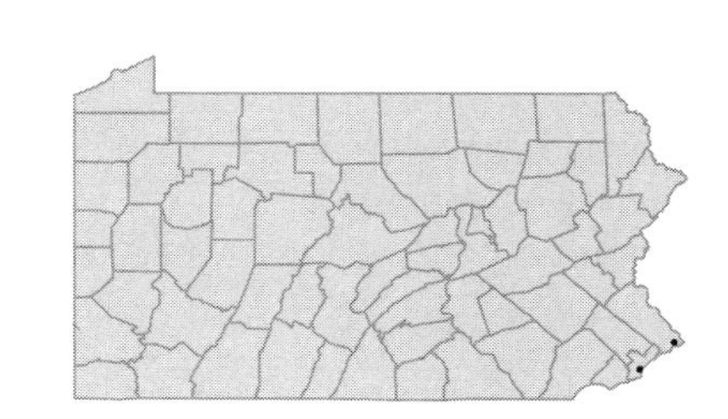

• ***Sanicula canadensis*** L. UPL
Canadian sanicle; Snake-root
Herbaceous biennial
Rich woods, rocky wooded slopes and roadsides.

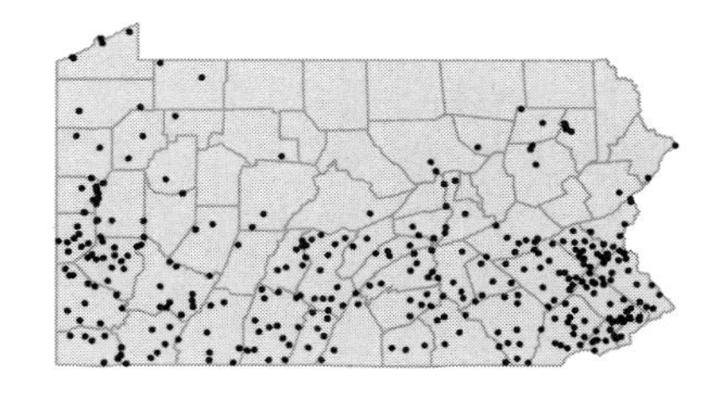

Sanicula odorata (Raf.) Pryer & Phillippe
Yellow-flowered sanicle; Fragrant snake-root
Herbaceous perennial
Moist, rich woods.
Sanicula gregaria Bickn. BFGKWC

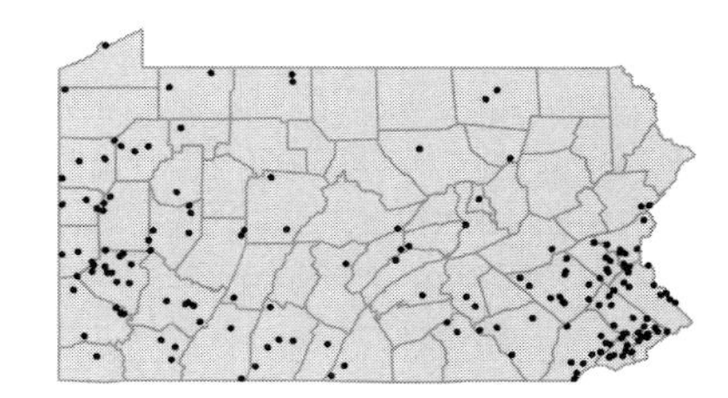

Sanicula marilandica L.

Black snake-root; Black sanicle
Herbaceous perennial
Moist woods, wooded limestone slopes, bogs and barrens.

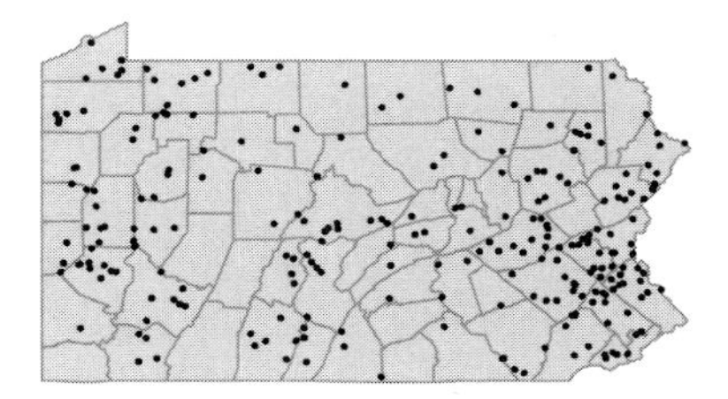

Sanicula trifoliata Bickn.
Large-fruited sanicle
Herbaceous biennial
Rich woods, wooded slopes and stream banks.

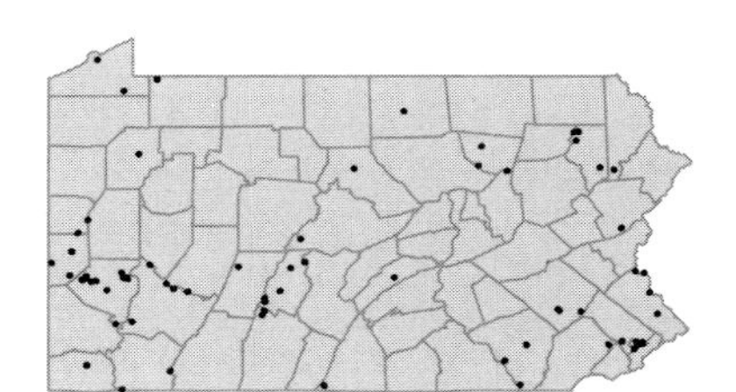

• ***Scandix pecten-veneris*** L. I
Lady's-comb; Venus'-comb
Herbaceous annual
Ballast. Represented by three collections from Philadelphia Co. 1878-1880.

• ***Sium suave*** Walt. OBL
Water-parsnip
Herbaceous perennial
Swamps, bogs, wet meadows and pond margins.
Sium cicutaefolium Gmel. P

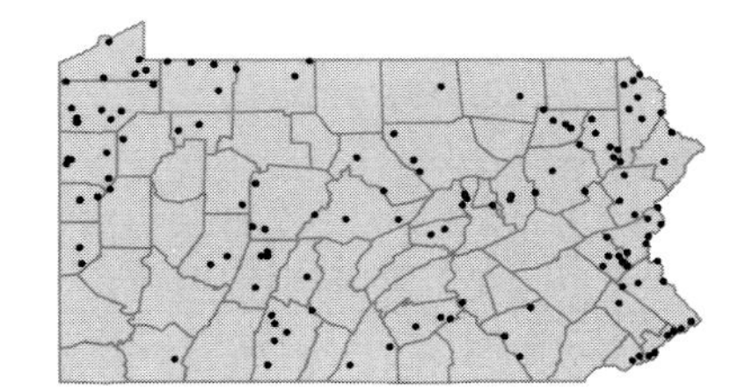

• ***Taenidia integerrima*** (L.) Drude
Yellow pimpernel
Herbaceous perennial
Rocky woods and slopes.

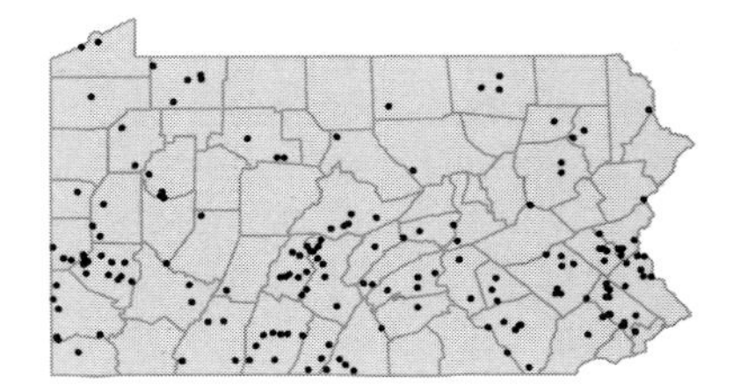

Taenidia montana (Mackenzie) Cronq.
Mountain pimpernel
Herbaceous perennial
Shale barrens and dry roadside banks.
Pseudotaenidia montana Mackenzie FGBW

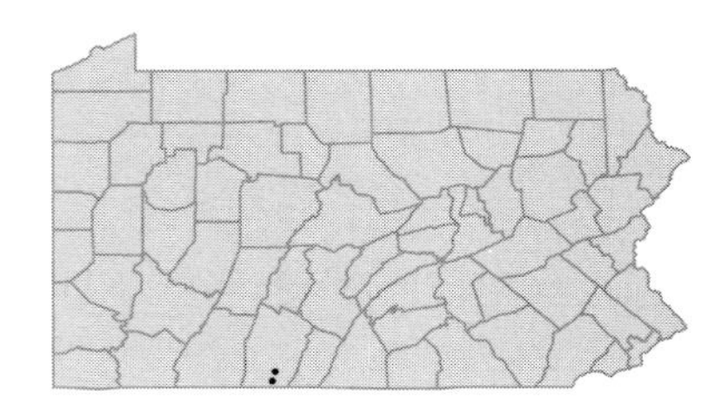

• ***Thaspium barbinode*** (Michx.) Nutt.
Meadow-parsnip
Herbaceous perennial
Open woods, wooded slopes and roadside banks.

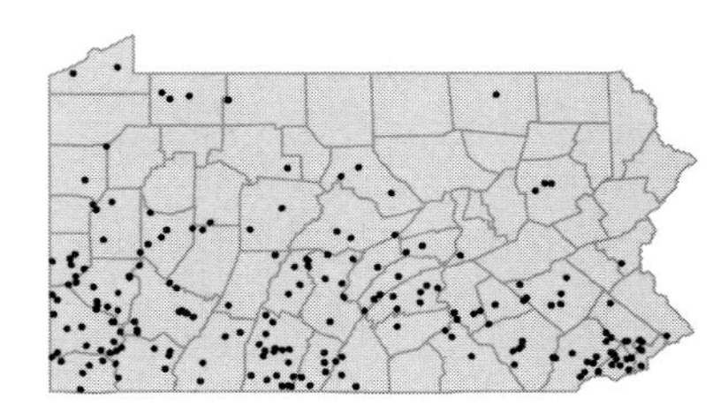

Thaspium trifoliatum (L.) A.Gray var. ***trifoliatum***
Meadow-parsnip
Herbaceous perennial
Woods, wooded slopes and edges.

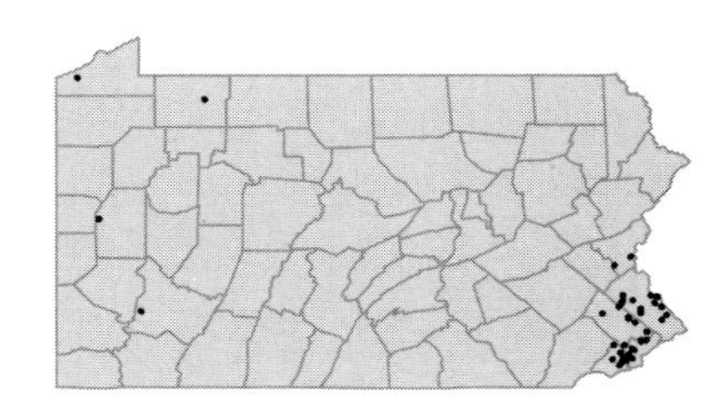

Thaspium trifoliatum (L.) A.Gray var. ***flavum*** S.F.Blake
Meadow-parsnip
Herbaceous perennial
Rich woods, ravines and roadsides.
Thaspium trifoliatum (L.) Britt. var. *aureum* (Nutt.) Britt. P

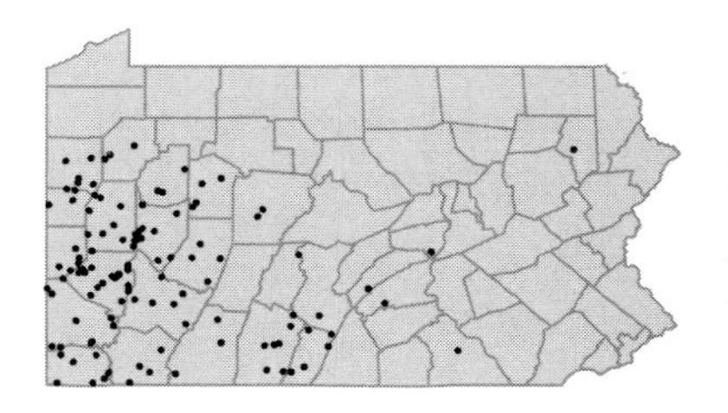

• ***Torilis arvensis*** (Huds.) Link [I]
Field hedge-parsley
Herbaceous annual
Thicket. Represented by a single collection from Bedford Co. in 1948.
Torilis anthriscus (L.) Gmel. W

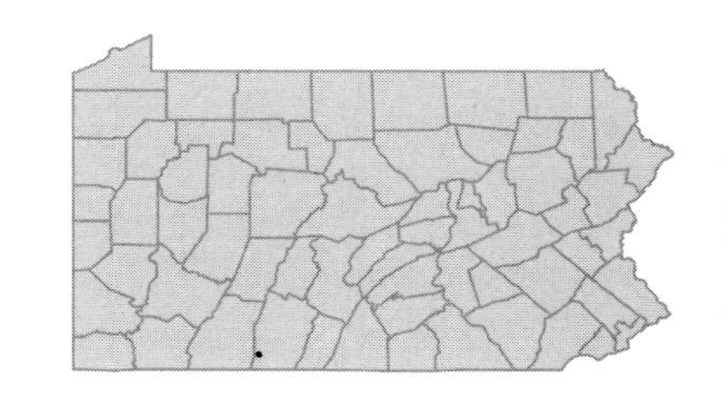

Torilis japonica (Houtt.) DC.
Japanese hedge-parsley
Herbaceous annual
Fields, thickets and waste ground.
Torilis anthriscus (L.) Gmel. P

I

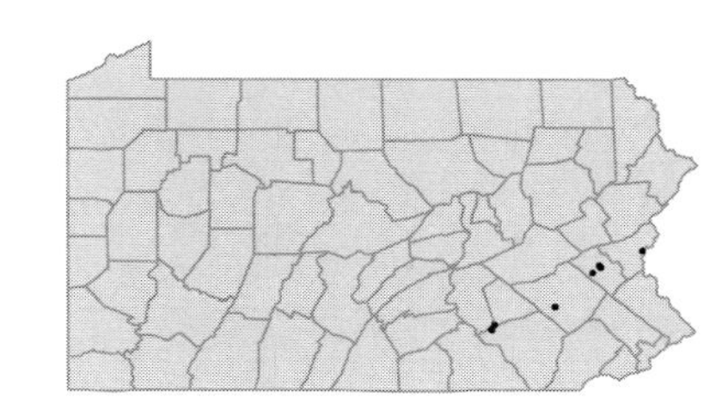

Torilis leptophylla (L.) Rchb.f.
Hedge-parsley
Herbaceous annual
Rubbish dump. Represented by two collections from Bucks Co. 1950-1961.

I

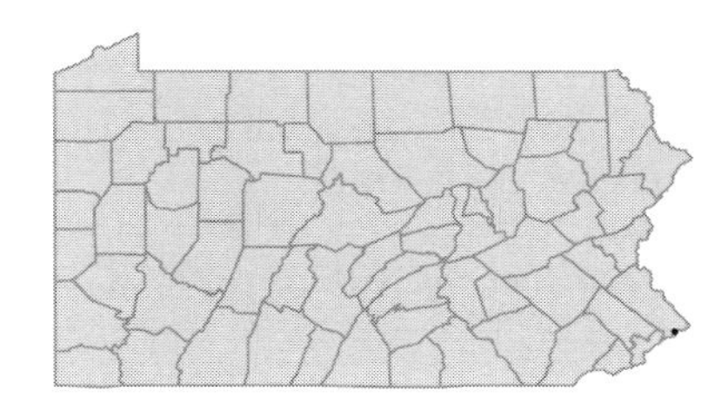

• ***Turgenia latifolia*** (L.) Hoffm.
Bur-parsley
Herbaceous annual
Ballast. Represented by three collections from Philadelphia Co. in the 1870's.
Caucalis latifolia L. BWK

I

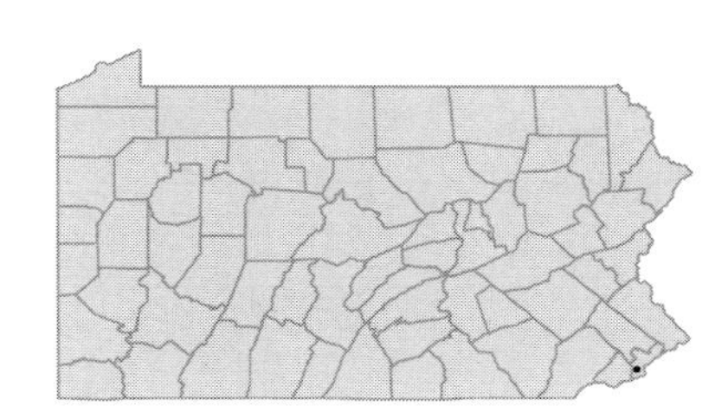

• ***Zizia aptera*** (A.Gray) Fern.
Golden-alexanders
Herbaceous perennial
Woods, wooded slopes, clearings and roadsides.
Zizia cordata (Walt.) DC. P

FAC

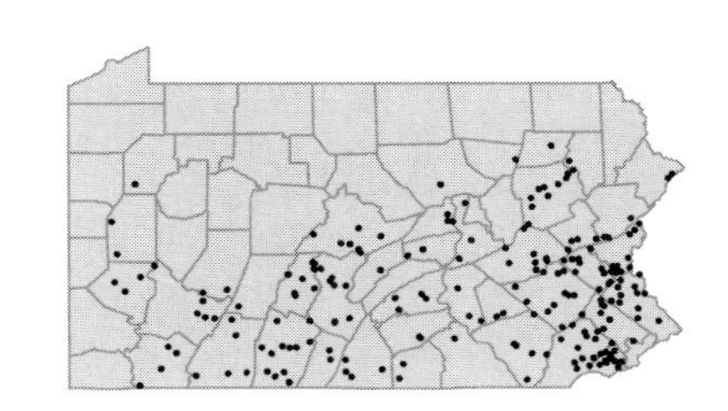

Zizia aurea (L.) W.D.J.Koch
Golden-alexanders
Herbaceous perennial
Wooded bottomland, stream banks, moist meadows and floodplains.

FAC

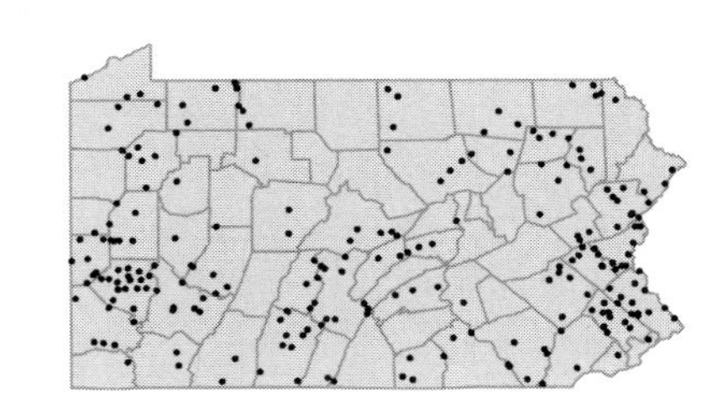

LOGANIACEAE

• ***Buddleja davidii*** Franch.
Butterfly-bush; Summer-lilac
Deciduous shrub
Cultivated and occasionally naturalized on roadsides, abandoned railroads and rubbish dumps.

I

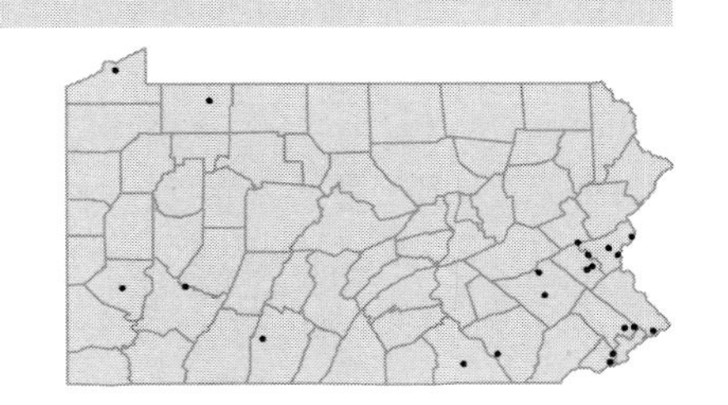

• ***Polypremum procumbens*** L.
Polypremum
Herbaceous annual
Ballast ground. Represented by several collections from Philadelphia Co. 1864-1865.

I

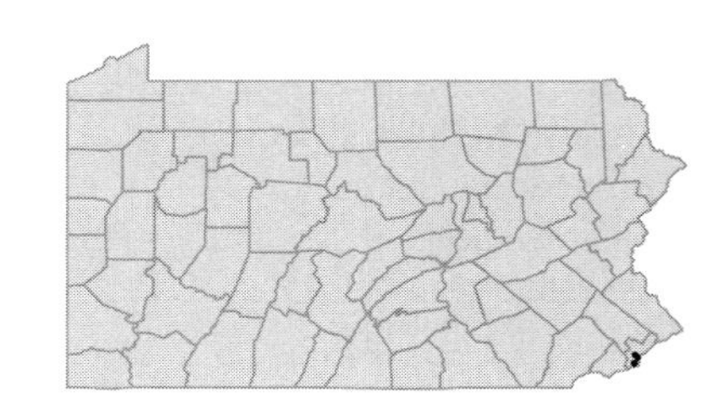

GENTIANACEAE

• ***Bartonia paniculata*** (Michx.) Muhl.
Screw-stem
Herbaceous annual
Hummocks in wet woods, wooded bogs and sphagnous pond margins.
Bartonia iodandra B.L.Robins. P

OBL

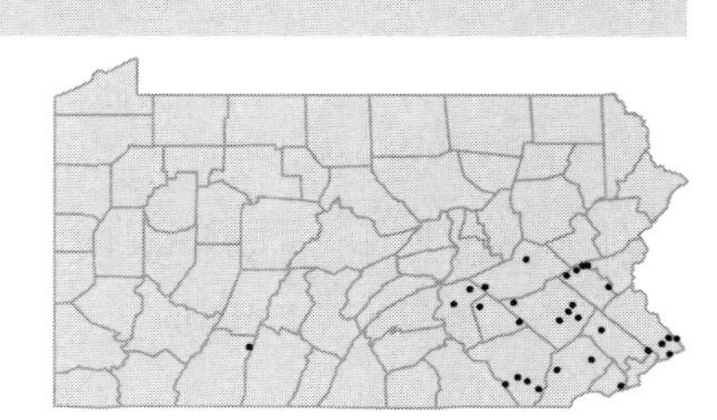

Bartonia virginica (L.) BSP
Bartonia
Herbaceous annual
Sphagnum bogs, swamps and moist fields.

FACW

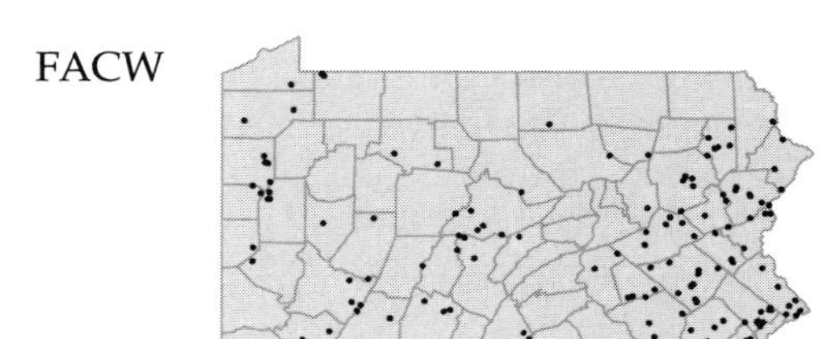

• ***Centaurium erythraea*** Raf.
Centaury
Herbaceous annual
Wet meadows and stream banks.
Centaurium umbellatum Gilib. FGBW

I

Centaurium pulchellum (Swartz) Druce
Centaury
Herbaceous annual
Fields, stream banks and waste ground.
Erythraea pulchella (Swartz) Fries P

I

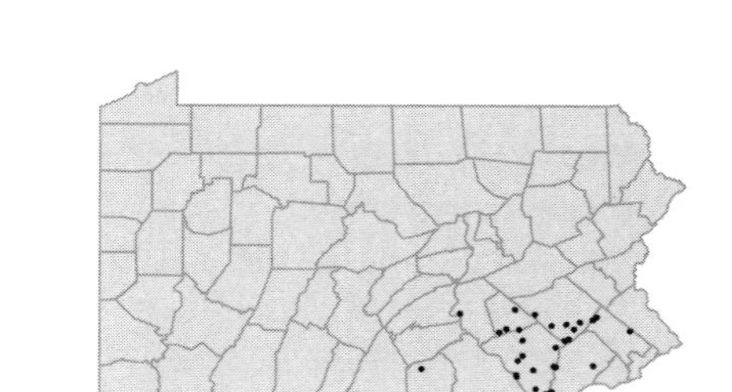

• ***Gentiana alba*** Muhl. ex Nutt.
Yellow gentian
Herbaceous perennial
Wooded hillsides.
Gentiana flavida A.Gray PFGBC

FACU

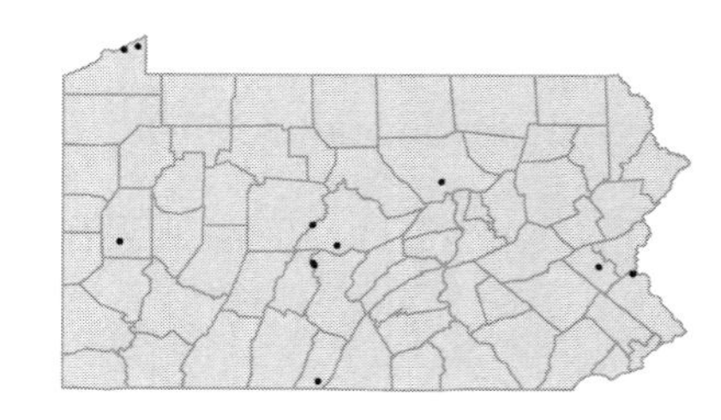

Gentiana andrewsii Griseb. var. ***andrewsii***
Bottle gentian; Closed gentian
Herbaceous perennial
Moist woods, swampy thickets, floodplains and low meadows.

FACW

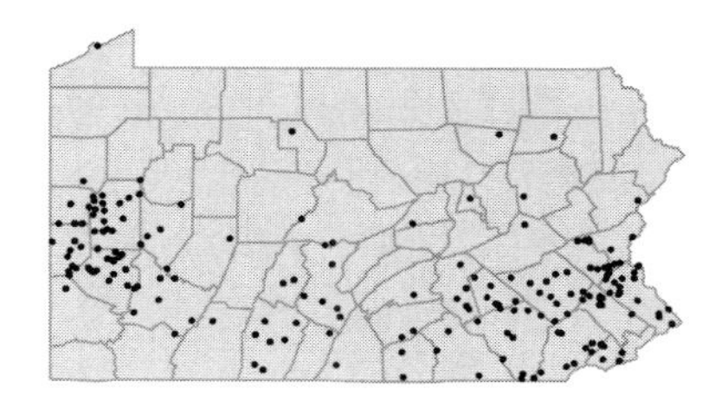

Gentiana catesbaei Walt.
Coastal Plain gentian
Herbaceous perennial
Open woods or clearings on seasonally wet sites. Believed to be extirpated, represented by a single collection from Delaware Co. in 1888.

OBL

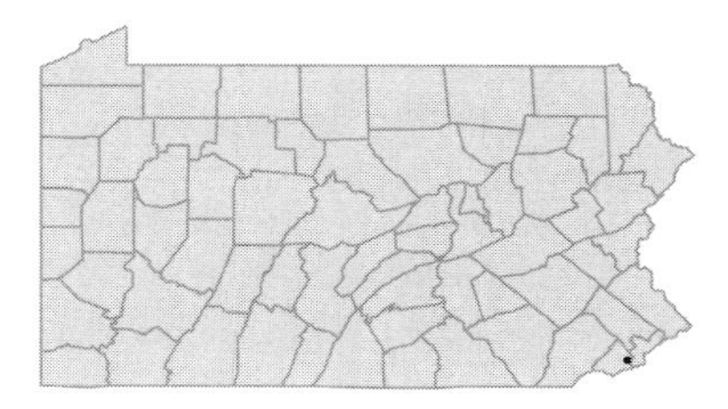

Gentiana clausa Raf.
Meadow bottle gentian
Herbaceous perennial
Moist woods, stream banks and meadows.

FACW

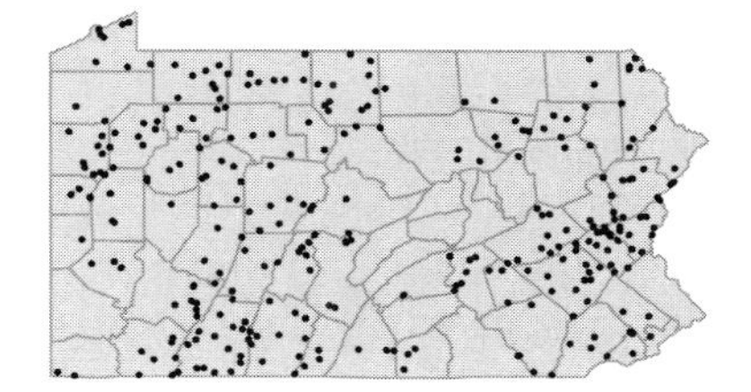

Gentiana linearis Froel.
Narrow-leaved gentian
Herbaceous perennial
Mountain bogs and stream banks.

OBL

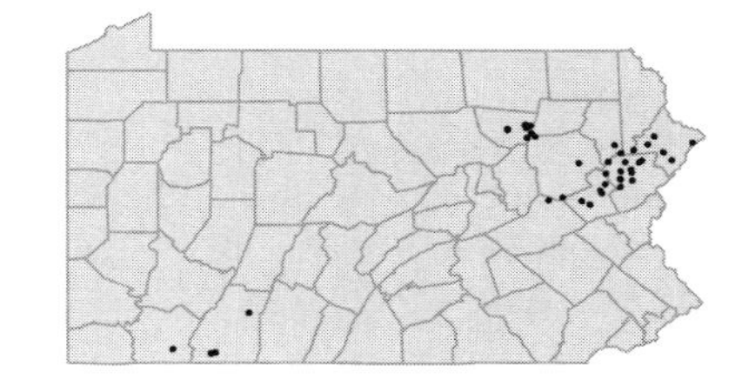

Gentiana saponaria L.
Soapwort gentian
Herbaceous perennial
Bogs, moist, open woods and roadsides.

FACW

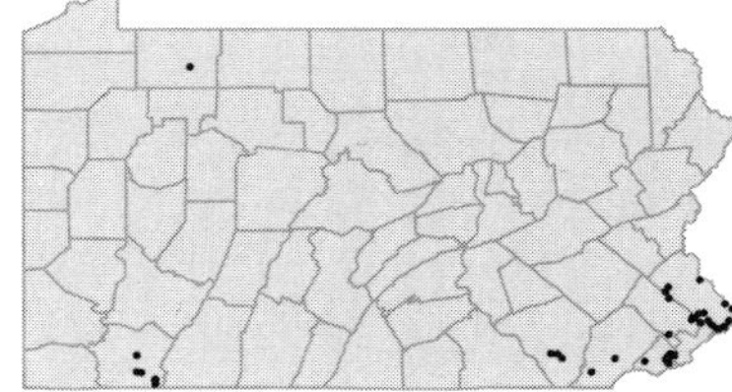

Gentiana villosa L.
Striped gentian
Herbaceous perennial
Dry, open woods and edges, also thickets on serpentine barrens.

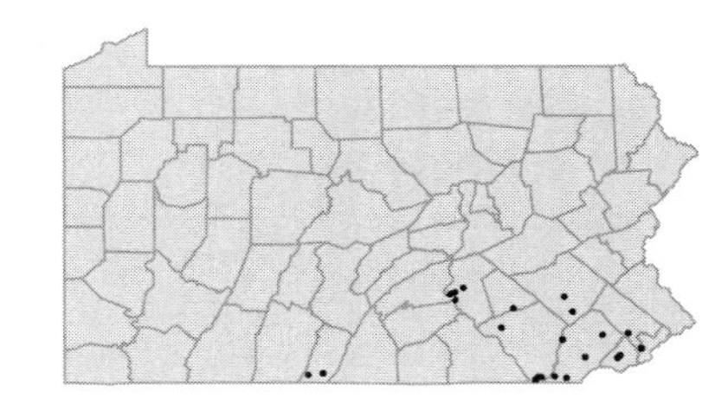

• ***Gentianella quinquefolia*** (L.) Small
Stiff gentian; Gall-of-the-earth; Ague-weed
Herbaceous annual
Woods, springy slopes, roadside banks and shaded limestone bluffs.
Gentiana quinquefolia L. PFGBW

FAC

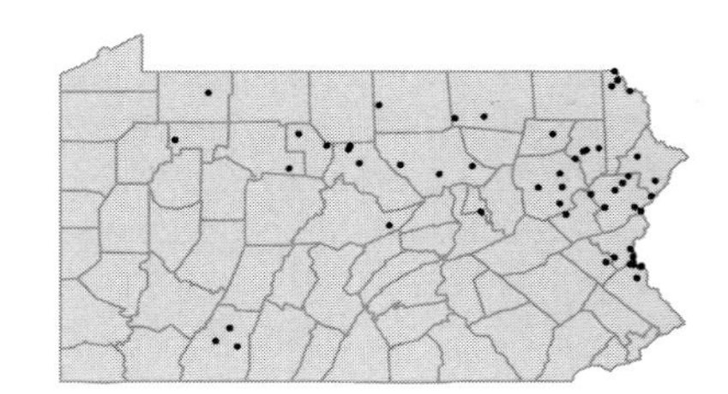

• ***Gentianopsis crinita*** (Froel.) Ma
Fringed gentian
Herbaceous annual
Wet meadows, boggy pastures, seepy slopes, swamps, fens and open slumps along creek valleys.
Gentiana crinita Froel. PFGBW

OBL

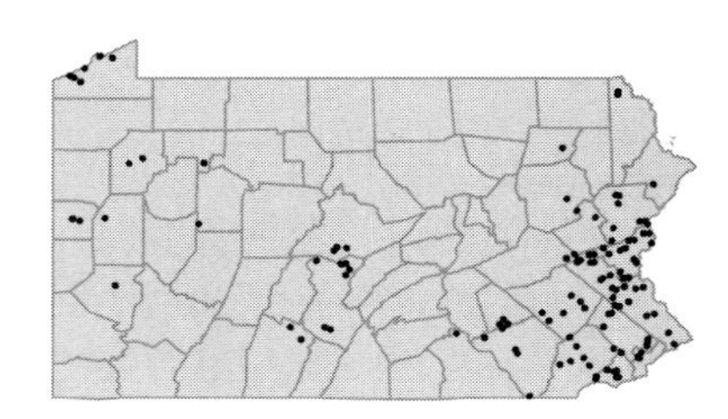

Gentianopsis procera (Holm) Ma
Lesser fringed gentian
Herbaceous annual
Moist shores. Believed to be extirpated, represented by a single collection from Erie Co. in 1900.
Gentiana procera Holm FB

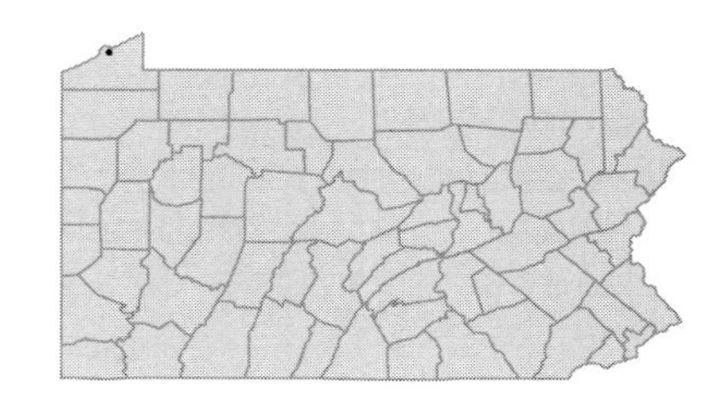

• ***Obolaria virginica*** L.
Pennywort
Herbaceous perennial
Rich, rocky woods.

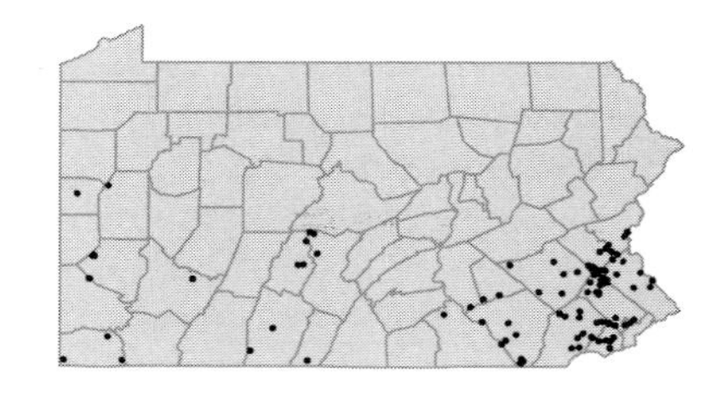

• ***Sabatia angularis*** (L.) Pursh
Marsh-pink
Herbaceous annual
Fields and meadows, dry woods, woodland edges and serpentine barrens.

FAC+

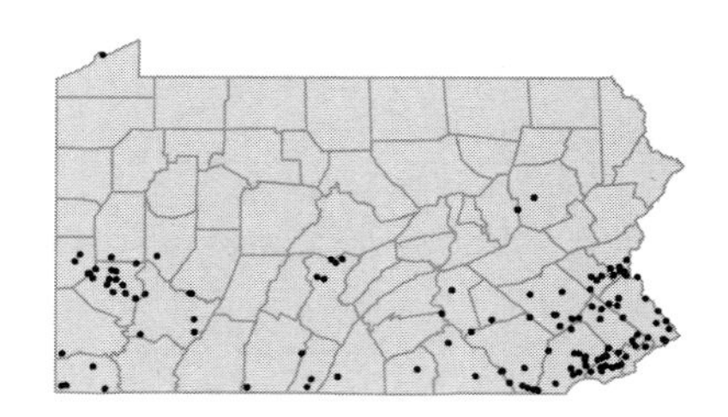

Sabatia campanulata (L.) Torr.
Slender marsh-pink
Herbaceous perennial
Coastal plain swamps. Believed to be extirpated, last collected in 1908.

FACW

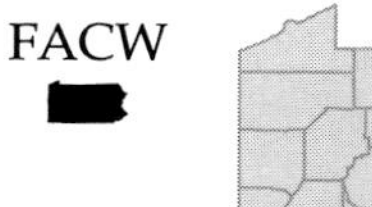

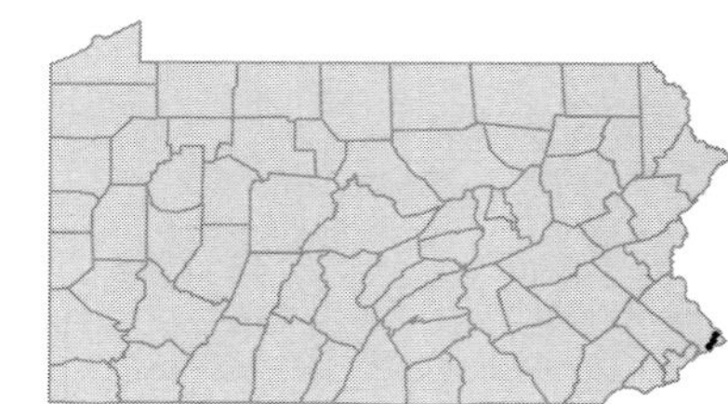

Sabatia stellaris Pursh

FACW+

Marsh-pink; Sea-pink
Herbaceous annual
Marshes or swamps. Believed to be extirpated, represented by a single collection from Bucks Co. in 1904.

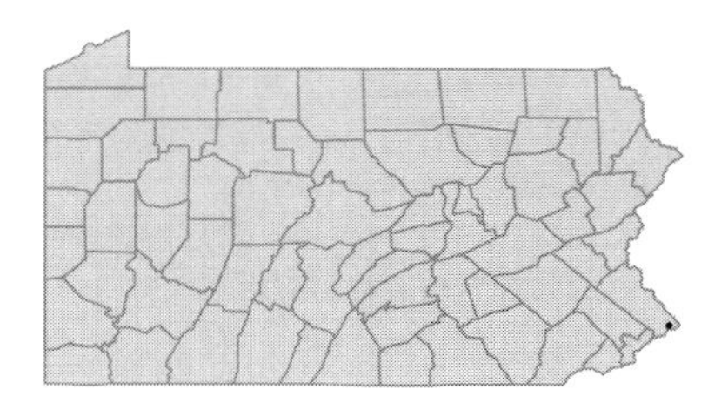

• ***Swertia caroliniensis*** (Walt.) Kuntze
American columbo
Herbaceous perennial
Woods, dry slopes and abandoned fields.
Frasera caroliniensis Walt. C

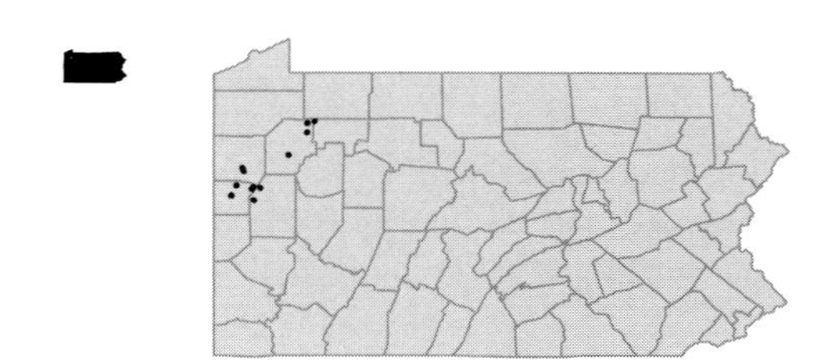

APOCYNACEAE

• ***Amsonia tabernaemontana*** Walt.

I

Blue-star
Herbaceous perennial
Cultivated and occasionally escaped to fields, woods or waste ground.
Amsonia amsonia (L.) Britt. P

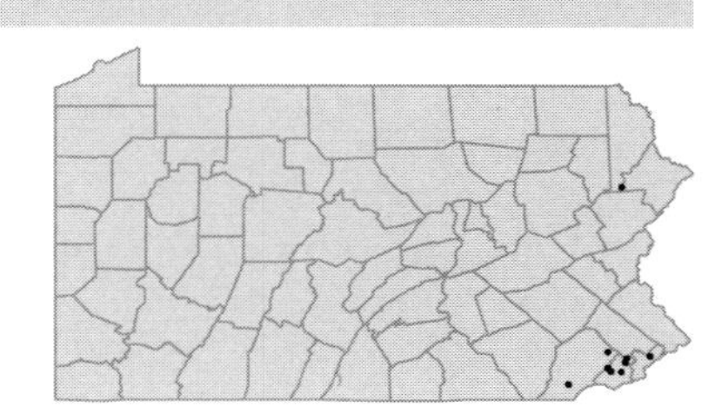

• ***Apocynum androsaemifolium*** L.
Pink dogbane; Spreading dogbane
Herbaceous perennial
Woods, roadsides, dry pastures, thickets and barrens.

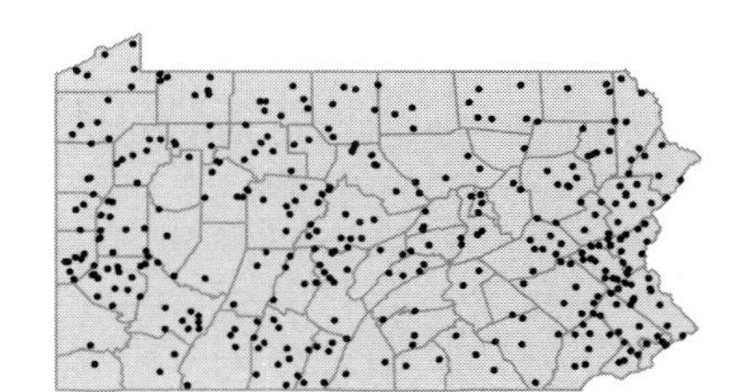

Apocynum androsaemifolium x cannabinum Greene
Dogbane
Herbaceous perennial
River banks, abandoned fields and waste ground.
Apocynum x medium Greene W

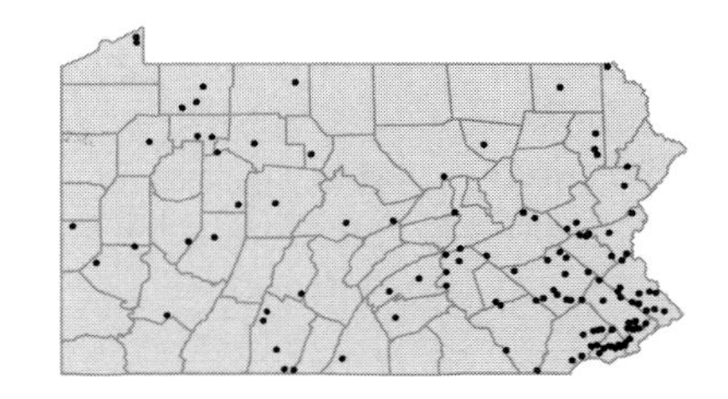

Apocynum cannabinum L. var. ***cannabinum***

FACU

Indian hemp
Herbaceous perennial
Woods, old fields, sandy flats, limestone bluffs and cindery waste ground.
Apocynum cannabinum L. var. *pubescens* (Mitchell) A.DC. FBW

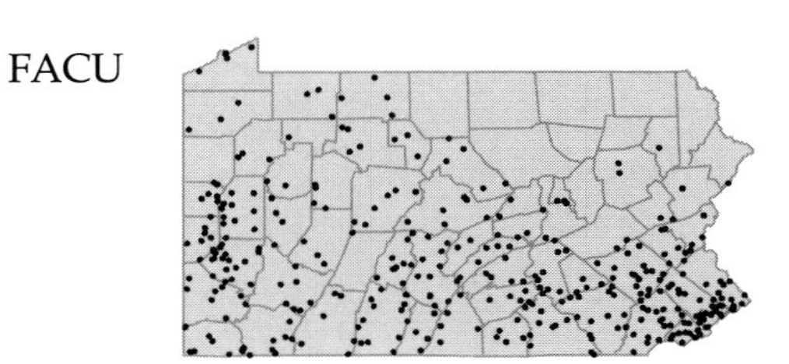

Apocynum cannabinum L. var. ***glaberrimum*** DC.
Indian hemp
Herbaceous perennial
Grassy meadows, fields, woods, dry railroad banks and river banks.
Apocynum album Greene P

FACU

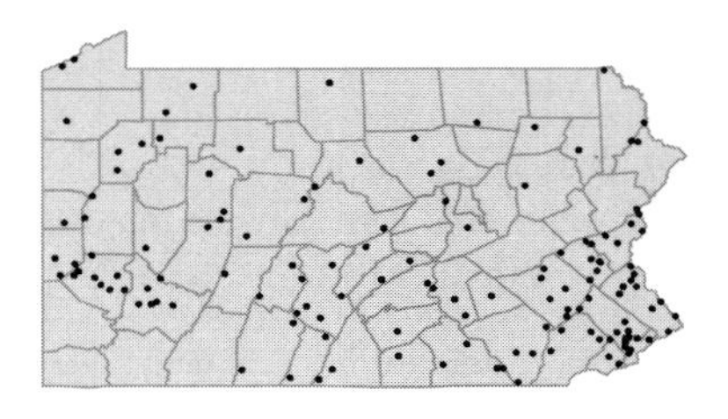

Apocynum cannabinum L. var. ***hypericifolium*** A.Gray
Indian hemp
Herbaceous perennial
Alluvial sand and cobble deposits, roadsides or dry banks.
Apocynum sibiricum Jacq. FGBWC

FACU

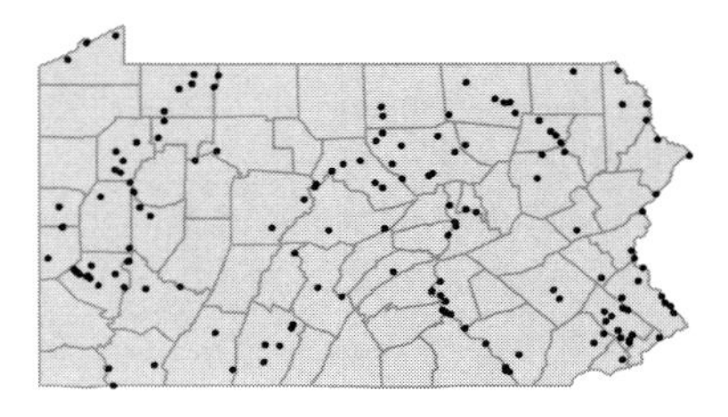

• ***Vinca major*** L.
Giant periwinkle
Herbaceous perennial
Cultivated and escaped at an abandoned greenhouse site.

[I]

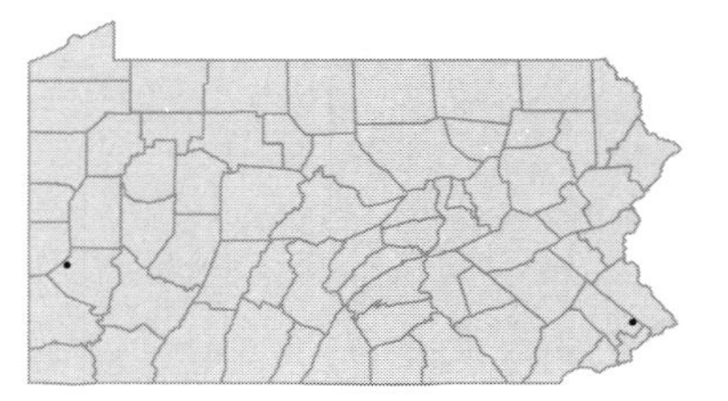

Vinca minor L.
Common periwinkle; Creeping myrtle
Herbaceous perennial
Cultivated and occasionally naturalized in woods, fields and on roadside banks.

[I]

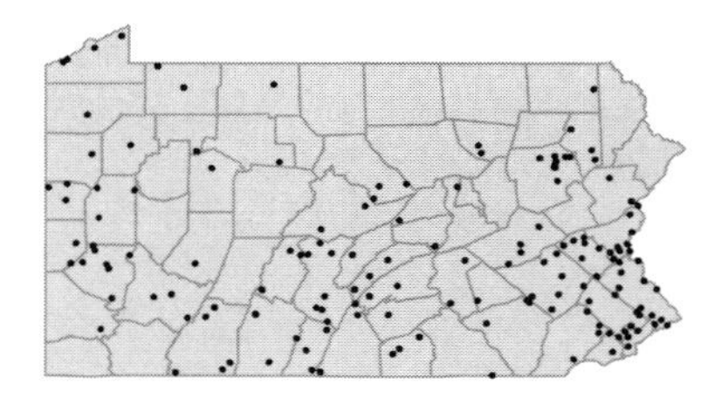

ASCLEPIADACEAE

• ***Asclepias amplexicaulis*** J.E.Smith
Blunt-leaved milkweed; Clasping milkweed
Herbaceous perennial
Dry pastures, sandy flats, barrens and roadsides.

Asclepias exaltata L.
Poke milkweed; Tall milkweed
Herbaceous perennial
Woods and wooded roadsides.

FACU

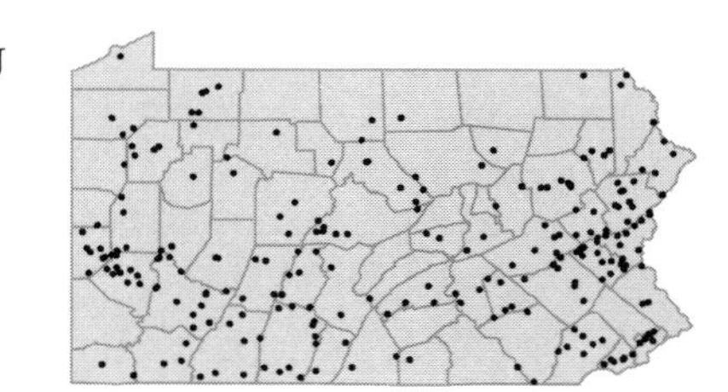

Asclepias incarnata L. ssp. ***incarnata***
Swamp milkweed
Herbaceous perennial
Swamps, floodplains and wet meadows.

OBL

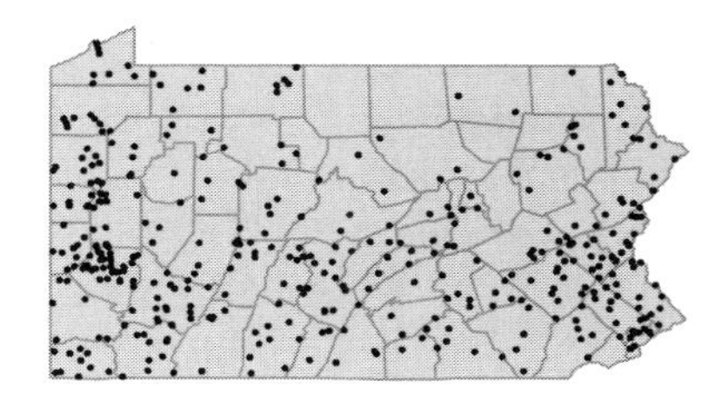

Asclepias incarnata L. ssp. ***pulchra*** (Ehrh. ex Willd.) Woods.
Swamp milkweed
Herbaceous perennial
Low, swampy ground, wet shores and sphagnum bogs.

OBL

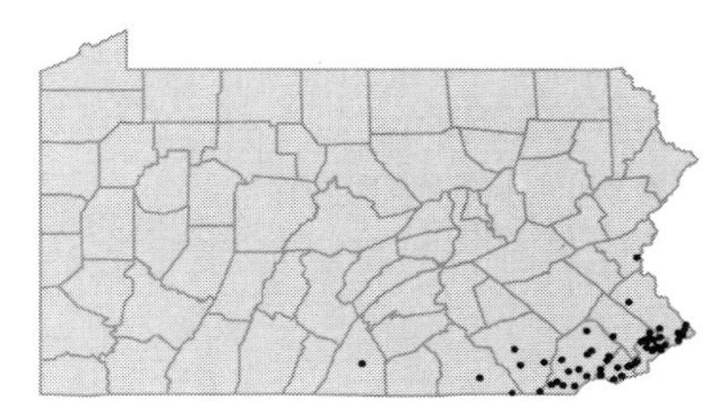

Asclepias purpurascens L.
Purple milkweed
Herbaceous perennial
Dry to moist woods, thickets, fields and roadsides.

FACU

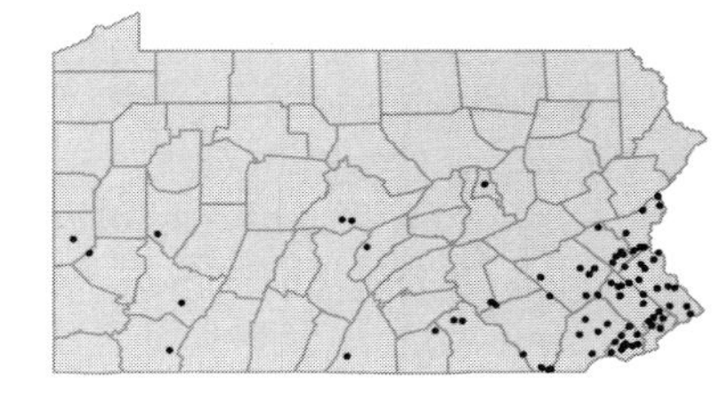

Asclepias quadrifolia Jacq.
Four-leaved milkweed
Herbaceous perennial
Dry woods and roadsides.

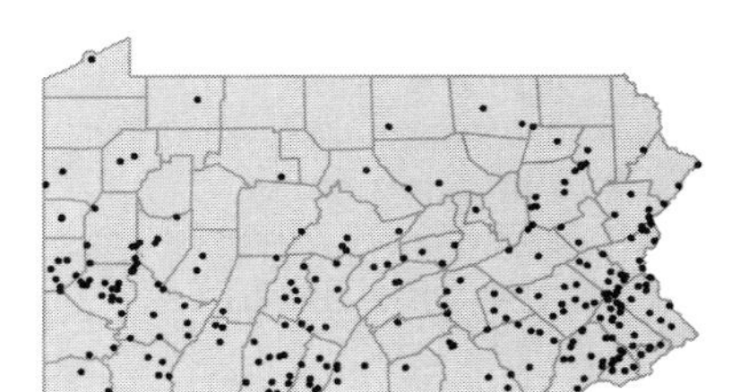

Asclepias rubra L.
Red milkweed
Herbaceous perennial
Sphagnum bogs. Believed to be extirpated, last collected in 1957.

OBL

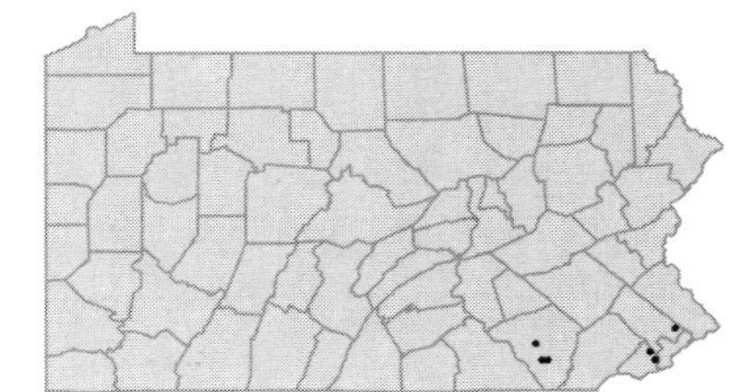

Asclepias syriaca L.
Common milkweed
Herbaceous perennial
Fields, roadsides and waste ground.

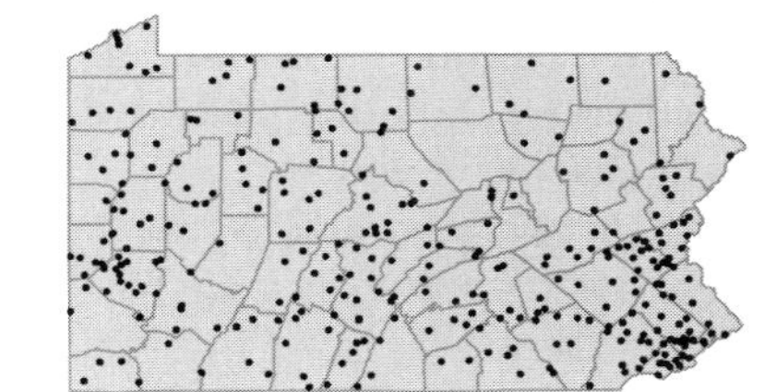

Asclepias tuberosa L.
Butterfly-weed
Herbaceous perennial
Dry woods, abandoned fields, roadsides and shale barrens.

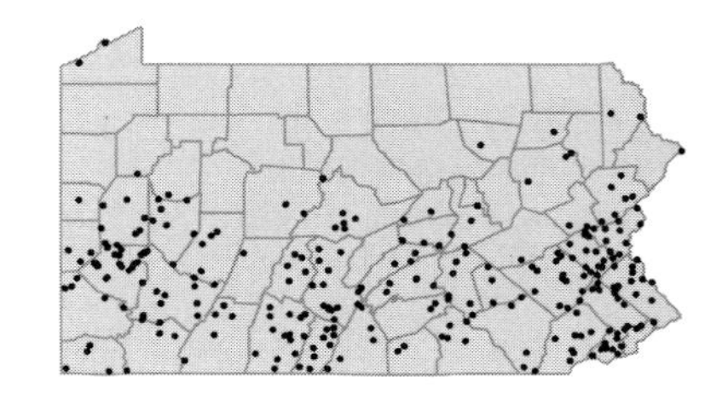

Asclepias variegata L.
White milkweed
Herbaceous perennial
Wooded slopes and roadside banks.

FACU

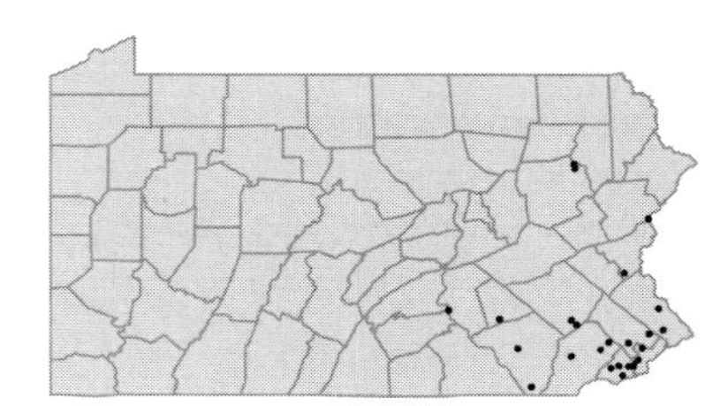

Asclepias verticillata L.
Whorled milkweed
Herbaceous perennial
Dry, sandy soil, serpentine barrens, shale barrens and calcareous slopes.

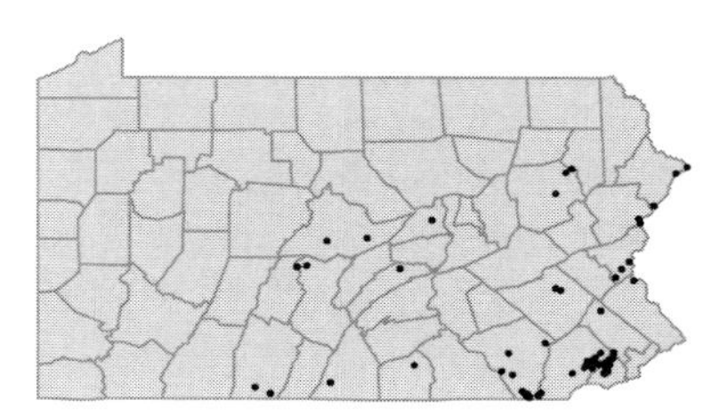

Asclepias viridiflora Raf.
Green milkweed
Herbaceous perennial
Dry fields and dry rocky slopes including limestone and serentine outcrops.
Acerates viridiflora (Raf.) Eat. P

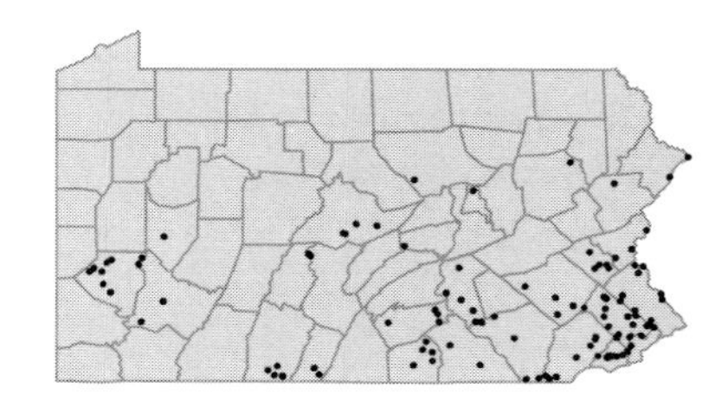

• ***Cynanchum laeve*** (Michx.) Pers.
Smooth sallow-wort
Herbaceous perennial vine
River bank. Represented by a single collection from York Co. in 1906.
Ampelamus albidus (Nutt.) Britt. FGBC

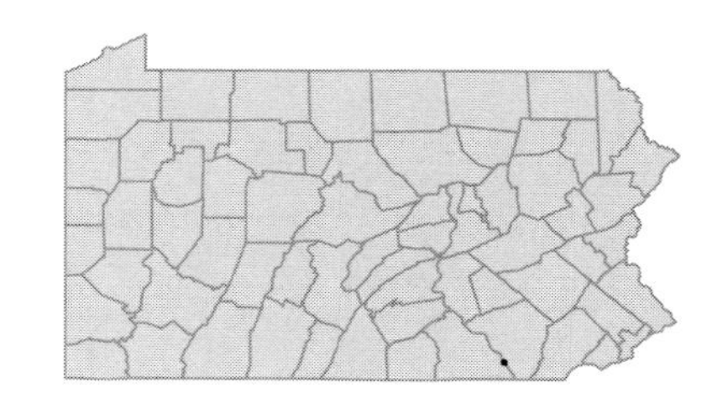

• ***Matelea obliqua*** (Jacq.) Woods.
Angle-pod; Oblique milkvine
Herbaceous perennial vine
Mesic woods, wooded edges and red cedar thickets on limestone.
Gonolobus obliquus (Jacq.) Schultes FGB; *Vincetoxicum obliquum* (Jacq.) Britt. P

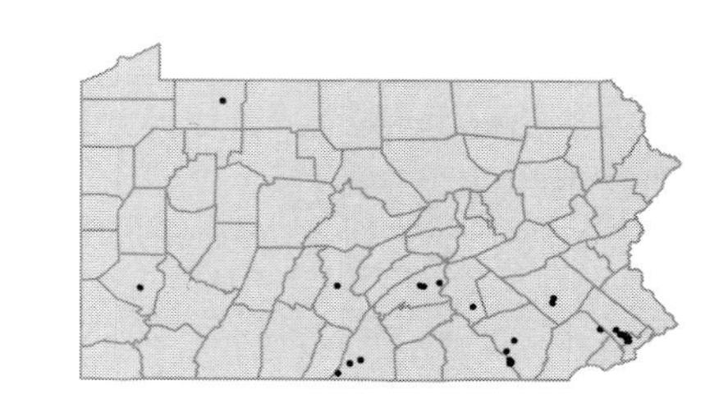

• ***Periploca graeca*** L.
Silkvine
Herbaceous perennial vine
River banks and urban waste ground.

I

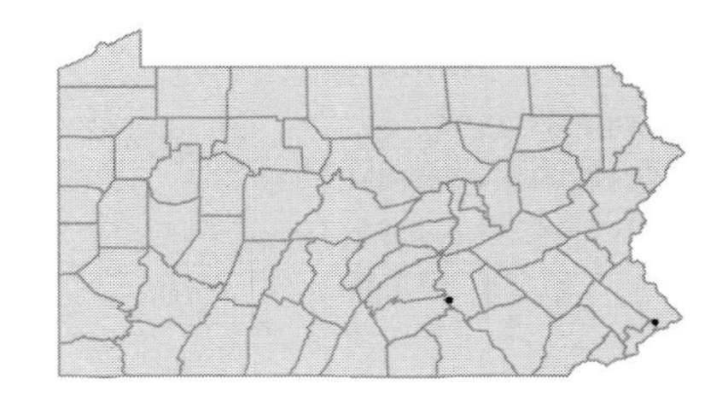

• ***Vincetoxicum nigrum*** (L.) Moench
Black swallow-wort
Herbaceous perennial vine
Roadsides, stream banks and woods edges.
Cynanchum nigrum (L.) Pers. FBGWK

I

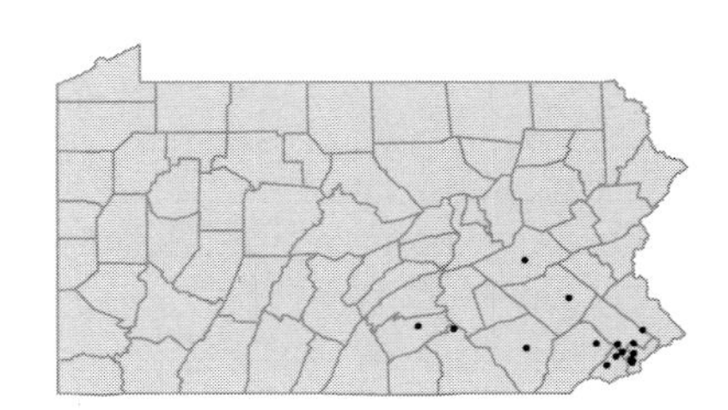

Vincetoxicum rossicum (Kleopow) Barbarich
Swallow-wort
Herbaceous perennial vine
Thickets and edges.
Vincetoxicum hirundinaria Medikus C

I

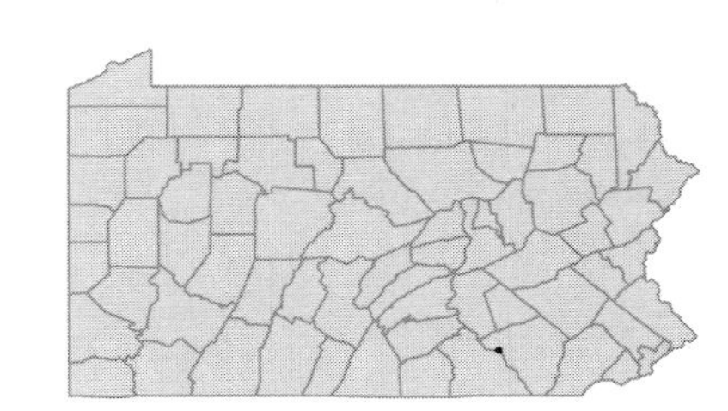

SOLANACEAE

• ***Capsicum annuum*** L.
Bell pepper
Herbaceous annual
Cultivated and occasionally occurring near rubbish dumps.
Capsicum frutescens L. BW

I

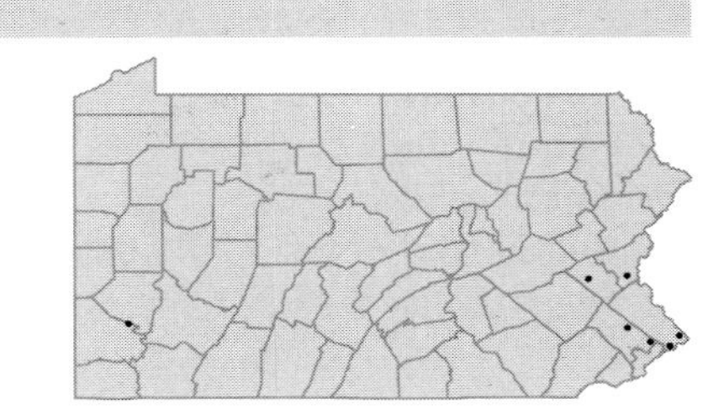

• ***Datura ferox*** L.
Jimsonweed; Thorn-apple
Herbaceous annual
Dump. Represented by a single collection from Philadelphia Co. in 1923.

I

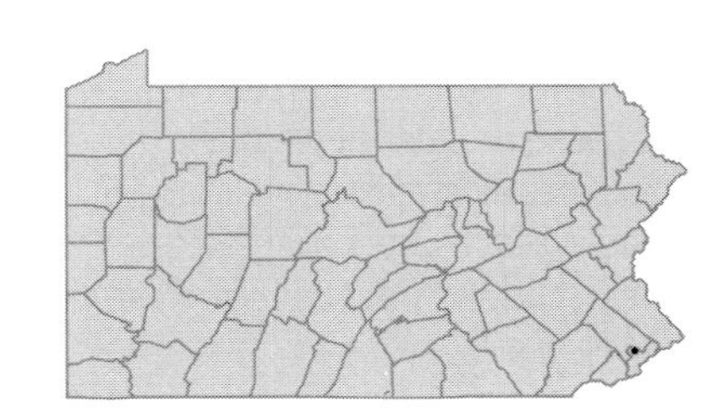

Datura meteloides DC. ex Dunal
Downy thorn-apple; Indian-apple
Herbaceous annual
Cultivated and occasionally escaping to ballast, railroad tracks, dumps or fallow ground.
Datura inoxia Miller C

I

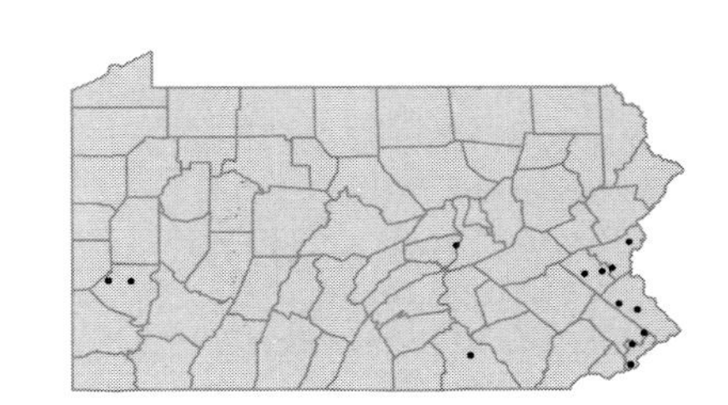

Datura stramonium L.
Jimsonweed
Herbaceous annual
Cultivated fields, roadsides and waste ground. Designated as a noxious weed in PA.

I

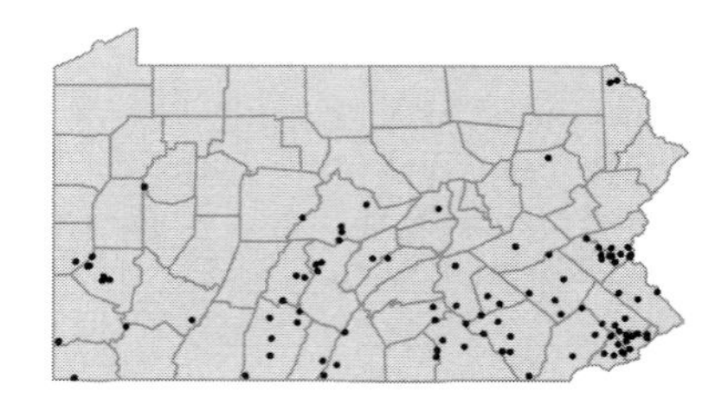

• ***Hyoscyamus albus*** L.
Henbane
Herbaceous annual
Ballast or waste ground. Represented by three collections from Philadelphia Co. 1879-1921.

I

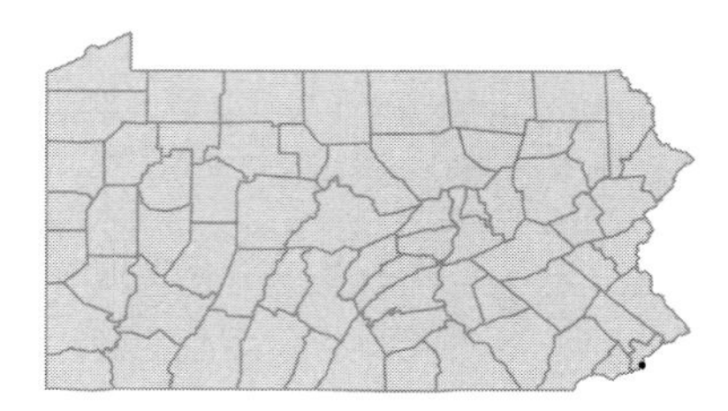

Hyoscyamus niger L.
Black henbane
Herbaceous annual
Ballast.

I

• ***Lycium barbarum*** L.
Matrimony-vine
Deciduous shrub
Cultivated and occasionally naturalized on roadsides or river banks.
Lycium halimifolium P.Mill. FGBW

I

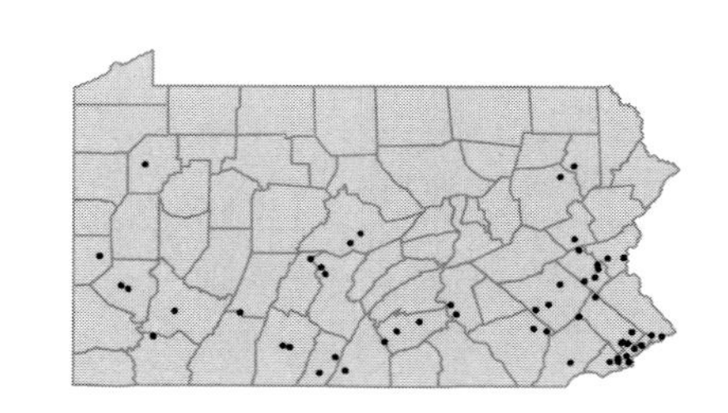

Lycium chinense P.Mill.
Chinese matrimory-vine
Deciduous shrub
Waste places and roadsides.
Lycium barbarum L. *pro parte* C

I

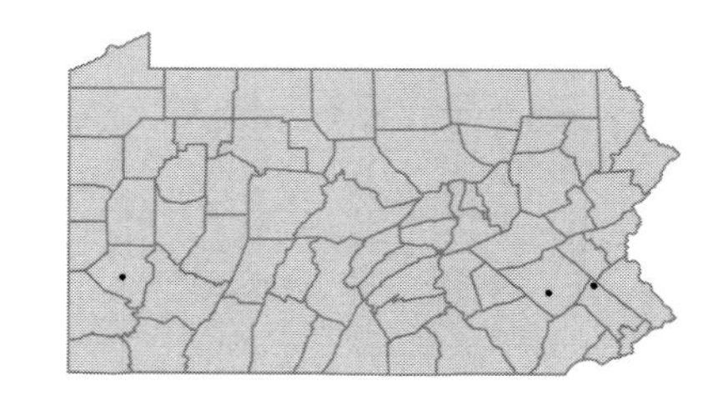

• ***Lycopersicon esculentum*** P.Mill.
Tomato
Herbaceous annual
Cultivated and occasionally occurring on alluvial shores, railroad banks or rubbish dumps.
Lycopersicon lycopersicum (L.) Karst. P

I

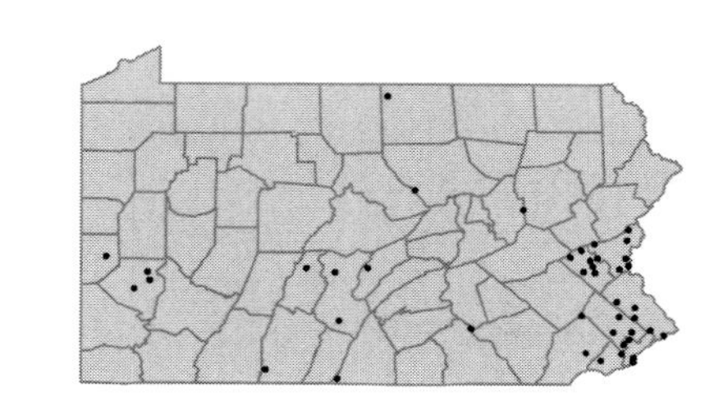

• ***Nicandra physalodes*** (L.) Gaertn.
Apple-of-peru; Shoofly-plant
Herbaceous annual
Cultivated and occasionally naturalized in old fields, or along railroads.
Physalodes physalodes (L.) Britt. P

I

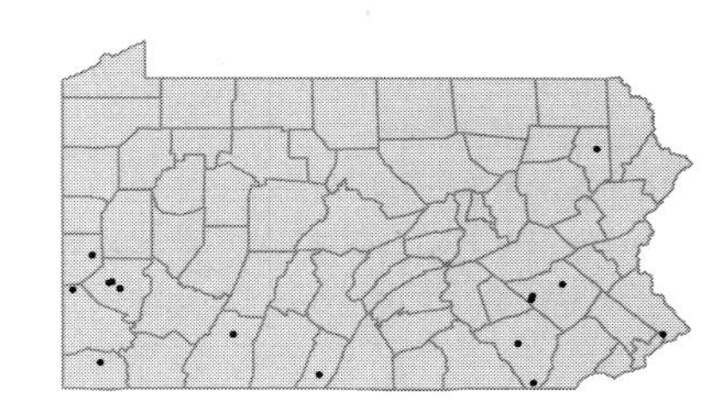

• ***Nicotiana alata*** Link & Otto
Flowering tobacco; Winged tobacco
Herbaceous annual
Rubbish dumps.

I

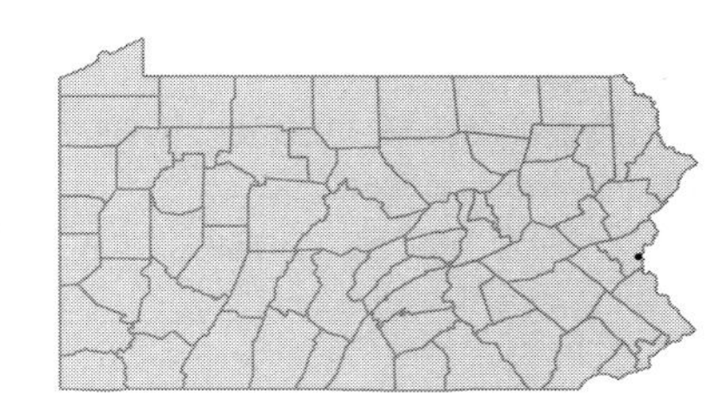

Nicotiana longiflora Cav.
Long-flowered tobacco
Herbaceous annual
Recently cleared woods and urban waste ground. Represented by four collections 1880-1890.

I

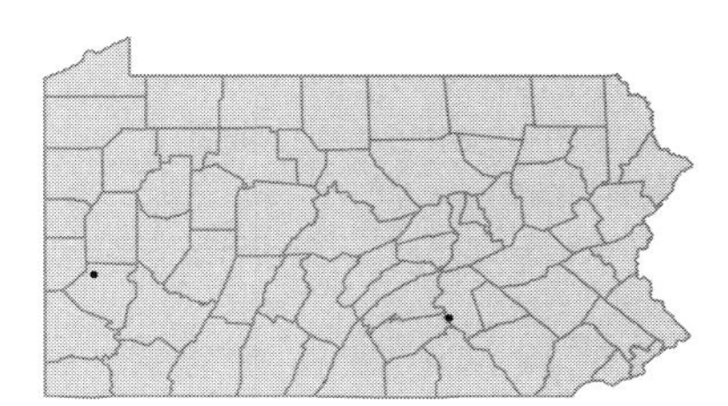

Nicotiana rustica L.
Wild tobacco
Herbaceous annual
Cultivated and rarely escaped to waste ground.

I

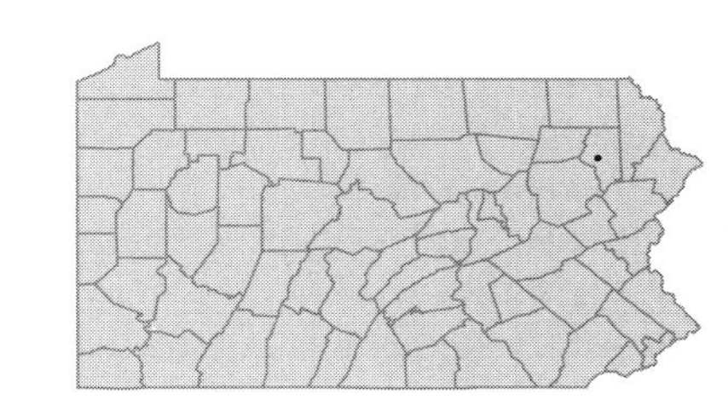

Nicotiana tabacum L.
Tobacco
Herbaceous annual
Cultivated and occasionally escaped near dumps or waste places.

I

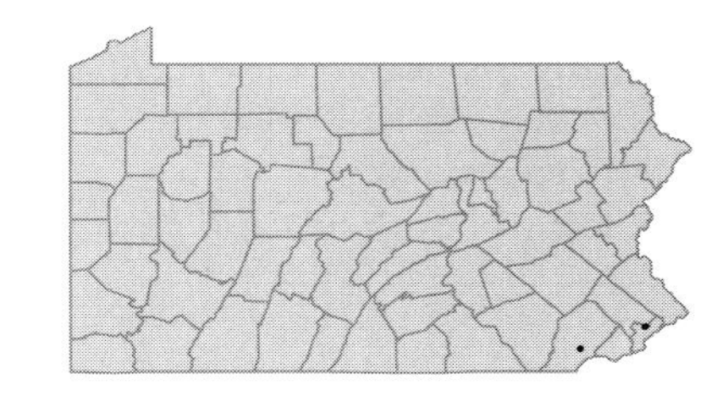

• ***Petunia x hybrida*** Vilm.
Petunia
Herbaceous annual
Cultivated and escaped to old fields, waste ground, rubbish dumps and paving cracks.
Petunia axillaris (Lam.) BSP PFBW; *Petunia integrifolia* (Hook.) Schinz & Thell. BW; *Petunia violacea* Lindl. PF

I

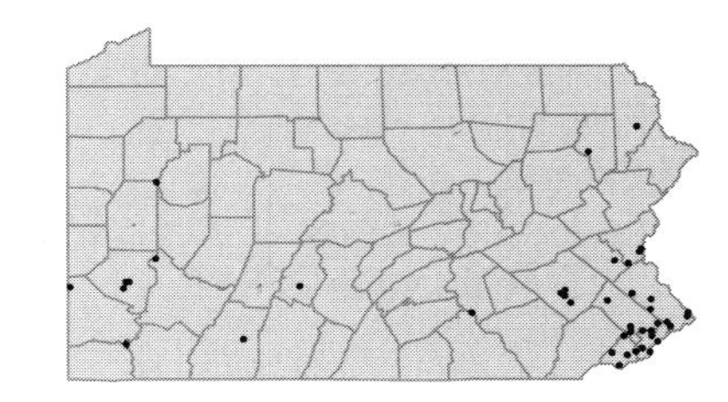

Petunia parviflora A.L.Juss.
Seaside petunia
Herbaceous annual
Waste ground or ballast. Represented by four collections from Philadelphia Co. 1864-1921.

FACW
[I]

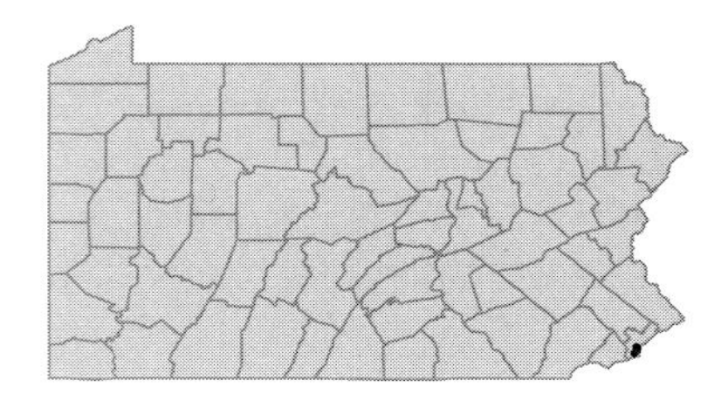

• ***Physalis alkekengi*** L.
Chinese-lantern; Winter-cherry
Herbaceous perennial
Cultivated and occasionally escaped to woods, roadsides, stream banks or vacant lots.

[I]

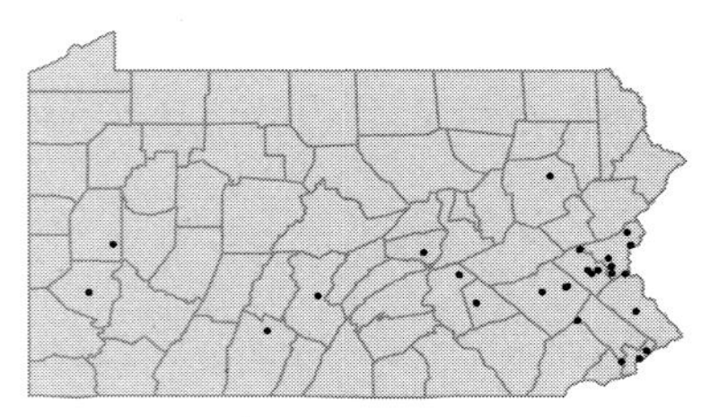

Physalis heterophylla Nees
Clammy ground-cherry
Herbaceous perennial
Fields, sandy or cindery open ground and cultivated areas.

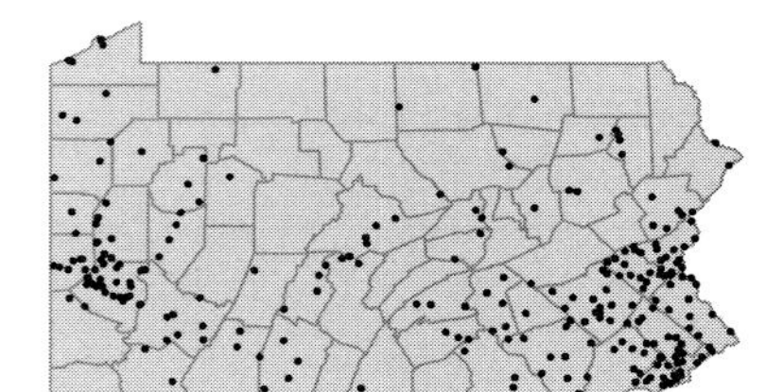

Physalis philadelphica Lam.
Tomatillo
Herbaceous annual
Cultivated and also occasionally occurring on ballast, wharves, and waste ground.
Physalis ixocarpa Hornem. FBGW

[I]

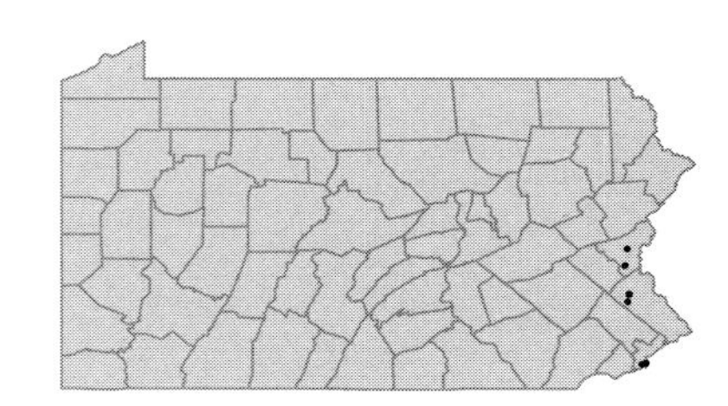

Physalis pubescens L. var. ***pubescens***
Hairy ground-cherry
Herbaceous annual
Hilltop woods, fields and gardens.

FACU-
[I]

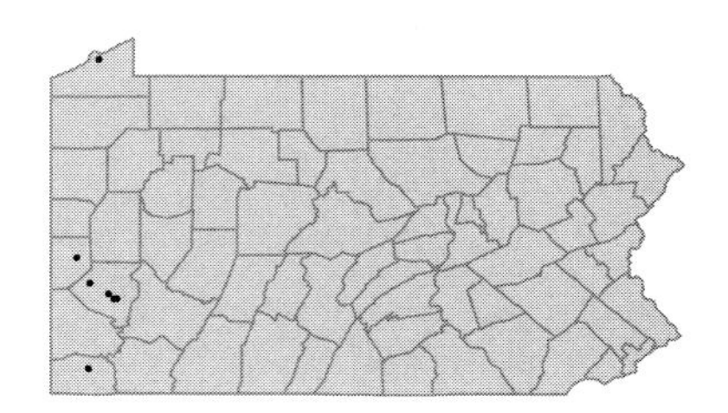

Physalis pubescens L. var. ***integrifolia*** (Dunal) Waterfall
Hairy ground-cherry
Herbaceous annual
Alluvial areas, fields and cultivated ground.
Physalis pruinosa L. PFBW

FACU-

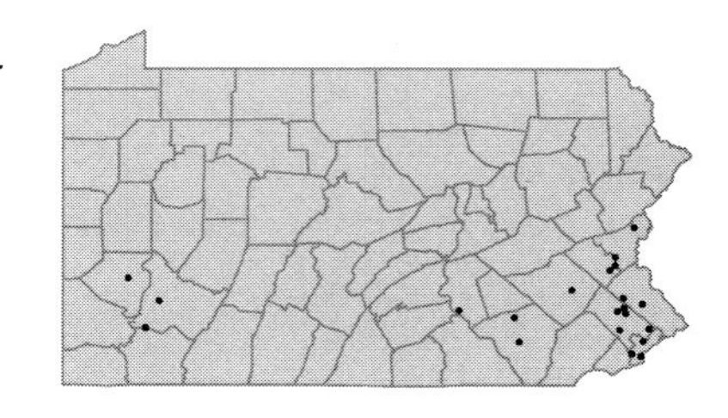

Physalis subglabrata Mackenzie & Bush
Ground-cherry
Herbaceous perennial
Fields, waste ground, hedgerows and limestone uplands.
Physalis longifolia Nutt. FBGC

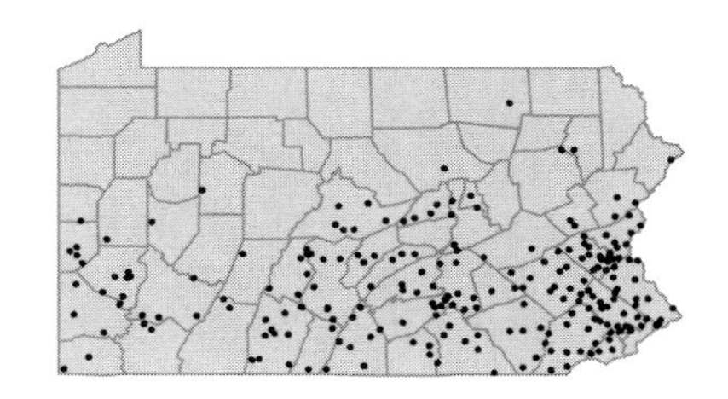

Physalis virginiana P.Mill.
Virginia ground-cherry
Herbaceous perennial
Ballast and waste ground.

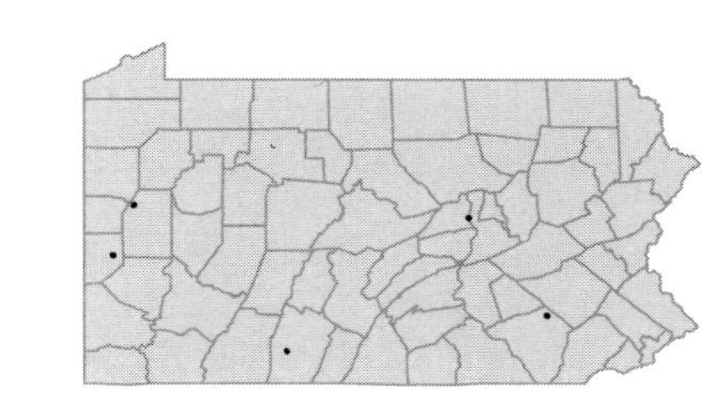

• ***Solanum carolinense*** L. UPL
Horse-nettle
Herbaceous perennial
Fields, roadsides and sandy river banks.

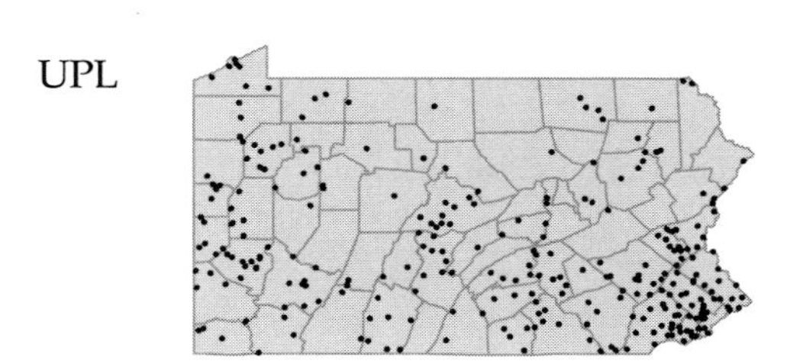

Solanum dulcamara L. var. ***dulcamara*** FAC- [I]
Trailing nightshade; Bittersweet
Woody vine
Wooded hillsides, bogs, wet thickets, roadsides and dumps.

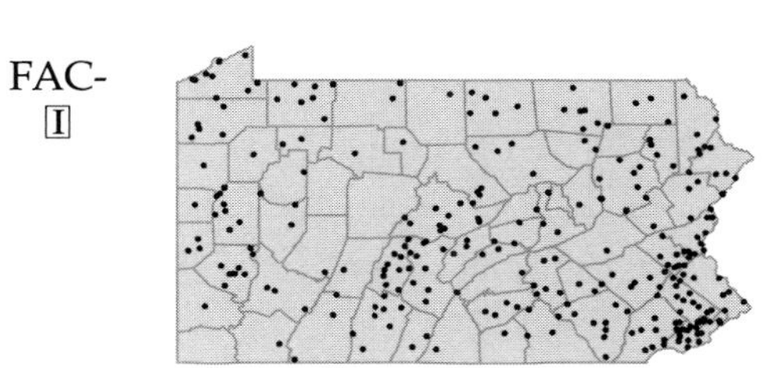

Solanum dulcamara L. var. ***villosissimum*** Desv. FAC- [I]
Trailing nightshade; Bittersweet
Woody vine
Rocky streambeds, thickets and waste ground.

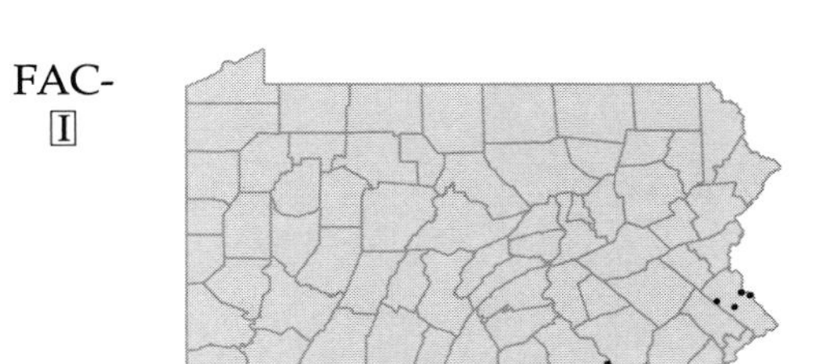

Solanum luteum P.Mill. [I]
Nightshade
Herbaceous annual
Ballast ground. Represented by a single collection from Philadelphia Co. in 1866.
Solanum nigrum L. var. *villosum* L. GB; *Solanum villosum* (L.) P.Mill. FWC

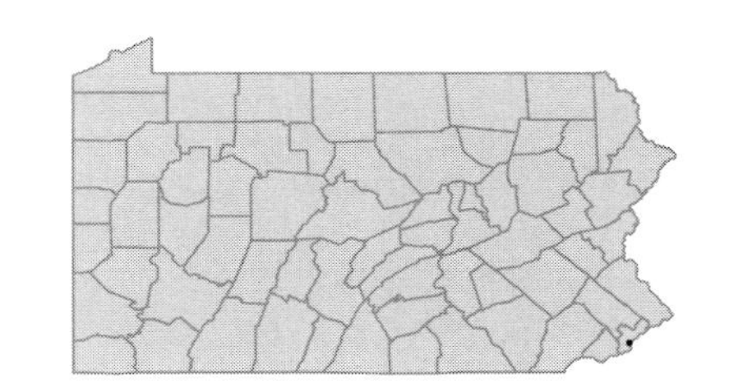

Solanum nigrum L.
Black nightshade
Herbaceous annual
Fields, woods and moist, disturbed ground. This taxon, apparently a mixture of the European *S. nigrum* var. *nigrum*, a hexaploid, and the American *S. nigrum* var. *virginicum* (*S. americanum*), a diploid, requires additional clarification.
Solanum americanum P.Mill. *pro parte*

FACU-
[I]

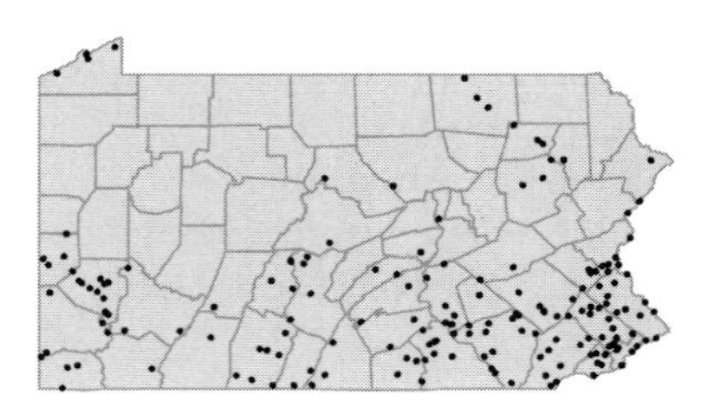

Solanum rostratum Dunal
Buffalo-bur
Herbaceous annual
Cultivated ground and river banks.
Solanum cornutum Lam. K

[I]

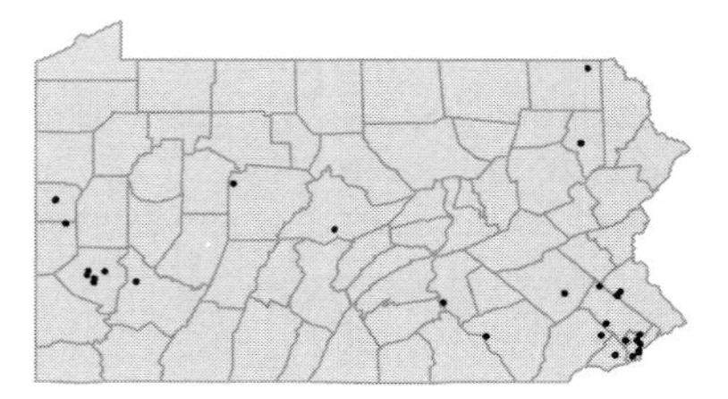

Solanum sarrachoides Sendtner
Nightshade
Herbaceous annual
Spontaneous in garden. Represented by a single collection from Allegheny Co. in 1948.

[I]

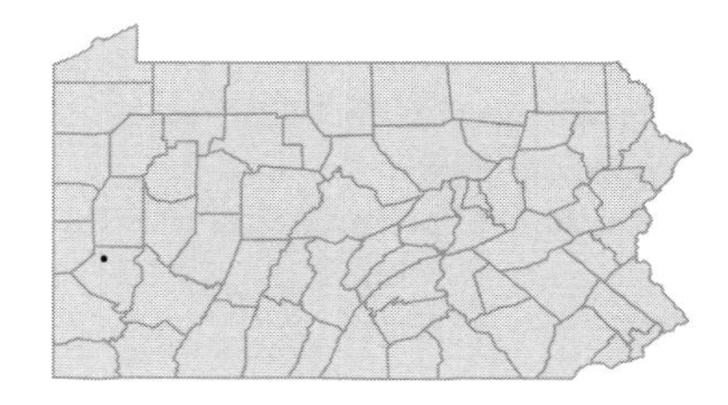

Solanum sisymbriifolium Lam.
Sticky nightshade
Herbaceous annual
Rubbish dump. Represented by a single collection from Philadelphia Co. in 1945.

[I]

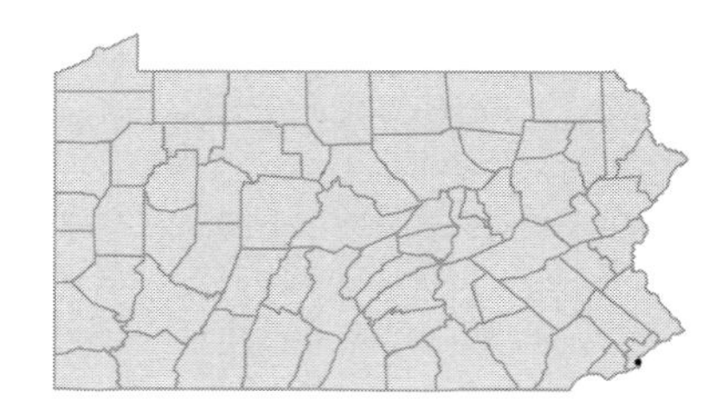

Solanum tuberosum L.
Potato
Herbaceous annual
Cultivated and occasionally occurring in abandoned fields, roadsides or rubbish dumps.

[I]

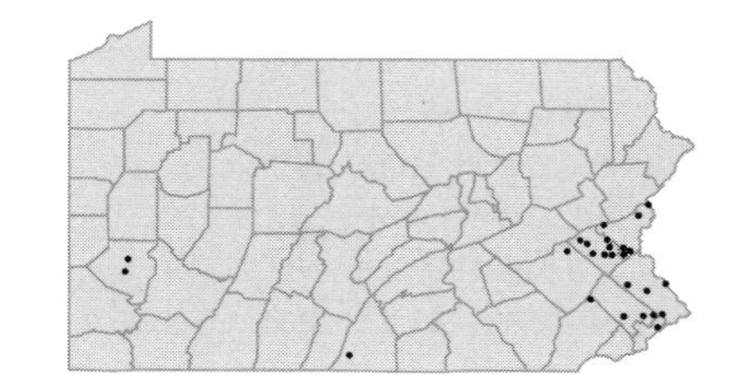

CONVOLVULACEAE

• ***Calystegia hederacea*** Wallich
Japanese bindweed
Herbaceous perennial vine
Hedges, flower beds, fields and rubbish dumps.
Convolvulus wallichianus Spreng. FGB

[I]

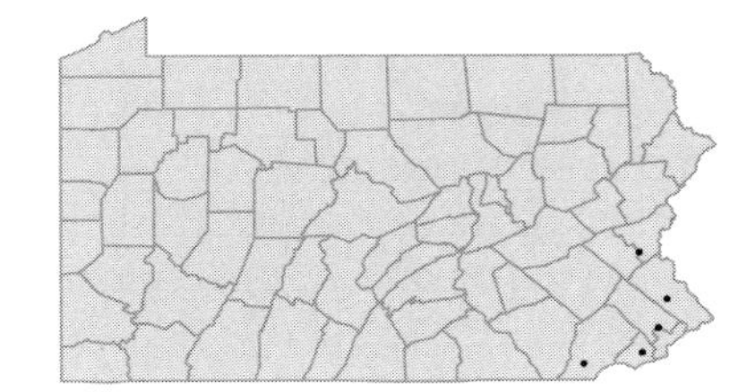

Calystegia pubescens Lindl.
Japanese bindweed
Herbaceous perennial vine
Fallow fields, vacant lots, roadsides, river banks and waste ground.
Convolvulus japonicus Thunb. PGB; *Convolvulus pellitus* Ledeb. *f. anestius* Fern. F; *Calystegia hederacea* Wallich *pro parte* C

[I]

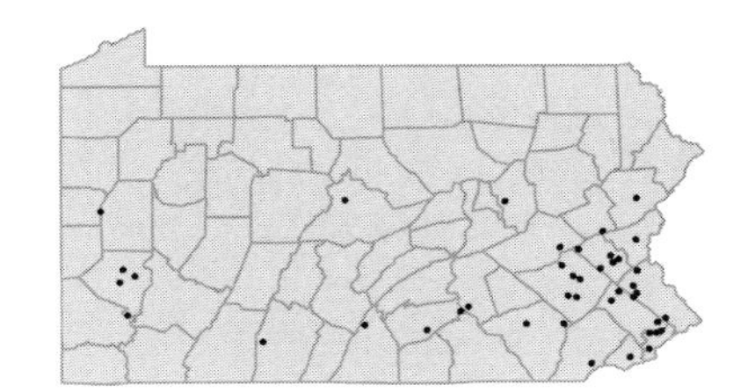

Calystegia sepium (L.) R.Br. *sensu lato*
Hedge bindweed; Wild morning-glory
Herbaceous perennial vine
Waste ground, fields and woods edges.

FAC-

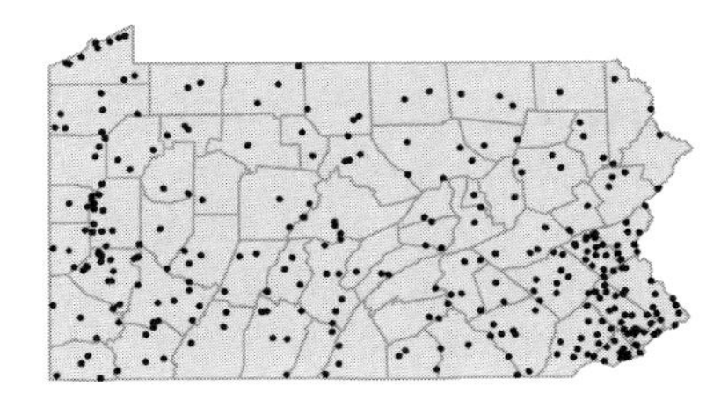

Calystegia silvatica (Kit.) Griseb. ssp. ***fraterniflora*** (Mackenzie & Bush) Brummitt
Bindweed; Morning-glory
Herbaceous perennial vine
Fields, stream banks, roadsides and moist waste ground.
Calystegia fraterniflora (Mackenzie & Bush) Brummitt W; *Convolvulus sepium* L. var. *fraterniflorus* Mackenzie & Bush FB

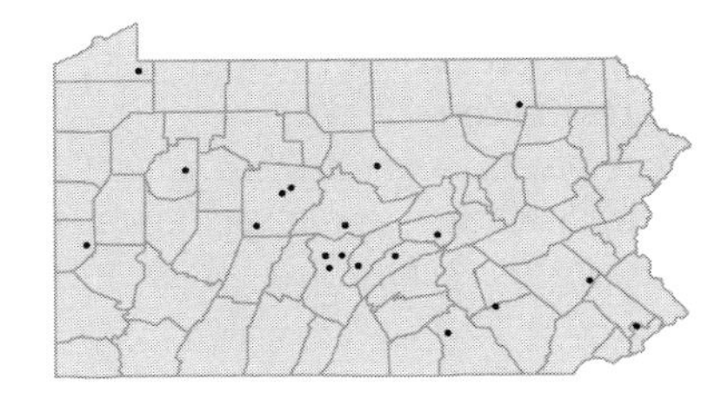

Calystegia spithamaea (L.) Pursh ssp. ***spithamaea***
Low bindweed
Herbaceous perennial vine
Fields, thickets, roadsides or waste ground.
Convolvulus spithamaeus L. *pro parte* PFGB

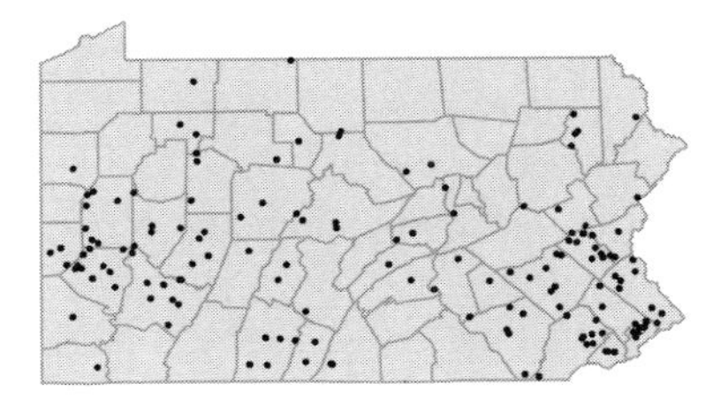

Calystegia spithamaea (L.) Pursh ssp. ***purshiana*** (Wherry) Brummitt
Low bindweed
Herbaceous perennial vine
Dry, open ground, dry slopes and shale barrens.
Convolvulus purshianus Wherry B; *Convolvulus spithamaeus* L. var. *pubescens* (A.Gray) Fern. FGC

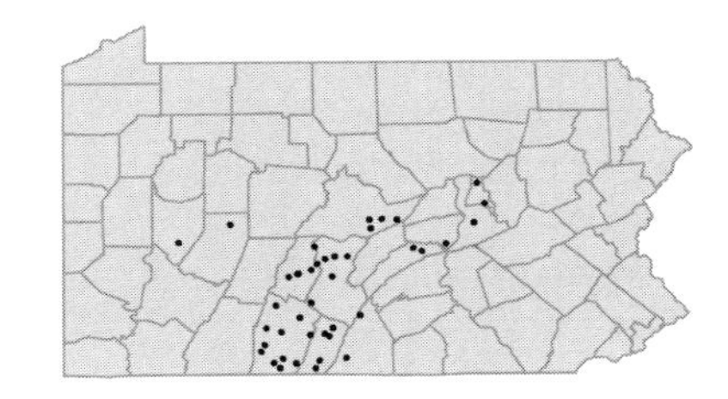

Calystegia spithamaea (L.) Pursh ssp. ***stans*** (Michx.) Brummitt
Low bindweed
Herbaceous perennial vine
Old fields, roadsides, railroad cuts and shaly banks.
Convolvulus spithamaeus L. *pro parte* PFGBC

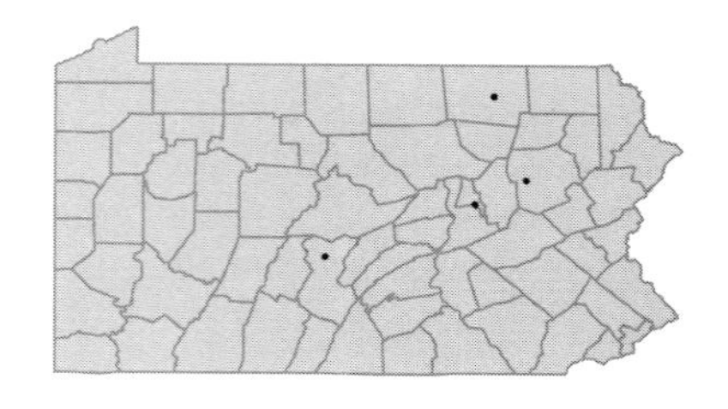

• ***Convolvulus arvensis*** L.
Field bindweed
Herbaceous perennial vine
Fields, roadsides, railroad banks and waste ground.

[I]

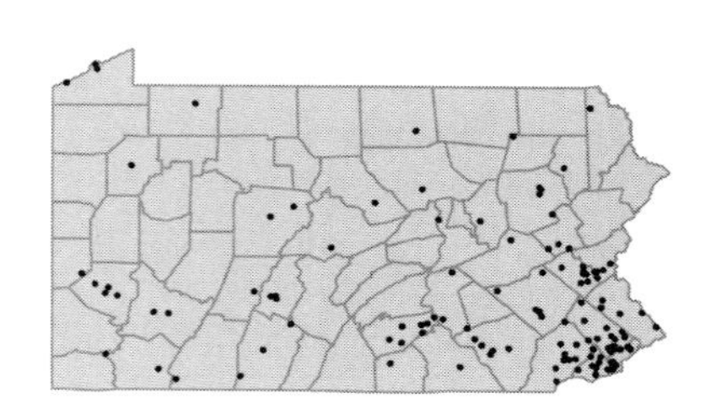

• ***Ipomoea batatas*** (L.) Lam.
Sweet potato
Herbaceous perennial vine
Cultivated and rarely escaped to waste places.

[I]

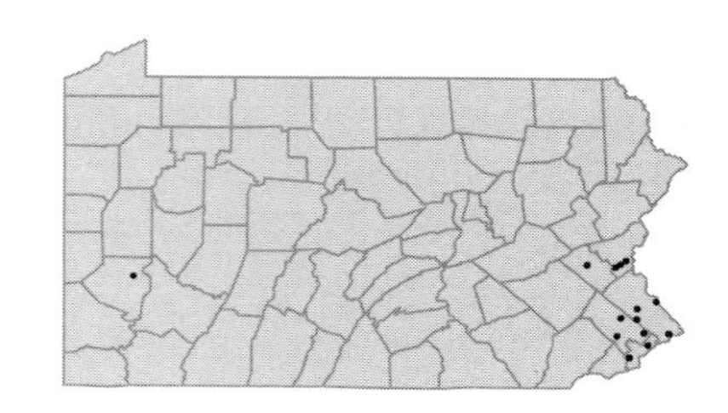

Ipomoea coccinea L.
Red morning-glory
Herbaceous annual vine
Roadsides, waste places, fencerows and old fields.
Quamoclit coccinea (L.) Moench PGB

FACU
[I]

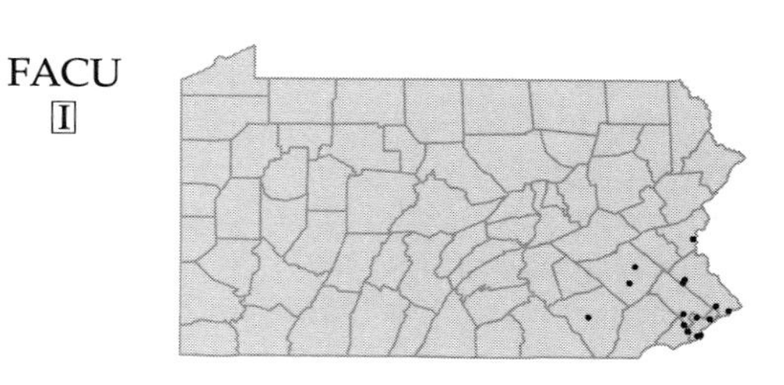

Ipomoea hederacea Jacq.
Ivy-leaved morning-glory
Herbaceous annual vine
Cultivated fields, gardens, shale barrens and ballast.

FACU
[I]

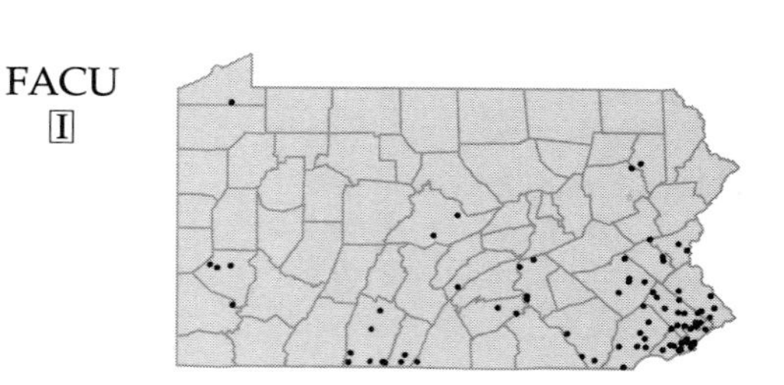

Ipomoea imperati (Vahl) Griseb.
Morning-glory
Herbaceous annual vine
Ballast. Represented by several collections from Philadelphia Co. in 1865.
Ipomoea littoralis (L.) Boissier *non* Blume W

[I]

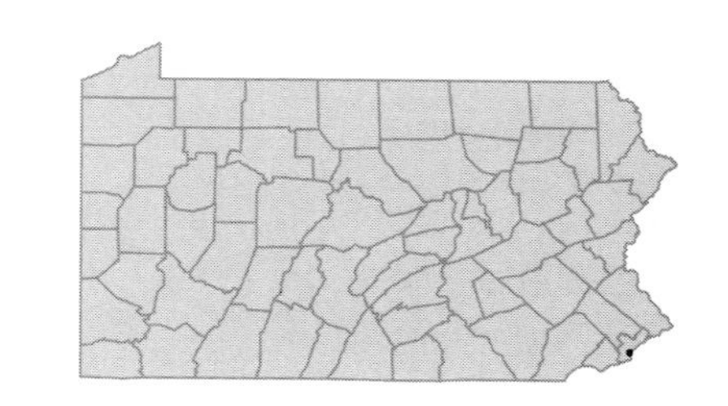

Ipomoea lacunosa L.
White morning-glory
Herbaceous annual vine
Fields, riverbanks, railroad sidings and rubbish dumps.

FACW

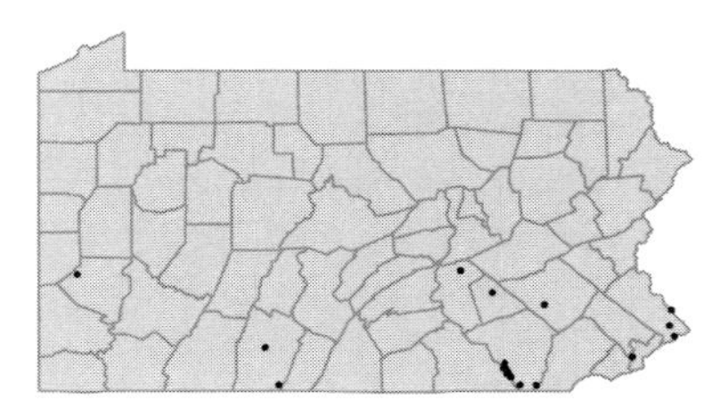

Ipomoea pandurata (L.) G.F.W.Mey.
Man-of-the-earth; Wild potato-vine
Herbaceous perennial vine
Calcareous uplands, thickets and roadsides.

FACU

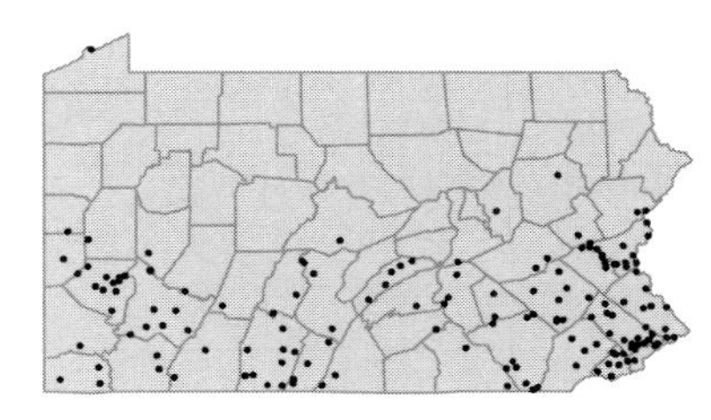

Ipomoea pes-caprae (L.) R.Br.
Railroad vine; Goat-foot morning-glory
Herbaceous perennial vine
Wet, open, sandy soil and beaches.

FAC
[I]

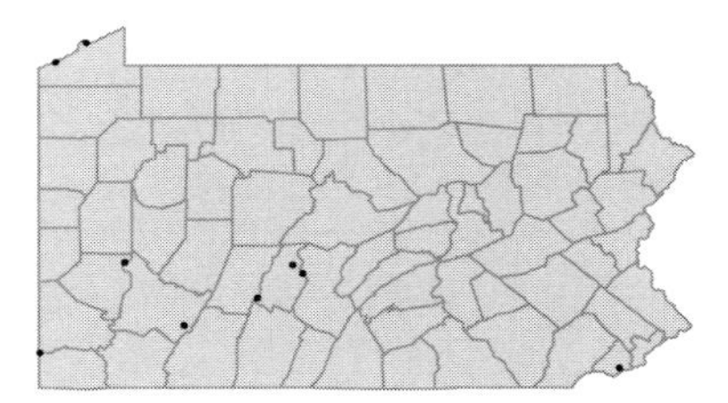

Ipomoea purpurea (L.) Roth
Common morning-glory
Herbaceous annual vine
Cultivated and occasionally escaped to hillsides, fields and roadsides.

UPL
[I]

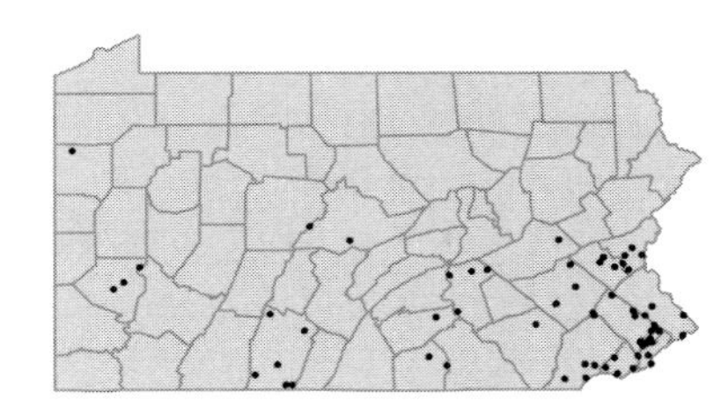

Ipomoea quamoclit L.
Cypress-vine
Herbaceous annual vine
Gardens, streets and waste ground.
Quamoclit quamoclit (L.) Britt. P; *Quamoclit vulgaris* Choisy GB

UPL
[I]

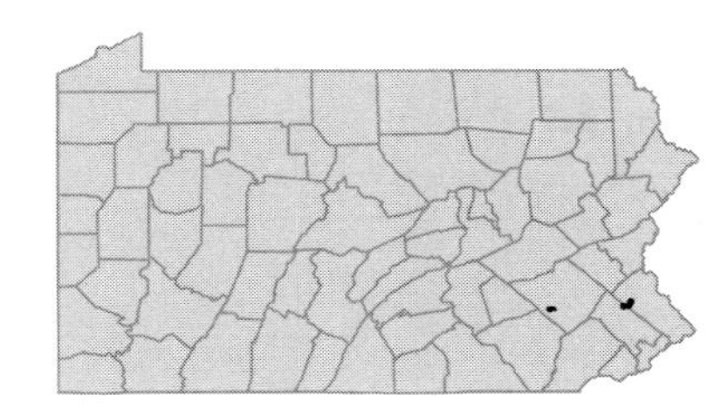

Ipomoea tricolor Cav.
Morning-glory
Herbaceous annual vine
Waste ground and rubbish dump.

[I]

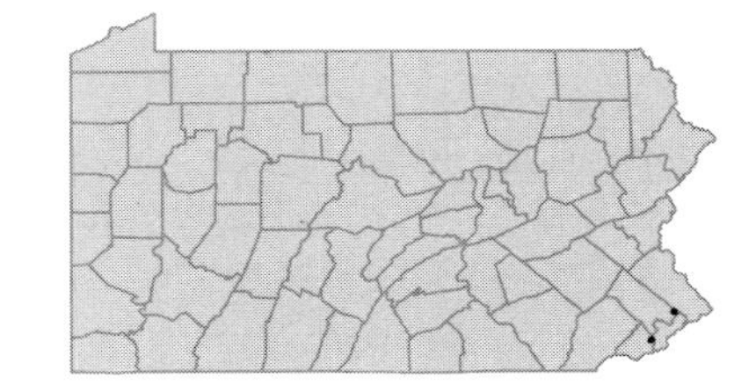

•***Jacquemontia tamnifolia*** (L.) Griseb. FACU- [I]
Bindweed
Herbaceous annual vine
Urban waste ground. Represented by a single collection from Erie Co. in 1914.

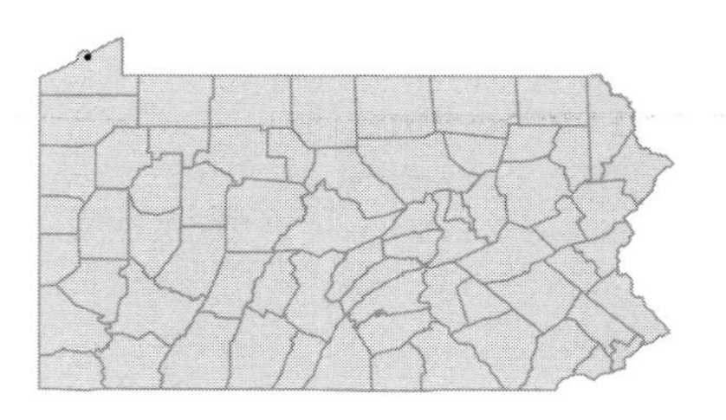

•***Merremia dissecta*** (Jacq.) Hallier f. [I]
Morning-glory
Herbaceous perennial vine
Ballast. Represented by a single collection from Philadelphia Co. ca. 1865.
Ipomoea sinuata Ortega W

CUSCUTACEAE

•***Cuscuta campestris*** Yuncker
Dodder
Herbaceous annual vine
Thickets and waste ground, parasitic on various hosts.
Cuscuta pentagona Engelm. *pro parte* GC

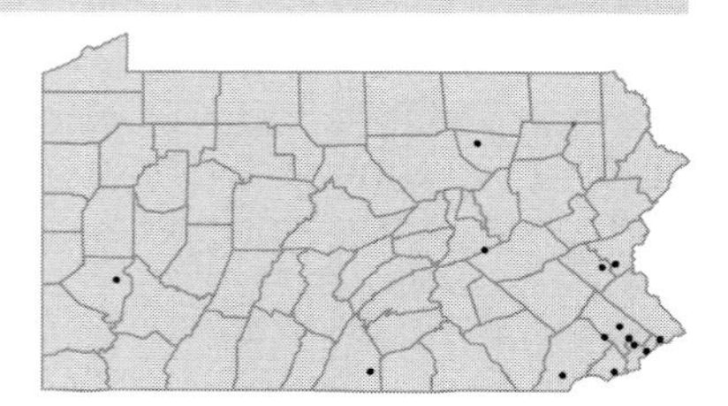

Cuscuta cephalanthii Engelm.
Buttonbush dodder
Herbaceous annual vine
Parasitic on a variety of shrubs and herbs of swamps and moist thickets.

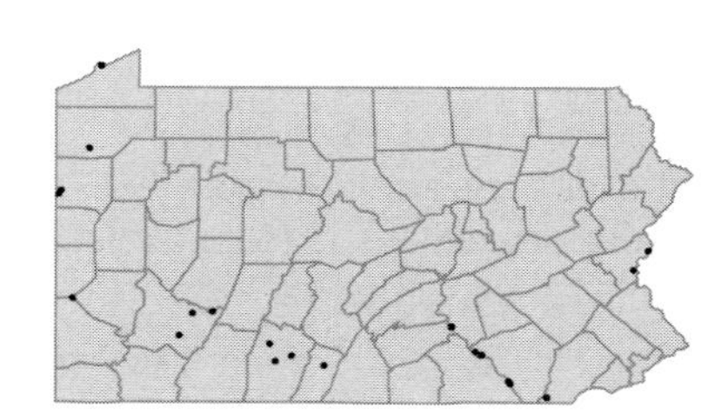

Cuscuta compacta Juss. ex Choisy
Dodder
Herbaceous annual vine
Parasitic on shrubs and herbs of moist thickets and stream banks.

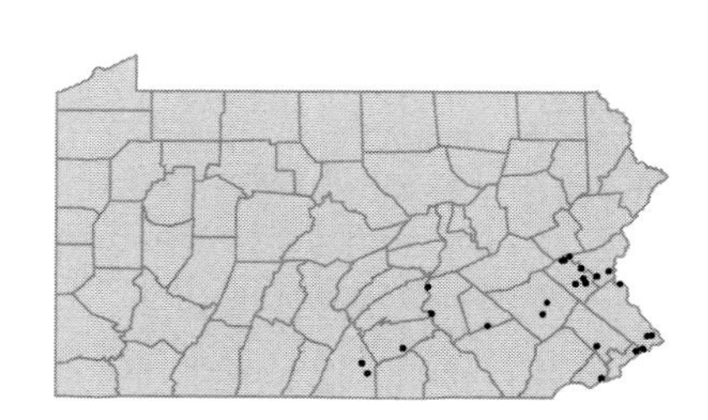

Cuscuta coryllii Engelm.
Hazel dodder
Herbaceous annual vine
Dry, rocky woods, clearings and hillsides, parasitic on various shrubs and herbs.

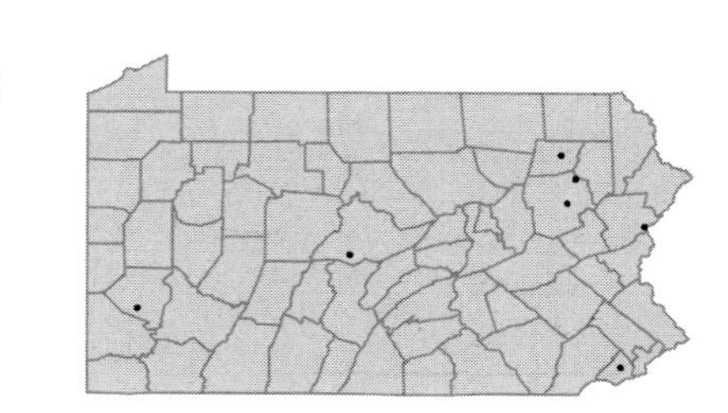

Cuscuta epilinum Weihe
Flax dodder
Herbaceous annual vine
Parasitic on flax. Represented by a single collection from Lancaster Co. in 1863. A federally designated noxious weed.

I

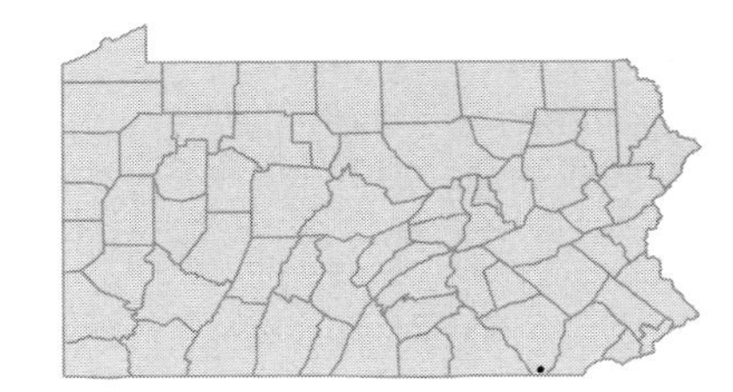

Cuscuta epithymum (L.) L.
Clover dodder
Herbaceous annual vine
Old fields and woods edges, parasitic on clover and other legumes. A federally designated noxious weed.

I

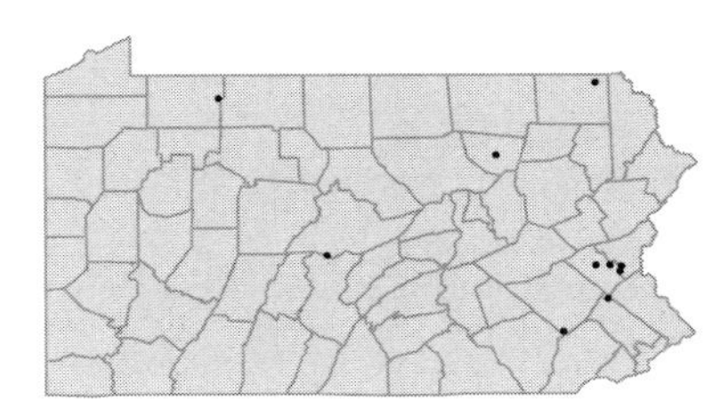

Cuscuta gronovii Willd. ex Schultz var. ***gronovii***
Common dodder
Herbaceous annual vine
Parasitic on a wide range of woody and herbaceous plants of low, wet habitats.

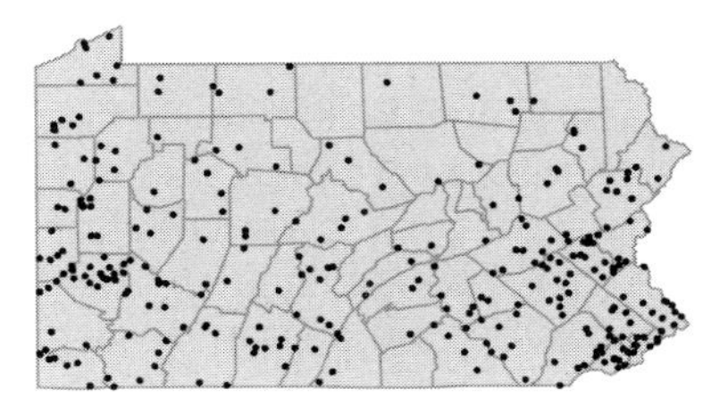

Cuscuta gronovii Willd. ex Schultz var. ***latiflora*** Engelm.
Dodder
Herbaceous annual vine
Moist thickets, river banks, roadsides and wet places, parasitic on many hosts.

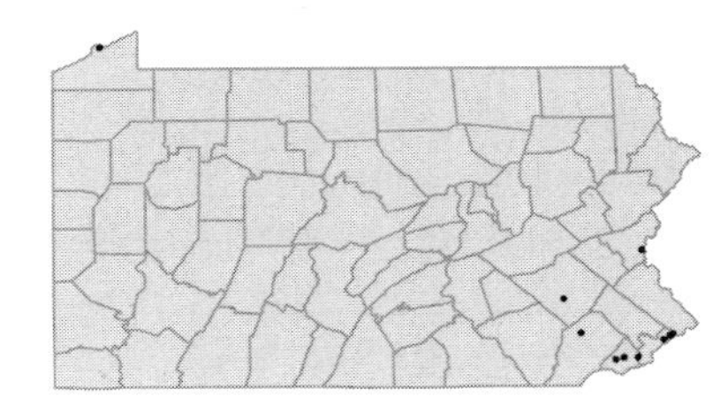

Cuscuta pentagona Engelm.
Field dodder
Herbaceous annual vine
Old fields, thickets and wet banks, parasitic on many hosts.
Cuscuta arvensis Beyrich P

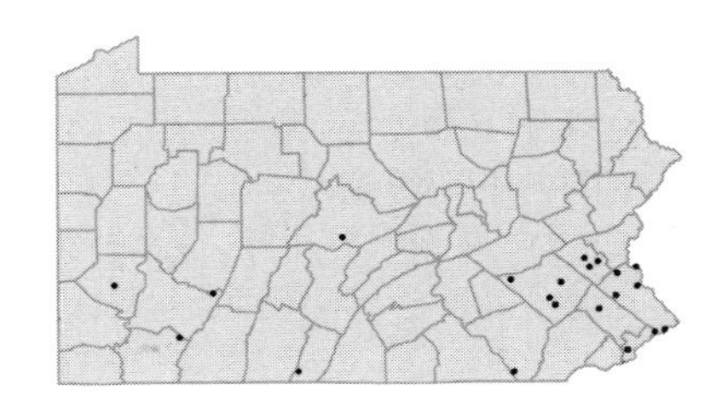

Cuscuta polygonorum Engelm.
Smartweed dodder
Herbaceous annual vine
Parasitic on *Polygonum* spp. and other plants of moist shores and river banks.

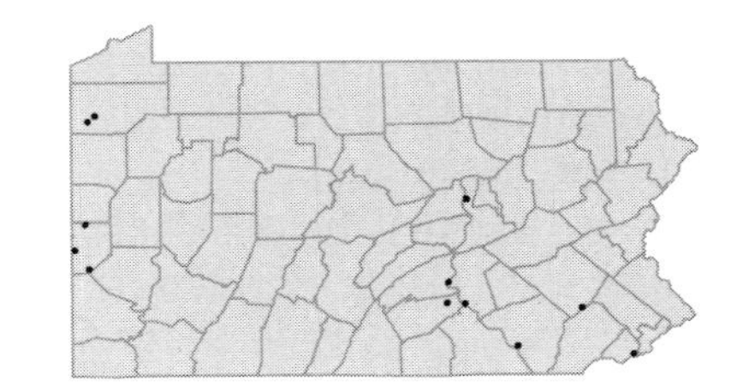

MENYANTHACEAE

• ***Menyanthes trifoliata*** L.
Bogbean; Buckbean
Herbaceous perennial, floating-leaf aquatic
Bogs, sphagnous swamps and pond margins.

OBL

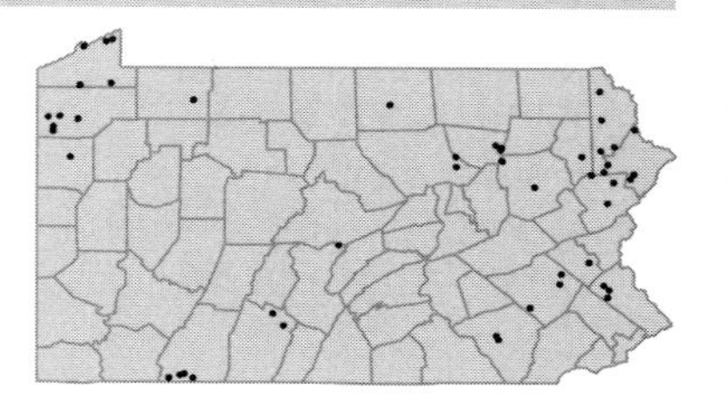

• ***Nymphoides cordata*** (Ell.) Fern.
Floating-heart
Herbaceous perennial, floating-leaf aquatic
Lakes and ponds.
Limnanthemum lacunosum (Vent.) Griseb. P

OBL

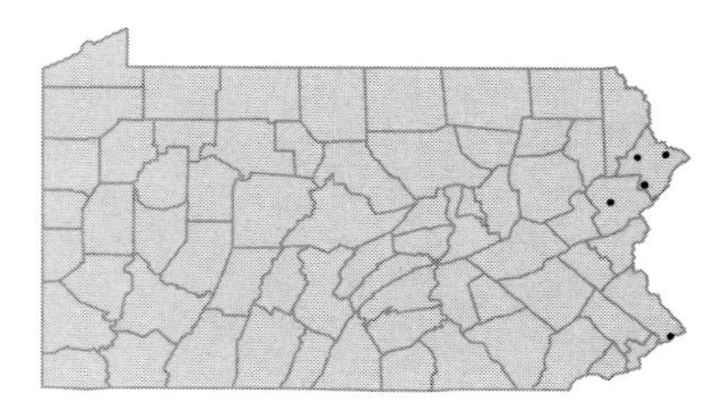

Nymphoides peltata (Gmel.) Kuntze
Water-fringe; Yellow floating-heart
Herbaceous perennial, floating-leaf aquatic
Abandoned quarry, small pond.

OBL
I

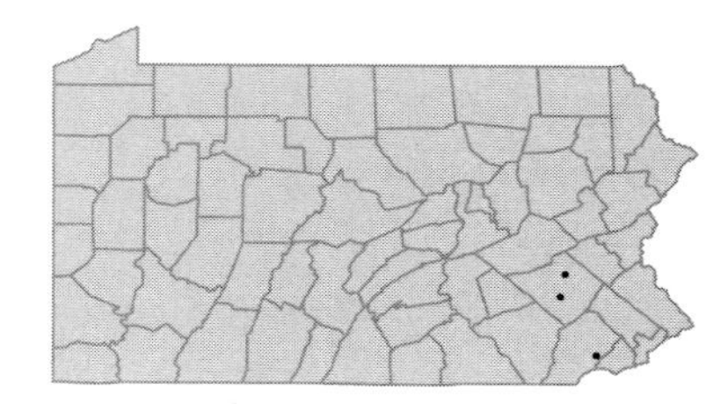

POLEMONIACEAE

• ***Collomia linearis*** Nutt.
Collomia
Herbaceous annual
Railroad tracks and waste places.

I

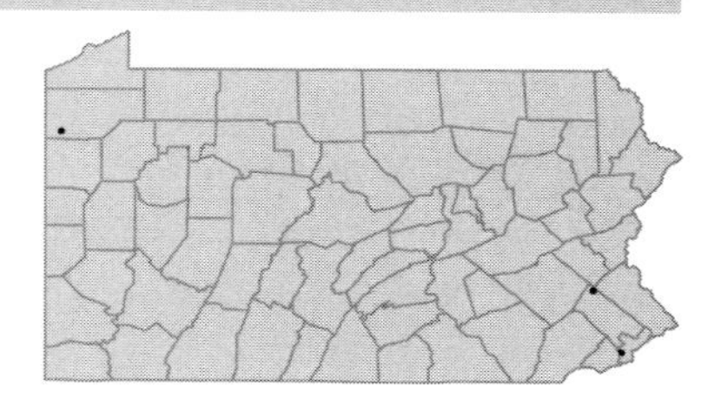

• ***Linanthus androsaceus*** (Benth.) Greene
Trumpet gilia
Herbaceous annual
Roadside. Represented by a single collection from Lehigh Co. in 1955.

I

• ***Navarretia squarrosa*** (Eschsch.) Hook. & Arn.
Navarretia
Herbaceous annual
Lawn. Represented by a single collection from Philadelphia Co. in 1955.

I

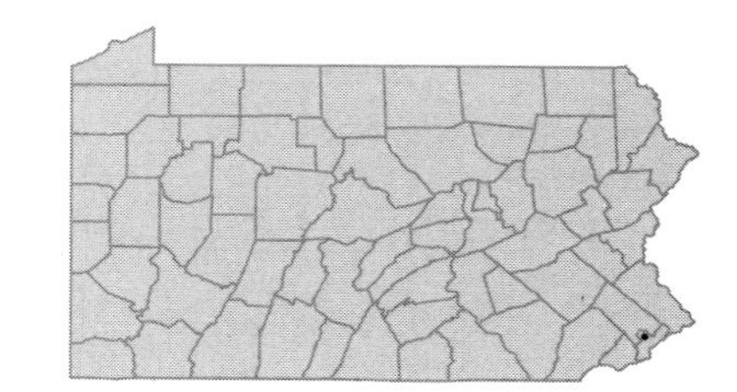

• ***Phlox divaricata*** L. ssp. ***divaricata*** FACU
Wild blue phlox; Wild sweet-william
Herbaceous perennial
Rich, deciduous woods.

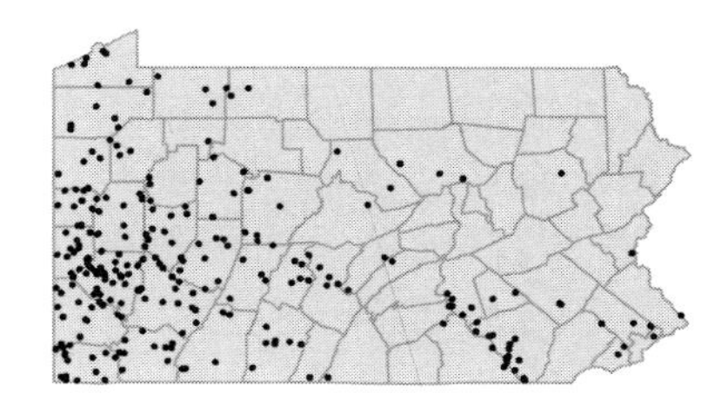

Phlox divaricata L. ssp. ***laphamii*** (A.Wood) Wherry [I]
Blue phlox
Herbaceous perennial
Cultivated and rarely escaped to wooded slopes and stream banks, this plant is a midwestern ssp. distinguished from the eastern ssp. *divaricata* by the un-notched petals.

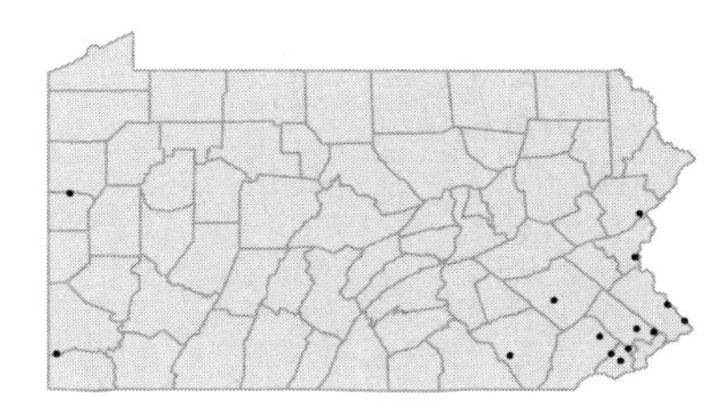

Phlox maculata L. ssp. ***maculata*** FACW
Wild sweet-william
Herbaceous perennial
Wet meadows, abandoned fields and open thickets.

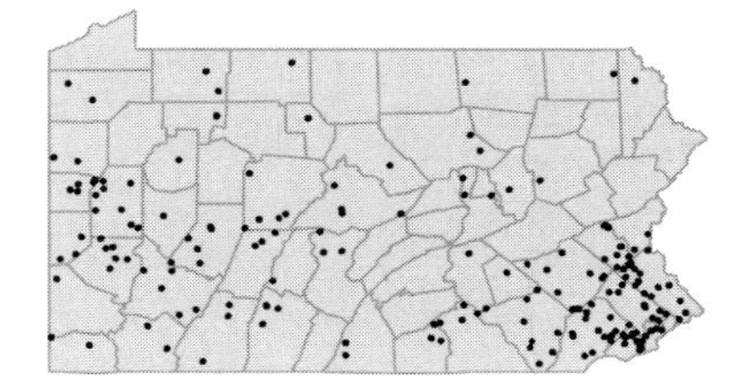

Phlox maculata L. ssp. ***pyramidalis*** (Small) Wherry FACW
Meadow phlox; Spotted phlox
Herbaceous perennial
Moist meadows and thickets, stream banks, swamps and ditches.
Phlox maculata L. var. *purpurea* Michx. F

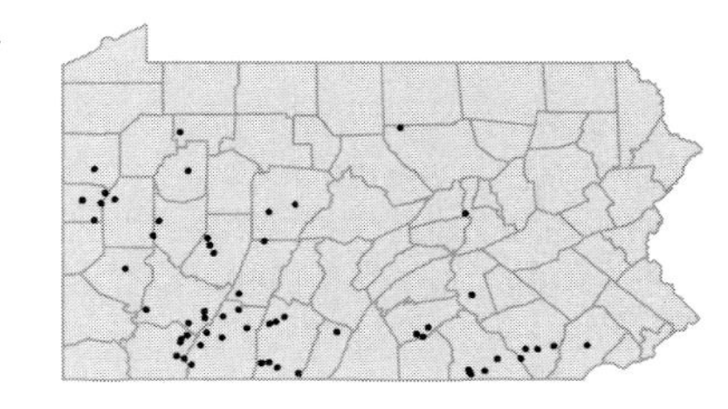

Phlox ovata L.
Mountain phlox
Herbaceous perennial
Openings and edges in dry, sandy woods.

Phlox paniculata L. FACU
Summer phlox
Herbaceous perennial
Thickets, hillsides and stream banks, often in calcareous soils, also cultivated and frequently escaped.

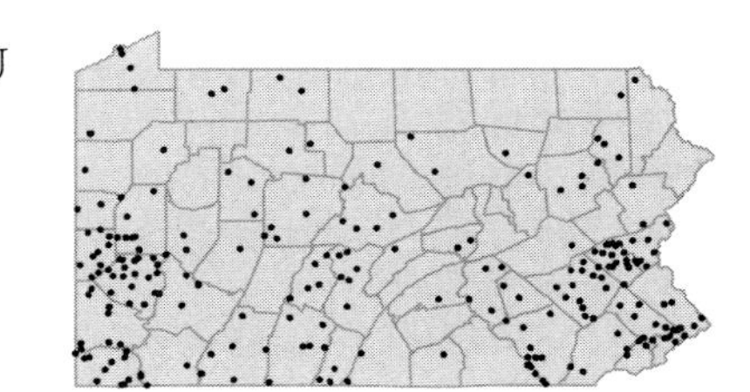

Phlox pilosa L.
Downy phlox
Herbaceous perennial
Open woods, moist slopes and wet meadows.

FACU

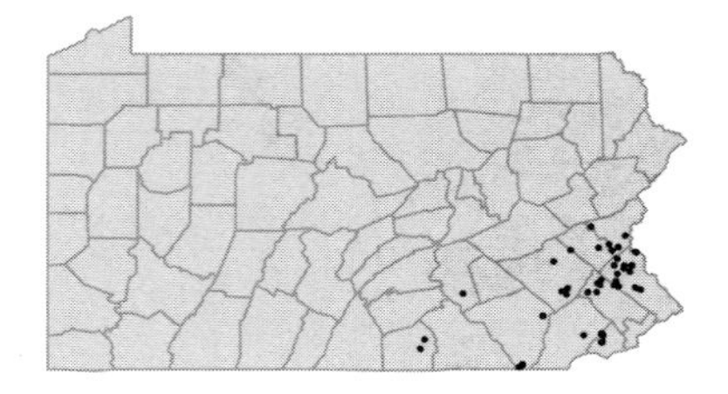

Phlox stolonifera Sims
Creeping phlox
Herbaceous perennial
Rich woods.
Phlox reptans Michx. P

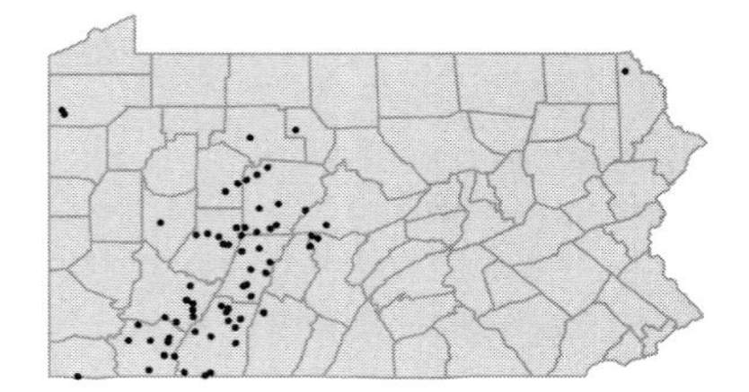

Phlox subulata L. ssp. **subulata**
Moss-pink
Herbaceous perennial
Dry slopes, rocky ledges and serpentine barrens.

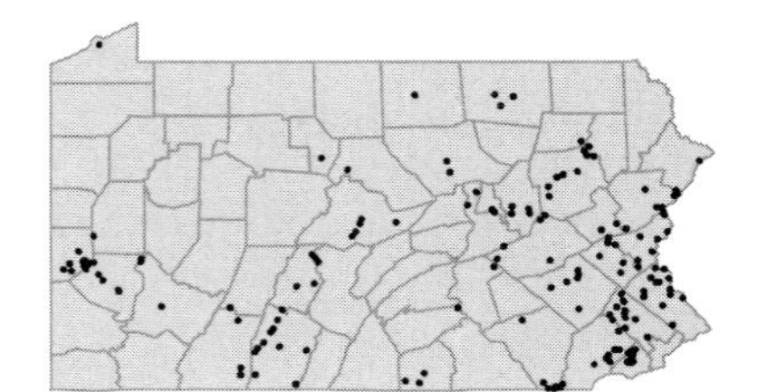

Phlox subulata L. ssp. **brittonii** (Small) Wherry
Moss-pink
Herbaceous perennial
Shale barrens.
Phlox subulata L. var. *setacea* (L.) A.Brand C

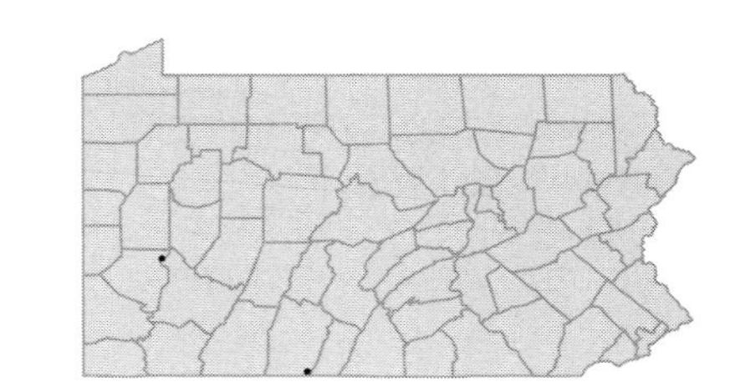

• **Polemonium caeruleum** L.
Jacob's-ladder
Herbaceous perennial
Cultivated and rarely escaped.

I

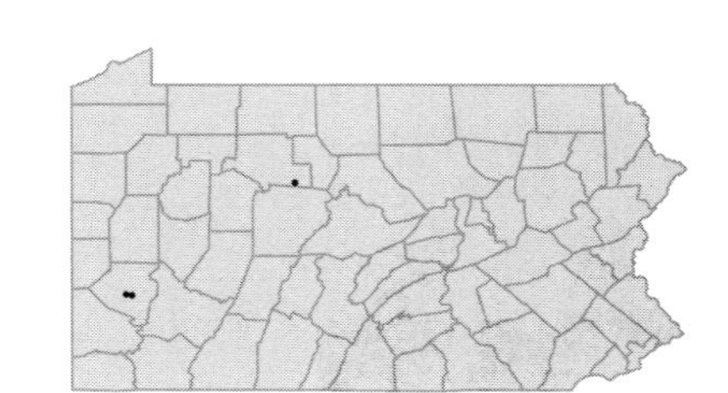

Polemonium micranthum Benth.
Polemony
Herbaceous perennial
Near shrubbery in yard. Represented by a single collection from Centre Co. in 1961.
Polemoniella micrantha (Benth.) Heller W

I

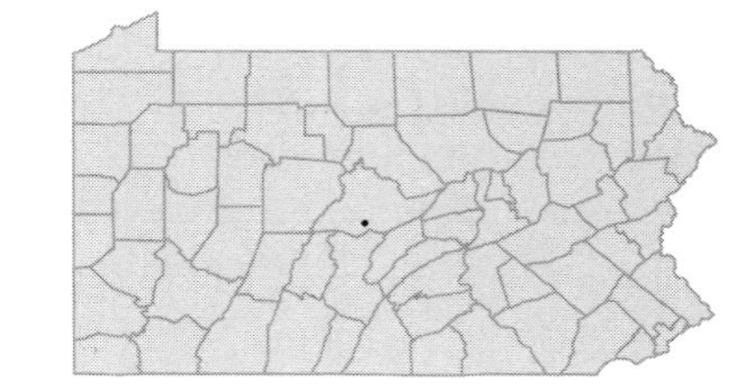

Polemonium reptans L.
Spreading Jacob's-ladder
Herbaceous perennial
Low, moist woods and floodplains.

FACU

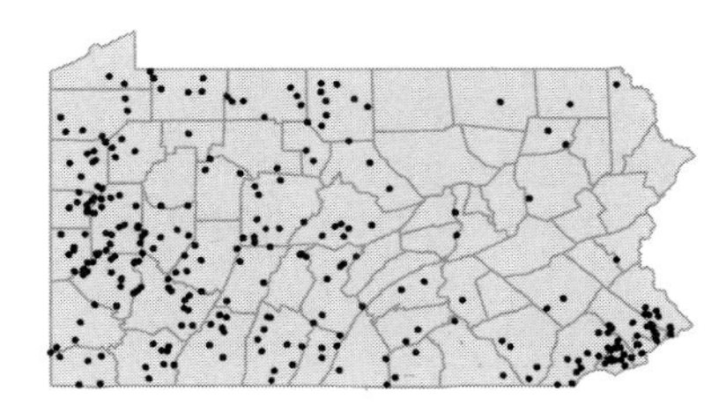

Polemonium van-bruntiae Britt.
Jacob's-ladder
Herbaceous perennial
Moist, sphagnous glades, swamps and marshes.

FACW

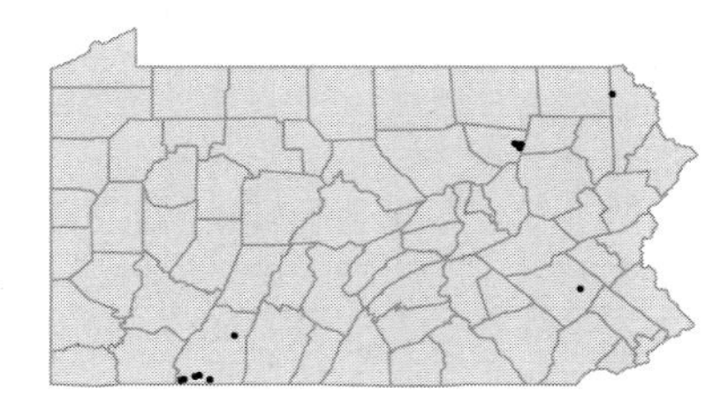

HYDROPHYLLACEAE

• ***Ellisia nyctelea*** L.
Aunt Lucy; Water-pod
Herbaceous annual
Damp, shady banks and rich, alluvial woods.
Macrocalyx nyctelea (L.) Kuntze P

FACU

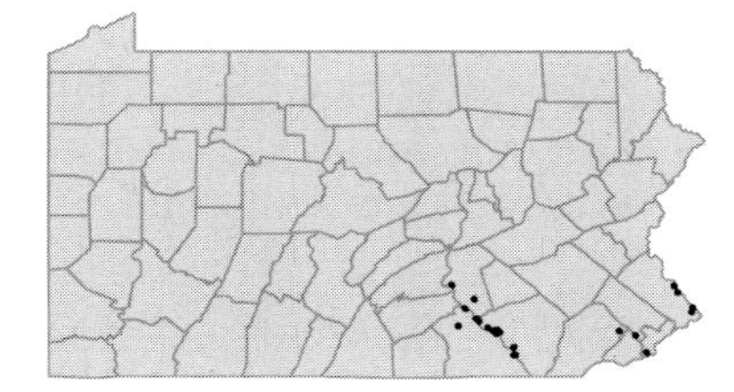

• ***Hydrophyllum appendiculatum*** Michx.
Waterleaf
Herbaceous biennial
Wooded slopes, thickets and stream banks.

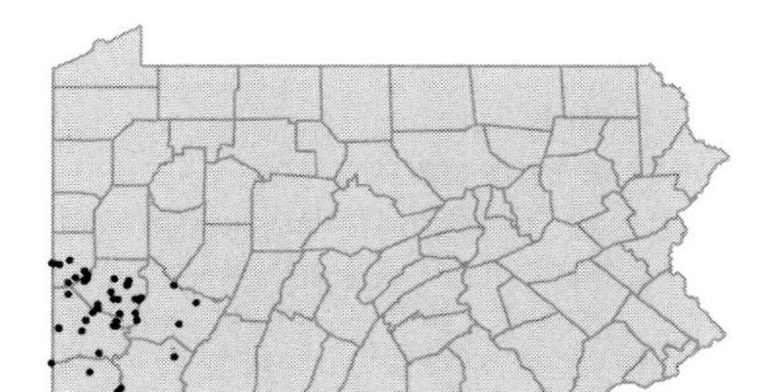

Hydrophyllum canadense L.
Canadian waterleaf
Herbaceous perennial
Rocky, wooded slopes, ravines and moist woods.

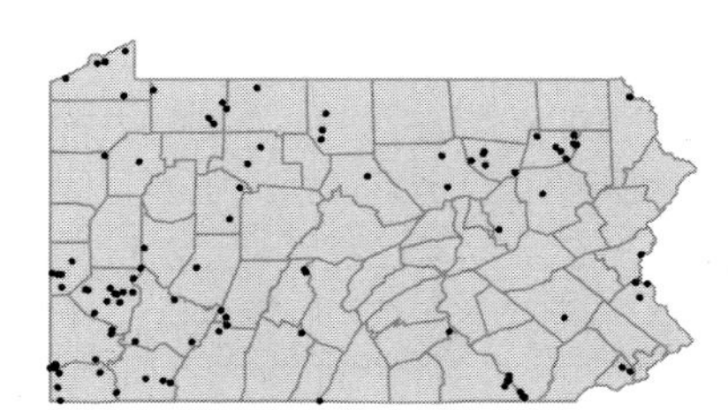

Hydrophyllum macrophyllum Nutt.
Large-leaved waterleaf
Herbaceous perennial
Mesic, calcareous woods.

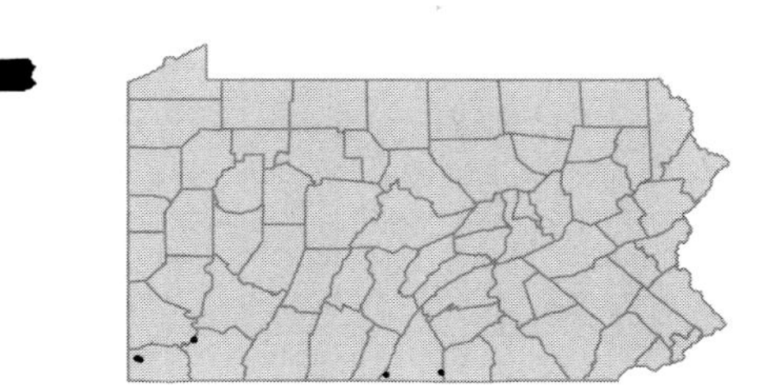

Hydrophyllum virginianum L. FAC
Virginia waterleaf
Herbaceous perennial
Moist woods, thickets and stream banks.

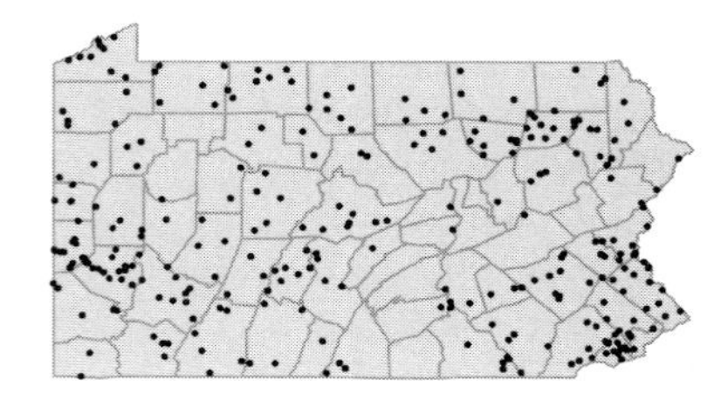

• *Phacelia bipinnatifida* Michx. [I]
Fern-leaf phacelia
Herbaceous biennial
Thicket. Represented by a single collection from Berks Co. in 1945.

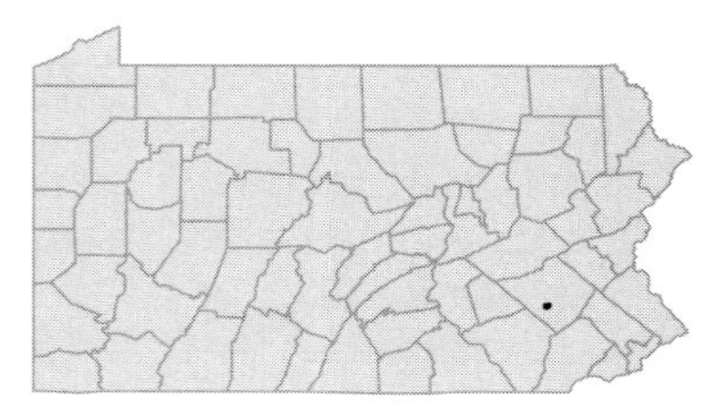

Phacelia dubia (L.) Trel.
Scorpion-weed; Small-flowered phacelia
Herbaceous annual
Shale barrens and shaly cliffs.

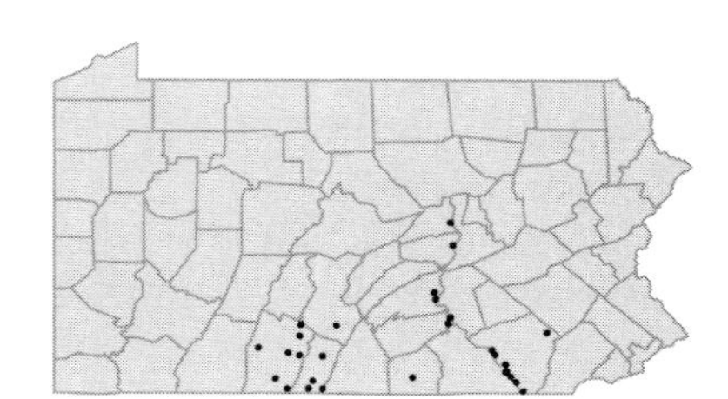

Phacelia hirsuta Nutt. [I]
Hairy phacelia
Herbaceous annual
Railroad tracks. Represented by a single collection from Bucks Co. in 1901.

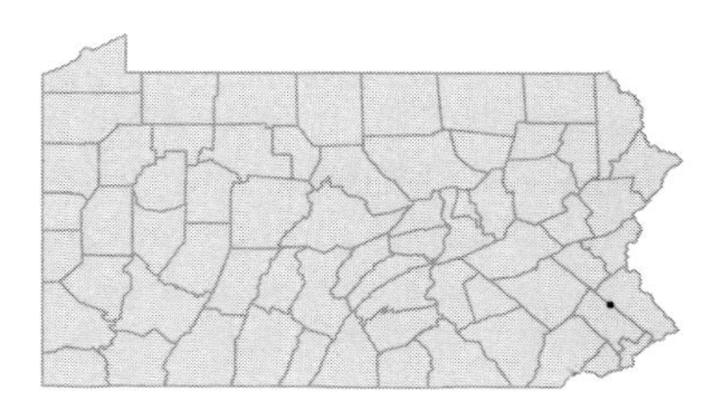

Phacelia purshii Buckl.
Miami-mist
Herbaceous annual
Open woods, along creeks and railroad cuts.

BORAGINACEAE

• *Amsinckia intermedia* Fisch. & Mey. [I]
Tarweed; Fiddle-neck
Herbaceous annual
Ballast. Represented by a single collection from Philadelphia Co. in 1877.

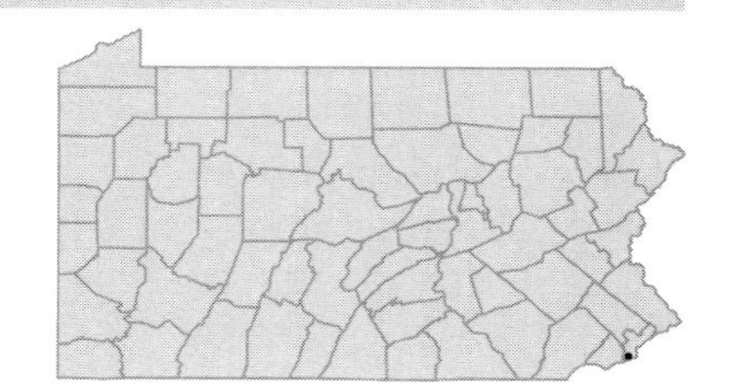

Amsinckia lycopsoides (Lehm.) Lehm.
Bearded tarweed; Fiddle-neck
Herbaceous annual
Ballast. Represented by a single collection from Philadelphia Co. in 1877.
Amsinckia barbata Greene FBW

I

• ***Anchusa arvensis*** (L.) Bieb.
Small bugloss
Herbaceous annual
Ballast and waste ground. Represented by two collections from Philadelphia Co. in 1878.
Lycopsis arvensis L. PFGBW

I

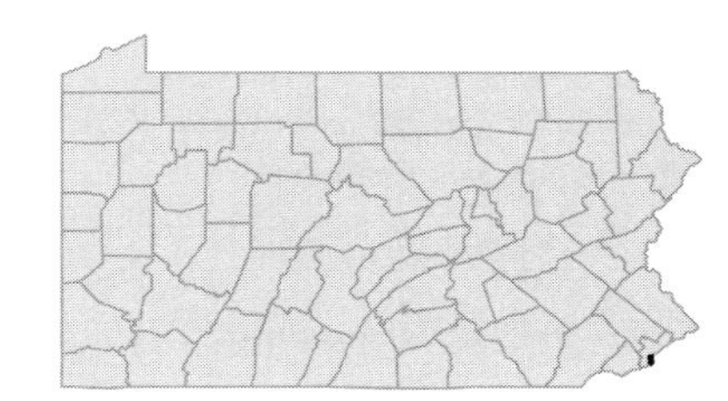

Anchusa azurea P.Mill.
Showy bugloss; Alkanet
Herbaceous biennial
Cultivated and occasionally escaped to waste ground.

I

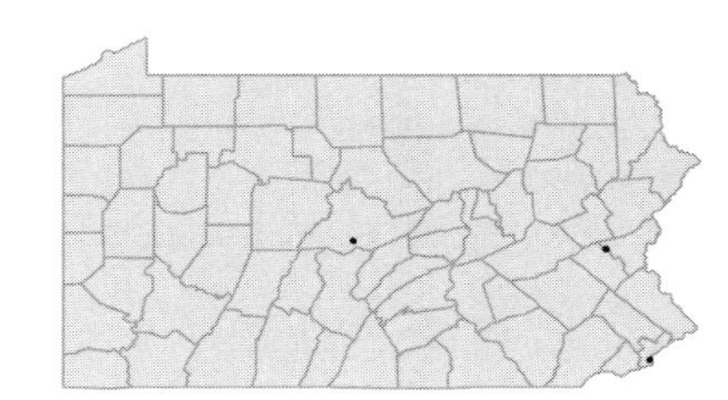

Anchusa officinalis L.
Common bugloss; Alkanet
Herbaceous perennial
Ballast and ore piles.

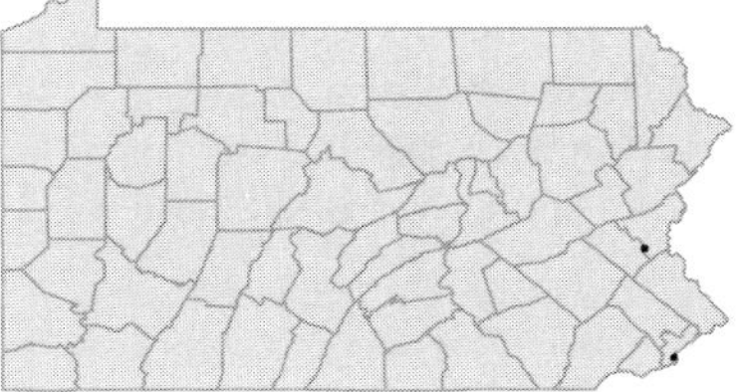

• ***Asperugo procumbens*** L.
Madwort
Herbaceous annual
Ballast and roadside ditches. Represented by three collections from Philadelphia and Chester Cos. 1877-1899.

I

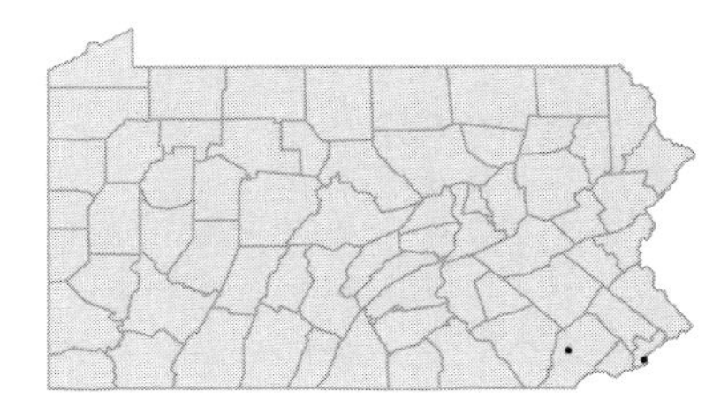

• ***Borago officinalis*** L.
Borage
Herbaceous annual
Cultivated and rarely escaped.

I

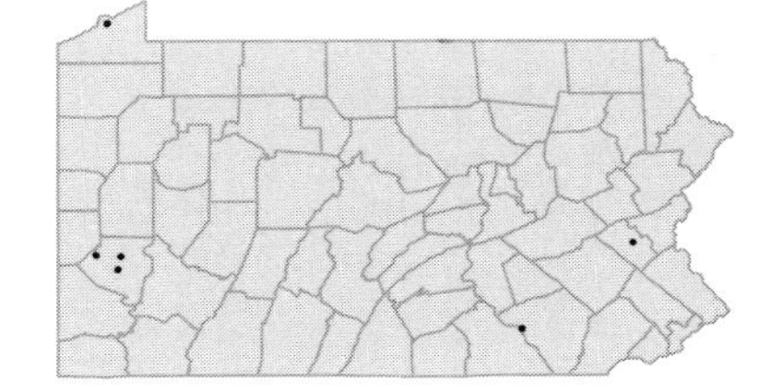

• ***Buglossoides arvense*** (L.) I.M.Johnston
Bastard alkanet; Corn gromwell
Herbaceous annual
Dry pastures, limestone slopes, roadsides and railroad beds.
Lithospermum arvense L. PFBWC

I

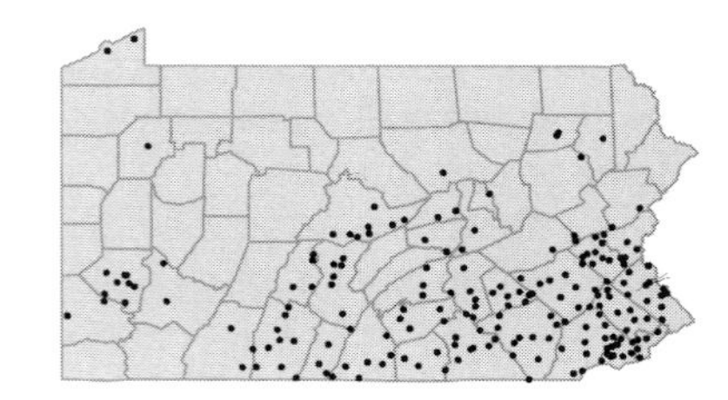

• ***Cynoglossum boreale*** Fern.
Northern hound's-tongue; Wild comfrey
Herbaceous perennial
Rich, open woods and roadsides. Believed to be extirpated, last collected in 1946.
Cynoglossum virginianum L. var. *boreale* (Fern.) Cooperrider C

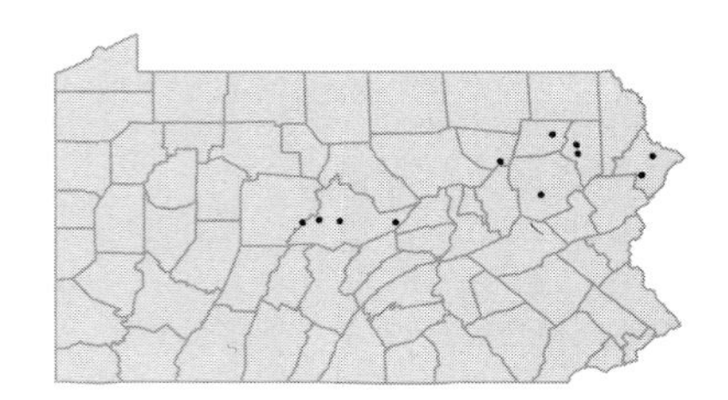

Cynoglossum officinale L.
Hound's-tongue
Herbaceous biennial
Limestone banks, pastures, railroad beds and waste ground.

I

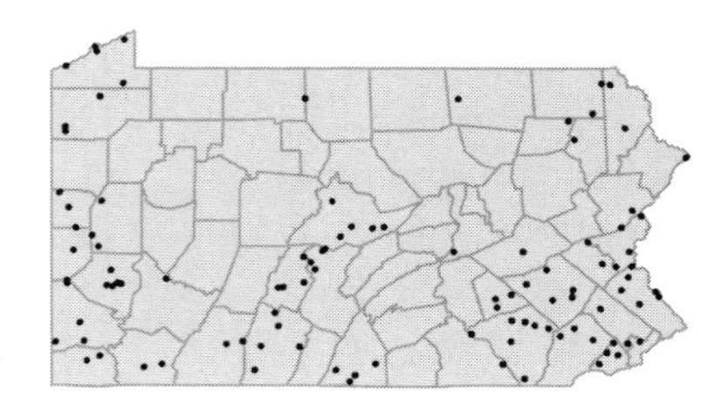

Cynoglossum virginianum L.
Wild comfrey
Herbaceous perennial
Rich, open woods and wooded slopes.

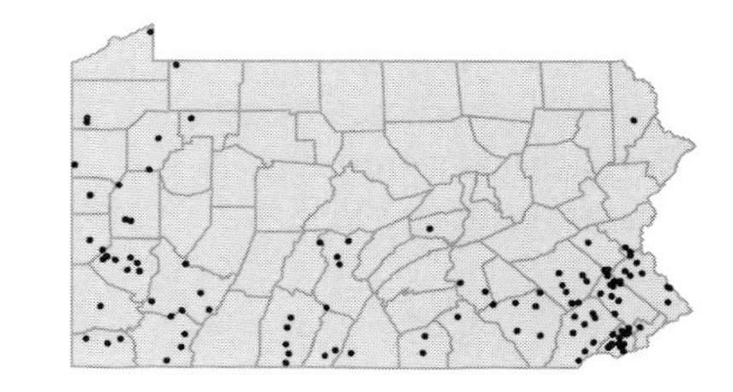

• ***Echium plantagineum*** L.
Viper's bugloss
Herbaceous biennial
Ballast and waste ground. Represented by a single collection from Philadelphia Co. in 1877.

I

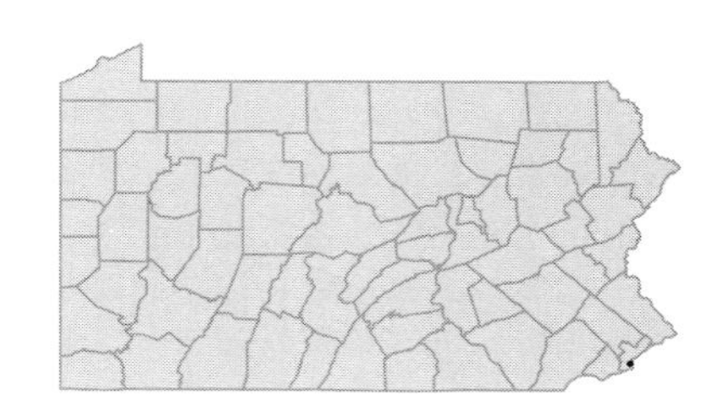

Echium vulgare L.
Viper's bugloss; Blueweed
Herbaceous biennial
Dry fields, roadsides, railroad banks and waste ground.

I

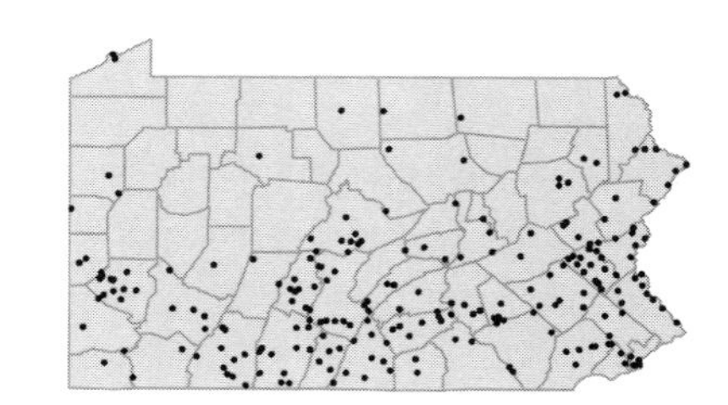

Hackelia virginiana (L.) I.M.Johnston FACU
Beggar's-lice; Stickseed
Herbaceous biennial
Dry to moist woods, wooded slopes and roadsides.
Lappula virginiana (L.) Greene P

• ***Heliotropium amplexicaule*** Vahl [I]
Wild heliotrope
Herbaceous perennial
Sandy, open ground. Represented by a single collection from Erie Co. in 1879.

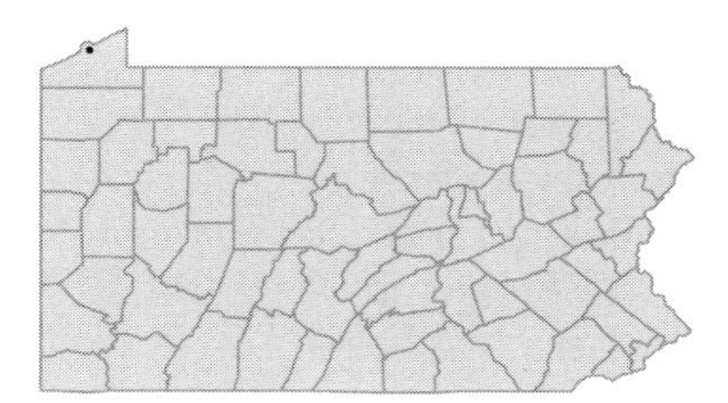

Heliotropium curassavicum L. OBL [I]
Seaside heliotrope
Herbaceous perennial
Ballast and waste ground. Represented by three collections from Philadelphia Co. 1865-1879.

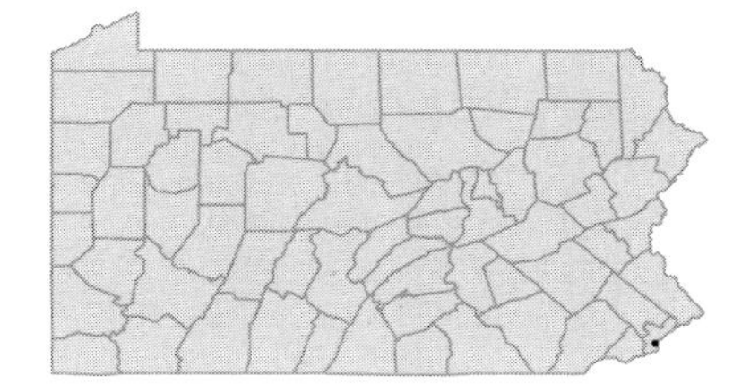

Heliotropium europaeum L. [I]
European heliotrope
Herbaceous annual
Waste ground, cinders and ballast.

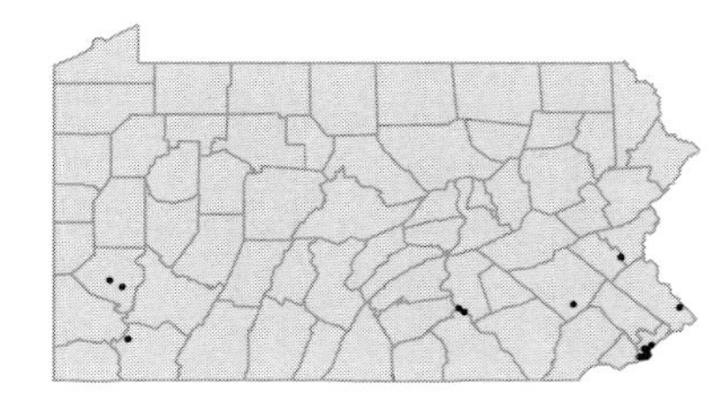

Heliotropium supinum L. [I]
Heliotrope
Herbaceous annual
Ballast. Represented by a single collection from Philadelphia Co. in 1879.

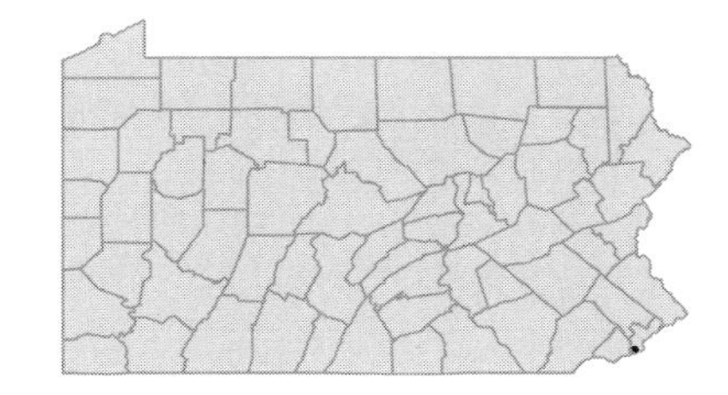

Heliotropium undulatum Vahl [I]
Heliotrope
Herbaceous annual
Ballast. Represented by three collections from Philadelphia Co. ca. 1879.

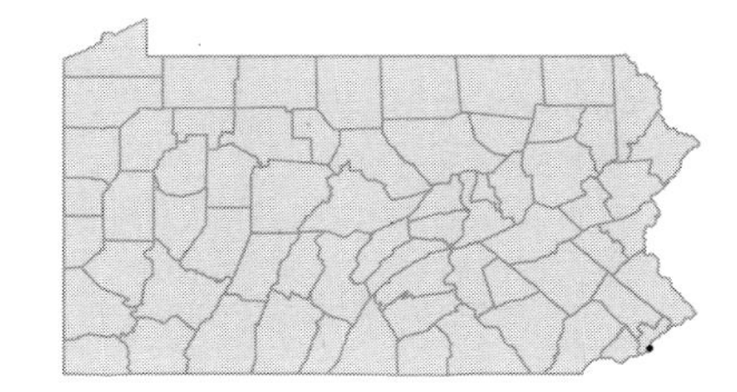

• ***Lappula marginata*** (Bieb.) Guerke
Stickseed
Herbaceous annual
Ballast. Represented by a single collection from Philadelphia Co. in 1878.
Lappula patula (L.) Ait. W

I

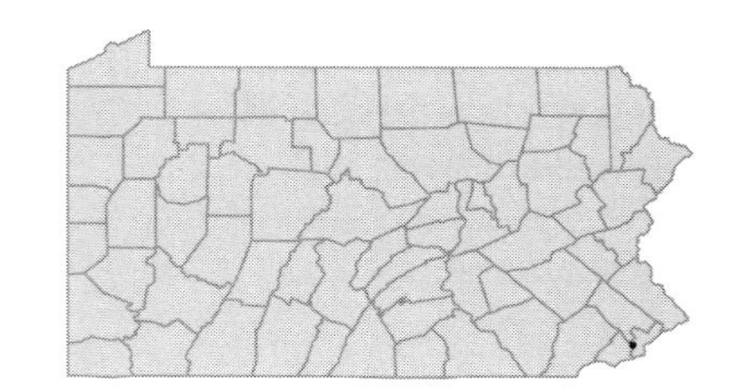

Lappula redowskii (Hornem.) Greene
Stickseed
Herbaceous annual
River bank and waste ground. Represented by two collections from Philadelphia Co. 1871-1877.

I

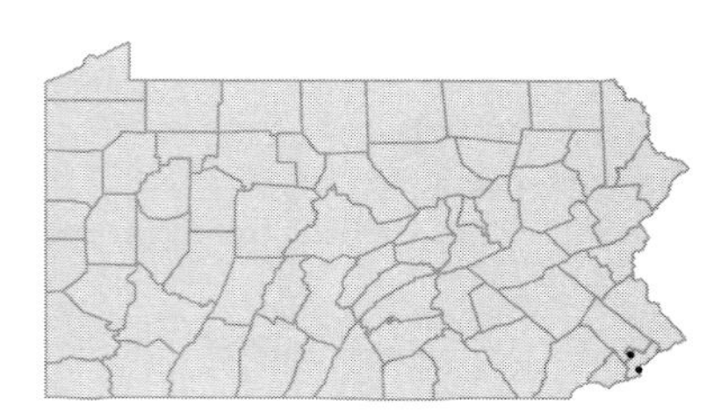

Lappula squarrosa (Retz.) Dumort.
Beggar's-lice; Stickseed
Herbaceous annual
Railroad ballast and cinders.
Lappula lappula (L.) Karst. P; *Lappula echinata* Gilib. FGBWK

I

• ***Lithospermum canescens*** (Michx.) Lehm.
Hoary puccoon; Indian-paint
Herbaceous perennial
River bluffs, dry rocky hillsides and barrens.

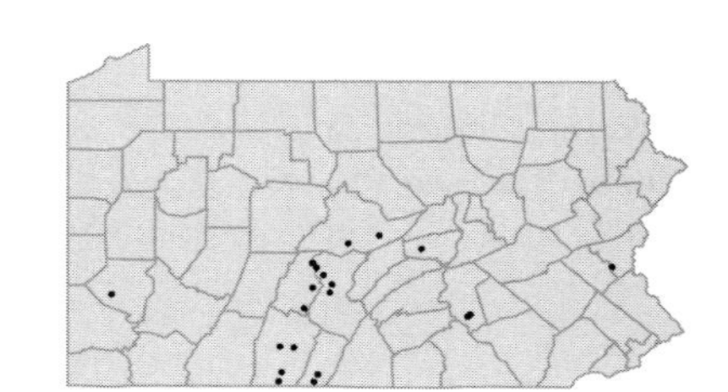

Lithospermum caroliniense (J.F.Gmel.) MacM.
Golden puccoon; Hispid gromwell
Herbaceous perennial
Open, sandy barrens and roadsides.
Lithospermum croceum Fern. FBW

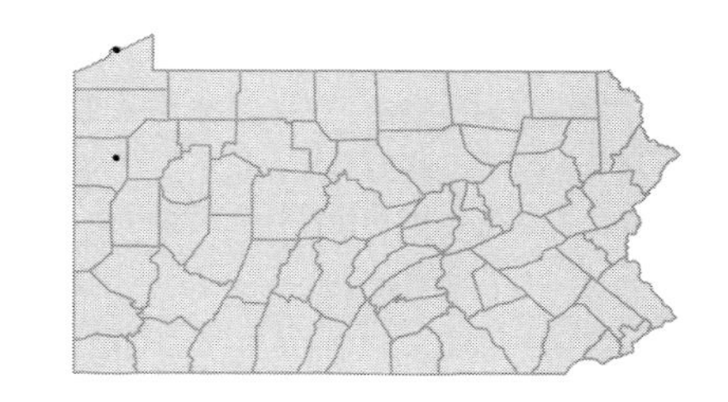

Lithospermum latifolium Michx.
American gromwell
Herbaceous perennial
Rich, wooded, limestone slopes and hilltop woods.

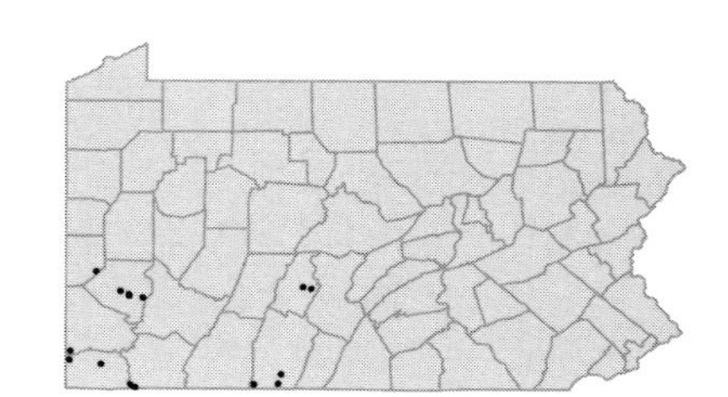

Lithospermum officinale L.
European gromwell
Herbaceous perennial
Field. Represented by a single collection from Lackawanna Co. in 1900.

[I]

• ***Mertensia virginica*** (L.) Pers. ex Link
Virginia bluebells; Virginia cowslip
Herbaceous perennial
Rich, wooded hillsides and bottomlands.

FACW

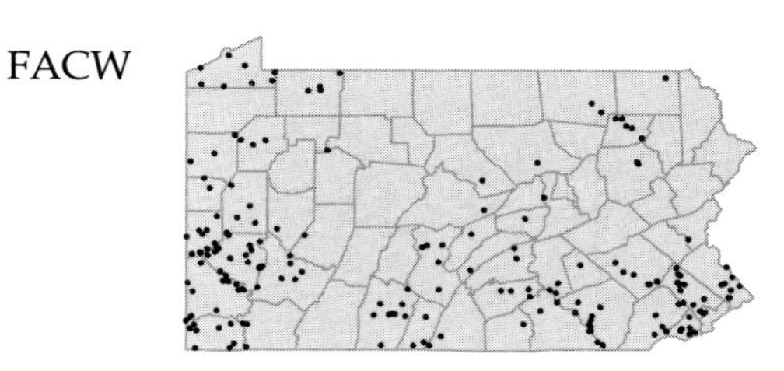

• ***Myosotis arvensis*** (L.) Hill
Forget-me-not; Field scorpion-grass
Herbaceous annual
Cultivated and occasionally escaped to roadsides or waste ground.

UPL
[I]

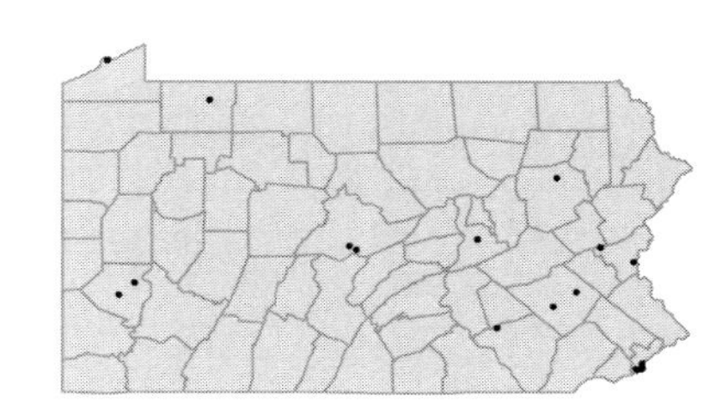

Myosotis discolor Pers.
Forget-me-not; Yellow and blue scorpion-grass
Herbaceous annual
Moist rocks.
Myosotis versicolor (Pers.) Smith PFW

UPL
[I]

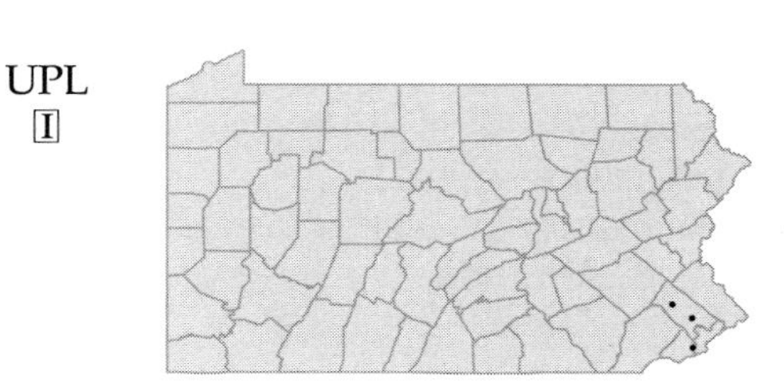

Myosotis laxa Lehm.
Wild forget-me-not
Herbaceous perennial
Wet, open ground and swamps.

OBL

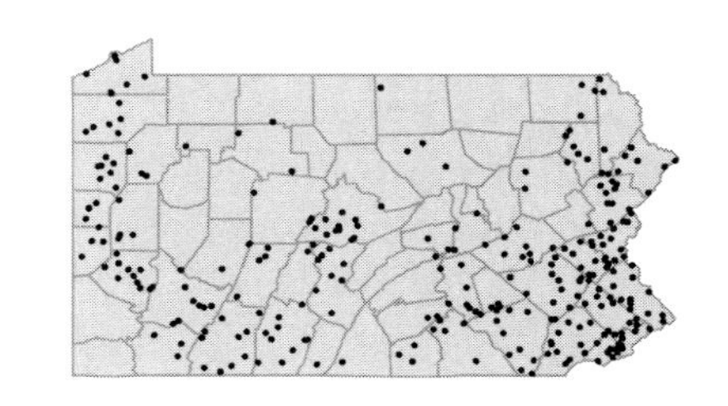

Myosotis scorpioides L.
Forget-me-not; Water scorpion-grass
Herbaceous perennial
Cultivated and frequently escaped to stream banks, floodplains and wet ditches.
Myosotis palustris (L.) Lam. P

OBL
[I]

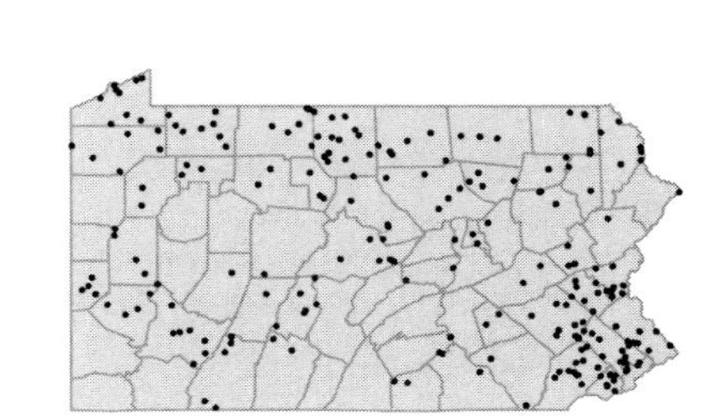

Myosotis stricta Link ex Roemer & Schultes
Forget-me-not
Herbaceous annual
Pastures, dry banks and lawns.
Myosotis micrantha Pall. GBC

[I]

Myosotis sylvatica Hoffm.
Garden forget-me-not
Herbaceous annual
Cultivated and occasionally naturalized.

UPL
[I]

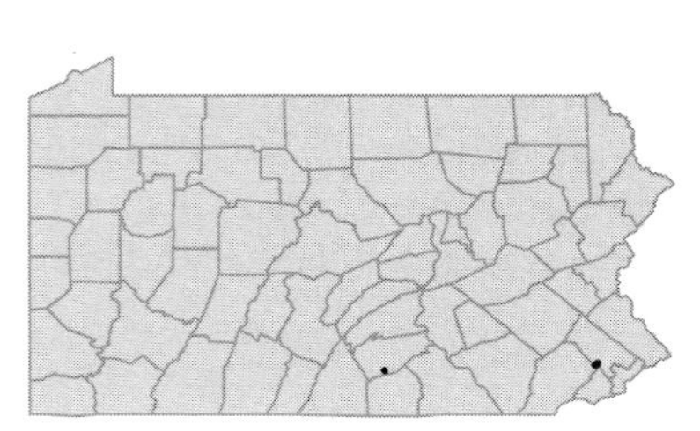

Myosotis verna Nutt.
Spring forget-me-not; Early scorpion-grass
Herbaceous annual
Dry, open woods, rocky ledges and roadside banks.
Myosotis virginica (L.) BSP PB

FAC-

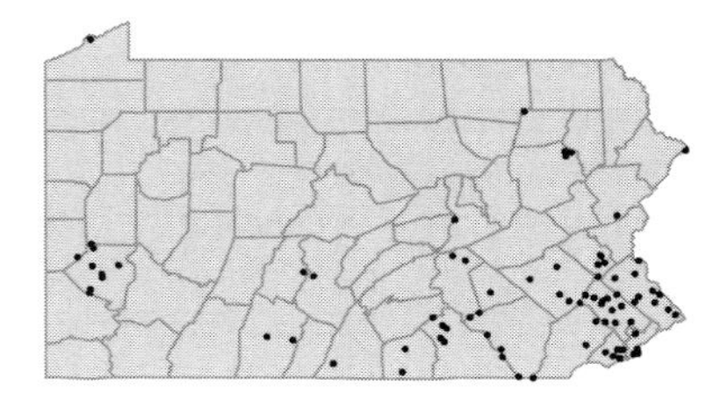

• ***Nonea rosea*** (Bieb.) Link
Nonea
Herbaceous annual
Urban waste ground. Represented by a single collection from Philadelphia Co. in 1889.

[I]

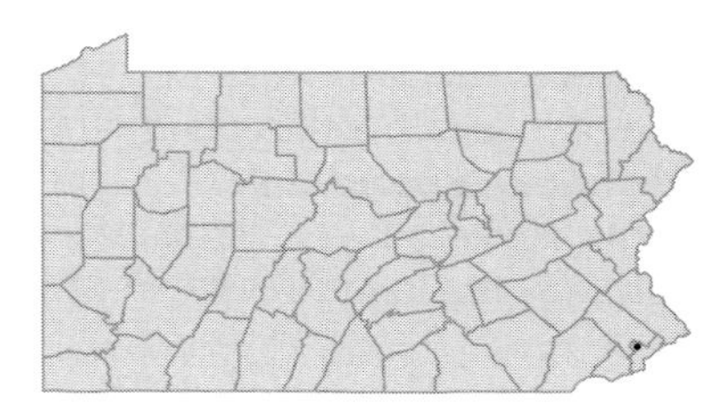

Nonea vesicaria (L.) Reichenb.
Nonea
Herbaceous annual
Ballast and waste ground along railroad tracks.

[I]

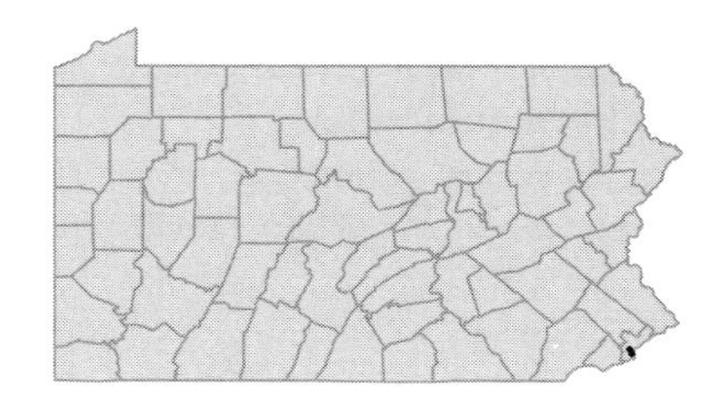

• ***Onosmodium molle*** Michx. ssp. ***hispidissimum*** (Mack.) Boivin
False gromwell; Marble-seed
Herbaceous perennial
Dry, calcareous hillsides and old pastures.
Onosmodium carolinianum (Lam.) DC. P; *Onosmodium hispidissimum* Mackenzie WFK

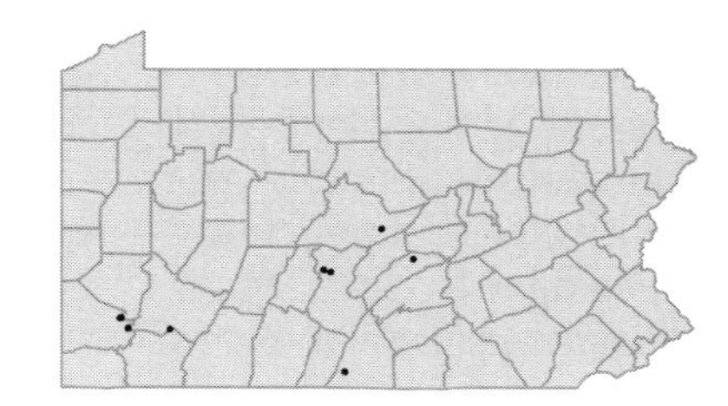

Onosmodium virginianum (L.) A.DC.
Virginia false gromwell
Herbaceous perennial
Sandy banks and limestone bluffs. Believed to be extirpated, last collected in 1908.

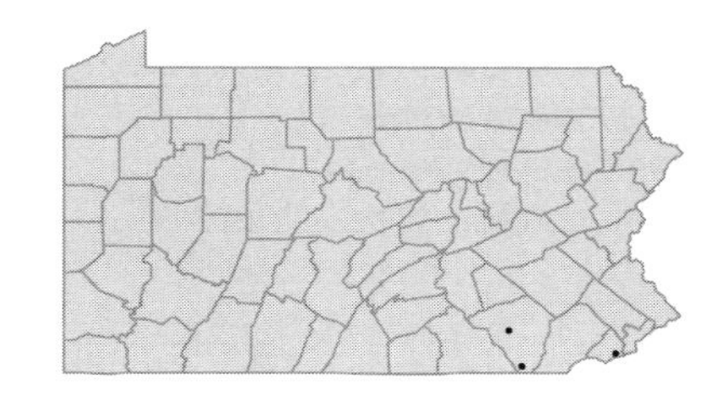

• ***Symphytum officinale*** L. — I
Comfrey
Herbaceous perennial
Cultivated and frequently naturalized in fields, roadsides and waste ground.

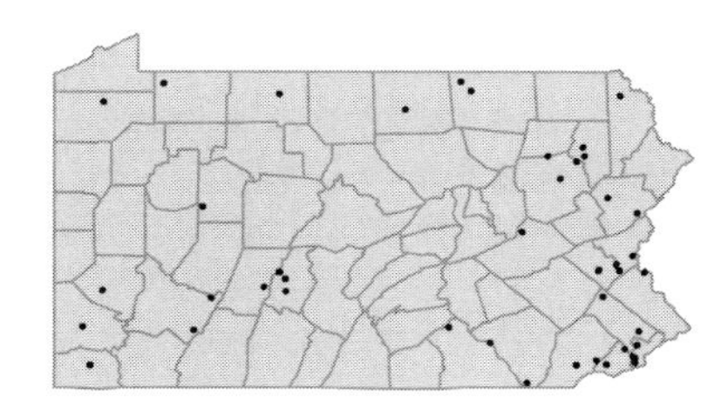

VERBENACEAE

• ***Clerodendrum trichotomum*** Thunb. — I
Glorybower
Deciduous shrub
River bank. Represented by a single collection from Venango Co. in 1964.

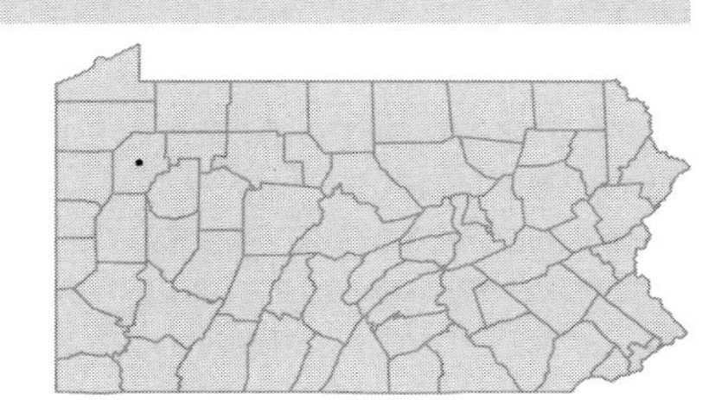

• ***Glandularia canadensis*** (L.) Nutt. — I
Rose verbena
Herbaceous perennial
Ballast, quarry waste, streets and sidewalks.
Verbena canadensis (L.) Britt. FGBWC

• ***Phryma leptostachya*** L.
Lopseed
Herbaceous perennial
Rich woods, rocky limestone slopes and swamps.

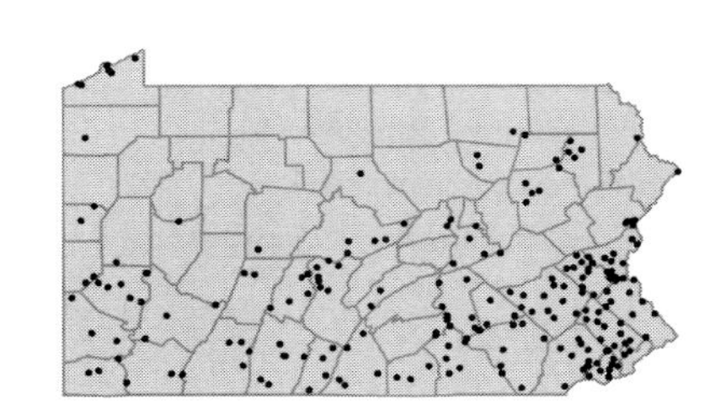

• ***Phyla lanceolata*** (Michx.) Greene — OBL
Fog-fruit
Herbaceous perennial
River and stream banks.
Lippia lanceolata Michx. PFW

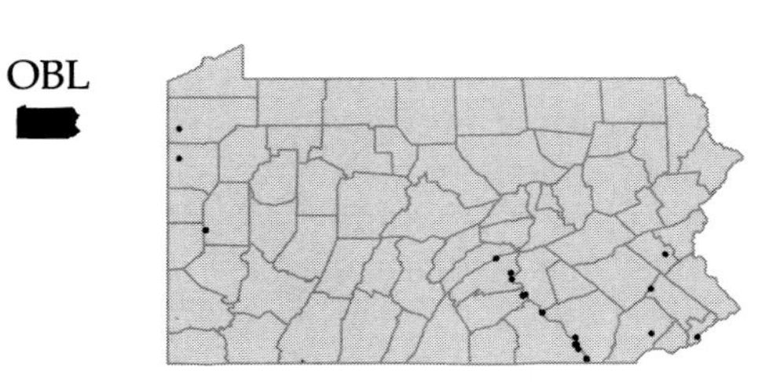

• ***Verbena bracteata*** Lag. & Rodr. I
Prostrate vervain
Herbaceous annual
Ballast, wharves, railroad banks and sidewalks.
Verbena bracteosa Michx. P

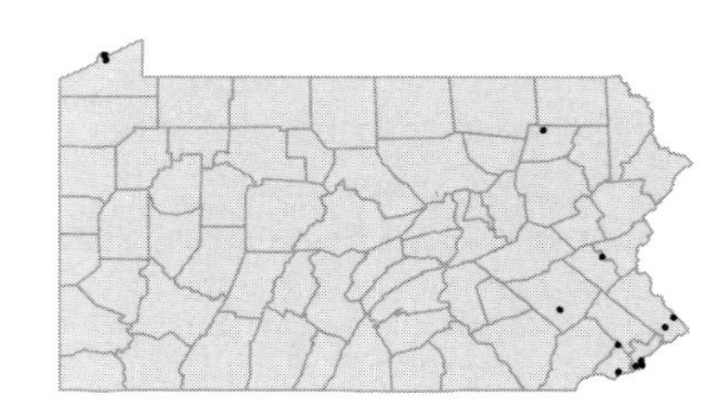

Verbena hastata L. FACW+
Blue vervain; Simpler's-joy
Herbaceous perennial
Moist thickets, floodplains, wet ditches and roadsides.

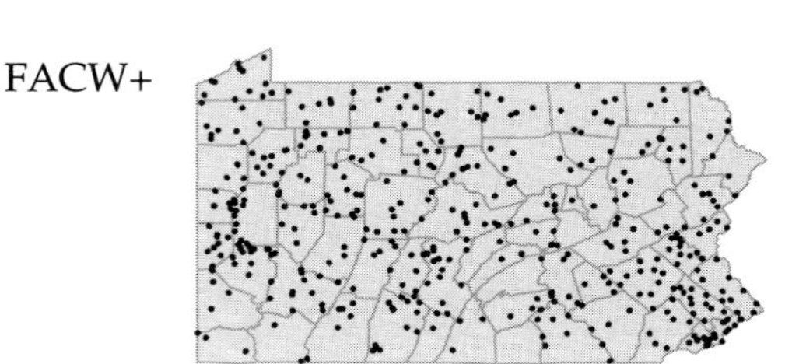

Verbena hastata x urticifolia
Vervain
Herbaceous perennial
Swamps, sandy river banks, wet meadows, fields and rubbish dumps.
Verbena x engelmannii Moldenke C

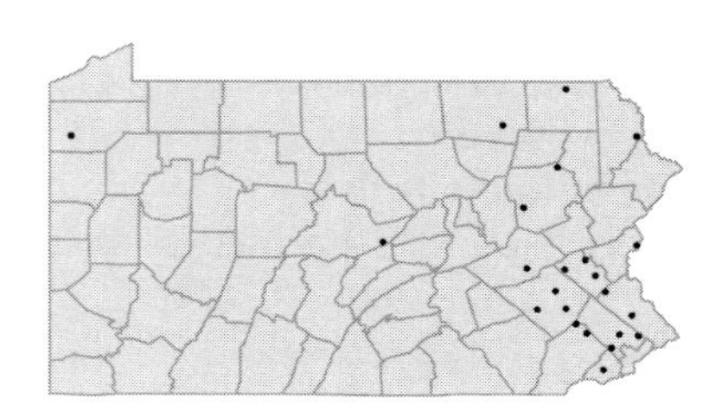

Verbena hybrida Voss ex Rumpl. I
Garden verbena
Herbaceous perennial
Cultivated and occasionally escaped at rubbish dumps.

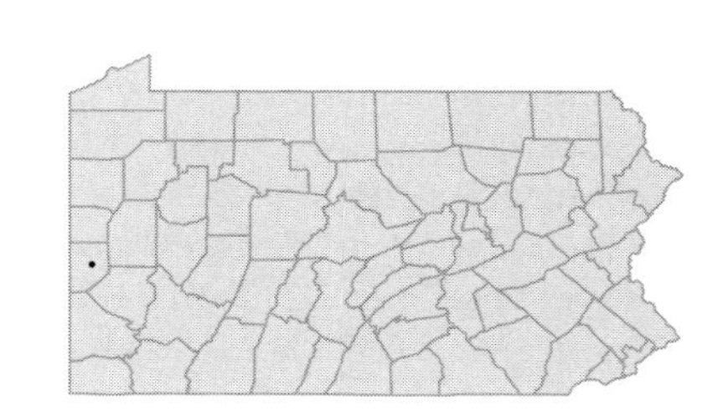

Verbena officinalis L. I
European vervain
Herbaceous annual
Ballast, wharves, railroad tracks and waste ground.

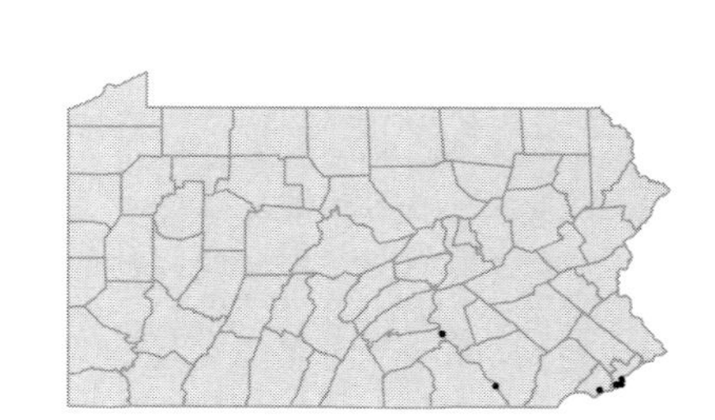

Verbena simplex Lehm.
Narrow-leaved vervain
Herbaceous perennial
Shale barrens, fields, railroad banks and roadside ditches.
Verbena angustifolia Michx. P

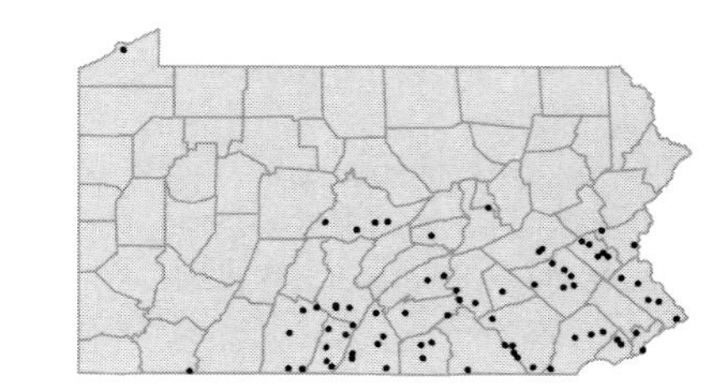

Verbena stricta Vent.
Hoary vervain
Herbaceous perennial
Old fields and abandoned lots.

I

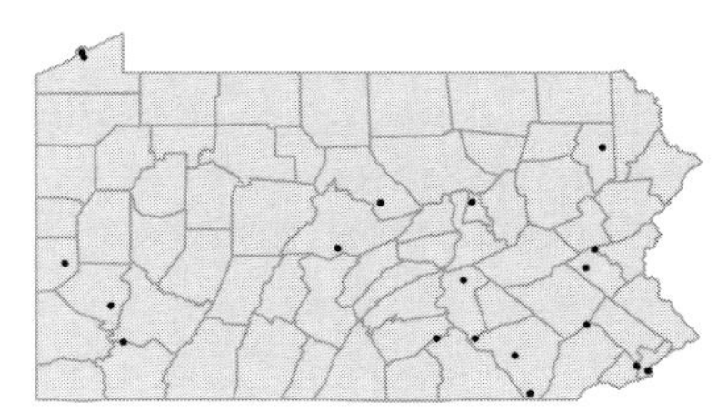

Verbena urticifolia L. var. ***urticifolia***
White vervain
Herbaceous annual
Moist fields, meadows and waste ground.

FACU

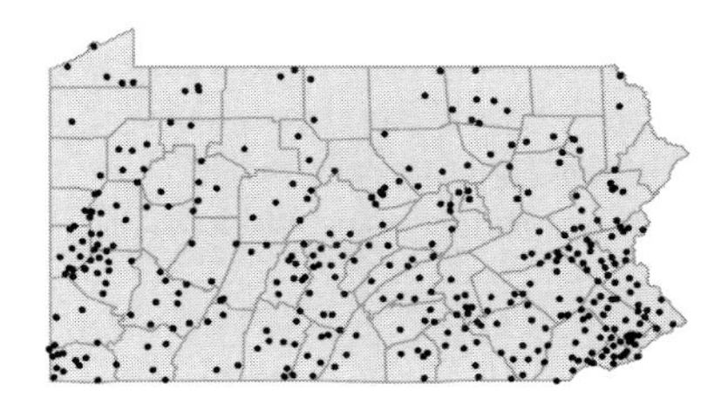

Verbena urticifolia L. var. ***leiocarpa*** Perry & Fern.
White vervain
Herbaceous annual
Moist woods.

FACU

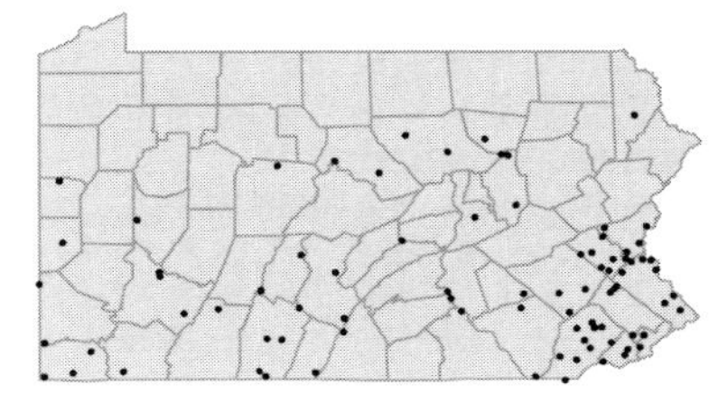

• ***Vitex agnus-castus*** L.
Chaste-tree
Deciduous shrub
Cultivated and rarely escaped to fencerows.

I

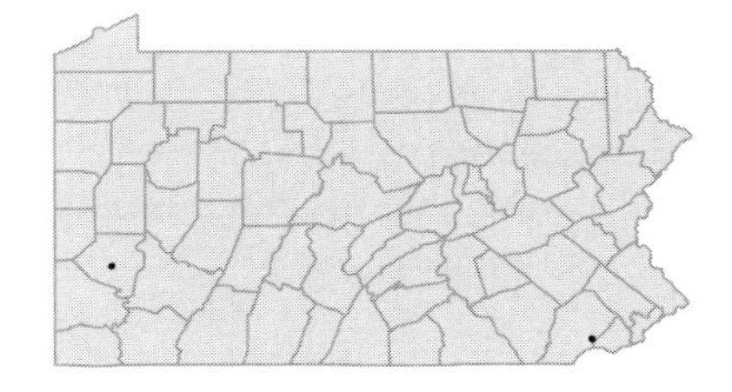

LAMIACEAE

• ***Acinos arvensis*** (Lam.) Dandy
Mother-of-thyme
Herbaceous annual
Calcareous hillsides and railroad banks.
Satureja acinos (L.) Scheele FGBWC

I

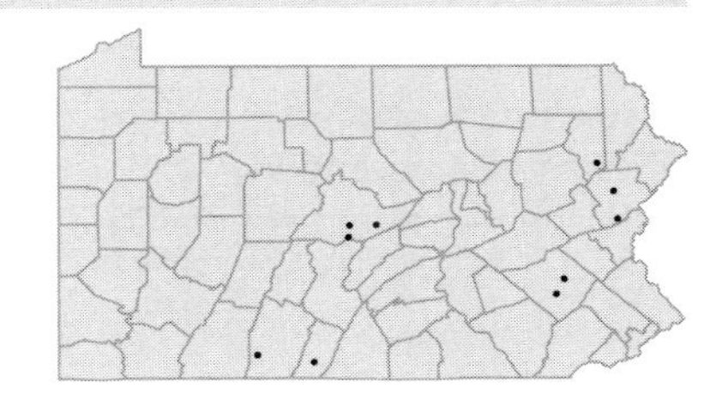

• ***Agastache foeniculum*** (Pursh) Kuntze
Anise giant-hyssop; Blue giant-hyssop
Herbaceous perennial
Cultivated and rarely escaped.

I

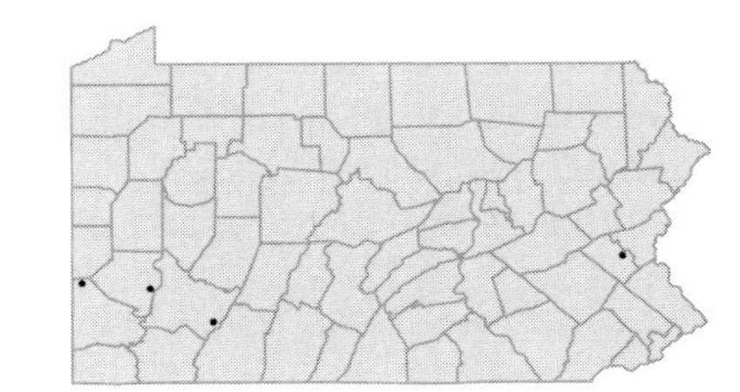

Agastache nepetoides (L.) Kuntze FACU
Yellow giant-hyssop
Herbaceous perennial
Rich woods, moist thickets, fields and roadsides.

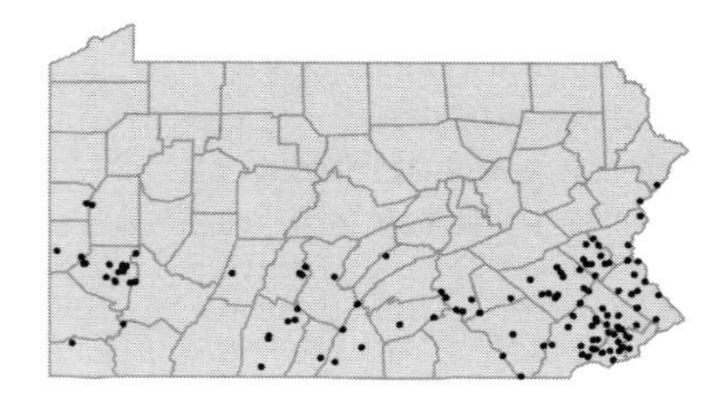

Agastache scrophulariifolia (Willd.) Kuntze
Purple giant-hyssop
Herbaceous perennial
Rich woods, moist thickets and roadsides.

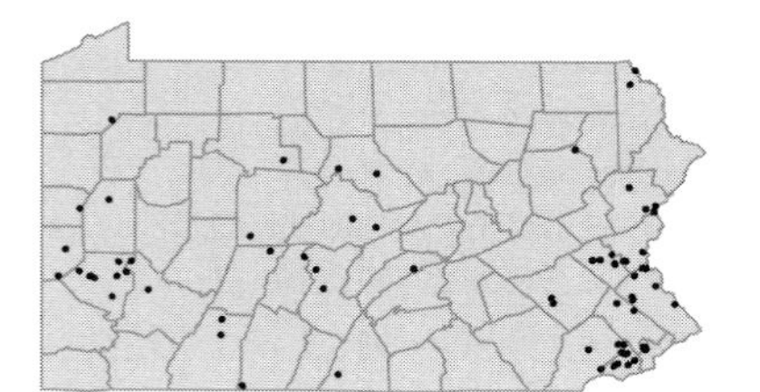

• *Ajuga genevensis* L. [I]
Bugleweed
Herbaceous perennial
Cultivated and rarely escaped to lawns and fallow fields.

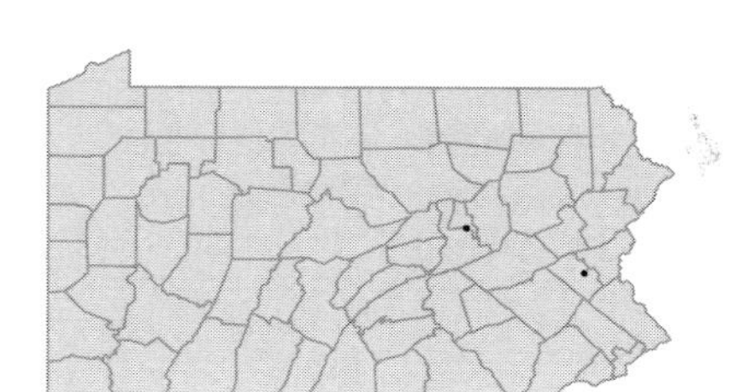

Ajuga reptans L. [I]
Carpet bugleweed
Herbaceous perennial
Escaped from cultivation and naturalized in fields and on wooded banks and roadsides.

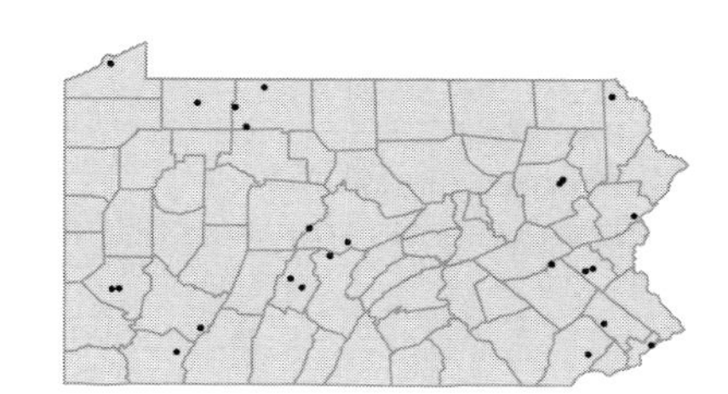

• *Ballota nigra* L. [I]
Black horehound
Herbaceous perennial
Ballast ground and rubbish dump.

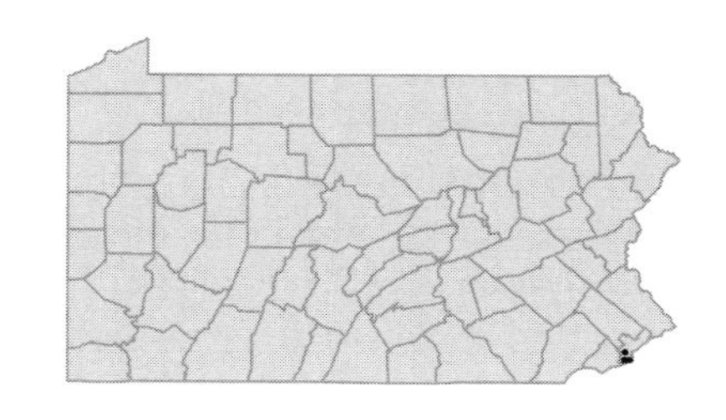

• *Blephilia ciliata* (L.) Benth.
Wood mint
Herbaceous perennial
Wooded slopes, swamps and calcareous hillsides.

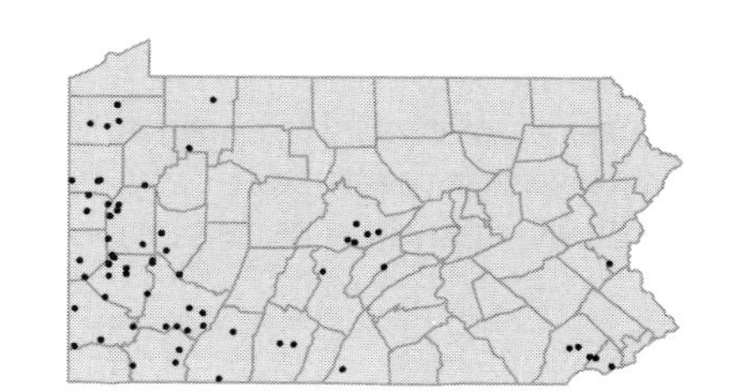

Blephilia hirsuta (Pursh) Benth.
Wood mint
Herbaceous perennial
Moist woods and swamps.

FACU-

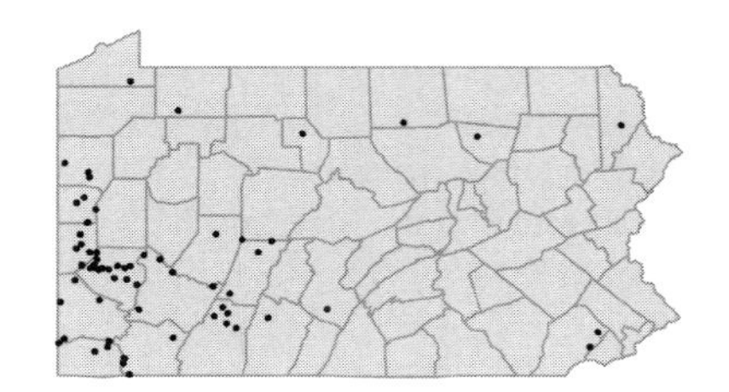

• ***Calamintha arkansana*** (Nutt.) Shinners
Calamint
Herbaceous perennial
Calcareous shores.
Satureja arkansana (Nutt.) Briq. F; *Satureja glabella* (Michx.) Briq. var. *angustifolia* (Torr.) Svenson C

FACU

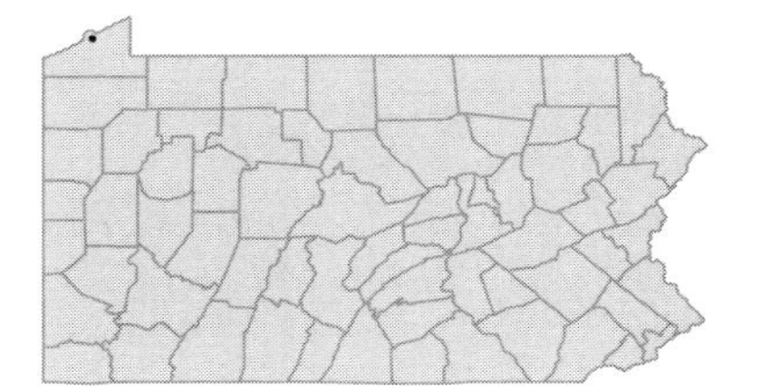

Calamintha nepeta (L.) Savi ssp. ***glandulosa*** (Requien) P.W.Ball
Basil-thyme; Calamint
Herbaceous perennial
Ballast. Represented by a single collection from Philadelphia Co. in 1865.
Satureja calamintha (L.) Scheele FGBWC

[I]

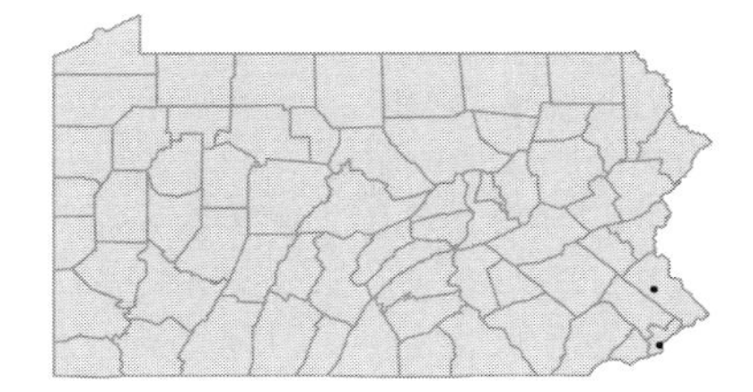

• ***Clinopodium vulgare*** L.
Wild basil
Herbaceous perennial
Open woods, fields and roadsides.
Satureja vulgaris (L.) Fritsch WFGBC

[I]

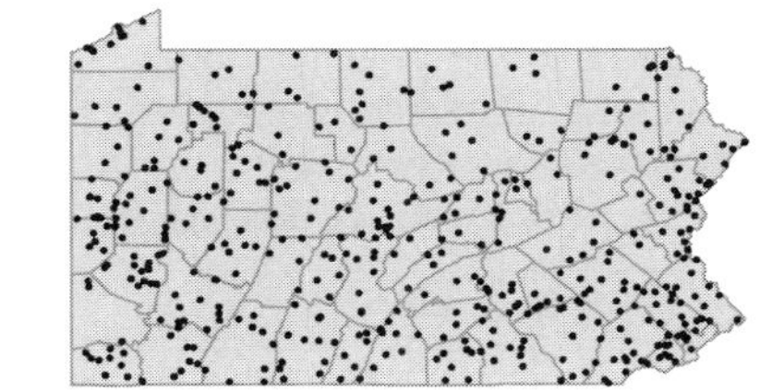

• ***Collinsonia canadensis*** L.
Horse-balm; Stoneroot
Herbaceous perennial
Rich woods and wooded floodplains.

FAC+

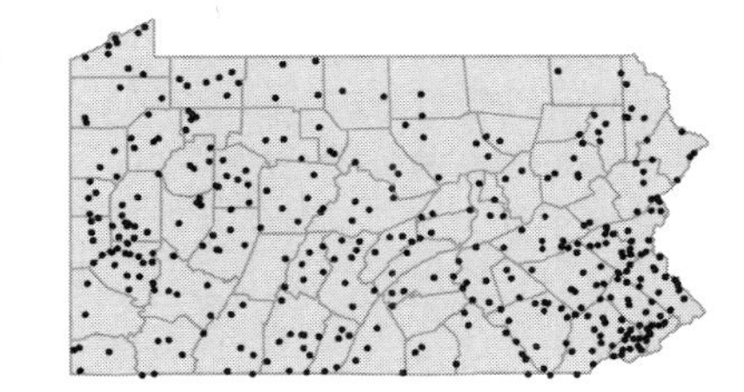

• ***Cunila origanoides*** (L.) Britt.
Common dittany; Stone-mint
Herbaceous perennial
Dry, open woods, shaly slopes and serpentine outcrops.

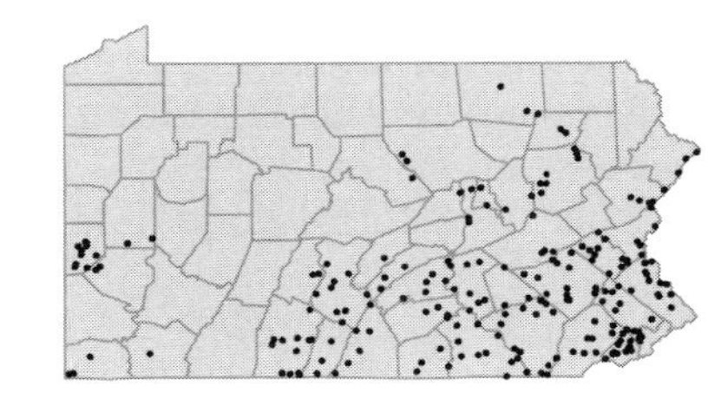

• ***Dracocephalum parviflorum*** Nutt.
Dragonhead
Herbaceous annual
Yards and waste ground.
Moldavica parviflora (Nutt.) Britt. B

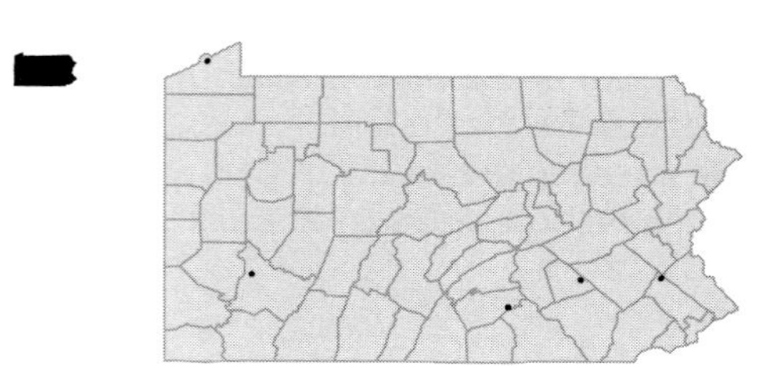

• ***Galeopsis bifida*** Boenn.
Hemp-nettle
Herbaceous annual
Moist or swampy woods, lake margins and roadsides.
Galeopsis tetrahit L. var. *bifida* (Boenn.) Lej. & Court. FGBWC

I

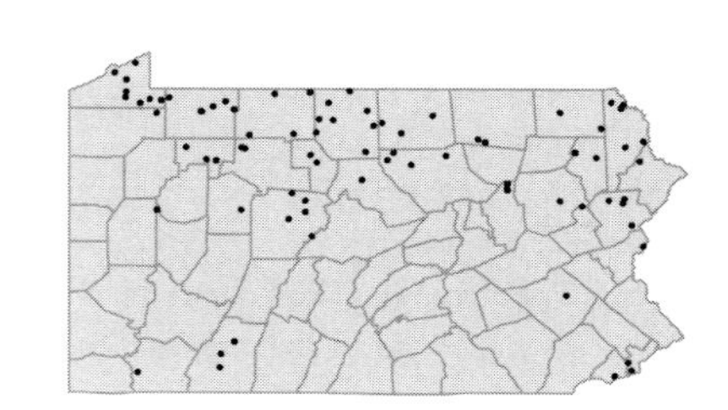

Galeopsis ladanum L.
Red hemp-nettle
Herbaceous annual
Ballast and waste ground.

I

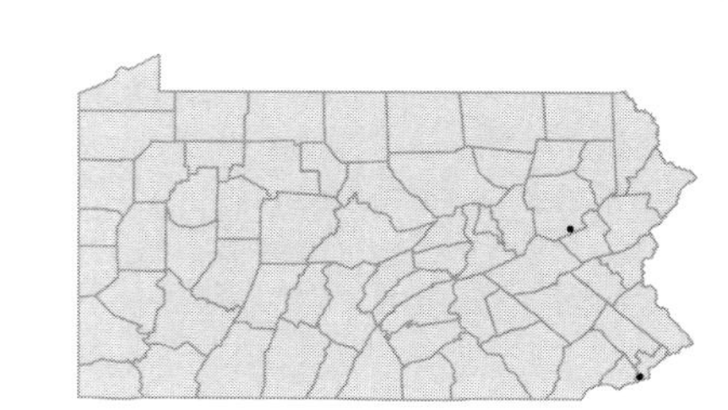

• ***Glechoma hederacea*** L.
Gill-over-the-ground; Ground-ivy
Herbaceous perennial
Fields, disturbed woods, roadsides, gardens and waste ground.

FACU
I

• ***Hedeoma hispidum*** Pursh
Mock pennyroyal
Herbaceous annual
Fallow field.

I

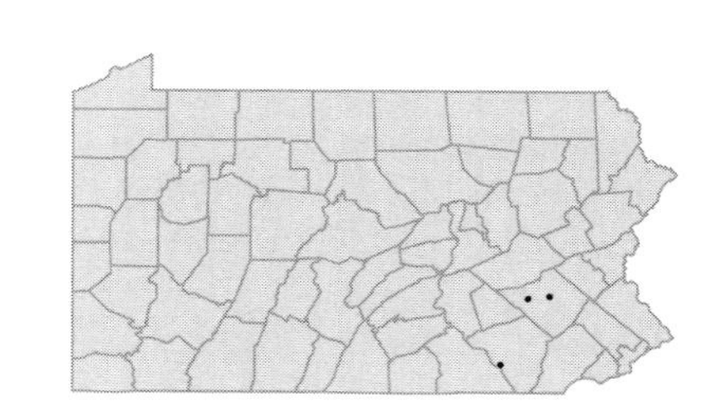

Hedeoma pulegioides (L.) Pers.
American pennyroyal; Pudding-grass
Herbaceous annual
Dry fields, pastures, woods and roadsides.

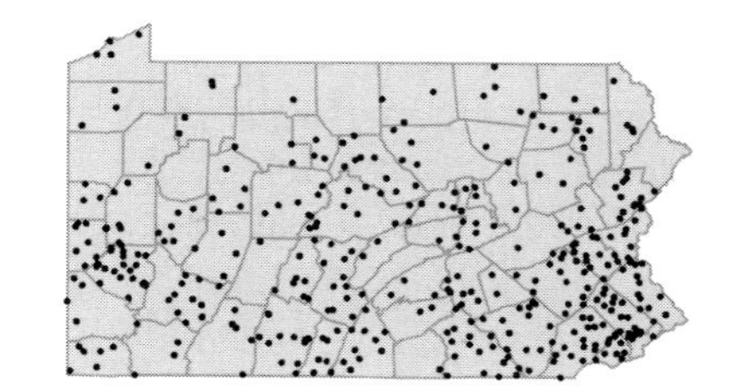

• ***Hyssopus officinalis*** L.
Hyssop
Herbaceous perennial
Cultivated and rarely escaped to streets and sidewalks.

I

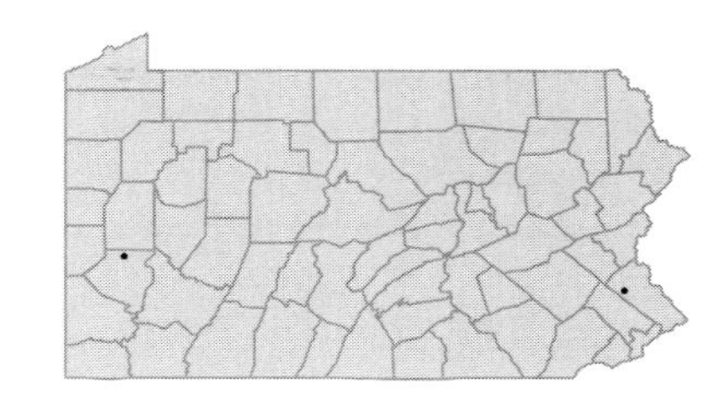

• ***Lamium album*** L.
Snowflake; White dead-nettle
Herbaceous perennial
Moist thickets.

I

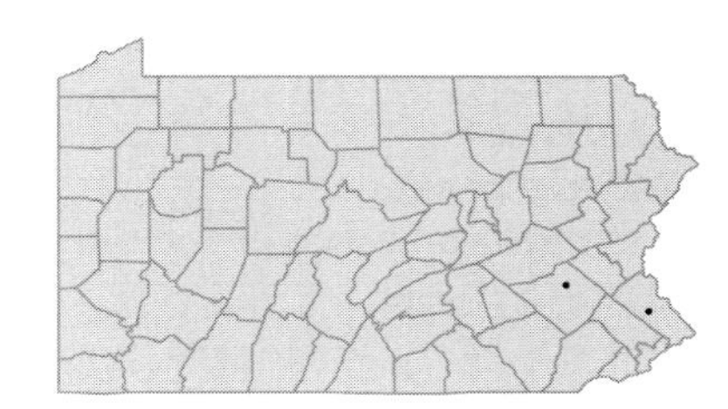

Lamium amplexicaule L.
Henbit
Herbaceous annual
Wooded slopes, fields, roadsides, fencerows and shale barrens.

I

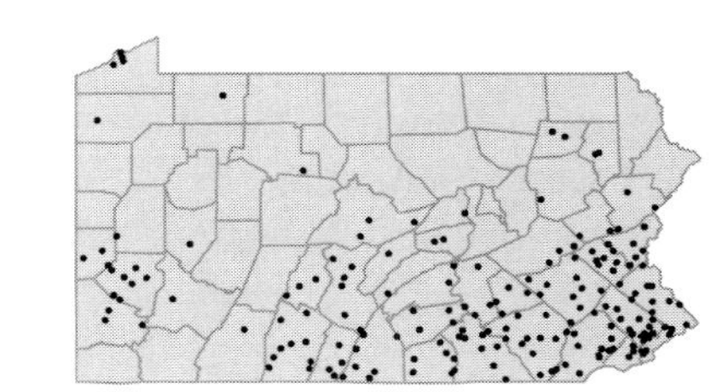

Lamium maculatum L.
Spotted dead-nettle
Herbaceous perennial
Wooded slopes and floodplains, cultivated fields, roadsides and waste ground.

I

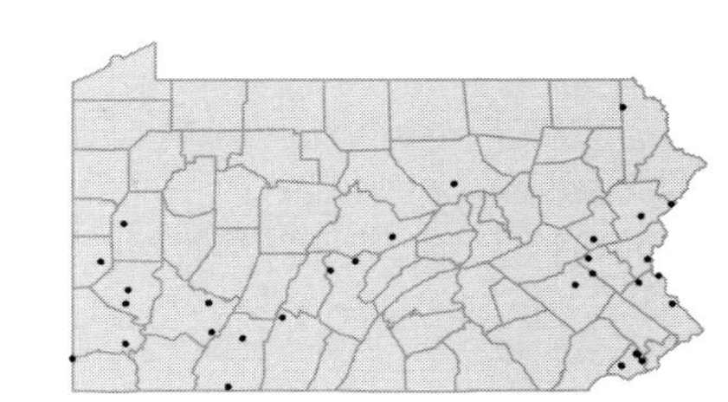

Lamium purpureum L.
Purple dead-nettle
Herbaceous annual
Wooded slopes, fields and roadsides.

I

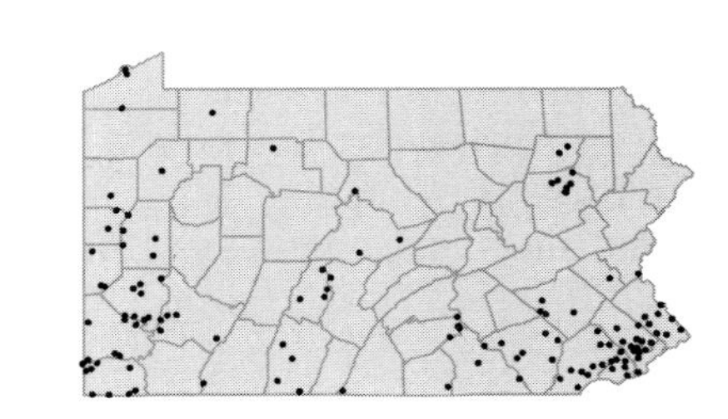

• ***Leonurus cardiaca*** L.
Common motherwort
Herbaceous perennial
Woods, stream banks, fields, roadsides and railroad cinders.

I

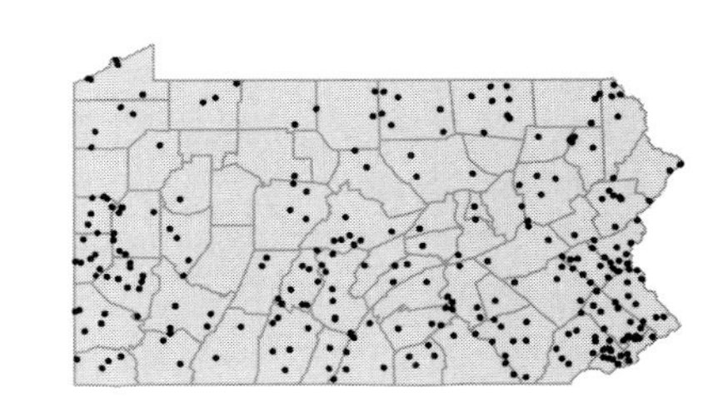

Leonurus marrubiastrum L.
Motherwort
Herbaceous biennial
Old fields, roadsides and vacant lots.

I

Leonurus sibiricus L.
Motherwort
Herbaceous annual
River banks and weedy ground along railroad tracks.

I

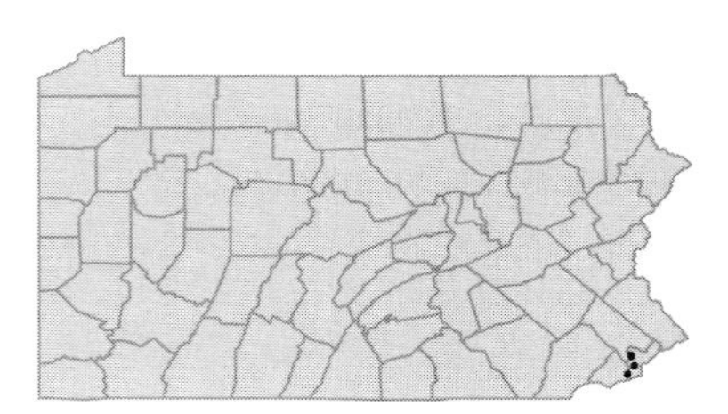

• ***Lycopus americanus*** Muhl. ex Bart.
Water-horehound
Herbaceous perennial
Shaded hillsides, fields, moist thickets, wet ditches and swamps.

OBL

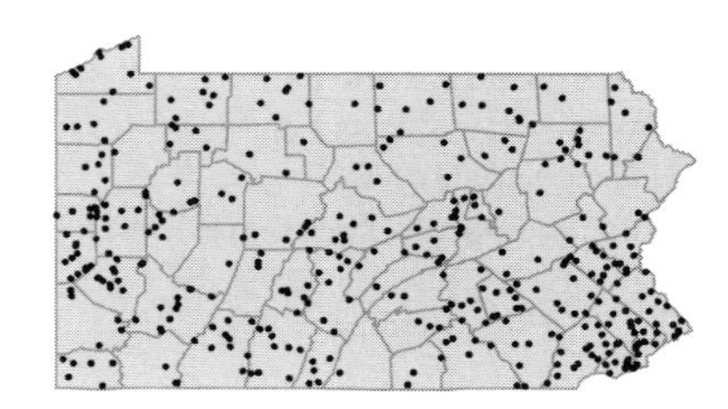

Lycopus europaeus L.
European water-horehound
Herbaceous perennial
Canal banks and moist waste ground.

OBL
I

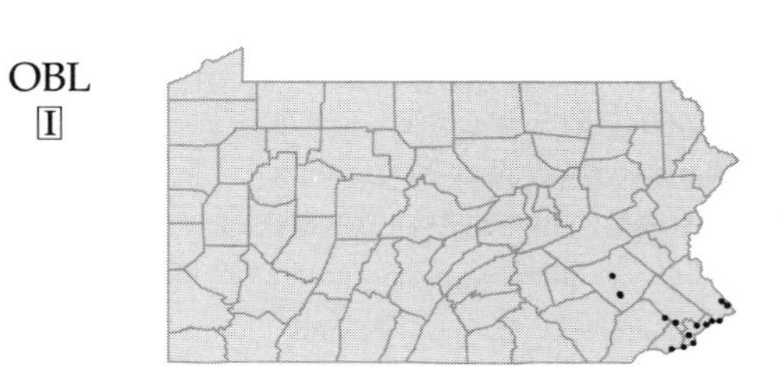

Lycopus rubellus Moench
Gypsy-wort; Water-horehound
Herbaceous perennial
Bogs, river banks, pond margins and wet ditches.

OBL

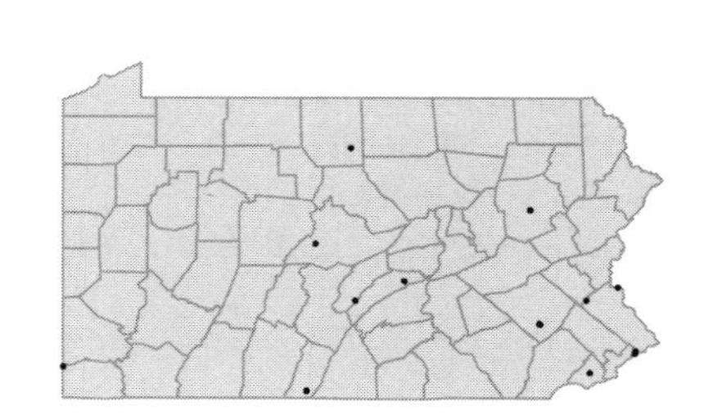

Lycopus uniflorus Michx.
Bugleweed; Water-horehound
Herbaceous perennial
Swampy meadows, bogs, lake margins and floodplains.
Lycopus communis Bickn. P

OBL

Lycopus uniflorus x virginicus
Water-horehound
Herbaceous perennial
Stream banks, swamps, canal or pond edges and moist thickets.
Lycopus x sherardii Steele C

OBL

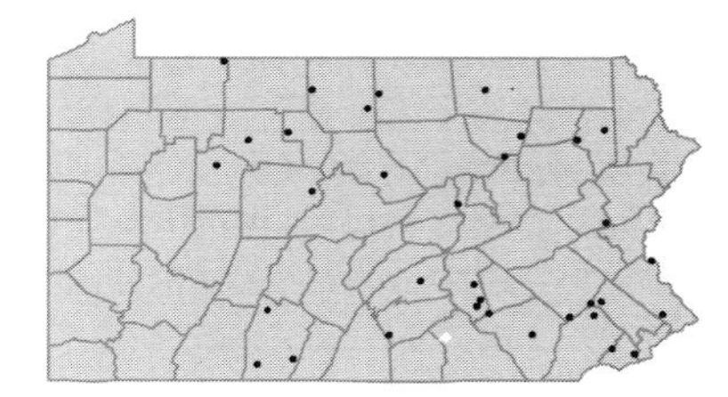

Lycopus virginicus L.
Bugleweed; Water-horehound
Herbaceous perennial
Moist woods, stream banks, swamps and wet ditches.

OBL

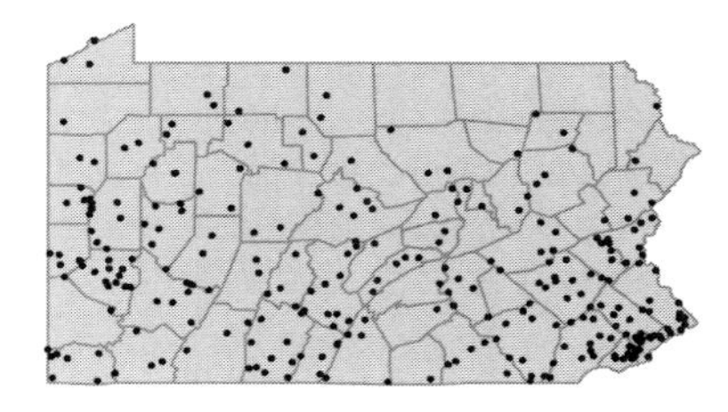

• ***Marrubium vulgare*** L.
Common horehound
Herbaceous perennial
Cultivated and occasionally escaped to fields and roadsides.

I

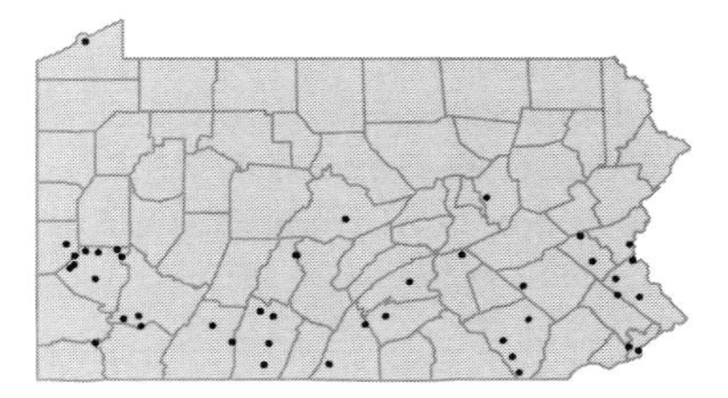

• ***Meehania cordata*** (Nutt.) Britt.
Heart-leafed meehania
Herbaceous perennial
Banks and wooded slopes.

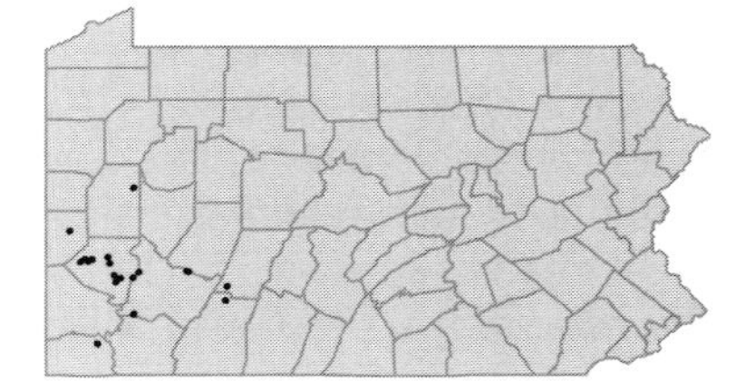

• ***Melissa officinalis*** L.
Lemon-balm
Herbaceous perennial
Cultivated and occasionally escaped to woods and roadsides.

I

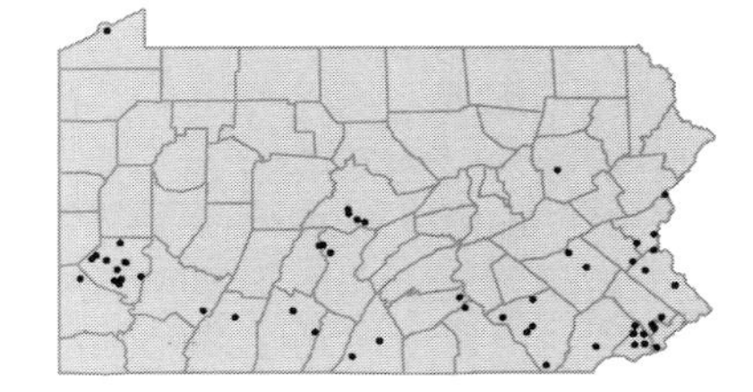

• ***Mentha aquatica*** L.
Water mint
Herbaceous perennial
River banks.

OBL
I

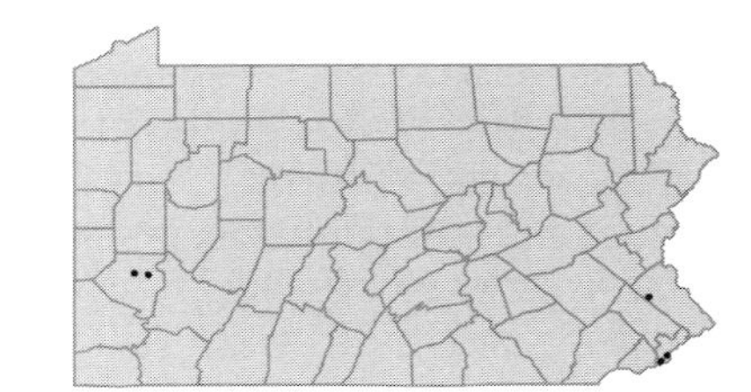

Mentha aquatica x arvensis
Mint
Herbaceous perennial
Ballast and waste ground.
Mentha sativa L. W; *Mentha x verticillata* L. C

I

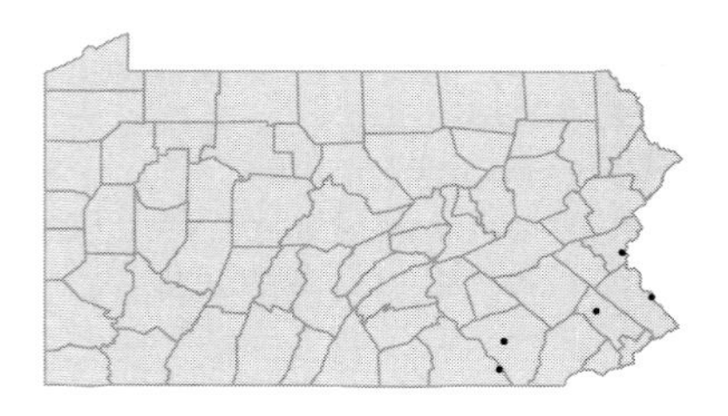

Mentha aquatica x spicata
Peppermint
Herbaceous perennial
Wet fields, moist thickets, swamps and stream banks.
Mentha x piperita L. CWFBG

FACW+
I

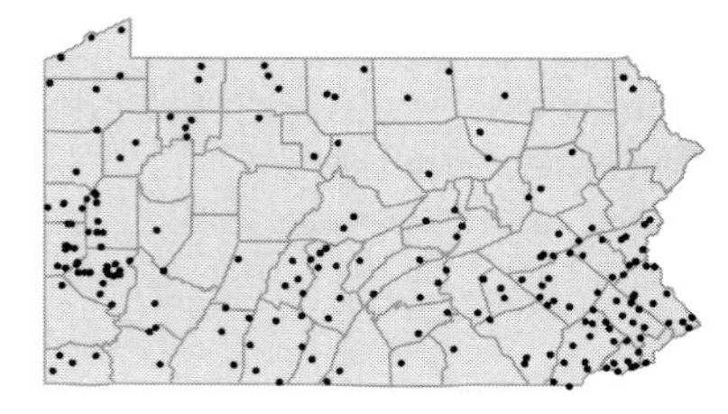

Mentha arvensis L.
Field mint
Herbaceous perennial
Moist banks, wet meadows and swamps.
Mentha canadensis L. K

FACW

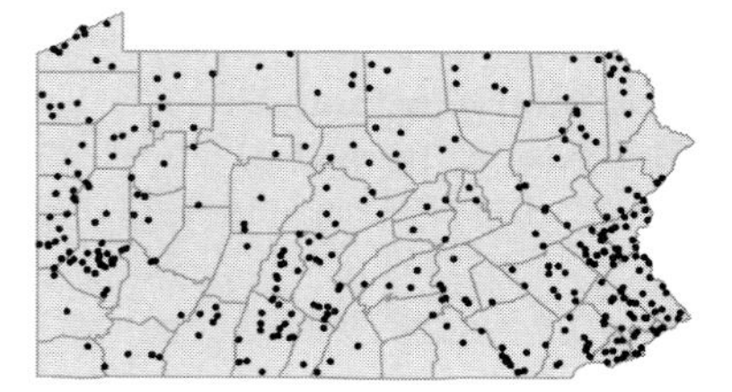

Mentha arvensis x spicata
Red mint; Scotch mint
Herbaceous perennial
Cultivated and frequently naturalized on gravelly shores, stream banks and roadsides.
Mentha cardiaca (S.F.Gray) Baker *pro parte* FGBW; *Mentha x gentilis* L. CW

FACW
I

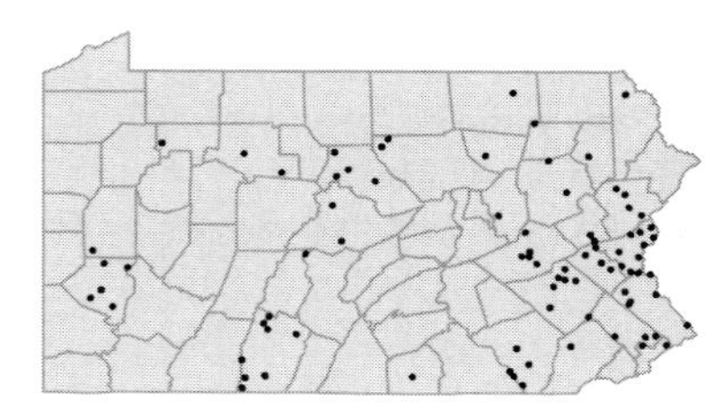

Mentha longifolia (L.) L.
Horse mint
Herbaceous perennial
Cultivated and rarely escaped to stream banks.

FACU
I

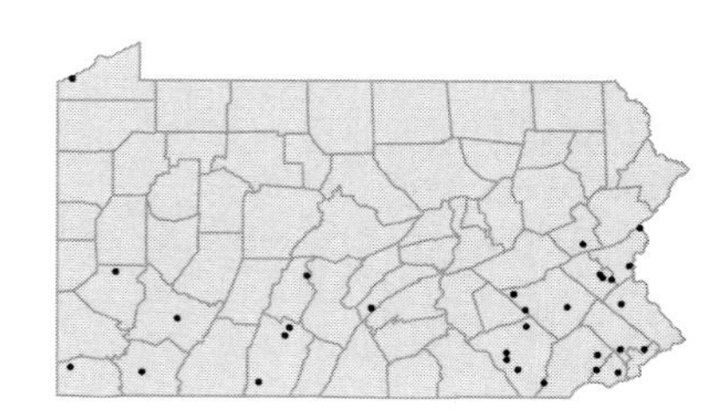

Mentha longifolia x suaveolens
Apple mint; Pineapple mint
Herbaceous perennial
Cultivated and rarely escaped to roadsides.
Mentha x rotundifolia (L.) Huds. CWFGB

FACW
I

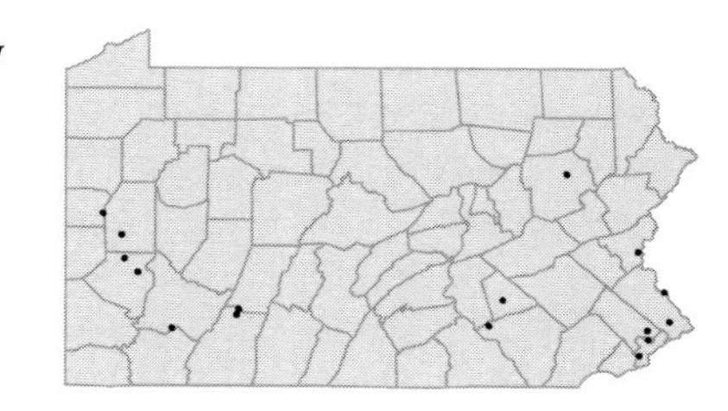

Mentha pulegium L.
Pennyroyal
Herbaceous perennial
Cultivated, also occasionally occurring on ballast or waste ground.

[I]

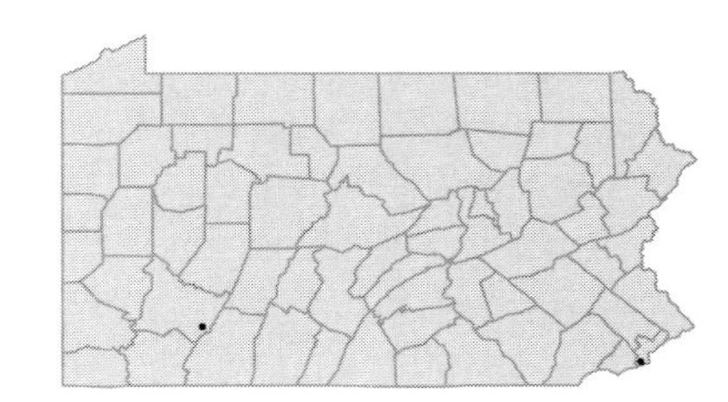

Mentha spicata L.
Spearmint
Herbaceous perennial
Naturalized on stream banks and in swamps and wet pastures.
Mentha crispa L. *pro parte* F

FACW+
[I]

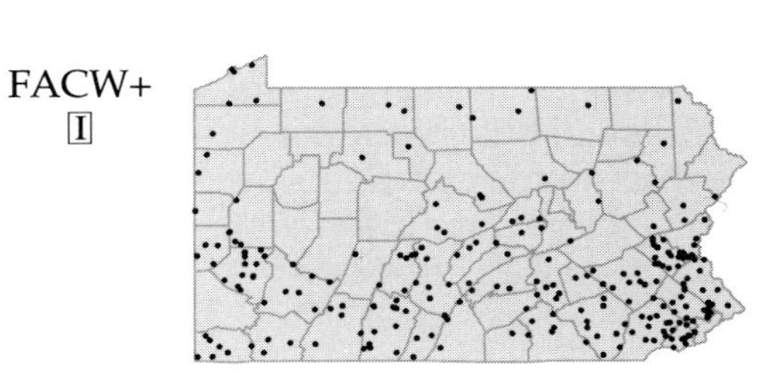

Mentha spicata x suaveolens
Apple mint; Wooly mint
Herbaceous perennial
Fields, springy slopes and roadsides.
Mentha alopecuroides (Hull) Briq. PFGBW; *Mentha x villosa* Huds. *pro parte* C

[I]

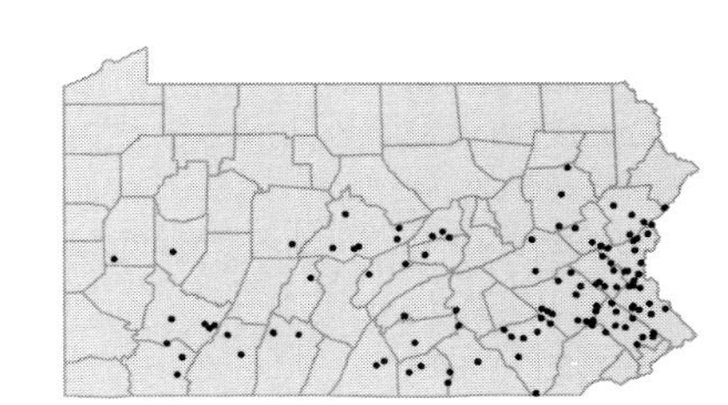

• ***Moluccella laevis*** L.
Bells-of-Ireland
Herbaceous annual
Cultivated and occasionally escaped to rubbish dumps.

[I]

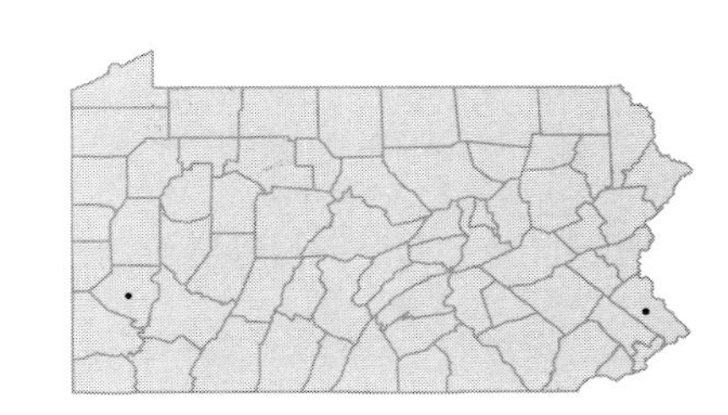

• ***Monarda clinopodia*** L.
Bee-balm
Herbaceous perennial
Moist woods, fields and floodplains.

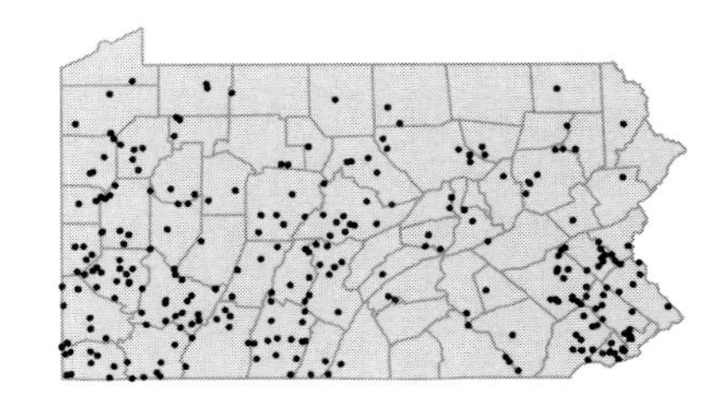

Monarda didyma L.
Bee-balm; Oswego-tea
Herbaceous perennial
Creek banks, floodplains, and moist woods, also cultivated.

FAC+

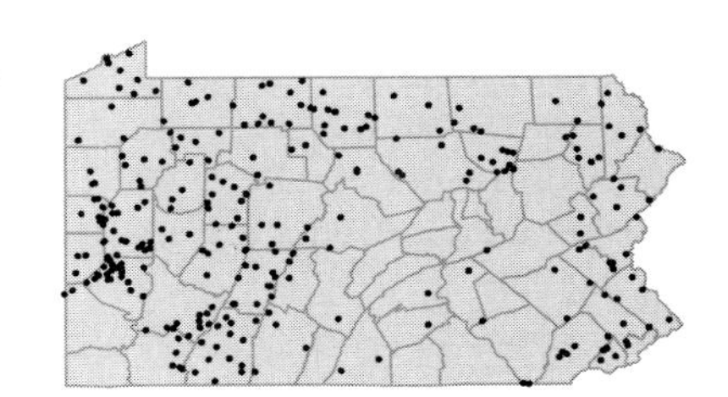

Monarda fistulosa L. var. ***fistulosa***
Horsemint; Wild bergamot
Herbaceous perennial
Fields, brushy thickets and roadsides.

UPL

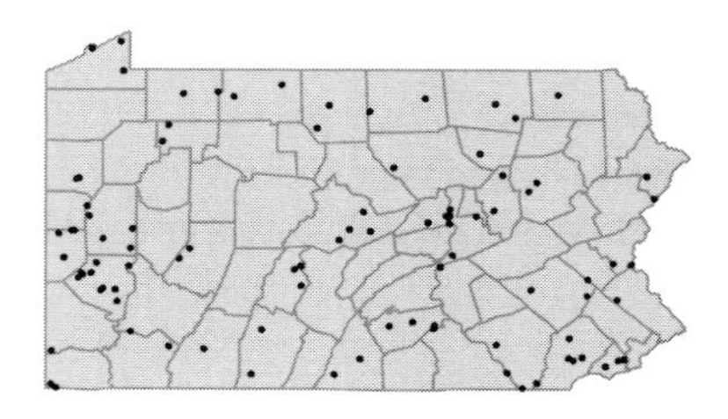

Monarda fistulosa L. var. ***mollis*** (L.) Benth.
Horsemint; Wild bergamot
Herbaceous perennial
Rocky thickets, meadows and roadsides.
Monarda mollis L. PGB

UPL

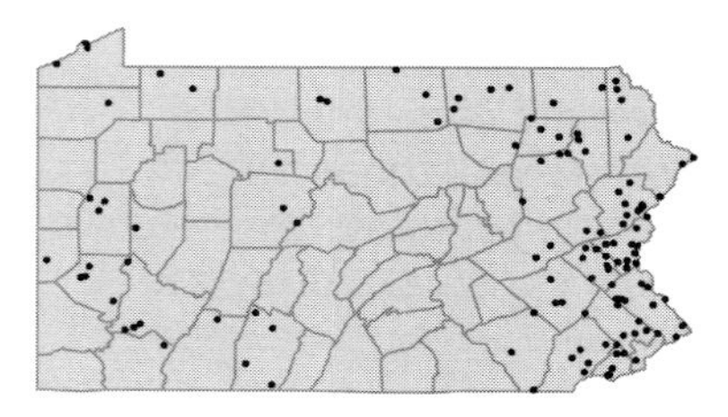

Monarda fistulosa L. var. ***rubra*** A.Gray
Horsemint; Wild bergamot
Herbaceous perennial
Alluvial ground along streams.

UPL

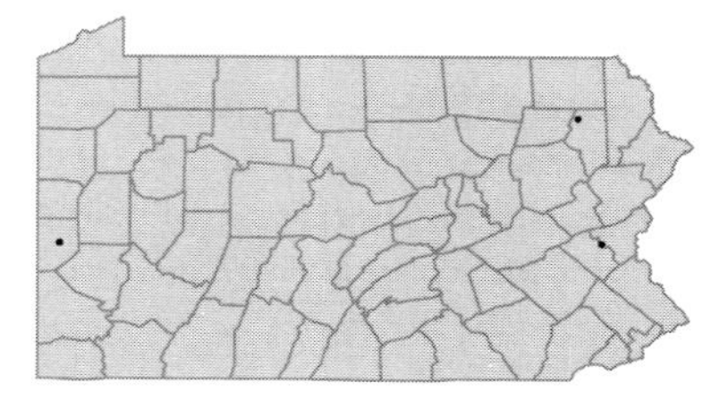

Monarda media Willd.
Bee-balm; Purple bergamot
Herbaceous perennial
Low woods, stream banks and floodplains.

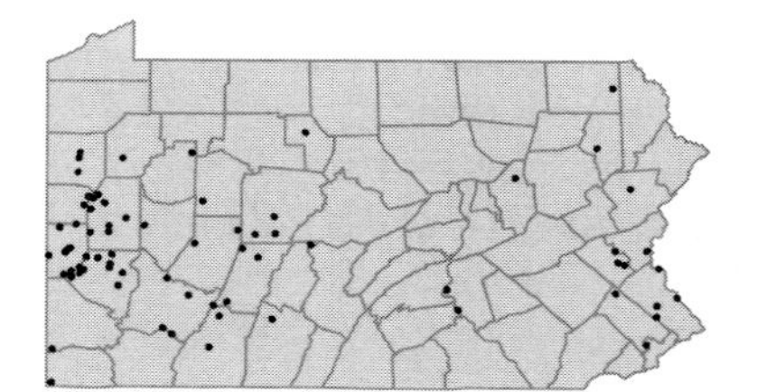

Monarda punctata L.
Spotted bee-balm
Herbaceous perennial
Dry, open, sandy fields.

UPL

• ***Nepeta cataria*** L.
Catnip
Herbaceous perennial
Cultivated and frequently escaped to woods, fields and roadsides.

FACU
[I]

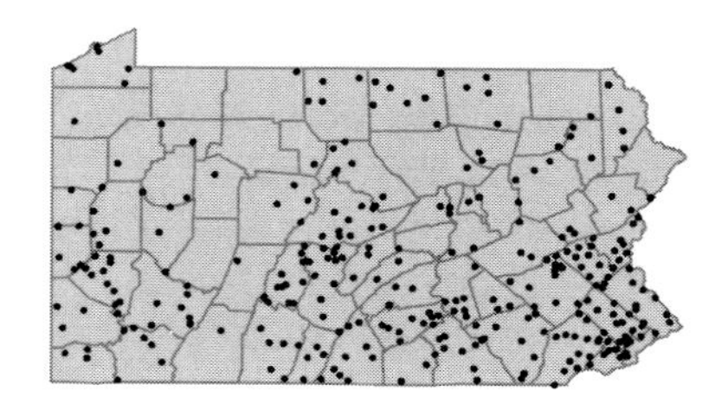

• ***Ocimum basilicum*** L. [I]
Basil
Herbaceous annual
Cultivated, occasionally escaped to fallow areas, river shores and rubbish dumps.

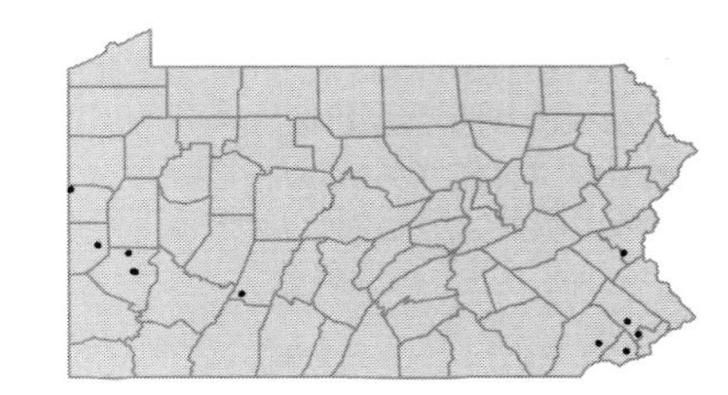

• ***Origanum majorana*** L. [I]
Sweet marjoram
Herbaceous annual
Cultivated and rarely escaped.
Majorana hortensis Moench W

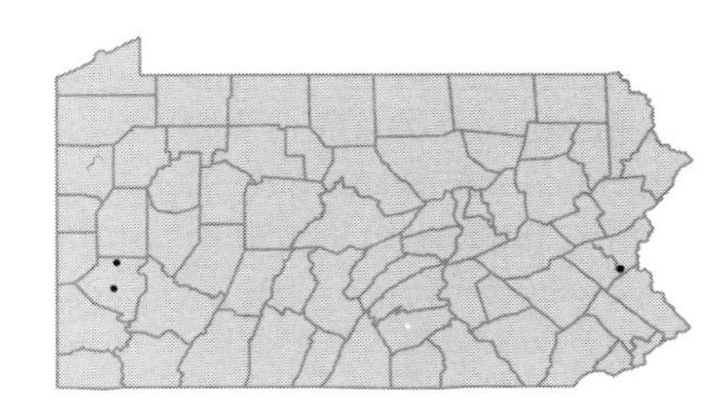

Origanum vulgare L. [I]
Oregano; Wild marjoram
Herbaceous perennial
Cultivated, occasionally escaped to roadsides.

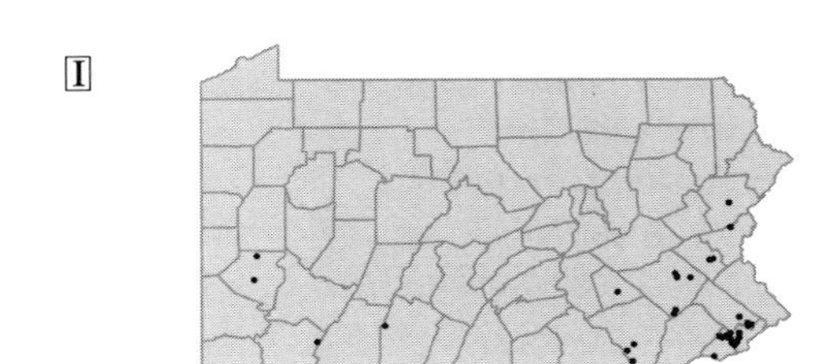

• ***Perilla frutescens*** (L.) Britt. FACU+ [I]
Perilla
Herbaceous annual
Cultivated and occasionally naturalized on moist or shaded roadsides or in disturbed woods.

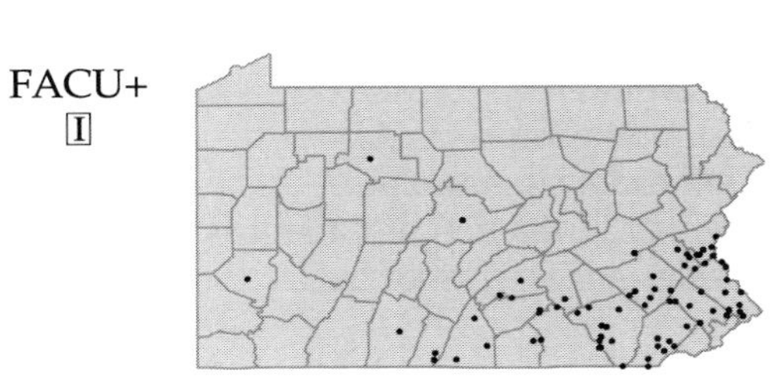

• ***Physostegia virginiana*** (L.) Benth. FAC+
False dragonhead
Herbaceous perennial
River banks and moist shorelines, also cultivated and occasionally escaped.
Dracocephalum virginianum L. B

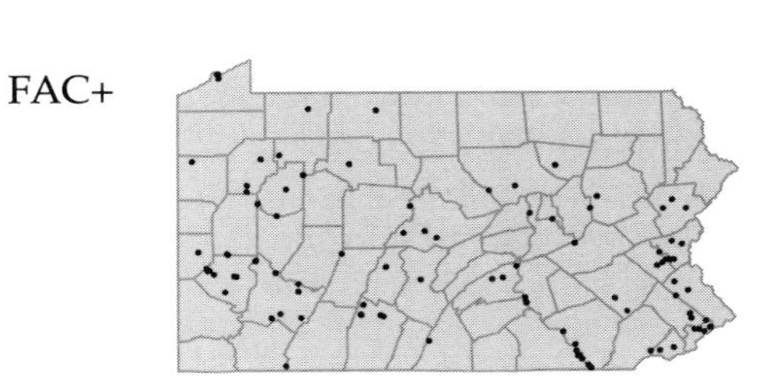

• ***Prunella laciniata*** L. [I]
Heal-all; Self-heal
Herbaceous perennial
Fields, open woods and weedy roadsides.

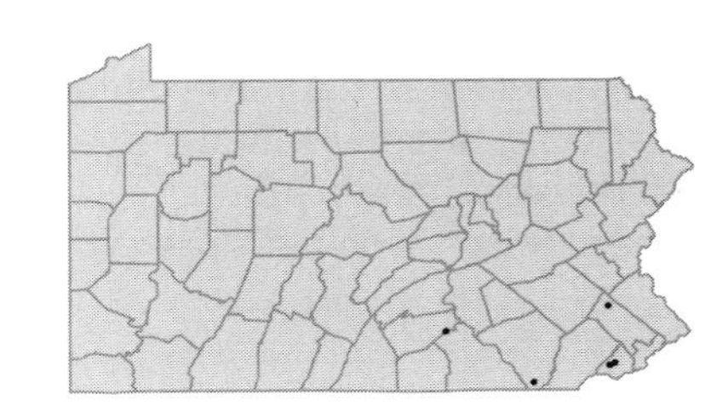

Prunella vulgaris L. ssp. ***vulgaris***
Heal-all; Self-heal
Herbaceous perennial
Moist fields and fencerows.

FACU+
[I]

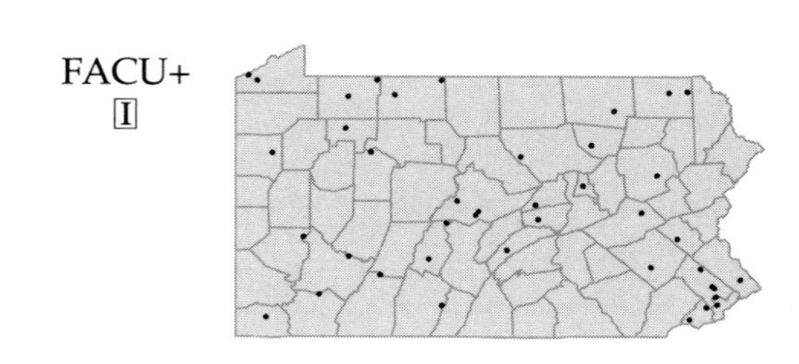

Prunella vulgaris L. ssp. ***lanceolata*** (Barton) Hulten
Heal-all; Self-heal
Herbaceous perennial
Fields, woods, floodplains and roadsides.

FACU+

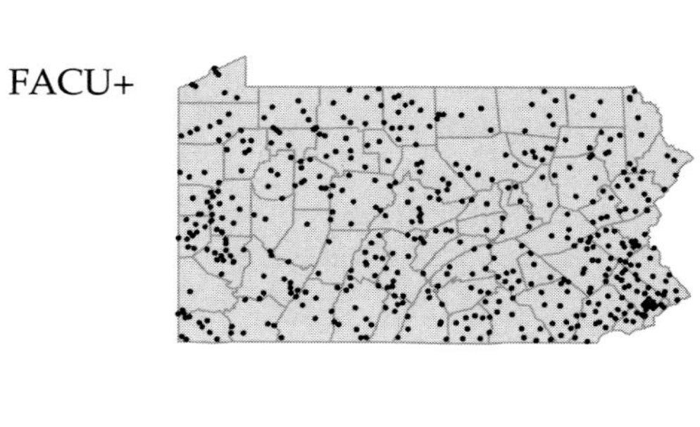

• ***Pycnanthemum clinopodioides*** Torr. & A.Gray
Mountain-mint
Herbaceous perennial
Dry slopes.
Koellia clinopodioides (Torr. & A.Gray) Kuntze P

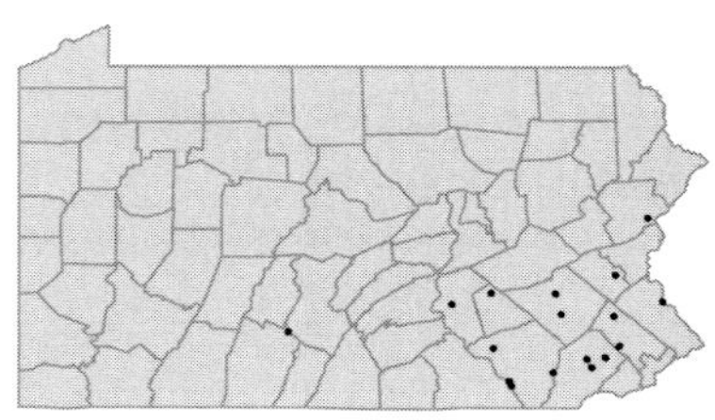

Pycnanthemum incanum (L.) Michx.
Mountain-mint
Herbaceous perennial
Old fields, thickets and barrens.
Koellia incana (L.) Kuntze P

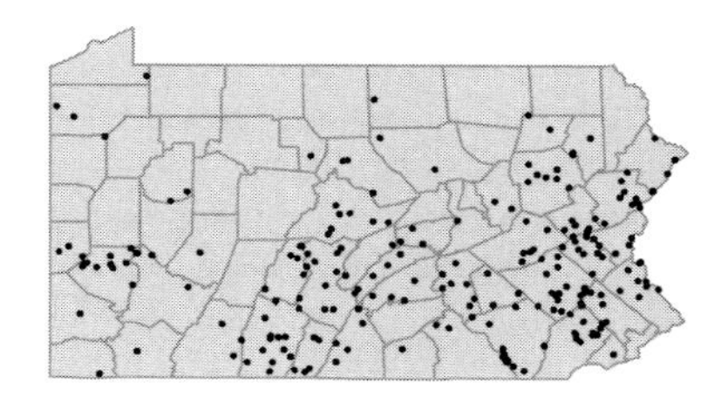

Pycnanthemum muticum (Michx.) Pers.
Mountain-mint
Herbaceous perennial
Moist woods, thickets, meadows and swales.
Koellia mutica (Michx.) Britt. P

FACW

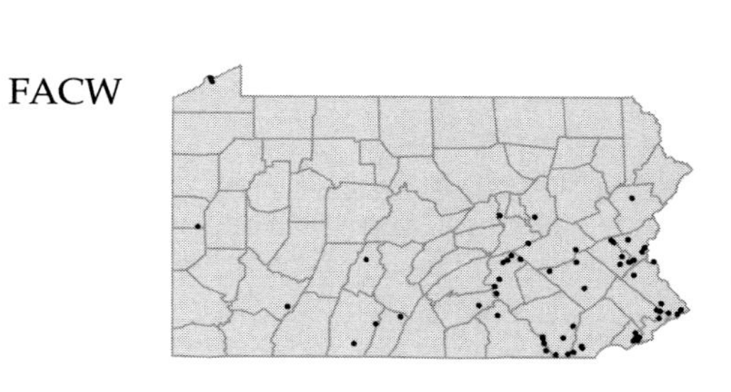

Pycnanthemum pycnanthemoides (Leavenw.) Fern.
Southern mountain-mint
Herbaceous perennial
Open, rocky, wooded slopes.

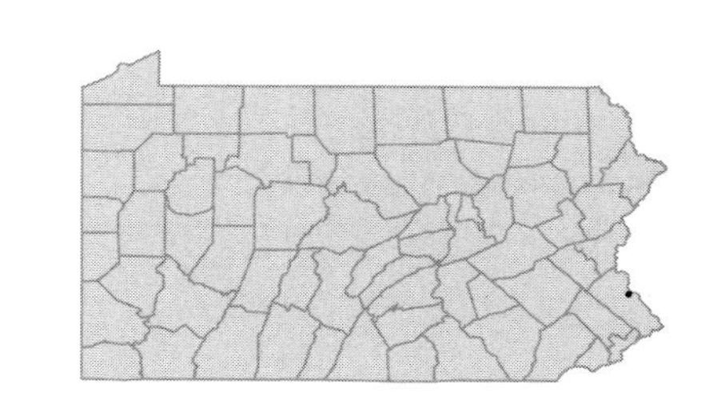

Pycnanthemum tenuifolium Schrad. FACW
Mountain-mint
Herbaceous perennial
Moist old fields, sandy river banks or floodplains.
Koellia flexuosa (Walt.) MacM. P; *Pycnanthemum flexuosum* (Walt.) BSP GB

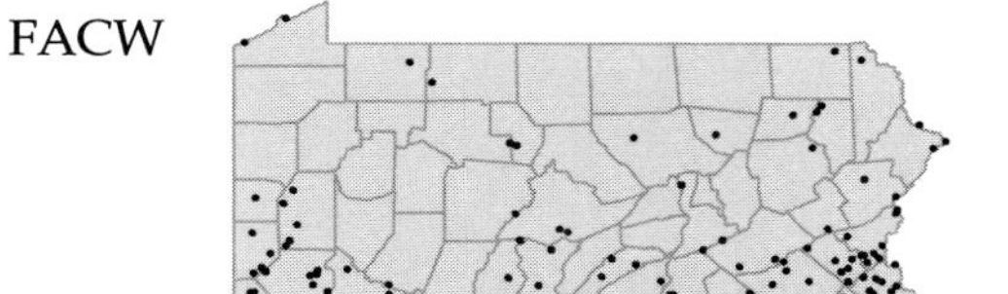

Pycnanthemum torrei Benth.
Torrey's mountain-mint
Herbaceous perennial
Upland woods and thickets.

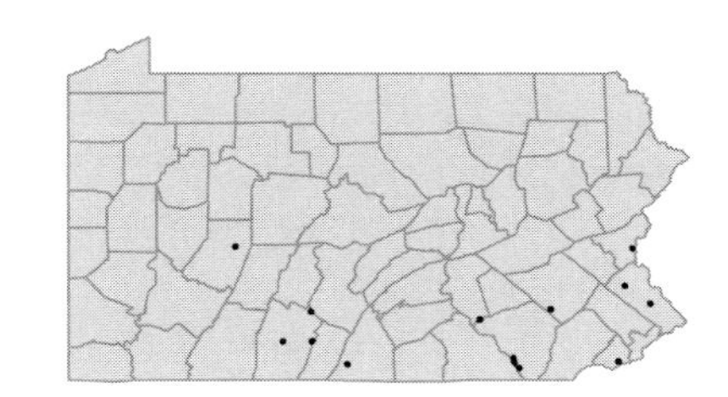

Pycnanthemum verticillatum (Michx.) Pers. var. ***verticillatum*** FAC
Mountain-mint
Herbaceous perennial
Abandoned fields, swampy meadows, marshes and woods.
Koellia verticillata (Michx.) Kuntze P

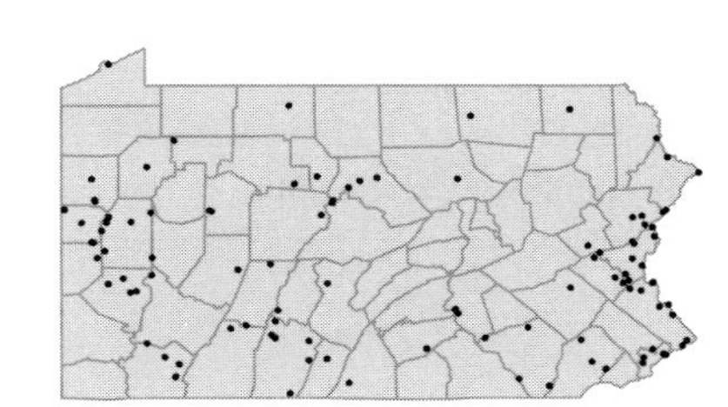

Pycnanthemum verticillatum (Michx.) Pers. var. ***pilosum*** (Nutt.) Cooperrider FAC
Mountain-mint
Herbaceous perennial
Rocky meadows.
Koellia pilosa (Nutt.) Britt. P; *Pycnanthemum pilosum* Nutt. FGBW

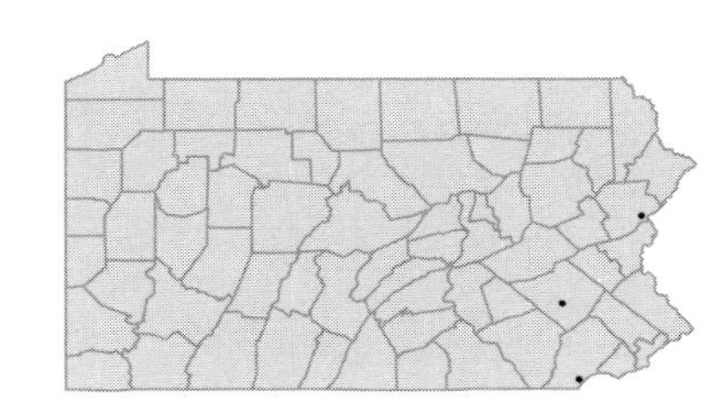

Pycnanthemum virginianum (L.) Durand & Jackson ex B.L.Robins. & Fern. FAC
Mountain-mint
Herbaceous perennial
Boggy fields, swamps and moist woods.
Koellia virginiana (L.) MacM. P

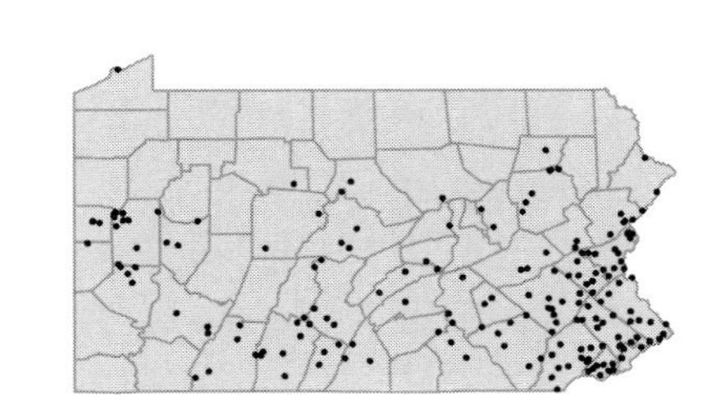

• ***Salvia lyrata*** L. UPL
Lyre-leaved sage
Herbaceous perennial
Moist pastures, thickets or woods.

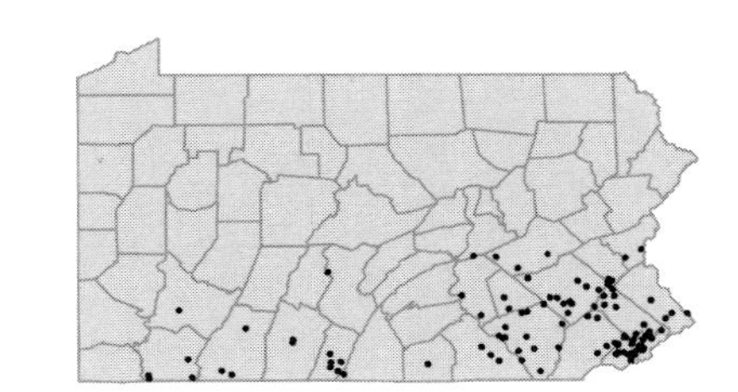

Salvia nemorosa L. I
Woodland sage
Herbaceous perennial
Railroad banks and cinders.
Salvia sylvestris L. FW

Salvia officinalis L. I
Garden sage
Herbaceous perennial
Cultivated and rarely escaped to railroad banks.

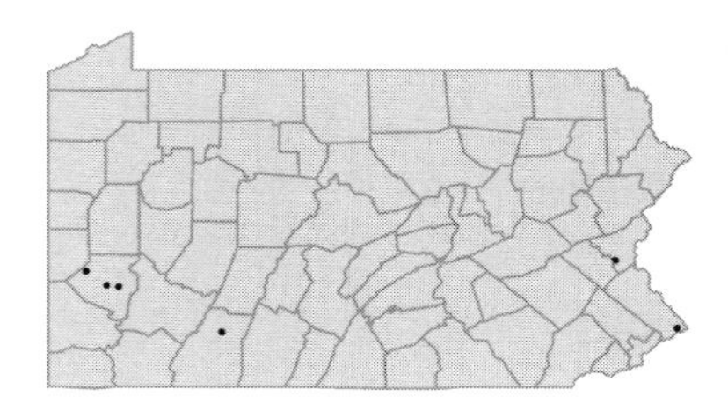

Salvia pratensis L. I
Meadow sage
Herbaceous perennial
Dry pastures and waste ground.

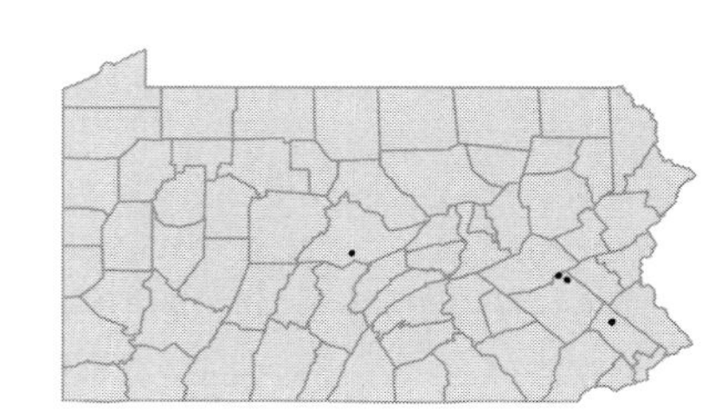

Salvia reflexa Hornem.
Lance-leaved sage
Herbaceous annual
River banks, old fields, roadsides, cinders and quarry waste.

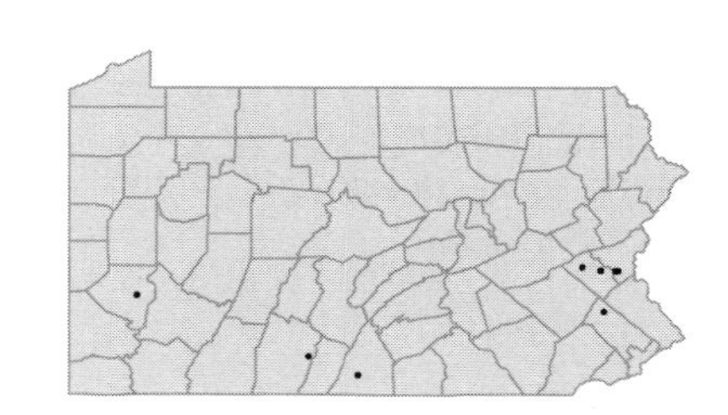

Salvia sclarea L. I
Clary
Herbaceous biennial
Cultivated, also occurring occasionally in fields or on ballast.

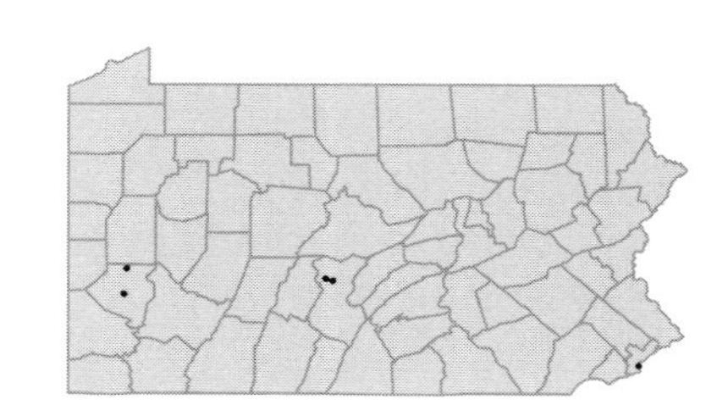

Salvia verbenacea L. I
Sage
Herbaceous perennial
Ballast. Represented by a single collection from Philadelphia Co. in 1879.

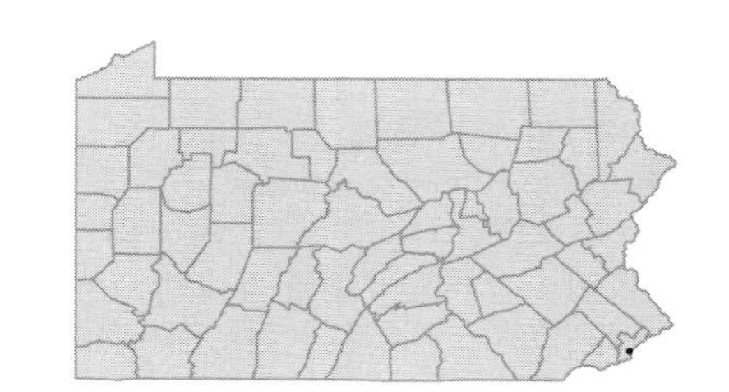

Salvia verticillata L. I
Sage
Herbaceous perennial
Cultivated and occasionally escaped to roadsides.

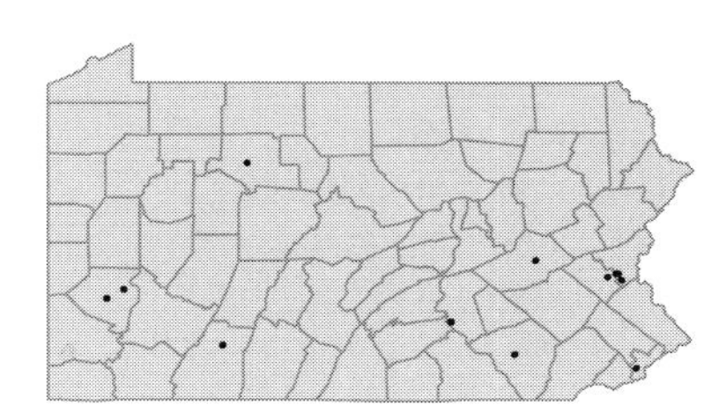

• ***Satureja hortensis*** L. I
Summer savory
Herbaceous annual
Cultivated and rarely escaped to rubbish dumps and waste ground.

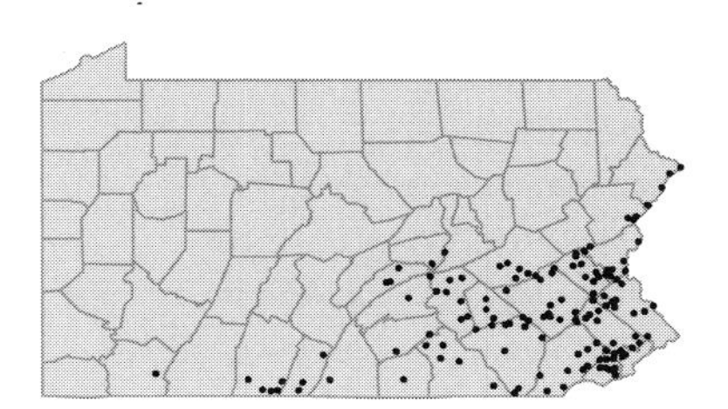

• ***Scutellaria elliptica*** Muhl. ex Spreng. var. ***elliptica***
Hairy skullcap
Herbaceous perennial
Open woods, wooded banks and shale barrens.
Scutellaria pilosa Michx. P

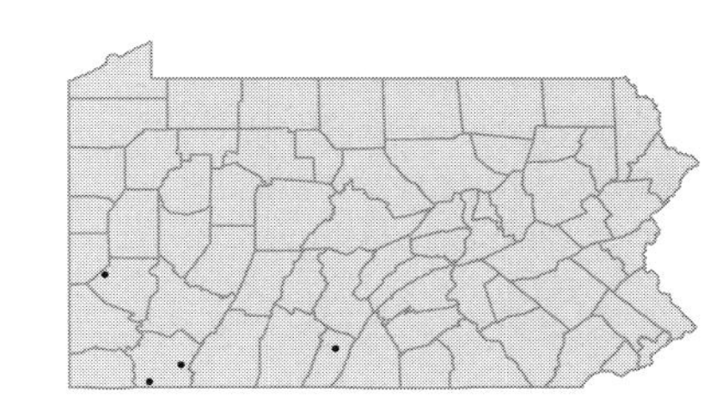

Scutellaria elliptica Muhl. ex Spreng. var. ***hirsuta*** (Short & Peter) Fern.
Hairy skullcap
Herbaceous perennial
Open woods.

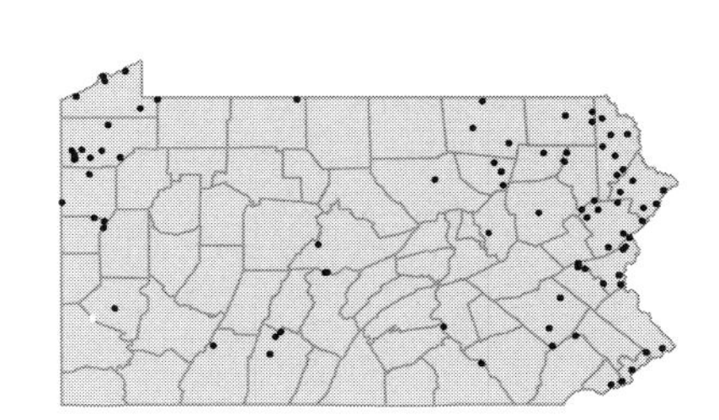

Scutellaria galericulata L. OBL
Common skullcap
Herbaceous perennial
Bogs, swamps and marshy meadows.
Scutellaria epilobiifolia A.Hamilton FW

Scutellaria incana Biehler
Downy skullcap
Herbaceous perennial
Rocky woods and roadsides.

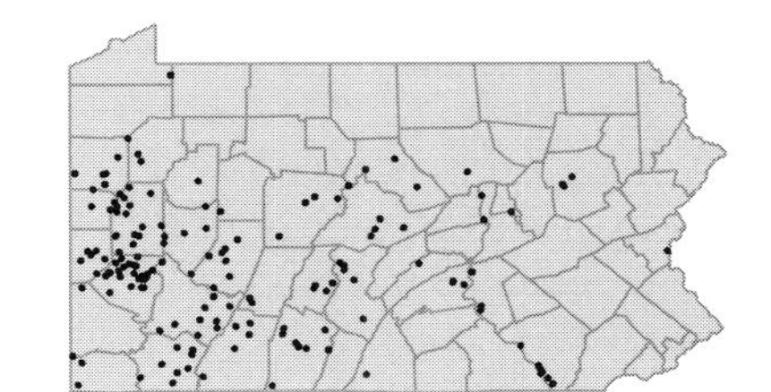

Scutellaria integrifolia L.
Hyssop skullcap
Herbaceous perennial
Swamps, bogs and moist woods or fields.

FACW

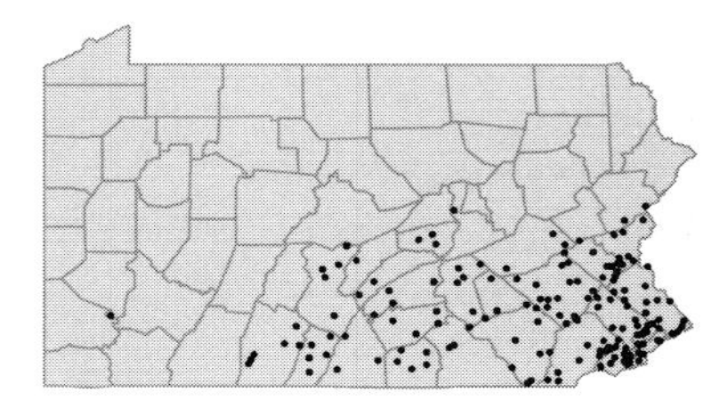

Scutellaria lateriflora L.
Mad-dog skullcap
Herbaceous perennial
Wet woods, bogs, lake margins, river banks, floodplains and swampy pastures.

FACW+

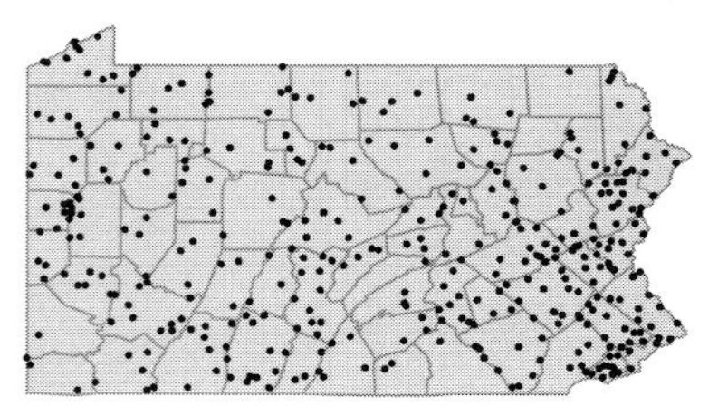

Scutellaria leonardii Epling
Small skullcap
Herbaceous perennial
Open woods and shores.
Scutellaria parvula Michx. var. *leonardii* (Epling) Fern. FW

I

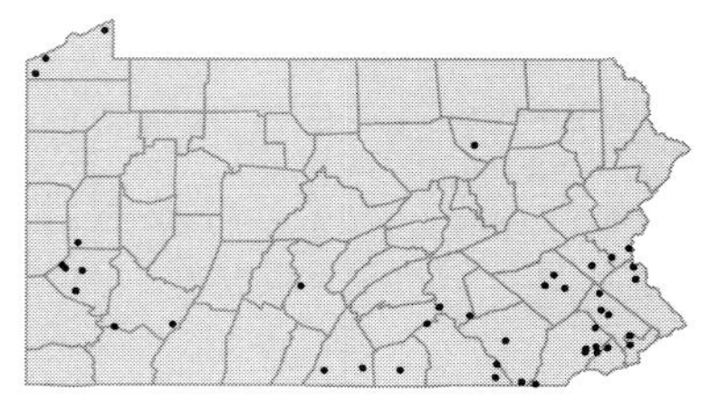

Scutellaria nervosa Pursh
Skullcap
Herbaceous perennial
Moist wooded slopes, stream banks and floodplains.

FAC

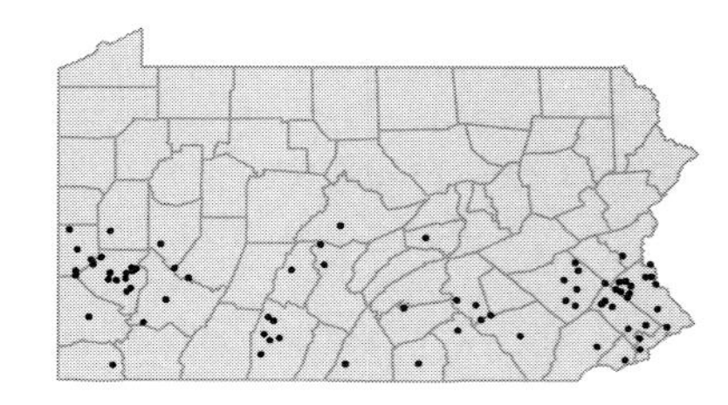

Scutellaria saxatilis Riddell
Rock skullcap
Herbaceous perennial
Low woods, rocky river banks and roadsides.

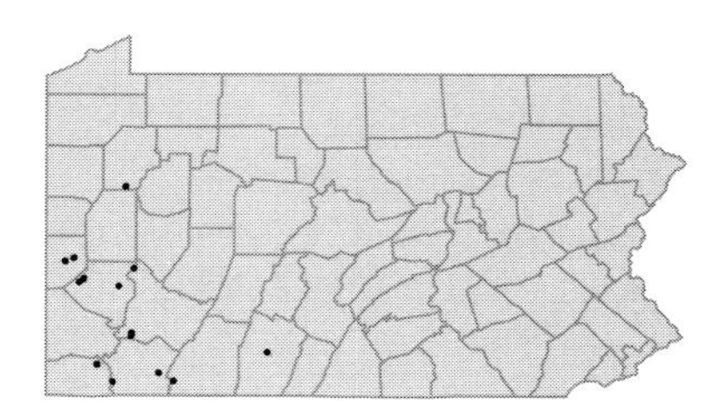

Scutellaria serrata Andr.
Showy skullcap
Herbaceous perennial
Rocky, humusy woods and floodplains. Believed to be extirpated, last collected in 1960.

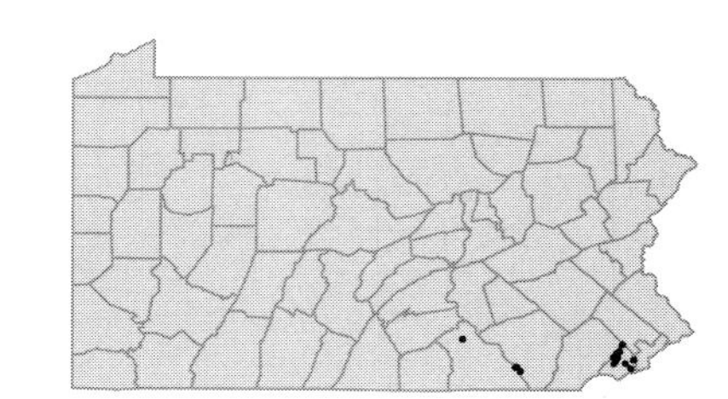

• ***Sideritis romana*** L.
Ironwort
Herbaceous annual
Fields, believed to have been introduced in imported clover seed.
Represented by a single collection from Lancaster Co. in 1909.

I

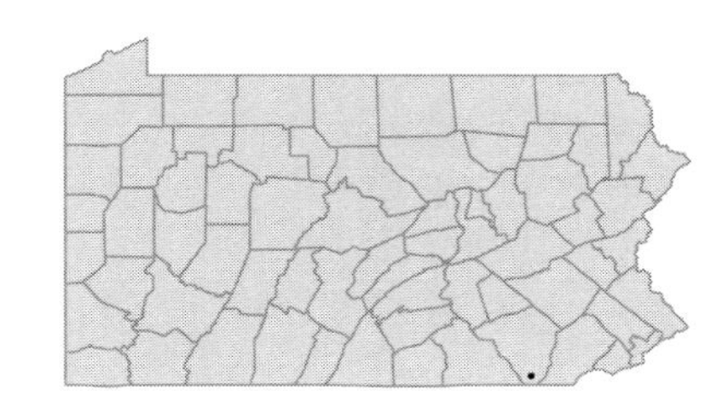

• ***Stachys annua*** L.
Hedge-nettle
Herbaceous annual
Ballast. Represented by several collections from Philadelphia Co. 1865-1920.

I

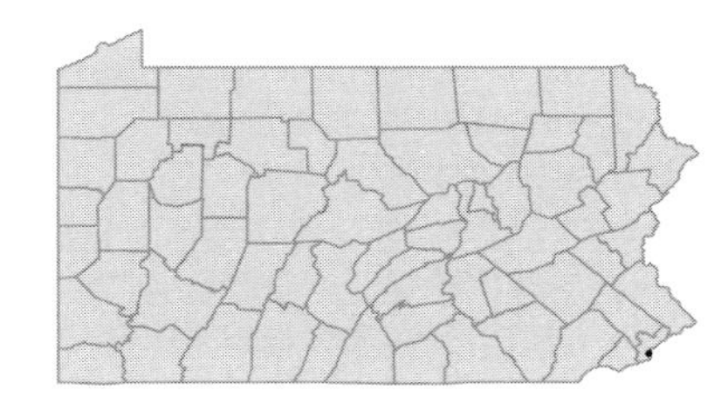

Stachys arvensis L.
Field hedge-nettle
Herbaceous annual
Ballast. Represented by two collections from Philadelphia Co. 1878-1883.

I

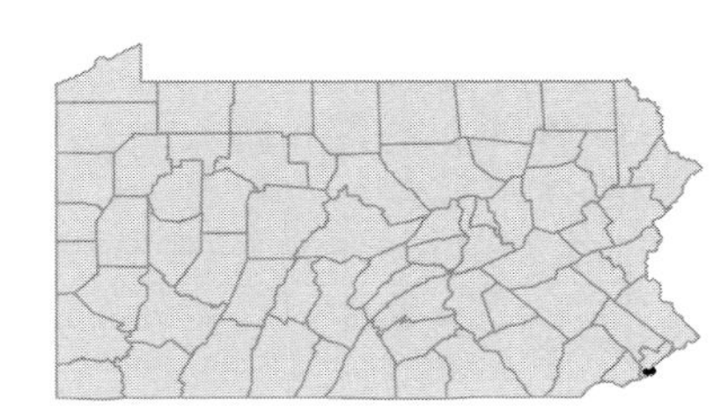

Stachys germanica L.
Hedge-nettle
Herbaceous biennial
Roadside and woods edge.

I

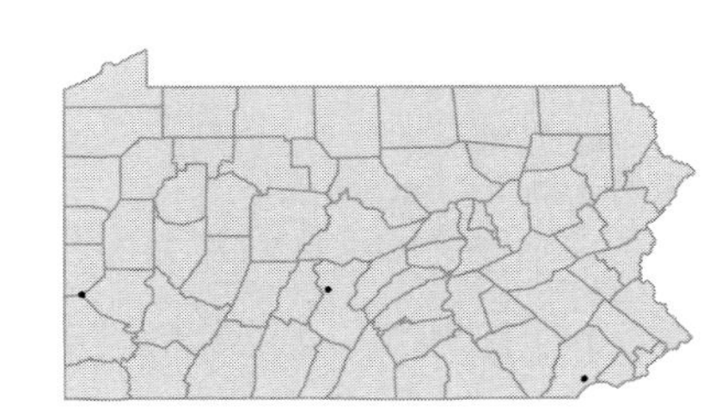

Stachys hyssopifolia Michx. var. ***hyssopifolia***
Hyssop hedge-nettle; Woundwort
Herbaceous perennial
Fallow fields, trolley tracks, ore pits and other low, waste ground.
Stachys atlantica Britt. P

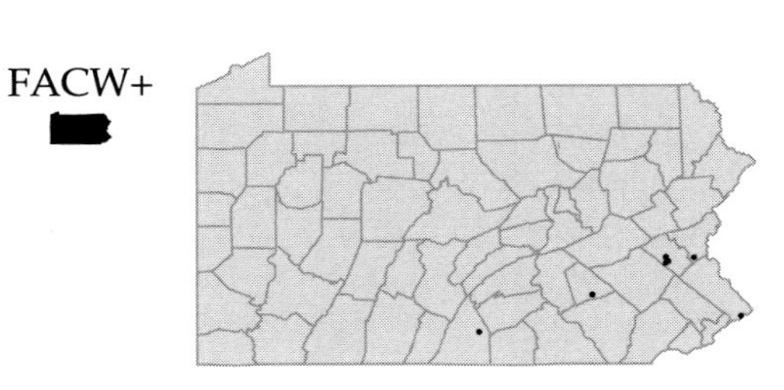

Stachys hyssopifolia Michx. var. ***ambigua*** A.Gray
Hedge-nettle
Herbaceous perennial
Fields and river banks.
Stachys aspera Michx. CG

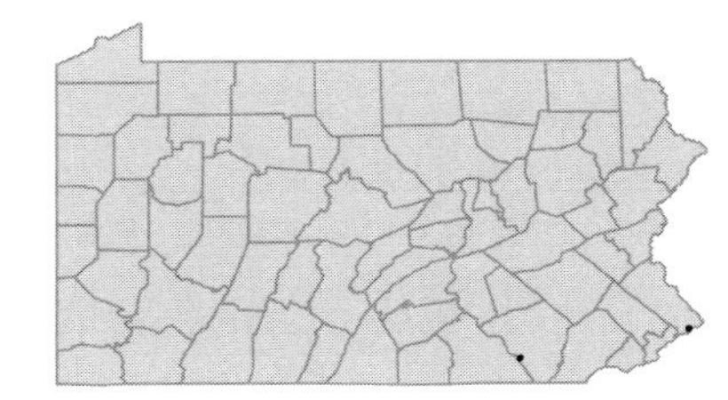

Stachys nuttallii Shuttlw. ex Benth.
Nuttall's hedge-nettle
Herbaceous perennial
Wooded mountain slopes.
Stachys cordata Riddell C

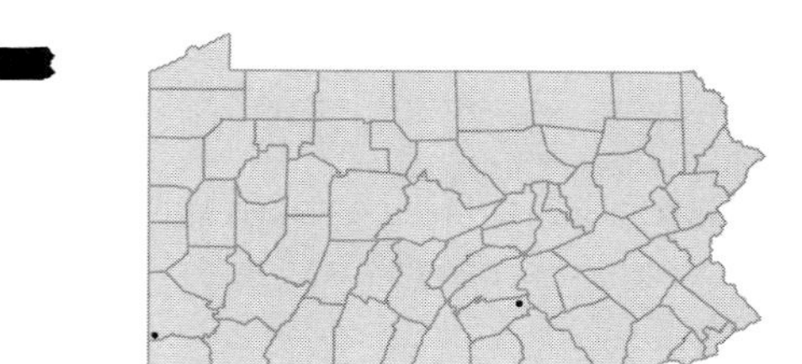

Stachys palustris L. ssp. ***palustris***
Hedge-nettle; Woundwort
Herbaceous perennial
Stream banks, moist fields, woods and edges.
Stachys palustris L. var. *homotrichia* Fern. *pro parte* FGBW; *Stachys palustris* L. var. *nipigonensis* Jennings *pro parte* FW

OBL
I

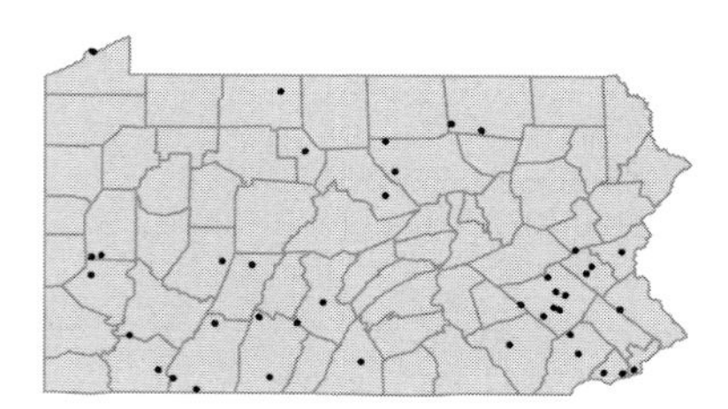

Stachys palustris L. ssp. ***pilosa*** (Nutt.) Fern.
Hedge-nettle; Woundwort
Herbaceous perennial
Campground. Represented by a single collection from Allegheny Co. in 1963.

OBL

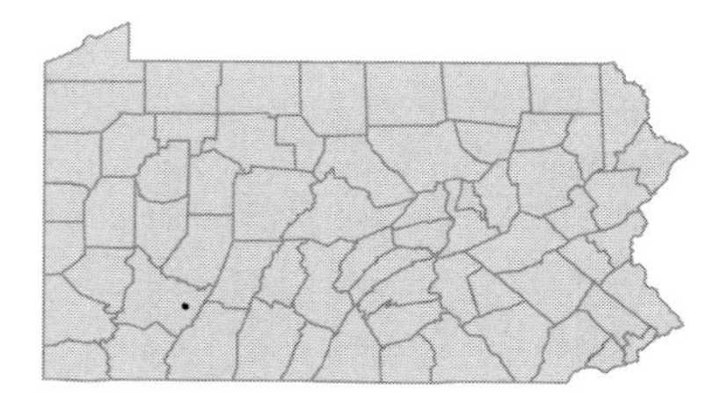

Stachys sylvatica L.
Hedge-nettle
Herbaceous perennial
Ballast. Represented by two collections from Philadelphia Co. 1879-1921.

I

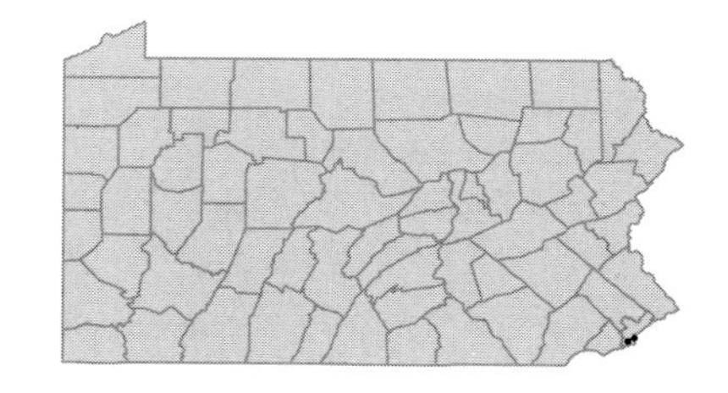

Stachys tenuifolia Willd.
Creeping hedge-nettle
Herbaceous perennial
Moist, wooded bottomland, river banks, wet meadows and fields.

FACW+

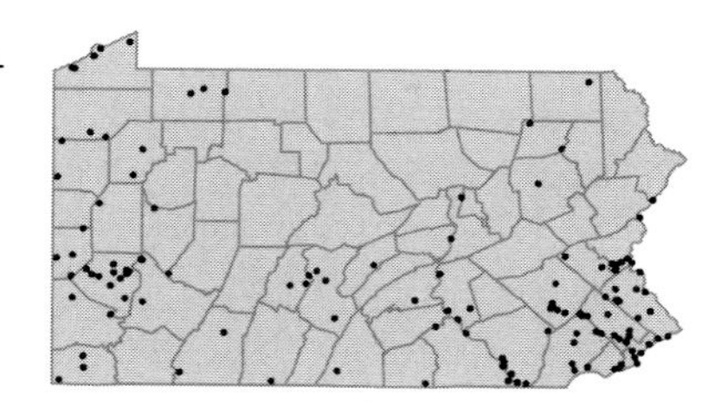

• ***Teucrium botrys*** L.
Cut-leaf germander
Herbaceous annual
Fallow slope. Represented by a single collection from Lehigh Co. in 1960.

I

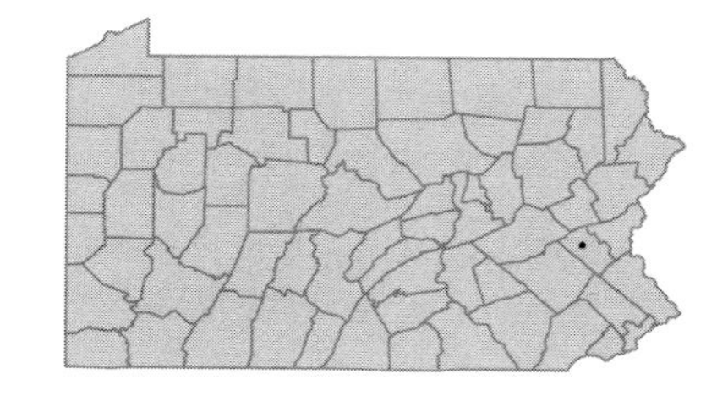

Teucrium canadense L. var. ***boreale*** (Bickn.) Shinners
Northern germander; Wood-sage
Herbaceous perennial
Ballast, wharves, swamps and cultivated ground.
Teucrium canadense L. var. *occidentale* (A.Gray) McClintock & Epling BC;
Teucrium occidentale A.Gray FW

FACW-
[I]

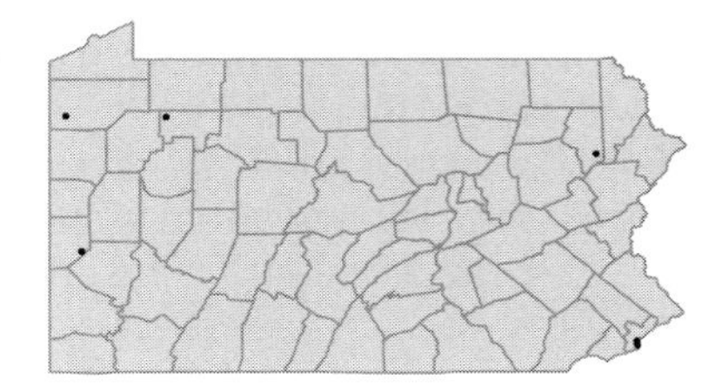

Teucrium canadense L. var. ***virginicum*** (L.) Eat.
Wild germander; Wood-sage
Herbaceous perennial
Floodplains, moist fields, fencerows and lake margins.

FACW
[I]

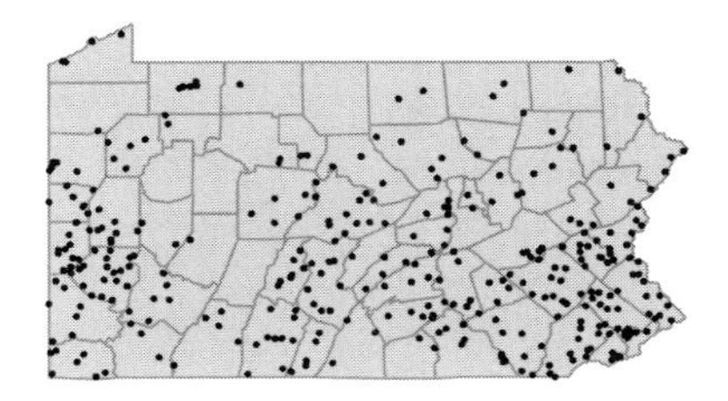

• ***Thymus pulegioides*** L.
Creeping thyme
Herbaceous perennial
Cultivated and occasionally escaped to roadside banks.
Thymus serpyllum sensu auctt., non L. FGBWC

[I]

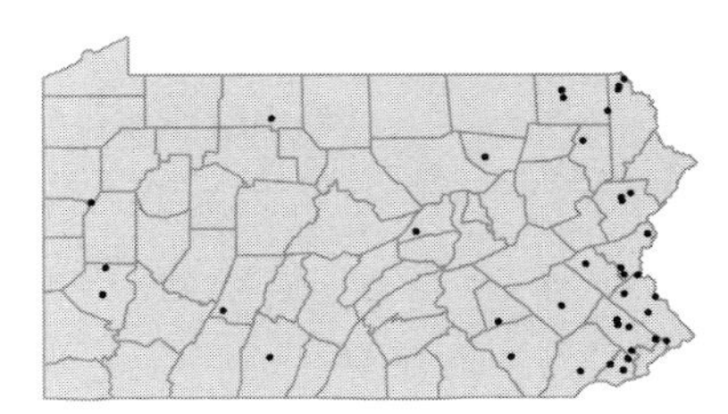

• ***Trichostema brachiatum*** L.
False pennyroyal
Herbaceous perennial
Open woods, rocky shores, shale barrens and dry slopes.
Isanthus brachiatus (L.) BSP PFGBC

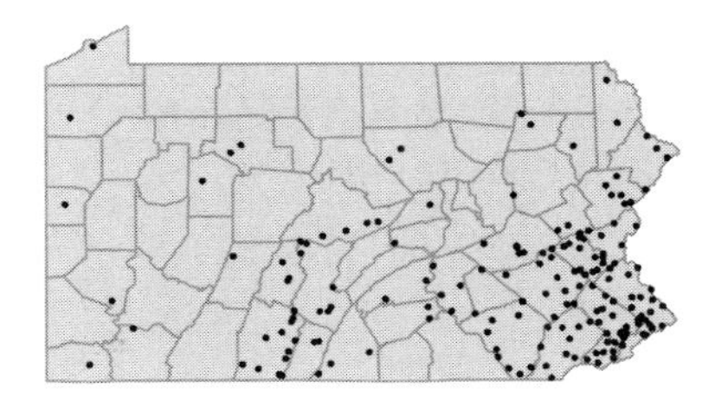

Trichostema dichotomum L.
Blue-curls
Herbaceous annual
Open woods, fields, rock outcrops, barrens, and dry roadsides.

Trichostema setaceum Houtt.
Narrow-leaved blue-curls
Herbaceous annual
Dry, sandy banks and shaly slopes.
Trichostema dichotomum L. var. *lineare* (Walt.) Pursh GB; *Trichostema lineare* Nutt. P

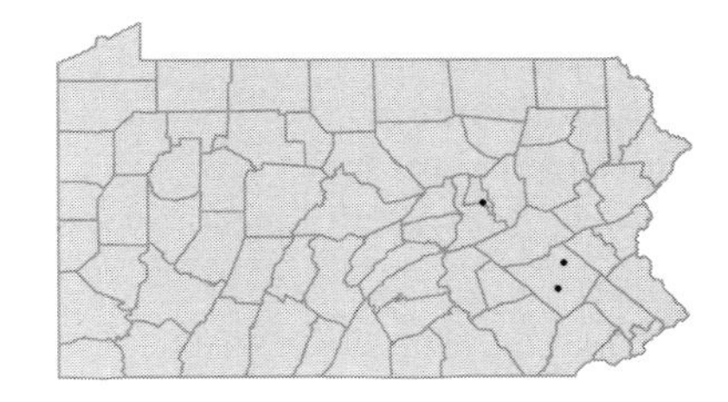

CALLITRICHACEAE

• ***Callitriche heterophylla*** Pursh *emend.* Darby
Water-starwort
Herbaceous perennial, rooted submergent aquatic
Ponds, slow-running streams and muddy shores.

OBL

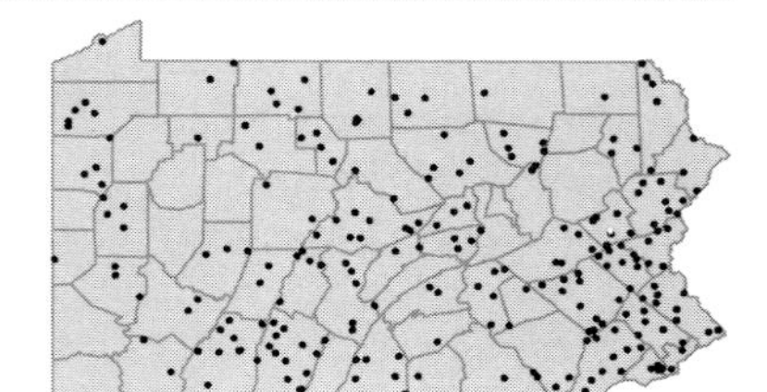

Callitriche palustris L.
Water-starwort
Herbaceous perennial, rooted submergent aquatic
Ponds, stream bottoms, swamps, springs and muddy shores.
Callitriche verna L. *emend.* Keutz. K

OBL

Callitriche stagnalis Scop.
Water-starwort; Water-chickweed
Herbaceous perennial, rooted submergent aquatic
Streams, swamps and ditches.

OBL
[I]

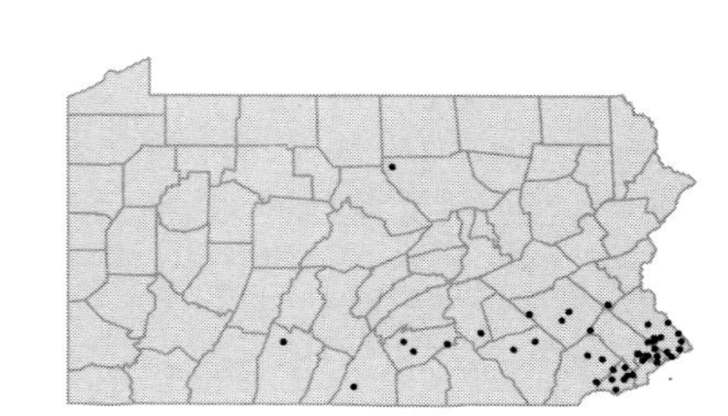

Callitriche terrestris Raf. *emend.* Torr.
Water-starwort
Herbaceous annual, rooted submergent aquatic
Pond edges, stream banks and damp woods.
Callitriche austinii Engelm. PW; *Callitriche deflexa* A.Braun var. *austini* (Engelm.) Hegelm. FGB

FACW+

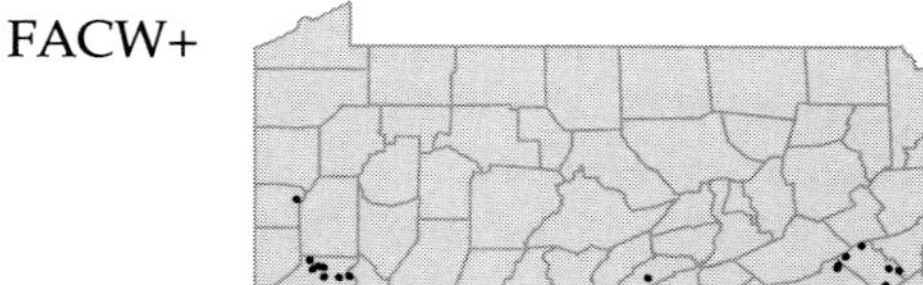

PLANTAGINACEAE

• ***Plantago aristata*** Michx.
Bristly plantain; Buckhorn
Herbaceous annual
Old fields, shale barrens, railroad cinders and other dry, open ground.

[I]

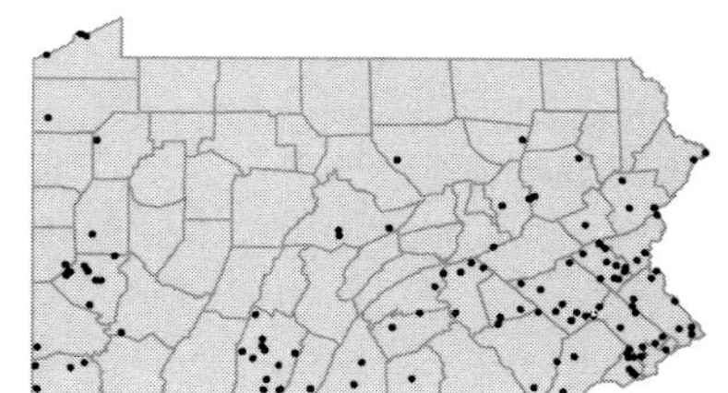

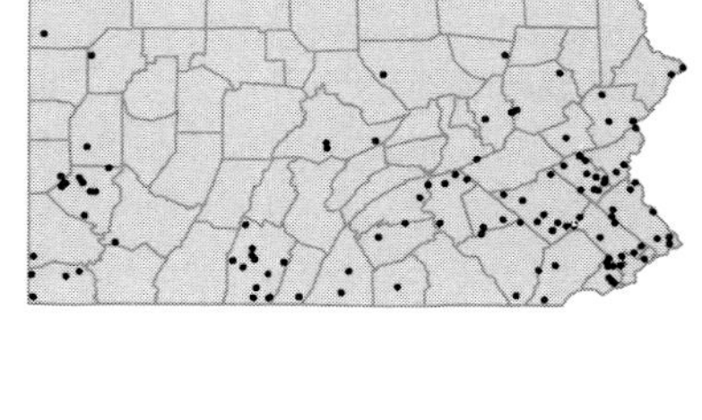

Plantago coronopus L.
Cutleaf plantain
Herbaceous annual
Ballast. Represented a single collection from Philadelphia Co. in 1878.

[I]

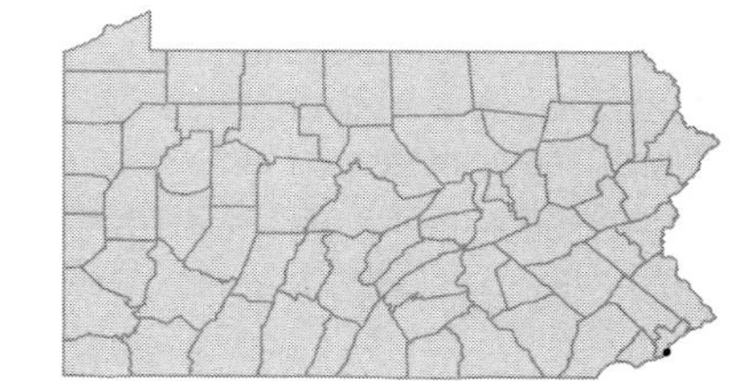

Plantago heterophylla Nutt.
Dwarf plantain
Herbaceous annual
Ballast. Represented by two collections from Philadelphia Co. 1864-1865.

FAC+
I

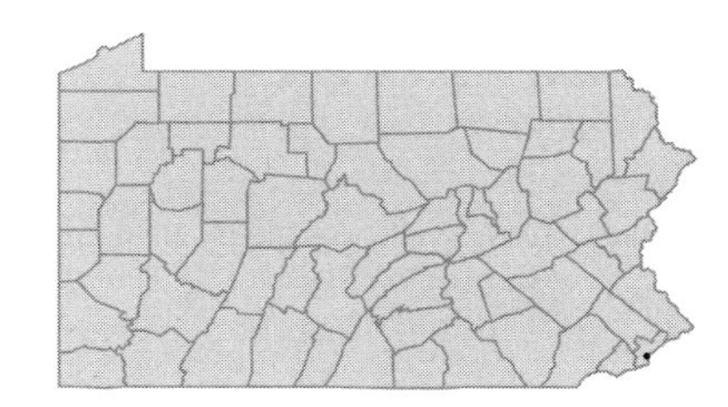

Plantago lanceolata L.
English plantain; Ribgrass
Herbaceous perennial
Lawns, roadsides, old fields, clearings and waste ground.

UPL
I

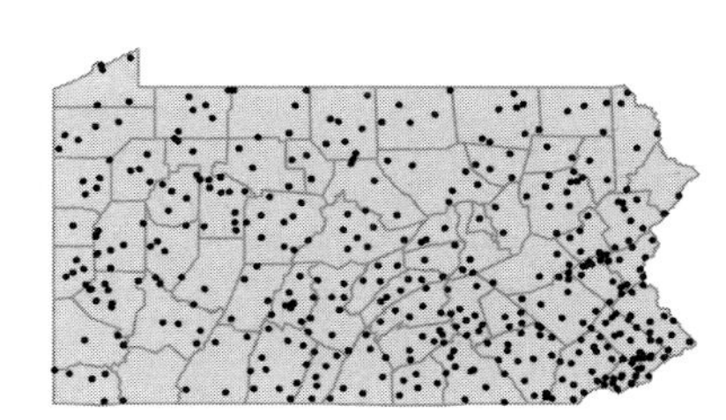

Plantago major L.
Broadleaf plantain; Whiteman's foot
Herbaceous perennial
Lawns, gardens, roadsides, railroad embankments and waste ground. Some authorities believe this to be a native species which increased dramatically in abundance with the clearing of the land.

FACU
I

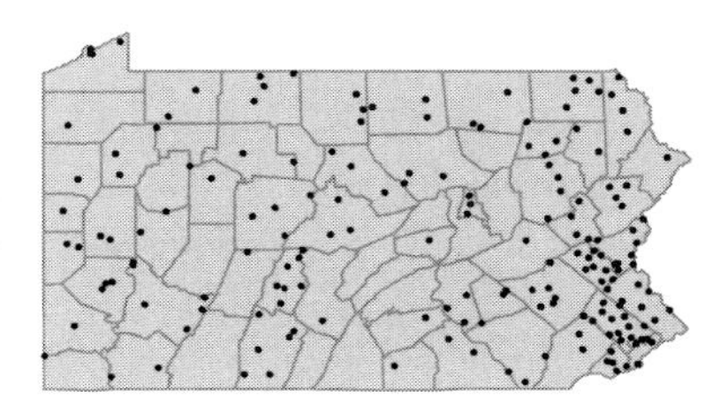

Plantago media L.
Hoary plantain; Lamb's-tongue
Herbaceous perennial
Ballast, waste ground and lawns.

I

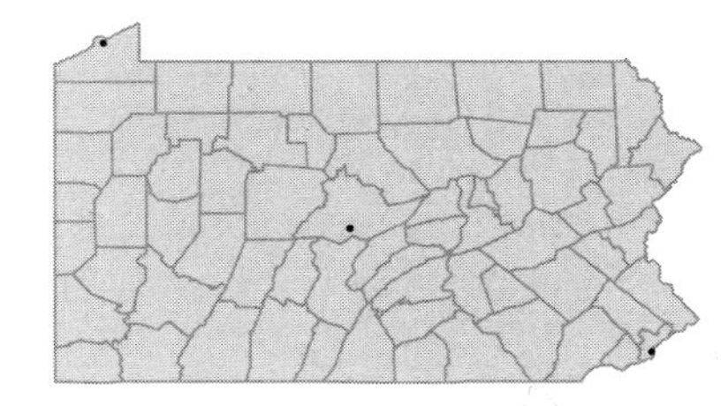

Plantago psyllium L.
Flaxseed plantain; Fleawort
Herbaceous annual
Sandy shores, railroad tracks and other dry waste ground.
Plantago arenaria Waldst. & Kit. G; *Plantago indica* L. BFW

I

Plantago pusilla Nutt.
Dwarf plantain
Herbaceous annual
Dry, sandy or rocky, open ground.

UPL
I

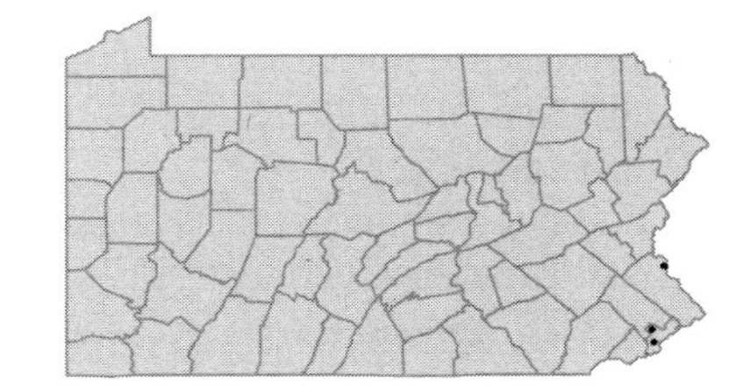

Plantago rugelii Decne. FACU
Rugel's plantain
Herbaceous perennial
Meadows, wet pastures, roadside banks and waste ground.

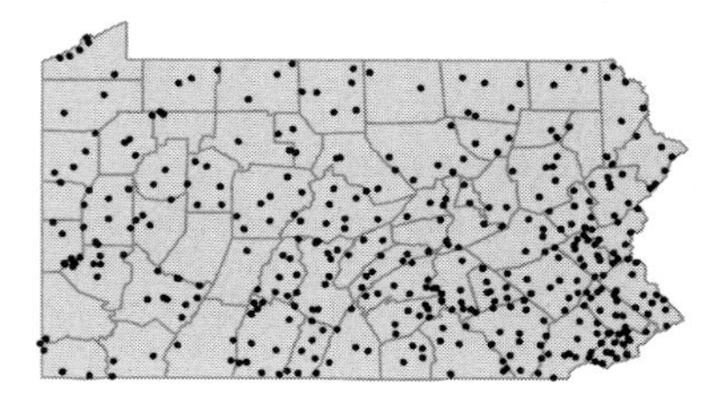

Plantago virginica L. UPL
Dwarf plantain; Hoary plantain; Pale-seeded plantain
Herbaceous annual
Fields, low meadows, open banks and waste ground along railroads.

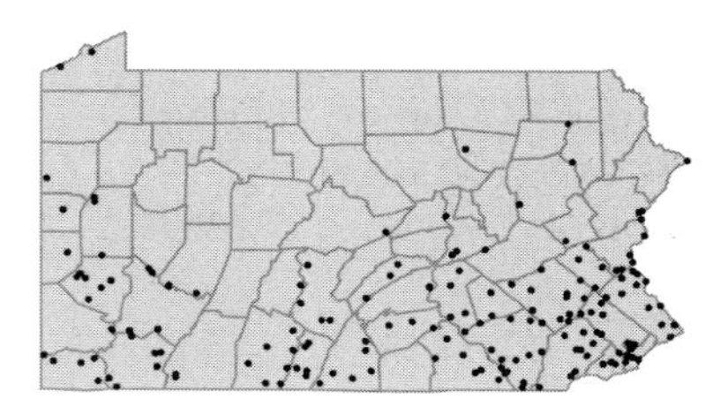

OLEACEAE

• ***Chionanthus virginicus*** L. FAC+
Fringe-tree
Deciduous tree
Moist, open woods and woods edges, also cultivated.

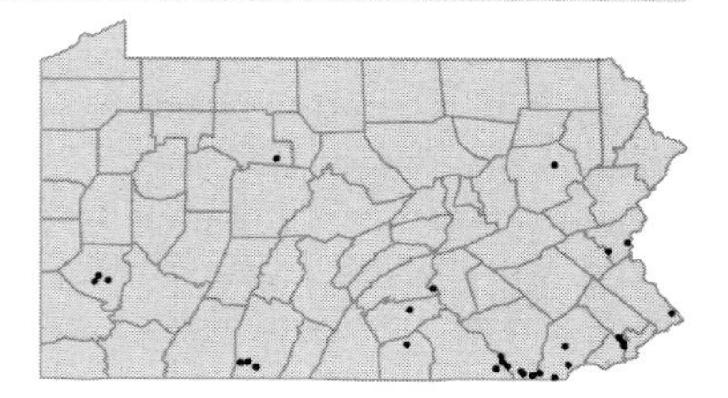

• ***Forsythia suspensa*** (Thunb.) Vahl [I]
Forsythia; Golden-bells
Deciduous shrub
Cultivated and occasionally escaped to roadsides or persisting at abandoned homesites.

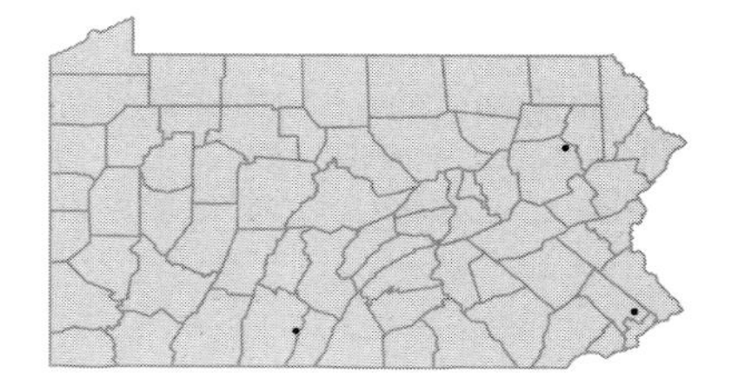

Forsythia viridissima Lindl. [I]
Forsythia; Golden-bells
Deciduous shrub
Cultivated and occasionally escaped to roadsides.

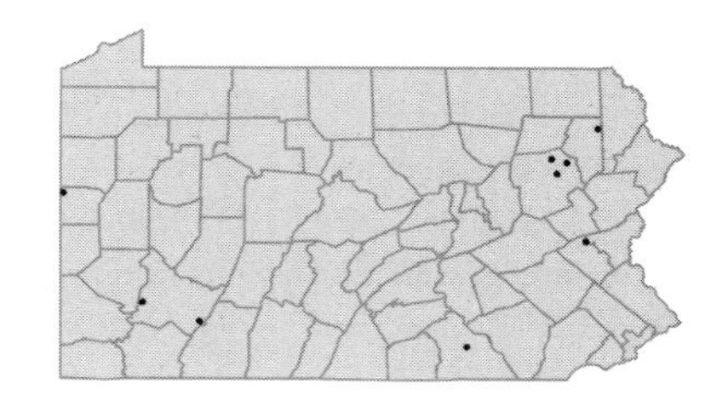

• ***Fraxinus americana*** L. var. ***americana*** FACU
White ash
Deciduous tree
Woods, fencerows and old fields.
Fraxinus americana L. var. *juglandifolia* (Lam.) Rehd. *pro parte* W; *Fraxinus iodocarpa* Fern. *pro parte* W

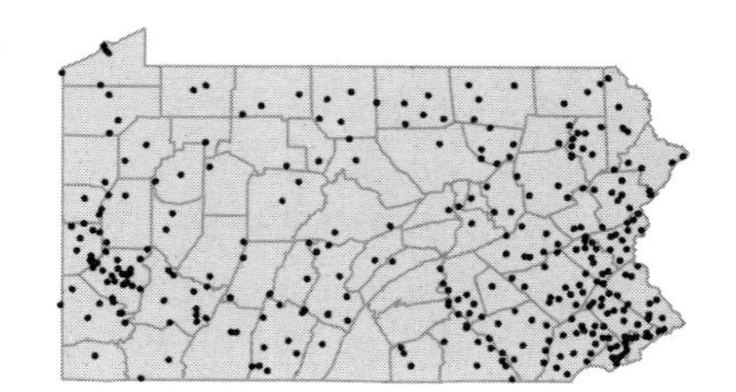

Fraxinus americana L. var. ***biltmoreana*** (Beadle) J.Wright — FACU
Biltmore ash
Deciduous tree
Rich, wooded slopes, river banks, fencerows and roadsides.
Fraxinus biltmoreana Beadle PG

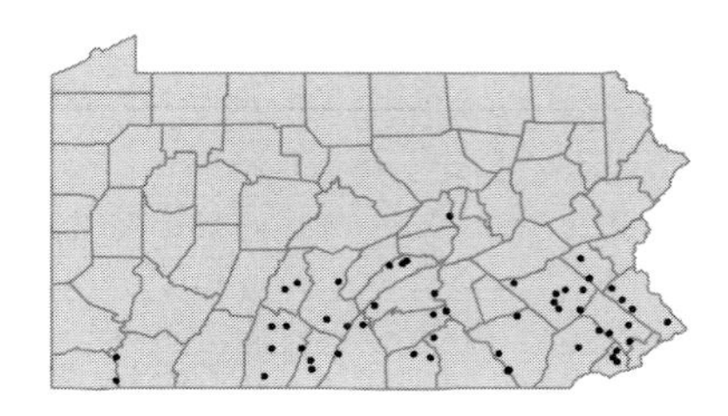

Fraxinus nigra Marshall — FACW
Black ash
Deciduous tree
Swamps, wet woods and bottomlands.

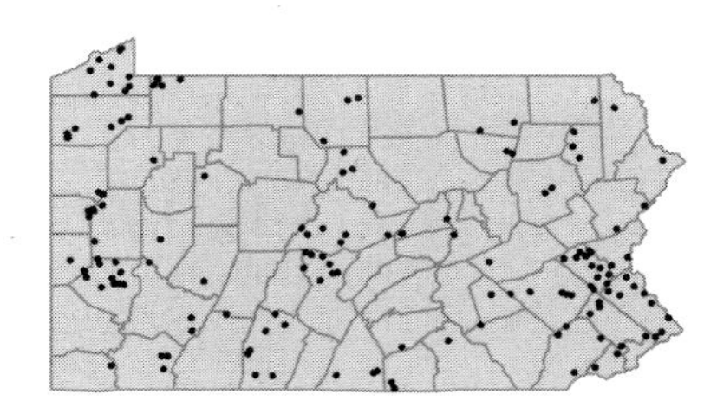

Fraxinus pennsylvanica Marshall — FACW
Red ash
Deciduous tree
Alluvial woods, stream banks and moist fields.

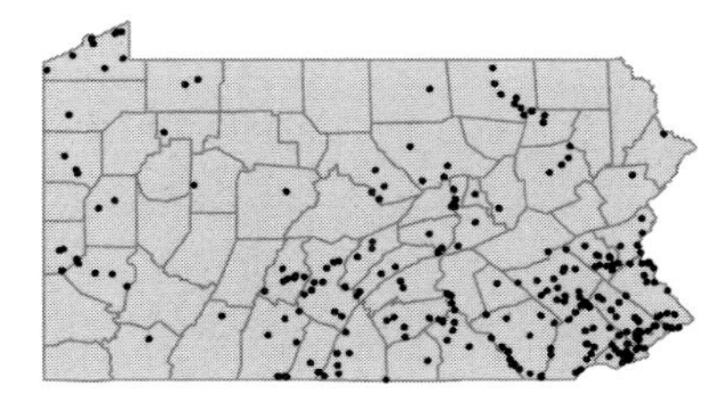

Fraxinus profunda (Bush) Bush — OBL
Pumpkin ash
Deciduous tree
Shallow woodland ponds and wet, wc oded flats.
Fraxinus tomentosa Michx. f. BGF

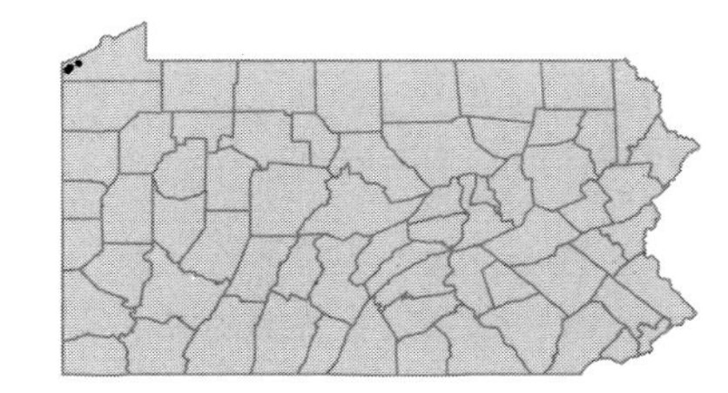

• ***Ligustrum amurense*** Carr. — [I]
Amur privet
Deciduous shrub
Cultivated and rarely spreading to thickets or rubbish dumps.

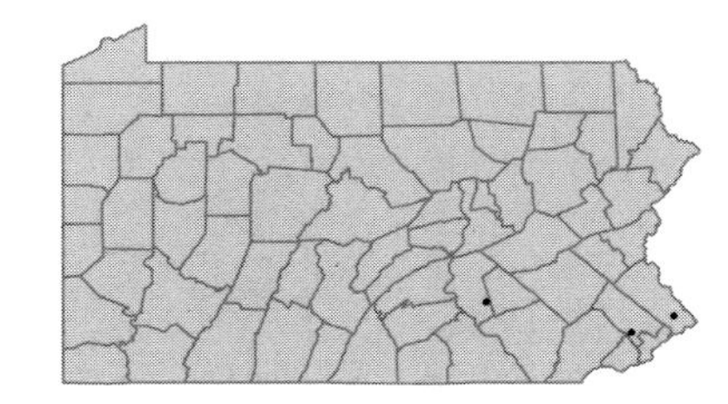

Ligustrum obtusifolium Sieb. & Zucc. — [I]
Privet
Deciduous shrub
Cultivated and frequently naturalized in disturbed woods, thickets, hedgerows and old fields.

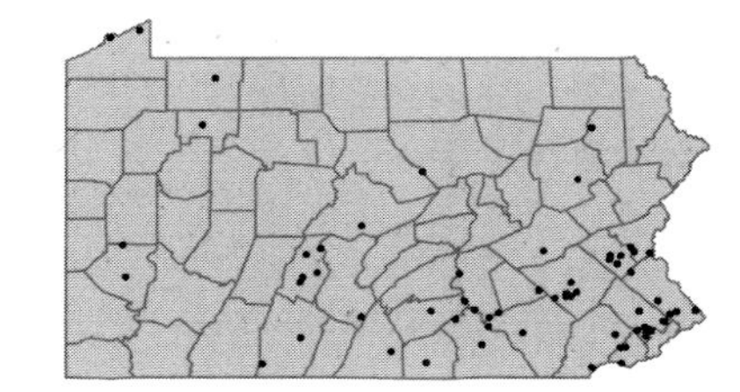

Ligustrum ovalifolium Hassk.
California privet
Deciduous shrub
Cultivated and occasionally escaped to roadsides and thickets.

I

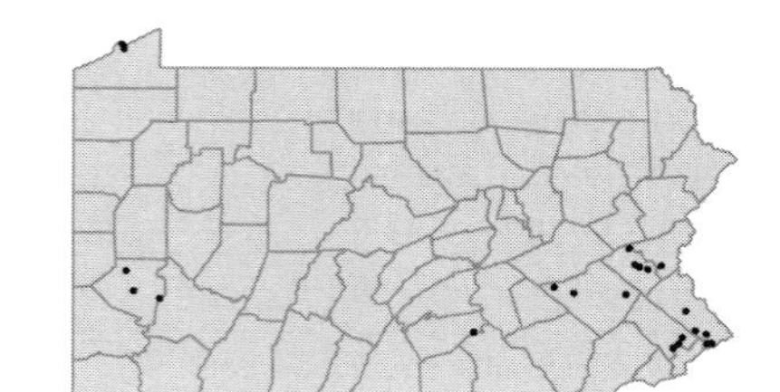

Ligustrum vulgare L.
Common privet
Deciduous shrub
Cultivated and frequently escaped to roadside banks, woods edges and waste ground.

I

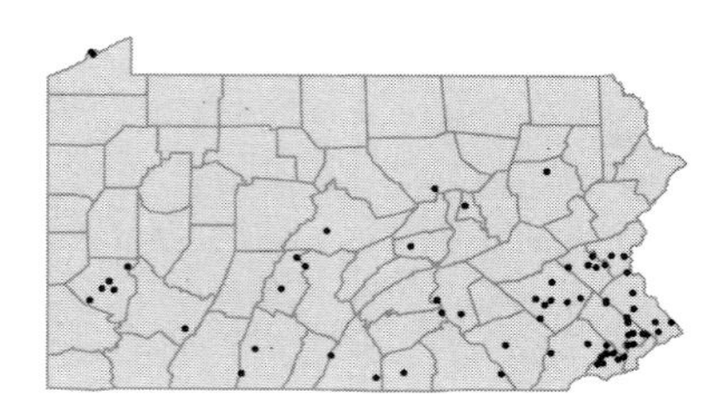

• ***Syringa vulgaris*** L.
Lilac
Deciduous shrub
Cultivated and occasionally persisting on roadsides or old home sites.

I

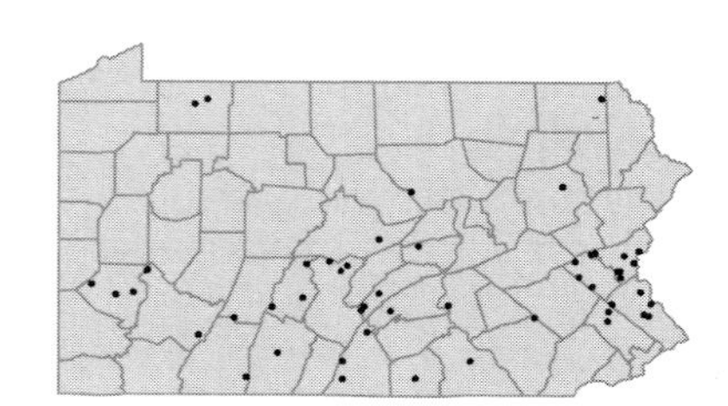

SCROPHULARIACEAE

• ***Agalinis auriculata*** (Michx.) S.F.Blake
Eared false-foxglove
Herbaceous annual
Wet meadows, fields, roadsides and waste ground along railroad tracks. Parasitic on the roots of several herbaceous species.
Gerardia auriculata Michx. F; *Tomanthera auriculata* (Michx.) Raf. WKBG

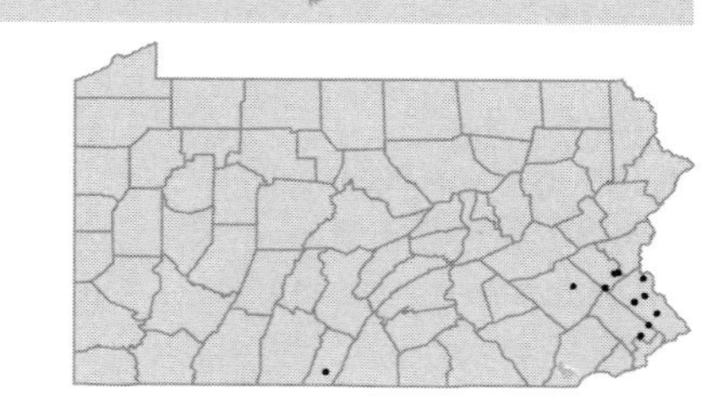

Agalinis decemloba (Greene) Pennell
Blue-ridge false-foxglove
Herbaceous annual
Dry roadside banks and serpentine barrens, parasitic on the roots of various woody and herbaceous species. Believed to be extirpated, last collected in 1907.
Gerardia decemloba Greene F; *Agalinis obtusifolia* Raf. *pro parte* C

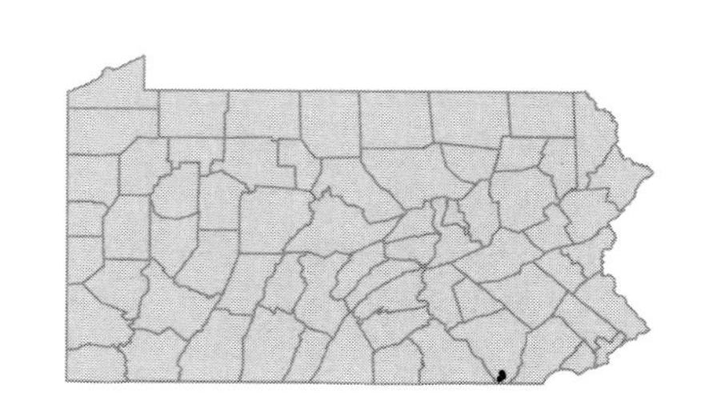

Agalinis fasciculata (Ell.) Raf.
False-foxglove
Herbaceous annual
Ballast. A root parasite of various plants. Represented by a single collection from Philadelphia Co. 1864.
Gerardia fasciculata Ell. FBG

FAC

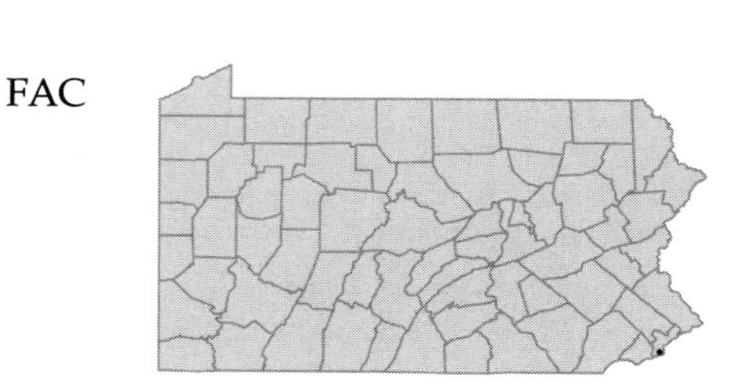

Agalinis paupercula (A.Gray) Britt.
Small-flowered false-foxglove
Herbaceous annual
Moist, open, sandy ground and pond shores. Parasitic on the roots of various plants.
Gerardia paupercula (A.Gray) Britt. PF; *Gerardia purpurea* L. var. *parviflora* Benth. GBC

FACW+

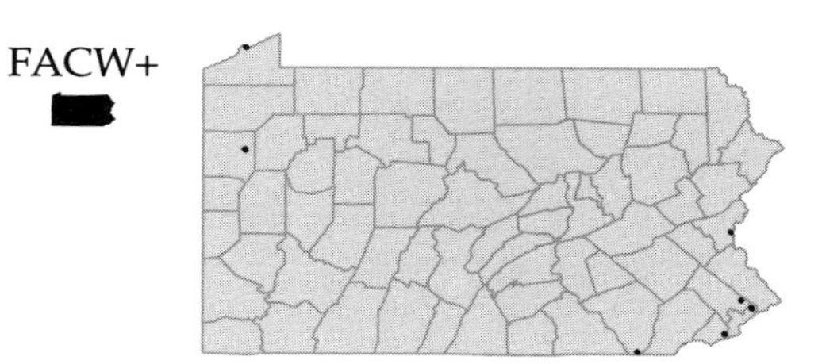

Agalinis purpurea (L.) Pennell
False-foxglove
Herbaceous annual
Moist, sandy fields, rocky shores and serpentine barrens. Parasitic on the roots of many woody and herbaceous plants.
Gerardia purpurea L. PFGB

FACW-

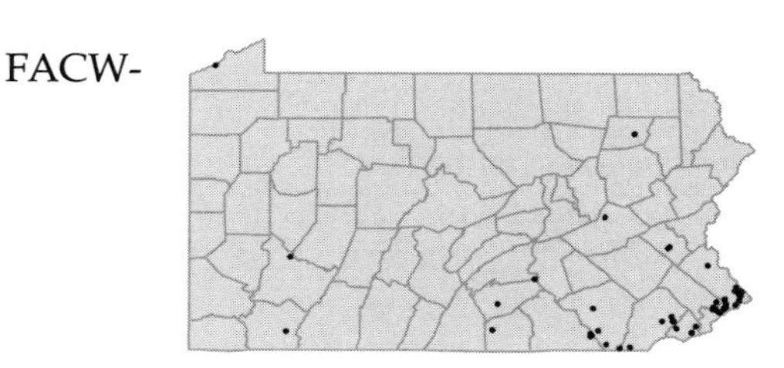

Agalinis tenuifolia (Vahl) Raf.
False-foxglove
Herbaceous annual
Dry fields, meadows and open woods. Parasitic on the roots of many woody and herbaceous plants.
Gerardia tenuifolia Vahl PFGB

FAC

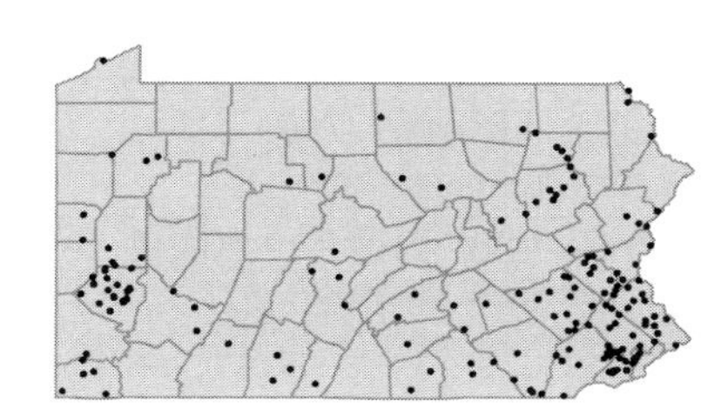

• ***Antirrhinum majus*** L.
Snapdragon
Herbaceous perennial
Cultivated and occasionally escaped to waste ground or rubbish dumps.

I

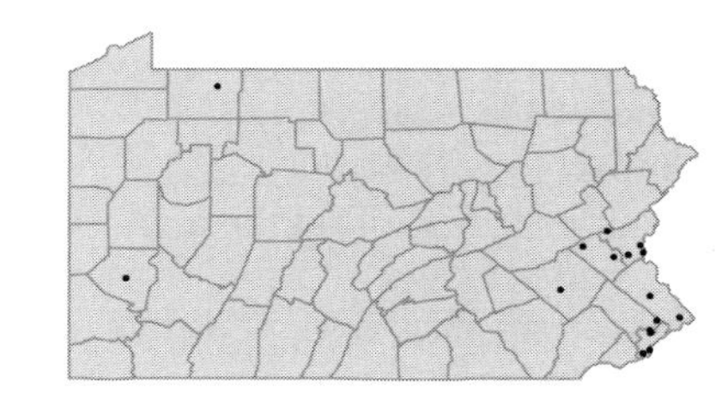

Antirrhinum orontium L.
Lesser snapdragon
Herbaceous annual
Ballast.

I

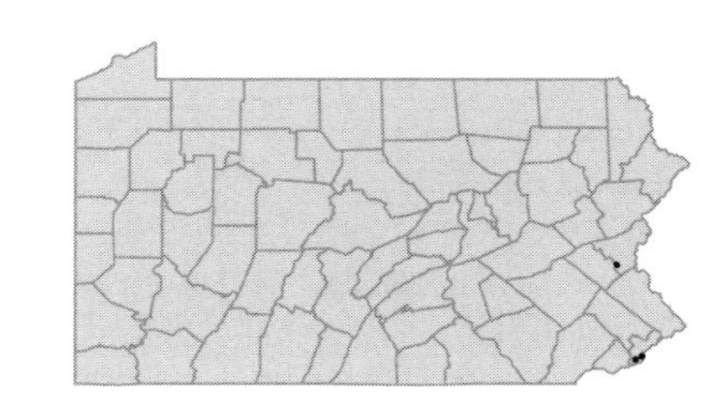

• ***Aureolaria flava*** (L.) Farw. var. ***flava***
Yellow false-foxglove
Herbaceous perennial
Dry, open woods and rocky thickets. Parasitic on the roots of the white oaks and other trees.
Dasystoma flava (L.) Wood P; *Gerardia flava* L. F

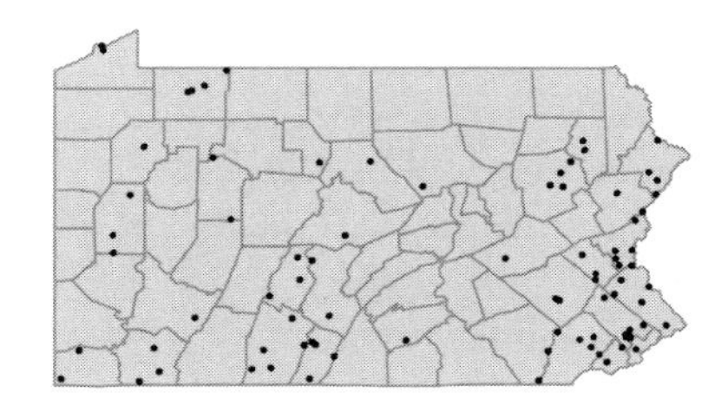

Aureolaria flava (L.) Farw. var. ***macrantha*** Pennell
Yellow false-foxglove
Herbaceous perennial
Deciduous woods. Parasitic on the roots of the white oaks and other trees.
Gerardia flava L. var. *macrantha* (Pennell) Fern. F

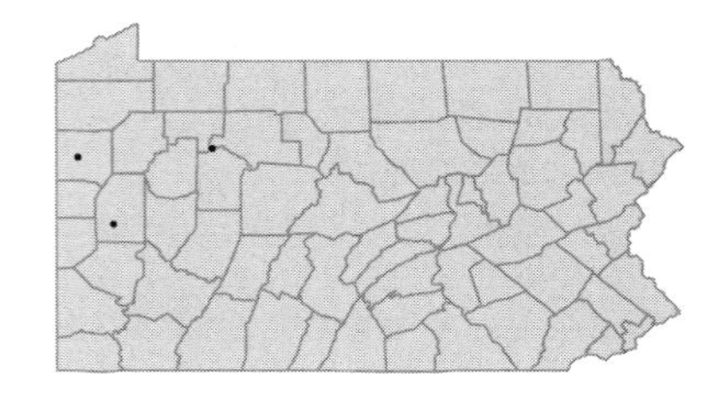

Aureolaria laevigata (Raf.) Raf.
False-foxglove
Herbaceous perennial
Rocky, open woods. Parasitic on the roots of the white oaks and other trees.
Dasystoma laevigata (Raf.) Chapman P; *Gerardia laevigata* Raf. F

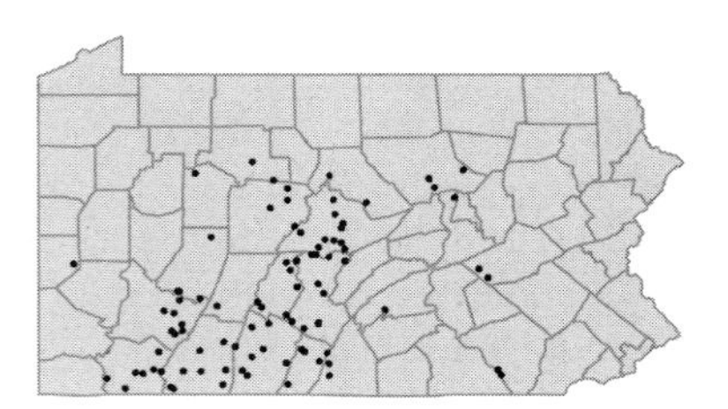

Aureolaria pedicularia (L.) Raf.
Cutleaf false-foxglove
Herbaceous annual
Dry, open woods and edges. Parasitic on the roots of the black oak group and occasionally other trees.
Dasystoma pedicularia (L.) Benth. P; *Gerardia pedicularia* L. F

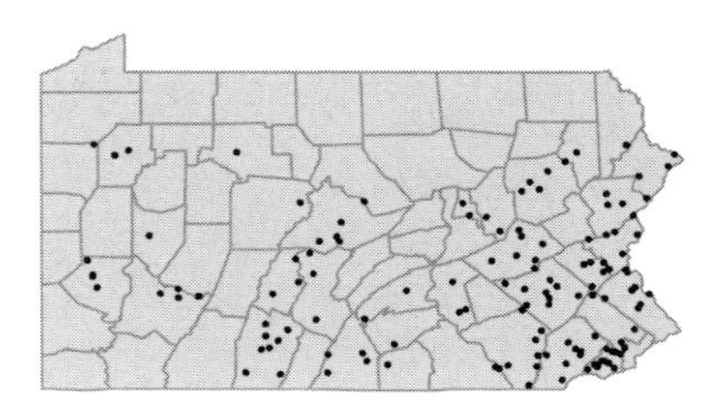

Aureolaria virginica (L.) Pennell
Downy false-foxglove
Herbaceous perennial
Dry, open, deciduous woods. Parasitic on the roots of the white oaks and other trees.
Dasystoma virginica (L.) Britt. P; *Gerardia virginica* (L.) BSP F

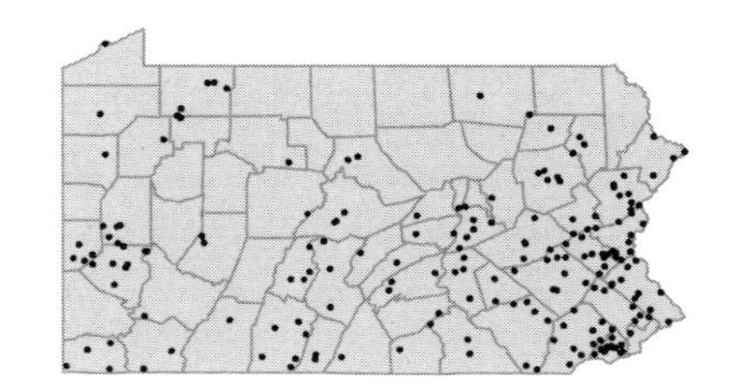

• ***Buchnera americana*** L.
Blue-hearts
Herbaceous biennial
Moist, sandy ground. Parasitic on the roots of other plants. Believed to be extirpated, last collected in 1905.

FACU

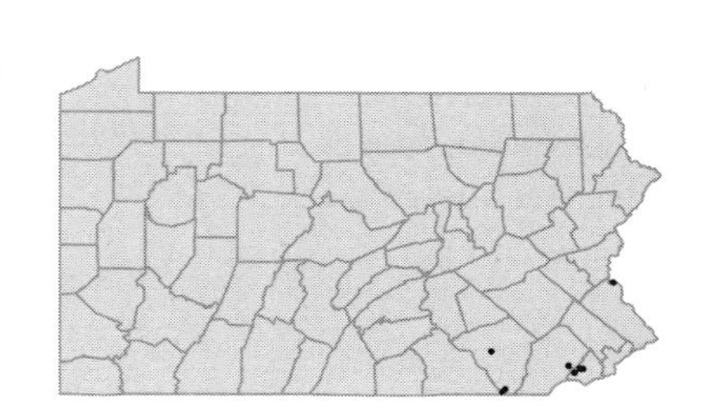

• ***Castilleja coccinea*** (L.) Spreng.
Indian paintbrush; Painted-cup
Herbaceous annual or biennial.
Moist meadows. Parasitic on the roots of other plants.

FAC

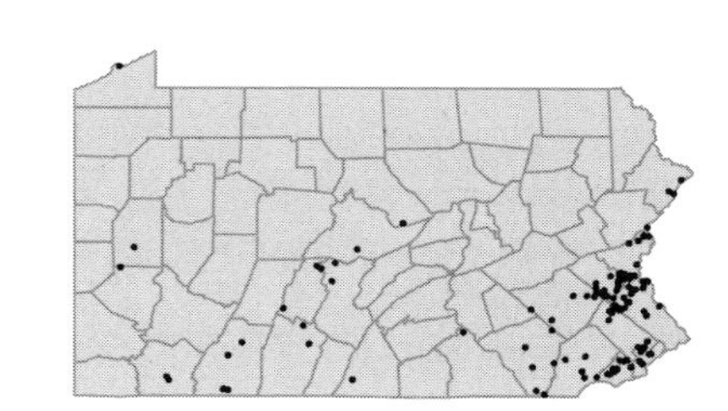

• ***Chaenorrhinum minus*** (L.) Lange
Dwarf snapdragon
Herbaceous annual
Railroad tracks, cinders and waste ground.

[I]

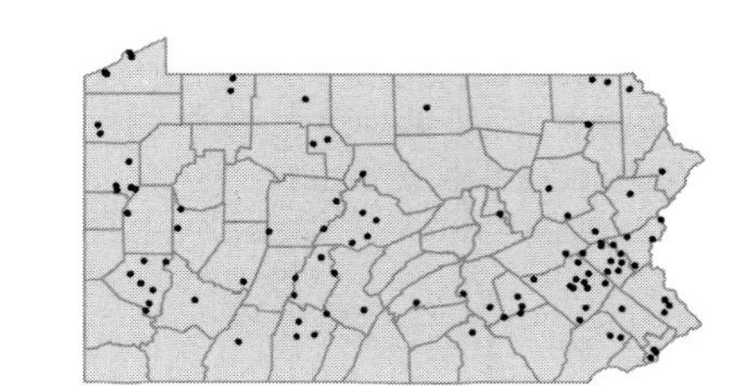

• ***Chelone glabra*** L.
Turtlehead
Herbaceous perennial
Stream banks, wet woods and swamps. Map includes vars. *elatior* and *linifolia* which have been collected in western counties primarily. The typical variety occurs throughout the state.

OBL

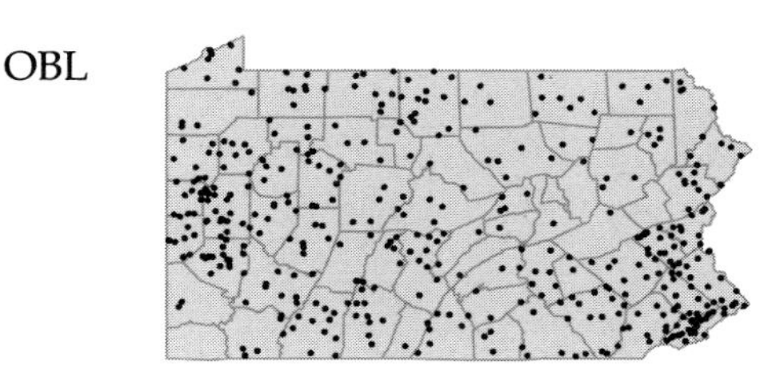

• ***Collinsia parviflora*** Dougl. ex Lindl.
Blue-eyed-Mary
Herbaceous annual
Waste ground and ballast. Represented by two collections from Philadelphia Co. in 1865.

[I]

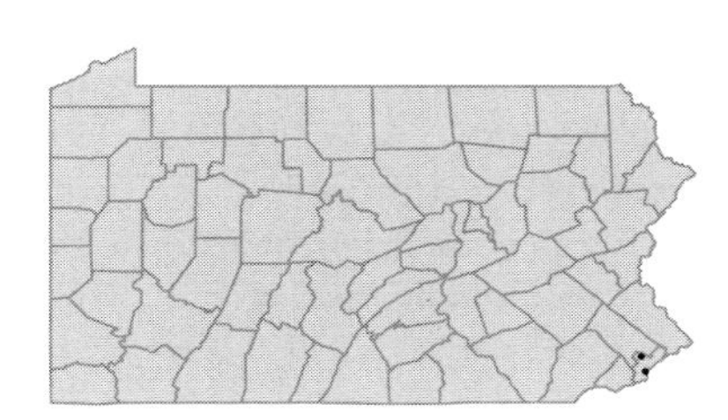

Collinsia verna Nutt.
Blue-eyed-Mary
Herbaceous annual
Wooded floodplains and alluvial thickets.

FAC-

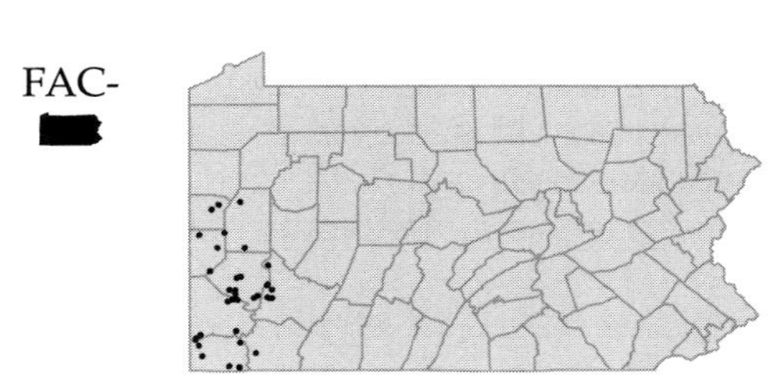

• ***Cymbalaria muralis*** Gaertn.,Mey. & Schreb.
Kenilworth-ivy
Herbaceous annual
Frequently naturalized in crevices in rock walls and other stone work.
Cymbalaria cymbalaria (L.) Wetts. P

[I]

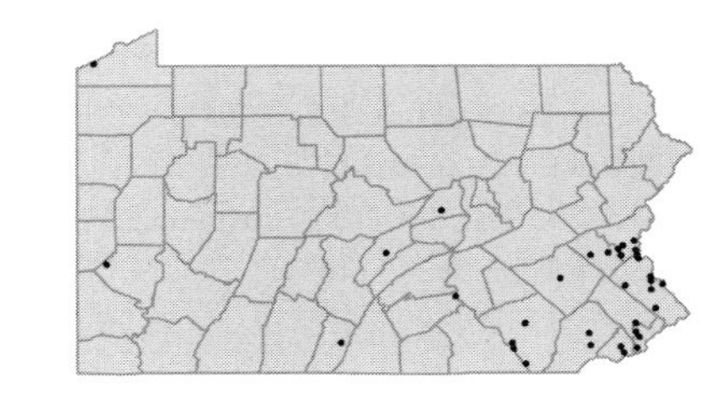

• ***Digitalis lanata*** Ehrh.
Wooly foxglove
Herbaceous perennial
Cultivated and rarely escaped to roadsides or wooded edges.

[I]

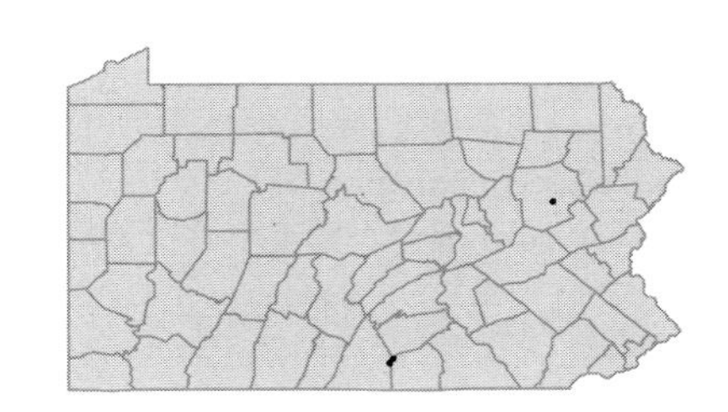

Digitalis lutea L.
Yellow foxglove
Herbaceous perennial
Cultivated and occasionally escaped to roadsides.

[I]

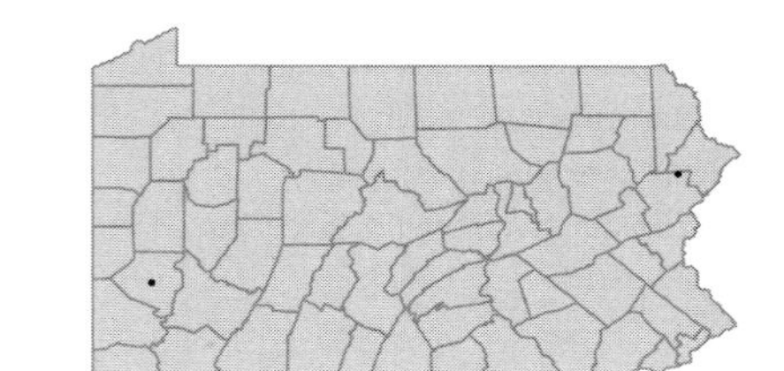

Digitalis purpurea L.
Foxglove
Herbaceous biennial
Cultivated and frequently naturalized in meadows, woods and roadsides.

[I]

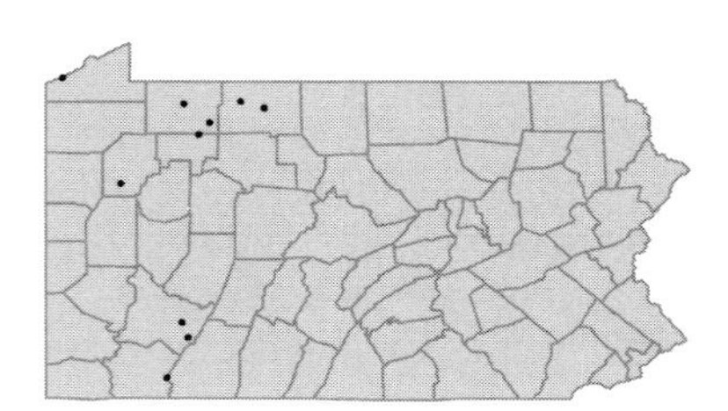

• ***Euphrasia stricta*** J.P.Wolff ex J.F.Lehm.
Eyebright
Herbaceous annual
Gravelly roadsides and sterile fields.
Euphrasia officinalis L. GC; *Euphrasia rigidula* Jord. FW

[I]

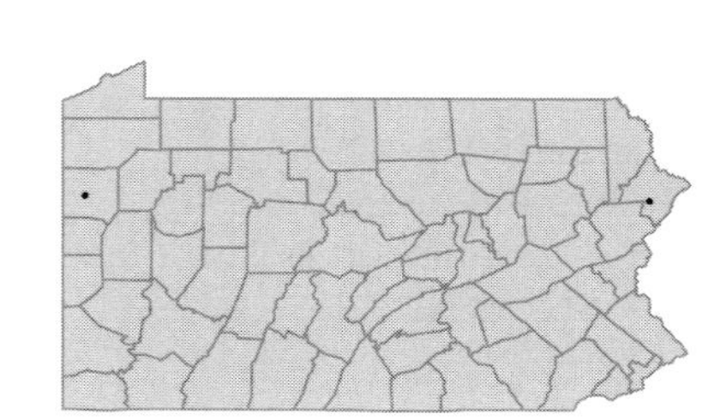

• ***Gratiola aurea*** Muhl. ex Pursh
Golden-pert; Hedge-hyssop
Herbaceous perennial
Moist river banks and sandy pond shores.

OBL

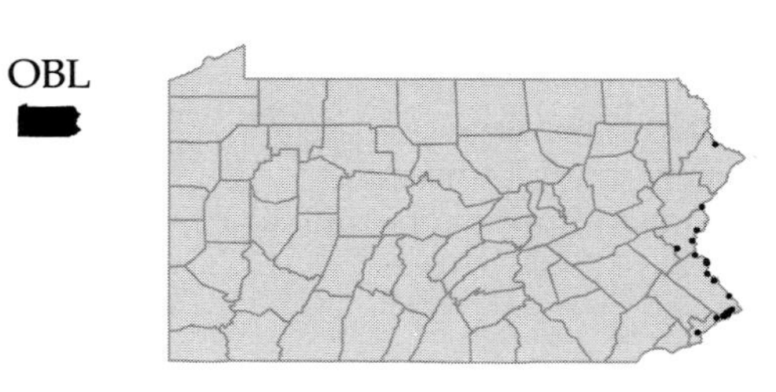

Gratiola neglecta Torr.
Hedge-hyssop; Mud-hyssop
Herbaceous annual
River banks, muddy shores, bogs, wet fields and ditches.

OBL

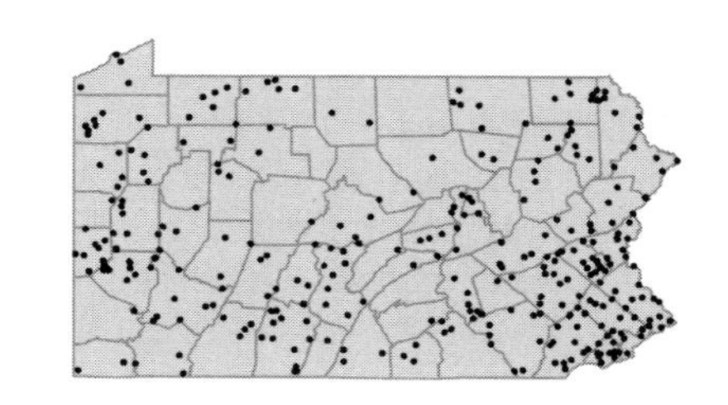

• ***Kickxia elatine*** (L.) Dumort.
Cancerwort
Herbaceous annual
Rocky slopes, waste ground and ballast.

FAC
[I]

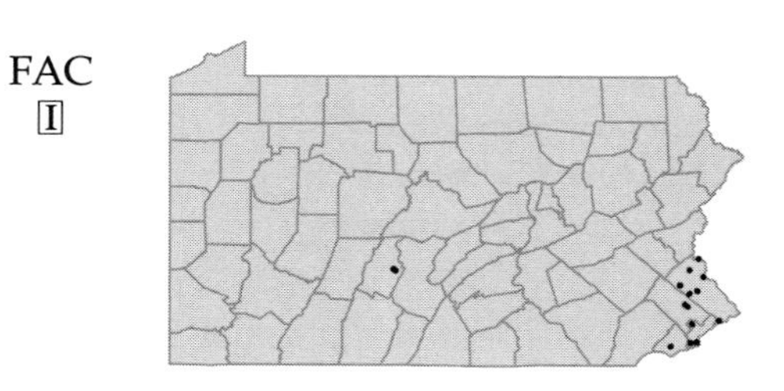

Kickxia spuria (L.) Dumort.
Cancerwort
Herbaceous annual
Waste ground and ballast.

I

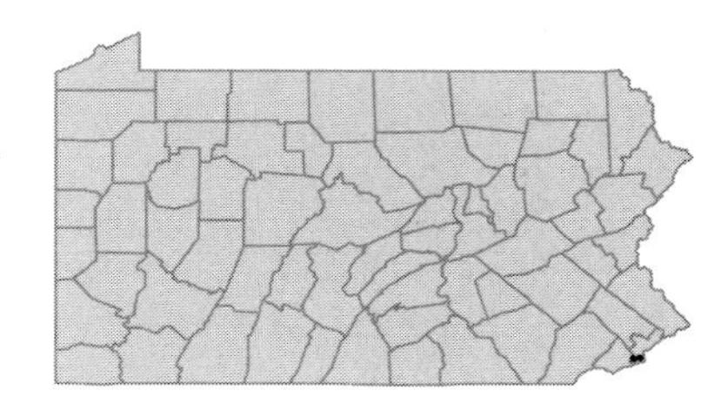

• ***Leucospora multifida*** (Michx.) Nutt.
Narrow-leaved paleseed
Herbaceous annual
Waste ground and ballast.
Conobea multifida (Michx.) Benth. FW

OBL
I

• ***Limosella australis*** R.Br.
Awl-shaped mudwort
Herbaceous annual
Tidal mudflats. Believed to be extirpated, last collected in 1917.
Limosella subulata Ives FGBWC

OBL

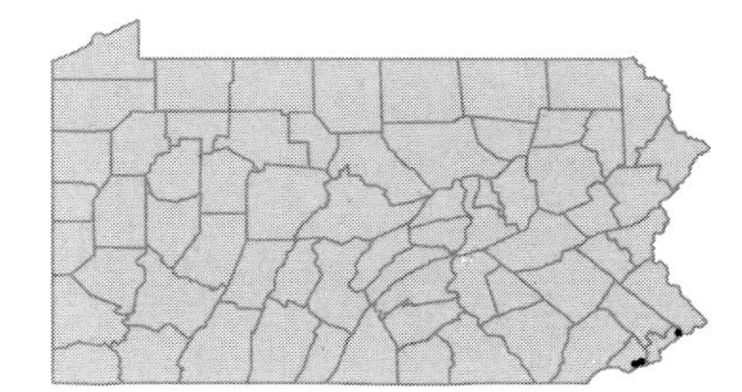

• ***Linaria canadensis*** (L.) Dum.-Cours.
Old-field toadflax
Herbaceous annual
River banks, sandy fields, serpentine barrens and railroad embankments.

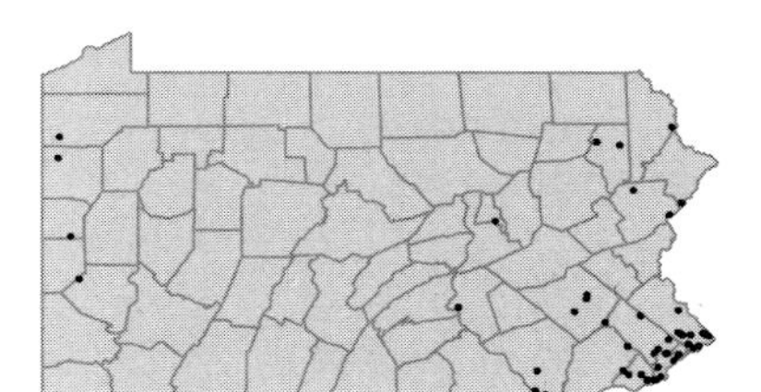

Linaria genistifolia (L.) P.Mill. ssp. ***dalmatica*** (L.) Maire & Petitmengen
Toadflax
Herbaceous perennial
Cultivated and occasionally escaped to railroad banks and roadsides.
Linaria dalmatica (L.) P.Mill. FGBWC

I

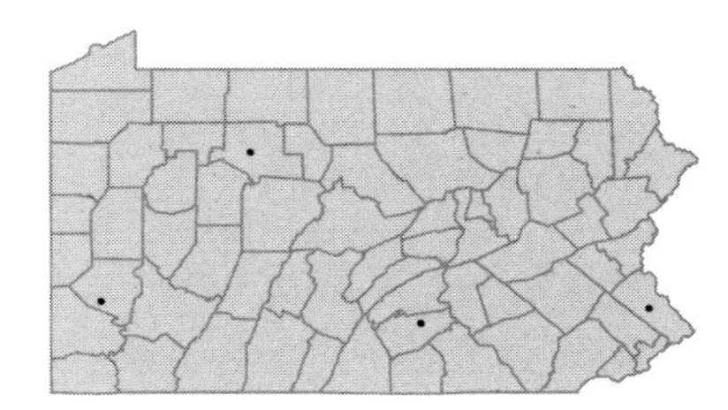

Linaria supina (L.) Chaz.
Toadflax
Herbaceous annual
Ballast. Represented by several collections from Philadelphia Co. 1879-1880.

I

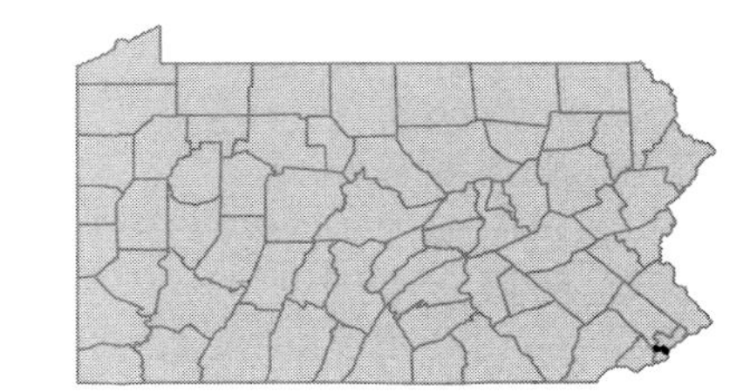

Linaria vulgaris Hill
Butter-and-eggs
Herbaceous perennial
Fields, roadsides, shale barrens, railroad tracks and waste ground.

[I]

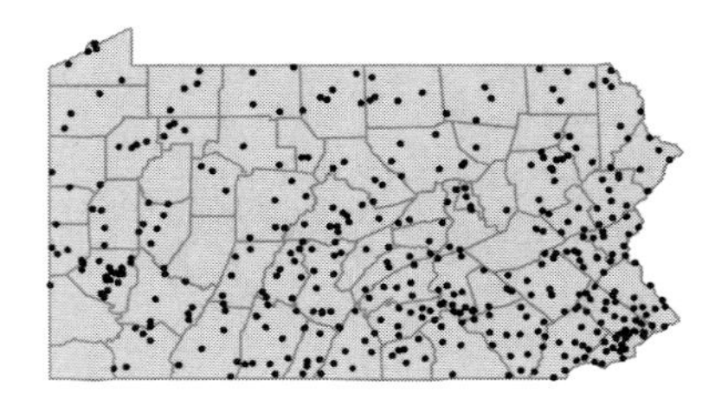

• ***Lindernia dubia*** (L.) Pennell var. ***dubia***
False pimpernel
Herbaceous annual
Swamps and muddy shores.
Lindernia dubia (L.) Pennell var. *riparia* (Raf.) Fern. FBW

OBL

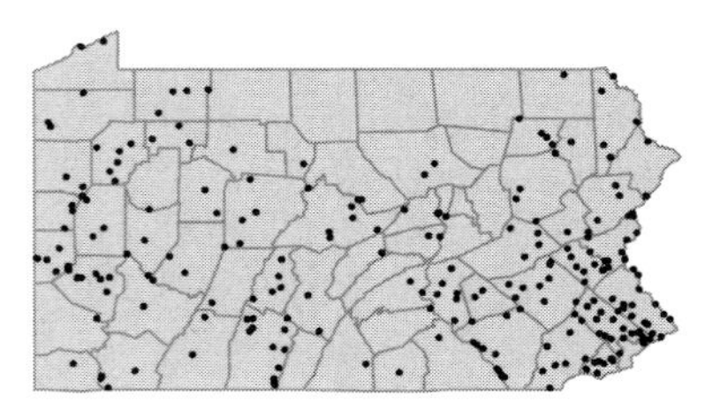

Lindernia dubia (L.) Pennell var. ***anagallidea*** (Michx.) Cooperrider
False pimpernel
Herbaceous annual
Moist meadows, stream banks, swampy thickets and ditches.
Lindernia anagallidea (Michx.) Pennell FGBW

OBL

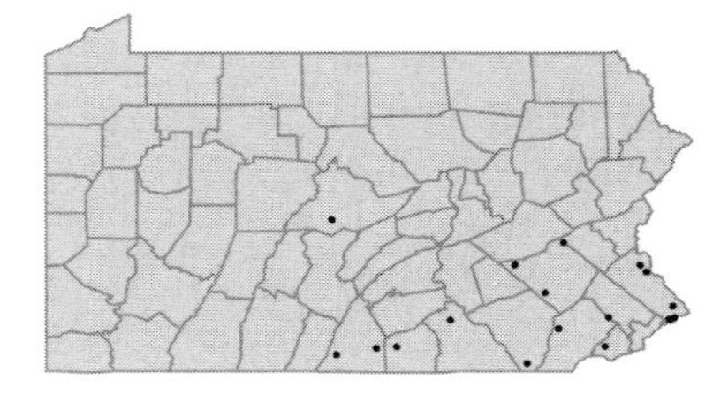

Lindernia dubia (L.) Pennell var. ***inundata*** (Pennell) Pennell
False pimpernel
Herbaceous annual
Wet shores and tidal mudflats.

OBL

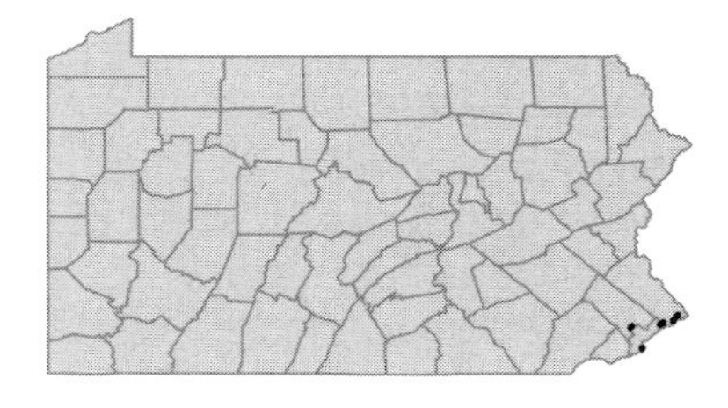

• ***Mazus miguelii*** Makino
Mazus
Herbaceous perennial
Lawns, gardens and roadsides, escaped from cultivation.
Mazus reptans N.E.Br. GW

[I]

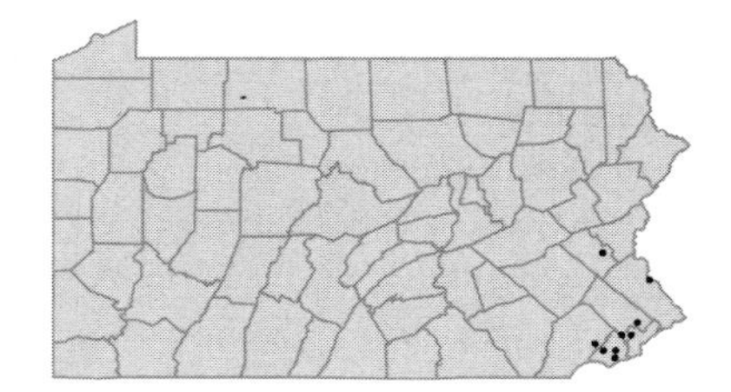

Mazus pumilus (Burm.f.) Steenis
Japanese mazus
Herbaceous perennial
Cultivated and occasionally escaped to lawns or moist, alluvial shores.
Mazus japonicus (Thunb.) Kuntze FGBW

[I]

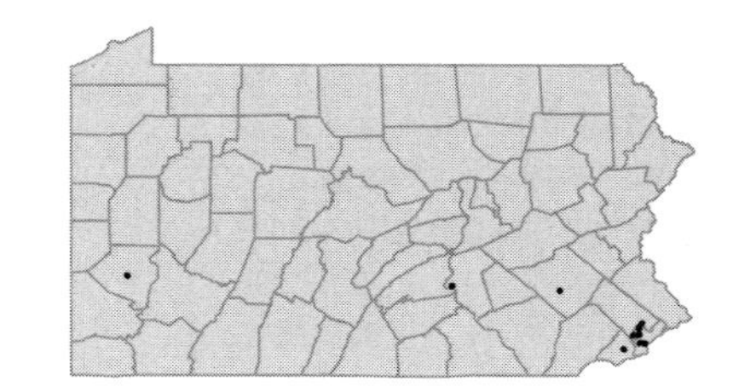

• ***Melampyrum lineare*** Desr. var. ***americanum*** (Michx.) Beauv.
Cow-wheat
Herbaceous annual
Dry, open woods and barrens. Parasitic on the roots of other plants.

FACU
I

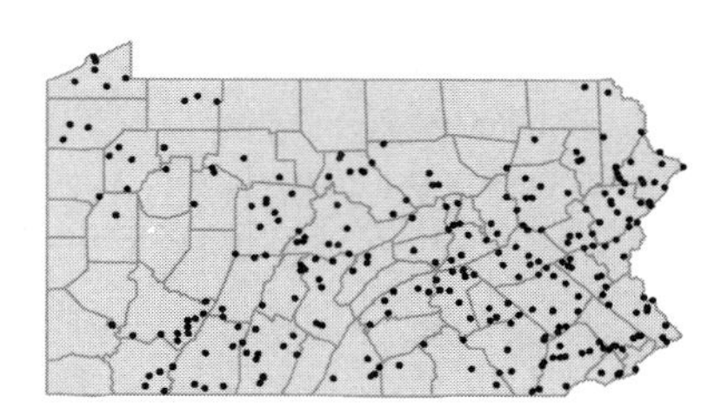

Melampyrum lineare Desr. var. ***pectinatum*** (Pennell) Fern.
Cow-wheat
Herbaceous annual
Dry woods, in sandy soil. Parasitic on the roots of other plants.

FACU

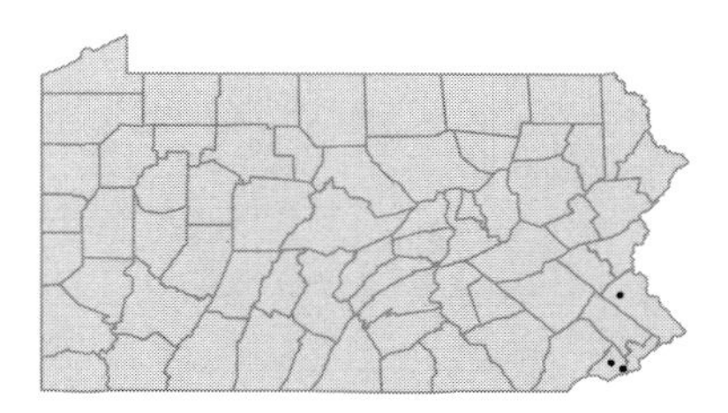

• ***Micranthemum micranthemoides*** (Nutt.) Wettst.
Nuttall's mud-flower
Herbaceous annual
Tidal mudflats. Believed to be extinct throughout its range, last collected in PA in 1932.
Hemianthus micranthoides Nutt. GBC

OBL

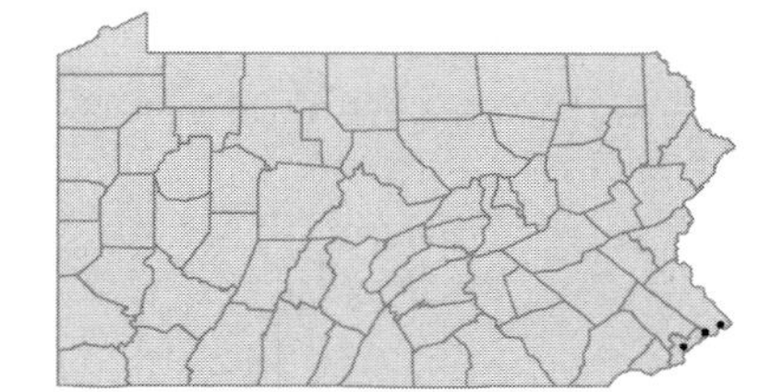

• ***Mimulus alatus*** Ait.
Winged monkey-flower
Herbaceous perennial
Swamps, wet meadows and shores.

OBL

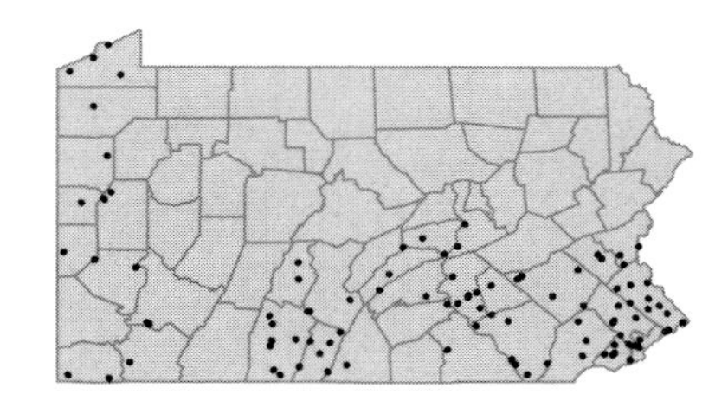

Mimulus guttatus Fisch. ex DC.
Monkey-flower
Herbaceous annual
Wet roadside ditch. Represented by a single collection from Clearfield Co. in 1955.

OBL
I

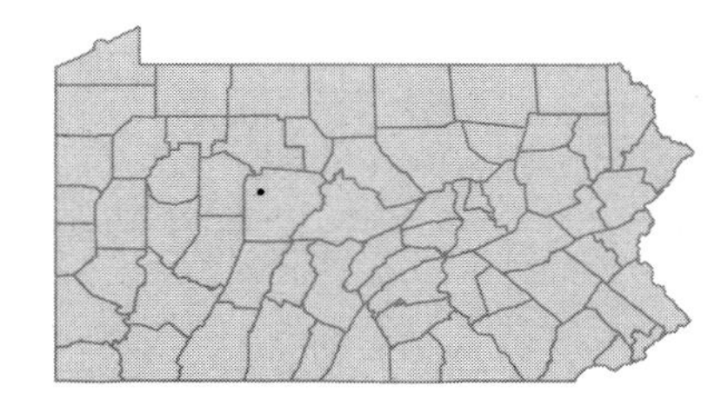

Mimulus moschatus Dougl. ex Lindl.
Muskflower
Herbaceous perennial
Wet shores, seeps and springy swales.

OBL
I

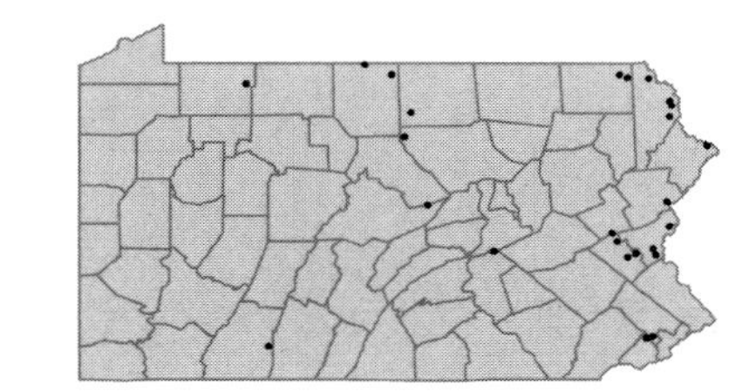

Mimulus ringens L. — OBL
Allegheny monkey-flower
Herbaceous perennial
Wet, open ground of swamps, meadows and shores.

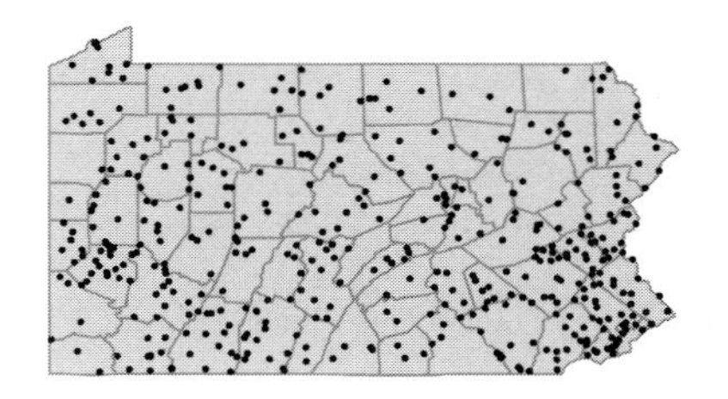

• *Pedicularis canadensis* L. — FACU
Forest lousewort; Wood-betony
Herbaceous perennial
Woods and edges. Parasitic on the roots of a wide variety of woody and herbaceous plants.

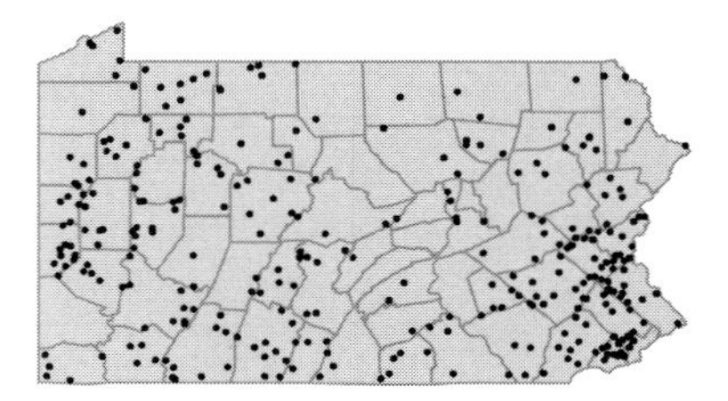

Pedicularis lanceolata Michx. — FACW
Swamp lousewort; Wood-betony
Herbaceous perennial
Swamps, boggy meadows and swales. Parasitic on the roots of other plants.

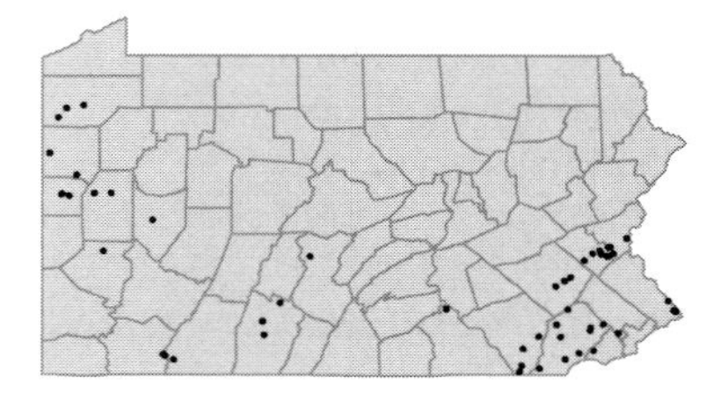

• *Penstemon calycosus* Small — I
Beard-tongue
Herbaceous perennial
Swampy woods, old fields and roadsides.
Penstemon laevigatus Ait. *pro parte* C

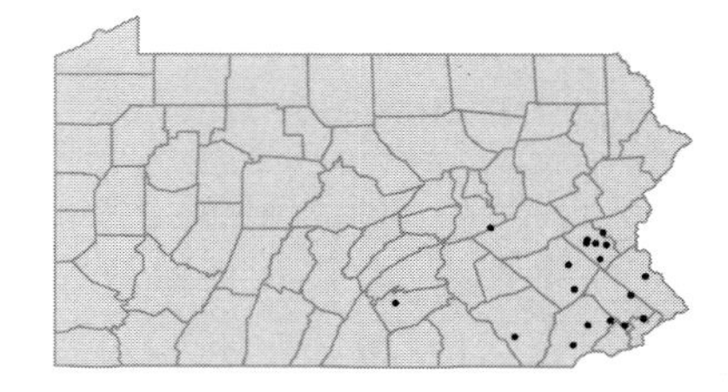

Penstemon canescens (Britt.) Britt.
Beard-tongue
Herbaceous perennial
Dry, rocky, wooded slopes and shale outcrops.

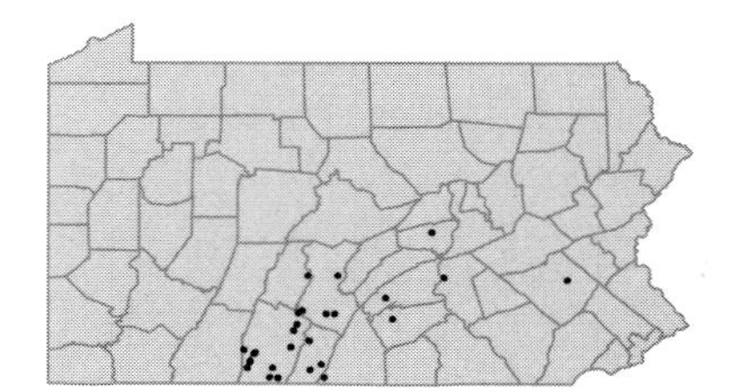

Penstemon digitalis Nutt. ex Sims — FAC
Beard-tongue
Herbaceous perennial
Meadows, old fields and roadsides.

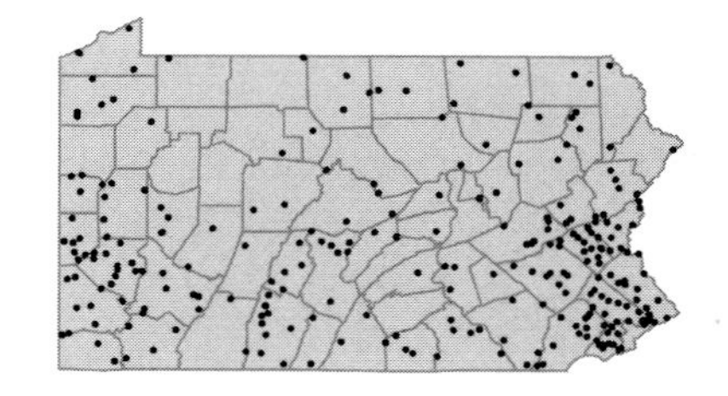

Penstemon hirsutus (L.) Willd.
Beard-tongue
Herbaceous perennial
Dry, open, rocky slopes and roadside banks.

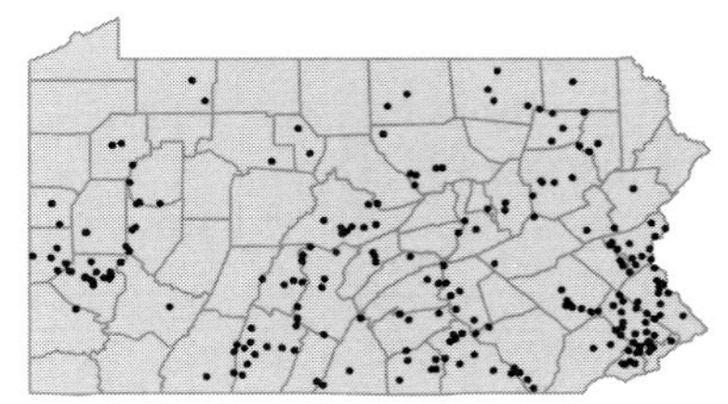

Penstemon laevigatus (L.) Ait.
Beard-tongue
Herbaceous perennial
Wooded hillsides, moist meadows and roadside banks.
Penstemon penstemon (L.) Britt. P

FACU

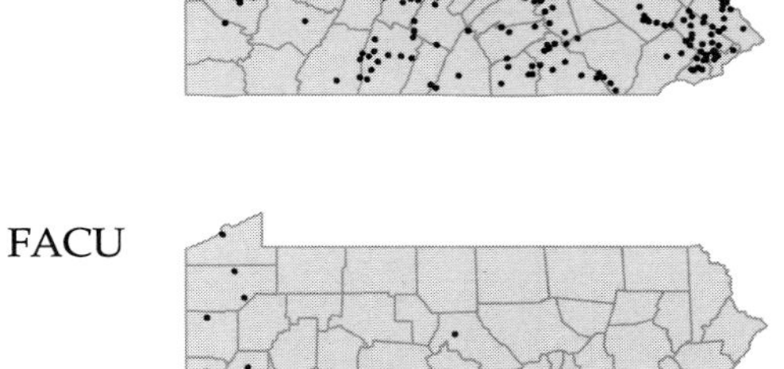

Penstemon pallidus Small
Beard-tongue
Herbaceous perennial
Wet meadows and old fields.

FACU

Penstemon tubiflorus Nutt.
Beard-tongue
Herbaceous perennial
Dry old fields and pastures. Represented by a single collection from Pike Co. in 1916.

I

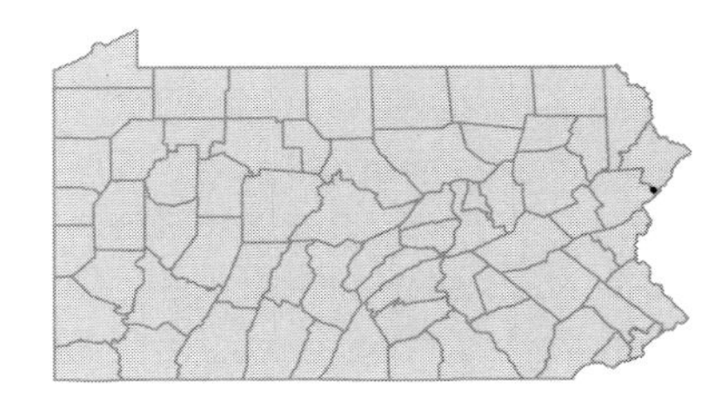

• ***Scrophularia aquatica*** L.
Figwort
Herbaceous perennial
Ballast. Represented by several collections from Philadelphia Co. 1878-1883.

I

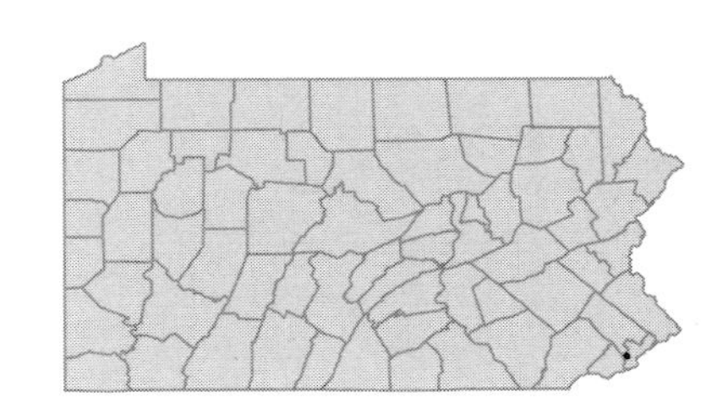

Scrophularia lanceolata Pursh
Lanceleaf figwort
Herbaceous perennial
Low woods, thickets, river banks and roadsides.
Scrophularia leporella Bicknell P

FACU+

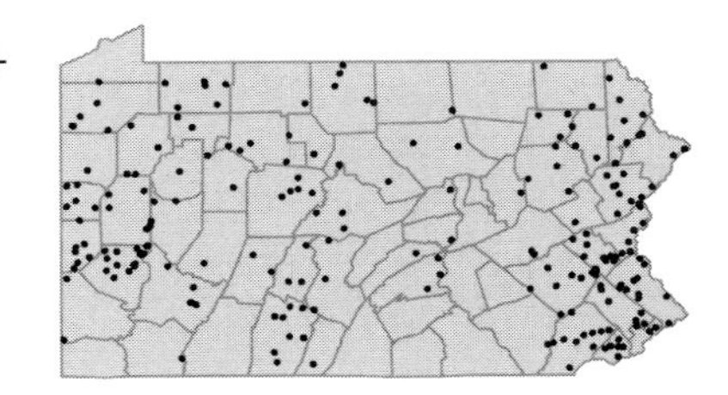

Scrophularia marilandica L. FACU-
Eastern figwort; Carpenter's-square
Herbaceous perennial
Alluvial woods, river banks, moist shores and roadsides.

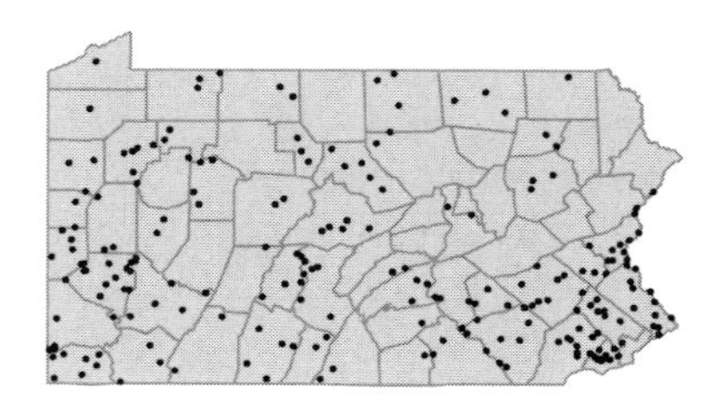

• ***Verbascum blattaria*** L. I
Moth mullein
Herbaceous biennial
Fields, roadsides, railroad embankments and waste ground.

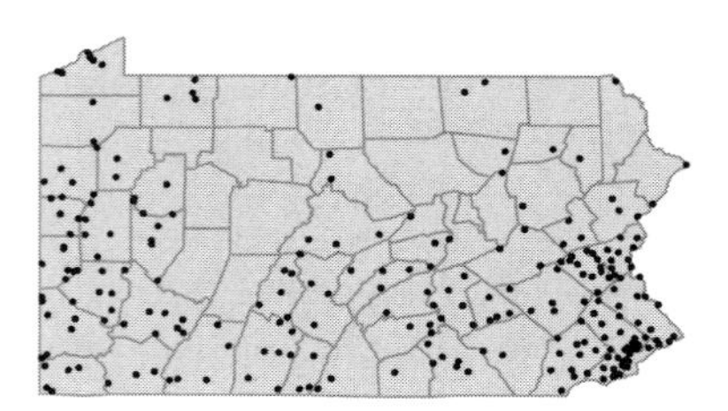

Verbascum lychnitis L. I
White mullein
Herbaceous biennial
Roadsides, shaly banks, railroad cuts and vacant lots.

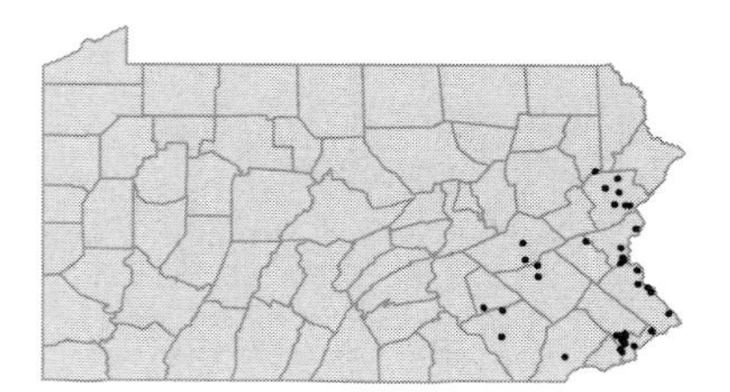

Verbascum nigrum L. I
Black mullein
Herbaceous biennial
Ballast. Represented by a single collection from Philadelphia Co. ca. 1865.

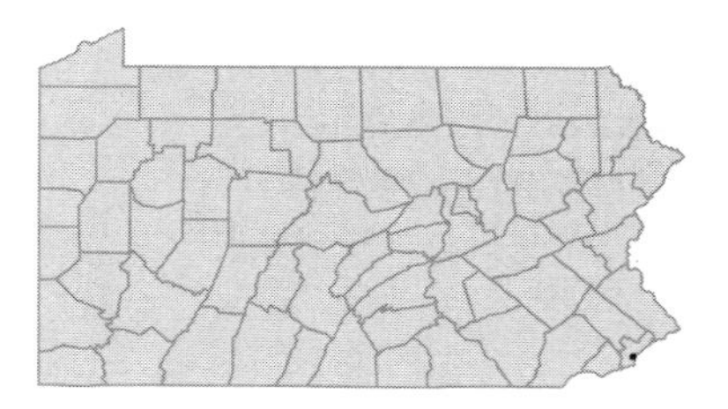

Verbascum phlomoides L. I
Mullein
Herbaceous biennial
Roadsides and disturbed, open ground.

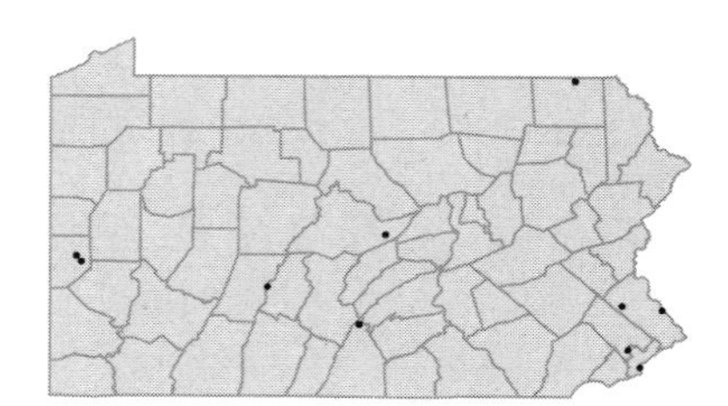

Verbascum sinuatum L. I
Mullein
Herbaceous biennial
Ballast.

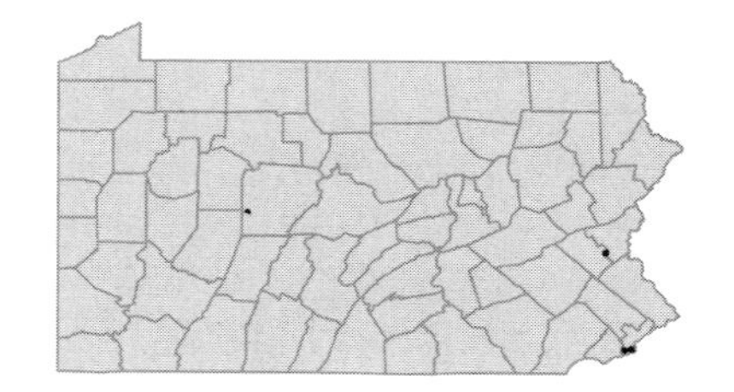

Verbascum thapsus L.
Common mullein; Flannel-plant
Herbaceous biennial
Fields, roadsides, shale barrens, railroad embankments and waste ground.

[I]

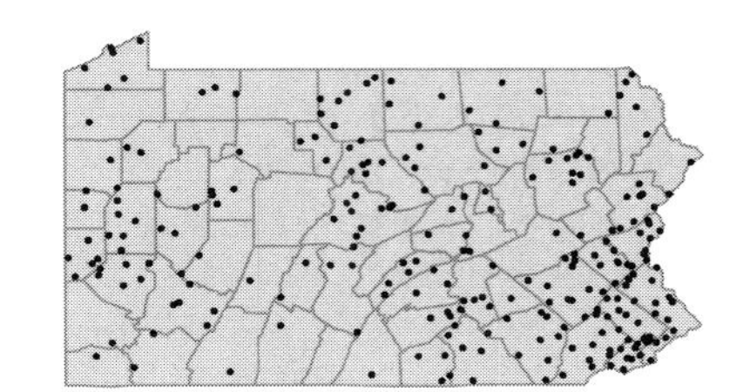

Verbascum virgatum Stokes
Mullein
Herbaceous biennial
Urban waste ground.

[I]

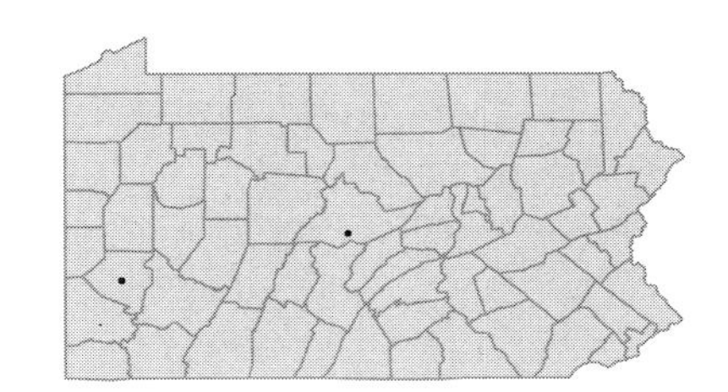

• ***Veronica agrestis*** L.
Field speedwell
Herbaceous annual
Roadsides, urban waste ground and ballast.

[I]

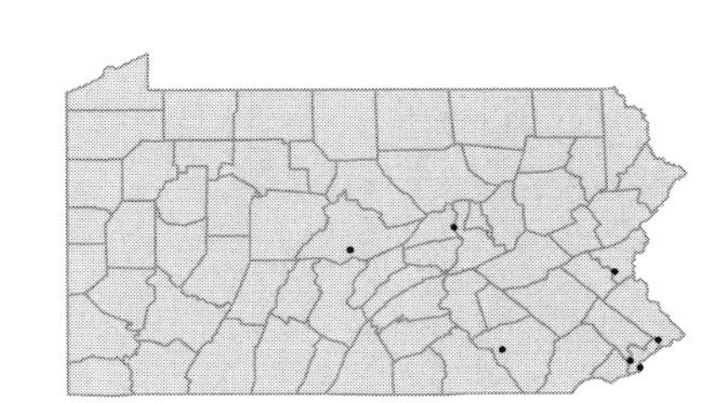

Veronica americana (Raf.) Schwein. ex Benth.
American brooklime; American speedwell
Herbaceous perennial
Moist banks, wet ditches and stream edges.

OBL

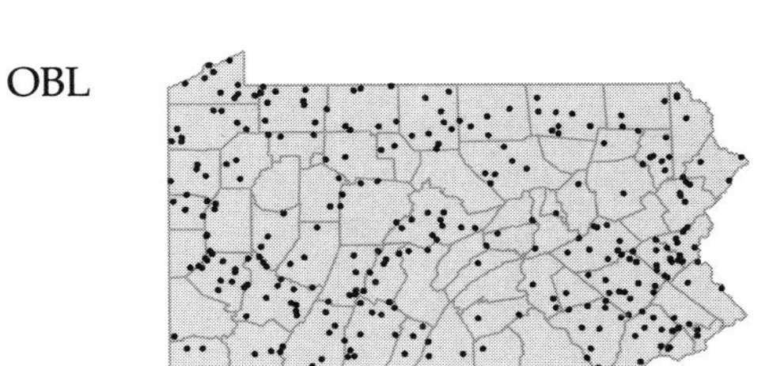

Veronica anagallis-aquatica L.
Brook-pimpernel; Water speedwell
Herbaceous biennial
Wet fields, ditches and stream edges in shallow water.
Veronica comosa Richter *pro parte* FW; *Veronica catenata* Pennell *pro parte* C

OBL
[I]

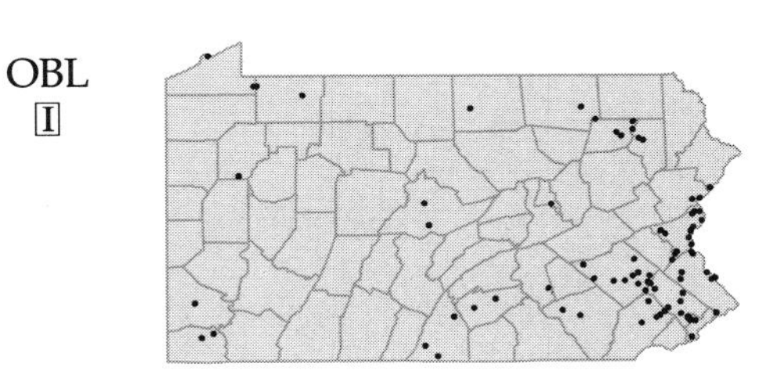

Veronica arvensis L.
Corn speedwell
Herbaceous annual
Woods, roadside banks, meadows and lawns.

[I]

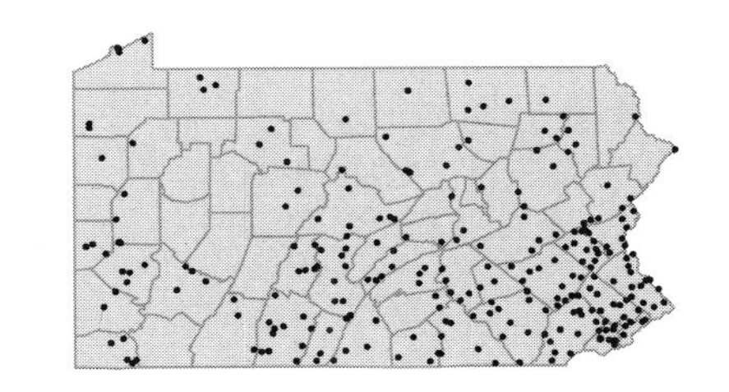

Veronica austriaca L.
Speedwell
Herbaceous perennial
Cultivated and rarely escaped to fields and roadsides.
Veronica latifolia L. FGBW

I

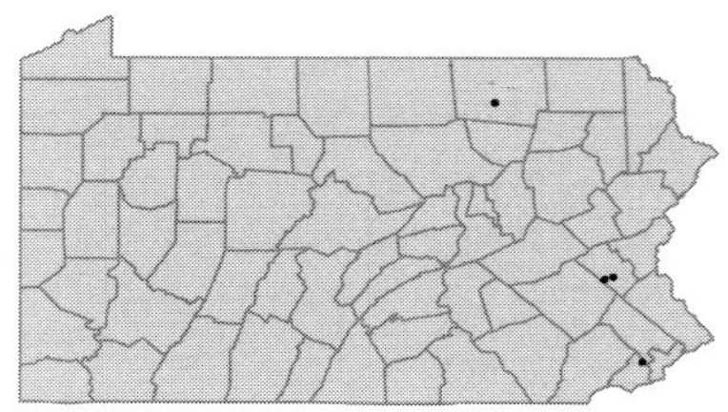

Veronica beccabunga L.
European brooklime
Herbaceous perennial
Wet shores and ditches.

OBL
I

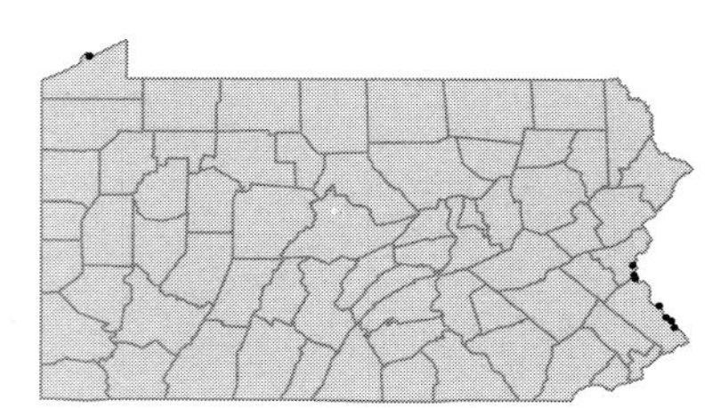

Veronica chamaedrys L.
Bird's-eye; Germander speedwell
Herbaceous perennial
Wooded slopes, roadsides and cultivated ground.

I

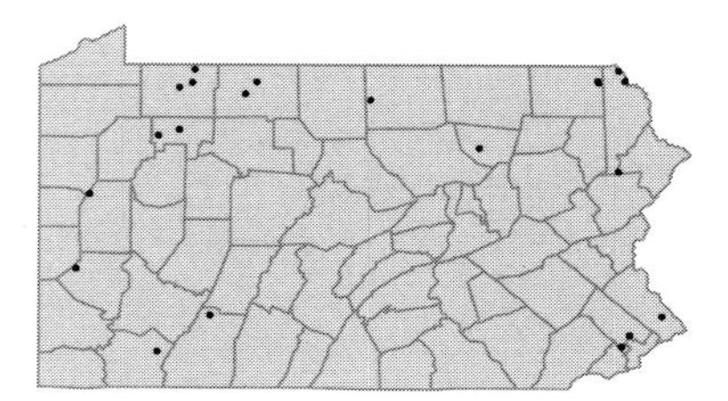

Veronica filiformis J.E.Smith
Creeping speedwell
Herbaceous perennial
Lawns and open ground.

I

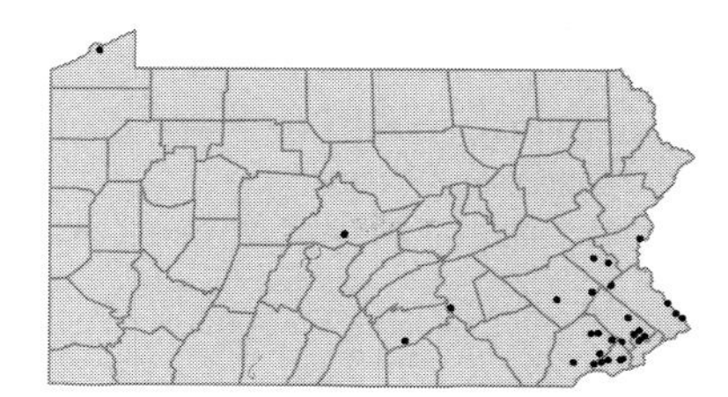

Veronica hederifolia L.
Ivy-leaved speedwell
Herbaceous annual
Open woods, roadside banks, lawns and ballast.

I

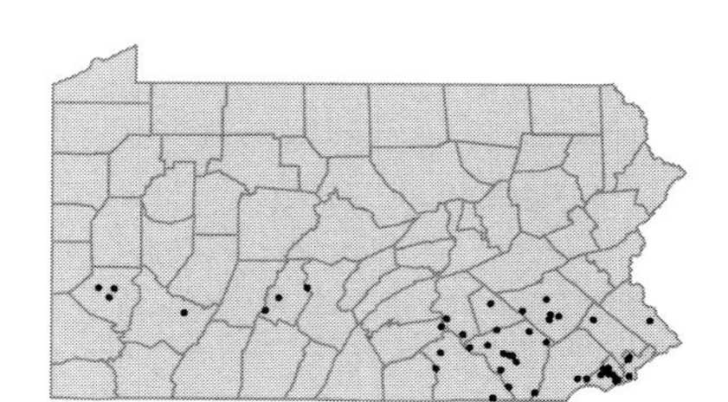

Veronica longifolia L.
Speedwell
Herbaceous perennial
Cultivated and occasionally escaped to roadsides and railroad banks.

I

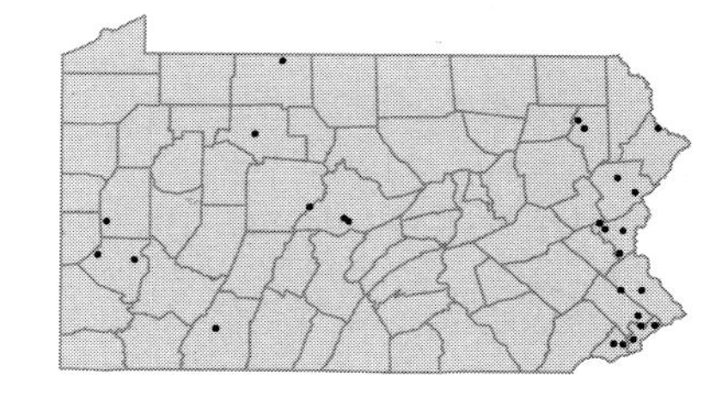

Veronica officinalis L.
Common speedwell; Gypsyweed
Herbaceous perennial
Woods, roadsides and shale barrens.

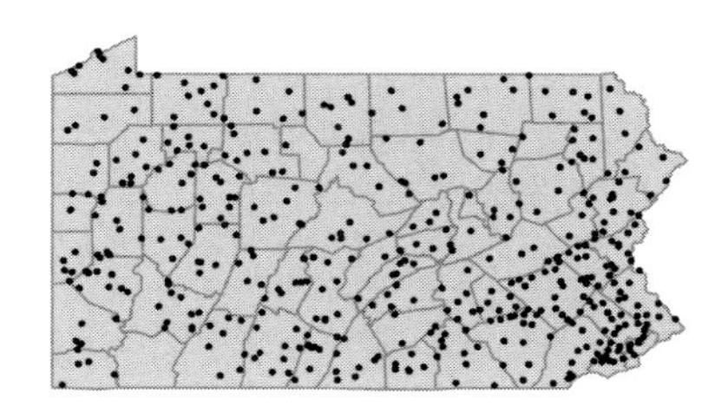

Veronica peregrina L. ssp. ***peregrina***
Neckweed; Purslane speedwell
Herbaceous annual
Open woods, alluvial banks, gardens and waste ground.

FACU-

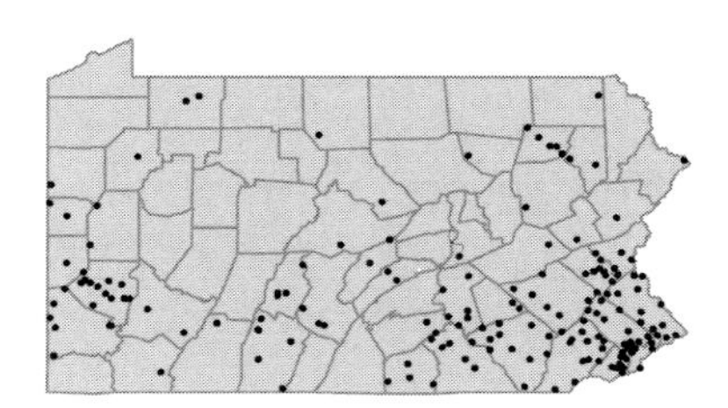

Veronica peregrina L. ssp. ***xalapensis*** (Kunth) Pennell
Neckweed; Purslane speedwell
Herbaceous annual
Cultivated or recently abandoned fields and moist ground along railroad tracks.

FACU-

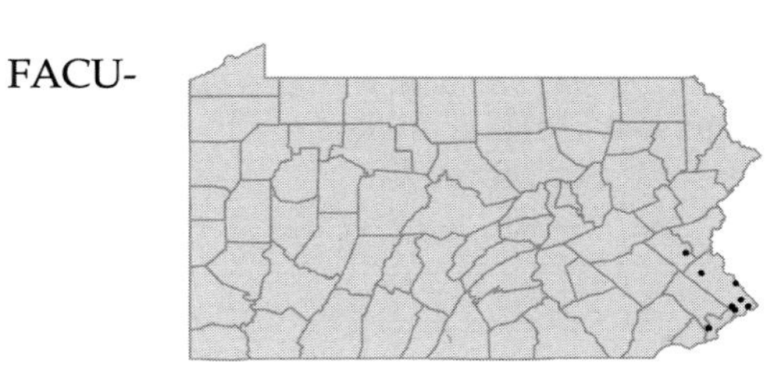

Veronica persica Poir.
Bird's-eye speedwell
Herbaceous annual
Lawns, roadsides, cultivated fields and waste ground.

[I]

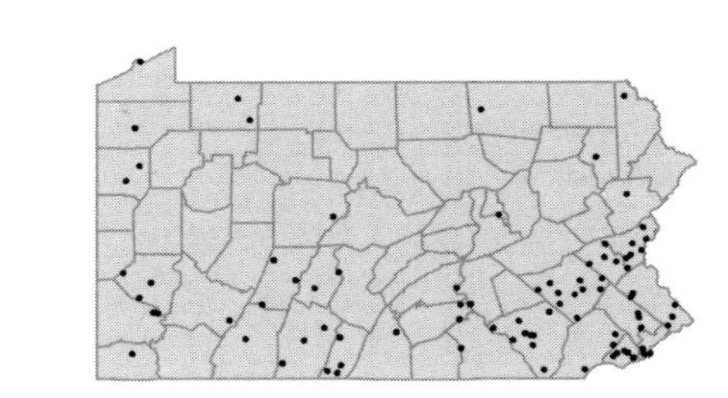

Veronica polita Fries
Speedwell
Herbaceous annual
Lawns, fields, roadsides, disturbed ground and ballast.
Veronica agrestis L. GB; *Veronica didyma* Tenore B

[I]

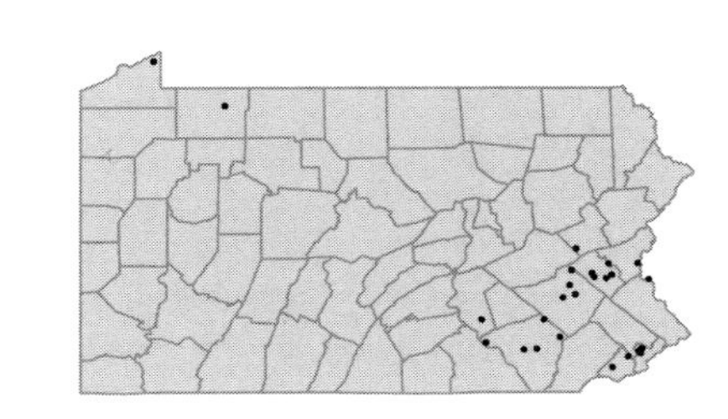

Veronica scutellata L.
Marsh speedwell; Narrow-leaved speedwell
Herbaceous perennial
Swamps, wet woods, ditches and swales.

OBL

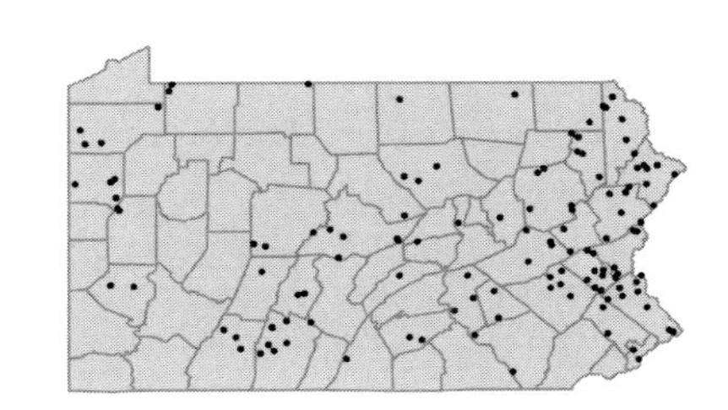

Veronica serpyllifolia L. FAC+ [I]
Thyme-leaved speedwell
Herbaceous perennial
Lawns, fields, meadows and open woods.

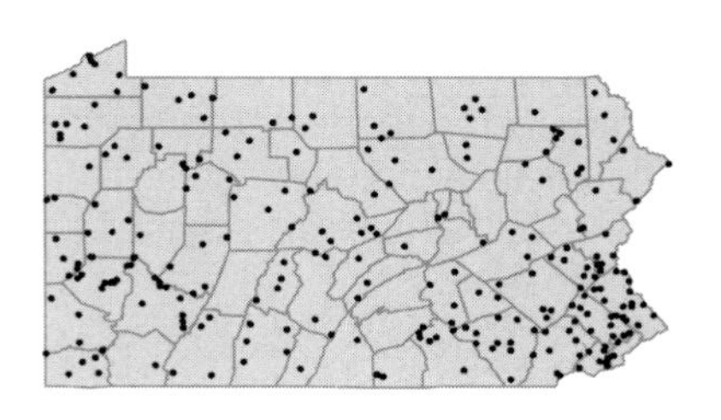

• ***Veronicastrum virginicum*** (L.) Farw. FACU
Culver's-root
Herbaceous perennial
Moist meadows, thickets and swamps.

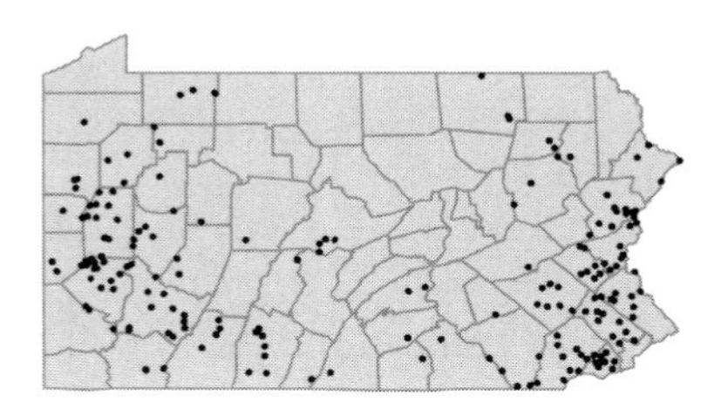

OROBANCHACEAE

• ***Conopholis americana*** (L.) Wallr.
Squaw-root; Cancer-root
Herbaceous perennial
Woods. Parasitic on the roots of oaks and other forest trees.

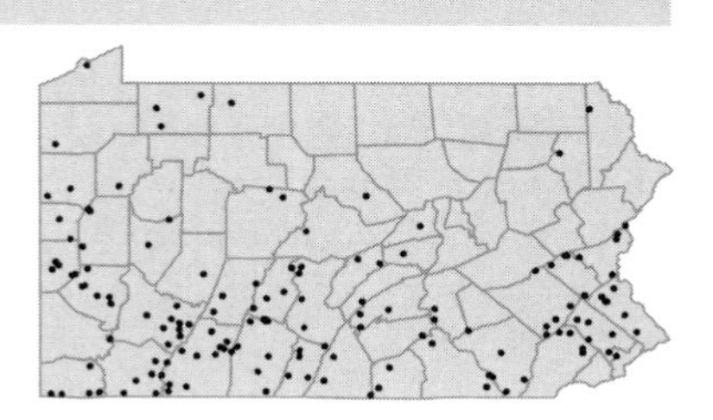

• ***Epifagus virginiana*** (L.) Bart.
Beech-drops
Herbaceous annual
Beech woods. Parasitic on the roots of American beech.
Leptamnium virginianum (L.) Raf. P

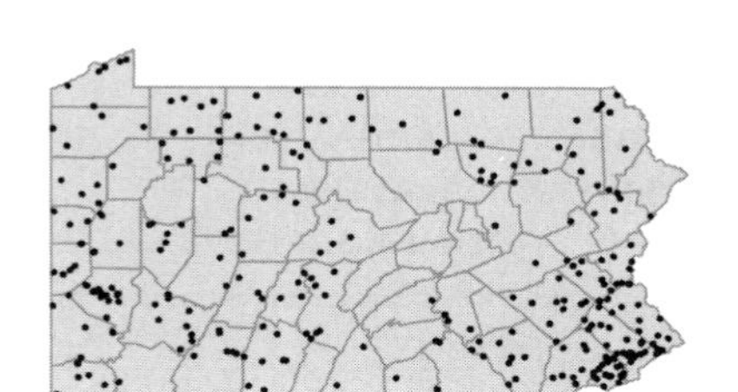

• ***Orobanche minor*** J.E.Smith ex Sowerby [I]
Small broom-rape
Herbaceous annual
Roadsides, yards and vacant lots. Parasitic on the roots of clover and other herbaceous plants and some ornamental shrubs. A federally designated noxious weed.

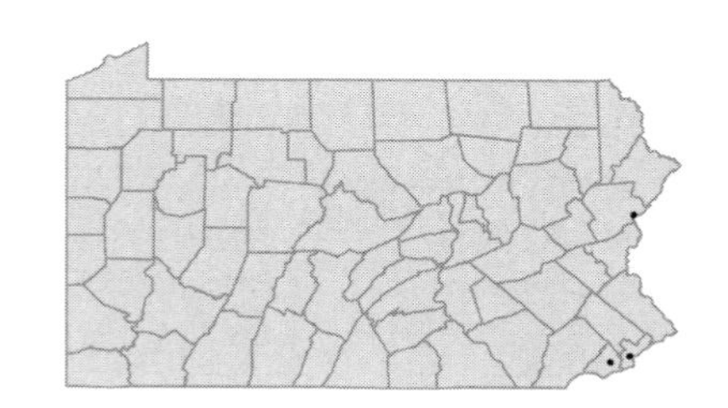

Orobanche uniflora L.
Broom-rape; Cancer-root
Herbaceous perennial
Woods. Parasitic on the roots of many plants.
Thalesia uniflora (L.) Britt. P

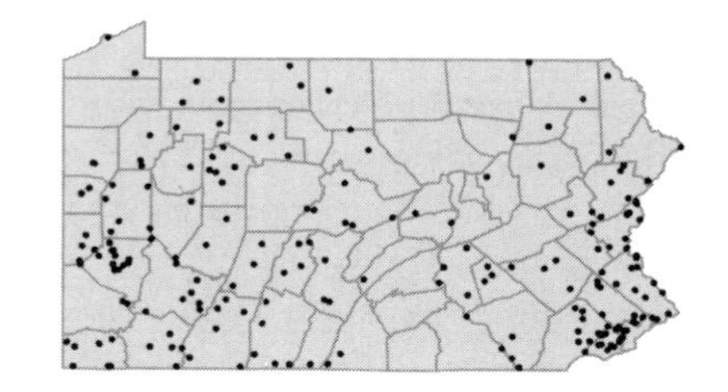

ACANTHACEAE

• ***Justicia americana*** (L.) Vahl
Water-willow
Herbaceous perennial, emergent aquatic
Marshy shorelines of lakes and rivers, in shallow water.

OBL

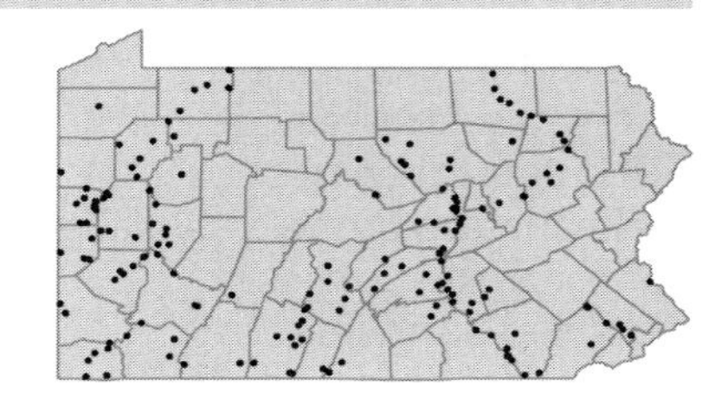

• ***Ruellia caroliniensis*** (Walt. ex J.F.Gmel.) Steud.
Carolina petunia
Herbaceous perennial
River bank. Believed to be extirpated, last collected in 1895.

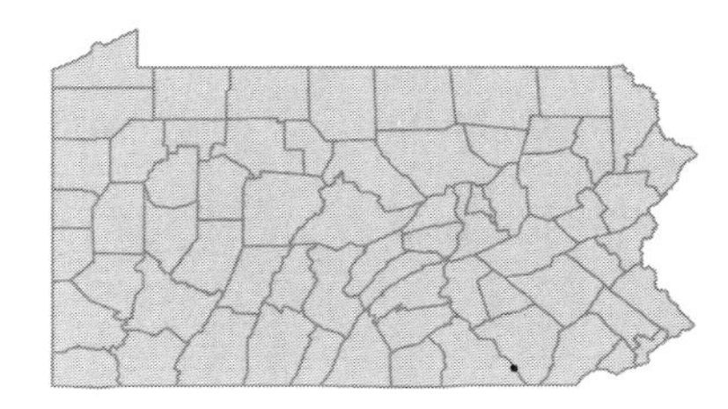

Ruellia humilis Nutt.
Fringed-leaved petunia
Herbaceous perennial
Limestone barrens and quarry waste.

UPL

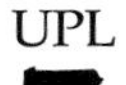

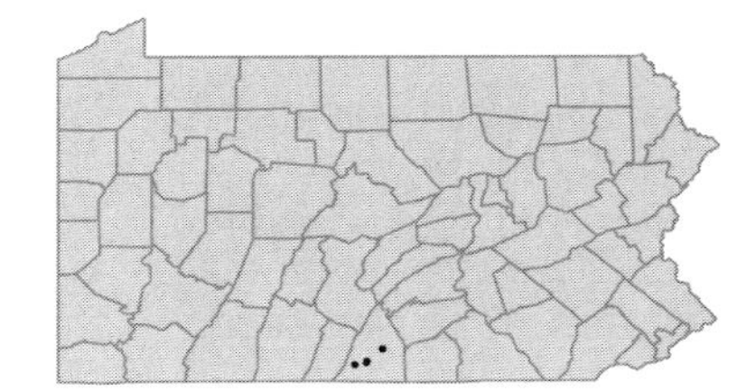

Ruellia strepens L.
Limestone petunia
Herbaceous perennial
Rich, wooded slopes, open woods, bluffs and roadsides, on limestone.
Ruellia strepens L. var. *micrantha* (Engelm. & Gray) Britt.

FAC

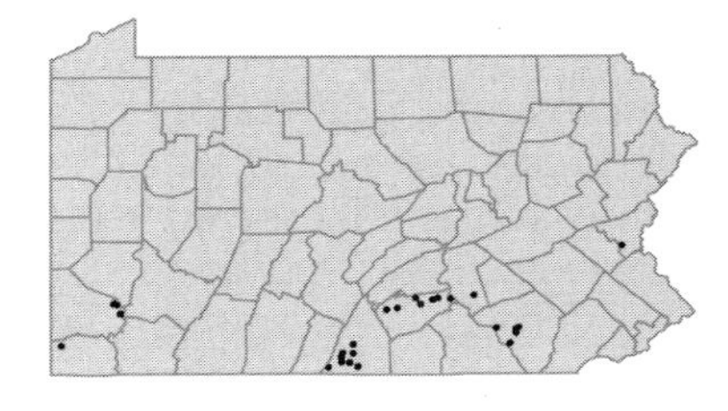

PEDALIACEAE

• ***Proboscidea louisianica*** (P.Mill.) Thell.
Unicorn-plant
Herbaceous annual
Cultivated and occasionally escaped to fields and rubbish dumps.
Martynia louisiana P.Mill. PC

FACU
[I]

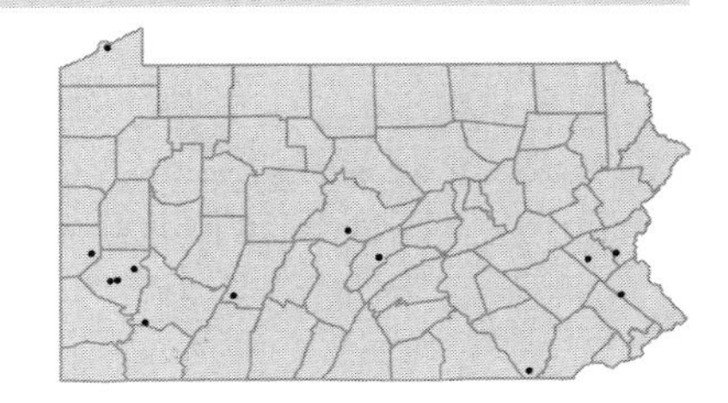

BIGNONIACEAE

• ***Campsis radicans*** (L.) Seem. ex Bureau
Trumpet-vine; Trumpet-creeper
Woody vine
River banks, roadside thickets and fencerows, also occasionally escaped from cultivation.
Tecoma radicans (L.) DC. P

FAC

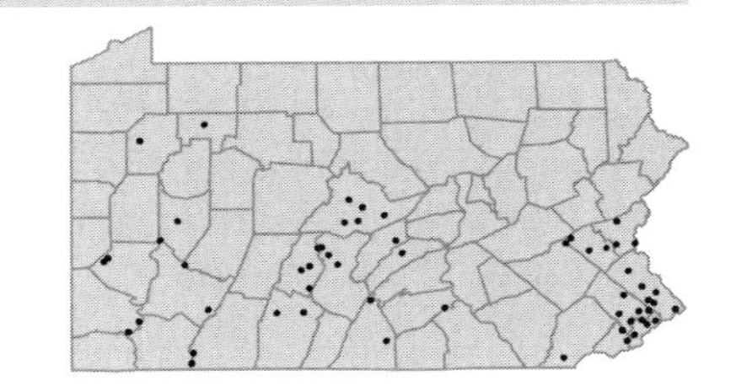

BIGNONIACEAE

• ***Catalpa bignonioides*** Walt. UPL [I]
Catalpa; Indian-bean
Deciduous tree
Cultivated and frequently naturalized in disturbed woods, floodplains, fields and waste ground.
Catalpa catalpa (L.) Karst. P

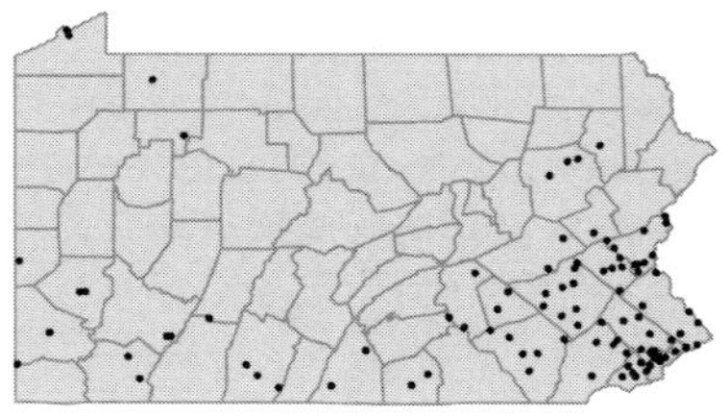

Catalpa ovata G.Don [I]
Chinese catalpa
Deciduous tree
Cultivated and rarely escaped to moist bottomland.

Catalpa speciosa (Warder ex Barney) Warder ex Engelm. FAC [I]
Catalpa; Cigar-tree
Deciduous tree
Cultivated, and occasionally escaped to low woods, roadsides and spoil banks.

• ***Paulownia tomentosa*** (Thunb.) Sieb. & Zucc. ex Steud. [I]
Empress-tree; Princess-tree
Deciduous tree
Cultivated, and frequently escaped to roadsides, railroad banks and woods edges.

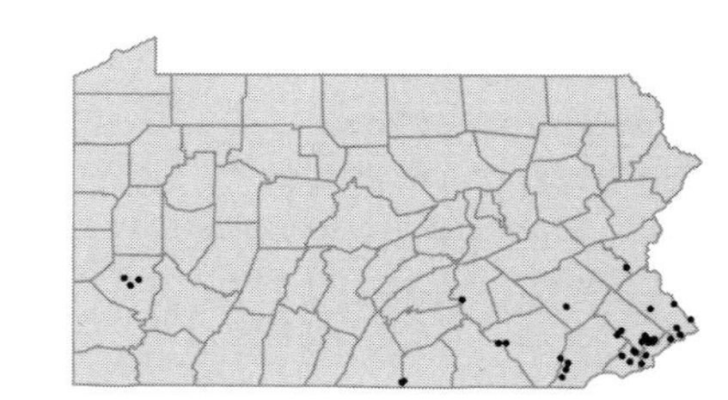

LENTIBULARIACEAE

• ***Utricularia cornuta*** Michx. OBL
Horned bladderwort
Herbaceous perennial, rooted submergent aquatic
Shallow water of marshes, ponds and ditches. Carnivorous.

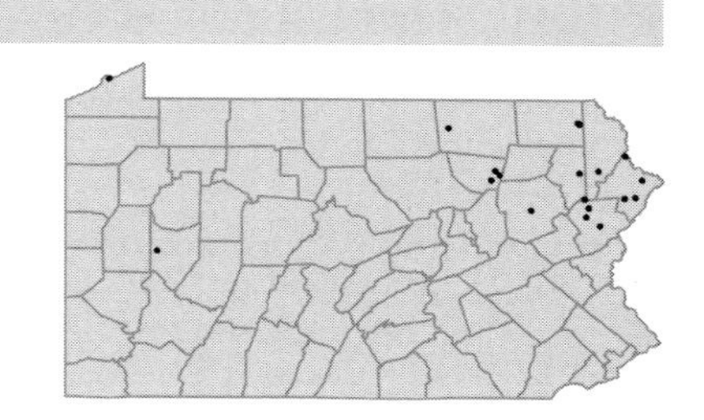

Utricularia geminiscapa Benj. OBL
Bladderwort
Herbaceous perennial, free-floating aquatic
Bogs, vernal ponds or river banks. Carnivorous.
Utricularia clandestina Nutt. P

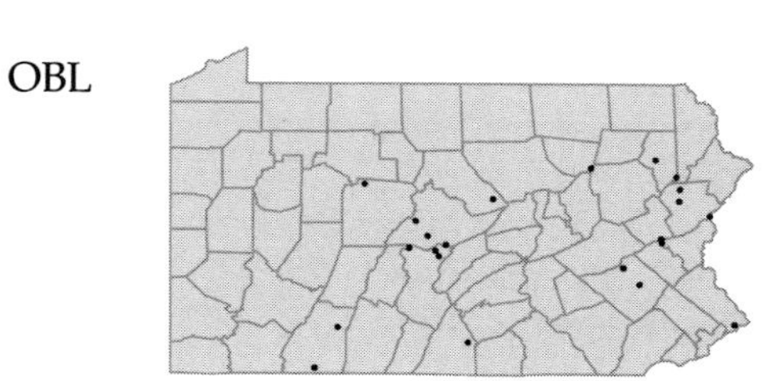

Utricularia gibba L.
Humped bladderwort; Fibrous bladderwort
Herbaceous annual, rooted submergent aquatic
Shallow water or exposed peat, sand or mud flats. Carnivorous.
Utricularia fibrosa Walt. *pro parte*

OBL

Utricularia intermedia Hayne
Flat-leaved bladderwort
Herbaceous perennial, rooted submergent aquatic
Lakes and wet edges of floating bog mats. Carnivorous.

OBL

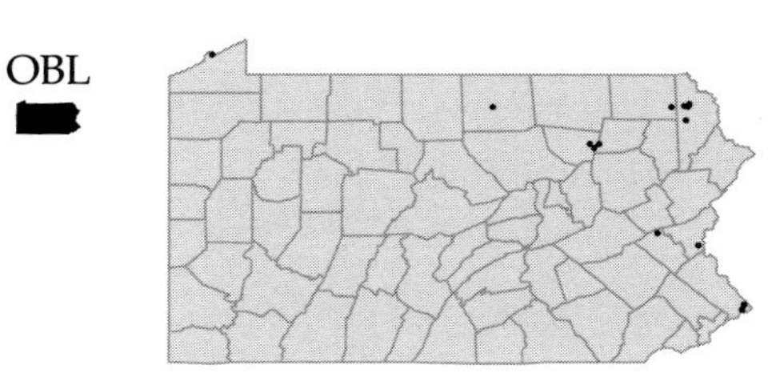

Utricularia macrorhiza LeConte
Common bladderwort
Herbaceous perennial, free-floating aquatic
Lakes, ponds, swamps, marshes and ditches. Carnivorous.
Utricularia vulgaris L. WFBGK

OBL

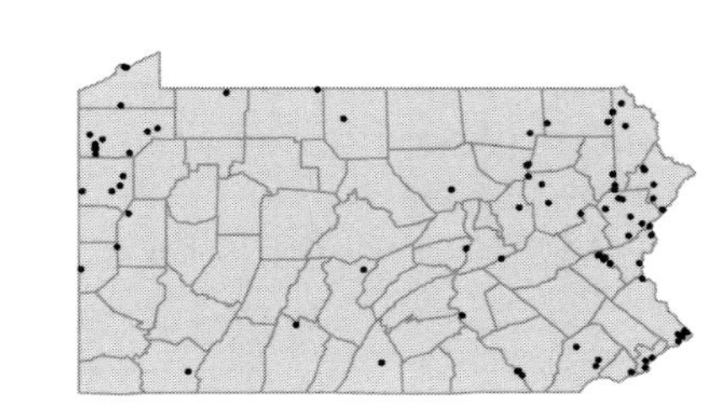

Utricularia minor L.
Lesser bladderwort
Herbaceous perennial, rooted submergent aquatic
Shallow water of lakes, ponds and swamps. Carnivorous.

OBL

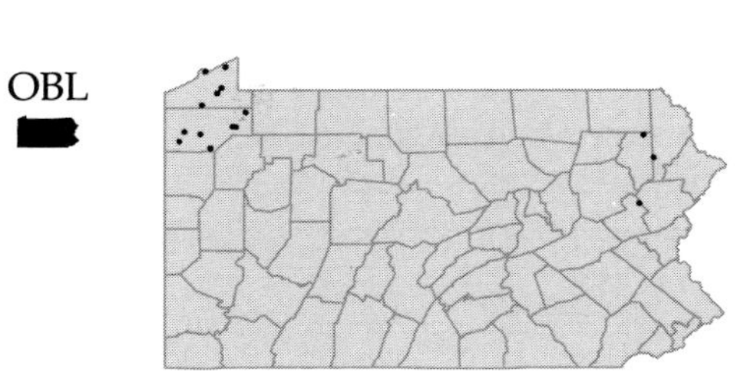

Utricularia purpurea Walt.
Purple bladderwort
Herbaceous perennial, free-floating aquatic
Lakes and ponds. Carnivorous.

OBL

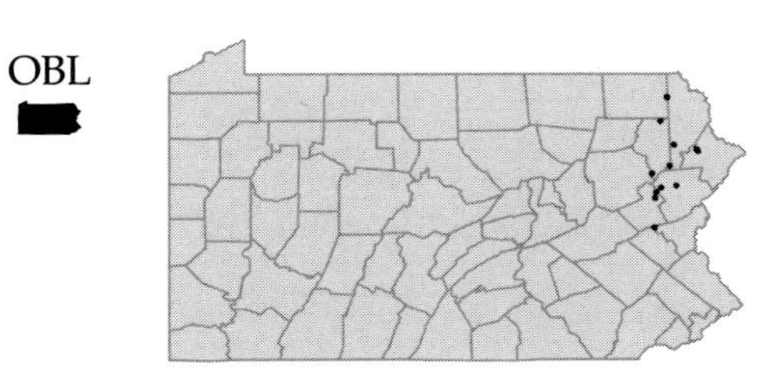

Utricularia radiata Small
Floating bladderwort
Herbaceous annual, free-floating aquatic
Shallow ponds and ditches. Believed to be extirpated, last collected in 1927. Carnivorous.
Utricularia inflata Walt. var. *minor* Chapm. FW

OBL

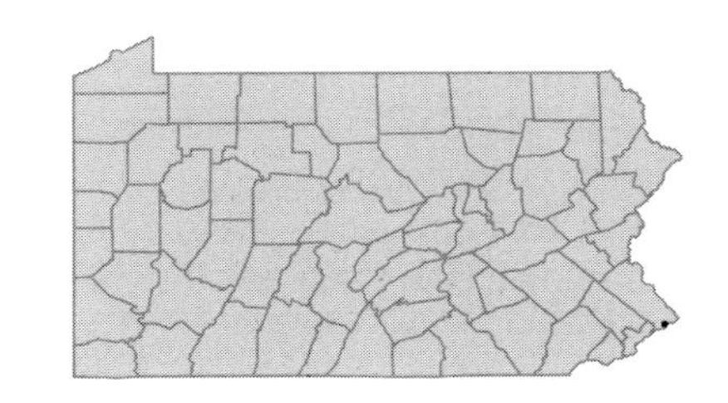

Utricularia resupinata B.D.Greene ex Bigelow
Northeastern bladderwort
Herbaceous perennial, rooted submergent aquatic
Swamp, in shallow water. Carnivorous. Believed to be extirpated, represented by a single collection from Erie Co. in 1879.

OBL

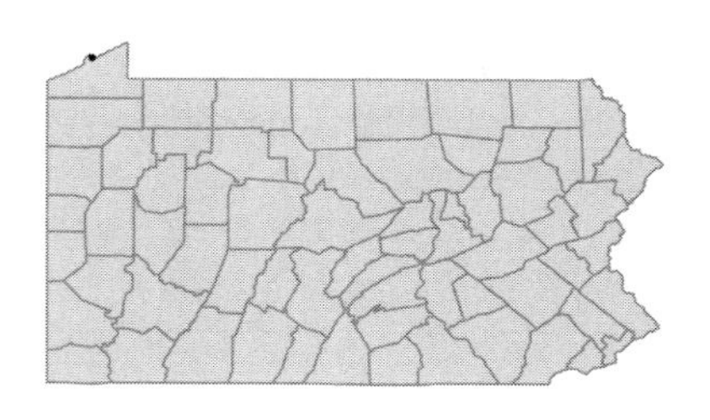

CAMPANULACEAE

• ***Campanula americana*** L.
Tall bellflower
Herbaceous annual
Moist woods, rocky wooded slopes and stream banks.

FAC

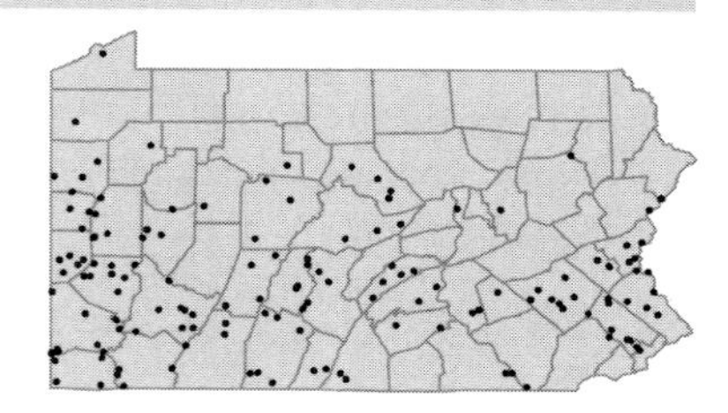

Campanula aparinoides Pursh
Marsh bellflower
Herbaceous perennial
Moist shores, swamps and seepy, open ground.

OBL

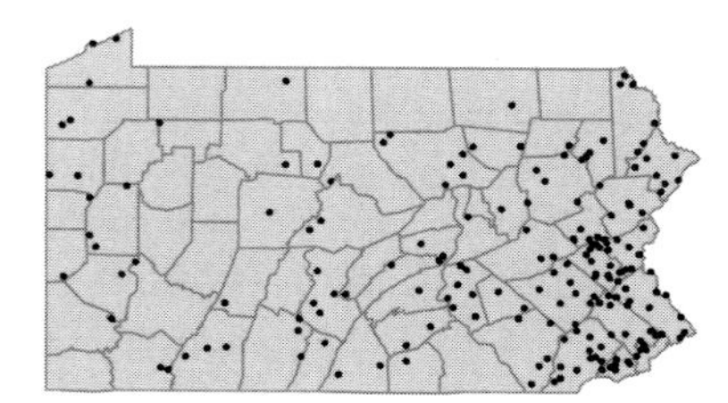

Campanula glomerata L.
Clustered bellflower
Herbaceous perennial
Roadside. Represented by a single collection from Bucks Co. in 1954.

I

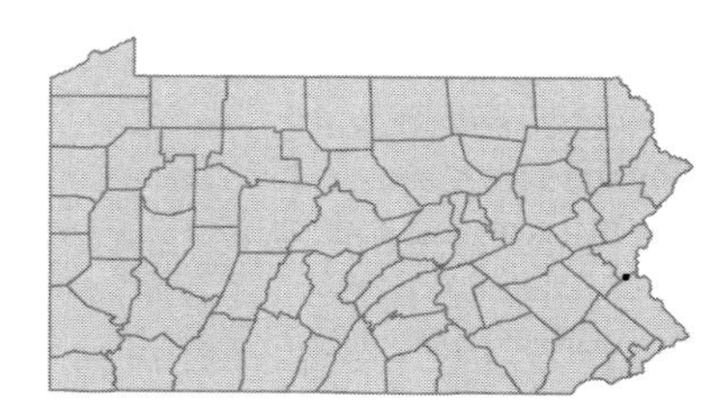

Campanula latifolia L.
Bellflower
Herbaceous perennial
Railroad bank. Represented by a single collection from Lancaster Co. in 1958.

I

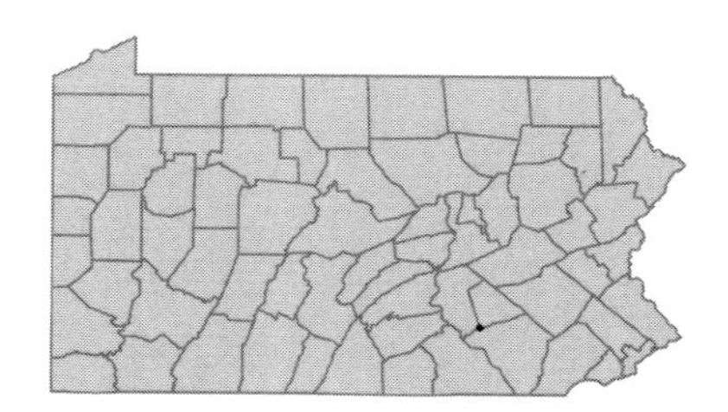

Campanula persicifolia L.
Peach-bells; Willow bellflower
Herbaceous perennial
Cultivated and rarely escaped to roadsides. Represented by a single collection from Bradford Co. in 1965.

I

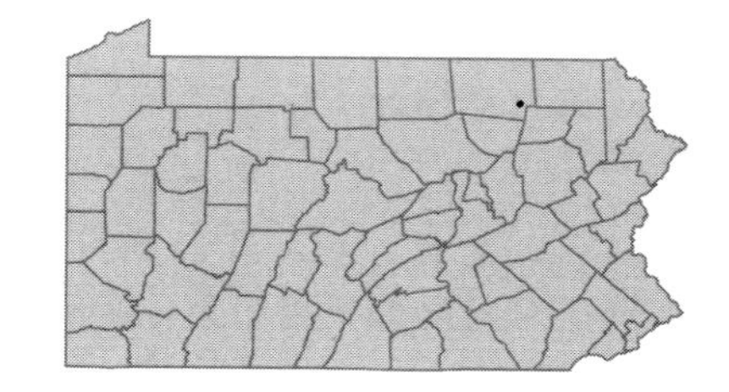

Campanula rapunculoides L.
Creeping bellflower; Roving bellflower
Herbaceous perennial
Roadsides and woods edges.

I

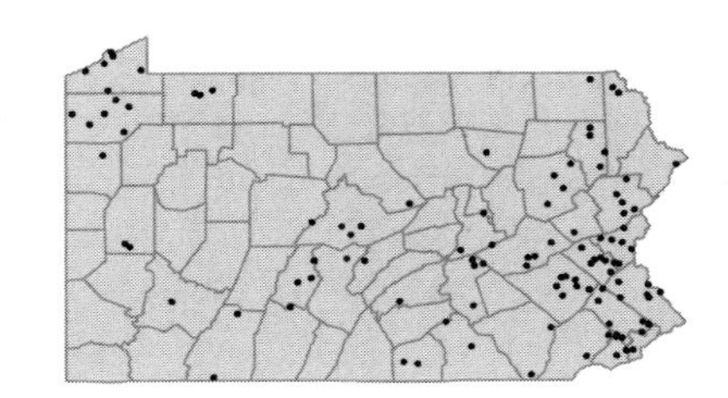

Campanula rotundifolia L.
Harebell
Herbaceous perennial
Dry, rocky slopes, bluffs and cliffs.

FACU

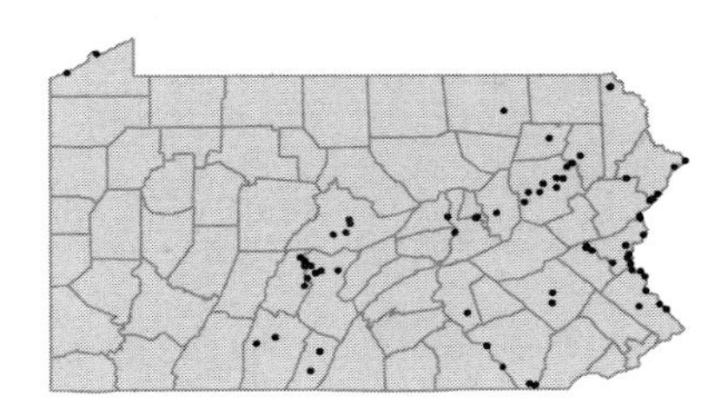

Campanula trachelium L.
Throatwort
Herbaceous perennial
Dry soil near cemetery. Represented by a single collection from Berks Co. in 1938.

I

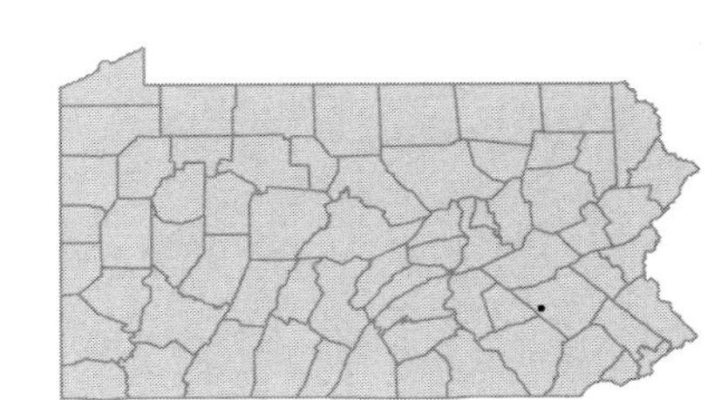

• ***Jasione montana*** L.
Sheep's-bit
Herbaceous annual
Cultivated and rarely occurring in abandoned pastures or on ballast.

I

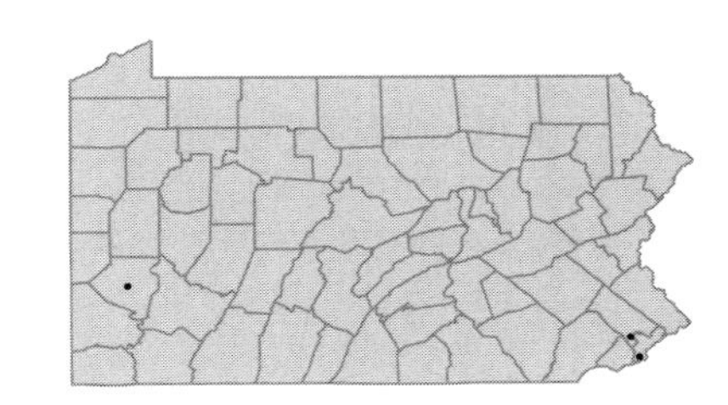

• ***Legousia speculum-veneris*** (L.) Fisch. ex A.DC.
Venus's-looking-glass
Herbaceous annual
Waste ground.
Specularia speculum-veneris (L.) Tanfani FGBWC

I

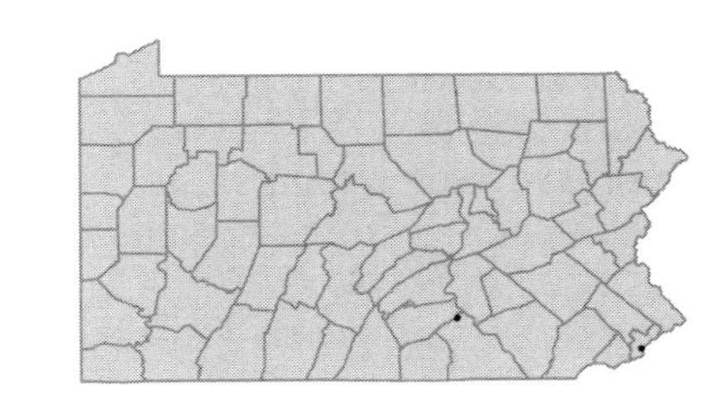

• ***Lobelia cardinalis*** L.
Cardinal-flower
Herbaceous perennial
Wet meadows, swamps, ditches, stream banks and lake shores.

FACW+

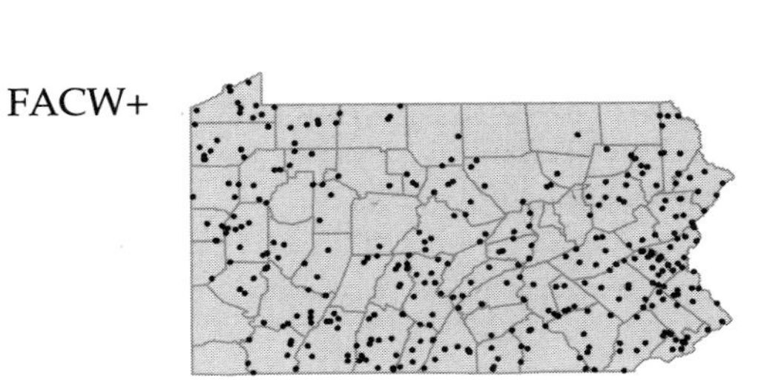

Lobelia chinensis Lour.
Chinese lobelia
Herbaceous perennial
Tidal river bank.

[I]

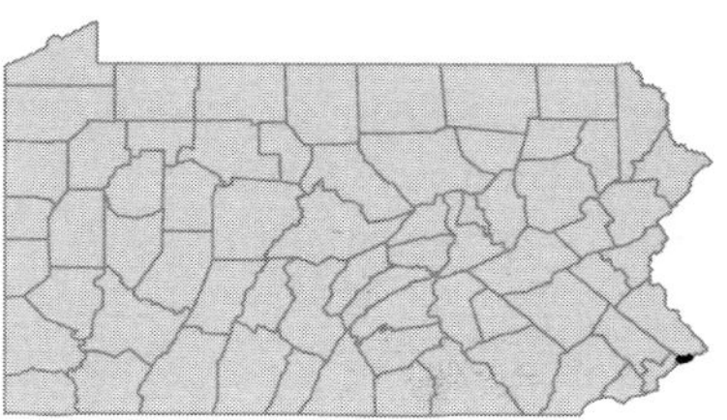

Lobelia dortmanna L.
Water lobelia
Herbaceous perennial, rooted submergent aquatic
Shallow water of ponds and lakes.

OBL

Lobelia erinus L.
Edging lobelia
Herbaceous annual
Cultivated and occasionally escaped to lawns.

[I]

Lobelia inflata L.
Indian-tobacco
Herbaceous annual
Woods, old fields, meadows and roadsides.

FACU

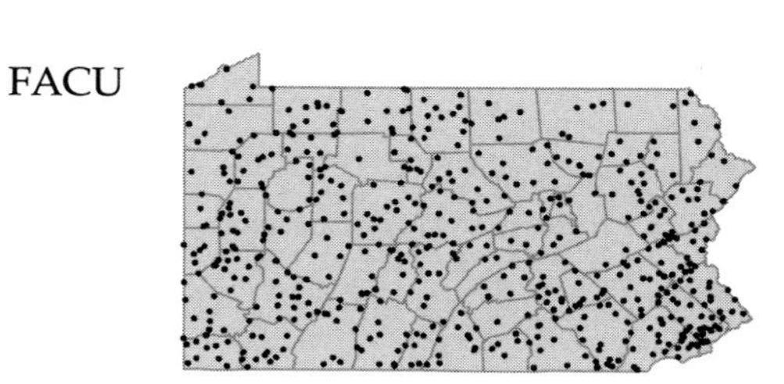

Lobelia kalmii L.
Brook lobelia
Herbaceous perennial
Calcareous swamps, moist pastures and fens.

OBL

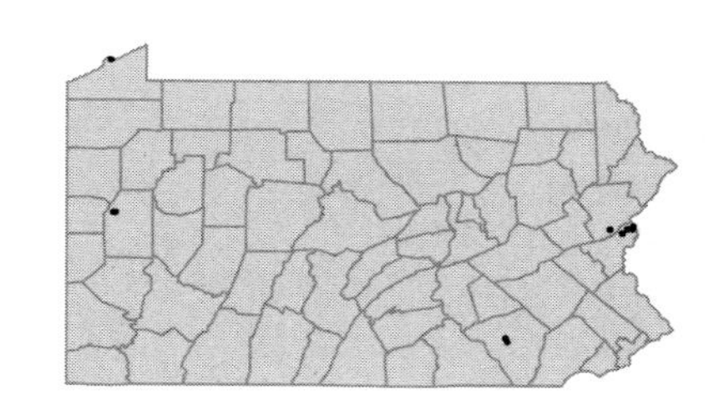

Lobelia nuttallii Roemer & Schultes
Nuttall's lobelia
Herbaceous perennial
Low woods, moist, sandy or peaty thickets and wet meadows. Believed to be extirpated, last collected in 1924.

FACW

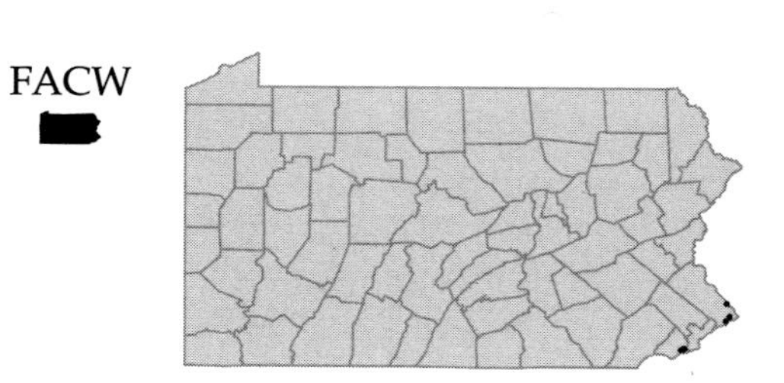

Lobelia puberula Michx.
Downy lobelia
Herbaceous perennial
Moist, sandy soil of old fields, gravel pits and serpentine barrens.

FACW-

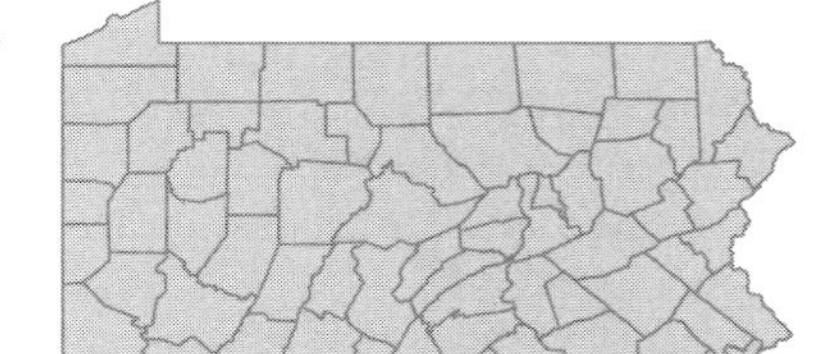

Lobelia siphilitica L.
Great lobelia
Herbaceous perennial
Swamps, moist meadows, stream banks and ditches.

FACW+

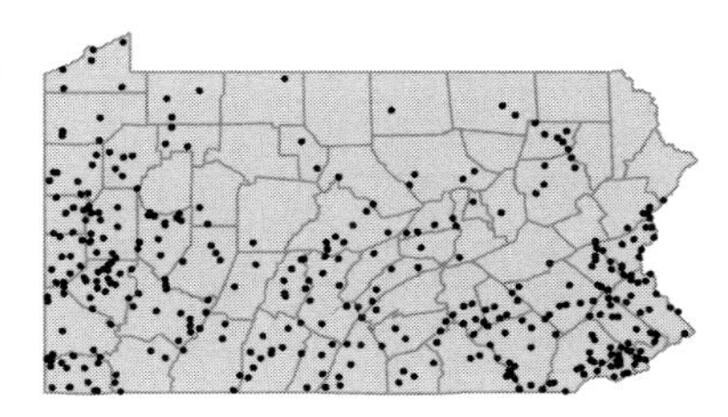

Lobelia spicata Lam. var. ***spicata***
Spiked lobelia
Herbaceous perennial
Dry ground of fields and woods.

FAC-

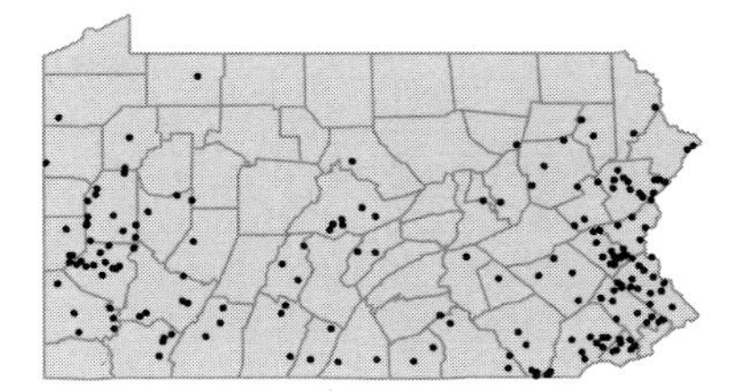

Lobelia spicata Lam. var. ***campanulata*** McVaugh
Spiked lobelia
Herbaceous perennial
Open woods, cedar glades and dry slopes.

FAC

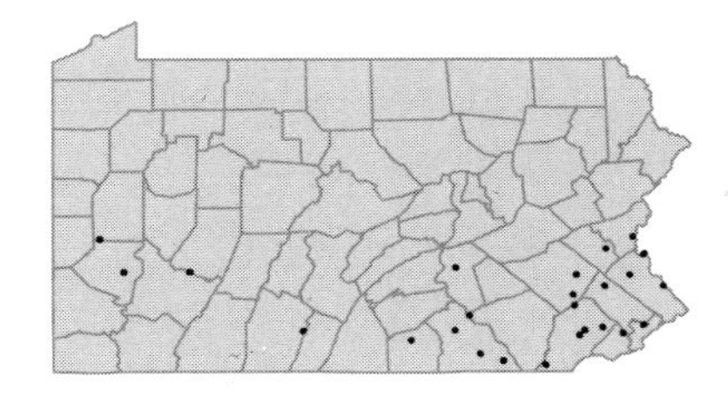

Lobelia spicata Lam. var. ***leptostachys*** (A.DC.) Mackenzie & Bush
Spiked lobelia
Herbaceous perennial
Abandoned field.

FAC

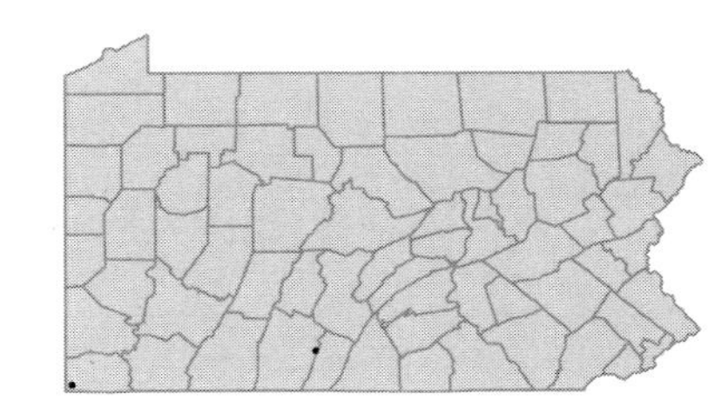

Lobelia spicata Lam. var. ***scaposa*** McVaugh
Spiked lobelia
Herbaceous perennial
Wooded slopes, shale barrens and moist meadows.

FAC-

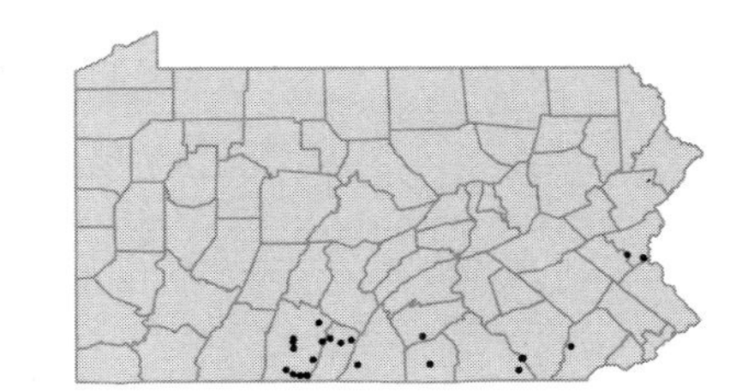

• ***Platycodon grandiflorum*** (Jacq.) A.DC. [I]
Balloon-flower
Herbaceous perennial
Cultivated and rarely occurring in hedgerows or waste ground.

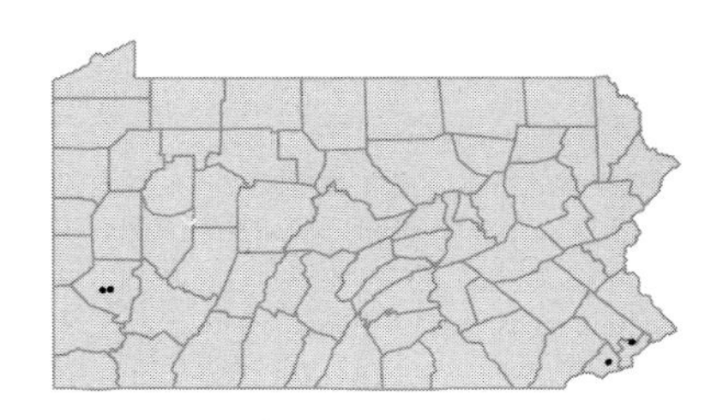

• ***Triodanis perfoliata*** (L.) Nieuwl. var. ***perfoliata*** FAC
Venus's-looking-glass
Herbaceous annual
Roadsides, woods edges and moist meadows.
Specularia perfoliata (L.) A.DC. PFB

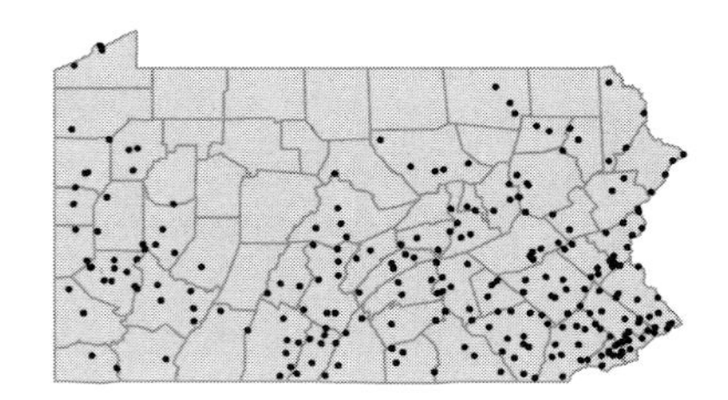

Triodanis perfoliata (L.) Nieuwl. var. ***biflora*** (Ruiz & Pavon) Bradley FAC [I]
Venus's-looking-glass
Herbaceous annual
Ballast and stream bank.
Specularia biflora (Ruiz & Pavon) Fisch. & Mey. FB; *Triodanis biflora* (Ruiz & Pavon) Greene GWC

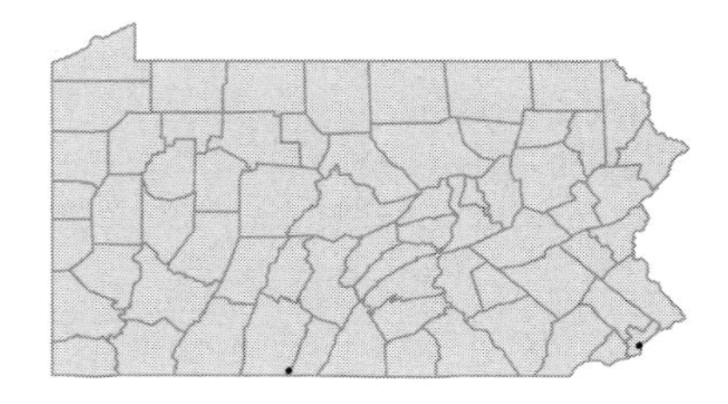

RUBIACEAE

• ***Asperula arvensis*** L. [I]
Woodruff
Herbaceous perennial
Ballast. Represented by a single collection from Philadelphia Co. in 1866.

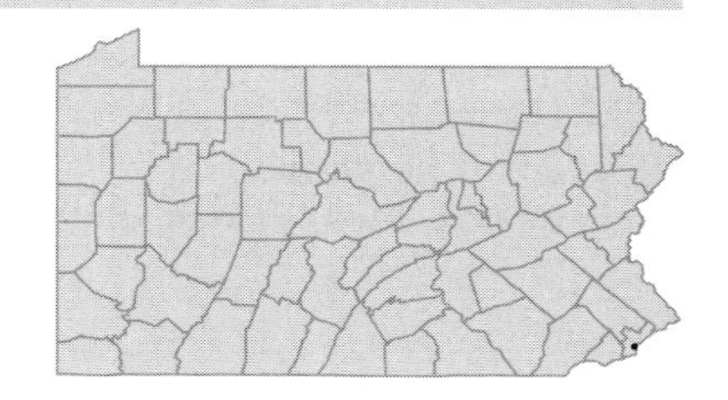

• ***Cephalanthus occidentalis*** L. OBL
Buttonbush
Deciduous shrub
Low wet ground, swamps, bogs and lake edges.

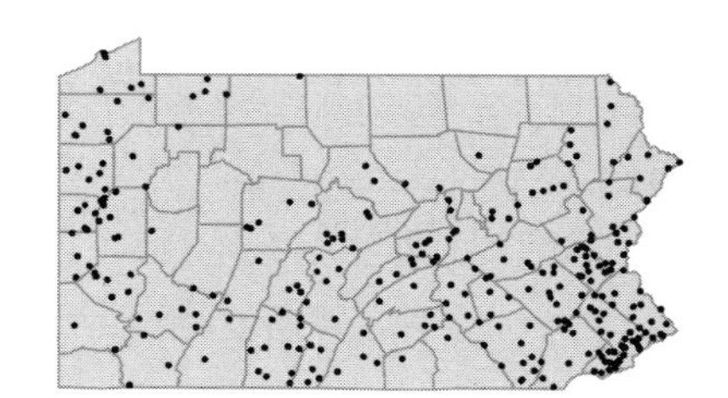

• ***Diodia teres*** Walt.
Rough buttonweed
Herbaceous annual
Railroad tracks, roadsides, old fields and sandy, open ground.

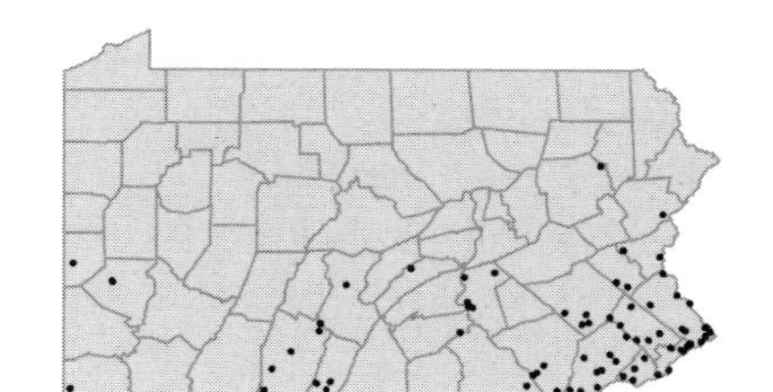

Diodia virginiana L. FACW
Virginia buttonweed
Herbaceous perennial
Ballast. Represented by several collections from Philadelphia Co. 1864-1868.

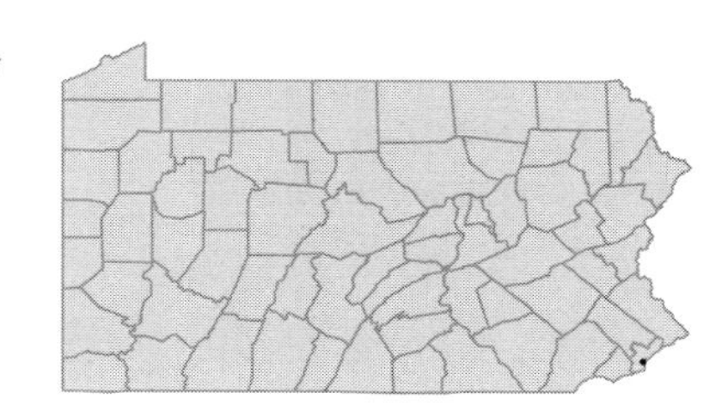

• *Galium aparine* L. FACU
Bedstraw; Cleavers; Goosegrass
Herbaceous annual
Woods, stream banks, wooded slopes and roadsides.

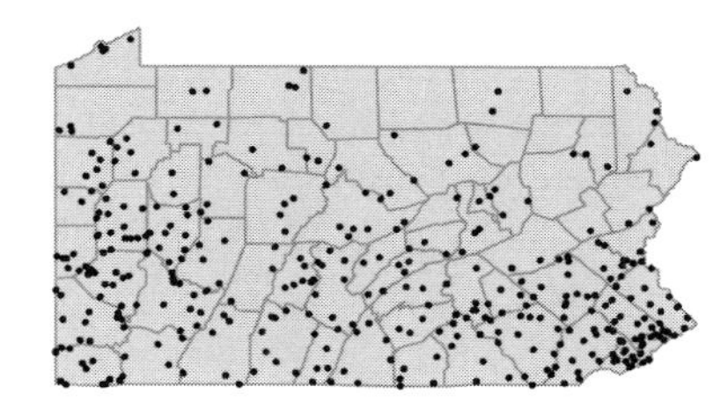

Galium asprellum Michx. OBL
Rough bedstraw
Herbaceous perennial
Swamps, bogs, stream banks and wet thickets.

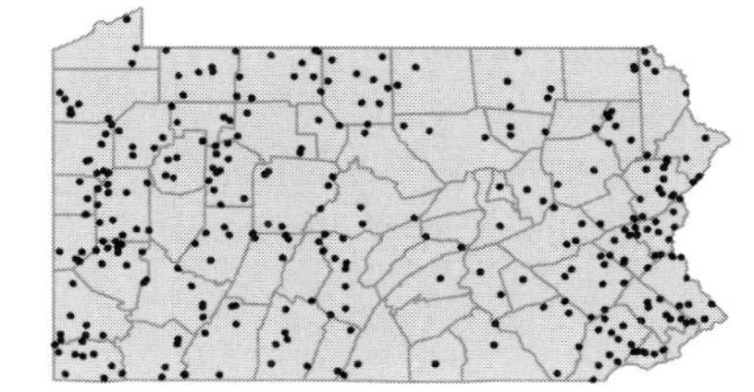

Galium boreale L. FACU
Northern bedstraw
Herbaceous perennial
Rocky slopes on limestone or serpentine, low fields, fens and roadside banks.

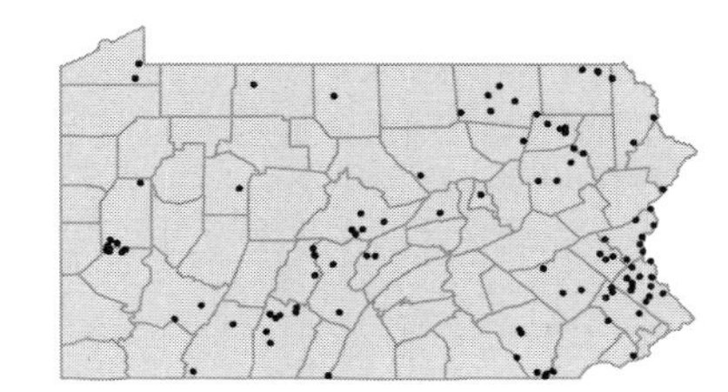

Galium circaezans Michx. var. *circaezans* UPL
Wild licorice
Herbaceous perennial
Rich woods and wooded, calcareous slopes.

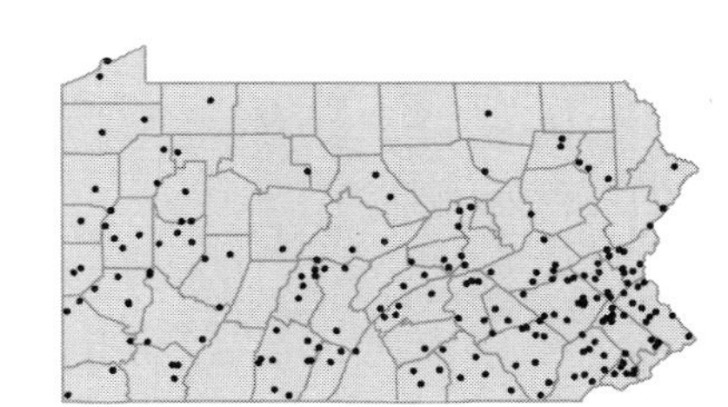

Galium circaezans Michx. var. *hypomalacum* Fern. UPL
Wild licorice
Herbaceous perennial
Woods, calcareous slopes and bluffs.

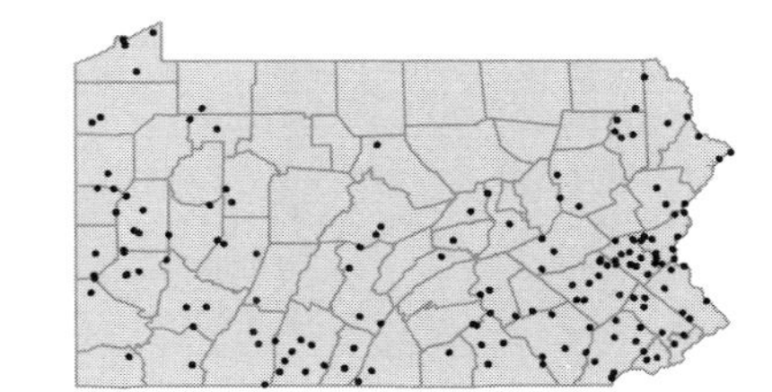

Galium concinnum Torr. & A.Gray
Shining bedstraw
Herbaceous perennial
Wooded slopes.

UPL

Galium labradoricum (Wieg.) Wieg.
Bog bedstraw; Labrador marsh bedstraw
Herbaceous perennial
Sphagnum bogs and moist banks.

OBL

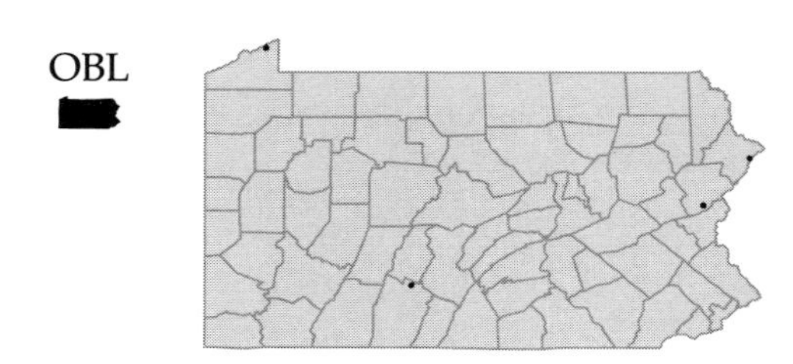

Galium lanceolatum Torr.
Wild licorice
Herbaceous perennial
Woods and wooded slopes.

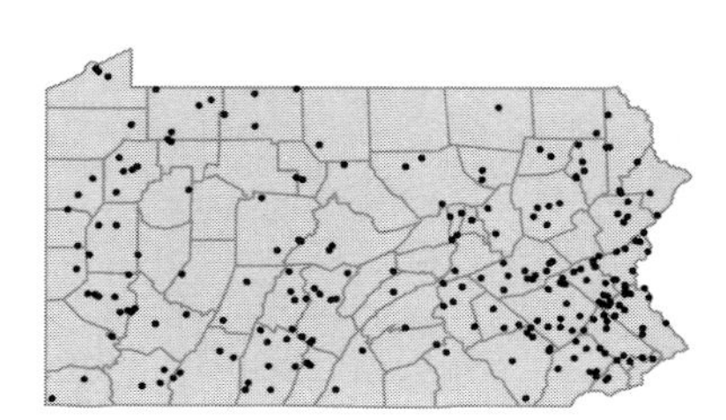

Galium latifolium Michx.
Purple bedstraw
Herbaceous perennial
Woods, rocky slopes and roadsides.

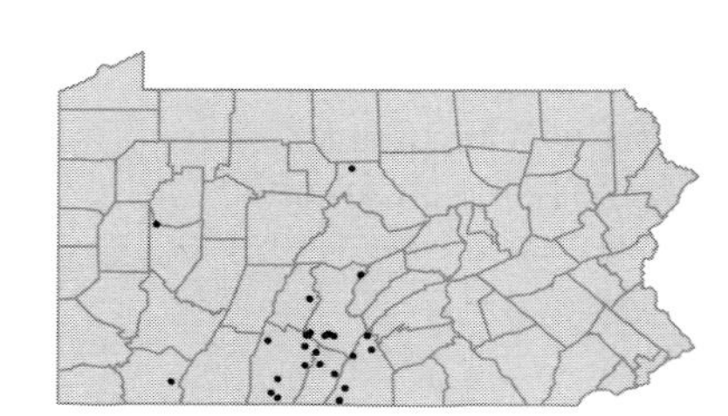

Galium mollugo L.
White bedstraw; Wild madder
Herbaceous perennial
Fields and roadsides.
Galium erectum Huds. FW

[I]

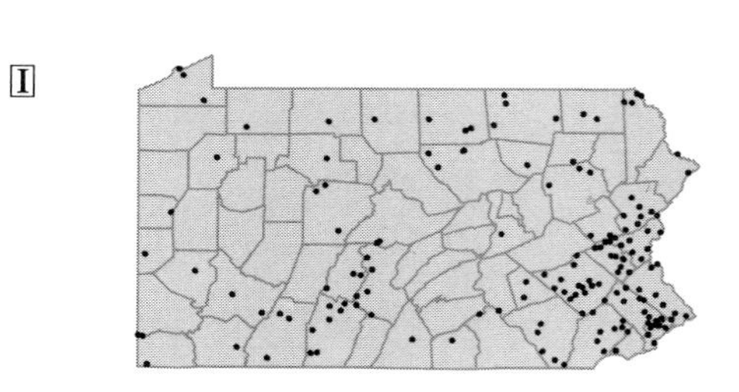

Galium obtusum Bigel.
Cleavers; Marsh bedstraw
Herbaceous perennial
Wet woods, bogs and swamps.

FACW+

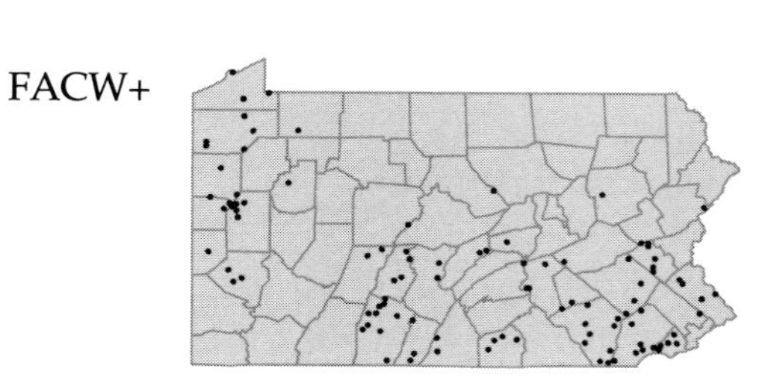

Galium odoratum (L.) Scop. [I]
Sweet woodruff
Herbaceous perennial
Cultivated and rarely escaped to rich, dry woods.
Asperula odorata L. FGBW

Galium palustre L. OBL
Ditch bedstraw; Marsh bedstraw
Herbaceous perennial
Stream banks, marshes, wet ditches and moist shores.

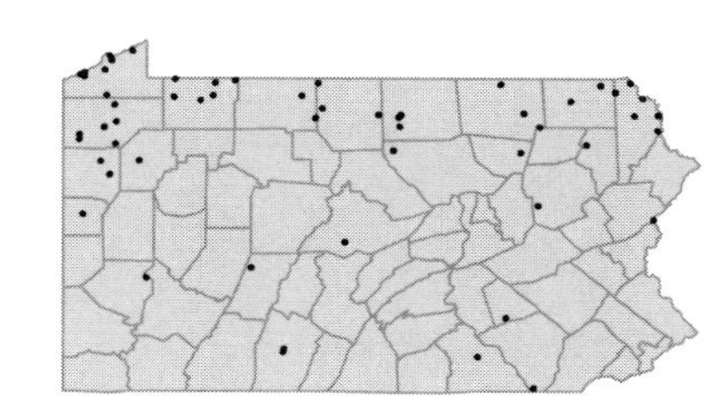

Galium pedemontanum All. [I]
Bedstraw
Herbaceous annual
Disturbed ground. Represented by a single collection from Franklin Co. in 1992.

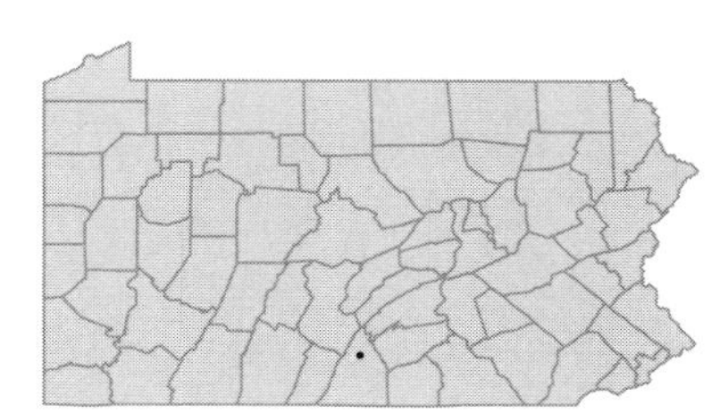

Galium pilosum Ait.
Bedstraw; Cleavers
Herbaceous perennial
Old fields, shale barrens and dry, sandy, open ground.

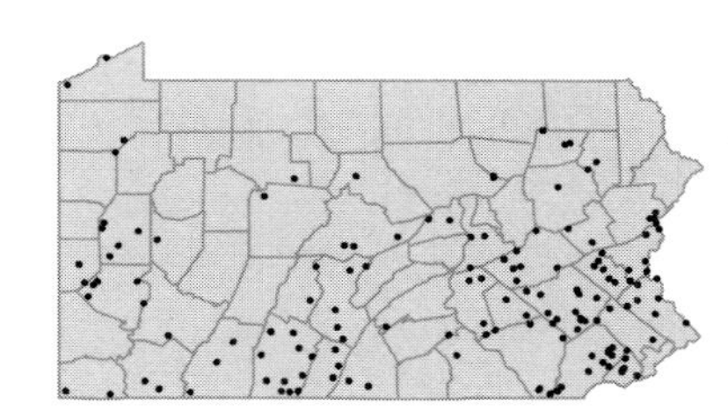

Galium rubioides L. [I]
Bedstraw
Herbaceous perennial
Open woods, site of abandoned garden. Represented by several collections from a single site in Northampton Co. 1941-1942.

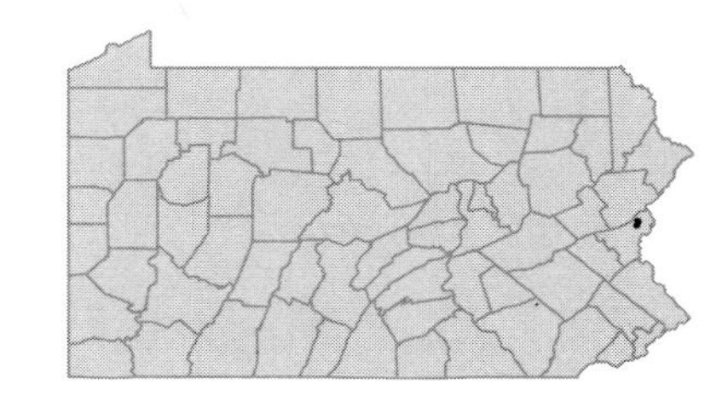

Galium tinctorium L. OBL
Bedstraw; Cleavers
Herbaceous perennial
Moist, wooded slopes, wooded floodplains, stream banks and swales.
Galium claytoni Michx. P; *Galium trifidum* L. var. *tinctorium* (L.) Torr. & A.Gray G

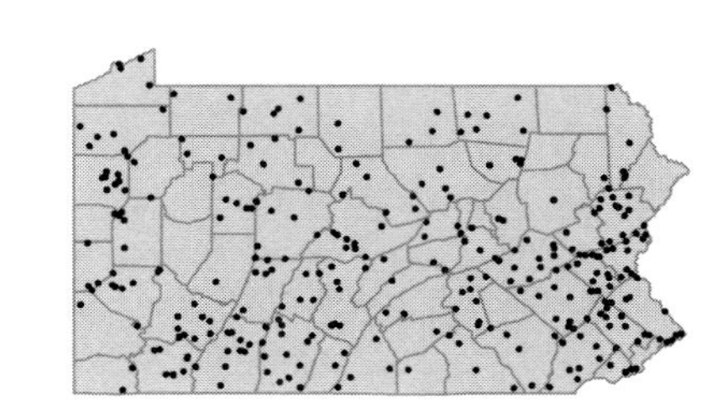

Galium tricornutum Dandy
Small bedstraw
Herbaceous annual
Ballast and waste ground.
Galium tricorne Stokes *pro parte* FBW

I

Galium trifidum L.
Cleavers; Small bedstraw
Herbaceous perennial
Moist woods, thickets and swales.

FACW+

Galium triflorum Michx.
Sweet-scented bedstraw
Herbaceous perennial
Rocky woods, shaded hillsides and roadside banks.

FACU

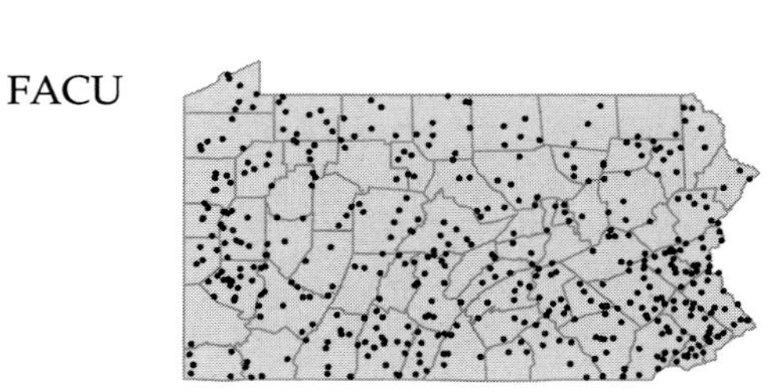

Galium verum L.
Our-lady's bedstraw; Yellow bedstraw
Herbaceous perennial
Fields, roadsides, waste ground and ballast.

I

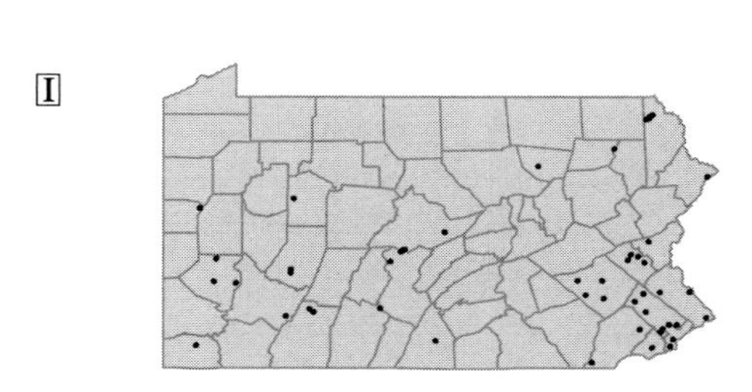

Galium wirtgenii F.W.Schultz
Bedstraw
Herbaceous perennial
Fields and meadows.

I

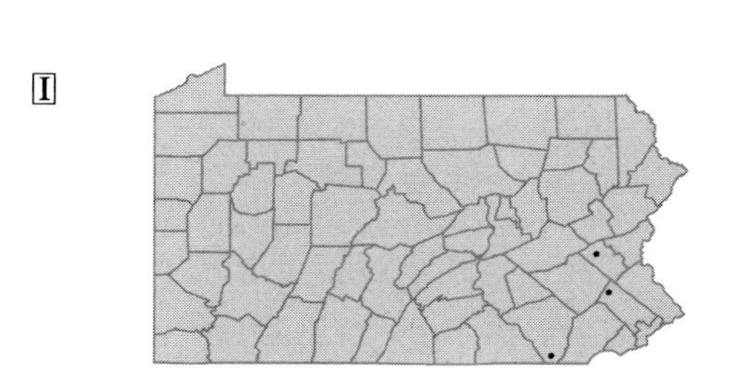

• ***Hedyotis caerulea*** (L.) Hook.
Bluets; Quaker-ladies
Herbaceous perennial
Meadows, open woods and edges.
Houstonia caerulea L. PFGB

FACU

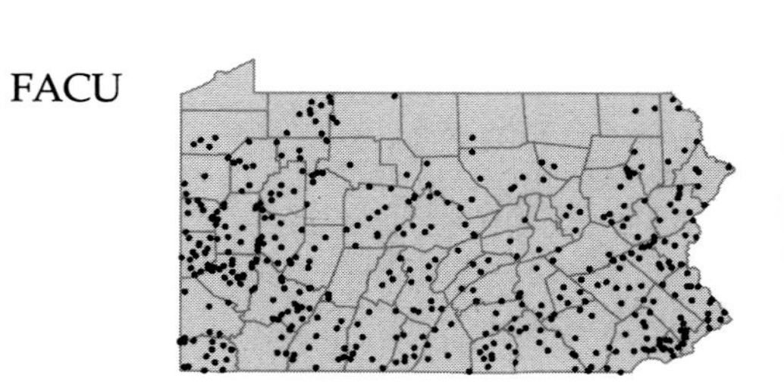

Hedyotis canadensis (Willd.) Fosberg
Purple bluets; Southern bluets
Herbaceous perennial
Dry, rocky slopes.
Houstonia canadensis Willd. FGB; *Houstonia ciliolata* Torr. P; *Hedyotis purpurea* (L.) Torr. & A.Gray var. *ciliata* (Torr.) Fosb. WK

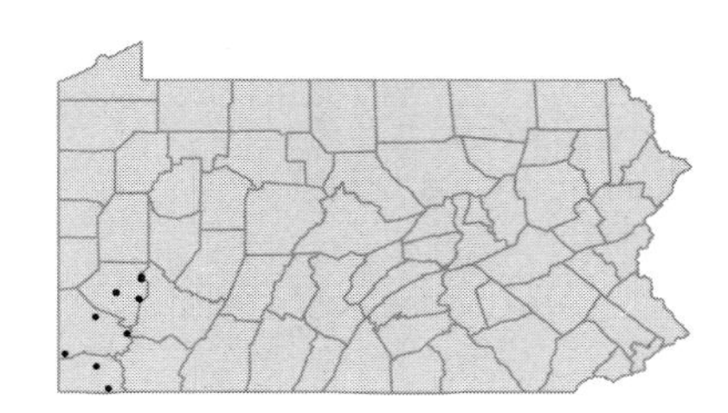

Hedyotis longifolia (Gaertn.) Hook.
Bluets
Herbaceous perennial
Dry sandy fields, shale barrens, dry wooded slopes and roadside banks.
Hedyotis purpurea (L.) Torr. & A.Gray var. *longifolia* (Gaertn.) Fosb. W; *Houstonia longifolia* Gaertn. FGB

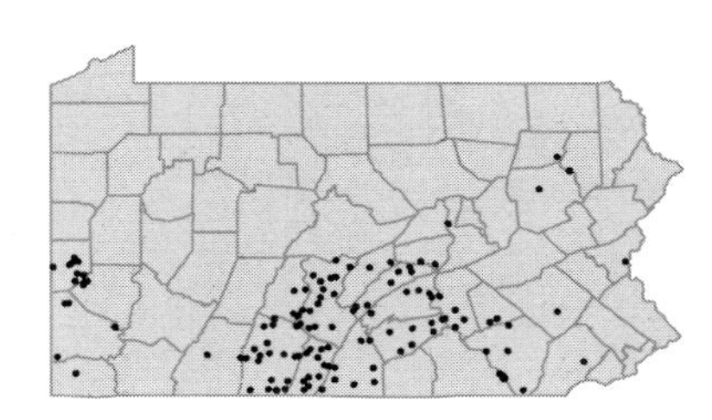

Hedyotis michauxii Fosberg
Creeping bluets; Thyme-leaved bluets
Herbaceous perennial
Moist stream border and pasture.
Houstonia serpyllifolia Michx. FGB

FAC

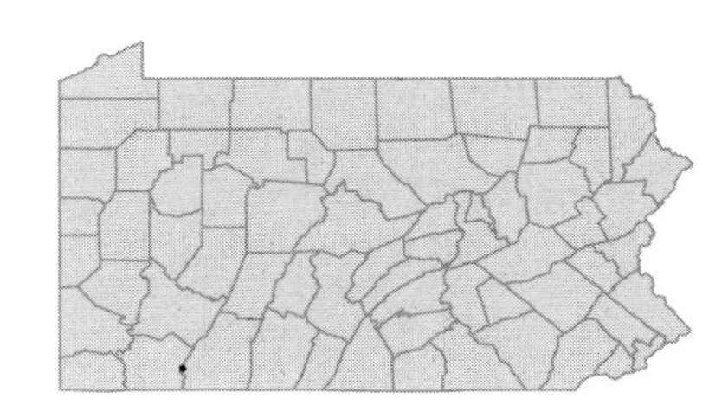

Hedyotis purpurea (L.) Torr. & A.Gray var. **purpurea**
Purple bluets; Southern bluets
Herbaceous perennial
Open woods and roadsides.
Houstonia purpurea L. FGB

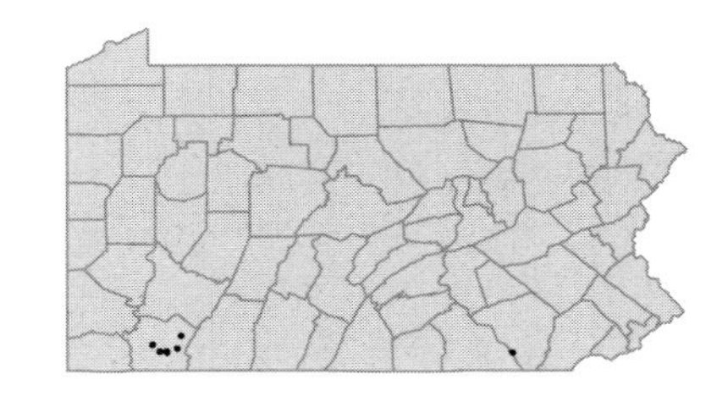

Hedyotis purpurea (L.) Torr. & A.Gray var. **calycosa** (A.Gray) Fosberg
Purple bluets; Southern bluets
Herbaceous perennial
Mountain woods.
Houstonia lanceolata (Poir.) Britt. F; *Houstonia purpurea* L. var. *calycosa* A.Gray GB

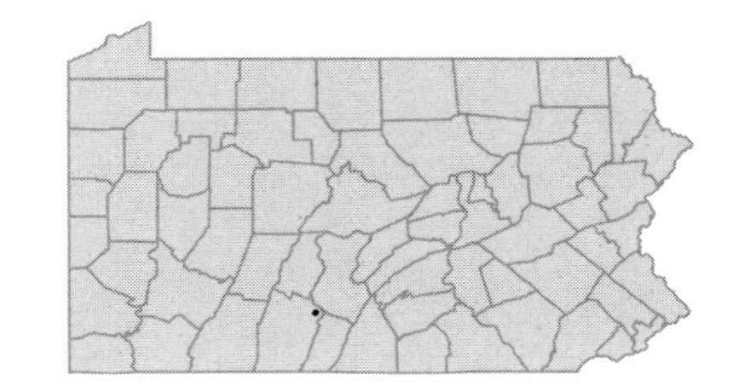

Hedyotis purpurea (L.) Torr. & A.Gray var. **tenuifolia** (Nutt.) Fosberg
Purple bluets; Southern bluets
Herbaceous perennial
Mesic, wooded, limestone slope.
Houstonia tenuifolia Nutt. FGB; *Hedyotis nuttalliana* Fosb. C

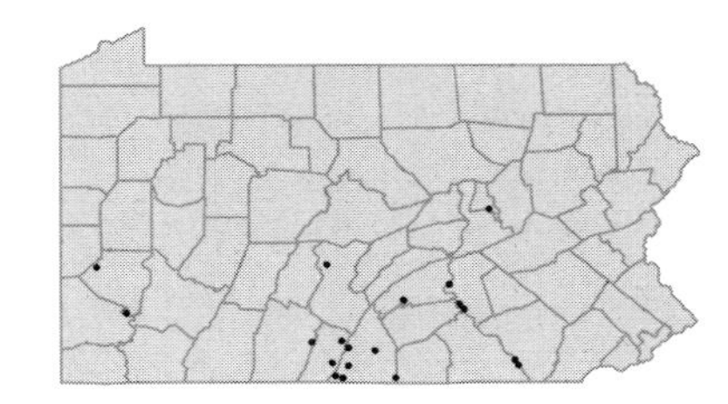

• ***Mitchella repens*** L.
Partridge-berry
Herbaceous perennial
Woods.

FACU

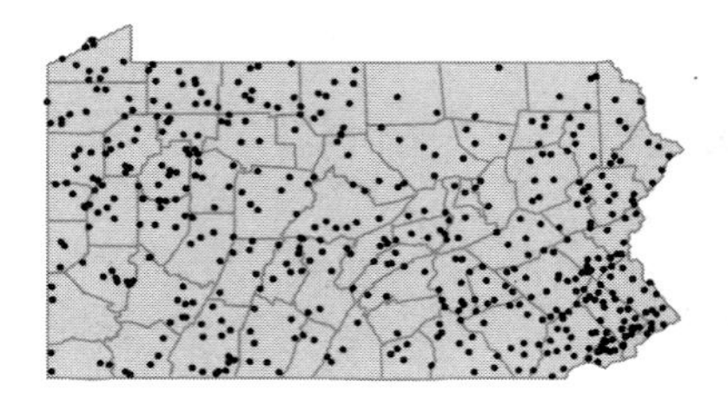

• ***Richardia brasiliensis*** (Moq.) Gomez
Mexican-clover
Herbaceous annual
Ballast. Represented by a several pre-1900 collections from Philadelphia Co.

I

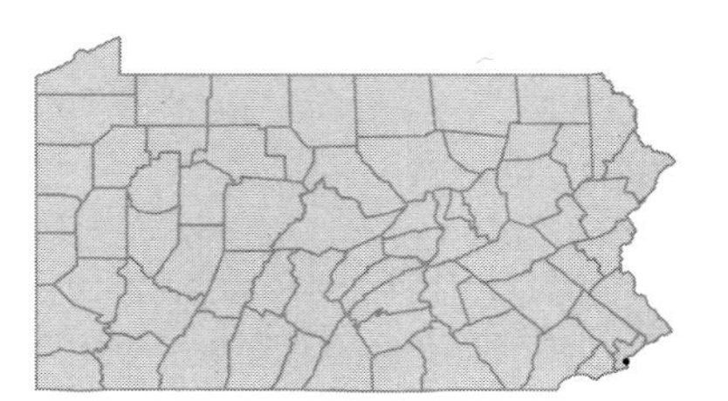

• ***Rubia tinctoria*** L.
Madder
Herbaceous perennial
Represented by a single collection from Franklin Co. in 1845.

I

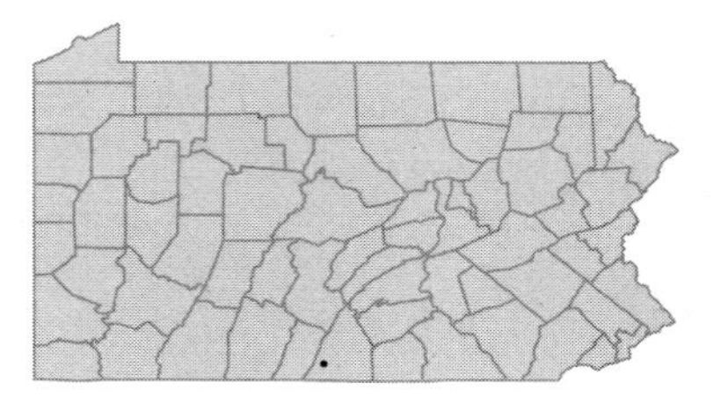

• ***Sherardia arvensis*** L.
Field-madder
Herbaceous annual
Cultivated fields, lawns and pavement.

I

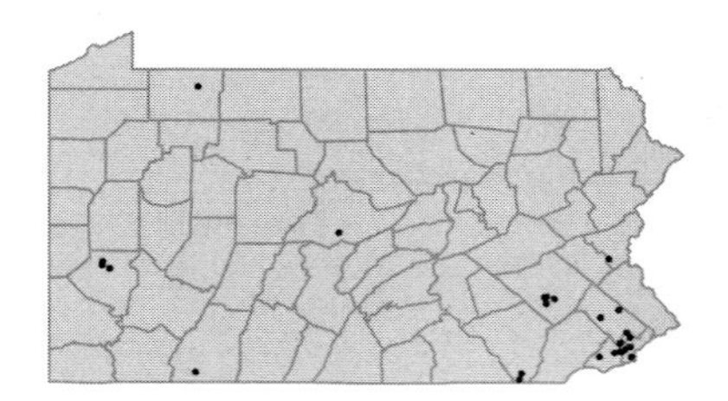

CAPRIFOLIACEAE

• ***Diervilla lonicera*** P.Mill.
Bush-honeysuckle
Deciduous shrub
Dry woods and rocky slopes.
Diervilla diervilla (L.) MacM. P

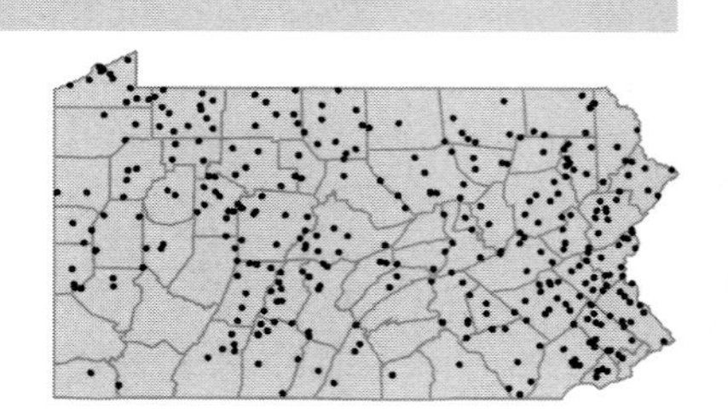

• ***Kolkwitzia amabilis*** Graebn.
Beautybush
Deciduous shrub
Cultivated and rarely escaped to alluvial woodlands.

I

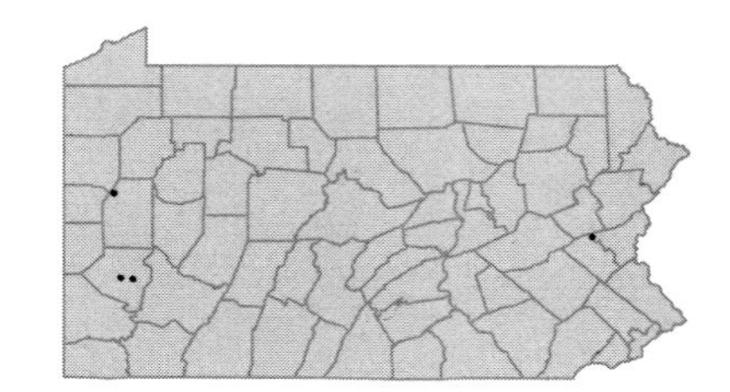

• ***Linnaea borealis*** L. var. ***americana*** (Forbes) Rehd.
Twinflower
Herbaceous perennial
Cool, moist woods.
Linnaea americana Forbes P; *Linnaea borealis* L. var. *longiflora* (Torr.) Hulten GBC

FAC

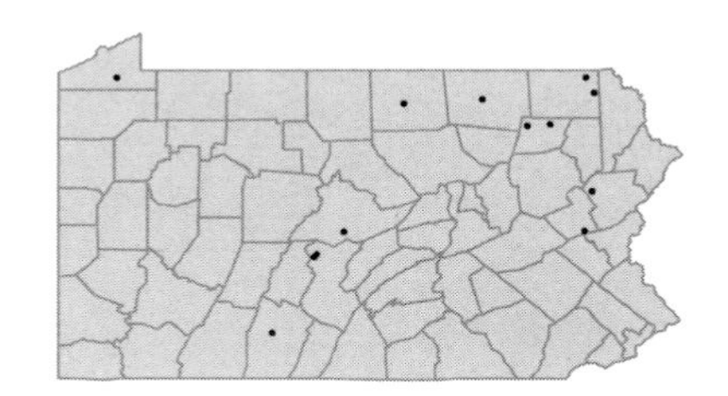

• ***Lonicera canadensis*** Bartr.
Fly honeysuckle
Deciduous shrub
Cool woods, ravines and rocky slopes.
Lonicera ciliata Muhl. P

FACU

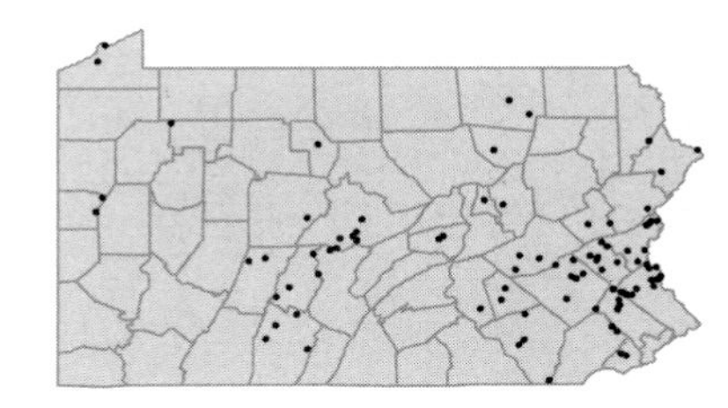

Lonicera dioica L. var. ***diocia***
Mountain honeysuckle
Deciduous shrub
Moist cliffs, rocky wooded banks and thickets.

FACU

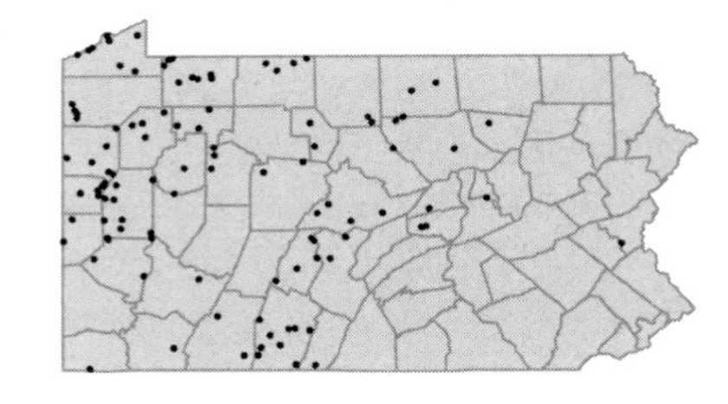

Lonicera dioica L. var. ***glaucescens*** (Rydb.) Butt.
Mountain honeysuckle
Deciduous shrub
Woods, bogs, mossy cliffs and wooded valleys.
Lonicera glaucescens Rydb. P

FACU

Lonicera dioica L. var. ***orientalis*** Gleason
Wild honeysuckle
Deciduous shrub
Clayey, rocky bank. Represented by a single collection from Butler Co. in 1976.

FACU

Lonicera fragrantissima Lindl. & Paxt.
Fragrant honeysuckle
Deciduous shrub
Cultivated and occasionally escaped to wooded slopes, thickets and rubbish dumps.

I

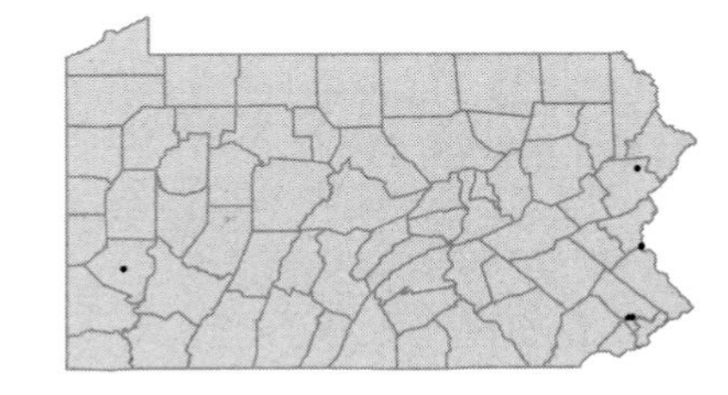

Lonicera hirsuta Eat.
Hairy honeysuckle
Deciduous shrub
Moist woods, swamps and rocky thickets.

FAC

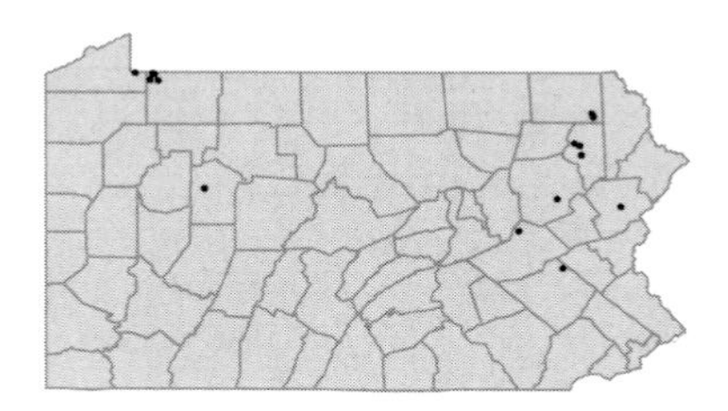

Lonicera japonica Thunb. var. ***japonica***
Japanese honeysuckle
Woody vine
Disturbed woods, fields, thickets, banks and roadsides.

FAC-
I

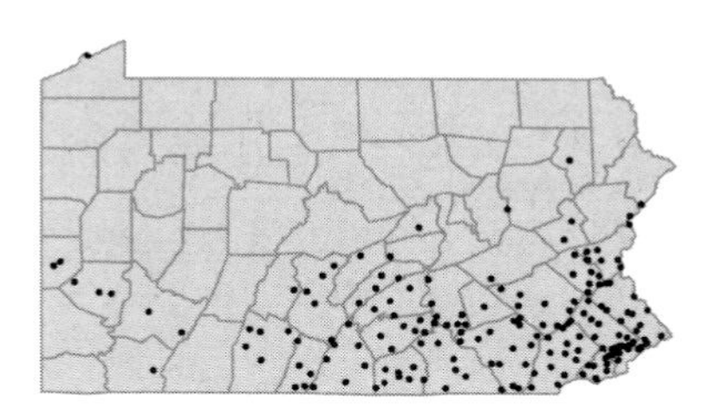

Lonicera japonica Thunb. var. ***chinensis*** (P.W.Watson) Baker
Japanese honeysuckle
Woody vine
Rocky, wooded slopes and moist, bottomland thickets.

FAC-
I

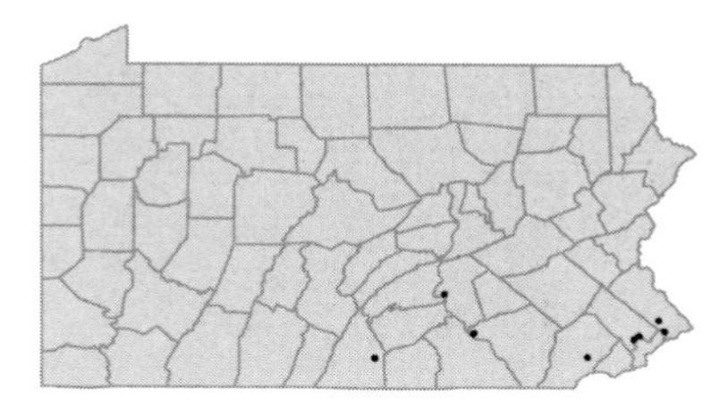

Lonicera maackii (Rupr.) Maxim.
Amur honeysuckle
Deciduous shrub
Cultivated and frequently naturalized in disturbed woods, thickets, old fields and roadsides.

I

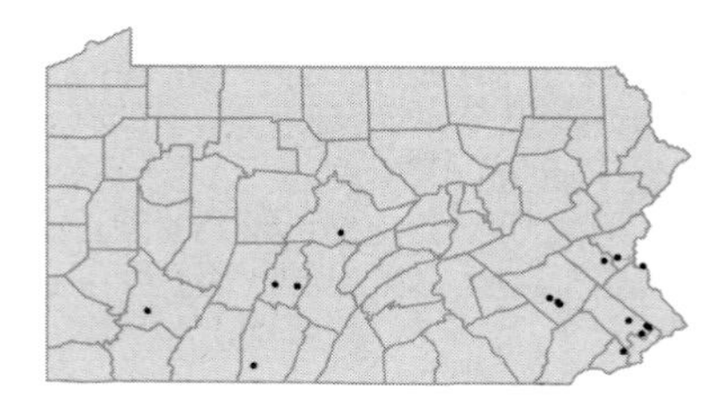

Lonicera morrowii A.Gray
Morrow's honeysuckle
Deciduous shrub
Cultivated and frequently naturalized in disturbed woods, old fields, roadsides and thickets.

I

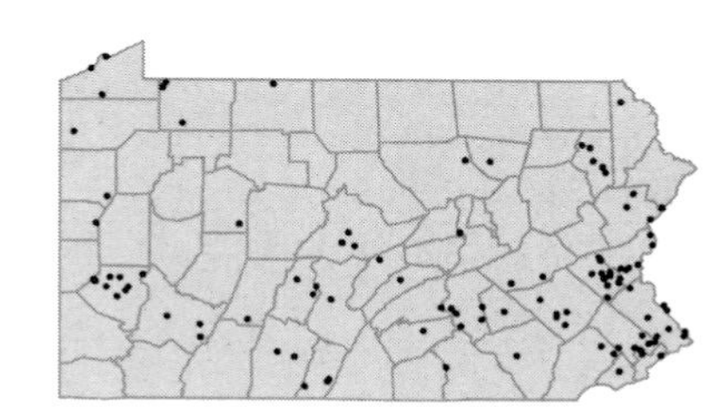

Lonicera morrowii x tatarica
Pretty honeysuckle
Deciduous shrub
Roadsides and stream banks, escaped from cultivation.
Lonicera x bella Zabel CGF

FACU-
I

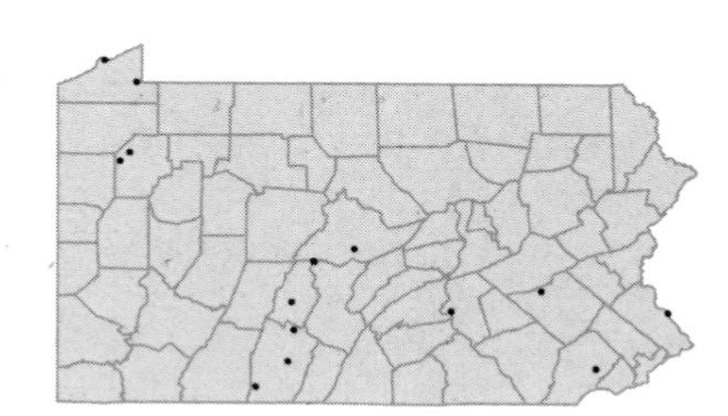

Lonicera oblongifolia (Goldie) Hook.
Swamp fly honeysuckle
Deciduous shrub
Bogs and swamps.

OBL

Lonicera sempervirens L.
Trumpet honeysuckle
Woody vine
Fencerows, thickets and roadsides.

FACU

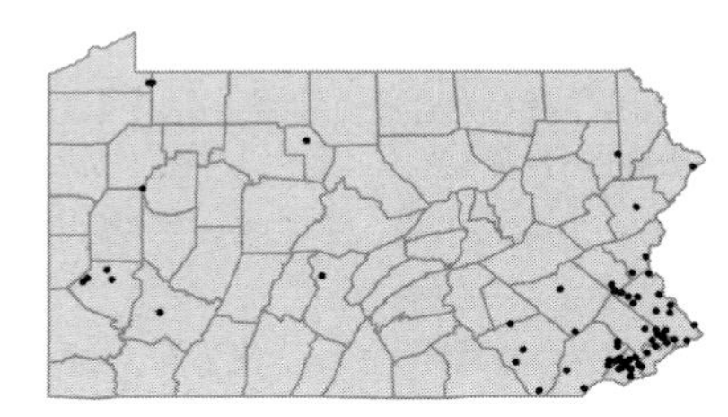

Lonicera standishii Jacques
Honeysuckle
Deciduous shrub
Cultivated and occasionally escaped to wooded slopes and edges.

[I]

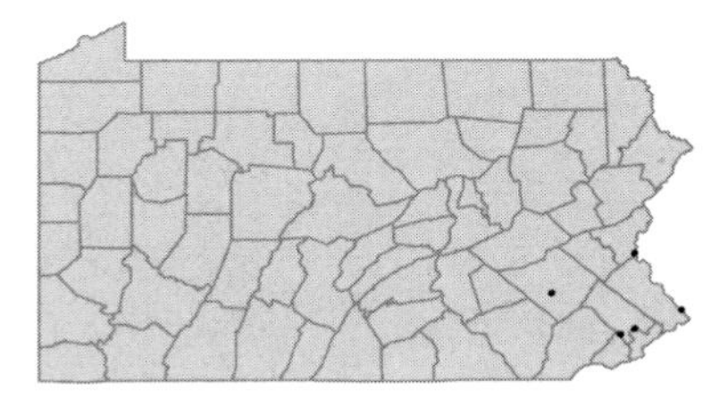

Lonicera tatarica L.
Tartarian honeysuckle
Deciduous shrub
Cultivated and frequently escaped to disturbed or waste ground, woods edges and roadsides.

FACU
[I]

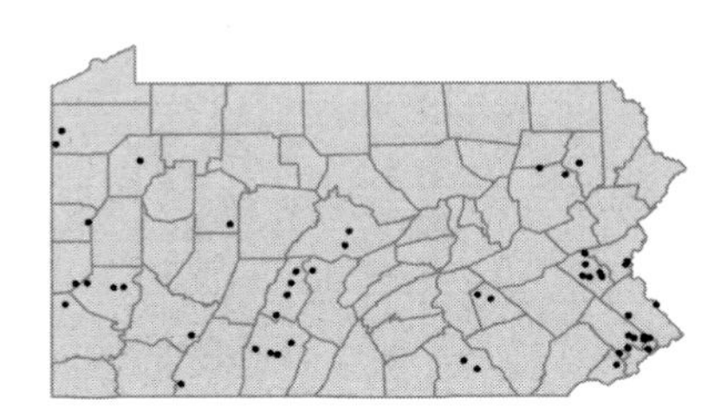

Lonicera villosa (Michx.) Roemer & Schultes
Waterberry; Mountain fly honeysuckle
Deciduous shrub
Bogs, swamps and wet thickets.
Lonicera caerulea L. C

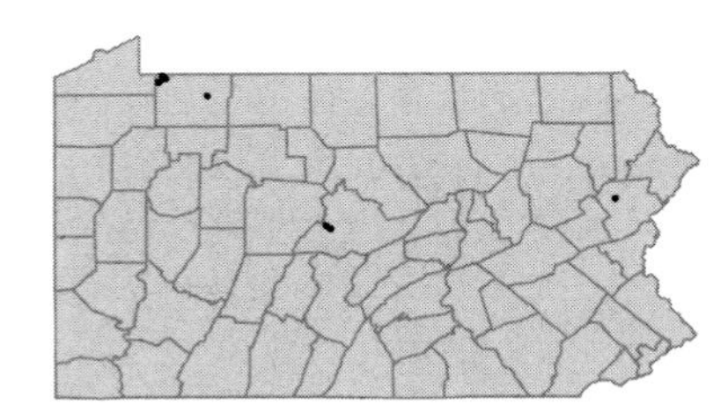

Lonicera xylosteum L.
European fly honeysuckle
Deciduous shrub
Cultivated and occasionally escaped to roadside thickets and stream banks.

[I]

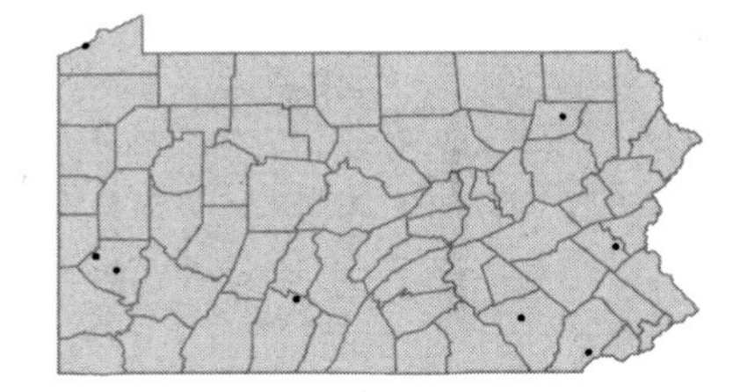

• ***Sambucus canadensis*** L.
American elder
Deciduous shrub
Woods, fields, stream banks and moist roadsides.

FACW-

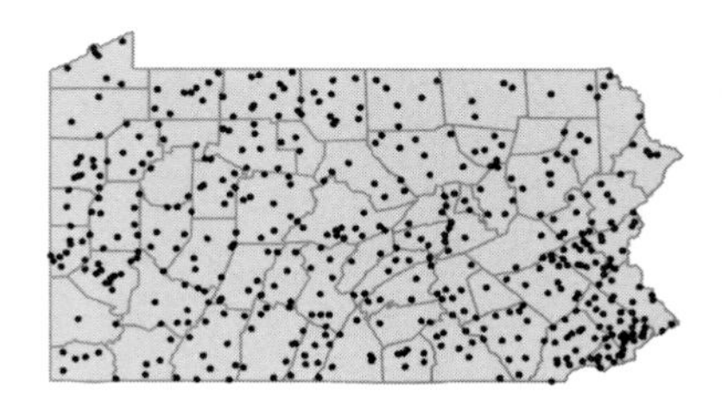

Sambucus nigra L.
Black elder; Elder-of-Europe
Deciduous tree
Floodplain. Reprsented by a single collection from Allegheny Co. in 1911.

[I]

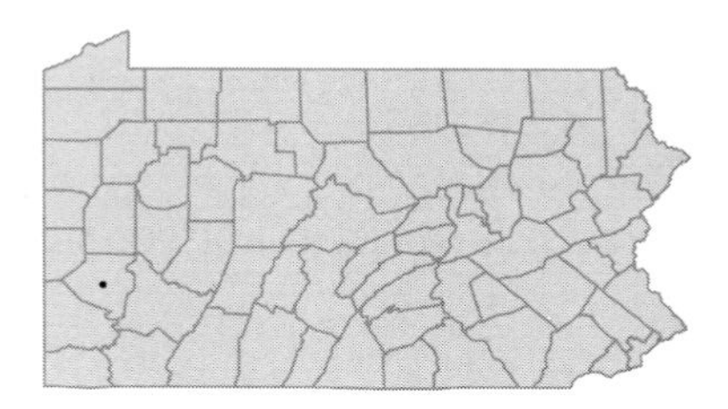

Sambucus racemosa L. var. ***pubens*** (Michx.) House
Red-berried elder
Deciduous shrub
Ravines, moist cliffs and rocky woods.
Sambucus pubens Michx. PFGBW

FACU

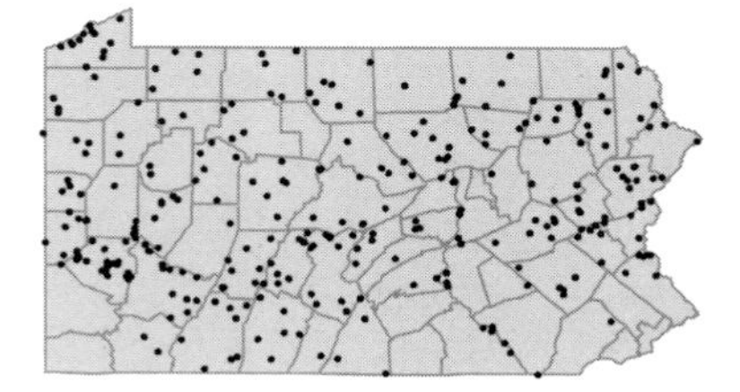

• ***Symphoricarpos albus*** (L.) S.F.Blake var. ***albus***
Snowberry
Deciduous shrub
Rocky, wooded, limestone slopes and barrens, also escaped from cultivation.
Symphoricarpos racemosus Michx. P

FACU-

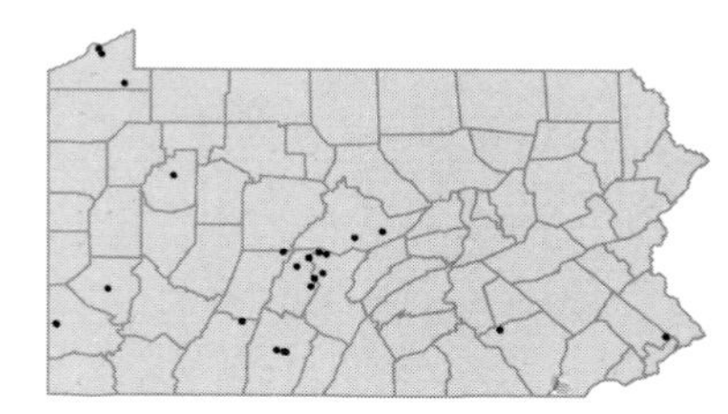

Symphoricarpos albus (L.) S.F.Blake var. ***laevigatus*** (Fern.) S.F.Blake
Snowberry
Deciduous shrub
Cultivated and occasionally escaped to roadsides and waste ground.

FACU-
[I]

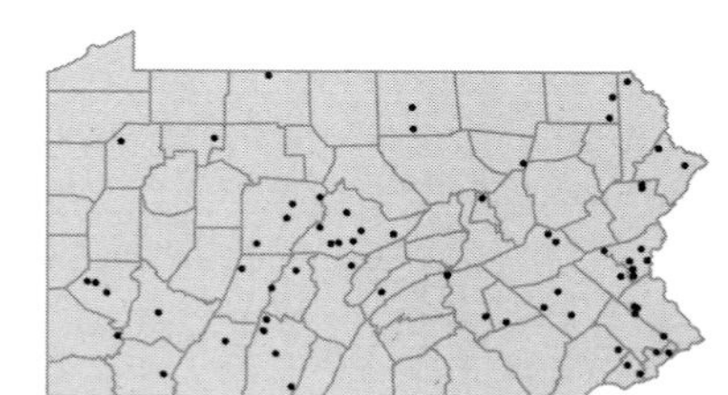

Symphoricarpos occidentalis Hook.
Wolfberry
Deciduous shrub
Cultivated and rarely escaped to railroad embankments and alluvium.

[I]

Symphoricarpos orbiculatus Moench UPL
Coralberry; Indian-currant
Deciduous shrub
Open woods, thickets, old fields, and dry banks, also escaped from cultivation.
Symphoricarpos symphoricarpos (L.) MacM. P

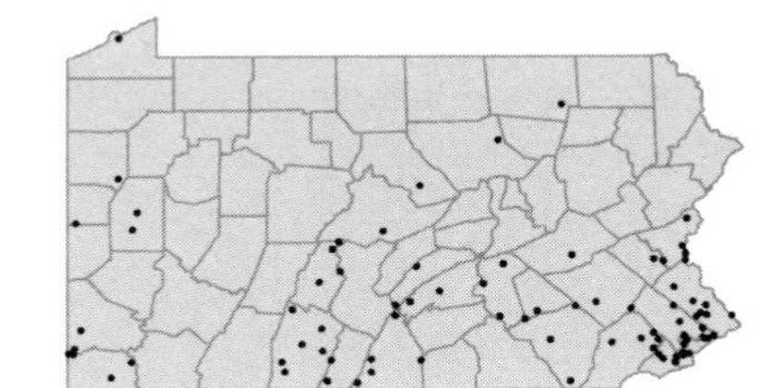

• ***Triosteum angustifolium*** L.
Horse-gentian; Feverwort
Herbaceous perennial
Woods and thickets.

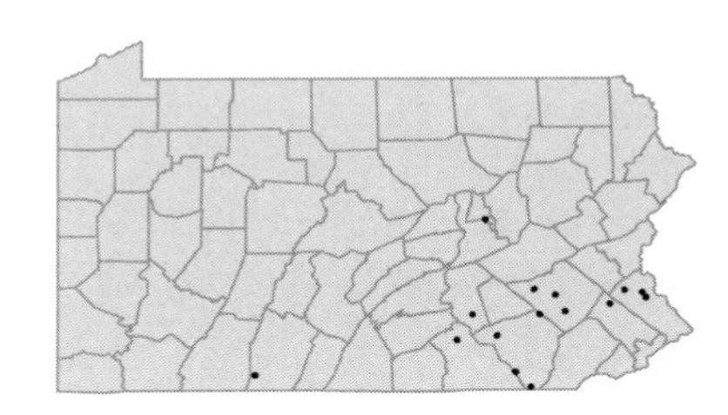

Triosteum aurantiacum Bickn. var. ***aurantiacum***
Wild coffee
Herbaceous perennial
Moist, rocky, limestone slopes and wooded ravines.
Triosteum perfoliatum L. var. *aurantiacum* (Bickn.) Wieg. GB

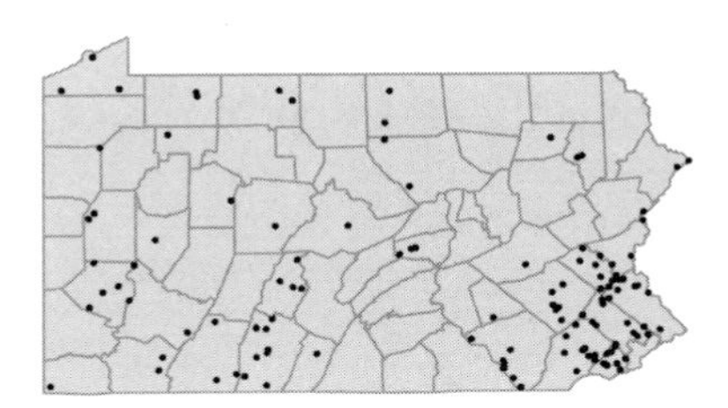

Triosteum aurantiacum Bickn. var. ***glaucescens*** Wieg.
Wild coffee
Herbaceous perennial
Rich woods.
Triosteum perfoliatum L. var. *glaucescens* Wieg. GB

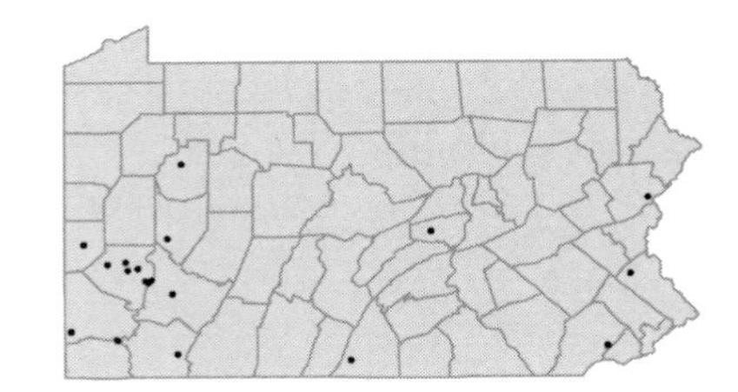

Triosteum aurantiacum Bickn. var. ***illinoense*** (Wieg.) Palmer & Steyermark
Wild coffee
Herbaceous perennial
Rich woods and wooded hillsides.
Triosteum illinoense (Wieg.) Rydb. W; *Triosteum perfoliatum* L. var. *illinoense* Wieg. GB

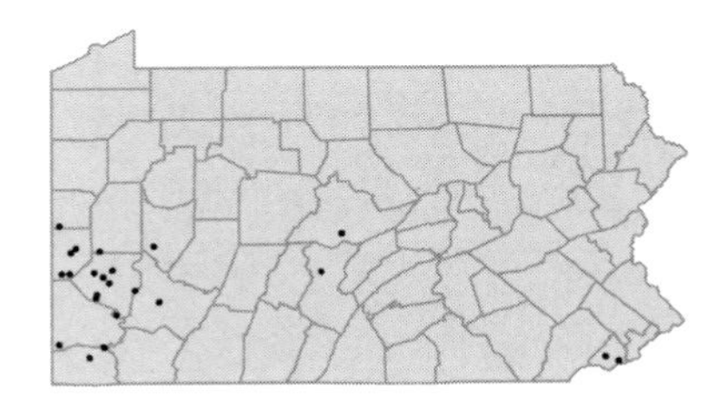

Triosteum perfoliatum L.
Horse-gentian
Herbaceous perennial
Moist woods, thickets, calcareous hillsides and roadsides.

• ***Viburnum acerifolium*** L.
Maple-leaved viburnum; Dockmackie
Deciduous shrub
Woods and thickets.

UPL

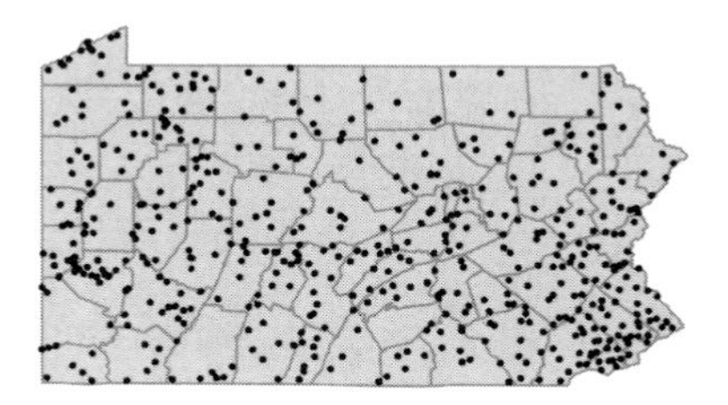

Viburnum cassinoides L.
Withe-rod
Deciduous shrub
Swamps, bogs and wet woods.
Viburnum nudum L. var. *cassinoides* (L.) Torr. & A.Gray C

FACW

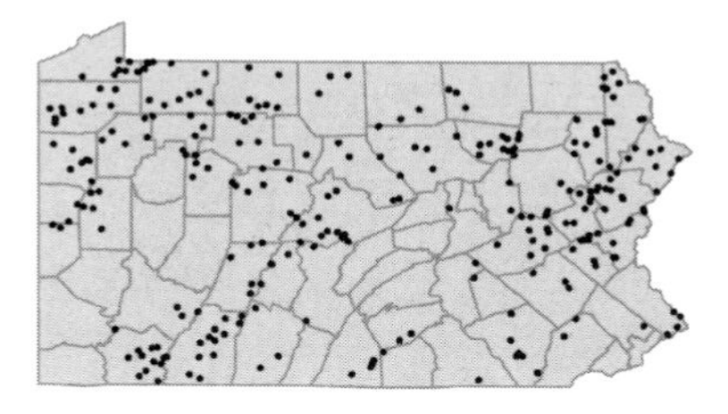

Viburnum dentatum L.
Southern arrow-wood
Deciduous shrub
Swamps and wet woods.
Viburnum pubescens (Ait.) Pursh P

FAC

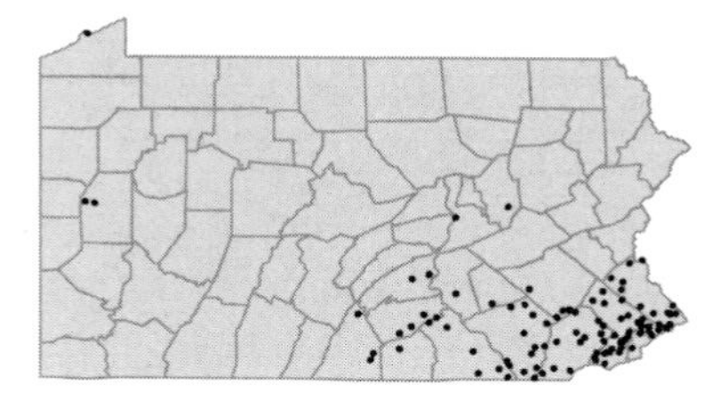

Viburnum lantana L.
Wayfaring-tree
Deciduous shrub
Cultivated and occasionally escaped.

[I]

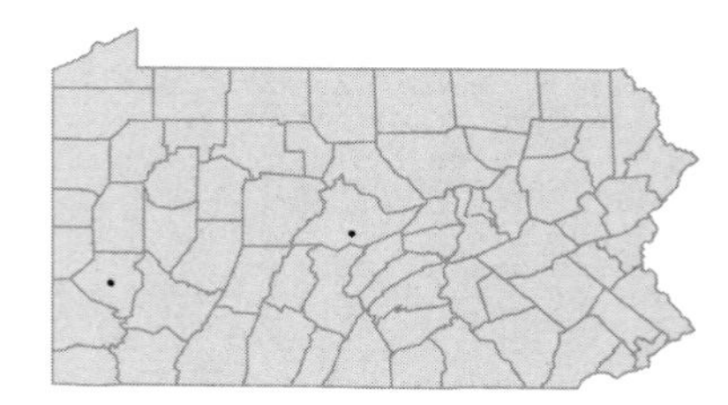

Viburnum lantanoides Michx.
Witch-hobble; Hobble-bush
Deciduous shrub
Cool, moist woods and ravines.
Viburnum alnifolium Marshall PFGBWC

FAC

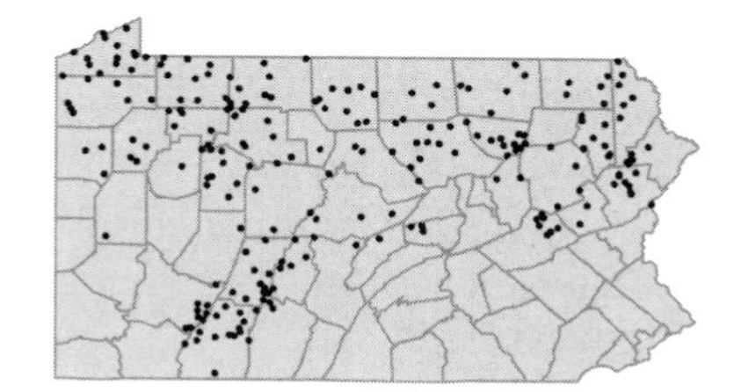

Viburnum lentago L.
Nannyberry; Sheepberry
Deciduous shrub
Woods, swamps and roadsides.

FAC

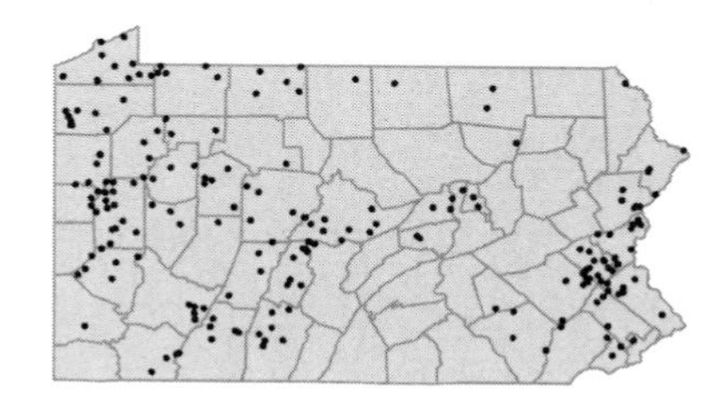

Viburnum nudum L.
Possum-haw; Swamp-haw
Deciduous shrub
Wet woods and swamps.

OBL

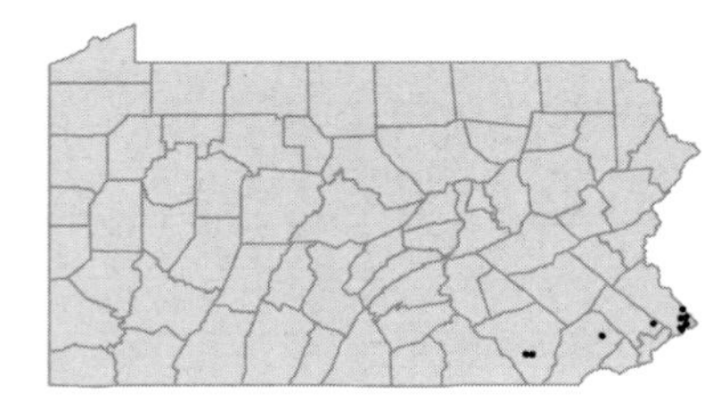

Viburnum opulus L.
Guelder-rose
Deciduous shrub
Cultivated and occasionally escaped to woods, fields and roadsides.

I

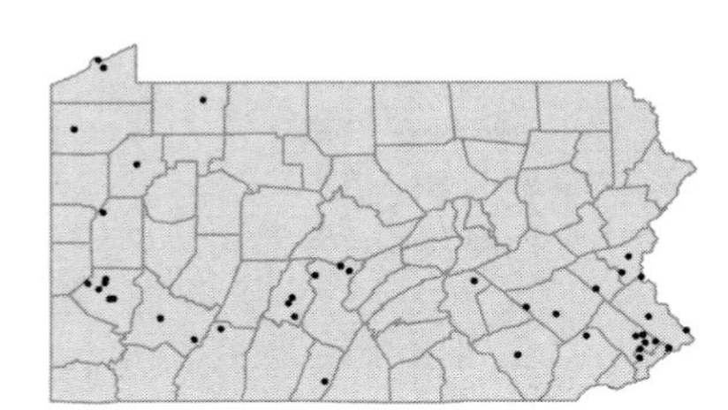

Viburnum plicatum Thunb.
Doublefile viburnum
Deciduous shrub
Cultivated and occasionally escaped.

I

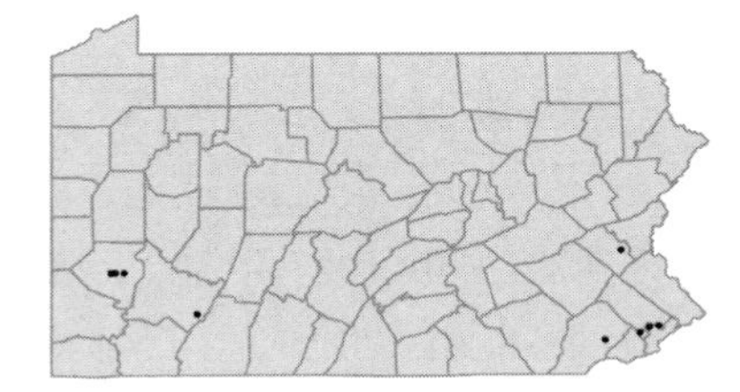

Viburnum prunifolium L.
Black-haw
Deciduous shrub
Woods, old fields, thickets and roadsides.

FACU

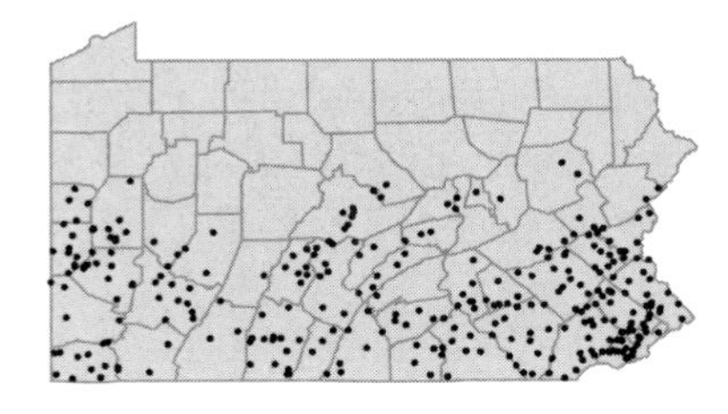

Viburnum rafinesquianum Schultes
Downy arrow-wood
Deciduous shrub
Dry slopes, open woods or barrens.

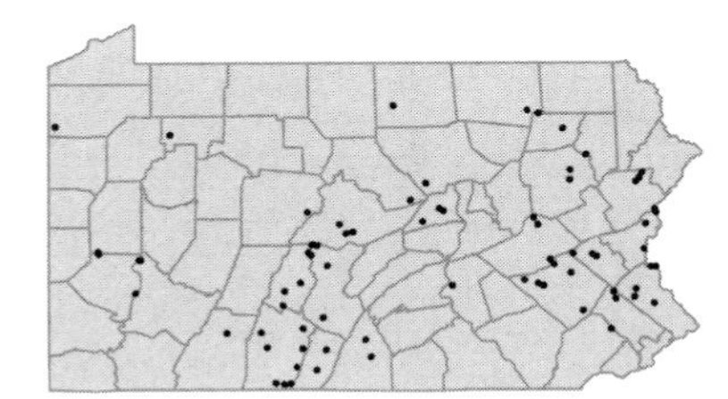

Viburnum recognitum Fern.
Arrow-wood
Deciduous shrub
Swamps, boggy woods, swampy pastures and stream banks.
Viburnum dentatum L. var. *lucidum* Ait. BC

FACW-

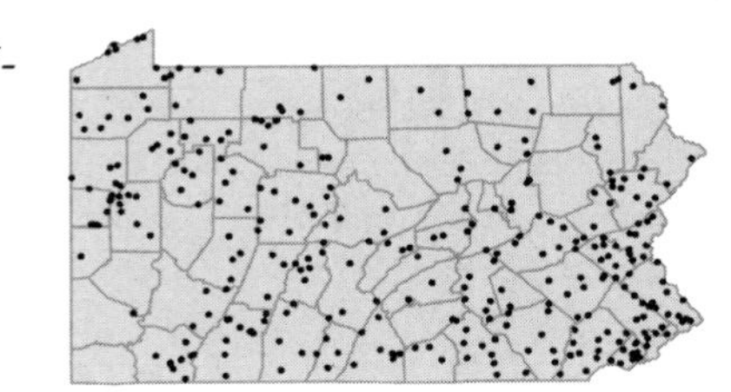

Viburnum sieboldii Miq.
Siebold viburnum
Deciduous shrub
Cultivated and occasionaly escaped to disturbed woods and stream banks.

I

Viburnum trilobum Marshall
Highbush-cranberry
Deciduous shrub
Swamps, fens and wet woods, also escaped from cultivation.
Viburnum opulus L. var. *americanum* Ait. GBC

FACW

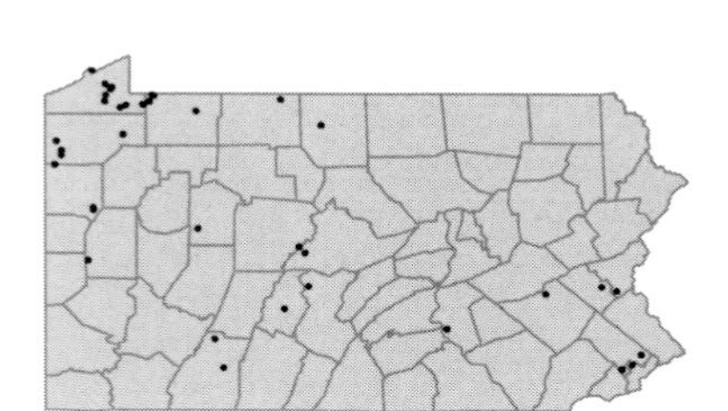

• ***Weigela florida*** (Sieb. & Zucc.) A.DC.
Crimson weigela
Deciduous shrub
Cultivated and rarely escaped. Represented by a single collection from Northampton Co.

I

VALERIANACEAE

• ***Valeriana officinalis*** L.
Garden heliotrope
Herbaceous perennial
Cultivated and occasionally escaped to woods, fields and roadsides.

I

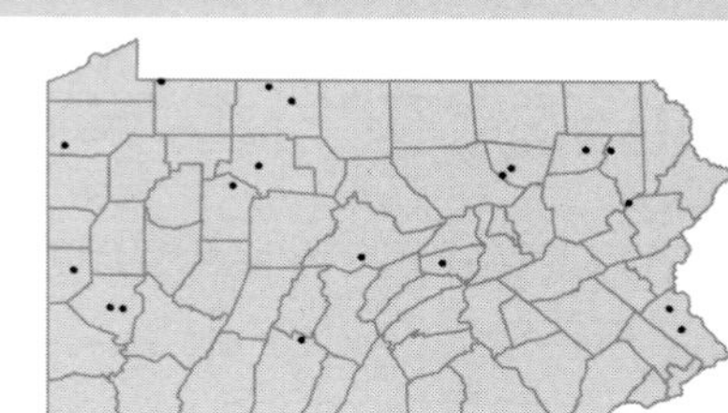

Valeriana pauciflora Michx.
Valerian
Herbaceous perennial
Wooded stream banks and floodplain forests.

FACW

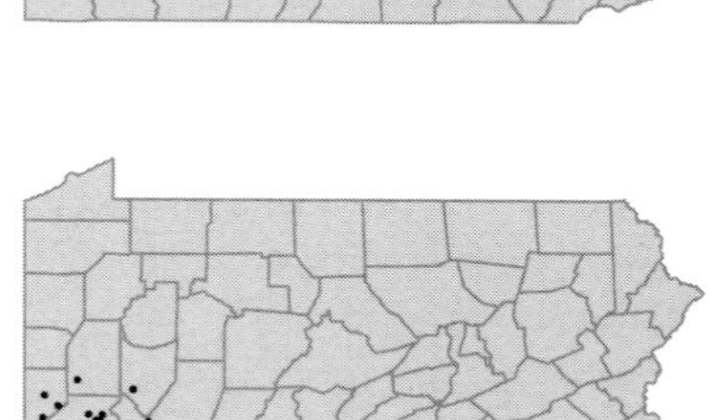

• ***Valerianella chenopodiifolia*** (Pursh) DC.
Goose-foot corn-salad
Herbaceous perennial
Mesic woods, floodplains, old fields and roadsides.

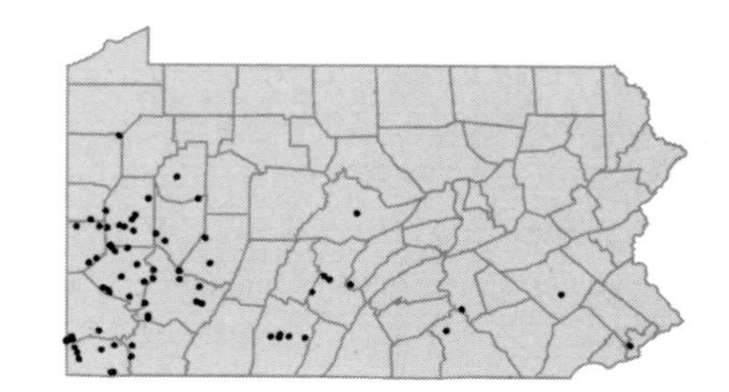

Valerianella locusta (L.) Betcke [I]
Corn-salad; Lamb's-lettuce
Herbaceous annual
Moist, open ground and roadside ditches.
Valerianella olitoria (L.) Pollich FBW

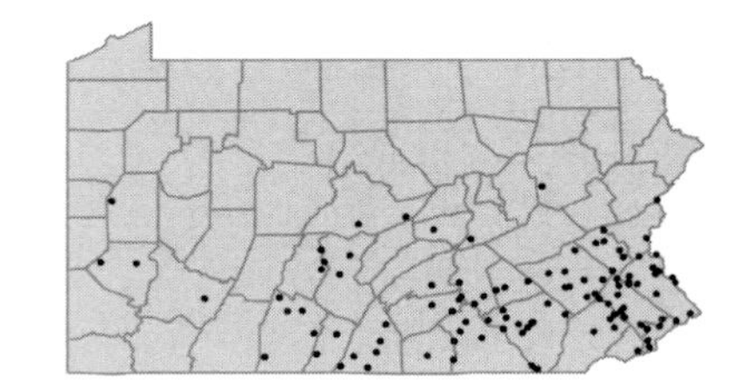

Valerianella umbilicata (Sullivant) Wood
Corn-salad
Herbaceous annual
Moist meadows, swampy woods and roadsides.
Valerianella intermedia Dyal FW; *Valerianella patellaria* (Sullivant) Wood PFW

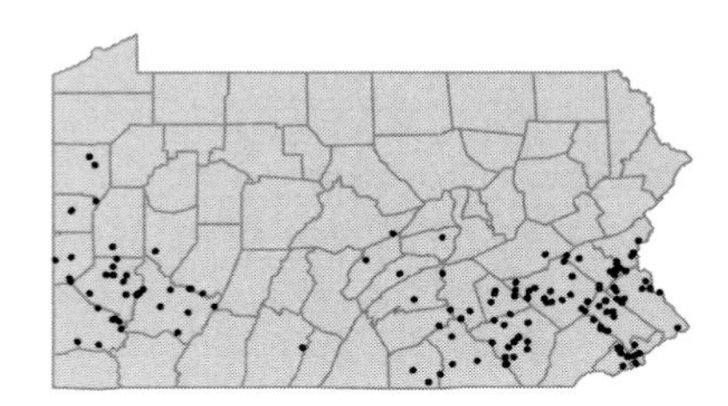

DIPSACACEAE

• ***Dipsacus laciniatus*** L. [I]
Cut-leaved teasel
Herbaceous biennial
Roadsides.

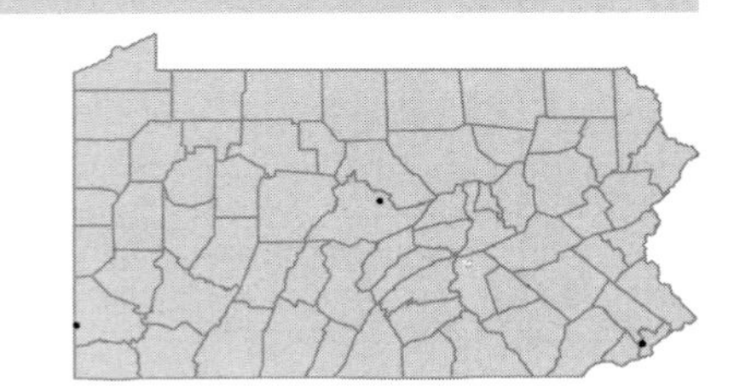

Dipsacus sativus (L.) Hockeny [I]
Teasel
Herbaceous biennial
Pastures and waste ground.
Dipsacus fullonum sensu Miller *non* L.

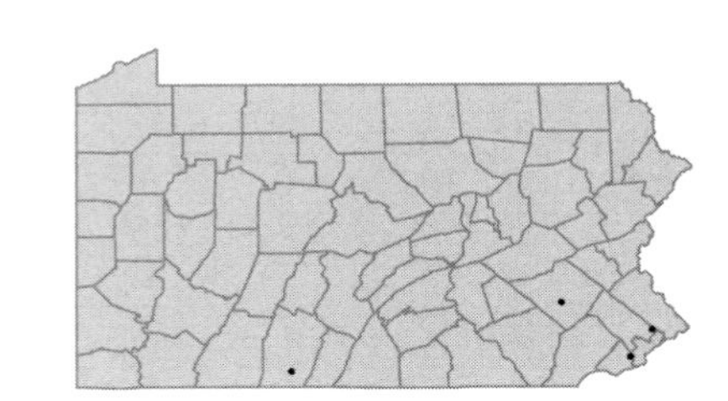

Dipsacus sylvestris Hudson [I]
Teasel
Herbaceous biennial
Roadsides, fields and waste ground.
Dipsacus fullonum L.

• ***Knautia arvensis*** (L.) Duby [I]
Blue-buttons
Herbaceous perennial
Fields and roadsides.

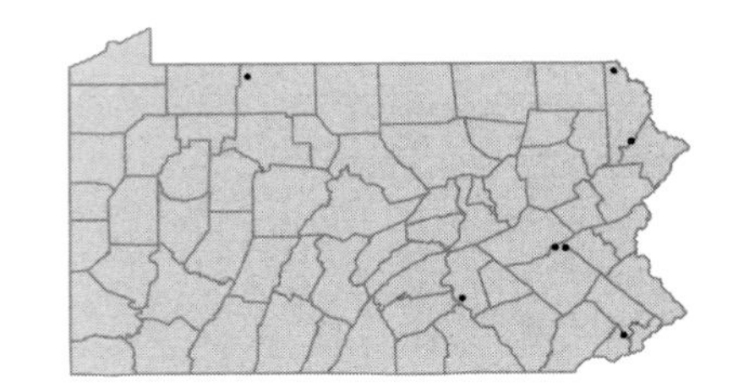

DIPSACACEAE

• ***Succisella inflexa*** (Kluk) G.Beck
Devil's-bit
Herbaceous annual
Moist meadows and pond margins.
Scabiosa australis Wulfen GB; *Succisa australis* (Wulfen) Reichenb. FW

[I]

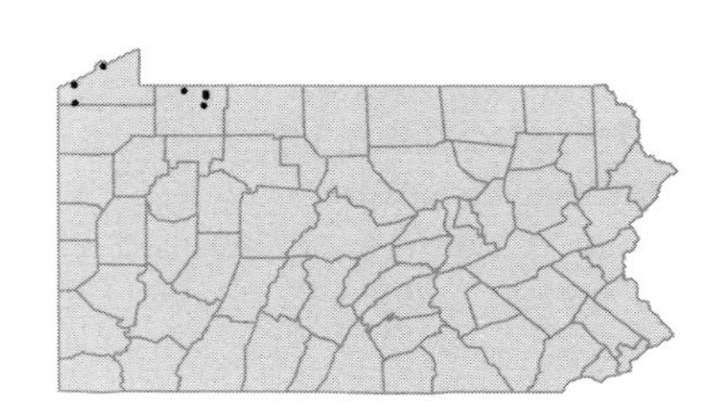

ASTERACEAE

• ***Acanthospermum australe*** (Loefl.) Kuntze.
Paraguay-bur
Herbaceous annual
Cinder fill. Represented by a single collection from Montgomery Co. in 1950.

[I]

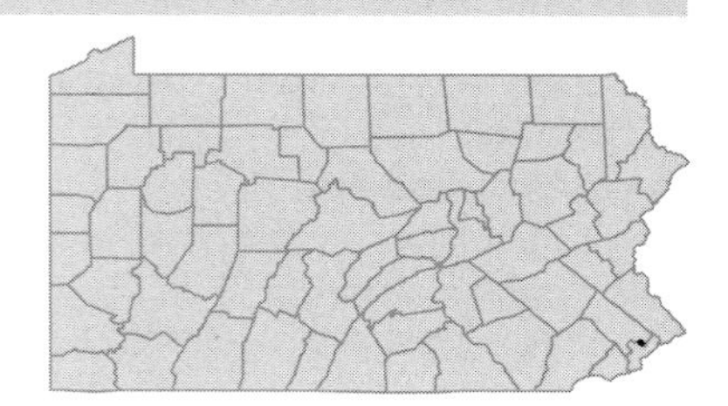

• ***Achillea millefolium*** L.
Common yarrow
Herbaceous perennial
Fields, roadsides and waste ground.

FACU
[I]

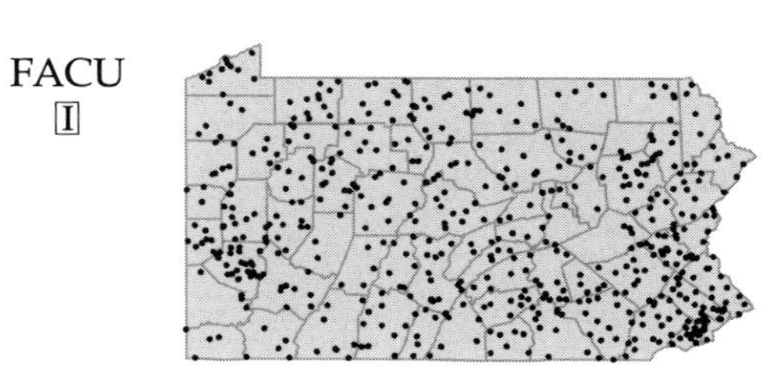

Achillea ptarmica L.
Sneezeweed
Herbaceous perennial
Cultivated and rarely escaped.

[I]

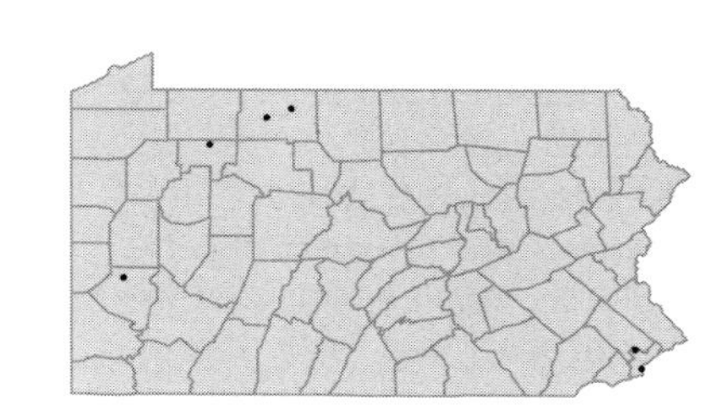

• ***Ambrosia artemisiifolia*** L.
Common ragweed
Herbaceous annual
Fields, meadows, cultivated areas, roadsides and waste ground.

FACU

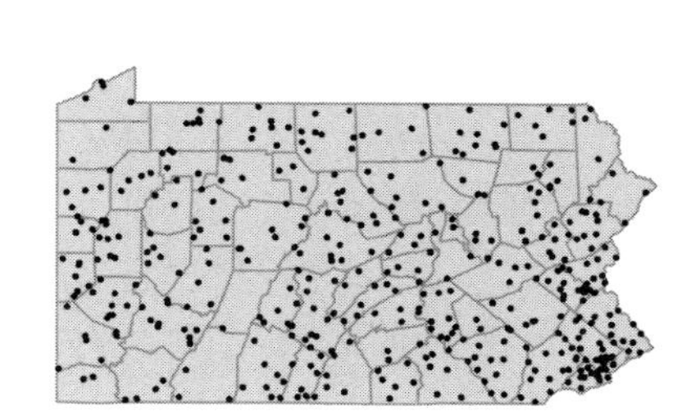

Ambrosia psilostachya DC.
Western ragweed
Herbaceous perennial
Sandy shores or meadows.

FACU-

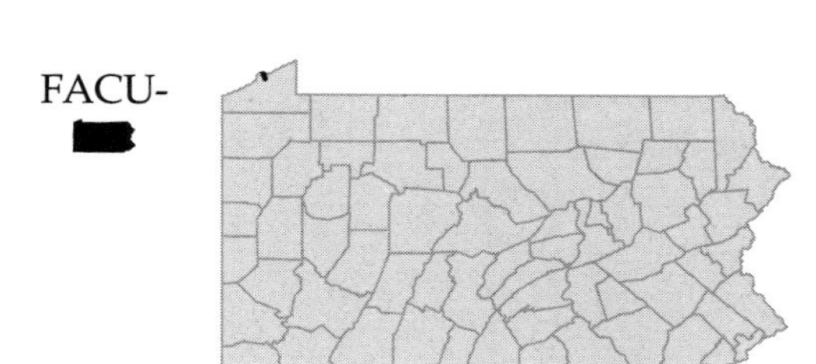

Ambrosia trifida L. FAC
Giant ragweed
Herbaceous annual
Fields, meadows, roadsides and floodplains.

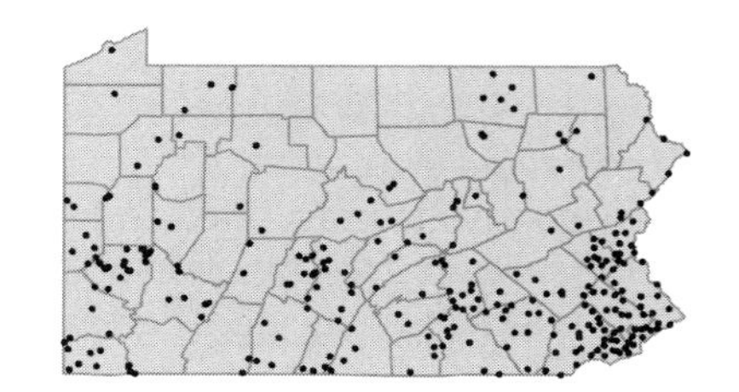

•***Amphiachyris dracunculoides*** (DC.) Nutt. [I]
Broom-snakeroot
Herbaceous annual
Urban grassy area. Represented by a single collection from Northampton Co. in 1884.
Gutierrezia dracunculoides (DC.) S.F.Blake FGBW

•***Anacyclus clavatus*** (Desf.) Pers. [I]
Anacylus
Herbaceous annual
Ballast ground. Represented by three collections from Philadelphia Co. ca. 1882.
Anacyclus tomentosus DC. W

•***Anaphalis margaritacea*** (L.) Benth. & Hook.f.
Pearly-everlasting
Herbaceous perennial
Dry, sandy or gravelly soil of fields, woods edges and roadsides.

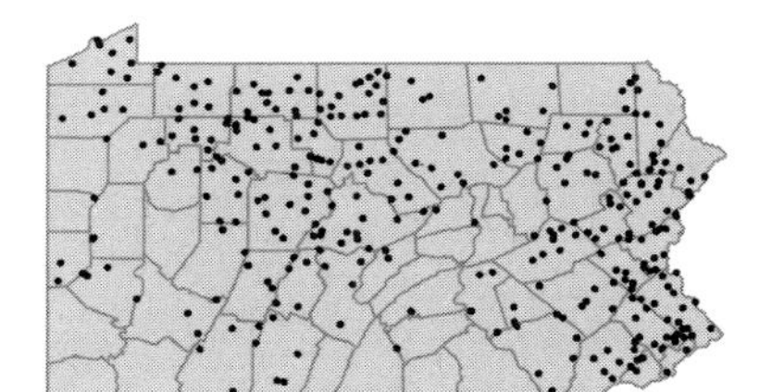

•***Antennaria neglecta*** Greene
Field pussytoes
Herbaceous perennial
Pastures, fields, woods borders and lawns.

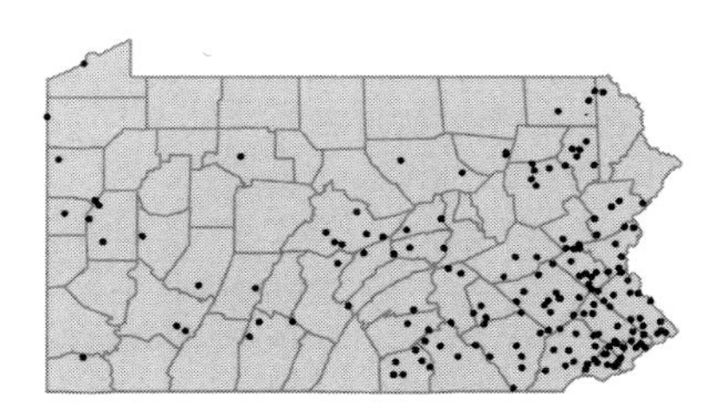

Antennaria neodioica Greene ssp. ***neodioica***
Pussytoes; Ladies'-tobacco
Herbaceous perennial
Old fields, woods, rocky slopes and roadsides.
Antennaria neglecta Greene var. *attenuata* (Fern.) Cronq. GB; *Antennaria neodioica* Greene var. *attenuata* Fern. FW; *Antennaria neglecta* Greene var. *neodioica* (Greene) Cronq. C

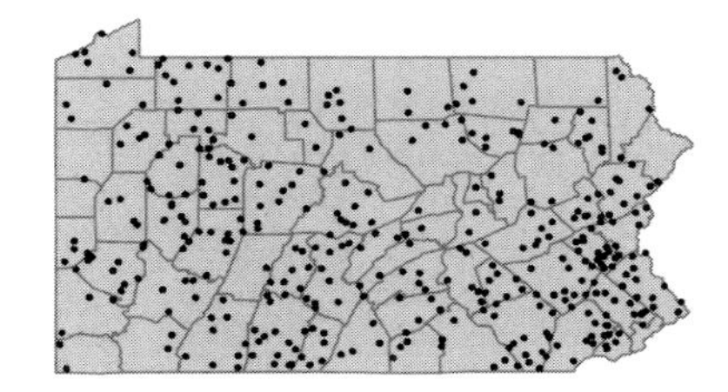

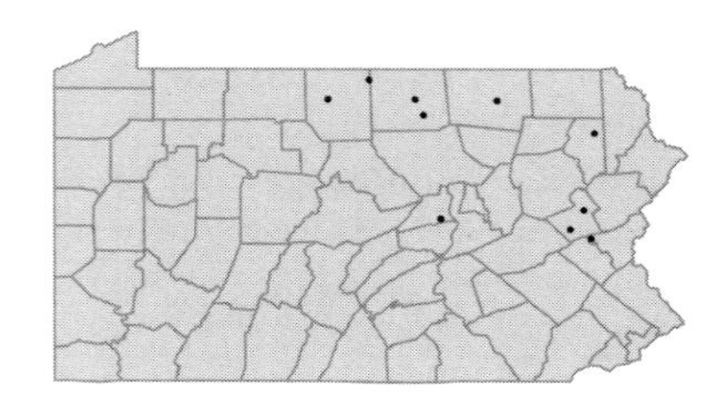

Antennaria neodioica Greene ssp. ***canadensis*** (Greene) Bayer & Stebbins
Pussytoes; Ladies'-tobacco
Herbaceous perennial
Dry woods, open wooded slopes, ridges, cliffs and roadside banks.
Antennaria canadensis Greene FW; *Antennaria neglecta* Greene var. *randii* (Fern.) Cronq. GB; *Antennaria neglecta* Greene var. *canadensis* (Greene) Cronq. C

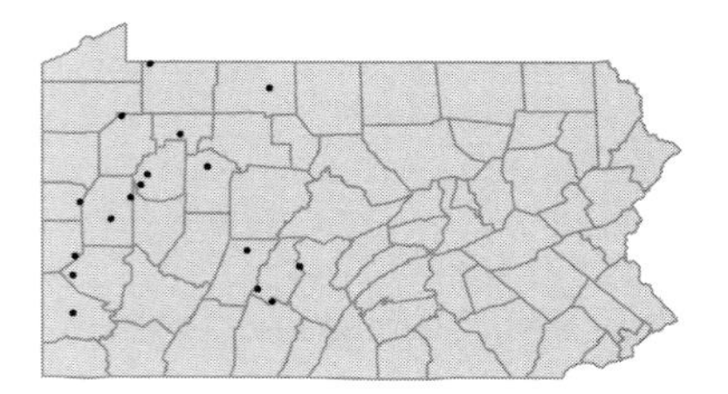

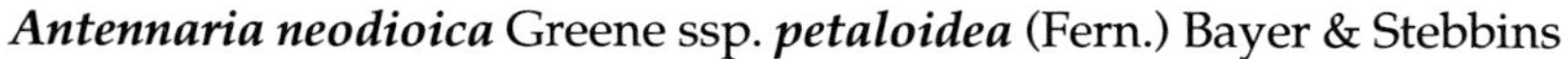

Antennaria neodioica Greene ssp. ***petaloidea*** (Fern.) Bayer & Stebbins
Pussytoes; Ladies'-tobacco
Herbaceous perennial
Woods, abandoned pasture and dry, shaly slopes.
Antennaria petaloidea Fern. FW; *Antennaria neglecta* Greene var. *petaloidea* (Fern.) Cronq. C

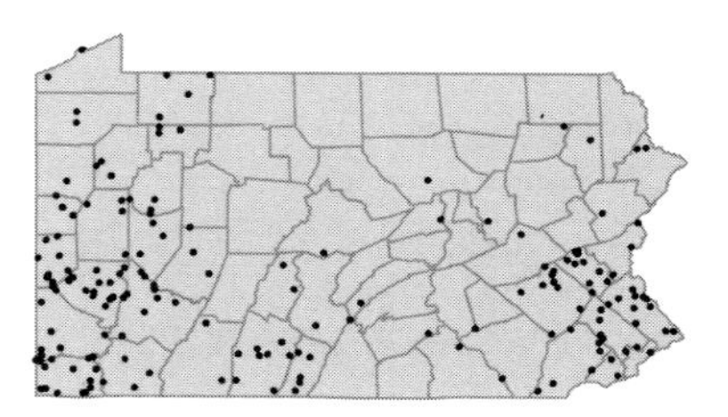

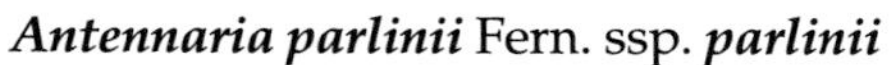

Antennaria parlinii Fern. ssp. ***parlinii***
Pussytoes; Ladies'-tobacco
Herbaceous perennial
Rocky woods, dry slopes, shaly hilltops and roadsides.
Antennaria parlinii Fern. var. *arnoglossa* (Greene) Fern. FW; *Antennaria plantaginifolia* (L.) Richardson var. *arnoglossa* (Greene) Cronq. GB; *Antennaria plantaginifolia* (L.) Richardson var. *parlinii* (Fern.) Cronq. C

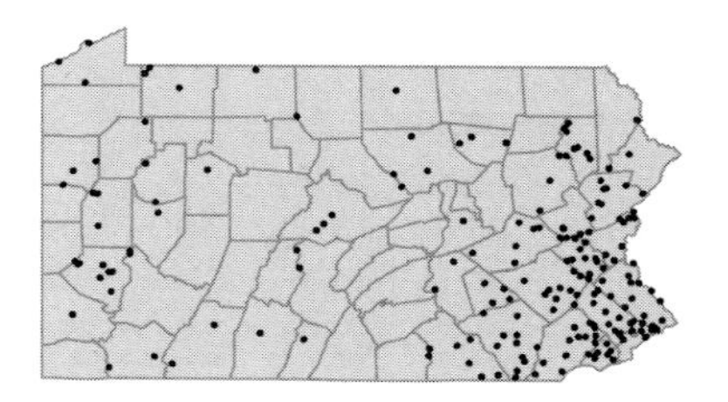

Antennaria parlinii Fern. ssp. ***fallax*** (Greene) Bayer & Stebbins
Pussytoes; Ladies'-tobacco
Herbaceous perennial
Dry, rocky slopes, fields, thickets, open woods and roadsides.
Antennaria brainerdii Fern. FW; *Antennaria fallax* Greene FW; *Antennaria munda* Fern. FW; *Antennaria plantaginifolia* (L.) Richardson var. *ambigens* (Greene) Cronq. GBC

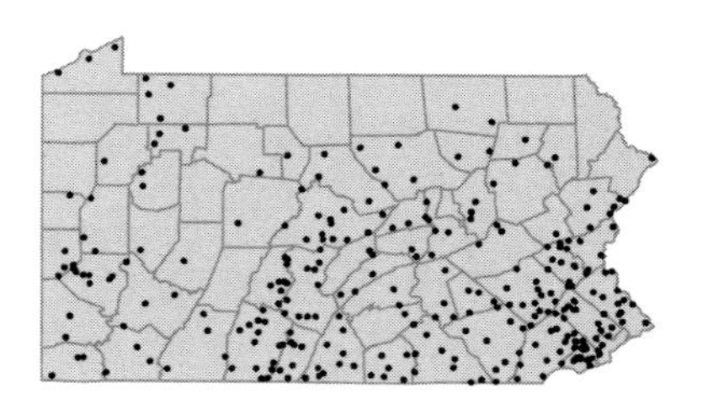

Antennaria plantaginifolia (L.) Richardson
Plantain pussytoes
Herbaceous perennial
Open woods, shaly banks, cliffs and old fields.
Antennaria plantaginifolia (L.) Richardson var. *petiolata* (Fern.) Heller *pro parte* FW; *Antennaria plantaginifolia* (L.) Richardson var. *plantaginifolia* C

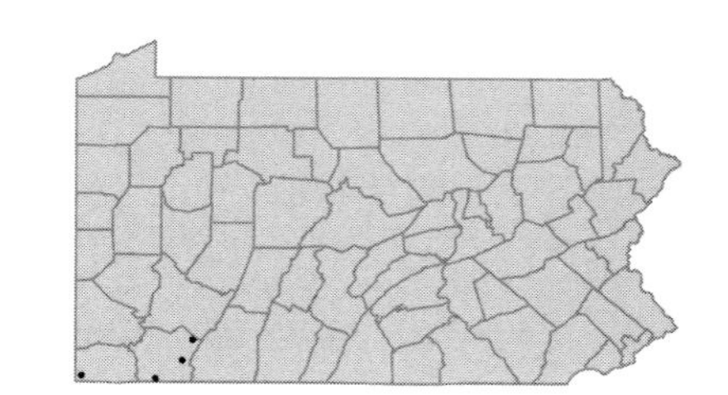

Antennaria solitaria Rydb.
Single-head pussytoes
Herbaceous perennial
Woods.

Antennaria virginica Stebbins
Shale-barren pussytoes
Herbaceous perennial
Old fields, woods, dry pastures and shale barrens.
Antennaria neglecta Greene var. *argillicola* (Stebbins) Cronq. GB

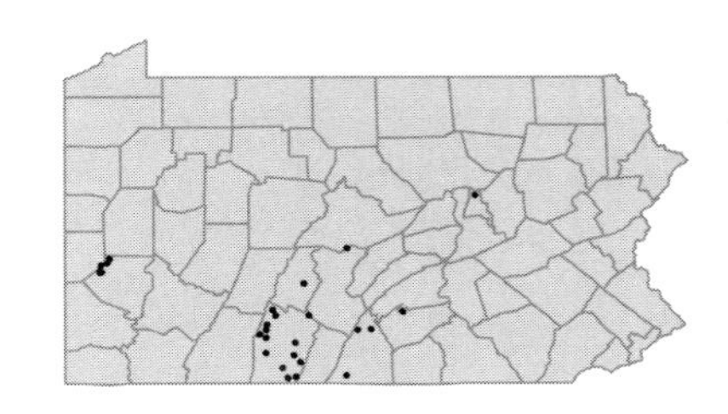

• ***Anthemis arvensis*** L.
Corn chamomile
Herbaceous annual
Woods, fields, roadsides and waste ground.

I

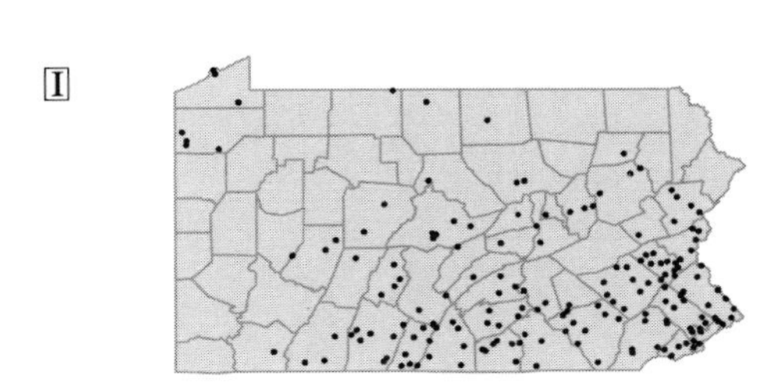

Anthemis cotula L.
Dog-fennel; Stinking chamomile
Herbaceous annual
Fields, roadsides and dry, waste ground.

FACU-
I

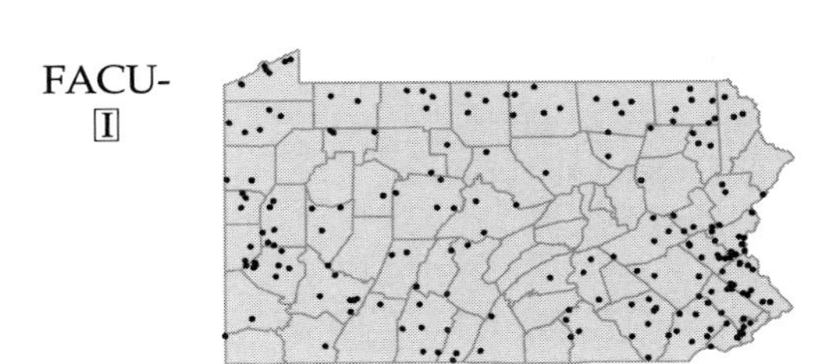

Anthemis tinctoria L.
Yellow chamomile
Herbaceous perennial
Abandoned fields and roadsides.

I

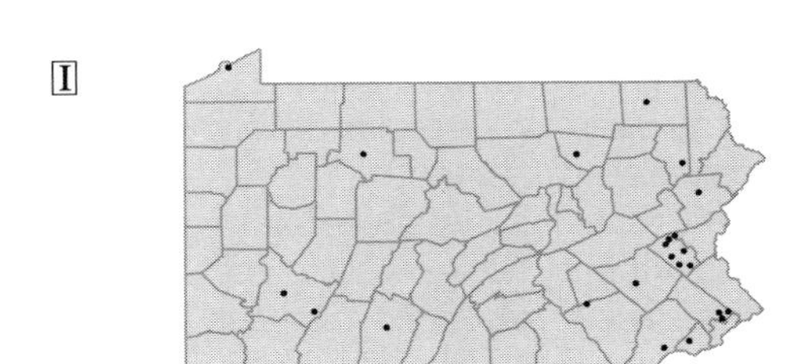

• ***Arctium lappa*** L.
Great burdock
Herbaceous biennial
Woods edges and roadsides.

I

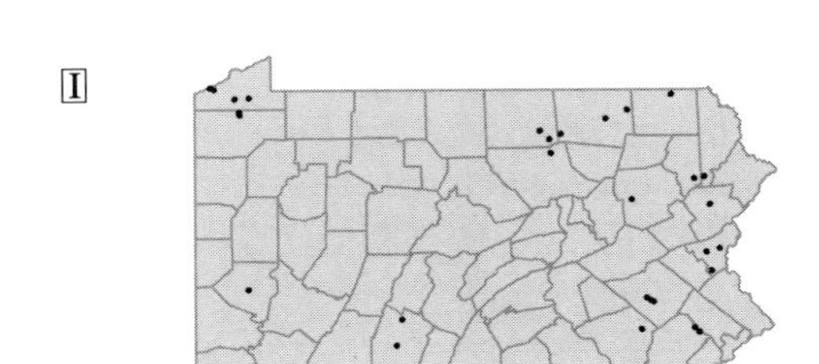

Arctium minus (Hill) Bernh.
Common burdock
Herbaceous biennial
Edges of fields and woods, railroad tracks and waste ground.

I

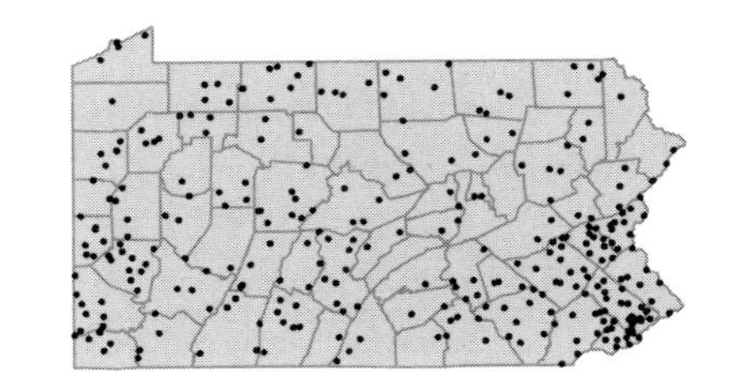

Arctium nemorosum Lej. & Court.
Woodland burdock
Herbaceous biennial
Waste ground.

I

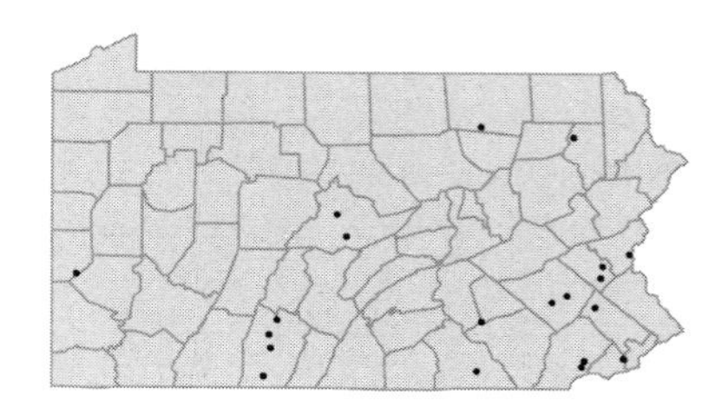

Arctium tomentosum P.Mill.
Wooly burdock
Herbaceous biennial
Roadsides, floodplains and waste ground.

I

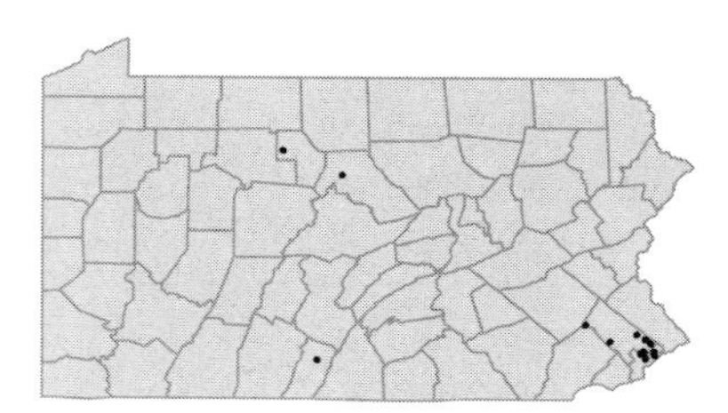

•***Arnica acaulis*** (Walt.) BSP
Leopard's-bane
Herbaceous perennial
Open woods and thickets of serpentine barrens.

FACU

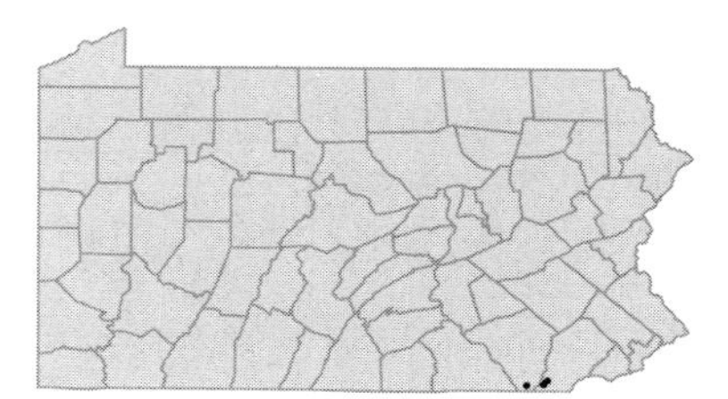

•***Arnoseris minima*** (L.) Schweig. & Koerte
Lamb-succory; Dwarf nipplewort
Herbaceous annual
Ballast ground. Represented by two collections from Philadelphia Co. 1877-1878.

I

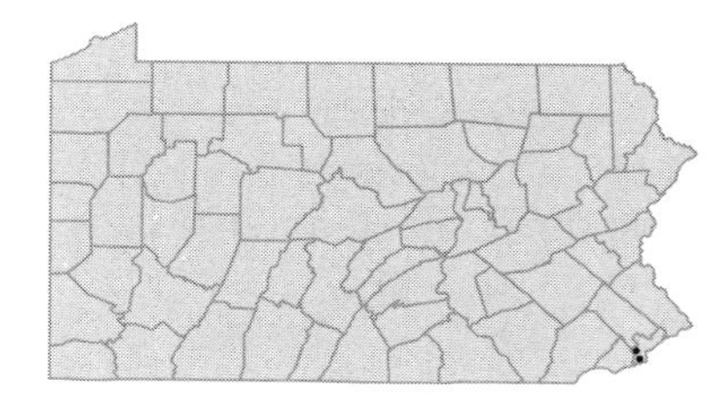

•***Artemisia absinthium*** L.
Common wormwood
Herbaceous perennial
Cultivated and occasionally escaped to waste ground or ballast.

I

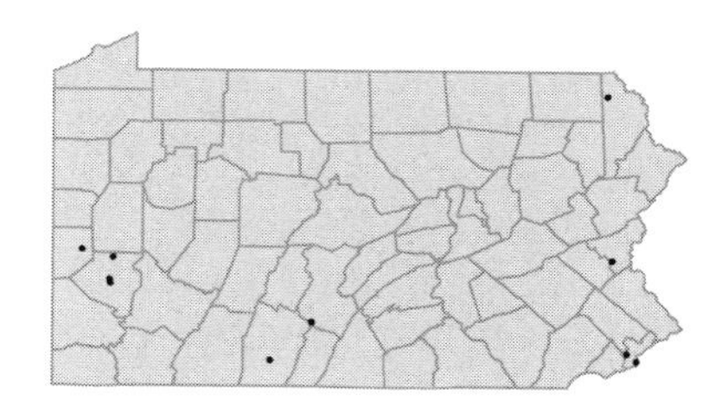

Artemisia annua L.
Sweet wormwood; Annual wormwood
Herbaceous annual
Cultivated fields, roadsides, dumps and urban waste ground.

FACU
I

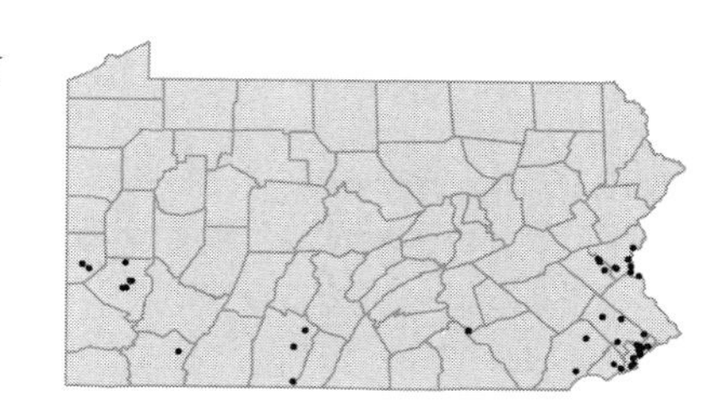

Artemisia biennis Willd.
Wormwood; Sage-weed
Herbaceous annual
Railroad tracks, wharves and ballast.

FACU-
[I]

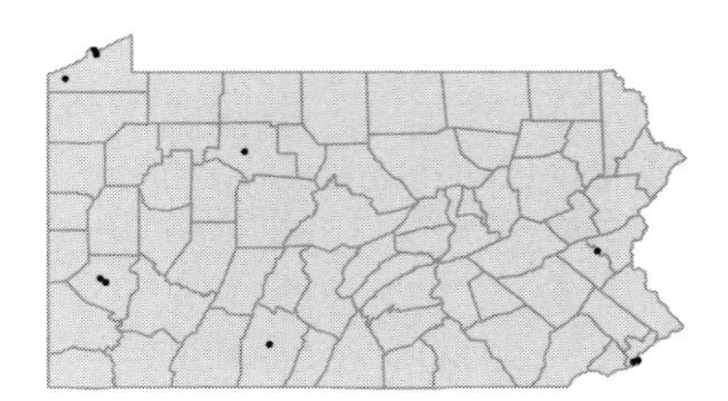

Artemisia campestris L. ssp. ***caudata*** (Michx.) Hall & Clements
Beach wormwood; Tall wormwood
Herbaceous biennial
Dry, sandy shores or sand flats.
Artemisia caudata Michx. PFBW

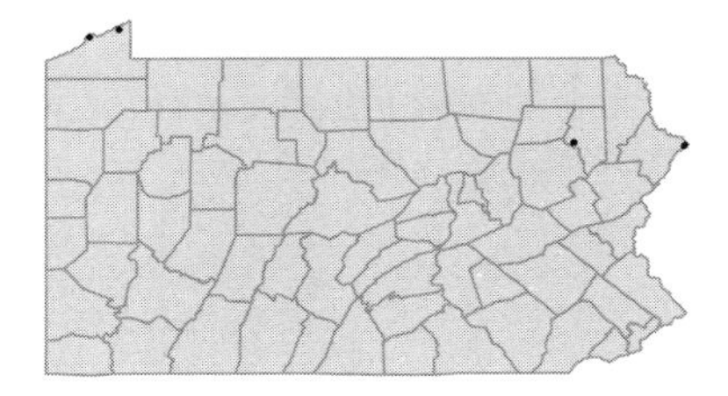

Artemisia frigida Willd.
Prairie sagewort
Herbaceous perennial
Railroad embankment. Represented by a single collection from Lehigh Co. in 1917.

[I]

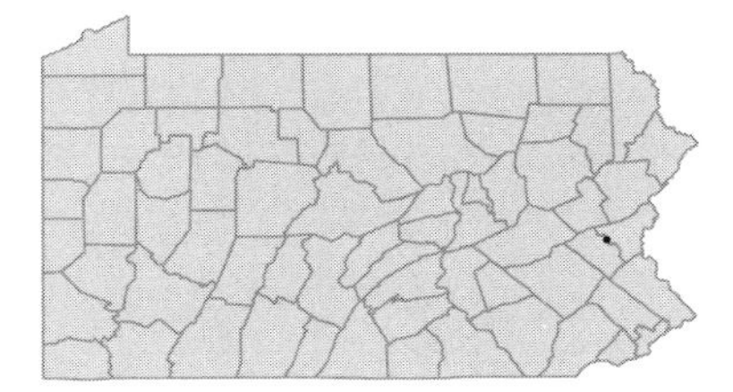

Artemisia ludoviciana Nutt.
Western mugwort
Herbaceous perennial
Vacant lots and waste ground.

UPL
[I]

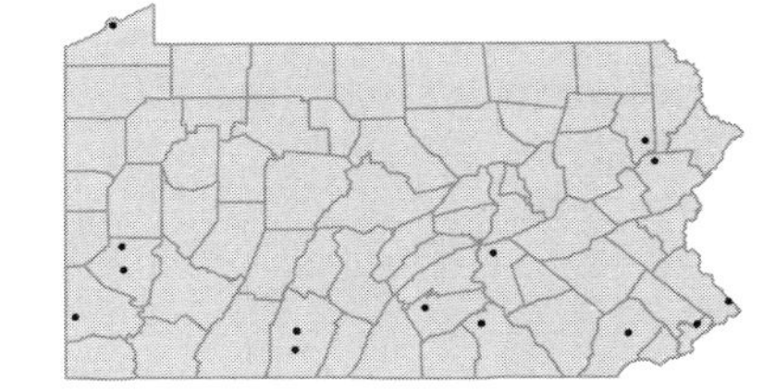

Artemisia pontica L.
Roman wormwood
Herbaceous perennial
Cultivated and rarely escaped.

[I]

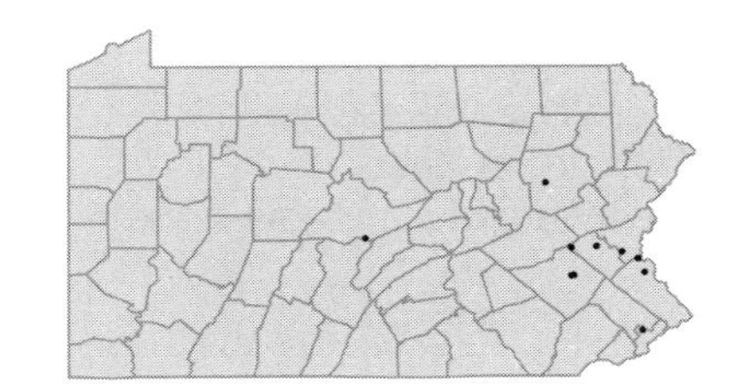

Artemisia stelleriana Bess.
Dusty-miller
Herbaceous perennial
Cultivated and rarely escaped to waste ground.

FACU
[I]

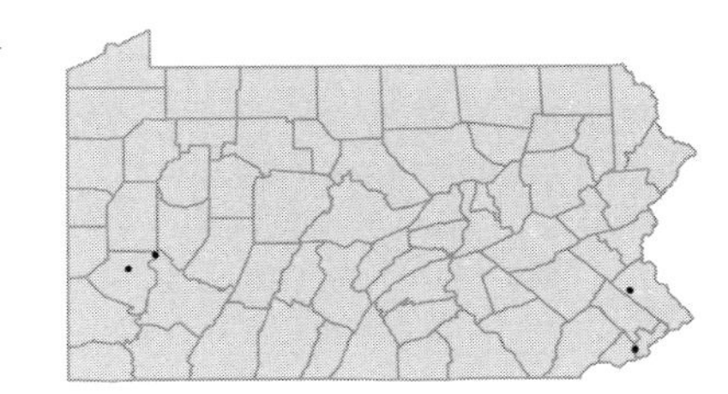

Artemisia vulgaris L.
Mugwort
Herbaceous perennial
Gardens, lawns, roadsides, thickets, waste ground and rubbish dumps.

I

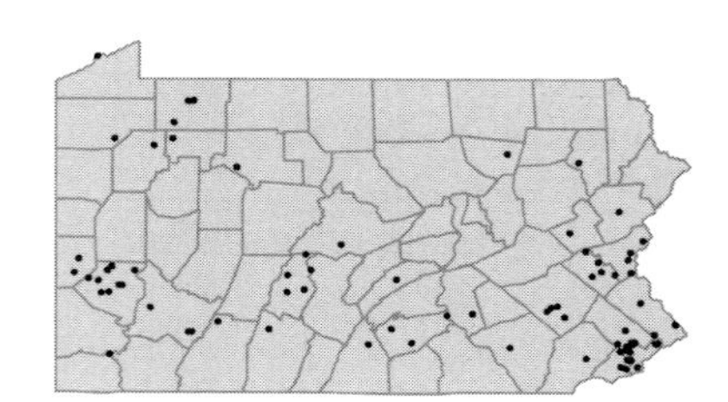

Aster—In addition to the taxa treated below, limited collections exist of the following hybrids: *acuminatus x nemoralis* (*A. x blakei* [Porter] House); *ericoides x novae-angliae* (*A. x amethystinus* Nutt.); *laevis x novi-belgii.*

• ***Aster acuminatus*** Michx.
Wood aster
Herbaceous perennial
Cool, moist woods.

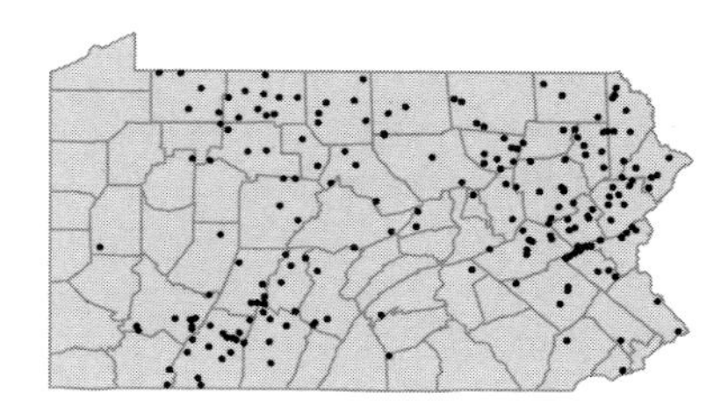

Aster borealis (Torr. & A.Gray) Prov.
Northern bog aster; Rush aster
Herbaceous perennial
Cold bogs.
Aster junceus sensu auct., non Aiton P; *Aster junciformis* Rydb. FGB

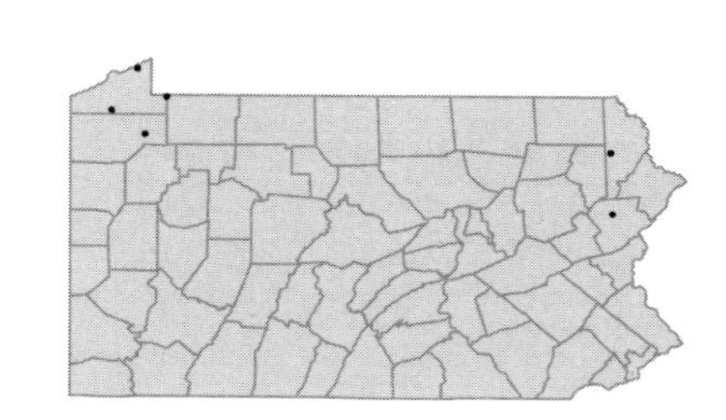

Aster brachyactis S.F.Blake
Western annual aster
Herbaceous annual
Open gravel along interstate highway.

FAC
I

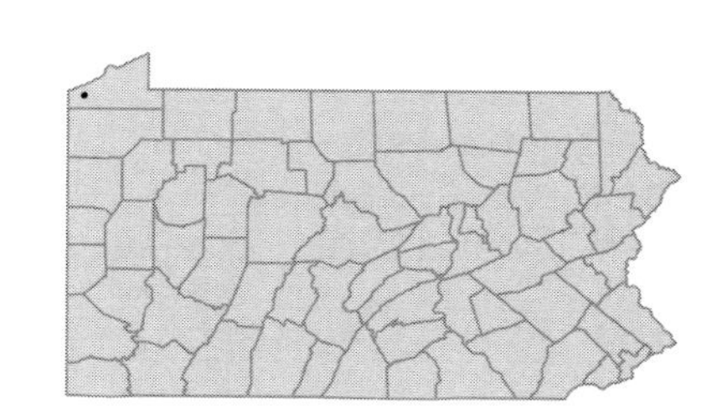

Aster cordifolius L. ssp. ***cordifolius***
Blue wood aster
Herbaceous perennial
Woods, meadows and roadsides. Plants with numerous small flowering heads have been segregated as var. *polycephalus*.

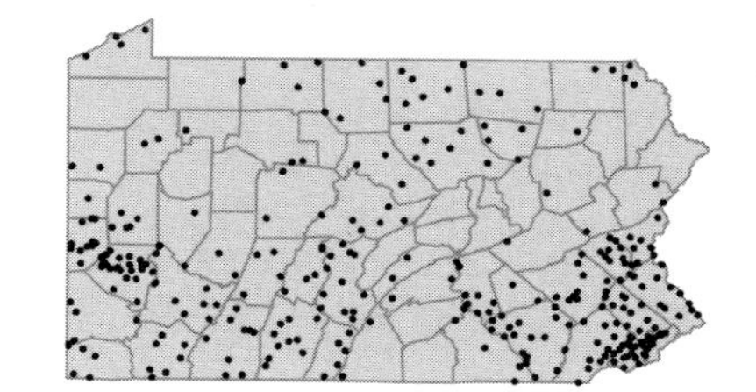

Aster cordifolius L. ssp. ***laevigatus*** (Porter) A.G.Jones
Smooth heart-leaved aster
Herbaceous perennial
Woods, shaly slopes, dry fields and roadsides.
Aster lowrieanus Porter WCBF

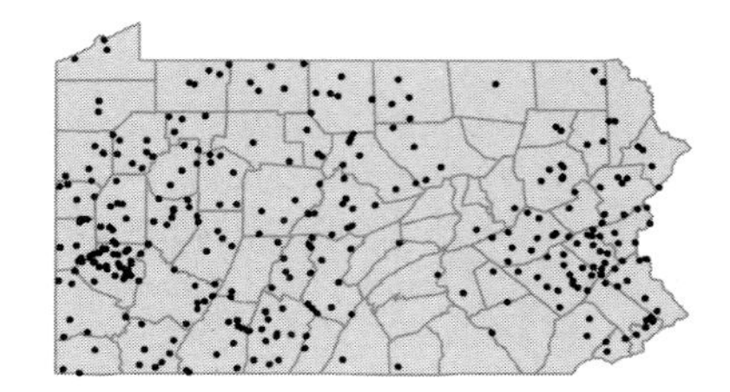

Aster cordifolius L. ssp. ***sagittifolius*** (Willd.) A.G.Jones
Blue wood aster
Herbaceous perennial
Rocky woods, upland fields, cliffs and roadsides.
Aster x sagittifolius Willd. CWGFBK

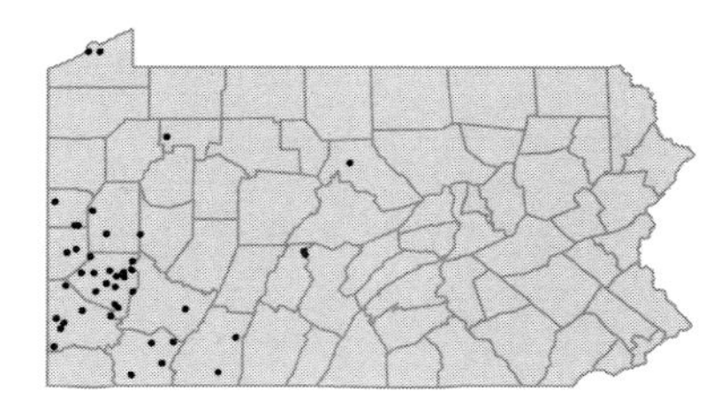

Aster depauperatus (Porter) Fern.
Serpentine aster
Herbaceous perennial
Open areas of serpentine barrens. There is mounting evidence that this aster should be considered conspecific with the midwestern *A. parviceps*.
Aster ericoides L. var. *depauperatus* Porter P; *Aster parviceps* (Burgess) Mackenzie & Bush var. *pusillus* (A.Gray) A.G.Jones; *Aster pilosis Willd.* ssp. *parviceps* A.G.Jones var. *pusillus* (A.Gray) A.G.Jones

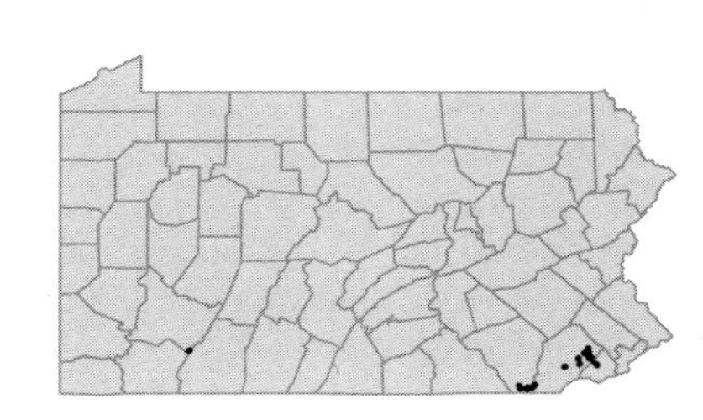

Aster divaricatus L. *sensu lato*
White wood aster
Herbaceous perennial
Woods.

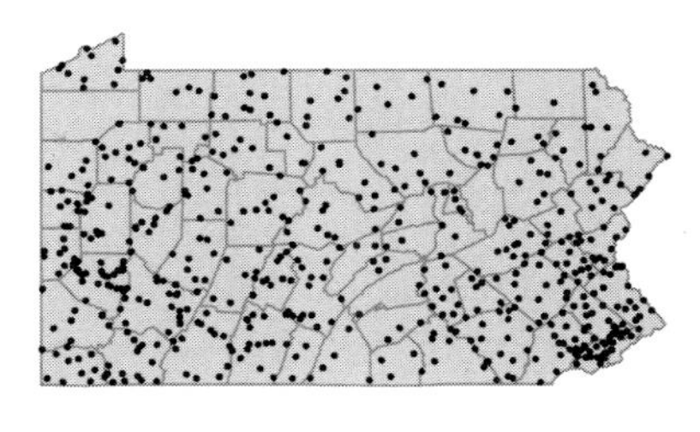

Aster drummondii Lindl.
Hairy heart-leaved aster
Herbaceous perennial
Stream valley.

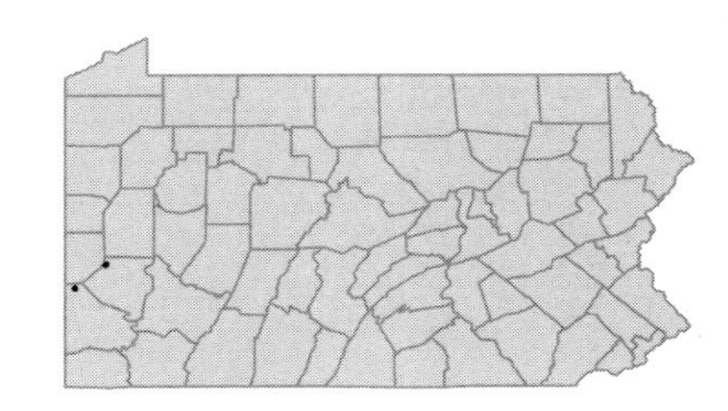

Aster dumosus L. *sensu lato*
Bushy aster
Herbaceous perennial
Open woods, moist fields, bogs and swales.

FAC

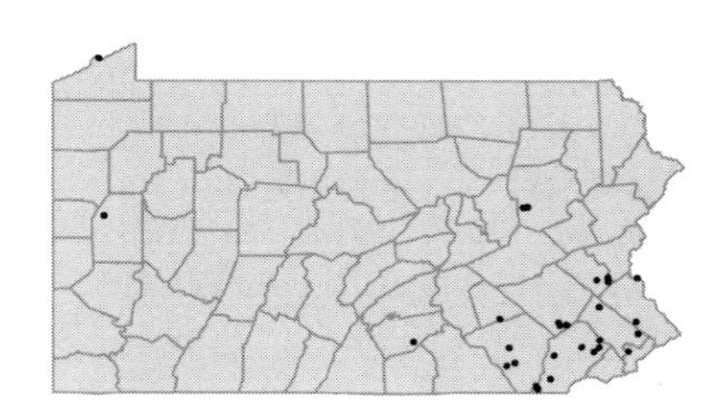

Aster ericoides L. ssp. ***ericoides***
White heath aster
Herbaceous perennial
Calcareous cliffs and outcrops.
Aster multiflorus Ait. P

FACU

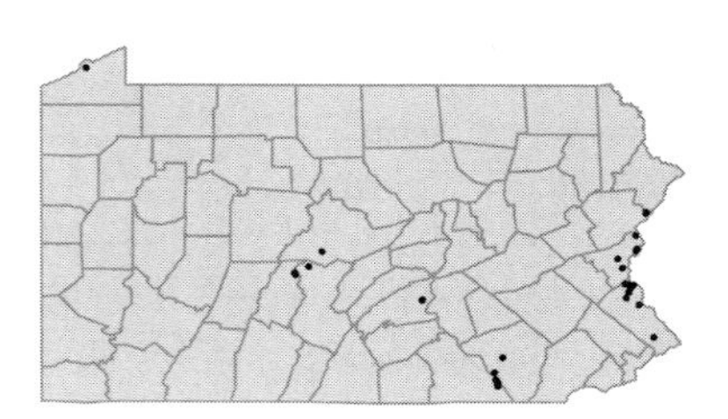

Aster fragilis Willd. var. ***fragilis***
Small white aster
Herbaceous perennial
Low meadows, floodplains and swamps.
Aster vimineus sensu auct., non Lam. var. *vimineus* BFG; *Aster racemosis* Elliot *pro parte* C

FAC

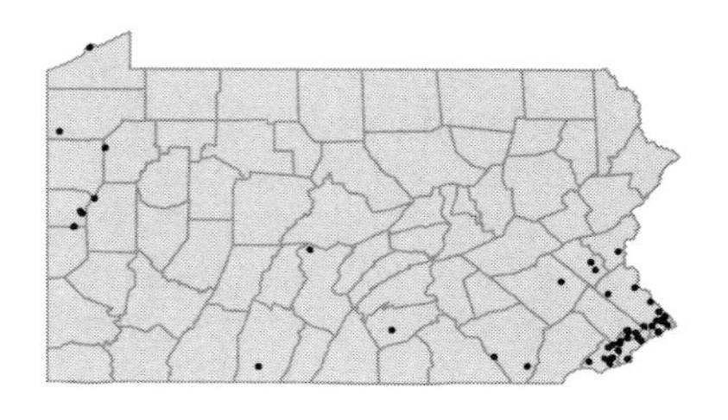

Aster fragilis Willd. var. ***subdumosus*** (Wieg.) A.G.Jones
Small white aster
Herbaceous perennial
Swampy woods, shores and barrens.
Aster vimineus sensu auct., non Lam. var. *subdumosus* Wieg. WF; *Aster racemosis* Elliot *pro parte* C

FAC

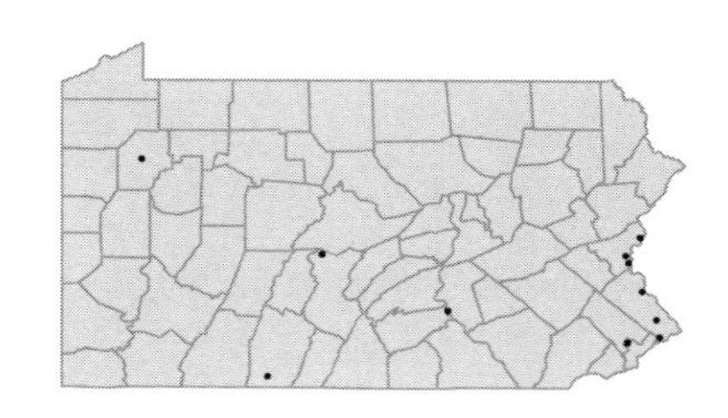

Aster infirmus Michx.
Flat-topped white aster
Herbaceous perennial
Rocky woods, thickets and barrens.
Doellingeria infirma (Michx.) Greene P

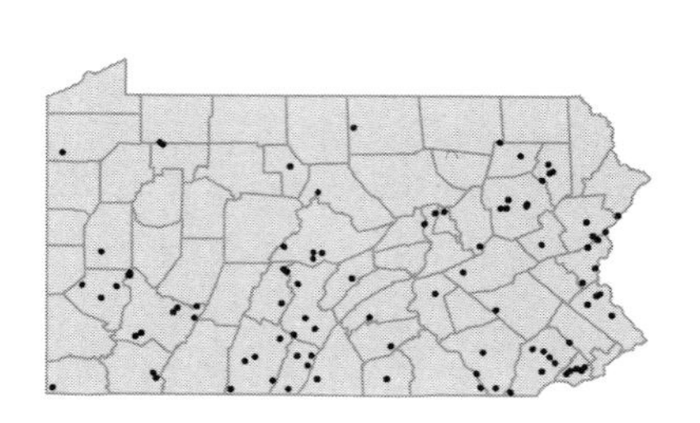

Aster laevis L. var. ***laevis***
Smooth blue aster
Herbaceous perennial
Dry woods, rocky ledges and roadsides.

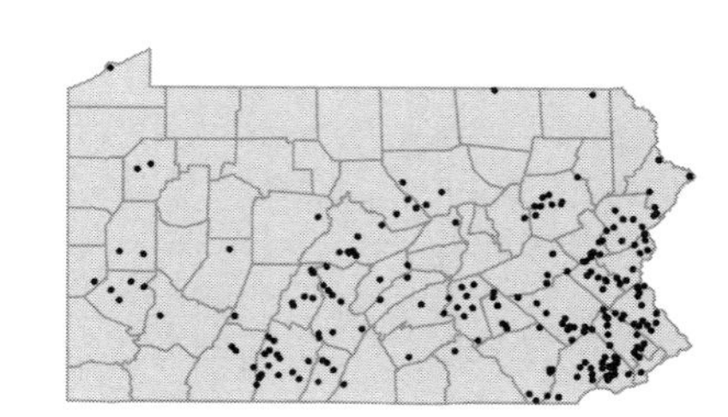

Aster laevis L. var. ***concinnus*** (Willd.) House
Smooth blue aster
Herbaceous perennial
Dry woods and serpentine barrens.
Aster concinnus Willd. C

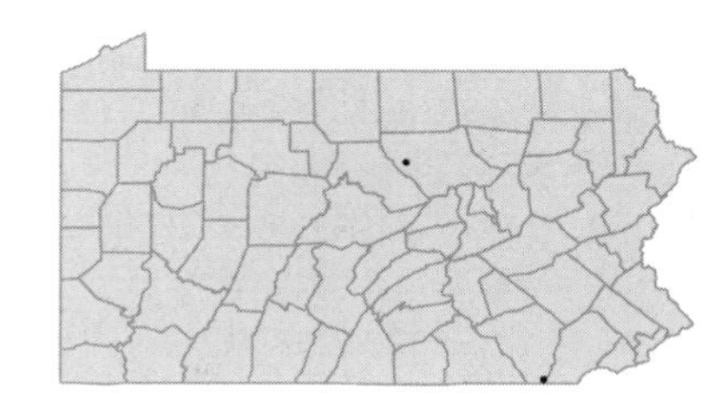

Aster lanceolatus Willd. ssp. ***lanceolatus***
Panicled aster
Herbaceous perennial
Open woods, old fields and moist roadsides.
Aster simplex Willd. var. *ramosissimus* (Torr. & A.Gray) Cronq. FBGW; *Aster paniculatus* Lam. P

FACW

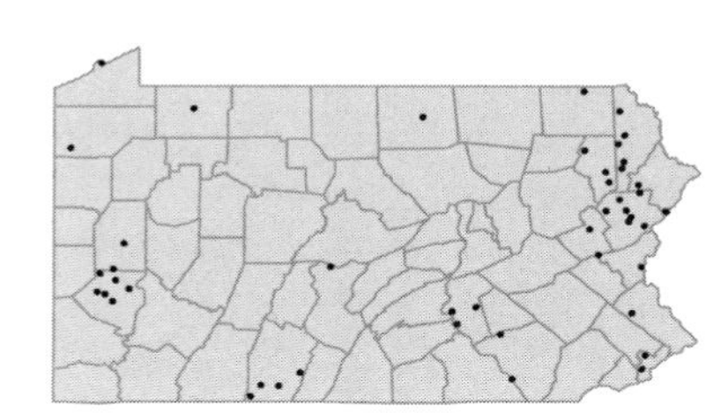

Aster lanceolatus Willd. ssp. ***interior*** (Wieg.) A.G.Jones
Eastern lined aster
Herbaceous perennial
Gravelly shores, moist meadows and bogs.
Aster simplex Willd. var. *interior* (Wieg.) Cronq. FGBW

FACW

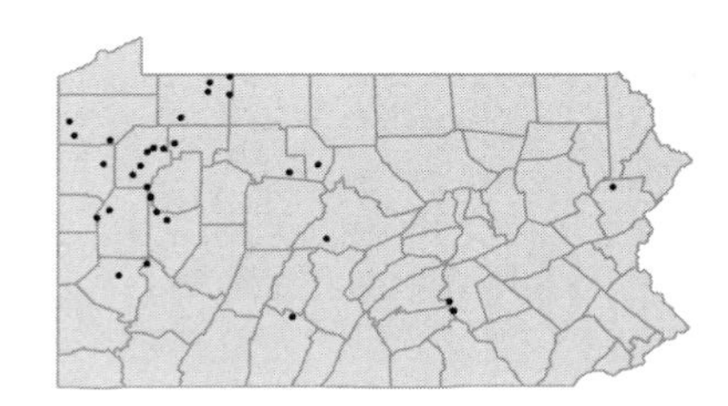

Aster lanceolatus Willd. ssp. ***simplex*** (Willd.) A.G.Jones
Simple aster
Herbaceous perennial
Moist fields, roadsides, floodplains or waste ground.
Aster simplex Willd. var. *simplex* GBFWK

FACW

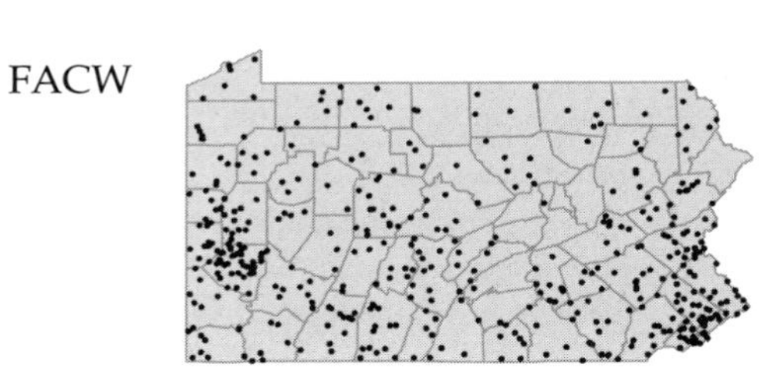

Aster lateriflorus (L.) Britt. *sensu lato*
Calico aster
Herbaceous perennial
Old fields, rocky woods, roadsides and waste ground.
Aster vimineus Lam.

FACW-

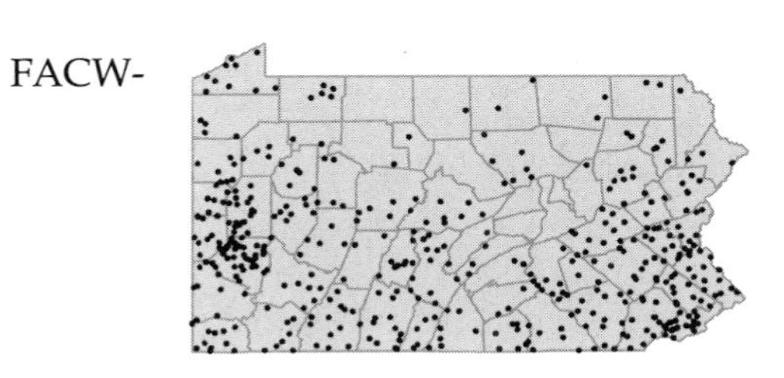

Aster linariifolius L.
Stiff-leaved aster
Herbaceous perennial
Dry, rocky woods or edges.
Ionactis linariifolius (L.) Greene P

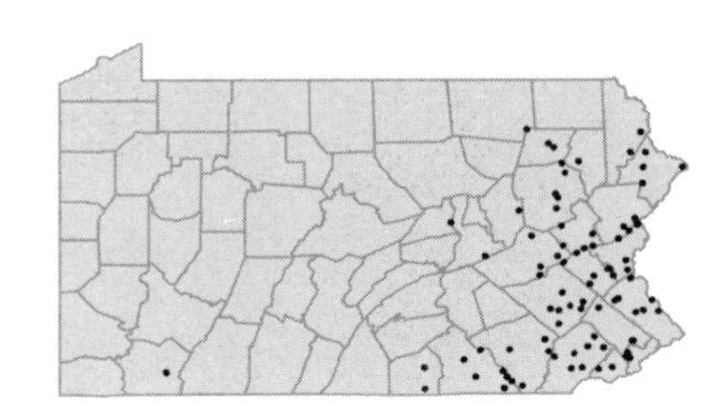

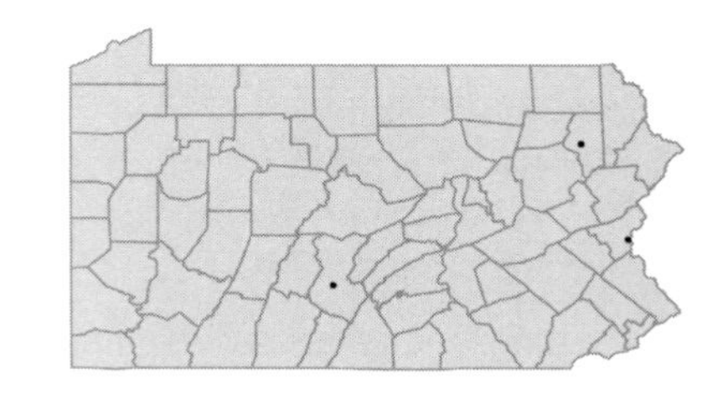

Aster longifolius Lam. *sensu lato*
Long-leaved aster
Herbaceous perennial
Moist, grassy meadows.
Aster salicifolius sensu auct., non Lam. P; *Aster nova-belgii sensu auct., non* L. C; *Aster praealtus sensu auct., non* Poir. K

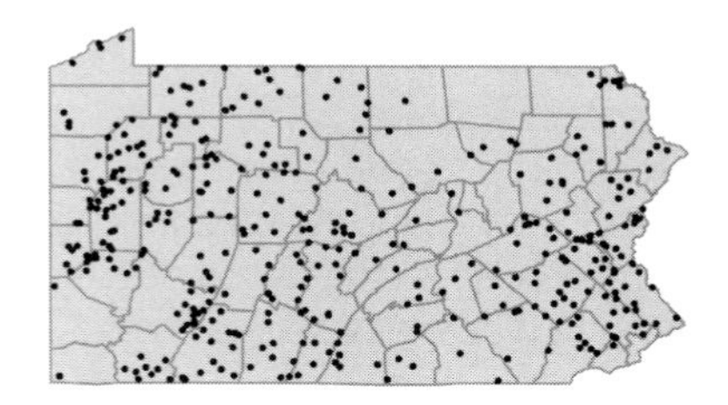

Aster macrophyllus L.
Bigleaf aster
Herbaceous perennial
Woods, rocky slopes and edges.

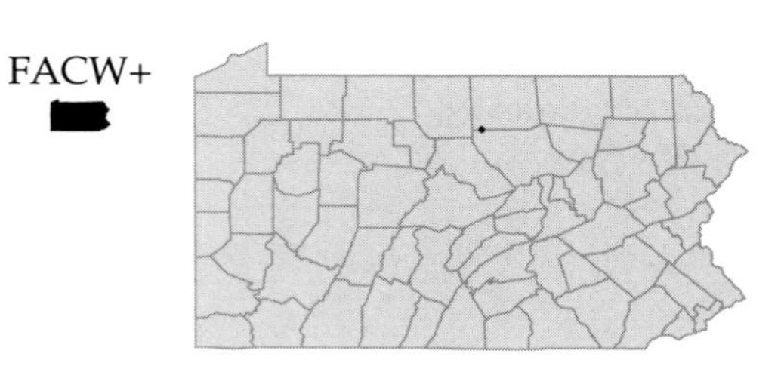

Aster nemoralis Ait.
Leafy bog aster
Herbaceous perennial
Cold bogs. This plant appears to hybridizing with *A. acuminatus* at both of its known sites in PA.

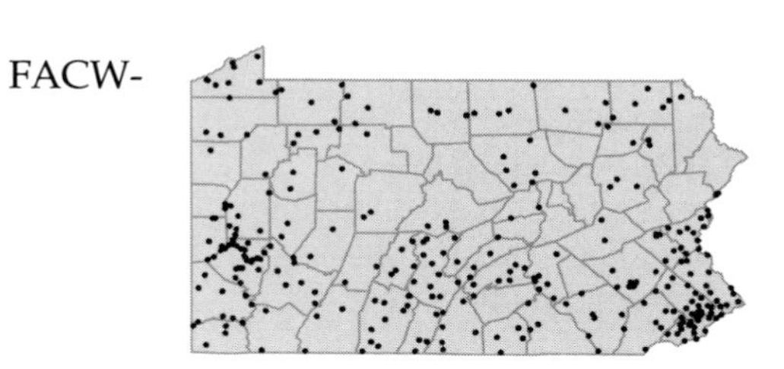

Aster novae-angliae L.
New England aster
Herbaceous perennial
Fields, roadsides, railroad tracks and waste ground.

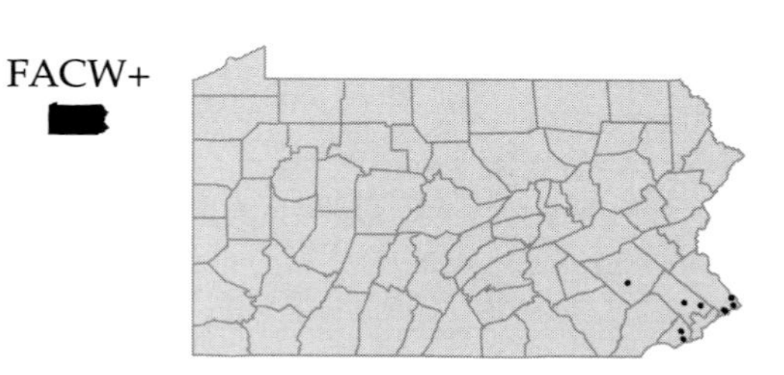

Aster novi-belgii L. *sensu lato*
New York aster
Herbaceous perennial
Swamps and moist meadows.

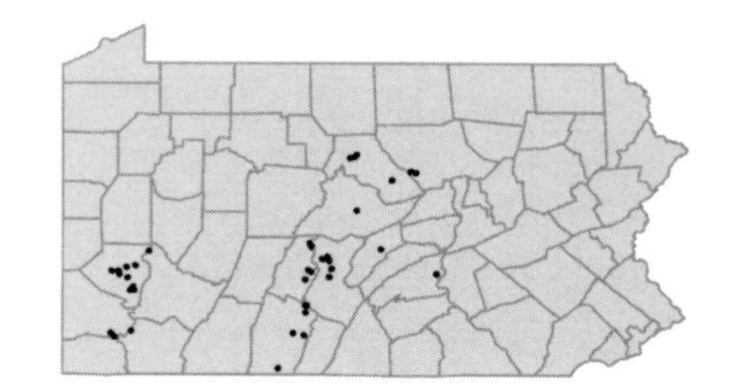

Aster oblongifolius Nutt.
Aromatic aster
Herbaceous perennial
Calcareous hillsides, cliffs and bluffs.

Aster patens Ait. var. ***patens***
Late purple aster; Clasping aster
Herbaceous perennial
Mesic woods and fields.

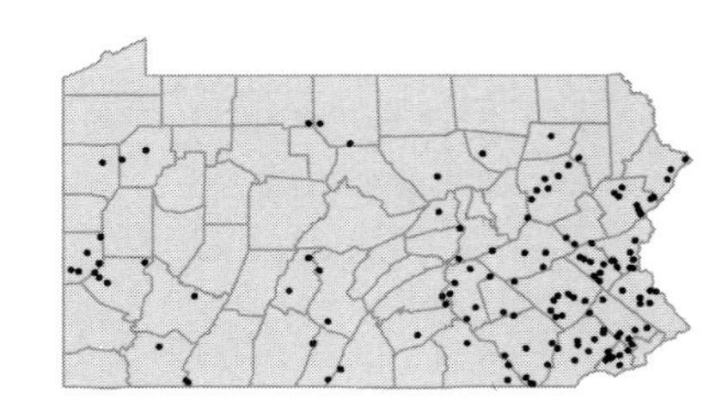

Aster paternus Cronq.
White-topped aster
Herbaceous perennial
Dry woods, fields and barrens.
Sericocarpus asteroides (L.) BSP PFW

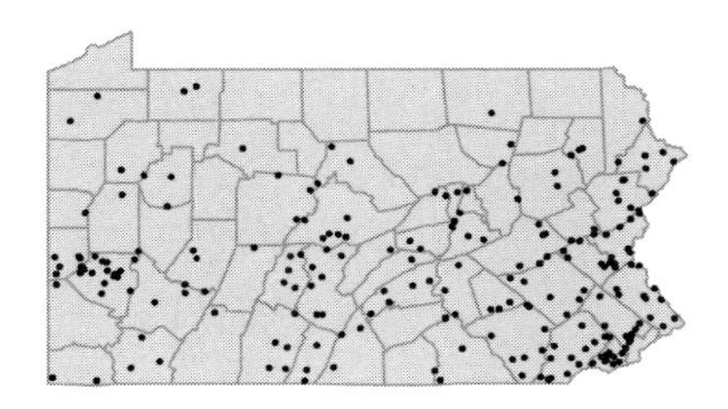

Aster phlogifolius Muhl. ex Willd.
Late purple aster
Herbaceous perennial
Open woods, old fields and rocky banks.
Aster patens Ait. var. *phlogifolius* (Muhl. ex Willd.) Nees WKCFGB

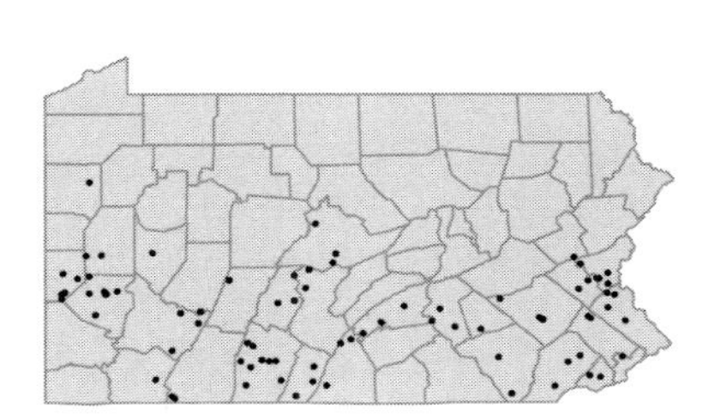

Aster pilosus Willd. var. ***pilosus*** UPL
Heath aster
Herbaceous perennial
Fields, open woods, vacant lots and roadsides.
Aster ericoides L. var. *pilosus* (Willd.) Porter P; *Aster pilosus* Willd. var. *platyphyllus* (Torr. & A.Gray) Blake FW

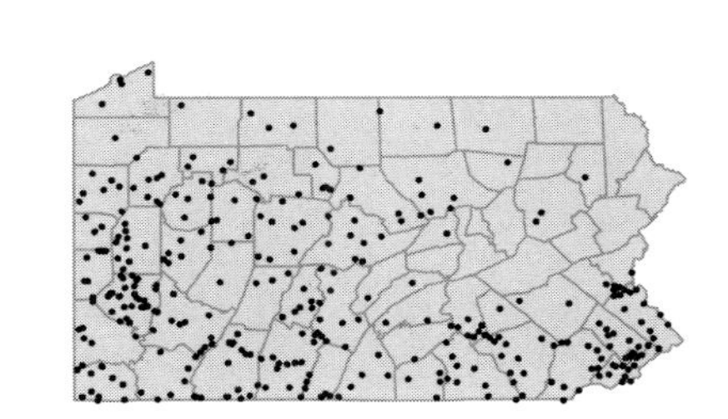

Aster pilosus Willd. var. ***demotus*** S.F.Blake UPL
Heath aster
Herbaceous perennial
Dry fields, open woods, vacant lots and roadsides.
Aster ericoides sensu auct., non L. P

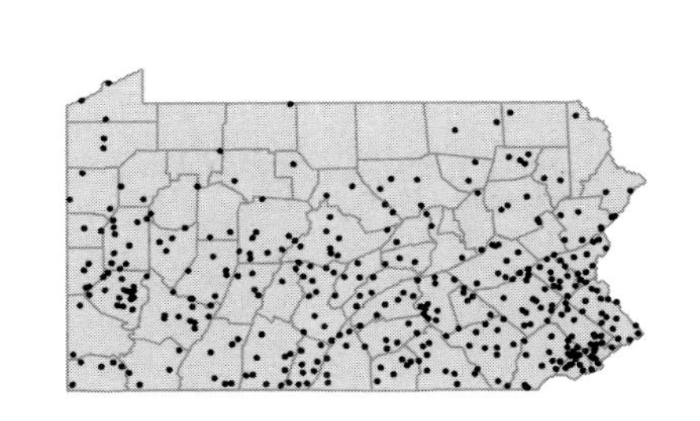

Aster pilosus Willd. var. ***pringlei*** (A.Gray) S.F.Blake UPL
Heath aster
Herbaceous perennial
Dry, gravelly or sandy soil.
Aster pringlei (A.Gray) Britt. FGBW

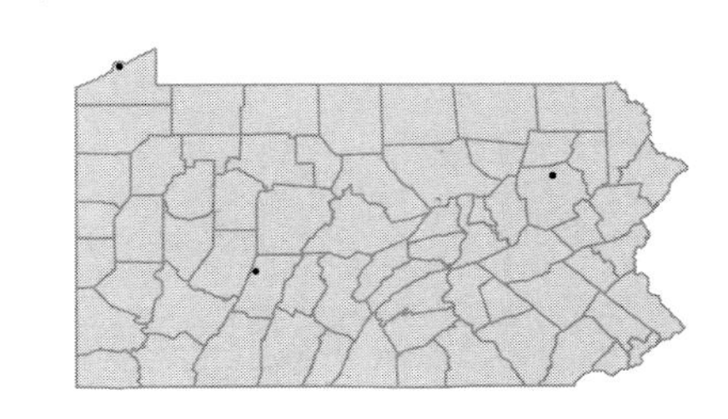

Aster praealtus Poir.
Veiny-lined aster
Herbaceous perennial
Woods, fields, thickets and roadsides.

FACW

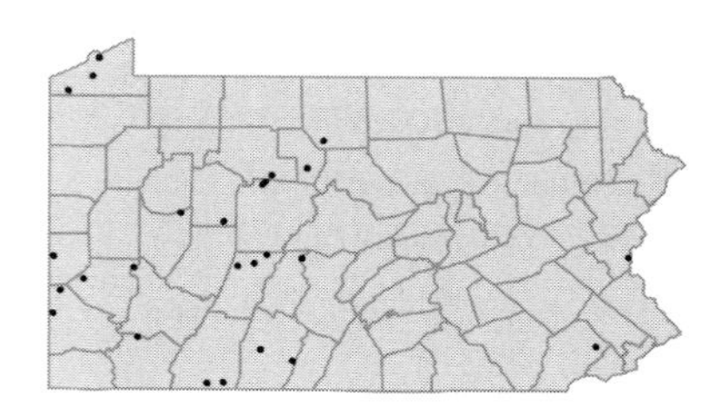

Aster prenanthoides Muhl. ex Willd.
Zig-zag aster
Herbaceous perennial
Swamps and low woods.

FAC

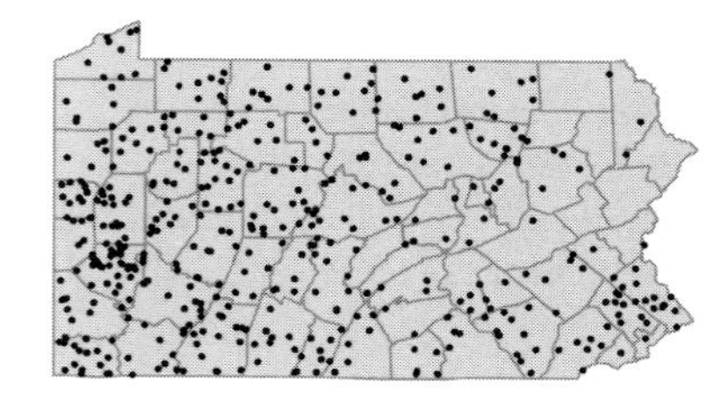

Aster puniceus L. ssp. ***puniceus***
Purple-stemmed aster
Herbaceous perennial
Wet meadows, river banks and moist roadsides.
Aster puniceus L. var. *calvus* Shinners W

OBL

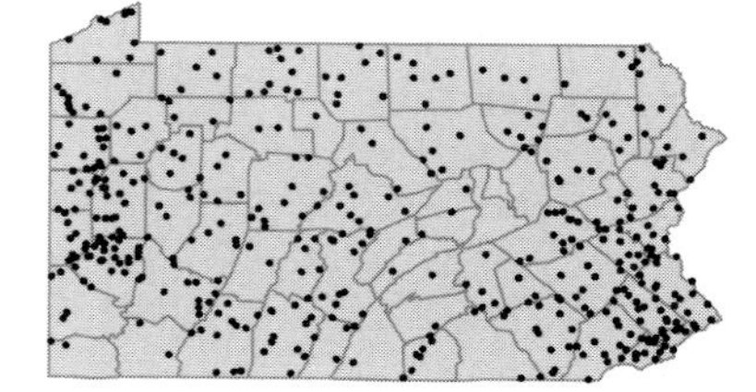

Aster puniceus L. ssp. ***firmus*** (Nees) A.G.Jones
Purple-stemmed aster
Herbaceous perennial
Swamps and moist, stream banks.
Aster lucidulus (A.Gray) Wieg. GBW; *Aster firmus* Nees C; *Aster puniceus* L. var. *lucidus* A.Gray

OBL

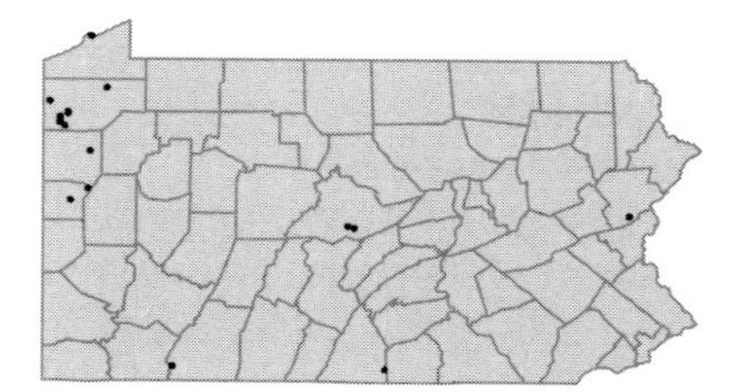

Aster radula Ait.
Rough aster; Swamp aster
Herbaceous perennial
Moist woods, swamps and bogs.

OBL

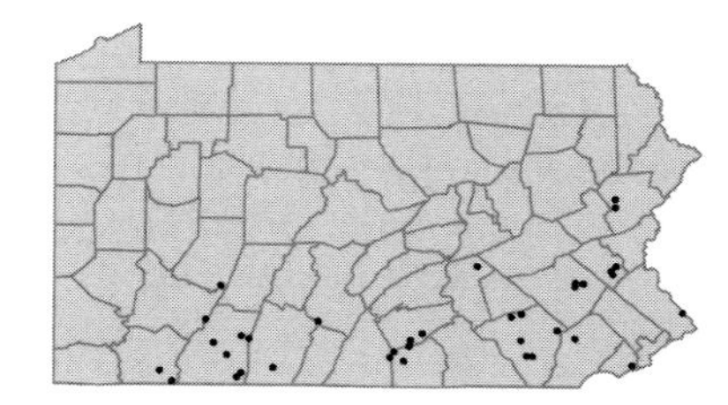

Aster schreberi Nees *sensu lato*
Schreber's aster
Herbaceous perennial
Woods, moist slopes and stream banks.

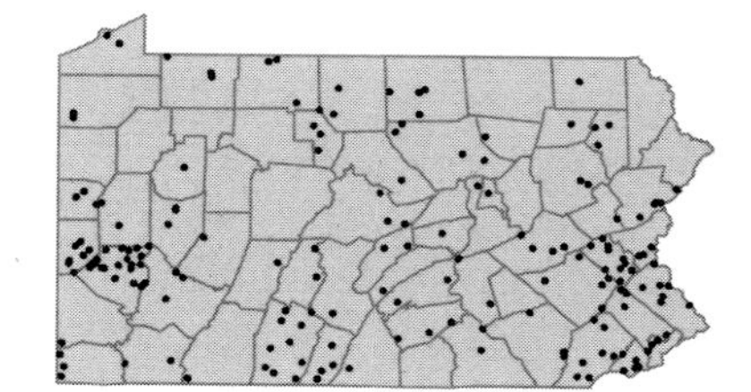

Aster shortii Lindl. in Hooker
Short's aster
Herbaceous perennial
Mesic, wooded slopes and ravines.

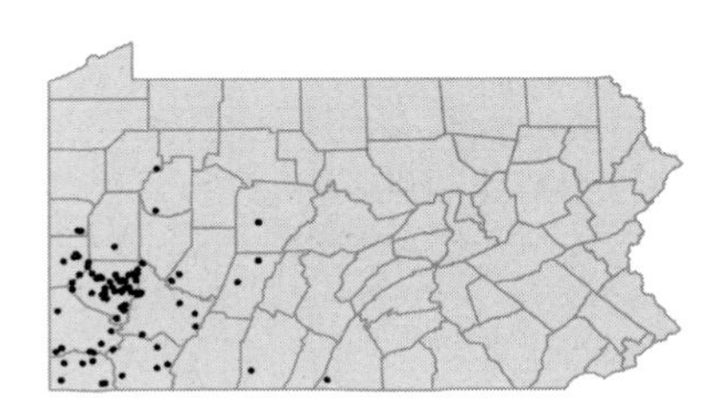

Aster solidagineus Michx.
Narrow-leaved white-topped aster
Herbaceous perennial
Swamps, bogs, serpentine barrens and sandy roadsides.
Sericocarpus linifolius (L.) BSP PFW

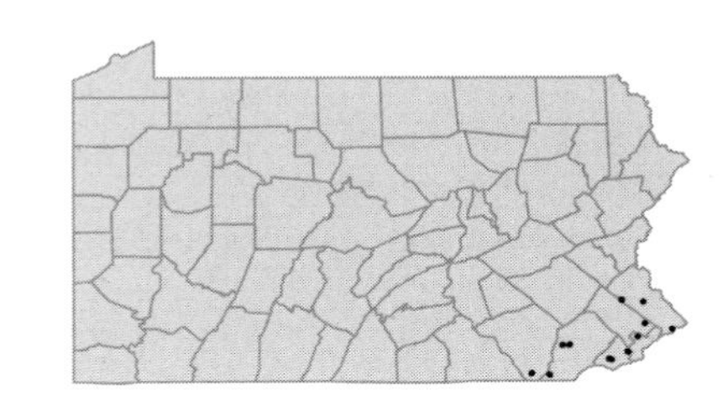

Aster spectabilis Ait.
Showy aster
Herbaceous perennial
Dry, sandy woods and rock outcrops.

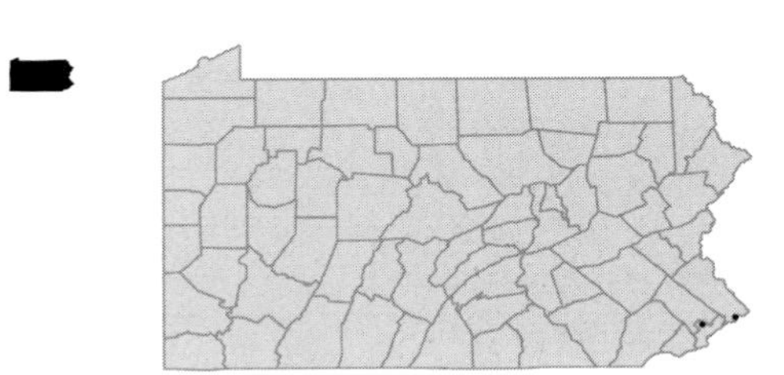

Aster subulatus Michx. var. ***subulatus***
Salt-marsh aster
Herbaceous annual
Ballast, sidewalks and marshy waste ground.

OBL

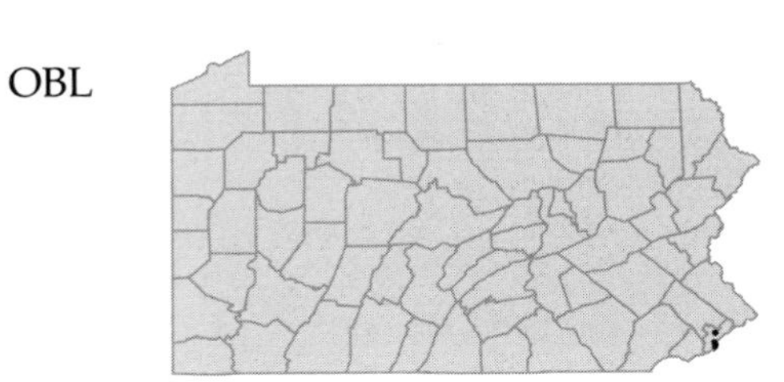

Aster tataricus L.f.
Tartarian aster
Herbaceous perennial
Railroad tracks, roadsides, old field and rubbish dump.

I

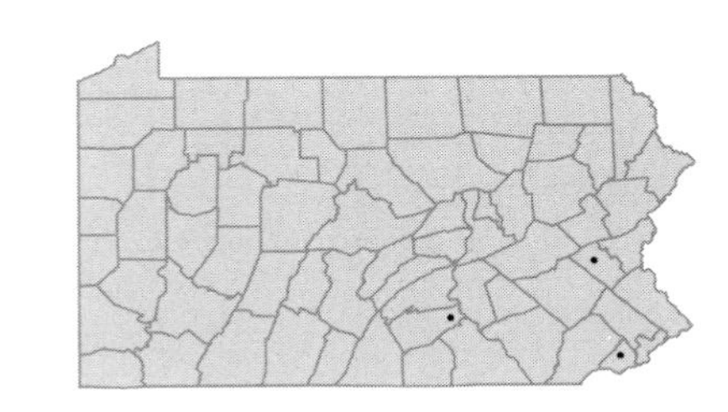

Aster tripolium L.
Aster
Herbaceous perennial
Ballast. Represented by two collections from Philadelphia Co. 1867-1880.

I

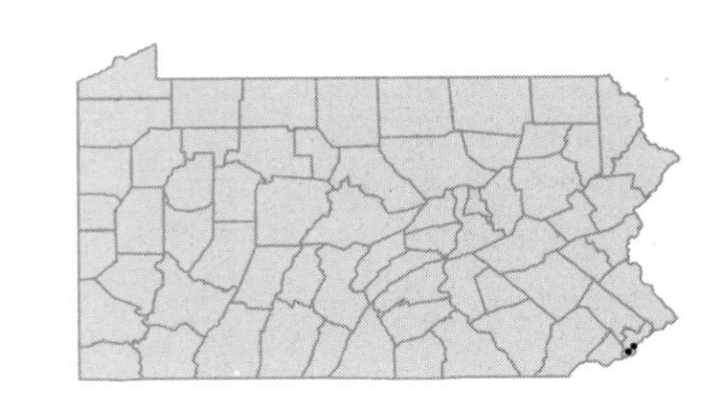

Aster umbellatus P.Mill.
Flat-topped white aster
Herbaceous perennial
Swampy woods and moist fields.
Doellingeria umbellata (P.Mill.) Nees P

FACW

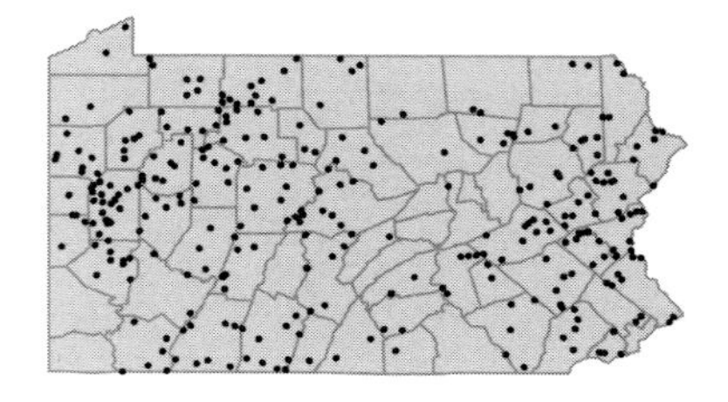

Aster undulatus L. *sensu lato*
Clasping heart-leaved aster
Herbaceous perennial
Dry woods, shaly slopes and fields.

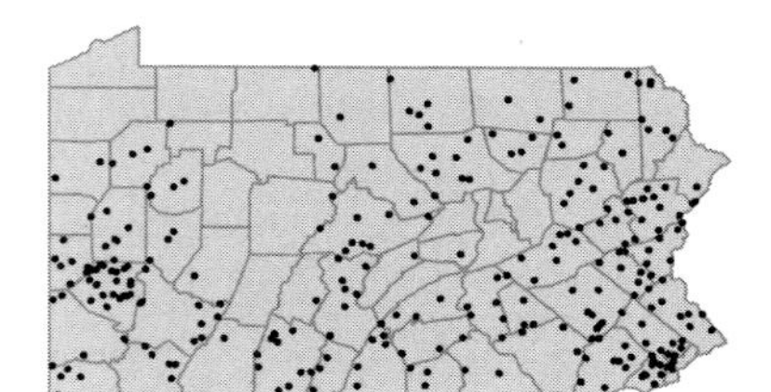

Aster urophyllus Lindley in DC.
Aster
Herbaceous perennial
Grassy roadside slopes, fields and woods edges.
Aster sagittifolius sensu auct., non Wedem. ex Willd. *pro parte* CF; *Aster hirtellus* Lindley in DC.

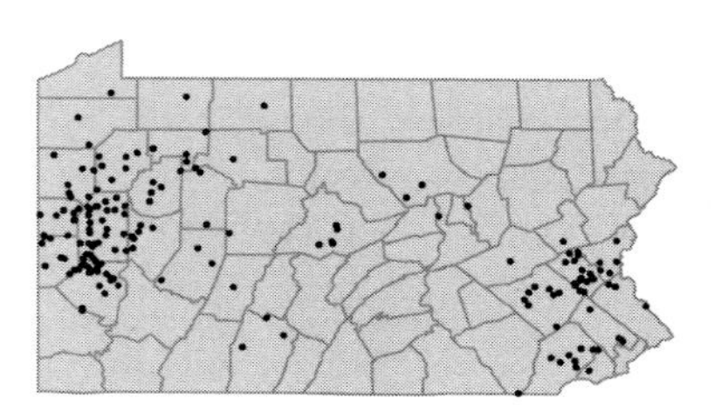

• ***Baccharis halimifolia*** L.
Groundsel-tree; Sea-myrtle
Deciduous shrub
Open woods, marshes and roadside ditches where de-icing salts are used.

FACW

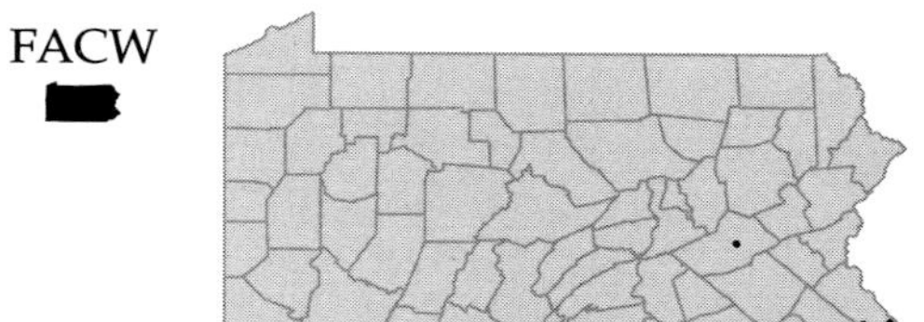

• ***Bellis perennis*** L.
English daisy
Herbaceous perennial
Naturalized in lawns and floodplains.

I

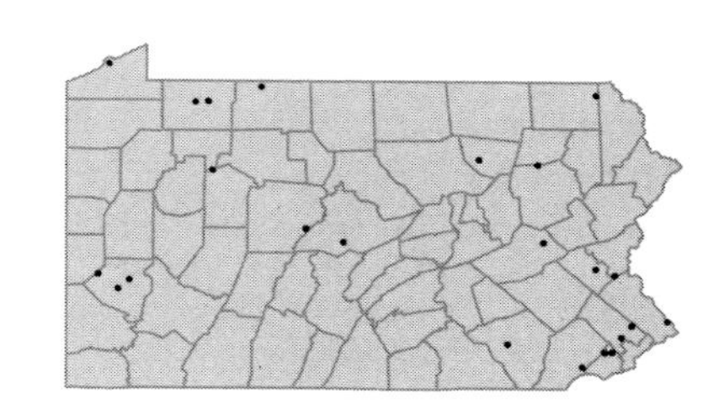

• ***Bidens aristosa*** (Michx.) Britt.
Tickseed-sunflower
Herbaceous annual
River banks.

FACW-
I

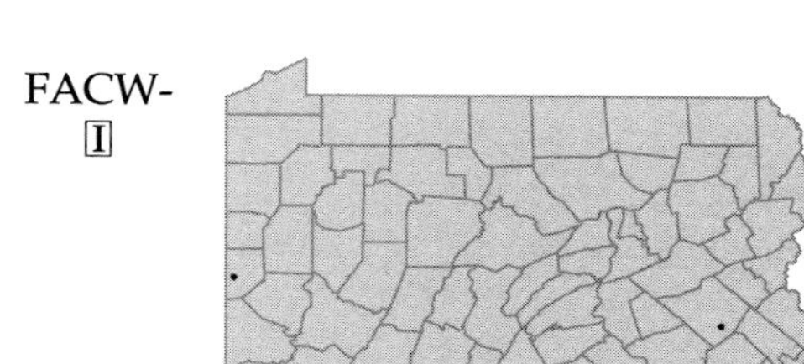

Bidens bidentoides (Nutt.) Britt.
Swamp beggar-ticks
Herbaceous annual
Tidal shores and mud flats.

FACW

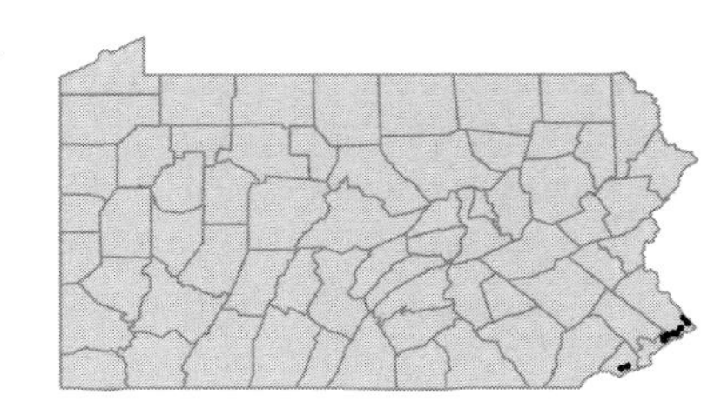

Bidens bipinnata L.
Spanish needles
Herbaceous annual
Dry rocky woods, shale barrens and roadside banks.

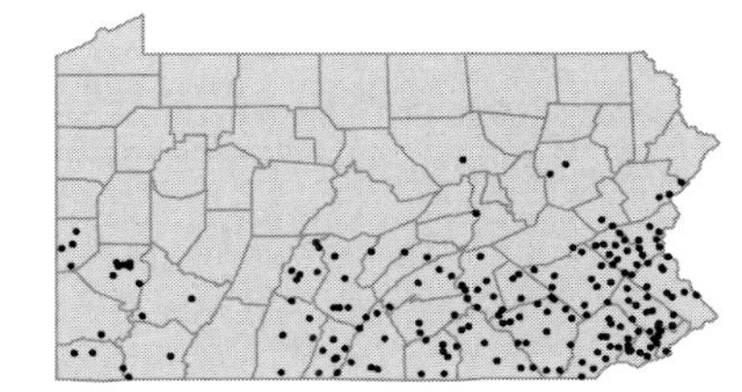

Bidens cernua L.
Bur-marigold; Stick-tight
Herbaceous annual
Swamps, wet shores and ditches.

OBL

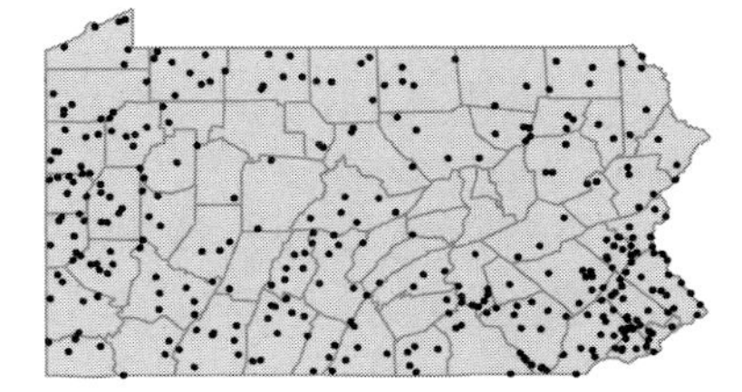

Bidens comosa (A.Gray) Wieg.
Beggar-ticks
Herbaceous annual
Stream banks, ditches and pond edges.

FACW

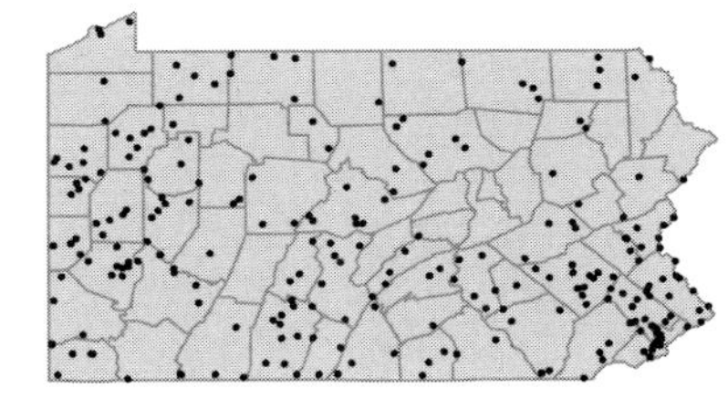

Bidens connata Muhl.
Beggar-ticks; Stick-tight
Herbaceous annual
Swamps, bogs, moist meadows and along streamlets.

FACW+

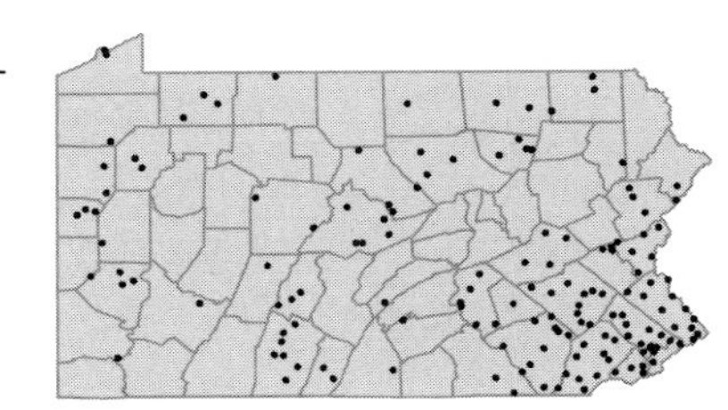

Bidens coronata (L.) Britt.
Tickseed-sunflower
Herbaceous annual
Bogs, swamps and wet ditches.
Bidens trichosperma (Michx.) Britt. P; *Bidens mitis* (Michx.) Sherff. C

OBL

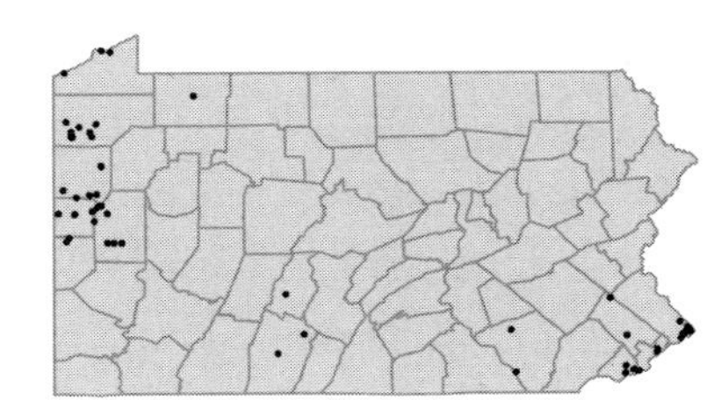

Bidens discoidea (Torr. & Gray) Britt.
Small beggar-ticks
Herbaceous annual
Bogs, vernal ponds and swampy ground.

FACW

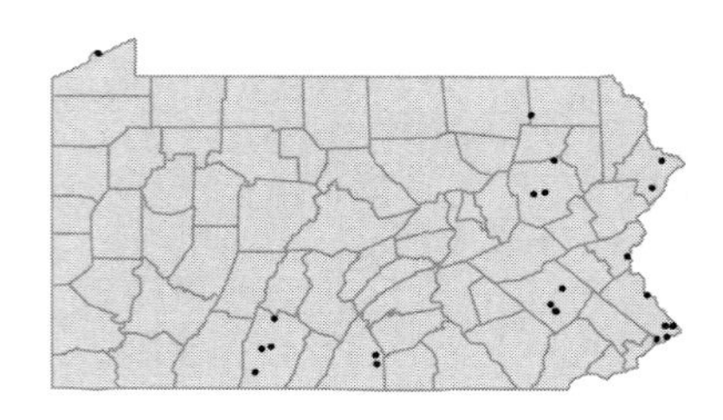

Bidens frondosa L.
Beggar-ticks; Stick-tights
Herbaceous annual
Moist, open ground, stream banks and roadsides.

FACW

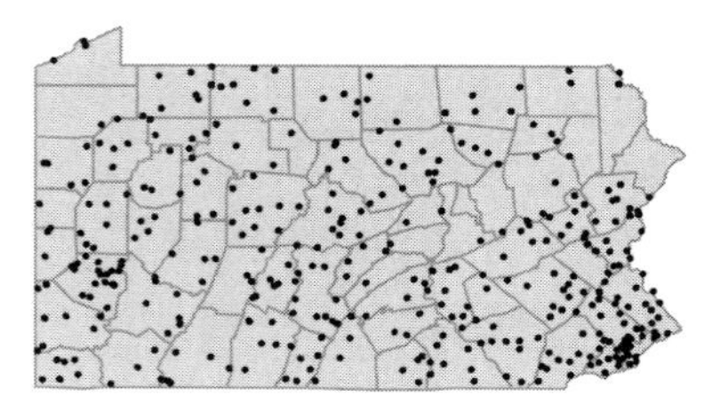

Bidens laevis (L.) BSP
Beggar-ticks; Bur-marigold
Herbaceous annual
Wet meadows and stream or pond edges.

OBL

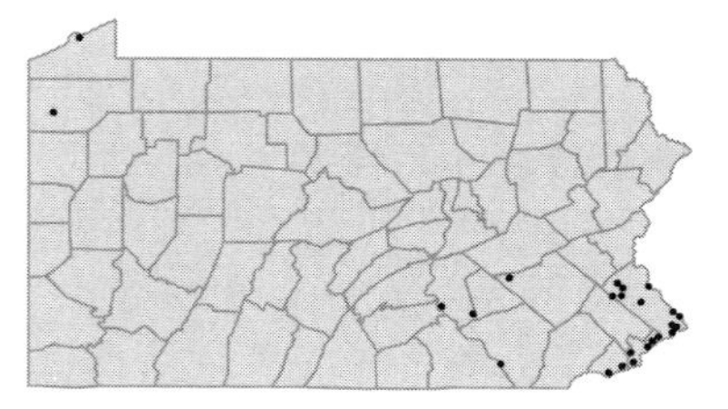

Bidens pilosa L. var. ***bimucronata*** (Turcz.) Sherff
Beggar-ticks
Herbaceous annual
Ballast. Represented by a single collection from Philadelphia Co. in 1879.

[I]

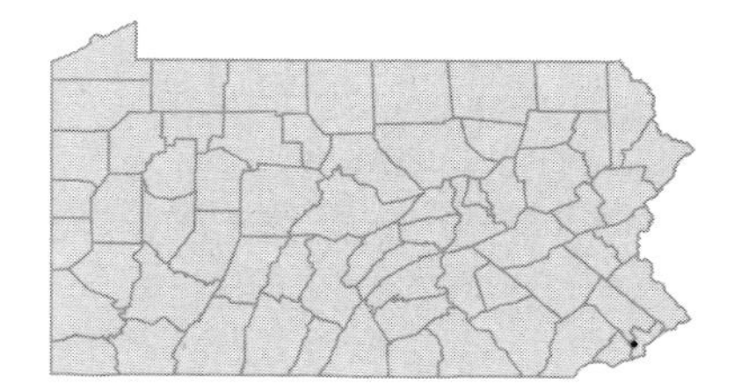

Bidens pilosa L. var. ***radiata*** Schultz-Bip.
Beggar-ticks
Herbaceous annual
Ballast ground. Represented by three collections from Philadelphia Co. 1868-1879.

[I]

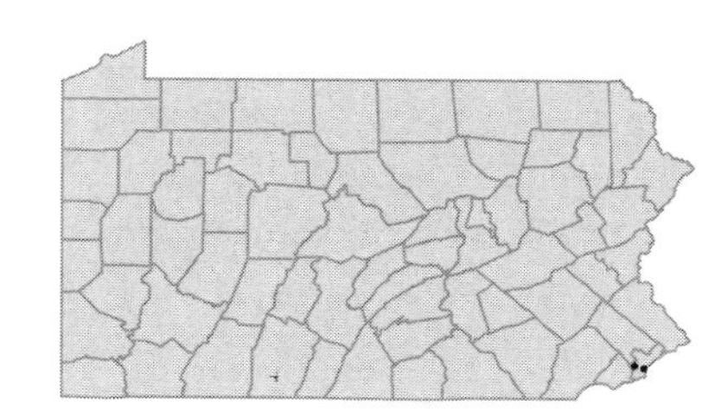

Bidens polylepis S.F.Blake
Tickseed-sunflower
Herbaceous annual
Moist fields, vacant lots and roadsides.

FACW
[I]

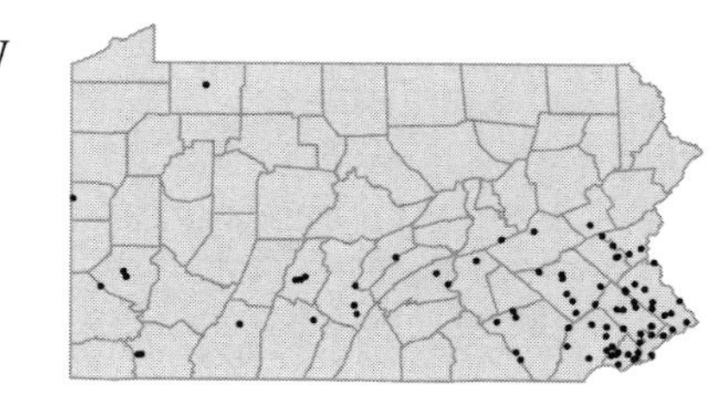

Bidens vulgata Greene
Beggar-ticks; Stick-tights
Herbaceous annual
Moist woods, wet fields, river banks and roadsides.

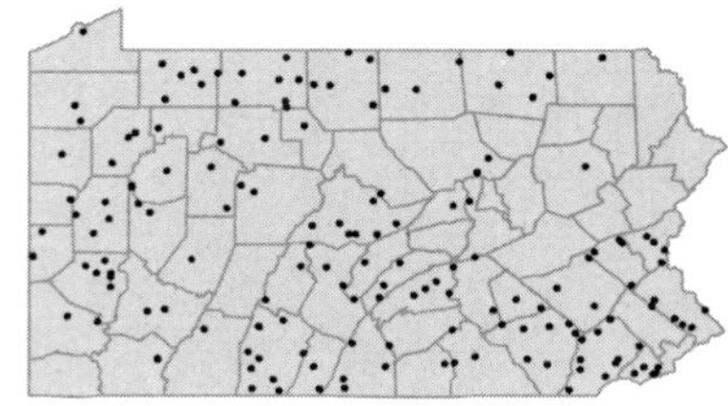

• ***Boltonia asteroides*** (L.) L'Her.
Aster-like boltonia
Herbaceous perennial
Rocky shores and exposed riverbed.

FACW

• ***Brickellia eupatorioides*** (L.) Shinners
False boneset
Herbaceous perennial
Shale barrens, wooded slopes and dry, rocky, limestone slopes.
Kuhnia eupatorioides L. PFGBWC

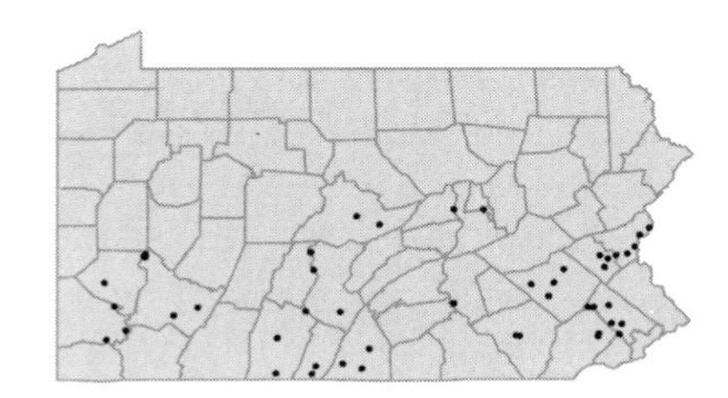

• ***Cacalia atriplicifolia*** L.
Pale Indian-plantain
Herbaceous perennial
Open woods, fields and moist banks.
Mesadenia atriplicifolia (L.) Raf. P

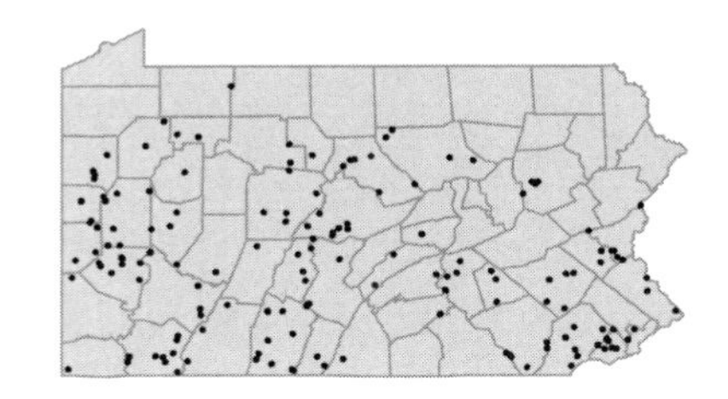

Cacalia muhlenbergii (Schultz-Bip.) Fern.
Great Indian-plantain
Herbaceous perennial
Woods and floodplains.
Mesadenia reniformis (Muhl.) Fern. P

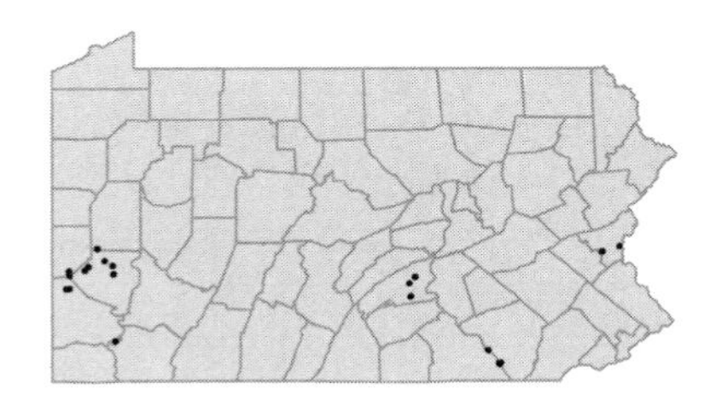

Cacalia suaveolens L.
Sweet-scented Indian-plantain
Herbaceous perennial
River banks, shaly slopes and meadows.
Synosma suaveolens (L.) Raf. P

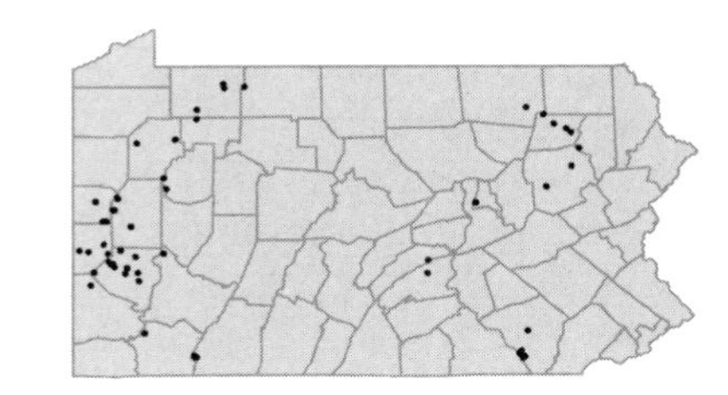

• ***Calendula officinalis*** L.
Pot marigold
Herbaceous annual
Cultivated and occasionally escaped to disturbed ground or roadsides.

[I]

• ***Callistephus chinensis*** (L.) Nees
China aster
Herbaceous annual
Cultivated and occasionally occurring spontaneously around rubbish dumps.

[I]

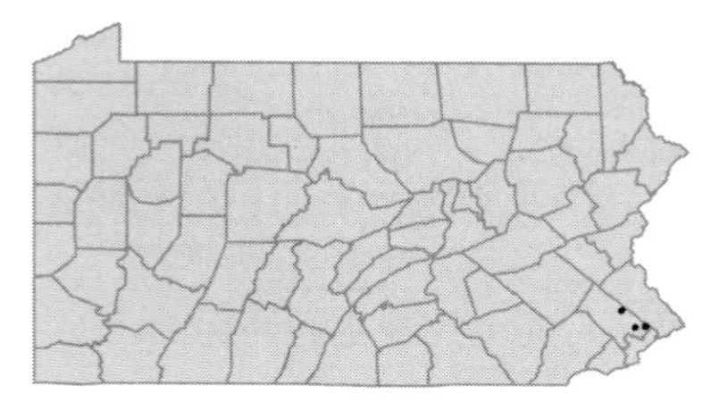

• ***Carduus acanthoides*** L.
Thistle
Herbaceous annual
Ballast, roadsides, stream banks and urban waste ground.

[I]

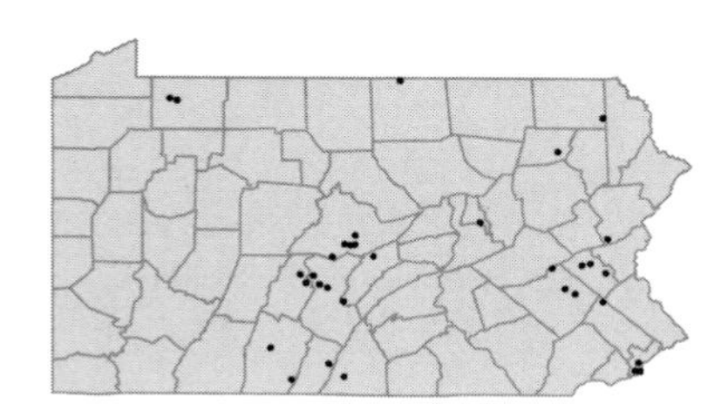

Carduus crispus L.
Welted thistle
Herbaceous annual
Fields. Represented by a single collection from Chester Co. in 1909.

[I]

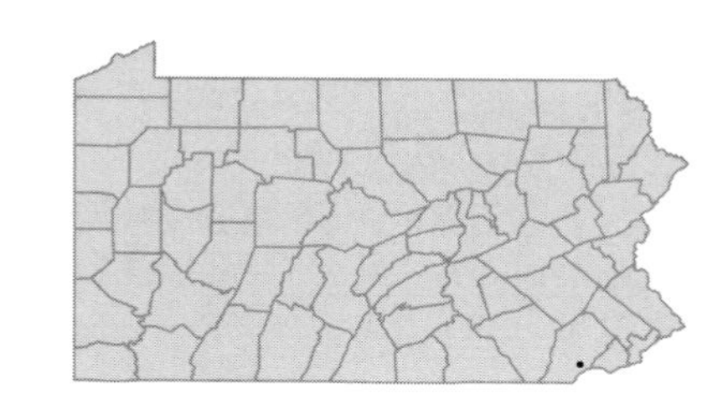

Carduus nutans L.
Nodding thistle; Musk thistle
Herbaceous biennial
Pastures, roadsides, waste ground and ballast. Designated as a noxious weed in PA.

[I]

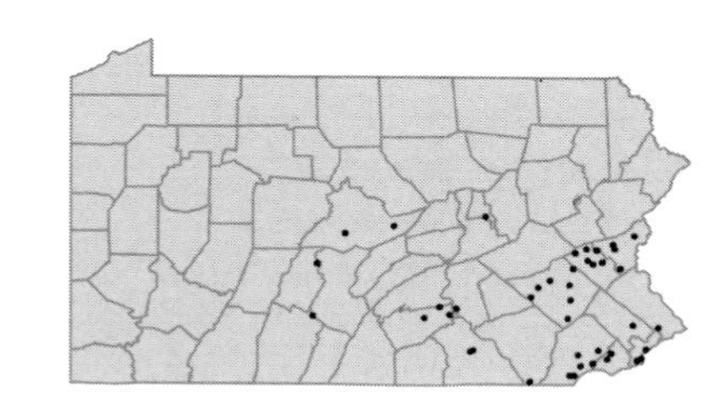

Carduus tenuiflorus Curtis
Plumeless thistle
Herbaceous annual
Ballast ground. Represented by several collections from Philadelphia Co. 1877-1879.

[I]

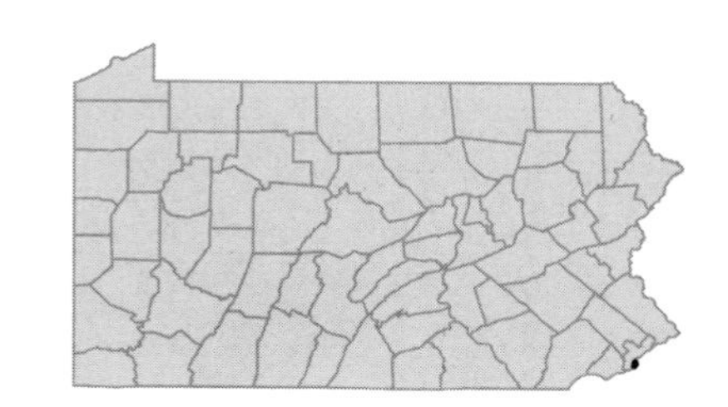

• ***Centaurea calcitrapa*** L.
Caltrops; Star-thistle
Herbaceous biennial
Ballast, rubbish dumps and waste ground.

I

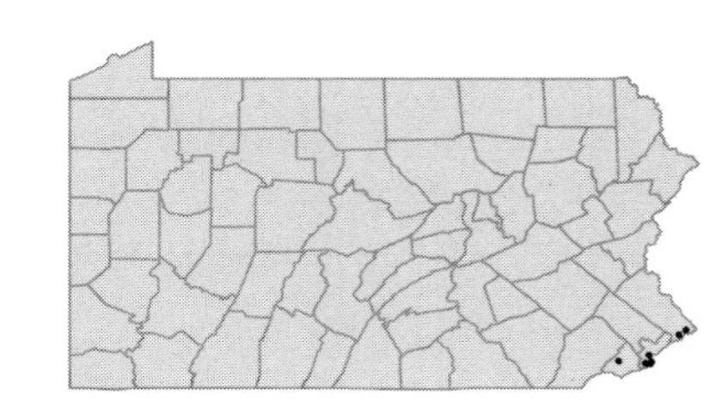

Centaurea cyanus L.
Bachelor's-button; Cornflower
Herbaceous annual
Cultivated and occasionally spreading to roadsides and fields, also a major component of seed mixtures for meadow establishment.

I

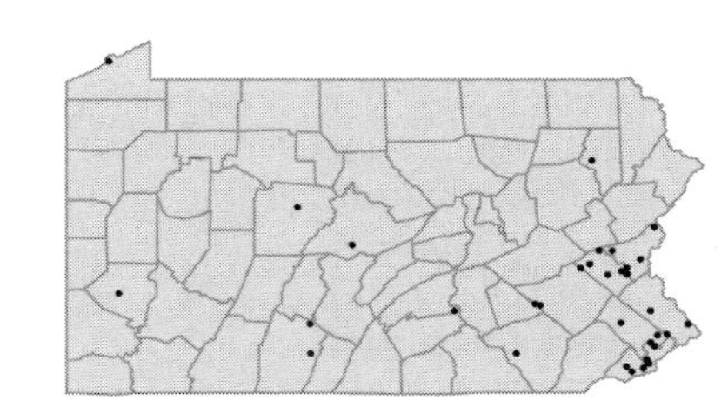

Centaurea jacea L.
Brown knapweed
Herbaceous perennial
Escaped to roadsides, fields and woods.

I

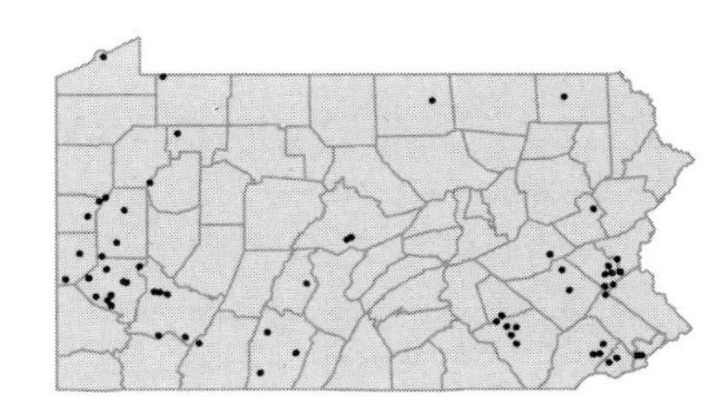

Centaurea maculosa Lam.
Bushy knapweed; Spotted knapweed
Herbaceous biennial
Woods, fields, roadsides and shale barrens.

I

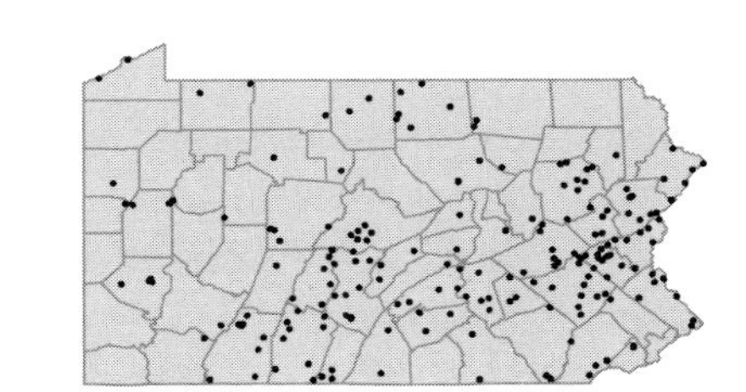

Centaurea melitensis L.
Star-thistle-of-Malta
Herbaceous annual
Ballast ground. Represented by a single collection from Philadelphia Co. in 1877.

I

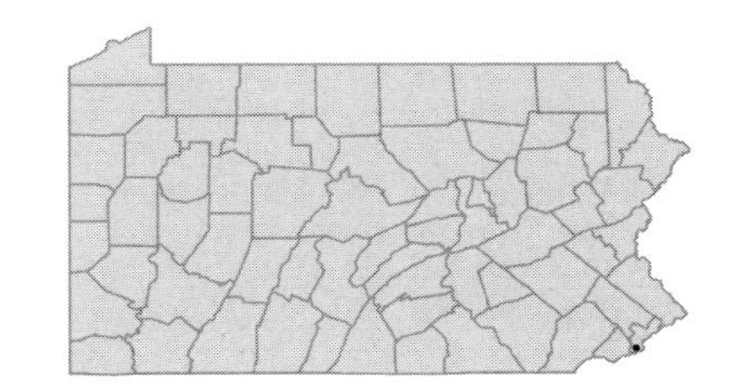

Centaurea montana L.
Mountain-bluet
Herbaceous perennial
Wooded roadsides. Represented by a single collection each from Elk Co. (1947) and Venango Co. (1964).

I

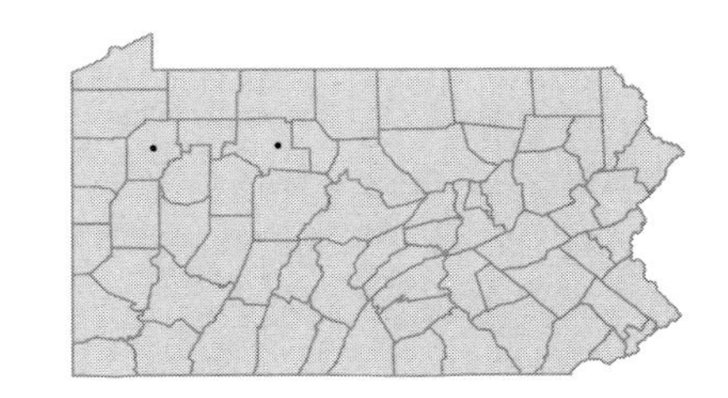

Centaurea nigra L. [I]
Black knapweed; Spanish-buttons
Herbaceous perennial
Dry, open fields, railroad banks, roadsides and waste ground.

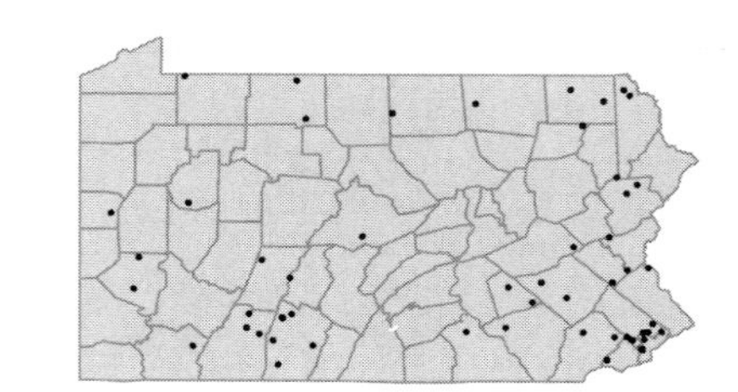

Centaurea nigrescens Willd. [I]
Knapweed
Herbaceous perennial
Fields and roadsides.
Centaurea dubia Suter ssp. *vochinensis* (Bernh.) Hayek GB; *Centaurea vochinensis* Bernh. FW

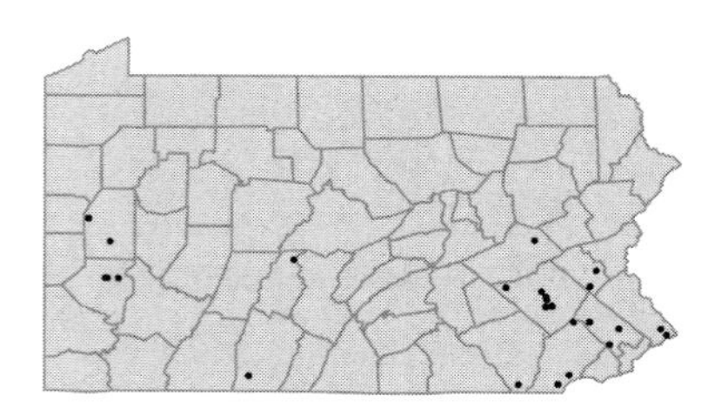

Centaurea solstitialis L. [I]
Barnaby's-thistle; Yellow star-thistle
Herbaceous annual
Wharves, ore heaps, ballast ground, cultivated fields, fencerows and gardens. Believed to have arrived in Delaware Co., PA in 1849 in seed of *Medicago sativa*.

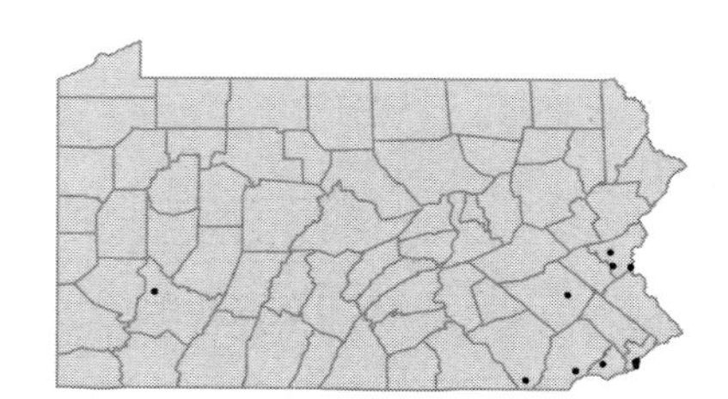

• ***Chondrilla juncea*** L. [I]
Gum-succory; Skeleton-weed
Herbaceous perennial
Fields, shale barrens and mine waste.

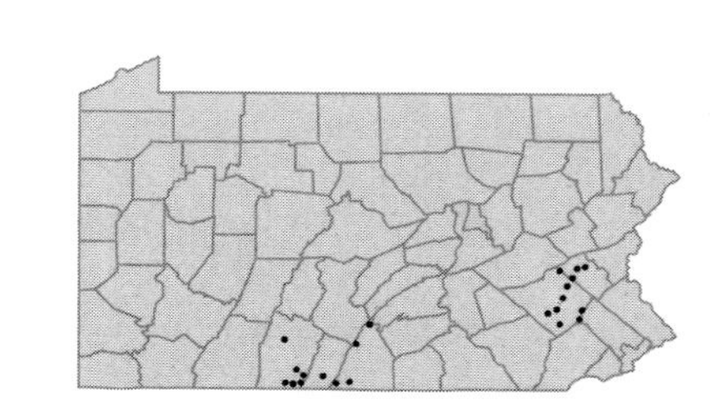

• ***Chrysanthemum balsamita*** L. [I]
Costmary
Herbaceous perennial
Cultivated and rarely escaped.

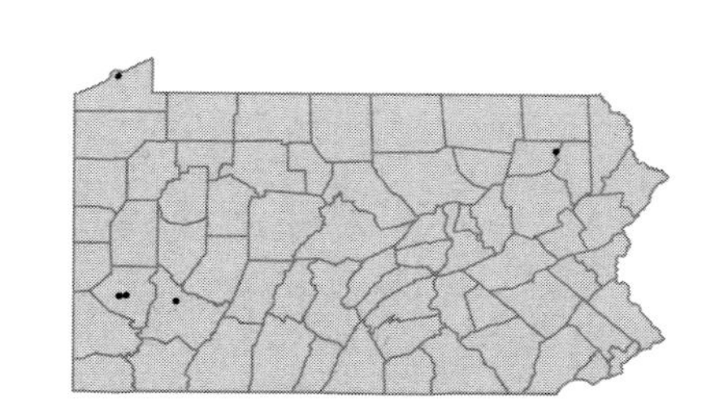

Chrysanthemum leucanthemum L. [I]
Ox-eye daisy
Herbaceous perennial
Fields, woods, meadows and roadsides.

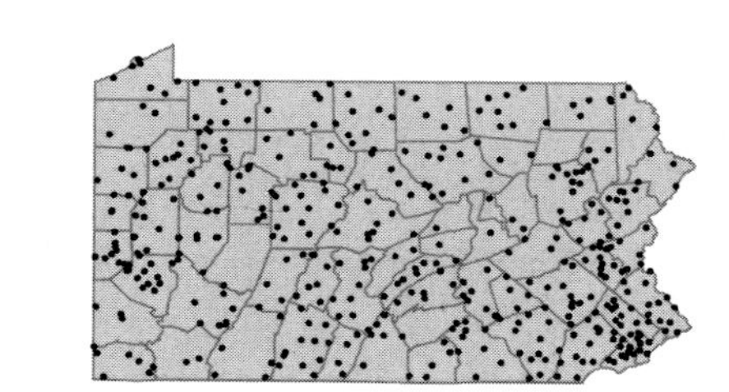

Chrysanthemum morifolium Ramat.
Garden chrysanthemum
Herbaceous perennial
Cultivated and occasionally escaped to fields, woods, railroad tracks and waste ground.

I

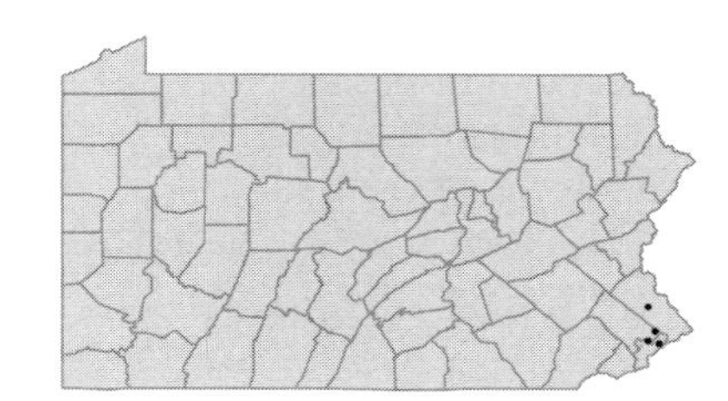

Chrysanthemum parthenium (L.) Bernh.
Feverfew
Herbaceous perennial
Cultivated and occasionally escaped to roadsides, waste ground and dumps.

I

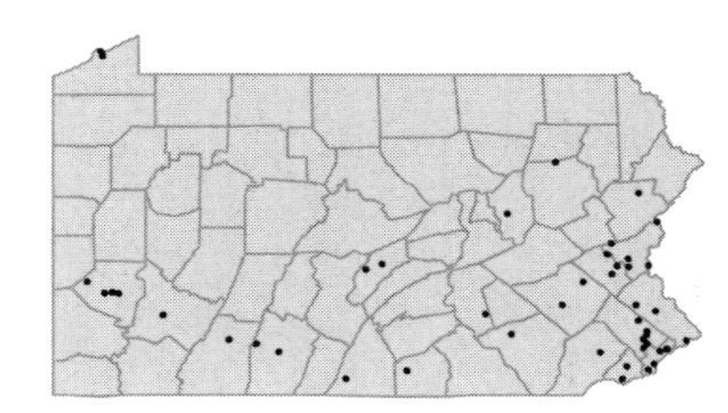

Chrysanthemum segetum L.
Corn-marigold
Herbaceous annual
Ballast. Represented by several collections from Philadelphia Co. 1877-1878.

I

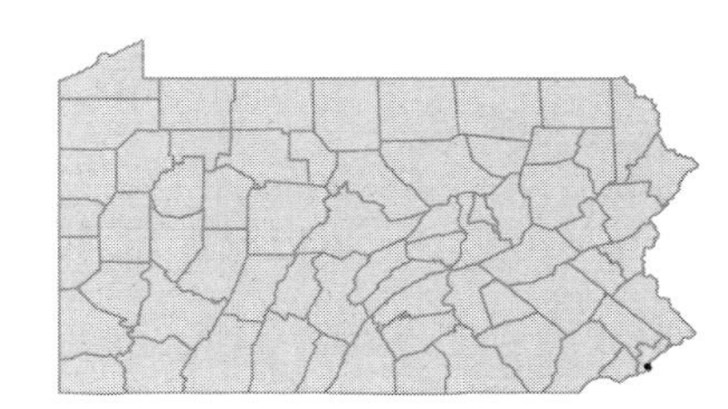

• ***Chrysogonum virginianum*** L.
Green-and-gold
Herbaceous perennial
Open woods, on limestone.

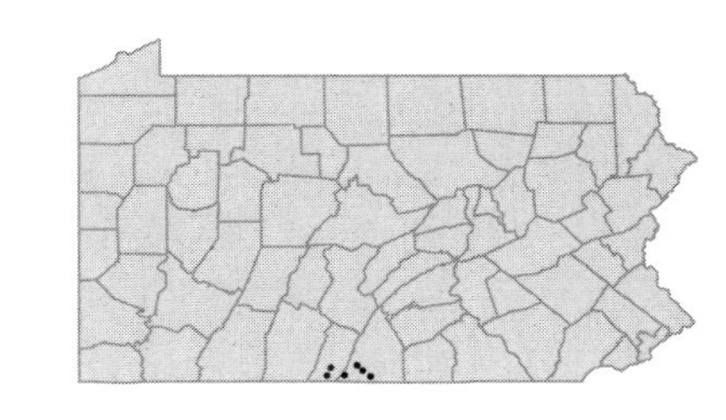

• ***Chrysopsis mariana*** (L.) Ell.
Golden aster
Herbaceous perennial
Dry, sandy woods or clearings, rock outcrops and serpentine barrens.
Heterotheca mariana (L.) Shinners W

UPL

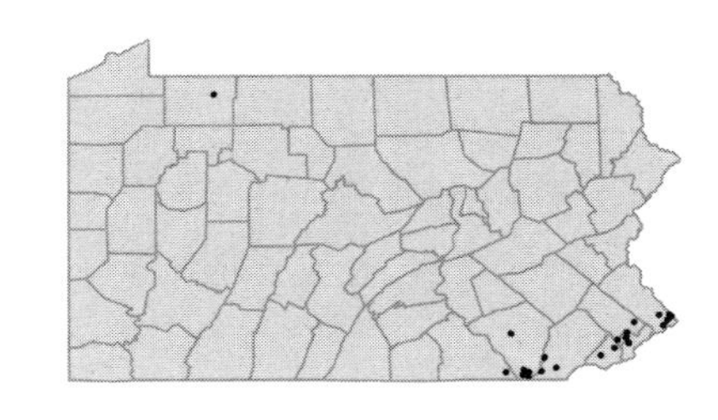

• ***Cichorium endiva*** L.
Endive
Herbaceous annual
Cultivated and occasionally escaped to alluvial shores and rubbish dumps.

I

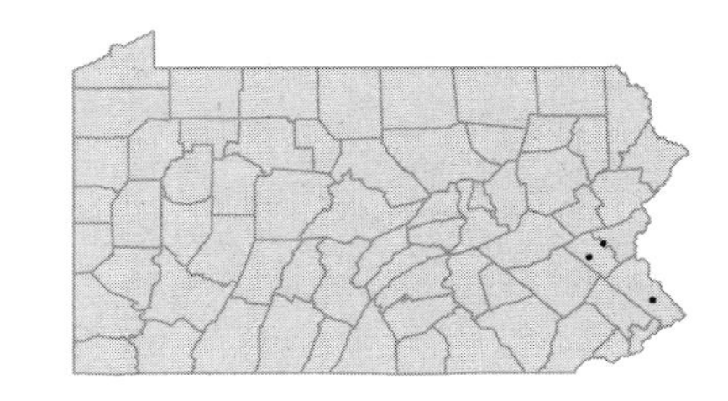

Cichorium intybus L. [I]
Blue-sailors; Blue chicory
Herbaceous perennial
Fields, roadsides and waste ground. Designated as a noxious weed in PA.

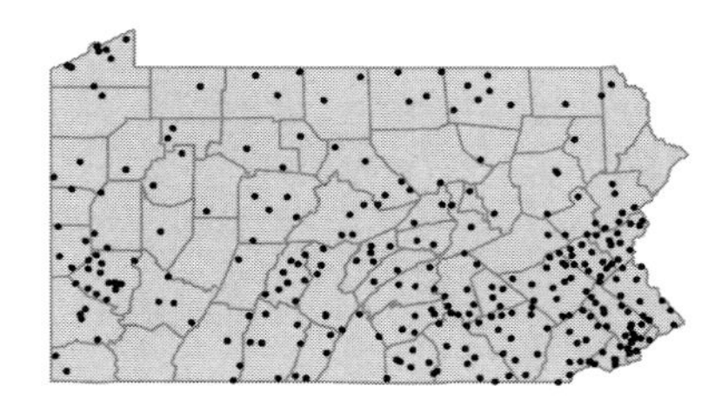

• ***Cirsium altissimum*** (L.) Spreng.
Tall thistle
Herbaceous biennial
Woods, river banks, fields and roadsides.
Carduus altissimus L. P

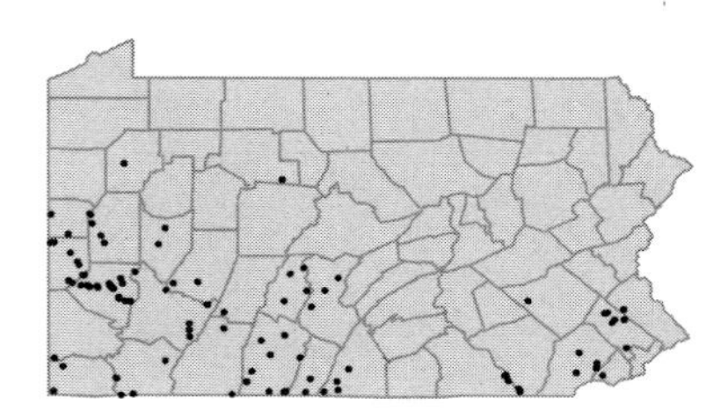

Cirsium arvense (L.) Scop. var. ***arvense*** FACU [I]
Canada thistle
Herbaceous perennial
Fields, pastures, roadsides and waste ground. Designated as a noxious weed in PA.
Carduus arvensis (L.) Robson P

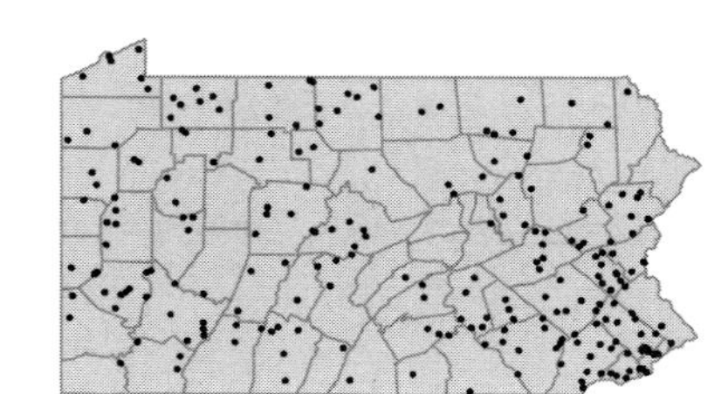

Cirsium arvense (L.) Scop. var. ***integrifolium*** Wimmer & Grab. FACU [I]
Canada thistle
Herbaceous perennial
Fields, roadsides and waste ground. Designated as a noxious weed in PA.

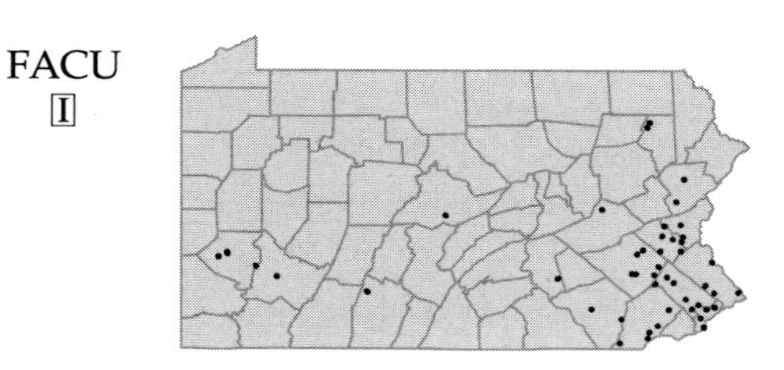

Cirsium arvense (L.) Scop. var. ***vestitum*** Wimmer & Grab. FACU [I]
Canada thistle
Herbaceous perennial
Moist pastures, fields and shores. Designated as a noxious weed in PA.

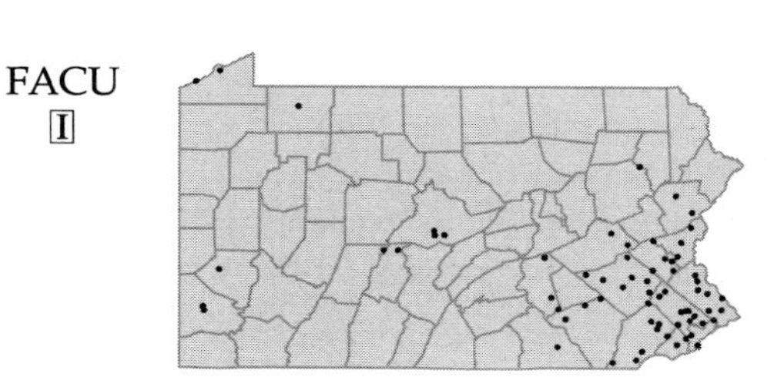

Cirsium discolor (Muhl.) Spreng.
Field thistle
Herbaceous biennial
Abandoned fields, open hillsides and roadside banks.
Carduus discolor (Muhl.) Nutt. P

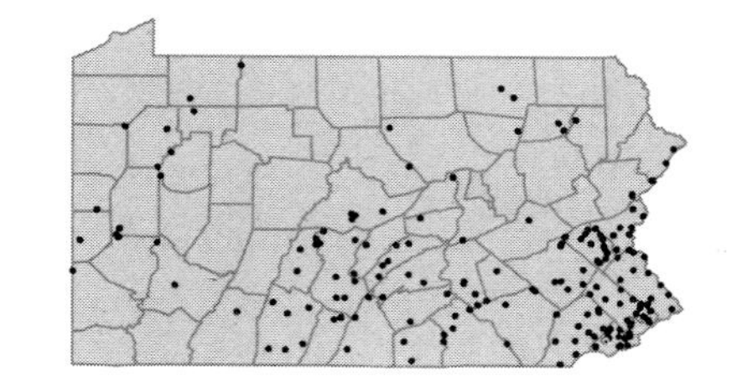

Cirsium horridulum Michx.
Yellow thistle; Horrible thistle
Herbaceous biennial
Moist, sandy or peaty meadows.
Carduus spinosissimus Walt. P

FACU-

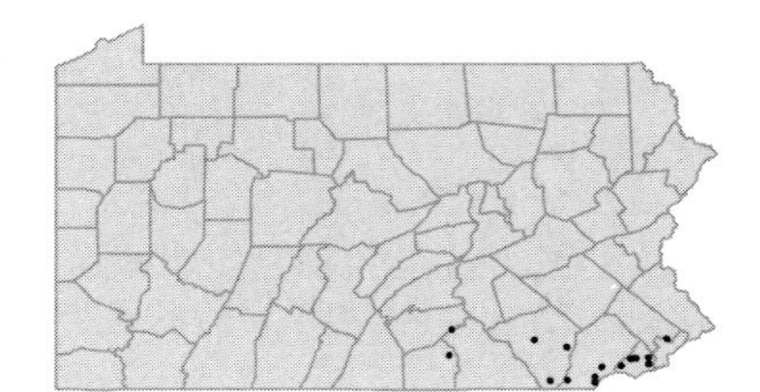

Cirsium muticum Michx.
Swamp thistle
Herbaceous biennial
Swamps, bogs, stream banks and wet meadows.
Carduus muticus (Michx.) Pers. P

OBL

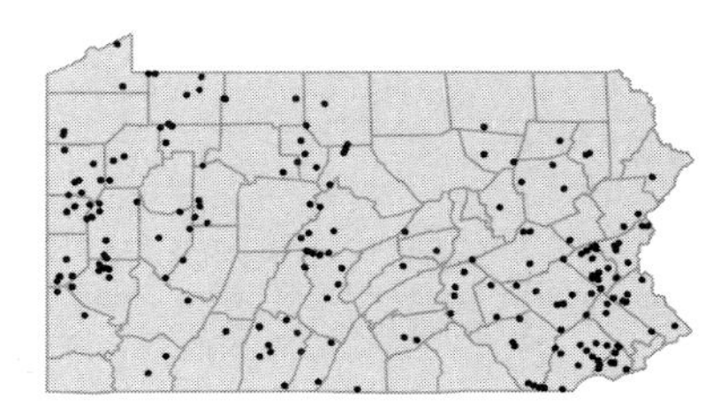

Cirsium pumilum (Nutt.) Spreng.
Pasture thistle
Herbaceous biennial
Dry fields, shaly hillsides, sandy floodplains, woods and roadsides.
Carduus odoratus (Muhl.) Porter P

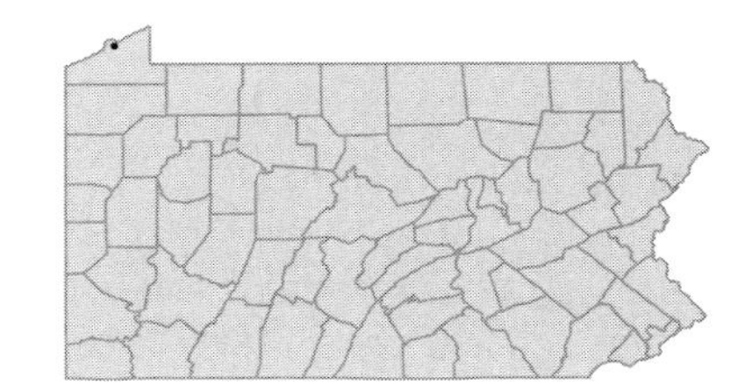

Cirsium undulatum (Nutt.) Spreng.
Wavy-leaved thistle
Herbaceous perennial
Wharves and docks. Represented by a single collection from Erie Co. in 1911.

I

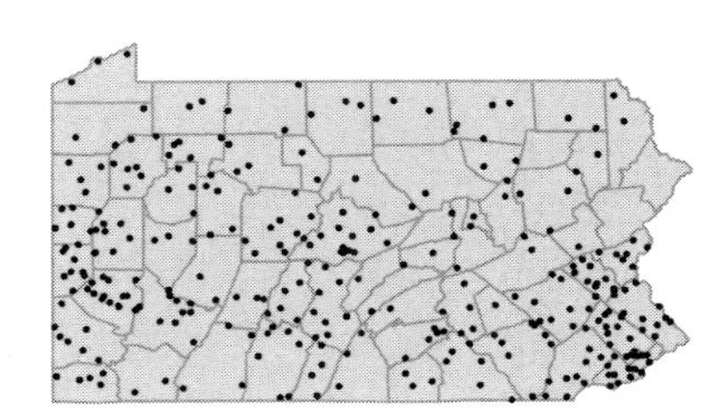

Cirsium vulgare (Savi) Tenore
Bull-thistle
Herbaceous biennial
Pastures, meadows and roadsides. Designated as a noxious weed in PA.
Carduus lanceolatus L. P

FACU-
I

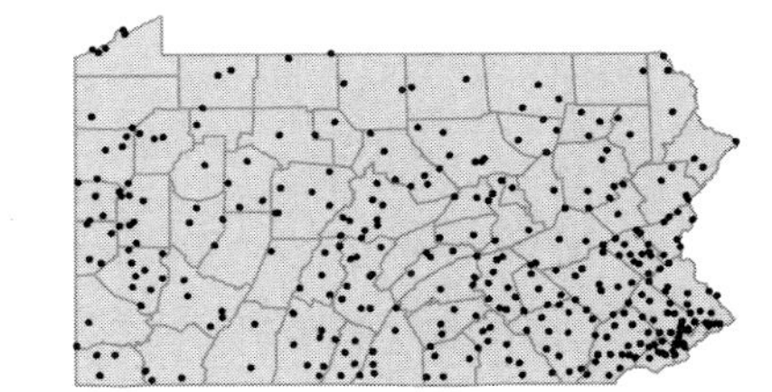

• ***Conyza canadensis*** (L.) Cronq. var. ***canadensis***
Horseweed
Herbaceous annual
Fields, roadsides, railroad tracks and waste ground.
Erigeron canadensis L. FW

UPL

Conyza canadensis (L.) Cronq. var. ***pusilla*** (Nutt.) Cronq.
Fleabane
Herbaceous annual
Sand, gravel, railroad cinders and roadsides.
Erigeron pusillus Nutt. WFB

[I]

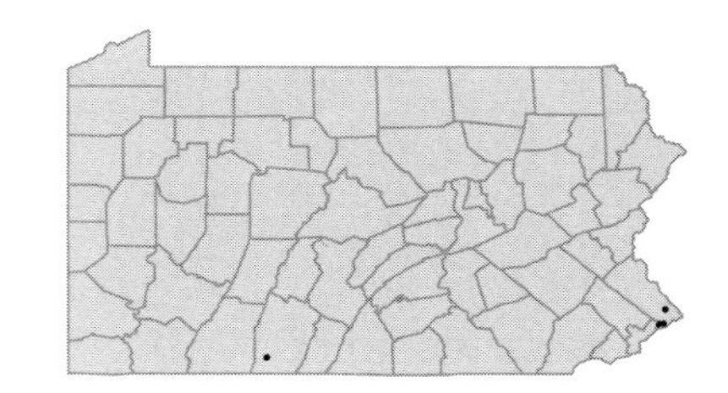

Conyza ramosissima Cronq.
Fleabane
Herbaceous annual
Garden. Represented by a single collection from Philadelphia Co. in 1877.
Erigeron divaricatus Michx. FW

[I]

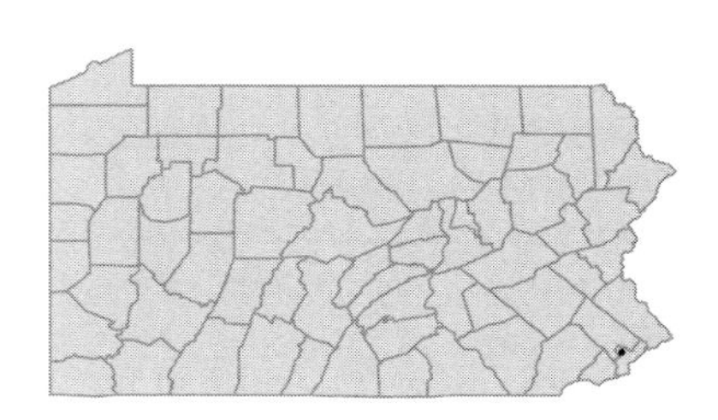

• ***Coreopsis lanceolata*** L.
Longstalk tickseed
Herbaceous perennial
Cultivated and frequently escaped to fields, shores and roadsides.

FACU
[I]

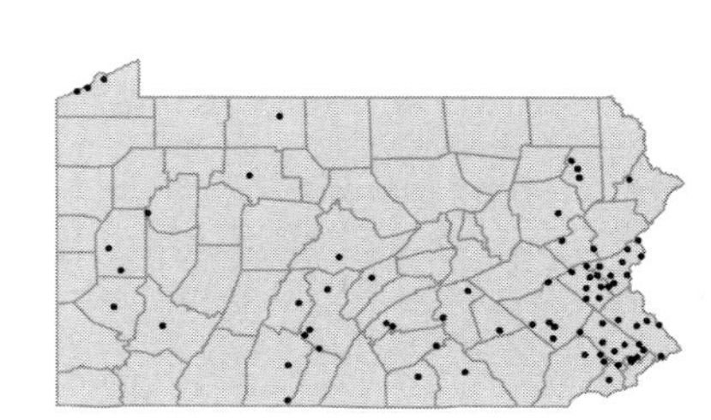

Coreopsis major Walt.
Wood tickseed
Herbaceous perennial
Although native as far north as VA, this species is only known from a single site in PA where it was growing as a weed in flowerbeds.

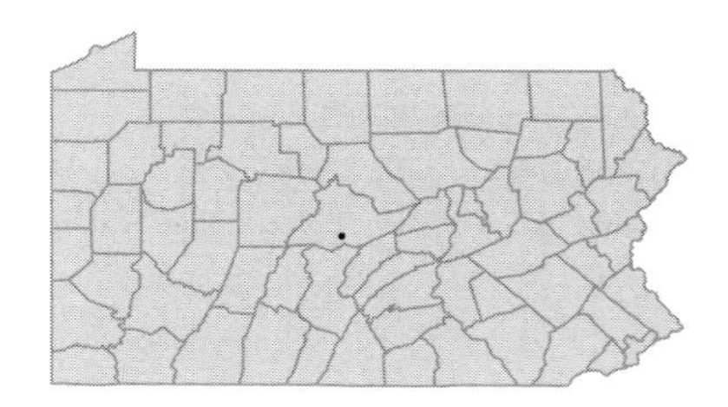

Coreopsis rosea Nutt.
Pink tickseed
Herbaceous perennial
Moist, open sandy soil. Believed to be extirpated, last collected in 1866.

FACW

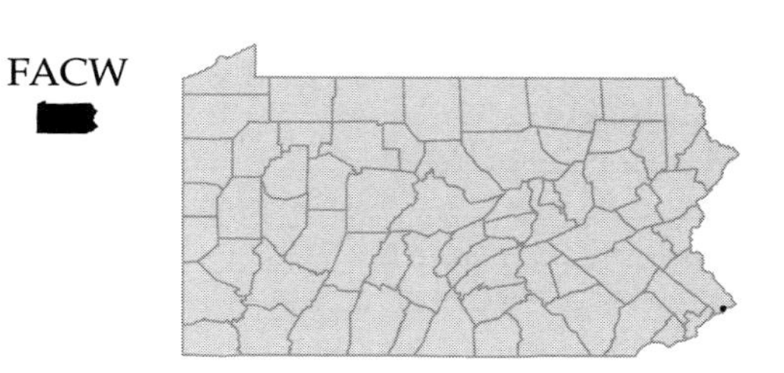

Coreopsis tinctoria Nutt.
Plains tickseed
Herbaceous annual
Cultivated and occasionally escaped to yards and river banks.

FAC-
[I]

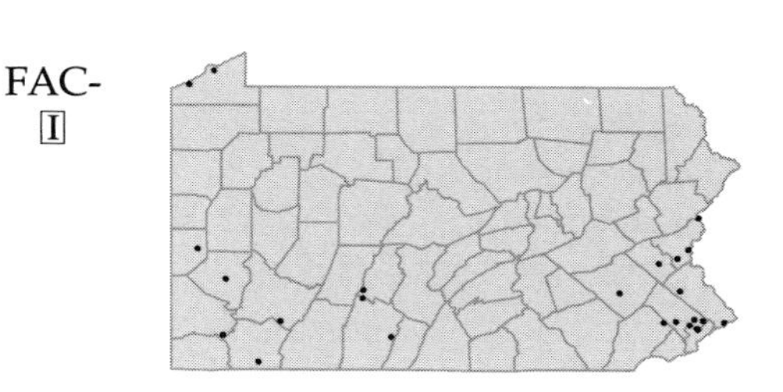

Coreopsis tripteris L.
Tall tickseed
Herbaceous perennial
Old fields, thickets, woods edges and roadsides.

FAC

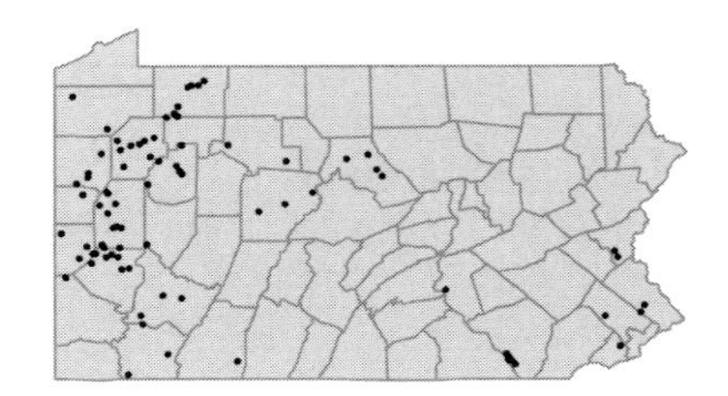

• *Cosmos bipinnatus* Cav.
Cosmos
Herbaceous annual
Cultivated and occasionally escaped to fields and roadsides, also a common component of wildflower seed mixtures for meadow establishment.

I

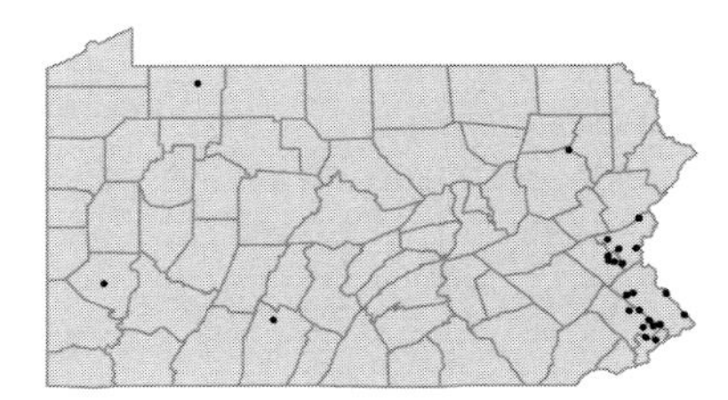

Cosmos sulphureus Cav.
Orange cosmos
Herbaceous annual
Cultivated and occasionally escaped to alluvium, vacant lots or rubbish dumps.

I

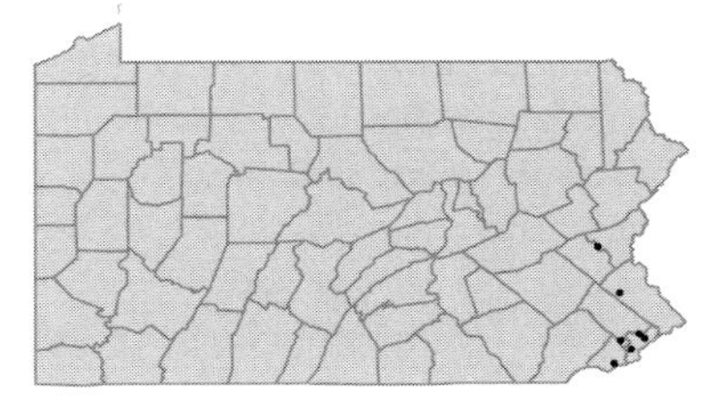

• *Crepis biennis* L.
Hawk's-beard
Herbaceous biennial
Represented by two early collections from Montour and Northampton Cos. (1863-1869).

I

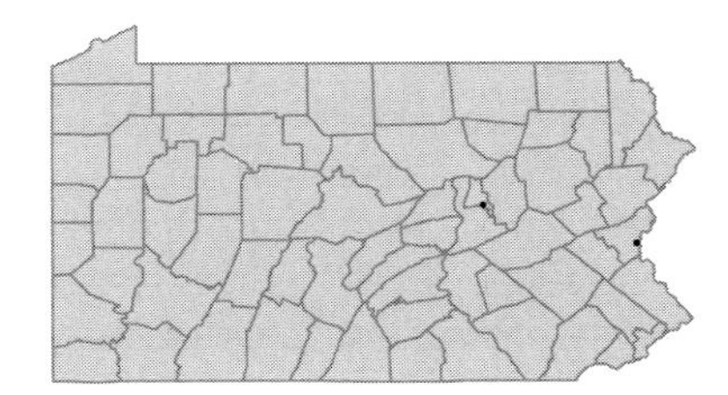

Crepis capillaris (L.) Wallr.
Hawk's-beard
Herbaceous annual
Fields, woods, lawns, roadsides, ballast and waste ground.
Crepis virens L. P

I

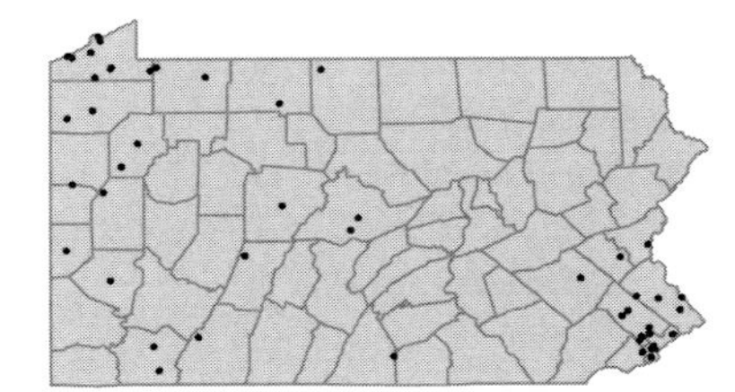

Crepis setosa Haller f.
Hawk's-beard
Herbaceous annual
Fields and waste ground.

I

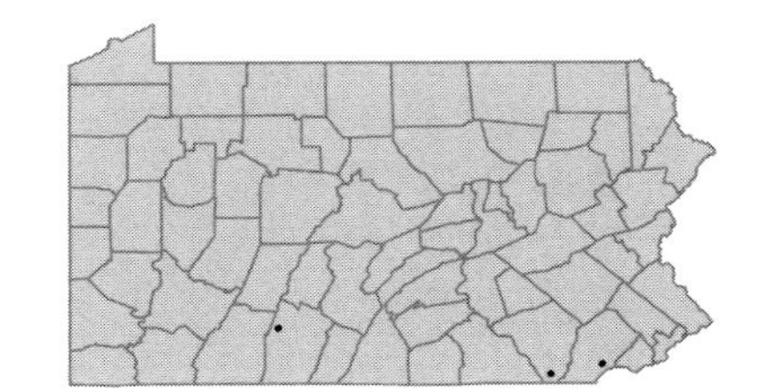

Crepis tectorum L.
Hawk's-beard
Herbaceous annual
Ballast, ore piles and waste ground.

[I]

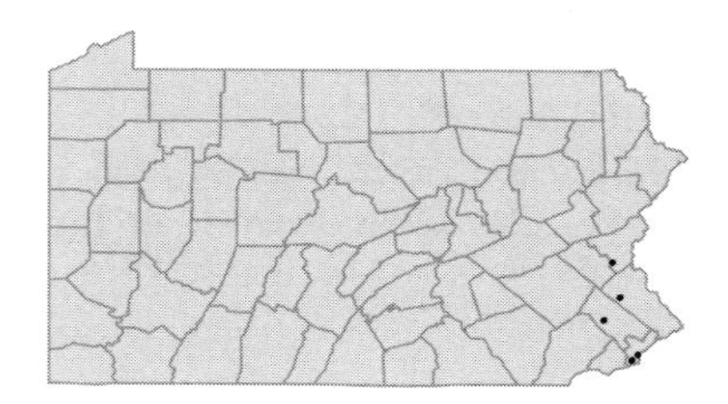

Crepis vesicaria L.
Hawk's-beard
Herbaceous annual
Roadsides.

[I]

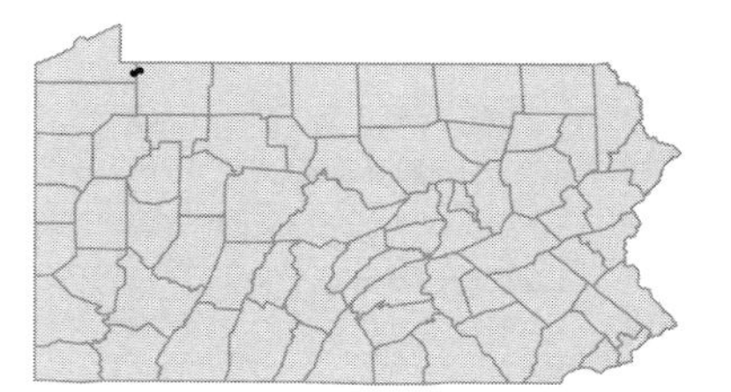

• ***Dyssodia papposa*** (Vent.) Hitchc.
Stinking-marigold
Herbaceous annual
Roadsides.

[I]

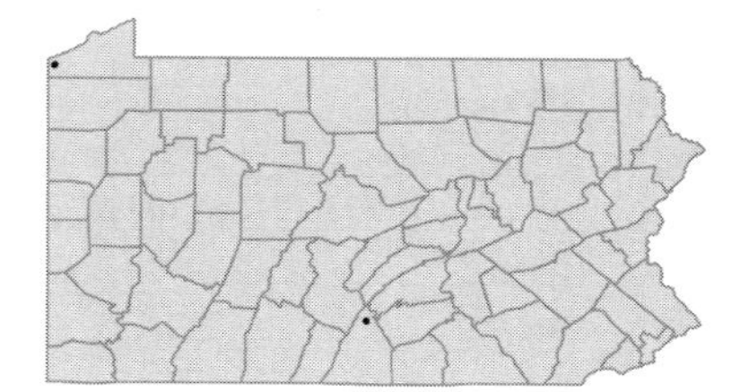

• ***Echinacea laevigata*** (C.L.Boynt. & Beadle) Blake
Appalachian coneflower; Smooth purple coneflower
Herbaceous perennial
Open woods and fields. Believed to be extirpated in PA, last collected in 1900.
Echinacea purpurea (L.) Moench var. *laevigata* (C.L.Boynt. & Beadle) Cronq. GB

Echinacea purpurea (L.) Moench
Purple coneflower
Herbaceous perennial
Cultivated and occasionally escaped to fields, abandoned nursery and rubbish dumps.

[I]

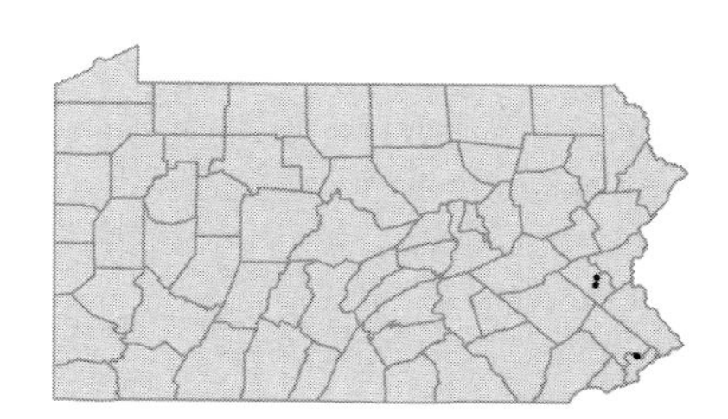

• ***Echinops sphaerocephalus*** L.
Globe-thistle
Herbaceous perennial
Cultivated and occasionally escaped to crevices in limestone rock.

[I]

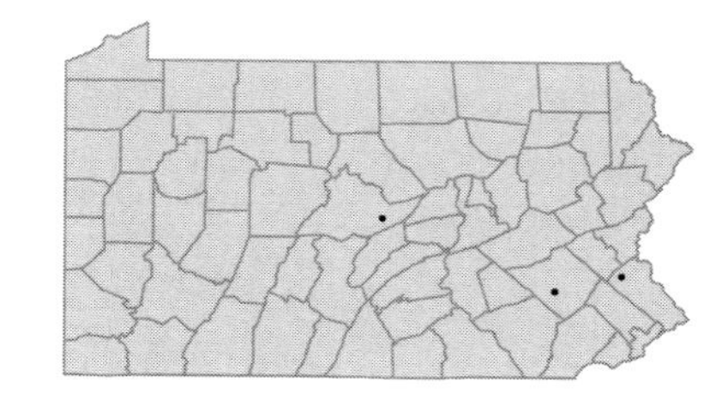

• ***Eclipta prostata*** (L.) L.
Yerba-de-tajo
Herbaceous annual
Wet shores, rocky river banks.
Eclipta alba (L.) Hassk. FGBW

FAC

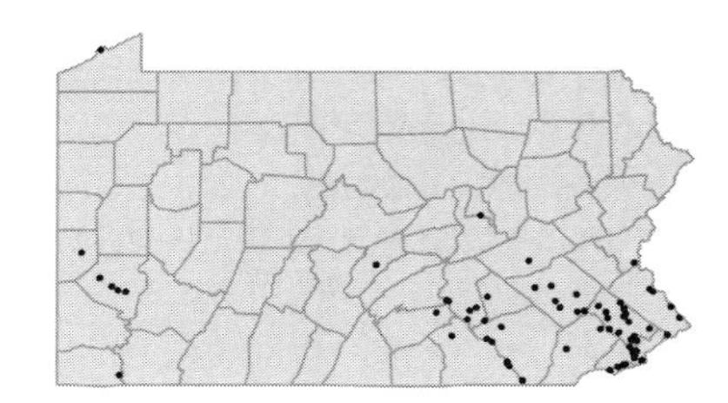

• ***Elephantopus carolinianus*** Willd.
Elephant's-foot
Herbaceous perennial
Open woods, serpentine barrens.

FACU

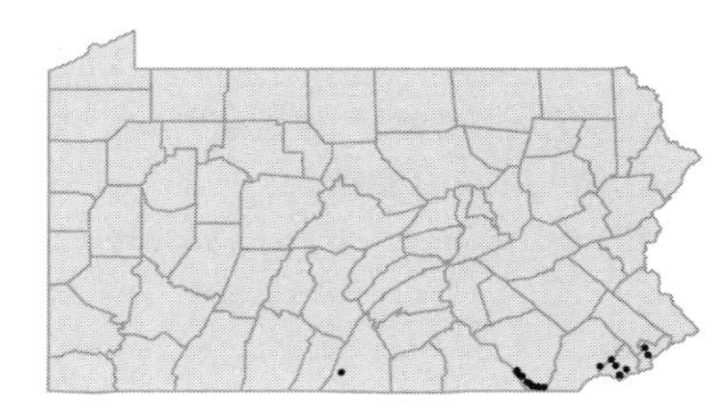

• ***Erechtites hieraciifolia*** (L.) Raf. ex DC.
Fireweed; Pilewort
Herbaceous annual
Fields, woods, clearings and waste ground.

FACU

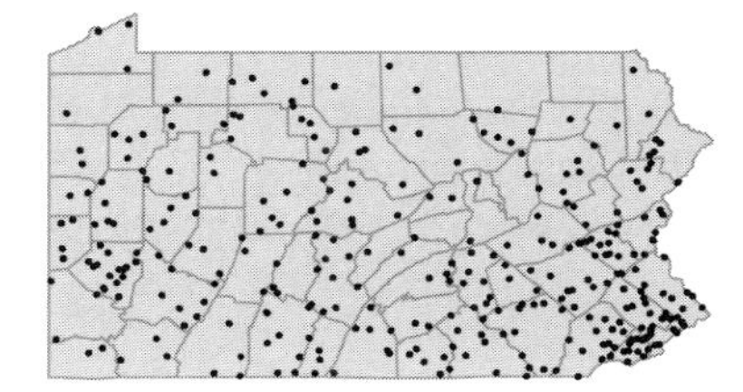

• ***Erigeron annuus*** (L.) Pers.
Daisy fleabane
Herbaceous annual
Fields, roadsides and waste ground.

FACU

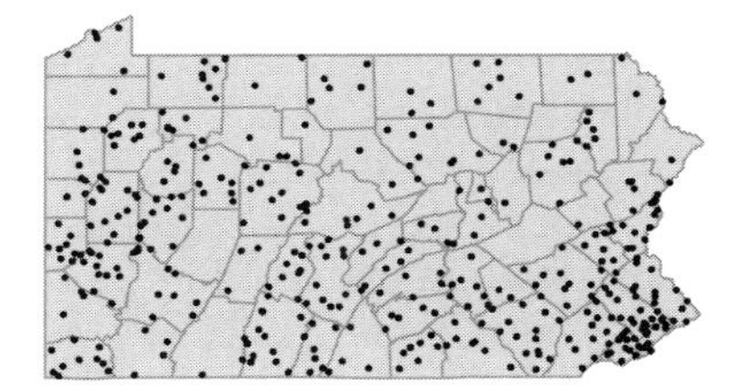

Erigeron philadelphicus L.
Daisy fleabane
Herbaceous perennial
Woods, edges, fields, roadsides and lawns.

FACU

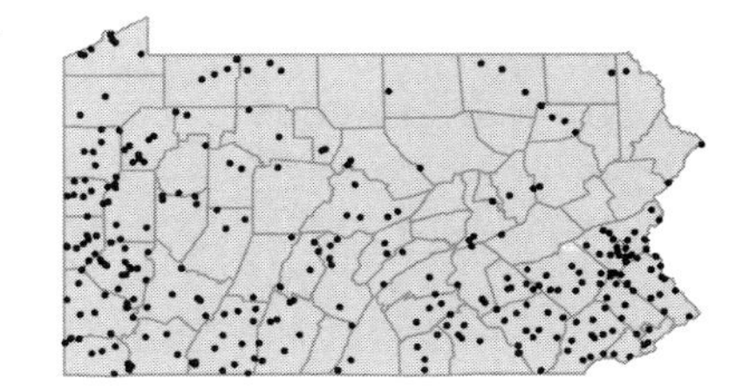

Erigeron pulchellus Michx.
Robin's-plantain
Herbaceous biennial
Meadows, wooded slopes, woodland edges and roadsides.

FACU

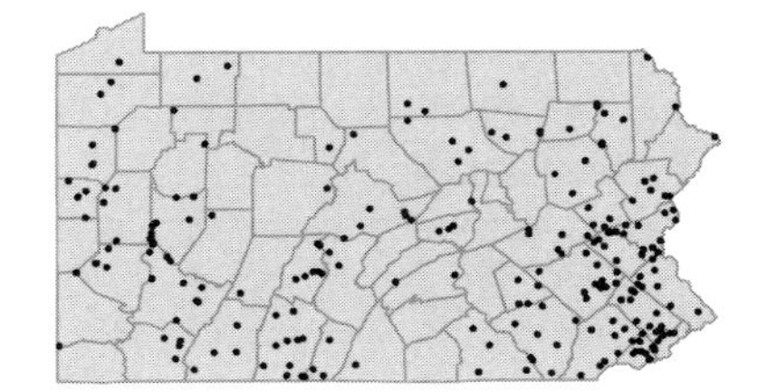

Erigeron strigosus Muhl. ex Willd. var. ***strigosus***
Daisy fleabane; White-top
Herbaceous annual
Fields, fencerows and dry, shaly slopes.
Erigeron ramosus (Walt.) BSP P

FACU+

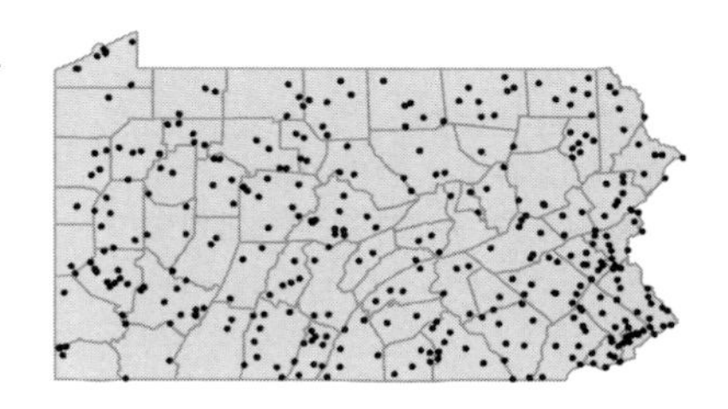

Erigeron strigosus Muhl. ex Willd. var. ***beyrichii*** (Fisch. & Mey.) A.Gray
Daisy fleabane; White-top
Herbaceous annual
Fallow fields, meadow swales and dry grasslands.

FACU+

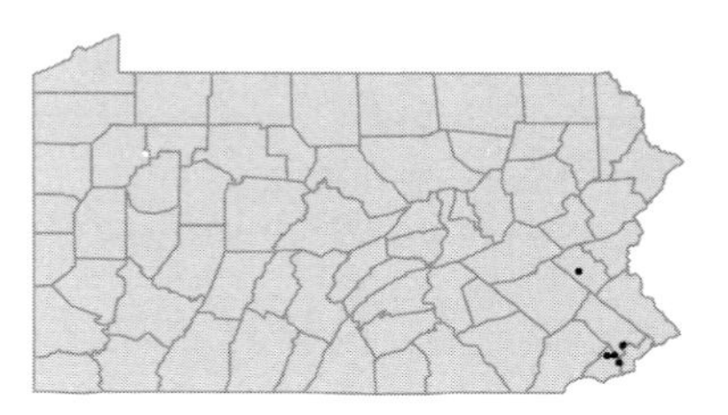

• ***Eupatorium album*** L.
White-bracted eupatorium; White thoroughwort
Herbaceous perennial
Sandy, open woods, dry slopes and serpentine barrens. Believed to be extirpated, last collected in 1964.

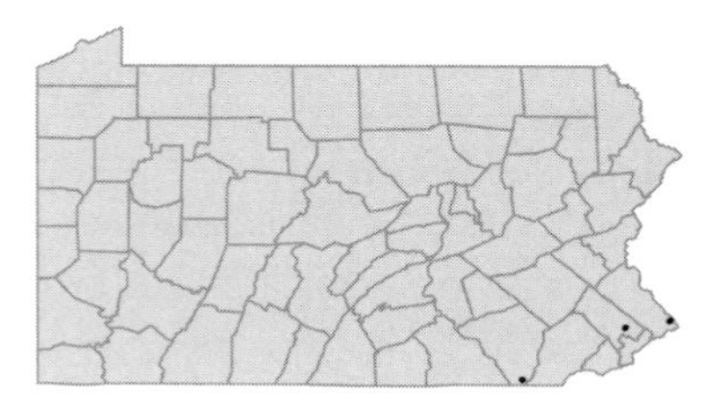

Eupatorium altissimum L.
Tall eupatorium
Herbaceous perennial
Dry, rocky slopes, bluffs, fields and roadsides.
Eupatorium rugosum Houtt. var. *tomentellum* (B.L.Robins.) Blake FW

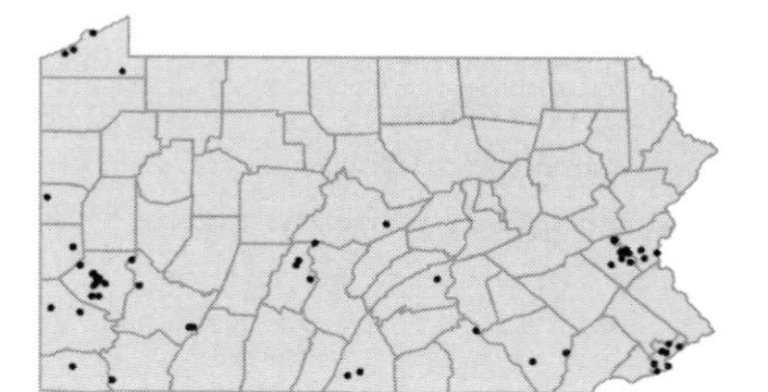

Eupatorium aromaticum L.
Small white-snakeroot
Herbaceous perennial
Dry woods and sandy, open areas.

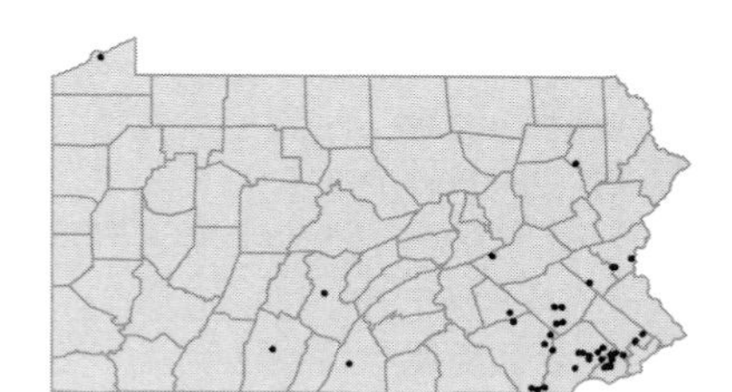

Eupatorium cannabinum L.
Eupatorium
Herbaceous perennial
Ballast. Represented by two collections from Philadelphia Co. in 1879.

I

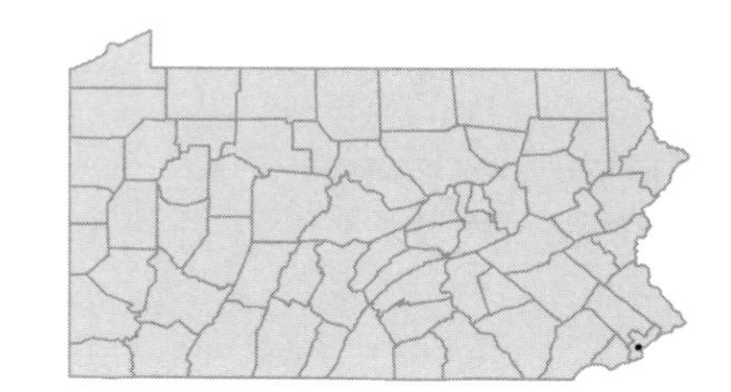

Eupatorium capillifolium (Lam.) Small
Dog-fennel
Herbaceous perennial
Ballast and rubbish dumps.

FACU-
I

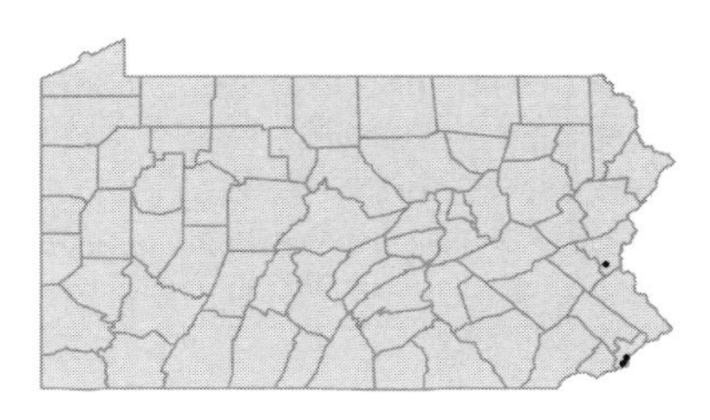

Eupatorium coelestinum L.
Mistflower; Wild ageratum
Herbaceous perennial
Old fields, meadows and stream banks, also cultivated and occasionally escaped.

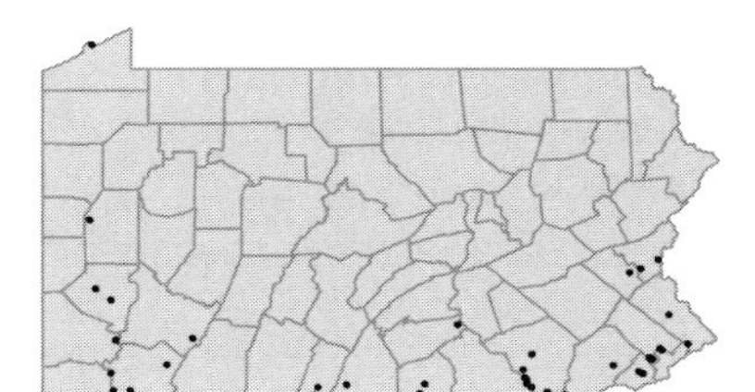

Eupatorium dubium Willd. ex Poir.
Joe-pye-weed
Herbaceous perennial
Swamps, bogs, calcareous marshes and swales.

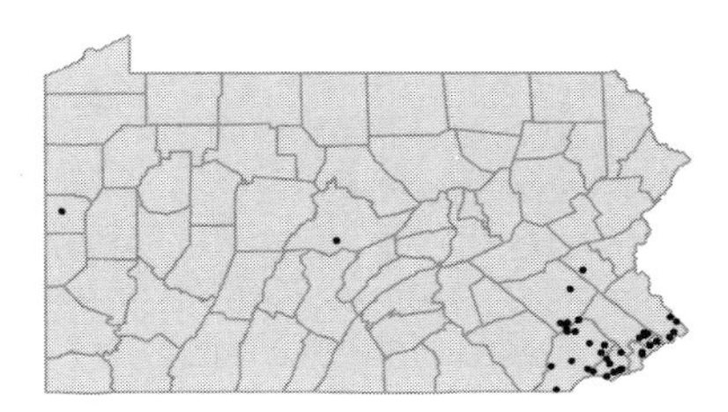

Eupatorium fistulosum Barratt
Joe-pye-weed
Herbaceous perennial
Floodplains, meadows, moist thickets and roadsides.

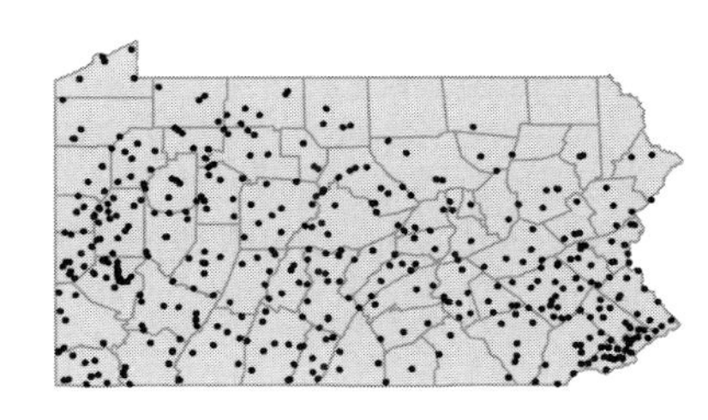

Eupatorium hyssopifolium L.
Hyssop-leaved eupatorium
Herbaceous perennial
Dry, sandy or gravelly fields, thickets and shores.

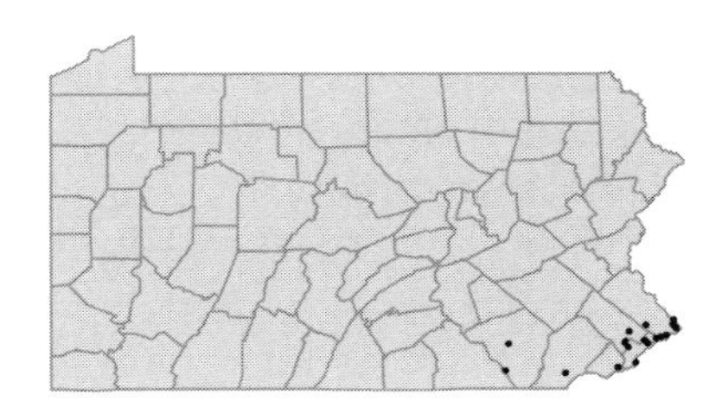

Eupatorium leucolepis (DC.) Torr. & A.Gray
White-bracted thoroughwort
Herbaceous perennial
Moist gravel pits. Believed to be extirpated, last collected in 1927.

FACW+

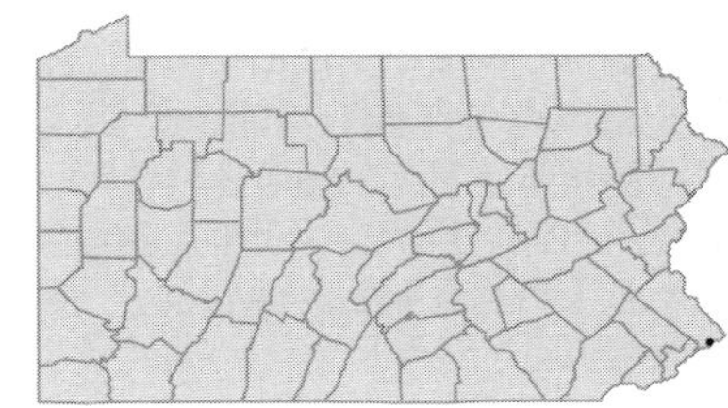

Eupatorium maculatum L.
Spotted joe-pye-weed
Herbaceous perennial
Floodplains, swamps and alluvial thickets.

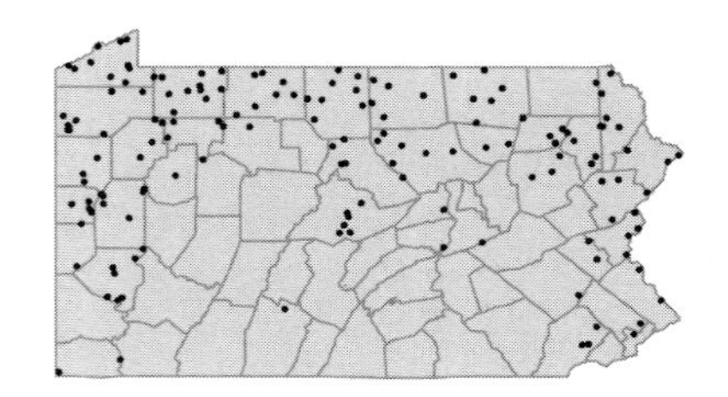

Eupatorium perfoliatum L.
Boneset
Herbaceous perennial
Floodplains, swamps, bogs, stream banks and wet meadows.

FACW+

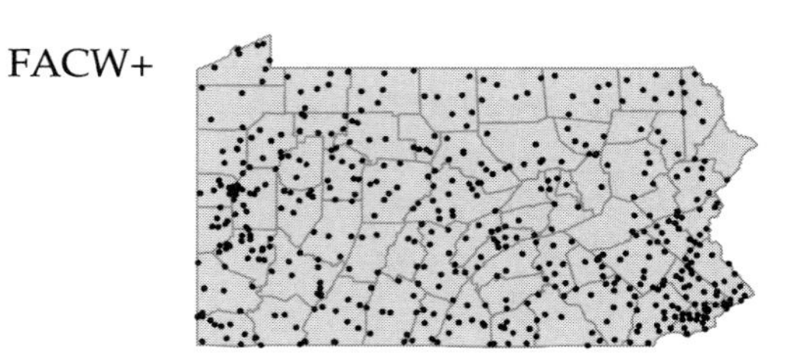

Eupatorium pilosum Walt.
Ragged eupatorium
Herbaceous perennial
Moist, rocky woods, sphagnum bogs or sandy-peaty openings.
Eupatorium rotundifolium L. var. *saundersii* (Porter) Cronq. K; *Eupatorium verbenaefolium* Michx. P

FAC-

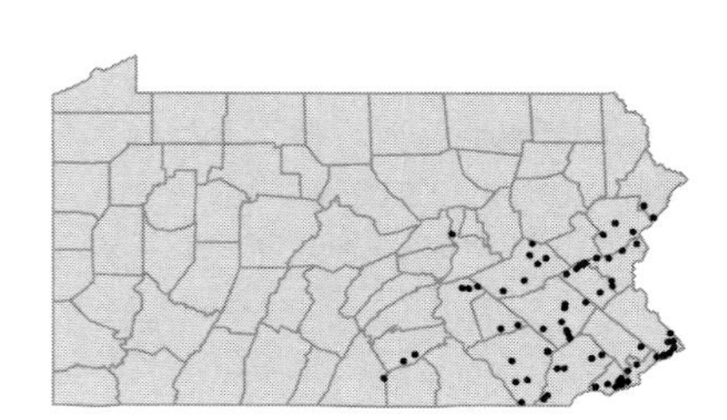

Eupatorium purpureum L.
Joe-pye-weed
Herbaceous perennial
Open woods, fields and floodplains.

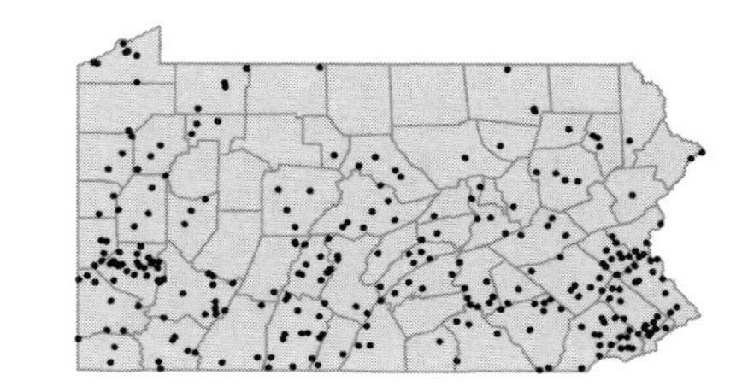

Eupatorium rotundifolium L. var. ***rotundifolium***
Round-leaved eupatorium
Herbaceous perennial
Sandy or clayey fields and open thickets.

FAC-

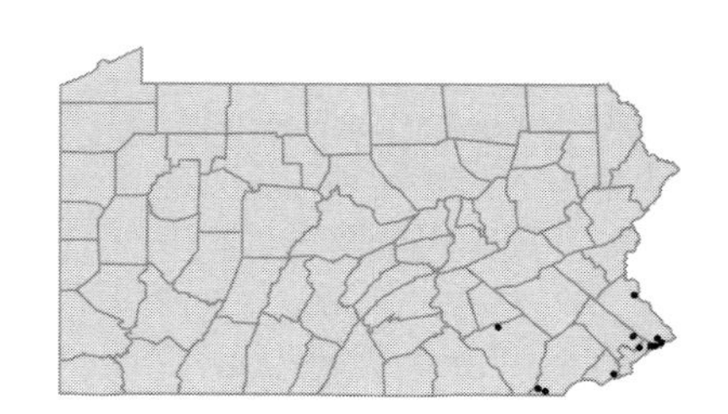

Eupatorium rotundifolium L. var. ***ovatum*** (Bigel.) Torr.
Round-leaved eupatorium
Herbaceous perennial
Dry sandy fields, exposed sandstone rocks and serpentine barrens.
Eupatorium pubescens Muhl. PFW

FAC-

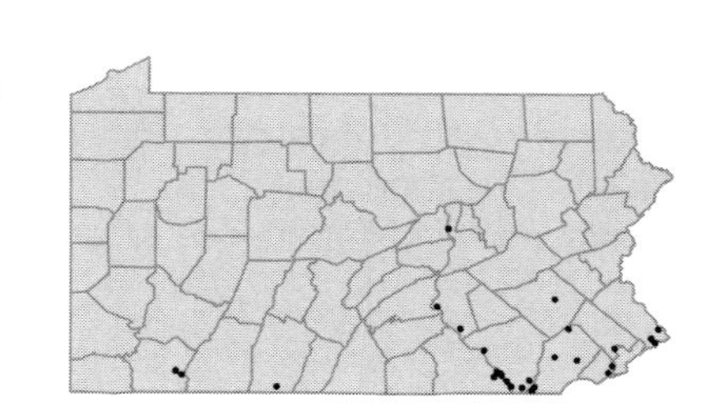

Eupatorium rugosum Houtt.
White-snakeroot
Herbaceous perennial
Woods, meadows and roadsides.

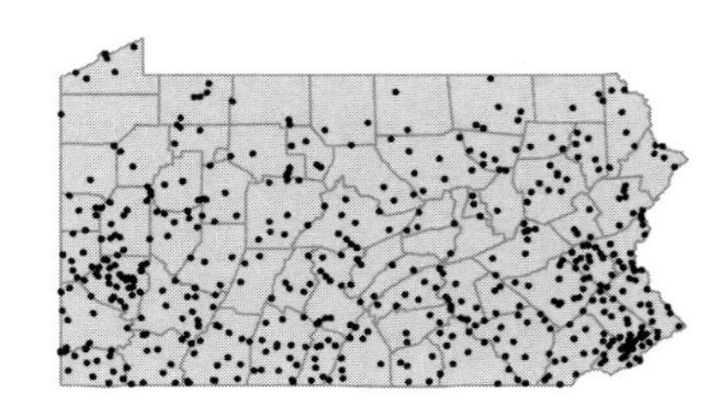

Eupatorium serotinum Michx.
Late eupatorium
Herbaceous perennial
Sandy fields, moist thickets, ditches, waste ground and ballast.

FAC-
[I]

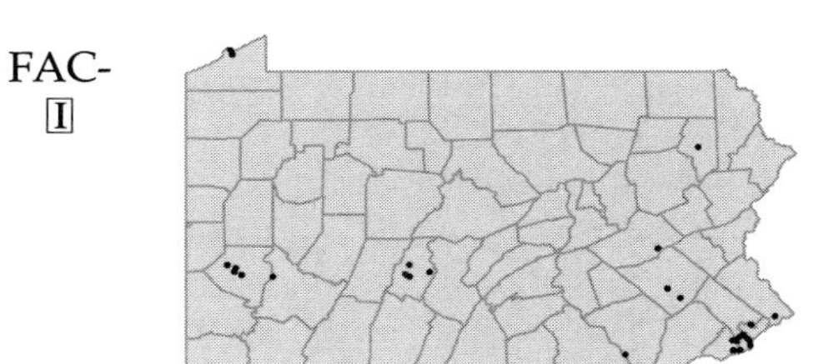

Eupatorium sessilifolium L.
Upland eupatorium
Herbaceous perennial
Dry wooded slopes, rocky banks and roadsides.

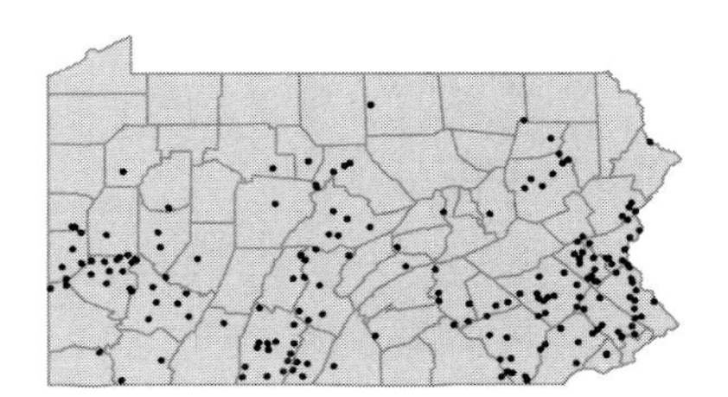

Eupatorium vaseyi Porter
Eupatorium
Herbaceous perennial
Stream banks, wooded roadside banks and shaley slopes.
Eupatorium sessilifolium L. var. *vaseyi* (Porter) Fern. & Grisc. FW; *Eupatorium album* L. var. *vaseyi* (Porter) Cronq. C

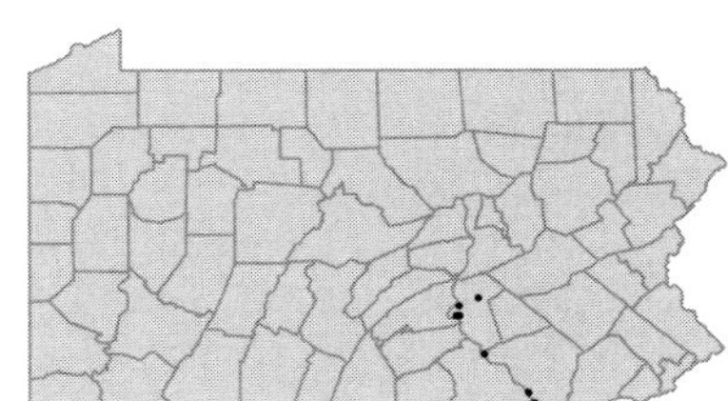

• ***Euthamia graminifolia*** (L.) Nutt. var. ***graminifolia***
Grass-leaved goldenrod; Flat-topped goldenrod
Herbaceous perennial
Fields, roadsides, moist ditches or shores.
Solidago graminifolia (L.) Salisb. var. *graminifolia* FGBW

FAC

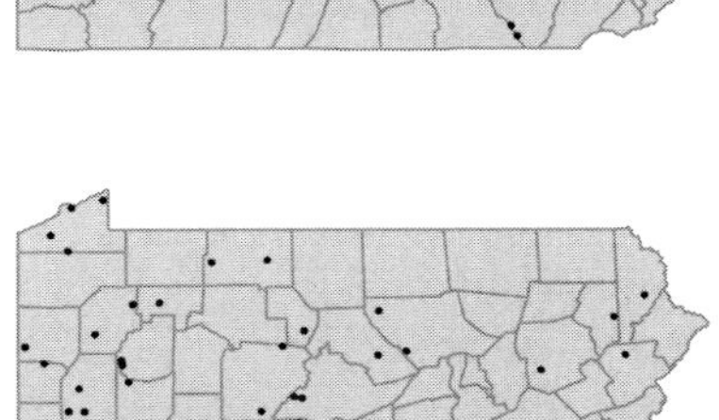

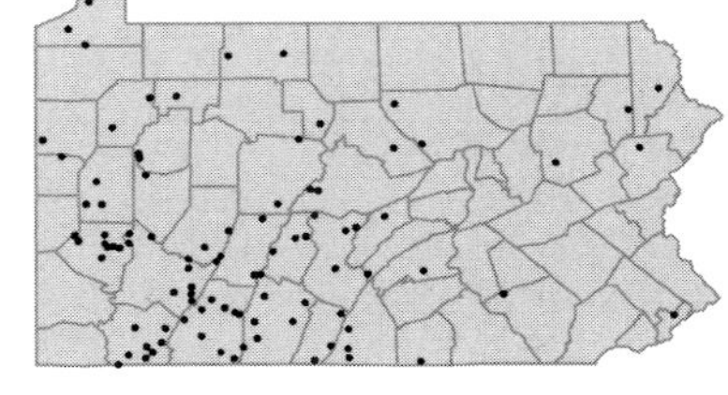

Euthamia graminifolia (L.) Nutt. var. ***nuttallii*** (Greene) W.Stone
Grass-leaved goldenrod; Flat-topped goldenrod
Herbaceous perennial
Fields, open woods, roadsides and waste ground.
Solidago graminifolia (L.) Salisb. var. *nuttallii* (Greene) Fern. FGBW

FAC

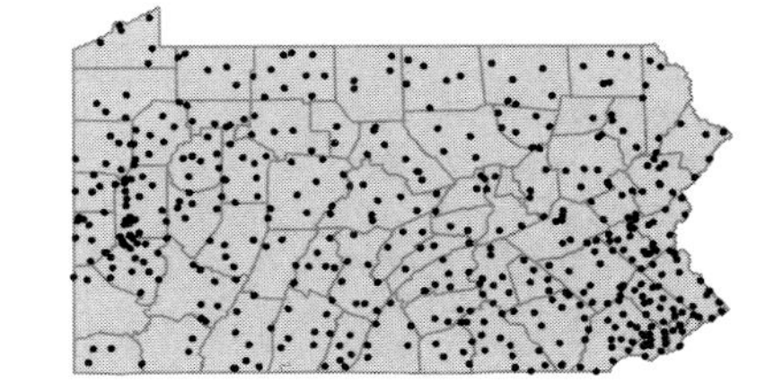

Euthamia tenuifolia (Pursh) Greene
Grass-leaved goldenrod; Coastal Plain flat-topped goldenrod
Herbaceous perennial
Moist, sandy or clayey fields or thickets.
Euthamia caroliniana (L.) Greene P; *Solidago tenuifolia* Pursh FGBW

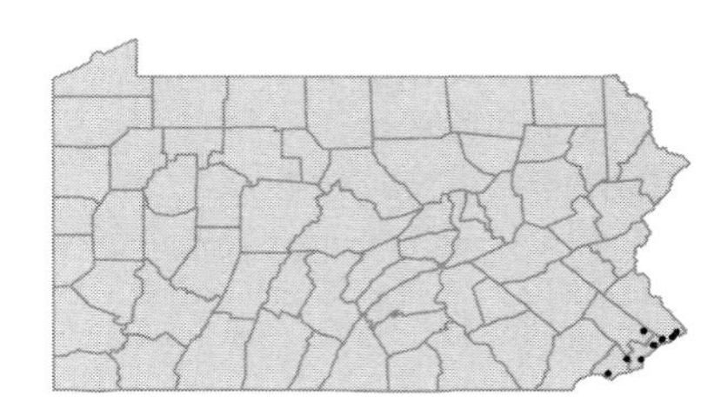

• ***Filago germanica*** (L.) Huds. I
Cotton-rose; Herba-impia
Herbaceous annual
River banks and fields. All collections are pre-1900.

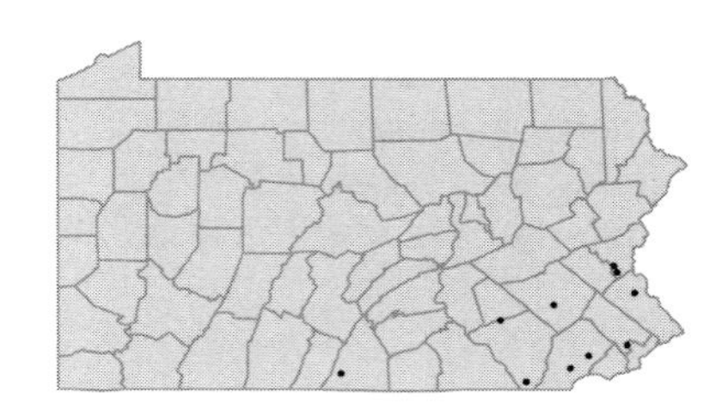

Filago minima Fries I
Cotton-rose
Herbaceous annual
Represented by a single collection from Philadelphia Co. in 1878.

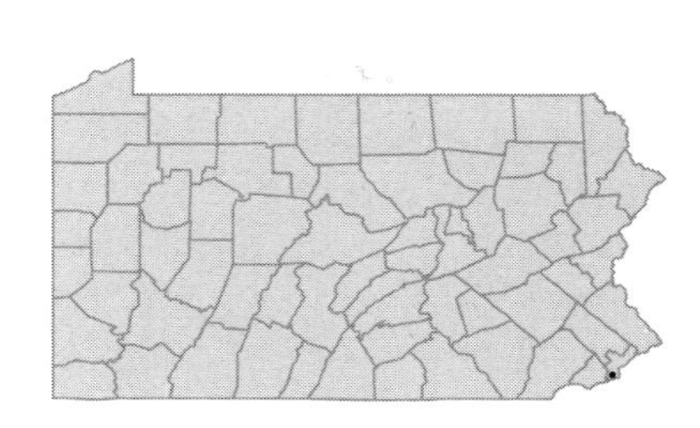

• ***Gaillardia grandiflora*** Van Houtte I
Blanket-flower
Herbaceous annual
Cultivated and occasionally naturalized in fallow fields, railroad yards or old nurseries.

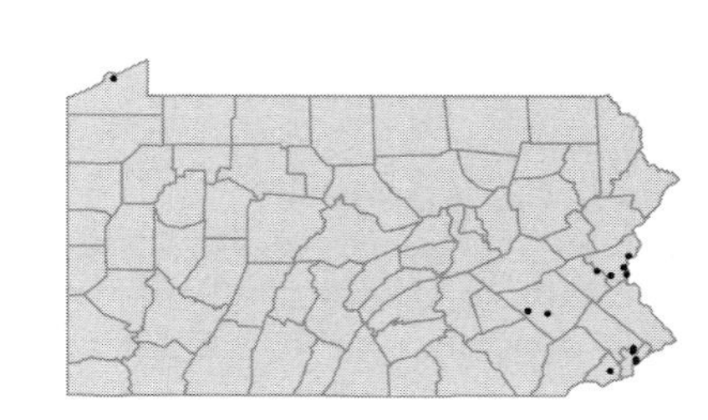

• ***Galinsoga parviflora*** Cav. I
Lesser quickweed
Herbaceous annual
Floodplains, streets, waste ground and ballast.

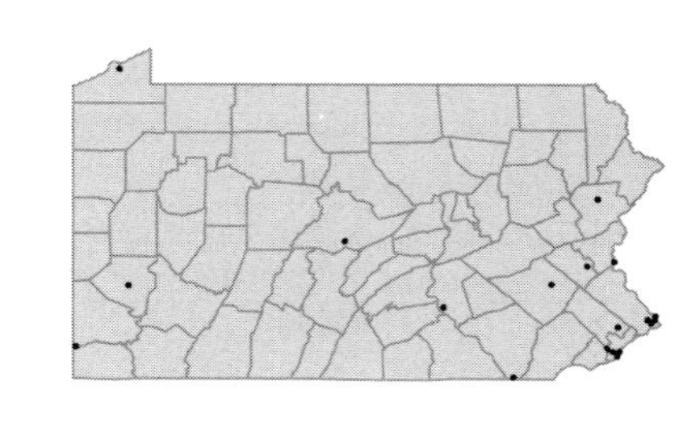

Galinsoga quadriradiata Ruiz & Pavon I
Quickweed
Herbaceous annual
Gardens, yards, roadsides and waste ground.
Galinsoga ciliata (Raf.) S.F.Blake FGBW

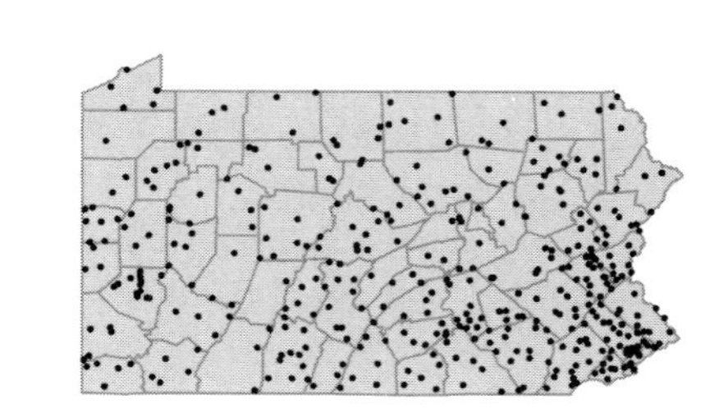

• ***Gnaphalium macounii*** Greene
Clammy cudweed; Everlasting
Herbaceous biennial
Old fields, cut-over woods, clearings and roadsides.
Gnaphalium viscosum HBK G

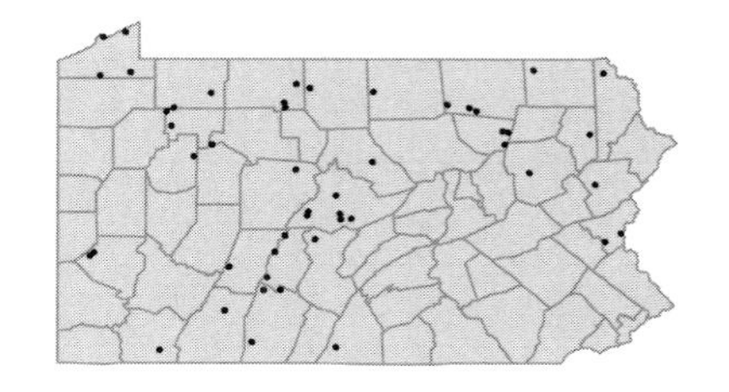

Gnaphalium obtusifolium L.
Fragrant cudweed; Catfoot
Herbaceous annual
Pastures, old fields, shale barrens and roadsides.

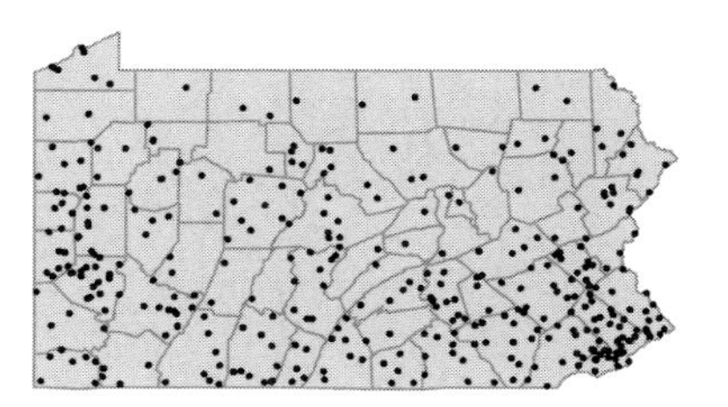

Gnaphalium purpureum L. var. ***purpureum***
Purple cudweed
Herbaceous annual
Dry, sandy or rocky fields.
Gamochaeta purpureum (L.) Cabrera K

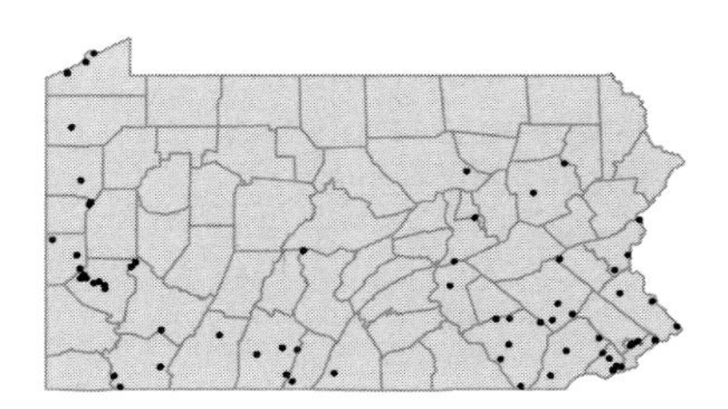

Gnaphalium sylvaticum L.
Cudweed; Everlasting
Herbaceous perennial
Dry, north-facing hillside. Represented by a single collection from Tioga Co. in 1941.

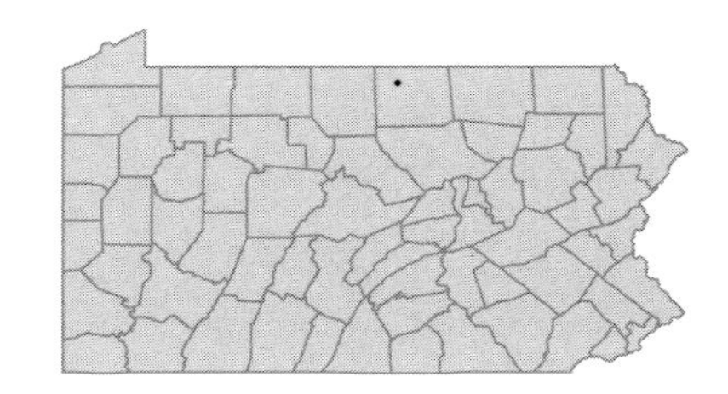

Gnaphalium uliginosum L.
Low cudweed
Herbaceous annual
Moist fields, river banks, woods and ditches.

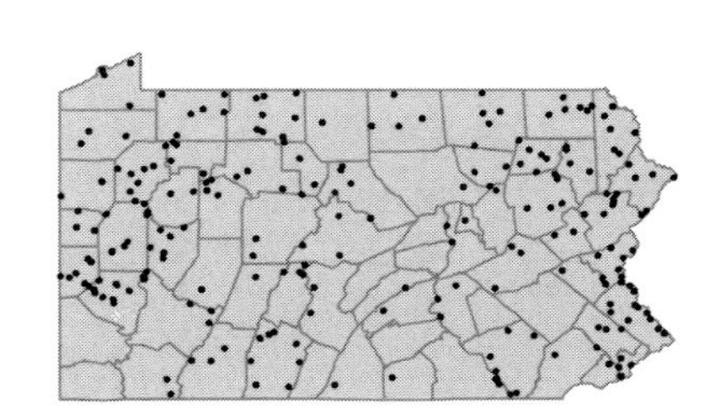

• ***Grindelia squarrosa*** (Pursh) Dunal
Curly-top gum-weed; Rosin-weed
Herbaceous biennial
Ballast, ore docks, sidewalks and lawns.

FACU
[I]

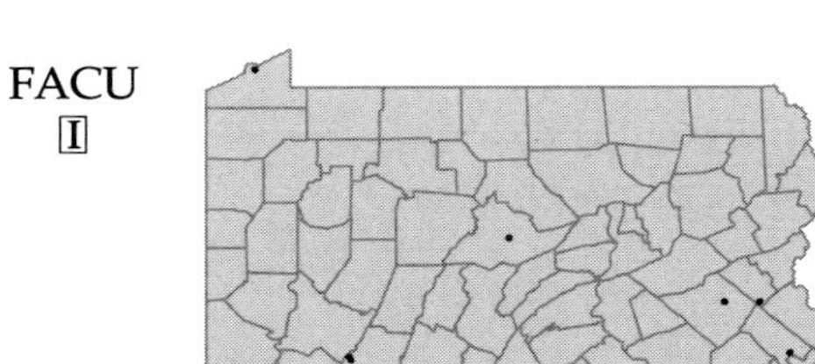

• ***Guizotia abyssinica*** (L.f.) Cass.
Ramtilla
Herbaceous annual
Waste ground, rubbish dumps and fill.

I

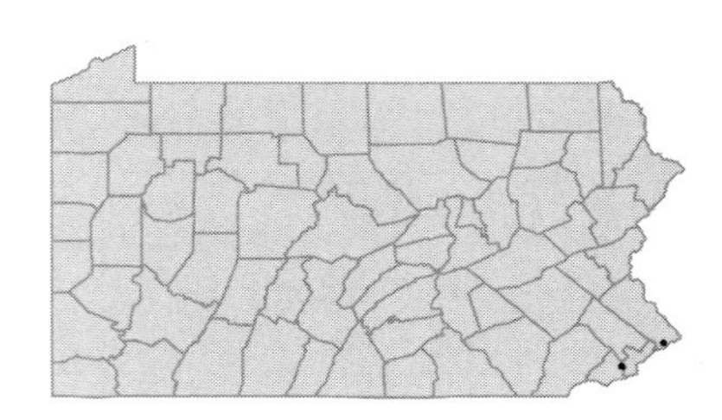

• ***Helenium amarum*** (Raf.) H.Rock
Sneezeweed
Herbaceous annual
Railroad tracks and river bank.
Helenium tenuifolium Nutt. FB

FACU-
I

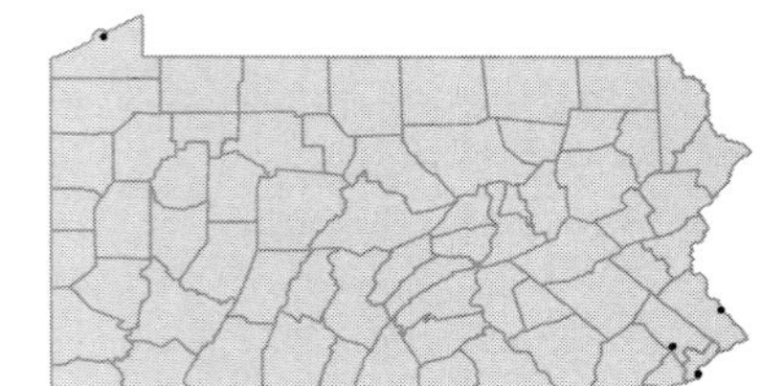

Helenium autumnale L.
Common sneezeweed
Herbaceous perennial
Swamps, moist river banks, alluvial thickets and wet fields.

FACW+

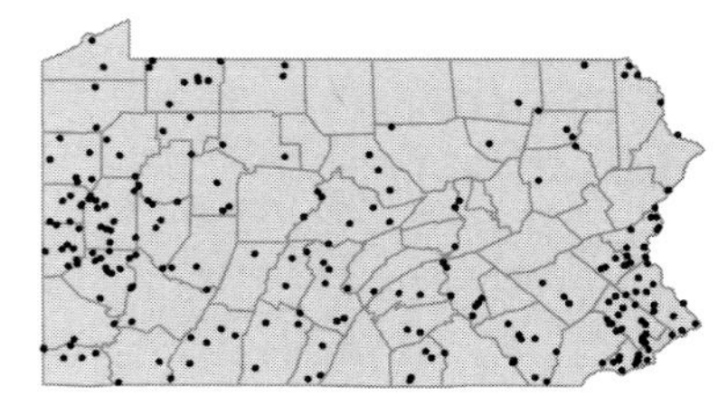

Helenium flexuosum Raf.
Southern sneezeweed
Herbaceous perennial
Moist fields, pastures shores and waste ground.
Helenium nudiflorum Nutt. PFB

FAC-
I

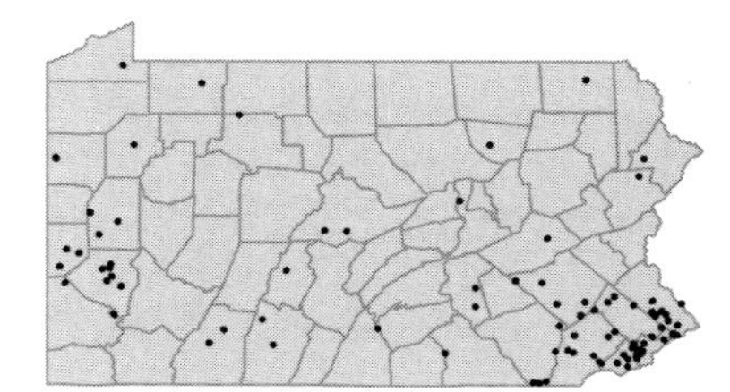

Helenium quadridentatum Labill.
Sneezeweed
Herbaceous annual
Meadows, ballast and waste ground. Represented by one collection from Philadelphia Co. (1864) and one from Chester Co. (1899).

I

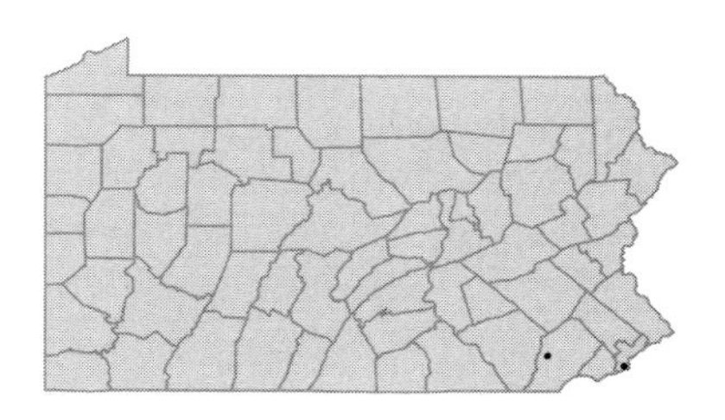

• ***Helianthus angustifolius*** L.
Swamp sunflower
Herbaceous perennial
Swamps and moist, sandy ground. Believed to be extirpated, last collected in 1934.

FACW

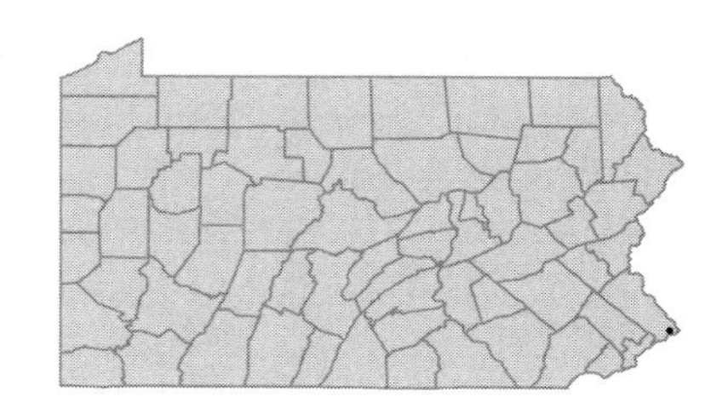

Helianthus annuus L.
Common sunflower
Herbaceous annual
Cultivated and frequently escaped to vacant lots, roadsides and rubbish dumps.

FAC-
[I]

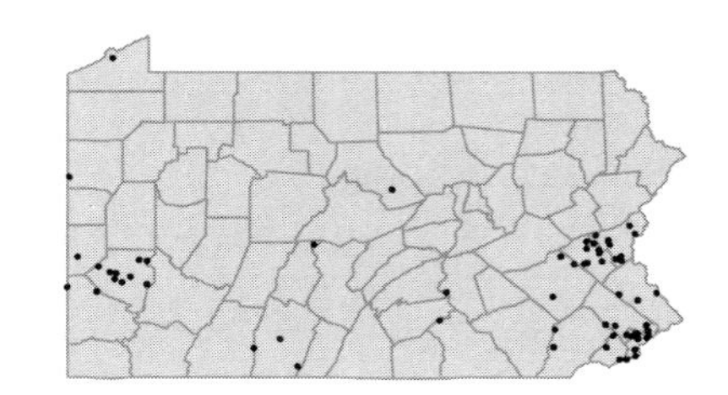

Helianthus debilis Nutt. ssp. ***cucumerifolius*** (Torr. & A.Gray) Heiser
Sunflower
Herbaceous annual
Waste ground and rubbish dumps. Represented by one collection from Philadelphia Co. (1901) and one from Montgomery Co. (1943).
Helianthus cucumerifolius Torr. & A.Gray B

UPL
[I]

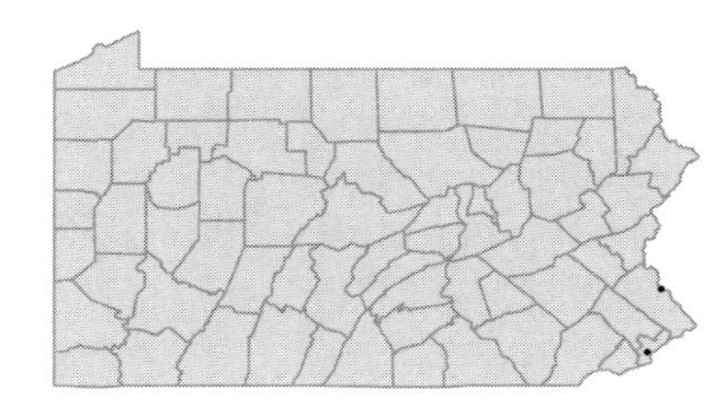

Helianthus decapetalus L.
Thin-leaved sunflower
Herbaceous perennial
Fields, moist bottomlands, stream banks and roadsides.
Helianthus trachelifolius P.Mill. P

FACU

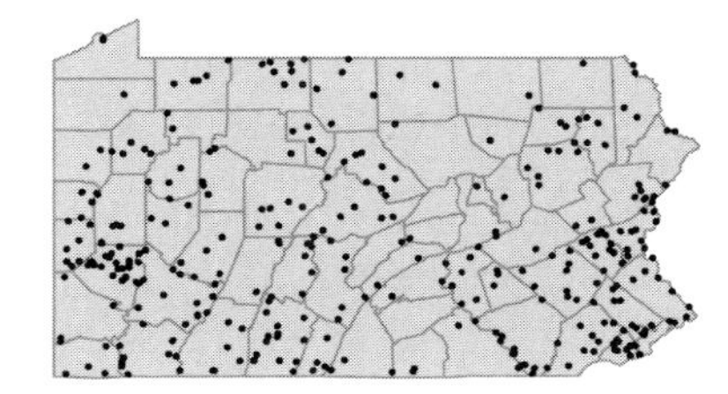

Helianthus divaricatus L.
Rough sunflower; Woodland sunflower
Herbaceous perennial
Dry, wooded slopes, shale barrens and roadsides.

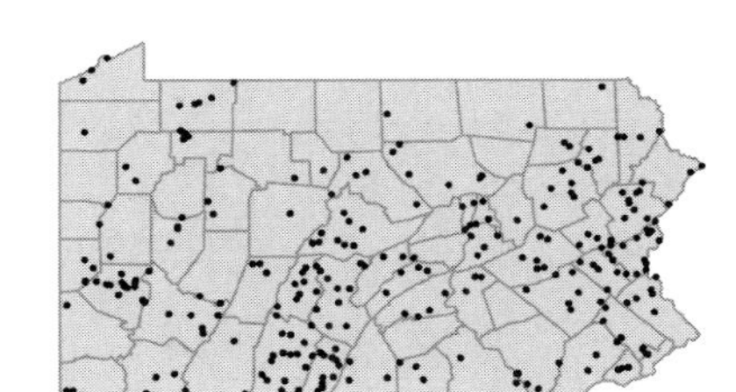

Helianthus giganteus L.
Swamp sunflower
Herbaceous perennial
Swamps, ditches and wet fields.

FACW

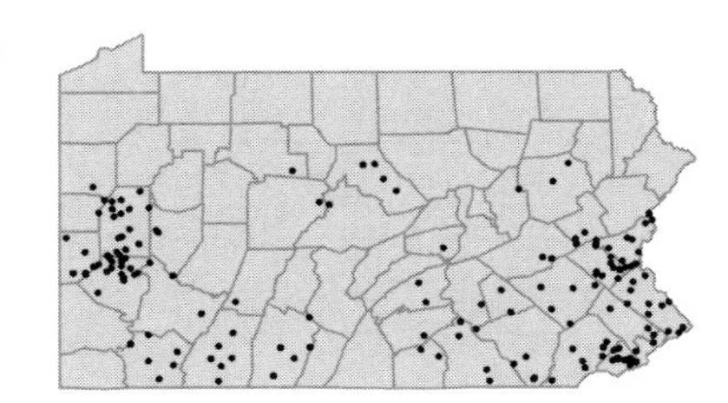

Helianthus grosseserratus Martens
Sawtooth sunflower
Herbaceous perennial
Fields, thickets and waste ground.

FACW
[I]

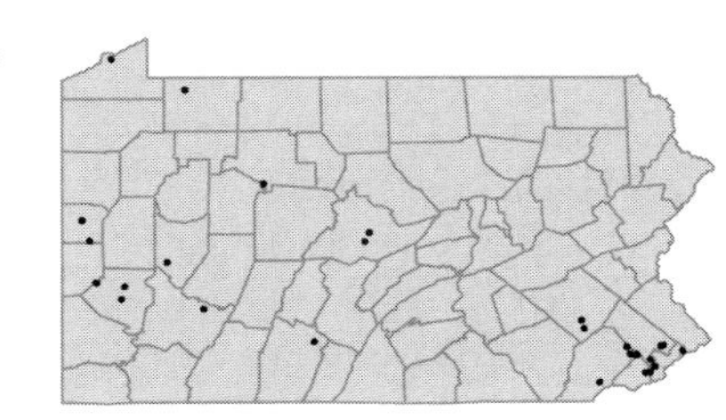

Helianthus hirsutus Raf.
Sunflower
Herbaceous perennial
Shaly slopes, upland meadows and dry roadside banks.

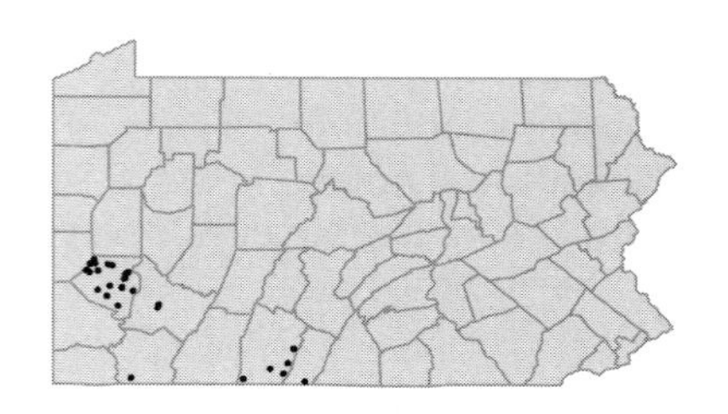

Helianthus laetiflorus Pers.
[I]
Showy sunflower
Herbaceous perennial
Old fields and stream banks. Considered by some to be a hybrid of *H. rigidus* and *H. tuberosus*.

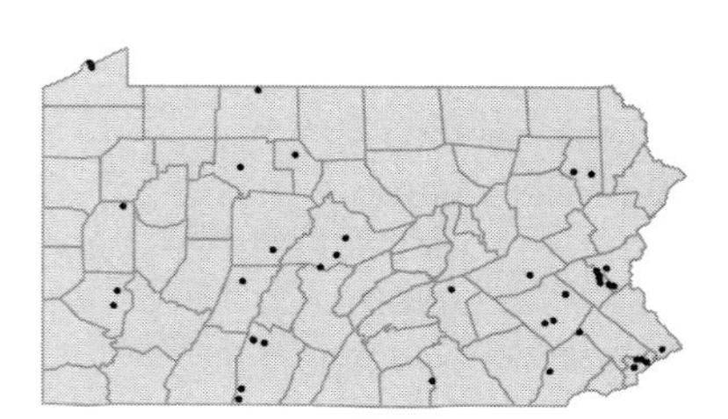

Helianthus maximilianii Schrad.
UPL
[I]
Maximilian's sunflower
Herbaceous perennial
Cultivated and occasionally escaped to old fields, railroad tracks or urban waste ground.

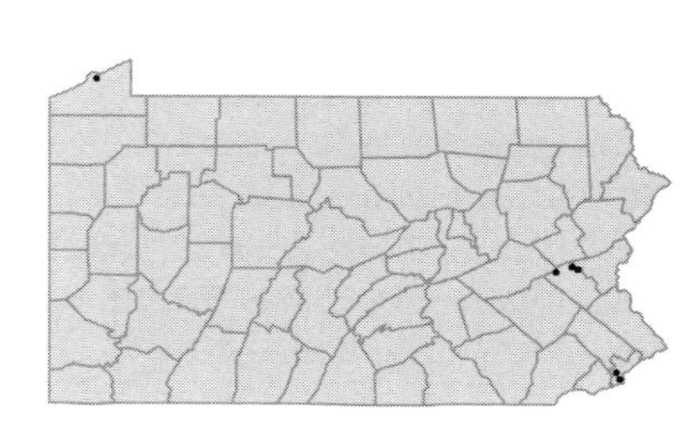

Helianthus microcephalus Torr. & A.Gray
Small wood sunflower
Herbaceous perennial
Upland woods, rocky slopes and dry roadside banks.

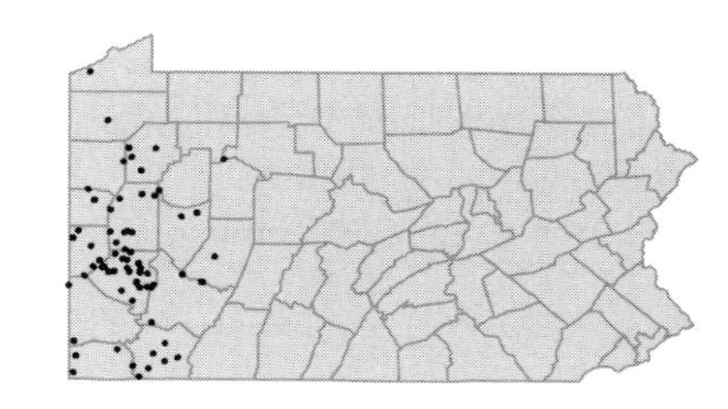

Helianthus mollis Lam.
[I]
Ashy sunflower
Herbaceous perennial
Clayey fields, shores, ballast and waste ground.

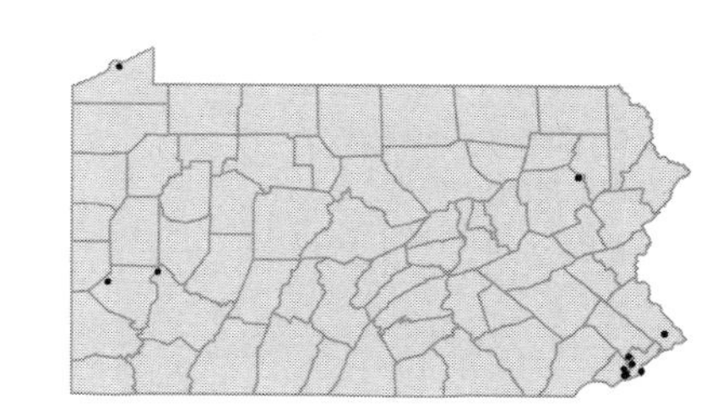

Helianthus occidentalis Riddell
UPL
Sunflower
Herbaceous perennial
Represented by a single collection from Warren Co. in 1836.

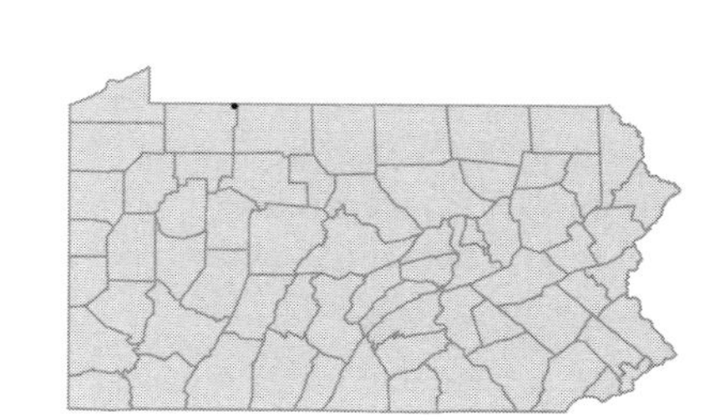

Helianthus petiolaris Nutt.
Sunflower
Herbaceous annual
Ballast, railroad tracks, rubbish dumps, wharves and vacant lots.

I

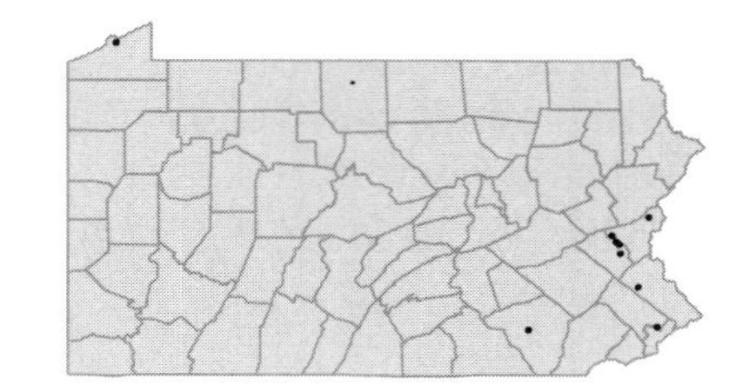

Helianthus strumosus L.
Rough-leaved sunflower
Herbaceous perennial
Fields, woods, stream banks and roadsides.

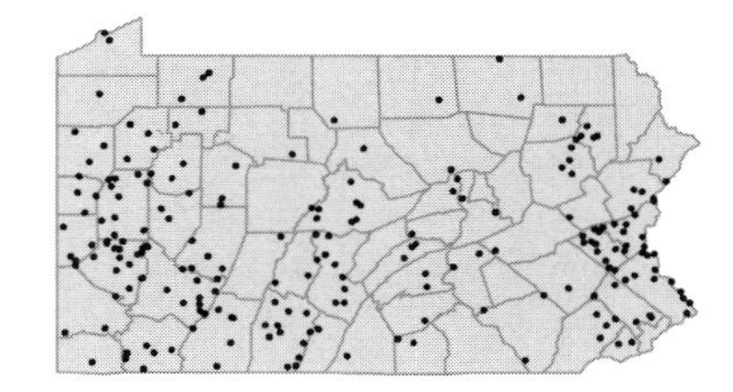

Helianthus tuberosus L.
Jerusalem artichoke
Herbaceous perennial
Cultivated and often naturalized in fields, woods and along railroad tracks and roads.

FAC
I

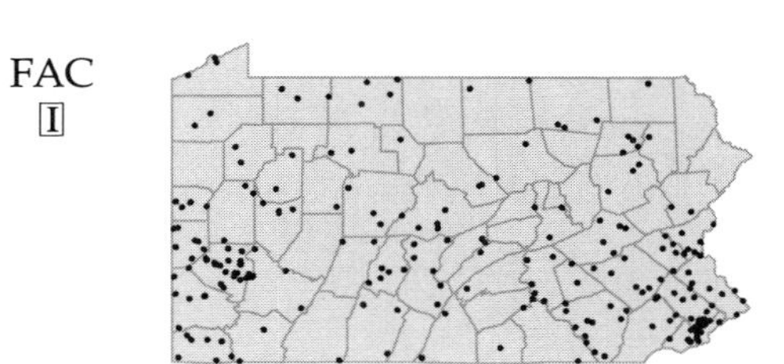

• ***Heliopsis helianthoides*** (L.) Sweet
Ox-eye
Herbaceous perennial
Fields, woods, floodplains and stream banks.

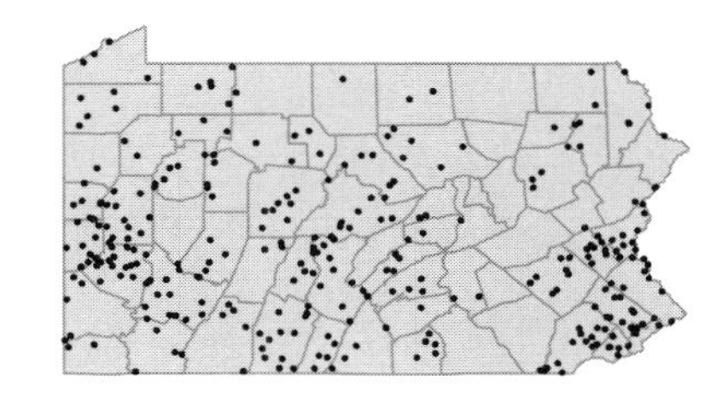

• ***Heterotheca subaxillaris*** (Lam.) Britt. & Rusby
Camphorweed
Herbaceous annual
Ballast, waste ground and sandy dredge spoil.

UPL
I

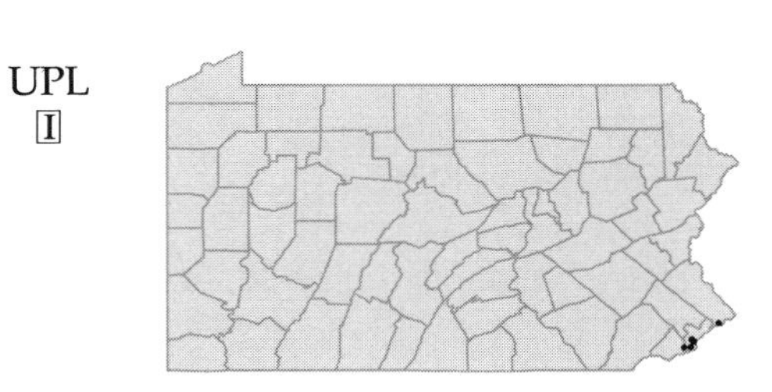

Hieracium—In addition to the taxa treated below, the following hybrids have been collected at least once in Pennsylvania: *kalmii x paniculatum* (*H. x mendicum* Lepage); *kalmii x scabrum* (*H. x fassettii* Lepage).

• ***Hieracium aurantiacum*** L.
Devil's paintbrush; Orange hawkweed
Herbaceous perennial
Abandoned fields, meadows and pastures.

I

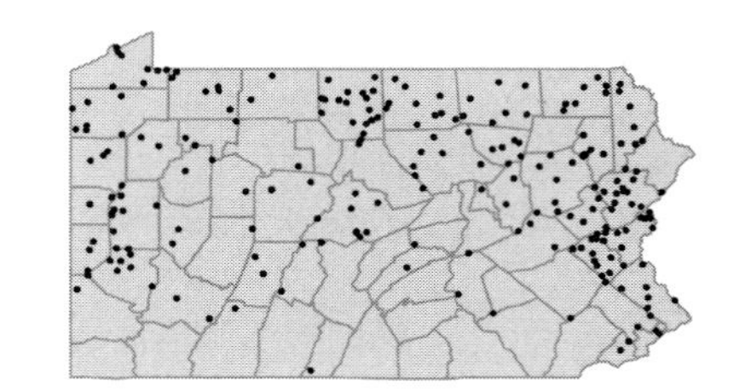

Hieracium caespitosum Dumort.
King-devil
Herbaceous perennial
Woods, fields and roadsides.
Hieracium pratense Tausch FGBW

I

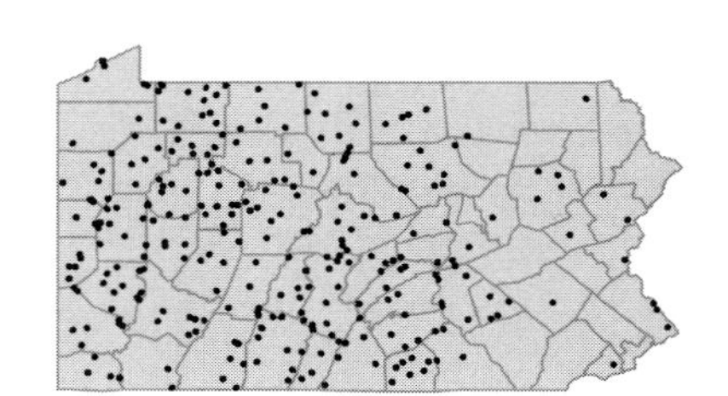

Hieracium flagellare Willd.
Hawkweed
Herbaceous perennial
Weedy fields and roadsides.
Hieracium x flagellare K

I

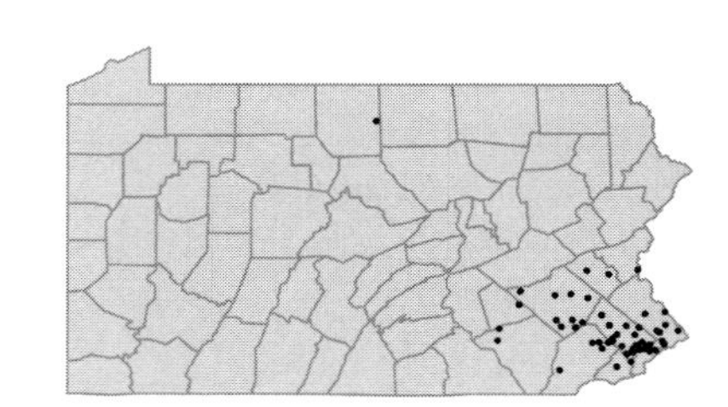

Hieracium gronovii L.
Hawkweed
Herbaceous perennial
Dry, open woods or thickets, usually in sandy soil.

UPL

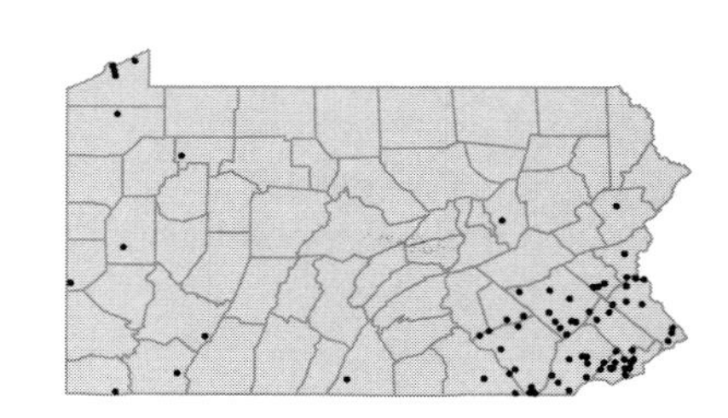

Hieracium gronovii x venosum
Hawkweed
Herbaceous perennial
Dry, open woods.
Hieracium x marianum Willd. FGBWC

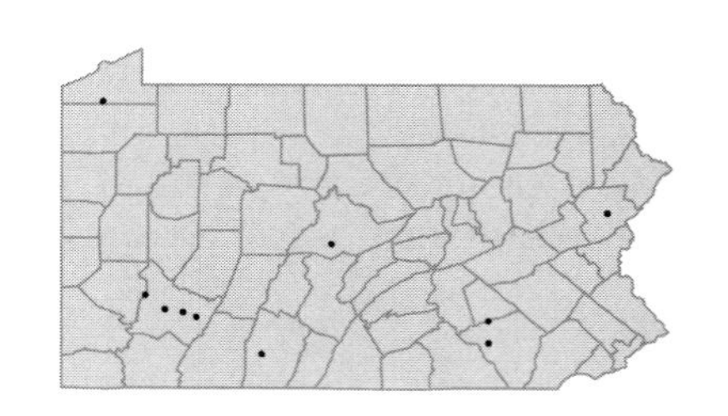

Hieracium kalmii L.
Canada hawkweed
Herbaceous perennial
Thickets, clearings and roadside banks.
Hieracium canadense sensu Torr. & A.Gray *non* Michx. *pro parte* WFB

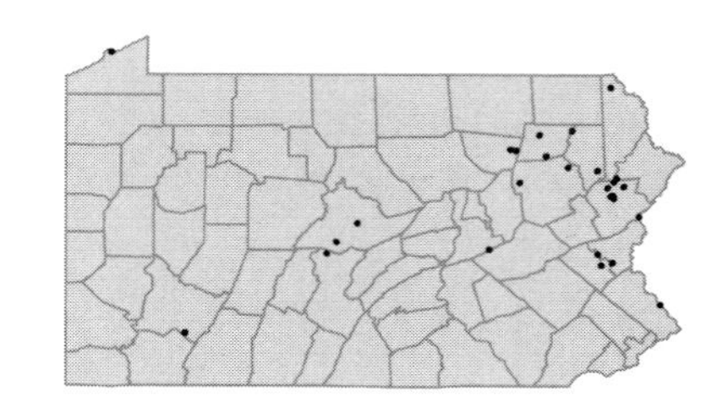

Hieracium lachenalii K.C.Gmel.
European hawkweed
Herbaceous perennial
Dry woods, grassy slopes, roadsides and lawns.
Hieracium vulgatum Fries FGBW

[I]

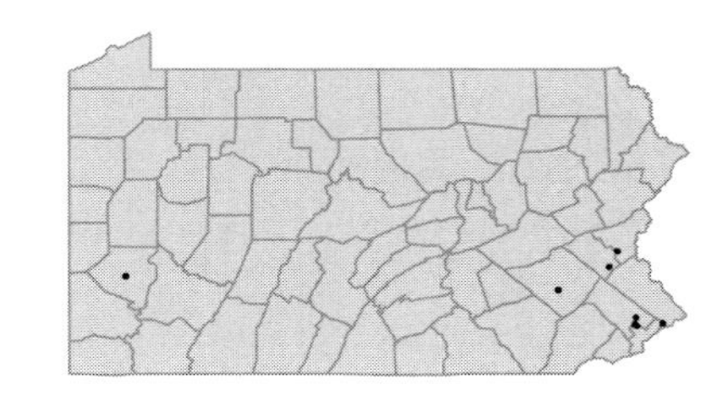

Hieracium murorum L.
Wall hawkweed; Golden lungwort
Herbaceous perennial
Rich, dry woods and shaded roadside banks.

[I]

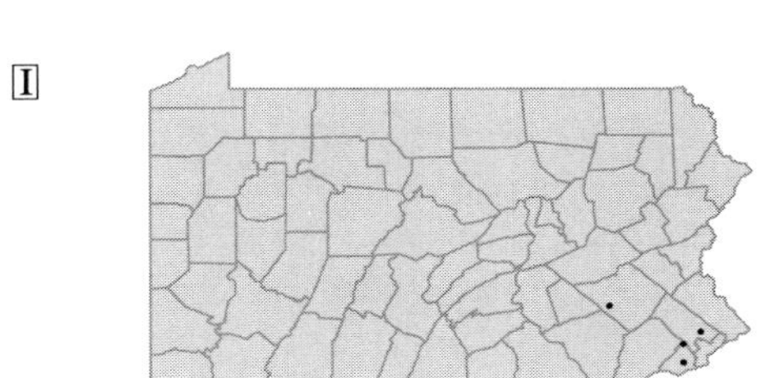

Hieracium paniculatum L.
Hawkweed
Herbaceous perennial
Dry, rocky or sandy woods or slopes.

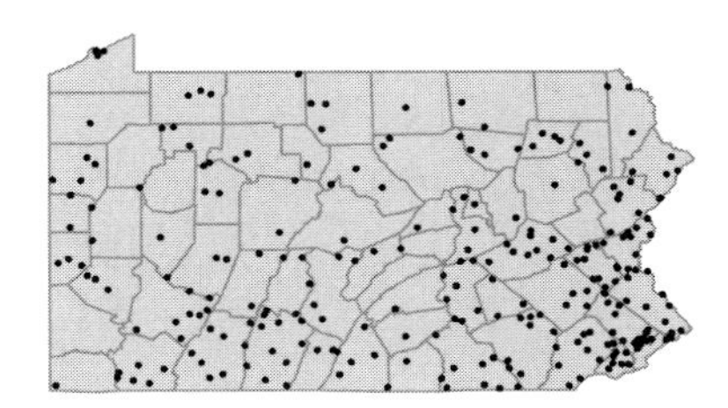

Hieracium pilosella L.
Mouse-ear hawkweed
Herbaceous perennial
Dry fields, pastures, roadsides and lawns.

[I]

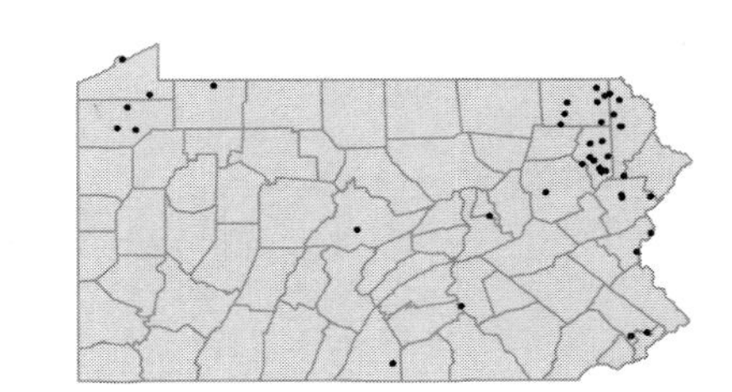

Hieracium piloselloides Vill.
King-devil
Herbaceous perennial
Dry fields, meadows and roadsides.
Hieracium florentinum All. FGBW

[I]

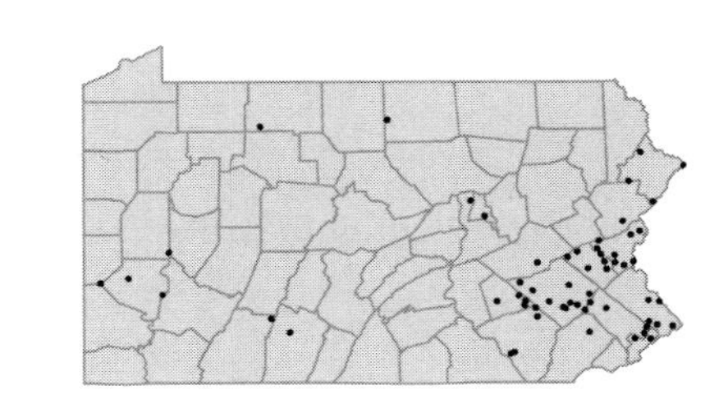

Hieracium sabaudum L.
Hawkweed
Herbaceous perennial
Alluvium and waste ground.

[I]

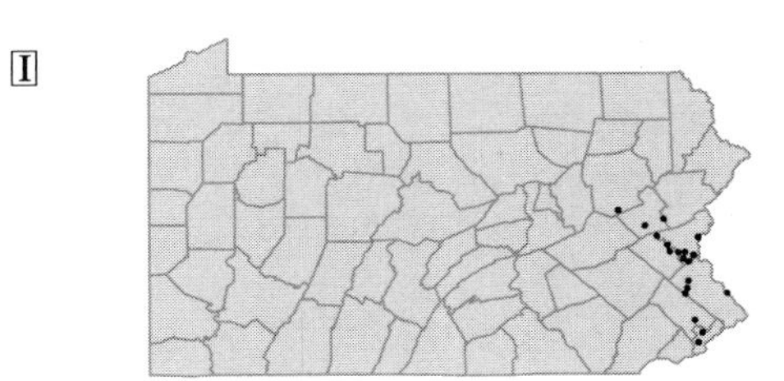

Hieracium scabrum Michx.
Hawkweed
Herbaceous perennial
Dry, open ground of fields, clearings and woods edges.

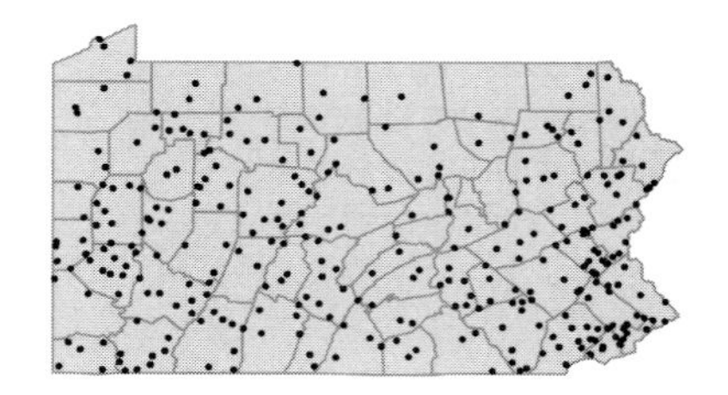

Hieracium traillii Greene
Green's hawkweed; Maryland hawkweed
Herbaceous perennial
Dry slopes, bluffs and shale barrens.
Hieracium greenii Porter & Britt. P

FACU

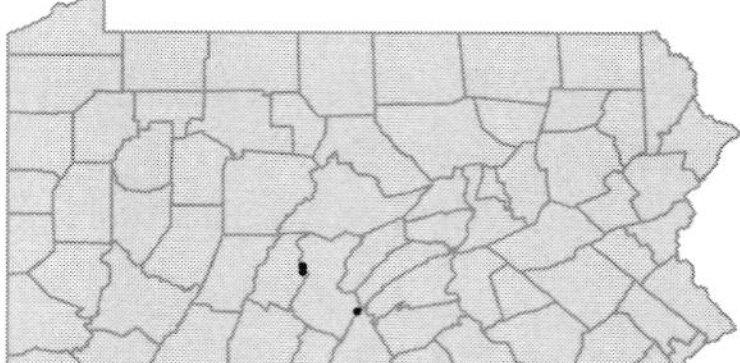

Hieracium venosum L.
Rattlesnake-weed
Herbaceous perennial
Upland woods, wooded slopes and edges.

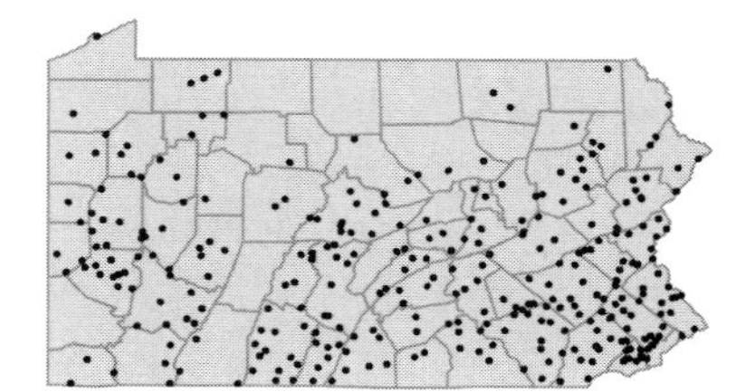

• ***Hypochoeris glabra*** L.
Cat's-ear
Herbaceous perennial
Ballast and waste ground. Represented by a single collection from Philadelphia Co. in 1865.

I

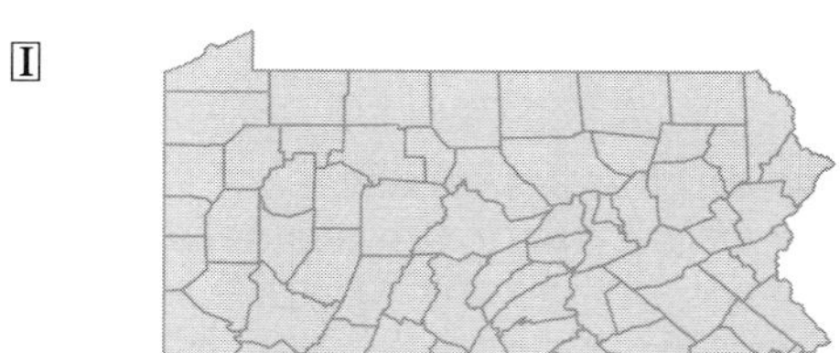

Hypochoeris radicata L.
Cat's-ear
Herbaceous perennial
Lawns, roadsides and urban waste ground.

I

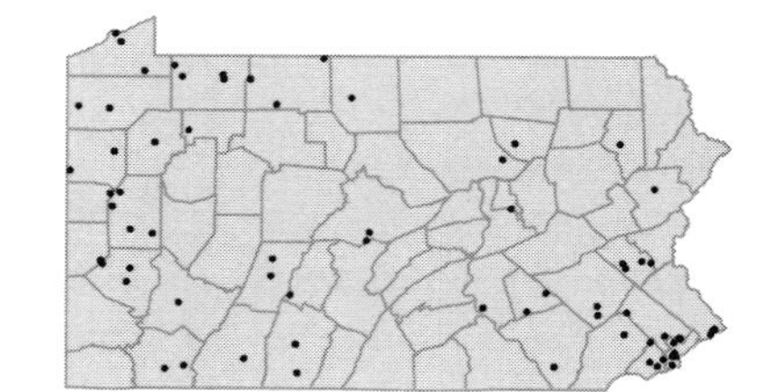

• ***Inula helenium*** L.
Elecampane
Herbaceous perennial
Cultivated and occasionally naturalized in pastures, roadsides and waste ground.

I

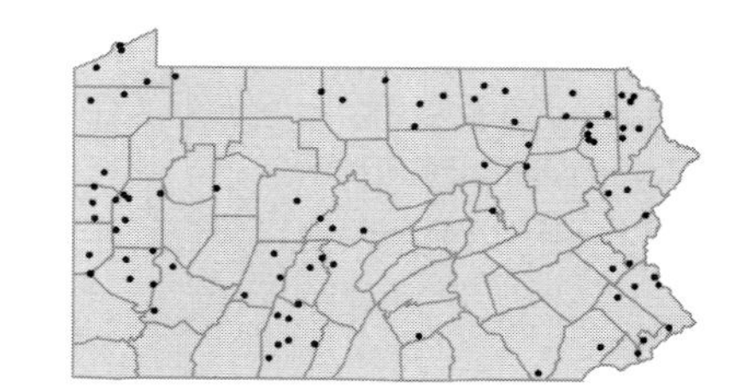

• ***Iva annua*** L.
Rough marsh-elder
Herbaceous annual
Moist, open ditch.

FAC
[I]

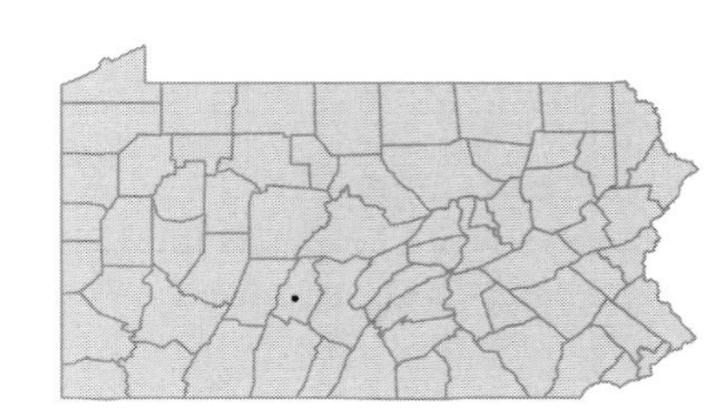

Iva frutescens L. ssp. ***oraria*** (Bartlett) R.C.Jackson
Marsh-elder; Highwater-shrub
Deciduous shrub
Ballast ground. Represented by a single collection from Philadelphia Co. in 1865.

FACW+

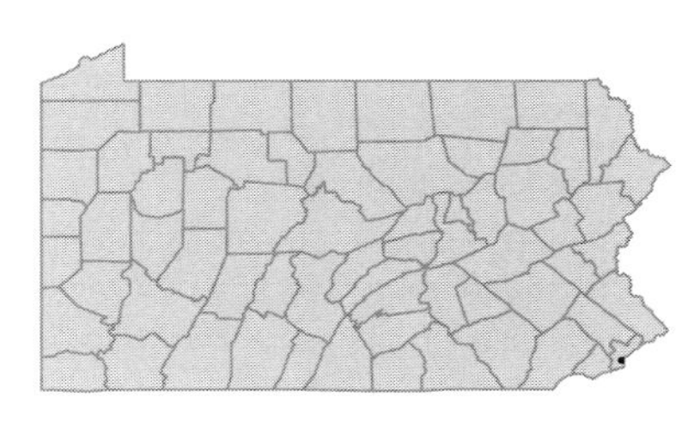

Iva xanthifolia Nutt.
Marsh-elder
Herbaceous annual
Docks, railroad embankments, rubbish dumps, vacant lots and fill.

FAC
[I]

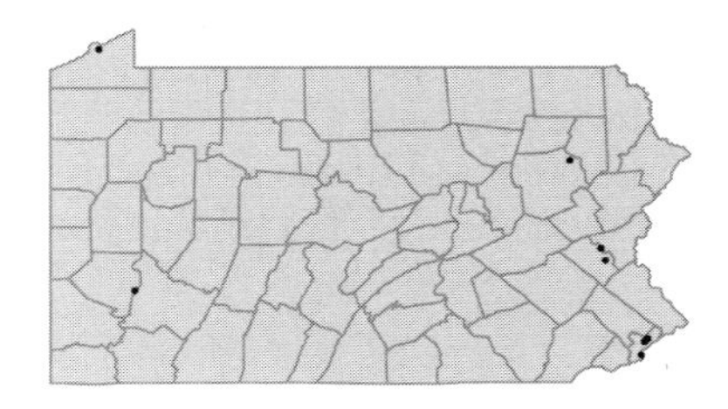

• ***Ixeris stolonifera*** A.Gray
Creeping lettuce
Herbaceous perennial
Lawns, gardens and nursery beds.
Lactuca stolonifera (A.Gray) Maxim. F

[I]

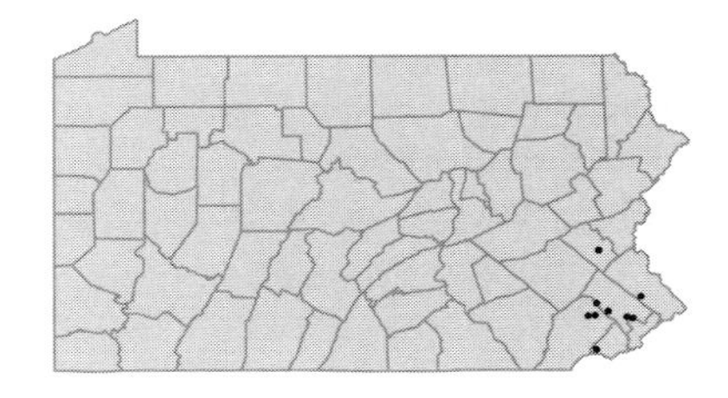

• ***Krigia biflora*** (Walt.) S.F.Blake
Dwarf dandelion; Two-flowered cynthia
Herbaceous perennial
Fields, meadows, woods and sandy banks.

FACW

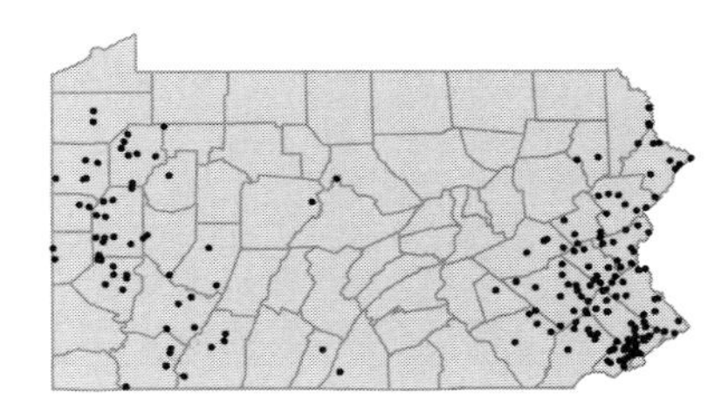

Krigia virginica (L.) Willd.
Dwarf dandelion
Herbaceous annual
Dry, rocky slopes and shale barrens.

UPL

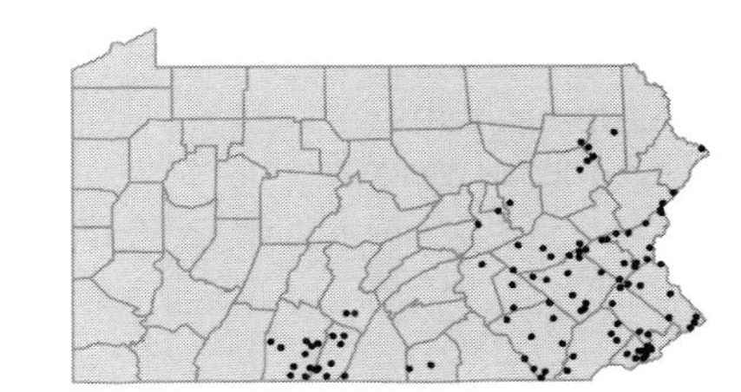

• ***Lactuca biennis*** (Moench) Fern. FACU
Blue lettuce
Herbaceous annual
Woods, stream banks, roadsides and vacant lots in moist, open soil.
Lactuca spicata (Lam.) A.S.Hitchc. P

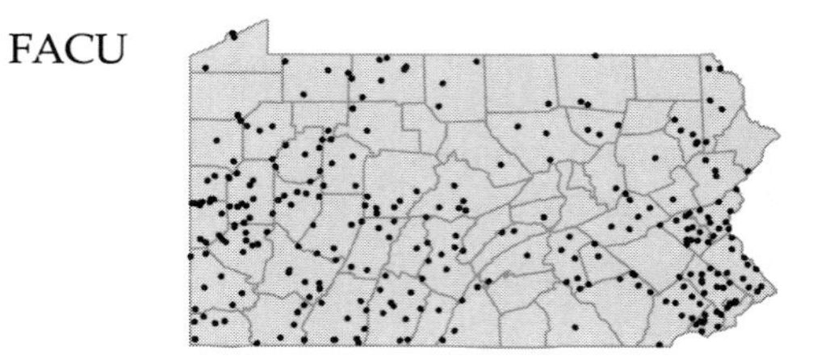

Lactuca canadensis L. var. ***canadensis*** FACU-
Wild lettuce
Herbaceous annual
Meadows, fields, rocky hillsides and roadside banks.
Lactuca canadensis L. var. *montana* Britt. *pro parte* P; *Lactuca sagittifolia* Ell. *pro parte* P

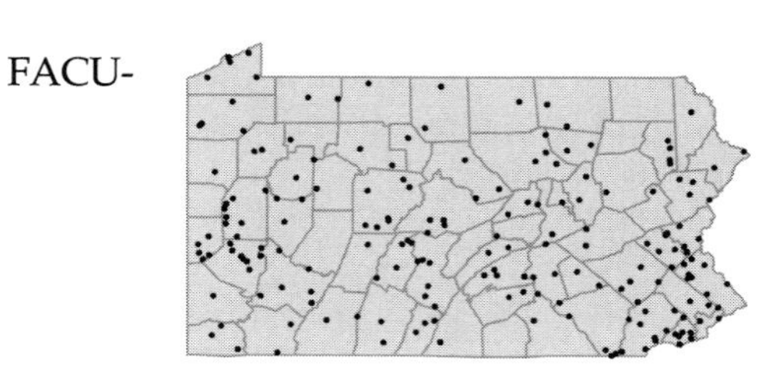

Lactuca canadensis L. var. ***latifolia*** Kuntze FACU-
Wild lettuce
Herbaceous annual
Woods, wooded slopes, thickets and roadsides.

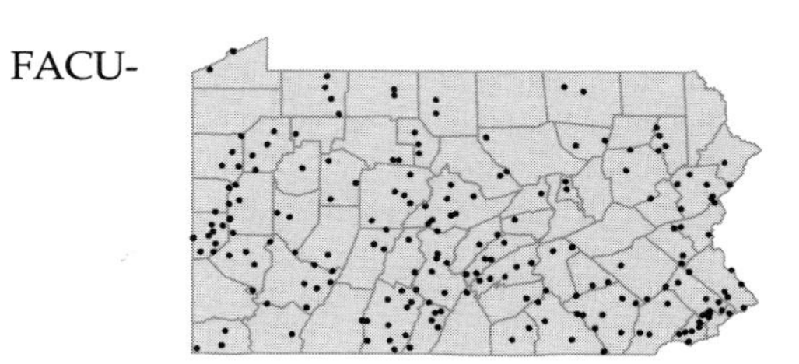

Lactuca canadensis L. var. ***longifolia*** (Michx.) Farw. FACU-
Wild lettuce
Herbaceous annual
Fields, woods, moist thickets, sandy barrens and roadsides.

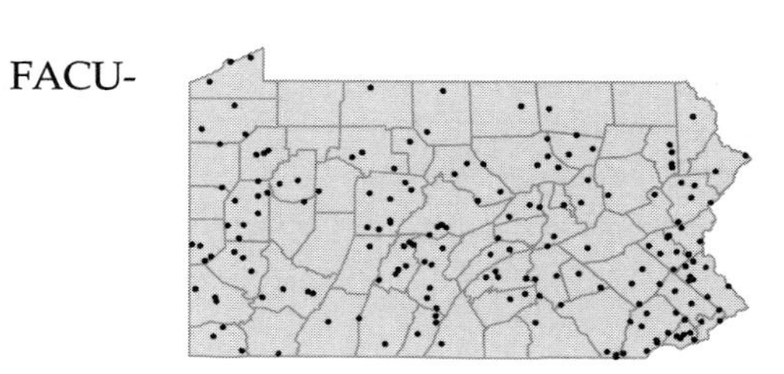

Lactuca canadensis L. var. ***obovata*** Wieg. FACU-
Wild lettuce
Herbaceous annual
Old fields, meadows, woods and roadside banks.

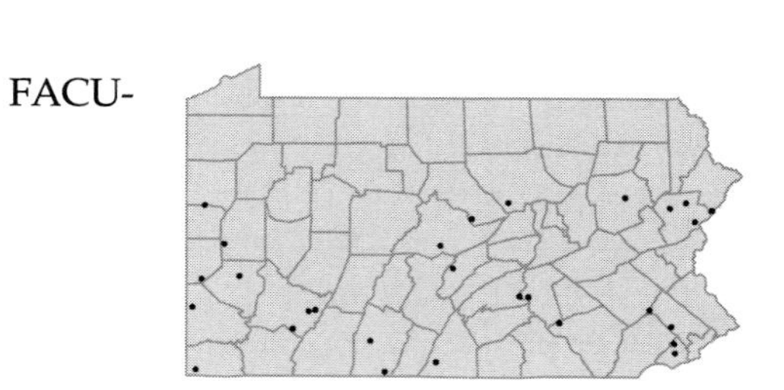

Lactuca floridana (L.) Gaertn. var. ***floridana*** FACU-
Woodland lettuce
Herbaceous annual
Rich, wooded slopes, meadows and roadsides.

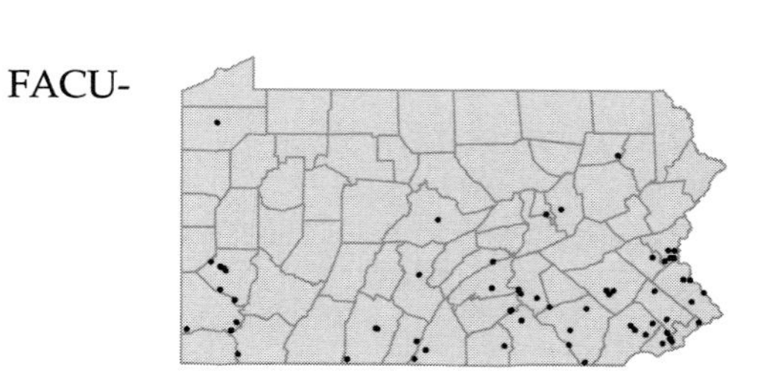

Lactuca floridana (L.) Gaertn. var. ***villosa*** (Jacq.) Cronq.
Woodland lettuce
Herbaceous annual
Rich woods and roadsides.
Lactuca villosa Jacq. PC

FACU-

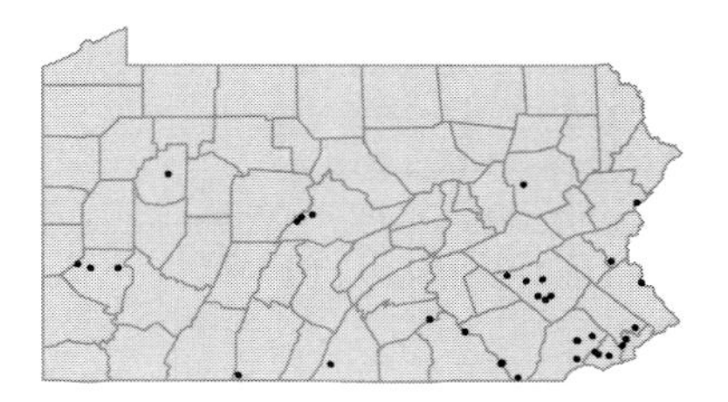

Lactuca hirsuta Muhl. var. ***hirsuta***
Downy lettuce
Herbaceous annual
Limestone woods, clearings and alluvial bottomlands.

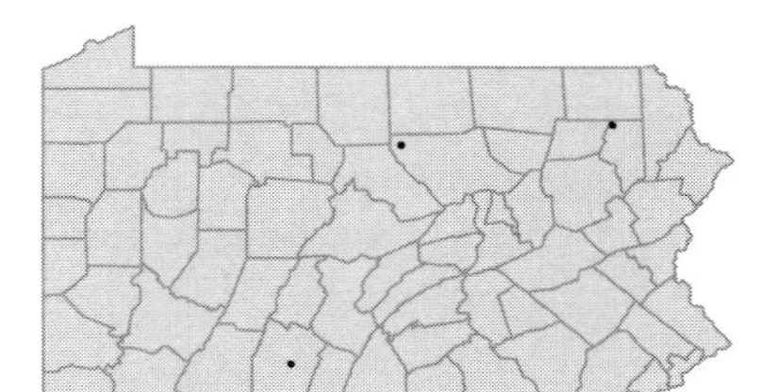

Lactuca hirsuta Muhl. var. ***sanguinea*** (Bigel.) Fern.
Downy lettuce
Herbaceous annual
Dry, open woods, thickets and rocky ledges.

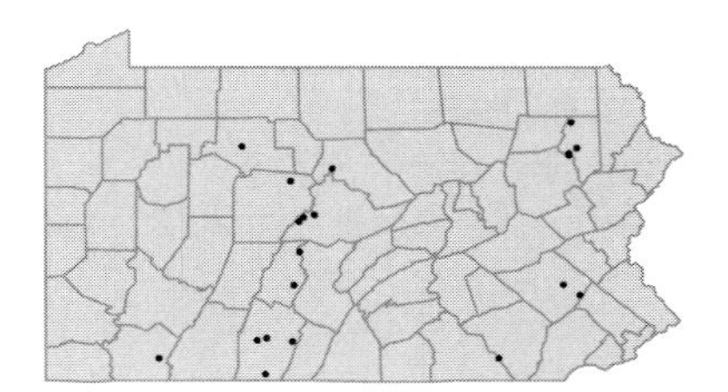

Lactuca pulchella (Pursh) DC.
Blue lettuce
Herbaceous perennial
Represented by a single collection from Berks Co. in 1888.
Lactuca tartarica (L.) C.A.Mey ssp. *pulchella* (Pursh) Stebbins

FAC
I

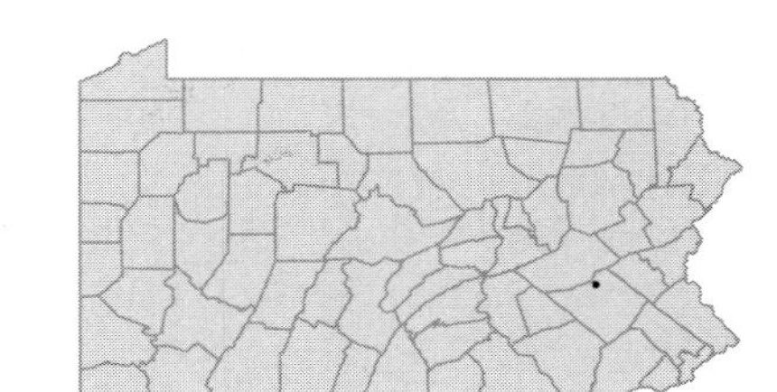

Lactuca saligna L.
Willow-leaf lettuce
Herbaceous annual
Woods, fields, strip mine areas, roadsides and a limestone quarry.

UPL
I

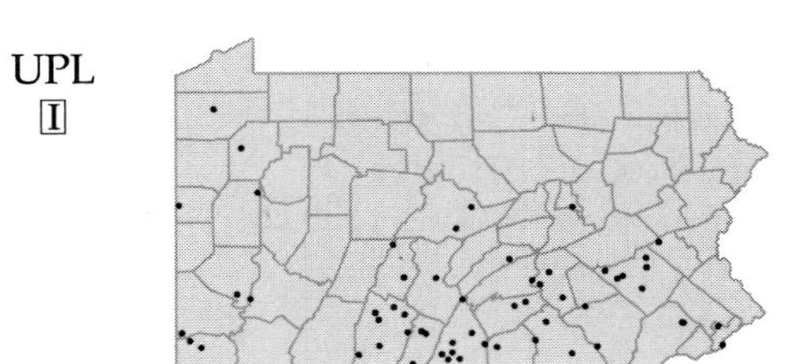

Lactuca sativa L.
Garden lettuce
Herbaceous annual
Cultivated and occasionally escaped to roadsides, railroad banks, dumps and waste ground.

I

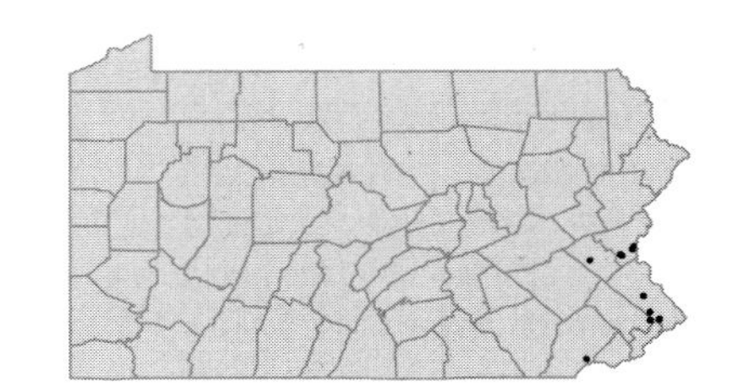

Lactuca serriola L.
Prickly lettuce
Herbaceous annual
Fields, woods, strip mines, roadsides and dumps.
Lactuca scariola L. PFW

FAC-
[I]

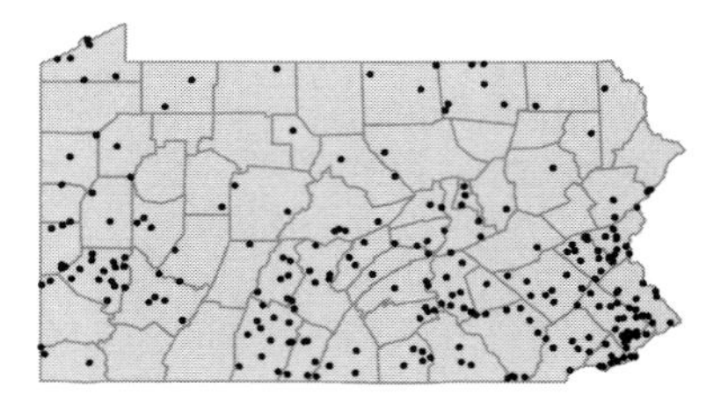

• ***Lapsana communis*** L.
Nipplewort
Herbaceous annual
Moist woods, roadsides, waste ground and ballast.

[I]

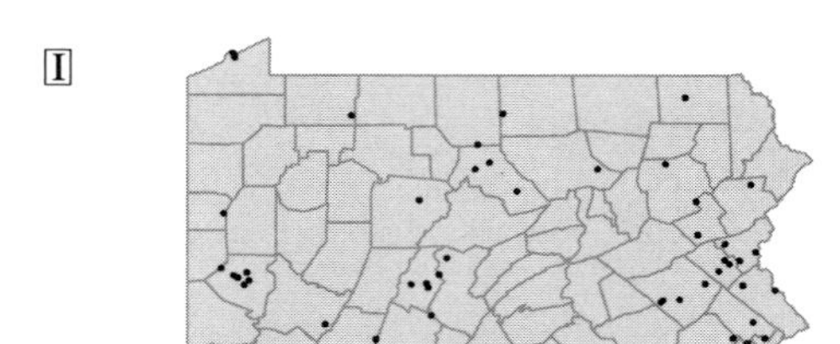

• ***Leontodon autumnalis*** L.
Fall-dandelion
Herbaceous perennial
Lawns, roadsides, waste ground and ballast.

[I]

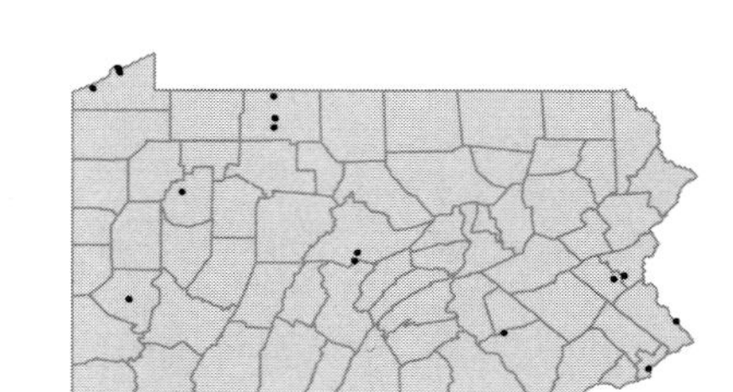

Leontodon taraxacoides (Vill.) Merat
Hawkbit
Herbaceous perennial
Lawns, grassy banks, dry yards, rubbish dumps and ballast.
Leontodon leysseri (Wallr.) G.Beck FGBW

FACU
[I]

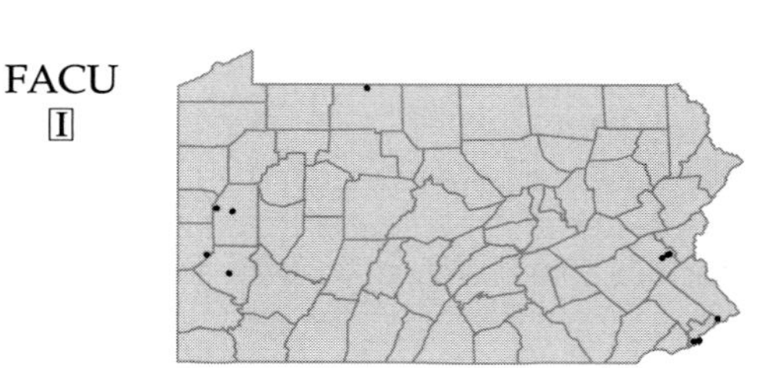

• ***Liatris pycnostachya*** Michx.
Blazing-star; Gay-feather
Herbaceous perennial
Moist ground along railroad tracks. Represented by a single collection from Chester Co. in 1929.

[I]

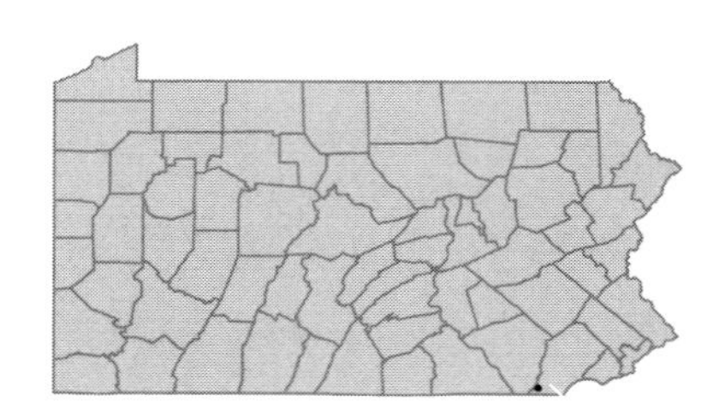

Liatris scariosa (L.) Willd. *sensu lato*
Northern blazing-star
Herbaceous perennial
Dry woods, shaly slopes and barrens. More work is needed to determine which varieties are present in the state.
Laciniaria scariosa (L.) Hill P; *Liatris borealis* Nutt. *pro parte* FW; *Liatris novae-angliae* (Lunell) Shinners var. *nieuwlandii* (Lunell) Shinners *pro parte* GB

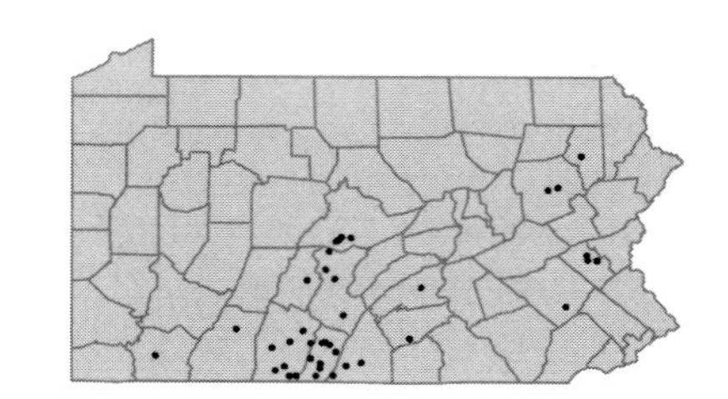

Liatris spicata (L.) Willd.
Blazing-star
Herbaceous perennial
Moist fields, fencerows and roadsides.
Laciniaria spicata (L.) Kuntze P

FAC+

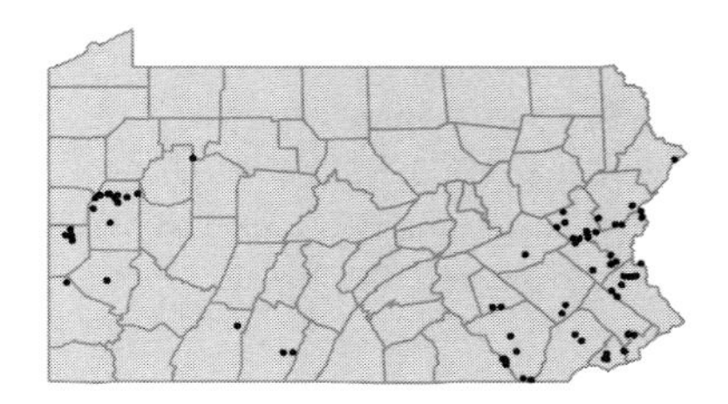

• ***Madia capitata*** Nutt.
Tarweed
Herbaceous annual
Urban waste ground. Represented by a single collection from Philadelphia Co. in 1931.
Madia sativa Molina var. *congesta* Torr. & A.Gray FGBWC

[I]

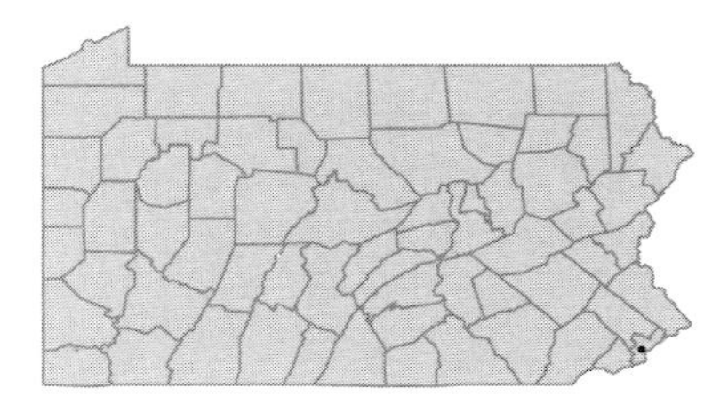

• ***Marshallia grandiflora*** Beadle & F.E.Boynt.
Large-flowered marshallia; Barbara's-buttons
Herbaceous perennial
Sandy or rocky river banks.

FAC

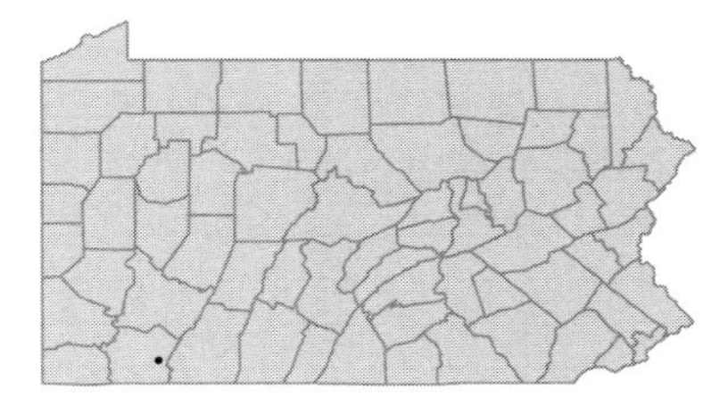

• ***Matricaria chamomilla*** L.
Wild camomile
Herbaceous annual
Ballast, wharves, dumps and waste ground.
Matricaria recrutita L. C

[I]

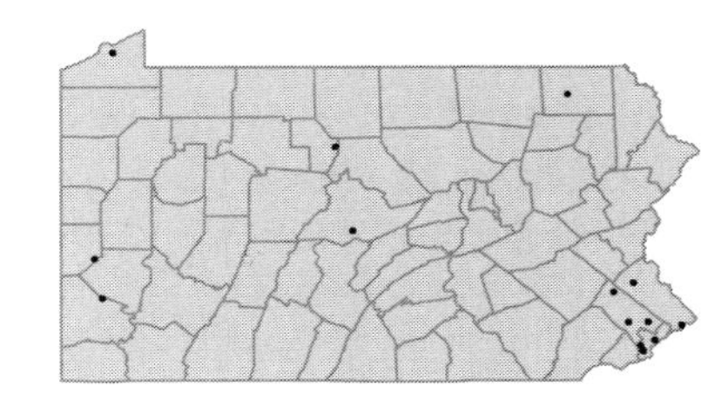

Matricaria matricarioides (Less.) Porter
Pineapple-weed
Herbaceous annual
Yards, railroad tracks, roadsides, waste ground and ballast.

FACU
[I]

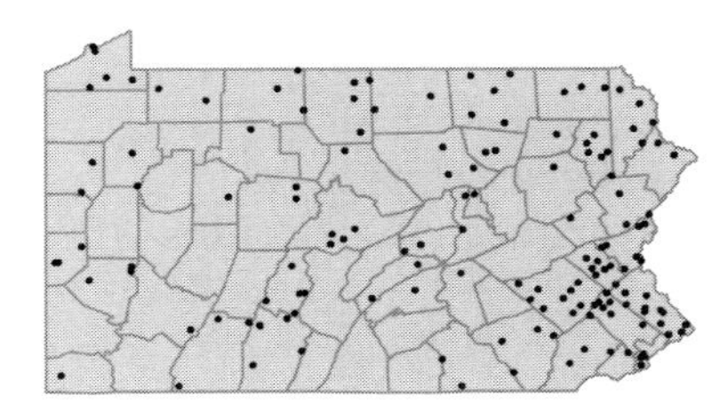

Matricaria perforata Merat
Wild chamomile
Herbaceous annual
Railroad tracks, streets, urban waste ground and ballast.
Matricaria inodora L. W; *Matricaria maritima* L. CF

UPL
[I]

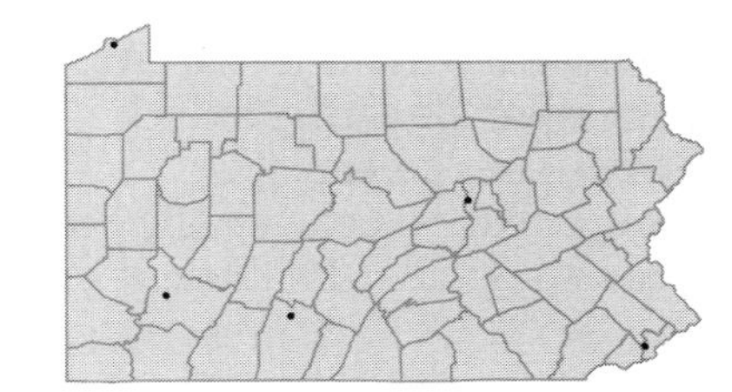

• ***Megalodonta beckii*** (Torr. ex Spreng.) Greene
Beck's water-marigold
Herbaceous perennial
Lakes and swamps, usually in calcareous water.
Bidens beckii Torr. PGBC

OBL

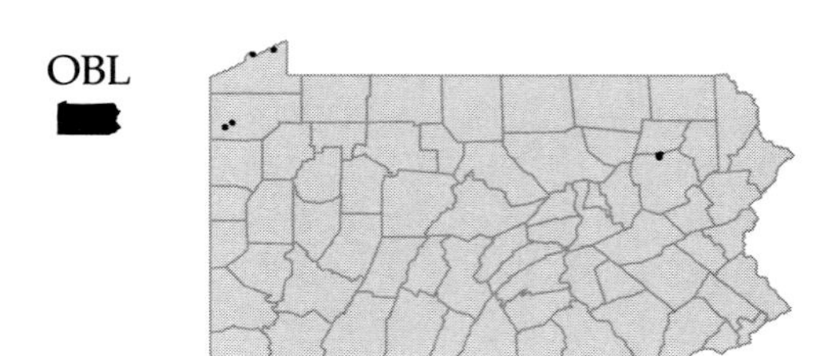

• ***Mikania scandens*** (L.) Willd.
Climbing hempweed
Herbaceous perennial vine
Swamps and moist thickets.

FACW+

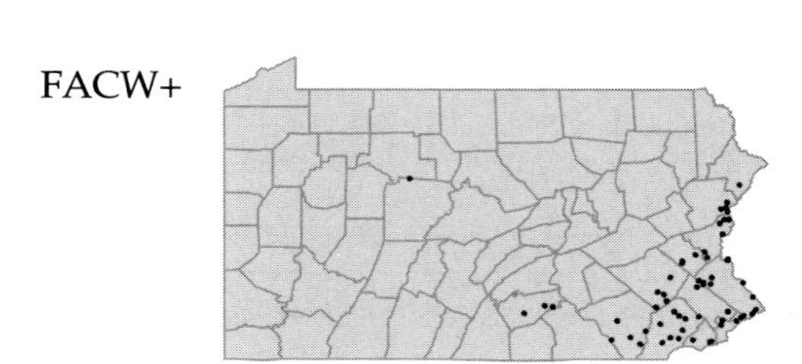

• ***Onopordum acanthium*** L.
Scotch-thistle
Herbaceous biennial
Streets, urban waste ground and ballast.

[I]

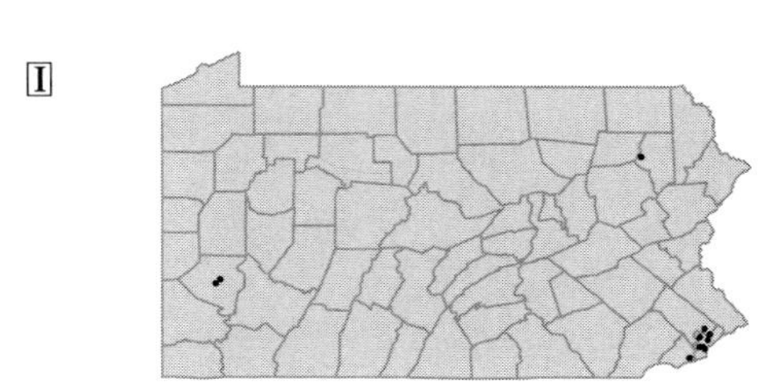

• ***Parthenium hysterophorus*** L.
Santa-Maria
Herbaceous annual
Ballast, waste ground and barnyards.

[I]

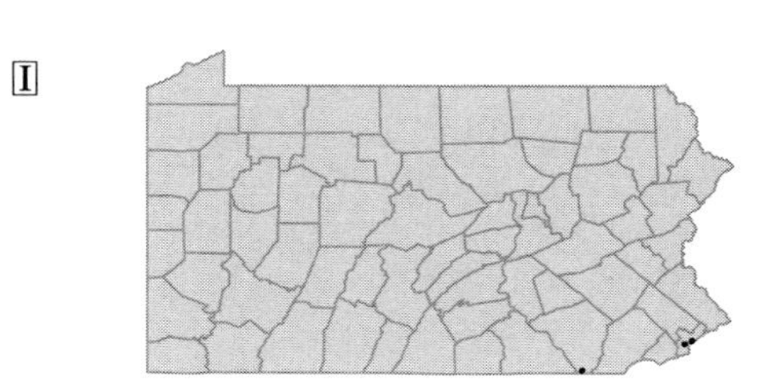

Parthenium integrifolium L.
American fever-few
Herbaceous perennial
Dry slopes and roadsides.

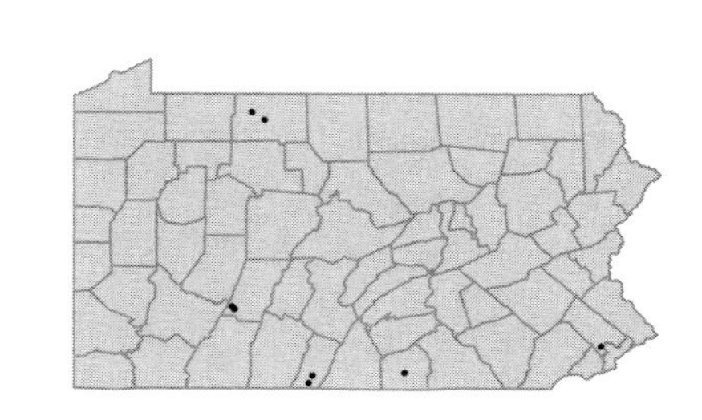

• ***Petasites hybridus*** (L.) Gaertn., Mey. & Scherb.
Butter-bur; Butterfly-dock
Herbaceous perennial
Cultivated and occasionally naturalized along creeks and in other moist sites.
Petasites petasites (L.) Karst. P

[I]

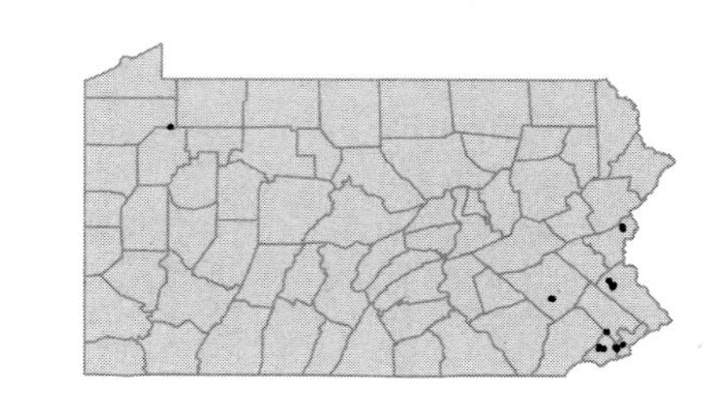

• ***Picris echioides*** L.
Bristly ox-tongue
Herbaceous annual
Gardens, waste ground and ballast.

I

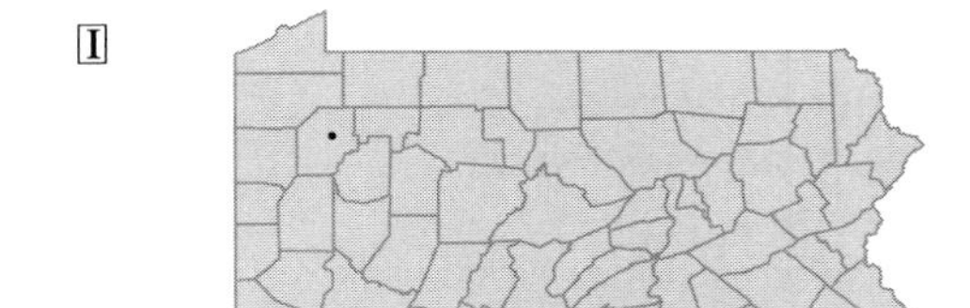

Picris hieracioides L.
Ox-tongue
Herbaceous biennial
Roadsides, fields and vacant lots.

I

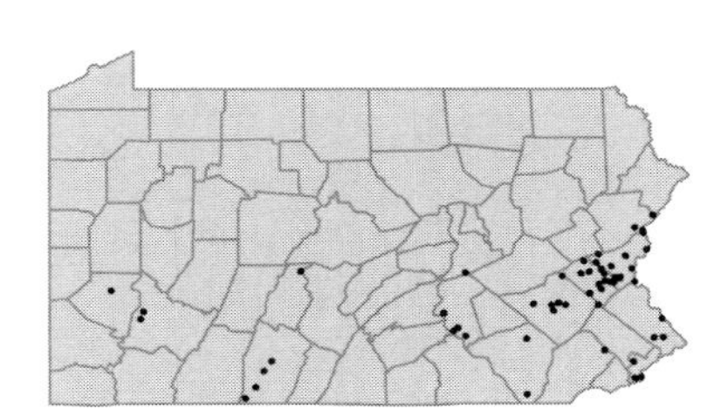

• ***Pluchea odorata*** (L.) Cass.
Marsh fleabane
Herbaceous annual
Tidal mudflats, wet ditches and railroad ballast, has also occurred locally in nursery beds where salt hay was used as mulch.
Pluchea purpurascens (Swartz) DC. var. *succulenta* Fern. F

FACW

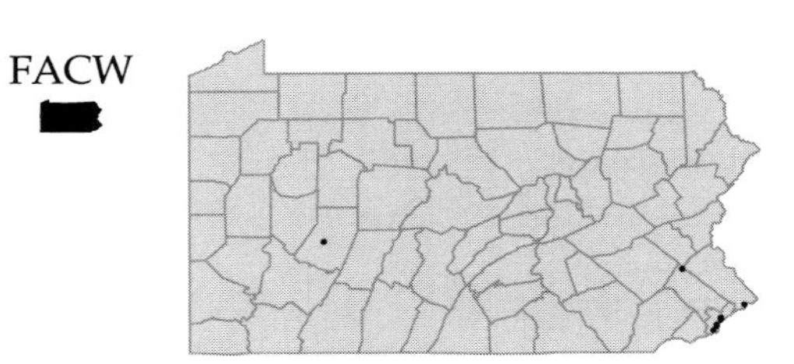

• ***Polymnia canadensis*** L.
Leaf-cup
Herbaceous perennial
Moist, rocky, wooded hillsides, floodplains and roadsides.

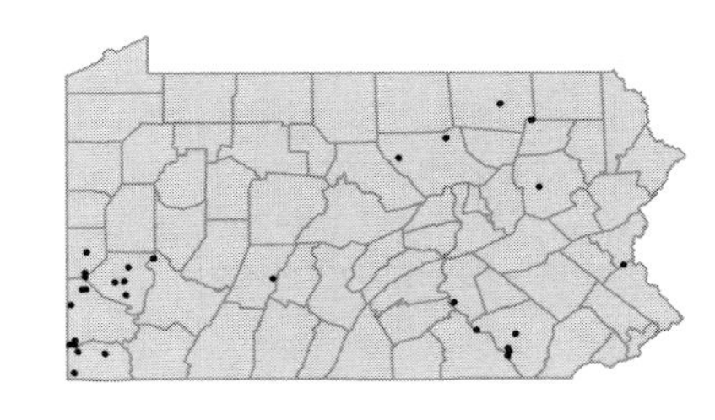

Polymnia uvedalia L.
Bear's-foot; Leaf-cup
Herbaceous perennial
Ravines, thickets and river or stream banks.

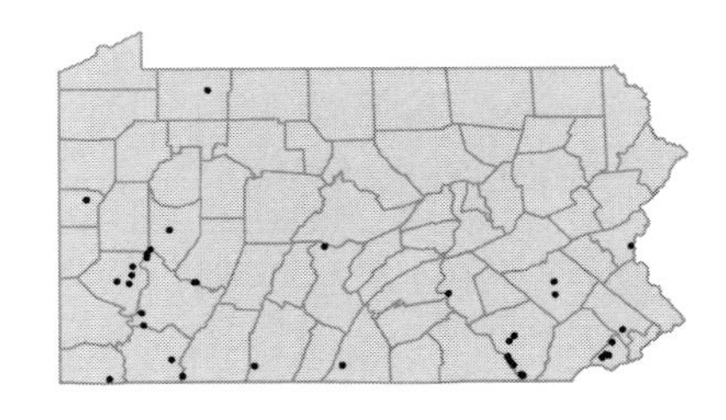

• ***Prenanthes alba*** L.
Rattlesnake-root
Herbaceous perennial
Rocky woods, barrens and roadsides.
Nabalus albus (L.) Hook. P

FACU

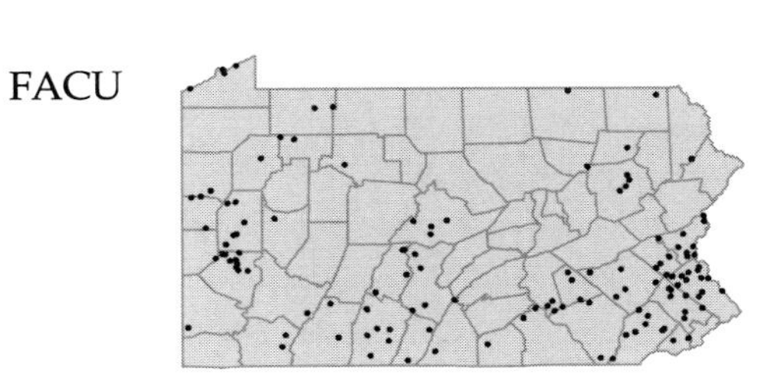

Prenanthes altissima L.
Rattlesnake-root
Herbaceous perennial
Woods.
Nabalus altissimus (L.) Hook. P

FACU-

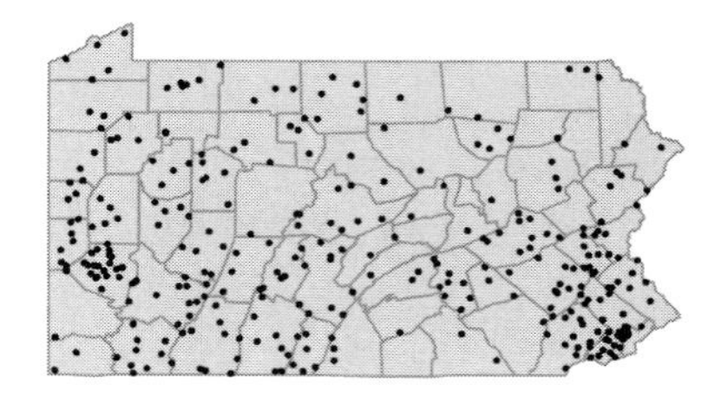

Prenanthes crepidinea Michx.
Rattlesnake-root
Herbaceous perennial
Moist, rocky woods and thickets.
Nabalus crepidineus (Michx.) DC. P

FACU

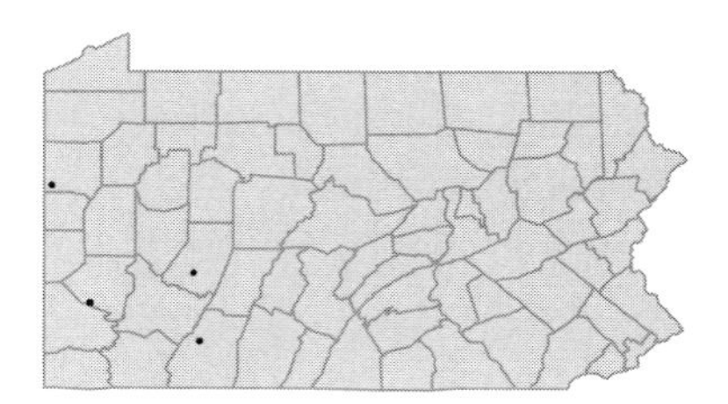

Prenanthes racemosa Michx.
Glaucous rattlesnake-root
Herbaceous perennial
Moist ground near bog, old quarry. Believed to be extirpated, last collected in 1964.
Nabalus racemosus (Michx.) DC. P

FACW-

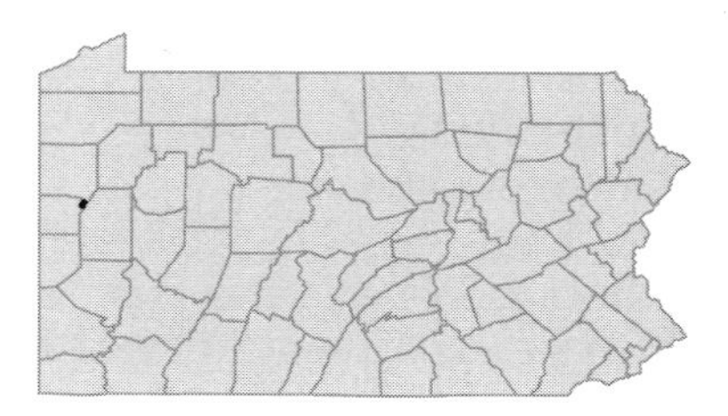

Prenanthes serpentaria Pursh
Lion's-foot
Herbaceous perennial
Dry woods, clearings and gravelly roadsides.
Nabalus serpentarius (Pursh) Hook. P

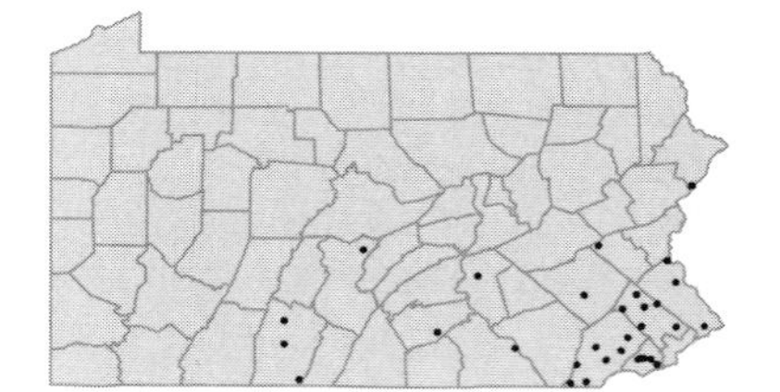

Prenanthes trifoliolata (Cass.) Fern.
Gall-of-the-earth
Herbaceous perennial
Sandy or rocky, open woods and shale barrens.
Nabalus trifoliatus Cass. P

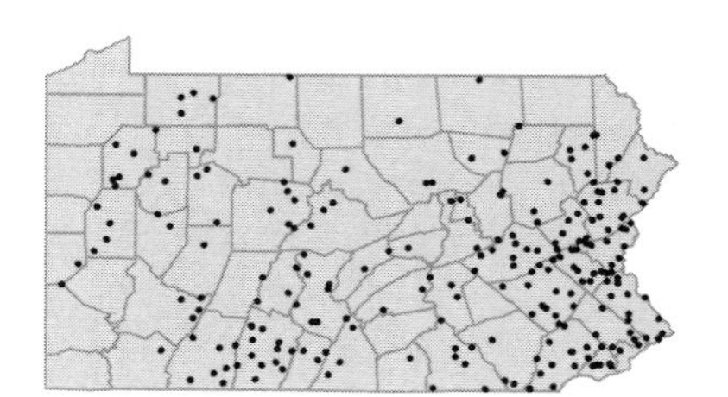

• ***Pyrrhopappus carolinianus*** (Walt.) DC.
False dandelion
Herbaceous annual
Ballast and waste ground. Represented by a single collection from Philadelphia Co. in 1864.

[I]

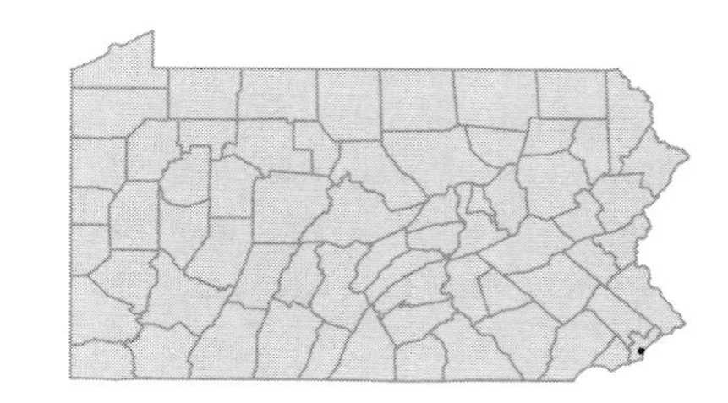

• ***Ratibida columnifera*** (Nutt.) Woot. & Standl. — I
Coneflower
Herbaceous perennial
Fields, wharves and docks.

Ratibida pinnata (Vent.) Barnh.
Prairie coneflower
Herbaceous perennial
Dry fields, limestone upland and open roadsides.

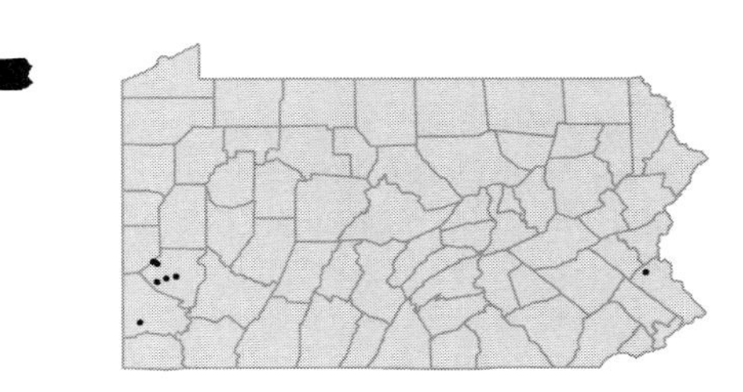

• ***Rudbeckia fulgida*** Ait. var. ***fulgida*** — FAC
Eastern coneflower
Herbaceous perennial
Moist fields and meadows.

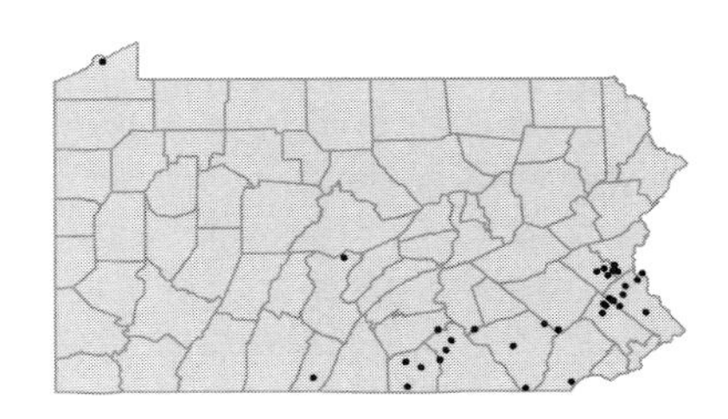

Rudbeckia fulgida Ait. var. ***speciosa*** (Wenderoth) Perdue — FAC
Coneflower
Herbaceous perennial
Fields.
Rudbeckia speciosa Wenderoth PFW

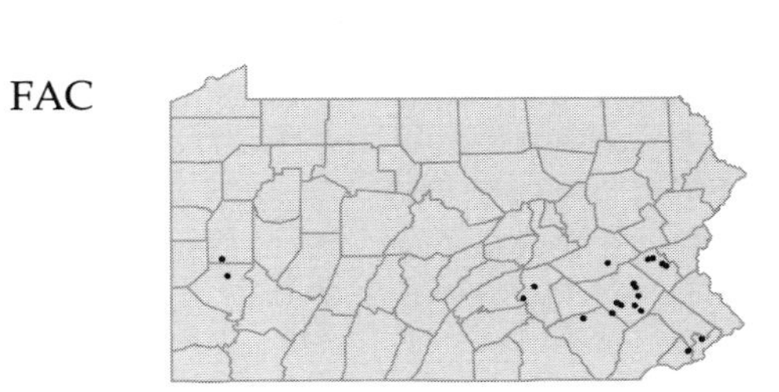

Rudbeckia hirta L. var. ***hirta*** — FACU-
Black-eyed-Susan
Herbaceous biennial
Fields, meadows and roadsides.

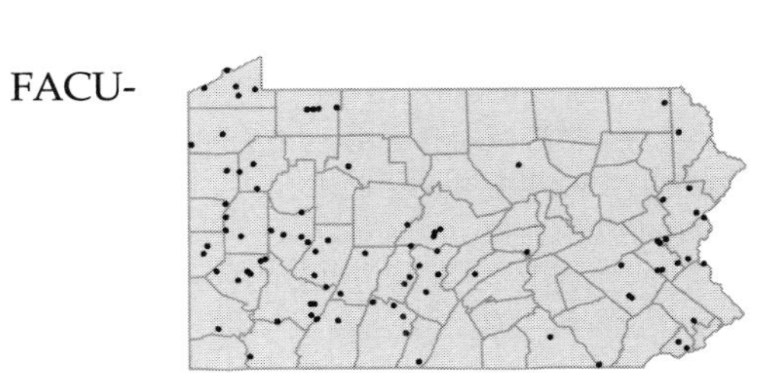

Rudbeckia hirta L. var. ***pulcherrima*** Farw. — FACU-
Black-eyed-Susan
Herbaceous biennial
Fields, woods, meadows and roadsides.

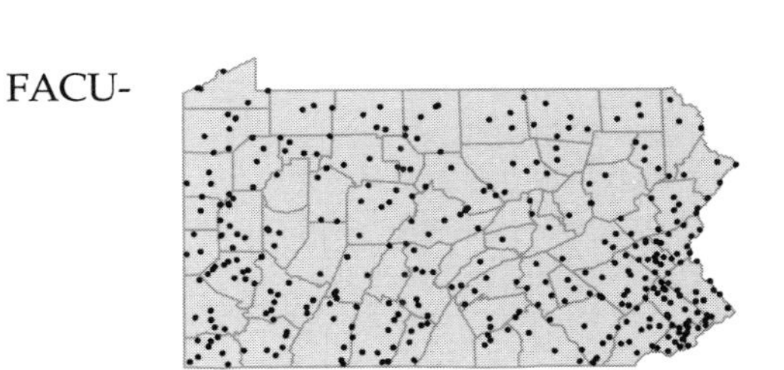

Rudbeckia laciniata L.
Cutleaf coneflower
Herbaceous perennial
Floodplains, stream banks and wet fields.

FACW

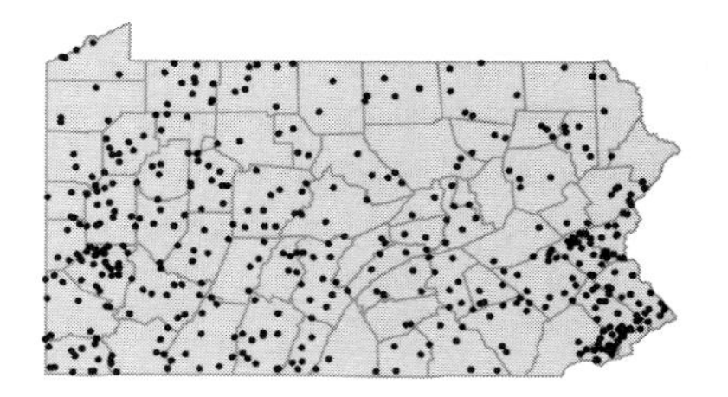

Rudbeckia triloba L.
Coneflower
Herbaceous perennial
Moist old fields, rocky woods and edges.

FACU

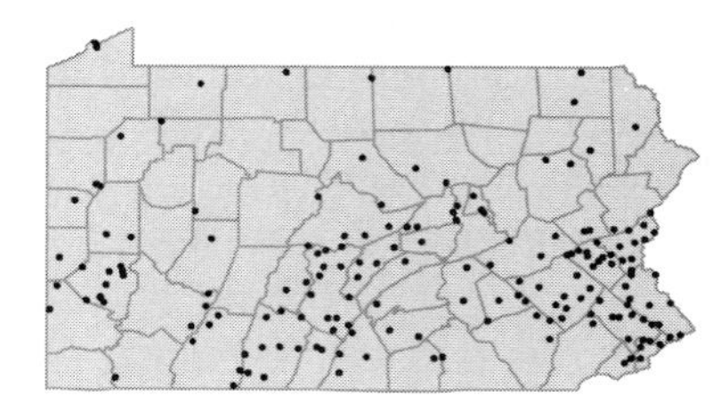

• ***Scolymus hispanicus*** L.
Golden thistle; Spanish oyster-plant
Herbaceous biennial
Ballast ground. Represented by two collections from Philadelphia Co. 1878-1880.

[I]

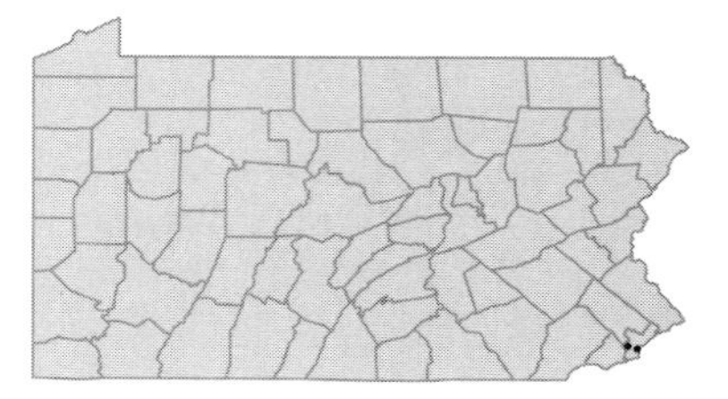

• ***Senecio anonymus*** A.Wood
Appalachian groundsel; Plain ragwort
Herbaceous perennial
Dry fields, open woods and serpentine barrens.
Senecio smallii Britt. FGBW

UPL

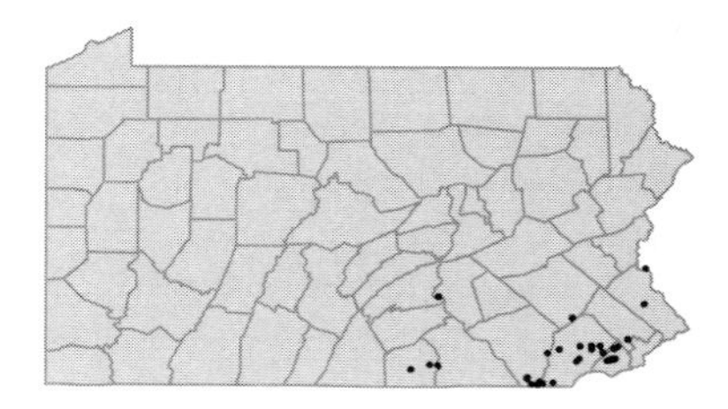

Senecio antennariifolius Britt.
Shale-barren ragwort; Cat's-paw ragwort
Herbaceous perennial
Shale barrens.

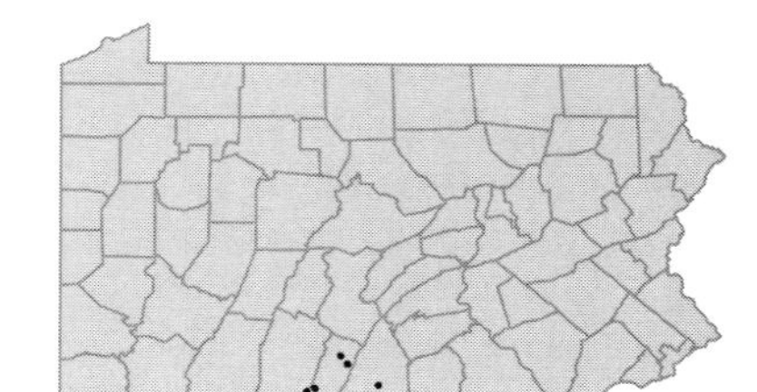

Senecio aureus L.
Golden ragwort
Herbaceous perennial
Moist fields, woods, floodplains and roadsides.

FACW

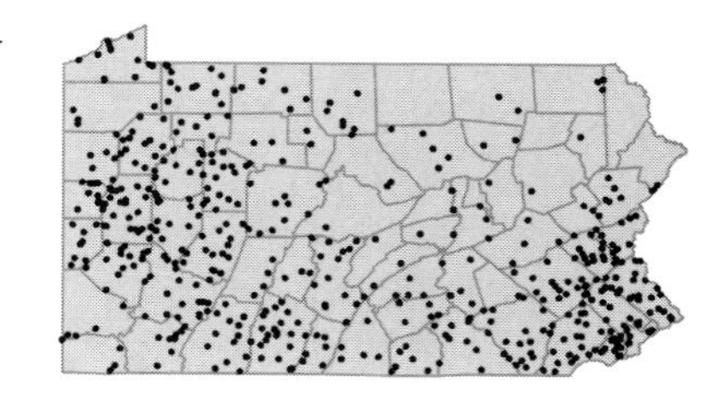

Senecio erucifolius L. I

Groundsel

Herbaceous perennial

Ballast ground. Represented by several collections from Philadelphia Co. 1865-1879.

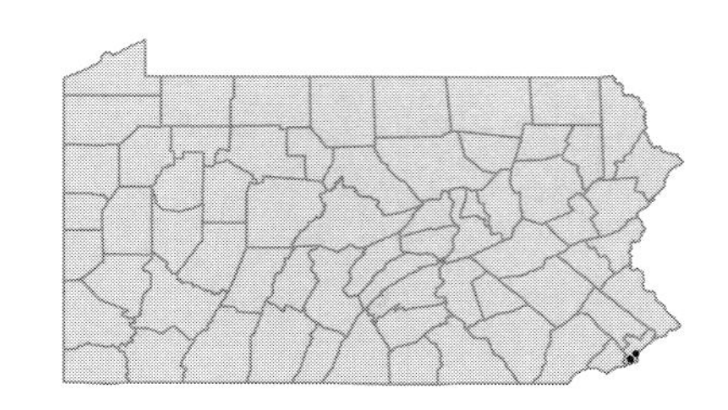

Senecio jacobaea L. I

Stinking-willie; Tansy ragwort

Herbaceous biennial

Clearings, urban waste ground and ballast.

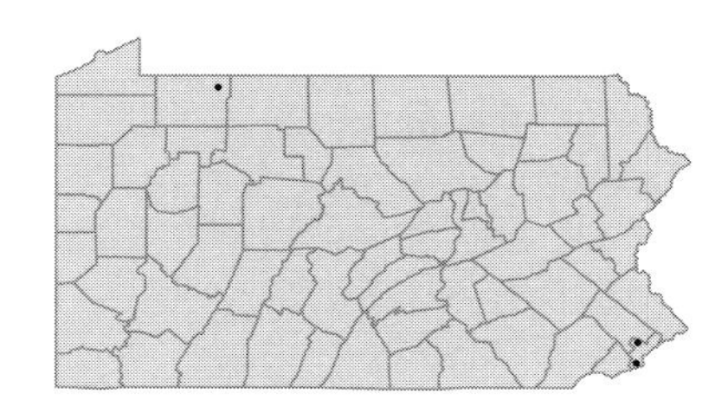

Senecio obovatus Muhl. FACU-

Groundsel; Ragwort

Herbaceous perennial

Moist fields, woods and calcareous slopes.

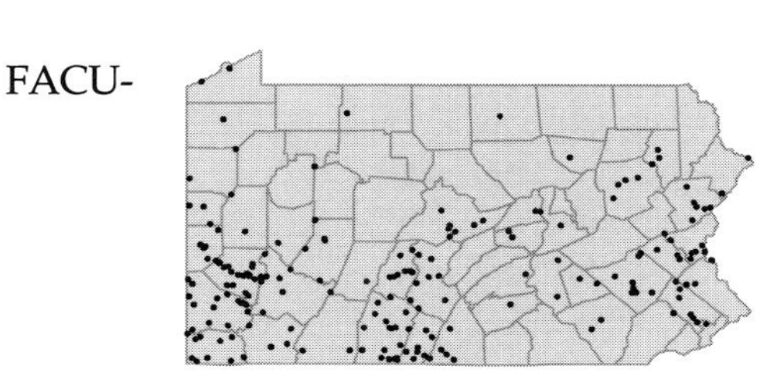

Senecio pauperculus Michx. FAC

Balsam groundsel; Balsam ragwort

Herbaceous perennial

Fields, meadows, peaty thickets, stream banks and roadsides.

Senecio crawfordii Britt. PF; *Senecio pauperculus* Michx. var. *crawfordii* (Britt.) T.M.Barkl. W

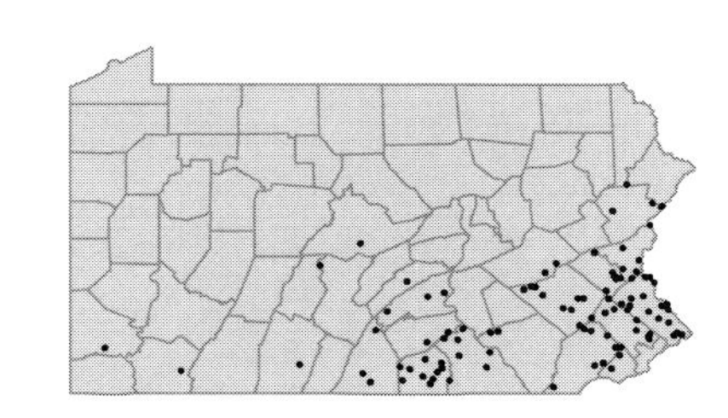

Senecio plattensis Nutt.

Prairie ragwort

Herbaceous biennial

Dry woods and abandoned field.

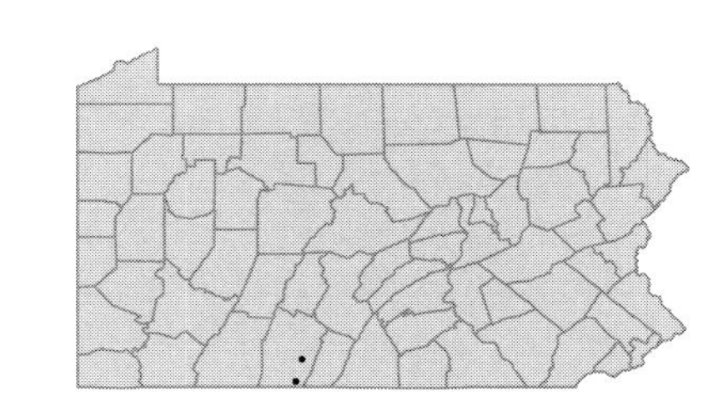

Senecio sylvaticus L. I

Groundsel

Herbaceous annual

Ballast ground. Represented by a single collection from Philadelphia Co. in 1878.

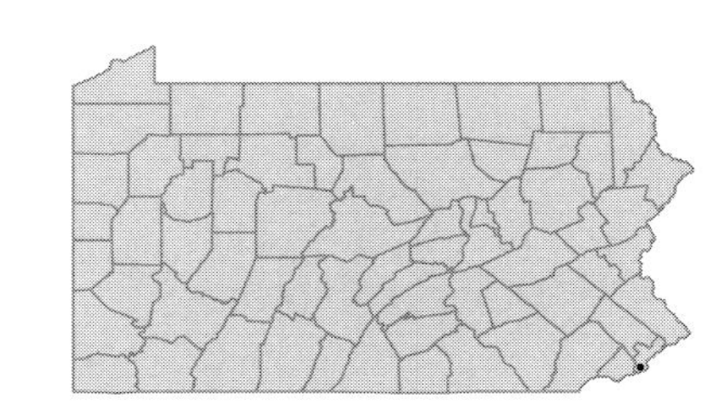

Senecio viscosus L.
Sticky groundsel
Herbaceous annual
Roadsides, floodplains, railroad tracks and waste ground.

I

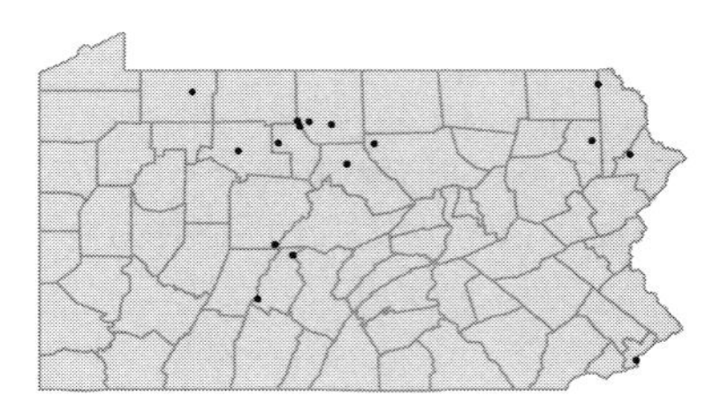

Senecio vulgaris L.
Common groundsel
Herbaceous annual
Streets, gardens, roadsides and vacant lots.

FACU
I

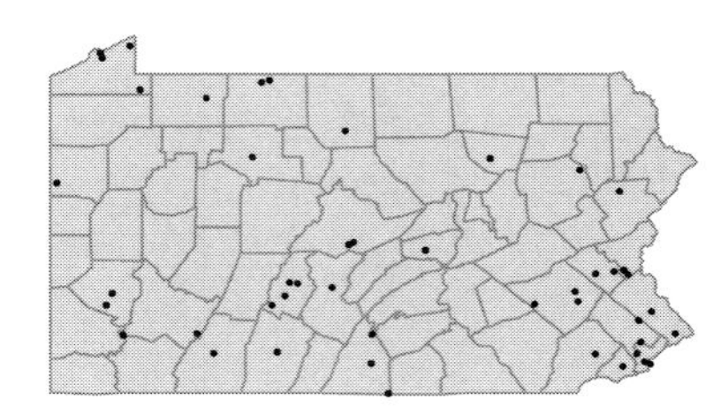

• ***Silphium laciniatum*** L.
Compass-plant
Herbaceous perennial
Cemetery grounds. Represented by several collections from Allegheny Co. 1885-1890.

I

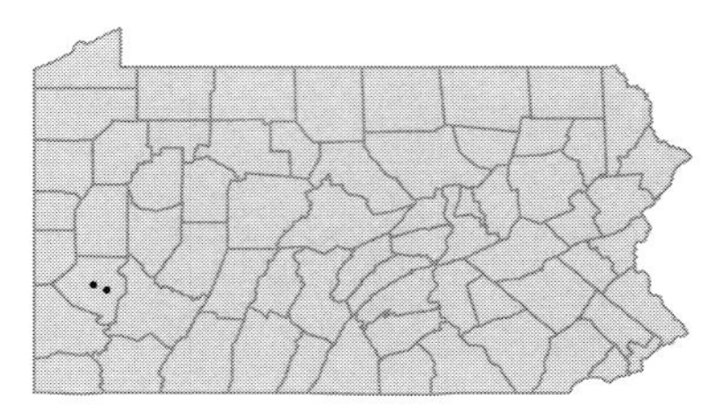

Silphium perfoliatum L.
Cup-plant
Herbaceous perennial
Floodplains, abandoned fields and moist meadows.

FACU
I

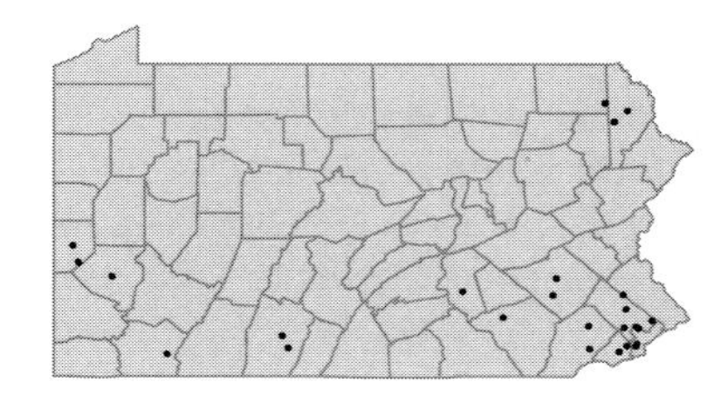

Silphium trifoliatum L. var. ***trifoliatum***
Whorled rosinweed
Herbaceous perennial
Roadsides, dry thickets and meadows.
Silphium asteriscus L. ssp. *trifoliatum* Weber, T.R.Fischer & Speer K

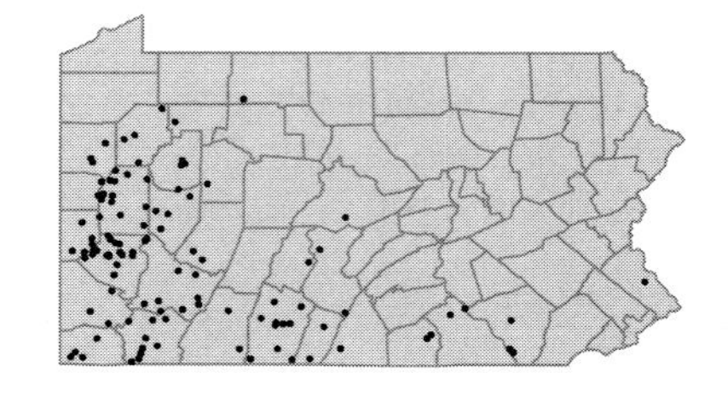

• ***Silybum marianum*** (L.) Gaertn.
Milk thistle
Herbaceous annual
Waste ground and ballast.
Mariana mariana (L.) Hill P

I

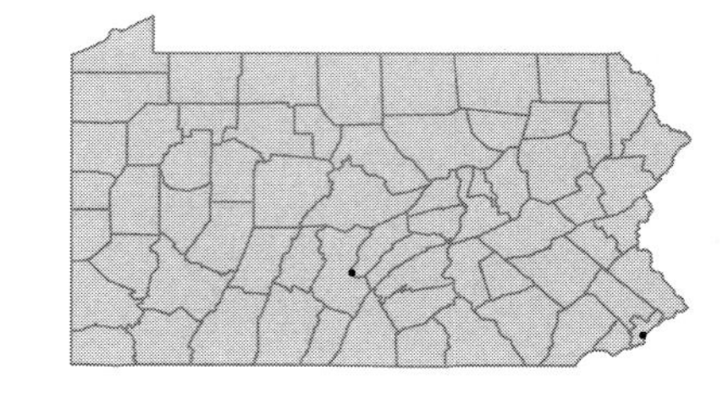

• ***Solidago arguta*** Ait. var. ***arguta***
Forest goldenrod
Herbaceous perennial
Rocky woods, dry thickets and roadsides.

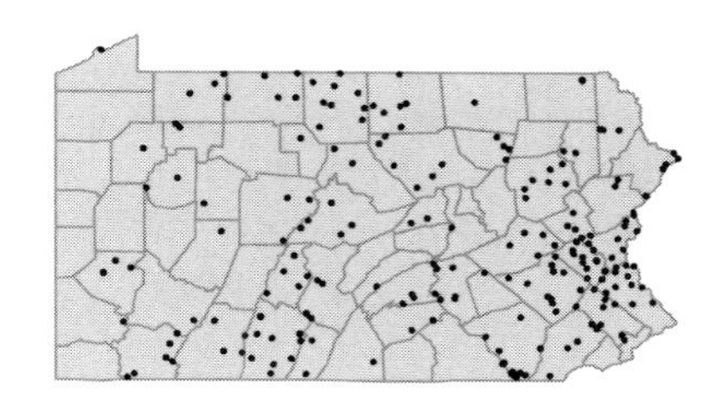

Solidago arguta Ait. ssp. ***harrisii*** (E.S.Steele) Cusick
Harris' goldenrod
Herbaceous perennial
Calcareous shales on steep slopes, shale barrrens.

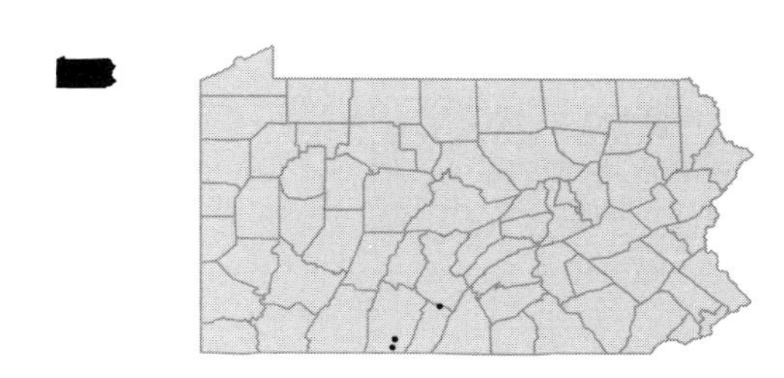

Solidago bicolor L.
Silver-rod; White goldenrod
Herbaceous perennial
Dry woods, wooded banks and shale barrens.

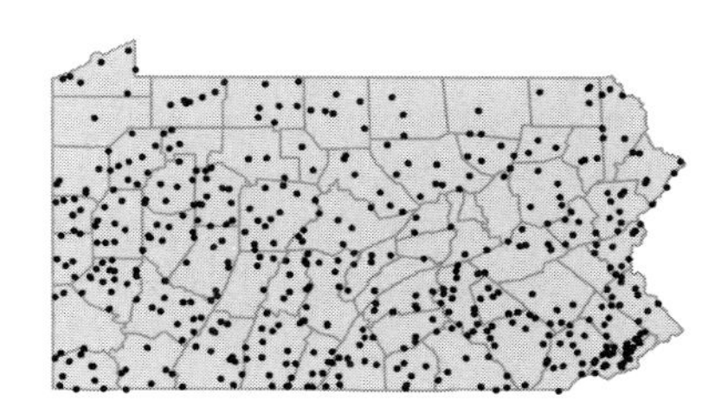

Solidago caesia L. var. ***caesia*** FACU
Blue-stem goldenrod; Wreath goldenrod
Herbaceous perennial
Rich woods.

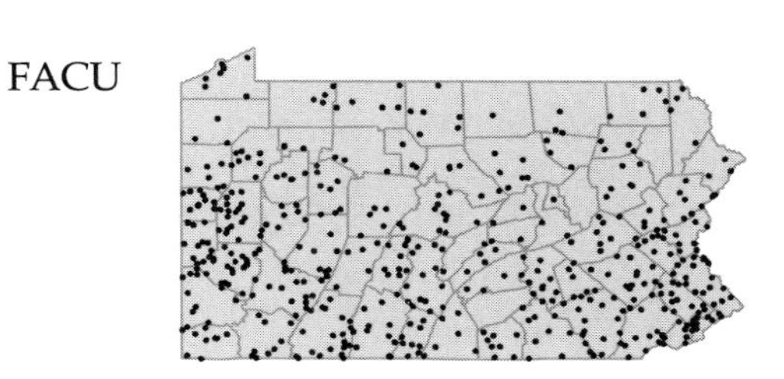

Solidago caesia L. var. ***curtisii*** (Torr. & A.Gray) A.Wood
Curtis's goldenrod
Herbaceous perennial
Floodplain forest.
Solidago curtisii Torr. & A.Gray FBGC

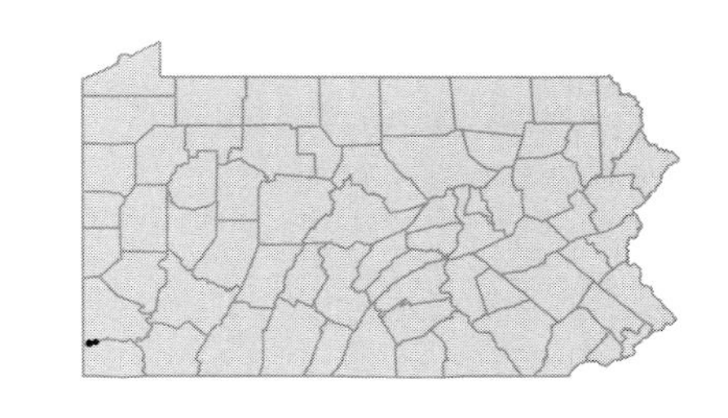

Solidago canadensis L. var. ***canadensis*** FACU
Canada goldenrod
Herbaceous perennial
Fields and roadsides.

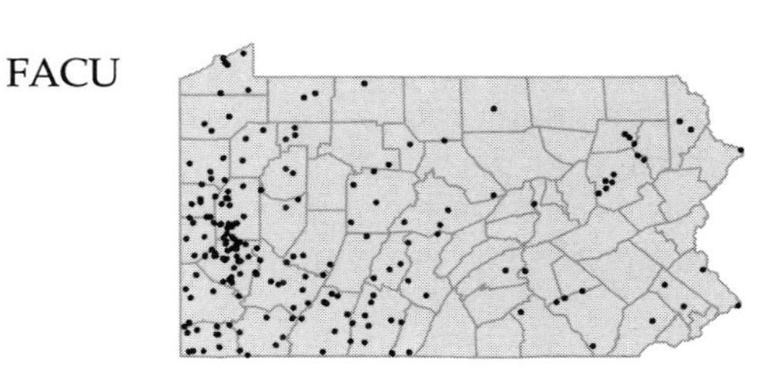

Solidago canadensis L. var. ***hargeri*** Fern. FACU
Canada goldenrod
Herbaceous perennial
Wooded slopes, meadows and roadsides.

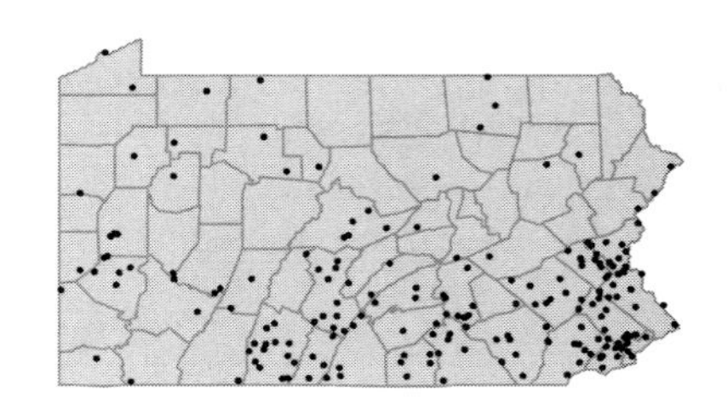

Solidago canadensis L. var. ***salebrosa*** (Piper) M.E.Jones FACU
Goldenrod
Herbaceous perennial
Open, valley-wall slumps. Represented by a single collection from Erie Co. in 1985.

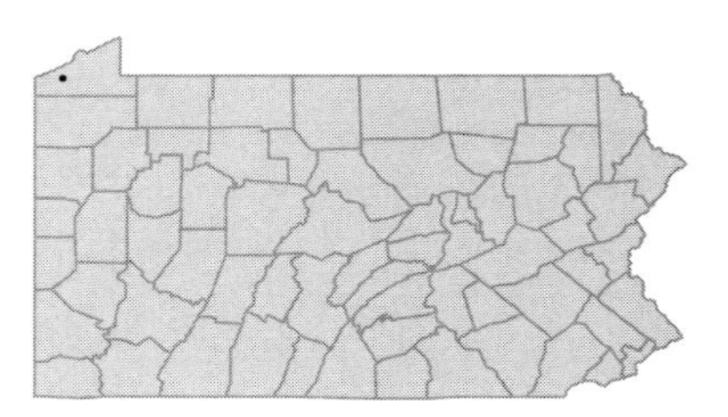

Solidago canadensis L. var. ***scabra*** Torr. & A.Gray FACU
Canada goldenrod
Herbaceous perennial
Wooded slopes, fields, river banks and roadsides.
Solidago altissima L. FBW

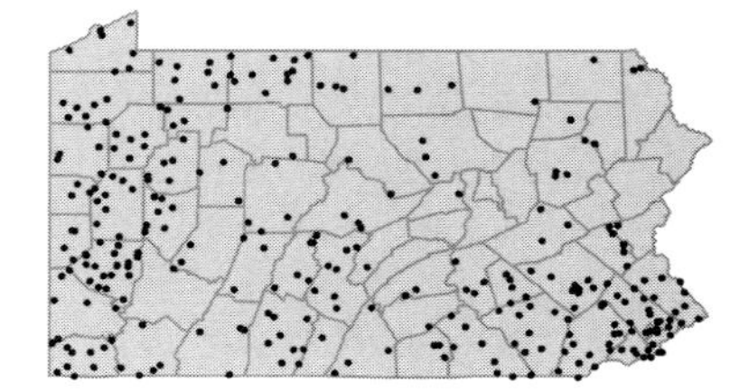

Solidago flexicaulis L. FACU
Zigzag goldenrod
Herbaceous perennial
Moist woods and rocky, wooded slopes.

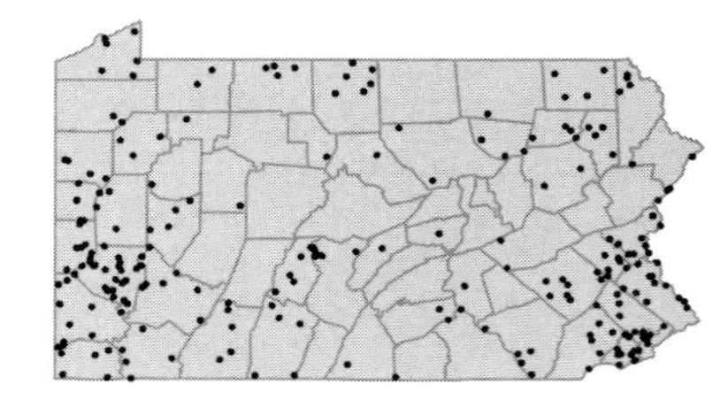

Solidago gigantea Ait. var. ***gigantea*** FACW
Smooth goldenrod
Herbaceous perennial
Moist fields, swamps, marshy shores and swales.

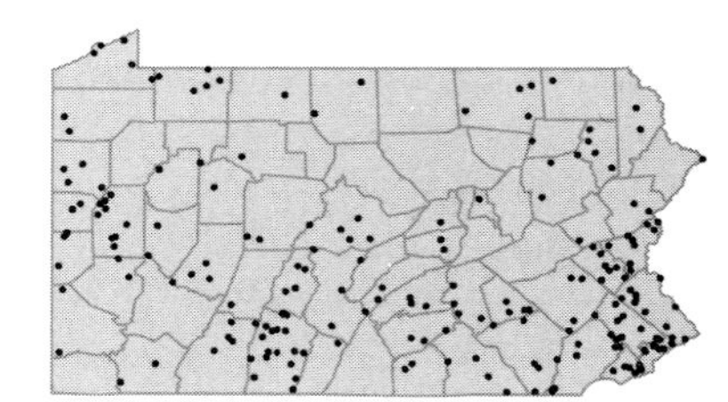

Solidago gigantea Ait. var. ***serotina*** (Ait.) Cronq. FACW
Smooth goldenrod
Herbaceous perennial
Moist meadows, banks and ditches.
Solidago gigantea Ait. var. *leiophylla* Fern. FBW; *Solidago serotina* Ait. P

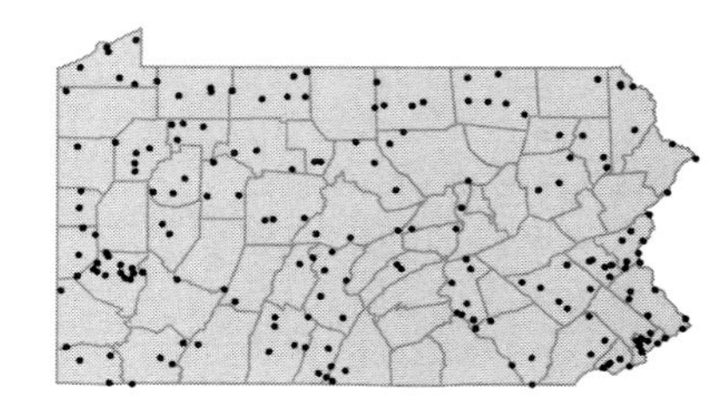

Solidago hispida Muhl.
Hairy goldenrod
Herbaceous perennial
Dry, rocky slopes and wooded roadside banks.

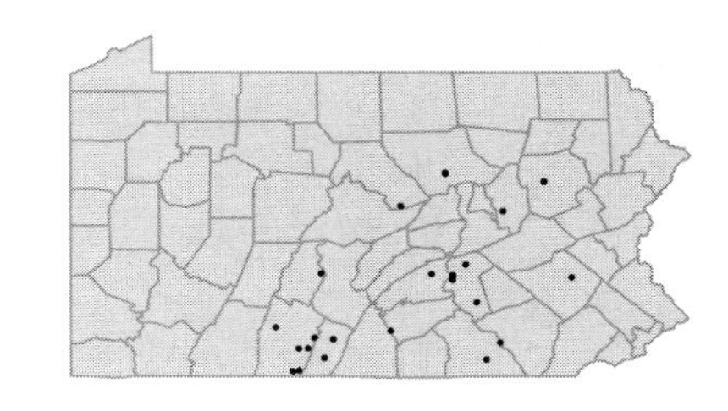

Solidago juncea Ait.
Early goldenrod
Herbaceous perennial
Fields, meadows, rocky slopes and roadsides.

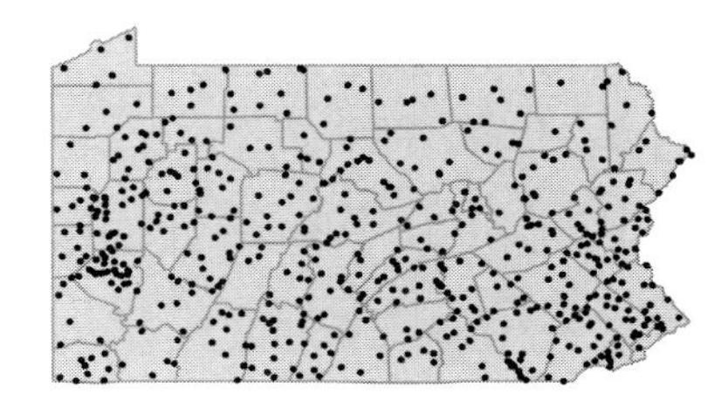

Solidago nemoralis Ait.
Gray goldenrod
Herbaceous perennial
Fields, woods and roadsides in dry, sterile soil.

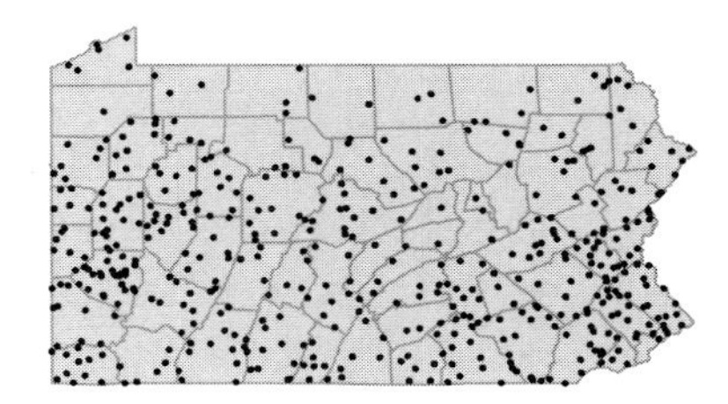

Solidago odora Ait.
Sweet goldenrod
Herbaceous perennial
Dry, open woods.

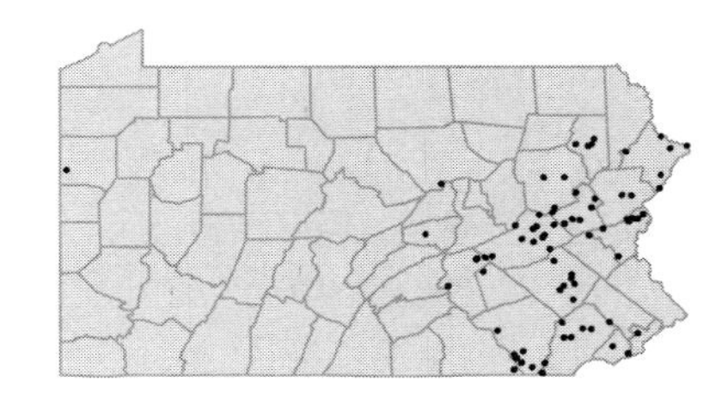

Solidago patula Muhl. ex Willd.
Spreading goldenrod
Herbaceous perennial
Swamps, floodplains and moist woods.

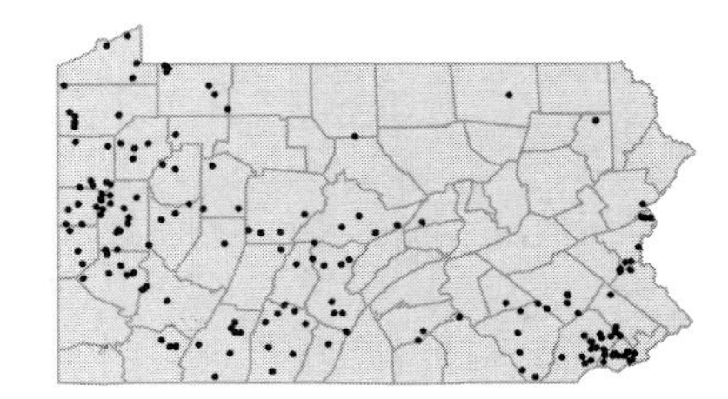

Solidago puberula Nutt.
Downy goldenrod
Herbaceous perennial
Rocky woods, thickets and roadsides.

FACU-

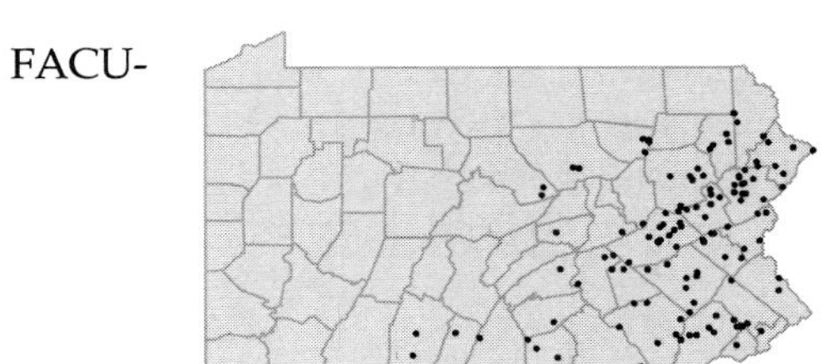

Solidago rigida L.
Stiff goldenrod
Herbaceous perennial
Moist fields or thickets, in rich soil.

UPL

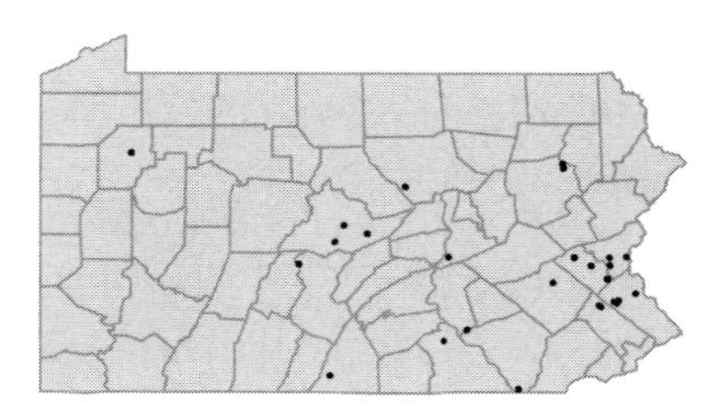

Solidago roanensis Porter
Mountain goldenrod
Herbaceous perennial
Rocky banks, roadsides, cut-over woods and woods edges.

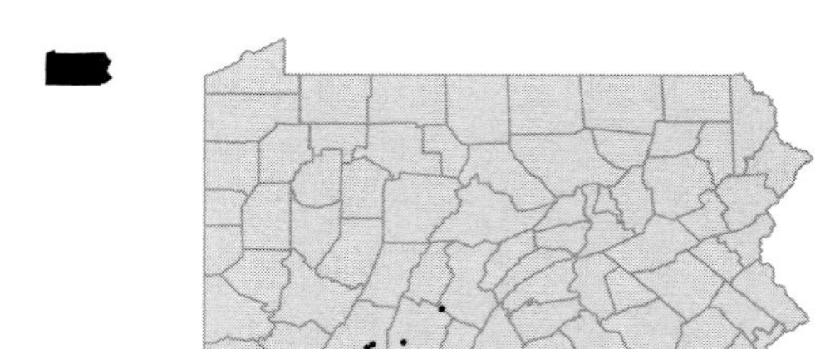

Solidago rugosa Ait. var. ***rugosa***
Wrinkle-leaf goldenrod
Herbaceous perennial
Fields, woods, floodplains, thickets, roadsides and waste ground.

FAC

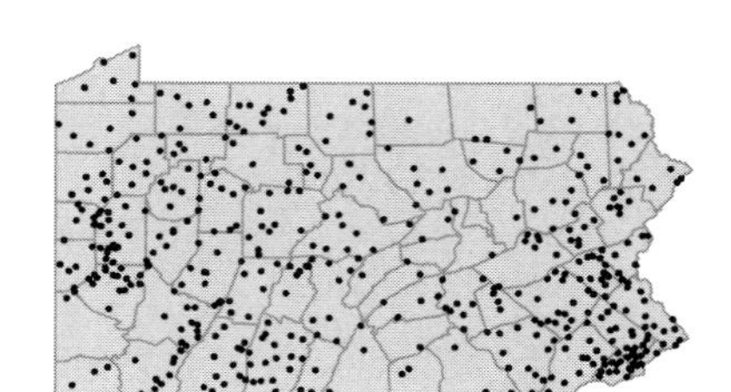

Solidago rugosa Ait. var. ***aspera*** (Ait.) Fern.
Wrinkle-leaf goldenrod
Herbaceous perennial
Fields, swamps and waste ground.
Solidago aspera (Aiton.) Cronq.

FAC

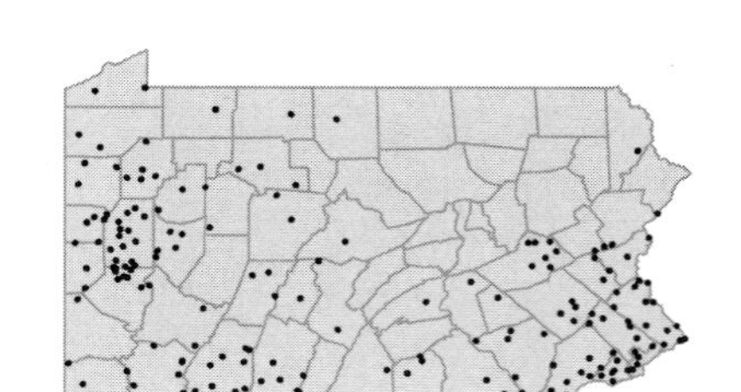

Solidago rugosa Ait. var. ***sphagnophila*** Graves
Wrinkle-leaf goldenrod
Herbaceous perennial
Wet meadows, low fields and bogs.

FAC

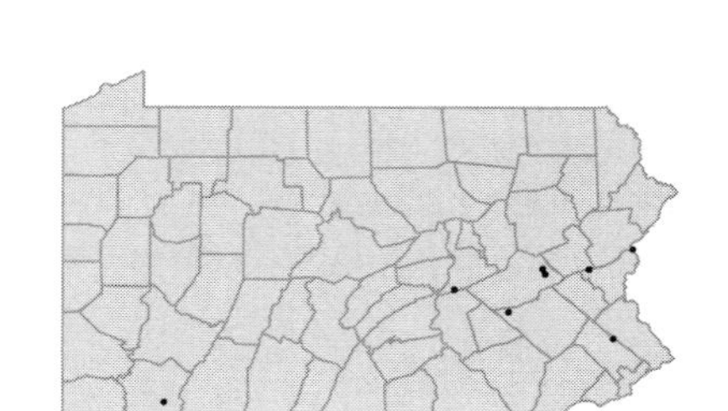

Solidago rugosa Ait. var. ***villosa*** (Pursh) Fern.
Wrinkle-leaf goldenrod
Herbaceous perennial
Dry, open fields and uplands.

FAC

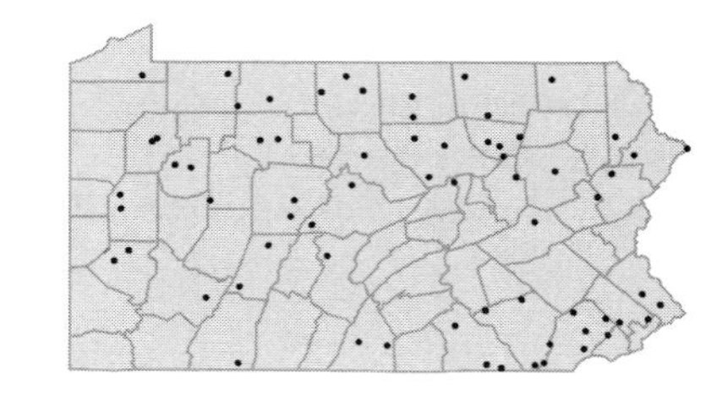

Solidago sempervirens L.
Seaside goldenrod
Herbaceous perennial
Waste ground, ballast and roadsides where de-icing salts are used. Native in coastal areas, but adventive in PA.

FACW

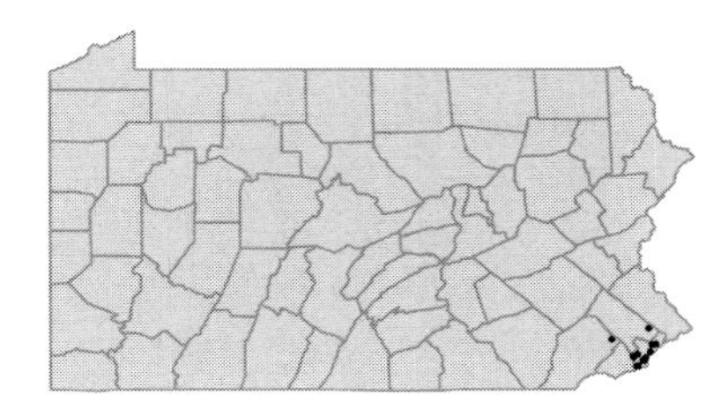

Solidago simplex HBK ssp. ***randii*** (Porter) Ringius var. ***racemosa*** (Greene) Ringius
Sticky goldenrod
Herbaceous perennial
Rock crevices and shores.
Solidago racemosa Greene FW; *Solidago spathulata* DC. ssp. *randii* (Porter) Cronq. var. *racemosa* (Greene) Cronq. G

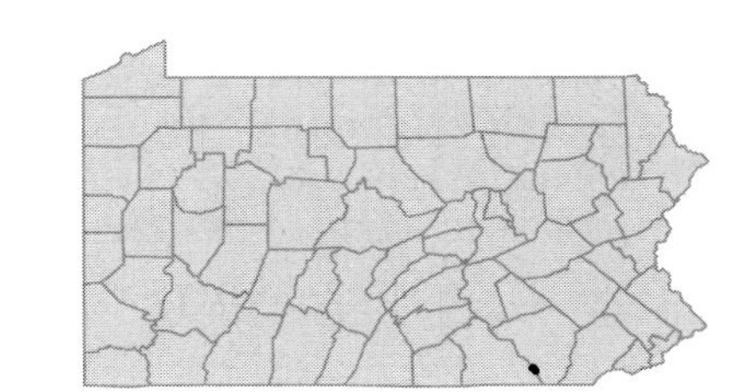

Solidago speciosa Nutt. var. ***speciosa***
Showy goldenrod
Herbaceous perennial
Moist meadows, rocky woods, thickets and rocky roadside banks.

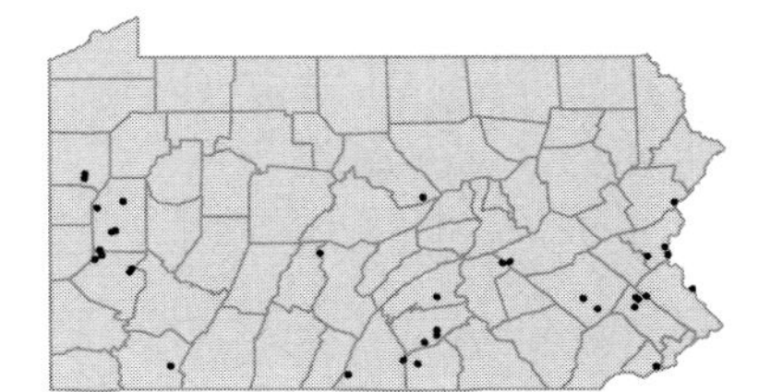

Solidago speciosa Nutt. var. ***erecta*** MacMillan
Slender goldenrod
Herbaceous perennial
Dry, acidic, shaly banks.
Solidago erecta Pursh FBCW

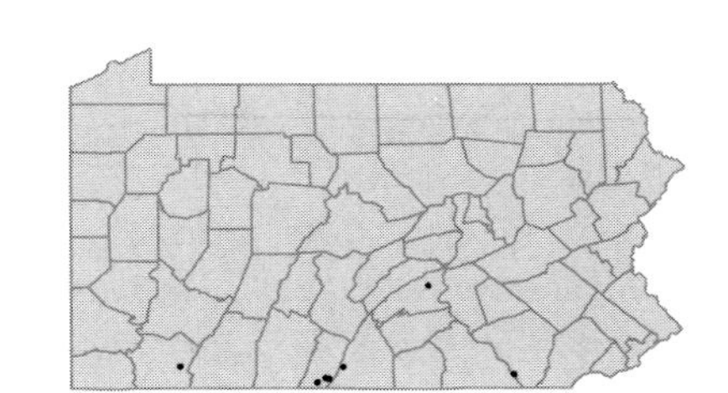

Solidago squarrosa Muhl.
Ragged goldenrod; Stout goldenrod
Herbaceous perennial
Mountain woods, rocky roadside banks and thickets.

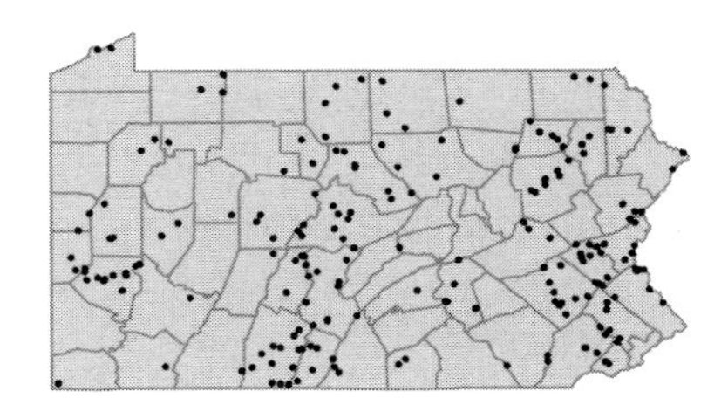

Solidago uliginosa Nutt. var. ***uliginosa***
Bog goldenrod
Herbaceous perennial
Bogs, swamps, sedge meadows and fens.
Solidago neglecta Torr. & A.Gray P

OBL

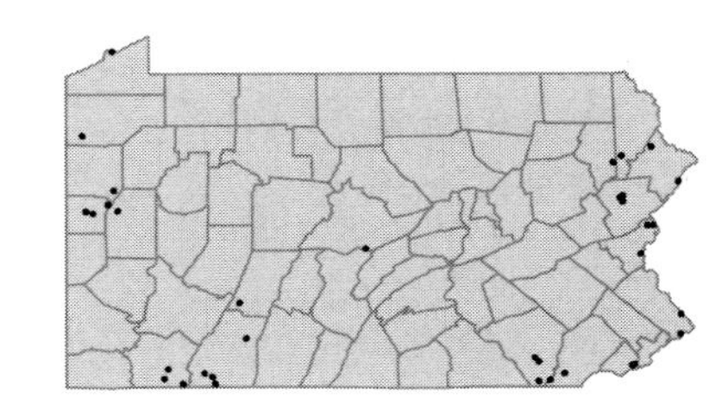

Solidago uliginosa Nutt. var. ***peracuta*** (Fern.) Friesner
Bog goldenrod
Herbaceous perennial
Bogs and swamps.
Solidago purshii Porter PFW

OBL

Solidago ulmifolia Muhl. ex Willd. var. ***ulmifolia***
Elm-leaved goldenrod
Herbaceous perennial
Woods, wooded slopes, roadside banks and shale barrens.

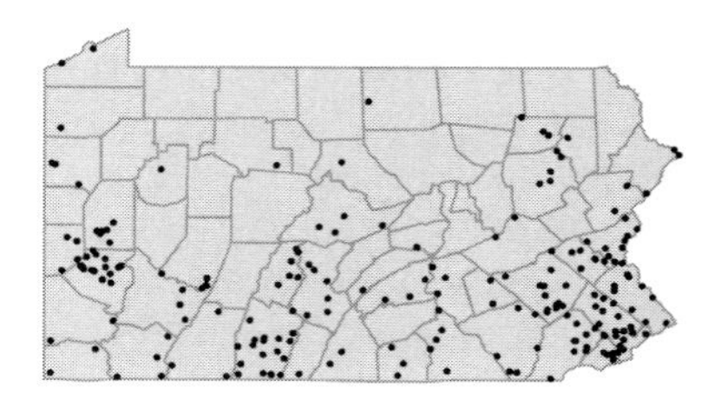

• ***Sonchus arvensis*** L. ssp. ***arvensis***
Field sow-thistle
Herbaceous perennial
Roadsides, fields, railroad tracks and vacant lots.

UPL
I

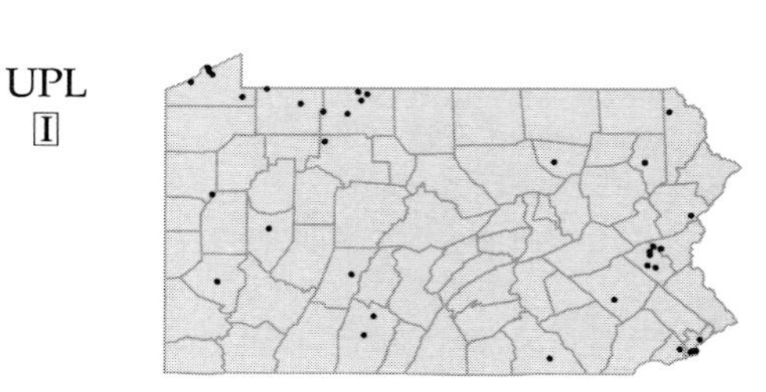

Sonchus arvensis L. ssp. ***uliginosus*** (Bieb.) Nyman
Field sow-thistle; Milk sow-thistle
Herbaceous perennial
Roadsides and waste ground.
Sonchus arvensis L. var. *glabrescens* (Guenther) Grab. & Wimmer C; *Sonchus uliginosus* Bieb. FW

UPL
I

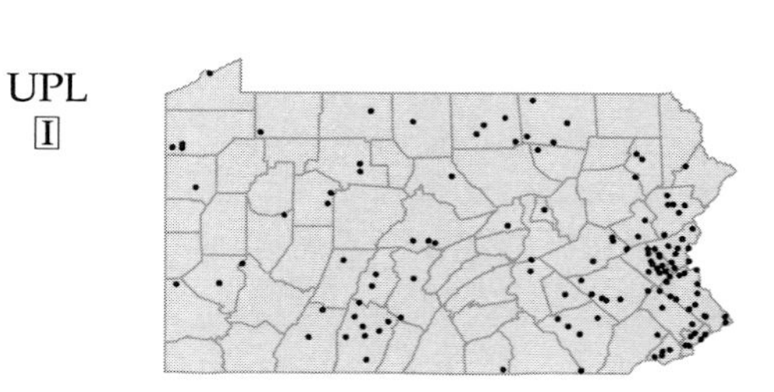

Sonchus asper (L.) Hill
Spiny-leaved sow-thistle
Herbaceous annual
Abandoned fields, roadsides and waste ground.

FAC
I

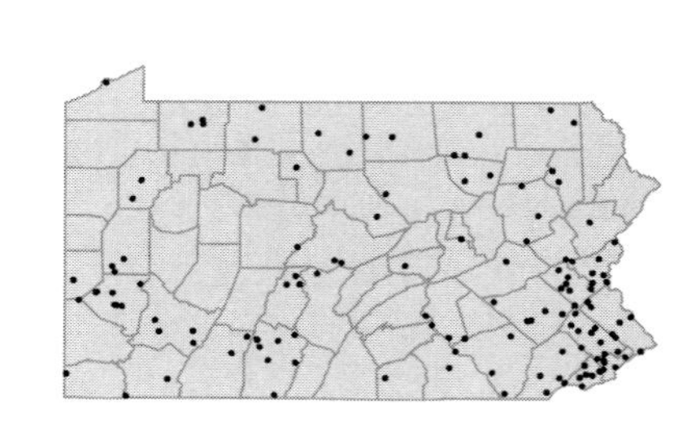

Sonchus oleraceus L.
Milk-thistle; Sow-thistle
Herbaceous annual
Roadsides, shores and urban waste ground.

UPL
I

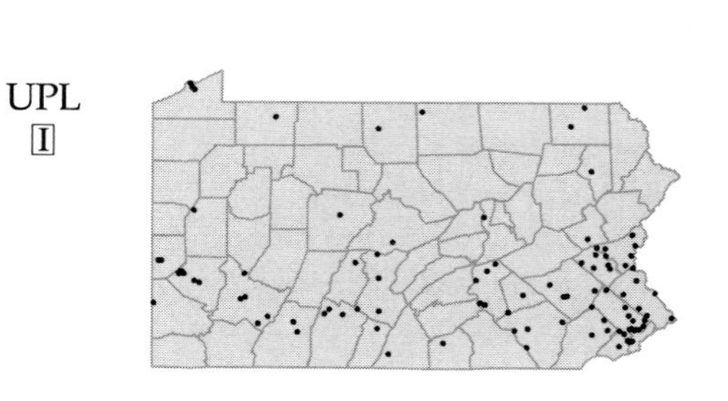

• ***Tagetes erecta*** L.
African marigold
Herbaceous annual
Cultivated and occasionally escaped to rubbish dumps and waste ground.

I

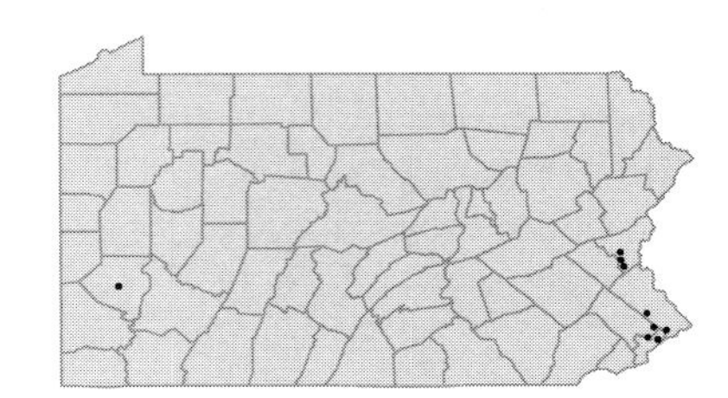

Tagetes minuta L.
Marigold
Herbaceous annual
Weed in cultivated ground. Represented by a single collection from Centre Co. in 1957.

I

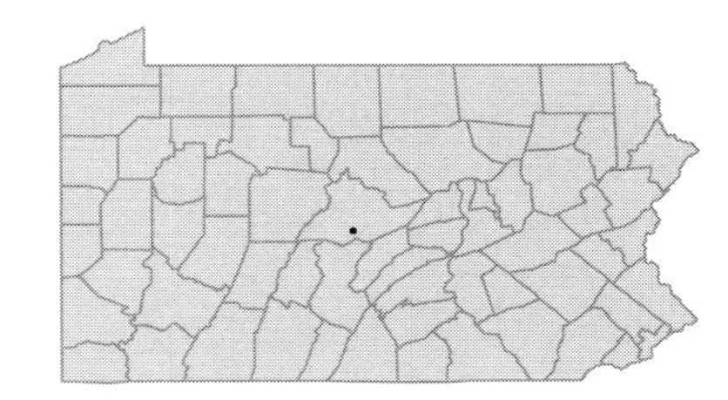

Tagetes patula L.
French marigold
Herbaceous annual
Cultivated and occasionally escaped to rubbish dumps and old fields.

I

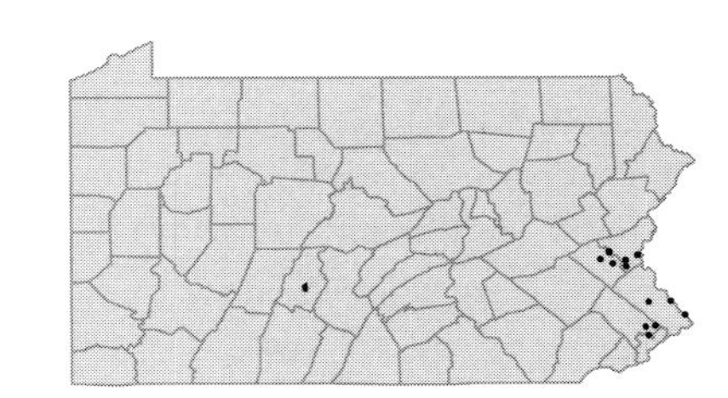

• ***Tanacetum vulgare*** L.
Tansy; Golden-buttons
Herbaceous perennial
Roadsides, fencerows, fields and pastures.

I

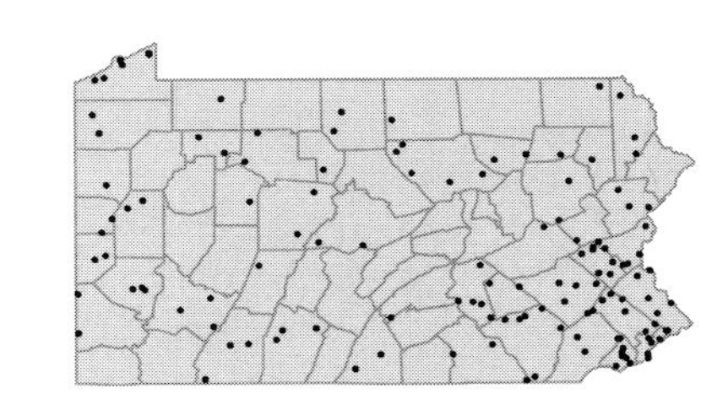

• ***Taraxacum laevigatum*** (Willd.) DC.
Red-seeded dandelion
Herbaceous perennial
Woods, rocky slopes, lawns and rubbish dumps.
Taraxacum erythrospermum Andrz. PFW

I

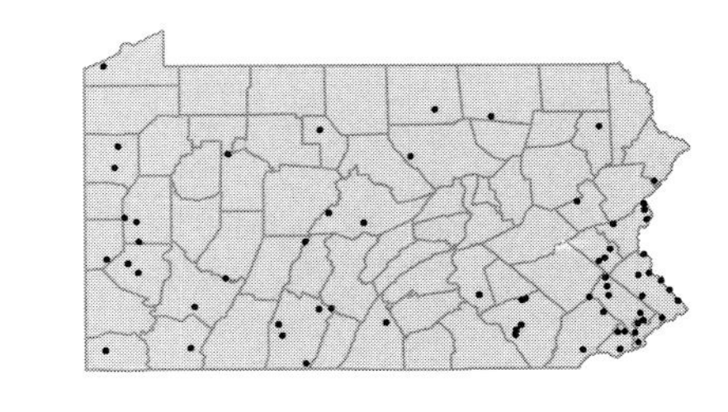

Taraxacum officinale Weber
Dandelion
Herbaceous perennial
Fields, roadsides, lawns, gardens and waste ground.
Taraxacum taraxacum (L.) Karst. P

FACU-
I

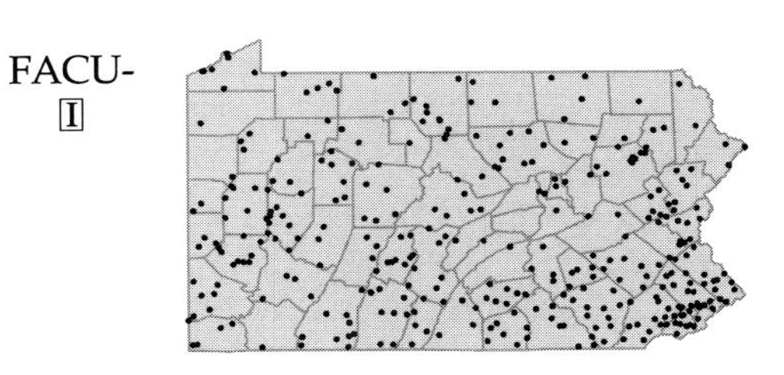

• ***Tragopogon dubius*** Scop.
Goat's-beard
Herbaceous biennial
Old fields, roadsides and railroad tracks.
Tragopogon major Jacq. FW

[I]

Tragopogon porrifolius L.
Salsify; Vegetable-oyster
Herbaceous biennial
Cultivated and occasionally escaped to railroad tracks, roadside banks or streets.

[I]

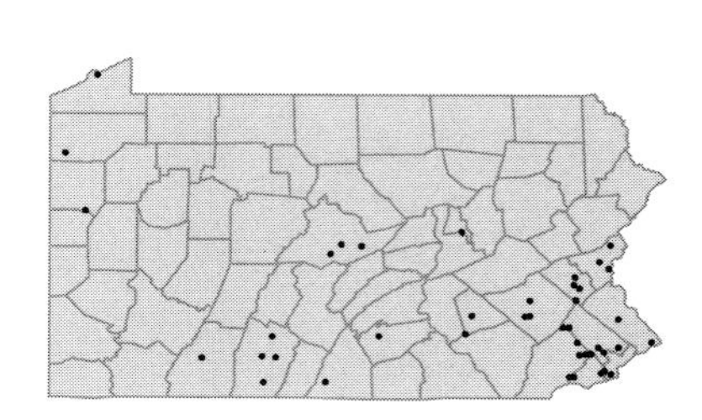

Tragopogon pratensis L.
Yellow goat's-beard
Herbaceous biennial
Fields, roadsides, railroad tracks and vacant lots.

[I]

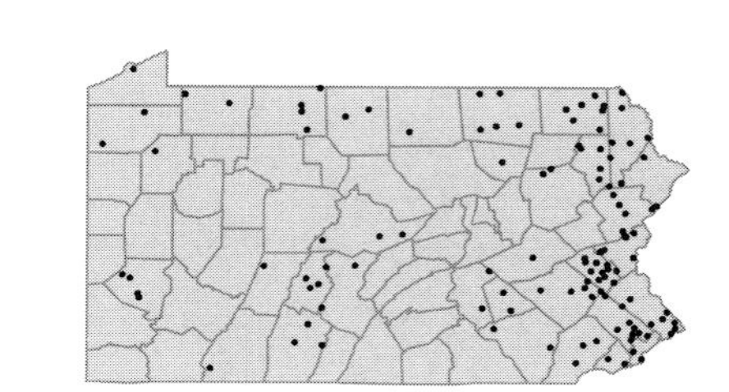

• ***Tussilago farfara*** L.
Coltsfoot
Herbaceous perennial
Woods, stream banks, roadsides, railroad tracks and waste ground.

FACU
[I]

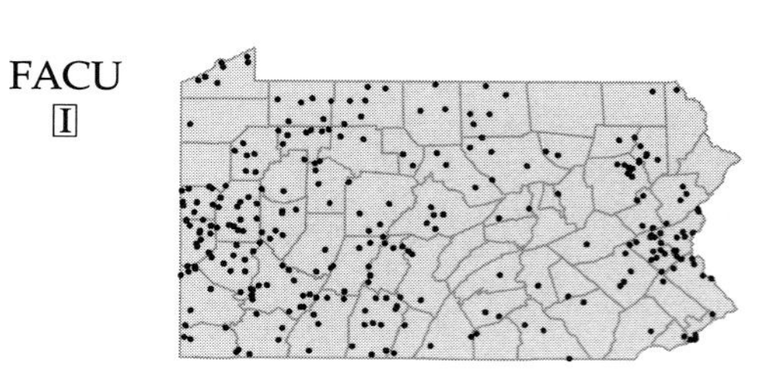

• ***Verbesina alternifolia*** (L.) Britt.
Wingstem
Herbaceous perennial
Moist, wooded slopes, shaded lowlands and roadsides.
Actinomeris alternifolia (L.) DC. F

FAC

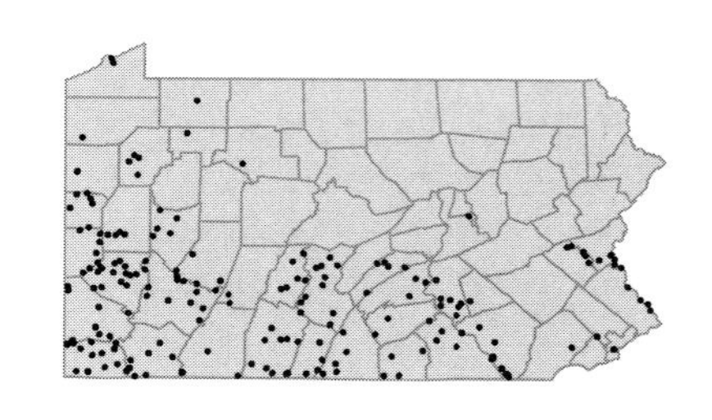

Verbesina encelioides (Cav.) Benth. & Hook.
Crownbeard
Herbaceous annual
Fields and waste ground.

FACU-
[I]

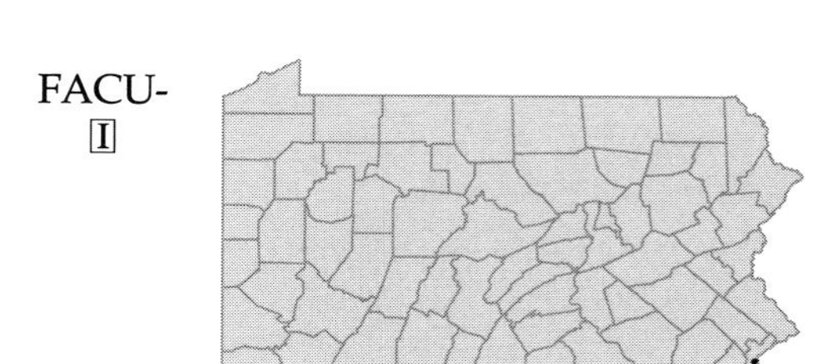

• ***Vernonia gigantea*** (Walt.) Trel. var. ***gigantea*** FAC
Ironweed
Herbaceous perennial
Moist fields, meadows, or floodplains.
Vernonia altissima Nutt. *pro parte* FBG

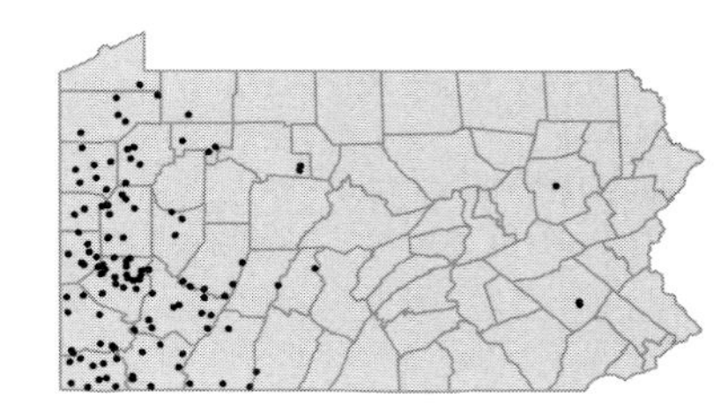

Vernonia glauca (L.) Willd.
Tawny ironweed; Appalachian ironweed
Herbaceous perennial
Dry fields, open slopes or clearings.

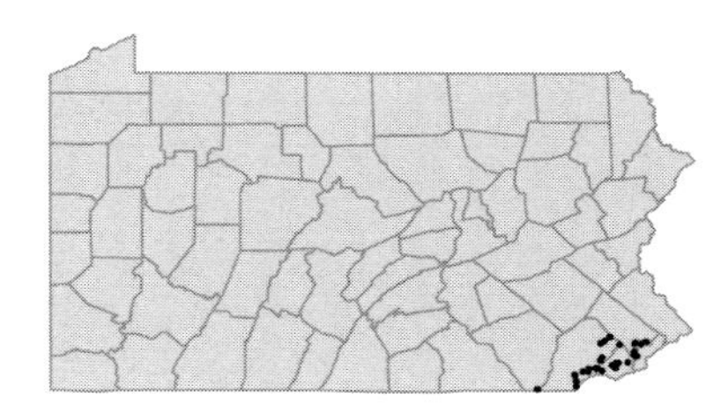

Vernonia noveboracensis (L.) Michx. FACW+
New York ironweed
Herbaceous perennial
Stream banks and wet fields, pastures or meadows.

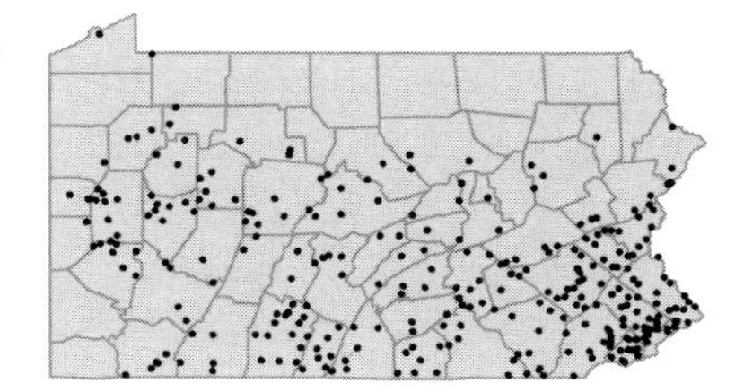

• ***Xanthium spinosum*** L. FACU I
Spiny cocklebur
Herbaceous annual
Waste ground, dump, cinders and ballast.

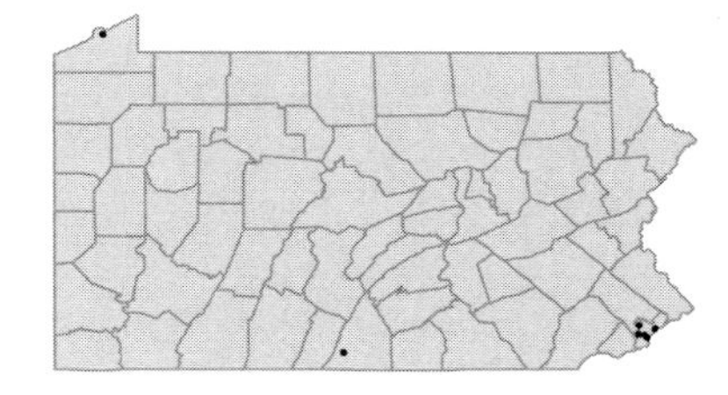

Xanthium strumarium L. var. ***canadense*** (P.Mill.) Torr. & A.Gray FAC
Common cocklebur
Herbaceous annual
Abandoned fields, cultivated ground and edges.
Xanthium echinatum Murr. PF; *Xanthium italicum* Moretti FW; *Xanthium oviforme* Wallr. FW; *Xanthium pensylvanicum* Wallr. FW

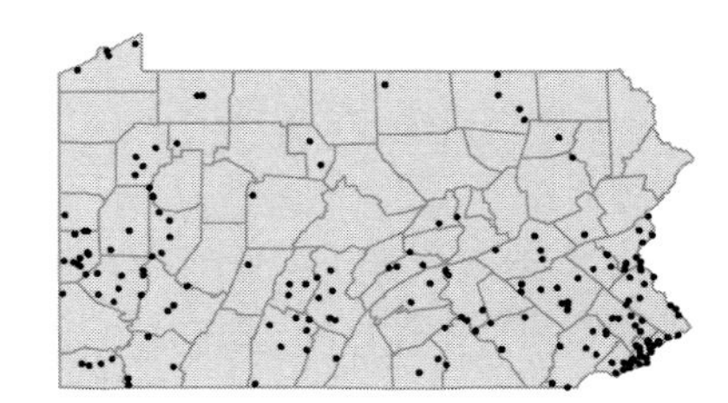

Xanthium strumarium L. var. ***glabratum*** (DC.) Cronq. FAC
Cocklebur
Herbaceous annual
Fields, stream banks and roadsides.
Xanthium chinense P.Mill. FW; *Xanthium glabratum* (DC.) Britt. P

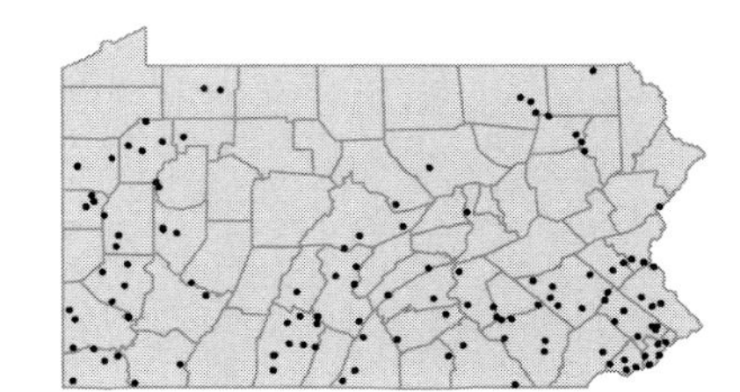

• ***Zinnia elegans*** Jacq. [I]
Zinnia
Herbaceous annual
Cultivated and occasionally escaped to roadsides, fallow fields, and dumps.

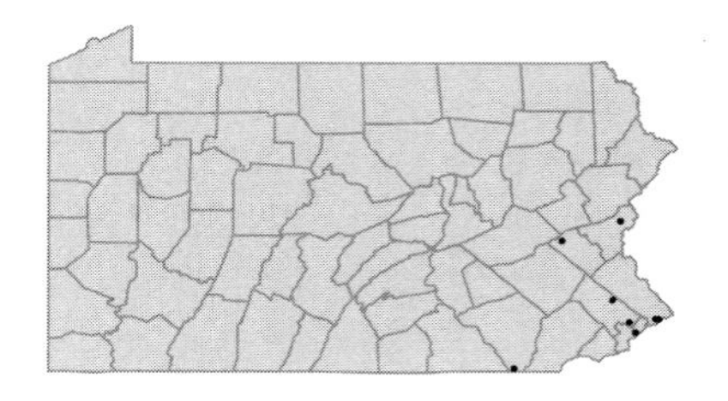

LILIOPSIDA
(Monocotyledons)

Jack-in-the-pulpit (*Arisaema triphyllum*) is a common member of the spring flora in moist woods.

BUTOMACEAE

• ***Butomus umbellatus*** L.
Flowering-rush
Herbaceous perennial, emergent aquatic
Pond margins and marshes.

OBL
I

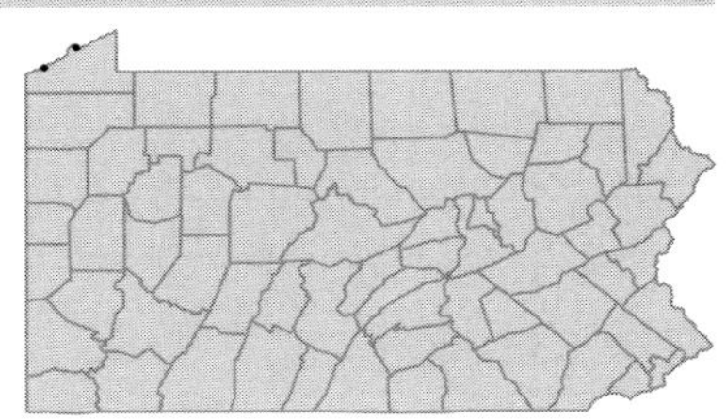

ALISMATACEAE

• ***Alisma plantago-aquatica*** L. var. ***americana*** Schultes & Schultes
Broad-leaved water-plantain; Northern water-plantain
Herbaceous perennial, emergent aquatic
Shallow water of ditches, lake margins and stream edges.
Alisma triviale Pursh FWC

OBL

Alisma plantago-aquatica L. var. ***parviflorum*** (Pursh) Torr.
Southern water-plantain
Herbaceous perennial, emergent aquatic
Marshes, stream and pond margins and muddy shores.
Alisma subcordatum Raf. FGBWC

OBL

• ***Sagittaria australis*** (J.G.Smith) Small
Appalachian arrowhead
Herbaceous perennial, emergent aquatic
Alluvial meadows, wet woods and backwater pools.
Sagittaria engelmanniana J.G.Smith ssp. *longirostra* (Micheli) Bogin G

OBL

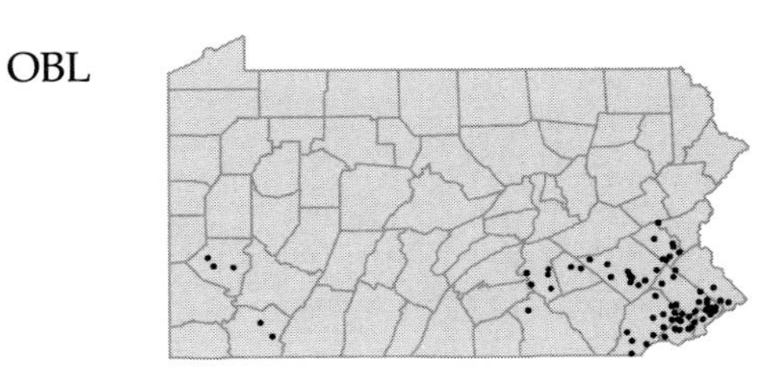

Sagittaria calycina Engelm.
Long-lobed arrowhead
Herbaceous annual, rooted submergent aquatic
Tidal mud flats.
Lophotocarpus spongiosus (Engelm.) J.G. Smith FW; *Sagittaria montevidensis* Cham. & Schlecht. B

OBL

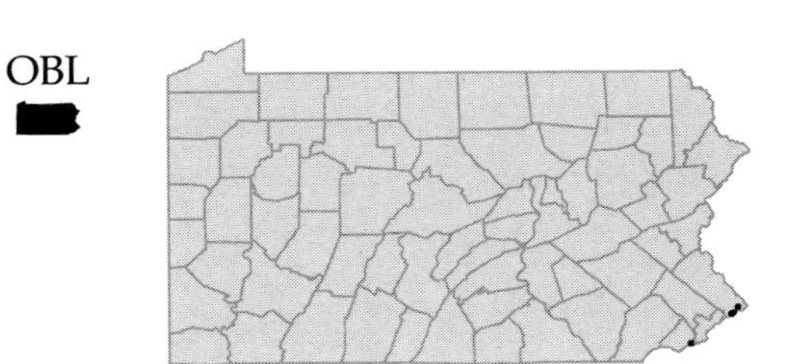

Sagittaria graminea Michx. var. ***graminea***
Grass-leaved sagittaria
Herbaceous perennial, emergent aquatic
Shallow water, mud flats and tidal shores.
Sagittaria eatonii J.G.Smith FW

OBL

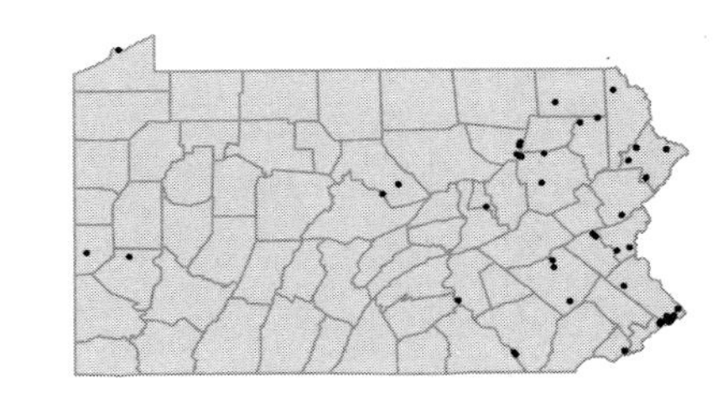

Sagittaria latifolia Willd. var. ***latifolia***
Arrowhead; Duck-potato; Wapato
Herbaceous perennial, emergent aquatic
Swamps, wet shores and shallow water of ponds and streams.

OBL

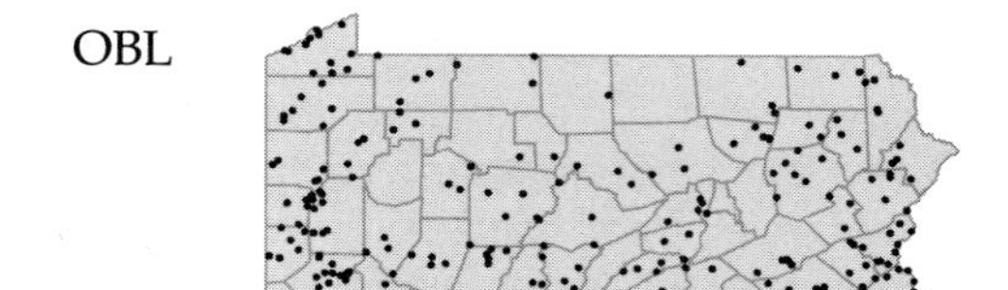

Sagittaria latifolia Willd. var. ***pubescens*** (Muhl. ex Nutt.) J.G.Smith
Arrowhead; Duck-potato; Wapato
Herbaceous perennial, emergent aquatic
Moist meadows and river banks.

OBL

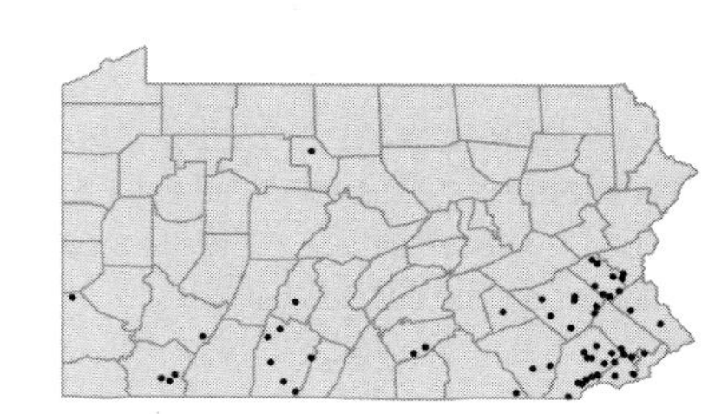

Sagittaria rigida Pursh
Arrowhead
Herbaceous perennial, emergent aquatic
Pond margins, mud flats and stream edges.

OBL

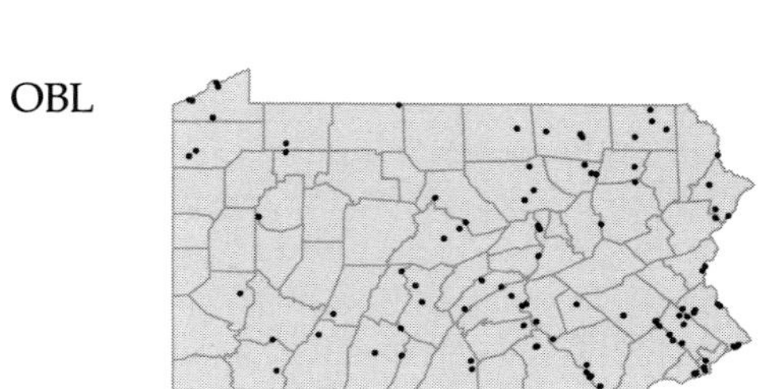

Sagittaria stagnorum Small
Arrowhead
Herbaceous perennial, rooted submergent aquatic
Lakes and ponds. Believed to be extirpated, represented by a single collection from Bucks Co. in the early 1900's .
Sagittaria subulata (L.) Buch. var. *gracillima* (S.Wats.) J.G.Smith PFGBWC

OBL

Sagittaria subulata (L.) Buch
Strap-leaf arrowhead; Subulate arrowhead
Herbaceous perennial, rooted submergent aquatic
Tidal shores and mud flats.

OBL

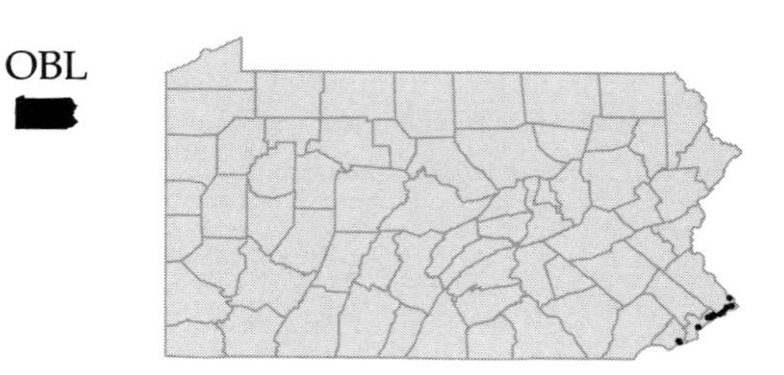

HYDROCHARITACEAE

• ***Egeria densa*** Planch.
Brazilian waterweed
Herbaceous perennial, rooted submergent aquatic
Naturalized in ponds and lakes.

OBL
I

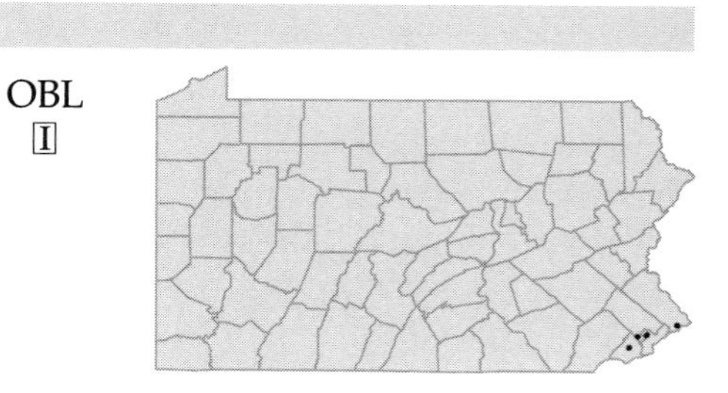

• ***Elodea canadensis*** L.C.Rich. ex Michx. OBL
Ditch-moss; Broad waterweed
Herbaceous perennial, rooted submergent aquatic
Shallow water of rivers, creeks, lakes and ponds. Pistillate plants are common, but staminate forms are known from only a few sites.
Anacharis canadensis (Michx.) L.C.Rich. GB; *Philotria canadensis* (Michx.) Britt. P

Elodea nuttallii (Planch.) St. John OBL
Waterweed
Herbaceous perennial, rooted submergent aquatic
Shallow water of rivers, streams and ponds, also tidal mud flats.
Anacharis nuttallii Planch. GB; *Philotria angustifolia* Muhl. P

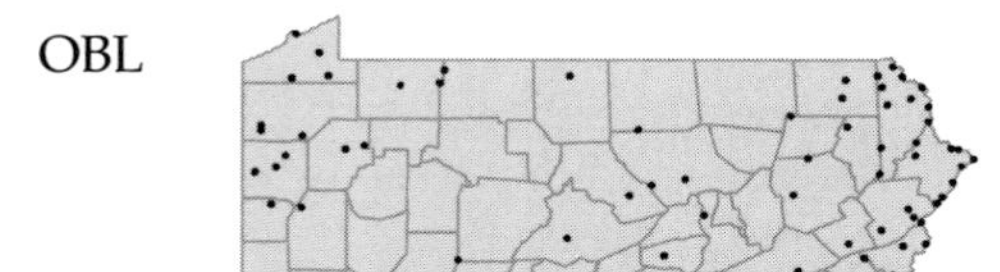

Elodea schweinitzii (Planch.) Caspary
Schweinitz's waterweed
Herbaceous perennial, rooted submergent aquatic
Calcareous water. Believed to be extirpated, this species, which has perfect flowers, has been collected only at or near the type locality in 1829-1832.

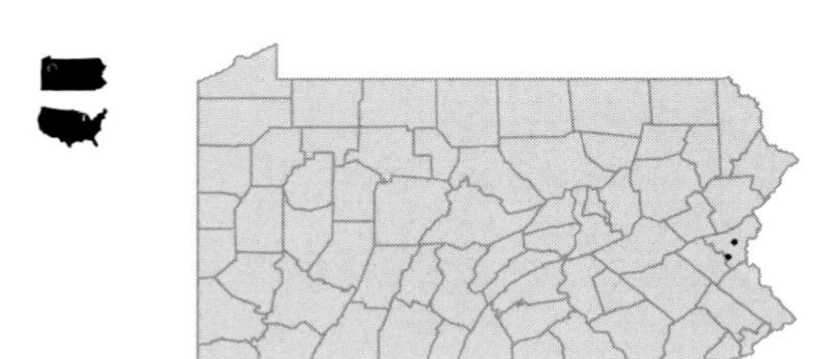

• ***Vallisneria americana*** Michx. var. ***americana*** OBL
Tape-grass; Water-celery
Herbaceous perennial, rooted submergent aquatic
Riverbeds, streams and lakes.

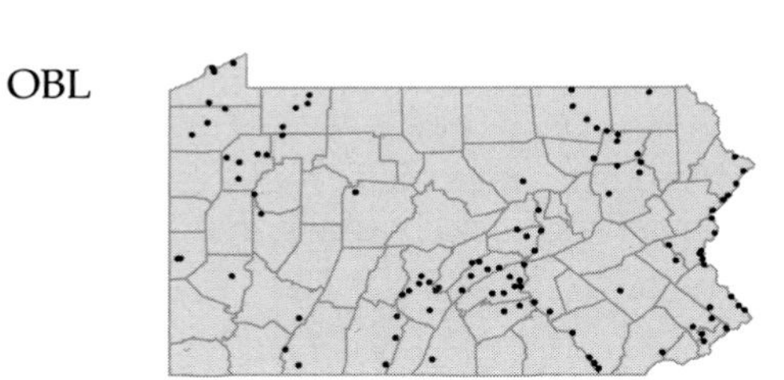

SCHEUCHZERIACEAE

• ***Scheuchzeria palustris*** L. OBL
Pod-grass
Herbaceous perennial, emergent aquatic
Sphagnum bogs.

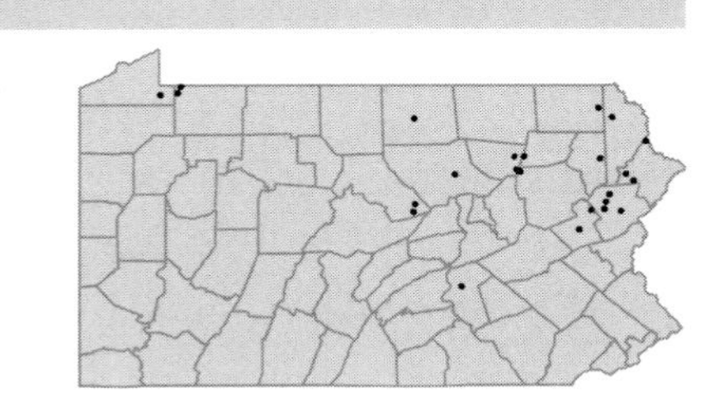

JUNCAGINACEAE

• ***Triglochin palustre*** L. OBL
Marsh arrow-grass; Slender arrow-grass
Herbaceous perennial, emergent aquatic
Moist, sandy shores. Believed to be extirpated, last collected in 1928.

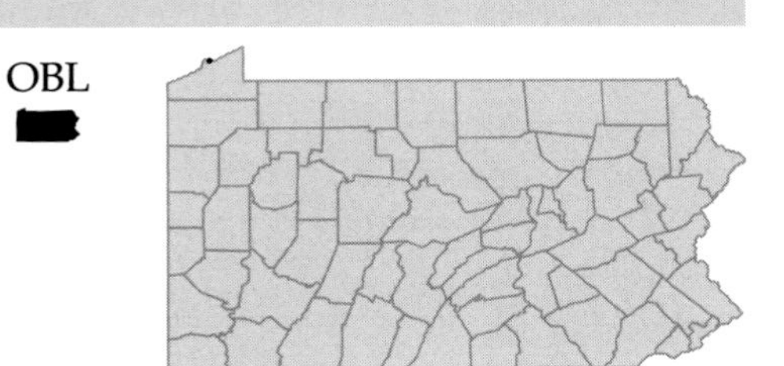

POTAMOGETONACEAE

• ***Potamogeton alpinus*** Balbis
Northern pondweed; Red pondweed
Herbaceous perennial, rooted submergent aquatic
Shallow river pools and backwater areas. Further study of plants resembling *P. alpinus* in the Delaware River basin is needed to determine whether they are true *P. alpinus* or hybrids of *P. gramineus* or *P. illinoensis* and *P. perfoliatus* has been suggested.

OBL

Potamogeton amplifolius Tuckerman
Bigleaf pondweed
Herbaceous perennial, rooted submergent aquatic
Lakes and streams.

OBL

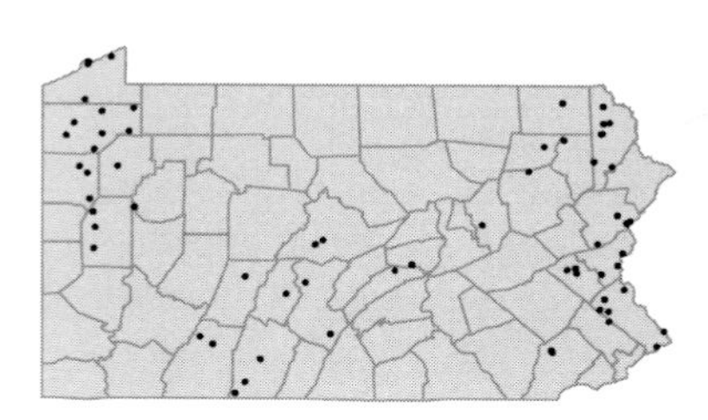

Potamogeton bicupulatus Fern.
Pondweed
Herbaceous perennial, rooted submergent aquatic
Shallow, quiet water.
Potamogeton diversifolius Raf. var. *trichophyllus* Morong GBC

OBL

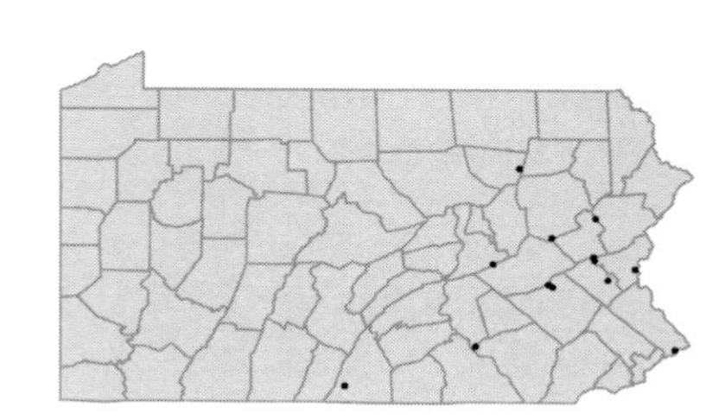

Potamogeton confervoides Reichenb.
Tuckerman's pondweed
Herbaceous perennial, rooted submergent aquatic
Glacial lakes and boggy ponds.

OBL

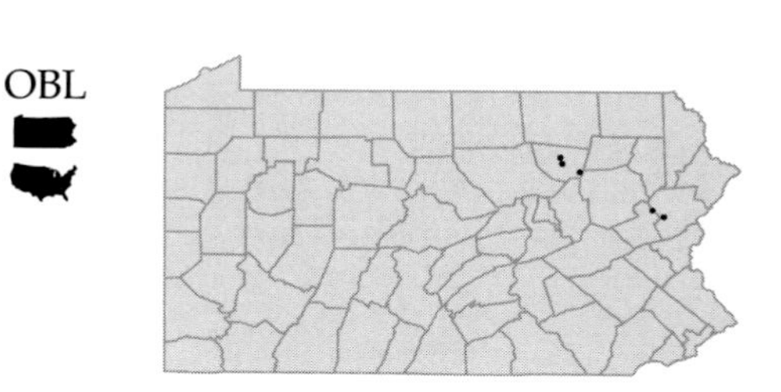

Potamogeton crispus L.
Curly pondweed
Herbaceous perennial, rooted submergent aquatic
Lakes, ponds, rivers and streams.

OBL
[I]

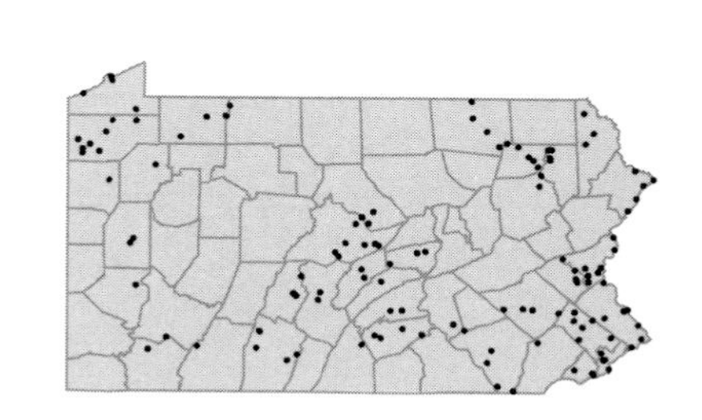

Potamogeton diversifolius Raf.
Snailseed pondweed
Herbaceous perennial, floating-leaf aquatic
Quiet, shallow water.

OBL

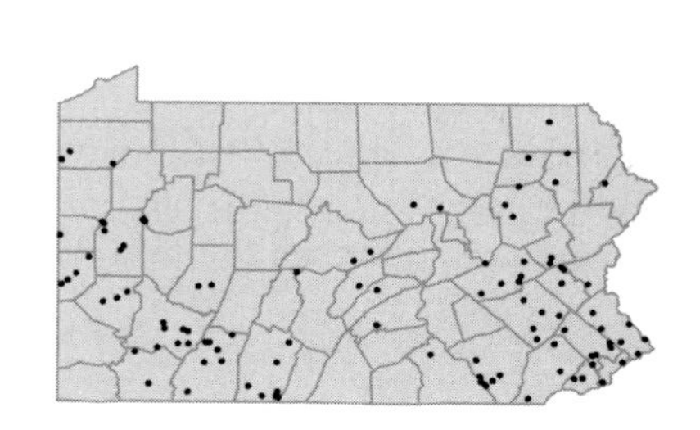

Potamogeton epihydrus Raf.
Ribbonleaf pondweed
Herbaceous perennial, floating-leaf aquatic
Ponds, lakes and streams.
Potamogeton nuttallii Cham. & Schlecht. P

OBL

Potamogeton filiformis Pers. var. ***borealis*** (Raf.) St.John
Threadleaf pondweed
Herbaceous perennial, rooted submergent aquatic
Shallow, calcareous water of streams and ponds.
Potamogeton filiformis Pers. var. *alpinus* (Blytt) Aschers. & Graebn. K

OBL

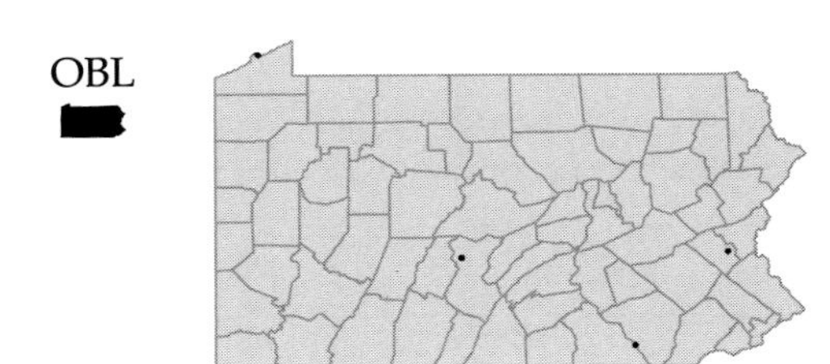

Potamogeton foliosus Raf.
Leafy pondweed
Herbaceous perennial, rooted submergent aquatic
Lakes and streams.

OBL

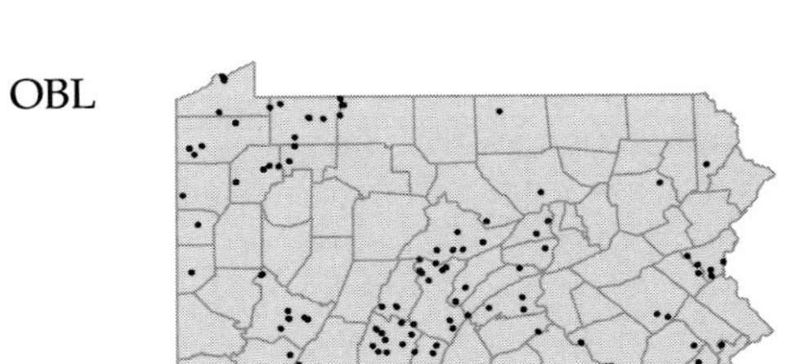

Potamogeton friesii Rupr.
Fries' pondweed
Herbaceous perennial, rooted submergent aquatic
Calcareous streams.

OBL

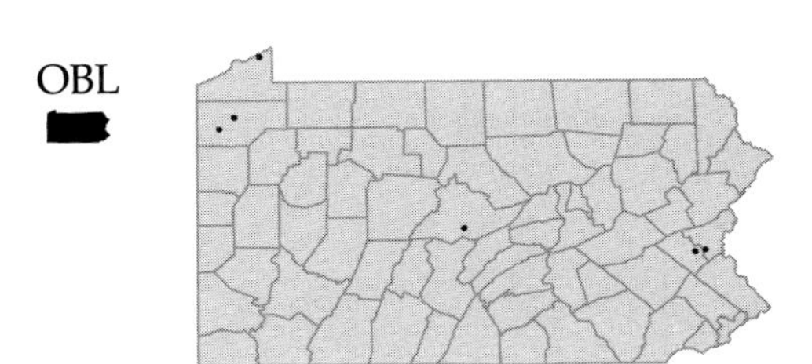

Potamogeton gramineus L.
Grassy pondweed; Variable pondweed
Herbaceous perennial, floating-leaf aquatic
Lakes and streams. A hybrid with *P. perfoliatus* (*P. x subnitens* Hagstr.) has been collected at two sites.

OBL

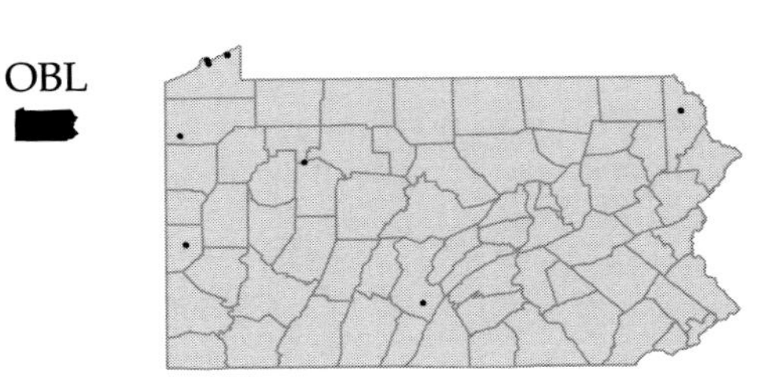

Potamogeton hillii Morong
Hill's pondweed
Herbaceous perennial, rooted submergent aquatic
Lakes and streams.
Potamogeton porteri Fern. FW

OBL

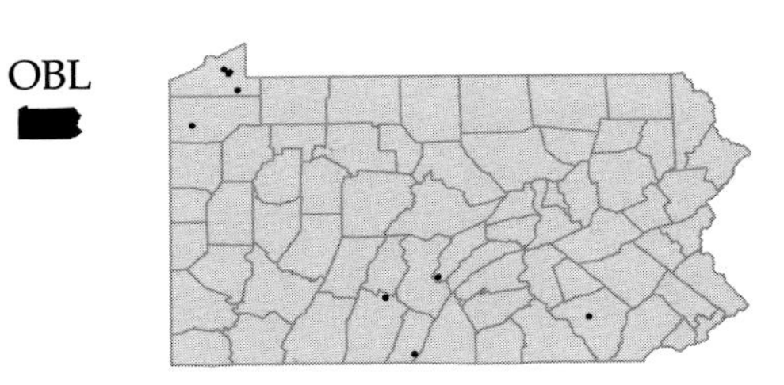

Potamogeton illinoensis Morong
Illinois pondweed
Herbaceous perennial, floating-leaf aquatic
Lakes and streams.
Potamogeton heterophyllus Schreb. P; *Potamogeton lucens* L. *pro parte* P;
Potamogeton zizii Roth *pro parte* P

OBL

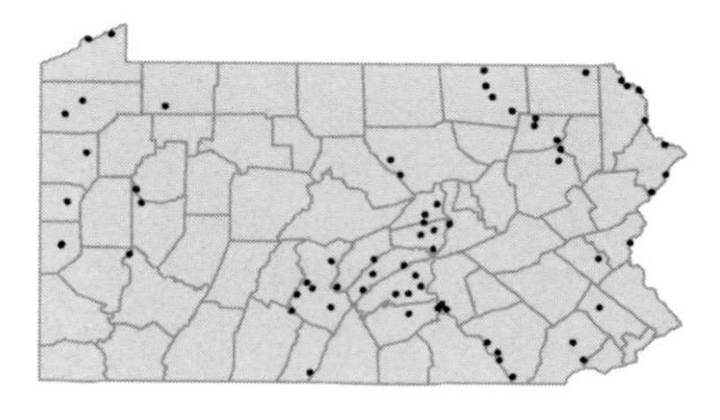

Potamogeton natans L.
Floating pondweed
Herbaceous perennial, rooted submergent aquatic
Lakes and streams.

OBL

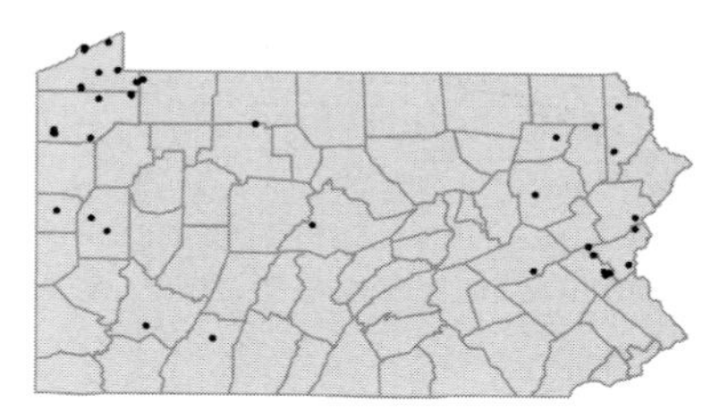

Potamogeton nodosus Poir.
Longleaf pondweed
Herbaceous perennial, floating-leaf aquatic
Lakes, ponds and streams.

OBL

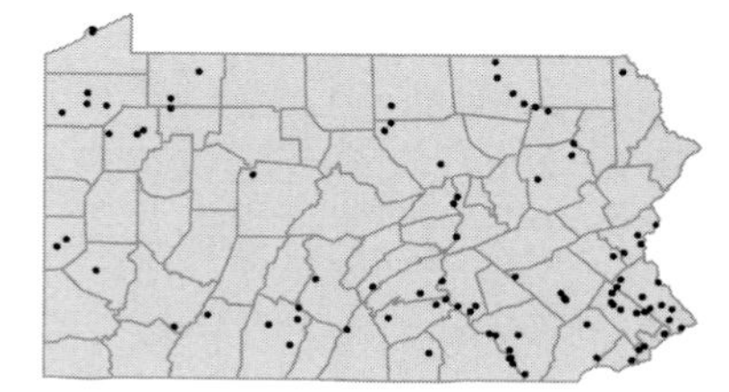

Potamogeton oakesianus J.W.Robbins
Oakes' pondweed
Herbaceous perennial, floating-leaf aquatic
Ponds and lakes.

OBL

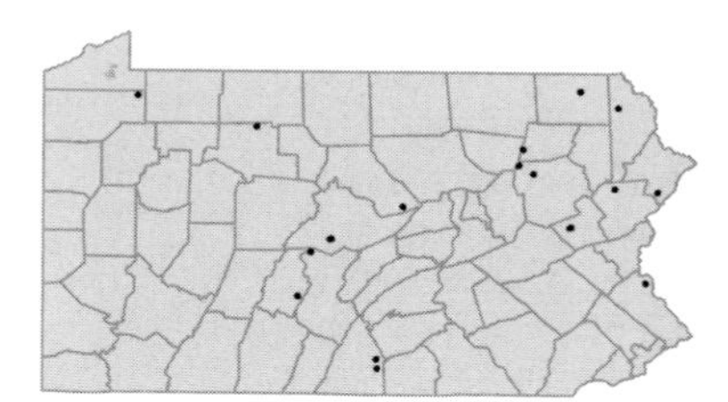

Potamogeton obtusifolius Mert. & Koch
Blunt-leaved pondweed
Herbaceous perennial, rooted submergent aquatic
Boggy ponds and lakes.

OBL

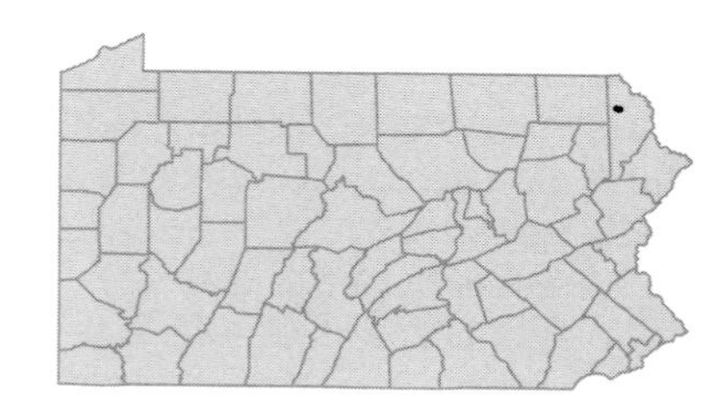

Potamogeton pectinatus L.
Sago pondweed
Herbaceous perennial, rooted submergent aquatic
Shallow, calcareous or brackish water.

OBL

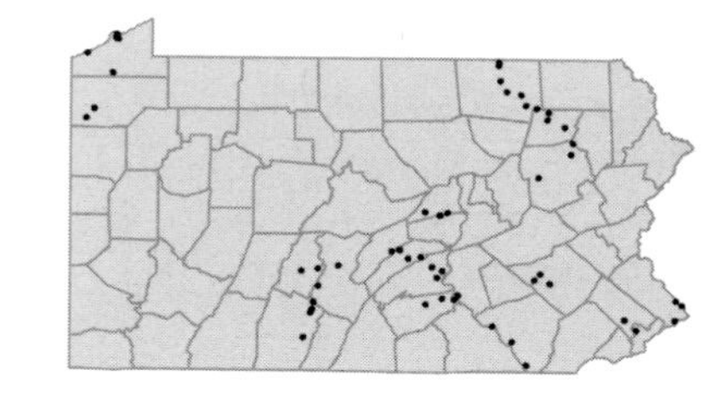

Potamogeton perfoliatus L.
Perfoliate pondweed; Redhead-grass
Herbaceous perennial, rooted submergent aquatic
Ponds, lakes and streams. Some PA material included here has previously been segregated as *P. richardsonii*, however, an analysis of leaf measurements of 105 specimens from throughout the state showed no basis for separation.

OBL

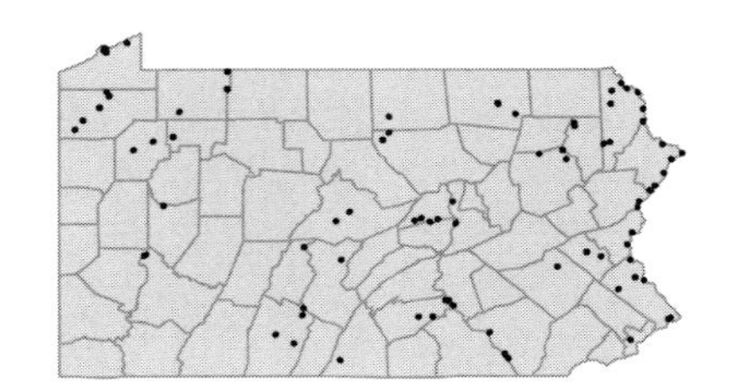

Potamogeton praelongus Wulfen
White-stem pondweed
Herbaceous perennial, rooted submergent aquatic
Lakes. Believed to be extirpated, last collected in the 1800's.

OBL

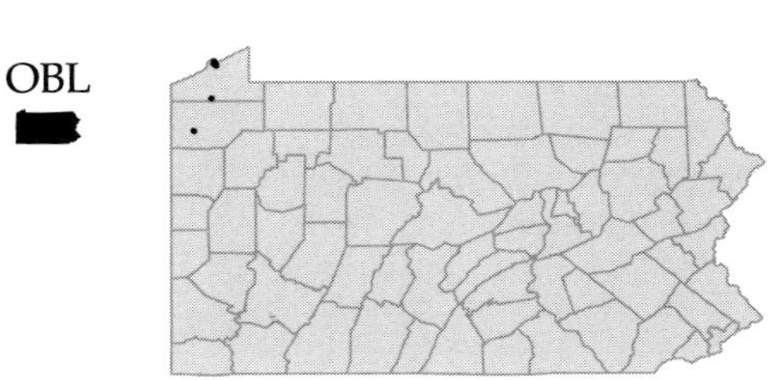

Potamogeton pulcher Tuckerman
Heartleaf pondweed; Spotted pondweed
Herbaceous perennial, floating-leaf aquatic
Shallow acidic water, swamps, and peaty or muddy shores.

OBL

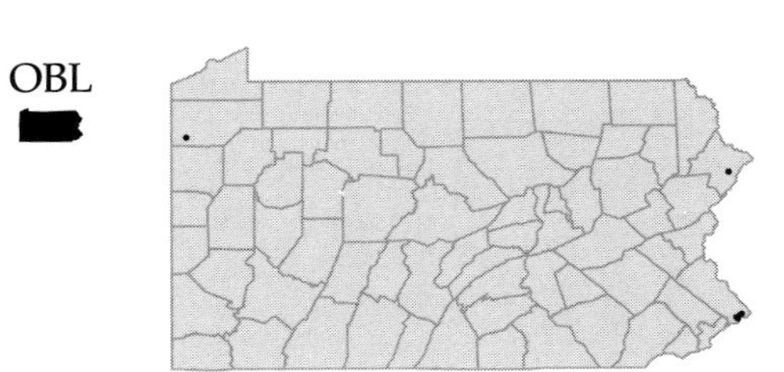

Potamogeton pusillus L.
Pondweed
Herbaceous perennial, rooted submergent aquatic
Lakes and ponds.
Potamogeton berchtoldii Fieber *pro parte* FW

OBL

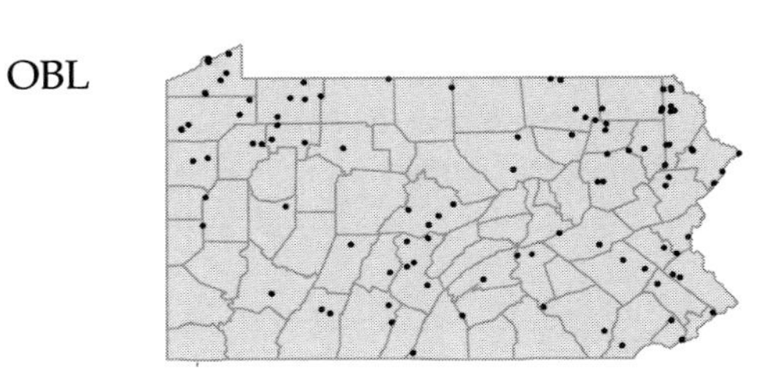

Potamogeton robbinsii Oakes
Flat-leaved pondweed; Fern pondweed
Herbaceous perennial, rooted submergent aquatic
Quiet water of lakes and ponds.

OBL

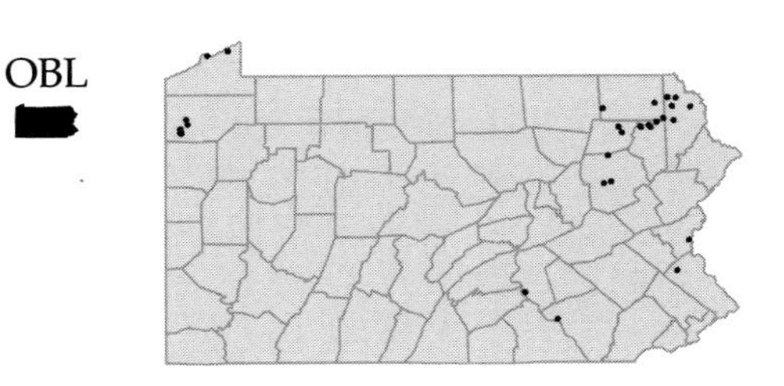

Potamogeton spirillus Tuckerman
Snailseed pondweed
Herbaceous perennial, floating-leaf aquatic
Shallow water of lakes, ponds and rivers.

OBL

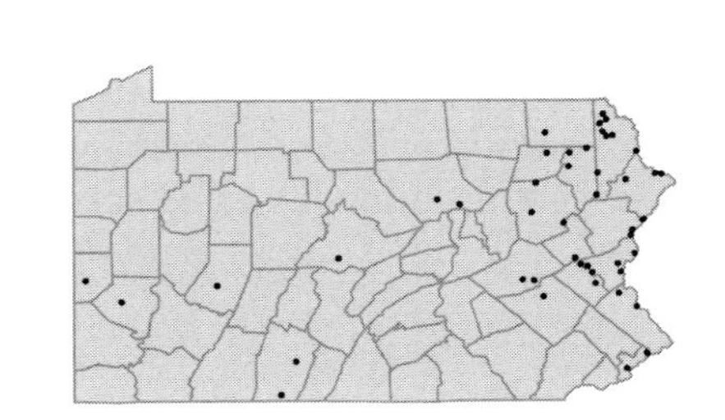

Potamogeton strictifolius Ar.Benn.
Narrow-leaved pondweed; Straight-leaved pondweed
Herbaceous perennial, rooted submergent aquatic
Calcareous ponds or streams.

OBL

Potamogeton tennesseensis Fern.
Tennessee pondweed
Herbaceous perennial, floating-leaf aquatic
Ponds or streams.

OBL

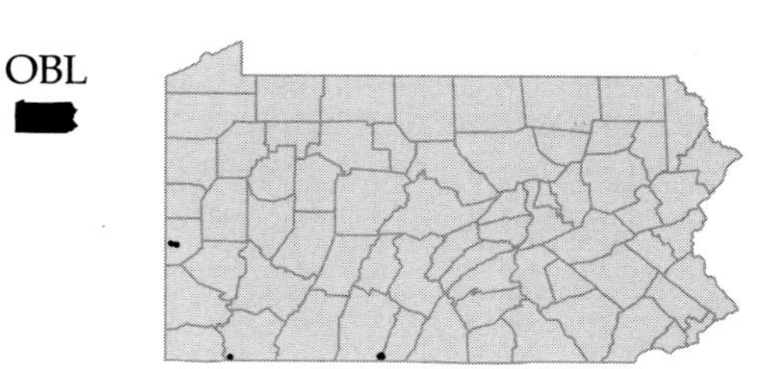

Potamogeton vaseyi J.W.Robbins
Vasey's pondweed
Herbaceous perennial, floating-leaf aquatic
Ponds, lagoons and other slow-moving water.

OBL

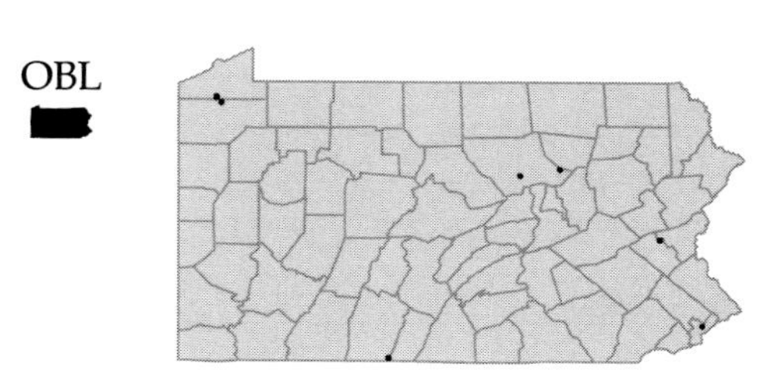

Potamogeton zosteriformis Fern.
Flat-stemmed pondweed
Herbaceous perennial, rooted submergent aquatic
Ponds and slow-moving streams.

OBL

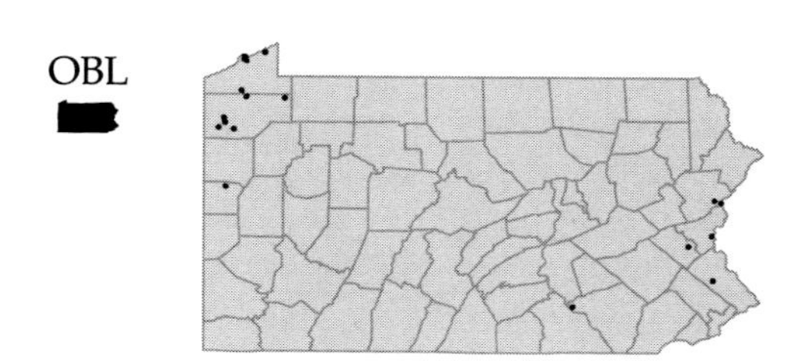

NAJADACEAE

• ***Najas flexilis*** (Willd.) Rostk. & Schmidt
Northern water-nymph; Naiad
Herbaceous annual, rooted submergent aquatic
Quiet water of lakes, ponds, creeks and canals.

OBL

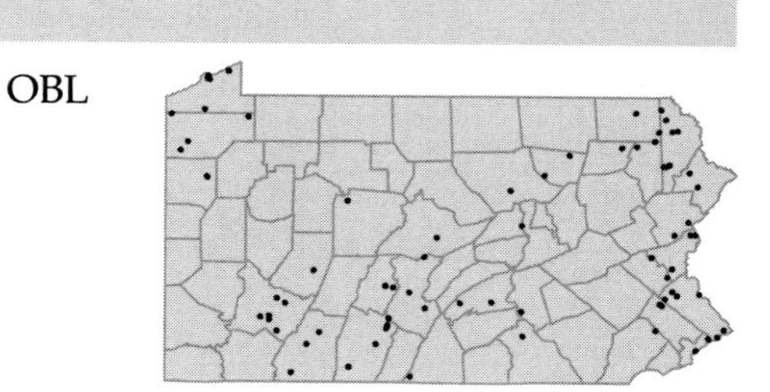

Najas gracillima (A.Braun) Magnus
Slender water-nymph; Bushy naiad
Herbaceous annual, rooted submergent aquatic
Shallow water of lakes, ponds and reservoirs.

OBL

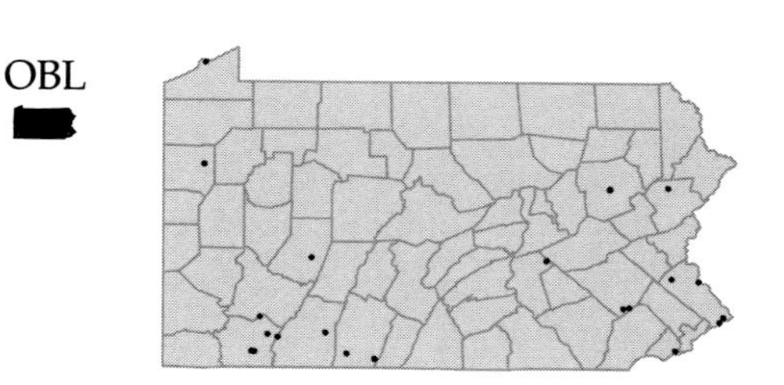

Najas guadalupensis (Spreng.) Magnus
Southern water-nymph; Naiad
Herbaceous annual, rooted submergent aquatic
Lakes, ponds, reserviors and tidal flats.

OBL

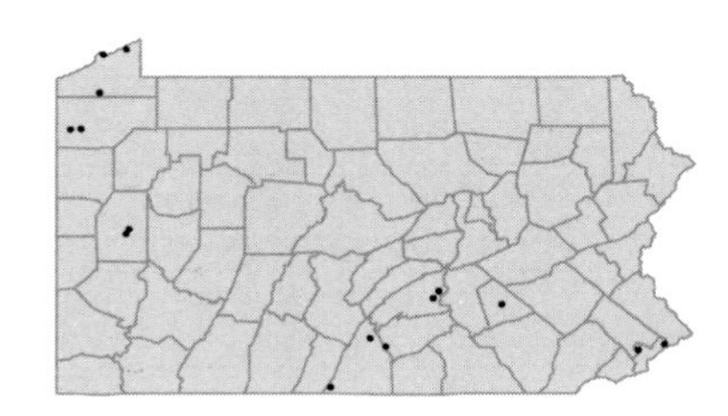

Najas marina L.
Holly-leaved naiad; Water-nymph
Herbaceous annual, rooted submergent aquatic
Shallow, calcareous water.

OBL

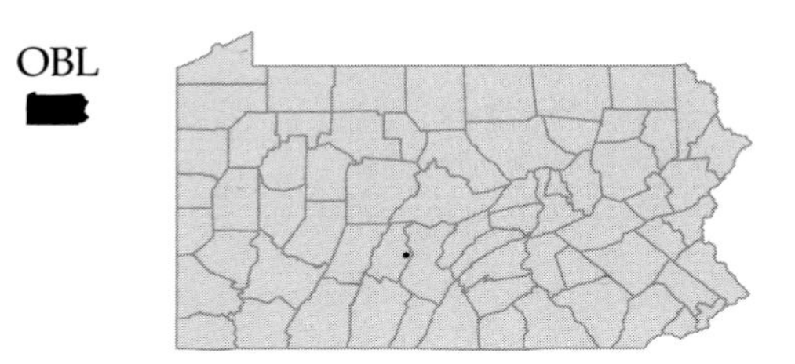

Najas minor All.
Water-nymph; Naiad
Herbaceous annual, rooted submergent aquatic
Shallow water of lakes, ponds and reserviors.

OBL
I

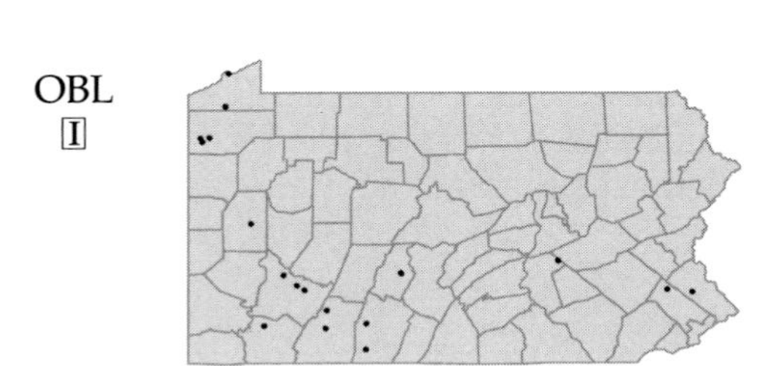

ZANNICHELLIACEAE

• ***Zannichellia palustris*** L.
Horned pondweed
Herbaceous perennial, rooted submergent aquatic
Streams, ponds, lakes, springs and tidal flats.

OBL

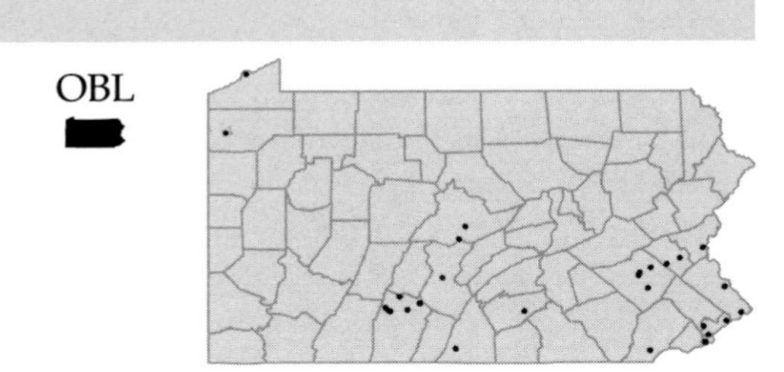

ACORACEAE

• ***Acorus americanus*** (Raf.) Raf.
Sweetflag
Herbaceous perennial, emergent aquatic
Shallow water of ponds and marshes.

OBL

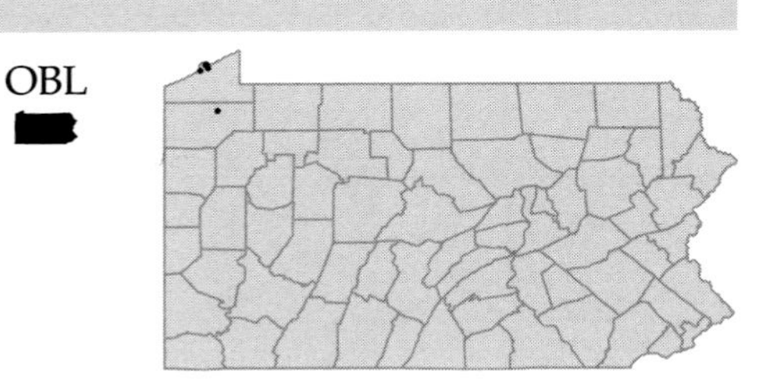

Acorus calamus L.
Sweetflag
Herbaceous perennial, emergent aquatic
Wet meadows, stream edges, ditches and swamps.

OBL
I

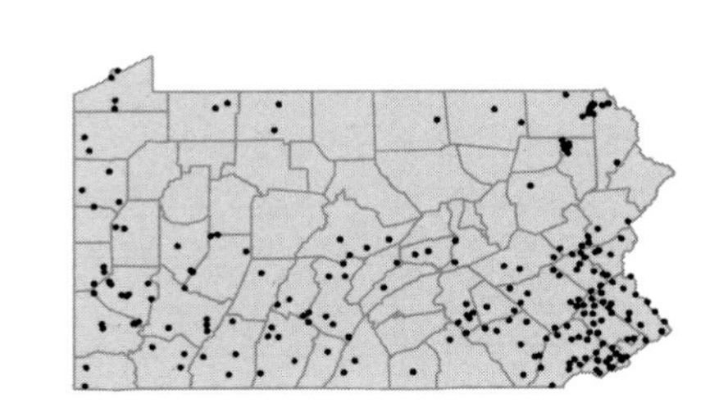

ARACEAE

• ***Arisaema dracontium*** (L.) Schott
Green-dragon
Herbaceous perennial
Low woods, floodplains and swamps.

FACW

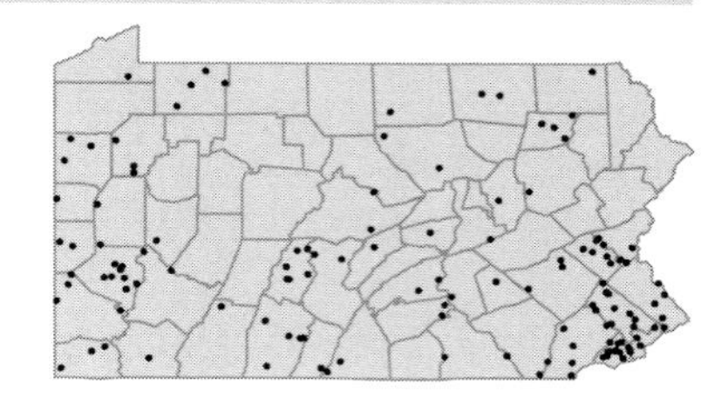

Arisaema triphyllum (L.) Schott ssp. ***triphyllum***
Jack-in-the-pulpit; Indian-turnip
Herbaceous perennial
Moist woods, swamps and bogs.

FACW-

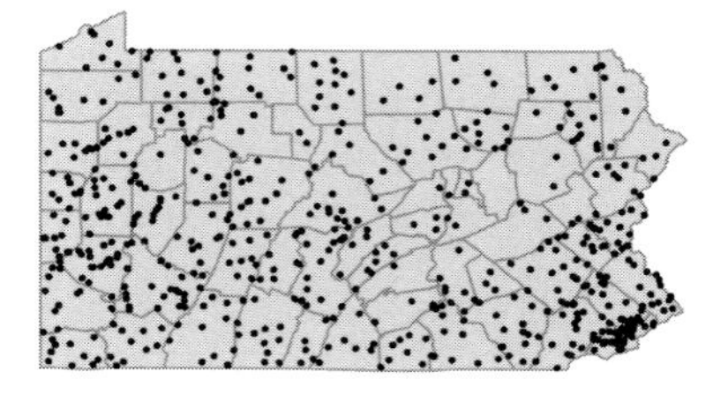

Arisaema triphyllum (L.) Schott ssp. ***pusillum*** (Peck) Huttleston
Small jack-in-the-pulpit
Herbaceous perennial
Moist woods, bogs and swamps.
Arisaema pusillum (Peck) Nash W

FACW-

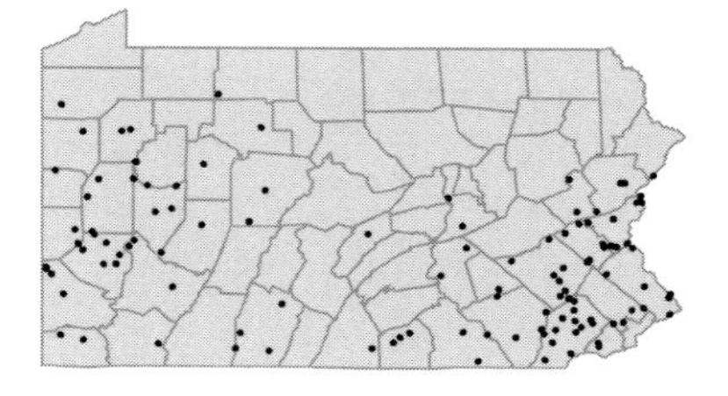

Arisaema triphyllum (L.) Schott ssp. ***stewardsonii*** (Britt.) Huttleston
Jack-in-the-pulpit; Indian-turnip
Herbaceous perennial
Swampy woods, wet thickets and bogs.
Arisaema stewardsonii Britt. PFW

FACW-

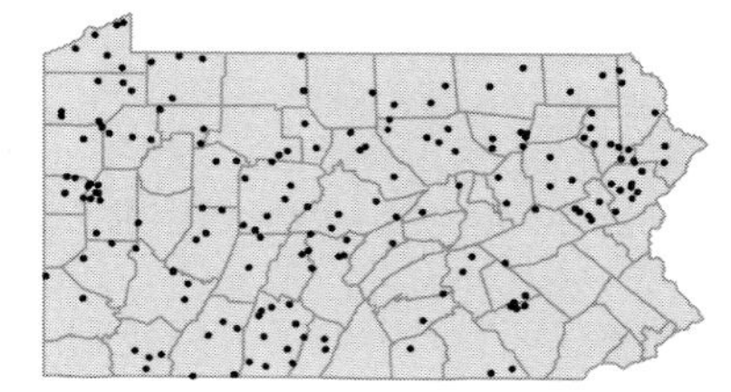

• ***Calla palustris*** L.
Wild calla
Herbaceous perennial, emergent aquatic
Bogs and swamps.

OBL

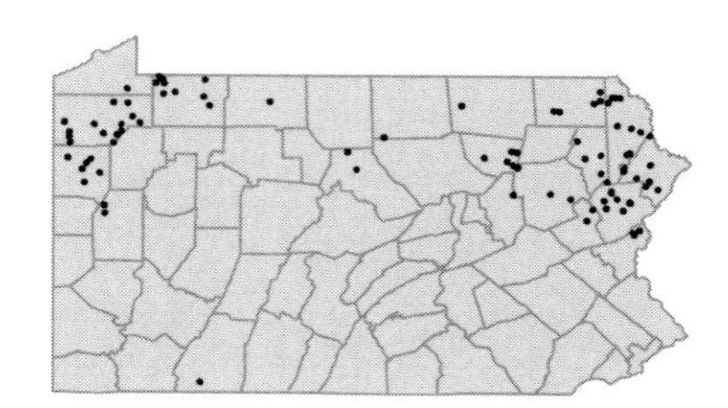

• ***Orontium aquaticum*** L.
Golden-club
Herbaceous perennial, emergent aquatic
Swamps, lakes, ponds, streams, ditches and wet shores.

OBL

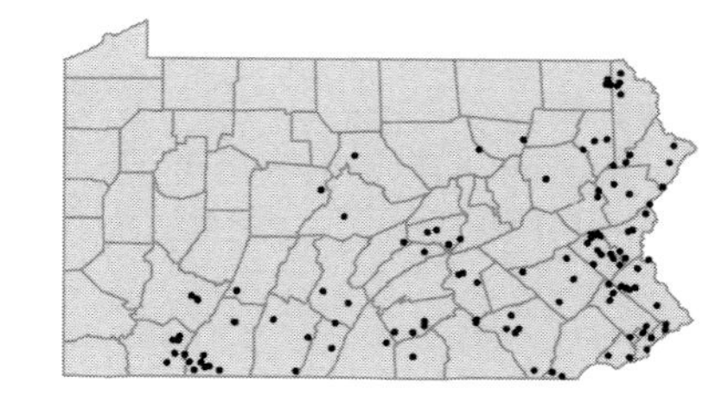

• ***Peltandra virginica*** (L.) Schott & Endl.
Arrow-arum; Tuckahoe
Herbaceous perennial, emergent aquatic
Swamps, stream or lake edges and tidal marshes.

OBL

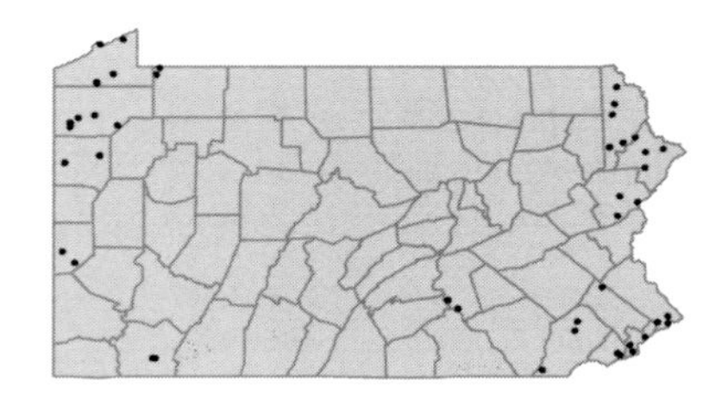

• ***Pinellia ternata*** (Thunb.) Tenore ex Breitenbach
Pinellia
Herbaceous perennial
A weed of gardens and planting beds.

I

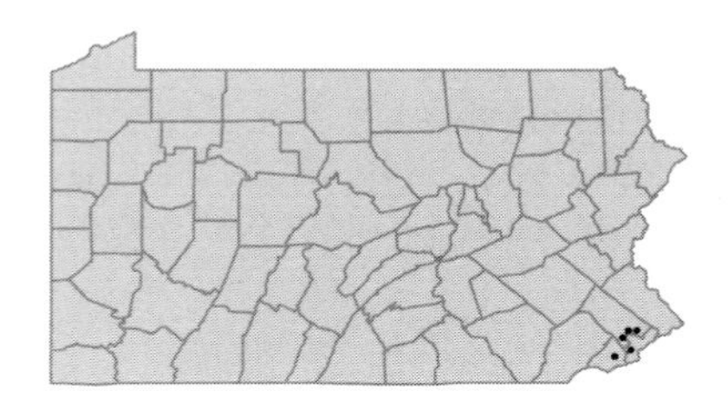

• ***Symplocarpus foetidus*** (L.) Salisb. ex Nutt.
Skunk-cabbage
Herbaceous perennial
Moist woods, swamps and bogs.
Spathyema foetida (L.) Raf. P

OBL

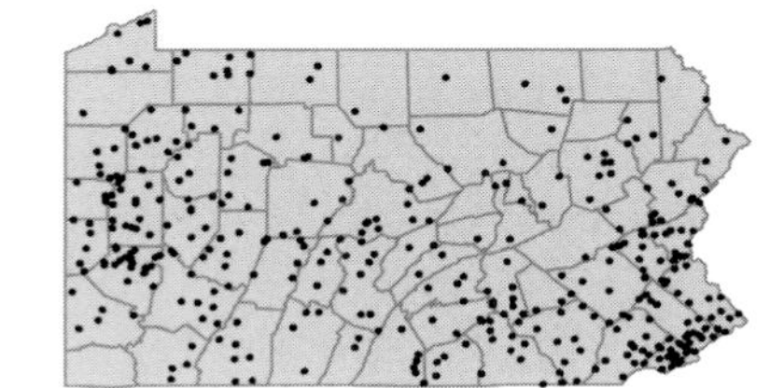

• ***Zantedeschia albomaculata*** (Hook.) Baill.
Black-throated calla; Spotted calla
Herbaceous perennial
Waste ground and rubbish dumps.

I

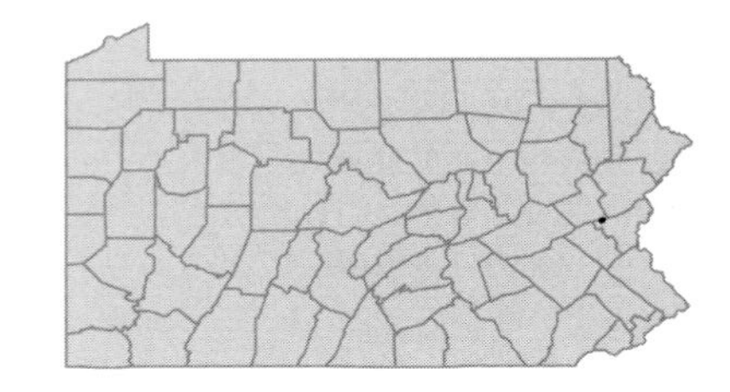

LEMNACEAE

• ***Lemna minor*** L.
Duckweed; Water-lentils
Herbaceous perennial, free-floating aquatic
Still water of lakes, ponds, streams, swamps and ditches.

OBL

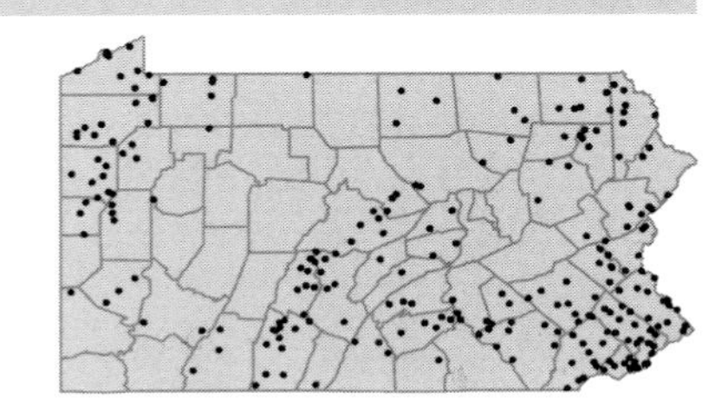

Lemna obscura (Austin) Daubs
Little water duckweed
Herbaceous perennial, free-floating aquatic
Shallow water. Believed to be extirpated, last collected in 1927.

OBL

Lemna perpusilla Torr.
Duckweed
Herbaceous perennial, free-floating aquatic
Ponds, bogs and marshes.

OBL

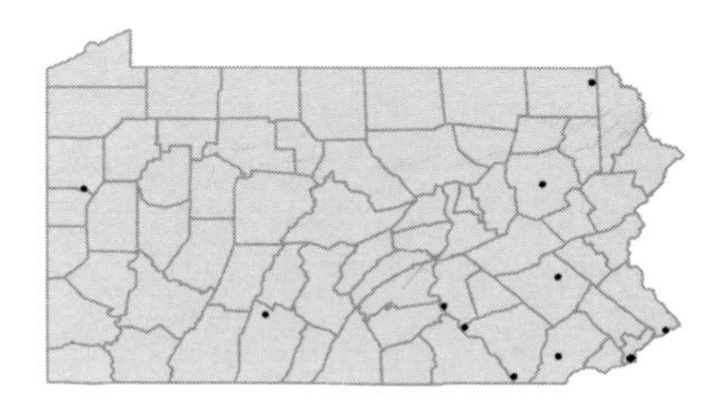

Lemna trisulca L.
Star duckweed
Herbaceous perennial, free-floating aquatic
Lakes, ponds, bogs, swamps, marshes and streams.

OBL

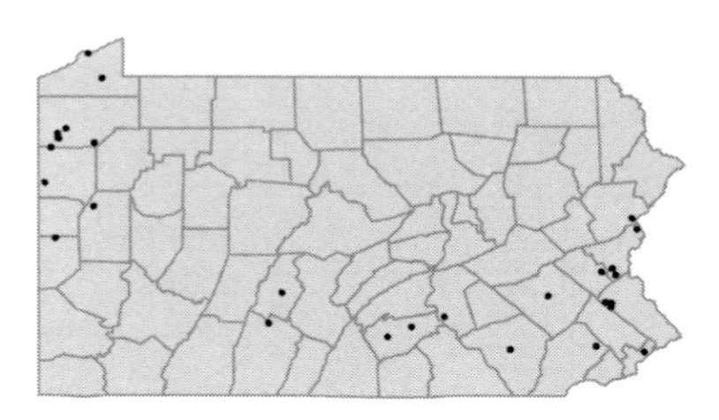

Lemna turionifera Landolt
Winter duckweed
Herbaceous perennial, free-floating aquatic
Lakes, ponds, swamps and marshes.

OBL

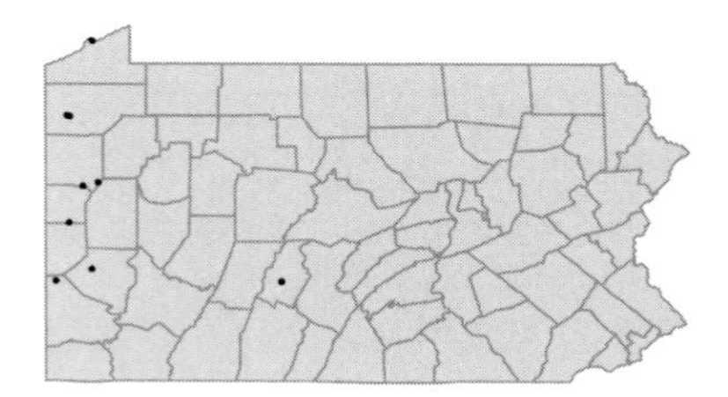

Lemna valdiviana Philippi
Pale duckweed
Herbaceous perennial, free-floating aquatic
In shallow water, floating or on submersed debris. Believed to be extirpated, last collected in 1953.

OBL

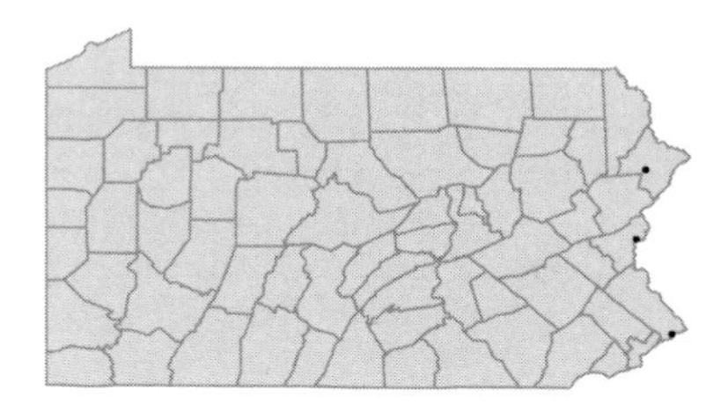

• ***Spirodela polyrhiza*** (L.) Schleid.
Greater duckweed; Water-flaxseed
Herbaceous perennial, free-floating aquatic
Ponds, lakes, swamps and margins of sluggish streams.

OBL

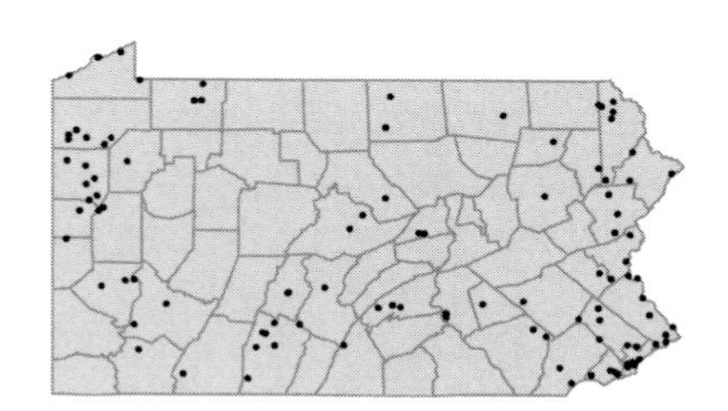

Spirodela punctata (G.F.W.Meyer) C.H.Thompson
Eastern water-flaxseed
Herbaceous perennial, free-floating aquatic
Pond.
Spirodela oligorrhiza (Kurtz) Hegelm. FW

OBL

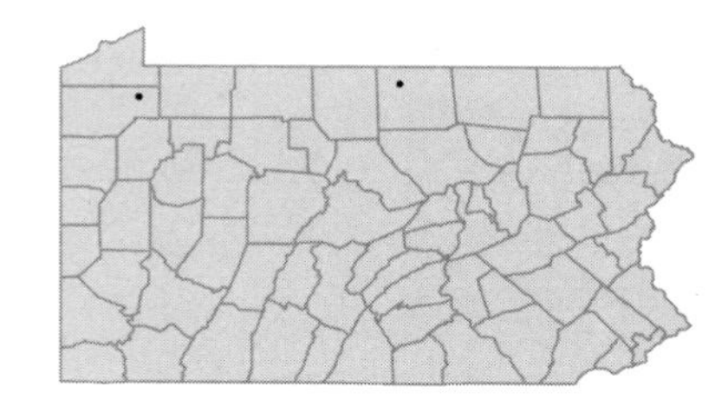

• ***Wolffia borealis*** (Engelm.) Landolt
Dotted water-meal
Herbaceous perennial, free-floating aquatic
Ponds and swamps. This taxon has been frequently confused with *W. punctata* Griseb. See Landolt (1980) for clarification.
Wolffia punctata auct. amer. nec Griseb.

OBL

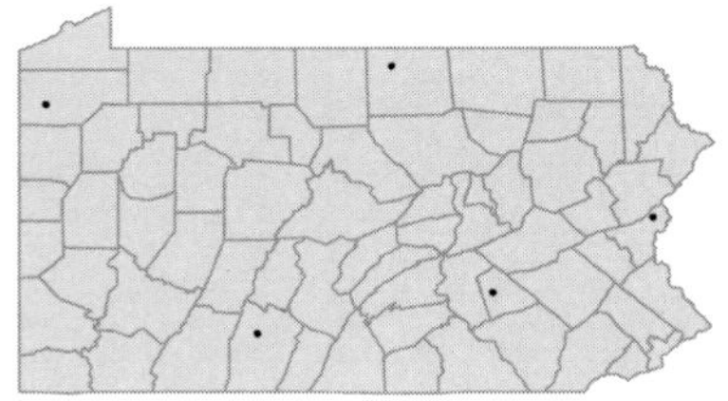

Wolffia brasiliensis Weddell
Pointed water-meal
Herbaceous perennial, free-floating aquatic
Lakes, ponds and margins of slow-moving streams.
Wolffia papulifera Thompson *pro parte* FGBC; *Wolffia punctata* Griseb. *pro parte* PFGBW

OBL

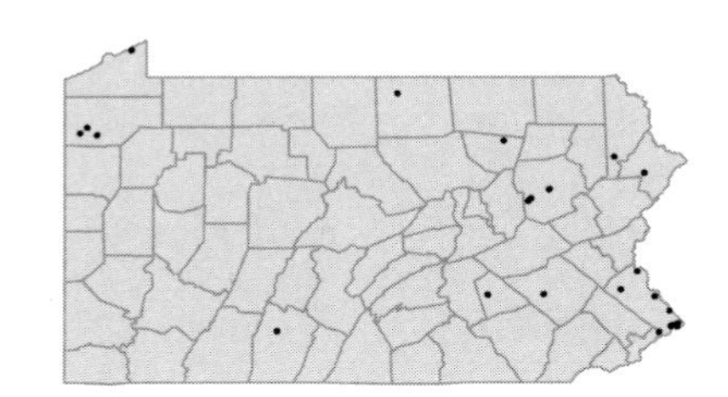

Wolffia columbiana Karst.
Water-meal
Herbaceous perennial, free-floating aquatic
Lakes, ponds, marshes, ditches, swamps and bogs.

OBL

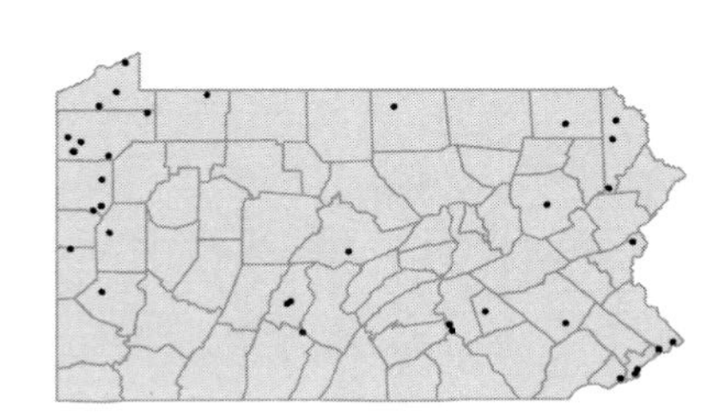

• ***Wolffiella gladiata*** (Hegelm.) Hegelm.
Bog-mat
Herbaceous perennial, free-floating aquatic
Open-water channel in swamp.
Wolffiella floridana (J.D.Smith) Thompson FGBC

OBL

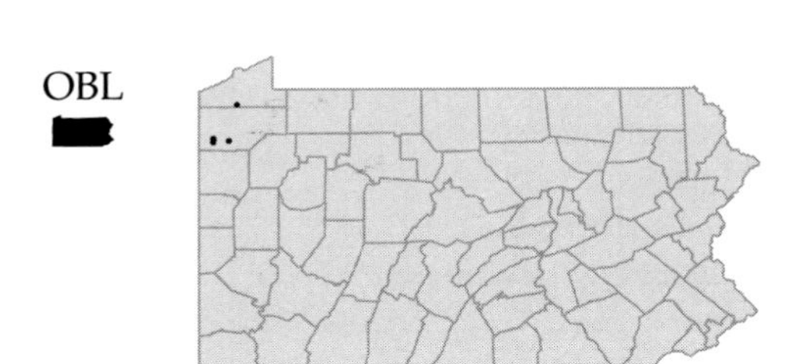

XYRIDACEAE

• ***Xyris difformis*** Chapm.
Yellow-eyed-grass
Herbaceous perennial, emergent aquatic
Damp, peaty or sandy soil of swamps or bogs.

OBL

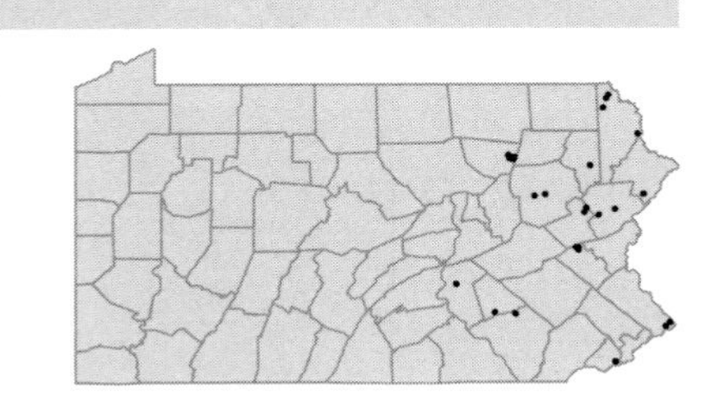

Xyris montana Ries
Yellow-eyed-grass
Herbaceous perennial, emergent aquatic
Exposed peat of floating bog mats.

OBL

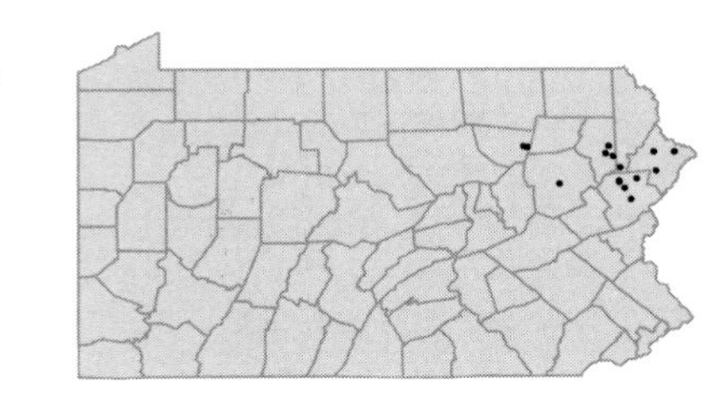

Xyris torta J.E.Smith
Yellow-eyed-grass
Herbaceous perennial, emergent aquatic
Sphagnum bogs and swampy meadows.

OBL

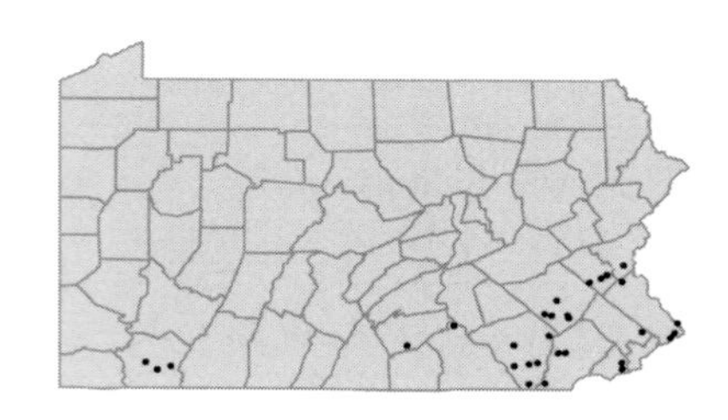

COMMELINACEAE

• ***Commelina communis*** L. var. ***communis***
Dayflower
Herbaceous annual
Gardens, woods, roadsides, stream banks and disturbed ground.

FAC-
[I]

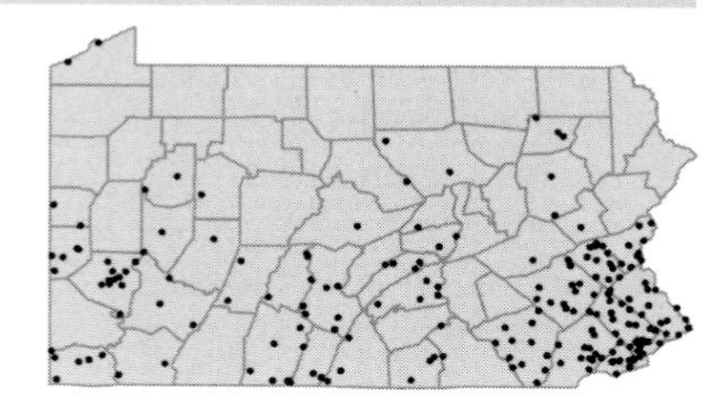

Commelina communis L. var. ***ludens*** (Miquel) C.B.Clarke
Dayflower
Herbaceous annual
Open woods, yards, dumps and waste ground.

FAC-
[I]

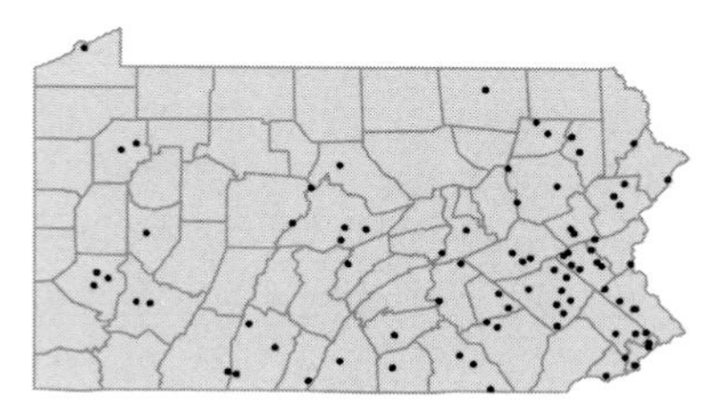

Commelina diffusa Burm.f.
Creeping dayflower
Herbaceous perennial
Waste ground.
Commelina nudiflora non L. W

FACW
[I]

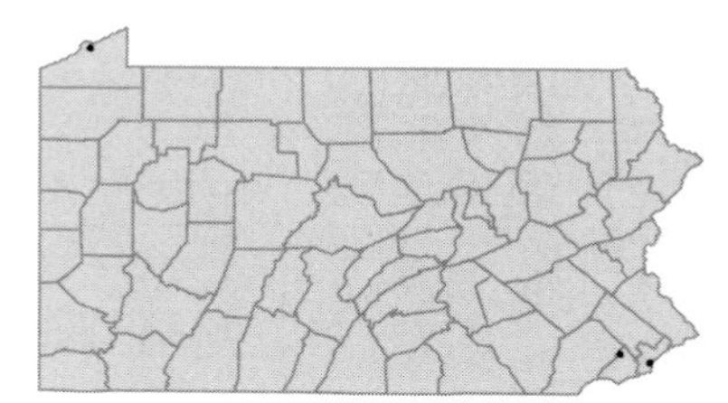

Commelina erecta L.
Erect dayflower; Slender dayflower
Herbaceous perennial
Dry, sandy soil and rocky banks. Believed to be extirpated, last collected in 1924.

Commelina virginica L.
Virginia dayflower
Herbaceous perennial
Moist shores, also cultivated. Believed to be extirpated, last collected in 1863.

FACW

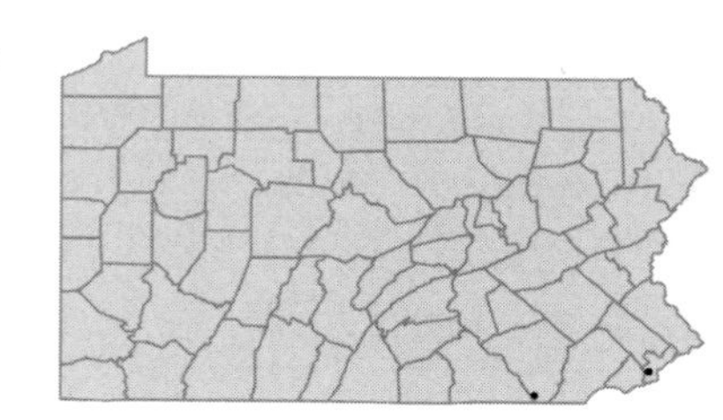

• ***Tradescantia ohiensis*** Raf.
Spiderwort
Herbaceous perennial
Moist fields, stream banks, alluvial woods and waste ground.

FAC

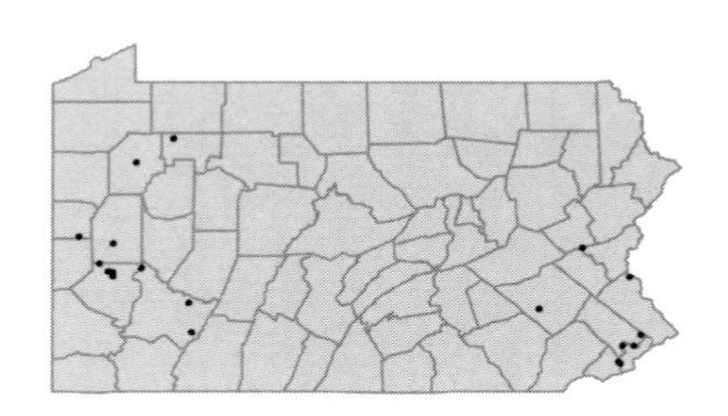

Tradescantia virginiana L.
Spiderwort; Widow's-tears
Herbaceous perennial
Wooded slopes, shale outcrops, moist fields and roadsides, also cultivated.

FACU

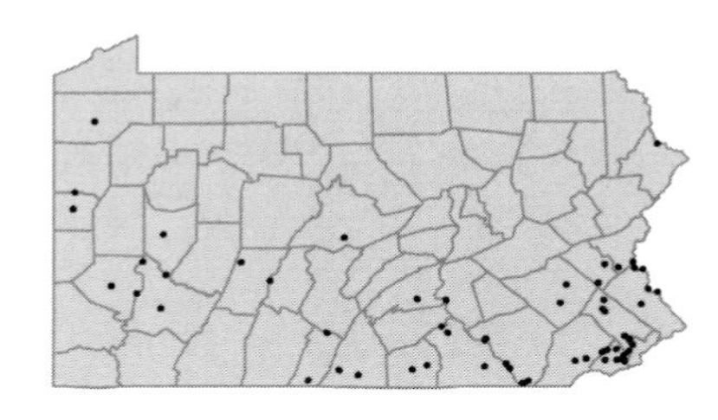

ERIOCAULACEAE

• ***Eriocaulon aquaticum*** (Hill) Druce
Seven-angle pipewort; White-buttons
Herbaceous perennial, rooted submergent aquatic
Shallow water and peaty shores of northern lakes.
Eriocaulon septangulare With. PFGBW

OBL

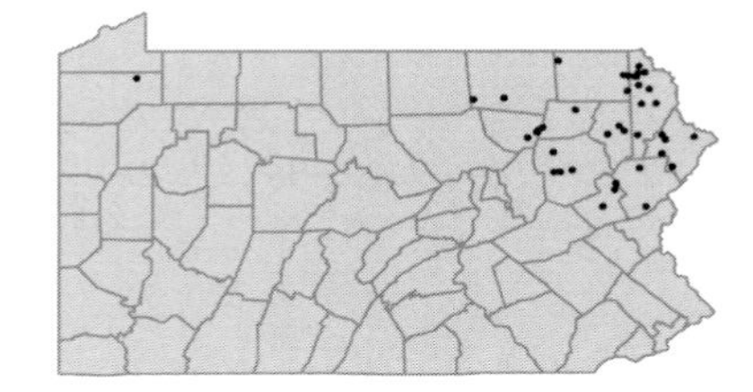

Eriocaulon decangulare L.
Ten-angle pipewort
Herbaceous perennial, rooted submergent aquatic
Glacial lake. Believed to be extirpated, represented by a single collection from Wayne Co. in 1907.

OBL

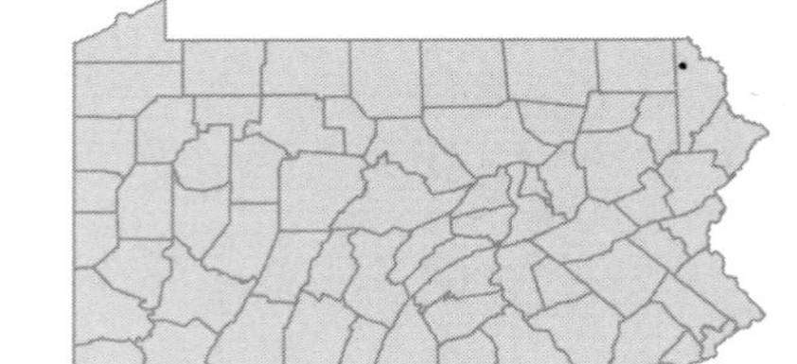

Eriocaulon parkeri B.L.Robins.
Parker's pipewort
Herbaceous perennial, rooted submergent aquatic
Tidal mud flats. Believed to be extirpated, last collected in 1927.

OBL

JUNCACEAE

• ***Juncus acuminatus*** Michx.
Sharp-fruited rush
Herbaceous perennial
Wet meadows, swamps, marshes and stream banks.

OBL

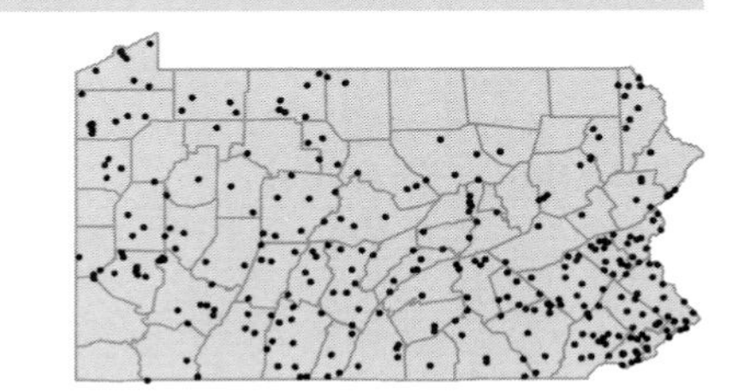

Juncus alpinoarticulatus Chaix in Vill. ssp. ***nodulosus*** (Wahlenb.) Hamet-Ahti
Alpine rush
Herbaceous perennial
Moist, sandy, calcareous shores and seeps. A hybrid with *J. articulatus* (*J. x alpiniformis* Fern.) has been collected at a site where both parent species also occur.
Juncus alpinus Vill. WFK; *Juncus alpinoarticulatus* Chaix in Vill. ssp. *americanus* (Farw.) Hamet-Ahti C

OBL

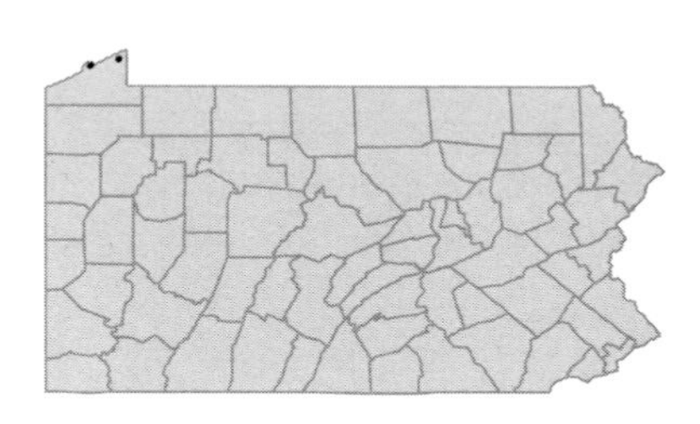

Juncus articulatus L.
Jointed rush
Herbaceous perennial, emergent aquatic
Bogs, swamps, swales and mud flats.

OBL

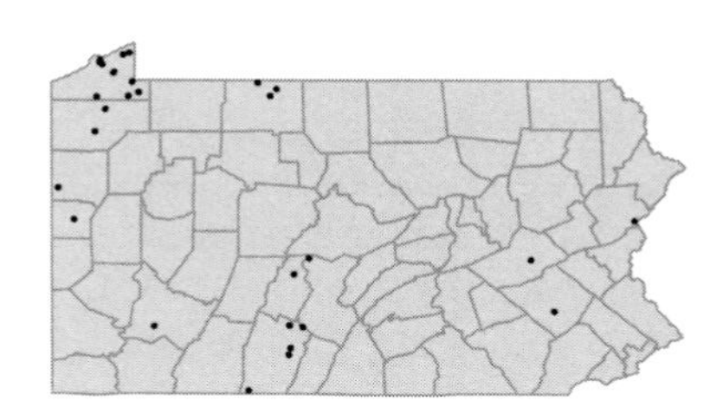

Juncus arcticus Willd. var. ***littoralis*** (Engelm.) Boivin
Baltic rush; Wire rush
Herbaceous perennial, emergent aquatic
Calcareous swamps and shores.
Juncus balticus Willd. var. *littoralis* Engelm. FWK

FACW+

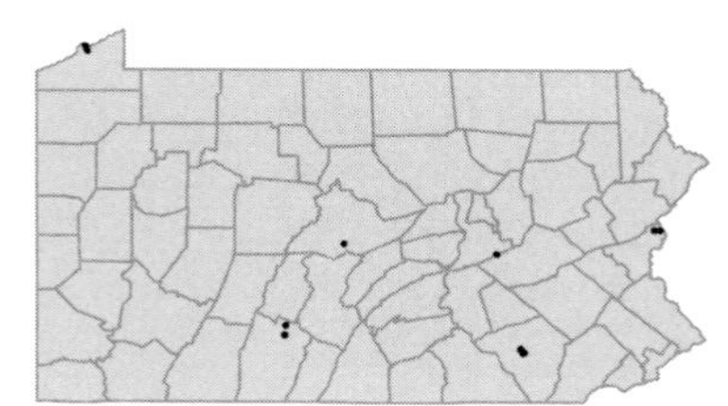

Juncus biflorus Ell.
Grass rush
Herbaceous perennial
Moist, open woods, boggy fields, gravel pits and ditches.
Juncus marginatus Rostk. var. *biflorus* (Elliot) Chapm. K

FACW

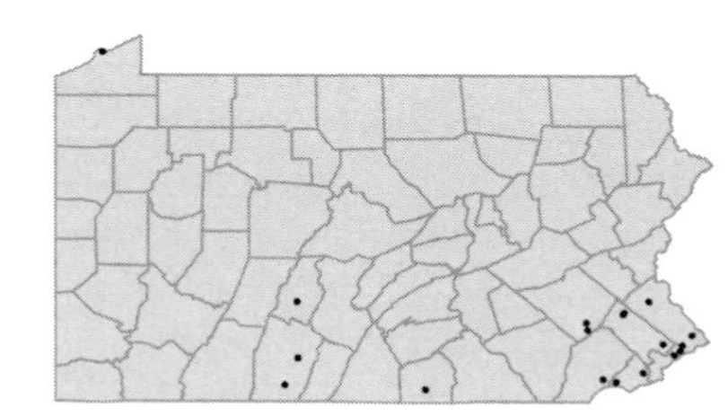

Juncus brachycarpus Engelm.
Short-fruited rush
Herbaceous perennial
Damp pasture. Represented by a single collection from Adams Co. in 1991.

FACW

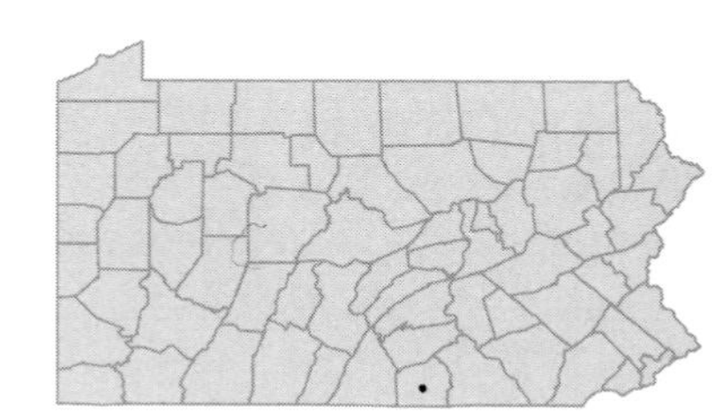

Juncus brachycephalus (Engelm.) Buch.
Small-headed rush
Herbaceous perennial, emergent aquatic
Muddy or sandy calcareous shores, clayey seeps and springy or boggy fields.

OBL

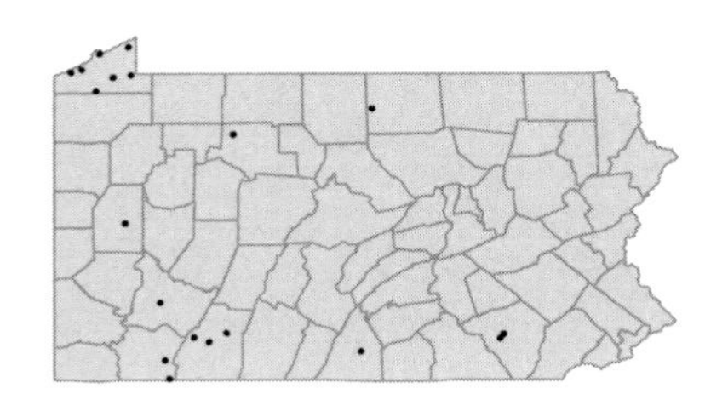

Juncus brevicaudatus (Engelm.) Fern.
Narrow-panicled rush
Herbaceous perennial, emergent aquatic
Bogs, swamps, moist shores, swales and ditches.
Juncus canadensis J.Gay var. *brevicaudatus* Engelm. P

OBL

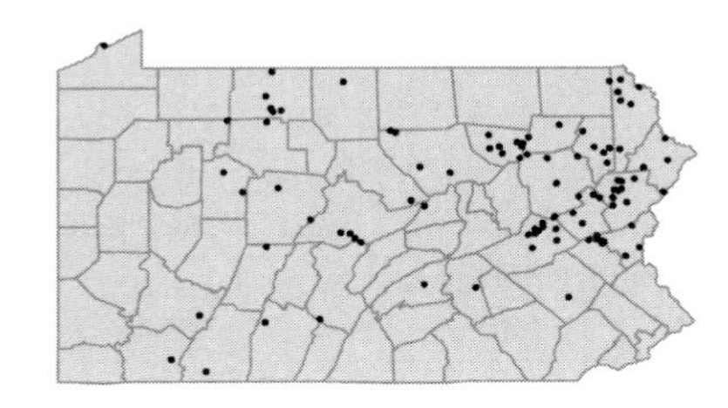

Juncus bufonius L.
Toad rush
Herbaceous annual, emergent aquatic
Muddy river banks, moist roadside ditches and other low ground.

FACW

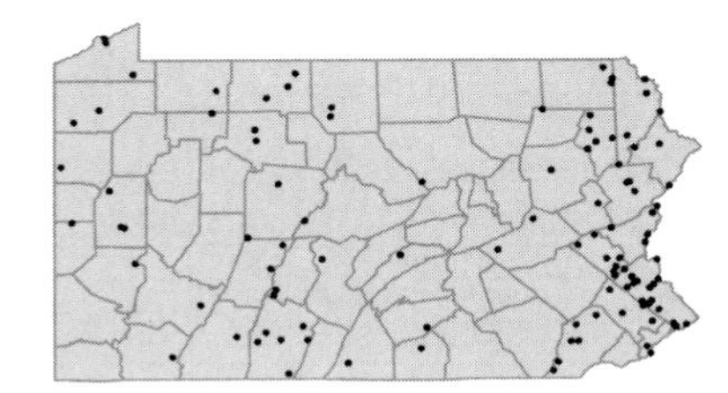

Juncus canadensis J.Gay ex Laharpe
Canada rush
Herbaceous perennial, emergent aquatic
Swamps, marshes, bogs, river banks, ponds and swales.

OBL

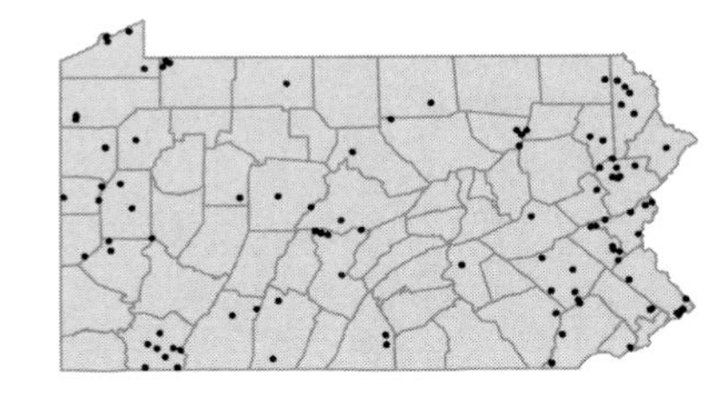

Juncus debilis A.Gray
Weak rush
Herbaceous perennial, emergent aquatic
River banks, mudflats, shores and ditches.
Juncus acuminatus Michx. var. *debilis* (A.Gray) Engelm. P

OBL

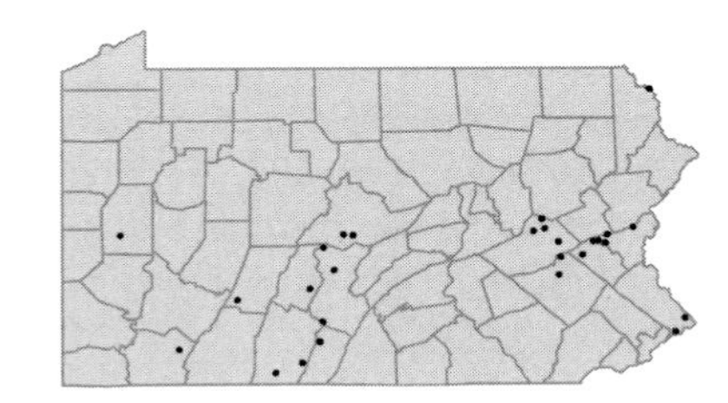

Juncus dichotomus Ell.
Forked rush
Herbaceous perennial
Moist, sandy old fields, open woods and gravel pits.
Juncus tenuis Willd. var. *dichotomus* (Elliot) A.Wood C; *Juncus platyphyllus* (Wieg.) Fern. KFW

FACW

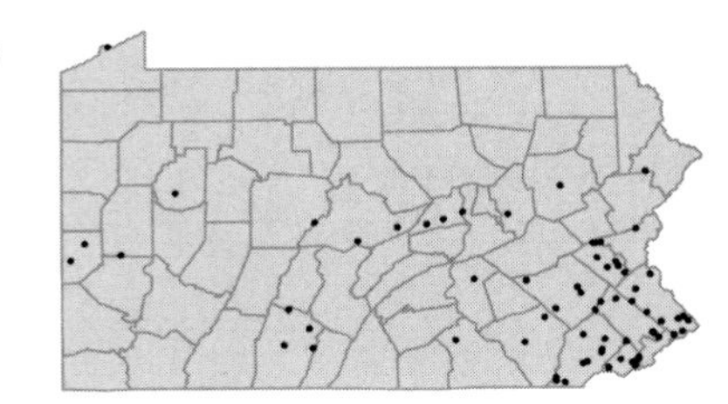

Juncus dudleyi Wieg.
Rush
Herbaceous perennial, emergent aquatic
Wet fields, river banks, swales and ditches.
Juncus tenuis Willd. var. *dudleyi* (Wieg.) F.J.Herm. C; *Juncus tenuis* Willd. var. *uniflorus* (Farw.) Farw.

FAC-

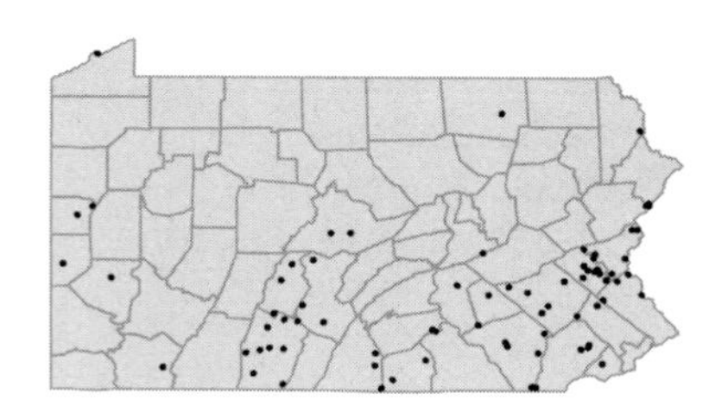

Juncus effusus L. var. ***pylaei*** (Laharpe) Fern. & Wieg.
Soft rush
Herbaceous perennial, emergent aquatic
Swamps, moist fields, floodplains, shores and ditches.
Juncus pylaei Laharpe C

FACW+

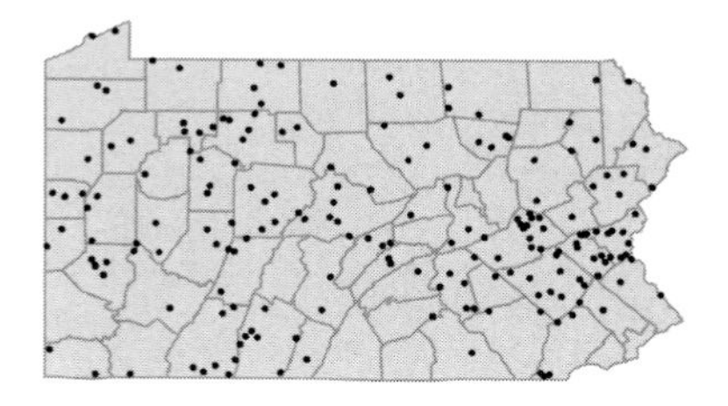

Juncus effusus L. var. ***solutus*** Fern. & Wieg.
Soft rush
Herbaceous perennial, emergent aquatic
Swamps, wet meadows, moist woods, shores and thickets.

FACW+

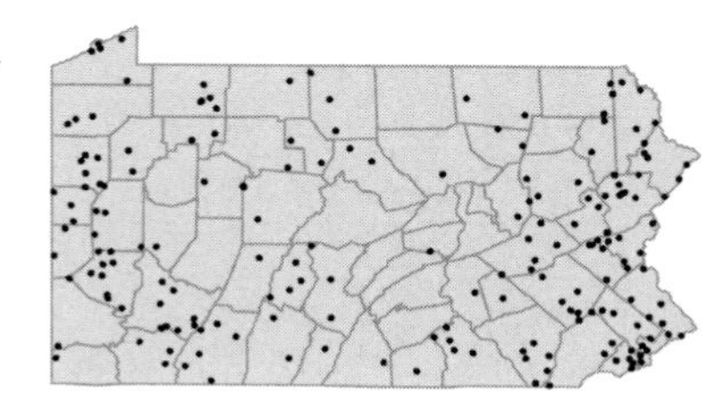

Juncus filiformis L.
Thread rush
Herbaceous perennial, emergent aquatic
Bogs and sandy shores.

FACW

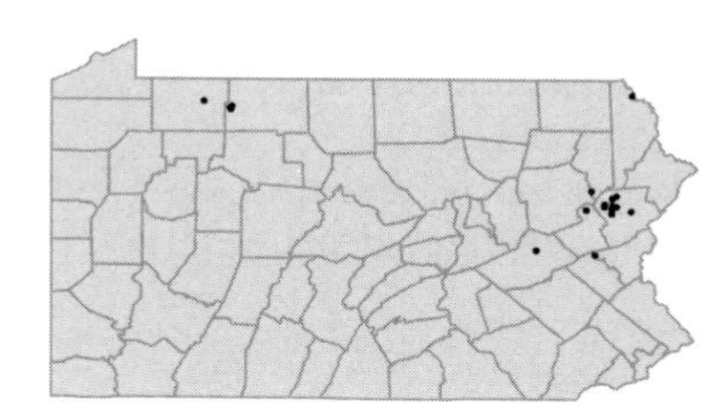

Juncus gerardii Loisel.
Blackfoot rush; Black-grass
Herbaceous perennial, emergent aquatic
Ballast, waste ground and moist roadsides, especially where de-icing salts are used. Native in coastal areas, but adventive in PA.

FACW+

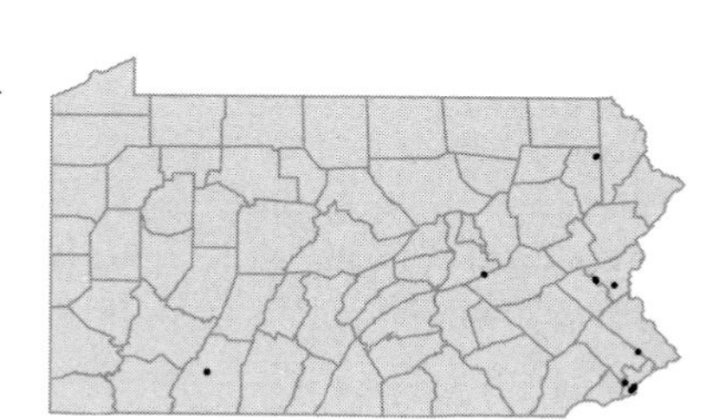

Juncus greenei Oakes & Tuckerman
Greene's rush
Herbaceous perennial
Sandstone cliffs and bluffs. Believed to be extirpated, last collected in 1938.

FAC

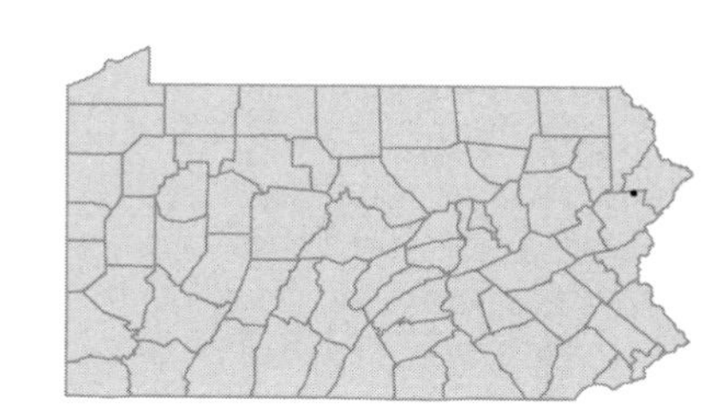

Juncus gymnocarpus Cov.
Coville's rush; Pennsylvania rush
Herbaceous perennial, emergent aquatic
Sphagnum swamps, seeps and springheads.

OBL

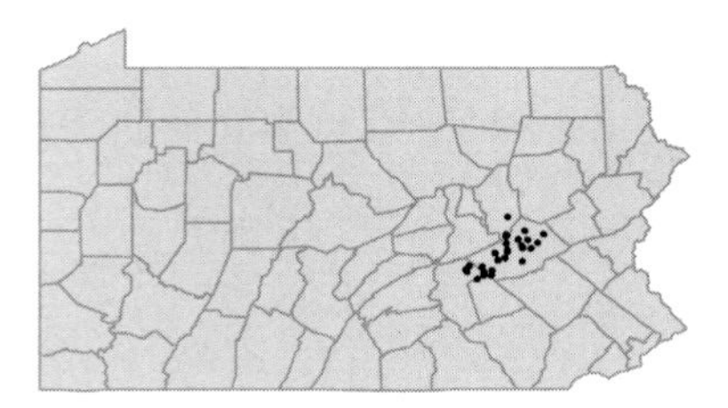

Juncus inflexus L.
Meadow rush
Herbaceous perennial, emergent aquatic
Wet bottomland. Represented by a single collection from Centre Co.

FACW
I

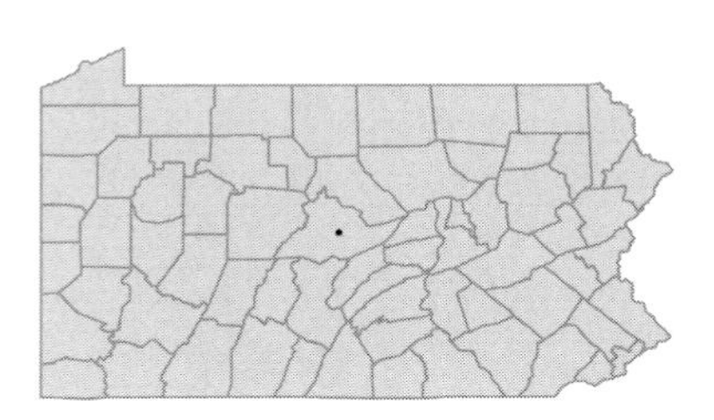

Juncus marginatus Rostk. var. ***marginatus***
Grass-leaved rush
Herbaceous perennial, emergent aquatic
Moist fields, ditches, swamps and roadsides.

FACW

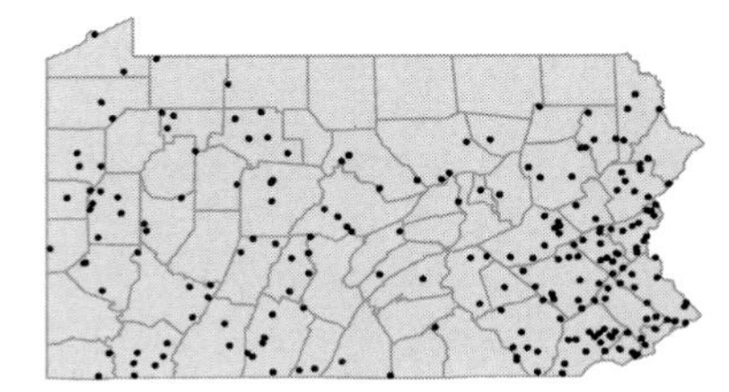

Juncus militaris Bigel.
Bayonet rush
Herbaceous perennial, emergent aquatic
Shallow water of lakes and ponds.

OBL

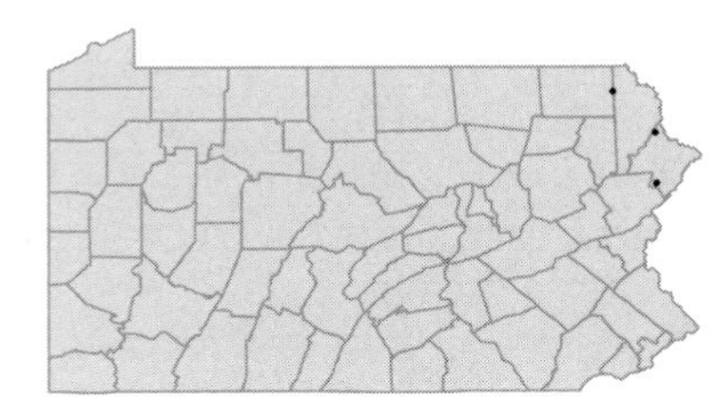

Juncus nodosus L.
Knotted rush
Herbaceous perennial, emergent aquatic
Moist fields, bogs, marshes, shores and swales; often on calcareous substrates.

OBL

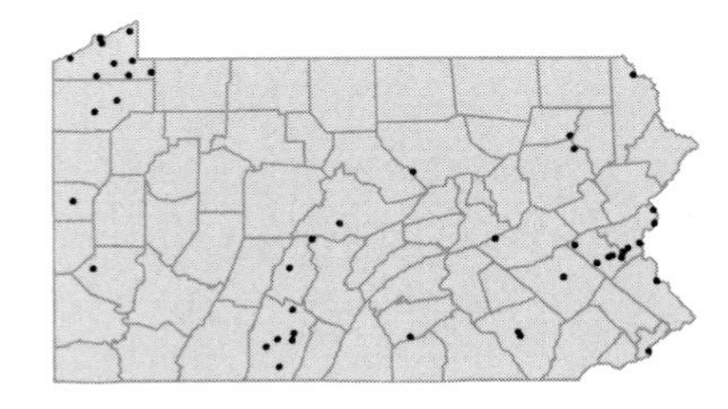

Juncus pelocarpus Mey.
Brown-fruited rush
Herbaceous perennial, emergent aquatic
Bogs and marshes.

OBL

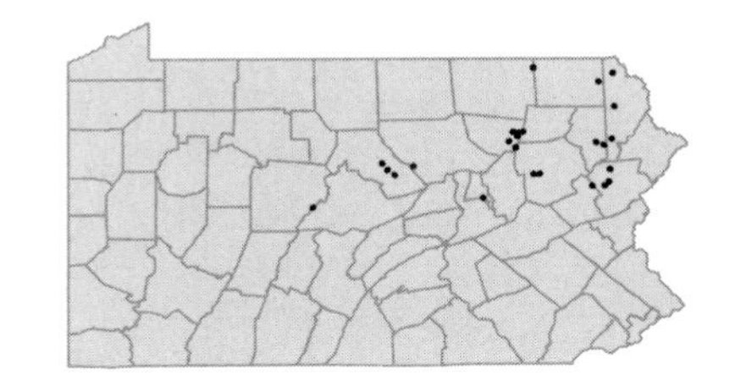

Juncus scirpoides Lam.
Sedge rush; Scirpus-like rush
Herbaceous perennial
Moist, sandy or peaty soil.

FACW

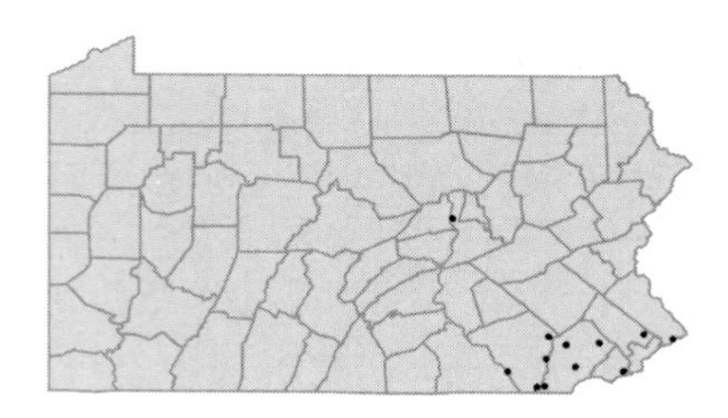

Juncus secundus Beauv. ex Poir.
Rush
Herbaceous perennial
Upland slopes, rocky ledges, serpentine barrens and roadside banks.

FACU

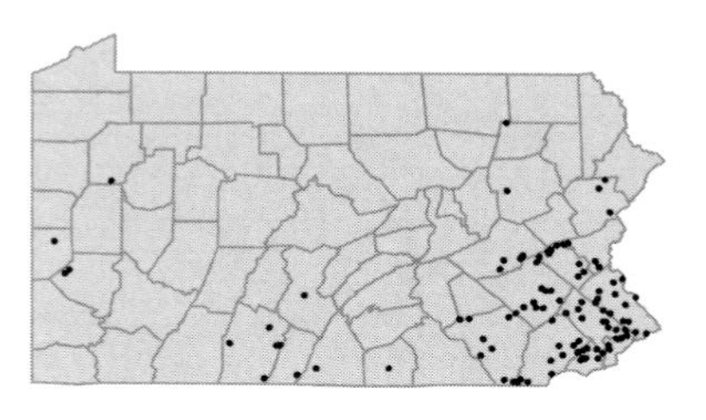

Juncus subcaudatus (Engelm.) Cov. & S.F.Blake
Rush
Herbaceous perennial, emergent aquatic
Swamps, wet fields, ditches, swales and bogs.
Juncus canadensis J.Gay var. *subcaudatus* Engelm. P

OBL

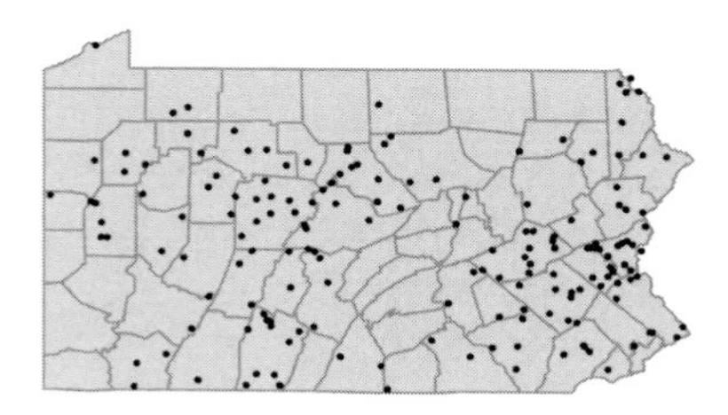

Juncus tenuis Willd. var. ***tenuis***
Path rush; Yard rush
Herbaceous perennial
Moist to dry, often heavily compacted soil of woods, fields, waste ground and paths.
Juncus tenuis Willd. var. *anthelatus* Wieg. *pro parte* FW; *Juncus tenuis* Willd. var. *williamsii* Fern. *pro parte* FW

FAC-

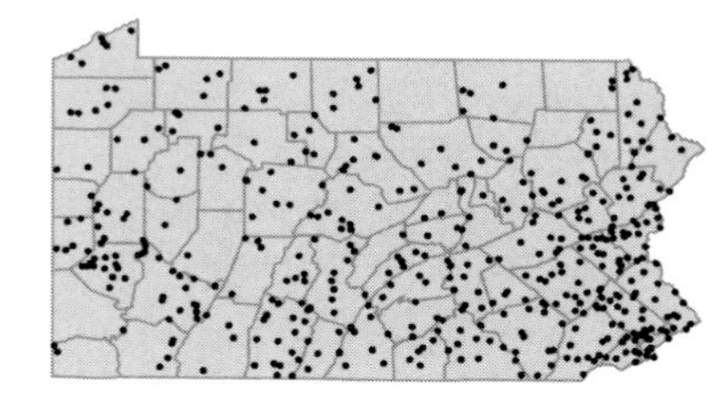

Juncus torreyi Cov.
Torrey's rush
Herbaceous perennial, emergent aquatic
Muddy or sandy shores, strip mine area, swales and ditches.

FACW

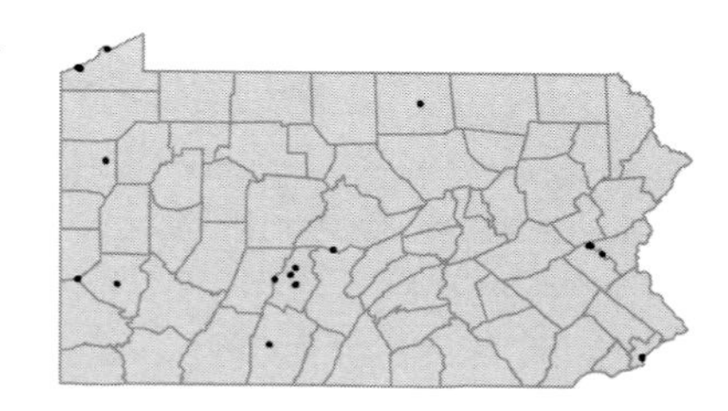

• ***Luzula acuminata*** Raf. var. ***acuminata***
Hairy wood-rush
Herbaceous perennial
Swampy woods and floodplains.

FAC

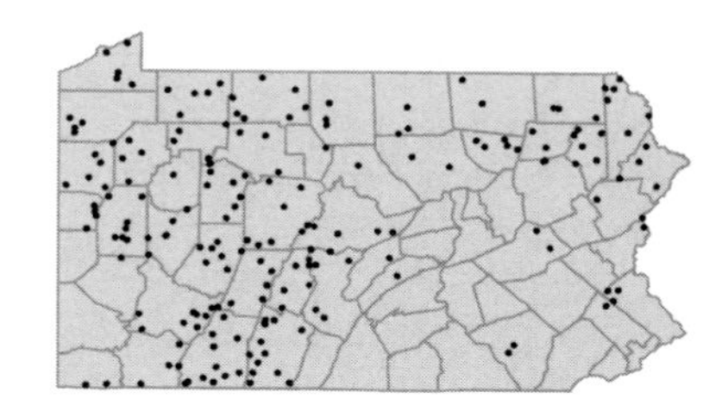

Luzula bulbosa (A.Wood) Rydb.
Wood-rush
Herbaceous perennial
Fields, woods borders and roadsides.
Luzula campestris (L.) DC. var. *bulbosa* A.Wood B

FACU

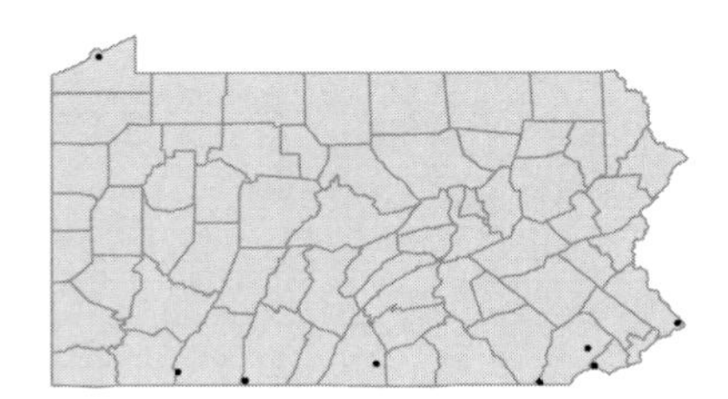

Luzula echinata (Small) F.J.Herm.
Common wood-rush
Herbaceous perennial
Moist, rocky woods and wet meadows.
Luzula campestris (L.) DC. var. *echinata* (Small) Fern. & Wieg. GB

FACU

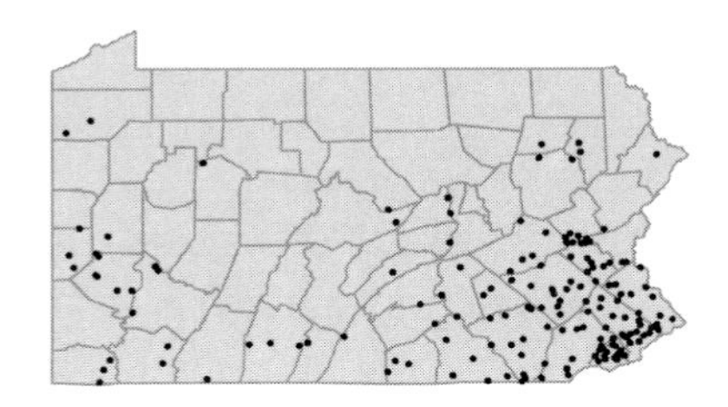

Luzula luzuloides (Lam.) Dandy & Wilmott
Forest wood-rush
Herbaceous perennial
Rich, dry woods. Represented by two collections from Berks Co. 1940-1942.

I

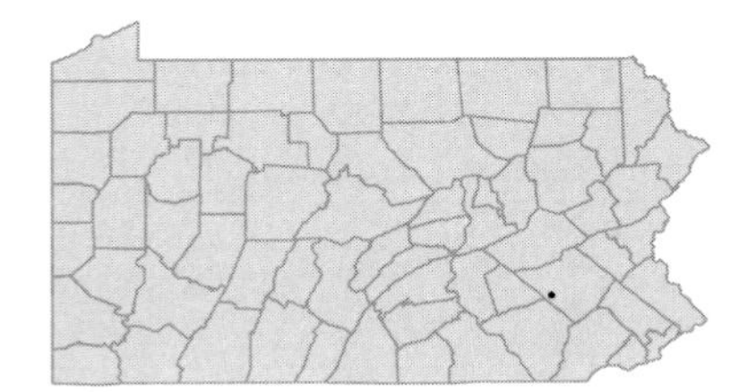

Luzula multiflora (Ehrh.) Lej.
Field wood-rush
Herbaceous perennial
Woods, swamps and floodplains.
Luzula campestris (L.) DC. var. *multiflora* (Ehrh.) Celak GB

FACU

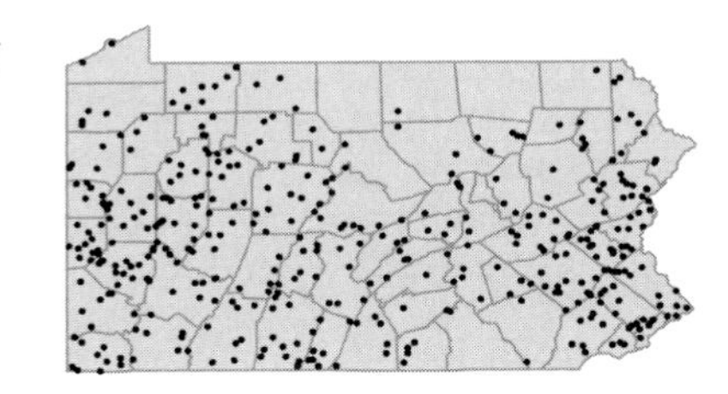

CYPERACEAE

• ***Bulbostylis capillaris*** (L.) C.B.Clarke
Sand-rush
Herbaceous annual
Dry or moist, open ground, often on railroad gravel.
Stenophyllus capillaris (L.) Britt. P

FACU

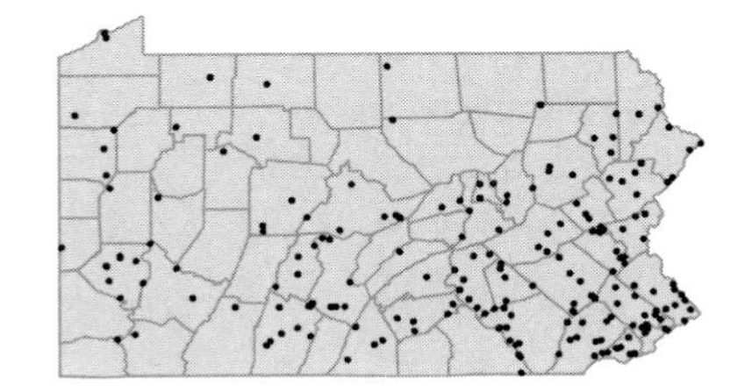

• ***Carex abscondita*** Mackenzie
Sedge
Herbaceous perennial
Moist to wet woods.

FAC

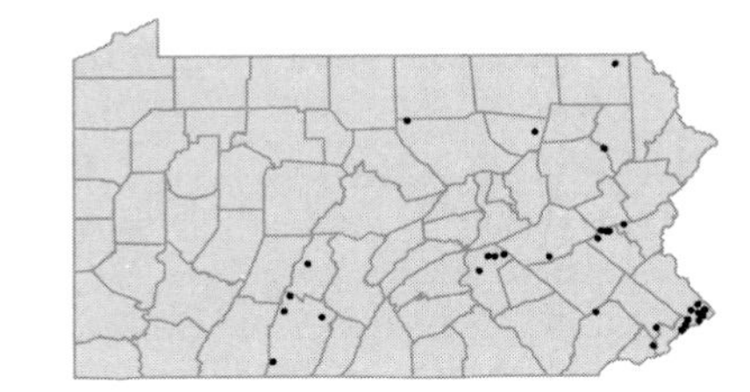

Carex adusta Boott
Crowded sedge
Herbaceous perennial
Dry, open woods and clearings. Believed to be extirpated, collected at a single site in Northampton Co. 1869-1870.

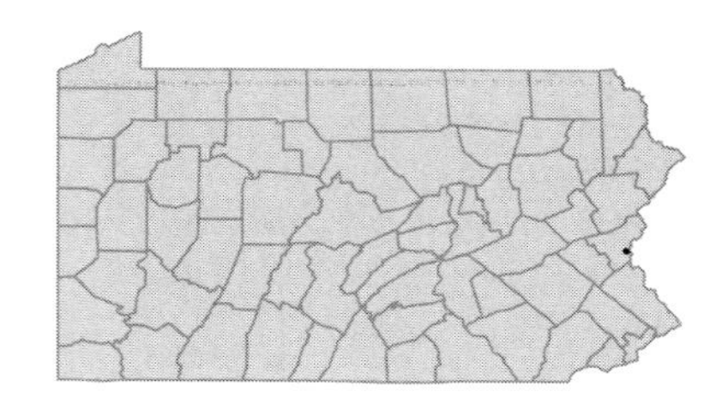

Carex aestivalis M.A.Curtis
Sedge
Herbaceous perennial
Dry to moist woods.

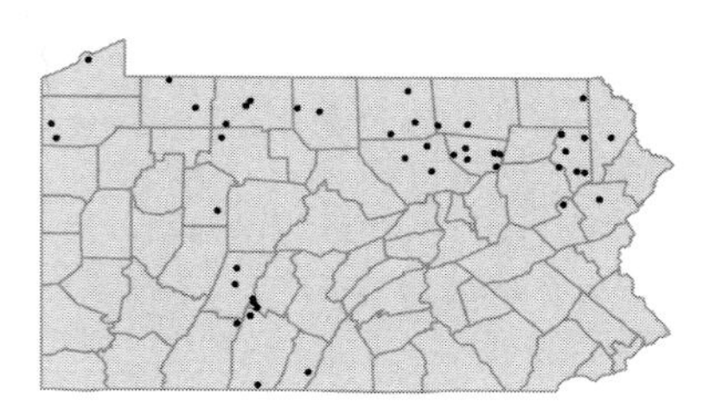

Carex aestivalis x gracillima
Sedge
Herbaceous perennial
Wooded slopes.
Carex x aestivaliformis Mackenzie C

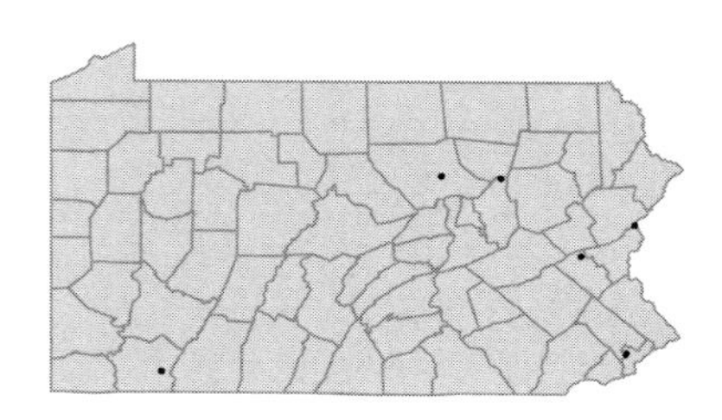

Carex aggregata Mackenzie
Sedge
Herbaceous perennial
Moist woods, meadows and ditches.
Carex sparganioides Muhl. var. *aggregata* (Mackenzie) Gleason GBC

FACU

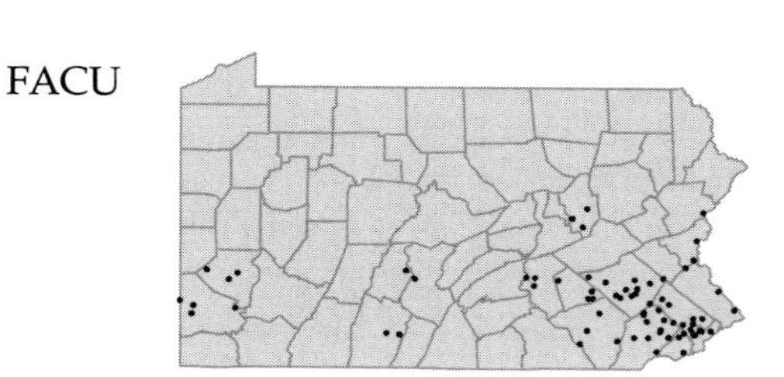

Carex alata Torr. & A.Gray
Broad-winged sedge
Herbaceous perennial
Swampy woods, meadows and marshes, usually on calcareous soils.

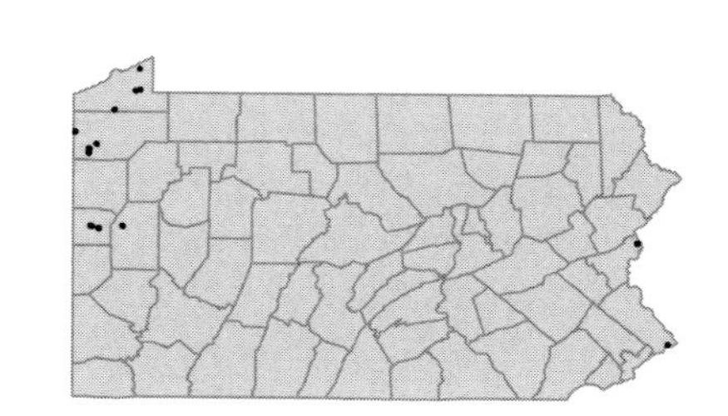

Carex albicans Willd. ex Sprengel
Sedge
Herbaceous perennial
Dry, wooded slopes.
Carex nigro-marginata Schwein. var. *muhlenbergii* (Gray) Gleason GB; *Carex varia* Muhl. P; *Carex artitecta* Mackenzie KWF

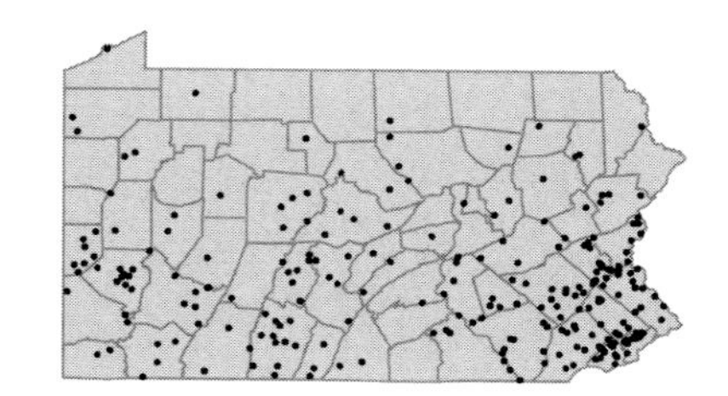

Carex albolutescens Schwein.
Sedge
Herbaceous perennial
Moist to wet woods and meadows.

FACW

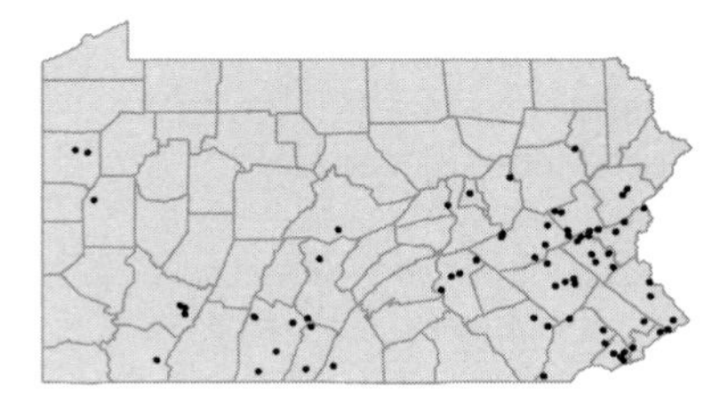

Carex albursina Sheldon
Sedge
Herbaceous perennial
Dry to moist, rich woods.
Carex laxiflora Lam. var. *latifolia* Boott GB

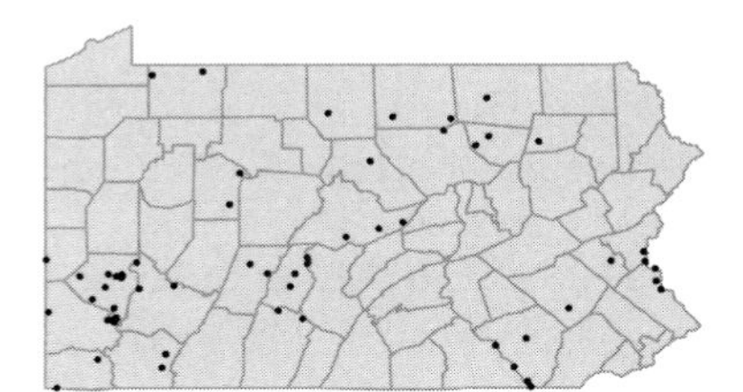

Carex alopecoidea Tuckerman
Foxtail sedge
Herbaceous perennial
Wet meadows. Believed to be extirpated, last collected in 1925.

FACW

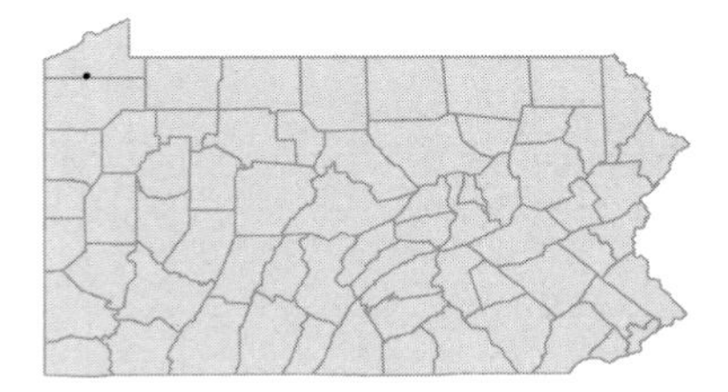

Carex amphibola Steud. var. ***amphibola*** *sensu* Fern.
Sedge
Herbaceous perennial
Dry to moist woods. Recent work has shown that this taxon should be named as new species distinct from true *C. amphibola*.

FAC

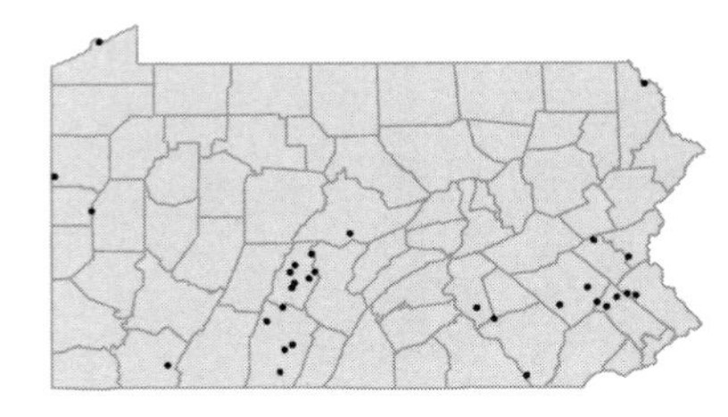

Carex amphibola Steud. var. ***rigida*** (Bailey) Fern.
Sedge
Herbaceous perennial
Dry to moist woods, meadows and swales. Shown above is the nomenclature as applied by Fernald (1950). However, recent work has shown that this taxon should be known as *C. amphibola* Steud. var. *amphibola*.

FAC

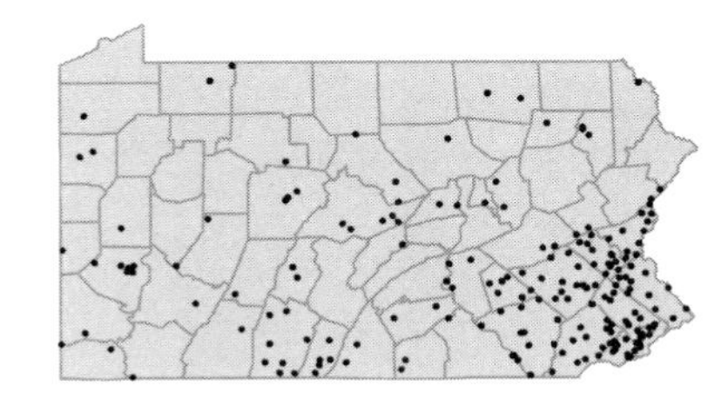

Carex annectens (Bickn.) Bickn.
Sedge
Herbaceous perennial
Dry to moist woods, fields and ditches.
Carex xanthocarpa Bickn. P; *Carex vulpinoidea* Michx. var. *ambigua* Boott C

FACW

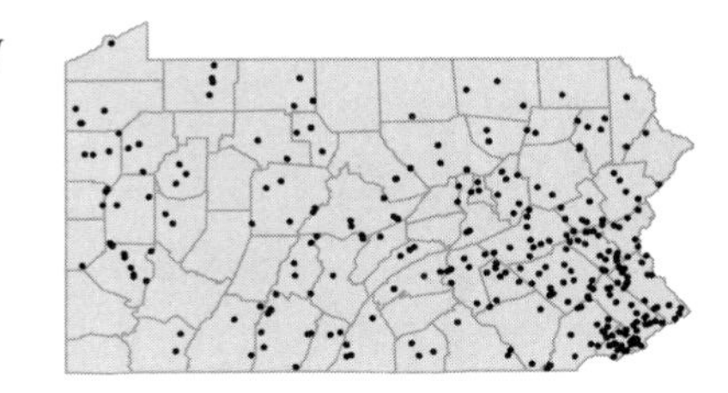

Carex appalachica Webber & Ball
Sedge
Herbaceous perennial
Moist, rocky woods.
Carex rosea Schkuhr *pro parte* GB; *Carex radiata sensu* Mackenzie *non* (Wallenb.) Small FW

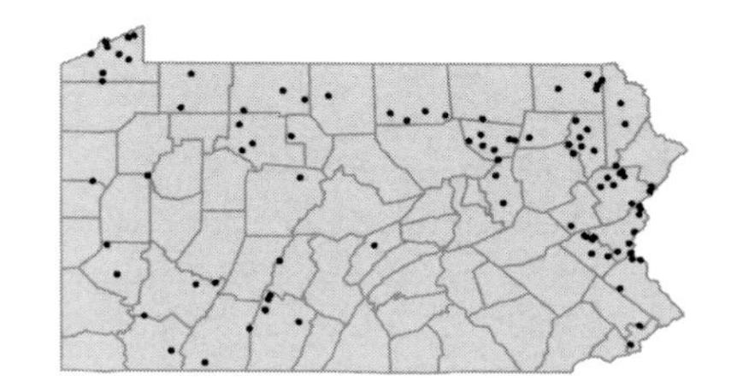

Carex aquatilis Wahlenb.
Water sedge
Herbaceous perennial, emergent aquatic
Marshy swales.
Carex substricta (Kukenth.) Mackenzie W

OBL

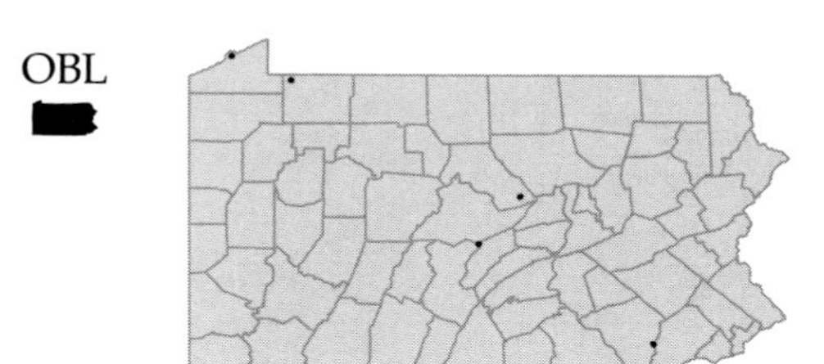

Carex arctata Boott
Sedge
Herbaceous perennial
Dry to moist woods and clearings.

OBL

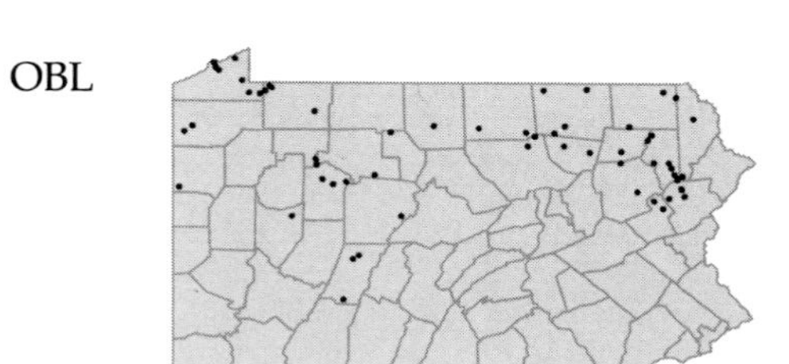

Carex argyrantha Tuckerman
Sedge
Herbaceous perennial
Dry, often rocky, woods and clearings.

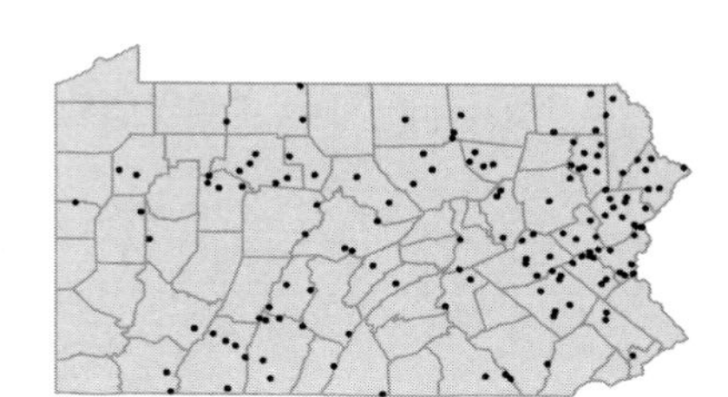

Carex atherodes Spreng.
Awned sedge
Herbaceous perennial
Open, seepy slope.

OBL

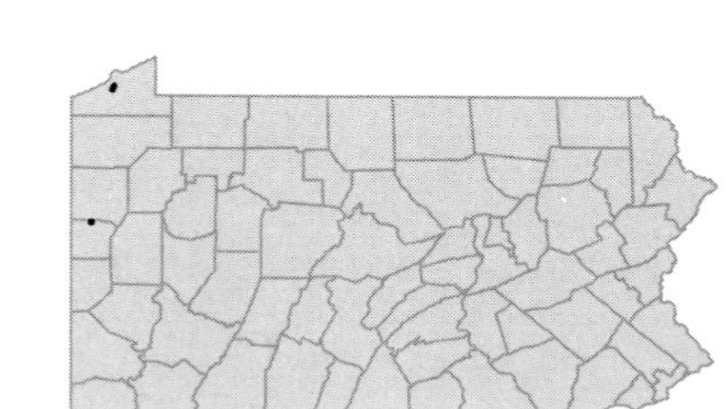
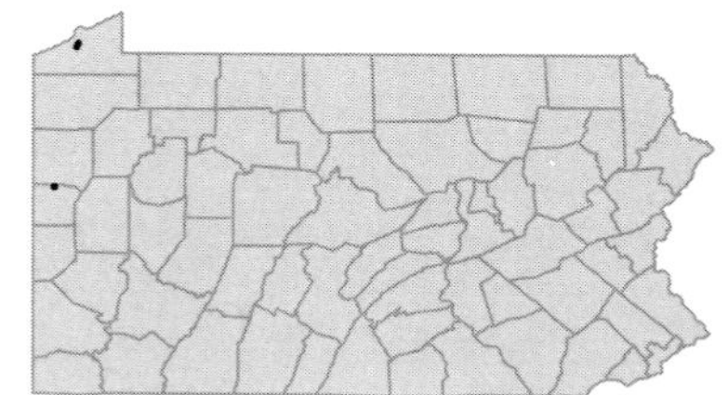

Carex atlantica Bailey ssp. ***atlantica***
Bog sedge
Herbaceous perennial
Swamps and bogs.
Carex incomperta Bickn. FGBW

FACW+

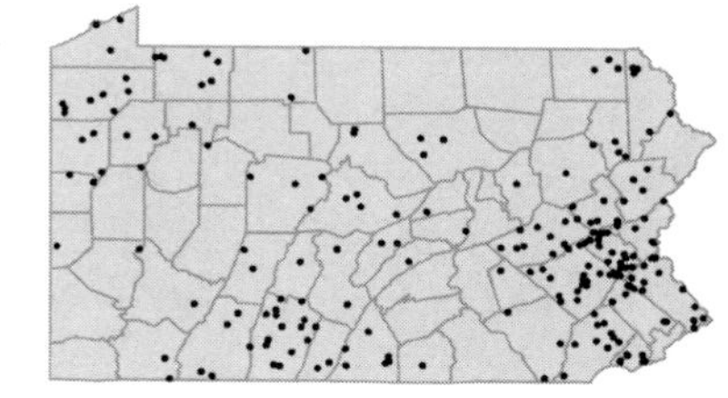

Carex atlantica Bailey ssp. ***capillacea*** (Bailey) Reznicek
Sedge
Herbaceous perennial
Swamps, bogs and moist shores.
Carex interior Bailey var. *capillacea* Bailey P; *Carex howei* Mackenzie WFB

OBL

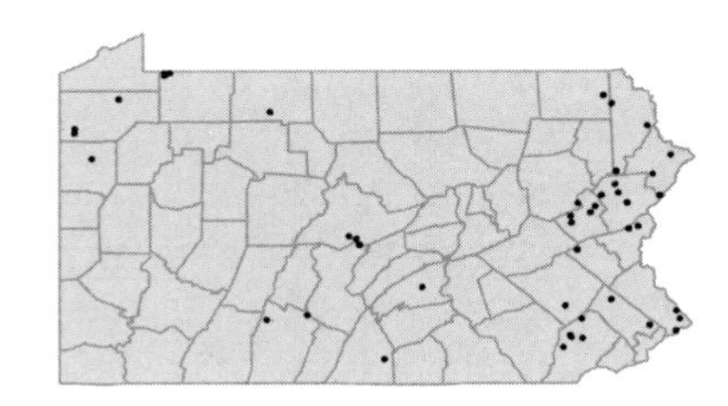

Carex aurea Nutt.
Golden-fruited sedge
Herbaceous perennial
Moist, calcareous slumps and seeps.

FACW

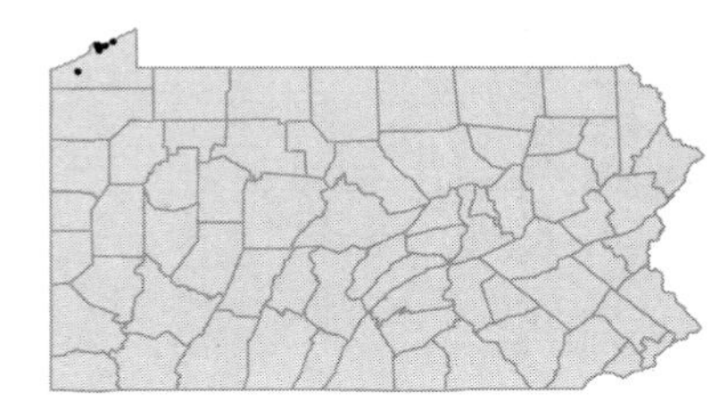

Carex backii Boott
Back's sedge; Rocky Mountain sedge
Herbaceous perennial
Rocky woods, base of sandstone ledge. Believed to be extirpated, represented by a single collection from Lackawanna Co. in 1951.

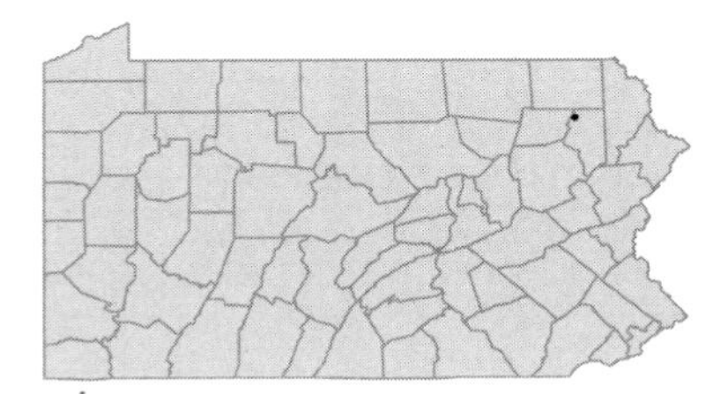

Carex baileyi Britt.
Sedge
Herbaceous perennial
Moist woods, swamps, meadows and shores.

OBL

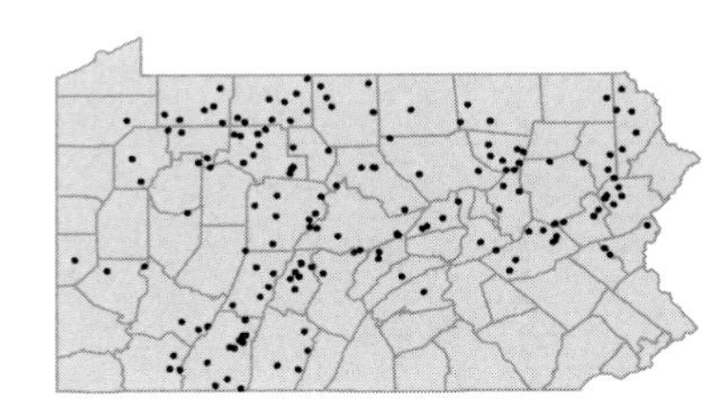

Carex barrattii Schwein. & Torr.
Barratt's sedge
Herbaceous perennial
Wet woods. Believed to be extirpated, last collected in 1914.
Carex littoralis Schwein. P

OBL

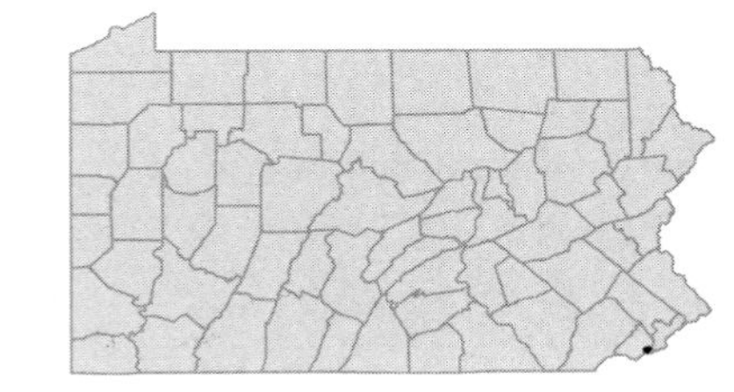

Carex bebbii (Bailey) Fern.
Bebb's sedge
Herbaceous perennial
Pond edges, boggy pastures and moist sand flats, usually on calcareous substrates.
Carex tribuloides Wahlenb. var. *bebbii* Bailey P

OBL

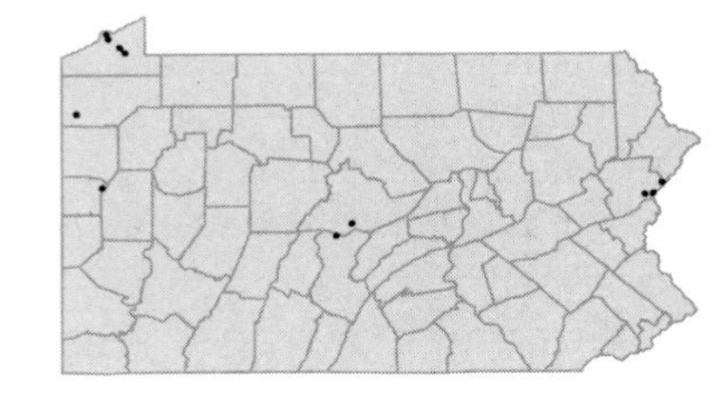

Carex bicknellii Britt.
Bicknell's sedge
Herbaceous perennial
Dry woods, thickets, fields and serpentine barrens.

Carex blanda Dewey
Sedge
Herbaceous perennial
Dry to moist woods, thickets and meadows.
Carex laxiflora Lam. var. *blanda* (Dewey) Boott PGB

FAC

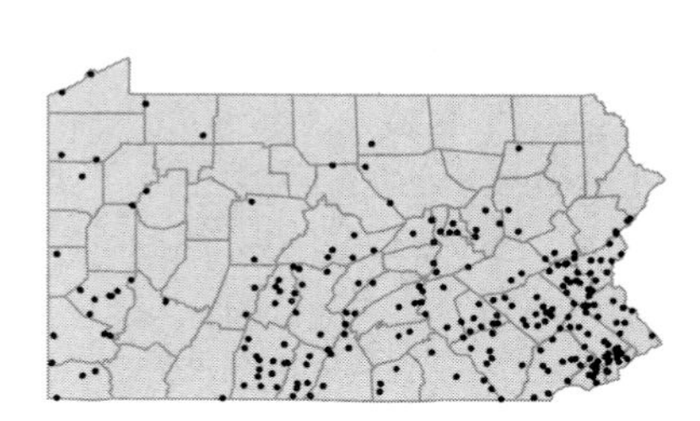

Carex brevior (Dewey) Mackenzie ex Lunell
Sedge
Herbaceous perennial
Dry, open thickets, banks, fields and roadsides.

UPL

Carex bromoides Willd.
Sedge
Herbaceous perennial
Wet woods, swamps, meadows and swales.

FACW

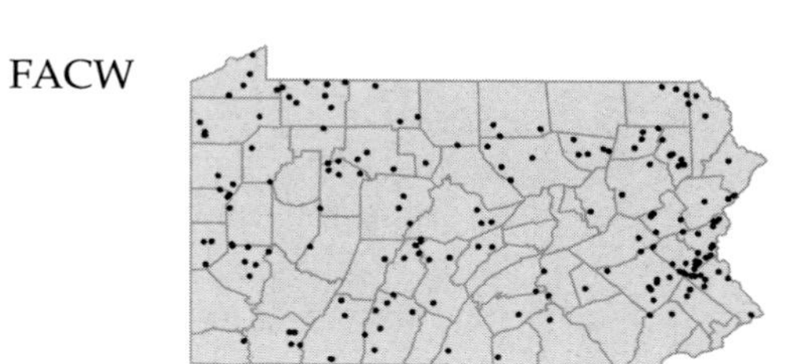

Carex brunnescens (Pers.) Poir.
Sedge
Herbaceous perennial
Dry to wet woods and clearings.

FACW

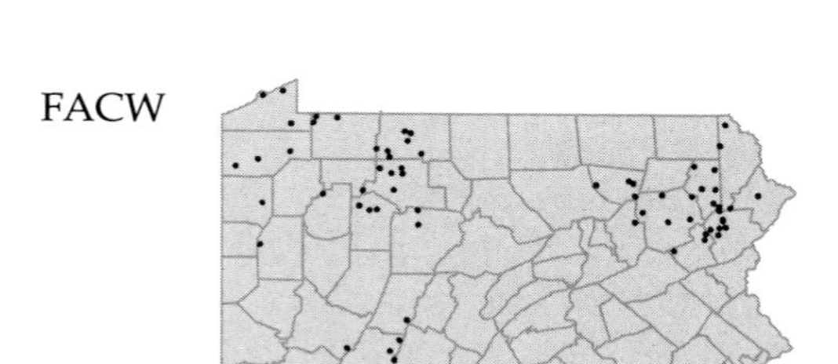

Carex bullata Schkuhr ex Willd.
Bull Sedge
Herbaceous perennial
Swales, meadows, openings in wet woods.

OBL

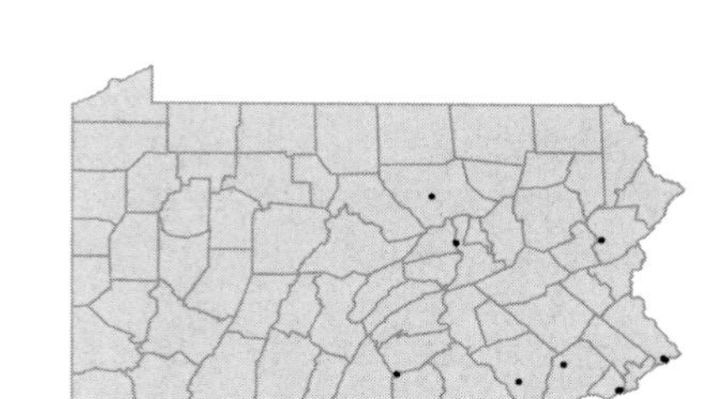

Carex bushii Mackenzie
Sedge
Herbaceous perennial
Dry to moist, open woods and fields.

FACW

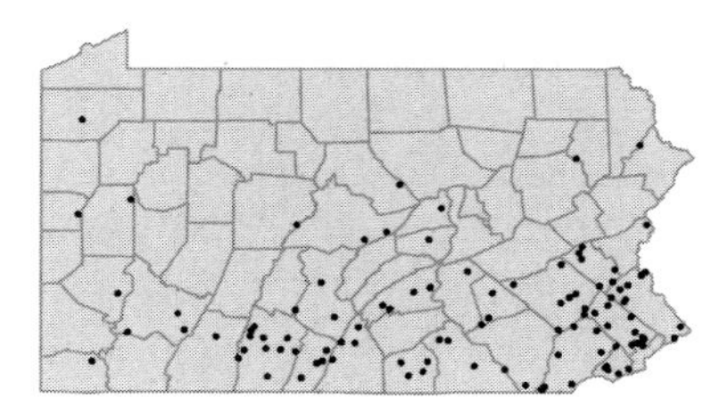

Carex buxbaumii Wahlenb.
Brown sedge
Herbaceous perennial
Open swamps, swales, meadows and boggy places.

OBL

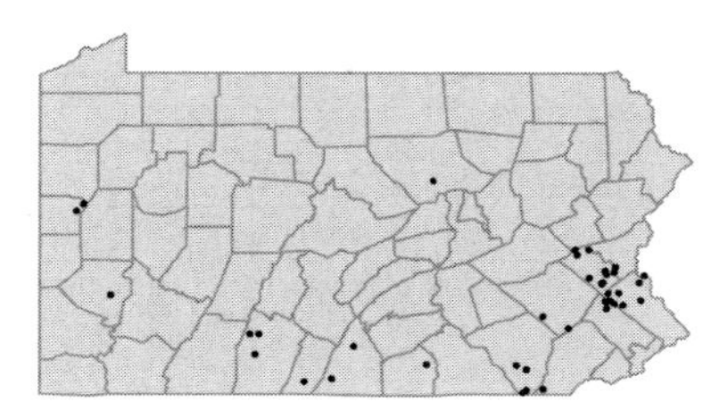

Carex canescens L. var. ***canescens***
Sedge
Herbaceous perennial
Swamps, marshes, meadows and shores.

OBL

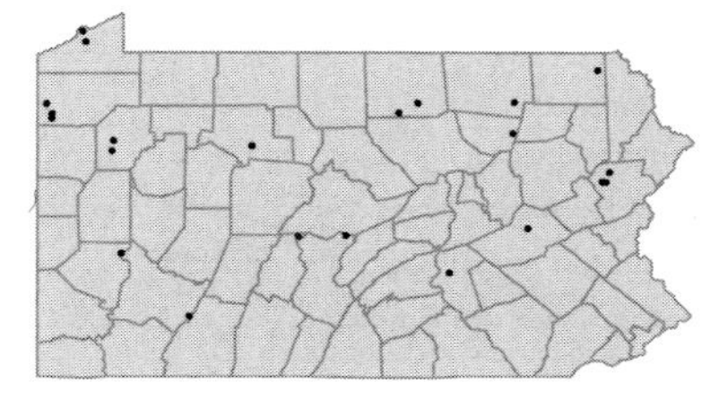

Carex canescens L. var. ***disjuncta*** Fern.
Sedge
Herbaceous perennial
Swamps, marshes, meadows and shores.

OBL

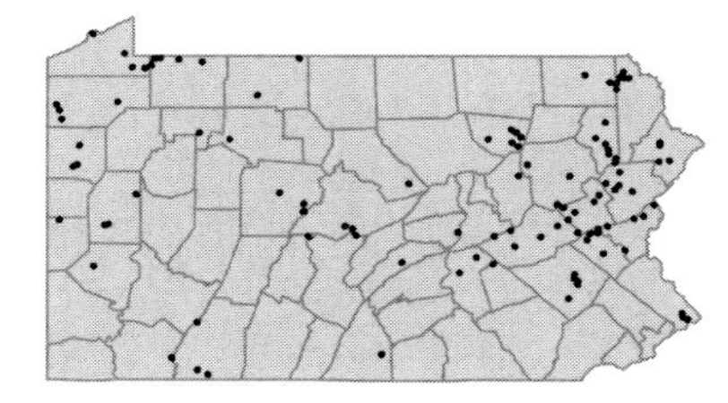

Carex careyana Torr. ex Dewey
Carey's sedge
Herbaceous perennial
Rich, calcareous woods.

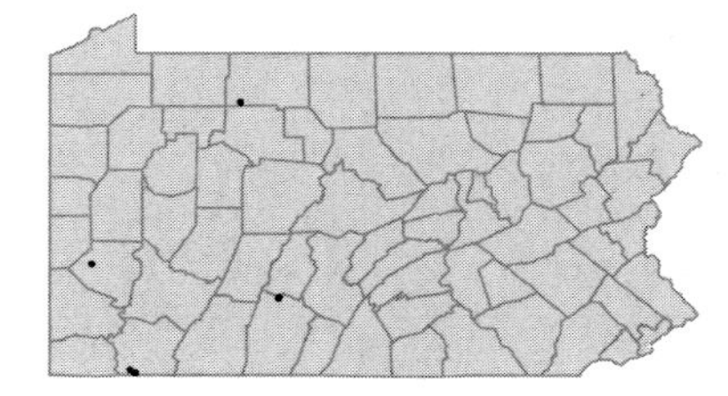

Carex caroliniana Schwein.
Sedge
Herbaceous perennial
Moist to wet woods, thickets, fields and meadows.

FACU

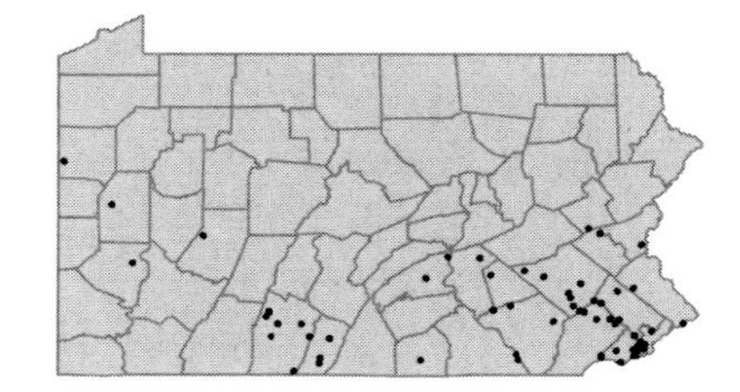

Carex cephaloidea (Dewey) Dewey
Sedge
Herbaceous perennial
Dry to moist woods, meadows and stream banks.
Carex sparganioides Muhl. var. *cephaloidea* (Dewey) Carey GBC

FACU

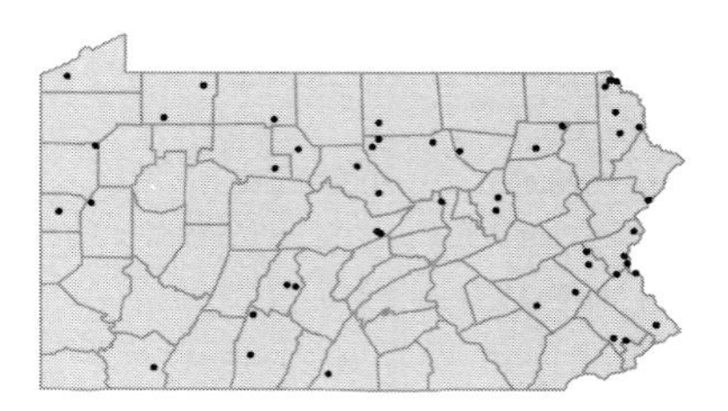

Carex cephalophora Muhl. ex Willd.
Sedge
Herbaceous perennial
Dry to moist woods, thickets and fields.

FACU

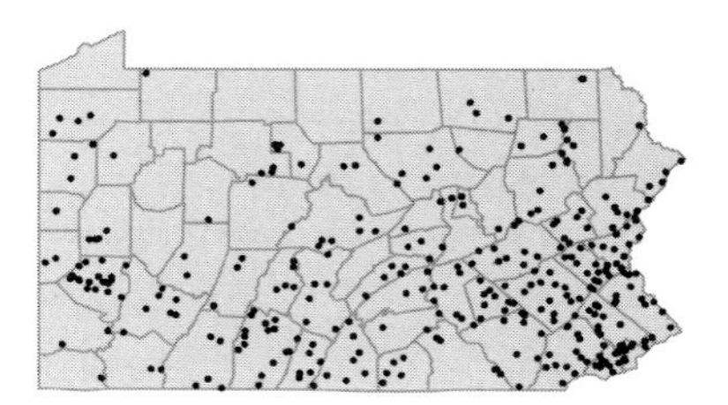

Carex chordorrhiza Ehrh. ex L.f.
Creeping sedge
Herbaceous perennial
Sphagnum bog. Believed to be extirpated, last collected in 1967.

OBL

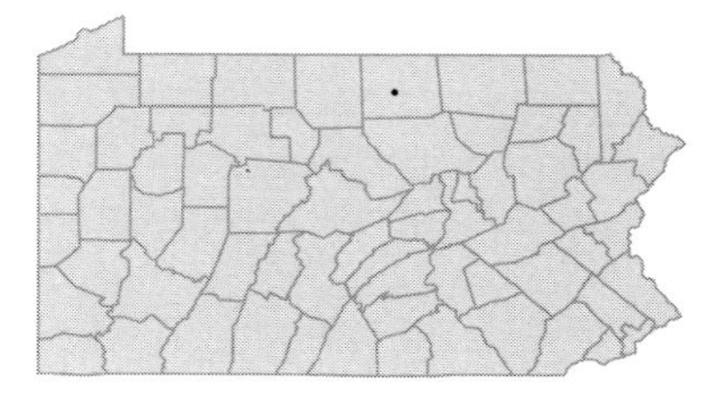

Carex collinsii Nutt.
Collin's sedge
Herbaceous perennial
Sphagnous swamps and swampy woods.

OBL

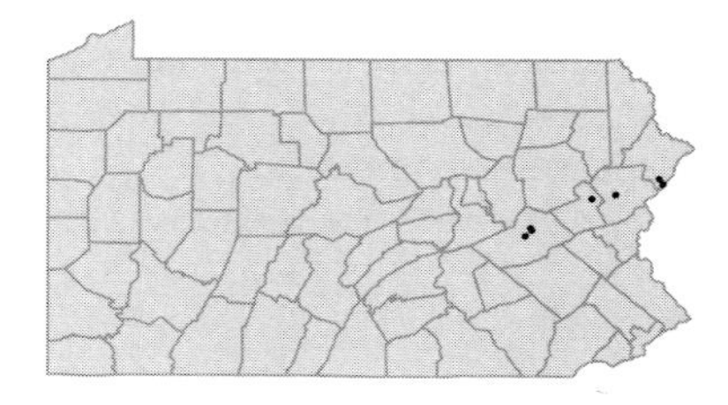

Carex communis Bailey
Sedge
Herbaceous perennial
Dry to moist woods and clearings.
Carex pedicellata (Dewey) Britt. P

Carex comosa Boott
Sedge
Herbaceous perennial
Swamps, marshes amd swales. A hybrid with *C. pseudocyperus* has been collected once.

OBL

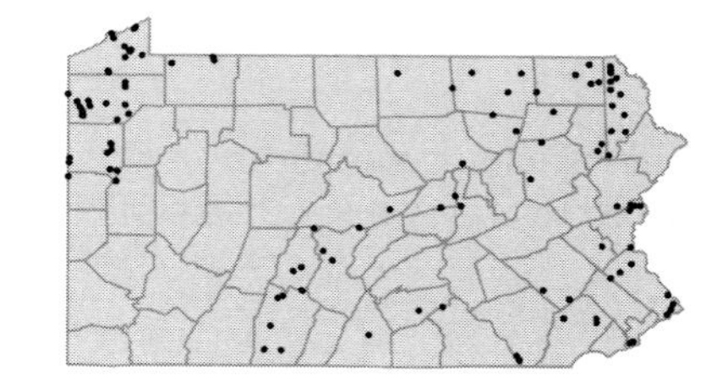

Carex conjuncta Boott
Sedge
Herbaceous perennial
Moist open woods, fields and meadows.

FACW

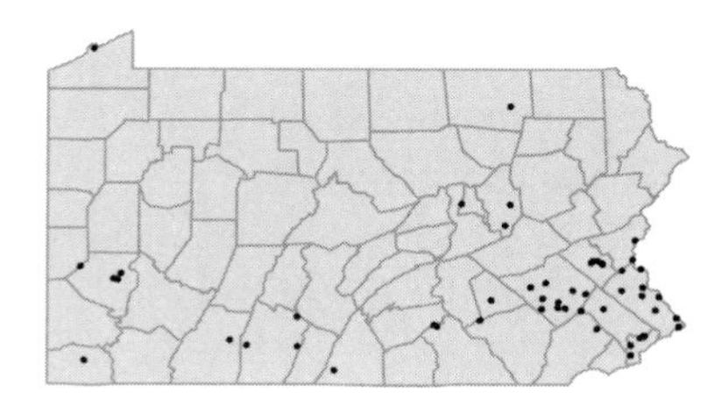

Carex conoidea Schkuhr ex Willd.
Sedge
Herbaceous perennial
Meadows and swales.

FACU

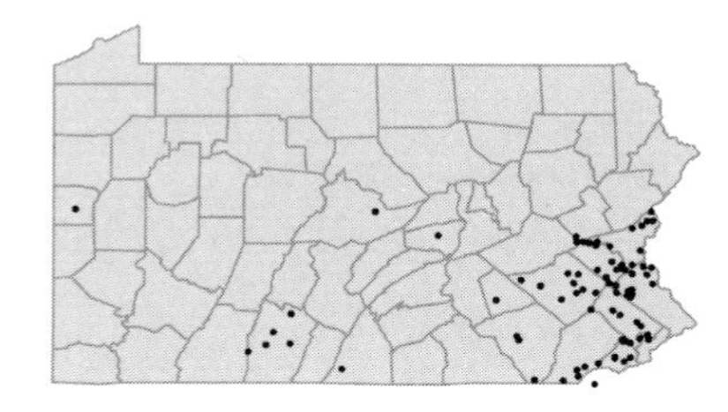

Carex crawfordii Fern.
Crawford's sedge
Herbaceous perennial
Wet or dry, open soil. Represented by a single collection from Pike Co. in 1944.

FAC

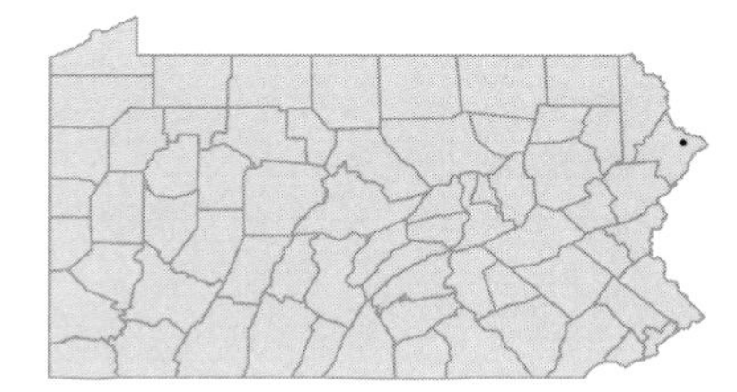

Carex crinita Lam. var. ***crinita***
Short hair sedge
Herbaceous perennial
Moist to wet woods, thickets, marshes, ditches and stream banks.

OBL

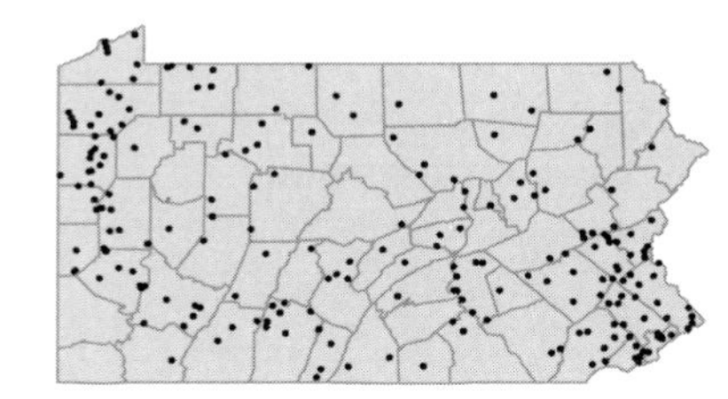

Carex crinita Lam. var. ***brevicrinis*** Fern.
Sedge
Herbaceous perennial
Moist to wet woods.

OBL

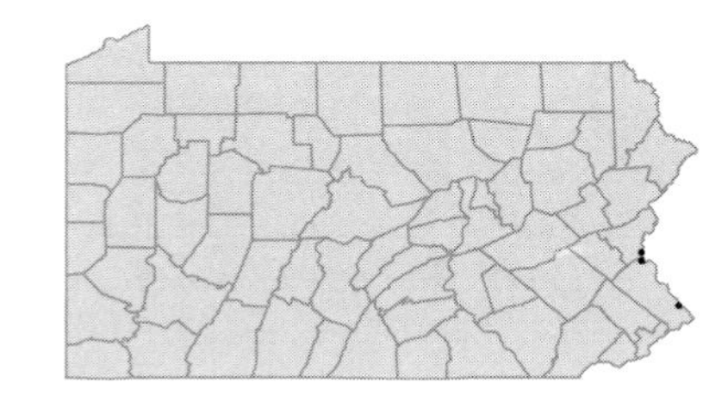

Carex cristatella Britt.
Sedge
Herbaceous perennial
Swamps, wet thickets, meadows, stream banks.

FACW

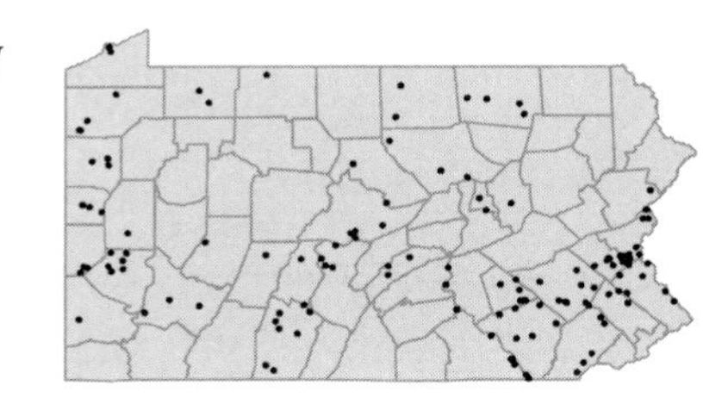

Carex cryptolepis Mackenzie
Northeastern sedge
Herbaceous perennial
Wet, calcareous meadows and marshy lake shore.
Carex flava L. var. *fertilis* Peck F

OBL

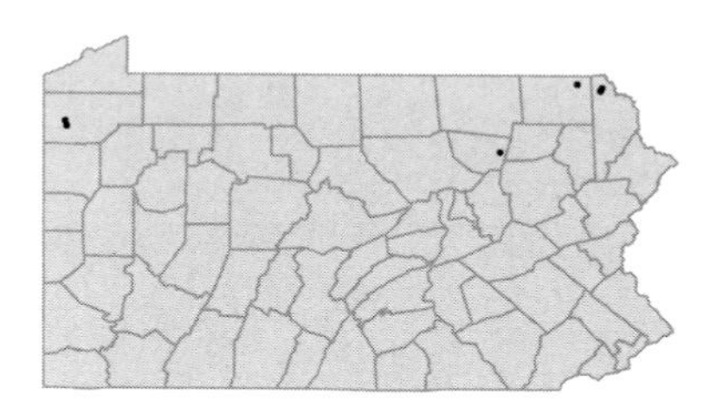

Carex cumulata (Bailey) Mackenzie
Sedge
Herbaceous perennial
Dry, rocky or sandy soil, abandoned railroad beds and alluvium.

FACU

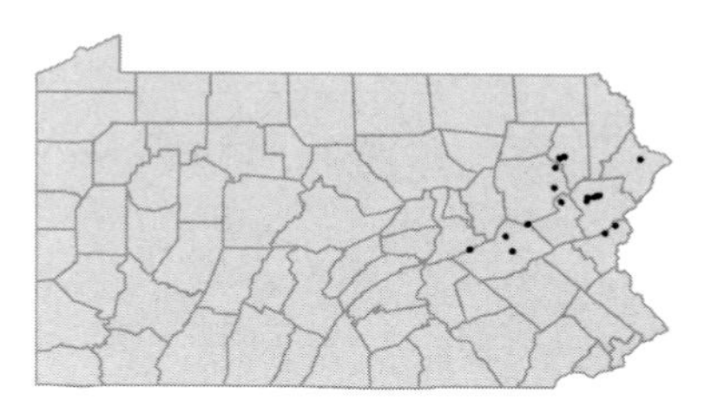

Carex davisii Schwein. & Torr.
Sedge
Herbaceous perennial
Rich woods and stream banks.

FAC-

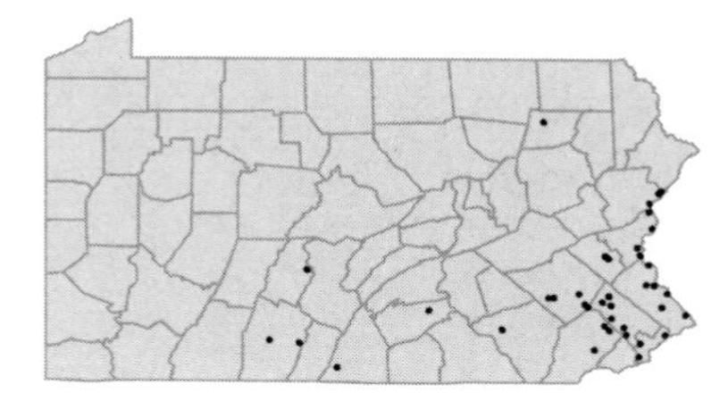

Carex debilis Michx. var. ***debilis***
Sedge
Herbaceous perennial
Swamps, thickets and low woods.

FAC

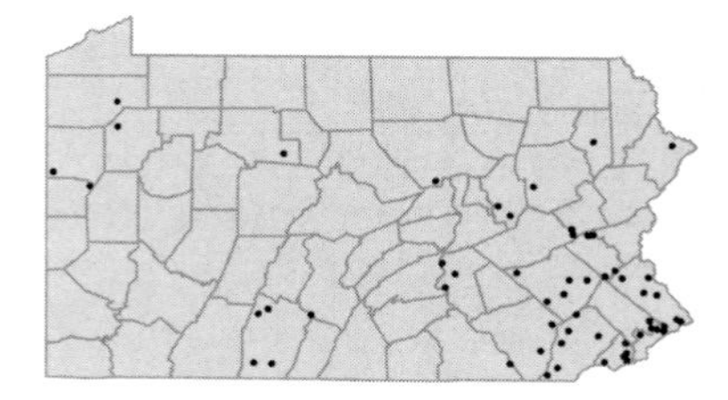

Carex debilis Michx. var. ***pubera*** A.Gray
Sedge
Herbaceous perennial
Wet woods, swamps and bogs.
Carex tenuis Rudge var. *pubera* A.Gray P

FAC

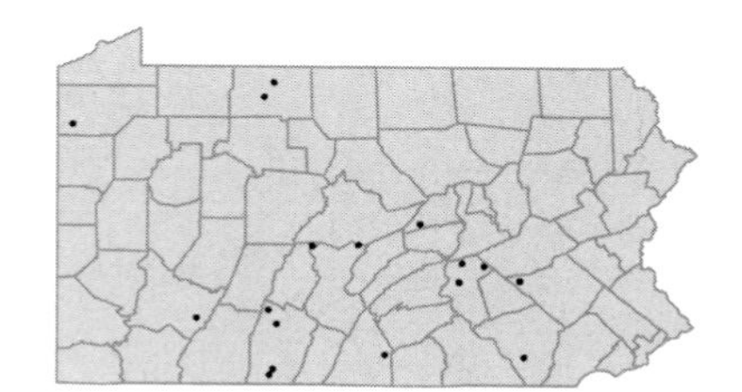

Carex debilis Michx. var. ***rudgei*** Bailey
Sedge
Herbaceous perennial
Moist, open or rocky woods.
Carex tenuis Rudge *pro parte* P

FAC

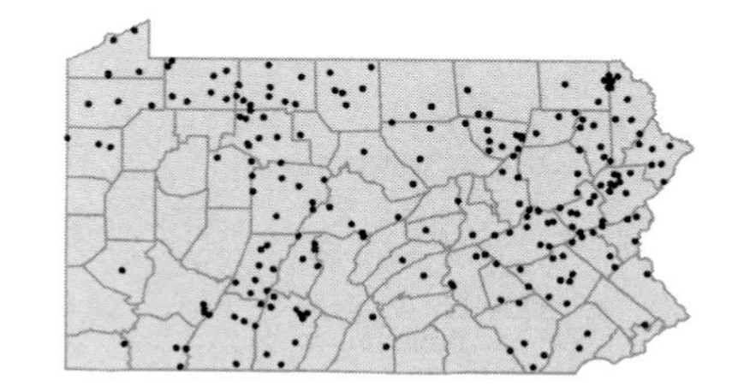

Carex deweyana Schwein. var. ***deweyana***
Sedge
Herbaceous perennial
Rocky woods and moist, wooded slopes.

FACU

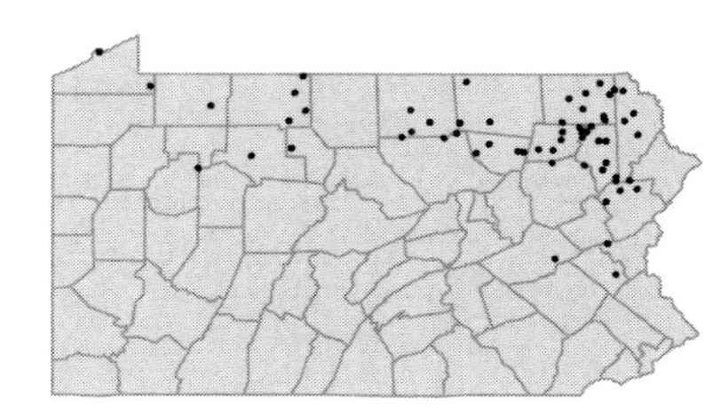

Carex diandra Schrank
Lesser panicled sedge
Herbaceous perennial
Bog hummocks, pond margins and marshy ground.
Carex teretiuscula Good. P

OBL

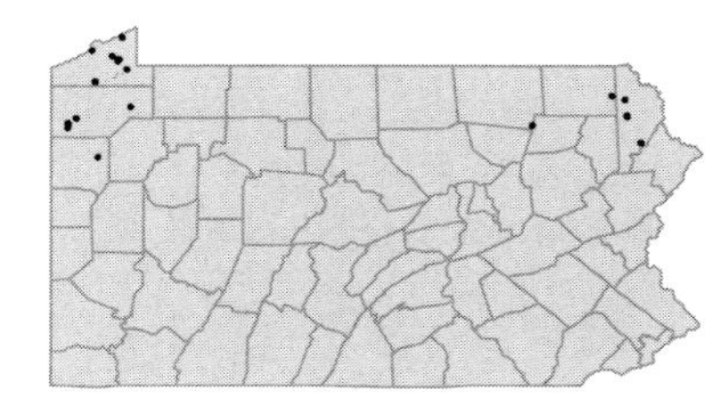

Carex digitalis Willd.
Sedge
Herbaceous perennial
Dry woods.

UPL

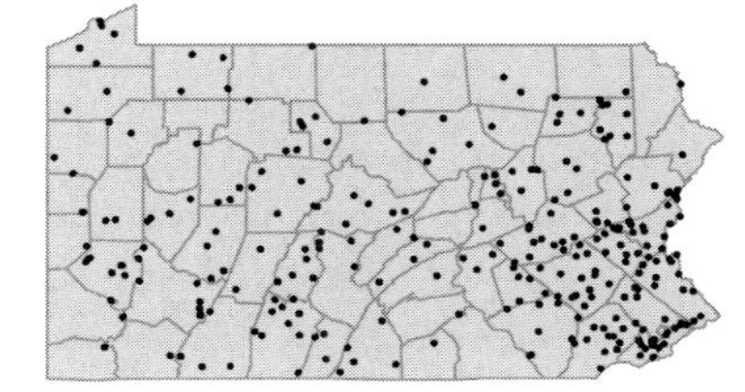

Carex disperma Dewey
Soft-leaved sedge
Herbaceous perennial
Swampy woods, bogs and rhododendron swamps.
Carex tenella Schkuhr P

FACW+

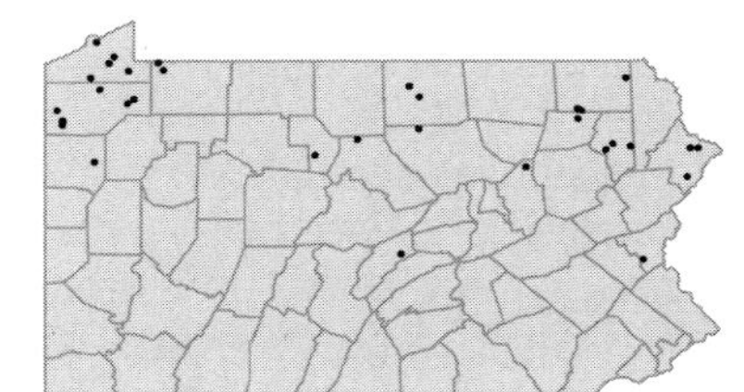

Carex distans L.
Sedge
Herbaceous perennial
Ballast. Represented by a single collection from Philadelphia Co. ca. 1865.

I

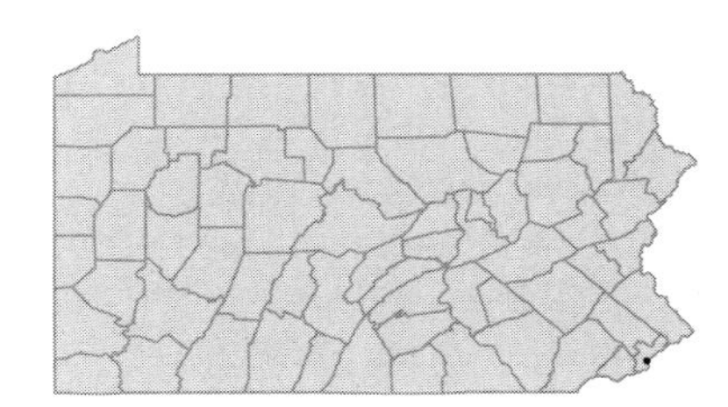

Carex distenta Kunze ex Kunth
Sedge
Herbaceous perennial
Represented by a single collection from Delaware Co. in 1901.

I

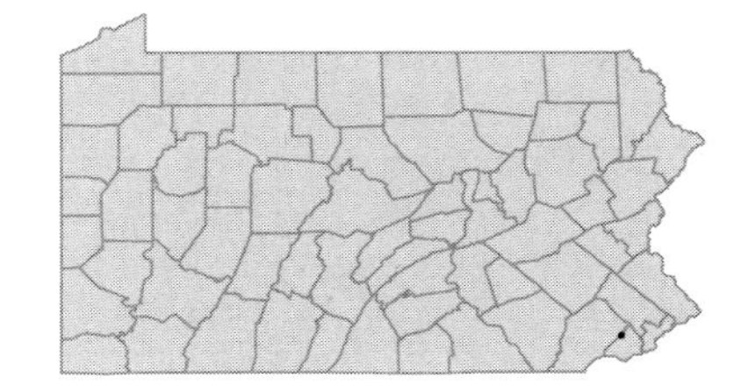

Carex divulsa Stokes
Sedge
Herbaceous perennial
Disturbed ground.
Carex virens Lam. FWGB

I

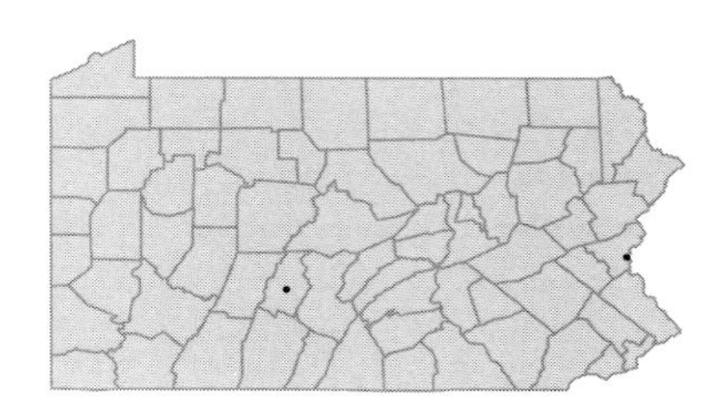

Carex eburnea Boott
Ebony sedge
Herbaceous perennial
Calcareous cliffs and slumps.
Carex setifolia (Dewey) Britt. P

FACU

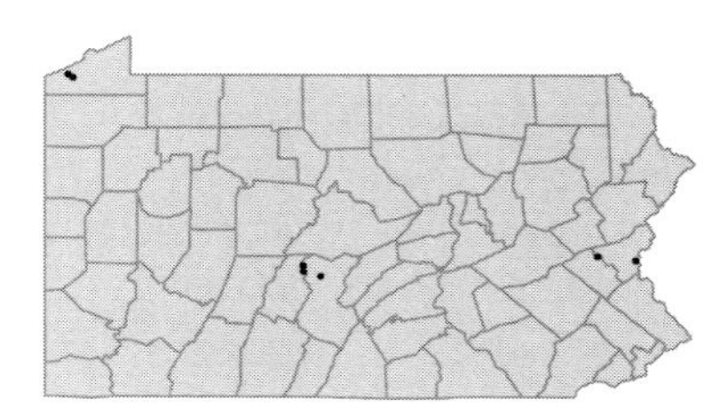

Carex echinata Murr.
Prickly sedge
Herbaceous perennial
Boggy woods, swamps, marshes, lake margins and swales.
Carex angustior Mackenzie *pro parte* FW; *Carex cephalantha* (Bailey) Bickn. *pro parte* FW; *Carex laricina* Mackenzie ex Bright *pro parte* W; *Carex muricata* L. *pro parte* GB

OBL

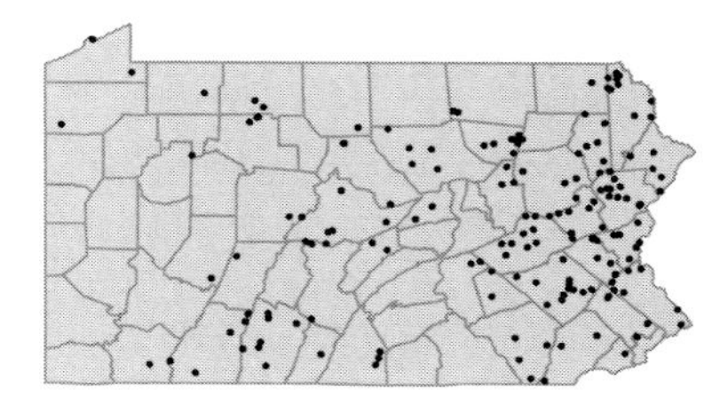

Carex emmonsii Dewey in Torr.
Sedge
Herbaceous perennial
Dry, acidic woods.
Carex nigromarginata Schwein. var. *minor* (Boott) Gleason GB; *Carex albicans* Willd. var. *emmonsii* (Dewey) Rettig. C

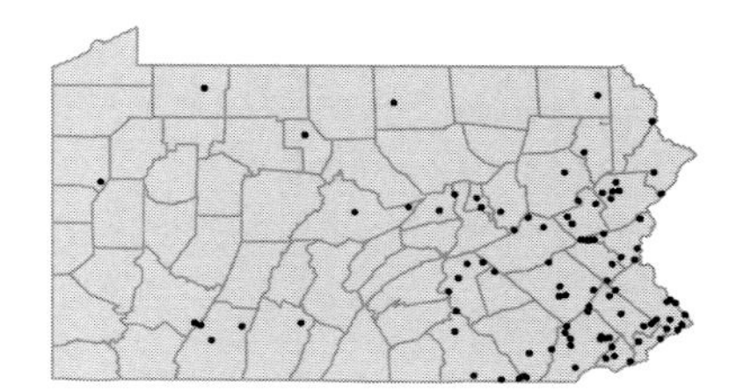

Carex emoryi Dewey in Torr.
Sedge
Herbaceous perennial
Marshes, stream banks and swales.
Carex stricta Lam. var. *elongata* (Boeckl.) Gleason GB

OBL

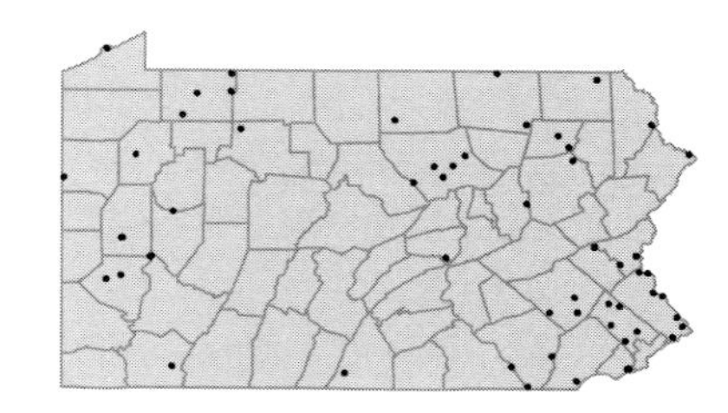

Carex festucacea Schkuhr ex Willd.
Sedge
Herbaceous perennial
Moist, open woods or thickets.

FAC

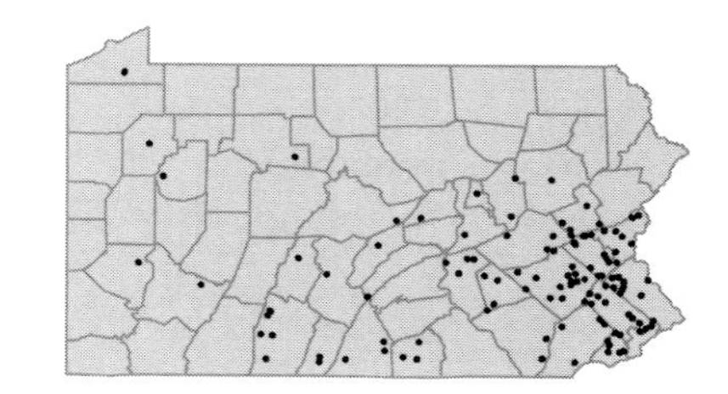

Carex flava L.
Yellow sedge
Herbaceous perennial
Calcareous bogs, fens and wet meadows.

OBL

Carex foenea Willd.
Fernald's hay sedge
Herbaceous perennial
Very dry, gravelly bank. Believed to be extirpated, last collected in 1931.
Carex aenea Fern. *pro parte* FGB

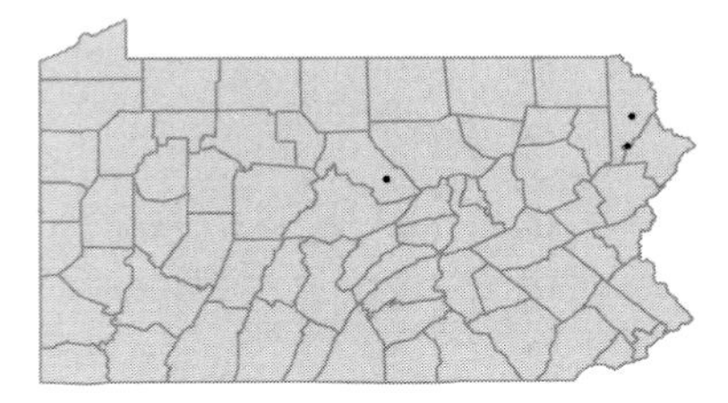

Carex folliculata L.
Sedge
Herbaceous perennial
Bogs, swamps and wet woods.

OBL

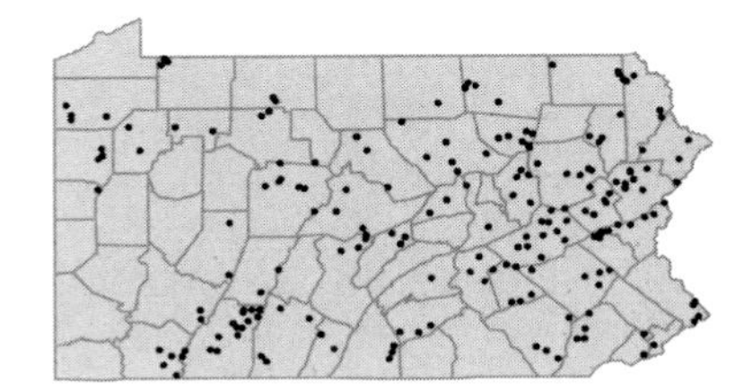

Carex formosa Dewey
Handsome sedge
Herbaceous perennial
Dry woods. Represented by a single collection from Centre Co. in 1975.

FAC

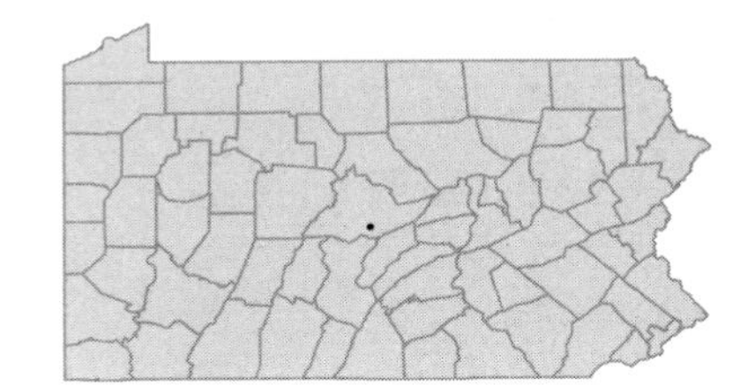

Carex frankii Kunth
Sedge
Herbaceous perennial
Moist woods, stream banks, low marshy ground and ditches.

OBL

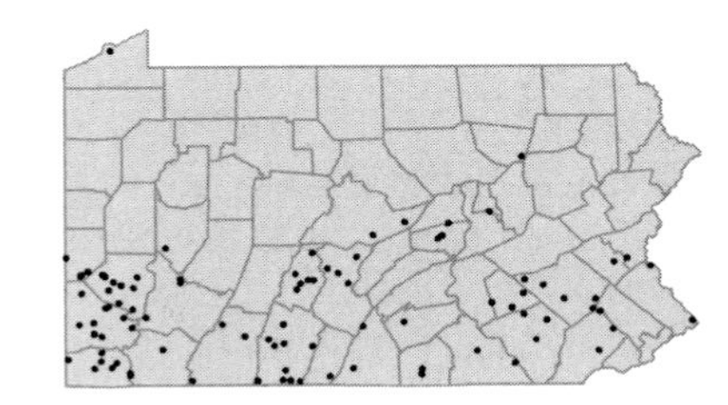

Carex garberi Fern.
Elk sedge
Herbaceous perennial
Sandy swales and calcareous gravel.
Carex aurea Nutt. *pro parte* C

FACW

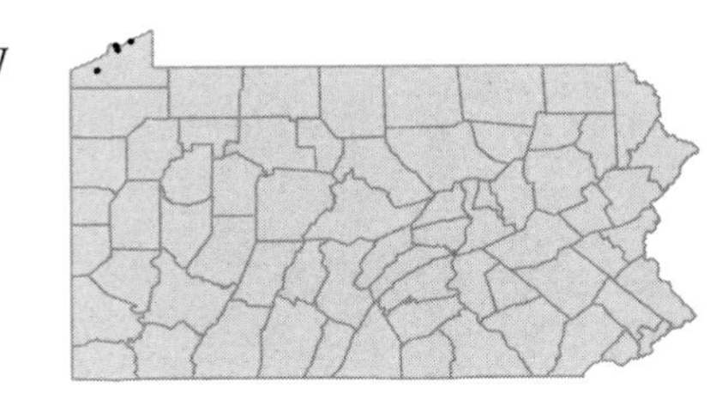

Carex geyeri Boott
Geyer's sedge
Herbaceous perennial
Wooded summit of limestone cliff.

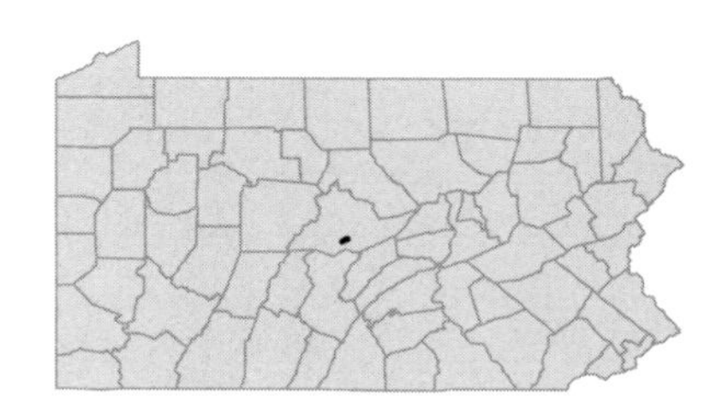

Carex glaucodea Tuckerman
Sedge
Herbaceous perennial
Dry to moist woods or fields.
Carex flaccosperma Dewey var. *glaucodea* (Tuckerman) Kukenth. C

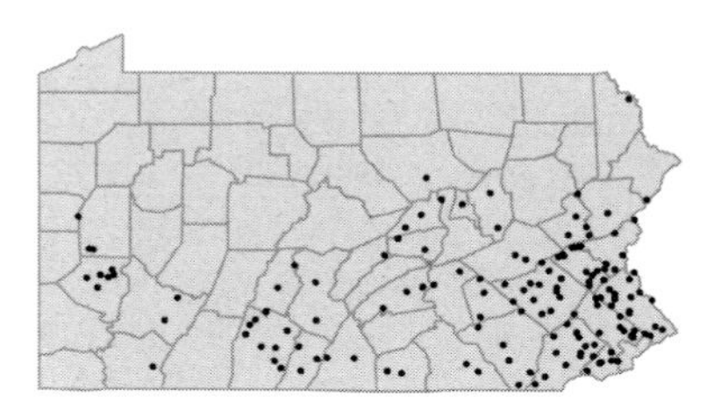

Carex gracilescens Steud.
Sedge
Herbaceous perennial
Woods and meadows.
Carex laxiflora Lam. var. *gracillima* (Boott) Robins. & Fern. GB

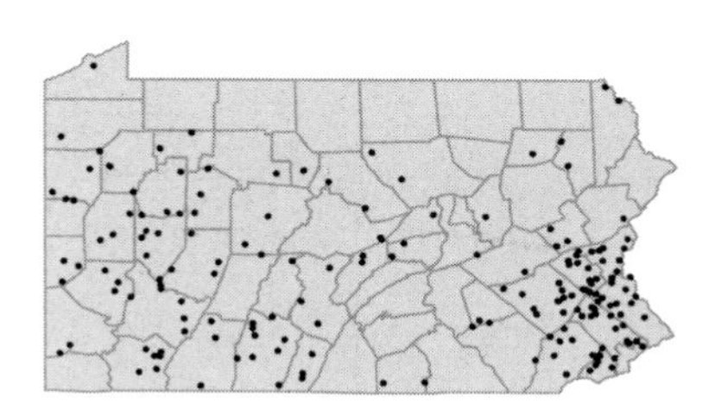

Carex gracillima Schwein.
Sedge
Herbaceous perennial
Dry to moist woods.

FACU

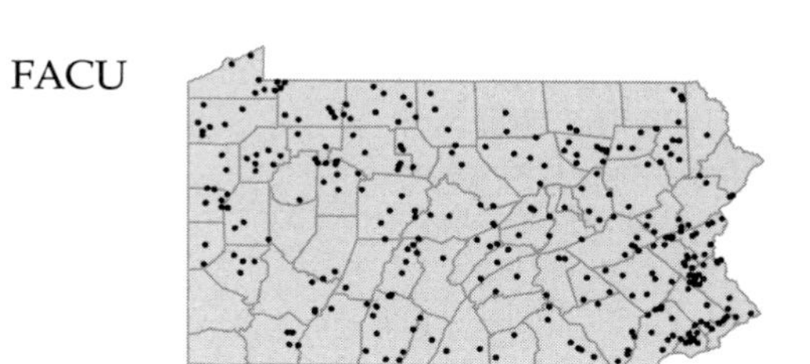

Carex granularis Muhl. ex Willd. var. ***granularis***
Sedge
Herbaceous perennial
Wet meadows, stream banks and wet woods.

FACW+

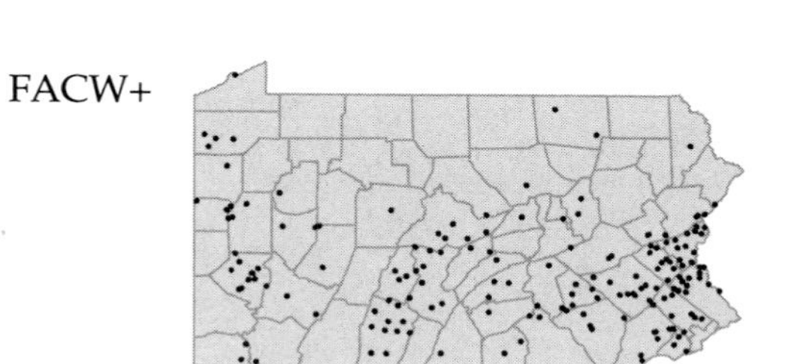

Carex granularis Muhl. ex Willd. var. ***haleana*** (Olney) Porter
Sedge
Herbaceous perennial
Wet meadows, swales or moist, limestone cliff.
Carex shriveri Britt. P

FACW+

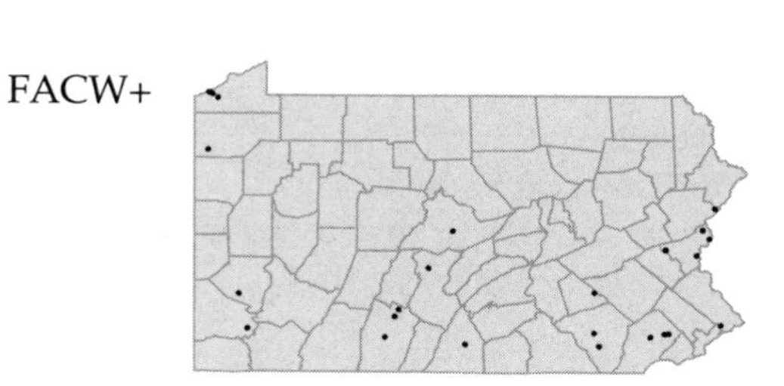

Carex gravida Bailey
Gravid sedge
Herbaceous perennial
Dry, open field.

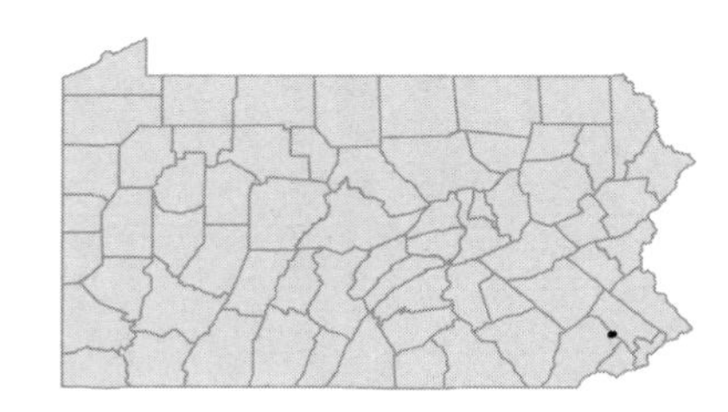

Carex grayi Carey
Sedge
Herbaceous perennial
Swamps and wet woods.
Carex asa-grayi Bailey P

FACW+

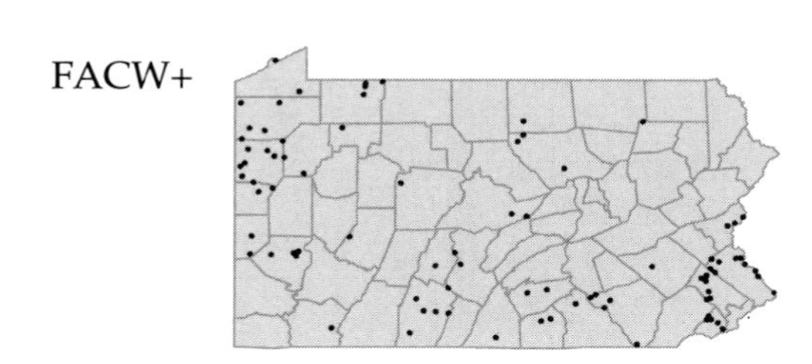

Carex grisea Wahlenb.
Sedge
Herbaceous perennial
Dry to moist woods, meadows and swales.
Carex amphibola Steud. var. *turgida* Fern. CFW

FAC

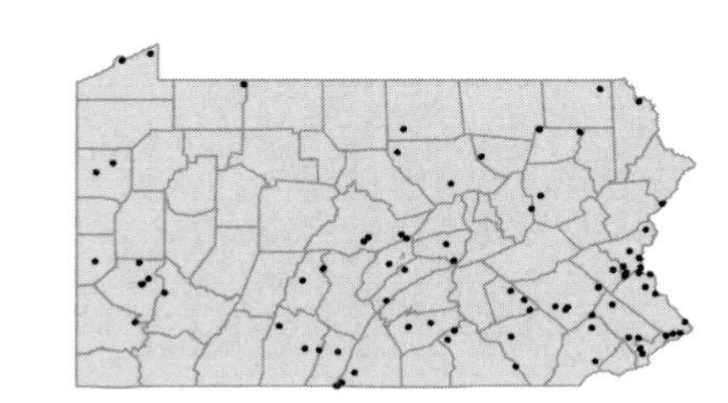

Carex gynandra Schwein.
Sedge
Herbaceous perennial
Swamps, swampy woods and lake margins.
Carex crinita Lam. var. *gynandra* (Schwein.) Schwein. & Torr. PFGB

OBL

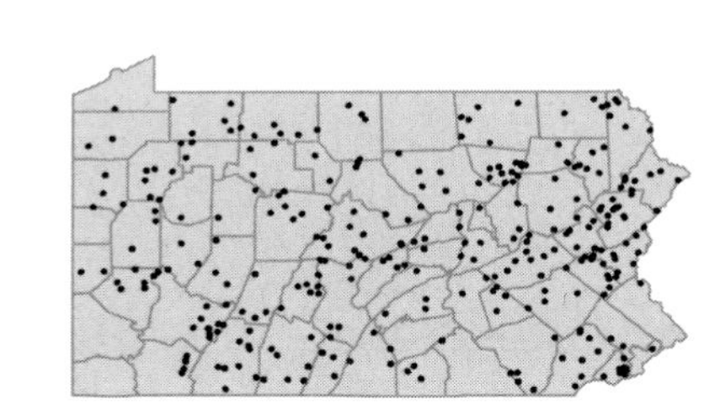

Carex haydenii Dewey
Cloud sedge
Herbaceous perennial
Swamps and wet meadows or woods.

OBL

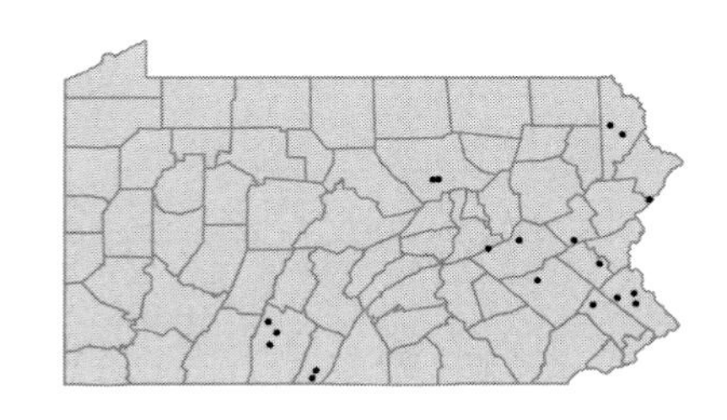

Carex hirsutella Mackenzie
Sedge
Herbaceous perennial
Dry fields, open woods and serpentine barrens.
Carex complanata Torr. & Hook. var. *hirsuta* (Bailey) Gleason GBC

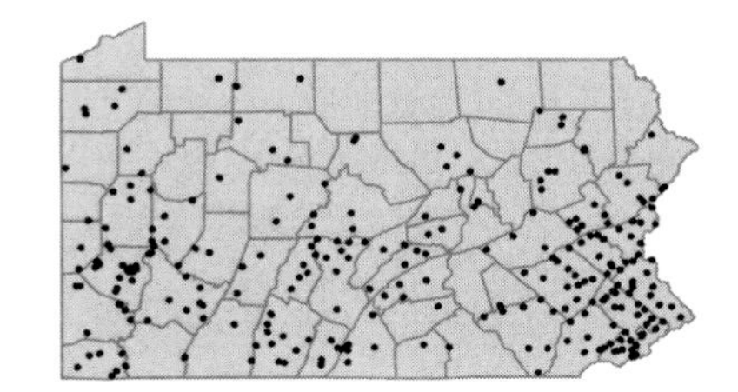

Carex hirta L.
I

Sedge
Herbaceous perennial
Ballast, meadows, fields and sandy alluvium.

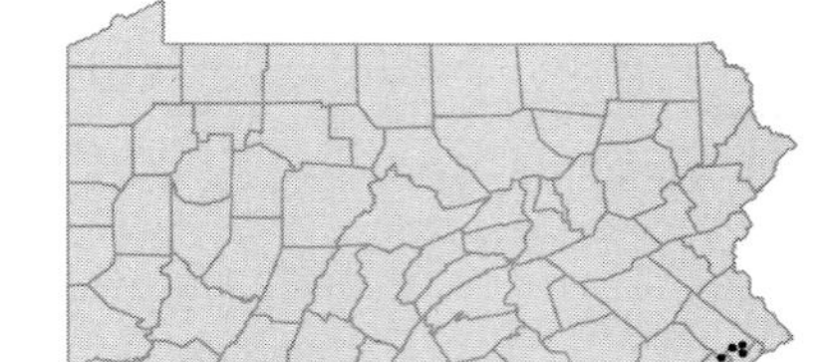

Carex hirtifolia Mackenzie
Sedge
Herbaceous perennial
Dry woods.
Carex pubescens Muhl. P

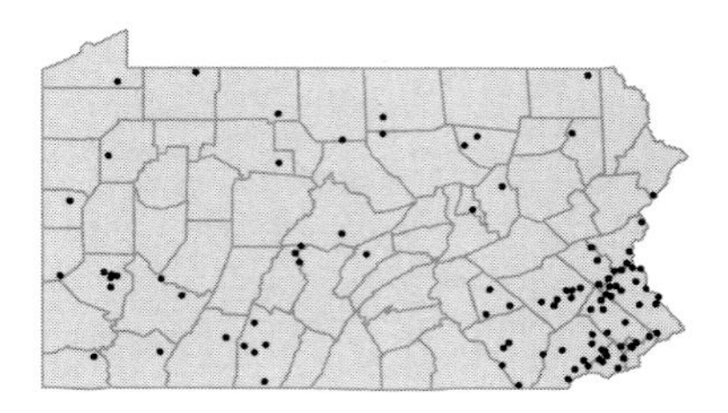

Carex hitchcockiana Dewey
Sedge
Herbaceous perennial
Moist, rocky, limestone woods and slopes.

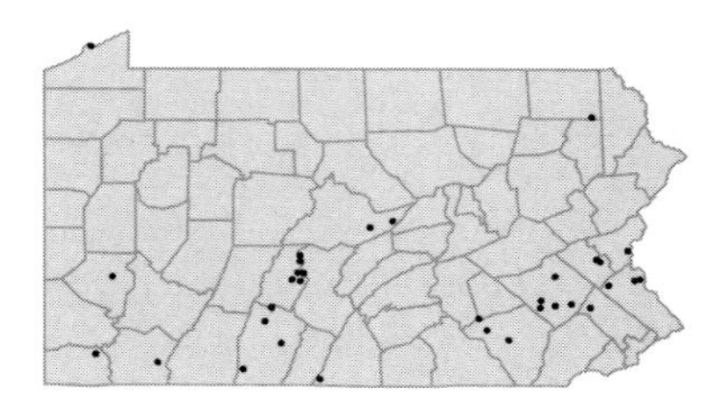

Carex hyalinolepis Steud.
OBL

Shoreline sedge
Herbaceous perennial
Damp banks. Believed to be extirpated, represented by a single collection from Philadelphia Co. in 1898.
Carex impressa (Wright) Mackenzie W; *Carex lacustris* Willd. var. *laxiflora* Dewey GB; *Carex riparia* M.A.Curtis P

Carex hystericina Muhl. ex Willd.
OBL

Sedge
Herbaceous perennial
Marshes, swamps, wet woods and swales.

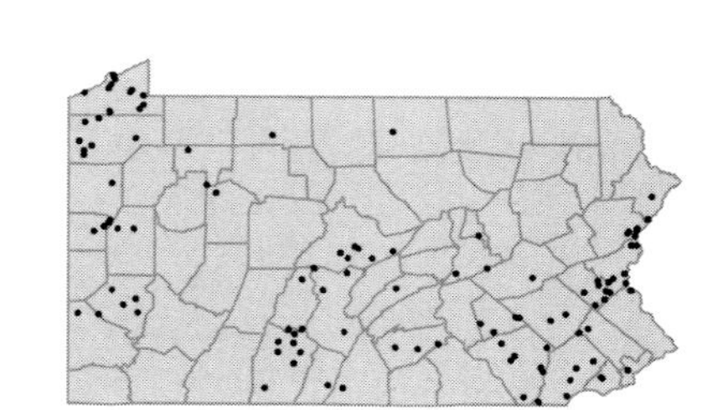

Carex interior Bailey
OBL

Sedge
Herbaceous perennial
Swamps, bogs, wet meadows and swales.

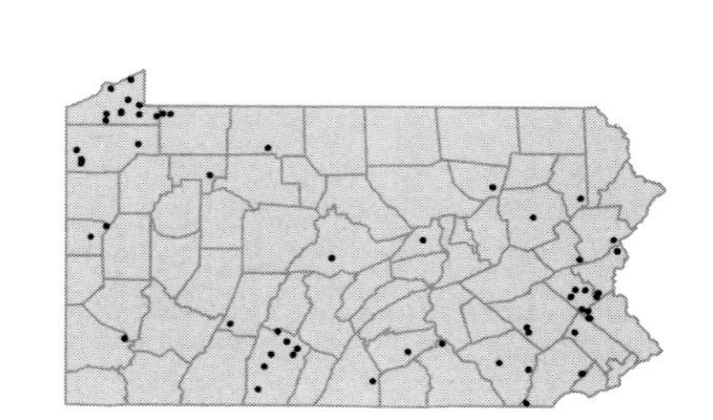

Carex intumescens Rudge
Sedge
Herbaceous perennial
Wet woods, meadows and swamps.

FACW+

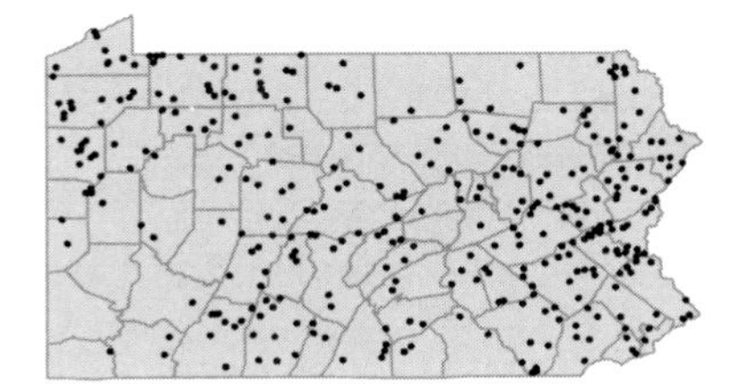

Carex jamesii Schwein.
Sedge
Herbaceous perennial
Rocky limestone woods.

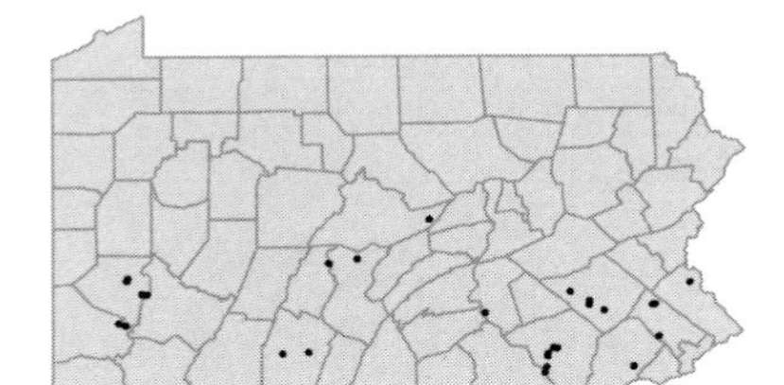

Carex lacustris Willd.
Sedge
Herbaceous perennial
Marshes, bogs and swamps.

OBL

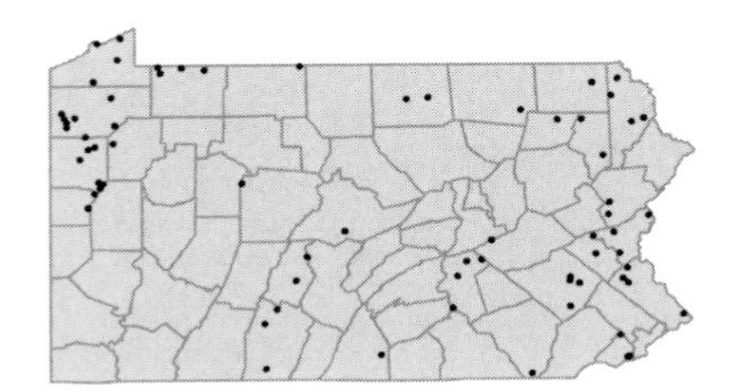

Carex laevivaginata (Kukenth.) Mackenzie
Sedge
Herbaceous perennial
Swamps and wet woods.

OBL

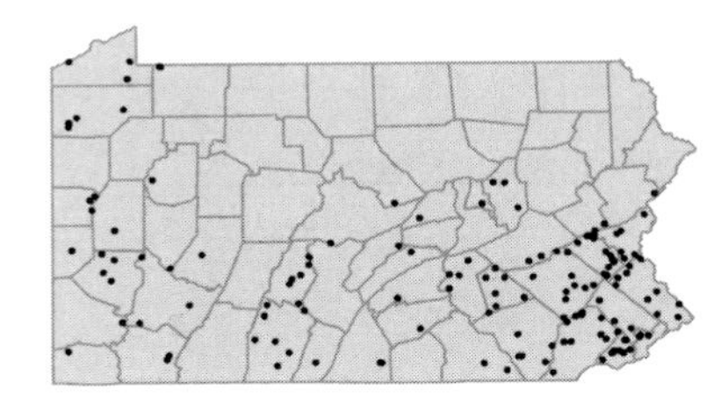

Carex lasiocarpa Ehrh. var. ***americana*** Fern.
Many-fruited sedge
Herbaceous perennial
Sphagnum bogs and boggy shores.
Carex filiformis L. P

OBL

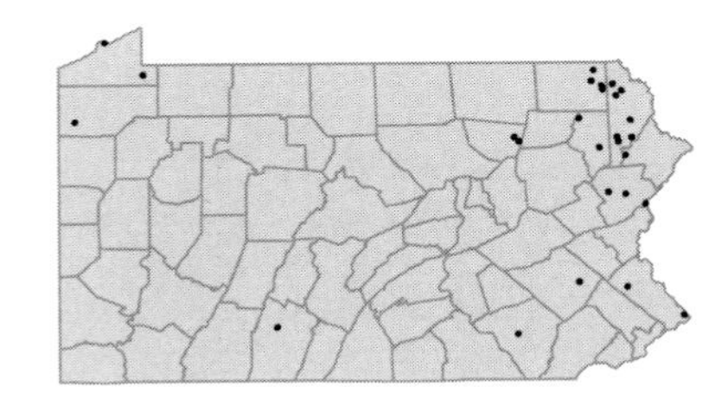

Carex laxiculmis Schwein. var. ***laxiculmis***
Sedge
Herbaceous perennial
Rich woods.

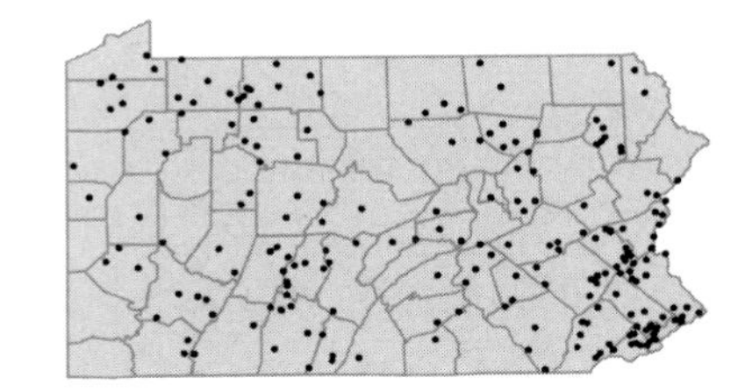

Carex laxiculmis Schwein. var. ***copulata*** (Bailey) Fern.
Sedge
Herbaceous perennial
Rich, calcareous woods.
Carex x copulata (Bailey) Mackenzie F

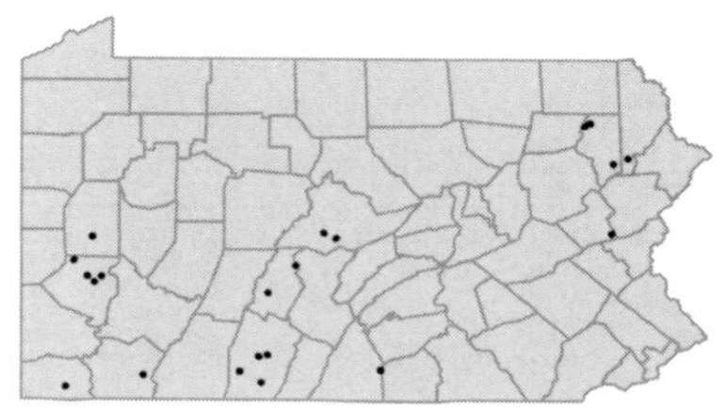

Carex laxiflora Lam.
Sedge
Herbaceous perennial
Low, rich woods.

FACU

Carex leavenworthii Dewey
Sedge
Herbaceous perennial
Fields, meadows, pastures and clearings.

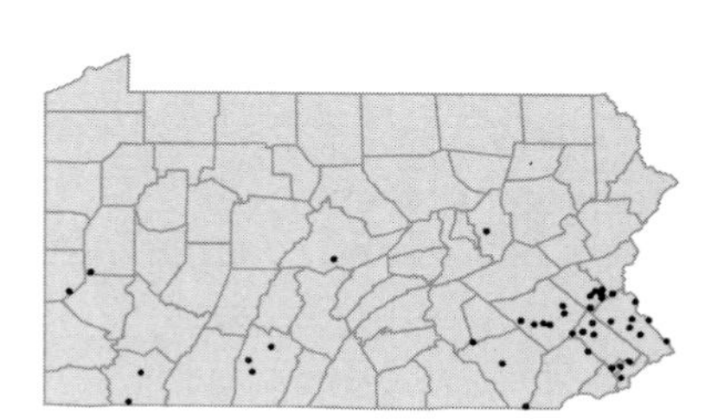

Carex leporina L.
Sedge
Herbaceous perennial
Abandoned lawn. Represented by a single collection from Philadelphia Co. in 1940.

[I]

Carex leptalea Wahlenb.
Sedge
Herbaceous perennial
Sphagnum bogs, wet woods, stream banks and swales.

OBL

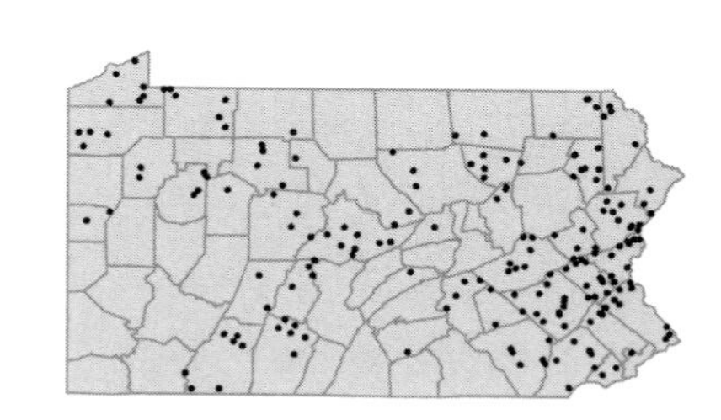

Carex leptonervia (Fern.) Fern.
Sedge
Herbaceous perennial
Moist woods.

FACW

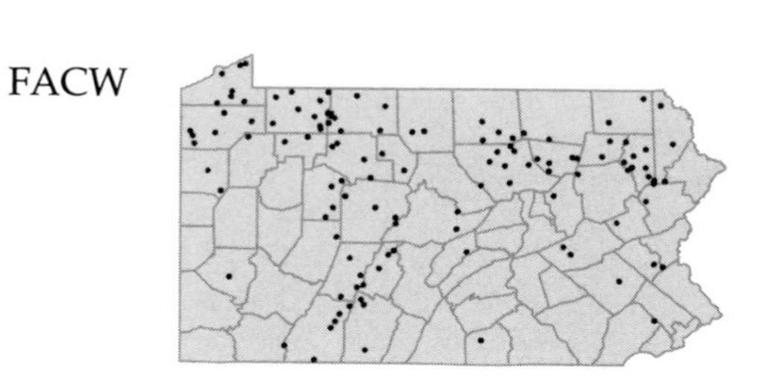

Carex limosa L.
Mud sedge
Herbaceous perennial
Sphagnum bog mats and hummocks.

OBL

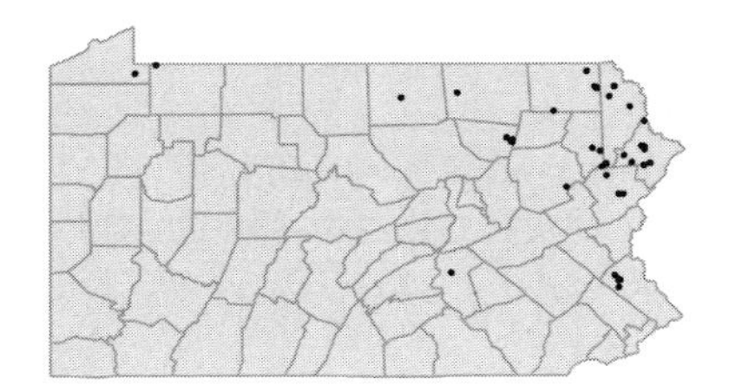

Carex longii Mackenzie
Long's sedge
Herbaceous perennial
Swamps, open thickets, moist meadows, old gravel pits and swales.
Carex albolutescens Schwein. *pro parte* BG

OBL

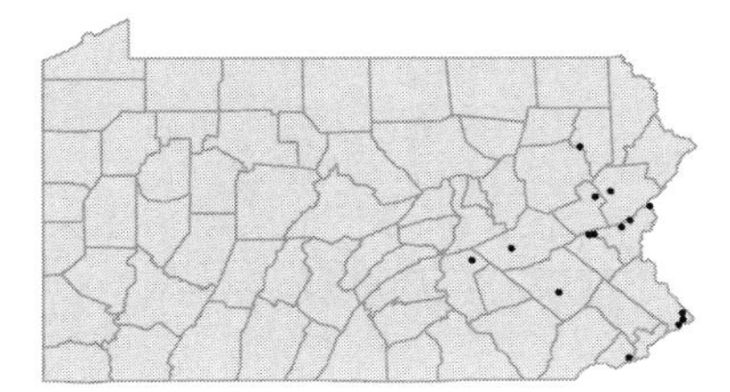

Carex lucorum Willd. ex Link
Sedge
Herbaceous perennial
Dry, open woods, clearings and slopes.
Carex pensylvanica Lam. var. *distans* Peck KFGBW

Carex lupuliformis Sartwell
False hop sedge
Herbaceous perennial
Hardwood swamp, wet meadow and railroad ditch. A hybrid with *C. retrorsa* has been collected at four locations.

FACW+

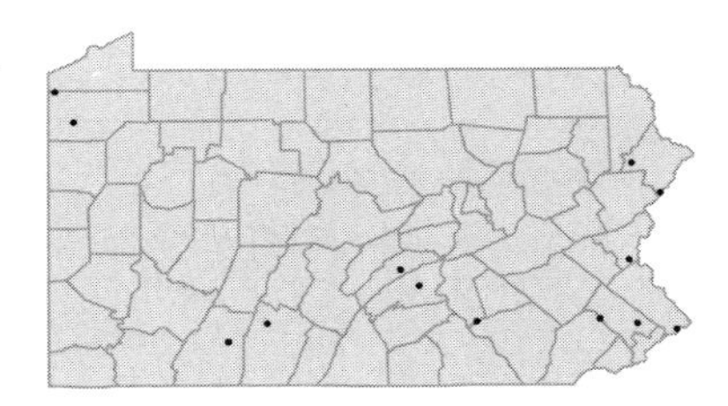

Carex lupulina Muhl. ex Willd.
Sedge
Herbaceous perennial
Swamps, bogs and wet woods.

OBL

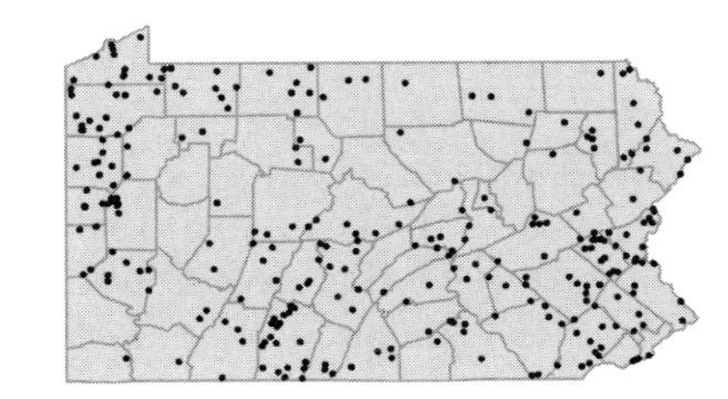

Carex lurida Wahlenb.
Sedge
Herbaceous perennial
Swamps, bogs and wet meadows.

OBL

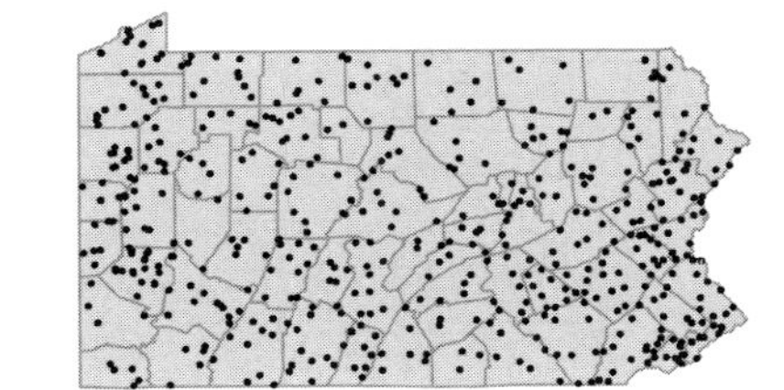

Carex meadii Dewey
Mead's sedge
Herbaceous perennial
Wet meadows, usually on diabase.

FAC

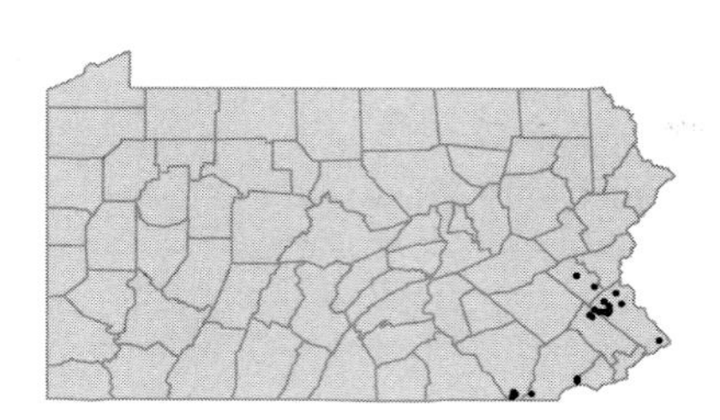

Carex mesochorea Mackenzie
Midland sedge
Herbaceous perennial
Dry, open woods, fallow fields, roadside banks and waste ground.
Carex cephalophora Muhl. var. *mesochorea* (Mackenzie) Gleason GBC

FACU

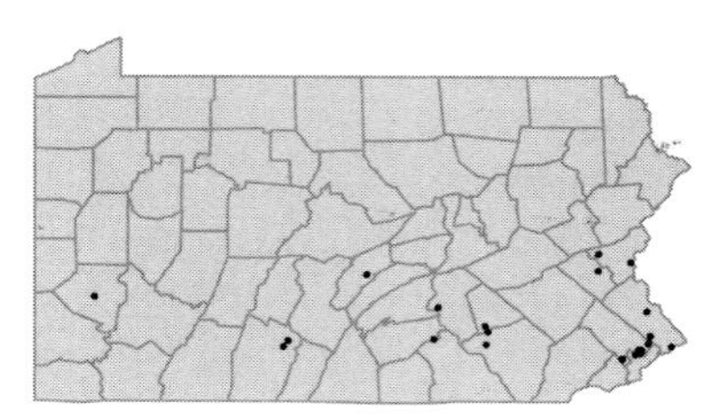

Carex mitchelliana M.A.Curtis
Mitchell's sedge
Herbaceous perennial
Swamps, wet meadows or stream banks.
Carex crinita Lam. var. *mitchelliana* (M.A.Curtis) Gleason GB

OBL

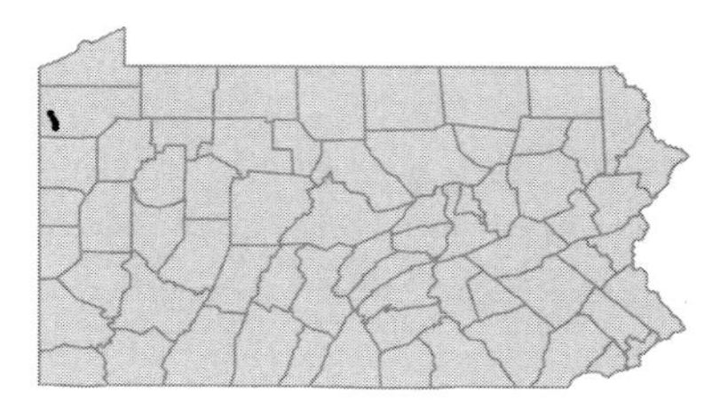

Carex molesta Mackenzie ex Bright
Sedge
Herbaceous perennial
Moist to dry, open ground and slopes.
Carex brevior (Dewey) Mackenzie *pro parte* C

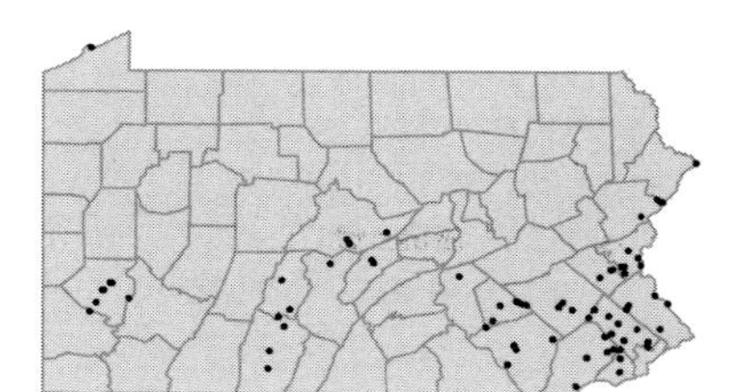

Carex muhlenbergii Schkuhr ex Willd.
Sedge
Herbaceous perennial
Dry woods, thickets, roadside banks and open sand.
Carex plana Mackenzie *pro parte* W

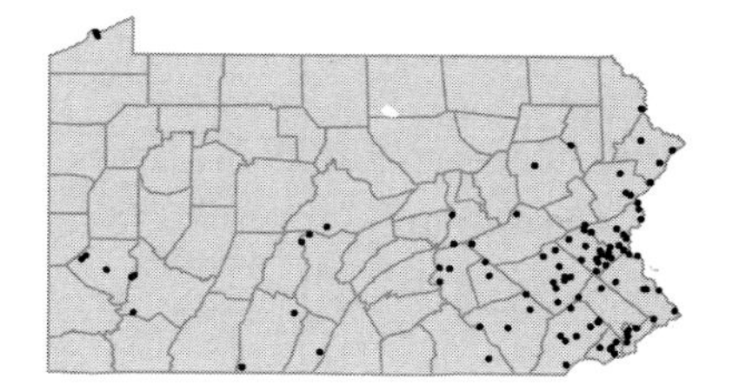

Carex muricata L.
Sedge
Herbaceous perennial
Represented by two collections from Northampton Co. 1891-1895.
Carex pairaei F.W.Schultz *pro parte* FWBG

[I]

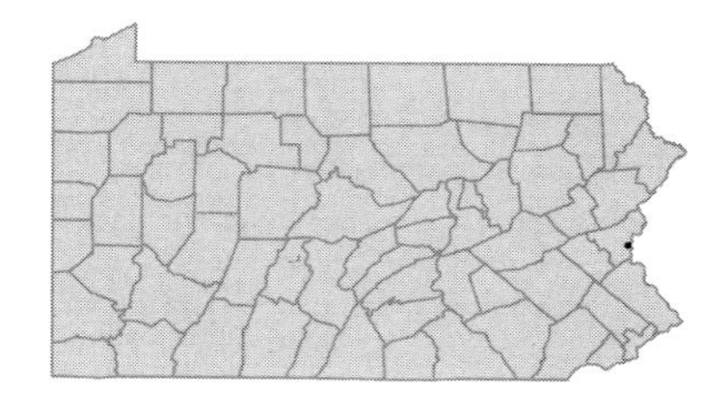

Carex nebraskensis Dewey
Sedge
Herbaceous perennial
Collected at a single site in Monroe Co. in 1920, probably an adventive associated with the railroad.

OBL
I

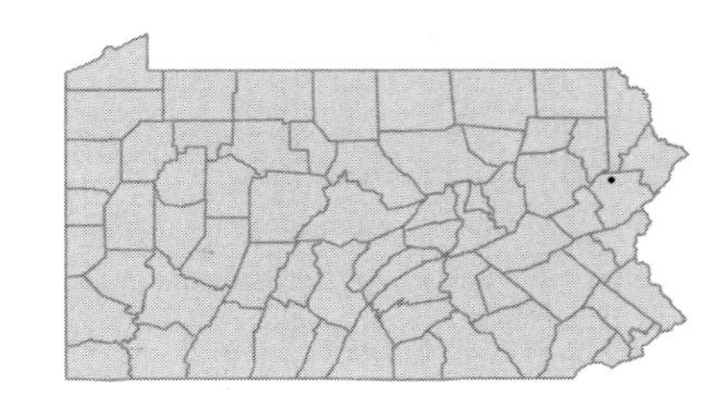

Carex nigromarginata Schwein.
Sedge
Herbaceous perennial
Dry woods and clearings.

UPL

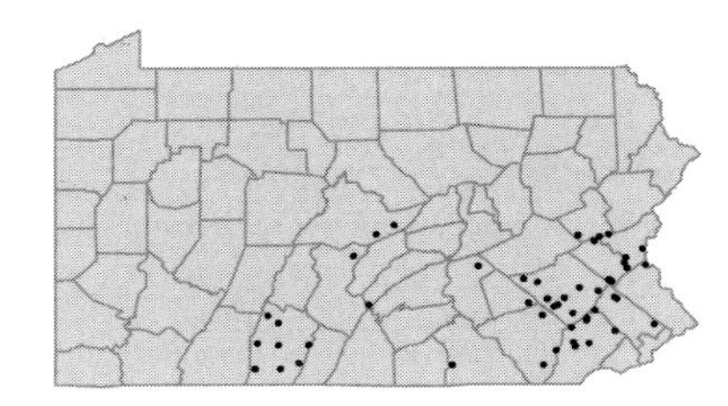

Carex normalis Mackenzie
Sedge
Herbaceous perennial
Dry to moist soil of open woods or meadows.

FACU

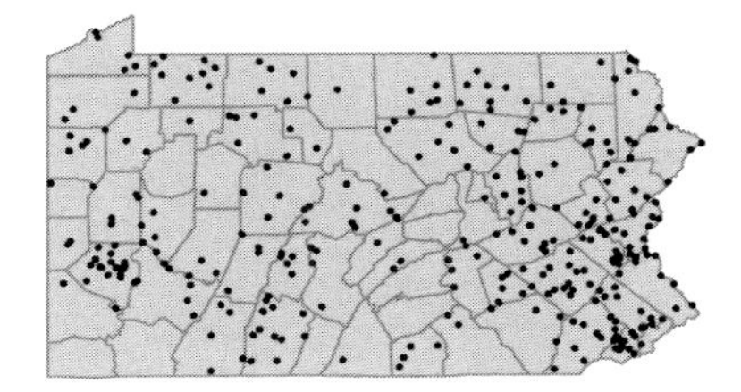

Carex novae-angliae Schwein.
Sedge
Herbaceous perennial
Wet woods.

FACU

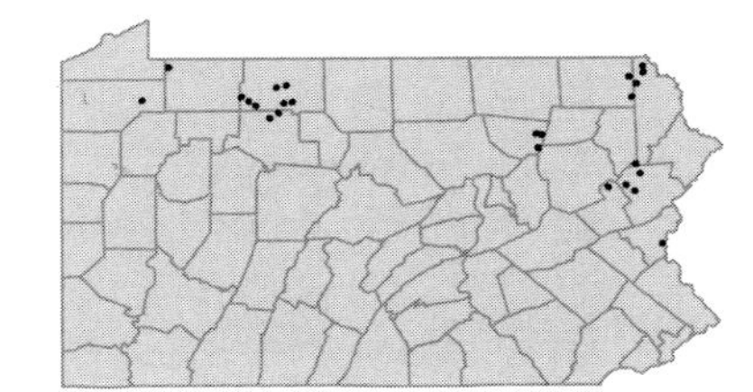

Carex oligocarpa Schkuhr ex Willd.
Sedge
Herbaceous perennial
Rich woods and limestone slopes.

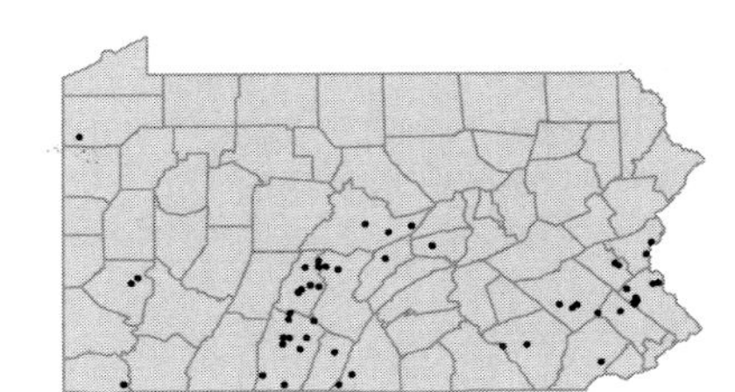

Carex oligosperma Michx.
Few-seeded sedge
Herbaceous perennial
Bogs.

OBL

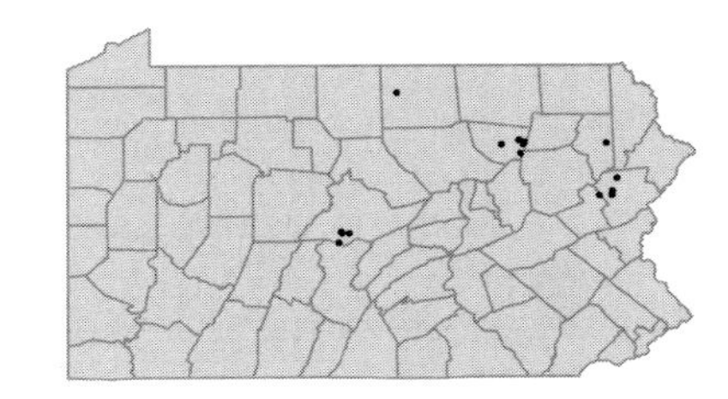

Carex ormostachya Wieg.
Spike sedge
Herbaceous perennial
Rocky woods and shale barrens.
Carex laxiflora Lam. var. *ormostachya* (Wieg.) Gleason GB; *Carex gracilescens* Steudel *pro parte* C

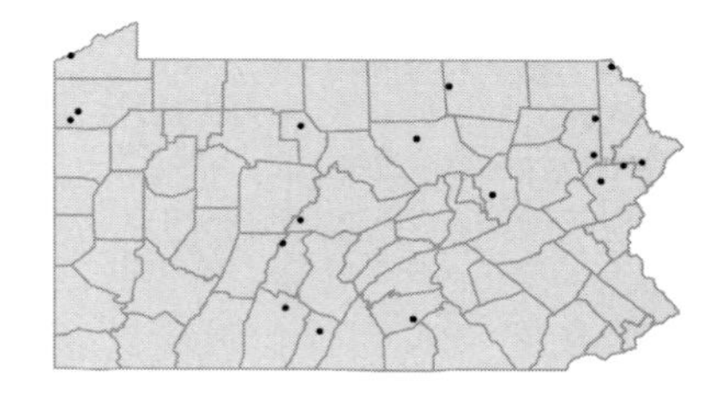

Carex pallescens L.
Sedge
Herbaceous perennial
Moist woods and meadows.

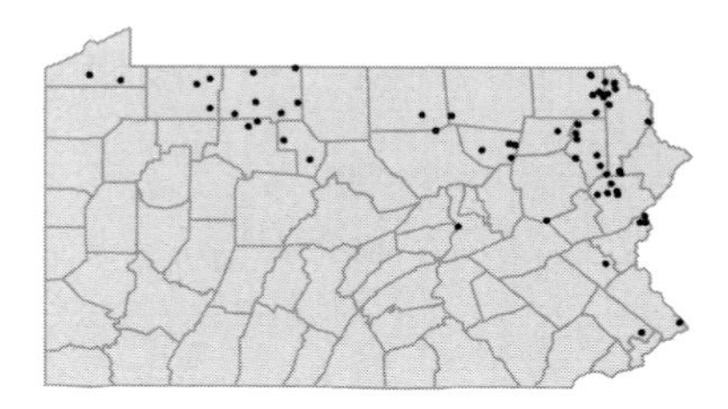

Carex pauciflora Lightfoot
Few-flowered sedge
Herbaceous perennial
Open sphagnum bogs.

OBL

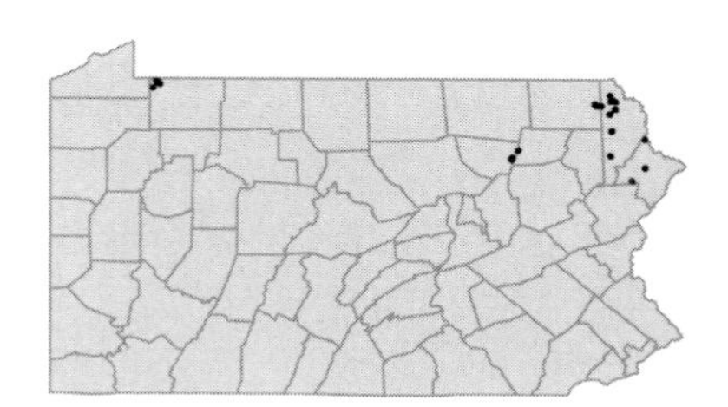

Carex paupercula Michx.
Bog sedge
Herbaceous perennial
Sphagnum bogs and boggy woods. Most, if not all, of the PA specimens are referable to var. *irrigua* (Wahlenb.) Fern.
Carex magellanica Lam. P

OBL

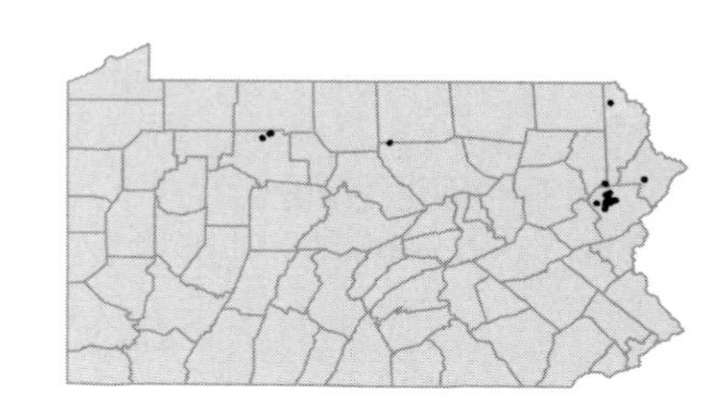

Carex pedunculata Muhl. ex Willd.
Sedge
Herbaceous perennial
Rocky, wooded slopes, swamps and floodplains, often on calcareous soils.

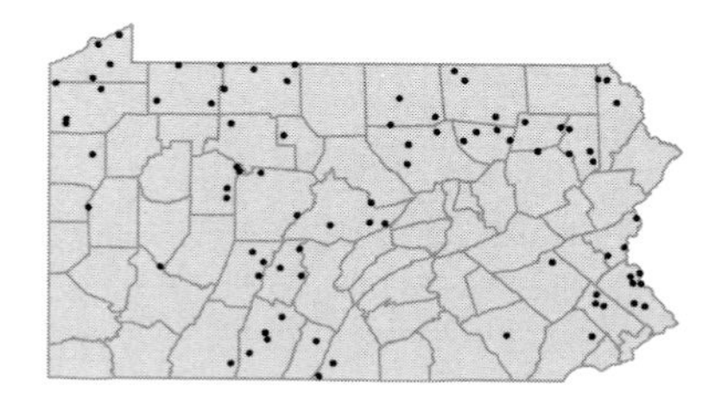

Carex pellita Muhl. ex Willd.
Sedge
Herbaceous perennial
Wet meadows and swamps. A hybrid with *C. trichocarpa* has been collected once.
Carex lasiocarpa Ehrh. var. *latifolia* (Boeckl.) Gleason GB; *Carex lanuginosa* Michx. WKF

OBL

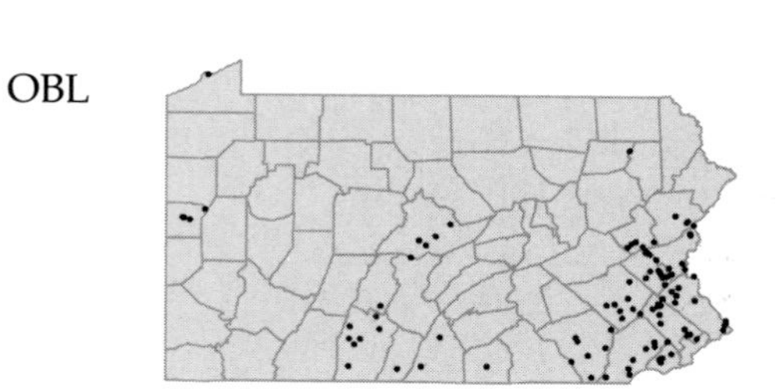

Carex pensylvanica Lam.
Sedge
Herbaceous perennial
Open woods and wooded slopes.

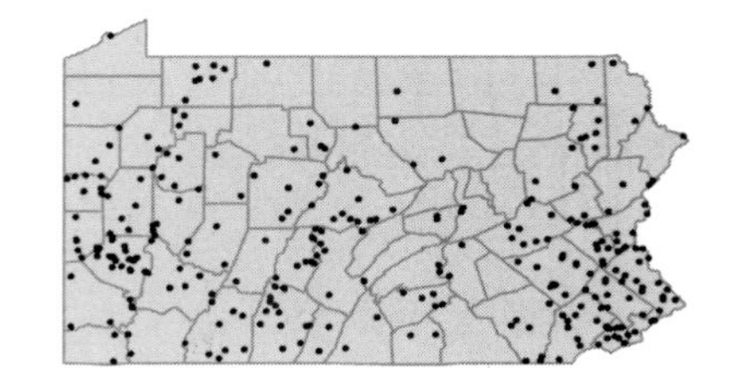

Carex plantaginea Lam.
Plantain sedge
Herbaceous perennial
Woods and wooded slopes.

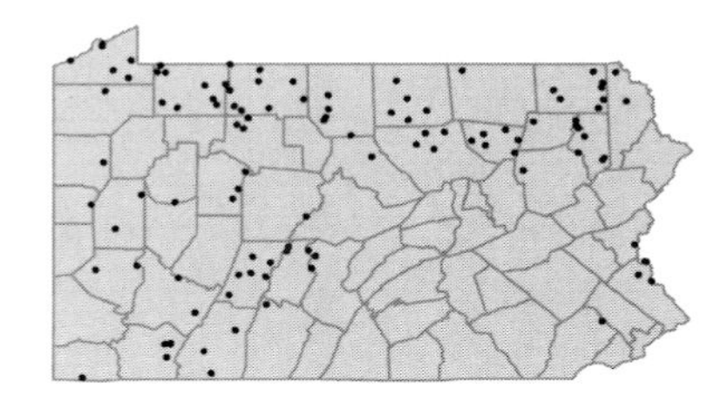

Carex platyphylla Carey
Broad-leaf sedge
Herbaceous perennial
Rich woods and wooded slopes.

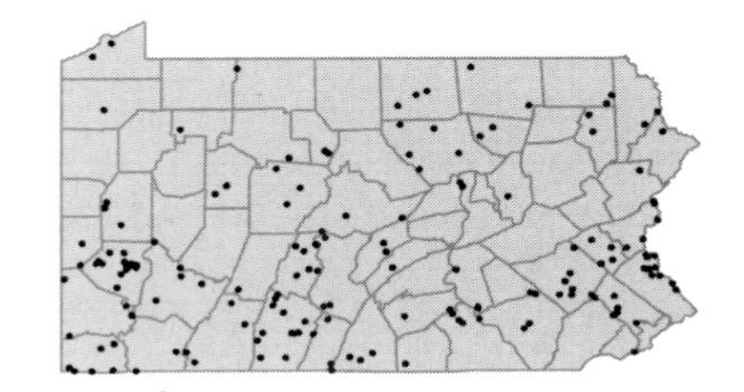

Carex polymorpha Muhl.
Variable sedge
Herbaceous perennial
Moist, open woods and barrens, in sandy-peaty soils.

FACU

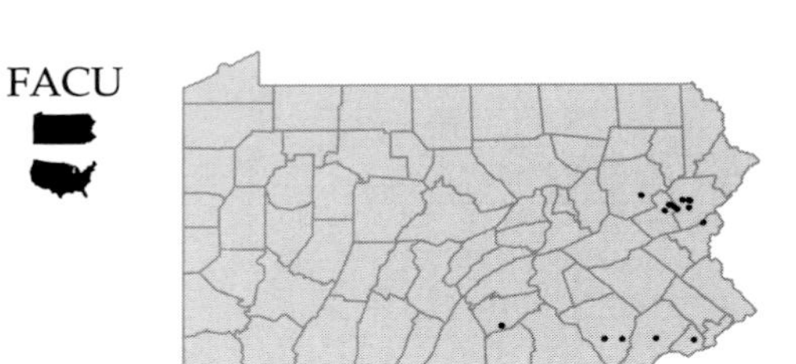

Carex praegracilis W. Boott.
Freeway sedge
Herbaceous perennial
Marshy floodplain. Represented by a single collection from Erie Co. in 1987.

I

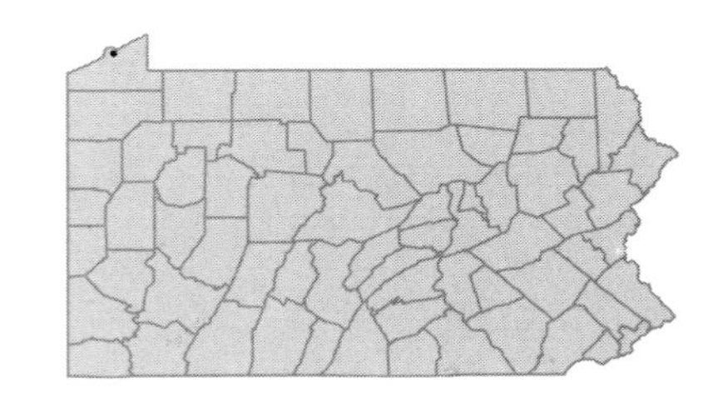

Carex prairea Dewey in A.Wood
Prairie sedge
Herbaceous perennial
Moist, calcareous meadows, marshes and fens.

FACW

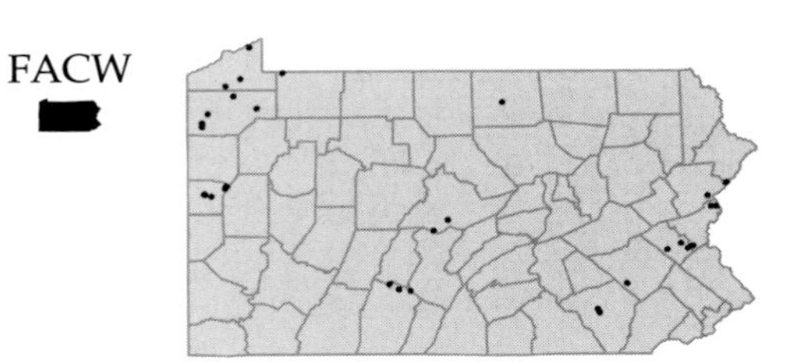

Carex prasina Wahlenb.
Sedge
Herbaceous perennial
Swampy or boggy woods, stream banks and ditches.

OBL

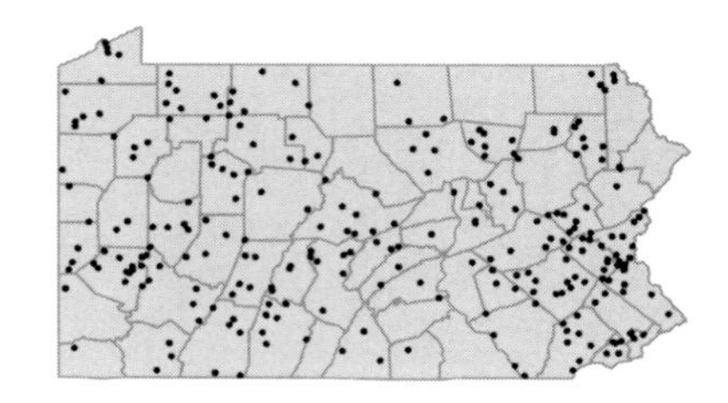

Carex projecta Mackenzie
Sedge
Herbaceous perennial
Wet woods and wooded swamps.

FACW

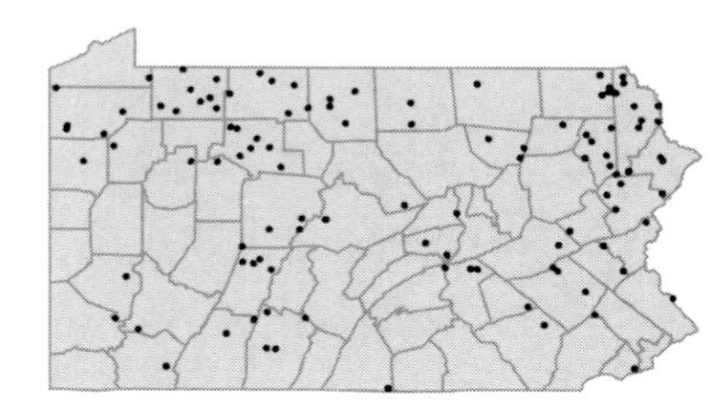

Carex pseudocyperus L.
Cyperus-like sedge
Herbaceous perennial
Calcareous swamps and swales.

OBL

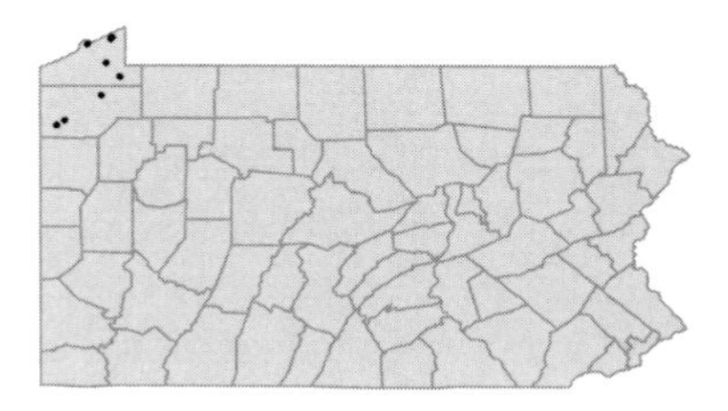

Carex radiata (Wahlenb.) Small
Sedge
Herbaceous perennial
Dry to moist woods.
Carex rosea sensu Mackenzie

Carex retroflexa Muhl. ex Schkuhr
Sedge
Herbaceous perennial
Dry, open, rocky woods.

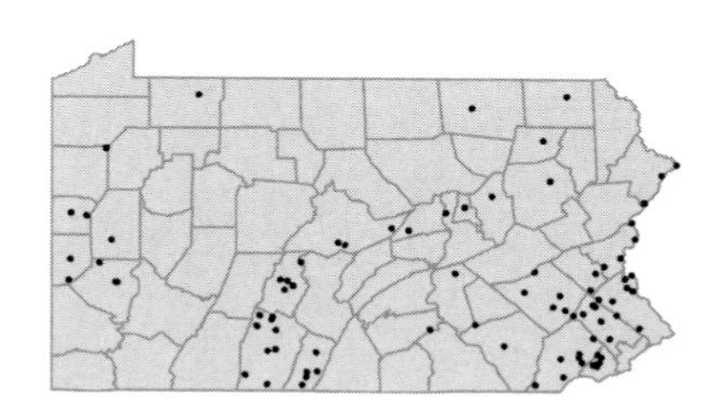

Carex retrorsa Schwein.
Backward sedge
Herbaceous perennial
Swamps, boggy swales, stream banks and open thickets.
Carex hartii Dewey P

FACW+

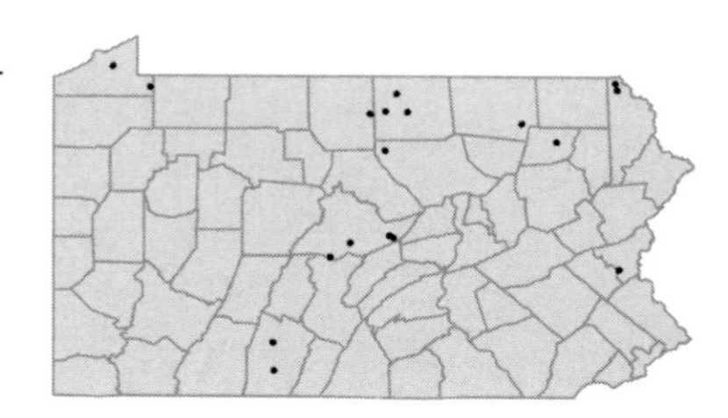

Carex rosea Schkuhr ex Willd.
Sedge
Herbaceous perennial
Dry to moist woods.
Carex convoluta Mackenzie GBFW

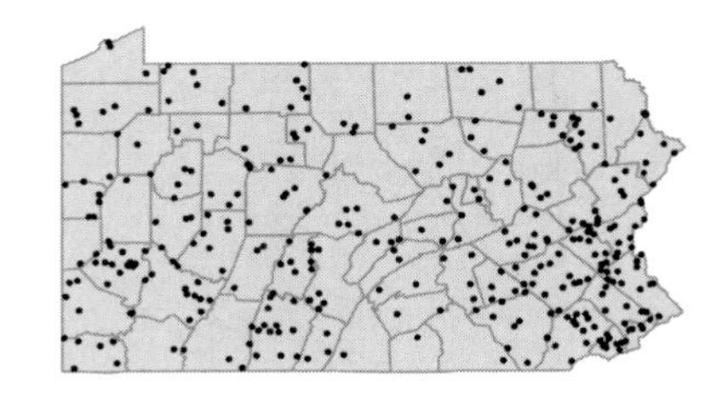

Carex sartwellii Dewey
Sartwell's sedge
Herbaceous perennial
Swamps and wet ditches. Believed to be extirpated, last collected in 1947.

OBL

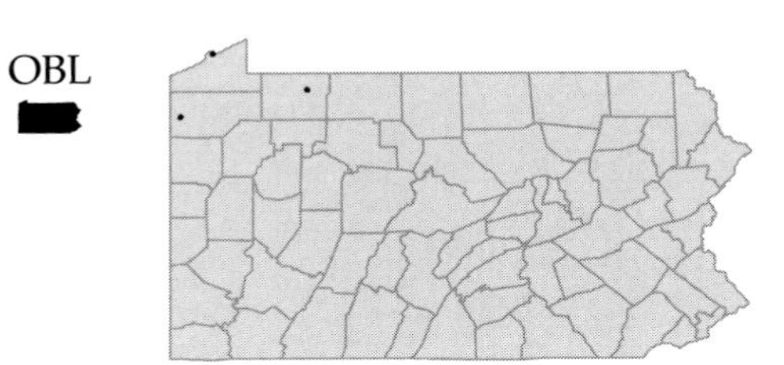

Carex scabrata Schwein.
Sedge
Herbaceous perennial
Bogs, swamps and moist woods.

OBL

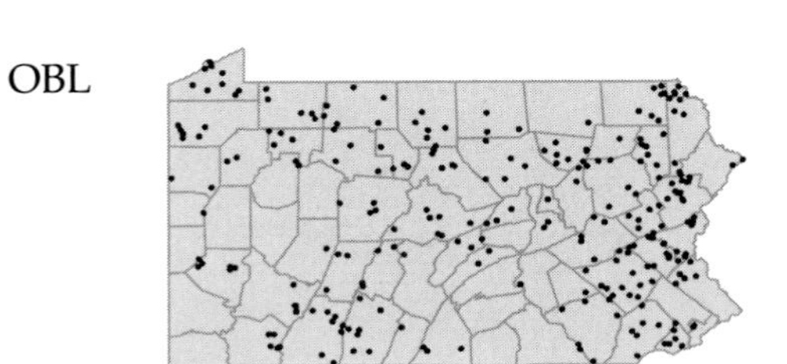

Carex schweinitzii Dewey ex Schwein.
Schweinitz' sedge
Herbaceous perennial
Calcareous marshes and stream banks.

OBL

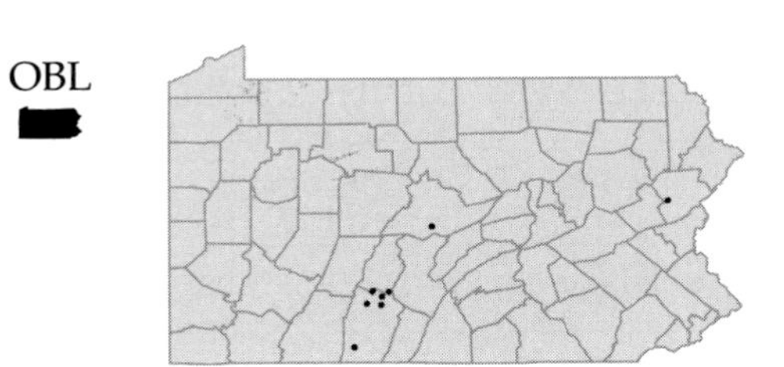

Carex scoparia Schkuhr ex Willd.
Broom sedge
Herbaceous perennial
Moist to dry, open ground.

FACW

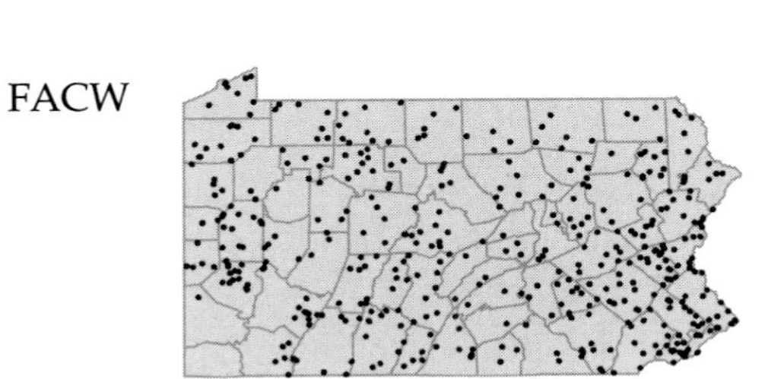

Carex seorsa Howe
Sedge
Herbaceous perennial
Wet woods and swamps.

FACW

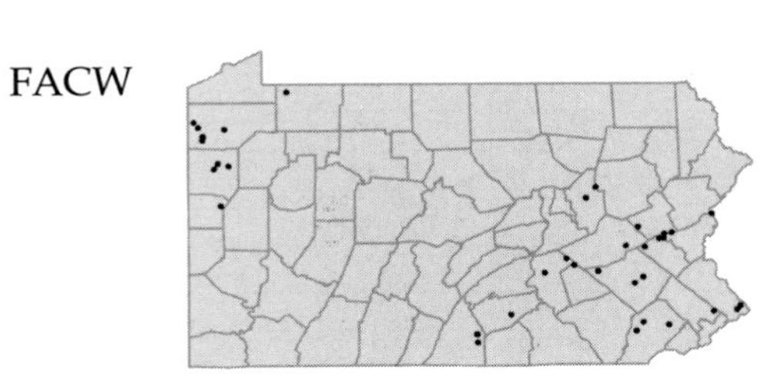

Carex shortiana Dewey
Sedge
Herbaceous perennial
Wet meadows and swamps.

FAC

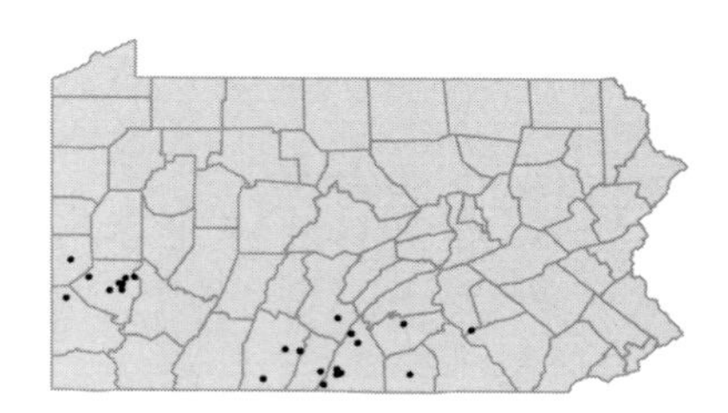

Carex siccata Dewey
Sedge
Herbaceous perennial
Sandy woods.
Carex foenea Willd. FGB

FAC+

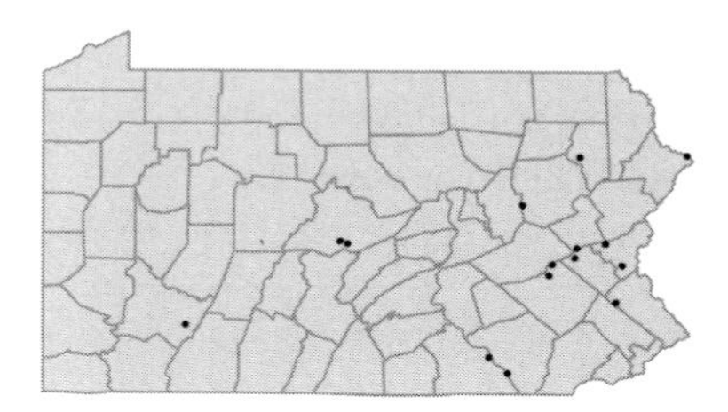

Carex sparganioides Muhl. ex Willd.
Sedge
Herbaceous perennial
Rich woods and meadows.

FACU

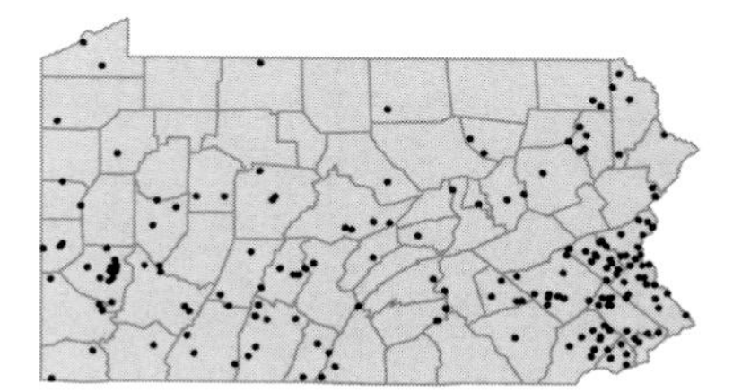

Carex spicata Huds.
Sedge
Herbaceous perennial
Fields and waste ground.

I

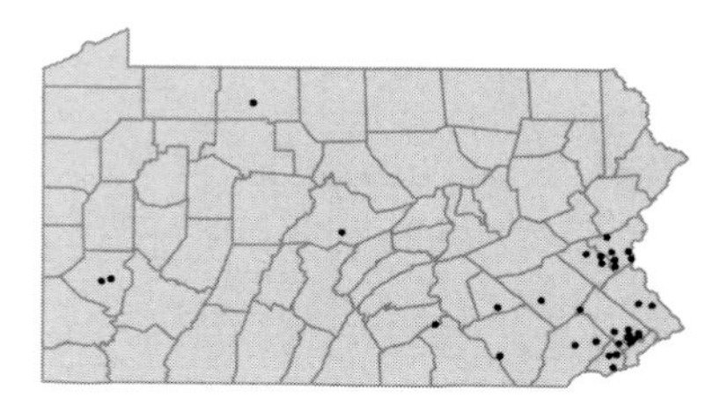

Carex sprengelii Dewey ex Spreng.
Sedge
Herbaceous perennial
Moist cliffs, thickets, woods and grassy banks.
Carex longirostris Torr. P

FACU

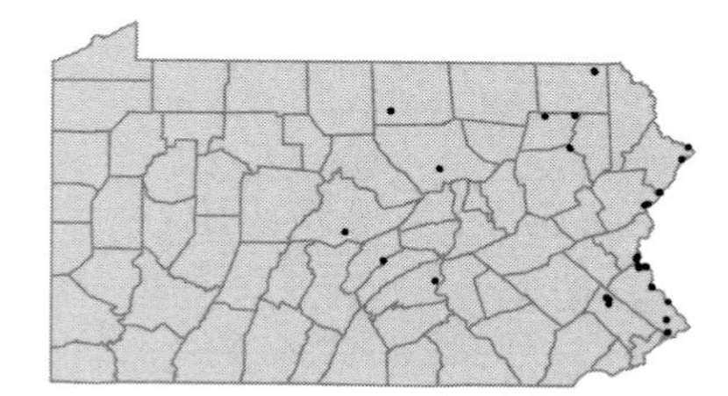

Carex squarrosa L.
Sedge
Herbaceous perennial
Swamps and wet woods.

FACW

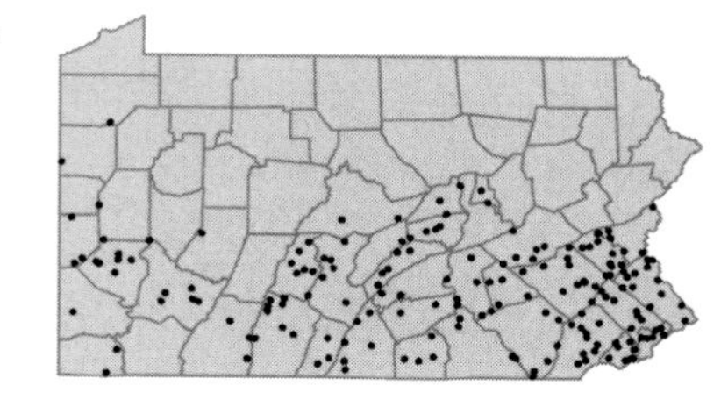

Carex sterilis Willd.
Atlantic sedge
Herbaceous perennial
Open, calcareous swamps and fens.
Carex muricata L. var. *sterilis* (Carey) Gleason GB

OBL

Carex stipata Muhl. ex Willd. var. ***stipata***
Sedge
Herbaceous perennial
Wet meadows and swampy woods.

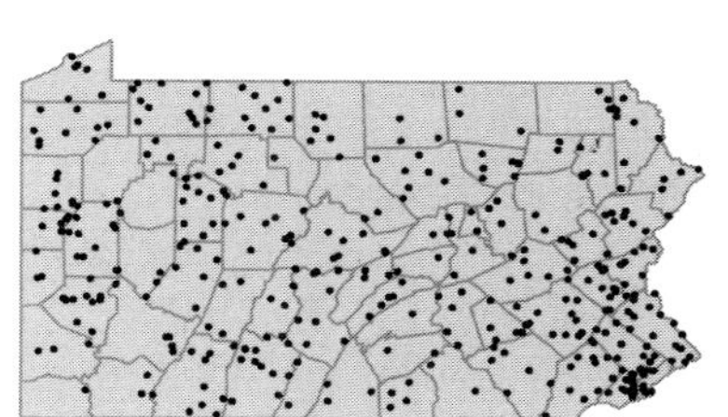

Carex stipata Muhl. ex Willd. var. ***maxima*** Chapm. in Boott
Sedge
Herbaceous perennial
Swamps and wet woods.

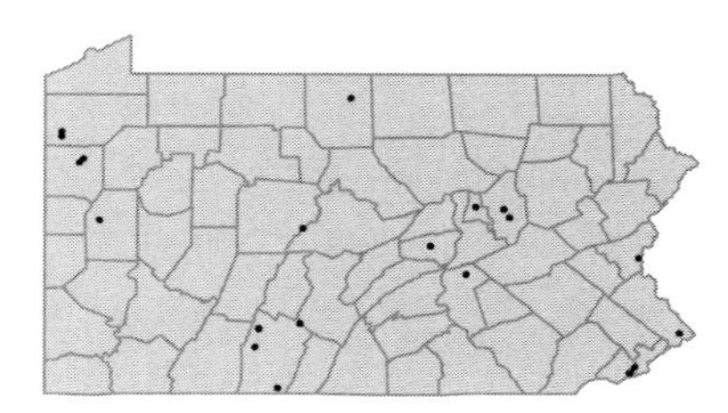

Carex straminea Willd. in Schkuhr
Sedge
Herbaceous perennial
Swamps, swales and marshy shores.

OBL

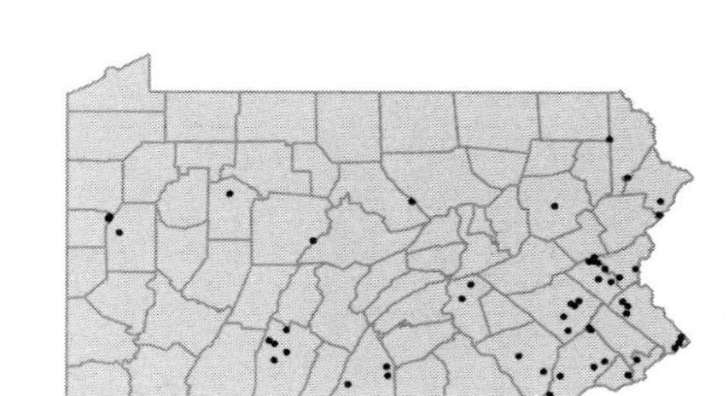

Carex striatula Michx.
Sedge
Herbaceous perennial
Rich, open woods.
Carex laxiflora Lam. var. *angustifolia* Dewey GB

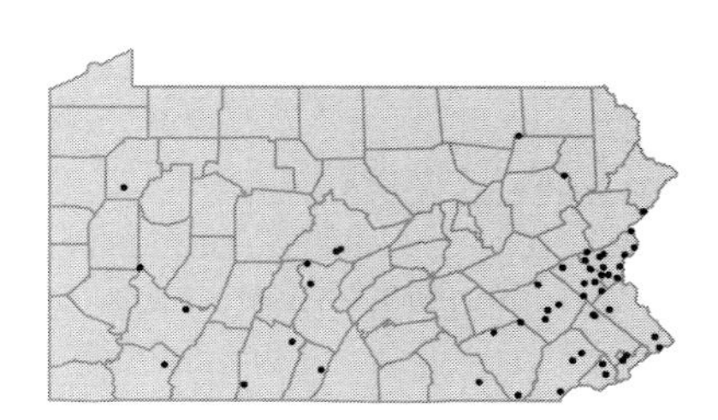

Carex stricta Lam.
Tussock sedge
Herbaceous perennial, emergent aquatic
Swamps, stream banks and wet meadows.
Carex strictior Dewey *pro parte* W

OBL

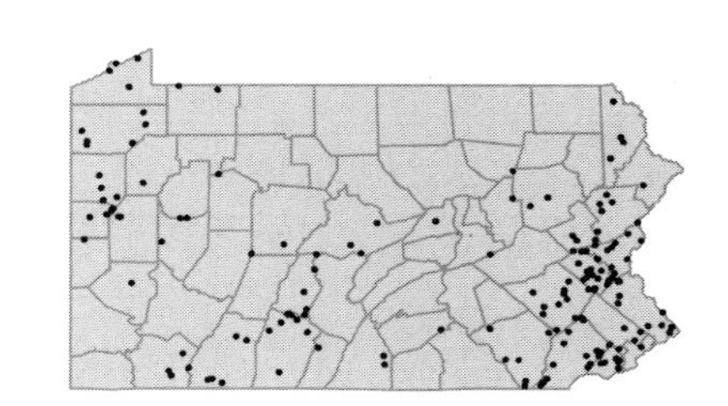

Carex styloflexa Buckl.
Sedge
Herbaceous perennial
Wet woods and bogs.

FACW-

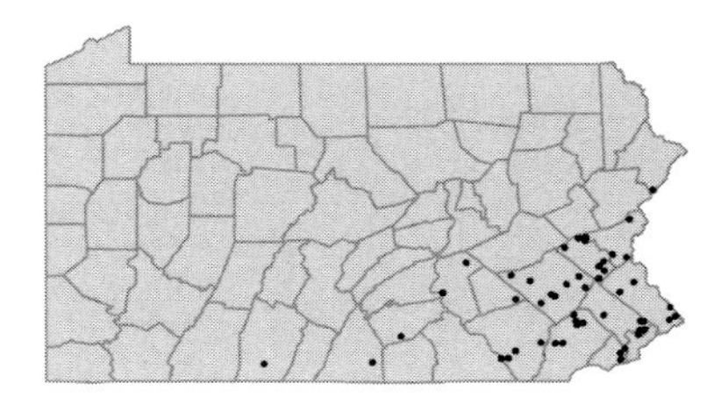

Carex swanii (Fern.) Mackenzie
Sedge
Herbaceous perennial
Dry woods, meadows and fields.

FACU

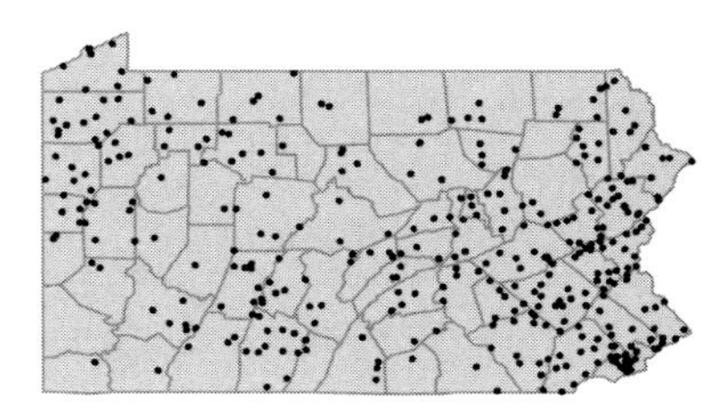

Carex tenera Dewey
Sedge
Herbaceous perennial
Moist meadows and marshes.

FAC

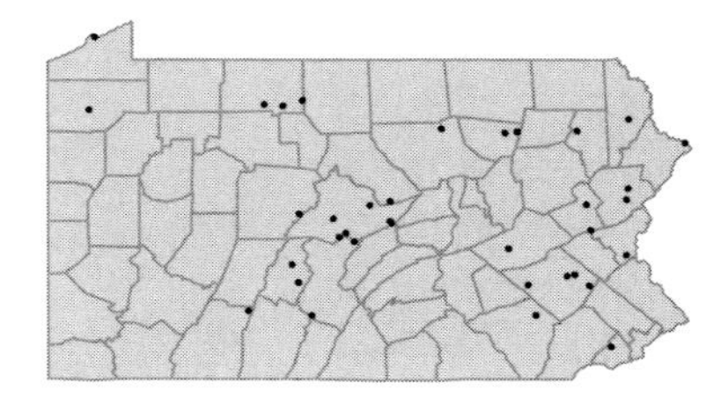

Carex tetanica Schkuhr
Wood's sedge
Herbaceous perennial
Moist, calcareous meadows, marshes, boggy swales and fens.

FACW

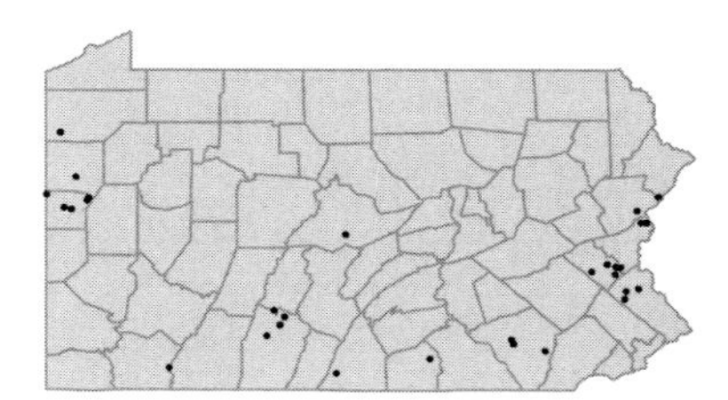

Carex tonsa (Fern.) Bickn.
Sedge
Herbaceous perennial
Rock ledges, roadside banks and abandoned fields.
Carex umbellata Schkuhr *pro parte* GBC

Carex torta Boott ex Tuckerman
Sedge
Herbaceous perennial
Stream banks and floodplains.

FACW

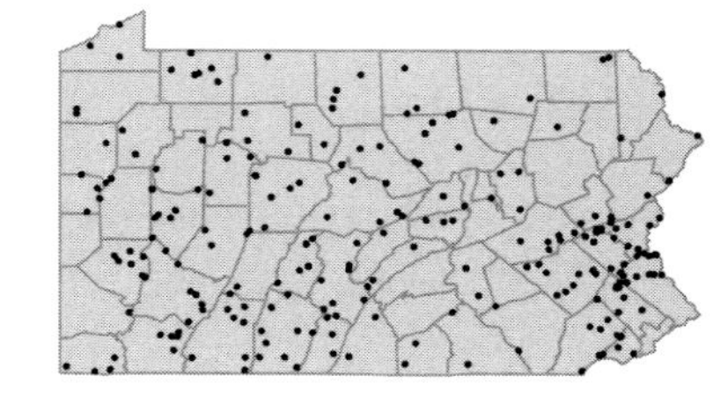

Carex tribuloides Wahlenb.
Sedge
Herbaceous perennial
Wet woods and meadows.

FACW+

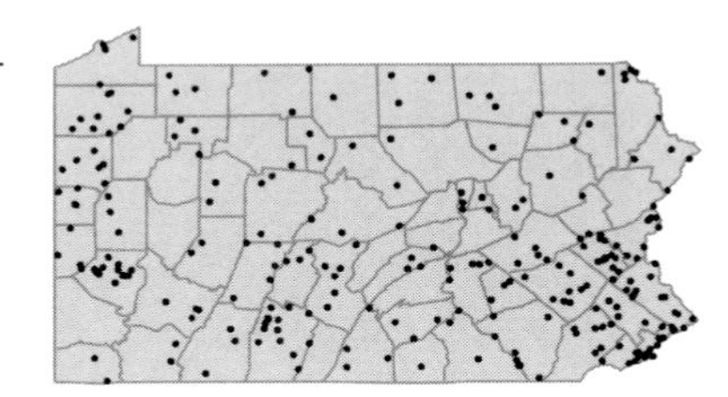

Carex trichocarpa Muhl. ex Willd.
Sedge
Herbaceous perennial
Marshes and wet meadows.

OBL

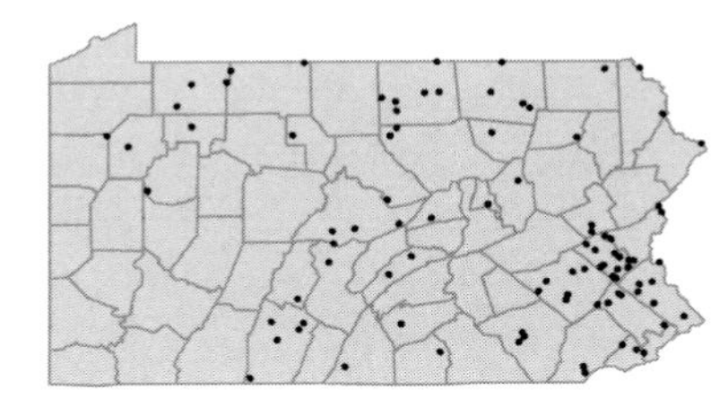

Carex trisperma Dewey
Sedge
Herbaceous perennial
Bogs, swamps and wet woods.

OBL

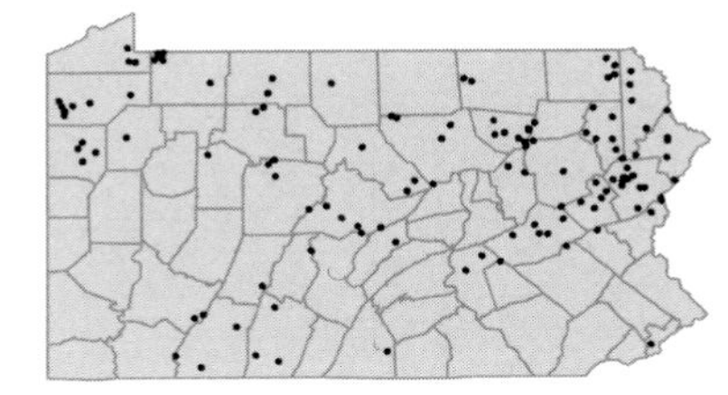

Carex tuckermanii Boott ex Dewey
Sedge
Herbaceous perennial
Bogs and swampy woods.

OBL

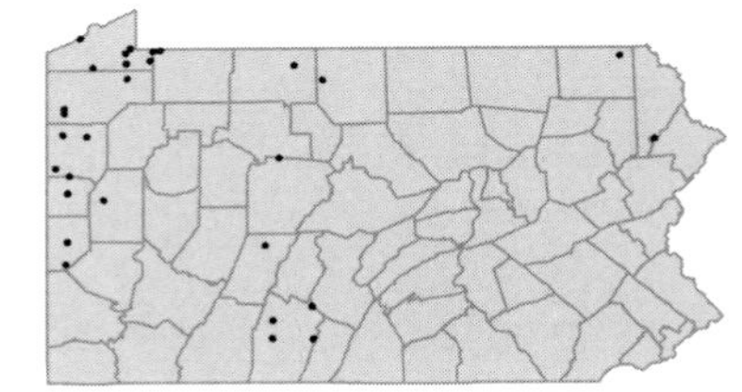

Carex typhina Michx.
Cat-tail sedge
Herbaceous perennial
Calcareous swamps, wet woods and swales.
Carex typhinoides Schwein. P

FACW+

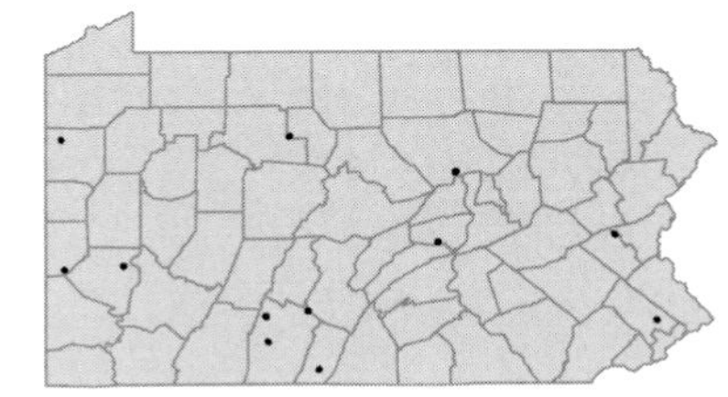

Carex umbellata Schkuhr ex Willd.
Sedge
Herbaceous perennial
Dry, open woods, roadside banks and railroad gravel.
Carex abdita Bickn. FW

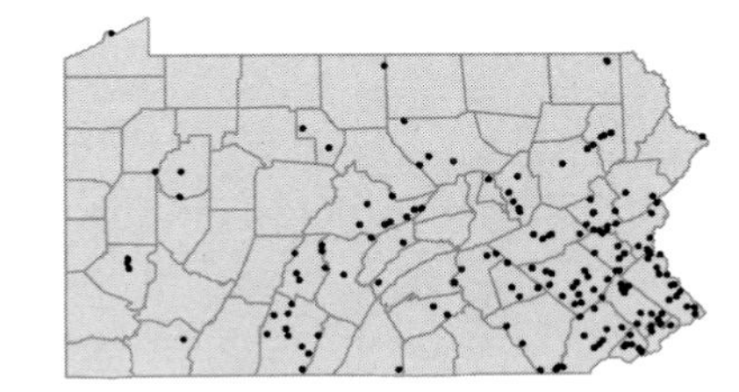

Carex utriculata Boott
Sedge
Herbaceous perennial
Bogs, swamps, fens and lake margins.
Carex rostrata Stokes var. *utriculata* (Boott) Bailey FBGWK

OBL

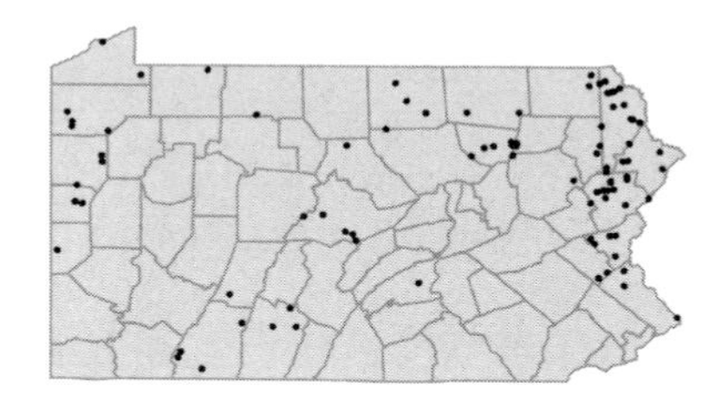

Carex vesicaria L.
Sedge
Herbaceous perennial, emergent aquatic
Swamps, marshes, vernal ponds, ditches and wet woods.
Carex monile Tuckerman P

OBL

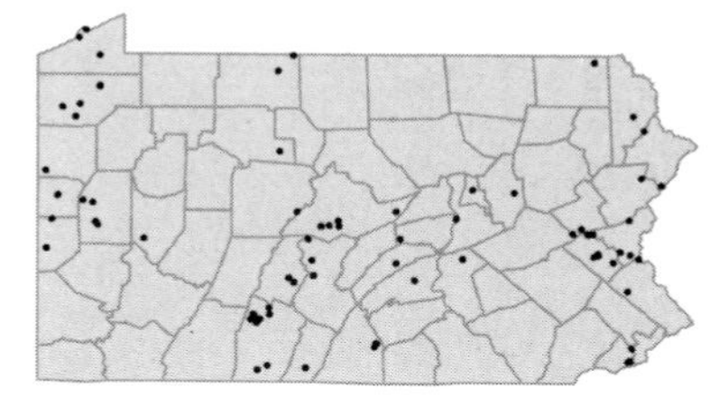

Carex vestita Willd.
Sedge
Herbaceous perennial
Dry or moist woods or sandy-peaty barrens.

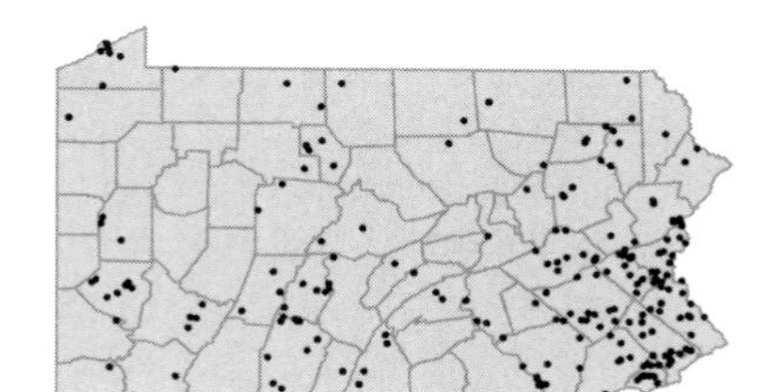

Carex virescens Muhl. ex Willd.
Sedge
Herbaceous perennial
Dry woods.
Carex costellata Britt. P

Carex viridula Michx. var. ***viridula***
Green sedge
Herbaceous perennial
Moist, open, calcareous sand flats.

OBL

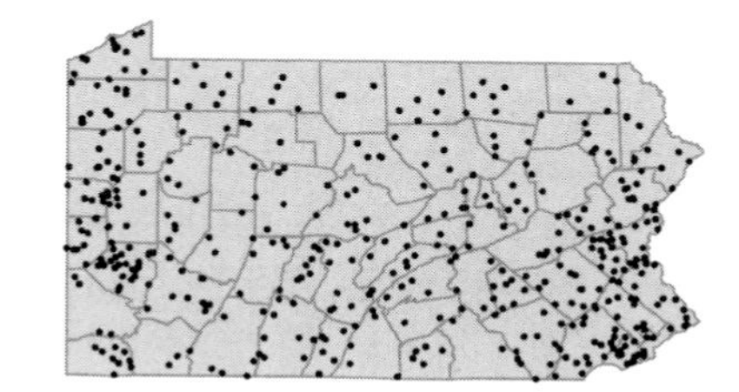

Carex vulpinoidea Michx. var. ***vulpinoidea***
Sedge
Herbaceous perennial
Moist meadows, fields and roadside ditches.

OBL

Carex wiegandii Mackenzie
Wiegand's sedge
Herbaceous perennial
Sphagnum openings in swamps.

OBL

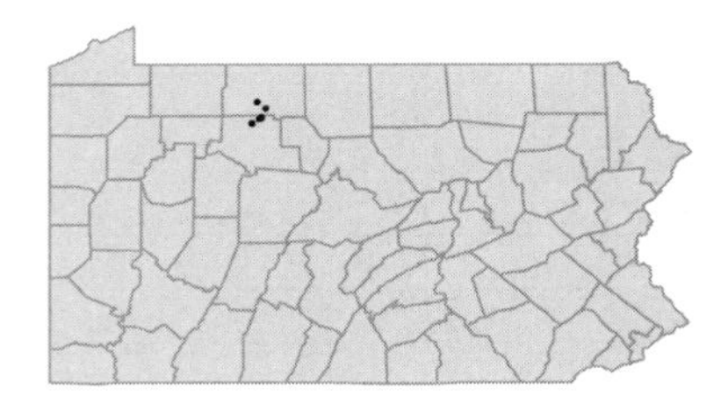

Carex willdenovii Schkuhr ex Willd.
Sedge
Herbaceous perennial
Open, rocky woods and slopes.

UPL
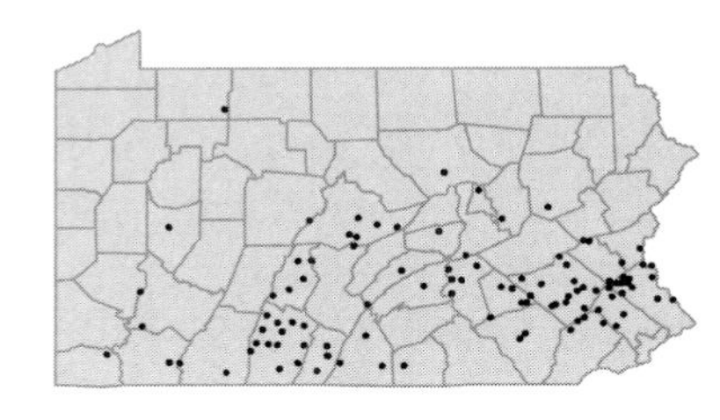

Carex woodii Dewey
Sedge
Herbaceous perennial
Rocky, limestone woods and rich, wooded slopes.
Carex tetanica Schkuhr var. *woodii* (Dewey) A.Wood GB

UPL
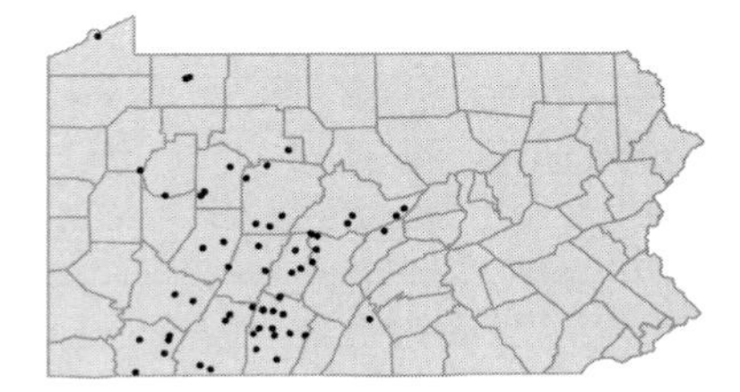

• ***Cladium mariscoides*** (Muhl.) Torr.
Twig-rush
Herbaceous perennial, emergent aquatic
Marshes, floating bog mats and shallow lake margins.

OBL
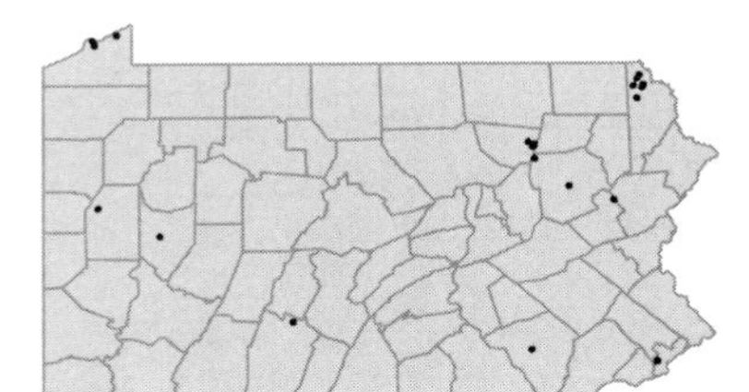

• ***Cymophyllus fraserianus*** (Ker-Gawl.) Kartesz & Gandhi
Fraser's sedge
Herbaceous perennial
Rich, wooded slopes.
Cymophyllus fraseri (Andr.) MacKenzie FBGWC

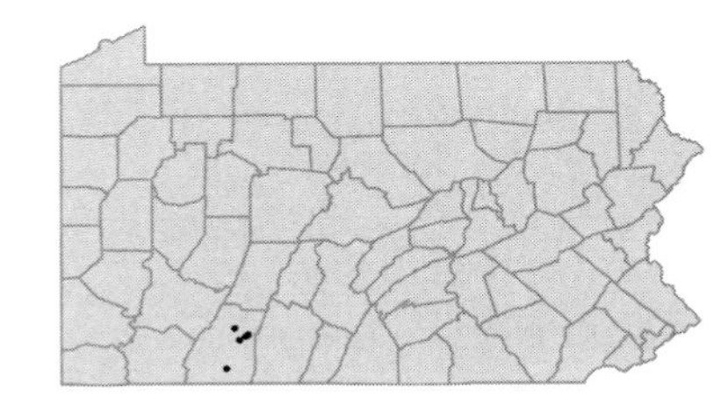

• ***Cyperus acuminatus*** Torr. & Hook.
Umbrella-sedge; Short-pointed flatsedge
Herbaceous annual
Wharves, railroad sidings or damp, sandy, disturbed open ground.

Although native as far east as VA, this species is known in PA only from very disturbed sites.

OBL
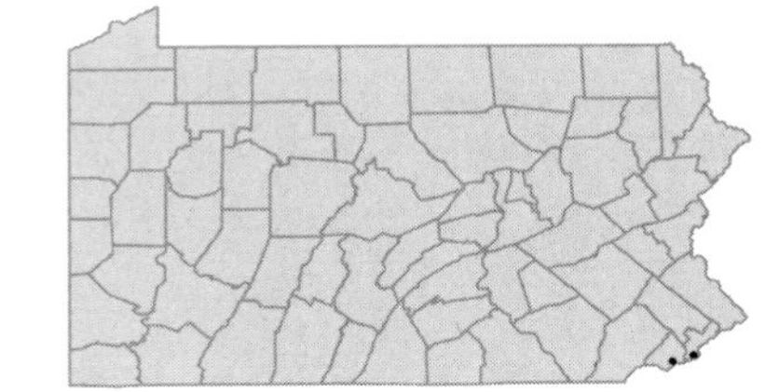

Cyperus aggregatus (Willd.) Endl.
Umbrella-sedge
Herbaceous annual
Ballast and urban waste ground. Represented by two collections from Philadelphia Co. 1877-1880.
Cyperus cayennensis (Lam.) Britt. F; *Cyperus flavus* (Vahl) Nees GC

FAC
I

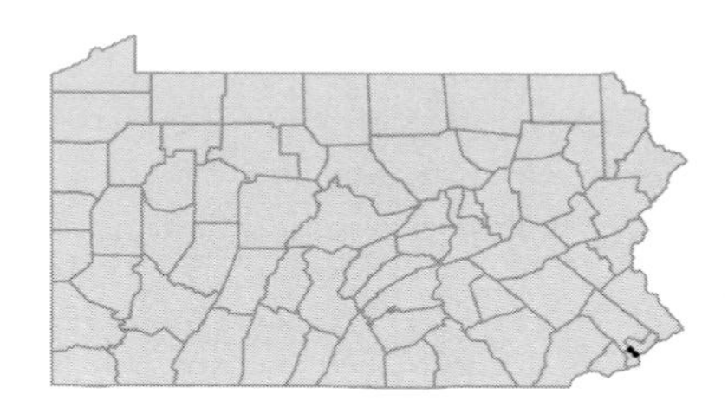

Cyperus bipartitus Torr.
Umbrella-sedge
Herbaceous annual
Wet sandy, gravelly or muddy shores. A hybrid with *C. flavescens* has been collected at a site where both parent species also occur.
Cyperus rivularis Kunth PFGBW

FACW+

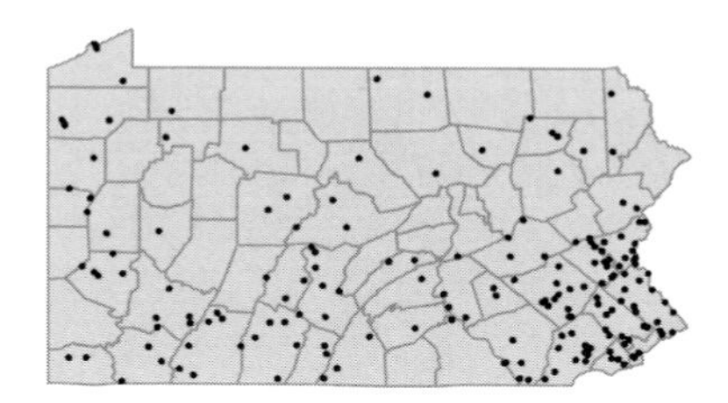

Cyperus brevifoliodes Thieret & Delahoussaye
Umbrella-sedge
Herbaceous perennial
Tidal mudflats, sandy dredge spoil and urban waste ground.
Kyllinga brevifolioides (Delahoussaye & Thieret) G.Tucker

FACW
I

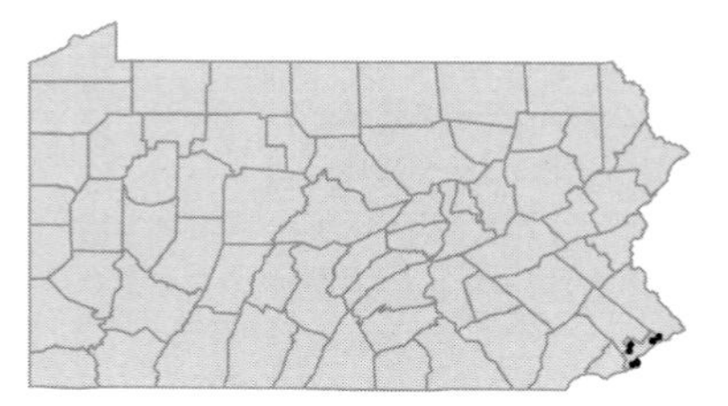

Cyperus compressus L.
Umbrella-sedge
Herbaceous annual
Sandy waste ground and ballast.

FAC+

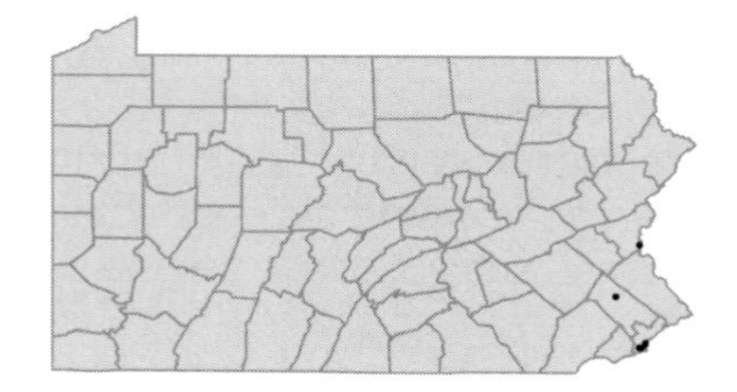

Cyperus croceus Vahl
Umbrella-sedge
Herbaceous perennial
Ballast and wharves. Represented by several collections from Philadelphia Co. 1864-1865.
Cyperus globulosus Aubl. FGBW

FACU

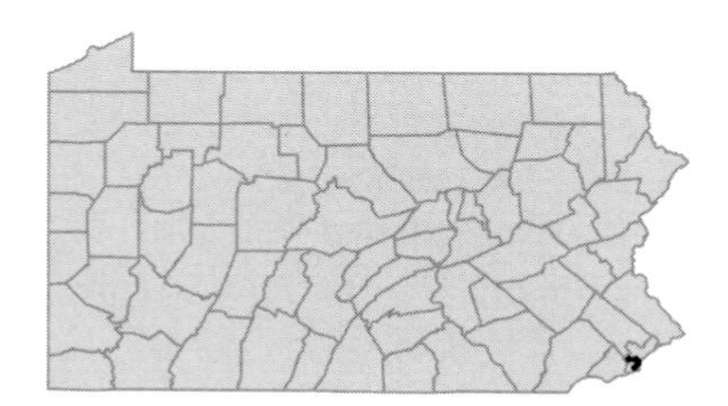

Cyperus dentatus Torr.
Umbrella-sedge
Herbaceous perennial
Moist, sandy river banks and dredge spoil.

FACW+

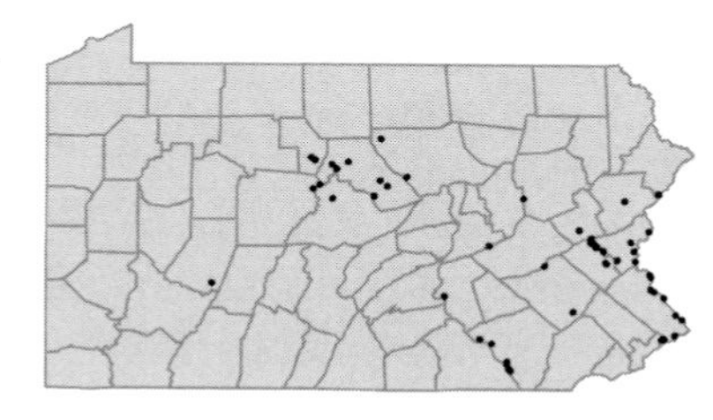

Cyperus diandrus Torr.
Umbrella-sedge
Herbaceous annual
Bogs, marshes and wet shores.

FACW

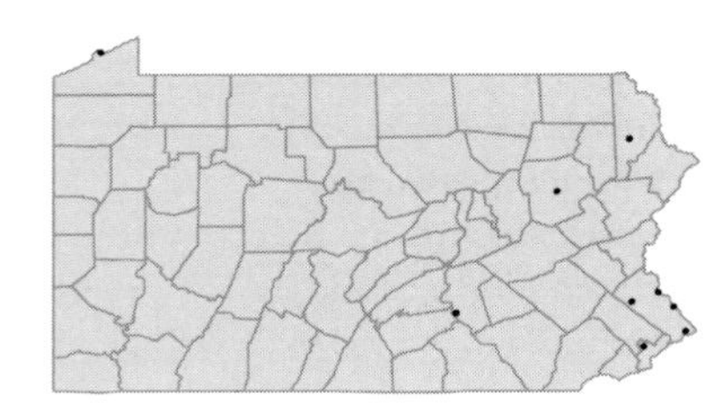

Cyperus difformis L.
Flatsedge
Herbaceous annual, emergent aquatic
Sandy pond margins and rock crevices in shallow water.

OBL
I

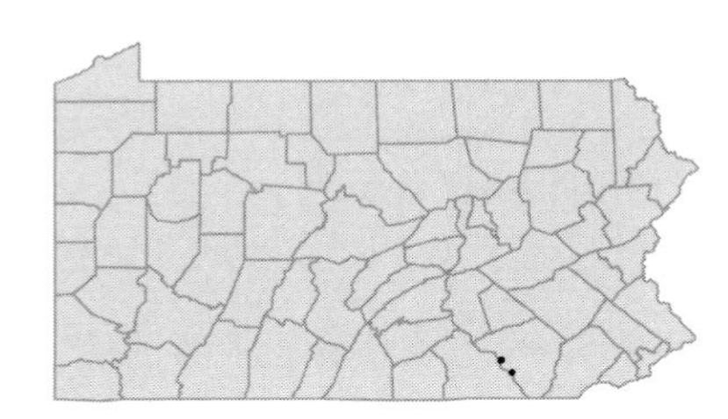

Cyperus echinatus (L.) A.Wood
Umbrella-sedge
Herbaceous perennial
Dry to moist soil of woods and fields.
Cyperus ovularis (Michx.) Torr. WFK

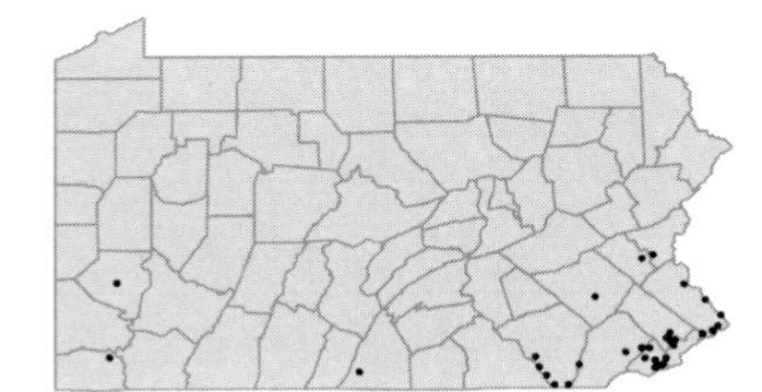

Cyperus engelmannii Steud.
Engelmann's flatsedge
Herbaceous annual
Mixed emergent marshes, shores and tidal mudflats.
Cyperus odoratus L. *pro parte* C

FACW+

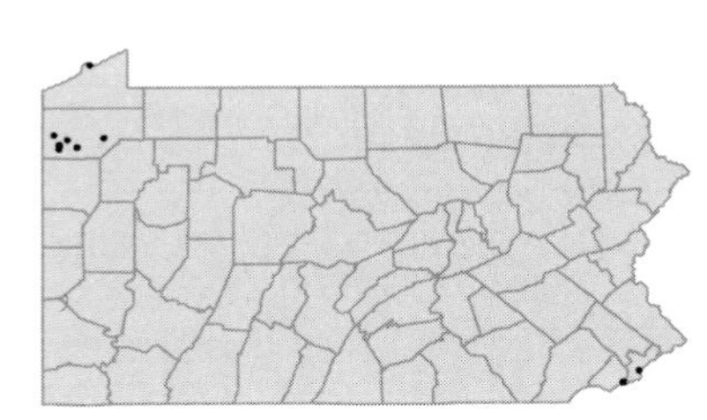

Cyperus erythrorhizos Muhl.
Redroot flatsedge
Herbaceous annual
Moist shores, swamps and wet ground.

FACW+

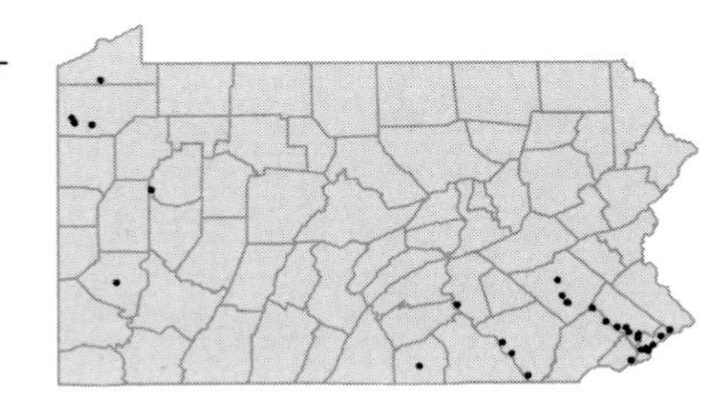

Cyperus esculentus L.
Yellow nutsedge
Herbaceous perennial
Moist ground of fields, meadows, lawns and gardens.

FACW

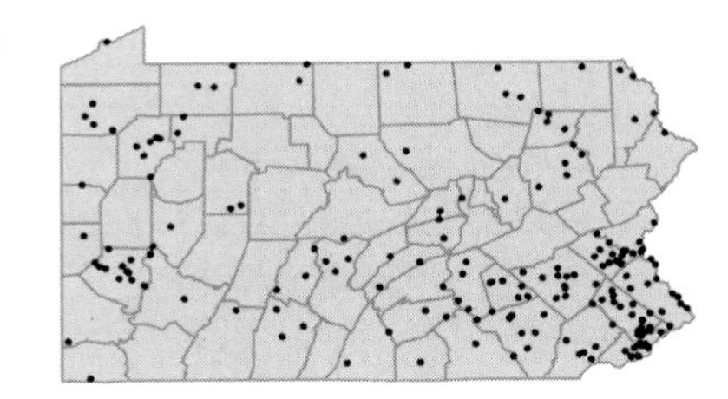

Cyperus filicinus Vahl
Umbrella-sedge
Herbaceous annual
Sandy stream banks and ballast.
Cyperus nuttallii Eddy P

OBL

Cyperus flavescens L.
Umbrella-sedge
Herbaceous annual
Fields, ditches and moist, open soil.

OBL

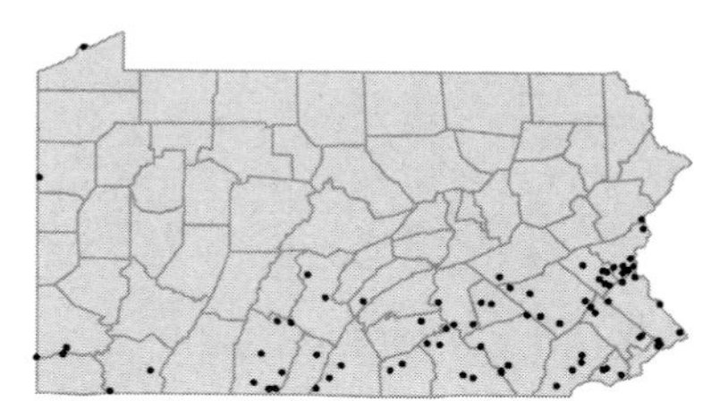

Cyperus flavicomus Michx.
Umbrella-sedge
Herbaceous annual
Tidal mudflats.
Cyperus albomarginatus Mart. & Schrad. FGBW

FAC

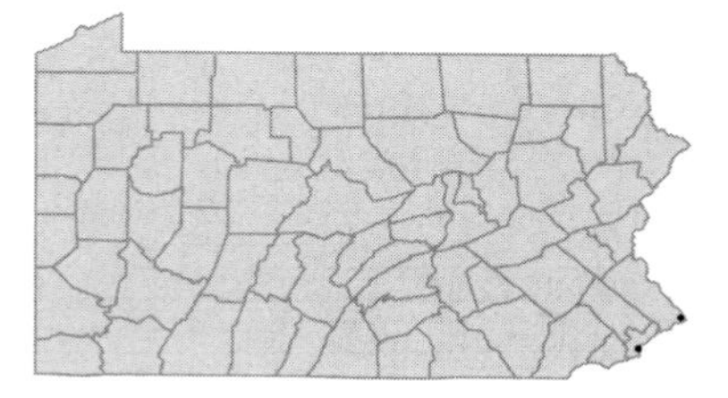

Cyperus fuscus L.
Umbrella-sedge
Herbaceous annual
Waste ground, ballast and wharf areas.

[I]

Cyperus grayi Torr.
Umbrella-sedge
Herbaceous perennial
Ballast and wharves. Represented by a single collection from Philadelphia Co. in 1865.

Cyperus houghtonii Torr.
Houghton's flatsedge
Herbaceous perennial
Railroad ballast and dry, sandy soil.
Cyperus lupulinus x schweinitzii

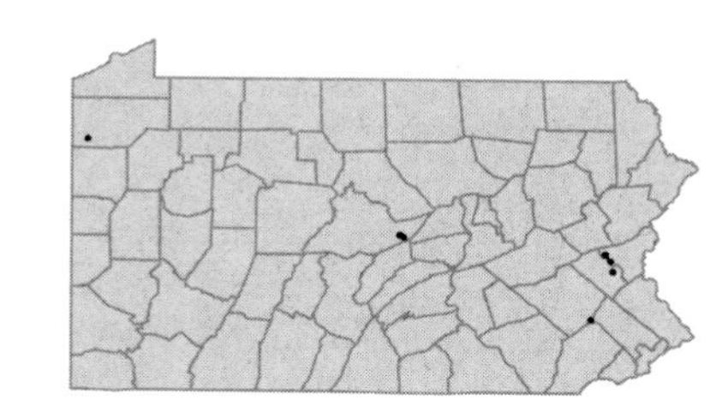

Cyperus iria L.
Umbrella-sedge
Herbaceous annual
Damp, sandy, disturbed ground.

FACW
I

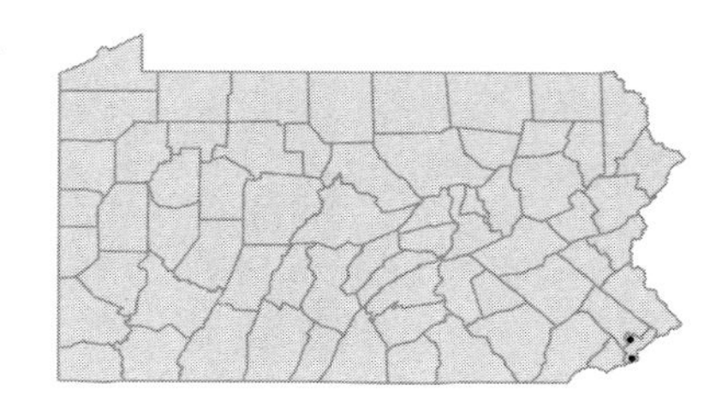

Cyperus lancastriensis Porter ex A.Gray
Umbrella-sedge
Herbaceous perennial
Dry slopes, open woods and fields.

FACU

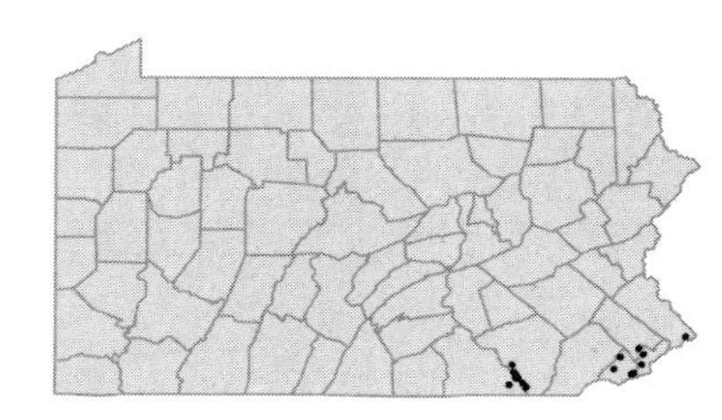

Cyperus lupulinus (Sprengel) Marcks
Umbrella-sedge
Herbaceous perennial
Dry woods, fields, roadsides, waste ground and railroad cinders. Although Marcks (1974) recognized two subspecies in Pennsylvania, we have chosen not to map them separately.
Cyperus filiculmis Vahl *pro parte* FWBG

UPL

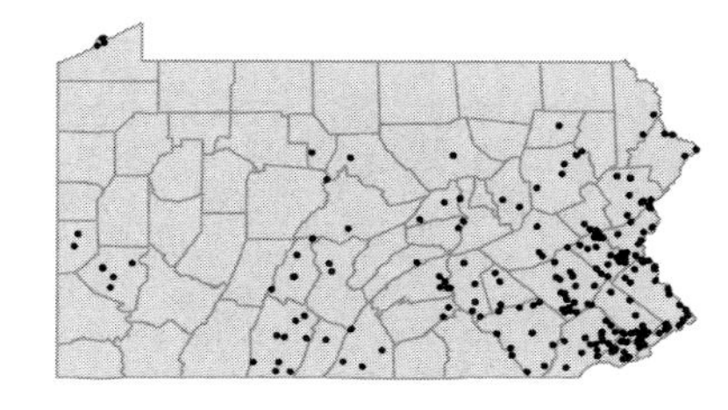

Cyperus microiria Steud.
Umbrella-sedge
Herbaceous annual
Damp, sandy, disturbed ground and rubbish dumps.
Cyperus amuricus Maxim. GB

I

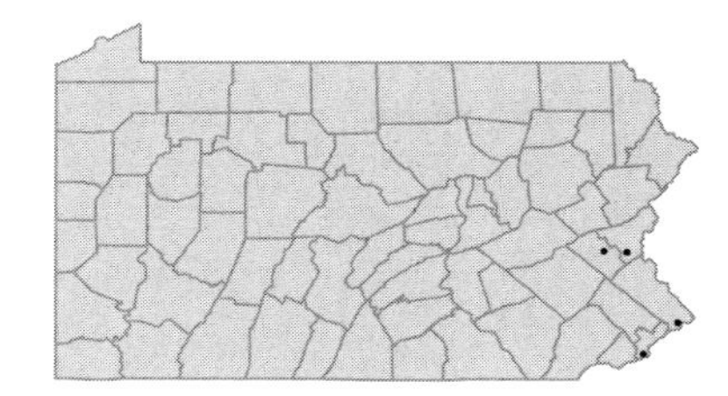

Cyperus ochraceus Vahl
Umbrella-sedge
Herbaceous perennial
Ballast and waste ground. Represented by two collections from Philadelphia Co. 1865-1877.

FACW
I

Cyperus odoratus L.
Umbrella-sedge; Rusty flatsedge
Herbaceous annual
Moist meadows, wet sand and gravel flats, ballast, wharves and railroad sidings.
Cyperus engelmanii Steud. *pro parte* FGBWC

FACW

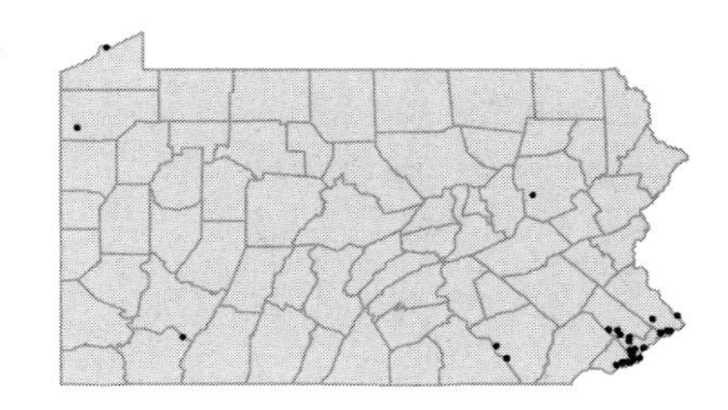

Cyperus polystachyos Rottb. var. ***texensis*** (Torr.) Fern.
Many-spiked flatsedge
Herbaceous annual
Swampy, tidal river bank. Believed to be extirpated, represented by a single collection from Philadelphia Co. in 1935.

FACW

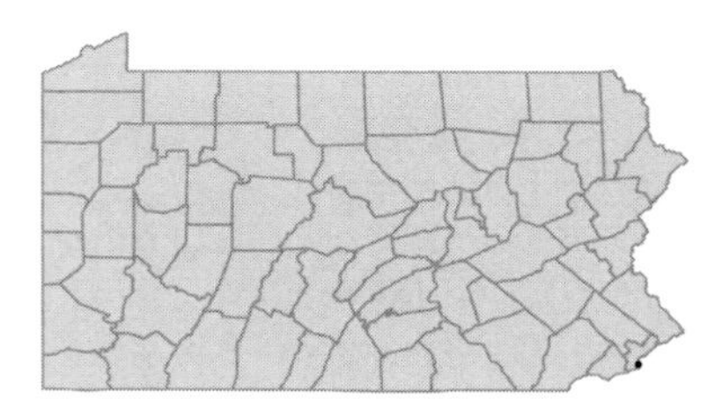

Cyperus refractus Engelm. ex Boeckl.
Reflexed flatsedge
Herbaceous perennial
Sandy, alluvial banks and dry woods.

FACU+

Cyperus retrorsus Chapm.
Retrorse flatsedge; Umbrella-sedge
Herbaceous perennial
Dry, sandy soil and ballast.

FAC-

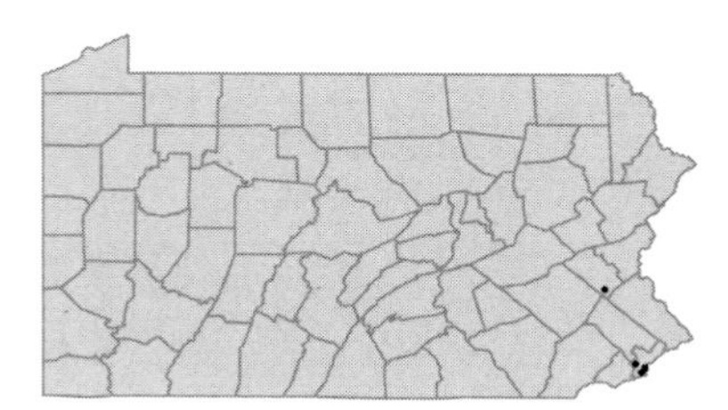

Cyperus rotundus L.
Purple nutsedge
Herbaceous perennial
Ballast and wharves.

FAC
[I]

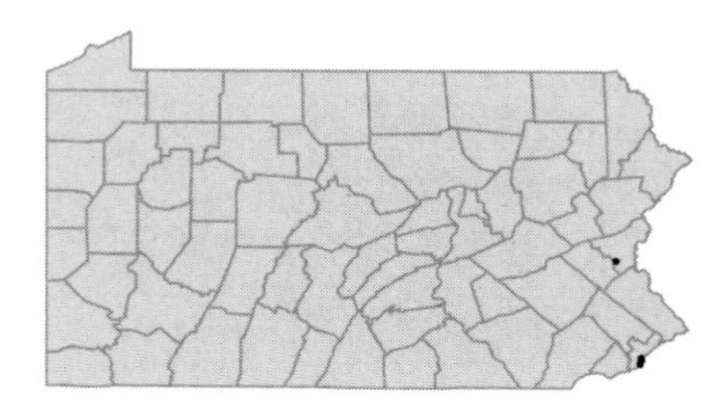

Cyperus schweinitzii Torr.
Schweinitz's flatsedge
Herbaceous perennial
Dry or moist sand flats or dunes, also railroad sidings.

FACU

Cyperus serotinus Rottb.
Umbrella-sedge
Herbaceous perennial
Tidal river bank.

OBL
[I]

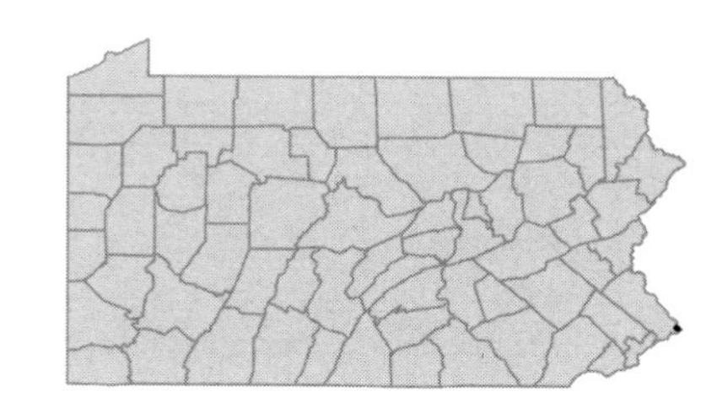

Cyperus squarrosus L.
Umbrella-sedge
Herbaceous annual
River banks, stream margins and wet ditches.
Cyperus aristatus Rottb. GBW; *Cyperus inflexus* Muhl. PF

FACW+

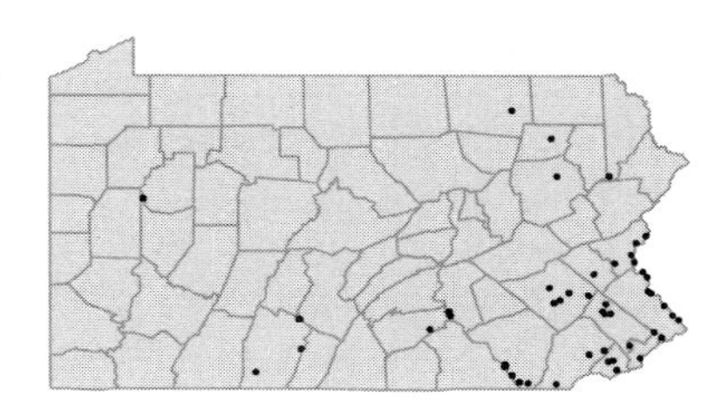

Cyperus strigosus L.
False nutsedge
Herbaceous perennial
Moist fields, woods, swamps and stream banks.

FACW

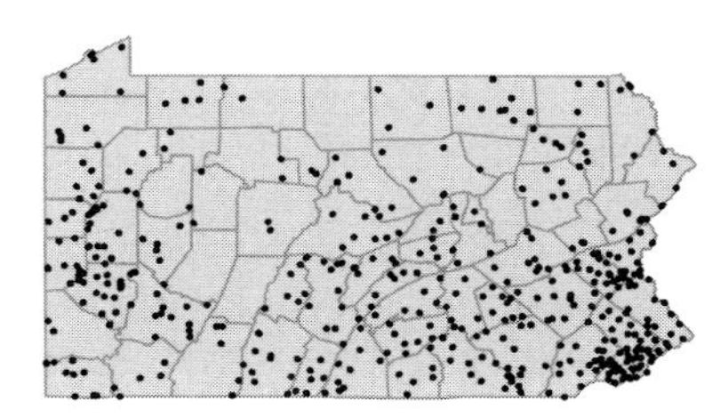

Cyperus tenuifolius (Steud.) Dandy
Thin-leaved flatsedge
Herbaceous annual
Moist, sandy or muddy shores.
Cyperus densicaespitosus Mattf. & Kukenth. B; *Kyllinga pumila* Michx.

FACW

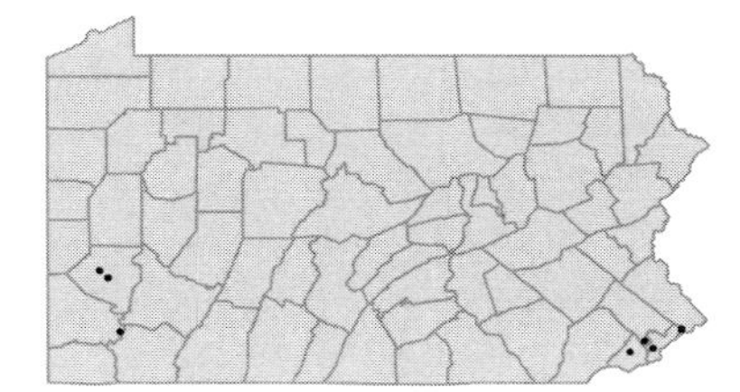

• ***Dulichium arundinaceum*** (L.) Britt.
Three-way sedge
Herbaceous perennial, emergent aquatic
Bogs, marshes, lake margins, swampy fields and ditches.

OBL

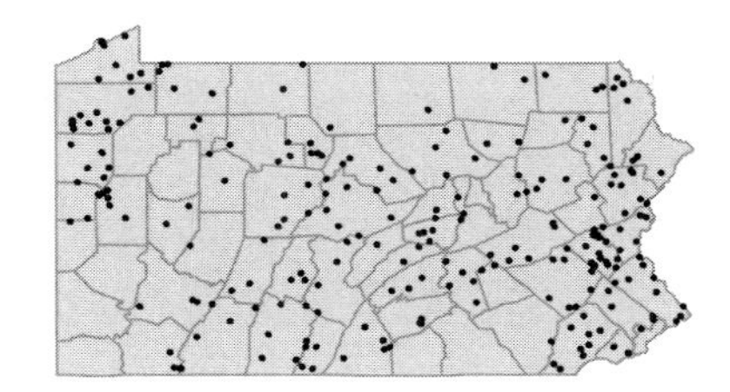

• ***Eleocharis acicularis*** (L.) Roemer & Schultes
Needle spike-rush
Herbaceous perennial, rooted submergent aquatic
Shallow water and moist shores.

OBL

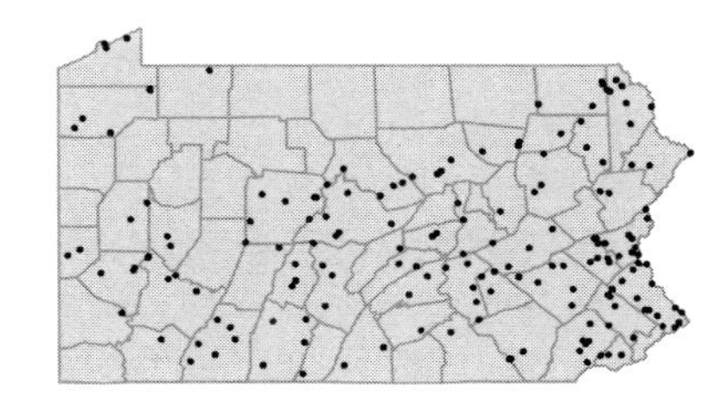

Eleocharis caribaea (Rottb.) S.F.Blake
Spike-rush
Herbaceous annual
Sand plains, lake shore.

FACW

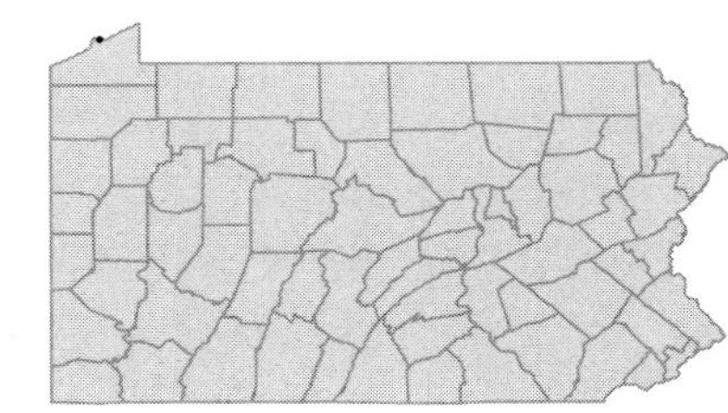

Eleocharis compressa Sullivant var. ***compressa***
Flat-stemmed spike-rush
Herbaceous perennial
Wet, sandy ground and river banks.
Eleocharis acuminata (Muhl.) Nees P

FACW+

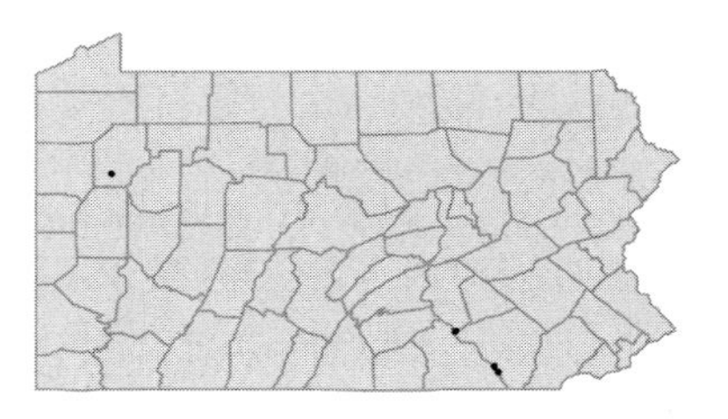

Eleocharis elliptica Kunth
Slender spike-rush
Herbaceous perennial
Moist sand flats and sandy swales.
Eleocharis tenuis (Willd.) Schultes var. *borealis* (Svenson) Gleason BGC;
Eleocharis compressa Sullivant var. *atrata* Svenson

FACW+

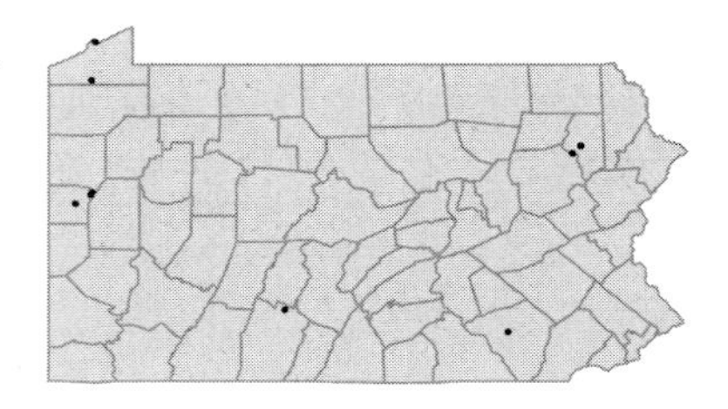

Eleocharis engelmannii Steud.
Spike-rush
Herbaceous annual
Vernal ponds, moist ditches and roadsides.
Eleocharis ovata (Roth) Roemer & Schultes *pro parte* GC

FACW+

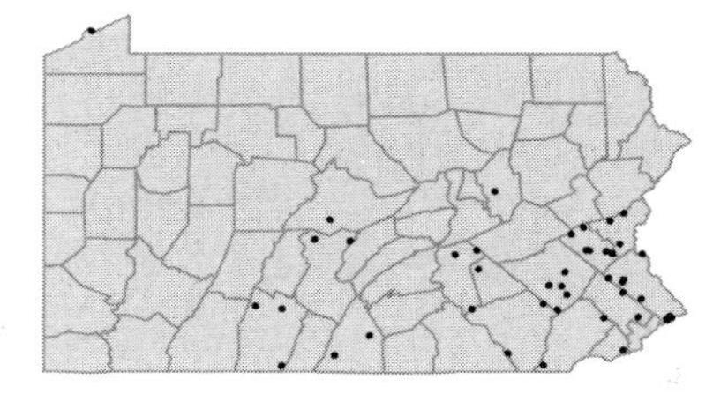

Eleocharis erythropoda Steud.
Spike-rush
Herbaceous perennial
River banks, stream banks, wet meadows and swales.
Eleocharis calva Torr. FBW

OBL

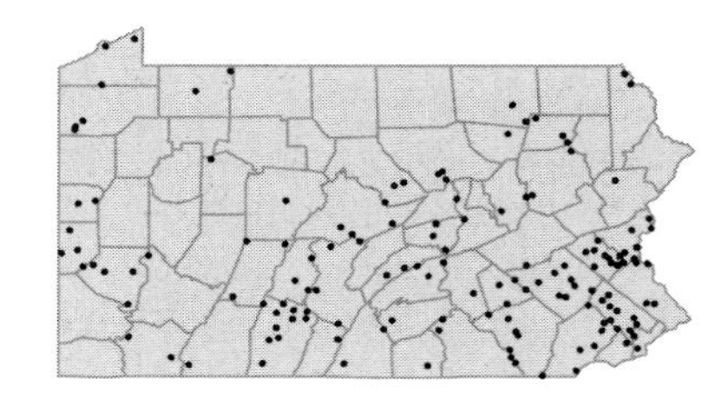

Eleocharis intermedia (Muhl.) Schultes
Matted spike-rush
Herbaceous annual
Calcareous swamps, bogs and lake shores.

FACW+

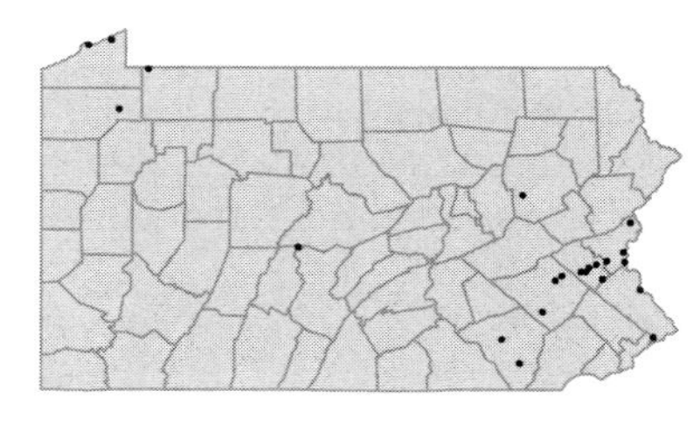

Eleocharis microcarpa Torr.
Spike-rush
Herbaceous annual
Moist ground. Represented by a single collection from Somerset Co. in 1898.

OBL

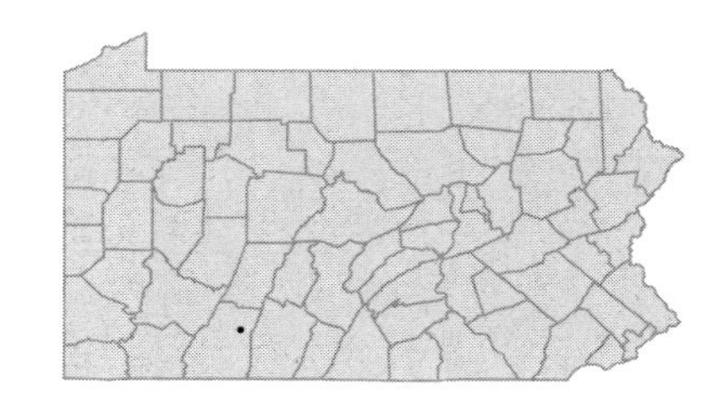

Eleocharis obtusa (Willd.) Schultes var. ***obtusa*** OBL
Wright's spike-rush
Herbaceous annual
Swampy ground, bogs, ditches and wet shores.
Eleocharis diandra C.Wright FBW; *Eleocharis obtusa* (Willd.) Schultes var. *ellipsoidalis* Fern. FW; *Eleocharis obtusa* (Willd.) Schultes var. *jejuna* Fern. PFW; *Eleocharis ovata* (Roth) Roemer & Schultes *pro parte* GBC

Eleocharis obtusa (Willd.) Schultes var. ***peasei*** Svenson OBL
Spike-rush
Herbaceous annual
Tidal mudflats.

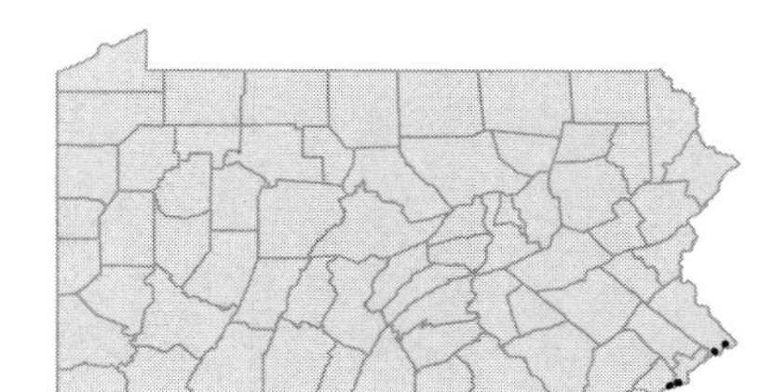

Eleocharis olivacea Torr. OBL
Capitate spike-rush
Herbaceous perennial
Wet pond margins and exposed peat.
Eleocharis flavescens (Poir.) Urban var. *olivacea* (Torr.) Gleason GBC

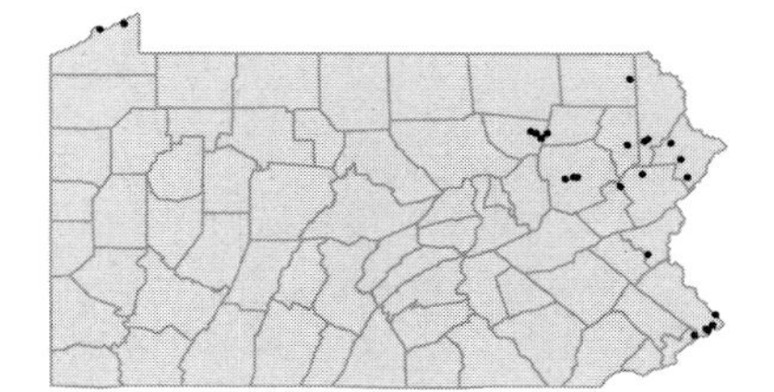

Eleocharis palustris (L.) Roemer & Schultes OBL
Creeping spike-rush
Herbaceous perennial, emergent aquatic
Lake margins, edges of streams, bogs, swamps and marshy fields. An 1879 collection from Erie Co. identified as *E. palustris* var. *major* requires further study.
Eleocharis smallii Britt. PFW

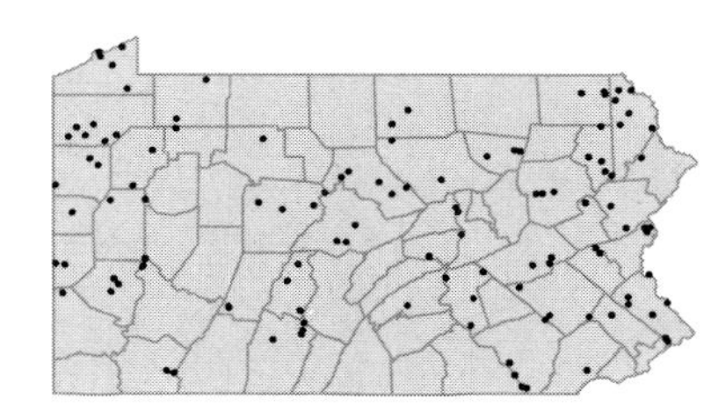

Eleocharis parvula (Roemer & Schultes) Link ex Buff. & Fingerh. OBL
Dwarf spike-rush
Herbaceous perennial
Tidal shores and mudflats.

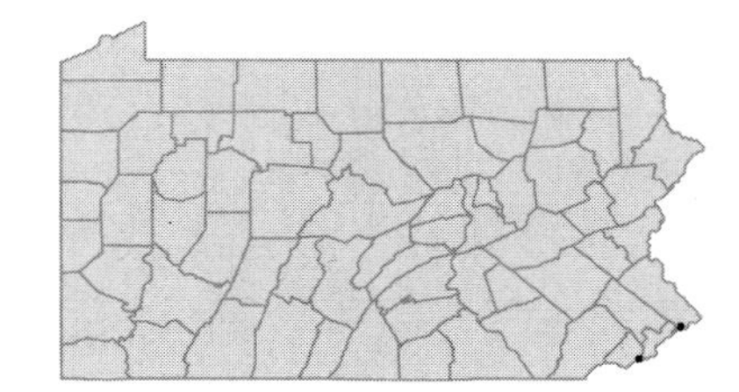

Eleocharis pauciflora (Lightfoot) Link var. ***fernaldii*** Svenson OBL
Spike-rush
Herbaceous perennial
Wet, calcareous sand, trail edges.
Eleocharis quinqueflora (F.X.Hartman) Schwarz K

Eleocharis quadrangulata (Michx.) Roemer & Schultes
Four-angled spike-rush
Herbaceous perennial, emergent aquatic
Lake margins, swamps and ponds.
Eleocharis mutata (L.) Roemer & Schultes P

OBL

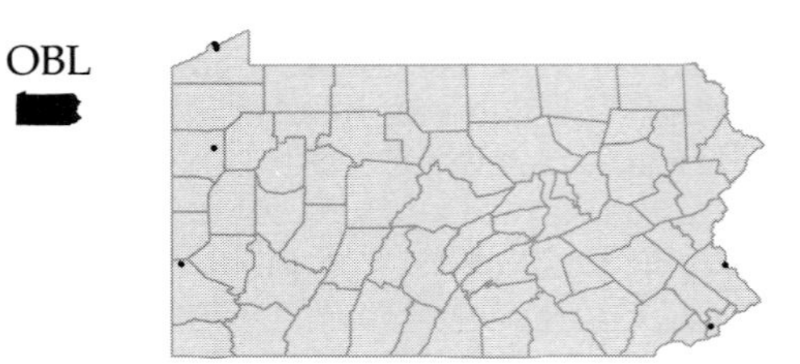

Eleocharis robbinsii Oakes
Robbins' spike-rush
Herbaceous perennial, emergent aquatic
Shallow water of lakes and ponds.

OBL

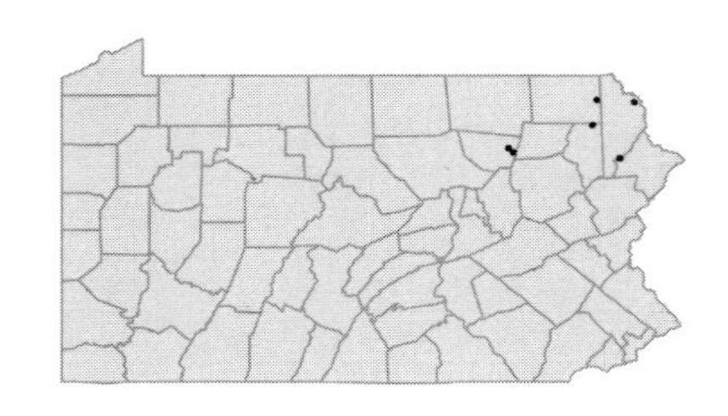

Eleocharis rostellata (Torr.) Torr.
Beaked spike-rush
Herbaceous perennial
Calcareous swamps.

OBL

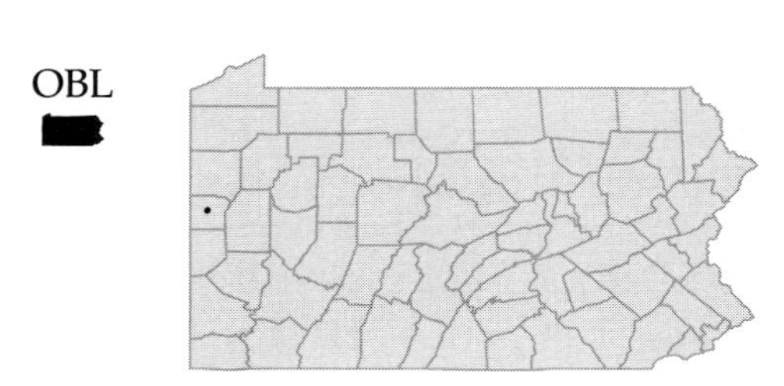

Eleocharis tenuis (Willd.) Schultes var. ***tenuis***
Spike-rush
Herbaceous perennial
Moist fields, swamps, bogs and wet ditches.

FACW+

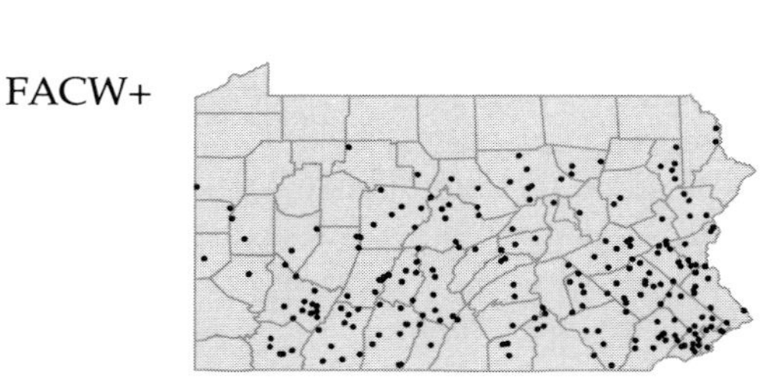

Eleocharis tenuis (Willd.) Schultes var. ***pseudoptera*** (Weatherby) Svenson
Slender spike-rush
Herbaceous perennial
Moist meadows, hayfields and damp areas of serpentine barrens.

FACW+

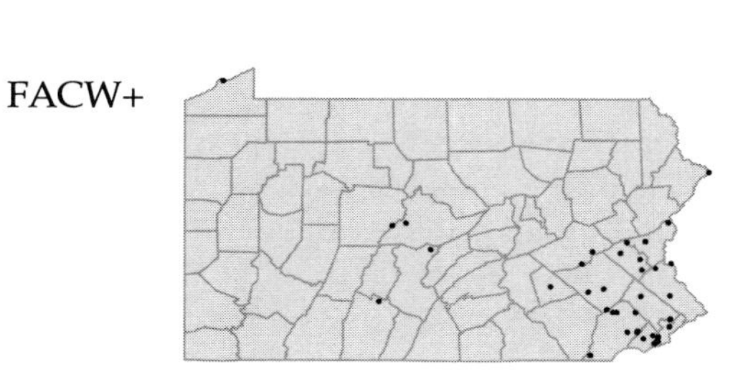

Eleocharis tenuis (Willd.) Schultes var. ***verrucosa*** (Svenson) Svenson
Slender spike-rush
Herbaceous perennial
Moist, open ground.

FACW+

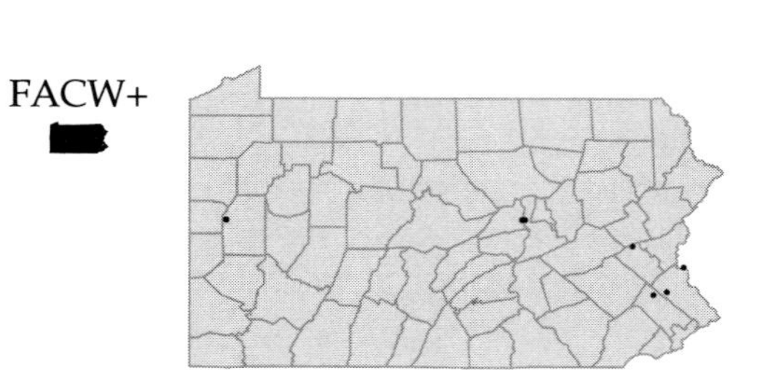

Eleocharis tricostata Torr.
Three-ribbed spike-rush
Herbaceous perennial
Wet, sandy-peaty meadows. Believed to be extirpated, represented by two collections from Delaware Co. 1864-1866.

OBL

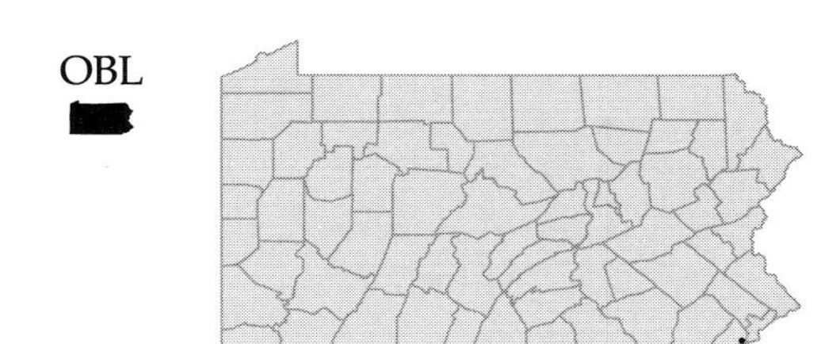

Eleocharis tuberculosa (Michx.) Roemer & Schultes
Long-tubercled spike-rush
Herbaceous perennial
Sphagnous bog. Believed to be extirpated, represented by a single collection from Montgomery Co. in 1898.

OBL

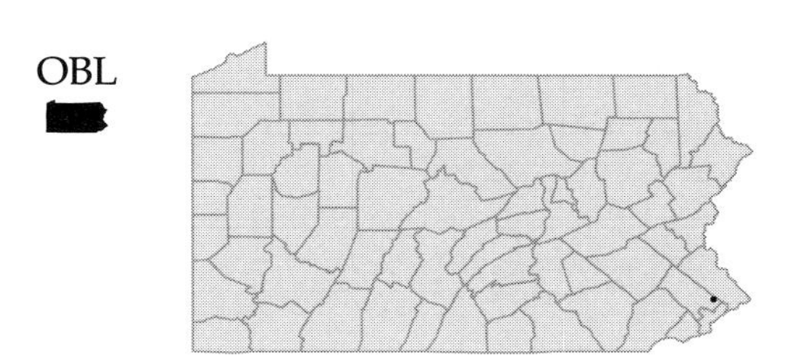

• ***Eriophorum gracile*** W.D.J.Koch ex Roth
Slender cotton-grass
Herbaceous perennial
Wet meadows, bogs and marshes.

OBL

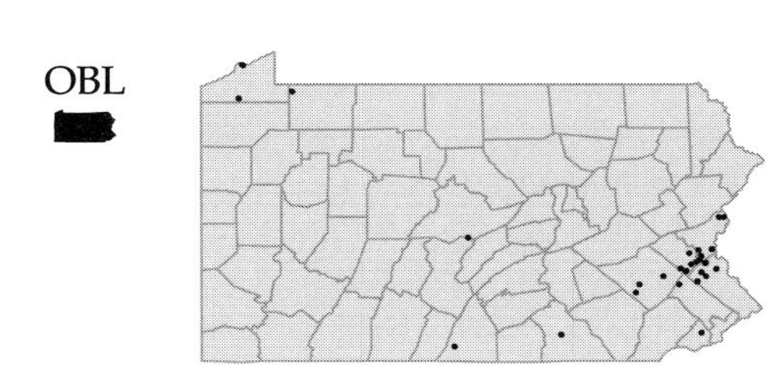

Eriophorum tenellum Nutt.
Rough cotton-grass
Herbaceous perennial, emergent aquatic
Bogs.

OBL

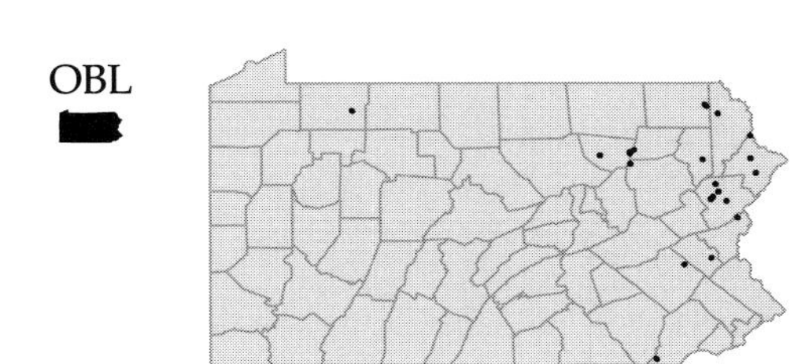

Eriophorum vaginatum L. ssp. ***spissum*** (Fern.) Hulten
Cotton-grass
Herbaceous perennial
Sphagnum bogs and swamps.
Eriophorum spissum Fern. FGBW

OBL

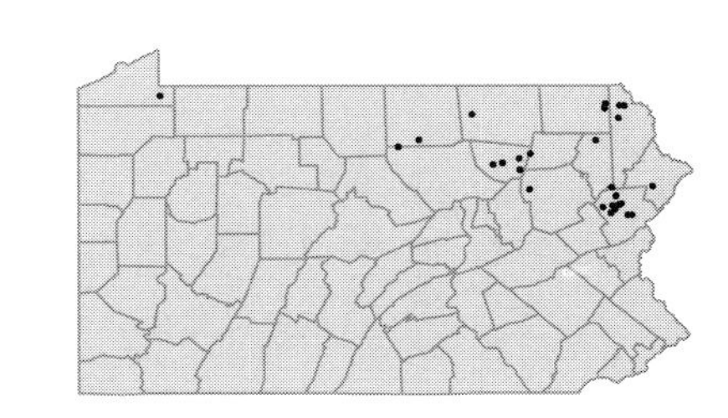

Eriophorum viridicarinatum (Engelm.) Fern.
Thin-leaved cotton-grass
Herbaceous perennial
Swamps, bogs, wet meadows and fens.

OBL

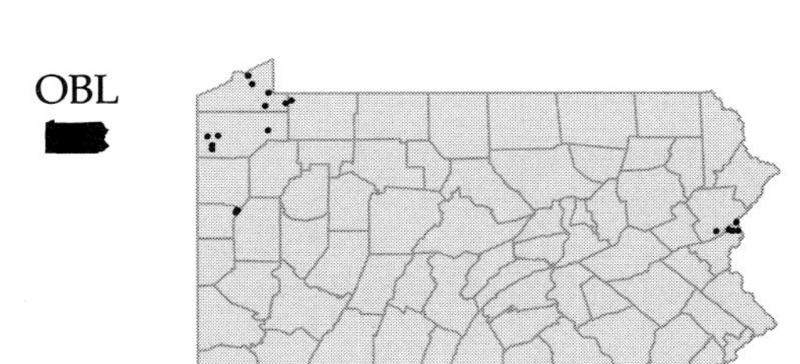

Eriophorum virginicum L.
Tawny cotton-grass
Herbaceous perennial
Swamps, bogs and swales.

OBL

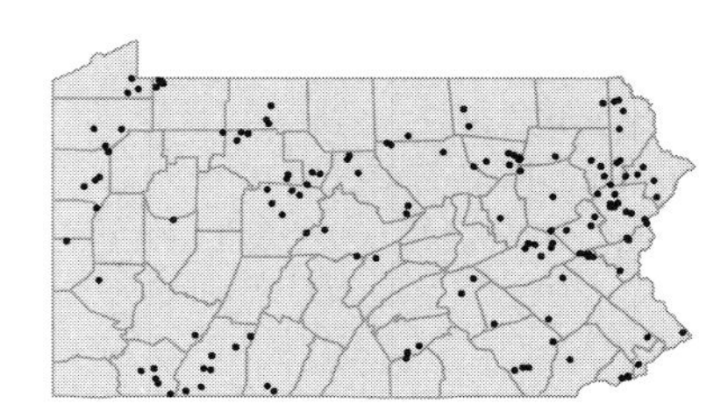

• ***Fimbristylis annua*** (All.) Roemer & Schultes
Annual fimbry
Herbaceous annual
Moist depressions on serpentine barrens, also wharves and waste ground.
Fimbristylis baldwiniana (Schultes) Torr. FW

FAC

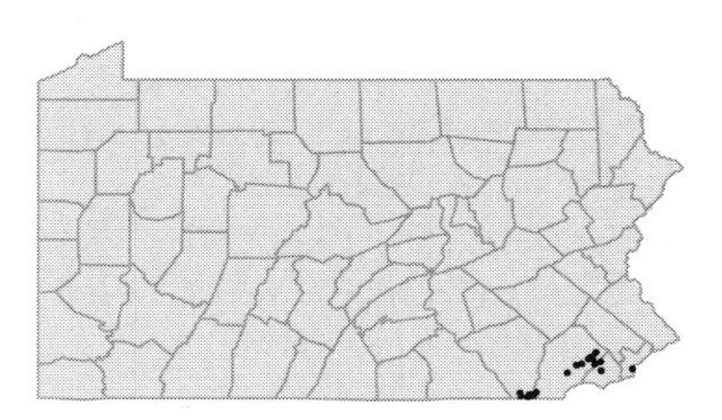

Fimbristylis autumnalis (L.) Roemer & Schultes
Slender fimbry
Herbaceous annual
Moist meadows, pond margins, river banks and wet ditches.

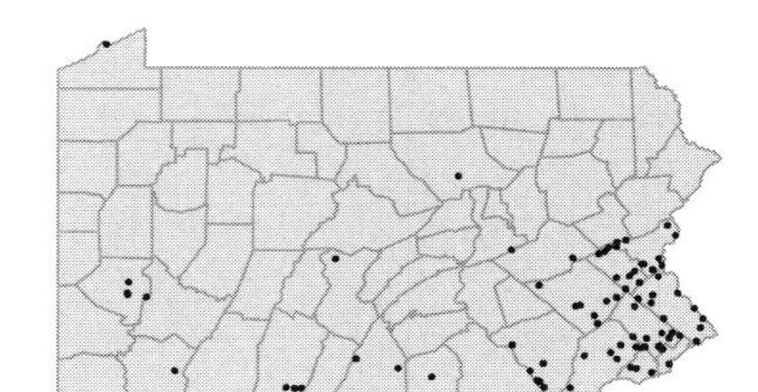

Fimbristylis castanea (Michx.) Vahl
Chestnut fimbry
Herbaceous perennial
Wharves and ballast ground.

OBL

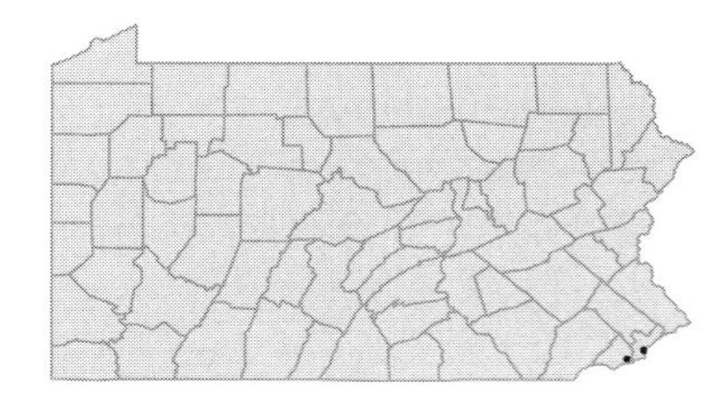

Fimbristylis miliacea (L.) Vahl
Fimbry
Herbaceous annual
Urban waste ground. Represented by a single collection from Philadelphia Co. in 1877.

I

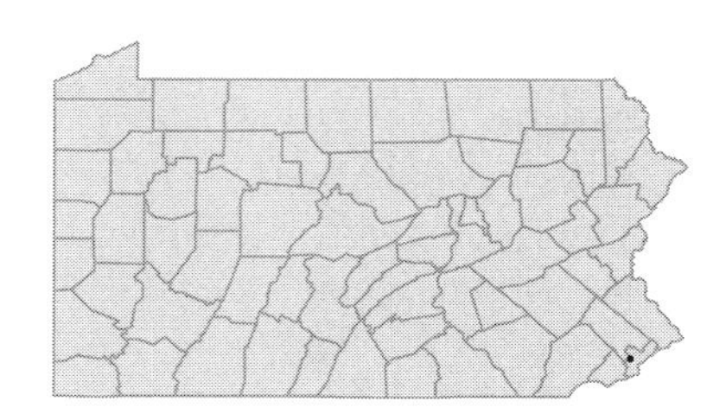

Fimbristylis puberula (Michx.) Vahl
Hairy fimbry
Herbaceous perennial
Edge of swamp, low ground. Believed to be extirpated, last collected in 1910.
Fimbristylis drummondii Boeckl. FW

OBL

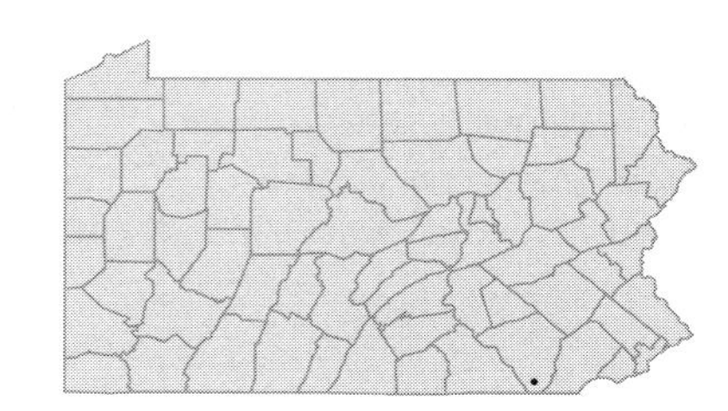

Fimbristylis vahlii (Lam.) Link
Fimbry
Herbaceous annual
Ballast and wharf areas. Represented by several collections from Philadelphia Co. 1864-1879.

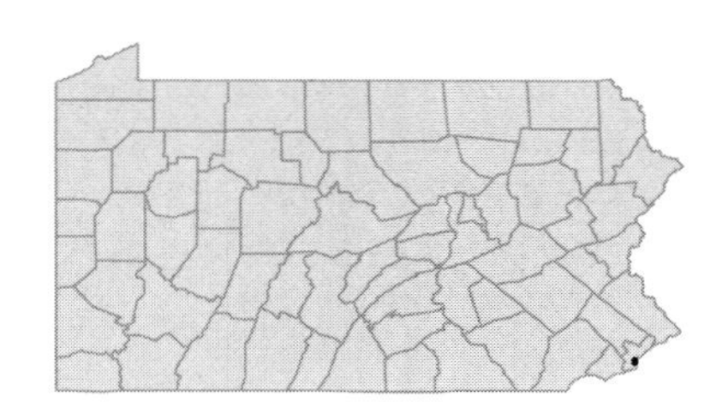

• ***Lipocarpha micrantha*** (Vahl) G.Tucker
Common hemicarpa
Herbaceous annual
Wet, sandy shores and barrens.
Hemicarpha micrantha (Vahl) Britt. FCBWK

FACW+

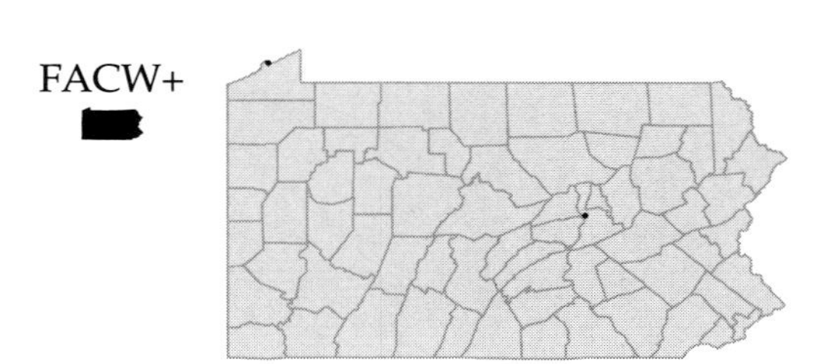

• ***Rhynchospora alba*** (L.) Vahl
White beak-rush
Herbaceous perennial
Bogs and swamps.

OBL

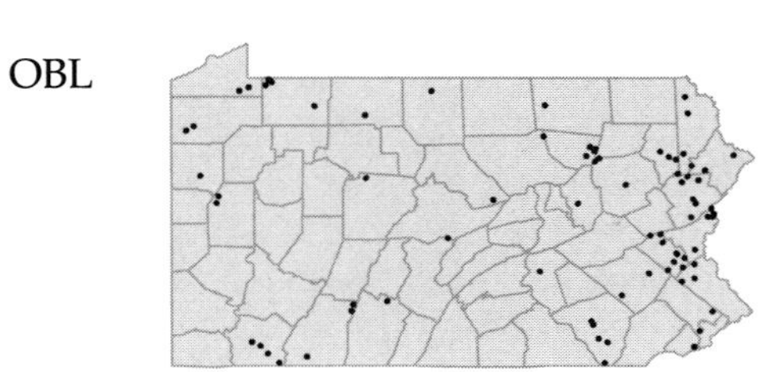

Rhynchospora capillacea Torr.
Capillary beak-rush
Herbaceous perennial
Calcareous swamps and fens.

OBL

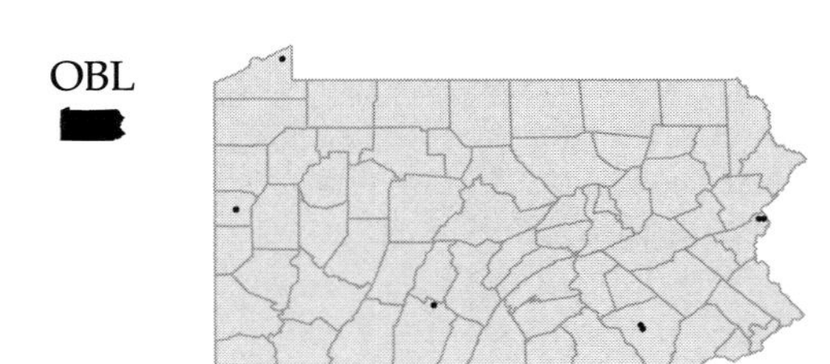

Rhynchospora capitellata (Michx.) Vahl
Beak-rush
Herbaceous perennial
Boggy meadows, low open woods, abandoned gravel pits and vernal ponds.

OBL

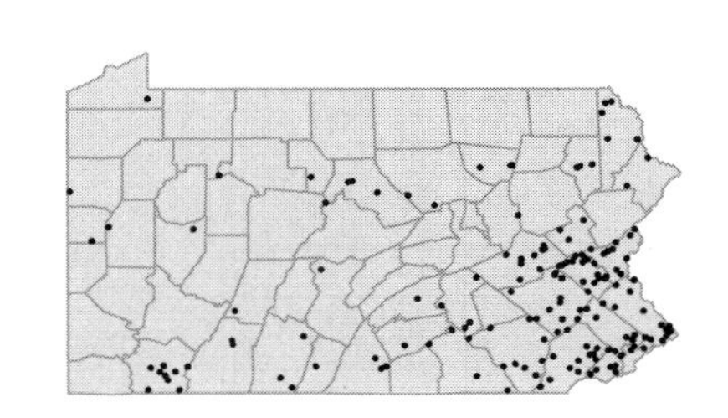

Rhynchospora fusca (L.) Ait.f.
Brown beak-rush
Herbaceous perennial
Bogs and glacial lakes. Believed to be extirpated, last collected in 1913.

OBL

Rhynchospora globularis (Chapm.) Small
Beak-rush
Herbaceous perennial
Sandy shores, swamps and sphagnum bogs.
Rhynchospora cymosa Ell. P

FACW

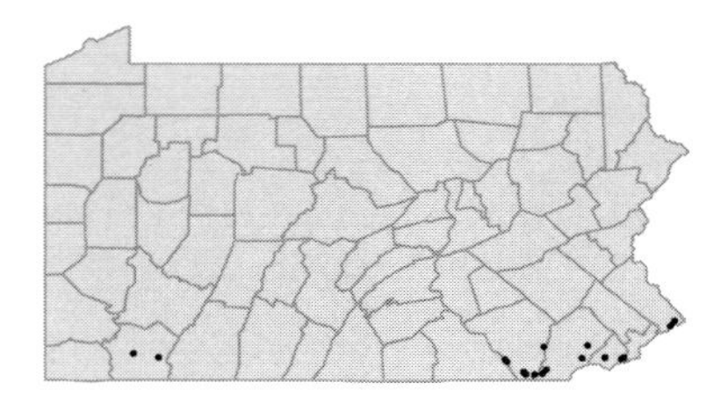

Rhynchospora gracilenta A.Gray
Beak-rush
Herbaceous perennial
Sphagnum bog. Believed to be extirpated, represented by single collection from Lancaster Co. in 1929.

OBL

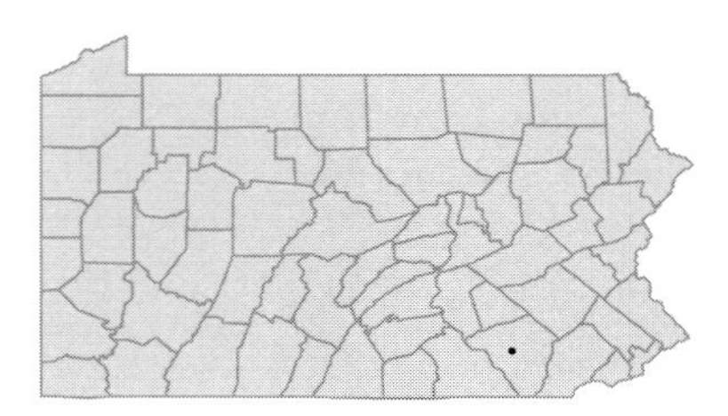

• ***Schoenoplectus acutus*** (Muhl. ex Bigel.) Love & Love
Great bulrush; Hard-stemmed bulrush
Herbaceous perennial, emergent aquatic
Shallow water of lake and pond margins.
Scirpus lacustris L. P; *Scirpus acutus* Muhl. ex Bigel. CFBGW

OBL

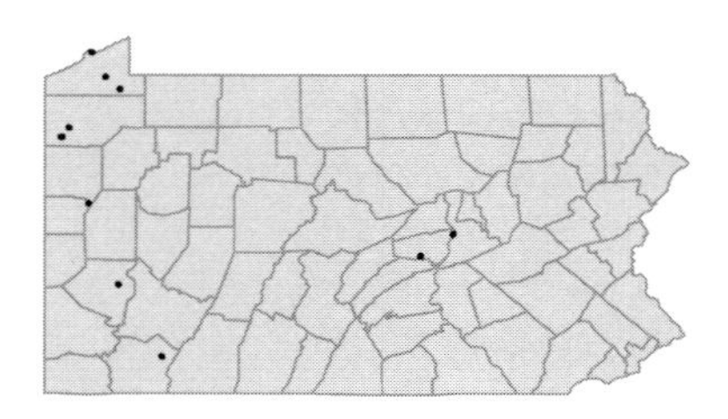

Schoenoplectus fluviatilis (Torr.) Strong
River bulrush
Herbaceous perennial, emergent aquatic
Moist, sandy shores and marshes, tidal and non-tidal.
Scirpus fluviatilis (Torr.) A.Gray. WBGKFC

OBL

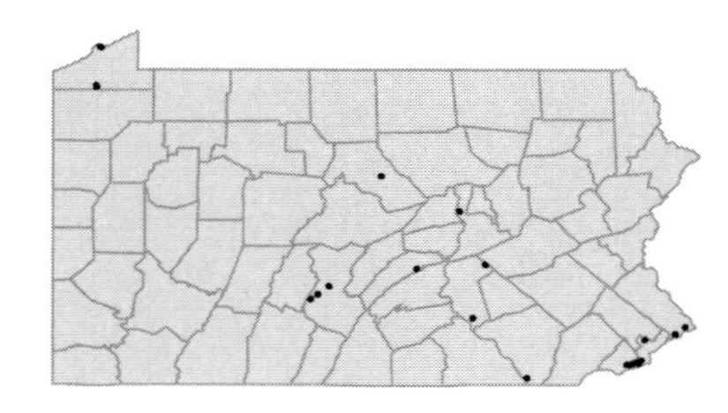

Schoenoplectus heterochaetus (Chase) Sojak
Slender bulrush
Herbaceous perennial, emergent aquatic
Believed to be extirpated, represented by a single collection from northeastern PA in 1897.
Scirpus heterochaetus Chase CWFBGK

OBL

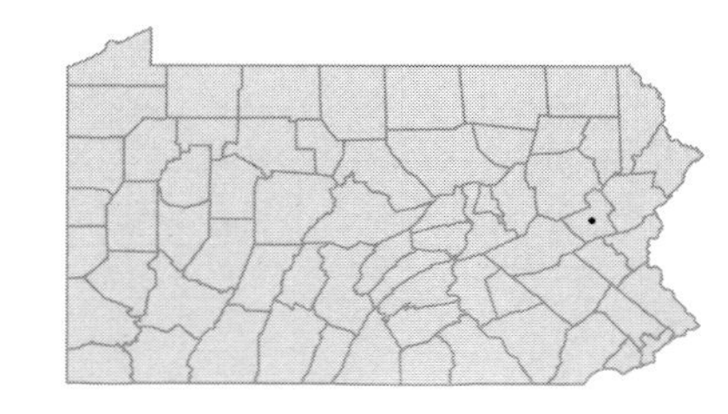

Schoenoplectus maritimus (L.) Lye
Salt-marsh bulrush
Herbaceous perennial, emergent aquatic
Ballast. Represented by two collections from Philadelphia Co. ca 1865-1877.
Scirpus maritimus L. CFBGW

OBL

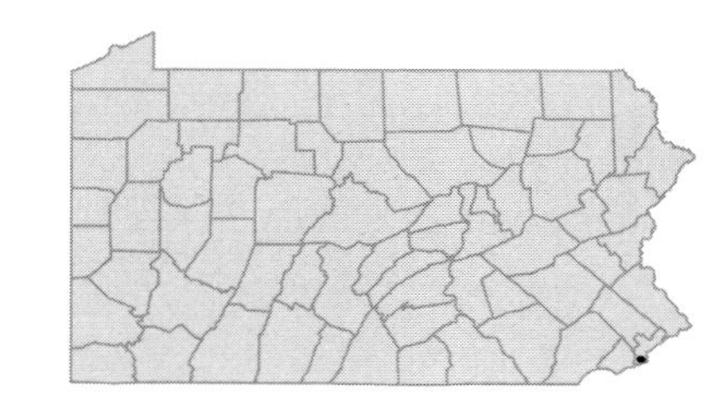

Schoenoplectus pungens (Vahl) Palla
Chairmaker's-rush; Three-square
Herbaceous perennial, emergent aquatic
Marshes, moist shores, river banks and tidal mudflats.
Scirpus americanus Pers. FBGW; *Scirpus pungens* Vahl CFGBW

FACW+

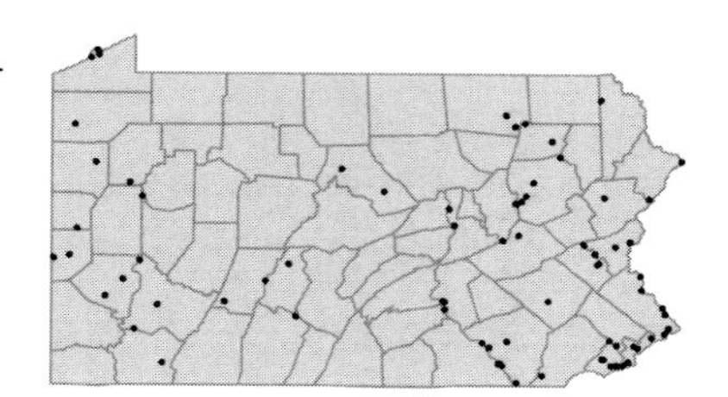

Schoenoplectus purshianus (Fern.) Strong
Bulrush
Herbaceous annual, emergent aquatic
River banks, stream margins and wet soil.
Scirpus debilis Pursh P; *Scirpus smithii* Gray var. *williamsii* (Fern.) Beetle GB; *Scirpus purshianus* Fern. CWF

OBL

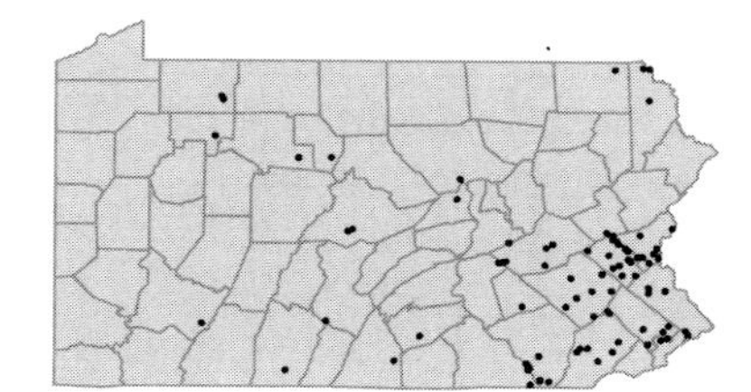

Schoenoplectus robustus (Pursh) Strong
Saltmarsh bullrush
Herbaceous perennial
Ballast and waste ground.
Scirpus robustus Pursh CFBGK

OBL

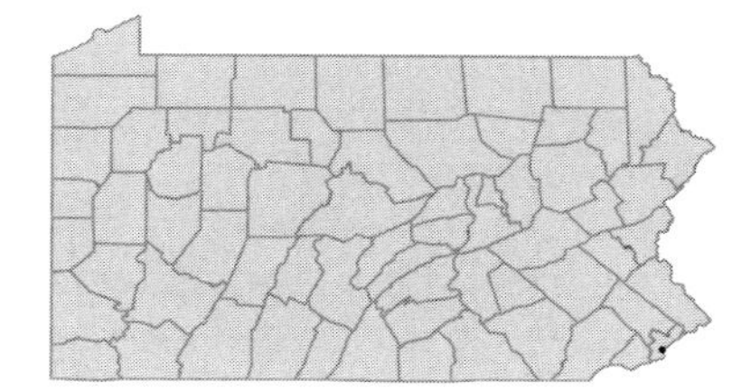

Schoenoplectus smithii (A.Gray) Sojak
Smith's bulrush
Herbaceous perennial, emergent aquatic
Moist shores and tidal mudflats.
Scirpus smithii A.Gray CFBGW

OBL

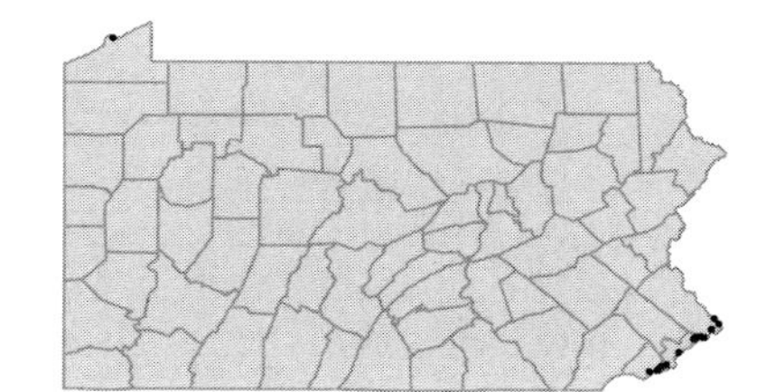

Schoenoplectus subterminalis (Torr.) Sojak
Water bulrush
Herbaceous perennial, rooted submergent aquatic
Quiet water of lakes and ponds.
Scirpus subterminalis Torr. CFGBW

OBL

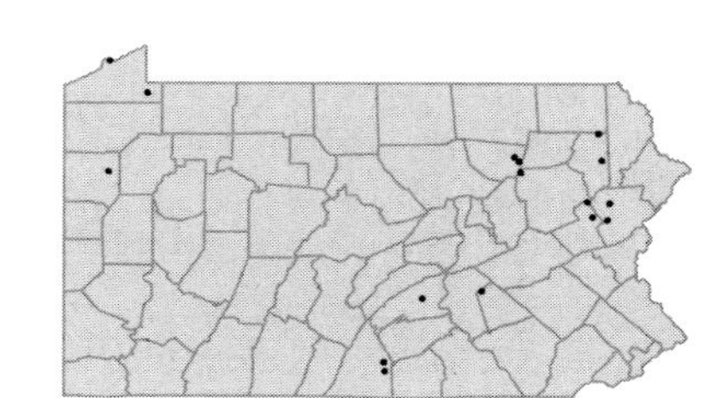

Schoenoplectus tabernaemontani (K.C.Gmel.) Palla
Great bulrush; Soft-stem bulrush
Herbaceous perennial, emergent aquatic
Swamps, lake and pond margins, wet ditches and mudflats.
Scirpus validus Vahl FGBCW; *Scirpus tabernaemontani* K.C.Gmel. K

OBL

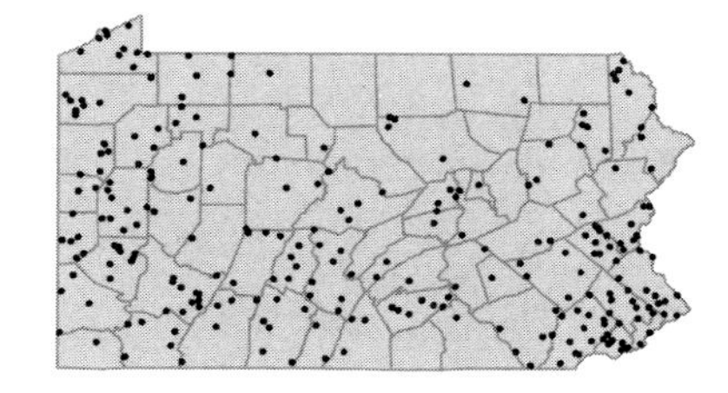

Schoenoplectus torreyi (Olney) Palla
Torrey's bulrush
Herbaceous perennial, emergent aquatic
Shallow water of lake and pond margins.
Scirpus torreyi Olney CFGBW

OBL

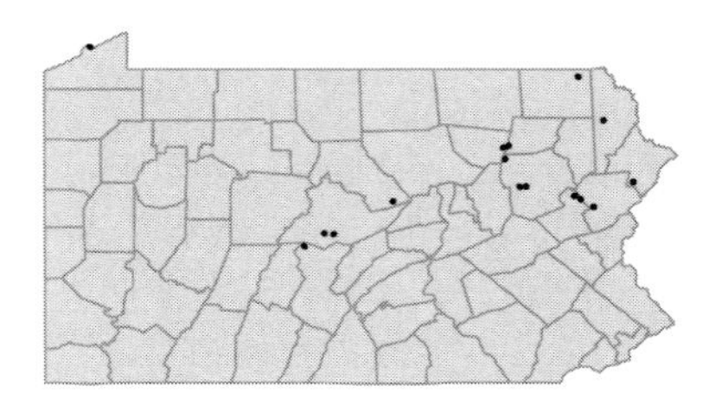

• ***Scirpus ancistrochaetus*** Schuyler
Northeastern bulrush
Herbaceous perennial
Vernal ponds and mudholes with fluctuating water levels.
Scirpus atrovirens Willd. var. *atrovirens pro parte* C

OBL

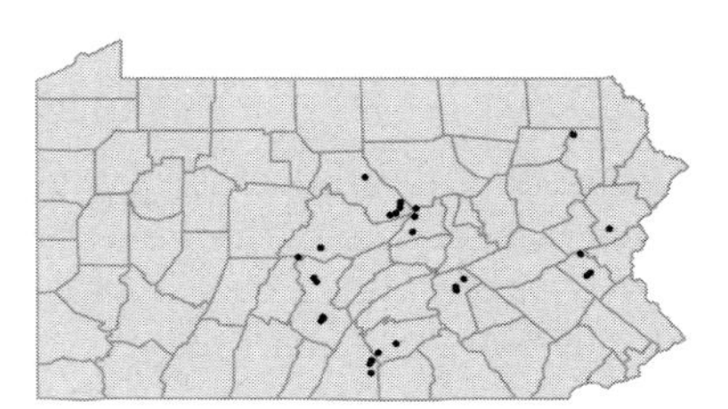

Scirpus atrocinctus Fern.
Blackish wool-grass
Herbaceous perennial
Marshes, bogs and swampy thickets.
Scirpus cyperinus (L.) Kunth *pro parte* GC

FACW+

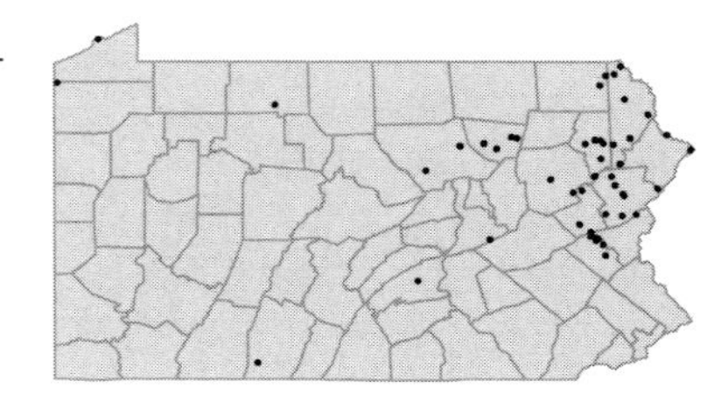

Scirpus atrovirens Willd.
Black bulrush
Herbaceous perennial
Moist meadows, marshes, bogs, pond and lake margins and stream banks.
Hybrids with *S. georgianus* and *S. hattorianus* have been collected at a few locations.

OBL

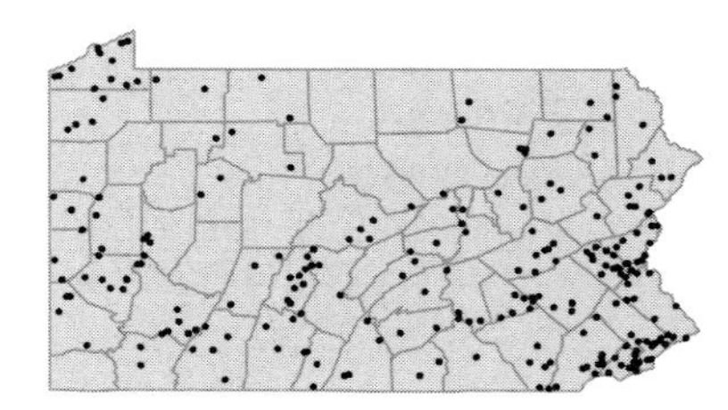

Scirpus cyperinus (L.) Kunth
Wool-grass
Herbaceous perennial
Marshes, moist meadows, swamps, shores and ditches.

FACW

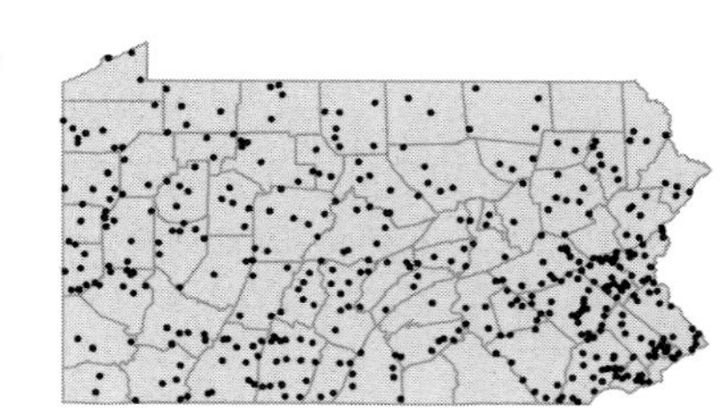

Scirpus expansus Fern.
Wood bulrush
Herbaceous perennial
Marshes and wet meadows.
Scirpus sylvaticus L. P

OBL

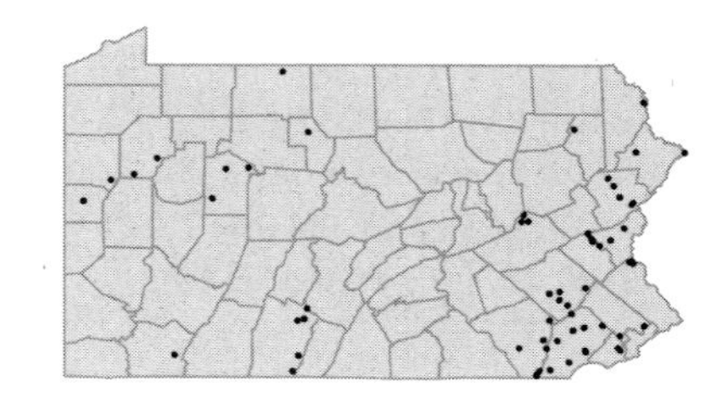

Scirpus georgianus Harper
Bulrush
Herbaceous perennial
Swamps, moist meadows and wet woods. A hybrid with *S. hattorianus* has been collected at a few locations. Fernald's concept of *S. atrovirens* var. *georgianus* also included some plants more properly identified as as *S. hattorianus.*
Scirpus atrovirens Willd. var. *atrovirens pro parte* GBC; *Scirpus atrovirens* Willd. var. *georgianus* (Harper) Fern. F

OBL

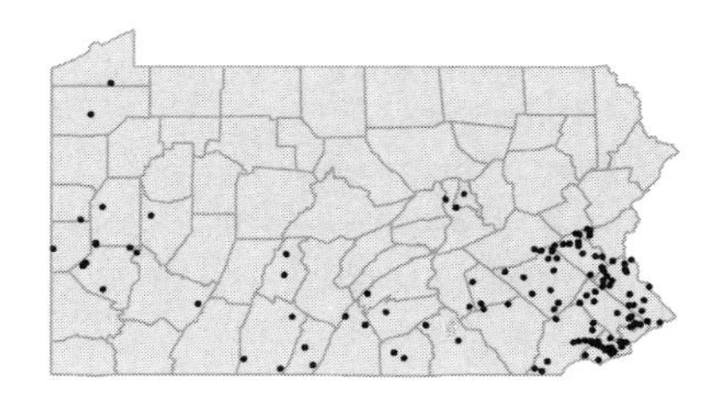

Scirpus hattorianus Makino
Bulrush
Herbaceous perennial
Moist meadows, swamps, bogs, river banks and ditches.
Scirpus atrovirens Willd. var. *atrovirens pro parte* C

OBL

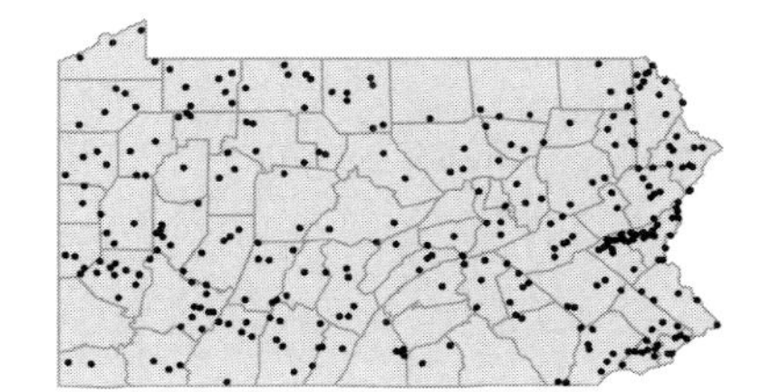

Scirpus microcarpus Presl
Bulrush
Herbaceous perennial
Marshes, swamps and wet meadows.

OBL

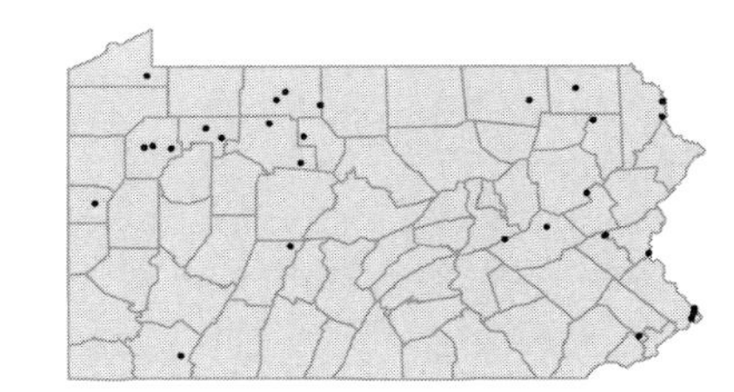

Scirpus pallidus (Britt.) Fern.
Pale bulrush
Herbaceous perennial
Moist depression along railroad, towpath.
Scirpus atrovirens Willd. var. *palidus* Britt. GBC

OBL
[I]

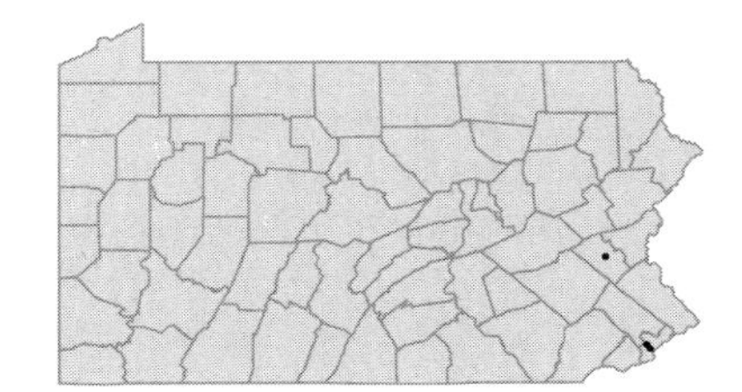

Scirpus pedicellatus Fern.
Wool-grass; Stalked bulrush
Herbaceous perennial
Wet shores and ditches.
Scirpus cyperinus (L.) Kunth *pro parte* GC

FACW+

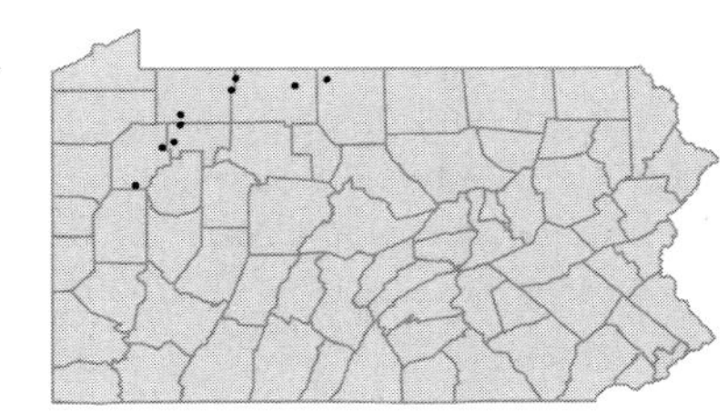

Scirpus pendulus Muhl.
Bulrush
Herbaceous perennial
Marshes, wet meadows, swampy floodplains, wet woods, moist sand flats and swales.

OBL

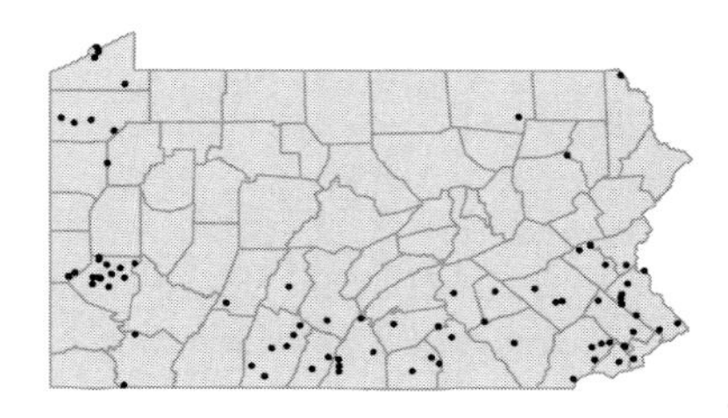

Scirpus polyphyllus Vahl
Bulrush
Herbaceous perennial
Swamps, woods and stream banks.

OBL

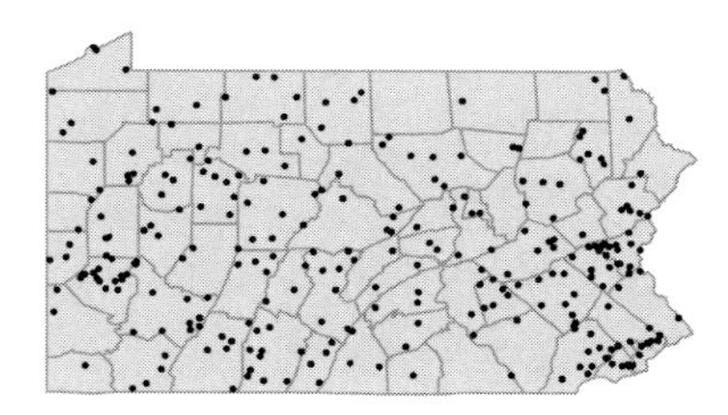

• ***Scleria minor*** W.Stone
Small nut-rush
Herbaceous perennial
Sphagnum bog, swamp.

FACW

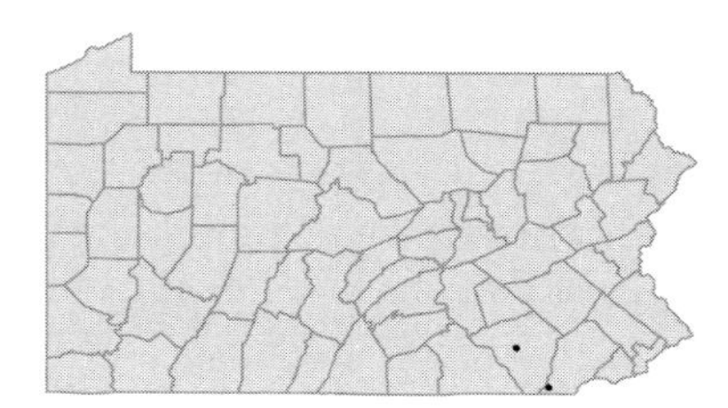

Scleria muhlenbergii Steud.
Reticulated nut-rush
Herbaceous annual
Moist, sandy meadows and boggy pastures.
Scleria reticularis Michx. GBC

OBL

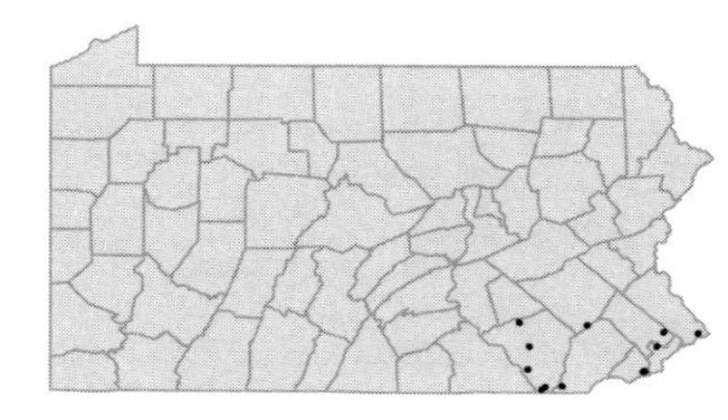

Scleria pauciflora Muhl. ex Willd.
Few-flowered nut-rush
Herbaceous perennial
Dry, open woods and serpentine barrens.

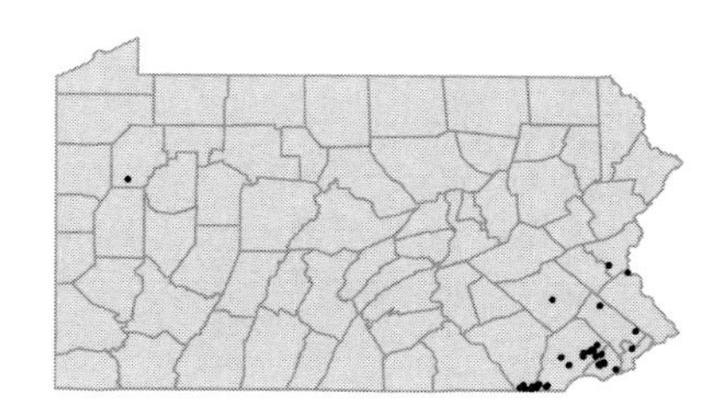

Scleria triglomerata Michx.
Whip-grass; Nut-rush
Herbaceous perennial
Sphagnum bogs, swampy meadows and moist serpentine barrens.

FAC

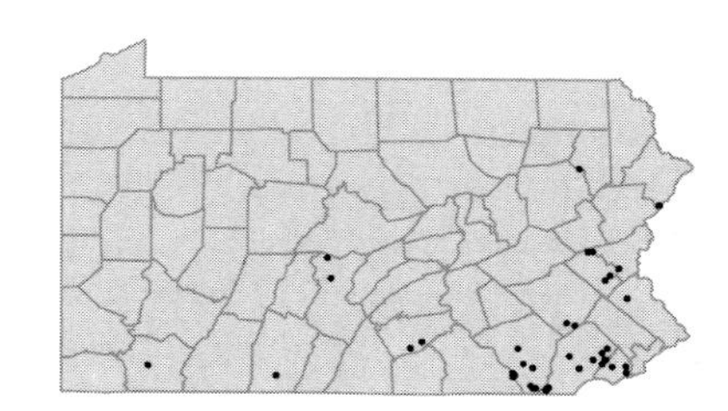

Scleria verticillata Muhl. ex Willd.
Whorled nut-rush
Herbaceous annual
Moist, calcareous meadows, bogs and fens.

OBL

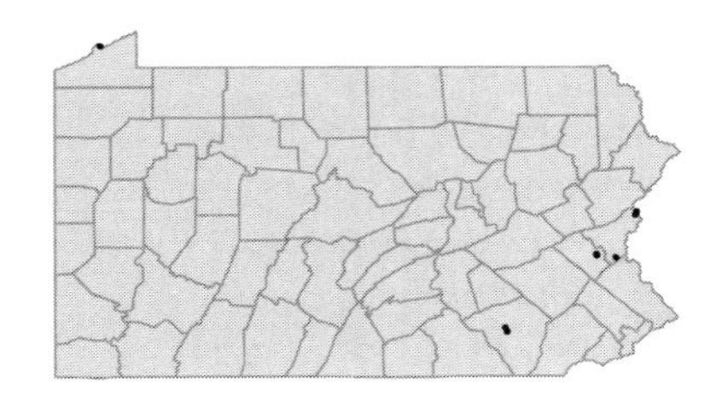

• ***Trichophorum planifolium*** (Sprengel) Palla
Club-rush
Herbaceous perennial
Woods and dry, rocky slopes.
Scirpus planifolius Muhl. P; *Scirpus verecundus* Fern. BCFGKW

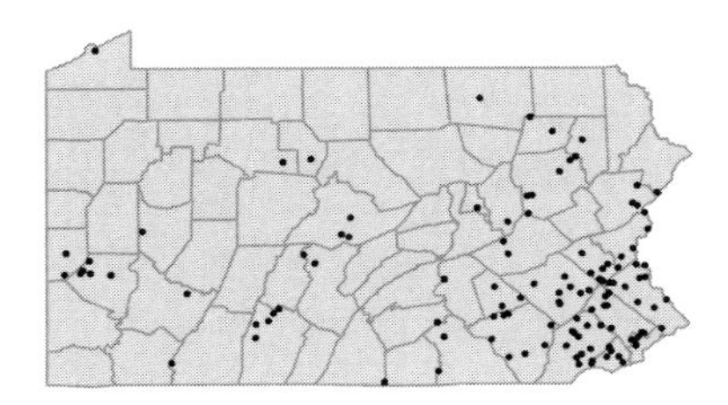

POACEAE

• ***Aegilops cylindrica*** Host
Jointed goatgrass
Herbaceous annual
Waste ground, rubbish dump and fill.

I

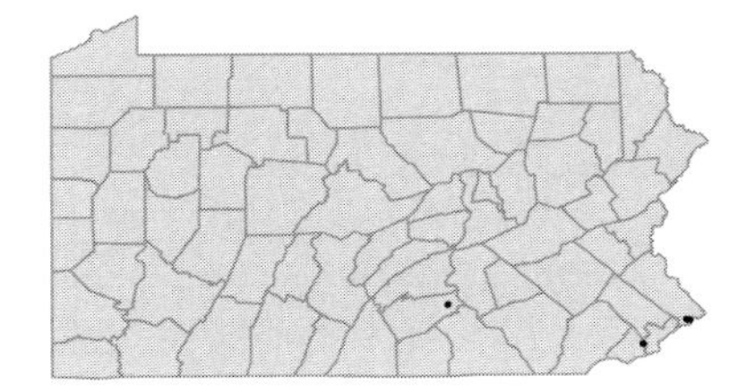

Aegilops triuncialis L.
Barbed goatgrass
Herbaceous annual
Waste ground near greenhouse. Represented by a single collection from Chester Co. in 1925.

I

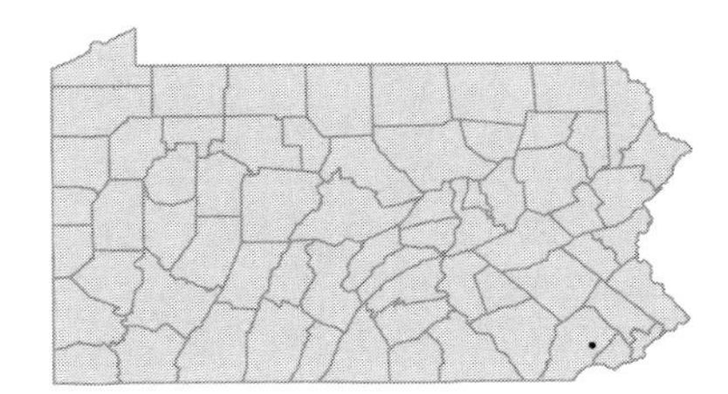

• ***Agrostis altissima*** (Walt.) Tuckerman
Tall bentgrass
Herbaceous perennial
Sphagnum bog, swale or swamp. Believed to be extirpated, last collected in 1946.
Agrostis perennans (Walt.) Tuckerman var. *elata* (Pursh) A.S.Hitchc. BGC

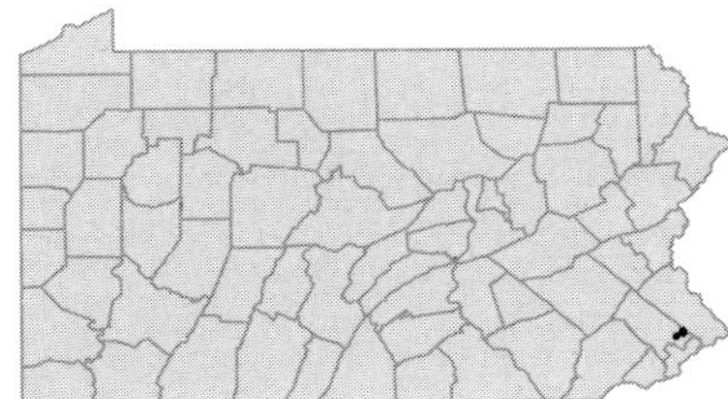

Agrostis canina L.
Brown bent; Velvet bent
Herbaceous perennial
Low, sandy or peaty soil.

I

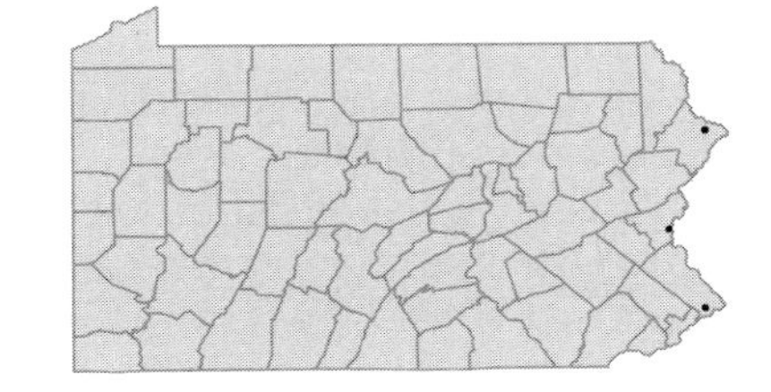

Agrostis capillaris L.
Rhode Island bent
Herbaceous perennial
Cultivated and occasionally established in dry, open ground or along roadsides.
Agrostis tenuis Sibth. FGBW

I

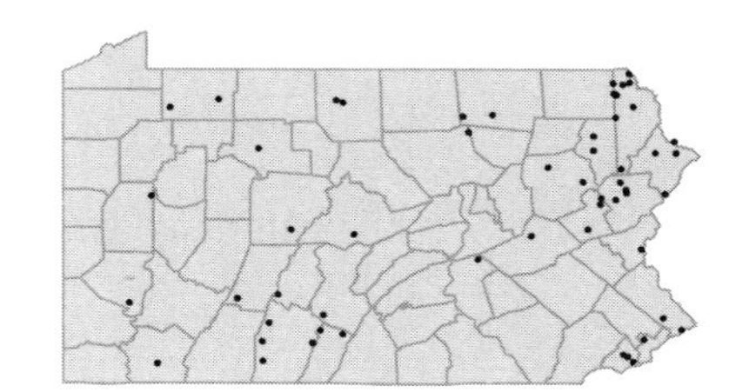

Agrostis elliottiana Schultes
Bentgrass
Herbaceous annual
Dry, open soil.

I

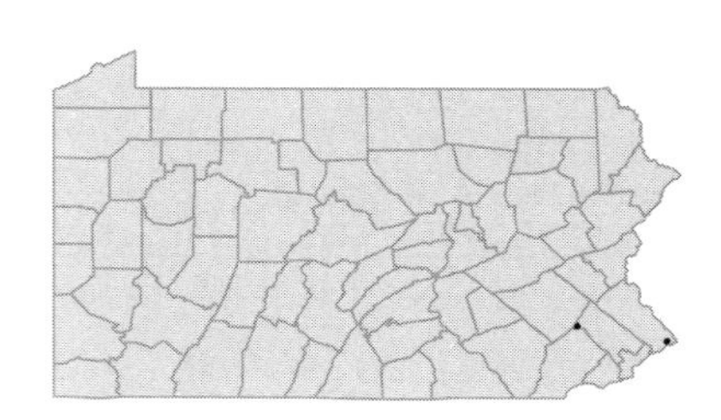

Agrostis gigantea Roth
Redtop
Herbaceous perennial
Cultivated and frequently established in moist soil of fields, roadsides and waste ground.
Agrostis alba L. PFW; *Agrostis stolonifera* L. var. *major* (Gaudin) Farw. GB

FACW
I

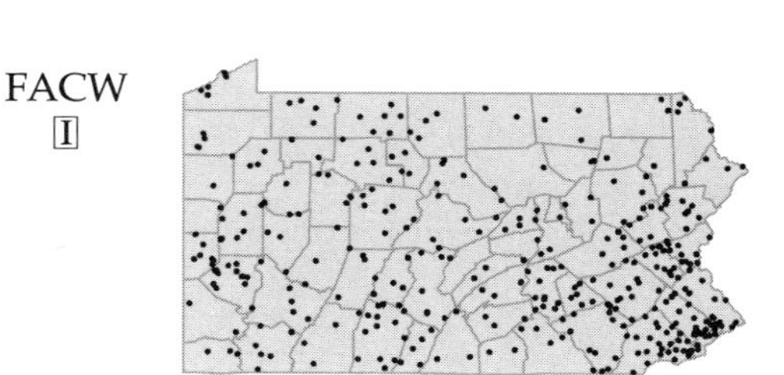

Agrostis hyemalis (Walt.) BSP
Hairgrass; Ticklegrass
Herbaceous perennial
Dry to moist, open, sterile soil.

FAC

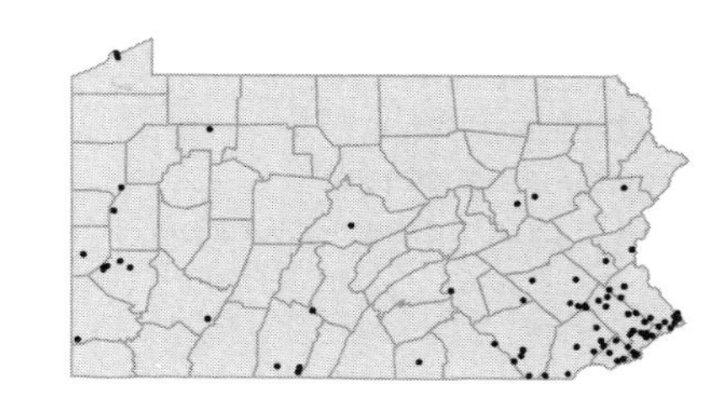

Agrostis perennans (Walt.) Tuckerman
Autumn bent; Upland bent
Herbaceous perennial
Dry, open ground or in light shade.

FACU

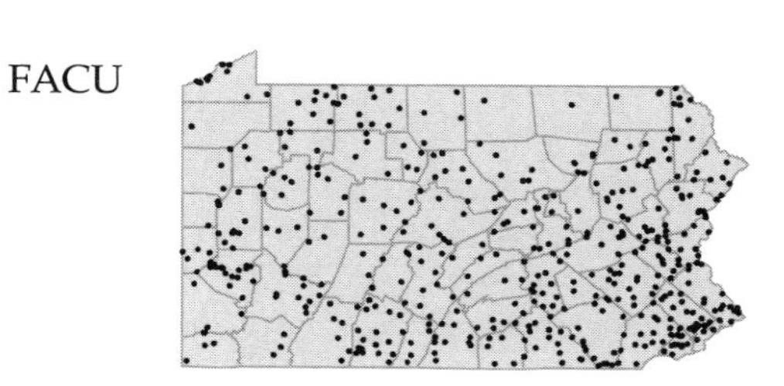

Agrostis scabra Willd.
Fly-away grass; Hairgrass; Ticklegrass
Herbaceous perennial
Moist, sandy-peaty ground and barrens.
Agrostis hyemalis (Walt.) BSP var. *scabra* (Willd.) Blomq. C

FAC

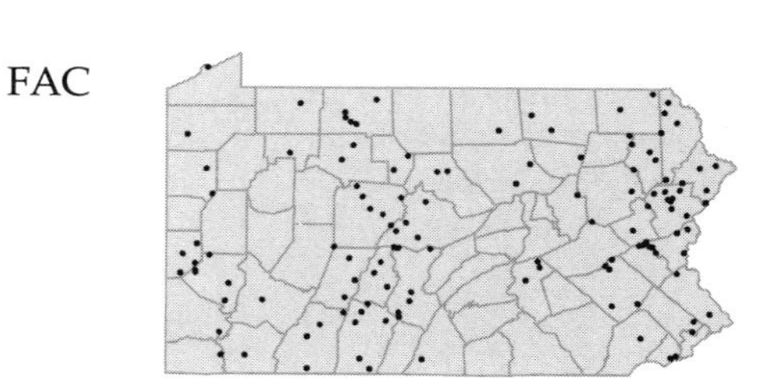

Agrostis stolonifera L. var. ***palustris*** (Huds.) Farw.
Carpet bentgrass; Creeping bentgrass
Herbaceous perennial
Cultivated and occasionally established in wet meadows or shores.
Agrostis alba L. var. *palustris* (Huds.) Pers. F; *Agrostis palustris* Huds. W; *Agrostis stolonifera* L. var. *compacta* Hartm. GB

FACW
[I]

• ***Aira caryophyllea*** L.
Silvery hairgrass
Herbaceous annual
Ballast. Represented by two collections from Philadelphia Co. 1878-1879.

[I]

Aira praecox L.
Hairgrass
Herbaceous annual
Ballast and waste ground. All specimens are pre-1900.

[I]

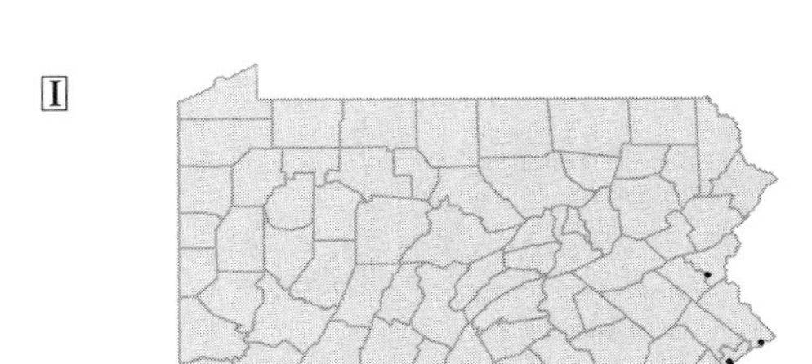

• ***Alopecurus aequalis*** Sobol.
Short-awned foxtail
Herbaceous perennial
Swamps, ditches and moist meadows.

OBL

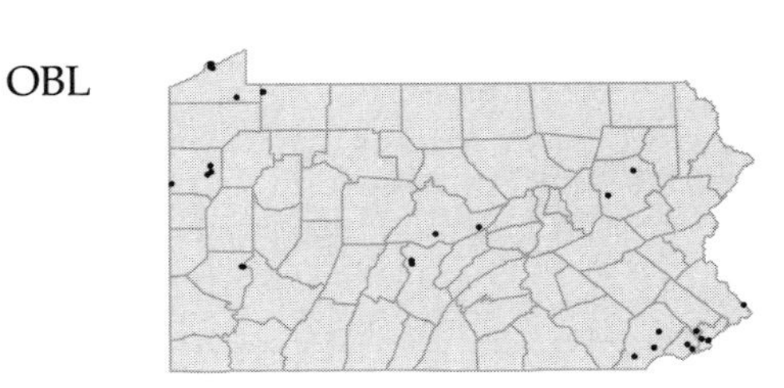

Alopecurus carolinianus Walt.
Carolina foxtail; Tufted foxtail
Herbaceous annual
Moist fields of the Coastal Plain or Piedmont.

FACW

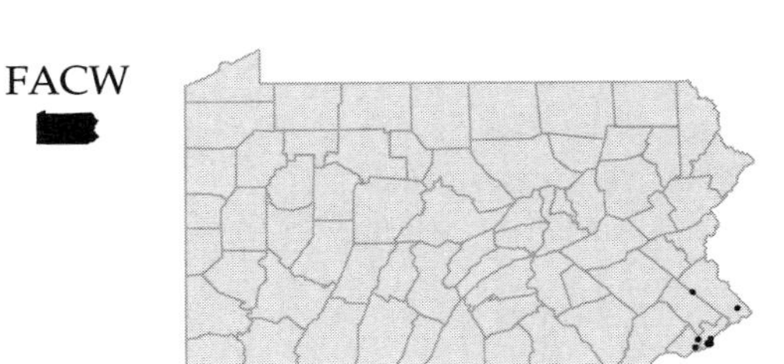

Alopecurus creticus Trin.
Foxtail
Herbaceous annual
Ballast. Represented by a single collection from Philadelphia Co. in 1877.

[I]

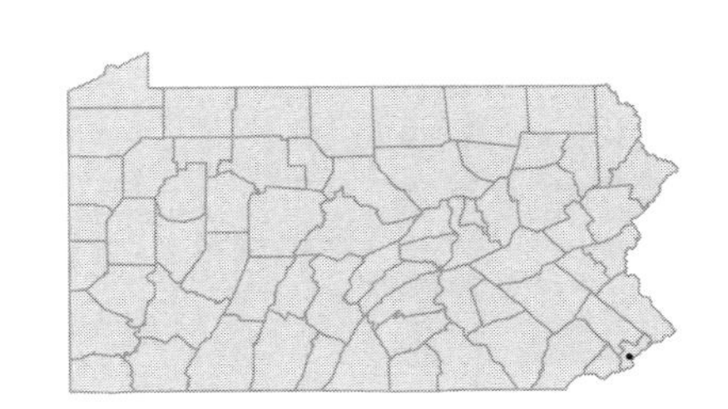

Alopecurus geniculatus L.
Marsh foxtail
Herbaceous perennial
Alluvial meadows and ballast. Represented by several specimens from Philadelphia Co. 1865-1879.

OBL
I

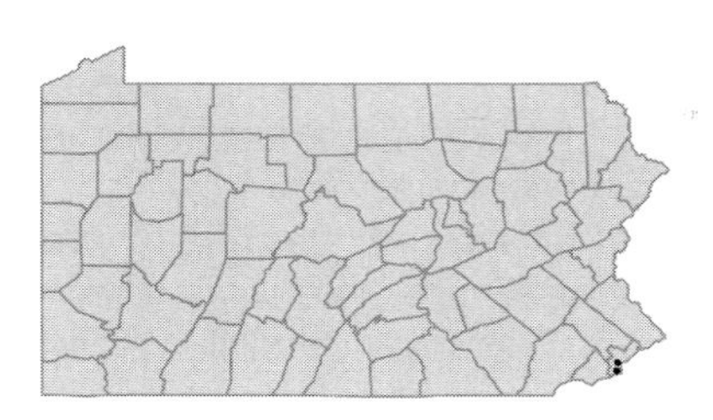

Alopecurus myosuroides Huds.
Slender foxtail
Herbaceous perennial
Fields, waste ground and ballast.
Alopecurus agrestis L. P

FACW
I

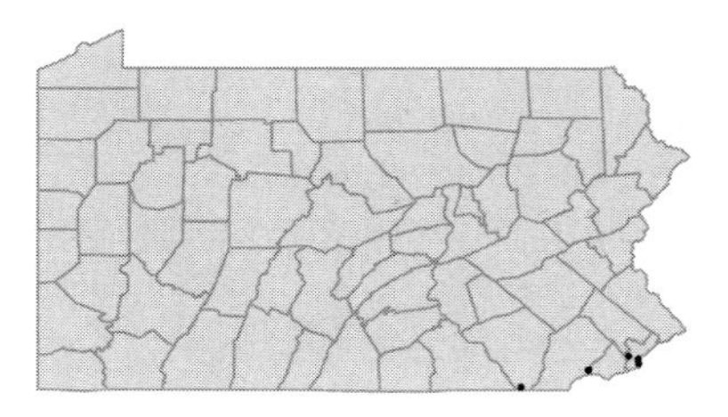

Alopecurus pratensis L.
Meadow foxtail
Herbaceous perennial
Meadows and waste ground.

FACW
I

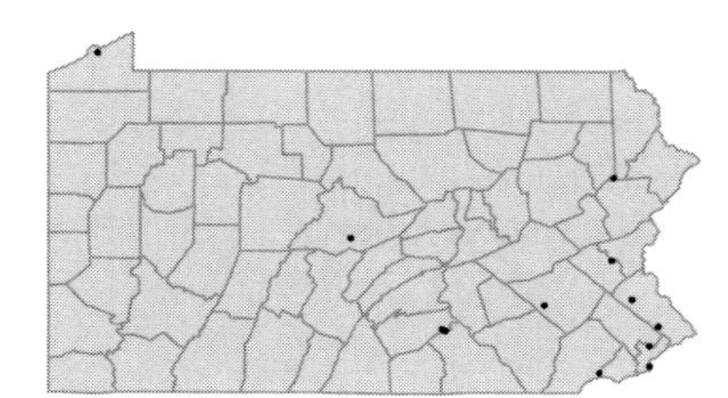

Alopecurus rendlei Eig.
Foxtail
Herbaceous annual
Ballast. Represented by a single collection from Philadelphia Co. in 1880.

I

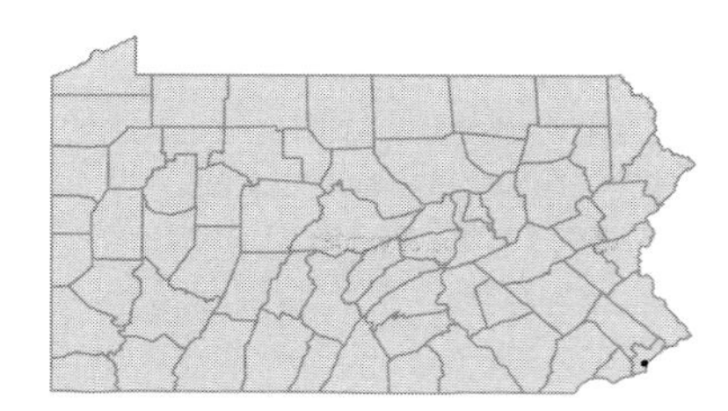

• ***Ammophila arenaria*** (L.) Link.
Marram grass
Herbaceous perennial
Sand dune. Represented by a single collection from Erie Co. in 1941.

FACU-
I

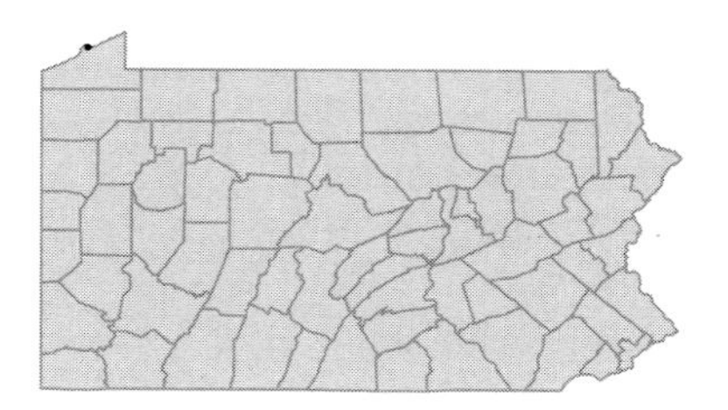

Ammophila breviligulata Fern.
American beachgrass
Herbaceous perennial
Sand dunes and beaches.

FACU–

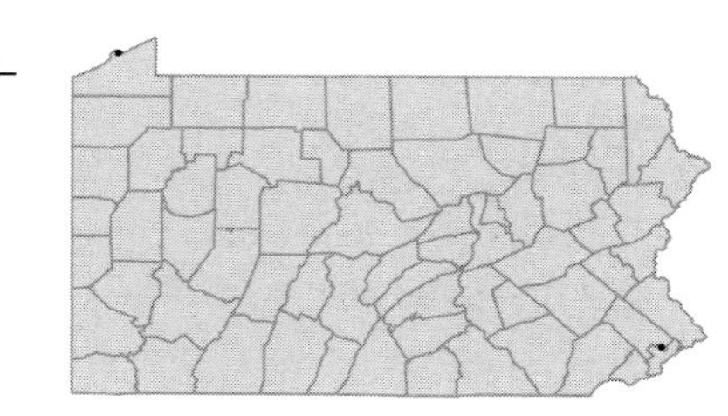

• ***Andropogon gerardii*** Vitman
Big bluestem; Turkey-foot
Herbaceous perennial
River banks, roadsides and moist meadows.
Andropogon furcatus Muhl. P

FAC

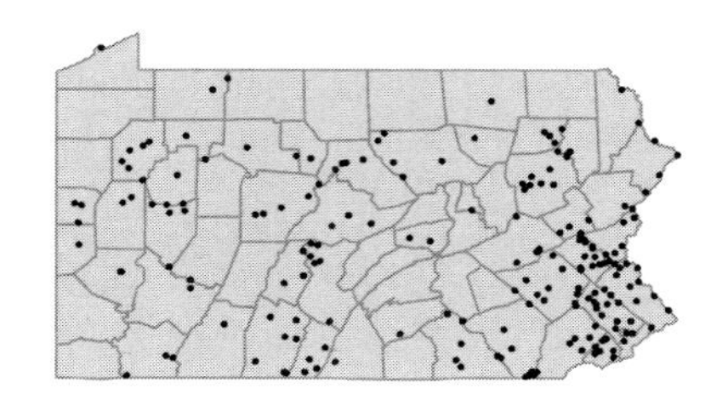

Andropogon glomeratus (Walt.) BSP
Broom-sedge
Herbaceous perennial
Swamps and moist meadows.
Andropogon corymbosus (Chapm.) Nash P; *Andropogon virginicus* L. var. *abbreviatus* (Hack.) Fern. & Grisc. FGBC

FACW+

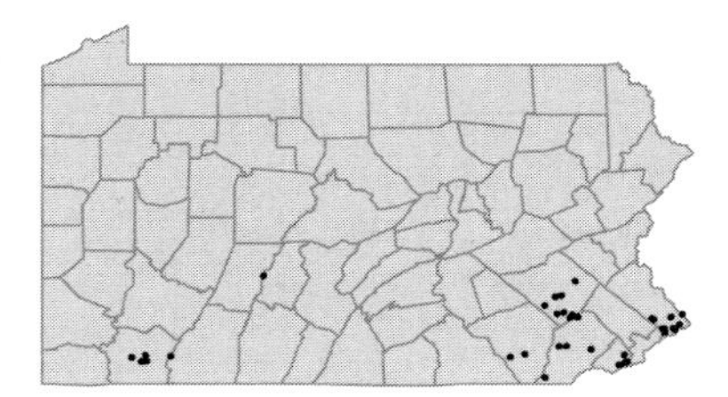

Andropogon gyrans Ashe
Elliott's beardgrass
Herbaceous perennial
Dry or moist fields or open woods.
Andropogon elliottii Chapm. FGBW

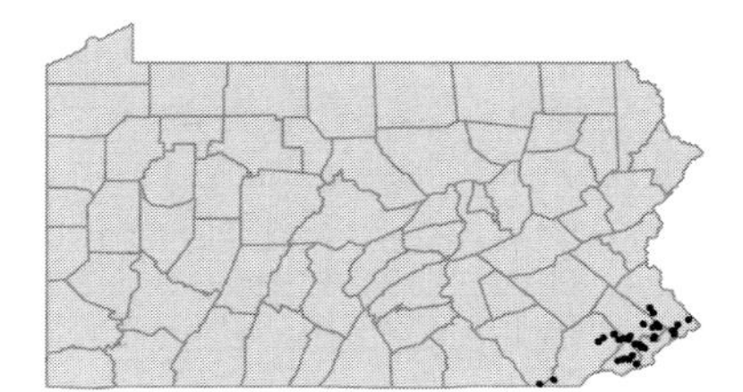

Andropogon virginicus L.
Broom-sedge
Herbaceous perennial
Old fields, hillsides and waste ground, in dry, sterile soil.

FACU

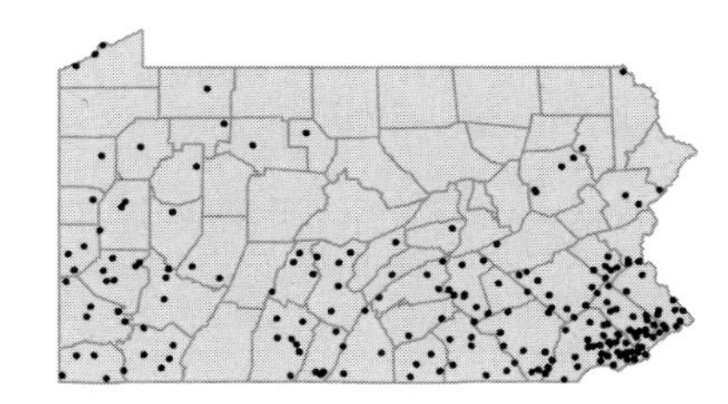

• ***Anthoxanthum aristatum*** Boiss.
Sweet vernal grass
Herbaceous annual
Cultivated and occasionally escaped.
Anthoxanthum puelii Lecoq & Lamotte F

I

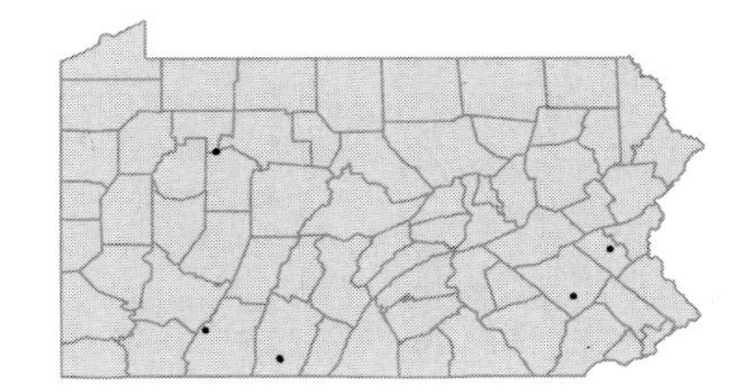

Anthoxanthum odoratum L.
Sweet vernal grass
Herbaceous perennial
Open fields, meadows and roadsides.

FACU
I

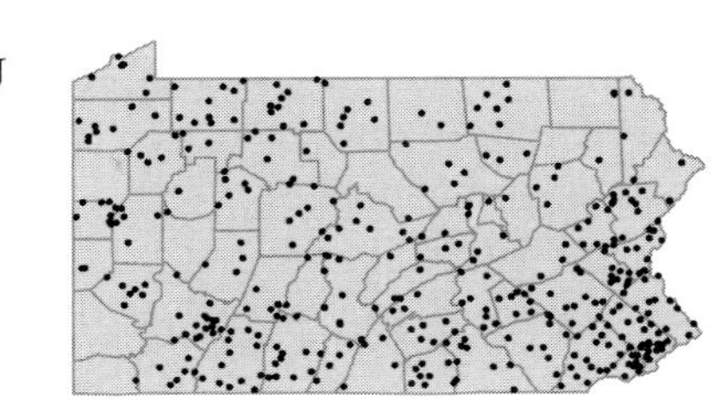

• ***Apera spica-venti*** (L.) Beauv. [I]
Silky bentgrass
Herbaceous annual
Ballast and waste ground.
Agrostis spica-venti L. GB

• ***Aristida dichotoma*** Michx. var. ***dichotoma*** UPL
Poverty grass
Herbaceous annual
Dry, sandy or sterile soil.

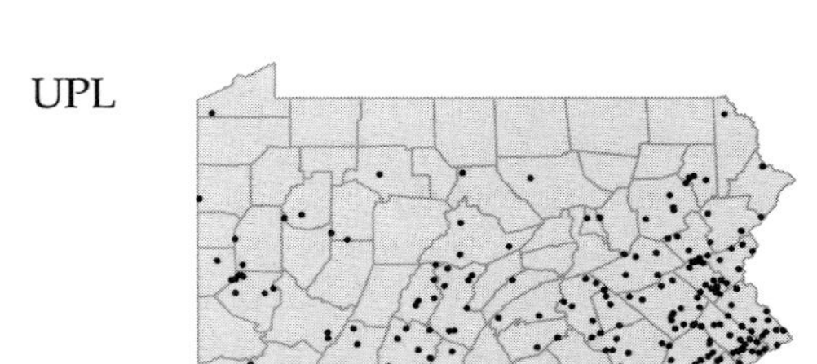

Aristida dichotoma Michx. var. ***curtissii*** A.Gray UPL
Three-awn; Poverty grass
Herbaceous annual
Dry, open ground.
Aristida curtissii (A.Gray) Nash K; *Aristida basiramea* Engel. var. *curtissii* (A.Gray) Shinners C

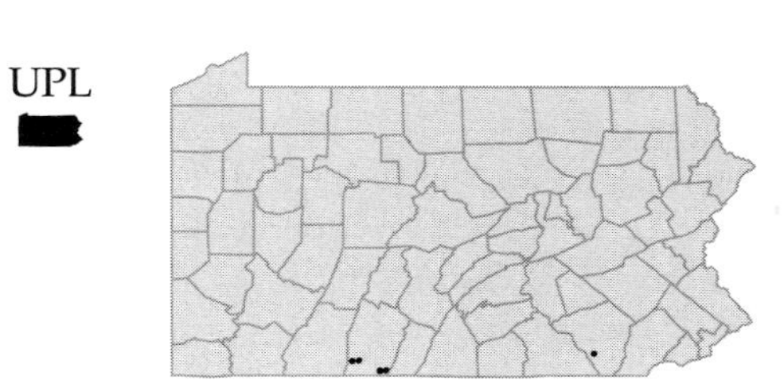

Aristida longispica Poir. var. ***longispica*** UPL
Slender three-awn
Herbaceous annual
Serpentine barrens and other dry, sandy soils.

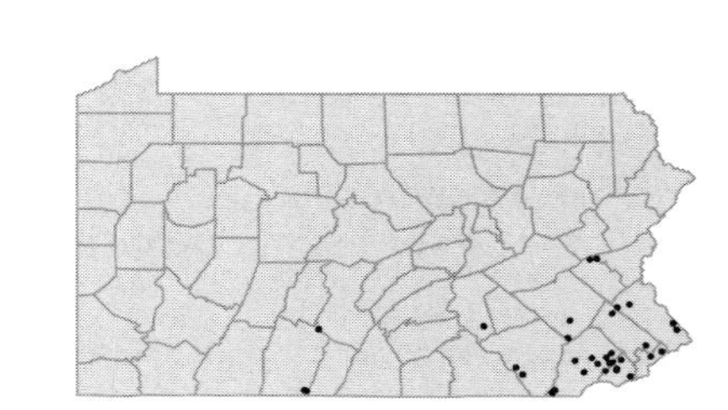

Aristida longispica Poir. var. ***geniculata*** (Raf.) Fern. UPL
Slender three-awn
Herbaceous annual
Dry, sterile soils.
Aristida gracilis Ell. P

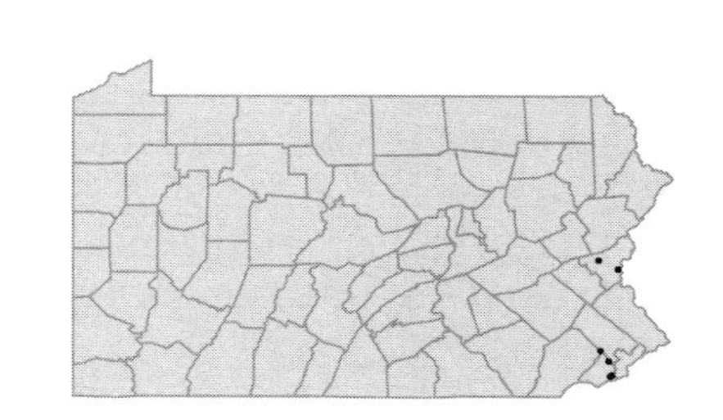

Aristida oligantha Michx.
Prairie three-awn
Herbaceous annual
Dry, open ground.

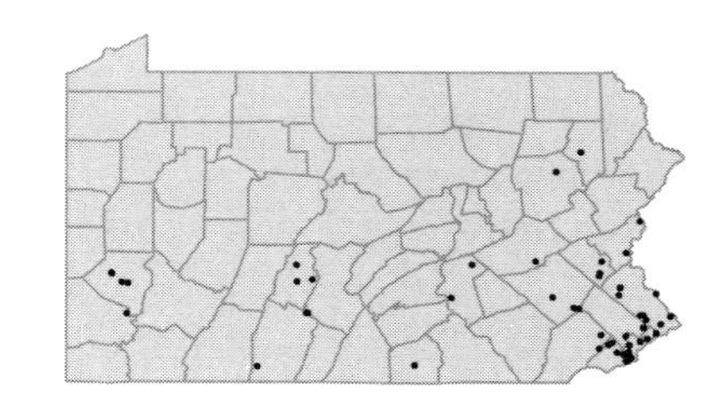

Aristida purpurascens Poir.
Arrow-feather; Three-awned grass
Herbaceous perennial
Serpentine barrens and other dry, sandy soils.

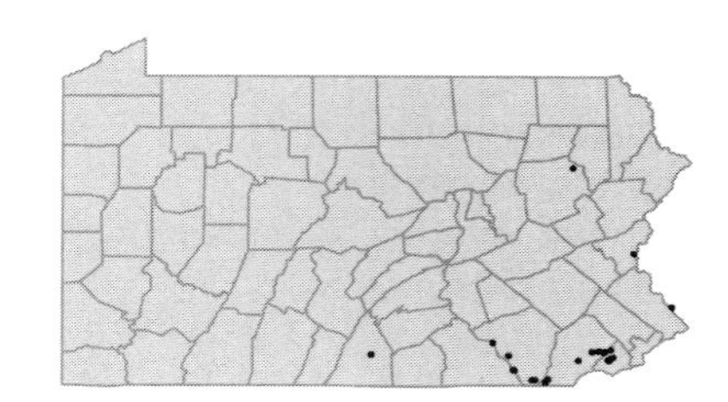

• ***Arrhenatherum elatius*** (L.) Beauv. var. ***elatius***
Tall oatgrass
Herbaceous perennial
Roadsides, fields and waste ground.

FACU
I

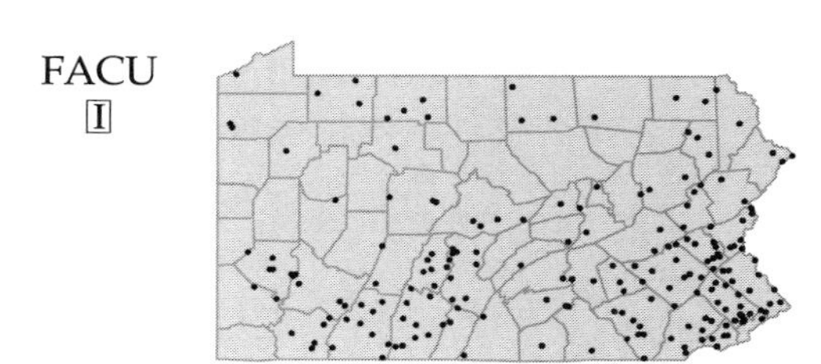

Arrhenatherum elatius (L.) Beauv. var. ***biaristatum*** (Peterm.) Peterm.
Tall oatgrass
Herbaceous perennial
Wooded slopes, roadsides and waste ground.

FACU
I

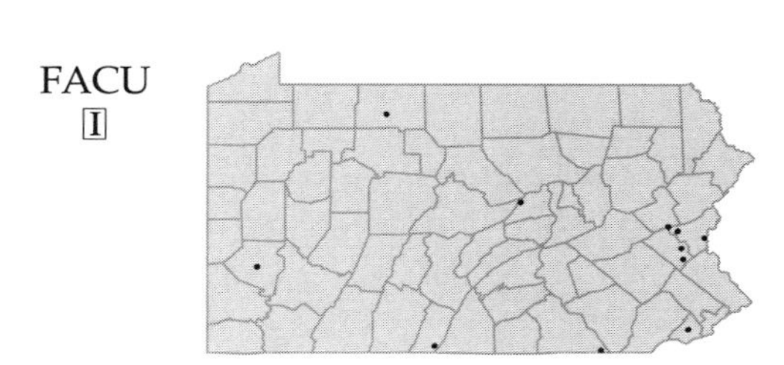

• ***Arthraxon hispidus*** (Thunb.) Makino
Grass
Herbaceous annual
Moist meadows and waste ground.

FACU+
I

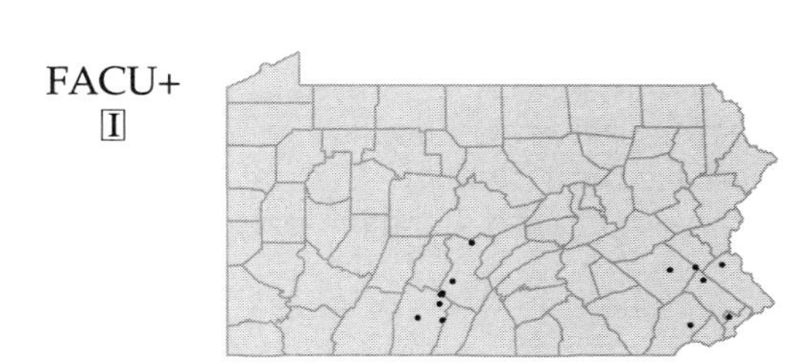

• ***Avena fatua*** L.
Wild oats
Herbaceous annual
Waste ground.

I

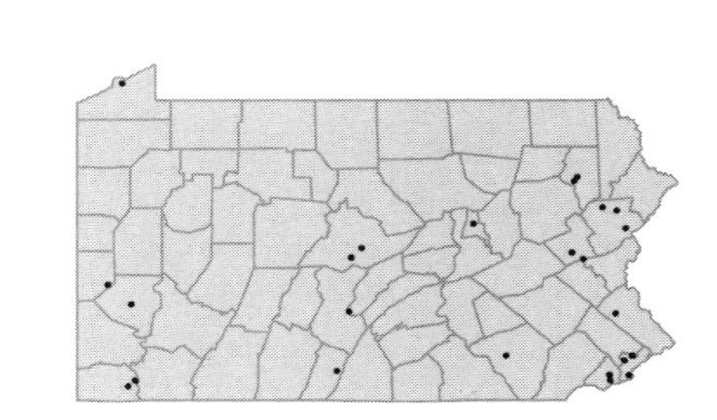

Avena sativa L.
Oats
Herbaceous annual
Cultivated and frequently escaped to fallow fields, roadsides and waste ground.

I

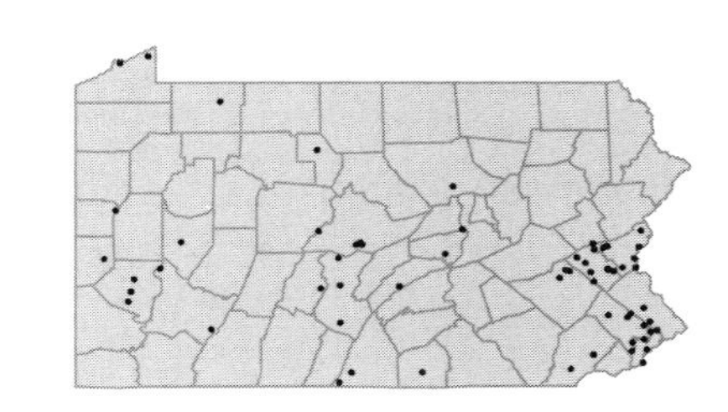

• ***Beckmannia syzigachne*** (Steud.) Fern.
American slough-grass
Herbaceous annual
Alluvial meadows. Represented by a single collection from Philadelphia Co. in 1921.

OBL
I

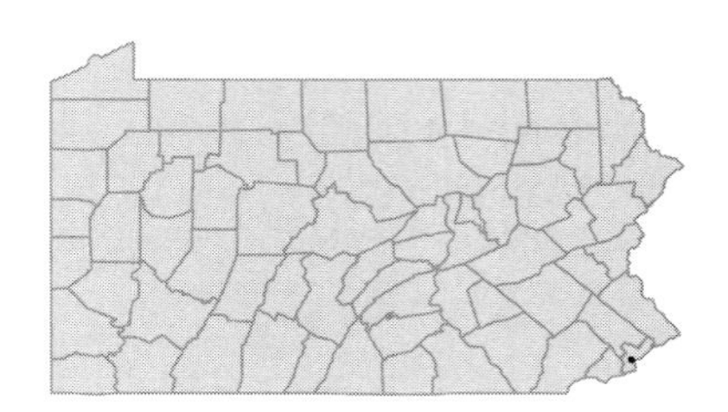

• ***Bouteloua curtipendula*** (Michx.) Torr.
Side-oats grama; Tall grama
Herbaceous perennial
Serpentine barrens, dry calcareous clearings and other dry, rocky or sandy sites.

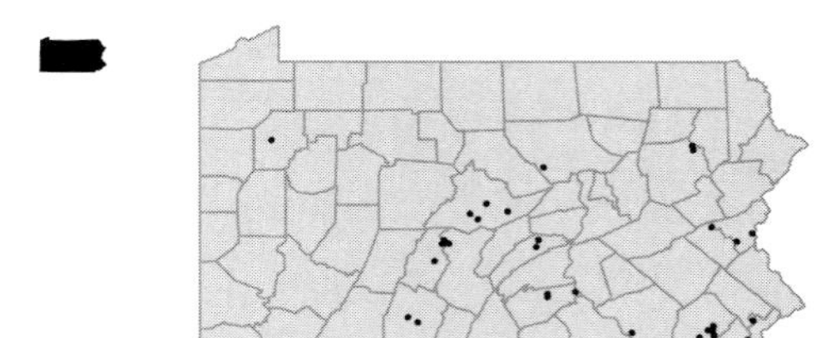

• ***Brachyelytrum erectum*** (Schreb.) Beauv.
Brachyelytrum
Herbaceous perennial
Moist, wooded hillsides, alluvial woods and moist thickets.

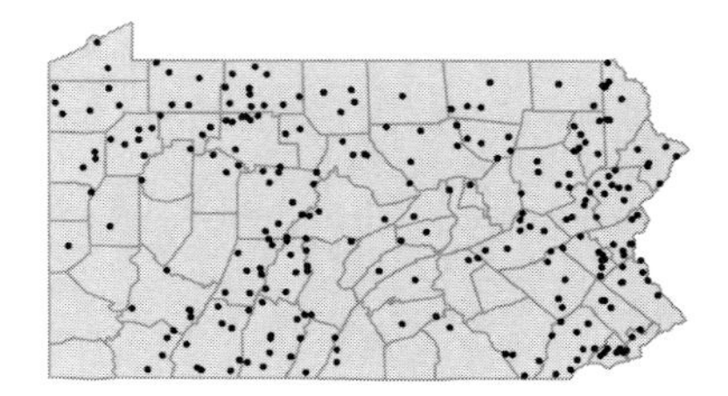

• ***Briza media*** L.
Quaking grass
Herbaceous perennial
Fallow fields and wet meadows.

FAC
I

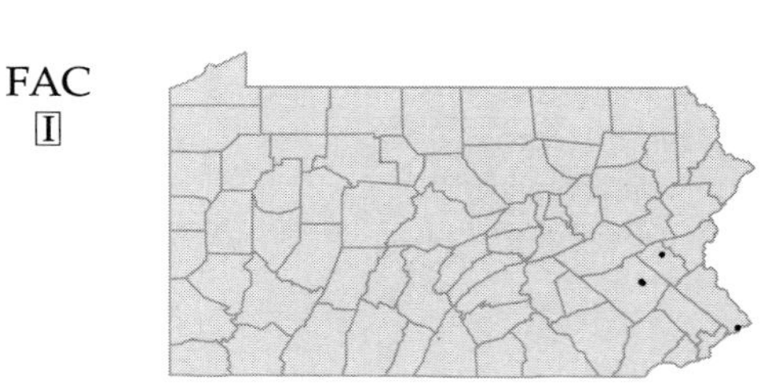

Briza minor L.
Quaking grass
Herbaceous annual
Ballast. Represented by two collections from Philadelphia Co. 1879-1889.

FACW
I

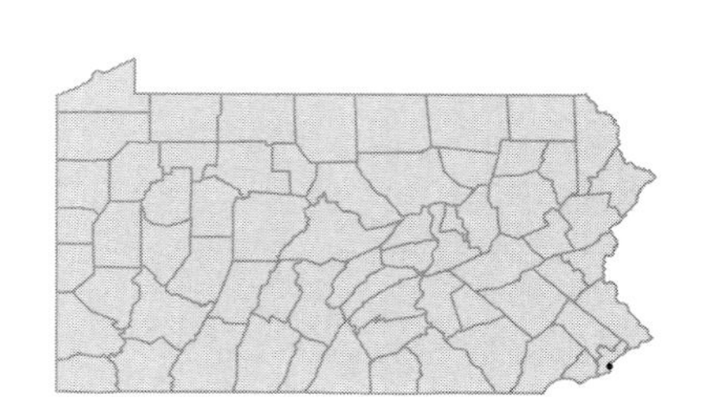

• ***Bromus altissimus*** Pursh
Brome grass
Herbaceous perennial
Rich, alluvial thickets or woods.
Bromus latiglumis (Shear) A.S.Hitchc. FGB

FACW

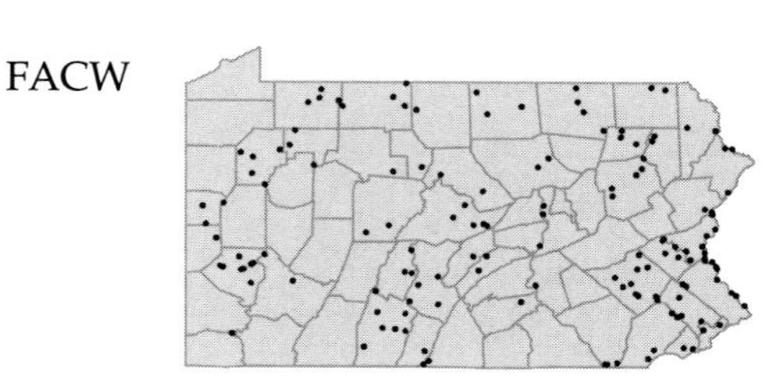

Bromus arenarius Labill.
Australian chess
Herbaceous annual
Ballast. Represented by a single collection from Philadelphia Co. in 1879.

I

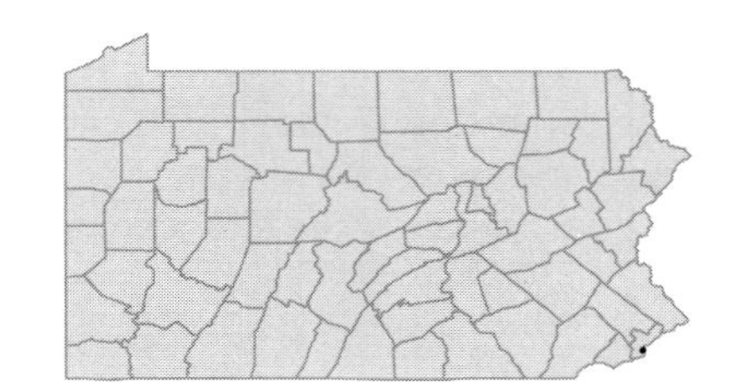

Bromus arvensis L.
Field chess
Herbaceous annual
Disturbed soil.

I

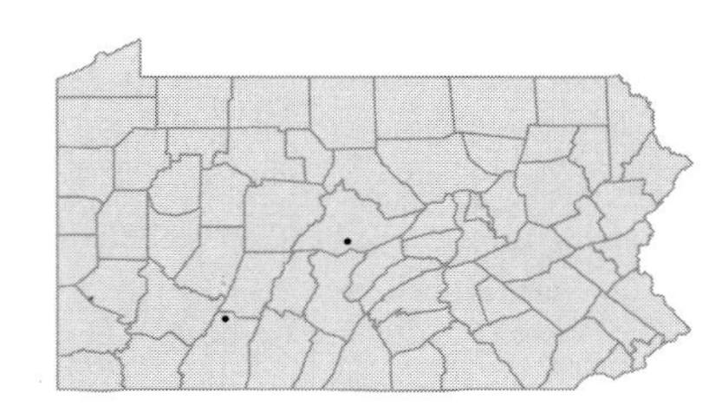

Bromus briziformis Fisch. & Mey.
Quake grass; Rattlesnake chess
Herbaceous annual
Dry, disturbed soil. Represented by a single collection from Berks Co. in 1933.

I

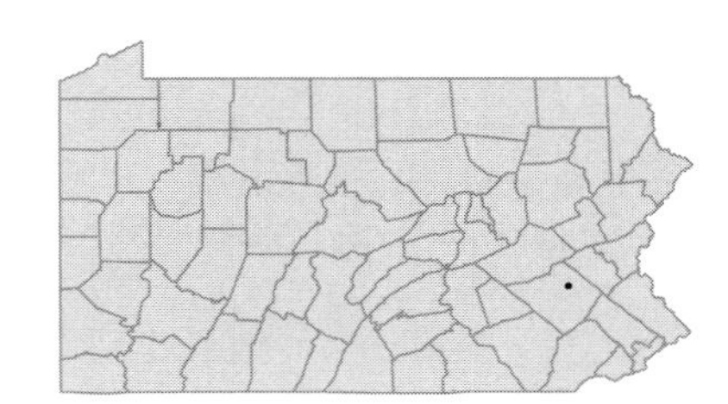

Bromus ciliatus L.
Fringed brome
Herbaceous perennial
Woods, clearings and meadows.

FACW
I

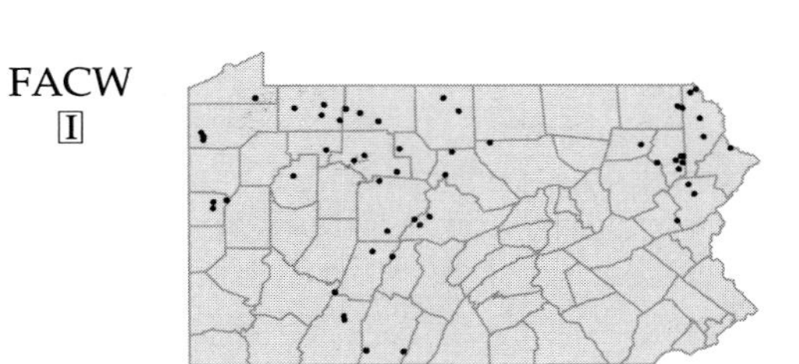

Bromus commutatus Schrad.
Hairy chess
Herbaceous annual
Fields, roadsides and waste ground.

I

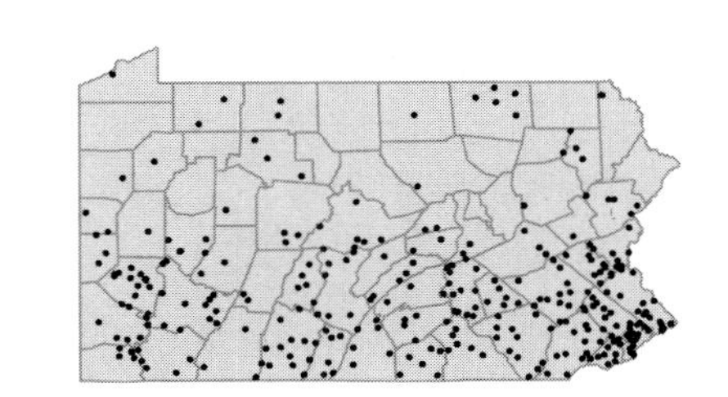

Bromus hordeaceus L.
Soft chess
Herbaceous annual
Fields and waste ground.
Bromus mollis L. FGBW

I

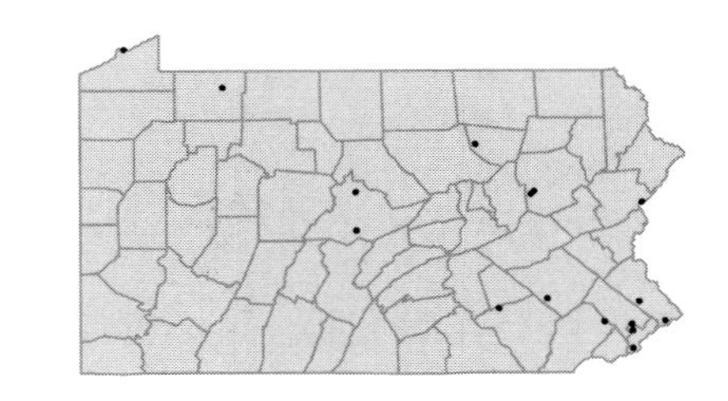

Bromus inermis Leyss.
Smooth brome
Herbaceous perennial
Fields, roadsides and waste ground.

I

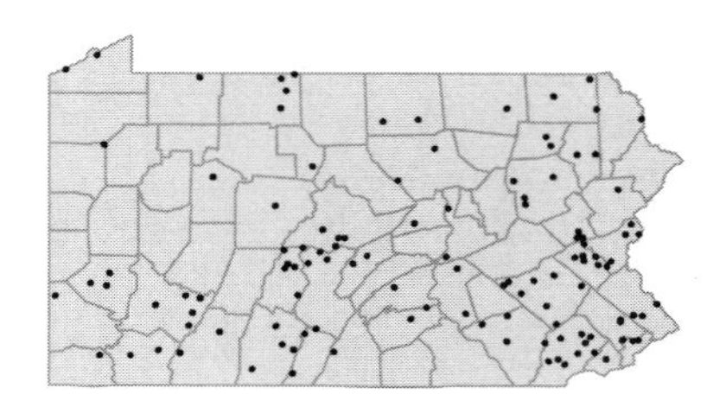

Bromus japonicus Thunb. ex Murr.
Japanese chess
Herbaceous annual
Roadsides and waste ground.

FACU-
I

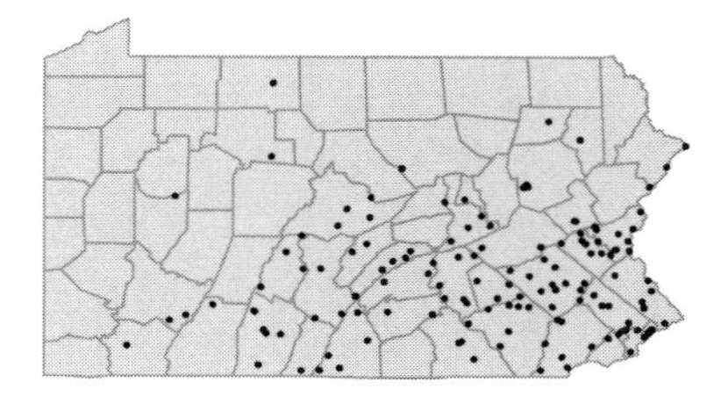

Bromus kalmii A.Gray
Brome grass
Herbaceous perennial
Dry or moist woods or rocky banks.

FAC-

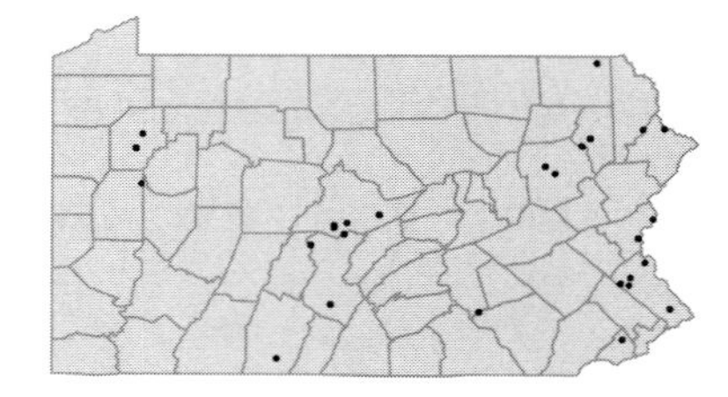

Bromus pubescens Muhl. ex Willd.
Canada brome
Herbaceous perennial
Dry to moist woods and thickets.
Bromus purgans sensu auctt., non L. PFGBW

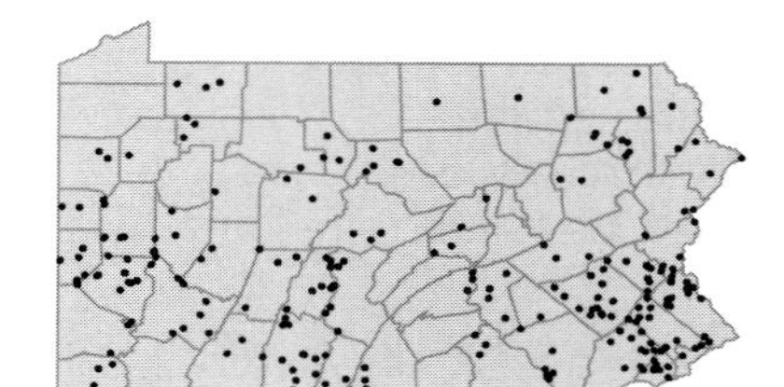

Bromus racemosus L.
Soft chess
Herbaceous annual
Waste ground and ballast.

I

Bromus secalinus L.
Cheat; Chess
Herbaceous annual
Grain fields and waste ground.

I

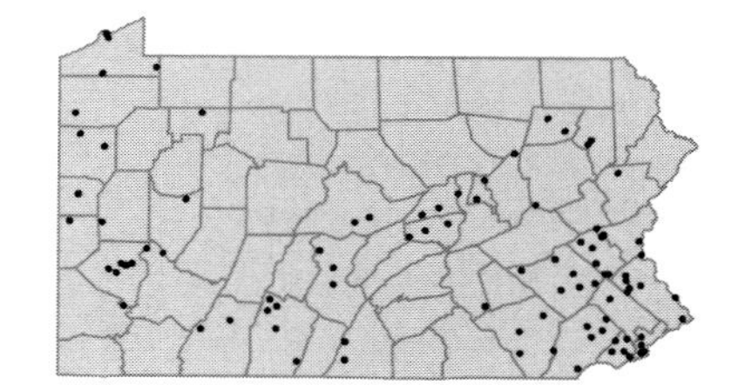

Bromus squarrosus L. I
Chess
Herbaceous annual
Railroad ballast. Represented by a single collection from Philadelphia Co. ca. 1898.

Bromus sterilis L. I
Barren brome
Herbaceous annual
Roadsides and waste ground.

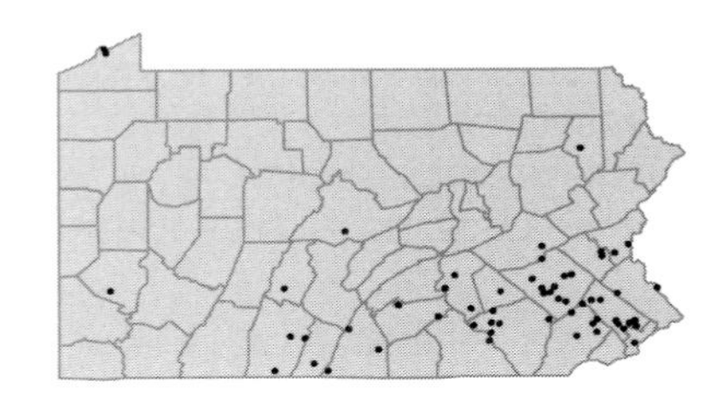

Bromus tectorum L. I
Downy chess
Herbaceous annual
Dry, fallow fields and waste ground.

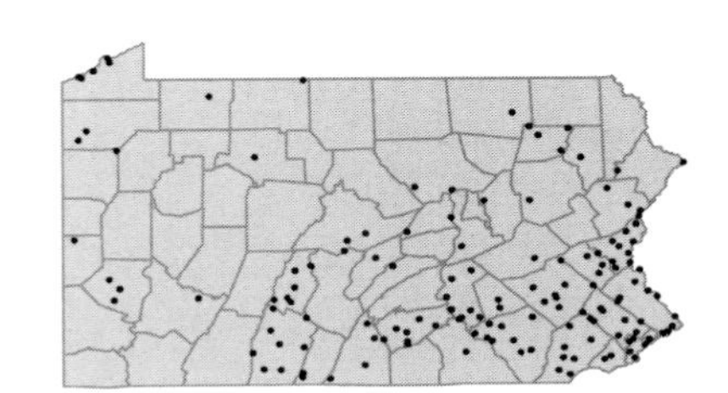

Bromus unioloides (Willd.) Kunth I
Rescue grass
Herbaceous annual
Agricultural fields. Represented by a single collection from Centre Co. in 1953.
Bromus catharticus Vahl FGBW

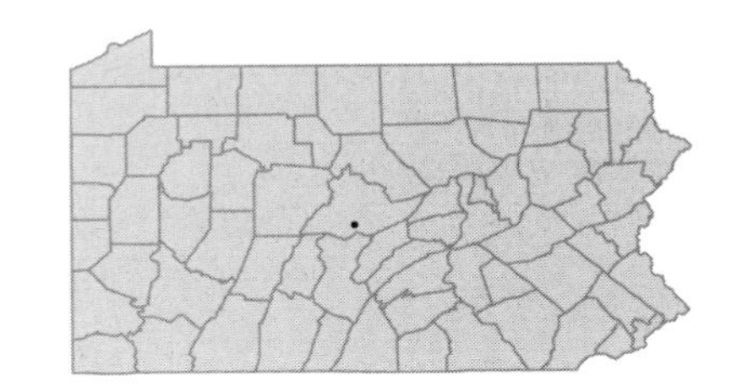

• ***Calamagrostis canadensis*** (Michx.) Beauv. var. ***canadensis*** FACW+
Canada bluejoint
Herbaceous perennial
Wet meadows, bogs and swamps.

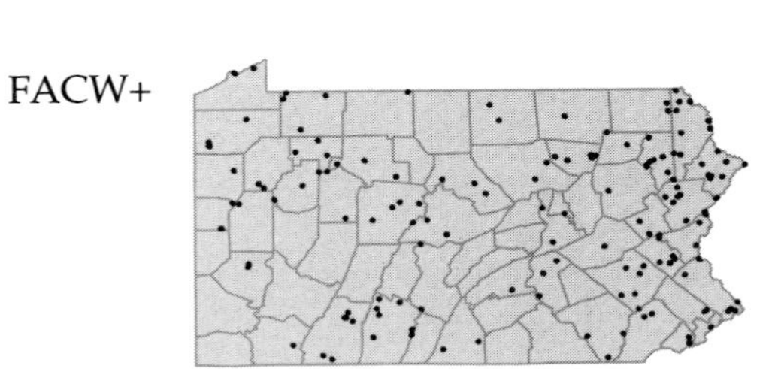

Calamagrostis canadensis (Michx.) Beauv. var. ***macouniana*** (Vasey) Stebbins FACW+
Canada bluejoint
Herbaceous perennial
Swamps and wet meadows.

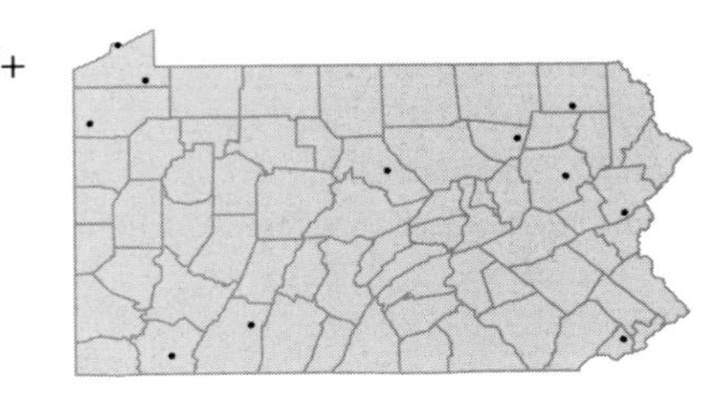

Calamagrostis cinnoides (Muhl.) Bart.
Reedgrass
Herbaceous perennial
Swamps or wet woods, in sandy or peaty soils.

OBL

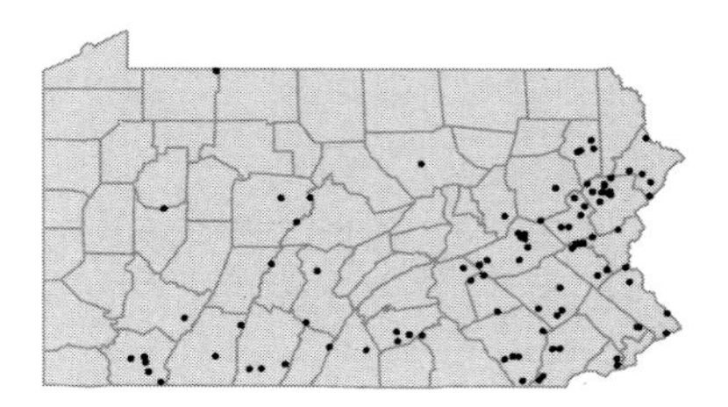

Calamagrostis epigejos (L.) Roth
Feather-top
Herbaceous perennial
Disturbed woods, roadsides and waste ground.

FAC
[I]

Calamagrostis porteri A.Gray
Porter's reedgrass
Herbaceous perennial
Roadsides and open woods.

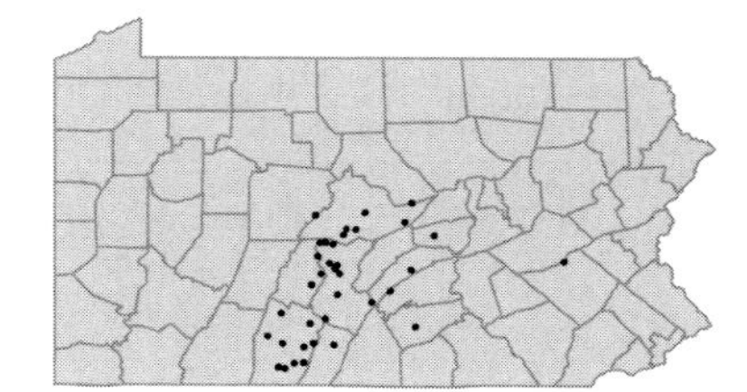

• ***Calamovilfa longifolia*** (Hook.) Scribn.
Sand-reed
Herbaceous perennial
Planted and naturalized on beaches.

[I]

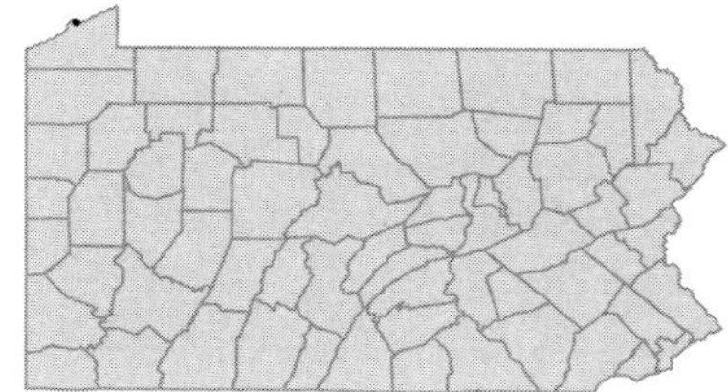

• ***Cenchrus longispinus*** (Hack.) Fern.
Sandbur
Herbaceous annual
Dry, sandy soil.
Cenchrus pauciflorus Benth. FBW

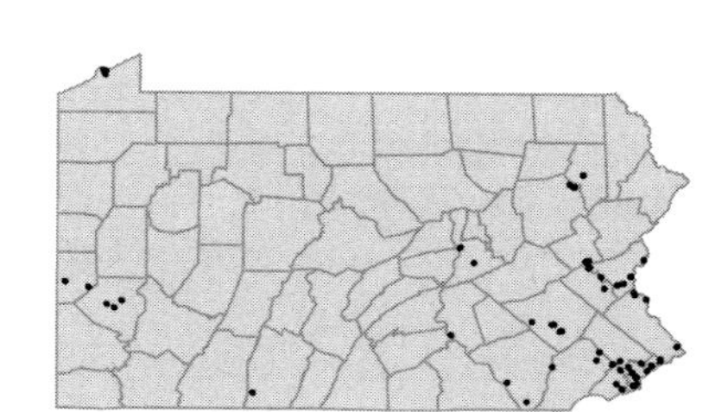

Cenchrus tribuloides L.
Dune sandbur
Herbaceous annual
Ballast ground, abandoned limestone quarry and other dry, open, sandy sites.

UPL
[I]

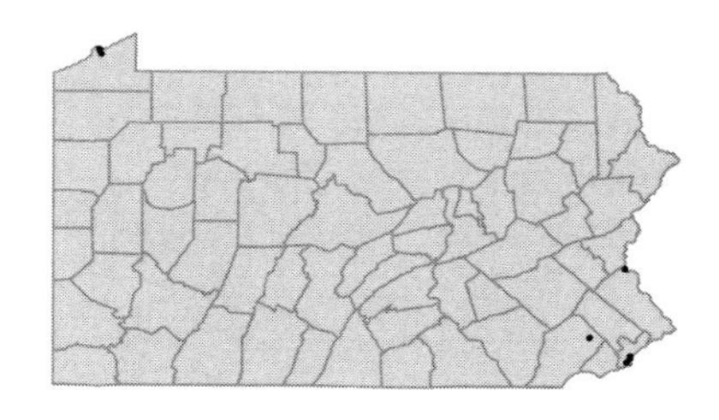

• ***Chasmanthium latifolium*** (Michx.) Yates
Sea-oats
Herbaceous perennial
River banks and alluvial woods.
Uniola latifolia Michx. PFGBW

FACU

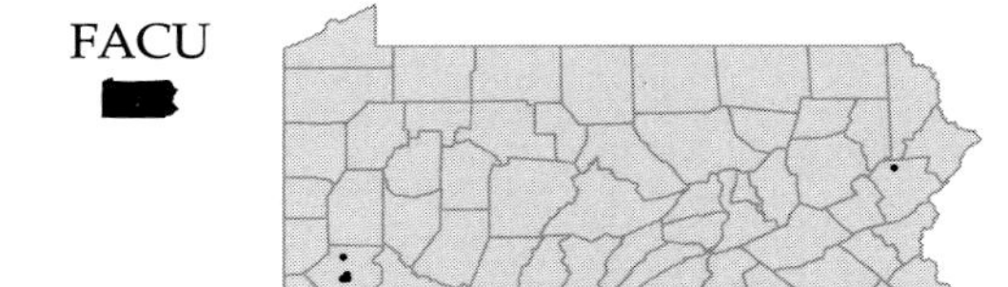

Chasmanthium laxum (L.) Yates
Slender sea-oats
Herbaceous perennial
Moist, sandy soils of the Coastal Plain.
Uniola laxa (L.) BSP PFGBW

FAC

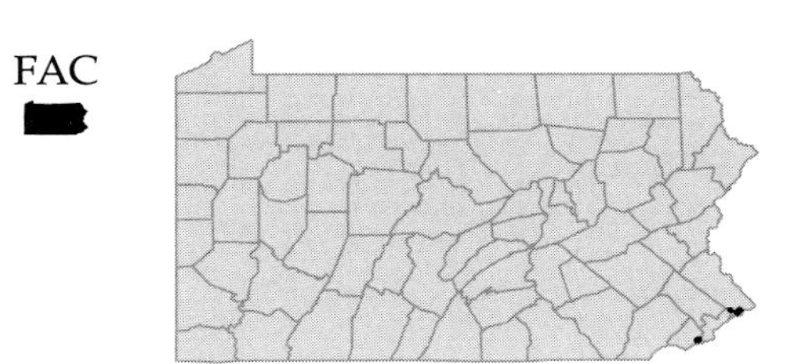

• ***Chloris verticillata*** Nutt.
Windmill grass
Herbaceous perennial
Fallow fields or lawns on sandy or droughty soils.

I

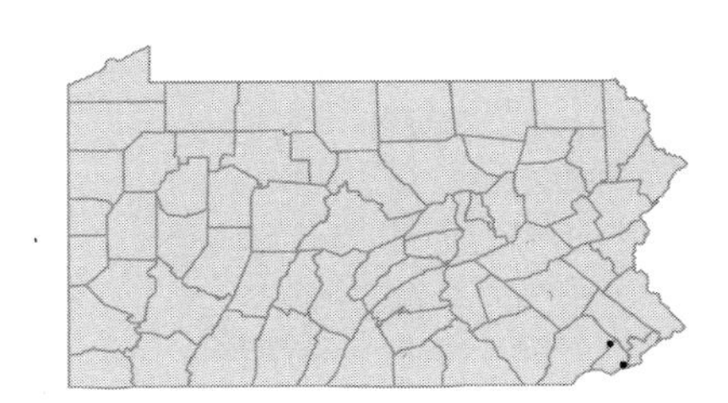

Chloris virgata Swartz
Feather fingergrass
Herbaceous annual
Waste ground, fill and rubbish heaps. Represented by a single collection from Philadelphia Co. in 1942.

I

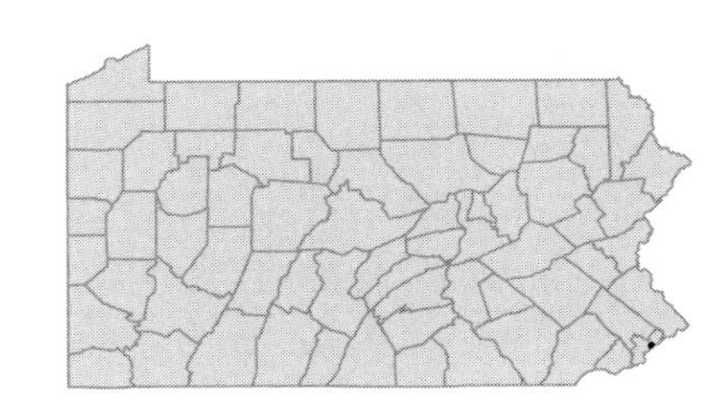

• ***Cinna arundinacea*** L.
Wood reedgrass
Herbaceous perennial
Swamps and wet woods.

FACW+

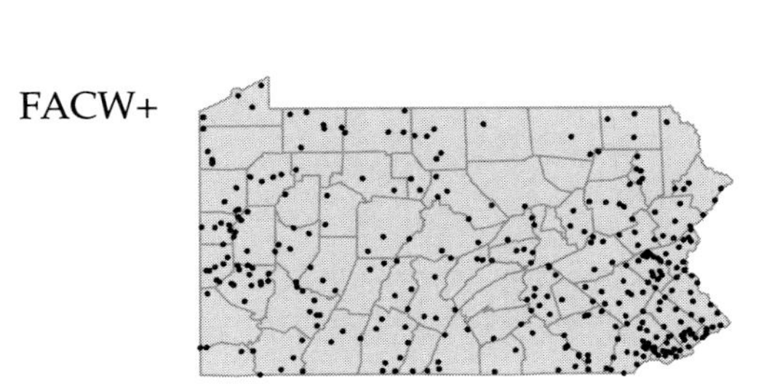

Cinna latifolia (Trev.) Griseb.
Drooping woodreed; Wood reedgrass
Herbaceous perennial
Moist shores, damp woods and bogs.

FACW

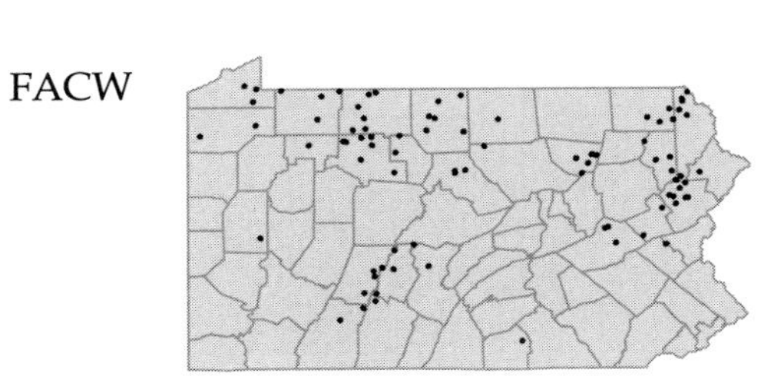

• ***Coix lacryma-jobi*** L.
Job's tears
Herbaceous annual
Cultivated and occasionally escaped in waste ground.

FACW
[I]

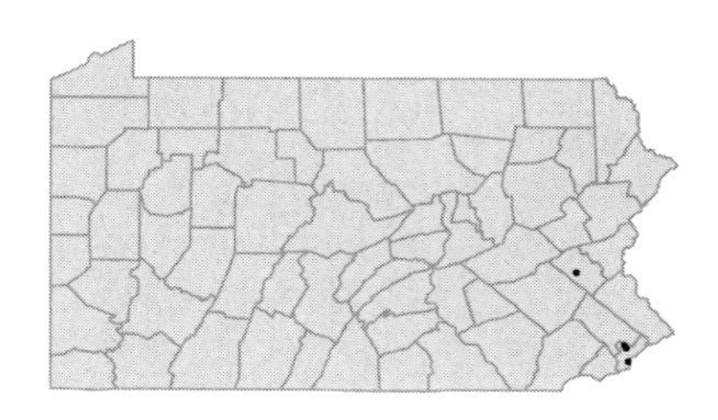

• ***Corynephorus canescens*** (L.) Beauv.
Gray hairgrass; Silver grass
Herbaceous perennial
Ballast. Represented by several collections from Philadelphia Co. 1878-1879.

[I]

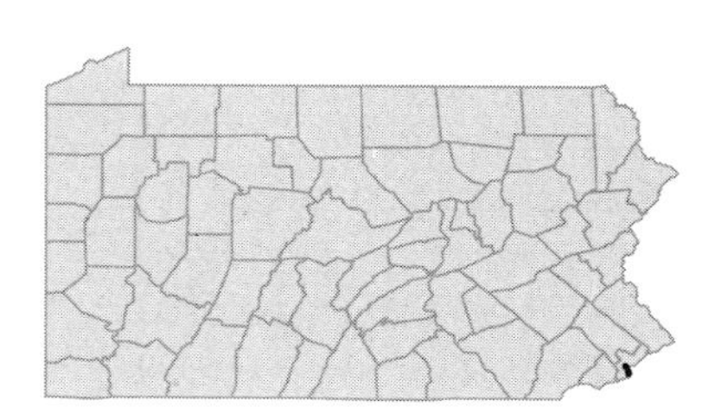

• ***Critesion brachyantherum*** (Nevski) Barkw. & D.R.Dewey
Meadow barley
Herbaceous perennial
Roadside ditch. Represented by a single collection from Bucks Co. in 1896.
Hordeum brachyantherum Nevski F; *Hordeum nodosum* L. PG

FAC+
[I]

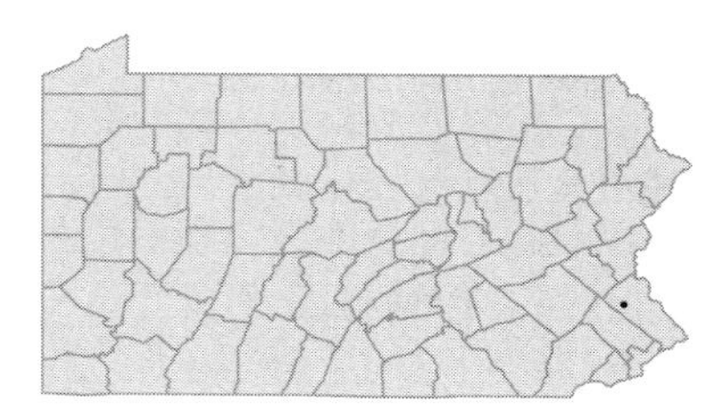

Critesion jubatum (L.) Nevski
Foxtail barley
Herbaceous perennial
Dry old fields, roadsides and waste ground.
Hordeum jubatum L. PFGBWKC

FAC

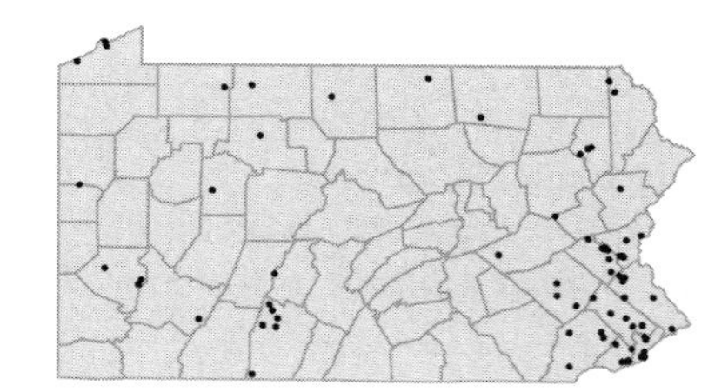

Critesion marinum (Huds.) A.Loeve ssp. ***gussoneanum*** (Parl.) Barkw. & D.R.Dewey
Squirrel-tail
Herbaceous perennial
Ballast and rubbish dumps.
Hordeum hystrix Nutt. GW; *Hordeum geniculatum* All. C

[I]

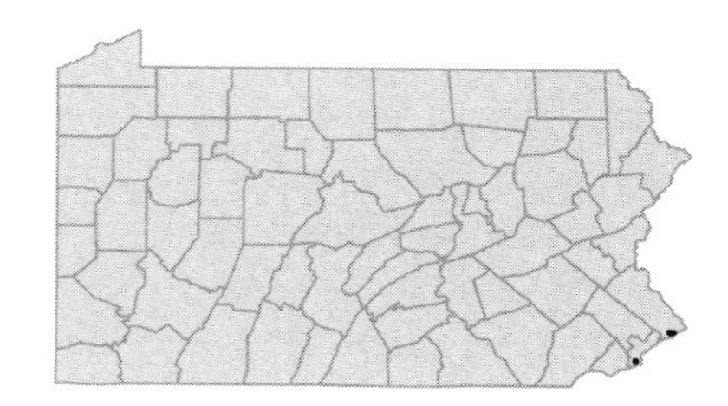

Critesion murinum (L.) A.Loeve ssp. ***murinum***
Wild barley
Herbaceous annual
Ballast. Represented by a single collection from Lehigh Co. in 1985.
Hordeum murinum L. ssp. *murinum* K

[I]

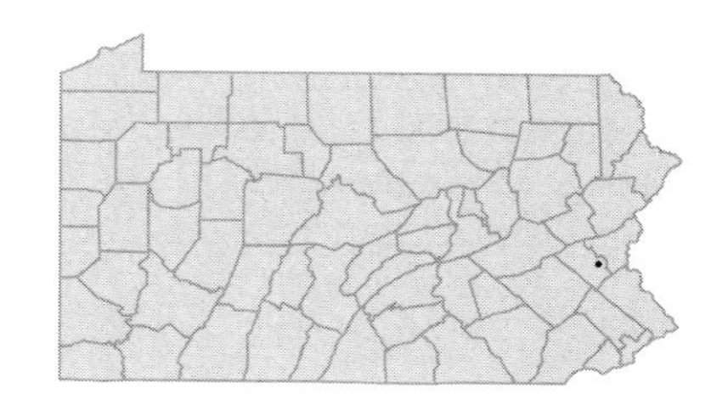

Critesion murinum (L.) A.Loeve ssp. **leporinum** (Link) A.Loeve
Wild barley
Herbaceous annual
Waste ground and ballast.
Hordeum leporinum Link GWC; *Hordeum murinum* L. ssp. *leporinum* (Link) Arcang. K

I

Critesion pusillum (Nutt.) A.Loeve
Little barley
Herbaceous annual
Roadside banks. Believed to be extirpated, last collected in 1952.
Hordeum pusillum Nutt. FGBWKC

FAC

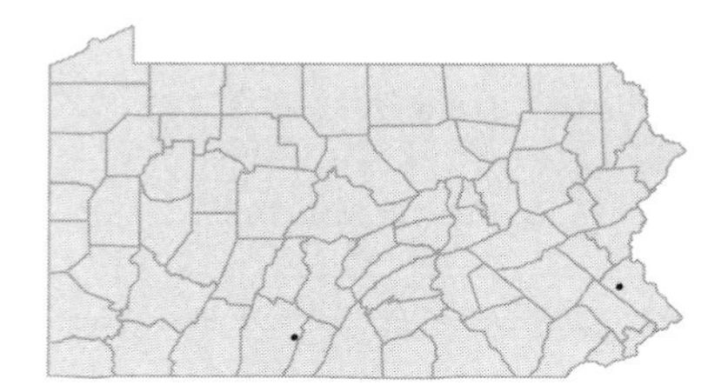

• **Crypsis alopecuroides** (Piller & Mitterp.) Schrad.
Grass
Herbaceous annual
Ballast. Represented by several collections from Philadelphia Co. in 1879.
Heleochloa alopecuroides (Piller & Mitterp.) Host BW

I

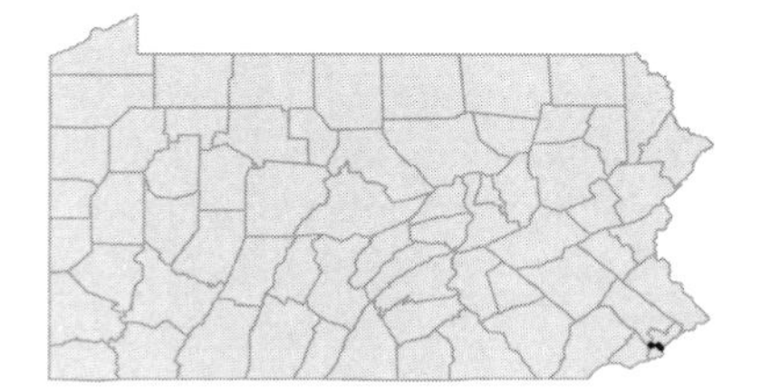

Crypsis schoenoides (L.) Lam.
Grass
Herbaceous annual
Waste ground and ballast.
Heleochloa schoenoides (L.) Host PFGBW

I

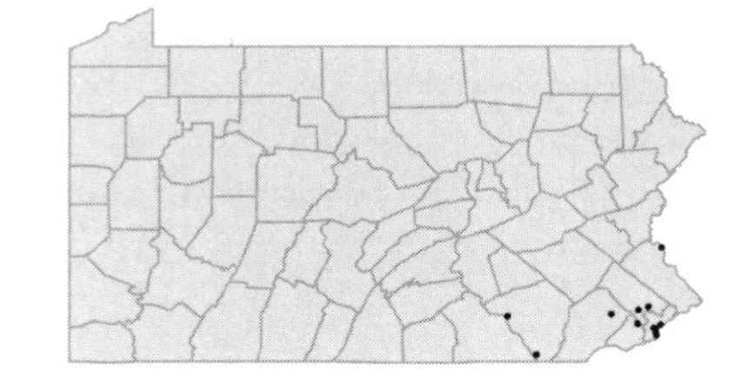

• **Cynodon dactylon** (L.) Pers.
Bermuda grass; Wire grass
Herbaceous perennial
Cultivated and occasionally escaped.

FACU
I

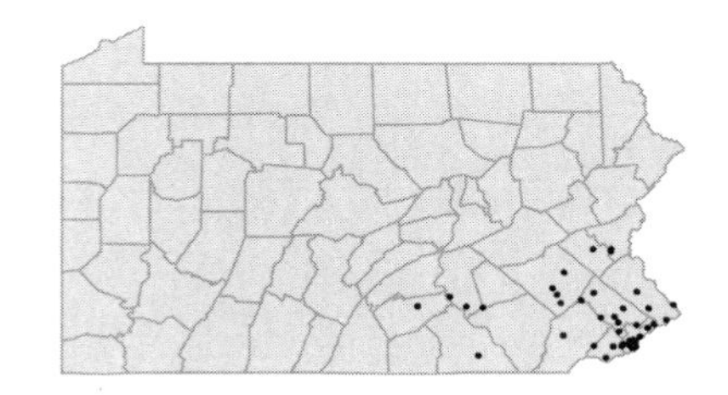

• **Cynosurus cristatus** L.
Crested dog's-tail
Herbaceous perennial
Lawns, roadsides and waste ground.

UPL
I

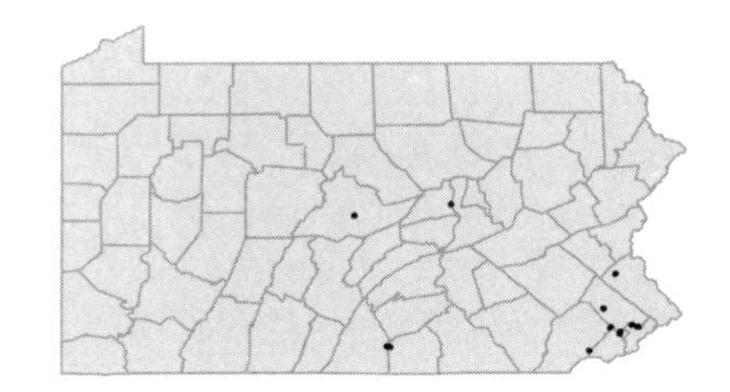

Cynosurus echinatus L.
Spiny dog's-tail
Herbaceous annual
Roadside banks, field margins and disturbed soil.

I

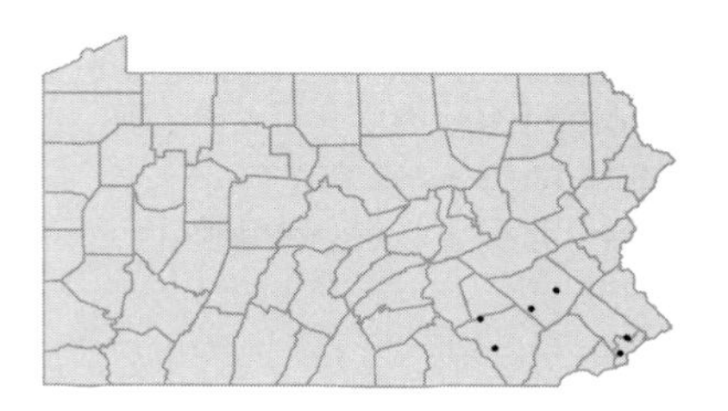

• *Dactylis glomerata* L.
Orchard grass
Herbaceous perennial
Fields, meadows and roadsides.

FACU
I

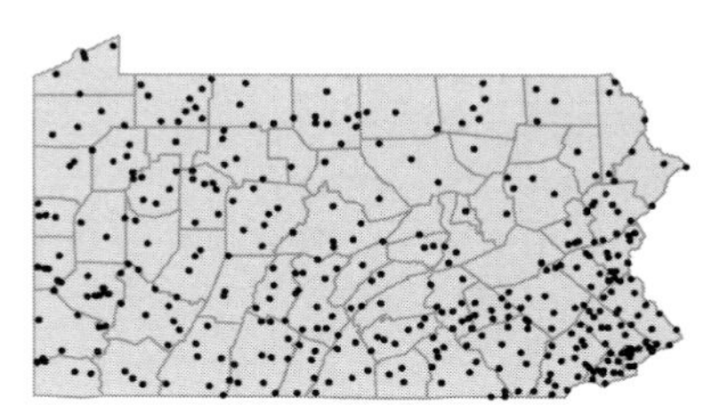

• *Dactyloctenium aegyptium* (L.) Willd.
Crowfoot grass
Herbaceous annual
Waste ground and ballast.

I

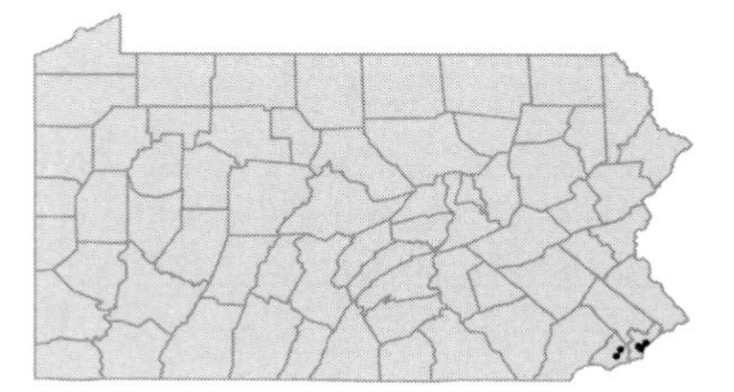

• *Danthonia compressa* Austin
Northern oatgrass
Herbaceous perennial
Dry, rocky woods and clearings.

FACU-

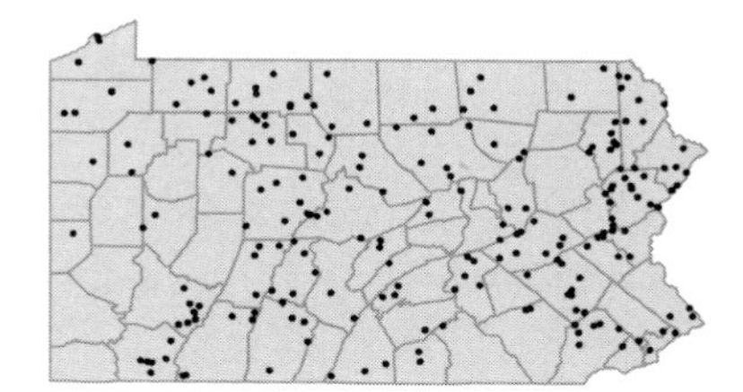

Danthonia spicata (L.) Beauv. ex Roemer & Schultes
Poverty grass
Herbaceous perennial
Dry, sandy or gravelly soil.

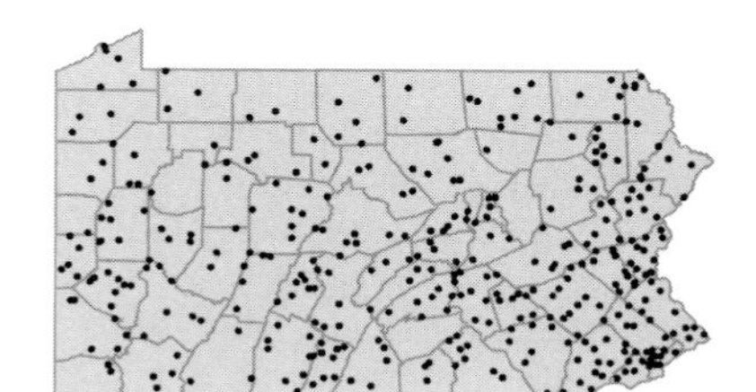

• *Dasypyrum villosum* (L.) Candargy
Grass
Herbaceous annual
Dry grassland. Represented by a single collection from Philadelphia Co. in 1877.
Haynaldia villosa (L.) Schur W

I

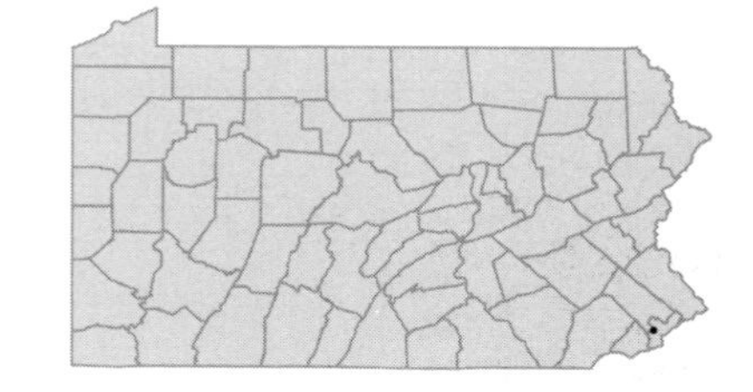

• ***Deschampsia cespitosa*** (L.) Beauv. FACW
Tufted hairgrass
Herbaceous perennial
Serpentine barrens, sandy shores and thickets.

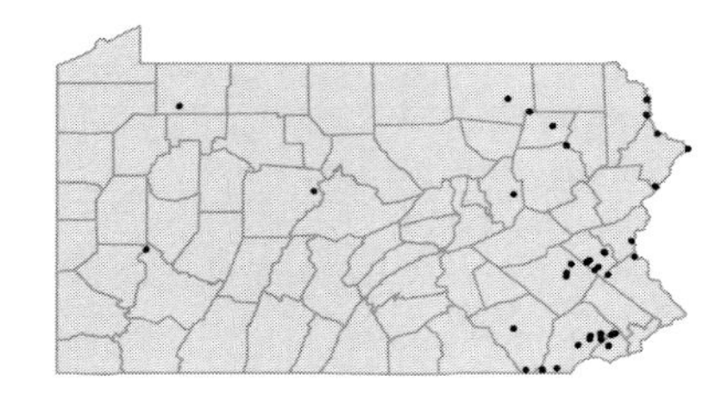

Deschampsia flexuosa (L.) Trin.
Common hairgrass
Herbaceous perennial
Dry woods or rocky slopes.

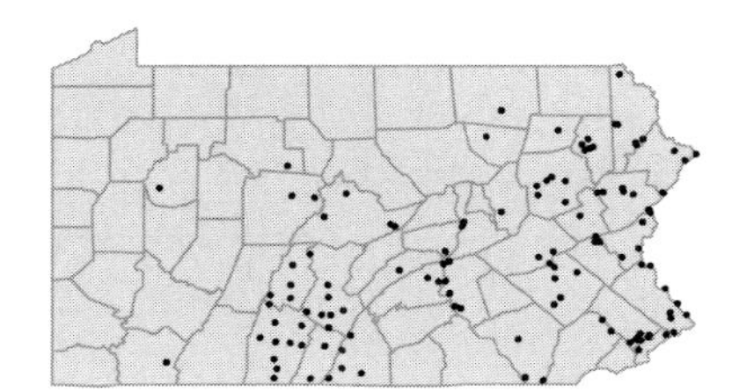

• ***Diarrhena obovata*** (Gleason) Brandenburg
American beakgrain
Herbaceous perennial
Rich woods.
Diarrhena americana Beauv. var. *obovata* Gleason C

• ***Digitaria ciliaris*** (Retz.) Koel. [I]
Southern crabgrass
Herbaceous annual
Dry, sandy, disturbed soil. Represented by several collections from a single site in Berks Co. 1931-1936.
Digitaria sanguinalis (L.) Scop. var. *ciliaris* (Retz.) Parl. FW

Digitaria filiformis (L.) Koel.
Slender crabgrass
Herbaceous annual
Serpentine barrens and other dry, open sites.
Syntherisma filiforme (L.) Nash P

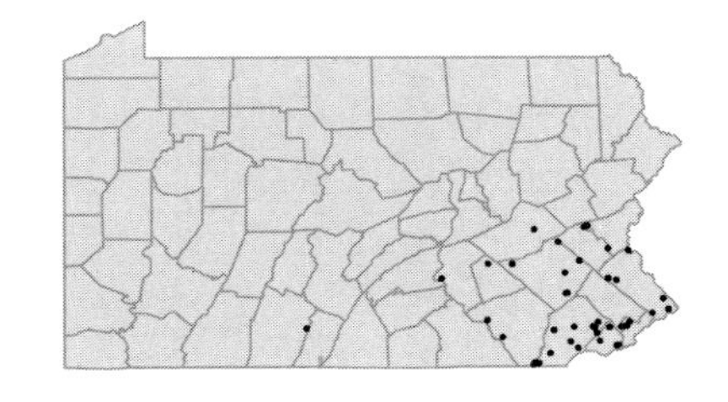

Digitaria ischaemum (Schreb. ex Schweig) Schreb. ex Muhl. UPL [I]
Smooth crabgrass
Herbaceous annual
Disturbed soil, roadsides and waste ground.
Syntherisma humifusa (Pers.) Rydb. P

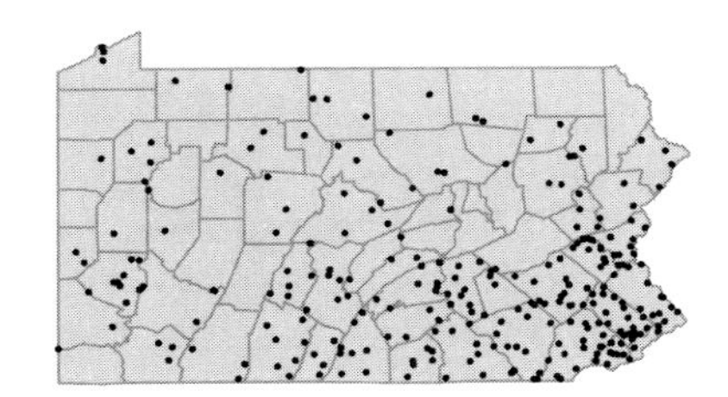

Digitaria sanguinalis (L.) Scop.
Northern crabgrass
Herbaceous annual
Disturbed soil, roadsides, lawns and waste ground.
Syntherisma sanguinale (L.) Dulac P

FACU
I

Digitaria serotina (Walt.) Michx.
Dwarf crabgrass
Herbaceous annual
Ballast. Represented by several collections from Philadelphia Co. in 1865.

FAC

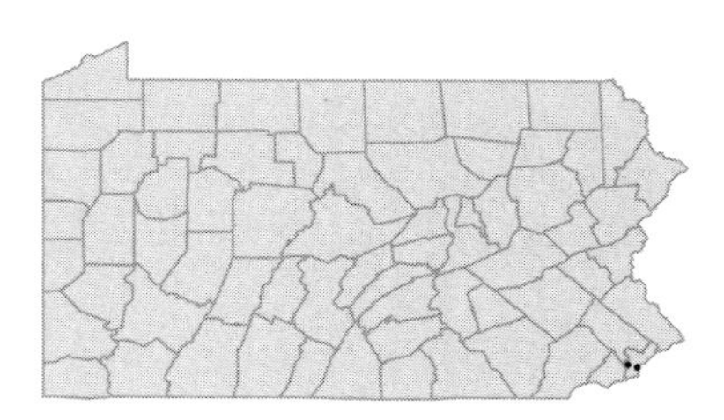

• ***Distichlis spicata*** (L.) Greene
Seashore salt grass
Herbaceous perennial
Ballast and waste ground. Believed to be extirpated, last collected in 1943.

FACW+

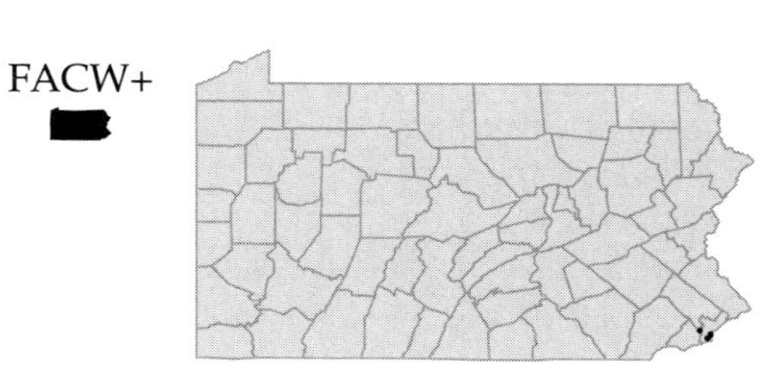

• ***Echinochloa colonum*** (L.) Link
Jungle-rice
Herbaceous annual
Dry, waste ground. Represented by a single collection from Berks Co. in 1956.

FACW
I

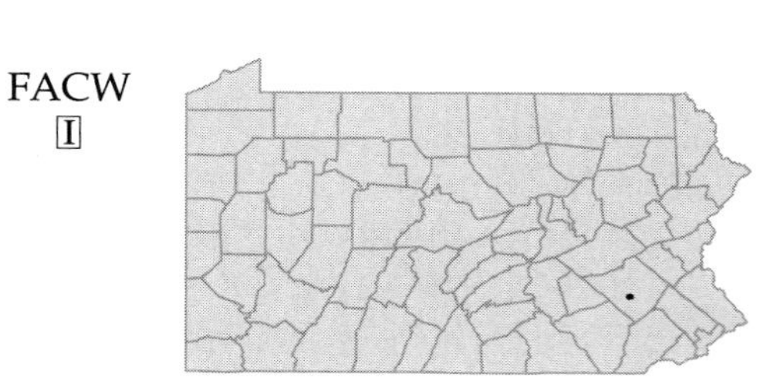

Echinochloa crusgalli (L.) Beauv. var. ***crusgalli***
Barnyard grass
Herbaceous annual
Fields, meadows, roadsides and waste ground.

FACU
I

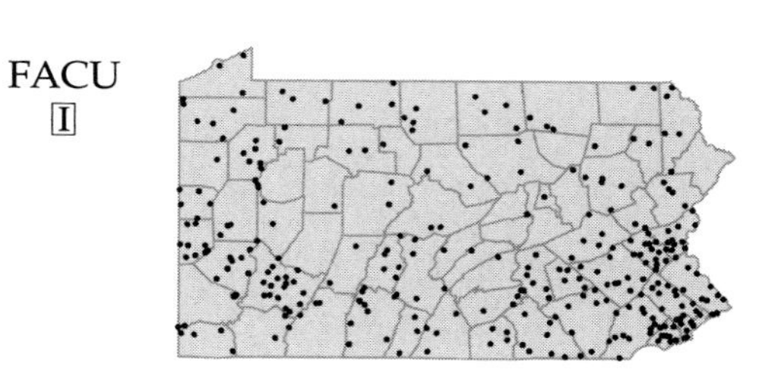

Echinochloa crusgalli (L.) Beauv. var. ***frumentacea*** (Roxb.) W.Wight
Billion-dollar grass; Japanese millet
Herbaceous annual
Cultivated and occasionally escaped.
Echinochloa frumentacea (Roxb.) Link FB

FACU
I

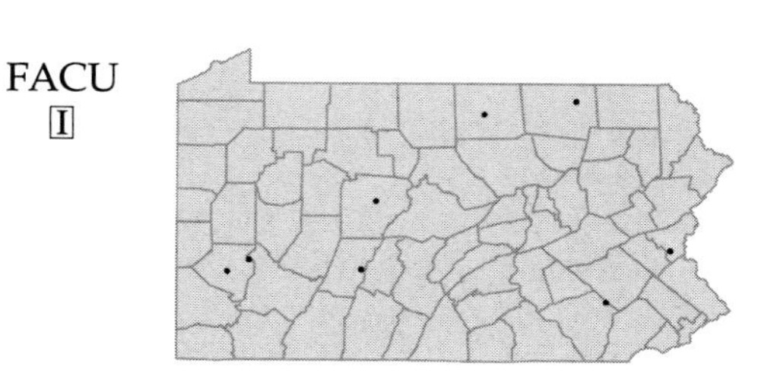

Echinochloa muricata (Beauv.) Fern.
Barnyard grass; Cockspur
Herbaceous annual
Moist ground and alluvial shores.
Echinochloa pungens (Poir.) Rydb. FBW

FACW+

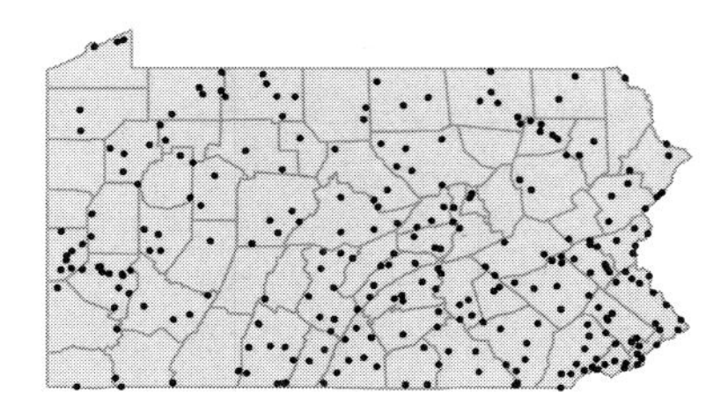

Echinochloa walteri (Pursh) Heller
Walter's barnyard grass
Herbaceous annual
Tidal marshes and mudflats.

FACW+

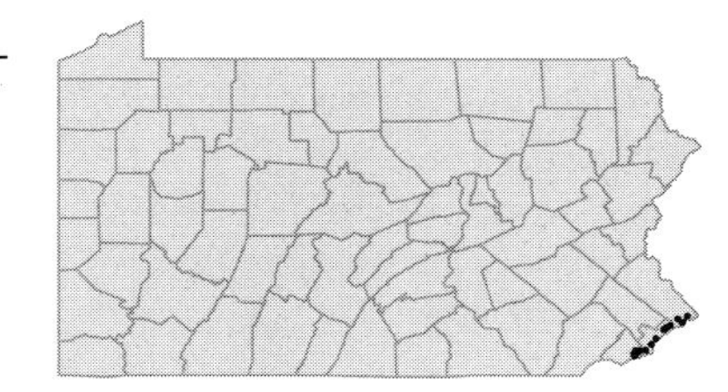

• ***Eleusine indica*** (L.) Gaertn.
Goosegrass; Wiregrass
Herbaceous annual
Gardens, disturbed soil and waste ground.

FACU-
I

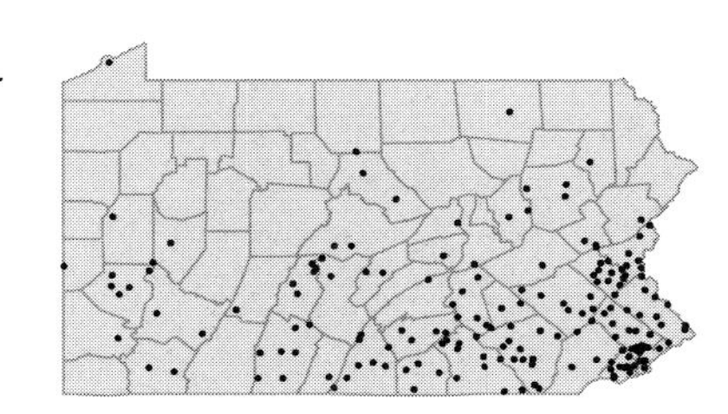

• ***Elymus canadensis*** L. var. ***canadensis***
Canada wild-rye
Herbaceous perennial
Alluvial shores and thickets.

FACU+

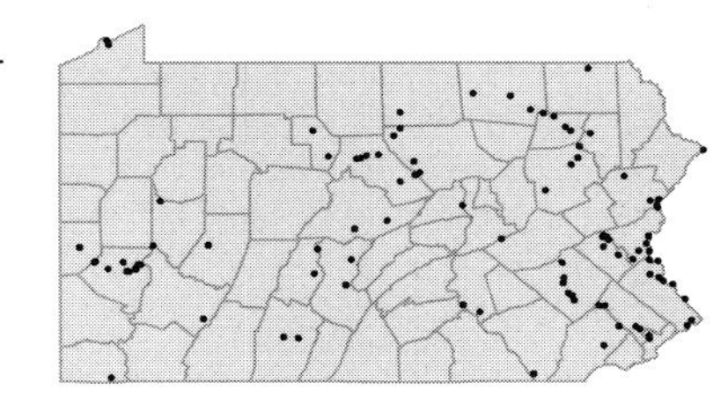

Elymus canadensis L. var. ***wiegandii*** (Fern.) Bowden
Canada wild-rye
Herbaceous perennial
Alluvial thickets.
Elymus wiegandii Fern. FGBWC

FAC

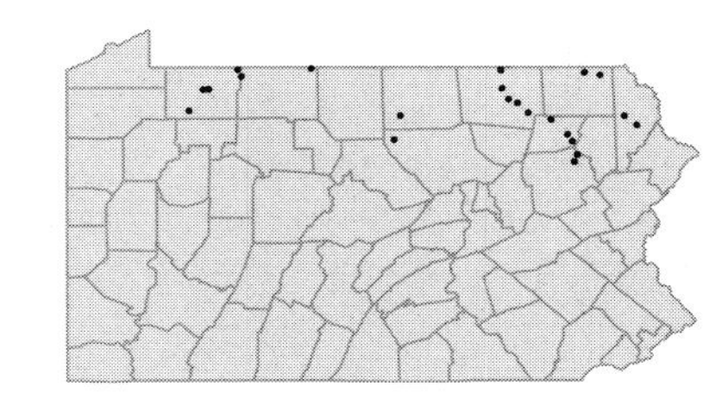

Elymus hystrix L.
Bottlebrush grass
Herbaceous perennial
Moist, alluvial woods.
Hystrix patula Moench FGBWK

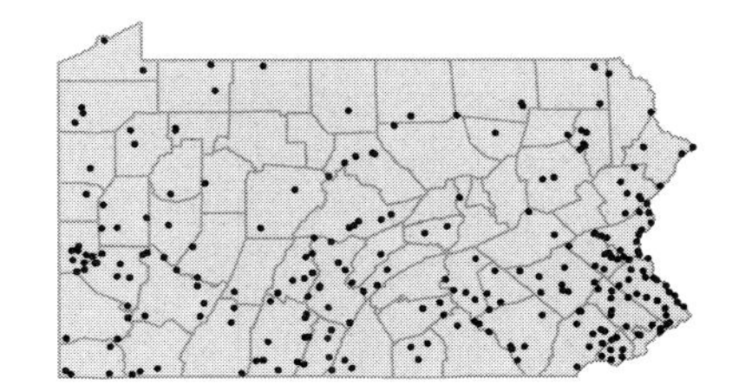

Elymus riparius Wieg.
Riverbank wild-rye
Herbaceous perennial
Alluvial flats, meadows and stream banks.

FACW

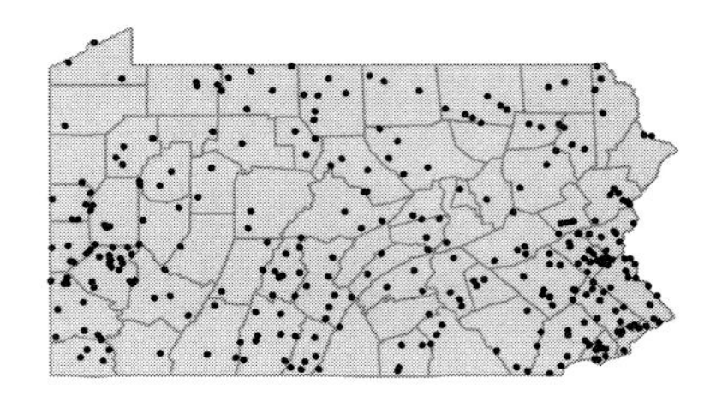

Elymus trachycaulus (Link) Gould ex Shinners
Slender wheatgrass
Herbaceous perennial
Open woods, barrens and banks.
Agropyron novae-angliae Scribn. P; *Agropyron trachycaulum* (Link) Malte FWGB

FACU

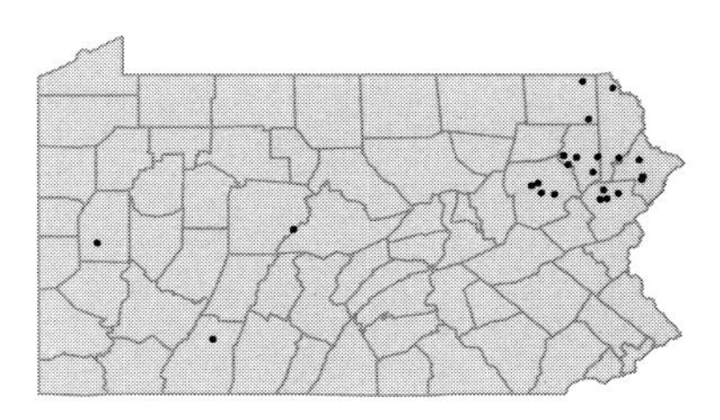

Elymus villosus Muhl. ex Willd.
Wild-rye
Herbaceous perennial
Stream banks, moist woods and marshes.
Elymus striatus Willd. P

FACU-

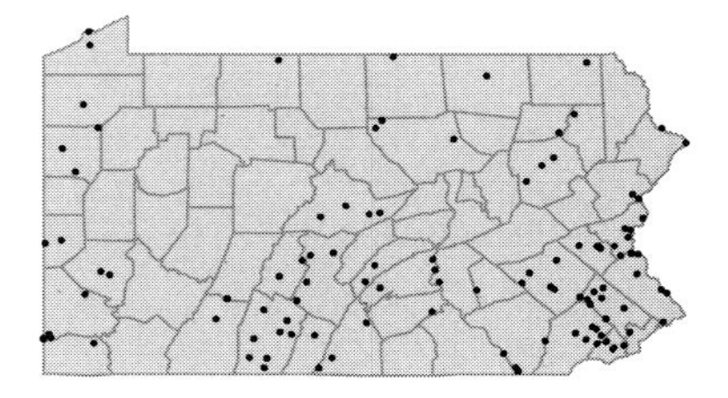

Elymus virginicus L.
Virginia wild-rye
Herbaceous perennial
Moist woods, meadows and river banks.

FACW-

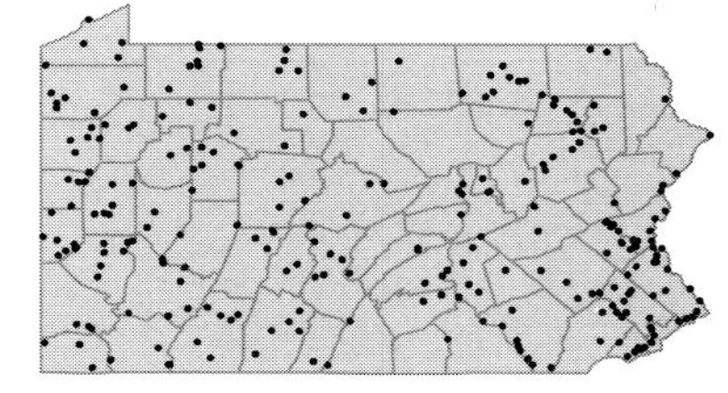

• **Elytrigia pungens** (Pers.) Tutin
Saltmarsh wheatgrass
Herbaceous perennial
Ballast, ditches and rubbish heaps.
Agropyron pungens (Pers.) Roemer & Schultes FGBW

FACW

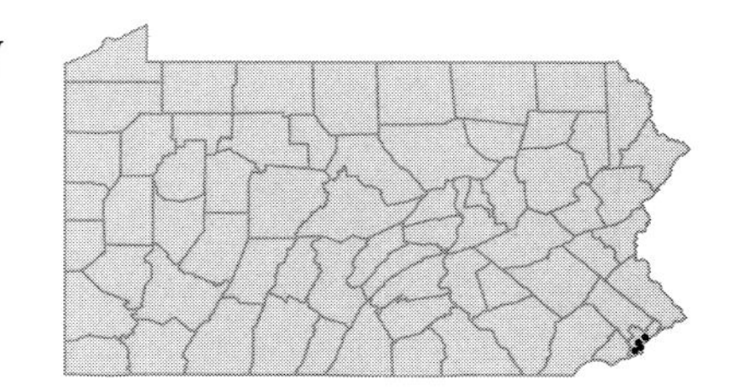

Elytrigia repens (L.) Desv. ex Nevski
Quackgrass; Witchgrass
Herbaceous perennial
Fields, roadsides and waste ground.
Agropyron repens (L.) Beauv. PFGBW

[I]

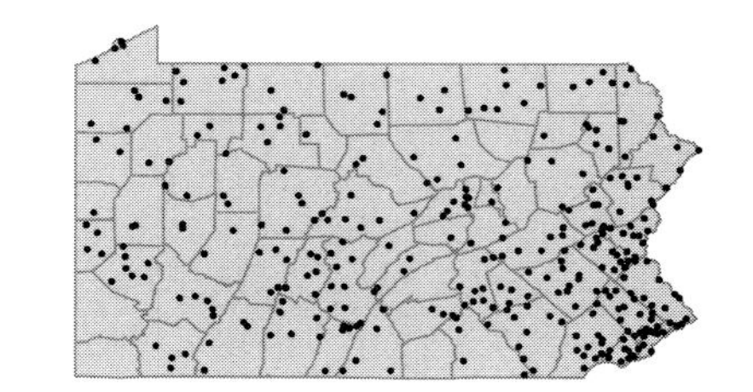

• ***Eragrostis capillaris*** (L.) Nees
Lacegrass
Herbaceous annual
Open woods, roadsides and waste ground, in dry sandy soil.

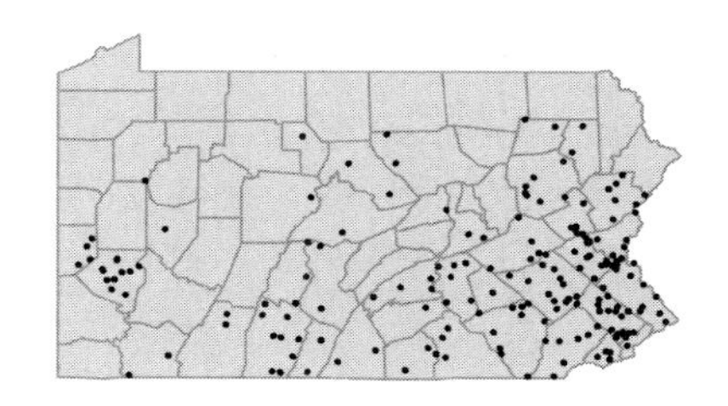

Eragrostis cilianensis (All.) Hubbard
Stink grass
Herbaceous annual
Fields, gardens and waste ground.
Eragrostis major Host P; *Eragrostis megastachya* (Koel.) Link F

FACU
[I]

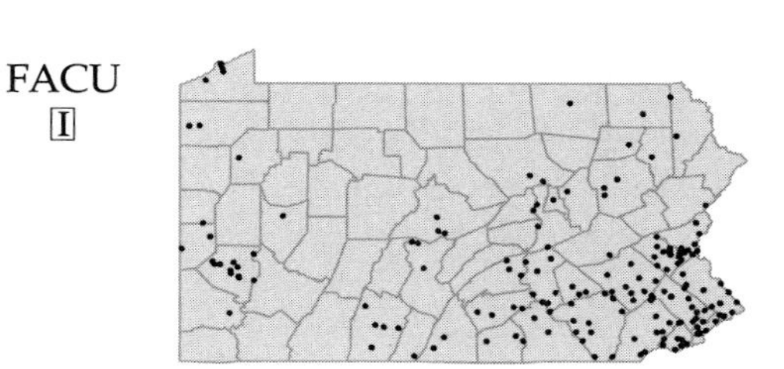

Eragrostis frankii C.A.Mey. ex Steud.
Lovegrass
Herbaceous annual
Moist river banks.

FACW

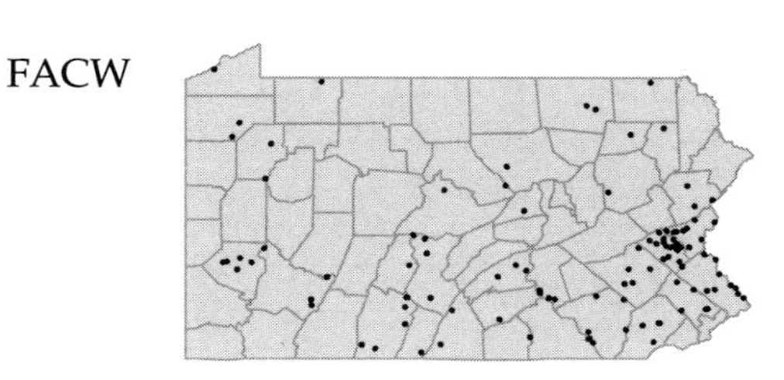

Eragrostis hypnoides (Lam.) BSP
Creeping lovegrass
Herbaceous annual
Wet shores and mudflats.

OBL

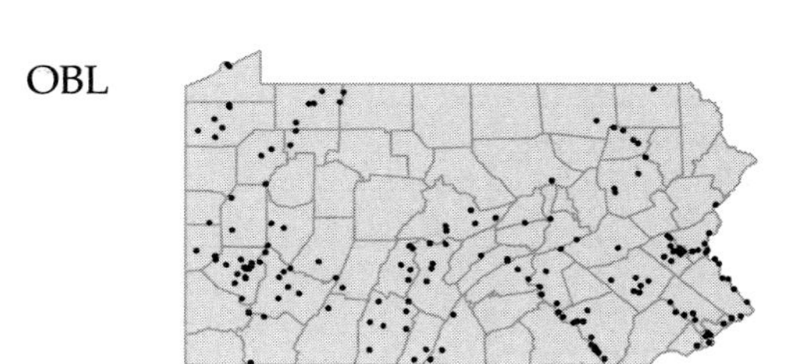

Eragrostis minor Host
Lovegrass
Herbaceous annual
Dry roadsides and railroad embankments.
Eragrostis poaeoides Beauv. FGBW

[I]

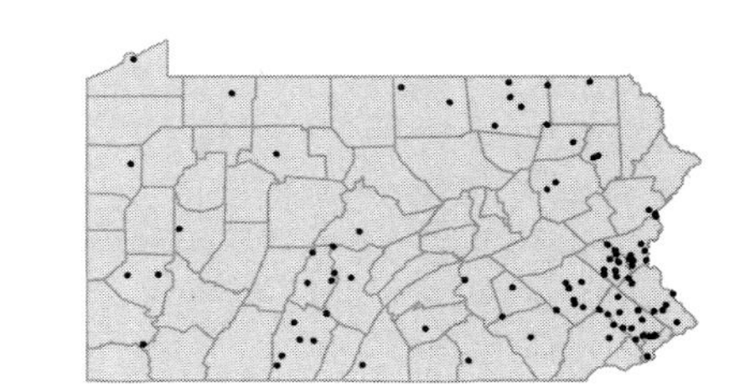

Eragrostis pectinacea (Michx.) Nees
Carolina lovegrass
Herbaceous annual
Gardens and waste ground.

FAC

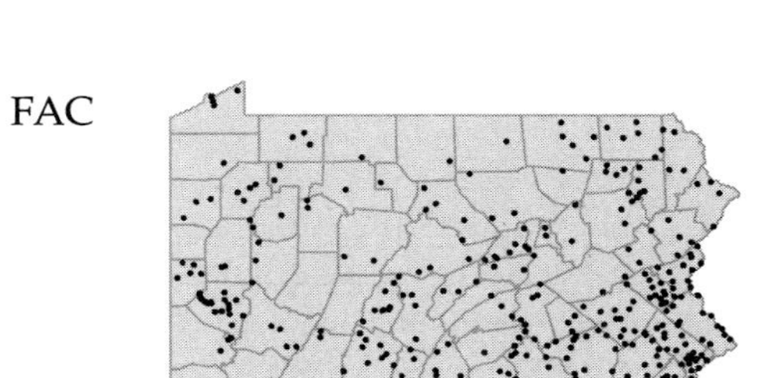

Eragrostis pilosa (L.) Beauv.
India lovegrass
Herbaceous annual
Gardens, roadsides and ballast.
Eragrostis multicaulis Steud. FGW

FACU
[I]

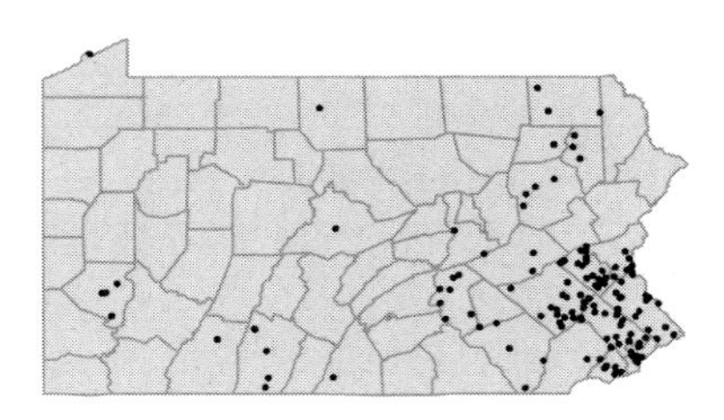

Eragrostis spectabilis (Pursh) Steud.
Purple lovegrass; Tumblegrass
Herbaceous perennial
Dry, sandy fields and roadsides.

UPL

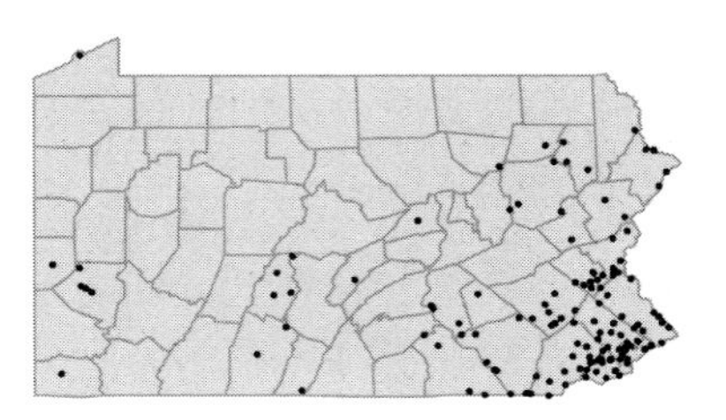

• ***Erianthus giganteus*** (Walt.) Muhl.
Giant beardgrass; Sugarcane plumegrass
Herbaceous perennial
Swamps and river banks. Believed to be extirpated, last collected in 1908.
Erianthus saccharoides Michx. W

FACW+

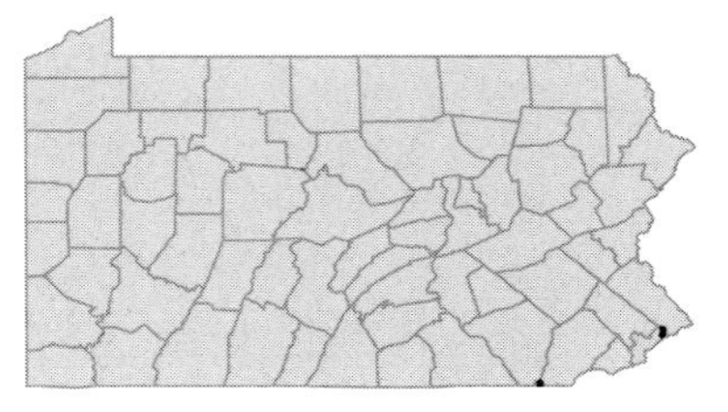

• ***Eriochloa villosa*** (Thunb.) Kunth
Chinese cupgrass
Herbaceous perennial
Open ground of powerline right-of-way.

[I]

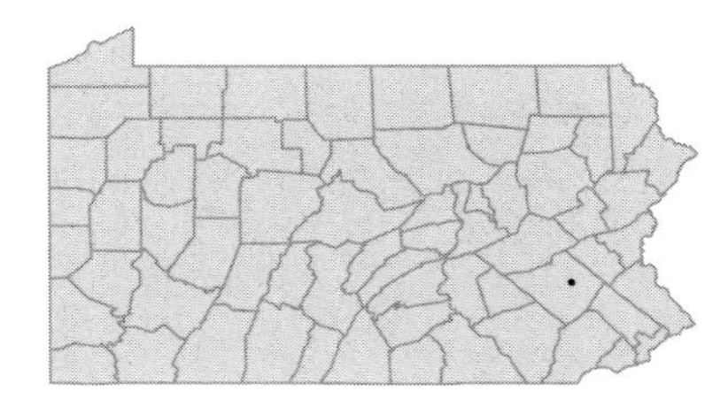

• ***Eustachys petraea*** (Swartz) Desv.
Fingergrass
Herbaceous perennial
Ballast. Represented by two collections from Philadelphia Co. in 1865.
Chloris petraea Swartz W

[I]

• ***Festuca ciliata*** Dumort.
Fescue
Herbaceous perennial
Ballast. Represented by a single collection from Philadelphia Co. in 1878.
Festuca danthonii Asch. & Graebn. W

[I]

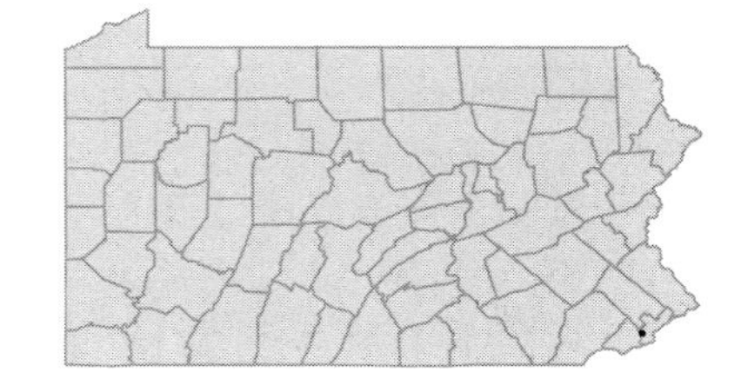

Festuca elatior L.
Fescue
Herbaceous perennial
Roadsides, fields and open ground.

FACU-
I

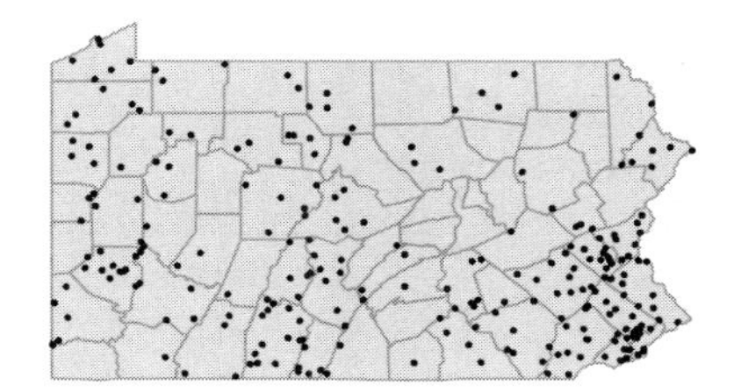

Festuca longifolia Thuill.
Hard fescue
Herbaceous perennial
Dry, open soil.
Festuca ovina L. var. *duriuscula* (L.) W.D.J.Koch PFGBW; *Festuca trachyphylla* (Hackel) Krajina C

I

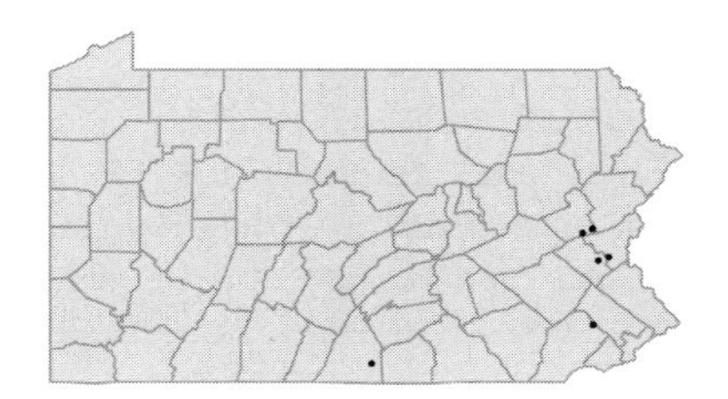

Festuca obtusa Biehler
Nodding fescue
Herbaceous perennial
Moist woods and clearings.
Festuca subverticillata (Pers.) E.Alexeev. C

FACU

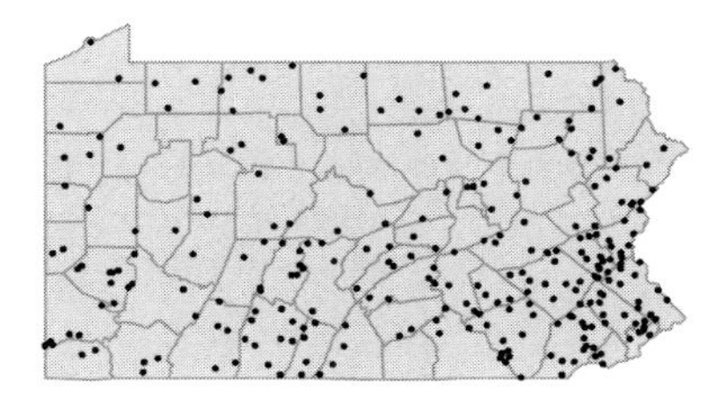

Festuca ovina L.
Sheep fescue
Herbaceous perennial
Open woods, dry fields and roadsides.

I

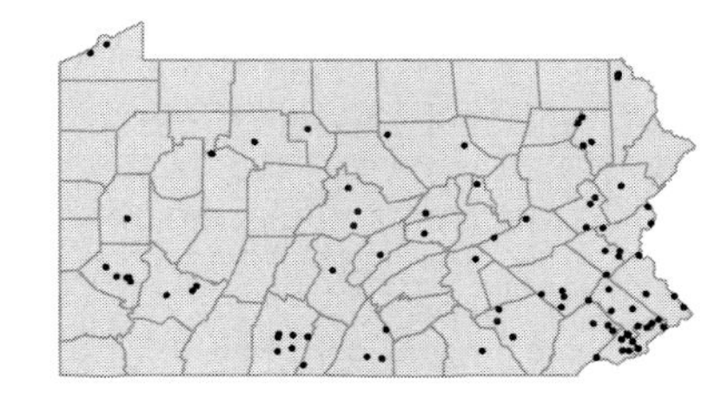

Festuca paradoxa Desv.
Cluster fescue
Herbaceous perennial
Moist, open ground and thickets.
Festuca nutans Willd. P

FAC

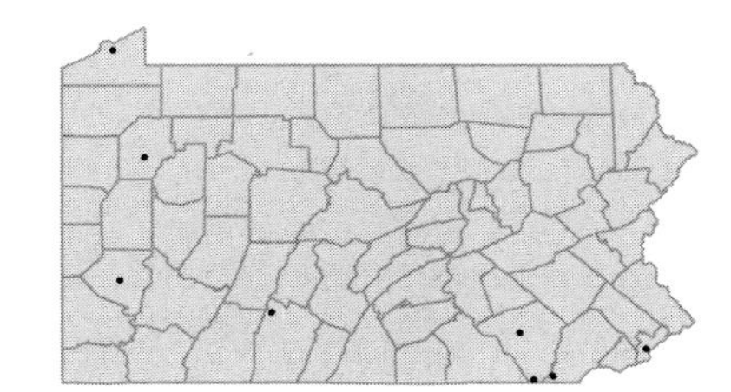

Festuca pratensis Huds.
Meadow fescue
Herbaceous perennial
Meadows, moist shores, roadsides and disturbed ground.

FACU-
I

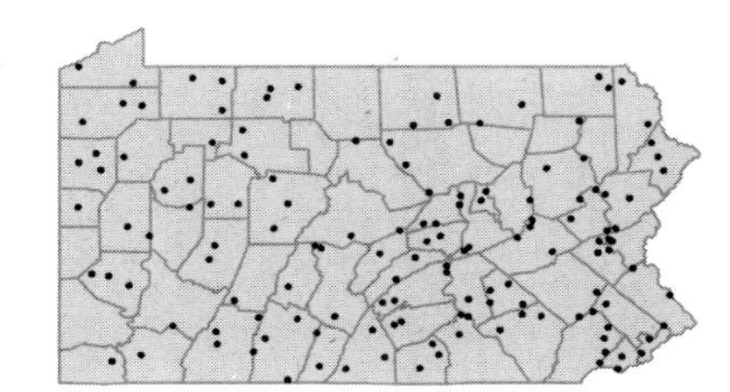

Festuca rubra L. *sensu lato*
Red fescue
Herbaceous perennial
Dry woods, roadsides, waste ground and ballast.

FACU
I

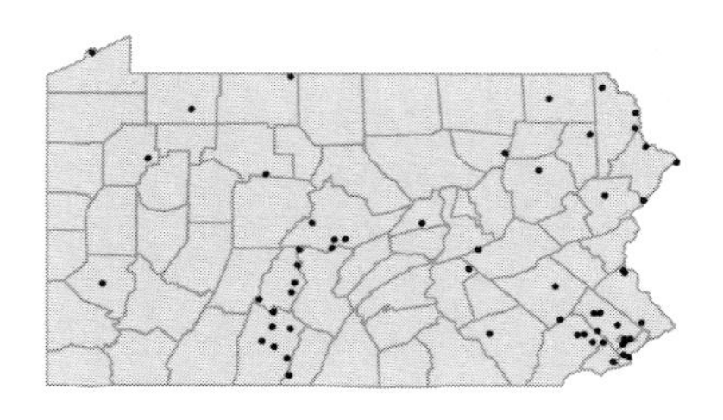

Festuca tenuifolia Sibth.
Hair fescue
Herbaceous perennial
Dry, open ground and oak woods.
Festuca capillata Lam. PFW; *Festuca ovina* L. var. *capillata* (Lam.) Alef. GB;
Festuca filiformis Pourret. C

I

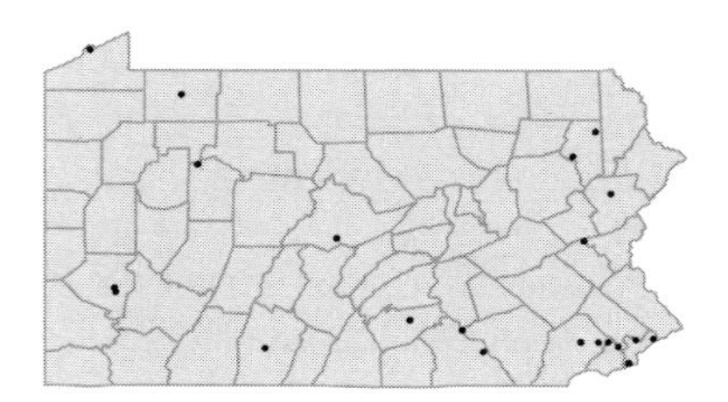

• ***Glyceria acutiflora*** Torr.
Mannagrass
Herbaceous perennial, emergent aquatic
Shallow water, muddy shores and swamps.
Panicularia acutiflora (Torr.) Kuntze P

OBL

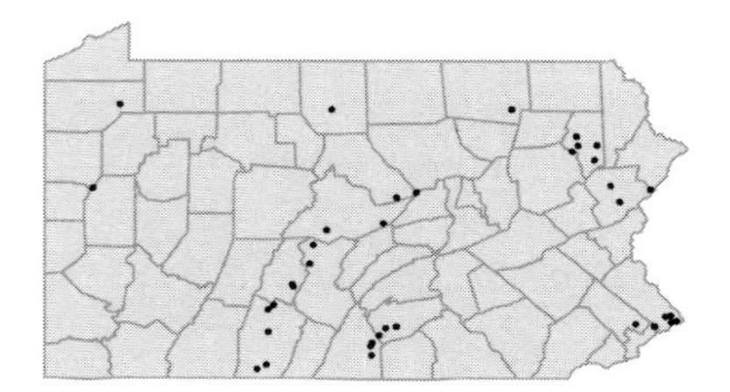

Glyceria borealis (Nash) Batchelder
Northern mannagrass; Small-floating mannagrass
Herbaceous perennial, emergent aquatic
Shallow water of lakes and streams.

OBL

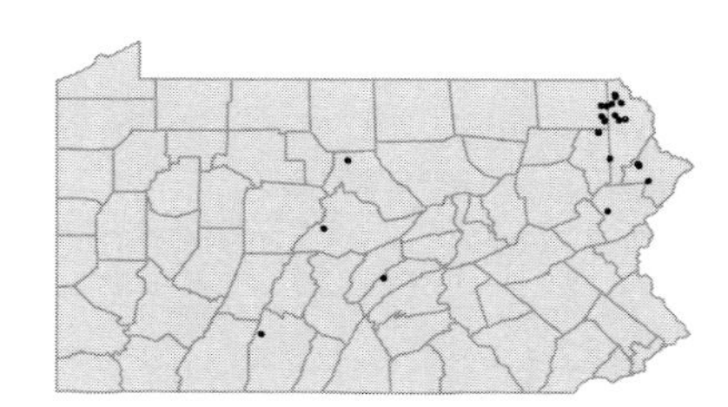

Glyceria canadensis (Michx.) Trin.
Rattlesnake mannagrass
Herbaceous perennial
Moist woods, marshes, swamps and wet shores.
Panicularia canadensis (Michx.) Kuntze P

OBL

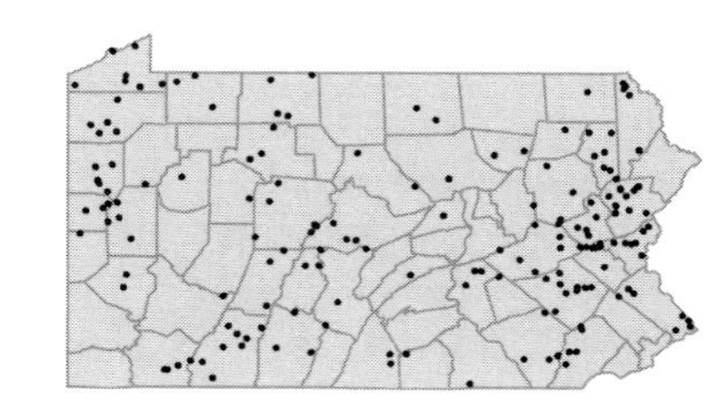

Glyceria canadensis x grandis
Rattlesnake grass
Herbaceous perennial
Marshes, swamps and wet woods.
Glyceria canadensis (Michx.) Trin. var. *laxa* (Scribn.) A.S.Hitchc. W; *Glyceria x laxa* (Scribn.) Scribn. *pro. sp.* FGBKC; *Panicularia laxa* Scribn. P

OBL

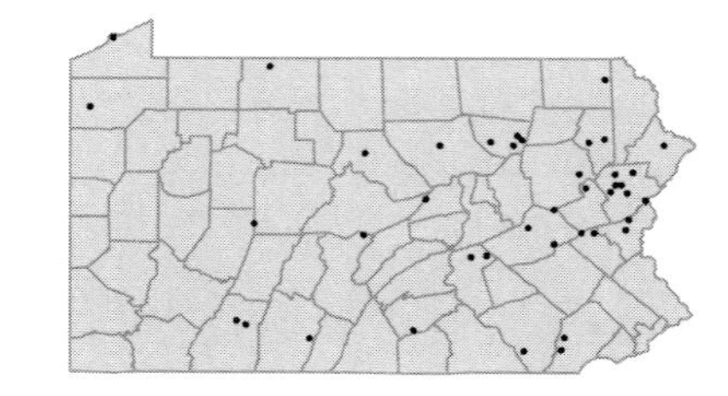

Glyceria fluitans (L.) R.Br.
Floating mannagrass
Herbaceous perennial, emergent aquatic
Shallow water and marshes.

OBL
I

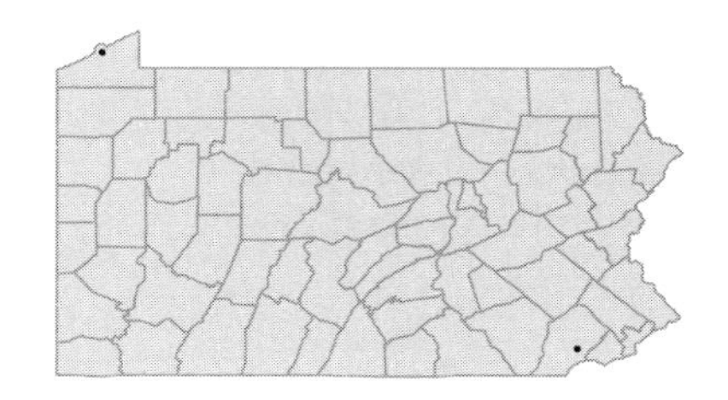

Glyceria grandis S.Wats.
American mannagrass
Herbaceous perennial, emergent aquatic
Shallow water or wet meadows.
Panicularia americana (Torr.) MacM. P

OBL

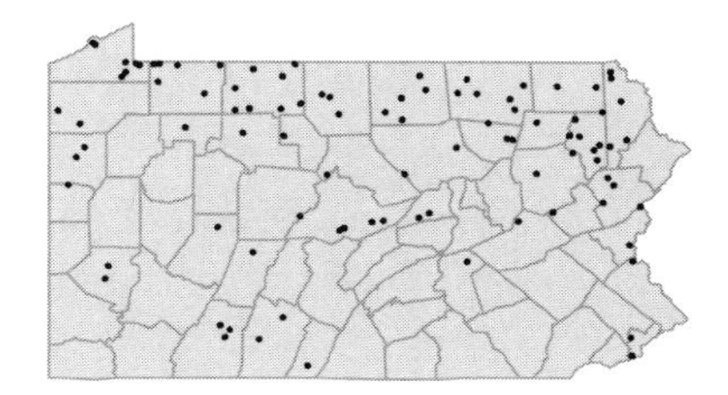

Glyceria melicaria (Michx.) F.T.Hubbard
Slender mannagrass
Herbaceous perennial
Wet woods, stream banks and swamps.

OBL

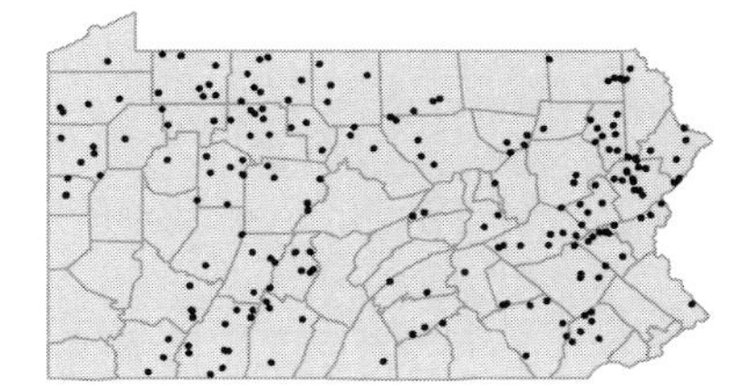

Glyceria obtusa (Muhl.) Trin.
Coastal mannagrass; Blunt mannagrass
Herbaceous perennial
Swamps, bogs and moist, sandy, peaty ground.
Panicularia obtusa (Muhl.) Kuntze P

OBL

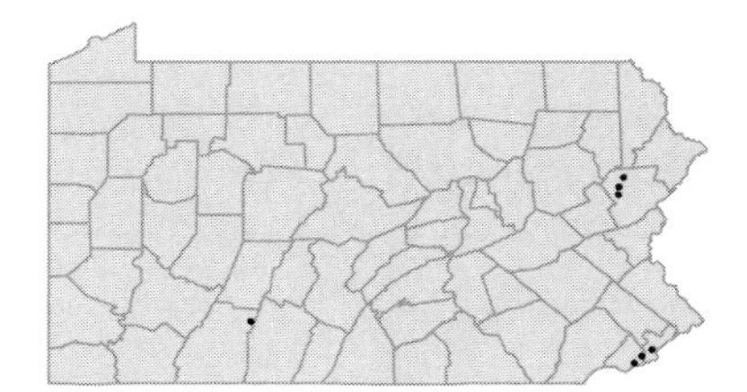

Glyceria septentrionalis A.S.Hitchc.
Floating mannagrass
Herbaceous perennial, emergent aquatic
Wet meadows and shallow water of stream margins.

OBL

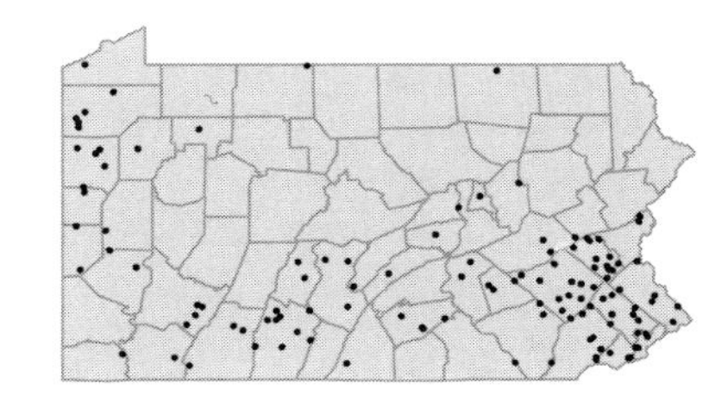

Glyceria striata (Lam.) A.S.Hitchc.
Fowl mannagrass
Herbaceous perennial
Wet woods, swamps and bogs.
Panicularia nervata (Willd.) Kuntze P

OBL

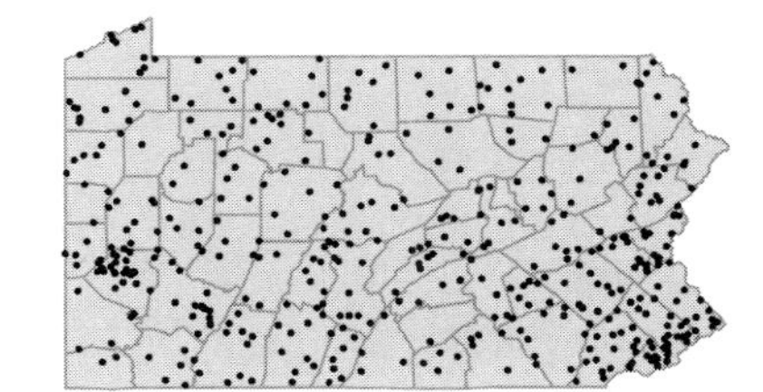

• ***Gymnopogon ambiguus*** (Michx.) BSP
Broad-leaved beardgrass
Herbaceous perennial
Serpentine barrens.

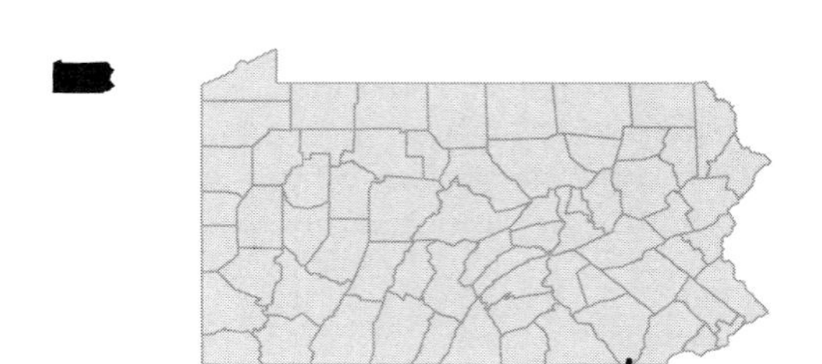

• ***Hemarthria altissima*** (Poir.) Stapf & C.E.Hubbard
Grass
Herbaceous perennial
Dry agricultural field and waste ground.
Manisuris altissima (Poir.) A.S.Hitchc. W

I

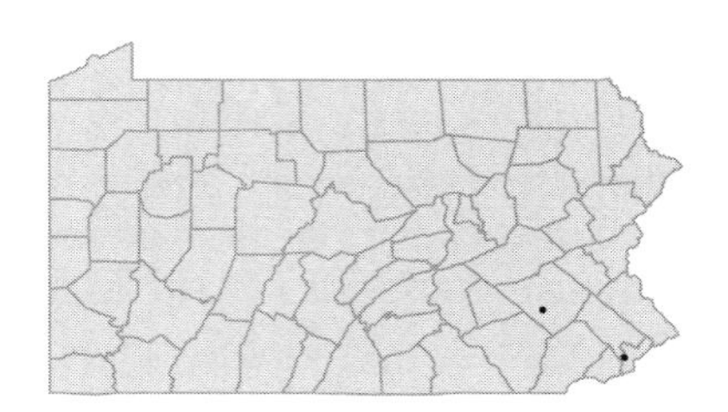

• ***Hierochloe odorata*** (L.) Beauv.
Vanilla sweet grass
Herbaceous perennial
Moist meadow or river shore.

FACW

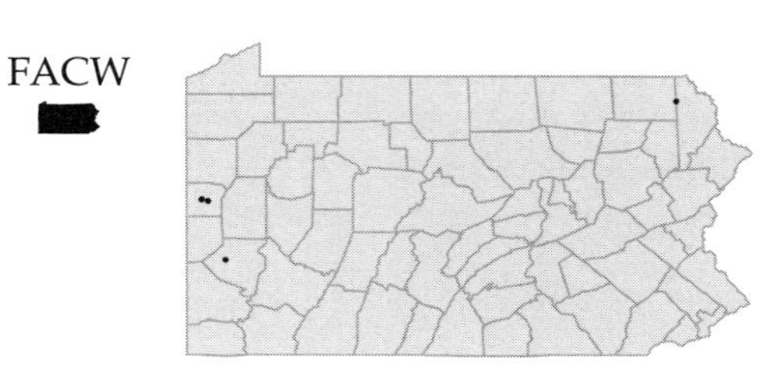

• ***Holcus lanatus*** L.
Velvet grass
Herbaceous perennial
Meadows, old fields, river shores and roadsides.

FACU
I

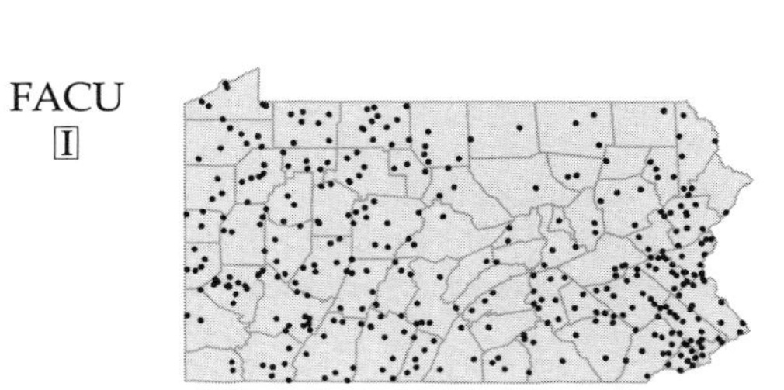

Holcus mollis L.
Velvet grass
Herbaceous perennial
Waste ground. Represented by a single collection from Philadelphia Co. ca. 1865.

I

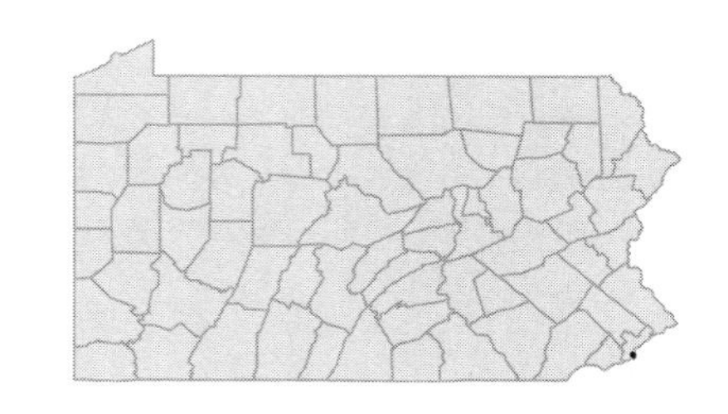

• ***Hordeum vulgare*** L.
Barley
Herbaceous annual
Cultivated and occasionally persisting on the edges of fields.

I

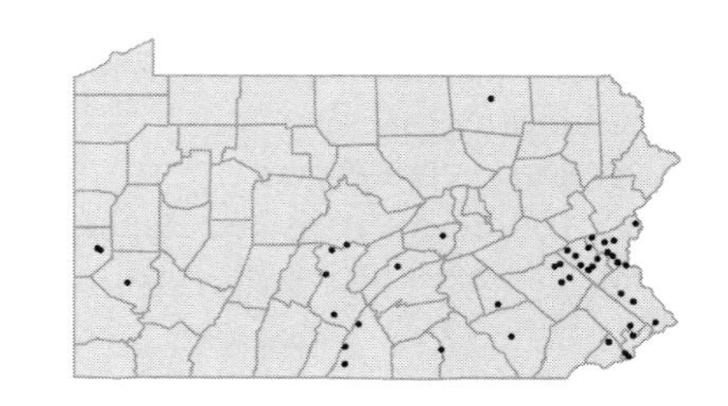

• ***Koeleria macrantha*** (Ledeb.) Schultes
June grass
Herbaceous perennial
Dry soil. Believed to be extirpated, last collected in 1912.
Koeleria cristata (L.) Pers. PFGBW; *Koeleria pyramidata* (Lam.) P.Beauv. *pro parte* C

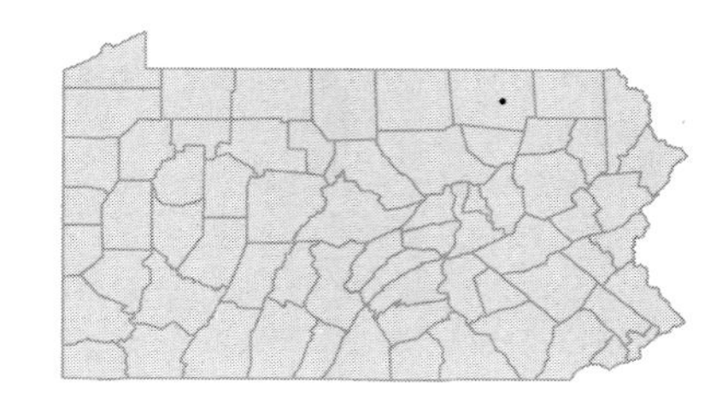

• ***Leersia oryzoides*** (L.) Swartz — OBL
Rice cutgrass
Herbaceous perennial
Marshes, bogs or wet meadows.
Homalocenchrus oryzoides (L.) Pollard P

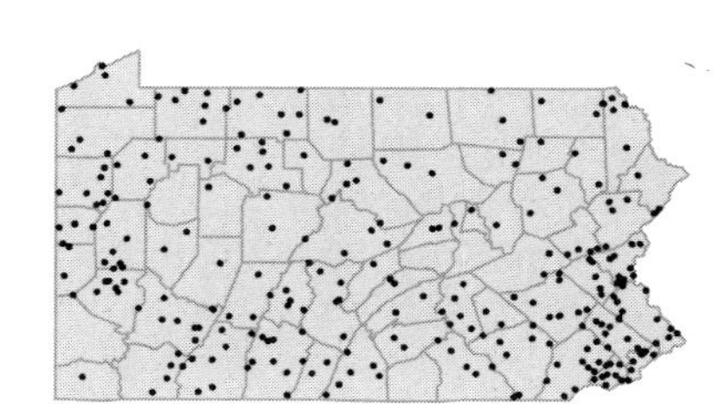

Leersia virginica Willd. — FACW
Cutgrass; White grass
Herbaceous perennial
Swamps or moist woods.
Homalocenchrus virginicus (Willd.) Britt. P

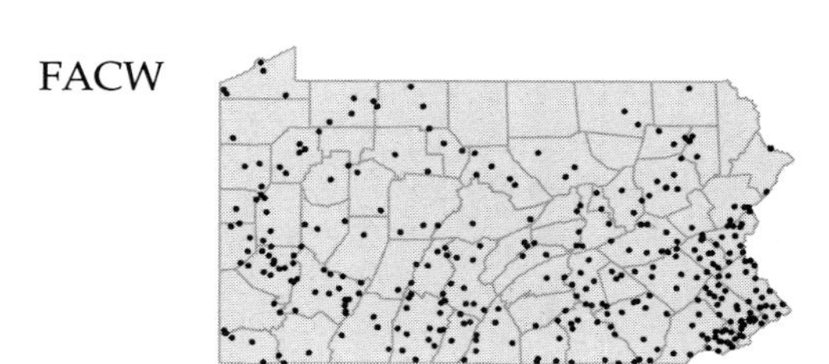

• ***Leptochloa fascicularis*** (Lam.) A.Gray var. ***acuminata*** (Nash) Gleason — FACW I
Sprangletop
Herbaceous annual
Roadsides, railroad sidings and waste ground.
Diplachne acuminata Nash FK

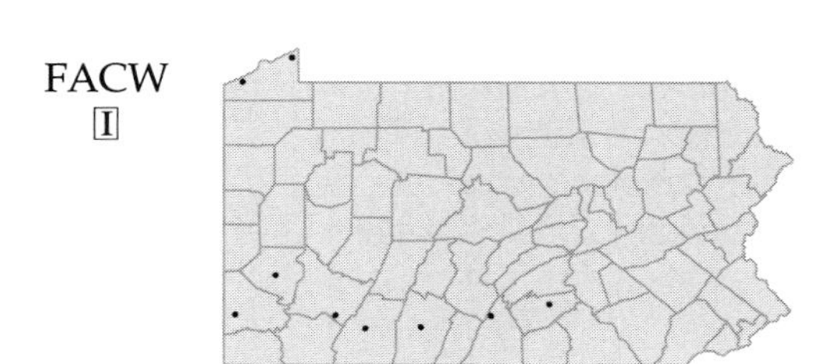

Leptochloa filiformis (Lam.) Beauv. — FACW I
Red sprangletop
Herbaceous annual
Wharves, railroad sidings and rubbish dumps.

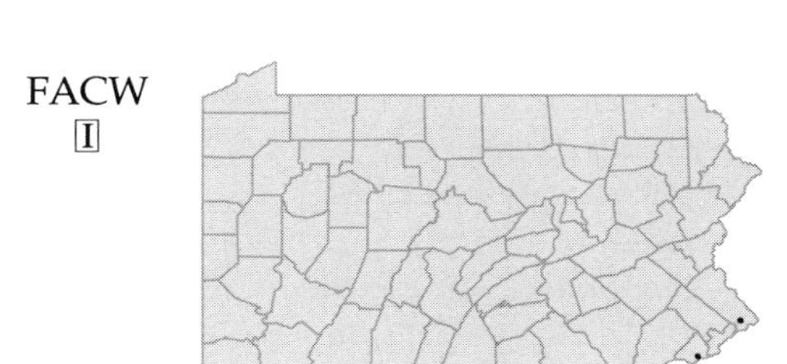

• ***Lophochloa cristata*** (L.) Hyl. — I
Grass
Herbaceous perennial
Ballast. Represented by a single collection from Philadelphia Co. ca. 1878.
Koeleria phleoides (Vill.) Pers. W

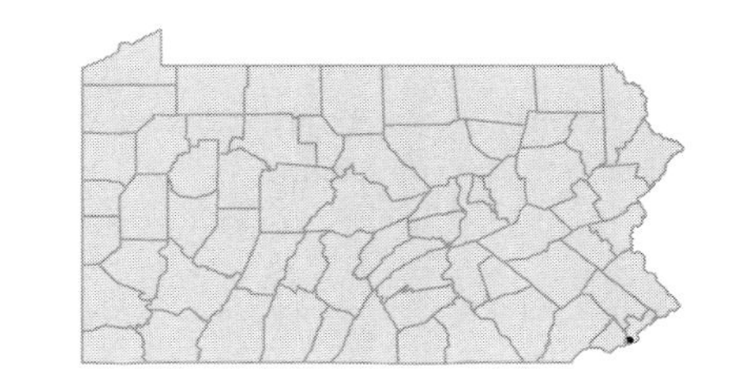

• ***Leptoloma cognatum*** (Schultes) Chase
Fall witchgrass
Herbaceous perennial
Sandy, open ground.
Digitaria cognatum (Schultes) Chase

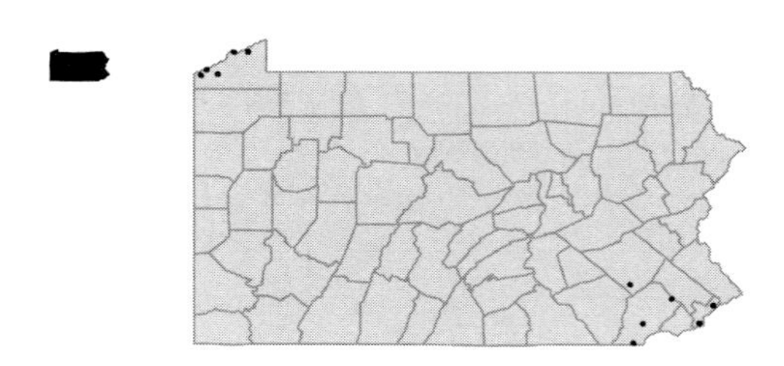

• ***Lolium multiflorum*** Lam.
Ryegrass
Herbaceous perennial
Cultivated and occasionally escaped.
Lolium perenne L. var. *aristatum* Willd. C

[I]

Lolium perenne L.
Perennial ryegrass
Herbaceous perennial
Cultivated and frequently escaped.

FACU-
[I]

Lolium temulentum L.
Darnel; Poison grass
Herbaceous annual
Waste ground and ballast.

[I]

• ***Leymus mollis*** (Trin.) Pilger ssp. ***mollis***
American dunegrass
Herbaceous perennial
Docks and wharves. Represented by a single collection from Erie Co. in 1882.
Elymus mollis Trin. var. *mollis*; *Elymus arenarius* L. var. *villosus* F

FACU-
[I]

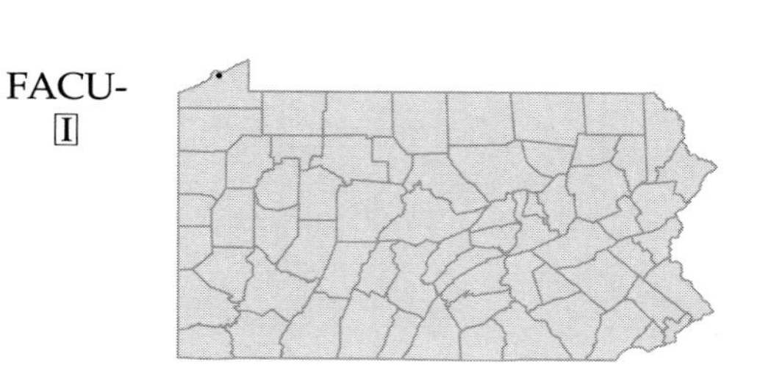

• ***Melica nitens*** Nutt.
Tall melic grass; Three-flowered melic grass
Herbaceous perennial
Steep, rocky slopes and river banks.

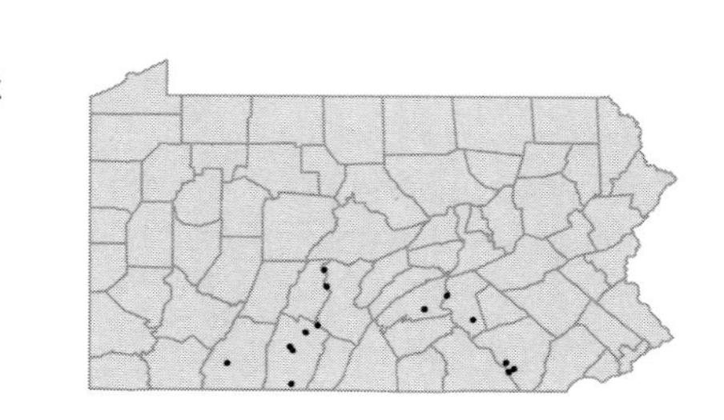

• ***Microstegium vimineum*** (Trin.) A.Camus.
Stilt grass
Herbaceous annual
Moist ground of open woods, thickets, paths, clearings, fields and gardens.
Eulalia vimineum (Trin.) Kuntze K

FAC
[I]

• ***Milium effusum*** L. var. ***cisatlanticum*** Fern.
Millet grass
Herbaceous perennial
Cool, rich woods.

[I]

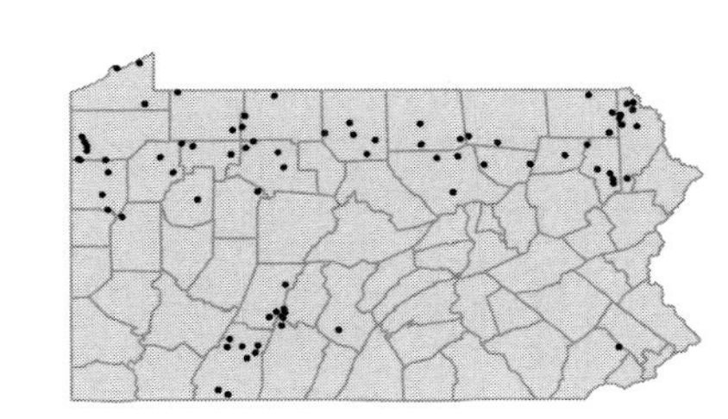

• ***Miscanthus sinensis*** Anderss. var. ***sinensis***
Japanese plumegrass; Eulalia
Herbaceous perennial
Cultivated and occasionally escaped.

FACU
[I]

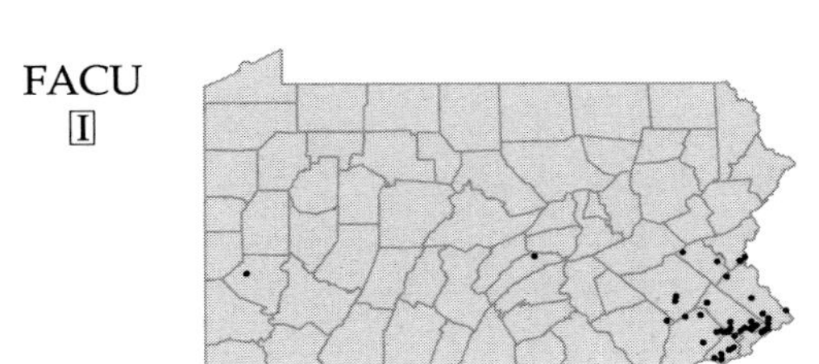

Miscanthus sinensis Anderss. var. ***zebrinus*** Beal
Zebra grass
Herbaceous perennial
Cultivated and occasionally established along banks and roadsides.

FACU
[I]

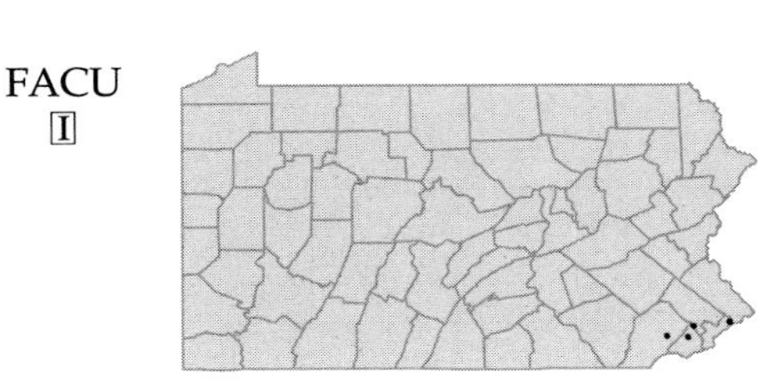

• ***Molinia caerulea*** (L.) Moench
Moor grass
Herbaceous perennial
Meadows. Represented by a single collection from Wayne Co. in 1944.

[I]

• ***Muhlenbergia asperifolia*** (Nees & Meyen) Parodi
Scratch grass
Herbaceous perennial
Waste ground.

FACW
[I]

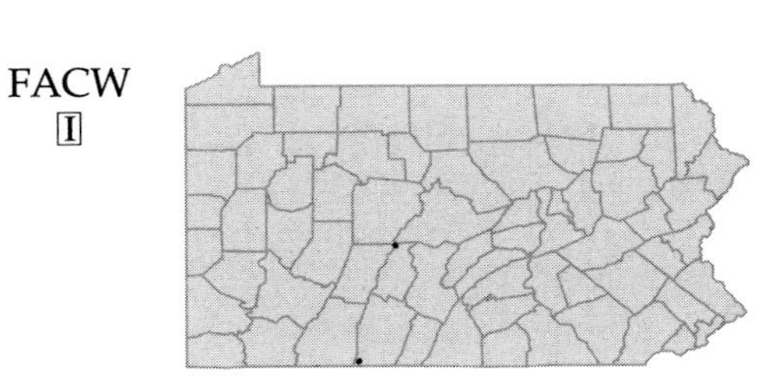

Muhlenbergia capillaris (Lam.) Trin.
Hairgrass; Short muhly
Herbaceous perennial
River shore. Believed to be extirpated, all collections from a single site in Lancaster Co. in 1864.

FACU-

Muhlenbergia cuspidata (Torr.)Rydb.
Sharp-pointed muhly
Herbaceous perennial
Alluvial shores and rock crevices. Represented by two collections from Northampton Co. 1946-1952.

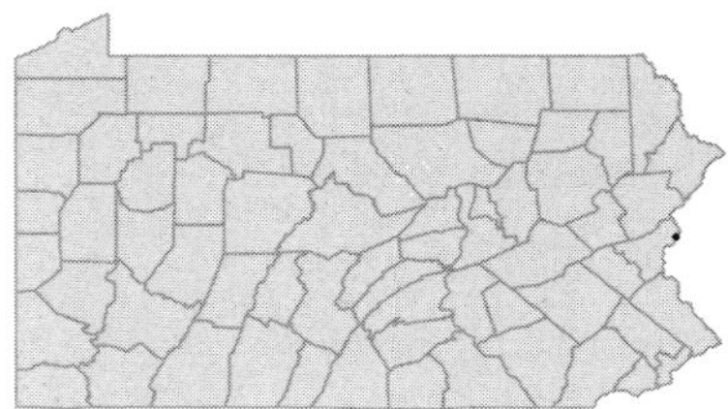

Muhlenbergia frondosa (Poir.) Fern.
Wirestem muhly
Herbaceous perennial
Moist, open woods and stream banks. A hybrid with *M. schreberi* (*M. x curtisetosa* [Scribn.] Pohl) has been collected at two sites.

FAC

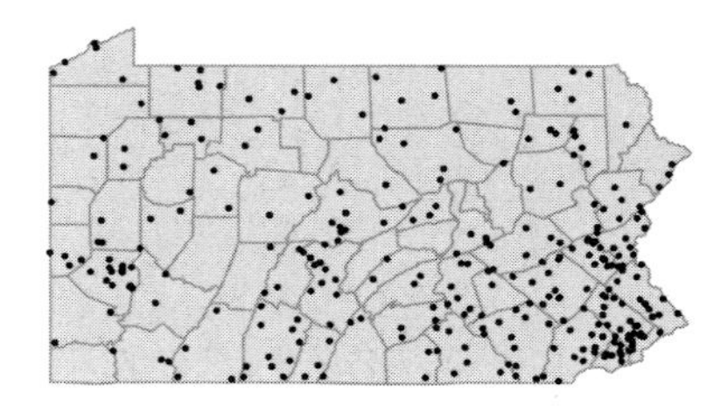

Muhlenbergia glomerata (Willd.) Trin.
Spike muhly
Herbaceous perennial
Rocky hillsides, sandy thickets and moist ground.

FACW

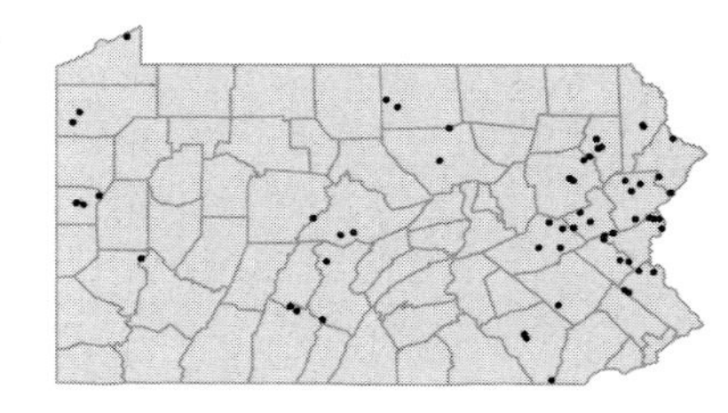

Muhlenbergia mexicana (L.) Trin.
Muhly; Satin grass
Herbaceous perennial
Woods, rocky shores, swamps and serpentine barrens.

FACW

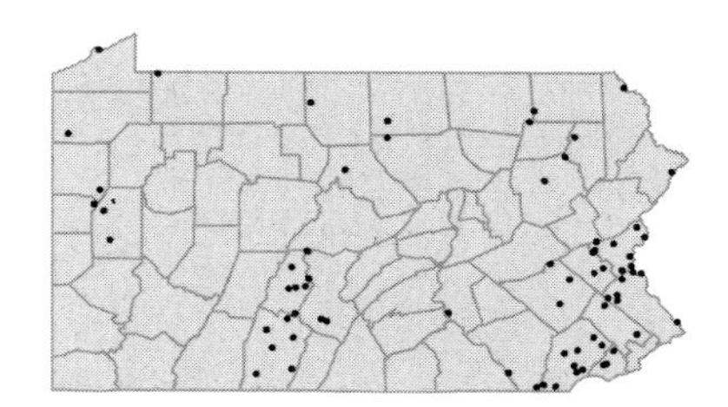

Muhlenbergia schreberi J.F.Gmel.
Dropseed; Nimble-will
Herbaceous perennial
Woods, thickets and waste ground.
Muhlenbergia diffusa Willd. P

FAC

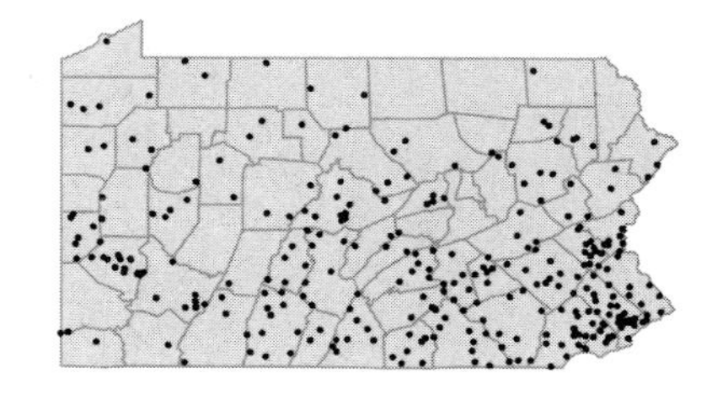

Muhlenbergia sobolifera (Muhl.) Trin.
Creeping muhly
Herbaceous perennial
Dry, rocky slopes.

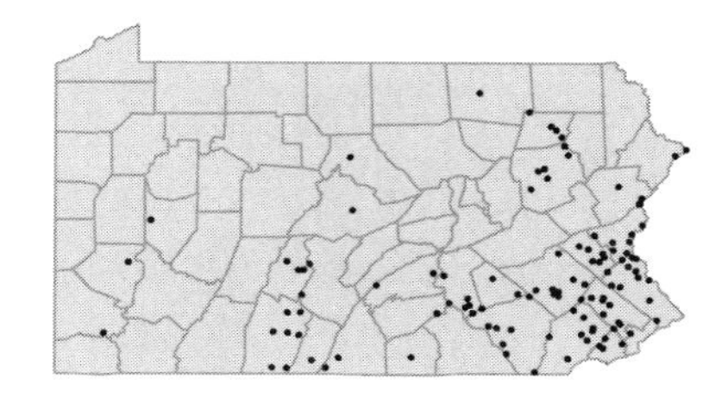

Muhlenbergia sylvatica (Torr.) Torr. ex A.Gray — FAC+
Muhly; Woodland dropseed
Herbaceous perennial
Moist woods and shaded banks.

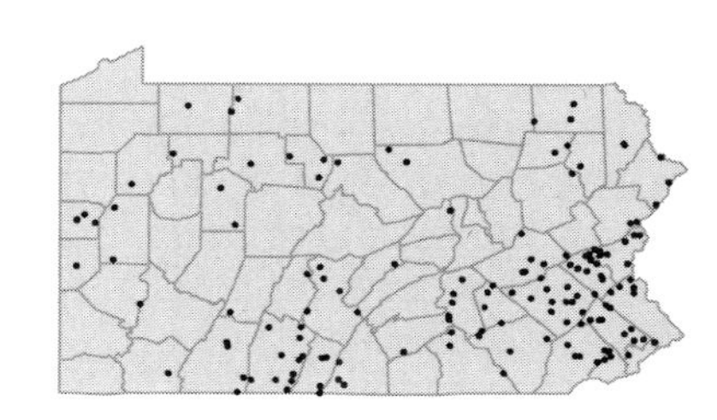

Muhlenbergia tenuiflora (Willd.) BSP
Muhly; Woodland dropseed
Herbaceous perennial
Rocky, wooded slopes along streams.

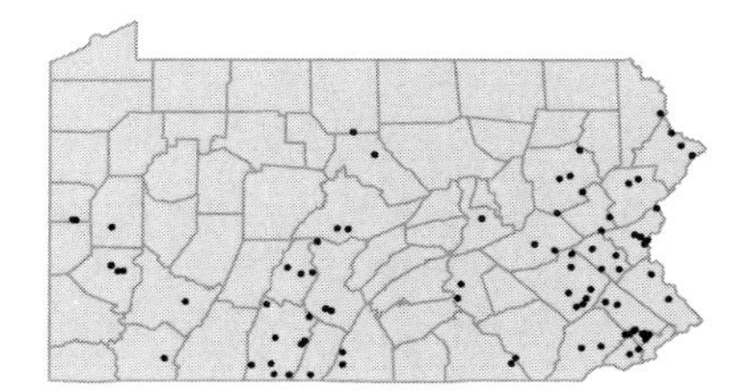

Muhlenbergia uniflora (Muhl.) Fern. — OBL
Fall dropseed muhly
Herbaceous perennial
Marshes, bogs and moist, sandy roadsides.

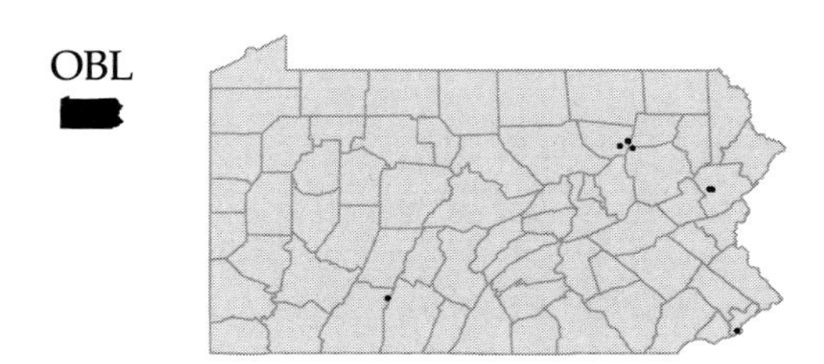

• ***Oryzopsis asperifolia*** Michx.
Spreading ricegrass
Herbaceous perennial
Dry, sandy soil or rocky woods.

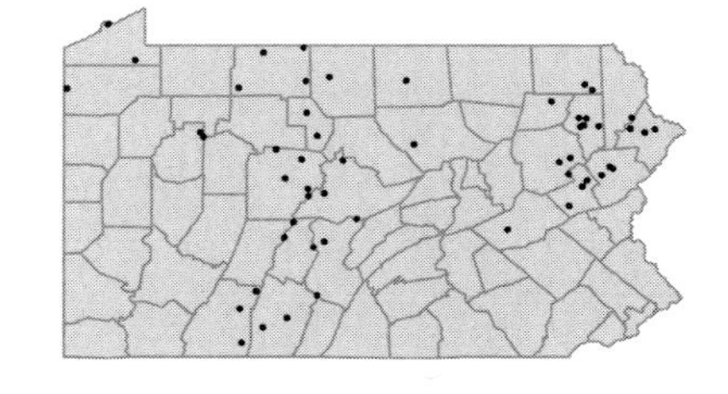

Oryzopsis pungens (Torr. ex Spreng.) A.S.Hitchc.
Slender mountain ricegrass
Herbaceous perennial
Dry, sandy thickets and barrens.
Oryzopsis juncea (Michx.) BSP P

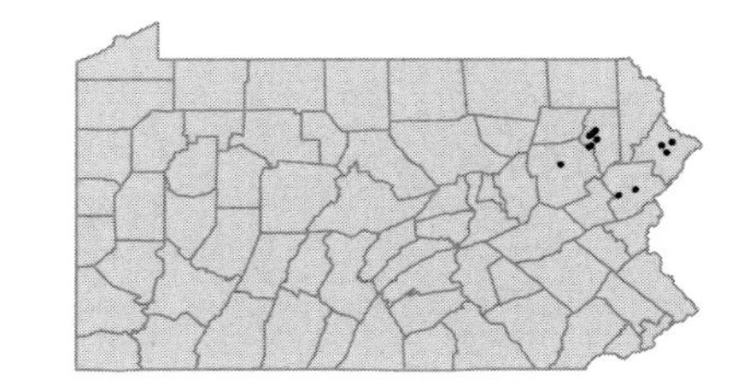

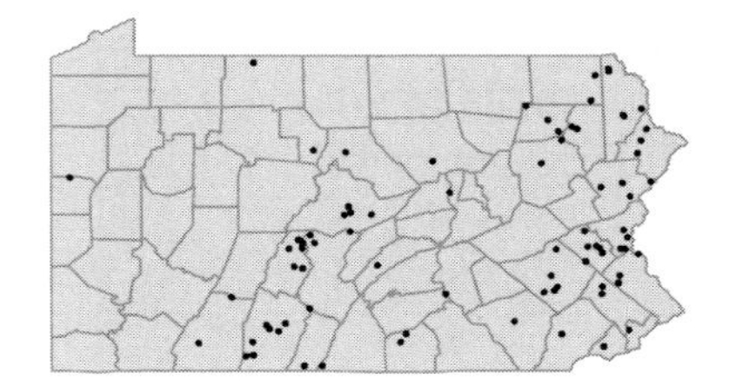

Oryzopsis racemosa (Smith) Ricker ex A.S.Hitchc.
Ricegrass
Herbaceous perennial
Open, rocky woods.
Oryzopsis melanocarpa Muhl. P

Panicum—We have chosen not to recognize the genus *Dichanthelium*, but rather place all species in *Panicum*. Examination of the type specimens convinces us that *P. acuminatum* Swartz is the correct name for the *acuminatum/lanuginosum* group. Within this group however, we find a continuous range of variation such that the segregation of subspecific taxa is not feasible.

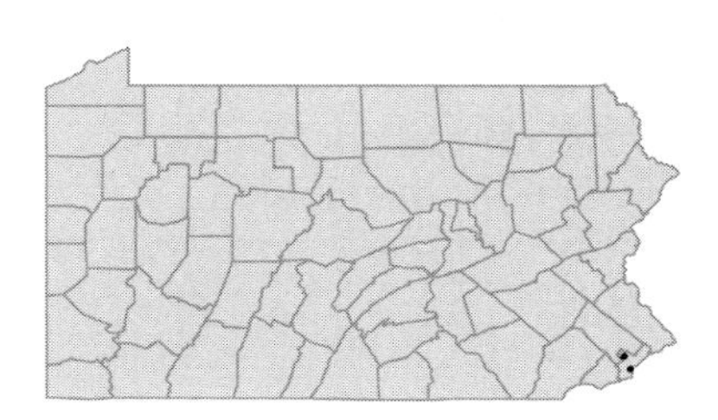

• ***Panicum acroanthum*** Steud. I
Panic-grass
Herbaceous annual
Ballast and waste ground. Represented by two collections from Philadelphia Co. 1865-1877.

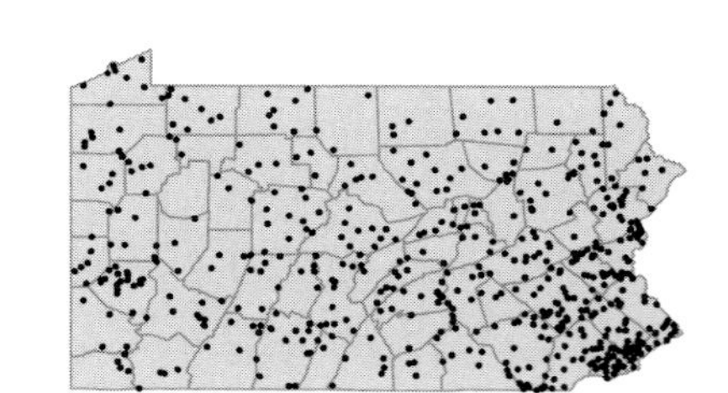

Panicum acuminatum Swartz *sensu lato* FAC
Panic-grass
Herbaceous perennial
Dry woods, slopes and clearings. Although four varieties are frequently recognized, we find the characters of spikelet size and pubescence to be highly variable and poorly correlated.
Panicum implicatum Scribner *pro parte* W; *Panicum lindheimeri* Nash *pro parte* W; *Dichanthelium longiligulatum* (Nash) Freckmann *pro parte* K; *Panicum longiligulatum* Nash *pro parte* W; *Panicum huachucae* Ashe *pro parte* ; *Panicum lanuginosum* Elliot CFB; *Dichanthelium lanuginosum* (Elliot) Gould *pro parte* ; *Dichanthelium acuminatum* (Swartz) Gould & Clark K

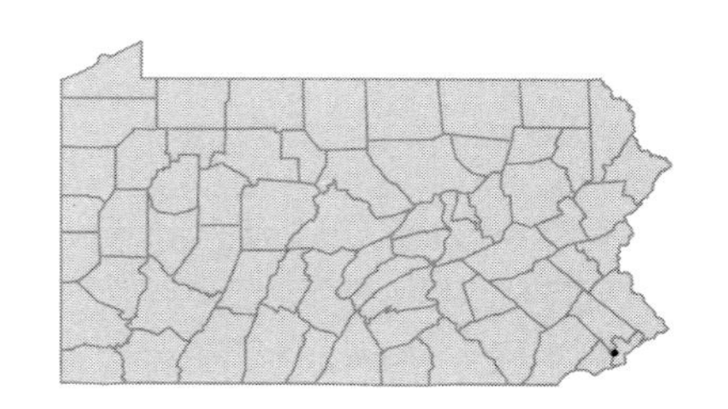

Panicum adspersum Trin. I
Panic-grass
Herbaceous annual
Ballast. Represented by two collections from Philadelphia Co. 1879-1881.

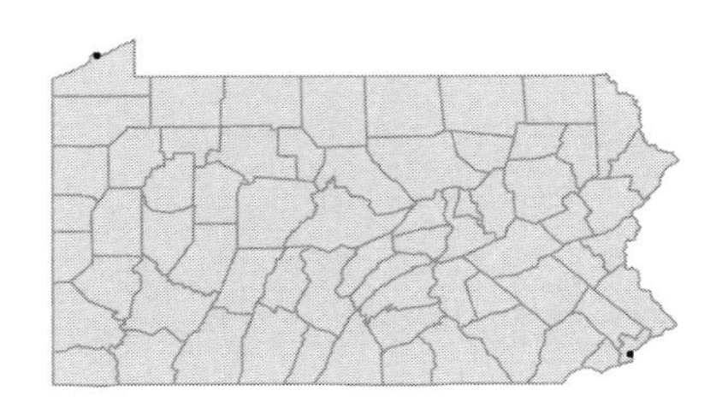

Panicum amarum Ell. var. ***amarum***
Panic-grass
Herbaceous perennial
Ballast and lakeshore.

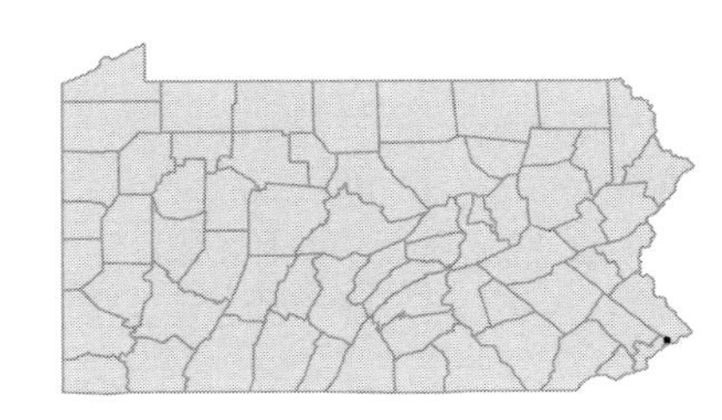

Panicum amarum Ell. var. ***amarulum*** (A.S.Hitchc. & Chase) P.G.Palmer
Beach grass
Herbaceous perennial
Sand and gravel fill along the Delaware River. Represented by a single collection from Bucks Co. in 1946.
Panicum amarulum A.S.Hitchc. & Chase FGBWC

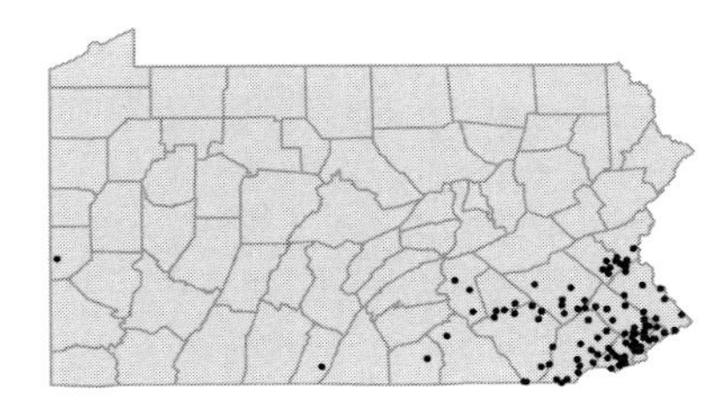

Panicum anceps Michx.
Panic-grass
Herbaceous perennial
Moist, open, sandy soil.

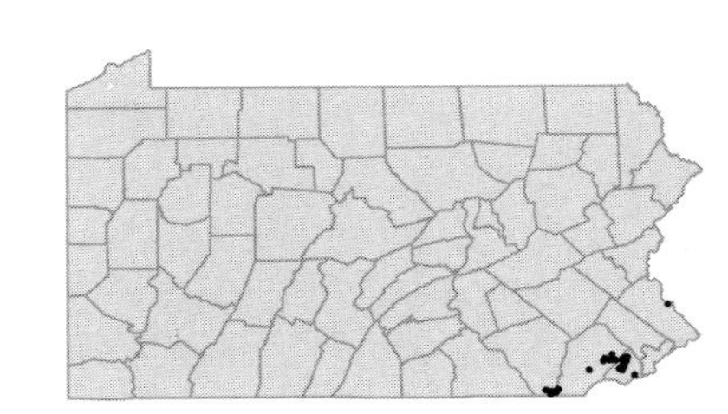

Panicum annulum Ashe
Annulus panic-grass
Herbaceous perennial
Dry soil of serpentine barrens.
Dichanthelium dichotomum (L.) Gould *pro parte* K; *Panicum dichotomum* L. *pro parte* C

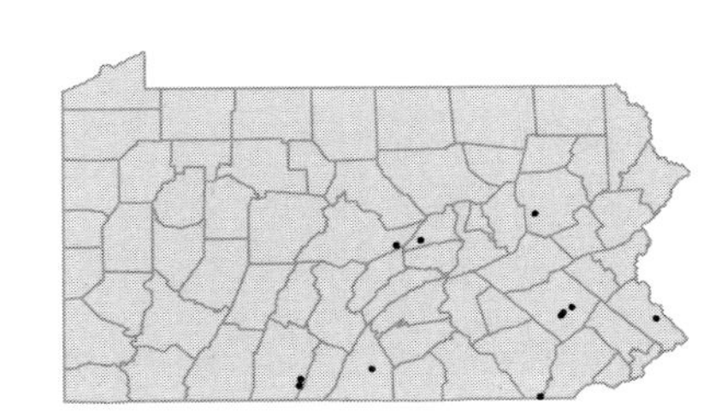

Panicum bicknellii Nash
Bicknell's panic-grass
Herbaceous perennial
Dry woods.
Dichanthelium boreale (Nash) Freckmann *pro parte* K; *Panicum boreale* Nash *pro parte* C

FACU

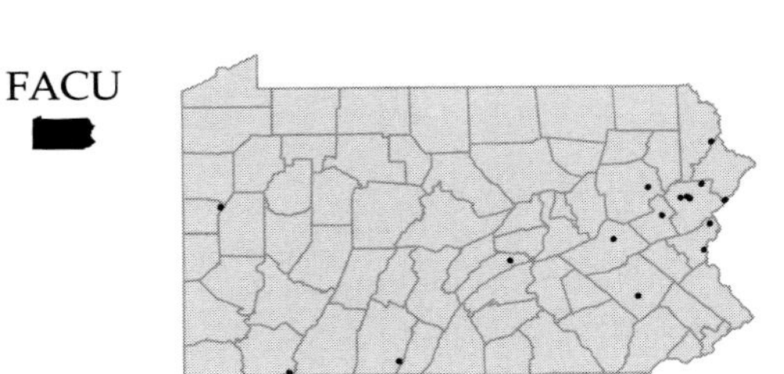

Panicum boreale Nash
Northern panic-grass
Herbaceous perennial
Bogs and fens.
Dichanthelium boreale (Nash) Freckmann *pro parte* K

Panicum boscii Poir.
Panic-grass
Herbaceous perennial
Moist woodlands, grassy slopes and river banks. A hybrid with *P. commutatum* has been collected at one Berks Co. site.
Dichanthelium boscii (Poir.) Gould and Clark K; *Panicum porterianum* Nash *pro parte* P; *Panicum pubifolium* Nash *pro parte* P

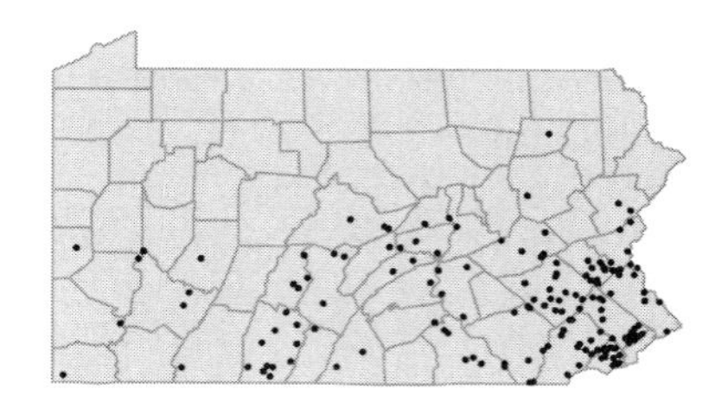

Panicum capillare L.
Witch grass
Herbaceous annual
Cultivated fields, shores and roadsides.
Panicum capillare L. var. *occidentale* Rydb. *pro parte* FW

FAC-

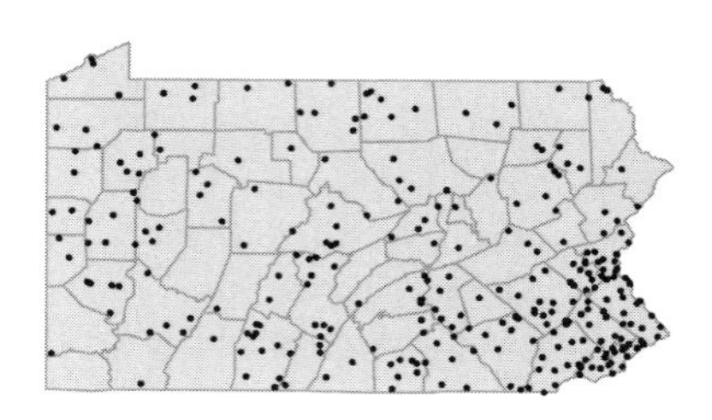

Panicum clandestinum L.
Deer-tongue grass
Herbaceous perennial
Moist, often sandy, soil of woods edges and clearings.
Dichanthelium clandestinum (L.) Gould K; *Panicum decoloratum* Nash P

FAC+

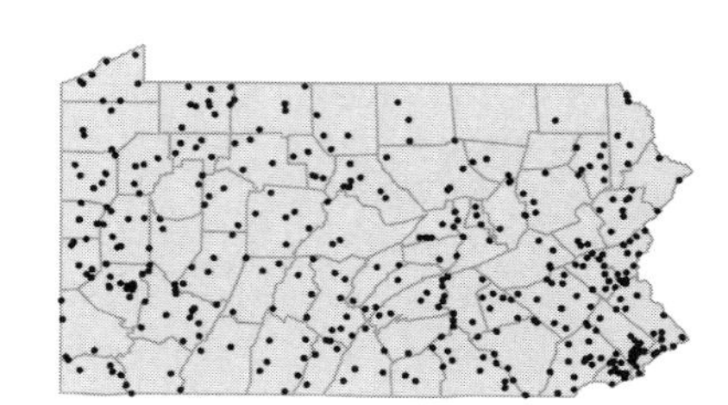

Panicum columbianum Scribn.
Panic-grass
Herbaceous perennial
Roadsides, waste ground and open woods, in dry sandy soil.
Dichanthelium sabulorum (Lam.) Gould & Clark var. *thinium* (A.S.Hitchc. & Chase) Gould & Clark K

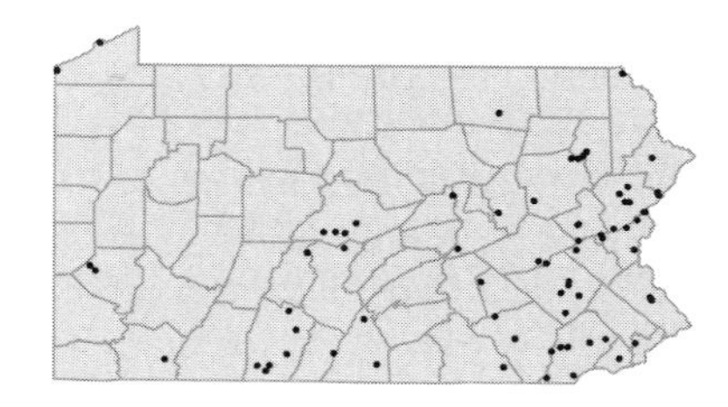

Panicum commonsianum Ashe var. ***commonsianum***
Cloaked panic-grass
Herbaceous perennial
Sandy, open ground. Represented by a single collection from Philadelphia Co. in 1937.
Dichanthelium ovale (Ell.) Gould & Clark var. *addisonii* (Nash) Gould & Clark K; *Panicum commonsianum* Ashe var. *addisonii* (Nash) Fern. WFB

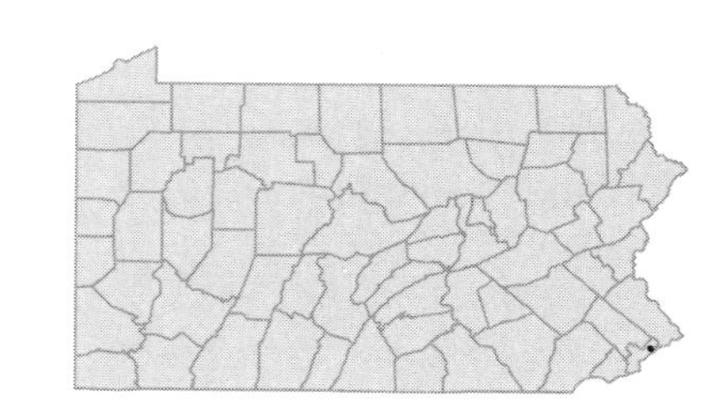

Panicum commonsianum Ashe var. ***euchlamydeum*** (Shinners) Pohl
Panic-grass
Herbaceous perennial
Sand plains and beaches.
Dichanthelium sabulorum (Lam.) Gould & Clark var. *patulum* (Scribn. & Merr.) Gould & Clark K

FACU

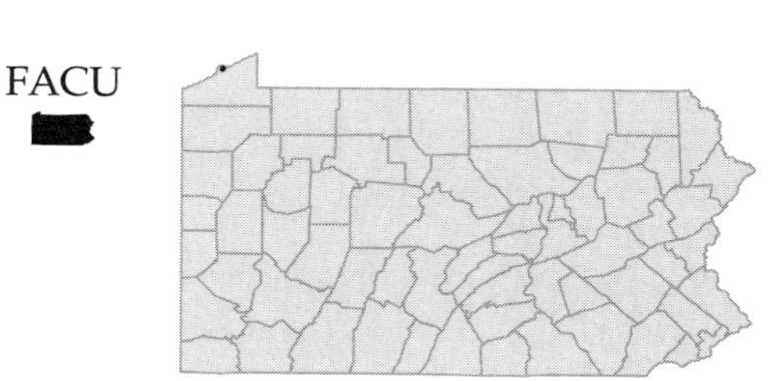

Panicum commutatum Schultes
Panic-grass
Herbaceous perennial
Dry woods, rocky slopes and barrens.
Dichanthelium commutatum (Schultes) Gould *pro parte* K; *Panicum ashei* Pearson *pro parte* P

FACU+

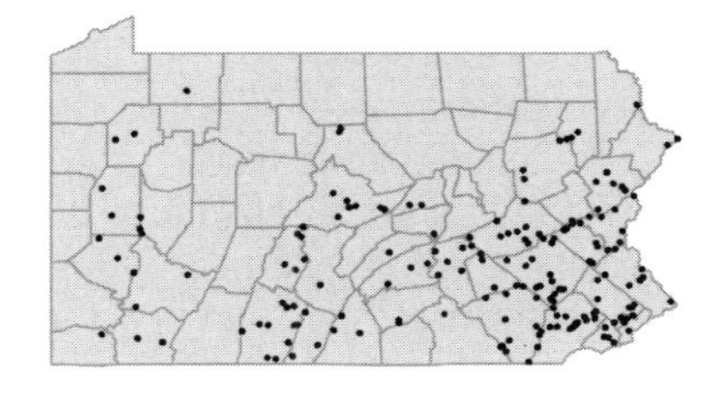

Panicum depauperatum Muhl.
Poverty panic-grass
Herbaceous perennial
Dry, open woods, roadside banks and serpentine barrens.
Dichanthelium depauperatum (Muhl.) Gould K

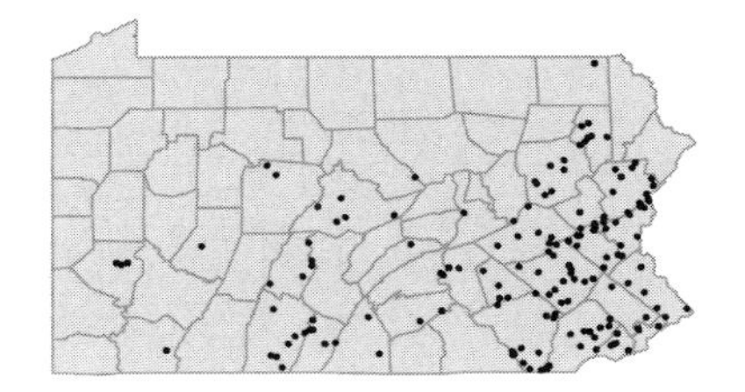

Panicum dichotomiflorum Michx.
Smooth panic-grass
Herbaceous annual
Dry to moist, open woods and meadows.
Panicum proliferum Lam. P

FACW-

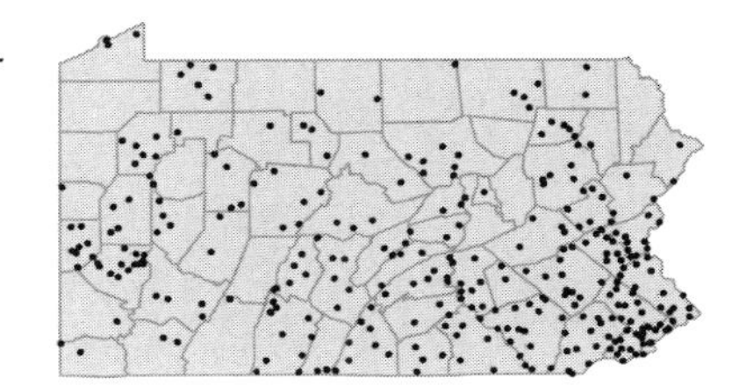

Panicum dichotomum L.
Panic-grass
Herbaceous perennial
Moist, sandy woods.
Dichanthelium dichotomum (L.) Gould *pro parte* K; *Panicum barbulatum* Michx. *pro parte* P; *Panicum nitidum* Lam. *pro parte* P

FAC

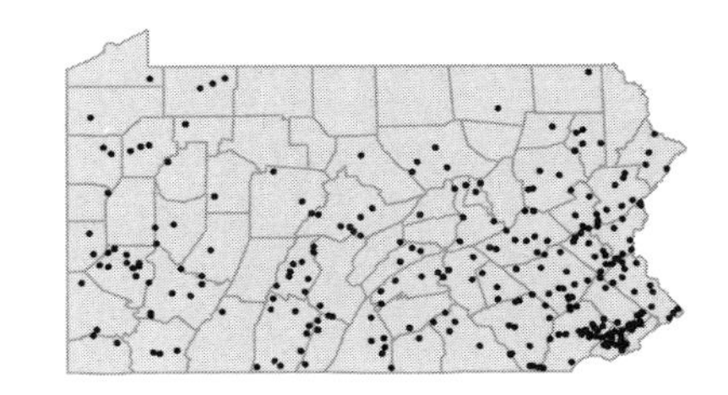

Panicum flexile (Gattinger) Scribn.
Old witch grass
Herbaceous annual
Dry fields, moist meadows, banks and swales.
Panicum capillare L. *pro parte* K

FACU

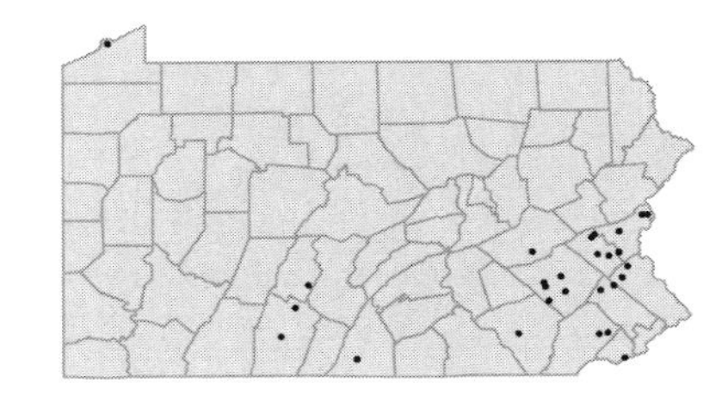

Panicum gattingeri Nash
Witch grass
Herbaceous annual
Cultivated fields, roadsides and waste ground in sandy soils.
Panicum capillare L. *pro parte* BKC

FAC-

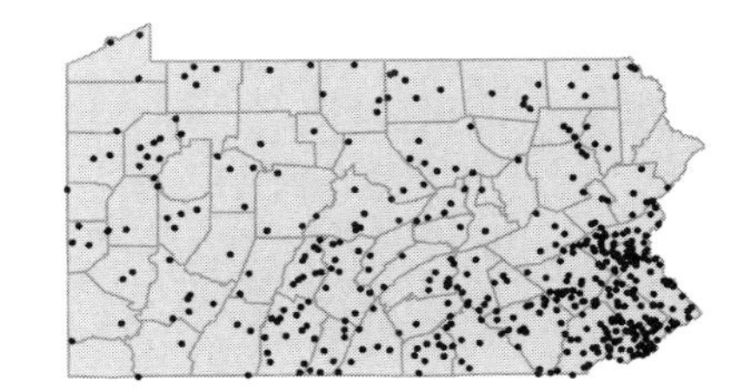

Panicum latifolium L.
Panic-grass
Herbaceous perennial
Roadsides, shores and thickets.
Dichanthelium latifolium (L.) Gould & Clark K; *Panicum macrocarpon* Leconte P

FACU-

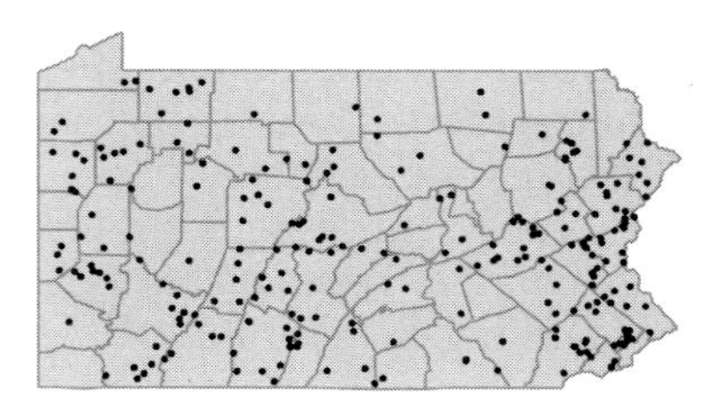

Panicum laxiflorum Lam.
Panic-grass
Herbaceous perennial
Sand barrens.
Dichanthelium laxiflorum (Lam.) Gould K

FACU

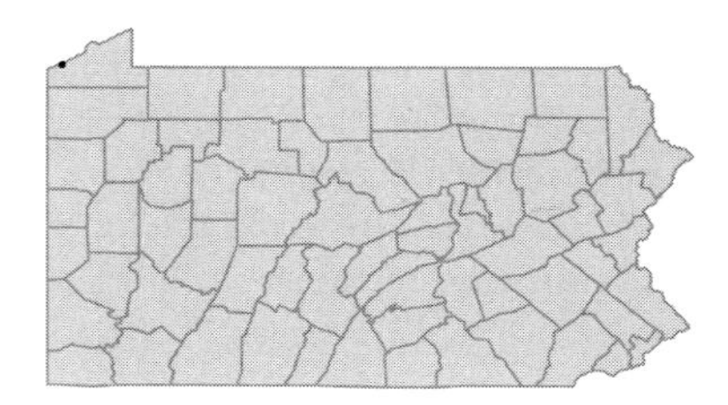

Panicum leibergii (Vasey) Scribn.
Leiberg's panic-grass
Herbaceous perennial
Limestone outcrop. Believed to be extirpated, represented by a single collection from Centre Co. in 1941.
Dichanthelium leibergii (Vasey) Freckmann K

FACU

Panicum linearifolium Scribn.
Panic-grass
Herbaceous perennial
Dry banks, open woods and fields.
Dichanthelium linearifolium (Scribn.) Gould K

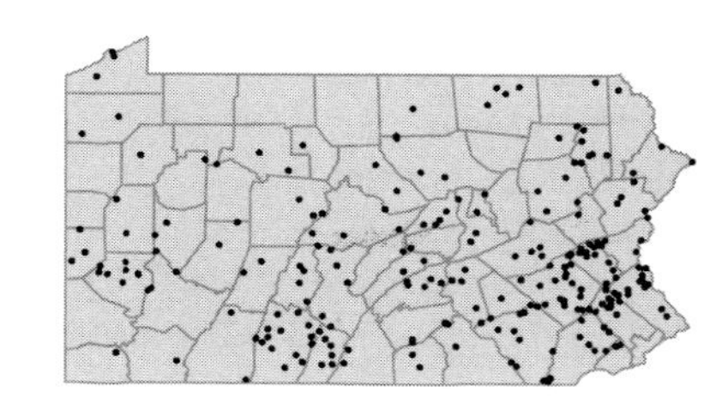

Panicum longifolium Torr.
Long-leaved panic-grass
Herbaceous perennial
Peaty and sandy bogs and shores.
Panicum rigidulum Bosc ex Nees var. *pubescens* (Vasey) Lelong K

OBL

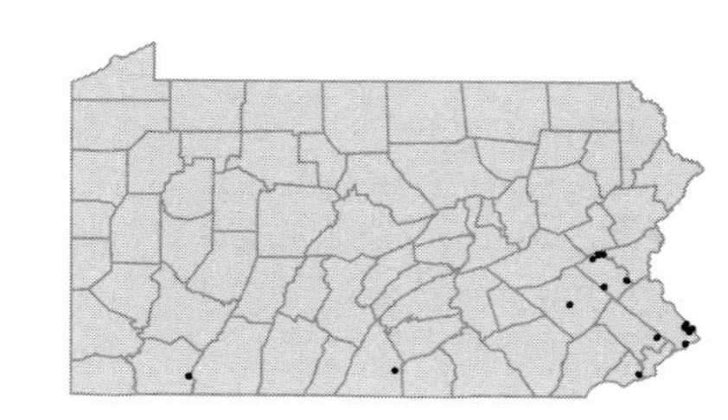

Panicum lucidum Ashe
Shining panic-grass
Herbaceous perennial
Bogs, swamps and boggy swales.
Dichanthelium dichotomum (L.) Gould *pro parte* K; *Panicum dichotomum* L. *pro parte* C

FAC

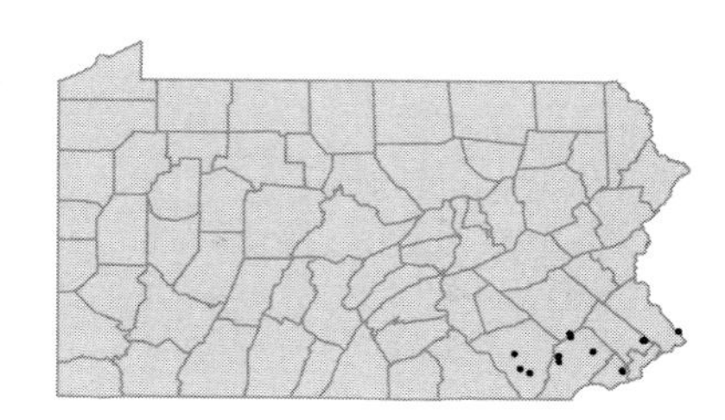

Panicum meridonale Ashe
Panic-grass
Herbaceous perennial
Dry, sandy soil.
Dichanthlium meridonale (Ashe) Freckmann K; *Panicum leucothrix* Nash C

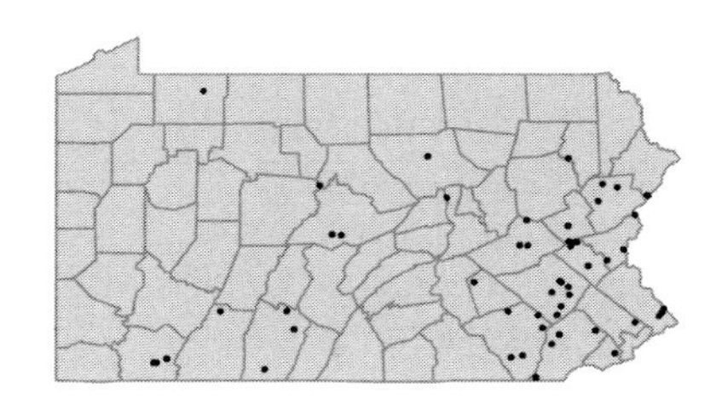

Panicum microcarpon Muhl. FACU
Panic-grass
Herbaceous perennial
Well-drained woods.
Dichanthelium sphaerocarpon (Ell.) Gould var. *isophyllum* (Scribn.) Gould & Clark *pro parte* K; *Panicum dichotomum* L. *pro parte* C

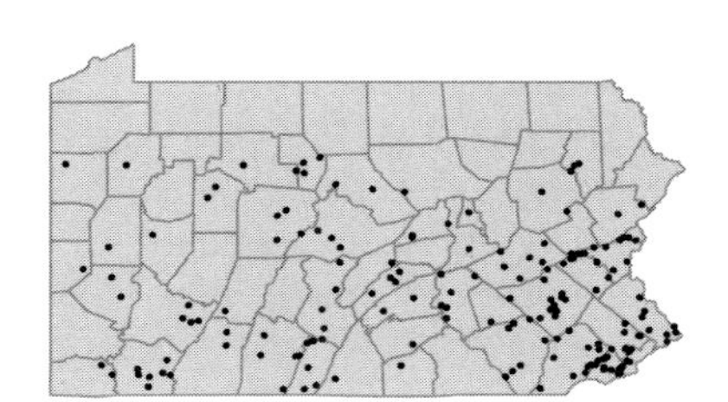

Panicum miliaceum L. I
Broomcorn millet
Herbaceous annual
Cultivated and occasionally escaped.
Panicum hirticaule Presl var. *miliaceum* (Vasey) Beetle K

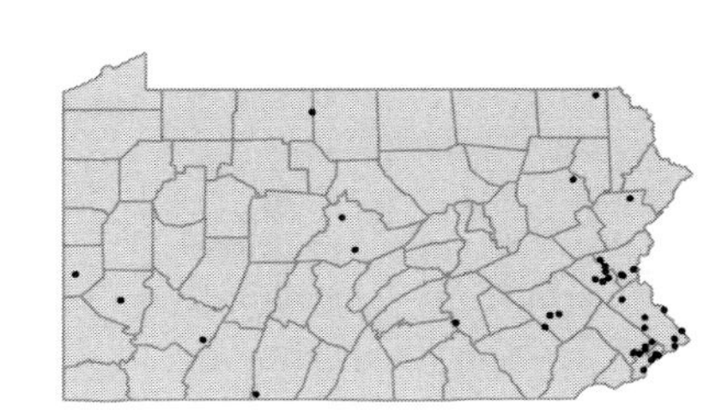

Panicum oligosanthes Schultes FACU
Panic-grass; Heller's witch grass
Herbaceous perennial
Thickets, in loamy or clayey soils.
Dichanthelium oligosanthes (Schultes) Gould K; *Panicum scribneranum* Nash W

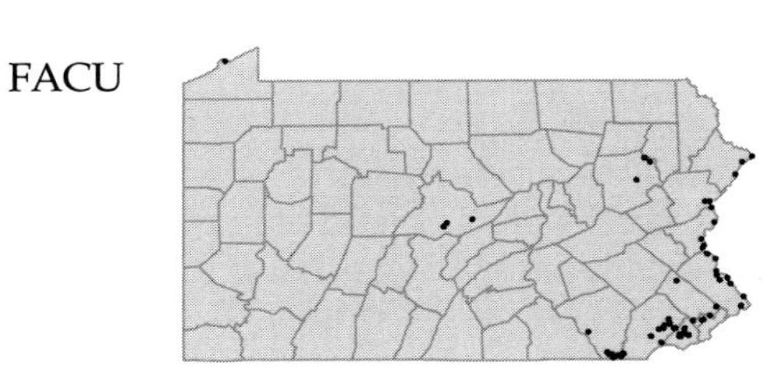

Panicum philadelphicum Bernh. ex Trin. FAC-
Panic-grass
Herbaceous annual
Dry, open woods, fields and roadsides.
Panicum capillare L. *pro parte* K

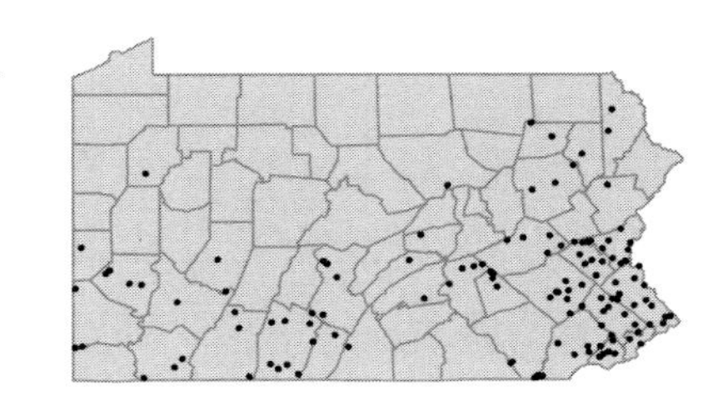

Panicum polyanthes Schultes FACU
Panic-grass
Herbaceous perennial
Roadsides and open woods.
Dichanthelium sphaerocarpon (Ell.) Gould var. *isophyllum* (Scribn.) Gould & Clark *pro parte* K

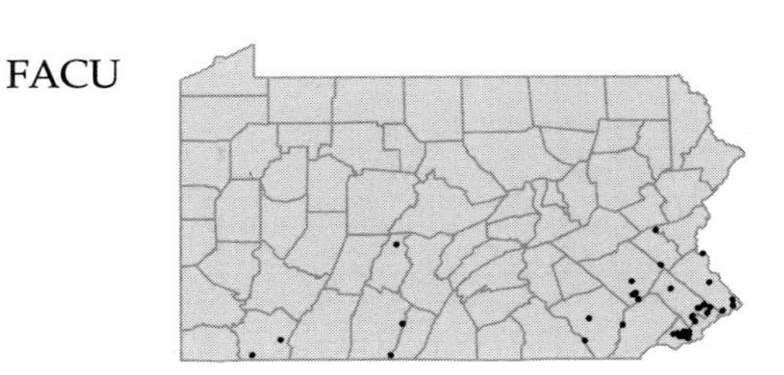

Panicum recognitum Fern.
Fernald's panic-grass
Herbaceous perennial
Woods or wooded slopes.
Dichanthelium scabriusculum (Ell.) Gould & Clark *pro parte* K; *Panicum scabriusculum* Elliot C

OBL

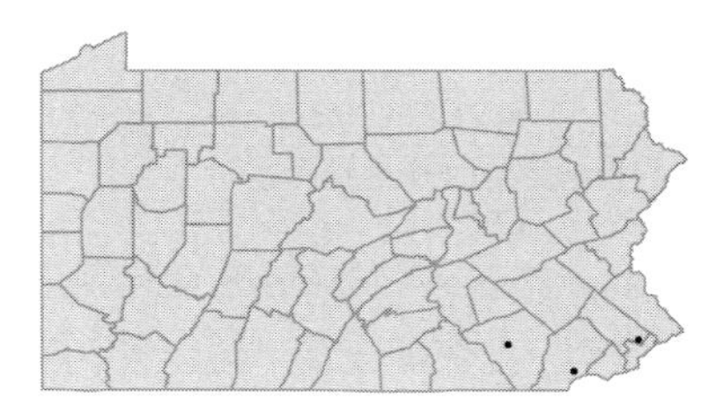

Panicum rigidulum Nees
Panic-grass
Herbaceous perennial
Moist, open, sandy or peaty ground.
Panicum agrostoides Spreng. FWGB; *Panicum condensum* Nash P

FACW+

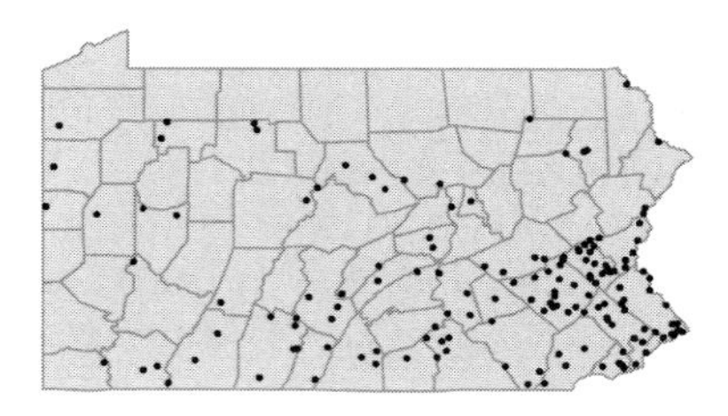

Panicum scoparium Lam.
Velvety panic-grass
Herbaceous perennial
Moist meadows and swales.
Dichanthelium scoparium (Lam.) Gould K

FACW

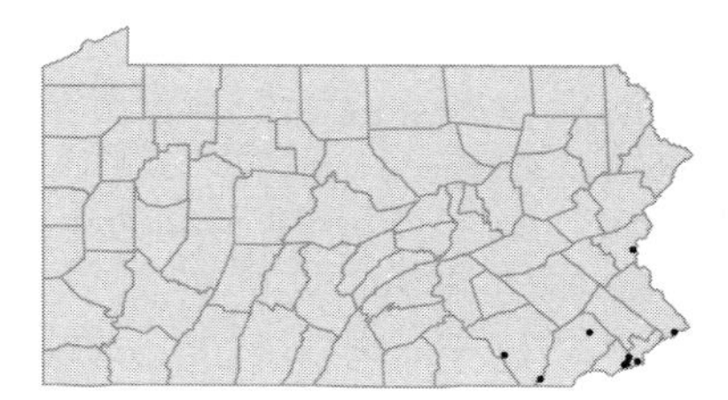

Panicum sphaerocarpon Ell.
Panic-grass
Herbaceous perennial
Dry woods, thickets and old fields.
Dichanthelium sphaerocarpon (Ell.) Gould var. *sphaerocarpon* K

FACU

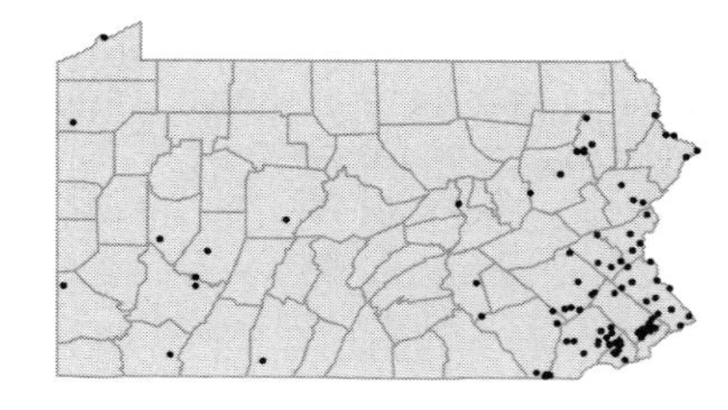

Panicum spretum Schultes
Panic-grass
Herbaceous perennial
Waste ground along railroad tracks. Believed to be extirpated, represented by a single collection from Bucks Co. in 1941.
Dichanthelium spretum (Schultes) Freckmann K

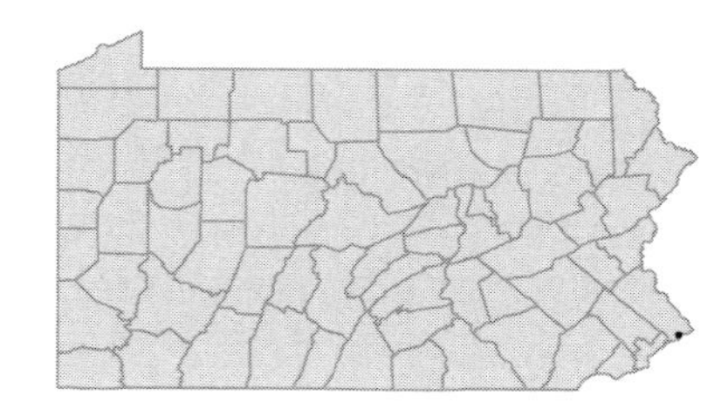

Panicum stipitatum Nash
Panic-grass
Herbaceous perennial
Wet meadows and sandy shores.
Panicum rigidulum Nees. *pro parte* C

FACW+

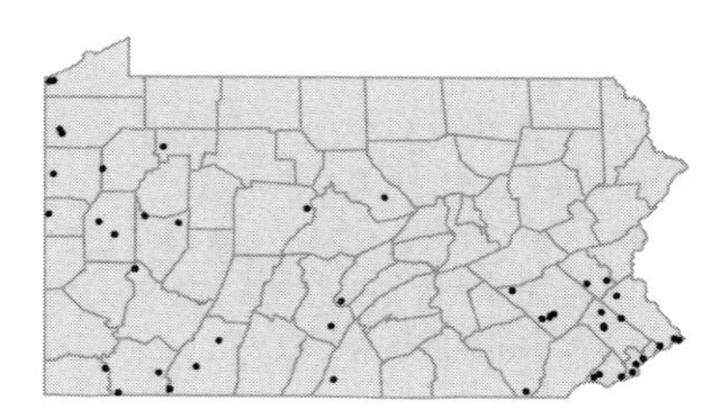

Panicum tuckermanii Fern.
Tuckerman's panic-grass
Herbaceous annual
Sandy flats.
Panicum philadelphicum Bernh. *pro parte* C

FAC-

Panicum verrucosum Muhl.
Panic-grass
Herbaceous annual
Moist, sandy or peaty soil.

FACW

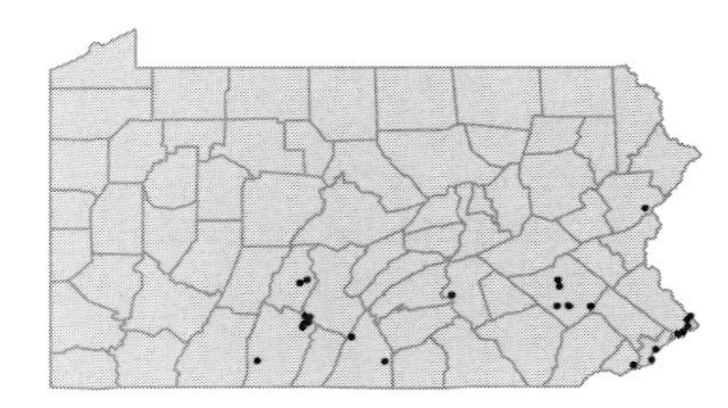

Panicum villosissimum Nash
Long-haired panic-grass
Herbaceous perennial
Dry woods and serpentine barrens.
Dichanthelium villosissimum (Nash) Freckmann K; *Panicum atlanticum* Nash P

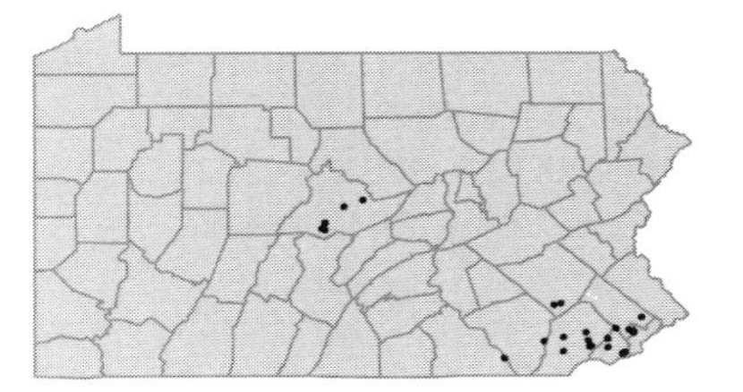

Panicum virgatum L.
Switch grass
Herbaceous perennial
Sandy shores, alluvium, fields and banks.

FAC

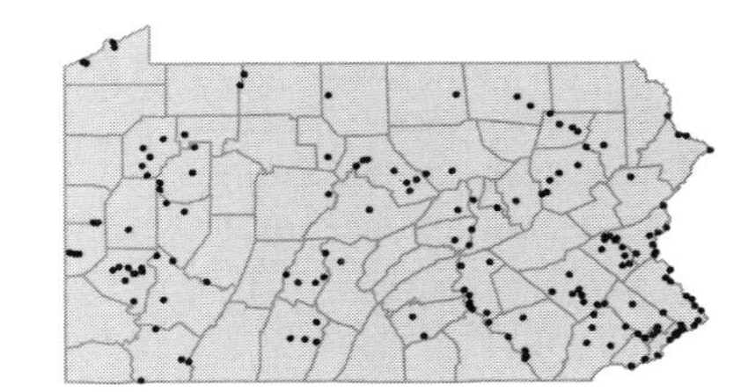

Panicum xanthophysum A.Gray
Slender panic-grass
Herbaceous perennial
Dry, rocky slopes or sandy, open woods.
Dichanthelium xanthophysum (A.Gray) Freckmann K

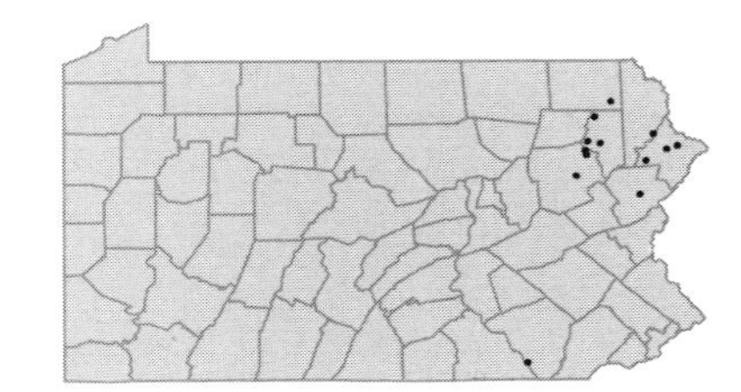

Panicum yadkinense Ashe
Yadkin River panic-grass
Herbaceous perennial
Dry woods.
Dichanthelium dichotomum (L.) Gould *pro parte* K

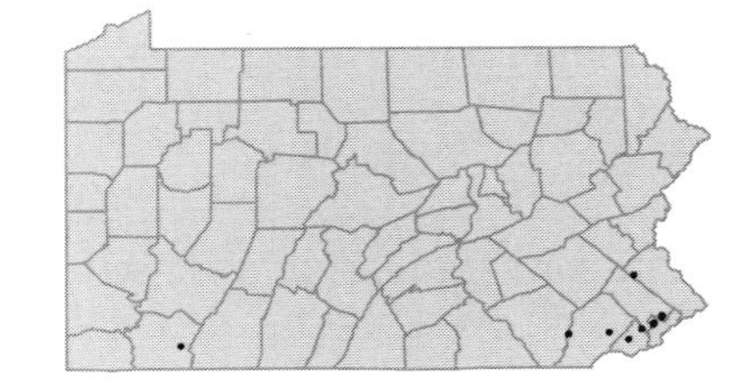

• ***Parapholis incurva*** (L.) C.E.Hubbard
Hard-grass; Thin-tail
Herbaceous annual
Ballast. Represented by collections from Philadelphia Co. ca 1878.
Pholiurus incurvus (L.) Schinz & Thell. FGB

OBL
I

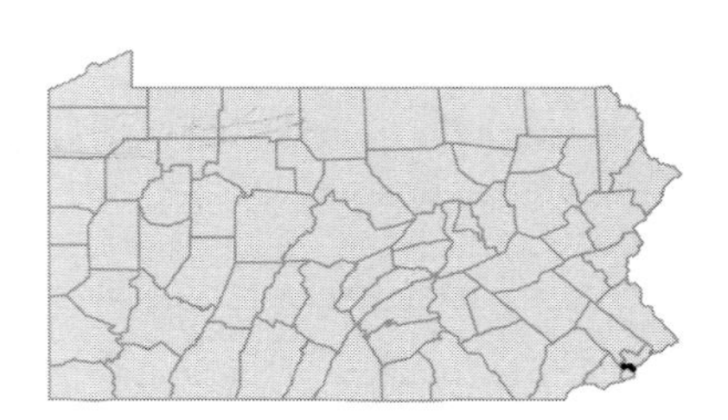

• ***Paspalum floridanum*** Michx. var. ***glabratum*** Engelm. ex Vasey
Florida beadgrass
Herbaceous perennial
Moist, sandy soils. Represented by a single collection from Lancaster Co. in 1933.

FACW

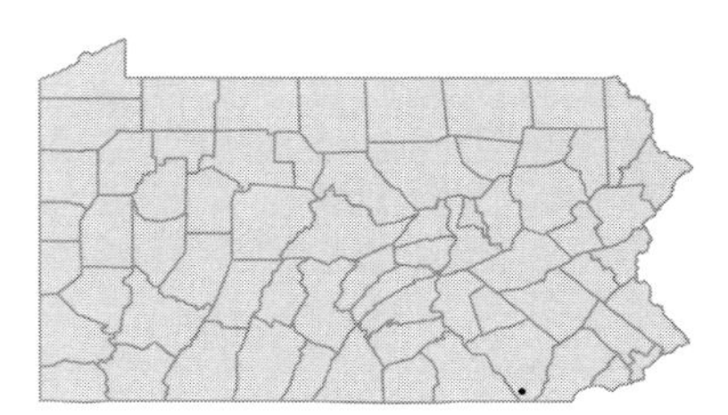

Paspalum laeve Michx. var. ***laeve***
Field beadgrass
Herbaceous perennial
Moist, sandy fields.

FAC+

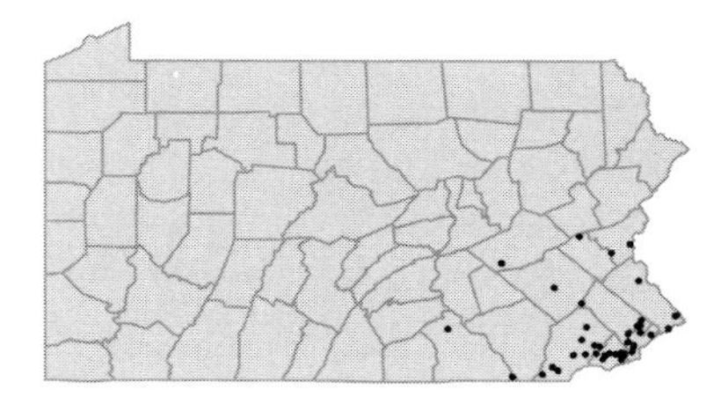

Paspalum laeve Michx. var. ***circulare*** (Nash) Fern.
Field beadgrass
Herbaceous perennial
Moist, sandy fields and shores.

FAC+

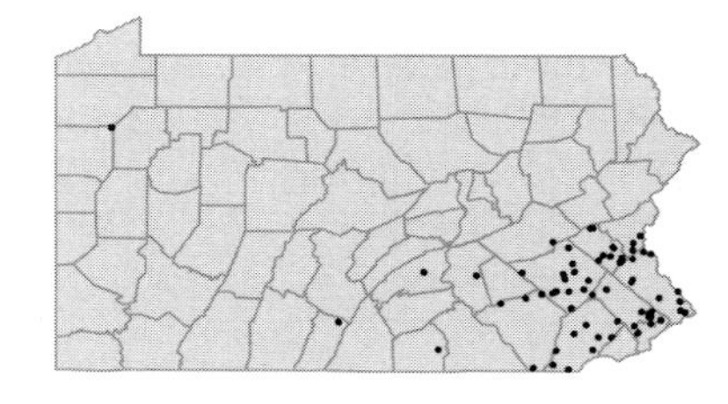

Paspalum laeve Michx. var. ***pilosum*** Scribn.
Field beadgrass
Herbaceous perennial
Moist, sandy fields.

FAC+

Paspalum paspalodes Michx.
Beadgrass
Herbaceous perennial
Ballast. All collections from Philadelphia Co. 1864-1879.
Paspalum distichum L. FGBWC

FACW+
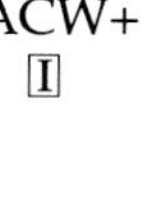

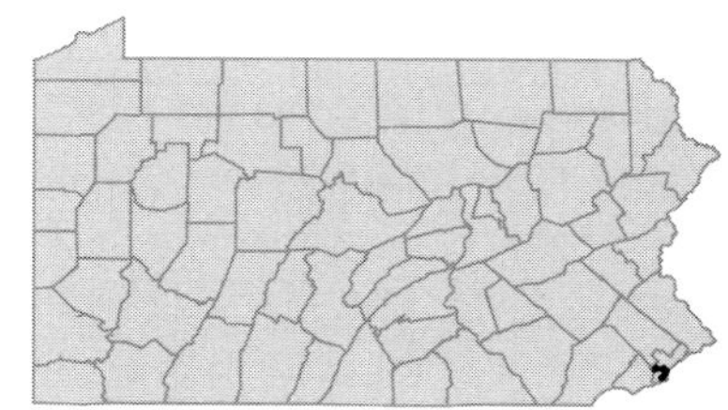

Paspalum pubiflorum Rupr. var. ***glabrum*** Vasey ex Scribn.
Beadgrass
Herbaceous perennial
Waste ground. Represented by a single collection from Delaware Co. in 1935.

FAC
I

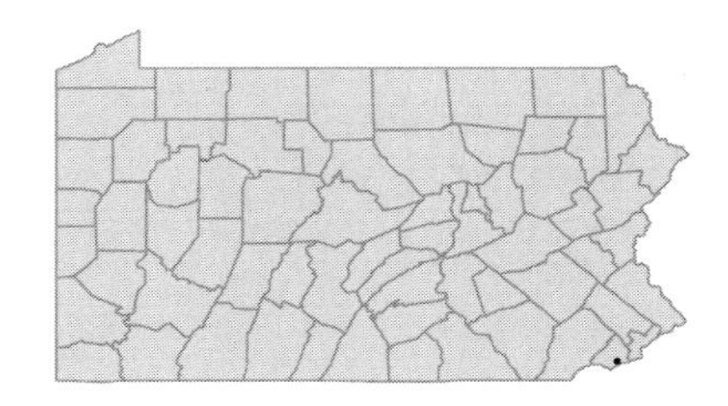

Paspalum setaceum Michx. var. ***setaceum***
Slender beadgrass
Herbaceous perennial
Dry, open ground.

FACU+

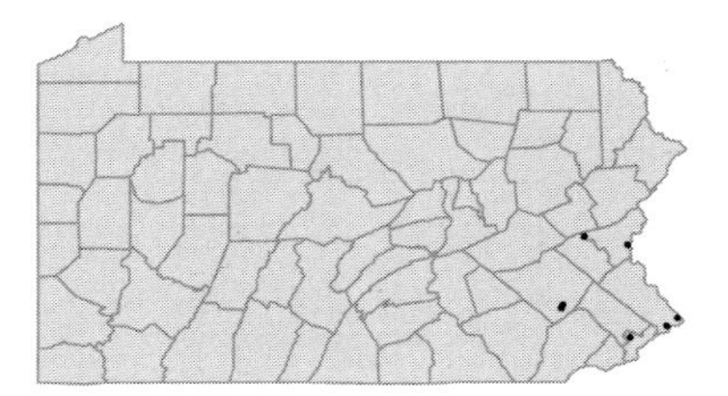

Paspalum setaceum Michx. var. ***muhlenbergii*** (Nash) D.Banks
Slender beadgrass
Herbaceous perennial
Dry to moist, open ground and serpentine barrens.
Paspalum ciliatifolium Michx. var. *muhlenbergii* (Nash) Fern. FB; *Paspalum muhlenbergii* Nash P; *Paspalum pubescens* Muhl. PW

FACU+

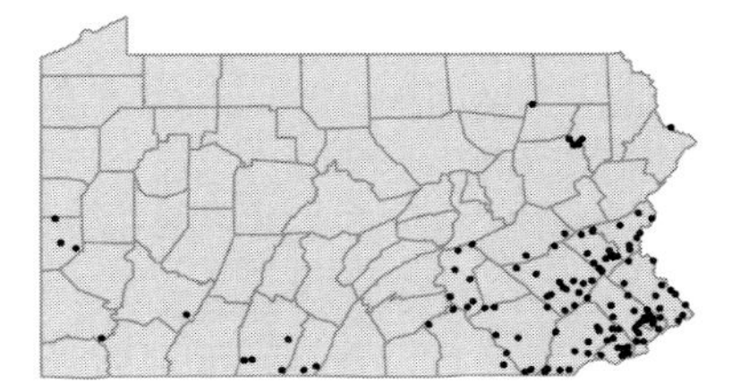

Paspalum setaceum Michx. var. ***psammophilum*** (Nash) D.Banks
Slender beadgrass
Herbaceous perennial
Dry, sandy soil.
Paspalum psammophilum Nash FGBW

FACU+

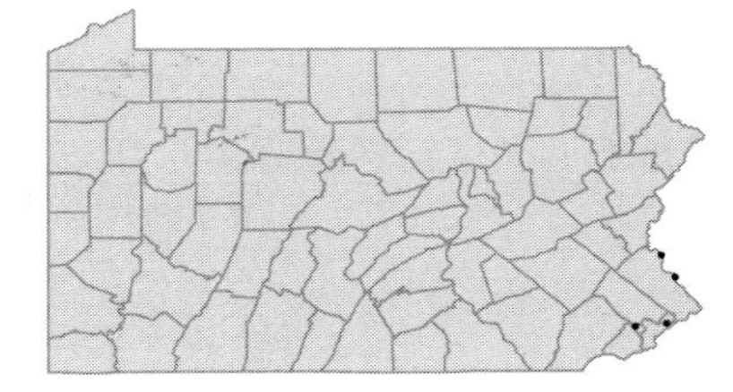

Paspalum setaceum Michx. var. ***supinum*** (Bosc & Poir.) Trin.
Slender beadgrass
Herbaceous perennial
Waste ground and ballast.
Paspalum supinum Bosc W

FACU+
I

• ***Pennisetum alopecuroides*** (L.) Spreng.
Fountain grass
Herbaceous perennial
Cultivted and rarely escaped to roadsides and abandoned fields.

I

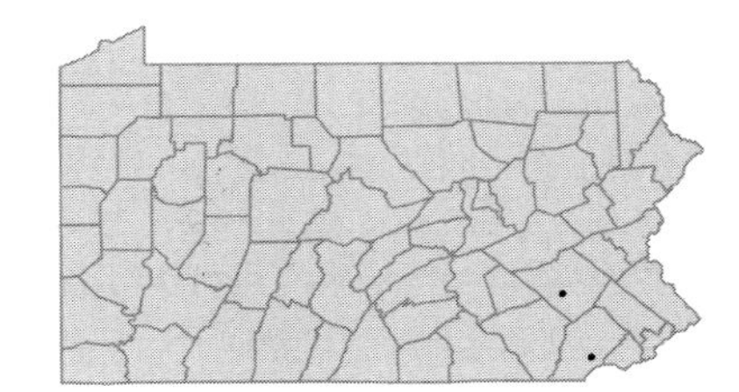

• ***Phalaris arundinacea*** L.
Reed canary-grass
Herbaceous perennial
Marshes, alluvial meadows, shores and ditches.

FACW+

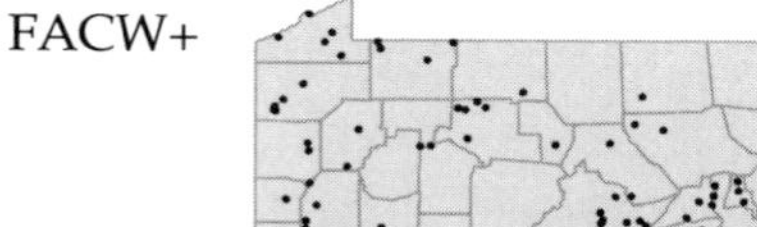

Phalaris canariensis L.
Canary-grass
Herbaceous annual
Waste ground.

FACU
[I]

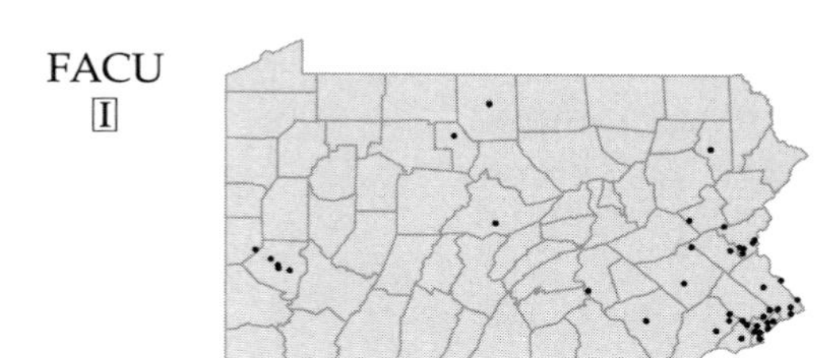

Phalaris paradoxa L.
Canary-grass
Herbaceous annual
Ballast, waste ground and rubbish dump.

[I]

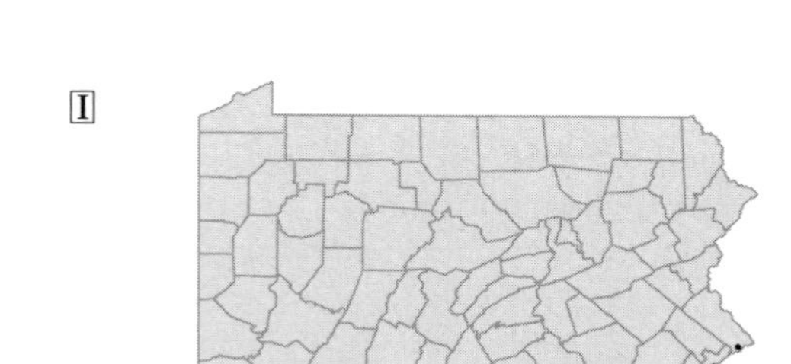

• ***Phleum pratense*** L.
Timothy
Herbaceous perennial
Fields, meadows and roadsides.

FACU
[I]

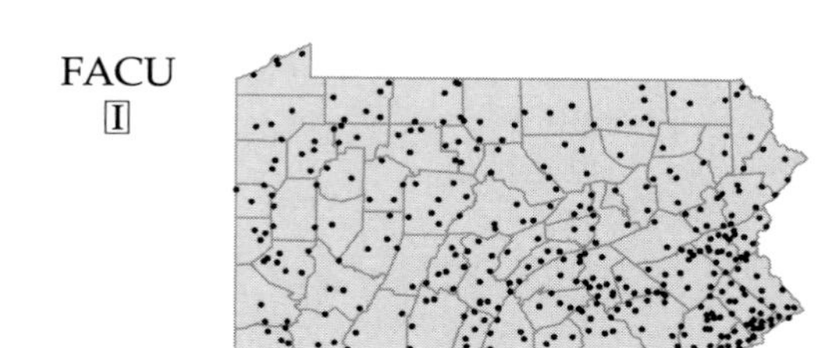

Phleum subulatum (Savi) Aschers. & Graebn.
Timothy
Herbaceous perennial
Ballast. All specimens collected at a single site in Philadelphia Co. in 1879.

[I]

• ***Phragmites australis*** (Cav.) Trin. ex Steud.
Common reed
Herbaceous perennial
Marshes, ditches and moist disturbed ground.
Phragmites communis Trin. FGB; *Phragmites phragmites* (L.) Karst. P

FACW

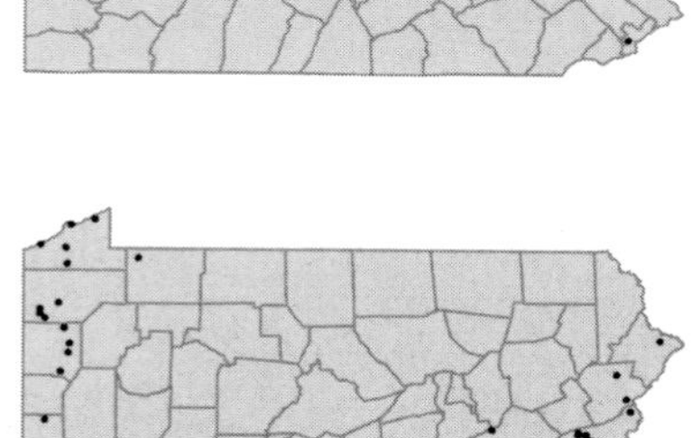

• ***Piptochaetium avenaceum*** (L.) Parodi
Black oatgrass
Herbaceous perennial
Thin, rocky woods and sandy, open ground.
Stipa avenacea L. PFGBW

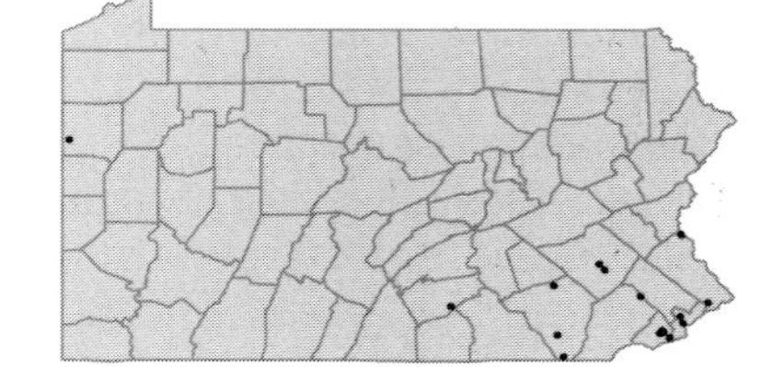

• ***Poa alsodes*** A.Gray
Woodland bluegrass
Herbaceous perennial
Cool, moist woods and thickets.

FACW-

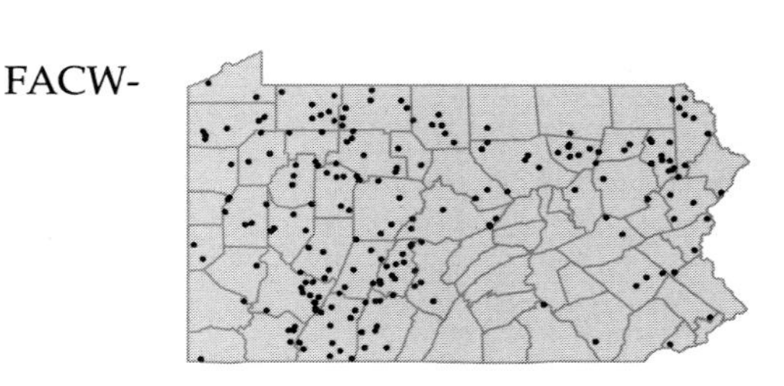

Poa annua L.
Annual bluegrass
Herbaceous annual
Roadsides, lawns, open woods and moist, alluvial soils.

FACU
[I]

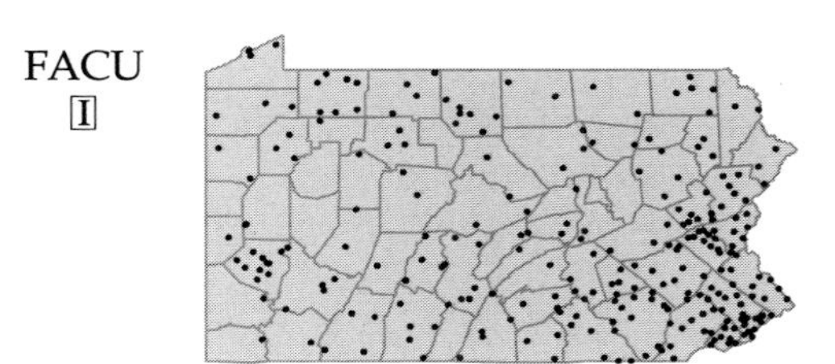

Poa autumnalis Muhl. ex Ell.
Autumn bluegrass
Herbaceous perennial
Moist woods.

FAC

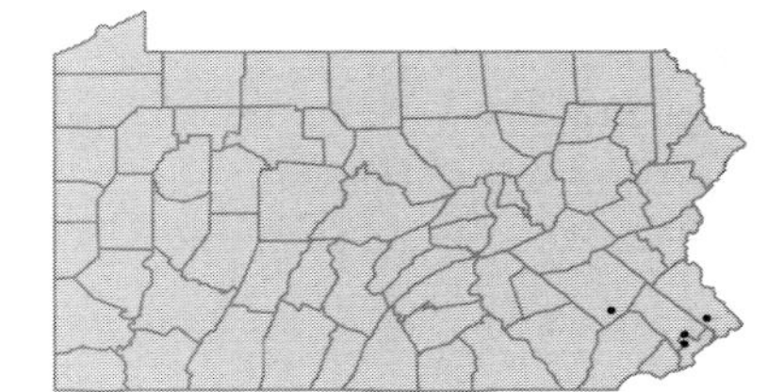

Poa bulbosa L.
Bulbous bluegrass
Herbaceous perennial
Waste ground.

[I]

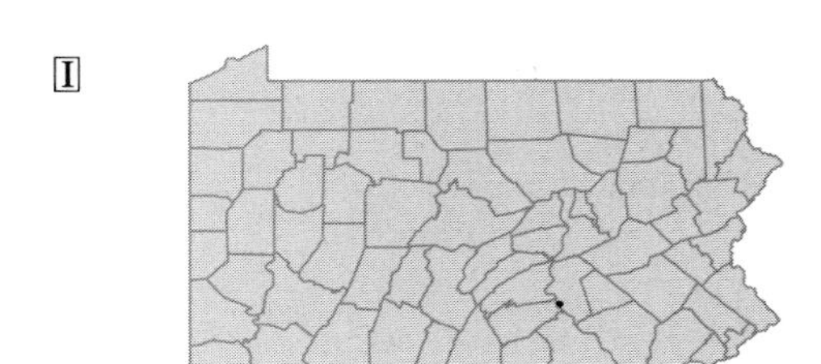

Poa compressa L.
Canada bluegrass
Herbaceous perennial
Dry woods, fields and rock outcrops.

FACU
[I]

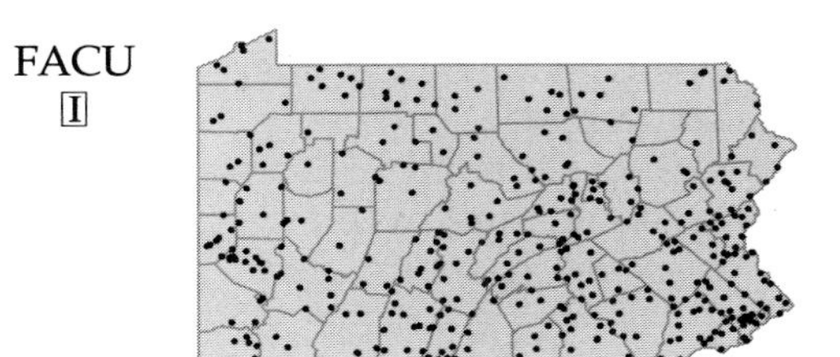

Poa cuspidata Nutt.
Bluegrass
Herbaceous perennial
Dry, wooded hillsides and banks.
Poa brevifolia Muhl. P

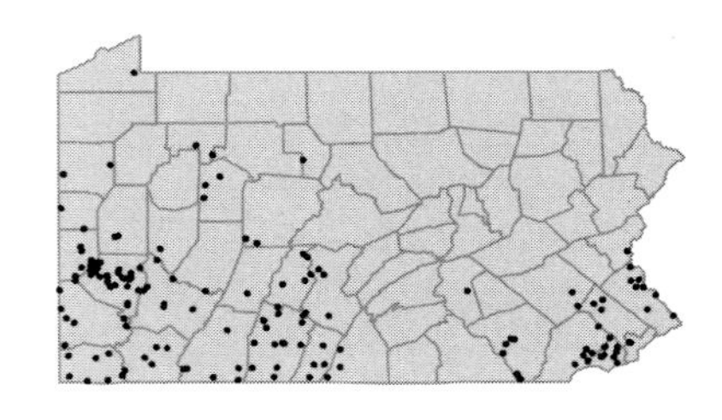

Poa glauca Vahl

I

Bluegrass
Herbaceous perennial
Dry soil. Represented by a single collection from Erie Co. in 1884.

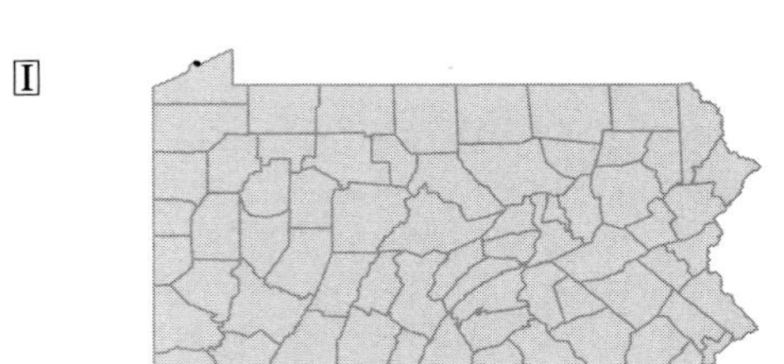

Poa languida A.S.Hitchc.
Woodland bluegrass; Drooping bluegrass
Herbaceous perennial
Moist woods and fen.

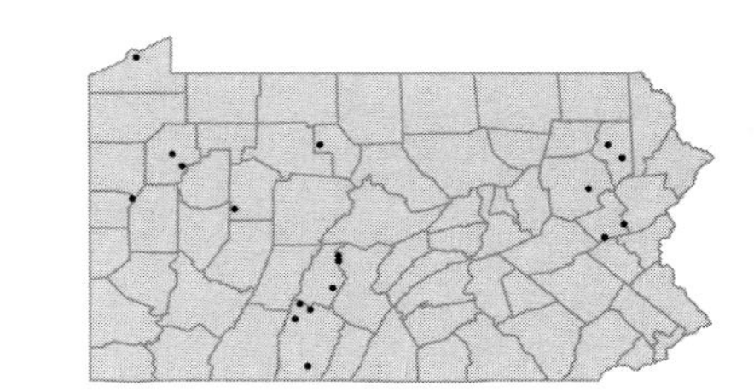

Poa laxa Haenke

I

Bluegrass
Herbaceous perennial
Calcareous fill. Represented by two collections from Erie Co. 1891-1900.

Poa nemoralis L.

FAC

Wood bluegrass
Herbaceous perennial
Dry woods and edges.

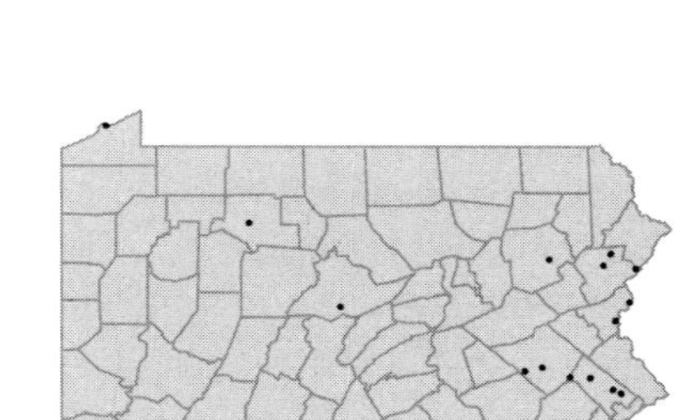

Poa paludigena Fern. & Wieg.

FACW+

Bog bluegrass
Herbaceous perennial
Boggy woods and swamps.

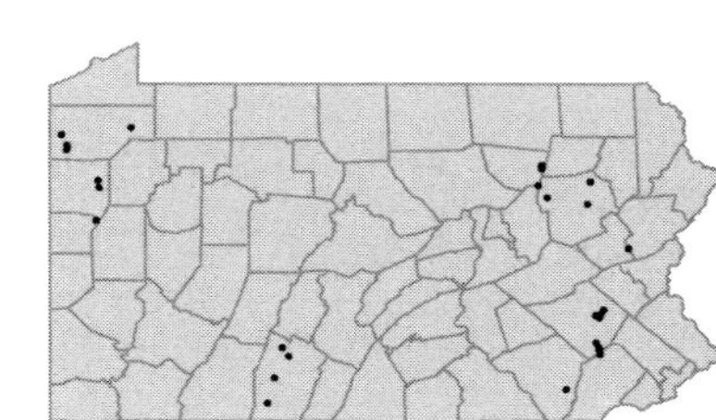

Poa palustris L.
Fowl bluegrass
Herbaceous perennial
Wet meadows, shores and thickets.
Poa serotina Ehrh. P

FACW

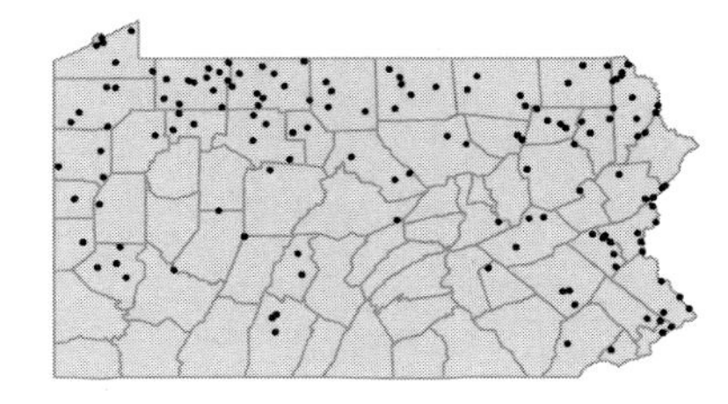

Poa pratensis L.
Kentucky bluegrass
Herbaceous perennial
Cultivated and widely naturalized in meadows, roadsides, open woods and waste ground.

FACU
I

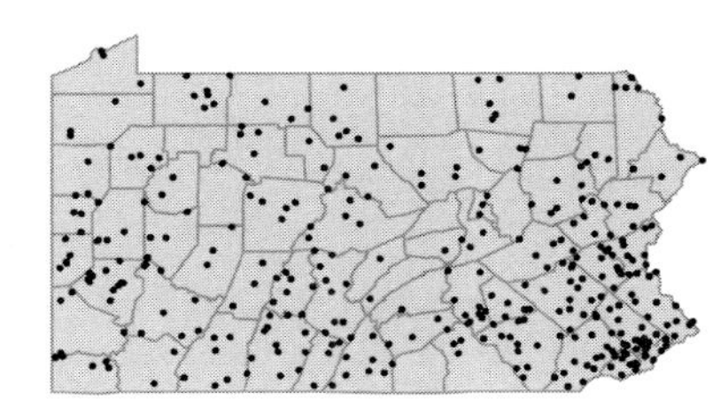

Poa saltuensis Fern. & Wieg.
Old-pasture bluegrass
Herbaceous perennial
Woods.

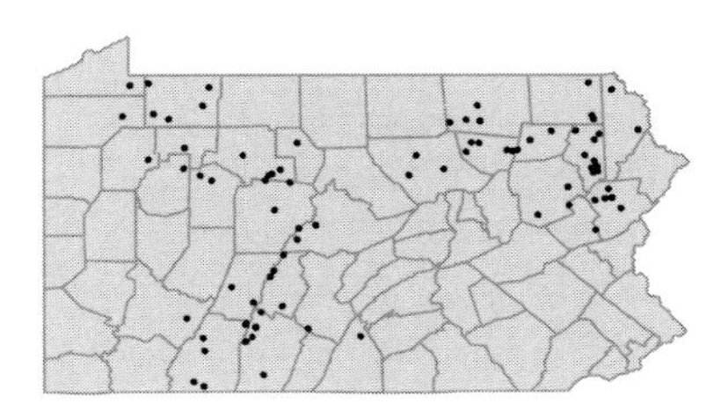

Poa sylvestris A.Gray
Woodland bluegrass
Herbaceous perennial
Rich woods.

FACW

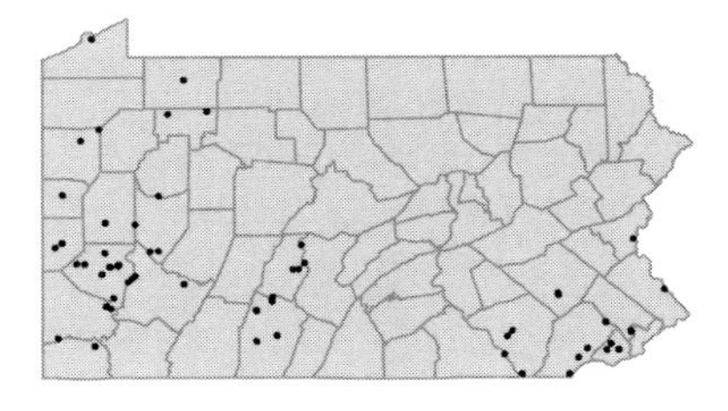

Poa trivialis L.
Rough bluegrass
Herbaceous perennial
Cultivated and frequently established in wet meadows, swamps and alluvial woods.

FACW
I

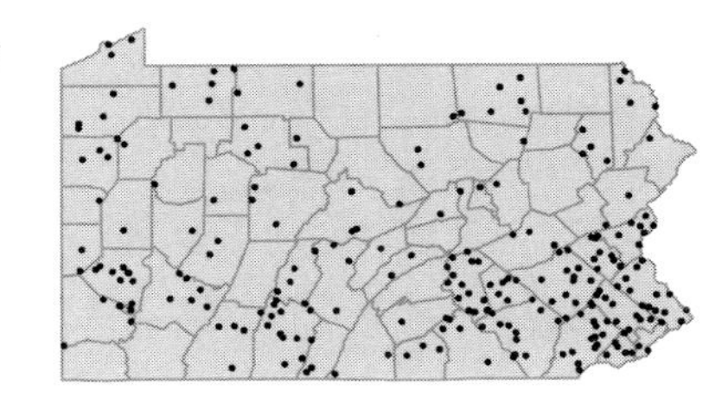

• ***Polypogon monspeliensis*** (L.) Desf.
Beardgrass; Rabbitfoot grass
Herbaceous annual
Ballast ground and rubbish dumps.

FACW+
I

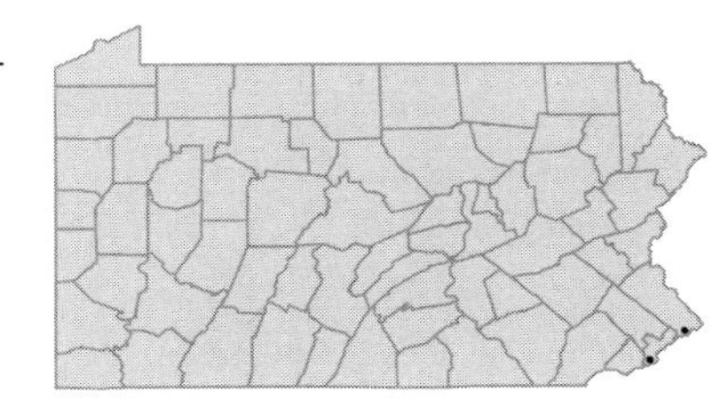

• ***Pseudosasa japonica*** (Sieb. & Zucc. ex Steud.) Makino
Bamboo
Herbaceous perennial
Cultivated and occasionally spreading to railroad embankments and waste ground.

[I]

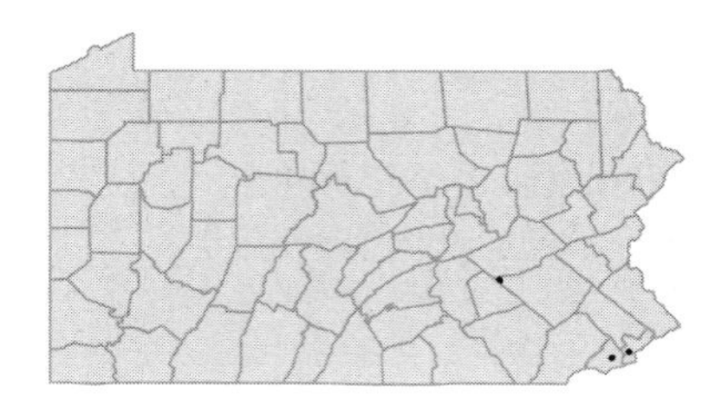

• ***Puccinellia distans*** (L.) Parl.
Alkali grass; Goosegrass
Herbaceous perennial
Roadsides, waste ground and ballast.

OBL
[I]

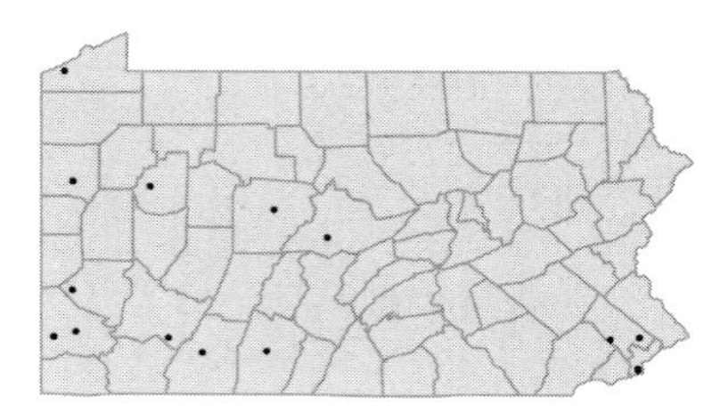

Puccinellia fasciculata (Torr.) Bickn.
Saltmarsh goosegrass
Herbaceous perennial
Waste ground and ballast. Represented by several collections from Philadelphia Co. 1862-1879.

OBL
[I]

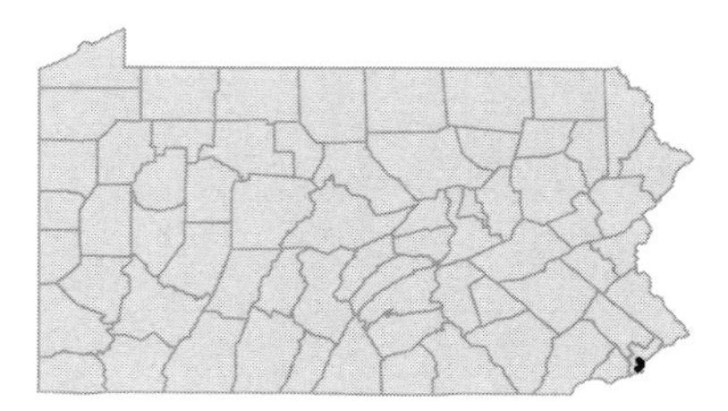

Puccinellia maritima (Huds.) Parl.
Seaside speargrass
Herbaceous perennial
Ballast and waste ground. Represented by several collections from Philadelphia Co. 1865-1870.

OBL
[I]

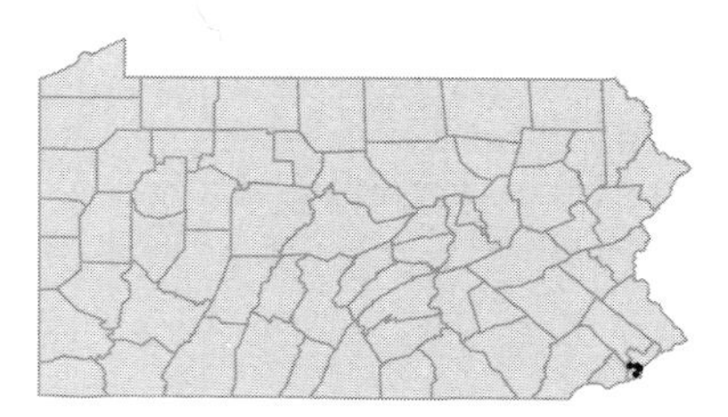

Puccinellia rupestris (With.) Fern. & Weatherby
Goosegrass
Herbaceous perennial
Ballast. Represented by a single collection from Philadelphia Co. in 1876.

[I]

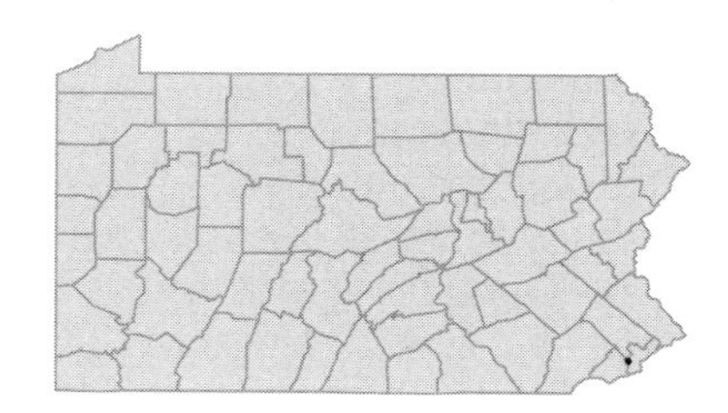

• ***Schizachne purpurascens*** (Torr.) Swallen
Grass
Herbaceous perennial
Rich, dry woods.

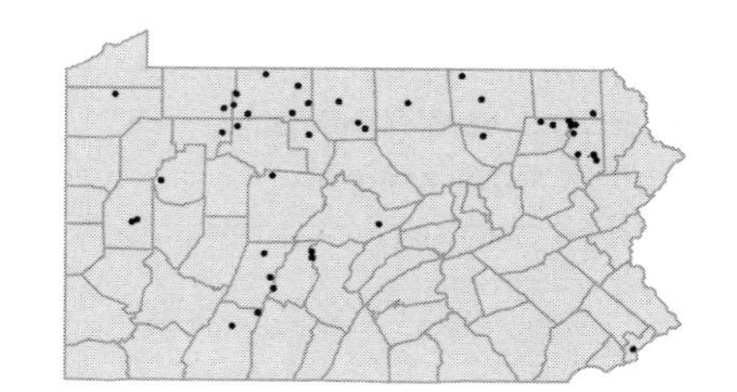

• ***Schizachyrium scoparium*** (Michx.) Nash var. ***scoparium***
Little bluestem
Herbaceous perennial
Old fields, roadsides and open woods.
Andropogon scoparius Michx. PFGBW

FACU

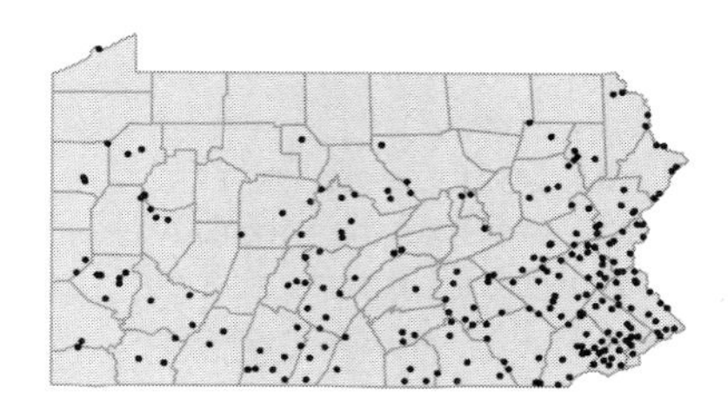

Schizachyrium scoparium (Michx.) Nash var. ***littorale*** (Nash) Gould
Seaside bluestem
Herbaceous perennial
Sandy shores or dunes.
Andropogon scoparius Michx. var. *littoralis* (Nash) A.S.Hitchc. FGBW

FACU

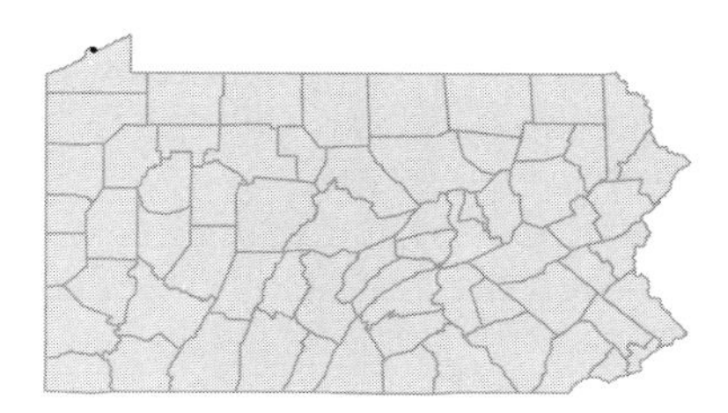

• ***Secale cereale*** L.
Rye
Herbaceous annual
Cultivated and occasionally occurring in fallow land or field margins.

I

• ***Setaria faberi*** Herrm.
Giant foxtail
Herbaceous annual
Cultivated fields, roadsides and waste ground.

UPL
I

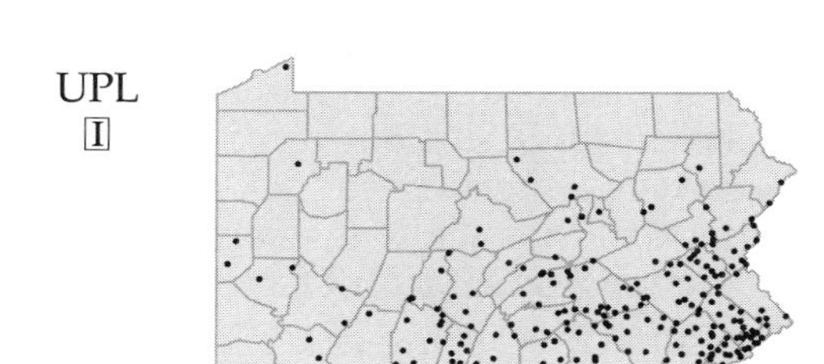

Setaria geniculata (Lam.) Beauv.
Perennial foxtail
Herbaceous perennial
Dry to moist, open soil.

FAC

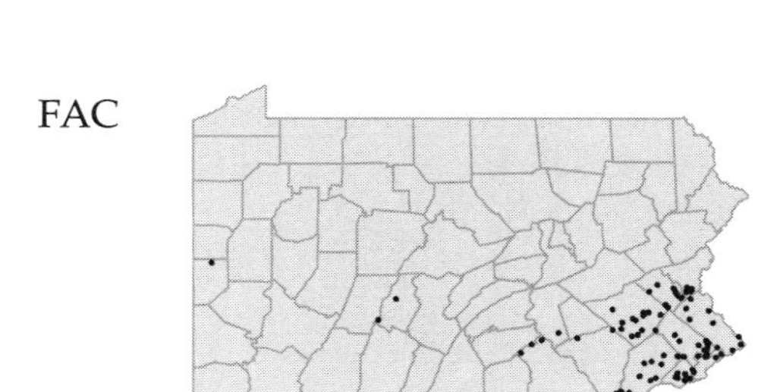

Setaria italica (L.) Beauv.
Foxtail millet; Italian millet
Herbaceous annual
Cultivated and occasionally escaping to roadsides and waste ground.
Chaetochloa italica (L.) Scribn. P; *Setaria viridis* (L.) Beauv. ssp. *italica* (L.) Briq. C

FACU
I

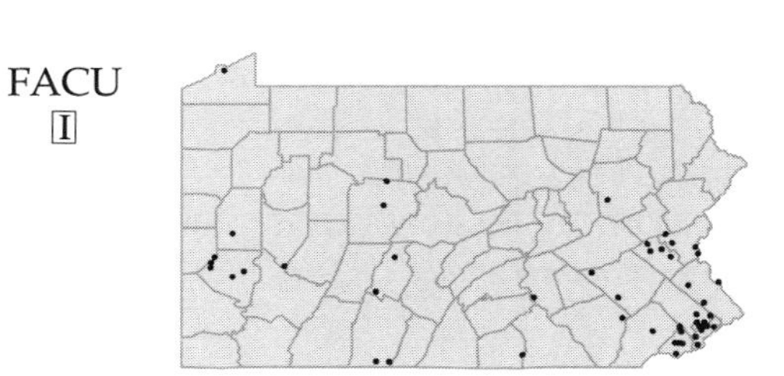

Setaria pumila (Poir.) Schultes
Yellow foxtail
Herbaceous annual
Fields, gardens and roadsides.
Chaetochloa glauca (L.) Scribn. P; *Setaria glauca sensu. auctt., non* (L.) Scribn. FGBC; *Setaria lutescens* (Weigel) F.T. Hubbard W

FAC
[I]

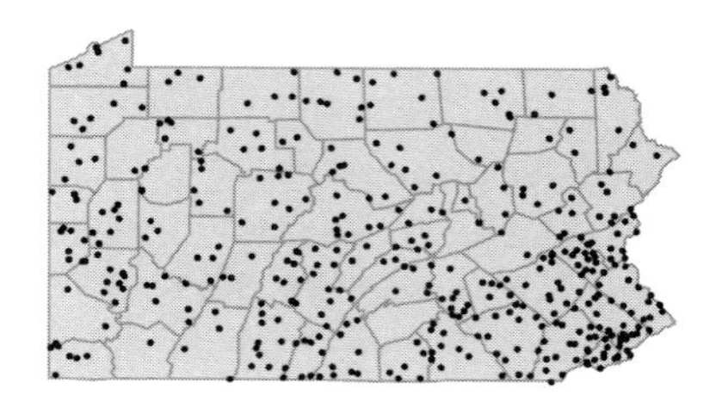

Setaria verticillata (L.) Beauv. var. ***verticillata***
Bristly foxtail
Herbaceous annual
Waste ground and roadsides.
Chaetochloa verticillata (L.) Scribn. P

FAC
[I]

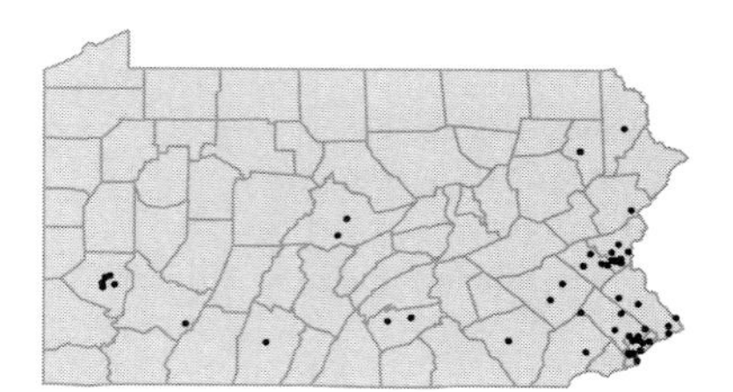

Setaria verticillata (L.) Beauv. var. ***ambigua*** (Guss.) Parl.
Bristly foxtail
Herbaceous annual
Urban waste ground.

FAC
[I]

Setaria viridis (L.) Beauv. var. ***viridis***
Green foxtail
Herbaceous annual
Cultivated fields, gardens and waste ground.
Chaetochloa viridis (L.) Scribn. P

[I]

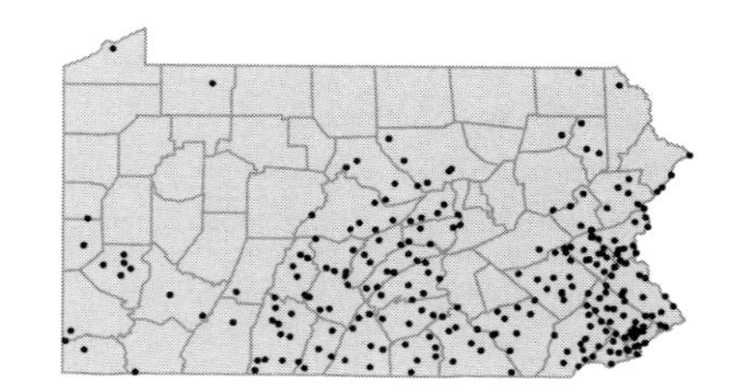

Setaria viridis (L.) Beauv. var. ***major*** (Gaudin) Pospichal
Green foxtail
Herbaceous annual
Fields and waste grounds.

[I]

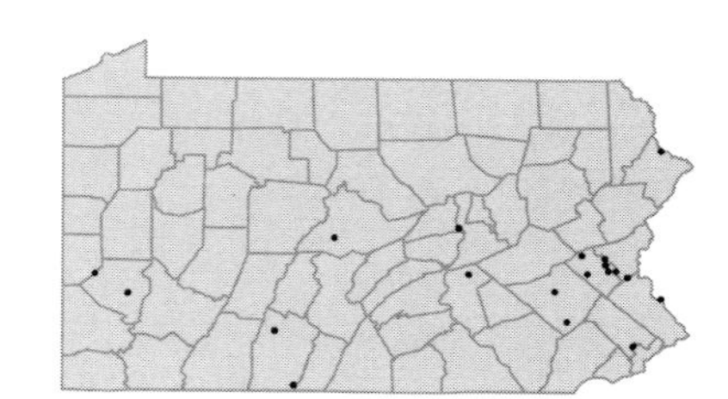

• ***Sorghastrum nutans*** (L.) Nash
Indian grass
Herbaceous perennial
Moist or dry fields, roadsides and serpentine barrens.
Sorghastrum avenaceum (Michx.) Nash P

UPL

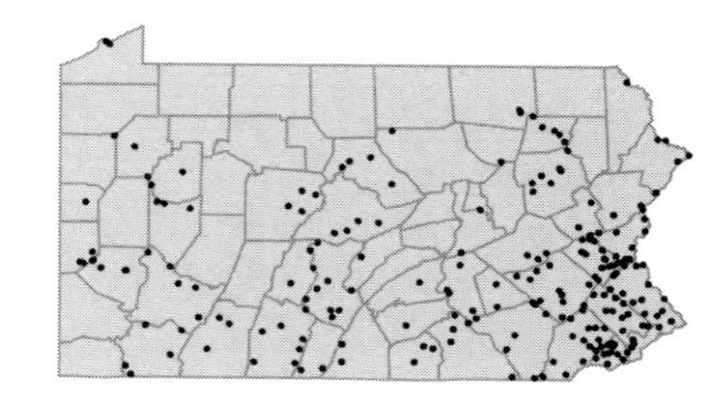

• ***Sorghum bicolor*** (L.) Moench ssp. ***bicolor***
Broom-corn
Herbaceous annual
Cultivated and occasionally spreading to waste ground.
Sorghum vulgare Pers. FBW

UPL
[I]

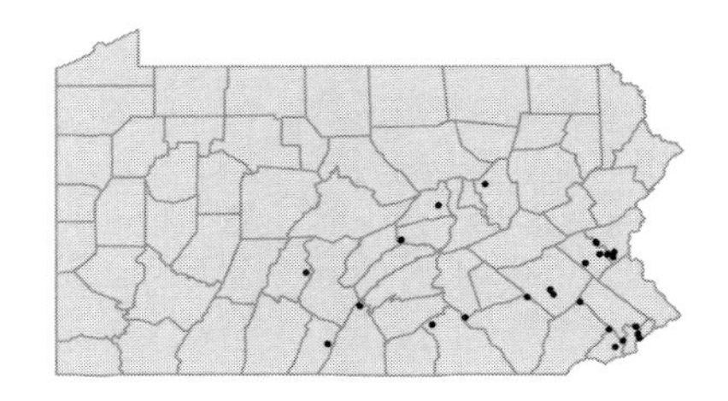

Sorghum bicolor (L.) Moench ssp. ***drummondii*** (Steud.) DeWet & Harlan
Shattercane
Herbaceous annual
Cultivated fields. Designated as a noxious weed in PA.
Sorghum sudanense (Piper) Stapf W

UPL
[I]

Sorghum halepense (L.) Pers.
Johnsongrass
Herbaceous perennial
Cultivated fields and waste ground. Designated as a noxious weed in PA.

FACU
[I]

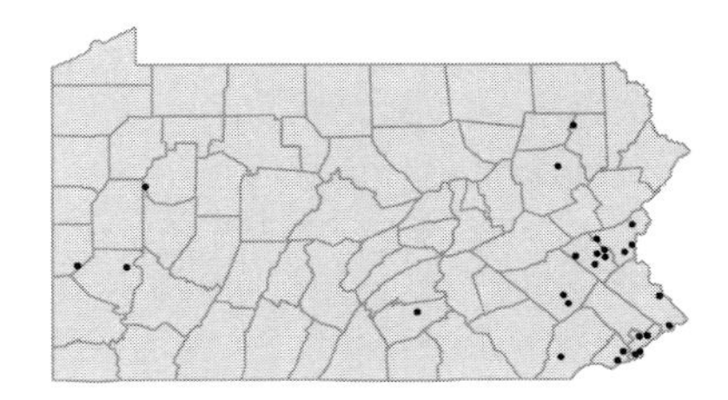

• ***Spartina patens*** (Ait.) Muhl.
Salt-meadow grass
Herbaceous perennial
Ballast. Represented by several pre-1900 collections from Philadelphia Co.

FACW+

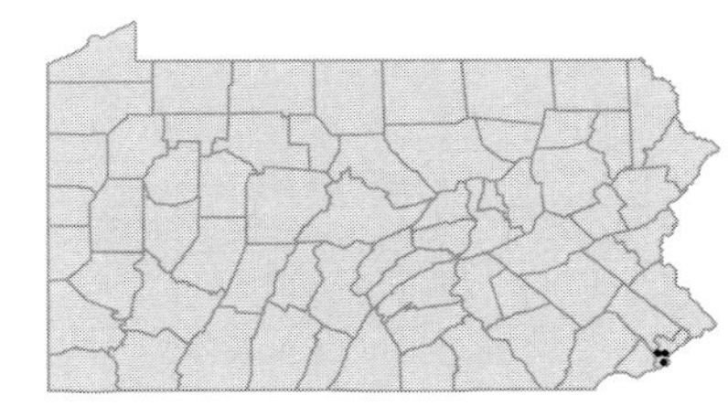

Spartina pectinata Link
Freshwater cordgrass
Herbaceous perennial
Sandy shores and alluvial flats.

OBL

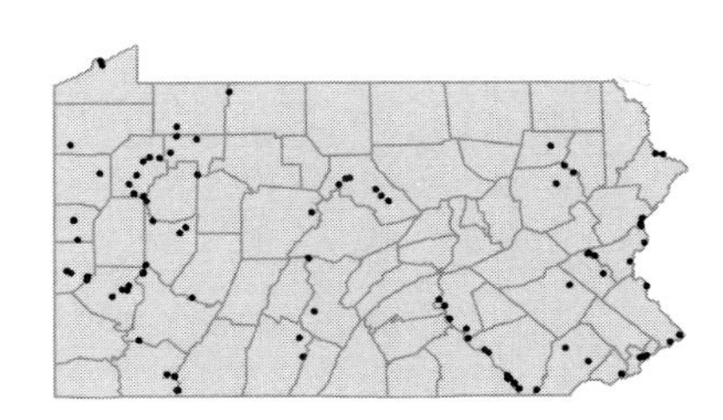

• ***Sphenopholis nitida*** (Biehler) Scribn.
Wedge-grass
Herbaceous perennial
Dry, rocky woods and roadside banks.

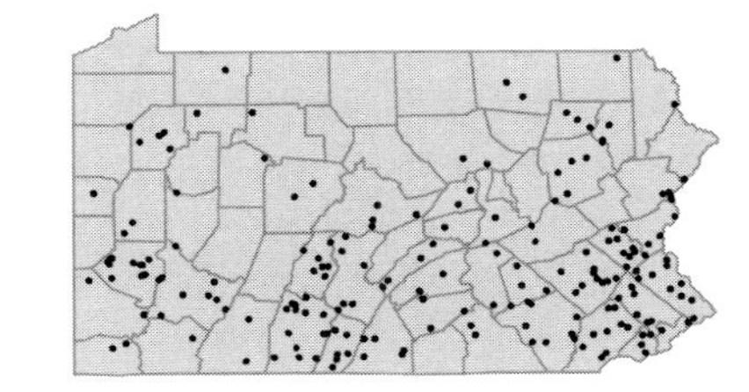

Sphenopholis obtusata (Michx.) Scribn. var. ***obtusata*** FAC-
Prairie wedge-grass
Herbaceous perennial
Serpentine barrens and other dry, open sites.
Sphenopholis obtusata (Michx.) Scribn. var. *pubescens* (Scribn. & Merr.) Scribn. FW

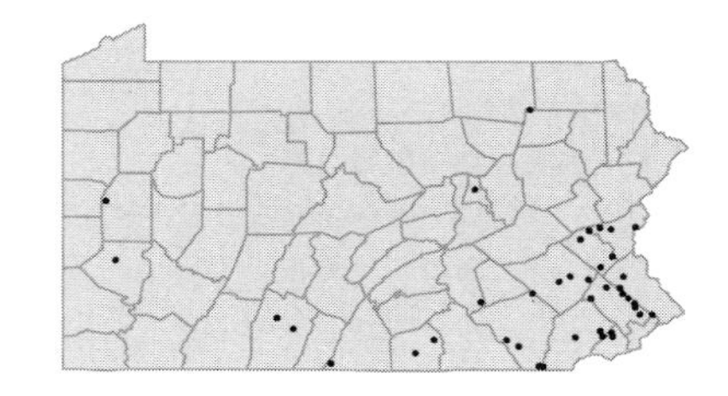

Sphenopholis obtusata (Michx.) Scribn. var. ***major*** (Torr.) K.S.Erdman FAC-
Slender wedge-grass
Herbaceous perennial
Stream banks, open woods and alluvial thickets.
Sphenopholis intermedia (Rydb.) Rydb. FGBW

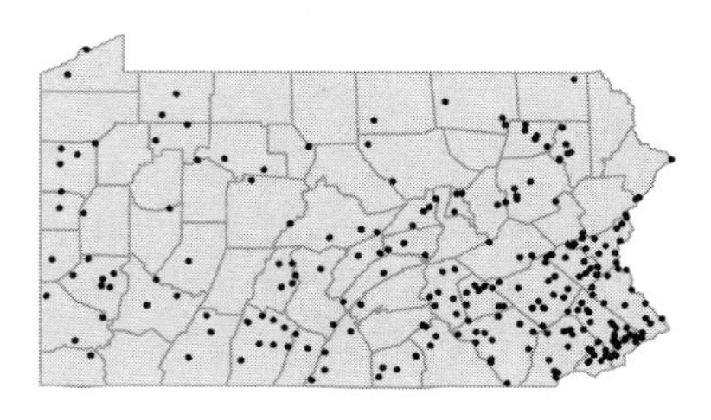

Sphenopholis obtusata x pensylvanica FAC
Wedge-grass
Herbaceous perennial
Wet woods, moist slopes and springheads.
Sphenopholis x pallens (Biehler) Scribn. FGBC; *Sphenopholis nitida x obtusata* C

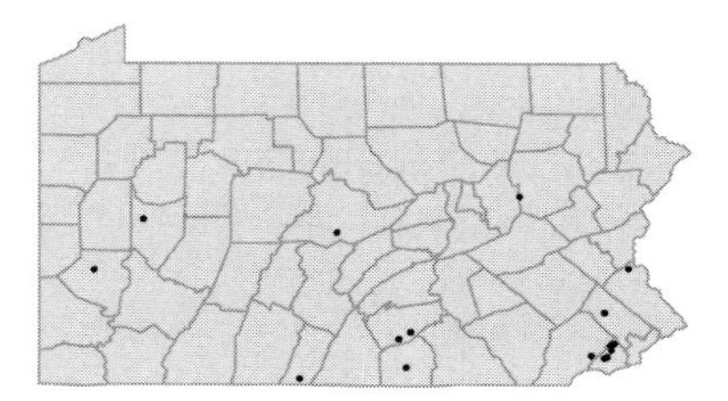

Sphenopholis pensylvanica (L.) A.S.Hitchc. OBL
Swamp oats
Herbaceous perennial
Swamps, wet woods or springy meadows.
Trisetum pensylvanicum (L.) Beauv. ex Roemer & Schultes PFGBW

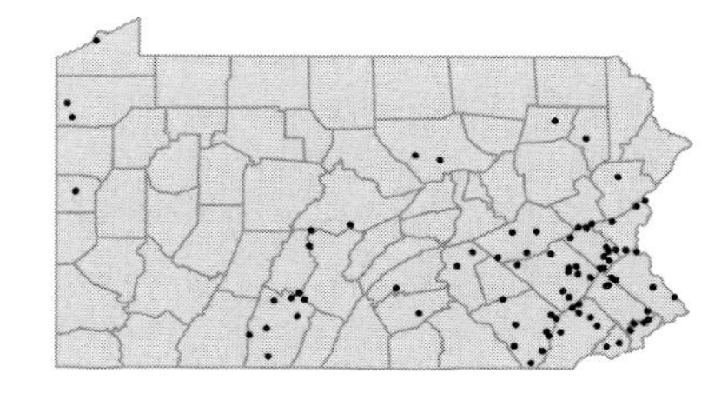

• ***Sporobolus asper*** (Michx.) Kunth UPL
Dropseed; Rough rush-grass
Herbaceous perennial
Dry, sandy banks.

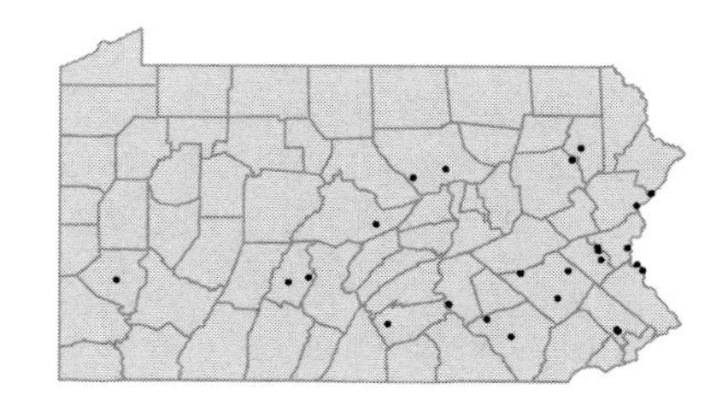

Sporobolus clandestinus (Biehler) A.S.Hitchc.

Rough dropseed; Rough rush-grass
Herbaceous perennial
Dry, sandy or rocky soil.
Sporobolus longifolius (Torr.) A.Wood P

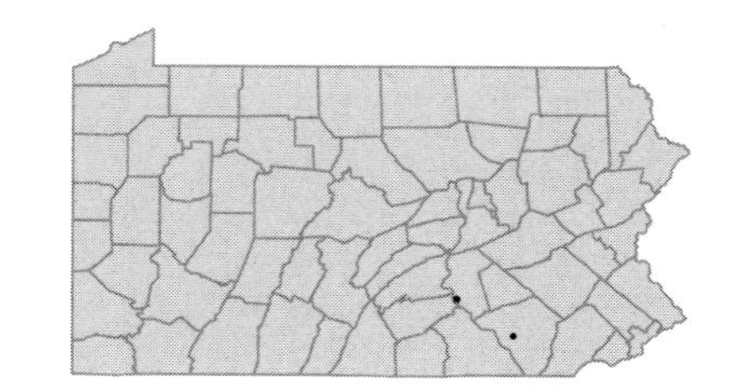

Sporobolus cryptandrus (Torr.) A.Gray
Sand dropseed
Herbaceous perennial
Dry waste ground and sandy shores.

UPL

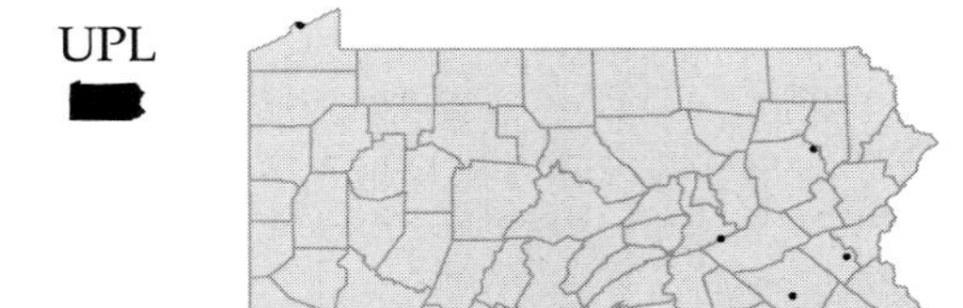

Sporobolus heterolepis (A.Gray) A.Gray
Prairie dropseed
Herbaceous perennial
Serpentine barrens.

UPL

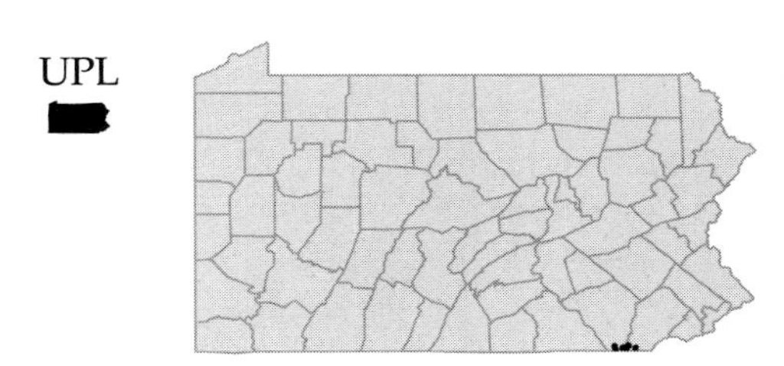

Sporobolus indicus (L.) R.Br.
Blackseed; Smutgrass
Herbaceous perennial
Ballast ground. Represented by several collections from Philadelphia Co. ca. 1865.
Sporobolis poiretii (Roemer & Schultes) A.S.Hitchc. FGBW

I

Sporobolus neglectus Nash
Small rush-grass
Herbaceous annual
Dry, sterile or sandy soil.

FACU-

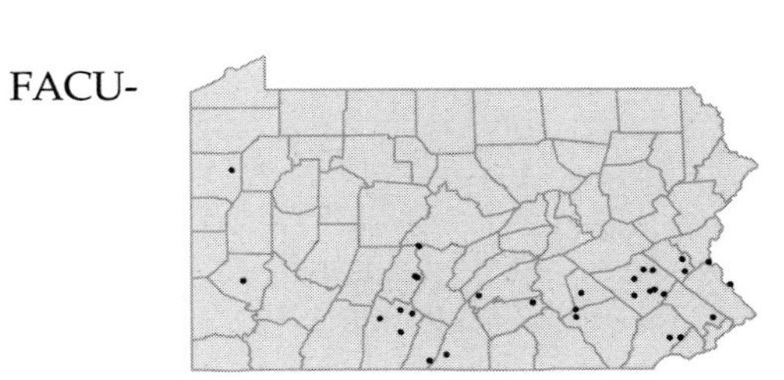

Sporobolus pyramidatus (Lam.) A.S.Hitchc.
Dropseed
Herbaceous perennial
Ballast. Represented by two pre-1900 collections from Philadelphia Co.

UPL
I

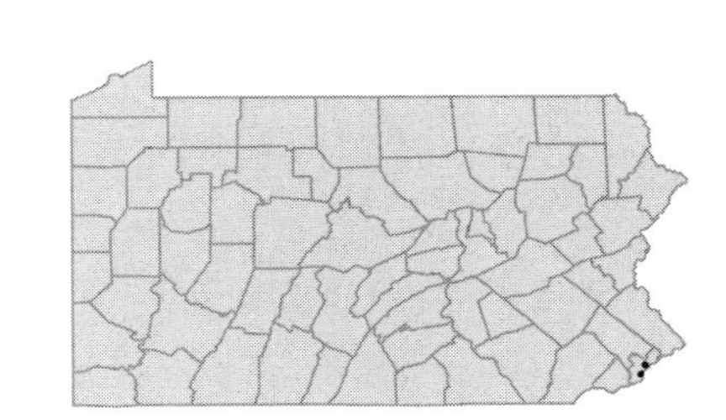

Sporobolus vaginiflorus (Torr. ex A.Gray) A.Wood
Poverty grass
Herbaceous annual
Dry, sandy or sterile soil including serpentine barrens.

UPL

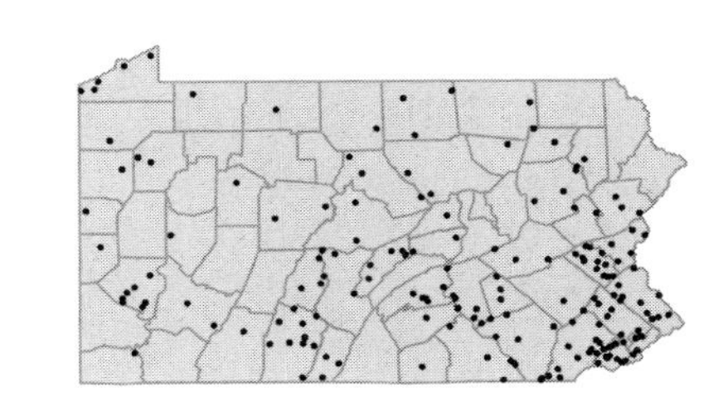

• ***Stipa spartea*** Trin.
Needle-grass; Porcupine grass
Herbaceous perennial
Roadside. Represented by a single collection from Lackawanna Co. in 1927.

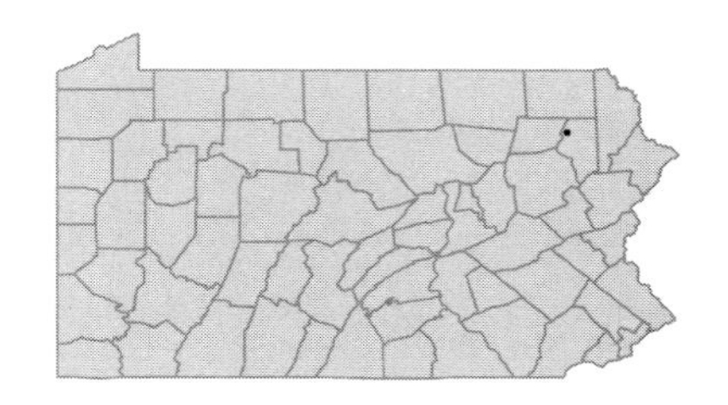

• ***Taeniatherum caput-medusae*** (L.) Nevski — I
Grass
Herbaceous annual
Rubbish dump and fill. Represented by two collections from a single site in Bucks Co. 1949-1950.

• ***Torreyochloa pallida*** (Torr.) Church var. ***pallida*** — OBL
Pale meadowgrass
Herbaceous perennial
Swamps, marshes and ditches.
Glyceria pallida (Torr.) Trin. FBW; *Puccinellia pallida* (Torr.) Clausen *pro parte* GC

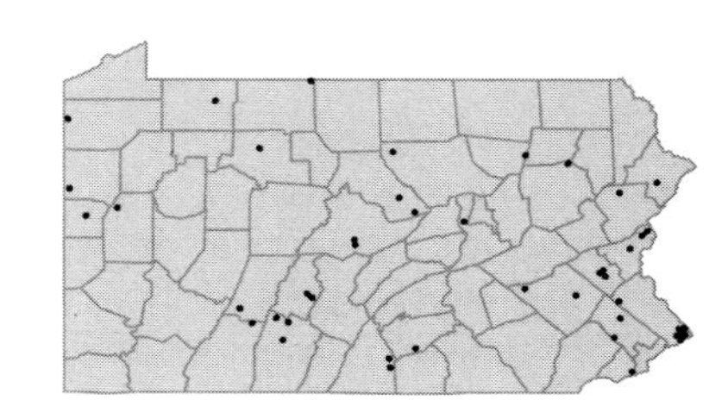

Torreyochloa pallida (Torr.) Church var. ***fernaldii*** (A.S.Hitchc.) Dore — OBL
Pale meadowgrass
Herbaceous perennial
Lake and stream edges.
Glyceria fernaldii (A.S.Hitchc.) St.John F; *Glyceria pallida* (Torr.) Trin. var. *fernaldii* A.S.Hitchc. BW; *Puccinellia pallida* (Torr.) Clausen *pro parte* GC

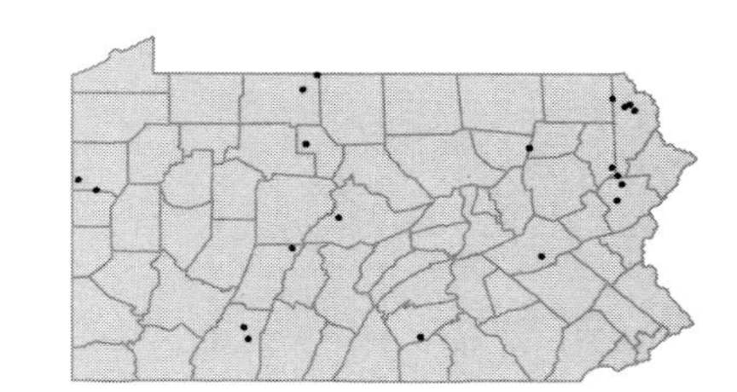

• ***Tragus racemosus*** (L.) All. — I
Texas bur
Herbaceous annual
Ballast.
Nazia racemosa (L.) Kuntze P

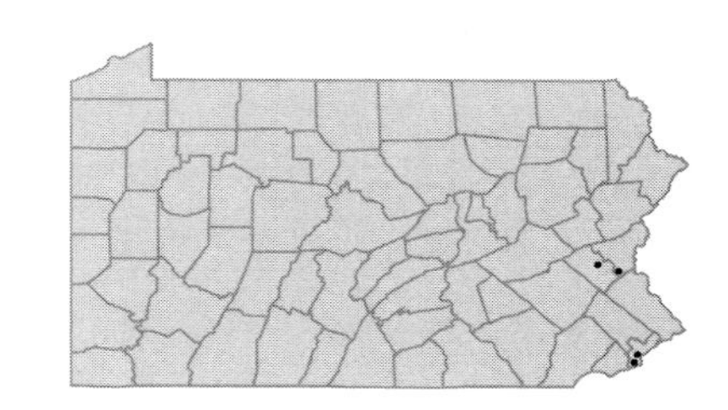

• ***Tridens flavus*** (L.) A.S.Hitchc. — FACU
Purple-top
Herbaceous perennial
Meadows, old fields and roadsides.
Tridens seslerioides (Michx.) Nash P; *Triodia flava* (L.) Smyth FGB

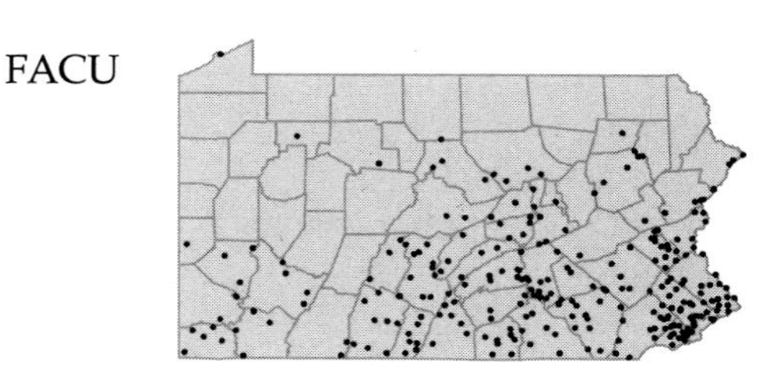

• ***Triodia stricta*** (Nutt.) Benth.
Grass
Herbaceous perennial
Moist soil. Represented by a single collection from Beaver Co. in 1885.

[I]

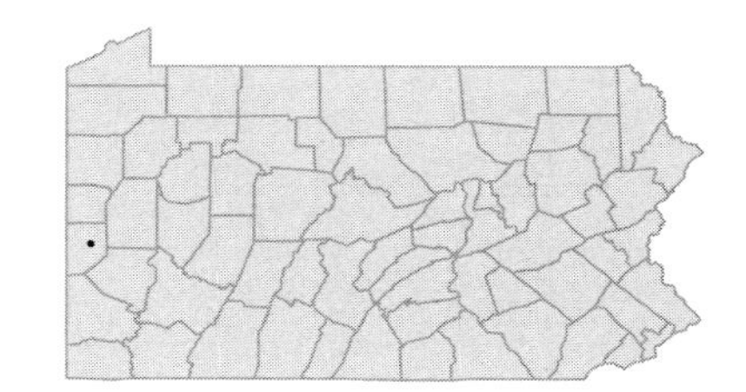

• ***Triplasis purpurea*** (Walt.) Chapm.
Purple sandgrass
Herbaceous annual
Dry, open, sandy soil.

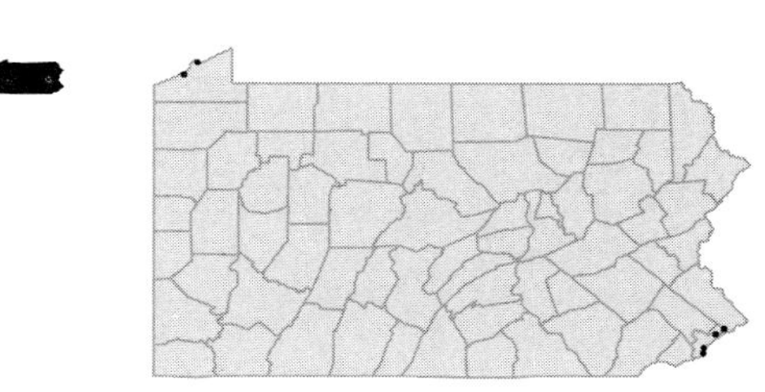

• ***Tripsacum dactyloides*** (L.) L.
Gamma grass
Herbaceous perennial
Swamps and wet shores.

FACW

• ***Trisetum spicatum*** (L.) Richter
Oatgrass
Herbaceous perennial
Open, shaly outcrops.

FACU

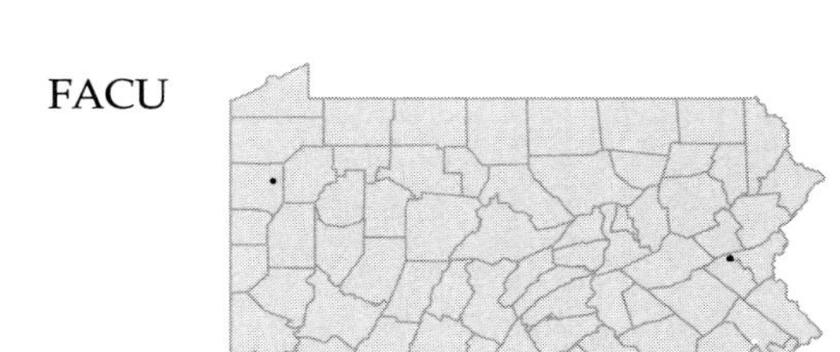

• ***Triticum aestivum*** L.
Wheat
Herbaceous annual
Cultivated and frequently occurring on roadsides and edges of fields.

[I]

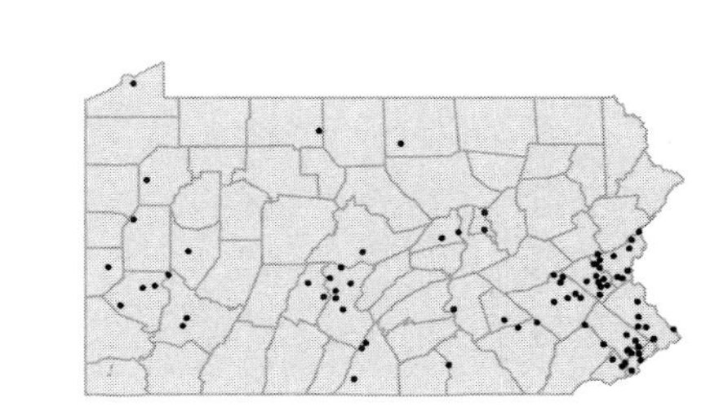

• ***Vulpia myuros*** (L.) K.C.Gmel. var. ***myuros***
Foxtail fescue
Herbaceous annual
Open fields and banks in sandy soil.
Festuca myuros L. PGBW

UPL
[I]

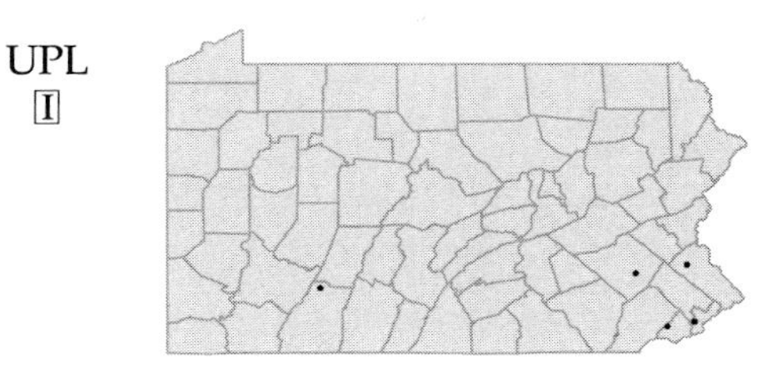

Vulpia myuros (L.) K.C.Gmel. var. ***hirsuta*** Hack.
Foxtail fescue
Herbaceous annual
Represented by a single collection from Philadelphia Co. ca. 1879.
Festuca megalura Nutt. W

UPL
[I]

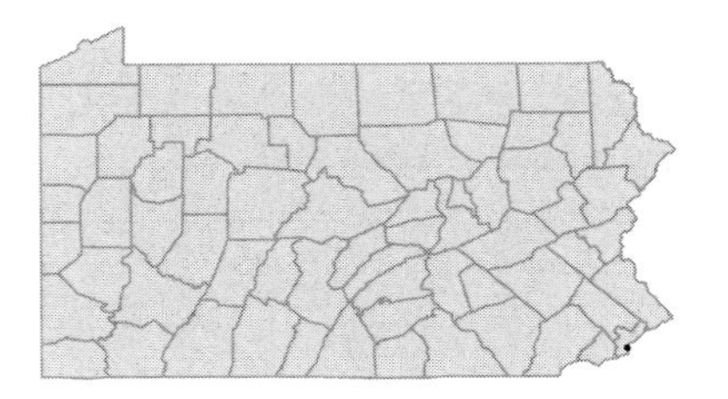

Vulpia octoflora (Walt.) Rydb. var. ***glauca*** (Nutt.) Fern.
Six-weeks fescue
Herbaceous annual
Dry, sterile soil.
Festuca octoflora Walt. var. *tenella* (Willd.) Fern. GBW

UPL

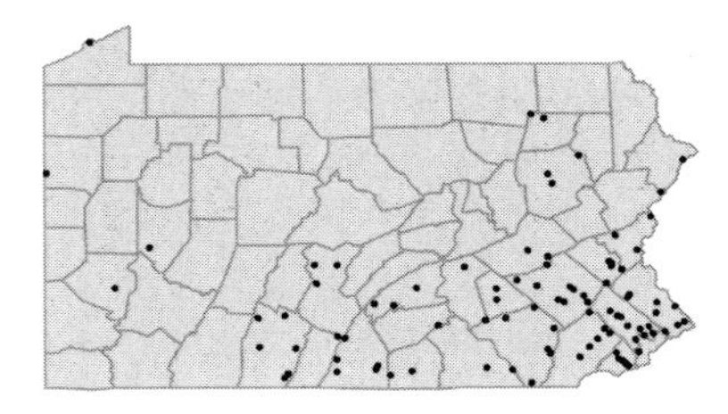

• ***Zea mays*** L.
Corn
Herbaceous annual
Cultivated and occasionally occurring in abandoned fields or roadsides.

[I]

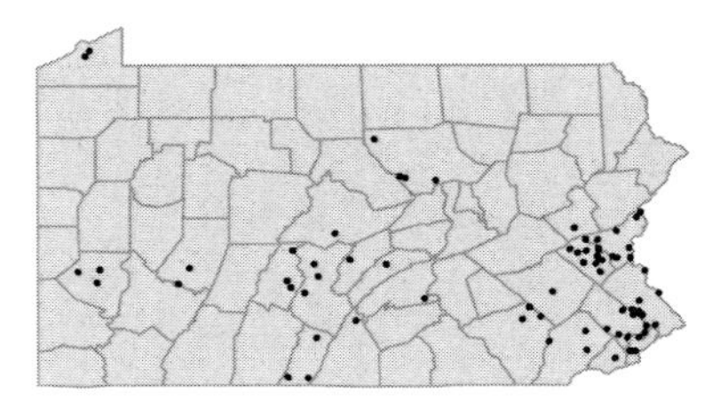

• ***Zizania aquatica*** L. var. ***aquatica***
Wild-rice
Herbaceous annual, emergent aquatic
Tidal and non-tidal marshes.

OBL

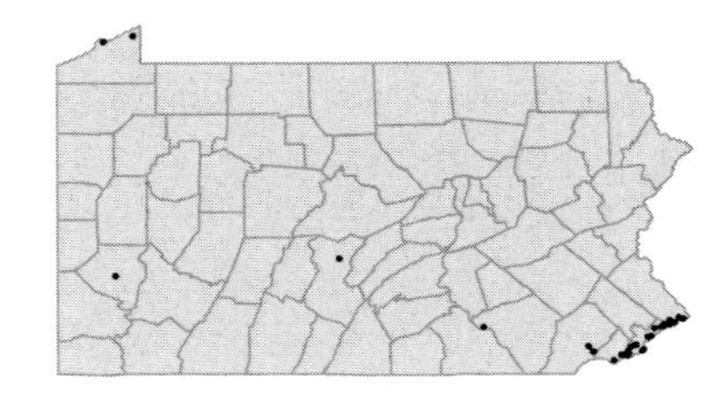

Zizania aquatica L. var. ***angustifolia*** A.S.Hitchc.
Wild-rice
Herbaceous annual, emergent aquatic
Shallow water. Represented by several collections from Erie Co. 1879-1928.
Zizania palustris L. var. *palustris* C

OBL

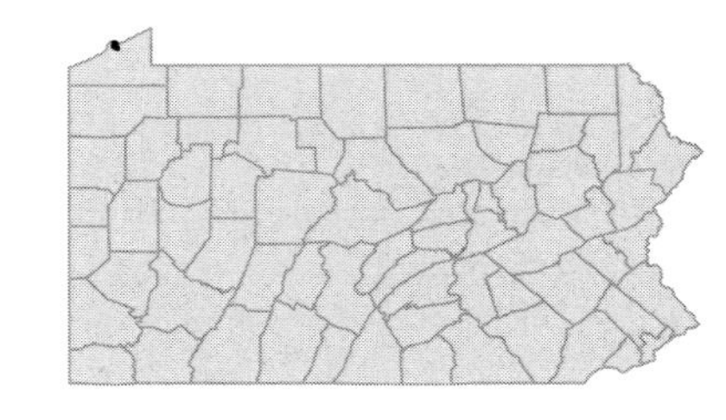

SPARGANIACEAE

• ***Sparganium americanum*** Nutt.
Bur-reed
Herbaceous perennial, emergent aquatic
Muddy shores and shallow water of rivers, streams, swamps or ponds.

OBL

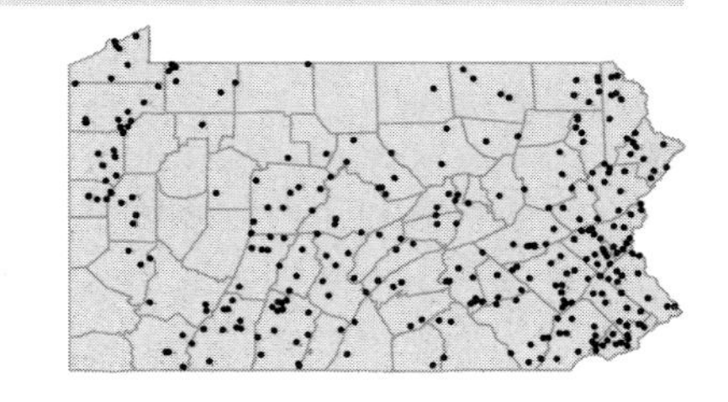

Sparganium androcladum (Engelm.) Morong
Branching bur-reed
Herbaceous perennial, emergent aquatic
Wet meadows, swales, stream banks and shallow water.

OBL

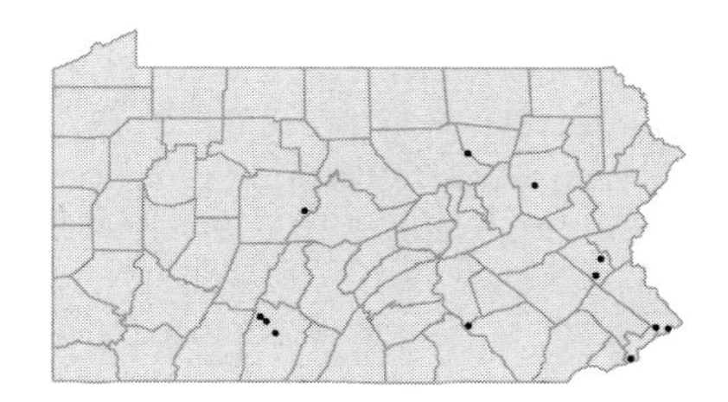

Sparganium angustifolium Michx.
Bur-reed
Herbaceous perennial, floating-leaf aquatic
Lakes and bogs.

OBL

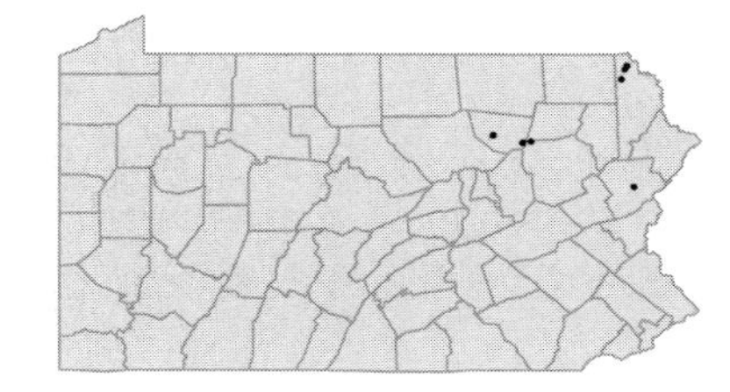

Sparganium chlorocarpum Rydb.
Bur-reed
Herbaceous perennial, floating-leaf aquatic
Slow-moving streams, lakes, bogs and wet meadows.

OBL

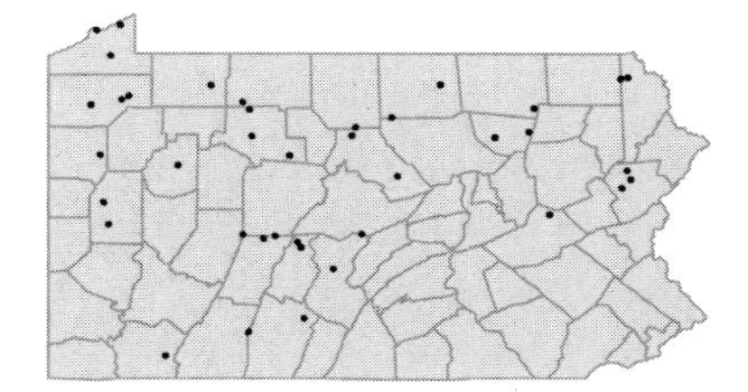

Sparganium eurycarpum Engelm.
Bur-reed
Herbaceous perennial, emergent aquatic
Bogs, swamps, lake margins, ditches and swampy meadows.

OBL

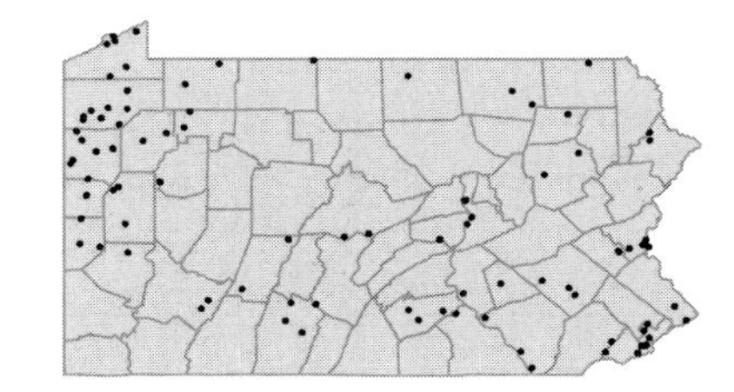

Sparganium fluctuans (Morong) B.L.Robins.
Bur-reed
Herbaceous perennial, floating-leaf aquatic
Lake margins, slow-moving streams and muddy shores.
Sparganium androcladum (Engelm.) Morong var. *fluctuans* Morong P

OBL

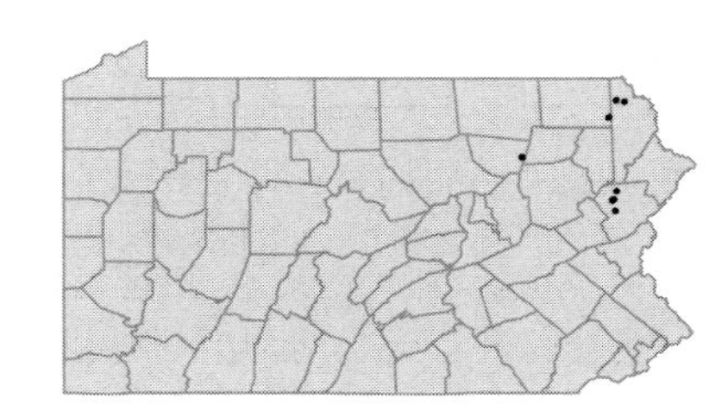

Sparganium minimum (Hartm.) Fries
Small bur-reed
Herbaceous perennial, floating-leaf aquatic
Shallow water. Believed to be extirpated, last collected in 1880.

OBL

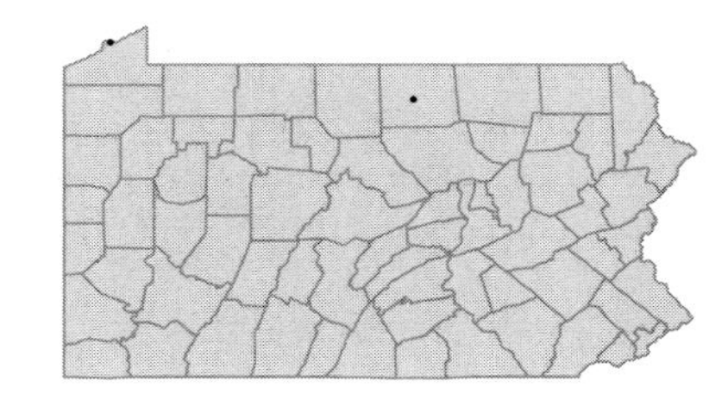

TYPHACEAE

• ***Typha angustifolia*** L.
Narrowleaf cat-tail
Herbaceous perennial, emergent aquatic
Wet meadows, marshes, shores and ditches, often in calcareous or brackish habitats.

OBL

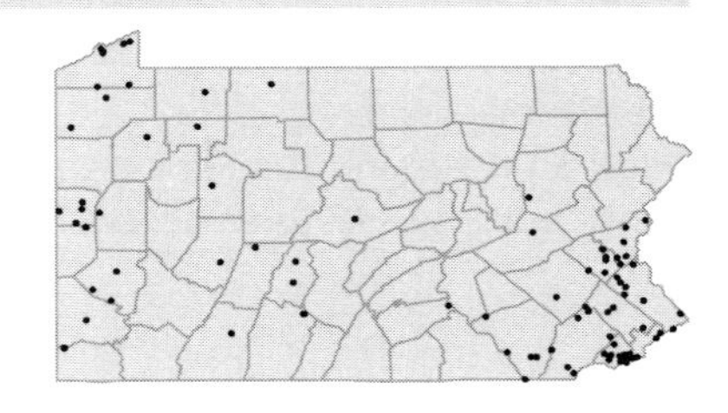

Typha angustifolia x latifolia
Cat-tail
Herbaceous perennial, emergent aquatic
Pond margins and marshes.
Typha x glauca Godr. FWC

OBL

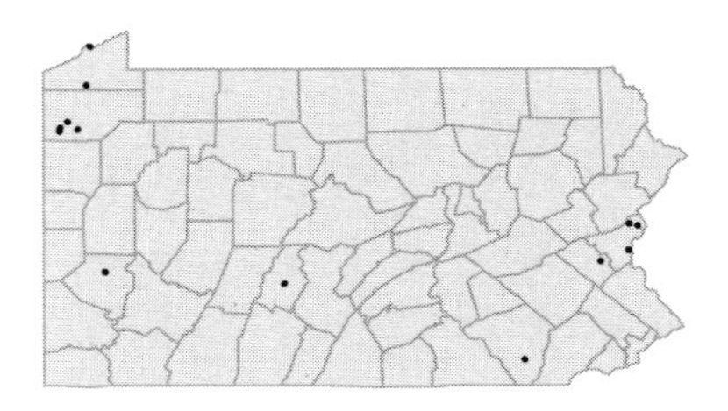

Typha latifolia L.
Common cat-tail
Herbaceous perennial, emergent aquatic
Swamps, marshes, wet shores and ditches.

OBL

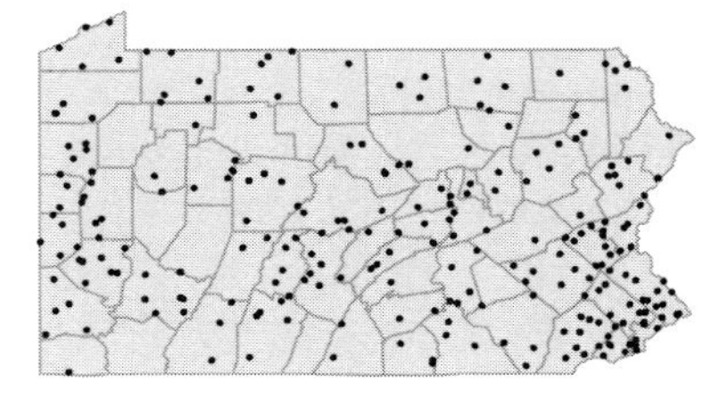

CANNACEAE

• ***Canna glauca x indica***
Canna
Herbaceous perennial
Cultivated and occasionally persisting.
Canna x generalis Bailey WK

[I]

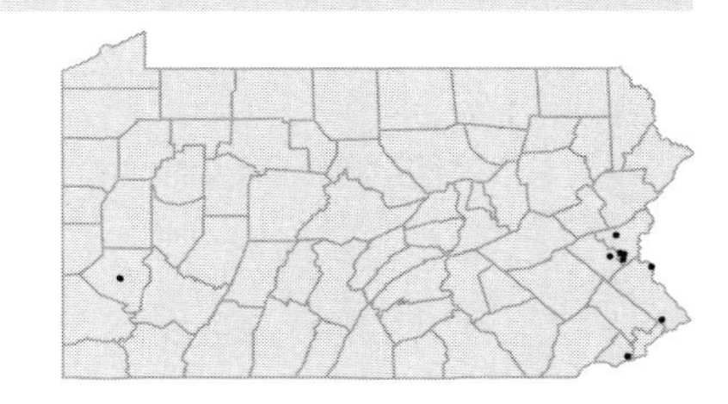

Canna neglecta Steud.
Canna
Herbaceous annual
Rubbish dump. Represented by a single collection from Northampton Co. in 1946.

[I]

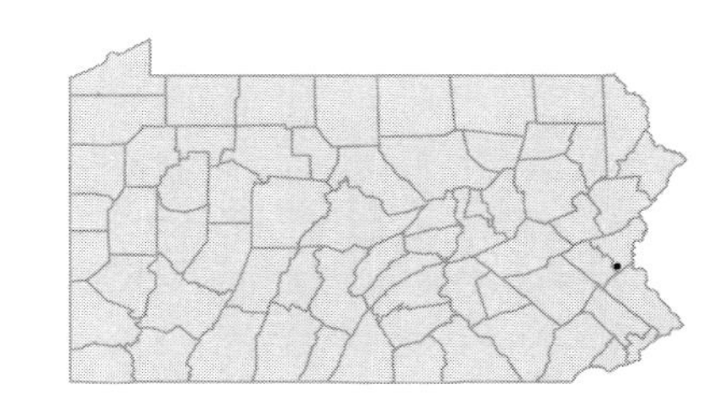

PONTEDERIACEAE

• ***Heteranthera multiflora*** (Griseb.) Horn
Mud-plantain
Herbaceous perennial, rooted submergent aquatic
Tidal shores and mud flats.

OBL

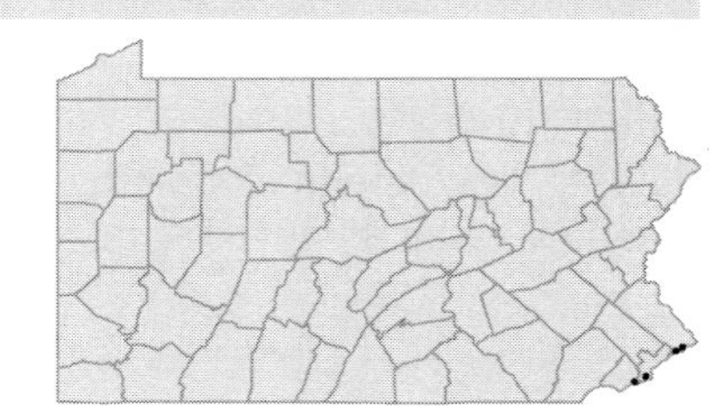

Heteranthera reniformis Ruiz & Pavon
Mud-plantain
Herbaceous perennial, floating-leaf aquatic
Shallow water or muddy shores.

OBL

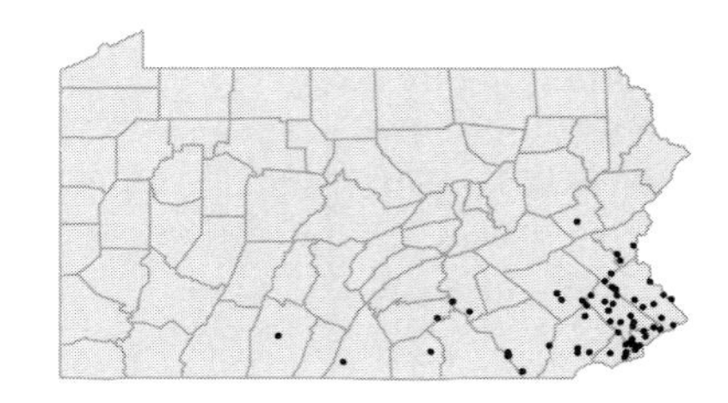

• ***Pontederia cordata*** L.
Pickerel-weed
Herbaceous perennial, emergent aquatic
Swampy edges of lakes and streams, also tidal shores.

OBL

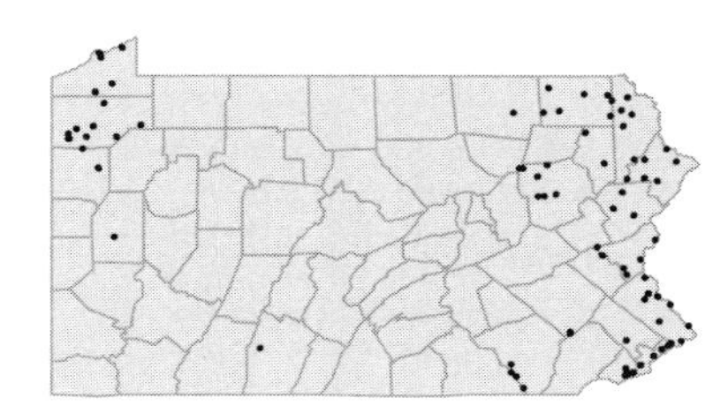

• ***Zosterella dubia*** (Jacq.) Small
Water-stargrass
Herbaceous perennial, rooted submergent aquatic
Lakes and streams.
Heteranthera dubia (Jacq.) MacM. FW

OBL

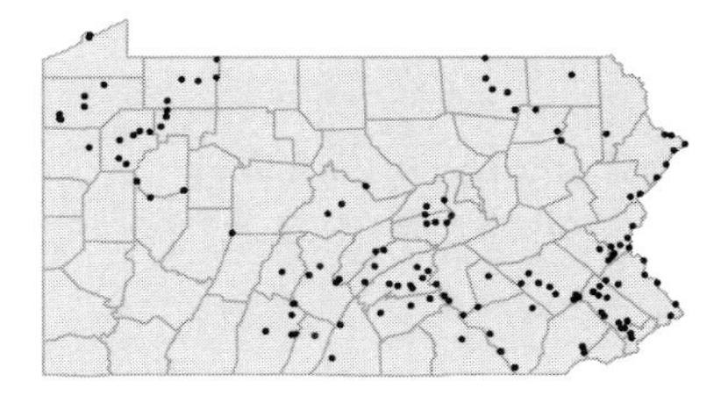

LILIACEAE

• ***Aletris farinosa*** L.
Colic-root
Herbaceous perennial
Moist, sandy soil of open woods and sphagnum bogs.

FAC

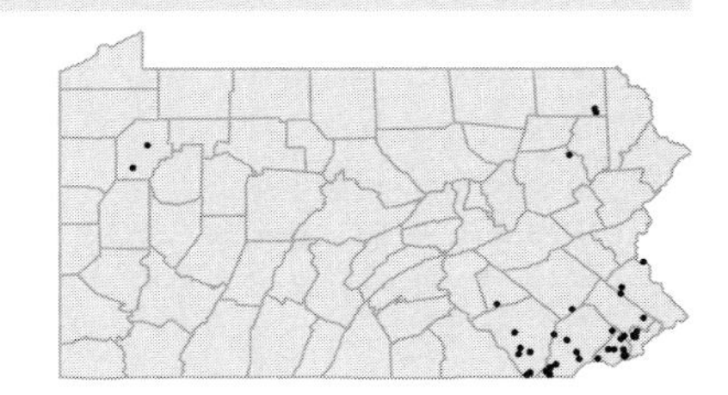

• ***Allium canadense*** L.
Wild onion
Herbaceous perennial
Low woods, stream banks and thickets.

FACU

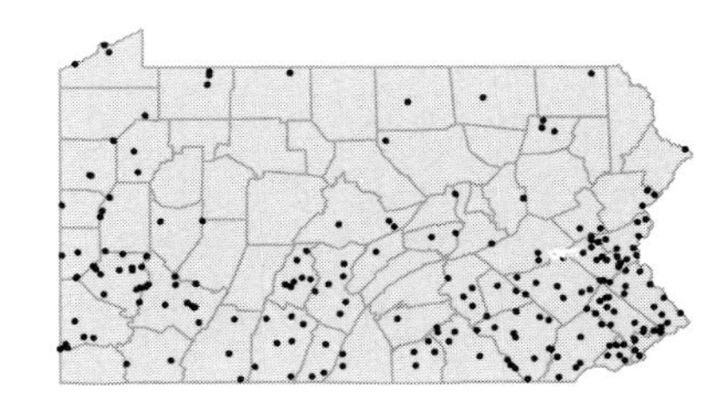

Allium cepa L.
Onion
Herbaceous perennial
Cultivated and occasionally occurring around rubbish dumps or waste ground.

I

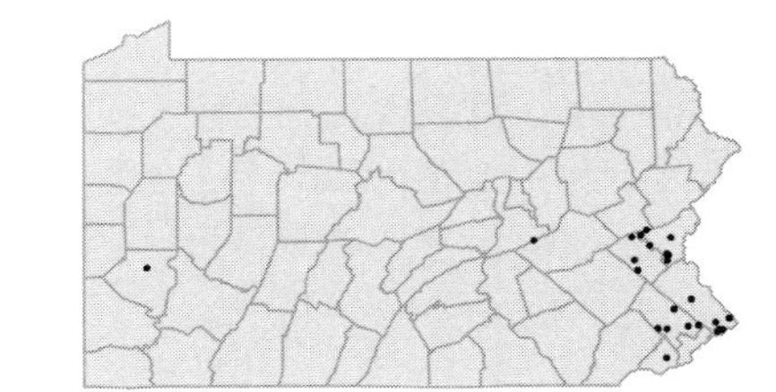

Allium cernuum Roth
Nodding onion
Herbaceous perennial
Dry, stony slopes and rocky shores, often on limestone.

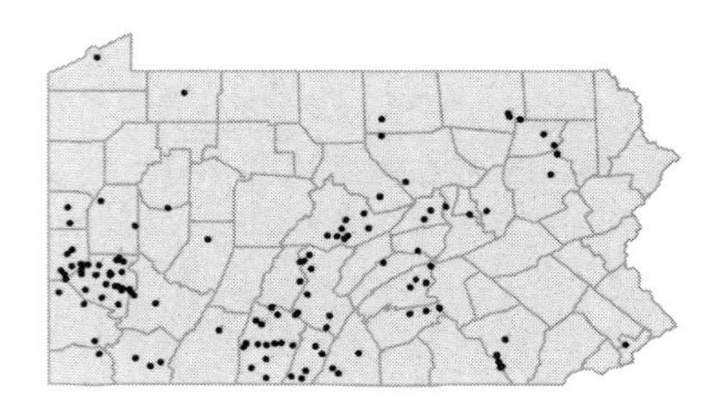

Allium oleraceum L.
Field garlic
Herbaceous perennial
Roadsides and alluvial thickets.

[I]

Allium sativum L.
Garlic
Herbaceous perennial
Cultivated and rarely escaped to roadsides and fields.

[I]

Allium schoenoprasum L. var. ***schoenoprasum***
Chives
Herbaceous perennial
Cultivated and rarely escaped to banks and roadsides.

FACU
[I]

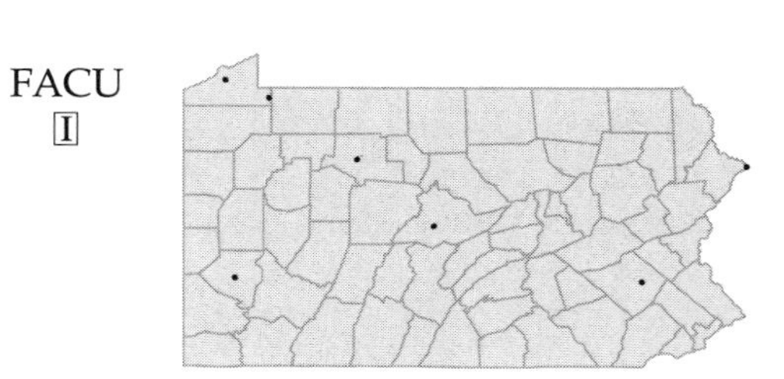

Allium tricoccum Ait.
Ramps; Wild leek
Herbaceous perennial
Rich woods and bottomlands.

FACU+

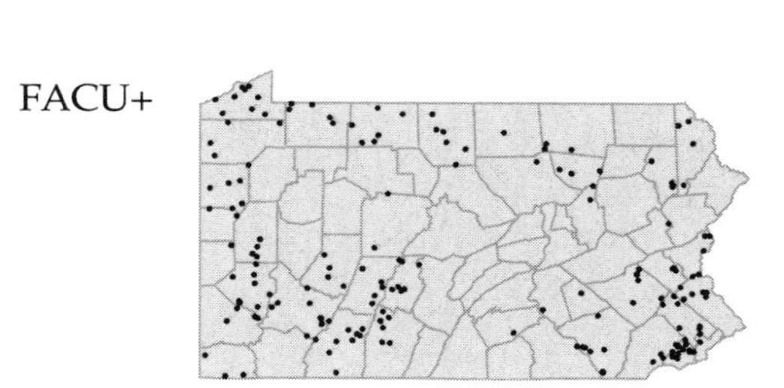

Allium vineale L.
Field garlic; Scallions
Herbaceous perennial
Disturbed woods, fields and lawns.

FACU-
[I]

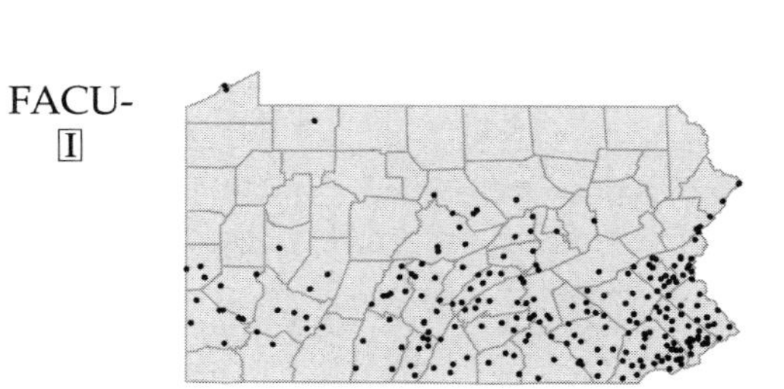

• ***Amianthium muscaetoxicum*** (Walt.) A.Gray
Fly-poison
Herbaceous perennial
Open woods, barrens and bogs, in sandy or peaty soils.
Chrosperma muscaetoxicum (Walt.) Kuntze P

FAC

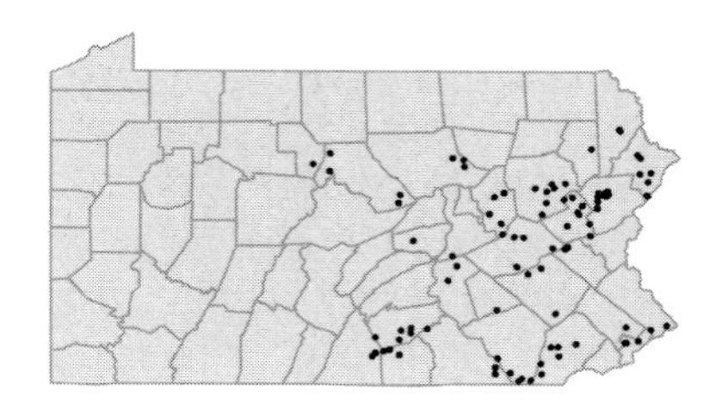

• ***Asparagus officinalis*** L.
Garden asparagus
Herbaceous perennial
Cultivated and frequently escaped to fields, roadsides and waste ground.

FACU
I

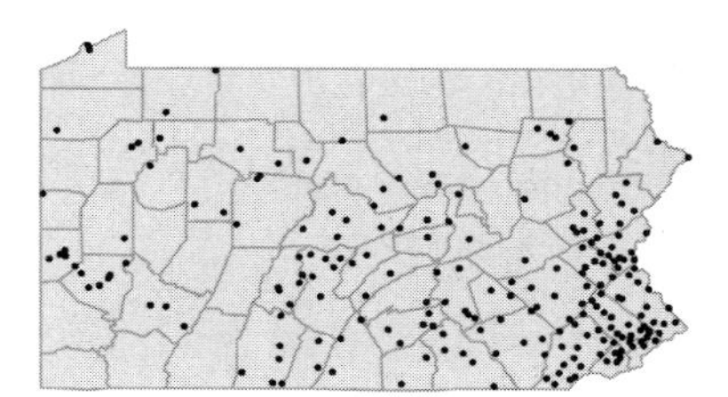

• ***Camassia scilloides*** (Raf.) Cory
Eastern camass; Wild hyacinth
Herbaceous perennial
Wooded floodplains and stream banks.
Quamasia hyacinthina (Raf.) Britt. P

FAC

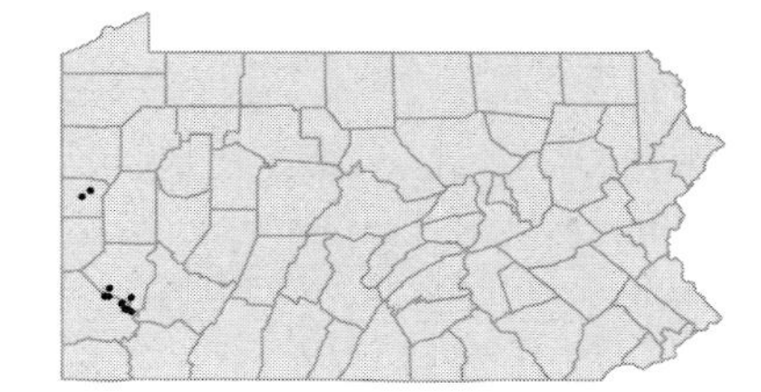

• ***Chamaelirium luteum*** (L.) A.Gray
Fairy-wand; Devil's-bit
Herbaceous perennial
Rich woods, rocky slopes, meadows and barrens.

FAC

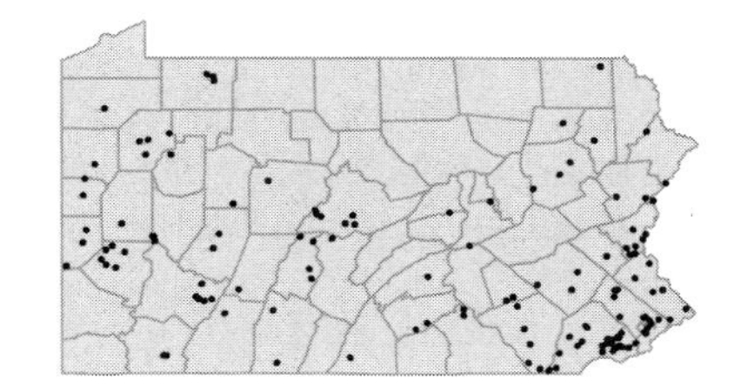

• ***Clintonia borealis*** (Ait.) Raf.
Bluebead lily
Herbaceous perennial
Cool, moist woods and bogs.

FAC

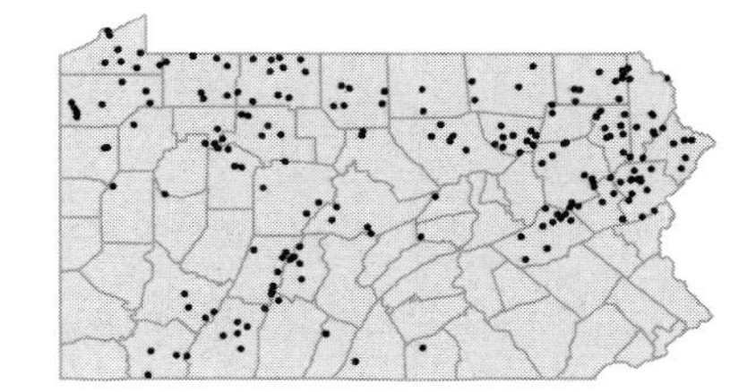

Clintonia umbellulata (Michx.) Morong
Speckled wood-lily; White wood-lily
Herbaceous perennial
Rich, moist woods, shaded ravines and banks.

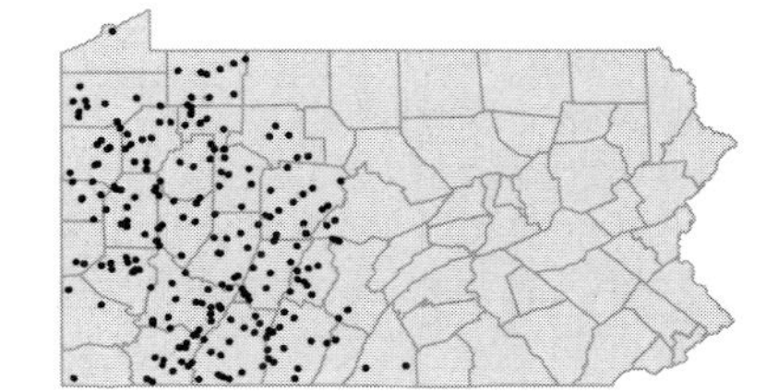

• ***Convallaria majalis*** L. I
Lily-of-the-valley
Herbaceous perennial
Cultivated and occasionally escaped to woods in the vicinity of old homesites.

• ***Disporum lanuginosum*** (Michx.) Nicholson
Yellow mandarin
Herbaceous perennial
Moist, rich woods.

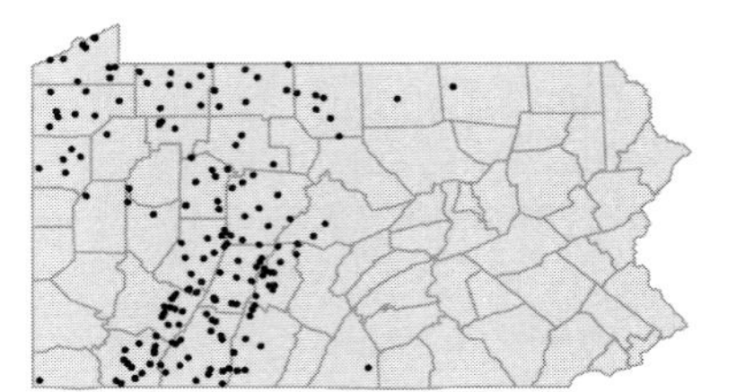

• ***Endymion nonscripta*** (L.) Garcke I
English bluebell
Herbaceous perennial
Cultivated and rarely escaped.
Scilla nonscripta (L.) Hoffmgg. & Link FBWC

• ***Erythronium albidum*** Nutt.
White trout-lily; White dog-tooth-violet
Herbaceous perennial
Wooded floodplains and low, moist ground.

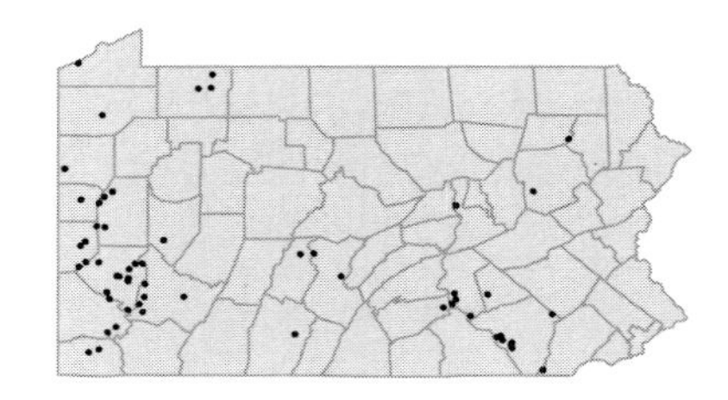

Erythronium americanum Ker-Gawl. FAC
Trout-lily; Dog-tooth-violet
Herbaceous perennial
Moist woods, bottomlands and meadows.

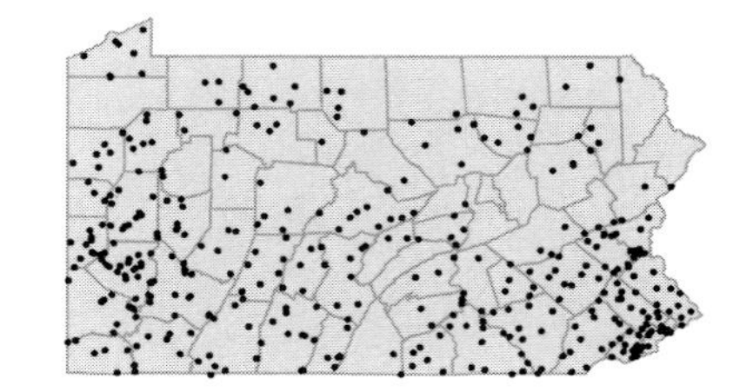

• ***Galanthus elwesii*** Hook.f. I
Giant snowdrop
Herbaceous perennial
Cultivated and rarely naturalized in woods. Represented by a single collection from Allegheny Co. in 1984.

Galanthus nivalis L.
Snowdrop
Herbaceous perennial
A rare garden escape.

I

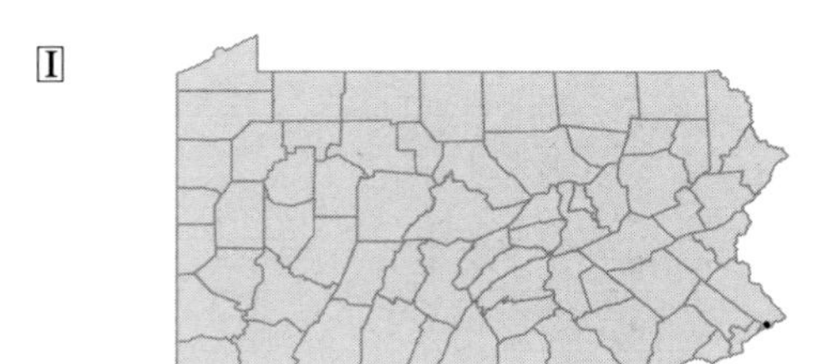

• ***Hemerocallis fulva*** (L.) L.
Orange day-lily
Herbaceous perennial
Fields, roadsides, stream banks, floodplains and woods edges.

I

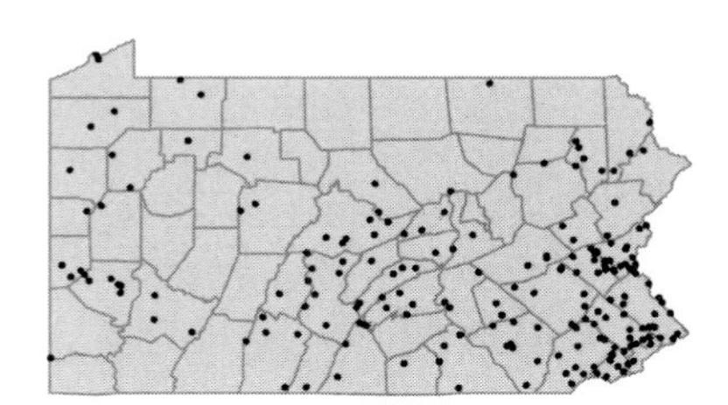

Hemerocallis lilioasphodelus L. *emend.* Hyl.
Yellow day-lily
Herbaceous perennial
Cultivated and occasionally escaped to fields and roadsides.
Hemerocallis flava (L.) L. FGBW

I

• ***Hosta lancifolia*** (Thunb. ex Houtt.) Engl.
Narrow-leaved plantain-lily
Herbaceous perennial
Cultivated and occasionally escaped at abandoned homesites, roadsides and waste areas.
Hosta japonica (Thunb.) Voss F

I

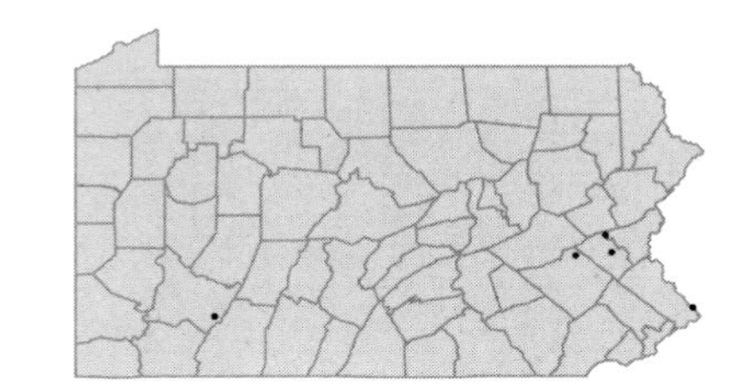

Hosta plantaginea (Lam.) Aschers.
Fragrant plantain-lily
Herbaceous perennial
Cultivated and rarely escaped.

I

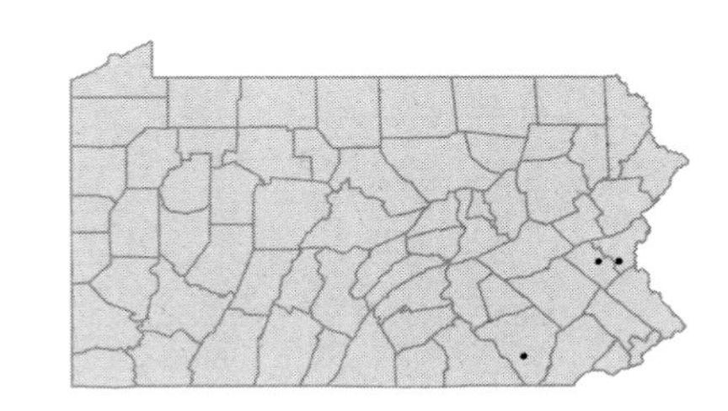

Hosta ventricosa (Salisb.) Stearn
Blue plantain-lily
Herbaceous perennial
Cultivated and occasionally escaped to moist woods and roadsides.

I

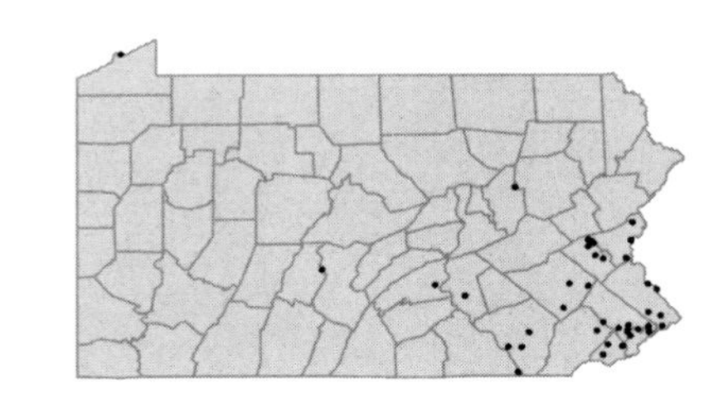

•***Hyacinthus orientalis*** L.
Hyacinth
Herbaceous perennial
Wooded floodplain, escaped from cultivation. Represented by single collection from Greene Co. in 1941.

[I]

•***Hypoxis hirsuta*** (L.) Cov.
Yellow star-grass
Herbaceous perennial
Dry, open woods and meadows.

FAC

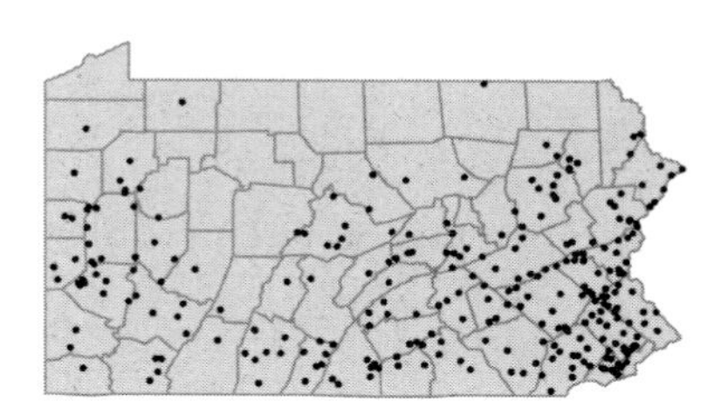

•***Leucojum aestivum*** L.
Summer snowflake
Herbaceous perennial
A rare garden escape.

[I]

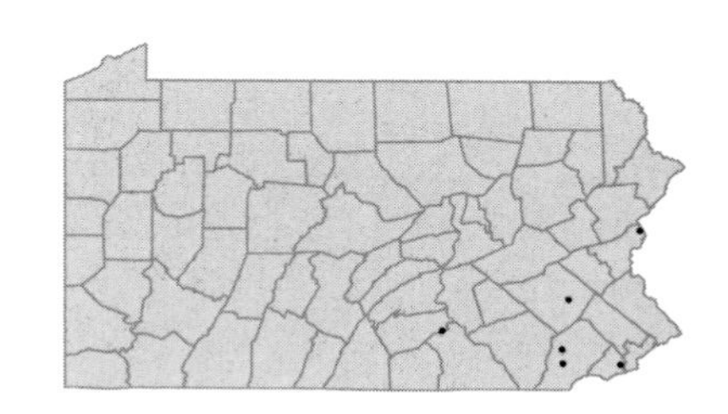

•***Lilium canadense*** L. ssp. ***canadense***
Canada lily
Herbaceous perennial
Wet meadows, moist woods and thickets.

FAC+

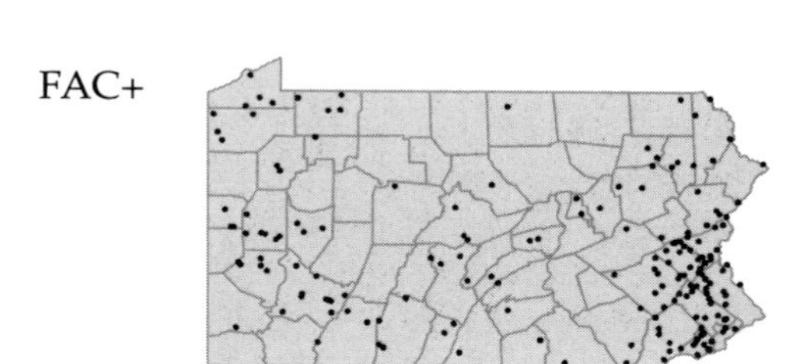

Lilium canadense L. ssp. ***editorum*** (Fern.) Wherry
Canada lily
Herbaceous perennial
Woods, swamps, bogs and stream banks.

FAC+

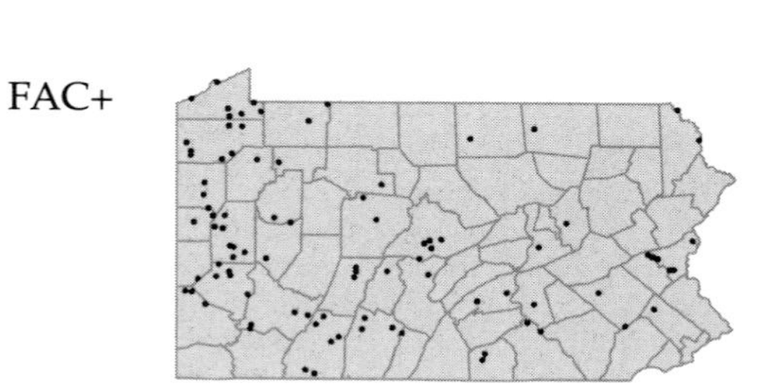

Lilium candidum L.
Madonna lily
Herbaceous perennial
Cultivated and rarely escaped.

[I]

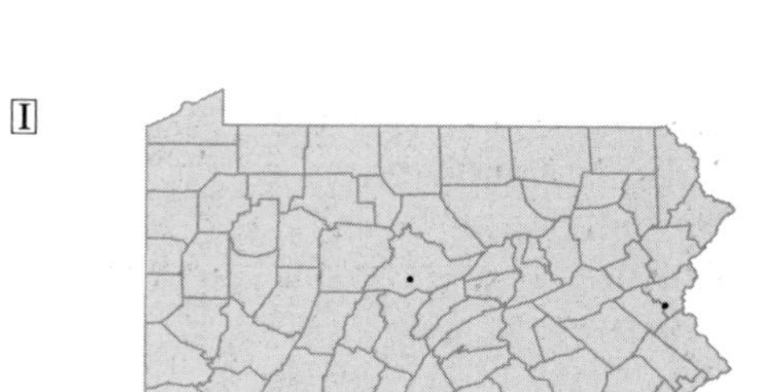

Lilium lancifolium Thunb.
Tiger lily
Herbaceous perennial
Roadsides and railroad banks, garden escape.
Lilium tigrinum L. FW

[I]

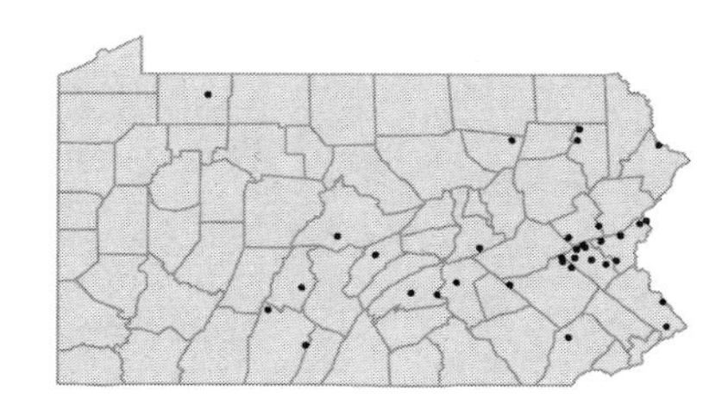

Lilium philadelphicum L.
Wood lily
Herbaceous perennial
Dry, open woods and barrens.

FACU+

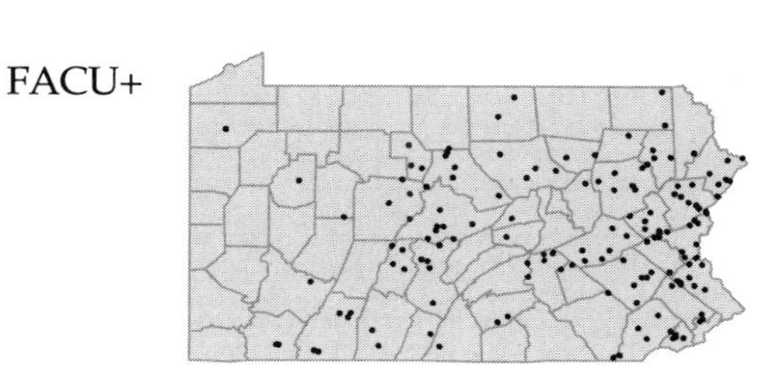

Lilium superbum L.
Turk's-cap lily
Herbaceous perennial
Moist meadows, low woods and swales in sandy-peaty soils.

FACW+

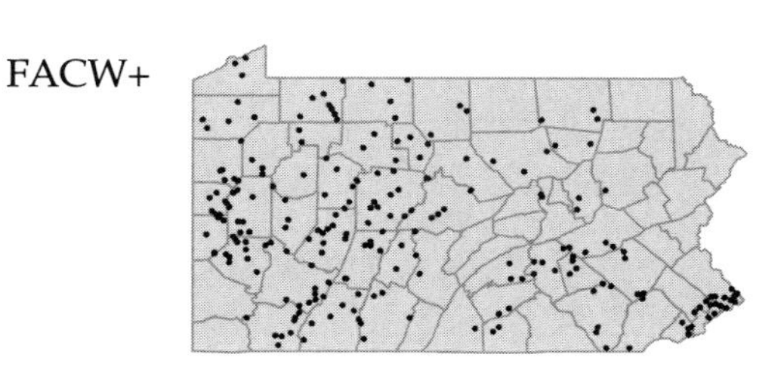

• ***Maianthemum canadense*** Desf.
Canada mayflower; False lily-of-the-valley
Herbaceous perennial
Moist woods.
Unifolium canadense (Desf.) Greene P

FAC-

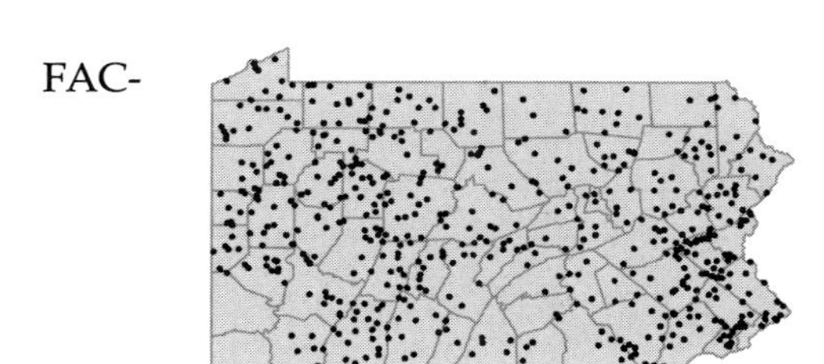

• ***Medeola virginiana*** L.
Indian cucumber-root
Herbaceous perennial
Moist woods.

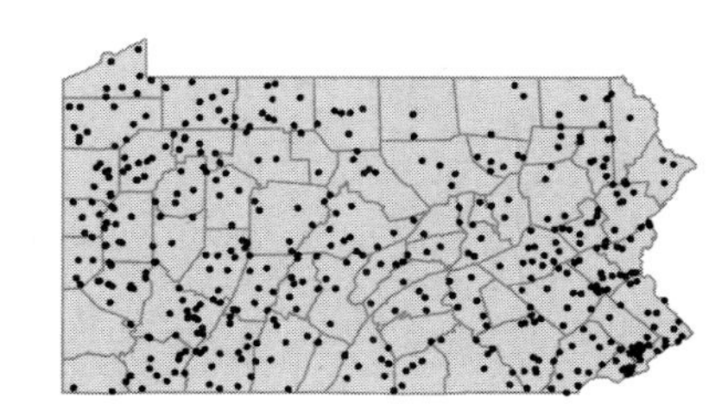

• ***Melanthium latifolium*** Desr.
Slender bunch-flower
Herbaceous perennial
Rich woods and slopes.
Melanthium hybridum Walt. FGBWC

FACU

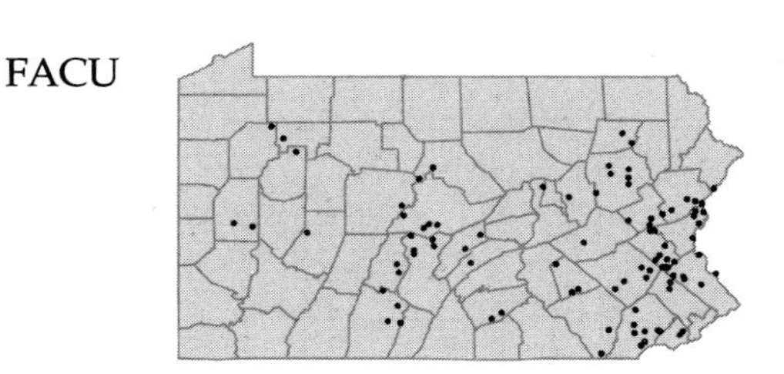

Melanthium virginicum L. FACW+
Bunch-flower
Herbaceous perennial
Rich woods, thickets and margins of swamps and bogs.

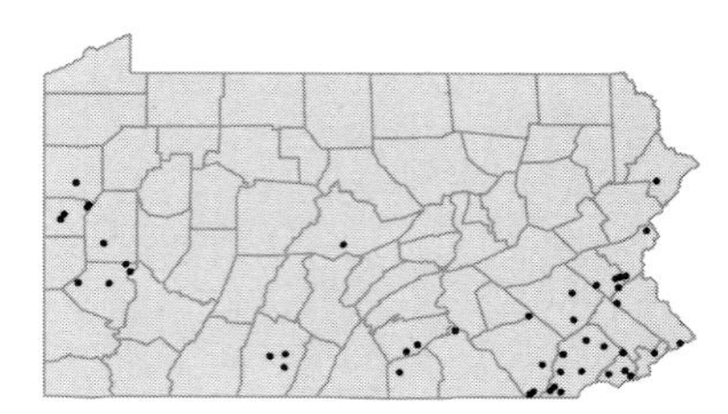

• ***Muscari botryoides*** (L.) P.Mill. [I]
Grape-hyacinth
Herbaceous perennial
Cultivated and occasionally naturalized on floodplains, stream banks and other disturbed areas.

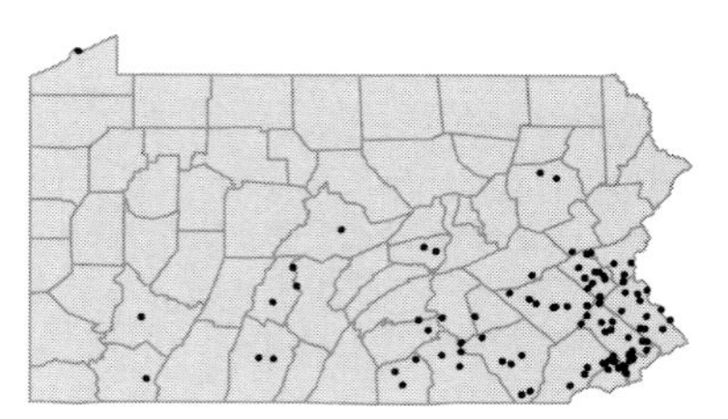

Muscari neglectum Guss. ex Ten. [I]
Grape-hyacinth
Herbaceous perennial
A rare garden escape.
Muscari atlanticum Boiss. & Reut. W; *Muscari racemosum* (L.) P.Mill. PFGBC

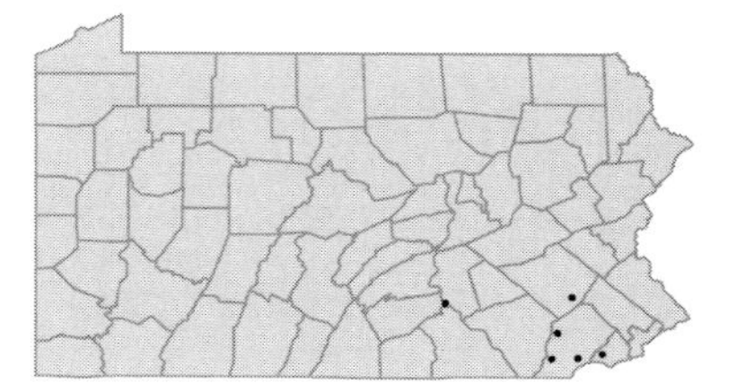

• ***Narcissus poeticus*** L. [I]
Poet's narcissus
Herbaceous perennial
Cultivated and occasionally persisting at old garden sites.

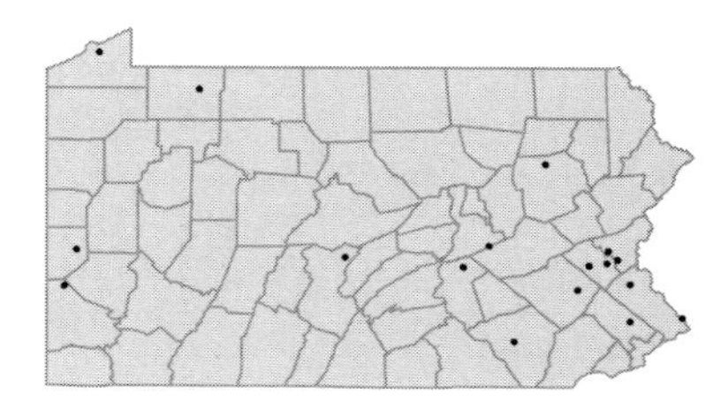

Narcissus pseudonarcissus L. [I]
Daffodil
Herbaceous perennial
Cultivated and occasionally escaped at abandoned gardens and home sites.

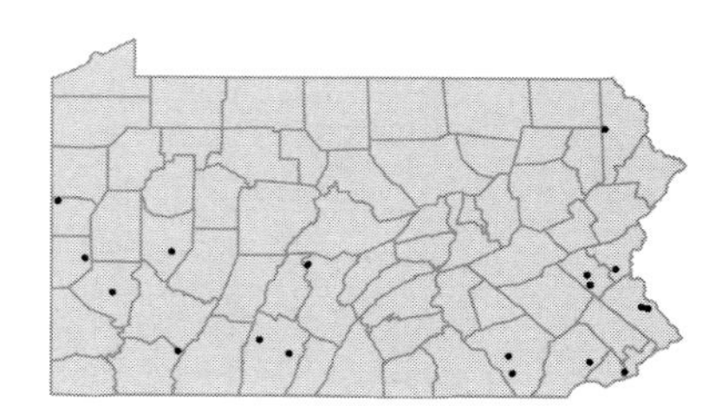

• ***Ornithogalum nutans*** L. [I]
Star-of-Bethlehem
Herbaceous perennial
Roadsides, fields and woods edges.

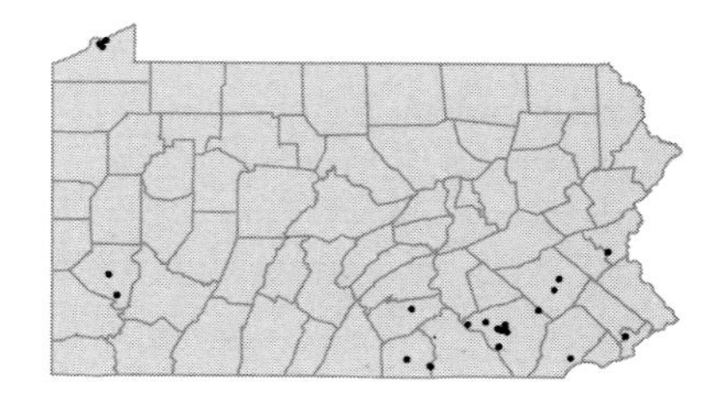

Ornithogalum umbellatum L.
Star-of-Bethlehem
Herbaceous perennial
Roadsides, lawns and meadows.

FACU
I

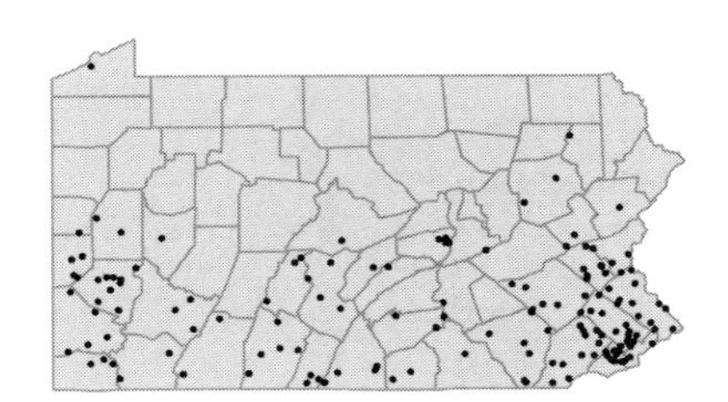

• ***Polygonatum biflorum*** (Walt.) Ell. var. ***biflorum***
Solomon's-seal
Herbaceous perennial
Deciduous woods, rocky slopes and roadside banks.

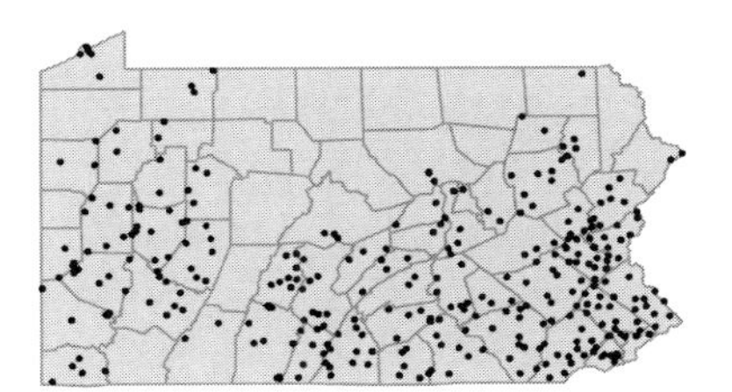

Polygonatum biflorum (Walt.) Ell. var. ***commutatum*** (Schultes f.) Morong
Solomon's-seal
Herbaceous perennial
Rich woods, stream banks and roadsides.
Polygonatum canaliculatum (Muhl.) Pursh FB; *Polygonatum commutatum* (Schultes f.) A.Dietr. GWC

FACU

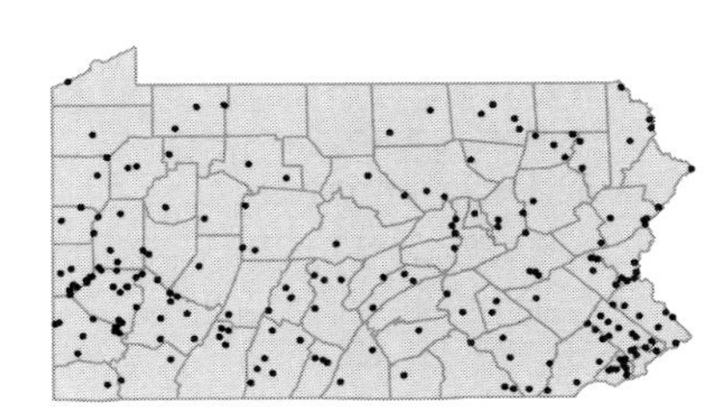

Polygonatum pubescens (Willd.) Pursh
Solomon's-seal
Herbaceous perennial
Wooded slopes, stream banks and fields.

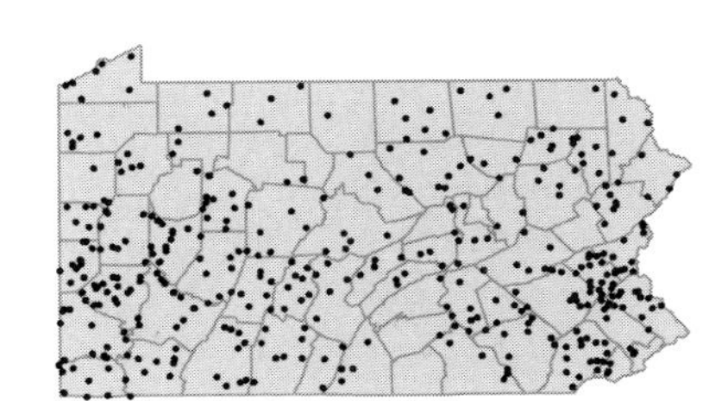

• ***Scilla sibirica*** Haw. & Andr.
Squill
Herbaceous perennial
Cultivated and rarely escaped.

I

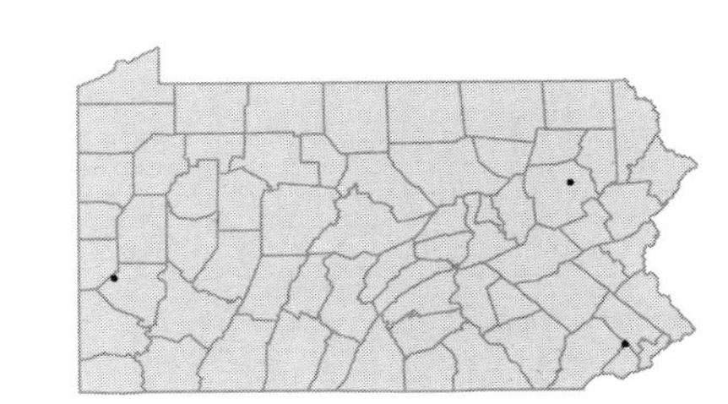

• ***Smilacina racemosa*** (L.) Desf.
False solomon's-seal
Herbaceous perennial
Woods and shaded edges.

FACU-

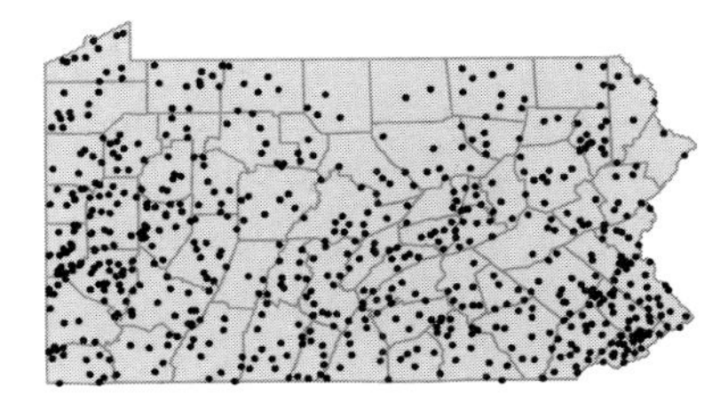

Smilacina stellata (L.) Desf.
Starflower
Herbaceous perennial
Swampy woods, bogs, stream banks and sandy shores.

FACW

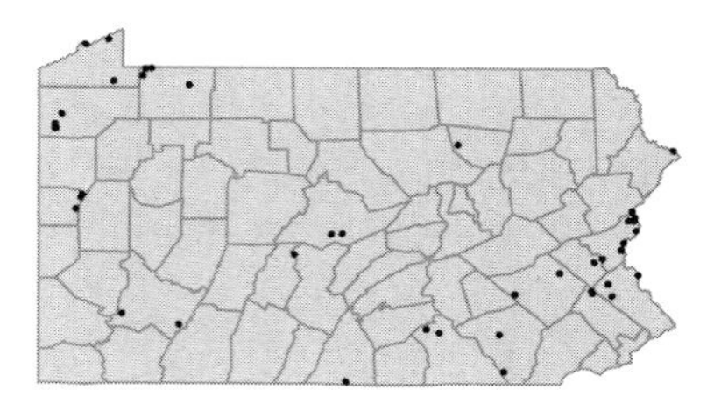

Smilacina trifolia (L.) Desf.
False solomon's-seal
Herbaceous perennial
Sphagnum bogs and swamps.

OBL

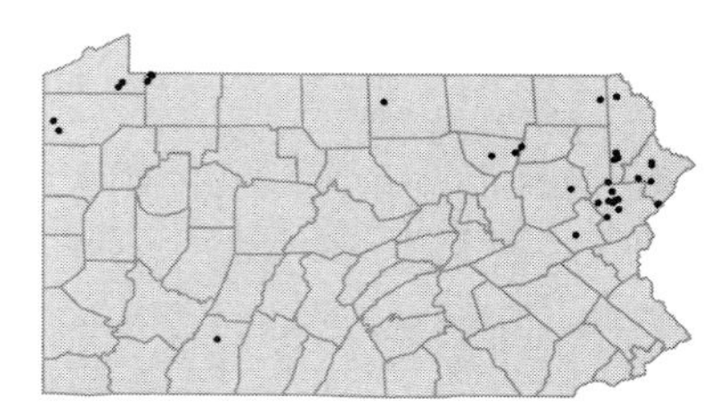

• ***Stenanthium gramineum*** (Ker-Gawl.) Morong
Featherbells
Herbaceous perennial
Moist woods, fields and floodplains.

FACW

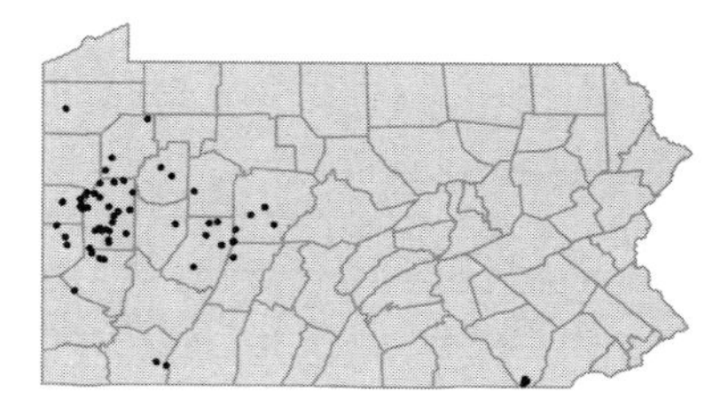

• ***Streptopus amplexifolius*** (L.) DC. var. ***americanus*** Schultes
White twisted-stalk
Herbaceous perennial
Cool ravines.

FAC+

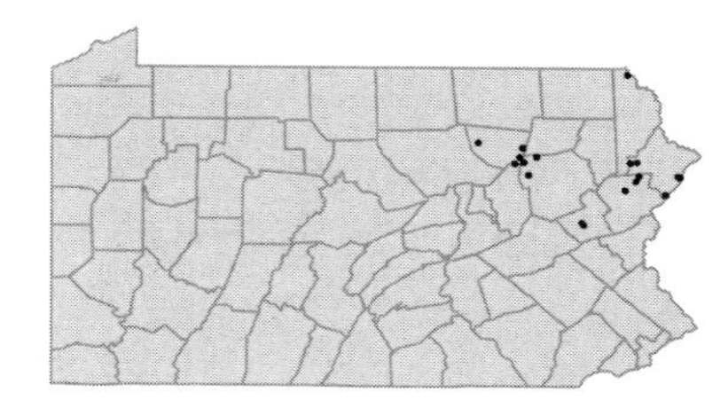

Streptopus roseus Michx. var. ***roseus***
Twisted-stalk
Herbaceous perennial
Moist woods and shaded ravines.

FAC-

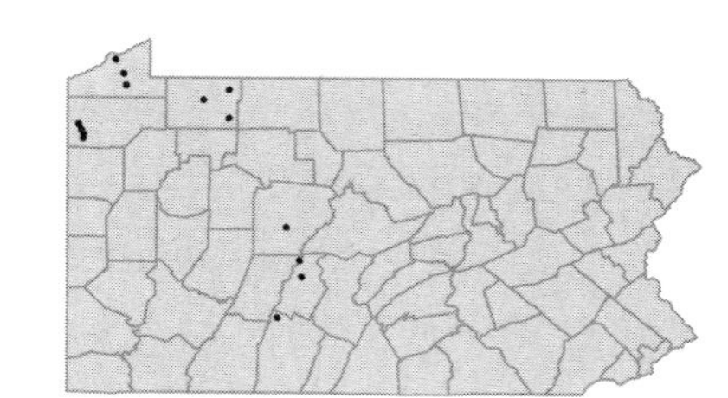

Streptopus roseus Michx. var. ***perspectus*** Fassett
Twisted-stalk
Herbaceous perennial
Cool, damp woods and slopes.

FAC-

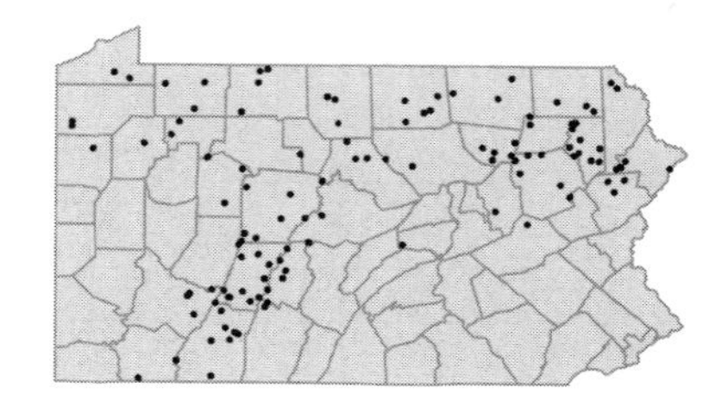

• ***Tricyrtis hirta*** (Thunb.) Hook.
Toadlily
Herbaceous perennial
Cultivated and rarely escaped to wooded slopes and edges.

I

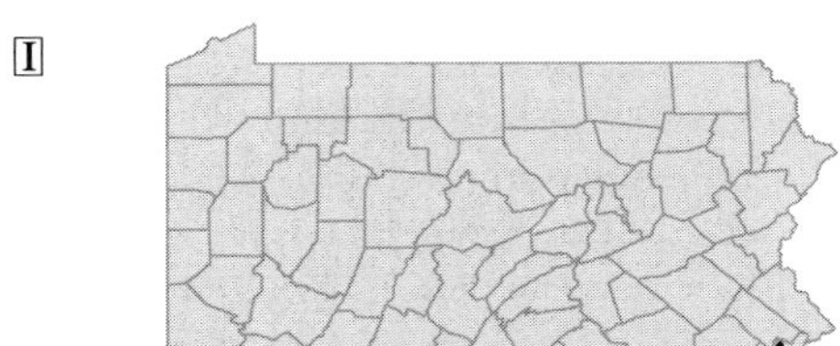
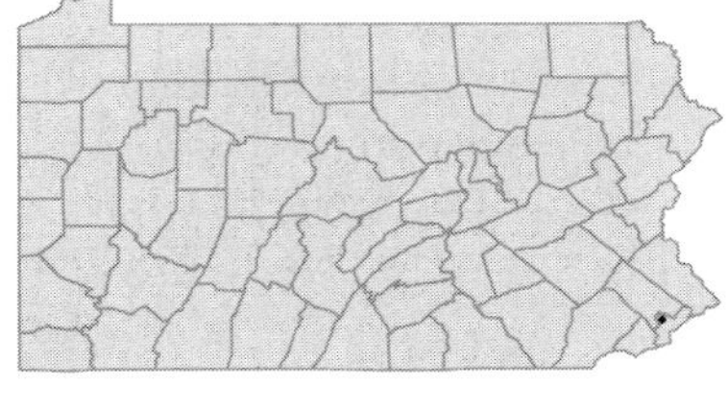

• ***Trillium cernuum*** L. var. ***cernuum***
Nodding trillium
Herbaceous perennial
Rich woods.

FACW

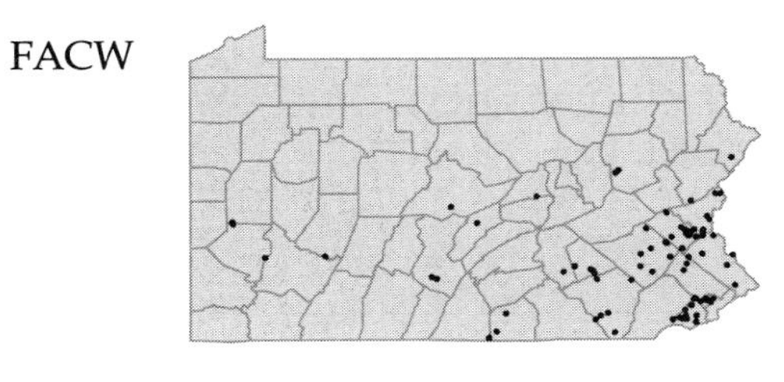

Trillium cernuum L. var. ***macranthum*** Wieg.
Nodding trillium
Herbaceous perennial
Moist, rich woods, often on calcareous soils.

FACW

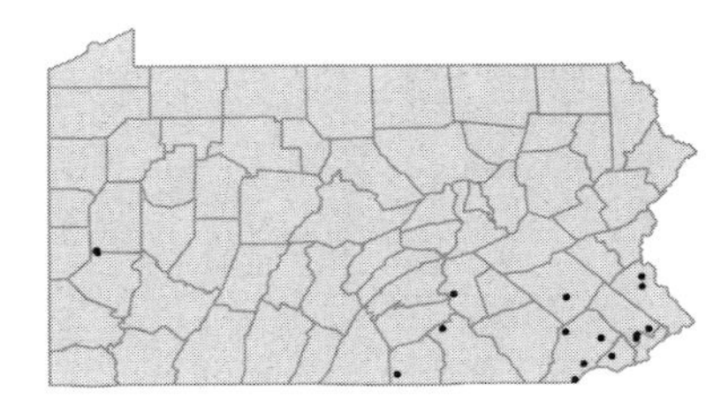

Trillium erectum L. var. ***erectum***
Purple trillium; Squawroot; Stinking benjamin
Herbaceous perennial
Moist, rocky, wooded slopes.

FACU-

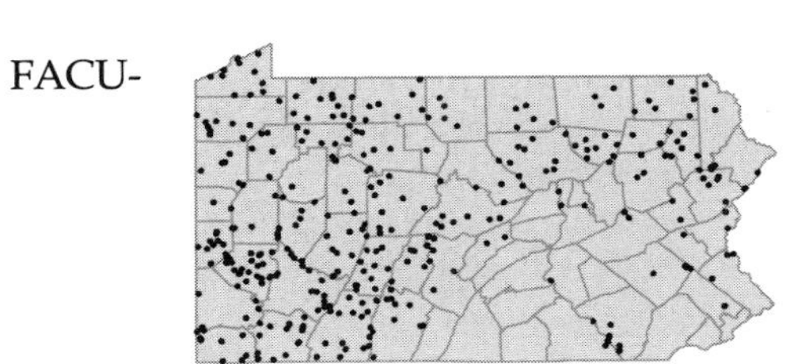

Trillium erectum L. var. ***album*** (Michx.) Pursh
Trillium
Herbaceous perennial
Moist, rich woods.

FACU-

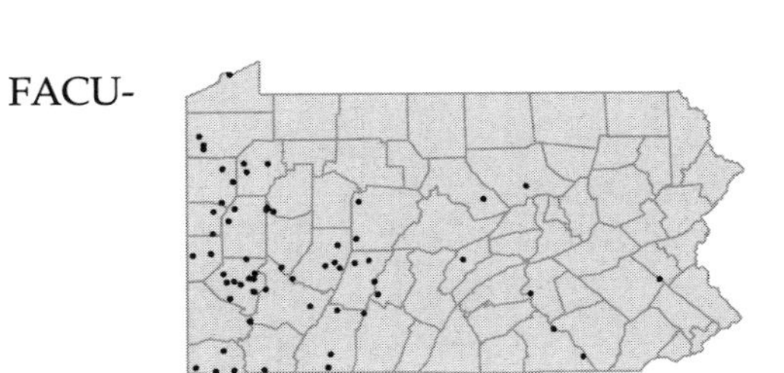

Trillium erectum x flexipes
Trillium
Herbaceous perennial
Rich woods.

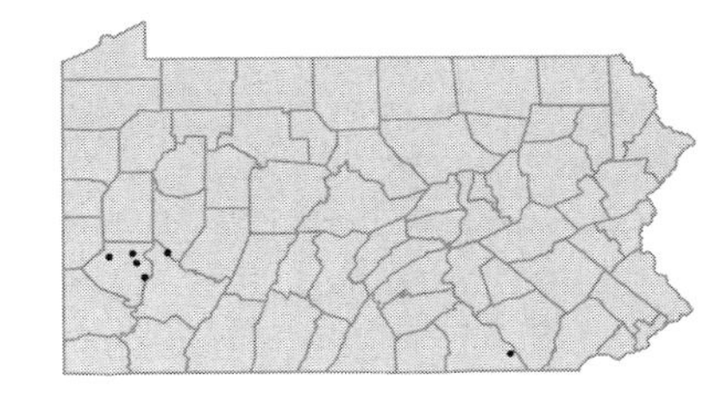

Trillium flexipes Raf.
Declined trillium; Nodding trillium
Herbaceous perennial
Wooded hillsides, swampy woods and floodplains.
Trillium erectum L. var. *blandum* Jenn. F; *Trillium gleasoni* Fern. GB

FAC

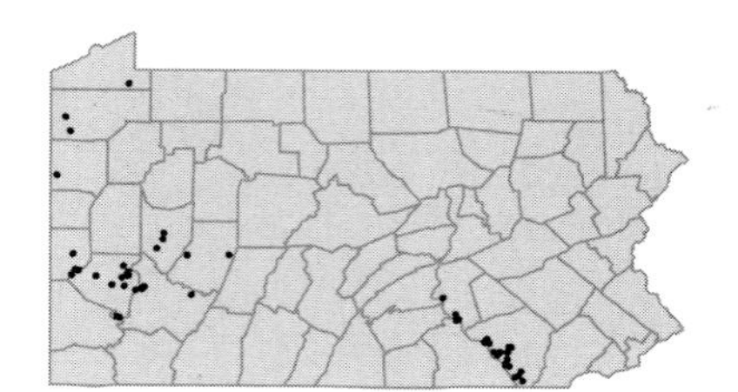

Trillium grandiflorum (Michx.) Salisb.
White trillium; Wake-robin
Herbaceous perennial
Rich woods and slopes.

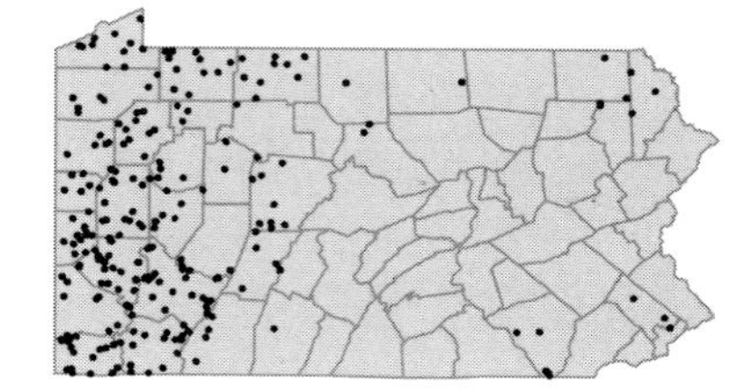

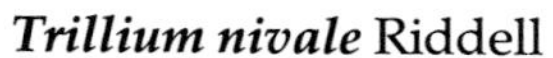

Trillium nivale Riddell
Snow trillium
Herbaceous perennial
Rocky, wooded slopes, frequently under hemlocks.

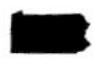

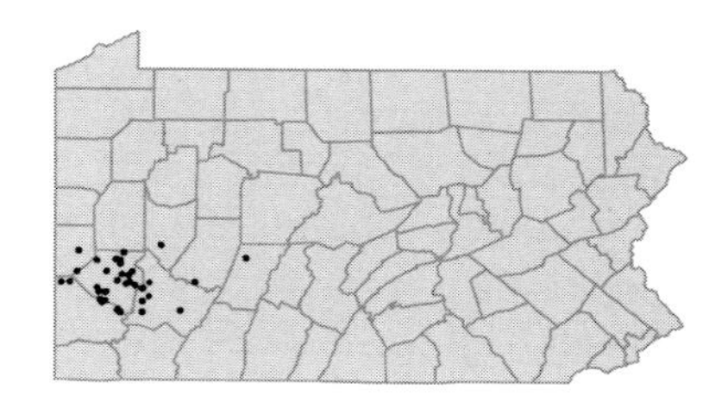

Trillium recurvatum Beck
Prairie trillium; Purple trillium
Herbaceous perennial
Apparently escaped from cultivation. Represented by a single collection from Lancaster Co. in 1955.

UPL
[I]

Trillium sessile L.
Toadshade
Herbaceous perennial
Rich, wooded hillsides and floodplains.

FACU-

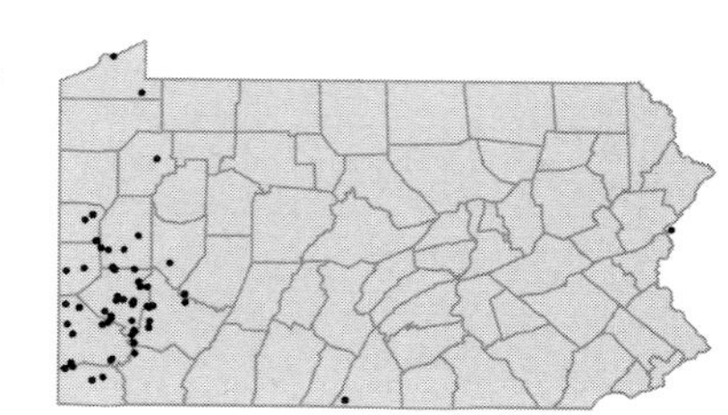

Trillium undulatum Willd.
Painted trillium
Herbaceous perennial
Woods, stream banks, bogs or swamps in moist, often peaty, soils.

FACU

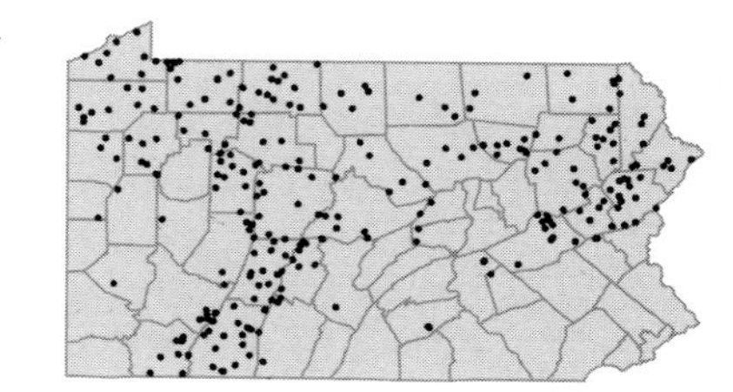

• ***Tulipa sylvestris*** L. [I]
Dutch-lily
Herbaceous perennial
Cultivated and occasionally naturalized in moist meadows, fields and roadsides.

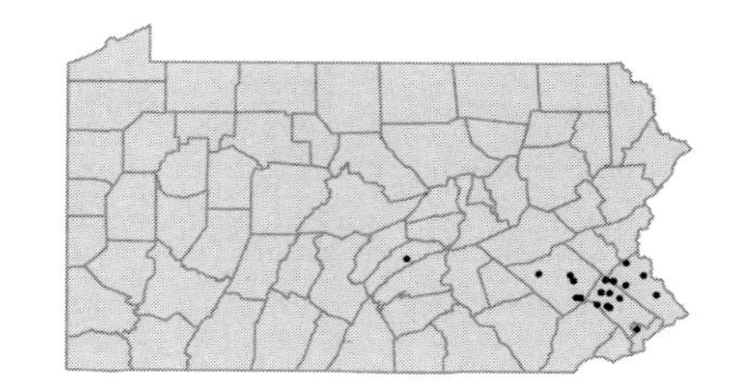

• ***Uvularia grandiflora*** J.E.Smith
Bellwort
Herbaceous perennial
Moist, rich woods and wooded slopes.

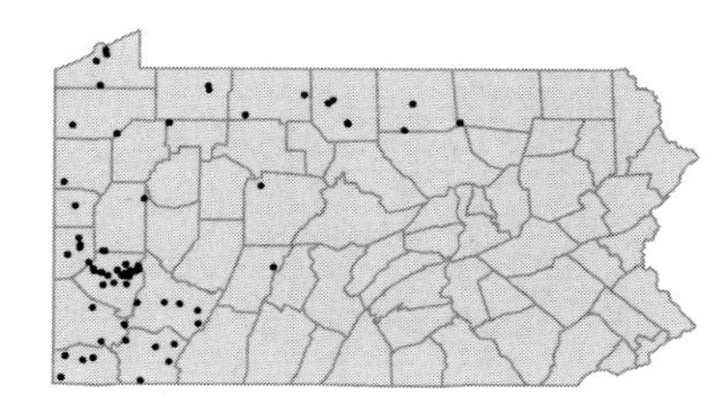

Uvularia perfoliata L. FACU
Bellwort
Herbaceous perennial
Moist, rich woods and slopes.

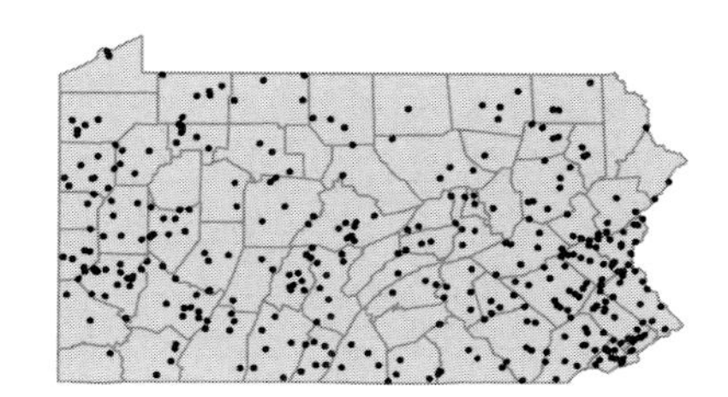

Uvularia puberula Michx. FACU
Mountain bellwort
Herbaceous perennial
Wooded slopes.
Uvularia pudica (Walt.) Fern. FGBW

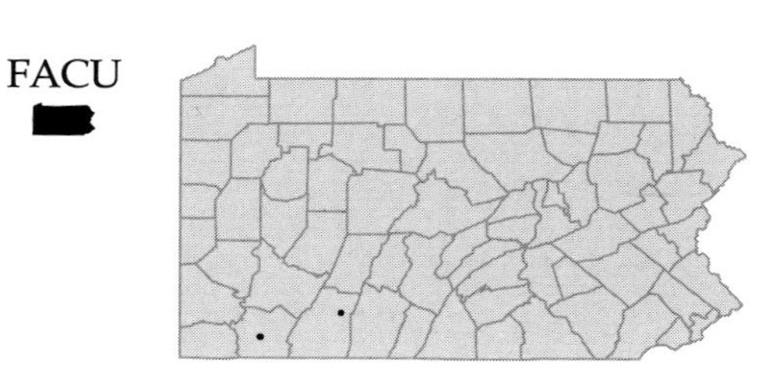

Uvularia sessilifolia L. FACU-
Bellwort; Wild-oats
Herbaceous perennial
Moist, rich woods.

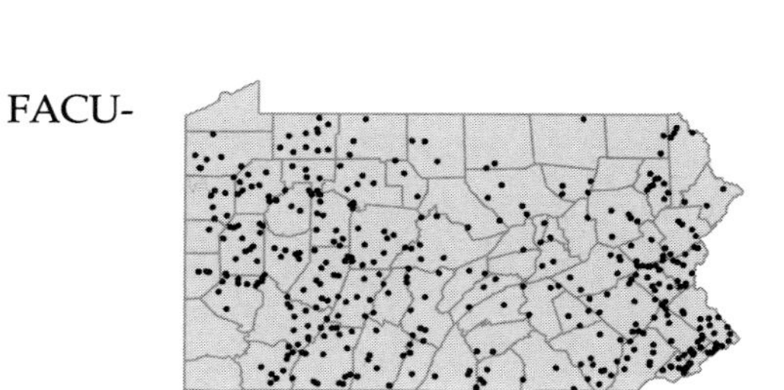

• ***Veratrum viride*** Ait. FACW+
False hellebore; Indian-poke
Herbaceous perennial
Wet woods and swamps.

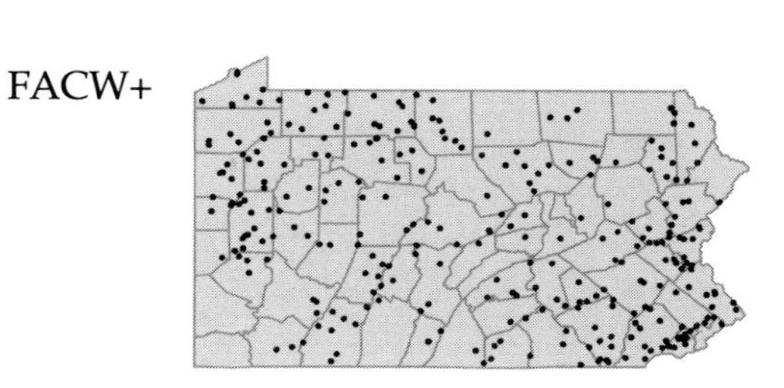

IRIDACEAE

• ***Belamcanda chinensis*** (L.) DC.
Blackberry-lily
Herbaceous perennial
Cultivated and occasionally escaped to roadside banks and open woods.

I

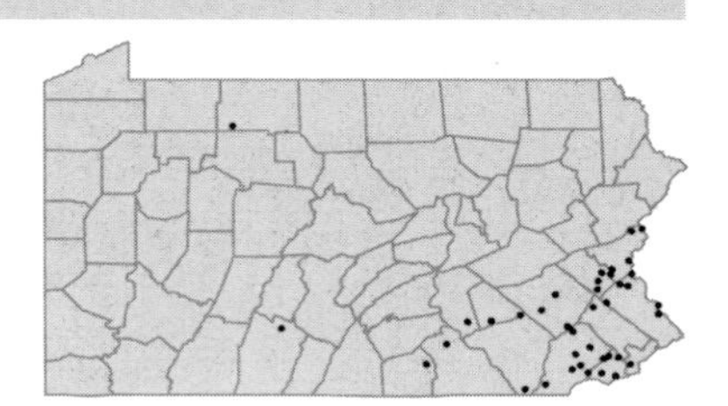

• ***Gladiolus gandavensis*** Van Houtte
Gladiolus
Herbaceous perennial
Cultivated and occasionally escaped to waste ground and abandoned dump sites.

I

• ***Iris cristata*** Soland.
Dwarf crested iris
Herbaceous perennial
Wooded slopes and stream banks.

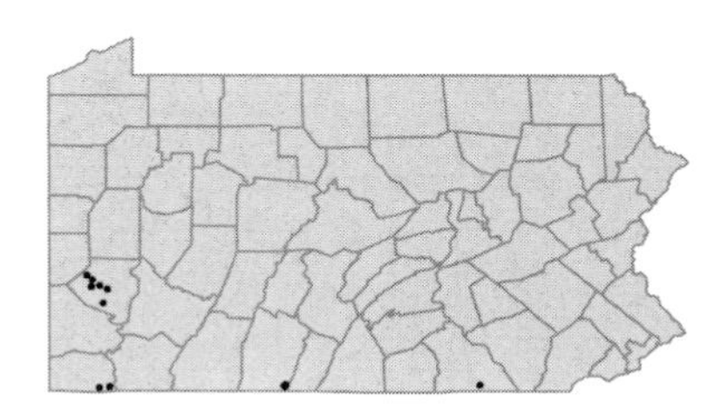

Iris ensata Thunb.
Japanese iris
Herbaceous perennial
Cultivated and rarely escaped at rubbish dumps and alluvial shores.
Iris kaempferi Maxim. W

I

Iris flavescens Delile
Iris
Herbaceous perennial
Cultivated and rarely persisting at abandoned homesites.

I

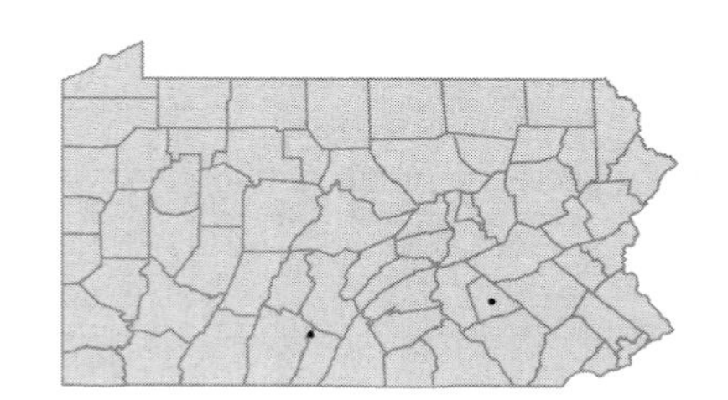

Iris germanica L.
German iris
Herbaceous perennial
Cultivated and occasionally persisting at abandoned homesites or gardens.

I

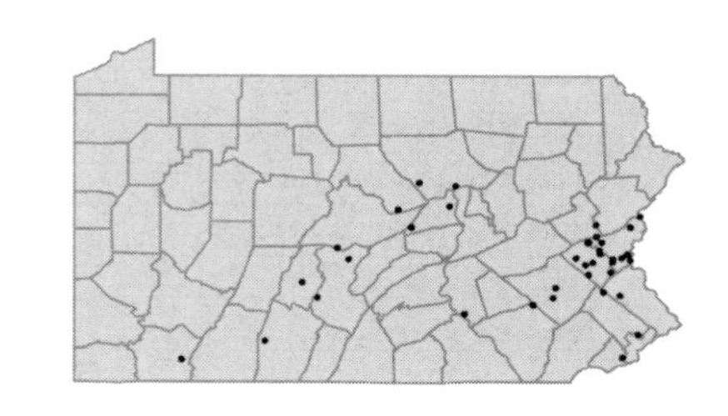

Iris prismatica Pursh
Slender blue flag
Herbaceous perennial
Moist meadows and sandy or gravelly shores.

OBL

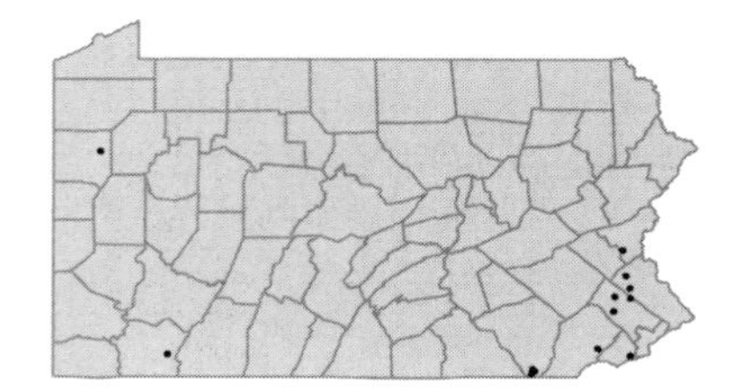

Iris pseudoacorus L.
Water flag; Yellow iris
Herbaceous perennial, emergent aquatic
Marshes, shallow water or wet shores.

OBL
I

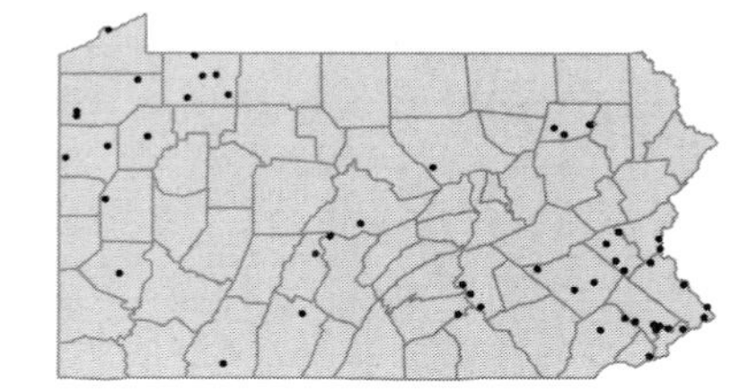

Iris sibirica L.
Siberian iris
Herbaceous perennial
A rare garden escape.

I

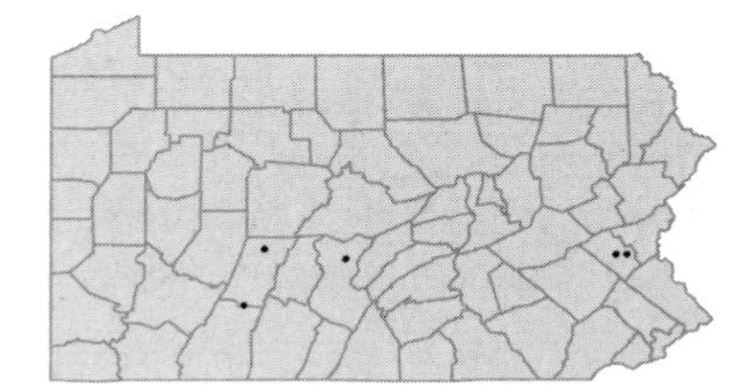

Iris verna L. var. ***smalliana*** Fern. ex M.E.Edwards
Dwarf iris
Herbaceous perennial
Dry to moist, acidic, sandy soils.

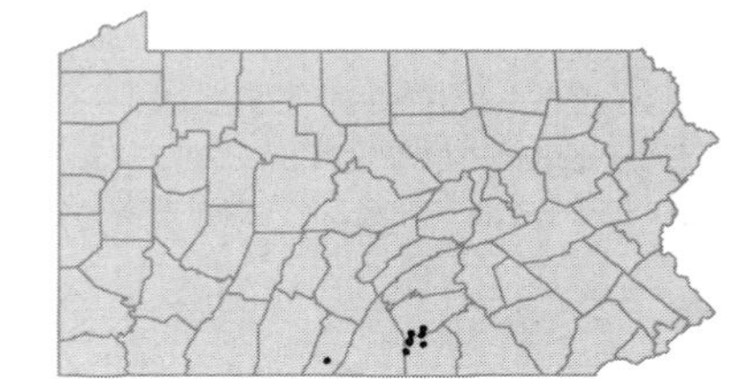

Iris versicolor L.
Northern blue flag
Herbaceous perennial, emergent aquatic
Wet meadows, bogs and marshes.

OBL

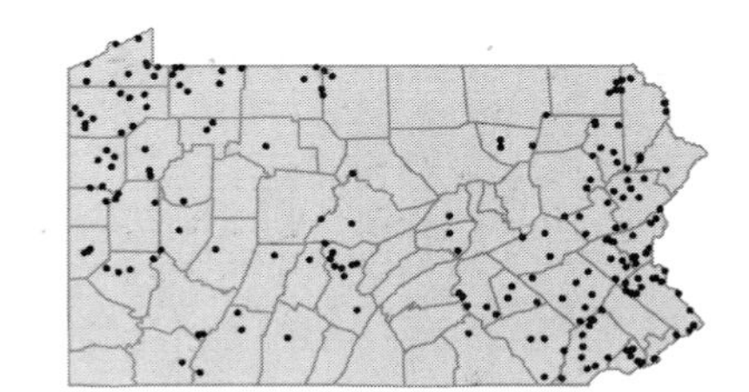

• ***Sisyrinchium albidum*** Raf.
Blue-eyed-grass
Herbaceous perennial
Dry, sandy, open soil, likely an adventive from further west. Represented by a single collection from Allegheny Co. in 1918, another specimen, reputedly from Dauphin Co., is questionable.

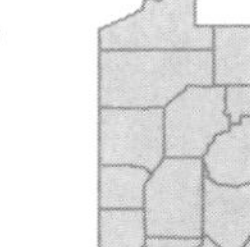
UPL

Sisyrinchium angustifolium P.Mill.
Blue-eyed-grass
Herbaceous perennial
Damp soil of meadows, floodplains, fields and open woods.
Sisyrinchium graminoides Bickn. PGB; *Sisyrinchium x intermedium* Bickn. W

FACW-

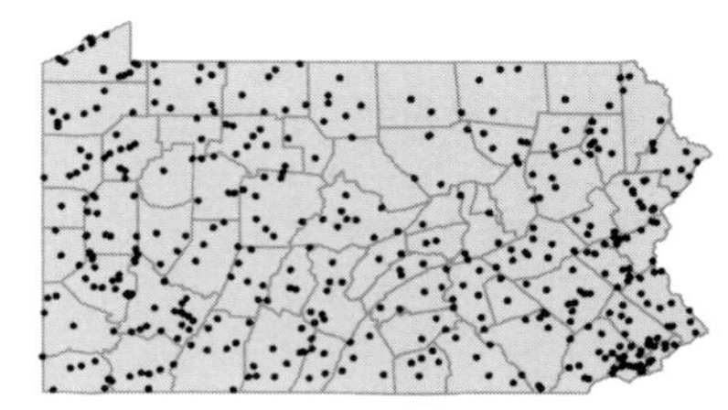

Sisyrinchium atlanticum Bickn.
Eastern blue-eyed-grass
Herbaceous perennial
Moist to dry, sandy, open ground of fields and thin woods.

FACW

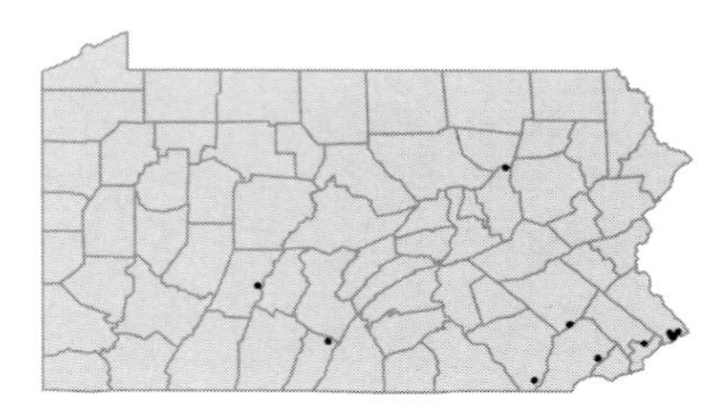

Sisyrinchium fuscatum Bickn.
Sand blue-eyed-grass
Herbaceous perennial
Dry, open, sandy soil. Believed to be extirpated, last collected in the 1930's.
Sisyrinchium arenicola Bickn. GFWKB

FACU

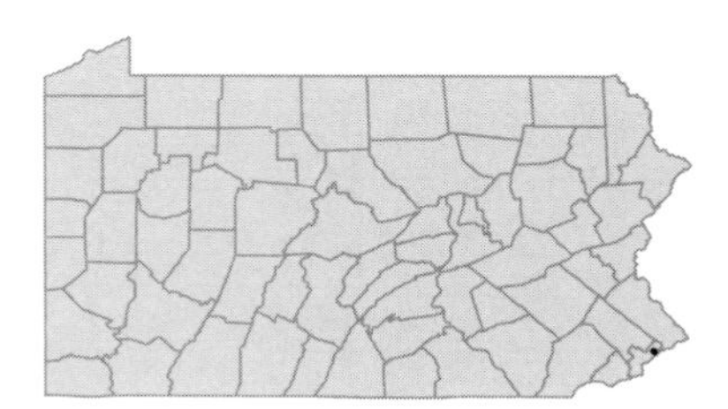

Sisyrinchium montanum Greene var. ***crebrum*** Fern.
Blue-eyed-grass
Herbaceous perennial
Stream banks, woods and old fields.

FAC

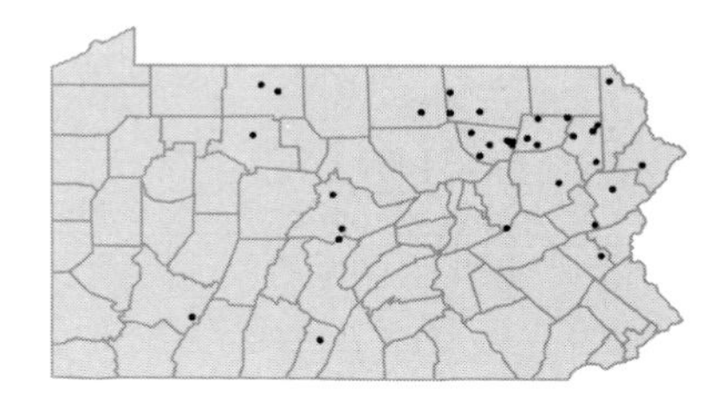

Sisyrinchium mucronatum Michx.
Blue-eyed-grass
Herbaceous perennial
Dry fields, roadsides and open woods.

FAC+

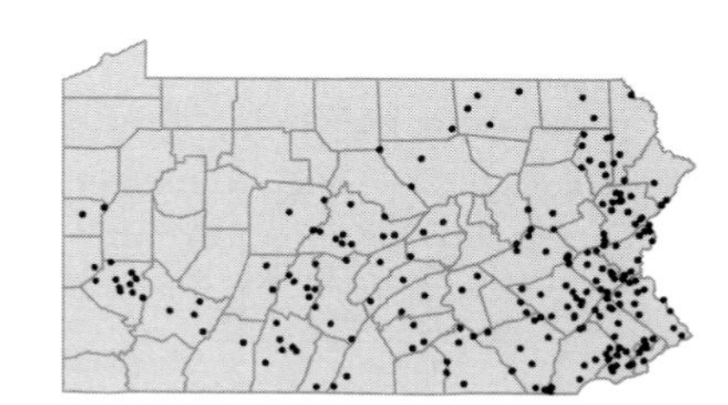

AGAVACEAE

• ***Yucca flaccida*** Haw.
Adam's-needle
Herbaceous perennial
Dry, sandy soil, old homesites and abandoned plantings.
Yucca smalliana Fern. FW; *Yucca filamentosa* L. C

[I]

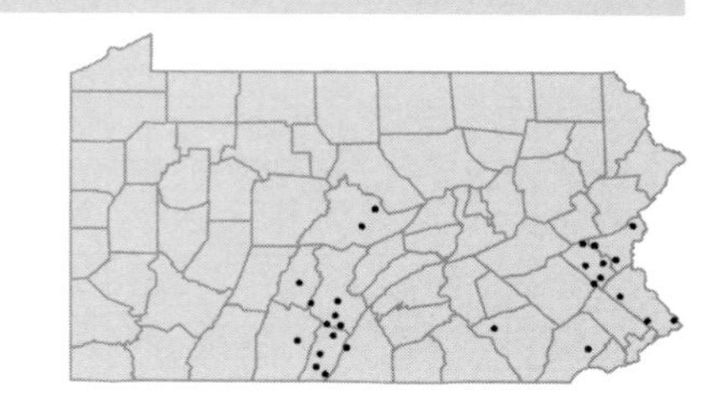

SMILACACEAE

• ***Smilax glauca*** Walt.
Catbrier; Greenbrier
Woody vine
Dry to moist, sandy soil of fields, woods, thickets, swamps and roadsides.

FACU

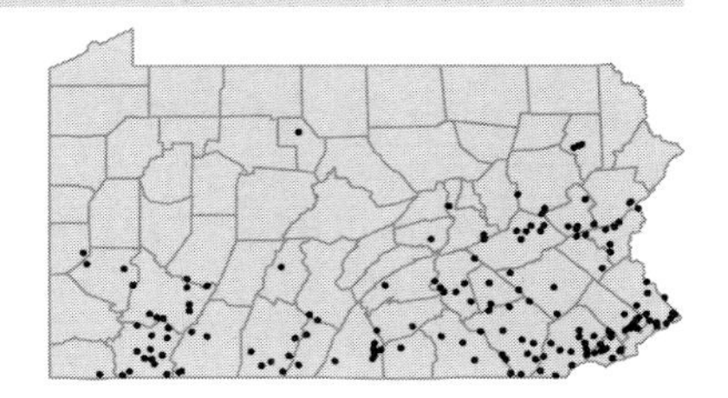

Smilax herbacea L.
Carrion-flower
Herbaceous perennial vine
Damp thickets, moist woods and floodplains.

FAC

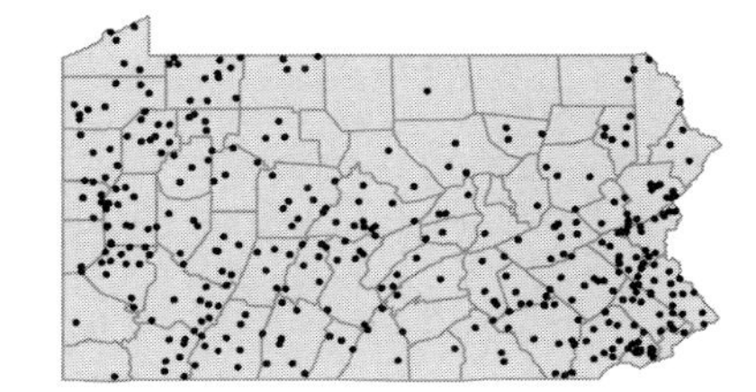

Smilax hispida Muhl. ex Torr.
Bristly greenbrier; Catbrier
Woody vine
Swamps, moist woods, thickets and roadsides.
Smilax tamnoides L. var. *hispida* (Muhl.) Fern. F

FAC

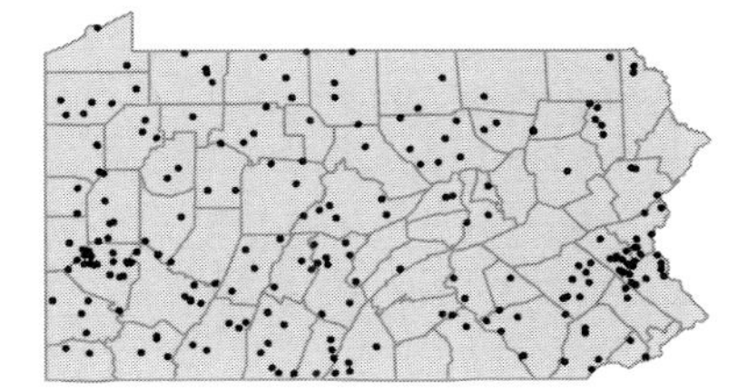

Smilax pseudochina L.
False china-root; Long-stalked greenbrier
Herbaceous perennial vine
Dry woods. Believed to be extirpated, last collected in 1906.
Smilax tamnifolia Michx. PGB

FAC+

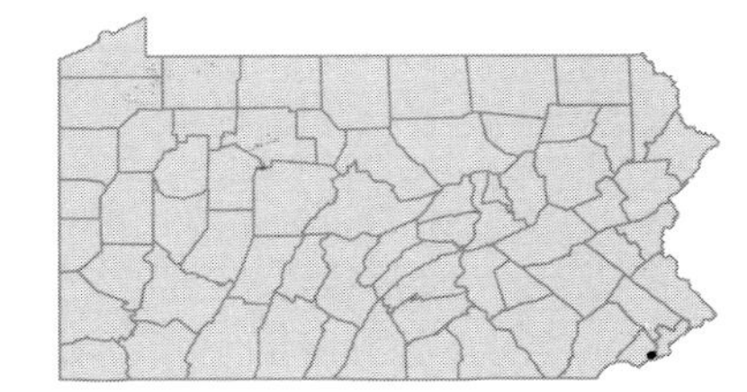

Smilax pulverulenta Michx.
Carrion-flower
Herbaceous perennial vine
Moist woods and thickets.
Smilax herbacea L. var. *pulverulenta* (Michx.) A.Gray GBC

FACW

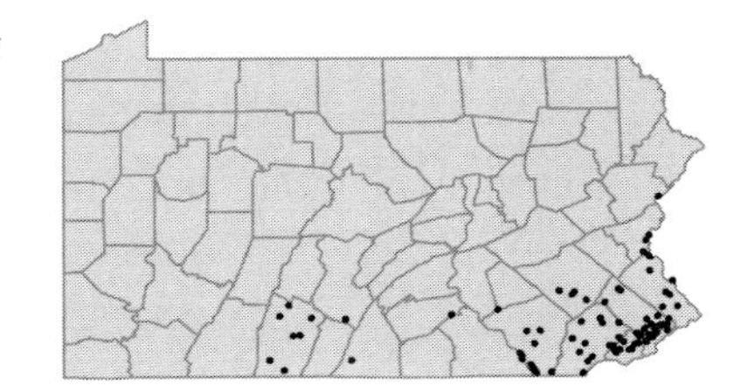

Smilax rotundifolia L.
Catbrier; Greenbrier
Woody vine
Moist to dry woods, thickets, roadsides, old fields and serpentine barrens.

FAC

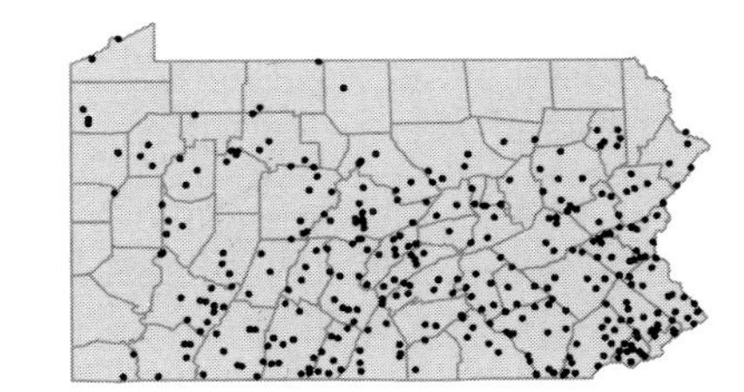

DIOSCOREACEAE

•***Dioscorea batatas*** Decne.
Chinese yam; Cinnamon vine
Herbaceous perennial vine
Old homesites, disturbed slopes and waste areas.

[I]

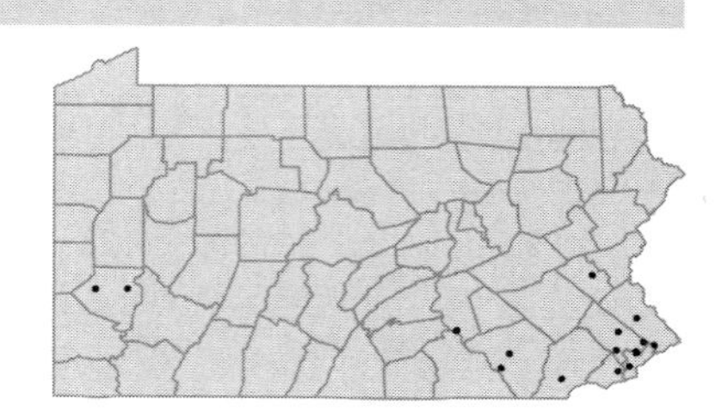

Dioscorea quaternata (Walt.) J.F.Gmel.
Wild yam
Herbaceous perennial vine
Dry, rocky, wooded slopes and thickets.

FACU

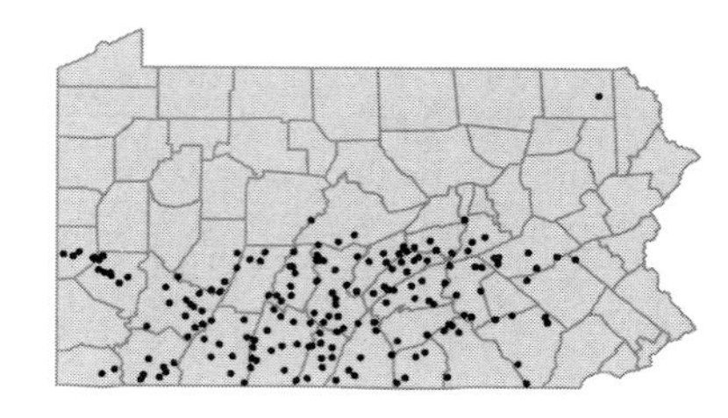

Dioscorea villosa L.
Wild yam; Colic-root
Herbaceous perennial vine
Woods, thickets and rocky slopes.

FAC+

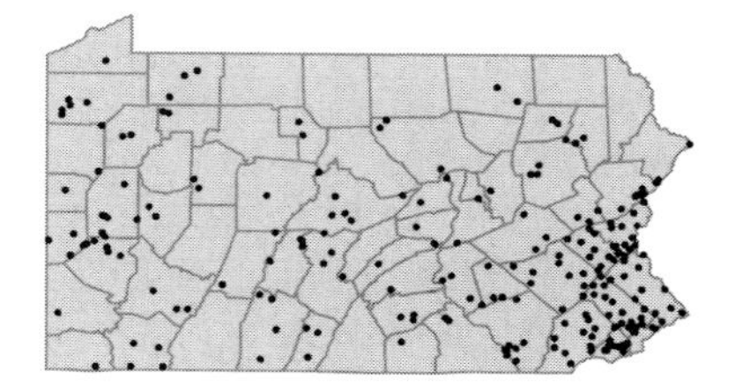

ORCHIDACEAE

•***Aplectrum hyemale*** (Muhl. ex Willd.) Nutt.
Puttyroot; Adam-and-Eve
Herbaceous perennial
Moist, rich wooded slopes.
Aplectrum spicatum (Walt.) BSP P

FAC

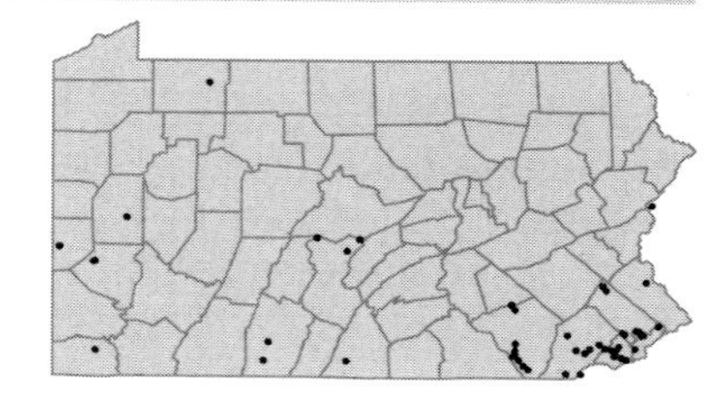

•***Arethusa bulbosa*** L.
Dragon's-mouth; Swamp-pink
Herbaceous perennial
Sphagnum bogs.

OBL

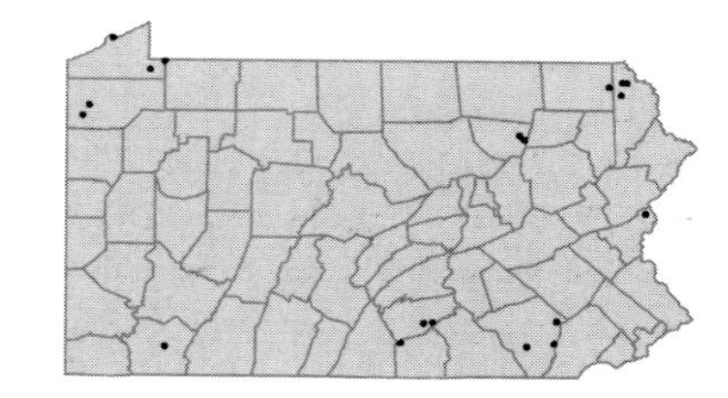

•***Calopogon tuberosus*** (L.) BSP
Grass-pink
Herbaceous perennial
Bogs, fens and wet meadows.
Calopogon pulchellus (Salisb.) R.Br. FGBW

FACW+

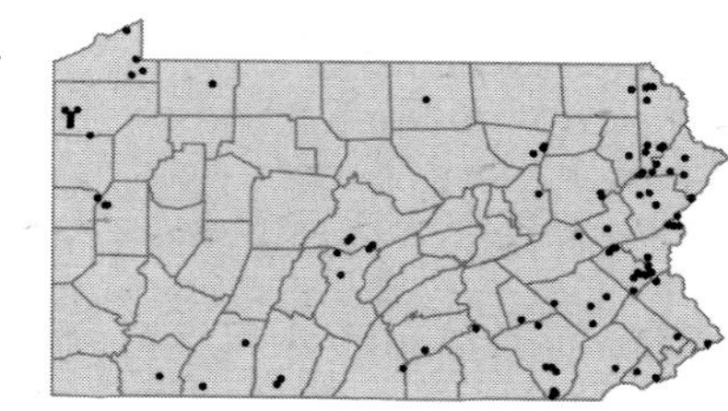

• ***Coeloglossum viride*** (L.) Hartm. var. ***virescens*** (Muhl. ex Willd.) Luer — FACU
Frog orchid; Long-bracted green orchis
Herbaceous perennial
Rich woods.
Coeloglossum bracteatum (Willd.) Parl. P; *Coeloglossum viride* (L.) Hartm. var. *bracteatum* (Muhl.) Reichard W; *Habenaria viridis* (L.) R.Br. var. *bracteata* (Muhl.) A.Gray FGBC

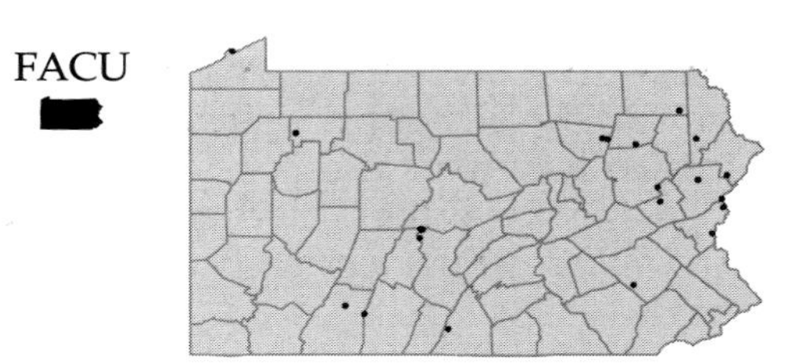

• ***Corallorhiza maculata*** (Raf.) Raf. — FACU
Spotted coral-root; Large coral-root
Herbaceous perennial, saprophytic
Dry, deciduous or coniferous woods in light soils.
Corallorhiza multiflora Nutt. P

Corallorhiza odontorhiza (Willd.) Nutt.
Autumn coral-root; Small-flowered coral-root
Herbaceous perennial, saprophytic
Rich woods on moist to dry soils.

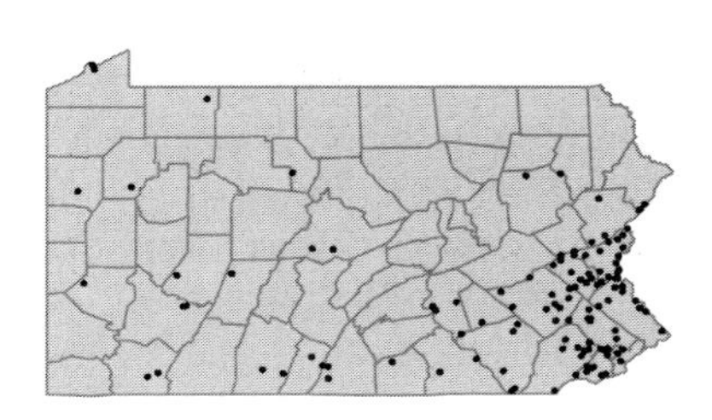

Corallorhiza trifida Chatelain — FACW
Early coral-root; Northern coral-root; Pale coral-root
Herbaceous perennial, saprophytic
Moist woods and bogs.
Corallorhiza corallorhiza (L.) Karst. P

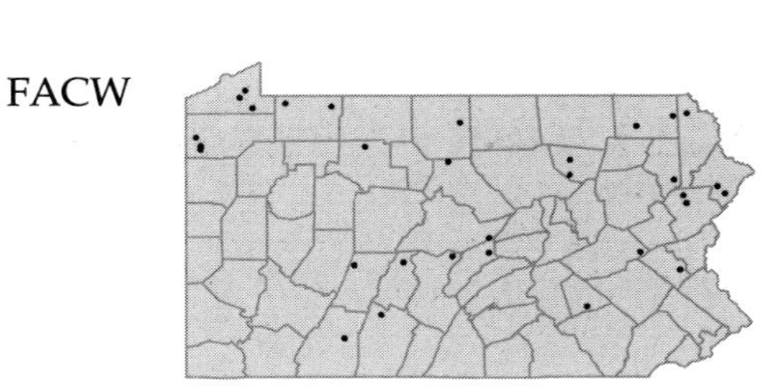

Corallorhiza wisteriana Conrad — FAC
Wister's coral-root
Herbaceous perennial, saprophytic
Rocky, wooded slopes, usually on limestone.

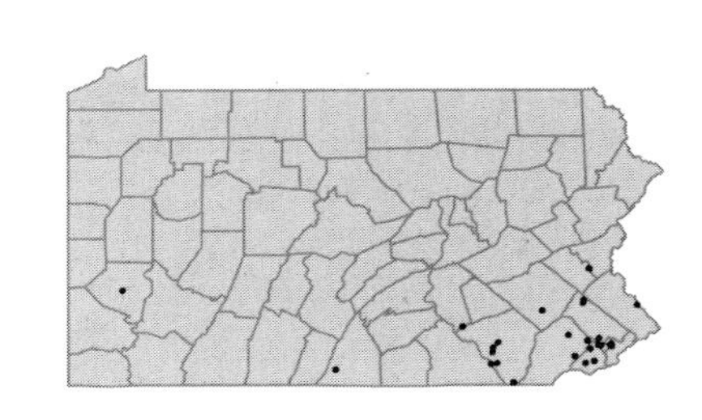

• ***Cypripedium acaule*** Ait. — FACU
Pink lady's-slipper; Pink moccasin-flower
Herbaceous perennial, saprophytic
Moist to dry acidic woods.

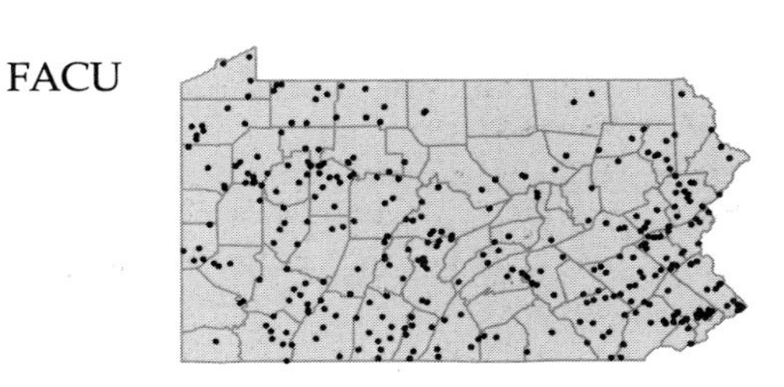

Cypripedium calceolus L. var. ***parviflorum*** (Salisb.) Fern.
Small yellow lady's-slipper; Small yellow moccasin-flower
Herbaceous perennial
Moist woods and bogs, often on limestone.
Cypripedium parviflorum Salisb. P

FAC+

Cypripedium calceolus L. var. ***pubescens*** (Willd.) Correll
Large yellow lady's-slipper; Large yellow moccasin-flower
Herbaceous perennial
Moist, rich, rocky woods and slopes.
Cypripedium pubescens Willd.

FAC+

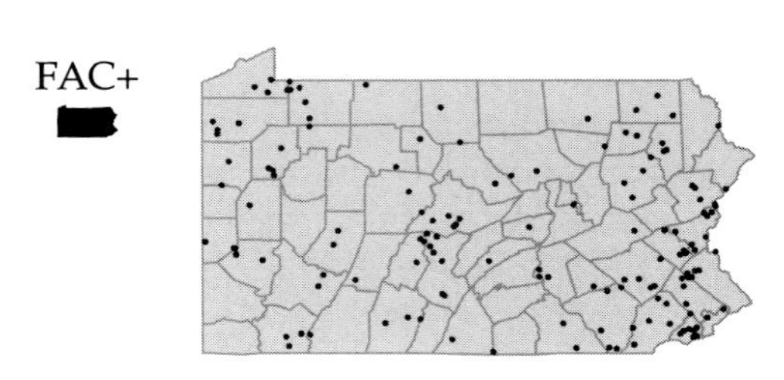

Cypripedium candidum Muhl. ex Willd.
Small white lady's-slipper
Herbaceous perennial
Damp, calcareous meadows and swamps. Believed to be extirpated, last collected in 1865.

OBL

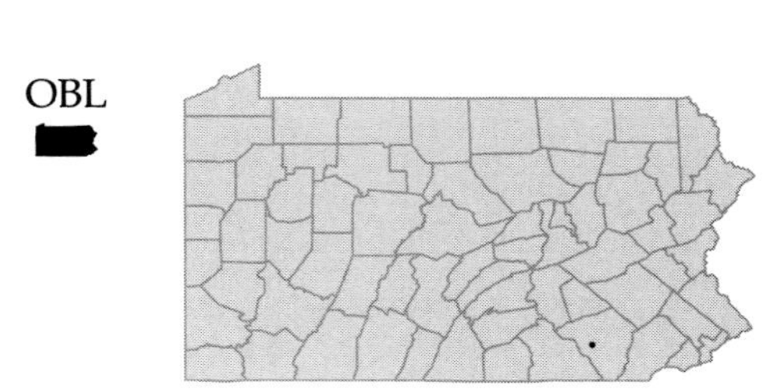

Cypripedium reginae Walt.
Large white lady's-slipper; Showy lady's-slipper
Herbaceous perennial
Bogs, fens and swampy woods.
Cypripedium hirsutum P.Mill. P

FACW

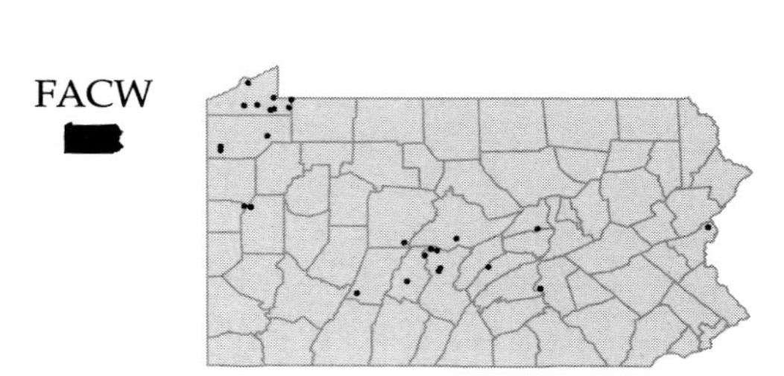

• ***Epipactis helleborine*** (L.) Crantz
Broad-leaved helleborine; Bastard hellebore
Herbaceous perennial
Woods, shaded roadsides and thickets.
Epipactis latifolia (L.) Crantz B

I

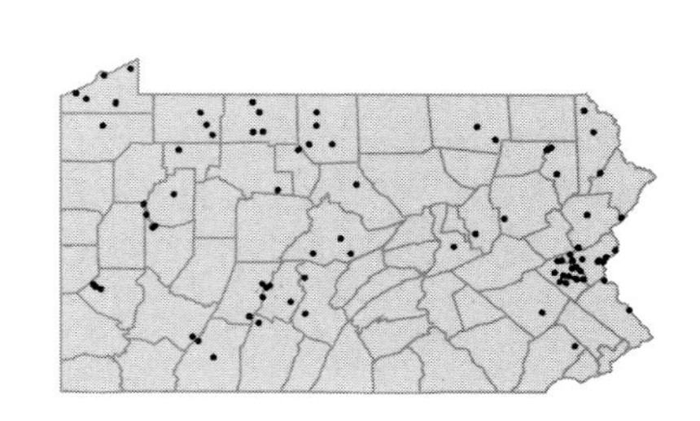

• ***Galearis spectabilis*** (L.) Raf.
Showy orchis
Herbaceous perennial
Rich, deciduous woods.
Galeorchis spectabilis (L.) Rydb. P; *Orchis spectabilis* L. FGBC

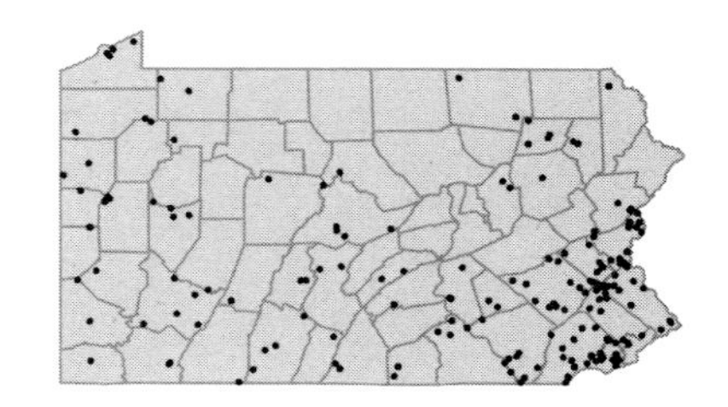

• ***Goodyera pubescens*** (Willd.) R.Br.
Downy rattlesnake-plantain
Herbaceous perennial
Dry to moist, deciduous or coniferous woods.

FACU-

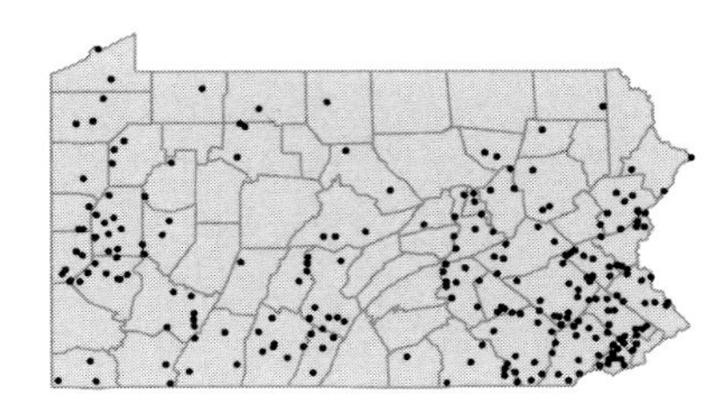

Goodyera repens (L.) R.Br.
Lesser rattlesnake-plantain; White-blotched rattlesnake-plantain
Herbaceous perennial
Moist, wooded slopes, mossy woods and swamps.
Peramium ophioides (Fern.) Rydb. P

FACU+

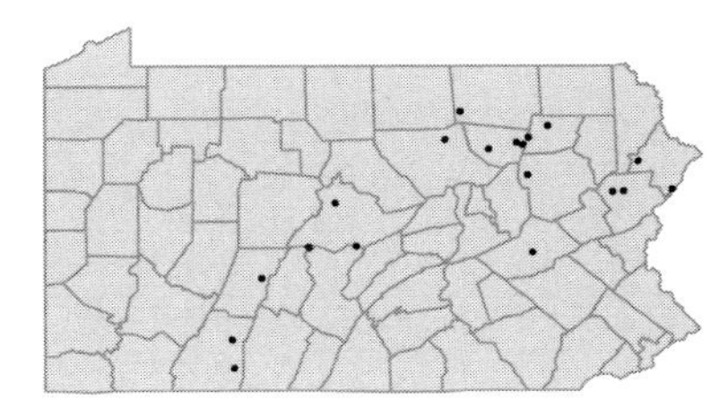

Goodyera tesselata Lodd.
Checkered rattlesnake-plantain
Herbaceous perennial
Moist, rich, coniferous or deciduous forests.

FACU-

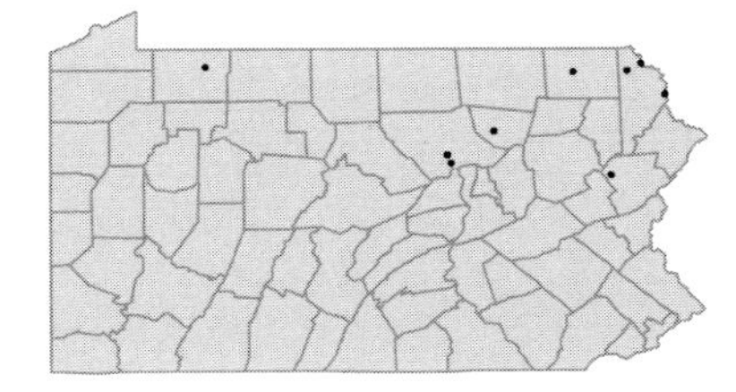

• ***Isotria medeoloides*** (Pursh) Raf.
Small whorled-pogonia
Herbaceous perennial
Dry, open deciduous woods on acidic soils.
Isotria affinis (Austin) Rydb. P

FACU

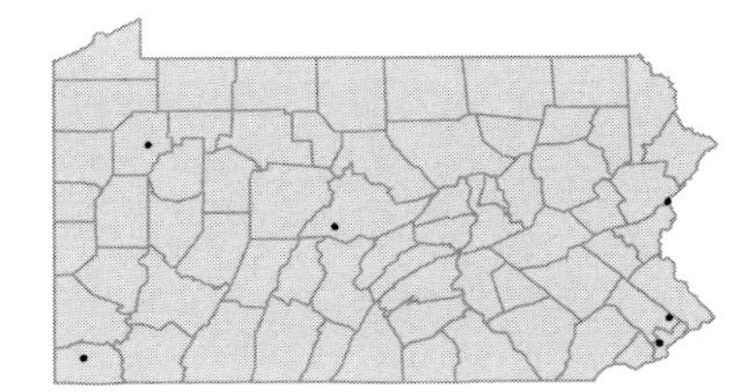

Isotria verticillata (Muhl. ex Willd.) Raf.
Whorled-pogonia
Herbaceous perennial
Dry to moist, moderately acidic woods.

FACU

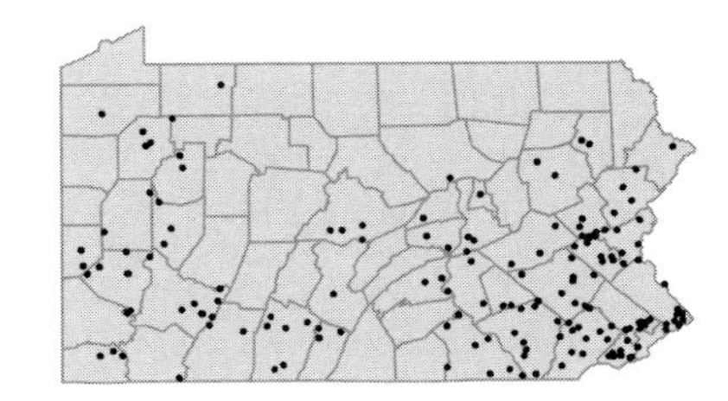

• ***Liparis liliifolia*** (L.) L.C.Rich. ex Lindl.
Large twayblade; Lily-leaved twayblade
Herbaceous perennial
Rich, rocky woods and slopes.

FACU-

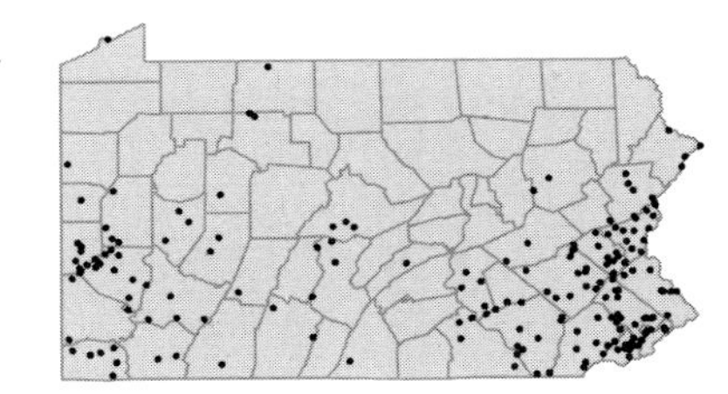

Liparis loeselii (L.) L.C.Rich.
Bog twayblade; Fen orchid; Yellow twayblade
Herbaceous perennial
Bogs, fens, wet meadows and cool ravines, especially on calcareous soils.

FACW

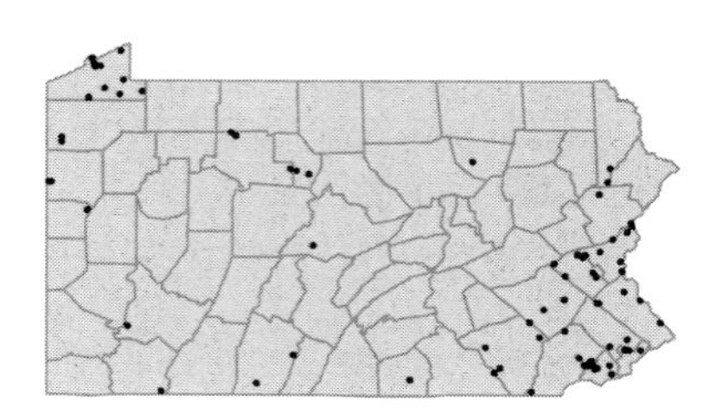

• ***Listera australis*** Lindl.
Southern twayblade
Herbaceous perennial
Bog edges, in sphagnum.

FACW

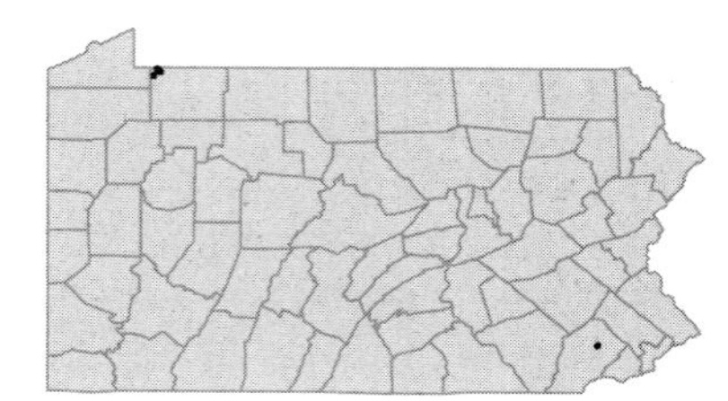

Listera cordata (L.) R.Br.
Heartleaf twayblade
Herbaceous perennial
Cool, wooded bogs or mossy woods.

FACW+

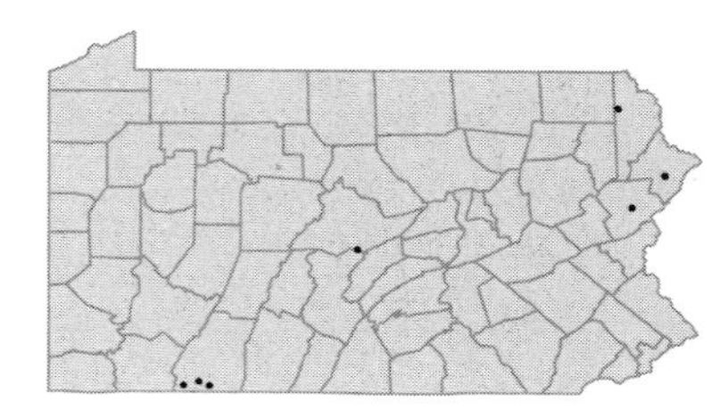

Listera smallii Wieg.
Appalachian twayblade; Small's twayblade; Kidney-leaved twayblade
Herbaceous perennial
Damp, shady woods or bogs.

FACW

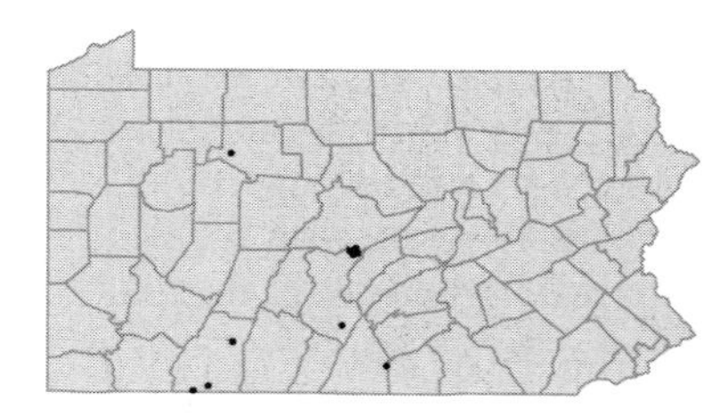

• ***Malaxis bayardii*** Fern.
Adder's-mouth
Herbaceous perennial
Dry, open, upland woods and shale barrens.

FAC

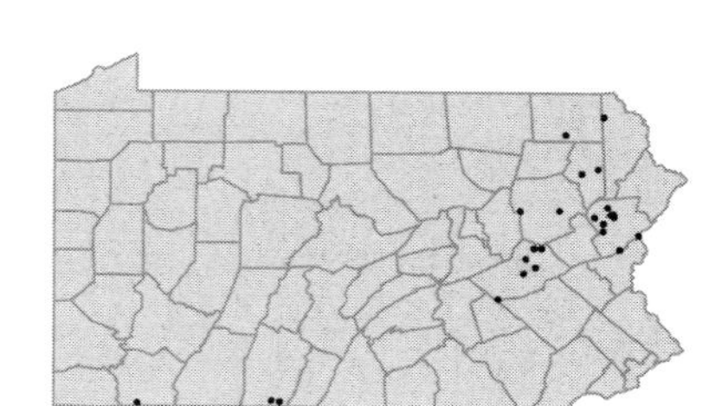

Malaxis monophyllos (L.) Swartz var. ***brachypoda*** (A.Gray) F.Morris
White adder's-mouth
Herbaceous perennial
Wooded swamps, wet woods and bogs.
Malaxis brachypoda (A.Gray) Fern. FW

FACW

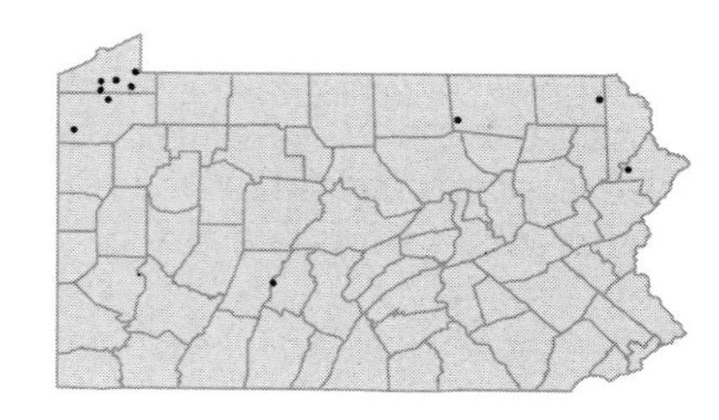

Malaxis unifolia Michx.
Green adder's-mouth
Herbaceous perennial
Dry to moist woods.

FAC

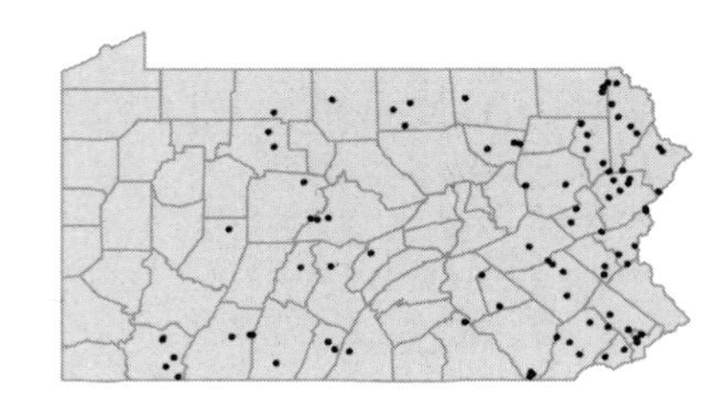

• ***Platanthera blephariglottis*** (Willd.) Lindl.
White fringed-orchid
Herbaceous perennial
Sphagnum bogs and swamps.
Blephariglottis blephariglottis (Willd.) Rydb. P; *Habenaria blephariglottis* (Willd.) Hook. FGBC

OBL

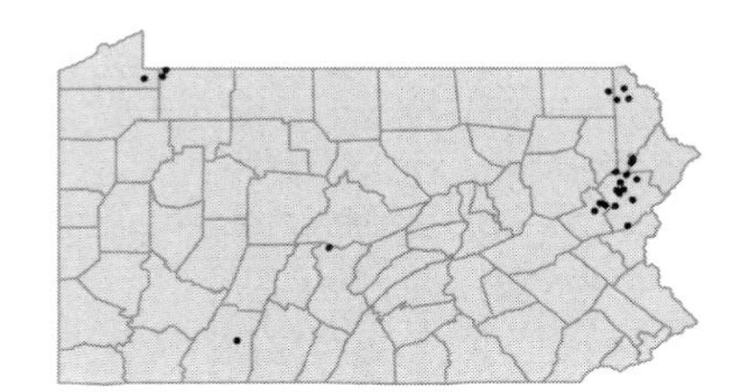

Platanthera ciliaris (L.) Lindl.
Yellow fringed-orchid
Herbaceous perennial
Bogs, moist meadows and woods.
Blephariglottis ciliaris (L.) Rydb. P; *Habenaria ciliaris* (L.) R.Br. FGBC

FACW

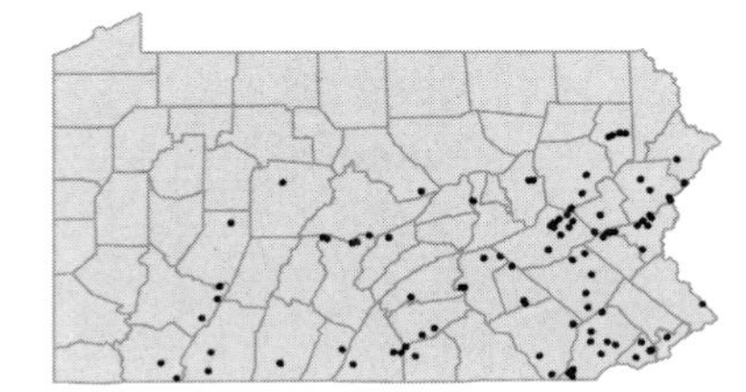

Platanthera clavellata (Michx.) Luer
Clubspur orchid; Small green woodland orchid
Herbaceous perennial
Bogs, wet woods and thickets.
Gymnadeniopsis clavellata (Michx.) Rydb. P; *Habenaria clavellata* (Michx.) Spreng. FGBC

FACW+

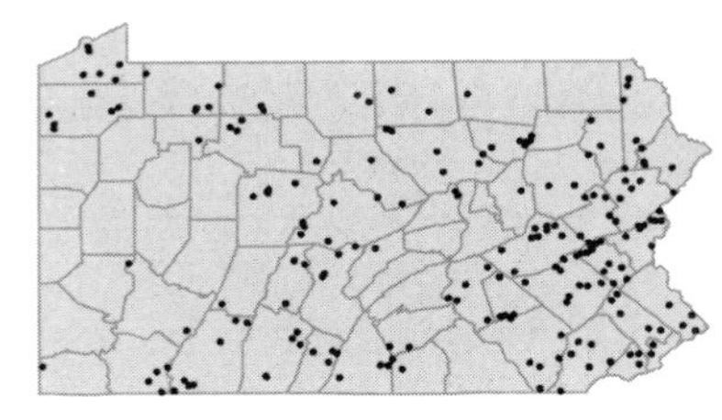

Platanthera cristata (Michx.) Lindl.
Crested fringed-orchid
Herbaceous perennial
Moist meadows, damp woods and swamps. Believed to be extirpated, last collected in 1902.
Blephariglottis cristata (Michx.) Raf. P; *Habenaria cristata* (Michx.) R.Br. FGBC

FACW+

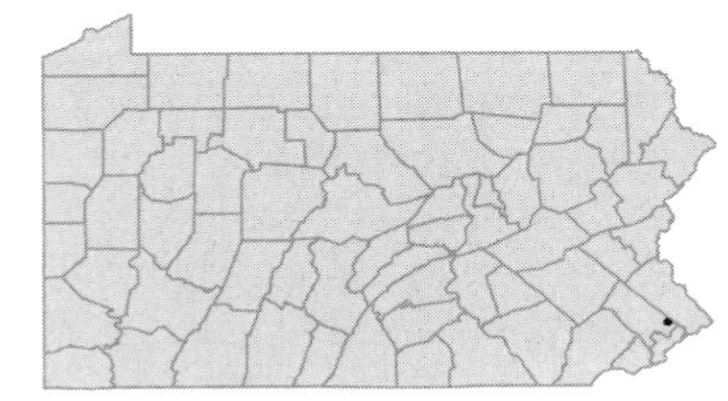

Platanthera dilatata (Pursh) Lindl. ex Beck
Bog-candles; Tall white bog orchid
Herbaceous perennial
Bogs, marshes or wet meadows. A hybrid with *P. hyperborea* (*P. x media* [Rydb.] Luer) has been collected at two locations.
Habenaria dilatata (Pursh) Hook. FGBC; *Limnorchis dilatata* (Pursh) Rydb. P

FACW

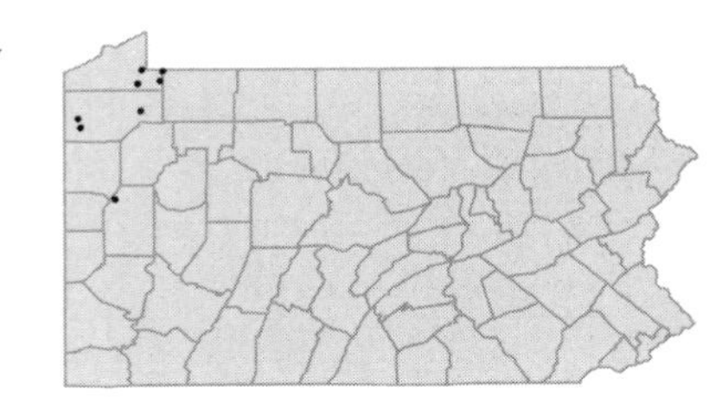

Platanthera flava (L.) Lindl. var. ***herbiola*** (R.Br.) Luer
Pale green orchid; Tubercled rein-orchid
Herbaceous perennial
Swamps, bogs, or wet, open woods.
Habenaria flava (L.) R.Br. var. *herbiola* (R.Br.) Ames & Correll FGBC

FACW

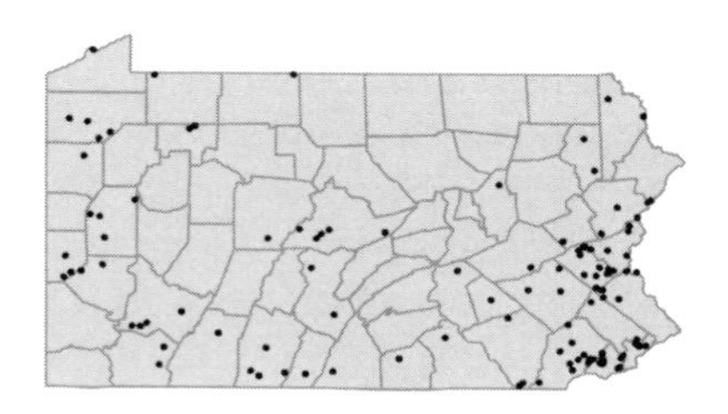

Platanthera grandiflora (Bigel.) Lindl.
Large purple fringed-orchid
Herbaceous perennial
Rich woods, thickets and meadows.
Blephariglottis grandiflora (Bigel.) Rydb. P; *Habenaria fimbriata* (Ait.) R.Br. F; *Habenaria psycodes* (L.) Spreng. var. *grandiflora* (Bigel.) A.Gray GBC

FACW

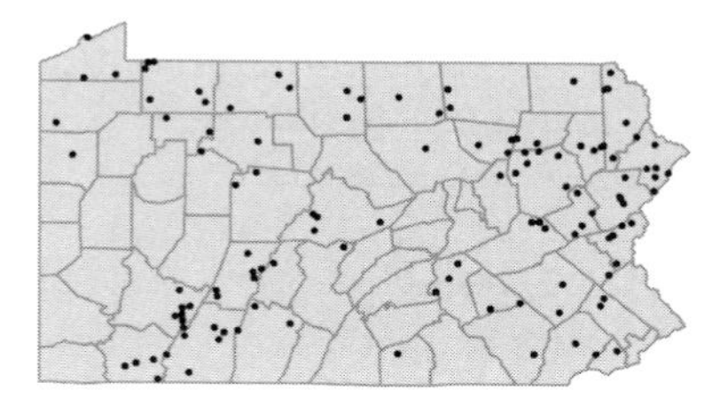

Platanthera hookeri (Torr. ex A.Gray) Lindl.
Hooker's orchid
Herbaceous perennial
Rich, well-drained, deciduous woods.
Habenaria hookeri Torr. FGBC; *Lysias hookeriana* (Torr.) Rydb. P

FAC

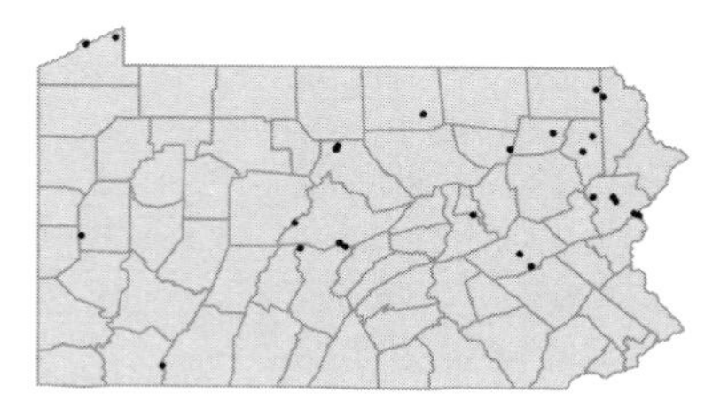

Platanthera hyperborea (L.) Lindl. var. ***huronensis*** (Nutt.) Luer
Northern green orchid; Tall green bog orchid
Herbaceous perennial
Wet meadows, bogs, woods and shores.
Habenaria hyperborea (L.) R.Br. var. *huronensis* (Nutt.) Farw. F; *Habenaria hyperborea* (L.) R.Br. var. *hyperborea* GBC

FACW

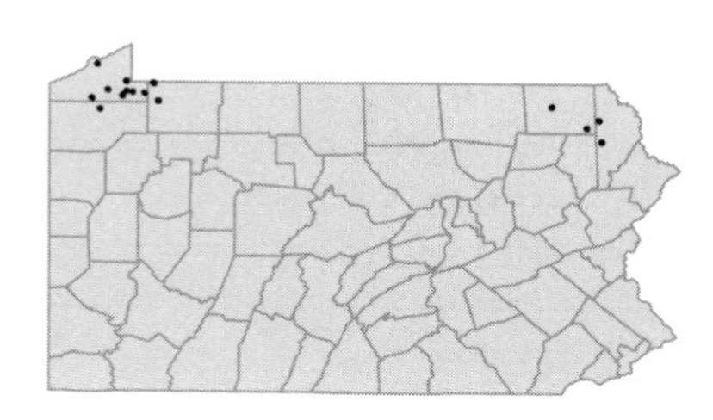

Platanthera lacera (Michx.) G.Don
Ragged fringed-orchid
Herbaceous perennial
Open woods, moist meadows, bogs and ditches. A hybrid with *P. psycodes* (*P. x andrewsii* M.White) has been collected at two locations.
Blephariglottis lacera (Michx.) Rydb. P; *Habenaria lacera* (Michx.) Lodd. FGBC

FACW

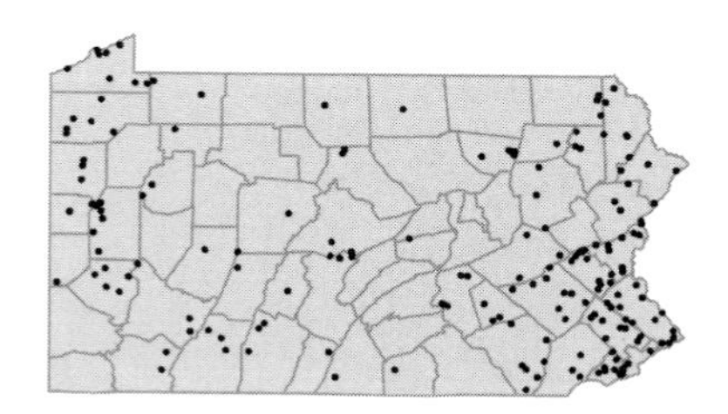

Platanthera leucophaea (Nutt.) Lindl.
Eastern prairie fringed-orchid
Herbaceous perennial
Damp, calcareous meadows and lake borders. Believed to be extirpated, last collected in 1881.
Habenaria leucophaea (Nutt.) A.Gray FGBC

FACW+

Platanthera orbiculata (Pursh) Lindl. var. ***orbiculata*** FAC
Large round-leaved orchid
Herbaceous perennial
Damp, rich humus in deep shade.
Habenaria orbiculata (Pursh) Torr. FGBC; *Lysias orbiculata* (Pursh) Rydb. P

Platanthera orbiculata (Pursh) Lindl. var. ***macrophylla*** (Goldie) Luer FAC
Large round-leaved orchid
Herbaceous perennial
Rich, shady woods.
Habenaria macrophylla Goldie F; *Habenaria orbiculata* A.Gray var. *macrophylla* (Goldie) Boivin C

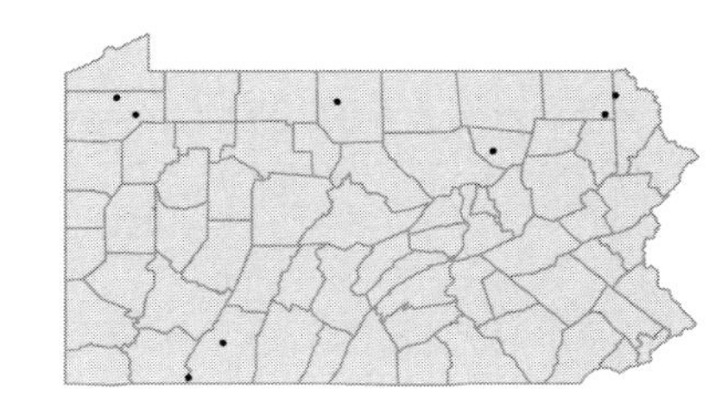

Platanthera peramoena (A.Gray) A.Gray FACW
Purple fringeless orchid
Herbaceous perennial
Moist meadows, low wet woods and ditches.
Blephariglottis peramoena (A.Gray) Rydb. P; *Habenaria peramoena* A.Gray FGBC

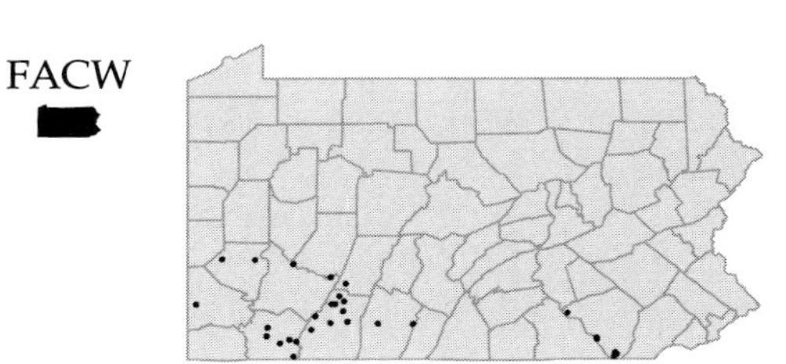

Platanthera psycodes (L.) Lindl. FACW
Purple fringed-orchid
Herbaceous perennial
Bogs, swamps, damp meadows and open woods.
Blephariglottis psycodes (L.) Rydb. P; *Habenaria psycodes* (L.) Spreng. var. *psycodes* FGBC

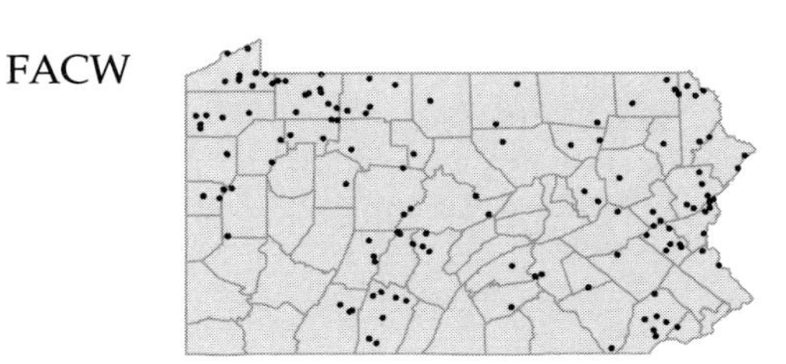

• ***Pogonia ophioglossoides*** (L.) Ker-Gawl. OBL
Rose pogonia; Snakemouth
Herbaceous perennial
Sphagnum bogs and boggy meadows.

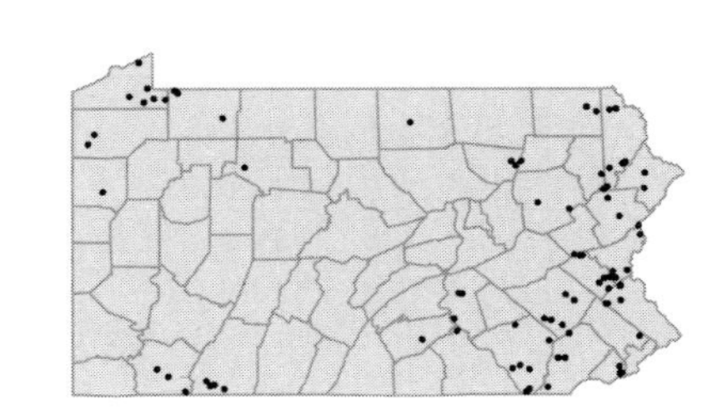

• ***Spiranthes casei*** Catling & Cruise
Case's ladies'-tresses
Herbaceous perennial
Dry, open, sandy soil.

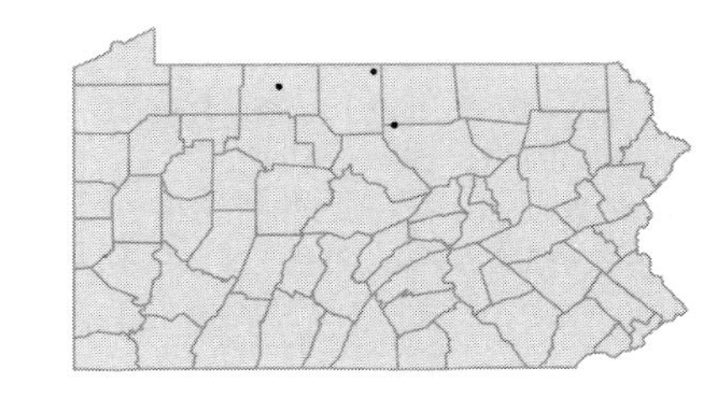

Spiranthes cernua (L.) L.C.Rich.
Nodding ladies'-tresses
Herbaceous perennial
Moist, acid soils of meadows, open woods and roadsides.

FACW

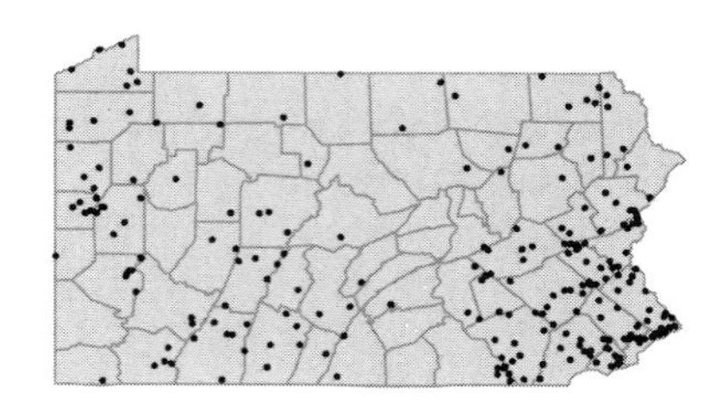

Spiranthes lacera (Raf.) Raf. var. ***lacera***
Northern slender ladies'-tresses
Herbaceous perennial
Open woods and grassy meadows.

FACU-

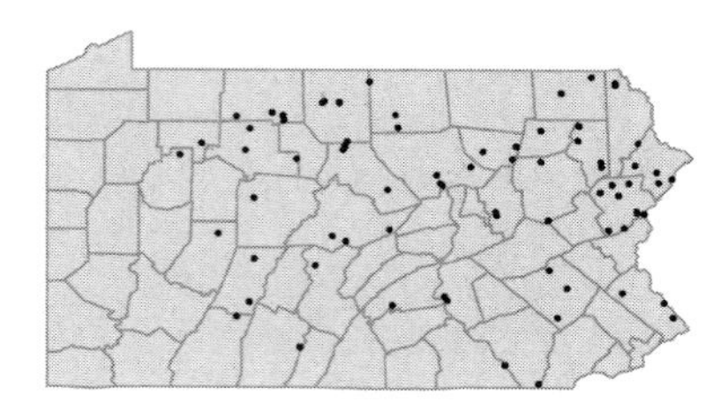

Spiranthes lacera (Raf.) Raf. var. ***gracilis*** (Bigel.) Luer
Southern slender ladies'-tresses
Herbaceous perennial
Dry, open woods, grassy meadows and roadside banks.
Spiranthes gracilis (Bigel.) Beck W

FACU-

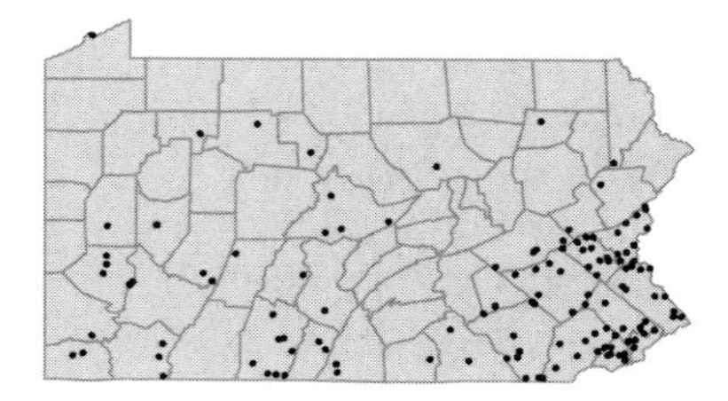

Spiranthes lucida (H.H.Eat.) Ames
Shining ladies'-tresses
Herbaceous perennial
Moist banks, lake shores and wet meadows, usually on calcareous soils.

FACW

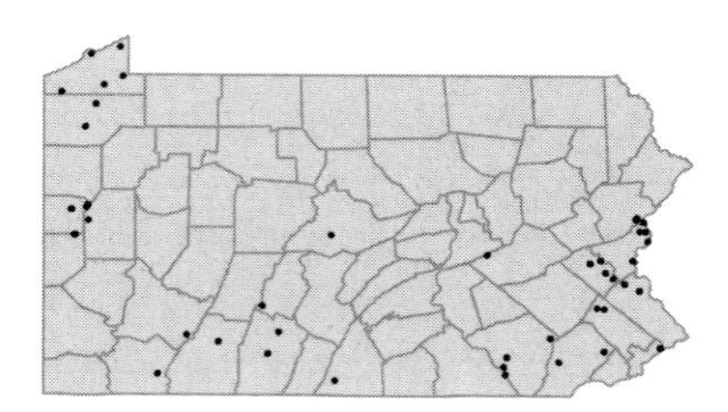

Spiranthes magnicamporum Sheviak
Great Plains ladies'-tresses
Herbaceous perennial
Wet meadows and fens on calcareous soils. Believed to be extirpated, last collected in 1892.

Spiranthes ochroleuca (Rydb.) Rydb.
Yellow nodding ladies'-tresses
Herbaceous perennial
Shady woods and roadsides.
Spiranthes cernua (L.) L.C.Rich. var. *ochroleuca* (Rydb.) Ames FC

FACW

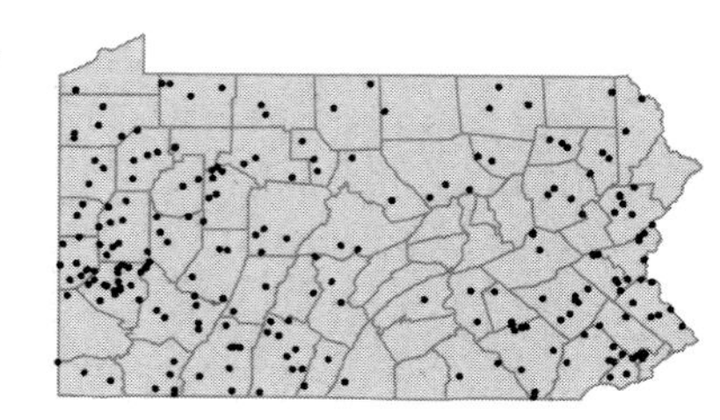

Spiranthes ovalis Lindl. var. ***erostellata*** Catling
October ladies'-tresses; Oval ladies'-tresses
Herbaceous perennial
Damp forests in rich humus.
Spiranthes ovalis Lindl. var. *ovalis pro parte* FGBW

FAC

Spiranthes romanzoffiana Cham.
Hooded ladies'-tresses
Herbaceous perennial
Bogs or rich, open woods.

OBL

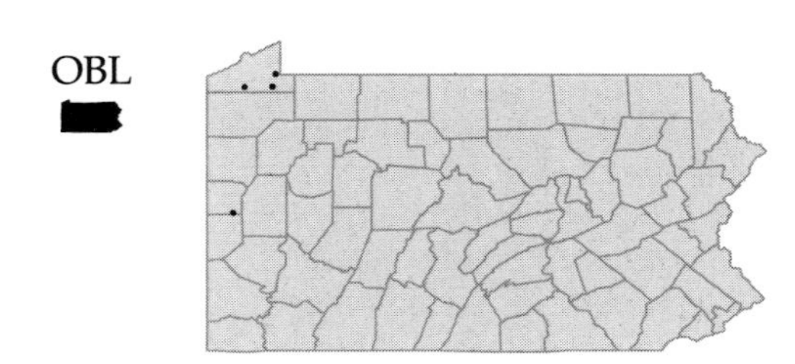

Spiranthes tuberosa Raf.
Slender ladies'-tresses
Herbaceous perennial
Grassy meadows, open woods and banks.

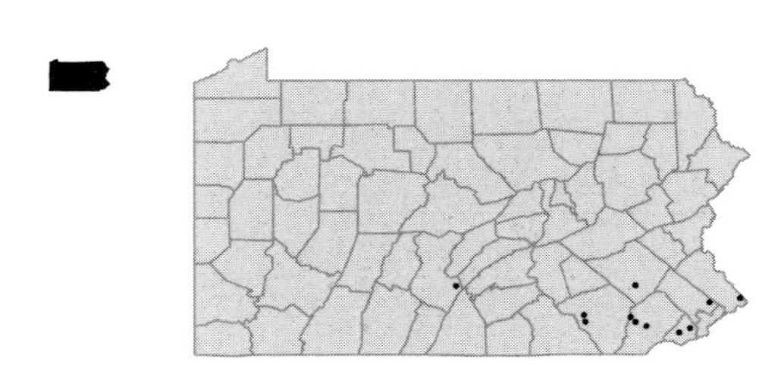

Spiranthes vernalis Engelm. & A.Gray
Spring ladies'-tresses
Herbaceous perennial
Moist, open, sandy soils.

FAC

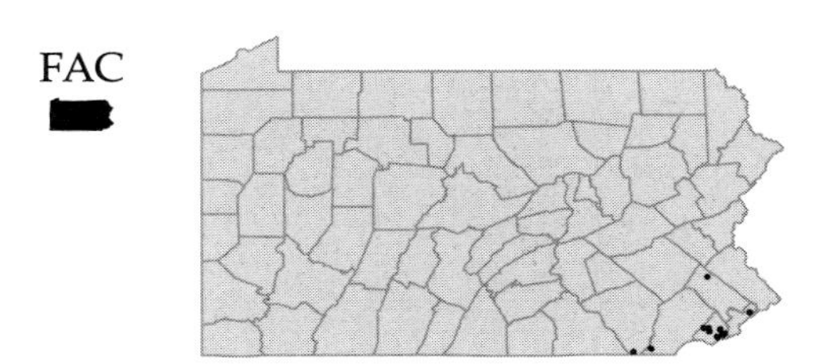

• ***Tipularia discolor*** (Pursh) Nutt.
Cranefly orchid; Tallow-root
Herbaceous perennial
Deciduous woods and stream banks.
Tipularia unifolia (Muhl.) BSP P

FACU

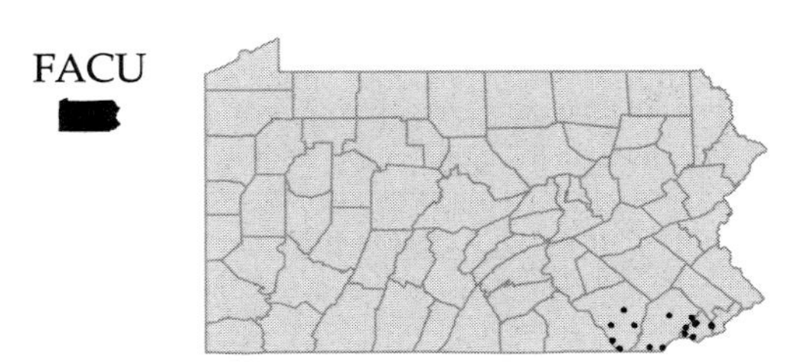

• ***Triphora trianthophora*** (Swartz) Rydb.
Nodding pogonia; Three-birds
Herbaceous perennial
Rich, moist woods, in humus.

UPL

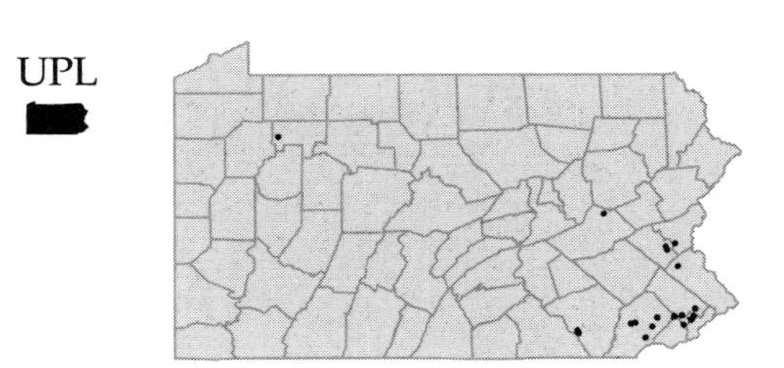

BIBLIOGRAPHY

GENERAL

Barton, William P.C. 1818. *Compendium Florae Philadelphicae: Containing a description of the indigenous and naturalized plants found within a circuit of ten miles.* Carey and Son, Philadelphia, PA.

Braun, E. Lucy. 1950. *Deciduous Forests of Eastern North America.* Hafner Press, New York, NY.

Bureau of Plant Industry. 1982. Act 1982-74 Pennsylvania Noxious Weed Control Act. PA Dept. Agriculture, Harrisburg, PA.

Burk, Isaac. 1877. List of plants recently collected on ships' ballast in the neighborhood of Philadelphia. Proc. Acad. Nat. Sci. Philadelphia 29: 105-109.

Cronquist, Arthur. 1981. *An Integrated System of Classification of Flowering Plants.* Columbia Univ. Press, New York, NY.

Department of Environmental Resources. 1987. Title 25, Chapter 82 Conservation of Pennsylvania Native Wild Plants. PA Bull. 17(49). Dec. 5, 1987.

Fernald, Merritt Lyndon. 1950. *Gray's Manual of Botany.* American Book Co., New York, NY.

Flora of North America Editorial Committee. *Flora of North America Vol 2.* Oxford Univ. Press, New York, NY (in press).

Fogg, John M., Jr. 1944. Some methods applied to a state flora survey. Contrib. Gray Herb. 165: 121-132.

Gleason, Henry A. 1952. *The New Britton and Brown Illustrated Flora of the Northeastern United States and Adjacent Canada.* Hafner Press of McMillan Pub. Co., New York, NY.

Gleason, Henry A. and Arthur Cronquist. 1963. *Manual of Vascular Plants of Northeastern United States and Adjacent Canada.* D. Van Nostrand Co., Inc., Princeton, NJ.

Gleason, Henry A. and Arthur Cronquist. 1991. *Manual of Vascular Plants of Northeastern United States and Adjacent Canada, 2nd ed.* NY Bot. Gard., Bronx, NY.

Godfrey, Robert K. and Jean W. Wooten. 1981. *Aquatic and Wetland Plants of the Southeastern United States - Dicotyledons.* Univ. GA Press, Athens, GA.

Godfrey, Robert K. and Jean W. Wooten. 1981. *Aquatic and Wetland Plants of the Southeastern United States - Monocotyledons.* Univ. GA Press, Athens, GA.

Kartesz, J.T. and R. Kartesz. 1980. *A Synonymized Checklist of the Vascular Flora of the United States, Canada, and Greenland.* Univ. NC Press, Chapel Hill, NC.

Kartesz, John T. and John W. Thieret. 1991. Common names for vascular plants: guidelines for use and application. Sida 14: 421-434.

Kuchler, A.W. 1964. *Potential Natural Vegetation of the Coterminous United States*, Special Pub. No. 36. Am. Geographical Soc., New York, NY.

Lellinger, D.B. 1985. *A Field Manual of the Ferns and Fern Allies of the US and Canada.* Smithsonian Inst. Press, Washington, DC.

Little, E.L. 1979. *Checklist of United States Trees (Native and Naturalized).* Agric. Handbook No. 541. USDA Forest Service, Washington, DC.

Marsh, Ben and Elizabeth R. Marsh. 1989. Landforms pp. 18-25 in: *The Atlas of Pennsylvania.* Temple Univ. Press, Philadelphia, PA.

Myer, George H. 1989. Geology pp. 12-17 in: *The Atlas of Pennsylvania.* Temple Univ. Press, Philadelphia, PA.

Porter, T.C. 1903. *Flora of Pennsylvania.* Ginn and Company, Boston, MA.

Reed, Porter B., Jr. 1988. National List of Plant Species That Occur in Wetlands: 1988 National Summary. US Dept. Interior, Fish and Wildlife Service, St. Petersburg, FL.

Smith, A.H. 1867. On colonies of plants observed near Philadelphia. Proc. Acad. Nat. Sci. Philadelphia 19: 1-10.

Smith, A.H. 1886. The railway cutting at Gray's Ferry Road. Proc. Acad. Nat. Sci. Philadelphia 52: 253-254.

Thomas, Joseph and I.S. Moyer. 1876. Plants, birds and mammals of Bucks County, Pennsylvania in: W.W.H. Davis. *History of Bucks County.* W.W.H. Davis, Doylestown, PA.

Thompson, Sue A., W.E. Buker and Myrta MacDonald. 1989. *Notes on the Distribution of Pennsylvania Plants based on*

Specimens in the Carnegie Museum Herbarium. Spec. Pub. No. 14, Carnegie Mus. Nat. Hist., Pittsburgh, PA.

Tutin, T.G., V.H. Heywood, N.A. Burgess, D.H. Valentine, S.M. Walters and D.A. Webb. 1964. *Flora Europaea Vol. 1, Lycopodiaceae to Plantanaceae.* Cambridge Univ. Press, Cambridge, England.

Tutin, T.G., V.H. Heywood, N.A. Burges, D.M. Moore, D.H. Valentine, S.M. Walters and D.A. Webb. 1968. *Flora Europaea Vol. 2, Rosaceae to Umbelliferae.* Cambridge Univ. Press, Cambridge, England.

Tutin, T.G., V.H. Heywood, N.A. Burges, D.M. Moore, D.H. Valentine, S.M. Walters and D.A. Webb. 1972. *Flora Europaea Vol. 3, Diapensiaceae to Myoporaceae.* Cambridge Univ. Press, Cambridge, England.

Tutin, T.G., V.H. Heywood, N.A. Burges, D.M. Moore, D.H. Valentine, S.M. Walters and D.A. Webb. 1976. *Flora Europaea Vol. 4, Plantaginaceae to Compositae.* Cambridge Univ. Press, Cambridge, England.

Tutin, T.G., V.H. Heywood, N.A. Burges, D.M. Moore, D.H. Valentine, S.M. Walters and D.A. Webb. 1980. *Flora Europaea Vol. 5, Alismataceae to Orchidaceae.* Cambridge Univ. Press, Cambridge, England.

United States Department of Interior, Fish and Wildlife Service. 1990. Endangered and threatened wildlife and plants; review of plant taxa for listing as endangered or threatened species. Combined Federal Register 55: 6184-6229. February 21, 1990.

United States Department of Interior, Fish and Wildlife Service. 1991. Title 50, Subpart B. Endangered and Threatened Wildlife and Plants. Combined Federal Register 17.11 and 17.12. July 15, 1991.

United States Geological Survey. 1988. Geographic Names Information System. National Cartographic Information Center, Reston, VA.

Voss, E.G. 1972. *Michigan Flora, Part I Gymnosperms and Monocots.* Cranbrook Inst. Sci., Bloomfield Hills, MI.

Voss, E.G. 1985. *Michigan Flora, Part II Dicots (Saururaceae -Cornaceae).* Cranbrook Inst. Sci., Bloomfield Hills, MI.

Wherry, Edgar T., John M. Fogg, Jr. and Herbert A. Wahl. 1979. *Atlas of the Flora of Pennsylvania.* Morris Arbor. Univ. of PA, Philadelphia, PA.

Whitney, G.G. 1990. The history and status of the hemlock-hardwood forests of the Allegheny Plateau. J. Ecol. 78: 443-458.

Yarnal, Brent. 1989. Climate pp 26-30 in: *The Atlas of Pennsylvania.* Temple Univ. Press., Philadelphia, PA.

PTERIDOPHYTA

Adiantaceae

Bartgis, Rodney L. and Michael Breiding. 1987. The status of *Cryptogramma stelleri, Matteuccia struthiopteris,* and *Ophioglossum pusillum* in West Virginia. Bartonia 53: 44-46.

Cody, W.J. 1983. *Adiantum pedatum* ssp. *calderi,* a new subspecies in northeastern North America. Rhodora 85: 93-96.

Gastony, G.J. 1988. The *Pellaea glabella* complex: electrophoretic evidence for the derivations of the agamosporous taxa and a revised taxonomy. Am. Fern J. 78: 44-67.

Paris, C.A. 1989. Maidenhair fern biogeography and M. L. Fernald's hypothesis of a Gaspe refugium. Rhodora 91: 143.

Paris, Cathy A. 1991. *Adiantum viridimontanum,* a new maidenhair fern in eastern North America. Rhodora 93: 105-122.

Paris, C.A. and M.D. Windham. 1988. A biosystematic investigation of the *Adiantum pedatum* complex in eastern North America. Syst. Bot. 13: 240-255.

Parks, J.C. 1984. New Pennsylvania locality for *Cryptogramma stelleri.* Bartonia 50: 63.

Aspleniaceae

Moran, R.C. 1982. The *Asplenium trichomanes* complex in the United States and adjacent Canada. Am. Fern J. 72: 5-11.

Dryopteridaceae

Barrington, D.S. 1986. The morphology and cytology of *Polystichum x potteri* Hybr. Nov. (= *P. acrostichoides x P. braunii*). Rhodora 88: 297-313.

Blasdell, R.F. 1963. A monographic study of the fern genus *Cystopteris.* Mem. Torrey Bot. Club 21: 1-102.

Haufler, C.H. and Michael D. Windham. 1991. New species of North American *Cystopteris* and *Polypodium* with comments on their reticulate relationships. Am. Fern J. 81: 7-23.

Haufler, C.H., M.D. Windham and T.A. Ranker. 1990. Biosystematic analysis of the *Cystopteris tennesseensis* (Dryopteridaceae) complex. Ann. Missouri Bot. Gard. 77: 314-329.

Kubitzki, K. (series editor). 1990. *The Families and Genera of Vascular Plants Vol. 1 Pteridophytes and Gymnosperms.* Kramer, K.U. and P.S. Green eds. Springer-Verlag, New York, NY.

Pryer, Kathleen M. 1992. The status of *Gymnocarpium heterosporum* and *G. robertianum* in Pennsylvania. Am. Fern J. 82: 34-39.

Tryon, R.F. and A.F. Tryon. 1982. *Ferns and Allied Plants with Special Reference to Tropical America.* Springer-Verlag, New York, NY.

Hymenophyllaceae

Farrar, Donald R. 1992. *Trichomanes intricatum*: the independent *Trichomanes* gametophyte in the eastern United States. Am. Fern J. 82: 68-74.

Farrar, D.R., J.C. Parks and B.W. McAlpin. 1983. The fern genera *Vittaria* and *Trichomanes* in the northeastern United States. Rhodora 85: 83-91.

Isoetaceae

Brunton, D.F. and W.C. Taylor. 1990. *Isoetes x brittonii* hyb. nov. (Isoetaceae): a naturally occurring hybrid (*I. engelmannii x I. riparia*) in the eastern United States. Am. Fern J. 80: 82-89.

Taylor, W. Carl and R. James Hickey. 1992. Habitat, evolution, and speciation in *Isoetes*. Ann. Missouri Bot. Gard. 79: 613-622.

Lycopodiacae

Bruce, J.G. 1975. Systematics and Morphology of Subgenus *Lepidotis* of the Genus *Lycopodium*, PhD Thesis. University of Michigan, Ann Arbor, MI.

Bruce, J.G., W.H. Wagner and J.M. Beitel. 1991. Two new species of bog clubmosses, *Lycopodiella* (Lycopodiaceae) from southwestern Michigan. Mich. Botanist 30: 3-10.

Hickey, R.J. 1977. The *Lycopodium obscurum* complex in North America. Am. Fern J. 67: 45-48.

Holub, Josef. 1975. *Diphasiastrum*, a new genus in Lycopodiaceae. Preslia 47: 97-110.

Ollgaard, B. 1987. A revised classification of the Lycopodiaceae *s. lat.* Opera Botanica 92: 153-178.

Ollgaard, B. 1989. Index of the Lycopodiaceae. Biologiske Skrifter 34: 1-135.

Wagner, W.H., Jr., J.M. Beitel and R.C. Moran. 1989. *Lycopodium hickeyi*: a new species of North American clubmoss. Am. Fern J. 79: 119-121.

Ophioglossaceae

Wagner, W.H., Jr., C.M. Allen and G.P. Landry. 1984. *Ophioglossum ellipticum* Hook. & Grev. in Louisiana and the taxonomy of *O. nudicaule* L.f. Castanea 49: 99-110.

Polypodiaceae

Haufler, C.H. and Michael D. Windham. 1991. New species of North American *Cystopteris* and *Polypodium* with comments on their reticulate relationships. Am. Fern J. 81: 7-23.

Haufler, C.H. and Wang Zhongren. 1991. Chromosomal analyses and the origin of allopolyploid *Polypodium virginianum* (Polypodiaceae). Am. J. Bot. 78: 624-629.

Vittariaceae

Farrar, D.R. 1990. Species and evolution in asexually reproducing independent fern gametophytes. Syst. Bot. 15: 98-111.

Farrar, Donald R. and John T. Mickel. 1991. *Vittaria appalachiana*: a name for the "Appalachian gametophyte". Am. Fern J. 81: 69-75.

Farrar, D.R., J.C. Parks and B.W. McAlpin. 1983. The fern genera *Vittaria* and *Trichomanes* in the northeastern United States. Rhodora 85: 83-91.

PINOPHYTA (GYMNOSPERMS)

Cupressaceae

Mickalitis, Albert B. 1949. The most unique tree native to Pennsylvania. Pennsylvania Forests and Waters 1: 179-188.

Taxaceae

Allison, Taber D. 1991. Variation in sex expression in Canada yew (*Taxus canadensis*). Am. J. Bot. 78: 569-578.

Allison, Taber D. 1992. The influence of deer browsing on the reproductive biology of Canada yew (*Taxus canadensis* Marsh.) III. Sex expression. Oecologia 89: 223-228.

Price, Robert A. 1990. The genera of Taxaceae in the southeastern United States. J. Arnold Arbor. 71: 69-91.

MAGNOLIOPSIDA (DICOTYLEDONS)

Aceraceae

Nowak, David J. and Rowan A. Rowntree. 1990. History and range of Norway maple. J. Arboric. 16: 291-296.

Nybom, Hilde and Steven H. Rogstad. 1990. DNA "fingerprints" detect genetic variation in *Acer negundo* (Aceraceae). Pl. Syst. Evol. 173: 49-56.

Amaranthaceae

Wheeler, Louis Cutter. 1940. *Amaranthus* of Philadelphia and vicinity. Bartonia 21: 2-4.

Anacardiaceae

Frankel, Edward. 1991. *Poison Ivy, Poison Oak, Poison Sumac and their Relatives, Pistacios, Mangoes, Cashews.* The Boxwood Press, Pacific Grove, CA.

Gillis, William T. 1971. The systematics and ecology of poison-ivy and the poison-oaks (I) (*Toxicodendron*, Anacardiaceae). Rhodora 73: 161-237.

Apiaceae

Buddell, G.F. and J.W. Thieret. 1985. Notes on *Erigenia bulbosa* (Apiaceae). Bartonia 51: 69-76.

Mountain, W.L. 1990. Poison hemlock, *Conium maculatum* L. Umbelliferae (Apiaceae). Regulatory Horticulture 16: 27-31.

Mulligan, G.A. 1980. The genus *Cicuta* in North America. Can. J. Bot. 58: 1755-1767.

Pryer, K.M. and L.R. Phillippe. 1989. A synopsis of the genus *Sanicula* (Apiaceae) in eastern Canada. Can. J. Bot. 67: 694-707.

Apocynaceae and Asclepiadaceae

Rosatti, Thomas J. 1989. The genera of suborder Apocynineae (Apocynaceae and Asclepiadaceae) in the southeastern United States. J. Arnold Arbor. 70: 307-514.

Asteraceae

Bayer, Randall J. 1990. A phylogenetic reconstruction of *Antennaria* (Asteraceae: Inuleae). Can. J. Bot. 68: 1389-1397.

Bayer, Randall J. and G. Ledyard Stebbins. 1982. A revised classification of *Antennaria* (Asteraceae: Inuleae) of the eastern United States. Syst. Bot. 7: 300-313.

Britton, N.L. 1901. A new *Senecio* from Pennsylvania. Torreya 1: 21.

Brouillet, Luc and Jean-Pierre Simon. 1981. An ecogeographical analysis of the distribution of *Aster acuminatus* Michx. and *A. nemoralis* Aiton (Asteraceae: Astereae). Rhodora 83: 521-550.

Chmielewski, J.G. and J.C. Semple. 1989. The cytogeography of *Aster pilosis* var. *pilosis* in southern Ontario revisited (Compositae: Astereae). Can. J. Bot. 67: 3517-3519.

Cronquist, Arthur. 1980. *Vascular Flora of the Southeastern United States Vol. I. Asteraceae.* Univ. NC Press, Chapel Hill, NC.

Cronquist, Arthur. 1985. The history of generic concepts in the Compositae. Taxon 34: 6-10.

Cusick, Allison W. 1990. Distribution and current status of three state-listed goldenrods (*Solidago*: Asteraceae) in Pennsylvania. Bartonia 56: 9-12.

Hart, Robin. 1990. *Aster depauperatus*: A midwestern migrant on eastern serpentine barrens? Bartonia 56: 23-28.

Hill, Steven R. 1990. *Hieracium trailii* or *Hieracium greenii*? (Asteraceae). Castanea 55: 211-212.

Hill, L. Michael and O.M. Rogers. 1970. Chromosome numbers of *Aster blakei* and *A. nemoralis*. Rhodora 72: 437-438.

Hill, Michael L. and O.M. Rogers. 1973. Chemical, cytological and genetic evidence for the hybrid origin of *Aster blakei* (Porter) House. Rhodora 75: 1-25.

Jones, Almut G. 1980. A classification of the New World species of *Aster* (Asteraceae). Brittonia 32: 230-239.

Jones, Almut G. 1980. Data on chromosome numbers in *Aster* (Asteraceae), with comments on the status and relationships of certain North American species. Brittonia 32: 240-261.

Jones, A.G. 1984. Nomenclatural notes on *Aster* (Asteraceae) - II. New combinations and some transfers. Phytologia 55: 373-388.

Jones, Almut G. and Paul Hiepko. 1981. The genus *Aster* s.l. (Asteraceae) in the Willdenow Herbarium at Berlin. Willdenowia 11: 343-360.

Jones, Ronald L. 1983. A systematic study of *Aster* section *Patentes* (Asteraceae). Sida 10: 41-81.

Kowal, Robert R. and T.M. Barkley. 1973. *Senecio anonymus* Wood, an earlier name for *Senecio smallii* Britton. Rhodora 75: 211-219.

Lamboy, Warren F. and Almut G. Jones. 1987. Lectotypifications and neotypifications in *Aster* section *Biotia* (Asteraceae) including a complete annotated synonymy. Brittonia 39: 286-297.

Lepage, Ernest. 1958. Etudes sur quelques Hieracia. Le Natur. Canadien 85: 81-93.

Lepage, Ernest. 1960. *Hieracium canandense* Michx. et ses allies en Amerique du Nord. Le Natural. Canadien 87: 59-107.

Lepage, Ernest. 1971. Les epervieres du Quebec. Le Natur. Canadien 98: 657-674.

McGregor, Ronald L. 1968. The taxonomy of the genus *Echinaceae* (Compositae). Univ. Kansas Sci. Bull. 48: 113-142.

Pike, Radcliffe B. 1970. Evidence for the hybrid status of *Aster x blakei* (Porter) House. Rhodora 72: 401-436.

Schrot, Edith Feuerstein. 1992. The first PA record of Bog aster (*Aster nemoralis* Aiton, Asteraceae) and other new county and site records for the Algerine Swamp Natural Area. Bartonia 57: 95-97.

Scott, Randall W. 1990. The genera of Cardueae (Compositeae; Asteraceae) in the southeastern United States. J. Arnold Arbor. 71: 391-451.

Semple, John C. 1978. The cytogeography of *Aster pilosis* (Compositae): Ontario and the adjacent United States. Can. J. Bot. 56: 1274-1279.

Semple, John C. 1978. The cytogeography of *Aster lanceolatus* (syn. *A. simplex, A. paniculatus*) in Ontario with additional counts from populations in the United States. Can. J. Bot. 57: 397-402.

Semple, John C. 1992. A geographic summary of chromosome number reports for North American asters and goldenrods (Asteraceae: Astereae). Ann. Missouri Bot. Gard. 79: 95-109.

Semple, John C. and C. C. Chinnappa 1986. The cytogeography of *Chrysopsis mariana* (Compositeae:Asteraceae): survey over the range of the species. Rhodora 88: 261-266.

Semple, John C. and Jerry G. Chmielewski. 1985. The cytogeography of *Aster pilosis* (Compositae-Asteraceae) II. Survey of the range with notes on *A. depauperatus, A. parviceps* and *A. porteri*. Rhodora 87: 367-379.

Semple, J.C., J.C. Chmielewski and C. Xiang. 1992. Chromosome number determinations in Fam. Compositae, Tribe Astereae. IV. Additional reports and comments on the cytogeography and status of some species. Rhodora 94: 48-62.

Stuessy, Tod F. 1977. Revision of *Chrysogonum* (Compositae, Heliantheae). Rhodora 79: 190-202.

Von Faasen, Paul and Fern Frank Sterk. 1973. Chromosome numbers in *Aster*. Rhodora 75: 26-33.

Voss, Edward G. and Mark W. Bohlke. 1978. The status of certain hawkweeds (*Hieracium* subgenus *Pilosella*) in Michigan. Mich. Bot. 17: 35-47.

Betulaceae

Furlow, John J. 1990. The genera of Betulaceae in the southeastern United States. J. Arnold Arbor. 71: 1-67.

Boraginaceae

Al-Shehbaz, Ihsan A. 1991. The genera of Boraginaceae in the southeastern United States. J.Arnold Arbor. Supplementary Series 1: 1-169.

Brassicaceae

Fogg, John M. Jr. 1931. Notes on a few introduced species in the Philadelphia local area. Bartonia 13: 48-49.

Hopkins, Milton. 1937. *Arabis* in eastern and central North America. Rhodora 39: 63-98, 106-148, 155-186.

Jacobson, H.A., J.B. Petersen and D.E. Putnam 1988. Evidence for pre-Columbian *Brassica* in the northeastern United States. Rhodora 90: 355-362.

Caesalpiniaceae

Irwin, H.S. and R.C. Barneby. 1982. The American Cassiinae. Mem. NY Bot. Gard. 35: 1-918.

Calycanthaceae

Nicely, K.A. 1965. A monographic study of the Calycanthaceae. Castanea 30: 38-81.

Capparaceae

Thieret, John W. 1984. *Cleome ornithopioides* (Capparaceae): adventive and spreading in North America. Bartonia 50: 25-26.

Caryophyllaceae

McNeill, J. 1978. *Silene alba* and *S. dioica* in North America and the generic delimitation of *Lychnis, Melandrium,* and *Silene* (Caryophyllaceae). Can. J. Bot. 56: 297-308.

Mitchell, Richard S. and Gordon C. Tucker. 1991. *Sagina japonica* (Sw.) Ohwi (Caryophyllaceae), an overlooked adventive in the northeastern United States. Rhodora 93: 192-194.

Morton, J.K. 1987. Caryophyllaceae, in: *Atlas of the Rare Vascular Plants of Ontario*. Argus. G.W., K.M. Pryer, D.J. White and C.J. Keddy eds. National Museum of Natural Sciences, Ottawa, Ontario, Canada.

Rabeler, Richard K. 1991. *Moenchia erecta* (Caryophyllaceae) in eastern North America. Castanea 56: 150-151.

Chenopodiaceae

Bassett, I.J., C.W. Crompton, J. McNeill and P.M. Tasche. 1983. The genus *Atriplex* (Chenopodiaceae) in Canada. Monograph No. 31. Agric. Canada, Ottawa, Ontario Canada.

Wahl, Herbert A. 1952. A preliminary study of the genus *Chenopodium* in North America. Bartonia 27: 1-46.

Clusiaceae

Adams, William P. 1957. A revision of the genus *Ascyrum* (Hypericaceae). Rhodora 59: 73-94.

Adams, Preston 1962. Studies in the Guttiferae II. Taxonomic and distributional observations on North American taxa. Rhodora 64: 231-242.

Cooperrider, Tom S. 1989. The Clusiaceae (or Guttiferae) of Ohio. Castanea 54: 1-11.

Gleason, H.A. 1947. Notes on some North American plants. Phytologia 2: 281-291.

Robson, N.K.B. 1980. The Linnean species of *Ascyrum*. Taxon 29: 267-274.

Utech, F.H. and H.H. Iltis 1970. Preliminary reports on the flora of Wisconsin No. 61 Hypericaceae (St. John's-wort Family). Trans. Wisc. Acad. Sci. 59: 325-351.

Wood, Carroll E. and Preston Adams. 1976. The genera of Guttiferae (Clusiaceae) in the southeastern United States. J. Arnold Arbor. 57: 74-90.

Convolvulaceae

Austin, Daniel F. 1986. Nomenclature of the *Ipomoea nil* complex (Convolvulaceae). Taxon 35: 355-358.

Brummitt, R.K. 1980. Further new names in the genus *Calystegia* (Convolvulaceae). Kew Bull. 35: 327-334.

Cornaceae

Eyde, Richard H. 1987. The case for keeping *Cornus* in the broad Linnaean sense. Syst. Bot. 12: 505-518.

Eyde, Richard H. 1988. Comprehending *Cornus*: Puzzles and progress in the systematics of the dogwoods. Bot. Rev. 54: 233-351.

Crassulaceae

Clausen, R.T. 1975. *Sedum of North America North of the Mexican Plateau.* Cornell Univ. Press, Ithaca, NY.

Ericaceae

Kron, Kathleen A. and Walter S. Judd. 1990. Phylogenetic relationships within the Rhodoreae (Ericaceae) with specific comments on the placement of *Ledum*. Syst. Bot. 15: 57-68.

Middleton, David J. 1991. Taxonomic studies in the *Gaultheria* group of genera of the tribe Andromedeae (Ericaceae). Edinb. J. Bot. 48: 283-306.

Vander Kloet, S.P. 1988. *The Genus* Vaccinium *in North America*, Pub. 1828. Res. Branch Agric. Canada, Ottawa, Canada.

Wherry, Edgar T. 1938. A northern and a southern plant in York County, Pennsylvania. Bartonia 20: 27.

Euphorbiaceae

Fogg, John M., Jr. 1933. *Euphorbia dentata* in the Philadelphia local area. Bartonia 15: 35-36.

Richardson, J.W., D. Burch and T.S. Cochrane. 1987. The flora of Wisconsin preliminary report No. 69 Euphorbiaceae - The Spurge Family. Trans. WI Acad. of Sci. 75: 97-129.

Fabaceae

Gillis, W.T. „1980. *Wisteria* in the Great Lakes region. Mich. Bot. 19: 79-83.

Isely, D. 1981. Leguminosae of the United States. III. Subfamily Papilionoideae: tribes Sophoreae, Podalyrieae, Loteae. Mem. NY Bot. Gard. 25: 1-264.

Isely, D. 1990. *Vascular Flora of the Southeastern United States Vol. 3, Part 2. Leguminosae (Fabaceae).* Univ. NC Press, Chapel Hill, NC.

Snyder, D.B. 1990. *Strophostyles leiosperma* in New Jersey: adventive or native? Bartonia 56: 34-37.

Fagaceae

Hardin, James W. 1975. Hybridization and introgression in *Quercus alba*. J. Arnold Arbor. 56: 336-363.

Pretz, Harold W. 1911. Some noteworthy plants of Lehigh County, PA. Bartonia 4: 6-10.

Gentianaceae

Pringle, James S. 1967. Taxonomy of *Gentiana*, section *Pneumonanthae,* in eastern North America. Brittonia 19: 1-32.

Grossulariaceae

Sinnott, Quinn P. 1985. A revision of *Ribes* L. subg. *Grossularia* (Mill.) Pers. Sec. *Grossularia* (Mill.) Nutt. (Grossulariaceae) in North America. Rhodora 87: 189-286.

Haloragaceae

Aiken, S.A., P.R. Newroth and I. Wile. 1979. The biology of Canadian weeds. 34. *Myriophyllum spicatum* L. Can. J. Plant Sci. 59: 201-215.

Ceska, A. and O. Ceska. 1986. Notes on *Myriophyllum* (Haloragaceae) in the Far East: The identity of *Myriophyllum sibiricum* Komarov. Taxon 35: 95-100.

Smith, Craig S. and J.W. Barko. 1990. Ecology of Eurasian watermilfoil. J. Aquatic Plant Manag. 28: 55-64.

Hydrophyllaceae

Klotz, Larry H. and Jeffrey L. Walck. 1990. Pennsylvania records of Large-leaf waterleaf (*Hydrophyllum macrophyllum* Hydrophyllaceae). Bartonia 56: 29-33.

Lamiaceae

Benner, Walter M. 1934. *Lycopus rubellus* Moench and its varieties. Bartonia 16: 47-50.

Marshall, H.H. and R.W. Scora. 1972. A new chemical race of *Monarda fistulosa* (Labiatae). Can. J. Bot. 50: 1845-1849.

Nelson, John B. 1981. *Stachys* (Labiatae) in the southeastern United States. Sida 9: 104-123.

Nelson, John B. and John E. Fairey, III. 1979. Misapplication of the name *Stachys nuttallii* (Lamiaceae) to a new southeastern species. Brittonia 31: 491-194.

Scora, Ranier W. 1967. Divergence in *Monarda* (Labiatae). Taxon 16: 499-505.

Tucker, Arthur O. 1986. Botanical Nomenclature of Culinary Herbs and Potherbs In: *Herbs, Spices and Medicinal Plants: Recent Advances in Bot. Vol. 1.* Eds. Lyle E. Craker and James E. Simon, Oryx Press.

Tucker, A.O., R.M. Harley and D.E. Fairbrothers,1980 The Linnean types of *Mentha* (Lamiaceae). Taxon 29: 233-255.

Lentibulariaceae

Taylor, Peter. 1989. The genus *Utricularia* - a taxonomic monograph. Kew Bull. Add'l. Series XIV. Royal Bot. Gard. Kew, London.

Linaceae

Rogers, C. Marvin. 1963. Yellow flowered species of *Linum* in eastern North America. Brittonia 15: 97-122.

Malvaceae

Bates, David M. 1968. Generic relationships in the Malvaceae, tribe Malvae. Gentes Herbarium 10: 117-135.

Clement, I.D. 1957. Studies in *Sida* (Malvaceae) 1. A review of the genus and monograph of the sections. Contrib. Gray Herb. 180: 5-91.

Hill, Steven Richard. 1982. A monograph of the genus *Malvastrum* A.Gray (Malvaceae: Malveae). Rhodora 84: 1-83.

Hill, Steven R. 1982. A monograph of *Malvastrum*, II. Taxonomic treatment. Rhodora 84: 159-264.

Hill, Steven R. 1982. A monograph of the genus *Malvastrum*, III. Rhodora 84: 317-409.

Nymphaeaceae

Beal, Ernest O. 1956. Taxonomic revision of the genus *Nuphar* Sm. of North America and Europe. J. Elisha Mitchell Sci. Soc. 72: 317-346.

Onagraceae

Raven, P.H. 1963. The old world species of *Ludwigia* (including *Jussiaea*) with a synopsis of the genus (Onagraceae). Reinwardtia 6: 327-427.

Raven, P.H. and D.P. Gregory. 1972. A revision of the genus *Gaura* (Onagraceae). Mem. Torrey Bot. Club 23: 1-96.

Raven, P.H. and W. Tai. 1979. Observations of chromosomes in *Ludwigia* (Onagraceae). Ann. Missouri Bot. Gard. 66: 862-879.

Walck, Jeffrey L. and Thomas L. Smith. 1988. The first Pennsylvania record of the Upright primrose-willow (*Ludwigia decurrens* Walt.). Bartonia 54: 24-25.

Zardini, Elsa M., Hongya Gu and Peter H. Raven. 1991. On the separation of two species within the *Ludwigia uruguayensis* complex (Onagraceae). Syst. Bot. 16: 242-244.

Papaveraceae

Buker, W.E. and Sue A. Thompson. 1986. Is *Stylophorum diphyllum* (Papaveraceae) native to Pennsylvania? Castanea 51: 66.

Plantaginaceae

Bassett, I. John. 1973. *The Plantains of Canada.* Monogr. No. 7. Canada Dept. Agric., Ottawa, Ont., Canada.

Polemoniaceae

Long, Bayard. 1927. Some noteworthy indigenous species new to the Philadelphia area. Bartonia 10: 30-52.

Wherry, Edgar T. 1955. *The Genus Phlox*. Monograph III. Morris Arbor. Univ. of PA, Philadelphia, PA.

Polygalaceae

Miller, Norton G. 1971. The Polygalaceae in the southeastern United States. J. Arnold Arbor. 52: 267-284.

Pennell, Francis W. 1931. *Polygala verticillata* in eastern North America. Bartonia 13: 7-17.

Polygonaceae

Hickman, James C. and Carole S. Hickman. 1978. *Polygonum perfoliatum*: a recent Asiatic adventive. Bartonia 45: 18-23.

Hill, Robert J., G. Springer, and L.B. Forer. 1951. Mile-a-minute, *Polygonum perfoliatum* L. (Polygonaceae), a new potential orchard and nursery weed. Weed Circular No.1. Regulatory Horticulture 7: 25-27.

Katz, Deborah S. and J. McNeill. 1987. The systematics of *Polygonum achoreum* Blake and *P. erectum* L. (section Polygonum). Abstract XIV International Botanical Congress 337.

Long, Bayard. 1910. Certain species becoming well established at Ashbourne and elsewhere near Philadelphia. Bartonia 3: 22-25.

Mitchell, Richard S. and J. Kenneth Dean. 1978. *Polygonaceae (Buckwheat Family) of New York State. Contrib. to a Flora of New York State I.*, Bull. No. 431, New York State Mus. Albany, NY.

Moul, Edwin T. 1948. A dangerous weedy *Polygonum* in Pennsylvania. Rhodora 50: 64-66.

Mountain, W.L. 1989. Mile-a-minute (*Polygonum perfoliatum* L.) update - distribution, biology, and control suggestions. Weed Circular No.15. Regulatory Horticulture 15: 21-24.

Portulacaceae

McAndrews, J.H. 1975. Pre-Columbian purslane (*Portulaca oleracea* L.) in the New World. Nature 253: 726-727.

Rhoads, A.F., A. Newbold, R.H. Mellon and R. Latham, 1985. *Montia chamissoi* rediscovered along the Delaware River in Wayne Co., PA. Bartonia 51: 77.

Wherry, Edgar T. 1964. The most disjunct species in Pennsylvania. Bartonia 34: 7.

Primulaceae

Channell, R.B. and C.E. Wood, Jr. 1959. Genera of Primulales 2. *Dodecatheon* L. J. Arnold Arbor. 40: 277-279.

Coffey, Vincent J. and Samuel B. Jones, Jr. 1980. Biosystematics of *Lysimachia* section Seleucia (Primulaceae). Brittonia 32: 309-322.

Fassett, Norman C. 1929. Notes from the herbarium of the University of Wisconsin IV. Rhodora 31: 49-53.

Fassett, Norman C. 1931. Notes from the herbarium of the University of Wisconsin VII. Rhodora 33: 224-228.

Fassett, Norman C. 1944. *Dodecatheon* in eastern North America. Am. Midland Natural. 81: 455-486.

Thompson, H.J. 1951. Experimental taxonomy: *Dodecatheon*. Carnegie Inst. Yrbk. 50: 120-121.

Thompson, Henry J. 1953. The biosystematics of *Dodecatheon*. Contrib. Dudley Herb. 4: 73-154.

Pyrolaceae

Haber, Erich and James E. Cruise. 1974. Generic limits in the Pyroloideae (Ericaceae). Can. J. Bot. 52: 877-883.

Ranunculaceae

Duncan, Thomas.1980. *A Taxonomic Study of the* Ranunculus hispidus *Michaux Complex in the Western Hemisphere*. Univ. of CA Press, Berkeley, CA.

Mitchell, R.S. and J.K. Dean. 1982. *Ranunculaceae (Crowfoot Family) of New York State. Contributions to a Flora of New York State IV*, Bulletin No.446. New York State Museum, Albany, NY,

Smit, P. 1973. A revision of *Caltha* (Ranunculaceae). Blumea 12: 119-150.

Voss, E.G. 1985. Nomenclatural notes on some Michigan dicots. Mich. Botanist 24: 117-124.

Rosaceae

Campbell, C.S., C.W. Greene, and T.A. Dickinson. 1991. Reproductive biology in subfam. Maloideae (Rosaceae). Syst. Bot. 16: 333-349.

Davis, H.A., A.M. Fuller and T. Davis. 1967. Contributions toward the revision of the *Eubati* of eastern North America. Castanea 32: 20-37.

Davis, H.A., A.M. Fuller and T. Davis. 1968. Contributions toward the revision of the *Eubati* of eastern North America II. *Setosi.* Castanea 33: 50-76.

Davis, H.A., A.M. Fuller and T. Davis. 1968. Contributions toward the revision of the *Eubati* of eastern North America III. *Flagellares.* Castanea 33: 206-241.

Davis, H.A., A.M. Fuller and T. Davis. 1969. Contributions toward the revision of the *Eubati* of eastern North America IV. Castanea 34: 157-179.

Davis, H.A., A.M. Fuller and T. Davis. 1969. Contributions toward the revision of the *Eubati* of eastern North America V. *Arguti*. Castanea 34: 235-266.

Davis, H.A., A.M. Fuller and T. Davis. 1970. Contributions toward the revision of the *Eubati* of eastern North America VI. *Cuneifolia*. Castanea 35: 176-194.

Dickinson, Timothy A. and Christopher S. Campbell. 1991. Evolution in the Maloideae (Rosaceae) - introduction. Syst. Bot. 16: 299-302.

Dickinson, Timothy A. and Christopher S. Campbell. 1991. Population struture and reproductive ecology in the Maloideae (Rosaceae). Syst. Bot. 16: 350-362.

Kruschke, Emil P. 1965. *Contributions to the Taxonomy of* Crataegus. Publication Botany No. 3. Milwaukee Public Mus., Milwaukee, WI.

Laughlin, Kendall. 1963. *Malus lancifolia* Rehder. Phytologia 9: 108-112.

Ogle, Douglas W. 1991. *Spiraea virginiana* Britton: I. Dilineation and distribution. Castanea 56 4: 287-296.

Palmer, Ernest J. 1946. *Crataegus* in the northeastern United States and adjacent Canada. Brittonia 5: 471-490.

Phipps, J.B. 1983. Biogeographic, taxonomic and cladistic relationships between East Asiatic and North American *Crataegus*. Ann. Missouri Bot. Gard. 70: 667-700.

Phipps, J.B. 1988. *Crataegus* (Maloideae, Rosaceae) of the southeastern United States, I. Introduction and series *Aestivales*. J. Arnold Arbor. 69: 401-431.

Phipps, J.B. and M. Muniyamma. 1980. A taxonomic revision of *Crataegus* (Rosaceae) in Ontario. Can. J. Bot. 58: 1621-1699.

Phipps, J.B., Kenneth R. Robertsom, Paul G. Smith and Joseph R. Rohrer. 1990. A checklist of the subfamily Maloideae (Rosaceae). Can. J. Bot. 68: 2209-269.

Robertson, Kenneth R. 1974. The genera of Rosaceae in the southeastern United States. J. Arnold Arbor. 55: 303-662.

Robertson, K.R., J.B. Phipps, J.R. Rohrer, P.G. Smith. 1991. A synopsis of genera in Maloideae (Rosaceae). Syst. Bot. 16: 376-394.

Robinson, A.W. and C.R. Partanen. 1980. Experimental taxonomy in the genus *Amelanchier*. Rhodora 82: 483-493.

Smith, P.G and J.B. Phipps. 1987. Studies in *Crataegus* (Rosaceae, Maloideae). XIV. Taxonomy of the C. series *Rotundifoliae* in Ontario. Can. J. Bot. 65: 2646-2667.

Smith, P.G. and J.B. Phipps. 1988. Studies in *Crataegus* (Rosaceae, Maloideae), XV. Morphometric variation in *Crataegus* series *Rotundifoliae* in Ontario. Syst. Bot. 13: 97-106.

Smith, P.G. and J.B. Phipps,,1988,,Studies in *Crataegus* (Rosaceae, Maloideae). XIX. Breeding behavior in Ontario *Crataegus* series *Rotundifoliae.* Can. J. Bot. 66: 1914-1920.

Thomas, Wm. Wayt, and David E. Boufford. 1986, C.S. Sargent's *Crataegus* (Rosaceae) types from western Pennsylvania. Brittonia 38: 27-31.

Rubiaceae

Lewis, Walter H. 1961. Merger of the North American *Houstonia* and *Oldenlandia* under *Hedyotis*. Rhodora 63: 216-223.

Puff, C. 1976. The *Galium trifidum* group (*Galium* sect. *Aparinoides*, Rubiaceae). Can. J. Bot. 54: 1911-1925.

Terrell, Edward E. 1959. A revision of the *Houstonia purpurea* group (Rubiaceae). Rhodora 61: 157-180, 188-207.

Salicaceae

Argus, George W. 1986. The genus *Salix* (Salicaceae) in the southeastern United States. Syst. Bot. Monogr. Vol. 9. Am. Soc. Plant Taxonomists, Ann Arbor, MI.

Sarraceniaceae

Wherry, Edgar T. 1933. The geographic relations of *Sarracenia purpurea*. Bartonia 15: 1-6.

Scrophulariaceae

Canne, Judith M. 1981. Chromosome counts in *Agalinis* and related taxa (Scrophulariaceae). Can. J. Bot. 59: 1111-1116.

Canne, Judith M. 1983. The taxonomic significance of seedling morphology in *Agalinis* (Scrophulariaceae). Can. J. Bot. 61: 1868-1874.

Canne, Judith M. 1984. Chromosome numbers and the taxonomy of North American *Agalinis* (Scrophulariaceae). Can. J. Bot. 62: 454-456.

Canne-Hilliker, Judith M. and Christine M. Kampny. 1991. Taxonomic significance of leaf and stem anatomy of *Agalinis* (Scrophulariaceae) from the U.S.A. and Canada. Can. J. Bot. 69: 1935-1950.

Musselman, Lytton J. and William F. Mann, Jr. 1978. *Root Parasites of Southern Forests*. USDA Forest Service Gen. Tech. Rep. SO-20. Southern For. Exp. Sta., New Orleans, LA.

Pennell, F.W. 1935. *The Scrophulariaceae of Eastern Temperate North America*. Monograph No. 1. Acad. Nat. Sci. Philadelphia, Philadelphia, PA.

Werth, Charles R. and James L. Riopel. 1979. A study of the host range of *Aureolaria pedicularia* (L.) Raf.

(Scrophulariaceae). Am. Midland Natural. 102: 300-306.

Solanaceae

Schilling, Edward E. and Qi-sheng Ma. 1992. Common names and species identification in black nightshades, *Solanum* (Solanaceae). Econ. Bot. 46: 223-225.

Stober, Spencer S. amd James C. Parks. 1985. Chromosomal studies of *Physalis virginiana* v *subglabrata* and *Physalis heterophylla*. Bartonia 51: 65-68.

Tiliaceae

Hardin, James W. 1990. Variation patterns and recognition of varieties of *Tilia americana* s.l. Syst. Bot. 15: 33-48.

Urticaceae

Townsend, C.C. 1968. *Parietaria officinalis* and *P. judaica*. Watsonia 6: 365-370.

Violaceae

Ballard, H.E., Jr. 1990. Hybrids among three caulescent violets with special reference to Michigan. Mich. Botanist 29: 43-54.

Cooperrider, T.S. 1984. Some species mergers and new combinations in the Ohio flora. Mich. Botanist 23: 165-168.

Levesque, Fr. L. and P. Dansereau. 1966. Etudes sur les violettes jaunes caulescentes de l'est de L'Amerique du Nord. I. Taxonomie, nomenclature, synonymie et bibliographie. Le Natur. Canadien 93: 489-569.

McKinney, L.E. 1986. The taxonomic status of *Viola appalachiensis* Henry. Bartonia 52: 42-43.

McKinney, Landon E. 1992. *A Taxonomic Revision of the Acaulescent Blue Violets* (Viola) *of North America*. Sida, Bot. Misc. No. 7. Botanical Research Institute of Texas, Fort Worth, TX.

McKinney, L.E. and K.E. Blum. 1978. Leaf variation in five species of acaulescent blue violets *(Viola)*. 1. Castanea 43: 95-107.

McKinney, L.E. and K.E. Blum. 1981. A preliminary study of a polyspecies complex in *Viola*. Castanea 46: 281-290.

Russell, N.H. 1955. The taxonomy of the North American acaulescent white violets. Am. Midland Natural. 54: 481-494.

Viscaceae

Reveal, James L. and Marshall C. Johnston. 1989. A new combination in *Phoradendron* (Viscaceae). Taxon 38: 107.

Vitaceae

Moore, Michael O. 1991. Classification and systematics of eastern North American *Vitis* L. (Vitaceae) north of Mexico. Sida 14: 339-367.

Zygophyllaceae

Porter, D.C.. 1970. *Kallstroemia* in the middle Atlantic states. Rhodora 72: 397-398.

LILIOPSIDA (MONOCOTYLEDONS)

Acoraceae

Grayum, Michael H. 1987. A summary of evidence and arguments supporting the removal of *Acorus* from the Araceae. Taxon 36: 723-729.

Grayum, Michael H. 1990. Evolution and phylogeny of the Araceae. Ann. Missouri Bot. Gard. 77: 628-697.

Hay, S.G., A. Bouchard and L. Brouillet. 1990. Additions to the flora of the island of Newfoundland. Rhodora 92: 277-293.

Packer, J.G. and G.S. Ringius. 1984. The distribution and status of *Acorus* (Araceae) in Canada. Can. J. Bot. 62: 2248-2252.

Alismataceae

Adams, Preston and R.K. Godfrey. 1961. Observations on the *Sagittaria subulata* complex. Rhodora 63: 247-266.

Bogin, Clifford. 1955. Revision of the genus *Sagittaria*. Mem. NY Bot. Gard. 9: 179-233.

Kaul, Robert B. 1991. Foliar and reproductive responses of *Sagittaria calycina* and *Sagittaria brevirostra* (Alismataceae) to varying natural conditions. Aquatic Botany 40: 47-59.

Araceae

Stone, Witmer. 1903. *Arisaema pusillum* in Pennsylvania and New Jersey. Torreya 3: 171-172.

Commelinaceae

Anderson, Edgar and Robert E. Woodson. 1935. The species of *Tradescantia* indigenous to the United States. Contrib. Arnold Arbor. 9: 1-132.

Hunt, D.R. 1981. Sections and series in *Tradescantia*, American Commelinaceae: IX. Kew Bull. 35: 437-442.

MacRoberts, D.T. 1979. Notes on Tradescantia III. *Tradescantia ohiensis* Raf. var. *paludosa* (Anderson & Woodson) MacRoberts. Phytologia 42: 380-382.

Cyperaceae

Britton, N.L. 1903. An undescribed *Eleocharis* from Pennsylvania: *Eleocharis smallii*. Torreya 3: 23-24.

Bruederle, Leo P. 1985. *Carex mitchelliana* in Crawford County, Pennsylvania. Bartonia 51: 112-113.

Bruederle, Leo P. and David E. Fairbrothers. 1986. Allozyme variation in populations of the *Carex crinita* complex (Cyperaceae). Syst. Bot. 11: 583-594.

Bruederle, L.P., D.E. Fairbrothers and S.L. Hanks. 1989. A systematic circumscription of *Carex mitchelliana* (Cyperaceae) with reference to taxonomic status. Am. J. Bot. 76: 124-132.

Bryson, Charles T. 1980 A Revision of the North American *Carex* Section *Laxiflorae* (Cyperaceae). PhD Thesis, Mississippi State University.

Carter, Richard and Robert Kral. 1990. *Cyperus echinatus* and *Cyperus croceus*, the correct names for North American *Cyperus ovularis* and *Cyperus globulosus* (Cyperaceae). Taxon 39: 322-327.

Crins, William J. and P.W. Ball. 1989. Taxonomy of the *Carex flava* complex (Cyperaceae) in North America and northern Eurasia, numerical taxonomy and character analysis. Can. J. Bot. 67: 1032-1047.

Crins, William J. and Peter W. Ball. 1989. Taxonomy of the *Carex flava* complex (Cyperaceae) in North America and northern Eurasia, taxonomic treatment. Can. J. Bot. 67: 1048-1065.

Crins, William J. and Peter W. Ball. 1983. The taxonomy of the *Carex pensylvanica* complex (Cyperaceae) in North America. Can. J. Bot. 61: 1692-1717.

Delahoussaye, A. James and John W. Thieret. 1967. *Cyperus* subgenus *Kyllinga* (Cyperaceae) in the continental United States. Sida 3: 128-136.

Denton, Melinda F. 1978. A taxonomic treatment of the *Luzulae* group of *Cyperus*. Contrib. Univ. Mich. Herb. 11: 197-271.

Fernald, M.L. and A.E. Brackett. 1929. The representatives of *Eleocharis palustris* in North America. Rhodora 31: 57-77.

Kartesz, John Y. and K.N. Gandhi. 1991. *Cymophyllus fraserianus* (Ker-Gawler) Kartesz & Gandhi (Cyperaceae), the correct name for Fraser's sedge. Rhodora 93: 136-140.

Lipscomb, Barney L. 1980. *Cyperus difformis* L. (Cyperaceae) in North America. Sida 8: 320-327.

Marcks, Brian G. 1974. Preliminary reports on the flora of Wisconsin No. 66. Cyperaceae II. (Sedge Family) the genus *Cyperus* - the umbrella sedges. Wisc. Acad. Sci., Arts and Letters 62: 261-284.

Menapace, Francis J. and Daniel E. Wujek. 1987. The systematic significance of achene micromorphology in *Carex retrorsa* (Cyperaceae). Brittonia 39: 278-283.

Naczi, Robert F. 1984. Rare sedges discovered and rediscovered in Delaware. Bartonia 50: 31-35.

Reznicek, Anton A. 1987. What is *Carex rostrata* Stokes? Abstract #474. Am. J. Bot. 72: 966.

Reznicek, A.A. 1989. New England *Carex* (Cyperaceae): Taxonomic problems and phytogeographical considerations. Rhodora 91: 144-152.

Reznicek, A.A. and P.W. Ball. 1980. The taxonomy of *Carex* sect. *Stellulatae* in North America north of Mexico. Contr. Univ. Mich. Herb. 14: 153-203.

Reznicek, A.A. and P.M. Catling. 1985. The status and identity of *Carex x caesariensis* (Cyperaceae). Rhodora 87: 529-537.

Reznicek, A.A. and P.M. Catling. 1986. *Carex striata*, the correct name for *C. walteriana* (Cyperaceae). Rhodora 88: 405-406.

Rothrock, Paul E. 1991. The identity of *Carex albolutescens*, *C. festucacea* and *C. longii* (Cyperaceae). Rhodora 93: 51-66.

Schuyler, Alfred E. 1962. A new species of *Scirpus* in the northeastern United States. Rhodora 64: 43-49.

Schuyler, Alfred E. 1964. A biosystematic study of the *Scirpus cyperinus* complex. Proc. Acad. Nat. Sci. Philadelphia 115: 283-311.

Schuyler, Alfred E. 1967. A taxonomic revision of North American leafy species of *Scirpus*. Proc. Acad. Nat. Sci. Philadelphia 119: 295-323.

Schuyler, A.E. 1972. Chromosome numbers of *Scirpus purshianus* and *S. smithii*. Rhodora 74: 398-402.

Schuyler, Alfred E. 1974. Typification and application of the names *Scirpus americanus* Pers., *S. olneyi* Gary, and *S. pungens* Vahl. Rhodora 76: 51-52.

Standley, Lisa A. 1989. Taxonomic revision of the *Carex stricta* (Cyperaceae) complex in eastern North America. Can. J. Bot. 67: 1-14.

Tucker, Gordon C. 1987. The genera of Cyperaceae in the southeastern United States. J. Arnold Arbor. 68: 361-445.

Tucker, Gordon C. 1985. The correct name for *Cyperus cayennensis* (*C. flavus*) Cyperaceae. Southwest. Natural. 30: 607-608.

Tucker, Gordon C. 1985. *Cyperus flavicomus*, the correct name for *Cyperus albomarginatus*. Rhodora 87: 539-541.

Tucker, Gordon C. 1983. The taxonomy of *Cyperus* (Cyperaceae) in Costa Rica and Panama. Syst. Bot. Monographs 2: 1-85.

Tucker, Gordon C. 1984. Taxonomic notes on two common neotropical species of *Cyperus* (Cyperaceae). Sida 10: 298-307.

Webb, David H., W. Michael Dennis and Thomas S. Patrick. 1981. Distribution and naturalization of *Cyperus brevifolioides* (Cyperaceae) in eastern United States. Sida 9: 188-190.

Webber, J.M. and P.W. Ball. 1979. Proposals to reject *Carex rosea* and *Carex radiata* of eastern North America (Cyperaceae). Taxon 28: 611-615.

Webber, J.M. and P.W. Ball. 1984. The taxonomy of the *Carex rosea* group (section *Phaestoglochin*) in Canada. Can. J. Bot. 62: 2058-2073.

Hydrocharitaceae

St. John, Harold. 1965. Monograph of the genus *Elodea*: Part 4 and summary. Rhodora 67: 1-35.

Iridaceae

Goldblatt, Peter. 1990. Phylogeny and classification of Iridaceae. Ann. Missouri Bot. Gard. 77: 607-627.

Hornberger, Kathleen L. 1991. The blue-eyed-grasses (*Sisyrinchium*: Iridaceae) of Arkansas. Sida 14: 597-604.

Juncaceae

Clemants, Steven E. 1990. *Juncaceae (Rush Family) of New York State. Contrib. to the Flora of New York State VII*, Bull. 475. NY State Mus., Albany, NY.

Kirschner, Jan. 1990. *Luzula multiflora* and allied species (Juncaceae): a nomenclatural study. Taxon 39: 106-114.

Juncaginaceae

Jennings, Otto E. 1906. A note on the occurrence of *Triglochin palustris* Linnaeus in Pennsylvania. Ann. Carnegie Mus. 3: 482.

Lemnaceae

Landolt, E. 1980. Key to the determination of taxa within the family of (Lemnaceae). Veroff. Geobot. Inst. ETH, Stiftung Rubel 70: 13-21.

Liliaceae

Long, Bayard. 1914. On the occurrence of the keeled garlic in America. Bartonia 7: 6-16.

Mathew, Brian. 1992. A taxonomic and horticultural review of *Erythronium* L. (Liliaceae). Bot. J. Linnean Soc. 109: 453-471.

Najadaceae

Haynes, R.R. 1977. The Najadaceae in the southeastern United States. J. Arnold Arbor. 58: 161-170.

Orchidaceae

Catling, Paul Miles. 1980. Systematics of *Spiranthes* L.C. Richard in northeastern North America. PhD thesis, Univ. Toronto, Toronto, Ont., Can.

Catling, P.M. 1983. *Spiranthes ovalis* var. *erostellata* (Orchidaceae), a new autogamous variety from the eastern United States. Brittonia 35: 120-125.

Catling, P.M. 1991. Systematics of *Malaxis bayardii* and *M. unifolia*. Lindleyana 6: 3-23.

Catling, P.M. and J.E. Cruise. 1974. *Spiranthes casei*, a new species from northeastern North America. Rhodora 76: 526-536.

Henry, L.K., W.E. Buker, and D.L. Pearth. 1975. Western Pennsylvania orchids. Castanea 40: 93-168.

Luer, C.A. 1975. *The Native Orchids of the United States and Canada Excluding Florida*. NY Bot. Gard., New York, NY.

Sheviak, C.J. 1982. *Biosystematic study of the* Spiranthes cernua *complex*. Bull. 448, NY State Mus., Albany, NY.,

Sheviak, C.J. 1973. A new *Spiranthes* from the grasslands of central North America. Bot. Mus. Leaflets, Harvard Univ, 23: 285-297.

Sheviak, C.J. and P.M. Catling. 1980. The identity and status of *Spiranthes ochroleuca* (Rydberg) Rydberg. Rhodora 82: 525-562.

Wander, Wade and Sharon Ann Wander. 1985. Range extension of *Goodyera tessalata* into northwestern New Jersey. Bartonia 51: 79.

Whiting, R.E. and P.M. Catling. 1986. *Orchids of Ontario, An Illustrated Guide*. The CanaColl Foundation, Ottawa, Canada.

Poaceae

Aiken, S.G., W.G. Dore, L.P. Lefkovitch, and K.C. Armstrong. 1989. *Calamagrostis epigejos* (Poaceae) in North America, especially Ontario. Can. J. Bot. 67: 3205-3218.

Alldred, Kelly W. 1986. Studies in *Aristida* (Gramineae) of the southeastern United States IV. key and conspectus. Rhodora 88: 367-387.

Anderson, D.E. 1974. Taxonomy of the genus *Chloris* (Gramineaea). Brigham Young Univ. Sci. Bull. 19: 1-133.

Barden, Lawrence S. 1987. Invasion of *Microstegium vimineum* (Poaceae), an exotic, annual, shade-tolerant, C4 grass, into a North Carolina floodplain. Am. Midland Natural. 118: 40-45.

Barkworth, M.E. and D.R. Dewey. 1985. Genomically based genera in the perennial Triticeae of North America: identification and membership. Am. J. Bot. 72: 767-776.

Baum, Bernard R. 1968. On some relationships between *Avena sativa* and *A. fatua* (Gramineae) as studied from Canadian material. Can. J. Bot. 46: 1013-

Baum, Bernard R. and L. Grant Bailey. 1989. Key and synopsis of North American *Hordeum* species. Can. J. Bot. 68: 2433-2442.

Brandenburg, D.M., J.R. Estes and S.L. Collins. 1991. A revision of *Diarrhena* (Poaceae) in the United States. Bull. Torrey Bot. Club 118: 128-136.

Crins, William J. 1991. The genera of Panicae (Gramineae: Panicoideae) in the southeastern United States. J.Arnold Arbor. Suppl. Series 1: 171-312.

DeLisle, Donald G. 1963. Taxonomy and distribution of the genus *Cenchrus*. Iowa State J. Science 37: 259-351.

Fernald, M.L. 1950. The North American variety of *Milium effusum*. Rhodora 52: 218-222.

Freckmann, Robert W. 1978. New combinations in *Dichanthelium* (Poaceae). Phytologia 39: 268-272.

Freckmann, Robert W. 1981. Realignments in the *Dichanthelium acuminatum* complex (Poaceae). Phytologia 48: 99-110.

Gould, F.W., M.A. Ali and D.E. Fairbrothers. 1972. A revision of *Echinochloa* in the United States. Am. Midland Natural. 87: 36-59.

Gould, Frank W. and Carolyn A. Clark. 1978. *Dichanthelium* (Poaceae) in the United States and Canada. Ann. Missouri Bot. Gard. 65: 1088-1132.

Hitchcock, A.S. 1971. *Manual of the Grasses of the United States, 2nd Ed.* 2 Vols. Dover Public., Inc., New York, NY.

Hunt, David M. and Robert E. Zaremba. 1992. The northeastward spread of *Microstegium vimineum* (Poaceae) into New York and adjacent states. Rhodora 94: 167-170.

Lelong, Michel G. 1984. New combinations for *Panicum* subgenus *Panicum* and subgenus *Dichanthelium* (Poaceae) of the southeastern United States. Brittonia 36: 262-273.

McCormick, J., D. Harper, L. Jones and W. Murray. 1970. Windmill grass (*Chloris verticillata* Nuttall) in Pennsylvania. Bartonia 40: 19-20.

McNeill, J. and W.G. Dore. 1976. Taxonomic and nomenclatural notes on Ontario grasses. Le Natur. Canadien 103: 553-567.

Palmer, Patricia G. 1975. A biosystematic study of the *Panicum amarum-P. amarulum* complex (Poaceae). Brittonia 27: 142-150.

Pohl, R.W. 1946. Notes on Pennsylvania and New Jersey grasses. Bartonia 24: 22-25.

Pohl, R.W. 1947. A taxonomic study on the grasses of Pennsylvania. Am. Midland Natural. 38: 513-604.

Sur, P.R. 1985. A revision of the genus *Microstegium* Nees (Poaceae) in India. J. Econ. Tax. Bot. 6: 167-176.

Thieret, John W. 1991. Revision of the genus *Cinna* (Poaceae). Sida 14: 581-596.

Tucker, Gordon C. 1990. The genera of Arundinoideae (Gramineae) in the southeastern United States. J. Arnold Arbor. 71: 145-177.

Webster, Robert D. 1988. Genera of the North American Panicae (Poaceae: Panicoideae). Syst. Bot. 13: 576-609.

Zuloaga, Fernando O. 1986. Systematics of New World species of *Panicum* (Poaceae: Paniceae). In: *Grass Systematics and Evolution*. Soderstrom,T.R., K.W.Hilu, C.S.Campbell and M.E.Barkworth. Eds. Smithsonian Instit. Press, Washington, DC.

Pontederiaceae

Horn, Charles N. 1985. A Systematic Study of the Genus *Heteranthera (sensu lato;* Pontederiaceae). PhD Thesis, Univ. Alabama, University, AL,

Horn, Charles N. 1986. Typifications and a new combination in *Heteranthera* (Pontederiaceae). Phytologia 59: 290.

Snyder, David B. 1988. *Heteranthera multiflora* in New Jersey: a first look. Bartonia 54: 21-23.

Potamogetonaceae

Haynes, R.R. 1974. A revision of North American *Potamogeton* subsection *Pusilli* (Potamogetonaceae). Rhodora 76: 564-649.

Haynes, R.R. 1985. A revision of the clasping-leaved Potamogetons (Potamogetonaceae). Sida 11: 173-188.

Ogden, E.C. 1974. Potamogeton *in New York*, Bull. 423. NY State Mus., Albany, NY.

Sparganiaceae

Beal, Ernest O. 1960. *Sparganium* (Sparganiaceae) in the southeastern United States. Brittonia 12: 176-181.

Typhaceae

Fassett, Norman C. and Barbara Calhoun. 1952. Introgression between *Typha latifolia* and *T. angustifolia*. Evolut. 6: 367-379.

Zannichelliaceae

Haynes, R.R. and L.B. Holm-Nielsen. 1987. The Zannichelliaceae in the southeastern United States. J. Arnold Arbor. 68: 259-268.

INDEX

Italics signifies names placed in synonymy.

A

B

C

D

E

F

G

H

I

J

K

L

M

N

S

T

U

V

W

X

Y

Z